Pro/ENGINEER® WILDFIRE™ 4.0

Louis Gary Lamit
with James Gee

De Anza College

CENGAGE Learning™

Australia • Brazil • Japan • Korea • Mexico • Singapore • Spain • United Kingdom • United States

Pro/ENGINEER® Wildfire™ 4.0

Louis Gary Lamit

Director, Global Engineering Program:
Chris Carson

Senior Developmental Editor:
Hilda Gowans

Marketing Services Coordinator:
Lauren Bestos

Content Project Manager:
Diane Bowdler

Senior Art Director:
Michelle Kunkler

Cover Design:
Johanna Liburd

Cover Design Modifications:
Andrew Adams

Cover Image:
J. Helgason/Shutterstock

Senior First Print Buyer:
Doug Wilke

Library of Congress Control Number: 2008928349

U.S. Student Edition:
ISBN-13: 978-0-495-41119-2
ISBN-10: 0-495-41119-1

Cengage Learning
200 First Stamford Place, Suite 400
Stamford, CT 06902
USA

Cengage Learning is a leading provider of customized learning solutions with office locations around the globe, including Singapore, the United Kingdom, Australia, Mexico, Brazil, and Japan. Locate your local office at: **international.cengage.com/region**.

Cengage Learning products are represented in Canada by Nelson Education Ltd.

For your course and learning solutions, visit **academic.cengage.com/engineering**.
Purchase any of our products at your local college store or at our preferred online store **www.ichapters.com**.

Printed in Canada
1 2 3 4 5 6 7 12 11 10 09 08

About the Authors

Louis Gary Lamit is currently a full time instructor at De Anza College in Cupertino, Ca, where he teaches Pro/ENGINEER, Pro/SURFACE, Pro/SHEETMETAL, Pro/NC, Expert Machinist, and Unigraphics NX. He is the founder of Scholarships for Veterans at www.scholarshipsforveterans.org.

Mr. Lamit has worked as a drafter, designer, numerical control (NC) programmer, technical illustrator, and engineer in the automotive, aircraft, and piping industries. A majority of his work experience is in the area of mechanical and piping design. He started as a drafter in Detroit (as a job shopper) in the automobile industry, doing tooling, dies, jigs and fixture layout, and detailing at Koltanbar Engineering, Tool Engineering, Time Engineering, and Premier Engineering for Chrysler, Ford, AMC, and Fisher Body. Mr. Lamit has worked at Remington Arms and Pratt & Whitney Aircraft as a designer, and at Boeing Aircraft and Kollmorgan Optics as an NC programmer and aircraft engineer. He also owns and operates his own consulting firm (CAD-Resources.com- Lamit and Associates), and has been involved with advertising, and patent illustration. He is the author of over 30 textbooks, workbooks, tutorials, and handbooks.

Mr. Lamit received a BS degree from Western Michigan University in 1970 and did Masters' work at Wayne State University and Michigan State University. He has also done graduate work at the University of California at Berkeley and holds an NC programming certificate from Boeing Aircraft.

Since leaving industry, Mr. Lamit has taught at all levels (Melby Junior High School, Warren, Mi.; Carroll County Vocational Technical School, Carrollton, Ga.; Heald Engineering College, San Francisco, Ca.; Cogswell Polytechnical College, San Francisco and Cupertino, Ca.; Mission College, Santa Clara, Ca.; Santa Rosa Junior College, Santa Rosa, Ca.; Northern Kentucky University, Highland Heights, Ky.; and De Anza College, Cupertino, Ca.). His textbooks include:

- ***Industrial Model Building,*** with Engineering Model Associates, Inc. (1981),
- ***Piping Drafting and Design*** (1981),
- ***Piping Drafting and Design Workbook*** (1981),
- ***Descriptive Geometry*** (1983),
- ***Descriptive Geometry Workbook*** (1983), and
- ***Pipe Fitting and Piping Handbook*** (1984), published by Prentice-Hall.
- ***Drafting for Electronics*** (3rd edition, 1998),
- ***Drafting for Electronics Workbook*** (2nd edition 1992), and
- ***CADD*** (1987), published by Charles Merrill (Macmillan-Prentice-Hall Publishing).
- ***Technical Drawing and Design*** (1994),
- ***Technical Drawing and Design Worksheets and Problem Sheets*** (1994),
- ***Principles of Engineering Drawing*** (1994),
- ***Fundamentals of Engineering Graphics and Design*** (1997),
- ***Engineering Graphics and Design with Graphical Analysis*** (1997), and
- ***Engineering Graphics and Design Worksheets and Problem Sheets*** (1997), published by West Publishing (ITP/Delmar).
- ***Basic Pro/ENGINEER in 20 Lessons*** (1998) (Revision 18) and
- ***Basic Pro/ENGINEER (with references to PT/Modeler)*** (1999) (Revision 19), published by PWS.
- ***Pro/ENGINEER 2000i*** (1999) (Revision 2000i), and
- ***Pro/E 2000i² (includes Pro/NC and Pro/SHEETMETAL)*** (2000) (Revision 2000i²), published by Brooks/Cole Publishing (ITP).
- ***Pro/ENGINEER Wildfire*** (2003) (Revision Wildfire) published by Brooks/Cole Publishing (ITP).
- ***Introduction to Pro/ENGINEER Wildfire 2.0*** (2004) (Revision Wildfire 2.0) published by SDC.
- ***Moving from 2D to 3D CAD for Engineering Design*** *(2007)* published by BookSurge, eBook by MobiPocket.
- ***Pro/ENGINEER Wildfire 3.0 Tutorial*** (2007) published by BookSurge, eBook by MobiPocket.
- ***Pro/ENGINEER Wildfire 4.0 Tutorial*** (2007) published by BookSurge eBook by MobiPocket.

James Gee is currently a part time instructor at De Anza College, where he teaches Pro/ENGINEER, Pro/MECHANICA, Pro/SURFACE, Pro/CABLE, and Pro/MOLD. Mr. Gee graduated from the University of Nevada- Reno with a BSME. He has worked in the Aerospace industry for Lockheed Missiles and Space Company, Sunnyvale, Ca.; Space Systems/Loral, Palo Alto, Ca.; and currently for BAE Systems in San Jose, Ca. Mr. Gee has assisted in checking and writing the Pro/ENGINEER series of textbooks.

Contents

Preface

Pro/ENGINEER® is one of the most widely used CAD/CAM software programs in the world today. Any aspiring engineer will benefit from the knowledge contained herein, while in school or upon graduation as a newly employed engineer.

The text involves creating a new part, an assembly, or a drawing, using a set of Pro/E commands that walk you through the process systematically.

Projects are not included in the text to keep the length and cost to the user down. For instructors and students wanting more material, compressive supplemental material can be downloaded at www.cad-resources.com.

Textbook Pro/E files are available from the author for **instructors** (not individuals) who adopt this text. The Pro/E files will open only on Academic and Commercial versions *(not Student Edition or Tryout Edition)* of Pro/ENGINEER software. *Please contact us from your academic email account.*

Lecture series for book is available at http://www.cad-resources.com/Lectures-On-DVD.html

Training CD, PTC's CADTRAIN ***COAch for Pro/ENGINEER®***, www.cadtrain.com, is a computer-based training (CBT) product designed to provide a comprehensive training program for Pro/E users in their actual CAD environment. COAch for Pro/ENGINEER, has been referenced in the book's figures with the authorized use of illustrations. COAch is one of the best ways available for expanding your knowledge of Pro/E software.

If you wish to contact the author concerning orders, questions, changes, additions, suggestions, comments, or to get on our email list, please send an email to one of the following:

Web Site: **www.cad-resources.com**
Email: **cad@cad-resources.com**

Downloads

Extra material and projects can be downloaded from ***www.cad-resources.com ⇒ Downloads***

- Part, Assembly, and Drawing Projects
- Lessons
- Tutorials
- Pro/E Files
- Pro/E related articles and information

Dedication

This book is dedicated to the men and women in the armed services who give their effort and sometimes their lives for freedom around the world.

Om Mani Padme Hum

Acknowledgments

I want to thank the following people and organizations for the support and materials granted the author:

Ken Louie	Part-time instructor, Pro/ENGINEER at De Anza College
Mike Campbell	Parametric Technology Corporation
Larry Fire	Parametric Technology Corporation
Dennis Stajic	CADTRAIN
Chris Carson	Thomson Engineering

Max Gilliland- is the instructional associate for CAD at De Anza College has been essential to the CAD program. Besides assisting in the classroom, he maintains the software and hardware for the program and in my home office.

Letha Jeanpierre- the dean of the Business and Computer Systems Division at De Anza College, has been the most supportive and professional administrator I have had the pleasure of working with out of the nine schools, colleges and universities at which I have taught during the last 37 years.

Donations and Scholarships for Veterans (SFV)

A portion of this text's profits go to my scholarship fund at Foothill-De Anza Community College District (FHDA Foundation). We fund one or more scholarships per year. Your contributions provide extra scholarships as funds are available.

Scholarships are available to any qualified veteran of the Army, Navy, Air Force, Marines, or Coast Guard. Scholarships are administered by the local college foundations. **No administration fees are taken by *Scholarships for Veterans*.** Committee members of Scholarships for Veterans make final selections. **All costs associated with Scholarships for Veterans are borne by Lamit and Associates and CAD-Resources.com.**

For more information, see Scholarships for Veterans at www.scholarshipsforveterans.org

Downloads

Extra material and projects can be downloaded from ***www.cad-resources.com ⇒ Downloads***

- Part, Assembly, and Drawing Projects
- Lessons
- Tutorials
- Pro/E Files
- Pro/E related articles and information

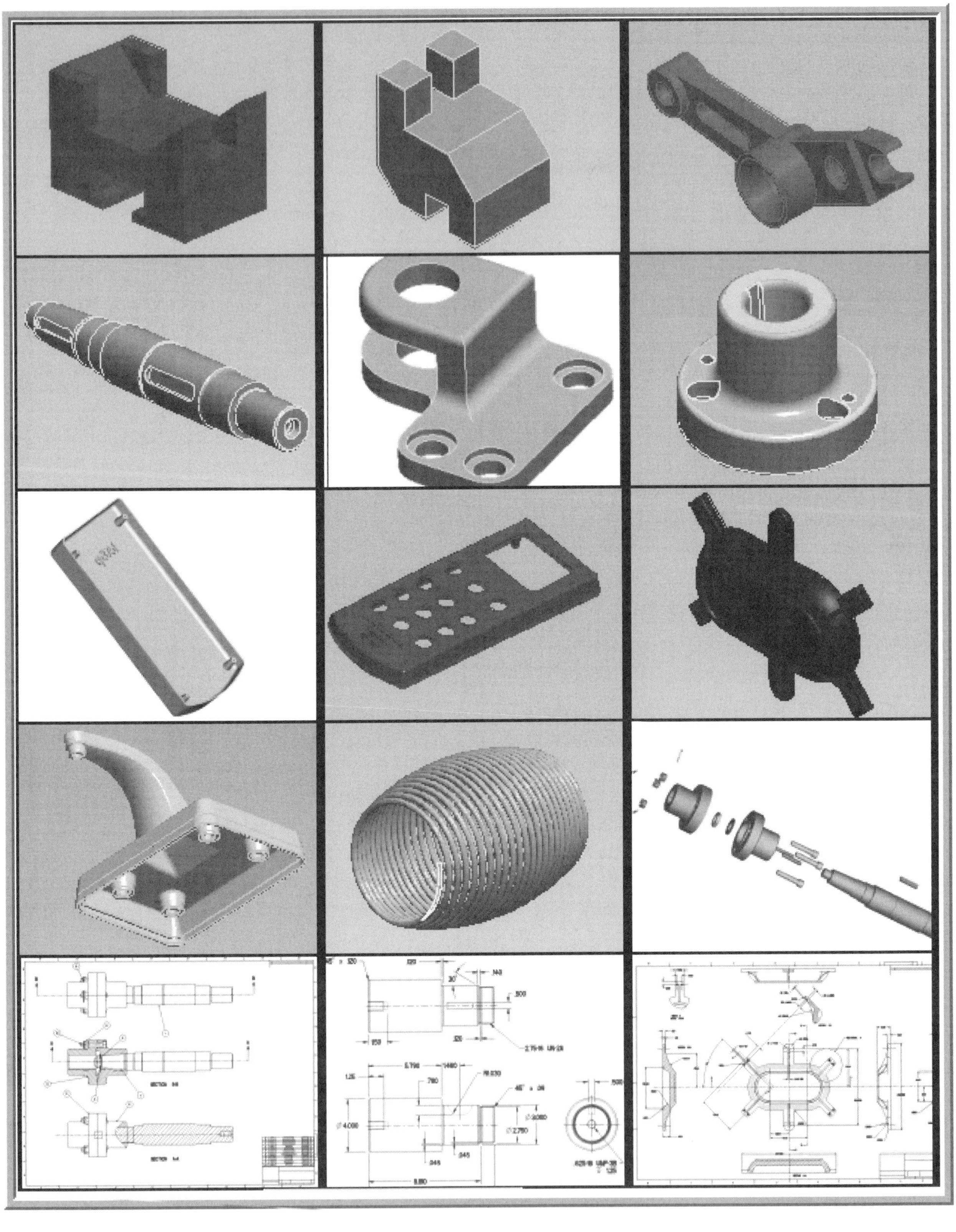

Introduction

This text guides you through parametric design using Pro/ENGINEER® Wildfire 4.0™. While using this text, you will create individual parts, assemblies, and drawings.

Parametric can be defined as *any set of physical properties whose values determine the characteristics or behavior of an object.* **Parametric design** enables you to generate a variety of information about your design: its mass properties, a drawing, or a base model. To get this information, you must first model your part design.

Parametric modeling philosophies used in Pro/E include the following:

Feature-Based Modeling Parametric design represents solid models as combinations of engineering features (Fig. 1).

Creation of Assemblies Just as features are combined into parts, parts may be combined into assemblies (Fig. 1).

Capturing Design Intent The ability to incorporate engineering knowledge successfully into the solid model is an essential aspect of parametric modeling.

Figure 1 Parts and Assembly Design

Parametric Design

Parametric design models are not drawn so much as they are *sculpted* from solid volumes of materials. To begin the design process, analyze your design. Before any work is started, take the time to ***tap*** into your own knowledge bank and others that are available. Think, Analyze, and Plan. These three steps are essential to any well-formulated engineering design process.

Break down your overall design into its basic components, building blocks, or primary features. Identify the most fundamental feature of the object to sketch as the first, or base, feature. Varieties of **base features** can be modeled using extrude, revolve, sweep, and blend tools.

Sketched features (*extrusions, sweeps, etc.)* and pick-and-place features called **referenced features** (*holes, rounds, chamfers, etc.*) are normally required to complete the design. With the SKETCHER, you use familiar 2D entities (points, lines, rectangles, circles, arcs, splines, and conics) (Fig. 2). There is no need to be concerned with the accuracy of the sketch. Lines can be at differing angles, arcs and circles can have unequal radii, and features can be sketched with no regard for the actual objects' dimensions. In fact, exaggerating the difference between entities that are similar but not exactly the same is actually a far better practice when using the SKETCHER.

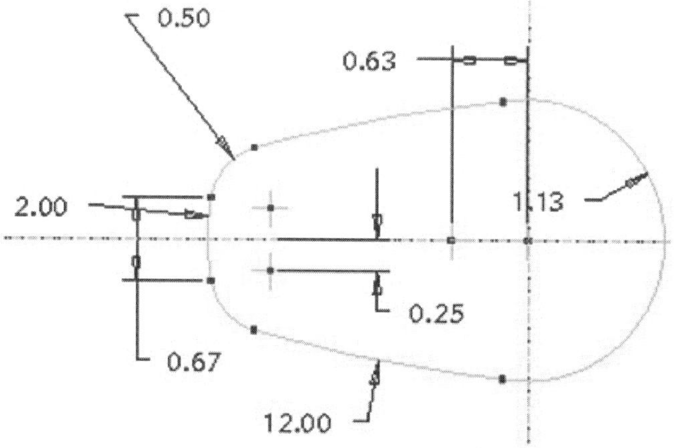

Figure 2 Sketching

Geometry assumptions and constraints will close ends of connected lines, align parallel lines, and snap sketched lines to horizontal and vertical (orthogonal) orientations. Additional constraints are added by means of **parametric dimensions** to control the size and shape of the sketch.

Features are the basic building blocks you use to create an object (Fig. 3). Features "understand" their fit and function as though "smarts" were built into the features themselves. For example, a hole or cut feature "knows" its shape and location and the fact that it has a negative volume. As you modify a feature, the entire object automatically updates after regeneration. The idea behind feature-based modeling is that the designer constructs an object so that it is composed of individual features that describe the way the geometry is supposed to behave if its dimensions change. This happens quite often in industry, as in the case of a design change. Feature-based modeling is diagramed in Figure 4.

Figure 3 Feature Design (Courtesy CADTRAIN)

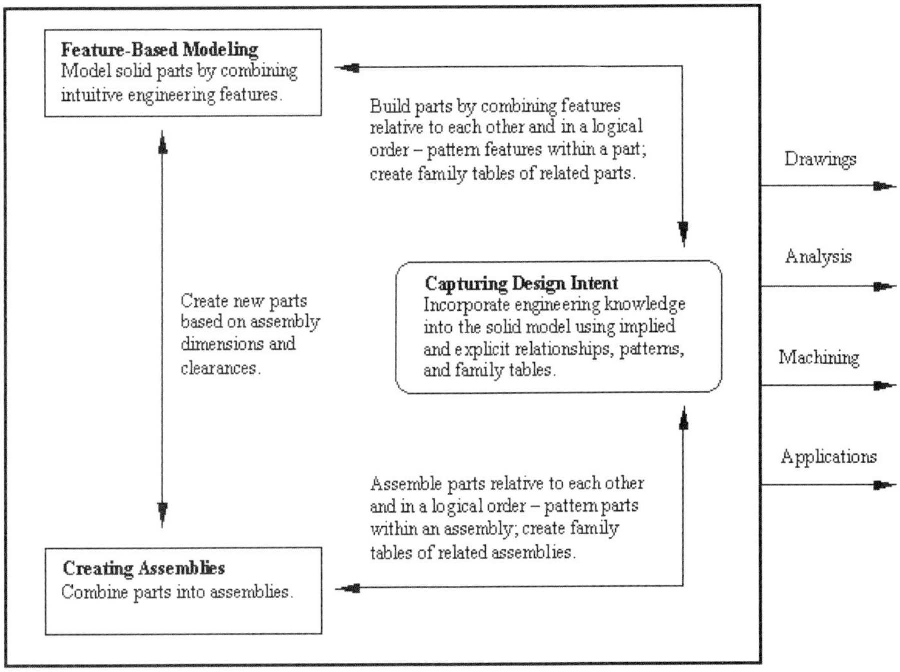

Figure 4 Feature-based Modeling

Parametric modeling is the term used to describe the capturing of design operations as they take place, as well as future modifications and editing of the design. The order of the design operations is significant. Suppose a designer specifies that two surfaces be parallel, such that surface two is parallel to surface one. Therefore, if surface one moves, surface two moves along with surface one to maintain the specified design relationship. The surface two is a **child** of surface one in this example. Parametric modeling software allows the designer to **reorder** the steps in the object's creation.

Various types of features are used as building blocks in the progressive creation of solid objects. Figures 5(a-c) illustrates base features, datum features, sketched features, and referenced features. The "chunks" of solid material from which parametric design models are constructed are called **features**.

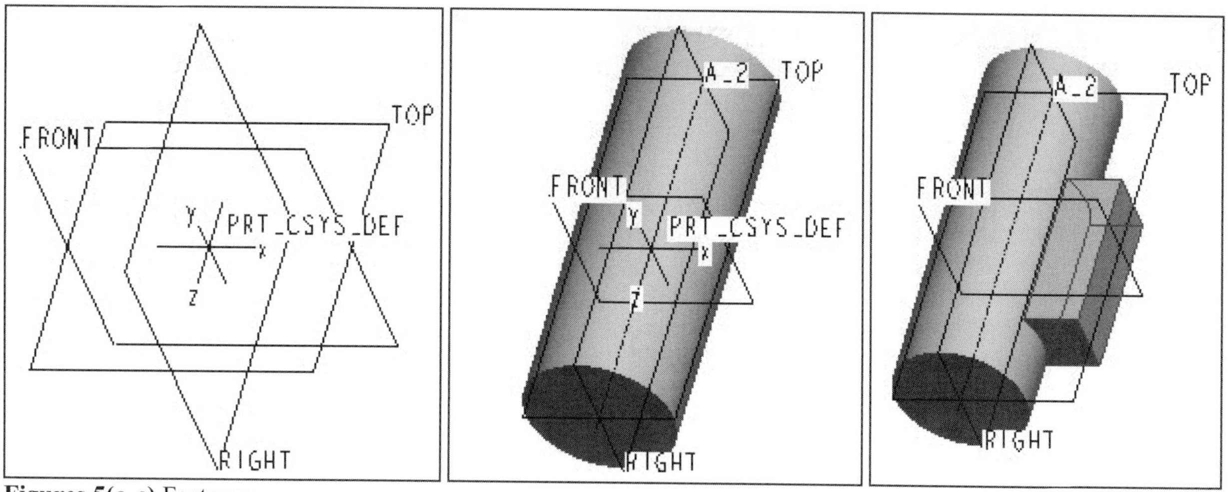

Figures 5(a-c) Features

Features generally fall into one of the following categories:

Base Feature The base feature is normally a set of datum planes referencing the default coordinate system. The base feature is important because all future model geometry will reference this feature directly or indirectly; it becomes the root feature. Changes to the base feature will affect the geometry of the entire model [Figs. 6(a-c)].

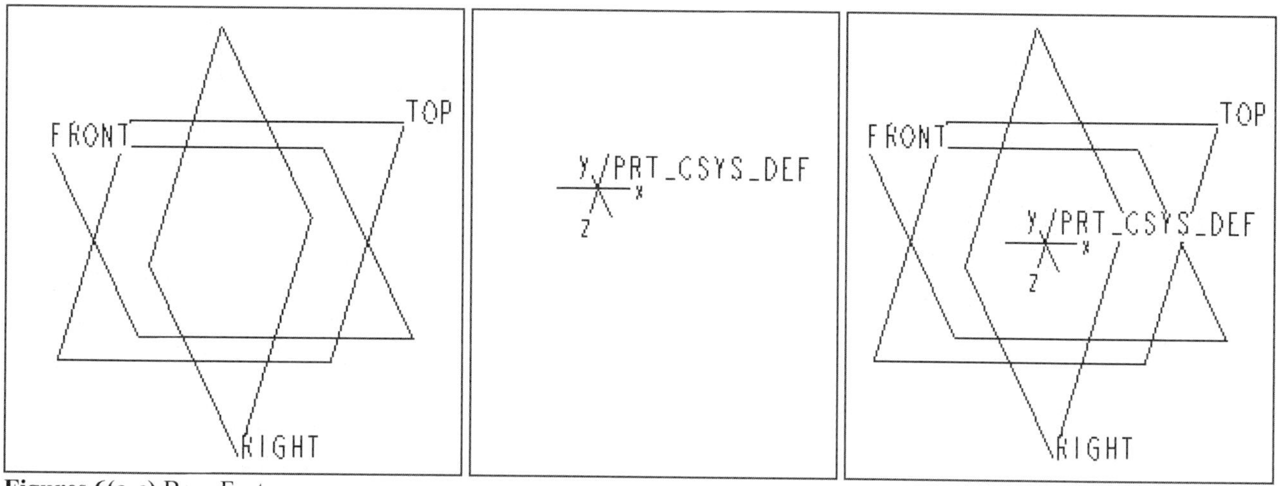

Figures 6(a-c) Base Features

Datum Features Datum features (lines, axes, curves, and points) are generally used to provide sketching planes and contour references for sketched and referenced features. Datum features do not have volume or mass and may be visually hidden without affecting solid geometry (Fig. 7).

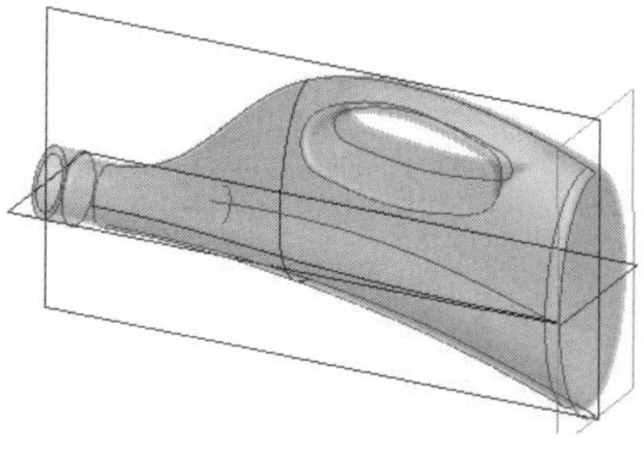

Figure 7 Datum Curves

Sketched Features Sketched features are created by extruding, revolving, blending, or sweeping a sketched cross section. Material may be added or removed by protruding or cutting the feature from the existing model (Fig. 8).

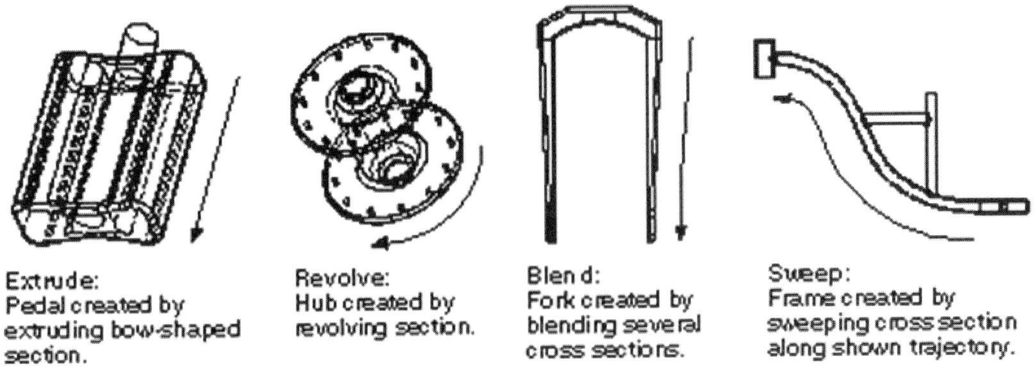

Figure 8 Sketched Features

Referenced Features Referenced features (rounds, holes, shells, and so on) utilize existing geometry for positioning and employ an inherent form; they do not need to be sketched [Figs. 9(a-b)].

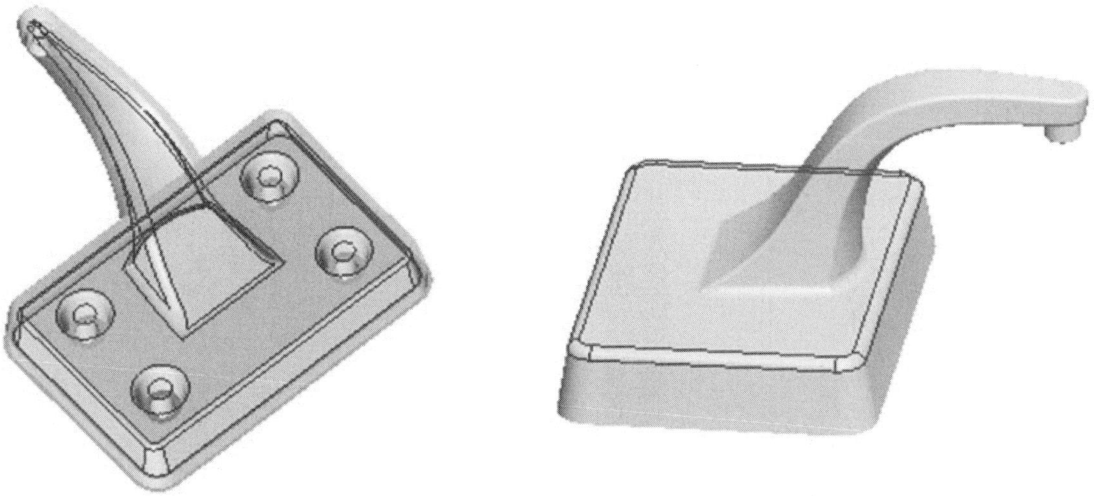

Figures 9(a-b) Referenced Features- Shell and Round (Spout is a Swept Blend Feature)

A wide variety of features are available. These tools enable the designer to make far fewer changes by capturing the engineer's design intent early in the development stage (Fig.10).

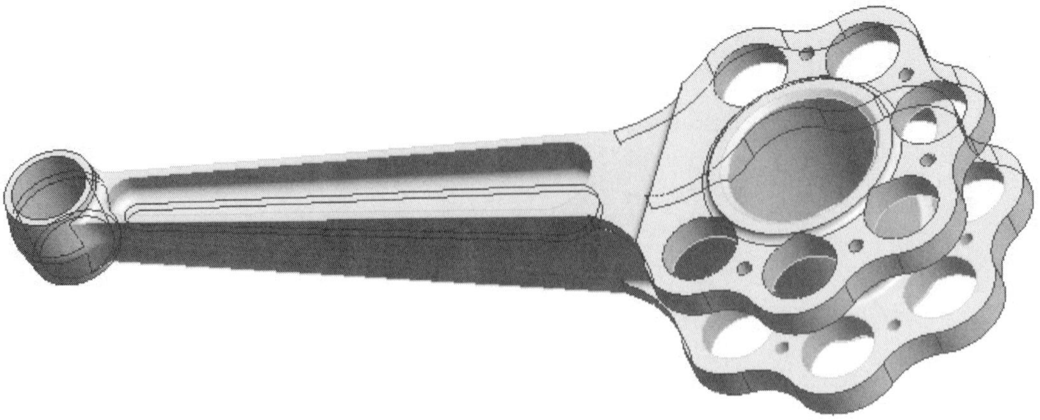

Figure 10 Parametric Designed Part

Fundamentals

The design of parts and assemblies, and the creation of related drawings, forms the foundation of engineering graphics. When designing with Pro/ENGINEER, many of the previous steps in the design process have been eliminated, streamlined, altered, refined, or expanded. The model you create as a part forms the basis for all engineering and design functions.

The part model contains the geometric data describing the part's features, but it also includes non-graphical information embedded in the design itself. The part, its associated assembly, and the graphical documentation (drawings) are parametric. The physical properties described in the part drive (determine) the characteristics and behavior of the assembly and drawing. Any data established in the assembly mode, in turn, determines that aspect of the part and, subsequently, the drawings of the part and the assembly. In other words, all the information contained in the part, the assembly, and the drawing is interrelated, interconnected, and parametric (Fig. 11).

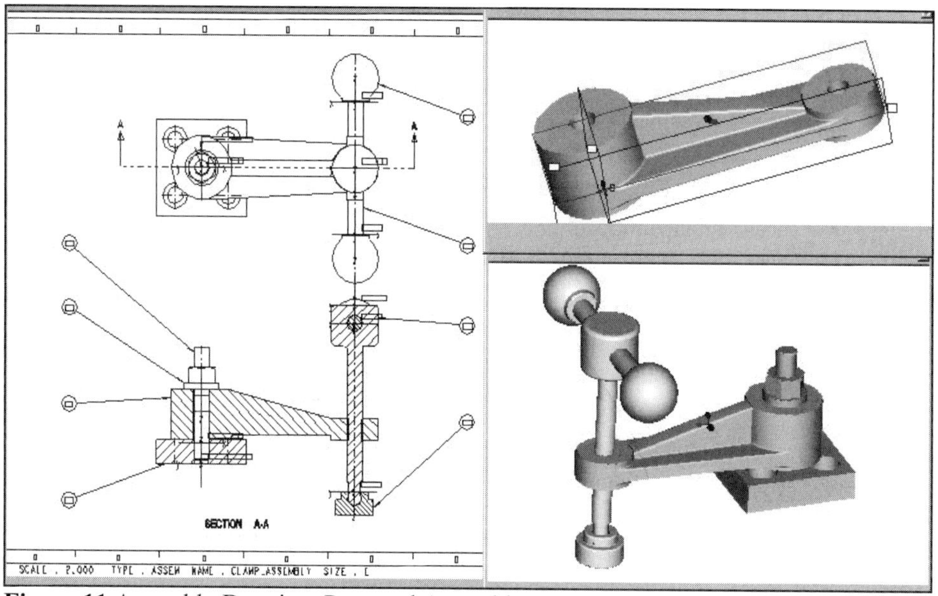

Figure 11 Assembly Drawing, Part, and Assembly

Part Design

In many cases, the part will be the first component of this interconnected process. The *part* function in Pro/E is used to design components.

During part design (Fig. 12), you can accomplish the following:

- Define the base feature
- Define and redefine construction features to the base feature
- Modify the dimensional values of part features (Fig. 13)
- Embed design intent into the model using tolerance specifications and dimensioning schemes
- Create pictorial and shaded views of the component
- Create part families (family tables)
- Perform mass properties analysis and clearance checks
- List part, feature, layer, and other model information
- Measure and calculate model features
- Create detail drawings of the part

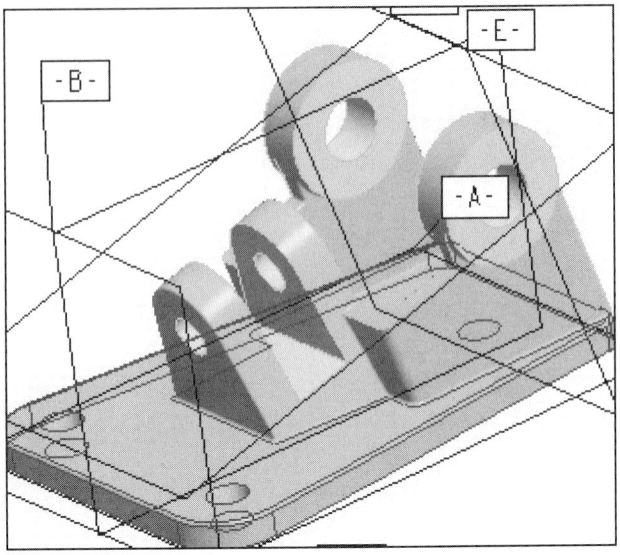

Figure 12 Part Design

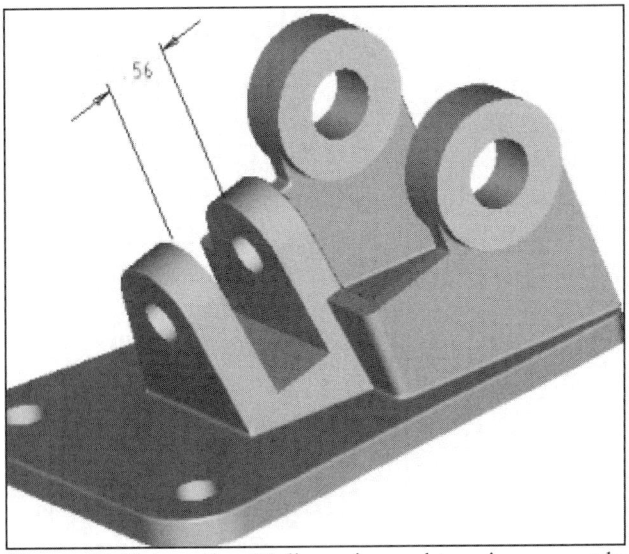

Figure 13 Pick on the **.56** dimension and type in a new value

Establishing Features

The design of any part requires that the part be *confined*, *restricted*, *constrained*, and *referenced*. In parametric design, the easiest method to establish and control the geometry of your part design is to use three datum planes. Pro/E automatically creates the three **primary datum planes**. The default datum planes (**RIGHT**, **TOP**, and **FRONT**) constrain your design in all three directions.

Datum planes are infinite planes located in 3D model mode and are associated with the object that was active at the time of their creation. To select a datum plane, you can pick on its name or anywhere on the perimeter edge. Datum planes are *parametric*--geometrically associated with the part. Parametric datum planes are associated with and dependent on the edges, surfaces, vertices, and axes of a part. Datum planes are used to create a reference on a part that does not already exist. For example, you can sketch or place features on a datum plane when there is no appropriate planar surface. You can also dimension to a datum plane as though it were an edge. In Figure 14, three **default datum planes** and a **default coordinate system**s were created when a new part (**PRT0001.PRT**) was started using the default template. Note that in the **Model Tree** window, they are the first four features of the part, which means that they will be the *parents* of the features that follow.

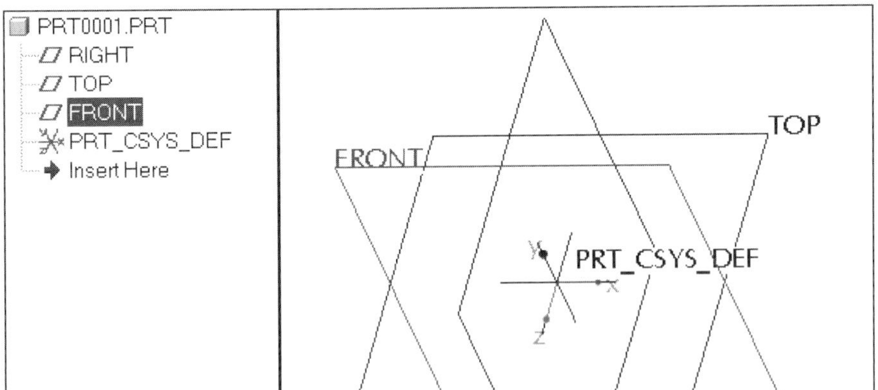

Figure 14 Default Datum Planes and Coordinate System

Datum Features

Datum features are planes, axes, and points you use to place geometric features on the active part. Datums other than defaults can be created at any time during the design process.

As we have discussed, there are three (primary) types of datum features (Fig. 14): **datum planes, datum axes**, and **datum points** (there are also ***datum curves*** and ***datum coordinate systems***). You can display all types of datum features, but they do not define the surfaces or edges of the part or add to its mass properties. In Figure 15, a variety of datum planes are used in the creation of the casting.

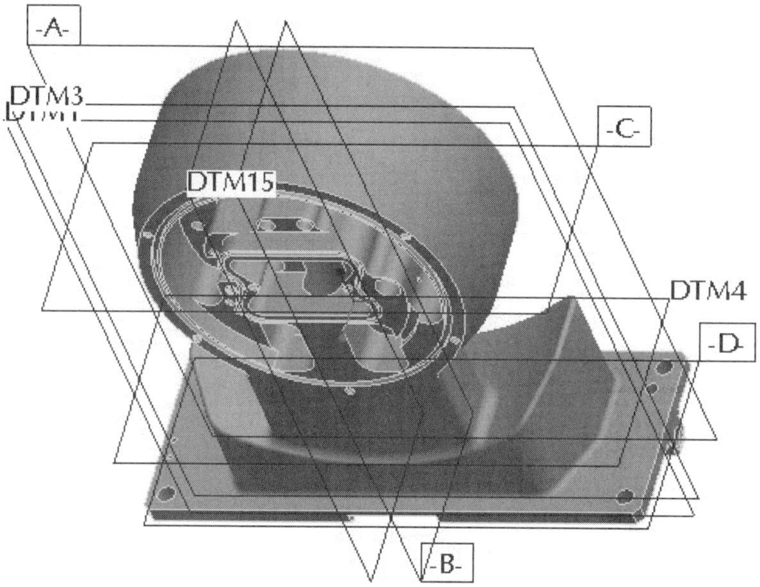

Figure 15 Datums in Part Design

Specifying constraints that locate it with respect to existing geometry creates a datum. For example, a datum plane might be made to pass through the axis of a hole and parallel to a planar surface. Chosen constraints must locate the datum plane relative to the model without ambiguity. You can also use and create datums in assembly mode.

Besides datum planes, datum axes and datum points can be created to assist in the design process. You can also automatically create datum axes through cylindrical features such as holes and solid round features by setting this as a default in your Pro/E configuration file. The part in Figure 16 shows the default axes for a variety of holes.

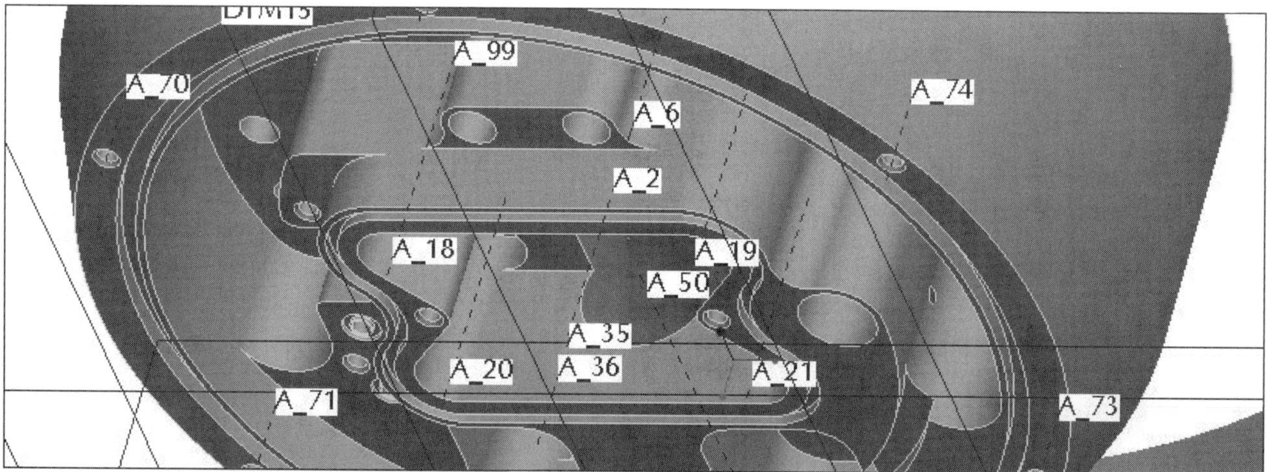

Figure 16 Feature Default Datum Axes

Parent-Child Relationships

Because solid modeling is a cumulative process, certain features must, by necessity precede others. Those that follow must rely on previously defined features for dimensional and geometric references. The relationships between features and those that reference them are termed ***parent-child relationships***. Because children reference parents, parent features can exist without children, but children cannot exist without their parents. This type of CAD modeler is called a history-based system. Using Pro/E's information command will list the models' information as shown in Figure 17.

Model Info : SLIDER_CRANK

ASSEMBLY NAME : SLIDER_CRANK

Name	ID
SLIDER_CRANK	1048576
BLOCK	1572864
CRANK_SHAFT	1572865
CON_ROD	1572866
END_CAP	1048579
PISTON_HEAD	1048580
BASE	1048581

Feature List

No.	ID	Name	Type
1	1	ASM_RIGHT	DATUM PLANE
2	3	ASM_TOP	DATUM PLANE
3	5	ASM_FRONT	DATUM PLANE
4	7	ASM_DEF_CSYS	COORDINATE S
5	39	BLOCK.PRT	COMPONENT
6	41	CRANK_SHAFT.PRT	COMPONENT
7	43	CON_ROD.PRT	COMPONENT
8	44	END_CAP.PRT	COMPONENT
9	45	PISTON_HEAD.PRT	COMPONENT
10	46	BASE.PRT	COMPONENT

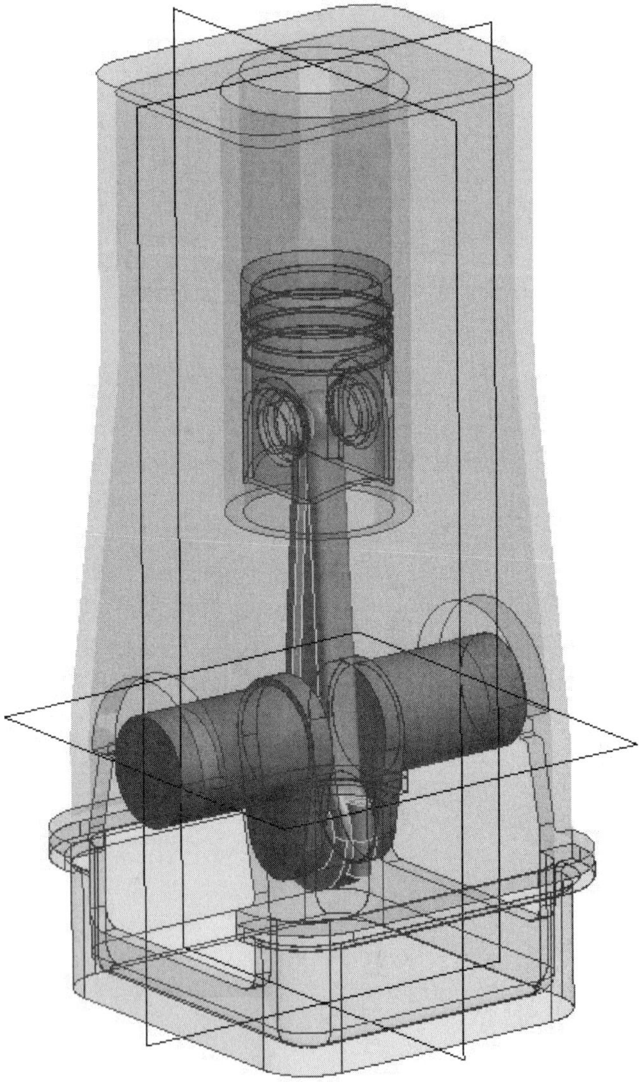

Figure 17 Model Information

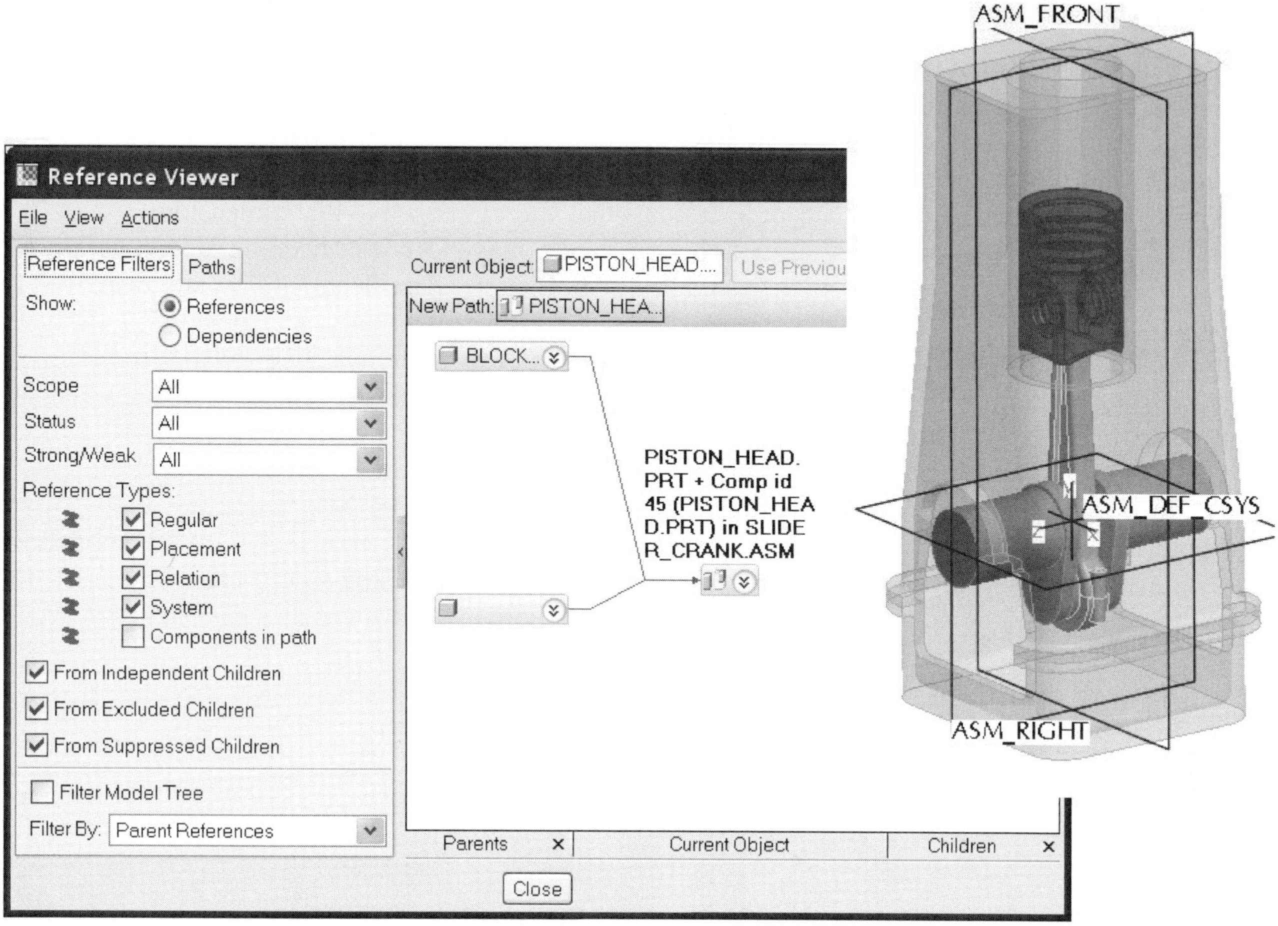

Figure 18 Parent-Child Information

The parent-child relationship (Fig. 18) is one of the most powerful aspects of parametric design. When a parent feature is modified, its children are automatically recreated to reflect the changes in the parent feature's geometry. It is essential to reference feature dimensions so that design modifications are correctly propagated through the model/part. Any modification to the part is automatically propagated throughout the model (Fig. 19) and will affect all children of the modified feature.

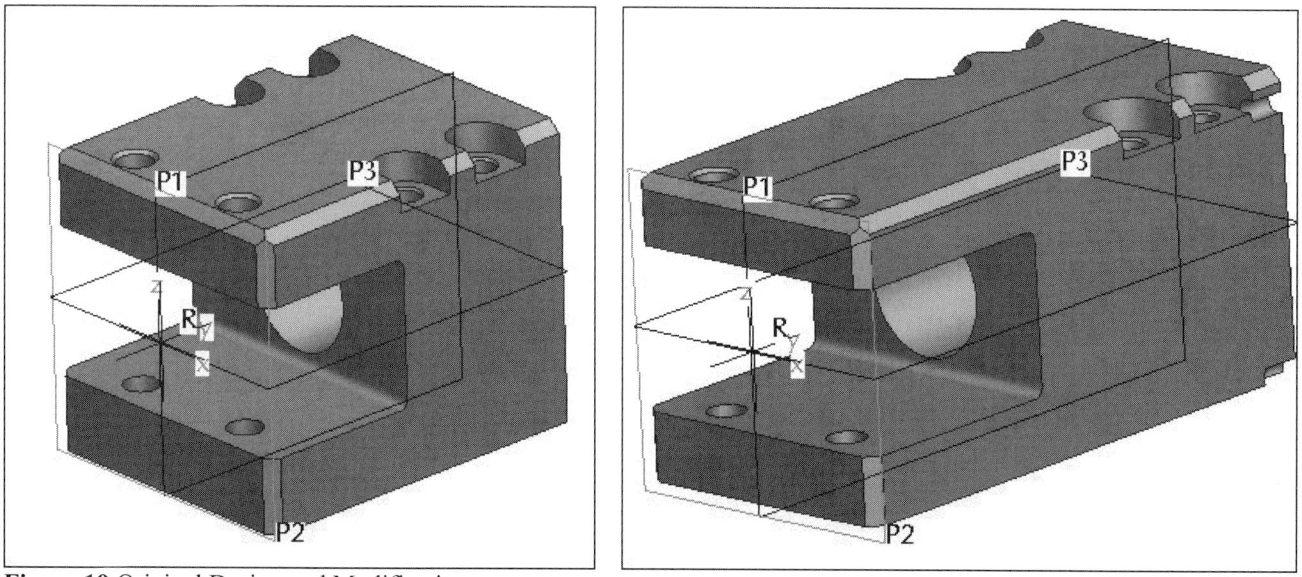

Figure 19 Original Design and Modification

Capturing Design Intent

A valuable characteristic of any design tool is its ability to *render* the design and at the same time capture its *intent* (Fig. 20). Parametric methods depend on the sequence of operations used to construct the design. The software maintains a *history of changes* the designer makes to specific parameters. The point of capturing this history is to keep track of operations that depend on each other. Whenever Pro/E is told to change a specific dimension, it can update all operations that are referenced to that dimension.

For example, a circle representing a bolt hole circle may be constructed so that it is always concentric to a circular slot. If the slot moves, so does the bolt circle. Parameters are usually displayed in terms of dimensions or labels and serve as the mechanism by which geometry is changed. The designer can change parameters manually by changing a dimension or can reference them to a variable in an equation (**relation**) that is solved either by the modeling program itself or by external programs such as spreadsheets.

Features can also store non-graphical information. This information can be used in activities such as drafting, numerical control (NC), finite-element analysis (FEA), and kinematics analysis.

Capturing design intent is based on incorporating engineering knowledge into a model by establishing and preserving certain geometric relationships. The wall thickness of a pressure vessel, for example, should be proportional to its surface area and should remain so, even as its size changes.

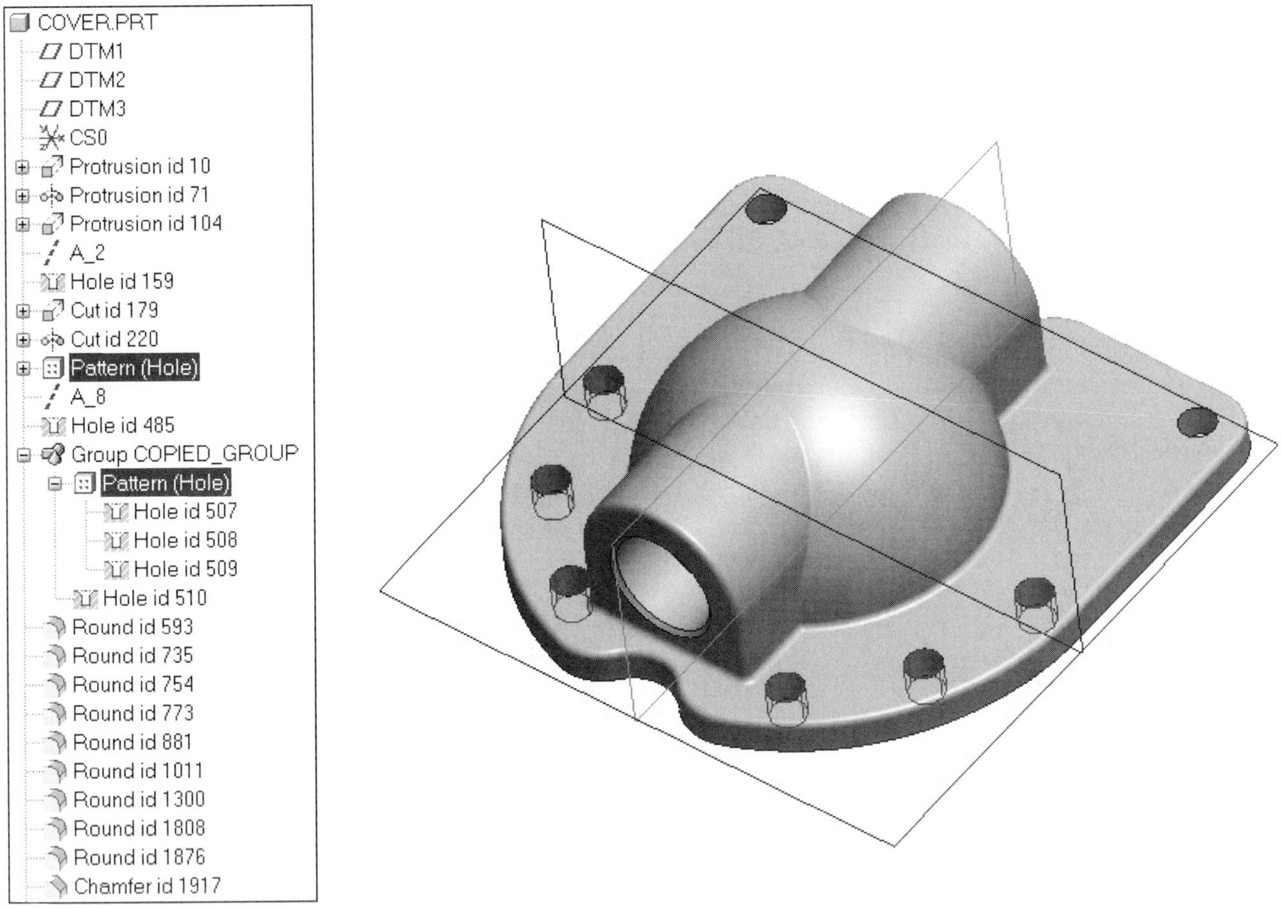

Figure 20 Capturing Design Intent

Parametric designs capture relationships in several ways:

Implicit Relationships Implicit relationships occur when new model geometry is sketched and dimensioned relative to existing features and parts. An implicit relationship is established, for instance, when the section sketch of a tire (Fig. 21) uses rim edges as a reference.

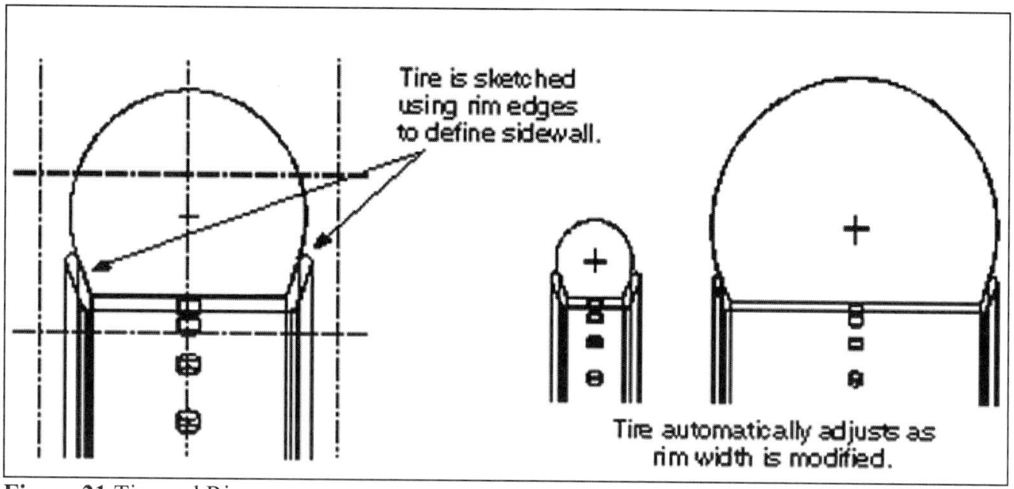

Figure 21 Tire and Rim

Patterns Design features often follow a geometrically predictable pattern. Features and parts are patterned in parametric design by referencing either construction dimensions or existing patterns. One example of patterning is a wheel hub with spokes (Fig. 22). First, the spoke holes are radially patterned. The spokes can then be strung by referencing this pattern.

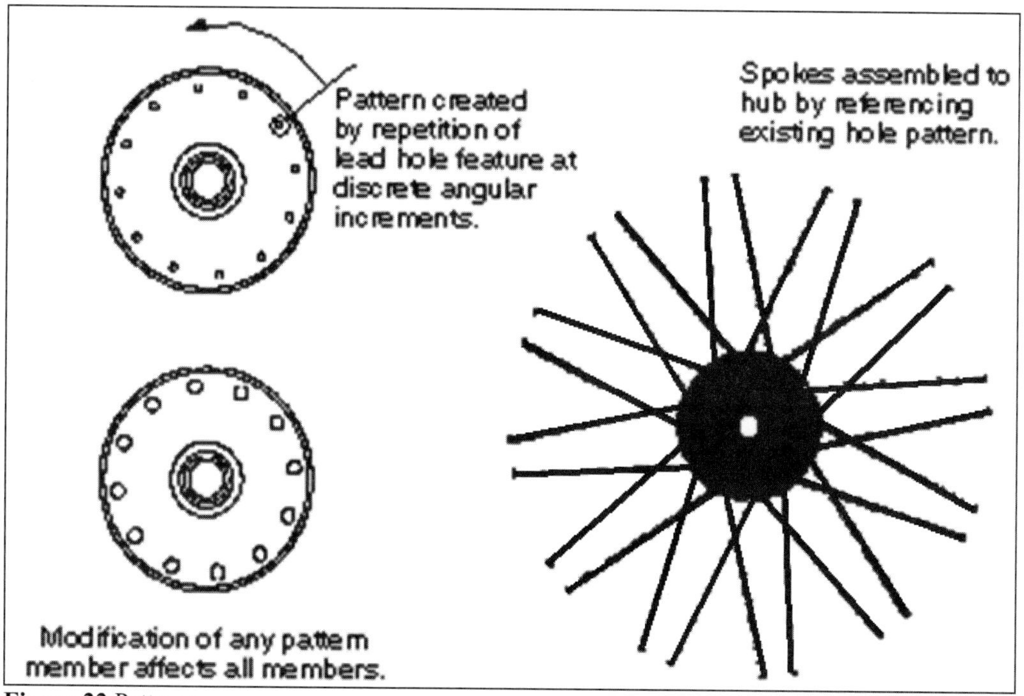

Figure 22 Patterns

Modification to a pattern member affects all members of that pattern. This helps capture design intent by preserving the duplicate geometry of pattern members.

The modeling task is to incorporate the features and parts of a complex design while properly capturing design intent to provide flexibility in modification. Parametric design modeling is a synthesis of physical and intellectual design (Fig. 23).

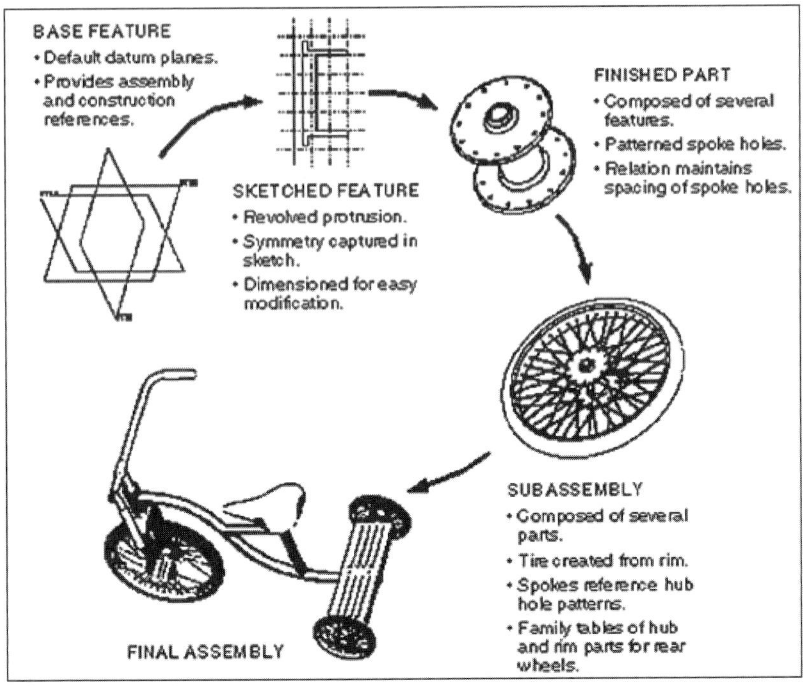

Figure 23 Relations

Explicit Relations Whereas implicit relationships are implied by the feature creation method, the user mathematically enters an explicit relation. This equation is used to relate feature and part dimensions in the desired manner. An explicit relation (Fig. 24) might be used, for example, to control sizes on a model.

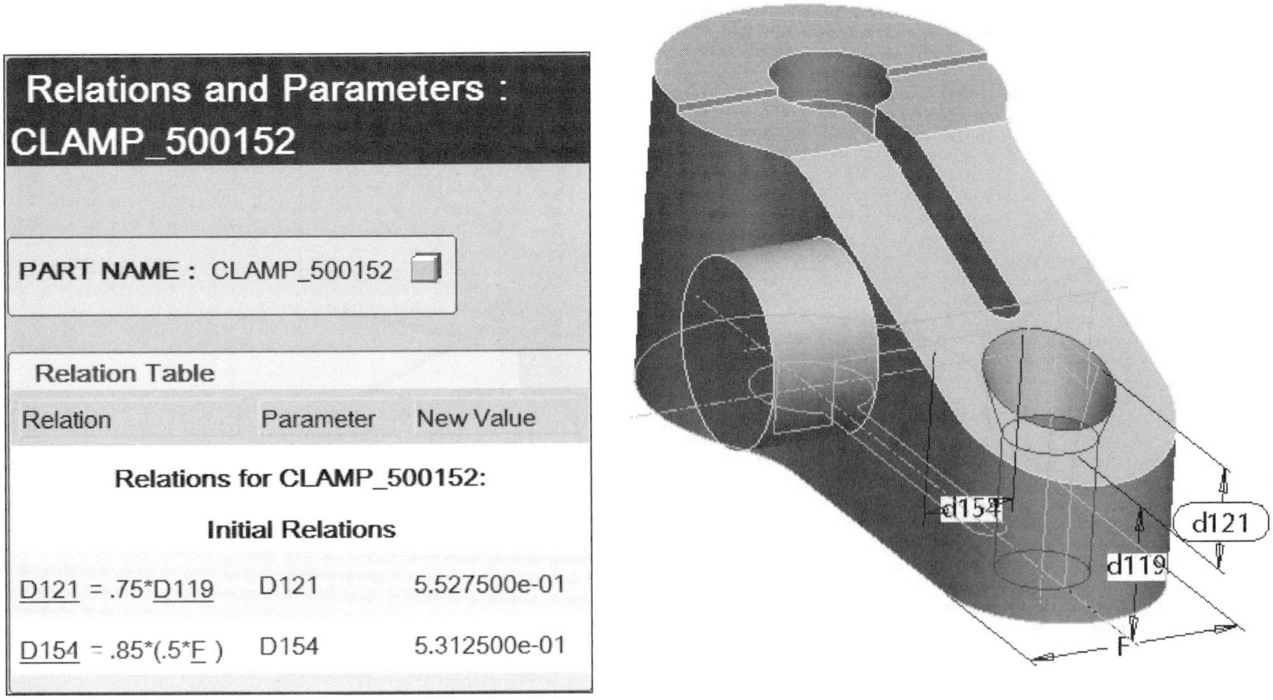

Figure 24 Adding Relations

Family Tables Family tables are used to create part families [Figs. 25(a-c)] from generic models by tabulating dimensions or the presence of certain features or parts. A family table might be used, for example, to catalog a series of couplings with varying width and diameter as shown in Figure 26.

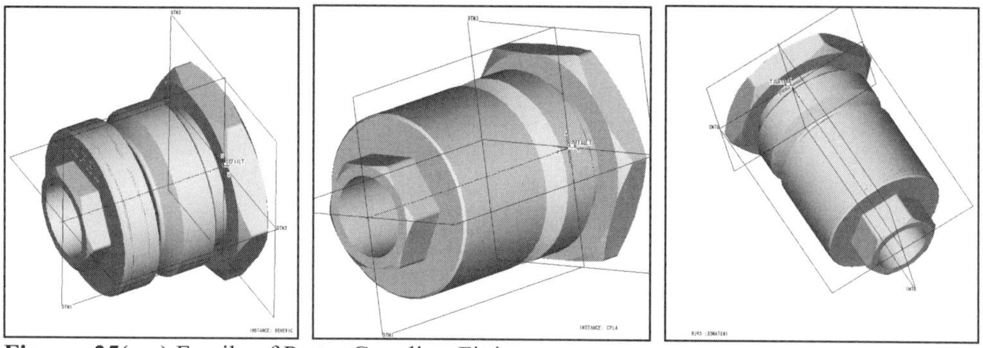

Figures 25(a-c) Family of Parts- Coupling-Fitting

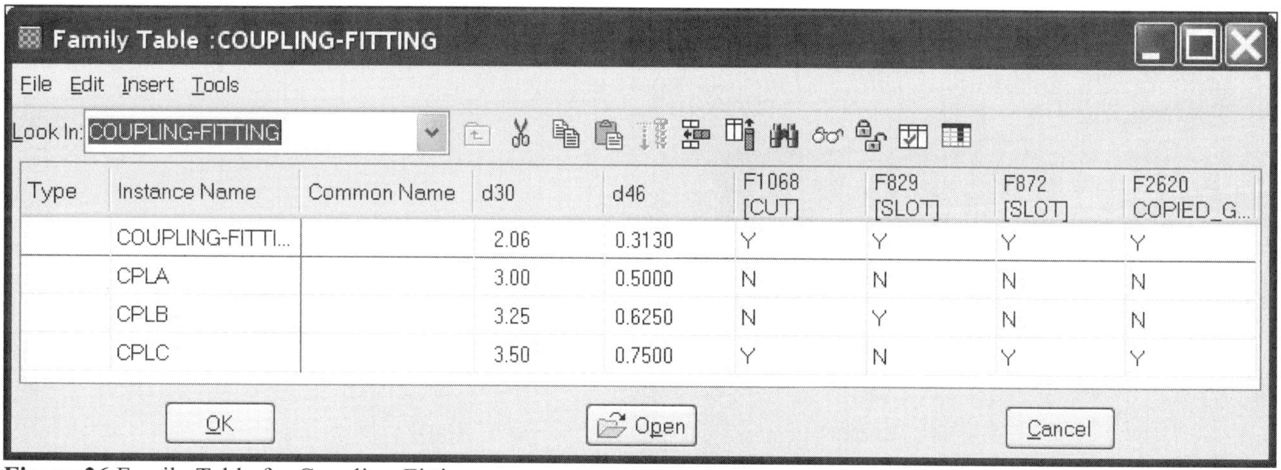

Type	Instance Name	Common Name	d30	d46	F1068 [CUT]	F829 [SLOT]	F872 [SLOT]	F2620 COPIED_G...
	COUPLING-FITTI...		2.06	0.3130	Y	Y	Y	Y
	CPLA		3.00	0.5000	N	N	N	N
	CPLB		3.25	0.6250	N	Y	N	N
	CPLC		3.50	0.7500	Y	N	Y	Y

Figure 26 Family Table for Coupling-Fitting

Assemblies

Just as parts are created from related features, **assemblies** are created from related parts. The progressive combination of subassemblies, parts, and features into an assembly creates parent-child relationships based on the references used to assemble each component (Fig. 27).

The *Assembly* functionality is used to assemble existing parts and subassemblies.

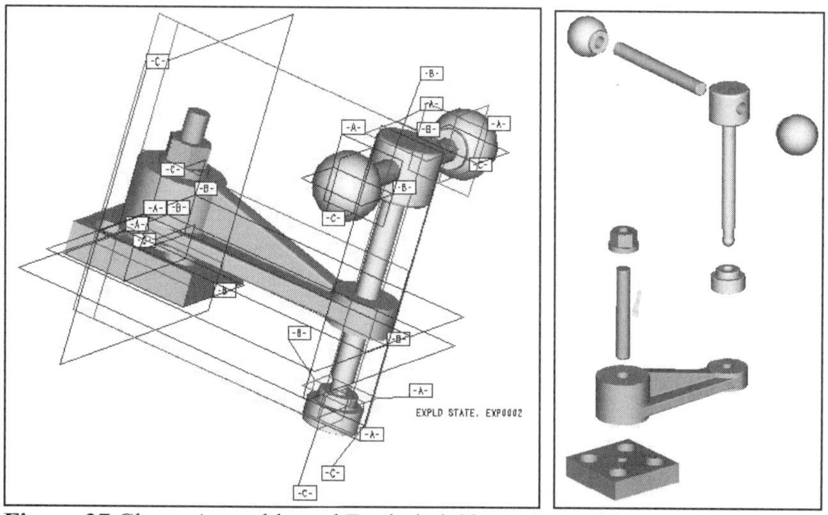

Figure 27 Clamp Assembly and Exploded Clamp Assembly

During assembly creation, you can:

- Simplify a view of a large assembly by creating a simplified representation
- Perform automatic or manual placement of component parts
- Create an exploded view of the component parts
- Perform analysis, such as mass properties and clearance checks
- Modify the dimensional values of component parts
- Define assembly relations between component parts
- Create assembly features
- Perform automatic interchange of component parts
- Create parts in Assembly mode
- Create documentation drawings of the assembly

Just as features can reference part geometry, parametric design also permits the creation of parts referencing assembly geometry. **Assembly mode** allows the designer both to fit parts together and to design parts based on how they should fit together.

In Figure 28, an assembly *Bill of Materials* report is generated.

Bom Report : CLAMP_ASSEMBLY

Assembly CLAMP_ASSEMBLY contains:

Quantity	Type	Name
1	Part	CLAMP_PLATE
1	Sub-Assembly	CLAMP_SUBASSEMBLY
1	Part	CLAMP_STUD35
1	Part	CLAMP_FLANGE_NUT

Sub-Assembly CLAMP_SUBASSEMBLY contains:

Quantity	Type	Name	Actions
1	Part	CLAMP_ARM	
1	Part	CLAMP_SWIVEL	
1	Part	CLAMP_FOOT	
1	Part	CLAMP_STUD5	
2	Part	CLAMP_BALL	

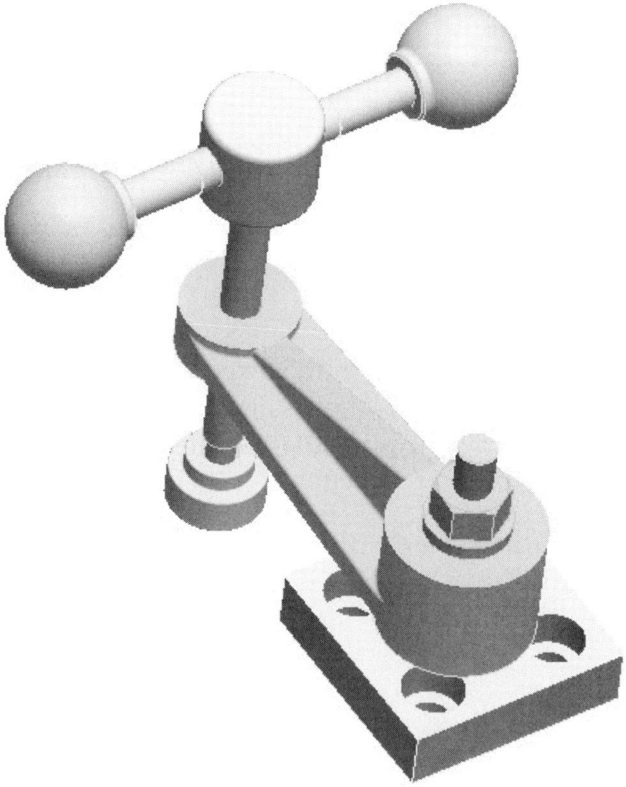

Figure 28 BOM Report

Drawings

You can create drawings of all parametric design models (Fig. 29). All model views in the drawing are *associative:* if you change a dimensional value in one view, other drawing views update accordingly. Moreover, drawings are associated with their parent models. Any dimensional changes made to a drawing are automatically reflected in the model. Any changes made to the model (e.g., addition of features, deletion of features, dimensional changes, and so on) in Part, Sheet Metal, Assembly, or Manufacturing modes are also automatically reflected in their corresponding drawings.

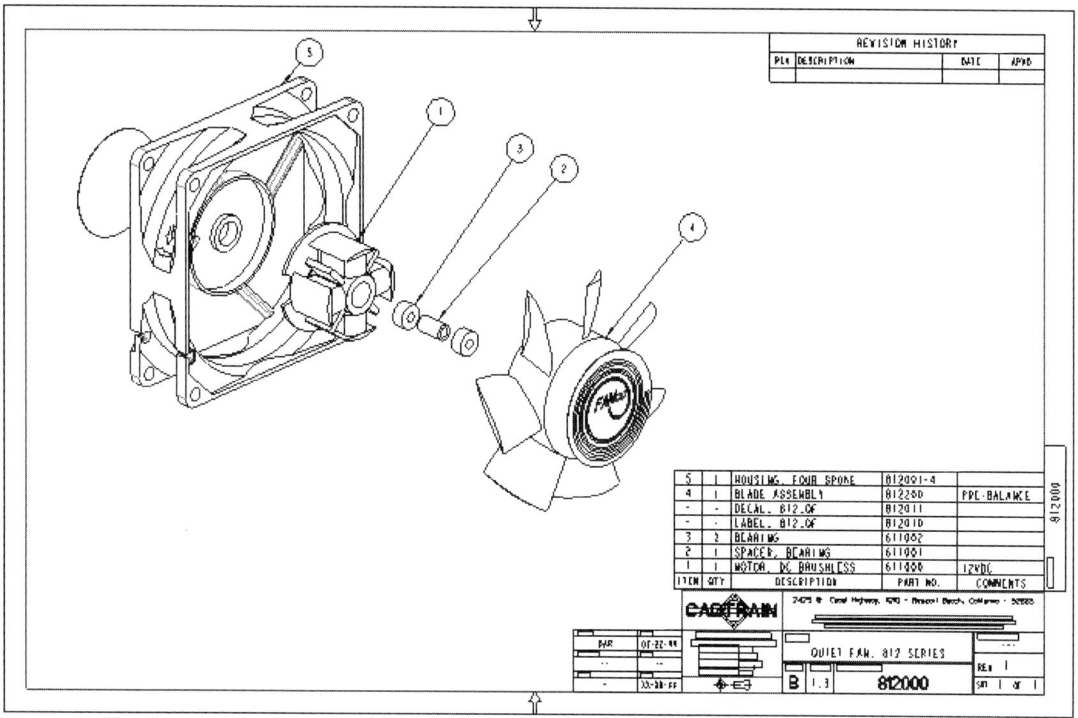

Figure 29 Ballooned Exploded View Assembly Drawing with Bill of Materials (BOM) (Courtesy CADTRAIN)

The **Drawing** functionality is used to create annotated drawings of parts and assemblies. During drawing creation, you can:

- Add views of the part or assembly
- Show existing dimensions
- Incorporate additional driven or reference dimensions
- Create notes on the drawing
- Display views of additional parts or assemblies
- Add sheets to the drawing
- Create draft entities on the drawing
- Balloon components on an assembly drawing (Fig. 30)
- Create an associative BOM

You can annotate the drawing with notes, manipulate the dimensions, and use layers to manage the display of different items on the drawing. The module **Pro/DETAIL** can be used to extend the drawing capability or as a stand-alone module allowing you to create, view, and annotate models and drawings. Pro/DETAIL supports additional view types and multi-sheets and offers commands for manipulating items in the drawing and for adding and modifying different kinds of textural and symbolic information. In addition, the abilities to customize engineering drawings with sketched geometry, create custom drawing formats, and make numerous cosmetic changes to the drawing are available.

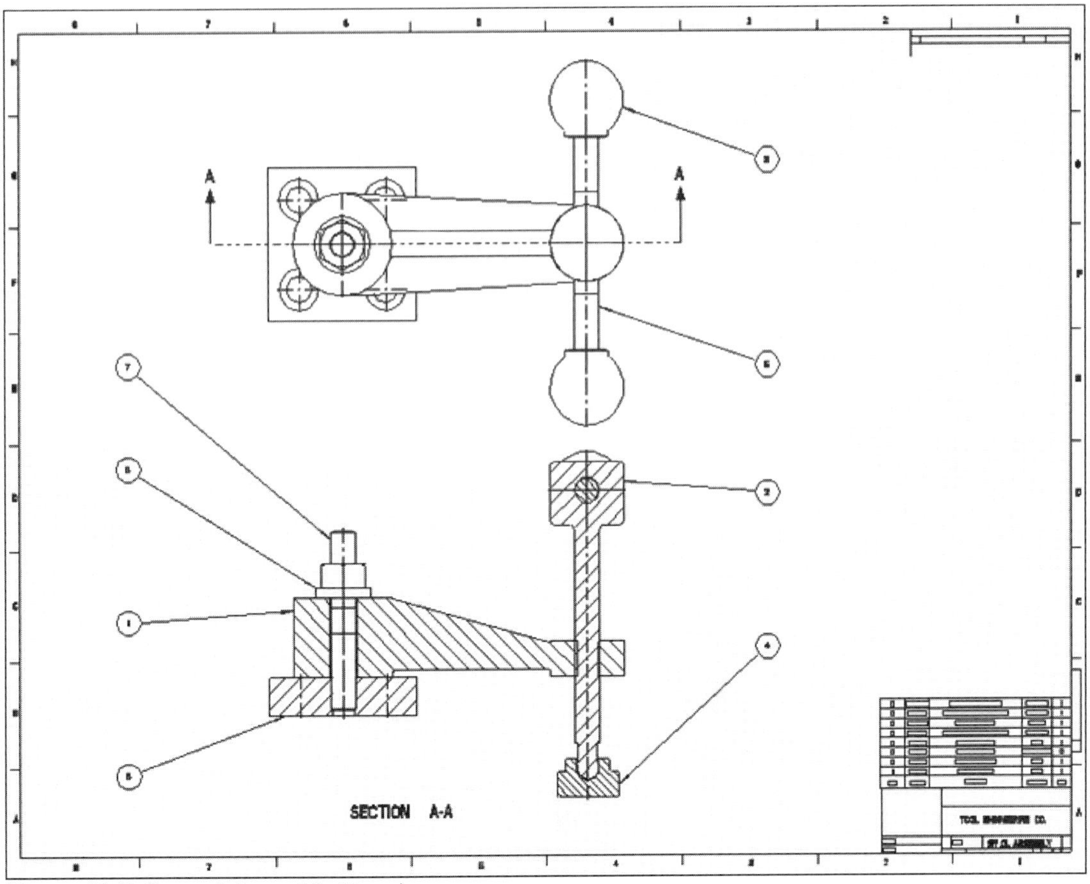

Figure 30 Ballooned Assembly Drawing

Drawing mode in parametric design provides you with the basic ability to document solid models in drawings that share a two-way associativity.

Changes that are made to the model in Part mode or Assembly mode will cause the drawing to update automatically and reflect the changes. Any changes made to the model in Drawing mode will be immediately visible on the model in Part and Assembly modes. The model shown in Figure 31 has been detailed in Figure 32.

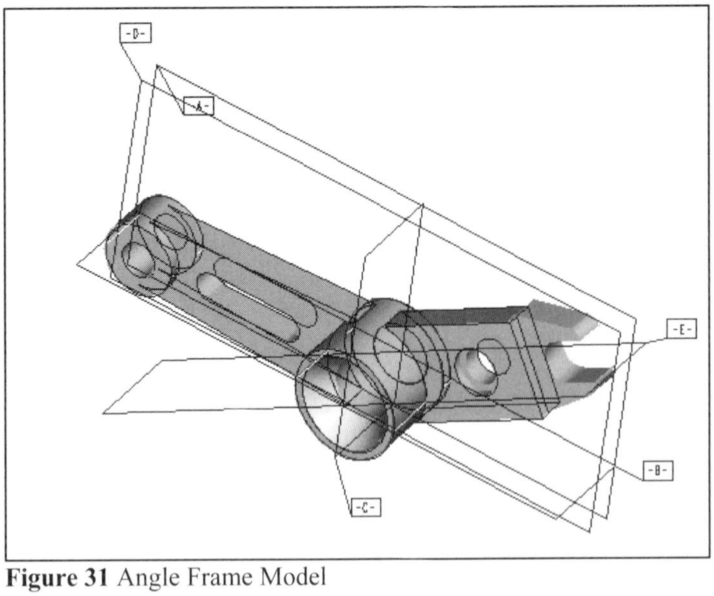

Figure 31 Angle Frame Model

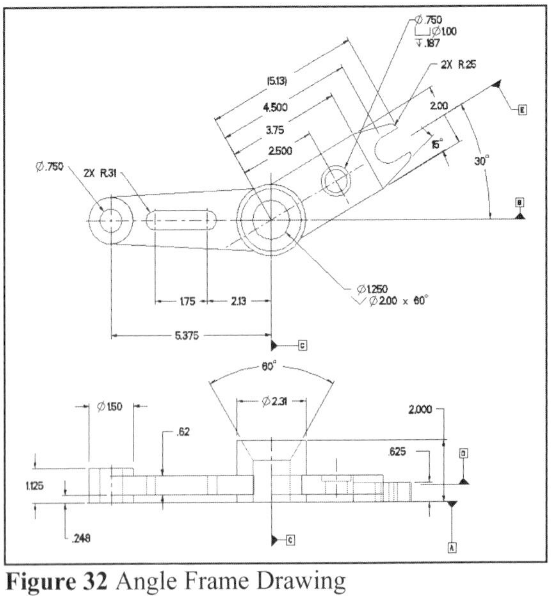

Figure 32 Angle Frame Drawing

Using the Text

The text utilizes a variety of command boxes and descriptions to lead you through construction sequences. Also, see the Pro/ENGINEER WILDFIRE 4.0 Quick Reference Cards at the end of this introduction. The following icons, symbols, shortcut keys, and conventions will be used *(command sequences are always in a box)*:

Click: **Tools** ⇒ Environment ⇒ Snap to Grid ⇒ Tangent Edges Dimmed ⇒ **Apply** ⇒ **OK** ⇒ **Revolve Tool** ⇒ **RMB** ⇒ **Define Internal Sketch** ⇒ Sketch Plane--- Plane: select **RIGHT** datum ⇒ **Reference: TOP** ⇒ Sketch ⇒ **Toggle the grid** on

Click: **Tools** ⇒ **Options** ⇒ Showing: Current Session ⇒ Show only options loaded from file ⇒ Slide the bar down to the option or type. Option: ***default_dec_places*** ⇒ Value: **3** ⇒ **Enter** ⇒ Slide the bar down to the option or type. Option: ***sketcher_dec_places*** ⇒ Value: **3** ⇒ **Enter** ⇒ Option: ***sketcher_starts_in_2d*** ⇒ Value: **no** ⇒ **Enter** ⇒ Option: (type) ***def_layer*** ⇒ Value: ⇒ **layer_axis datum_axes** ⇒ **Add/Change** ⇒ **Apply** ⇒ Show only options loaded from file ⇒ **Save a copy of the currently displayed configuration file** ⇒ Name **clamp.pro** ⇒ **Ok** ⇒ **Close** ⇒ **Create a new object** ⇒ Part ⇒ Name CLAMP_FOOT ⇒ Use default template ⇒ **OK** ⇒ **Edit** ⇒ **Setup** ⇒ **Units** ⇒ Inch lbm Second (Pro/E Default) ⇒ **Close** ⇒ **MMB**

Commands:

- ⇒ — Continue with command sequence or screen picks using **LMB**
- **Create lines** — icon (with description) indicates command to pick using **LMB**

Mouse or keyboard terms used in this text:

- **LMB** — **L**eft **M**ouse **B**utton
  - or "**Pick**" — term used to direct an action (i.e., "Pick the surface")
  - or "**Click**" — term used to direct an action (i.e., "Click on the icon")
  - or "**Select**" — term used to direct an action (i.e., "Select the feature")
- **MMB** — **M**iddle **M**ouse **B**utton (accept the current selection or value)
  - or **Enter** — press **Enter** key to accept entry
  - or ✓ — Click on this icon to accept entry
- **RMB** — **R**ight **M**ouse **B**utton (toggles to next selection, or provides a list of available commands)

Shortcut Keys:

Ctrl+A (Activate)	**Ctrl+C** (Copy)	**Ctrl+D** (Standard Orientation)
Ctrl+F (Find)	**Ctrl+G** (Regenerate)	**Ctrl+N** (New)
Ctrl+O (Open)	**Ctrl+P** (Print)	**Ctrl+R** (Repaint)
Ctrl+S (Save)	**Ctrl+V** (Paste)	**Ctrl+Y** (Redo)
Ctrl+Z (Undo)		

Text Organization

Text Lessons (Parts, Assemblies, and Drawings)

- **Lesson 1** has you complete two simple parts, an assembly, and a drawing using default settings.
- **Lesson 2** introduces Pro/ENGINEER's Wildfire 4.0 interface and embedded Browser.
- **Lesson 3** provides uncomplicated instructions to model a variety of simple-shaped parts.
- **Lessons 4** and **5** involve part modeling, using a variety of commands and tools.
- **Lessons 6** through **12** involve modeling the parts, creating the assembly [Fig. 33(a)], and documenting a design with detail and assembly drawings [Fig. 33(b)].
- **Lessons 13-18** provide instructions to model various parts using other features.

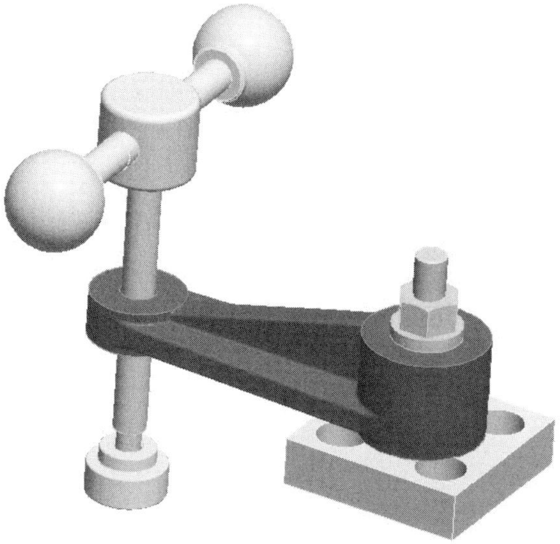

Figure 33(a) Clamp Assembly

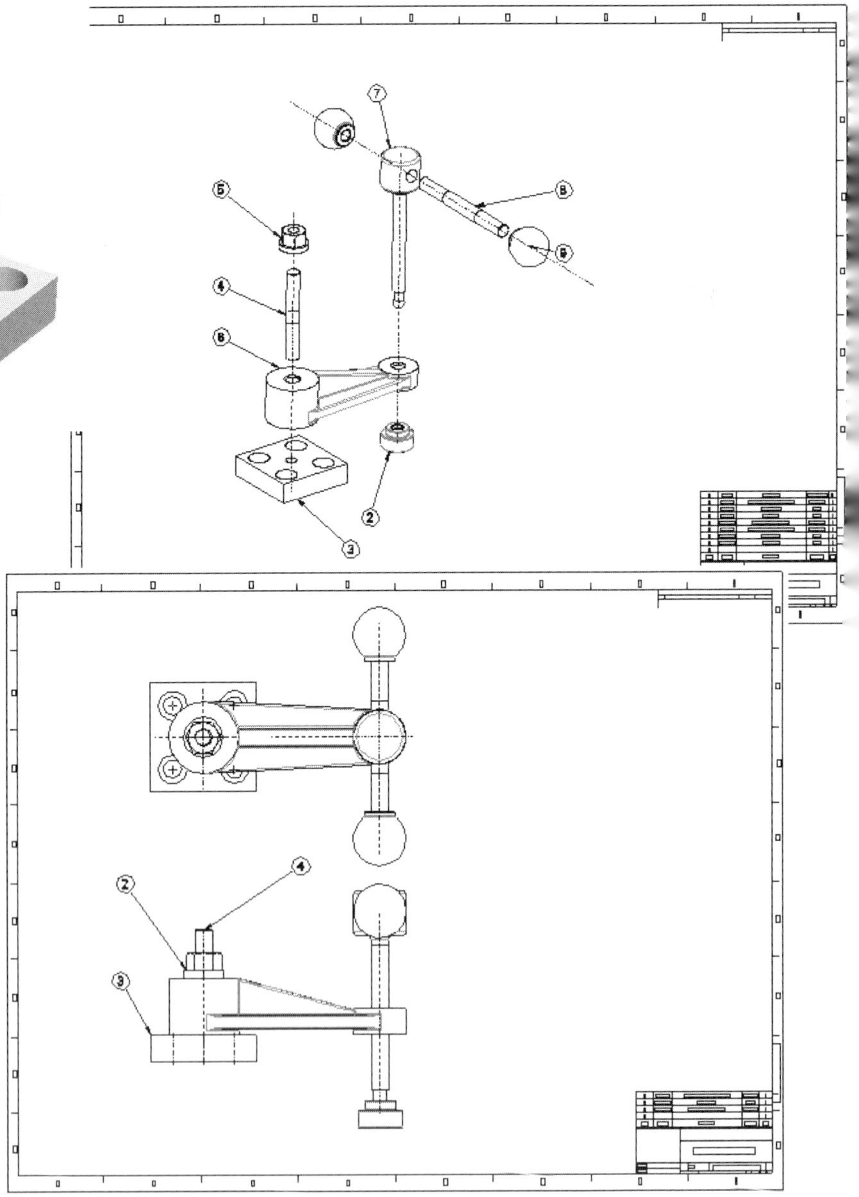

Figure 33(b) Clamp Assembly Drawing

Pro/ENGINEER Wildfire 4.0 Quick Reference Cards

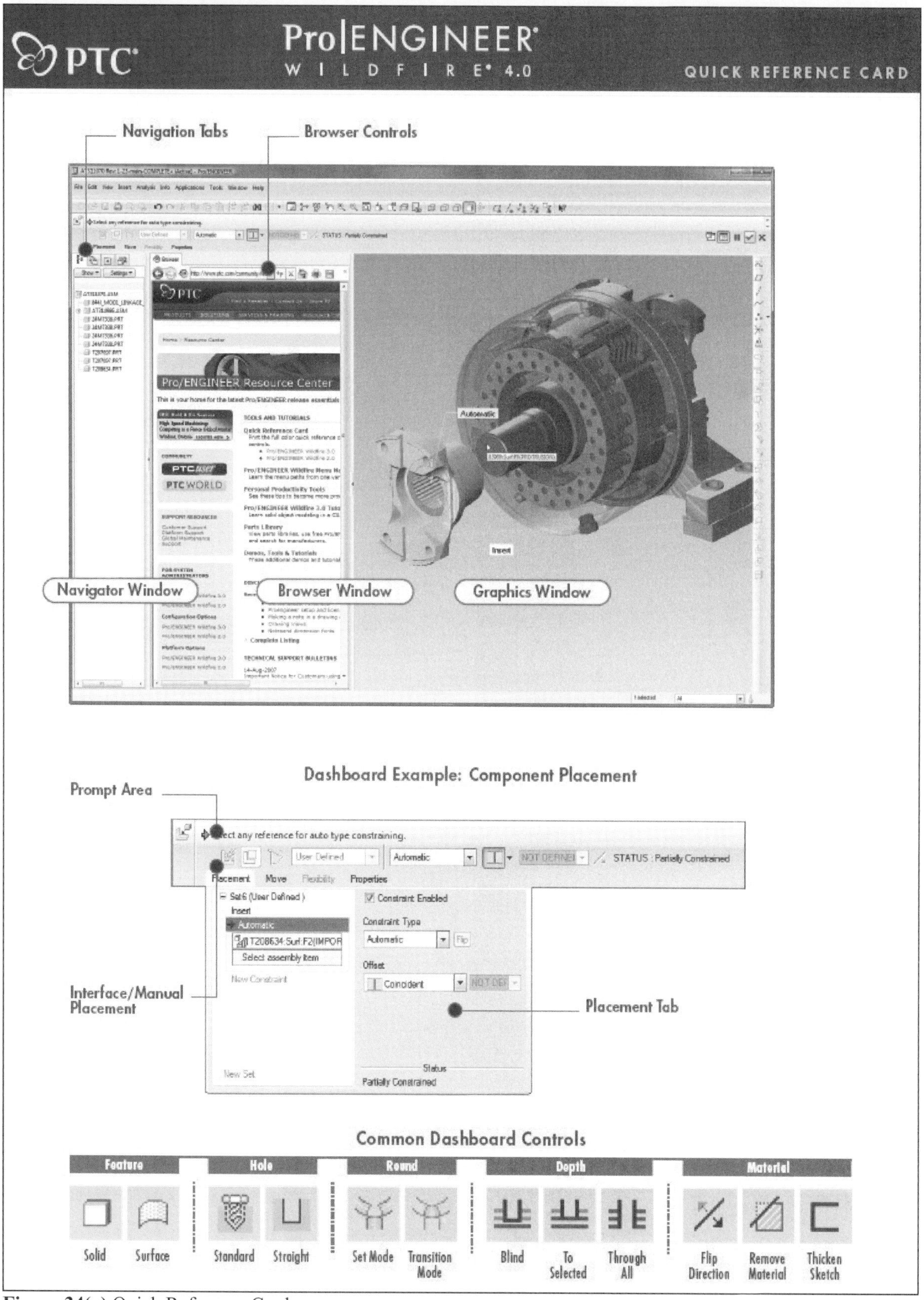

Figure 34(a) Quick Reference Card

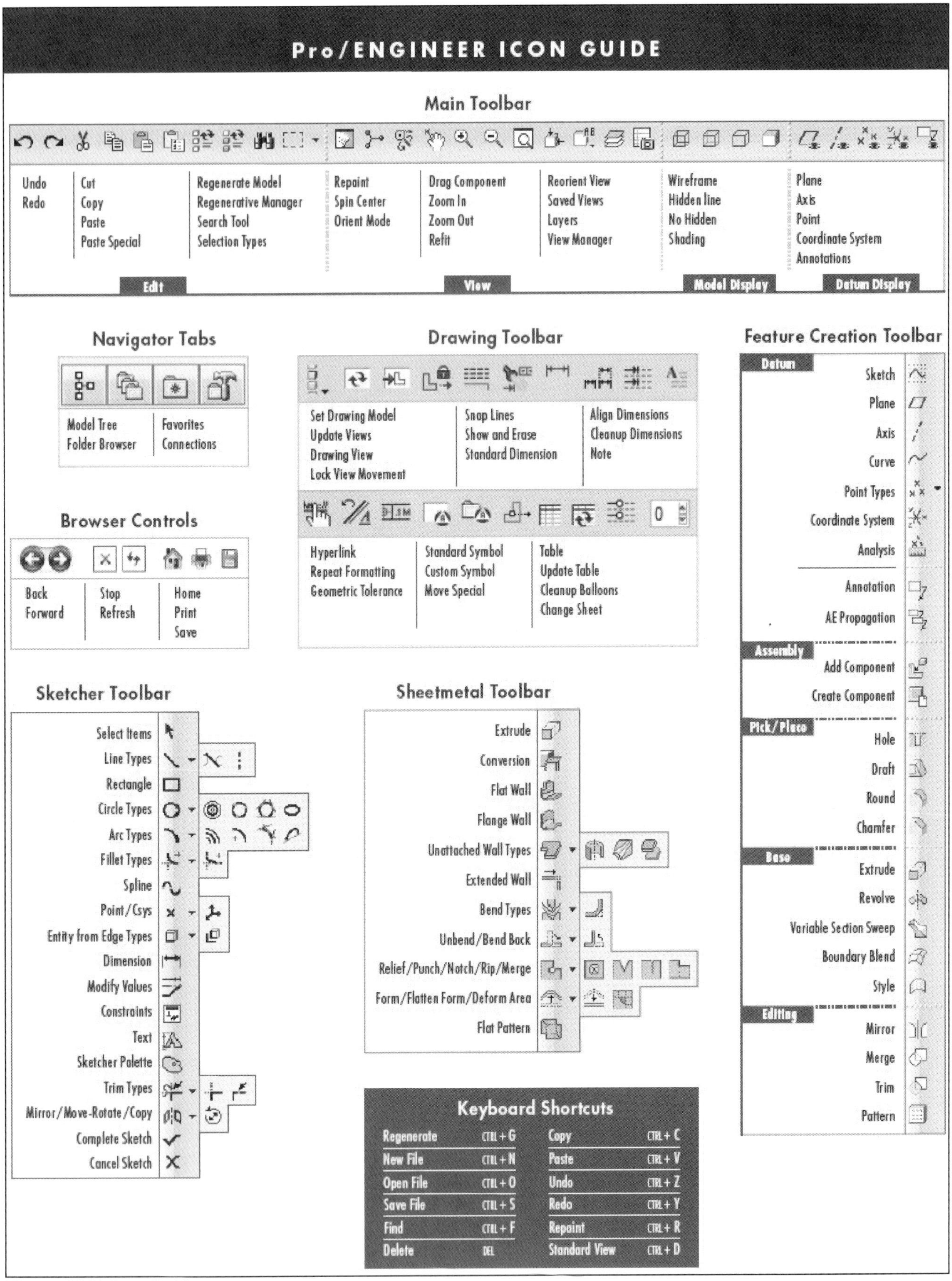

Figure 34(b) Icon Guide

MAKING SELECTIONS

System Color Assignments

Cyan	**Preselection Highlight**	Item will be added to or removed from the set of selected items
Red	**Selected Geometry**	Items currently selected
Yellow	**Preview Geometry**	Results of the current operation when complete

Making Selections

Mouse Controls	
Highlight Geometry	Over Geometry
Query to Next Item	Until Highlighted
Select Highlighted Geometry	
Add or Remove Items from Selection	CTRL +
Construct Chains or Surface Sets	SHIFT +
Clear Selection	On Background

USING FILTERS

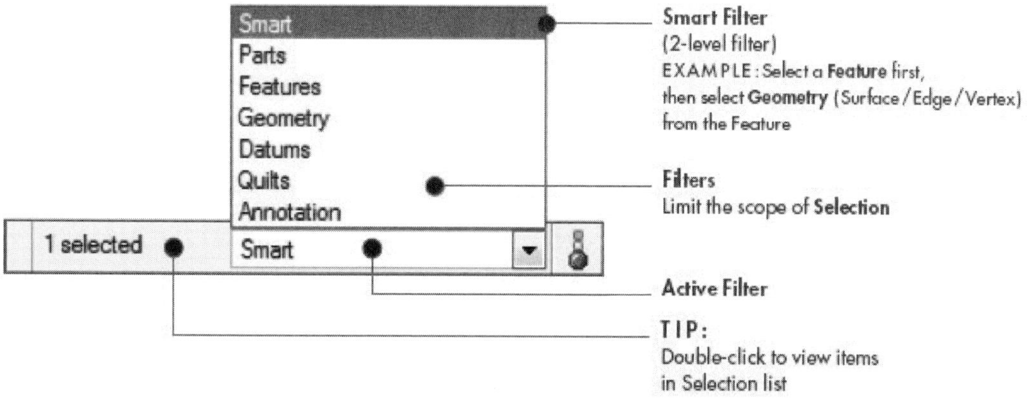

Figure 34(c) Making Selections

ADVANCED SELECTION: Chain and Surface Set Construction

DEFINITIONS

General Definitions

Chain
A collection of adjacent edges and curves that share common endpoints. Chains can be open-ended or closed-loop, but they are always defined by two ends.

Surface Set
A collection of surface patches from solids or quilts. The patches do not need to be adjacent.

Methods of Construction

Individual
Constructed by selecting individual entities (edges, curves, or surface patches) one at a time. This is also called the One-by-One method.

Rule-Based
Constructed by first selecting an anchor entity (edge, curve, or surface patch), and then automatically selecting its neighbors (a range of additional edges, curves, or surface patches) based on a rule. This is also called the Anchor/Neighbor method.

CONSTRUCTING CHAINS

Individual Chains

One-by-One
To select adjacent edges one at a time along a continuous path:

1 Select an edge
2 Hold down SHIFT
3 Select adjacent edges
4 Release SHIFT

Rule-Based Chains

Tangent
To select all the edges that are tangent to an anchor edge:

1 Select an edge
2 Hold down SHIFT
3 Highlight **Tangent** chain (Query may be required)
4 Select tangent chain
5 Release SHIFT

Boundary
To select the outermost boundary edges of a quilt:

1 Select a one-sided edge of a quilt
2 Hold down SHIFT
3 Highlight **Boundary** chain (Query may be required)
4 Select boundary chain
5 Release SHIFT

Surface Loop
To select a loop of edges on a surface patch:

1 Select an edge
2 Hold down SHIFT
3 Highlight **Surface** chain (Query may be required)
4 Select surface loop
5 Release SHIFT

From-To
To select a range of edges from a surface patch or a quilt:

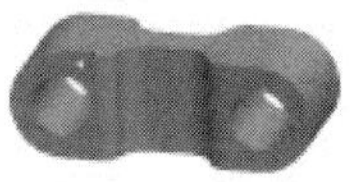

1 Select the **From** edge

2 Hold down SHIFT

3 Query to highlight the desired **From-To** chain

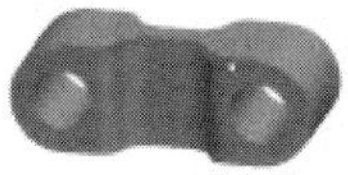

4 Select From-To chain

5 Release SHIFT

Multiple Chains

1 Construct initial chain
2 Hold down CTRL
3 Select an edge for new chain
4 Release CTRL
5 Hold down SHIFT
6 Complete new chain from selected edge

Figure 34(d) Advanced Selection

Pro|ENGINEER®
WILDFIRE® 4.0

CONSTRUCTING SURFACE SETS

Individual Surface Sets

Single Surfaces
To select multiple surface patches from solids or quilts one at a time:

1 Select a surface patch
2 Hold down CTRL
3 Select additional patches (Query may be required)
4 Release CTRL

Rule-Based Surface Sets

Solid Surfaces
To select all the surface patches of solid geometry in a model:

1 Select a surface patch on solid geometry
2 Right-click and select **Solid Surfaces**

Quilt Surfaces
To select all the surface patches of a quilt:

1 Select a surface feature
2 Select the corresponding quilt

Loop Surfaces
To select all the surface patches that are adjacent to the edges of a surface patch:

1 Select a surface patch
2 Hold down SHIFT
3 Place the pointer over an edge of the patch to highlight the **Loop Surfaces**
4 Select the Loop Surfaces (The initial surface patch is de-selected)
5 Release SHIFT

Seed and Boundary Surfaces
To select all surface patches, from a **Seed** surface patch up to a set of **Boundary** surface patches:

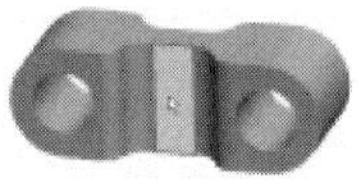

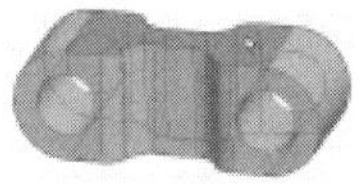

1 Select the **Seed** surface patch
2 Hold down SHIFT
3 Select one or more surface patches to be used as boundaries
4 Release SHIFT (All surfaces from the Seed up to the Boundaries are selected)

Excluding Surface Patches from Surface Sets

To exclude surface patches during or after construction of a surface set:

1 Construct a surface set
2 Hold down CTRL
3 Highlight a patch from the surface set
4 Select the patch to de-select it
5 Release CTRL

CONSTRUCTING CHAINS AND SURFACE SETS USING DIALOG BOXES

To explicitly construct and edit Chains and Surface Sets, click **Details** next to a collector:

Chain Dialog Box

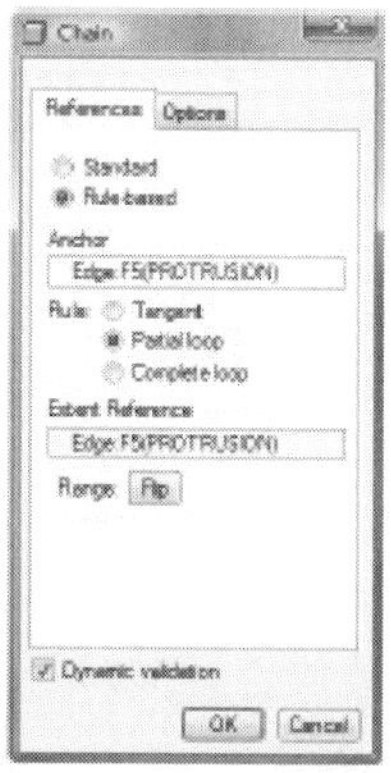

Dashboard Collector

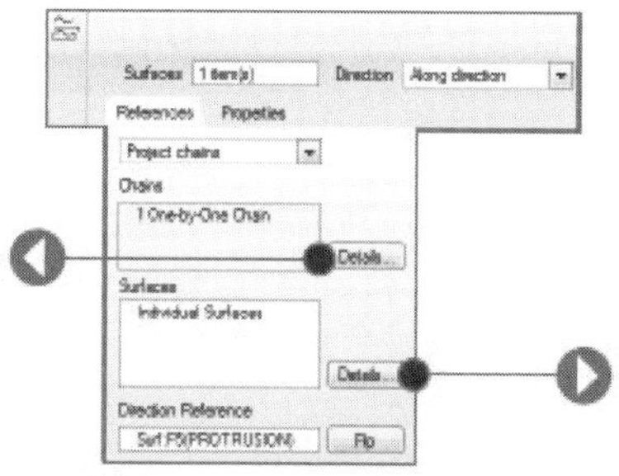

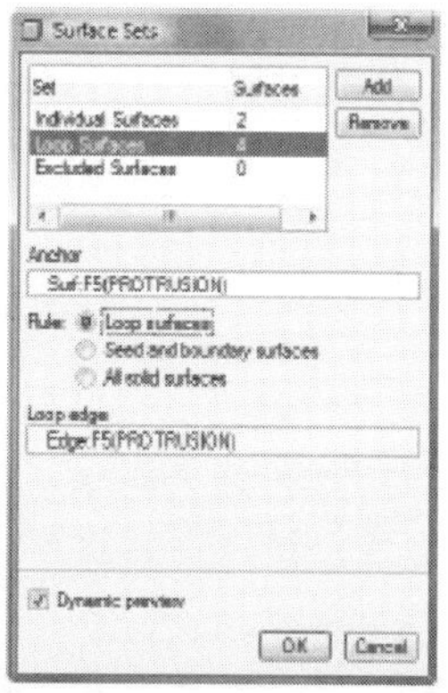

Surface Set Dialog Box

Figure 34(e) Constructing Surface Sets

ORIENTING THE MODEL

DYNAMIC VIEWING

3D Mode

Hold down the key and button. Drag the mouse.

SPIN	
PAN	SHIFT +
ZOOM	CTRL +
TURN	CTRL +

2D Mode

PAN	
ZOOM	CTRL +

2D and 3D Modes

Hold down the key and roll the mouse wheel.

ZOOM	
FINE ZOOM	SHIFT +
COARSE ZOOM	CTRL +

Using the Spin Center

Click the icon in the Main Toolbar to enable the Spin Center.

- Enabled – The model spins about the location of the spin center
- Disabled – The model spins about the location of the mouse pointer

Using Orient Mode

Click the icon in the Main Toolbar to enable Orient mode.

- Provides enhanced Spin/Pan/Zoom Control
- Disables selection and highlighting
- Right-click to access additional orient options
- Use the shortcut: CTRL + SHIFT + Middle-click

Using Component Drag Mode in an Assembly

Click the icon in the Main Toolbar to enable Component Drag mode.

- Allows movement of components based on their kinematic constraints or connections
- Click a location on a component, move the mouse, click again to stop motion.
- Middle-click to disable Component Drag mode

COMPONENT PLACEMENT CONTROLS

Allows reorientation of components during placement

COMPONENT DRAG	CTRL + ALT +
SPIN	CTRL + ALT +
MOVE	CTRL + ALT +

Object Mode

Provides enhanced Spin/Pan/Zoom Control:

1 Enable Orient mode
2 Right-click to enable Orient Object mode
3 Use Dynamic Viewing controls to orient the component
4 Right-click and select Exit Orient mode

DOC-RC60388-EN-360

Figure 34(f) Orienting the Model

Lesson 1 Pro/ENGINEER Wildfire 4.0 Overview

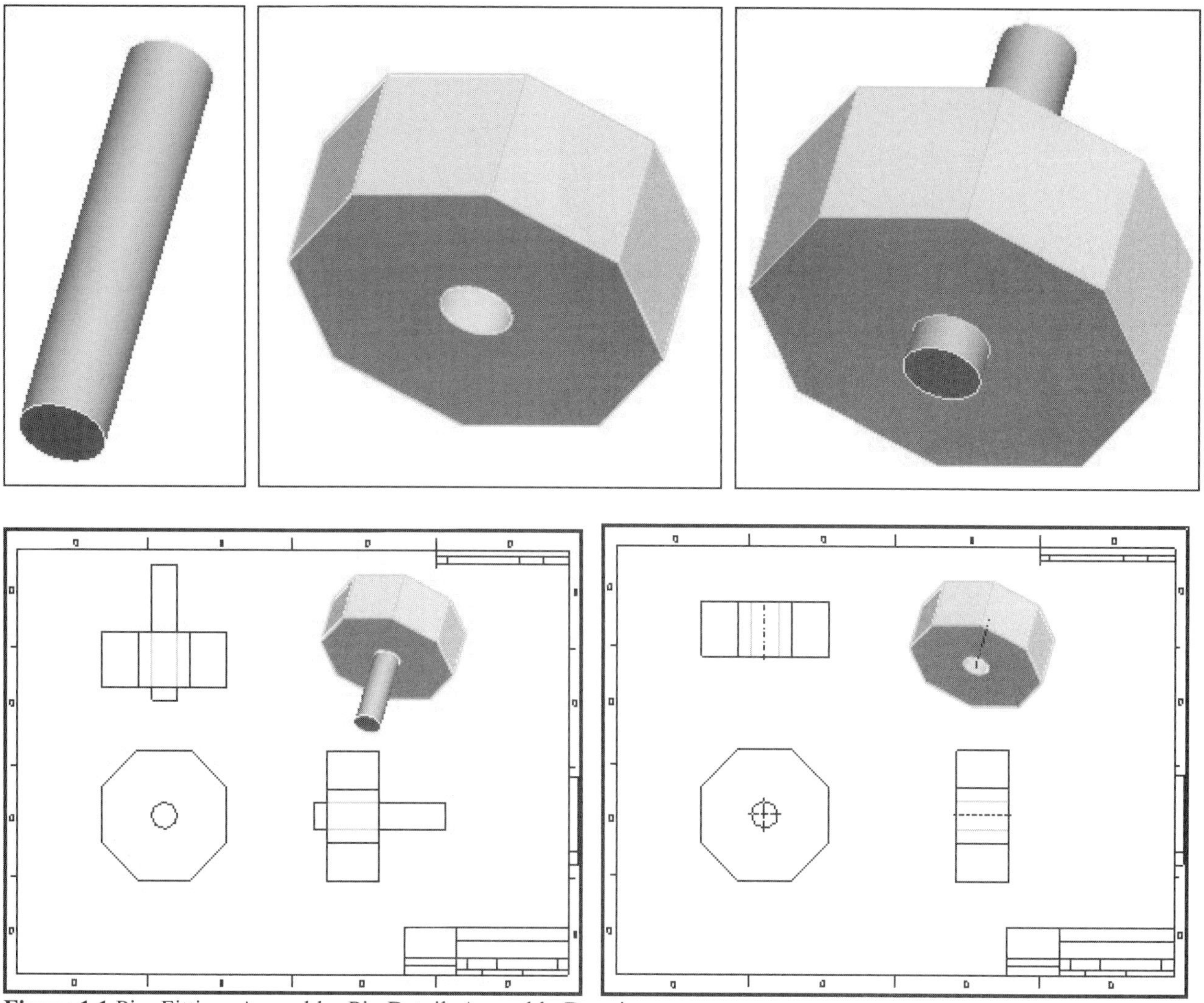

Figure 1.1 Pin, Fitting, Assembly, Pin Detail, Assembly Drawing

OBJECTIVES

- Create two parts
- Assemble parts to create an assembly
- Create a part drawing
- Create an assembly drawing

Pro/ENGINEER Wildfire 4.0 Overview

This lesson will allow you to experience the part, assembly, and drawing modes (Fig. 1.1) of Pro/ENGINEER Wildfire 4.0 by creating two simple parts (Figs. 1.2 and 1.5), assembling them, and creating drawings.

A minimum of explanation and instruction is provided here. This lesson will quickly get you up and running on Pro/ENGINEER Wildfire 4.0.

Lesson 1 STEPS

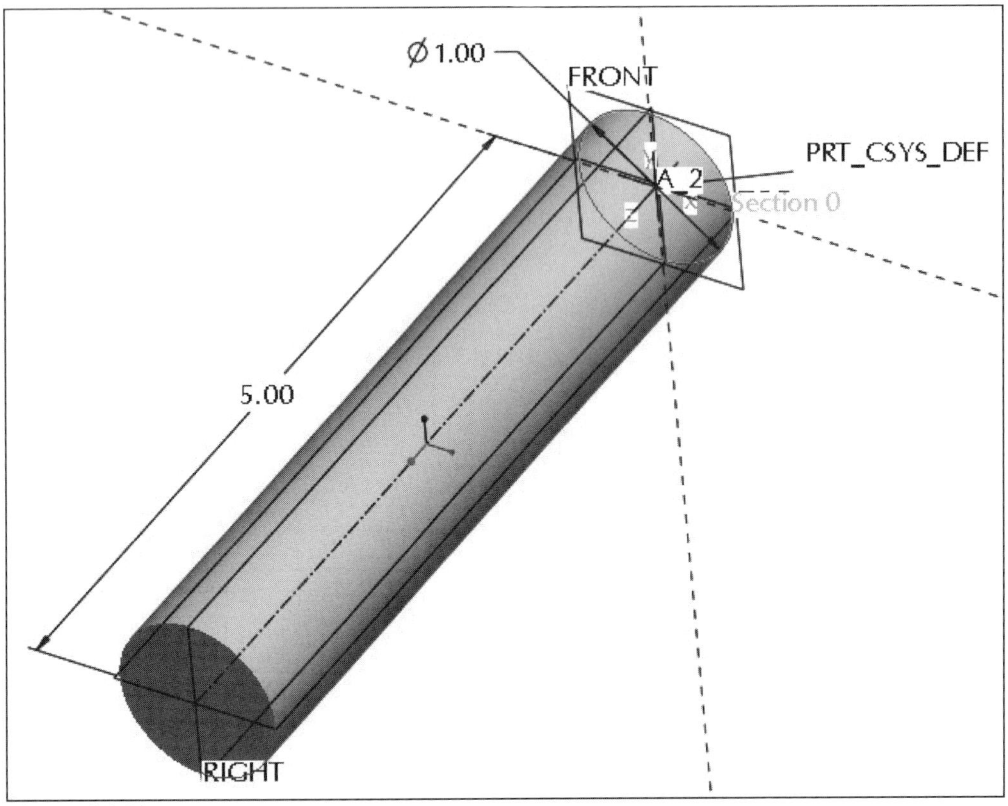

Figure 1.2 Pin

Creating the Pin Part

In this lesson most commands and picks will be accompanied by the window or dialog box that will open as the command is initiated. After this lesson, the commands will not show every window and dialog box, as this would make the text extremely long. Appropriate illustrations will be provided.

Throughout the text, a box surrounds all commands and menu selections.

Open **Pro/ENGINEER Wildfire 4.0** using a shortcut icon on your Desktop 

*(or with WINDOWS, click: **Start ⇒ Programs ⇒ proewildfire4.0**)* Pro/E will open on your computer.

File ⇒ Set Working Directory 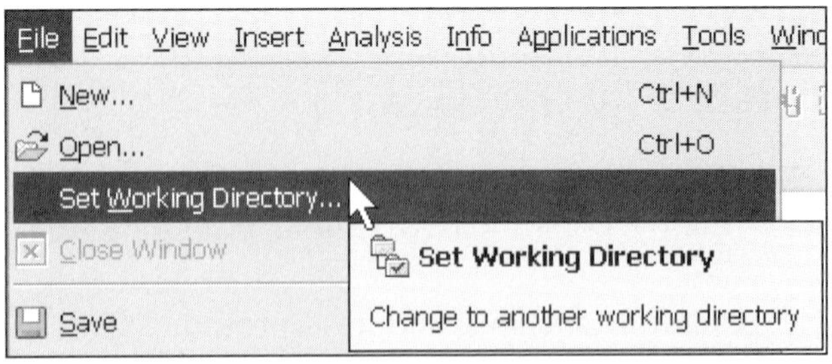

⇒ select the desired working directory or accept the default directory [Fig. 1.3(a)] ⇒ **OK** ⇒  **Create a new object** from Top Toolchest ⇒ Part ⇒ Name **PIN** ⇒ Use default template [Fig. 1.3(b)] ⇒ **OK** [Fig. 1.3(c)]

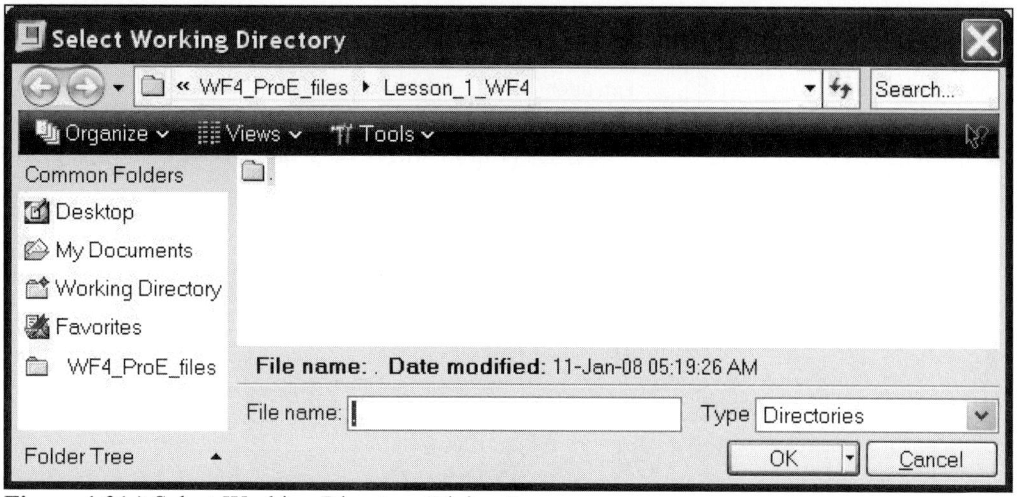

Figure 1.3(a) Select Working Directory Dialog Box

Figure 1.3(b) New Dialog Box, Part

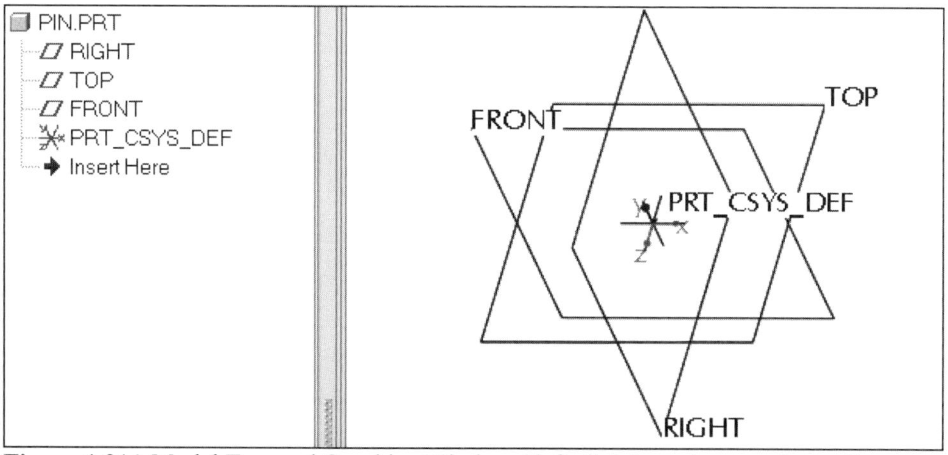

Figure 1.3(c) Model Tree and Graphics Window (default datum planes and coordinate system)

Click: **Extrude Tool** from Right Toolchest 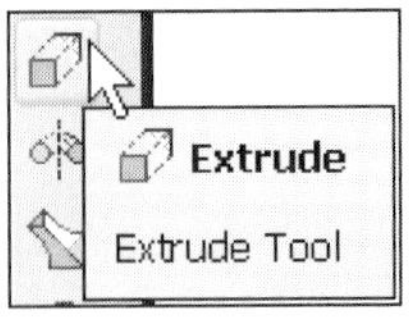

[Fig. 1.4(a)] ⇒ **Placement** tab ⇒ **Define** [Fig. 1.4(b)]

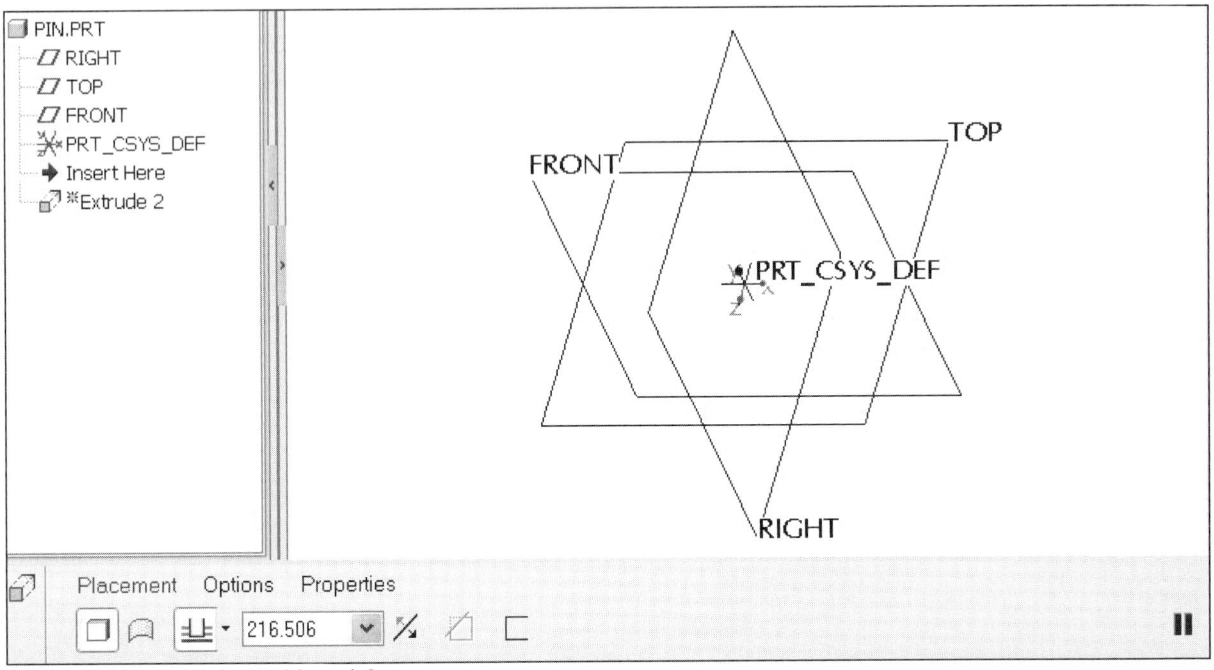

Figure 1.4(a) Extrude Dashboard Opens

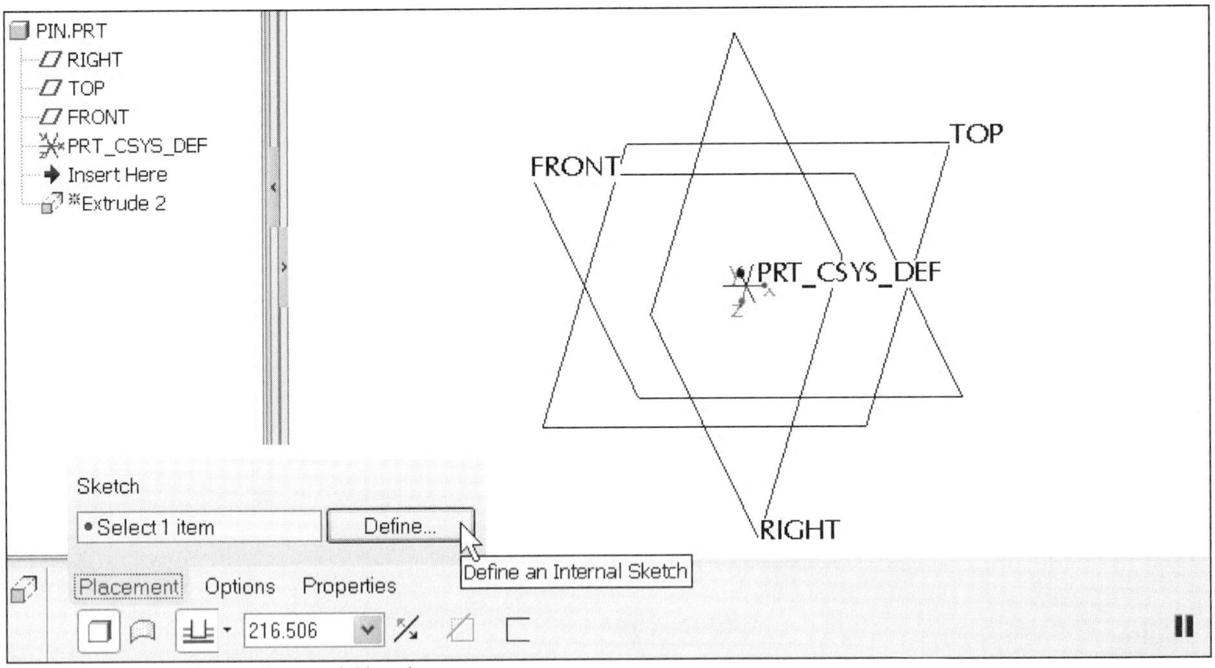

Figure 1.4(b) Define an Internal Sketch

Select the **FRONT** datum plane from the graphics window (or Model Tree) [Fig. 1.4(c)]. The FRONT datum plane is the Sketch Plane and the RIGHT datum plane is automatically selected as the Sketch Orientation Reference [Fig. 1.4(d)].

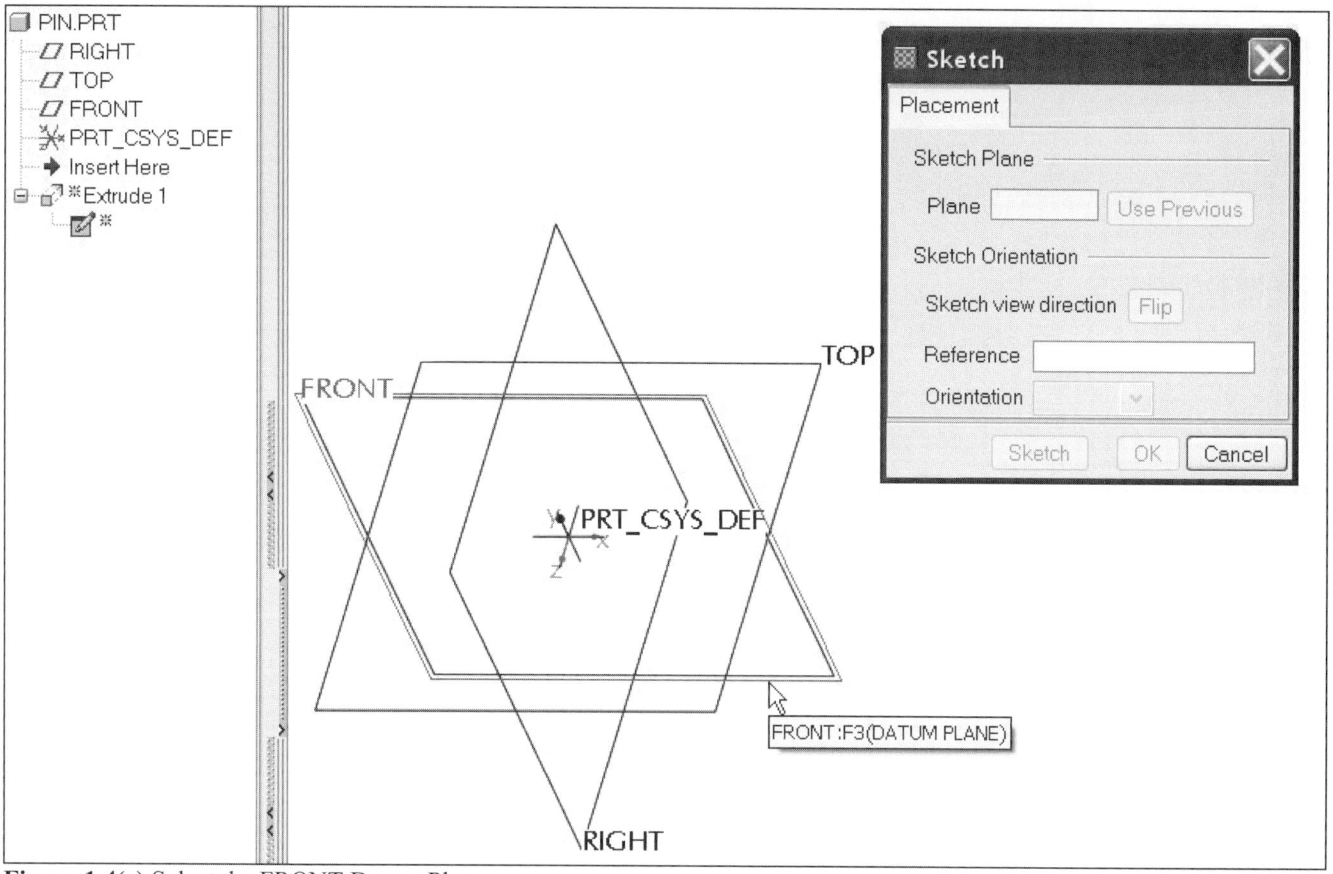

Figure 1.4(c) Select the FRONT Datum Plane

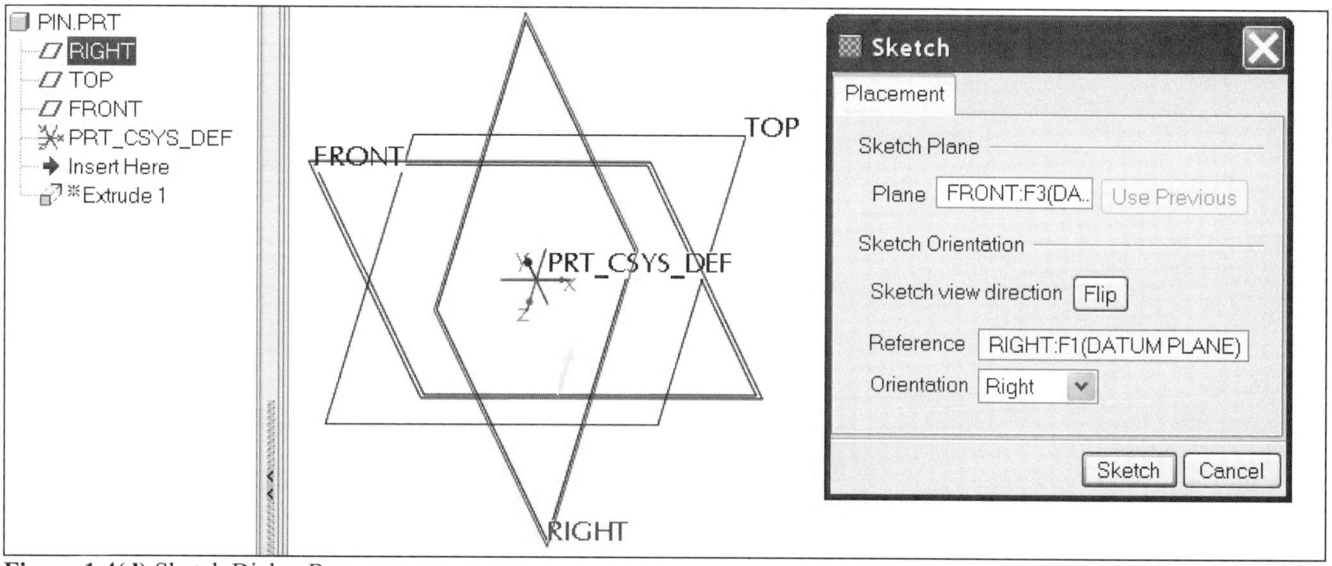

Figure 1.4(d) Sketch Dialog Box

Click: **Sketch** [Fig. 1.4(e)] ⇒  **Create circle by picking the center and a point on the circle** from Right Toolchest [Fig. 1.4(f)]

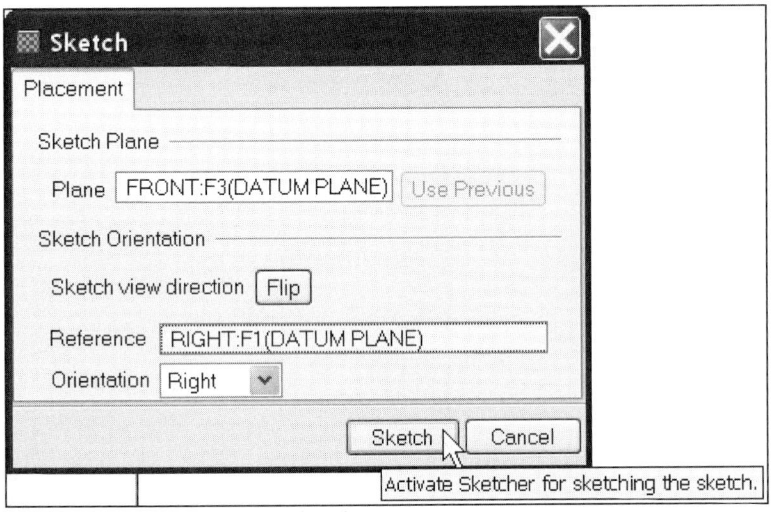

Figure 1.4(e) Activate the Sketcher

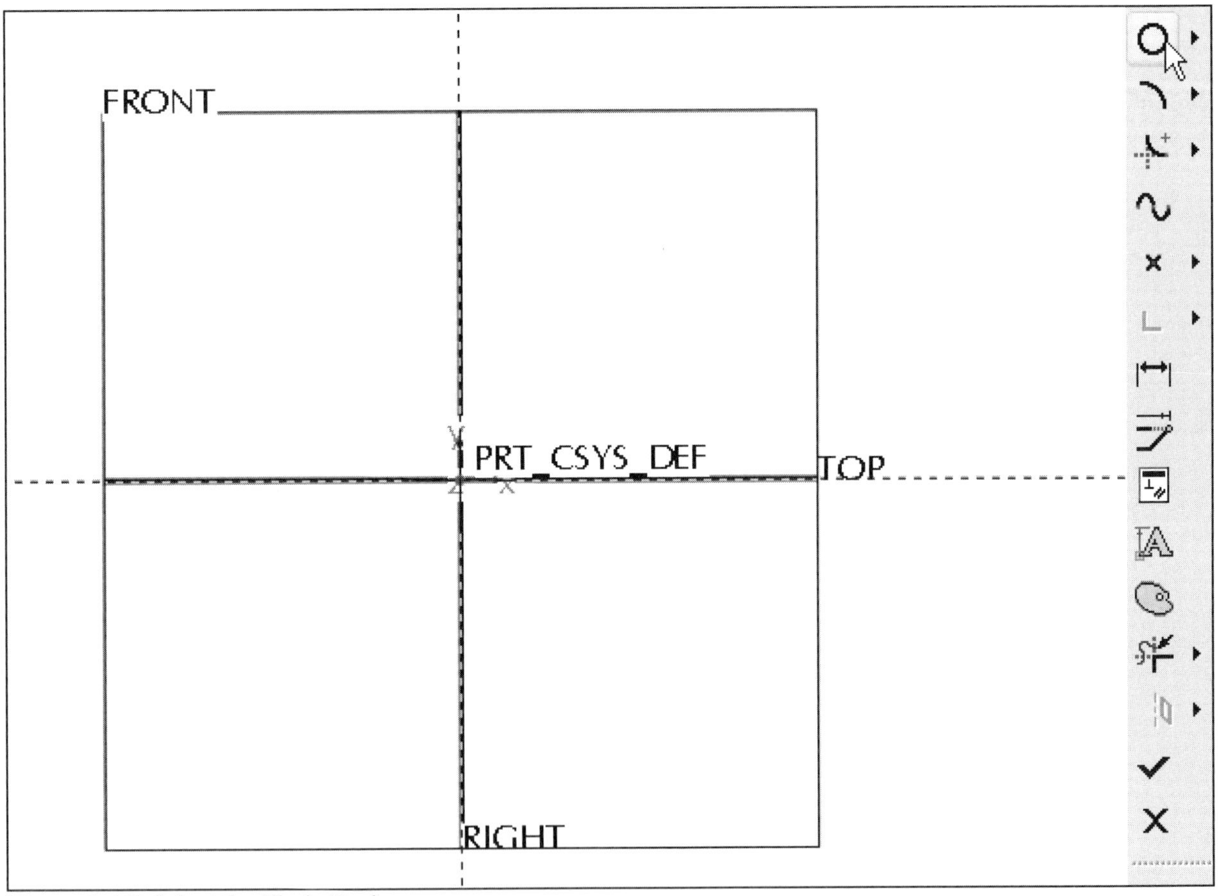

Figure 1.4(f) Select the Circle Tool

Pick the origin for the circle's center [Fig. 1.4(g)] ⇒ pick a point on the circle's edge [Fig. 1.4(h)] ⇒ **MMB** (**M**iddle **M**ouse **B**utton) [Figs. 1.4(i-l)] ⇒ double-click on the dimension ⇒ type **1.00** ⇒ **Enter**

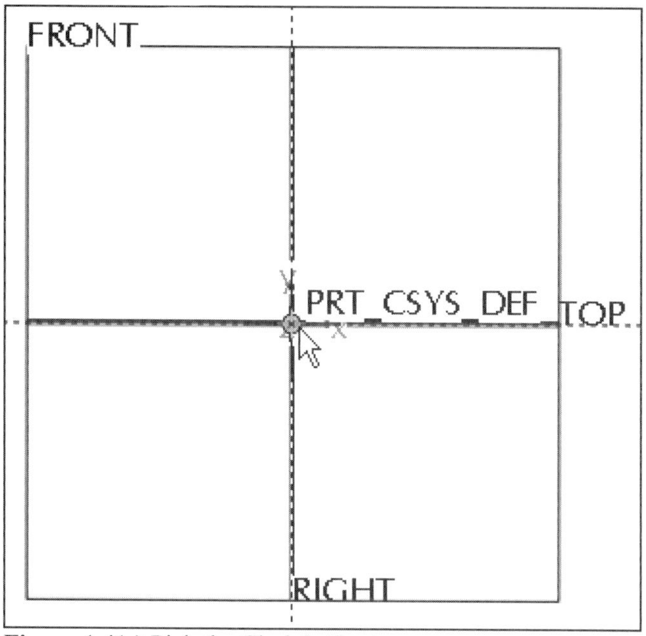

Figure 1.4(g) Pick the Circle's Center

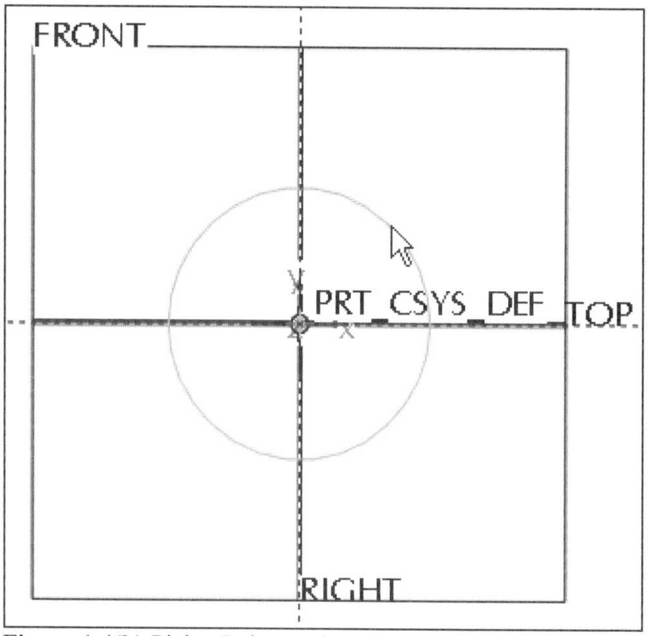

Figure 1.4(h) Pick a Point on the Circle's Edge

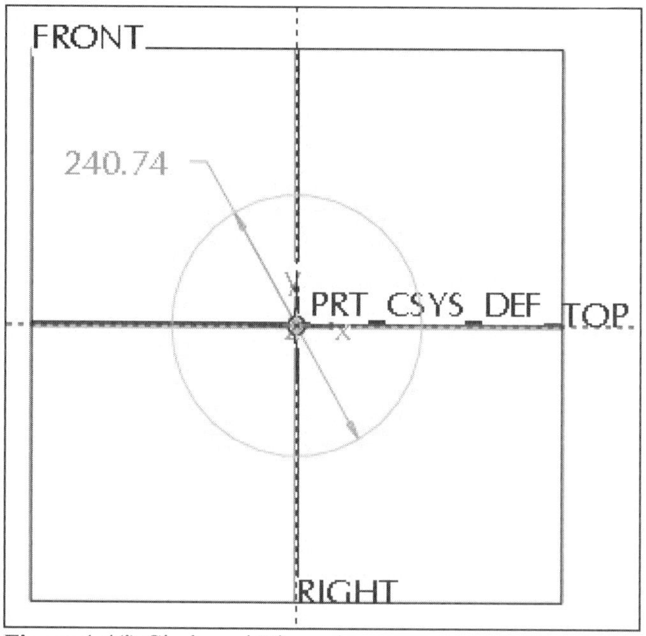

Figure 1.4(i) Circle and Dimension

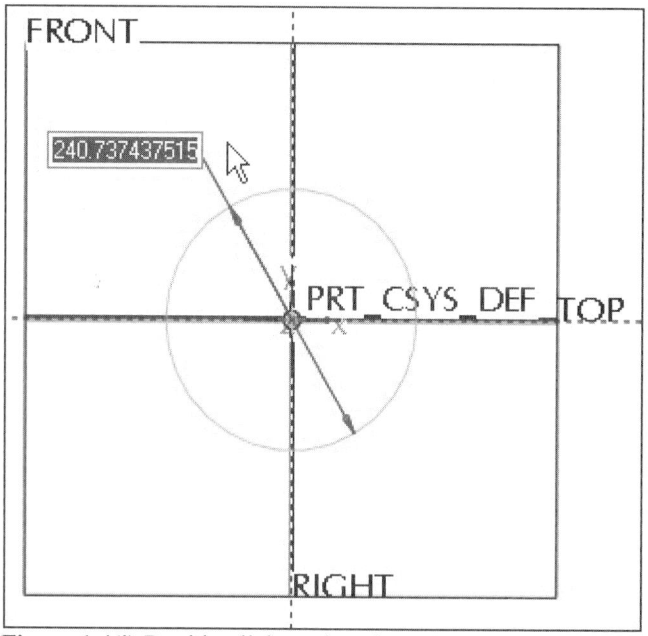

Figure 1.4(j) Double-click on the Dimension

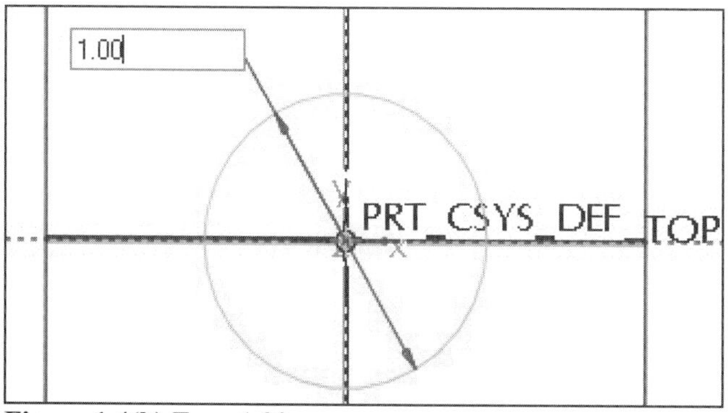

Figure 1.4(k) Type **1.00**

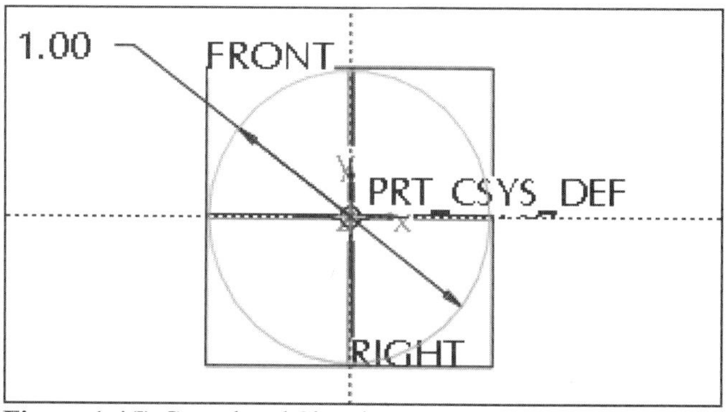

Figure 1.4(l) Completed Sketch

Click: Continue with the current section (✓ icon) from Right Toolchest [Fig. 1.4(m)] ⇒ click in value field [Fig. 1.4(n)] ⇒ type **5.00** [Fig. 1.4(o)] ⇒ **Enter** ⇒ (Saved View List icon) ⇒ **Standard Orientation** [Fig. 1.4(p)]

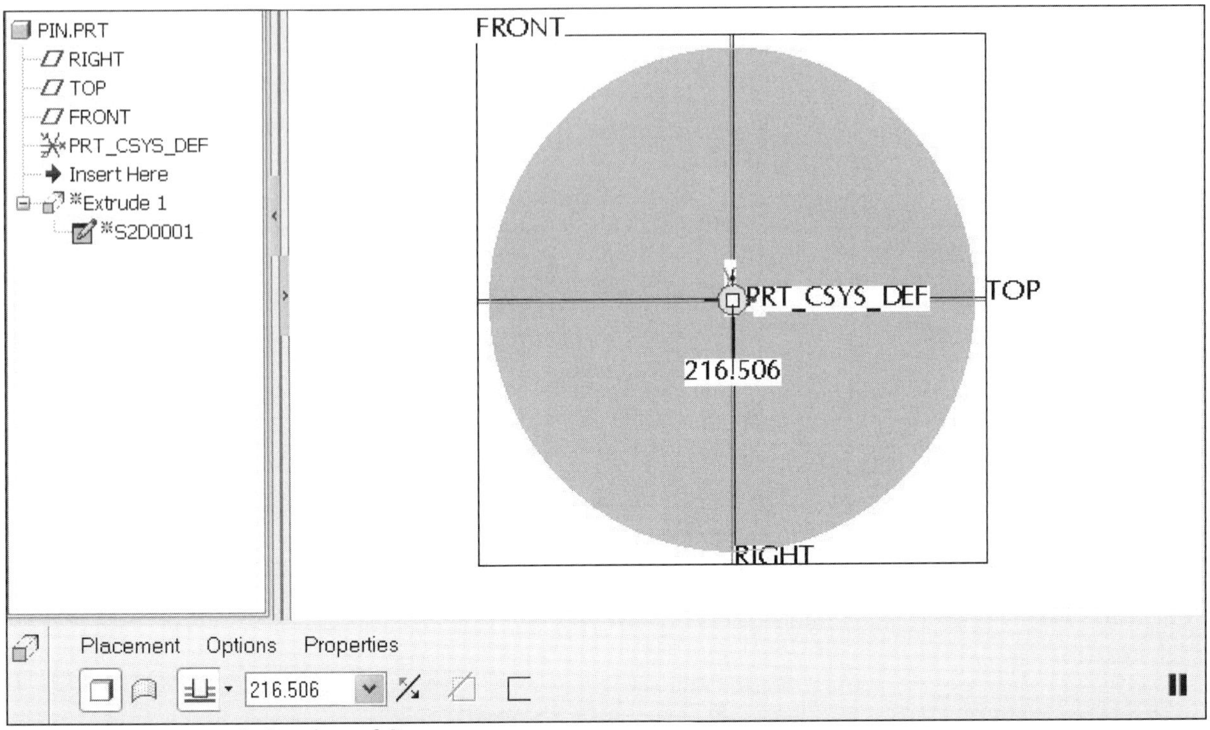

Figure 1.4(m) Dynamic Preview of Geometry

Figure 1.4(n) Extrude Dashboard Depth Value Field

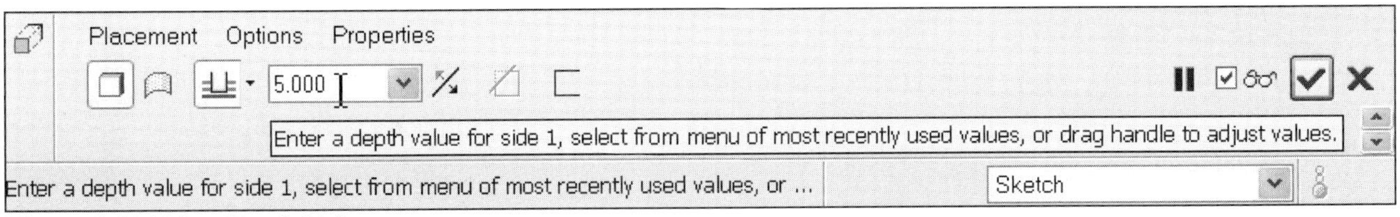

Figure 1.4(o) New Depth Value

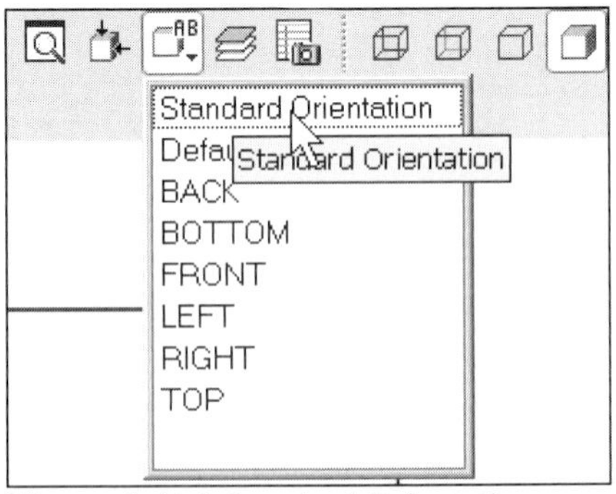

Figure 1.4(p) Standard Orientation

Click: **Refit object to fully display it on the screen** from Top Toolchest ⇒ hold down the **Ctrl** key and your **MMB**, move your mouse downward to zoom in [Fig. 1.4(q)] *[if you have a three-button mouse with a thumb wheel, simply rotate (not click) the thumb wheel to zoom]* ⇒ click ⇒ ⇒ **Shading** [Fig. 1.4(r)] ⇒ **File** ⇒ **Save** [Fig. 1.4(s)] ⇒ **MMB** *(or Enter)* The Model Tree displays the Extrude feature [Fig. 1.4(t)]

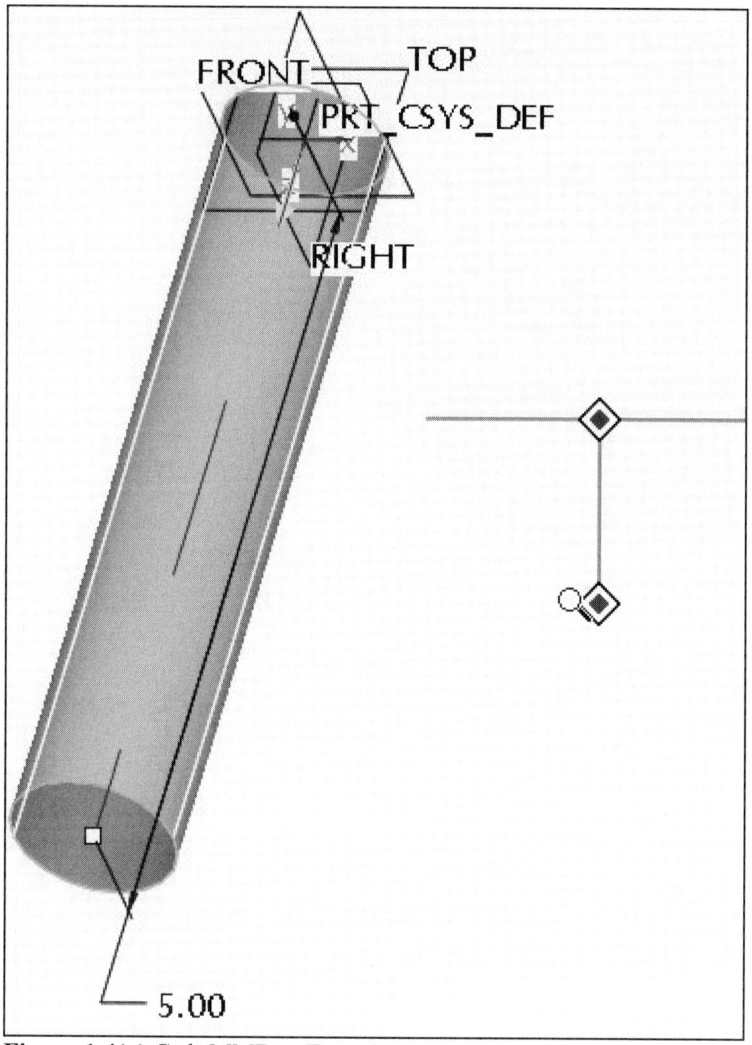

Figure 1.4(q) Ctrl+MMB to Zoom In

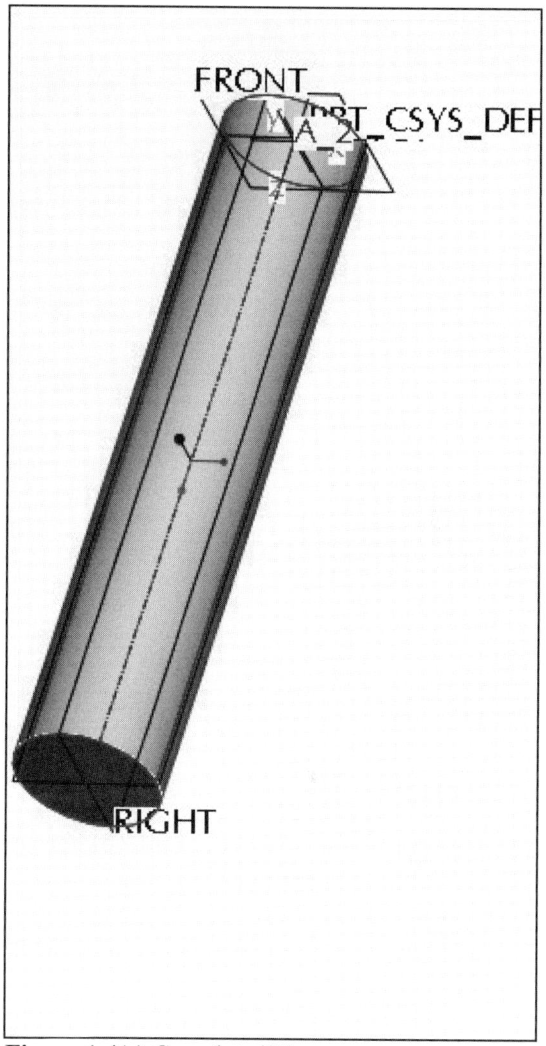

Figure 1.4(r) Completed Pin

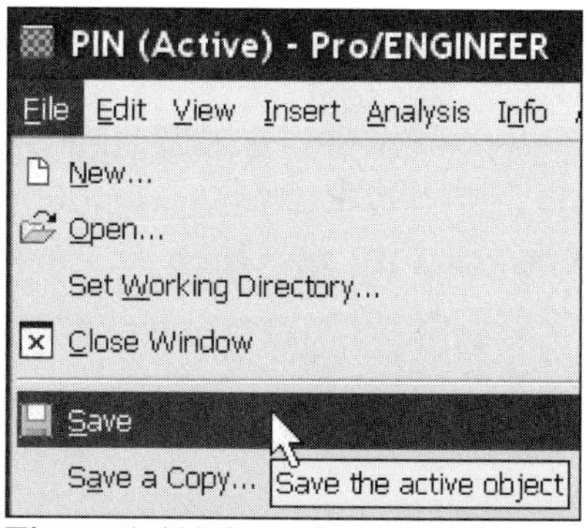

Figure 1.4(s) Save the active object

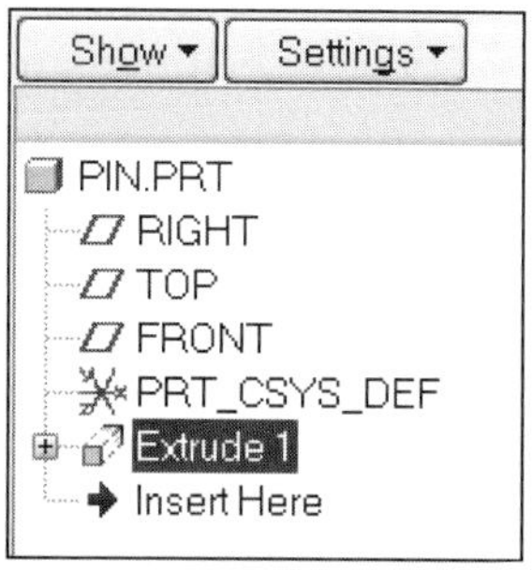

Figure 1.4(t) Model Tree

You have completed your first Pro/ENGINEER component.

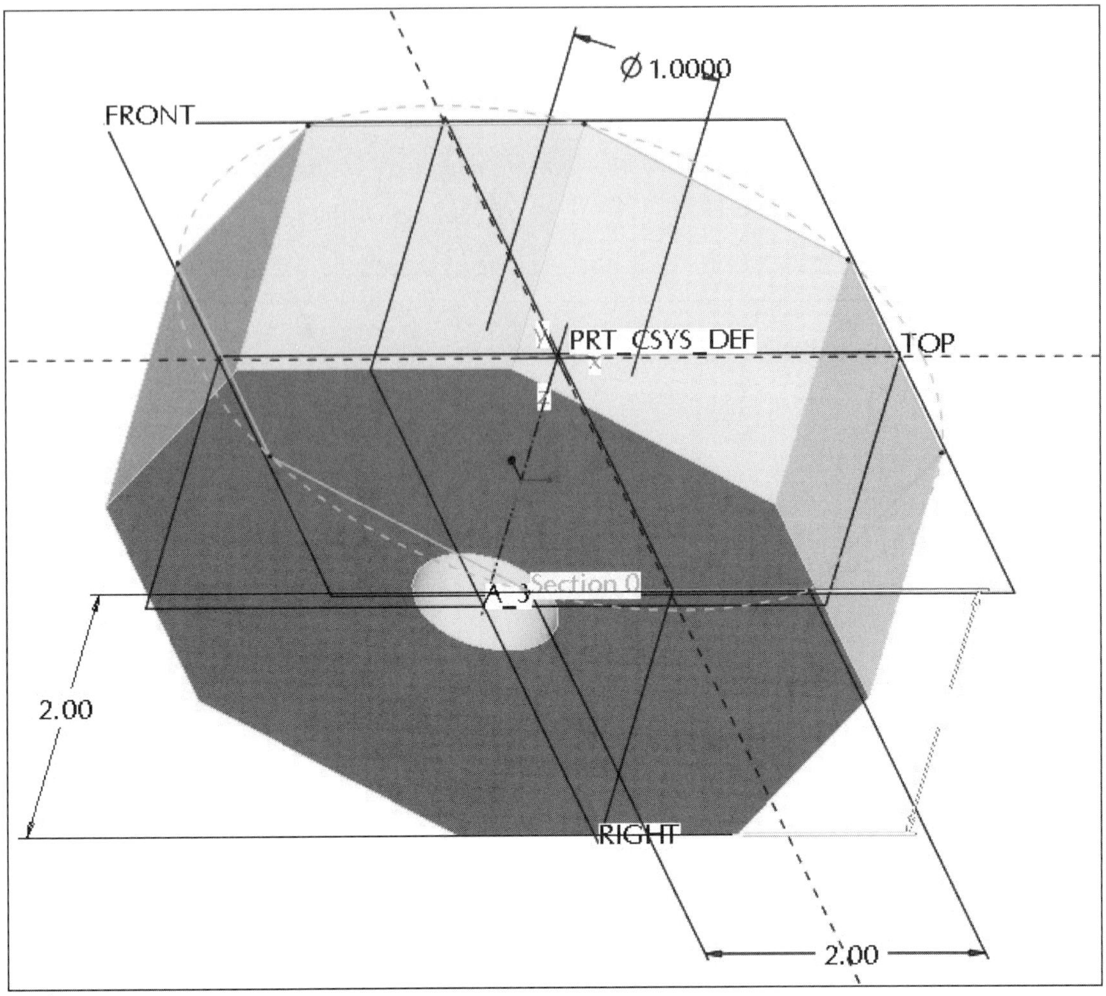

Figure 1.5 Plate

Creating the Plate Part

In this lesson most commands and picks will be accompanied by the window or dialog box that will open as the command is initiated.

After this lesson, the commands will not show every window and dialog box, as this would make the text extremely long. Appropriate illustrations will be provided.

If you are continuing from the last section, then skip this set of commands:

Open **Pro/ENGINEER Wildfire 4.0** using a shortcut icon on your Desktop 

⇒ **File** ⇒ **Set Working Directory** ⇒ select the working directory or accept the default directory (make sure this working directory is the same directory where the PIN was saved) ⇒ **OK** ⇒ **File** ⇒ **Open** ⇒ pick **pin.prt** ⇒ **Open** ⇒  **Shading**

Click: **Create a new object** ⇒ Part ⇒ Name **PLATE** ⇒ Use default template [Fig. 1.6(a)] ⇒ **OK** ⇒ **Window** from menu bar [Fig. 1.6(b)] (note: you now have two parts *"in session")* ⇒ **Activate** ⇒ **Save the active object** *(Pro/ENGINEER does NOT save for you. So save after the completion of each feature.)* ⇒ **OK** *(or MMB)* ⇒ [Fig. 1.6(c)]

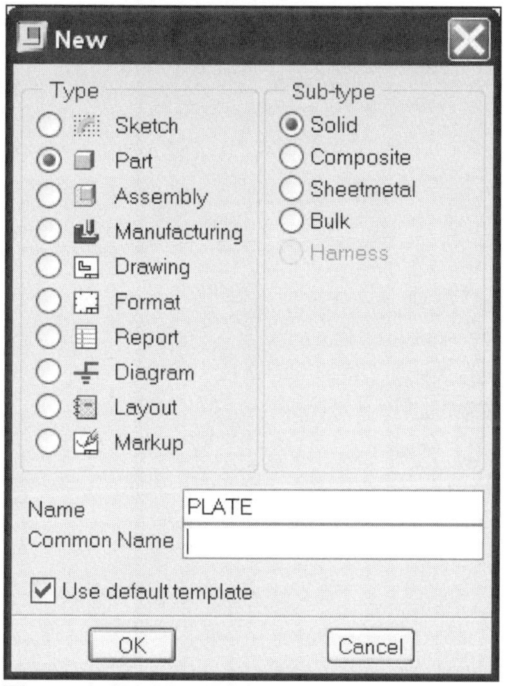

Figure 1.6(a) New Dialog Box, Part

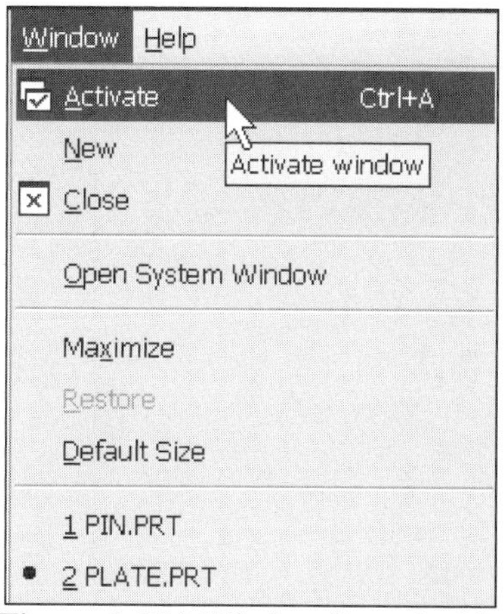

Figure 1.6(b) Window from the Menu Bar

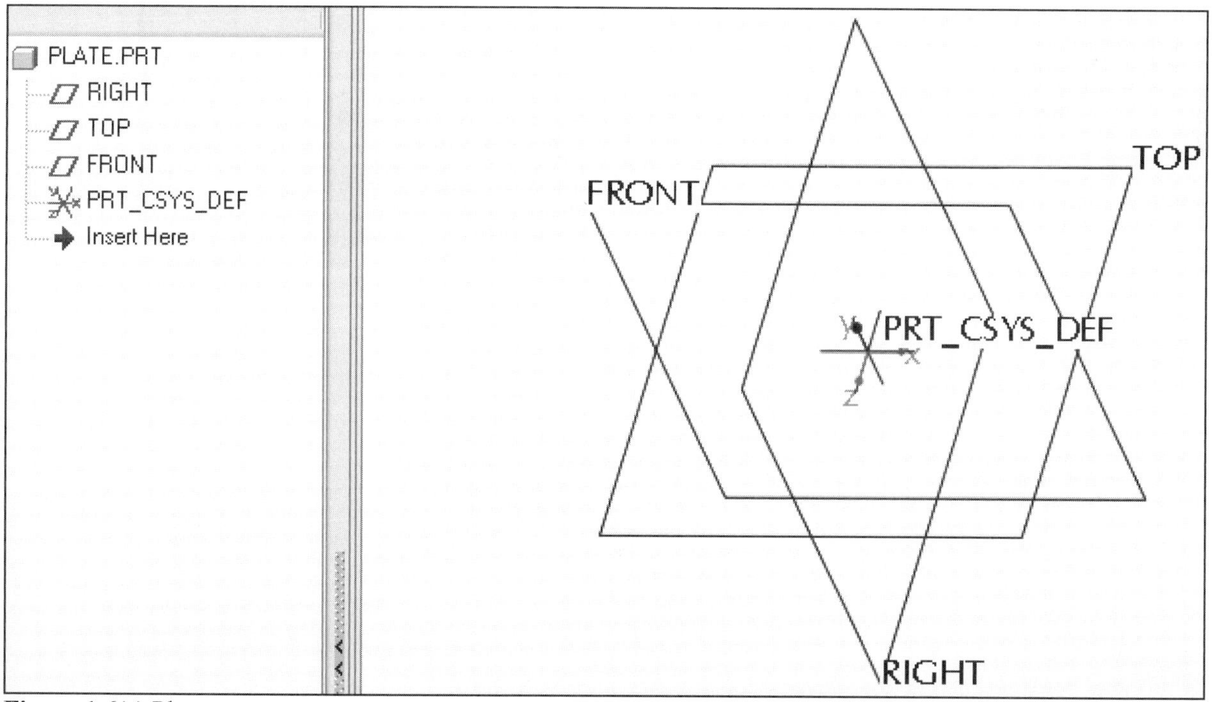

Figure 1.6(c) Plate

Click: **Tools** from menu bar [Fig. 1.6(d)] ⇒ **Environment** [Fig. 1.6(e)] ⇒ ☑ Snap to Grid *(your system may have this already set as the default)* ⇒ **Apply** ⇒ **OK**

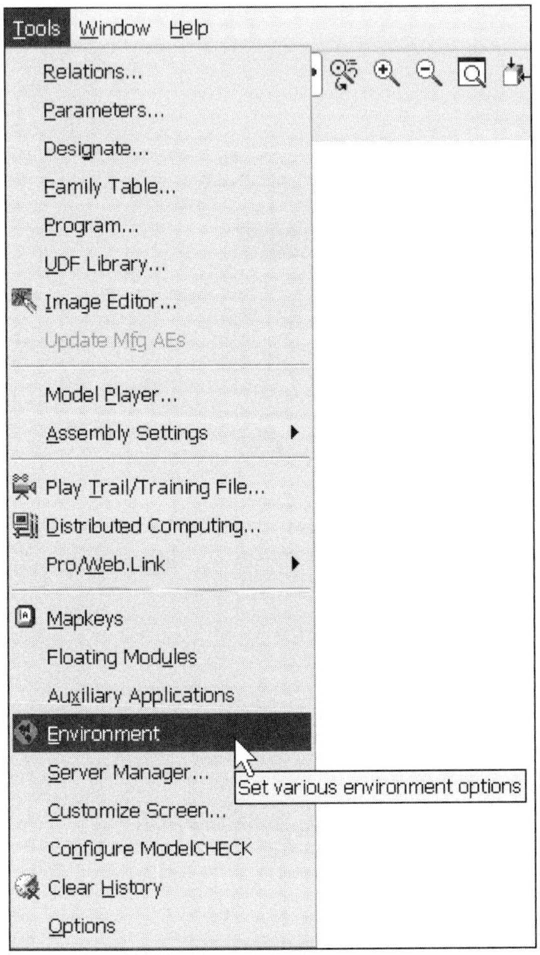

Figure 1.6(d) Tools ⇒ Environment

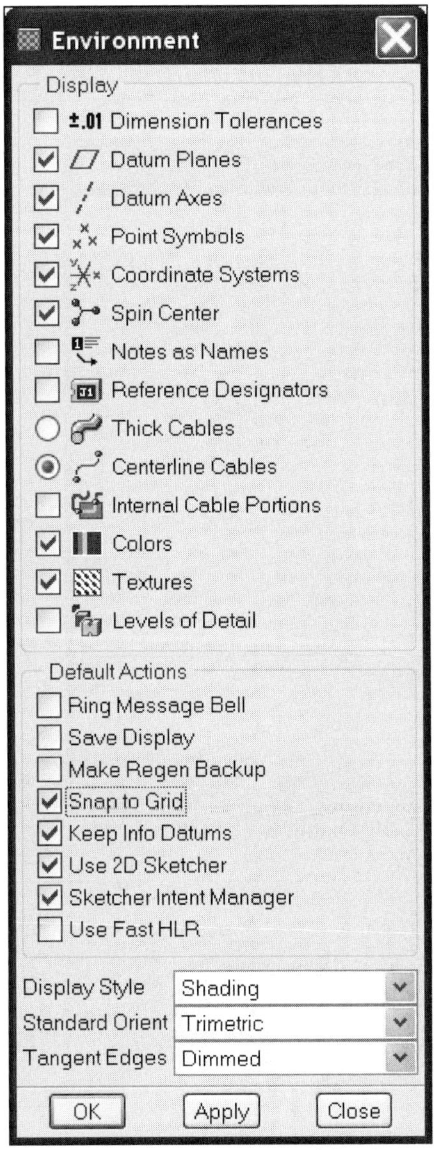

Figure 1.6(e) Environment Dialog Box

Pick the **FRONT** datum plane [Fig. 1.7(a)] ⇒ **Extrude Tool** ⇒ place cursor inside graphics area and press **RMB** (**R**ight **M**ouse **B**utton) ⇒ **Define Internal Sketch** [Fig. 1.7(b)] ⇒ **Sketch** [Fig. 1.7(c)]

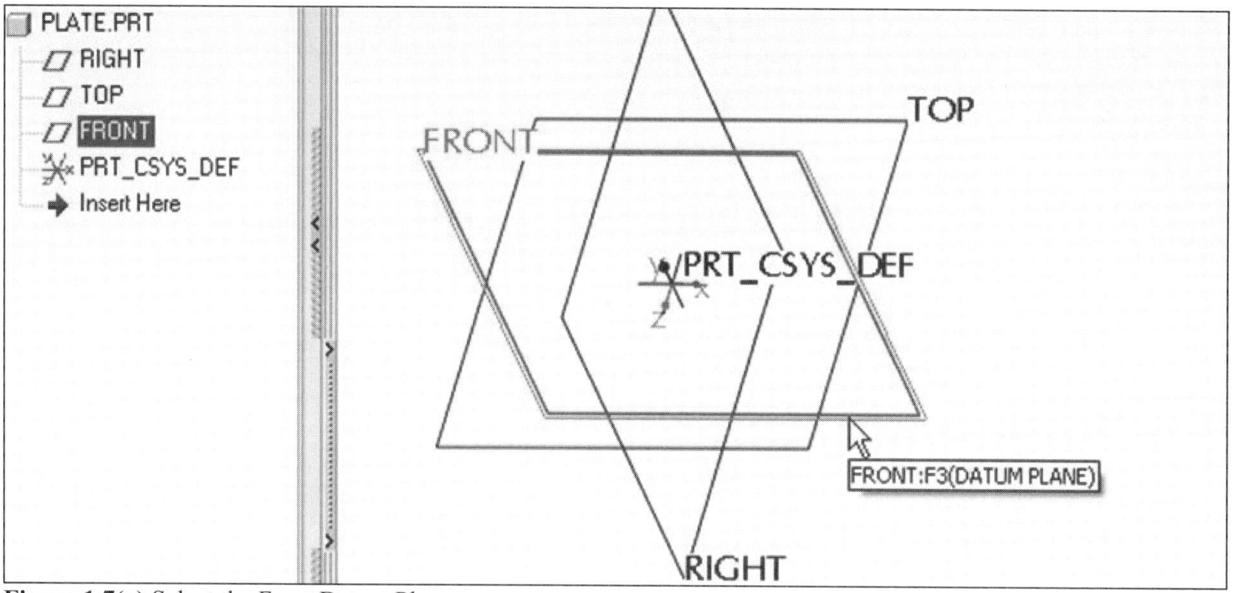

Figure 1.7(a) Select the Front Datum Plane

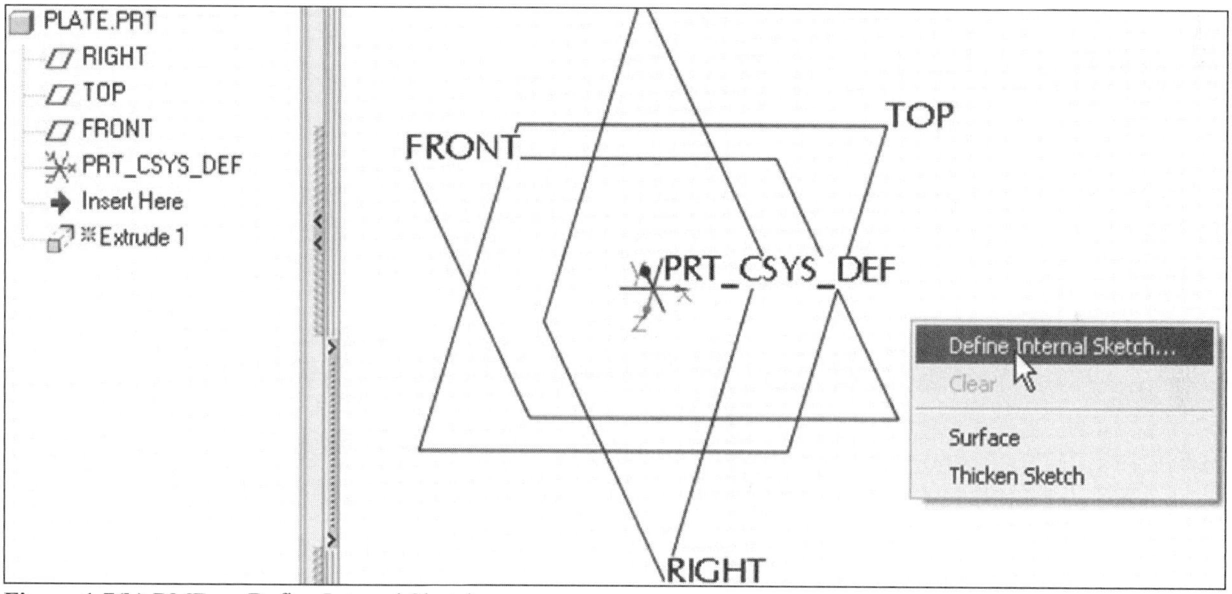

Figure 1.7(b) RMB ⇒ Define Internal Sketch

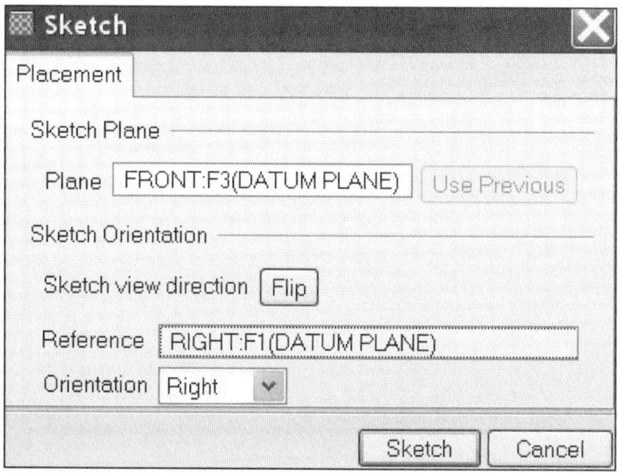

Figure 1.7(c) Sketch Dialog Box

Click:  **Insert foreign data from Palette into active object** [Fig. 1.7(d)] ⇒ scroll down to see the octagon 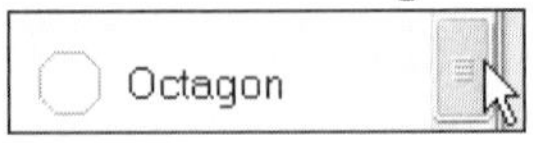 ⇒ double-click **Octagon** [Figs. 1.7(e-f)]

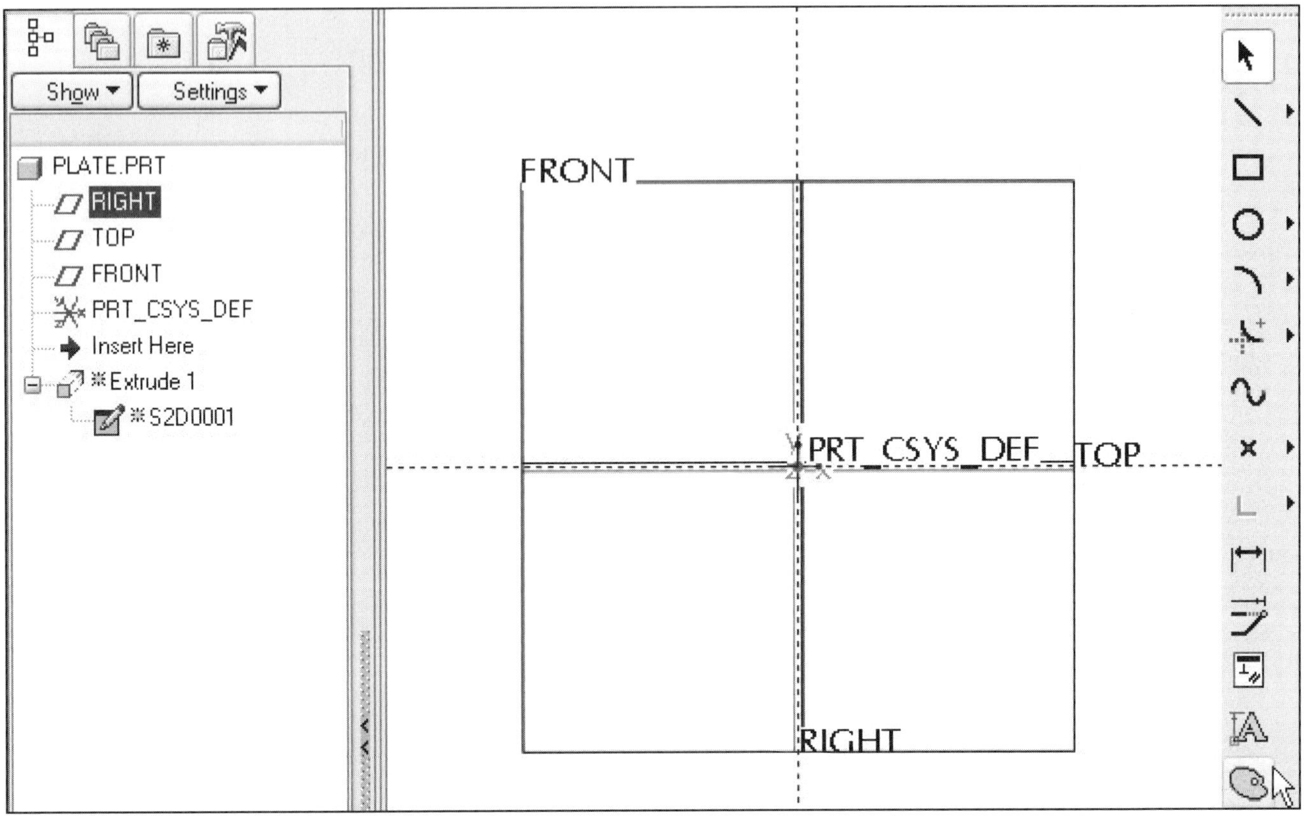

Figure 1.7(d)  Insert Foreign Data from Palette into Active Object

Figure 1.7(e) Sketcher Palette

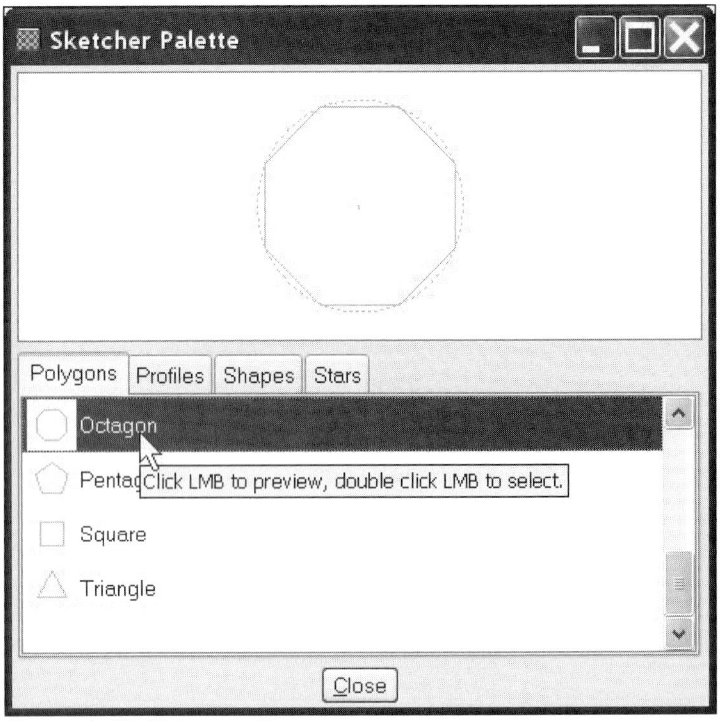

Figure 1.7(f) Polygons: Octagon

Place the octagon on the sketch by picking a position [Fig. 1.7(g)] ⇒ with the **LMB,** drag and drop the

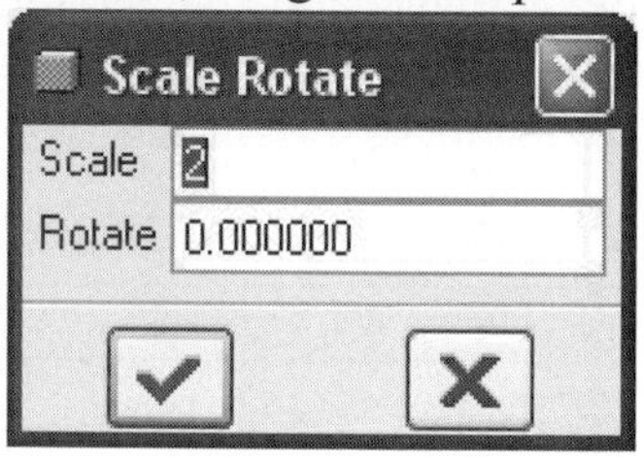

center of the octagon at the origin [Fig. 1.7(h)] ⇒ modify Scale value to **2** ⇒ **Enter** ⇒  from the Scale Rotate dialog box

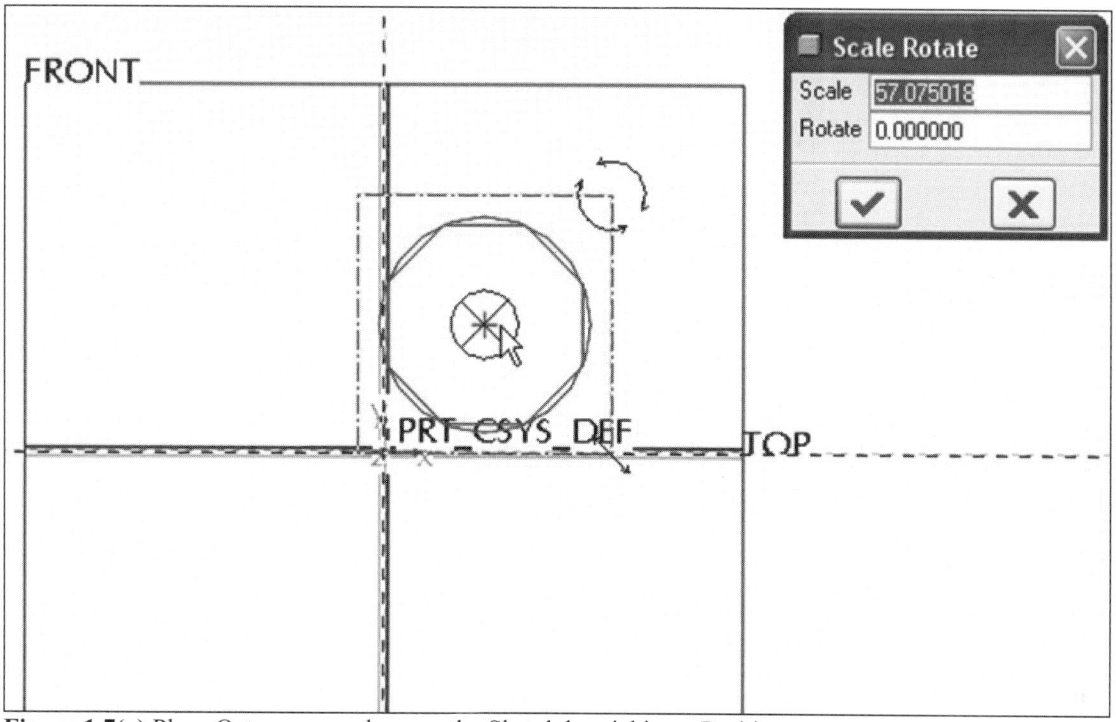

Figure 1.7(g) Place Octagon anywhere on the Sketch by picking a Position

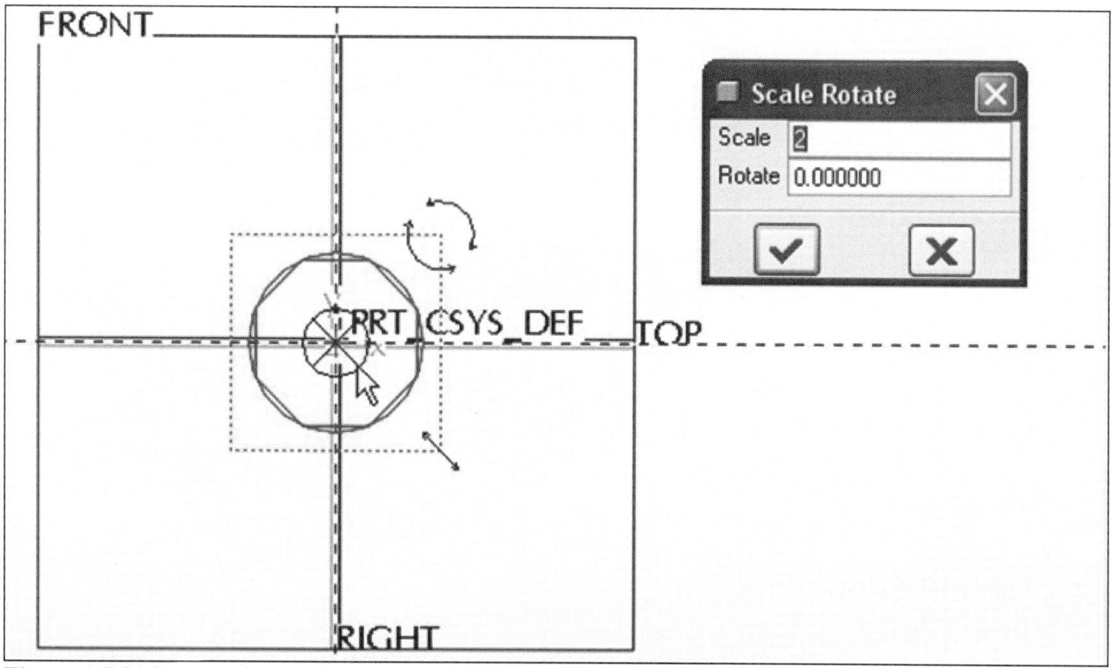

Figure 1.7(h) Drag and Drop the Center of the Octagon at the Origin of the Part (PRT_CSYS_DEF)

Click: **Close** from the Sketcher Palette ⇒ [Fig. 1.7(i)] ⇒ from Right Toolchest [Fig. 1.7(j)] ⇒ modify the depth in the dashboard to **2.00** [Fig. 1.7(k)] ⇒ **Enter** ⇒ ⇒ **Standard Orientation** from Top Toolchest [Fig. 1.7(l)] ⇒ from dashboard [Fig. 1.7(m)] ⇒ **LMB** to deselect

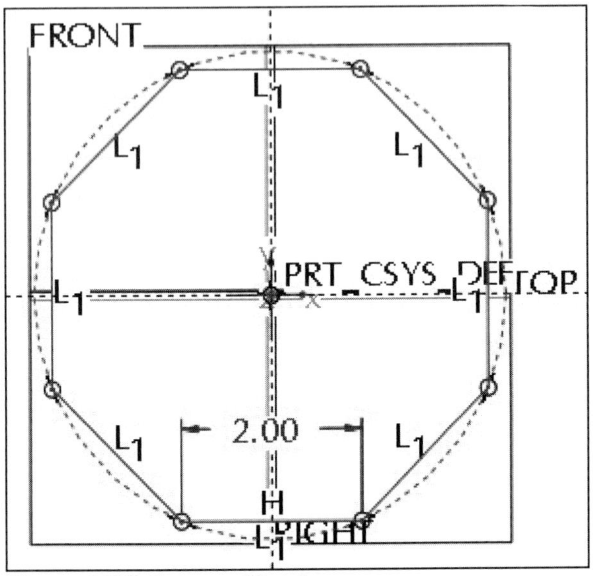

Figure 1.7(i) Octagon Sketch

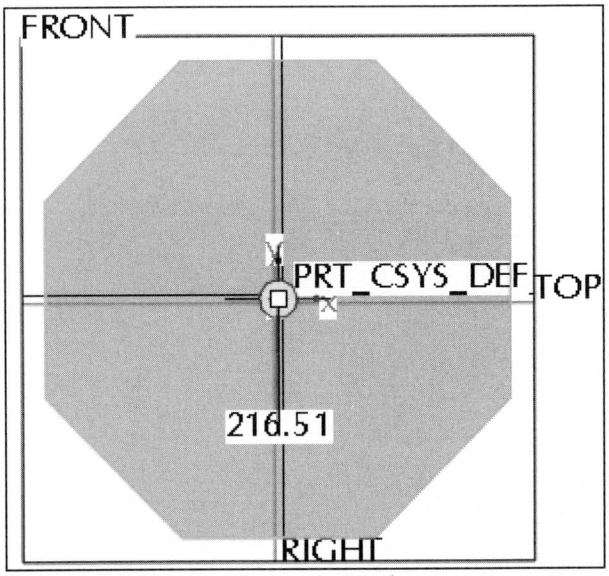

Figure 1.7(j) Octagon Dynamic Preview

Figure 1.7(k) Modify Depth to **2.00**

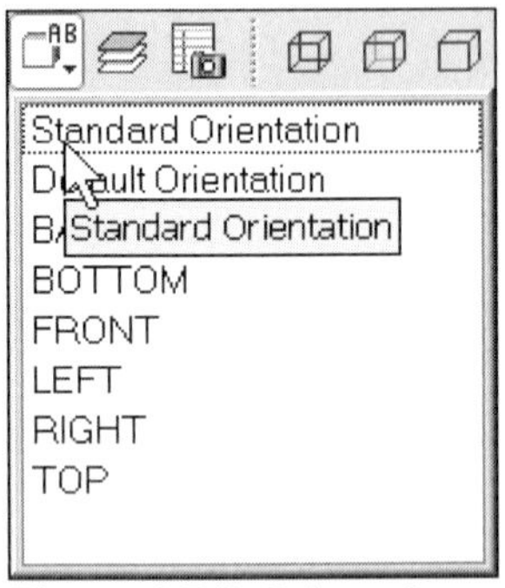

Figure 1.7(l) Orientation

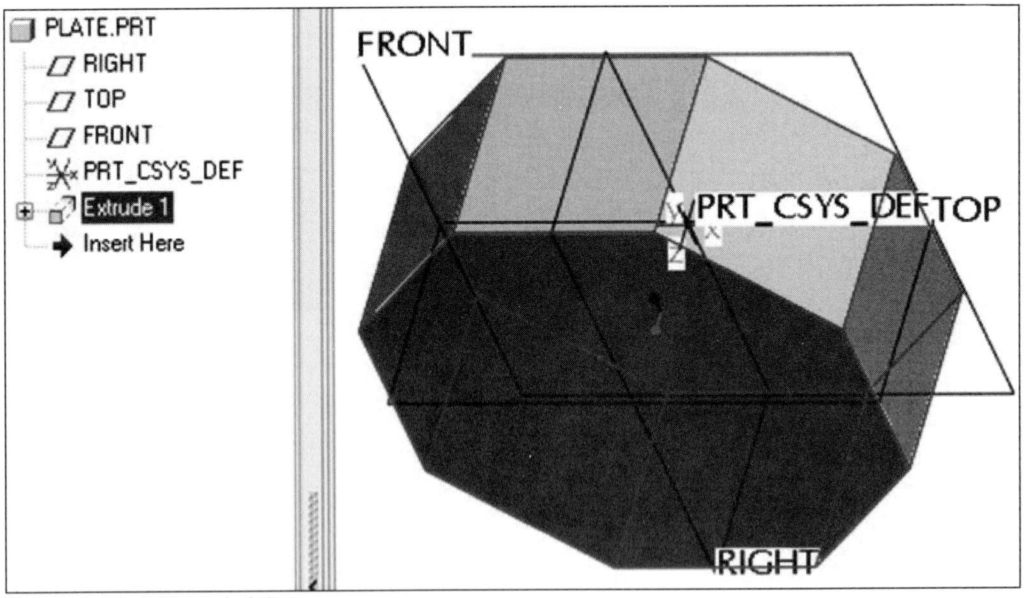

Figure 1.7(m) Extruded Octagon

Click: Hole Tool from Right Toolchest ⇒ pick on the top surface [Fig. 1.8(a)] ⇒ **Hidden line** from Top Toolchest [Fig. 1.8(b)]

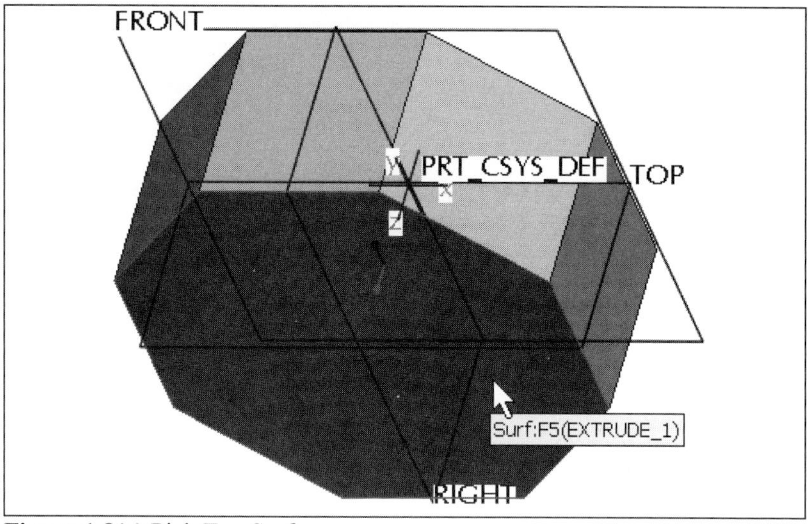

Figure 1.8(a) Pick Top Surface

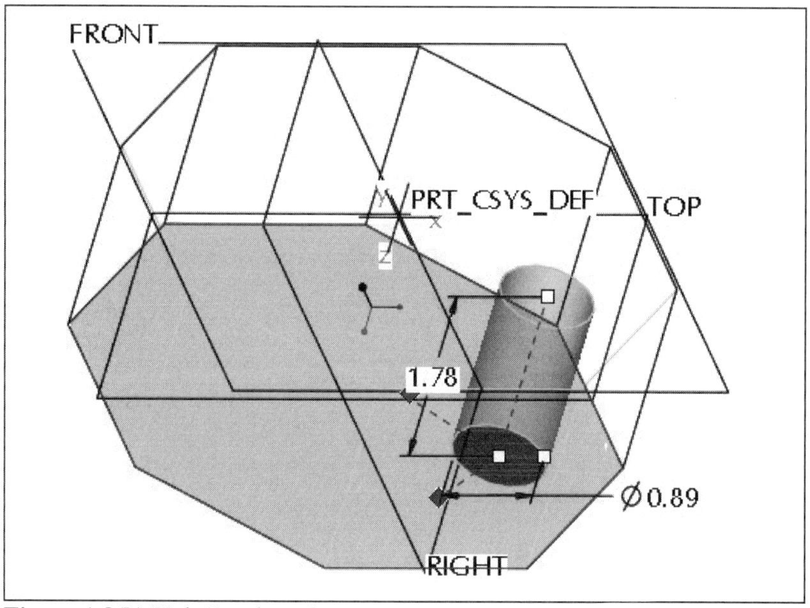

Figure 1.8(b) Hole Previewed

Click: **RMB** [Fig. 1.8(c)] ⇒ **Offset References Collector** ⇒ press and hold the **Ctrl** key ⇒ select the **RIGHT** datum plane from the Model Tree ⇒ with the **Ctrl** key still pressed, select the **TOP** datum plane from the Model Tree [Fig. 1.8(d)] ⇒ **Placement** tab from the dashboard [Fig. 1.8(e)]

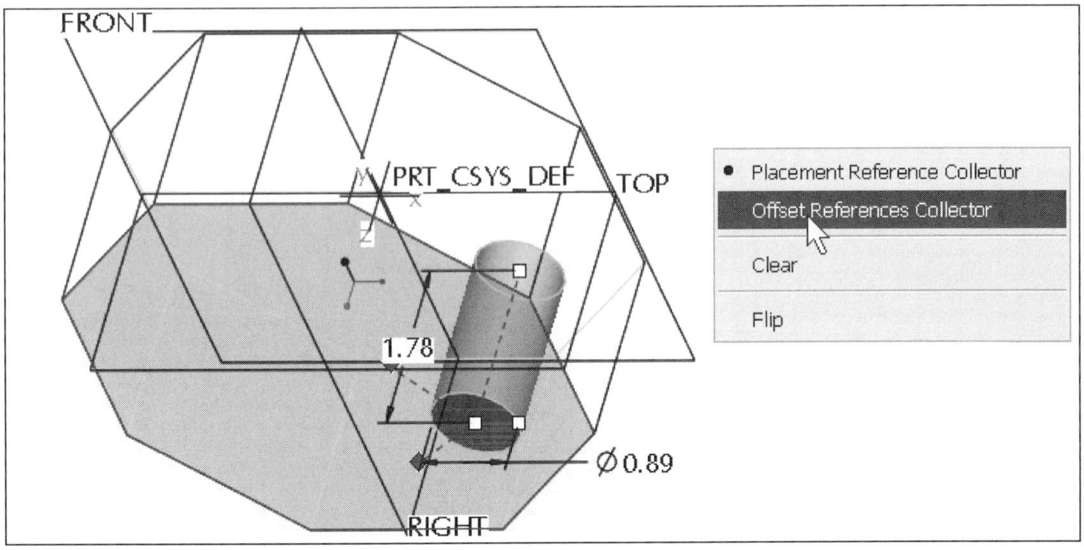

Figure 1.8(c) RMB ⇒ Secondary References Collector

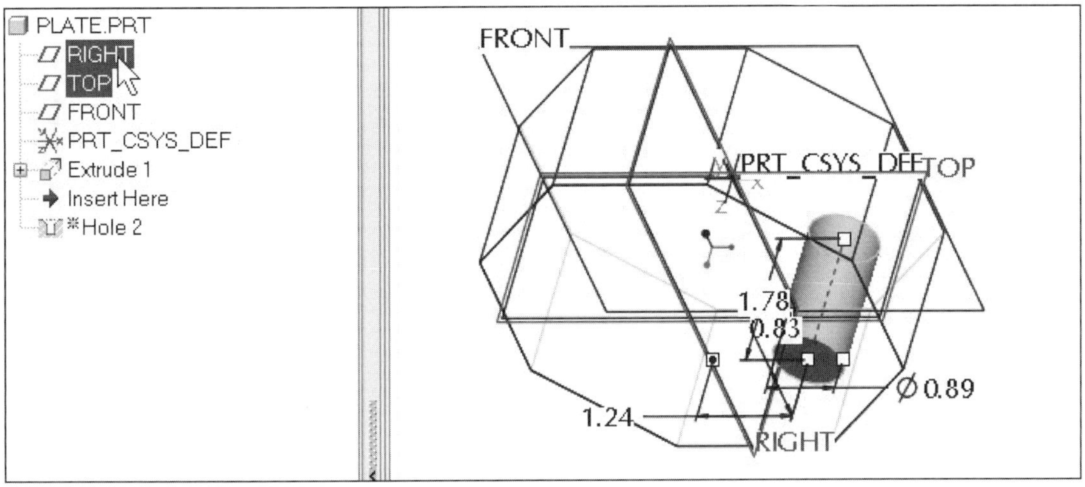

Figure 1.8(d) Select the RIGHT and the TOP Datum Planes as Secondary References

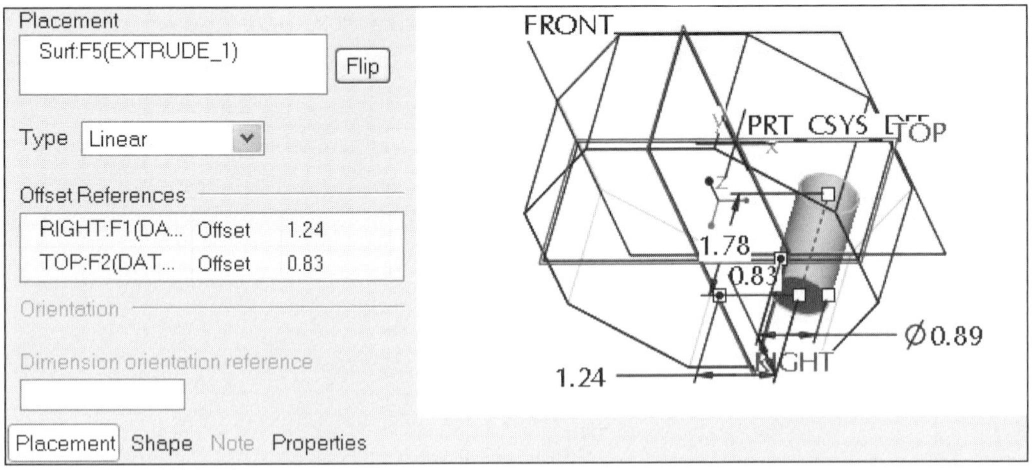

Figure 1.8(e) Click Placement

In the Offset References collector, click on **RIGHT** [Fig. 1.8(f)] ⇒ click on Offset ⇒ **Align** [Fig. 1.8(g)] ⇒ click on **TOP** [Fig. 1.8(h)] ⇒ click on Offset ⇒ **Align** [Figs. 1.8(i-j)]

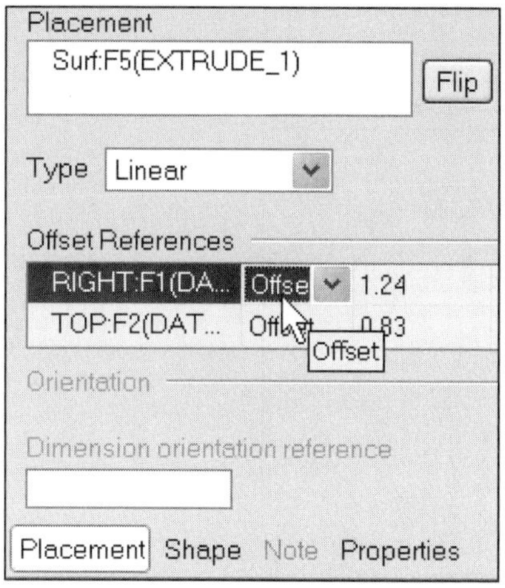

Figure 1.8(f) Click on RIGHT ⇒ Offset

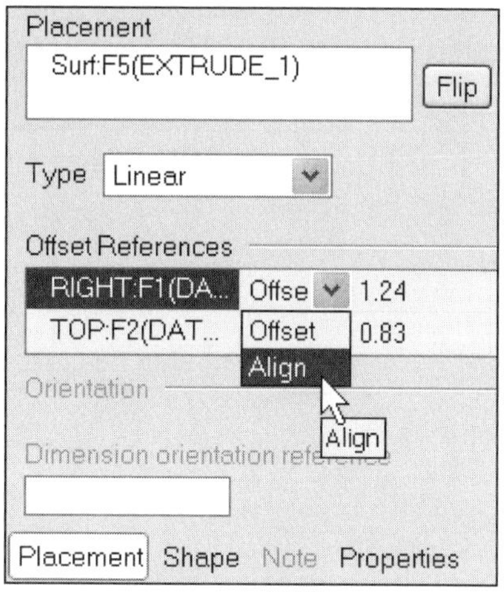

Figure 1.8(g) Click ⇒ Align

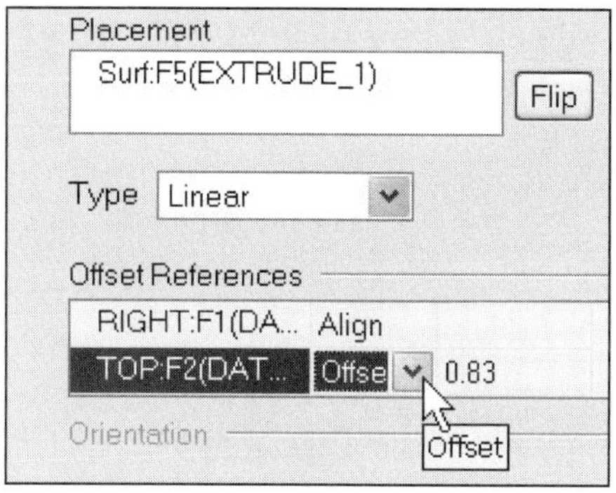

Figure 1.8(h) Click on TOP ⇒ Offset

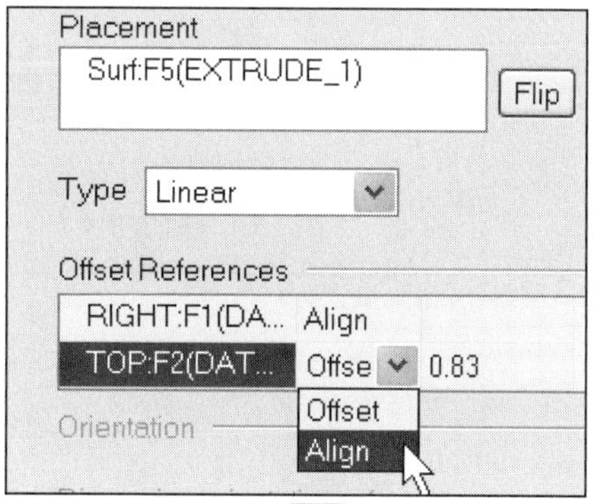

Figure 1.8(i) Click ⇒ Align

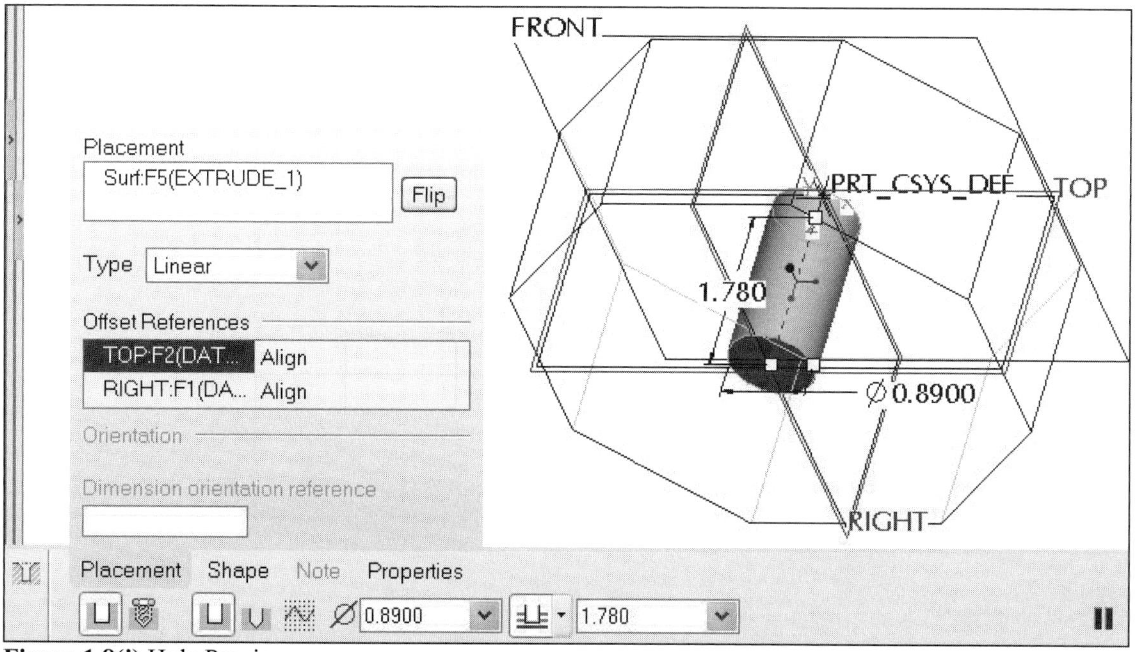

Figure 1.8(j) Hole Preview

Click: **Placement** tab to close the Placement panel ⇒ 1.78 ⇒ **Drill to intersect with all surfaces** [Fig. 1.8(k)] ⇒ modify the hole's diameter, type **1.00** 1.00 ⇒ **Enter** [Fig. 1.8(l)] ⇒ [Fig. 1.8(m)] ⇒ **Ctrl+S** to save the part ⇒ **OK**

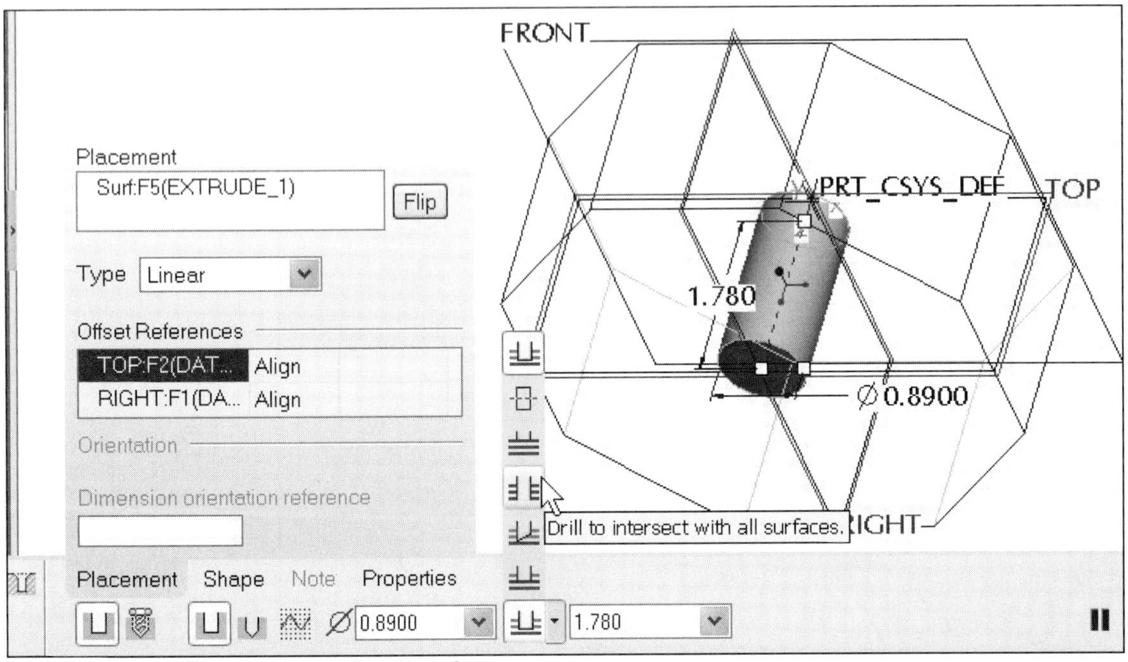

Figure 1.8(k) Drill to intersect with all surfaces

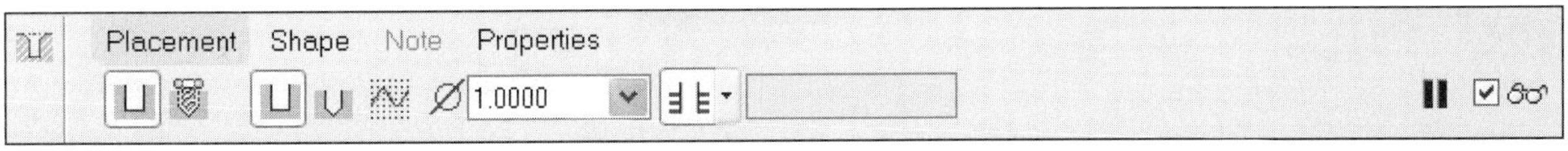

Figure 1.8(l) Diameter **1.00**

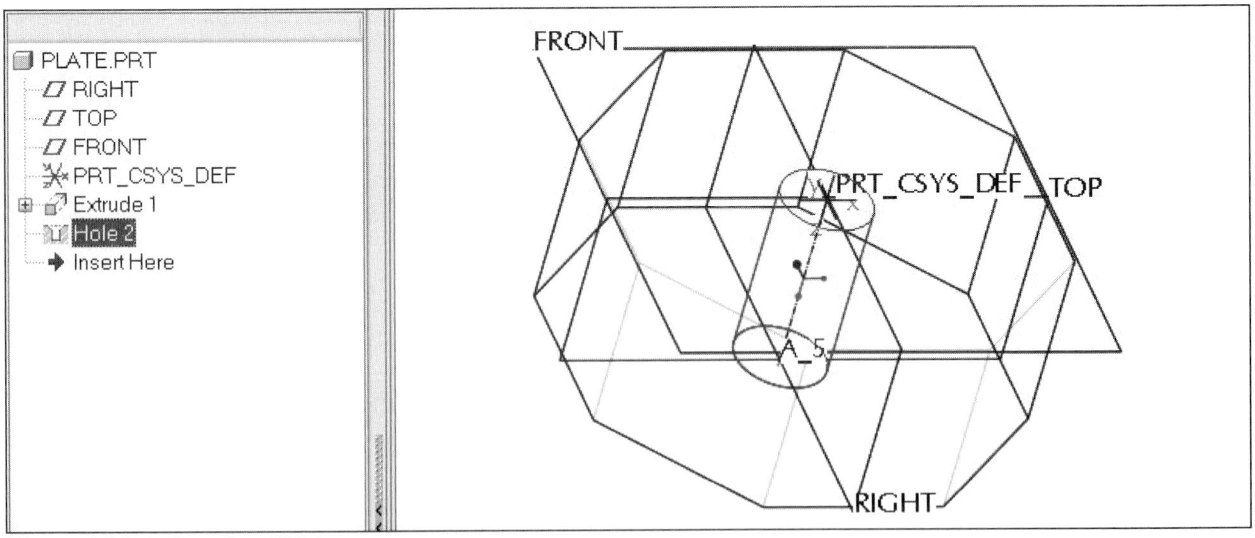

Figure 1.8(m) Completed Hole

Click:  **Shading** ⇒ **View** from menu bar [Fig. 1.9(a)] ⇒ **Color and Appearance** ⇒  **ptc-metallic-gold** [Fig. 1.9(b)] ⇒ **Apply** ⇒ **Close** the dialog box

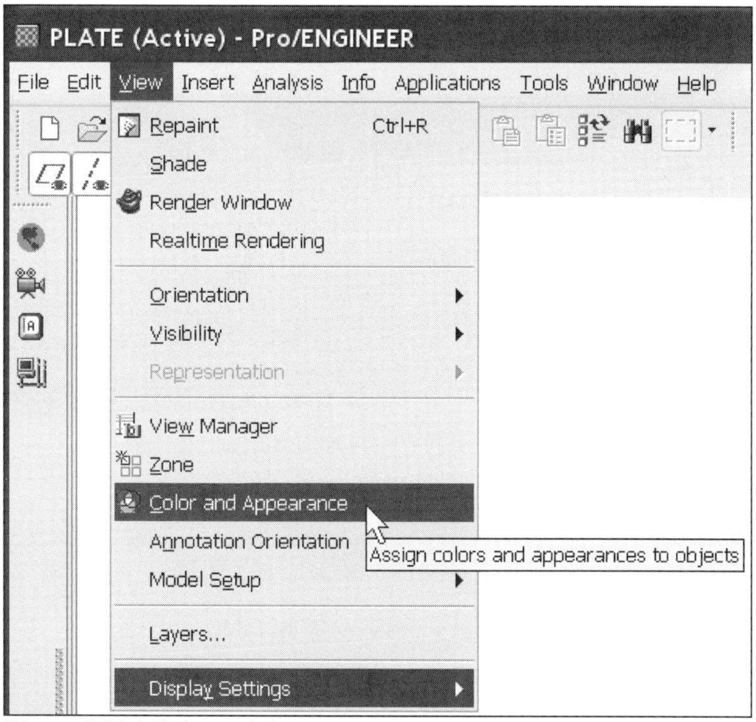

Figure 1.9(a) Color and Appearance

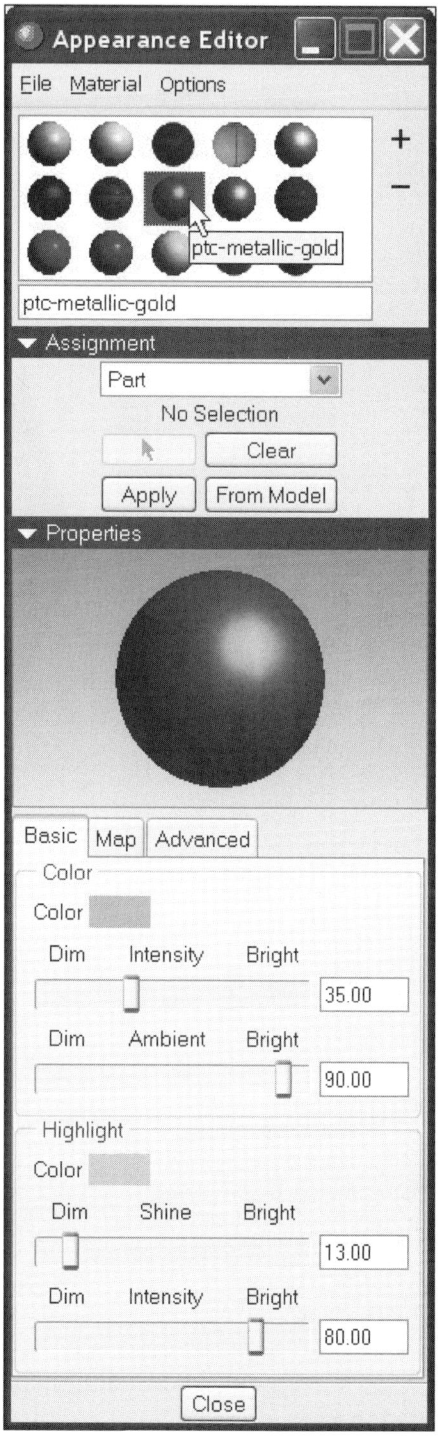

Figure 1.9(b) Appearance Editor Dialog Box

Click: ⇒ ⇒ **OK** (Fig. 1.10)

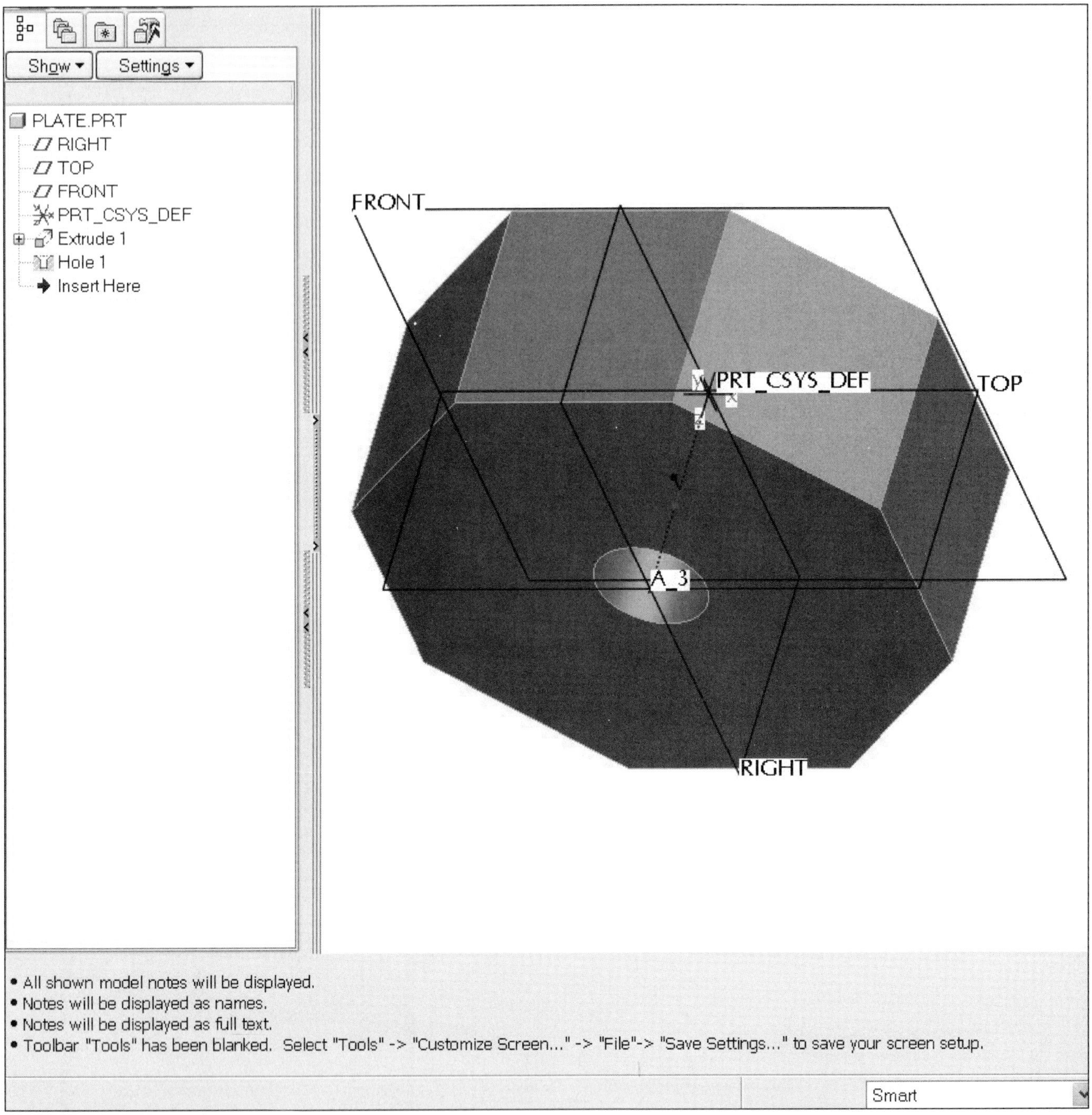

Figure 1.10 Completed Plate

You have completed your second Pro/ENGINEER component. In the next section, you will put the parts together to create an assembly.

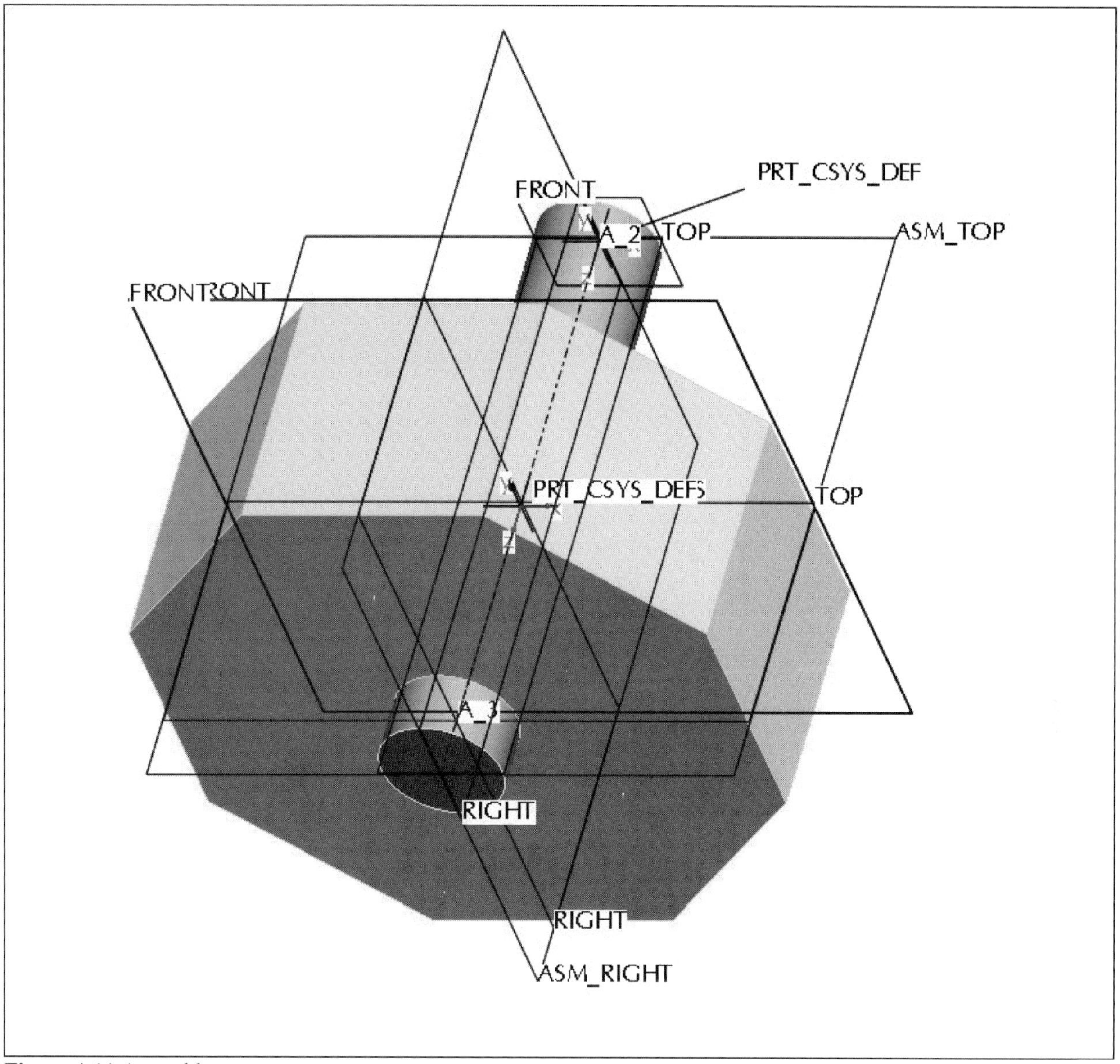

Figure 1.11 Assembly

Creating the Assembly

Using the two simple parts just created you will now create an assembly (Fig. 1.11). Just as you can combine features into parts, you can also combine parts into assemblies. The Assembly mode in Pro/E enables you to place component parts and subassemblies together to form assemblies.

As with a part, an assembly starts with default datum planes and a default coordinate system. To create a subassembly or an assembly, you must first create datum features. You then create or assemble additional components to the existing component(s) and datum features.

***If you are continuing from the last section, then skip this set of commands*:**

Open **Pro/ENGINEER Wildfire 4.0** using a shortcut icon on your Desktop ProE 4.0 ⇒ **File** ⇒ **Set Working Directory** ⇒ select the working directory or accept the default directory (make sure you set the same working directory as where your parts were saved) ⇒ **OK** ⇒ **File** ⇒ **Open** ⇒ pick **pin.prt** ⇒ **Open** ⇒ **File** ⇒ **Open** ⇒ pick **plate.prt** ⇒ **Open** ⇒ **Shading**

Click: **Create a new object** ⇒ Assembly ⇒ Design ⇒ Name **CONNECTOR** ⇒ Use default template [Fig. 1.12(a)] ⇒ **OK** ⇒ **Settings** from Navigator [Fig. 1.12(b)] ⇒ **Tree Filters** ⇒ toggle all on [Fig. 1.12(c)] ⇒ **Apply** ⇒ **OK** [Fig. 1.12(d)] ⇒ **File** ⇒ **Save** ⇒ **OK**

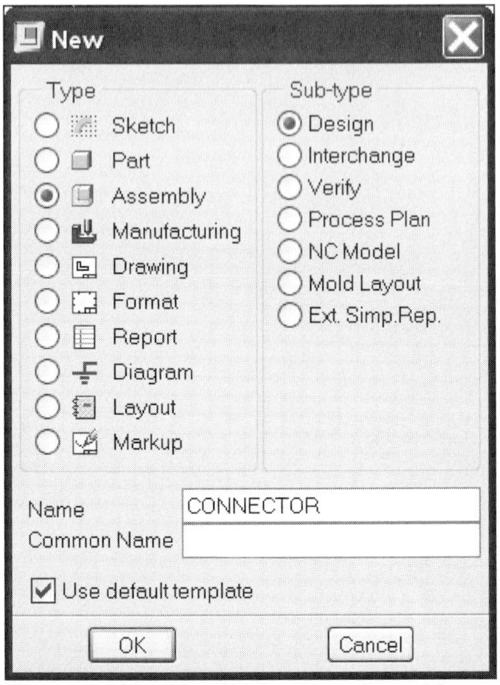

Figure 1.12(a) New Dialog Box, Assembly

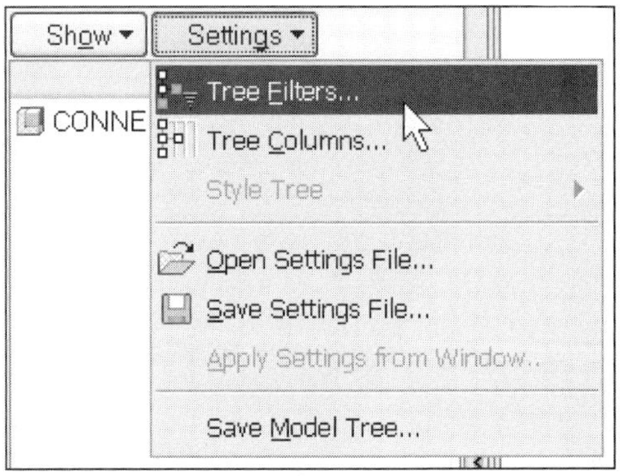

Figure 1.12(b) Tree Filters

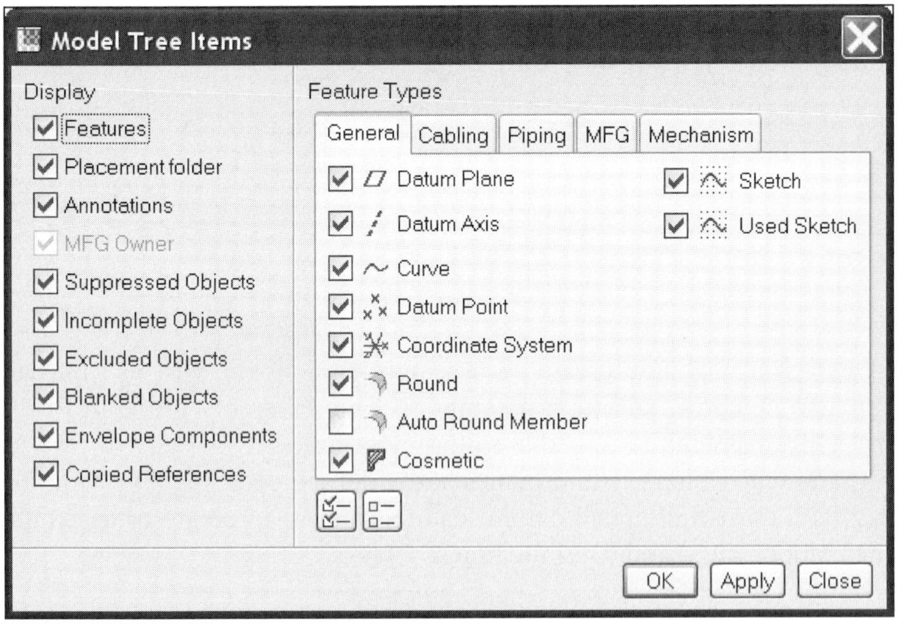

Figure 1.12(c) Model Tree Items Dialog Box

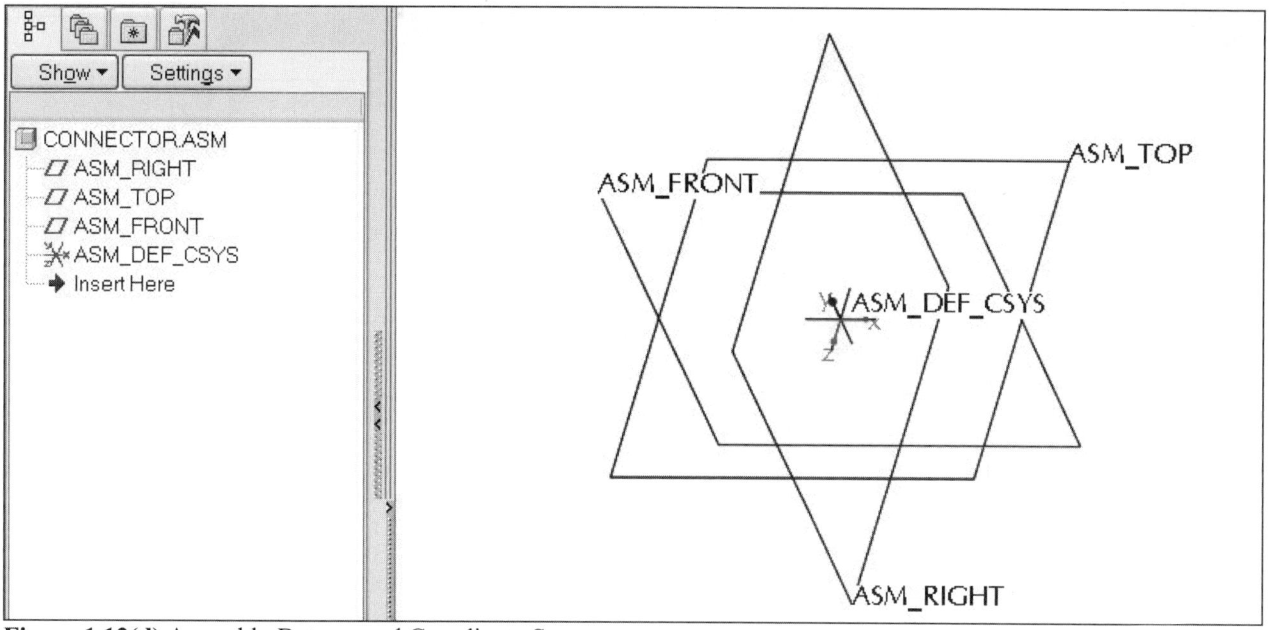

Figure 1.12(d) Assembly Datums and Coordinate System

Click: **Add component to the assembly** from Right Toolchest ⇒ Type **Part (*.prt)** ⇒ **plate.prt** [Fig. 1.13(a)] ⇒ **Open**

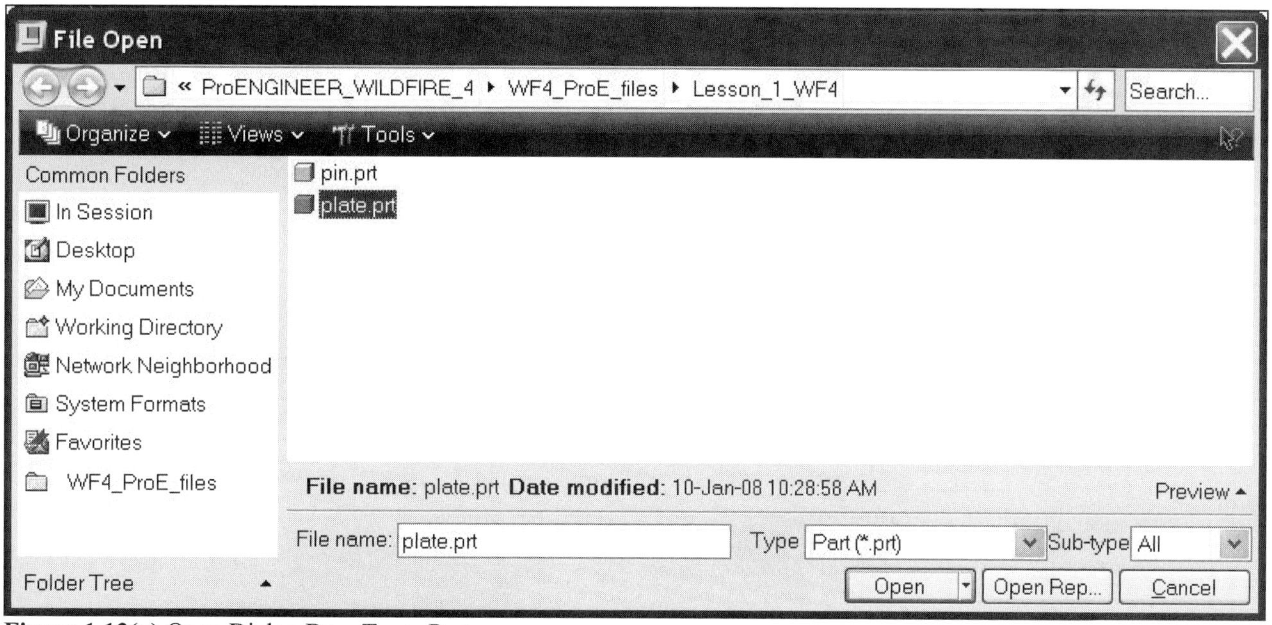

Figure 1.13(a) Open Dialog Box, Type: Part

Click:  from Top Toolchest [Fig. 1.13(b)] ⇒ 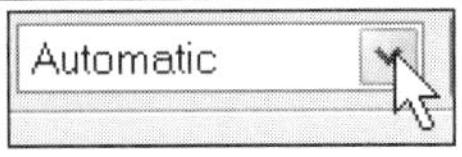

⇒ 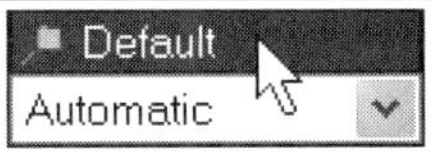

[Fig. 1.13(c)] ⇒ 

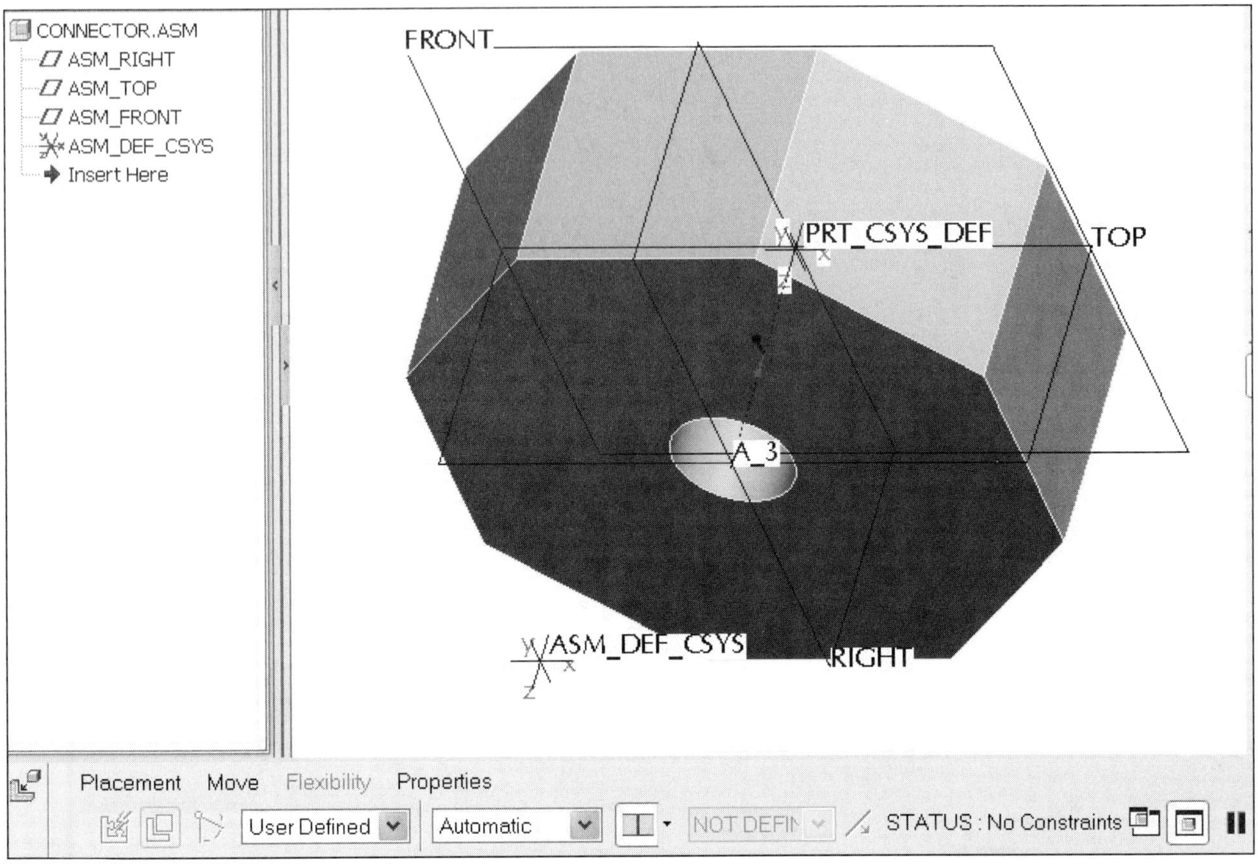

Figure 1.13(b) Assembly Dashboard

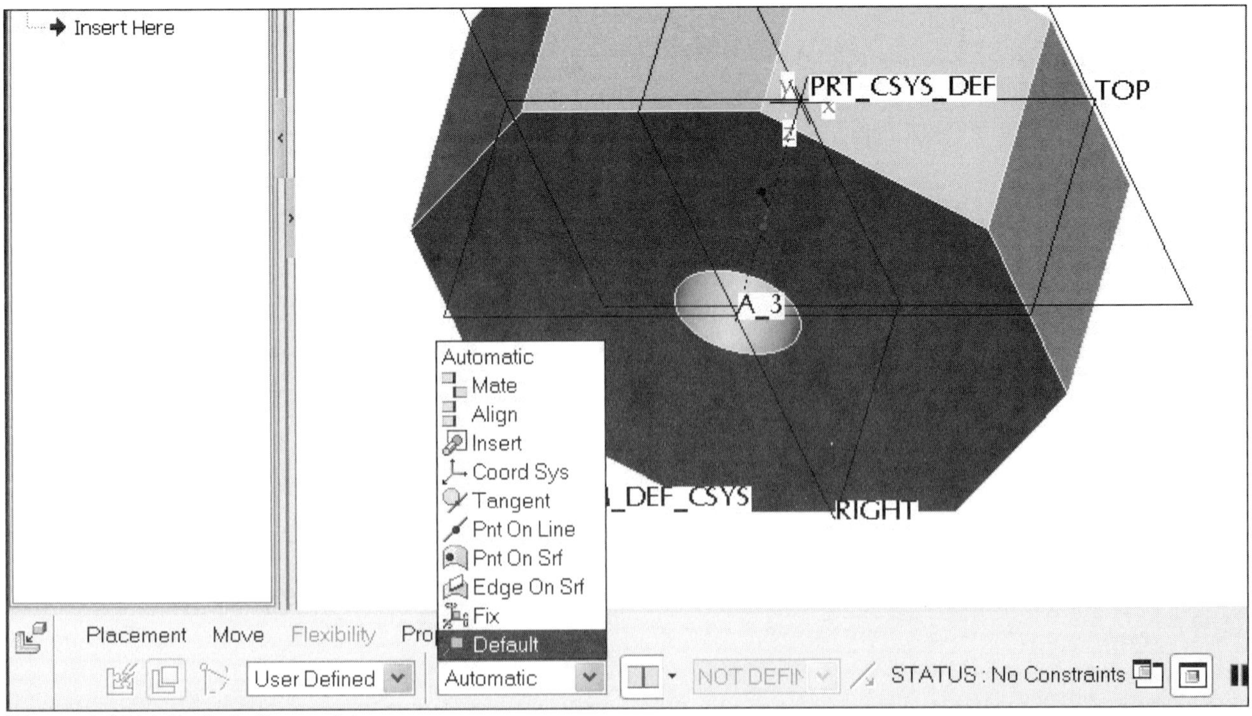

Figure 1.13(c) Default Constraint

Click:  from dashboard [Fig. 1.13(d)] ⇒  ⇒ **Ctrl+S** ⇒ **OK** ⇒ in the Model Tree expand PLATE.PRT ⇒ expand Placement ⇒ expand Set1 [Fig. 1.13(e)]

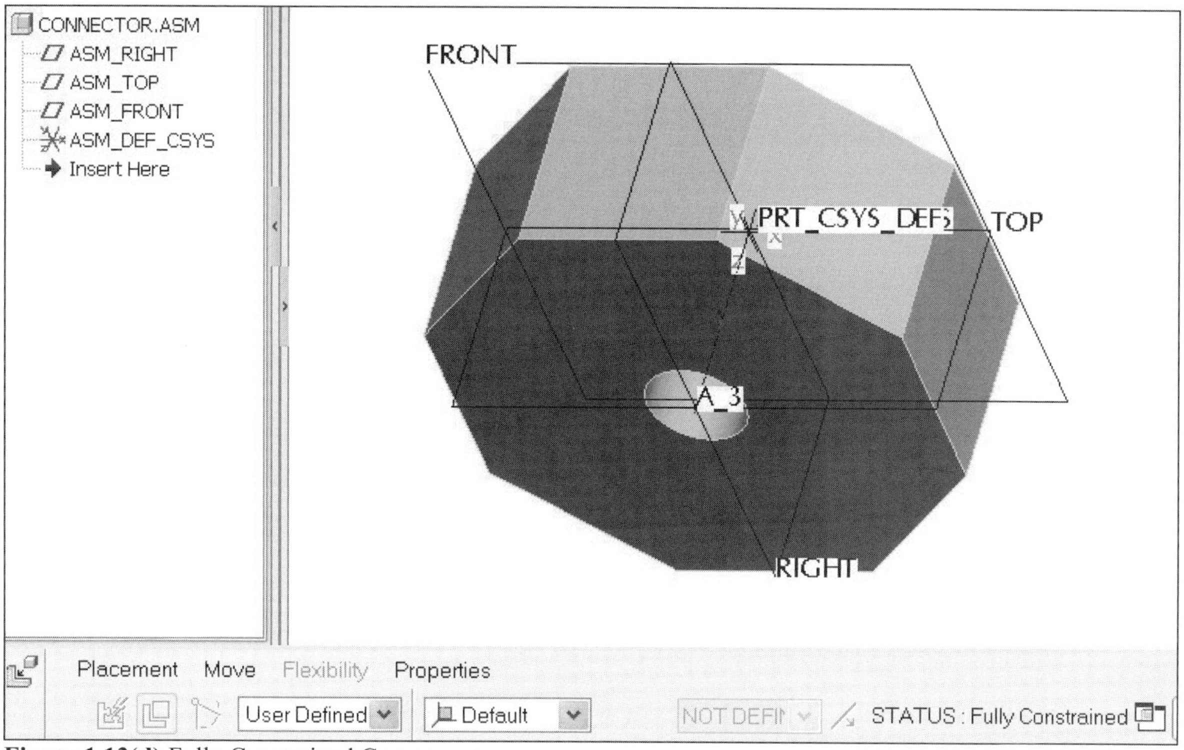

Figure 1.13(d) Fully Constrained Component

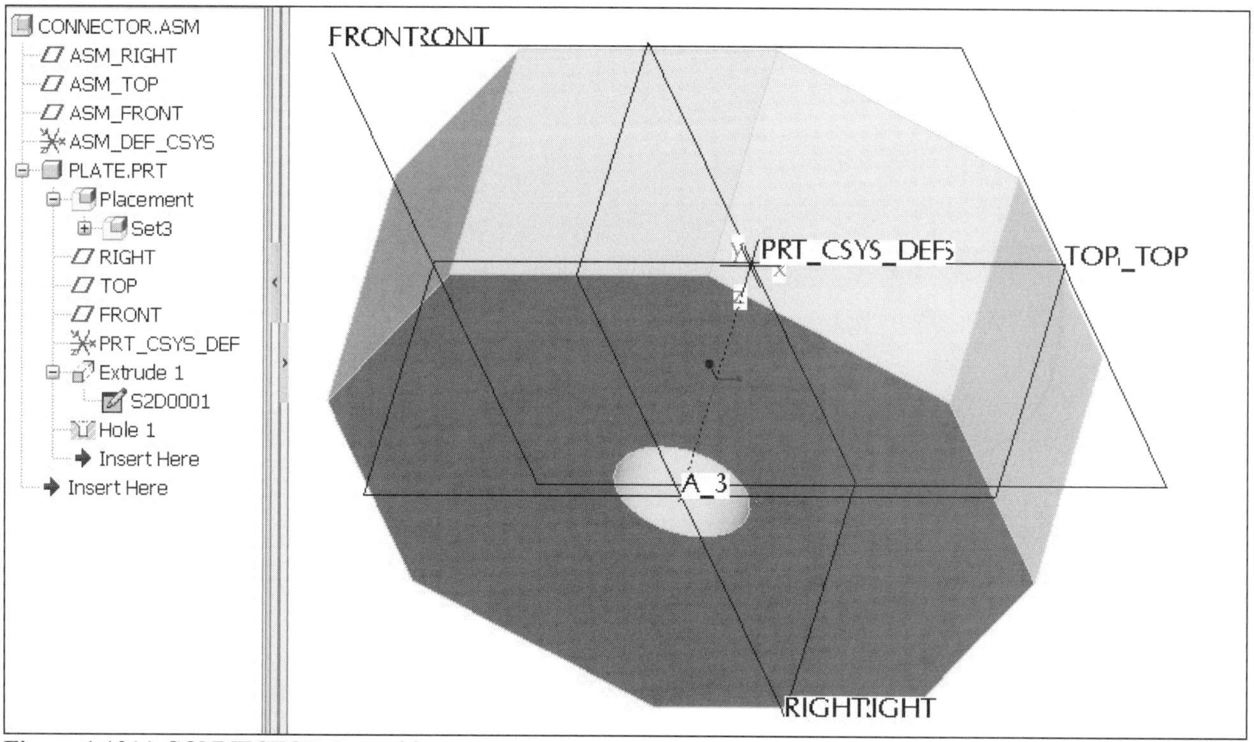

Figure 1.13(e) CONNECTOR Assembly

Click: Add component to the assembly from Right Toolchest ⇒ **pin.prt** ⇒ **Preview** [Fig. 1.14(a)] ⇒ **Open** [Fig. 1.14(b)]

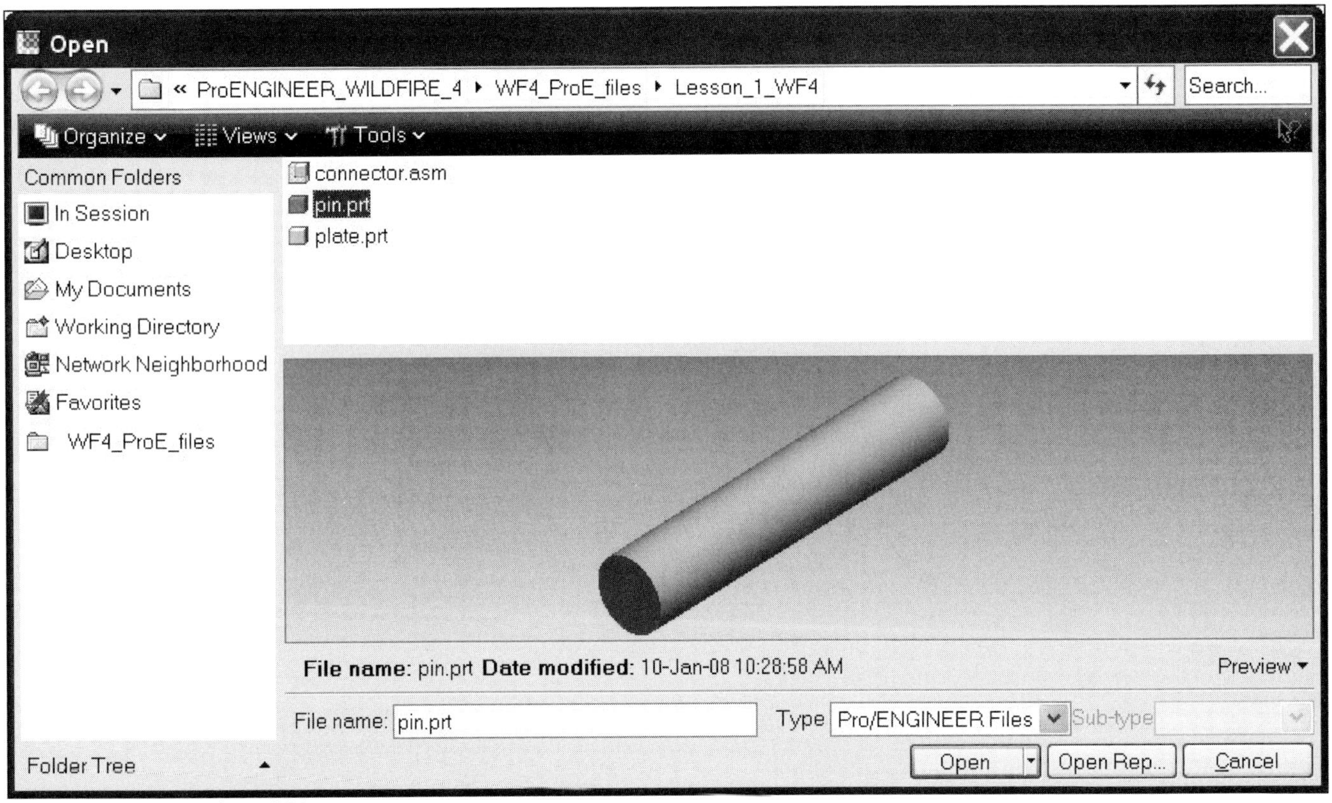

Figure 1.14(a) Previewed Pin

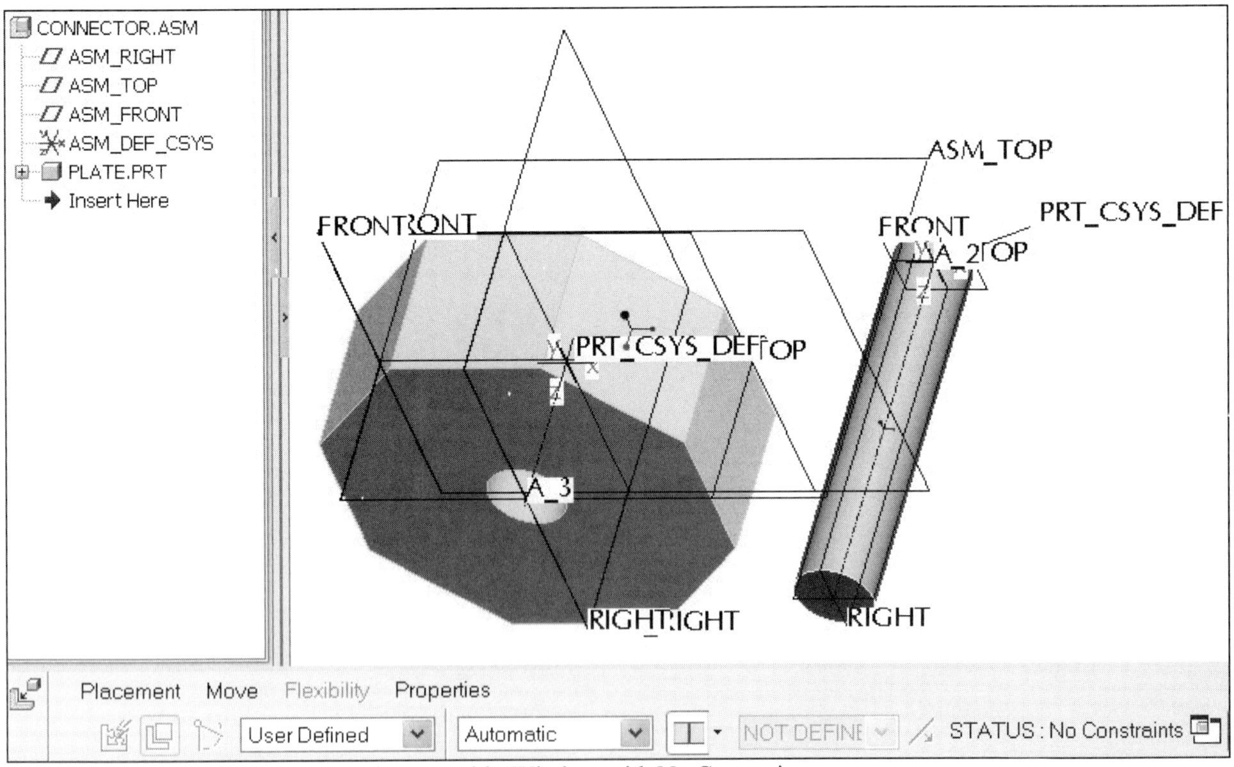

Figure 1.14(b) Component Shown in Assembly Window with No Constraints

Pick on the Pin's cylindrical surface [Fig. 1.14(c)] ⇒ 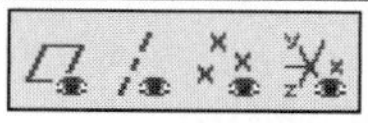 off ⇒ pick on the Plate's cylindrical hole surface [Fig. 1.14(d)] ⇒ **Placement** tab from dashboard 

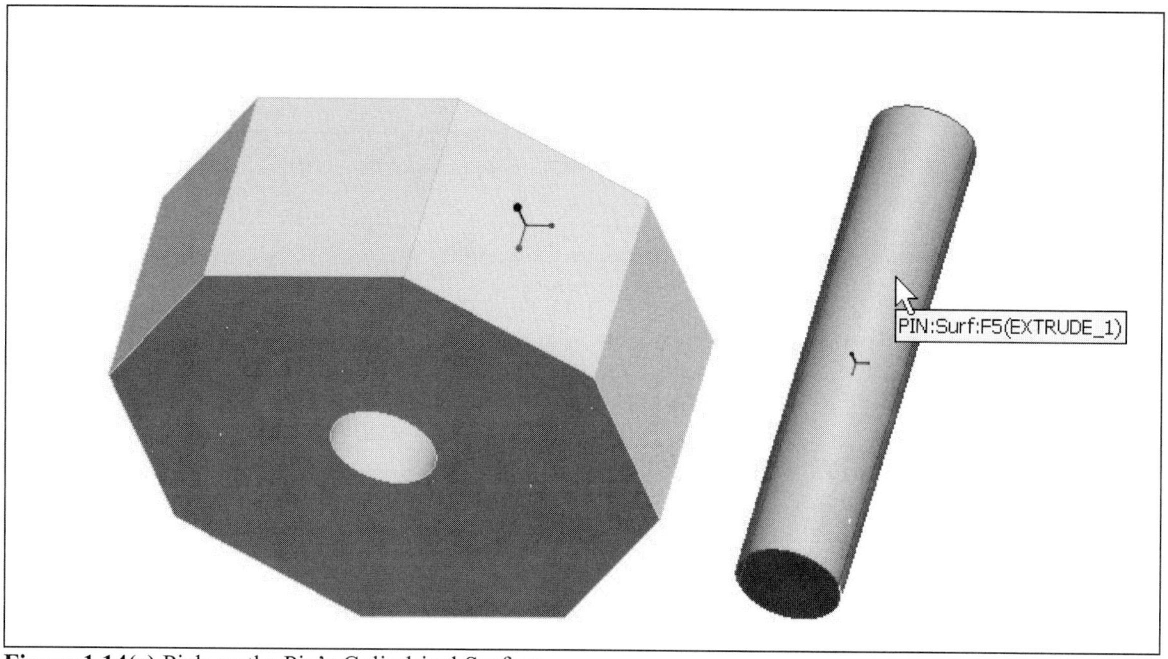

Figure 1.14(c) Pick on the Pin's Cylindrical Surface

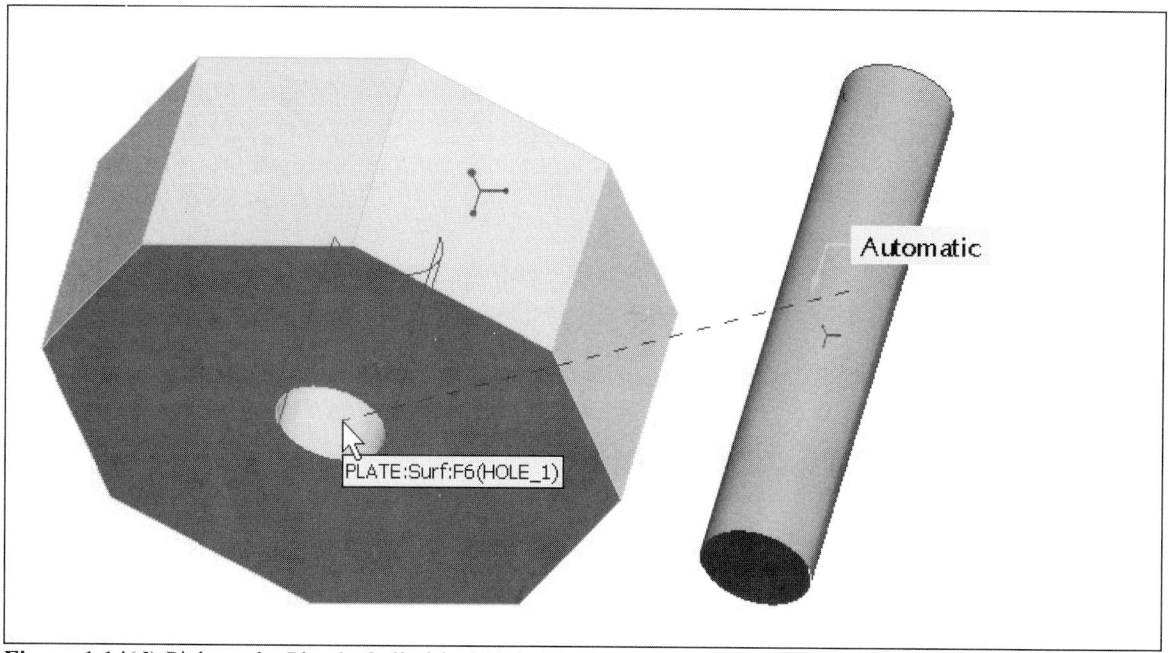

Figure 1.14(d) Pick on the Plate's Cylindrical Hole (Surface)

Click: 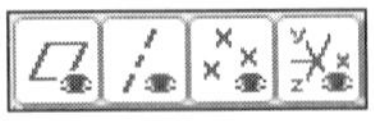 on ⇒ **New Constraint** 

[Fig. 1.14(e)]

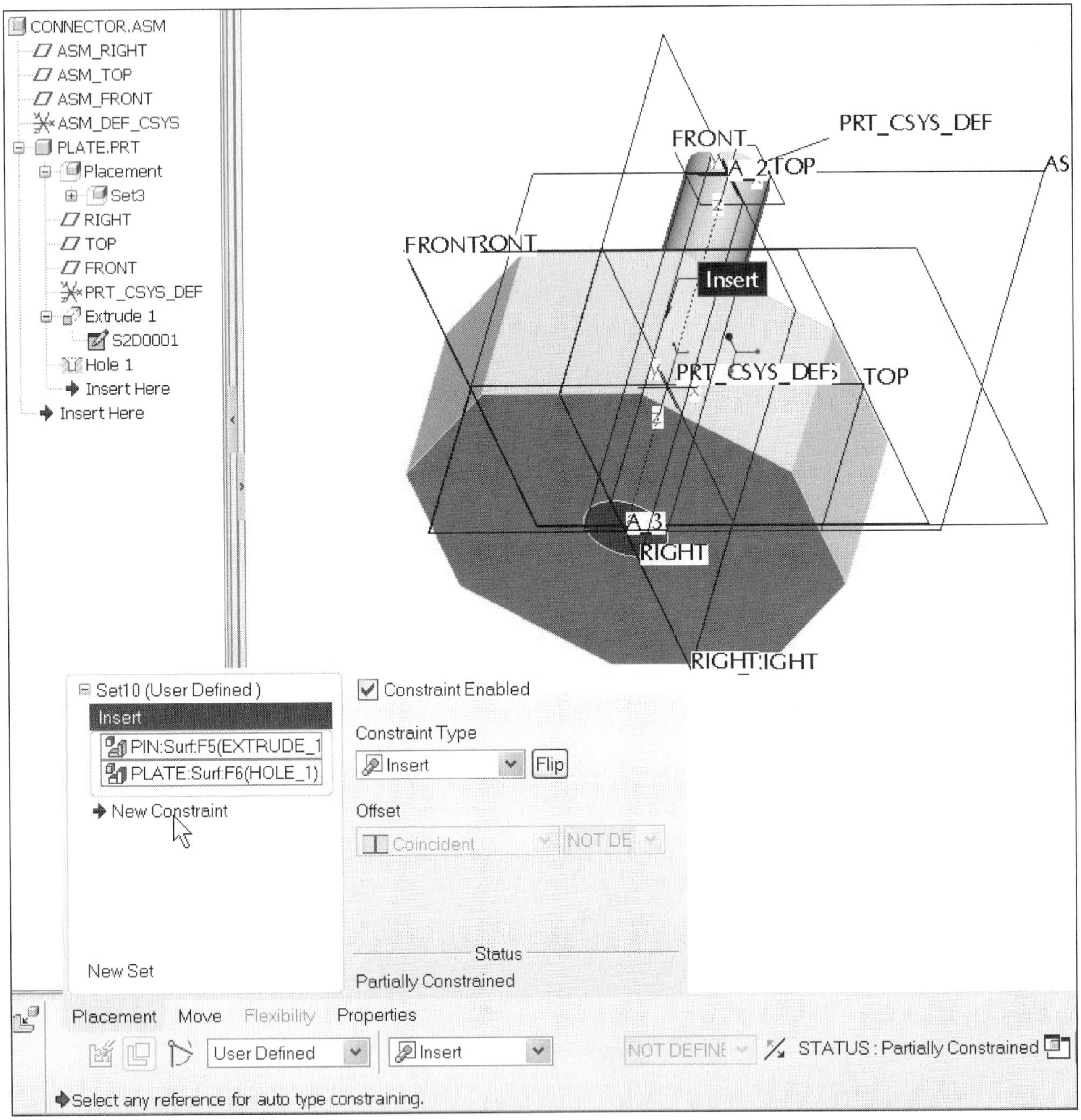

Figure 1.14(e) New Constraint

Pick on the Plate's front surface [Fig. 1.14(f)]

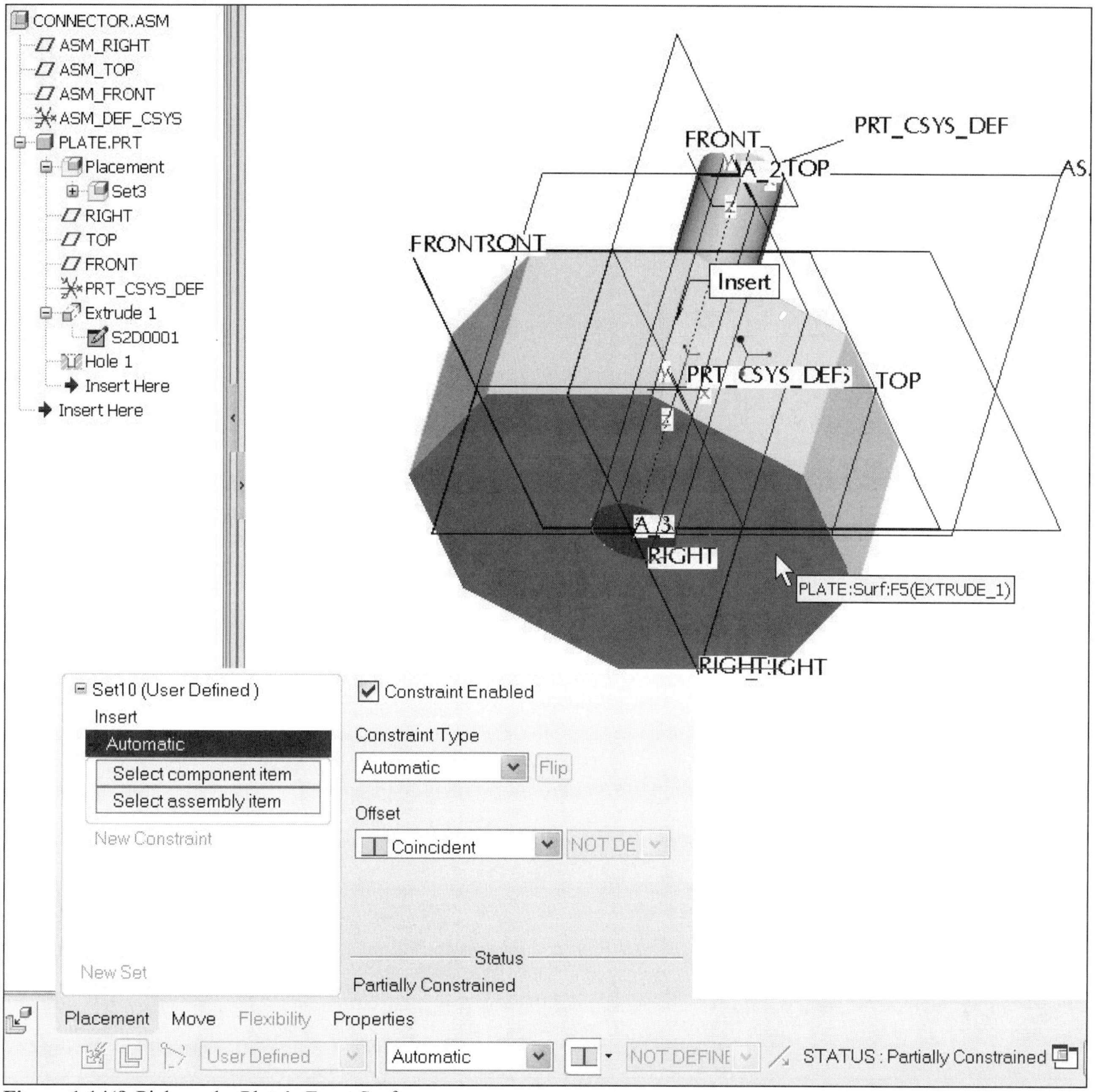

Figure 1.14(f) Pick on the Plate's Front Surface

Pick on the Pin's end face surface [Fig. 1.14(g)]

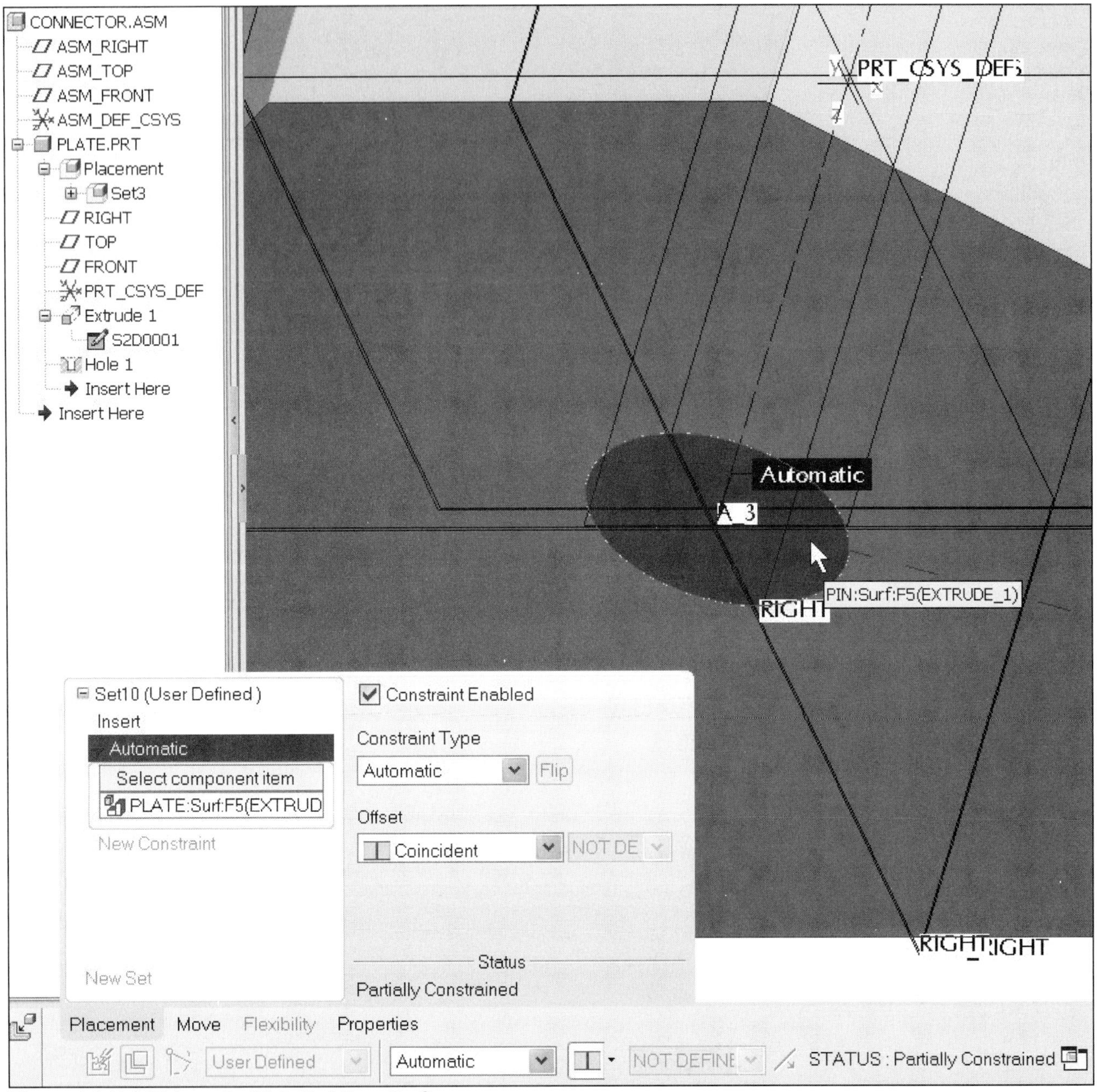

Figure 1.14(g) Pick on the Pin's End Face Surface

Click: Coincident [Fig. 1.14(h)] ⇒ Offset [Fig. 1.14(i)] ⇒ click in value field next to Offset, type **.50** Offset 0.50 [Fig. 1.14(j)] ⇒ **Enter**

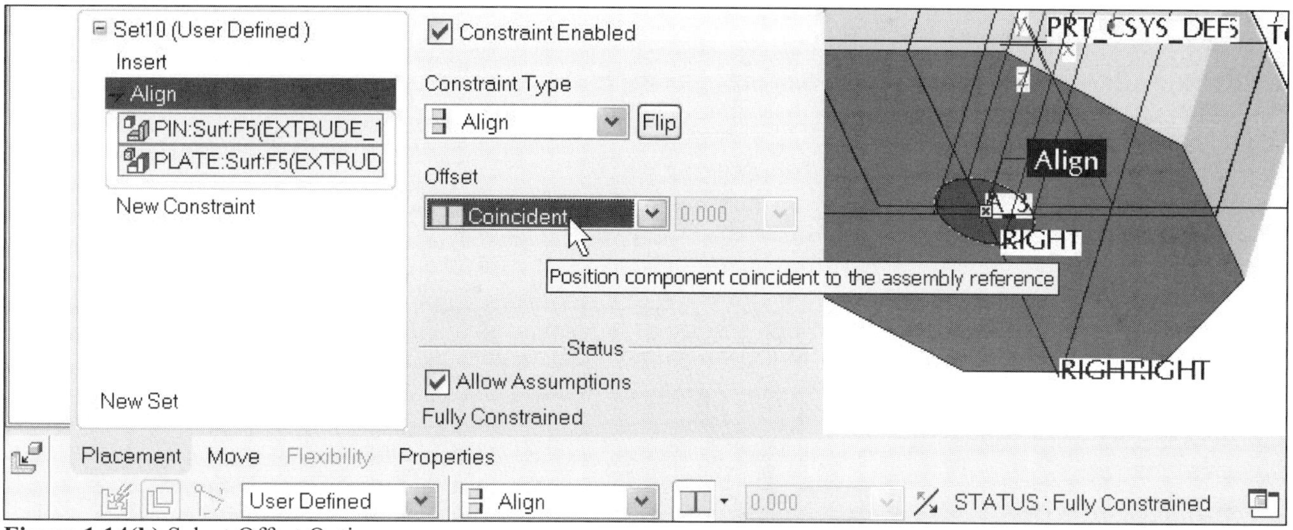

Figure 1.14(h) Select Offset Option

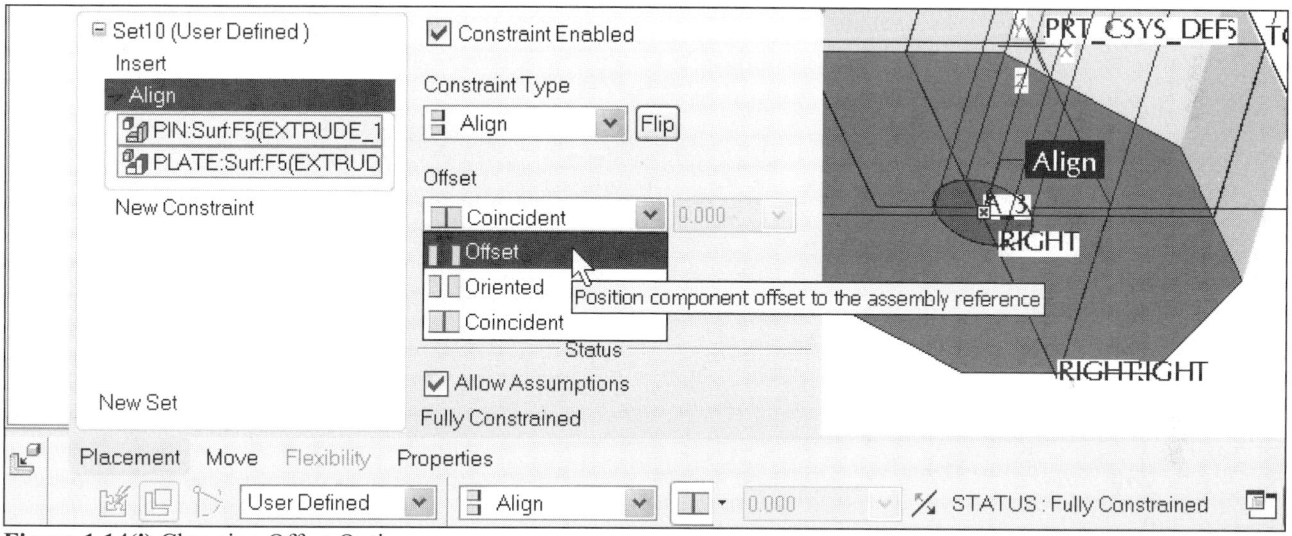

Figure 1.14(i) Changing Offset Option

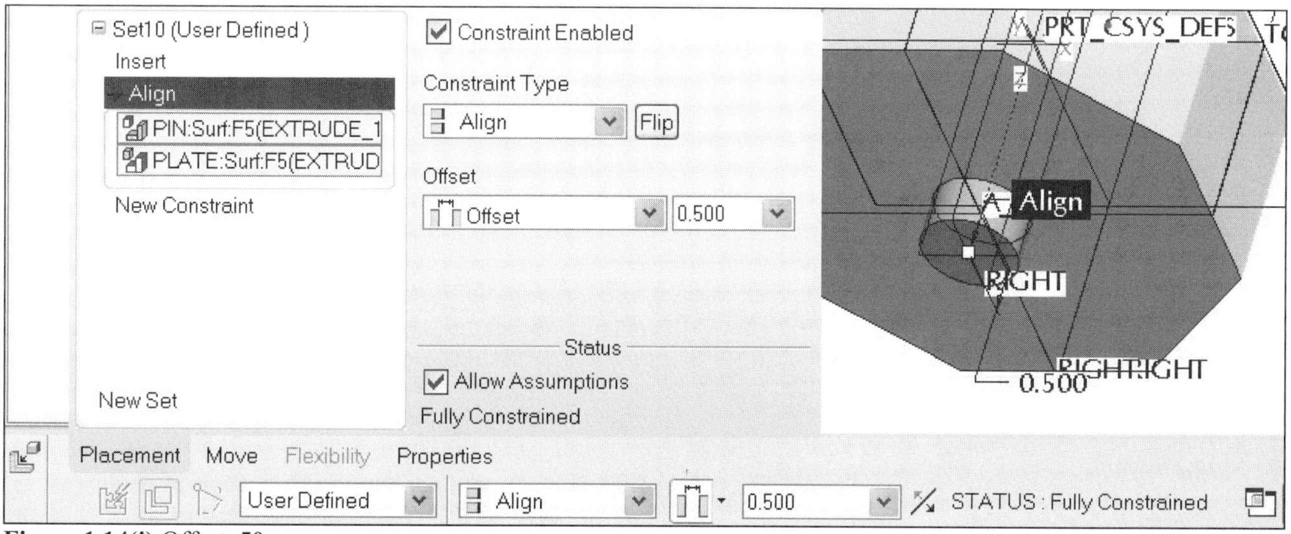

Figure 1.14(j) Offset **.50**

Click: ☑ from the dashboard [Fig. 1.14(k)] ⇒ **Ctrl+S** ⇒ **OK** [Fig. 1.14(l)]

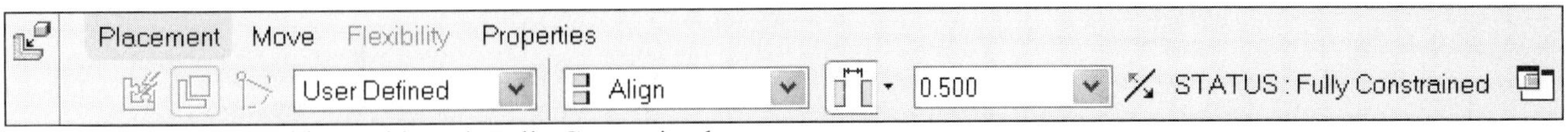

Figure 1.14(k) Assembly Dashboard: Fully Constrained

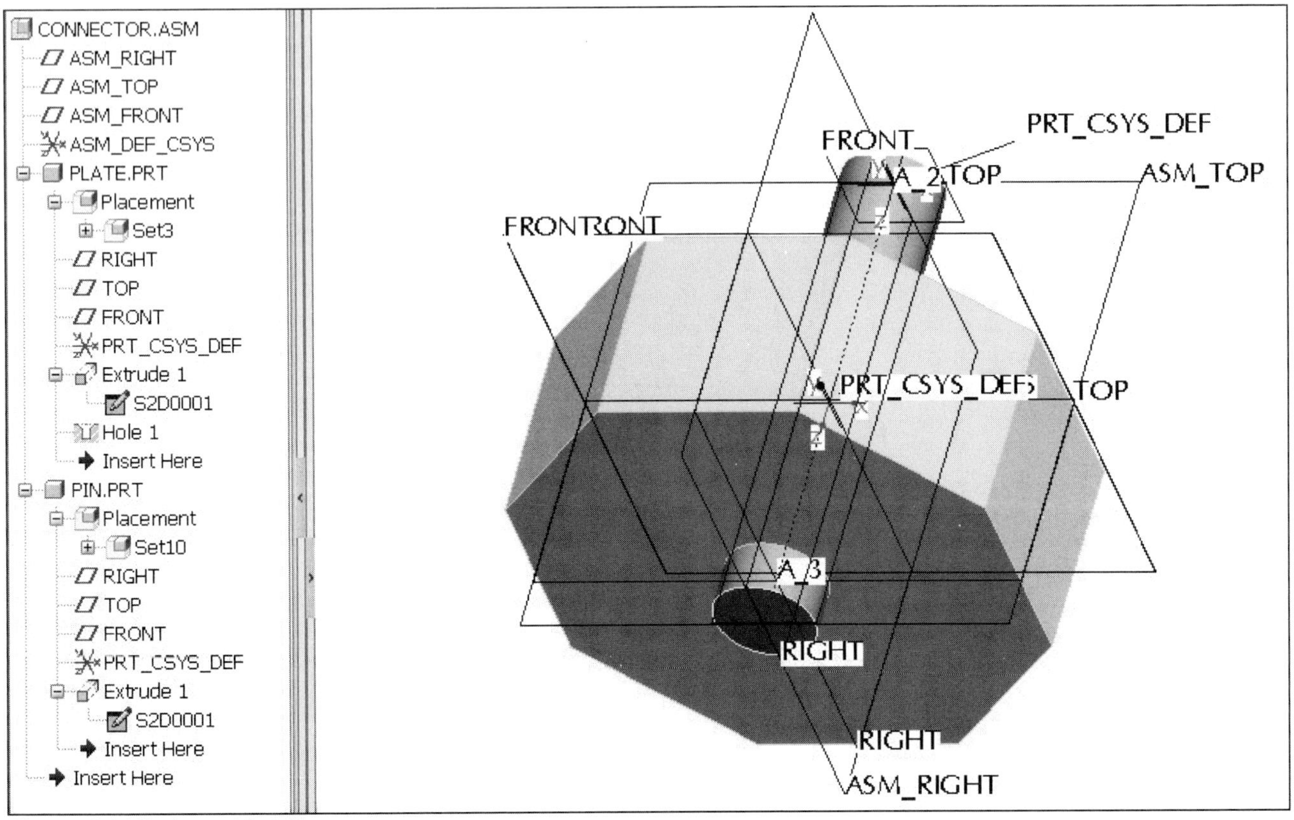

Figure 1.14(l) Assembled Pin

You have completed your first Pro/ENGINEER assembly. In the next section you will create a set of drawings for the Plate component and the Connector assembly.

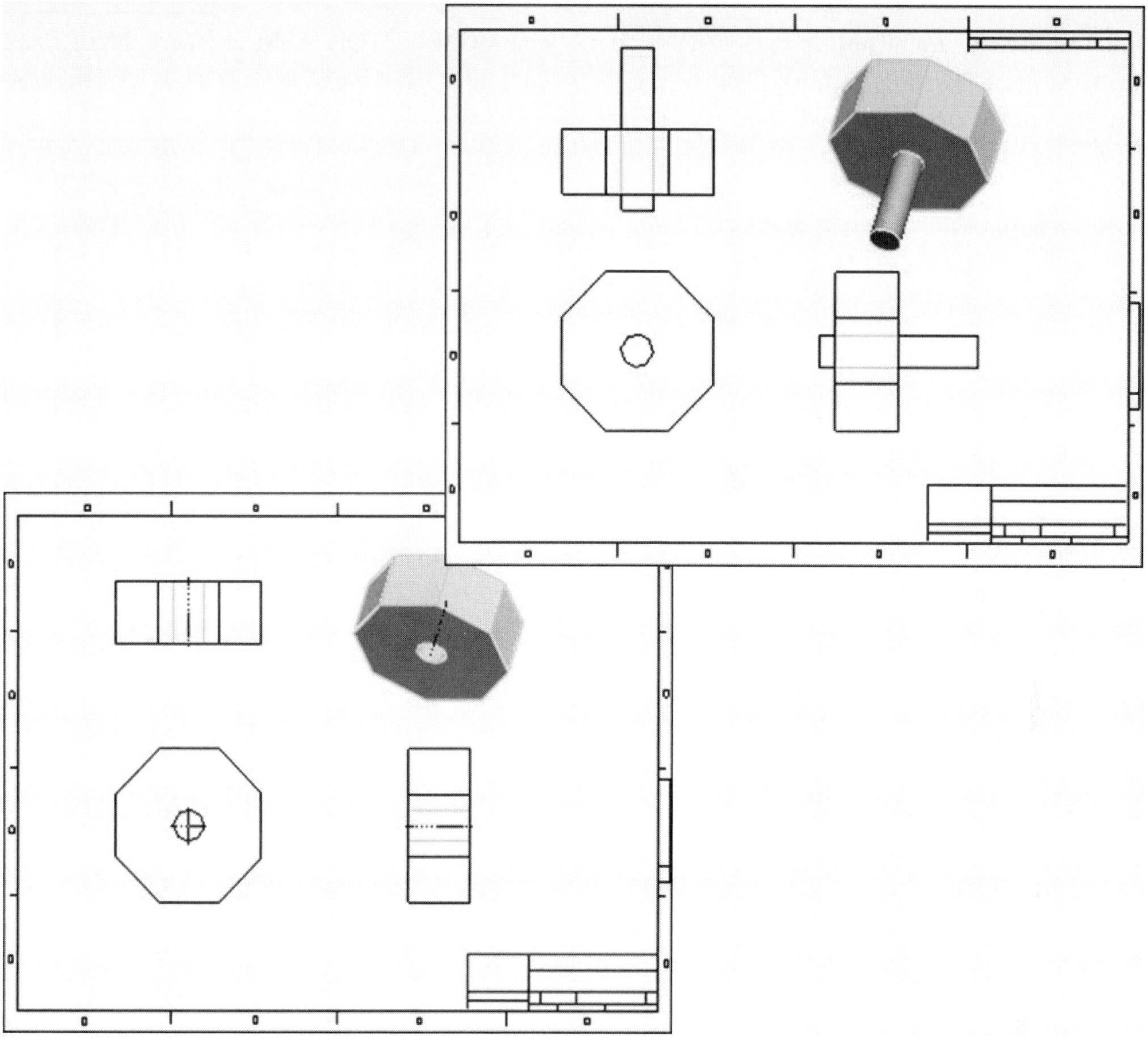

Figure 1.15 Drawings

Creating Drawings

Using the parts and the assembly just created, you will now create an assembly drawing of the Connector and a detail drawing of the Plate (Fig. 1.15).

The Drawing mode in Pro/E enables you to create and manipulate engineering drawings that use the 3D model (part or assembly) as a geometry source. You can pass dimensions, notes, and other elements of design between the 3D model and its views on a drawing.

If you are continuing from the last section, then skip this set of commands:

Open **Pro/ENGINEER Wildfire 4.0** using a shortcut icon on your Desktop ⇒ **File** ⇒ **Set Working Directory** ⇒ select the working directory or accept the default directory (make sure you set the same working directory as where your parts and assembly models were saved) ⇒ **OK** ⇒ **File** ⇒ **Open** ⇒ **pin.prt** ⇒ **Open** ⇒ **File** ⇒ **Open** ⇒ **plate.prt** ⇒ **Open** ⇒ **File** ⇒ **Open** ⇒ **connector.asm** ⇒ **Open** *[you now have all three objects (the two parts and the assembly) "in session"]* ⇒ **Shading**

Click: **Create a new object** ⇒ Drawing ⇒ Name CONNECTOR_ASM ⇒ Use default template [Fig. 1.16(a)] ⇒ **OK** [Fig. 1.16(b)] *[Note: if the Default Model is not the CONNECTOR.ASM, then click: Browse ⇒ connector.asm [Fig. 1.16(c)] ⇒ Open]*

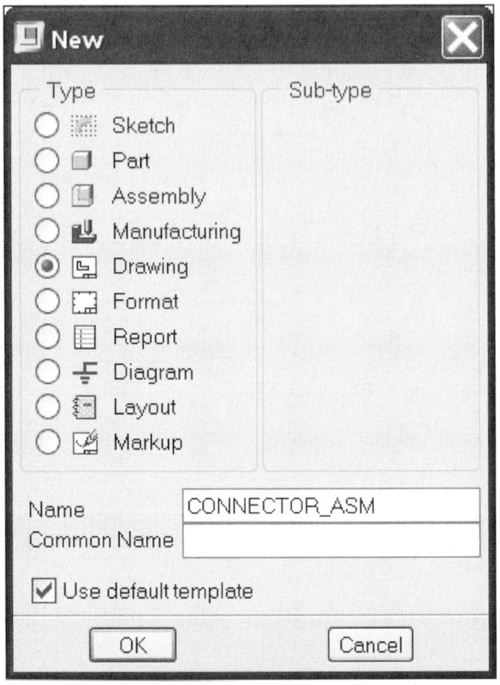

Figure 1.16(a) New Dialog Box, Drawing

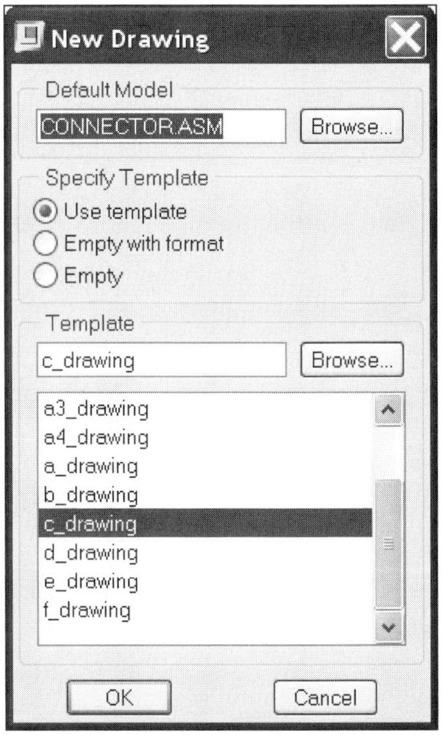

Figure 1.16(b) New Drawing Dialog Box

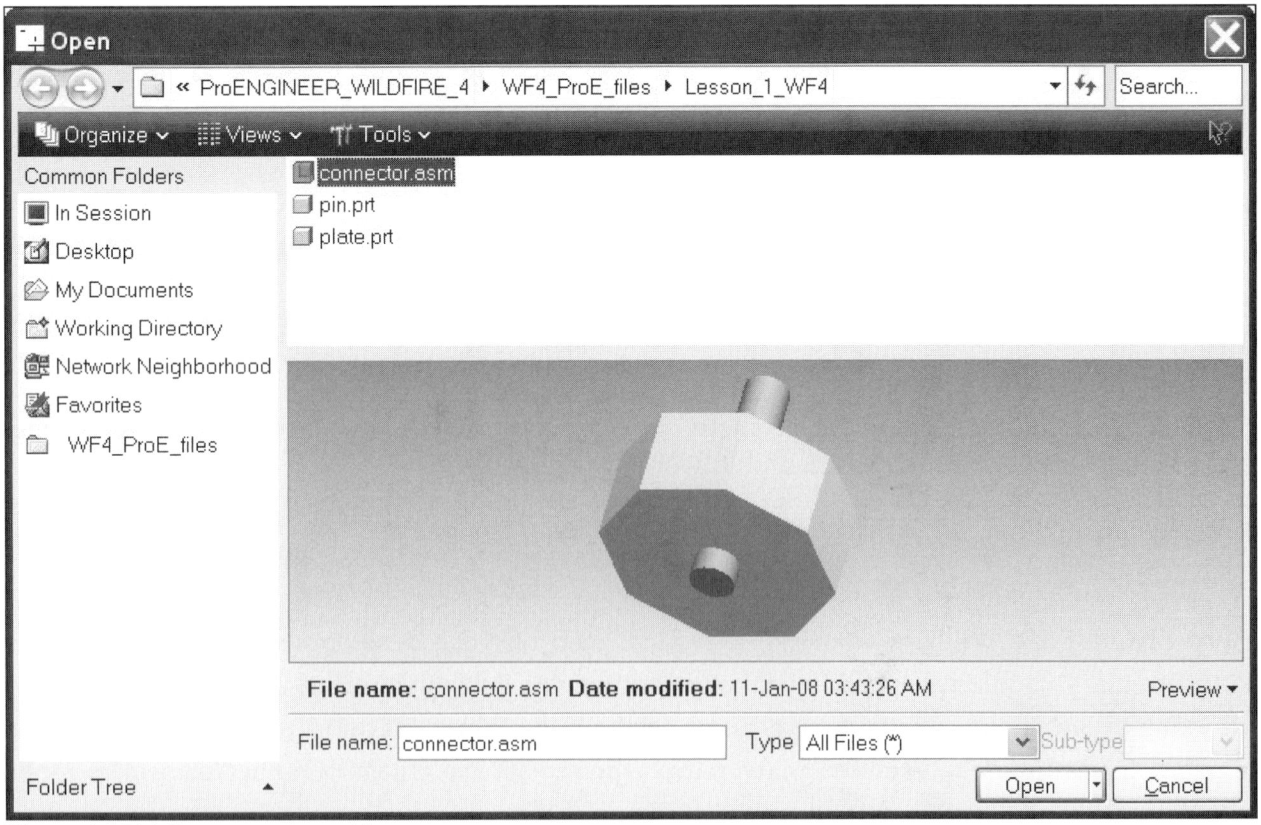

Figure 1.16(c) Connector Assembly

Click: **OK** with the default model set as 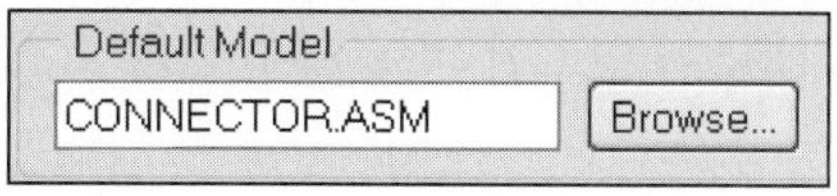

the drawing will display [Fig. 1.16(d)] ⇒ **RMB** ⇒ **Page Setup** [Fig. 1.16(e)]

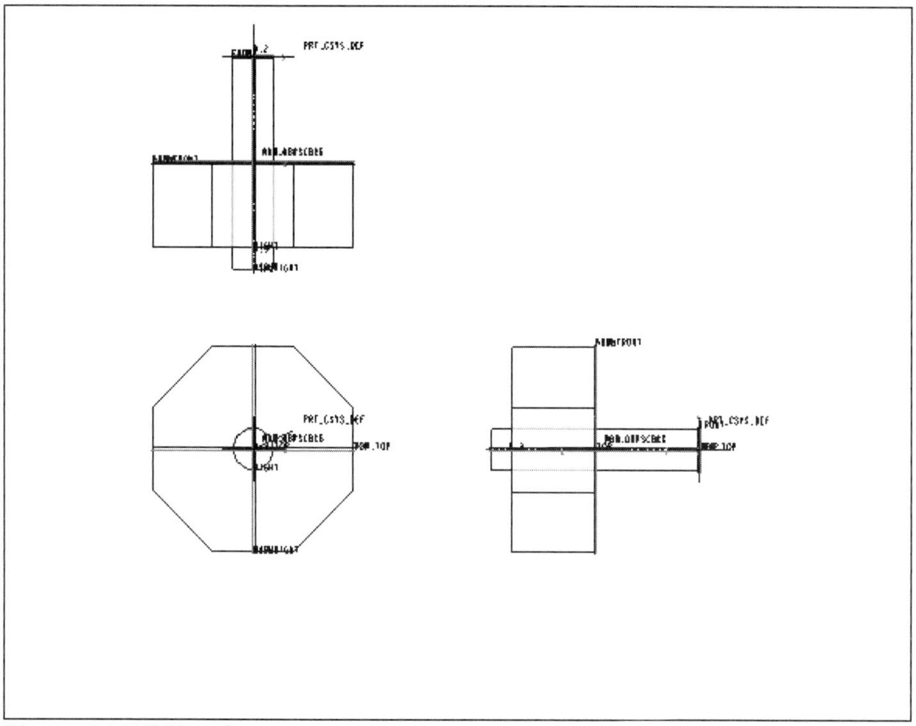

Figure 1.16(d) Drawing

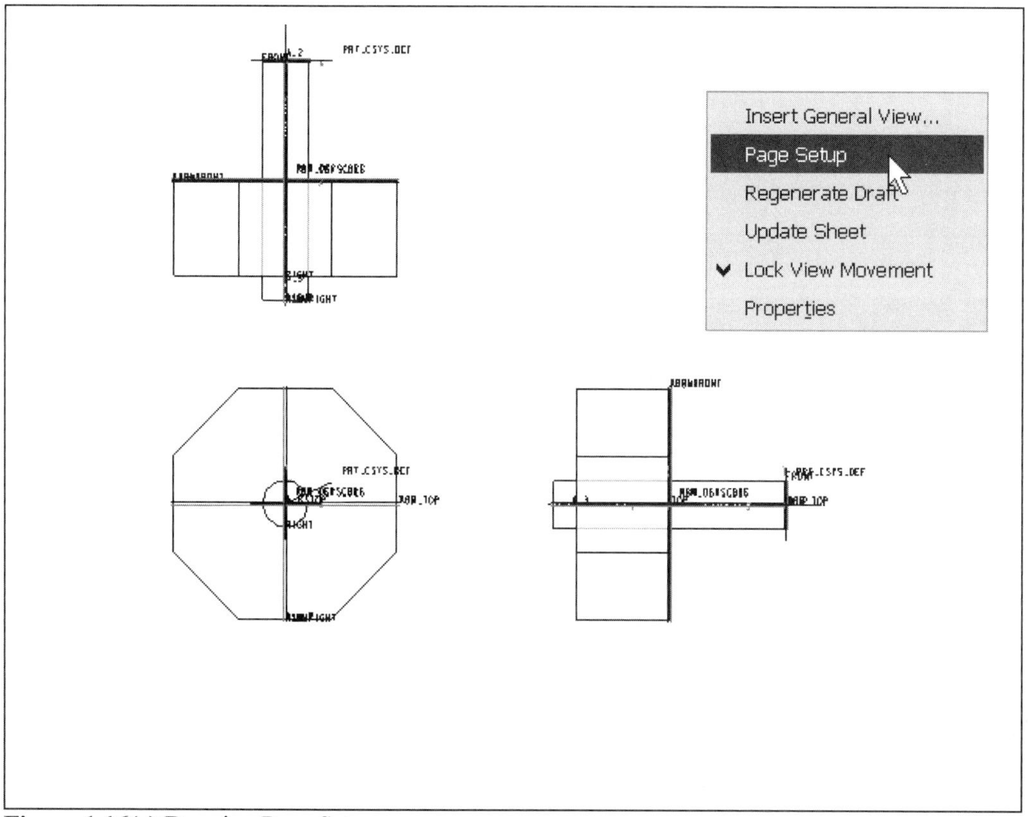

Figure 1.16(e) Drawing Page Setup

Page Setup dialog box displays [Fig. 1.16(f)] ⇒ click in the **C Size** field [Fig. 1.16(g)] ⇒ to see options 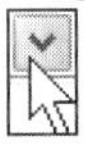 [Fig. 1.16(h)] ⇒ **Browse** [Fig. 1.16(i)] ⇒ **c.frm** [Fig. 1.16(j)] ⇒ **Open** [Fig. 1.16(k)] ⇒ **OK**

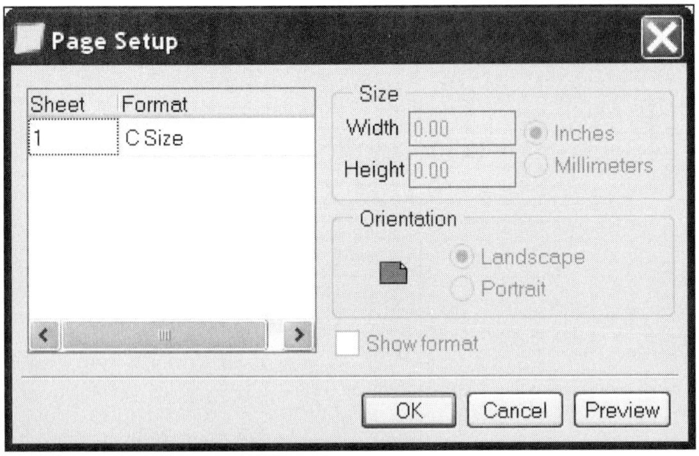

Figure 1.16(f) Page Setup Dialog Box

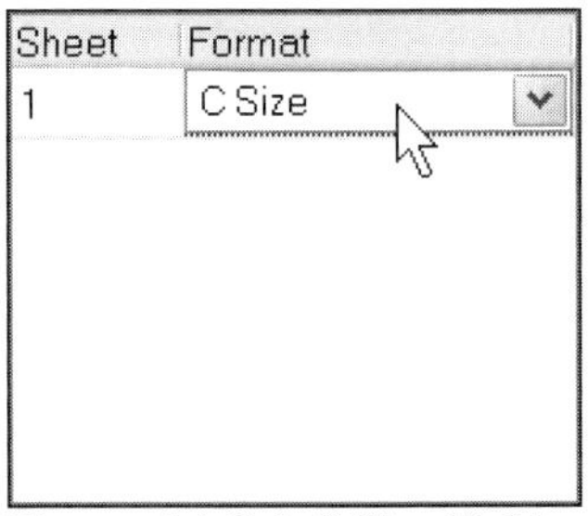

Figure 1.16(g) C Size Field

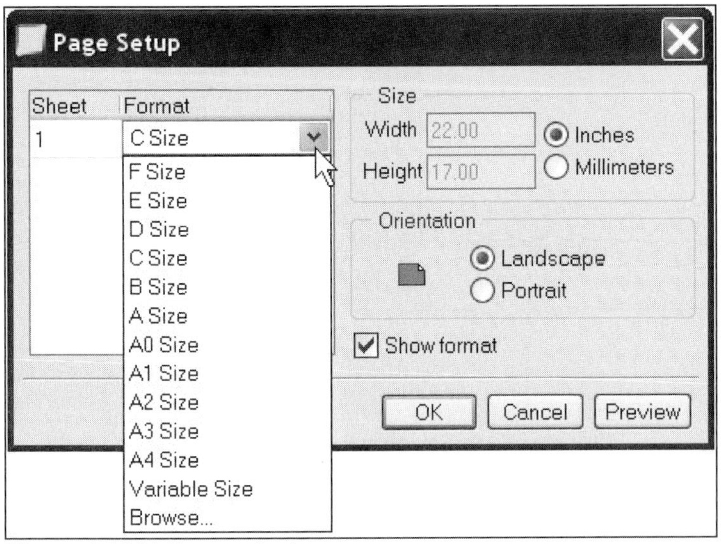

Figure 1.16(h) Format Options

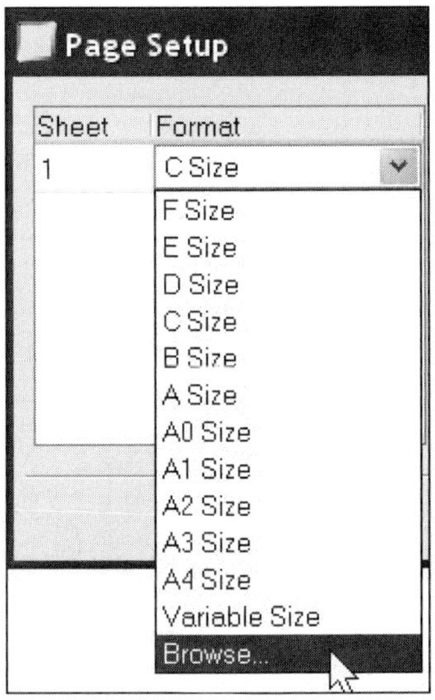

Figure 1.16(i) Browse

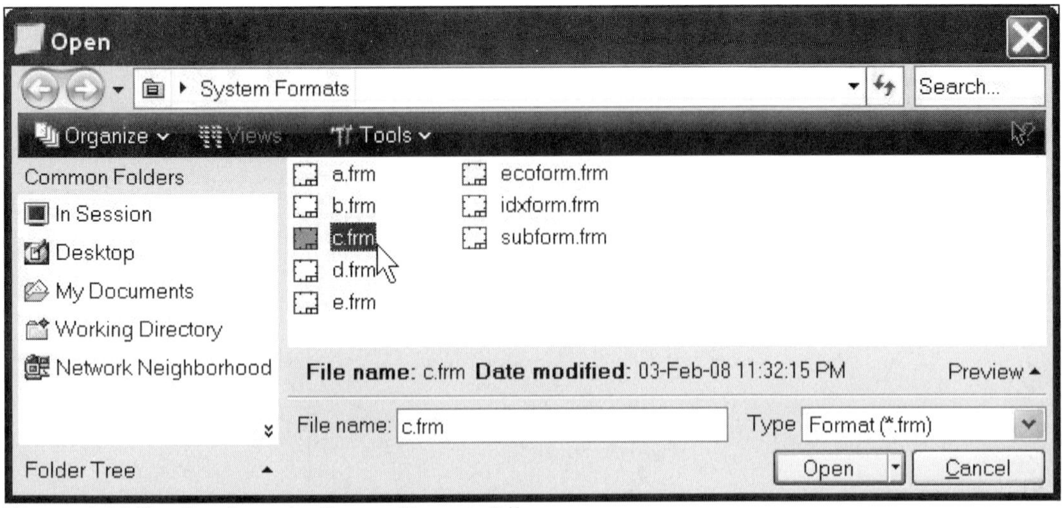

Figure 1.16(j) c.frm from the System Formats Library
C

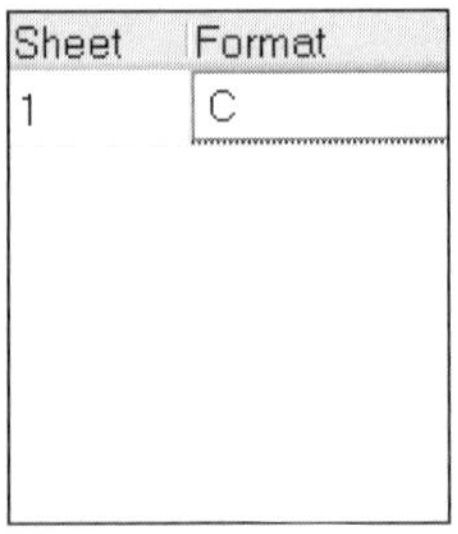

Figure 1.16(k) Format

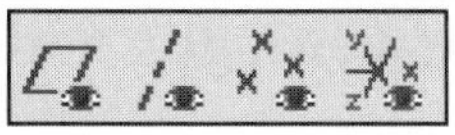 **off** ⇒  ⇒ **RMB** ⇒ **Insert General View** [Fig. 1.16(l)] ⇒ **No Combine State** [Fig. 1.16(m)] ⇒ **OK** ⇒ pick a position for the new view [Fig. 1.16(n)]

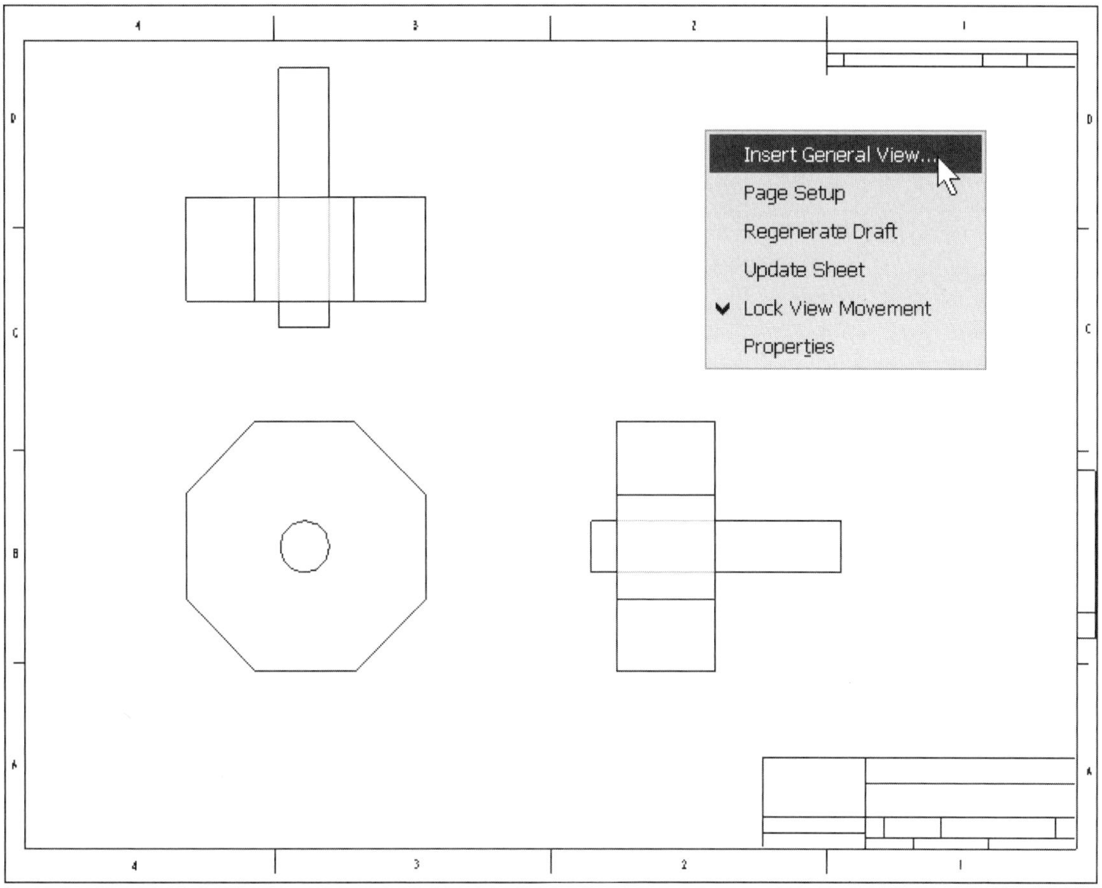

Figure 1.16(l) Insert General View

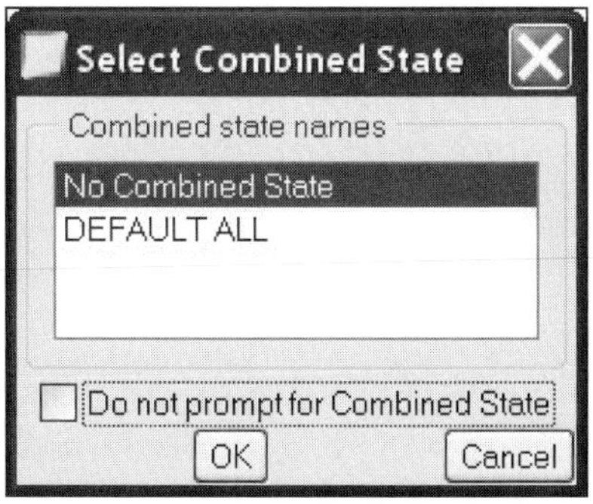

Figure 1.16(m) Select No Combined State

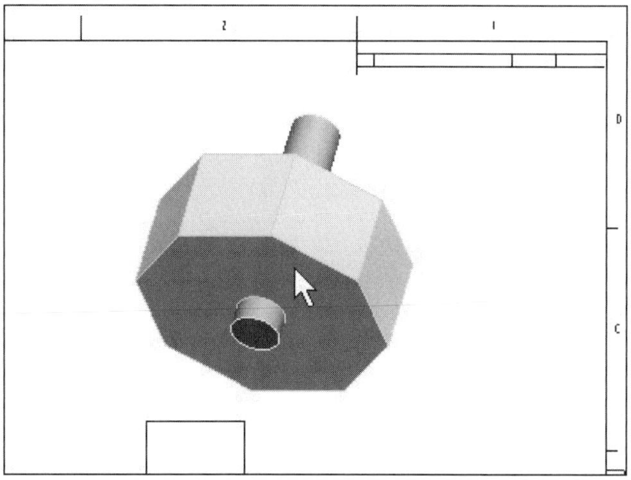

Figure 1.16(n) Pick a Position for the New View

Click: **OK** from the Drawing View dialog box [Fig. 1.16(o)] ⇒ **Ctrl+S** ⇒ **OK** the drawing is now complete [Fig. 1.16(p)] ⇒  ⇒ **LMB** (to deselect)

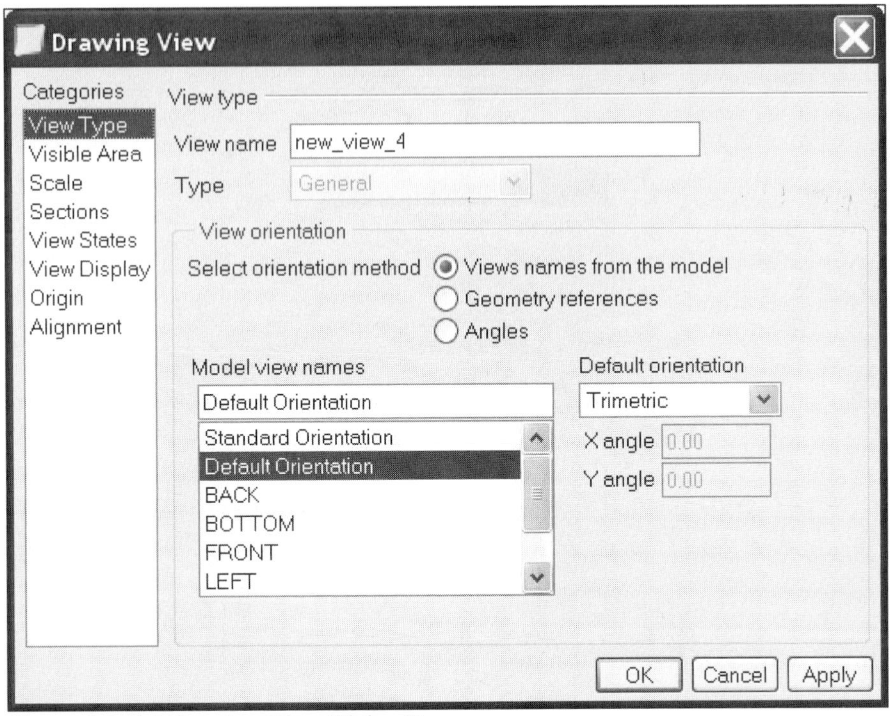

Figure 1.16(o) Drawing View Dialog Box

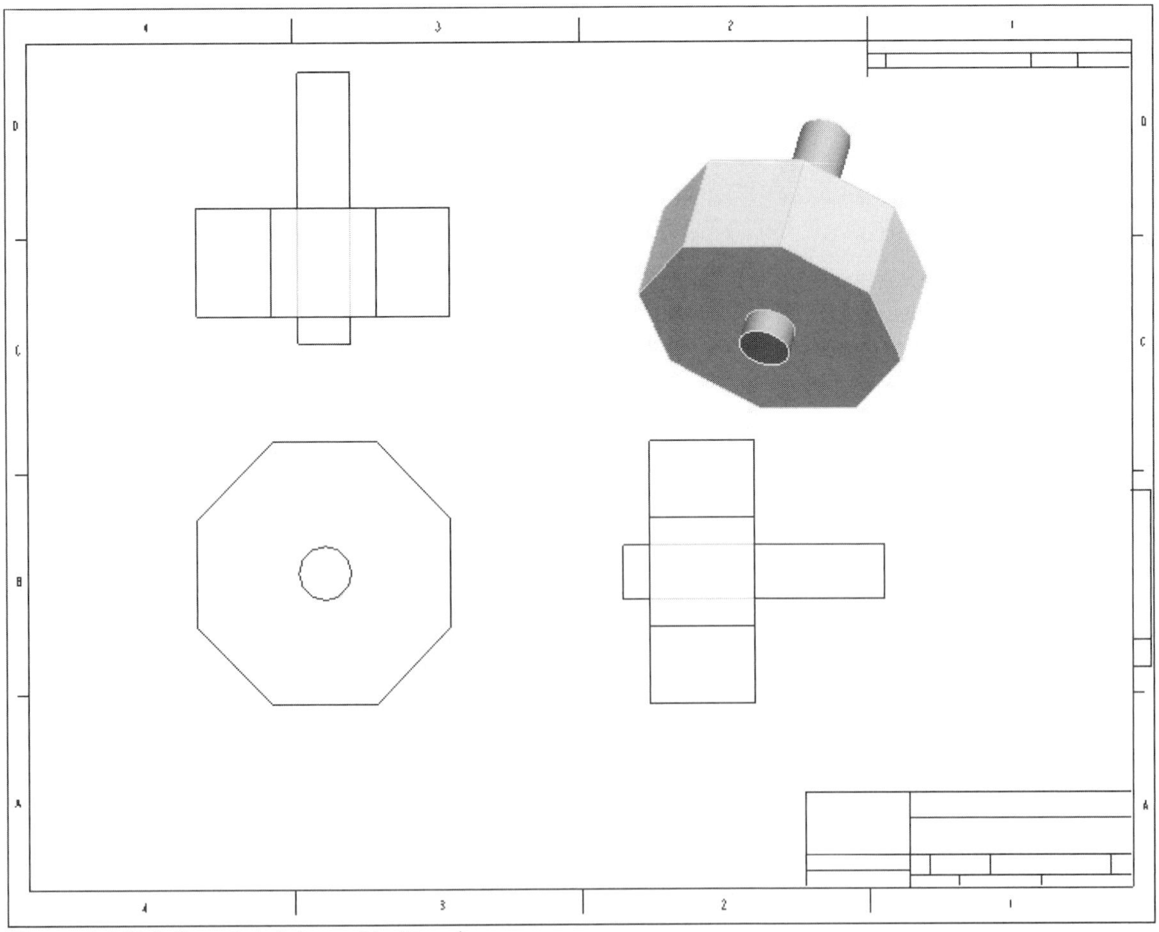

Figure 1.16(p) Completed Assembly Drawing

Click: Create a new object ⇒ Drawing ⇒ Name PLATE ⇒ Use default template [Fig. 1.17(a)] ⇒ **OK** ⇒ Default Model **Browse** [Fig. 1.17(b)] ⇒ Preview ⇒ **plate.prt** [Fig. 1.17(c)] ⇒ **Open**

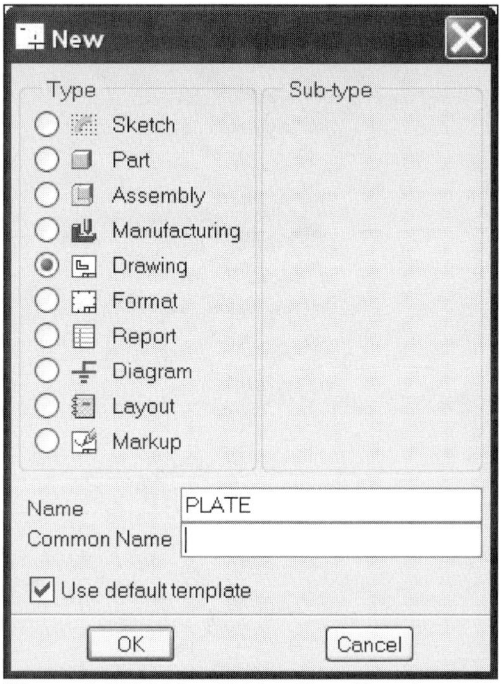

Figure 1.17(a) New Dialog Box, Drawing

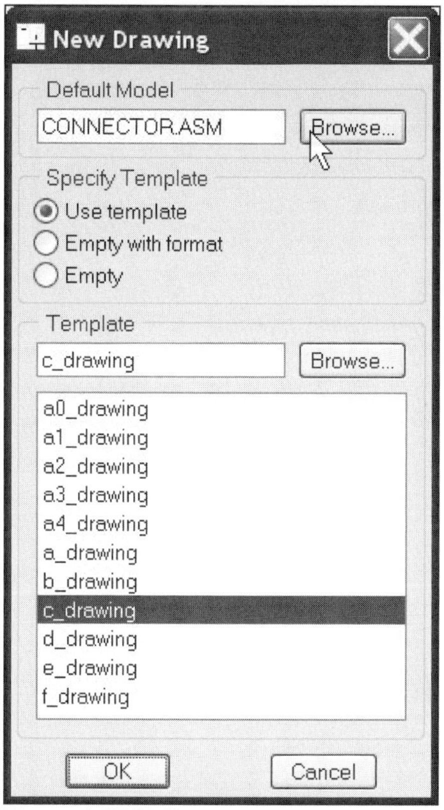

Figure 1.17(b) Browse

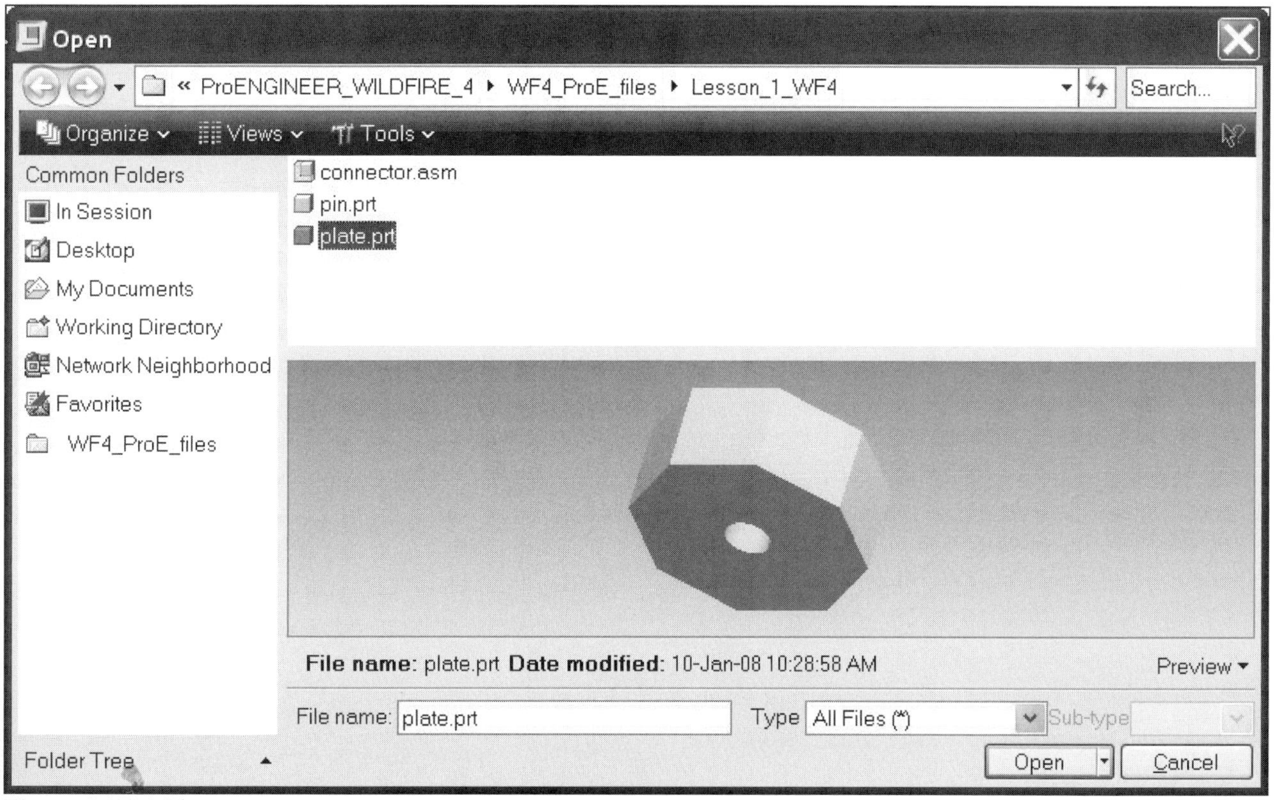

Figure 1.17(c) Plate

Click: **OK** [Fig. 1.17(d)] drawing opens ⇒ **RMB** ⇒ **Page Setup** [Figs. 1.17(e-f)] ⇒ click in the **C Size** field [Fig. 1.17(g)] ⇒  to see options [Fig. 1.17(h)] ⇒ **Browse** [Fig. 1.17(i)] ⇒ **c.frm** [Fig. 1.17(j)] ⇒ **Open** ⇒ **OK** formatted drawing is displayed [Fig. 1.17(k)]

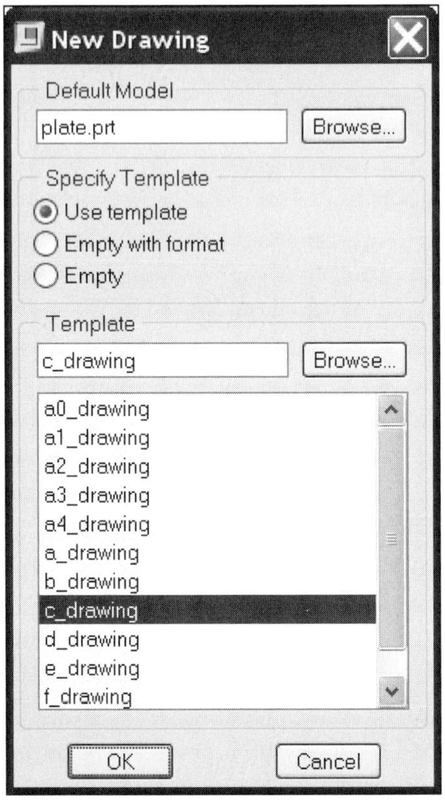

Figure 1.17(d) Browse for a different Template

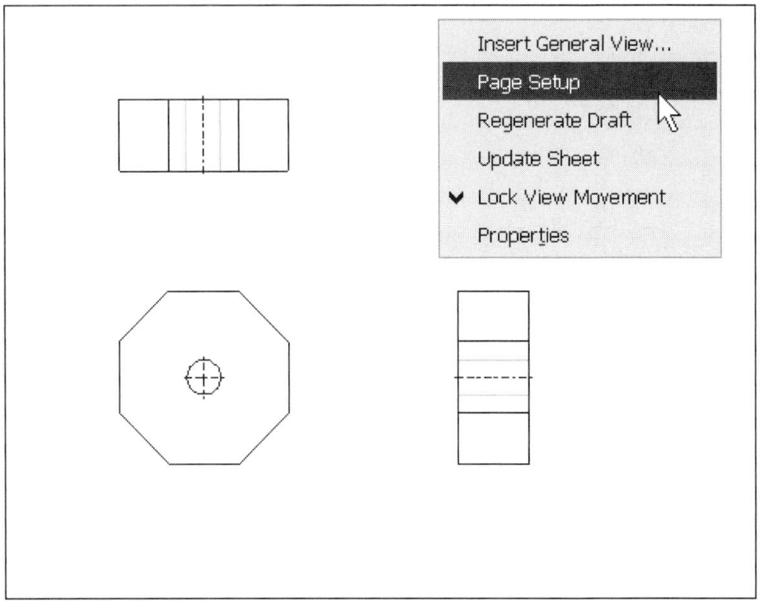

Figure 1.17(e) Plate Drawing, RMB ⇒ Page Setup

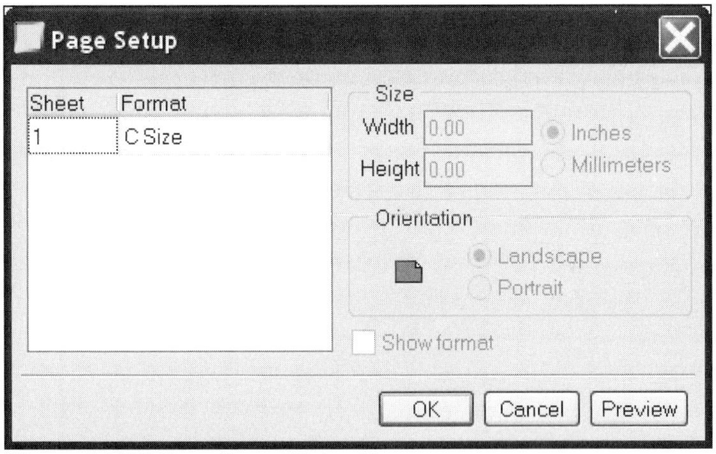

Figure 1.17(f) Page Setup Dialog Box

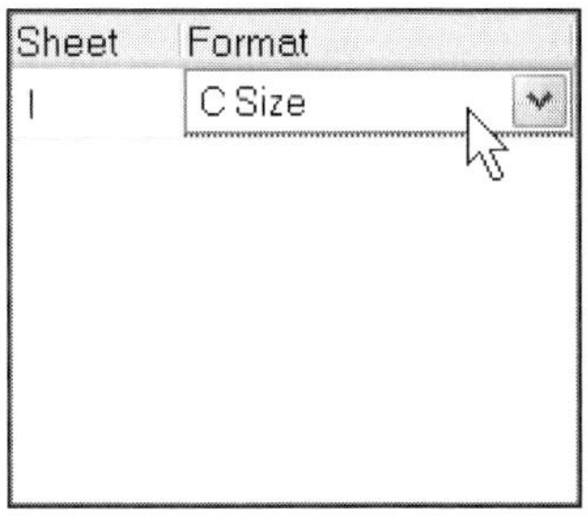

Figure 1.17(g) C Size Field

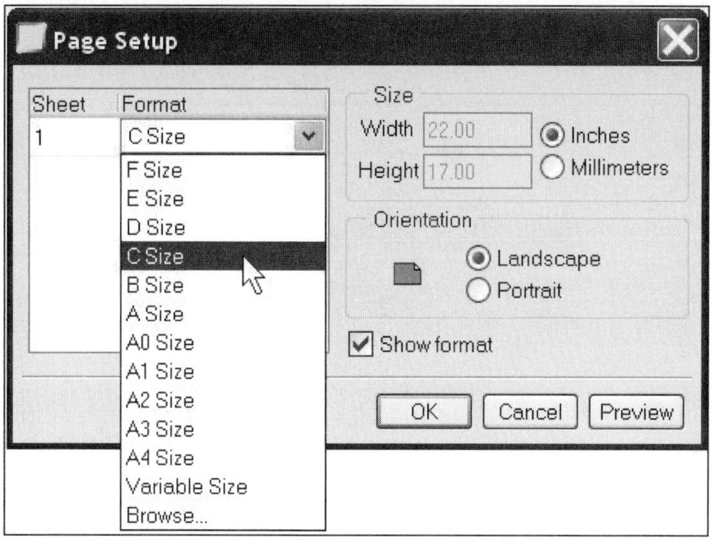

Figure 1.17(h) Page Setup Dialog Box, Format Options

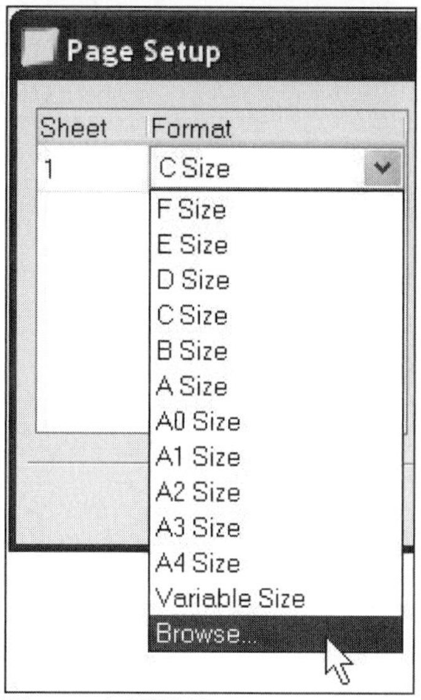

Figure 1.17(i) Browse

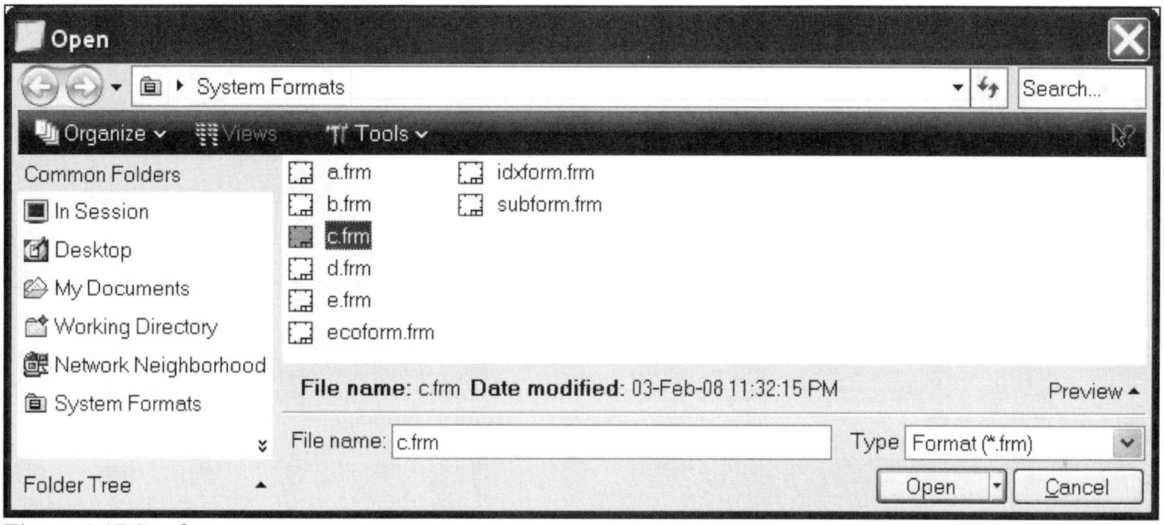

Figure 1.17(j) c.frm

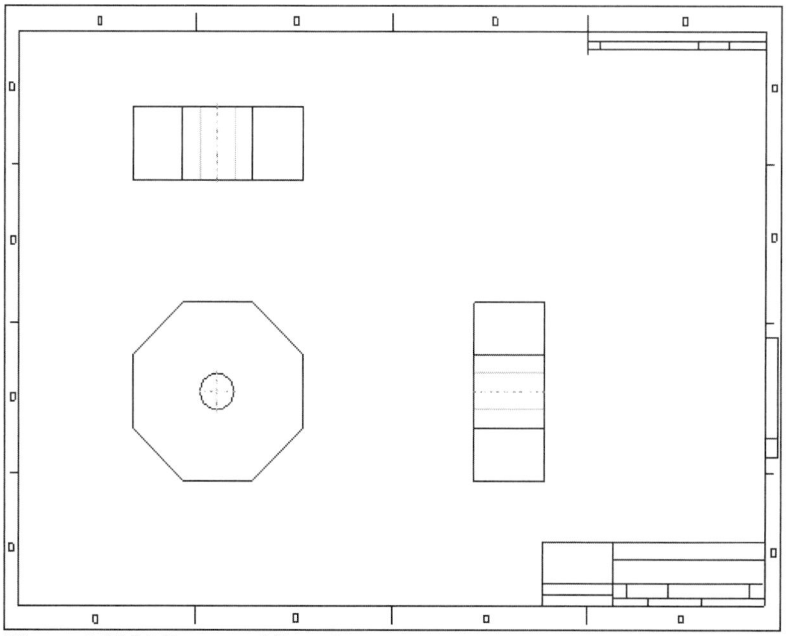

Figure 1.17(k) Formatted Drawing

Click: **RMB** ⇒ **Insert General View** [Fig. 1.17(l)] ⇒ pick a position for the new view ⇒ **OK** from the Drawing View dialog box [Fig. 1.17(m)]

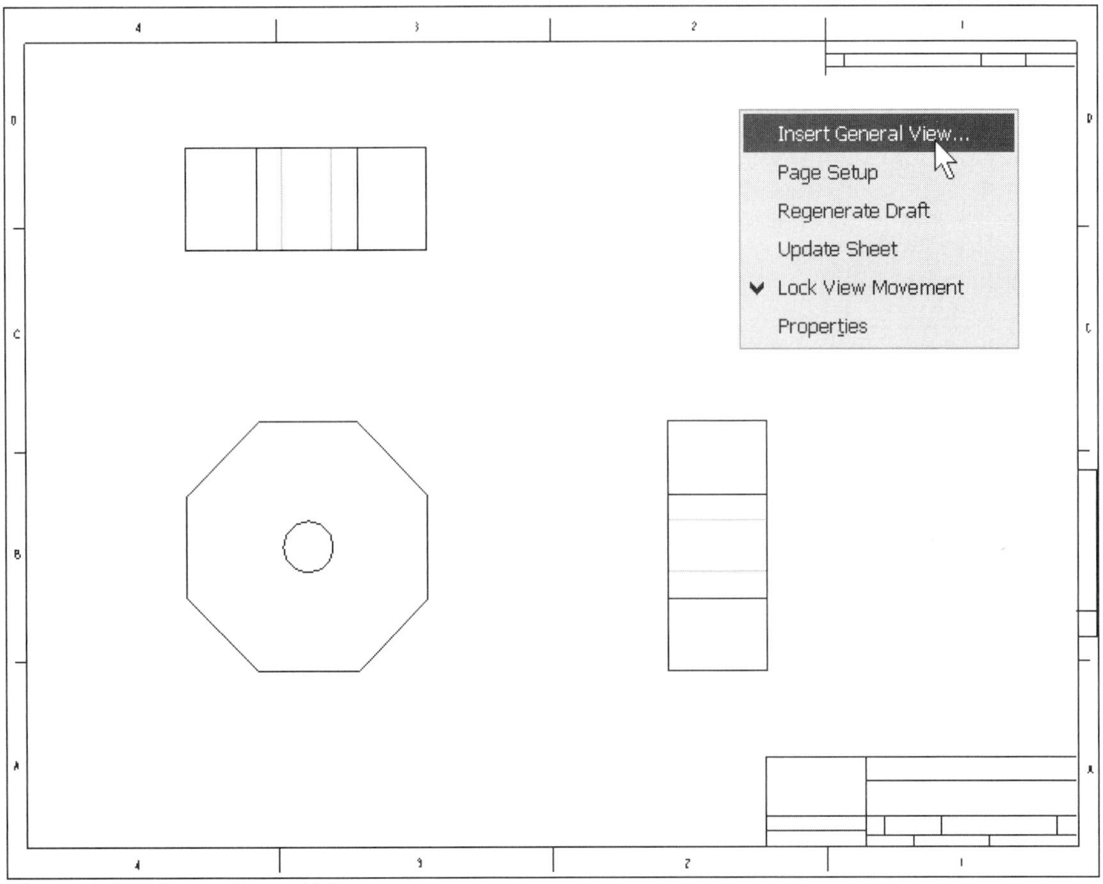

Figure 1.17(l) Insert General View

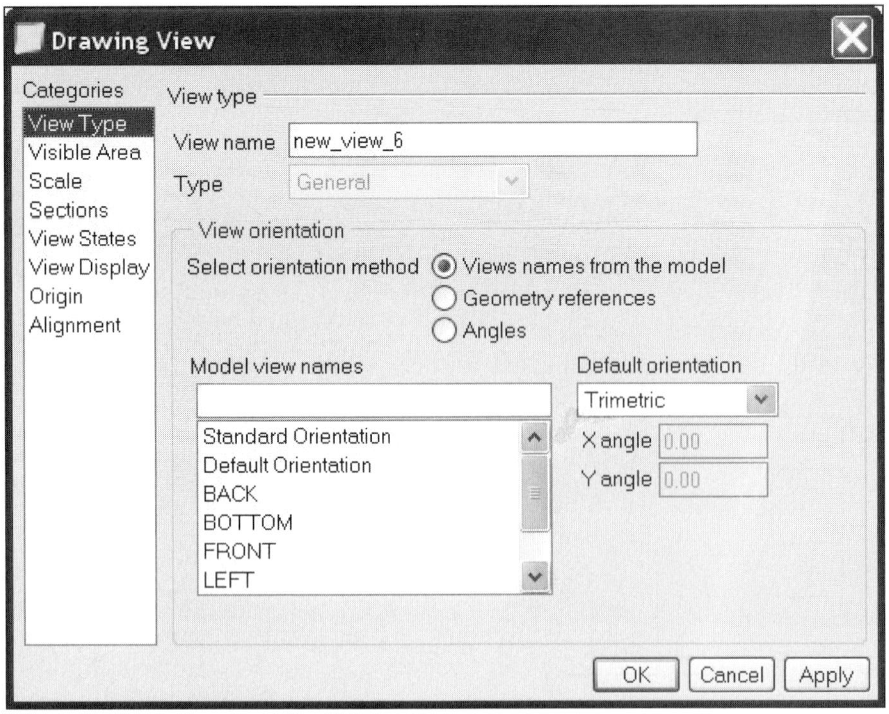

Figure 1.17(m) Drawing View Dialog Box

Click: **LMB** ⇒ ⇒ **Open the Show/Erase dialog box** [Fig. 1.18(a)] ⇒ **Show** ⇒ **Axis** [Fig. 1.18(b)] ⇒ Show All **Show all items** [Fig. 1.18(c)] ⇒ **Yes** [Fig. 1.18(d)] ⇒ **Accept All** [Fig. 1.18(e)] ⇒ **Close** ⇒ **LMB** ⇒ [Fig. 1.18(f)] ⇒ ⇒ **MMB**

Figure 1.18(a) Show/Erase

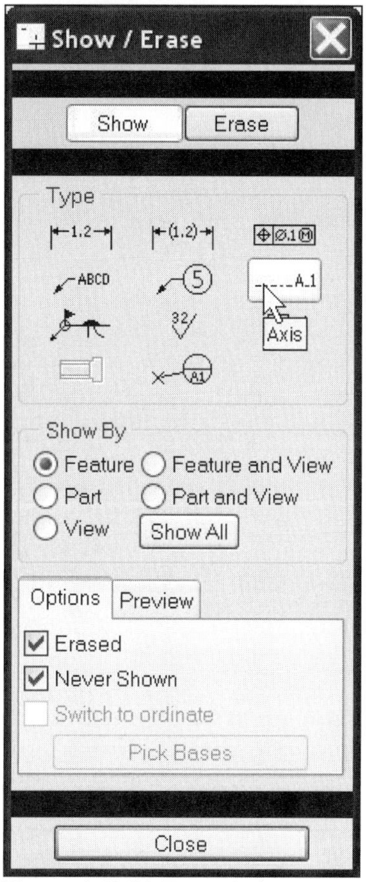

Figure 1.18(b) Type: Axis

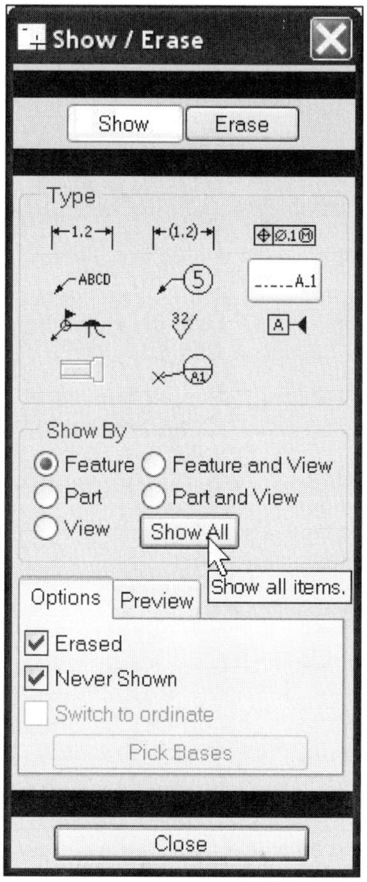

Figure 1.18(c) Show All

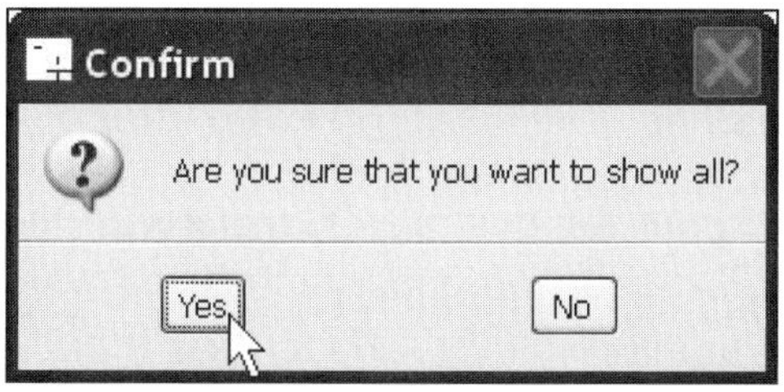

Figure 1.18(d) Confirm Dialog Box ⇒ Yes

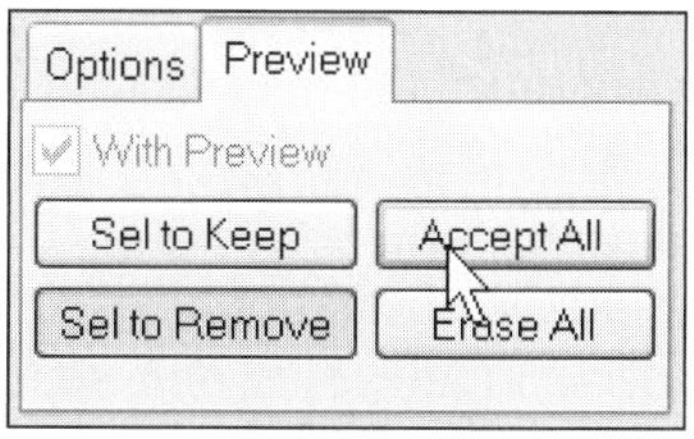

Figure 1.18(e) Accept All

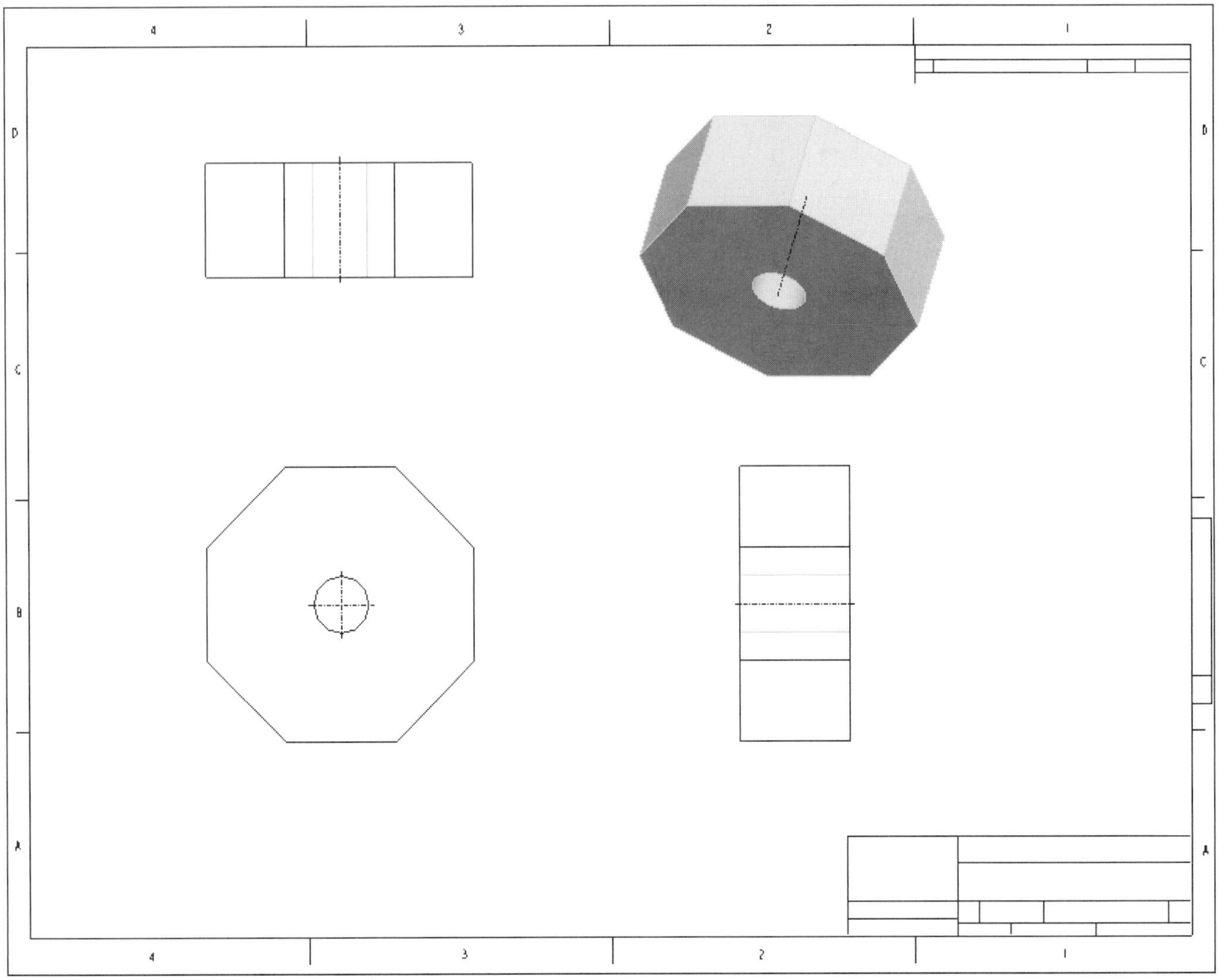

Figure 1.18(f) Completed Plate Drawing

Click: **Window ⇒ Close** *(five times)* ⇒ **File ⇒ Erase ⇒ Not Displayed ⇒ OK**

This concludes the overview of Pro/ENGINEER Wildfire 4.0.

Spend some time exploring the website *www.cad-resources.com* for downloads, materials, and other CAD related information. To get on the CAD-Resources email list for job opportunities, software updates, tutorials, and new lessons, send an email to *cad@cad-resources.com*.

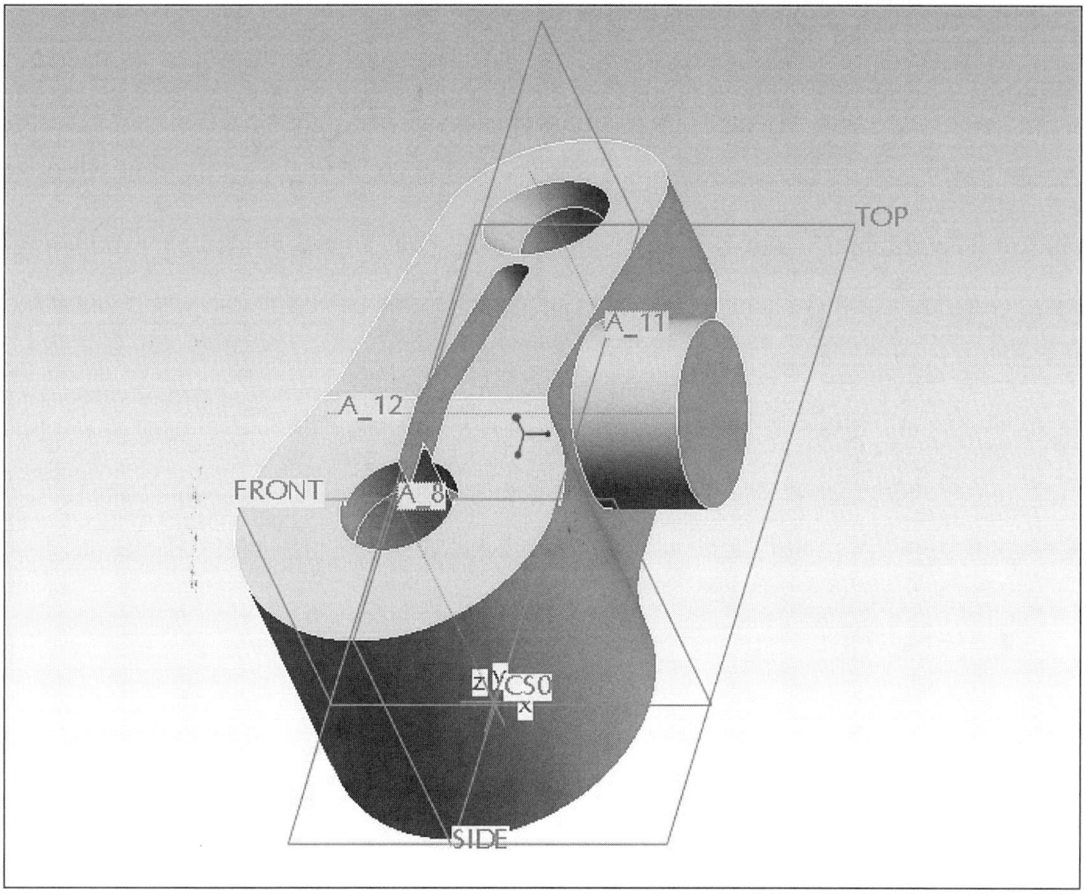

Figure 2.1 Part 500152 - Swing/Pull Clamp Arm (SPX Fluid Power Part)

OBJECTIVES

- Understand the **User Interface (UI)**
- **Download 3D CAD models** from the internet
- Master the **File Functions**
- Learn how to **Email** active Pro/ENGINEER objects
- Become familiar with the **Help** center
- Be introduced to the **Display** and **View** capabilities
- Use **Mouse Buttons** to **Pan**, **Zoom**, and **Rotate** the object
- Change **Display Settings**
- Investigate an object with **Information Tools**
- Experience the **Model Tree** functionality

Pro/ENGINEER Wildfire 4.0

This lesson will introduce you to Pro/ENGINEER's working environment. An existing 3D CAD model (Fig. 2.1) will be downloaded and used to demonstrate the UI (user interface) and the general interaction required to master Pro/E.

You will be using a part available on the web using the **Browser**. If you are not connected to the Internet, your instructor will provide you with a simple start part. In addition, if you have Pro/Library installed, you may use any library part that you wish.

Pro/ENGINEER's Interface

The Pro/ENGINEER user interface consists of a navigation window, an embedded Web browser, the menu bar, toolbars, information areas, and the graphics window (Fig. 2.2).

- **Navigation Window** Located on the left side of the Pro/E main window, it includes tabs for the Model Tree and Layer Tree, Folder Browser, Favorites, History and Connections.
- **Browser Window** An embedded Web browser is located to the right of the navigation window. This browser provides you access to internal or external Web sites.
- **Graphics Window** The graphics window is the main working space (main window) for modeling.

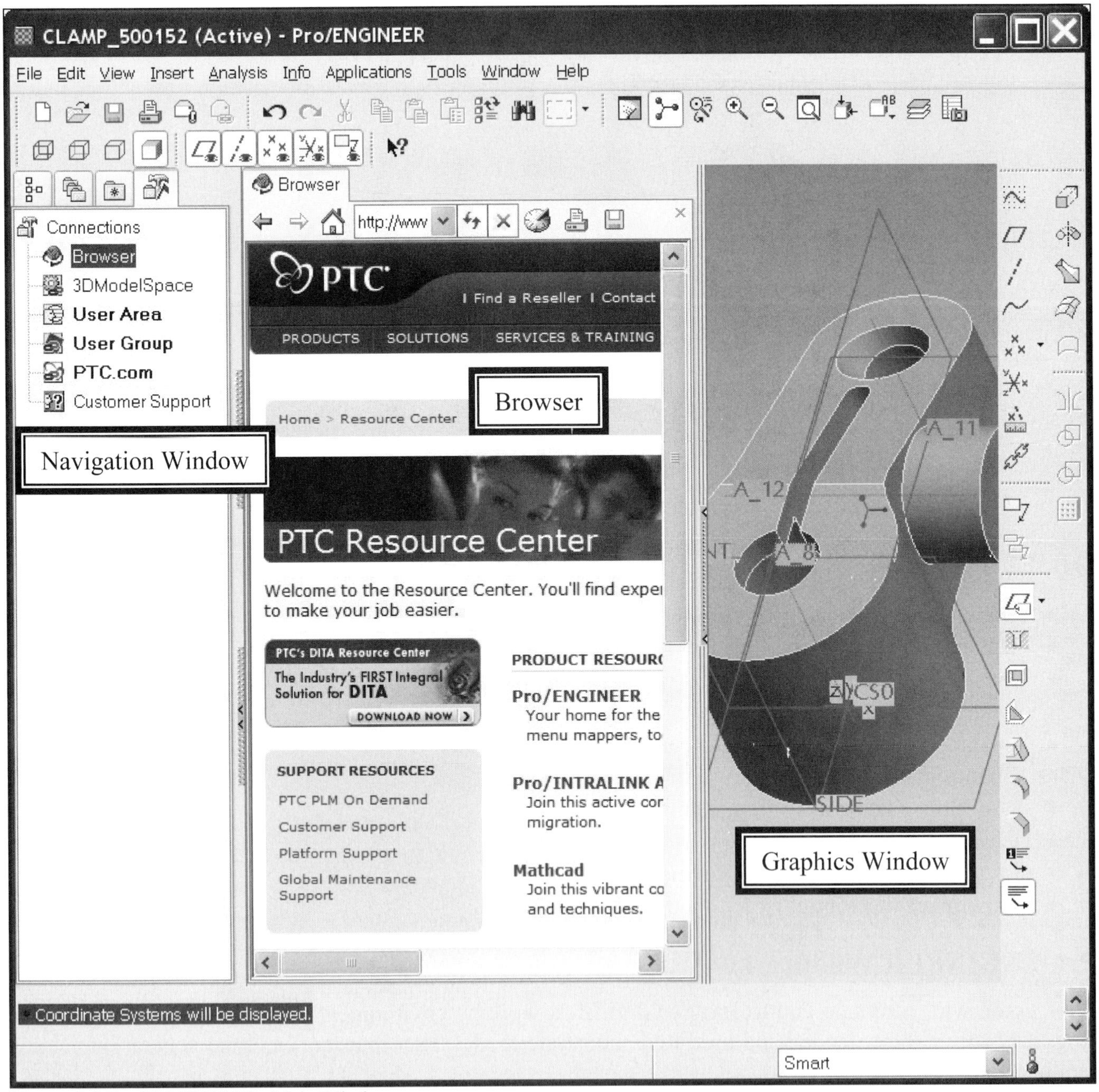

Figure 2.2 Pro/ENGINEER's Main Window

- **Menu Bar** The menu bar contains menus with options for creating, saving, and modifying models, and also contains menus with options for setting environment and configuration options.

- **Information Areas** Each Pro/E window has a message area near the bottom of the window for displaying one-line Help messages.
    - **Message Area** Messages related to work performed are displayed here.

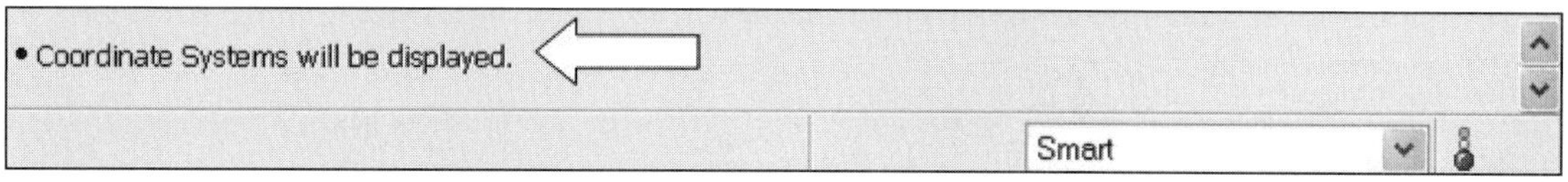

    - **Status Bar** At the bottom of the window is a one-line Help area that dynamically displays one-line context-sensitive Help messages. If you move your mouse pointer over a menu command or dialog box option, a one-line description appears in this area.

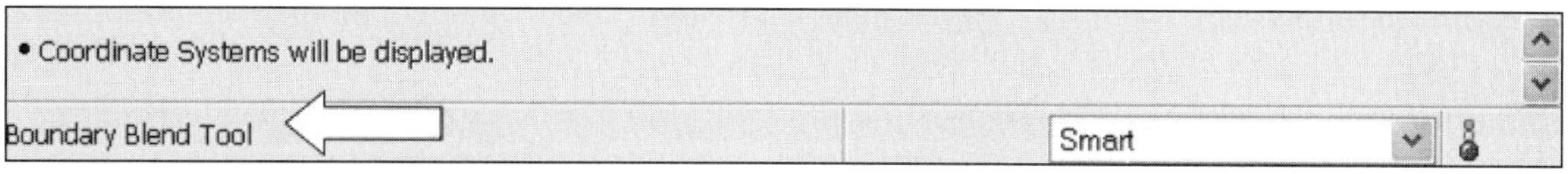

    - **Screen Tips** The status bar messages also appear in small yellow boxes near the menu option or dialog box item or toolbar button that the mouse pointer is passing over.

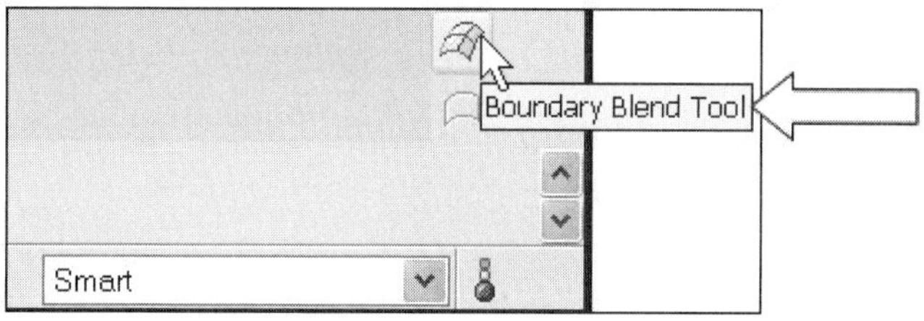

- **Toolbars** The toolbars contain icons to speed up access to commonly used menu commands. By default, the toolbars consist of a row of buttons located directly under the main menu bar. Toolbar buttons can be positioned on the top, left, and right of the window. Toolbar buttons can be added or removed from the Toolchest by customizing the layout.

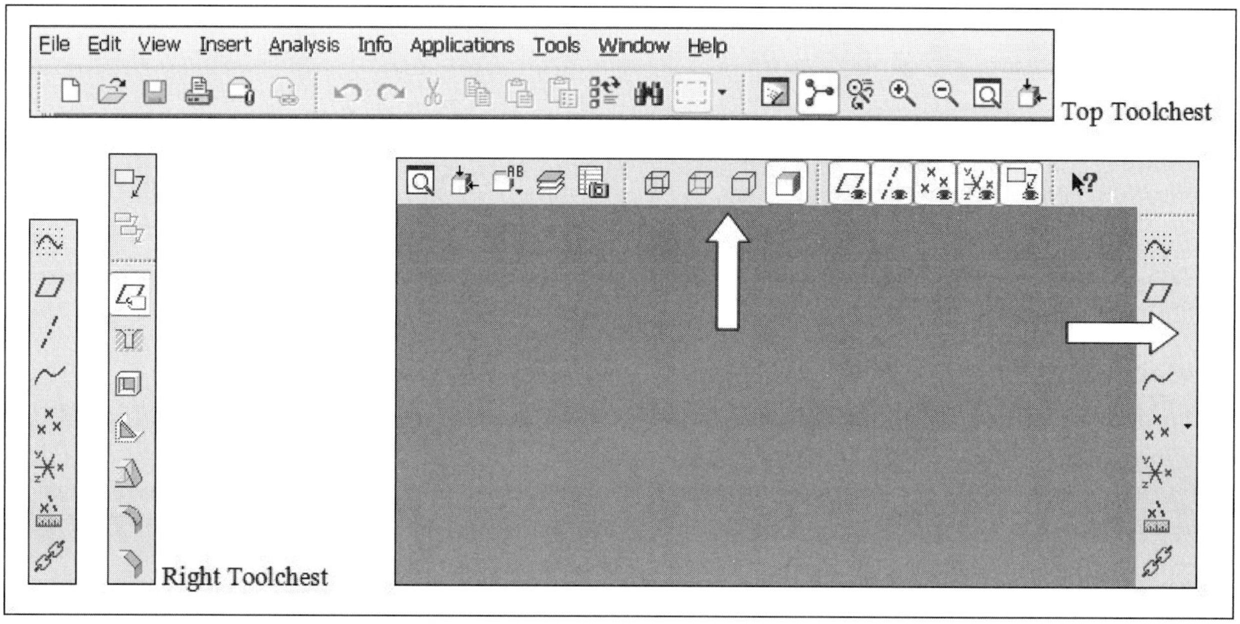

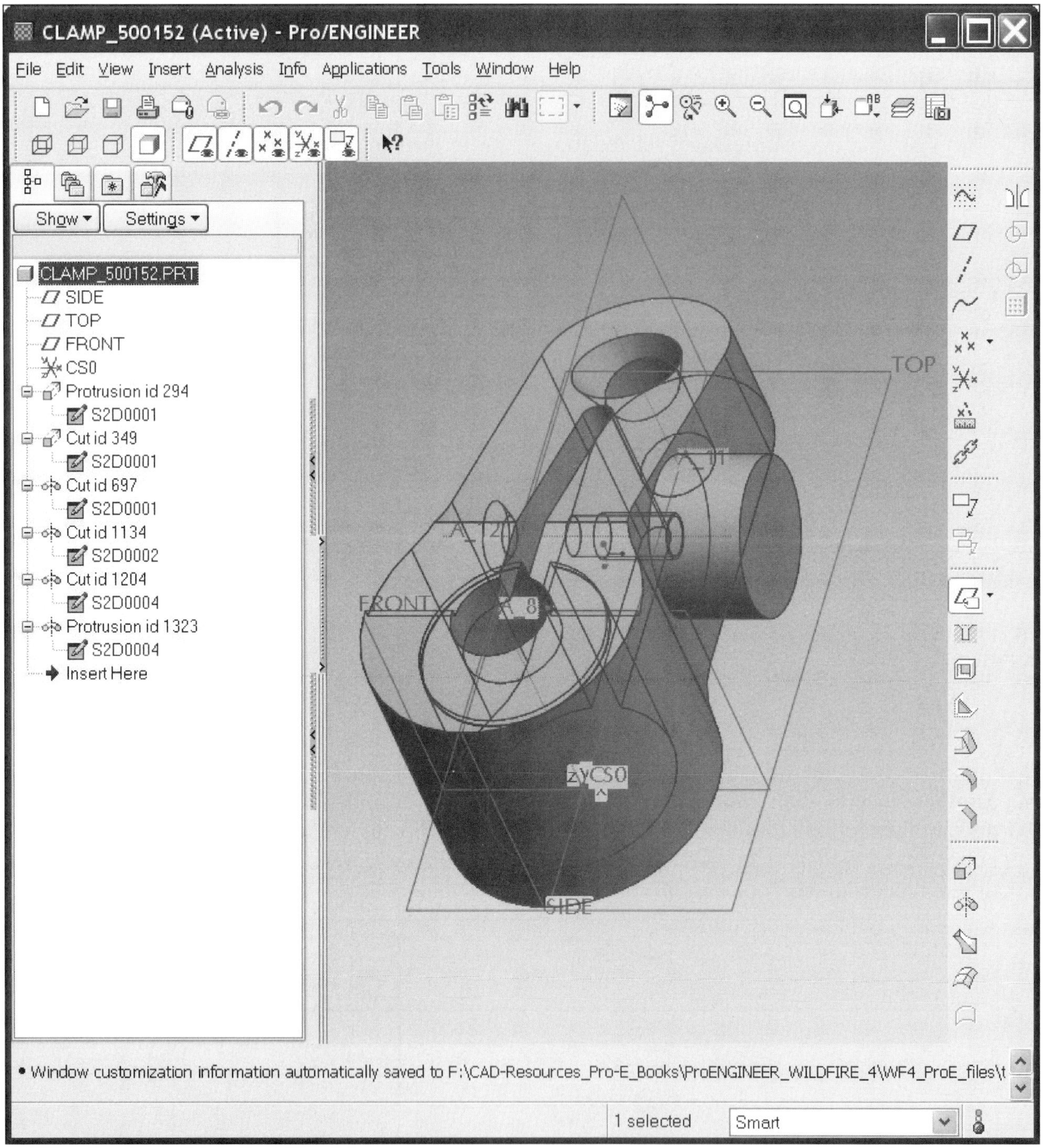

Figure 2.3 Swing/Pull Clamp Arm

Catalog Parts

In order to see and use Pro/ENGINEER's UI, we must have an active object (Fig. 2.3). You will download an existing model from the Internet using Pro/ENGINEER's integrated Browser. You will also experience using the Catalog of parts from PTC. *If you are using an Academic or Commercial version of Pro/ENGINEER Wildfire 4.0, you can download catalog parts directly. The Student Edition (SE) and Tryout Edition (TE) do not have access to these selections.*

Throughout the text, a box surrounds all commands and menu selections.

START HERE: Open **Pro/ENGINEER Wildfire 4.0** using a shortcut icon on your Desktop 

*(or with WINDOWS, click: **Start ⇒ Programs ⇒ proewildfire4.0**)* ⇒ Pro/E will open on your computer ⇒ **Connections** tab 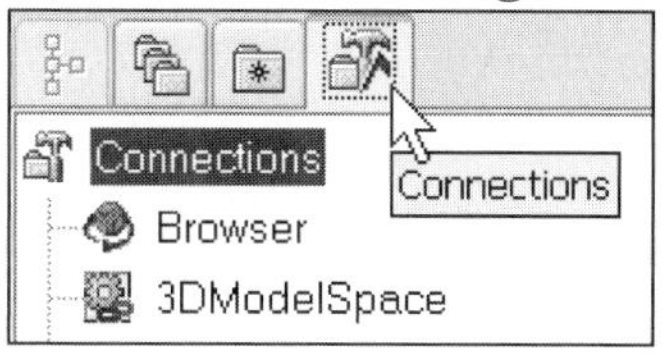

⇒ **3DModelSpace** 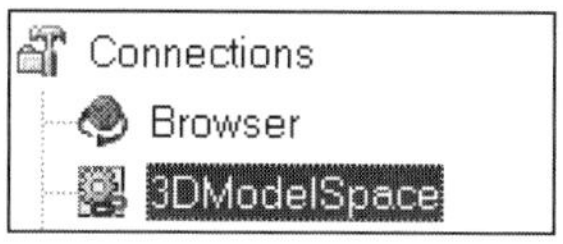

[Fig. 2.4(a)] ⇒ **Manufacturer List** [Fig. 2.4(b)]

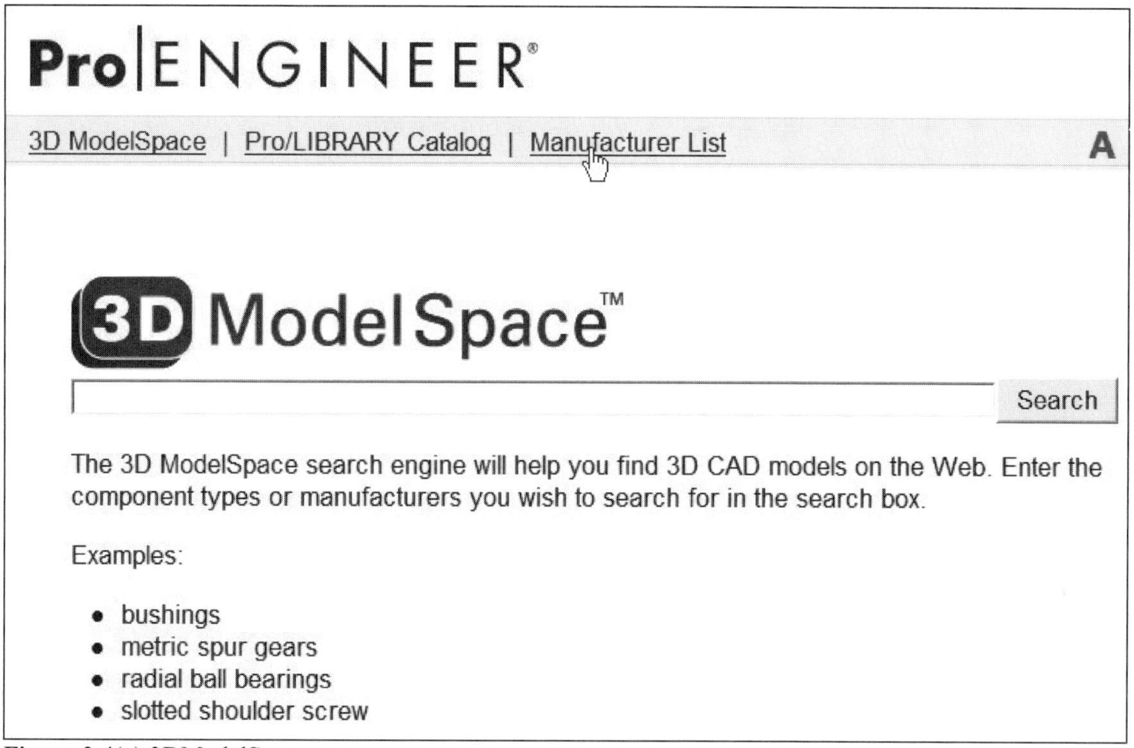

Figure 2.4(a) 3DModelSpace

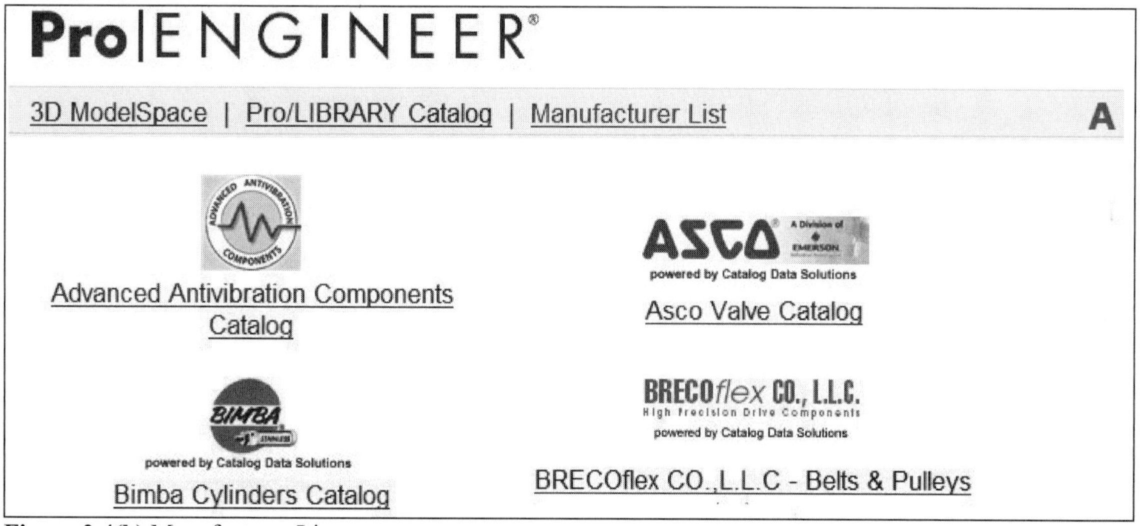

Figure 2.4(b) Manufacturer List

Scroll down the list to see **SPX Hytec** [Fig. 2.4(c)] ⇒ Click: **SPX Hytec – Workholding Catalog** [Fig. 2.4(d)]

Figure 2.4(c) Manufacturer List

Figure 2.4(d) SPX Hytec Workholding Catalog

Click: **Products** on the Hytec web page [Fig. 2.4(e)] ⇒ type **500152** Search field [Fig. 2.4(f)] ⇒ **Search**

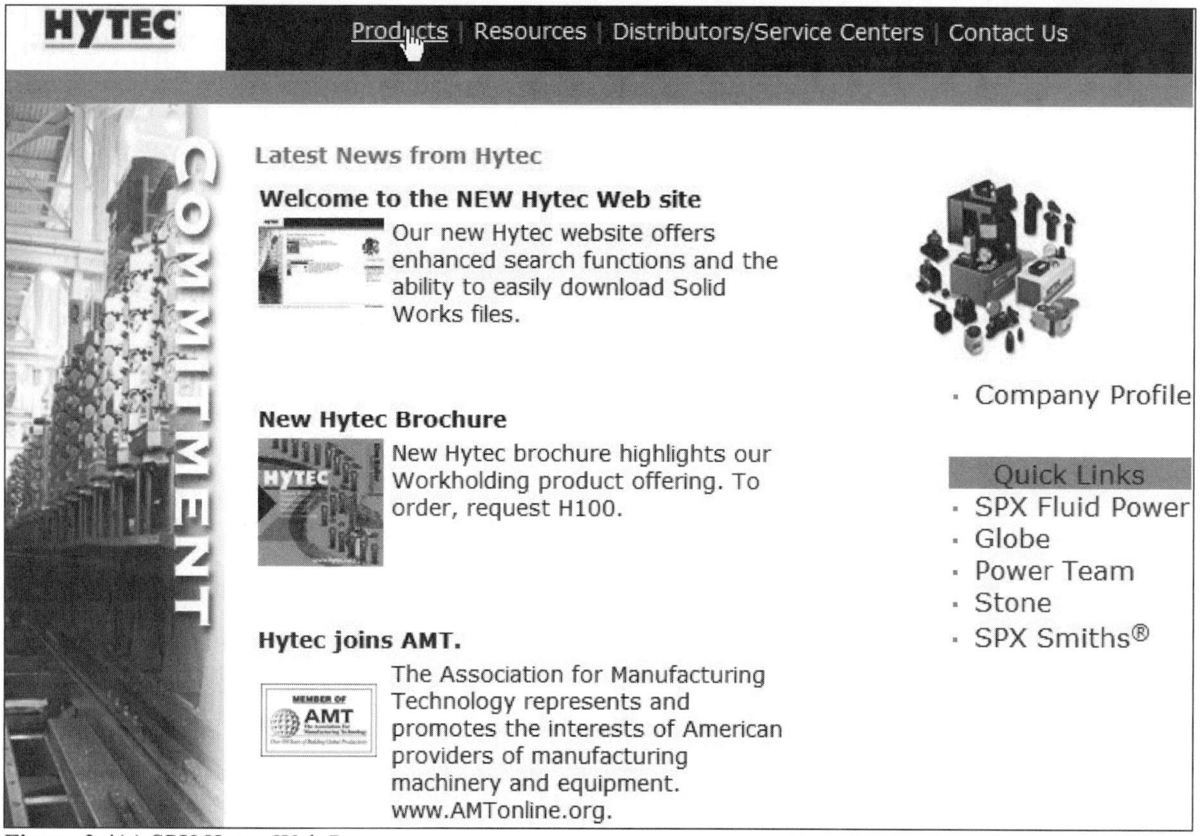

Figure 2.4(e) SPX Hytec Web Page

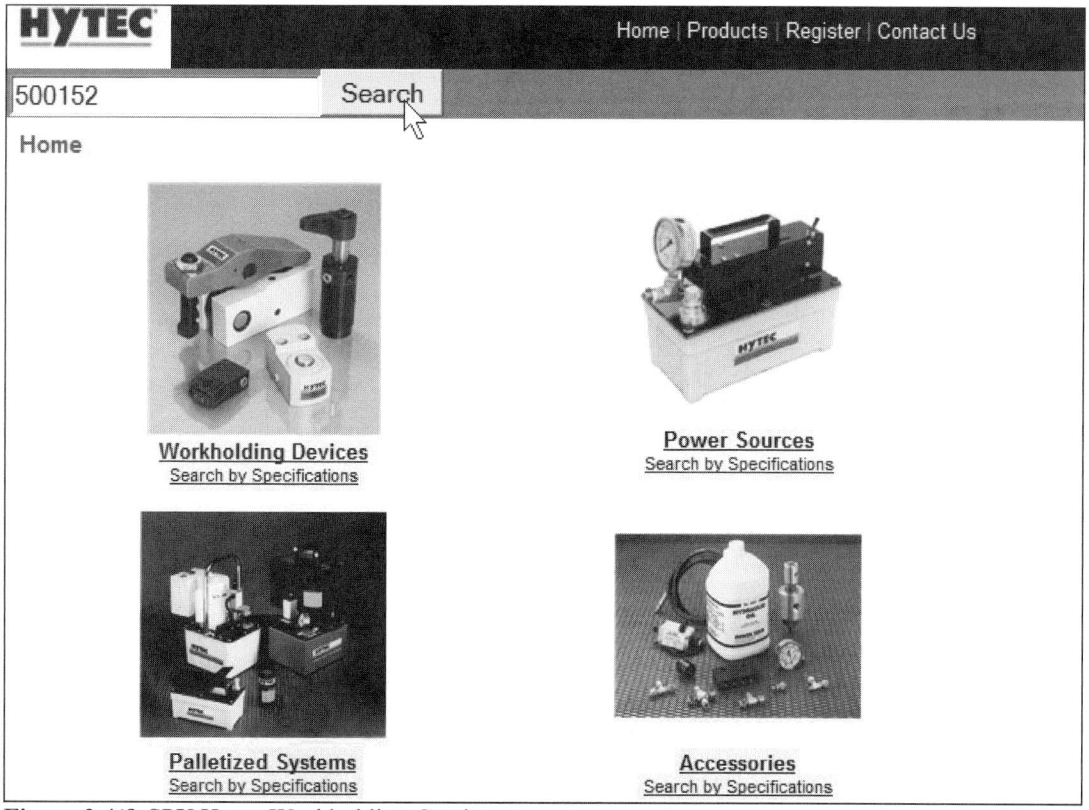

Figure 2.4(f) SPX Hytec Workholding Catalog

Click: **500152** 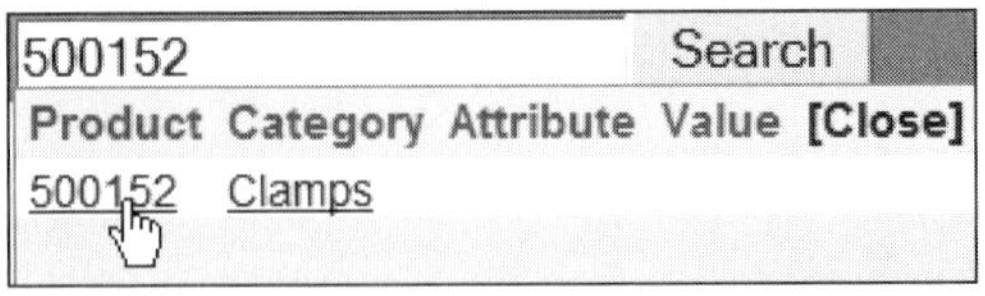

⇒ **Choose Format** ⇒ **Pro/E Part/Assembly**

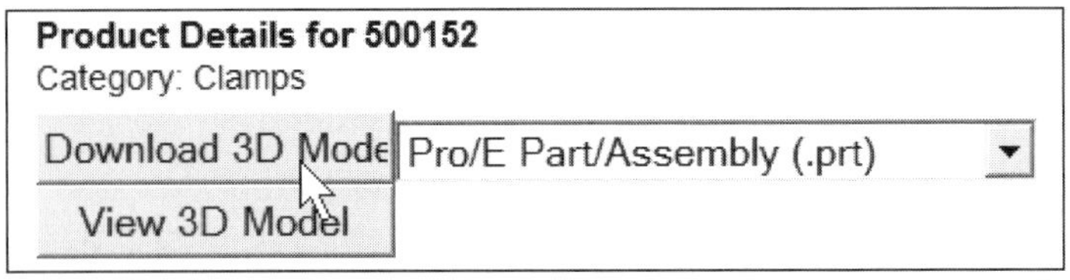

(.prt) [Fig. 2.5(g)] ⇒ **Download 3D Model** ⇒

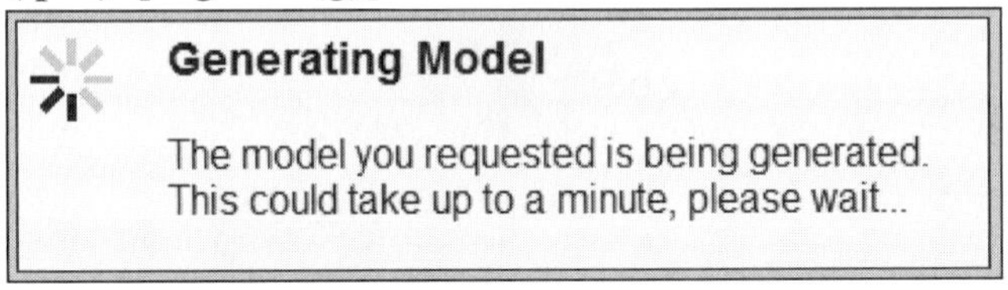

⇒

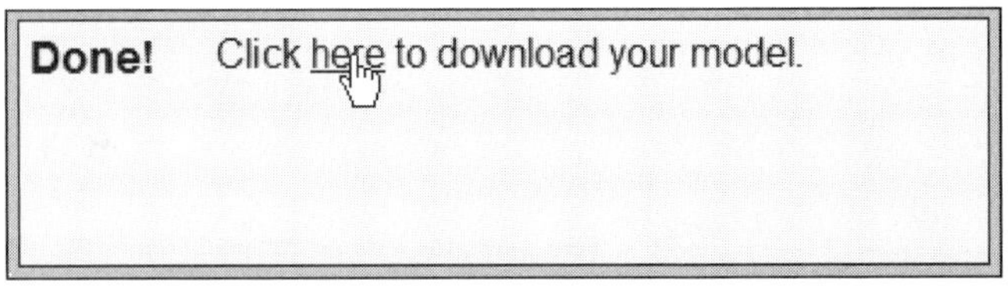

⇒ **Open** the zipped file *(depending on your zip program, you may have to respond to various licensing messages)*

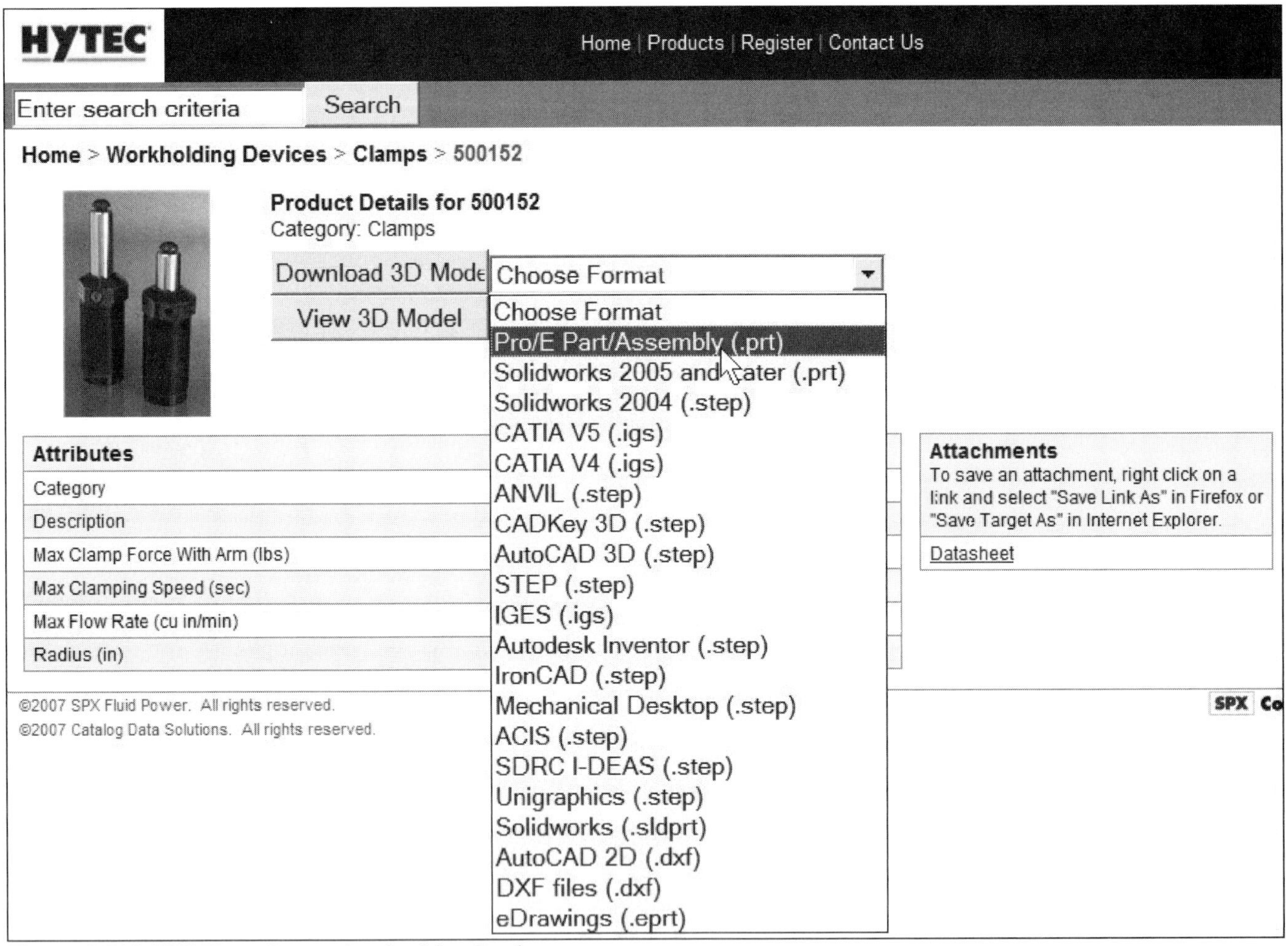

Figure 2.5(g) SPX Hytec Interactive On-Line Catalog

Click on **500152** [Fig. 2.5(h)] ⇒ *(Depending on your zip program, you may have to extract the file.)* ⇒ drag and drop the part in the graphics window [Fig. 2.5(i)] ⇒ close the zip dialog window

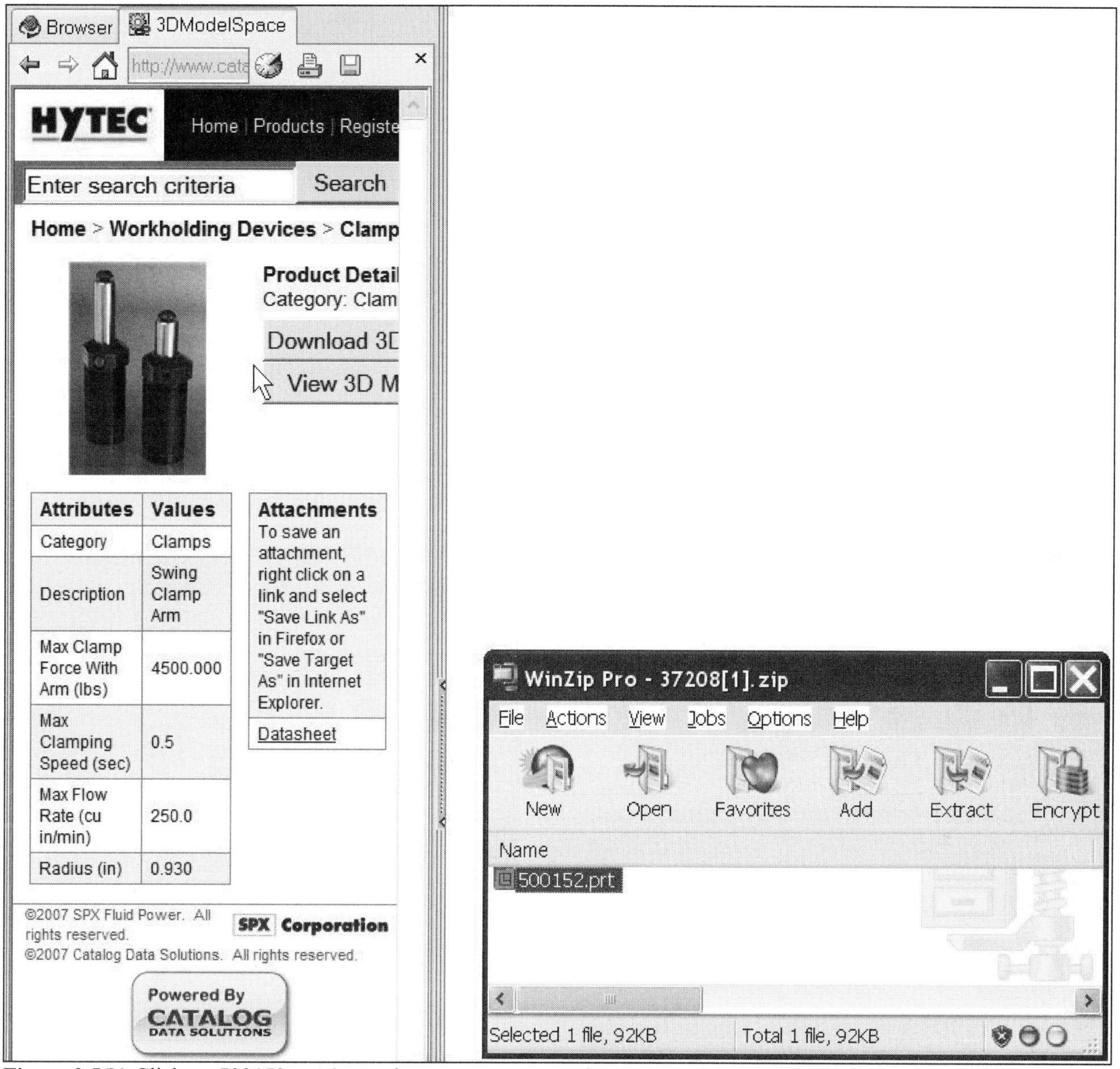

Figure 2.5(h) Click on **500152.prt** (your zip program may require you to extract the file first)

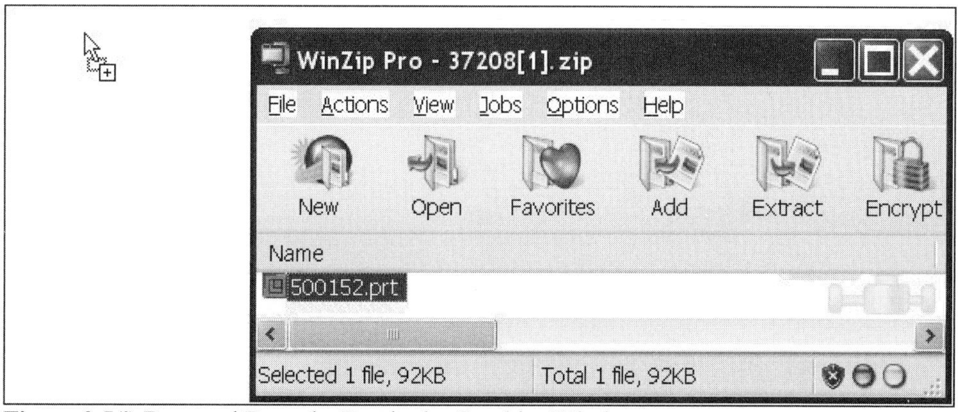

Figure 2.5(i) Drag and Drop the Part in the Graphics Window

The part is an imported feature and is not composed of Pro/E features [Fig. 2.5(j)]. It is also has colored *surfaces* instead of *part* colors. Click: **View** from the menu bar ⇒ **Shade**⇒ **View** ⇒ **Color and Appearance** ⇒ in the Assignment block select **All Surfaces** [Fig. 2.5(k)] ⇒ **Clear** [Fig. 2.5(l)] ⇒ **Close** ⇒ **Ctrl+S** ⇒ **Enter** [Fig. 2.5(m)] ⇒ **File** ⇒ **Close Window** ⇒ **File** ⇒ **Erase** ⇒ **Current** ⇒ **Yes**

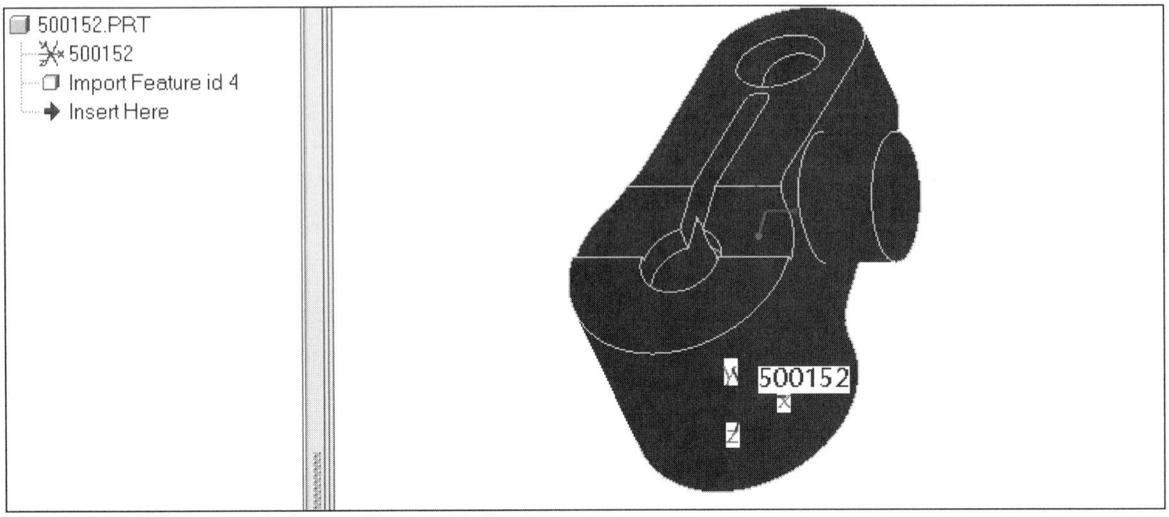

Figure 2.5(j) Part 500152 (Import Feature id 4)

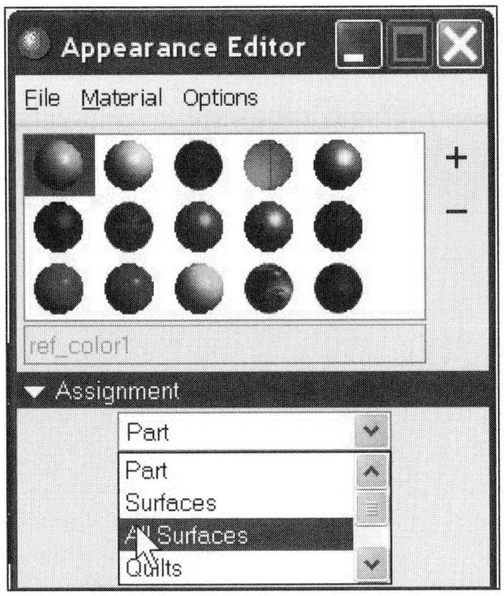

Figure 2.5(k) All Surfaccs

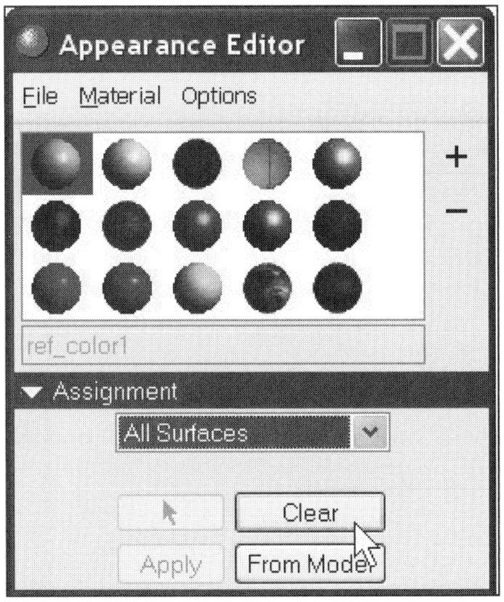

Figure 2.5(l) Clear

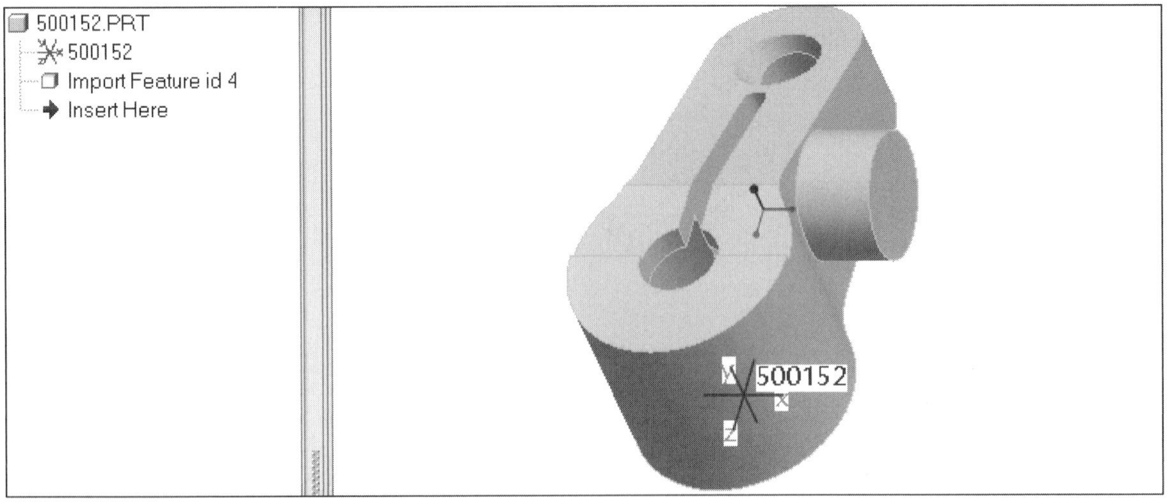

Figure 2.5(m) 500152 Hytec Part

A Pro/E version of this component is also available on the www.cad-resources.com web site. In the next sequence of commands we will download this part and drag and drop it into the graphics window. *[If you are using the* ***Tryout Edition*** *or the* ***Student Edition of Pro/ENGINEER Wildfire 4.0****]*

Click: **Connections** tab in the Navigator panel ⇒ **Browser** [Fig. 2.6(a)] ⇒ type **www.cad-resources.com** in the Browser Address Bar [Fig. 2.6(b)] 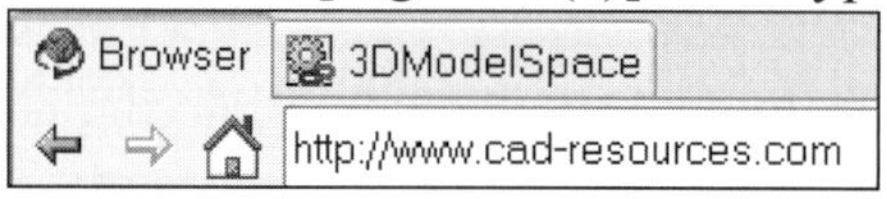

⇒ **Enter** ⇒ ⇒ **Downloads**

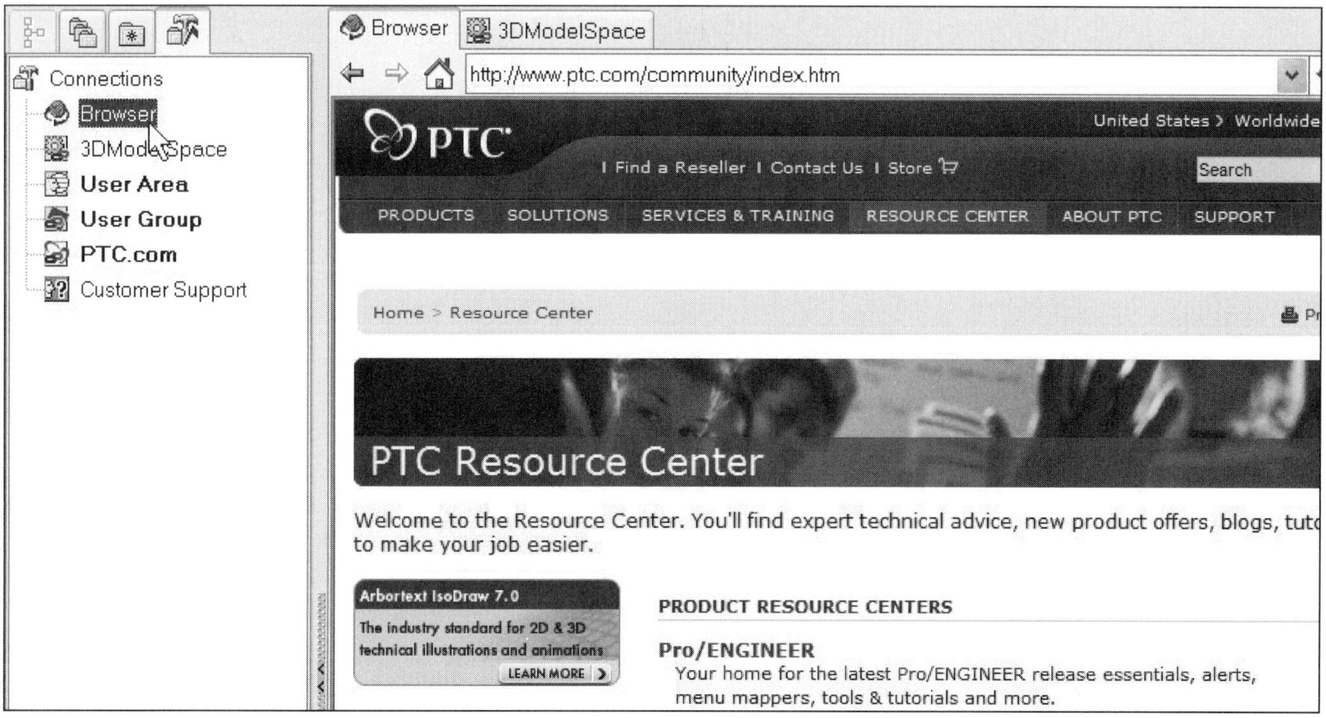

Figure 2.6(a) www.cad-resources.com

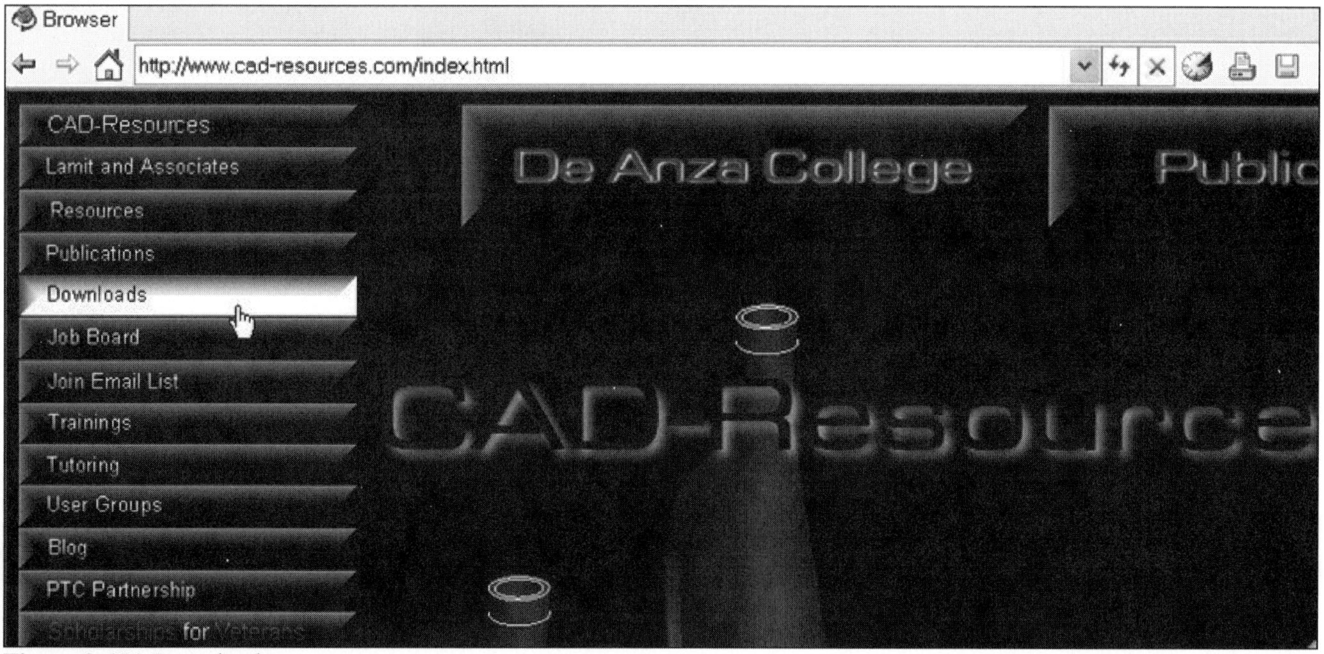

Figure 2.6(b) Downloads

Click: **TE SE Version** or **Commercial Version** depending on your Pro/E license

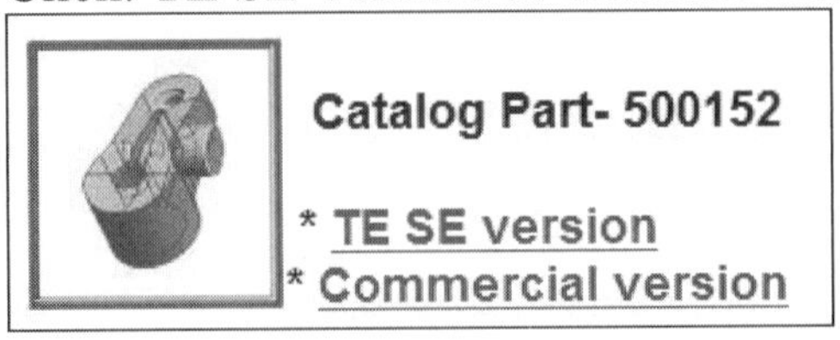

⇒ Extract & open the file if necessary, otherwise simply drag and drop the part from the Zip dialog box into the Graphics Window. [Fig. 2.6(c)] ⇒ close the zip dialog box

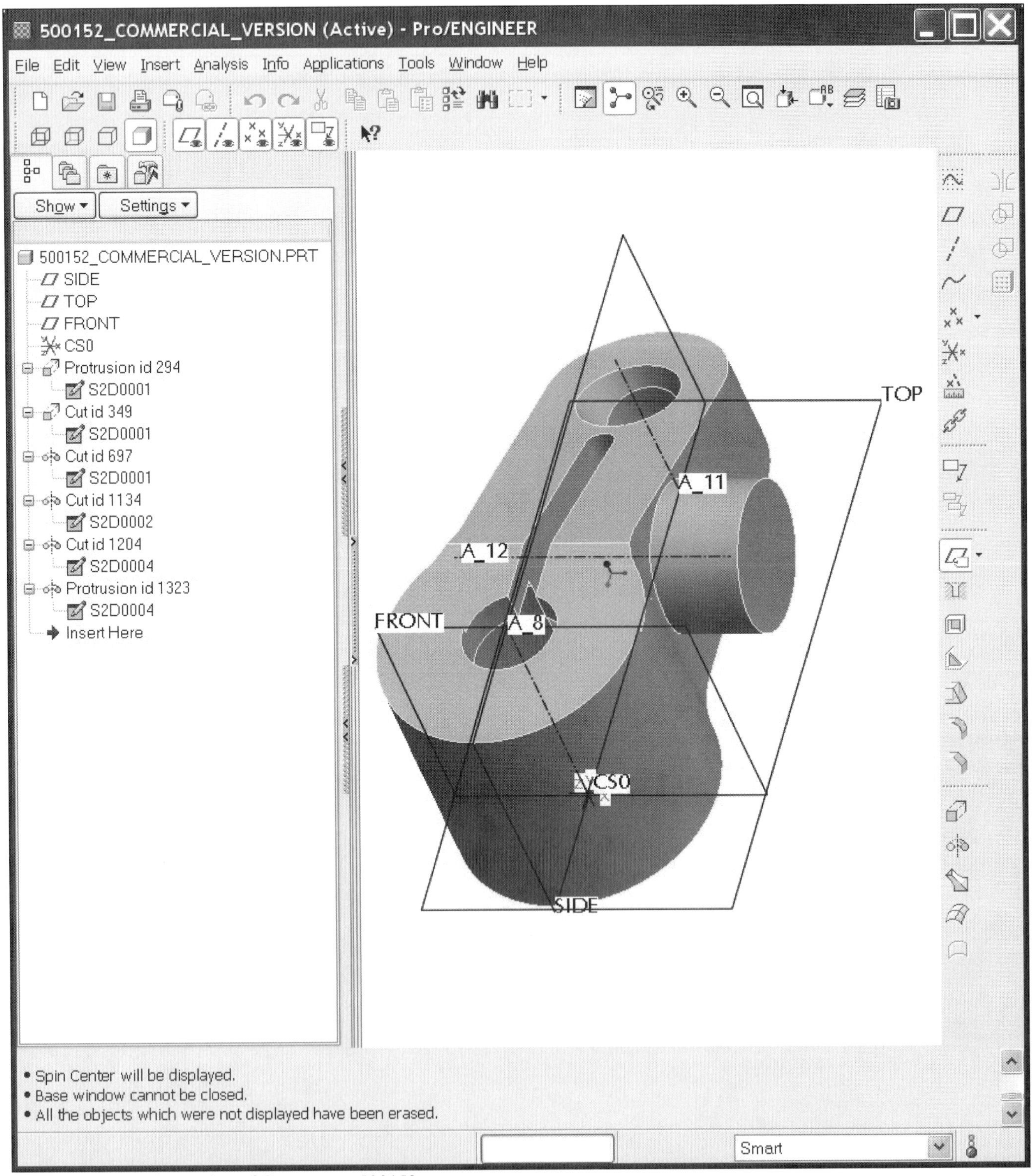

Figure 2.6(c) Pro/E version of Component 500152

File Functions

The **File** menu, from the menu bar, provides options for opening, creating, saving, renaming, backing up files, and printing. File functions include options for importing files from and exporting files to external formats, setting your working directory and performing operations on instances. Before using any File tool, make sure you have set your working directory to the folder where you wish to save objects for the project on which you are working. The Working Directory is a directory that you set up to contain Pro/E files. You can save a Pro/E file using *Save* or *Save a Copy* from the File menu. The Save a Copy dialog box allows the exportation of Pro/E files to different formats, and to save files as images. Since the name already exists in session, you cannot save or Rename a file using the same name as the original file name, even if you save the file in a different directory. Pro/E forces you to enter a unique file name by displaying the message: *An object with that name exists in session. Choose a different name.* Names for all Pro/E files are restricted to a maximum of 31 characters and must contain no spaces. A File can be a part, assembly, or drawing. Each is considered an "object".

Click: **File** ⇒ **Set Working Directory** [Fig. 2.7(a)] ⇒ select the directory you wish all objects to be saved to for this Pro/E project ⇒ **OK** ⇒ **File** ⇒ **Save a Copy** ⇒ type in the New Name field: **CLAMP_500152** [Fig. 2.7(b)] ⇒ **OK** ⇒ **File** ⇒ **Close Window**

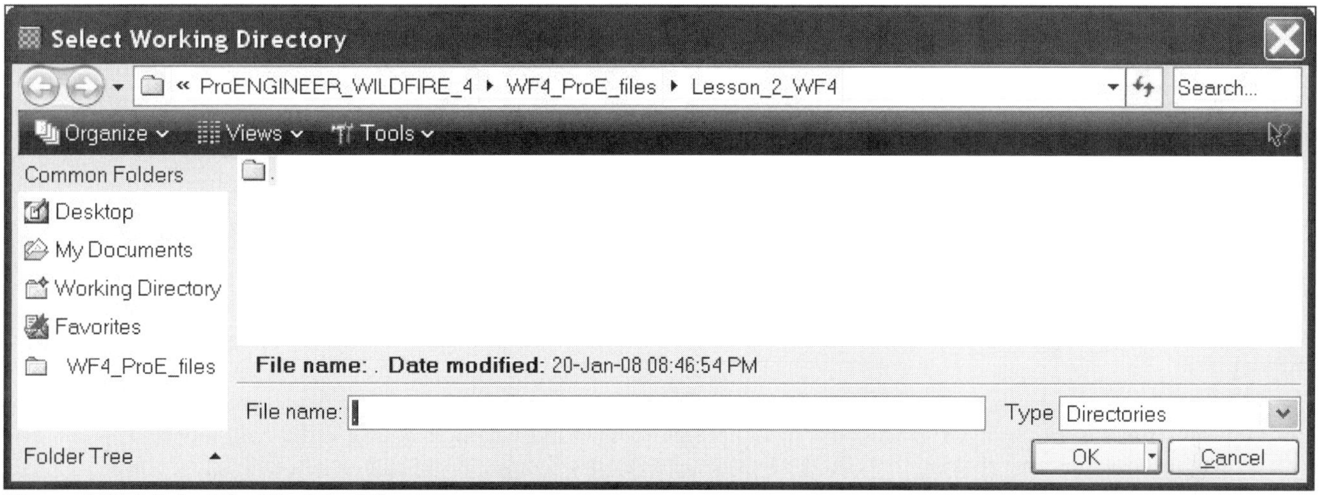

Figure 2.7(a) Select the Working Directory

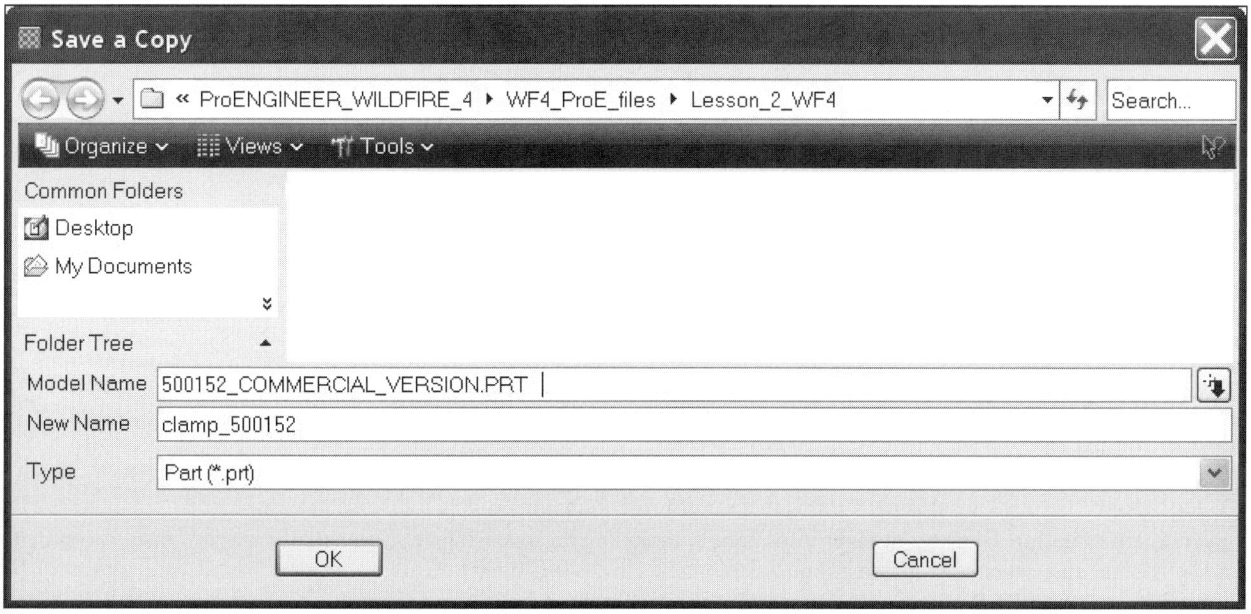

Figure 2.7(b) Save a Copy

Click: **File ⇒ Open ⇒ clamp_500152 ⇒ Open**

[READ THIS - Depending on the computer you are using (yours or an organization's) the following may not work. If you have your mail server configured as the default then proceed, otherwise skip this step. ***(Again- If you do not own the computer you are working on, then do not do this step)]***

 ***Send email with object in active window*** *⇒ email the object to yourself (or a friend with Pro/E software) ⇒ check on* ***Create a ZIP file ⇒ OK*** *(Fig. 2.8) ⇒ follow your email procedure as required*

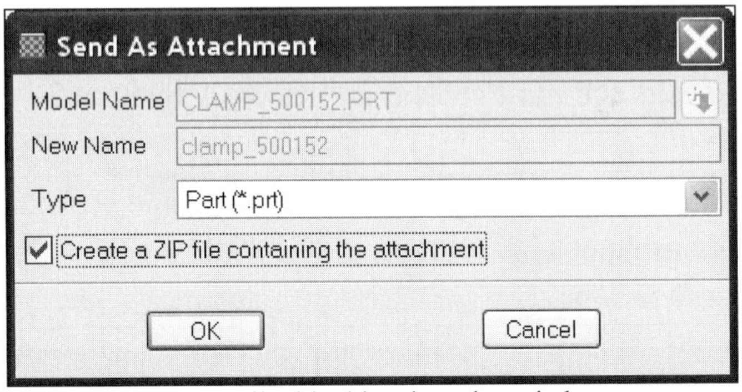

Figure 2.8 Send email with object in active window

Help

Accessing the Help function is one of the best ways to learn CAD software. Use the Help tool as often as possible to understand the tool or command you are using at the time and to expand your knowledge of the other capabilities provided by Pro/ENGINEER. Use the **Help** menu to gain access to online information, Pro/E release information, and customer service information. The following commands are available on the Pro/E **Help** menu in standard Part and Assembly modes.

- **Help Center** Displays the context-sensitive online help system. When you select this, your supported network browser opens to display a navigation tree and search tools to aid you in finding specific help topics. You can also access these topics by clicking for context-sensitive Help from windows, menu commands, and dialog boxes.

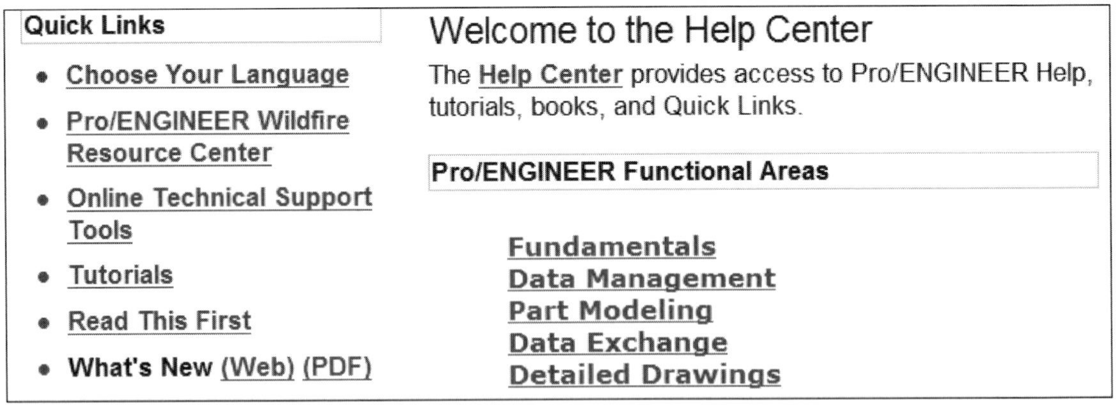

- **What's This?** Enables context-sensitive Help mode.
- **Menu Mapper** To help show you where older release selections have been moved.
- **Search Online Knowledge Base** Useful tutorials, tools and various help guides.
- **What's New?** Displays what is new in this release of Pro/E software.
- **System Information** Displays product information, including the release level, license information, installation date, and customer support contact information.
- **About Pro/ENGINEER** Displays Pro/E copyright and release information.

Click: **Help** ⇒ **Help Center** [Fig. 2.9(a)] ⇒ **Fundamentals** ⇒ **Pro/ENGINEER Fundamentals** ⇒ Inside the Contents tab choose (to expand or to collapse) ⇒ explore the Help documentation before continuing ⇒ click: to close Help Center

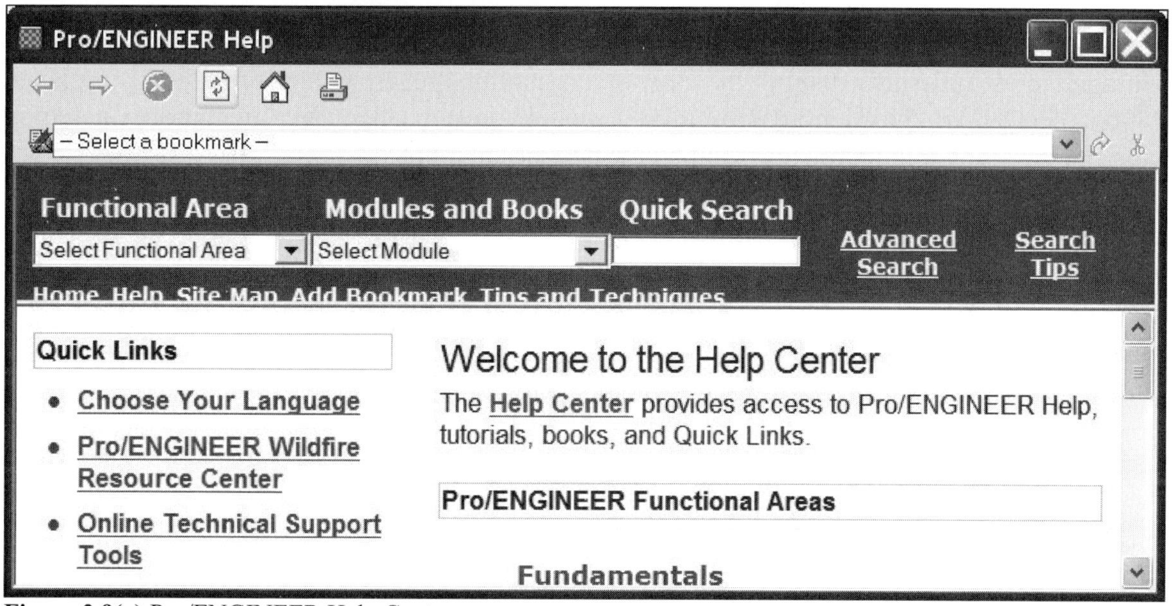

Figure 2.9(a) Pro/ENGINEER Help Center

Click: **Context sensitive help** 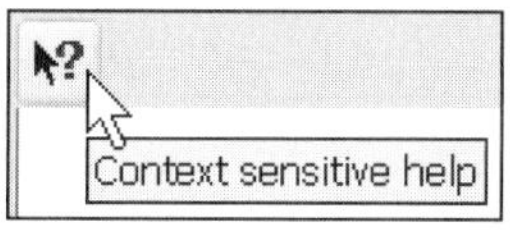

from Top Toolchest ⇒ then click on the **Orient Mode on/off** button from the Top Toolchest **Orient Mode on/off** ⇒ read the documentation [Fig. 2.9(b)] ⇒ close the window ⇒ repeat, and choose some other buttons

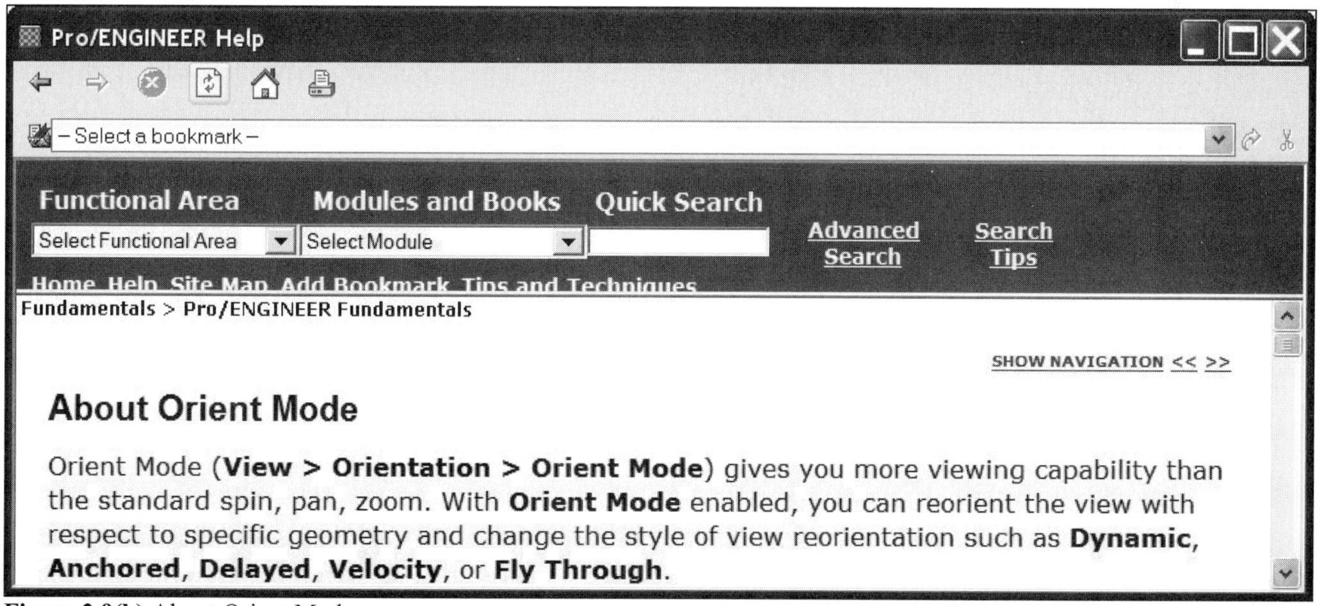

Figure 2.9(b) About Orient Mode

View and Display Functions

Using the Pro/E **View** menu, you can adjust the model view, orient the view, hide and show entities, create and use advanced views, and set various model display options. The following list includes some of the View operations you can perform:

- Orient the model view in the following ways, using the **Orientation** dialog box: spin, pan, and zoom models and drawings, display the default orientation, revert to the previously displayed orientation, change the position or size of the model view, change the orientation (including changing the view angle in a drawing), and create new orientations
- Temporarily shade a model by using cosmetic shading
- Show, dim, or remove hidden lines
- Highlight items in the graphics window when you select them in the Model Tree
- Explode or unexplode an assembly view
- Repaint the Pro/E graphics window
- Refit the model to the Pro/E window after zooming in or out on the model
- Update drawings of model geometry
- Hide and unhide entities, and hide or show items during spin or animation
- Use advanced views
- Add perspective to the model view

Click: 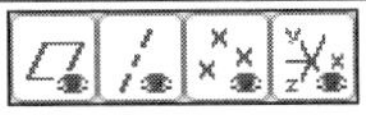 on ⇒  **Wireframe** [Fig. 2.10(a)] ⇒  **Hidden line** [Fig. 2.10(b)] ⇒  **No hidden** [Fig. 2.10(c)] ⇒ 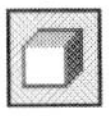 **Shading** ⇒ **View** ⇒ **Shade** ⇒ **View** ⇒ **Repaint** (Ctrl+R)

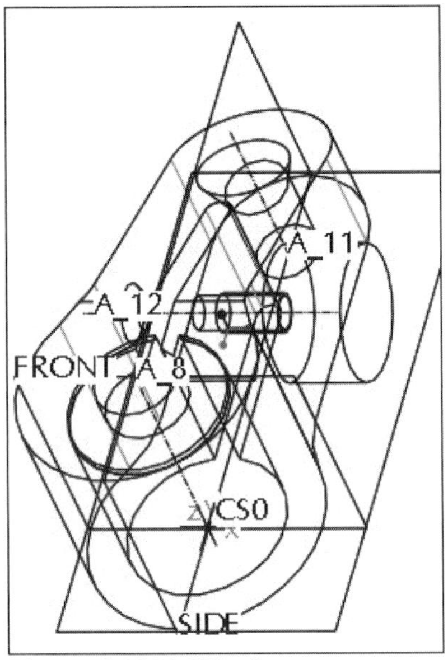

Figure 2.10(a) Wireframe

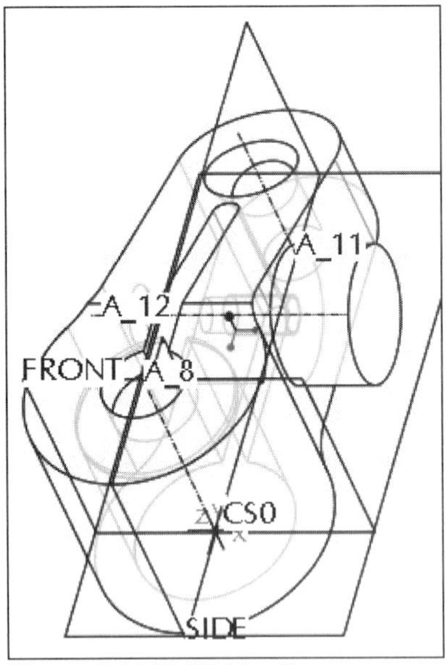

Figure 2.10(b) Hidden line

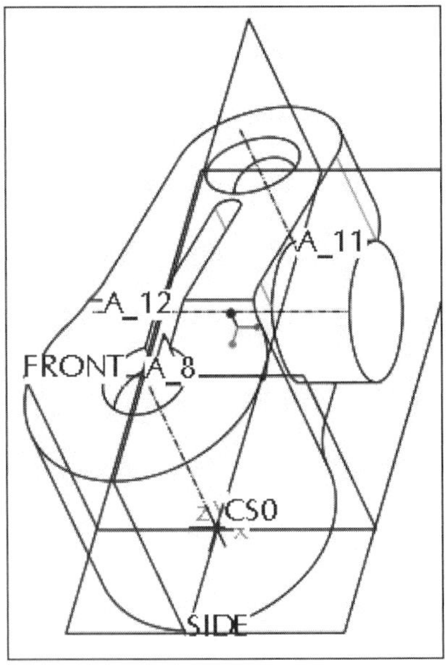

Figure 2.10(c) No hidden

To see the standard views, click: **Saved view list** ⇒ **FRONT** [Fig. 2.11(a)] ⇒ **Saved view list** ⇒ **RIGHT** [Fig. 2.11(b)] ⇒ try all the variations ⇒ **Saved view list** ⇒ **Standard Orientation**

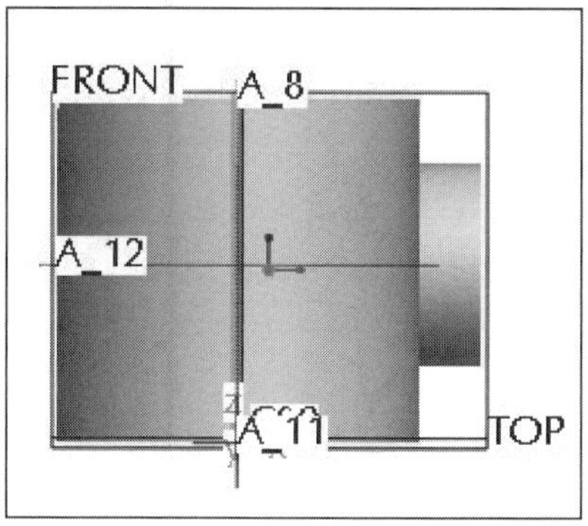

Figure 2.11(a) FRONT View

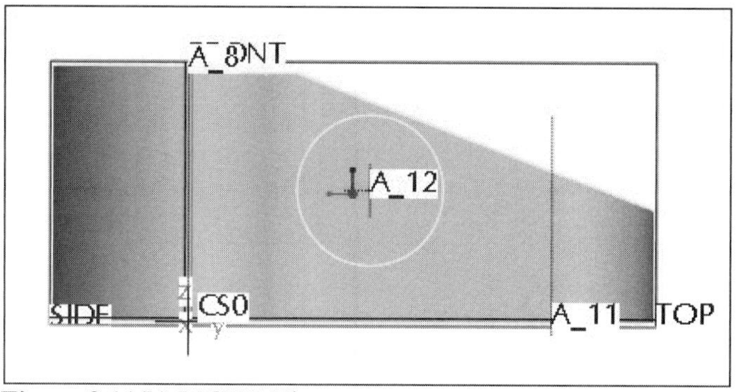

Figure 2.11(b) RIGHT View

View Tools

As with all CAD systems, Pro/E provides the typical view tools associated with CAD:

- **Zoom In** Use this tool to zoom in on a specific portion of the model. Pick two positions of a rectangular zoom box.
- **Zoom Out** Use this tool to reduce the view size of the model on the screen by 50%.
- **Refit object to fully display it on the screen** Use this tool to refit the model to the screen so that you can view the entire model. A refitted model fills 80% of the graphics window.

Click: **Zoom In** [Fig. 2.12(a)] ⇒ pick two positions about an area you wish to enlarge [Fig. 2.12(b)] ⇒ **Zoom Out** ⇒ **Refit** [Fig. 2.12(c)] ⇒ **Redraw the current view**

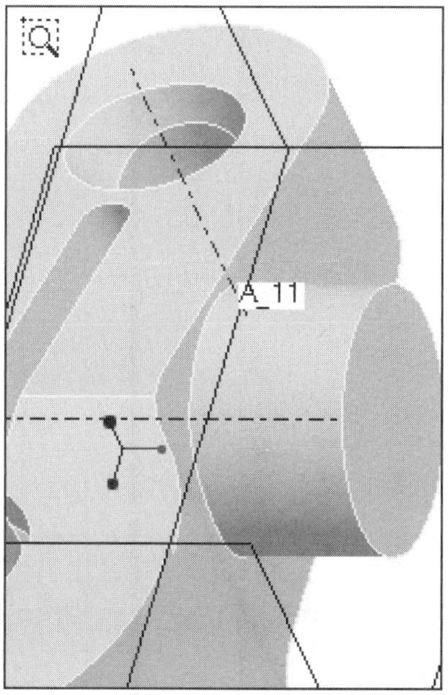

Figure 2.12(a) View

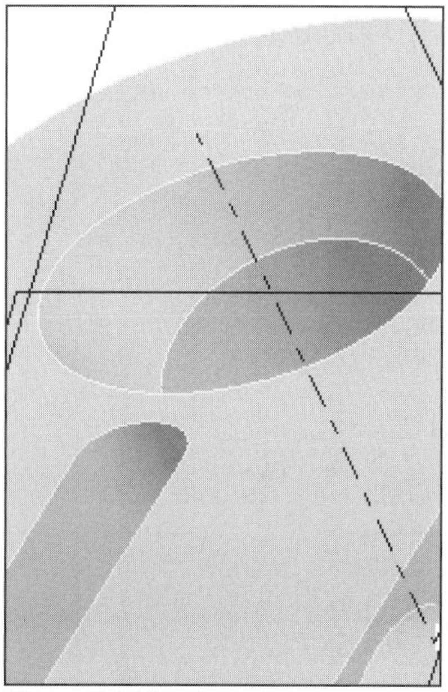
Figure 2.12(b) Zoom In

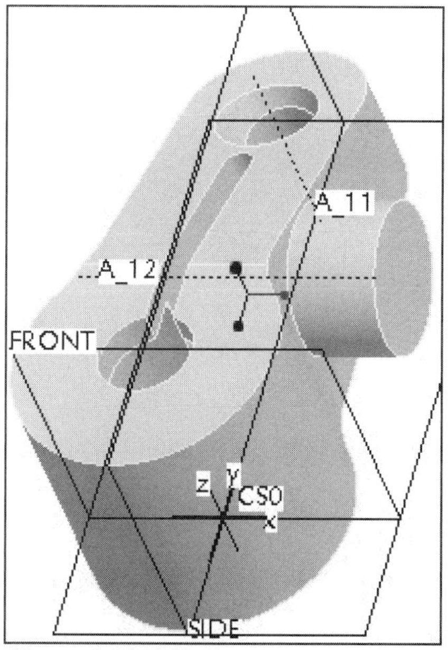

Figure 2.12(c) Refit

Using Mouse Buttons to Manipulate the Model

You can also dynamically reorient the model using the **MMB** by itself (**Spin**) or in conjunction with the **Shift** key (**Pan**) or **Ctrl** key (**Zoom, Turn**).

Hold down **Ctrl** key and **MMB** in the graphics area near the model and move the cursor up (zoom out) [Fig. 2.13(a)] ⇒ hold down **Shift** key and **MMB** in the graphics area near the model and move the cursor about the screen (pan) [Fig. 2.13(b)] ⇒ hold down **MMB** in the graphics area near the model and move the cursor around (spin) [Figs. 2.13(c-d)] ⇒  ⇒ **Standard Orientation**

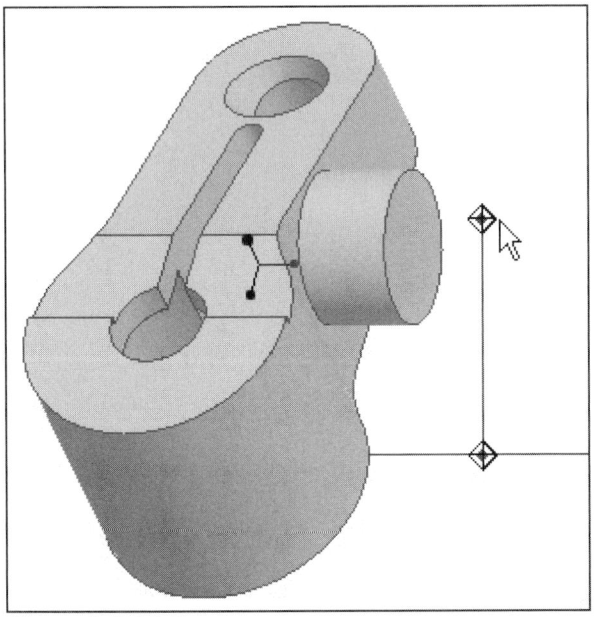

Figure 2.13(a) Zoom

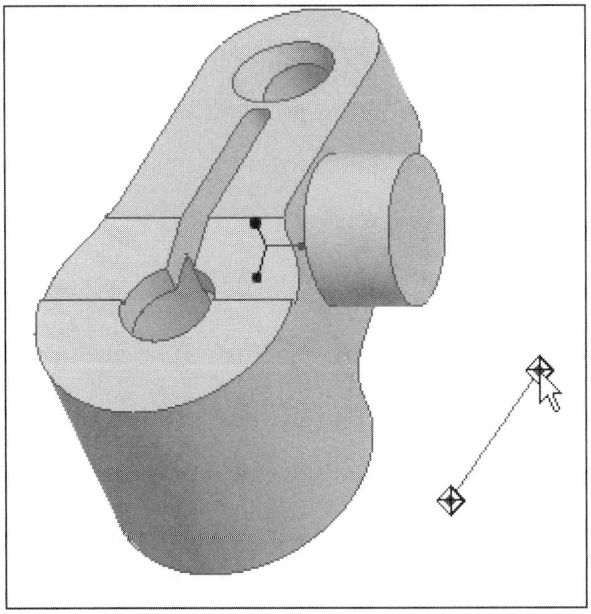

Figure 2.13(b) Pan

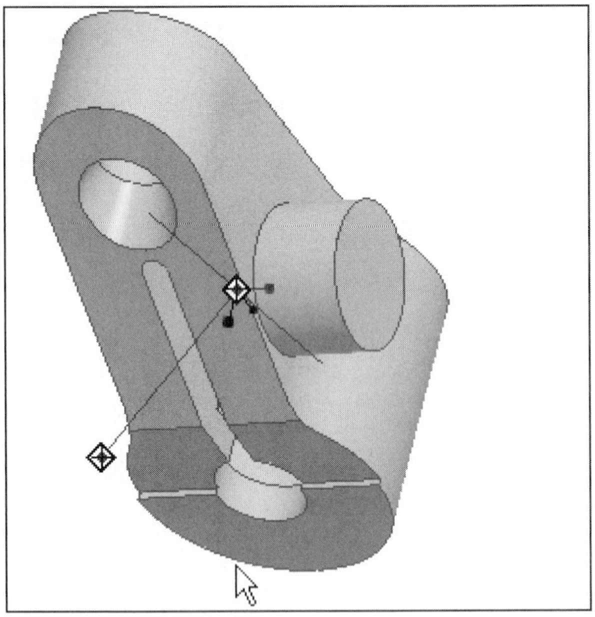

Figure 2.13(c) Spin

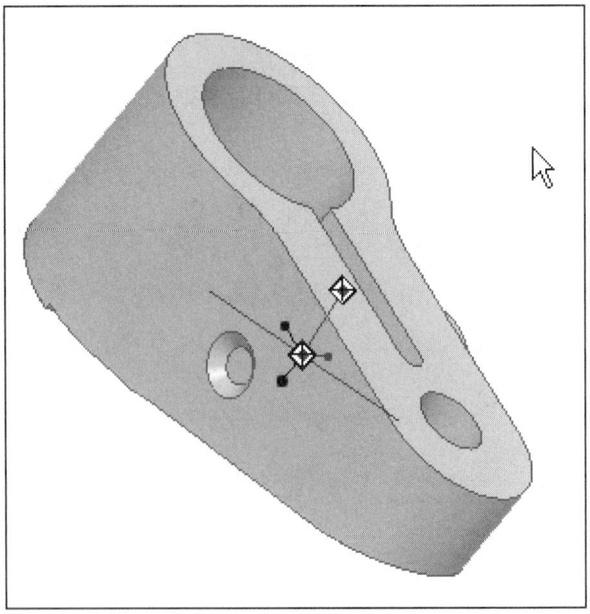

Figure 2.13(d) Spin Again

Please note that the default for the background and geometry colors has been changed so that the illustrations will capture and print clearer. In the next section, you will learn how to change the system display settings.

System Display Settings

You can make a number of changes to the default colors furnished by Pro/E, customizing them for your own use:

- Define, save, and open color schemes
- Customize colors used in the user interface
- Change your entire color scheme to a predefined color scheme (such as black on white)
- Change the top or bottom background colors
- Redefine basic colors used in models
- Assign colors to be used by an entity
- Store a color scheme so you can reuse it
- Open a previously used color scheme

The **Scheme** menu includes the following color schemes (text uses the **Black on White** selection):

- **Black on White** Black entities shown on a white background
- **White on Black** White entities shown on a black background
- **White on Green** White entities shown on a dark-green background
- **Initial** Reset the color scheme to the one defined by the configuration file settings
- **Default** Reset the color scheme to the system default

Click: **View** ⇒ **Display Settings** ⇒ **System Colors** [Figs. 2.14(a-c)] ⇒ uncheck Blended Background ⇒ **Scheme** ⇒ **Black on White** ⇒ try other color schemes and system colors ⇒ **Scheme** ⇒ **Default** ⇒ **OK**

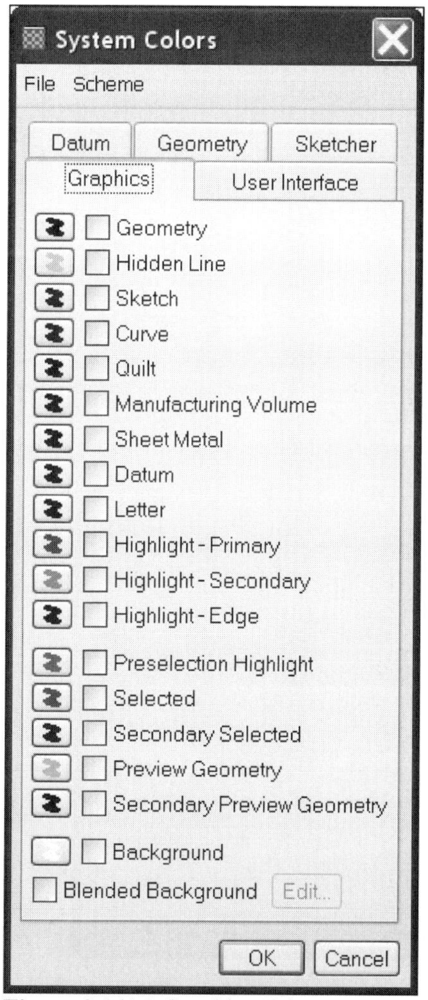

Figure 2.14(a) Graphics Tab

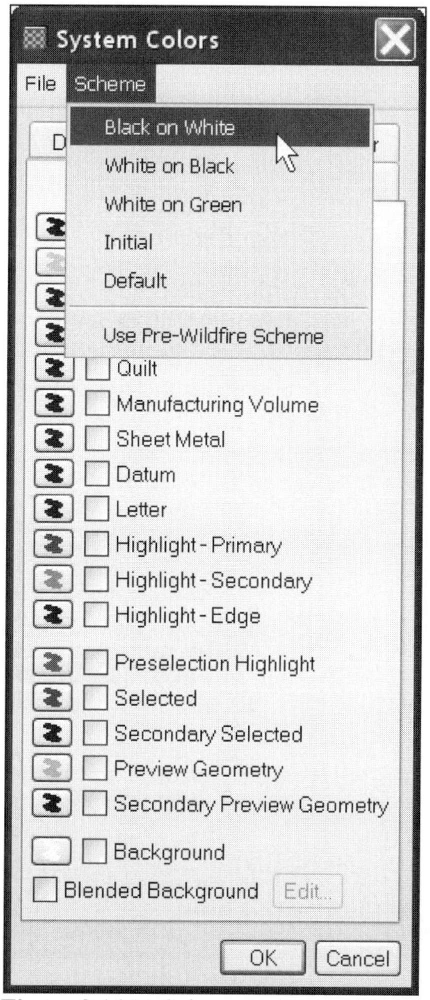

Figure 2.14(b) Scheme Tab

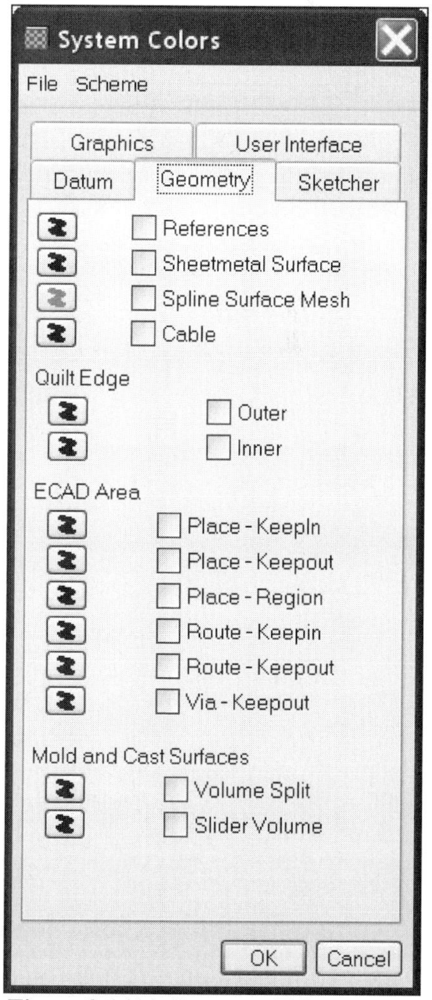

Figure 2.14(c) Geometry Tab

Information Tools

At any time during the design process you can request model, feature, or other information. Picking on a feature in the graphics window and then **RMB ⇒ Info** will provide information about that feature in the Browser. This can also be accomplished by clicking on the feature name in the Model Tree and then **RMB ⇒ Info**. Both feature and model information can be obtained using this method. A variety of information can also be extracted using the **Info** tool from the menu bar.

Pick once on the revolved protrusion ⇒ **RMB** ⇒ **Info** ⇒ **Feature** [Fig. 2.15(a-b)]

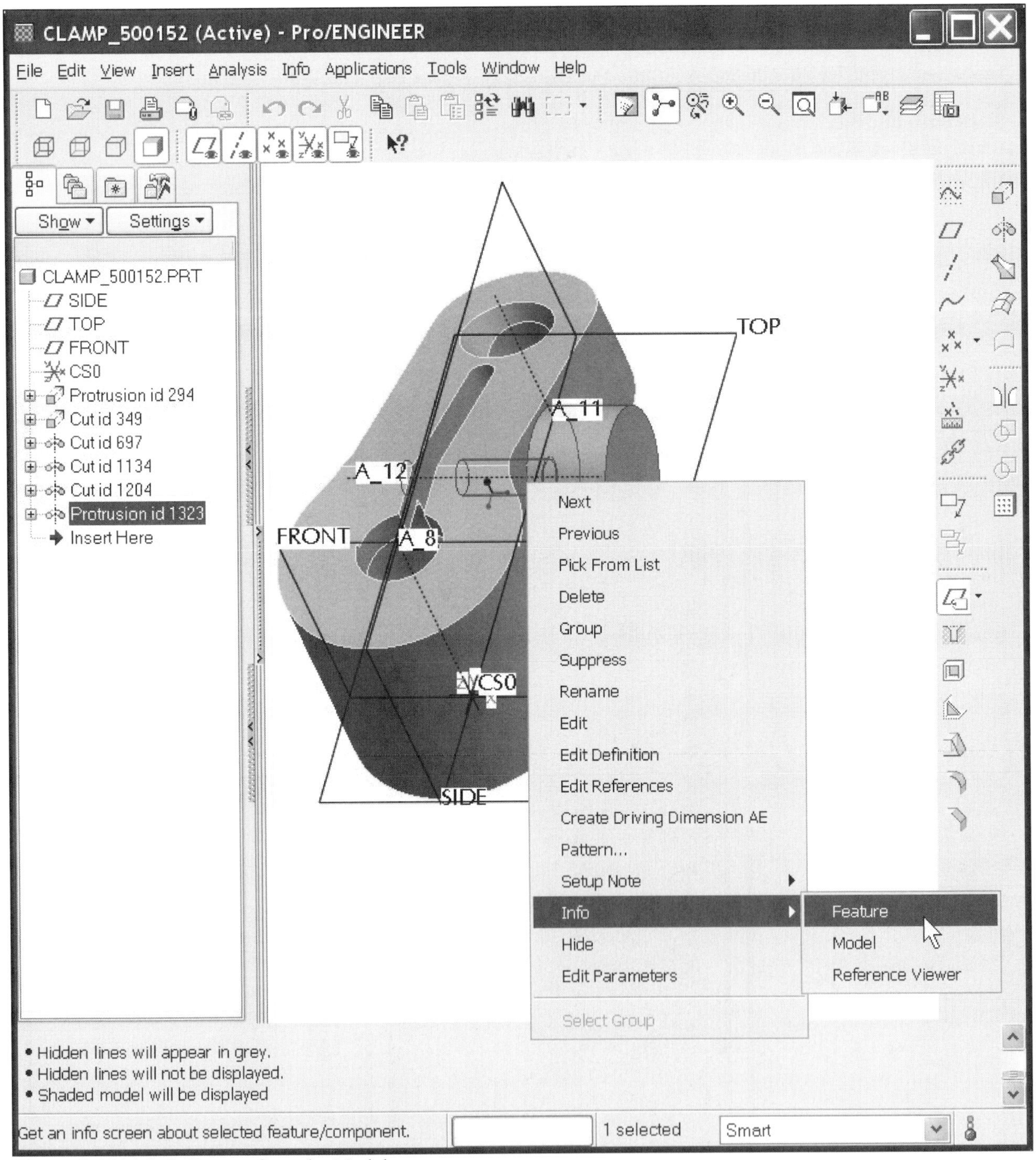

Figure 2.15(a) Info Feature from the Model

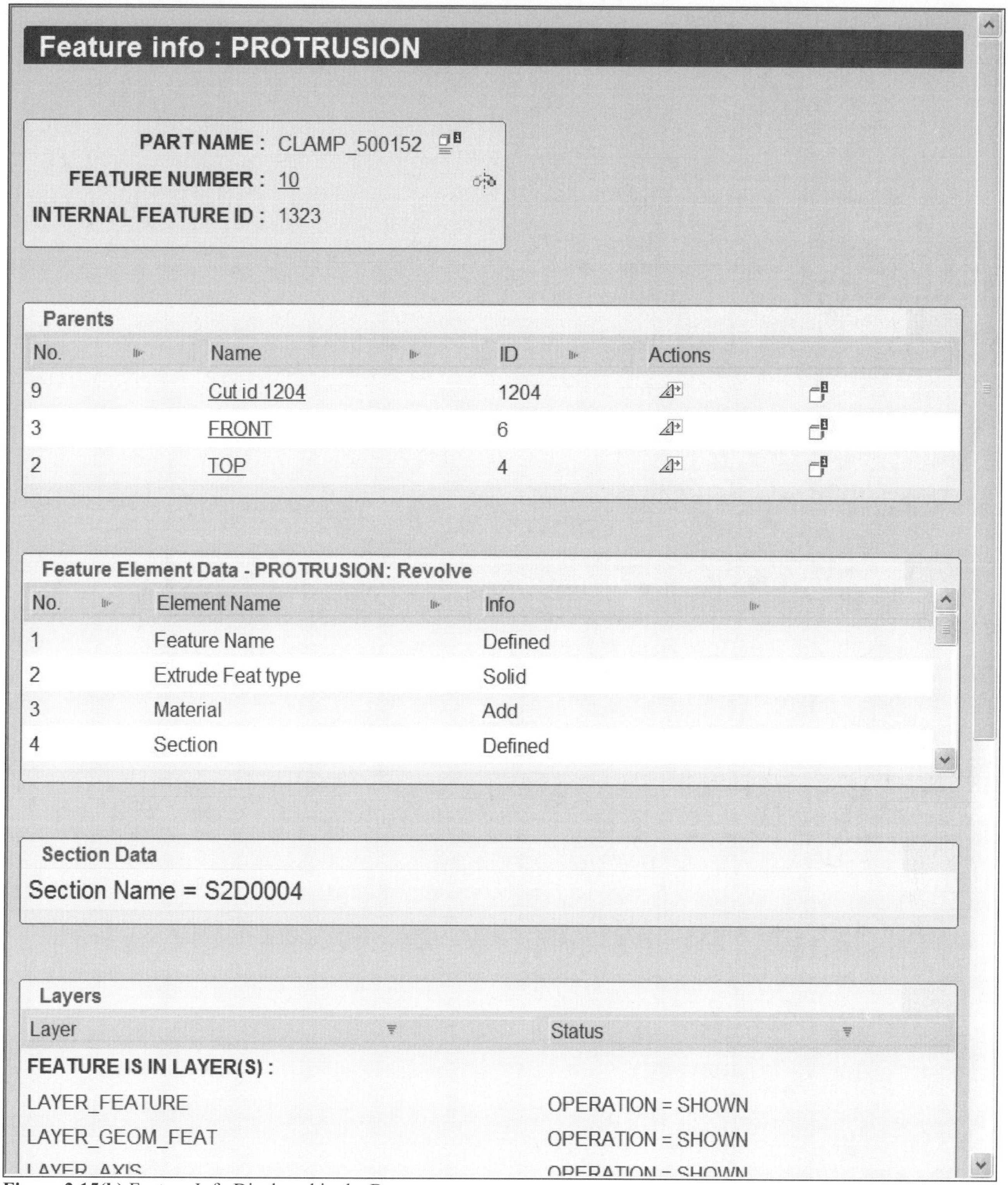

Figure 2.15(b) Feature Info Displayed in the Browser

Click on the "quick sash" to collapse the Browser 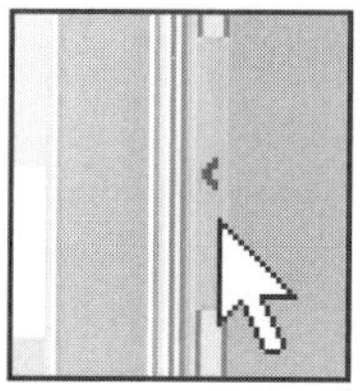

In the graphics window click: **RMB ⇒ Info ⇒ Model** [Fig. 2.15(c-d)] ⇒ collapse the Browser again

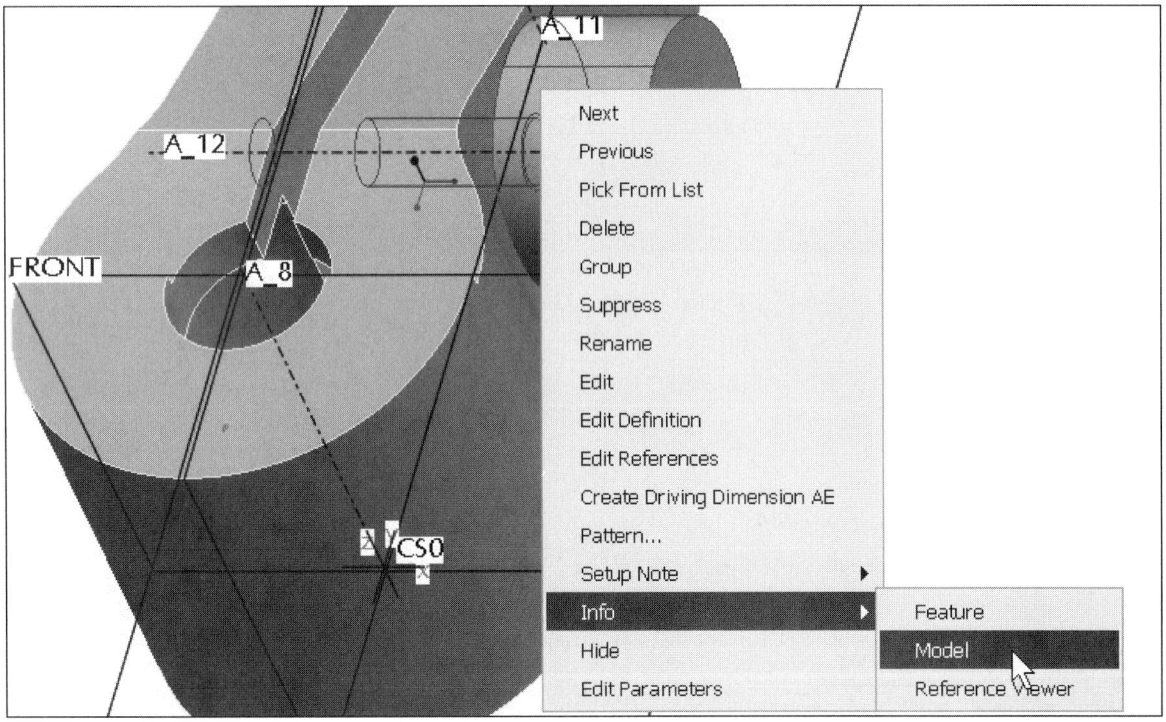

Figure 2.15(c) Feature Dimension Information

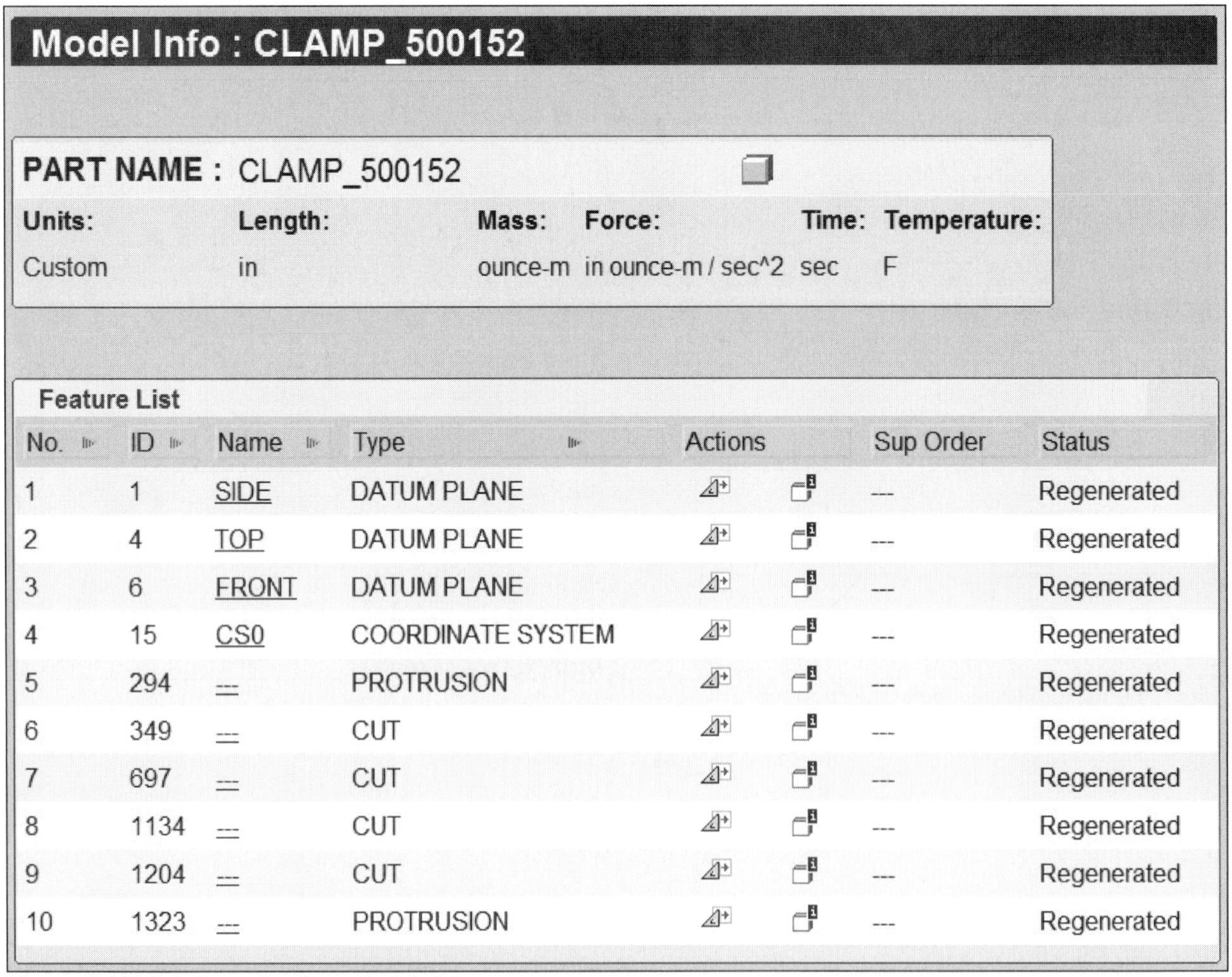

Model Info : CLAMP_500152

PART NAME : CLAMP_500152

Units:	Length:	Mass:	Force:	Time:	Temperature:
Custom	in	ounce-m	in ounce-m / sec^2	sec	F

Feature List

No.	ID	Name	Type	Actions		Sup Order	Status
1	1	SIDE	DATUM PLANE			---	Regenerated
2	4	TOP	DATUM PLANE			---	Regenerated
3	6	FRONT	DATUM PLANE			---	Regenerated
4	15	CS0	COORDINATE SYSTEM			---	Regenerated
5	294	---	PROTRUSION			---	Regenerated
6	349	---	CUT			---	Regenerated
7	697	---	CUT			---	Regenerated
8	1134	---	CUT			---	Regenerated
9	1204	---	CUT			---	Regenerated
10	1323	---	PROTRUSION			---	Regenerated

Figure 2.15(d) Information from the Model

The Model Tree

The **Model Tree** is a tabbed feature on the Pro/E navigator that displays a list of every feature or part in the current part, assembly, or drawing.

The model structure is displayed in hierarchical (tree) format with the root object (the current part or assembly) at the top of its tree and the subordinate objects (parts or features) below. If you have multiple Pro/E windows open, the Model Tree contents reflect the file in the current active window.

The Model Tree lists only the related feature and part level items in a current object and does not list the entities (such as edges, surfaces, curves, and so forth) that comprise the features.

Each Model Tree item contains an icon that reflects its object type, for example, hidden, assembly, part, feature, or datum plane (also a feature). The icon can also show the display status for a feature, part, or assembly, for example, suppressed.

Selection in the Model Tree is object-action oriented; you select objects in the Model Tree without first specifying what you intend to do with them. You can select components, parts, or features using the Model Tree. You cannot select the individual geometry that makes up a feature (entities). To select an entity, you must select it in the graphics window.

With the **Settings** tab you can control what is displayed in the Model Tree.

You can add informational columns to the Model Tree window, such as **Tree Columns** containing parameters and values, assigned layers, or feature name for each item. You can use the cells in the columns to perform context-sensitive edits and deletions. These options will be covered elsewhere in the text, as they are needed in the design process.

Click: **Settings** 

tab ⇒ **Tree Filters** 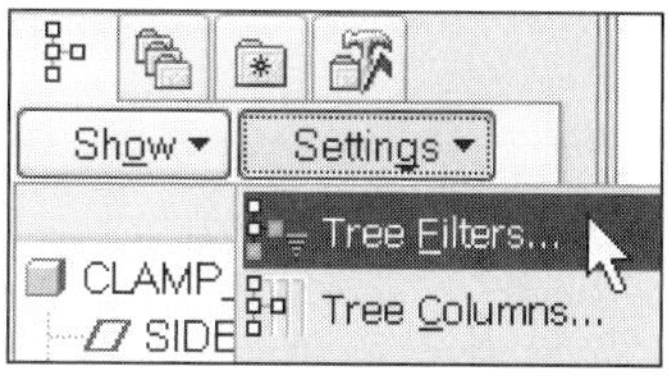

⇒ Display (toggle on Display options [Fig. 2.16(a)] ⇒ **Apply** ⇒ **OK**

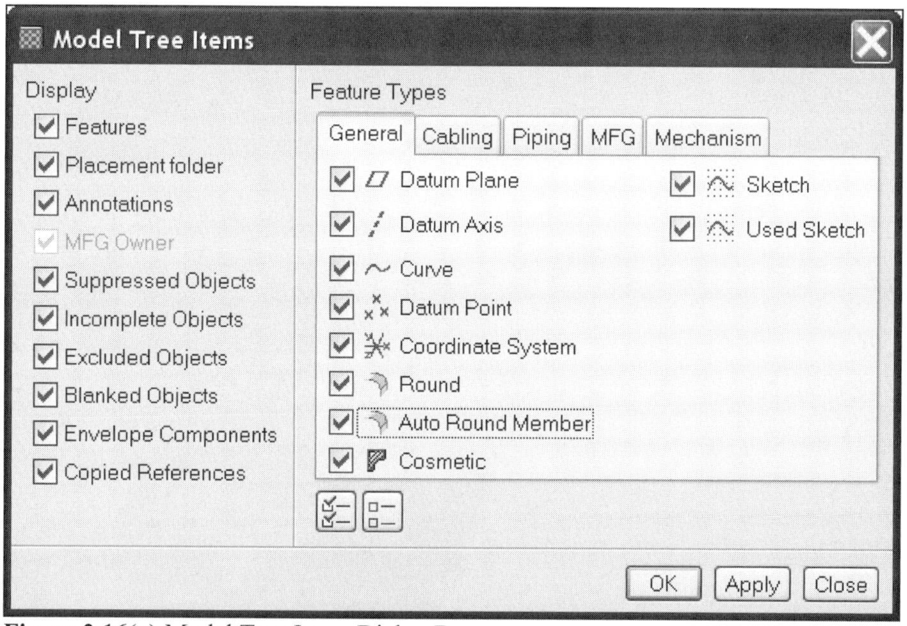

Figure 2.16(a) Model Tree Items Dialog Box

Working on the Model

In Pro/E, you can select objects to work on from within the graphics window or in the Model Tree by using the mouse or the keyboard. The object types that are available for selection vary depending on whether you select an object from within the graphics window or in the Model Tree. You can select any type of object, including features, 3-D notes (model Annotations), parts, datum objects (planes, axes, curves, points, and coordinate systems), and geometry (edges and surfaces) from within the graphics window. Additionally, since the Model Tree displays only parts, components, and features, you can select only those object types from within the Model Tree.

Selection in both the graphics window and the Model Tree can be action-object or object-action oriented, depending on the process you choose within Pro/E to build your model. You can specify the action you want to perform on an object before you select the object, or you can select the object before you specify the action.

You can *dynamically* modify certain features from within the graphics window as you work. Features can be edited by selecting them from the graphics window directly, or from the Model Tree. Dimensions of the following features can be modified dynamically:

- **Protrusions** Extruded protrusions (adding or removing material) with variable depth, and revolved protrusions of variable angle
- **Surfaces** Extruded surfaces with variable depth, and revolved surfaces with variable angle
- **Rounds** Simple, constant, and edge chain

Pick on the cylindrical protrusion ⇒ **RMB** [Fig. 2.17(a)] ⇒ **Delete** ⇒ **OK** in the Delete dialog box [Fig. 2.17(b)]

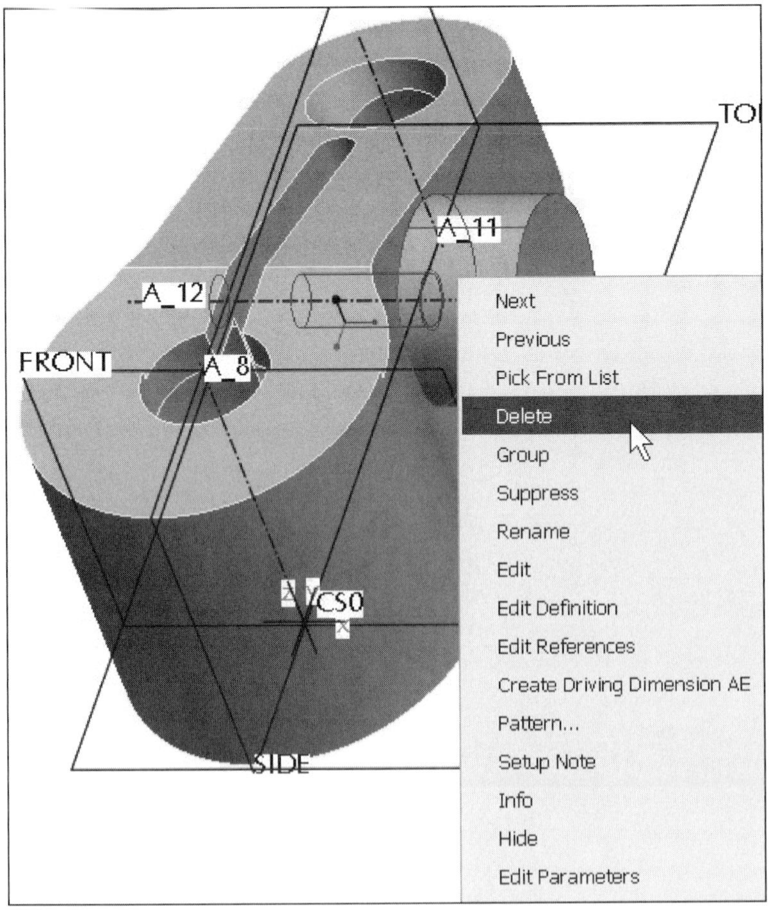

Figure 2.17(a) Delete

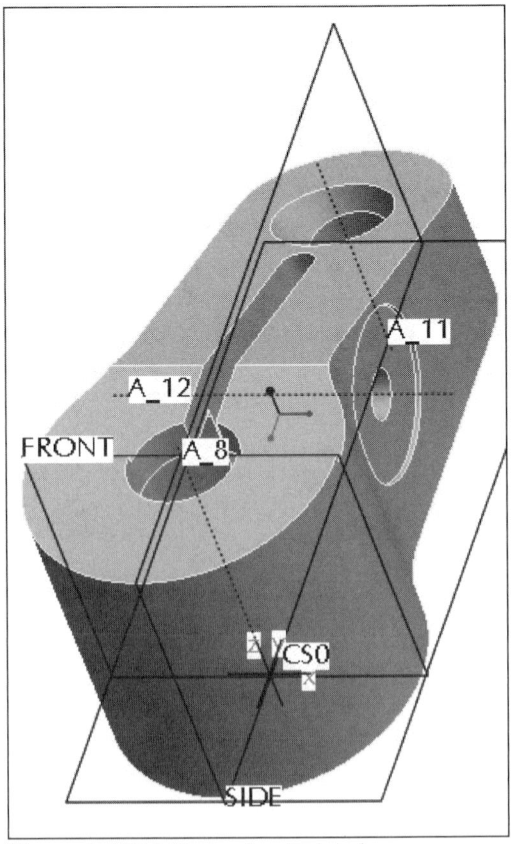

Figure 2.17(b) Protrusion Deleted

Click on the protrusion in the Model Tree [Fig. 2.18(a)] ⇒ **RMB** ⇒ **Edit Definition** [Fig. 2.18(b)]

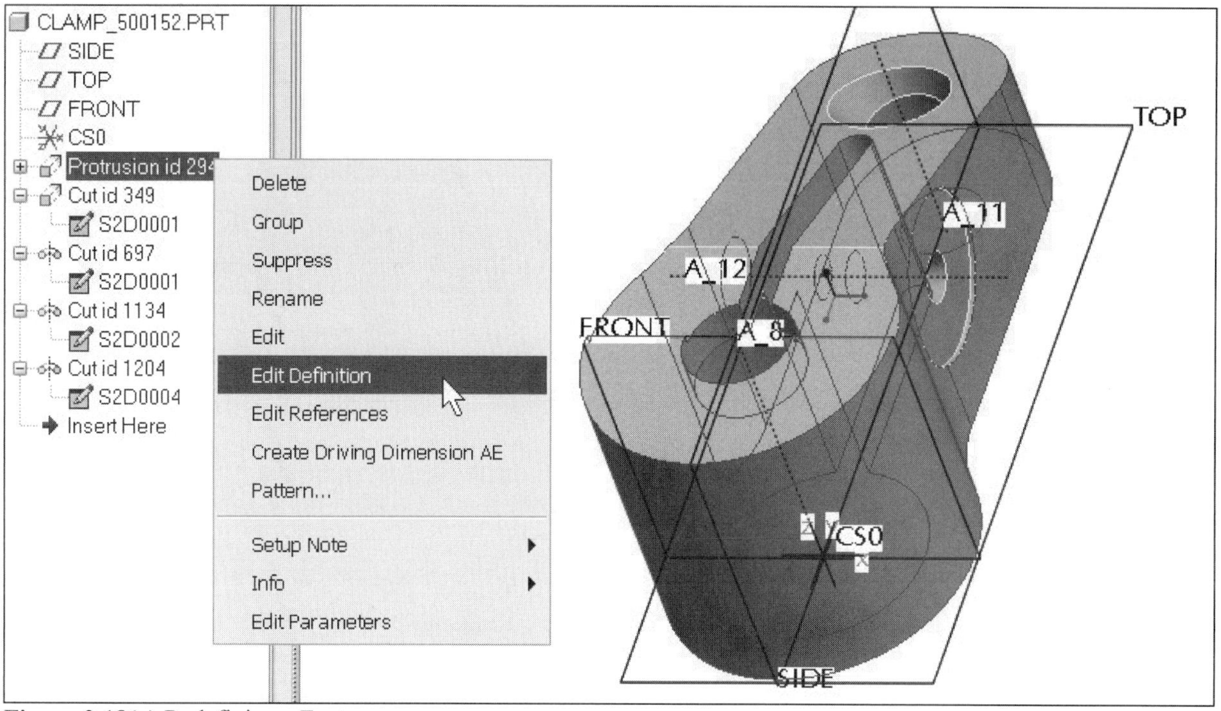

Figure 2.18(a) Redefining a Feature

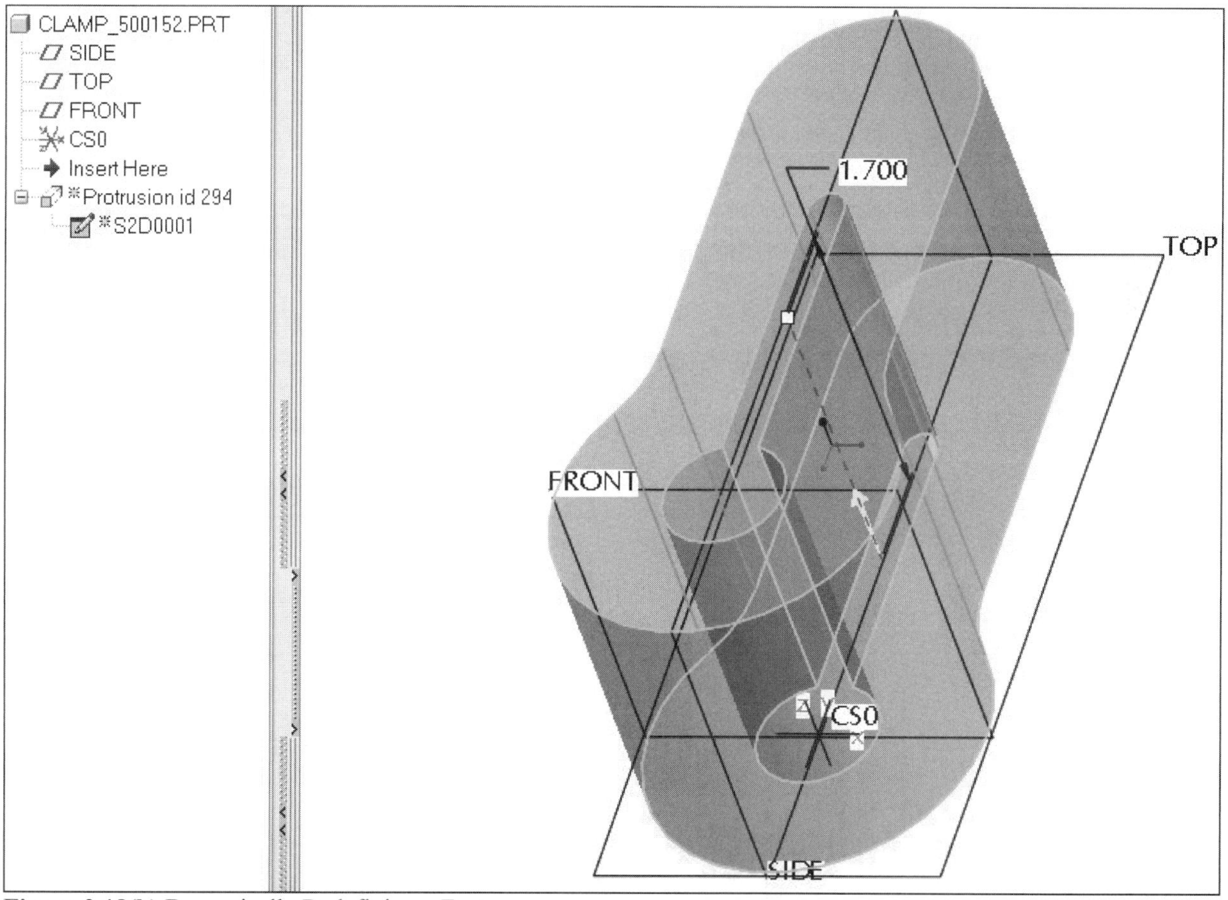

Figure 2.18(b) Dynamically Redefining a Feature

Pick on and drag the white nodal handle to change the protrusion's height from **1.700** [Fig. 2.18(c)] to **2.000** [Fig. 2.18(d)] ⇒ **MMB** *(or press the Enter key, or click on the green check mark in the dashboard)* to regenerate the model ⇒ **LMB** to deselect

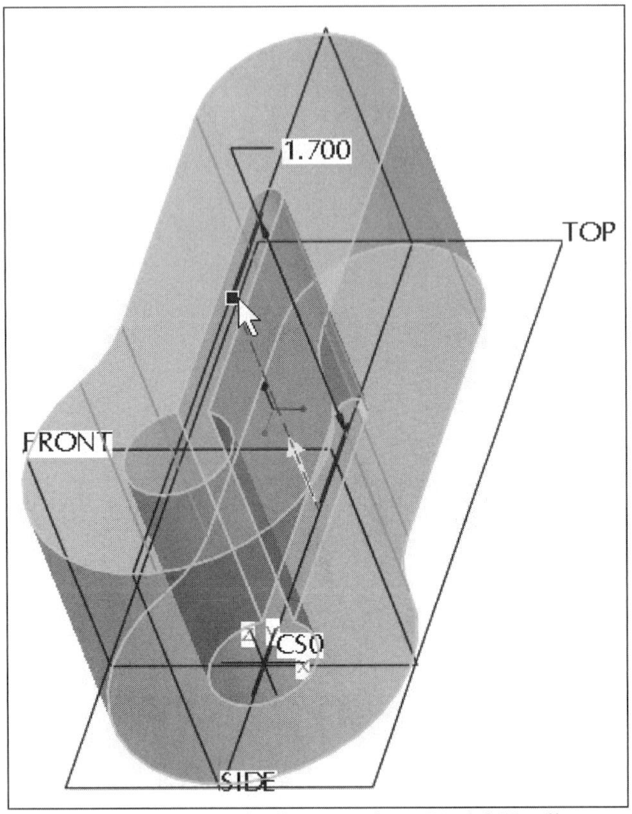

Figure 2.18(c) Dynamically Dragging a Nodal Handle

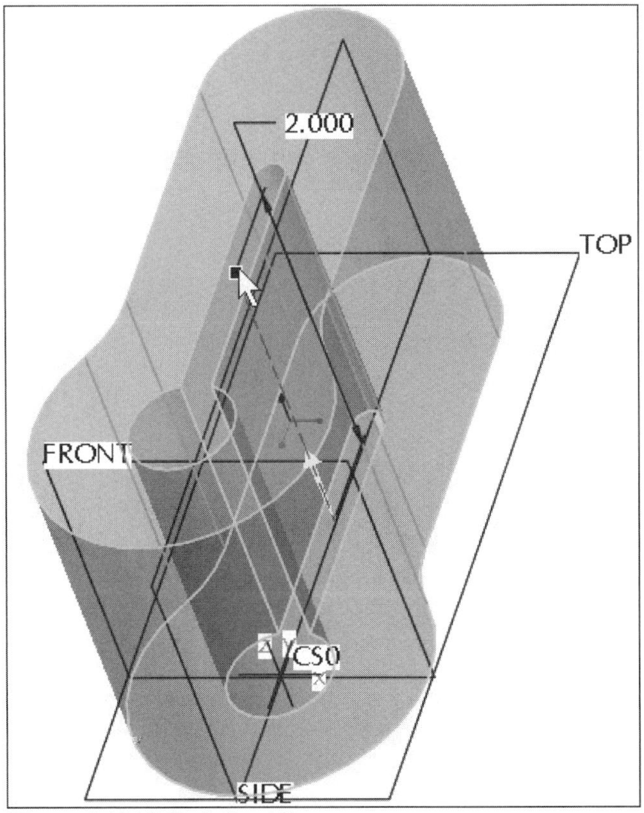

Figure 2.18(d) Redefined Feature

Although you may have not noticed it, there was other information and editing available during the redefine process. Click on the protrusion in the Model Tree ⇒ **RMB** ⇒ **Edit Definition** ⇒ look at the "dashboard" on the lower part of the screen [Fig. 2.19(a)] ⇒ **Options** tab [Fig. 2.19(b)] ⇒ change the **2.000** to **4.000** ⇒ **Enter** ⇒ ✓ [Figs. 2.19(c-d)] ⇒ **Window** ⇒ **Close**

Note that *RMB ⇒ Edit Definition* is the solution to 75% of all questions-problems in Pro/E, regardless of whether you are working on a component part or an assembly model.

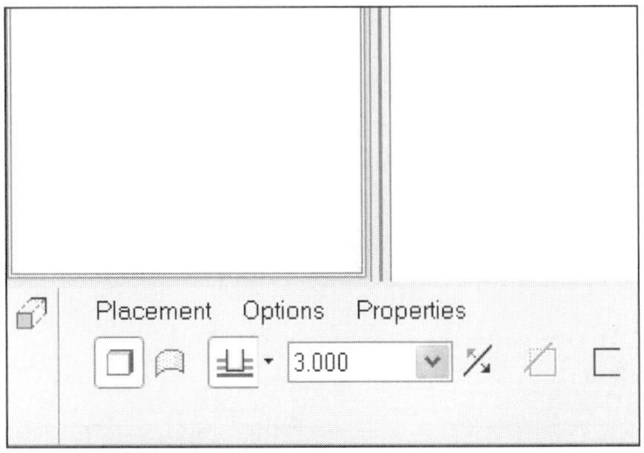

Figure 2.19(a) Dashboard

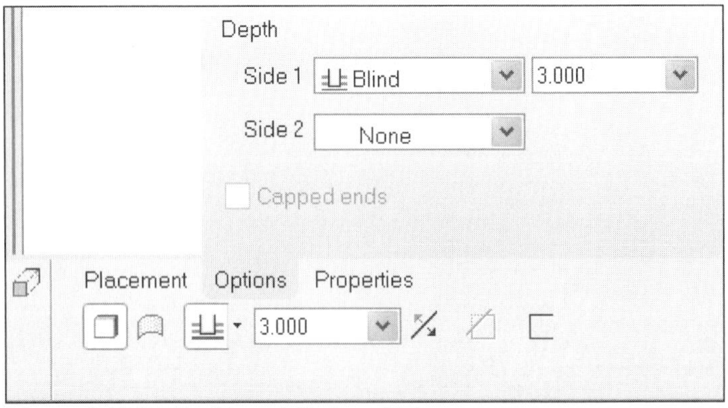

Figure 2.19(b) Dashboard Options Tab

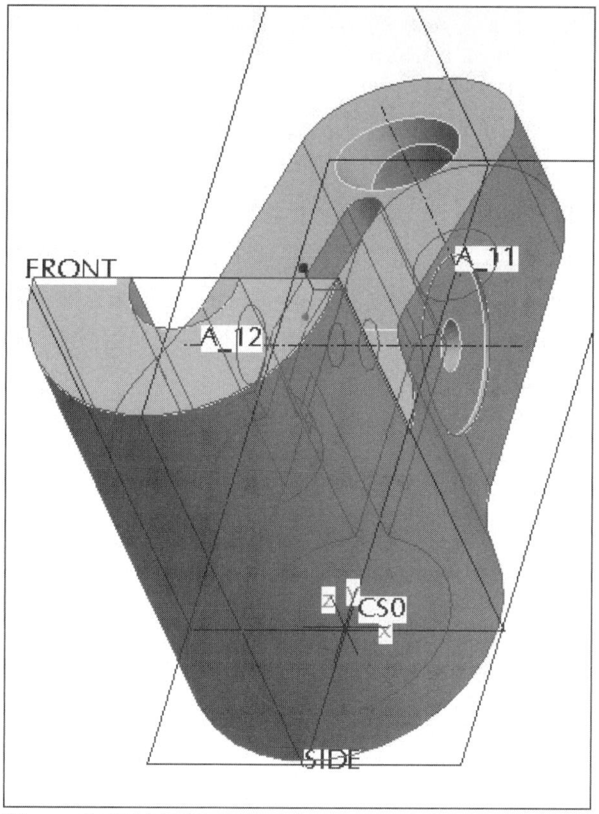

Figure 2.19(c) Edited Model

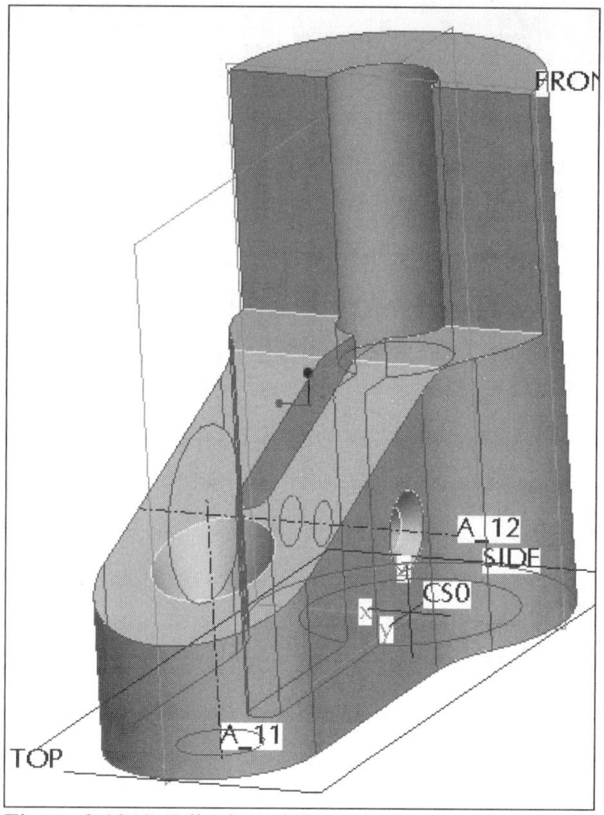

Figure 2.19(d) Edited Model Reoriented

About the Dashboard

As you create and modify your models using direct graphical manipulation in the graphics window, the **Dashboard** guides you throughout the modeling process. This context sensitive interface monitors your actions in the current tool and provides you with basic design requirements that need to be satisfied to complete your feature. As you select individual geometry, the Dashboard narrows the available options enabling you to make only targeted modeling decisions. For advanced modeling, separate slide-up panels provide all relevant advanced options for your current modeling action. These Dashboard panels remain hidden until needed. This enables you to remain focused on successfully capturing the design intent of your model. The Dashboard for a revolve feature is different from an extrude feature [Figs. 2.20(a-c)]. Figure 2.20(d) shows Edit Definition applied to the revolved cut.

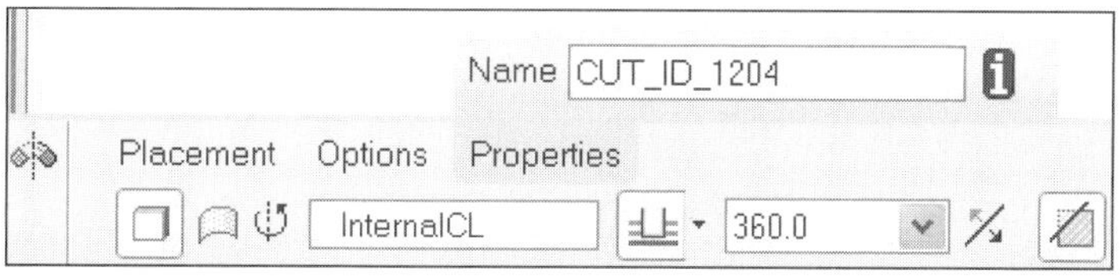

Figure 2.20(a) Revolved Feature Properties

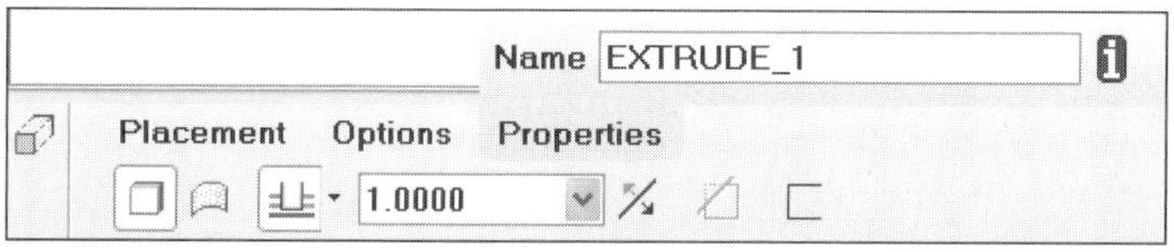

Figure 2.20(b) Extruded Feature Properties

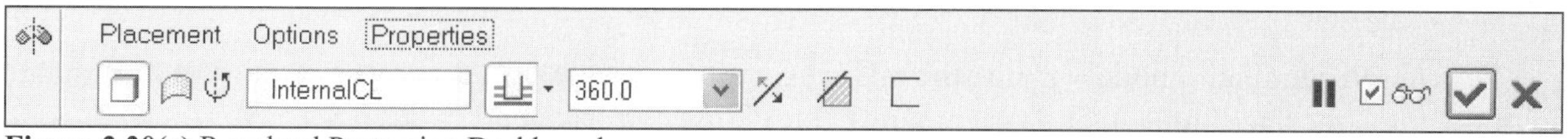

Figure 2.20(c) Revolved Protrusion Dashboard

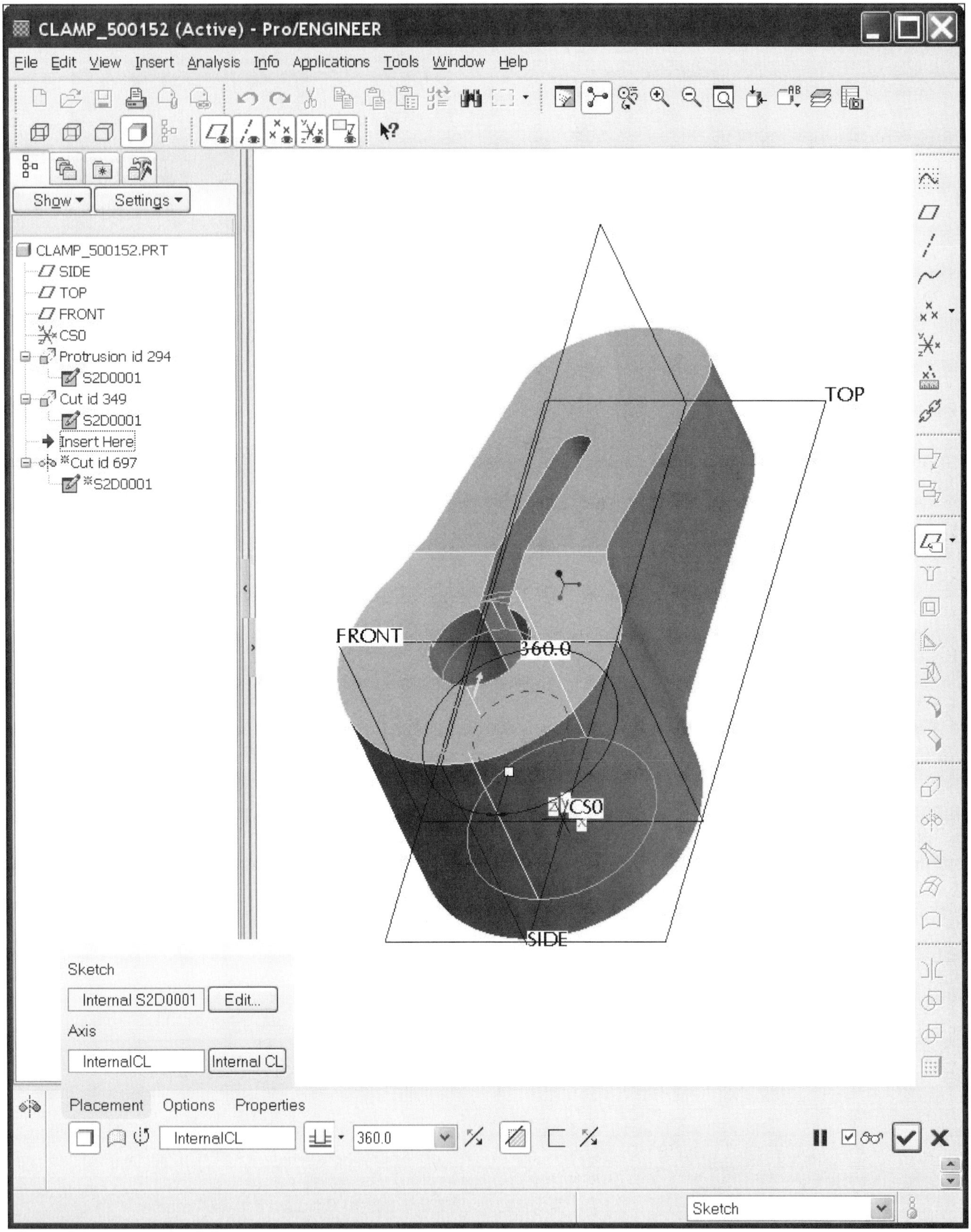

Figure 2.20(d) Revolved Cut

Download another model from the PTC Catalog or www.cad-resources.com and practice navigating the interface and using commands previously introduced. ***Changes, corrections, and suggestions can be found at www.cad-resources.com on the Pro/ENGINEER Wildfire 4.0 book page.***

Lesson 3 Direct Modeling

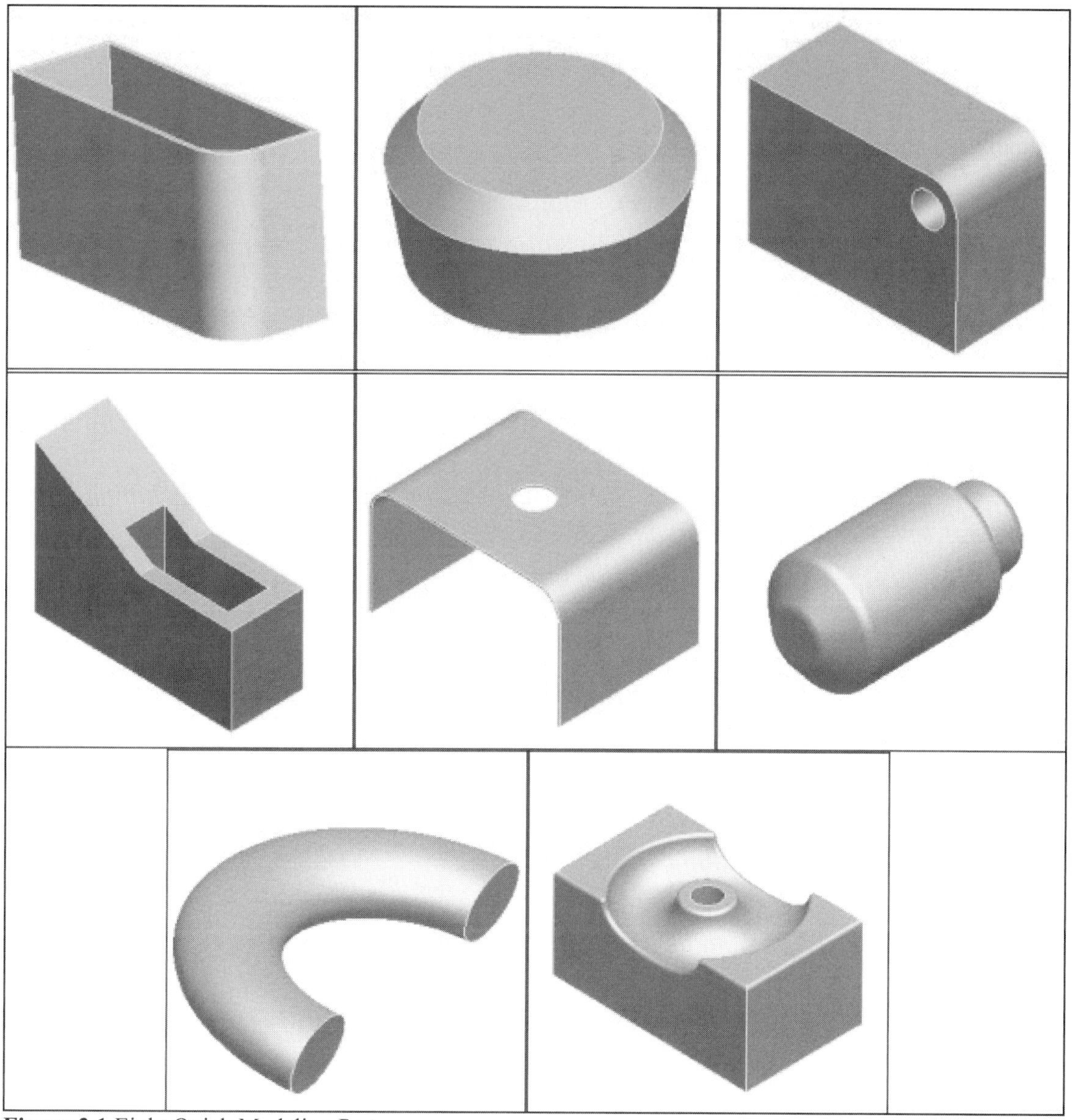

Figure 3.1 Eight Quick Modeling Parts

OBJECTIVES

- Modeling simple parts quickly using **default selections**
- Sample the **Extrude Tool** and the **Revolve Tool**
- Try out a variety of **Engineering Tools** including: **Hole**, **Shell**, **Round**, **Chamfer**, and **Draft**
- Sketch simple **sections**

Modeling

The purpose of this lesson is to quickly introduce you to a variety of **Feature Tools** and **Engineering Tools**. You will model a variety of very simple parts (Fig. 3.1). Little or no explanation of the methodology or theory of the Tool or process will accompany the instructions. By using mainly default selections; you will create models that will display the power and capability of Pro/E Wildfire 4.0.

In the next lesson, a detailed step-by-step systematic description will accompany all commands. Here, we hope to get you up and running on Pro/E without any belabored explanations.

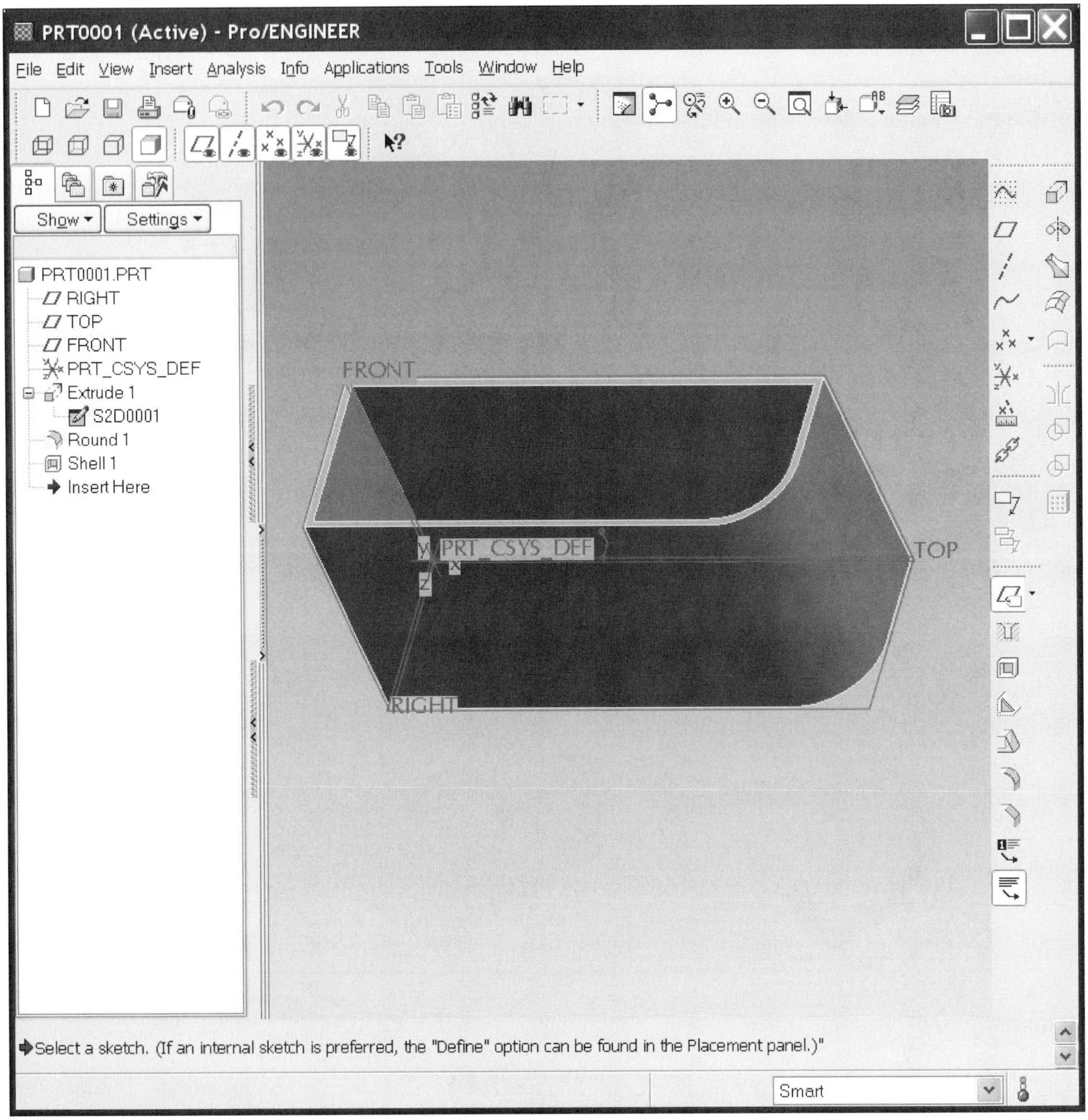

Figure 3.2(a) Part One

Part Model One (PRT0001.PRT) (Extrude)

This model [Fig. 3.2(a)] will introduce the **Extrude Tool** to create a simple box-shape, the **Round Tool** to add a round to one edge, and the **Shell Tool** to remove one surface and make the part walls a consistent thickness. In this direct modeling example, we will provide step-by-step illustrations. For subsequent parts, only important steps or illustrations displaying aspects of the command sequence that represent new material will be provided. The same applies to ***Tool Tips*** that appear as you pass your cursor over a button or icon. All parts have been created with out-of-the-box system settings for Pro/E Wildfire 4.0 including default templates, grid settings, and so forth.

Click: **Launch Pro/E** ⇒ **File** ⇒ **Set Working Directory** ⇒ select the working directory ⇒ **OK** ⇒ **Create a new object** ⇒ Part ⇒ **OK** ⇒ pick on the **FRONT** datum plane from the Model Tree [Fig. 3.2(b)]

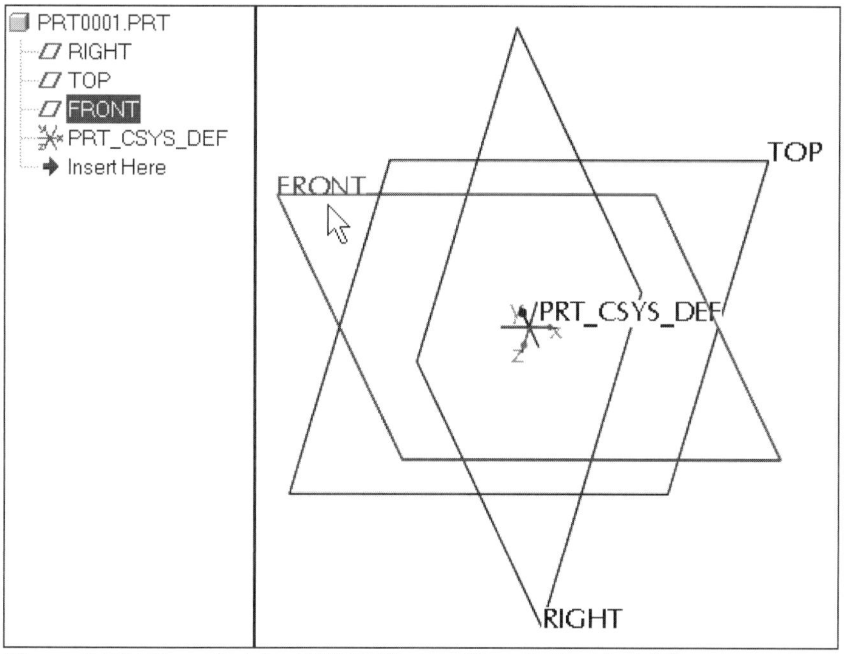

Figure 3.2(b) Pre-select the FRONT Datum Plane in the Model Tree

Click: **Extrude Tool** ⇒ **RMB** ⇒ **Define Internal Sketch** [Fig. 3.2(c)] ⇒ Sketch dialog box opens. Pre-selected FRONT datum is the Sketch Plane; RIGHT datum is automatically selected as the Sketch Orientation Reference [Fig. 3.2(d)]

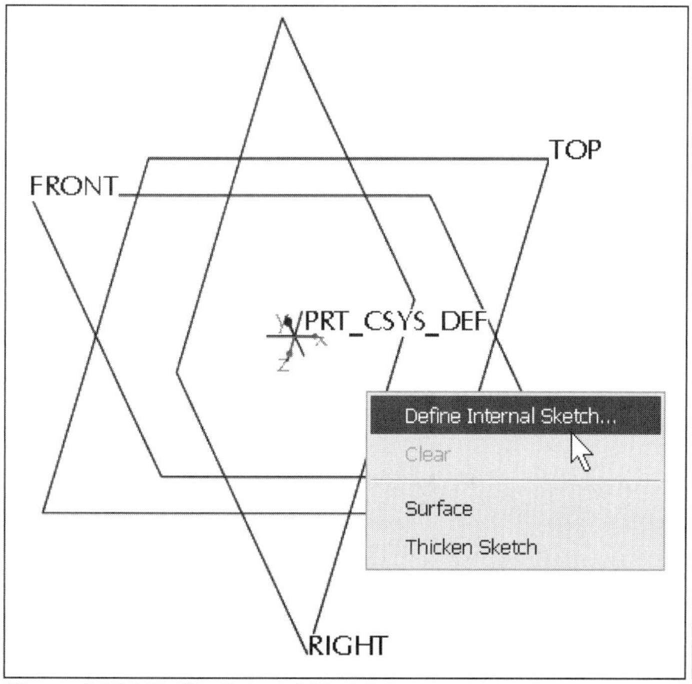

Figure 3.2(c) Define Internal Sketch

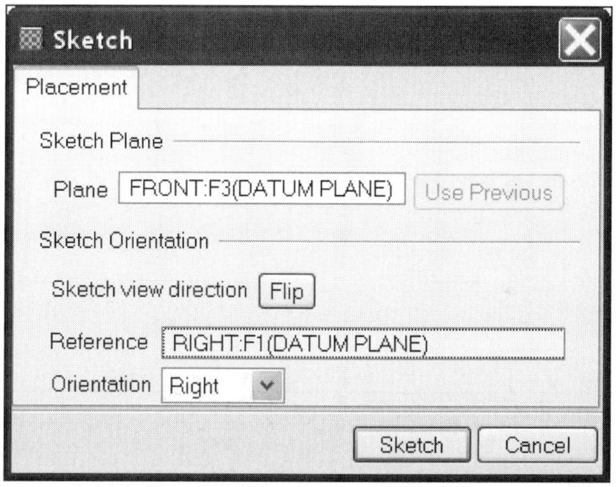

Figure 3.2(d) Sketch Dialog Box

With your mouse pointer in the Pro/E graphics window, click: **MMB** the Sketcher activates in order to define an independent section [Fig. 3.2(e)]

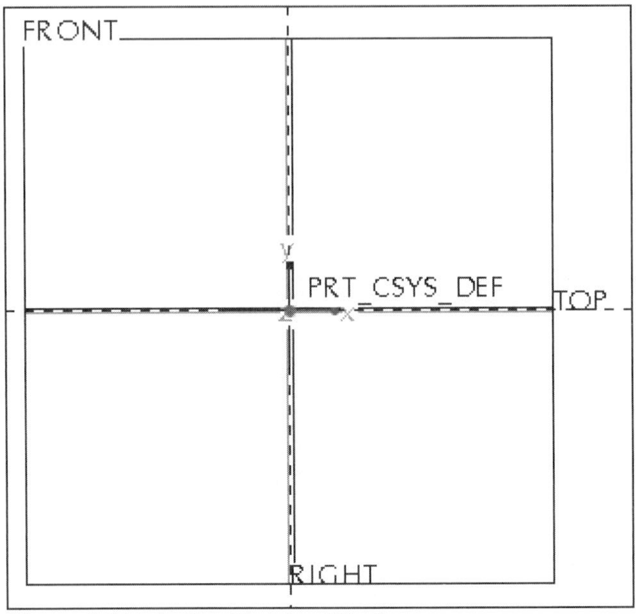

Figure 3.2(e) Active Sketcher Displays

Click: 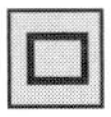 **Create rectangle** ⇒ sketch a rectangle by picking two corners [Fig. 3.2(f)] ⇒  **Continue with the current section**

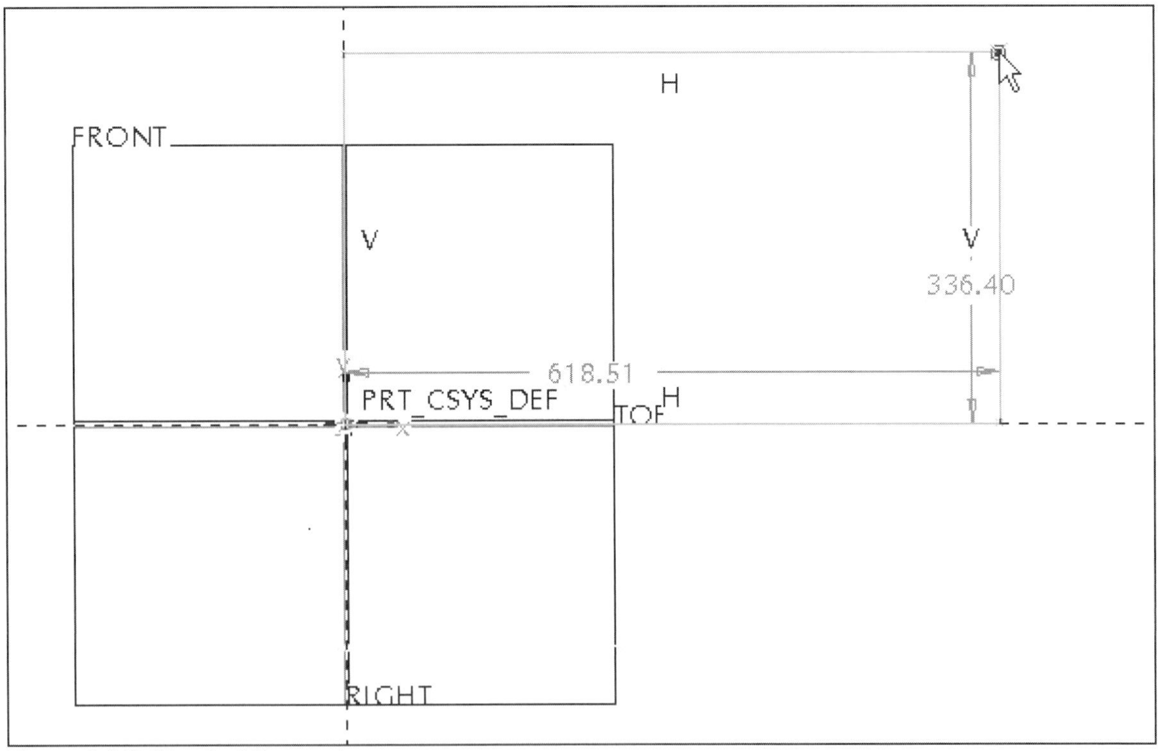

Figure 3.2(f) Sketched Rectangle

Click: ⇒ **Standard Orientation** [Fig. 3.2(g)] ⇒ **Shading** ⇒ **Enter** [Fig. 3.2(h)]

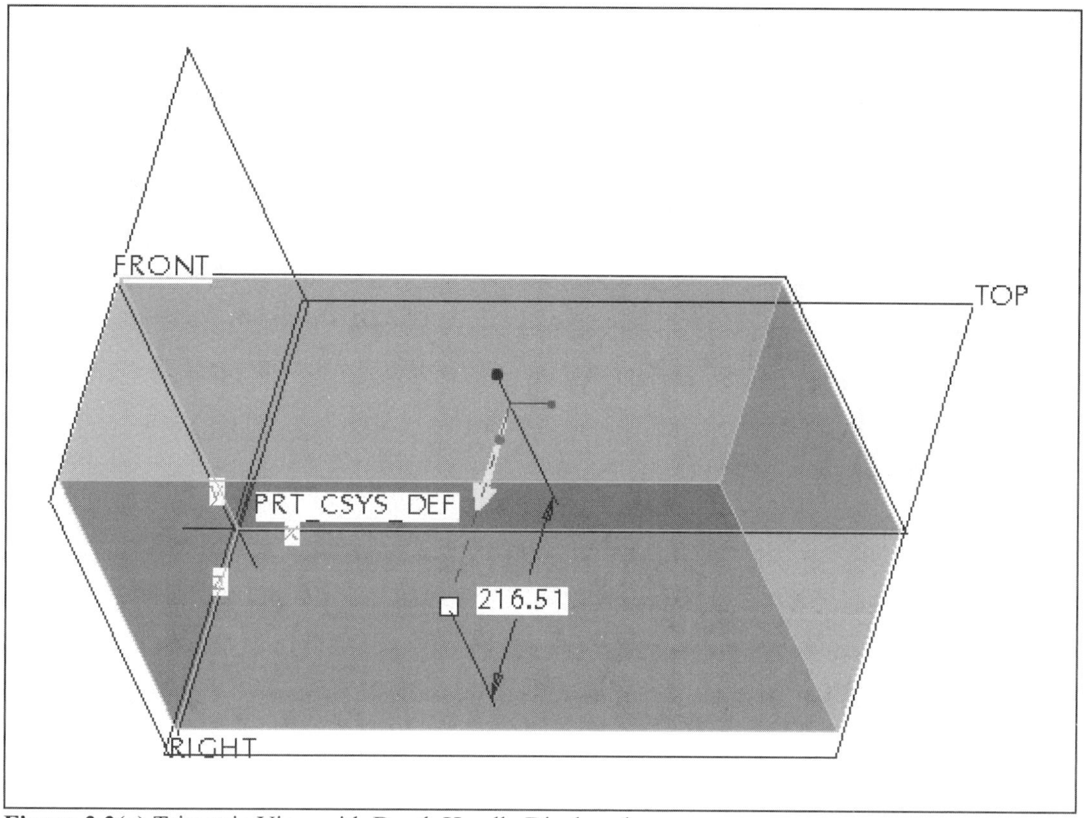

Figure 3.2(g) Trimetric View with Depth Handle Displayed

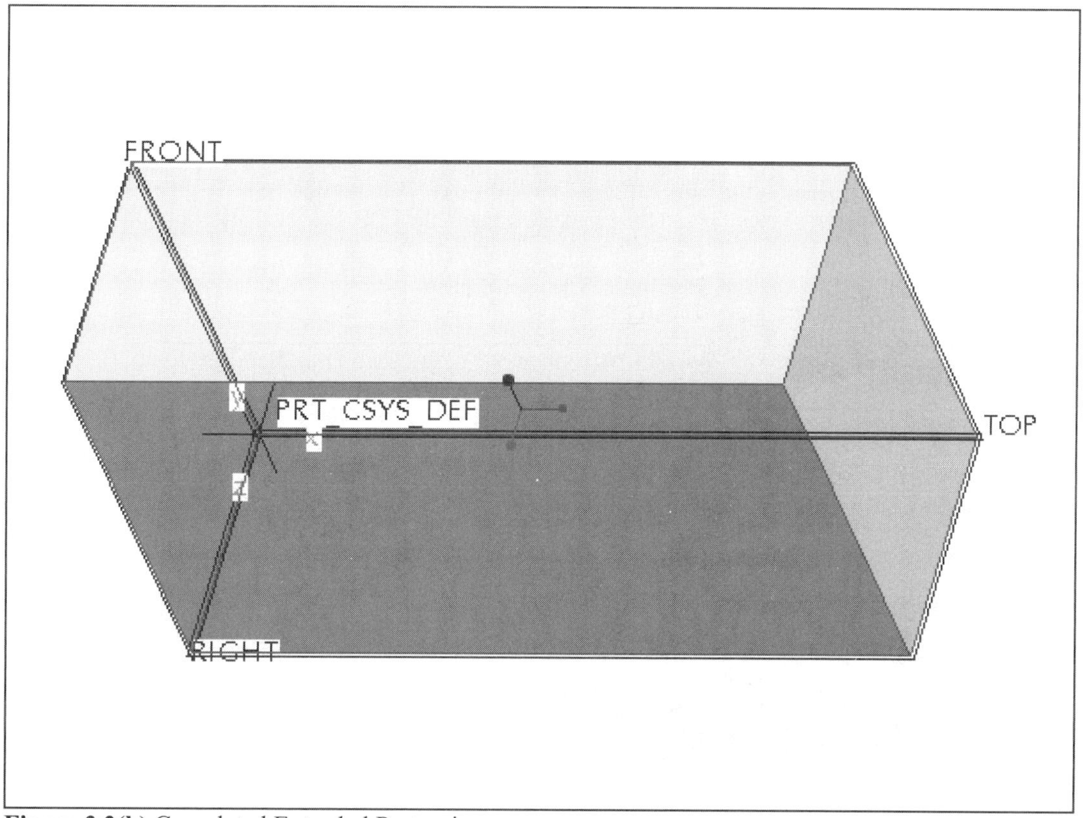

Figure 3.2(h) Completed Extruded Protrusion

Slowly pick on the right front edge of the part until it highlights [Fig. 3.2(i)] ⇒ **RMB** ⇒ **Round Edges** [Fig. 3.2(j)] ⇒ pick on and move a drag handle [Fig. 3.2(k)] ⇒ **MMB** (or Enter) [Fig. 3.2(l)]

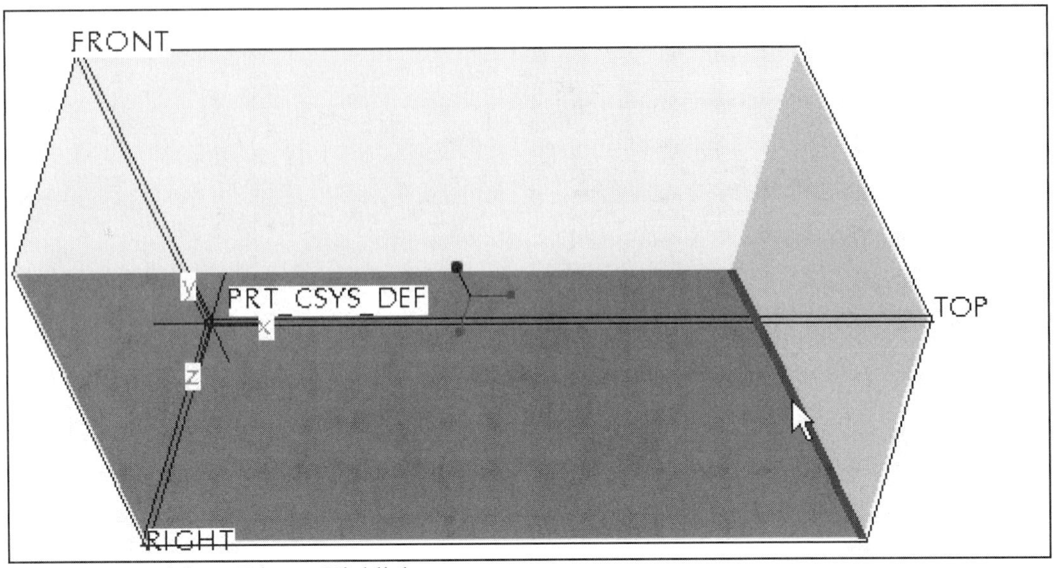

Figure 3.2(i) Pick on Edge to Highlight

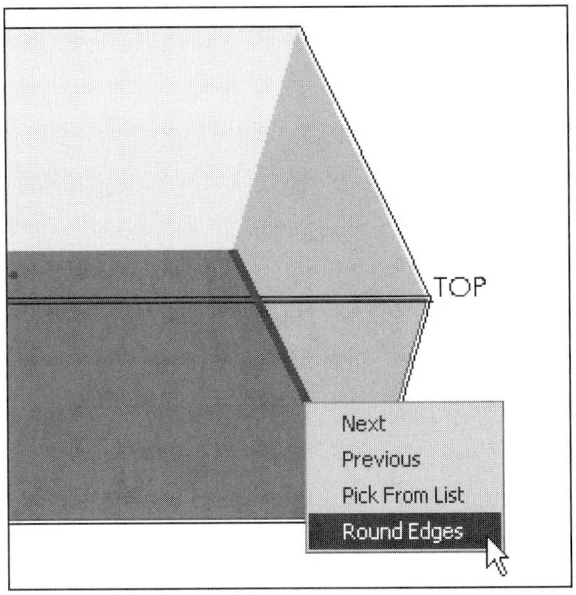

Figure 3.2(j) Round Edges

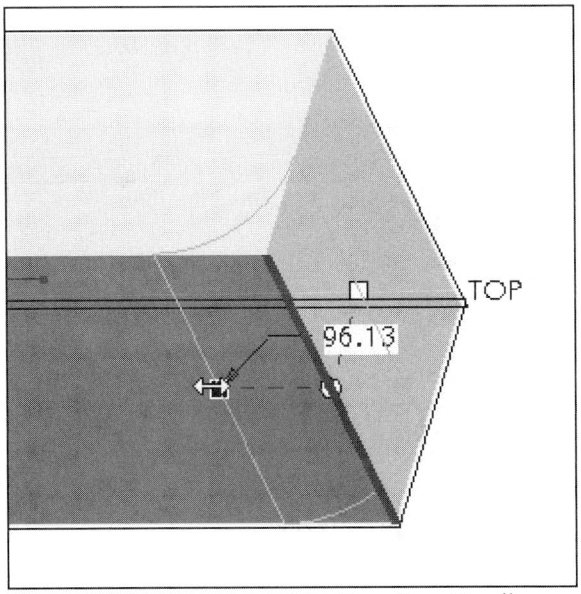

Figure 3.2(k) Move one of the Two Drag Handles

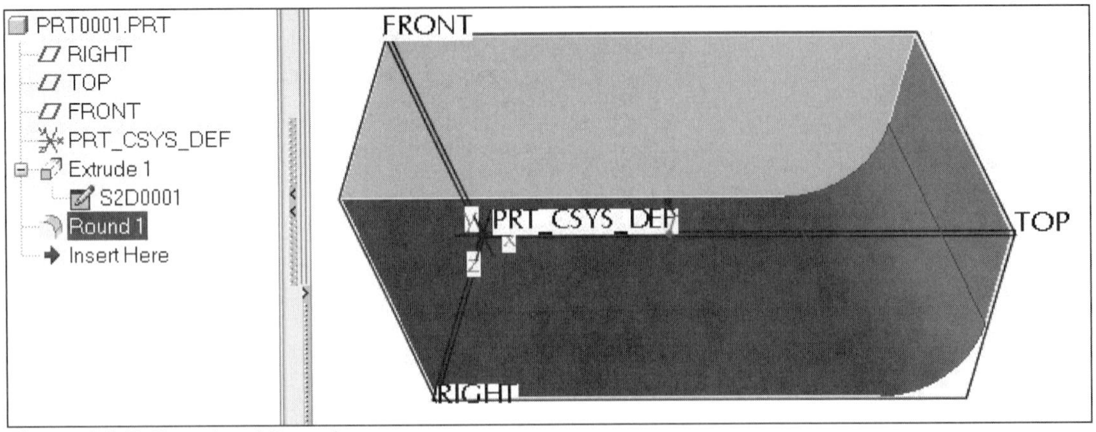

Figure 3.2(l) Completed Round

Pick on the top surface ⇒ slightly move the mouse and pick again (it will highlight in red) [Fig. 3.2(m)] ⇒ **Shell Tool** [Fig. 3.2(n)] ⇒ **MMB** [Fig. 3.2(o)] ⇒ ⇒ **Enter** ⇒ **File** ⇒ **Close Window**

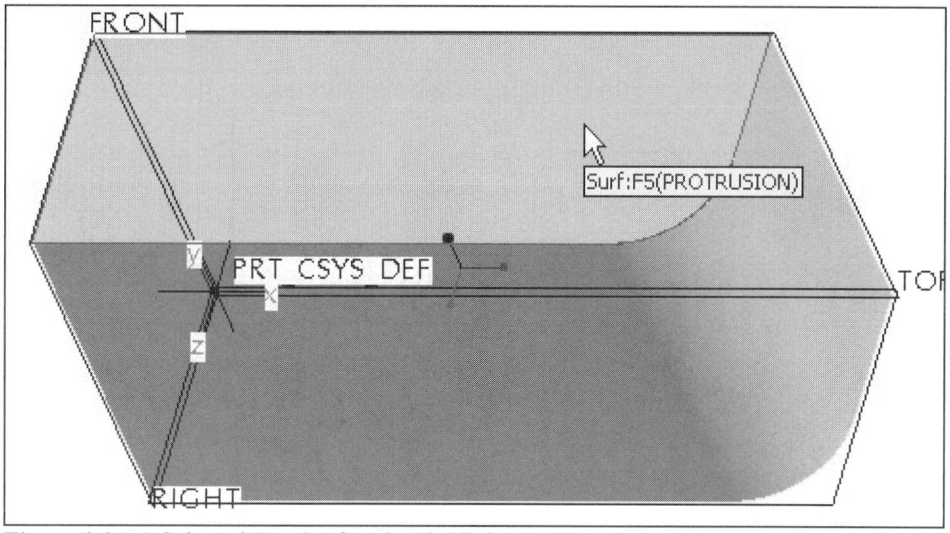

Figure 3.2(m) Selected Top Surface is Highlighted

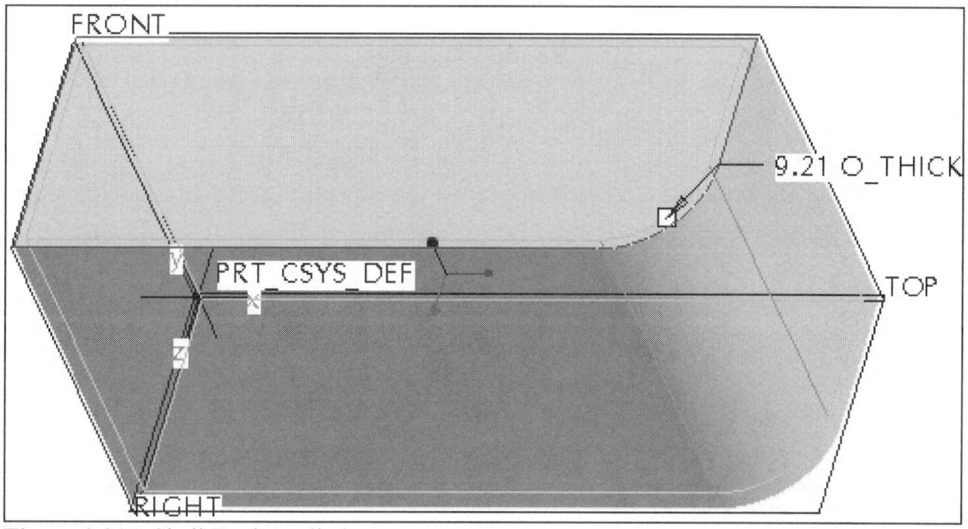

Figure 3.2(n) Shell Tool Applied

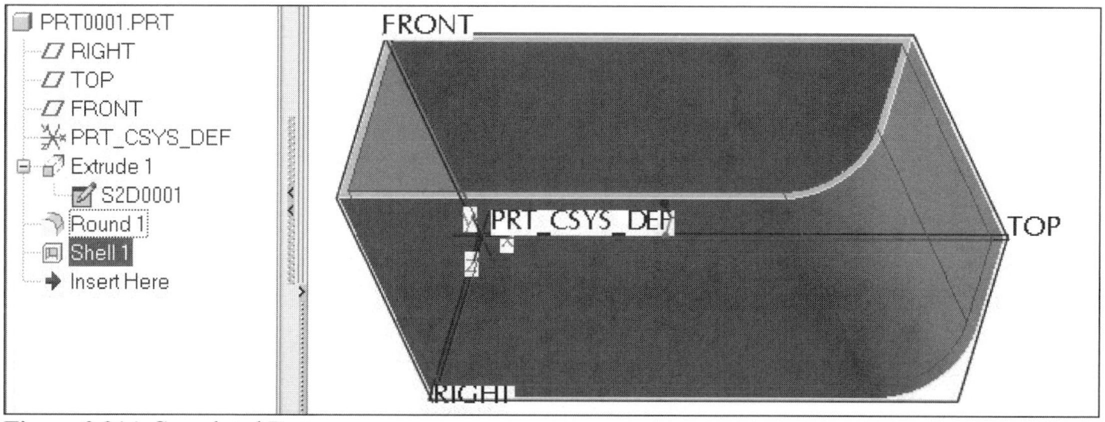

Figure 3.2(o) Completed Part

Part Model Two (PRT0002.PRT) (Draft)

Click: ⇒ **(PRT0002.PRT)** ⇒ **MMB** ⇒ pick on the **TOP** datum plane from the Model Tree ⇒ **Extrude Tool** ⇒ **RMB** ⇒ **Define Internal Sketch** Sketch dialog box opens with Sketch Plane and Sketch Orientation selected [Fig. 3.3(a)] ⇒ **MMB** in the graphics window

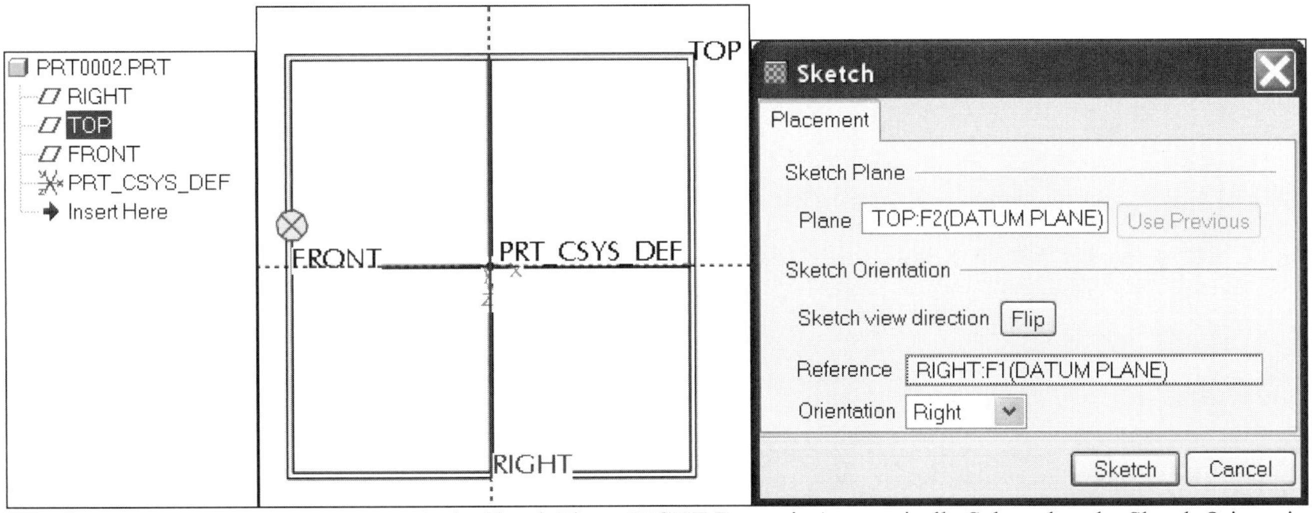

Figure 3.3(a) Pre-selected TOP Datum is the Sketch Plane, RIGHT Datum is Automatically Selected as the Sketch Orientation Reference

Click: **RMB** ⇒ **Circle** ⇒ sketch a circle by picking at the origin and then stretching the circle's diameter [Fig. 3.3(b)] to a convenient size ⇒ pick again to establish the circle's diameter [Fig. 3.3(c)] ⇒ **MMB** to deselect the circle command

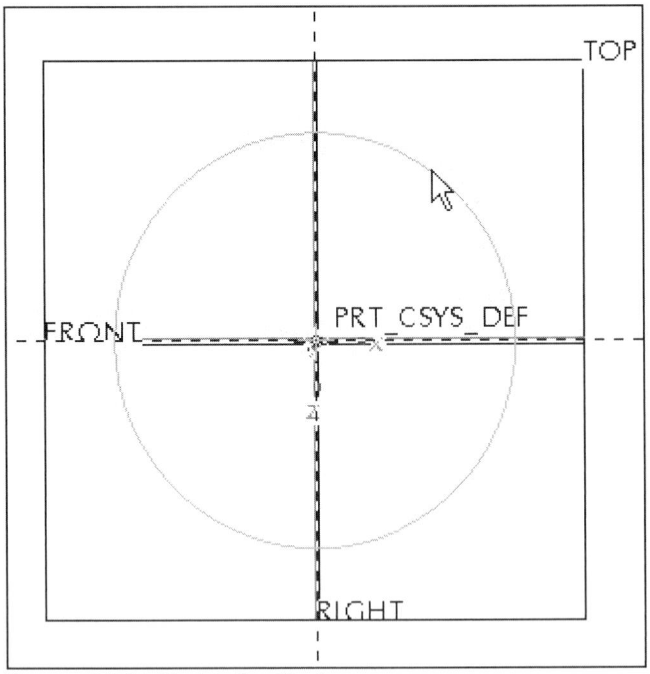

Figure 3.3(b) Sketch the Circle

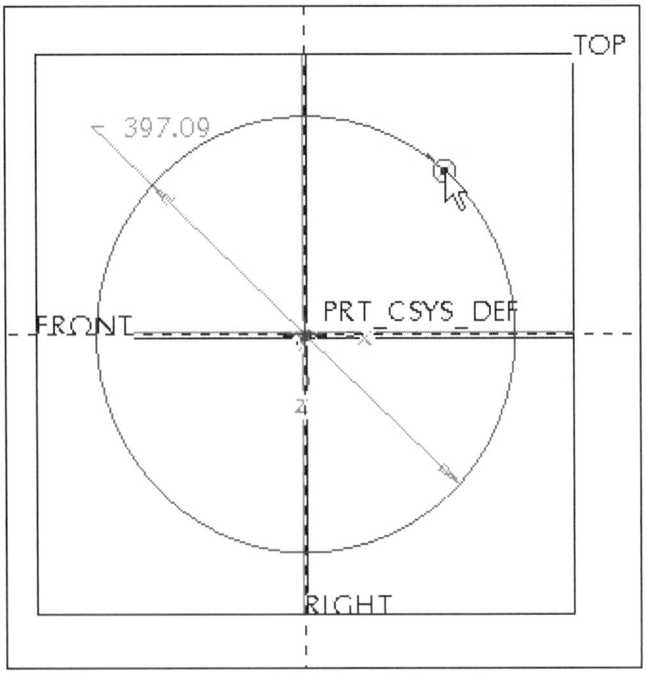

Figure 3.3(c) Pick to Establish the Circle's Diameter

Click: ⇒ **Standard Orientation** [Fig. 3.3(d)] ⇒ [Fig. 3.3(e)] ⇒ **MMB** [Fig. 3.3(f)] ⇒ **Hidden line** ⇒ **LMB** (in the graphics window) to deselect ⇒ pick on the cylindrical surface ⇒ pick on the surface again (it will highlight in red) [Fig. 3.3(g)] ⇒ **Draft Tool**

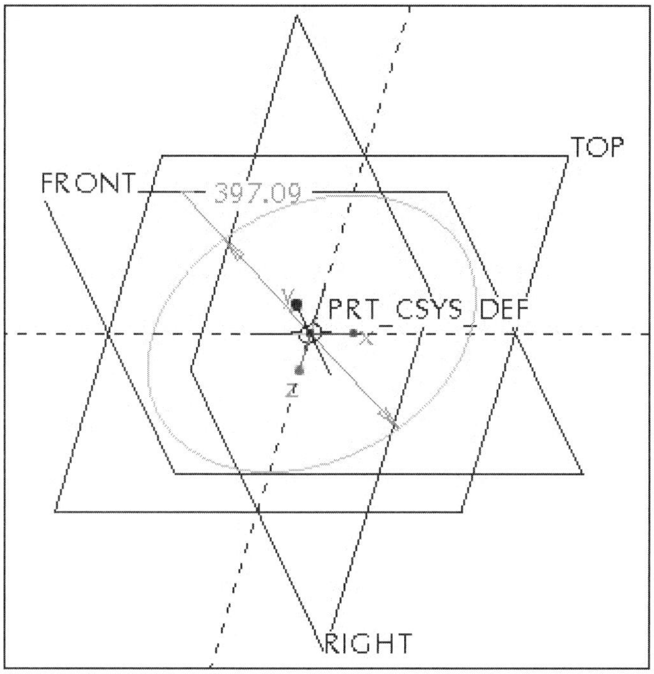

Figure 3.3(d) Sketch in Standard Orientation

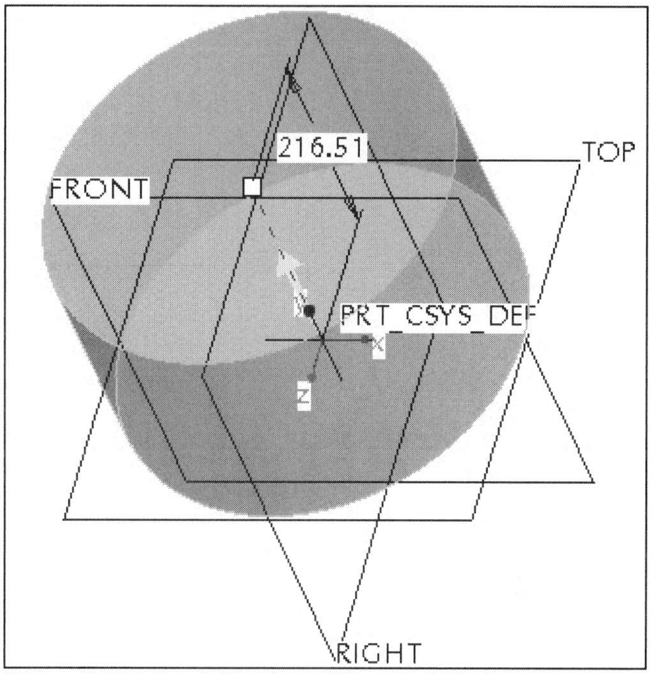

Figure 3.3(e) Depth Preview of Extruded Circle

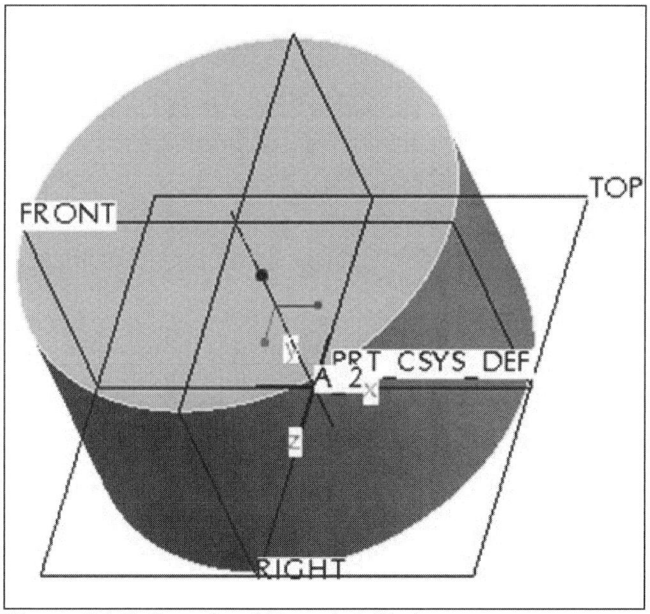

Figure 3.3(f) Completed Extruded Circle

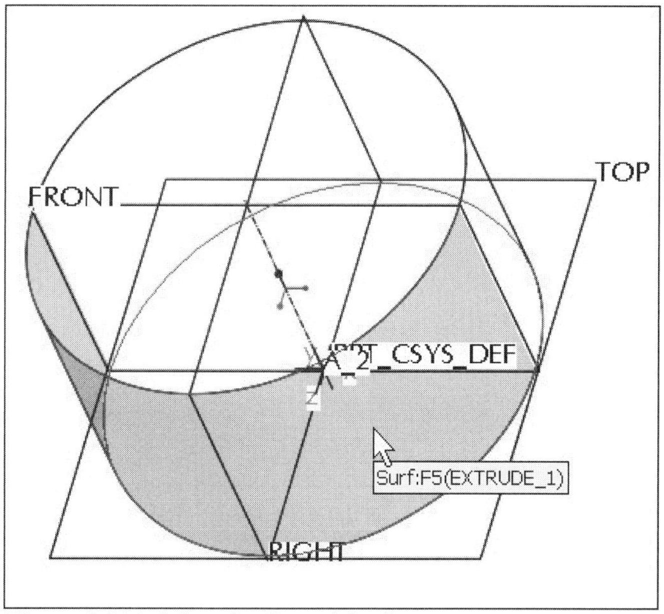

Figure 3.3(g) Select the Curved Surface

Click: References ⇒ pick the **TOP** datum plane as the Draft hinges ⇒ click in the Angle dimension box and type **10** 10.00 ⇒ **Enter** [Fig. 3.3(h)] ⇒ **Shading** ⇒ **MMB** [Fig. 3.3(i)] ⇒ pick on the top surface ⇒ pick on the top edge [Fig. 3.3(j)] ⇒ **Chamfer Tool** ⇒ type **50** D 50.00 ⇒ **Enter** ⇒ **MMB** [Fig. 3.3(k)] ⇒ press and hold **MMB** and spin the part to display the bottom surface ⇒ pick on the bottom surface ⇒ slightly move your cursor and pick again ⇒ **Shell Tool** ⇒ **MMB** [Fig. 3.3(l)] ⇒ ⇒ **Enter** (or MMB) ⇒ **Window** ⇒ **Close**

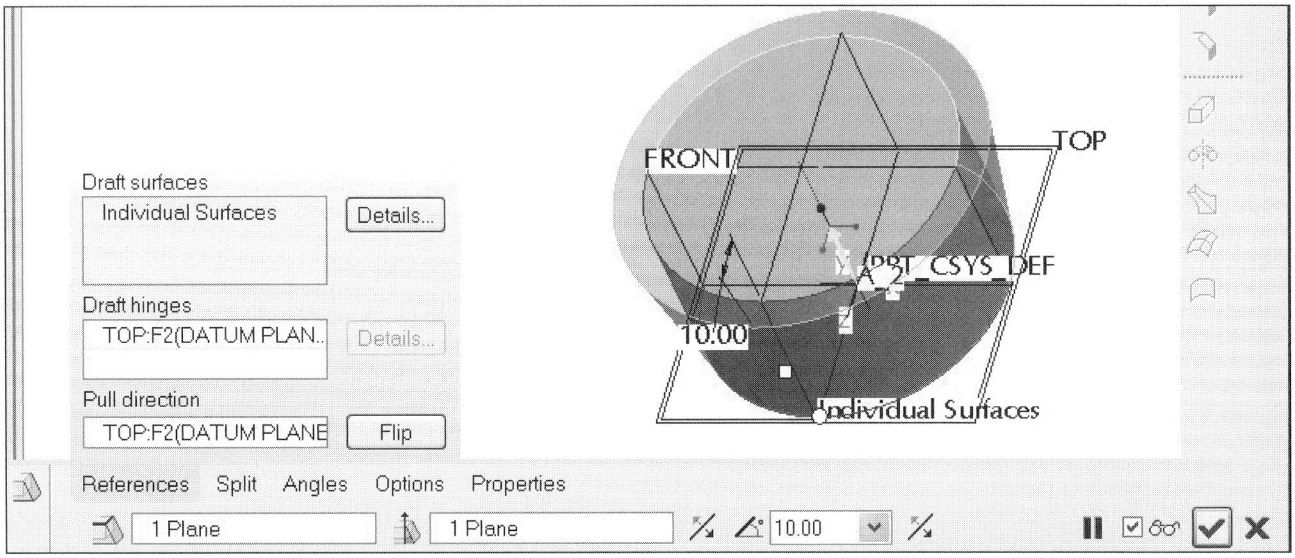

Figure 3.3(h) TOP Datum Plane Selected as the Draft Hinges

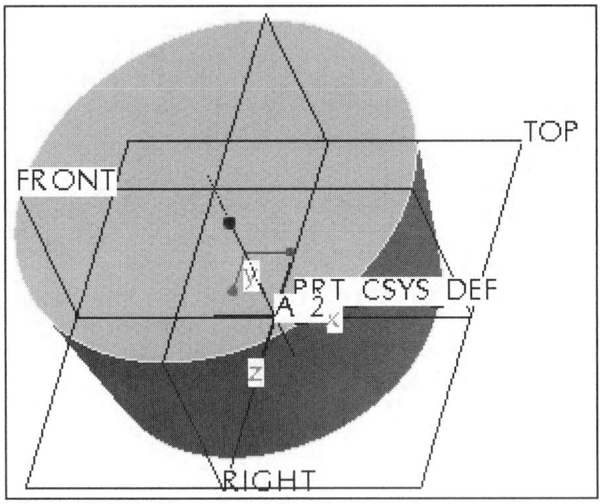

Figure 3.3(i) Completed Draft

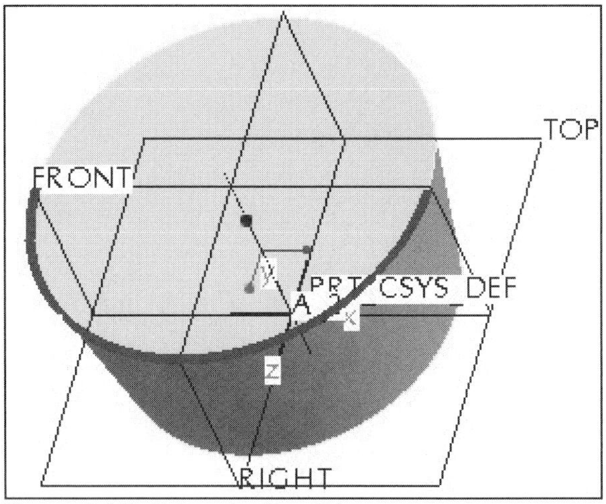

Figure 3.3(j) Select the Top Edge

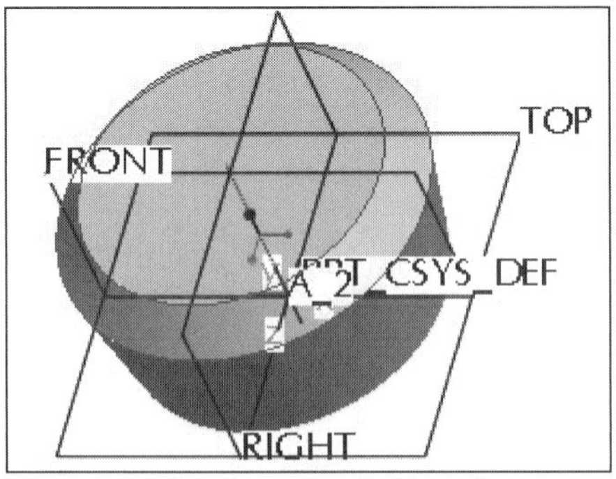

Figure 3.3(k) Completed Chamfer

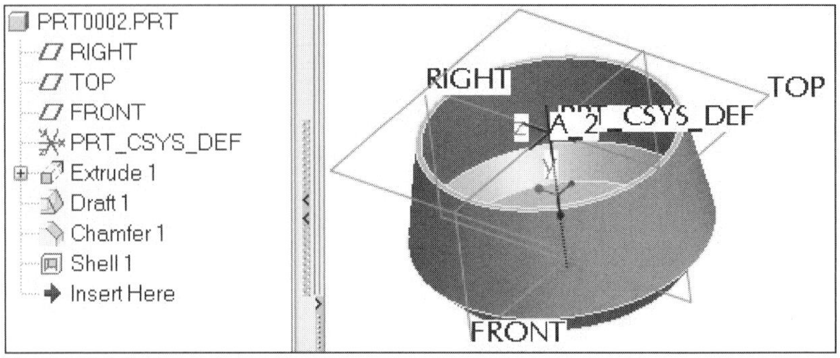

Figure 3.3(l) Completed Part

Part Model Three (PRT0003.PRT) (Hole)

Click: ⇒ Part (PRT0003.PRT) ⇒ **MMB** ⇒ pick the **FRONT** datum ⇒ **Extrude Tool** ⇒ **RMB** ⇒ **Define Internal Sketch** ⇒ **MMB** ⇒ **Create rectangle** ⇒ sketch a rectangle [Fig. 3.4(a)] ⇒ **Create a circular fillet between two entities** ⇒ pick the two lines that form the upper right-hand corner [Fig. 3.4(b)] ⇒ ⇒ ⇒ **Standard Orientation** [Fig. 3.4(c)] ⇒ **MMB** [Fig. 3.4(d)] ⇒ **Datum Axis Tool** ⇒ pick on the cylindrical surface [Fig. 3.4(e)] ⇒ **MMB**

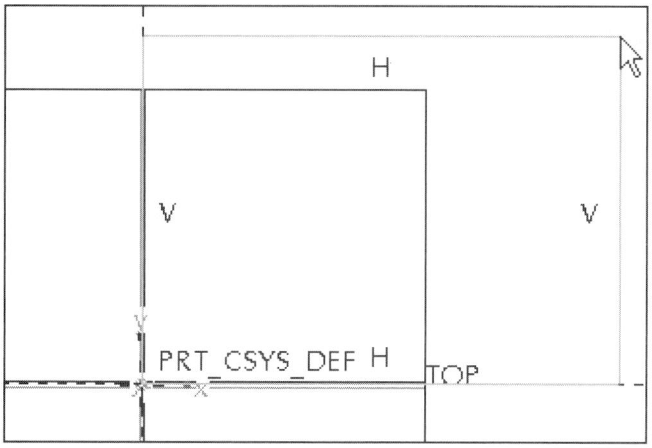

Figure 3.4(a) Sketch a Rectangle

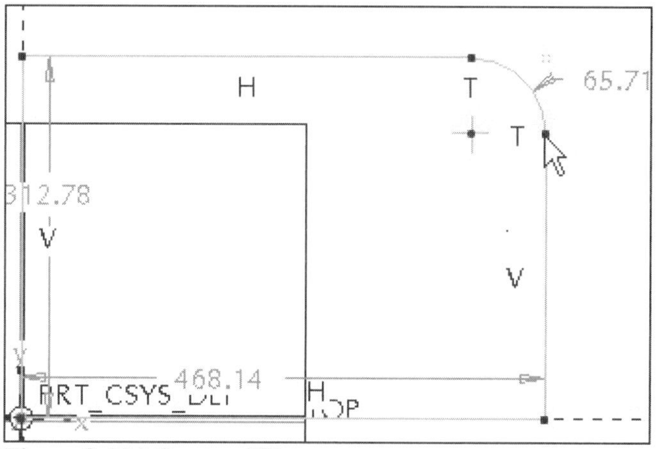

Figure 3.4(b) Create a Fillet

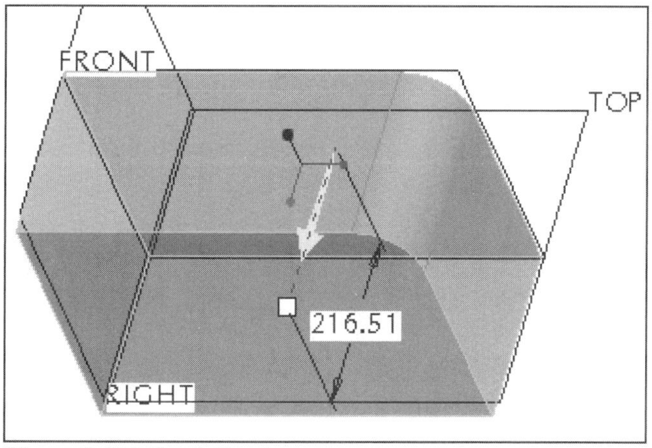

Figure 3.4(c) Depth Preview

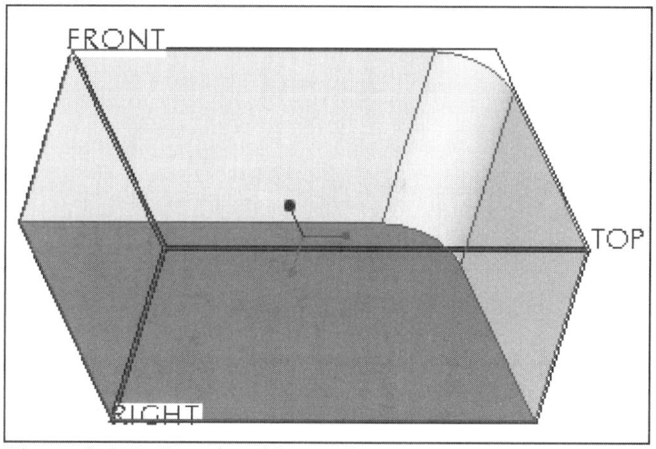

Figure 3.4(d) Completed Protrusion

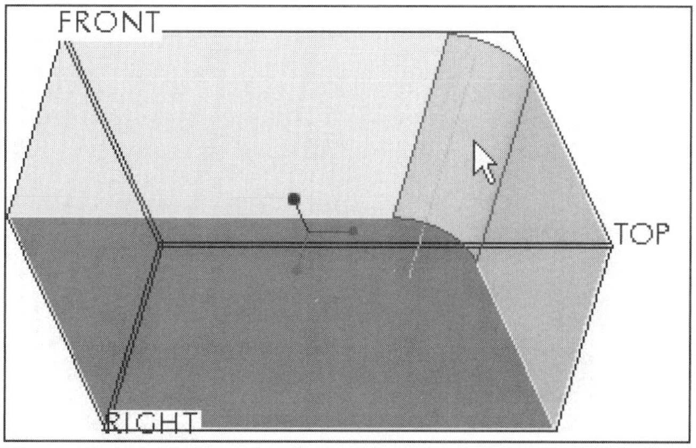

Figure 3.4(e) Pick on the Cylindrical Surface to Establish the References for the Datum Axis

With the datum axis still highlighted, click: **Hole Tool** [Fig. 3.4(f)] ⇒ [Fig. 3.4(g)] ⇒ while holding down the **Ctrl** key, pick on the front face [Fig. 3.4(h)] ⇒ expand depth options by opening slide-up panel ⇒ **Drill to intersect with all surfaces** [Fig. 3.4(i)] ⇒ [Fig. 3.4(j)] ⇒ ⇒ **Enter** ⇒ **Window** ⇒ **Close**

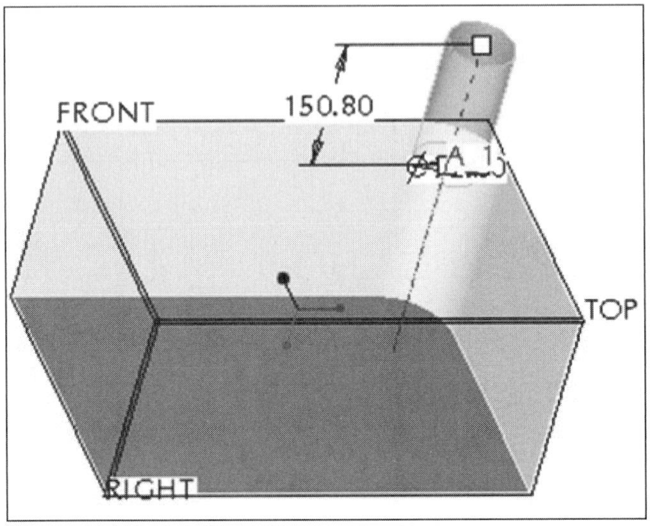

Figure 3.4(f) Hole Tool Display Preview

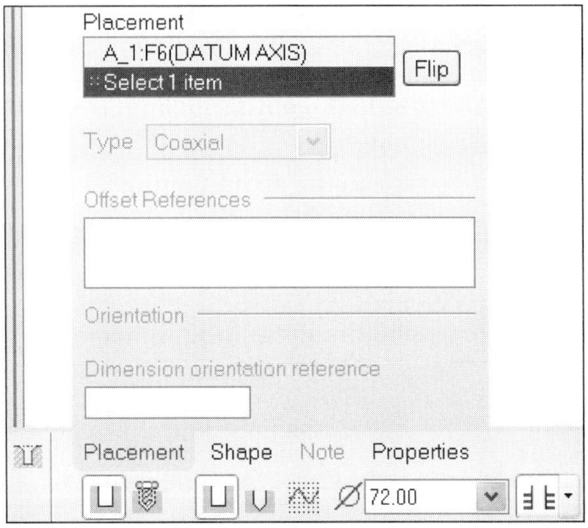

Figure 3.4(g) Placement tab

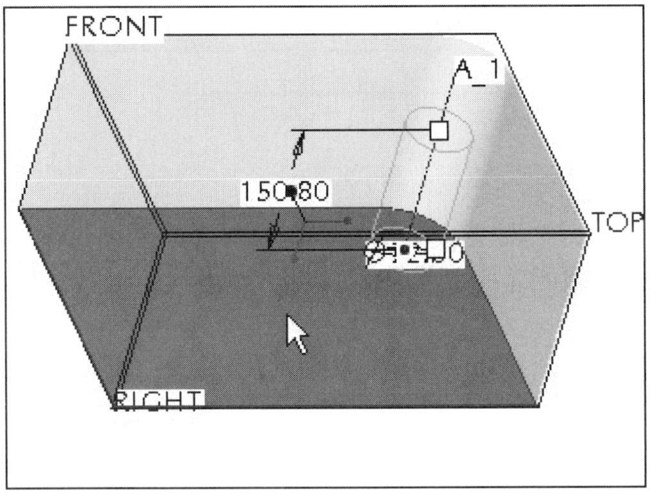

Figure 3.4(h) Pick on the Front Face

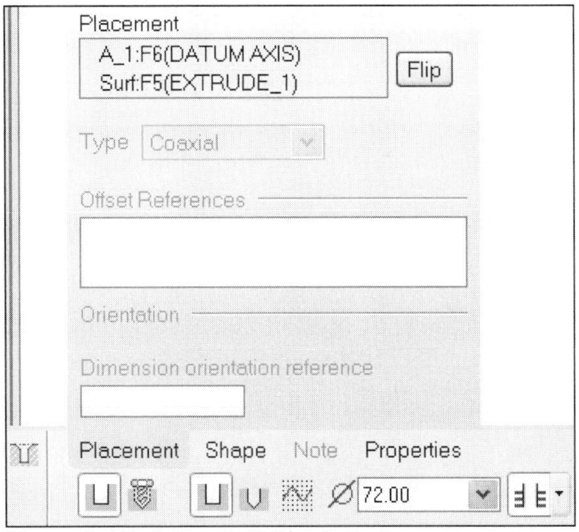

Figure 3.4(i) Select Drill Depth Option

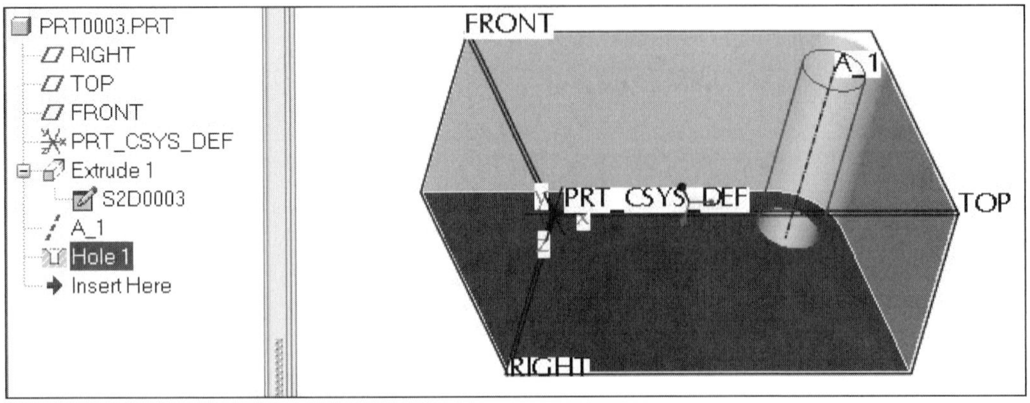

Figure 3.4(j) Completed Part

Part Model Four (PRT0004.PRT) (Cut)

Click: ⇒ Part **(PRT0004.PRT)** ⇒ **Enter** *(or MMB)* ⇒ pick the **FRONT** datum ⇒ ⇒ Placement from the dashboard ⇒ Define.. ⇒ **MMB** ⇒ **Create 2 point lines** ⇒ sketch the outline using five lines forming a closed section [Fig. 3.5(a)] ⇒ **MMB** to end the current tool ⇒ ⇒ ⇒ **Standard Orientation** [Fig. 3.5(b)] ⇒ **MMB** *(or Enter)* [Fig. 3.5(c)]

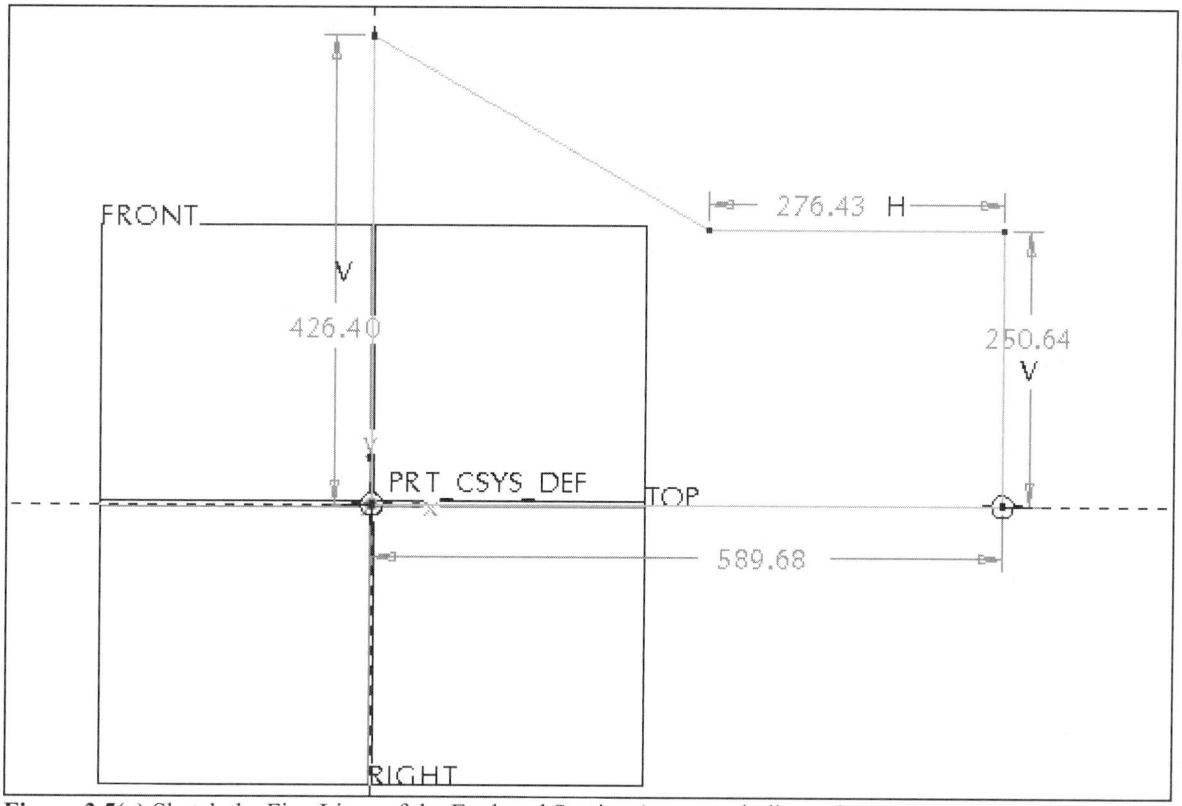

Figure 3.5(a) Sketch the Five Lines of the Enclosed Section (your weak dimensions will be different)

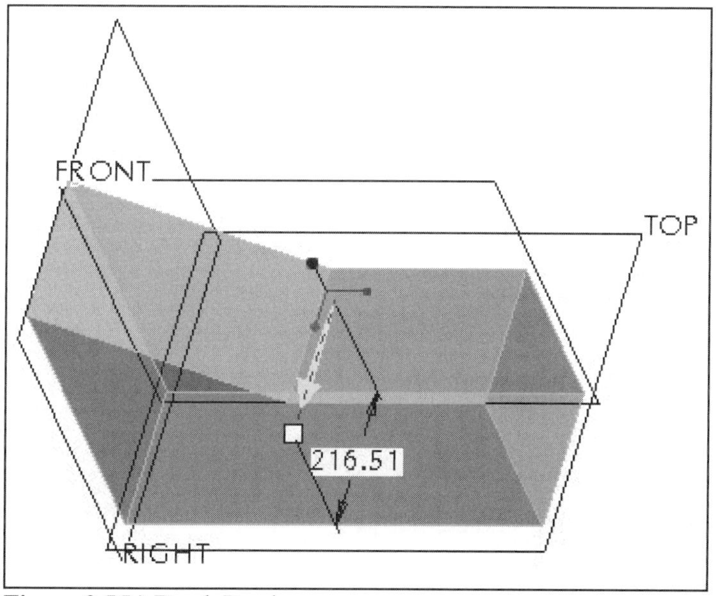

Figure 3.5(b) Depth Preview

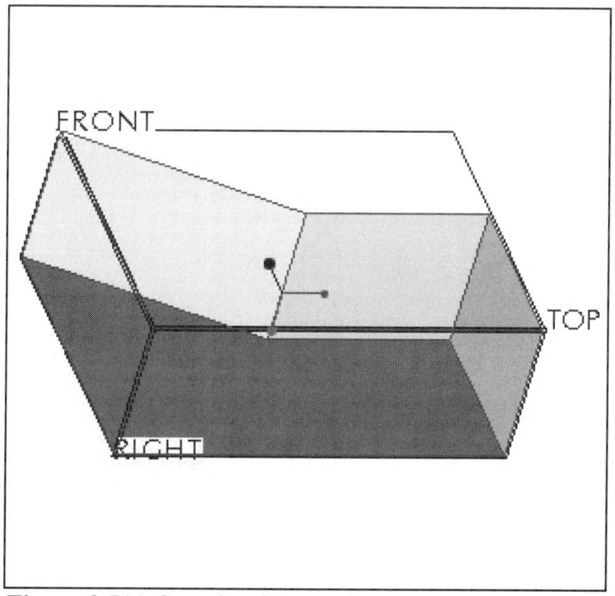

Figure 3.5(c) Completed Protrusion

Pick on the top horizontal surface of the part until it highlights [Fig. 3.5(d)] ⇒ **Extrude Tool** ⇒ **Remove Material** ⇒ Placement ⇒ Define.. ⇒ **MMB** ⇒ **Hidden line** ⇒ **Create rectangle** ⇒ sketch a rectangle [Fig. 3.5(e)] ⇒ **Shading** ⇒ ⇒ **Standard Orientation** [Fig. 3.5(f)] ⇒ [Fig. 3.5(g)] ⇒ expand ⇒ **Extrude on both sides** ⇒ move a drag handle so that the cut passes thru the part [Fig. 3.5(h)] ⇒ **MMB** [Fig. 3.5(i)] ⇒ ⇒ **Enter** ⇒ **Window** ⇒ **Close**

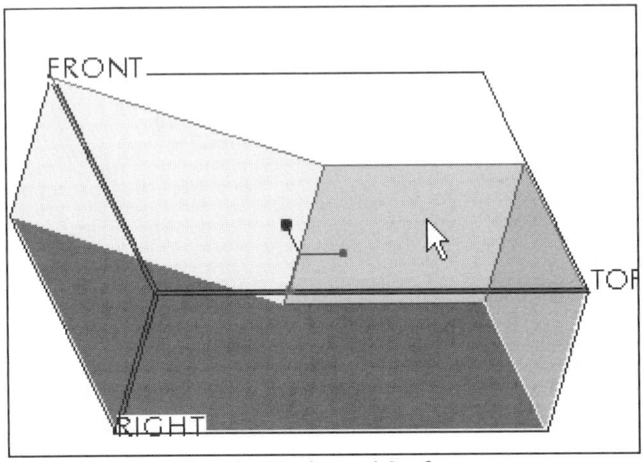

Figure 3.5(d) Select the Horizontal Surface

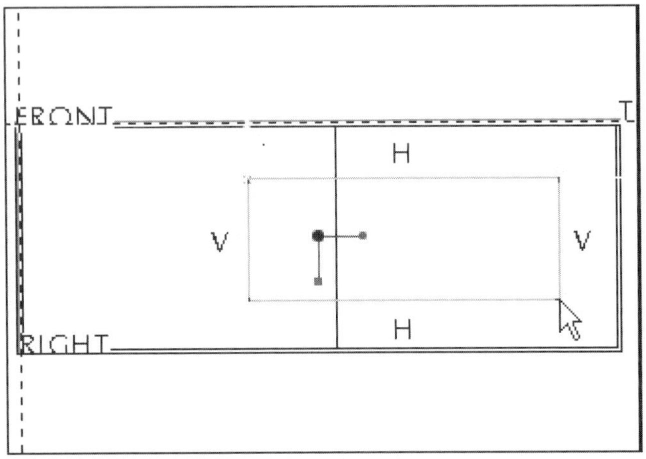

Figure 3.5(e) Sketch a Rectangle

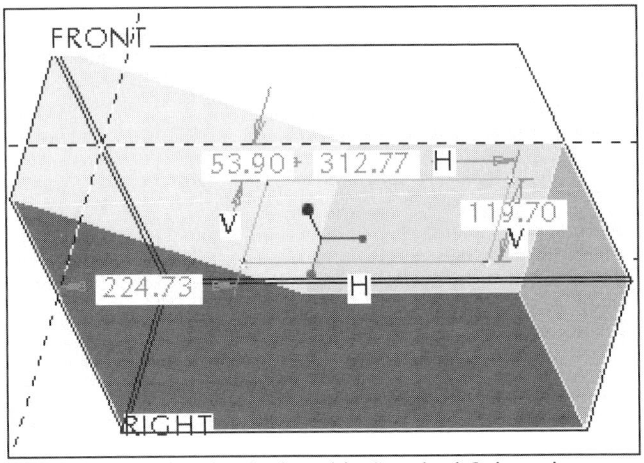

Figure 3.5(f) Sketch Displayed in Standard Orientation

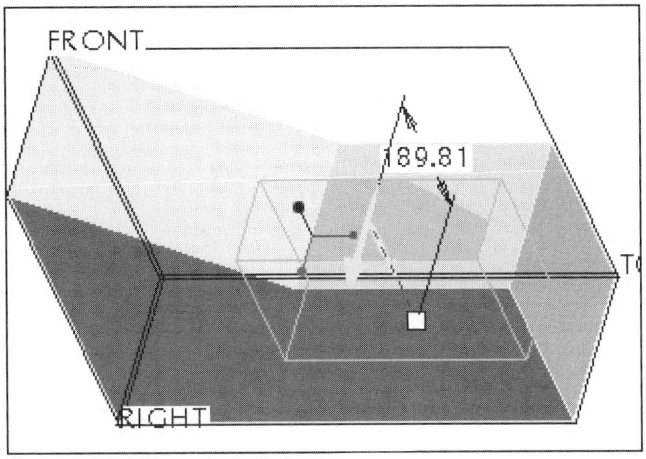

Figure 3.5(g) Previewed Cut

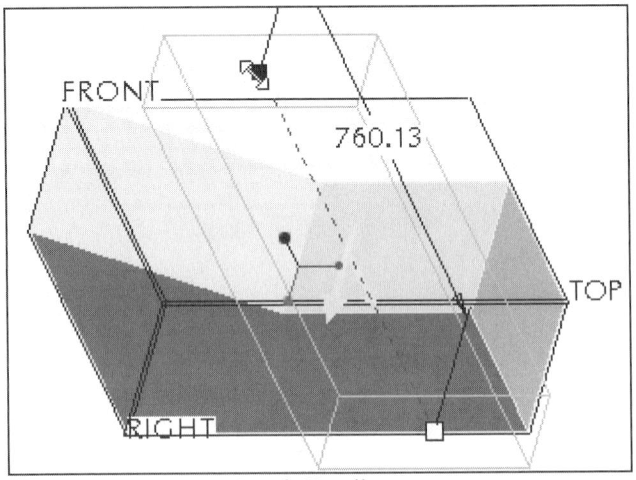

Figure 3.5(h) Drag a Depth Handle

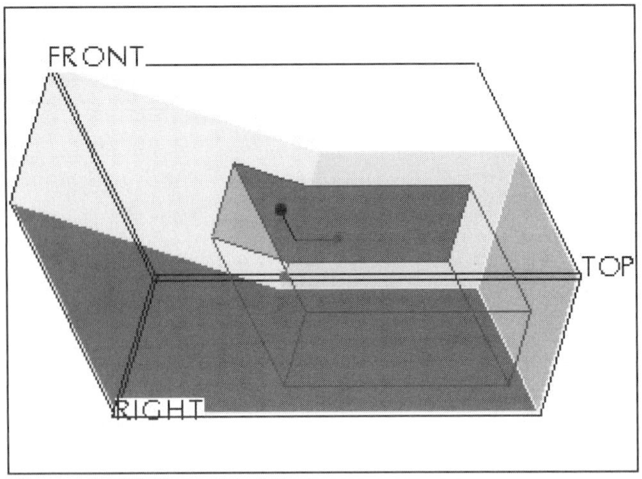

Figure 3.5(i) Completed Part

Part Model Five (PRT0005.PRT) (Mirror)

Click: ⇒ Part **(PRT0005.PRT)** ⇒ **MMB** ⇒ pick the **FRONT** datum ⇒ ⇒ **Thicken Sketch** from the dashboard ⇒ Placement ⇒ Define.. ⇒ **MMB** ⇒ **RMB** ⇒ **Centerline** ⇒ pick two points vertically on the edge of the RIGHT datum plane [Figs. 3.6(a-b)] ⇒ **MMB** to end the current tool ⇒ **RMB** ⇒ **Line** [Fig. 3.6(c)] ⇒ sketch the vertical and horizontal lines [Fig. 3.6(d)] ⇒ **MMB** to end the current line

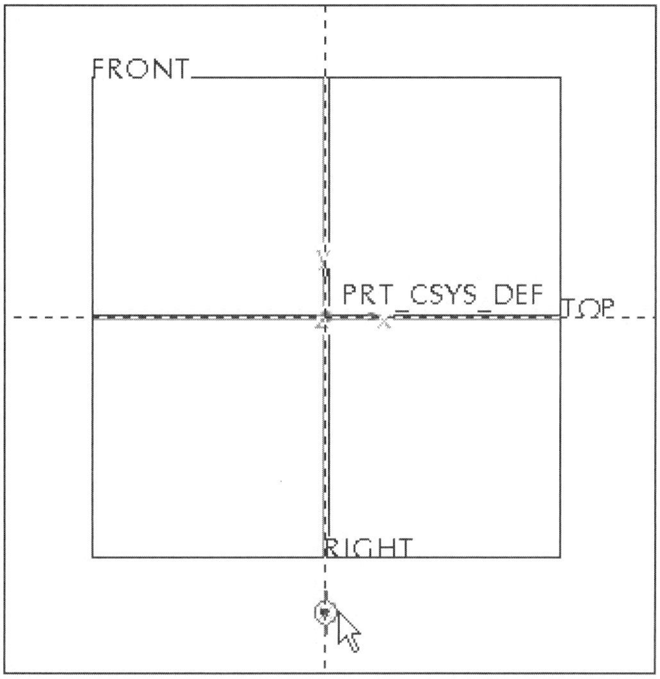

Figure 3.6(a) Pick the First Point of the Vertical Centerline

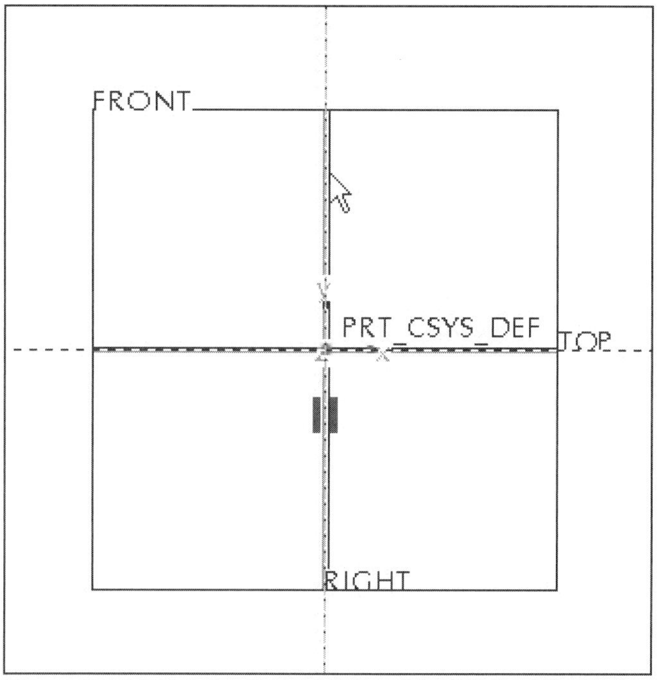

Figure 3.6(b) Pick the Second Point of the Vertical Centerline

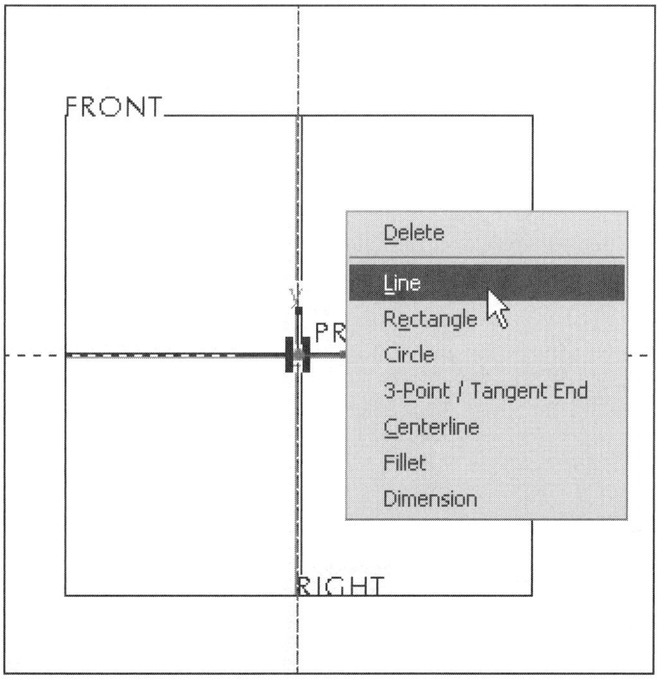

Figure 3.6(c) RMB ⇒ Line

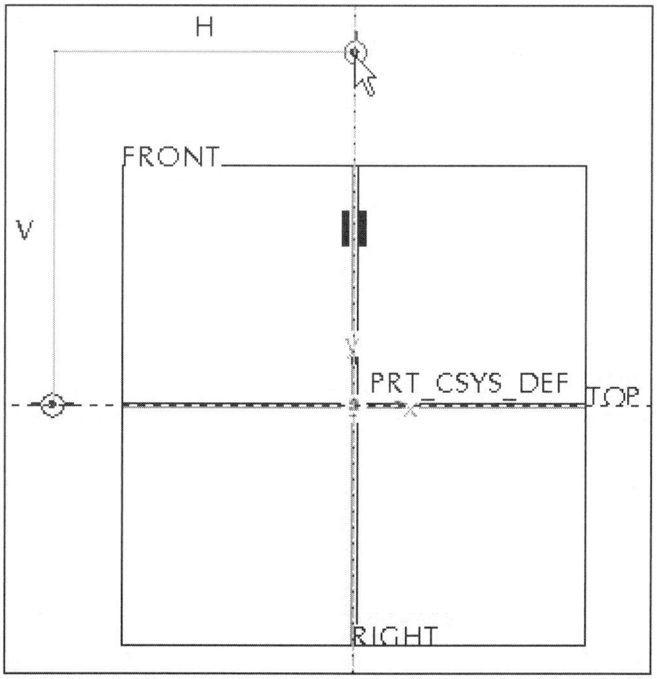

Figure 3.6(d) Sketch Two Lines

Click: **RMB** ⇒ **Fillet** [Fig. 3.6(e)] ⇒ pick the lines near the corner ⇒ **MMB** ⇒ press **LMB** and hold while dragging a window until it incorporates the two lines and fillet [Fig. 3.6(f)] ⇒ **Mirror selected entities** ⇒ pick the centerline [Fig. 3.6(g)] ⇒ ⇒ **Standard Orientation** ⇒ [Fig. 3.6(h)]

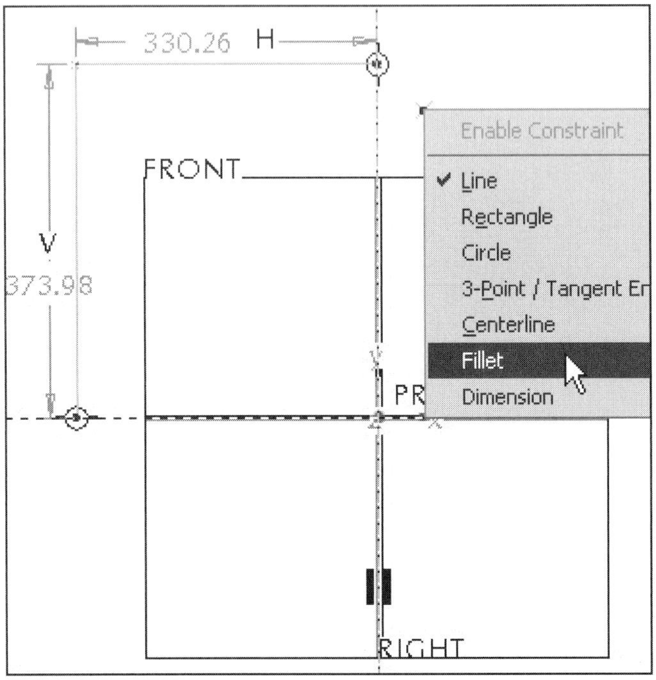

Figure 3.6(e) RMB ⇒ Fillet

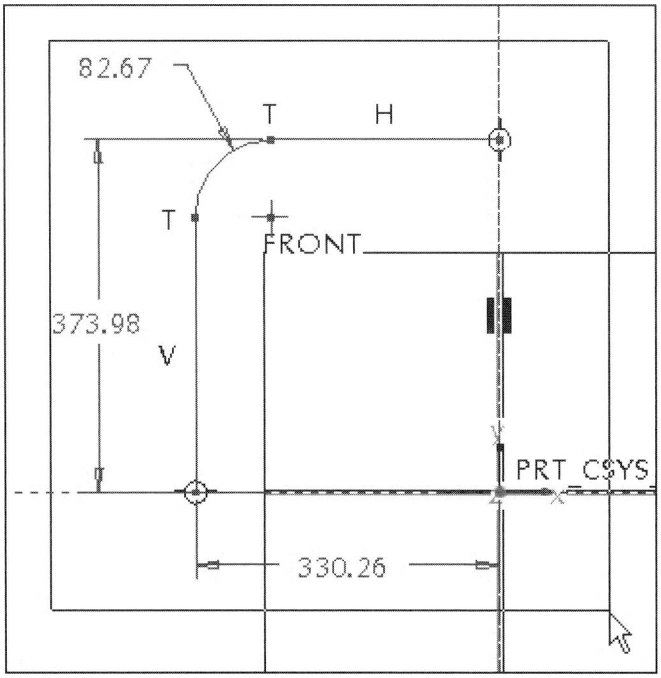

Figure 3.6(f) Select Sketch Entities by Windowing

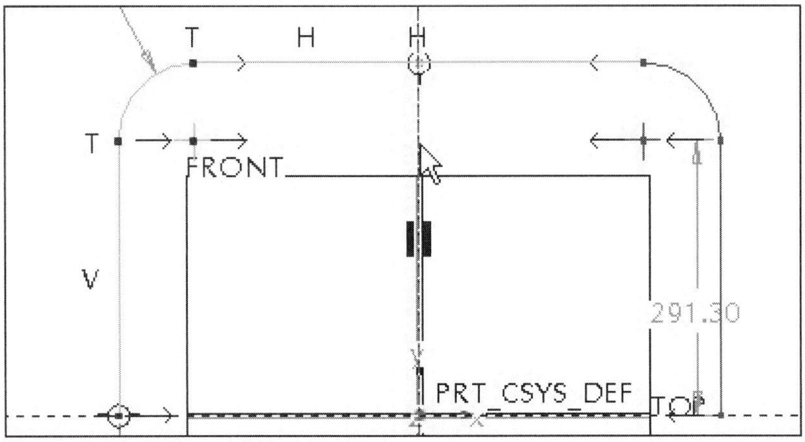

Figure 3.6(g) Mirrored Sketch

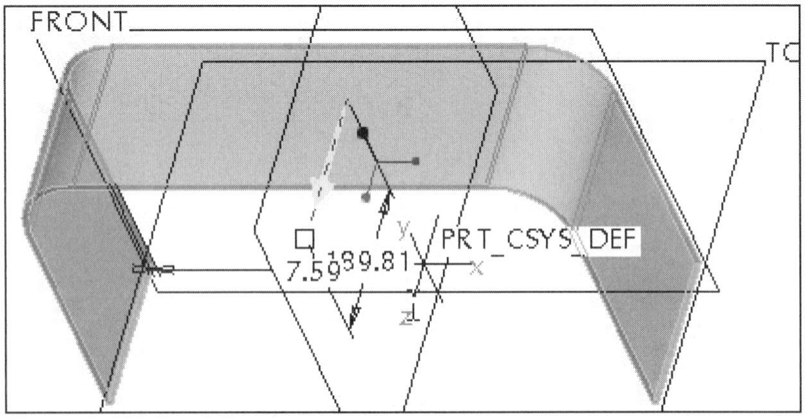

Figure 3.6(h) Depth Preview

Click: ⇒ **Extrude on both sides** ⇒ drag a depth handle [Fig. 3.6(i)] ⇒ **MMB** ⇒ **LMB** ⇒ press and hold the **Ctrl** key and select the **RIGHT** and **FRONT** datum planes in the Model Tree [Fig. 3.6(j)] ⇒ **Datum Axis Tool** [Fig. 3.6(k)] ⇒ **Hole Tool** ⇒ Placement ⇒ ⇒ ⇒ while holding down the **Ctrl** key, pick on the top face [Fig. 3.6(l)] ⇒ **MMB** ⇒ ⇒ **MMB** ⇒ **Window** ⇒ **Close**

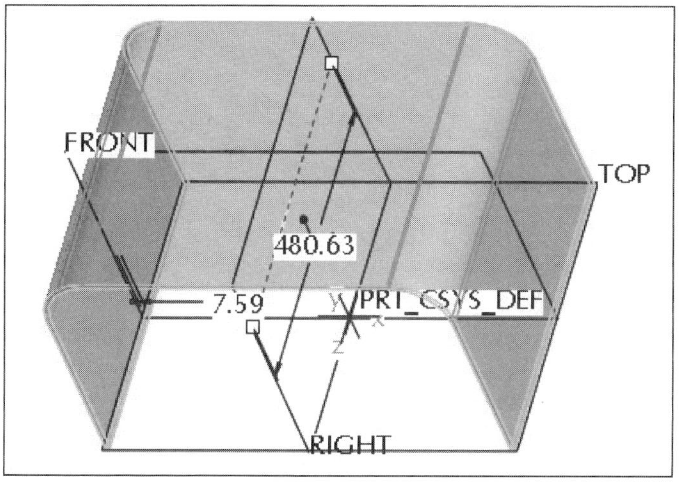

Figure 3.6(i) Drag Depth Handle

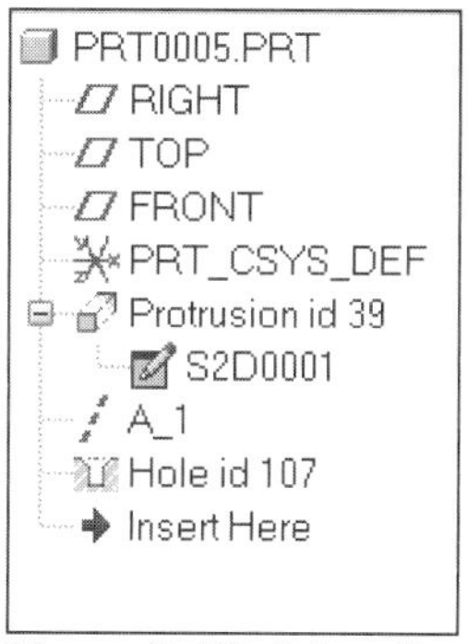

Figure 3.6(j) Select Datums

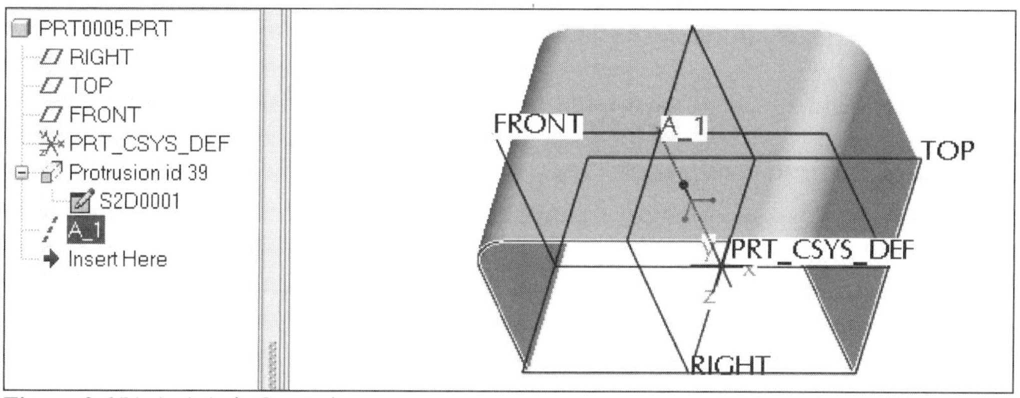

Figure 3.6(k) A_1 Axis Created

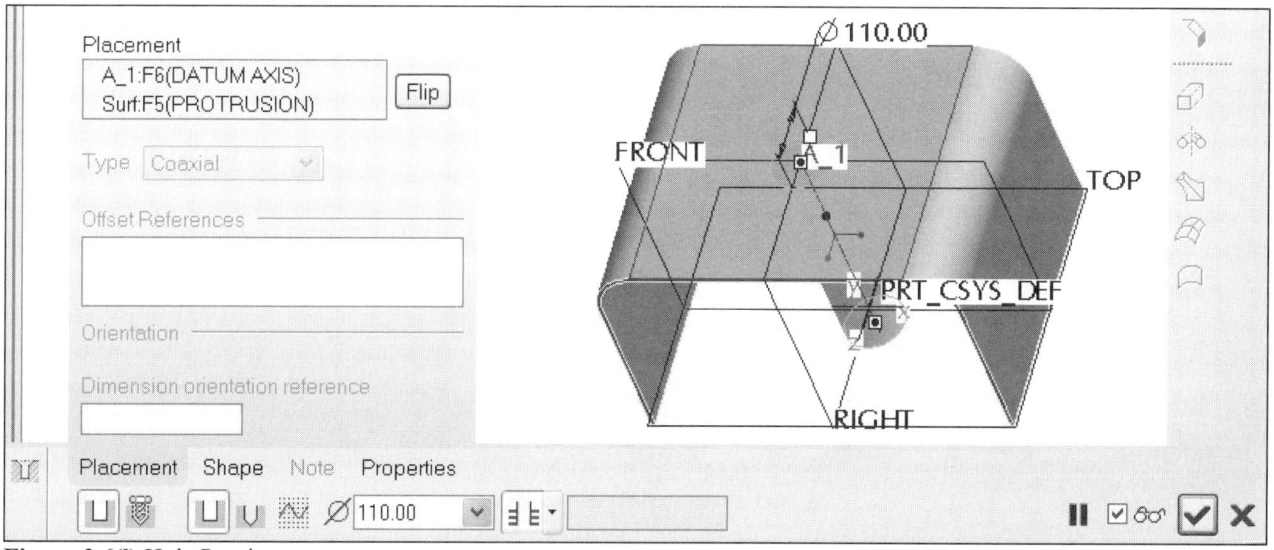

Figure 3.6(l) Hole Preview

Part Model Six (PRT0006.PRT) (Revolve)

Click:  ⇒ 

(PRT0006.PRT) ⇒ **Enter** ⇒ pick the **RIGHT** datum ⇒  **Revolve Tool** ⇒ **RMB** ⇒ **Define Internal Sketch** ⇒ Orientation **Top** [Fig. 3.7(a)] ⇒ 

⇒ **RMB** ⇒ **Centerline** ⇒ create a horizontal centerline on the edge of the TOP datum ⇒ **MMB** ⇒ **RMB** ⇒ **Line** ⇒ sketch the outline of the closed section [Fig. 3.7(b)] ⇒ **MMB** to end the current line ⇒  ⇒  ⇒ **Standard Orientation** [Fig. 3.7(c)] ⇒ **MMB** ⇒ **LMB** [Fig. 3.7(d)]

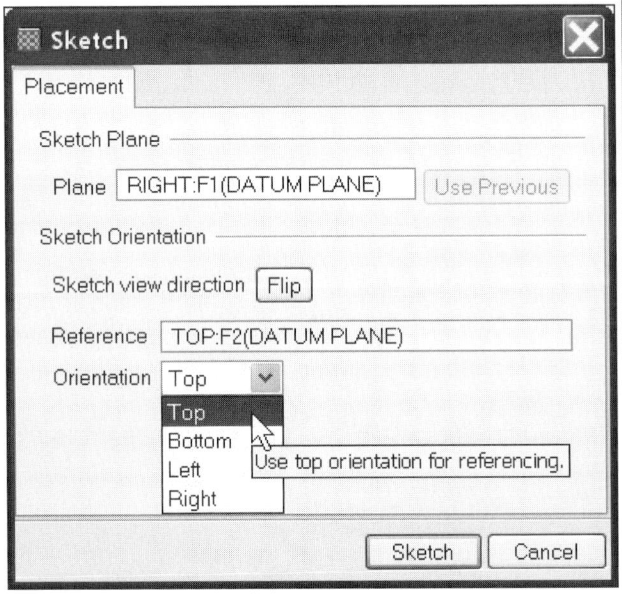

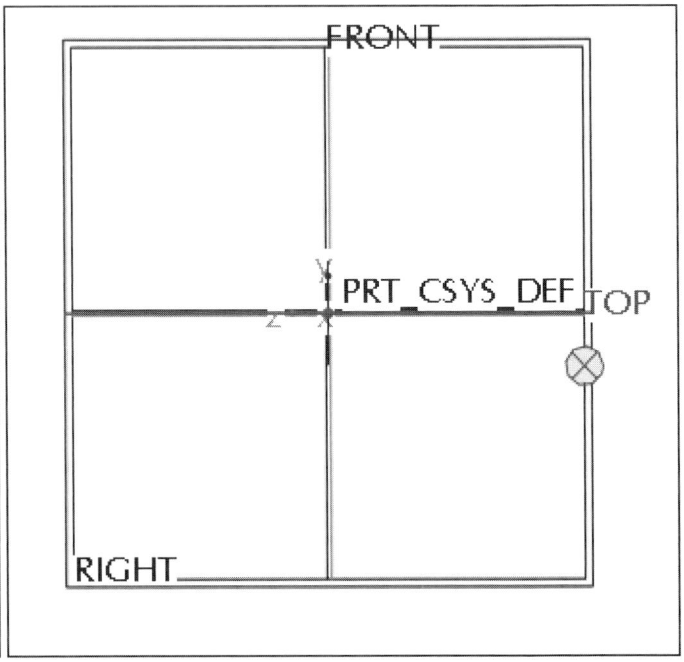

Figure 3.7(a) Sketch Dialog Box

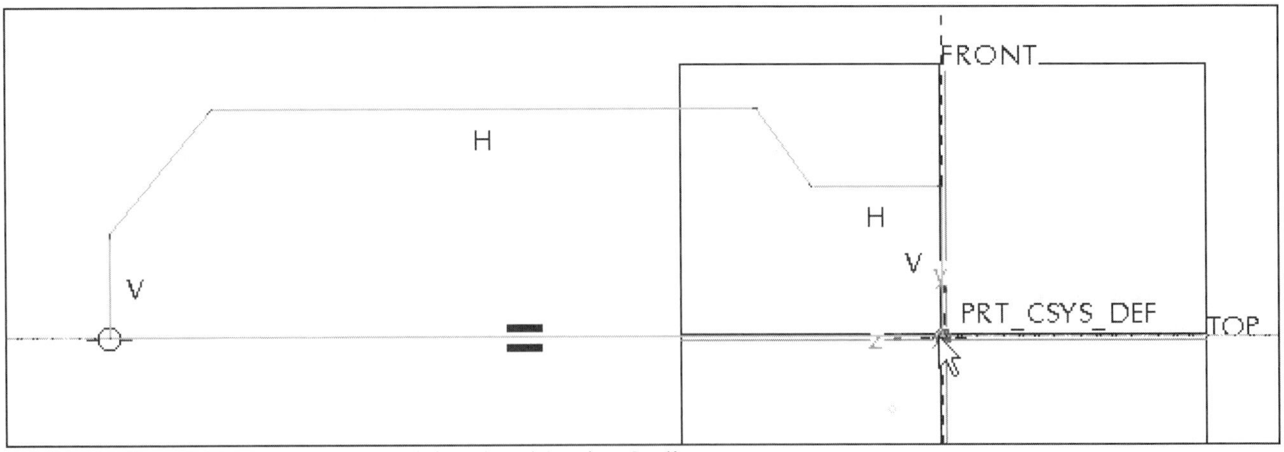

Figure 3.7(b) Sketch the Centerline and the Closed Section Outline

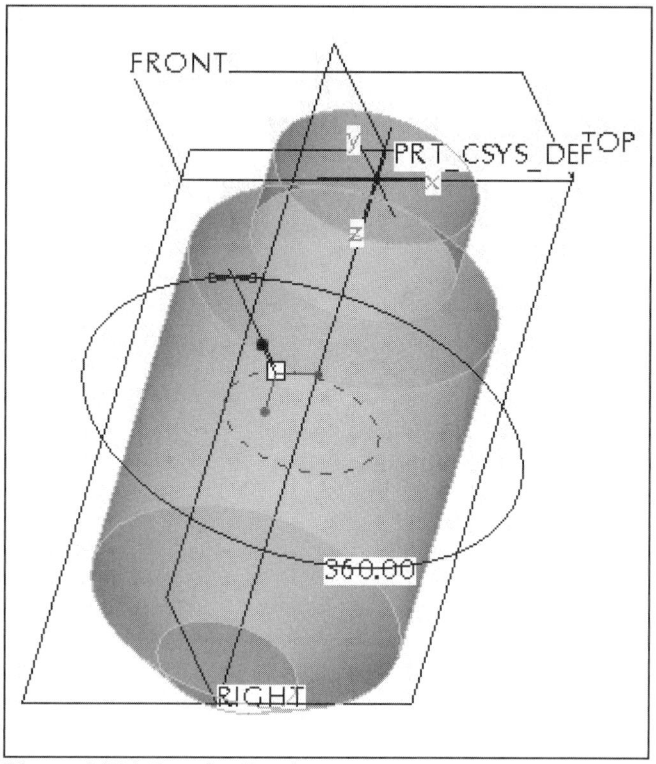

Figure 3.7(c) Revolve Preview

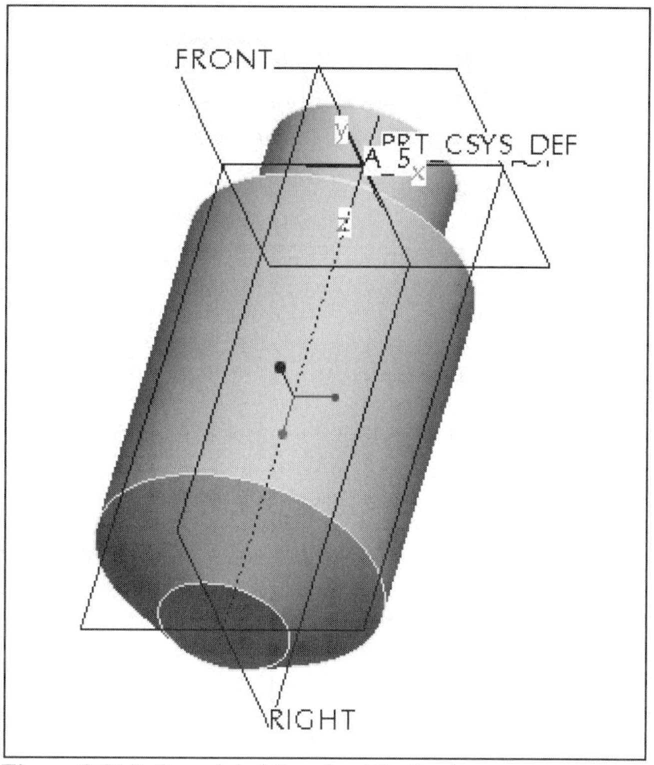

Figure 3.7(d) Completed Revolved Protrusion

Slowly pick on the front edge of the part until it highlights ⇒ press and hold the **Ctrl** key ⇒ pick the other visible edges ⇒ **RMB** ⇒ **Round Edges** [Fig. 3.7(e)] ⇒ drag a handle [Fig. 3.7(f)] ⇒ **MMB** ⇒ **LMB**

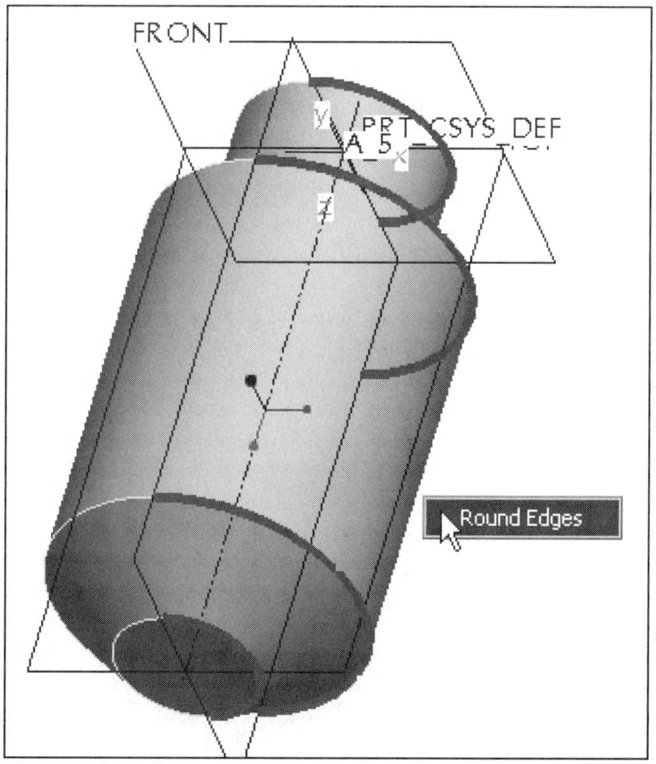

Figure 3.7(e) Select the Edges

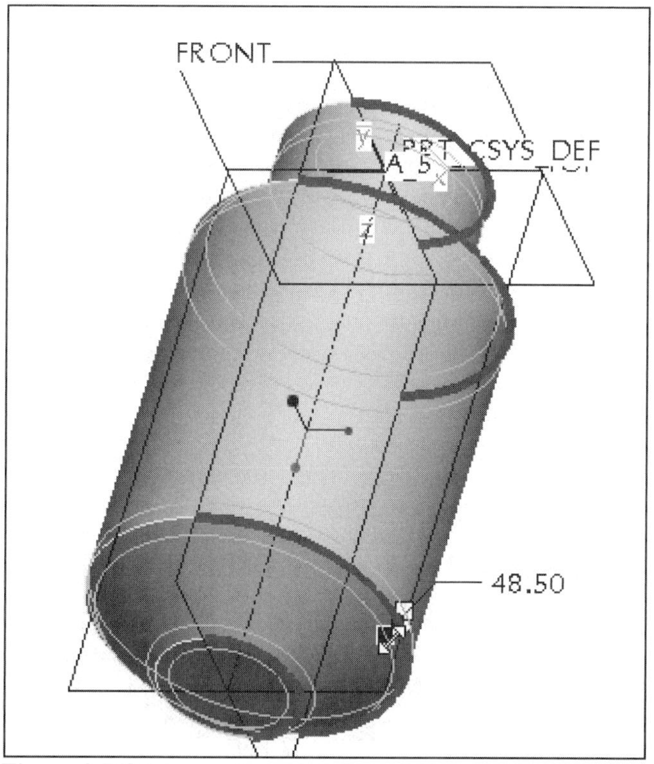

Figure 3.7(f) Move a Drag Handle to Adjust the Size

Click: **View** from the menu bar ⇒ View Manager [Fig. 3.7(g)] ⇒ Xsec tab [Fig. 3.7(h)] ⇒ New [Fig. 3.7(i)] ⇒ type **A** ⇒ **Enter** [Fig. 3.7(j)] ⇒ **MMB** ⇒ select **RIGHT** from the Model Tree [Fig. 3.7(k)] ⇒ Display ⇒ **Set Active** ⇒ Display ⇒ **Visibility** ⇒ Display ⇒ **Flip** ⇒ Display ⇒ **Flip** ⇒ Display ⇒ **Visibility** ⇒ select **No Cross Section** ⇒ Display ⇒ **Set Active** ⇒ double click on **A** ⇒ double click on **No Cross Section** ⇒ Display ⇒ **Normal**⇒ Display ⇒ Xhatching ⇒ select **A** [Fig. 3.7(l)] ⇒ Close ⇒ **LMB** ⇒ [save icon] ⇒ **MMB** ⇒ **Window** ⇒ **Close**

Figure 3.7(g) View Manager

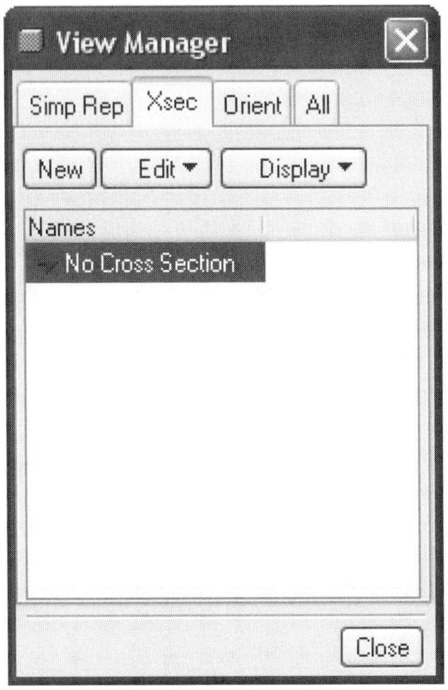

Figure 3.7(h) Select Xsec Tab

Figure 3.7(i) New Xsec

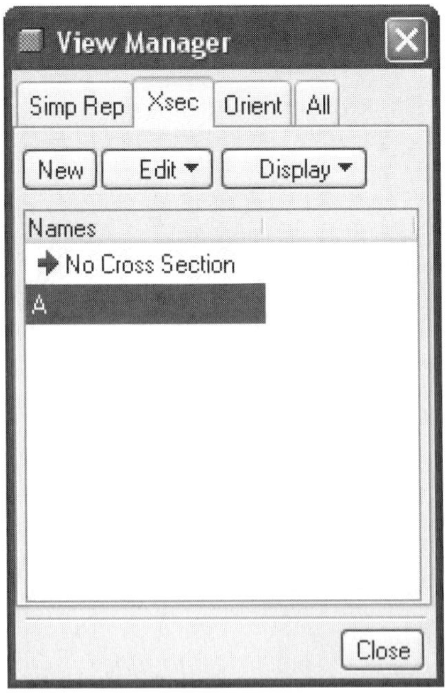

Figure 3.7(j) Section A

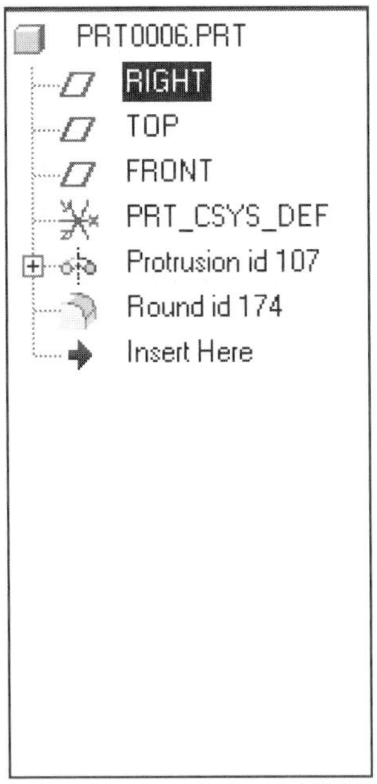

Figure 3.7(k) Select RIGHT

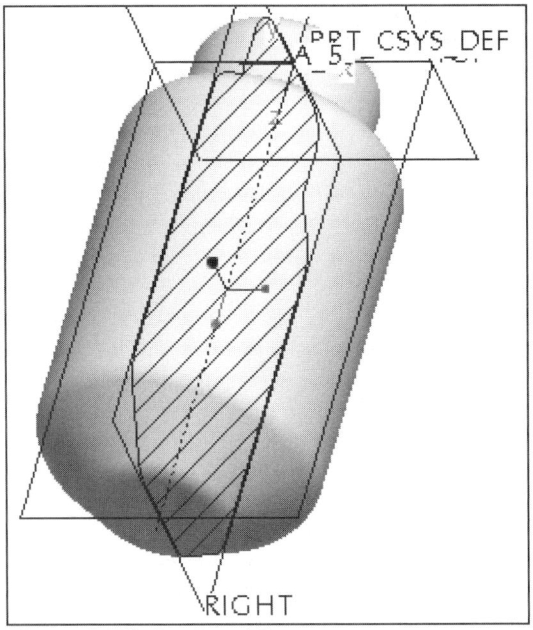

Figure 3.7(l) Sectioned Part

Part Model Seven (PRT0007.PRT) (Revolve Ellipse)

Click: ⇒ Part (**PRT0007.PRT**) ⇒ **MMB** ⇒ pick the **FRONT** datum ⇒ **Revolve Tool** ⇒ **RMB** ⇒ **Define Internal Sketch** ⇒ **MMB** ⇒ **RMB** ⇒ **Centerline** ⇒ create a vertical centerline on the edge of the RIGHT datum ⇒ **MMB** ⇒ ⇒ **Create a full ellipse** ⇒ pick one point to locate the center and the second to determine the shape of the ellipse [Fig. 3.8(a)] ⇒ **MMB** to end the current tool ⇒ ⇒ ⇒ **Standard Orientation** ⇒ Placement ⇒ 180.00 [Fig. 3.8(b)] ⇒ **MMB** [Fig. 3.8(c)] ⇒ **LMB** to deselect

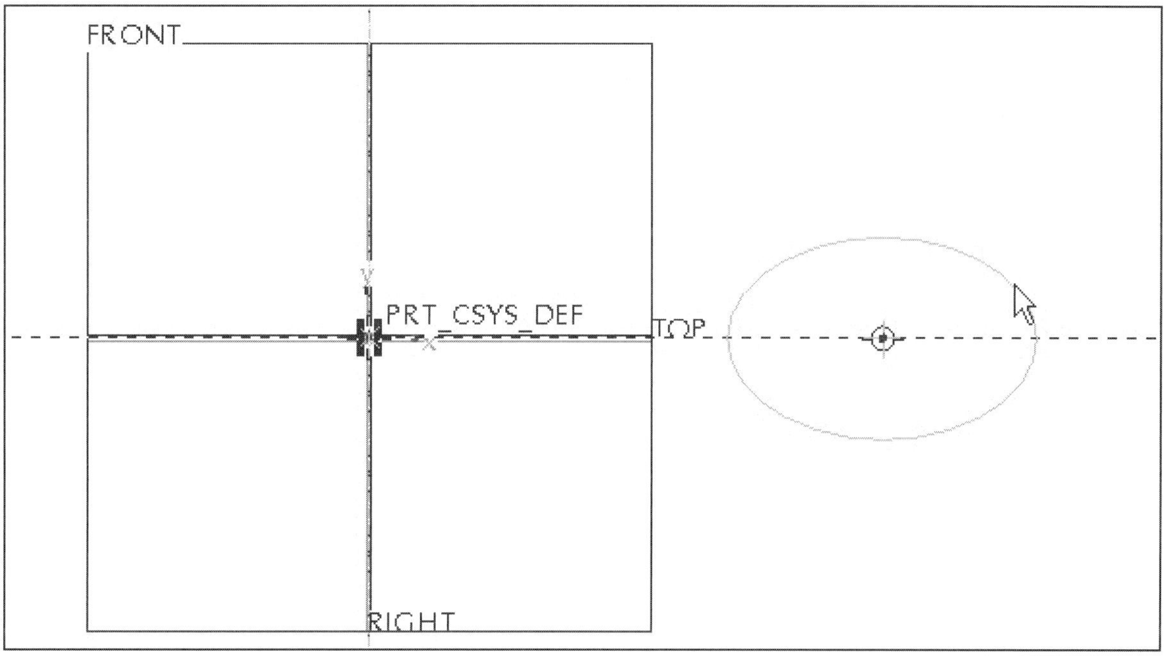

Figure 3.8(a) Sketch a Vertical Centerline and an Ellipse

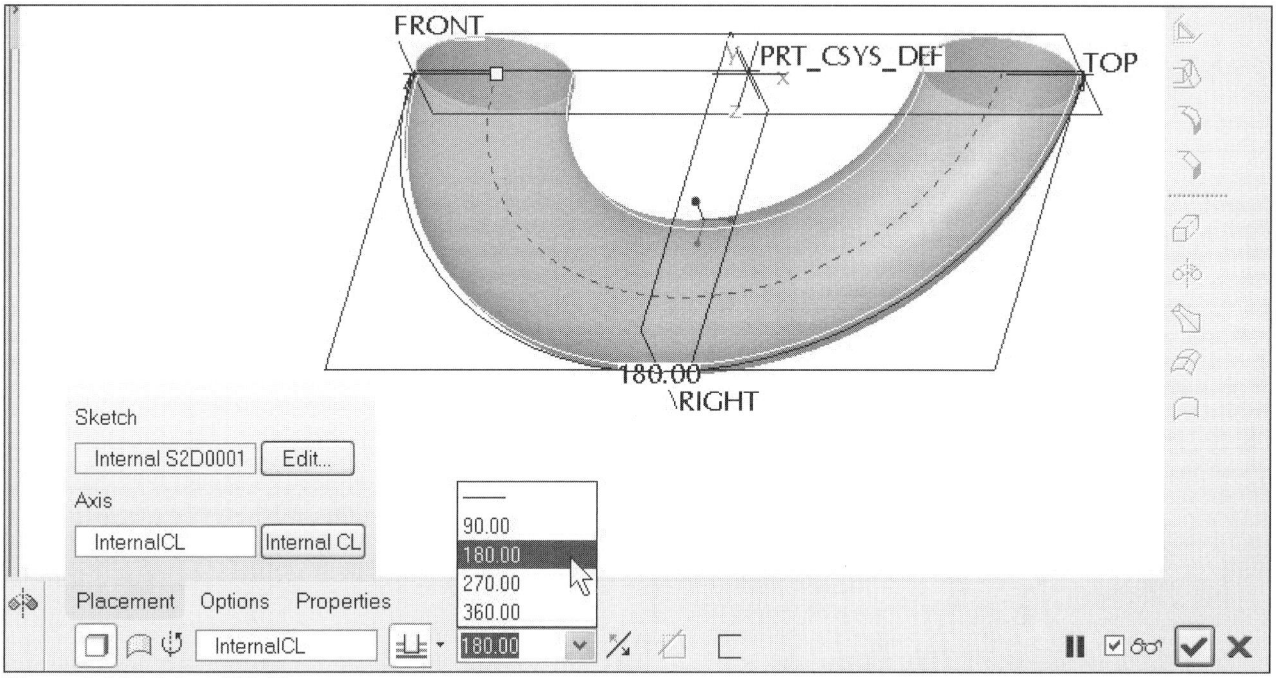

Figure 3.8(b) Select **180.00**

Click: **Info** from the menu bar ⇒ **Model** [Fig. 3.8(d)] ⇒ collapse the Browser shade 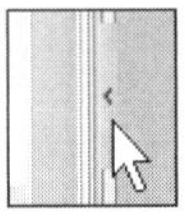 ⇒  ⇒ **MMB** ⇒ **Window** ⇒ **Close**

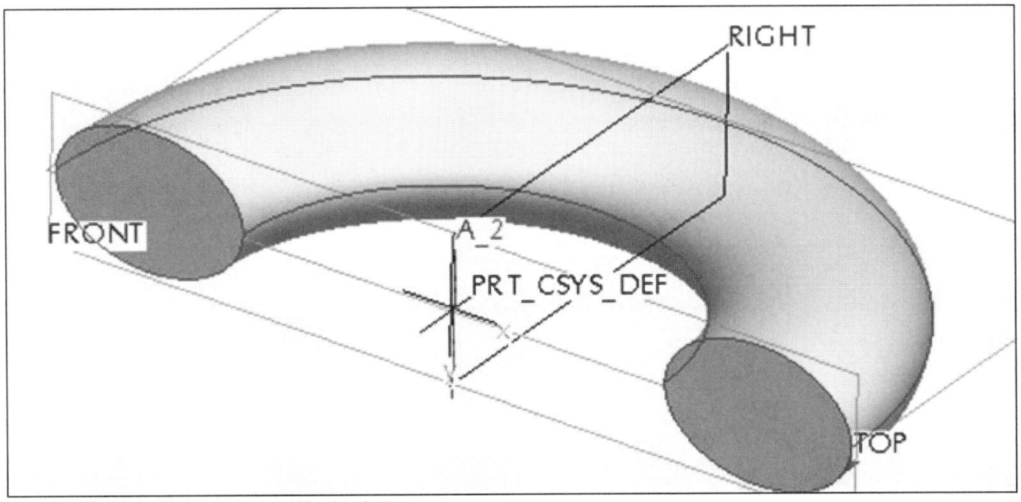

Figure 3.8(c) Completed Elliptical Torus

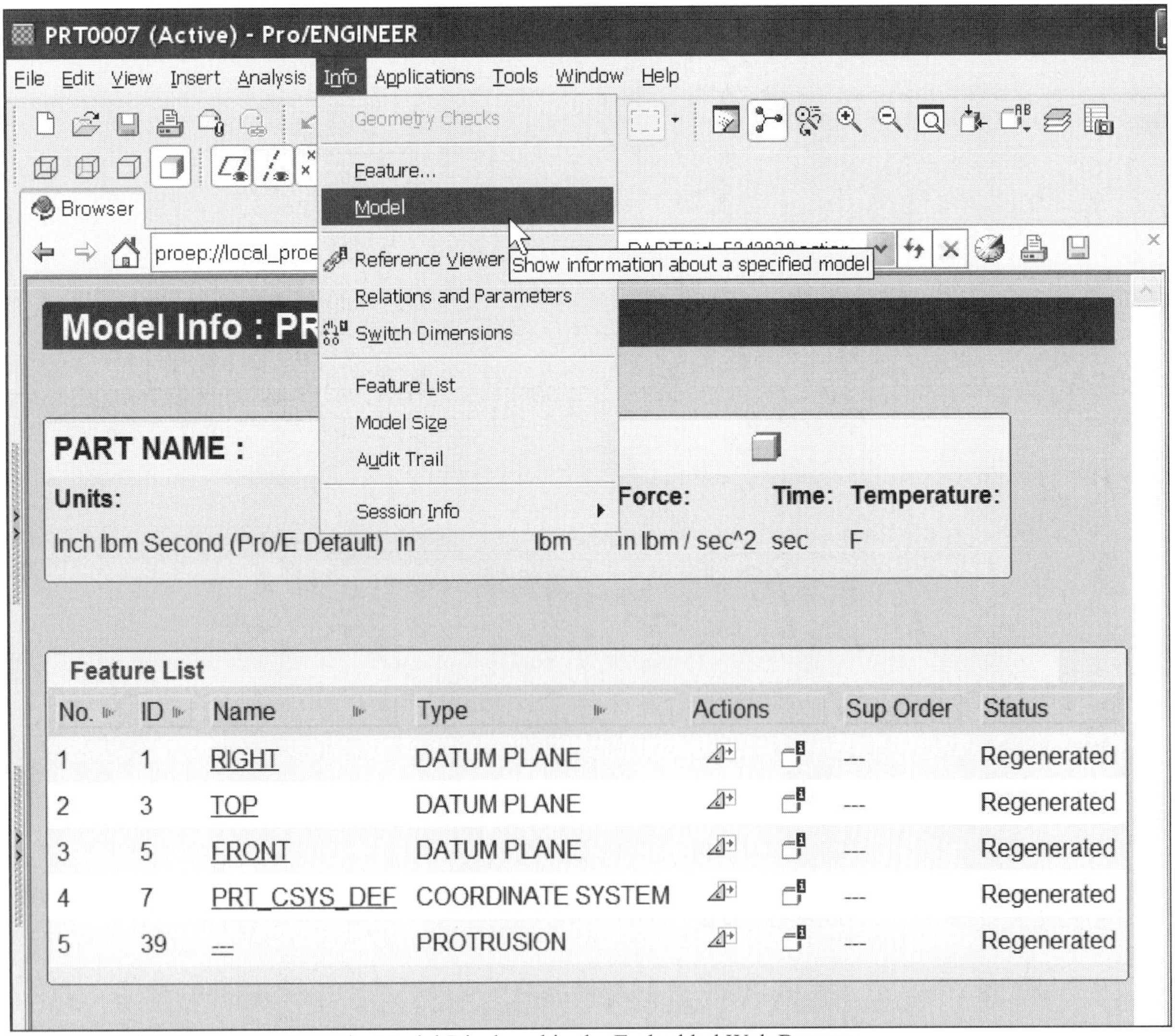

Figure 3.8(d) Information about the Model Displayed in the Embedded Web Browser

Part Model Eight (PRT0008.PRT) (Revolve Cut)

Click: ⇒ Part (PRT0008.PRT) ⇒ **MMB** ⇒ pick the **FRONT** datum ⇒ ⇒ ⇒ **Extrude on both sides** ⇒ Placement from the dashboard ⇒ Define.. ⇒ **MMB** ⇒ **RMB** ⇒ **Centerline** ⇒ create a vertical centerline on the edge of the RIGHT datum plane ⇒ **MMB** ⇒ **RMB** ⇒ **Rectangle** ⇒ sketch a rectangle [Fig. 3.9(a)] ⇒ **MMB** ⇒ ⇒ ⇒ **Standard Orientation** ⇒ **MMB** ⇒ **LMB**

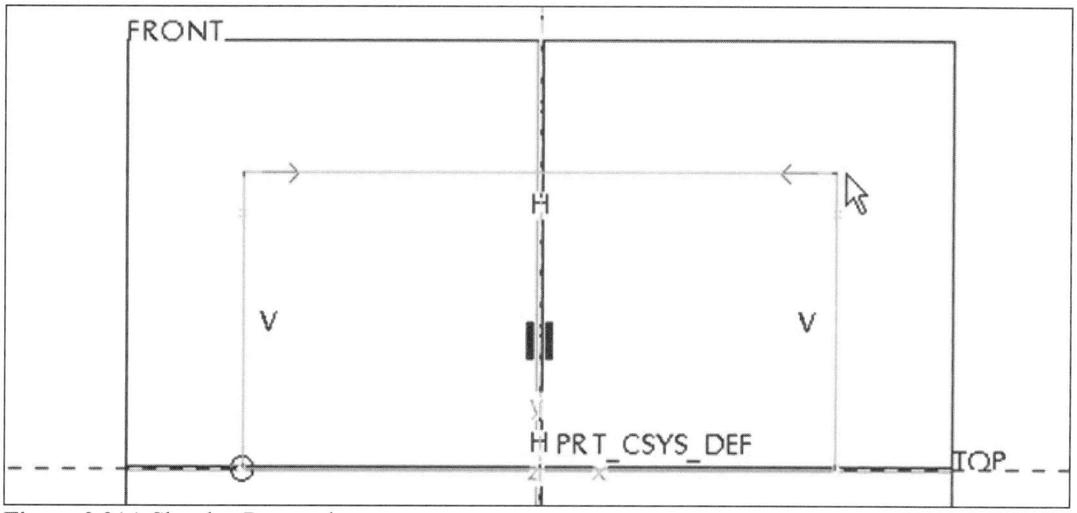

Figure 3.9(a) Sketch a Rectangle

Click: ⇒ **Remove Material** ⇒ Placement ⇒ Define.. ⇒ Use Previous ⇒ **Sketch** from the menu bar ⇒ **References** ⇒ **Select references for dimensioning and constraining** ⇒ pick the top edge of the part [Fig. 3.9(b)] ⇒ **MMB** ⇒ **MMB**

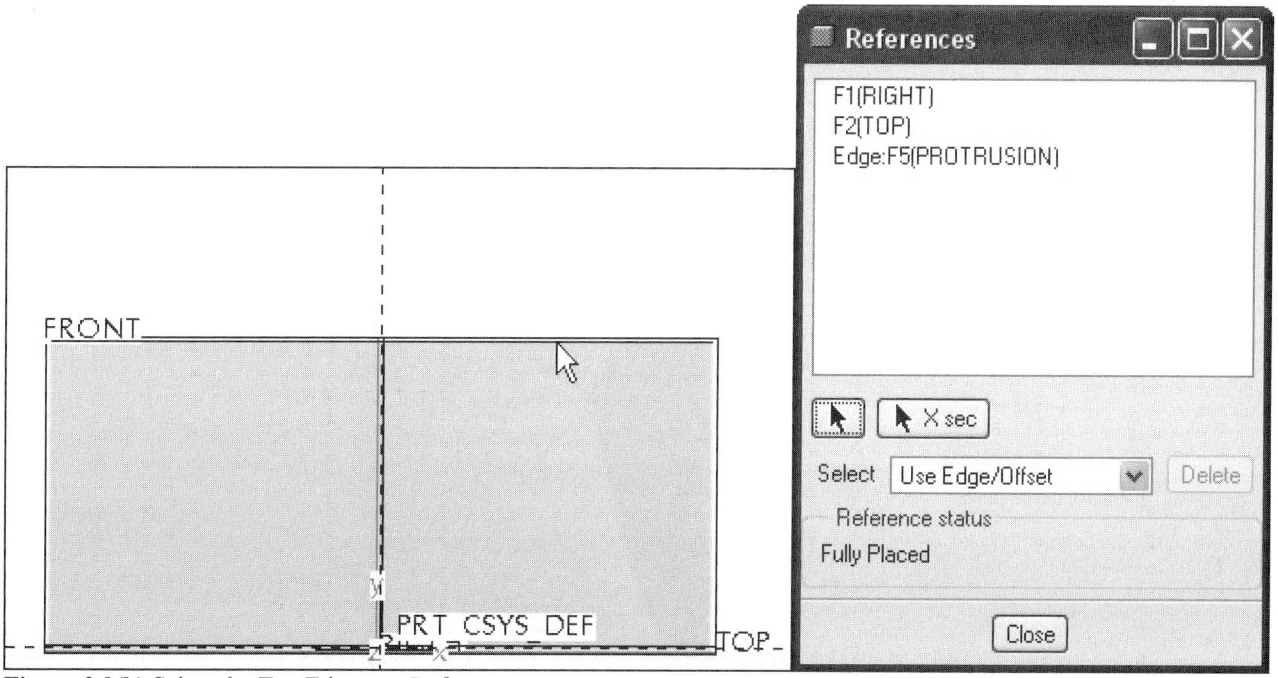

Figure 3.9(b) Select the Top Edge as a Reference

Click: **RMB** ⇒ **Centerline** ⇒ create a vertical centerline through the middle of the part ⇒ **MMB** ⇒ **RMB** ⇒ **Circle** ⇒ sketch a circle [Figs. 3.9(c-d)] ⇒ **MMB** ⇒ ✓ ⇒ **Ctrl+D** [Fig. 3.9(e)] ⇒ **MMB** [Fig. 3.9(f)] ⇒ **LMB**

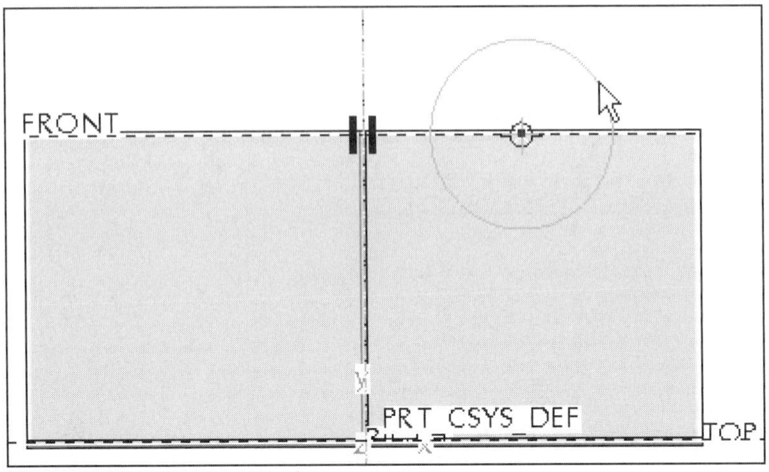

Figure 3.9(c) Sketch a Circle

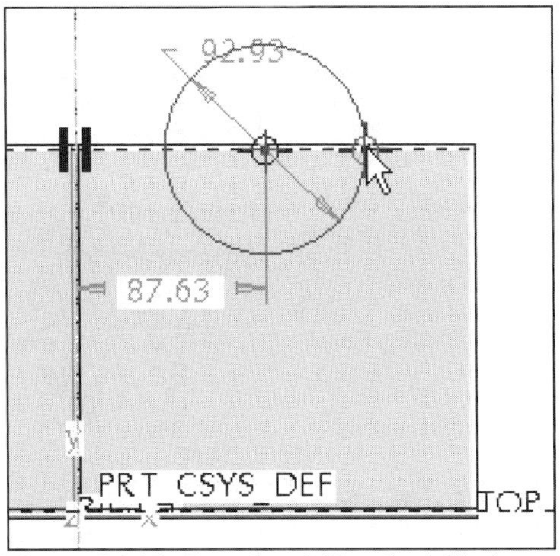

Figure 3.9(d) Completed Sketched Circle

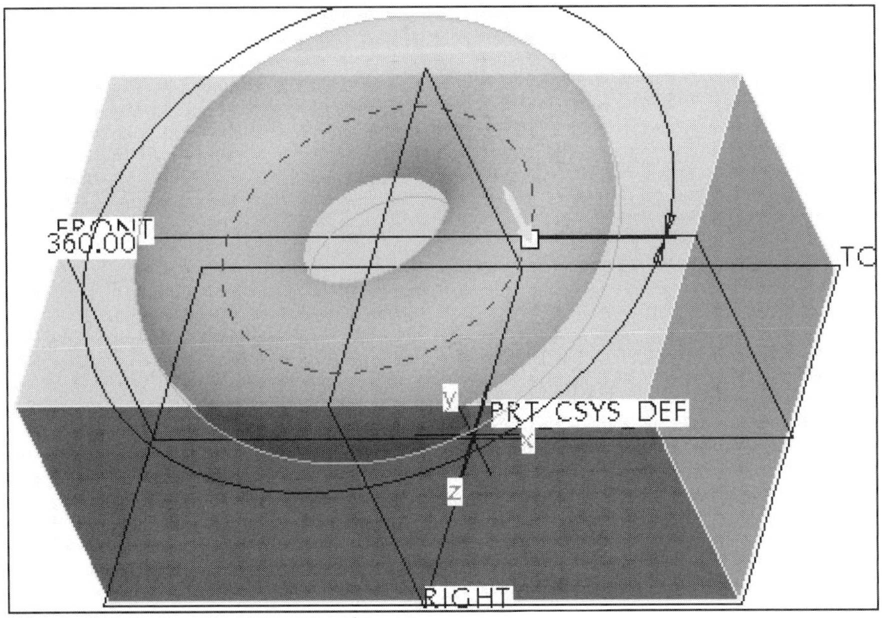

Figure 3.9(e) Revolved Cut Preview

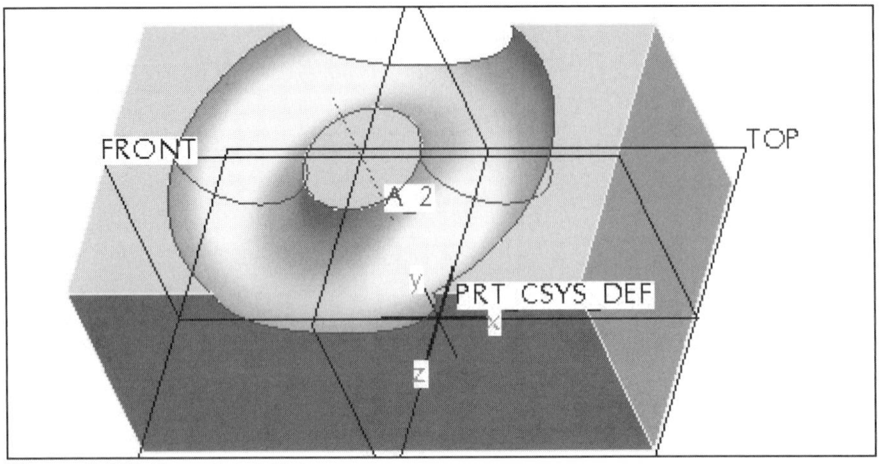

Figure 3.9(f) Completed Cut

Slowly pick on the edge of the part until it highlights ⇒ press and hold the **Ctrl** key ⇒ pick the other edges [Fig. 3.9(g)] ⇒ **RMB** ⇒ Round Edges [Fig. 3.9(h)] ⇒ **MMB** [Fig. 3.9(i)] ⇒ **LMB**

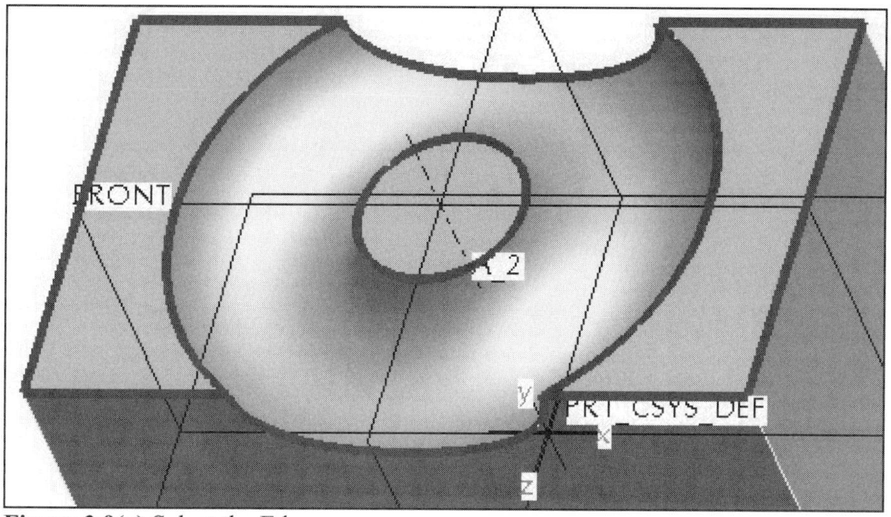

Figure 3.9(g) Select the Edges

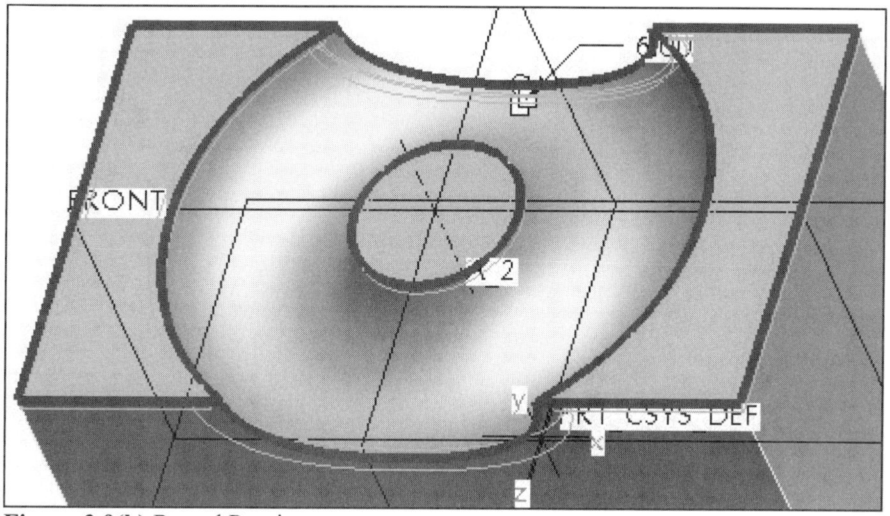

Figure 3.9(h) Round Preview

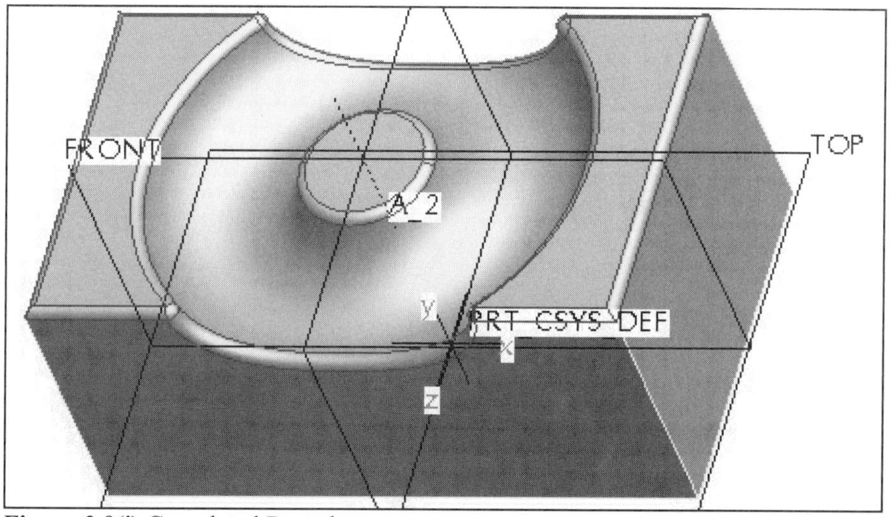

Figure 3.9(i) Completed Round

Smart *(lower right-hand corner of the screen)* ⇒ ⇒ Datums ⇒ pick Axis **A_2** *(your axis label may be different)* [Fig. 3.9(j)] ⇒ **Hole Tool** ⇒ Placement ⇒ while holding down the **Ctrl** key, pick on the top face of the part [Fig. 3.9(k)] ⇒ ⇒ ⇒ **MMB** [Fig. 3.9(l)] ⇒ **LMB** ⇒ ⇒ **MMB** ⇒ **Window** ⇒ **Close**

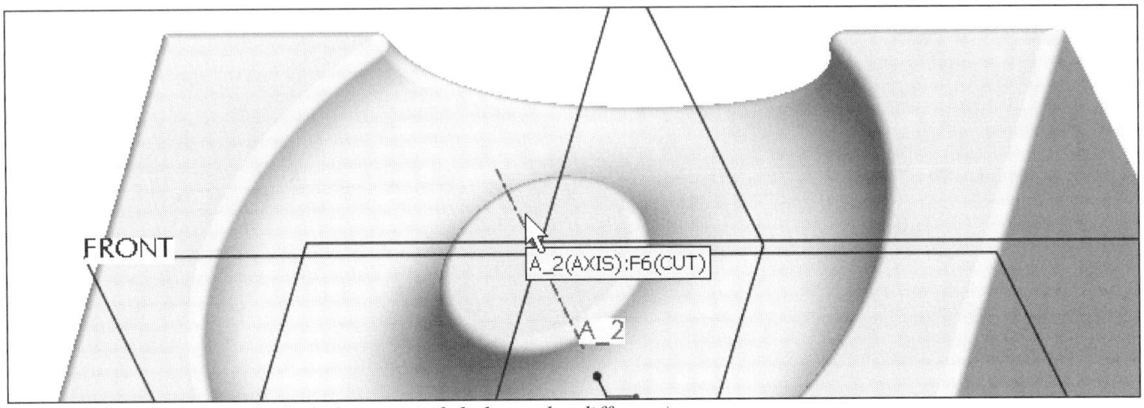

Figure 3.9(j) Select Axis **A_2** *(your axis label may be different)*

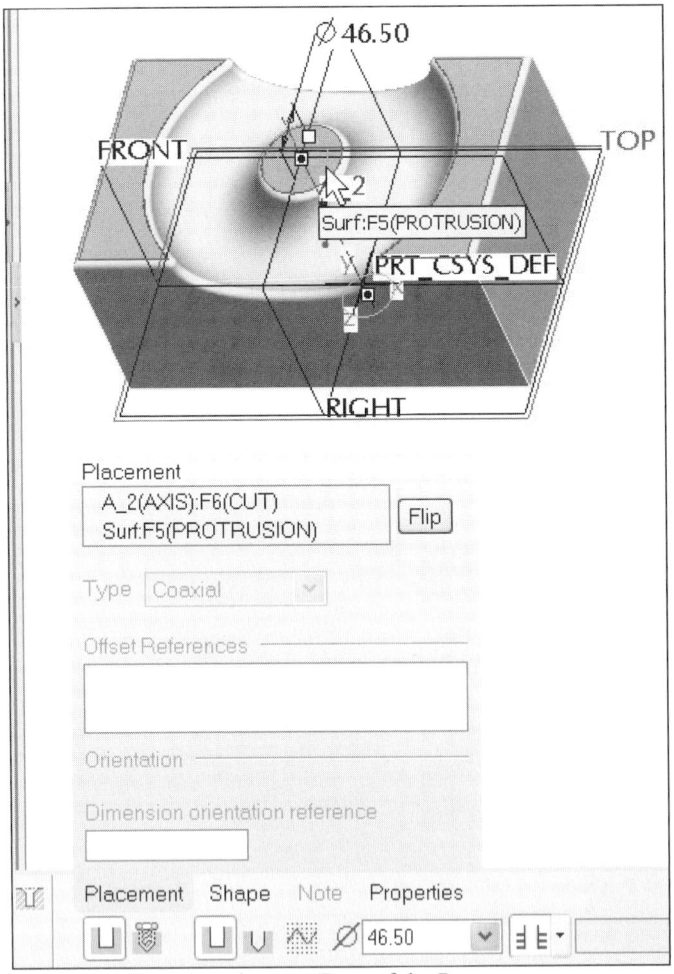

Figure 3.9(k) Select the Top Face of the Part

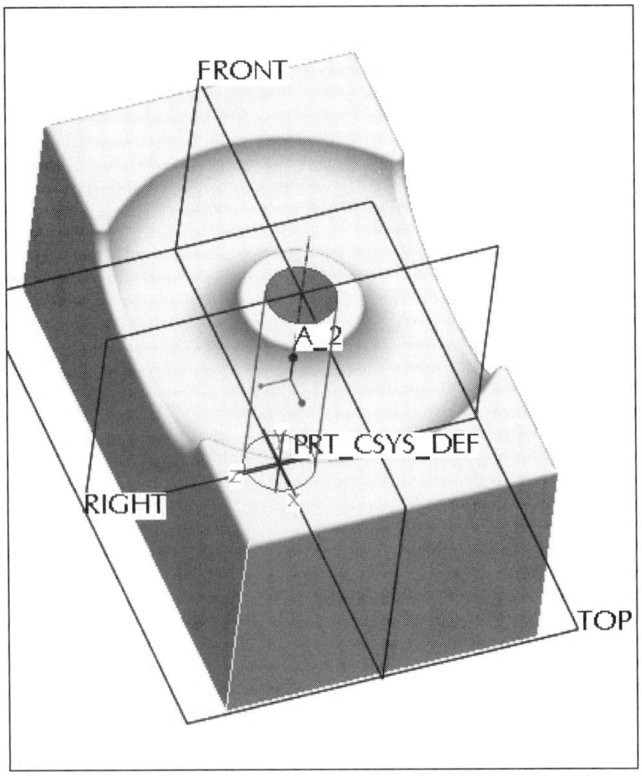

Figure 3.9(l) Completed Part

See ***www.cad-resources.com*** ***⇒ Downloads***, for extra Lessons and projects.

Lesson 4 Extrusions

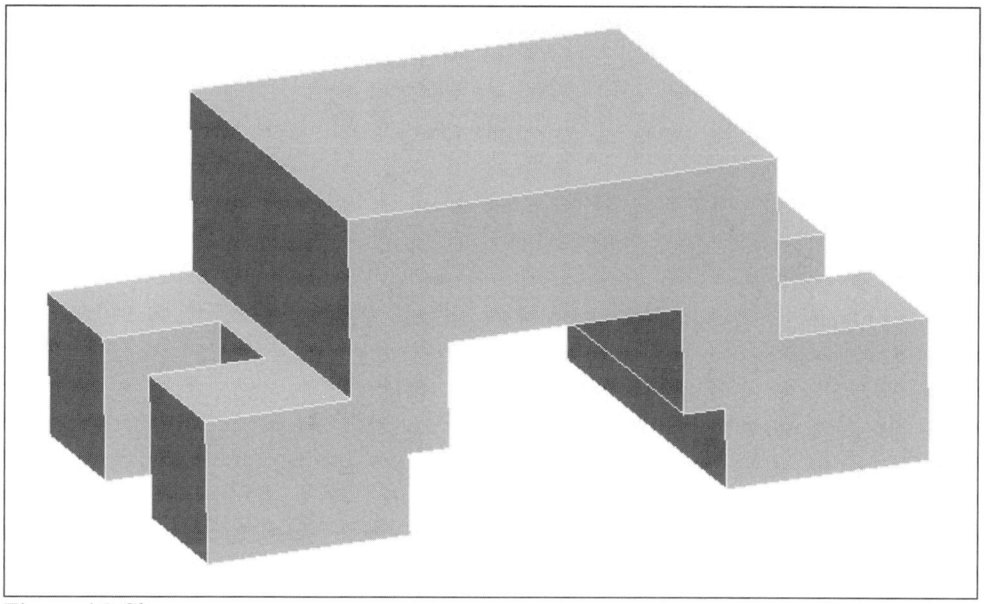

Figure 4.1 Clamp

OBJECTIVES

- Create a feature using an **Extruded** protrusion
- Understand **Setup** and **Environment** settings
- Define and set a **Material** type
- Create and use **Datum** features
- Sketch protrusion and cut feature geometry using the **Sketcher**
- Understand the feature **Dashboard**
- **Copy** a feature
- **Save** and **Delete Old Versions** of an object

Extrusions

The design of a part using Pro/E starts with the creation of base features (normally datum planes), and a solid protrusion. Other protrusions and cuts are then added in sequence as required by the design. You can use various types of Pro/E features as building blocks in the progressive creation of solid parts (Fig. 4.1). Certain features, by necessity, precede other more dependent features in the design process. Those dependent features rely on the previously defined features for dimensional and geometric references.

The progressive design of features creates these dependent feature relationships known as *parent-child relationships*. The actual sequential history of the design is displayed in the Model Tree. The parent-child relationship is one of the most powerful aspects of Pro/E and parametric modeling in general. It is also very important after you modify a part. After a parent feature in a part is modified, all children are automatically modified to reflect the changes in the parent feature. It is therefore essential to reference feature dimensions so that Pro/E can correctly propagate design modifications throughout the model.

An **extrusion** is a part feature that adds or removes material. A protrusion is *always the first solid feature created.* This is usually the first feature created after a base feature of datum planes. The **Extrude Tool** is used to create both protrusions and cuts. A toolchest button is available for this command or it can be initiated using Insert ⇒ Extrude from the menu bar. Figure 4.2 shows four different types of basic protrusions.

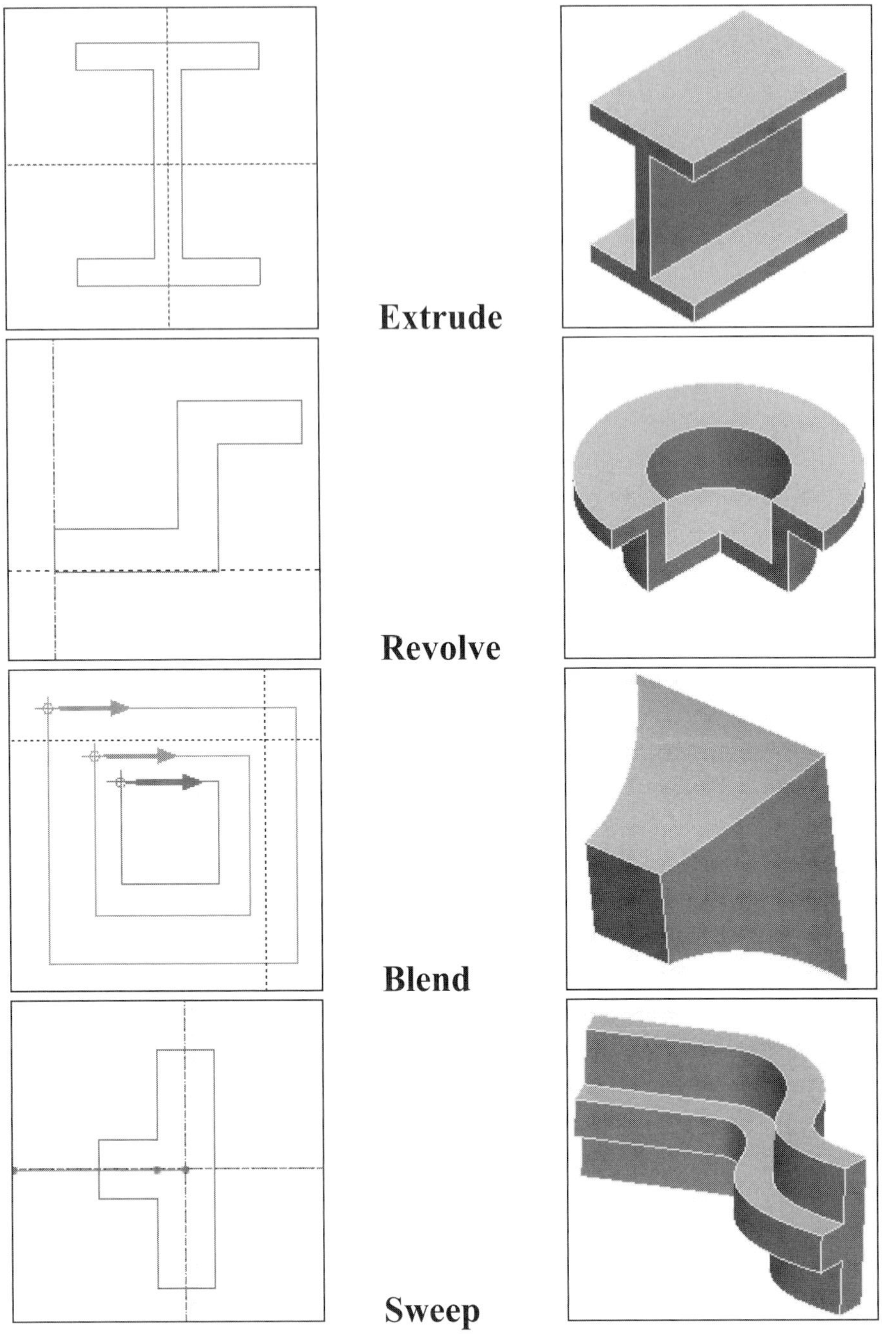

Figure 4.2 Basic Protrusions

The Design Process

It is tempting to directly start creating models. Nevertheless, in order to build value into a design, you need to create a product that can keep up with the constant design changes associated with the design-through-manufacturing process. Flexibility must be integral to the design. Flexibility is the key to a friendly and robust product design while maintaining design intent, and you can accomplish it through planning. To plan a design, you need to understand the overall function, form, and fit of the product. This understanding includes the following points:

- Overall size of the part
- Basic part characteristics
- The way in which the part can be assembled
- Approximate number of assembly components
- The manufacturing processes required to produce the part

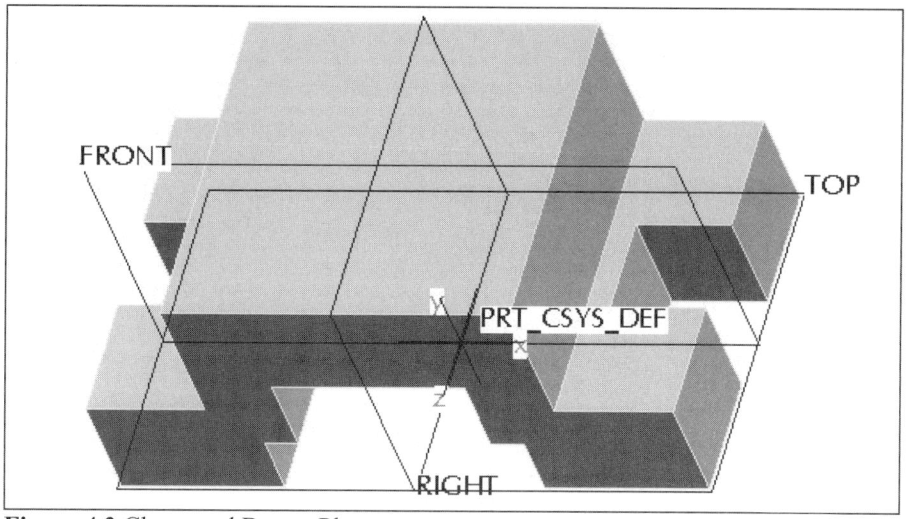

Figure 4.3 Clamp and Datum Planes

Clamp

The clamp in Figure 4.3 is composed of a protrusion and two cuts. A number of things need to be established before you actually start modeling. These include setting up the *environment*, selecting the *units*, and establishing the *material* for the part.

Before you begin any part using Pro/E, you must plan the design. The **design intent** will depend on a number of things that are out of your control and many that you can establish. Asking yourself a few questions will clear up the design intent you will follow: Is the part a component of an assembly? If so, what surfaces or features are used to connect one part to another? Will geometric tolerancing be used on the part and assembly? What units are being used in the design, SI or decimal inch? What is the part's material? What is the primary part feature? How should I model the part, and what features are best used for the primary protrusion (the first solid mass)? On what datum plane should I sketch to model the first protrusion? These and many other questions will be answered as you follow the systematic lesson part. However, you must answer many of the questions on your own when completing the *lesson project*, which does not come with systematic instructions.

Launch **Pro/ENGINEER WILDFIRE 4.0** ⇒ **File** ⇒ **Set Working Directory** ⇒ select the working directory ⇒ **OK** ⇒  **Create a new object** ⇒ 

⇒ Name **CLAMP** ⇒ 

⇒ **OK** ⇒ **Edit** ⇒ **Setup** ⇒ **Units** (Units Manager dialog box opens) [Fig. 4.4(a)] ⇒

millimeter Newton Second (mmNs) ⇒ **Set** ⇒ 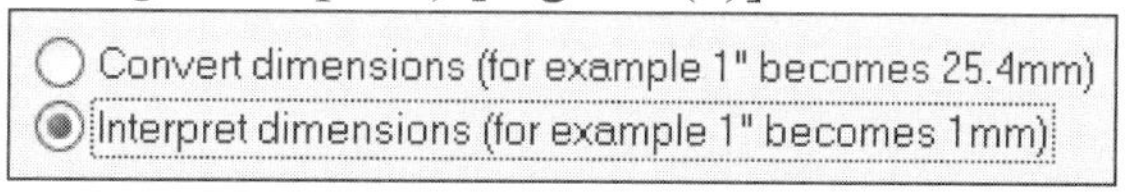

[Fig. 4.4(b)]

⇒ **OK** 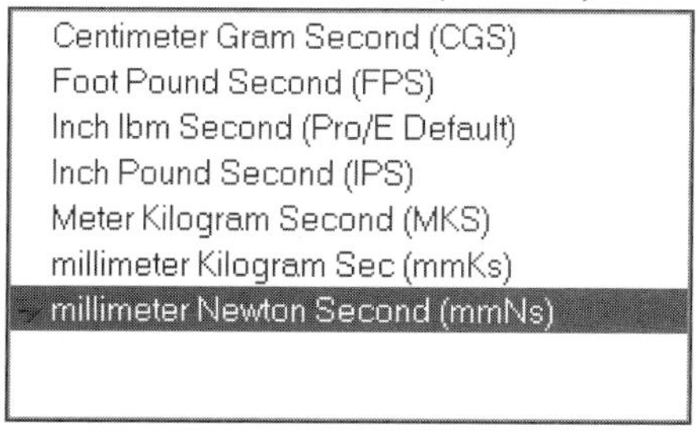

⇒ **Close**

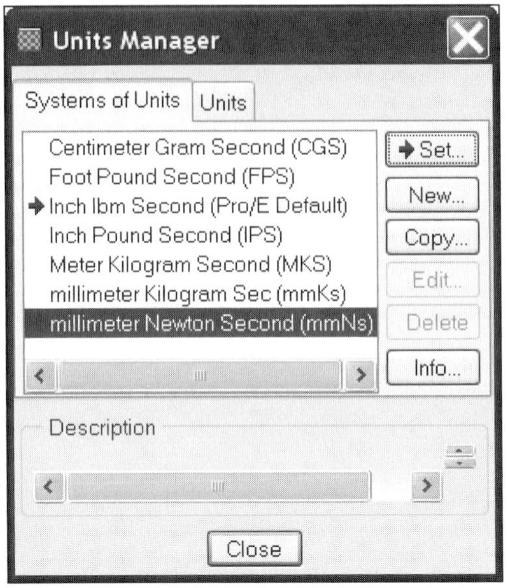

Figure 4.4(a) Units Manager Dialog Box

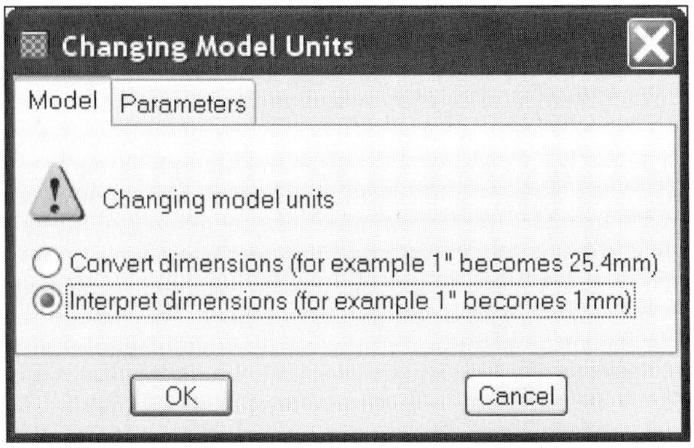

Figure 4.4(b) Changing Model Units Dialog Box

Click: **Material** ⇒ **steel.mtl** ⇒  [Fig. 4.4(c)] ⇒ 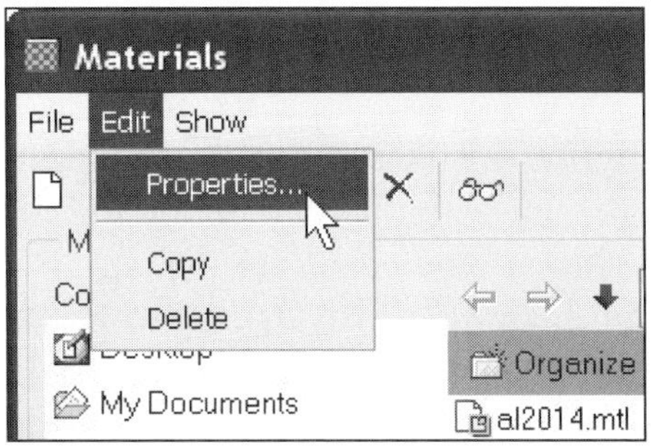

[Fig. 4.4(d)]

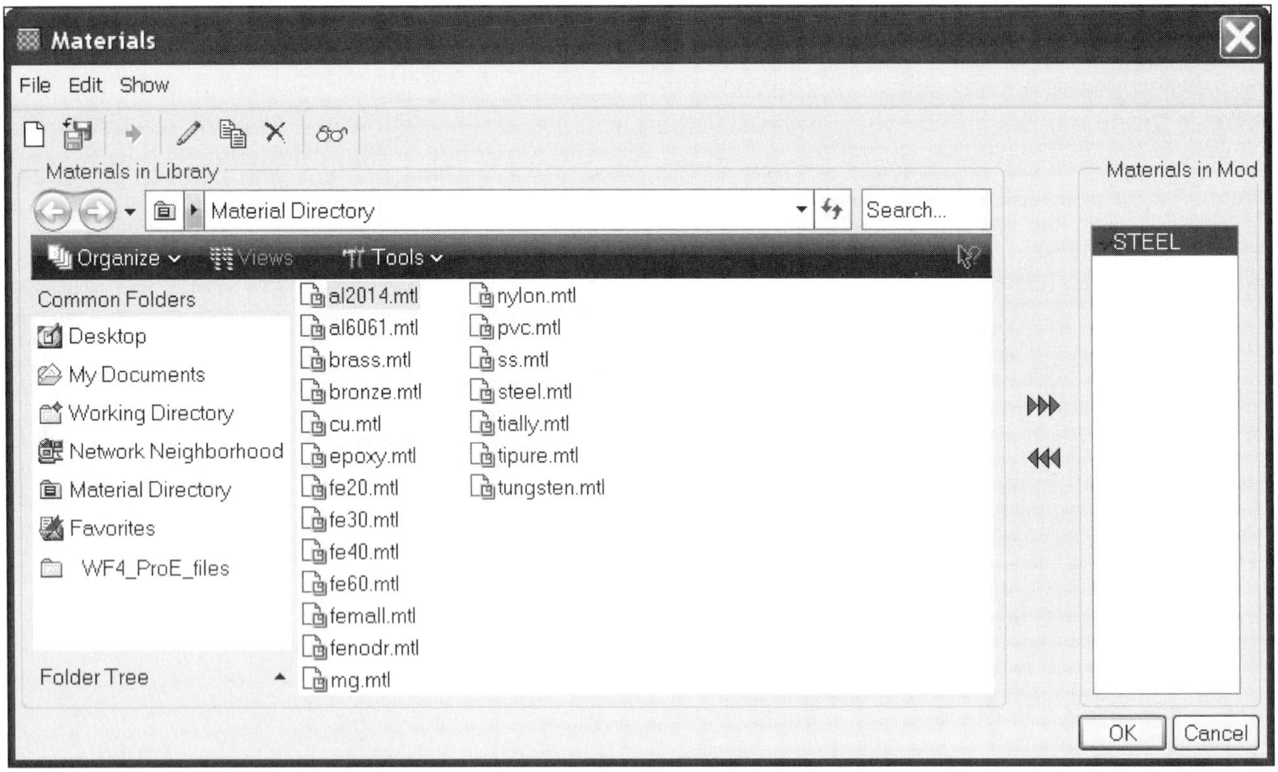

Figure 4.4(c) Material File

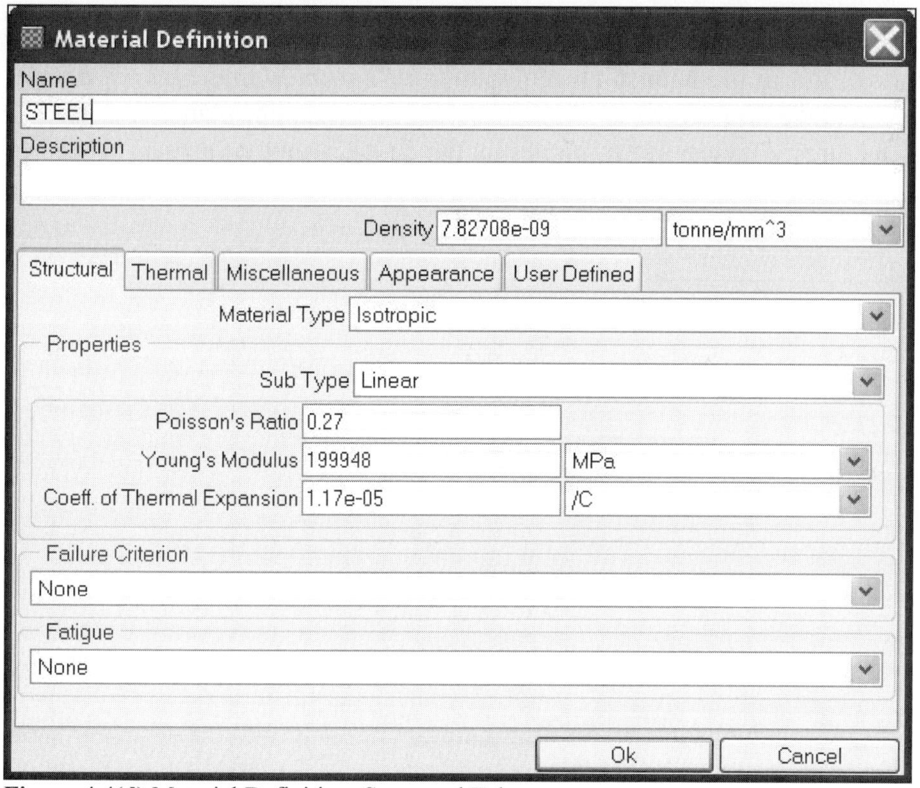

Figure 4.4(d) Material Definition, Structural Tab

Click: **Thermal** tab [Fig. 4.4(e)] ⇒ investigate other options and tabs ⇒ **Ok** ⇒ **OK** ⇒ **MMB** ⇒ 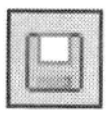 ⇒ **MMB** *(or Enter or OK)*

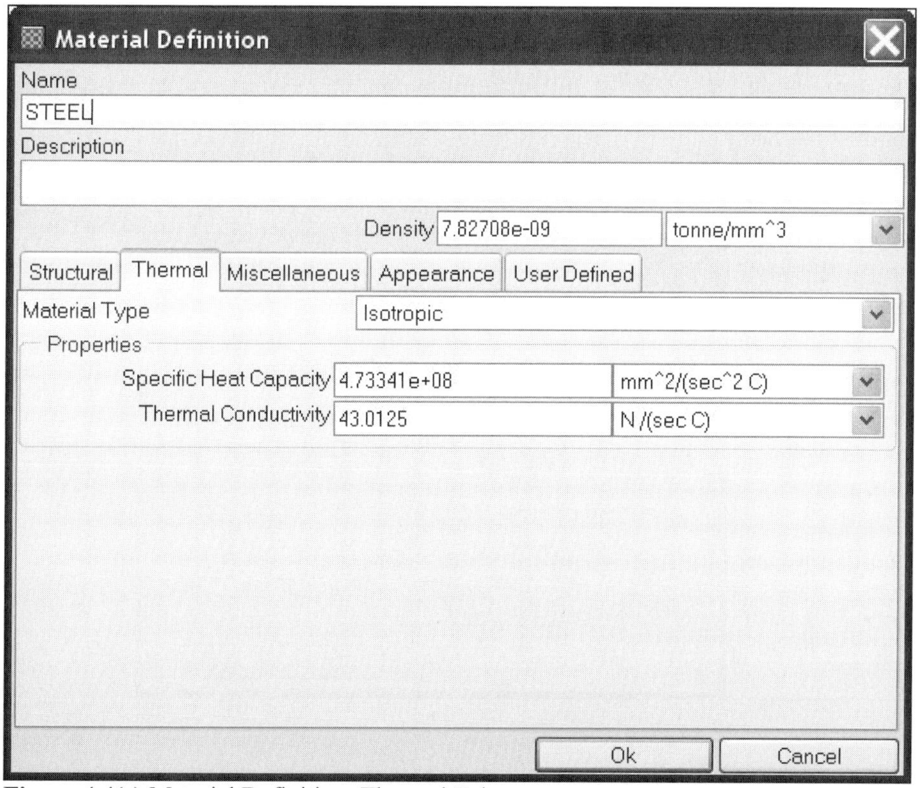

Figure 4.4(e) Material Definition, Thermal Tab

Since [✓ Use default template] was selected, the default datum planes and the default coordinate system are displayed in the graphics window and in the Model Tree. *The **default datum planes** and the **default coordinate system** will be the first features on all parts and assemblies.* The datum planes are used to sketch on and to orient the part's features. Having datum planes as the first features of a part, instead of the first protrusion, gives the designer more flexibility during the design process. Picking on an item in the Model Tree will highlight that item on the model (Fig. 4.5).

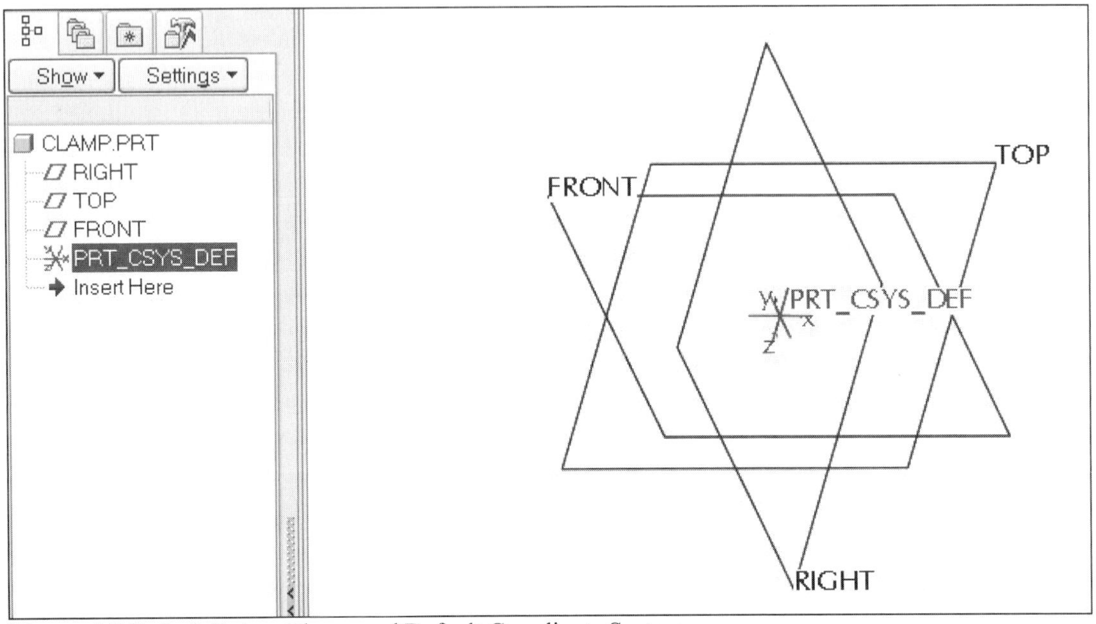

Figure 4.5 Default Datum Planes and Default Coordinate System

Select on the **FRONT** datum plane in the Model Tree ⇒ **Sketch Tool** from Right Toolchest ⇒ Sketch dialog box opens [Fig. 4.6(a)] ⇒ *accept the default selections,* click: [Sketch]

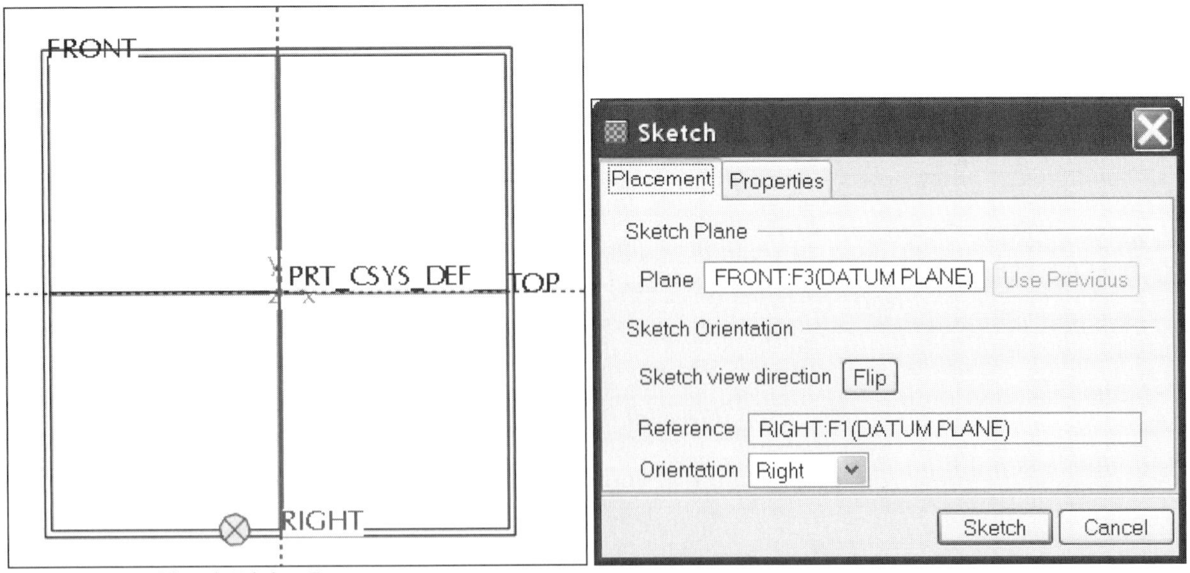

Figure 4.6(a) Sketch Dialog Box

From menu bar, click: **Sketch ⇒ References** [Fig. 4.6(b)] *(the RIGHT and TOP datum planes are the positional/dimensional references)* ⇒ **Close** ⇒ **Toggle the grid on** from Top Toolchest [Fig. 4.6(c)]

Figure 4.6(b) References Dialog Box

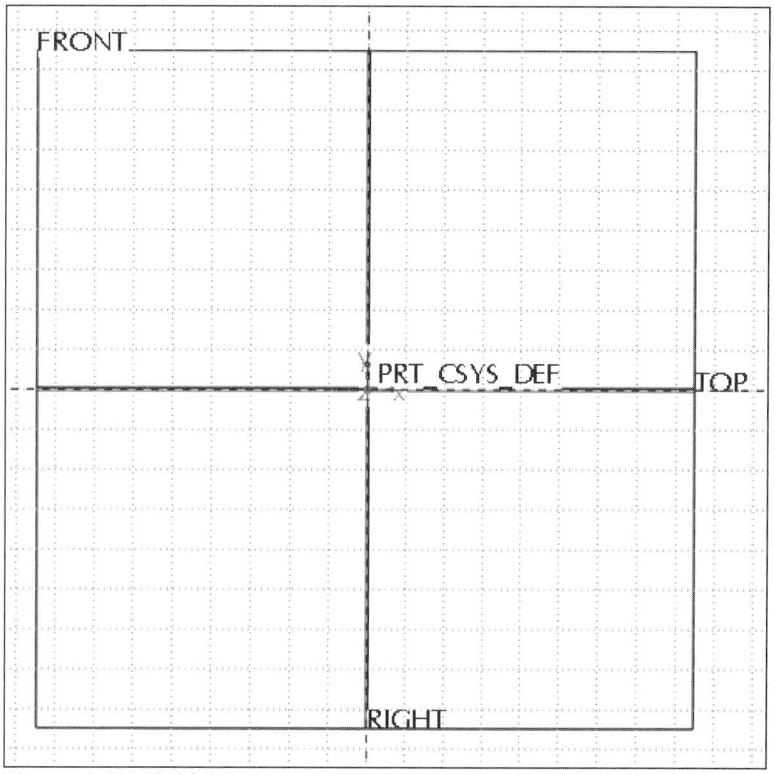

Figure 4.6(c) Grid On

The sketch is now displayed and oriented in 2D [Fig. 4.6(c)]. The coordinate system is at the middle of the sketch, where datum RIGHT and datum TOP intersect. The X coordinate arrow points to the right and the Y coordinate arrow points up. The Z arrow is pointing toward you (out from the screen). The square box you see is the limited display of datum FRONT. This is similar to sketching on a piece of graph paper. Pro/E is not coordinate-based software, so you need not enter geometry with X, Y, and Z coordinates.

Use **Shift+MMB** and **Ctrl+MMB** to reposition and resize the sketch as needed. Since you now have a visible grid, turn on the grid snap to have your sketch picks lock to the grid position. Click: **Tools** from Top Toolchest ⇒ Environment ⇒ Snap to Grid [Fig. 4.6(d)] ⇒ **Apply** ⇒ **OK**

You can control many aspects of the environment in which Pro/E runs with the Environment dialog box. To open the Environment dialog box, click Tools ⇒ Environment on the menu bar or click the appropriate icon in the toolbar. When you make a change in the Environment dialog box, it takes effect for the current session only. When you start Pro/E, the environment settings are defined by Pro/E configuration defaults. *Config* settings can also be set using: Tools ⇒ Options.

Depending on which Pro/E Mode is active (here it is the Part Mode), some or all of the following options may be available in the **Environment** dialog box:

Display:

Dimension Tolerances Display model dimensions with tolerances
Datum Planes Display the datum planes and their names
Datum Axes Display the datum axes and their names
Point Symbols Display the datum points and their names
Coordinate Systems Display the coordinate systems and their names
Spin Center Display the spin center for the model
Reference Designators Display reference designation of Cabling, ECAD, and Piping components
Thick Cables Display a cable with 3-D thickness
Centerline Cables Display the centerline of a cable with location points
Internal Cable Portions Display cable portions that are hidden from view
Colors Display colors assigned to model surfaces
Textures Display textures on shaded models
Levels of Detail Controls levels of detail available in a shaded model during dynamic orientation

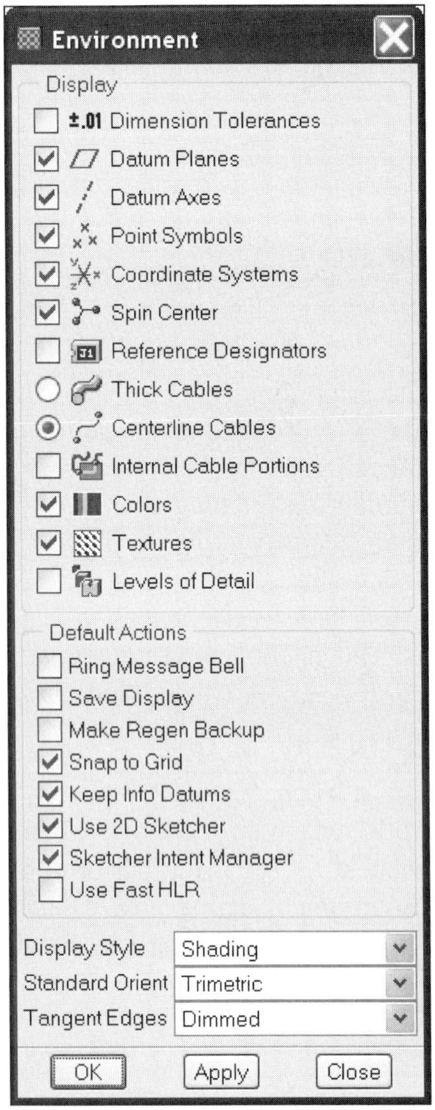

Figure 4.6(d) Environment Dialog Box

Default Actions:

Ring Message Bell Ring bell (beep) after each prompt or system message
Save Display Save objects with their most recent screen display
Make Regen Backup Backs up the current model before every regeneration
Snap to Grid Make points you select on the Sketcher screen snap to a grid
Keep Info Datums Control how Pro/E treats datum planes, datum points, datum axes, and coordinate systems created on the fly under the Info functionality
Use 2D Sketcher Control the initial model orientation in Sketcher mode
Sketcher Intent Manager Use the Intent Manager when in Sketcher
Use Fast HLR Make possible the hardware acceleration of dynamic spinning with hidden lines, datums, and axes

Display Style:

- **Wireframe** Model is displayed with no distinction between visible and hidden lines
- **Hidden Line** Hidden lines are shown in gray
- **No Hidden** Hidden lines are not shown
- **Shading** All surfaces and solids are displayed as shaded

Standard Orient:

- **Isometric** Standard isometric orientation
- **Trimetric** Standard trimetric orientation
- **User Defined** User-defined orientation

Tangent Edges:

- **Solid** Display tangent edges as solid lines
- **No Display** Blank tangent edges
- **Phantom** Display tangent edges in phantom font
- **Centerline** Display tangent edges in centerline font
- **Dimmed** Display tangent edges in the Dimmed Menu system

Because you checked [Snap to Grid], you can now sketch by simply picking grid points representing the part's geometry (outline). Because this is a sketch in the true sense of the word, you need only create geometry that *approximates* the shape of the feature; the sketch does not have to be accurate as far as size or dimensions are concerned. No two sketches will be the same between those using these steps, unless you count each grid space (which is unnecessary). Even with the grid snap off, Pro/E constrains the geometry according to rules, which include but are not limited to the following:

- **RULE:** Symmetry
  DESCRIPTION: Entities sketched symmetrically about a centerline are assigned equal values with respect to the centerline
- **RULE:** Horizontal and vertical lines
  DESCRIPTION: Lines that are approximately horizontal or vertical are considered exactly horizontal or vertical
- **RULE:** Parallel and perpendicular lines
  DESCRIPTION: Lines that are sketched approximately parallel or perpendicular are considered exactly parallel or perpendicular
- **RULE:** Tangency
  DESCRIPTION: Entities sketched approximately tangent to arcs or circles are assumed to be exactly tangent

The outline of the part's primary feature is sketched using a set of connected lines. The part's dimensions and general shape are provided in Figure 4.6(e). The cut on the front and sides will be created with separate sketched features. Sketch only one series of lines (8 lines in this sketch). ***Do not sketch lines on top of lines.***

It is important not to create any unintended constraints while sketching. Therefore, remember to exaggerate the sketch geometry and not to align geometric items that have no relationship. Pro/E is very smart: if you draw two lines at the same horizontal level, Pro/E assumes they are horizontally aligned. Two lines the same length will be constrained as so.

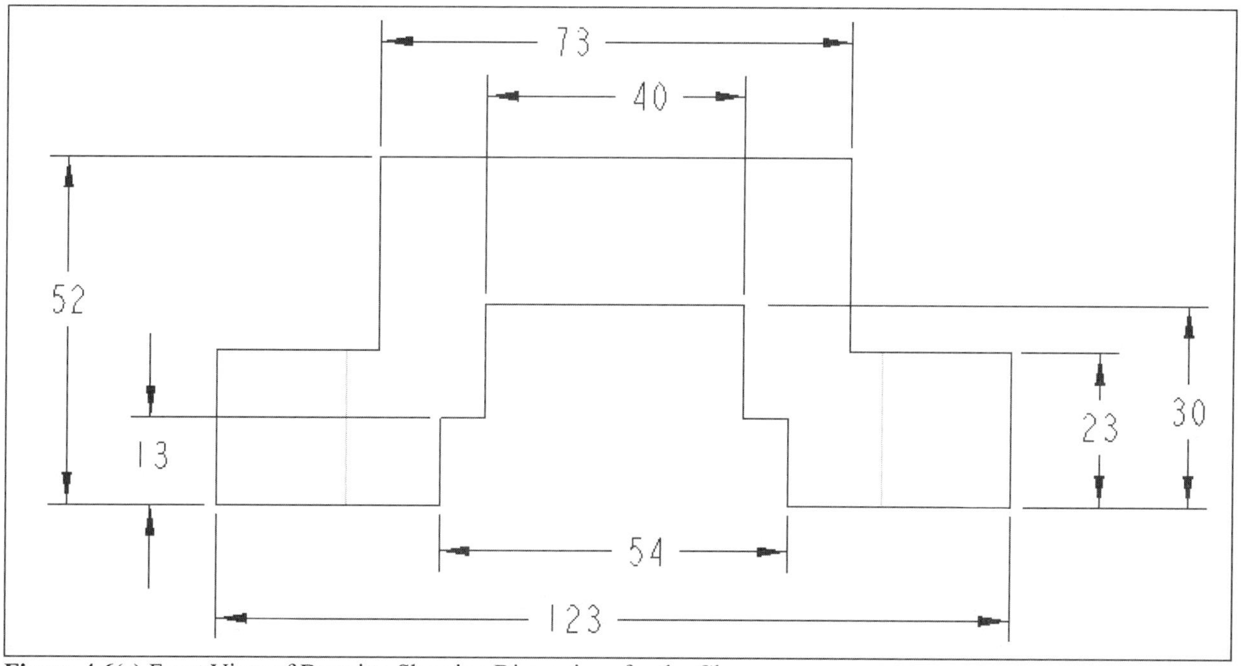

Figure 4.6(e) Front View of Drawing Showing Dimensions for the Clamp

With your cursor anywhere in the graphics window, but not on an object, click: **RMB** [Fig. 4.6(f)] ⇒ **Centerline** ⇒ pick two vertical positions on the RIGHT datum plane to create the centerline [Fig. 4.6(g)]

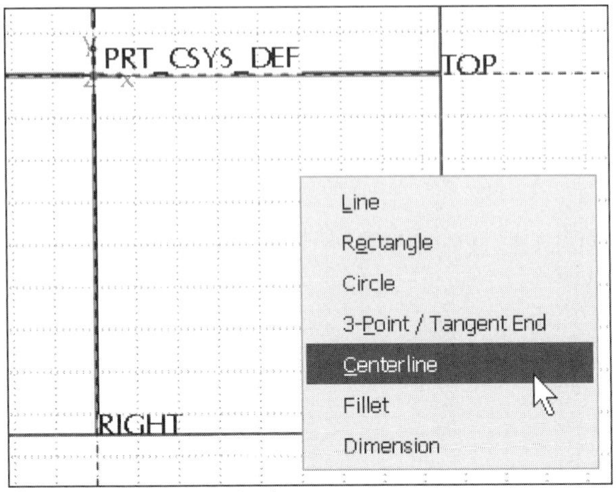

Figure 4.6(f) RMB Options

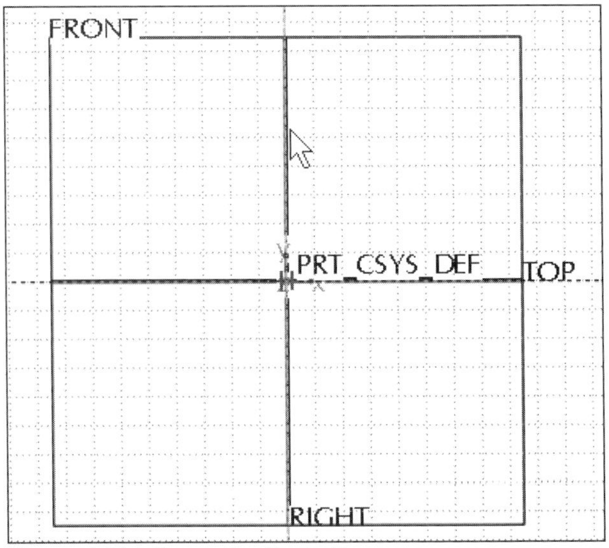

Figure 4.6(g) Create the Centerline

Click: **RMB** ⇒ **Line** ⇒ sketch the eight lines of the closed outline [Fig. 4.6(h)] ⇒ **MMB** to end the line sequence [Fig. 4.6(i)] ⇒ **MMB** to end the current tool

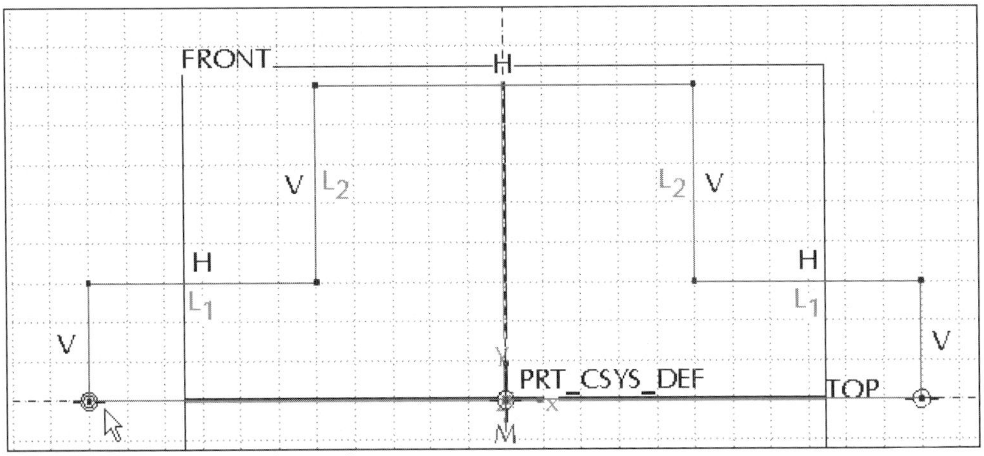

Figure 4.6(h) Sketching the Outline

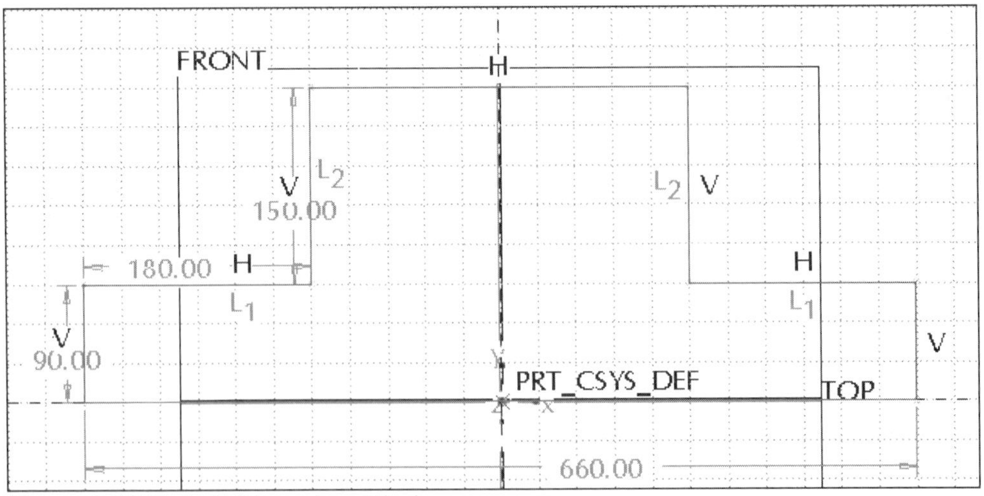

Figure 4.6(i) Default Dimensions Display

A sketcher *constraint symbol* 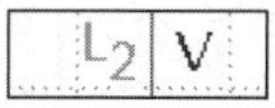 appears next to the entity that is controlled by that constraint. Sketcher constraints can be turned on or off (enabled or disabled) while sketching. Simply click your RMB as you sketch- before picking the position- and the constraint that is displaying will have a slash imposed over it. This will disable it for that entity. An **H** next to a line means horizontal; a **T** means tangent. Dimensions display, as they are needed according to the references selected and the constraints. Seldom are they the same as the required dimensioning scheme needed to manufacture the part. You can add, delete, and move dimensions as required. *The dimensioning scheme is important,* ***not the dimension value****, which can be modified now or later.*

Place and create the dimensions as required [Fig. 4.6(j)]. Do not be concerned with the perfect positioning of the dimensions, but in general, follow the spacing and positioning standards found in the **ASME Geometric Tolerancing and Dimensioning** standards. This saves you time when you create a drawing of the part. Dimensions placed at this stage of the design process are displayed on the drawing document by simply showing all the dimensions.

To dimension between two lines, simply pick the lines with the left mouse button (LMB) and place the dimension value with the middle mouse button (MMB). To dimension a single line, pick on the line (LMB), and then place the dimension with MMB.

Click: **Tools** ⇒ Environment ⇒ Snap to Grid *(it is easier to position the dimensions with Snap to Grid off)* ⇒ **Apply** ⇒ **OK** ⇒ **RMB** ⇒ **Dimension** ⇒ add and reposition dimensions [Fig. 4.6(j)]

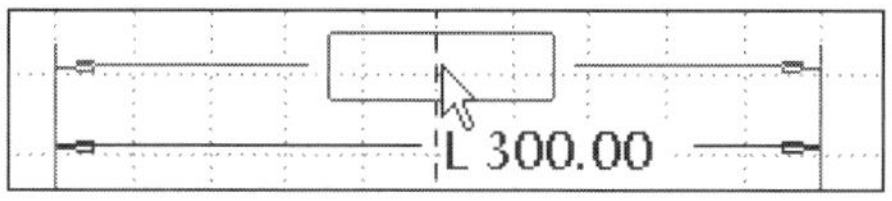

(To move a dimension—click:  *⇒ pick a dimension ⇒ hold down the LMB ⇒ move it to a new position ⇒ release the LMB)*

If any of the dimension values are *light gray* in color, they are called *weak* dimensions. If a *weak* dimension matches your dimensioning scheme, make them *strong—click:*  ⇒ pick on a *weak* dimension value *(will highlight in Red)* ⇒ **RMB** ⇒ **Strong** [Fig. 4.6(j)]

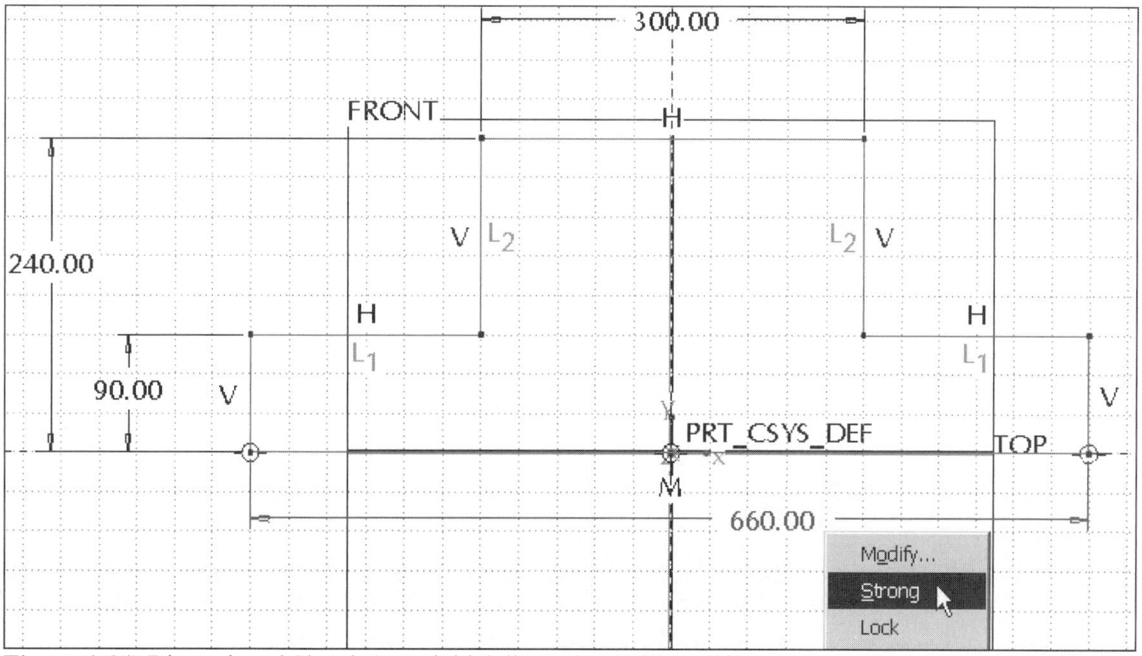

Figure 4.6(j) Dimensioned Sketch (your initial dimensions will be different)

Next, control the sketch by adding symmetry constraints, click:  **Impose sketcher constraints on the section** ⇒  **Make two points or vertices symmetric about a centerline** [Fig. 4.6(k)] ⇒ pick the centerline and then pick two vertices (endpoints) to be symmetric 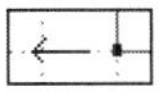 [Fig. 4.6(l)] ⇒ repeat the process and make the sketch symmetrical if needed [Fig. 4.6(m)] ⇒ **Close**

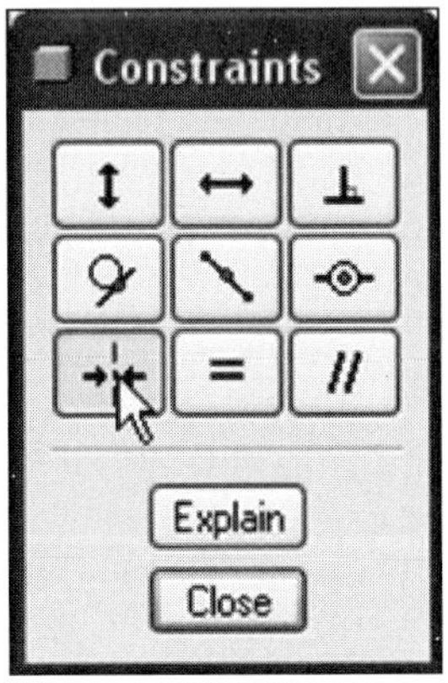

Figure 4.6(k) Constraints Dialog

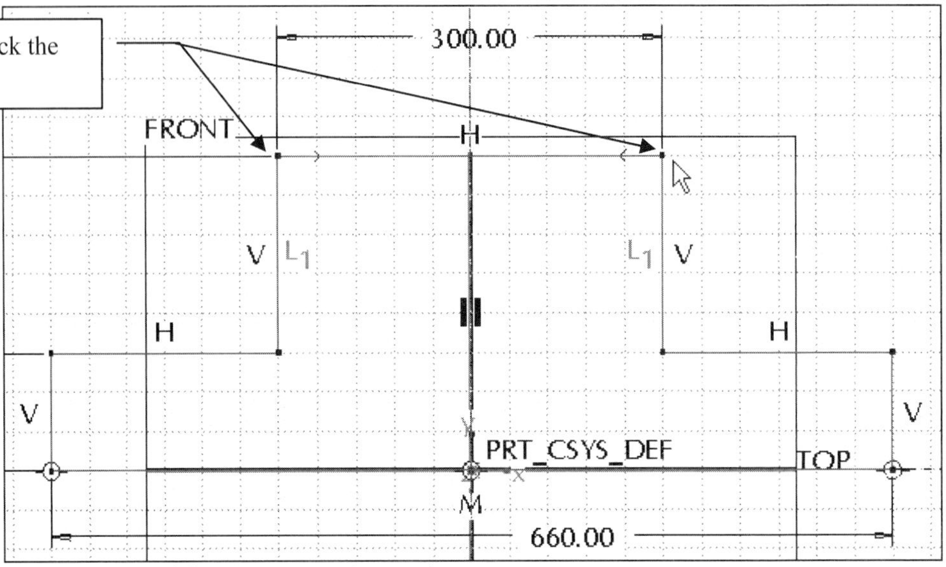

Figure 4.6(l) Adding Symmetry Constraint

Your original sketch values will be different from the example, but the final design values will be the same. **DO NOT CHANGE YOUR SKETCH DIMENSION VALUES TO THOSE IN FIGURE 4.6(m).**

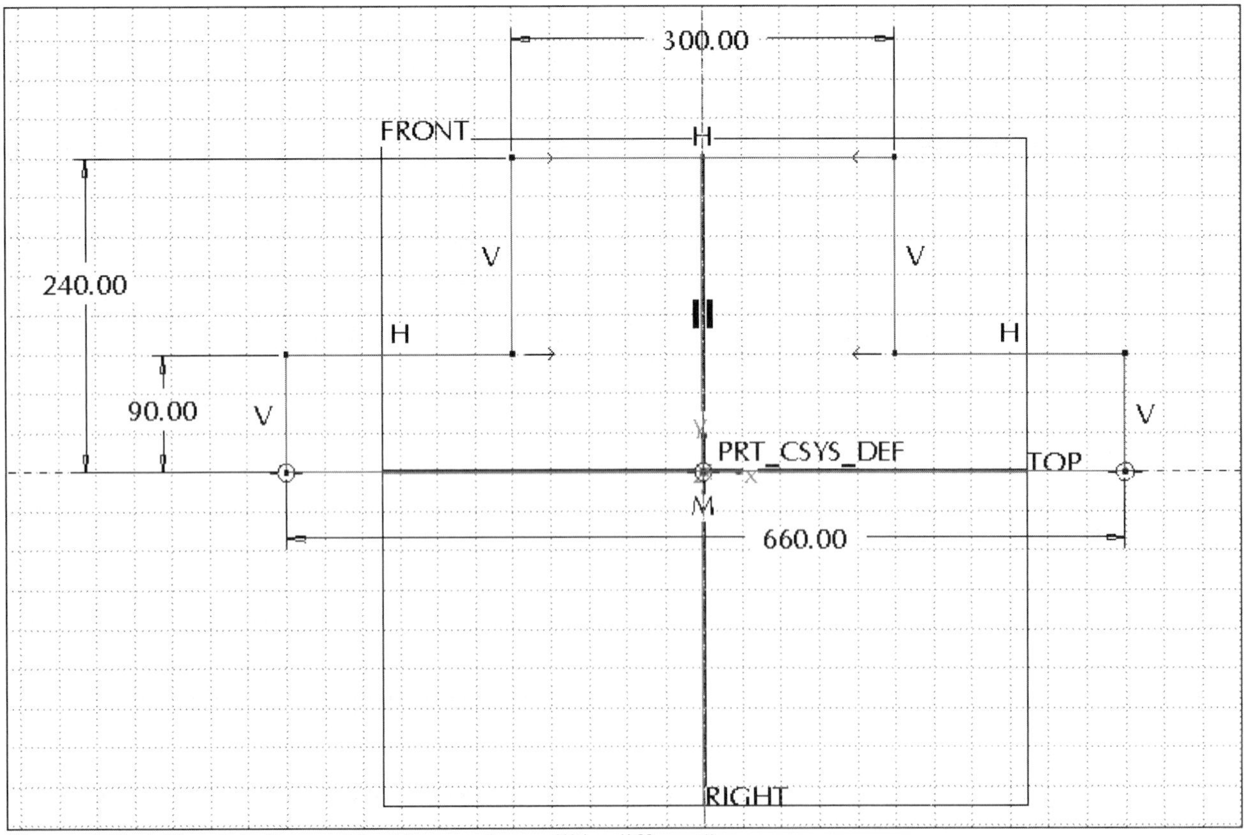

Figure 4.6(m) Sketch is Symmetrical (your values will be different!)

Click:  ⇒ Window-in the sketch (place the cursor at one corner of the window with the **LMB** depressed, drag the cursor to the opposite corner of the window and release the **LMB**) to capture all four dimensions. They will turn red. ⇒ **RMB** ⇒ **Modify** ⇒ ☑ Lock Scale ⇒ ☑ Regenerate [Fig. 4.6(n)] ⇒ double-click on length dimension (*here it is 660, but your dimension may be different*) in the Modify Dimensions dialog box and type the design value at the prompt (**123**) [Fig. 4.6(o)] ⇒ **Enter** ⇒  **Regenerate the section and close the dialog** ⇒ double-click on another dimension and modify the value [Fig. 4.6(p)] ⇒ **Enter** ⇒ continue until all of the values are changed to the design sizes [Fig. 4.6(q)]

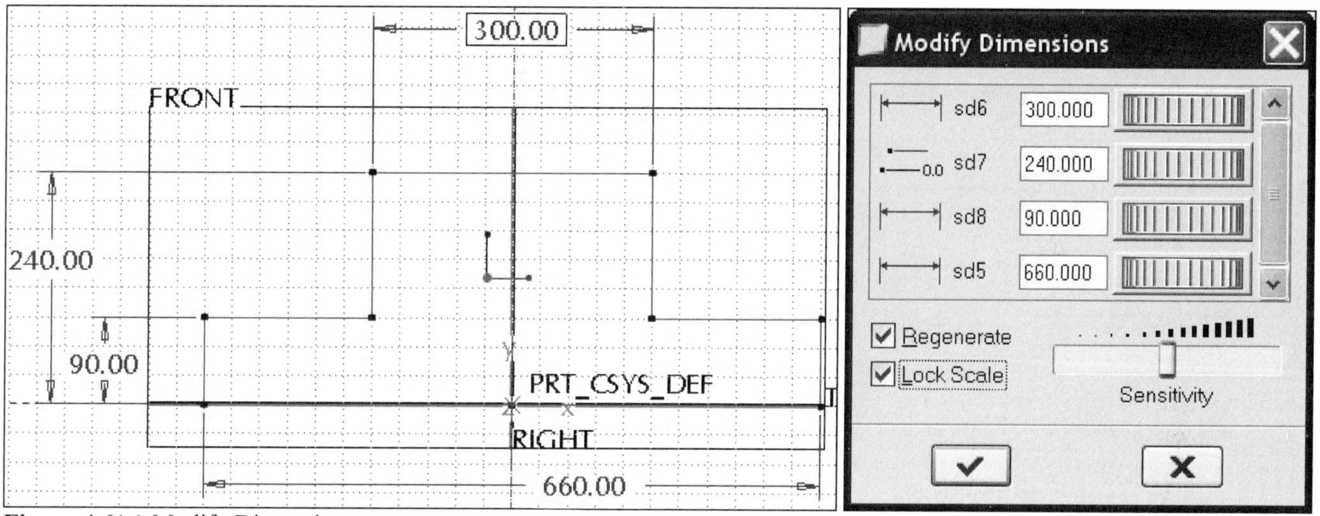

Figure 4.6(n) Modify Dimensions

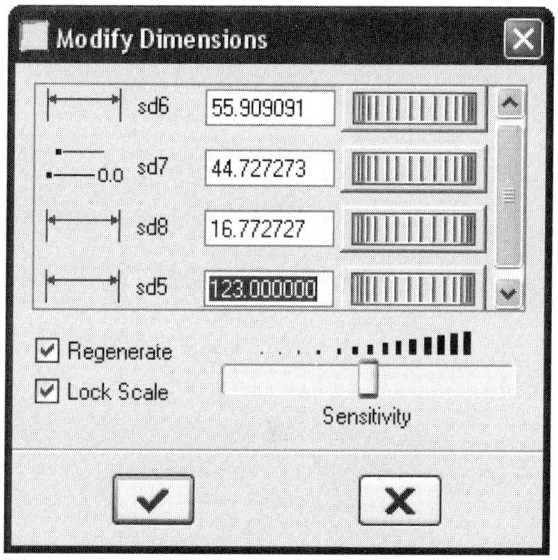

Figure 4.6(o) Modify the Dimension

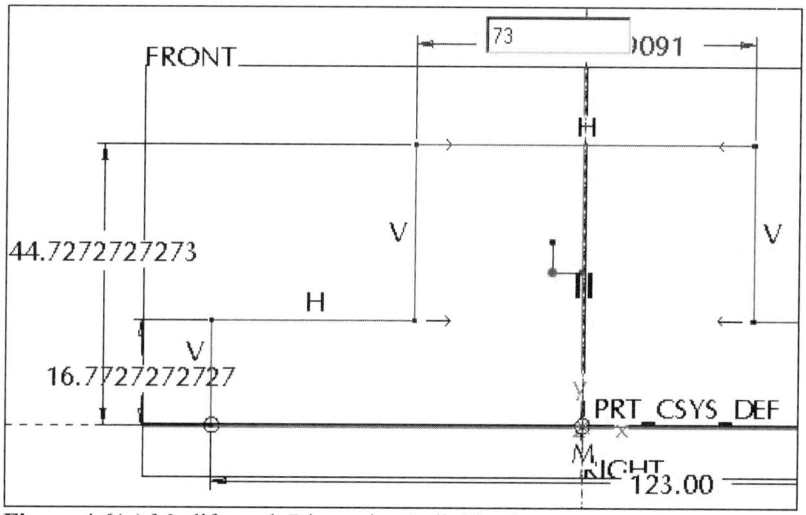

Figure 4.6(p) Modify each Dimension Individually

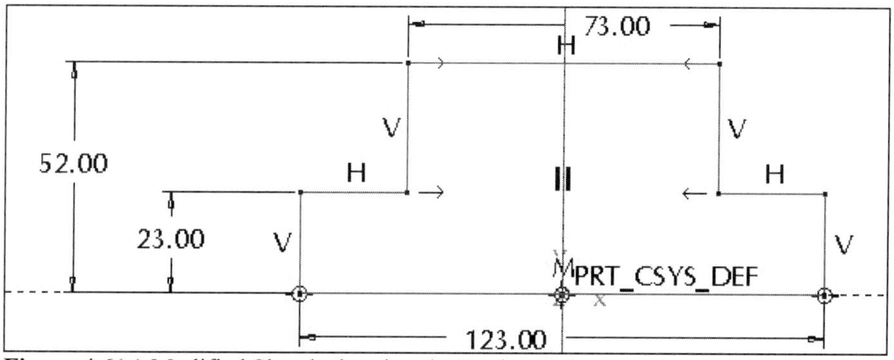

Figure 4.6(q) Modified Sketch showing the Design Values

In the previous two figures, the grid visibility has been turned off so that the illustrations are clearer. From the Top Toolchest, click: **Toggle the grid** off ⇒ **Color the inside of closed chains of sketched entities** ⇒ ⇒ **Standard Orientation** [Fig. 4.6(r)] ⇒ **Continue with the current section** from the Right Toolchest [Fig. 4.6(s)] ⇒ **Refit** ⇒ ⇒ OK

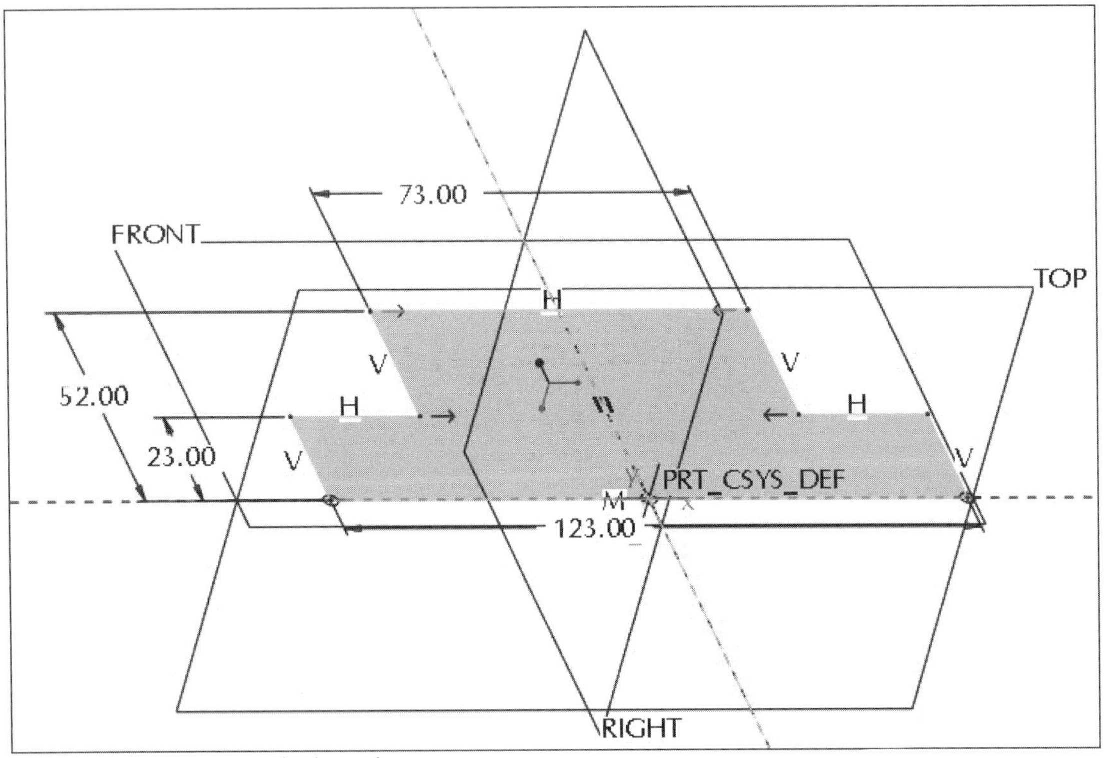

Figure 4.6(r) Regenerated Dimensions

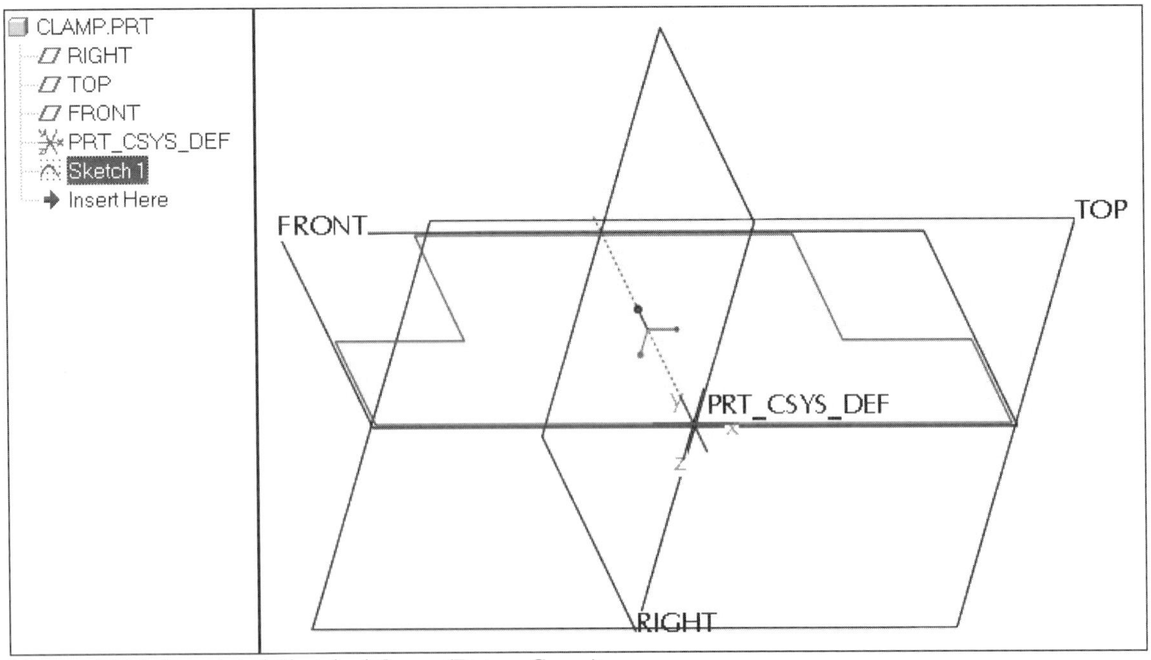

Figure 4.6(s) Completed Sketched Curve (Datum Curve)

The datum curve (Sketch1) will remain highlighted, active and therefore selected.

Click:  **Extrude Tool** [Fig. 4.7(a)] ⇒ double-click on the depth value on the model ⇒ type **70** [Fig. 4.7(b)] ⇒ **Enter** ⇒ place your pointer over the square *white* drag handle  (it will turn *black*) ⇒ **RMB** ⇒ **Symmetric** [Fig. 4.7(c)] ⇒  [Fig. 4.7(d)] ⇒  ⇒ **Enter**

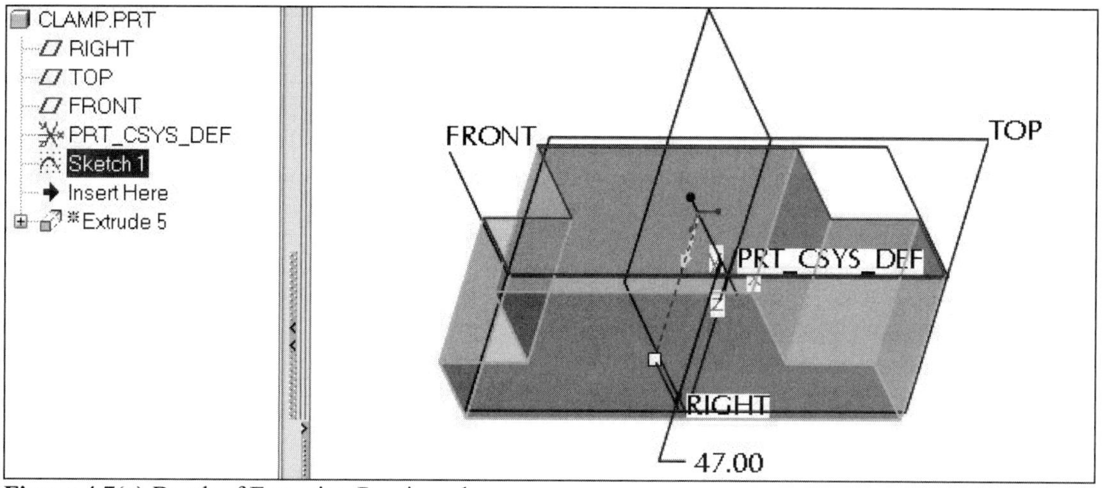

Figure 4.7(a) Depth of Extrusion Previewed

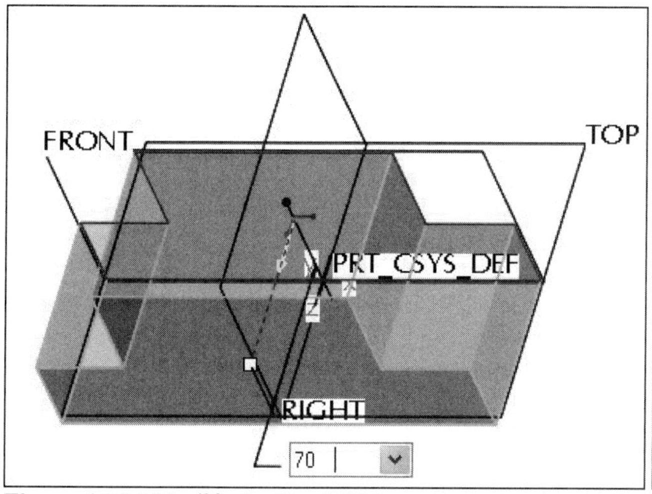

Figure 4.7(b) Modify the Depth Value

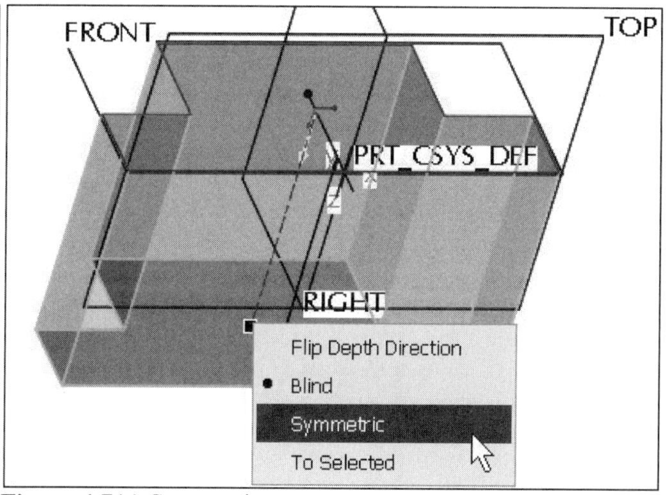

Figure 4.7(c) Symmetric

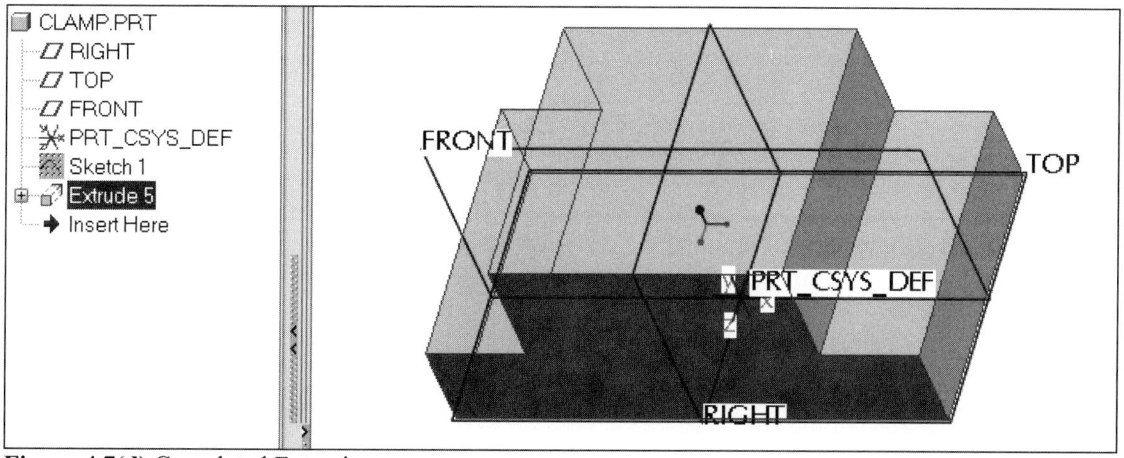

Figure 4.7(d) Completed Extrusion

Click: **Tools** ⇒ 

⇒ 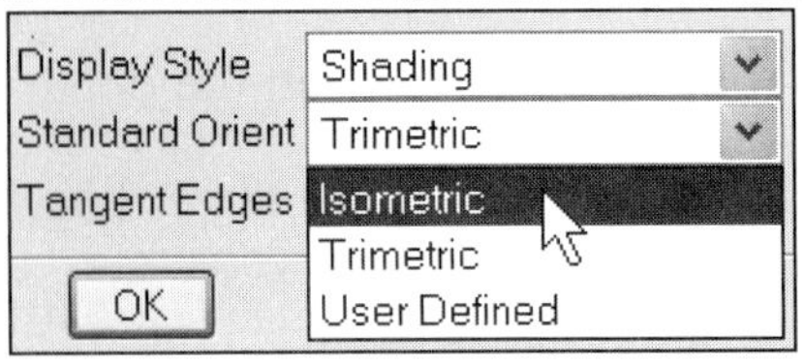

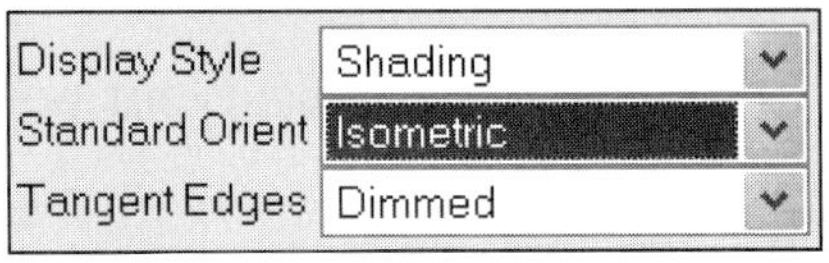

⇒ **Apply** ⇒ **OK** [Fig. 4.7(e)] ⇒  ⇒ **Standard Orientation** ⇒  ⇒  ⇒ **MMB** ⇒ **File** ⇒ **Delete** ⇒ **Old Versions** ⇒ **MMB** ⇒ **LMB** to deselect

Storing an object on the disk does not overwrite an existing object file. To preserve earlier versions, Pro/E saves the object to a new file with the same object name but with an updated version number. Every time you store an object using Save, you create a new version of the object in memory, and write the previous version to disk. Pro/E numbers each version of an object storage file consecutively (for example, box.sec.1, box.sec.2, box.sec.3). If you save 25 times, you have 25 versions of the object, all at different stages of completion. You can use *File ⇒ Delete ⇒ Old Versions* after the *Save* command to eliminate previous versions of the object that may have been stored.

When opening an existing object file, you can open any version that is saved. Although Pro/E automatically retrieves the latest saved version of an object, you can retrieve any previous version by entering the full file name with extension and version number (for example, **partname.prt.5**). If you do not know the specific version number, you can enter a number relative to the latest version. For example, to retrieve a part from two versions ago, enter **partname.prt.3** *(or **partname.prt.-2**)*.

You use *File ⇒ Erase* to remove the object and its associated objects from memory. If you close a window before erasing it, the object is still in memory. In this case, you use *File ⇒ Erase ⇒ Not Displayed* to remove the object and its associated objects from memory. This does not delete the object. It just removes it from active memory. *File ⇒ Delete ⇒ All Versions* removes the file from memory and from disk completely. You are prompted with a Delete All Confirm dialog box when choosing this command. Be careful not to delete needed files.

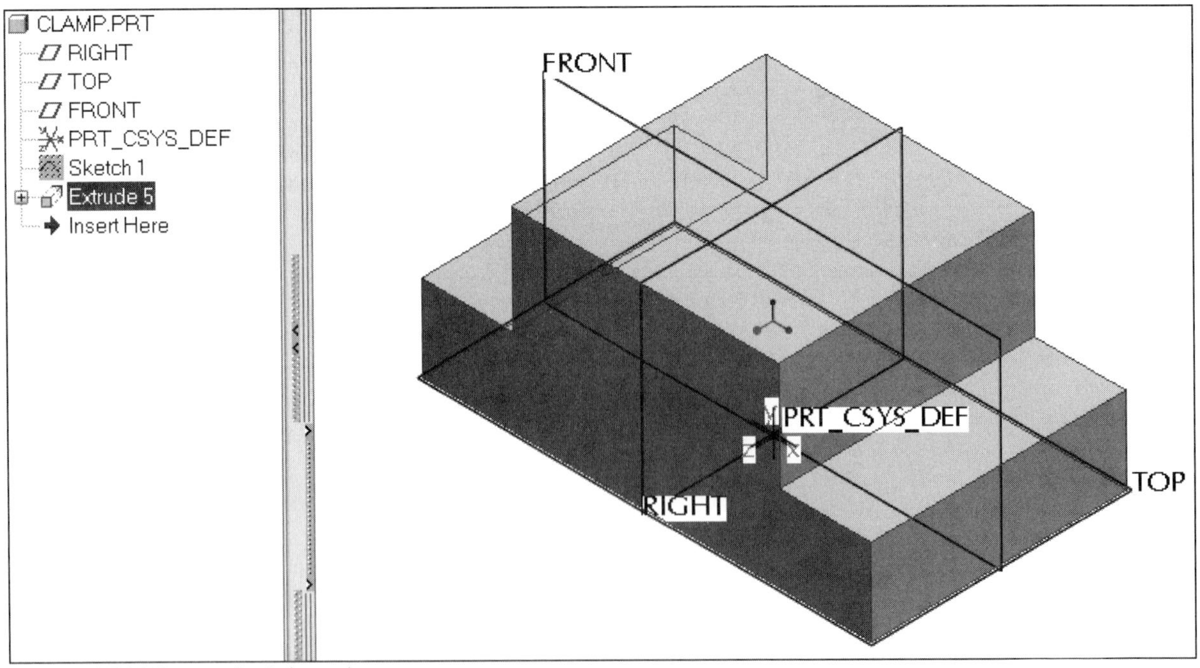

Figure 4.7(e) Isometric Orientation

Next, the cut through the middle of the part will be modeled.

Click: **Extrude Tool** ⇒ click on **Sketch 1** [Sketch 1] in the Model Tree [Fig. 4.8(a)] ⇒ [Placement] from the dashboard ⇒ [Unlink] [Fig. 4.8(b)] ⇒ **OK** [Fig. 4.8(c)] ⇒ [Edit...] [Figs. 4.8(d-e)]

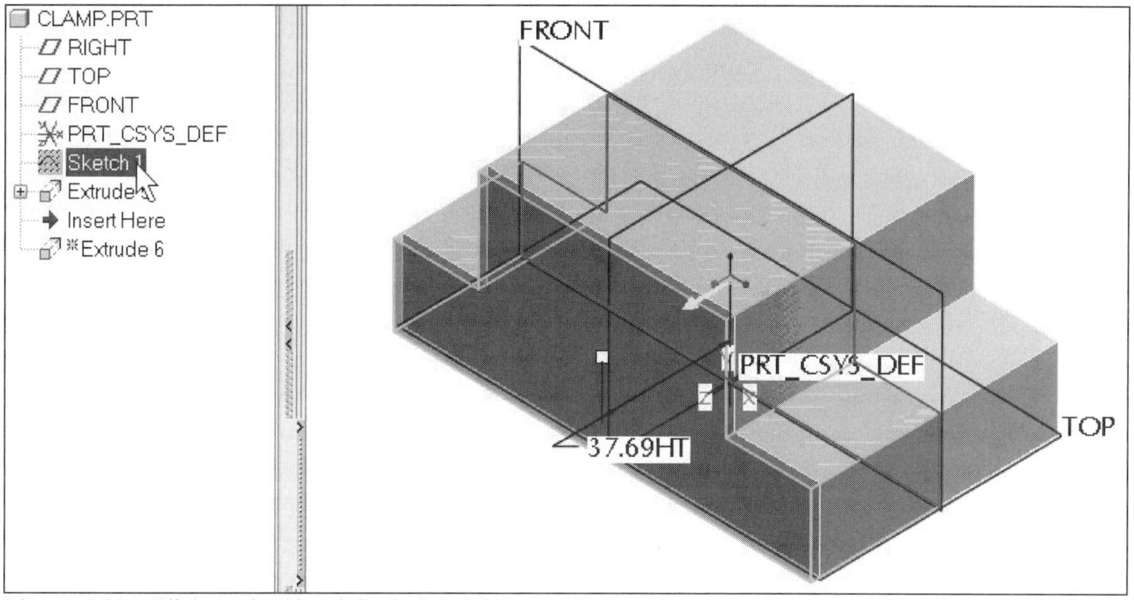

Figure 4.8(a) Click on the Sketch in the Model Tree

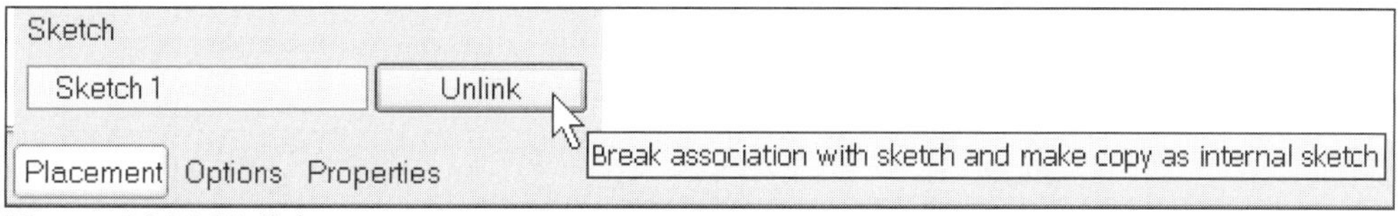

Figure 4.8(b) Unlink

Figure 4.8(c) Unlink Dialog Box

Figure 4.8(d) Edit the Internal Sketch

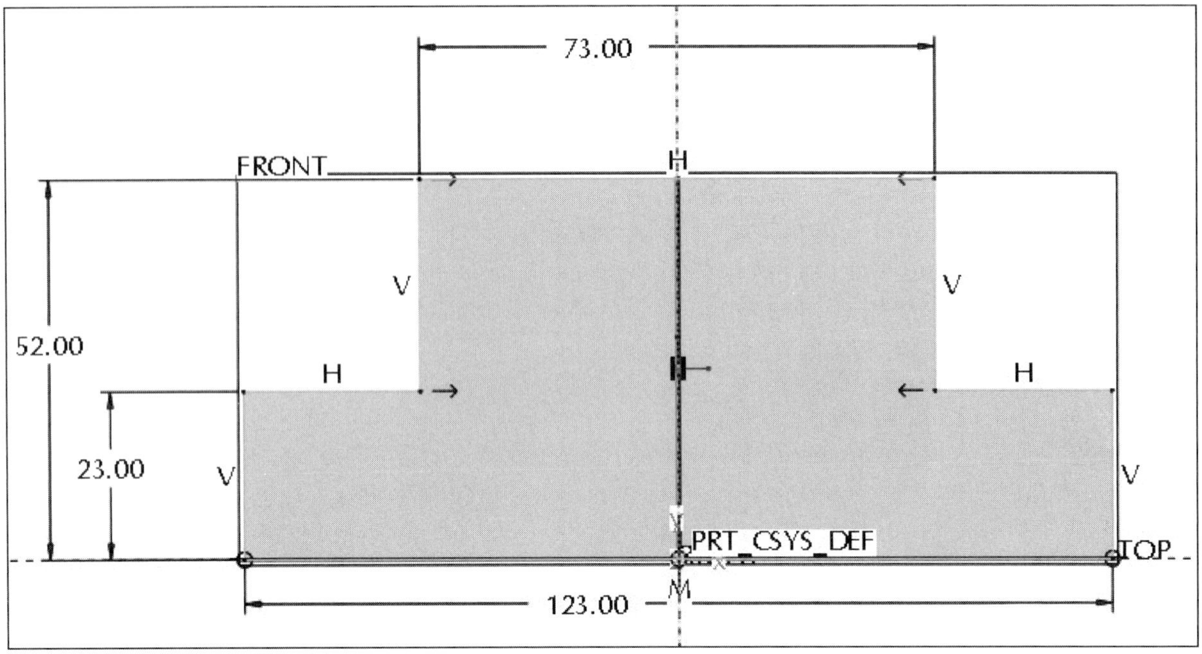

Figure 4.8(e) Outline of Sketch 1

Click: **Hidden line** ⇒ double-click on each value and modify to the design size [Fig. 4.8(f)]

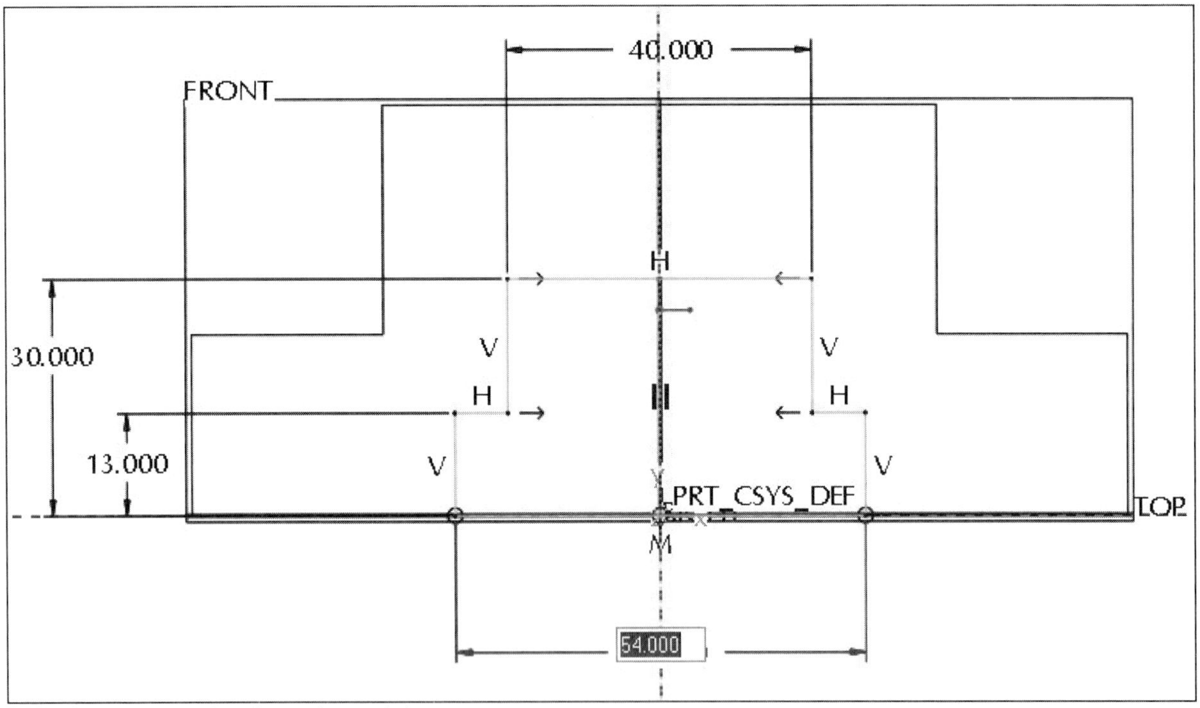

Figure 4.8(f) Modify Dimensions Dialog Box

Click: **Shading** ⇒ ⇒ **Standard Orientation**

Click: ☑ from the Right Toolchest ⇒ **RMB** ⇒ **Remove Material** ⇒ note the yellow direction arrow [Fig. 4.8(g)] ⇒ **Options** from the dashboard ⇒ **Side 1** ⇒ ⌄ ⇒ **Through All** ⇒ **Side 2** ⇒ ⌄ ⇒ **Through All** [Fig. 4.8(h)] ⇒ ☑ from dashboard ⇒ 💾 ⇒ **MMB** [Fig. 4.8(i)]⇒ **LMB**

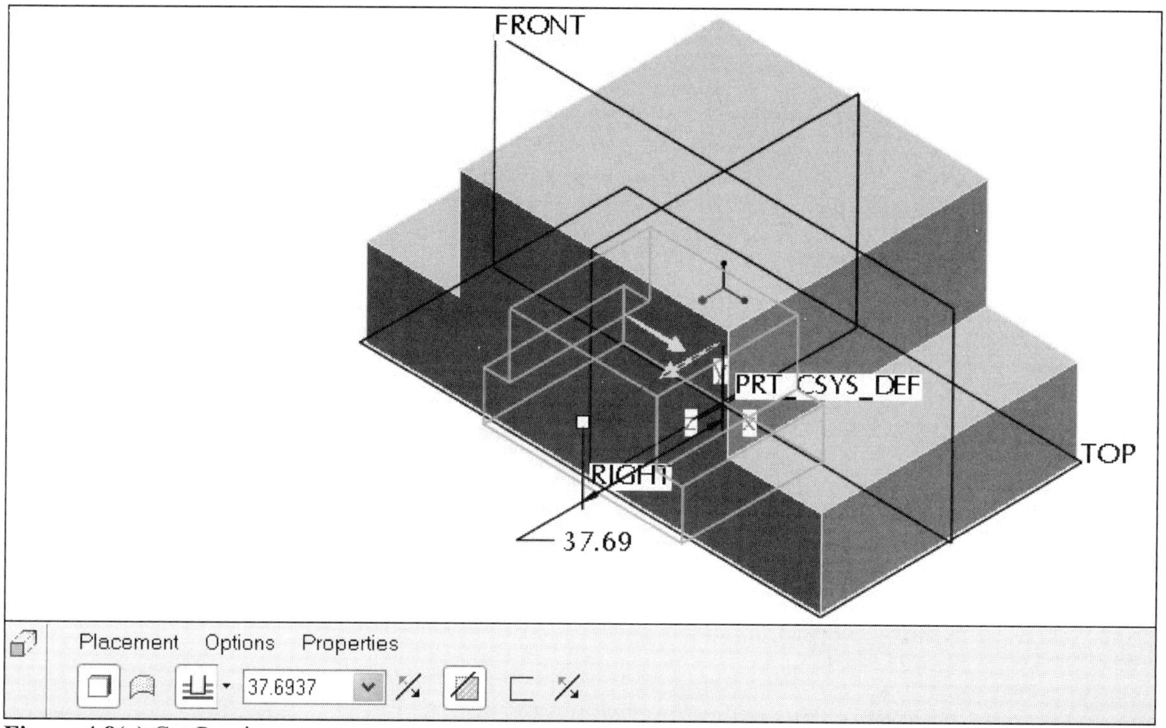

Figure 4.8(g) Cut Preview

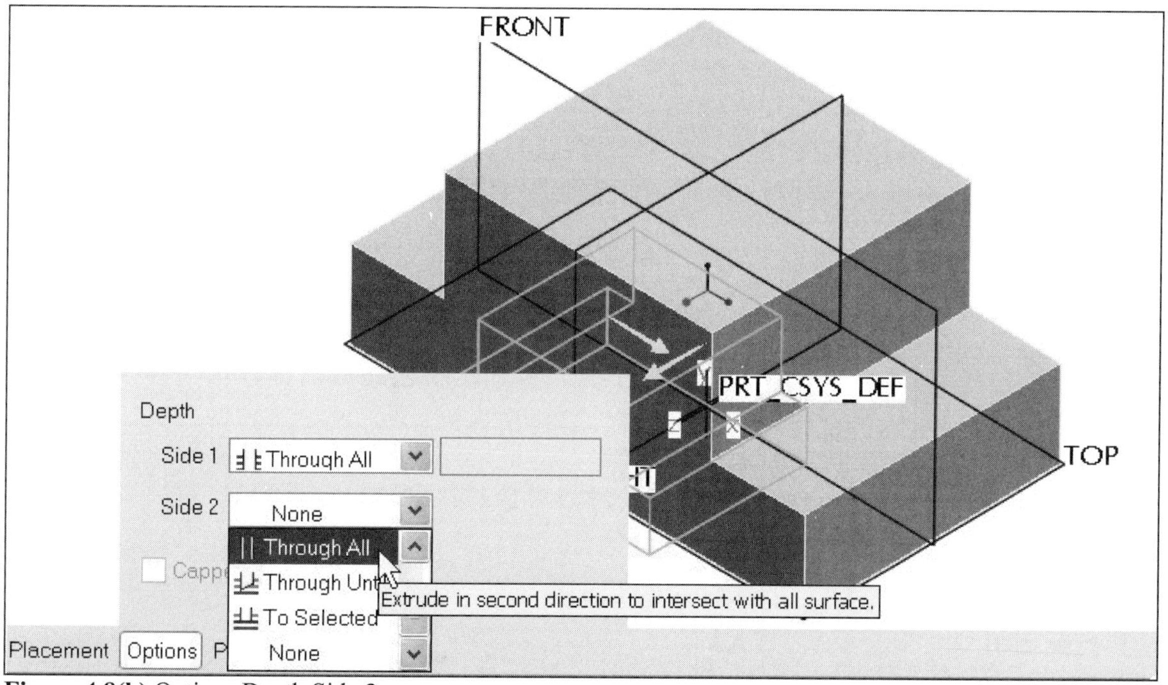

Figure 4.8(h) Options Depth Side 2

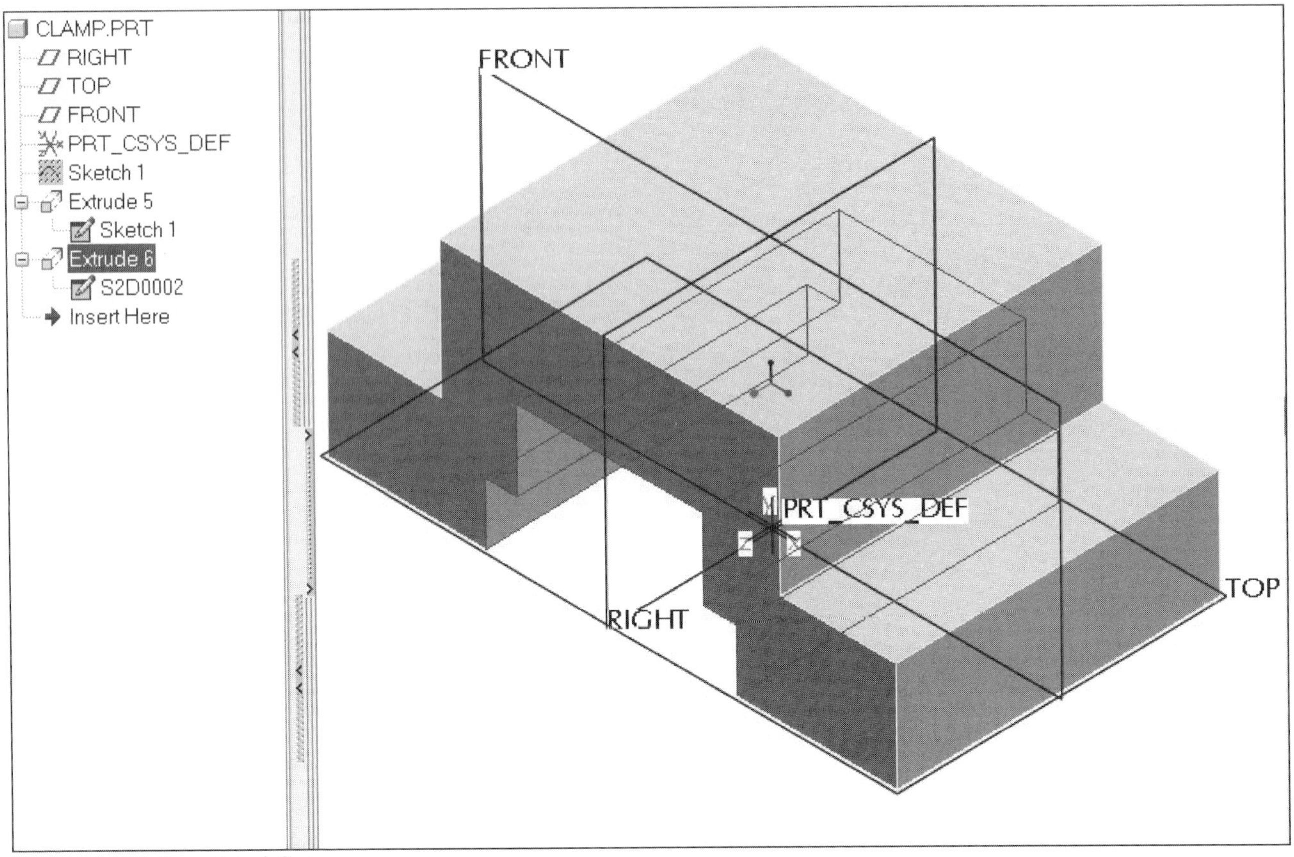

Figure 4.8(i) Completed Cut

The next feature will be a **20 X 20** centered cut (Fig. 4.9). Because the cut feature is identical on both sides of the part, you can mirror and copy the cut after it has been created.

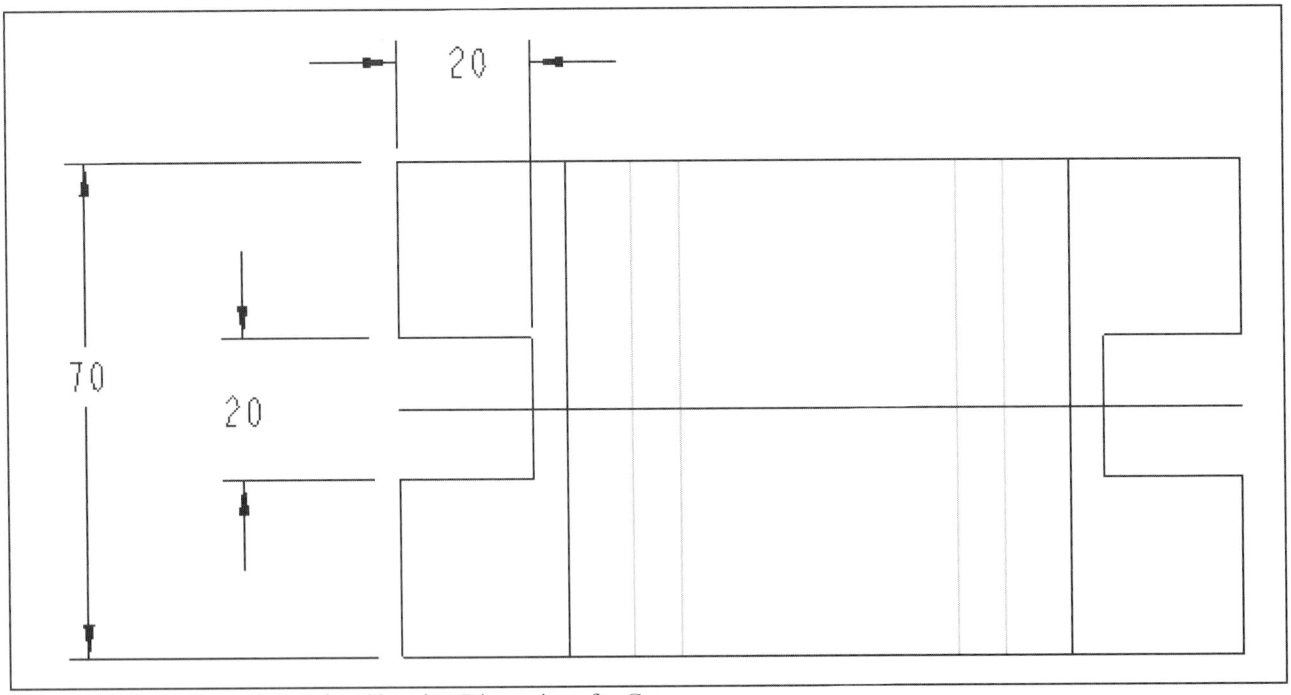

Figure 4.9 Top View of Drawing Showing Dimensions for Cut

Click: **Extrude Tool** ⇒ **Remove Material** ⇒ Placement from the dashboard [Fig. 4.10(a)] ⇒ Define... Sketch dialog box opens ⇒ Sketch Plane--- Plane: select **TOP** datum from the model as the sketch plane [Fig. 4.10(b)] ⇒ Sketch from the Sketch dialog box [Fig. 4.10(c)]

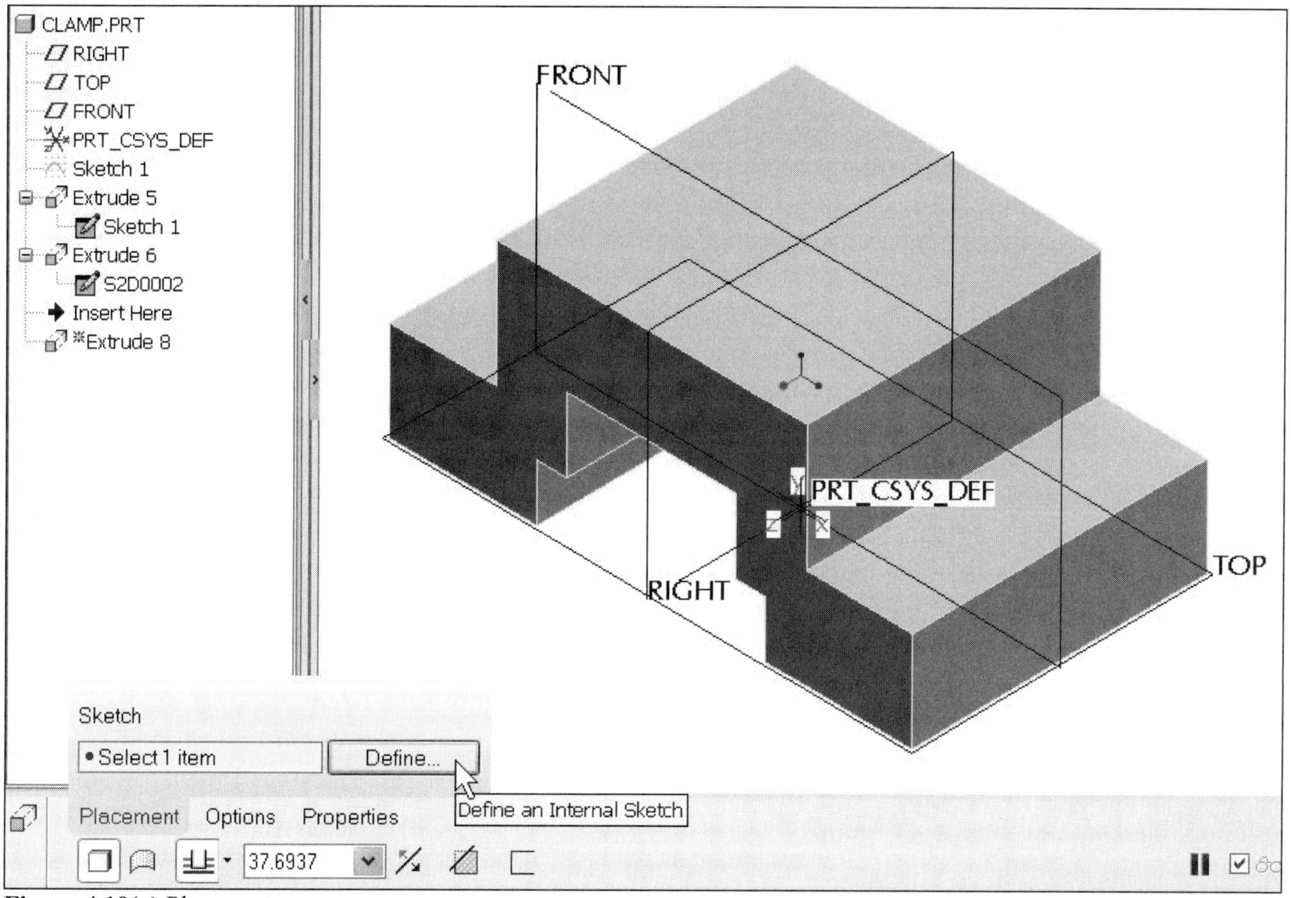

Figure 4.10(a) Placement

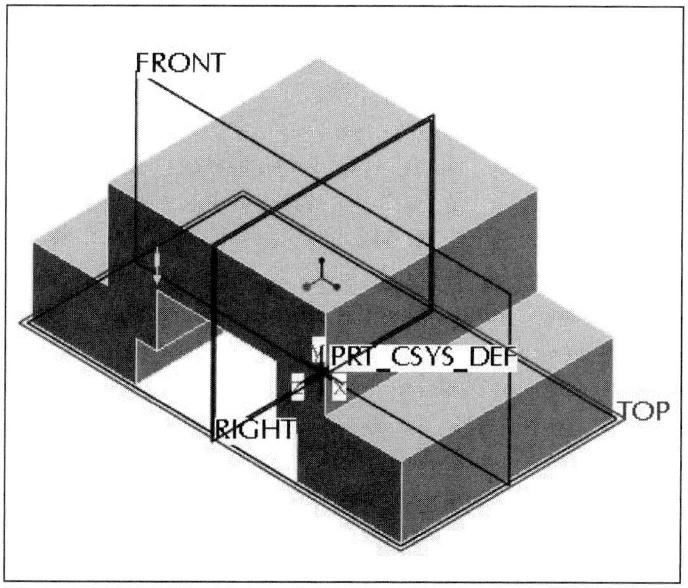

Figure 4.10(b) Top Datum Selected as Sketch Plane

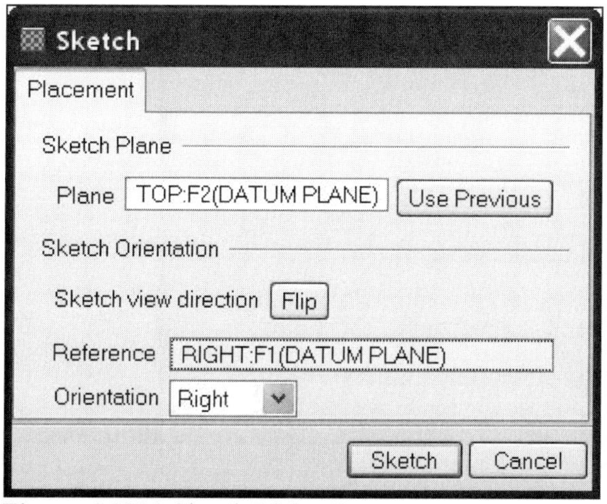

Figure 4.10(c) Sketch Dialog Box

Click: **Sketch** from the menu bar ⇒ **References** 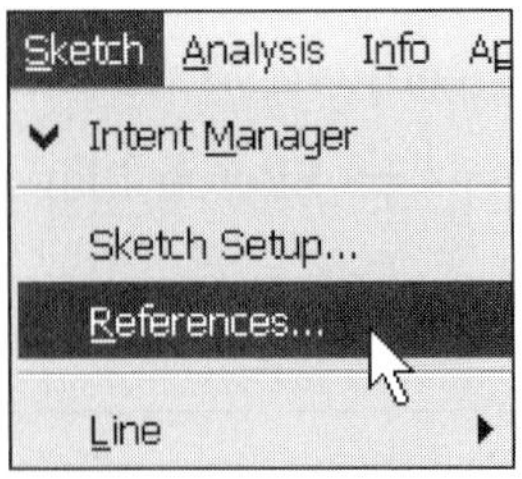

⇒ pick the left edge of the part [Fig. 4.10(d)] to add it to the References dialog box [Fig. 4.10(e)] ⇒ **Close** ⇒ **Tools** from Top Toolchest ⇒ **Environment** ⇒ 

⇒ **Apply** ⇒ **OK**

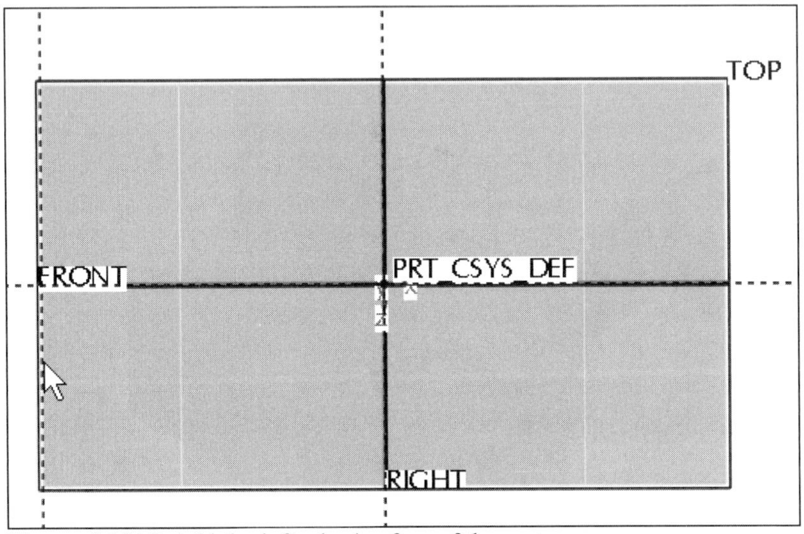

Figure 4.10(d) Add the left edge/surface of the part

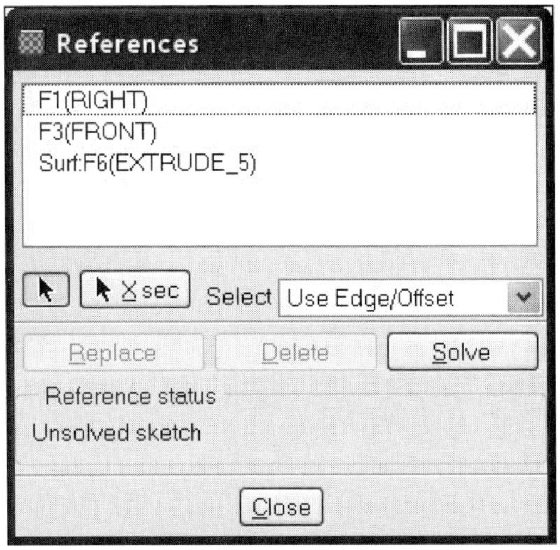

Figure 4.10(e) References Dialog Box

Click:  **Hidden line** ⇒ **RMB** ⇒ **Centerline** ⇒ create a horizontal centerline through the center of the part by picking two positions along the FRONT datum plane [Fig. 4.10(f)] ⇒ **MMB**

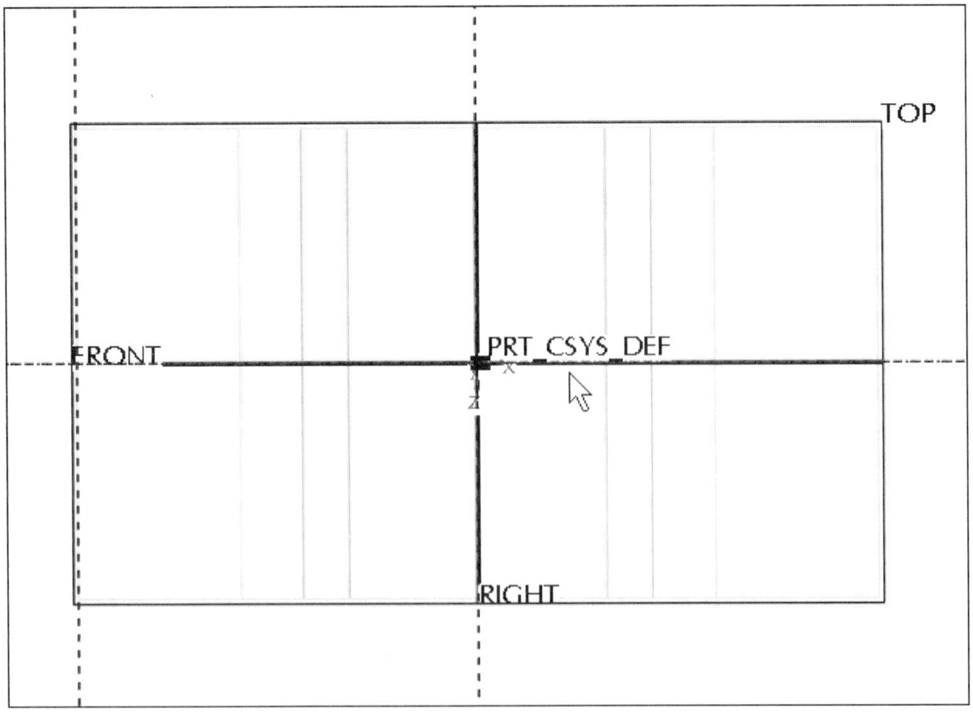

Figure 4.10(f) Horizontal Centerline

Click:  **Create two point lines** from the Right Toolchest ⇒ place the mouse on the left edge and create an open section with *three* lines [Fig. 4.10(g)] ⇒ **MMB** to end the line sequence ⇒ **MMB** to end the current tool [Fig. 4.10(h)] ⇒  **Impose sketcher constraints on the section** from the Right Toolchest [Fig. 4.10(i)] ⇒ 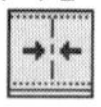 ⇒ pick the centerline [Fig. 4.10(j)] ⇒ pick a vertex (endpoint) [Fig. 4.10(k)] ⇒ pick a second vertex [Fig. 4.10(l)] to be symmetric ⇒ **Close** [Fig. 4.10(m)]

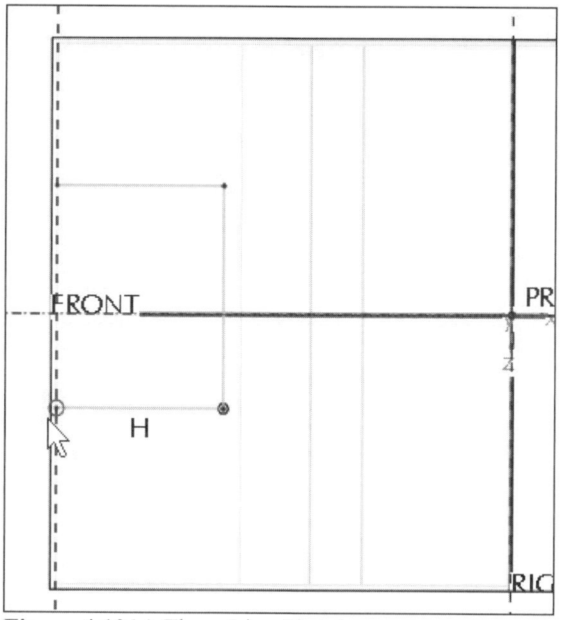

Figure 4.10(g) Three Line Sketch

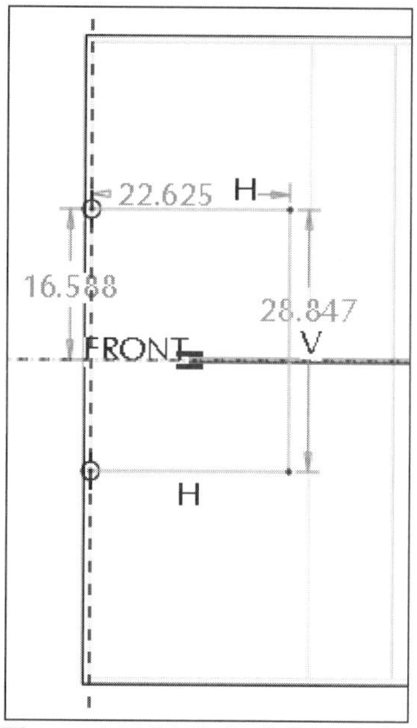

Figure 4.10(h) Default Dimension

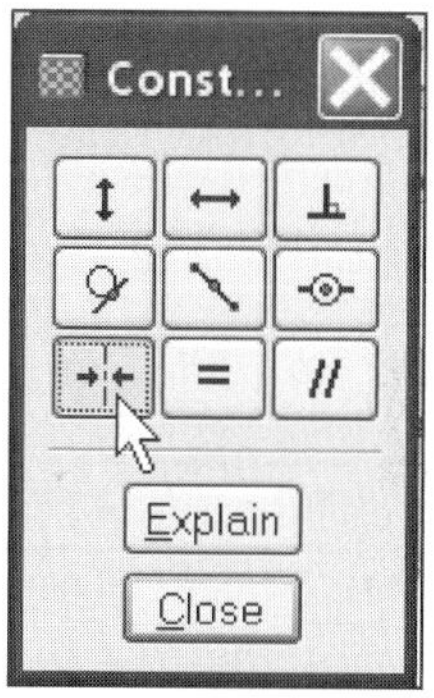

Figure 4.10(i) Constraints

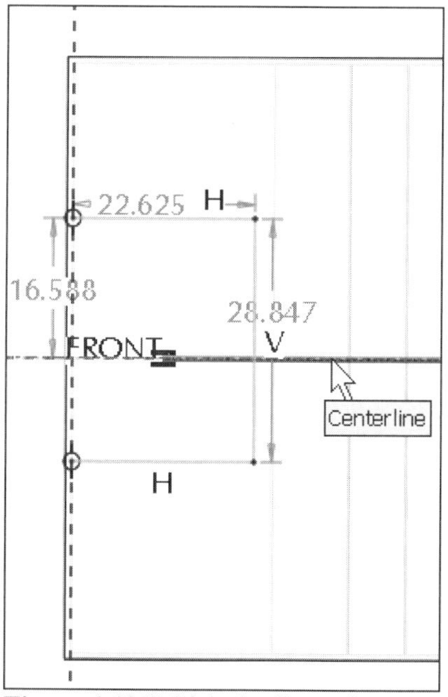

Figure 4.10(j) Pick the Centerline

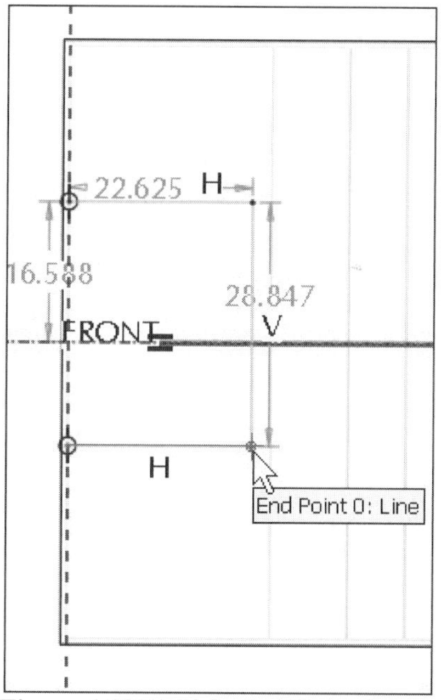

Figure 4.10(k) Pick First Endpoint

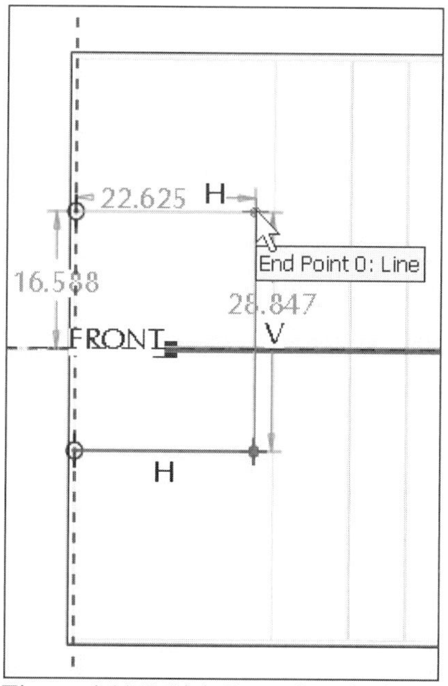

Figure 4.10(l) Pick Second Endpoint

Modify and reposition the values for the two dimensions (**20 X 20**) [Fig. 4.10(n)] ⇒ ⇒ ⇒ **Standard Orientation** ⇒ **Change depth direction** ⇒ **Options** tab ⇒ ⇒ Side 1 Through All [Fig. 4.10(o)]

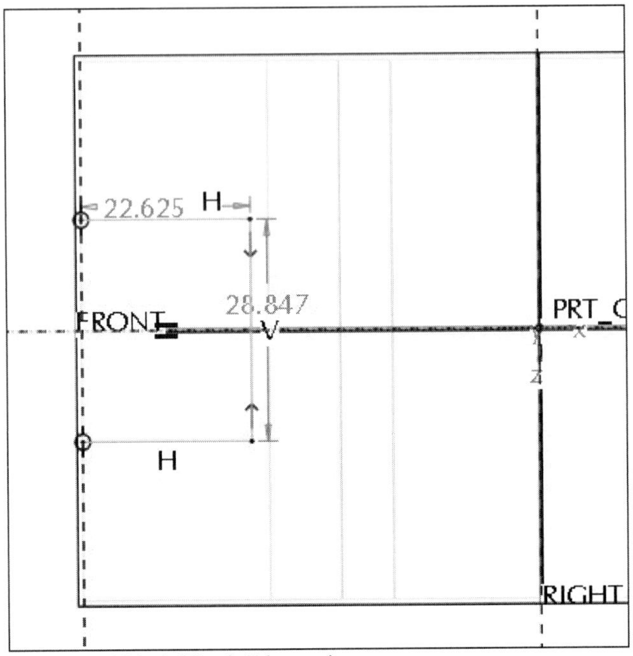

Figure 4.10(m) Weak Dimensions

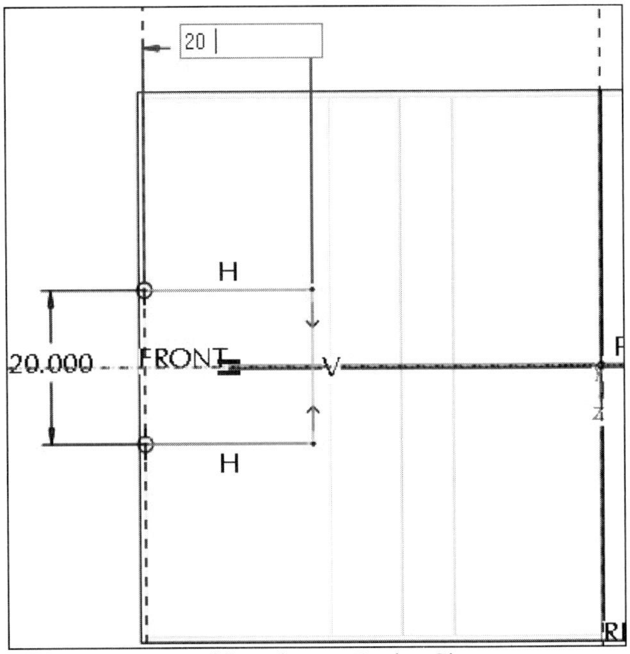

Figure 4.10(n) Modify Values to Design Sizes

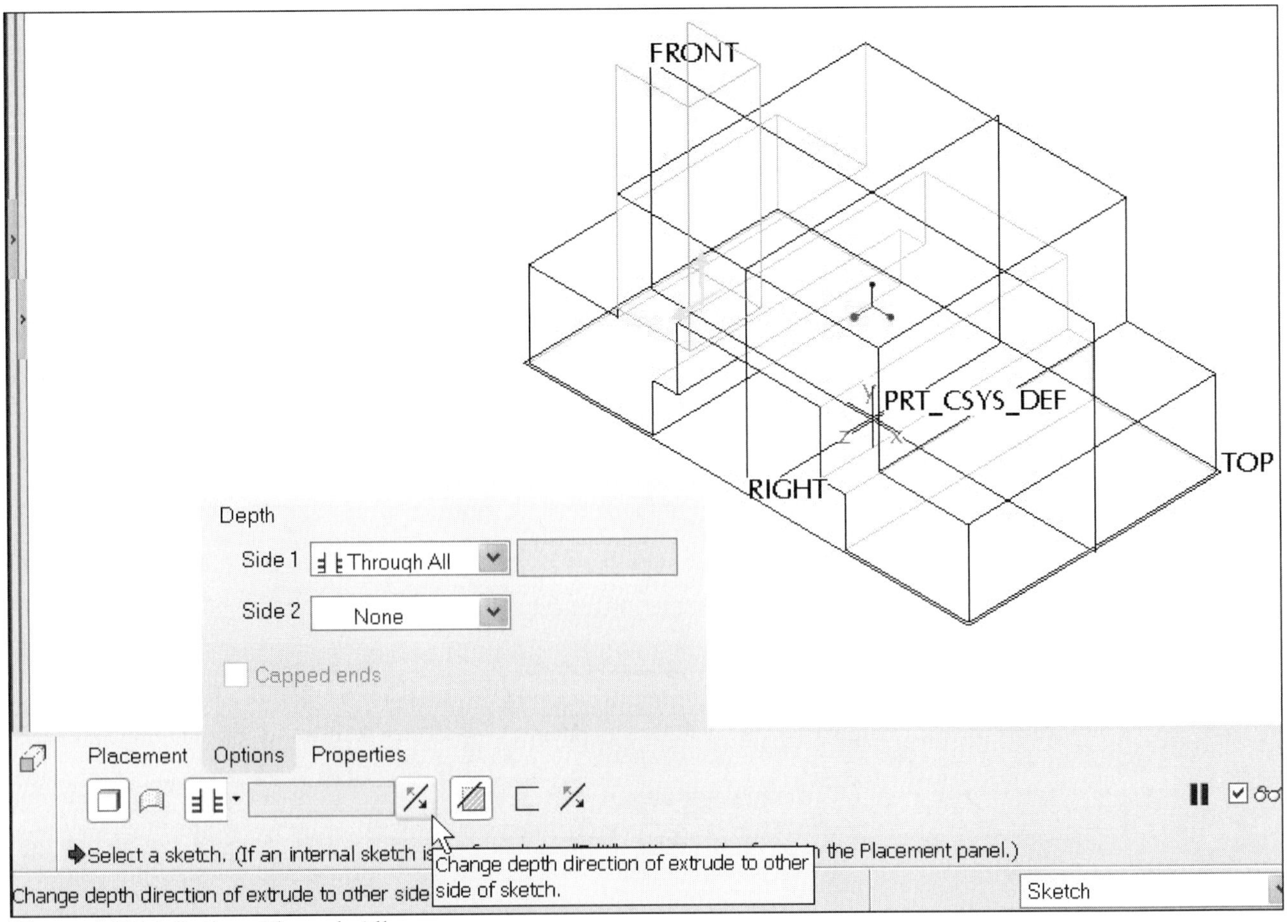

Figure 4.10(o) Options Through All

Click: from the dashboard ⇒ ⇒ **Shading** [Fig. 4.10(p)] ⇒ ⇒ **OK**

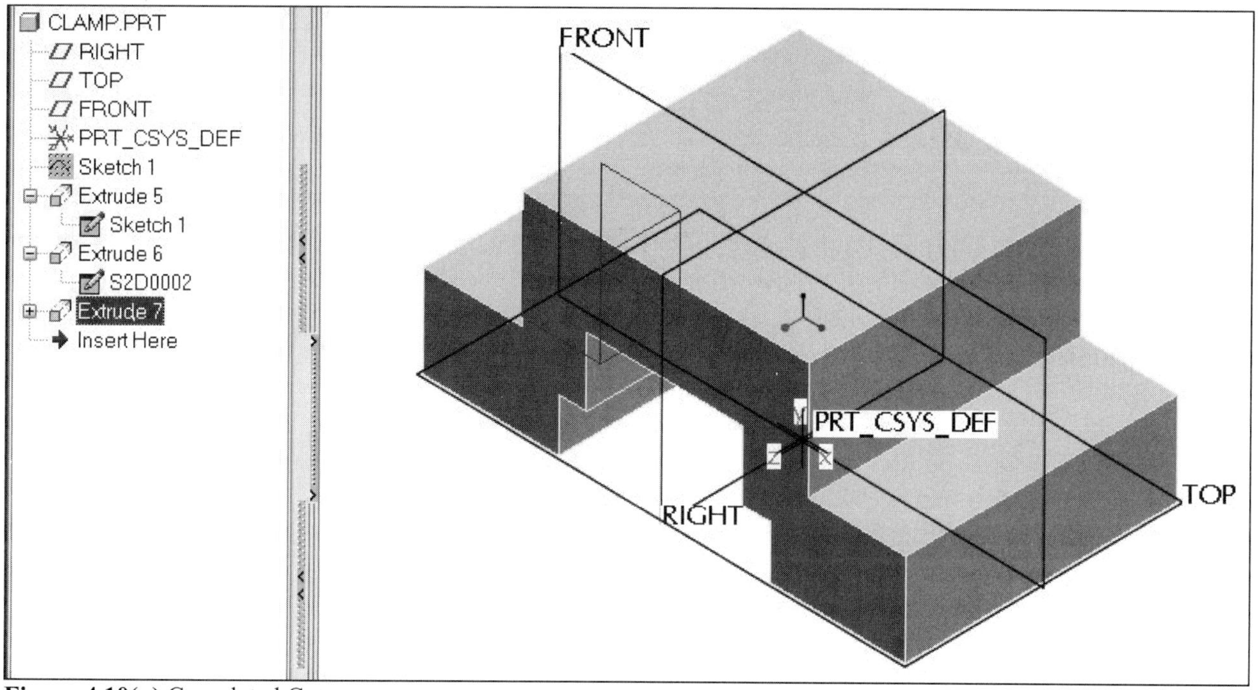

Figure 4.10(p) Completed Cut

With the cut still highlighted (selected), click: **Mirror Tool** from the Right Toolchest *(the extrude cut must be selected-highlighted for this tool to become active)* Mirror Tool ⇒ select the **RIGHT** datum plane from the model *or in the Model Tree* (Fig. 4.11) ⇒ or MMB ⇒ **File** ⇒ **Save** ⇒ **Enter**

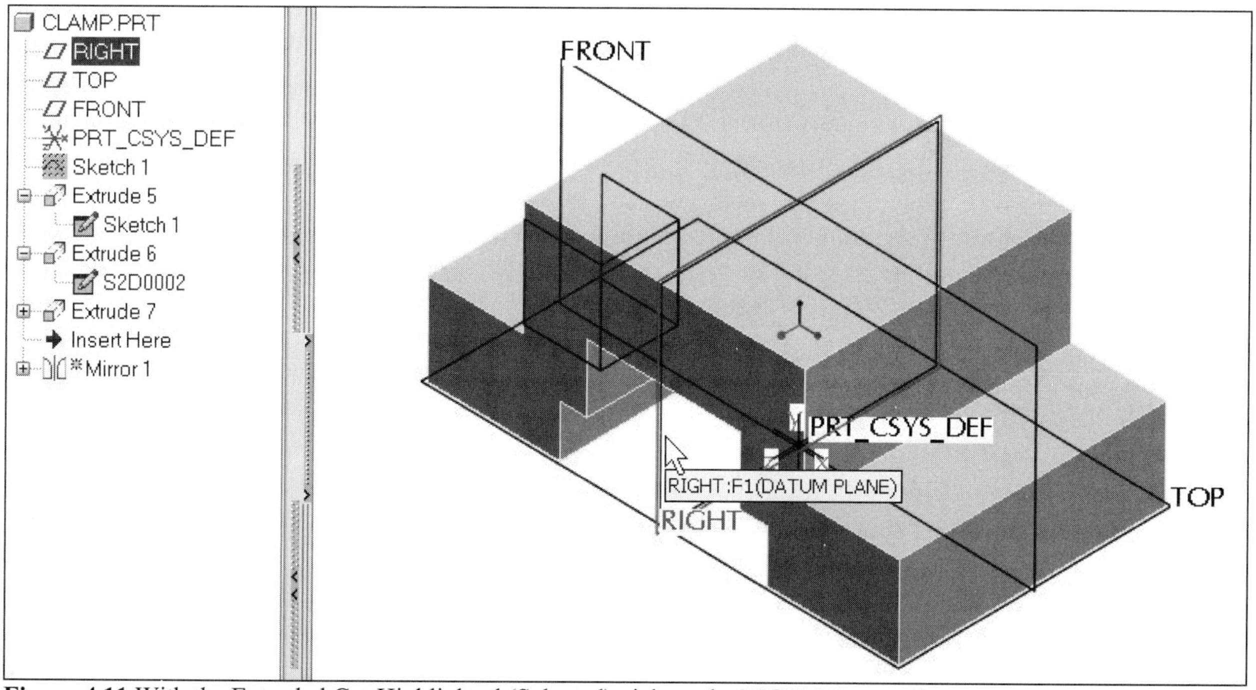

Figure 4.11 With the Extruded Cut Highlighted (Selected) pick on the RIGHT Datum Plane

Rotate the model [Fig. 4.12(a)] ⇒  ⇒ **Standard Orientation** [Fig. 4.12(b)] ⇒ **File** ⇒ **Exit** ⇒ **No**

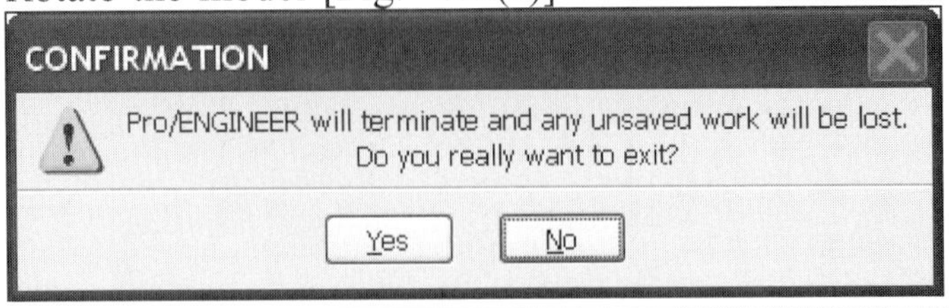

⇒ **File** ⇒ **Save** ⇒ **OK** ⇒ **File** ⇒ **Exit** ⇒ **Yes**

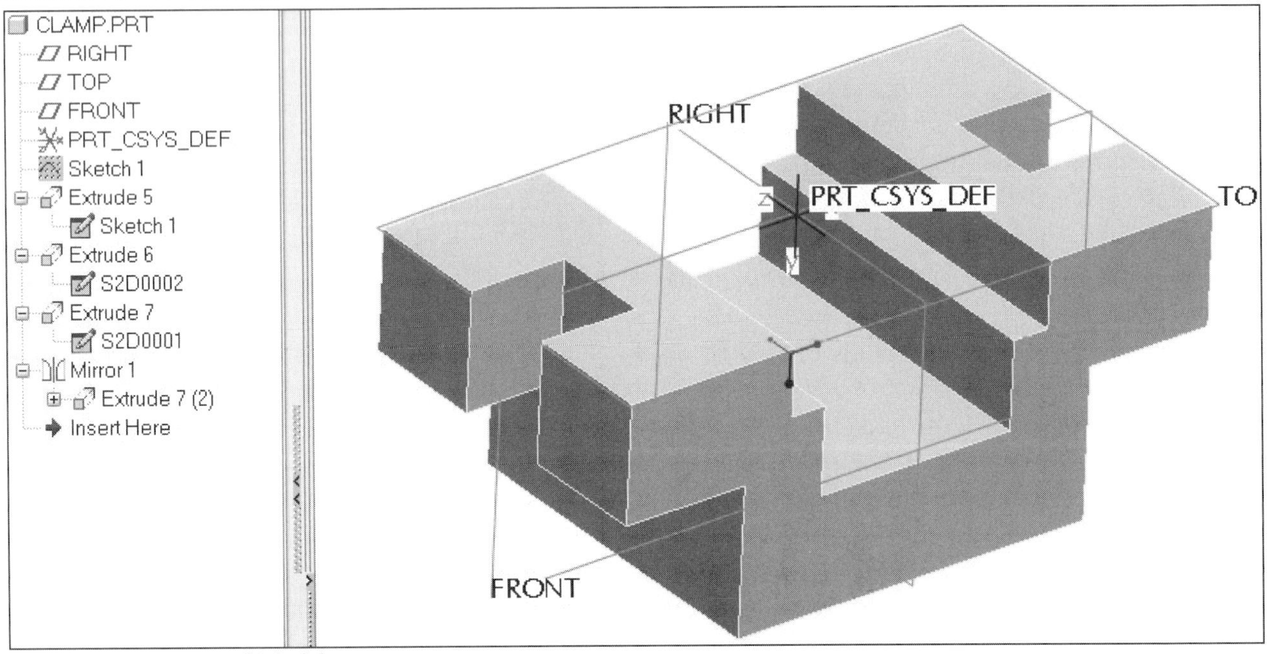

Figure 4.12(a) Rotated Model

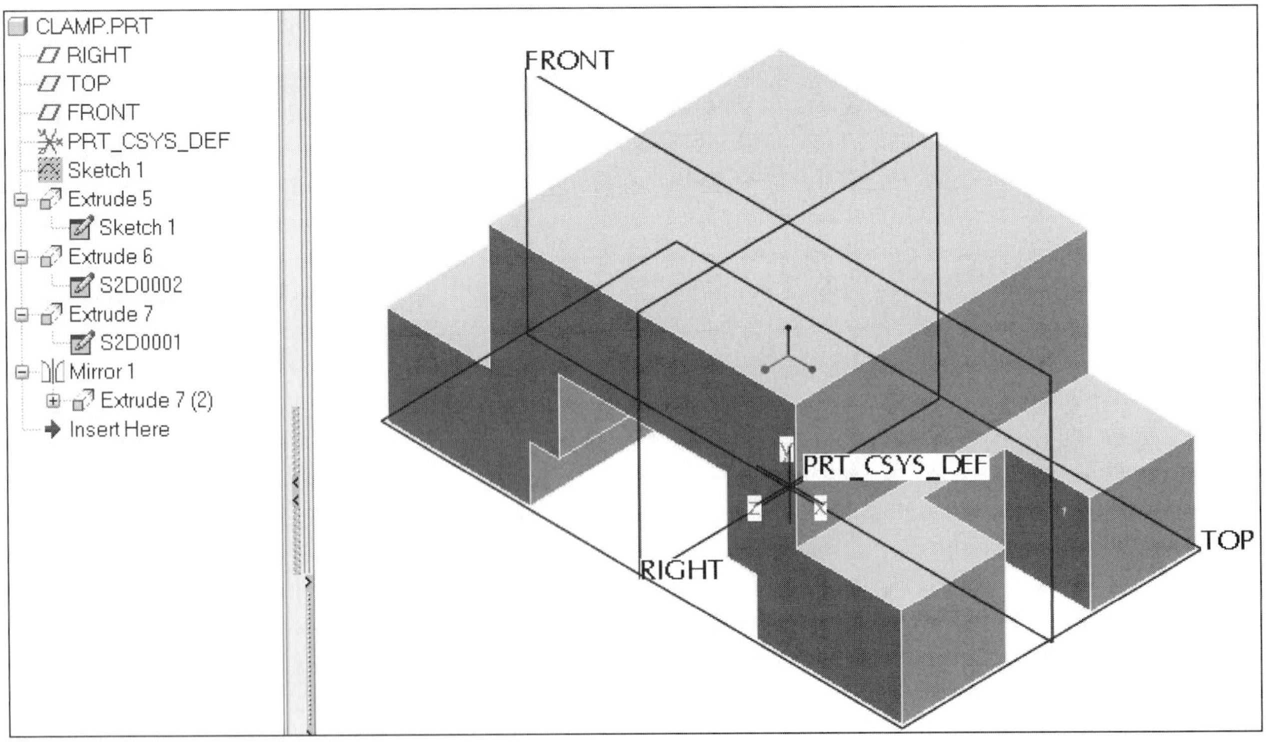

Figure 4.12(b) Completed Part

The lesson is now complete. If you wish to model a project without instructions, a complete set of projects and illustrations are available at www.cad-resources.com ⇒ Downloads.

Lesson 5 Datums, Layers, and Sections

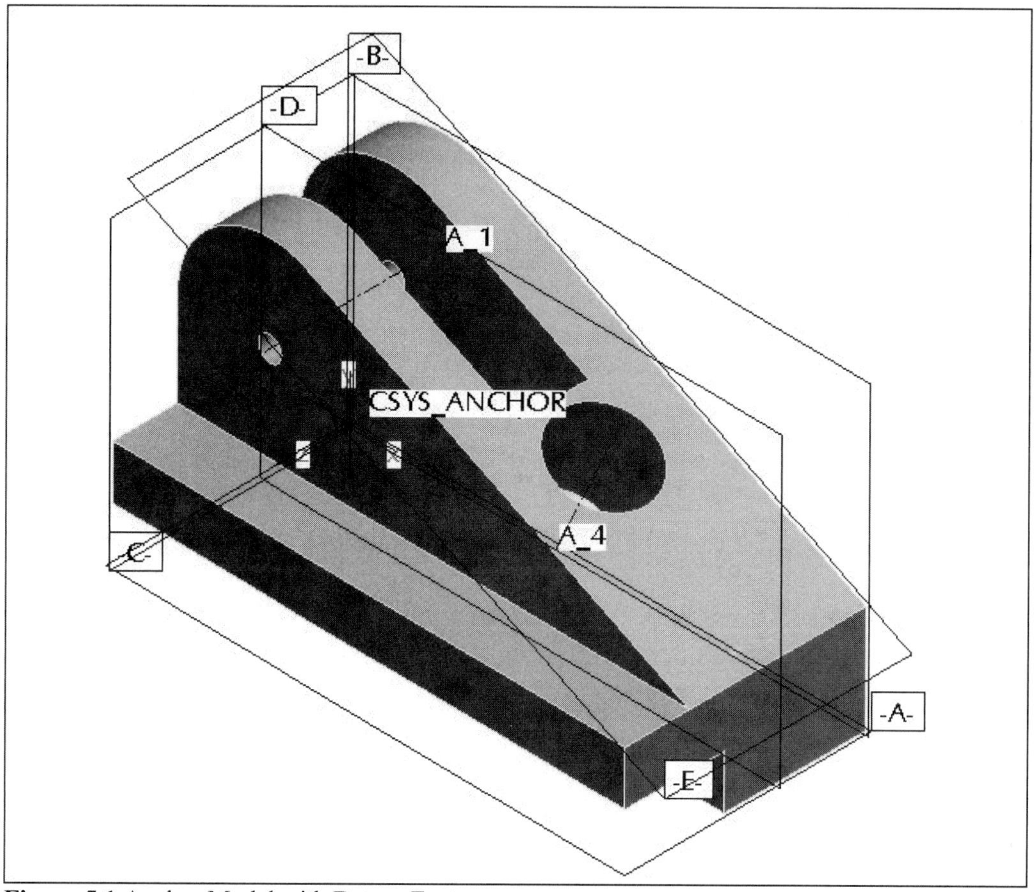

Figure 5.1 Anchor Model with Datum Features

OBJECTIVES

- Create **Datums** to locate features
- Set datum planes for **geometric tolerancing**
- Learn how to change the **Color** and **Shading** of models
- Use **Layers** to organize features
- Use datum planes to establish **Model Sectioning**
- Add a simple **Relation** to control a feature
- Use **Info** command to extract **Relations** information

DATUMS, LAYERS, and SECTIONS

Datums and **layers** are two of the most useful mechanisms for creating and organizing your design (Fig. 5.1). Features such as *datum planes* and *datum axes* are essential for the creation of all parts, assemblies, and drawings using Pro/E.

Layers are an essential tool for grouping items and performing operations on them, such as selecting, hide/unhide, plotting, and suppressing. Any number of layers can be created. User-defined names are available, so layer names can be easily recognized.

Most companies have a layering scheme that serves as a *default standard* so that all projects follow the same naming conventions and objects/items are easily located by anyone with access. Layer information, such as display status, is stored with each individual part, assembly, or drawing.

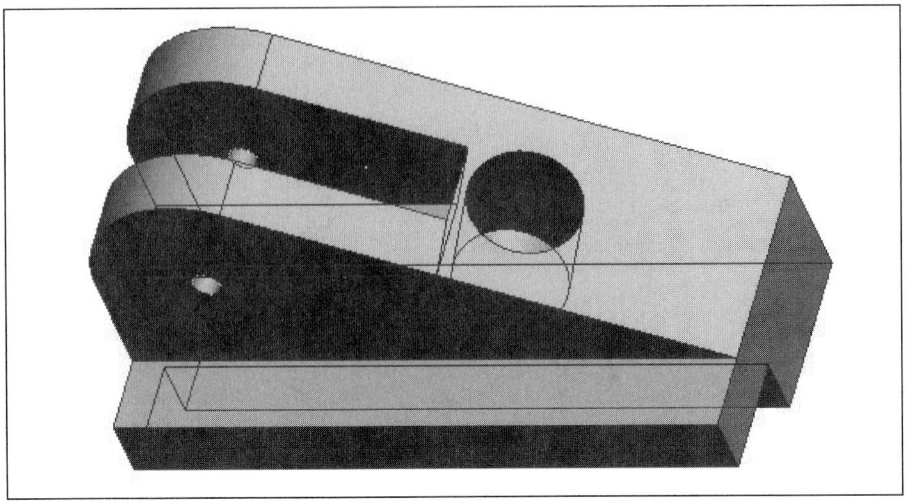

Figure 5.2 Anchor

Anchor

Though default datum planes have been sufficient in previous lessons, the Anchor (Fig. 5.2) incorporates the creation of user-defined datums and the assignment of datums to layers. The datum planes will be set as geometric tolerance features and put on a separate layer.

Launch **Pro/ENGINEER WILDFIRE 4.0** ⇒ **File** ⇒ **Set Working Directory** ⇒ navigate to your directory if necessary ⇒ **OK** ⇒ [icon] **Create a new object** ⇒ [Part] ⇒ Name **anchor** ⇒ [Use default template] ⇒ **OK** ⇒ **Edit** ⇒ **Setup** ⇒ **Units** ⇒ Units Manager **Inch lbm Second (Pro/E Default)** ⇒ **Close** ⇒ **Material** ⇒ **steel.mtl** ⇒ [icon] ⇒ **OK** ⇒ **Done** ⇒ select the coordinate system on the model or in the Model Tree-- **PRT_CSYS_DEF** ⇒ **RMB** [Fig. 5.3(a)] ⇒ **Rename** ⇒ type **CSYS_ANCHOR** in the Model Tree [Fig. 5.3(b)] ⇒ **Enter**

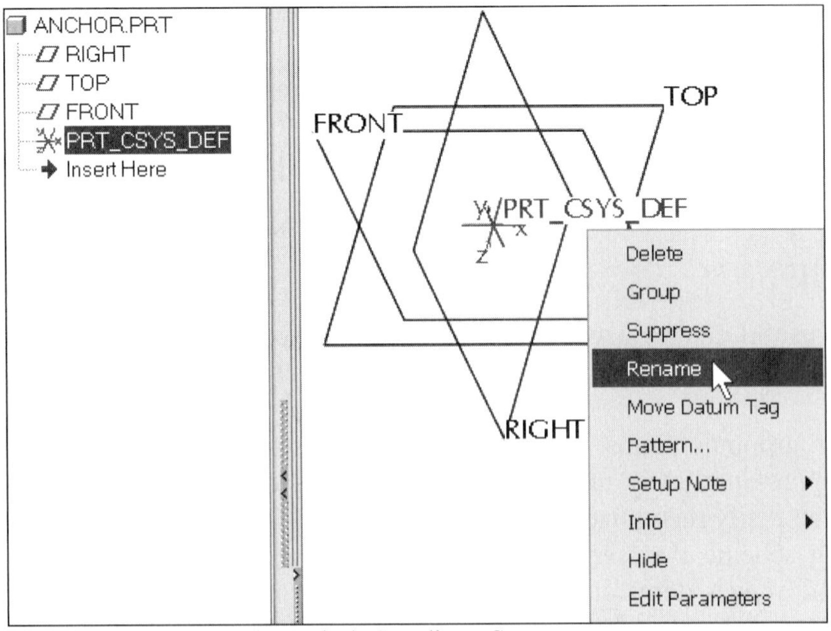

Figure 5.3(a) Rename the Default Coordinate System

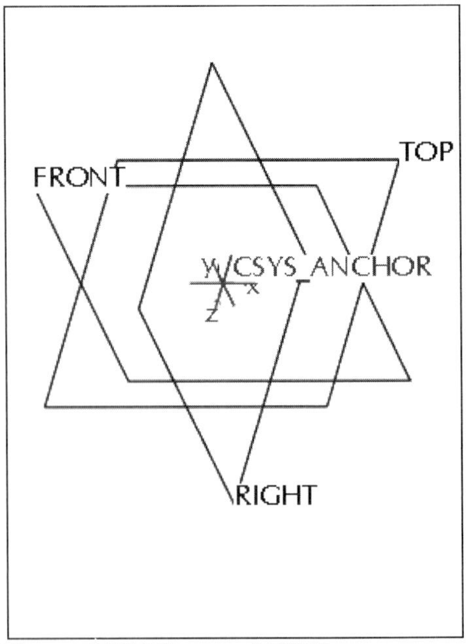

Figure 5.3(b) Renamed CSYS

It is considered good practice to rename the default coordinate system to something similar to the components name. If an assembly has 25 components, it will have 25 default coordinate systems. Renaming the coordinate system to the component name will make it easier to identify.

In the next set of steps, you will show the Layer Tree and see the default layering that automatically takes place as you model. The default datum planes and the default coordinate system are automatically placed on two layers each. For the datum planes; the *part default datum plane layer* and the *part all datum plane layer* [Fig 5.3(d)].

The coordinate system will also be layered in a similar fashion. The *part all datum layer* for the coordinate system displays the new name created for the coordinate system.

Click: **Show** in the Navigator ⇒ **Layer Tree** Layer Tree displays in place of the Model Tree [Fig. 5.3(c)] *(or click* *on)* ⇒ 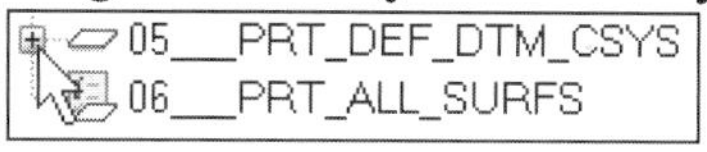

⇒ **Settings** ⇒ **Setup File** ⇒ **Save** [Fig. 5.3(d)] ⇒ **OK** ⇒ **Save** from the Top Toolchest ⇒ **OK** ⇒ **Show** in the Navigator ⇒ **Model Tree** *(or click* *off)*

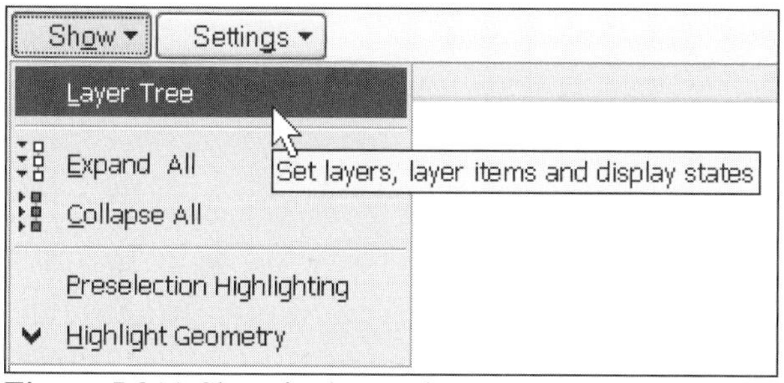

Figure 5.3(c) Show in the Navigator

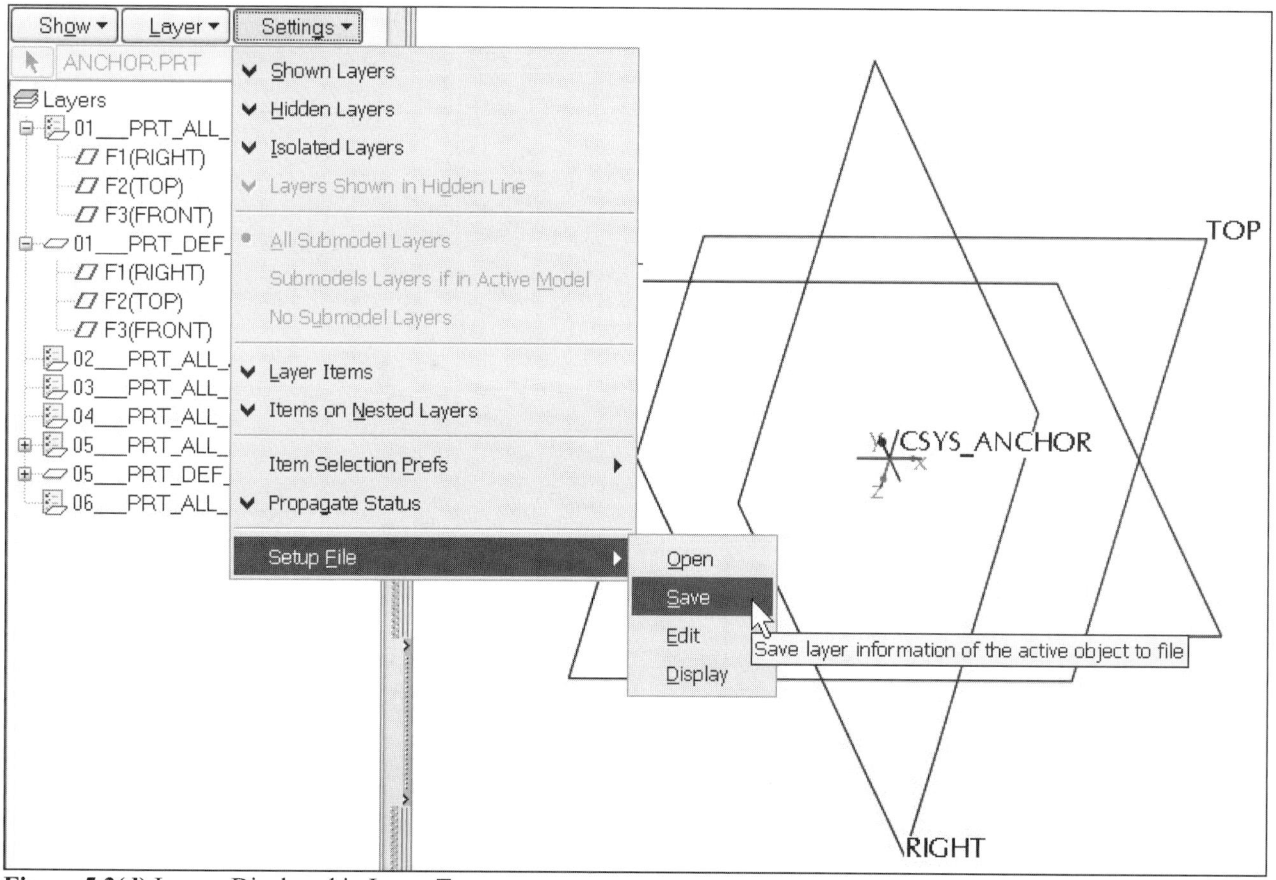

Figure 5.3(d) Layers Displayed in Layer Tree

As in previous lessons, the first protrusion will be sketched on FRONT datum plane. Use Figure 5.4 for the protrusion dimensions. Use only the **5.50**, **R1.00**, **1.125** (vertical), and **25°** dimensions.

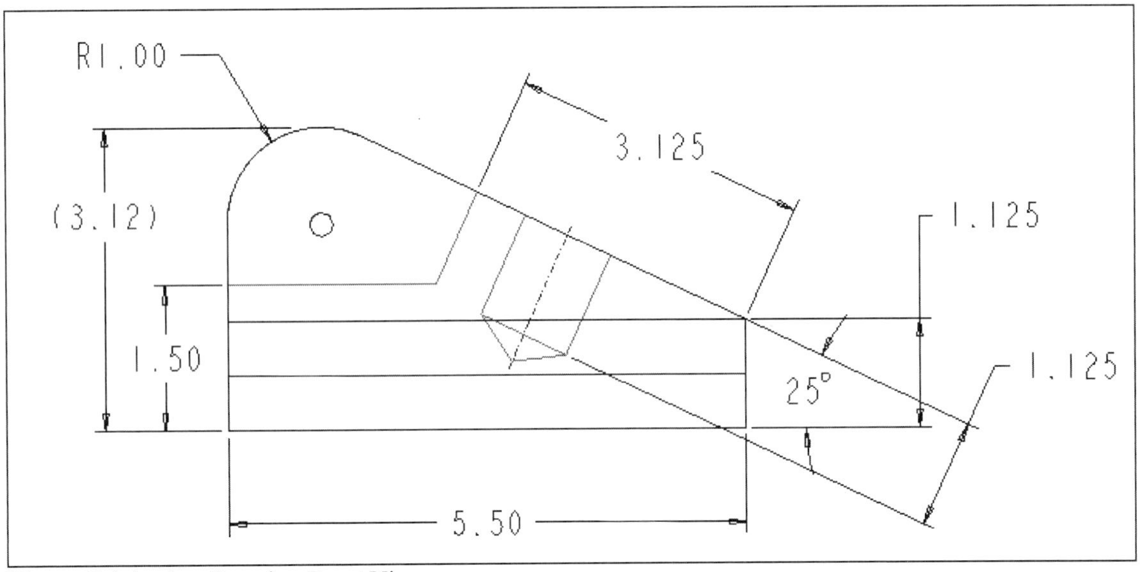

Figure 5.4 Anchor Drawing Front View

Click: **Extrude Tool** ⇒ **Placement** from dashboard ⇒ **Define** [Fig. 5.5(a)] ⇒ *in the Sketch Dialog box:* Sketch Plane--- Plane: select **FRONT** datum from the model or Model Tree ⇒ **Sketch** to close the Sketch dialog box ⇒ **Tools** ⇒ Environment ⇒ Snap to Grid ⇒ **OK** ⇒ **Toggle the grid** on ⇒ **RMB** anywhere in the graphics window ⇒ **Line** [Fig. 5.5(b)]

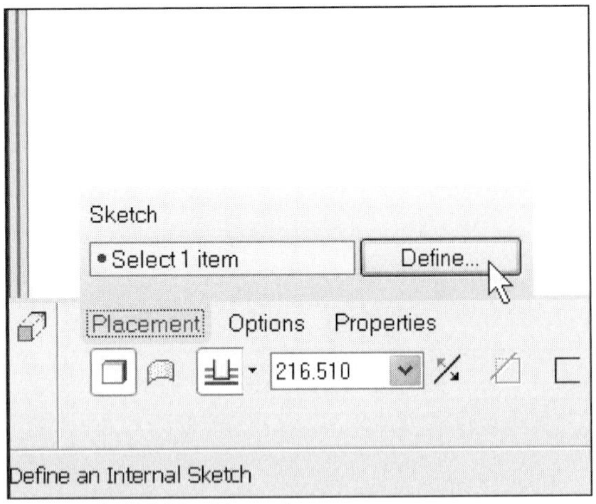

Figure 5.5(a) Dashboard

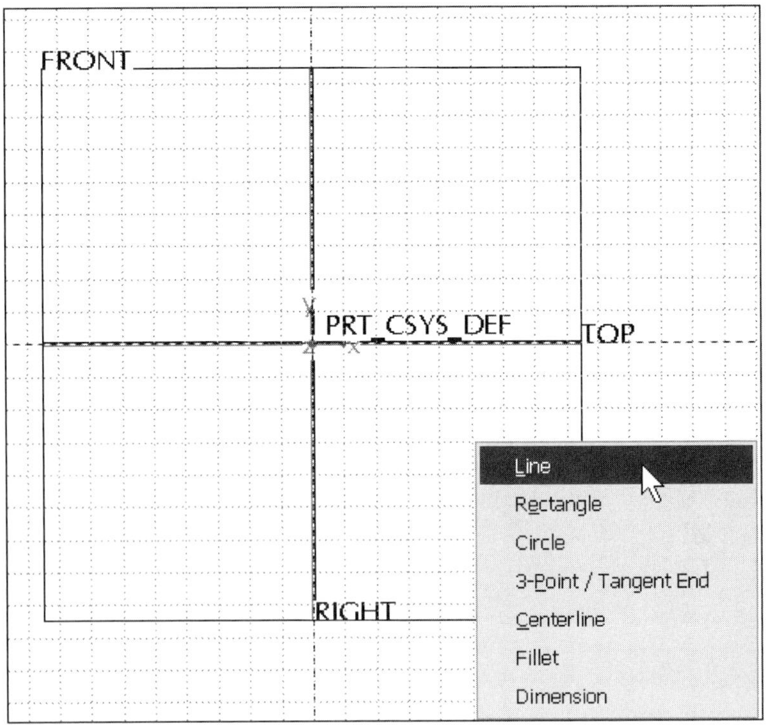

Figure 5.5(b) RMB ⇒ Line

Starting at the coordinate system, sketch the four lines ⇒ **MMB** ⇒ **MMB** ⇒ **RMB** anywhere in the graphics window ⇒ **Fillet** [Fig. 5.5(c)] ⇒ pick the left vertical line and the angled line to create a fillet [Fig. 5.5(d)] ⇒ **Tools** from menu bar ⇒ Environment ⇒ Snap to Grid toggle off ⇒ **OK** ⇒ from the Top Toolchest **Toggle the grid** off ⇒ **Ctrl+R** repaints the screen

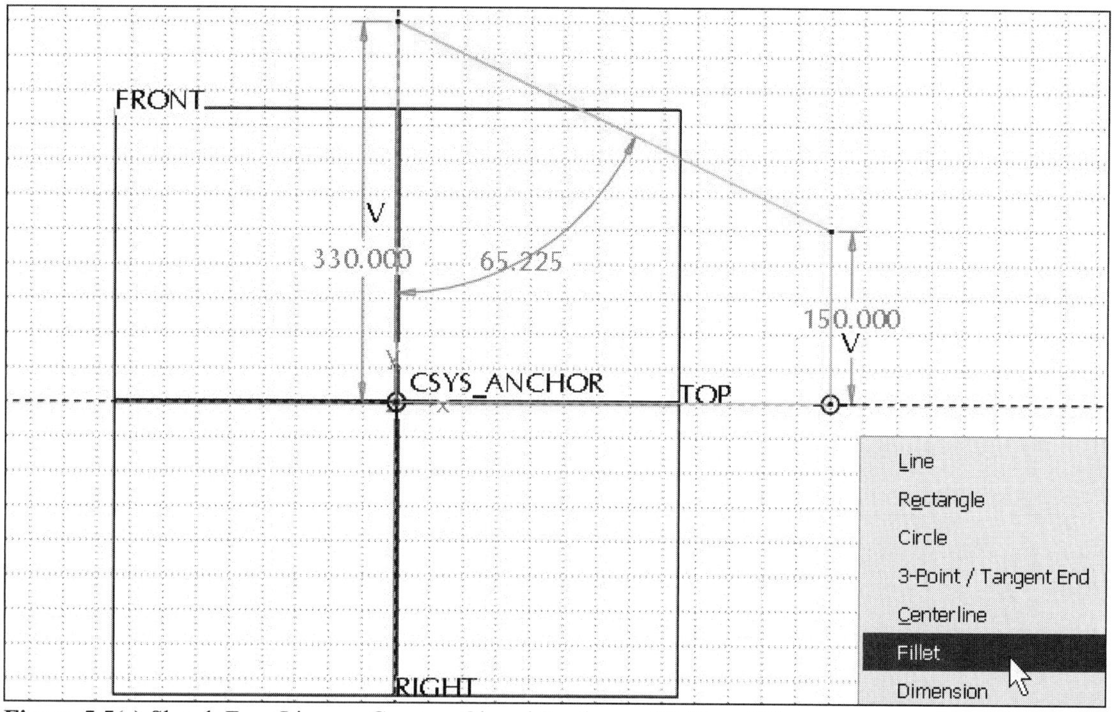

Figure 5.5(c) Sketch Four Lines to Create a Closed Section ⇒ RMB ⇒ Fillet

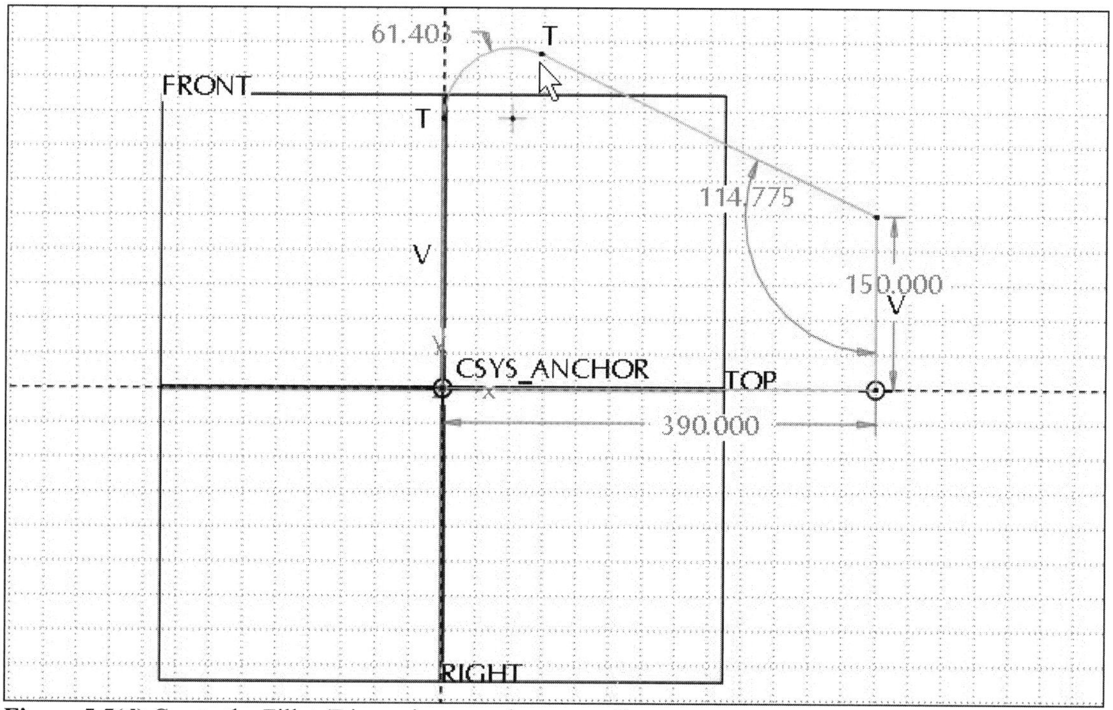

Figure 5.5(d) Create the Fillet (Dimensions are shown *weak*)

Click: **Create defining dimension** ⇒ pick on the horizontal line and then the angled line and click **MMB** to position the dimension value ⇒ **Select items** from the Right Toolchest ⇒ reposition the default dimensions [Fig. 5.5(e)] ⇒ Window-in dimensions *(or select all dimensions while pressing the Ctrl key)* ⇒ **Modify the values of dimensions** [Fig. 5.5(f)] ⇒ modify each dimension, hitting Enter after each edit ⇒ **Refit object** ⇒ **Regenerate the section and close the dialog** ⇒ **Continue** from Right Toolchest

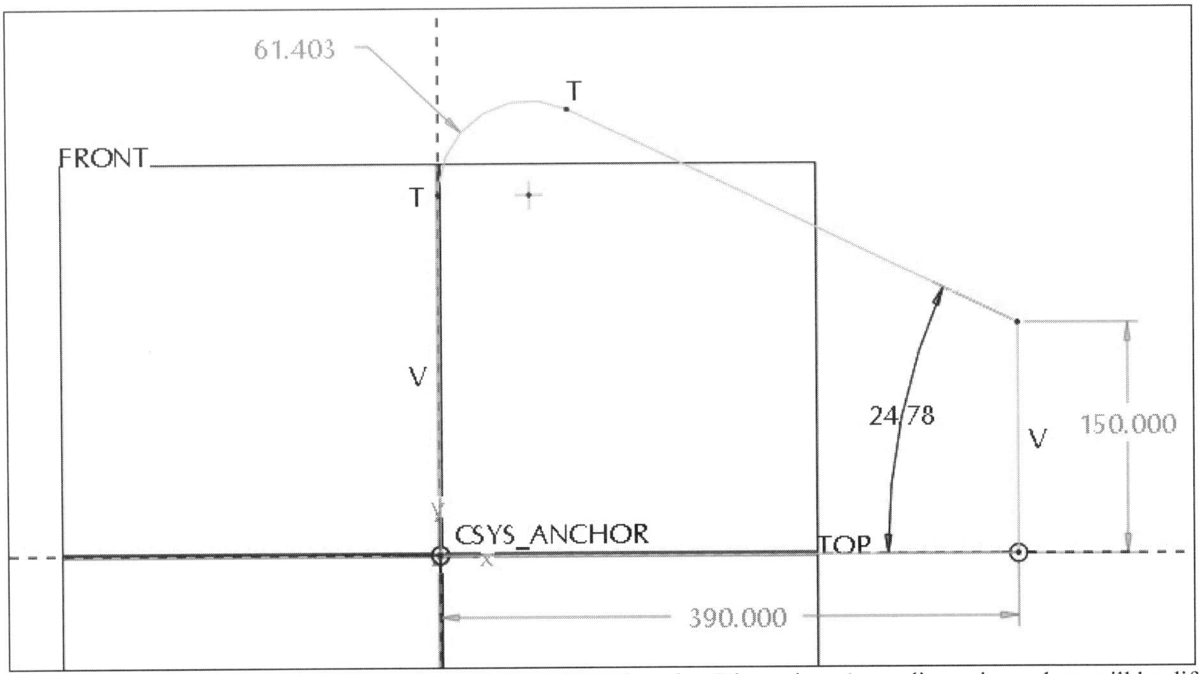

Figure 5.5(e) Create the Angle Dimension and Reposition the other Dimensions (your dimension values will be different)

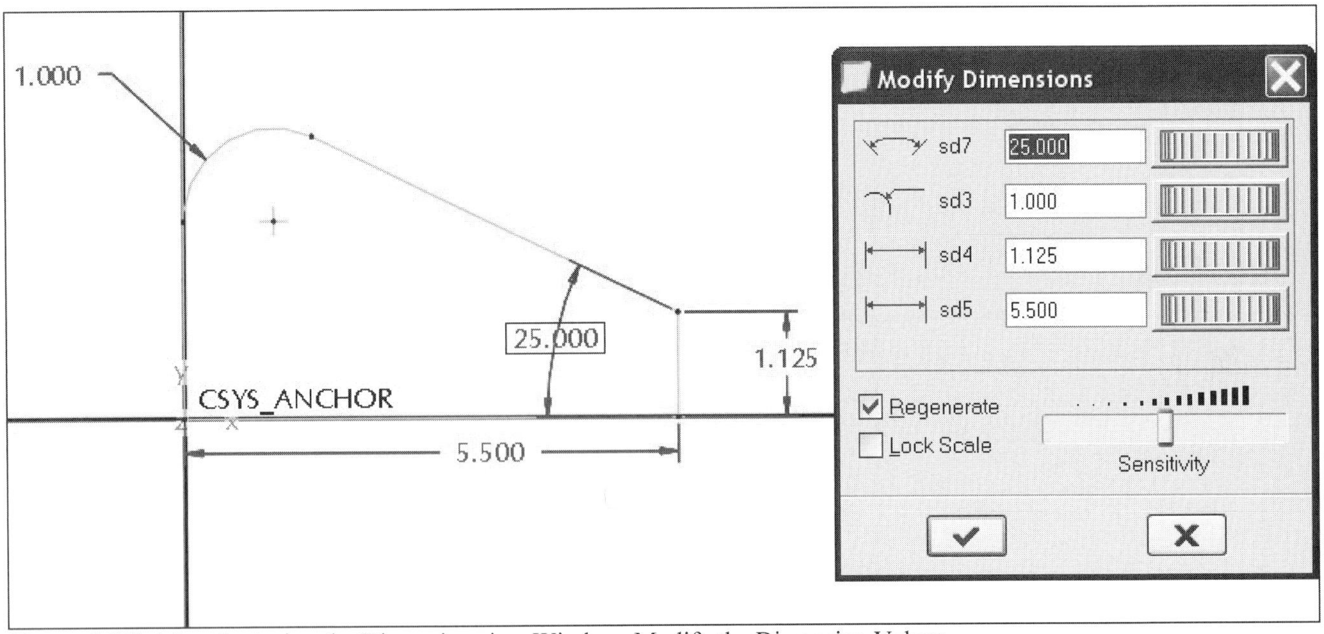

Figure 5.5(f) After Capturing the Dimensions in a Window, Modify the Dimension Values

Type: **Ctrl+D** Standard Orientation ⇒ in the dashboard, type: **2.5625** [Fig. 5.5(g)] ⇒ **Enter** ⇒ **Coordinate systems** off [Fig. 5.5(h)] ⇒ **Shading** from the Top Toolchest ⇒ in the dashboard [Fig. 5.5(i)] (if necessary, use your **MMB** to zoom in) ⇒ **Resumes the previously paused tool** *(or RMB ⇒ Exit Verify)* ⇒ *(or MMB, or Enter)* ⇒ **Ctrl+D** *(or Refit object)* ⇒ ⇒ **MMB**

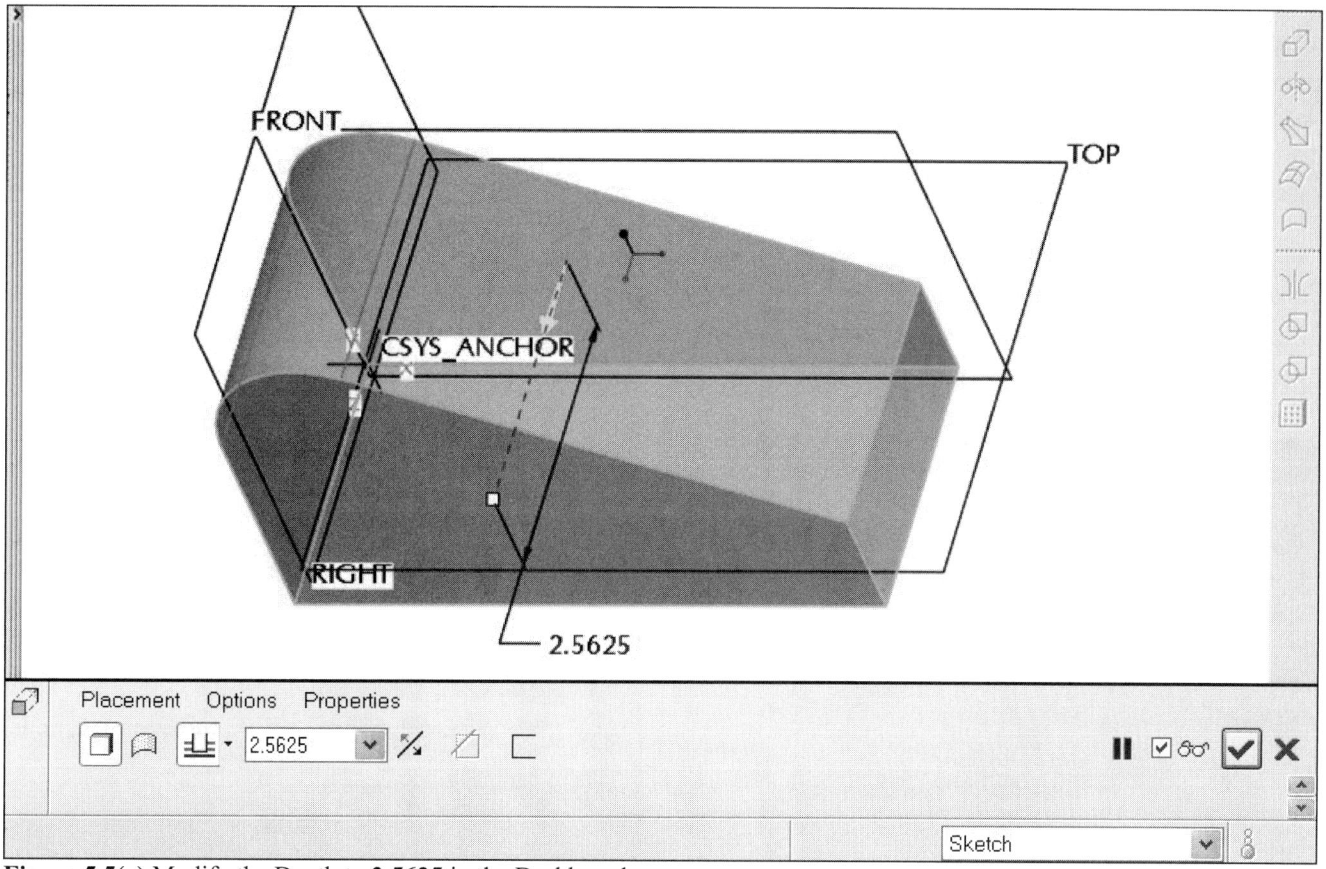

Figure 5.5(g) Modify the Depth to **2.5625** in the Dashboard

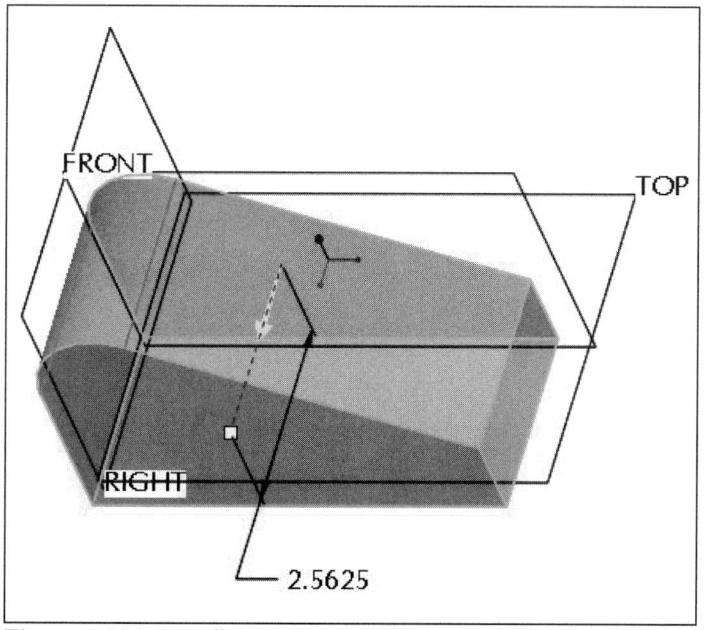

Figure 5.5(h) Coordinate systems off

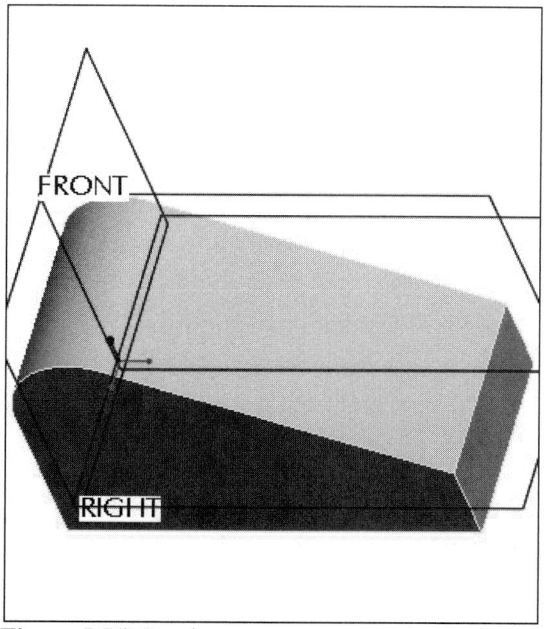

Figure 5.5(i) Preview

Colors

Colors are used to define material and light properties. Use the **Color Editor** to create and modify colors and to customize the display of geometric items. You can access the Color Editor by clicking a color swatch to set the **Appearance Color** or the **Highlight Color**. Avoid the colors that are used as Pro/E defaults. Because feature and entity highlighting is defaulted to *red*, colors similar to *red* should be avoided. Datum planes have *tan* and *gray* sides; therefore, they should be avoided. It is up to you to select colors that work with the type of project you are modeling. If the parts will be used in an assembly, each component should have a unique color scheme.

Click: **Coordinate systems** on ⇒ **View** from menu bar ⇒ Color and Appearance [Fig. 5.6(a)] ⇒ click: + in the Appearance Editor dialog box [Fig. 5.6(b)] ⇒ **Properties** [Fig. 5.6(c)] ⇒ **Basic** tab ⇒ in the Color region, click: Color [Fig. 5.6(d)] *(color swatch)* to open the Color Editor dialog box [Fig. 5.6(e)] ⇒ slide the RGB/HSV bars to create a new color ⇒ you can also use the **Color Wheel** [Fig. 5.6(f)] or the **Blending Palette** [Fig. 5.6(g)] to adjust your colors⇒ **Close** ⇒ **Apply** [Fig. 5.6(h)]

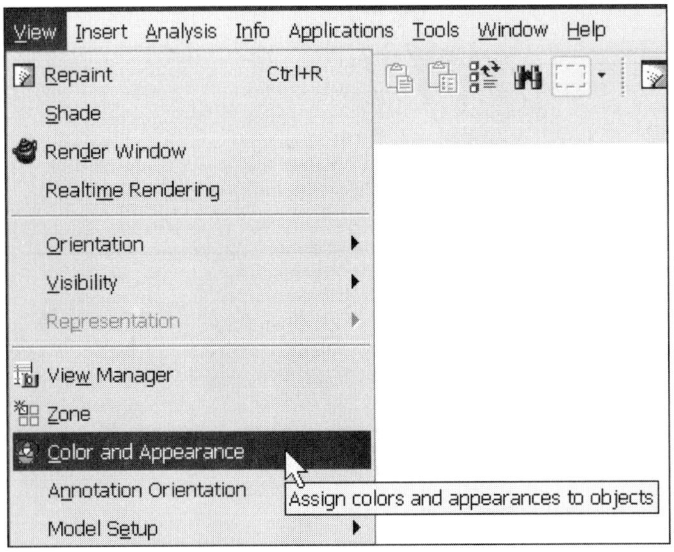

Figure 5.6(a) Color and Appearance

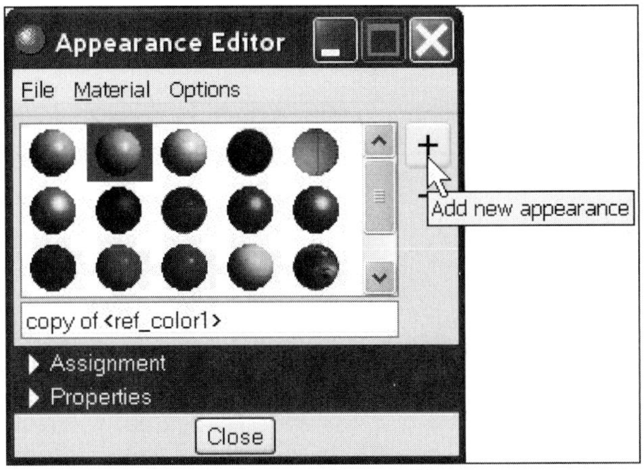

Figure 5.6(b) Appearance Editor, Add new appearance

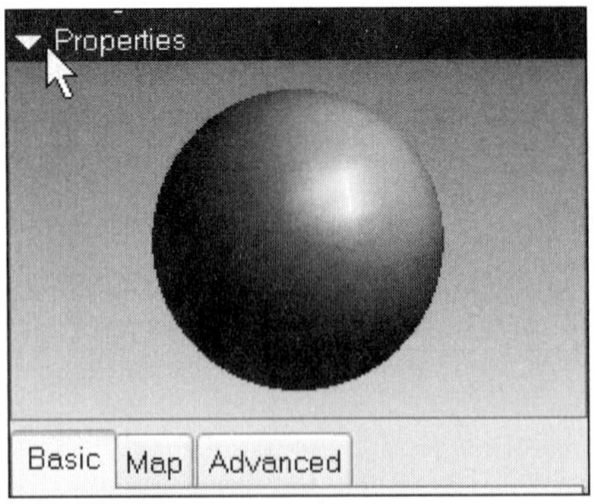

Figure 5.6(c) Properties

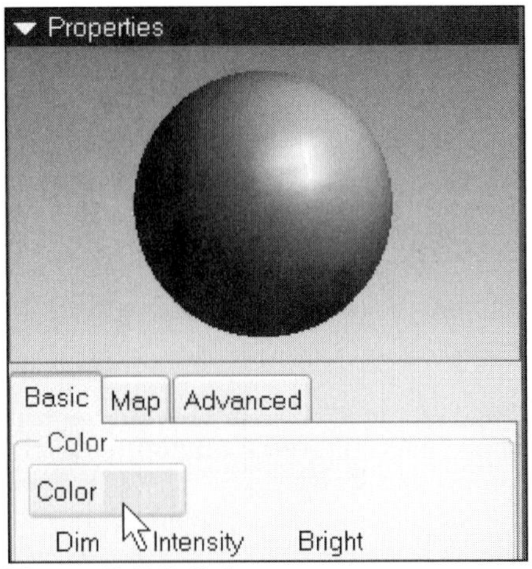

Figure 5.6(d) Color

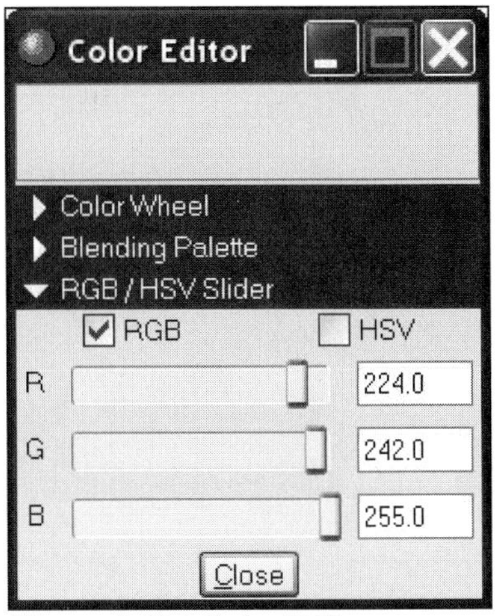

Figure 5.6(e) Color Editor

Repeat to make a few more colors, adding each to the palette [Figs. 5.6(i-j)]

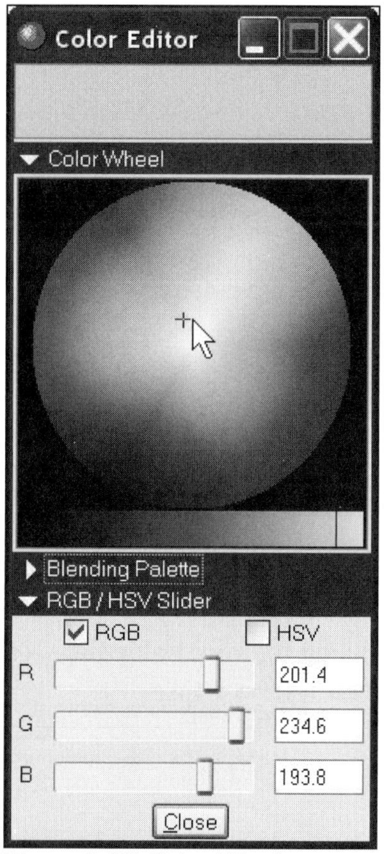

Figure 5.6(f) Color Wheel

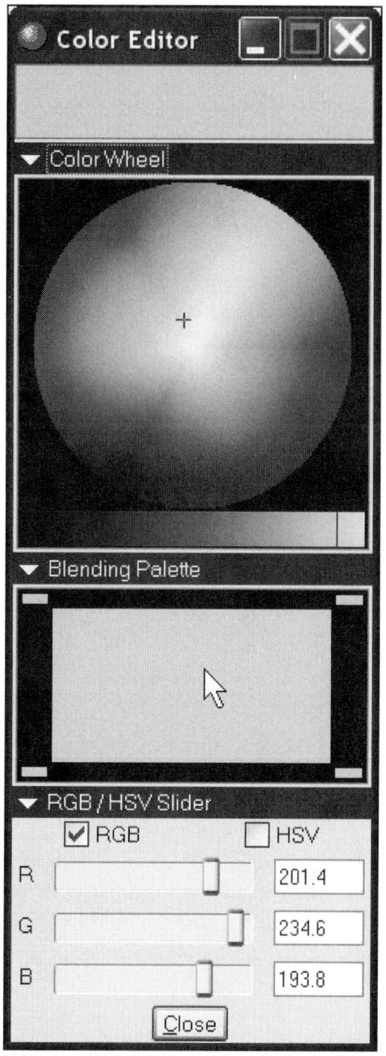

Figure 5.6(g) Blending Palette

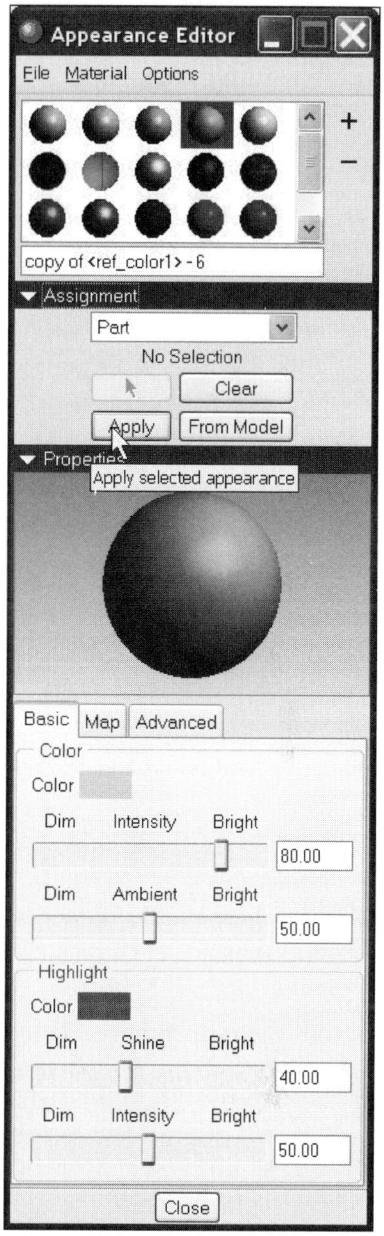

Figure 5.6(h) Apply New Color to Part

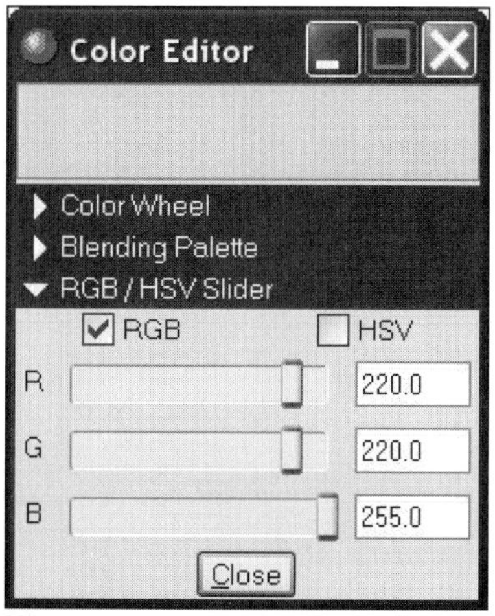

Figure 5.6(i) New Color

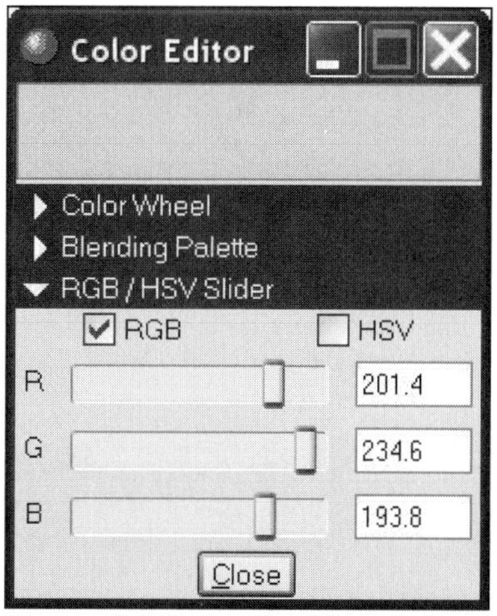

Figure 5.6(j) Another New Color

Also, add some Highlight colors to each of your colors using the same basic method [Fig. 5.6(k)

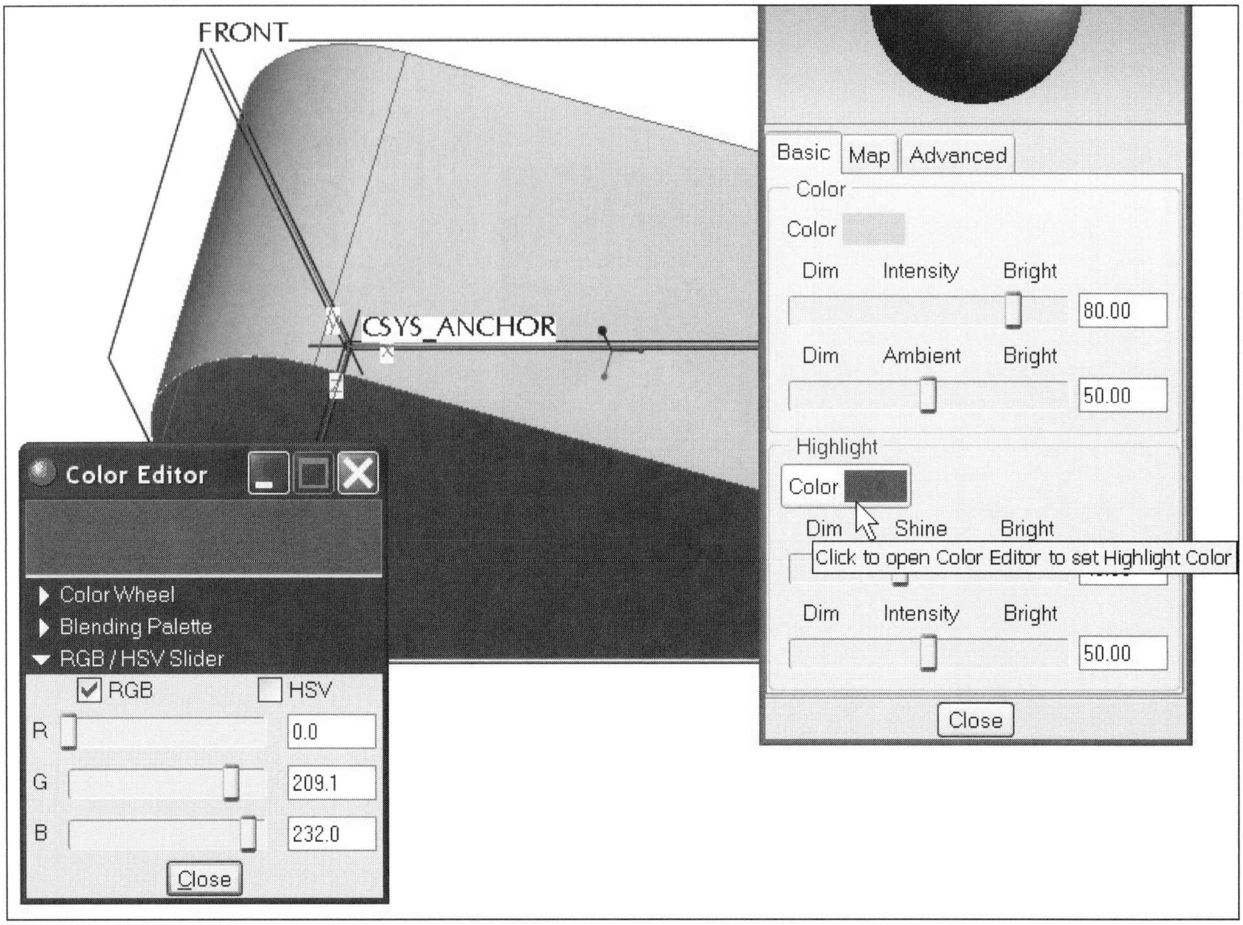

Figure 5.6(k) Highlight Color Editor

File (from the Appearance Editor dialog box) [Fig. 5.6(l)]⇒ **Save** ⇒ type a name for your color file *(you can open and use this file for other projects instead of recreating new colors)* ⇒ **OK** ⇒ **Close**

Figure 5.6(l) Saving Appearance File

The next two features will remove material. For both of the cuts, use the RIGHT datum plane as the sketching plane, and TOP datum plane as the reference plane. Each cut requires just two lines, and two dimensions. Use the first cut's sketching/placement plane and reference/orientation (**Use Previous**) for the second cut. Create the cuts separately, as open sections. Figure 5.7 shows the cuts dimensions.

Click: **Tools** ⇒ Environment ⇒ Snap to Grid ⇒

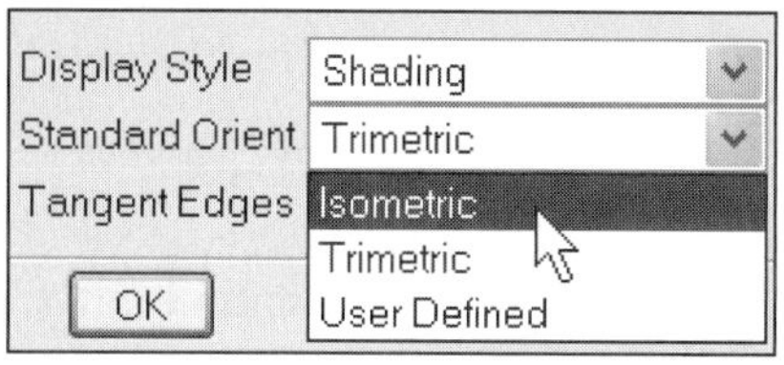

⇒

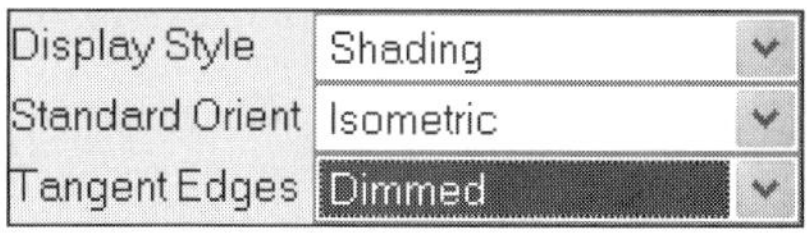

⇒ **Apply** ⇒ **OK** [Fig. 5.8(a)] ⇒ **Refit object to fully display**

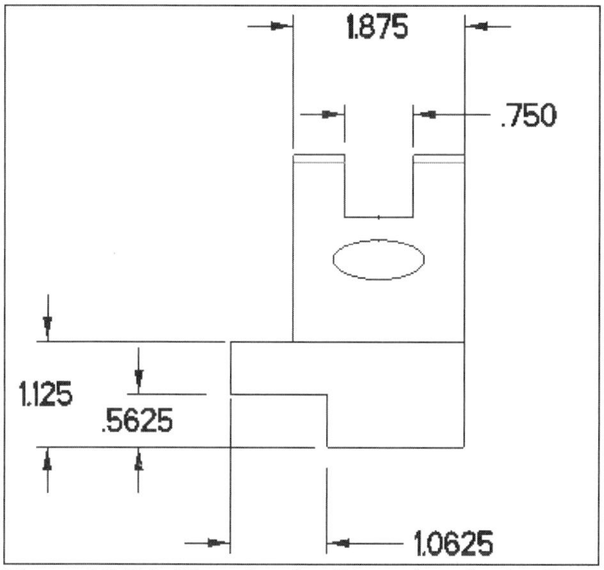

Figure 5.7 Side View showing Dimensions for the Cuts

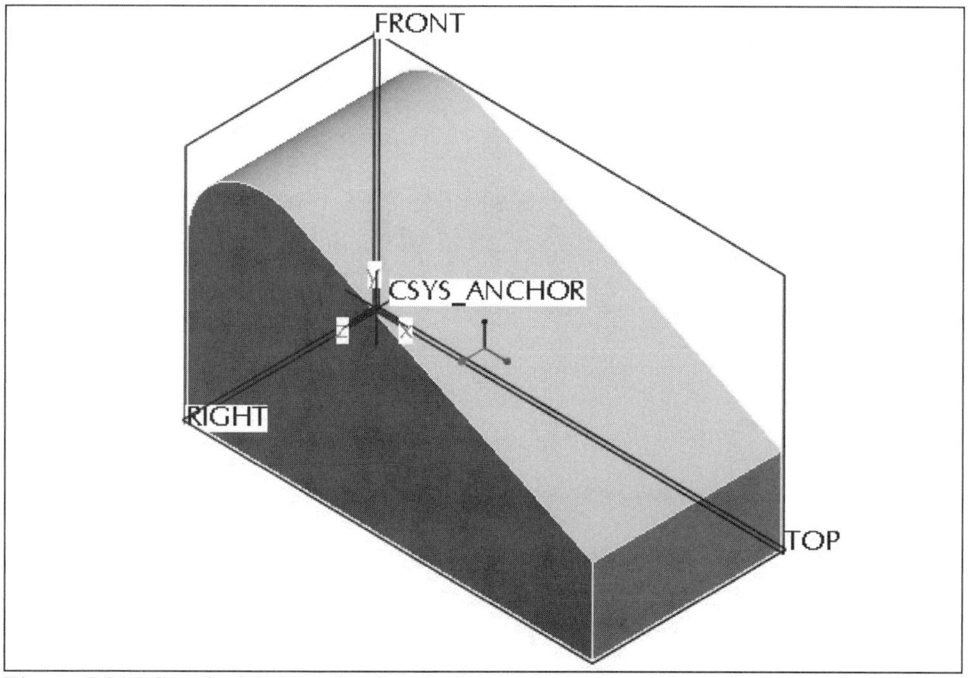

Figure 5.8(a) Standard Orientation Isometric

Create the first cut, click:  **Extrude Tool** ⇒ 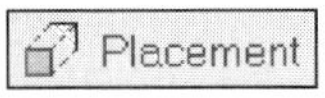

⇒ 

⇒ Plane: select **RIGHT** datum [Fig. 5.8(b)] ⇒ Sketch Orientation--- Reference: **TOP** datum [Fig. 5.8(c)] ⇒ Orientation: **Top** [Fig. 5.8(d)] ⇒ **Sketch** ⇒  **Hidden line** ⇒ add a reference to align the cut with the left edge and the top edge of the part, click: **Sketch** from the menu bar ⇒ **References** 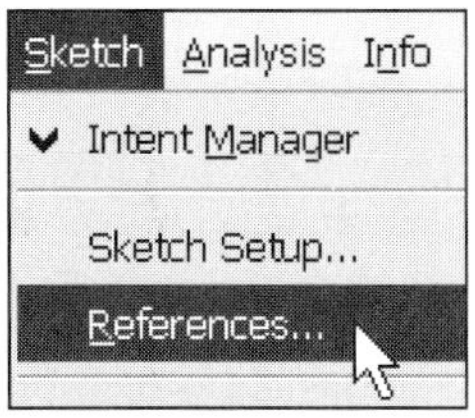

[Fig. 5.8(e)] ⇒ pick on the *left* edge of the part ⇒ pick on the *top* edge of the part [Fig. 5.8(f)]. *If you select the wrong edge, pick the incorrect reference (in the dialog box) ⇒ Delete and try again.* ⇒ **Close**

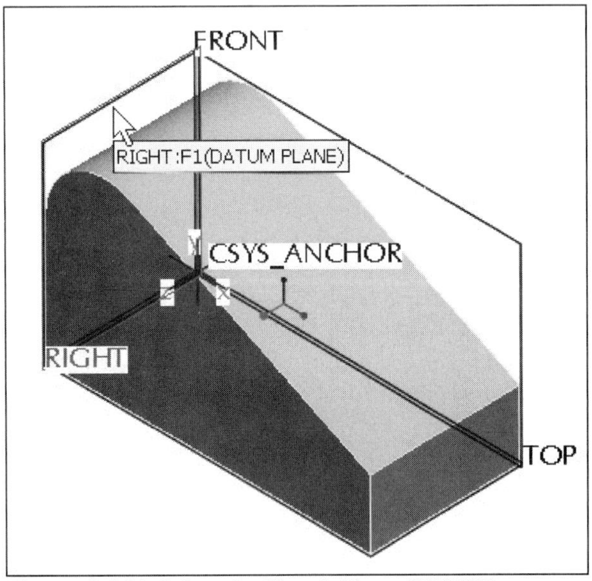

Figure 5.8(b) Select the Sketch Plane (RIGHT)

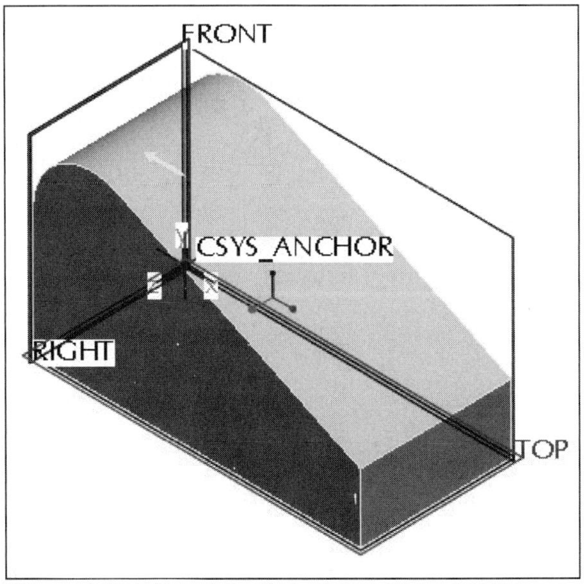

Figure 5.8(c) Select the Orientation Plane (TOP)

Figure 5.8(d) Sketch Dialog Box

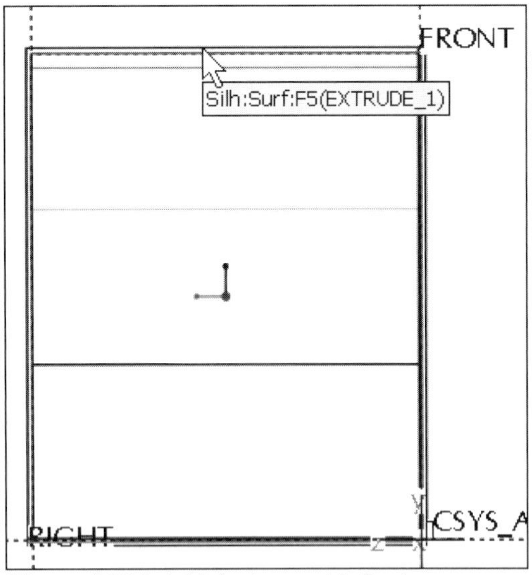

Figure 5.8(e) Add the top and left edges

Figure 5.8(f) References Dialog Box

Click: Create 2 point lines ⇒ sketch the lines- start from the top with a vertical line [Fig. 5.8(g)] ⇒ Select items from the Right Toolchest ⇒ Toggle display of dimensions on ⇒ Highlights vertices of sketched entities which are not common to more than one entity [Fig. 5.8(h)] ⇒ Redraw the current view ⇒ RMB ⇒ Dimension ⇒ add dimensions between the horizontal line and the Top datum, and another between the vertical line and the Front datum plane [Fig. 5.8(i)]

Note that the "weak" (light gray) dimensions disappear as new dimensions are added. Created dimensions are "strong" dimensions (yellow) and represent the design intent of the feature. These dimensions are displayed on the drawing as driving-design dimensions to be used in manufacturing.

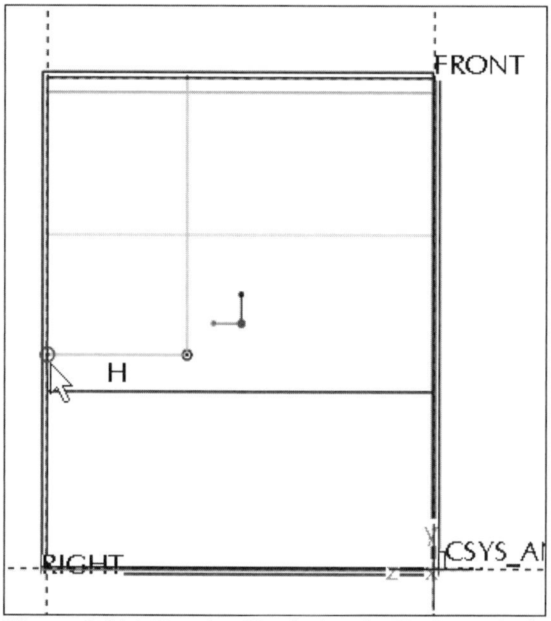

Figure 5.8(g) Sketch a Vertical and a Horizontal Line

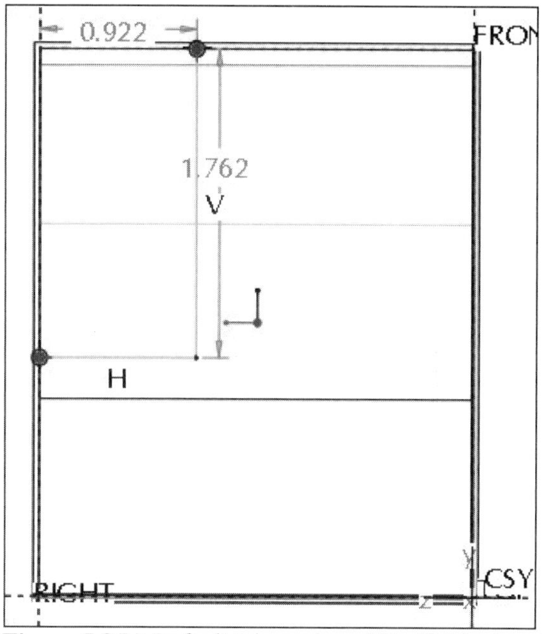

Figure 5.8(h) Default Dimensions (*Weak*) Displayed

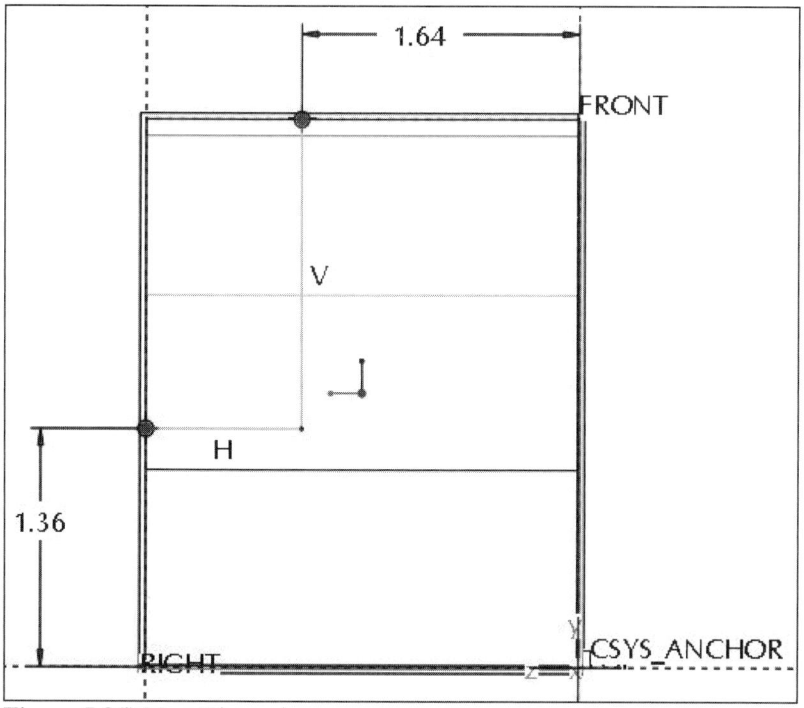

Figure 5.8(i) New Dimensions

The dimensional *design scheme* is correct for the cut, but the values are not the correct *design value*. Click: **Select items** from the Right Toolchest ⇒ double-click on the horizontal dimension and change to the design value [Fig. 5.8(j)] ⇒ **Enter** ⇒ repeat for the vertical dimension [Fig. 5.8(k)] from the Right Toolchest ⇒ **Ctrl+D** ⇒ **Shading** ⇒ use your **MMB** to zoom in [Fig. 5.8(l)] ⇒ flip the direction arrows if necessary

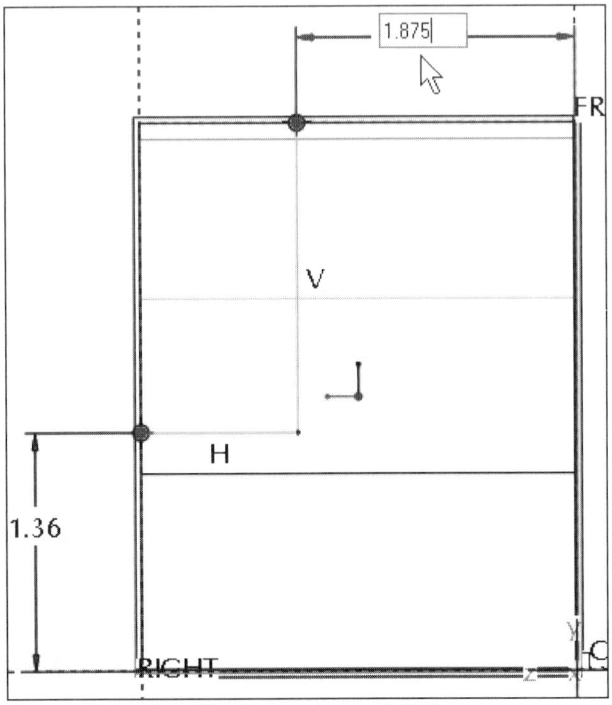

Figure 5.8(j) Modify Dimension to **1.875**

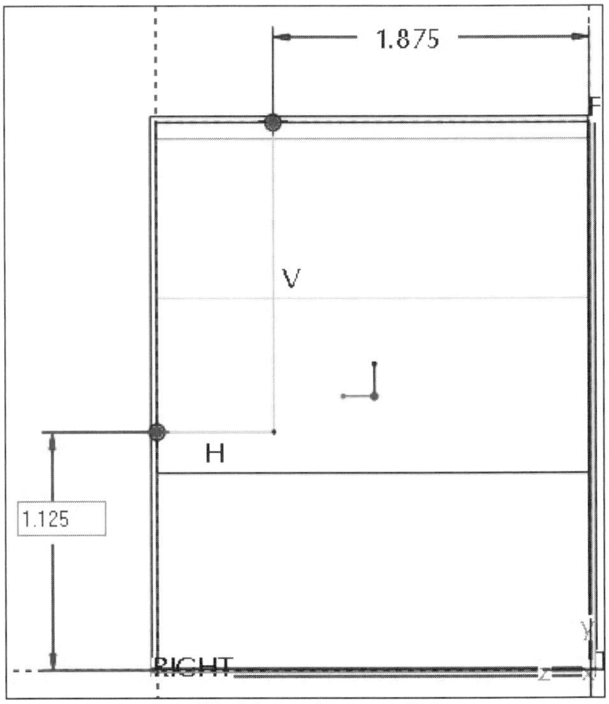

Figure 5.8(k) Modify Dimension to **1.125**

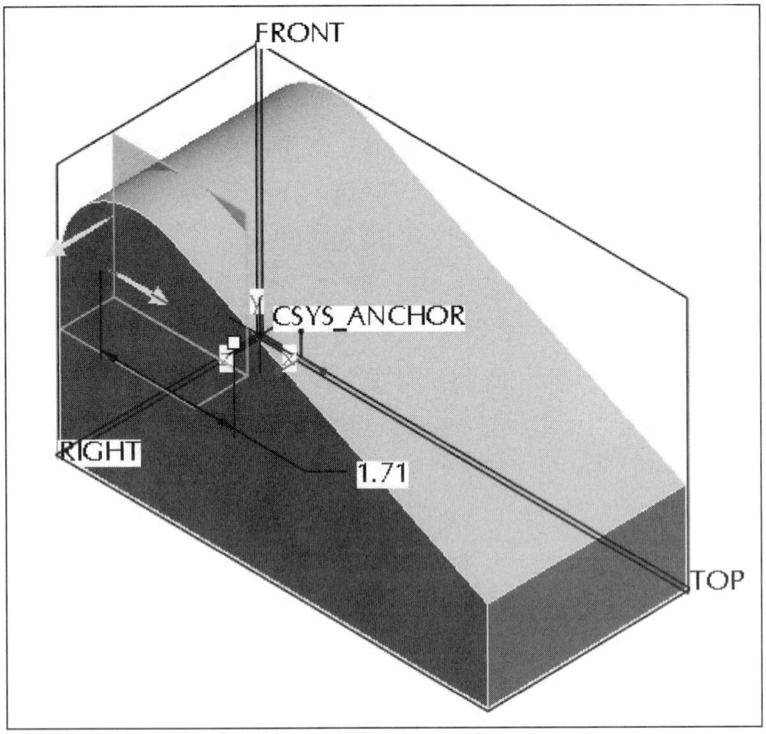

Figure 5.8(l) Extrusion Preview (flip arrow if necessary)

From the dashboard click: off ⇒ **Remove Material** ⇒ **Extrude to intersect with all surface** ⇒ **Options** tab [Fig. 5.8(m)] ⇒ ⇒ on ⇒ **Refit object** [Fig. 5.8(n)] ⇒ **Ctrl+S** ⇒ **Enter**

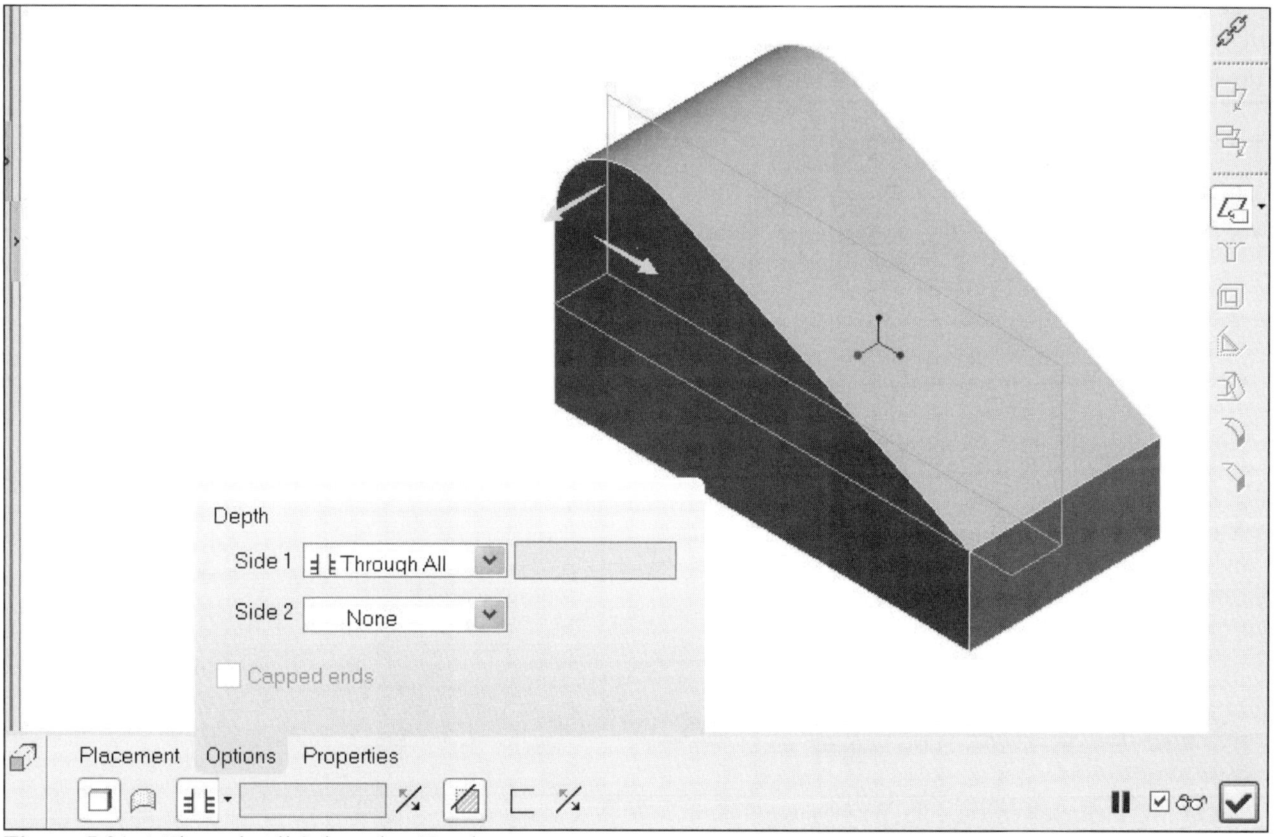

Figure 5.8(m) Through All Selected as Depth

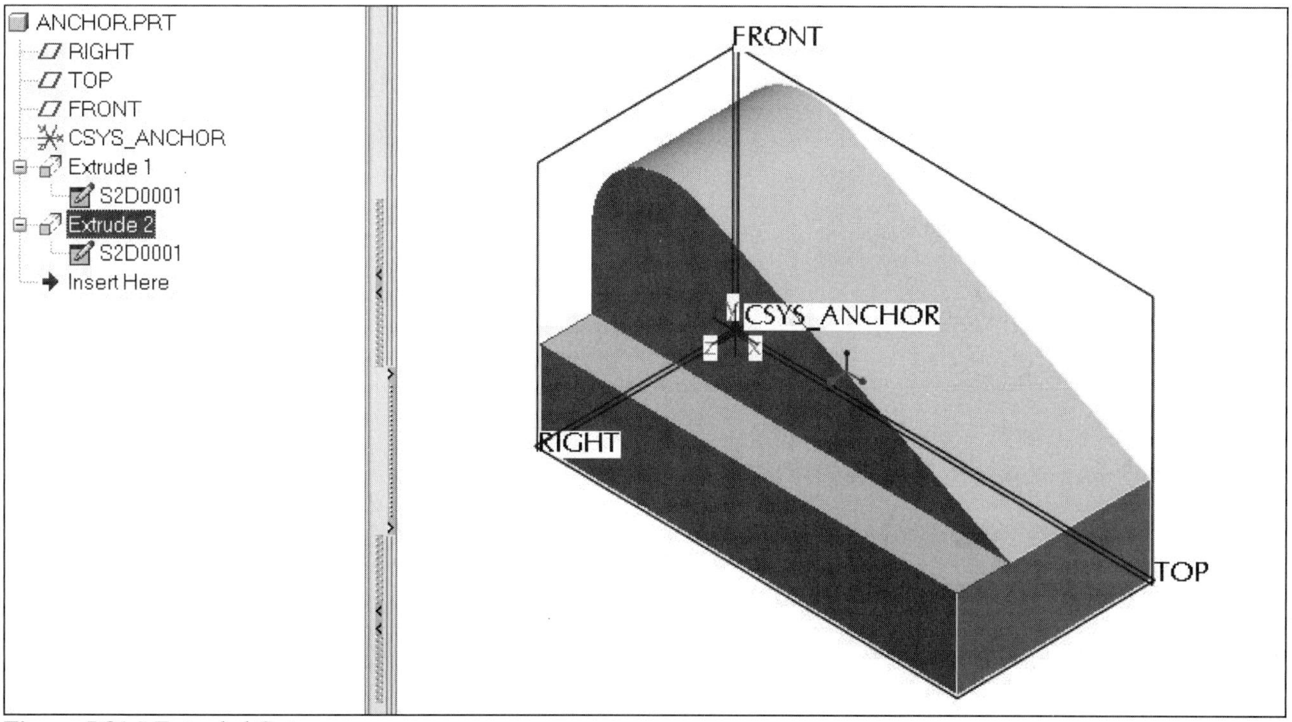

Figure 5.8(n) Extruded Cut

Complete the second cut, click: **Extrude Tool** ⇒ **RMB** ⇒ **Define Internal Sketch** ⇒ **Use Previous** ⇒ **No hidden** ⇒ **Sketch** ⇒ **References** ⇒ add the left vertical edge/surface as a reference [Fig. 5.9(a)] ⇒ **Close** ⇒ **RMB** ⇒ **Line** ⇒ sketch the vertical and horizontal lines [Fig. 5.9(b)] ⇒ **MMB** ⇒ **MMB** ⇒ **Toggle display of dimensions** on ⇒ **Highlights vertices** [Fig. 5.9(c)] ⇒ reposition the default weak dimension scheme by moving the mouse pointer on top of each dimension value, press and hold down the **LMB**, move the pointer to a new position and release the **LMB** to place the dimension

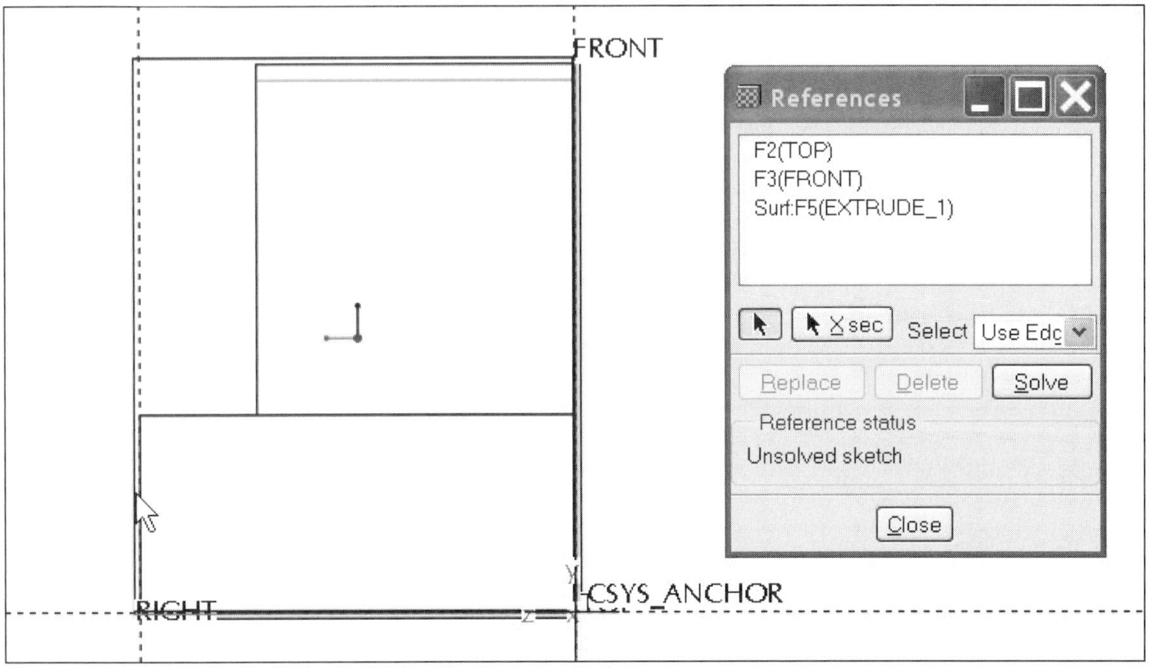

Figure 5.9(a) Add the Left Vertical Edge/Surface as a Reference

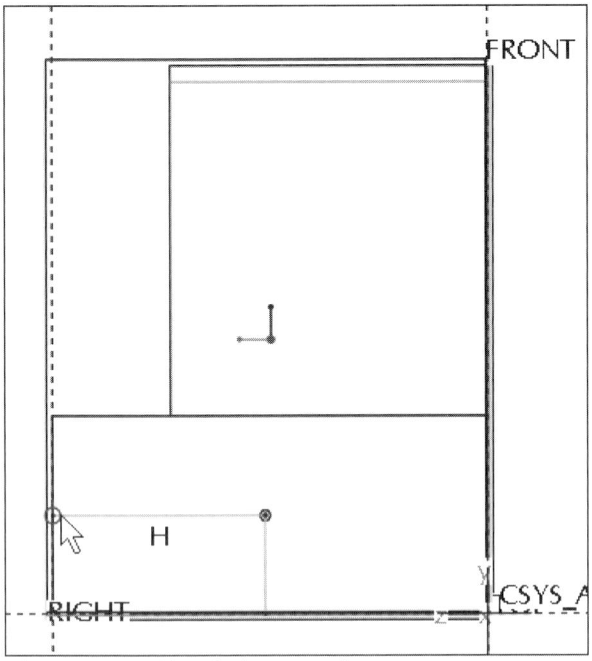

Figure 5.9(b) Sketch the Two Lines

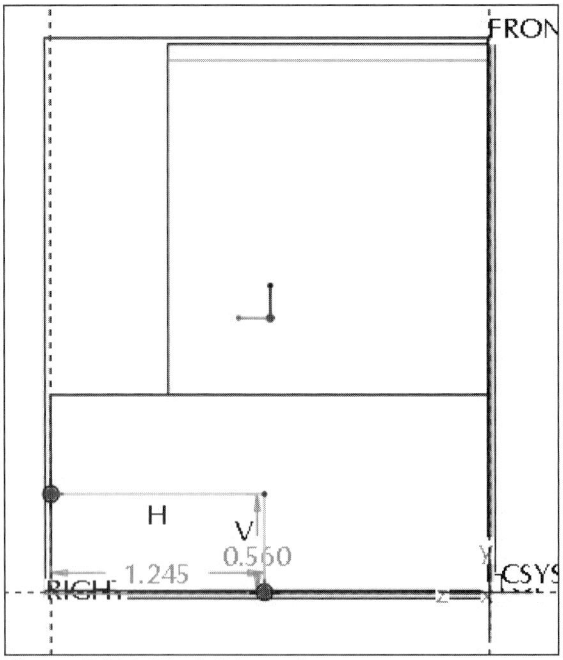

Figure 5.9(c) Weak Dimensions

Double-click on the vertical dimension and type **.5625** [Fig. 5.9(d)] ⇒ **Enter** ⇒ double-click on the horizontal dimension and type **1.0625** [Fig. 5.9(e)] ⇒ **Enter** ⇒ from Right Toolchest ⇒ ⇒ **Standard Orientation** ⇒ **Shading** ⇒ **Remove Material** ⇒ **Through All** [Fig. 5.9(f)] ⇒ **Ctrl+MMB** to zoom in if necessary ⇒ flip the direction arrows if necessary ⇒ from dashboard ⇒ [Fig. 5.9(g)] ⇒ ⇒ **MMB** ⇒ **LMB**

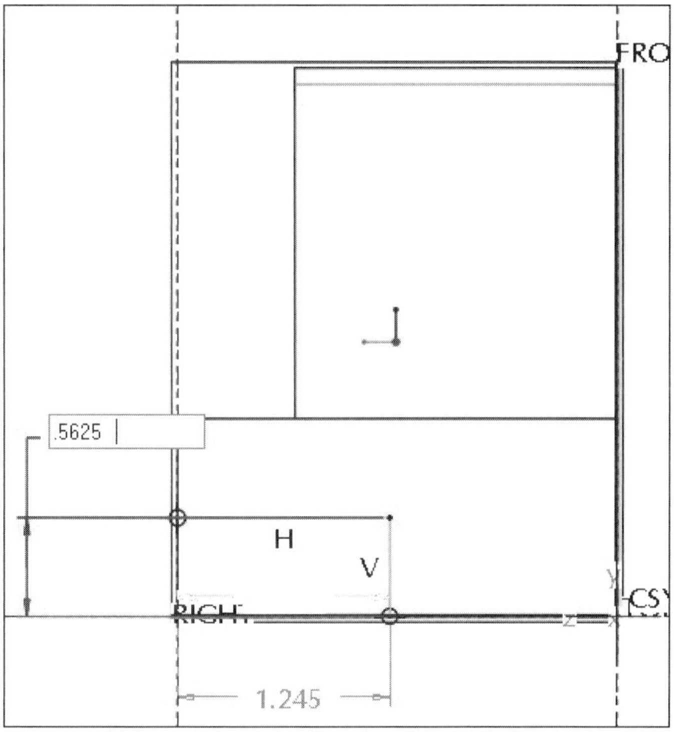

Figure 5.9(d) Modify Vertical Dimension to **.5625**

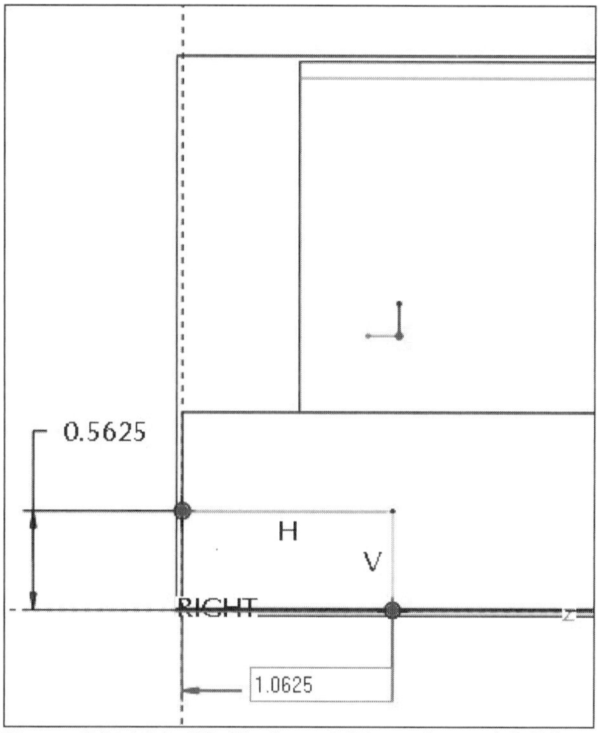

Figure 5.9(e) Modify Horizontal Dimension to **1.0625**

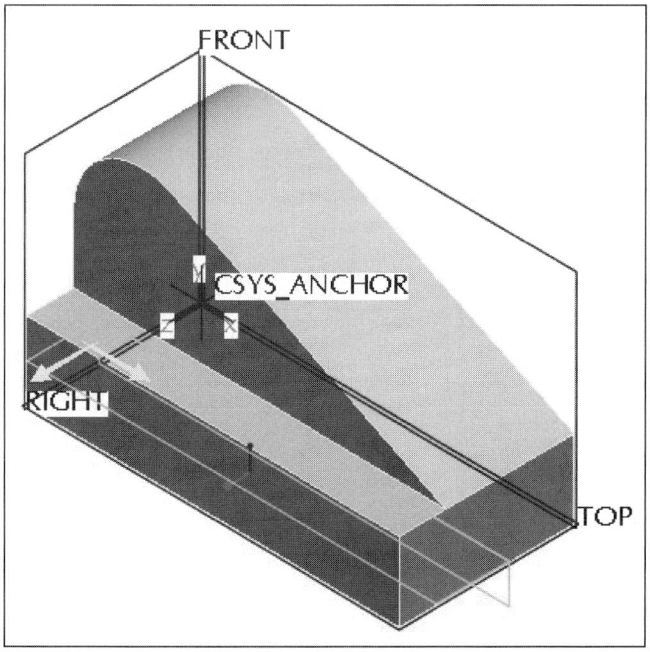

Figure 5.9(f) Dynamic Preview of the Cut

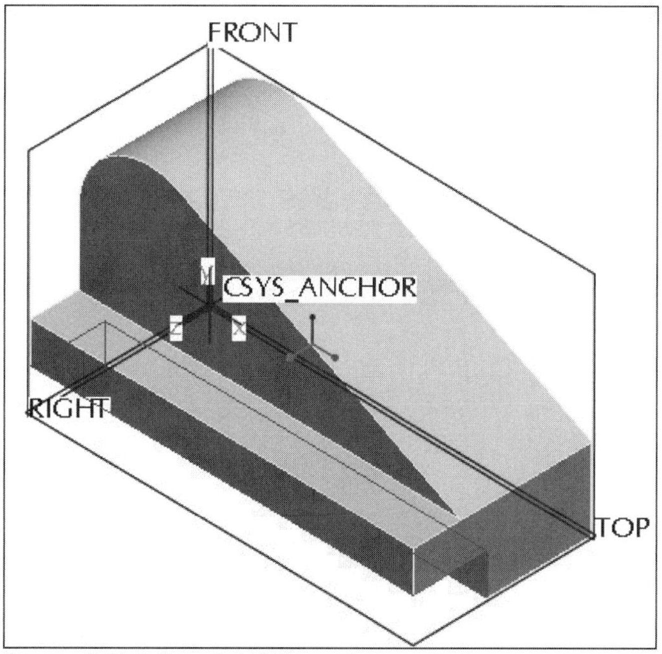

Figure 5.9(g) Completed Geometry

For the next feature, you will need to create a new datum plane on which to sketch. New part datum planes are by default numbered sequentially, DTM1, DTM2, and so on. Also, a datum axis (A_1) will be created through the curved top of the part and used later for the axial location of a small hole. Holes and circular features automatically have axes when they are created. Features created with arcs, fillets and so on need to have an axis added if needed as a design reference.

Click: **Coordinate systems** off ⇒ **Datum Plane Tool** from Right Toolchest ⇒ References: pick datum **FRONT** as the plane to offset from [Fig. 5.10(a)] ⇒ in the DATUM PLANE dialog box, Offset: Translation, type the distance **1.875/2** *(1.875/2 = .9375, always have Pro/E do the math!)* ⇒ **Enter** *(make sure that the value is .9375, if not type it in – your default number of decimal places may be set to low, changing this value will be shown later in the text)* ⇒ **OK** ⇒ **LMB** in the graphics area to deselect ⇒ **Datum Axis Tool** ⇒ References: pick the curved surface [Fig. 5.10(b)] ⇒ **OK** ⇒ **LMB** to deselect

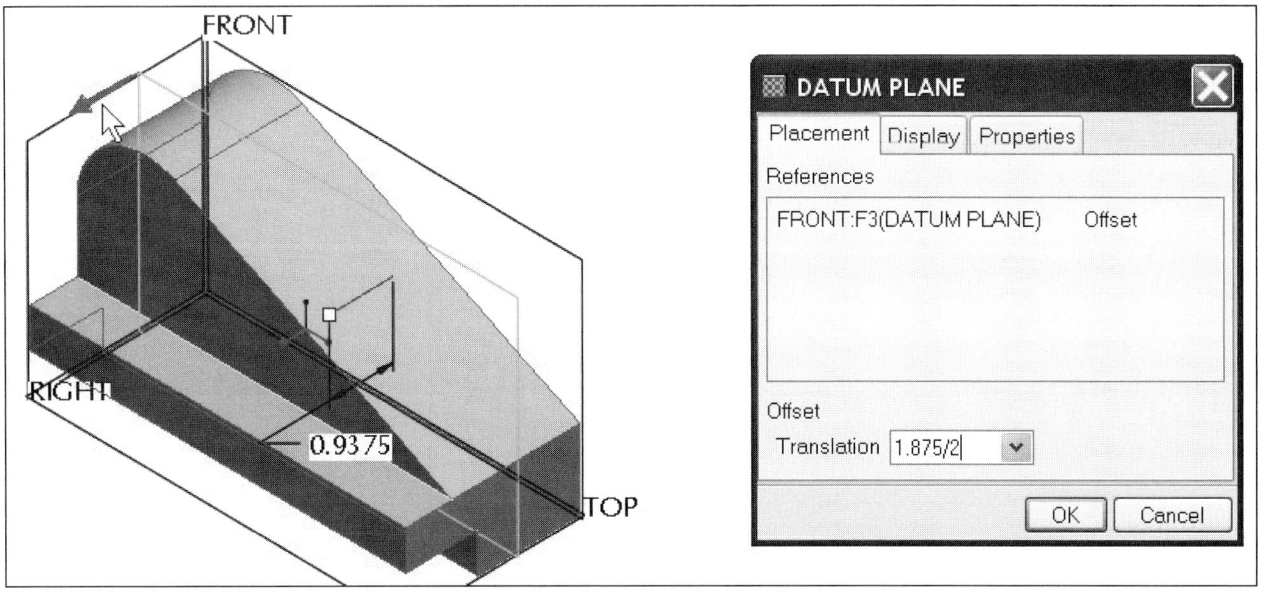

Figure 5.10(a) Offset Datum

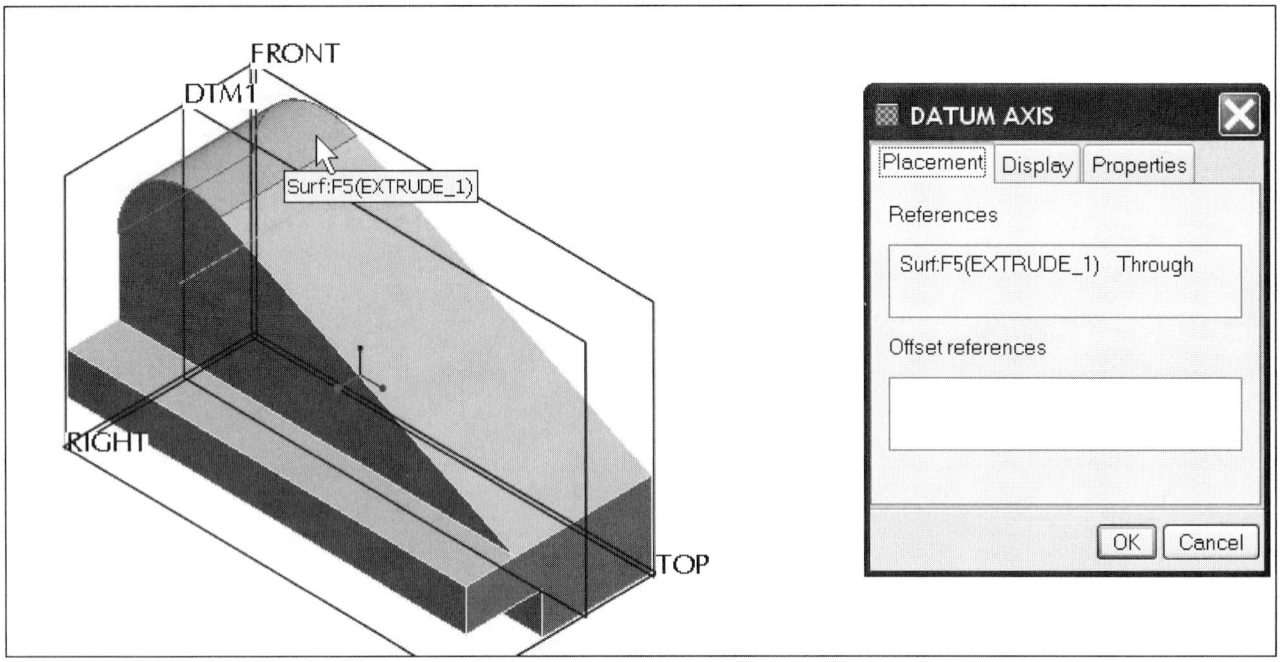

Figure 5.10(b) Datum Axis A_1 through the Center of the Curved Surface

Datum Plane Tool ⇒ References: pick the angled surface ⇒ click **Offset** ⇒ click **Offset** again to see options ⇒ click **Through** [Fig. 5.10(c)] ⇒ **OK** [Fig. 5.10(d)] ⇒ **LMB** to deselect ⇒ **Coordinate systems** on ⇒ ⇒ **OK** ⇒ **File** ⇒ **Delete** ⇒ **Old Versions** ⇒ **Enter** *(or MMB)*

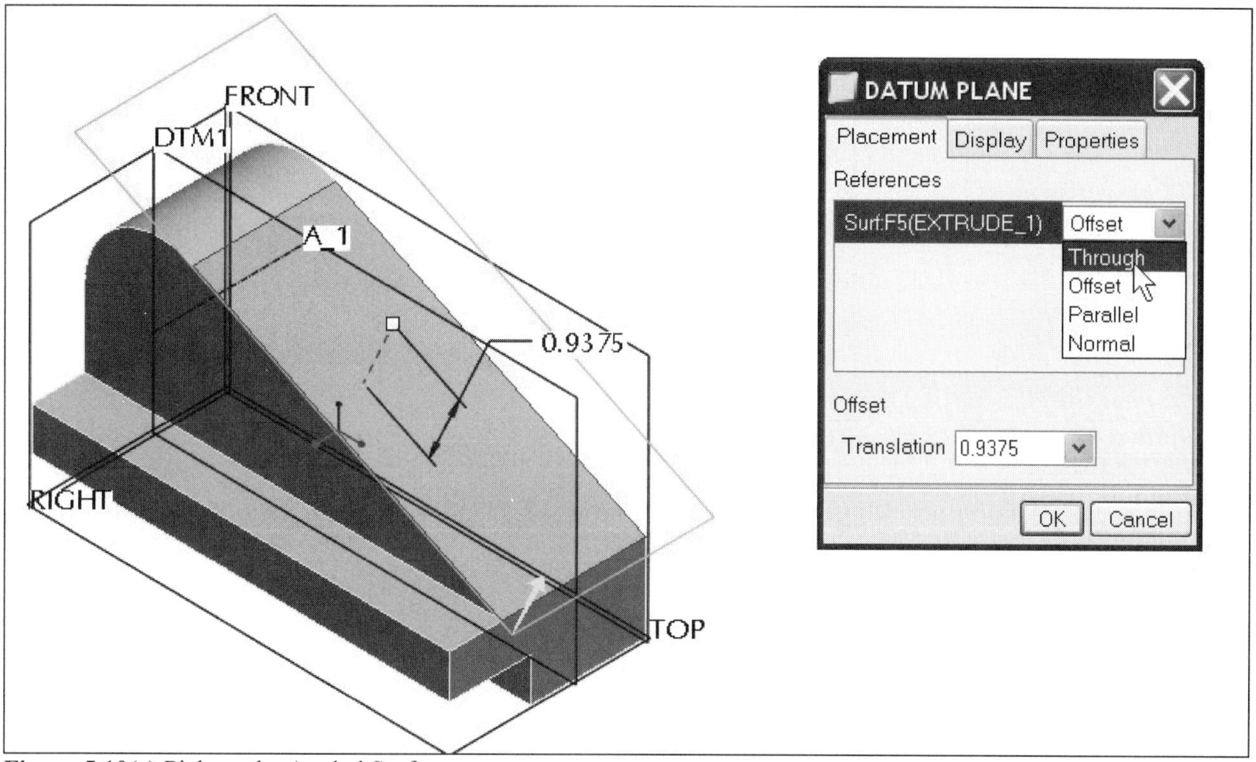

Figure 5.10(c) Pick on the Angled Surface

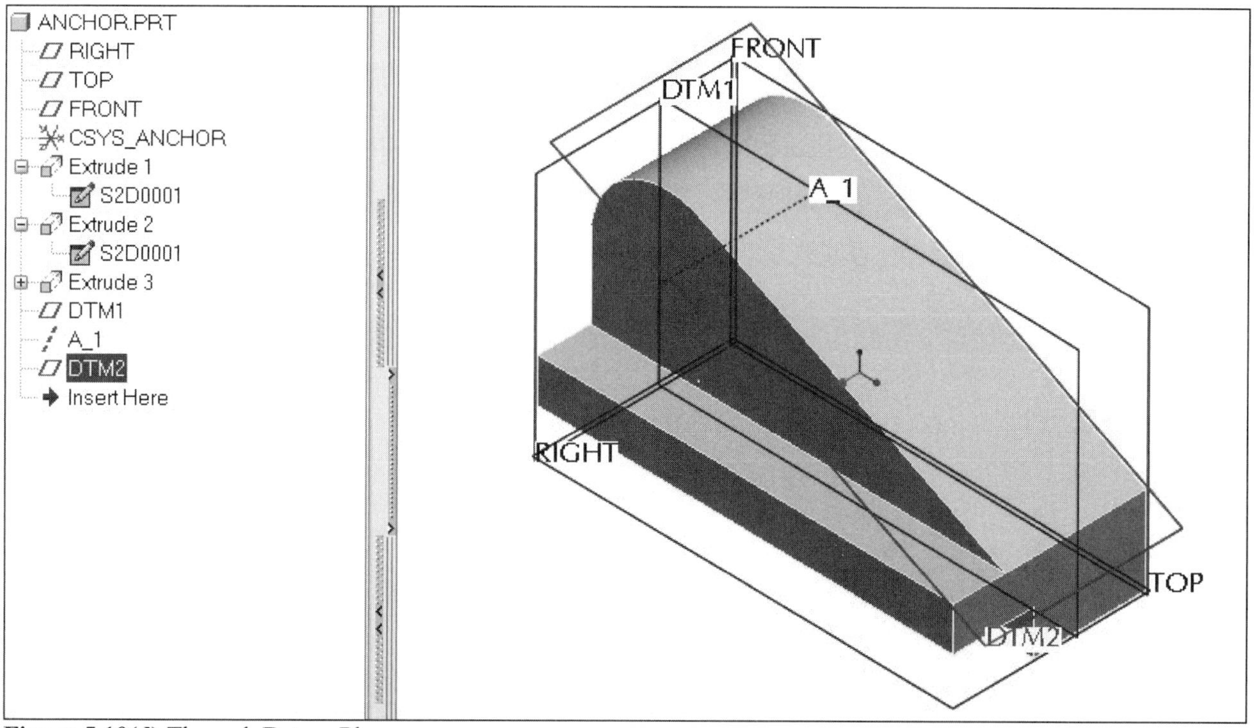

Figure 5.10(d) Through Datum Plane

Create a new layer and add the new datum planes and axis to it, click: **Set layers** from Top Toolchest, Layer Tree displays in place of the Model Tree [Fig. 5.11(a)] ⇒ **Layer** ⇒ **New Layer** [Fig. 5.11(b)]

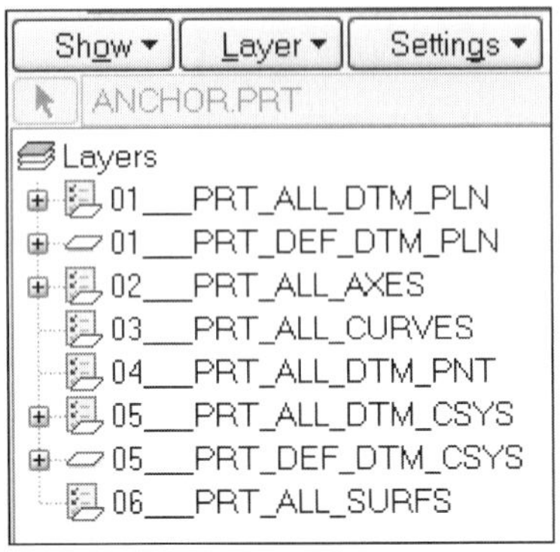

Figure 5.11(a) Layer Tree

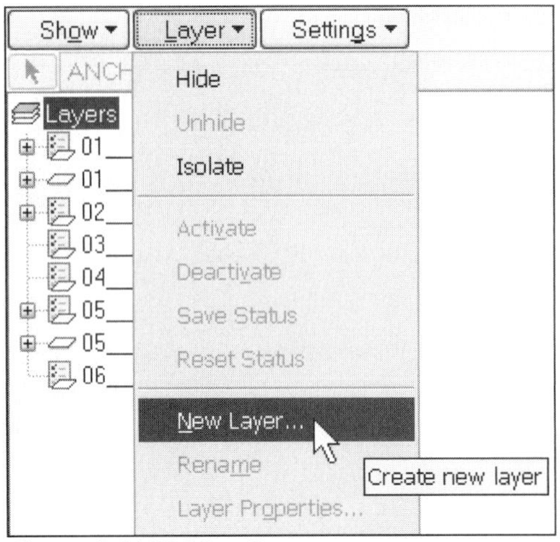

Figure 5.11(b) New Layer

In the Layer Properties dialog box [Fig. 5.11(c)]-Name: type **DATUM_FEATURES** *(do not press Enter)* as the name for new layer [Fig. 5.11(d)]

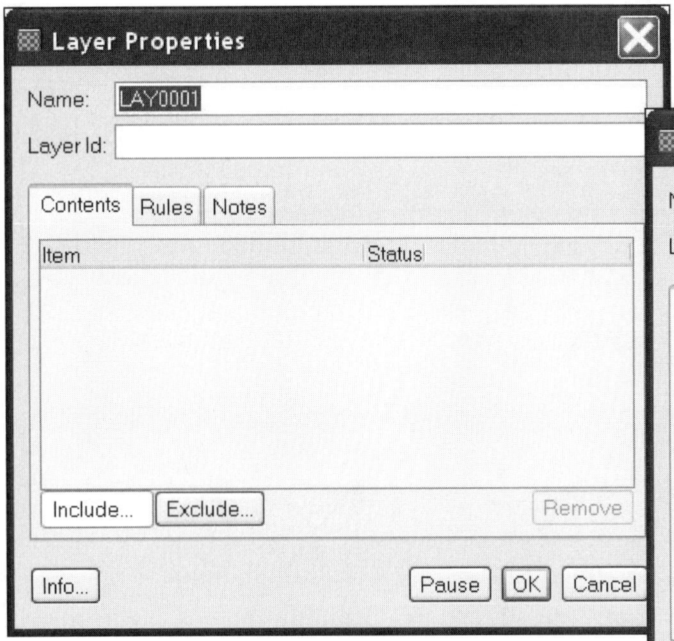

Figure 5.11(c) Layer Properties Dialog Box

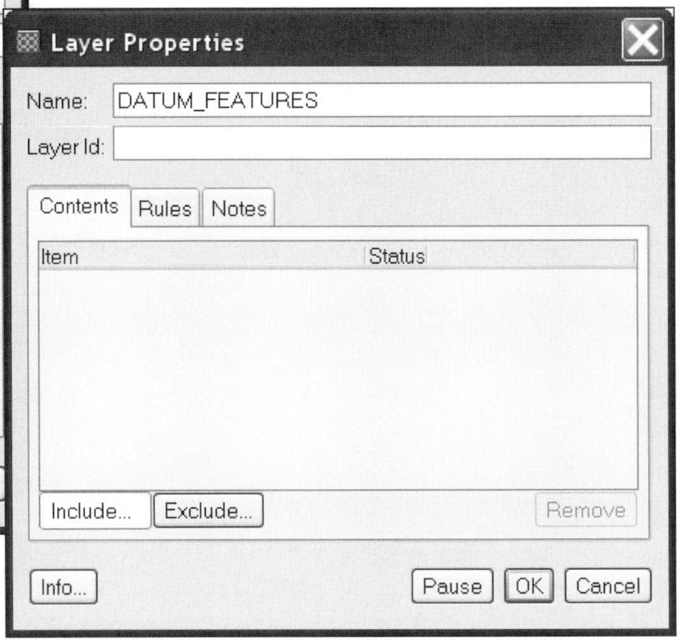

Figure 5.11(d) Layer Name - DATUM_FEATURES

Select the two *new* datum planes from the model ⇒ select the axis from the model, with your cursor over the axis on the model, click **RMB** to toggle to the correct selection ⇒ select the axis with **LMB** [Fig. 5.11(e)] ⇒ **OK** [Fig. 5.11(f)] ⇒ expand layer to see items 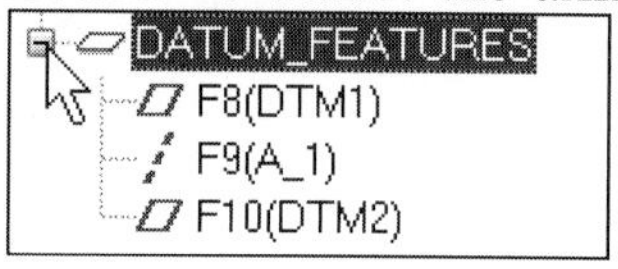

[Fig. 5.11(g)] ⇒ **Show** in the Navigator ⇒ **Model Tree** ⇒ **Ctrl+R**

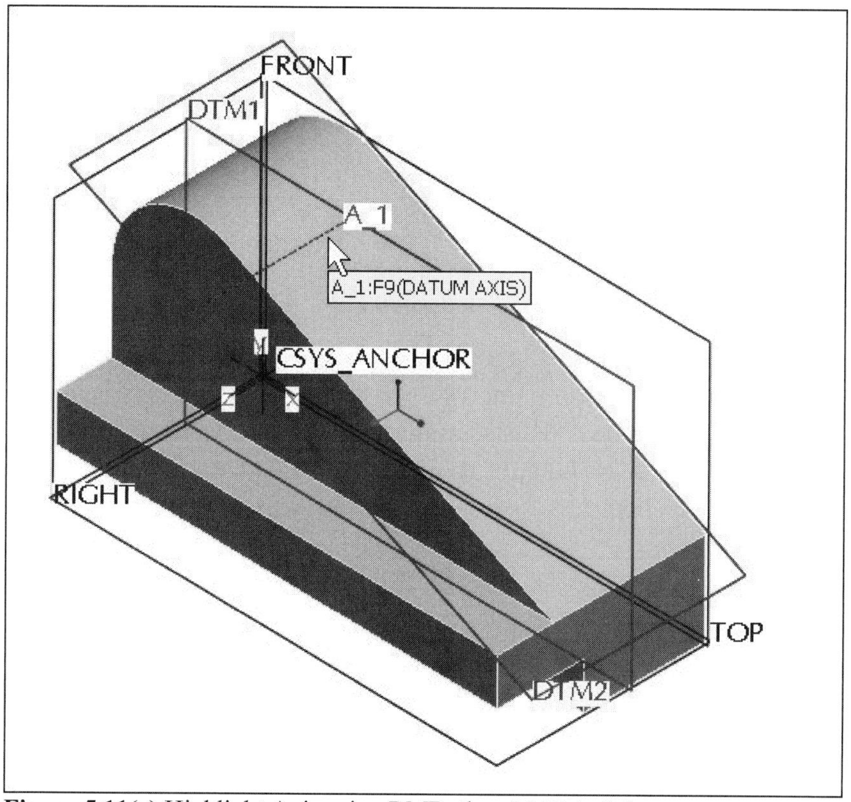

Figure 5.11(e) Highlight Axis using RMB, then LMB to Select

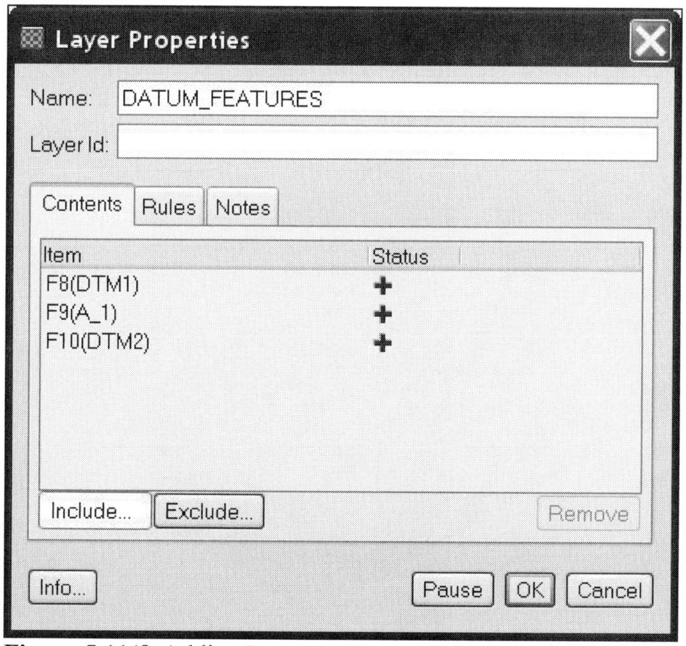

Figure 5.11(f) Adding Items to a Layer

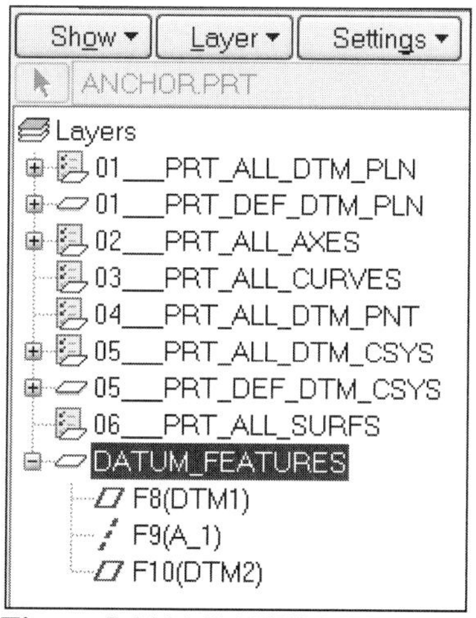

Figure 5.11(g) DATUM_FEATURES Layer

Geometric Tolerances

Before continuing with the modeling of the part, the datums used and created thus far will be "set" for geometric tolerancing. Geometric tolerances (GTOLs) provide a comprehensive method of specifying where on a part the critical surfaces are, how they relate to one another, and how the part must be inspected to determine if it is acceptable. They provide a method for controlling the location, form, profile, orientation, and run out of features. When you store a Pro/E GTOL in a solid model, it contains parametric references to the geometry or feature it controls—its reference entity—and parametric references to referenced datums and axes. In Assembly mode, you can create a GTOL in a subassembly or a part. A GTOL that you create in Part or Assembly mode automatically belongs to the part or assembly that occupies the window; however, it can refer only to set datums belonging to that model itself, or to components within it. It cannot refer to datums outside of its model in some encompassing assembly, unlike assembly created features. You can add GTOLs in Part or Drawing mode, but they are reflected in all other modes. Pro/E treats them as annotations, and they are always associated with the model. Unlike dimensional tolerances, though, GTOLs do not affect part geometry. Before you can reference a datum plane or axis in a GTOL, you must set it as a reference. Pro/E encloses its name using the set datum symbol. After you have set a datum, you can use it in the usual way to create features and assemble parts. You enter the set datum command by picking on the datum plane or axis in the Model Tree or on the model itself ⇒ RMB ⇒ Properties. Pro/E encloses the datum name in a feature control frame. If needed, type a new name in the Name field of the Datum dialog box. Most datums will follow the alphabet, A, B, C, D, and so on. You can hide (not display) a set datum by placing it on a layer and then hiding the layer *(or clicking on the item in the Model Tree ⇒ RMB ⇒ Hide)*.

Pick **TOP** in the Model Tree ⇒ **RMB** ⇒ **Properties** [Figs. 5.12(a-b)] ⇒ double-click in the Name field ⇒ type **A** [Fig. 5.12(c)] ⇒ **Set** [Fig. 5.12(d)] ⇒ **OK** in the Datum dialog box [Fig. 5.12(d)] ⇒ repeat process and set the Front datum plane as **B** and the Right datum plane as **C**.

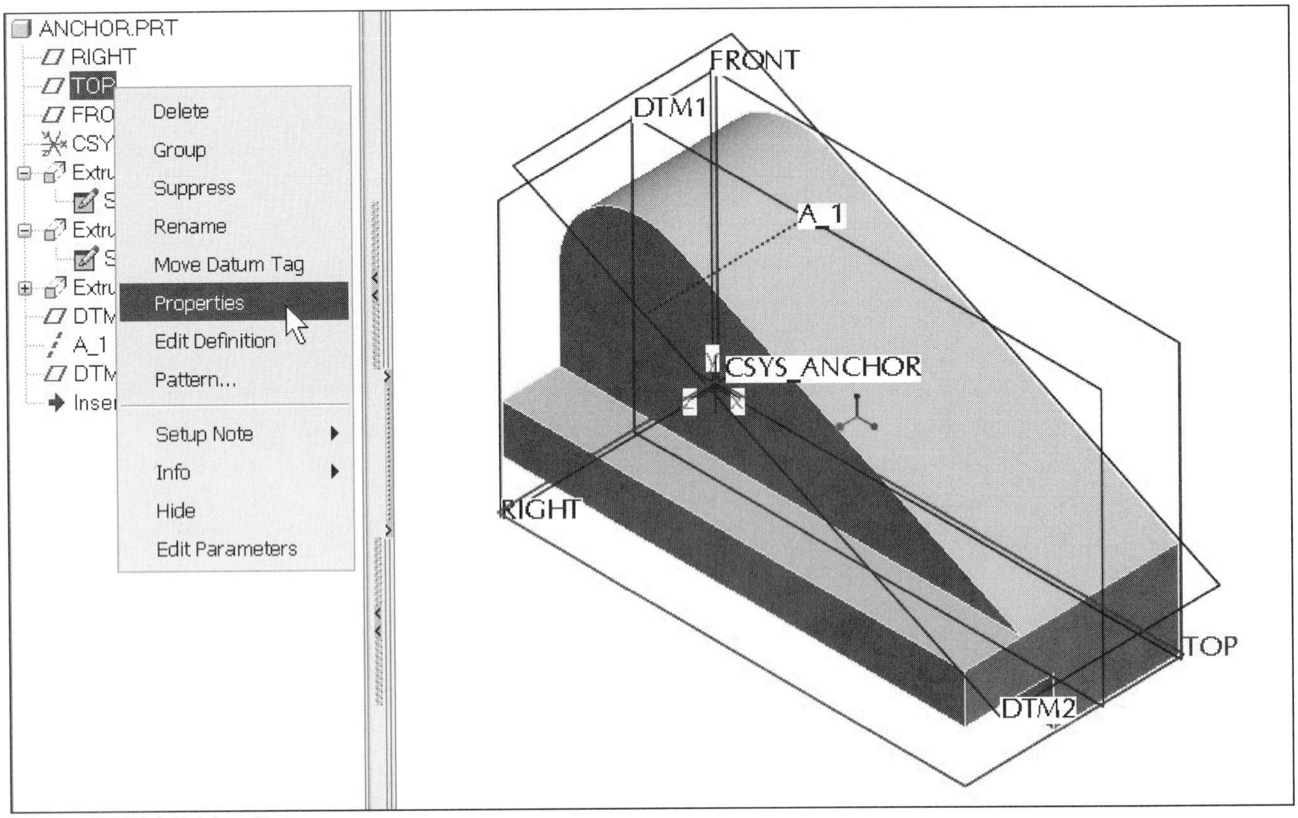

Figure 5.12(a) Setting Datums

Figure 5.12(b) Datum Dialog Box

Figure 5.12(c) Type A

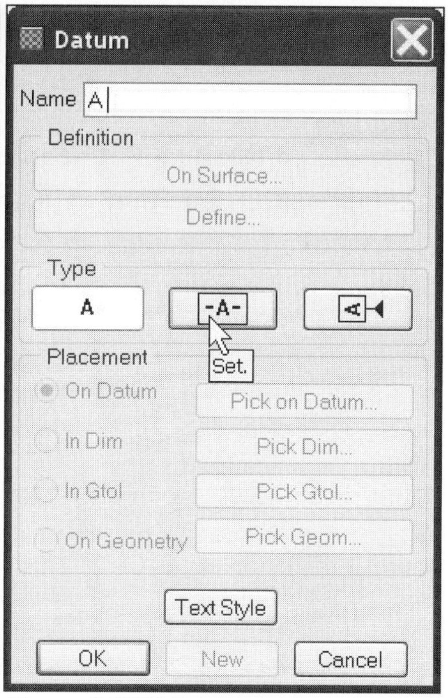

Figure 5.12(d) Set

You can also set datums directly on the model or using the Model Tree. Pick **DTM1** on the model ⇒ **RMB** ⇒ **Properties** [Fig. 5.12(e)] ⇒ Name- type **D** ⇒ **Set** ⇒ **OK** ⇒ pick **DTM2** ⇒ **RMB** ⇒ **Properties** ⇒ Name- type **E** ⇒ **Set** ⇒ **OK** [Fig. 5.12(f)] ⇒ ⇒ **Save** ⇒ **MMB** *(or Enter)* ⇒ **LMB** to deselect

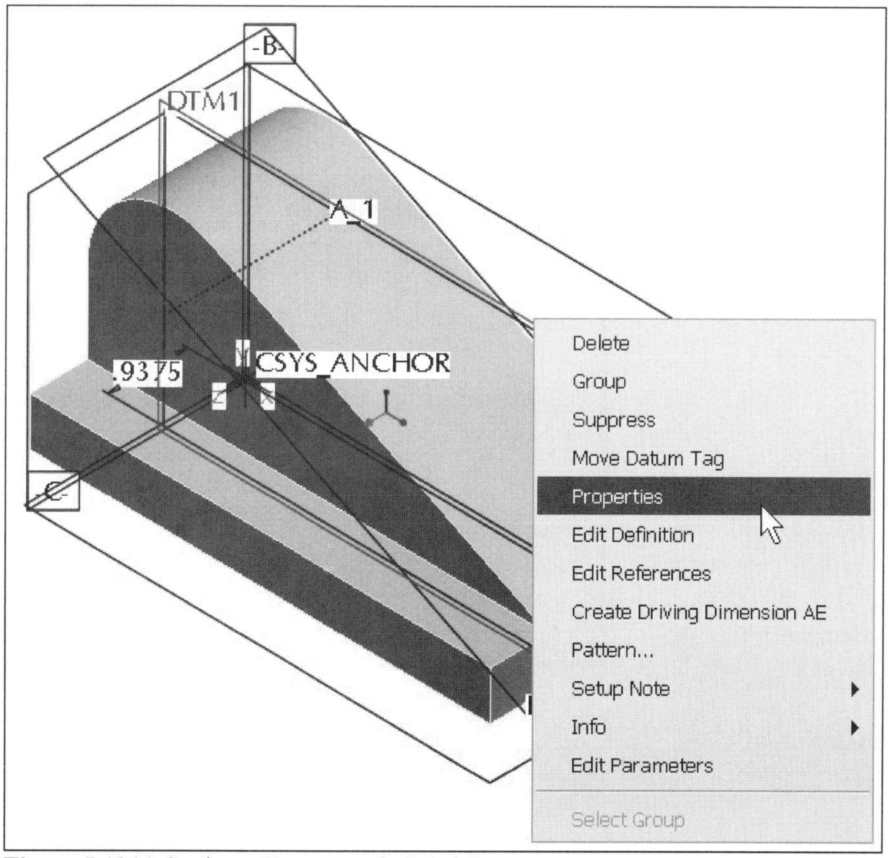

Figure 5.12(e) Setting a Datum on the Model

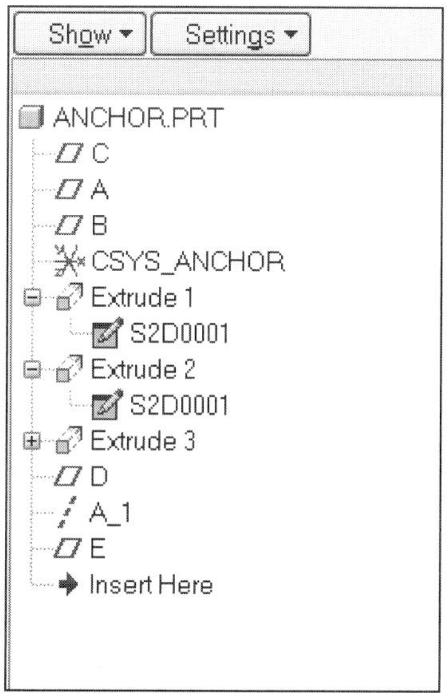

Figure 5.12(f) Model Tree

Create the slot on the top of the part by sketching on datum D and projecting the cut toward both sides. Use the Model Tree to select the appropriate datum planes.

Click: ⇒ **RMB** ⇒ **Define Internal Sketch** ⇒ Plane: select datum **D** [Fig. 5.13(a)] ⇒ **Sketch** ⇒ ⇒ **Standard Orientation** ⇒ add a reference to align the cut with the angled edge of the part, click: **Sketch** from the menu bar ⇒ **References** ⇒ pick on datum **E** ⇒ pick the small vertical surface on the right side of the part as the fourth reference [Fig. 5.13(b)] ⇒ **Close**

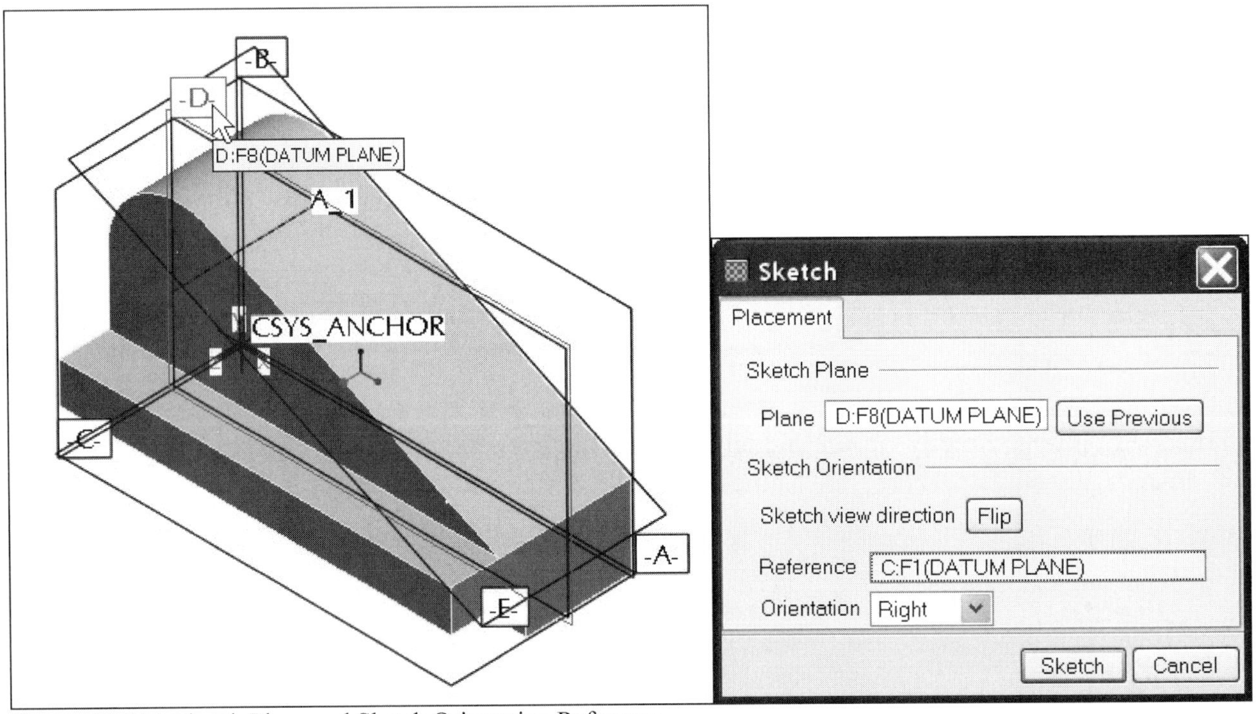

Figure 5.13(a) Sketch Plane and Sketch Orientation Reference

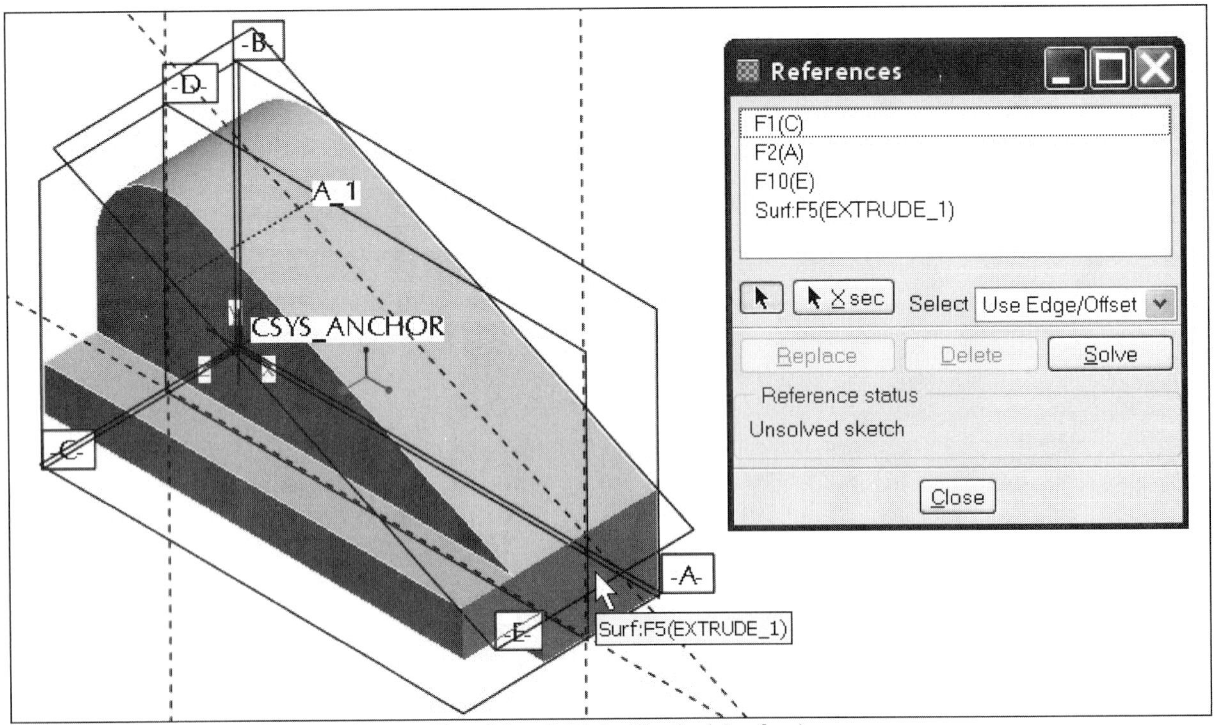

Figure 5.13(b) New References (Datum E and Small Right Vertical Surface)

Click: Hidden line ⇒ **Orient the sketching plane parallel to the screen** from the Top Toolchest ⇒ **Create points** pick at the corner of the angled datum **E** and the right vertical reference [Fig. 5.13(c)] ⇒ **MMB**

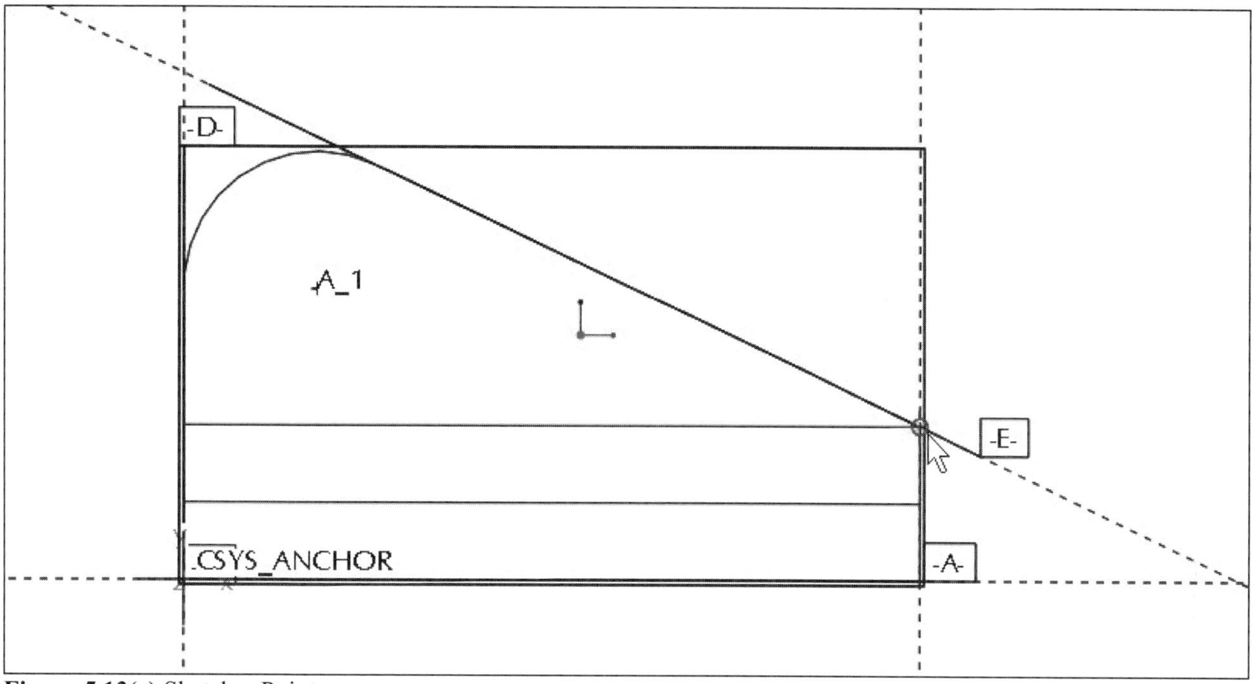

Figure 5.13(c) Sketcher Point

Click: **Spin center** off ⇒ **Create 2 point lines** ⇒ sketch the first line from and perpendicular to the angled edge [Fig. 5.13(d)]

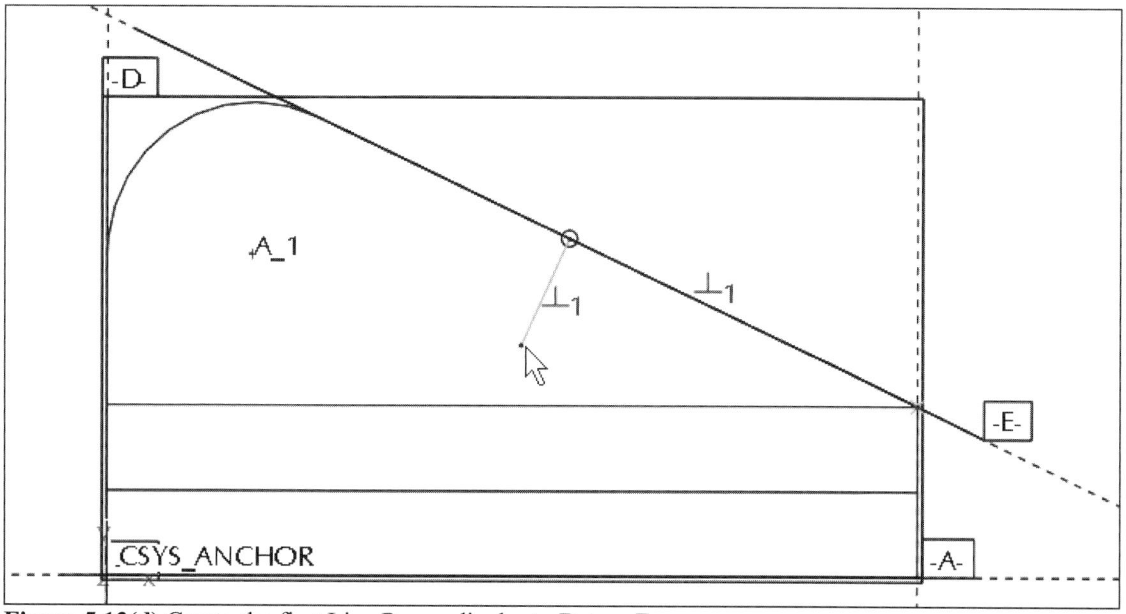

Figure 5.13(d) Create the first Line Perpendicular to Datum E

Sketch the next line horizontal [Fig. 5.13(e)] ⇒ **MMB** ⇒ **MMB** to end the line command [Fig. 5.13(f)] ⇒ move the dimension off the object [Fig. 5.13(g)]

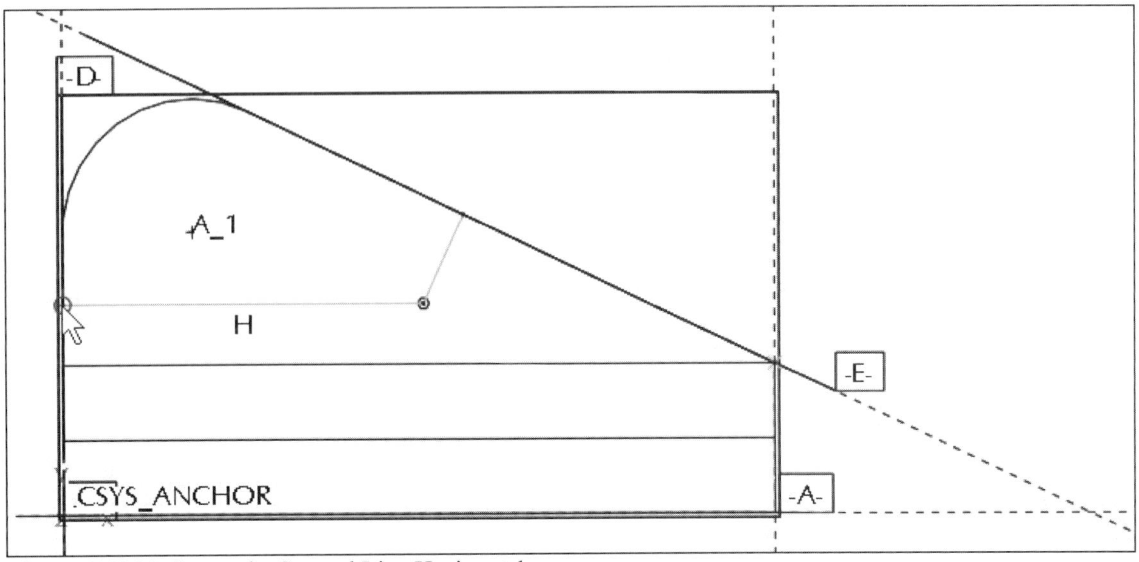

Figure 5.13(e) Create the Second Line Horizontal

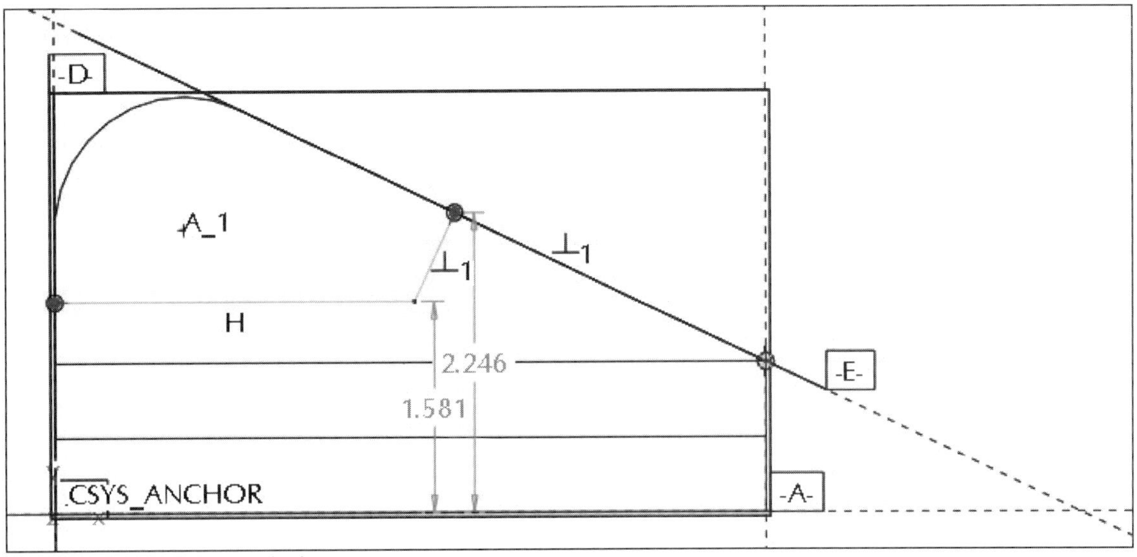

Figure 5.13(f) Completed Lines

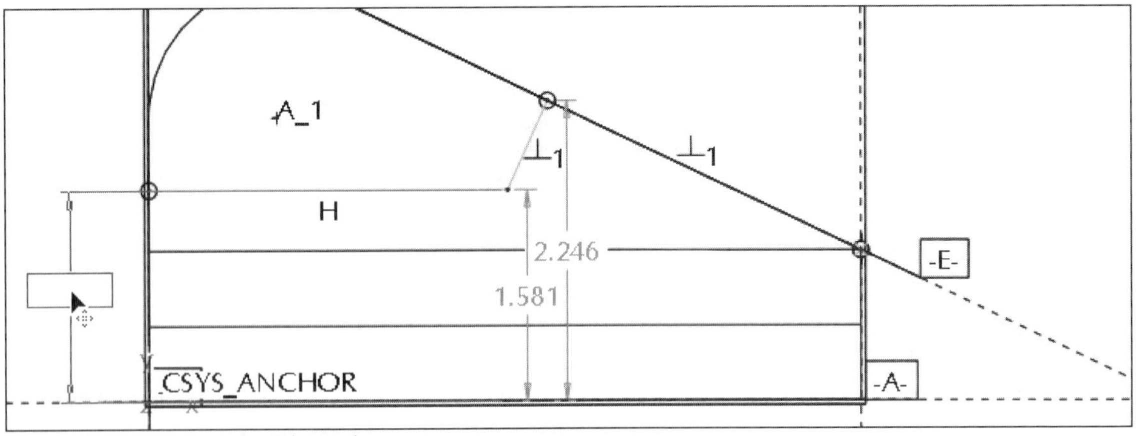

Figure 5.13(g) Move the Dimension

Click: double-click on the moved dimension ⇒ type the design value of **1.50** [Fig. 5.13(g)] ⇒ **Enter** ⇒ **RMB** ⇒ **Dimension** ⇒ add a dimension by selecting angled line and then the point *(do not pick endpoint to point)* ⇒ **MMB** to place the dimension [Fig. 5.13(h)] ⇒ **MMB** ⇒ modify the value to **2.652** ⇒ **Enter**

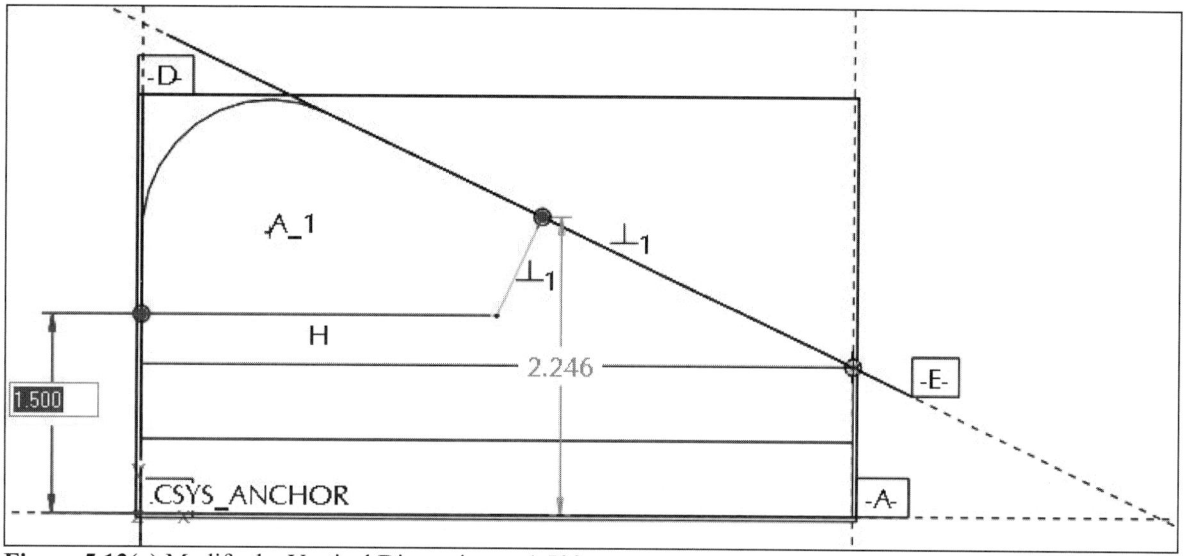

Figure 5.13(g) Modify the Vertical Dimension to **1.500**

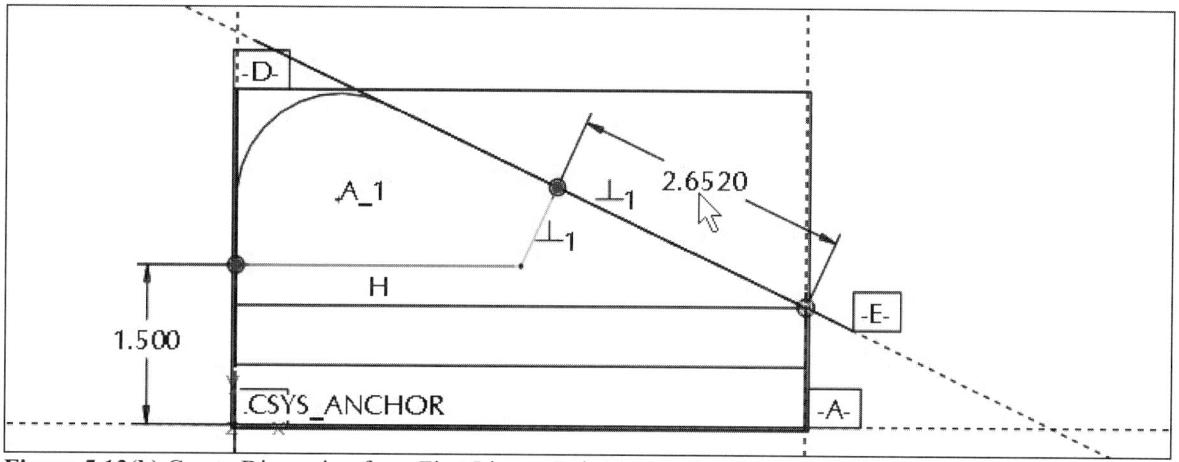

Figure 5.13(h) Create Dimension from First Line to Point (not end point to point!)

The **2.652** value is different than that shown in Figure 5.4. When this part is used in Lesson 11, the detail will show the dimension modeled here. You will be instructed on how to modify the value at the drawing level instead of on the model. Since the part, assembly and drawing are associative, the component will regenerate with the new size.

Click: ✓ from the Right Toolchest ⇒ from Top Toolchest, click: [datum display icon] off ⇒ [saved views icon] ⇒ **Standard Orientation** ⇒ from the dashboard, click: [remove material icon] **Remove Material** ⇒ **Options** ⇒ [-⊟- Symmetric] [Fig. 5.13(i)] ⇒ depth value **.75** ⇒ **Enter** ⇒ ✓ ⇒ **LMB** to deselect ⇒ **Ctrl+D** ⇒ [shading icon] **Shading** ⇒ [datum display icon] on ⇒ **Ctrl+S** ⇒ **Enter** [Fig. 5.13(j)]

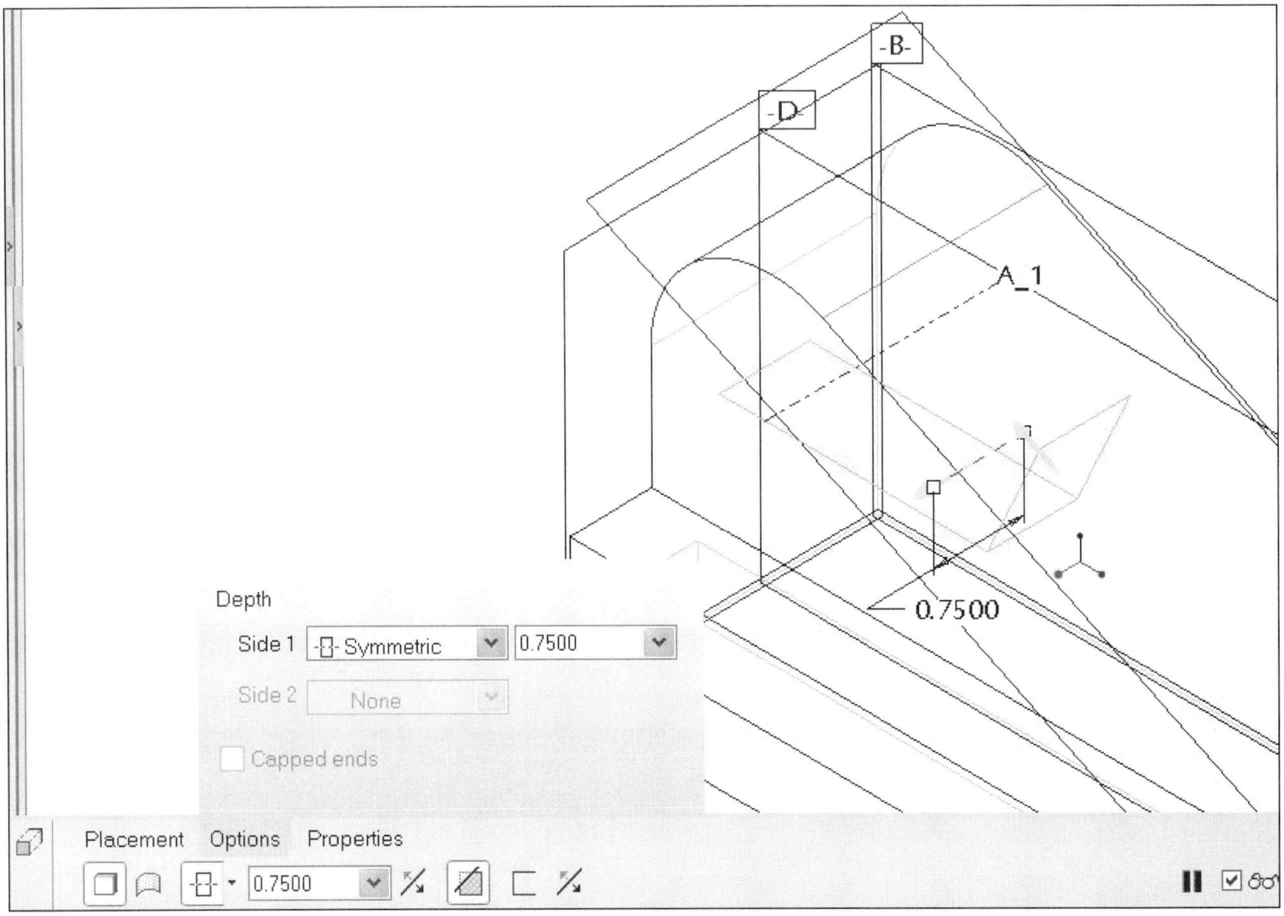

Figure 5.13(i) Symmetric Cut

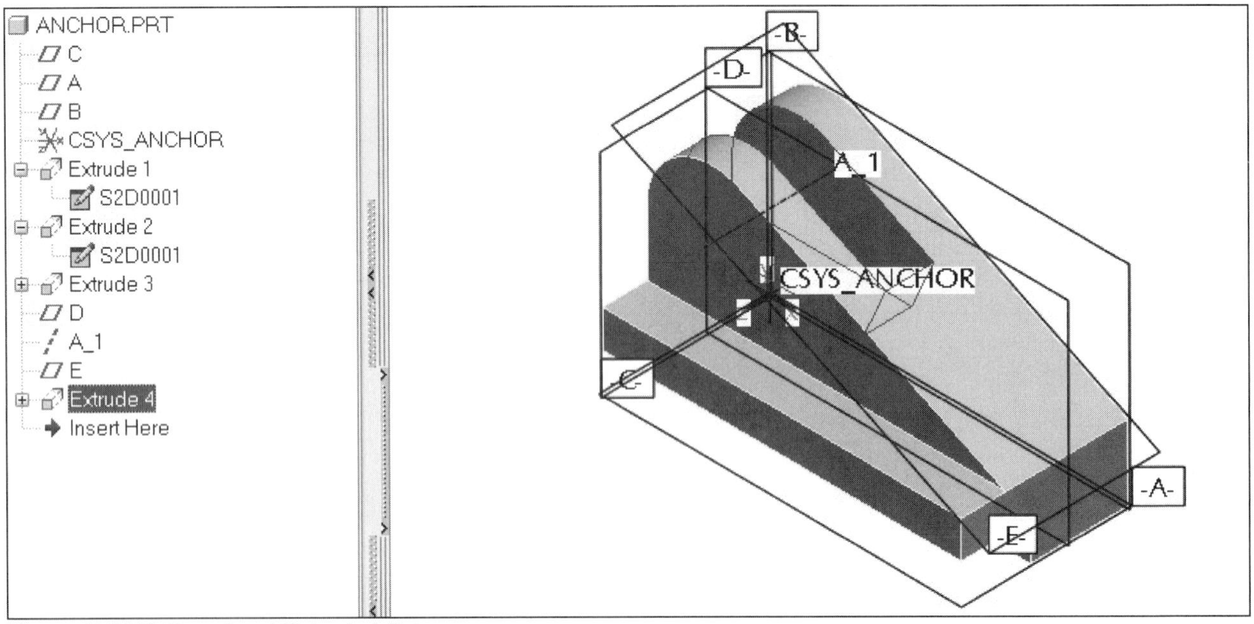

Figure 5.13(j) Completed Extruded Cut

The hole drilled in the angled surface appears to be aligned with datum **D**. Upon inspection (Fig. 5.7), the hole is at a different distance and is not in line with the slot and datum plane.

Click: **Hole Tool** ⇒ **Hidden line** ⇒ pick on the angled face and a hole will display with green drag handles [Fig. 5.14(a)] ⇒ drag one handle to datum **B** [Fig. 5.14(b)] and the other to the edge between the angled surface and the right vertical face [Fig. 5.14(c)]

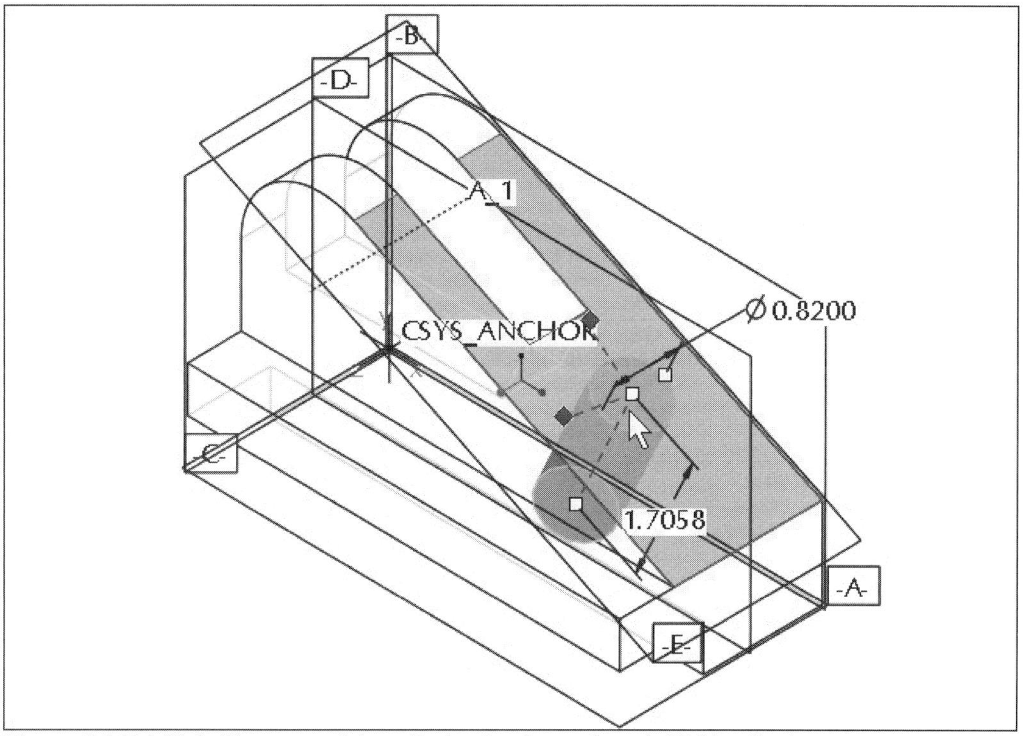

Figure 5.14(a) Hole Placement (view display has been altered- Hidden line)

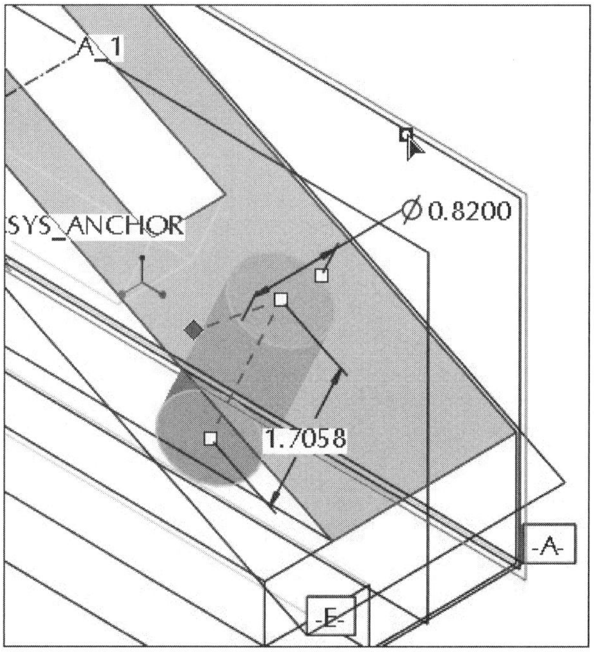

Figure 5.14(b) Drag Handle to Datum B

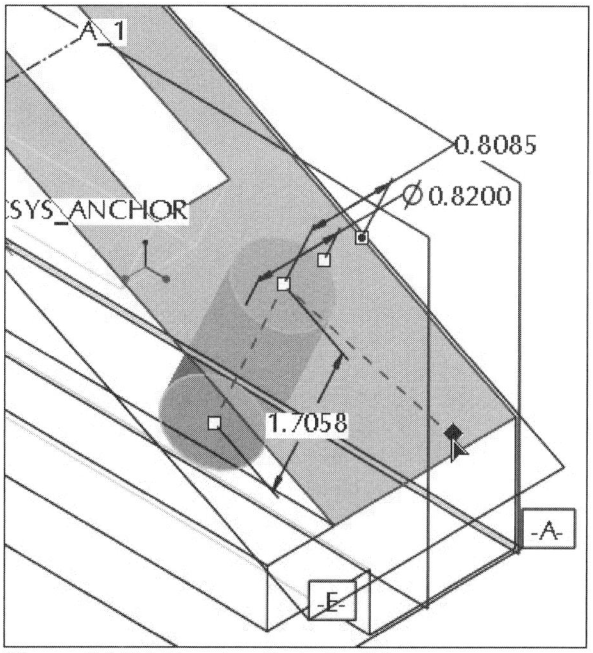

Figure 5.14(c) Drag handle to the Edge

Double-click on the dimension from the hole's center to datum **B** and modify the value to **.875** [Fig. 5.14(d)] ⇒ **Enter** ⇒ double-click on the dimension from the hole's center to the edge and modify the value to **2.0625** [Fig. 5.14(e)] ⇒ **Enter** ⇒ **Use standard hole profile as drill hole profile** in the dashboard ⇒ ⌀ 1.0000 ⇒ 1.1250 [Fig. 5.14(f)] ⇒ **Shape** tab ⇒ ⇒ ⇒ **MMB**

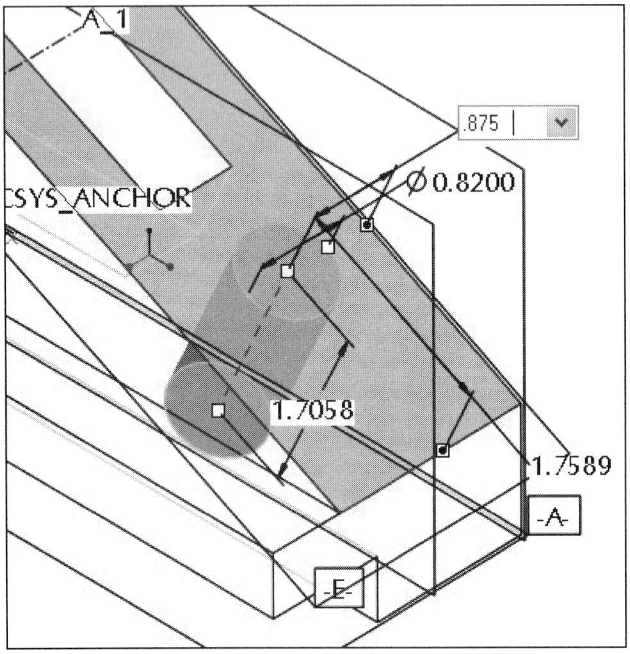

Figure 5.14(d) Modify the Distance to Datum B

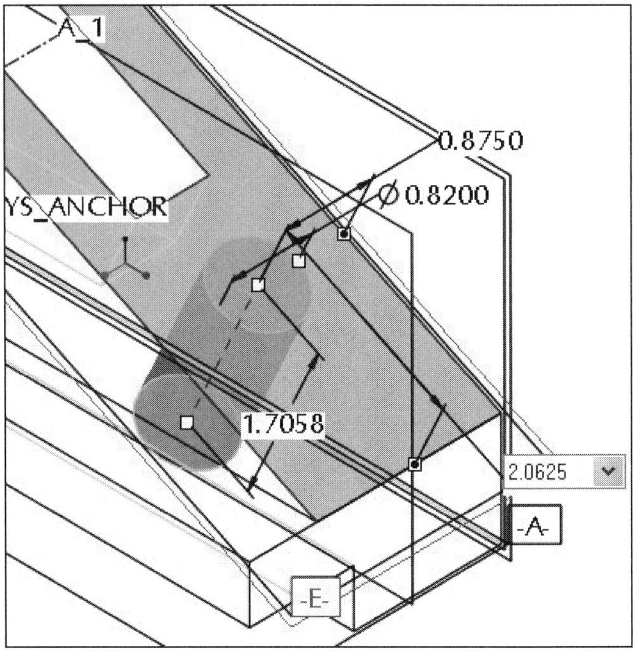

Figure 5.14(e) Modify the Distance to the Part Edge

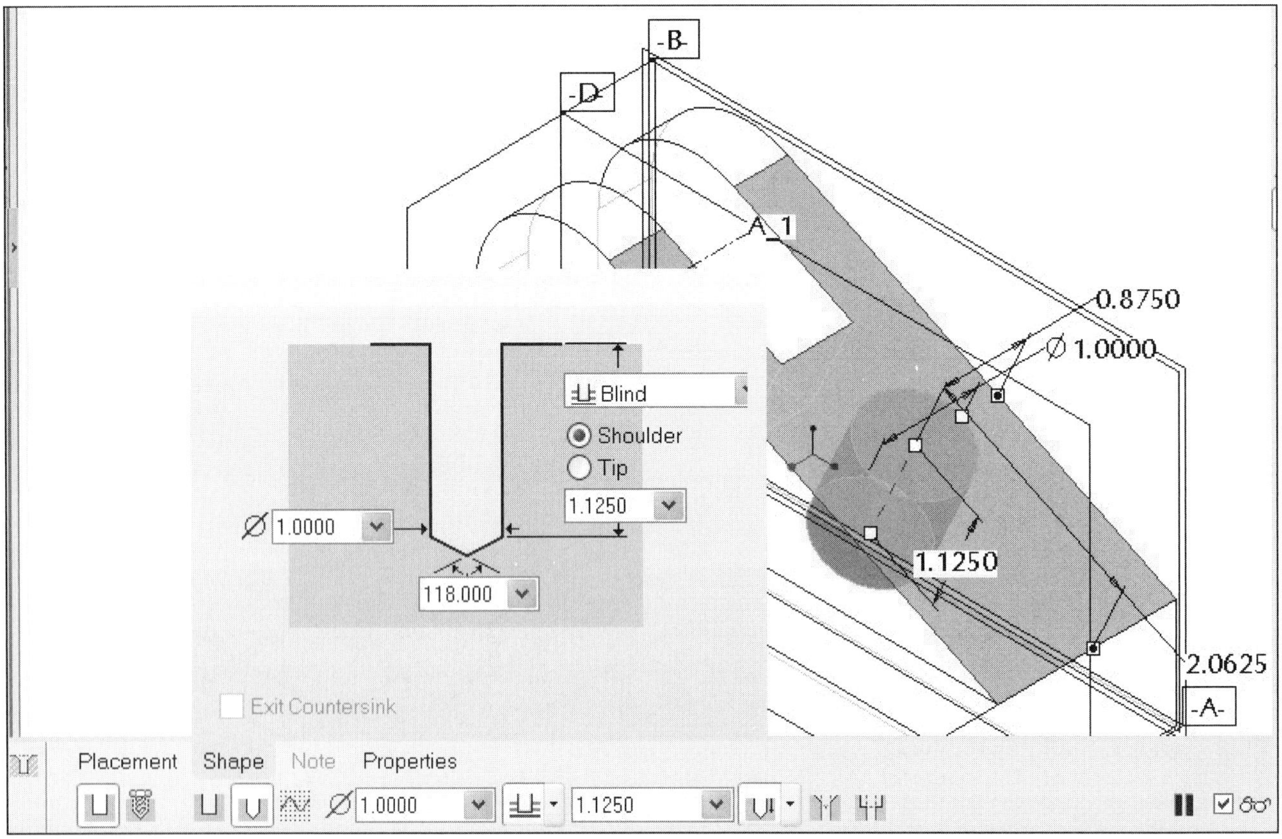

Figure 5.14(f) Shape Tab

Click: Hole Tool ⇒ pick **A_1** from the model or the Model Tree ⇒ hold down the **Ctrl** key and pick datum **D** from the model or the Model Tree ⇒ **Placement** tab [Fig. 5.15(a)] ⇒ **Shape** tab ⇒ diameter **.250** ∅0.2500 ⇒ **Through All** ⇒ (Side 2) **Through All** [Fig. 5.15(b)] ⇒ ✓ ⇒ **Ctrl+S** ⇒ **Enter**

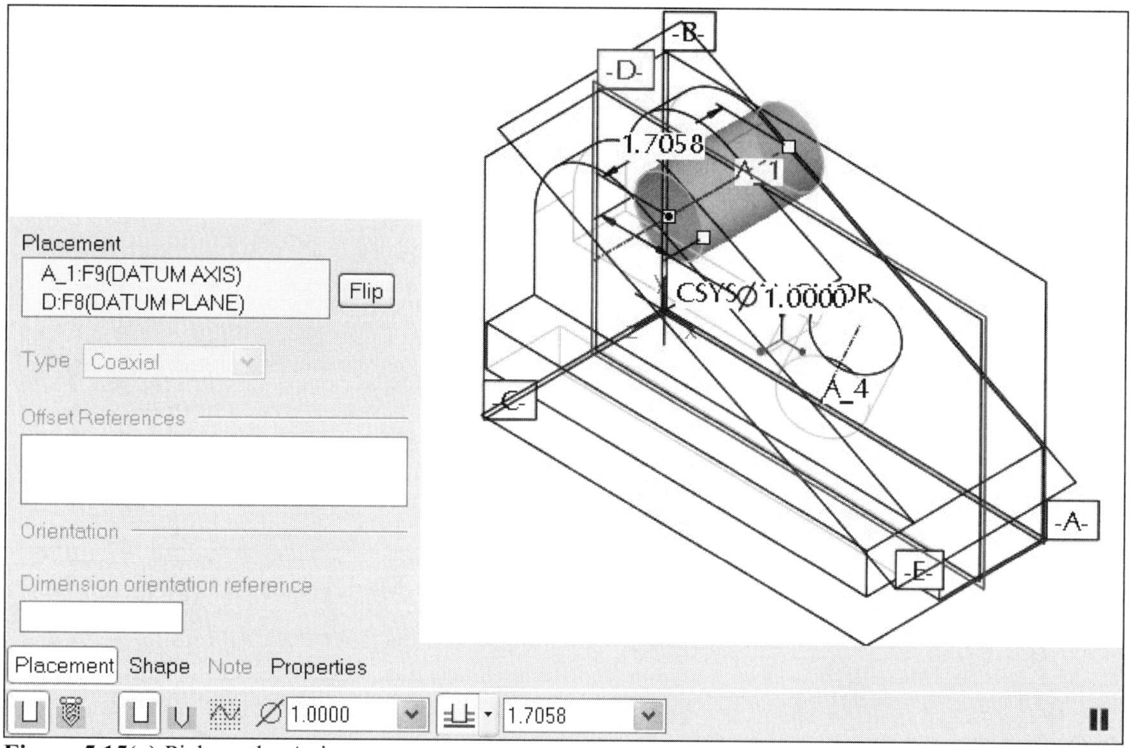

Figure 5.15(a) Pick on the Axis

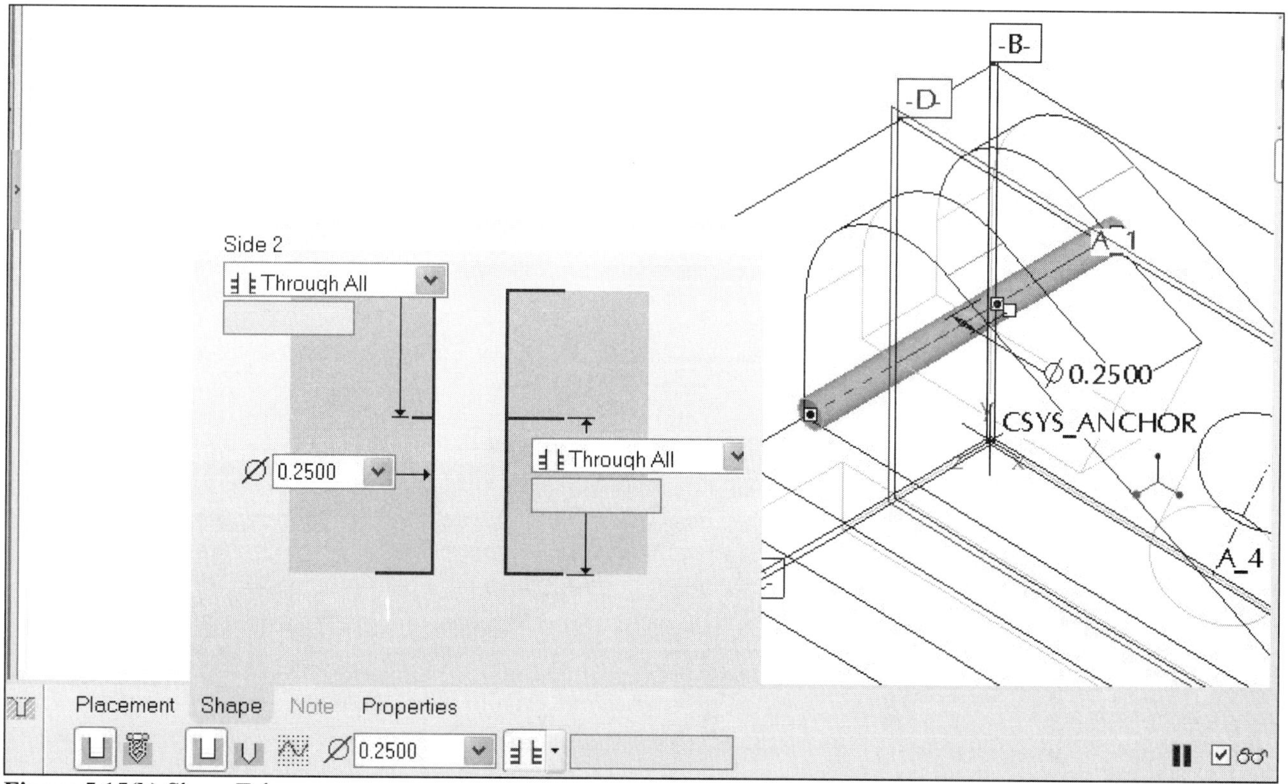

Figure 5.15(b) Shape Tab

Suppressing and Resuming Features using Layers

Next, you will create a new layer and add the two holes to it. The holes will then be selected in the Layer Tree and suppressed. Suppressed features are temporarily removed from the model along with their children (if any). In the example, you will notice that the holes and one axis (A_4) will be suppressed. Since axis A_1 is the parent of the small hole, it will not be suppressed. Suppressing features is like removing them from regeneration temporarily. However, you can "unsuppress" (resume) suppressed features at any time.

You can suppress features on a part to simplify the part model and decrease regeneration time. For example, while you work on one end of a shaft, it may be desirable to suppress features on the other end of the shaft. Similarly, while working on a complex assembly, you can suppress some of the features and components for which the detail is not essential to the current assembly process. Suppress features to do the following:

- Concentrate on the current working area by suppressing other areas
- Speed up a modification process because there is less to update
- Speed up the display process because there is less to display
- Temporarily remove features to try different design iterations

Unlike other features, the base feature cannot be suppressed. If you are not satisfied with your base feature, you can redefine the section of the feature, or start another part.

Click: **Shade** ⇒ **Set layers** (Layer Tree displays in place of the Model Tree) ⇒ in the Layer Tree to expand and show the layer items [Fig. 5.16(a)]

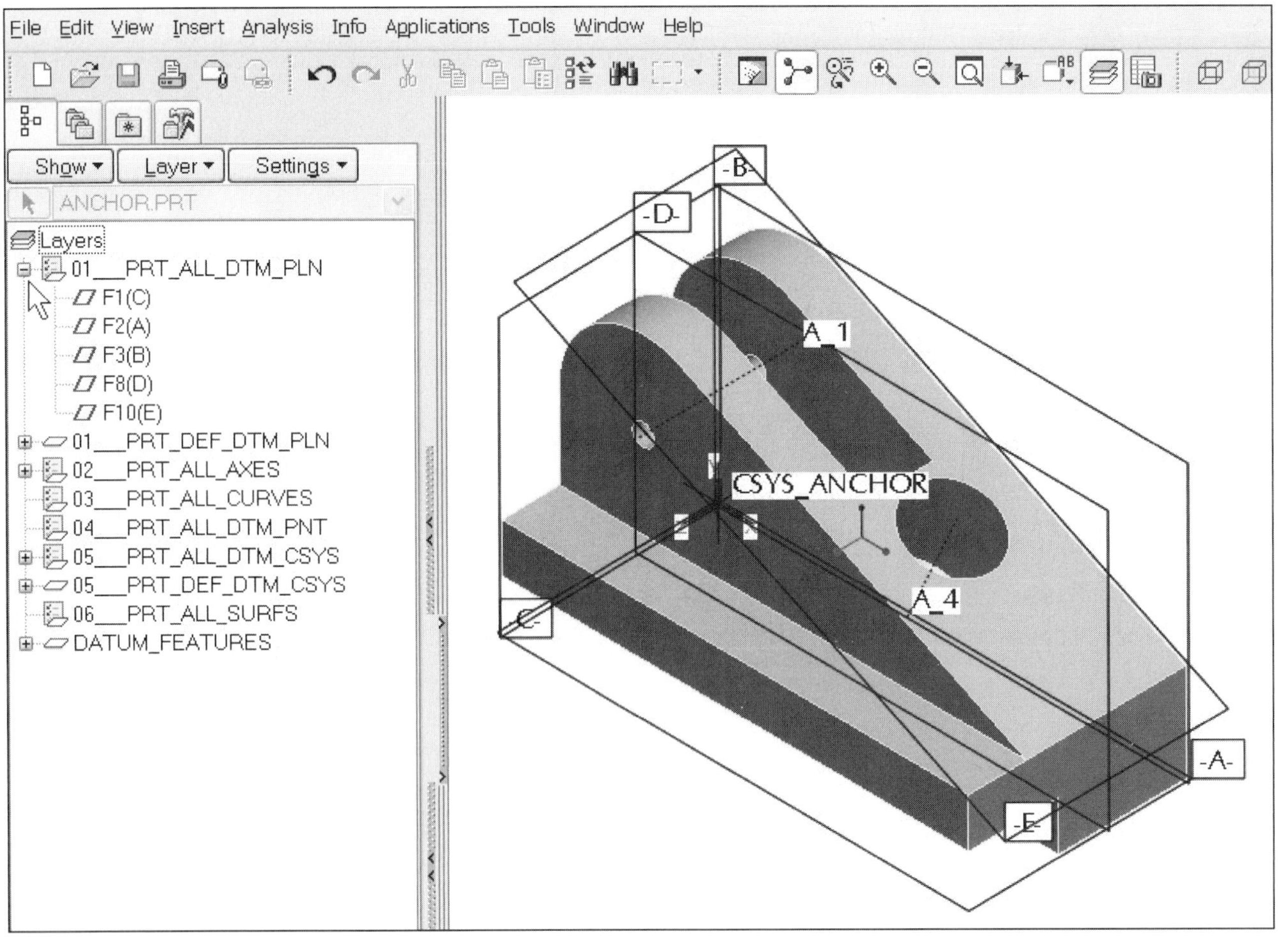

Figure 5.16(a) Layer Tree

Click **Layer** 

⇒ **New Layer** [Fig. 5.16(b)] ⇒ in the Layer Properties dialog box**,** Name: type **HOLES** as the name for the new layer *(do not press Enter)* [Fig. 5.16(c)] ⇒ Selection Filter (bottom right-hand side below graphics window), click: **Feature** [Fig. 5.16(d)] ⇒ select the two holes from the model *(no need to hold down the Ctrl key)* [Fig. 5.16(e)] ⇒ **OK** ⇒ expand the **HOLES** layer ⇒ press and hold down the **Ctrl** key and pick on the two holes in the **HOLES** layer [Fig. 5.16(f)]

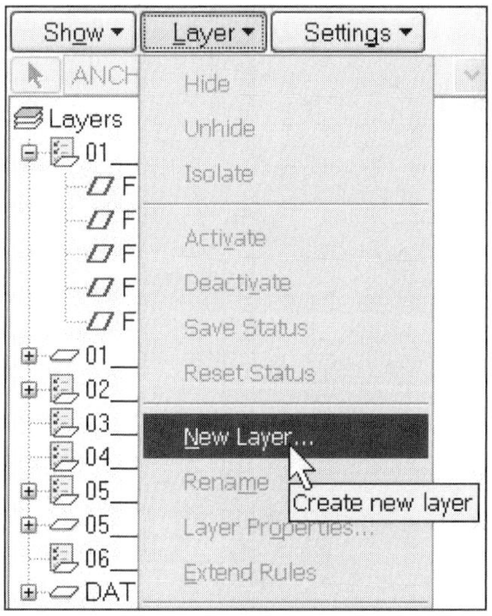

Figure 5.16(b) New Layer

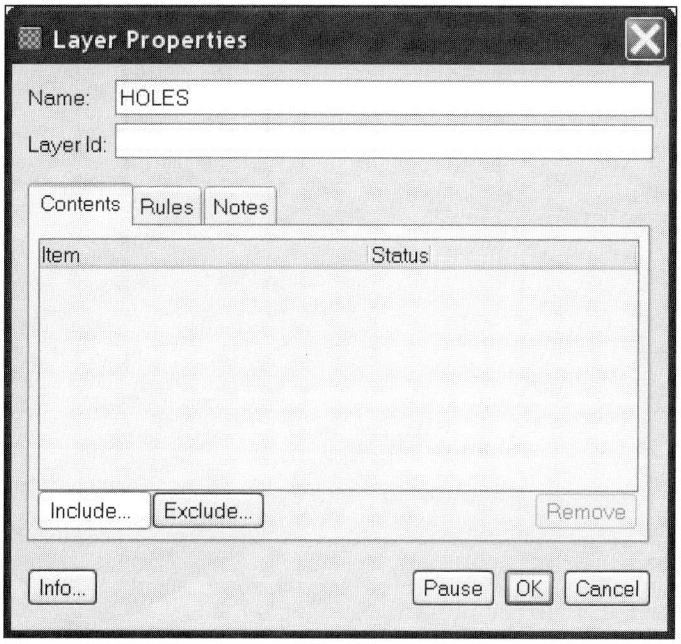

Figure 5.16(c) New Layer: HOLES

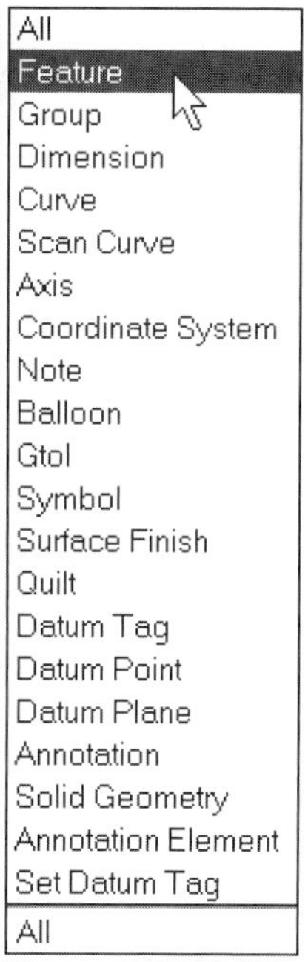

Figure 5.16(d) Filter

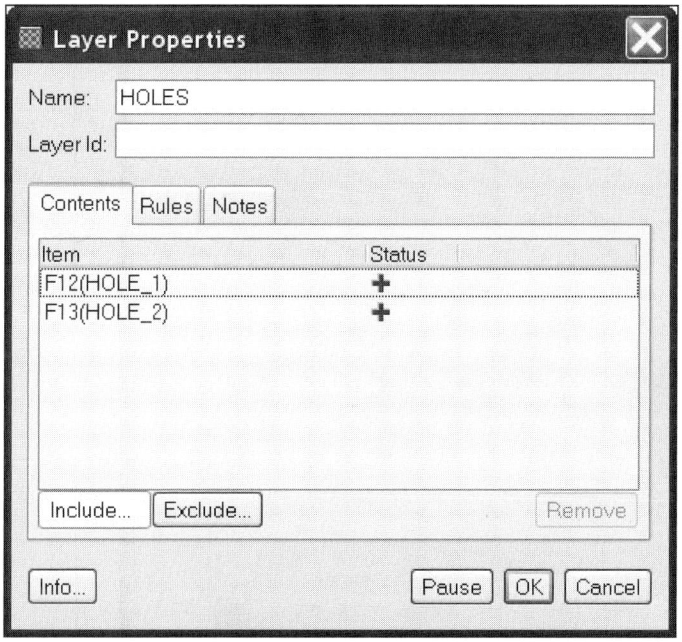

Figure 5.16(e) Holes Layer Items

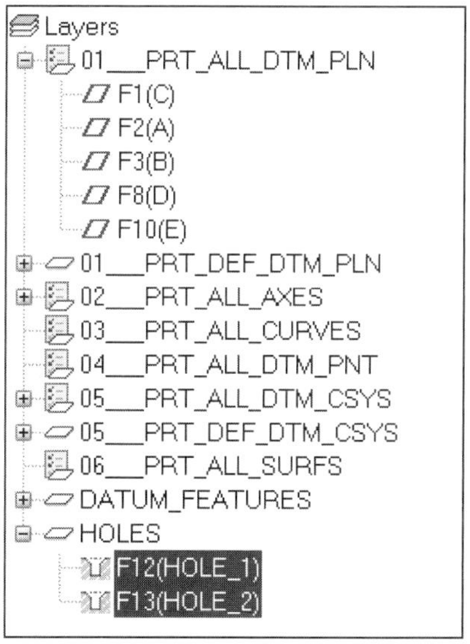

Figure 5.16(f) Layer Tree Items Selected

Click: **Edit** from the menu bar ⇒ **Suppress** ⇒ **Suppress** [Fig. 5.16(g)] ⇒ **OK** from the Suppress dialog box ⇒ **LMB** to deselect *(look at the Layer Tree and the Model to see that the holes are gone)* [Fig. 5.16(h)] ⇒ **Edit** from the menu bar ⇒ **Resume** ⇒ **Resume All** ⇒ **Show** in the Navigator ⇒ **Model Tree** ⇒ [save icon] ⇒ **OK** ⇒ **LMB** to deselect

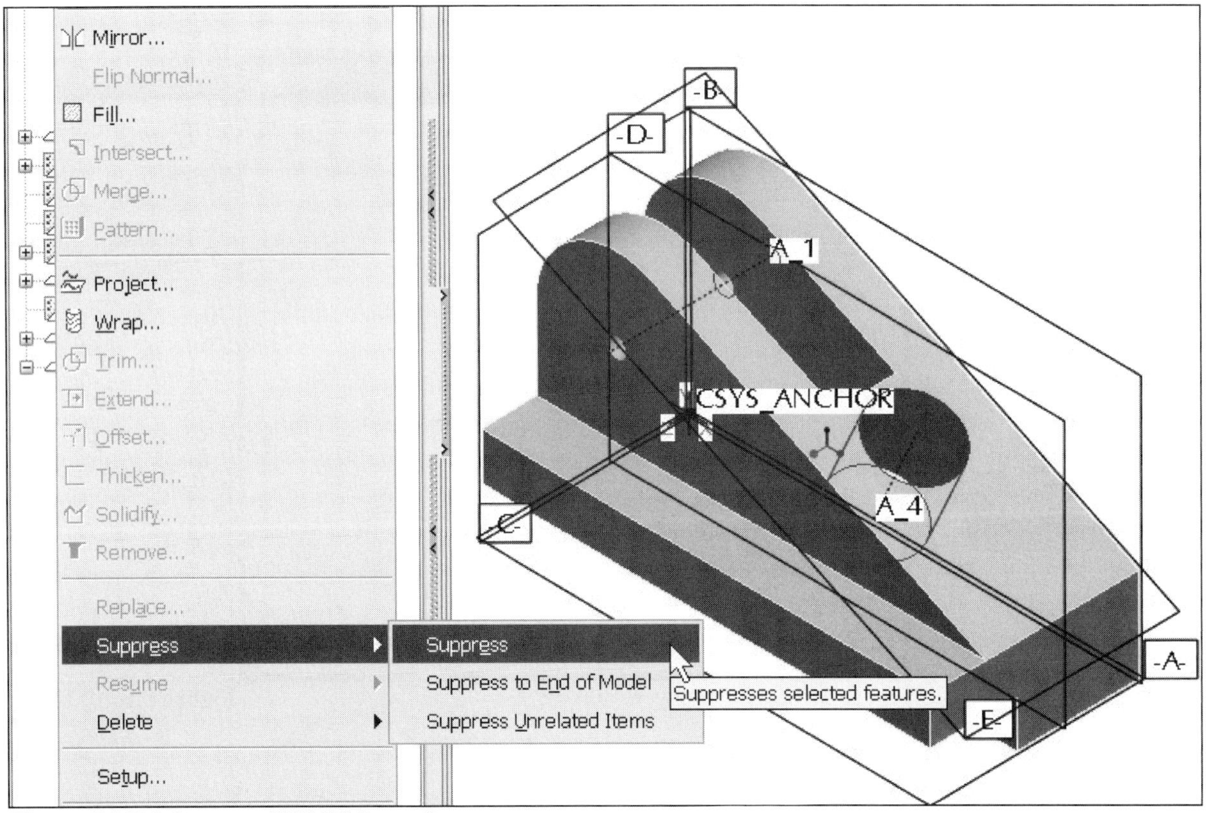

Figure 5.16(g) Suppress HOLES Layer Items

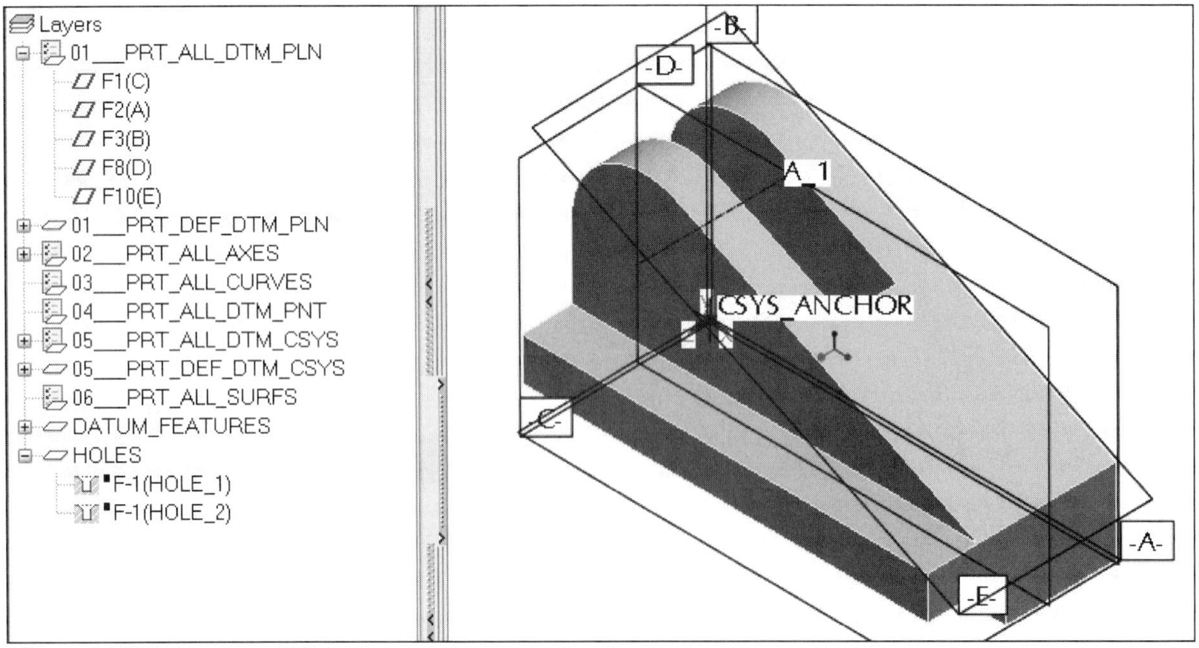

Figure 5.16(h) Suppressed Holes

Cross Sections

There are two types of cross sections: **planar** and **offset**. Planar cross sections can be crosshatched or filled, while offset cross sections can be crosshatched but not filled. You will be creating a planar cross section. Pro/E can create standard planar cross sections of models (parts or assemblies), offset cross sections of models (parts or assemblies), planar cross sections of datum surfaces or quilts (Part mode only) and planar cross sections that automatically intersect all quilts and all geometry in the current model.

Click: **Start the view manager** from Top Toolchest ⇒ View Manager dialog box displays [Fig. 5.17(a)] ⇒ **Xsec** tab [Fig. 5.17(b)] ⇒ **New** [Fig. 5.17(c)] ⇒ type name **A** [Fig. 5.17(d)] ⇒ **Enter** ⇒ **Planar** [Fig. 5.17(e)] ⇒ **Single** [Fig. 5.17(f)] ⇒ **Done** ⇒ **Plane** [Fig. 5.17(g)]

Figure 5.17(a) View Manager

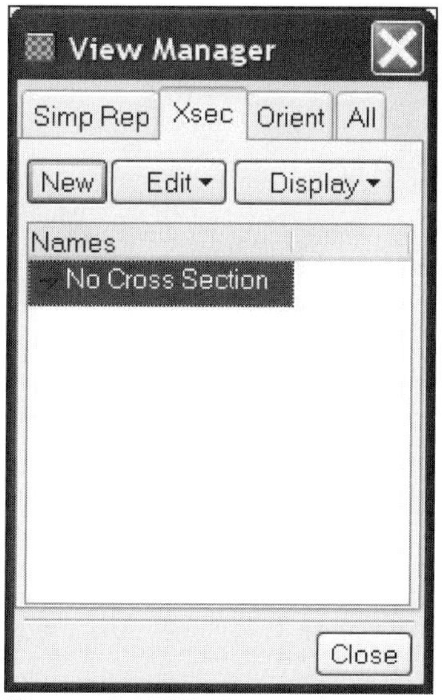

Figure 5.17(b) Xsec Tab

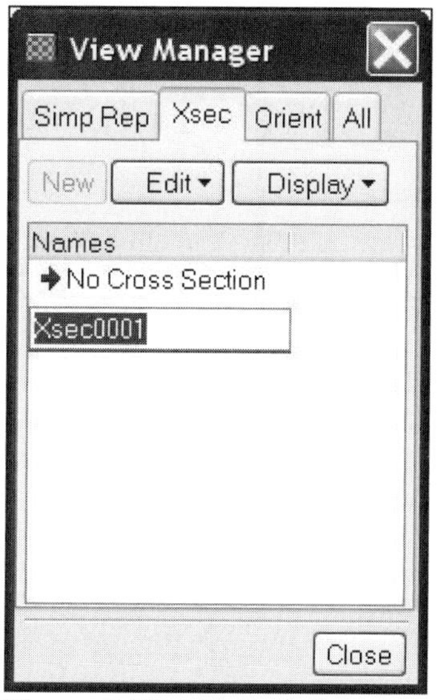

Figure 5.17(c) Default Xsec

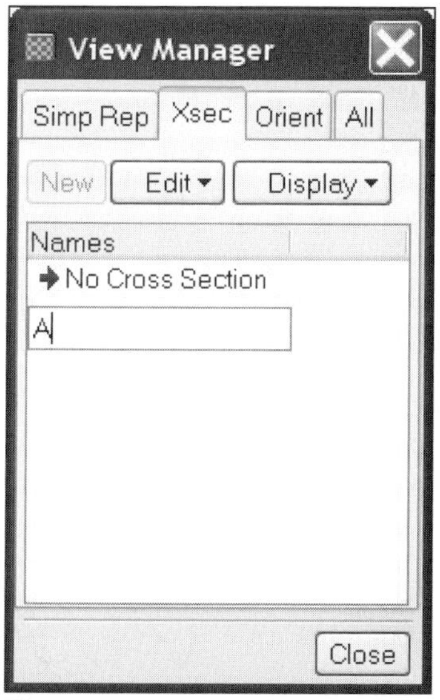

Figure 5.17(d) Section A

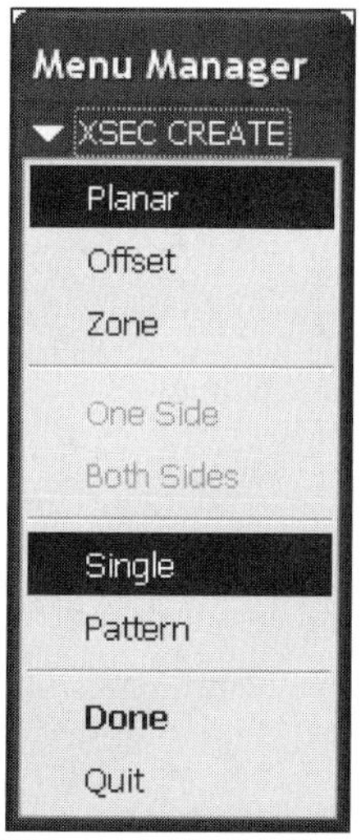

Figure 5.17(e) Planar

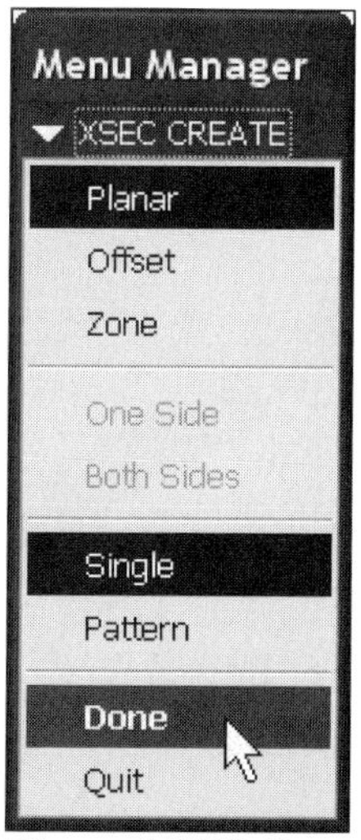

Figure 5.17(f) Done

Figure 5.17(g) Plane

Pick datum **D** ⇒ **Display** ⇒ **Visibility** ⇒ **Edit** ⇒ **Redefine** [Fig. 5.17(h)] ⇒ **Hatching** ⇒ **Fill** [Fig. 5.17(i)] ⇒ **Hatch** ⇒ **Spacing** ⇒ **Half** [Fig. 5.17(j)] ⇒ **Done** *or MMB* ⇒ **Done/Return** *or MMB* ⇒ **Display** ⇒ **Visibility** off ⇒ **Close** ⇒ ⇒ **OK** *or MMB*

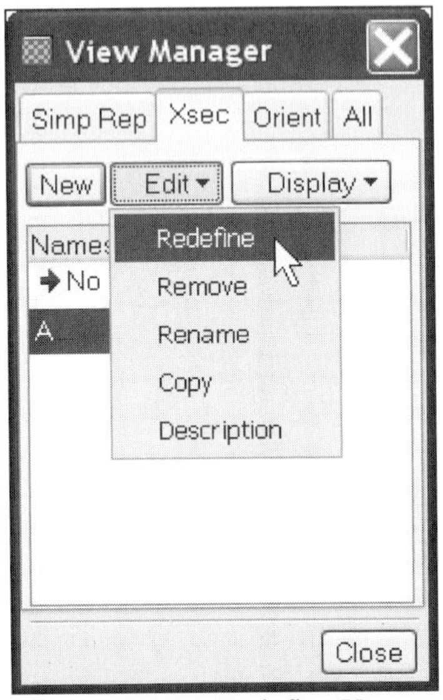

Figure 5.17(h) Redefine Xsec

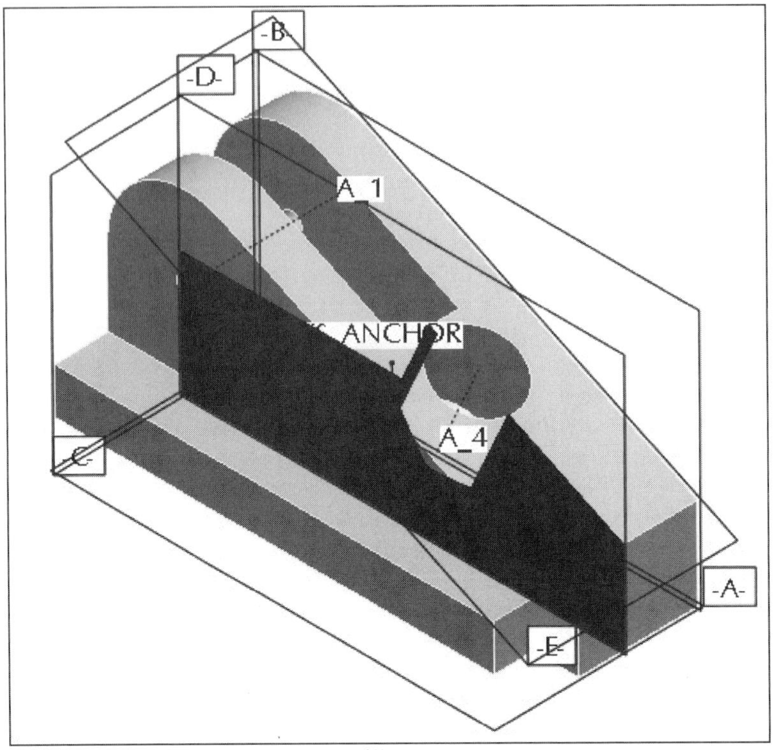

Figure 5.17(i) Model X-Section Fill

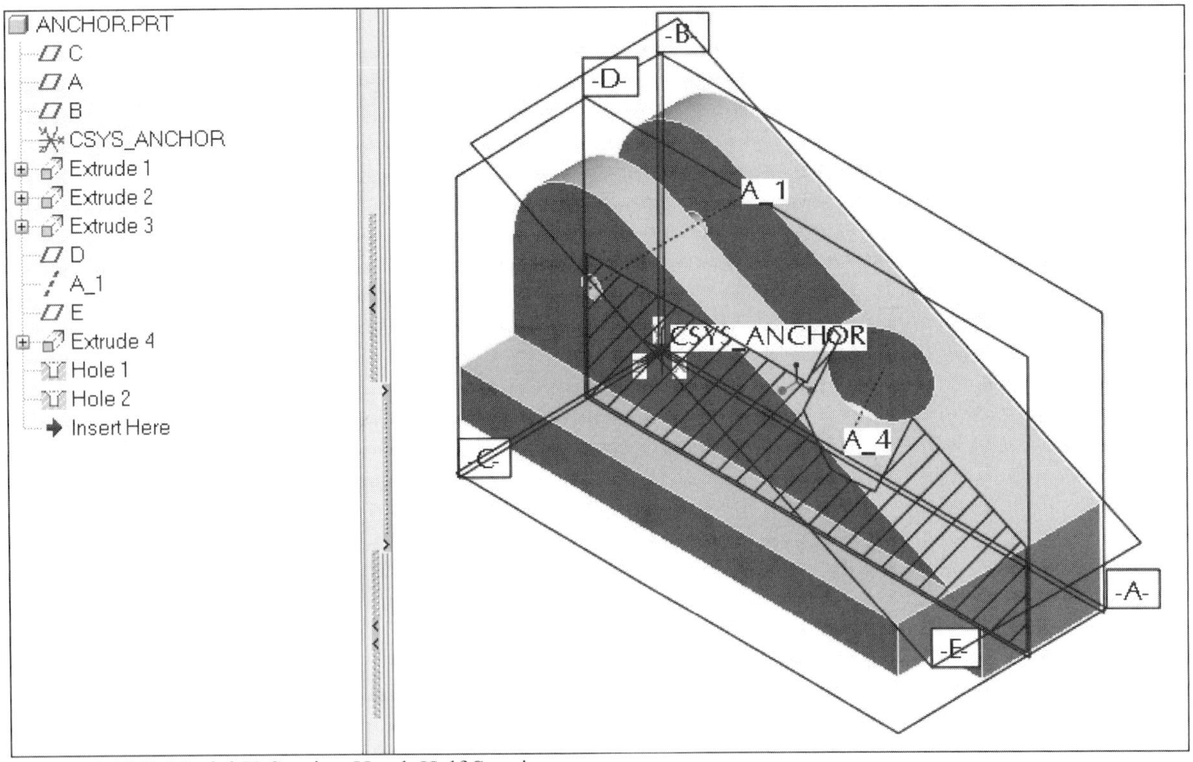

Figure 5.17(j) Model X-Section Hatch Half Spacing

The section passes through datum D. The slot is the *child* of datum D. If datum D moves, so will the slot and the small hole (and X-Section A). In order to ensure that the datum D stays centered on the upper portion of the part, you will need to create a relation to control the location of datum D. Relations will be covered in-depth in a later lesson. The first cut used a dimension from datum B for location (**1.875**). The relation should state that the distance from datum B to datum D will be one-half the value of the distance from datum B to the first cut. If the thickness of the upper portion of the part (**1.875**) changes datum D will remain centered as will the slot and the X-Section. To start, you must first find out the feature dimension symbols (**d#**) required for the relation.

(**READ!** - Almost everyone will have a different **d#** numbers than the ones displayed here)

Select the first cut from the model or the Model Tree ⇒ **RMB** ⇒ **Edit** [Figs. 5.18(a-b)] ⇒ **Info** from the menu bar ⇒ Switch Dimensions note the **d** symbol **d6** [Fig. 5.18(c)] *(Note: your "**d**" values may be different)* ⇒ select datum **D** in the Model Tree ⇒ **RMB** ⇒ **Edit** ⇒ note the **d** symbol **d11** [Fig. 5.18(d)] *(Note: your "**d**" values may be different. If you have the incorrect symbol, your relation will not work. Use your **d** symbols not the ones in the text!)*

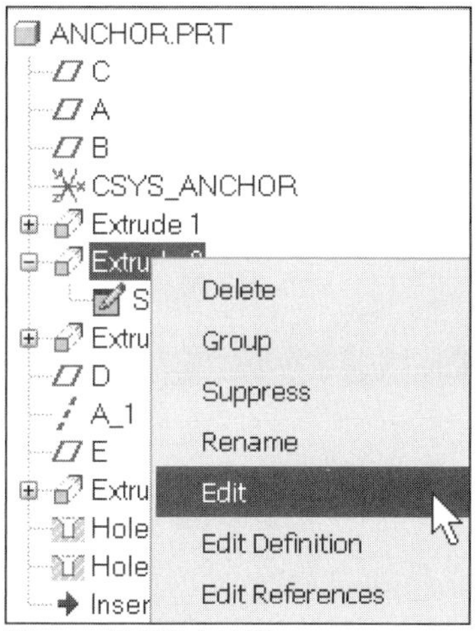

Figure 5.18(a) Edit

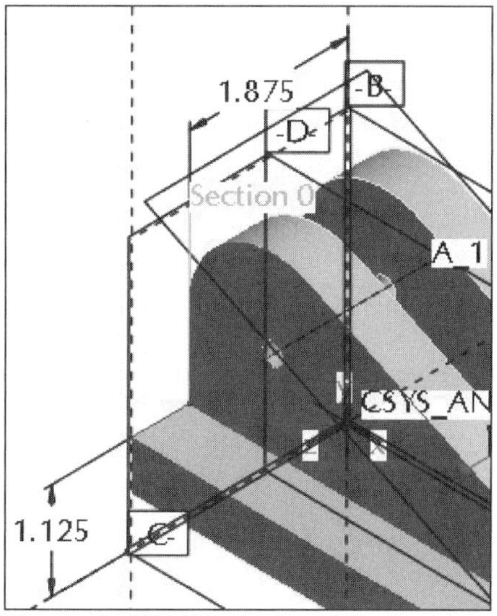

Figure 5.18(b) Dimensions Displayed

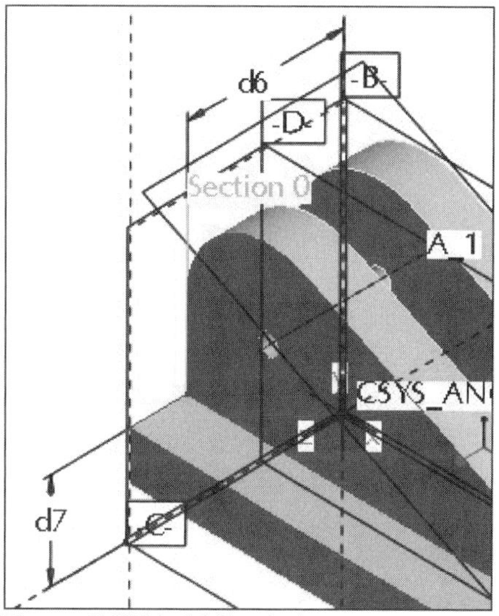

Figure 5.18(c) d6 Symbol

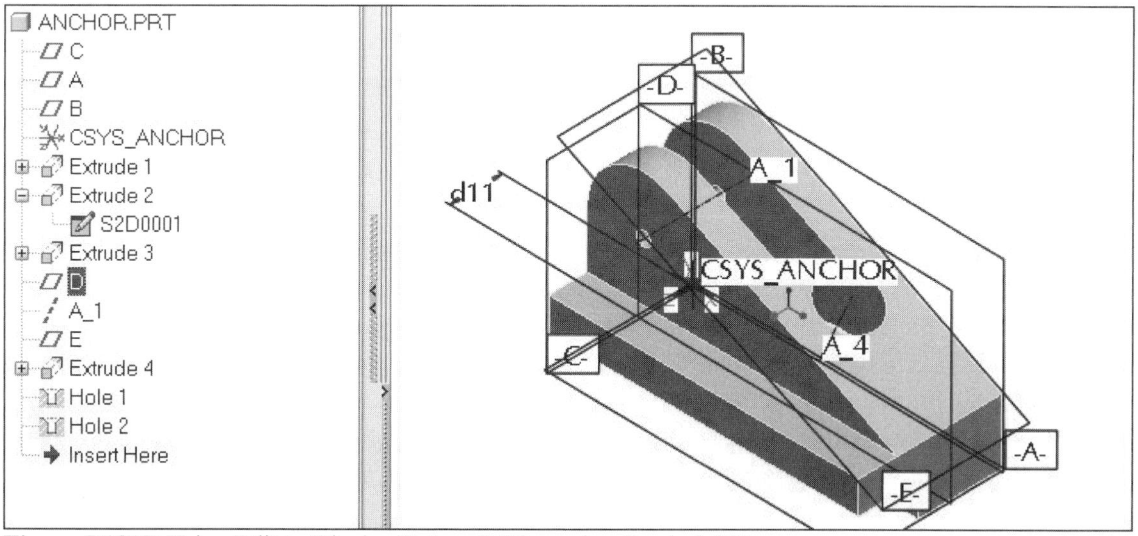

Figure 5.18(d) Using Edit to Display Datum D Dimension Symbol **d11**

Click: **Tools** from the menu bar ⇒ **Relations** Relations dialog box displays ⇒ **Local Parameters** to see the parameters of the part ⇒ type *(or pick from the model)* **d11=d6/2** in the Relations field *(your "d" values may be different)* ⇒ **Execute/Verify** [Fig. 5.18(e)] ⇒

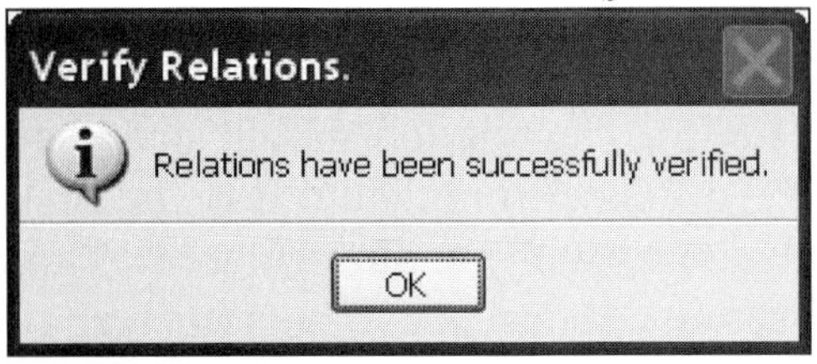

OK ⇒ **Ok** ⇒ **Info** ⇒ **Switch Dimensions** ⇒ **Edit** ⇒ **Regenerate** ⇒ ⇒ **MMB**

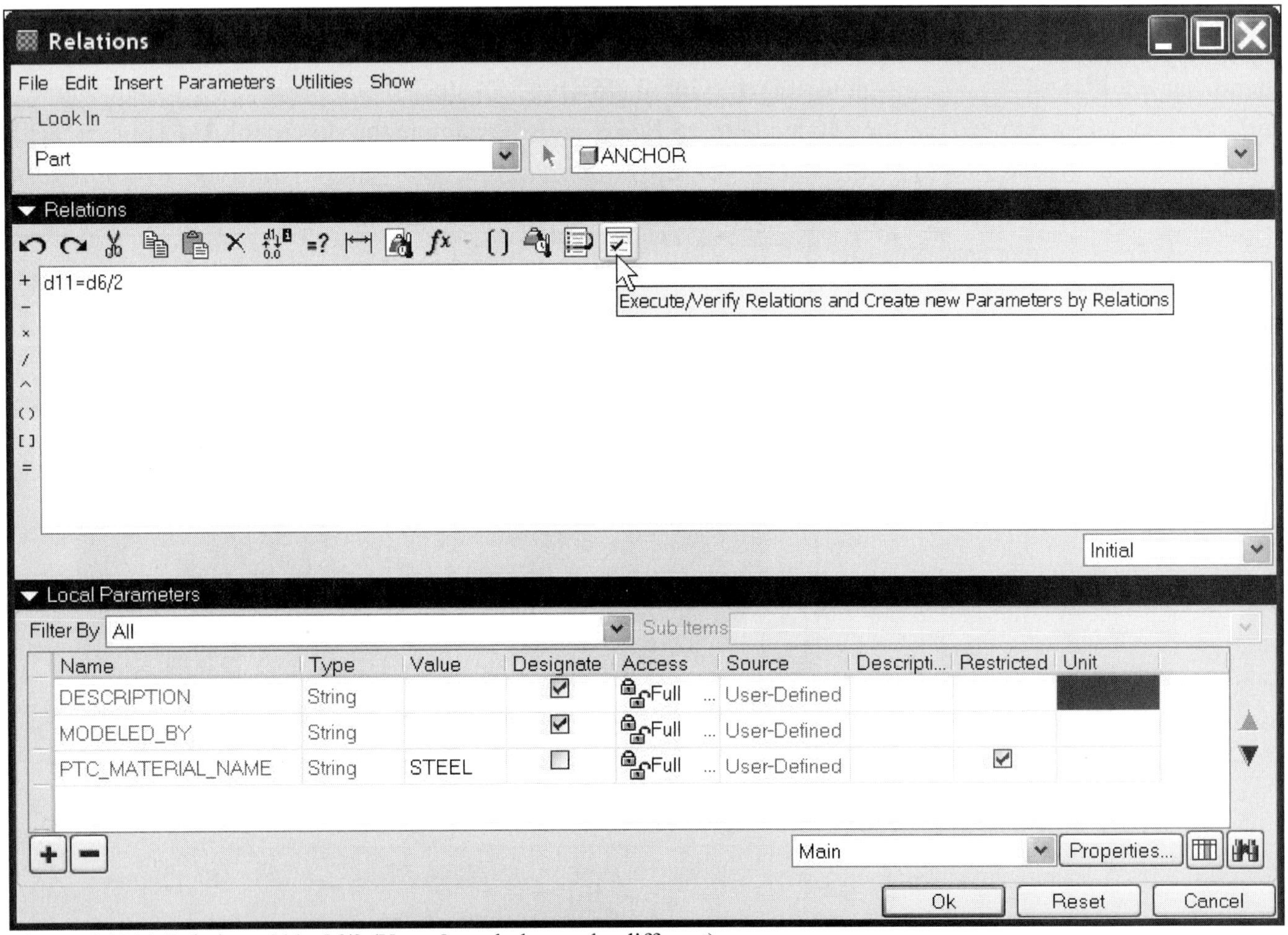

Figure 5.18(e) Relations **d11=d6/2** (Your **d** symbols may be different)

The slot will now remain in the center of the upper protrusion regardless of changes in the protrusions width.

In the real world, you will seldom encounter a situation where the project is designed and modeled without a "design change" or **ECO** (Engineering Change Order). Therefore, let us assume that an ECO has been "issued" that states: *the location of the hole on the angled surface must be aligned with the center of the slot at all times.*

A relation could be established to control the hole as was done with the slot, but instead of referencing the hole from Datum B with a dimension we will *align* the holes offset reference from the D datum.

Select the large hole in the Model Tree or on the Model ⇒ **RMB** ⇒ **Edit Definition** [Fig. 5.18(f)] ⇒ **Placement** ⇒ pick on datum **B (DATUM)** in the Offset References collector ⇒ **RMB** ⇒ **Remove** [Fig. 5.18(g)]

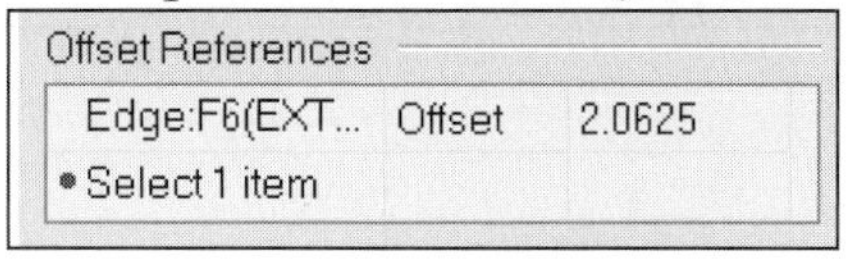

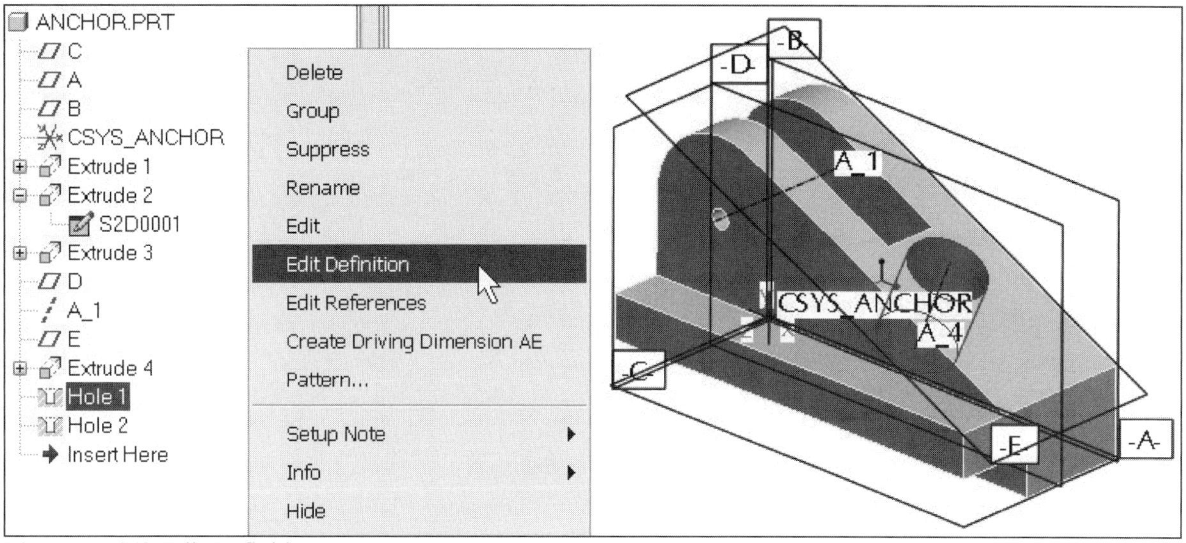

Figure 5.18(f) Edit Definition

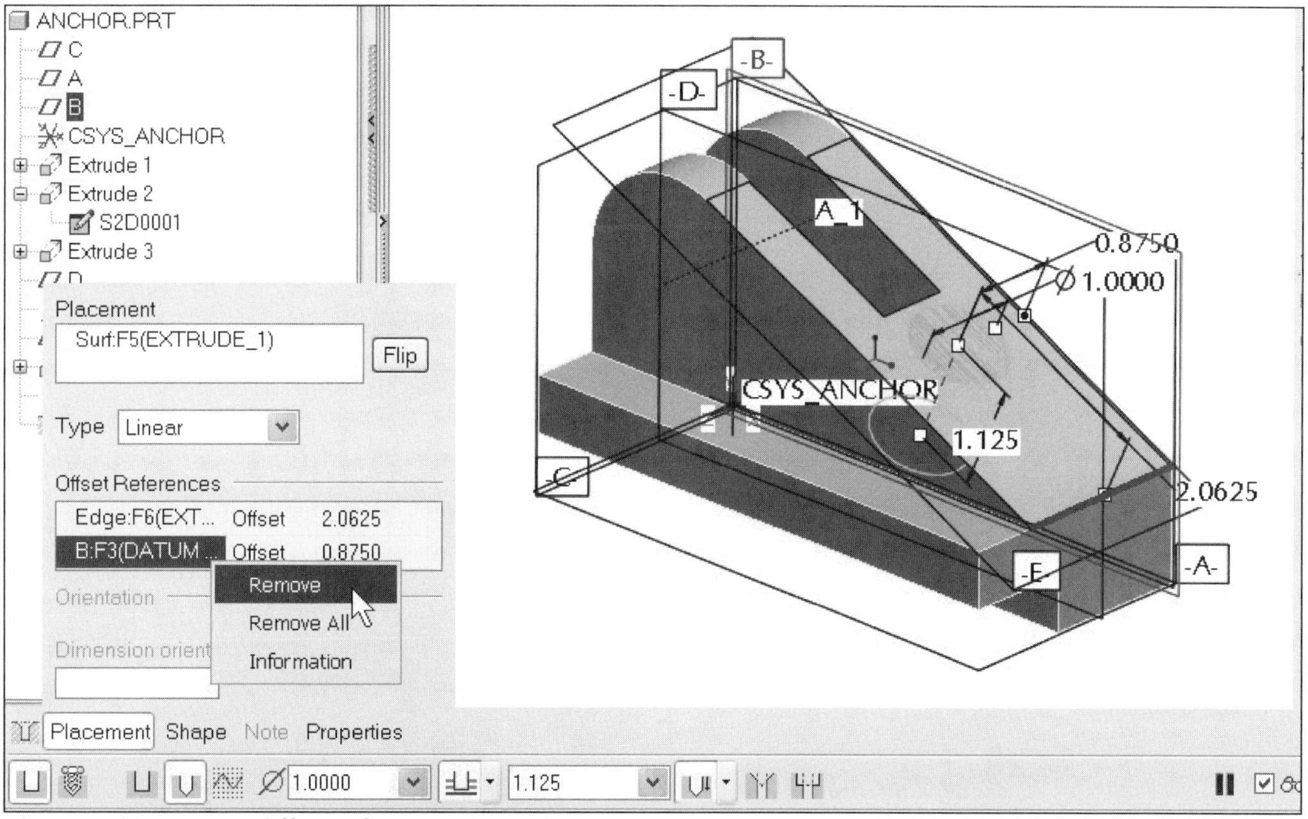

Figure 5.18(g) Remove Offset Reference B (Datum)

Press and hold down **Ctrl** key and select **D (DATUM)** from the model or Model Tree [Fig. 5.18(h)] ⇒ **Offset** in the Offset References collector ⇒ **Align** [Fig. 5.18(i)] ⇒  ⇒ **Ctrl+S** ⇒ **Enter**

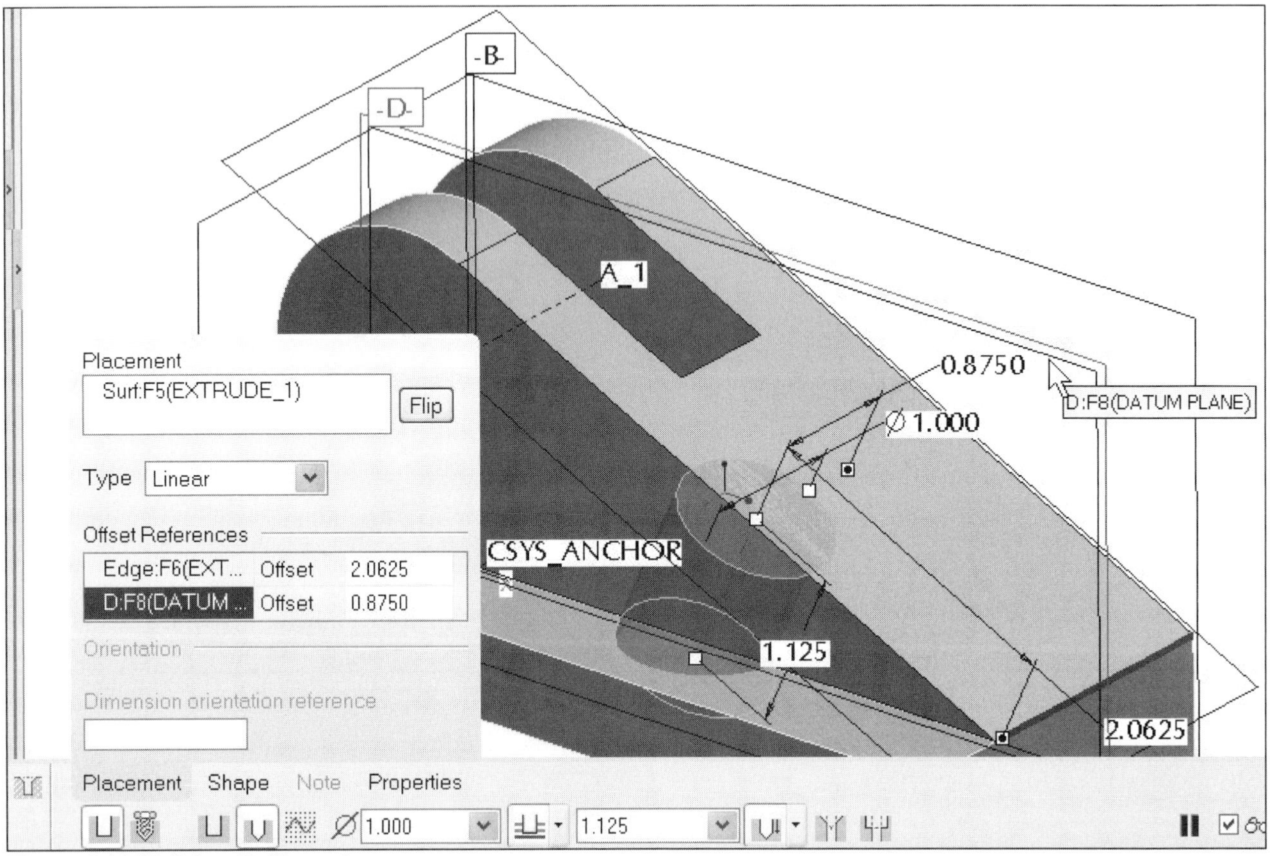

Figure 5.18(h) Select D (Datum)

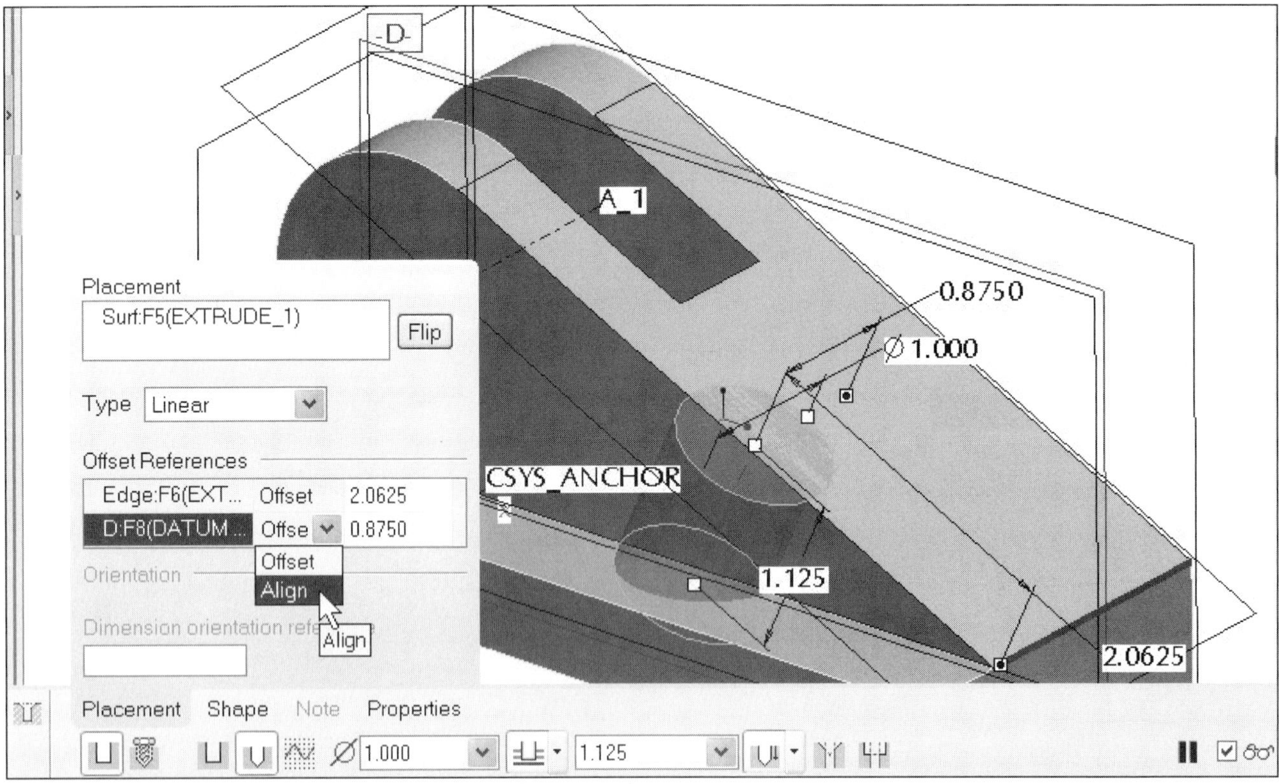

Figure 5.18(i) Align (Use 2.0625 as the distance from the edge)

Click: ⇒ **Standard Orientation** ⇒ **Info** ⇒ **Relations and Parameters** [Fig. 5.18(j)]

Relations and Parameters : ANCHOR

PART NAME : ANCHOR

Relation Table

Relation	Parameter	New Value
	Relations for ANCHOR:	
	Initial Relations	
d11 =d6 /2	D11	9.375000e-01

Local Parameters

Symbolic constant	Current value	TYPE	SOURCE	ACCESS	DESIGNATED	DESCRIP
DESCRIPTION	---	String	User-Defined	Full	YES	
MODELED_BY	---	String	User-Defined	Full	YES	

Alternative Mass Property Parameters

Symbolic constant	Current value	TYPE	SOURCE	ACCESS	DESIGNA
MP_DENSITY	2.827713e-01	Real Number	Alternate Mass Properties	Full	NO
PRO_MP_SOURCE	GEOMETRY	String	Alternate Mass Properties	Full	NO
PRO_MP_ALT_MASS	---	Unknown	Alternate Mass Properties	Full	NO
PRO_MP_ALT_VOLUME	---	Unknown	Alternate Mass Properties	Full	NO
PRO_MP_ALT_AREA	---	Unknown	Alternate Mass Properties	Full	NO
PRO_MP_ALT_COGX	---	Unknown	Alternate Mass Properties	Full	NO
PRO_MP_ALT_COGY	---	Unknown	Alternate Mass Properties	Full	NO
PRO_MP_ALT_COGZ	---	Unknown	Alternate Mass Properties	Full	NO
PRO_MP_ALT_IXX	---	Unknown	Alternate Mass Properties	Full	NO
PRO_MP_ALT_IYY	---	Unknown	Alternate Mass Properties	Full	NO
PRO_MP_ALT_IZZ	---	Unknown	Alternate Mass Properties	Full	NO
PRO_MP_ALT_IXZ	---	Unknown	Alternate Mass Properties	Full	NO
PRO_MP_ALT_IYZ	---	Unknown	Alternate Mass Properties	Full	NO
PRO_MP_ALT_IXY	---	Unknown	Alternate Mass Properties	Full	NO

Figure 5.18(j) Relations and Parameters in the Browser

Click on **d6**, and **d11** (your **d** values may be different) in the Browser [Fig. 5.18(k)] to display them in the graphics window⇒ close the Browser with the quick sash 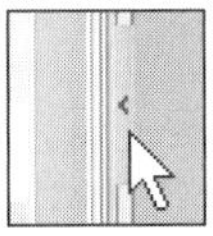 ⇒ **Ctrl+S** ⇒ **Enter** ⇒ **File** ⇒ **Delete** ⇒ **Old Versions** ⇒ **MMB** ⇒ **Info** ⇒ **Feature List** ⇒ **File** ⇒ **Close Window**

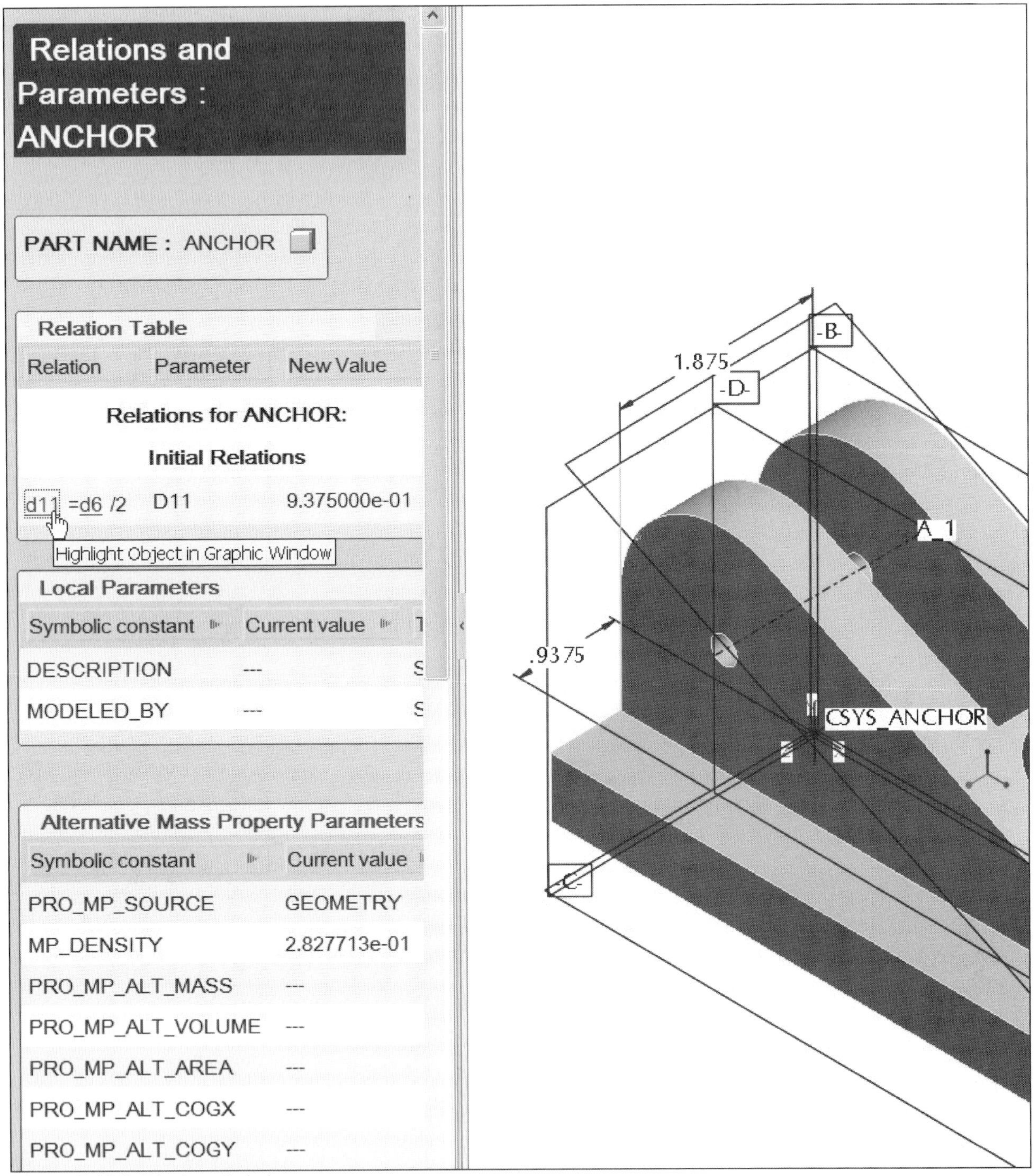

Figure 5.18(j) Displays Relation

The lesson is now complete. If you wish to model a project without instructions, a complete set of projects and illustrations are available at ***www.cad-resources.com*** ⇒ ***Downloads***.

Lesson 6 Revolved Features

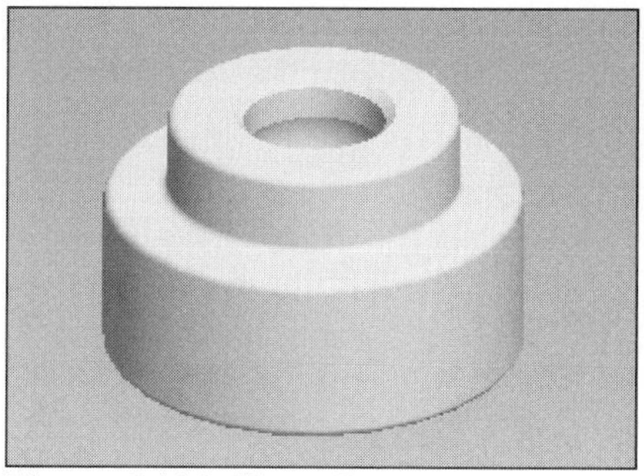

Figure 6.1(a) Clamp Foot

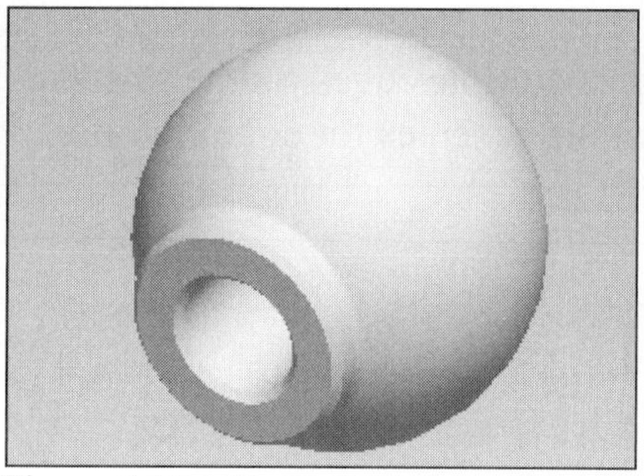

Figure 6.1(b) Clamp Swivel

Figure 6.1(c) Clamp Ball

OBJECTIVES

- Master the **Revolve Tool** [Figs. 6.1(a-c)]
- Create **Chamfers** along part edges
- Learn how to **Sketch in 3D**
- Understand and use the **Navigation browser**
- Alter and set the **Items** and **Columns** displayed in the **Model Tree**
- Create standard **Tapped Holes**
- Create **Cosmetic Threads** and complete **tabular information** for threads
- Edit **Dimension Properties**
- Use the **Model Player** to extract information and dimensions
- Get a hard copy using the **Print** command

REVOLVED FEATURES

The **Revolve Tool** creates a *revolved solid* or a *revolved cut* by revolving a sketched section around a centerline from the sketching plane (Fig. 6.2). You can have any number of centerlines in your sketch/section, but only one will be used to rotate your section geometry. Rules for sketching a revolved feature include:

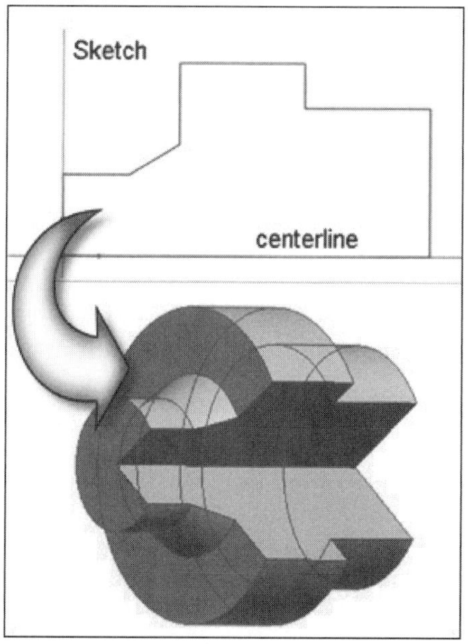

- The revolved section must have a centerline
- By default Pro/E uses the first centerline sketched as the *axis of revolution* (you may select a different axis of revolution)
- The geometry must be sketched on only one side of the *axis of revolution*
- The section must be closed for a solid (Fig. 6.2) but can be open for a cut

A variety of geometric shapes and constructions are used on revolved features. For instance, **chamfers** are created at selected edges of the part. Chamfers are *pick-and-place* features.

Threads can be a *cosmetic feature* representing the *nominal diameter* or the *root diameter* of the thread. Information can be embedded in the feature. Threads show as a unique color. By putting cosmetic threads on a separate layer, you can hide and unhide them.

Figure 6.2 Revolved Protrusion (CADTRAIN, COAch for Pro/ENGINEER)

Chamfers

Chamfers are created between abutting edges of two surfaces on the solid model. An edge chamfer removes a flat section of material from a selected edge to create a beveled surface between the two original surfaces common to that edge. Multiple edges can be selected.

There are four basic dimensioning schemes for edge chamfers: (Fig. 6.3).

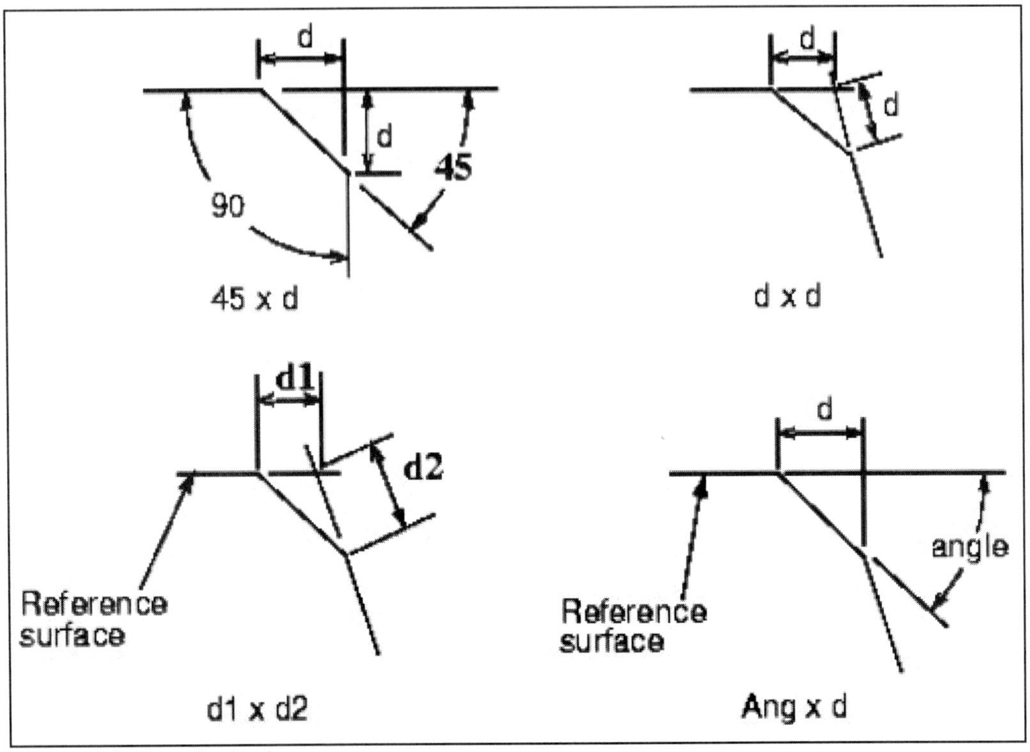

Figure 6.3 Chamfer Options

- **45 x d** Creates a chamfer that is at an angle of **45°** to both surfaces and a distance **d** from the edge along each surface. The distance is the only dimension to appear when edited. **45 x d** chamfers can be created only on an edge formed by the intersection of two *perpendicular* surfaces.
- **d x d** Creates a chamfer that is a distance **d** from the edge along each surface. The distance is the only dimension to appear when edited.
- **d1 x d2** Creates a chamfer at a distance **d1** from the selected edge along one surface and a distance **d2** from the selected edge along the other surface. Both distances appear along their respective surfaces.
- **Ang x d** Creates a chamfer at a distance **d** from the selected edge along one adjacent surface at an **Angle** to that surface.

Threads

Cosmetic threads are displayed with *magenta/purple* lines and circles. Cosmetic threads can be external or internal, blind or through. A cosmetic thread has a set of embedded parameters that can be defined at its creation or later, when the thread is added.

Standard Holes

Standard holes are a combination of sketched and extruded geometry. It is based on industry-standard fastener tables. You can calculate either the tapped or clearance diameter appropriate to the selected fastener. You can use Pro/E supplied standard lookup tables for these diameters or create your own. Besides threads, standard holes can be created with chamfers.

Navigation Window

Besides using the File command and corresponding options, (File ⇒ Set Working Directory), the *Navigation window* can be used to directly access other functions.

As previously mentioned, the working directory is a directory that you set up to contain Pro/E files. You must have read/write access to this directory. You usually start Pro/E from your working directory. A new working directory setting is not saved when you exit the current Pro/E session. By default, if you retrieve a file from a non-working directory, rename the file and then save it, the renamed file is saved to the directory from which it was originally retrieved, if you have read/write access to that directory.

The navigation area is located on the left side of the Pro/E main window. It includes tabs for the Model Tree and Layer Tree, Folder Browser, Favorites, and Connections:

- **Model Tree** (default)
- **Layer Tree (Show ⇒ Layer Tree)**
- **Folder Browser**
- **Favorites**
- **Connections**

Folder Browser

The Folder browser (**Folder Browser**) is an expandable tree that lets you browse the file systems and other locations that are accessible from your computer.

As you navigate the folders, the contents of the selected folder appear in the Pro/E browser as a Contents page.

The Folder browser contains top-level nodes for accessing file systems and other locations that are known to Pro/E:

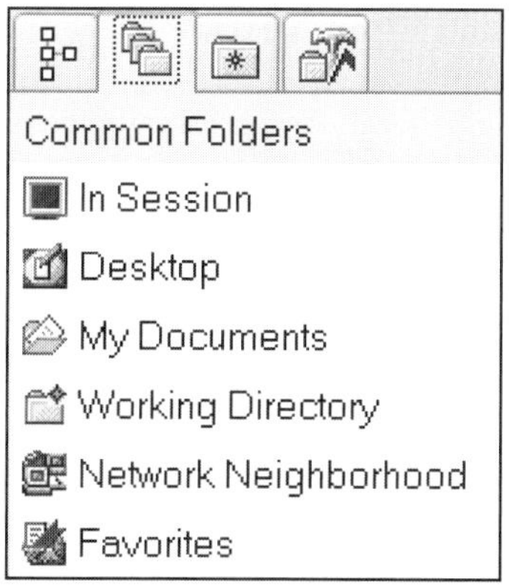

Figure 6.4 Folder Browser

- **In Session** Pro/E objects that have been retrieved into local memory.
- **Desktop** Files and programs on the Desktop. Only Pro/E items can be opened.
- **My Documents** Files created and saved in Windows *My Documents* folder.
- **Working Directory** Directory linked with the Set Working directory command. All work will be accessed and saved in this location.
- **Network Neighborhood** *(only for Windows)* The navigator shows computers on the networks to which you have access. The operations you can perform depend on your permissions to the remote computers.
- **Favorites** Saved folder locations for fast retrieval.

Manipulating Folders

To work with folders, you can use the **Browser** to accomplish most file management requirements:

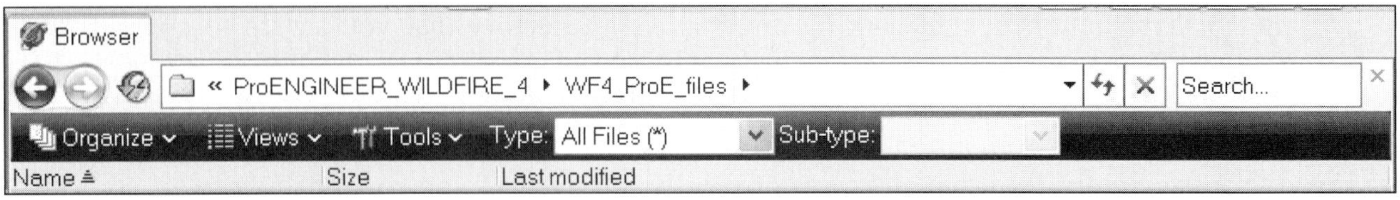

You can perform a variety of tasks from the **Organize** drop-down menu in the Browser including **New Folder**, **Rename**, **Cut**, **Copy**, **Paste**, **Delete**, and **Add to common folders**:

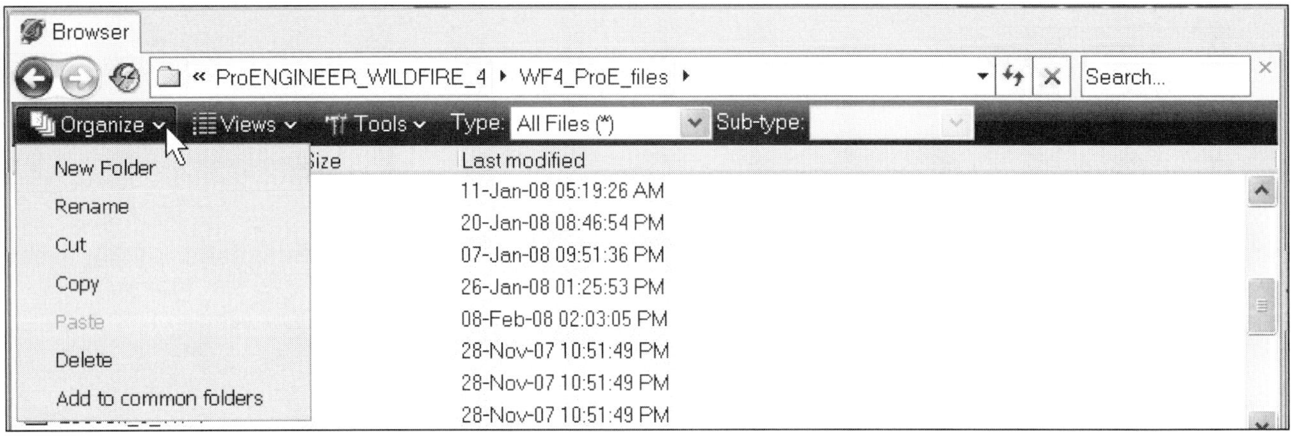

From the Browser's **Views** drop-down menu you can change between a simple **List** and **Details**:

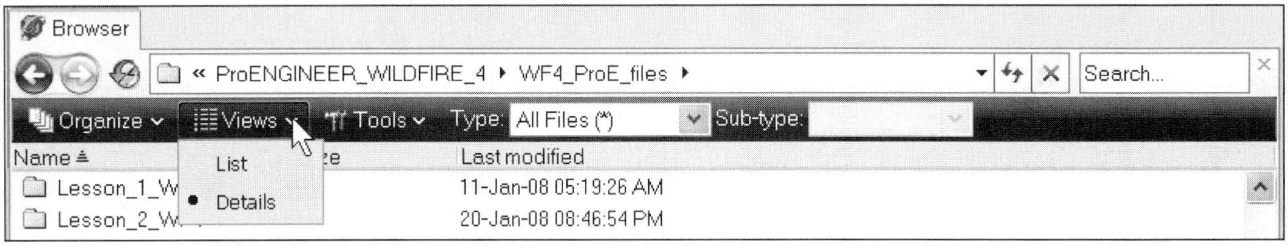

From the Browser's **Tools** drop-down menu you can change **Sort By**, go **Up One Level**, **Add to Favorites**, **Remove from Favorites**, **Organize Favorites**, show **All Versions**, and **Show Instances**:

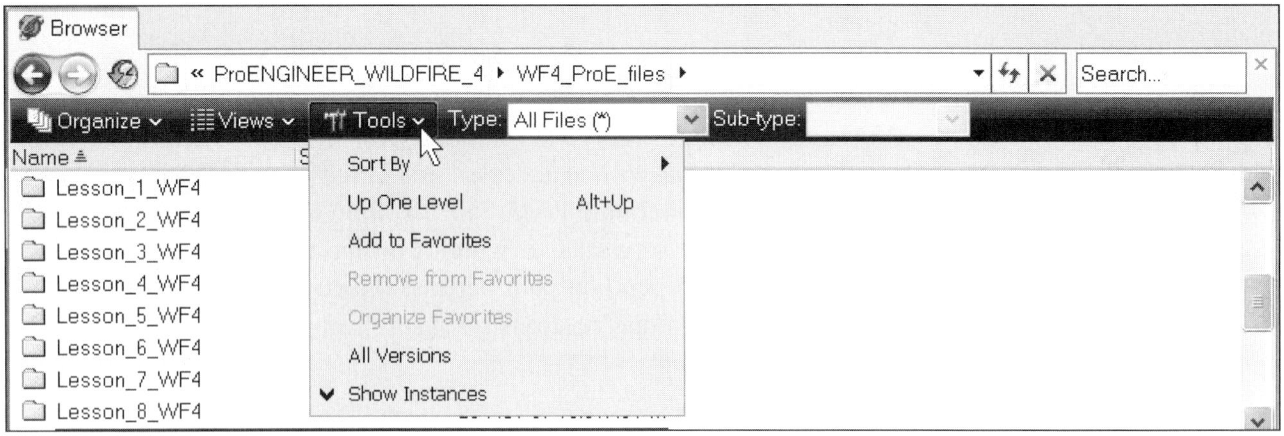

Do not change your directory from your default system work folder unless instructed to do so.

In the Browser's **Type** field, you limit the search to one of Pro/E's file types:

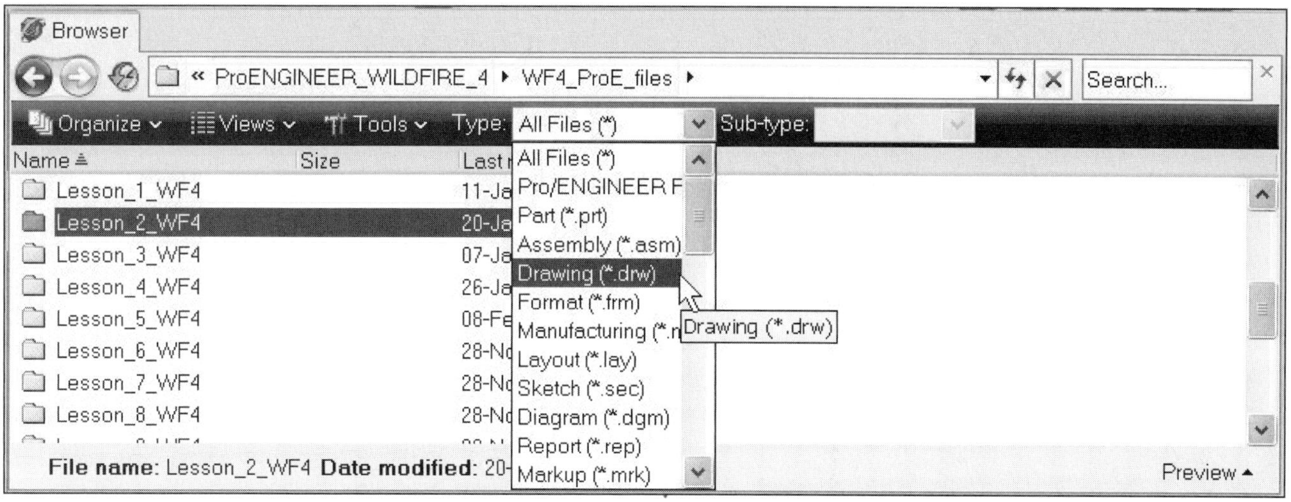

To set a working directory using the Browser, click on the desired folder ⇒ ***RMB*** ⇒ ***Set Working Directory***:

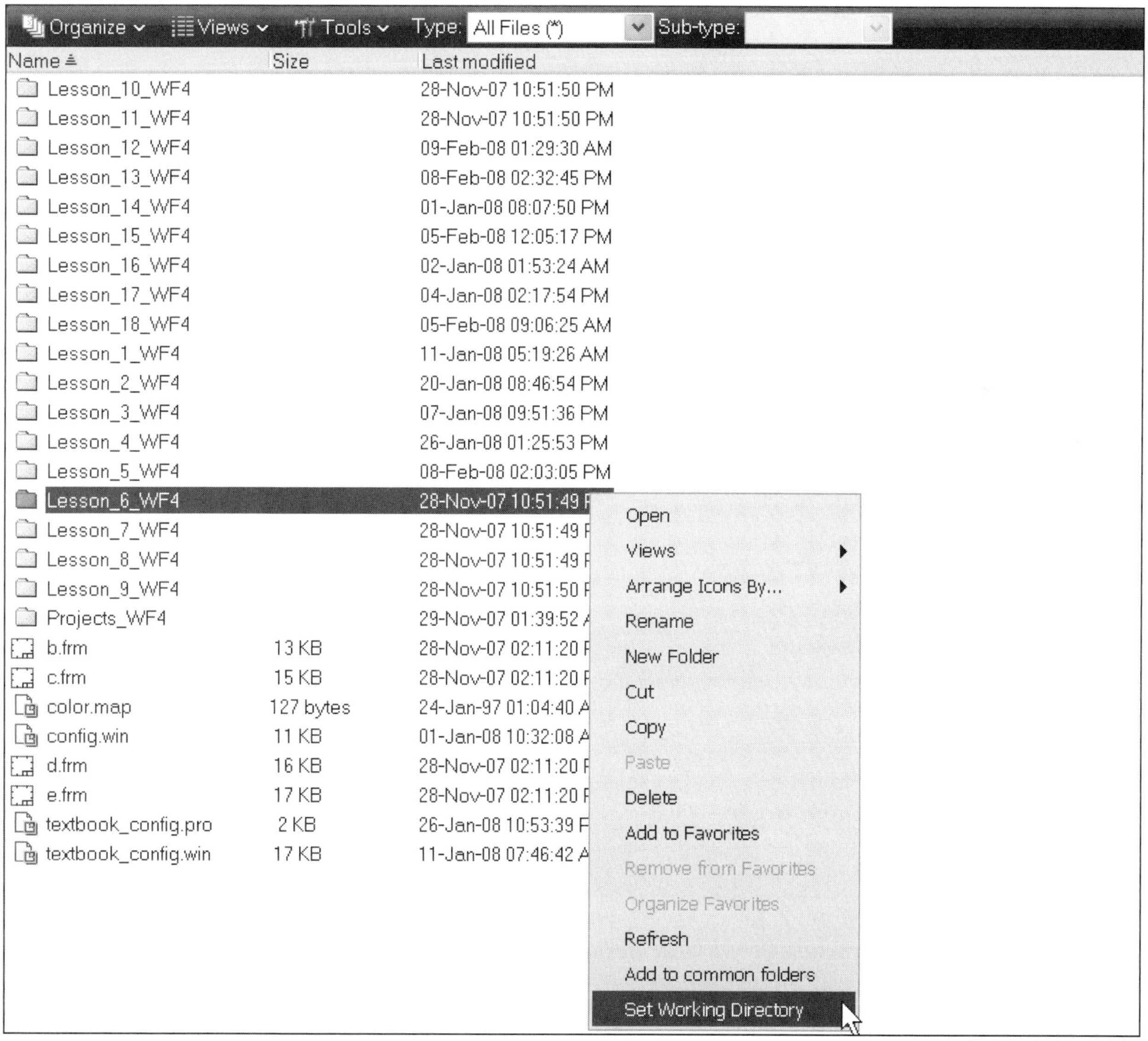

You can set your working directory and then click on the Pro/E object that you wish to preview:

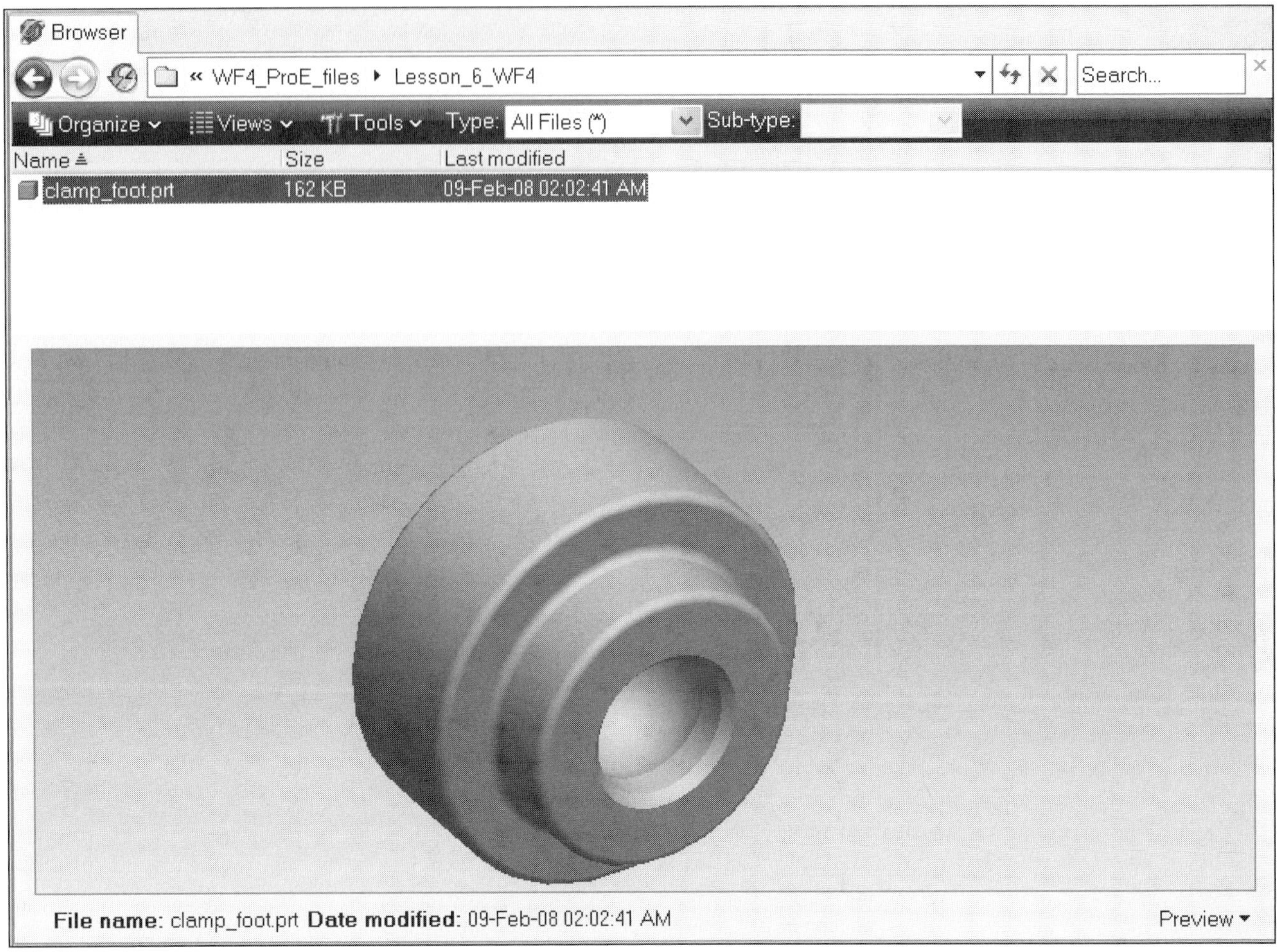

Use the MMB (press and hold as you move the mouse) to rotate or (roll the middle mouse button) to zoom the object in the preview window. Double-clicking on the file name will open the part. After you complete this lesson you will have this same part in your working directory.

From the Browser toolbar you can also navigate to other directories or computer areas:

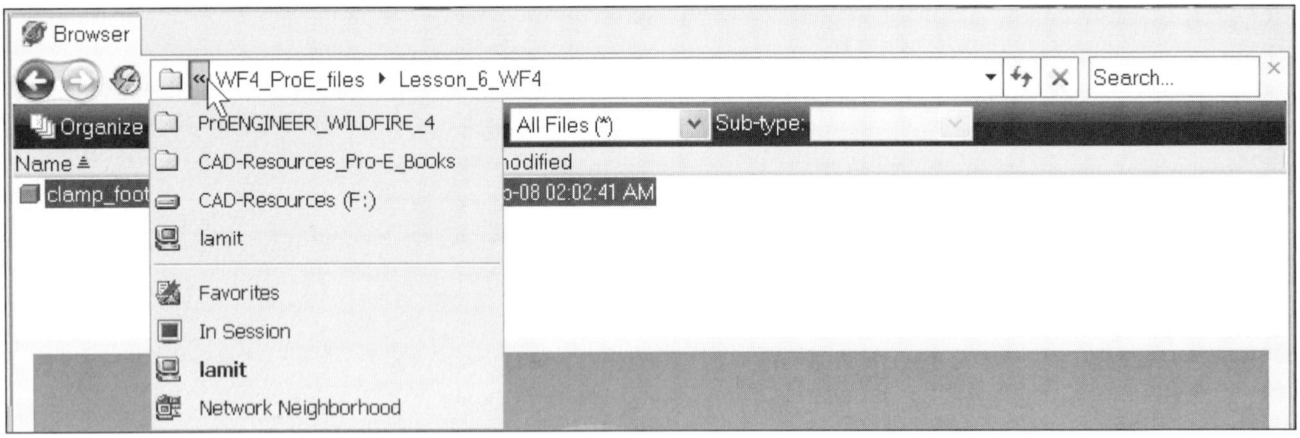

Do not change your directory from your default work folder unless you are instructed to do so by your teacher or you are on your own computer.

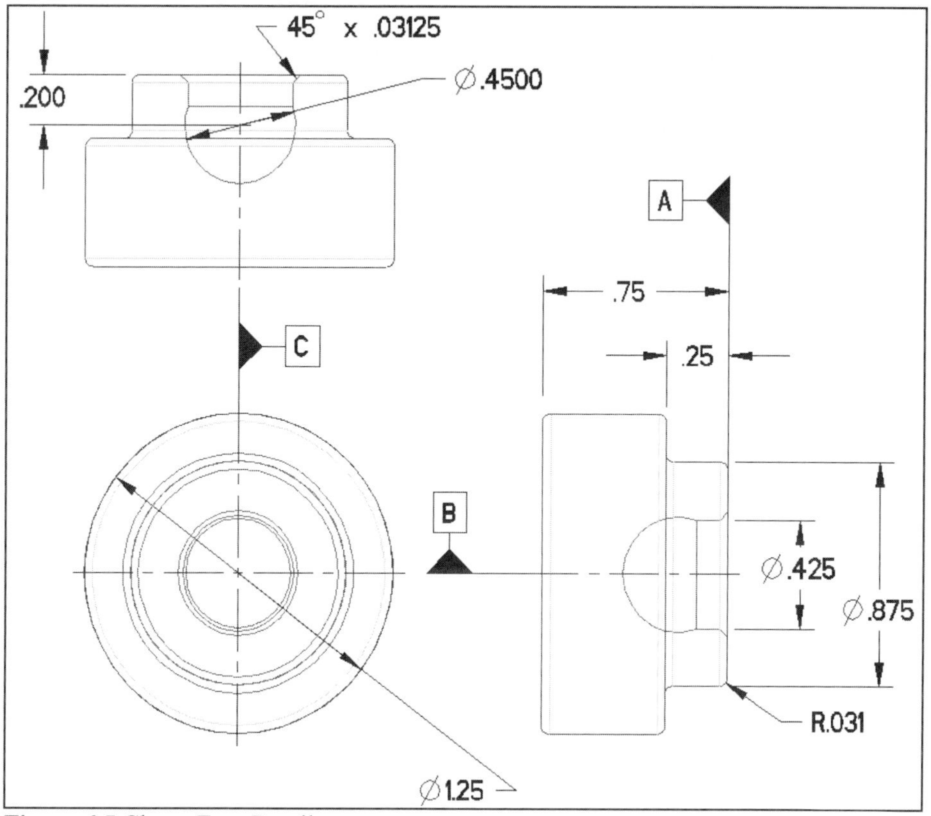

Figure 6.5 Clamp Foot Detail

Clamp Foot

The Clamp Foot (Fig. 6.5) is the first of three revolved parts created for Lesson 6. The Clamp Foot, Clamp Ball, and Clamp Swivel are three revolved parts needed for the Clamp assembly and drawings later in the text. This part requires a revolved extrusion (protrusion), a revolved cut, a chamfer and rounds.

Launch **Pro/ENGINEER WILDFIRE 4.0** ⇒ **File** ⇒ **Set Working Directory** ⇒ navigate to your directory if necessary ⇒ **OK** ⇒ **Create a new object** ⇒ Part ⇒ Name CLAMP_FOOT ⇒ Use default template ⇒ **OK** ⇒ **Edit** ⇒ **Setup** ⇒ **Units** ⇒ Inch lbm Second (Pro/E Default) ⇒ **Close** ⇒ **Done**

or MMB ⇒ **Tools** ⇒ **Options** ⇒ Showing: Current Session ⇒ Show only options loaded from file [Fig. 6.6(a)]

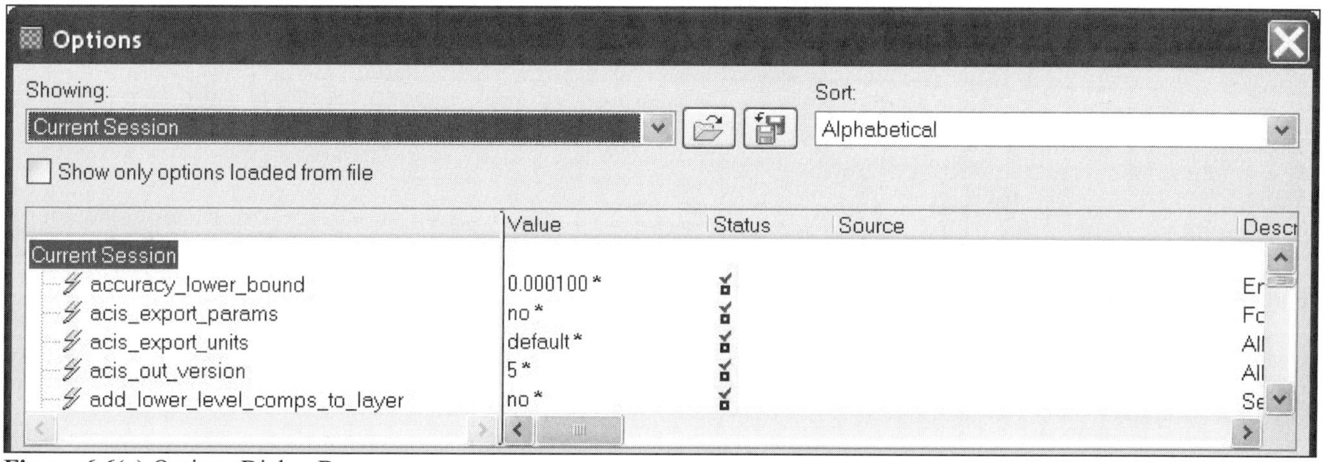

Figure 6.6(a) Options Dialog Box

Slide the vertical scroll bar *(right side of Options dialog box)* down to the option or type the option. Option: ***default_dec_places*** ⇒ Value: **3** [Fig. 6.6(b)] ⇒ **Enter** [Fig. 6.6(c)]

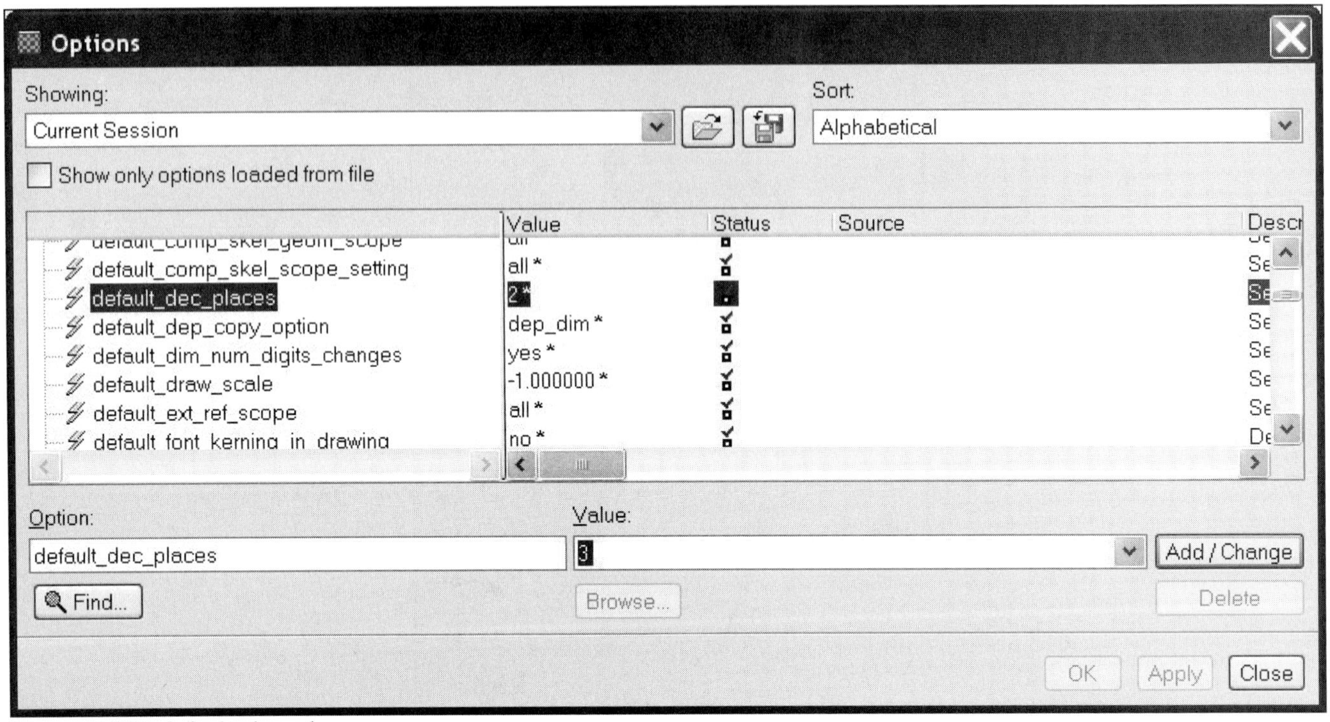

Figure 6.6(b) default_dec_places

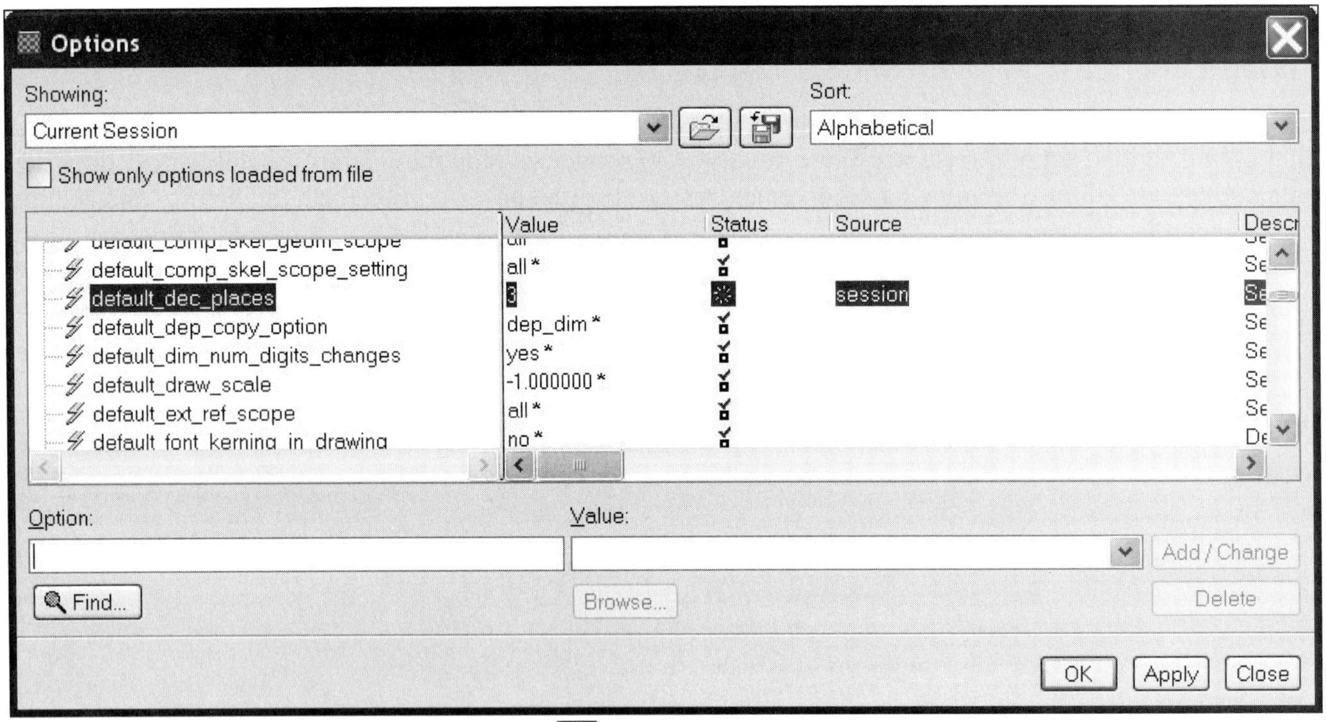

Figure 6.6(c) default_dec_places Value: 3 Status: Source: session

Slide the bar down to the option or type. Option: ***sketcher_dec_places*** ⇒ Value: **3** [Fig. 6.6(d)] ⇒ **Enter** ⇒ Option: ***sketcher_starts_in_2d*** ⇒ Value: **no** [Fig. 6.6(e)] ⇒ Add / Change *(same as Enter)* ⇒ Option: *(you must type this with your keyboard)* ***def_layer*** ⇒ Value: ▾ ⇒ **layer_axis** *(space here)* **datum_axes** *(after you select layer_axis, type datum_axes)* [Figs. 6.6(f-g)] ⇒ **Add/Change** ⇒ **Apply**

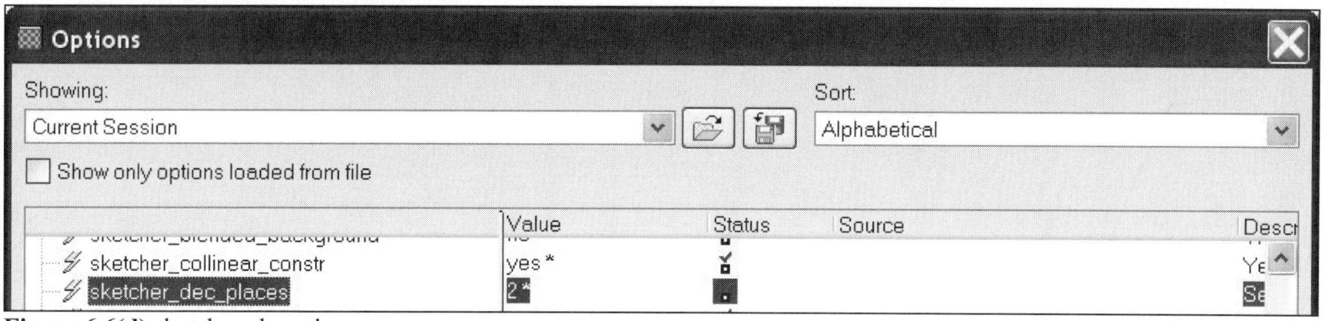

Figure 6.6(d) sketcher_dec_places

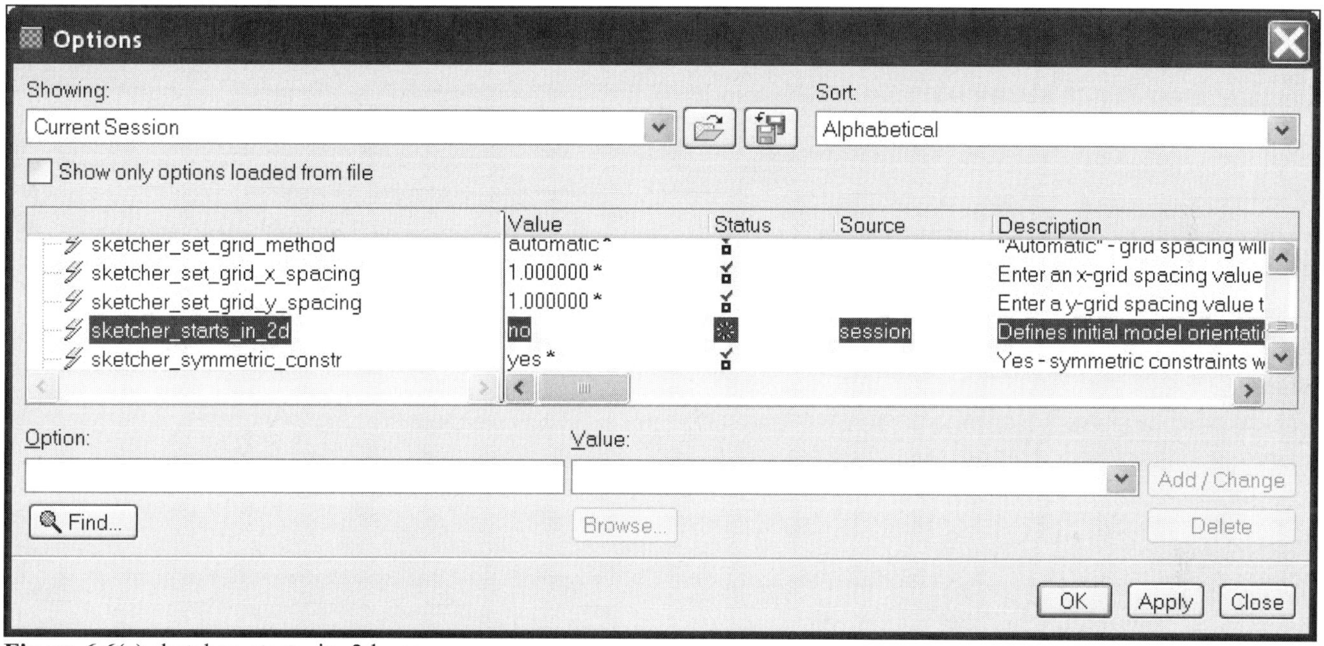

Figure 6.6(e) sketcher_starts_in_2d no

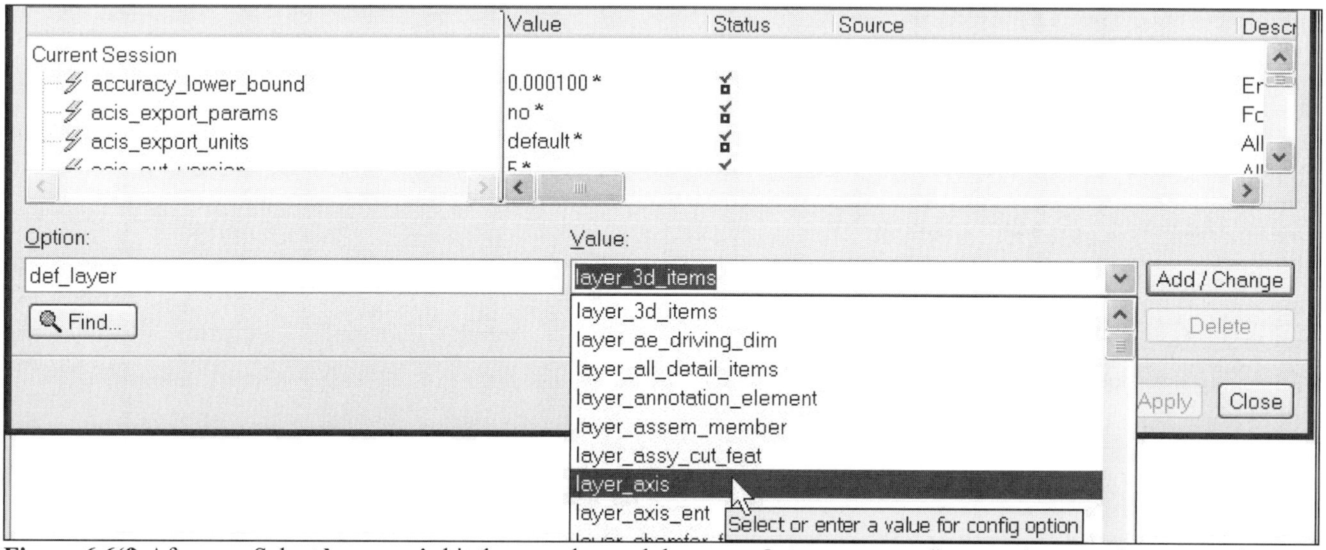

Figure 6.6(f) After you Select **layer_axis** hit the spacebar and then type **datum_axes** ***(layer_axis*** *space* ***datum_axes)***

Click: Show only options loaded from file ⇒ from the Options dialog box **Save a copy of the currently displayed configuration file** [Fig. 6.6(h)] ***(to use this configuration file in other sessions- it MUST be saved here)*** ⇒ File name: **clamp.pro** [Fig. 6.6(i)] ⇒ **Ok** ⇒ **Close** ⇒ **Ctrl+S** ⇒ **Enter**

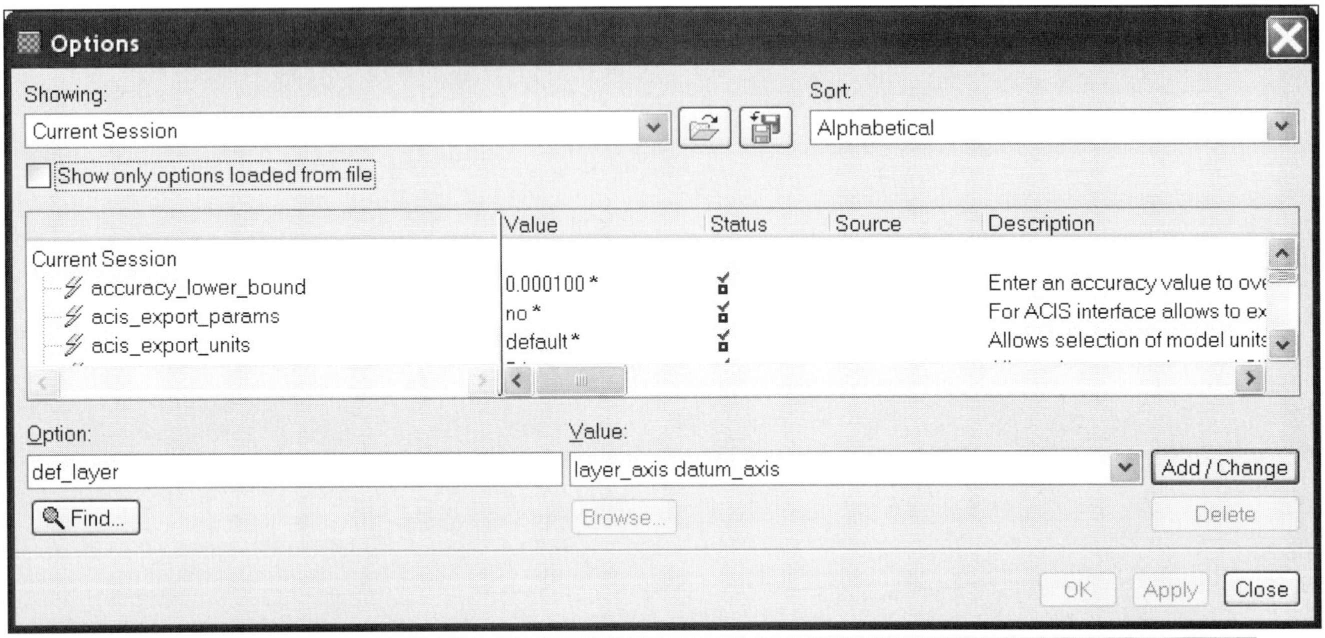

Figure 6.6(g) def_layer

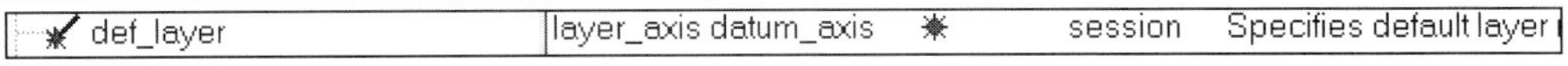

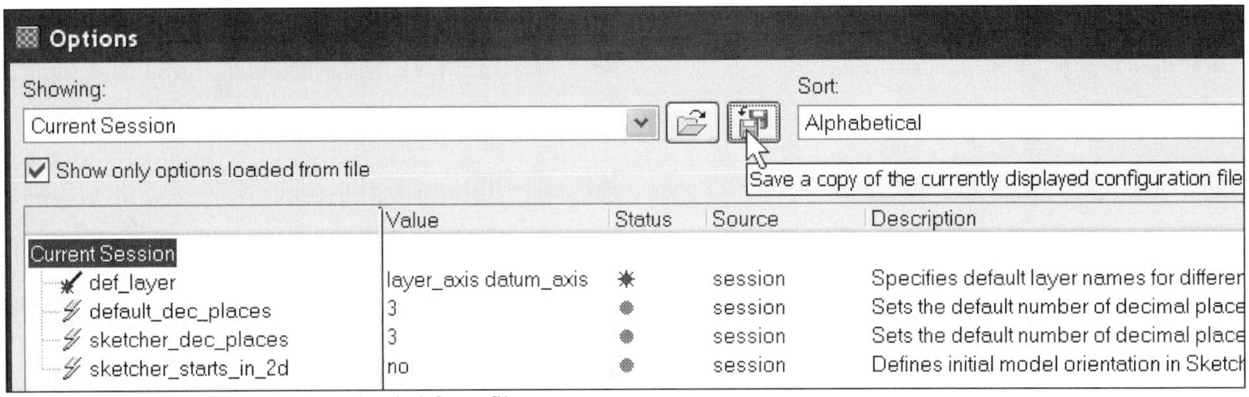

Figure 6.6(h) Show only options loaded from file

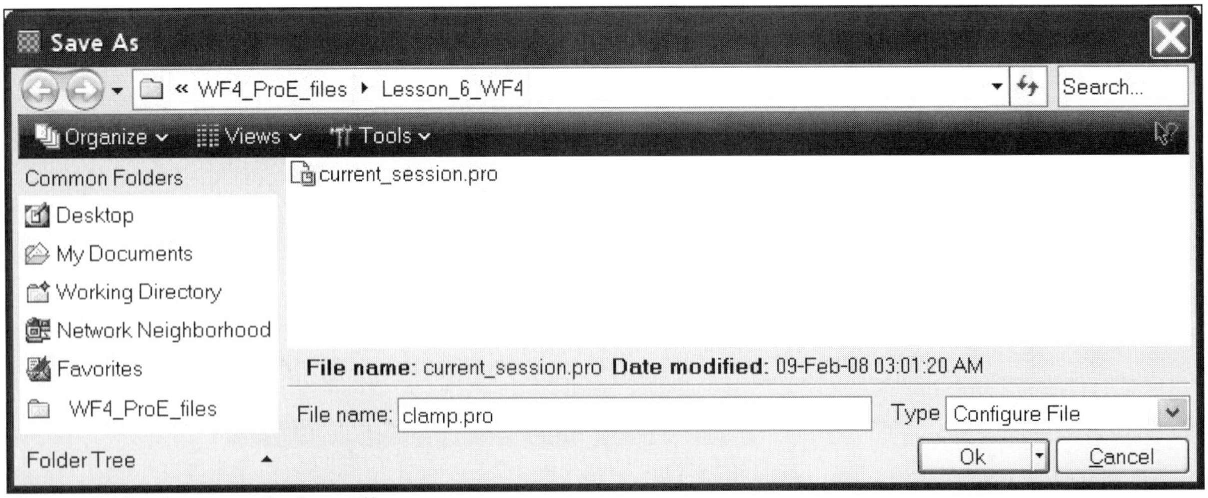

Figure 6.6(i) clamp.pro Options File

The first protrusion is a revolved protrusion created with the Revolve Tool, you will be sketching the section in 3D since the configuration option; *sketcher_starts_in_2d,* was previously set to *no.* In the Environment dialog box, you can see that [Use 2D Sketcher] is deactivated.

Click: **Tools** ⇒ [Environment] ⇒ [Snap to Grid] ⇒ [Tangent Edges: Dimmed] ⇒ [Standard Orient: Trimetric] ⇒ **Apply** ⇒ **OK** ⇒ **Revolve Tool** ⇒ **RMB** ⇒ **Define Internal Sketch** ⇒ Sketch Plane--- Plane: select the **RIGHT** datum ⇒ Reference: **TOP:(DATUM PLANE)** [Fig. 6.7(a)] ⇒ Orientation **Top** ⇒ [Sketch] ⇒ on [Fig. 6.7(b)]

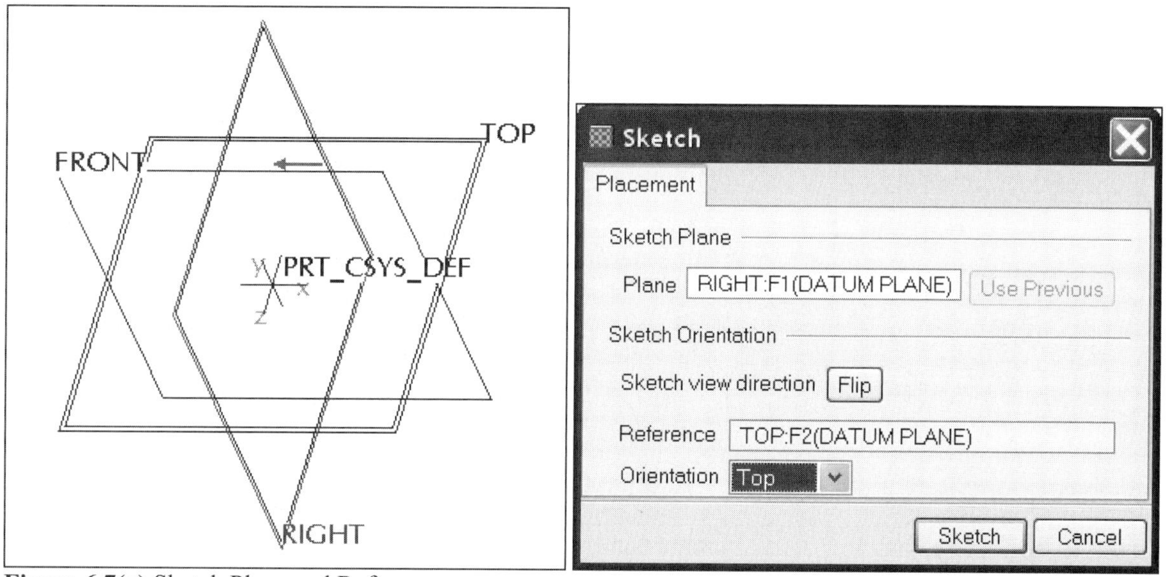

Figure 6.7(a) Sketch Plane and Reference

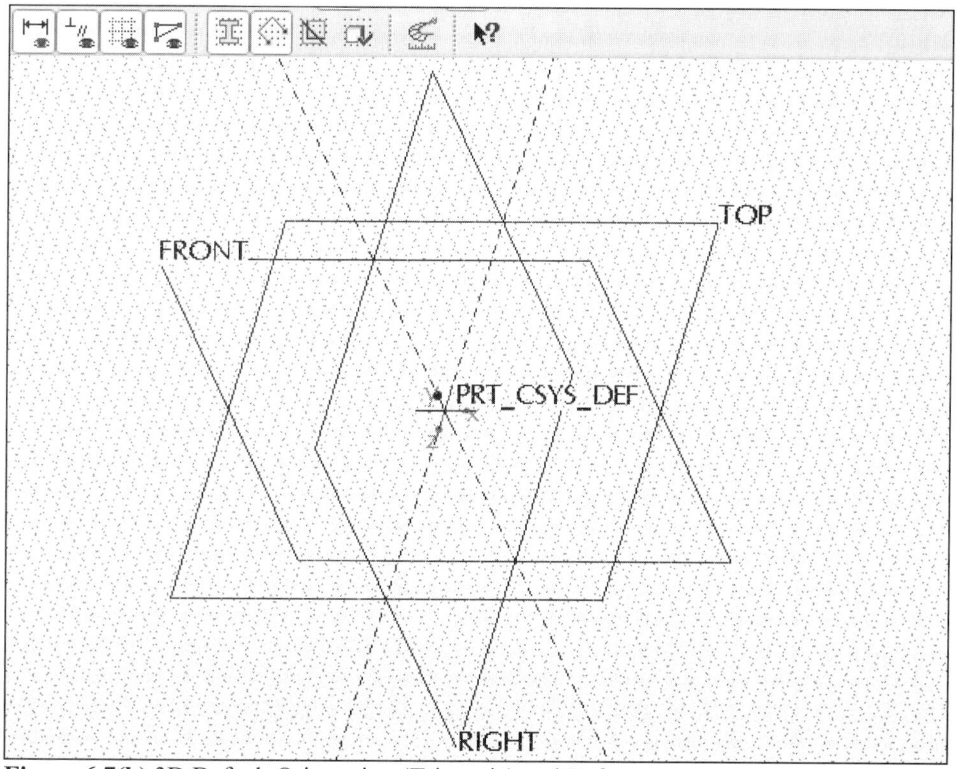

Figure 6.7(b) 3D Default Orientation (Trimetric) and References

Click: **RMB** ⇒ **Centerline** sketch a *vertical* centerline through the default coordinate system to be used as the axis of revolution [Fig. 6.7(c)] ⇒ Rotate the part/sketch with your **MMB** to see the sketching plane clearly [Fig. 6.7(d)] ⇒ **RMB** ⇒ **Line** sketch the six lines on the RIGHT datum [Fig. 6.7(e)] ⇒ **MMB** ⇒ **MMB** *(Note: if you have difficulty, you can delete the lines and start again after clicking*  ***Orient the sketching plane parallel to the screen*** *and sketch the lines in 2D as in previous lessons)*

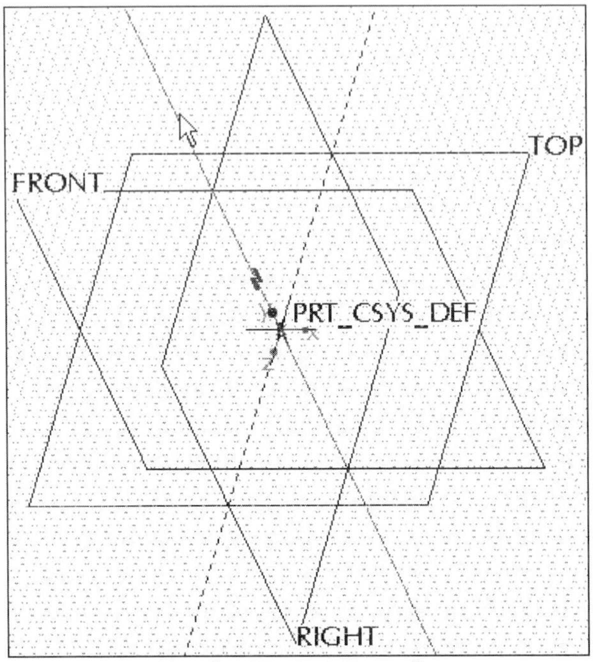

Figure 6.7(c) Sketch a Vertical Centerline

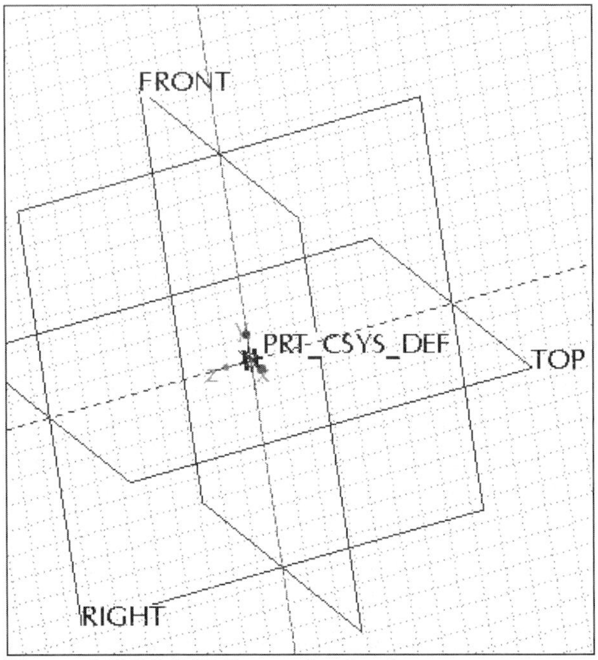

Figure 6.7(d) MMB Rotate the sketch

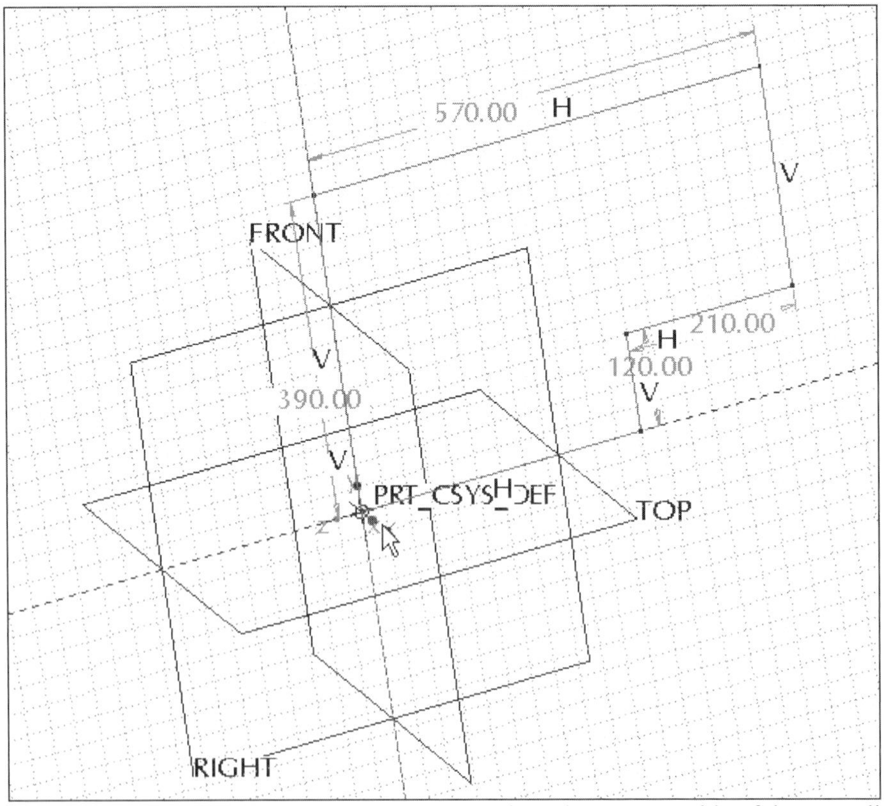

Figure 6.7(e) Sketch Six Lines to Form a Closed Section (on one side of the centerline)

Click: **Sketch** from menu bar ⇒ **Options** ⇒ Snap To Grid [Fig. 6.7(f)] ⇒ ⇒ off ⇒ **Ctrl+R** ⇒ **RMB** ⇒ **Dimension** add the two vertical dimensions [Fig. 6.7(g)] ⇒ add a diameter dimension by picking the centerline, then pick the outer vertical edge line, and then pick the centerline again ⇒ **MMB** to place the dimension ⇒ repeat to dimension the other diameter [Fig. 6.7(h)] ⇒ **Ctrl+R** repaint

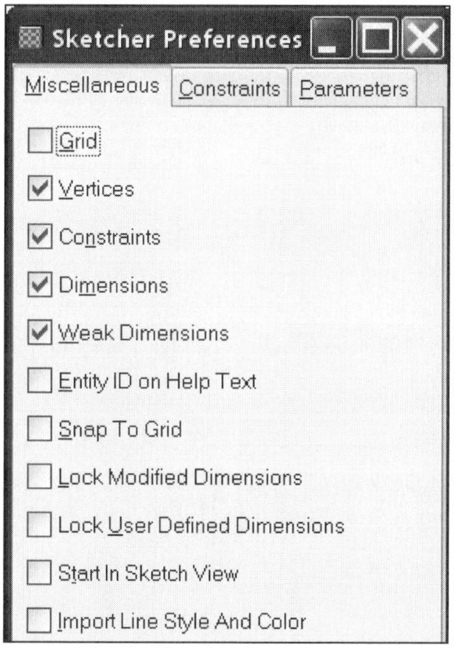

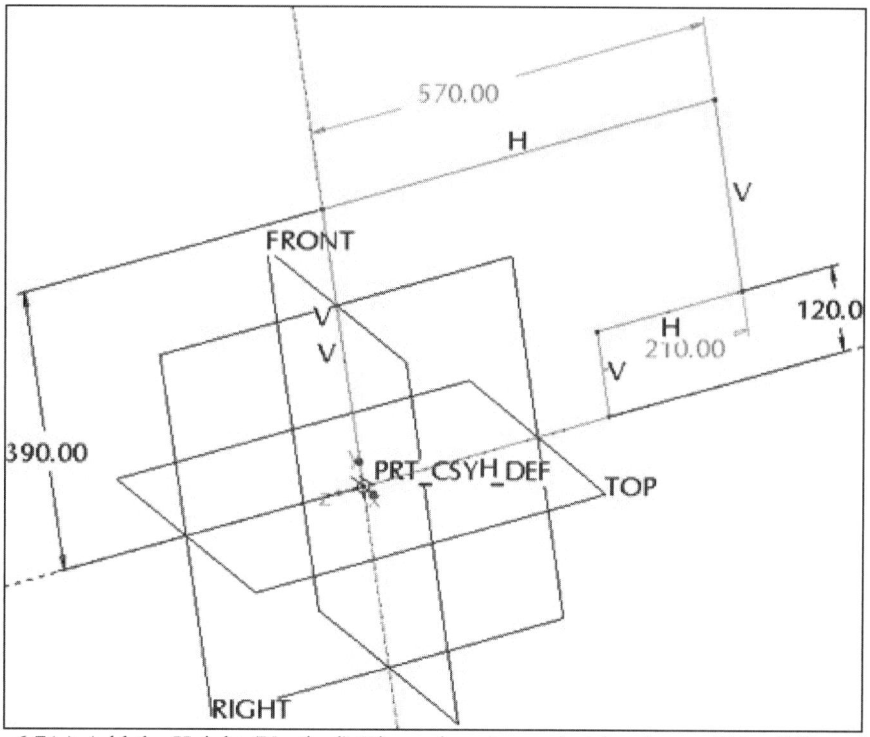

Figure 6.7(f) Sketcher Preferences **Figure 6.7(g)** Add the Height (Vertical) Dimensions (your sketch values will be different)

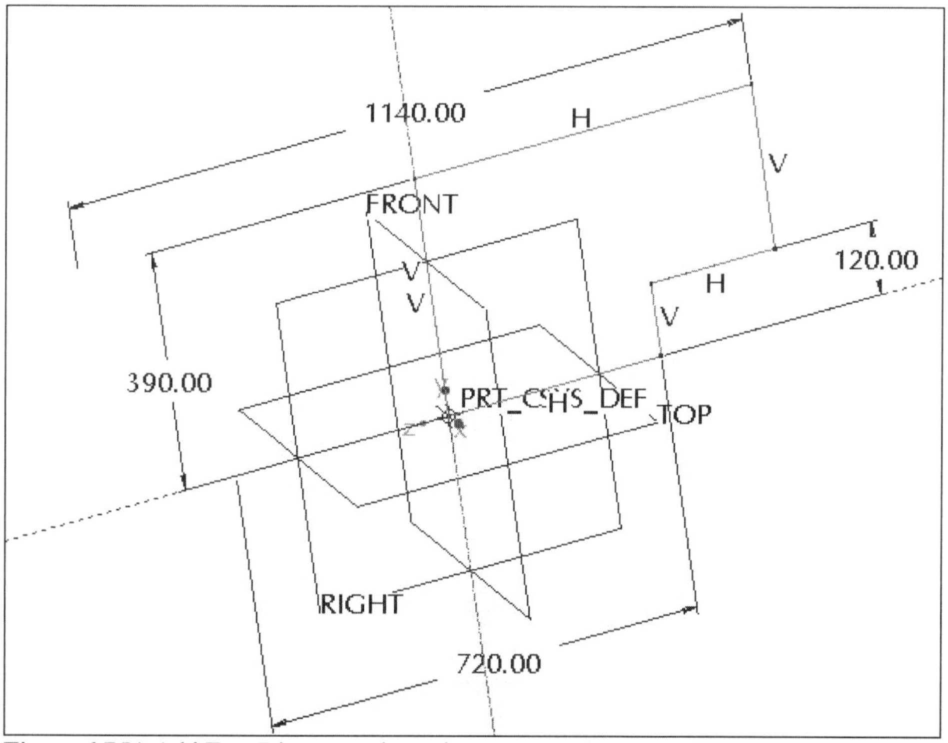

Figure 6.7(h) Add Two Diameter Dimensions (Your values should be different, DO NOT change them to these values!)

Modify the dimensions, click: **Select items** ⇒ window-in the sketch to capture all four dimensions ⇒ **RMB** ⇒ **Modify** ⇒ Lock Scale ⇒ Regenerate [Fig. 6.7(i)] ⇒ modify **only** the overall height to the design value (see Fig. 6.5) of **.750** ⇒ **Enter** ⇒ [Fig. 6.7(j)] ⇒ **Ctrl+R** repaint ⇒ **LMB** to deselect

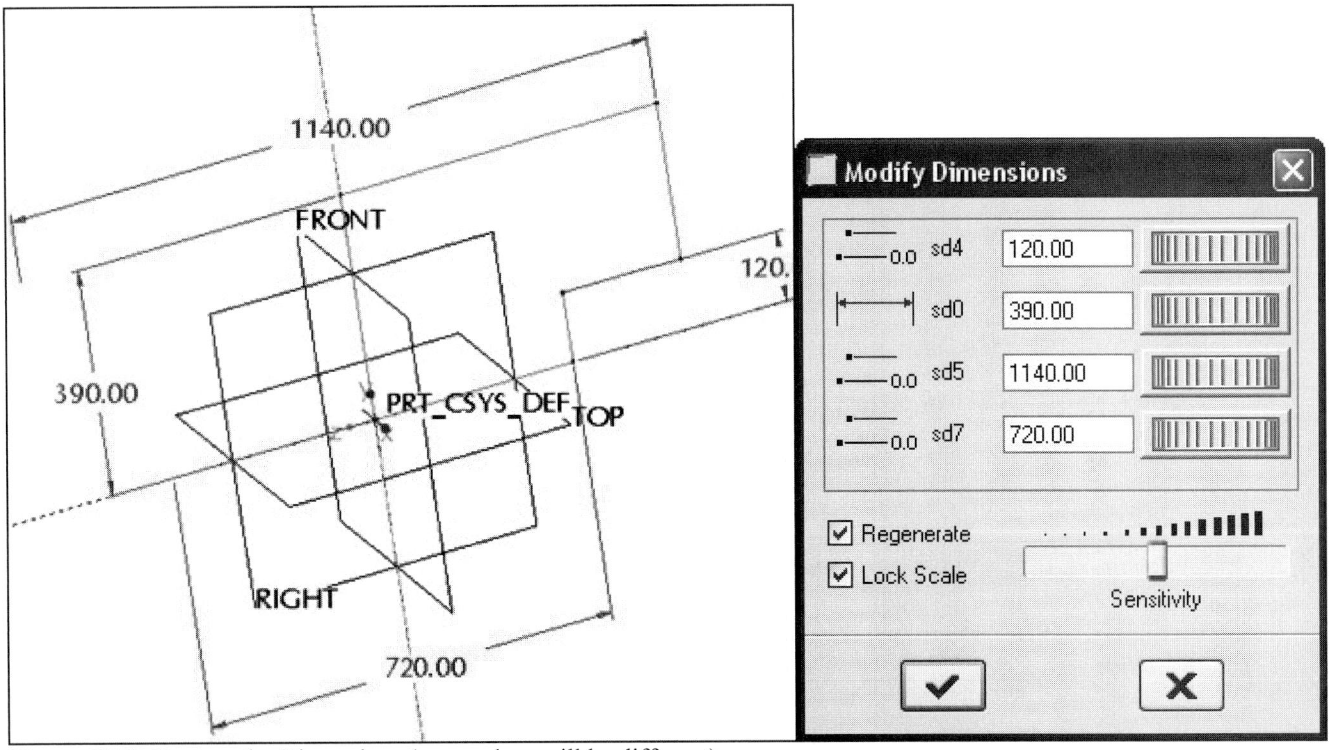

Figure 6.7(i) Capture the Dimensions (your values will be different)

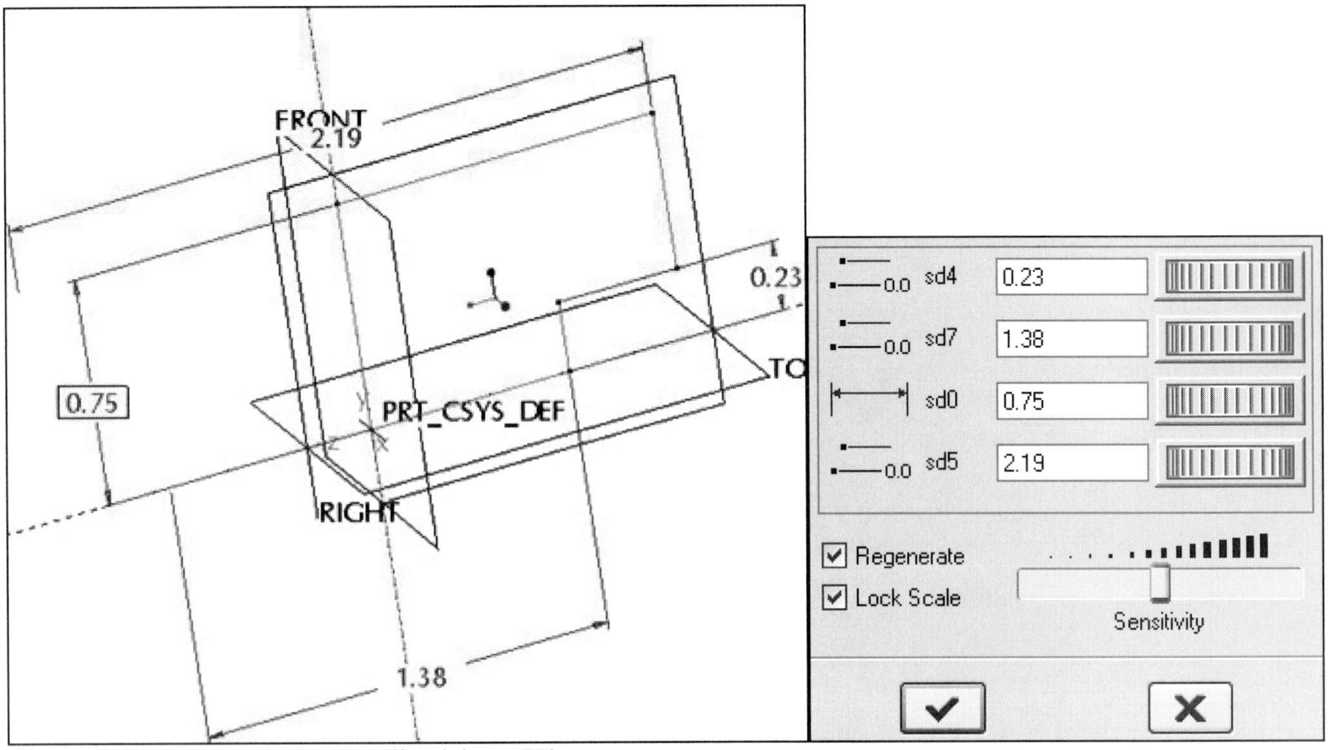

Figure 6.7(j) Modify **only** the Overall Height to **.750**

Double-click on each dimension individually and change to the design value ⇒ [icon] **Color the inside of closed chains** ⇒ [icon] [Fig. 6.7(k)] ⇒ [dashboard: InternalCL, 360.000] ⇒ [icon] **Shading** ⇒ [icon] [Fig. 6.7(l)] ⇒ [icon] ⇒ **Standard Orientation** ⇒ [icon] ⇒ **MMB** ⇒ **LMB** to deselect

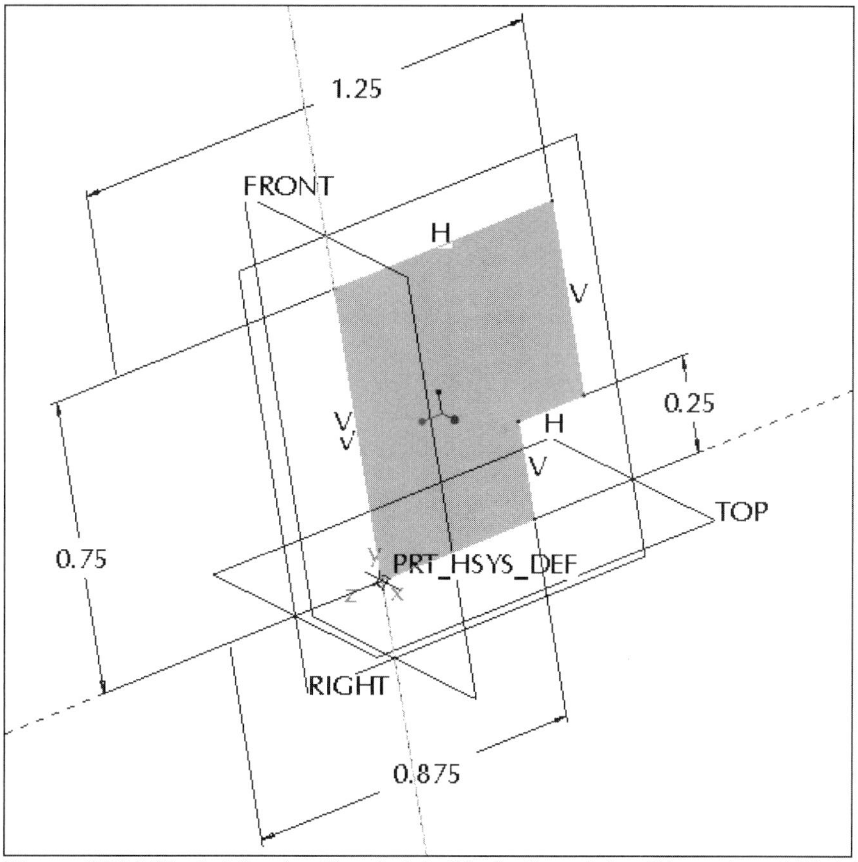

Figure 6.7(k) Shaded Section Sketch. Modify the Remaining Dimensions (**.75**, **1.25**, **.25**, **.875**)

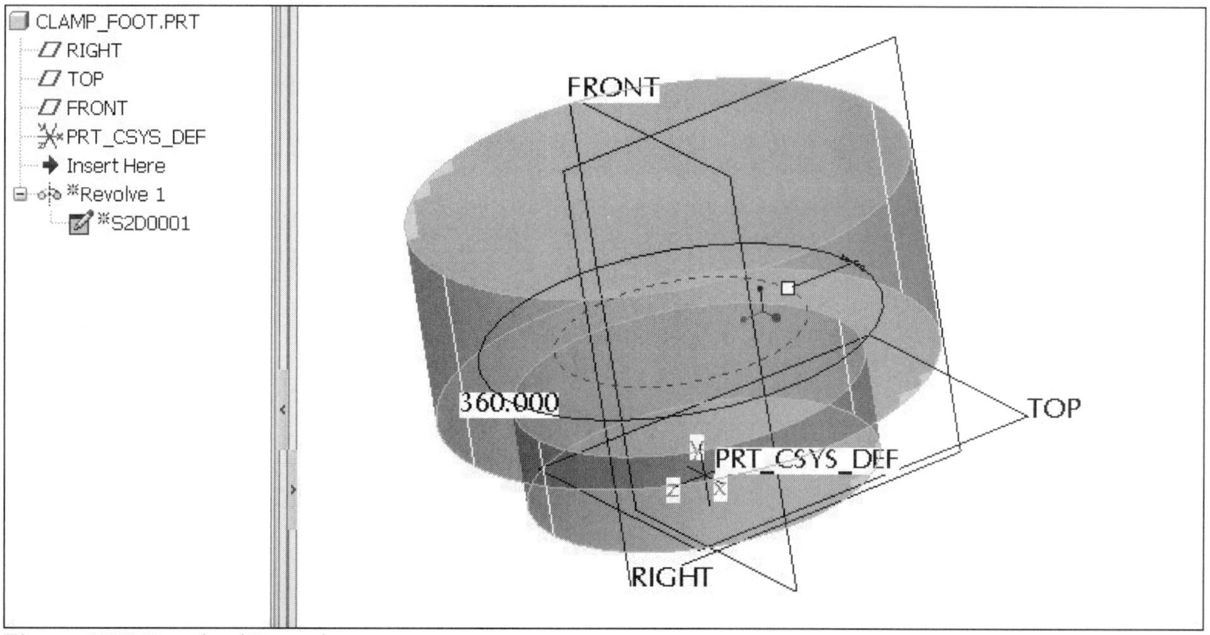

Figure 6.7(l) Revolved Extrusion

Click: **Revolve Tool** ⇒ **Remove Material** ⇒ **RMB** ⇒ **Define Internal Sketch** ⇒ Use Previous [Figs. 6.8(a-b)] ⇒ **Orient the sketching plane parallel to the screen** [Fig. 6.8(c)]

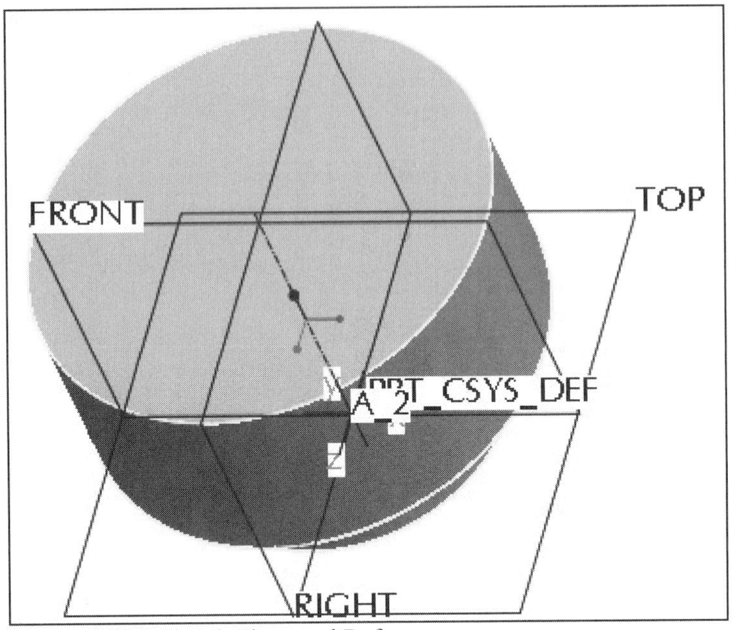

Figure 6.8(a) Sketch Plane and Reference

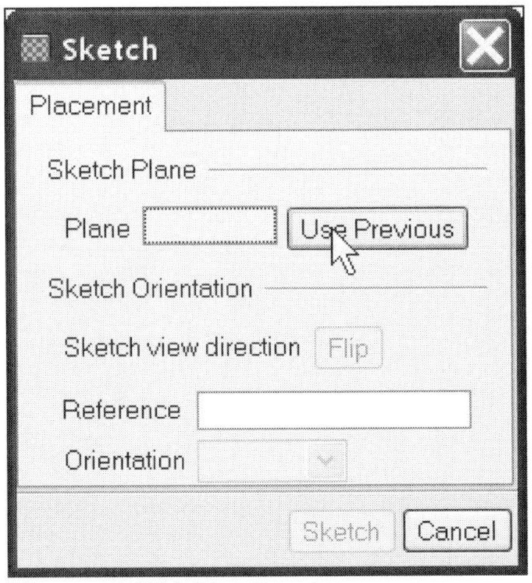

Figure 6.8(b) Use Previous

Click: **RMB** ⇒ **Centerline** sketch a *vertical* centerline through the default coordinate system to be used as the axis of revolution [Fig. 6.8(d)] ⇒ **MMB** ⇒ **Hidden line** [Fig. 6.8(e)]

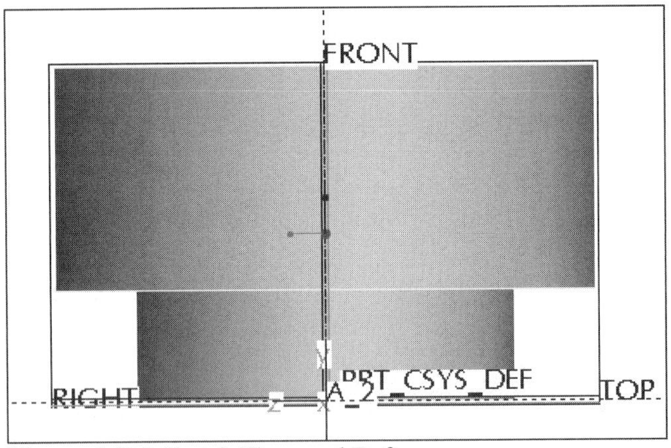

Figure 6.8(c) Sketch Plane and Reference

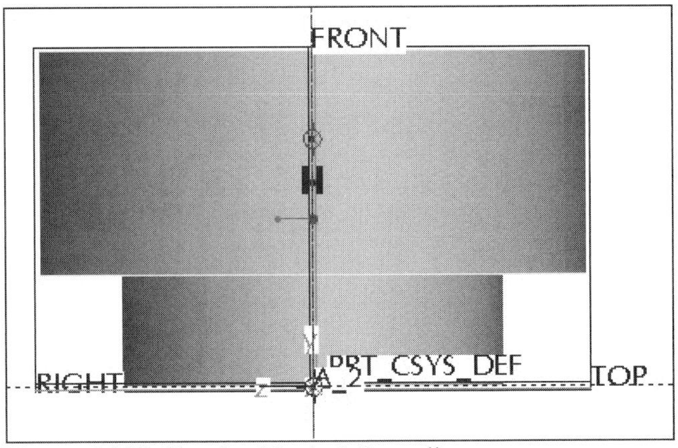

Figure 6.8(d) Sketch a Vertical Centerline

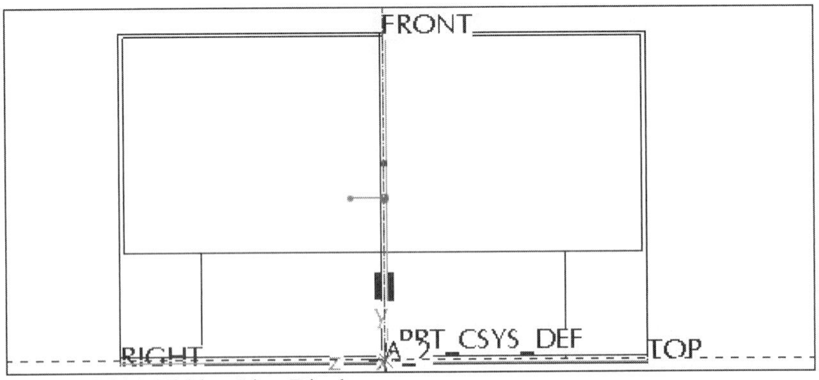

Figure 6.8(e) Hidden Line Display

From the menu bar, click: **Sketch** ⇒ **Arc** ⇒ **Center and Ends** ⇒ sketch the arc [Fig. 6.8(f)] *(LMB to place the center of the arc, LMB to place the start point of the arc, LMB to place the end point of the arc)* ⇒ **MMB** ⇒

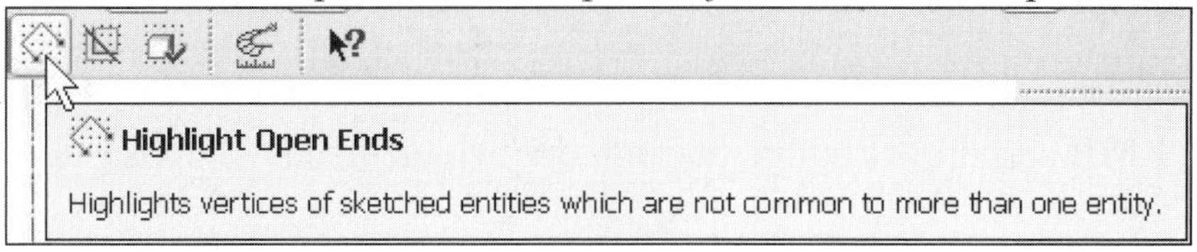

[Fig. 6.8(g)] ⇒ 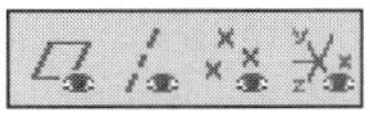 off ⇒ **RMB** ⇒ **Line** sketch the vertical and horizontal lines [Figs. 6.8(h-i)] ⇒ **MMB** ⇒ **MMB** to end the current tool

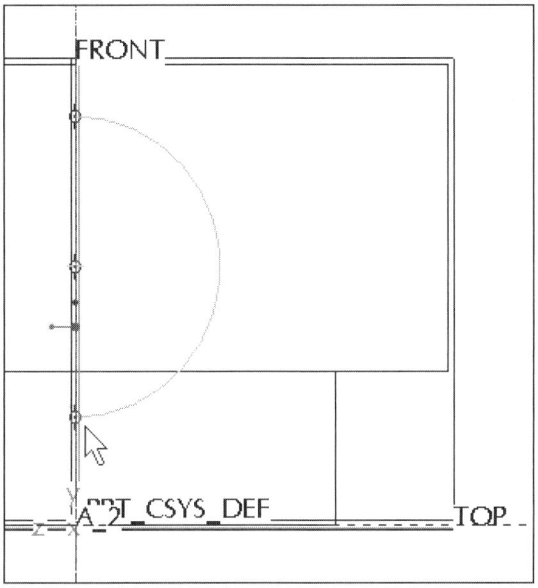

Figure 6.8(f) Sketch the Arc

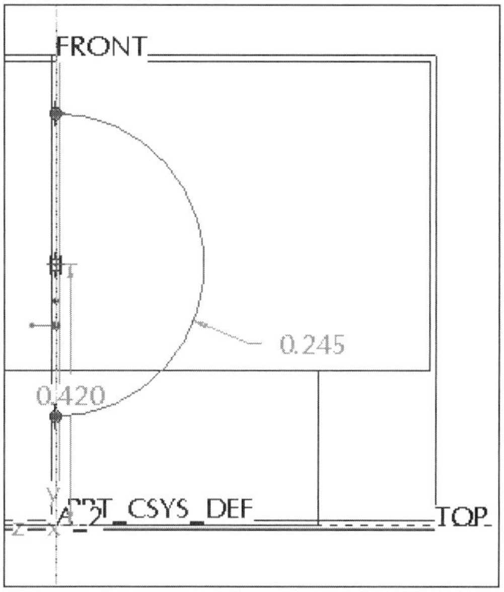

Figure 6.8(g) Completed Arc

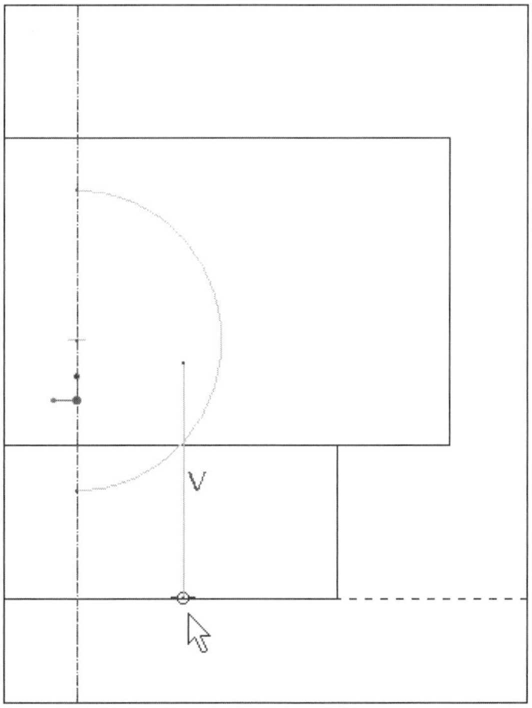

Figure 6.8(h) Sketch the First Line

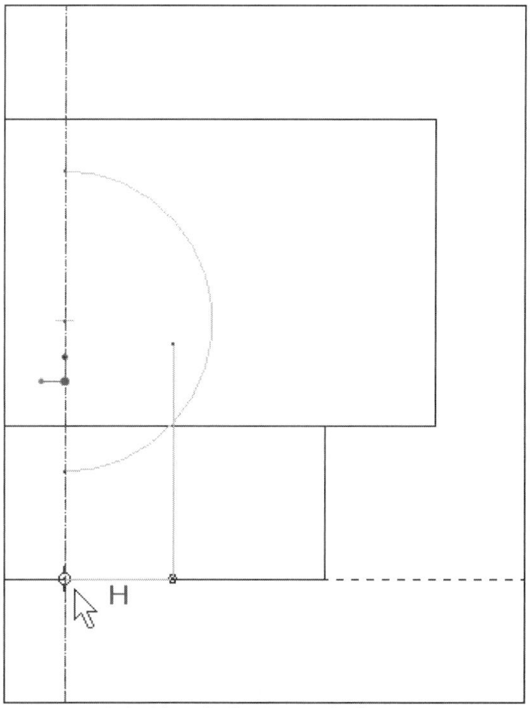

Figure 6.8(i) Sketch the Second Line

Click: 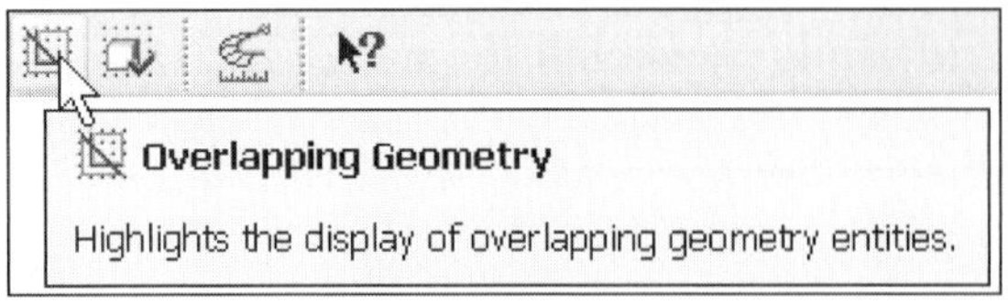

[Fig. 6.8(j)] ⇒  **Dynamically trim section entities** ⇒ press and hold the **LMB** and draw a spline through the two elements you wish to remove [Fig. 6.8(k)] ⇒ release the **LMB** to complete the trim ⇒ 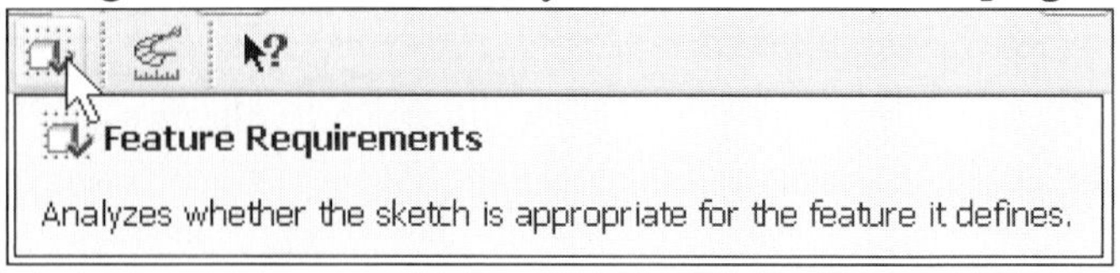

[Fig. 6.8(l)] ⇒ **Close** *(Note: if you delete the wrong items, click:*  ***Undo sketcher operations** and try again)*

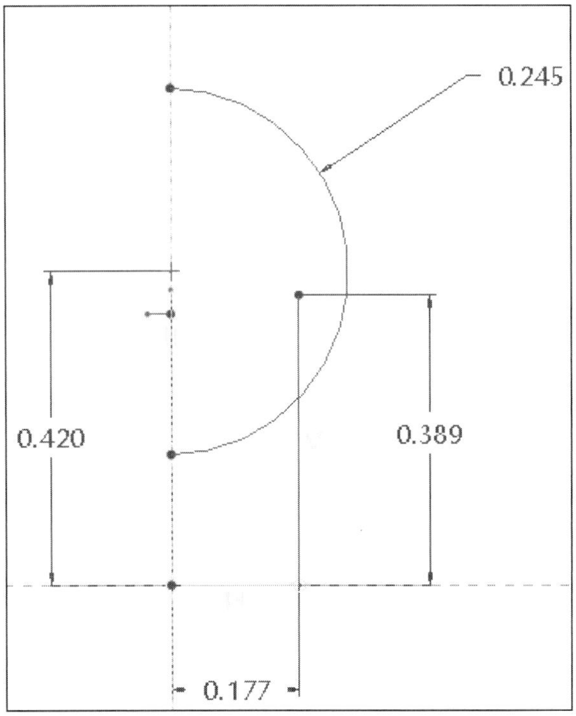

Figure 6.8(j) Highlighted Overlapping Geometry

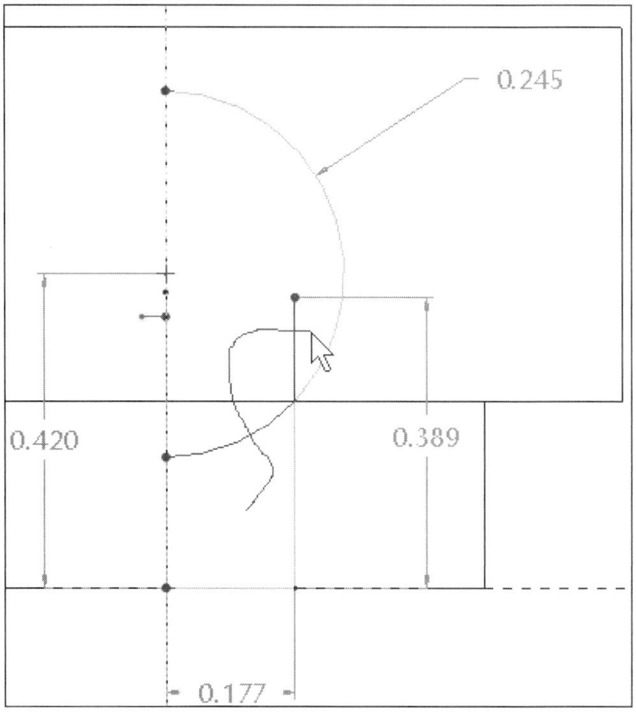

Figure 6.8(k) Dynamically Trim the Arc and Line

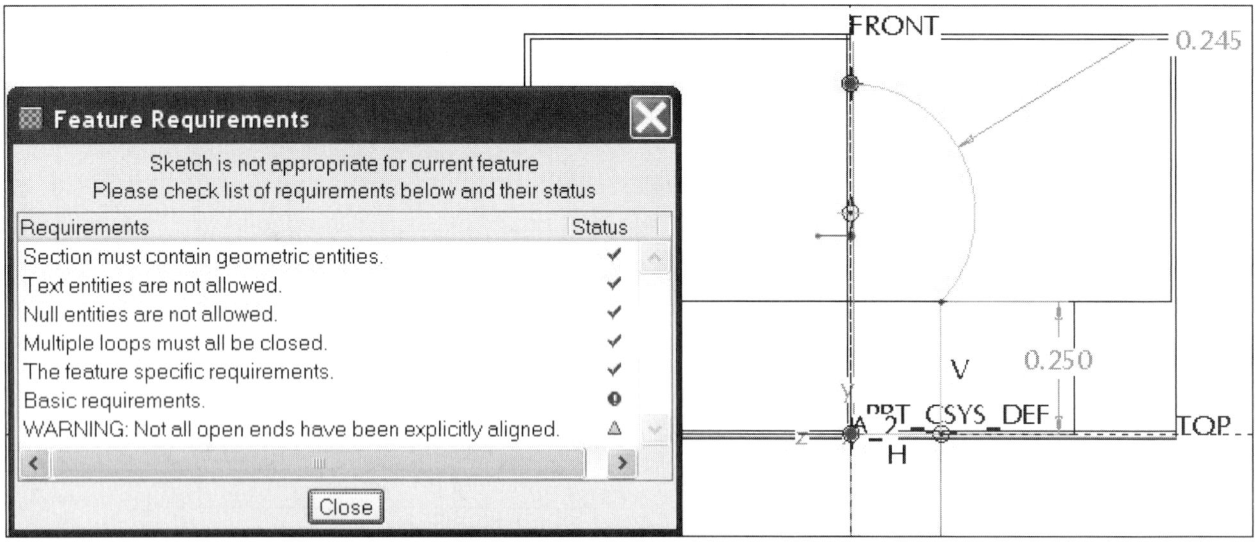

Figure 6.8(l) Feature Requirements

Click: on ⇒ **RMB** ⇒ **Dimension** ⇒ *double-click* on the arc edge ⇒ **MMB** to place the diameter dimension ⇒ add a dimension to the center of the arc from the bottom of the part ⇒ pick the centerline, then pick the vertical line, and then pick the centerline again ⇒ **MMB** to place the diameter dimension [Fig. 6.8(m)] ⇒ **MMB** to end the current tool *(or click:* ***Select items)*** ⇒ *(if you have an extra weak dimension, click:* ***Impose sketcher constraints on the section*** ⇒ ***Create same points*** ⇒ *pick the centerline, then pick the FRONT datum from the model tree* ⇒ ***Close)*** ⇒ window–in all three dimensions ⇒ **RMB** ⇒ **Modify** modify the dimensions to the design values (Fig. 6.5) [Fig. 6.8(n)] *(Note: sketch is displayed without datums and coordinate system for clarity)* ⇒ ⇒ **LMB** to deselect

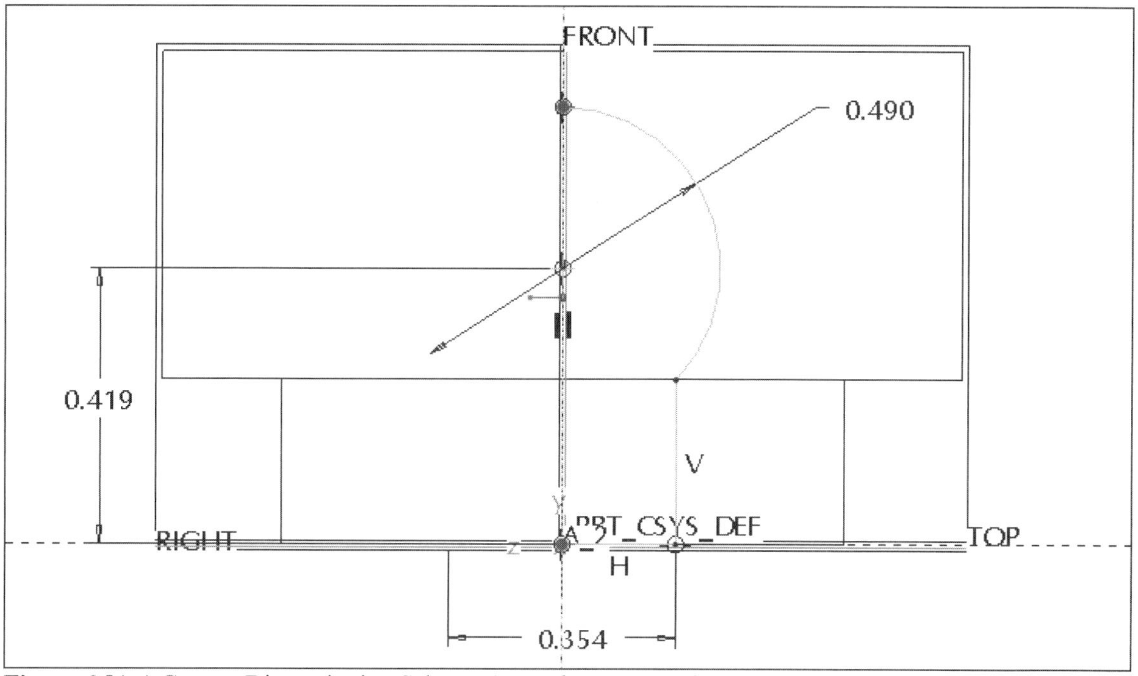

Figure 6.8(m) Correct Dimensioning Scheme (*your dimension values* ***will*** *be different*)

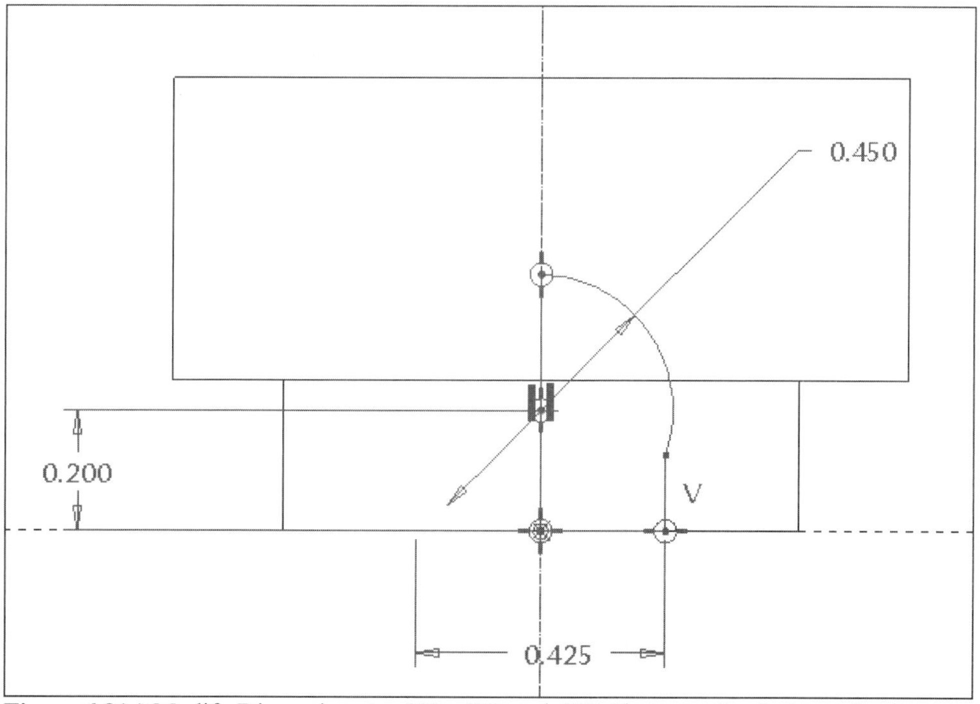

Figure 6.8(n) Modify Dimensions to **.450**, **.425**, and **.200** (these are the design values)

Click: **File** ⇒ **Save a Copy** ⇒ New Name- type **ball_end** ⇒ **OK** ⇒  ⇒ 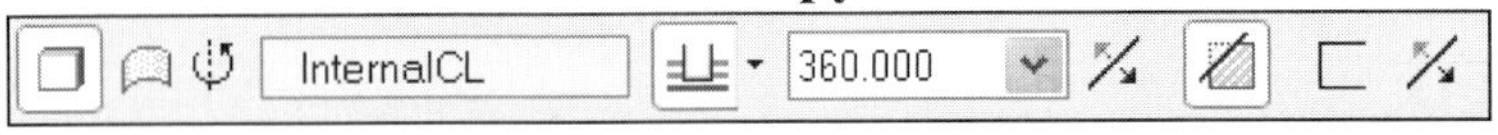

[Fig. 6.8(o)] ⇒ **Enter** ⇒ **MMB** rotate the part to view the hole/cut [Fig. 6.8(p)] ⇒  **Shading** [Fig. 6.8(q)] ⇒  ⇒ **Enter**

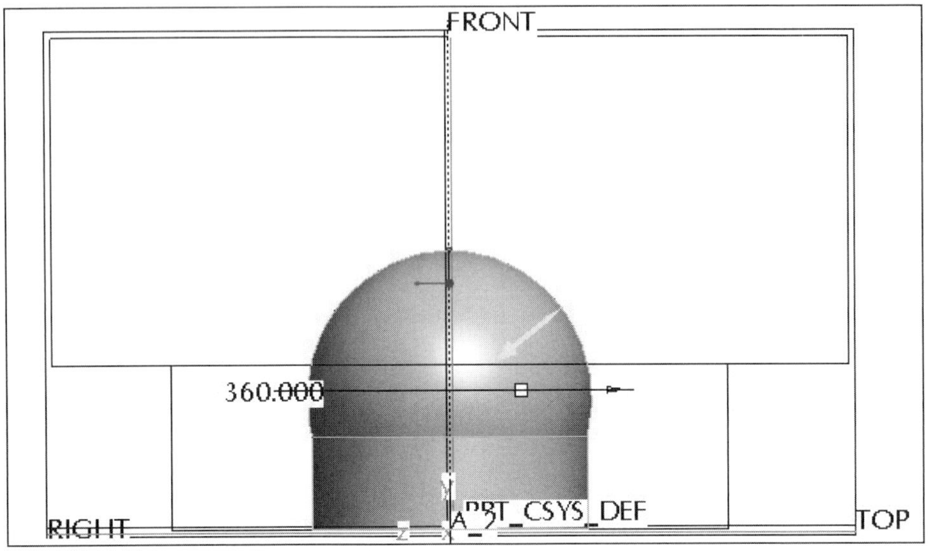

Figure 6.8(o) Previewed Revolved Cut

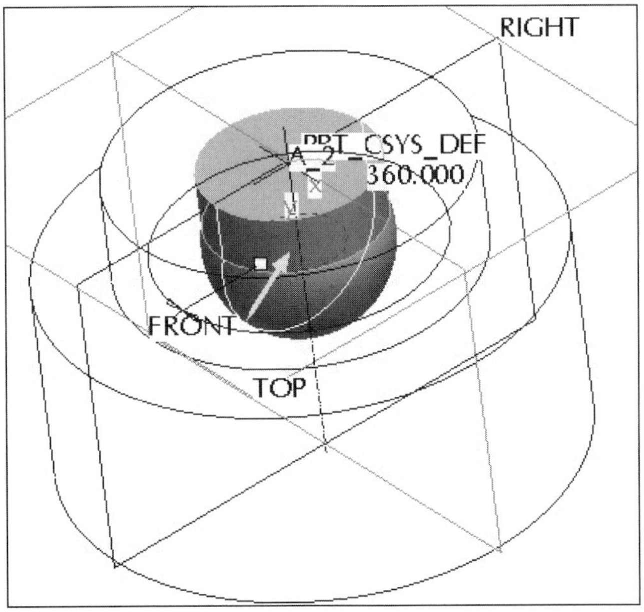

Figure 6.8(p) Revolved Cut

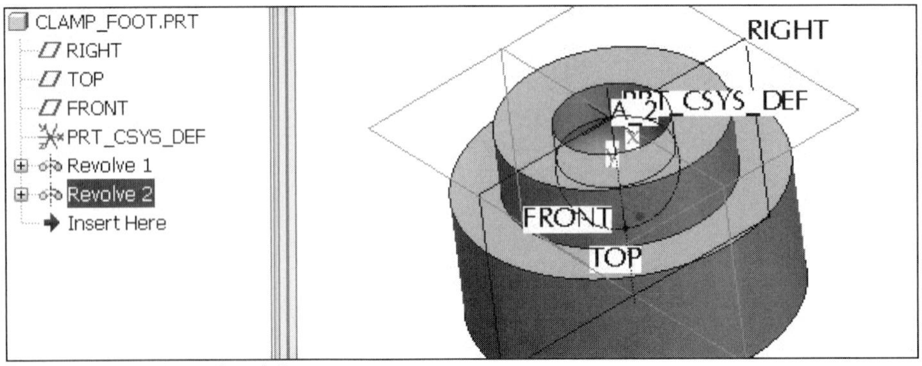

Figure 6.8(q) Completed Cut

Pick on the revolved protrusion then pick on the edge of the cut until selected (highlights) [Fig. 6.8(r)] ⇒  **Chamfer Tool** [Fig. 6.8(s)] ⇒ double-click on the dimension and modify to the design value of **.03125** [Fig. 6.8(t)] ⇒ **Enter** ⇒ **MMB** [Fig. 6.8(u)] ⇒ **LMB** to deselect ⇒ 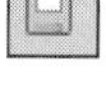 ⇒ **MMB** ⇒ **File** ⇒ **Delete** ⇒ **Old Versions** ⇒ **MMB**

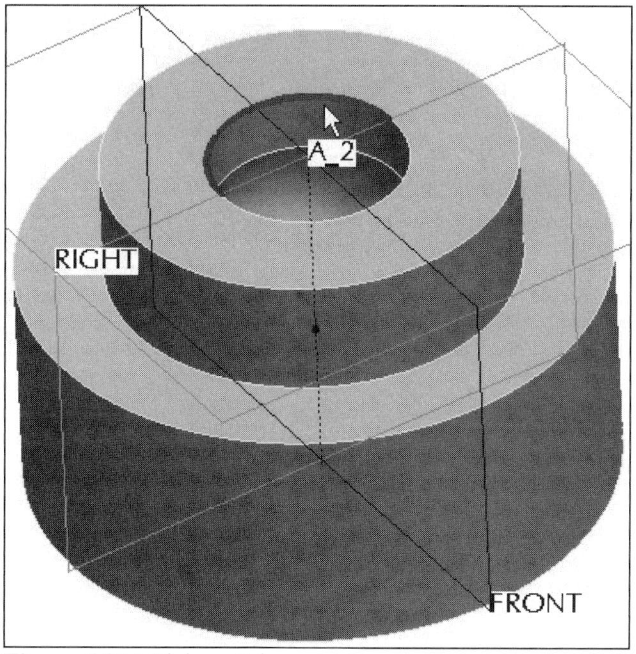

Figure 6.8(r) Pick on Edge

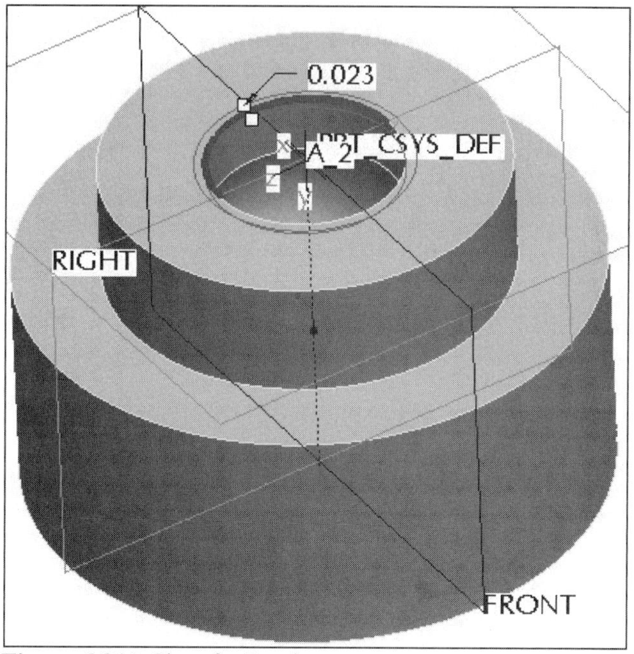

Figure 6.8(s) Chamfer Preview

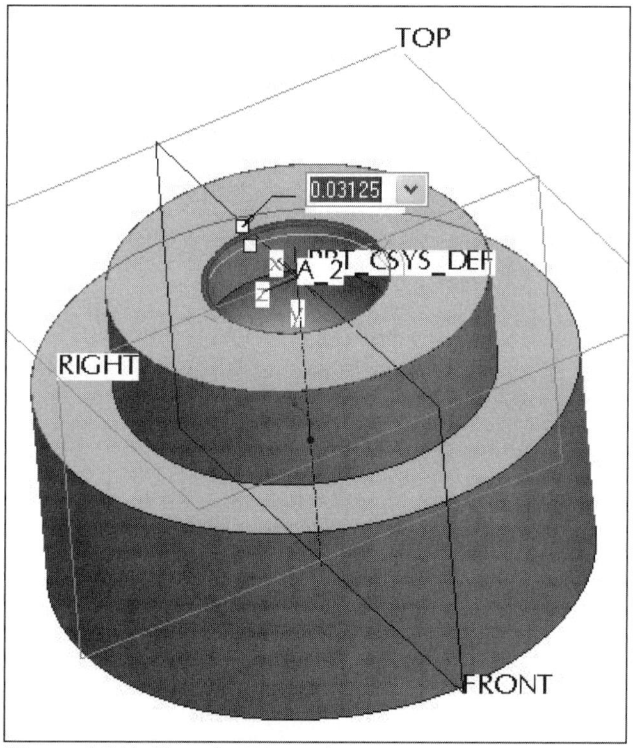

Figure 6.8(t) Chamfer Dimension

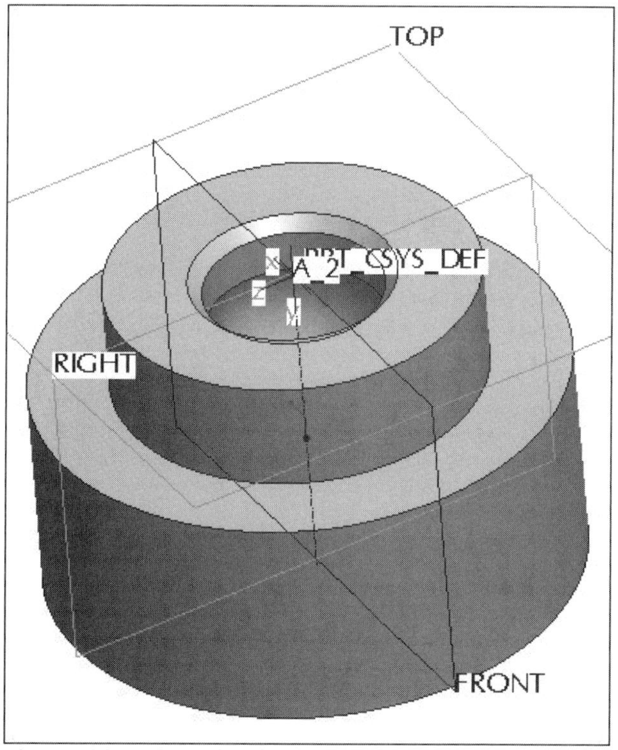

Figure 6.8(u) Completed Chamfer

Select one edge, then press and hold the **Ctrl** key and pick on the remaining three edges ⇒ **RMB** ⇒ **Round Edges** [Fig. 6.8(v)] ⇒ double-click on the dimension and modify to the design value of **.03125** [Fig. 6.8(w)] ⇒ **Enter** ⇒ **MMB** ⇒ [save icon] ⇒ **Enter** ⇒ **LMB** to deselect ⇒ **Ctrl+D**⇒ **Ctrl+R**

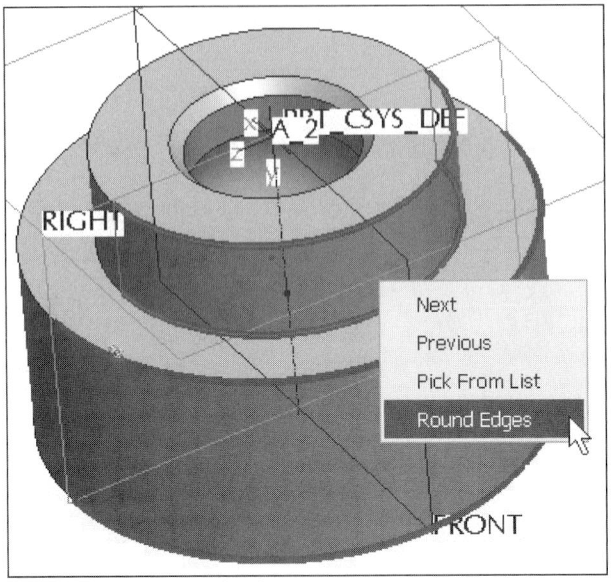

Figure 6.8(v) Round Edges

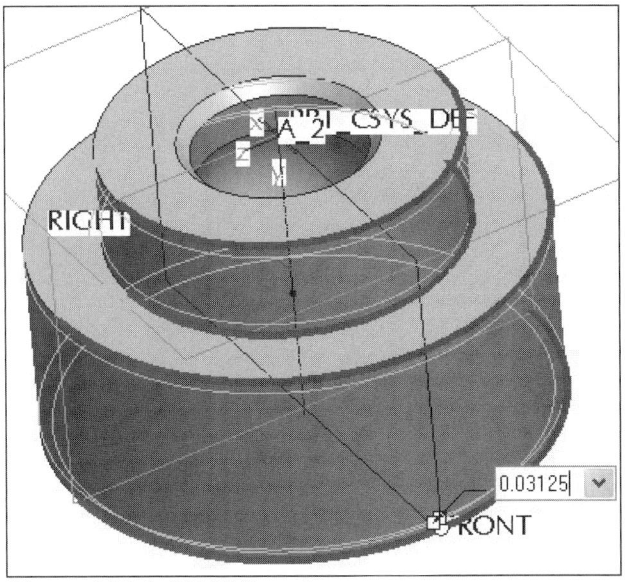

Figure 6.8(w) .03125 Round

Pick: **TOP** from the model or Model Tree ⇒ **RMB** ⇒ **Properties** ⇒ Name- type **A** ⇒ [icon] ⇒ **OK** ⇒ pick **RIGHT** ⇒ **RMB** ⇒ **Properties** ⇒ Name- type **B** ⇒ [icon] ⇒ **OK** ⇒ pick **FRONT** ⇒ **RMB** ⇒ **Properties** ⇒ Name- type **C** ⇒ [icon] ⇒ **OK** ⇒ **Edit** ⇒ **Setup** ⇒ **Material** ⇒ steel.mtl ⇒ [icon] ⇒ **OK** ⇒ **Done** ⇒ pick twice on the coordinate system in the Model Tree and rename CLAMP_FOOT_CSY [Fig. 6.8(x)] ⇒ **Ctrl+S** ⇒ **OK** ⇒ **File** ⇒ **Delete** ⇒ **Old Versions** ⇒ **MMB** ⇒ **Window** ⇒ **Close**

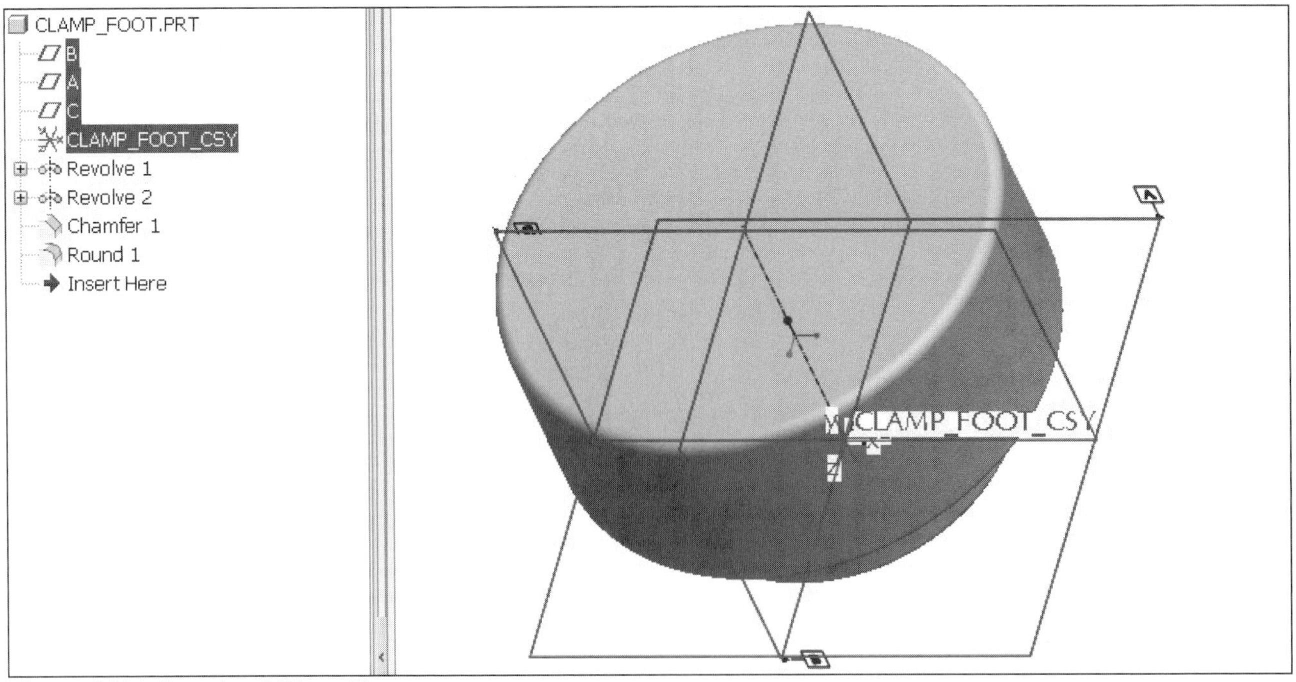

Figure 6.8(x) Completed Clamp Foot

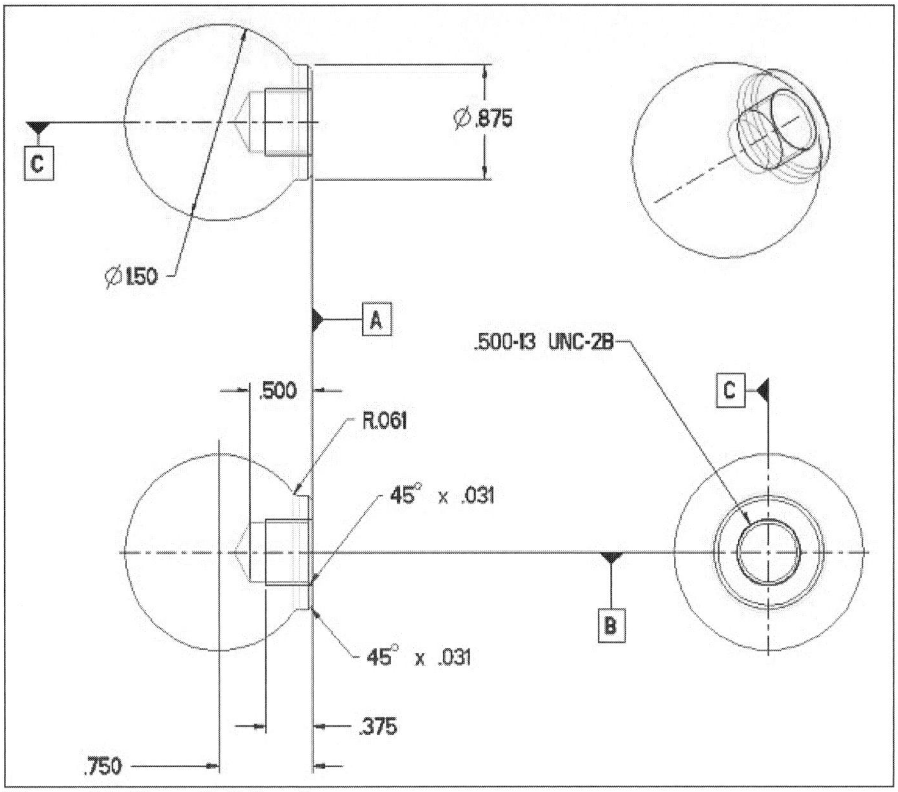

Figure 6.9 Clamp Ball Detail

Clamp Ball

The second part for this lesson is the Clamp Ball (Fig. 6.9). Much of the sketching is the same for this part as was done for the first revolved feature of the Clamp Foot. Instead of an internal revolved cut, you will create a standard hole. The Clamp Ball is made of *nylon*.

Click: **Create a new object** ⇒ Part ⇒ Name CLAMP_BALL ⇒ Use default template ⇒ **OK**

Tools ⇒ **Options** ⇒ **Open a configuration file** ⇒ click on **clamp.pro** [Fig. 6.10(a)] from your working directory ⇒ **Open** ⇒ **Apply** ⇒ **Close** ⇒ **Edit** from menu bar ⇒ **Setup** ⇒ **Units** ⇒ Inch lbm Second (Pro/E Default) ⇒ **Close** ⇒ **Material** ⇒ nylon.mtl ⇒ ▶▶▶ [Fig. 6.10(b)] ⇒ **OK** ⇒ **Done**

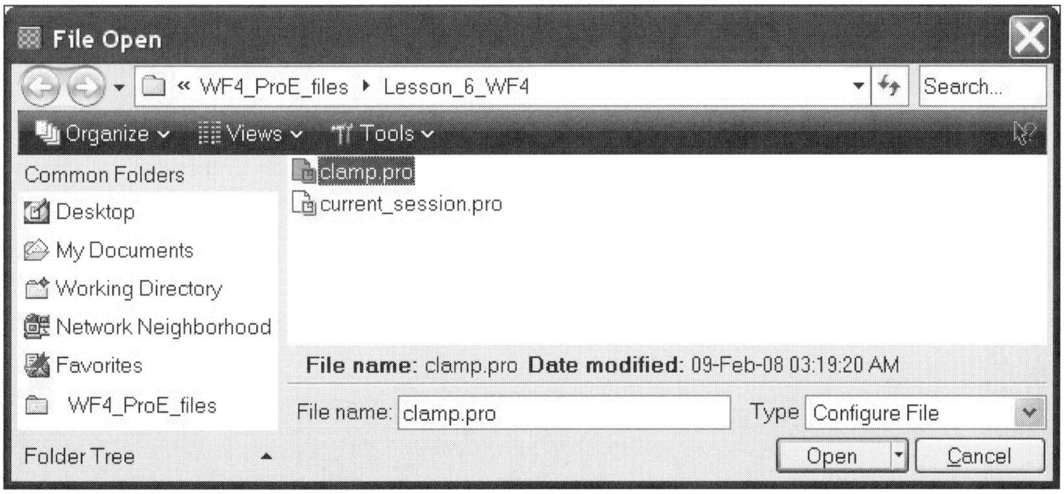

Figure 6.10(a) Configuration File

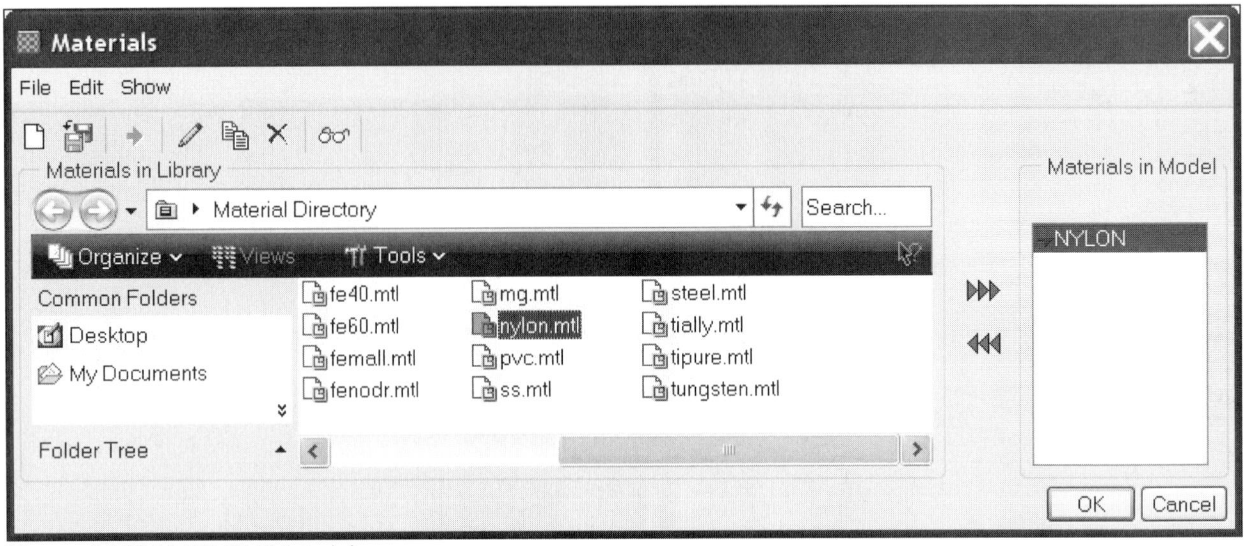

Figure 6.10(b) NYLON Material

Model Tree

The Model Tree is a tabbed feature on the Pro/E navigator that contains a list of every feature (or component) in the current part, assembly, or drawing. The model structure is displayed in hierarchical (tree) format with the root object (the current feature, part or assembly) at the top of its tree and the subordinate objects (features, parts, or assemblies) below.

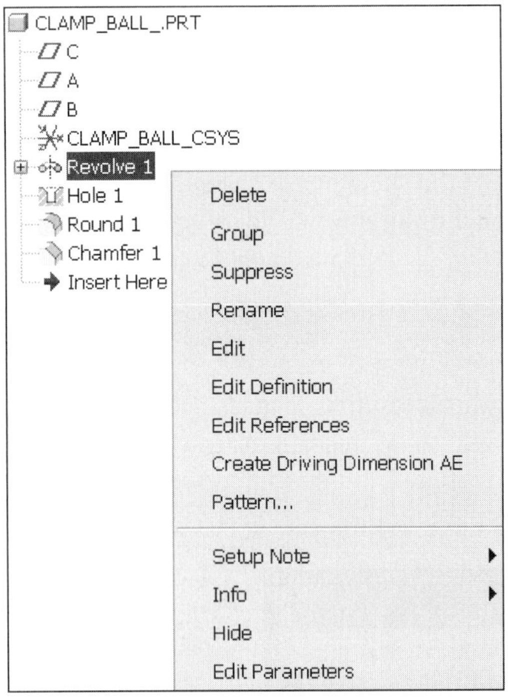

If you have multiple Pro/E windows open, the Model Tree contents reflect the file in the current "active" window. The Model Tree lists only the related feature- and part-level objects in a current file and does not list the entities (such as edges, surfaces, curves, and so forth) that comprise the features.

Each Model Tree item displays an icon that reflects its object type, for example, assembly, part, feature, or datum plane (also a feature). The icon can also show the display status for a feature, part, or assembly, for example, suppressed or hidden.

You can save the Model Tree as a .txt file. Selection in the Model Tree is object-action oriented; you select objects in the Model Tree without first specifying what you intend to do with them. Items can be added or removed from the Model Tree column display using Settings in the Navigator:

- Select objects, and perform object-specific operations on them using the shortcut menu.
- Filter the display by item type, for example, hiding or unhiding datum features.
- Open a part within an assembly file by right-clicking the part in the Model Tree.
- Create or modify features and perform other operations such as deleting and redefining.
- Search the Model Tree for model properties or other feature information.
- Show the display status for an object, for example, suppressed or hidden.

Click: **Settings** in the Navigator [Fig. 6.11(a)] ⇒ **Tree Filters** ⇒ check all Display options on ⇒ [Fig. 6.11(b)] ⇒ **Apply** ⇒ **OK**

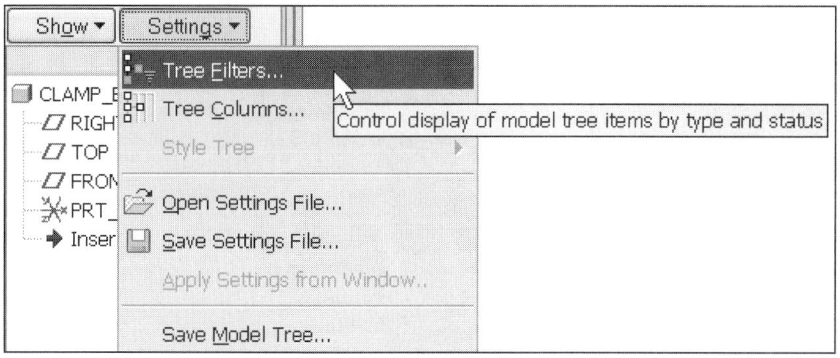

Figure 6.11(a) Tree Filters

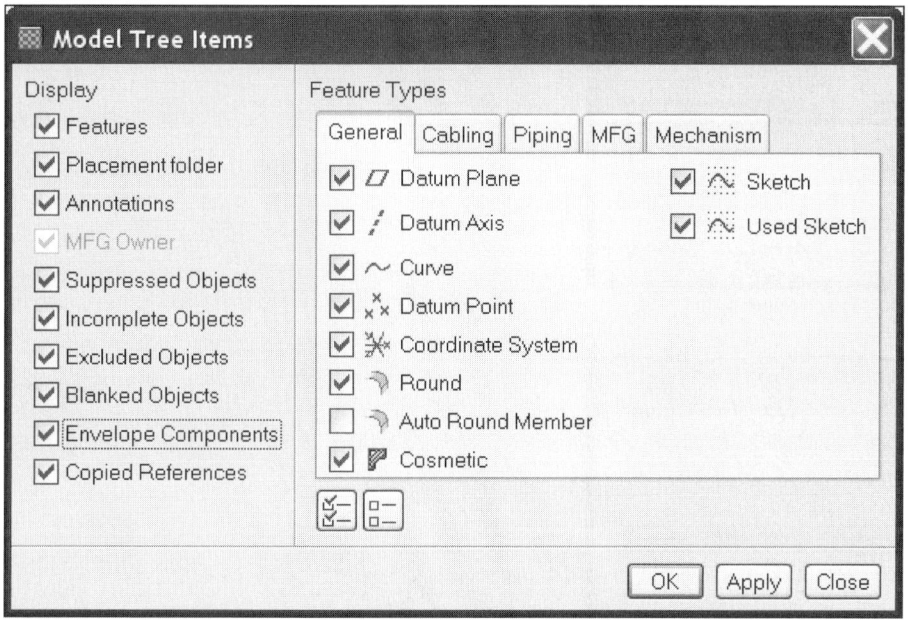

Figure 6.11(b) Model Tree Items Dialog Box

Click: **Settings** in the Navigator ⇒ 

⇒ the Model Tree Columns dialog box opens [Fig. 6.12(a)]

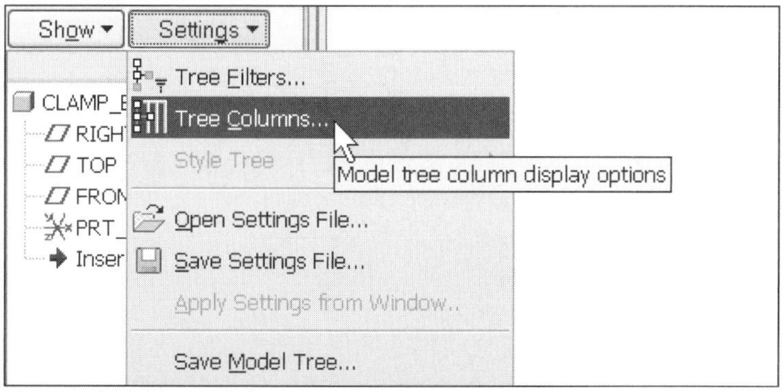

Figure 6.12(a) Model Tree Columns

Type **Info** [Fig. 6.12(b)], add types to the Displayed list, click: **Feat #** ⇒ >> ⇒ **Feat ID** ⇒ >> ⇒ Type ⌄ ⇒ select **Layer** [Fig. 6.12(c)] ⇒ **Layer Names** ⇒ >> ⇒ **Layer Status** ⇒ >> [Fig. 6.12(d)] ⇒ **Apply** ⇒ **OK**

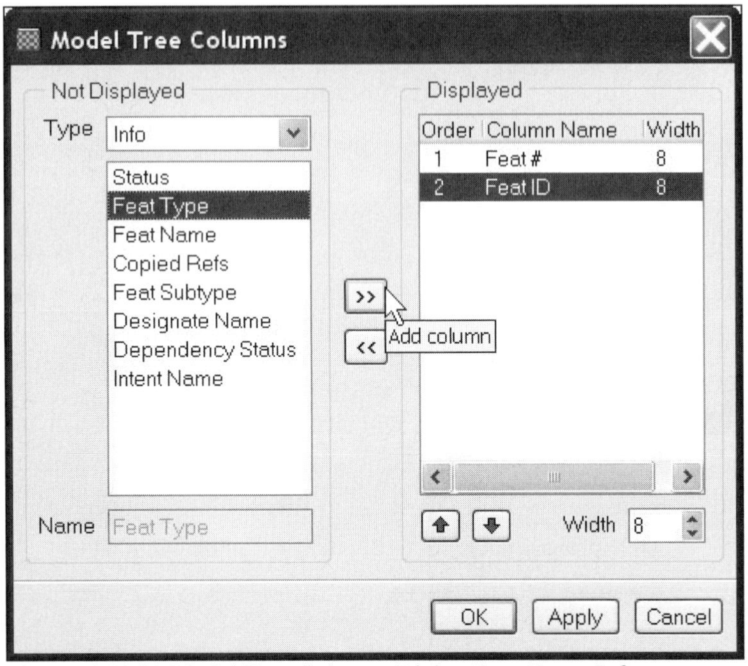

Figure 6.12(b) Model Tree Columns Dialog Box, Type: Info

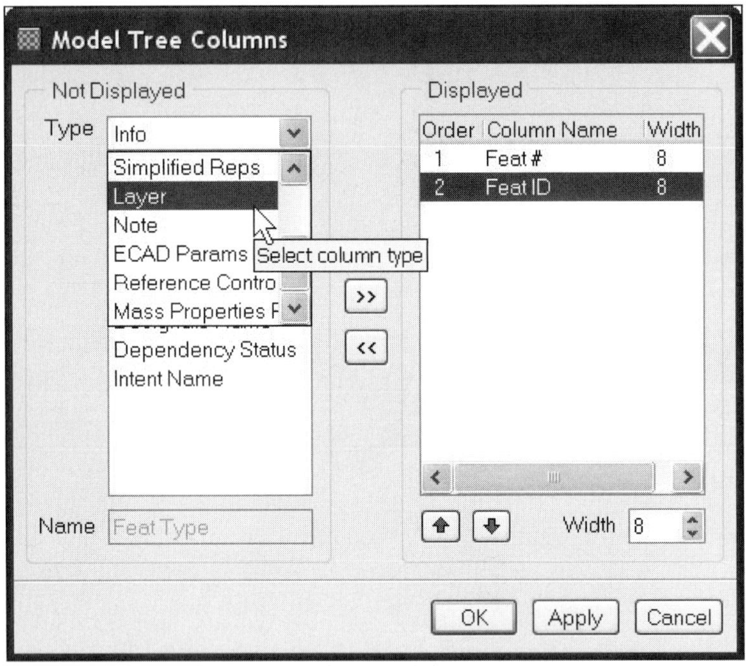

Figure 6.12(c) Model Tree Columns Dialog Box, Type: Layer

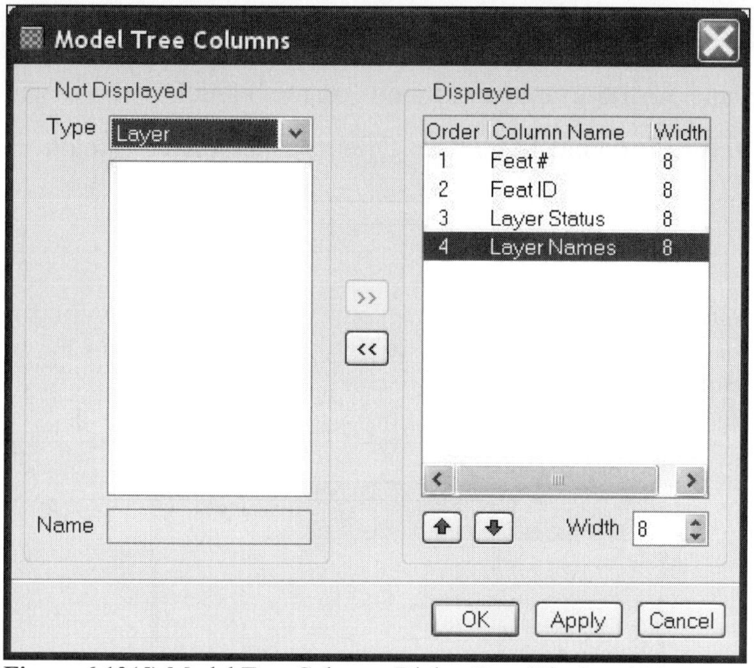

Figure 6.12(d) Model Tree Columns Dialog Box, Displayed List

Click on the sash 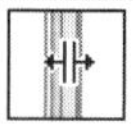 and drag to expand Model Tree [Fig. 6.12(e)] ⇒ adjust the width of each column by dragging the column divider ⇒ 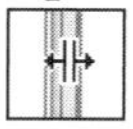 drag the sash to the left to decrease the Model Tree size [Fig. 6.12(f)]

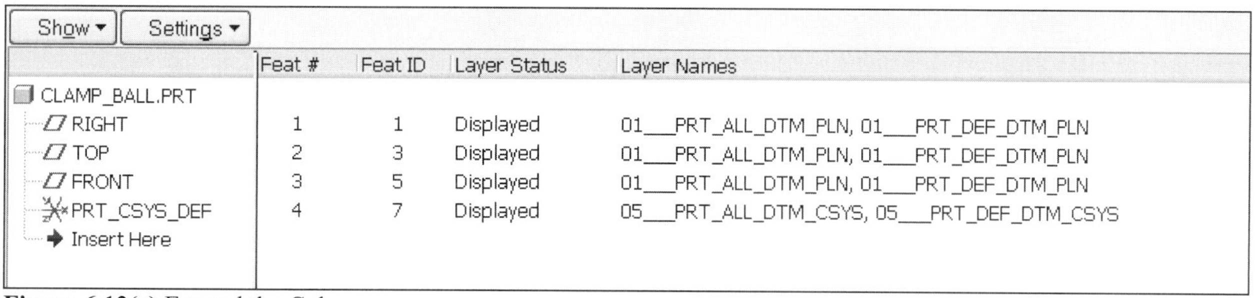

Figure 6.12(e) Expand the Columns

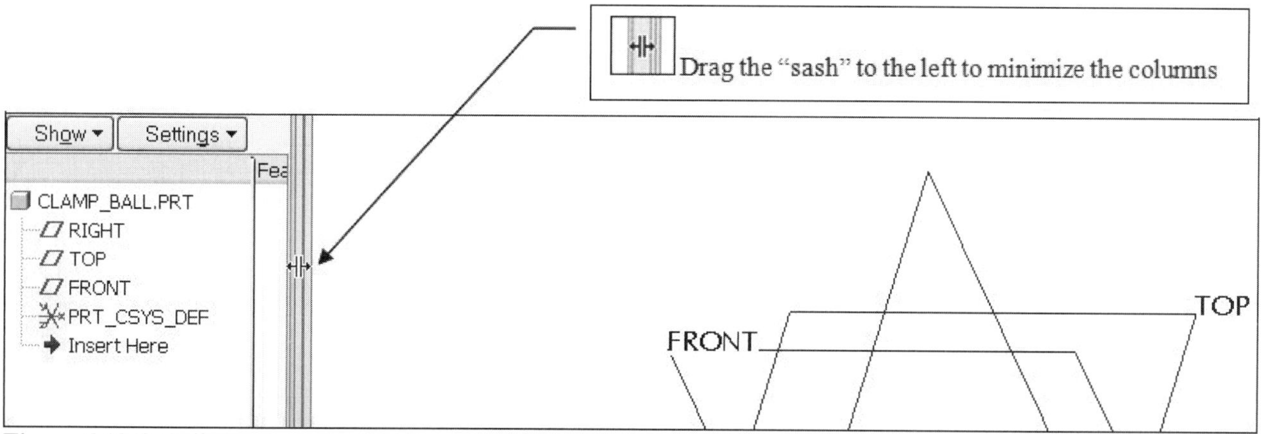

Figure 6.12(f) Collapse the Columns

Click: **Tools** ⇒ Environment ⇒ Snap to Grid ⇒ Use 2D Sketcher ⇒ **Apply** ⇒ **OK** ⇒ **Spin Center** off ⇒ pick on the **RIGHT** datum plane in the graphics window or the Model Tree (it will highlight) ⇒ **Revolve Tool** ⇒ **RMB** ⇒ **Define Internal Sketch** ⇒ Orientation: **Left** ⇒ **Sketch** ⇒

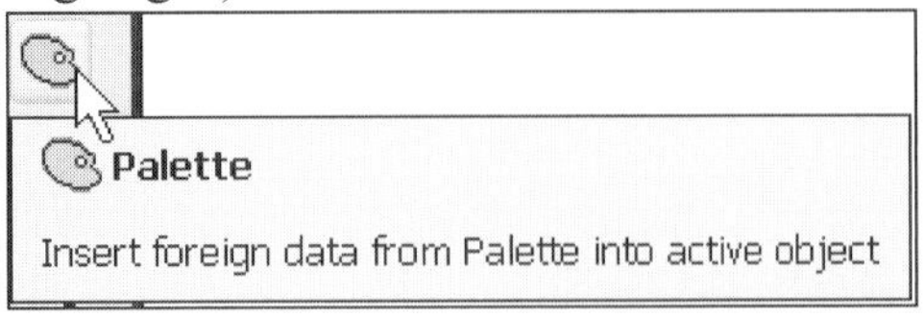

[Fig. 6.13(a)] ⇒ pick the tab with your Working Directory name ⇒ double-click

[Fig. 6.13(b)] ⇒ place the section [Figs. 6.13(c-d)]

Figure 6.13(a) Sketcher Palette

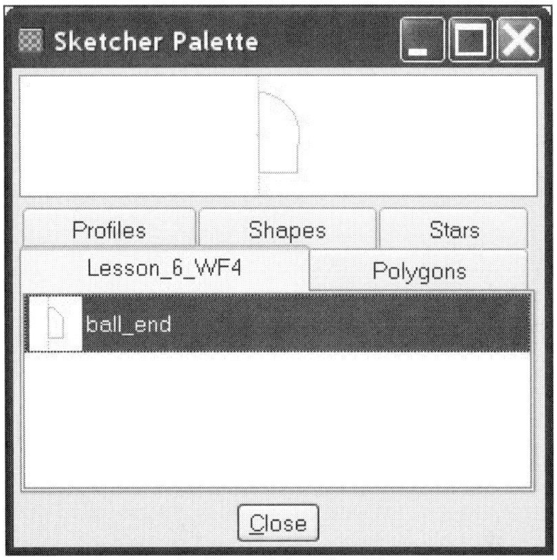

Figure 6.13(b) ball_end for the Sketcher Palette

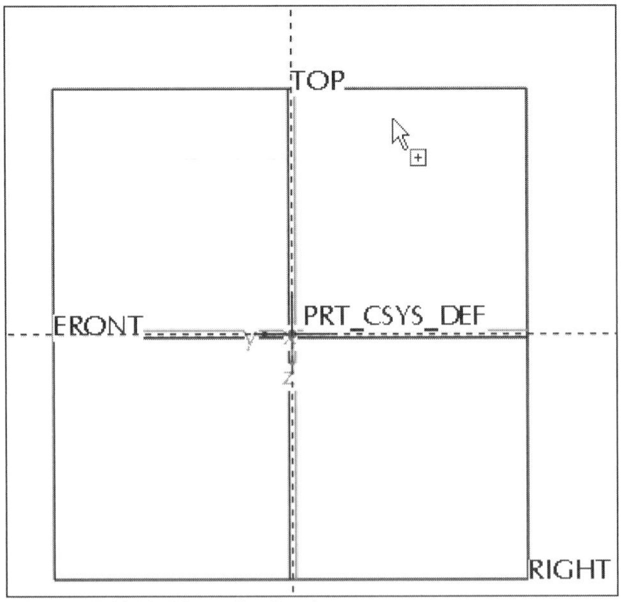

Figure 6.13(c) Place the Section

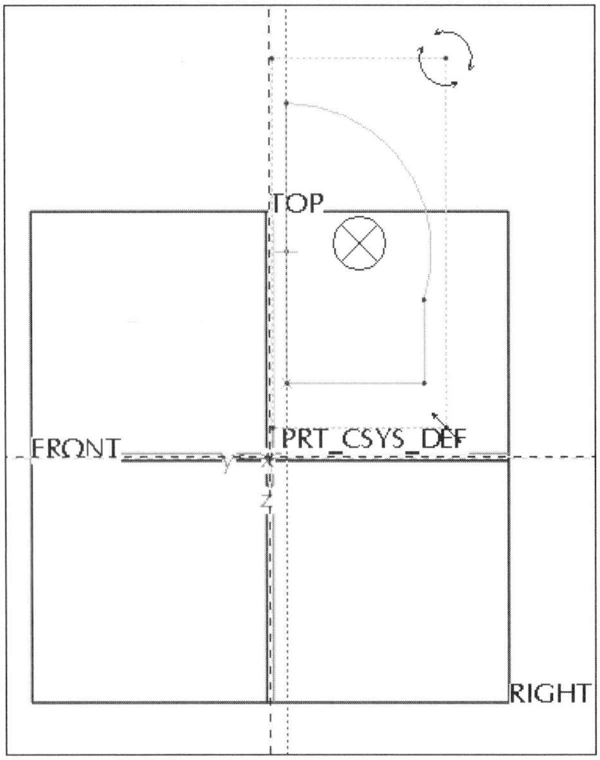

Figure 6.13(d) Placed Section

With your pointer on the move handle , click and hold the **RMB**, move it to the lower left-hand corner of the ball_end sketch [Fig. 6.13(e)] ⇒ with your **LMB** move the sketch to the part's origin [Fig. 6.13(f)] ⇒ Scale **2** ⇒ **Enter** 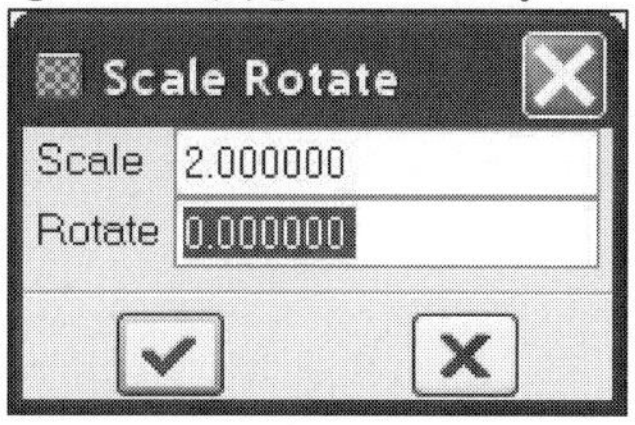

⇒  [Fig. 6.13(g)] ⇒  ⇒ 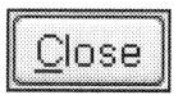

the Sketcher Palette ⇒ add the dimension from the bottom of the section to the center of the arc [Fig. 6.13(h)]

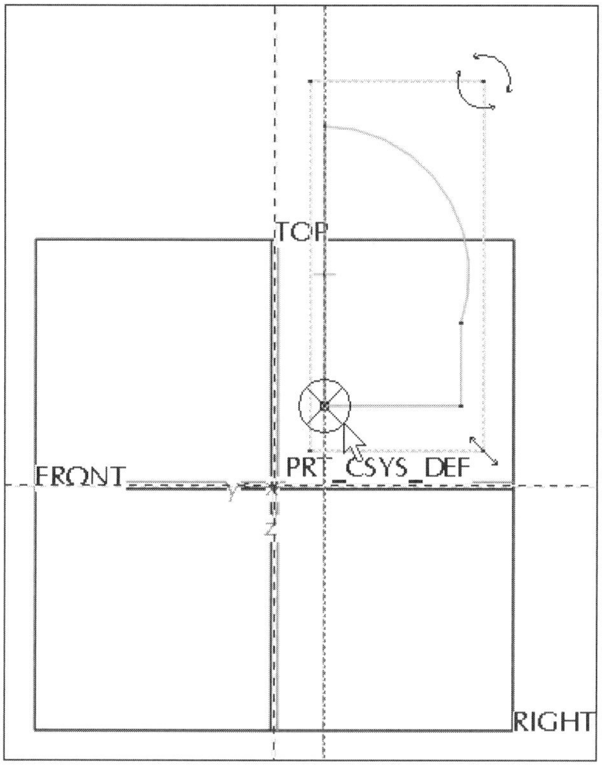

Figure 6.13(e) Repositioned Move Handle

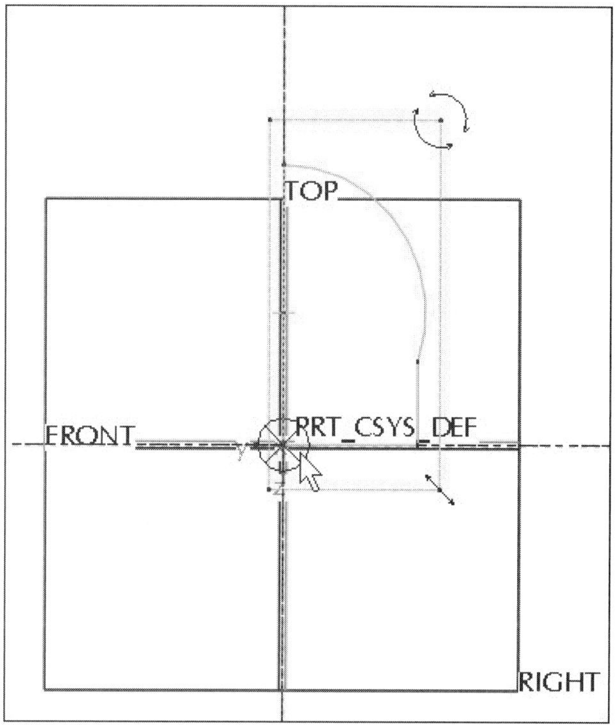

Figure 6.13(f) Placed Section

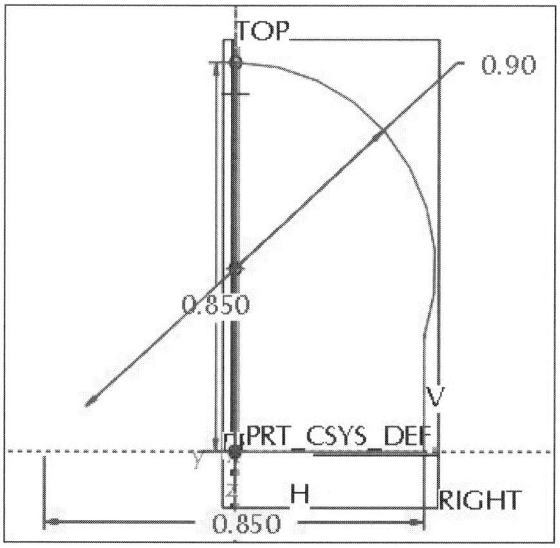

Figure 6.13(g) Section

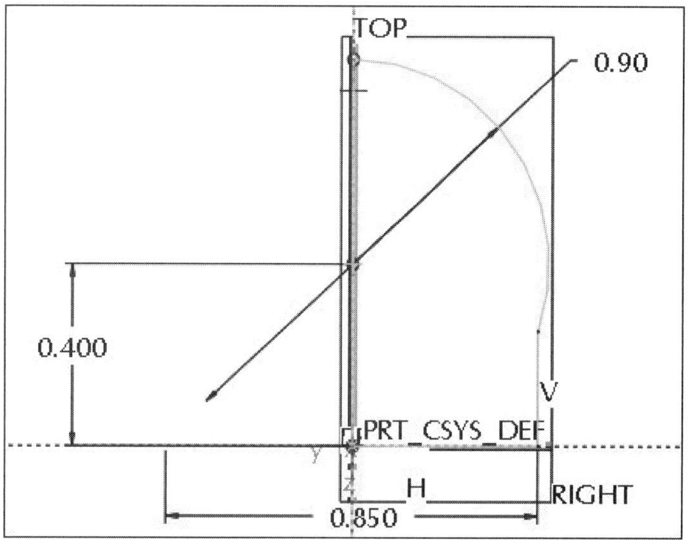

Figure 6.13(h) Add Dimension

Since you are adding material, the section must be closed. Click: **Tools** ⇒ **Environment** ⇒ Snap to Grid *make sure snap is off* ⇒ **Apply** ⇒ **OK** ⇒ **RMB** ⇒ **Line** sketch a line from the bottom of the section to the top of the arc ⇒ **MMB** ⇒ ⇒ window–in all three dimensions ⇒ **RMB** ⇒ **Modify** modify the dimensions to the design values (Fig. 6.9) [Fig. 6.13(e)] ⇒ ⇒ ⇒ ⇒ **Standard Orientation** ⇒ ⇒ InternalCL 360.000 [Fig. 6.13(f)] ⇒ ⇒ drag the Model Tree sash open [Fig. 6.13(g)] ⇒ drag the sash shut ⇒ ⇒ **Enter** ⇒ **Spin Center** on

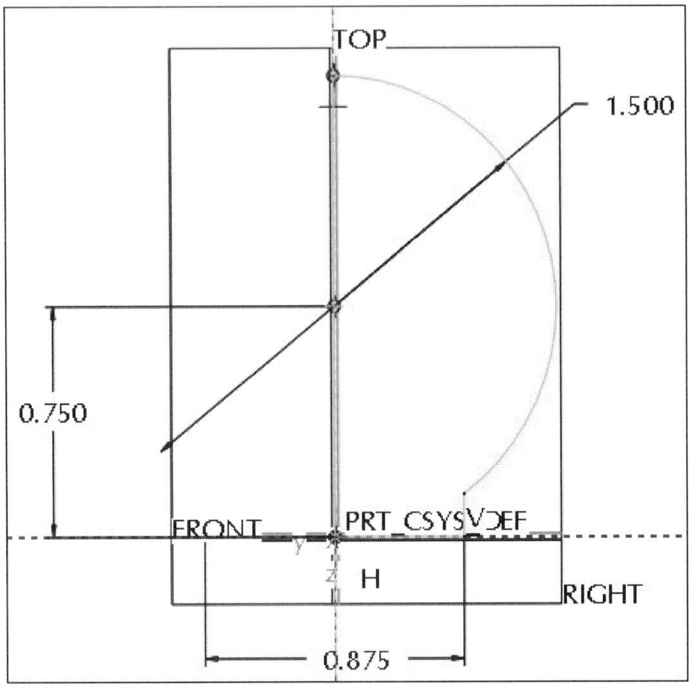

Figure 6.13(e) Modified Dimensions

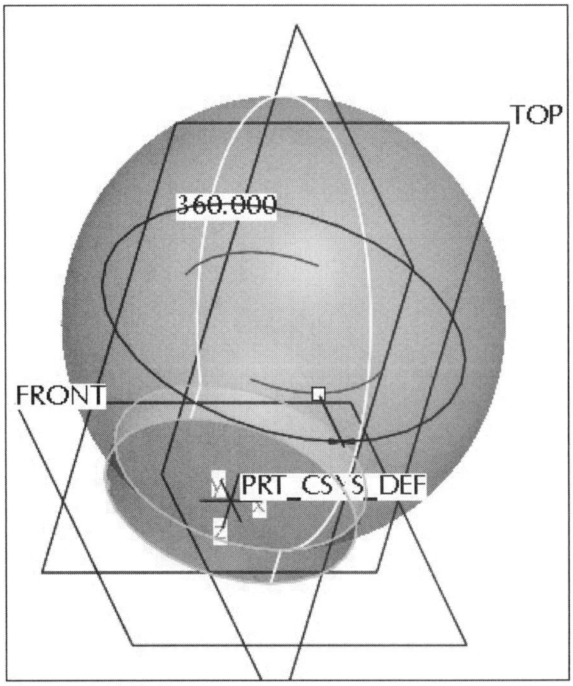

Figure 6.13(f) Feature Preview

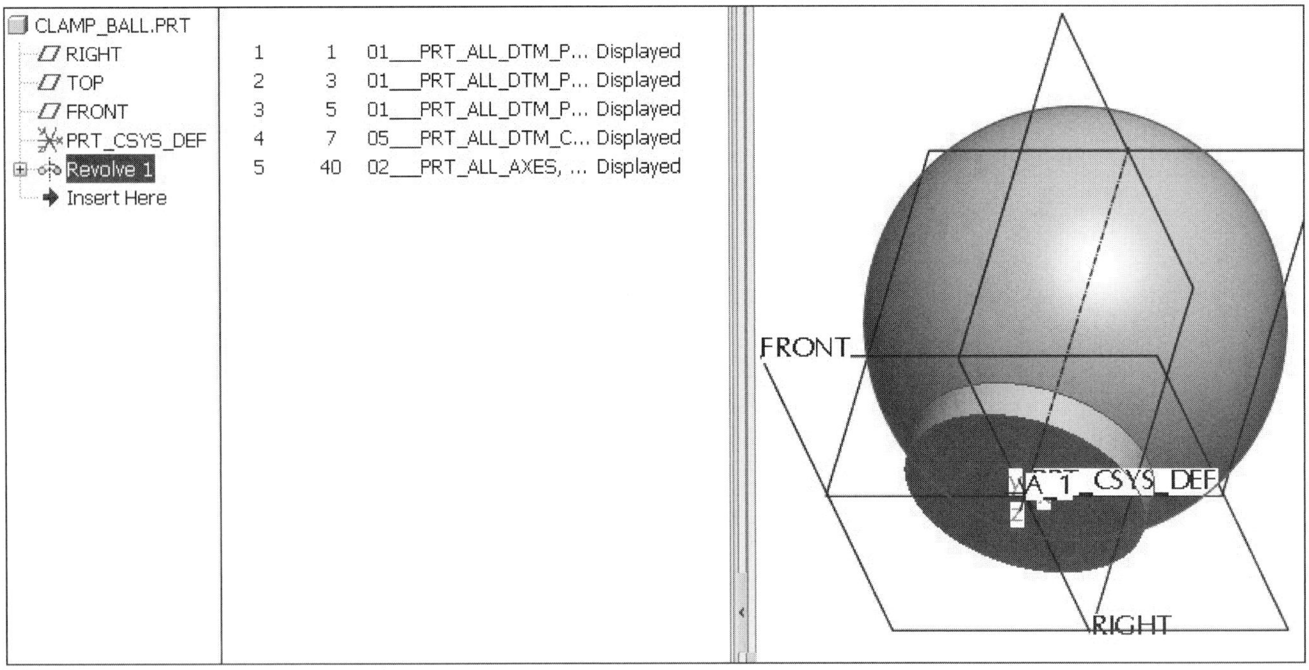

Figure 6.13(g) Completed Revolved Feature

Holes

Hole charts are used to lookup diameters for a given fastener size. You can create custom hole charts and specify their directory location with the configuration file option *hole_parameter_file_path.* UNC, UNF and ISO hole charts are supplied with Pro/E. Create a standard **.500-13 UNC-2B** hole, **.375** thread depth, **.50** tap drill. Include a standard chamfer.

Click: **Hole Tool** from Right Toolchest ⇒ ⇒ 1/2-13 ⇒ **Adds countersink ⇒ Shape** tab ⇒ 

⇒ countersink (.0625 x 2 + .4219 = **.5469**) ⇒ **.500** depth

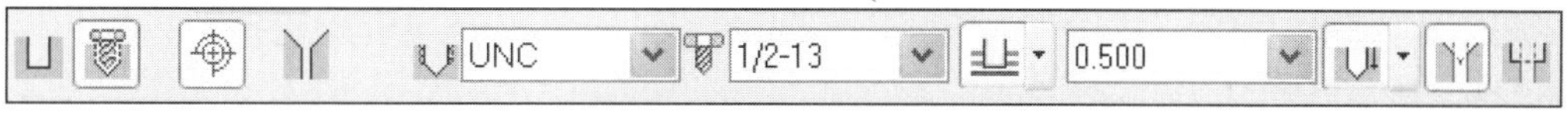

[Fig. 6.14(a)] ⇒ using your **MMB**, rotate the part ⇒ **Hidden line ⇒ Placement** tab ⇒ pick on axis **A_2** *(your axis name may be different)* ⇒ press and hold your **Ctrl** key and pick on the flat end surface [Fig. 6.14(b)]

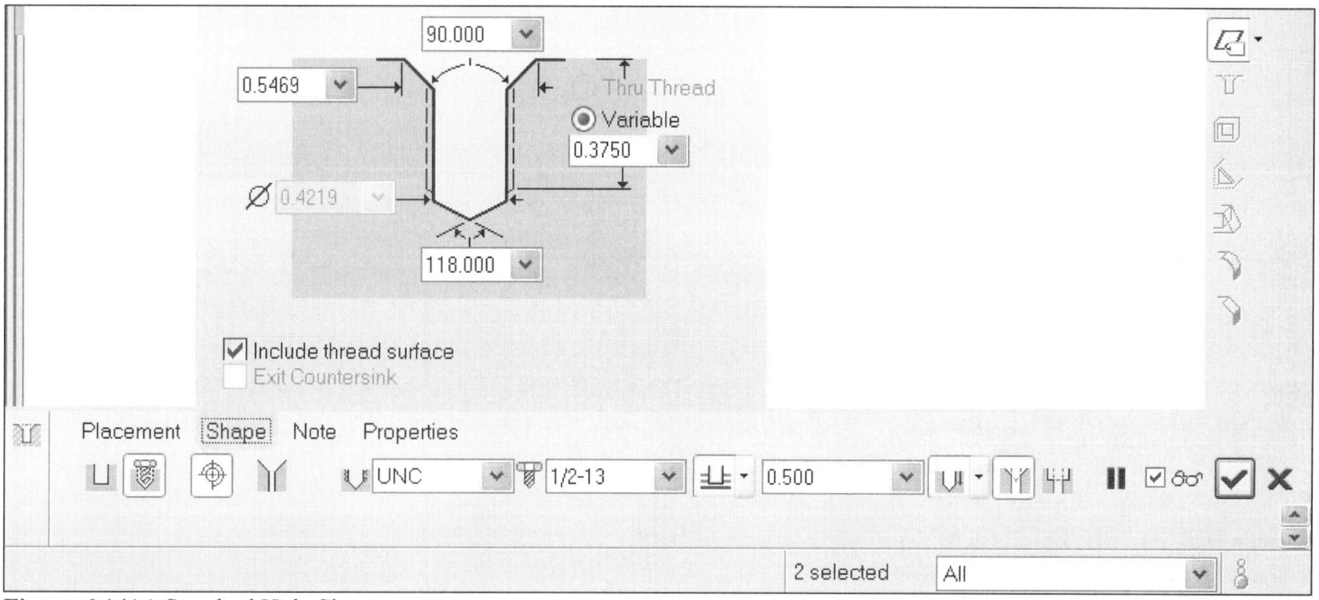

Figure 6.14(a) Standard Hole Shape

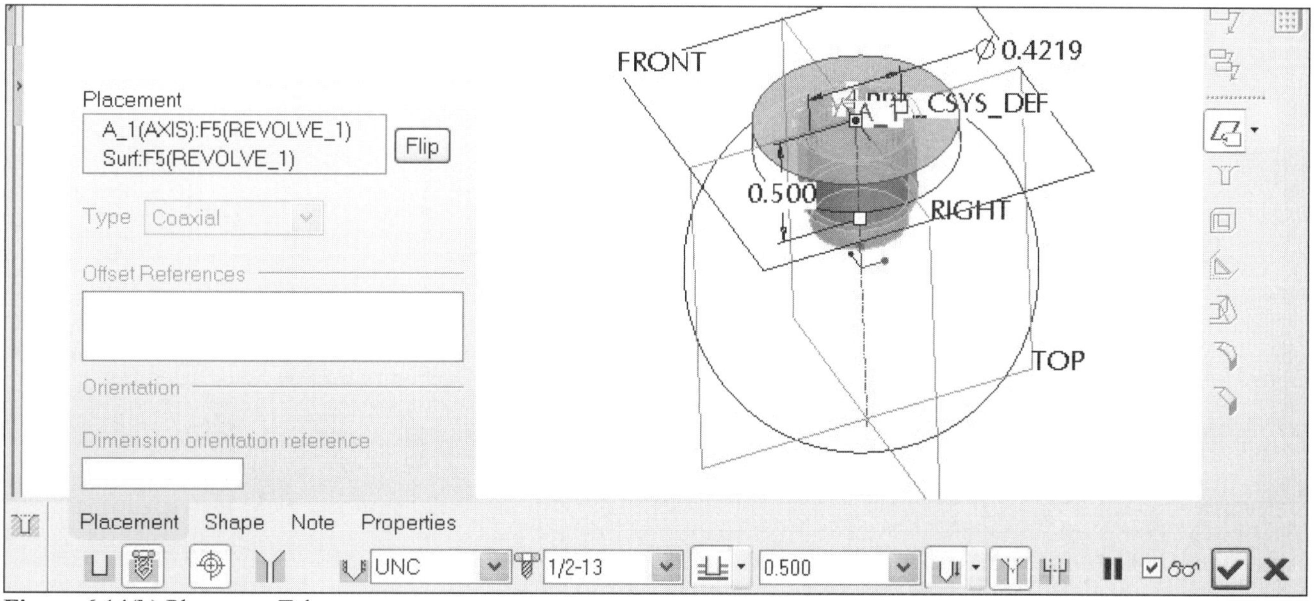

Figure 6.14(b) Placement Tab

Click: **Note** tab [Fig. 6.14(d)] ⇒ ⇒ ⇒ pop-up filters in the lower right-hand corner below the graphics window ⇒ **Annotation**

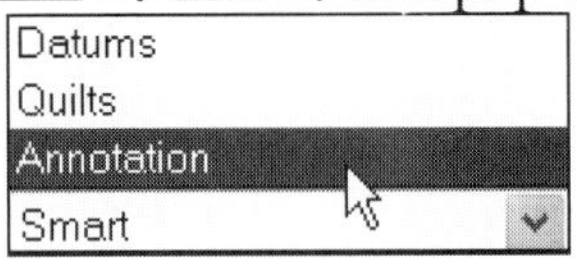

⇒ pick on the note ⇒ **RMB** ⇒ **Properties** [Fig. 6.15(a)] ⇒ Note dialog box opens [Fig. 6.15(b)] ⇒ **Ctrl+D**

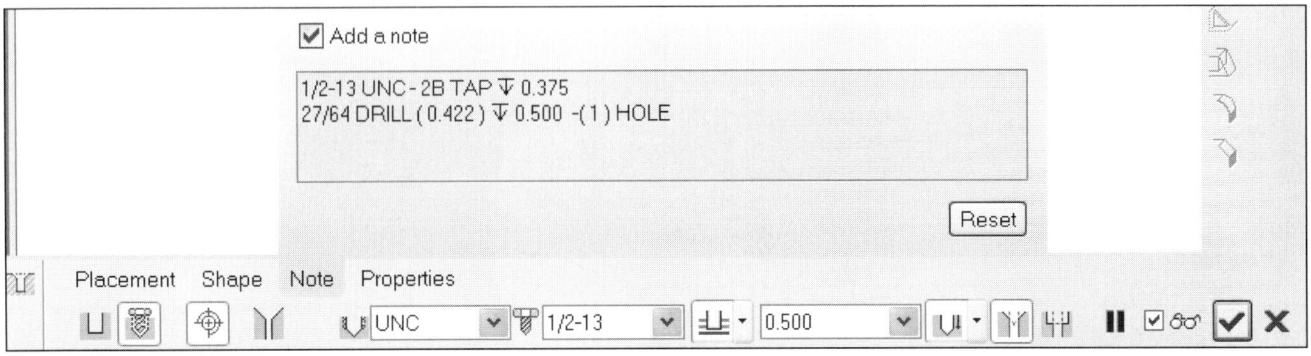

Figure 6.14(d) Hole Note

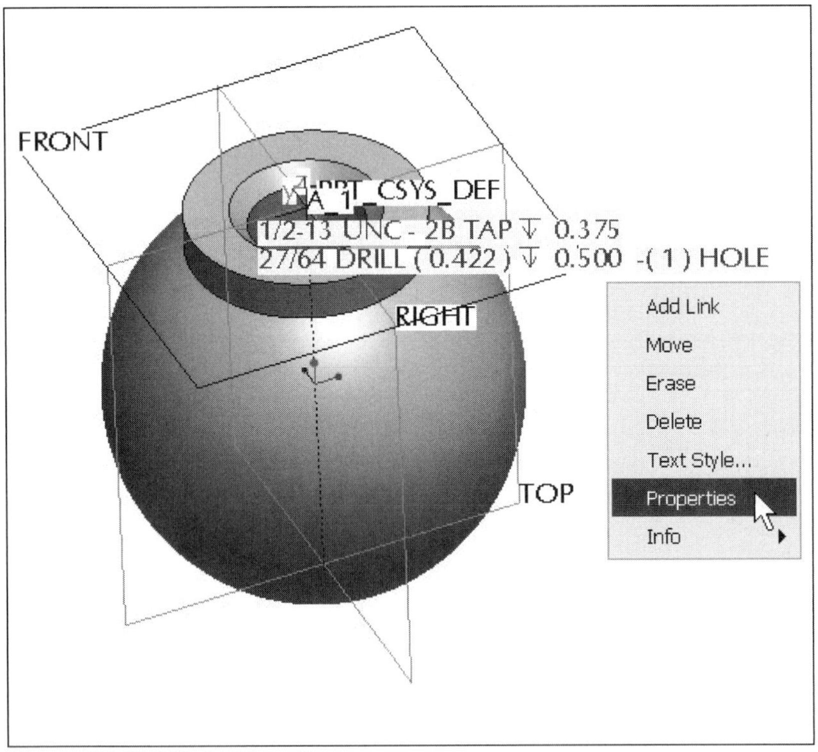

Figure 6.15(a) Note Properties

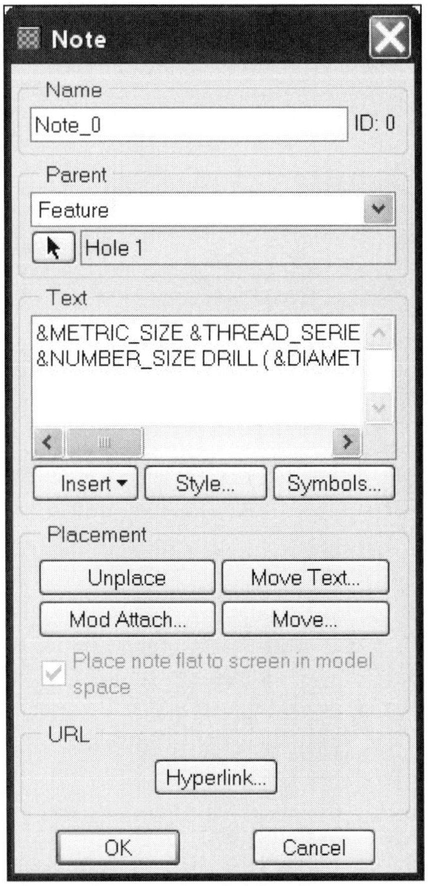

Figure 6.15(b) Note Dialog Box

Modify the note [Fig. 6.15(c)]. Click on the Symbols button ⇒ ⌀ add the diameter symbol [Fig. 6.15(d)] ⇒ **OK** ⇒ with the note still selected, click: **RMB** ⇒ **Move** ⇒ reposition the note [Fig. 6.15(e)] ⇒ ⌄ ⇒ Smart ⌄ (lower right corner below graphics window)

Figure 6.15(c) Modified Note

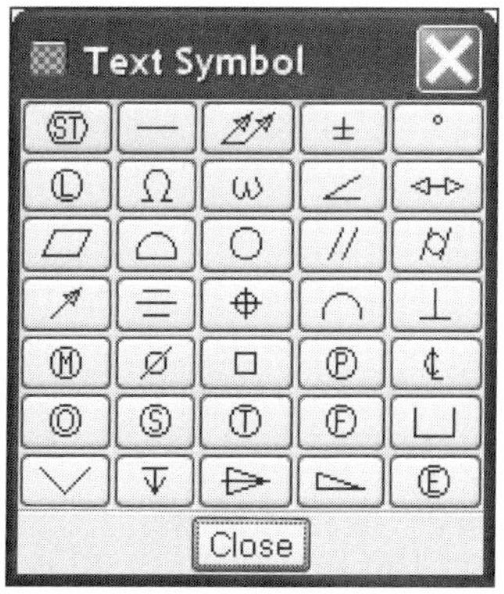

Figure 6.15(d) Text Symbol Dialog Box

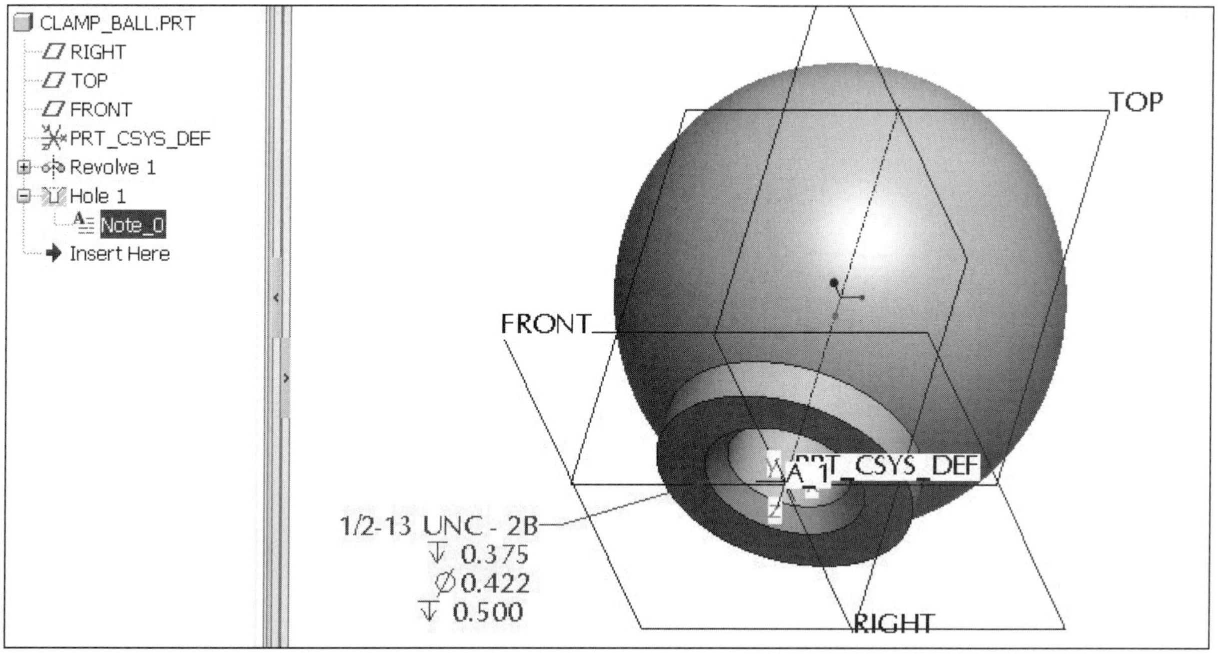

Figure 6.15(e) Annotation

Pick: **FRONT** from the model or Model Tree ⇒ **RMB** ⇒ **Properties** ⇒ Name- type **A** ⇒ [icon] ⇒ **OK** ⇒ pick **TOP** ⇒ **RMB** ⇒ **Properties** ⇒ Name- type **B** ⇒ [icon] ⇒ **OK** ⇒ pick **RIGHT** ⇒ **RMB** ⇒ **Properties** ⇒ Name- type **C** ⇒ [icon] ⇒ **OK** ⇒ pick twice on the coordinate system in the Model Tree and rename [CLAMP_BALL_CSYS] [Fig. 6.15(f)] ⇒ **LMB** to deselect ⇒ [icon] ⇒ **Enter**

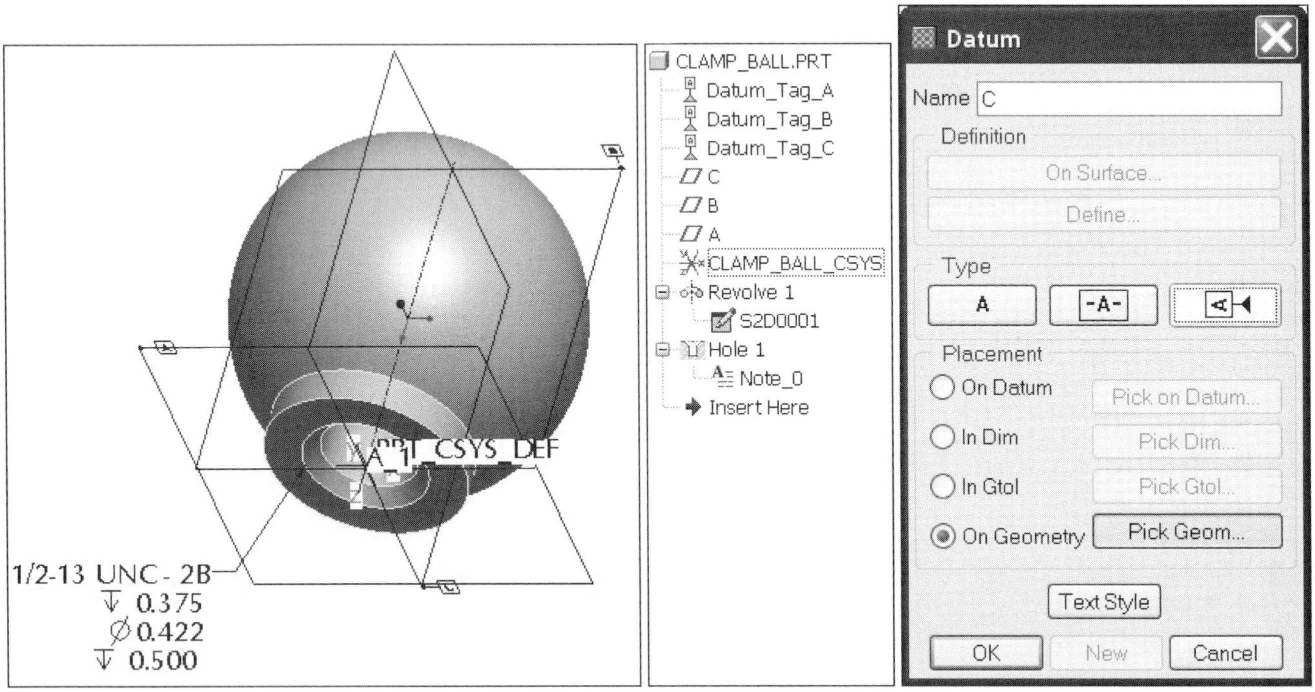

Figure 6.15(f) Set Datums

Double-click on the model to display the dimensions for the protrusion ⇒ pick on the **∅.875** dimension ⇒ **RMB** [Fig. 6.16(a)] ⇒ **Properties** Dimension Properties dialog box displays [Fig. 6.16(b)]

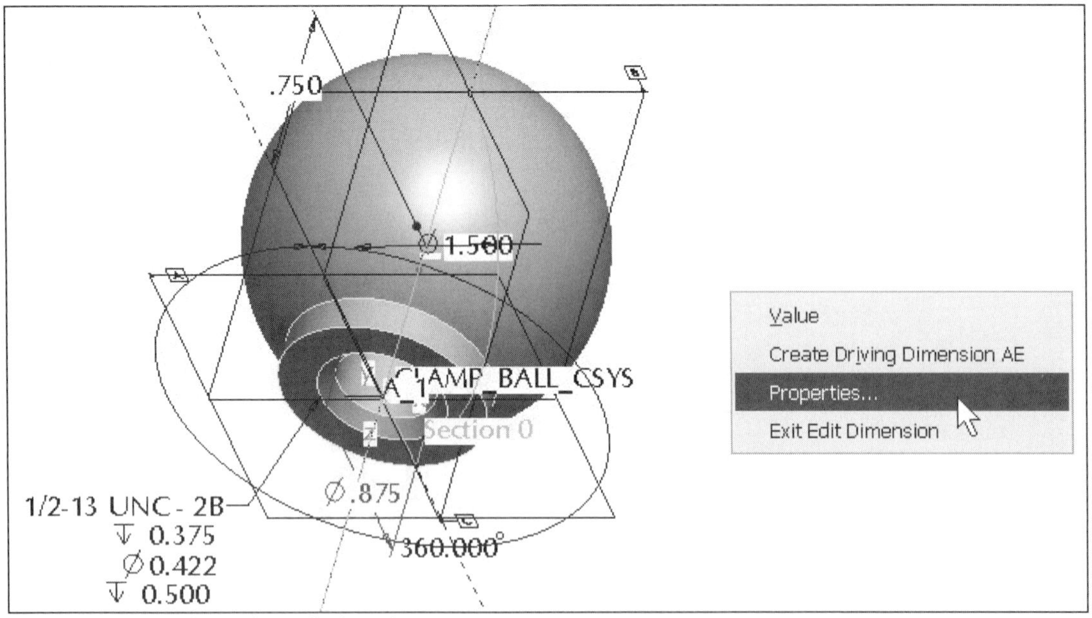

Figure 6.16(a) Dimension Displayed

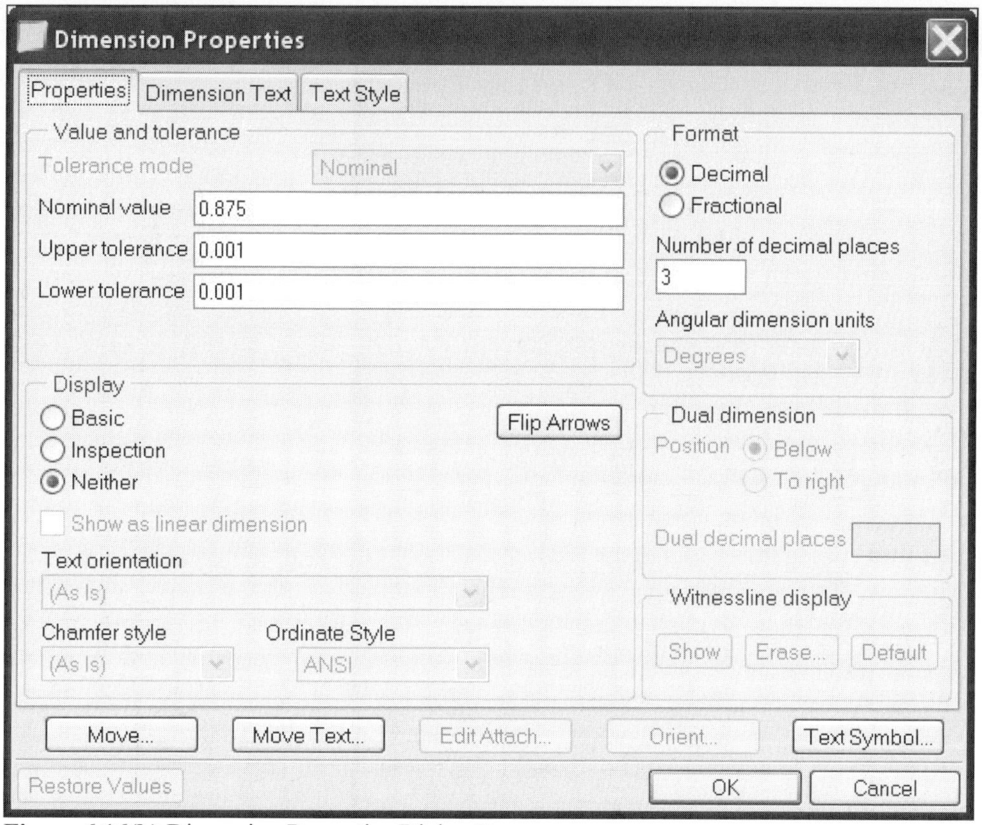

Figure 6.16(b) Dimension Properties Dialog Box

The Dimension Text tab [Fig. 6.16(c)] shows the parametric dimension symbol. The Text Style tab [Fig. 6.16(d)] provides options for Character, and Note/Dimension variations.

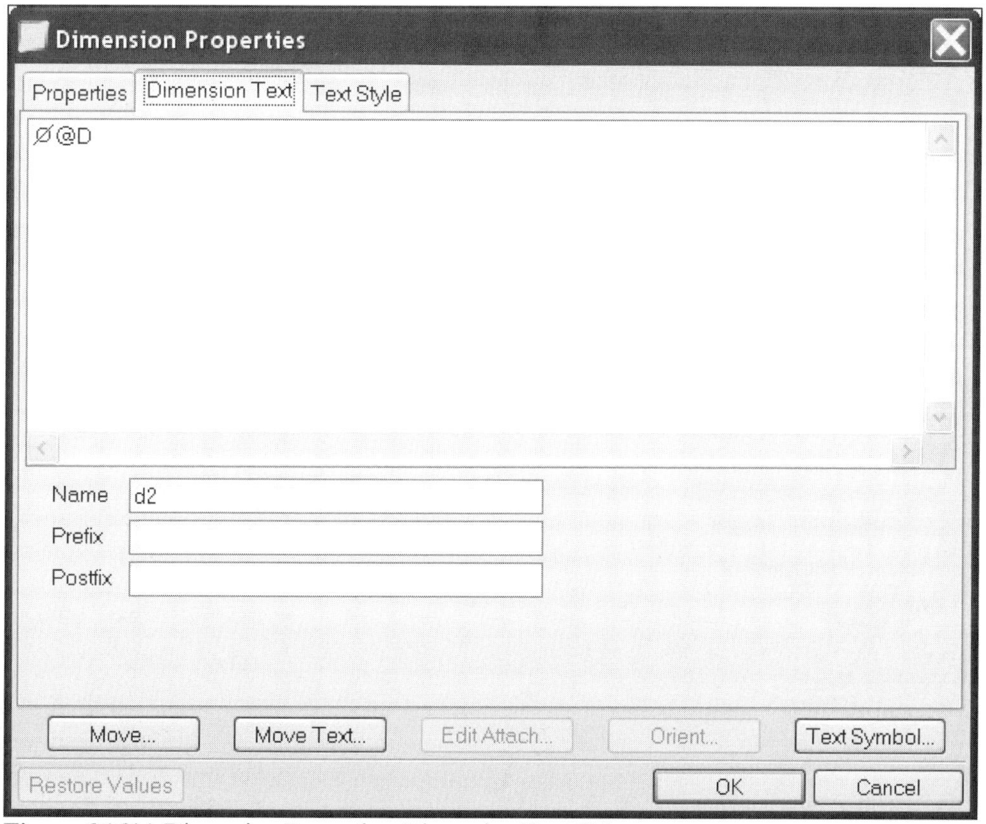

Figure 6.16(c) Dimension Properties, Dimension Text Tab

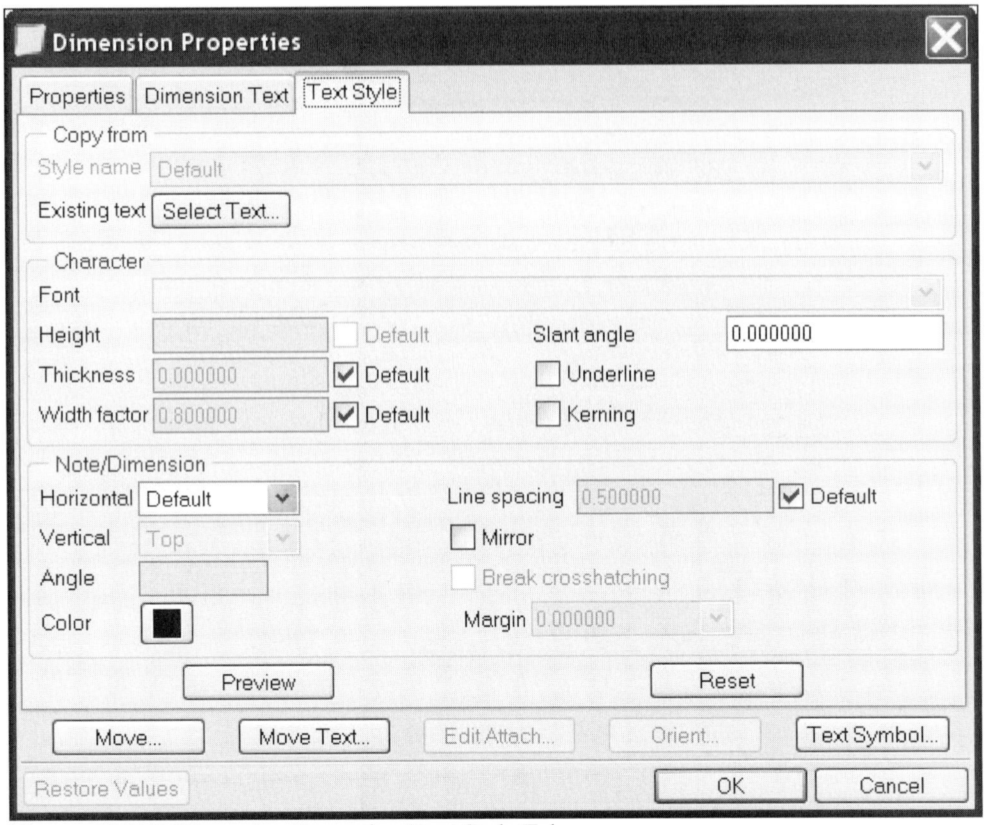

Figure 6.16(d) Dimension Properties, Text Style Tab

Click: **Move** ⇒ select a new position for the **∅.875** dimension [Fig. 6.16(e)] ⇒ **OK** ⇒ repeat to move the other dimensions as required *(pick on a dimension ⇒* ***RMB*** *⇒* ***Properties*** *⇒* ***Move*** *⇒ select a new position ⇒* ***OK****)* ⇒ **LMB** to deselect the last dimension ⇒ **LMB** to deselect the protrusion

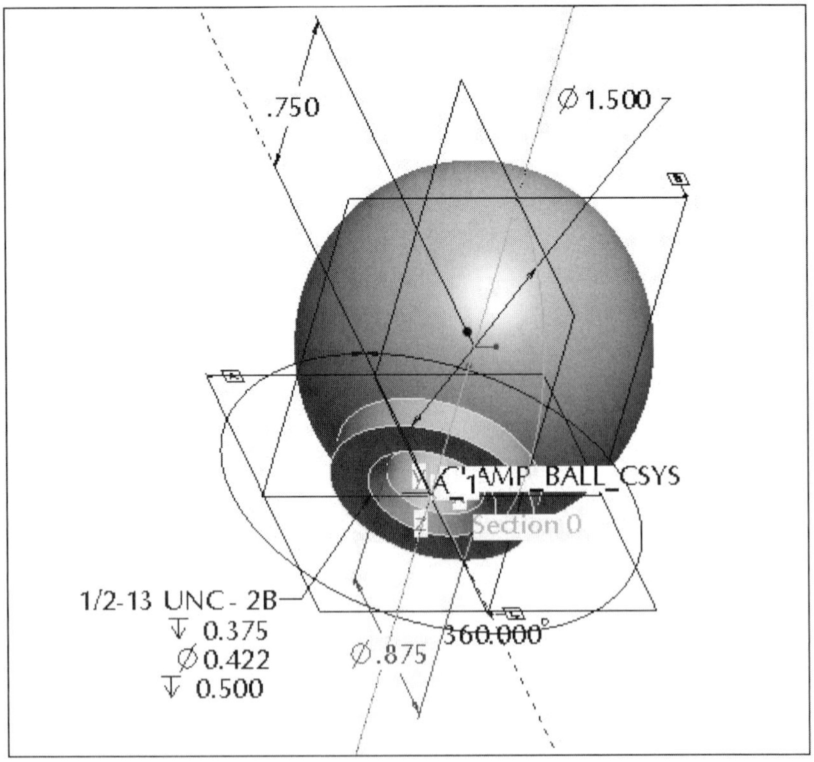

Figure 6.16(e) Moved Dimension

Click: **Annotations** off ⇒ ⇒ Geometry ⇒ pick on the edge [Fig. 6.17(a)] ⇒ **RMB** ⇒ **Round Edges** ⇒ double-click on the dimension and modify to the design value of **.06125** [Fig. 6.17(b)] ⇒ **Enter** ⇒ **Enter** ⇒ **Ctrl+S** ⇒ **Enter** ⇒ **LMB** to deselect

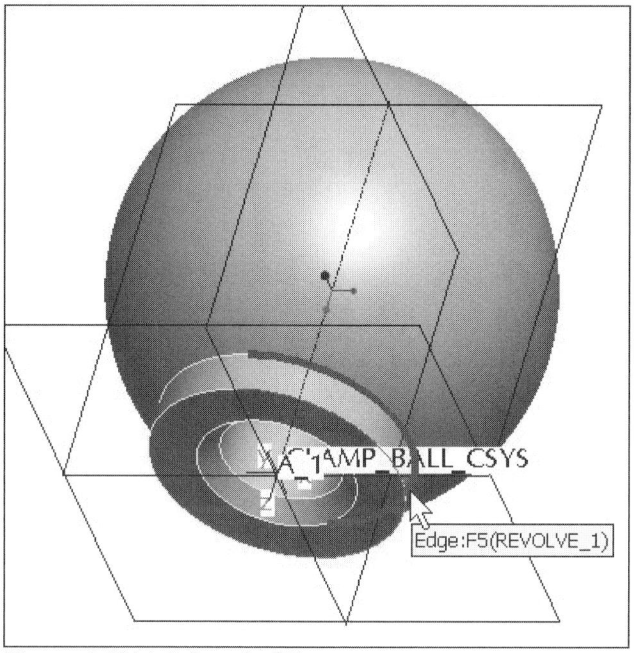

Figure 6.17(a) Select the Edge

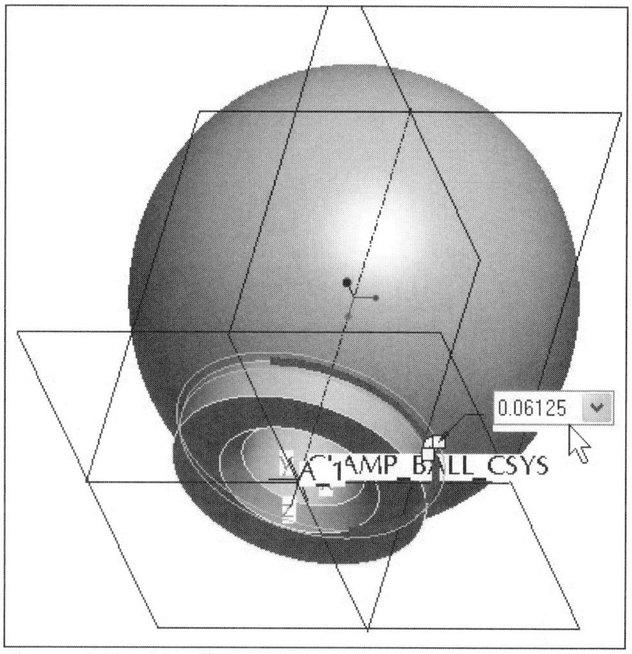

Figure 6.17(b) Radius **.06125**

Click: off ⇒ pick on the edge of the protrusion [Fig. 6.18(a)] ⇒ **Chamfer Tool** ⇒

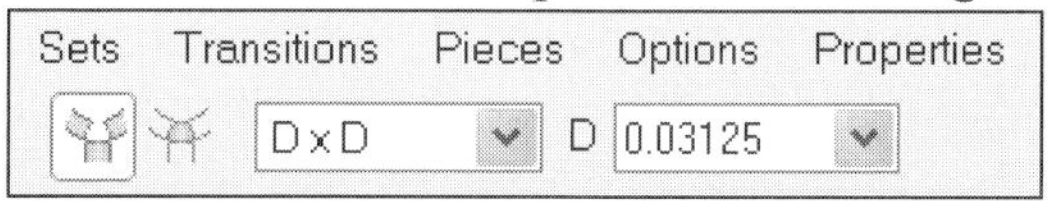

⇒ **Enter** ⇒ ⇒ **Ctrl+S** ⇒ **Enter** ⇒ **LMB** to deselect ⇒ ⇒ Smart ⇒ **Ctrl+D** ⇒ on [Fig. 6.18(b)] ⇒ **File** ⇒ **Save** ⇒ **Enter** ⇒ **File** ⇒ **Delete** ⇒ **Old Versions** ⇒ **MMB** ⇒ **File** ⇒ **Close Window**

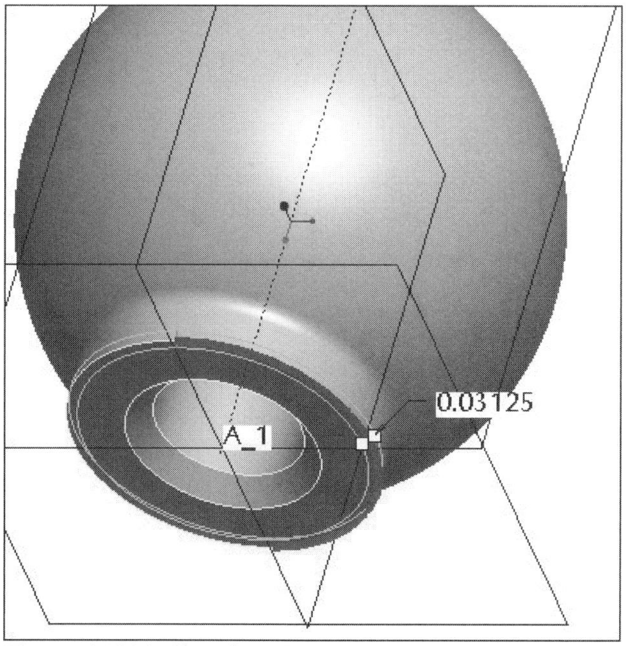

Figure 6.18(a) Chamfer **.03125**

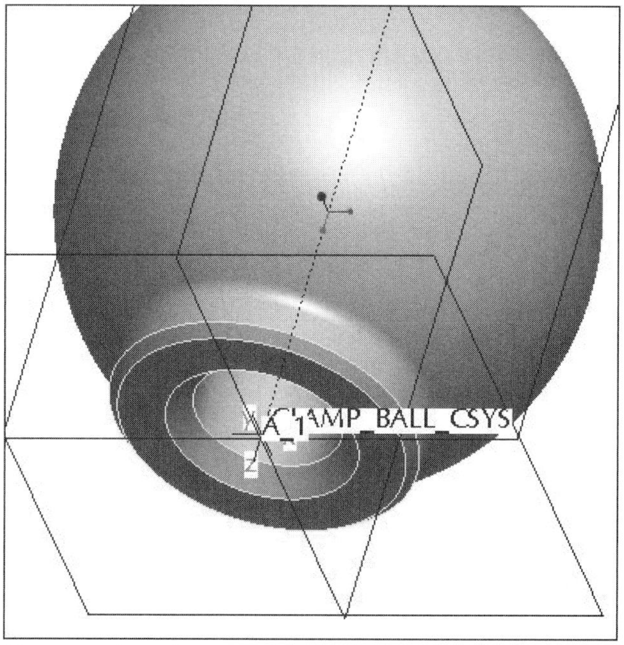

Figure 6.18(b) Completed Chamfer

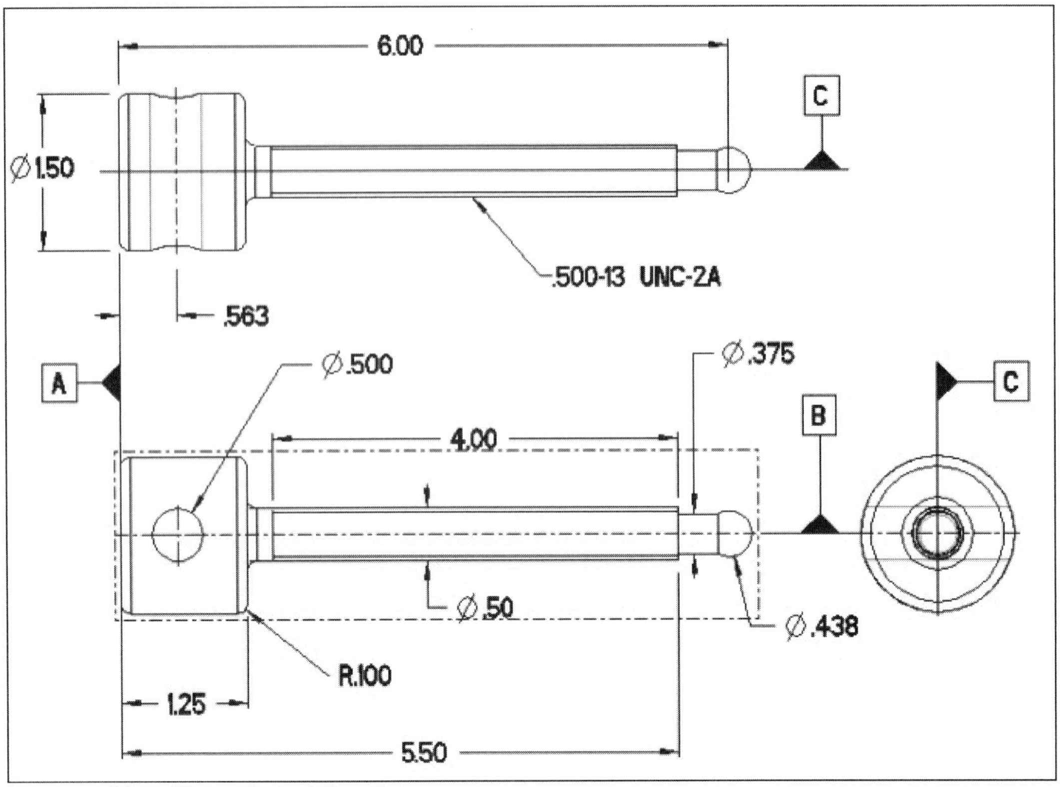

Figure 6.19(a) Clamp Swivel Detail

Clamp Swivel

The Clamp Swivel is the third part created by revolving one section about a centerline [Fig. 6.19(a)]. The Clamp Swivel is a component of the Clamp Assembly [Fig. 6.19(b)].

Click: **Create a new object** ⇒ Part ⇒ Name CLAMP_SWIVEL ⇒ Use default template ⇒ **OK** ⇒ *(if you have just started a new session of Pro/ENGINEER WILDFIRE 4.0, remember to load your configuration file* ⇒ ***Tools*** ⇒ ***Options*** ⇒ ***Open a configuration file*** ⇒ *click on* ***clamp.pro*** *which was previously created and saved* ⇒ ***Open*** ⇒ ***Apply*** ⇒ ***Close)*** ⇒ **Edit** ⇒ **Setup** ⇒ **Units** ⇒ Inch lbm Second (Pro/E Default) ⇒ **Close**

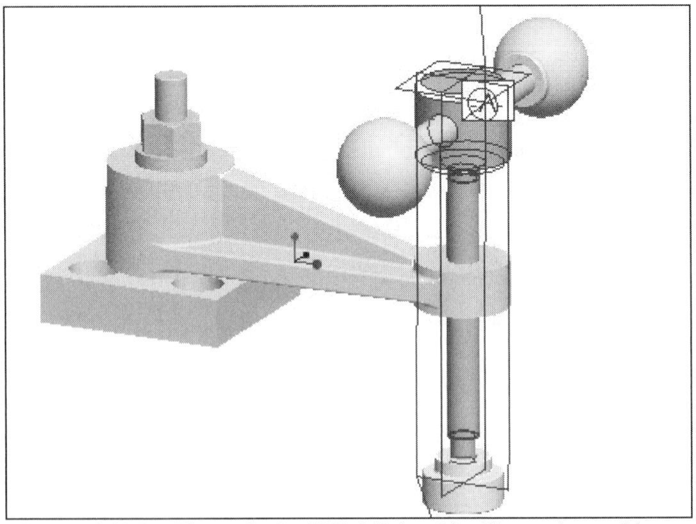

Figure 6.19(b) Clamp Swivel Highlighted in Clamp Assembly

Click: **Material** ⇒ **Create new material** [Fig. 6.21(a)] ⇒ type **STEEL_1066** ⇒ fill in the form [Fig. 6.21(b)] ⇒ **Save To Model** [Fig. 6.21(c)] ⇒ **OK** ⇒ **Done**

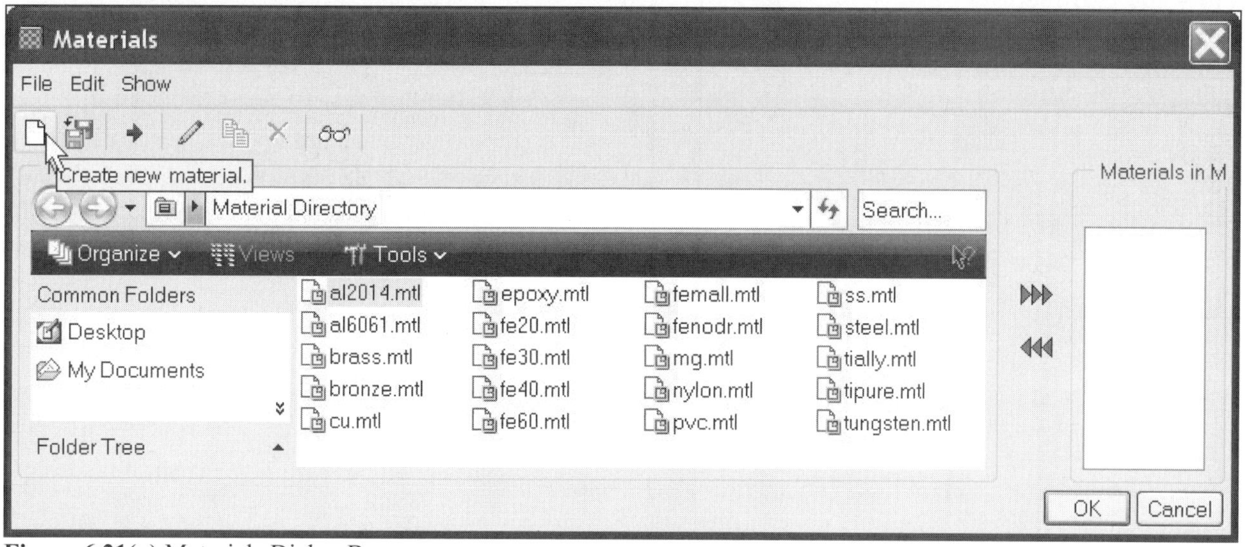

Figure 6.21(a) Materials Dialog Box

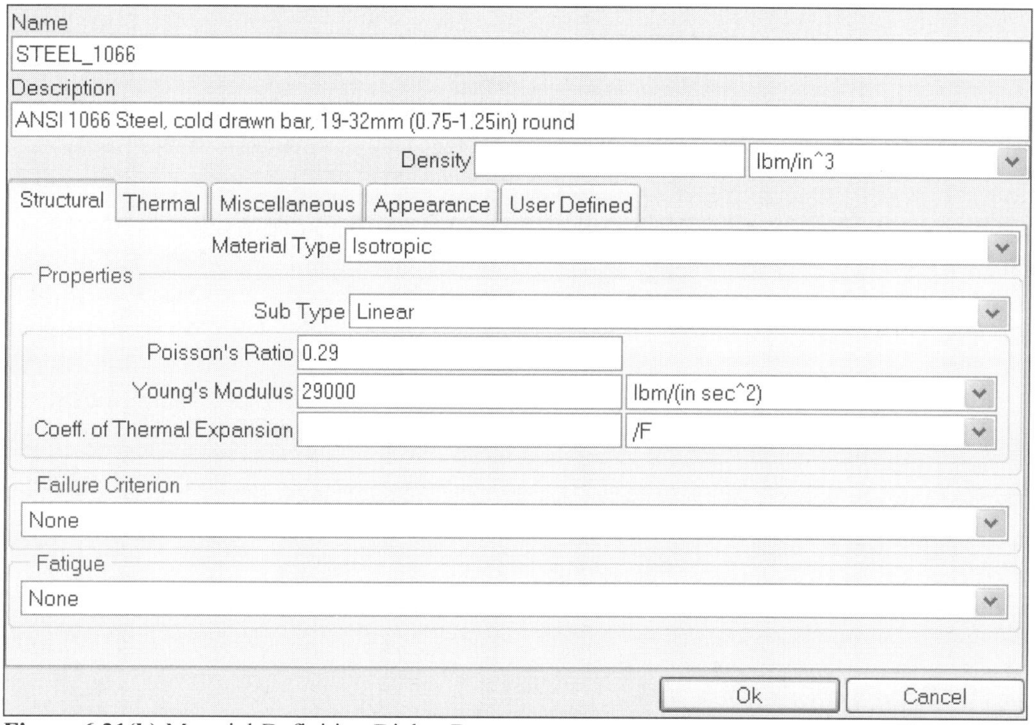

Figure 6.21(b) Material Definition Dialog Box

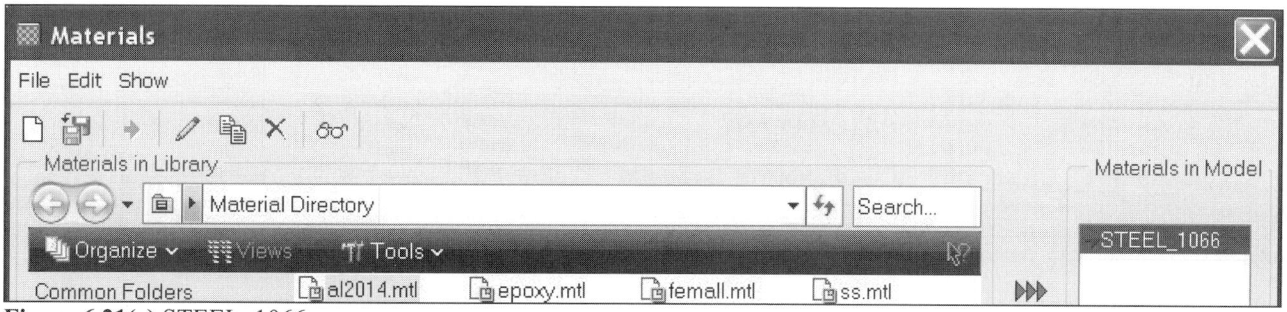

Figure 6.21(c) STEEL_1066

Click: **Save** ⇒ **OK** ⇒ **Info** from menu bar ⇒ **Model** [Figs. 6.21(d-e)] ⇒  click the quick sash to close the Browser

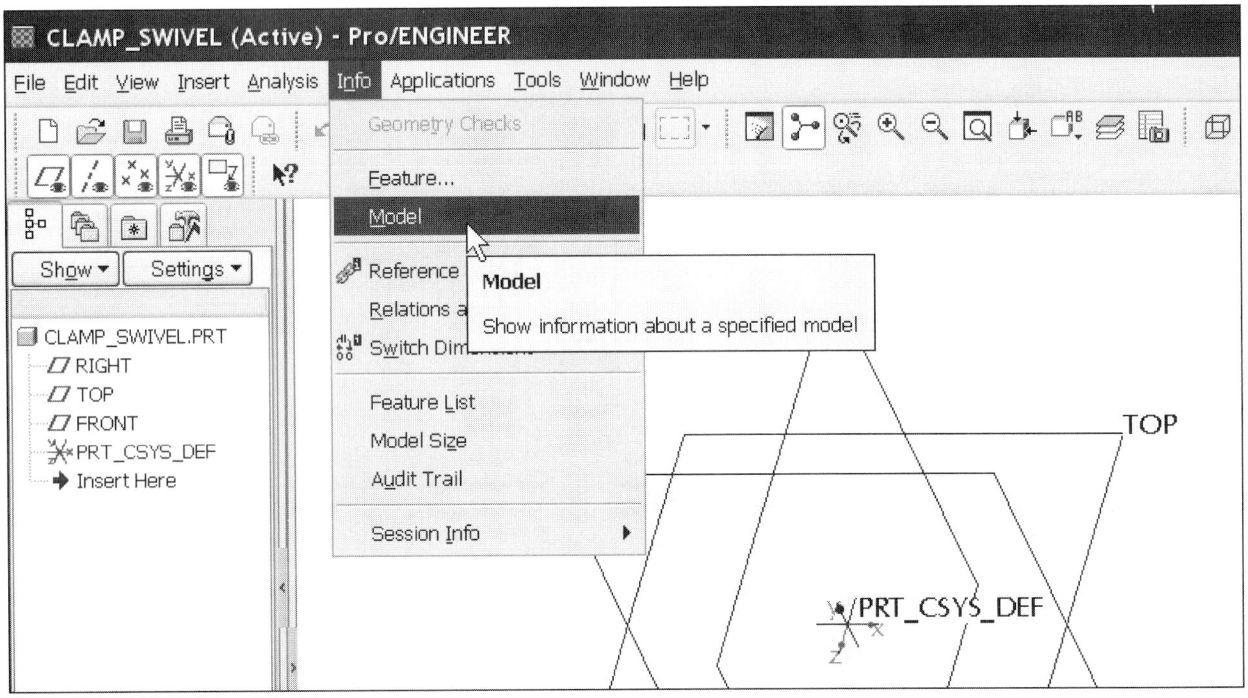

Figure 6.21(d) Info ⇒ Model

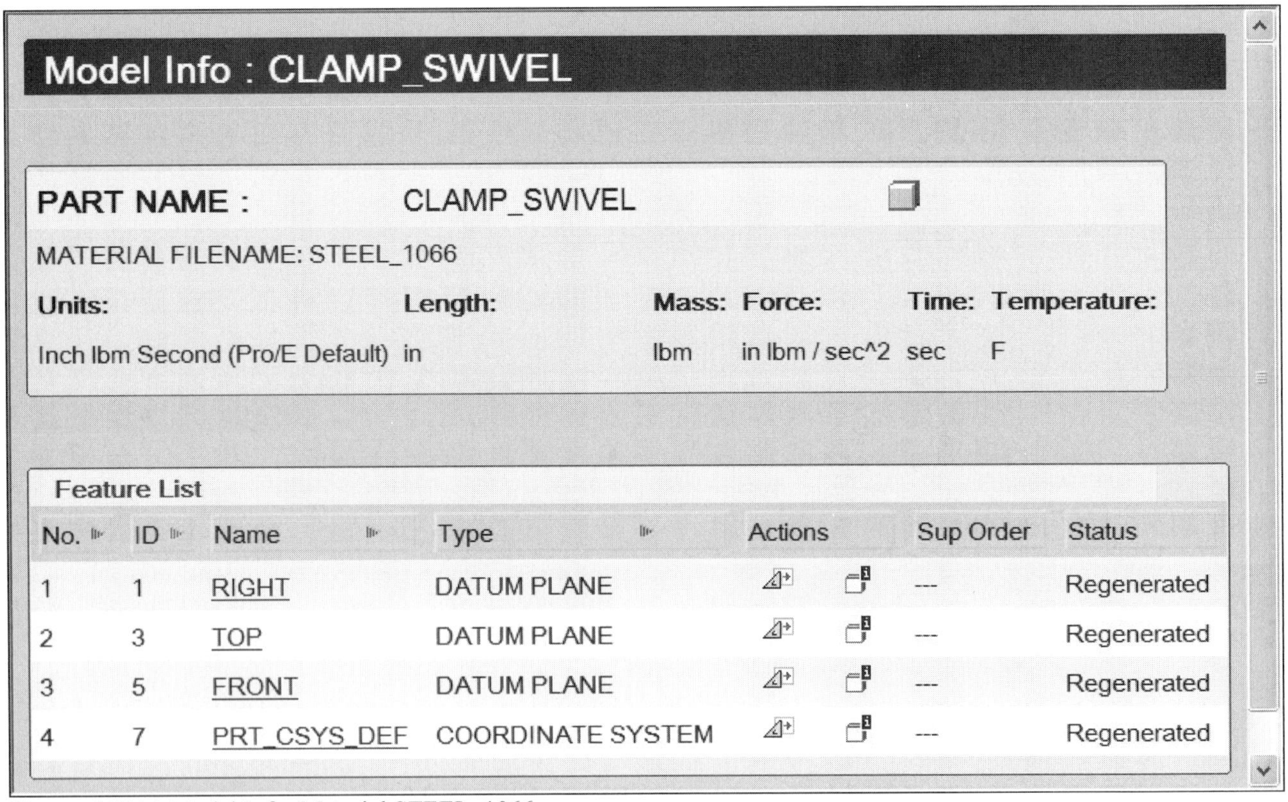

Figure 6.21(e) Model Info: Material STEEL_1066

Click: **Tools** ⇒ Environment ⇒ Snap to Grid ⇒ Tangent Edges Dimmed ⇒ **Apply** ⇒ **OK** ⇒ pick on the **RIGHT** datum plane ⇒ **Revolve Tool** ⇒ **RMB** ⇒ Define Internal Sketch... ⇒ Orientation: **Right** ⇒ **Sketch** ⇒ ⇒ **Toggle the grid** on ⇒ off ⇒ on ⇒ on ⇒ ⇒ **RMB** ⇒ Centerline sketch a *vertical* centerline through the default coordinate system to be used as the axis of revolution ⇒ sketch the section [Fig. 6.22(a)] ⇒ **Tools** ⇒ **Environment** ⇒ Snap to Grid ⇒ **Apply** ⇒ **Close** ⇒ off ⇒ **Toggle the grid** off ⇒ dimension and modify values as per the design requirements [Fig. 6.22(b)] ⇒ on ⇒ ⇒ ⇒ **File** ⇒ **Save** ⇒ **OK** ⇒ **Ctrl+D**

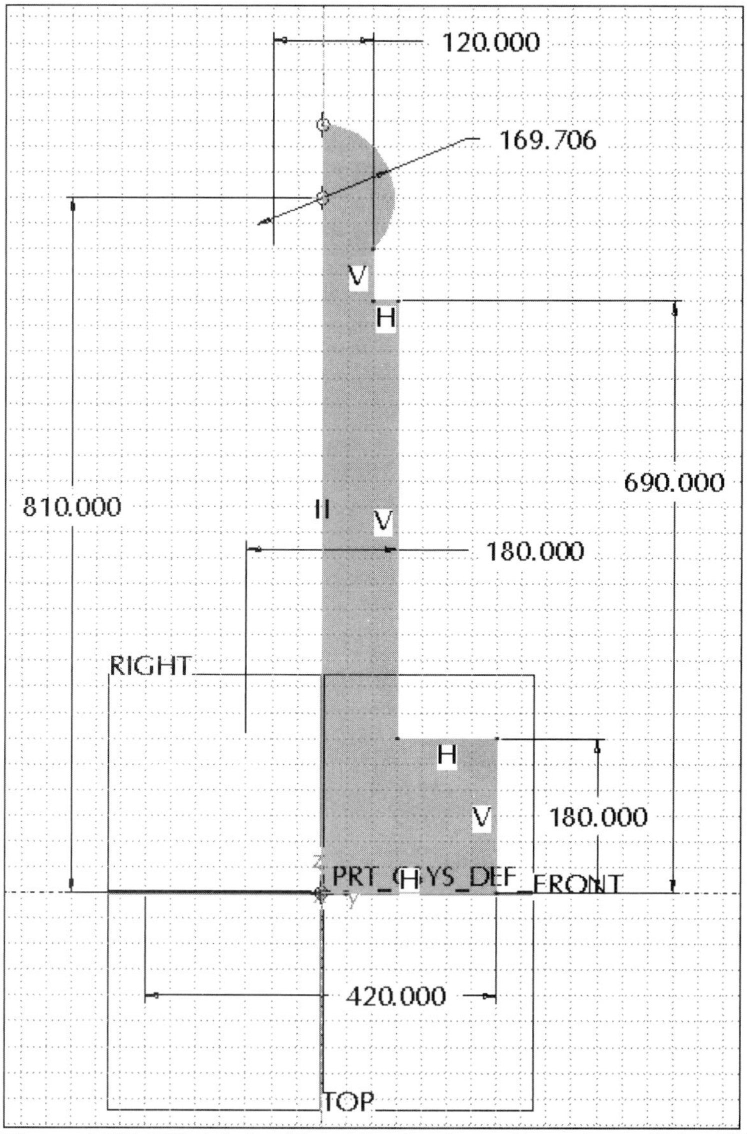

Figure 6.22(a) Sketch the Closed Section (your dimensions will be different)

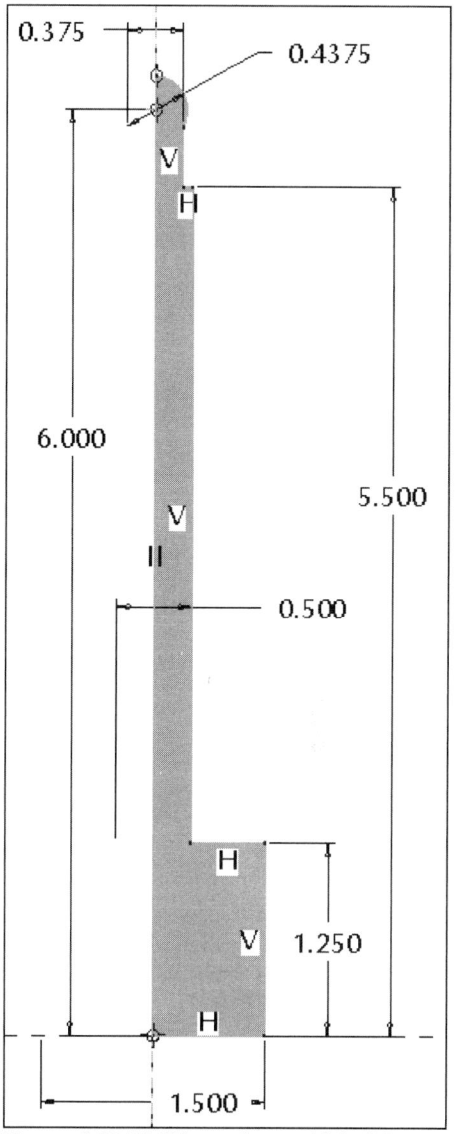

Figure 6.22(b) Dimension and Modify

Click: on ⇒ pick: **RIGHT** ⇒ **RMB** ⇒ **Properties** ⇒ Name- type **C** ⇒ ⇒ **OK** ⇒ pick **FRONT** ⇒ **RMB** ⇒ **Properties** ⇒ Name- type **A** ⇒ ⇒ **OK** ⇒ pick **TOP** ⇒ **RMB** ⇒ **Properties** ⇒ Name- type **B** ⇒ ⇒ **OK** ⇒ **Tools** ⇒ **Environment** ⇒ **Isometric** ⇒ **OK**

Rename the coordinate system CLAMP_SWIVEL_CSYS ⇒ **Ctrl+S** ⇒ **Enter** ⇒ **Shading** (Fig. 6.23)

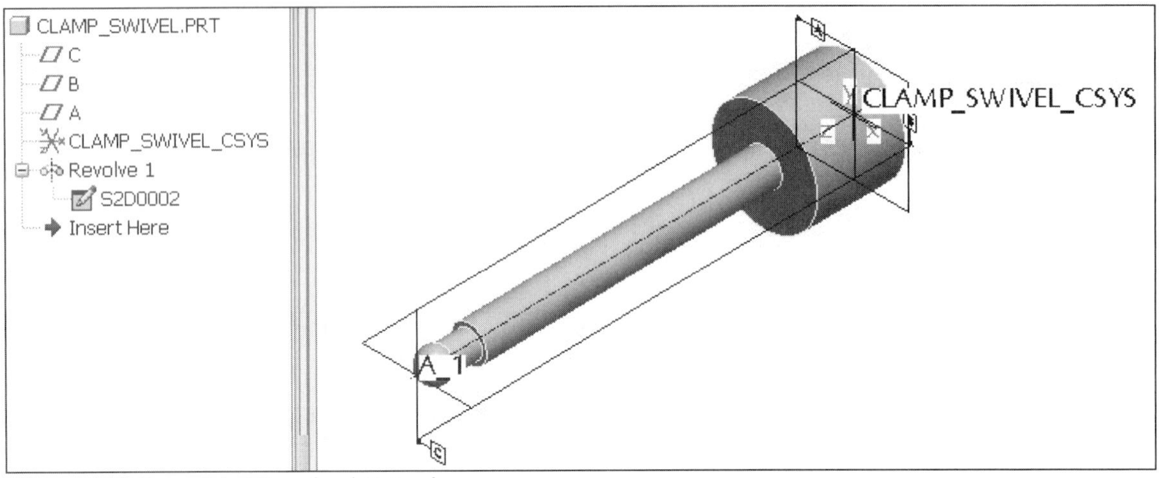

Figure 6.23 Completed Revolved Extrusion

Pick on datum plane **B** ⇒ **Hole Tool** ⇒ 0.500 ⇒ **Enter** ⇒ ⇒ **RMB** ⇒ Offset References Collector ⇒ **Placement** tab ⇒ pick datum **A** ⇒ Offset **.5625** ⇒ **Enter** ⇒ press and hold the **Ctrl** key and pick on datum **C** ⇒ **Align** [Fig. 6.24(a)] ⇒ **Shape** tab [Fig. 6.24(b)] ⇒ Side 2: Through All ⇒ **Enter** ⇒ **Ctrl+S** ⇒ **OK** ⇒ **Ctrl+D** ⇒ **LMB** to deselect

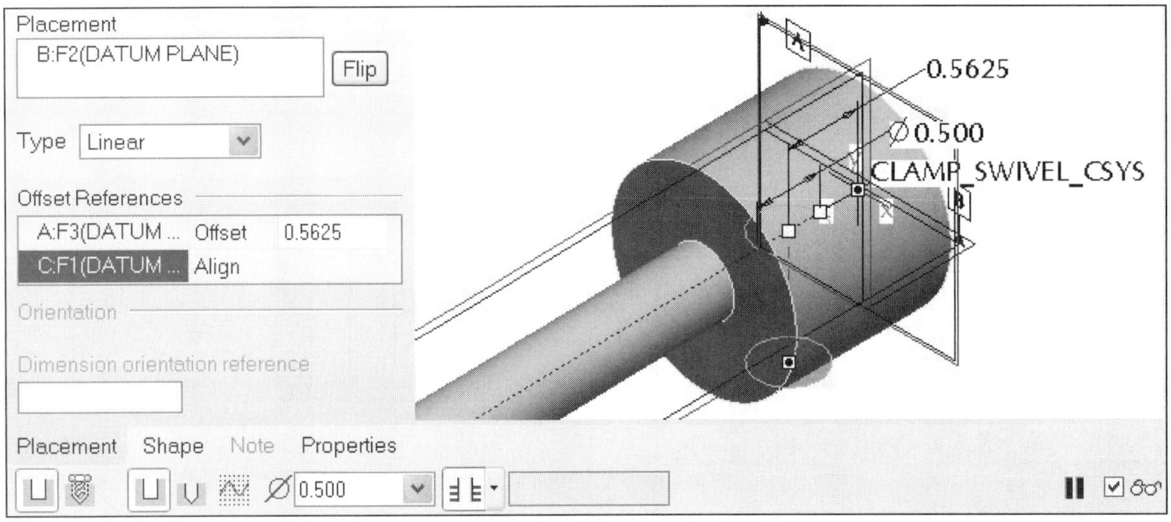

Figure 6.24(a) Placement Tab

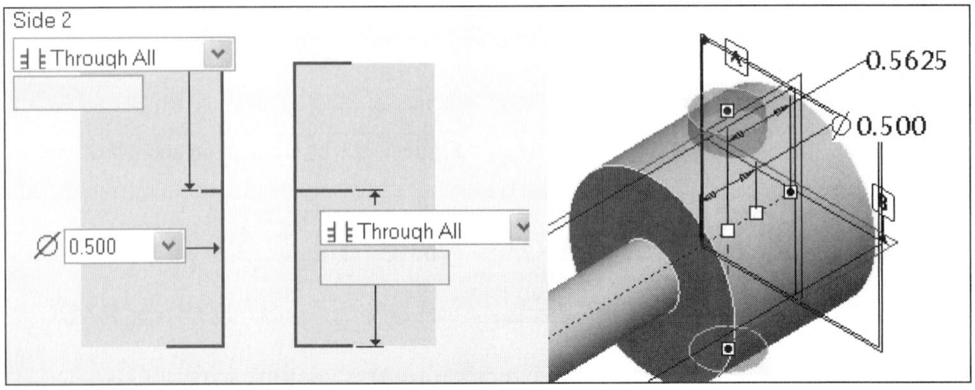

Figure 6.24(b) Shape Tab

Click: Round Tool ⇒ hold down your **Ctrl** key and select the three edges ⇒ modify the dimension to **.100** (Fig. 6.25) ⇒ **Enter** ⇒ ⇒ **File** ⇒ **Save** ⇒ **OK** ⇒ **Ctrl+D** ⇒ **LMB** to deselect

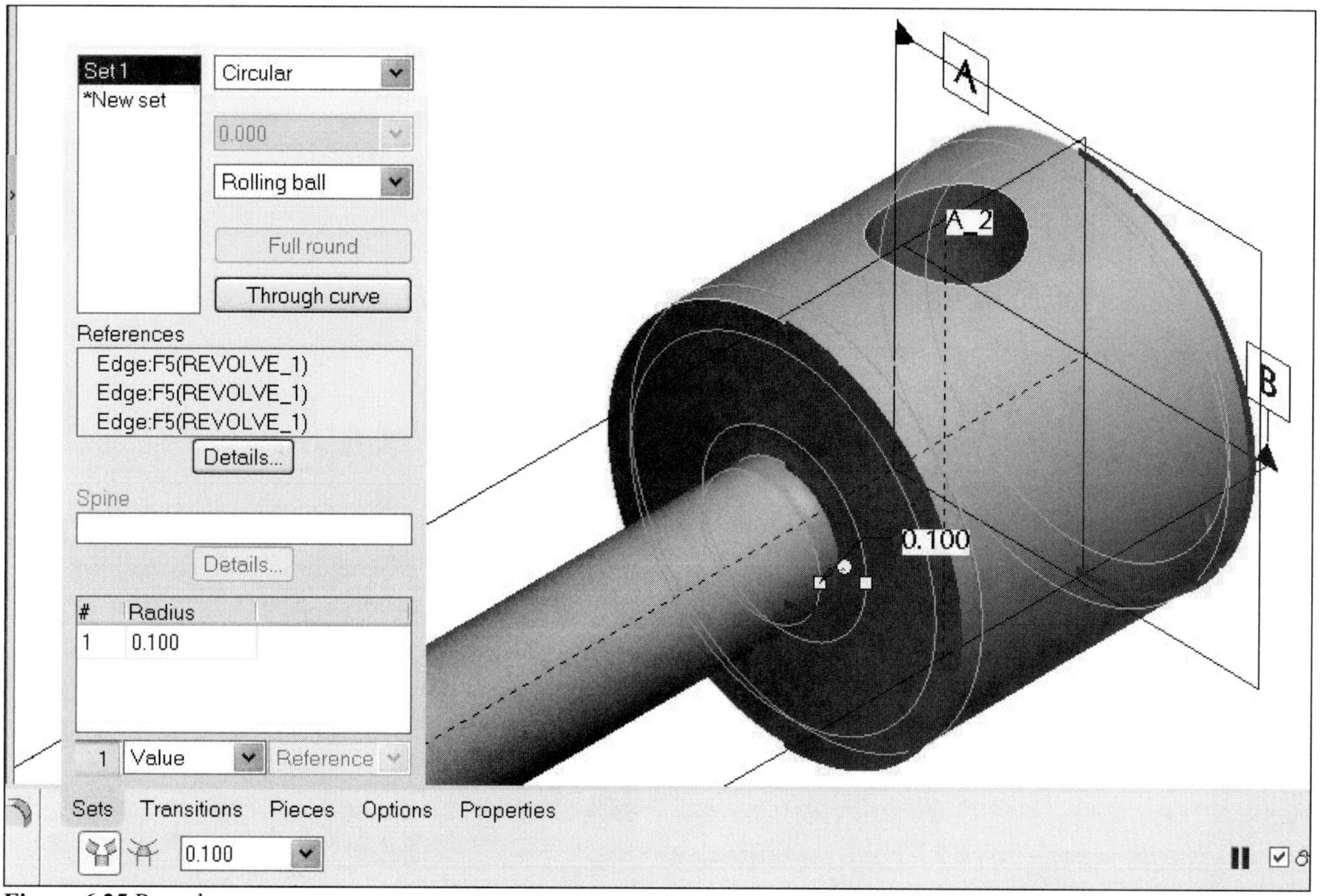

Figure 6.25 Rounds

Cosmetic Threads

A **cosmetic thread** is a feature that "represents" the diameter of a thread without having to show the actual threaded feature. Since a threaded feature is memory intensive, using cosmetic threads can save an enormous amount of visual memory on your computer. It is displayed in a unique default color. **Internal cosmetic threads** are created automatically when holes are created using the Hole Tool (Standard-Tapped). In situations where the internal thread is unique and the Hole Tool cannot be used, internal cosmetic threads can be added to the hole. **External cosmetic threads** represent the *root diameter.* For threaded shafts you must create the external cosmetic threads. The following table lists the parameters that can be defined for a cosmetic thread. In this table, "pitch" is the distance between two threads.

PARAMETER NAME	**PARAMETER TYPE**	**PARAMETER DESCRIPTION**
MAJOR_DIAMETER	Real Number	Thread major diameter
THREADS_PER_INCH	Real Number	Threads per inch (1/pitch)
THREAD FORM	String	Thread form
CLASS	String	Thread class
PLACEMENT	Character	Thread placement (A-external, B-internal)
METRIC	YesNo (True/False)	Thread is metric

Click: **Insert** ⇒ **Cosmetic** ⇒ **Thread** ⇒ pick the cylindrical surface [Fig. 6.26(a)] ⇒ pick the thread start surface--the edge lip surface [Fig. 6.26(b)] ⇒ **Okay** [Fig. 6.26(c)] ⇒ **Blind** ⇒ **Done** ⇒ **4.00** Enter depth 4.00 ⇒ **Enter** ⇒ type the value of the cosmetic thread root diameter: **.4485** Enter DIAMETER .4485 ⇒ **Enter** [Fig. 6.26(d)] ⇒ **Mod Params**

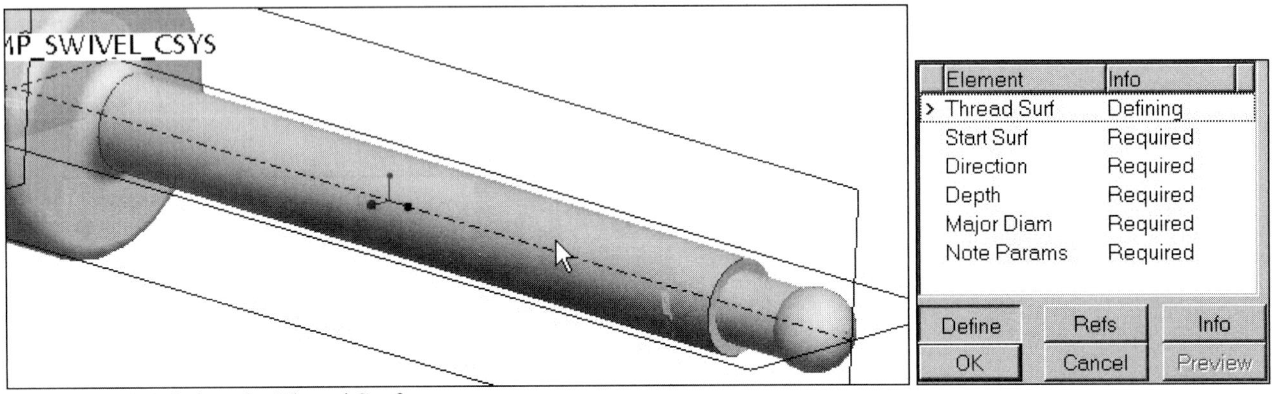

Figure 6.26(a) Select the Thread Surface

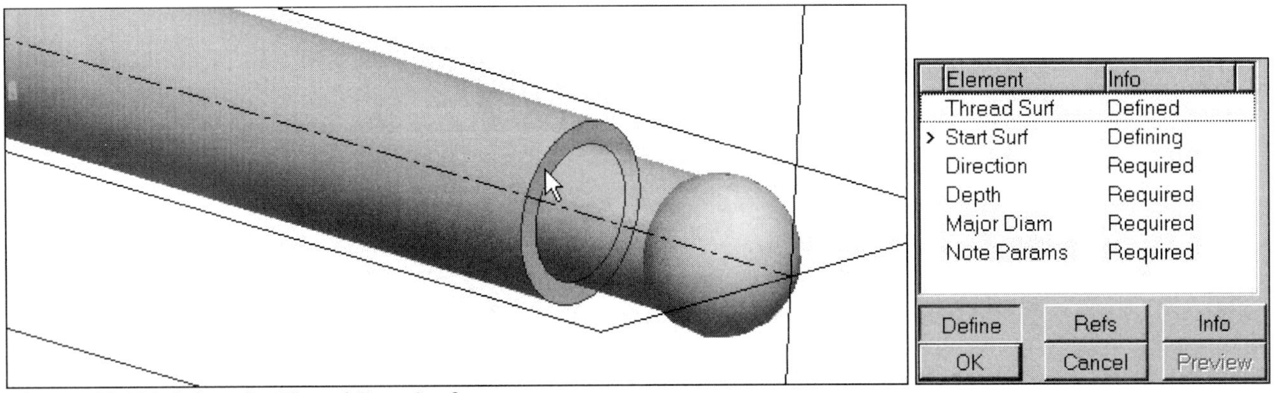

Figure 6.26(b) Select the Thread Start Surface

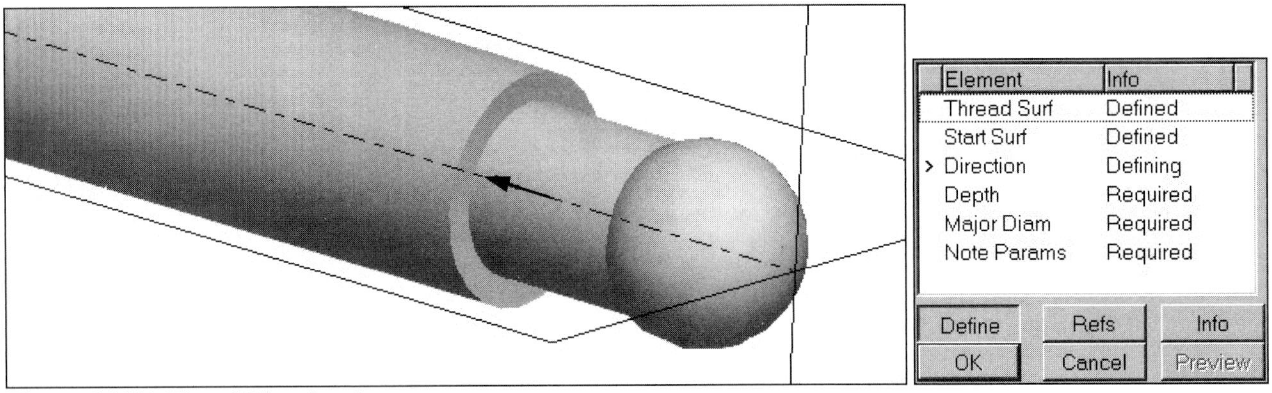

Figure 6.26(c) Thread Direction Arrow

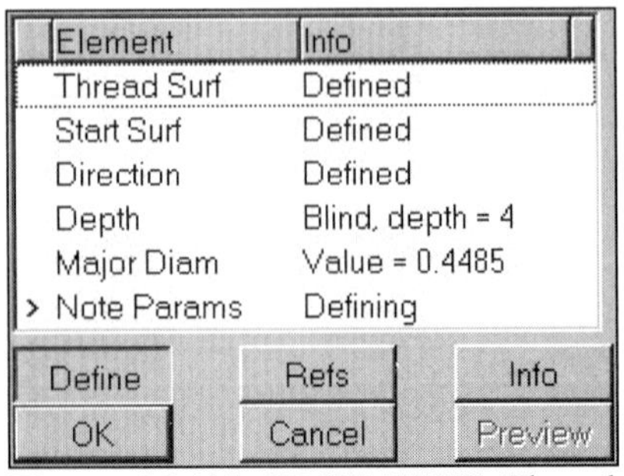

Figure 6.26(d) COSEMETIC: Thread Dialog Box

After creating the cosmetic thread, edit the thread table. The thread size of an external thread must be changed to the nominal size from the root diameter defaulted on the thread table. Create an external cosmetic thread (**.500-13 UNC-2A**) using the **∅.500** surface [Fig. 6.26(a)]. The thread starts at the "neck" and extends **4.00** along the swivel's shaft [Fig. 6.26(b)]. The Pro/TABLE thread parameter [Fig. 6.26(e)] shows the major diameter as **0.4485**. Since you are cosmetically representing the *root diameter (.4485)* of the external thread on the model, the *thread diameter* is *smaller* than the *nominal (.50)* thread size.

Click in each table field and change the values [Fig. 6.26(f)] ⇒ MAJOR_DIAMETER **.5** ⇒ THREADS_PER_INCH **13** ⇒ FORM **UNC** ⇒ CLASS **2** ⇒ PLACEMENT **A** ⇒ METRIC **FALSE** ⇒ from Pro/TABLE, click: **File** ⇒ **Save** ⇒ **File** ⇒ **Exit** ⇒ **Show** [Fig. 6.26(g)] ⇒ **Close** ⇒ **MMB** [Fig. 6.26(h)] ⇒ **OK** ⇒ **LMB** ⇒ **Hidden line** (to see the cosmetic surface) ⇒ **Shading** ⇒ off ⇒ pick on the cosmetic feature from the model tree ⇒ **RMB** ⇒ **Info** ⇒ **Feature** [Figs. 6.26(i-j)]

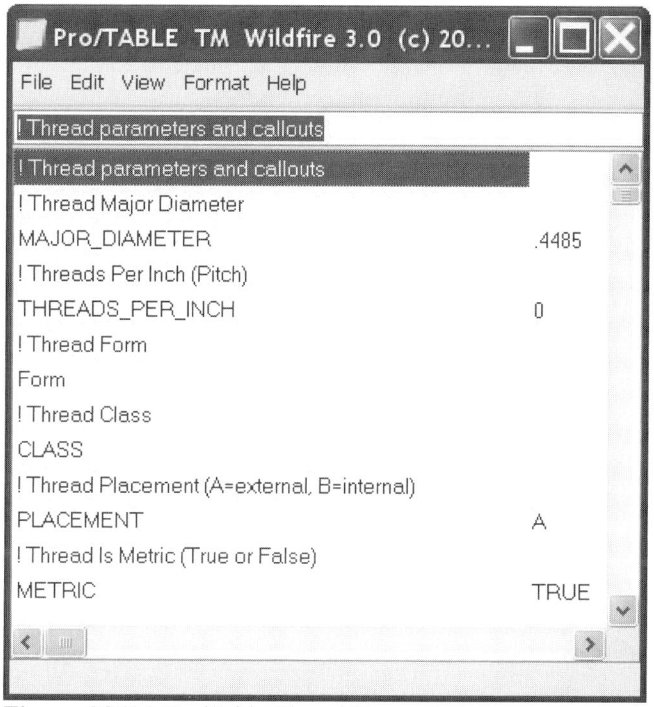

Figure 6.26(e) Pro/Table Thread Parameters

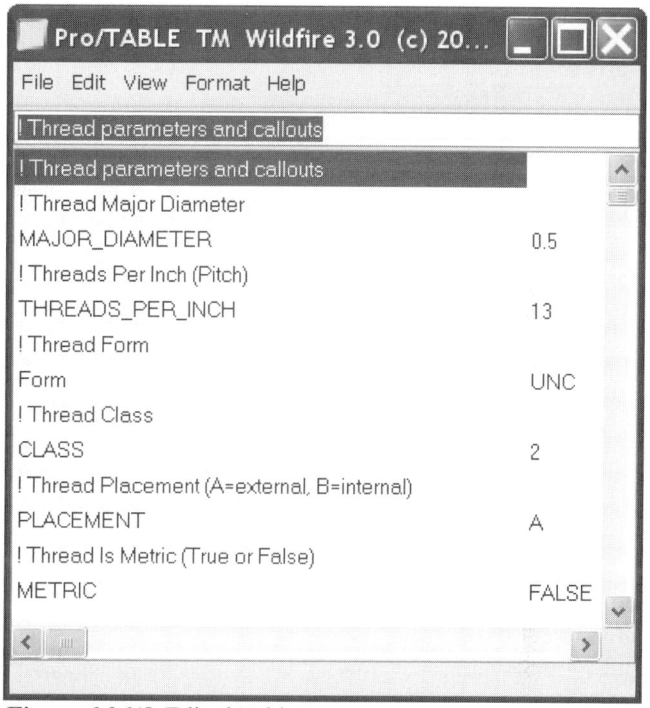

Figure 6.26(f) Edited Table Parameters

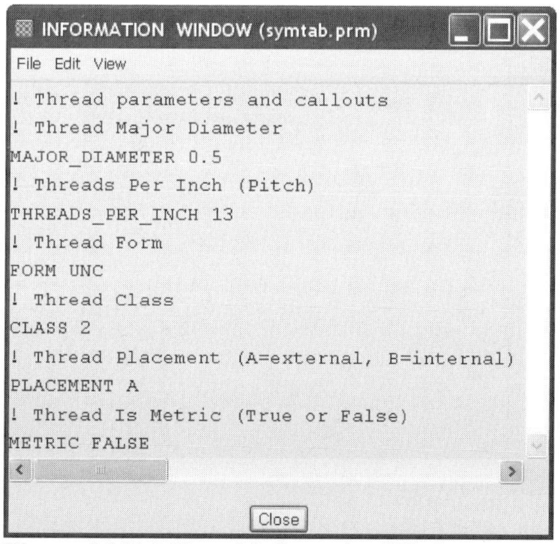

Figure 6.26(g) Thread Information Window

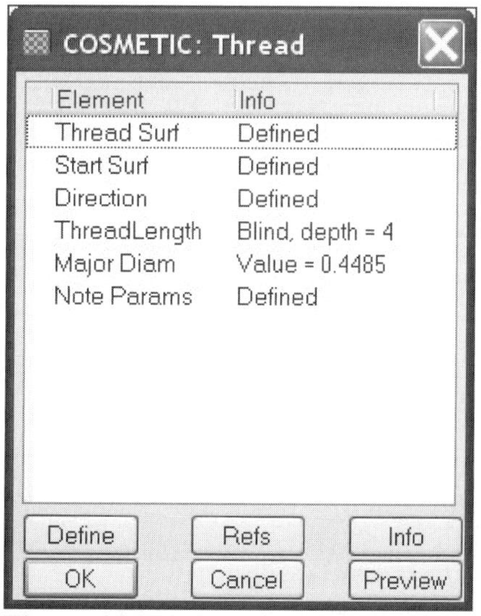

Figure 6.26(h) Completed Thread Elements

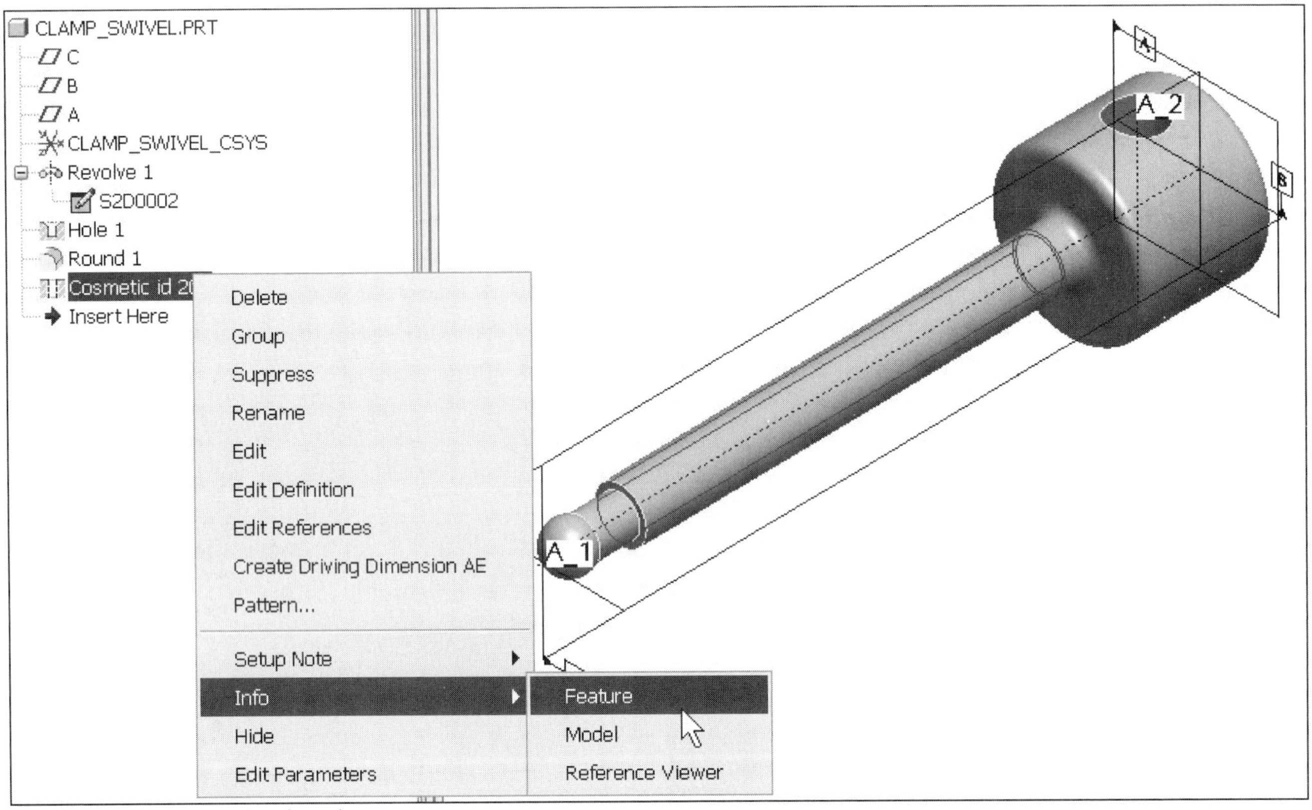

Figure 6.26(i) Cosmetic Thread

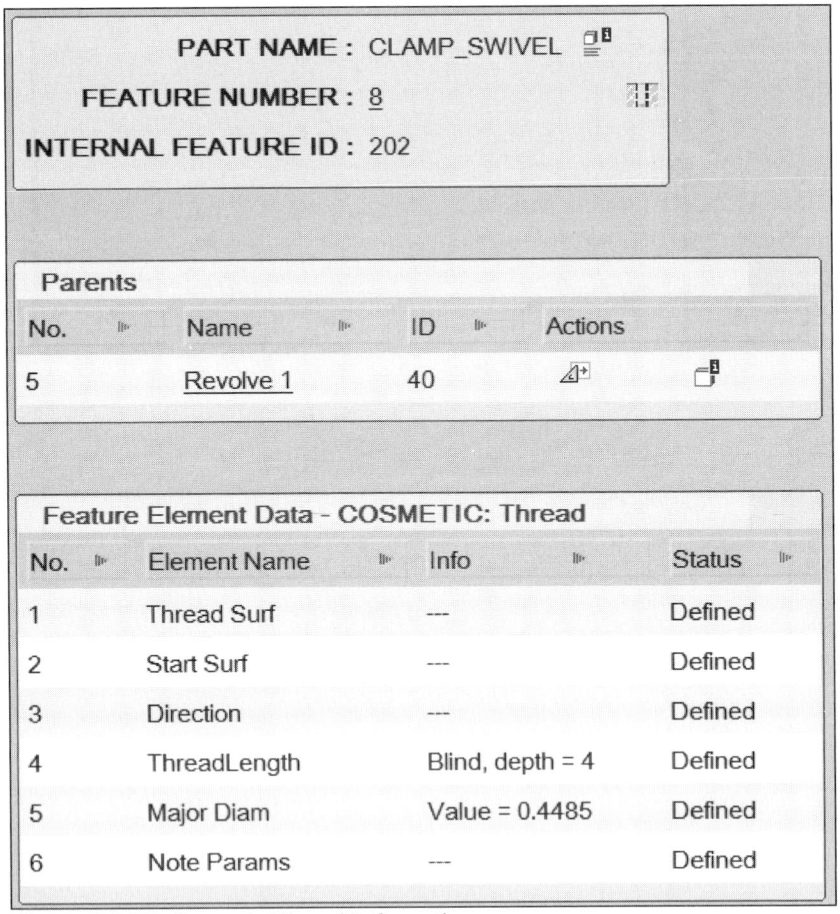

PART NAME : CLAMP_SWIVEL

FEATURE NUMBER : 8

INTERNAL FEATURE ID : 202

Parents

No.	Name	ID	Actions
5	Revolve 1	40	

Feature Element Data - COSMETIC: Thread

No.	Element Name	Info	Status
1	Thread Surf	---	Defined
2	Start Surf	---	Defined
3	Direction	---	Defined
4	ThreadLength	Blind, depth = 4	Defined
5	Major Diam	Value = 0.4485	Defined
6	Note Params	---	Defined

Figure 6.26(j) Cosmetic Thread Information

Using the Model Player

The **Model Player** (Fig. 6.27) option on the **Tools** menu lets you observe how a part is built. You can:

- Move backward or forward through the feature-creation history of the model in order to observe how the model was created. You can start the model playback at any point in its creation history
- Regenerate each feature in sequence, starting the specified feature
- Display each feature as it is regenerated or rolled forward
- Update (regenerate all the features in) the entire display when you reach the desired feature
- Obtain information about the current feature (you can show dimensions, obtain regular feature information, investigate geometry errors, and enter Fix Model mode)

To open the **Model Player** dialog box, click Tools ⇒ Model Player. You can select one of the following:

Figure 6.27 Model Player

- **Regenerate features** Regenerates each feature in sequence, starting from the specified feature
- **Display each feature** Displays each feature in the graphics window as it is being regenerated
- **Compute CL** (Available in Manufacturing mode only) When selected, the CL data is recalculated for each NC sequence during regeneration

Select one of the following commands:

Go to the beginning of the model moves immediately to the beginning of the model

Step backward through the model one feature at-a-time and regenerates the preceding feature

Stop play

Step forward through the model one feature at-a-time and regenerates the next feature

Go to the last feature in the model moves immediately to the end of the model (resume all)

Slider Bar Drag the slider handle to the feature at which you want model playback to begin. The features are highlighted in the graphics window as you move through their position with the slider handle. The feature number and type are displayed in the selection panel [such as #4 (COORDINATE SYSTEM)] [Fig. 6.28(b)], and the feature number is displayed in the **Feat #** box.

Select feature from screen or model tree Lets you select a starting feature from the graphics window or the Model Tree. Opens the **SELECT FEAT** and **SELECT** menus. After you select a starting feature, its number and ID are displayed in the selection panel, and the feature number is displayed in the **Feat #** box.

Feat # 5 of 25 Lets you specify a starting feature by typing the feature number in the box. After you enter the feature number, the model immediately rolls or regenerates to that feature.

To stop playback, click the **Stop play** button. Use the following commands for information:

- **Show Dims** Displays the dimensions of the current feature
- **Feat Info** Provides regular feature information about the current feature in an Information window
- **Geom Check** Investigates the geometry error for the current feature
- **Fix Model** Activates Resolve mode by forcing the current feature to abort regeneration
- **Close** Closes the Model Player and enters Insert mode at the current feature
- **Finish** Closes the Model Player and returns to the last feature in the model

Click on the quick sash to close the Browser ⇒ on ⇒ **Tools** ⇒ **Model Player** ⇒ Regenerate features ⇒ Display each feature ⇒ **Go to the beginning of the model** ⇒ [Fig. 6.28(a)] ⇒ ⇒ ⇒ [Fig. 6.28(b)] ⇒ [Fig. 6.28(c)] ⇒ **View** ⇒ **Shade** ⇒ ⇒ Show Dims [Fig. 6.28(d)]

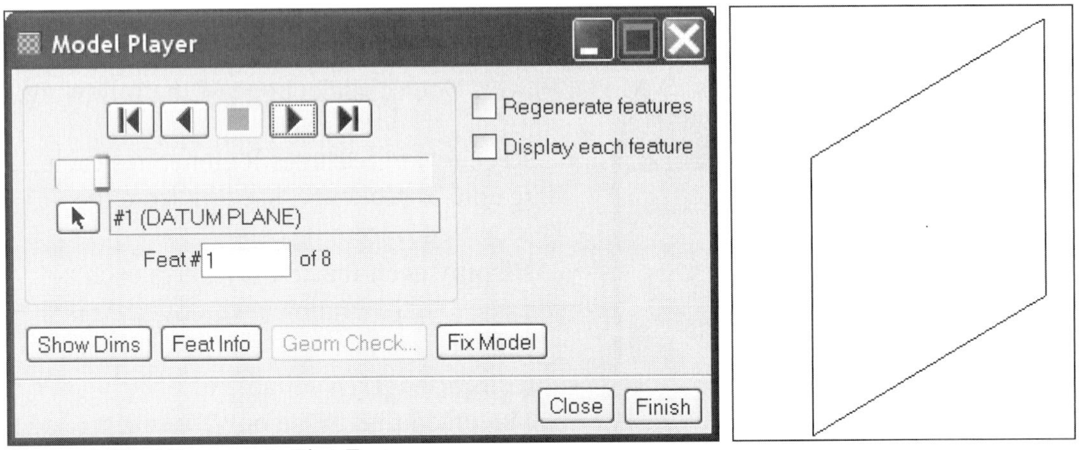

Figure 6.28(a) Regenerate First Feature

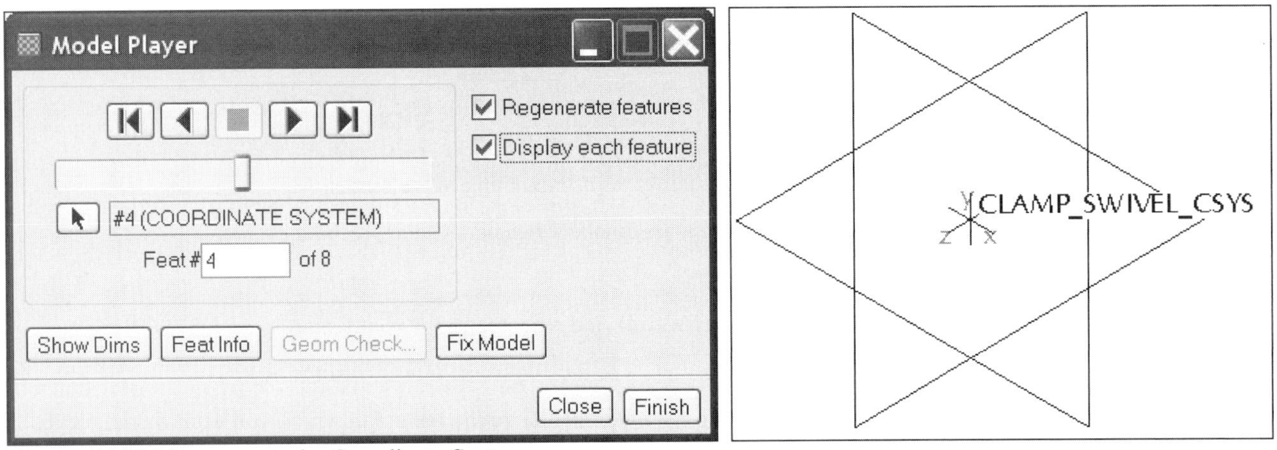

Figure 6.28(b) Regenerate the Coordinate System

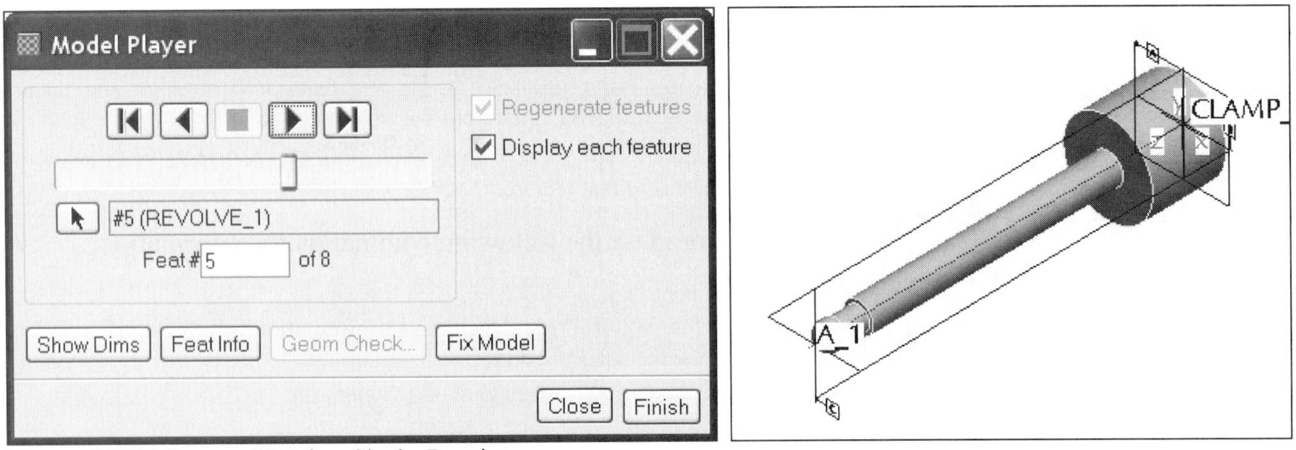

Figure 6.28(c) Feature #5, View Shade, Repaint

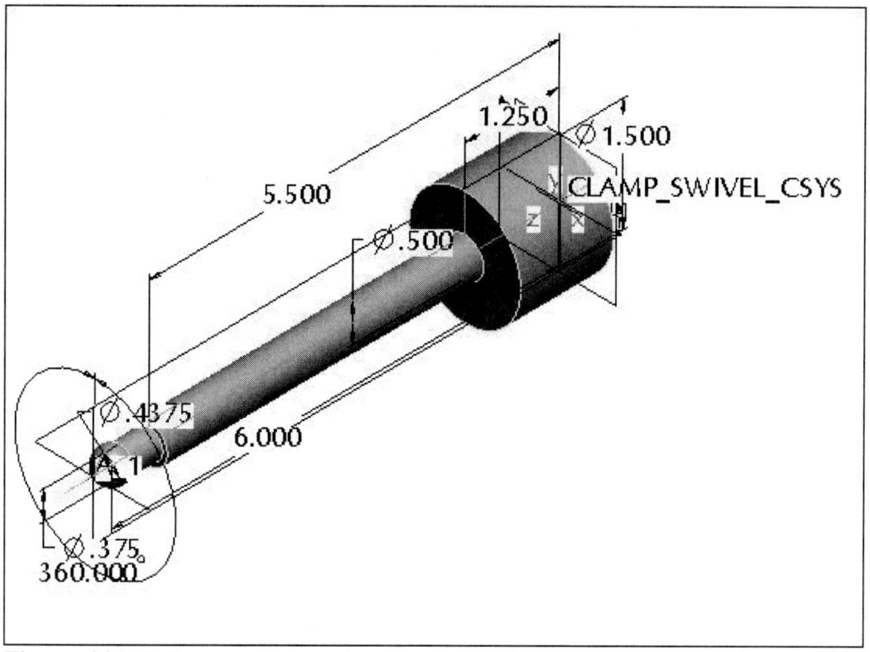

Figure 6.28(d) Feature #5, Revolve Dimensions

Click: ▶ ⇒ Feat Info [Fig. 6.28(e)] ⇒ close the **Browser** by clicking the quick sash ⇒ ▶ ⇒ **Finish** ⇒ ⇒ ⇒ ⇒ **MMB**

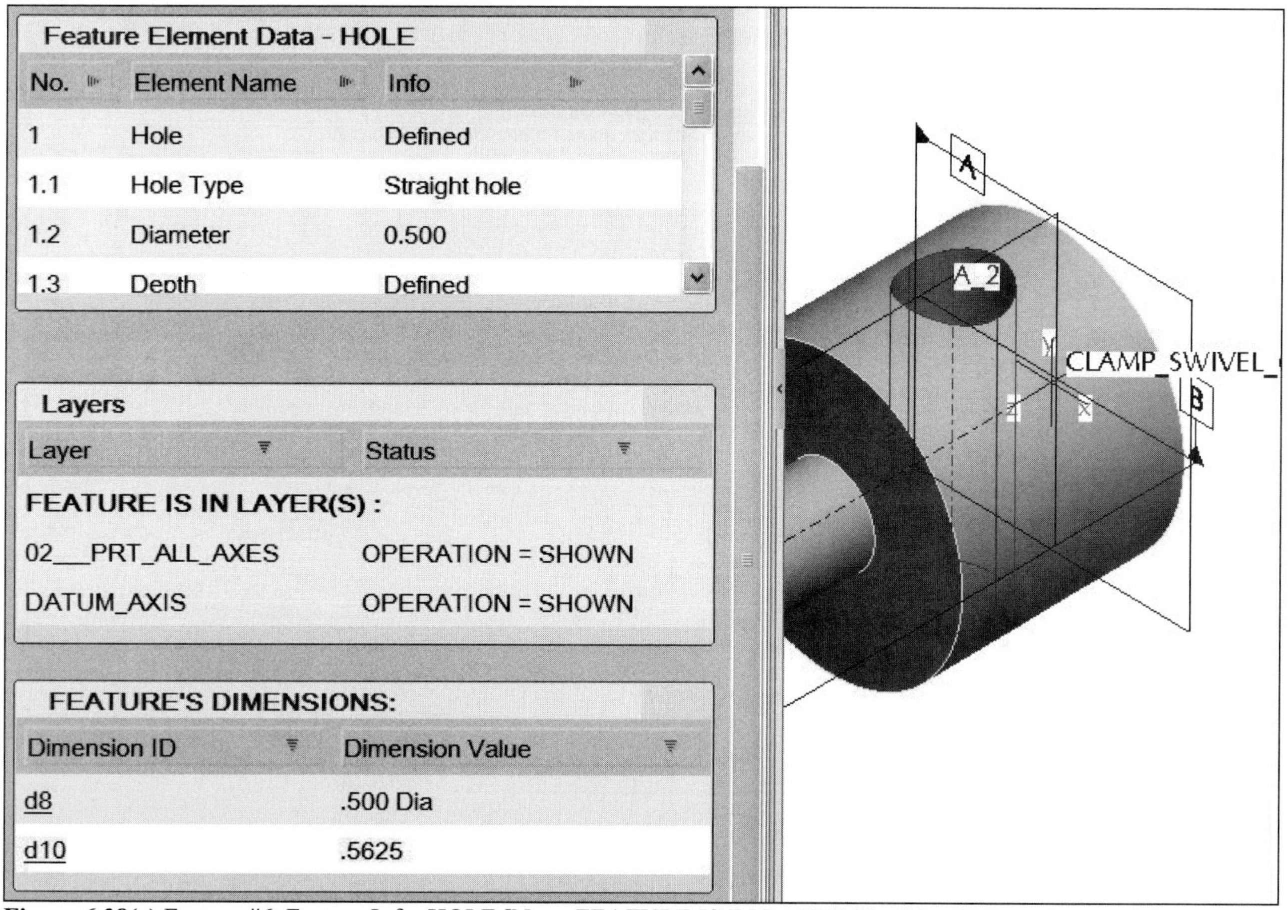

Figure 6.28(e) Feature **#6**, Feature Info: HOLE [Note: FEATURE IS IN LAYER(S): (02__PRT_ALL_AXES)]

Printing and Plotting

From the File menu, you can print with the following options: scaling, clipping, displaying the plot on the screen, or sending the plot directly to the printer. Shaded images can also be printed from this menu. You can create plot files of the current object (sketch, part, assembly, drawing, or layout) and send them to the print queue of a plotter. The plotting interface to HPGL and PostScript formats are standard.

You can configure your printer using the MS Printer Manager Print dialog box, available from the Pro/E Print dialog box. If you are printing a shaded image, the Printer Configuration dialog box opens.

The following applies for plotting:

- Hidden lines appear as gray for a screen plot, but as dashed lines on paper.
- When Pro/E plots Pro/E line fonts, it scales them to the size of a sheet. It does not scale the user-defined line fonts, which do not plot as defined.
- You can use the configuration file option *use_software_linefonts* to make sure that the plotter plots a user-defined font exactly as it appears in Pro/E.
- You can plot a cross section from Part or Assembly mode.
- With a Pro/PLOT license, you can write plot files in a variety of formats.

Print your object, click: **Print the active object** ⇒ **OK** [Fig. 6.29(a)] ⇒ **OK** [Fig. 6.29(b)] ⇒ ⇒ **MMB** ⇒ **File** ⇒ **Delete** ⇒ **Old Versions** ⇒ **MMB** ⇒ **Window** ⇒ **Close**

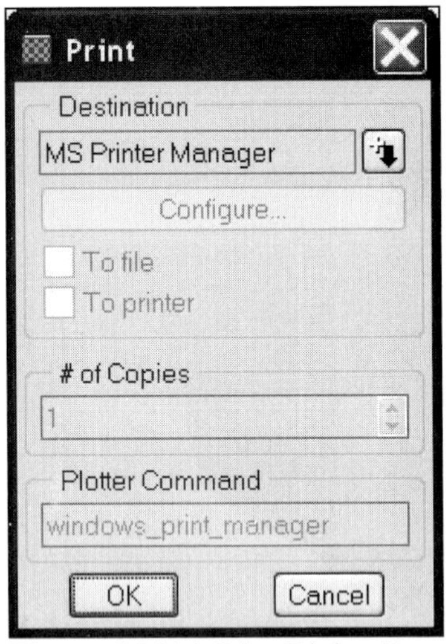

Figure 6.29(a) Print Dialog Box

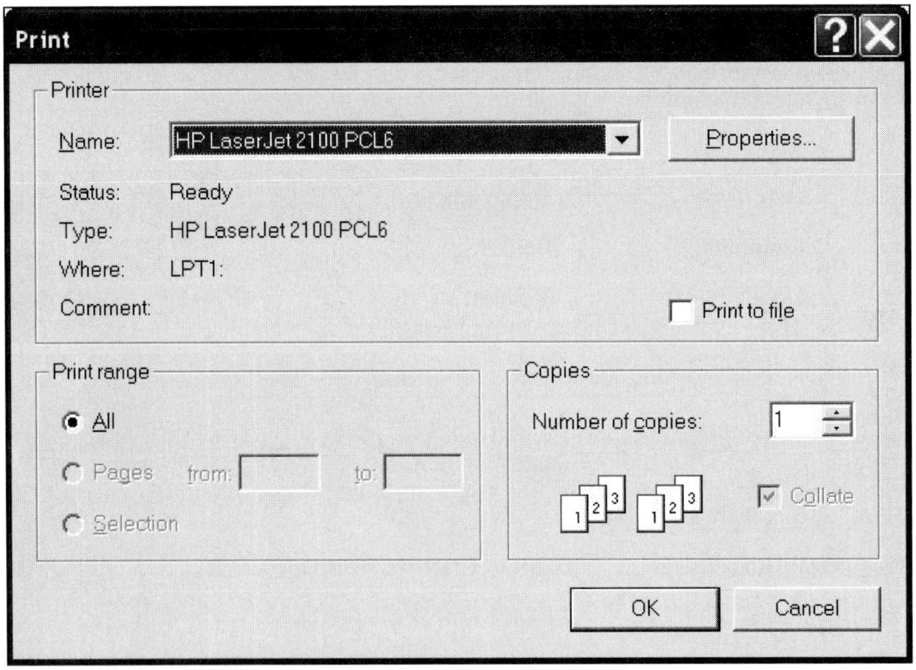

Figure 6.29(b) MS Printer Manager Print Dialog Box

The lesson is now complete. If you wish to model a project without instructions, a complete set of projects and illustrations are available at ***www.cad-resources.com*** ***⇒ Downloads.***

Lesson 7 Feature Operations

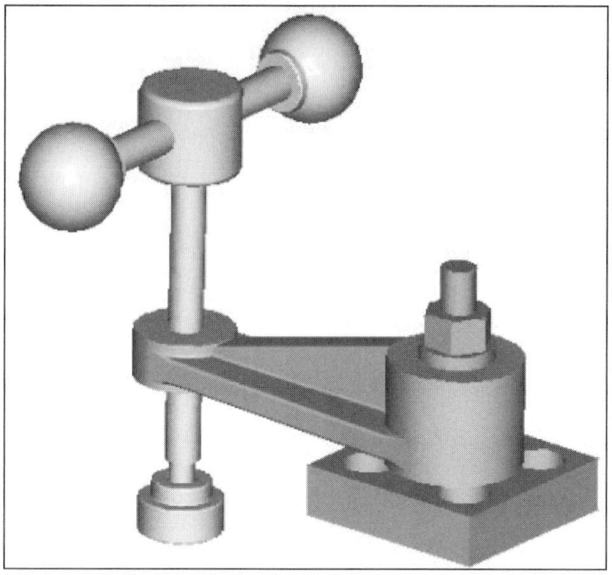

Figure 7.1 Swing Clamp Assembly

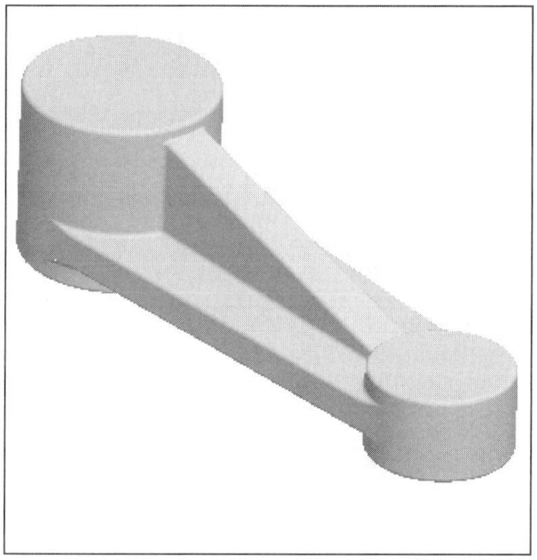

Figure 7.2(a) Clamp Arm (Casting- Workpiece)

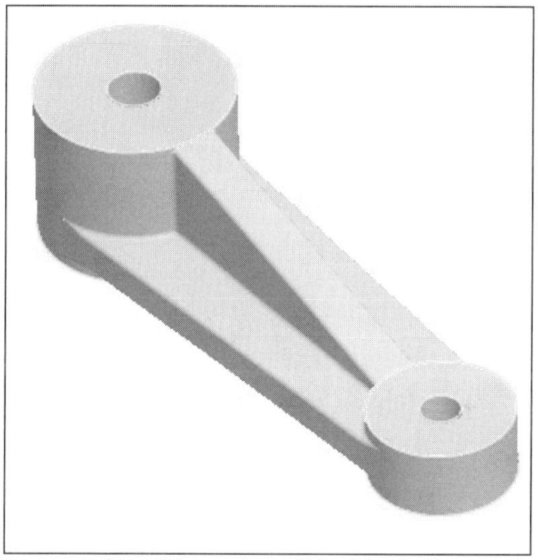

Figure 7.2(b) Clamp Arm (Machined- Design Part)

OBJECTIVES

- Use **Copy**, **Paste**, and **Paste Special**
- Create **Ribs**
- Understand **Parameters** and **Relations**
- **Measure** geometry
- Solve **Failures**
- Create a workpiece using a **Family Table**

FEATURE OPERATIONS

The Clamp Arm is used in the Swing Clamp Assembly (Fig. 7.1). This lesson will cover a wide range of Pro/E capabilities including: **Feature Operations**, **Copy** and **Paste Special**, **Relations**, **Parameters**, **Failures**, **Family Tables**, and the **Rib Tool**. You will create two versions of the Clamp Arm; one with all cast surfaces [Fig. 7.2(a)] and the other with machined ends [Fig. 7.2(b)] (using a Family Table).

Ribs

A rib is a special type of protrusion designed to create a thin fin or web that is attached to a part. You always sketch a rib from a side view, and it grows about the sketching plane symmetrically or to either side. Because of the way ribs are attached to the parent geometry, they are always sketched as open sections.

When sketching an open section, Pro/E may be uncertain about the side to which to add the rib. Pro/E adds all material in the direction of the arrow. If the incorrect choice is made, toggle the arrow direction by clicking on the direction arrow on the screen.

A rib must "see" material everywhere it attaches to the part; otherwise, it becomes an unattached feature. There are two types of ribs: straight [Fig. 7.3(a)] and rotational [Fig. 7.3(b)]. The type is automatically set according to the attaching geometry (planar or curved).

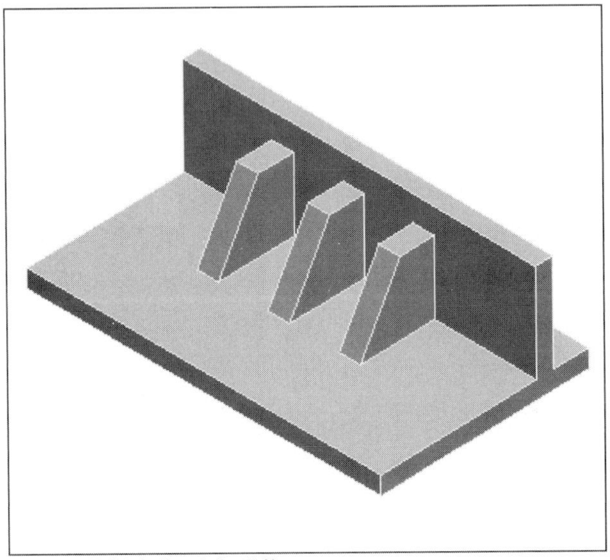

Figure 7.3(a) Straight Rib

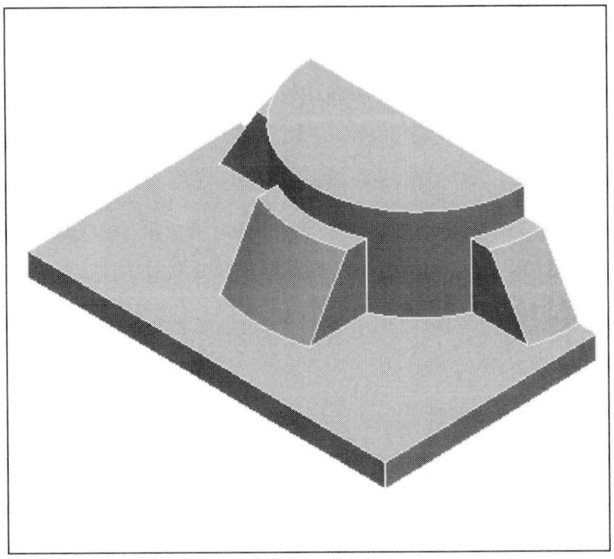

Figure 7.3(b) Rotational Rib

Relations

Relations (also known as parametric relations) are user-defined equations written between symbolic dimensions and parameters. Relations capture design relationships within features or parts, or among assembly components, thereby allowing users to control the effects of modifications on models.

Relations can be used to control the effects of modifications on models, to define values for dimensions in parts and assemblies, and to act as constraints for design conditions (for example, specifying the location of a hole in relation to the edge of a part). They are used in the design process to describe conditional relationships between different features of a part or an assembly.

Relations can be used to provide a value for a dimension. However, they can also be used to notify you when a condition has been violated, such as when a dimension exceeds a certain value. There are two basic types of relations, equality and comparison.

An equality relation equates a parameter on the left side of the equation to an expression on the right side. This type of relation is used for assigning values to dimensions and parameters. The following are a few examples of equality relations:

d2 = 25.500   **d8 = d4/2**   **d7 = d1+d6/2**   **d6 = d2*(sqrt(d7/4.0+d4))**

A comparison relation compares an expression on the left side of the equation to an expression on the right side. This type of relation is commonly used as a constraint or as a conditional statement for logical branching. The following are examples of comparison relations:

d1 + d2 > (d3 + 5.5)	Used as a constraint
IF (d1 + 5.5) > = d7	Used in a conditional statement

Parameter Symbols

Four types of parameter symbols are used in relations:

- **Dimensions** These are dimension symbols, such as **d8**, **d12**.
- **Tolerances** These are parameters associated with ± symmetrical and plus-minus tolerance formats. These symbols appear when dimensions are switched from numeric to symbolic.
- **Number of Instances** These are integer parameters for the number of instances in a direction of a pattern.
- **User Parameter** These can be parameters defined by adding a parameter or a relation (e.g., **Volume** = **d3** * **d4** * **d5**).

Operators and Functions

The following operators and functions can be used in equations and conditional statements:

Arithmetic Operators

+	**Addition**
–	**Subtraction**
/	**Division**
*	**Multiplication**
^	**Exponentiation**
( )	**Parentheses for grouping** [for example, **(d0 = (d1–d2)*d3)**]

Assignment Operators

=	**Equal to**

The = (equals) sign is an assignment operator that equates the two sides of an equation or relation. When it is used, the equation can have only a single parameter on the left side.

Comparison Operators

Comparison operators are used whenever a TRUE/FALSE value can be returned. For example, the relation **d1** >= **3.5** returns TRUE whenever d1 is greater than or equal to **3.5**. It returns FALSE whenever **d1** is less than **3.5**. The following comparison operators are supported:

==	**Equal to**
>	**Greater than**
>=	**Greater than or equal to**
!=, <>,~=	**Not equal to**
<	**Less than**
<=	**Less than or equal to**

|	**Or**
&	**And**
~, !	**Not**

Mathematical Functions

The following operators can be used in relations, both in equations and in conditional statements. Relations may include the following mathematical functions:

cos ( )	**cosine**
tan ( )	**tangent**
sin ( )	**sine**
sqrt ( )	**square root**
asin ( )	**arc sine**
acos ( )	**arc cosine**
atan ( )	**arc tangent**
sinh ( )	**hyperbolic sine**
cosh ( )	**hyperbolic cosine**
tanh ( )	**hyperbolic tangent**

Failures

Sometimes model geometry cannot be constructed because features that have been modified or created conflict with or invalidate other features. This can happen when the following occurs:

- A protrusion is created that is unattached and has a one-sided edge.
- New features are created that are unattached and have one-sided edges.
- A feature is resumed that now conflicts with another feature (i.e. two chamfers on the same edge).
- The intersection of features is no longer valid because dimensional changes have moved the intersecting surfaces.
- A relation constraint has been violated.

Resolve Feature

After a feature fails, Pro/E enters Resolve Feature mode. Use the commands in the RESOLVE FEAT menu to fix the failed feature:

- **Undo Changes** Undo the changes that caused the failed regeneration attempt, and return to the last successfully regenerated model.
- **Investigate** Investigate the cause of the regeneration failure using the Investigate submenu.
- **Fix Model** Roll the model back to the state before failure and select commands to fix the problem.
- **Quick Fix** Choose an option from the **QUICK FIX** menu, the options are as follows:
    - **Redefine** Redefine the failed feature.
    - **Reroute** Reroute the failed feature.
    - **Suppress** Suppress the failed feature and its children.
    - **Clip Supp** Suppress the failed feature and all the features after it.
    - **Delete** Delete the failed feature.

Failed Features

If a feature fails during creation and it does not use the dialog box interface, Pro/E displays the FEAT FAILED menu with the following options:

- **Redefine** Redefine the feature
- **Show Ref** Display the SHOW REF menu so you can see the references of the failed feature. Pro/E displays the reference number in the Message Window.
- **Geom Check** Check for problems with overlapping geometry, misalignment, and so on
- **Feat Info** Get information about the feature

If a feature fails, you can redisplay the part with all failed geometry highlighted in different colors. Pro/E displays the corresponding error messages in an Information Window. Features can fail during creation for the following reasons:

- **Overlapping geometry** A surface intersects itself. If Pro/E finds a self-intersecting surface, it does not perform any further surface checks.
- **Surface has edges that coincide** The surface has no area. Pro/E highlights the surface in red.
- **Inverted geometry** Pro/E highlights the inverted geometry in purple and displays an error message.
- **Bad edges** Pro/E highlights bad edges in blue and displays an error message.

Family Tables

Family Tables are effective for two main reasons: they provide a beneficial tool, and they are easy to use. You need to understand the functionality of Family Tables, and you must understand when a Family Table is required and what circumstances should promote its use.

To determine whether a model is a candidate for a Family Table: establish whether the original and the variation would ever have to co-exist at the same time (both in the same assembly, both shown in the same drawing, both with an independent Bill of Materials) and whether they should be tied together (most of the same dimensions, features, and parameters). If so, the component is a candidate for the creation of a Family Table [Figs. 7.4(a-b)], otherwise, the model may be a candidate for copying to an independent model.

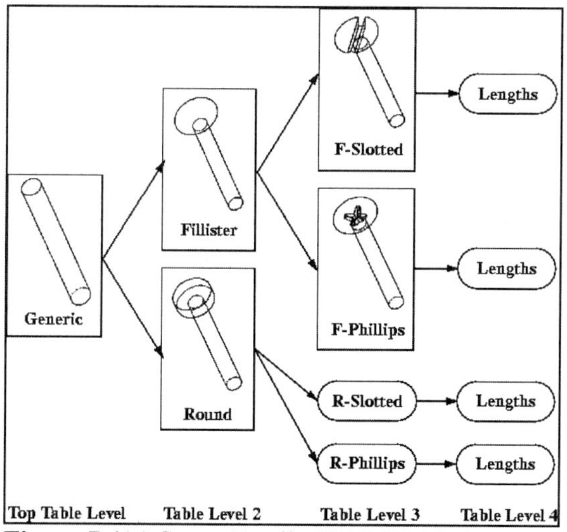

Figure 7.4(a) Screw Family (CADTRAIN)

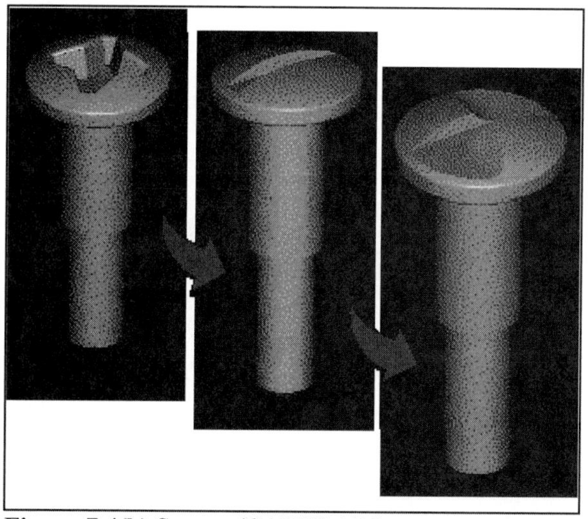

Figure 7.4(b) Screws (CADTRAIN)

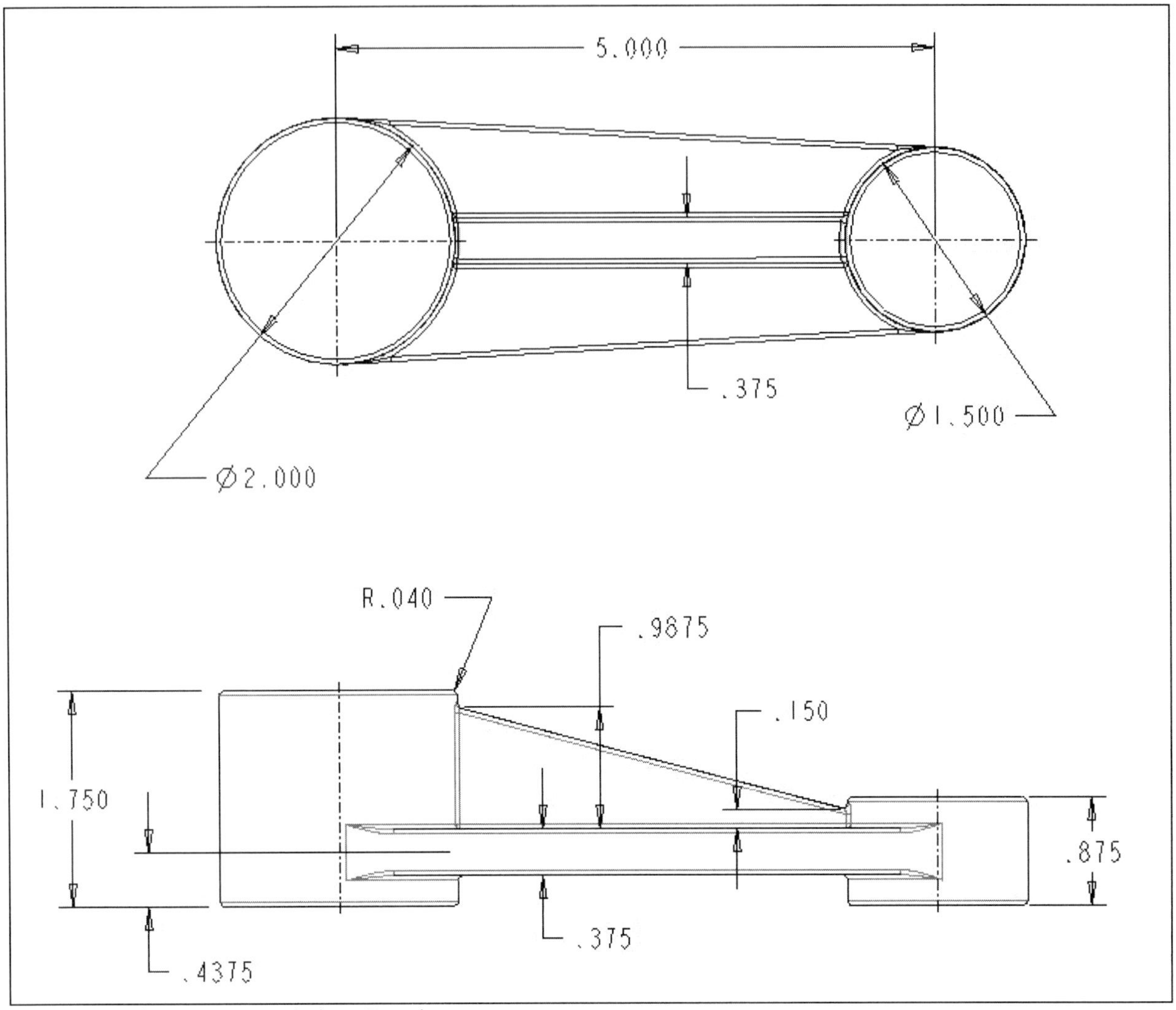

Figure 7.5 Clamp Arm Workpiece Drawing

Clamp Arm

The Clamp_Arm is modeled in two stages. Model the casting (Fig. 7.5) using the casting detail, and then use the machine detail to complete the last (machined) features. The last step will be to create a Family Table with an instance that suppresses the machined features. By having a *casting part* (which is called a ***workpiece*** in **Pro/NC**) and a separate but almost identical *machined part* (which is called a ***design part*** in **Pro/NC**), you can create an operation for machining and an NC sequence. During the manufacturing process, *you merge the workpiece into the design part* and create a ***manufacturing model***.

The difference between the two files is the difference between the volume of the workpiece/casting part and the volume of the design part/machined part. The removed volume can be seen as *material removal* when you are performing an **NC Check** operation on the manufacturing model. If the machining process gouges the part, the gouge will display the interference. The cutter location can also be displayed as an animated machining process.

For additional tutorials on Pro/MANUFACTURING: Pro/NC, and Expert Machinist see: ***www.cad-resources.com*** *and click* ***CDI72 Pro/ENGINEER Advanced***.

Start a new part, click: **File** ⇒ **Set Working Directory** ⇒ select your working directory ⇒ **OK** ⇒ **File** ⇒ **New** ⇒ [Part] ⇒ [Name CLAMP_ARM] ⇒ [Use default template] ⇒ **OK** ⇒ **Tools** ⇒ **Options** ⇒ Showing: **Current Session** ⇒ **Open a configuration file** ⇒ **clamp.pro** *(open your previously created option file from Lesson 6)* ⇒ **Open** ⇒ **Apply** ⇒ **Close** ⇒ **Tools** ⇒ [Environment] ⇒ [Snap to Grid] ⇒ [Use 2D Sketcher] ⇒ [Tangent Edges Dimmed] ⇒ **OK** ⇒ **Edit** ⇒ **Setup** ⇒ **Material** ⇒ [steel.mtl] ⇒ ⇒ **OK** ⇒ **Done** ⇒ change the coordinate system name to [CLAMP_ARM_CSYS] ⇒ ⇒ **Enter** ⇒ pick on datum **FRONT** to pre-select it ⇒ ⇒ **RMB** ⇒ [Define Internal Sketch...] [Fig. 7.6(a)] ⇒ **Sketch** ⇒ **RMB** ⇒ [Circle] sketch a circle ⇒ **MMB** ⇒ **Ctrl+D** ⇒ double-click on the diameter dimension and change to **2.00** [Fig. 7.6(b)] ⇒ **Enter** ⇒ [Fig. 7.6(c)] ⇒ double-click on the height dimension and change to **1.75** [Fig. 7.6(c)] ⇒ **Enter** ⇒ **Enter** ⇒ **Refit** ⇒ ⇒ **Ctrl+S** ⇒ **MMB**

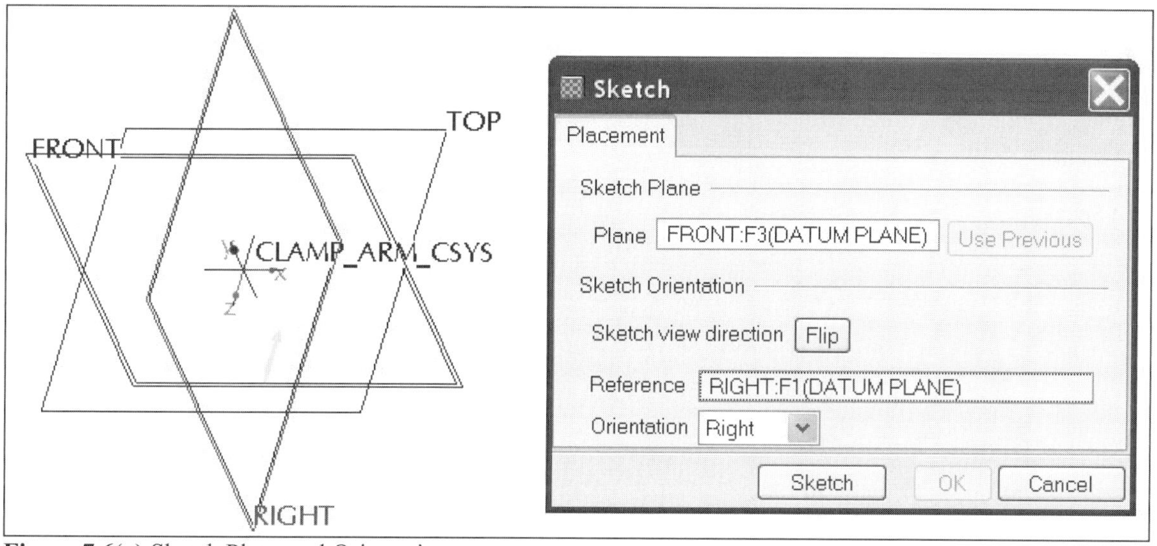

Figure 7.6(a) Sketch Plane and Orientation

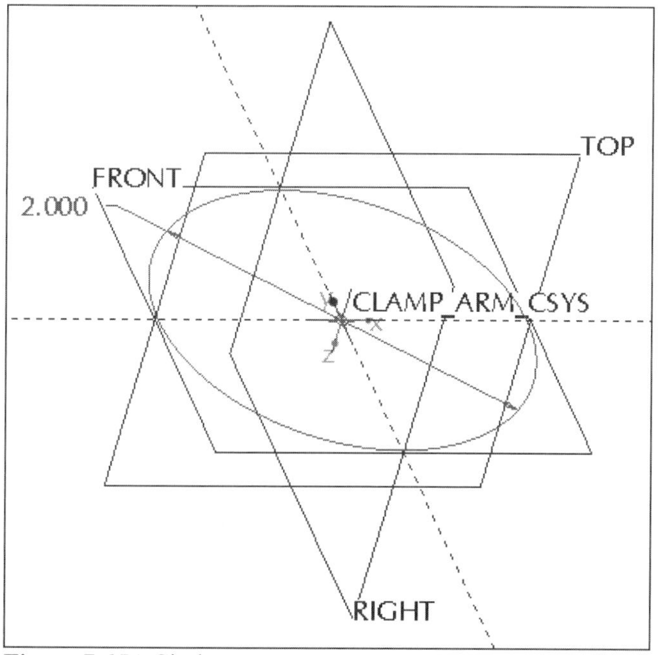

Figure 7.6(b) Circle

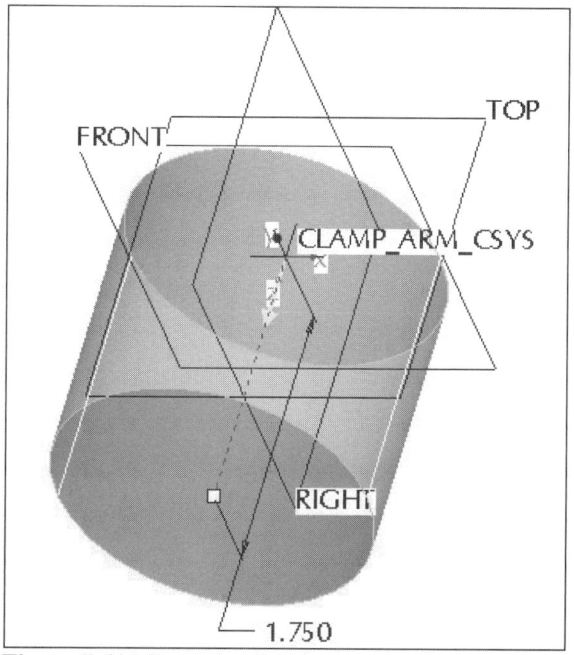

Figure 7.6(c) Extrusion Preview

With the extrusion selected/highlighted, click: **Ctrl+C** ⇒ **Edit** ⇒ Paste Special... [Fig. 7.7(a)] ⇒ Dependent copy ⇒ Apply Move/Rotate transformations to copies ⇒ **OK** ⇒ Transformations ⇒ pick datum **Right** as the **Direction reference** ⇒ move the drag handle to the right until **5.00** [Fig. 7.7(b)] ⇒ **Enter** ⇒ **Ctrl+S** ⇒ **Enter** ⇒ **Ctrl+D** ⇒ **LMB** to deselect

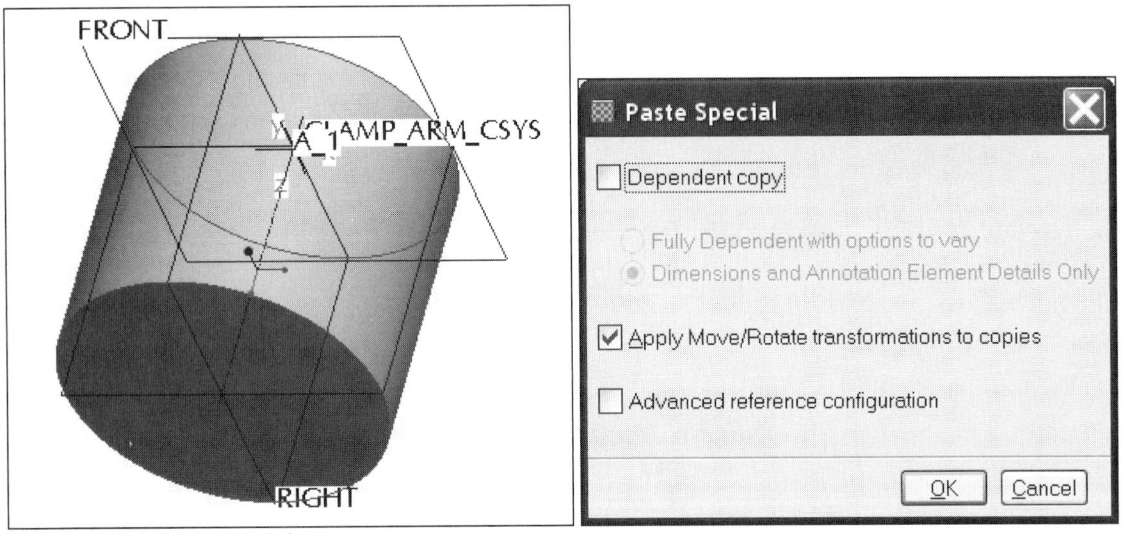

Figure 7.7(a) Paste Special

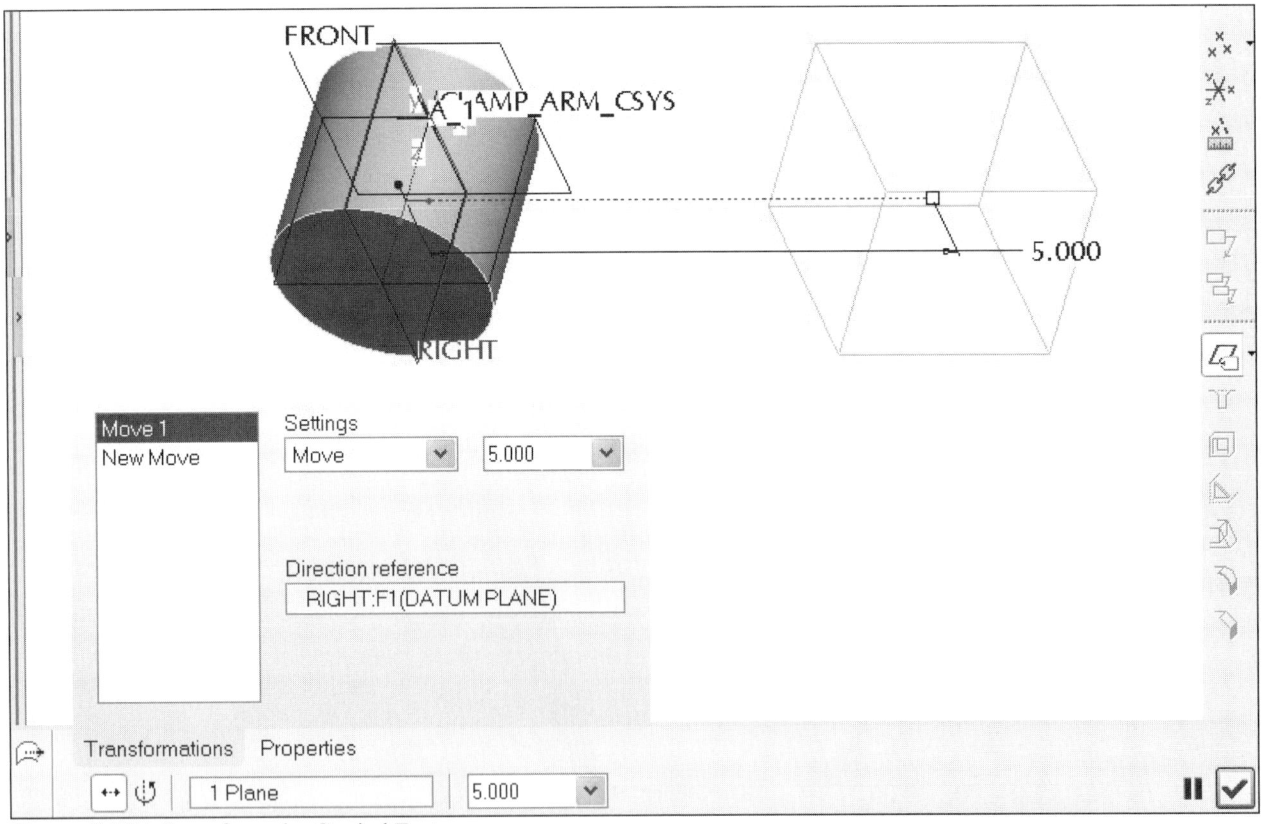

Figure 7.7(b) Transform the Copied Feature

Double-click on the copied feature and modify the diameter dimension to **1.50** and the height dimension to **.875** [Fig. 7.7(c)] ⇒ **Regenerates Model** [Fig. 7.7(d)] ⇒ **Ctrl+S** ⇒ **MMB**

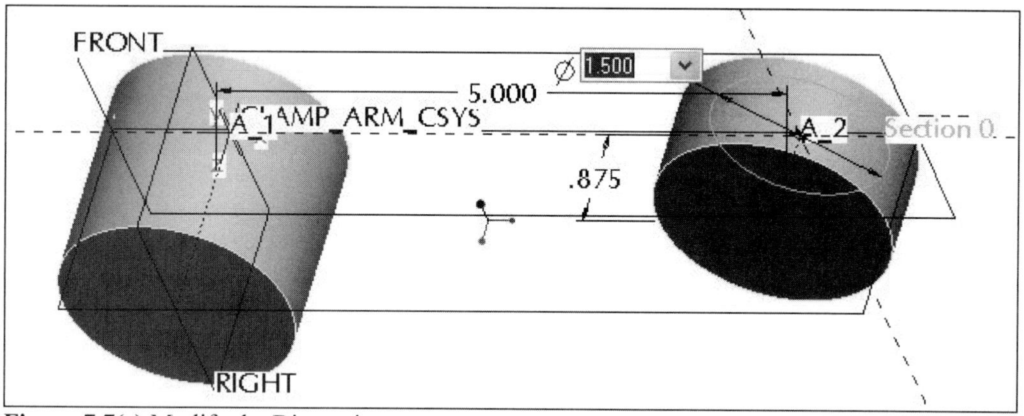

Figure 7.7(c) Modify the Dimensions

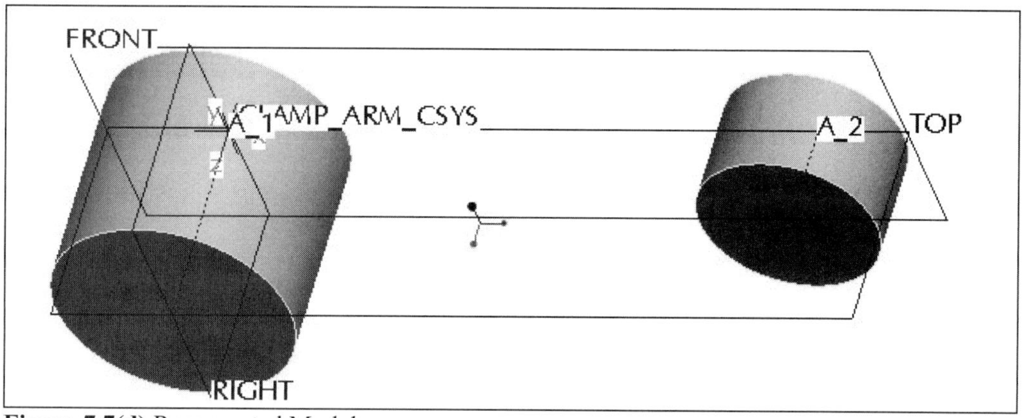

Figure 7.7(d) Regenerated Model

Pick on the **FRONT** datum plane ⇒ **Datum Plane Tool** ⇒ Translation 0.4375 (Fig. 7.8) ⇒ **OK**

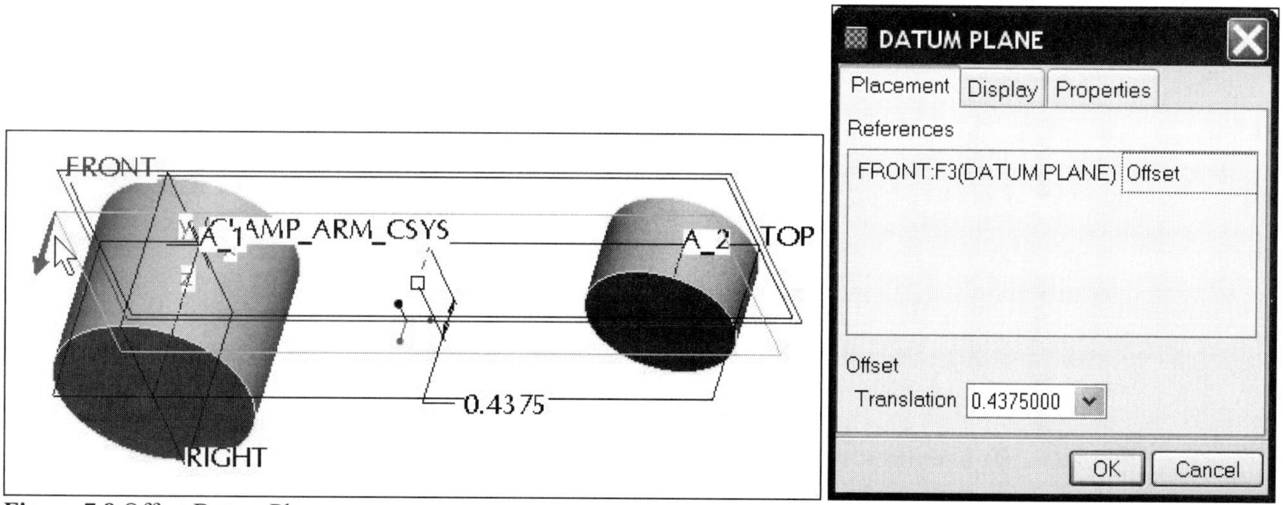

Figure 7.8 Offset Datum Plane

With datum **DTM1** selected, click:  **Extrude Tool** ⇒ **Placement** tab ⇒ **Define** ⇒ **Sketch** to close Sketch dialog box ⇒ **Sketch** from the menu bar ⇒ **References** [Figs. 7.9(a-b)] ⇒ add one arc from both circles [Fig. 7.9(c)] ⇒ select all four References *(use the Ctrl key – Note that as you select each reference, its corresponding feature highlights in the window)* ⇒ **Ctrl+D** [Fig. 7.9(d)] ⇒  ⇒ **Close**

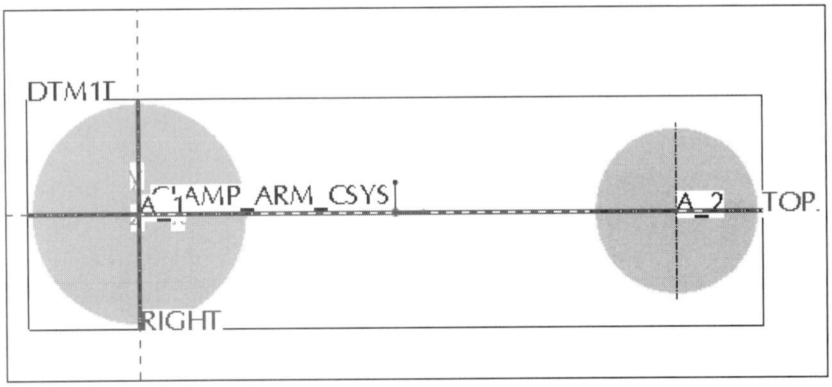

Figure 7.9(a) References

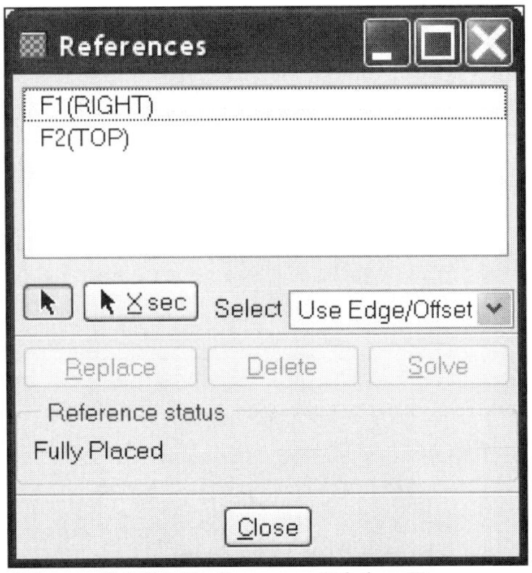

Figure 7.9(b) References Dialog Box

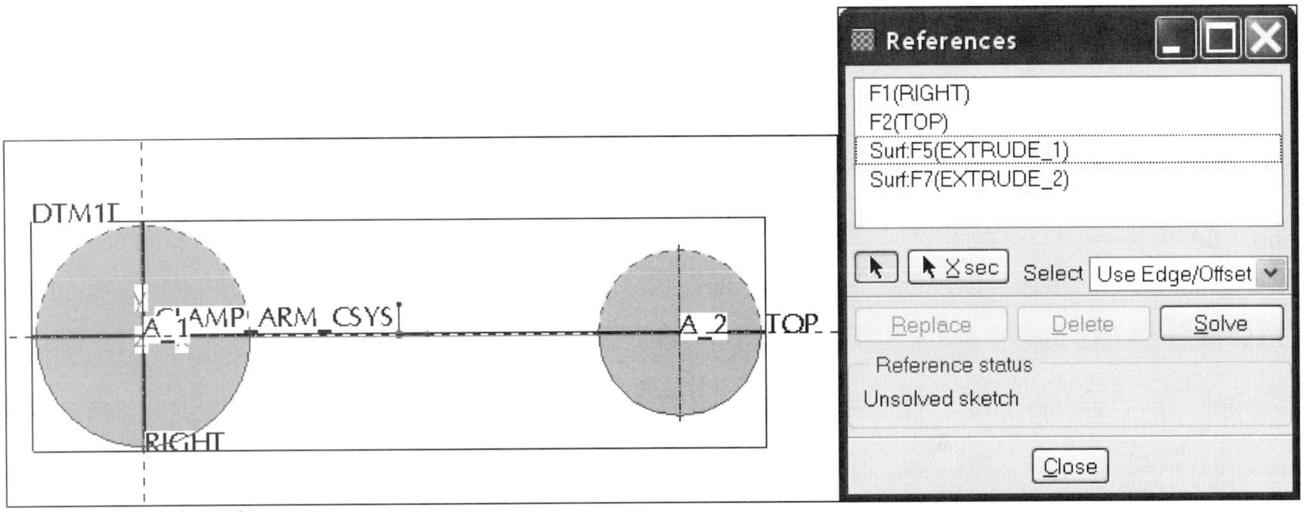

Figure 7.9(c) Add Reference

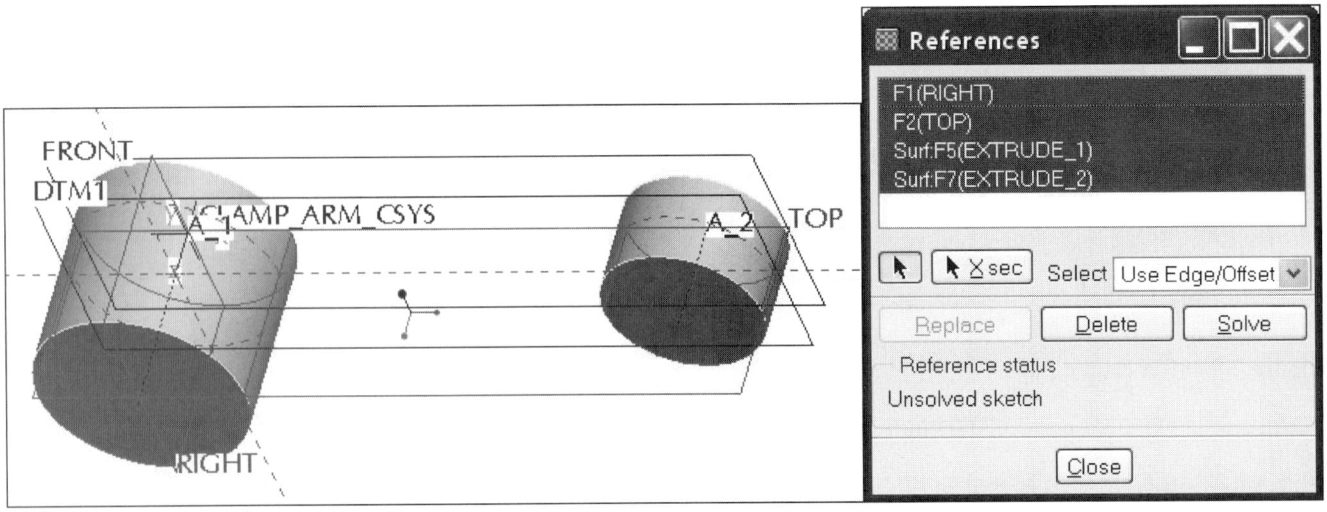

Figure 7.9(d) Four References

Click: **RMB** ⇒ **Centerline** ⇒ create a horizontal centerline [Fig. 7.9(e)] ⇒  **Hidden line** ⇒ **RMB** ⇒ **Circle** ⇒ pick on the center of a circular reference and then on its circular reference [Fig. 7.9(f)] ⇒ repeat on the opposite end [Fig. 7.9(g)] ⇒ **MMB** to end the Circle Tool ⇒  off ⇒ **Ctrl+D** [Fig. 7.9(h)]

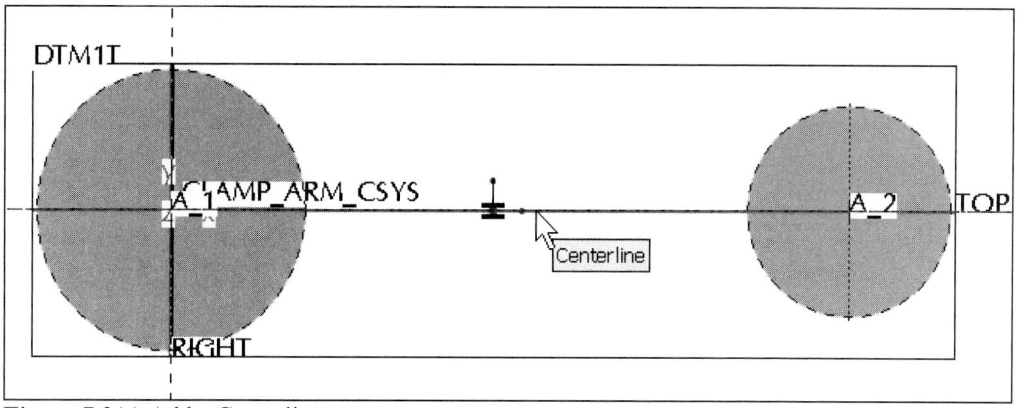

Figure 7.9(e) Add a Centerline

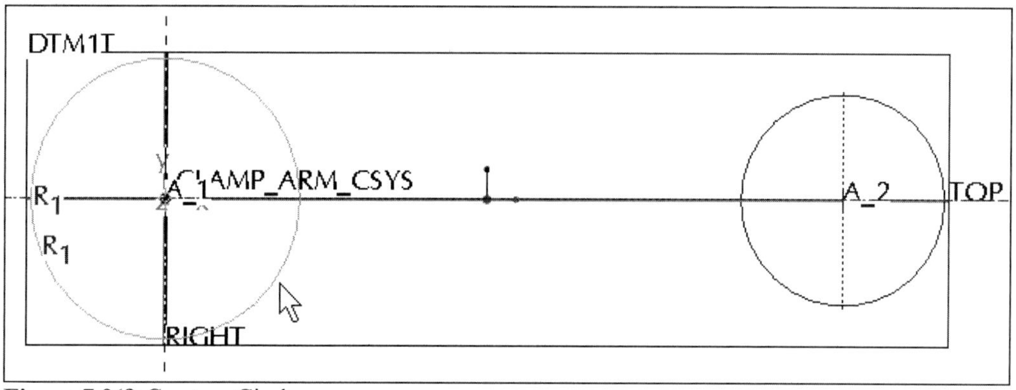

Figure 7.9(f) Create a Circle

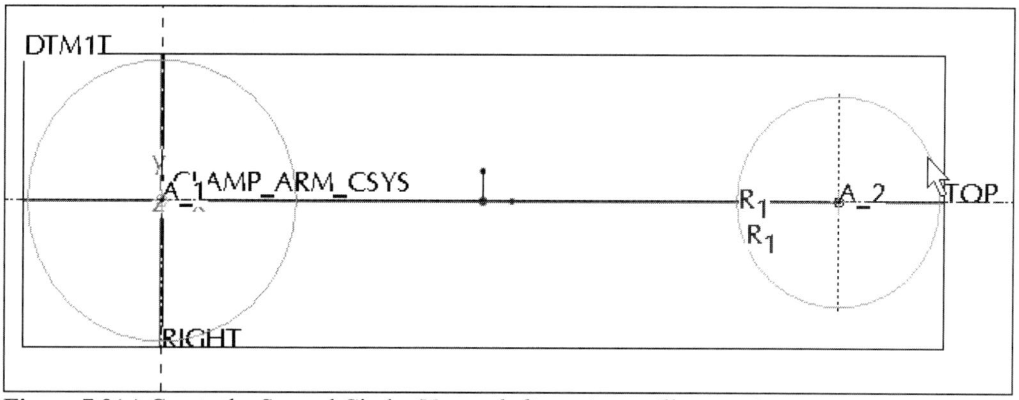

Figure 7.9(g) Create the Second Circle *(No weak dimensions will appear)*

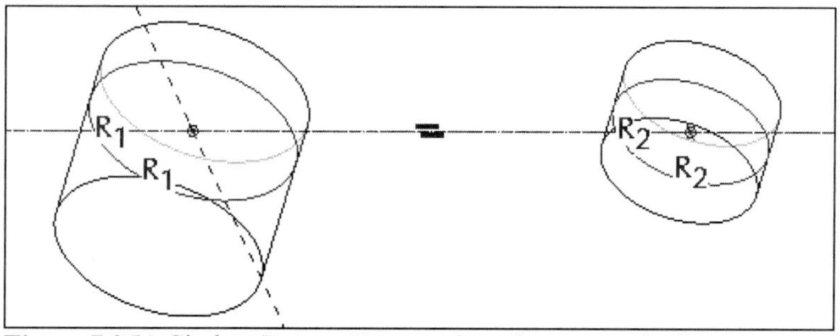

Figure 7.9(h) Circles *(Datum Features have been turned off)*

Click:  ⇒ 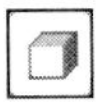 **Shading** ⇒ 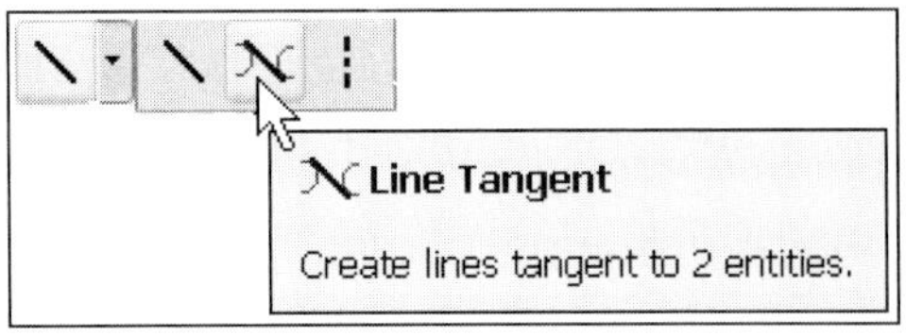

create a line tangent between the circular extrusions [Fig. 7.9(i)] ⇒ create the bottom tangent line [Fig. 7.9(j)] ⇒ **MMB** ⇒ **Ctrl+D** ⇒ 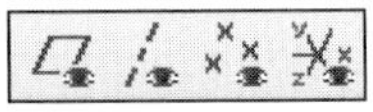 off [Fig. 7.9(k)] ⇒ **Orient the sketching plane parallel to the screen**

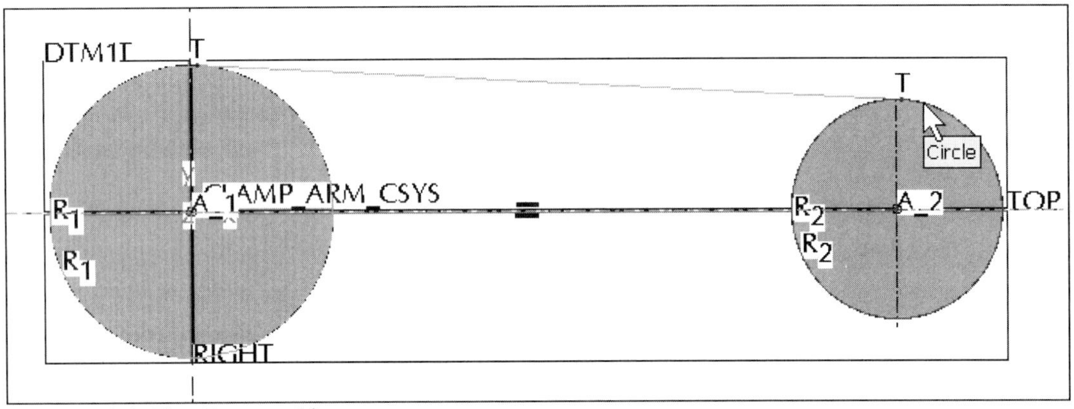

Figure 7.9(i) First Tangent Line

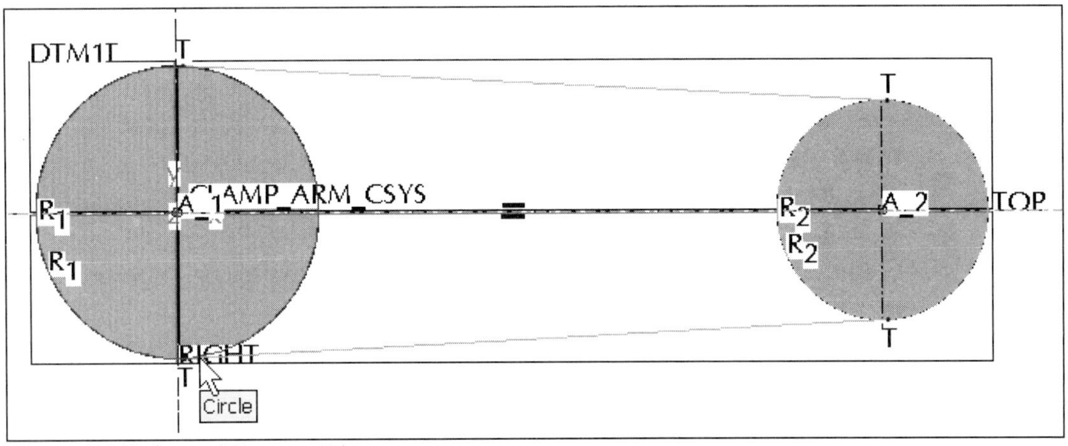

Figure 7.9(j) Second Tangent Line

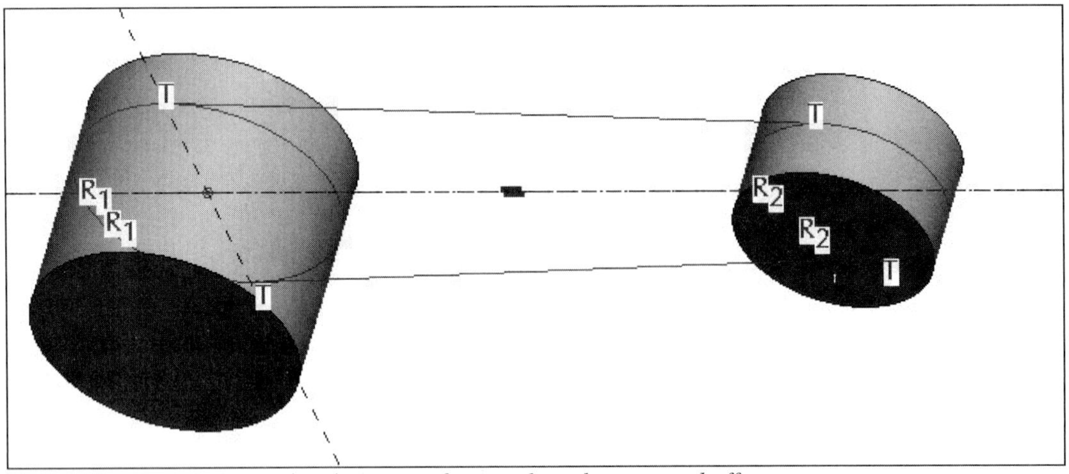

Figure 7.9(k) Completed Sketch *(Datum features have been turned off)*

Click: on ⇒ on ⇒ **Delete Segment** 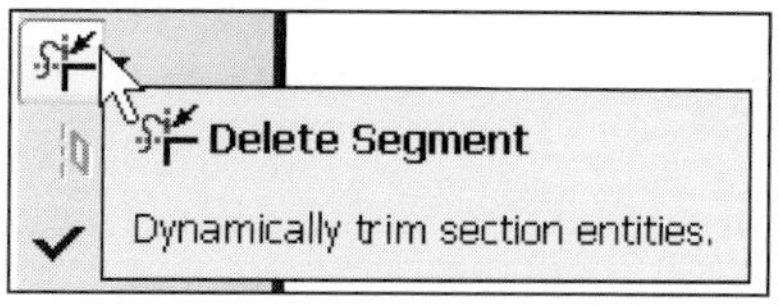

⇒ draw a spline through the unwanted entities on the right side [Fig. 7.9(l)] ⇒ draw a spline through the unwanted entities on the left side [Fig. 7.9(m)] ⇒ draw a spline through the tiny leftover piece on the lower-left side between the tangent position and the vertical reference [Fig. 7.9(n)] ⇒ repeat for the upper-left side [Fig. 7.9(o)]

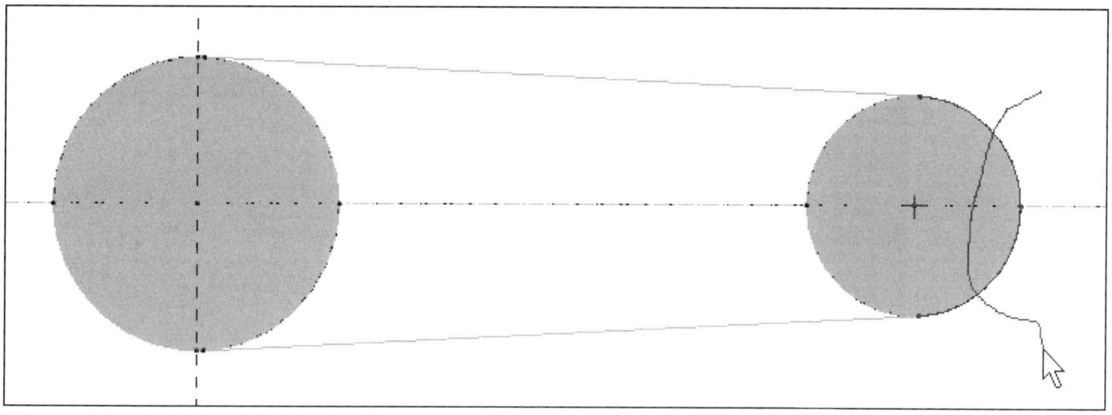

Figure 7.9(l) Trim Unwanted Entities on Right Side

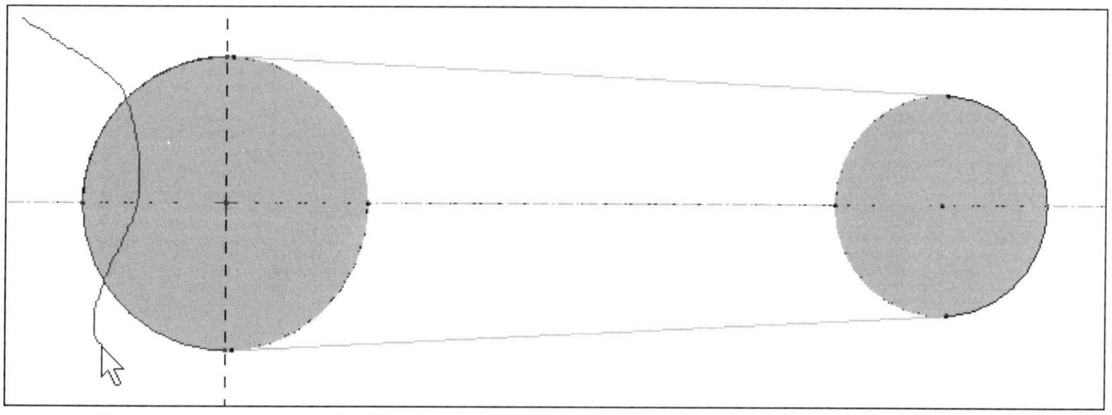

Figure 7.9(m) Trim Unwanted Entities on Left Side

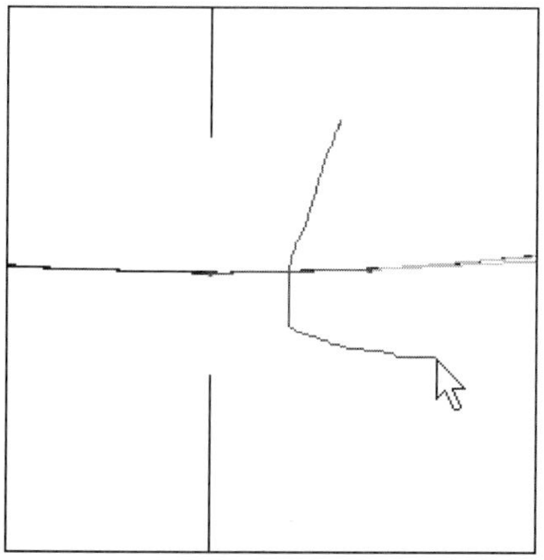

Figure 7.9(n) Trim Leftover on Bottom

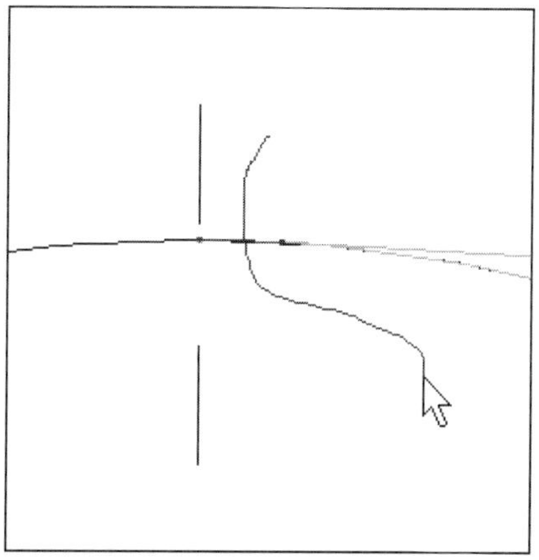

Figure 7.9(o) Trim Leftover on Top

Click: **LMB** ⇒ **MMB** [Fig. 7.9(p)] ⇒ **Ctrl+D** ⇒ on [Fig. 7.9(q)] ⇒ [Fig. 7.9(r)] ⇒ ⇒ modify depth value to **.375** ⇒ **Enter** [Fig. 7.9(s)]

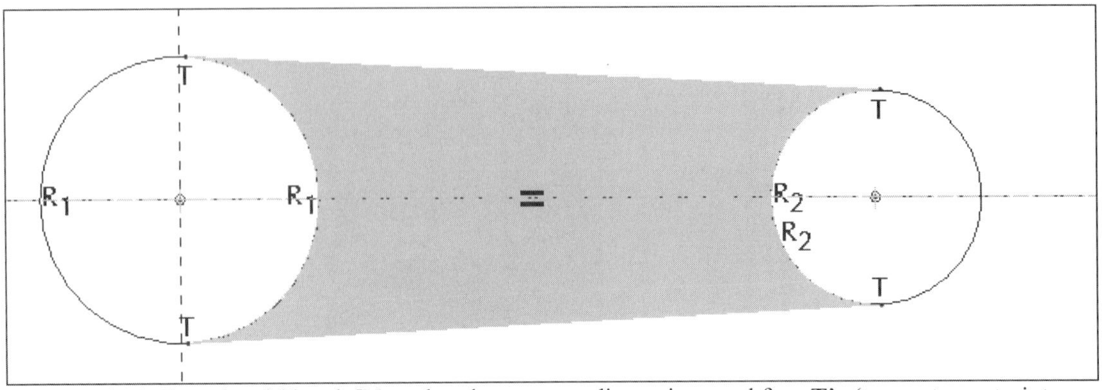

Figure 7.9(p) Completed Sketch [Note that there are no dimensions and four **T**'s (tangent constraints are displayed)]

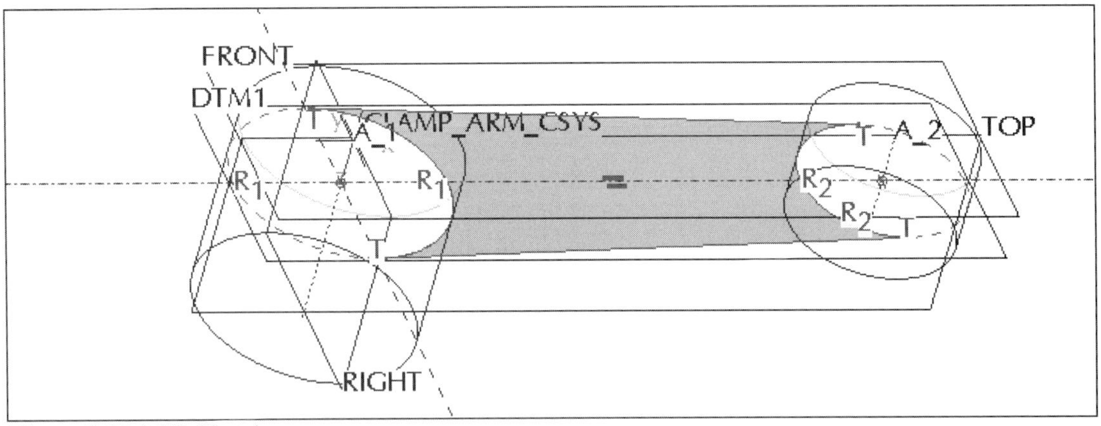

Figure 7.9(q) 3D Sketch

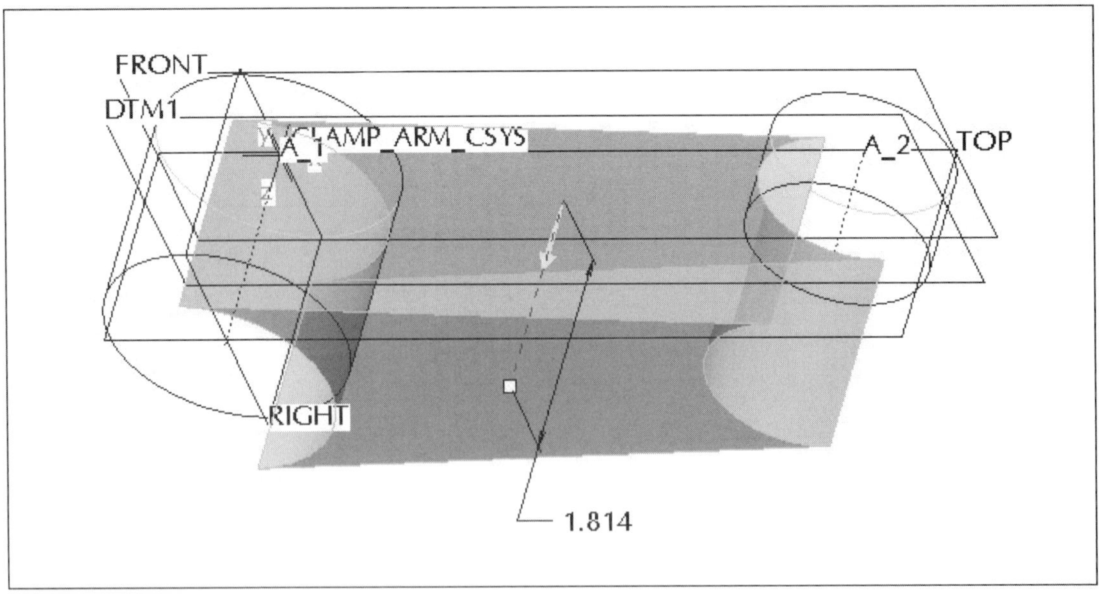

Figure 7.9(r) Preview

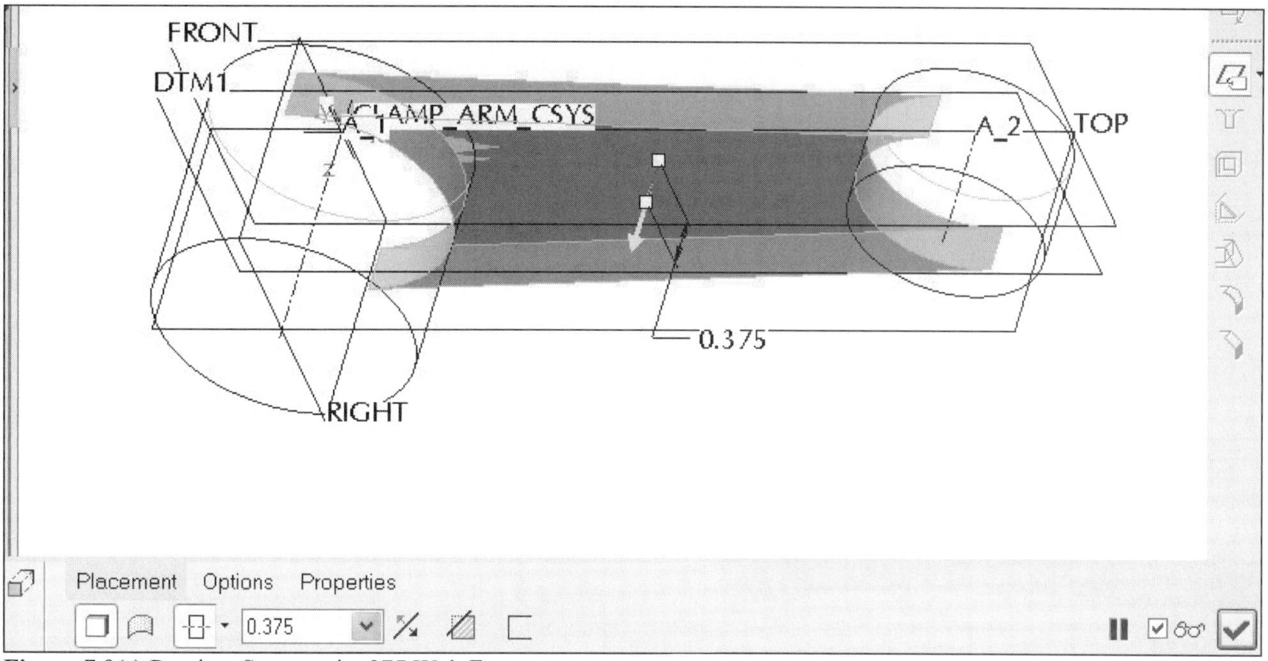

Figure 7.9(s) Preview Symmetric **.375** Web Feature

Click: [checkmark icon] [Fig. 7.9(t)] ⇒ [icon] ⇒ **Ctrl+S** ⇒ **Enter** ⇒ in the Model Tree, click [⊞] next to each feature ⇒ select the sketch **S2D0002** *(your name may be different)* for Extrude 3 [Fig. 7.9(u)] ⇒ **LMB** to deselect

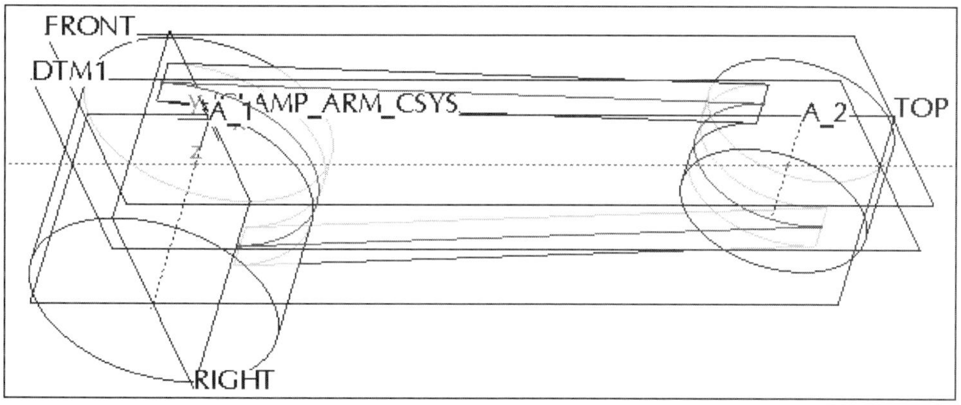

Figure 7.9(t) Completed Feature

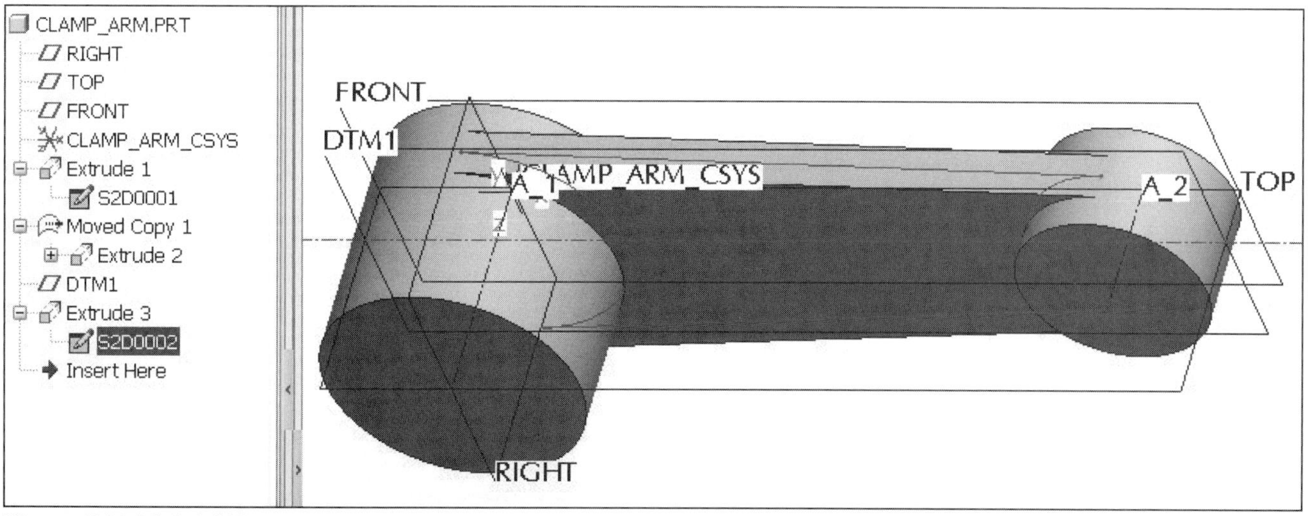

Figure 7.9(u) Model Tree

Select **TOP** datum ⇒ 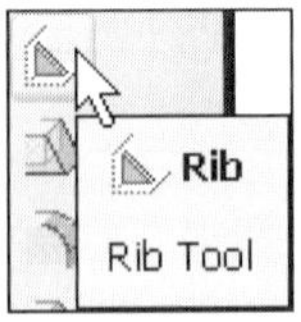

Rib Tool [Fig. 7.10(a)] ⇒ **RMB** ⇒ Define Internal Sketch... ⇒ **Flip** Sketch view direction Flip [Fig. 7.10(b)] ⇒ **Sketch** from dialog box ⇒ 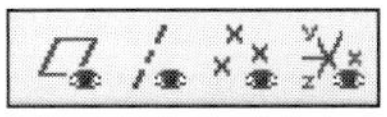 off ⇒ **Sketch** from menu bar ⇒ **References** ⇒ delete the two existing references [Fig. 7.10(c)] ⇒  [Fig. 7.10(d)]

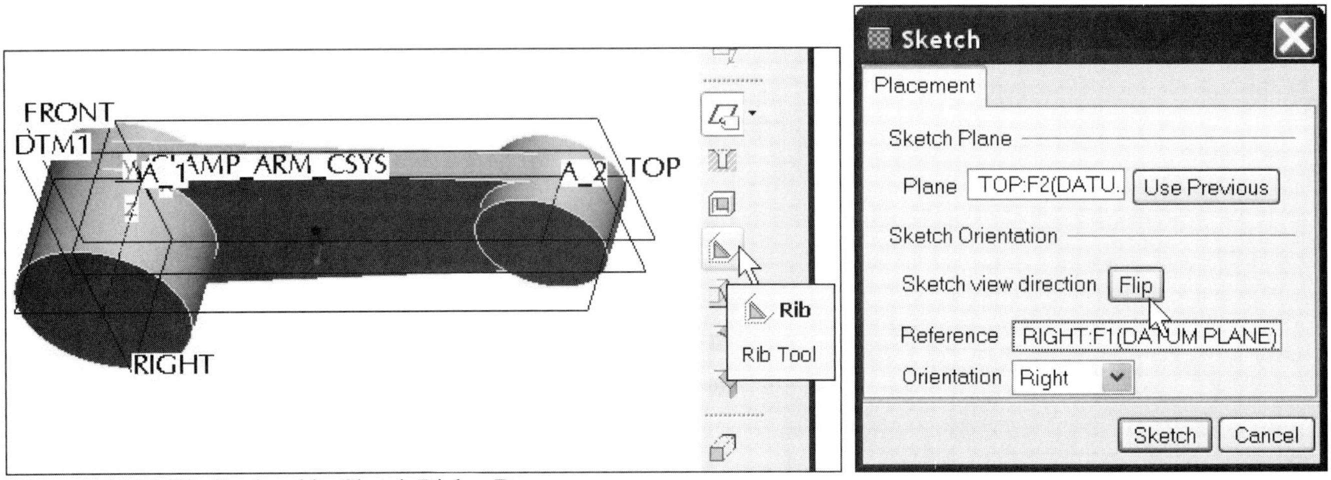

Figure 7.10(a) Rib Tool and its Sketch Dialog Box

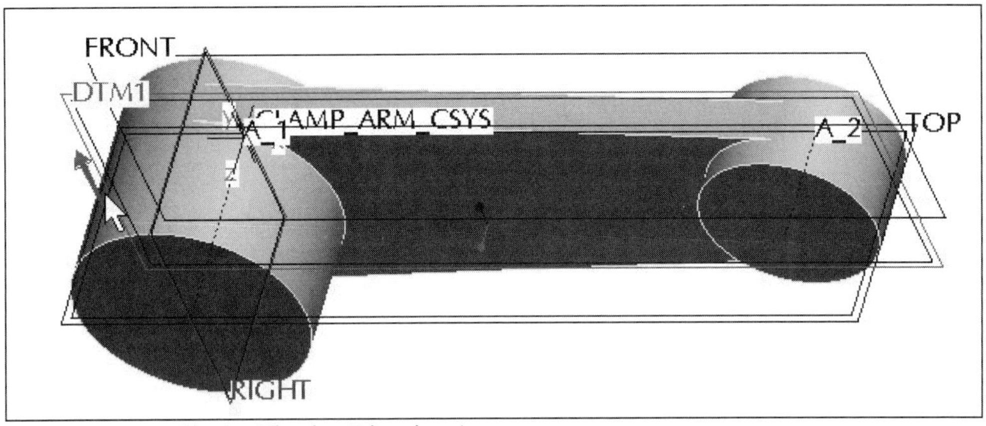

Figure 7.10(b) Flip the Viewing Direction Arrow

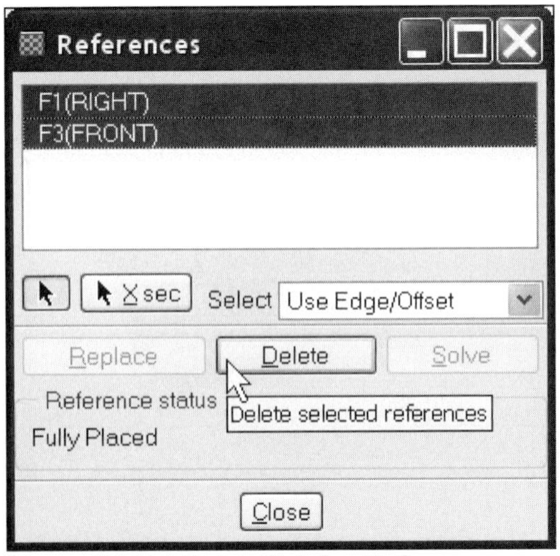

Figure 7.10(c) Delete Existing References

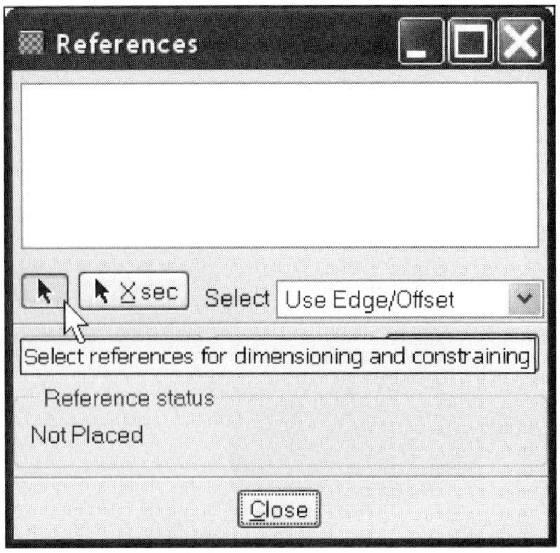

Figure 7.10(d) Select Button

Add three new references, pick on the inside vertical edge of both cylindrical extrusions and the top edge of the web extrusion [Figs. 7.10(e-f)] ⇒ **Close** ⇒ 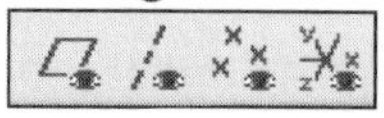 on ⇒ **RMB** ⇒ **Line** ⇒ **Ctrl+MMB** to zoom in ⇒ draw one angled line from one vertical reference to the other [Fig. 7.10(g)] (zoom in if necessary to pick the vertical surface/edge, not the endpoint) ⇒ **MMB** ⇒ **MMB** [Fig. 7.10(h)]

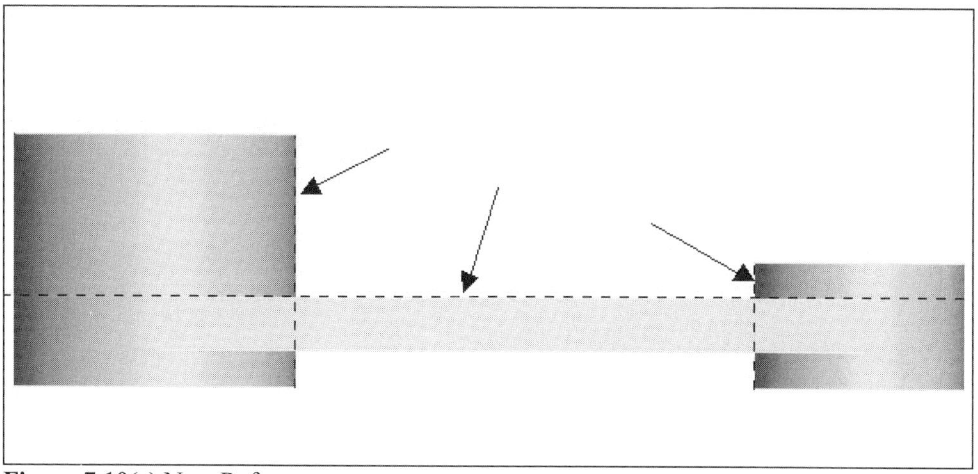

Figure 7.10(e) New References

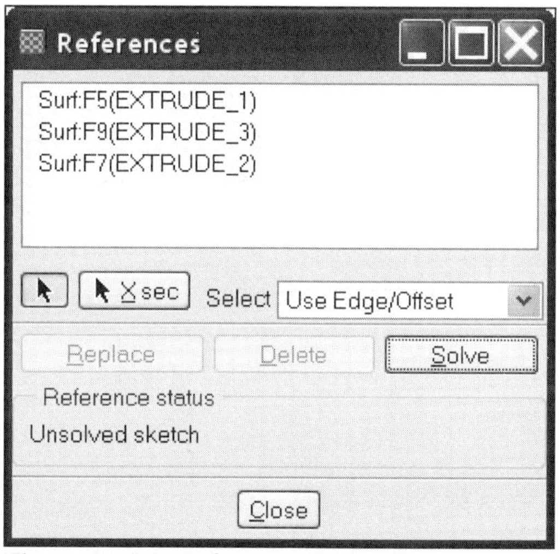

Figure 7.10(f) References Dialog Box

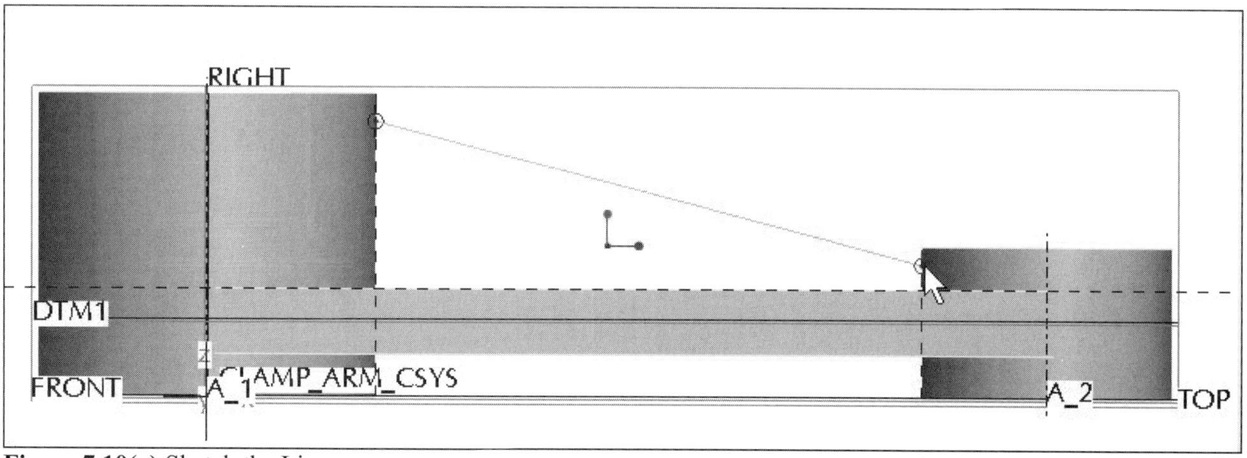

Figure 7.10(g) Sketch the Line

Click: **RMB** ⇒ **Dimension** add the vertical dimensions [Fig. 7.10(i)] ⇒ **MMB** ⇒ double-click on a dimension ⇒ change the dimension value ⇒ **Enter** ⇒ repeat for the other dimension (**.9875** and **.15** are the design values) [Fig. 7.10(j)] ⇒ **Ctrl+D** [Fig. 7.10(k)]

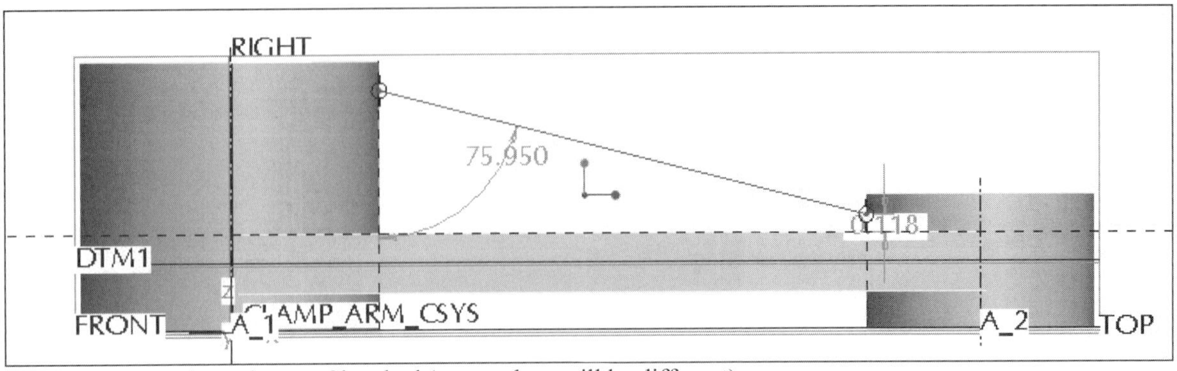

Figure 7.10(h) Dimensions as Sketched (your values will be different)

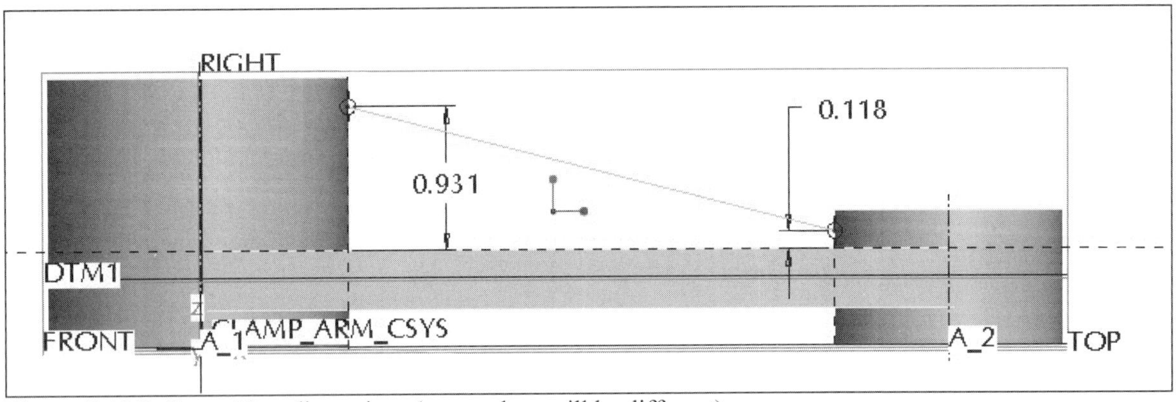

Figure 7.10(i) Create New dimensions (your values will be different)

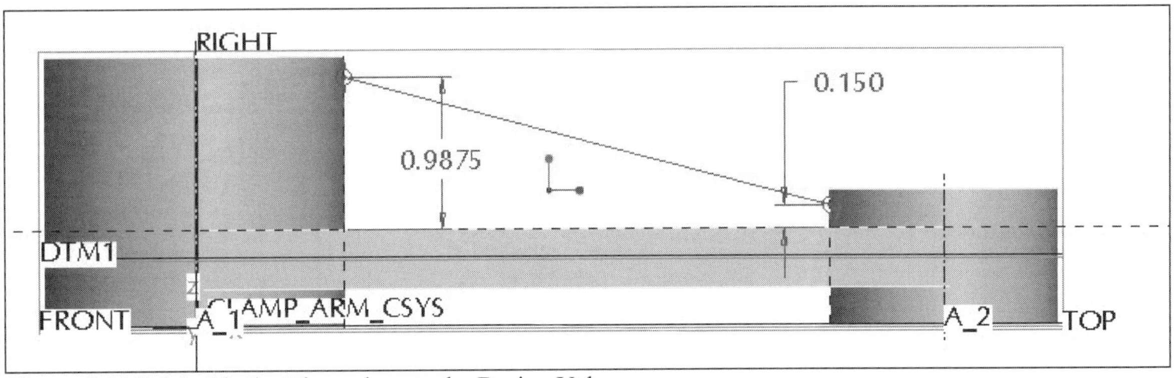

Figure 7.10(j) Modify the Dimensions to the Design Values

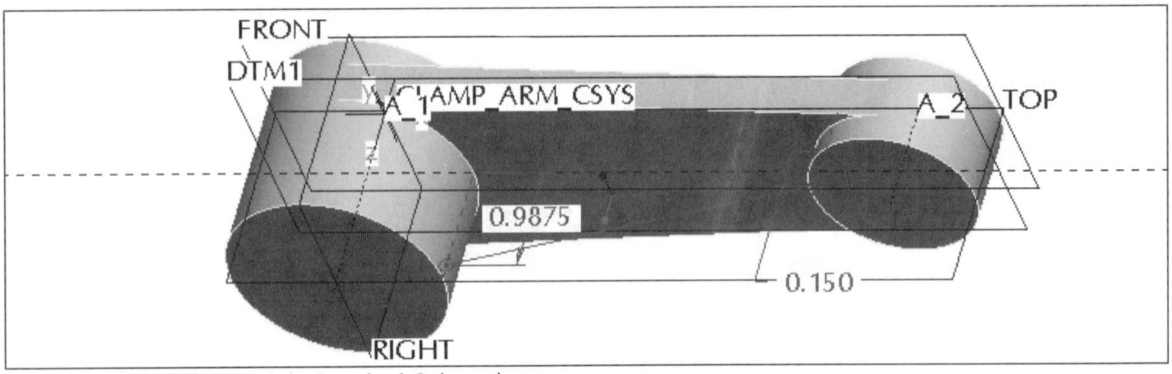

Figure 7.10(k) Rib Sketch in Standard Orientation

Click: [✓] ⇒ pick on the yellow arrow to flip the direction of the rib creation towards the part [Fig. 7.10(l)] ⇒ move the drag handle until a rib thickness of **.375** is displayed [Fig. 7.10(m)] [0.375] *(or type the value and press Enter)* ⇒ [✓] ⇒ **Ctrl+S** ⇒ **MMB** ⇒ **LMB** in graphics area to deselect [Fig. 7.10(n)] *(Note: clicking* [icon] *in the* ***dashboard*** *of the* ***Rib Tool*** *toggles the rib from centered, to the right side, or to the left side of the sketch plane.)*

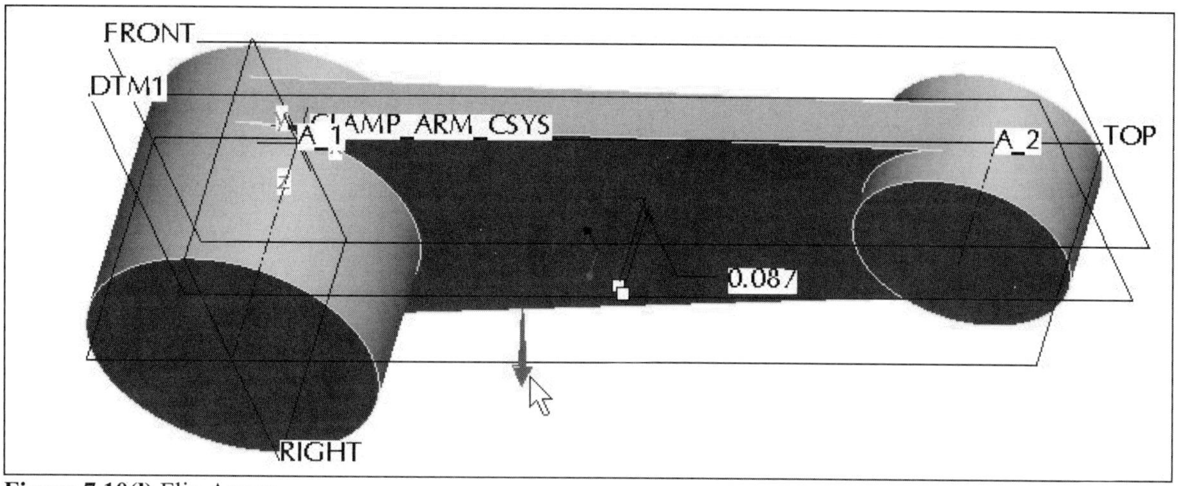

Figure 7.10(l) Flip Arrow

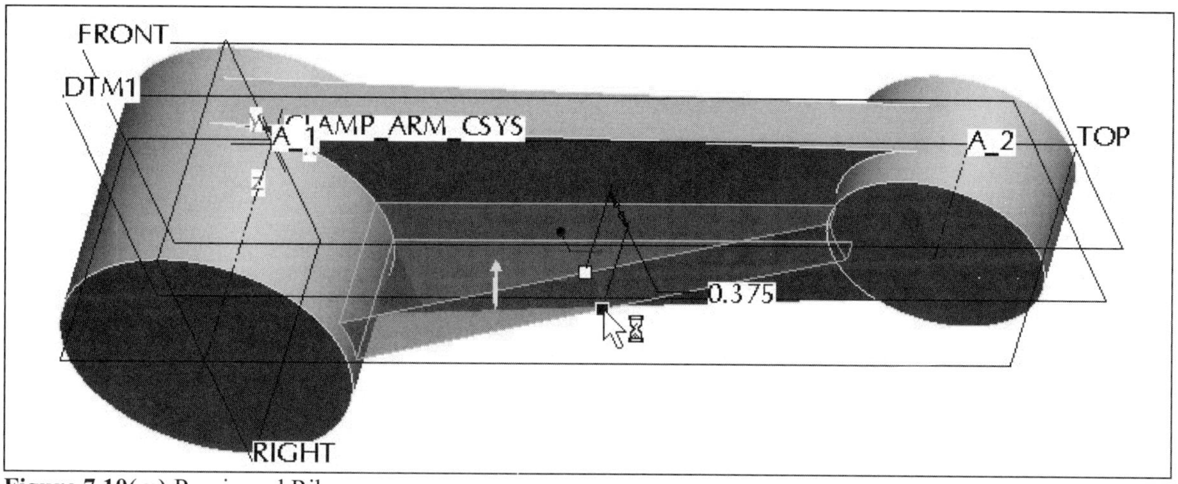

Figure 7.10(m) Previewed Rib

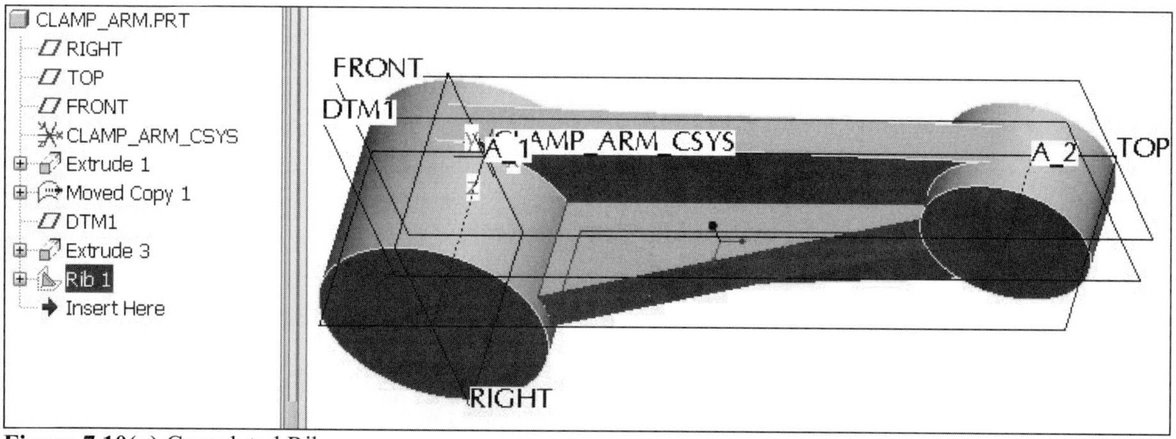

Figure 7.10(n) Completed Rib

Create the rounds, pick on the left extrusion and then on an edge until it highlights in red ⇒ press and hold the **Ctrl** key and pick on the other edges until they highlight ⇒ **RMB** ⇒ **Round Edges** [Fig. 7.11(a)] ⇒ modify the radius to **.04** 

⇒ **Enter** [Fig. 7.11(b)] ⇒ **MMB** ⇒ **LMB** ⇒ rotate the part ⇒ **View** ⇒ **Shade** ⇒ select the six edges [Fig. 7.11(c)] ⇒ **RMB** ⇒ **Round Edges** ⇒ modify the radius to **.04** ⇒ **Enter** ⇒ **MMB** ⇒ **LMB**

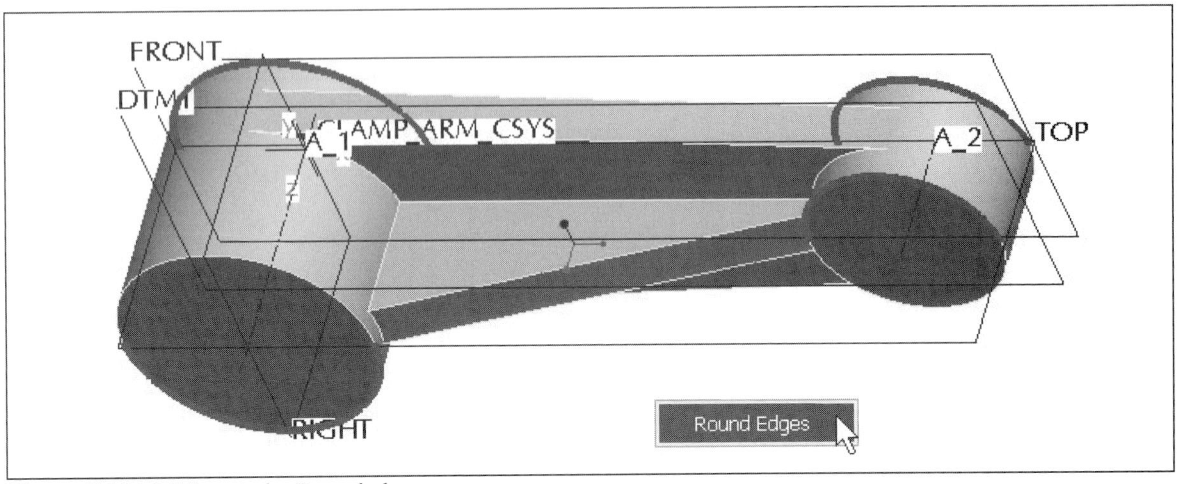

Figure 7.11(a) Edges to be Rounded

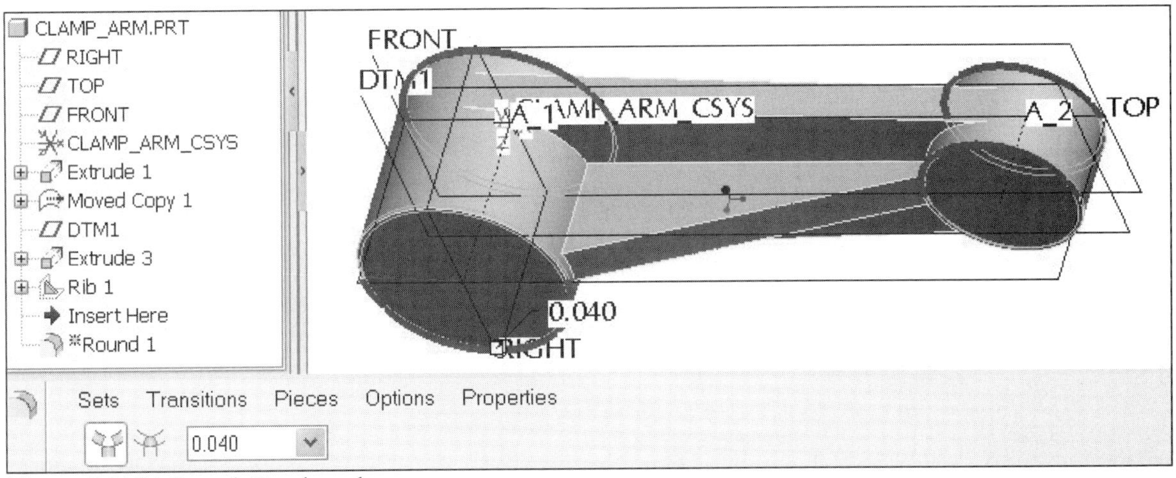

Figure 7.11(b) Rounds Previewed

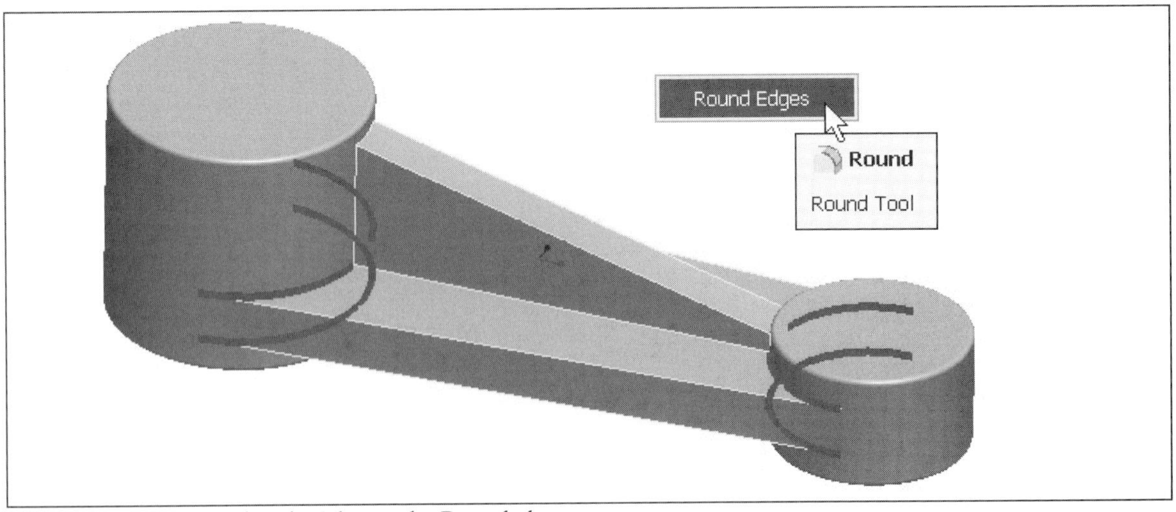

Figure 7.11(c) Select the Six Edges to be Rounded

Click: off ⇒ pick on the four edges ⇒ **RMB** ⇒ **Round Edges** [Fig. 7.11(d)] ⇒ modify the radius to **.04** ⇒ **Enter** ⇒ **RMB** ⇒ **Add set** [Fig. 7.11(e)] ⇒ press and hold the **Ctrl** key and pick on the remaining edges [Fig. 7.11(f)] (Note: the tangent edges will automatically be selected) ⇒ **Sets** tab [Fig. 7.11(g)] ⇒ **MMB** ⇒ on ⇒ **Ctrl+D** ⇒ **Ctrl+S** ⇒ **OK** ⇒ **LMB**

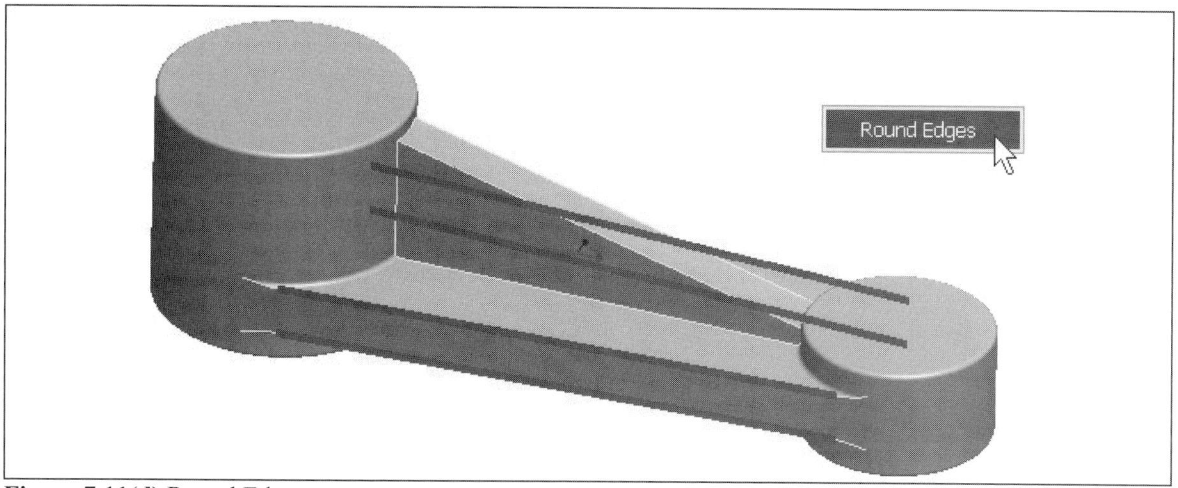

Figure 7.11(d) Round Edges

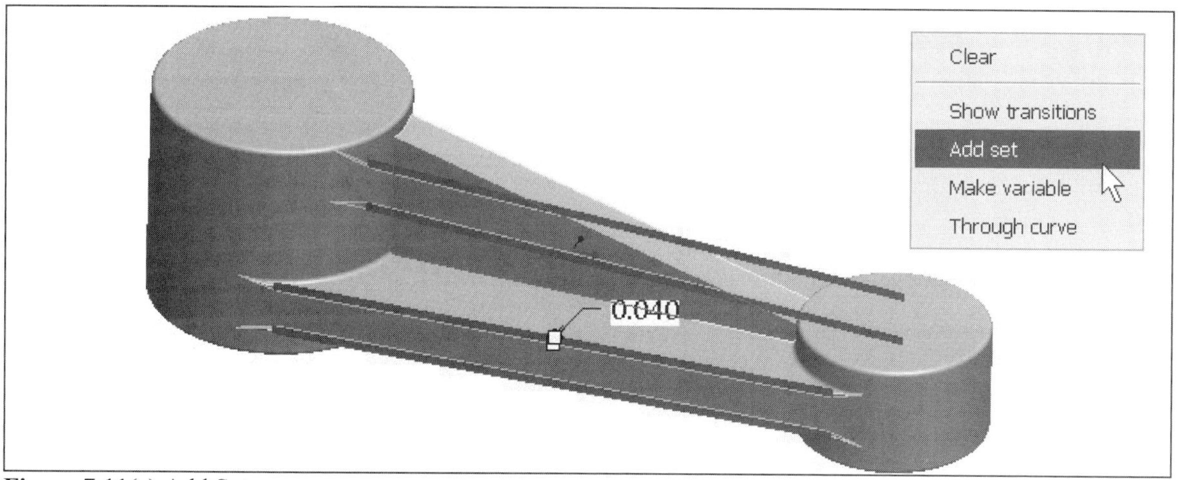

Figure 7.11(e) Add Set

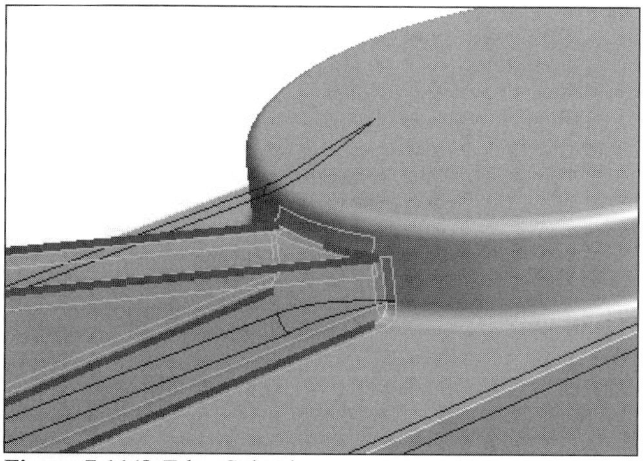

Figure 7.11(f) Edge Selection

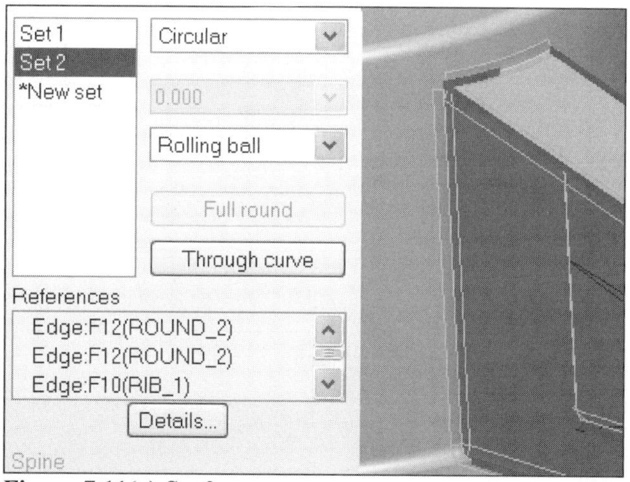

Figure 7.11(g) Set 2

The "machined" features of the part can be created with cuts to "face" the ends of the cylindrical protrusions. A standard tapped hole, and a thru hole with two countersinks are also added as "machined" features. You will be using the dimensions from the machine drawing (Fig. 7.12).

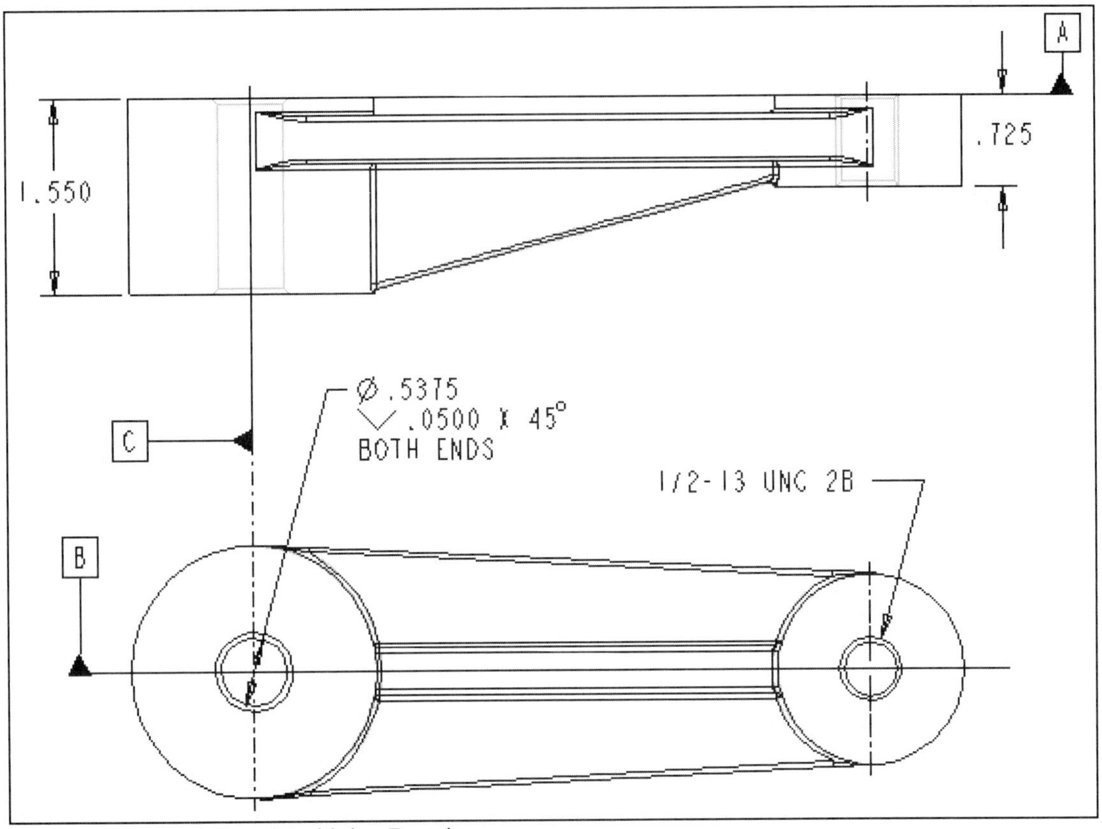

Figure 7.12 Clamp Arm Machining Drawing

Pick on datum **TOP** to pre-select it ⇒ ⇒ **RMB** ⇒ **Remove Material** ⇒ **RMB** ⇒ **Define Internal Sketch** ⇒ **Flip** [Sketch view direction Flip] [Fig. 7.13(a)] ⇒ **Sketch**

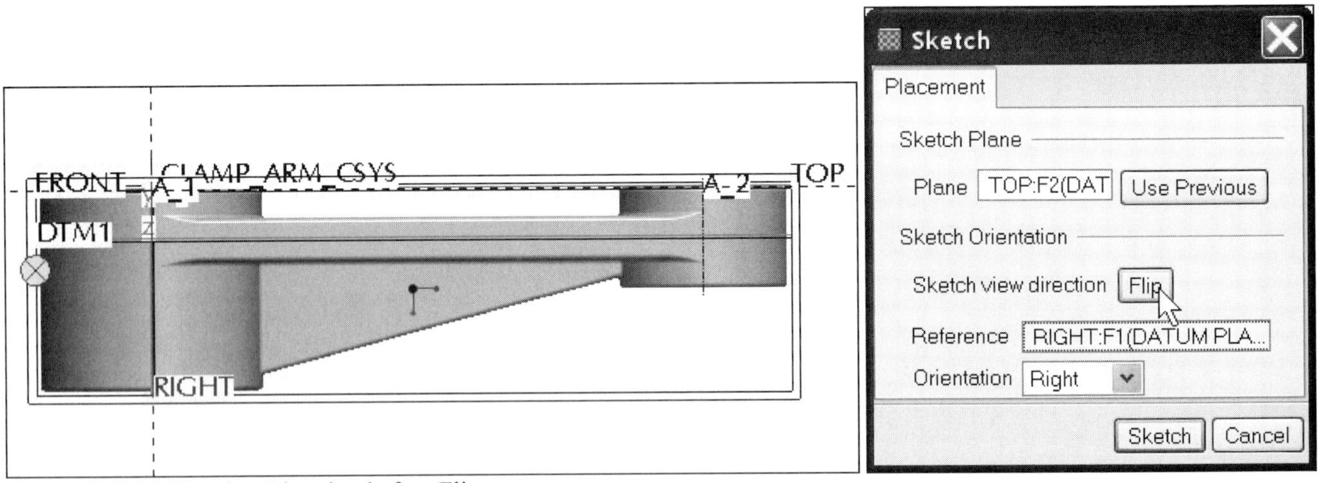

Figure 7.13(a) Viewing Direction before Flip

Click: off ⇒ **No hidden** ⇒ **Highlights vertices** ⇒ **Sketch** ⇒ **References** ⇒ ⇒ pick on the outside vertical edge of both cylindrical extrusions as references [Fig. 7.13(b)] ⇒ **RMB** ⇒ **Line** ⇒ zoom in as needed ⇒ draw a horizontal line from one vertical reference to the other ⇒ **MMB** ⇒ **MMB** ⇒ modify the dimension to **.100** [Fig. 7.13(c)] ⇒ **Enter** ⇒ on ⇒ ⇒ **Ctrl+D** ⇒ Options ⇒ Side 1 Through All ⇒ Side 2 Through All ⇒ flip material direction arrow [Fig. 7.13(d)]

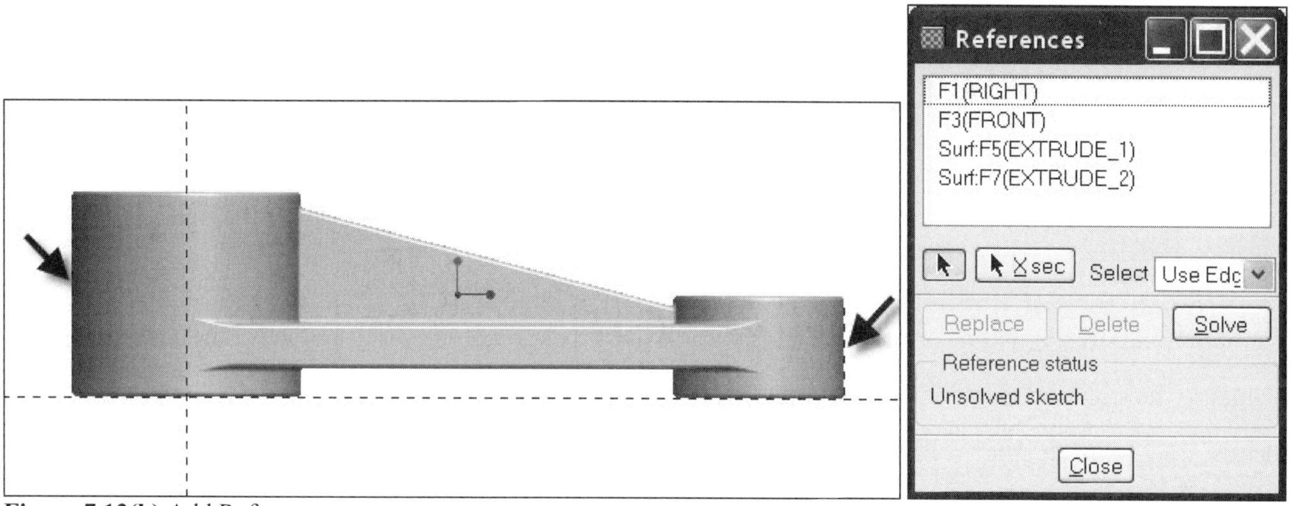

Figure 7.13(b) Add References

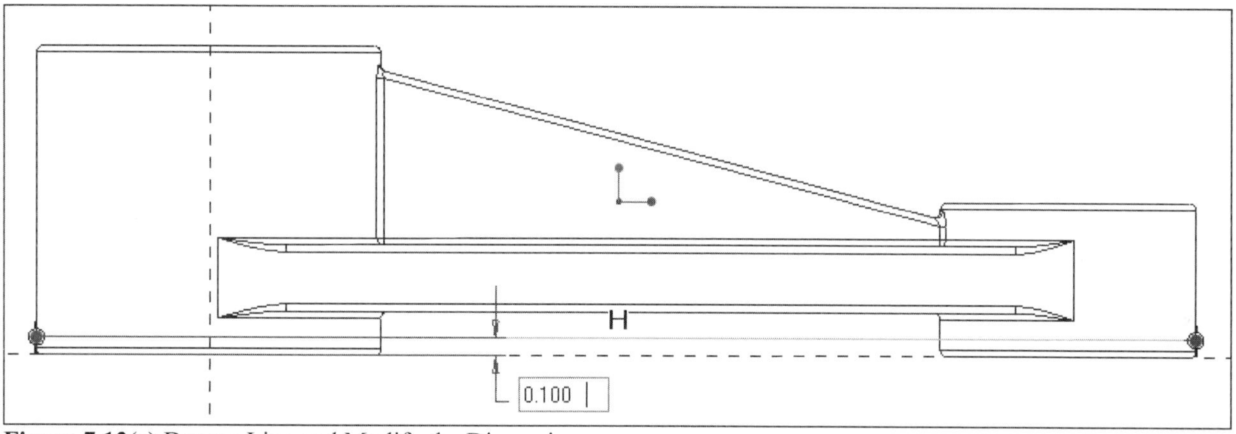

Figure 7.13(c) Draw a Line and Modify the Dimension

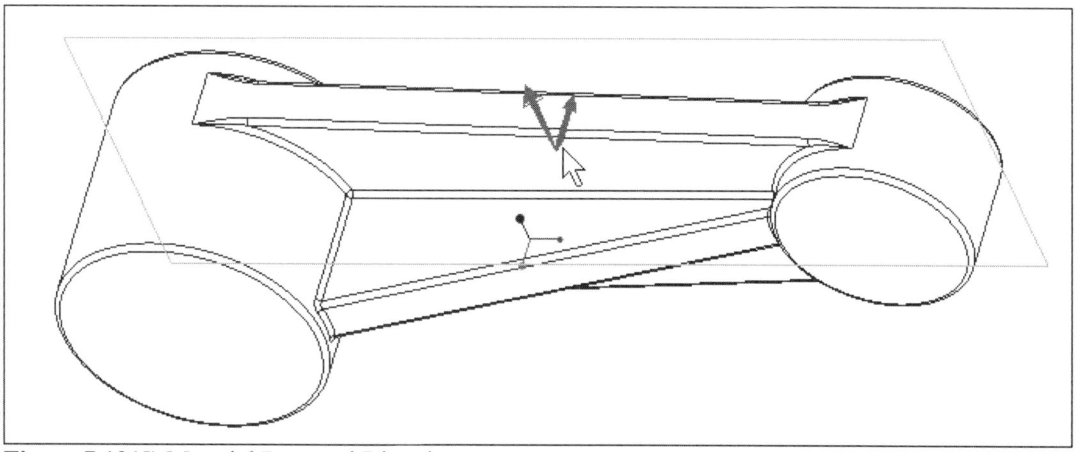

Figure 7.13(d) Material Removal Direction

Click: ⇒ ⇒ ⇒ **RMB** ⇒ **Edit** [Figs. 7.13(e-f)] ⇒ **LMB** to deselect

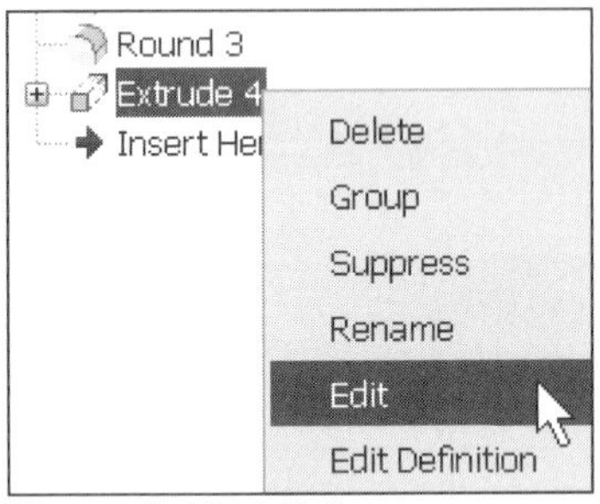

Figure 7.13(e) Edit

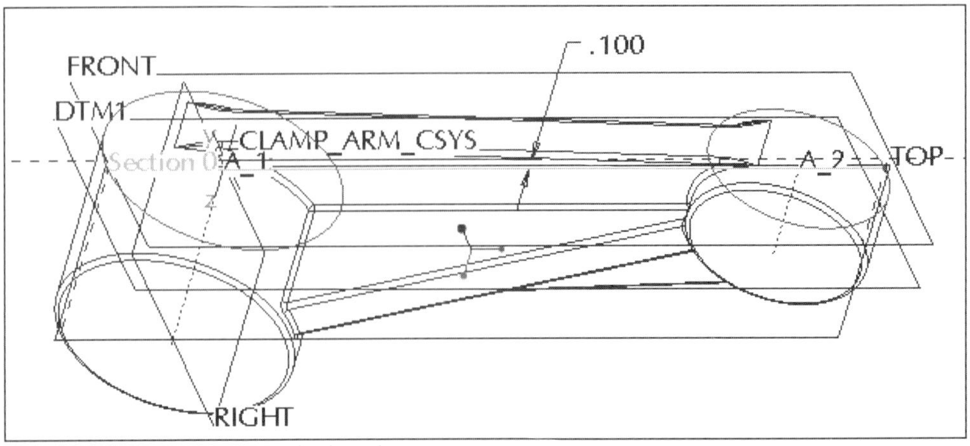

Figure 7.13(f) Dimension Displayed

Rotate the model to see its bottom surface cut ⇒ pick on the cut ⇒ move your cursor a small distance and pick again (the cut *surface* will highlight in pink) [Fig. 7.14(a)] ⇒ **Datum Plane Tool** from Right Toolchest ⇒ **Through** [Fig. 7.14(b)] ⇒ **MMB** ⇒ **Shading** ⇒ **Ctrl+S** ⇒ **OK** [Fig. 7.14(c)] ⇒ **LMB**

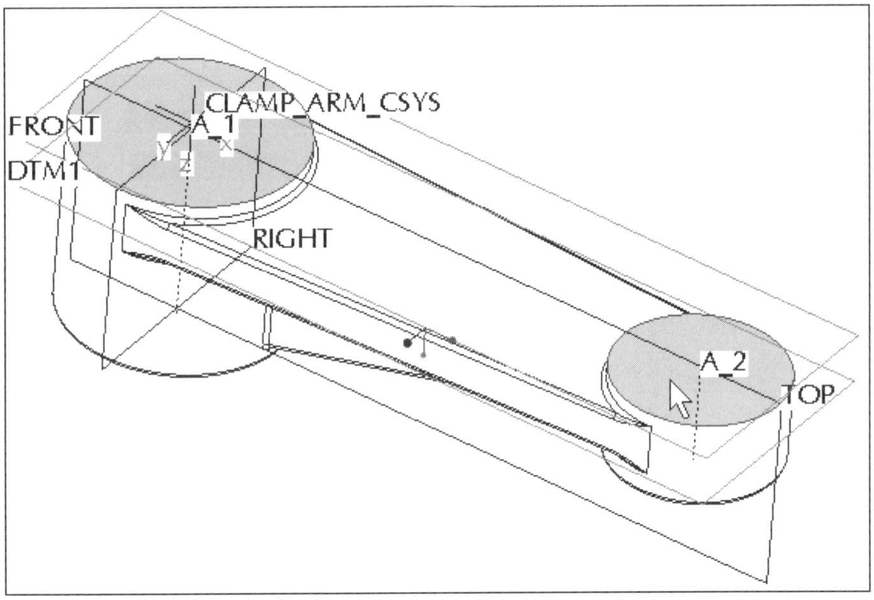

Figure 7.14(a) Select the Surface

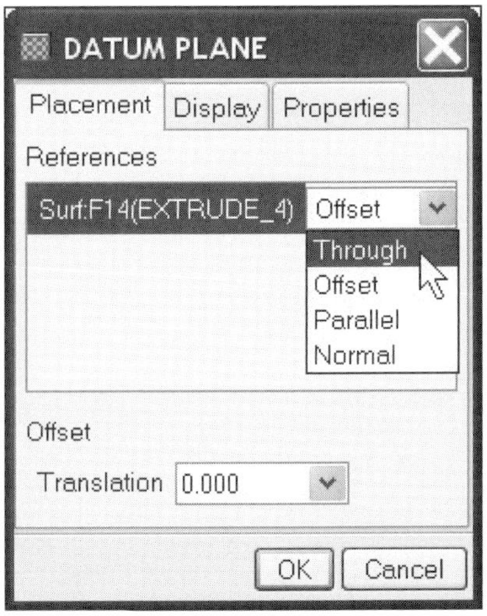

Figure 7.14(b) Datum Plane Through

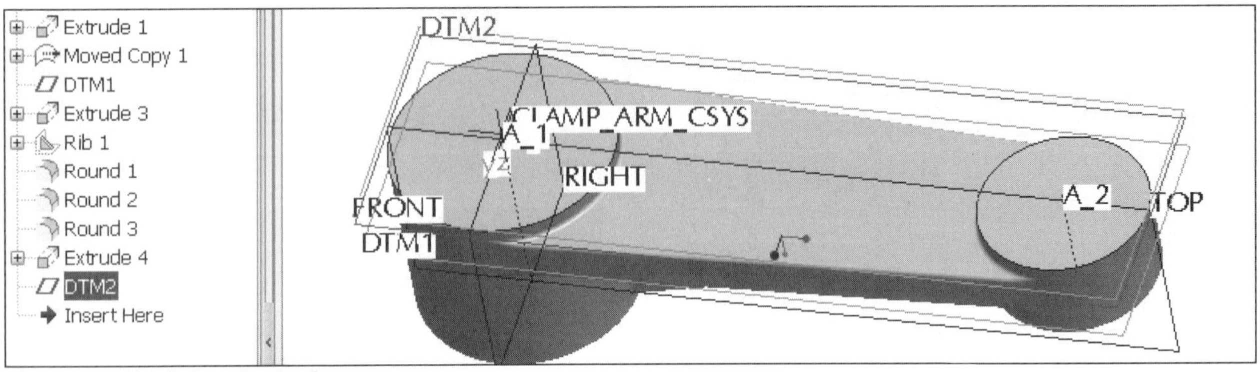

Figure 7.14(c) Select the Surface

Set the datums for geometric tolerancing, select **DTM2** from the model or Model Tree ⇒ **RMB** ⇒ **Properties** ⇒ Name- type **A** ⇒ ⇒ **OK** ⇒ pick **TOP** ⇒ **RMB** ⇒ **Properties** ⇒ Name- type **B** ⇒ ⇒ **OK** ⇒ pick **RIGHT** ⇒ **RMB** ⇒ **Properties** ⇒ Name- type **C** ⇒ ⇒ **OK** ⇒ off [Fig. 7.15] ⇒ **LMB** ⇒ pick datum **A** ⇒ **Datum Plane Tool** from Right Toolchest ⇒ move the drag handle towards the top of the rib ⇒ **Offset** ⇒ Offset Translation **1.550** [Fig. 7.16(a)] ⇒ **OK** ⇒ on [Fig. 7.16(b)]

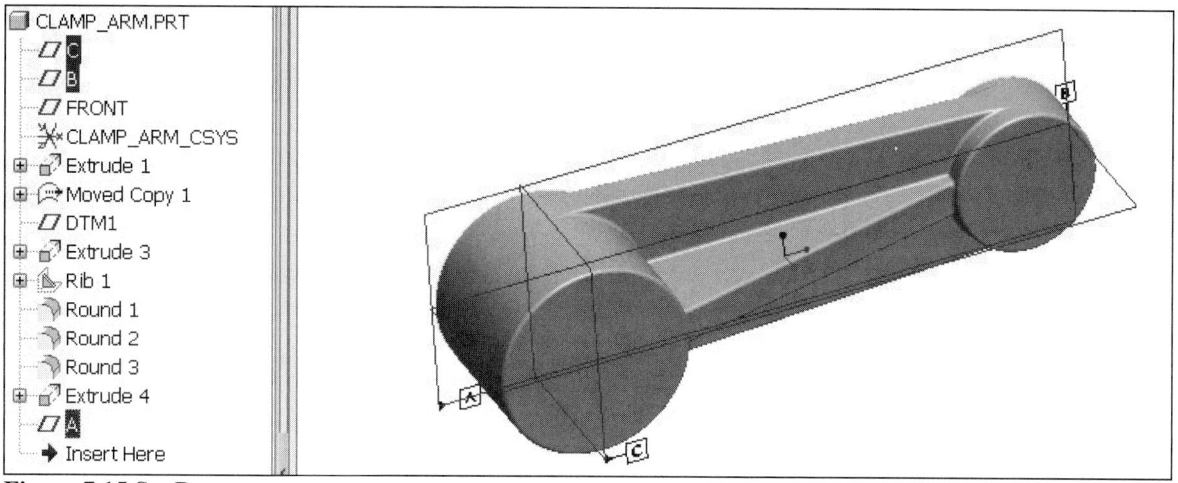

Figure 7.15 Set Datums

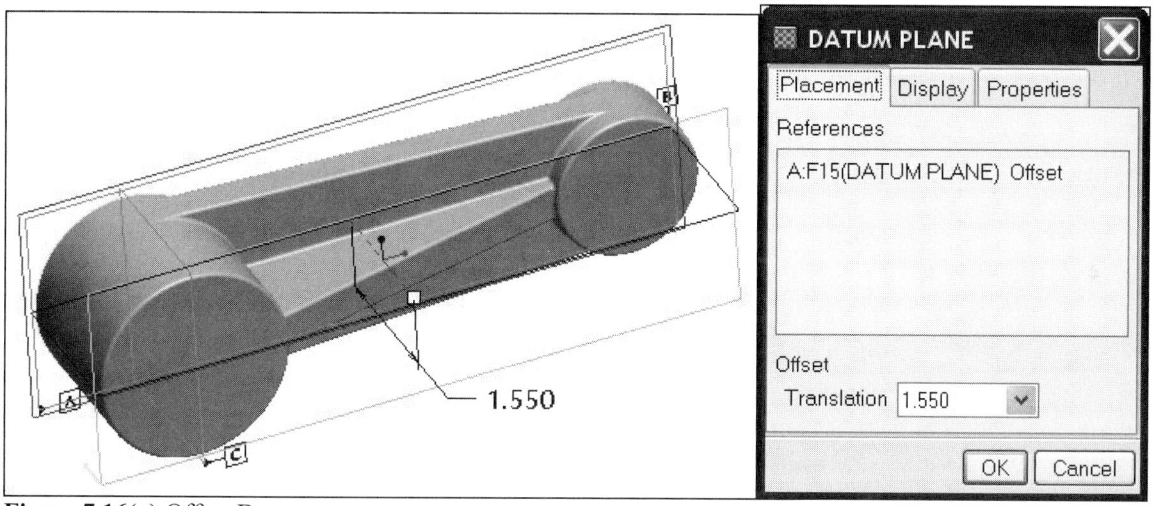

Figure 7.16(a) Offset Datum

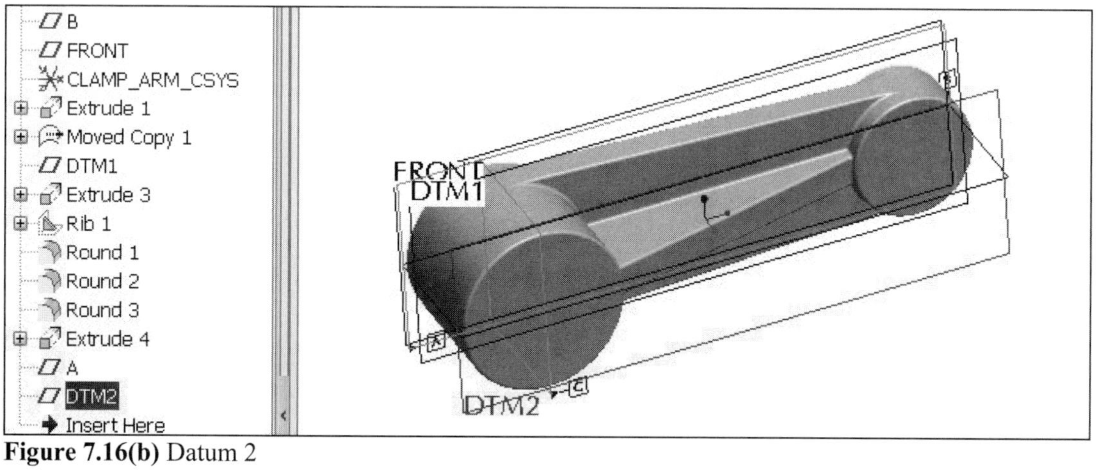

Figure 7.16(b) Datum 2

With **DTM2** selected, click: ⇒ **RMB** ⇒ **Define Internal Sketch** ⇒ Reference **B (Datum Plane)** ⇒ Orientation **Bottom** [Fig. 7.17(a)] ⇒ **Sketch** to close the Sketch dialog box [Fig. 7.17(b)] ⇒ **Sketch** from menu bar ⇒ **References** ⇒ **Ctrl+D** ⇒ pick on the circular surface of the large protrusion to add as a reference [Fig. 7.17(c)] ⇒ **Close** [Figs. 7.17(d-e)] ⇒ **No hidden**

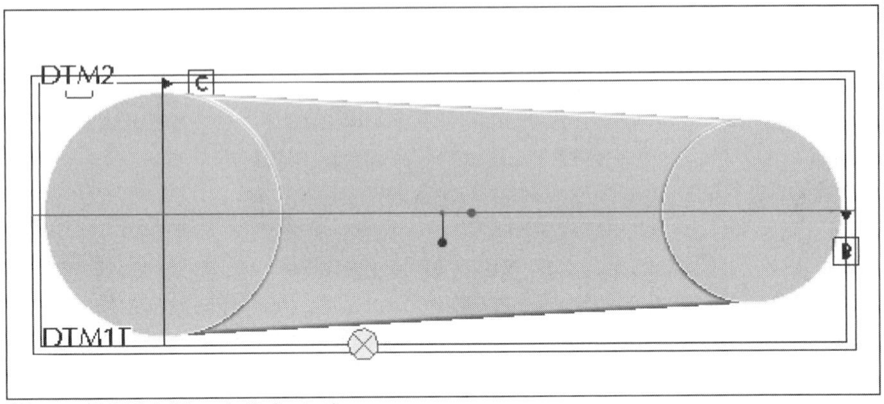

Figure 7.17(a) Sketch Orientation

Figure 7.17(b) Sketch Dialog Box

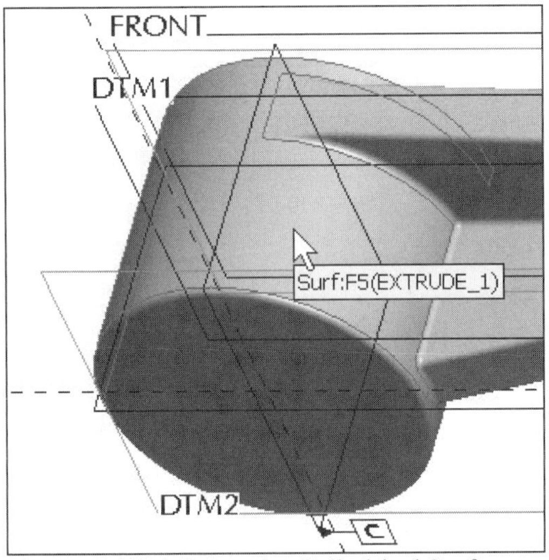

Figure 7.17(c) Select the Cylindrical Surface

Figure 7.17(d) References

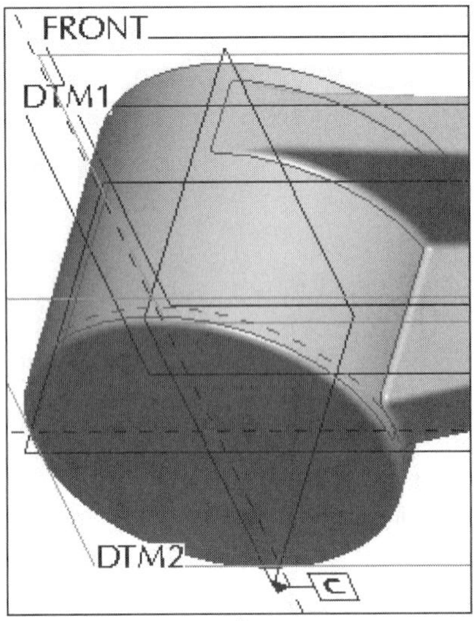

Figure 7.17(e) Reference

Click: [icon] ⇒ **RMB ⇒ Circle ⇒** sketch a circle by picking the center (at references) [Fig. 7.17(f)] and one edge (on surface reference) [Fig. 7.17(g)] *(the circle is locked into references, no dimensions will be displayed, if you get dimensions, redo the circle)* [Fig. 7.17(h)] ⇒ [icon] ⇒ **RMB ⇒ Remove Material ⇒ RMB ⇒ Flip Depth Direction** [Fig. 7.17(i)] ⇒ [icon] ⇒ [icon] Fig. 7.17(j)] ⇒ **Enter** [Fig. 7.17(k)] ⇒ **LMB**

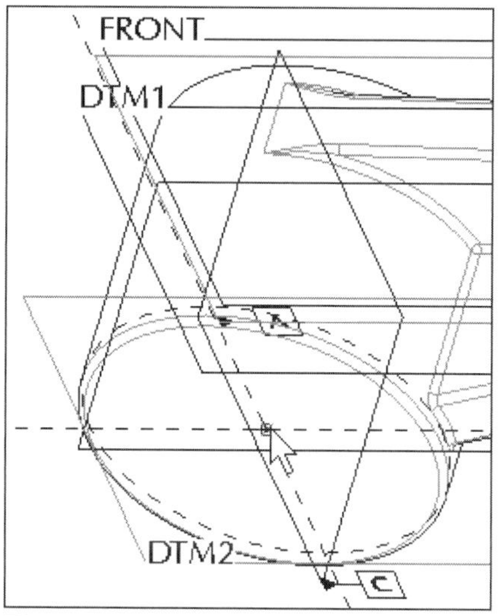

Figure 7.17(f) Pick the Center

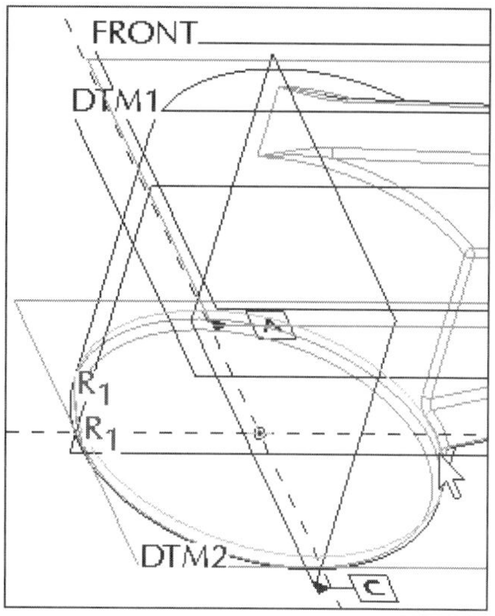

Figure 7.17(g) Pick the Edge References

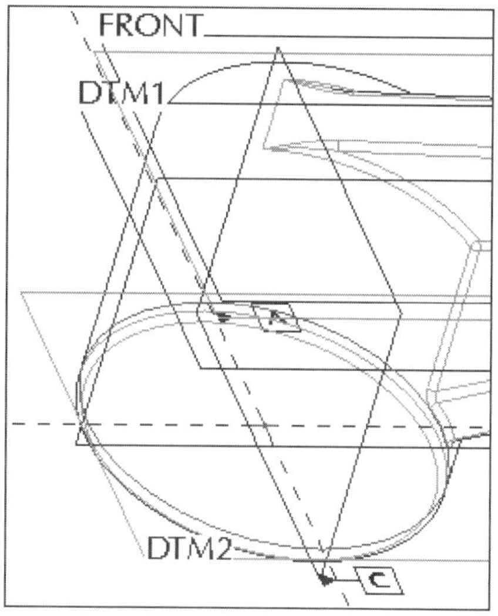

Figure 7.17(h) Circle

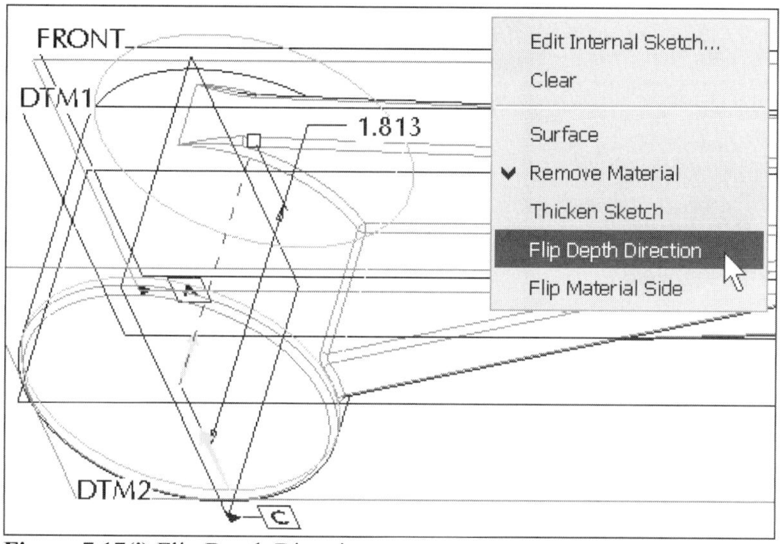

Figure 7.17(i) Flip Depth Direction

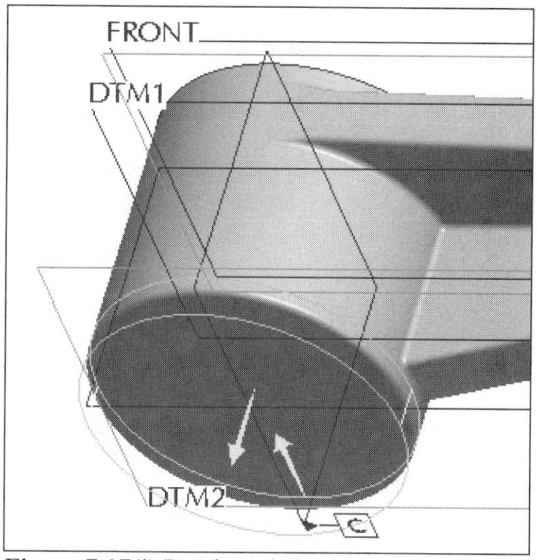

Figure 7.17(j) Previewed Cut

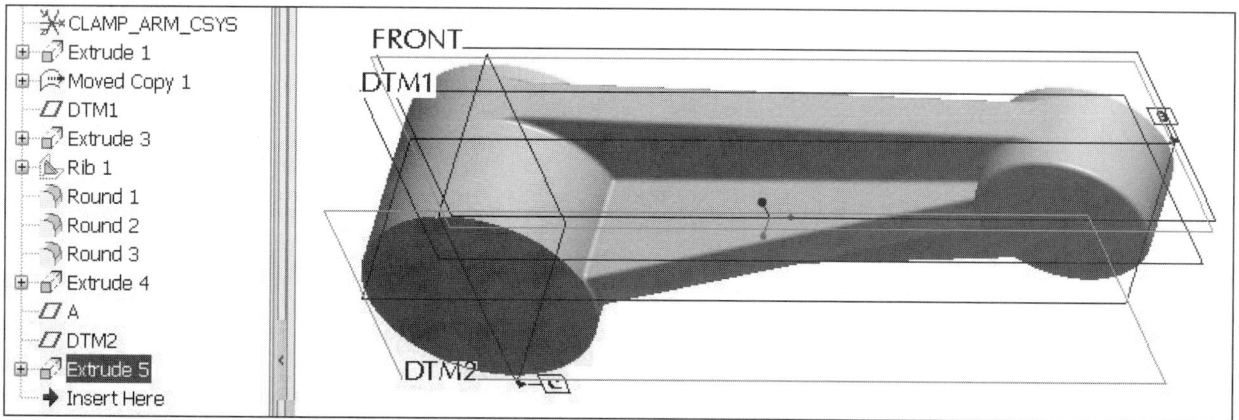

Figure 7.17(k) Completed Cut

Click: ⇒ ⇒ **RMB** ⇒ **Remove Material** ⇒ **RMB** ⇒ **Define Internal Sketch** ⇒ **Datum Plane Tool** ⇒ pick Datum **A** ⇒ move the drag handle to the opposite side ⇒ double-click on the dimension ⇒ type **.725** ⇒ **Enter** [Fig. 7.17(l)] ⇒ **OK** ⇒ Reference **B (Datum Plane)** ⇒ Orientation **Top** [Fig. 7.17(m)] ⇒ **Sketch** to close dialog ⇒ ⇒ **Sketch** from menu bar ⇒ **References** ⇒ rotate and zoom in on the feature ⇒ pick on the circular surface of the small protrusion [Fig. 7.17(n)] ⇒ **Close**

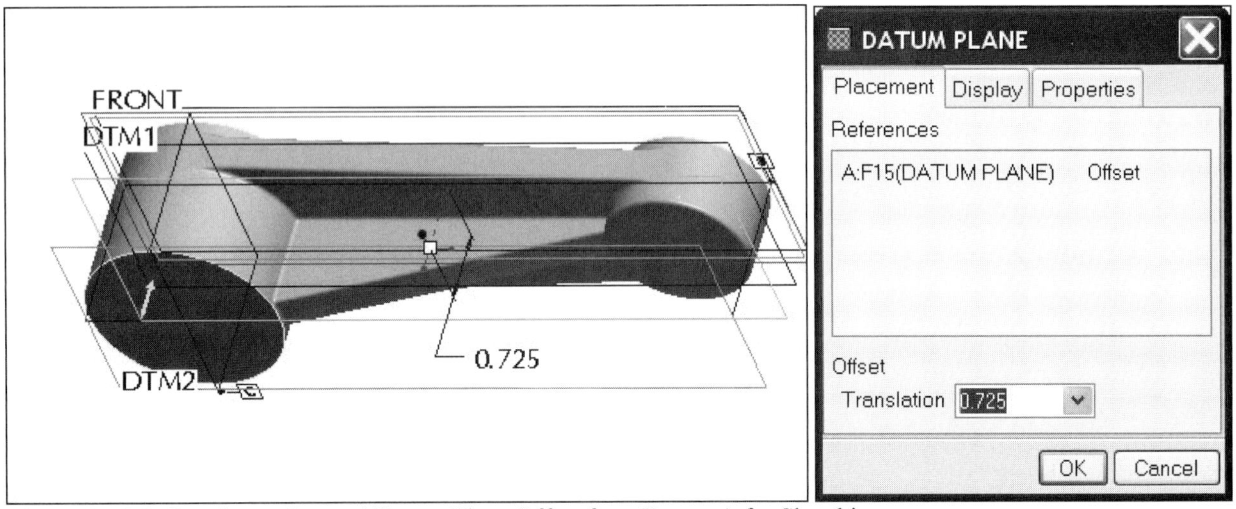

Figure 7.17(l) Creating an Internal Datum Plane Offset from Datum A for Sketching

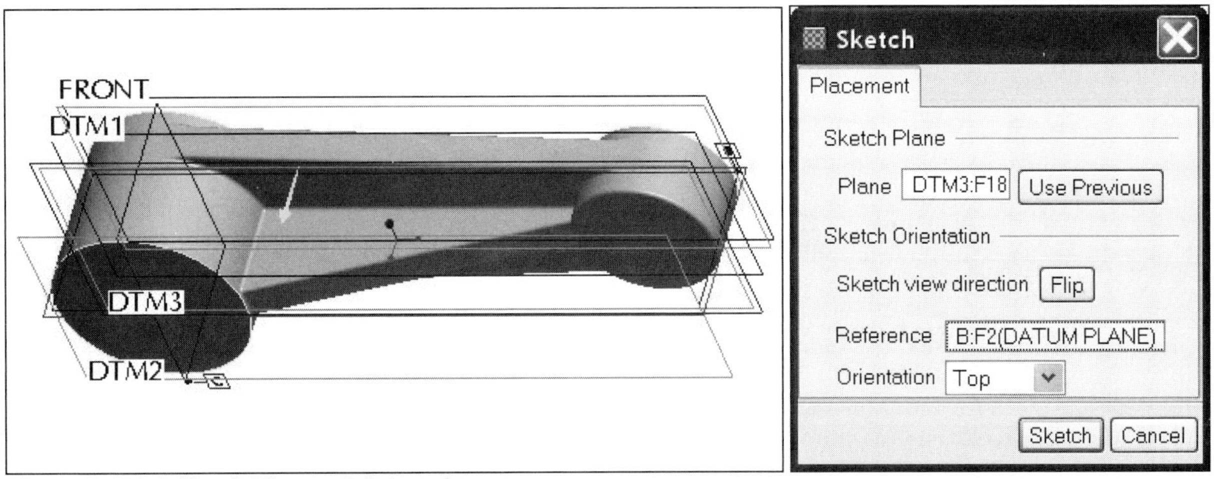

Figure 7.17(m) Sketch Plane and Orientation

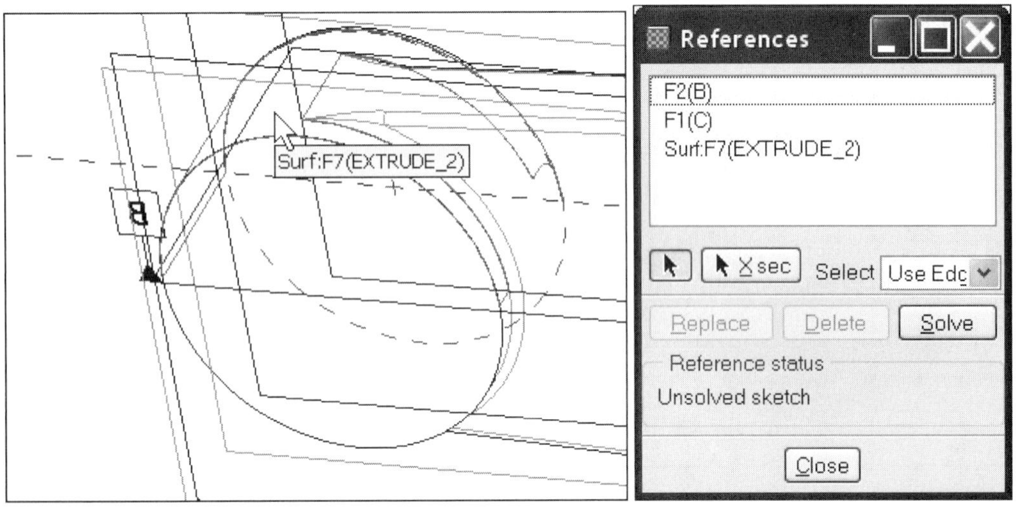

Figure 7.17(n) References

Click: ⇒ **RMB** ⇒ **Circle** ⇒ sketch a circle by picking the center (at references) and one edge (on surface reference) [Fig. 7.17(o)] *(the circle is locked into references, no dimensions will be displayed, if you get dimensions, redo the circle)* [Fig. 7.17(p)] ⇒ **Ctrl+D** ⇒ **MMB** [Fig. 7.17(q)] ⇒ [Fig. 7.17(r)] ⇒ ⇒ [Fig. 7.17(s)] ⇒ **LMB** ⇒ ⇒ **OK** ⇒ **File** ⇒ **Delete** ⇒ **Old Versions** ⇒

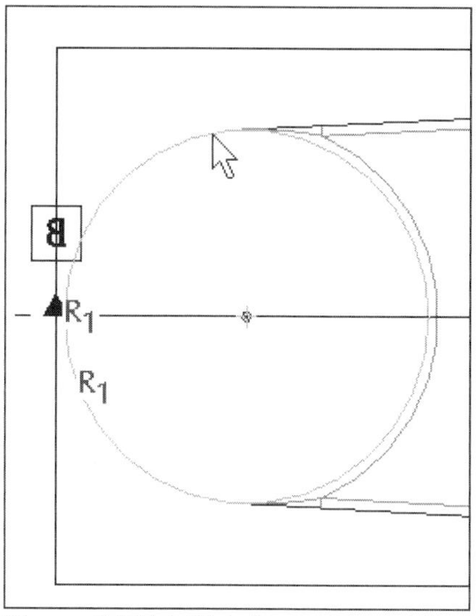

Figure 7.17(o) Pick the center and the Edge

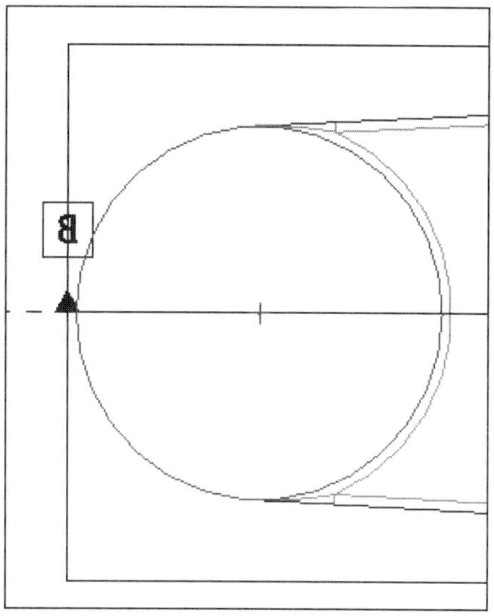

Figure 7.17(p) Circle

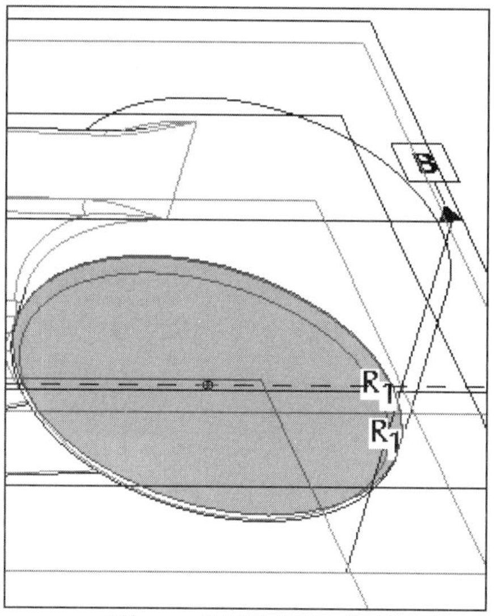

Figure 7.17(q) Shaded Section

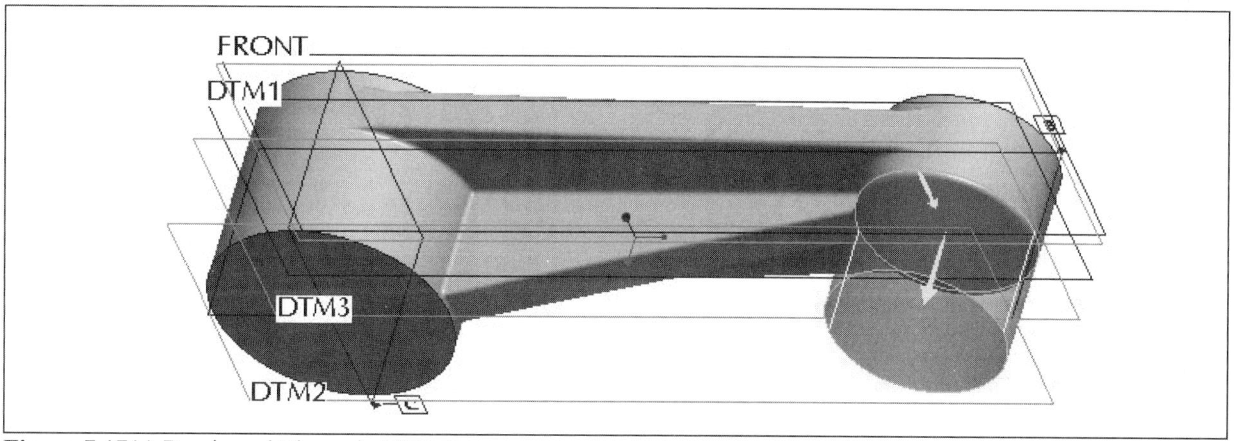

Figure 7.17(r) Previewed Through All Cut

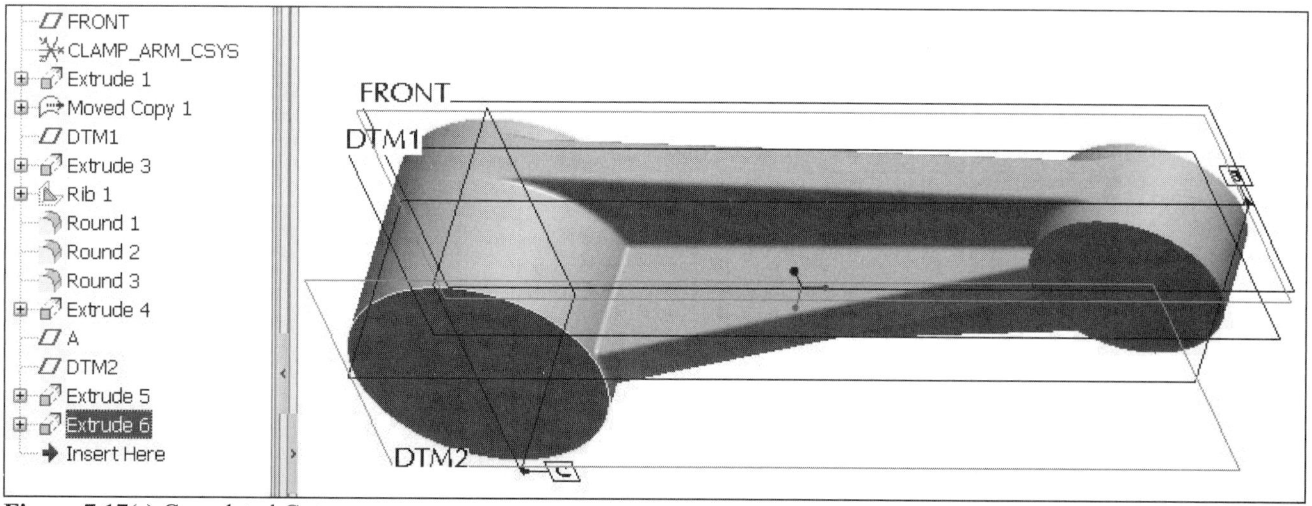

Figure 7.17(s) Completed Cut

Measuring Geometry

Using analysis measure, you can measure model geometry with one of the following commands:

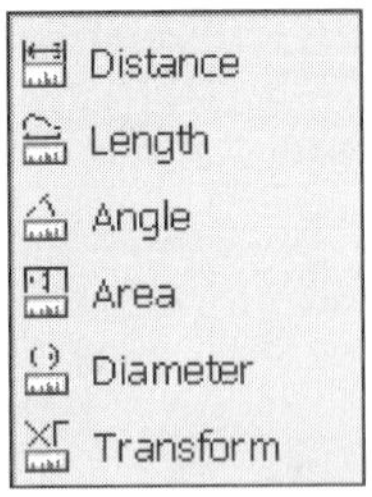

- **Distance** Displays the distance between two entities
- **Length** Displays the length of the curve or edge
- **Angle** Displays the angle between two entities
- **Area** Displays the area of the selected surface, quilt, facets, or an entire model
- **Diameter** Displays the diameter of the surface
- **Transform** Displays a note showing the transformation matrix values between two coordinate systems

Click: **Analysis** ⇒ **Measure** ⇒ **Distance** ⇒ pick the top of the large circular protrusion [Fig. 7.18(a)] ⇒ place your cursor over the bottom and click **RMB** until it highlights [Fig. 7.18(b)] ⇒ **LMB** to select *(measures the height- 1.550)* [Figs. 7.18(c-d)] ⇒ ⇒ Repeat the process to measure the shorter end ⇒ (The dimensions are the design sizes specified in Figure 7.12, **1.550** and **.725** respectively)

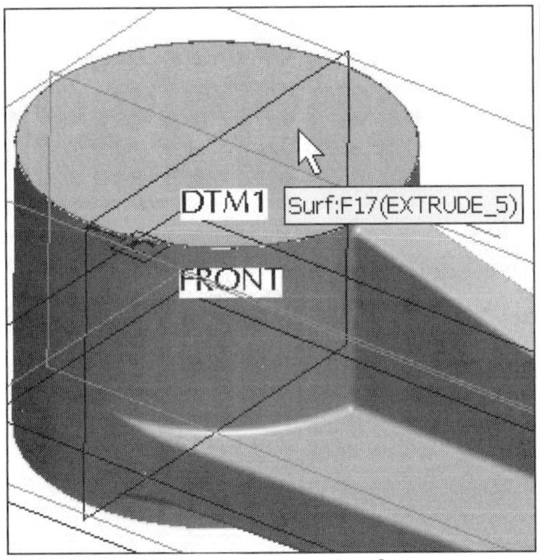

Figure 7.18(a) Select Top Surface

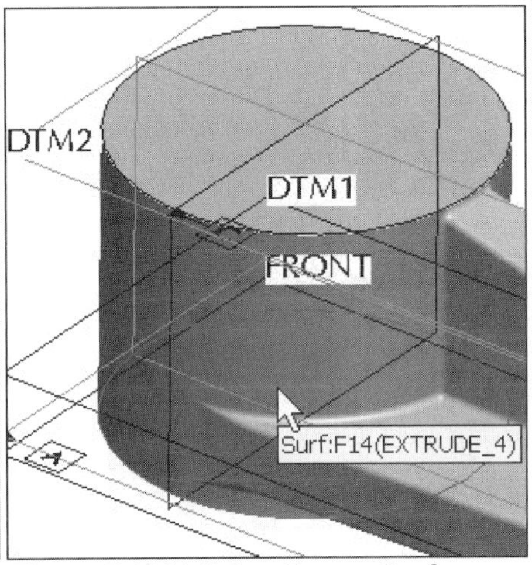

Figure 7.18(b) Select Bottom Surface

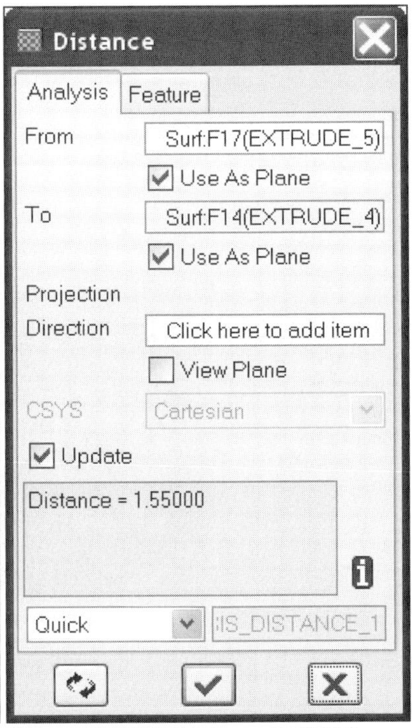

Figure 7.18(c) Distance **1.55000**

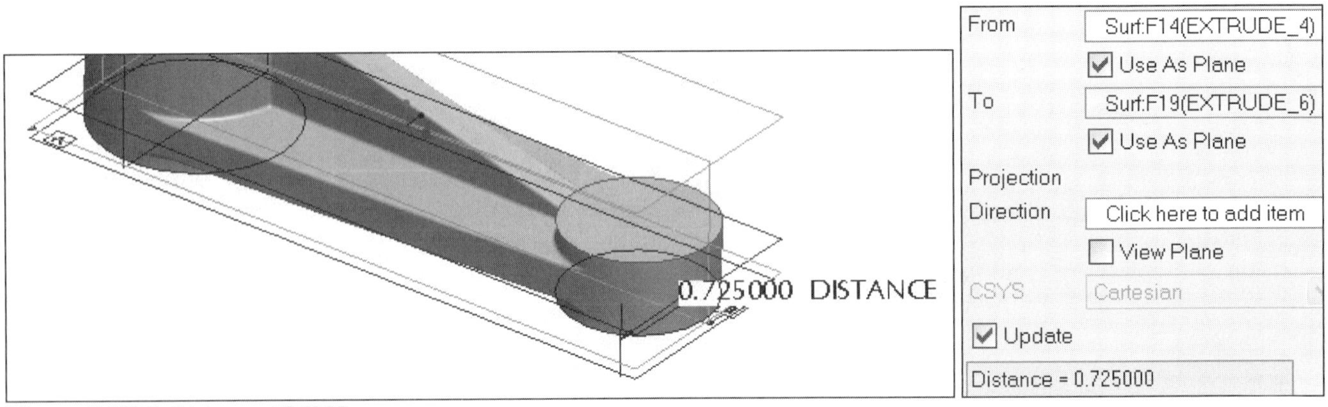

Figure 7.18(d) Distance **.725000**

Flexing the Model

During the design of a component there are modifications made to the design. The ability to make changes without causing failures is important. "Flexing" the model; changing and editing dimension values to see if the model integrity withstands these modifications, establishes your designs robustness. Modify the part's geometry and observe the change.

Click: **MMB** to rotate the part ⇒ pick on the rib ⇒ **RMB** ⇒ **Edit** ⇒ double-click on the **.375** rib width dimension [Fig. 7.19(a)] ⇒ modify to **1.00** [Fig. 7.19(b)] ⇒ **Enter** ⇒ 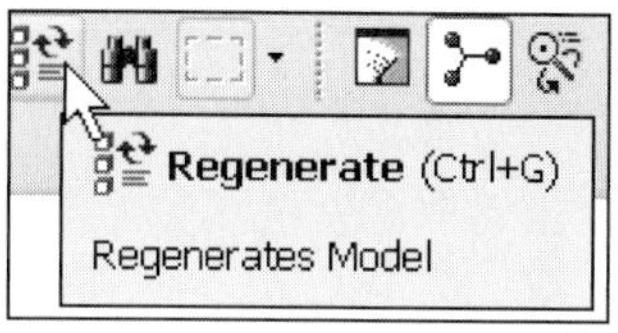

[Figs. 7.19(c-d)] ⇒  **Undo** ⇒ Try other dimensional changes to see how your model reacts, If you get a failure, use **Undo Changes** ⇒ **Confirm**.

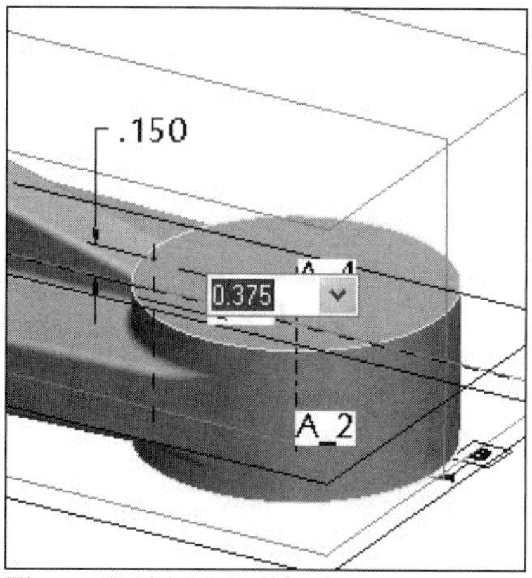

Figure 7.19(a) Modify the **.375** Dimension

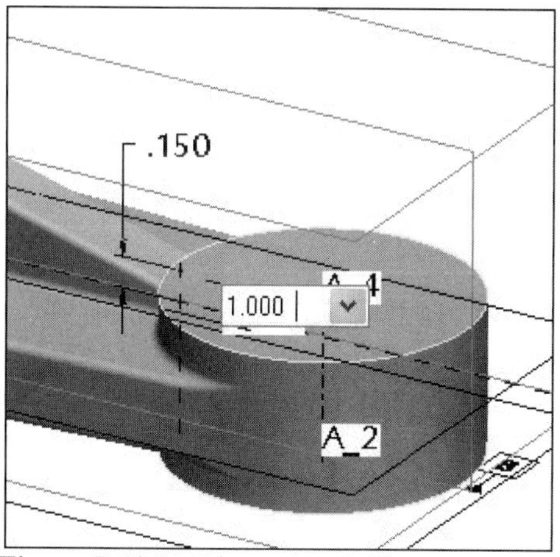

Figure 7.19(b) Dimension **1.00**

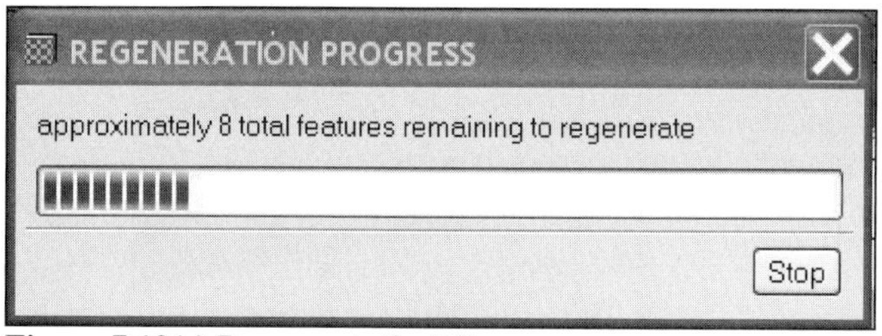

Figure 7.19(c) Regenerate the Rib

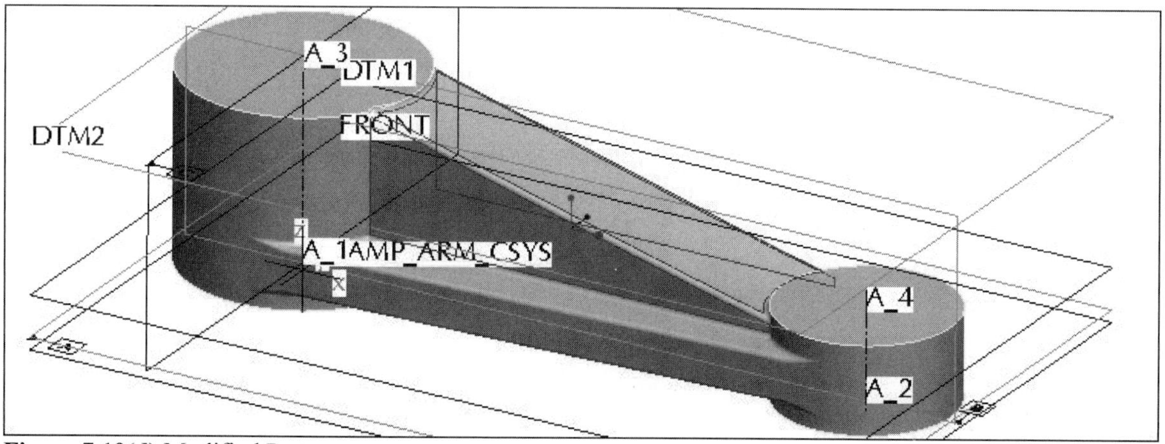

Figure 7.19(d) Modified Part

Click: **Ctrl+S** ⇒ **Enter** ⇒ ⇒ ⇒ ⌀ 0.5375 ⇒ **Placement** ⇒ Primary: pick axis A_1 [Fig. 7.20(a)] ⇒ hold down **Ctrl** key ⇒ pick top surface [Fig. 7.20(b)] ⇒ **Shape** [Fig. 7.20(c)] ⇒ **Enter**

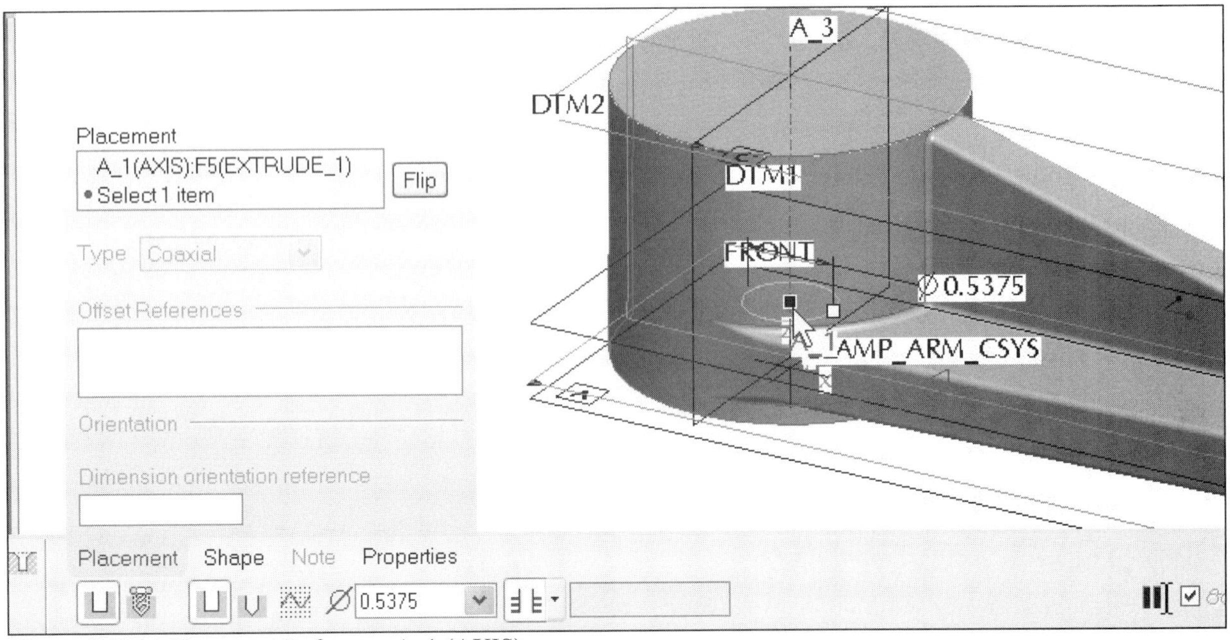

Figure 7.20(a) Placement Reference A_1 (AXIS)

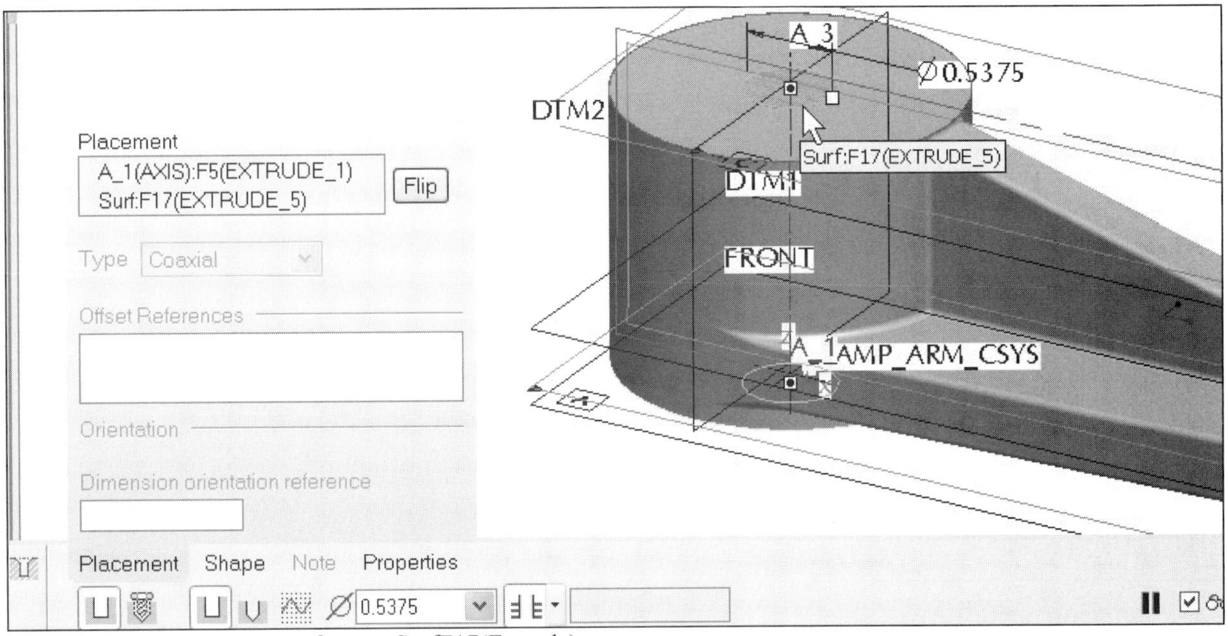

Figure 7.20(b) Placement Reference SurfF17(Extrude)

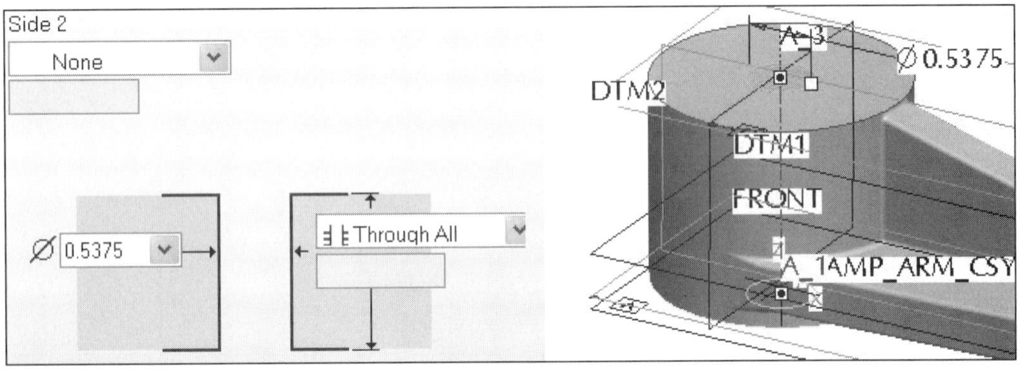

Figure 7.20(c) Shape

Click: **Ctrl+S** ⇒ **OK** ⇒ 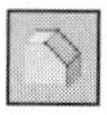 **Chamfer Tool** ⇒ 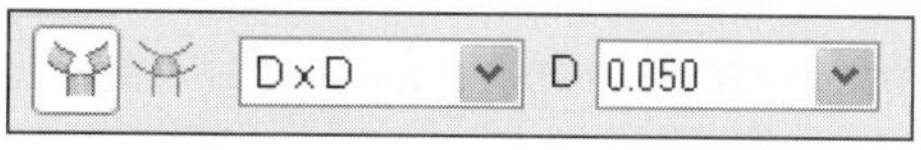

⇒ **Sets** tab ⇒ press and hold **Ctrl** and pick the top and bottom edges of the hole [Fig. 7.20(d)] ⇒  ⇒ **Ctrl+S** ⇒ **Enter**

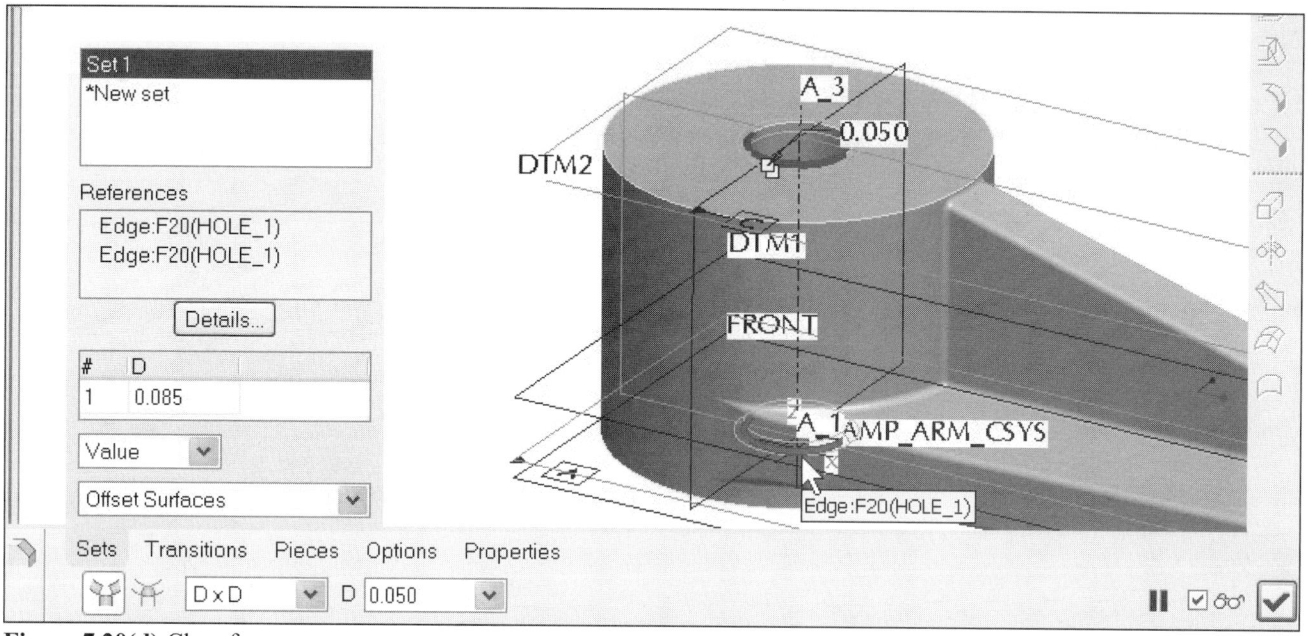

Figure 7.20(d) Chamfers

Click:  **Hole Tool** ⇒ pick axis **A_2** [Fig. 7.21(a)]

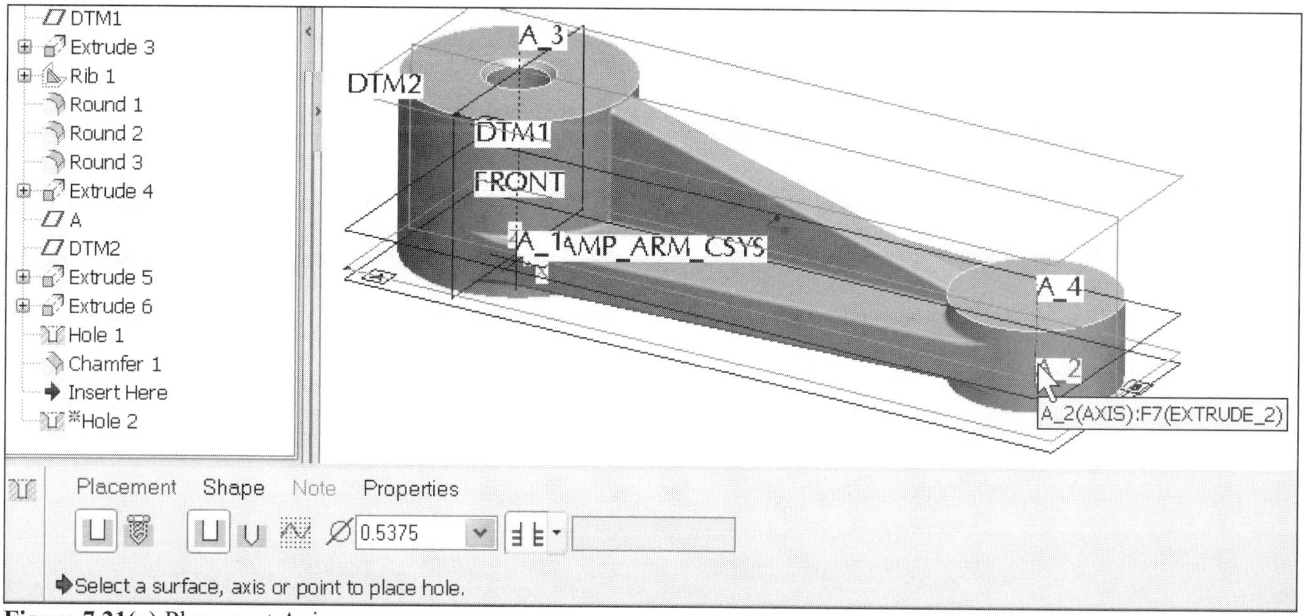

Figure 7.21(a) Placement Axis

Press and hold **Ctrl** and pick the top surface of the small cylinder ⇒ **Placement** tab [Fig. 7.21(b)] ⇒ **Create standard hole** ⇒ 1/2-13 ⇒ ⇒ **Adds countersink** ⇒ UNC 1/2-13 ⇒ **Shape** tab [Fig. 7.21(c)] ⇒ Thru Thread ⇒ Exit Countersink ⇒ 0.5625 top and bottom chamfer [Fig. 7.21(c)]

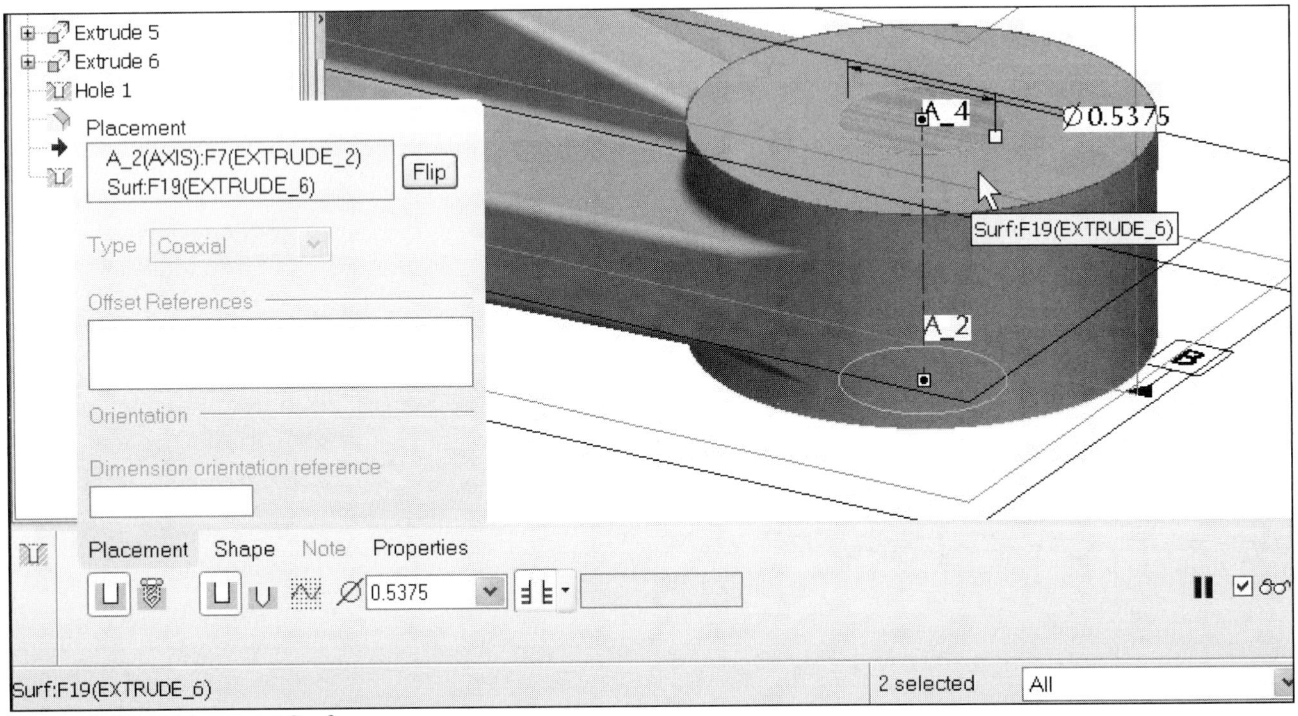

Figure 7.21(b) Placement Surface

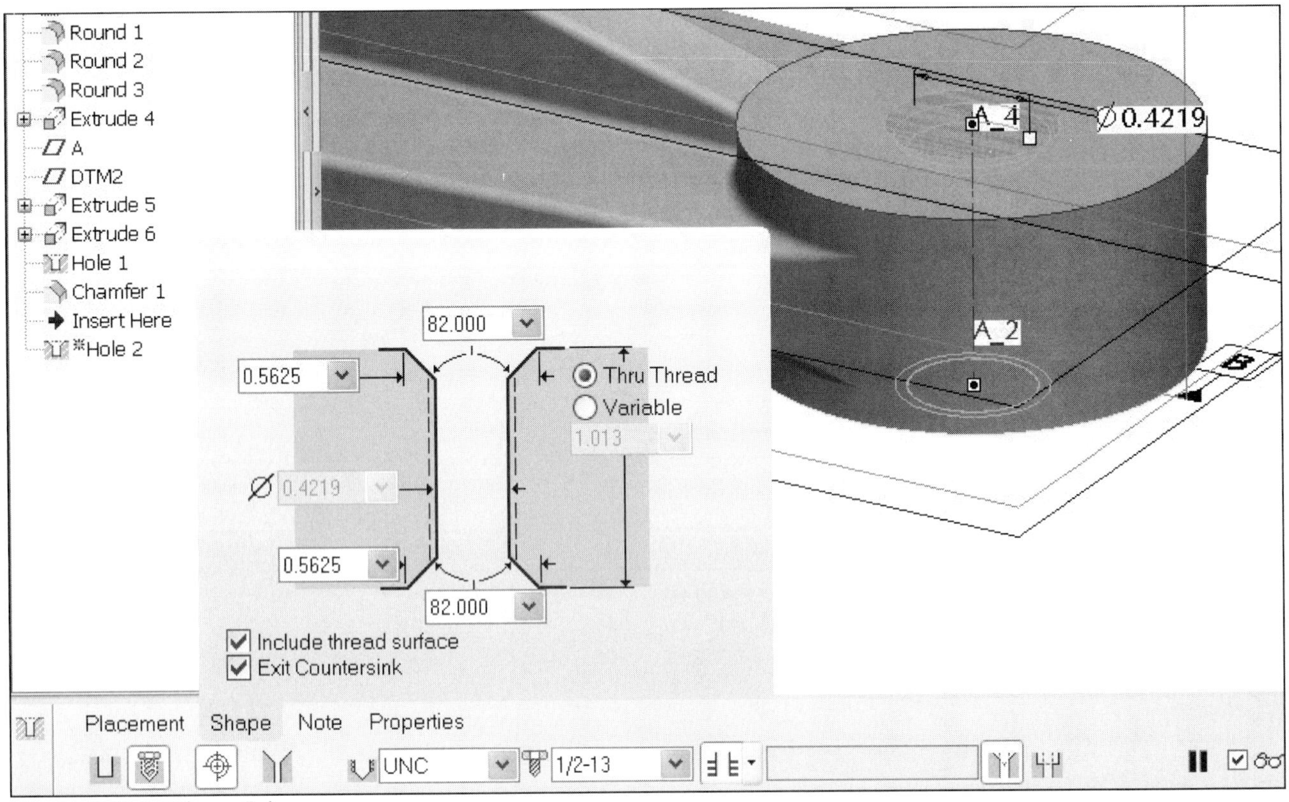

Figure 7.21(c) Shape Tab

Click: **Note** tab [Fig. 7.21(d)] ⇒ (if you get a failure, open the Shape tab and check your options and dimensions) ⇒ **Settings** ⇒ **Tree Filters** ⇒ Annotations [Fig. 7.21(e)] ⇒ **OK** ⇒ Hole 2 Note_0 [Fig. 7.21(f)] ⇒ **Datum planes** off ⇒ **Turn off 3D Annotations** [the note and the datum plane tags (set datums) will not display] ⇒ **Ctrl+D** ⇒ **Ctrl+S** ⇒ **OK** ⇒ **LMB** to deselect

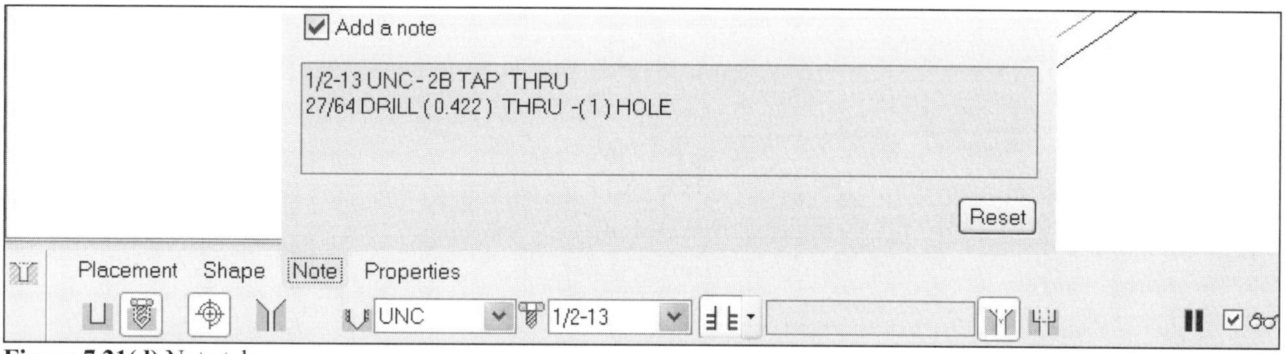

Figure 7.21(d) Note tab

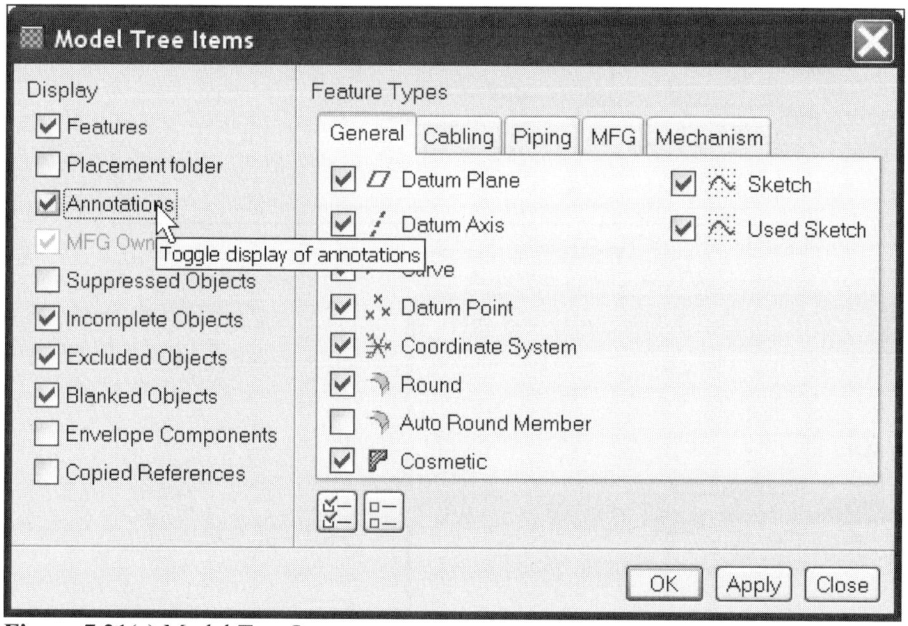

Figure 7.21(e) Model Tree Items

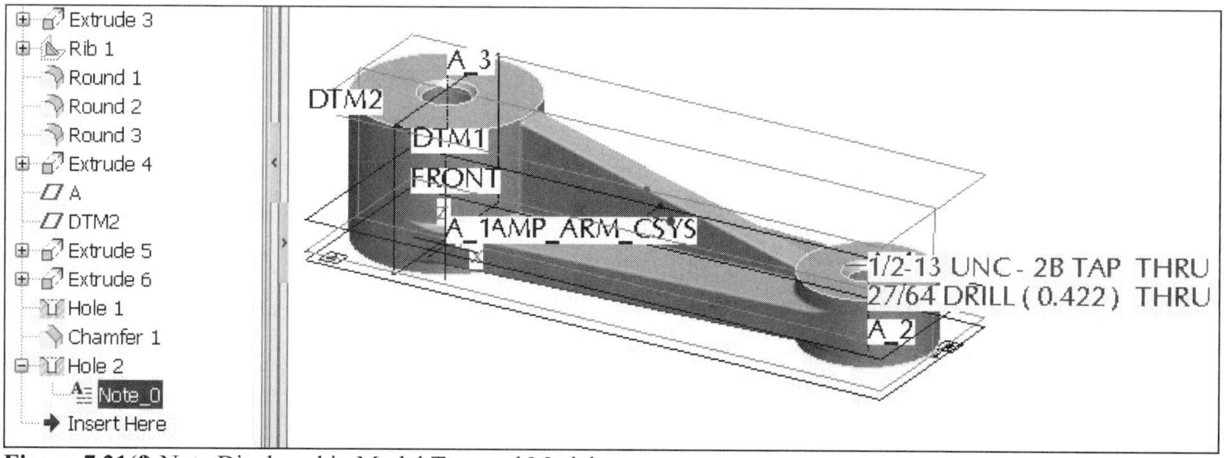

Figure 7.21(f) Note Displayed in Model Tree and Model

Write a relation to keep the thickness of the rib the same as that for the "web". Click: **Hidden line** ⇒ with the **Ctrl** key pressed, pick on the "web" extrusion and the rib in the Model Tree or on the model ⇒ **RMB** ⇒ **Edit** ⇒ **Info** ⇒ Switch Dimensions [Fig. 7.22(a)] ⇒ **Tools** ⇒ **Relations** ⇒ type **d8=d7** [Fig. 7.22(b)] *(your "**d**" values may be different)* ⇒ **Execute/Verify** ⇒ **OK** ⇒ **Ok** ⇒ **Info** ⇒ Switch Dimensions ⇒ ⇒ **LMB** to deselect ⇒ **Regenerates Model** ⇒ **Ctrl+S** ⇒ **Enter**

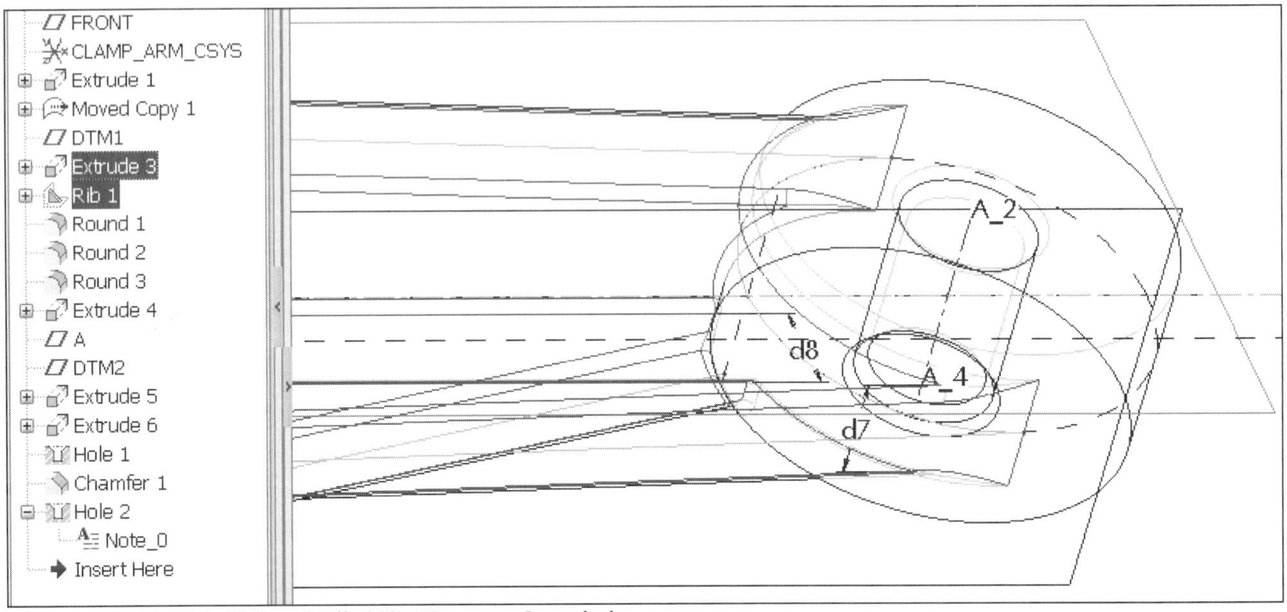

Figure 7.22(a) Web (**d7**) and Rib (**d8**). Use your **d** symbols.

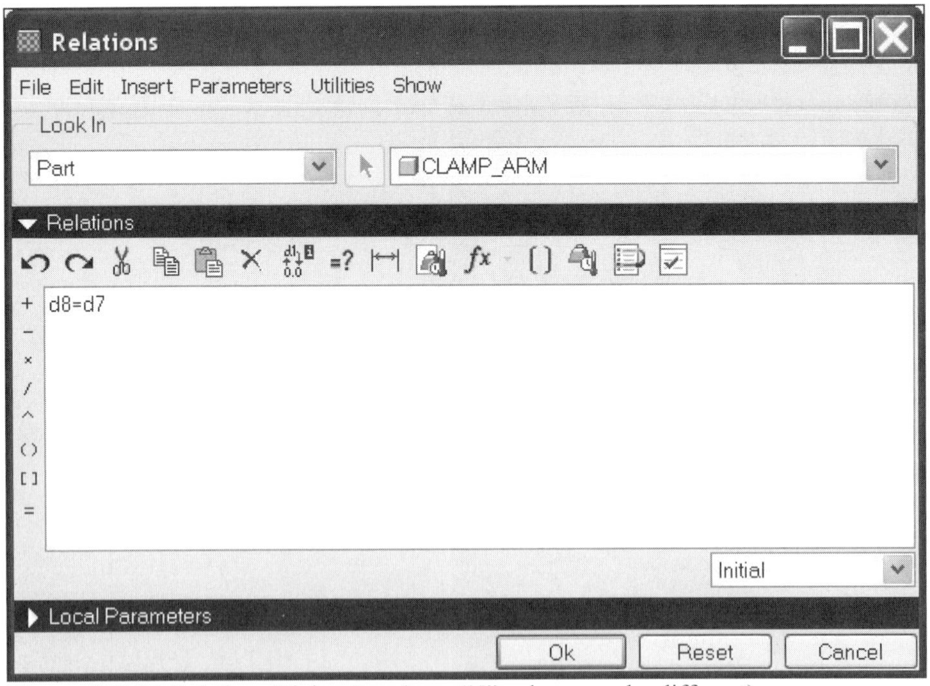

Figure 7.22(b) Relations Dialog Box (your "**d**" values may be different)

Click: **Ctrl+D** ⇒ pick on the "web" ⇒ **RMB** ⇒ **Edit** ⇒ double-click on the **.375** dimension [Fig. 7.22(c)] and change to **.60** [Fig. 7.22(d)] ⇒ **Enter** ⇒ **Edit** ⇒ **Regenerate** [Fig. 7.22(e)] ⇒ **Undo Changes** [Fig. 7.22(f)] ⇒ **Confirm** ⇒ modify the value to **.20** ⇒ **Enter** ⇒ **Ctrl+G** regenerates model [Fig. 7.22(g)] ⇒ **Undo** ⇒ **Ctrl+D** ⇒ **Ctrl+S** ⇒ **Enter**

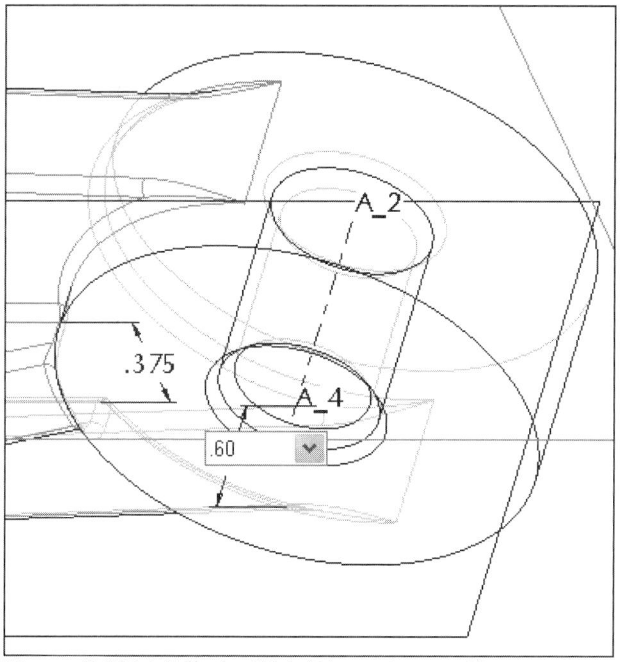

Figure 7.22(c) Edit the Web Dimension to **.60**

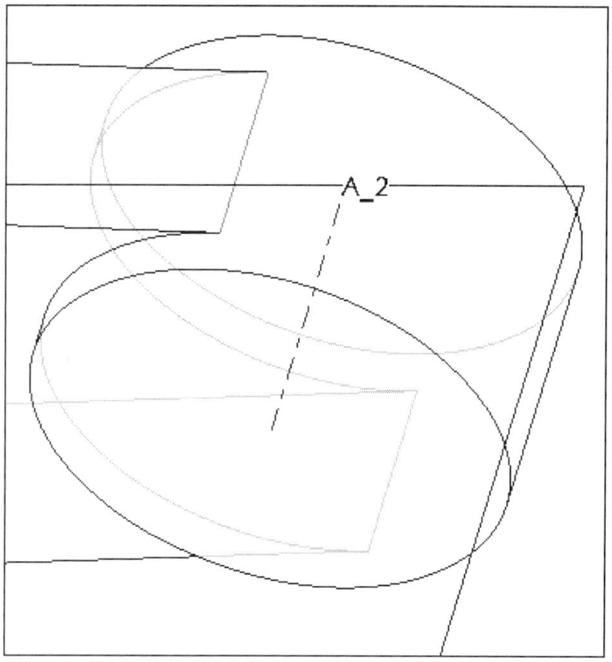

Figure 7.22(d) Failed feature

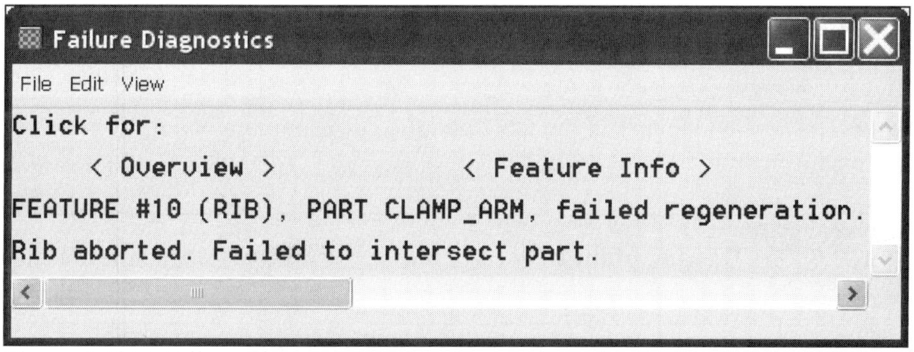

Figure 7.22(e) Failure Diagnostics Dialog Box

Figure 7.22(f) Resolve

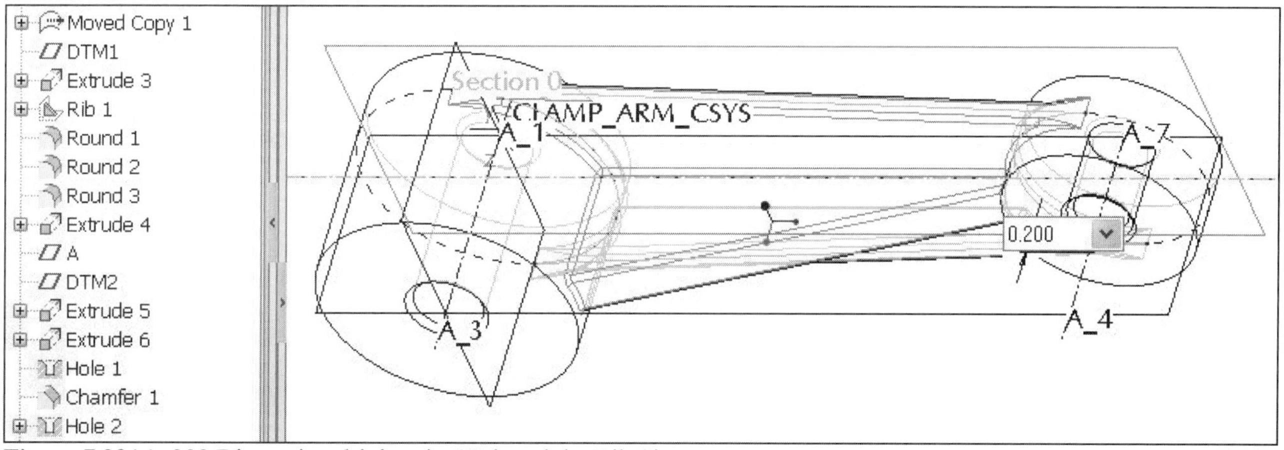

Figure 7.22(g) .200 Dimension driving the Web and the Rib Size

Family Tables

Family Tables are used any time a part or assembly has several unique iterations developed from the original model. The iterations must be considered as separate models, not just iterations of the original model. In this lesson, you will add instances of a family table to create a machined part and a casting of the Clamp Arm (a version without machined cuts or holes).

You will be creating a Family Table from the generic model. The (base) model is the **Generic**. Each variation is referred to as an **Instance**. When you create a Family Table, Pro/E allows you to *select dimensions,* which can vary between instances. You can also *select features* to add to the Family Table. Features can vary by being suppressed or resumed in an instance. When you are finished selecting items (e.g., dimensions, features, and parameters), the Family Table is automatically generated.

When adding features to the table, enter an **N** to suppress the feature, or a **Y** to resume the feature. Each instance must have a unique name.

Family tables are spreadsheets, consisting of columns and rows. *Rows* contain instances and their corresponding values; *columns* are used for items. The column headings include the *instance name* and the names of all of the *dimensions, parameters, features, members,* and *groups* that were selected to be in the table. The Family Table dialog box is used to create and modify family tables.

Family tables include:

- The base object (generic object or *generic*) on which all members of the family are based.
- Dimensions, parameters, feature numbers, user-defined feature names, and assembly member names that are selected to be table-driven (*items*).
    - **Dimensions** are listed by name (for example, **d125**) with the associated symbol name (if any) on the line below it (for example, depth).
    - **Parameters** are listed by name (dim symbol).
    - **Features** are listed by feature number with the associated feature type (for example, [cut]) or feature name on the line below it. The generic model is the first row in the table. Only modifying the actual part, suppressing, or resuming features can change the table entries belonging to the generic; *you cannot change the generic model by editing its row entries in the family table.*
- Names of all family members (*instances*) created in the table and the corresponding values for each of the table-driven items

Click: **Turn on 3D Annotations** ⇒ **Datum planes** on ⇒ **Tools** ⇒ **Family Table**--the Family Table: dialog box opens [Fig. 7.23(a)]

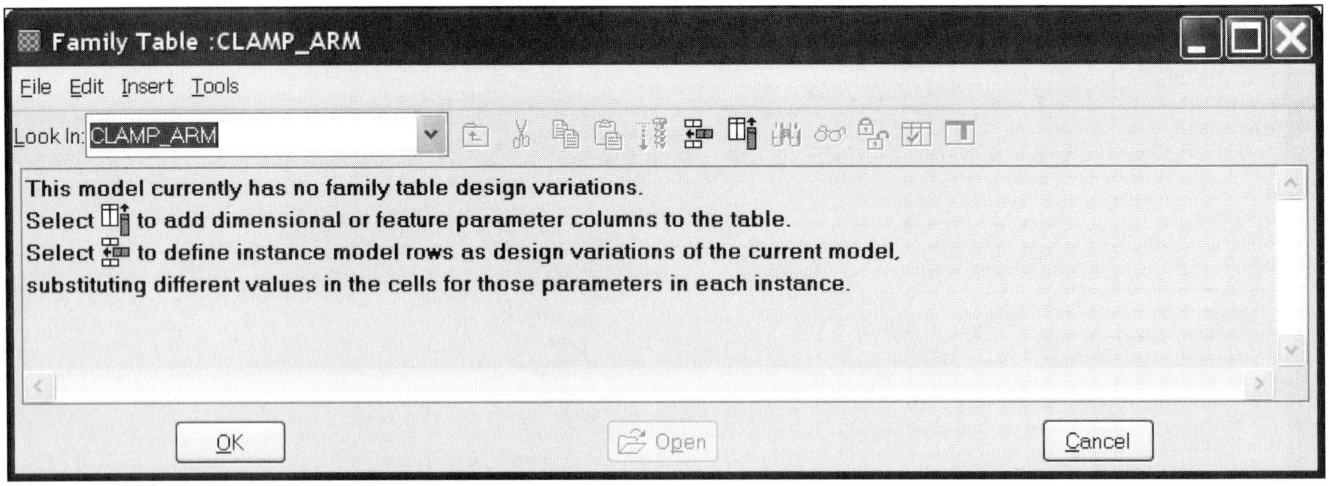

Figure 7.23(a) Family Table Dialog Box

Click: **Add/delete the table columns** ⇒ Feature from the Add Item options ⇒ select the cuts and holes from the model or the Model Tree [Fig. 7.23(b)] *(the order in which you select items will determine the default order in which they will appear in the table – default column order)* ⇒ **OK** [Fig. 7.23(c)] ⇒ **Insert a new instance at the selected row** ⇒ click on the name of the new instance **CLAMP_ARM_INST** ⇒ type **CLAMP_ARM_DESIGN** [Fig. 7.23(d)] ⇒ **Enter** *(adds a new instance)*

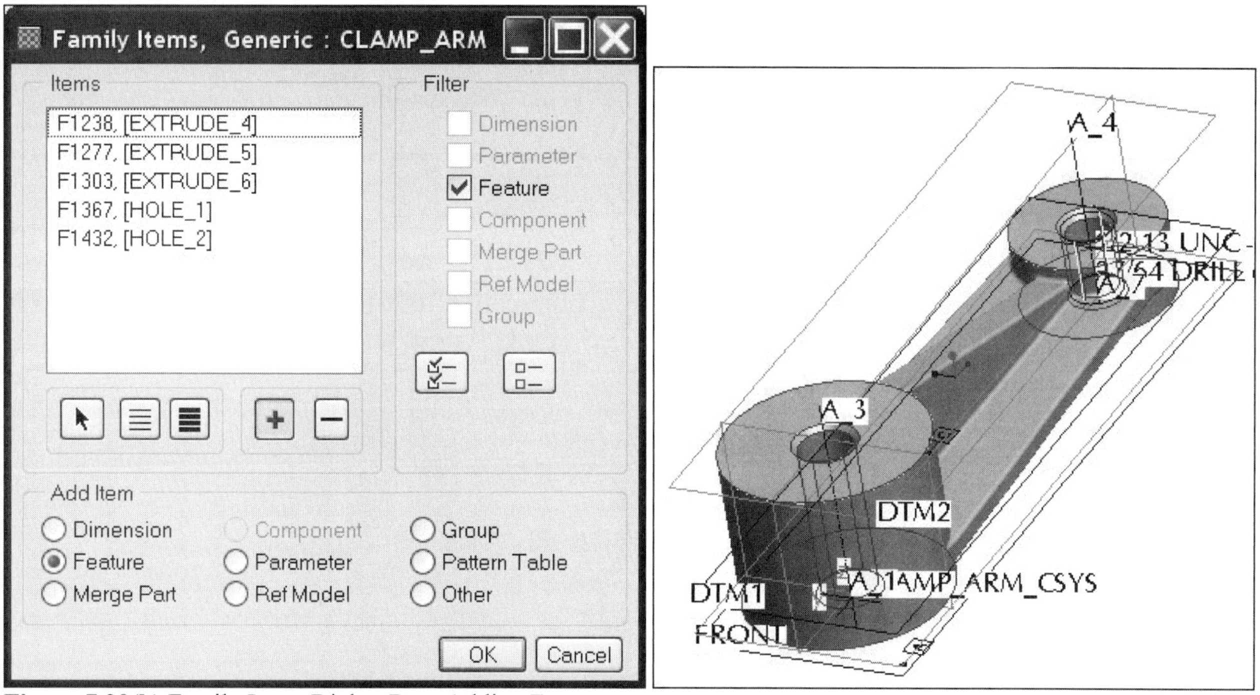

Figure 7.23(b) Family Items Dialog Box, Adding Features

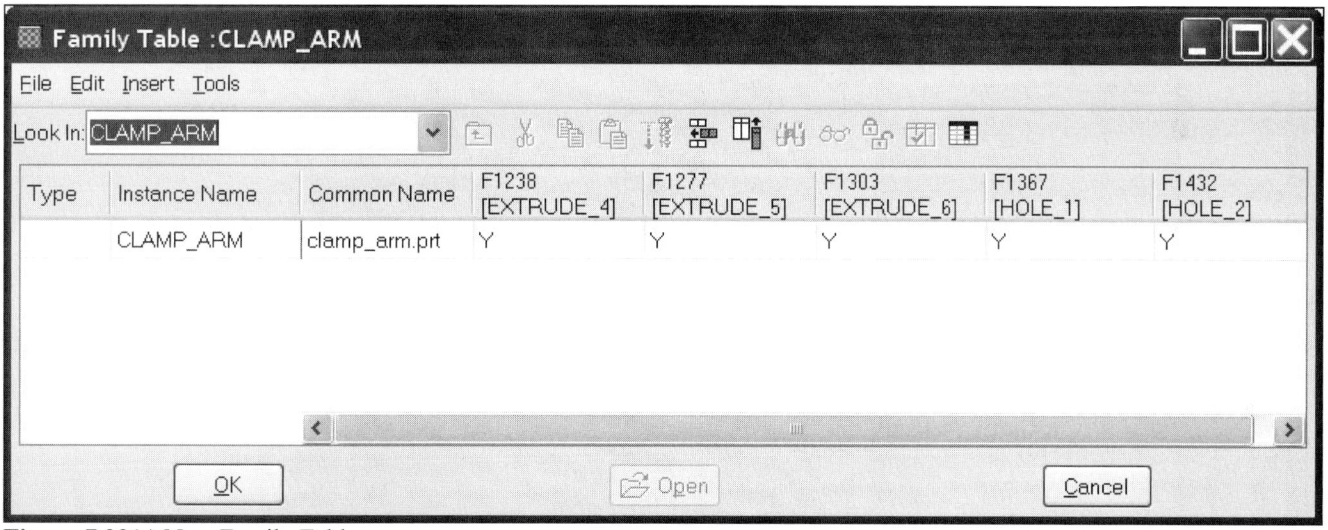

Figure 7.23(c) New Family Table

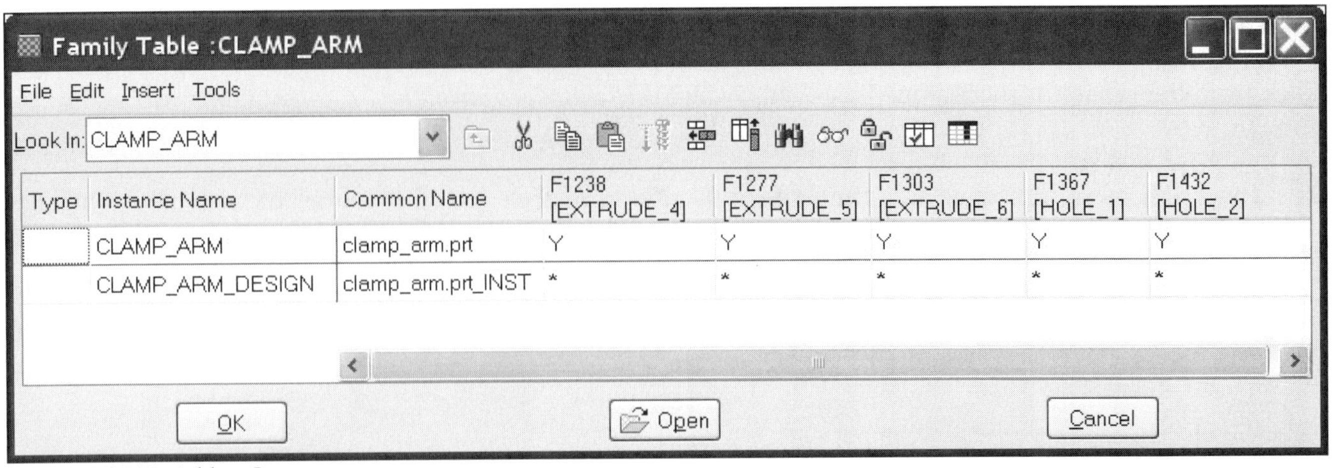

Figure 7.23(d) Add an Instance

The second instance **CLAMP_ARM_INST** should be highlighted ⇒ type **CLAMP_ARM_WORKPIECE** [Fig. 7.23(e)] ⇒ click in the cell of the first feature and change to **N** (not used) [Fig. 7.23(f)] ⇒ change all cells for the **CLAMP_ARM_WORKPIECE** to **N** [Fig. 7.23(g)]

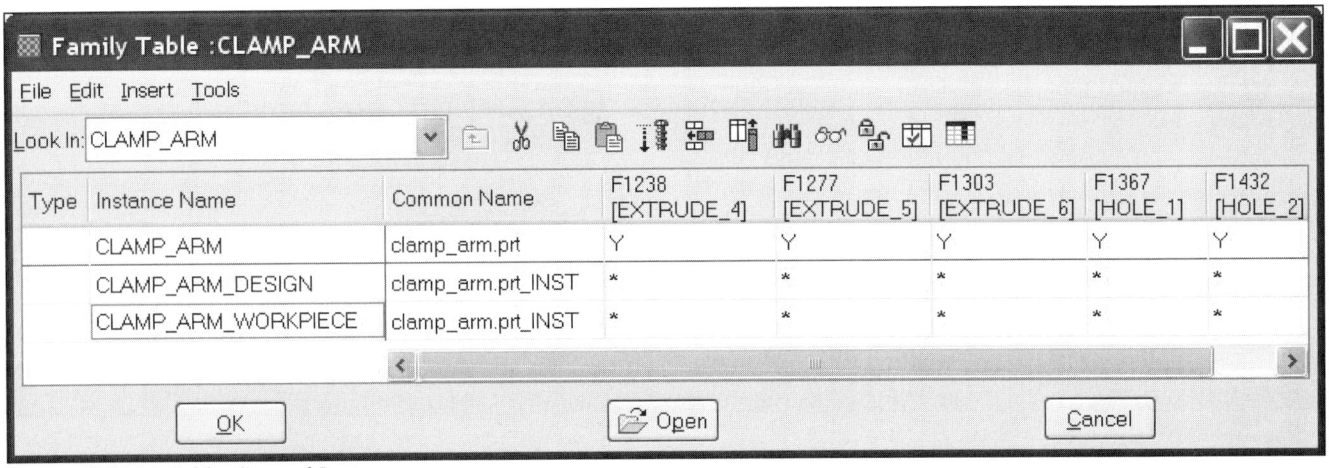

Figure 7.23(e) Add a Second Instance

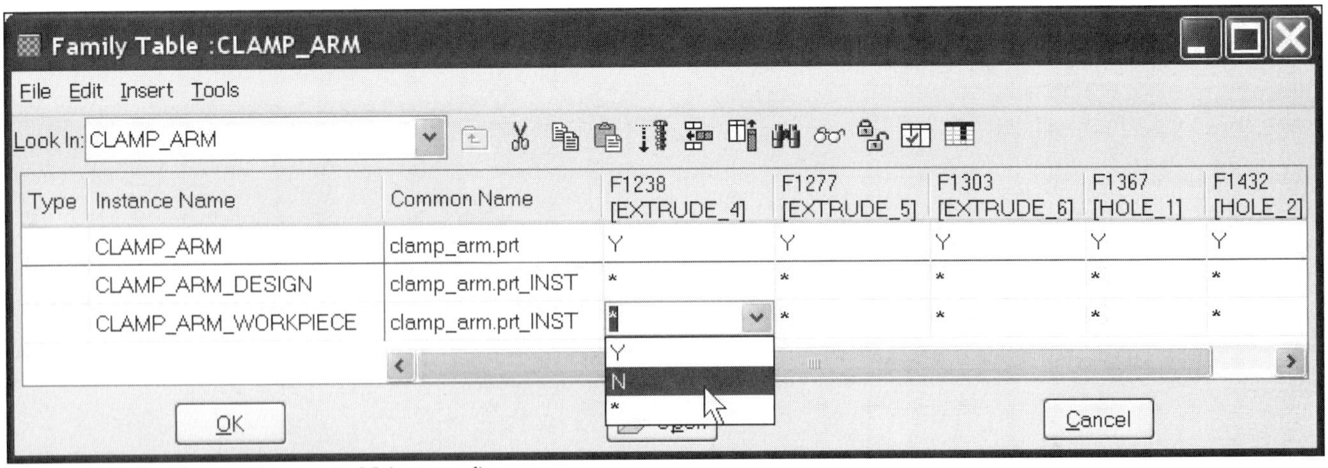

Figure 7.23(f) Change Feature to N (not used)

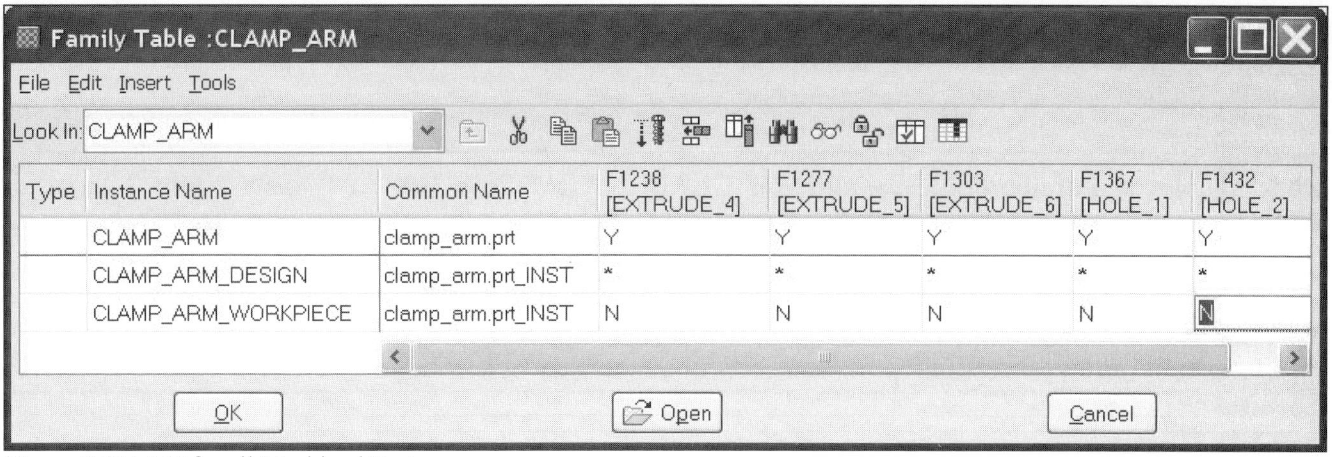

Figure 7.23(g) N for all Machined Features

Click: Verify instances of the family [Fig. 7.23(h)] ⇒ **VERIFY** [Fig. 7.23(i)] ⇒ **Close** [Fig. 7.23(j)] ⇒ **OK** from Family Table dialog box ⇒ **Ctrl+S** ⇒ **Enter** ⇒ **Ctrl+D** ⇒ ⇒ **LMB** to deselect ⇒

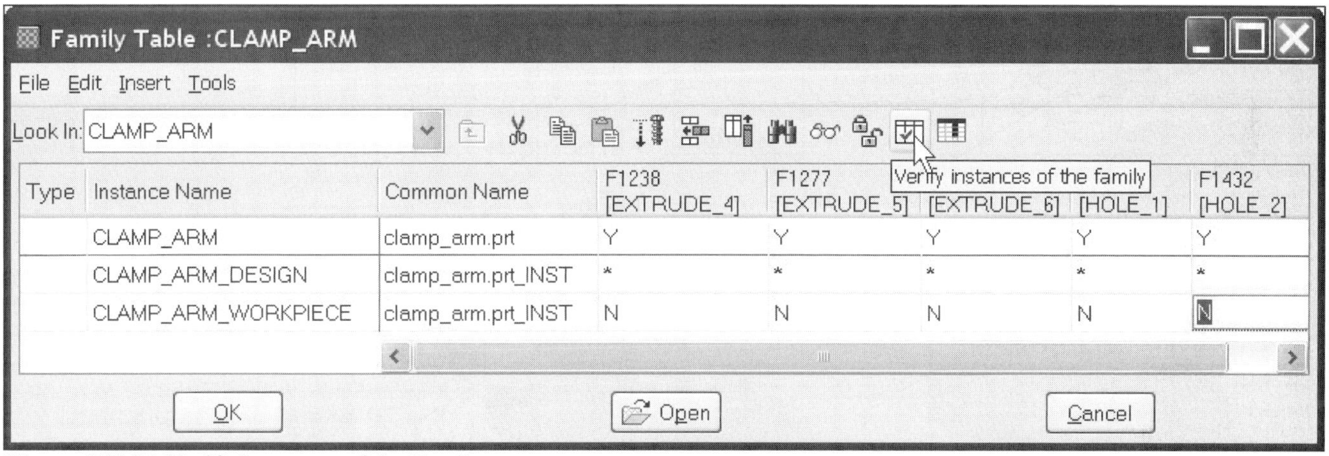

Figure 7.23(h) Verify Instances of the Family

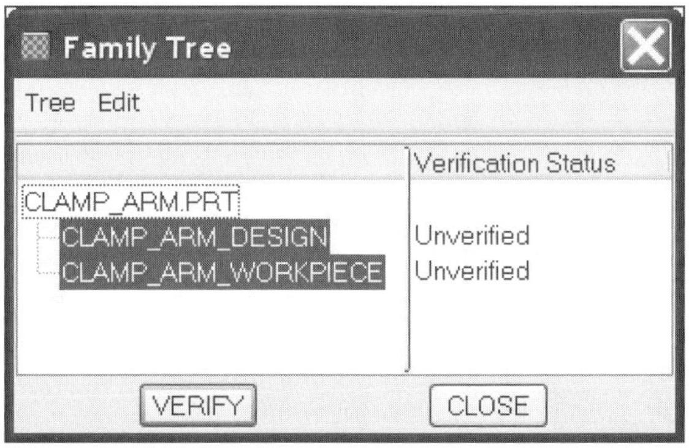

Figure 7.23(i) Verify

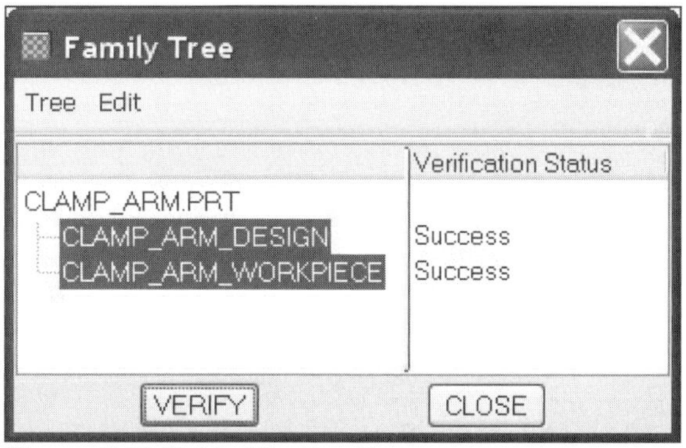

Figure 7.23(j) Verification Status

A Family Table controls whether a feature is present or not for a given design instance, not whether a feature is displayed. The Generic is the base model [Fig. 7.23(k)] and is typically not a member of an assembly.

Click: **Tools** ⇒ **Family Table** ⇒ click on **CLAMP_ARM_DESIGN** ⇒ Open [Fig. 7.23(l)] ⇒ **Window** from the menu bar of **CLAMP_ARM_DESIGN** ⇒ click: **1 CLAMP_ARM.PRT** ⇒ **Tools** ⇒ **Family Table** ⇒ click on **CLAMP_ARM_WORKPIECE** ⇒ **RMB** ⇒ Open [Fig. 7.23(m)] ⇒ **Window** ⇒ **Close** ⇒ **Window** from the menu bar of **CLAMP_ARM_DESIGN** ⇒ **Activate** ⇒ **Window** ⇒ **Close** ⇒ **Window** ⇒ **Activate** (the Generic will be the only object on your screen) ⇒ **Ctrl+S** ⇒ **OK**

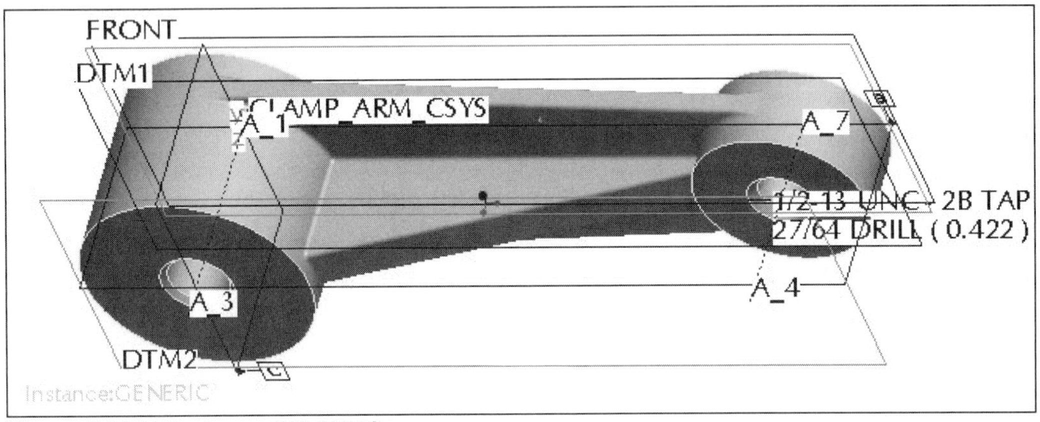

Figure 7.23(k) Instance: GENERIC

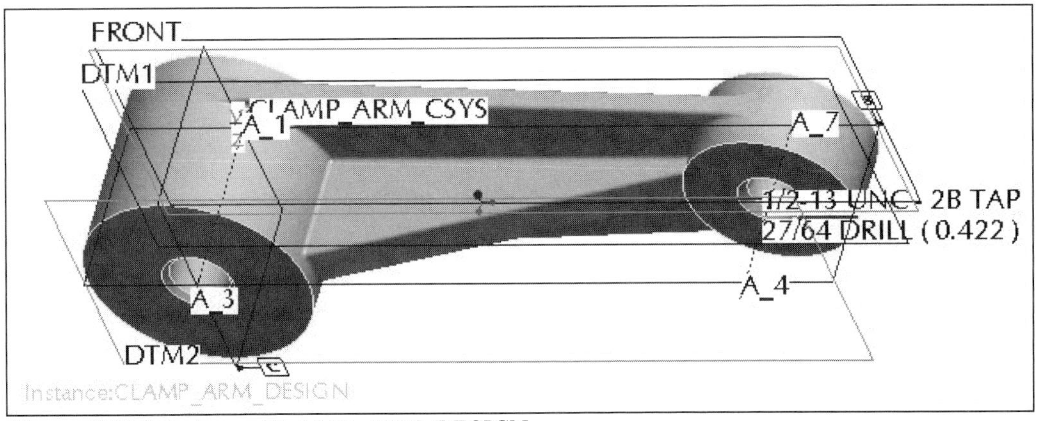

Figure 7.23(l) Instance: CLAMP_ARM_DESIGN

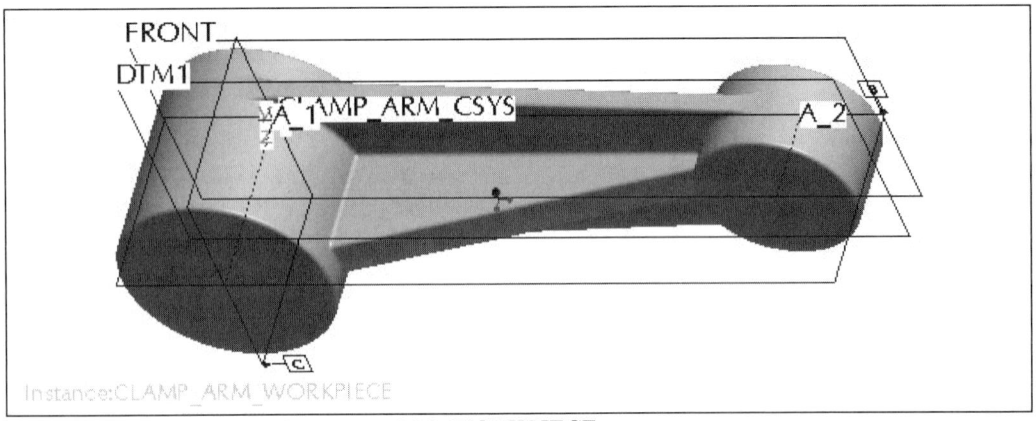

Figure 7.23(m) Instance: CLAMP_ARM_WORKPIECE

Click: 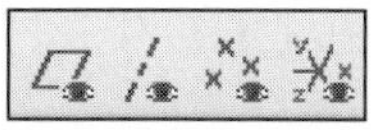 off ⇒ rotate the model ⇒ highlight the part's features in the Model Tree⇒ **RMB** ⇒ **Create Driving Dimensions AE** [Fig. 7.24(a)] ⇒ from selection filters, click: **Smart** ⇒ **Annotation** ⇒ pick on a dimension ⇒ **RMB** ⇒ **Move** ⇒ pick a new position ⇒ continue to move dimensions to more visible locations ⇒ pick on 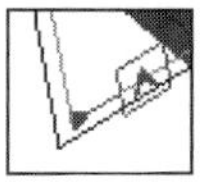 ⇒ **RMB** ⇒ **Flip** ⇒ 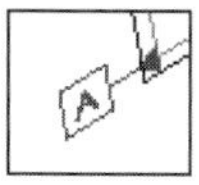 [Fig. 7.24(b)]

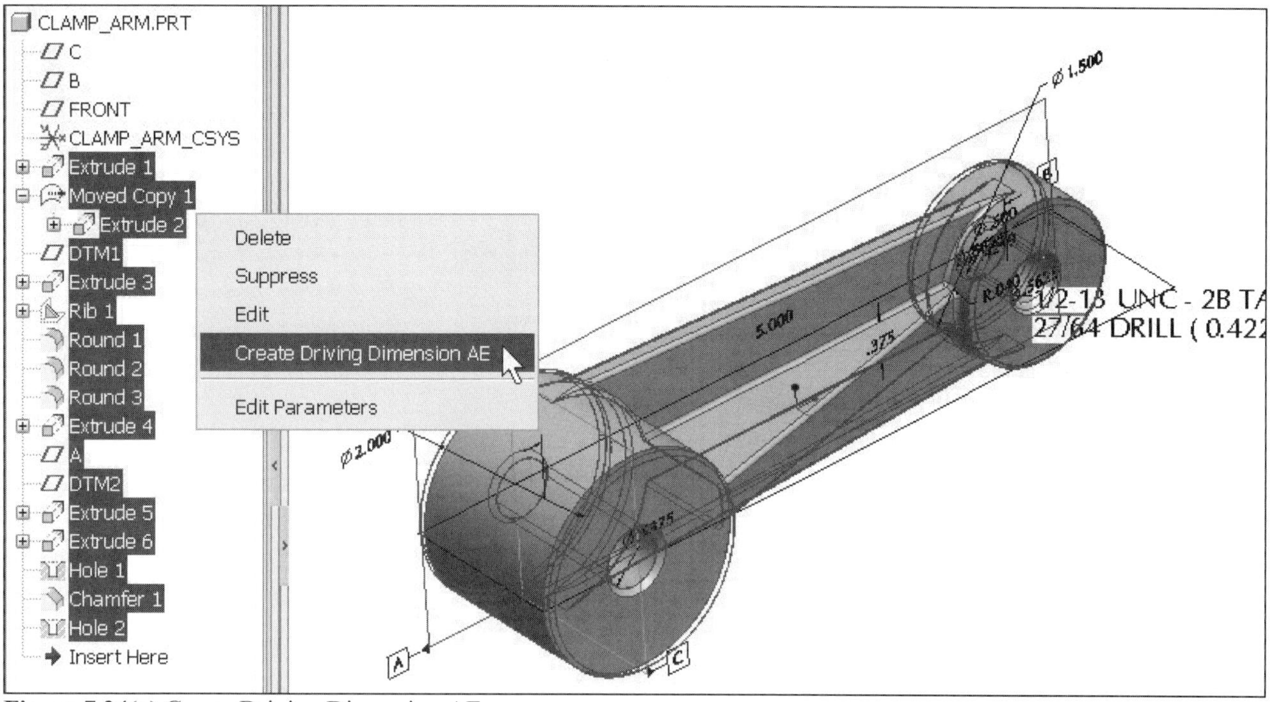

Figure 7.24(a) Create Driving Dimension AE

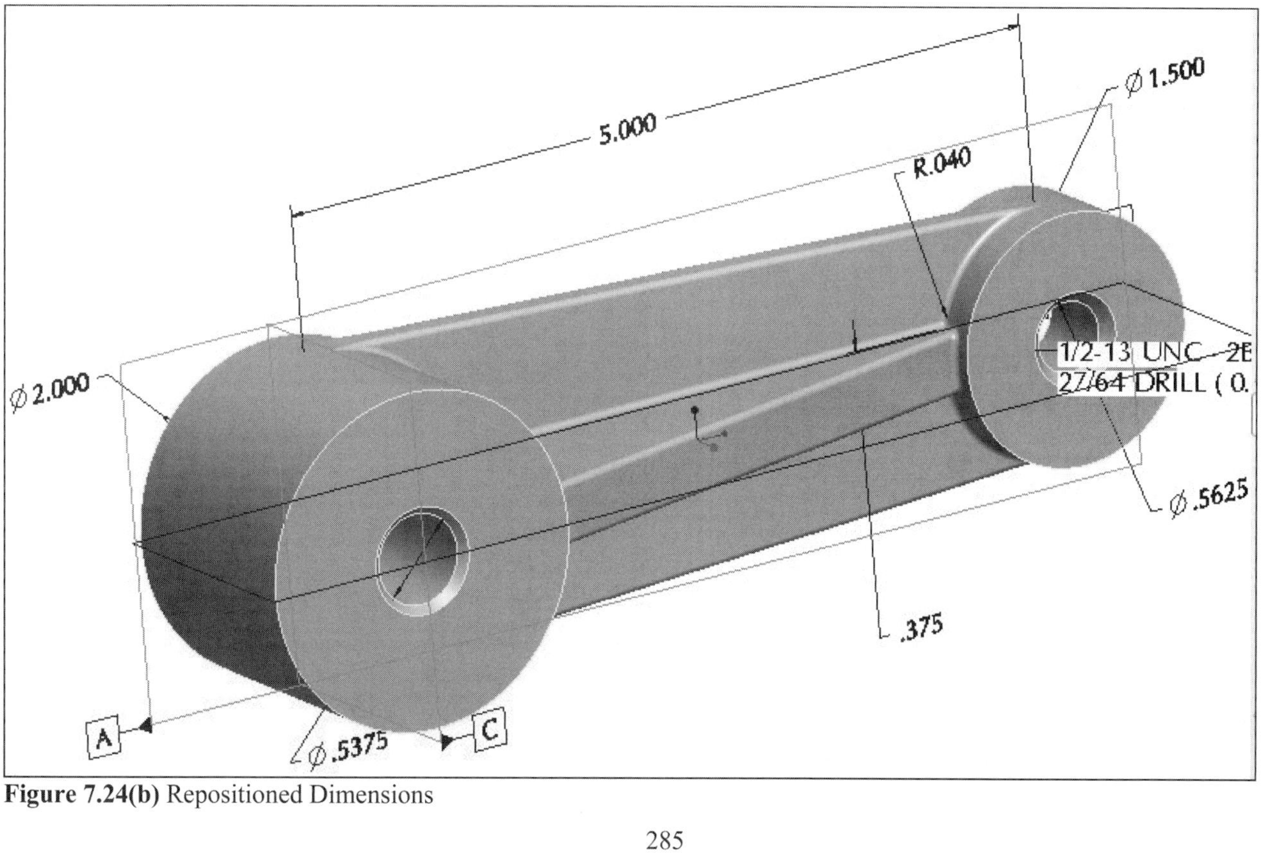

Figure 7.24(b) Repositioned Dimensions

Click: **View ⇒ Annotation Orientation ⇒ Front** [Fig. 7.24(c)] **⇒ OK ⇒ Ctrl+D ⇒ Ctrl+S ⇒ Enter**

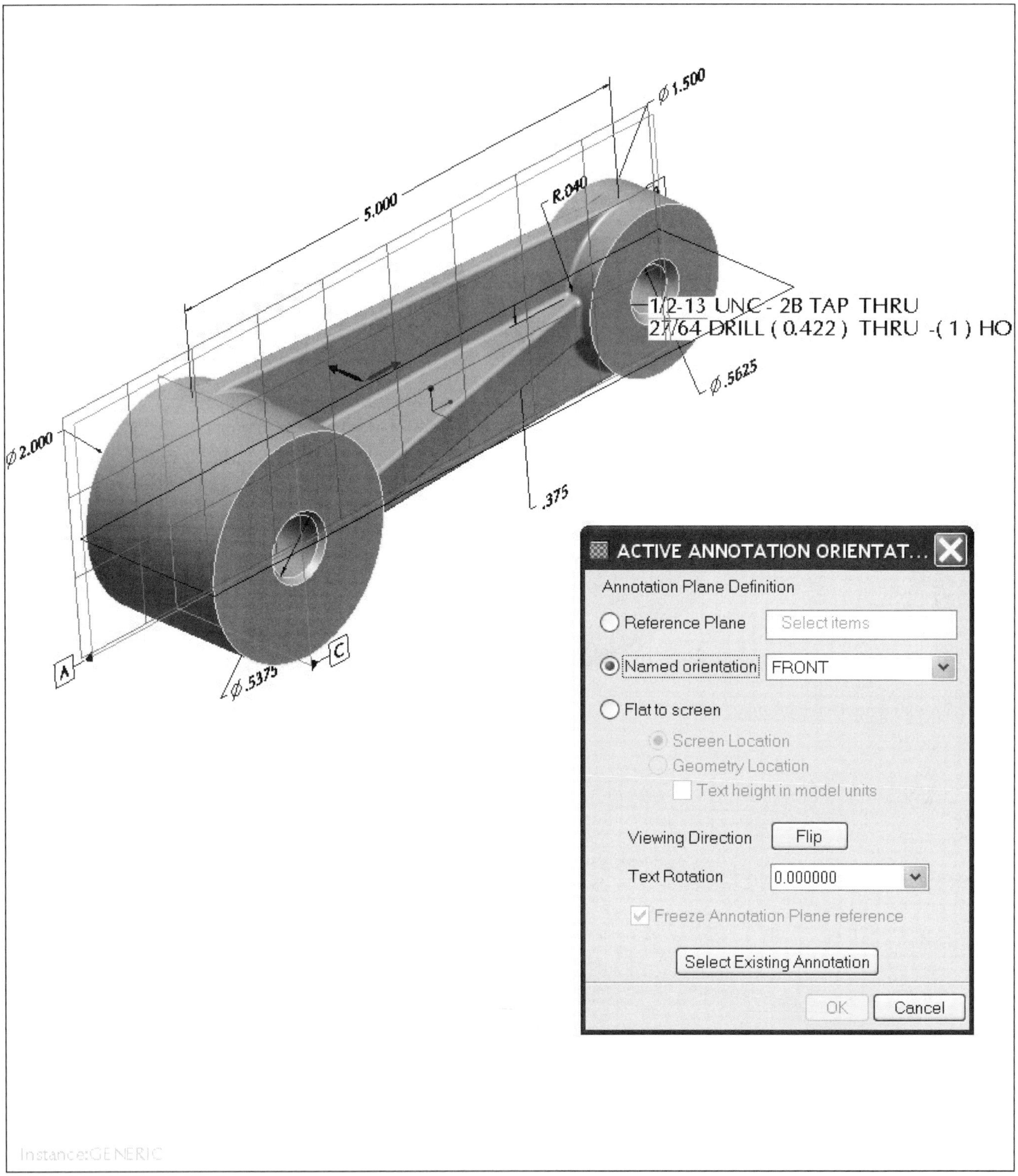

Figure 7.24(c) Annotation Orientation

Click: **Info** ⇒ **Relations and Parameters** [Fig. 7.25] ⇒ **File** ⇒ **Delete** ⇒ **Old Versions** ⇒ **Enter**

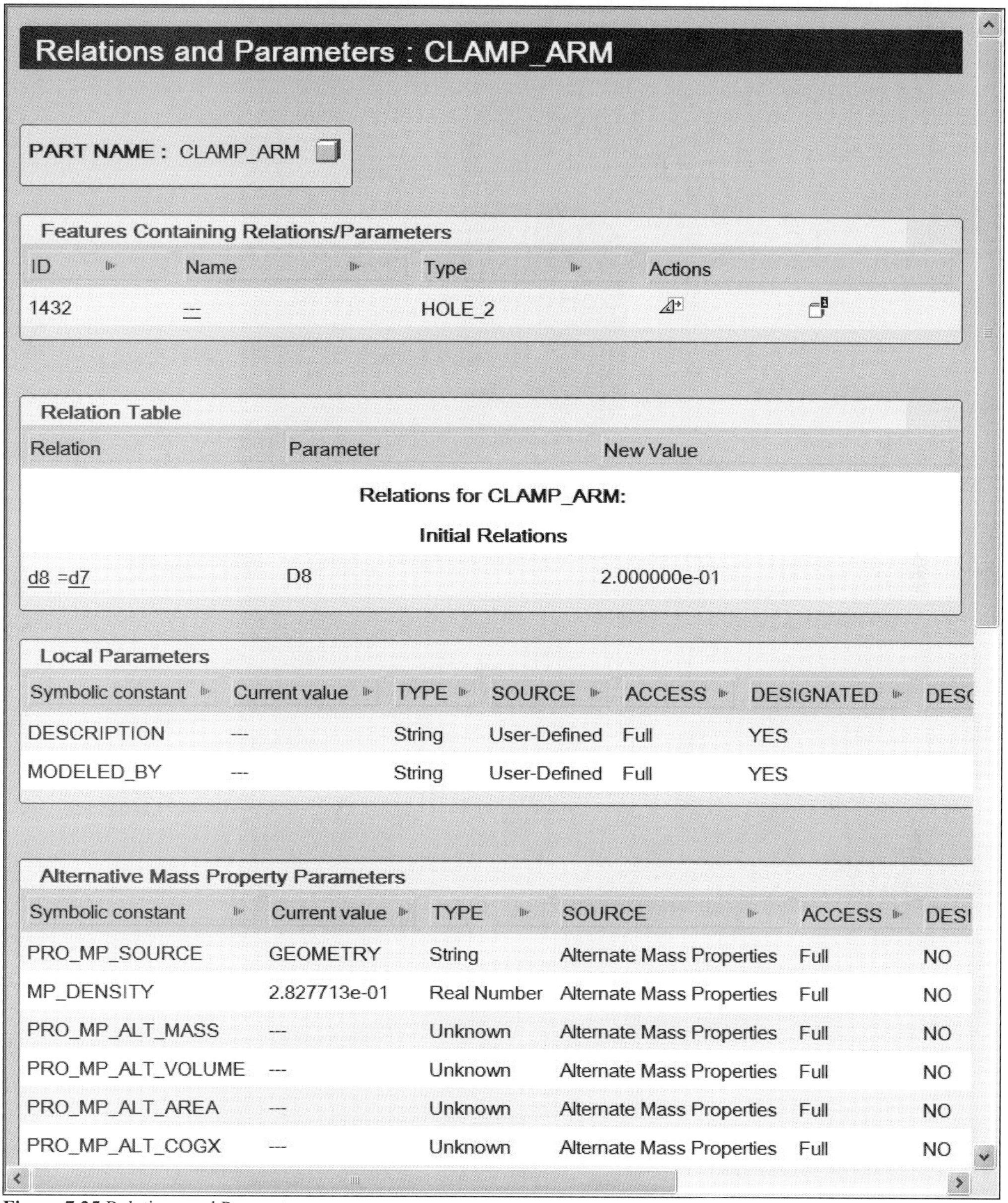

Figure 7.25 Relations and Parameters

Pro/MANUFACTURING

You now have two separate models, a casting (workpiece) and a machined part (design part). During the manufacturing process, the workpiece is merged (assembled) into the design part thereby creating a manufacturing model [Fig. 7.26(a)]. The difference between the two objects is the difference between the volume of the casting and the volume of the machined part. The manufacturing model is used to machine the part [Fig. 7.26(b)]. If you have the Pro/MANUFACTURING module, Pro/NC, and or Expert Machinist; you could now machine the part.

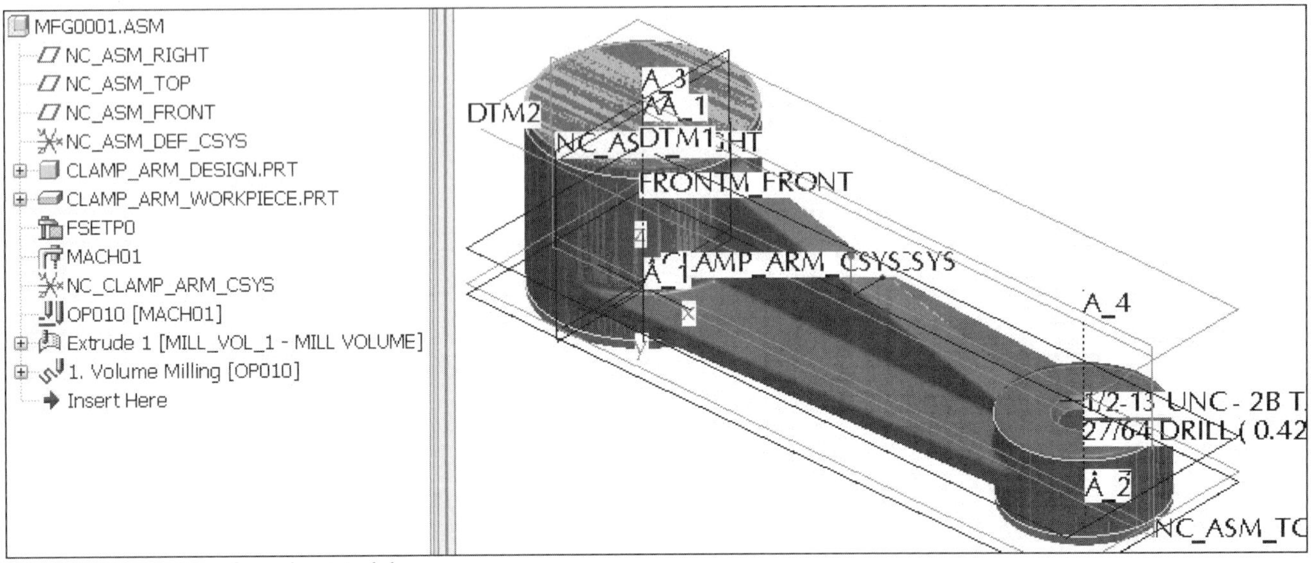

Figure 7.26(a) Manufacturing Model

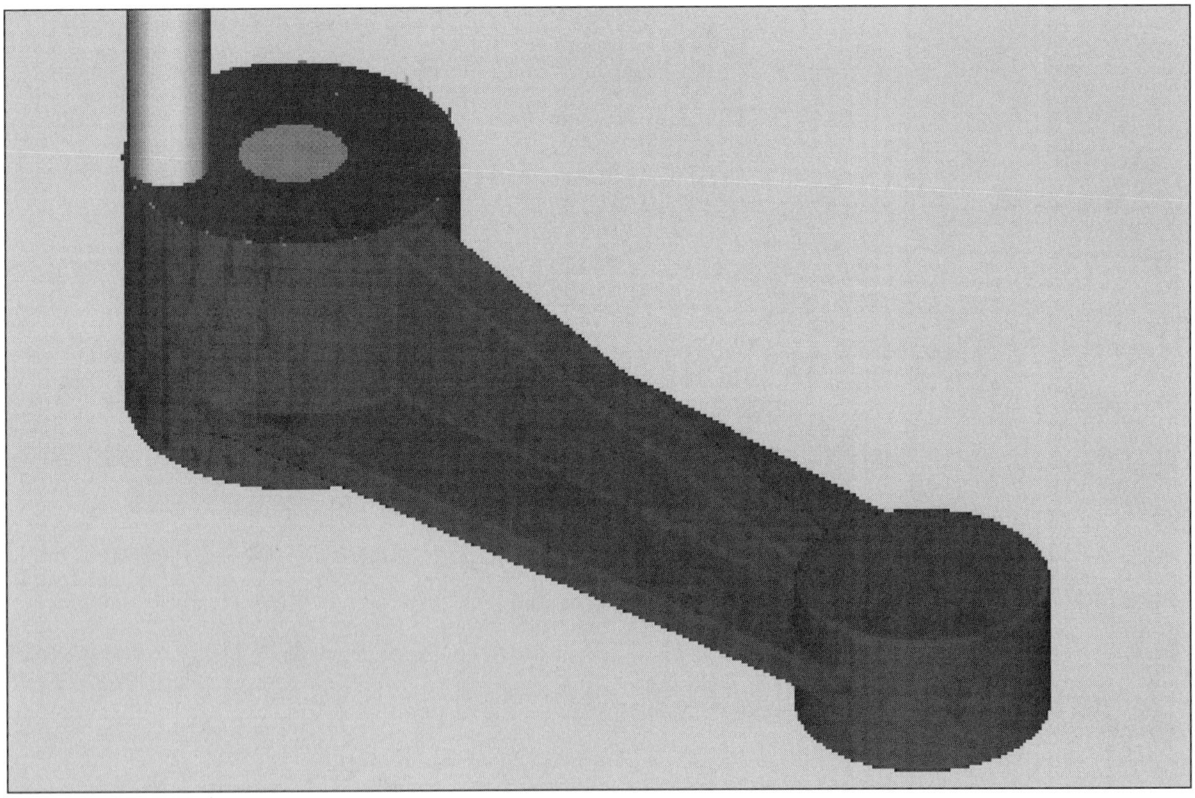

Figure 7.26(b) Facing

This Lesson is now complete. If you wish to model a project without instructions, a complete set of projects and illustrations are available at ***<u>www.cad-resources.com</u>* ⇒ *Downloads.***

Lesson 8 Assemblies

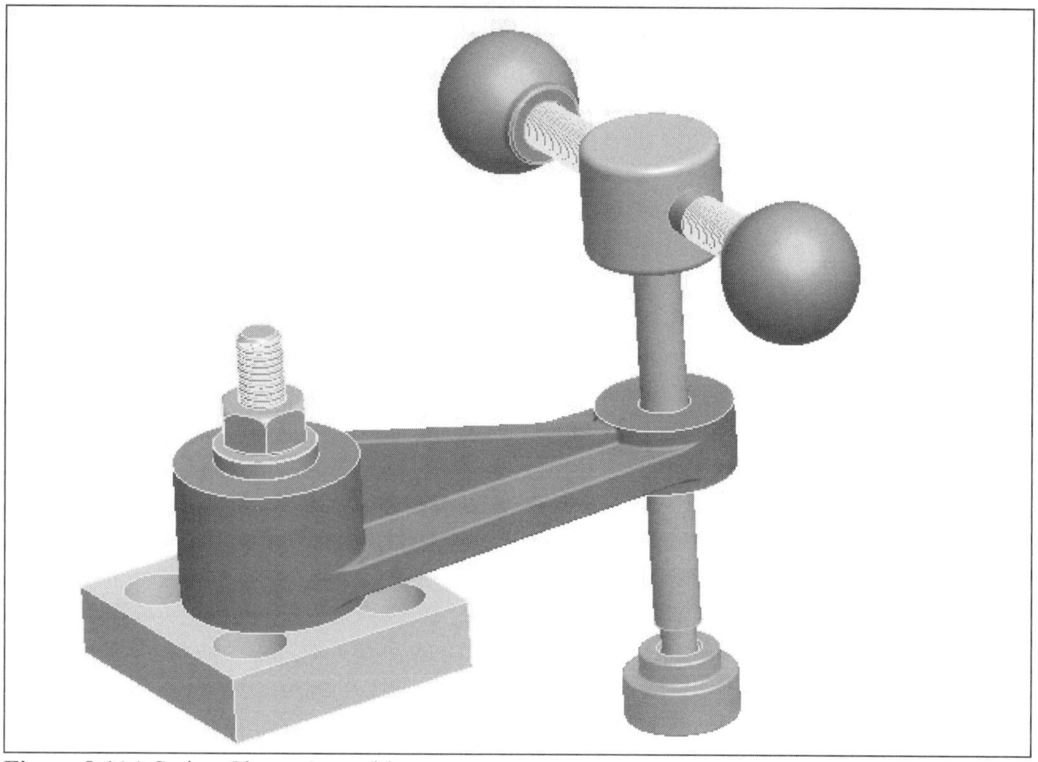

Figure 8.1(a) Swing Clamp Assembly

OBJECTIVES

- **Assemble** components to create an assembly
- Create a **subassembly**
- Understand and use a variety of **Assembly Constraints**
- **Modify** a component constraint
- **Edit** a constraint value
- Check for **clearance** and **interference**

ASSEMBLY CONSTRAINTS

Assembly mode allows you to place together components and subassemblies to create an assembly [Fig. 8.1(a)]. Assemblies [Fig. 8.1(b)] can be modified, reoriented, documented, or analyzed. An assembly can be assembled into another assembly, thereby becoming a subassembly.

Figure 8.1(b) Swing Clamp (CARRLANE at **www.carrlane.com**)

Placing Components

To assemble components, use: **Add component to the assembly** or **Insert ⇒ Component ⇒ Assemble**. After selecting a component from the Open dialog box, the dashboard [Fig. 8.2(a)] opens and the component appears in the assembly window. Alternatively, you can select a component from a browser window and drag it into the Pro/E window. If there is an assembly in the window, Pro/E will begin to assemble the component into the current assembly.

Using icons in the dashboard, you can specify the screen window in which the component is displayed while you position it. You can change window options at any time using: **Show component in a separate window while specifying constraints** or **Show component in the assembly window while specifying constraints**.

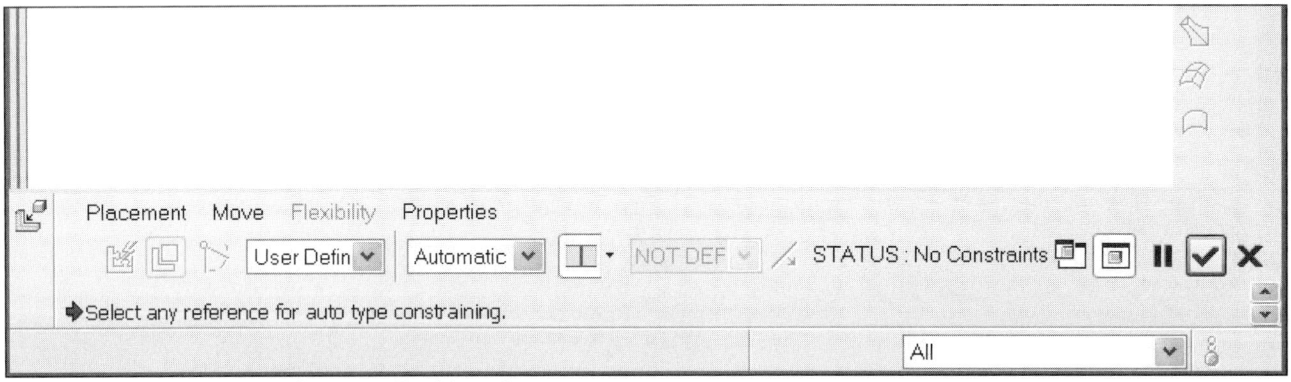

Figure 8.2(a) Component Placement Dashboard

The Automatic placement constraint Automatic is selected by default when a new component is introduced into an assembly for placement. After you select a pair of valid references from the component and the assembly, Pro/E automatically selects a constraint type appropriate to the specified references.

Before selecting references, you may also change the type of constraint. Clicking on the current constraint in the Constraint Type box [Fig. 8.2(b)] shows the list of constraints.

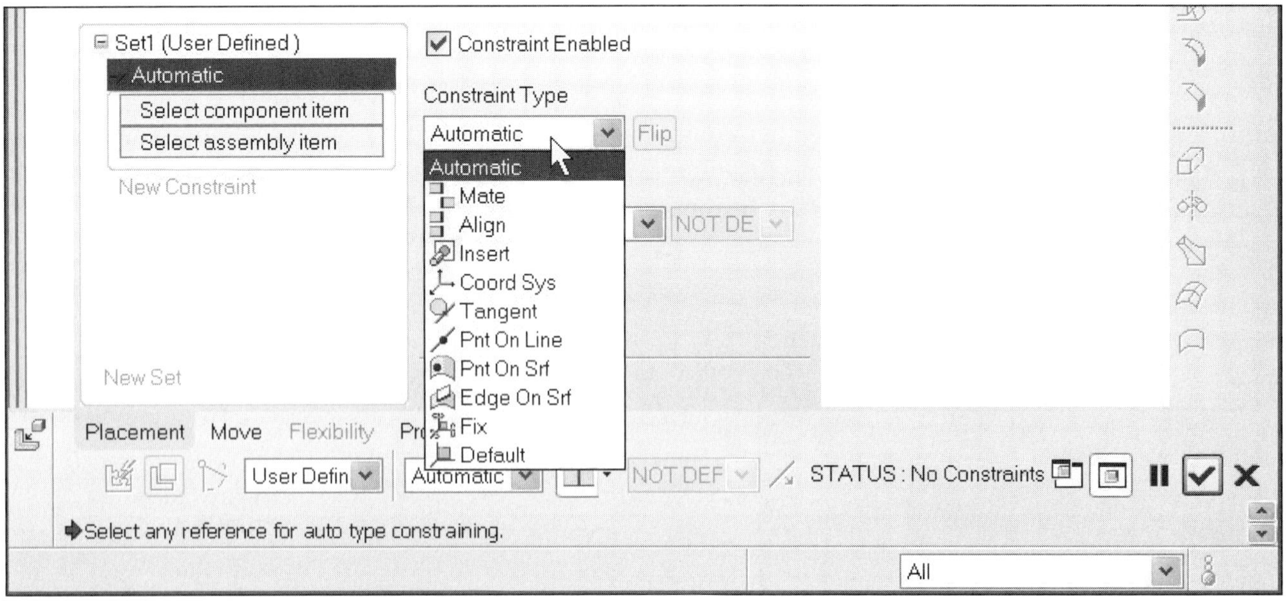

Figure 8.2(b) Component Placement Dashboard, Constraint Type List

Lesson 8 STEPS- Part One Bottom-Up Design

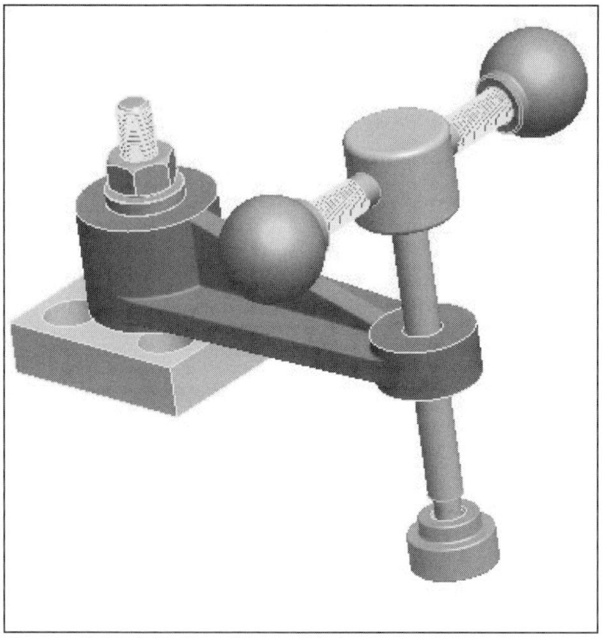

Figure 8.3(a) Swing Clamp Main Assembly

Figure 8.3(b) Swing Clamp Sub-Assembly

Swing Clamp Assembly

The parts required in this lesson are from this text. *If you have not modeled these parts previously, please do so before you start the following systematic instructions.* The other components required for the assembly are standard *off-the-shelf* hardware items that you can get from Pro/E by accessing the Pro/Library. If your system does not have a Pro/LIBRARY license for the Basic and Manufacturing libraries, model the parts using the detail drawings provided in this text or download them from the online PTC Catalog *(the Student Edition and the Tryout Edition do not allow access to the catalog parts).* The **Flange Nut**, the **3.50 Double-ended Stud**, and the **5.00 Double-ended Stud** are standard items. The **Clamp Plate** component is the first component of the main assembly and will be modeled later when completing the main assembly [Fig. 8.3(a)] using the *top-down design* approach.

Because you will be creating the sub-assembly [Fig. 8.3(b)] using the *bottom-up design* approach, all the components must be available before any assembling starts. *Bottom-up design* means that existing parts are assembled, one by one, until the assembly is complete. The assembly starts with a set of default datum planes and a coordinate system. The parts are constrained to the datum features of the assembly. The sequence of assembly will determine the parent-child relationships between components.

Top-down design is the design of an assembly where one or more component parts are created in Assembly mode as the design unfolds. Some existing parts are available, such as standard components and a few modeled parts. The remaining design evolves during the assembly process. The main assembly will involve creating one part using the *top-down design* approach.

Regardless of the design method, the assembly default datum planes and coordinate system should be on their own separate *assembly layer*. Each part should also be placed on separate assembly layers; the part's datum features should already be on *part layers*.

Before starting the assembly, you will be modeling each part or retrieving *standard parts* from the library and saving them under unique names into *your* working directory. *Unless instructed to do so,* ***do not use the library parts directly in the assembly***. Start this process by retrieving the standard parts [Figs. 8.3(c-e)].

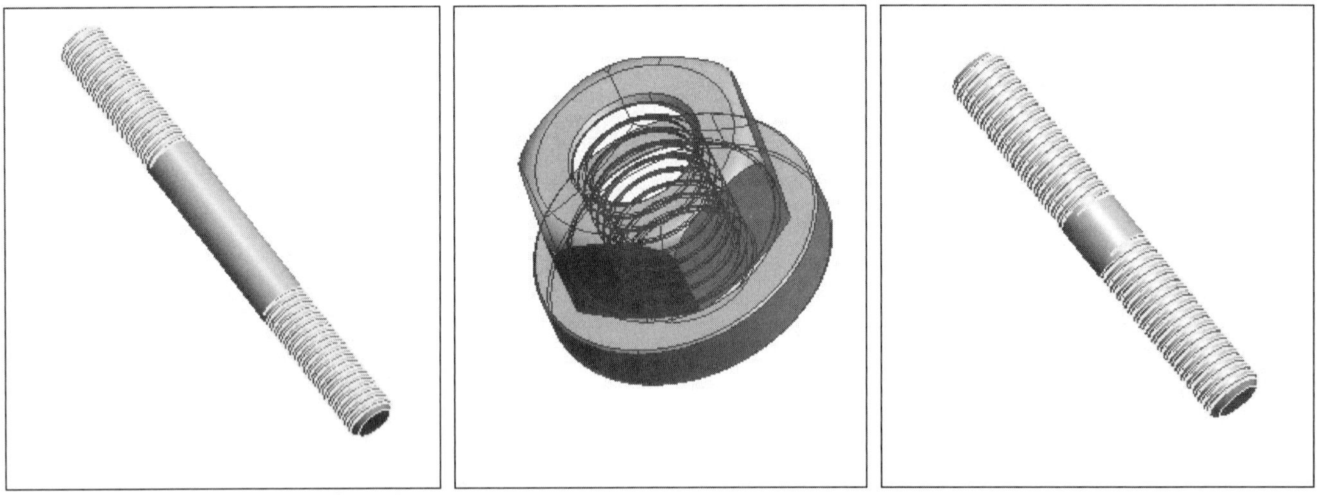

Figure 8.3(c-e) Standard Parts from Carr Lane

Click: **File** ⇒ **Set Working Directory** ⇒ select your working directory ⇒ **OK** ⇒ Click  **Connections** 

⇒ 

⇒ open the Browser shade 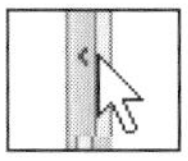 ⇒ Type **www.carrlane.com** in the address bar 

⇒ **Enter** [Fig. 8.4(a)]

*(**Alternative method 1:** model the 3.50 stud) [for the unix os, all (new) file names should be lower case]*
*(**Alternative method 2:** download from http://www.cad-resources.com/services-WF-4.html)*

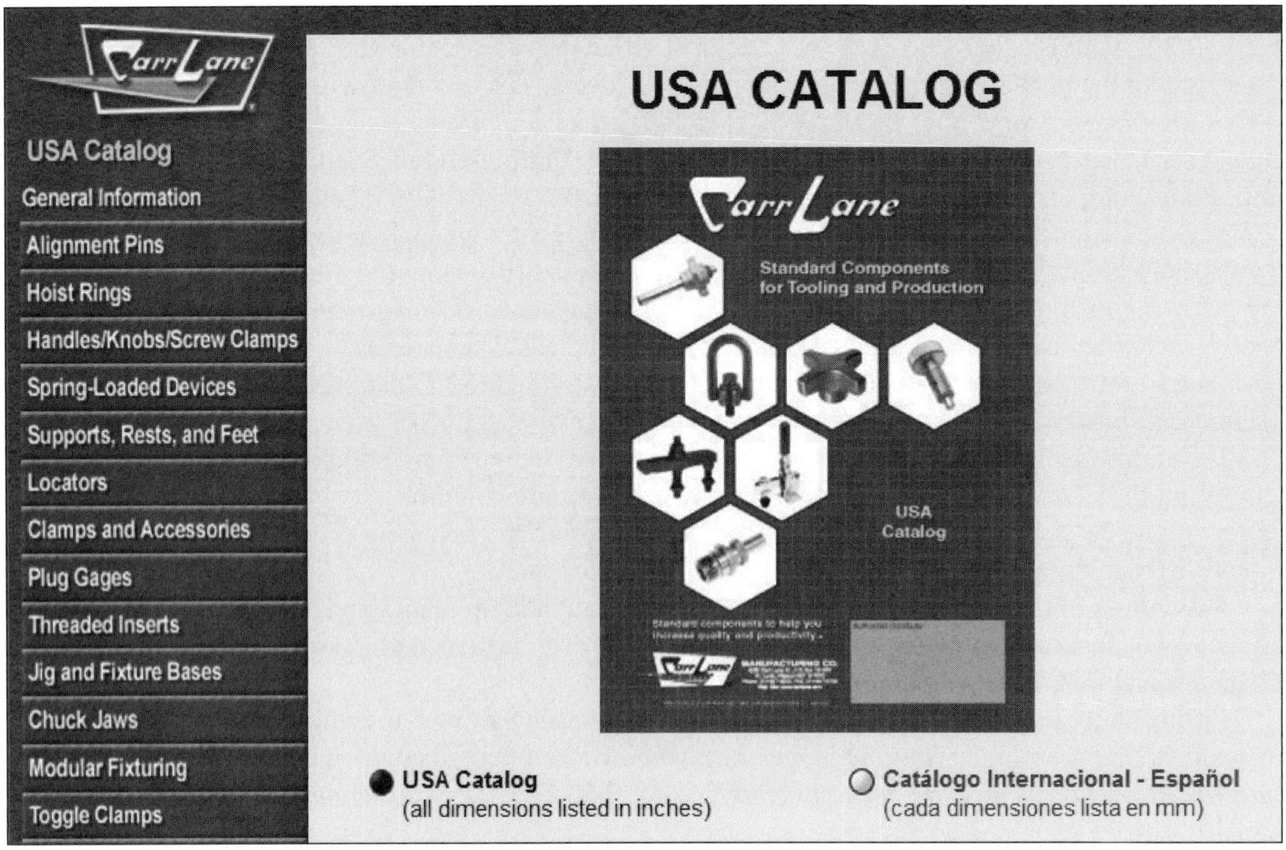

Figure 8.4(a) Carr Lane

Click: **Online Catalog** [Online Catalog] ⇒ **Clamps and Accessories** [Clamps and Accessories] ⇒

Studs, Bolts, and T Nuts 

⇒ **Clamping Studs** 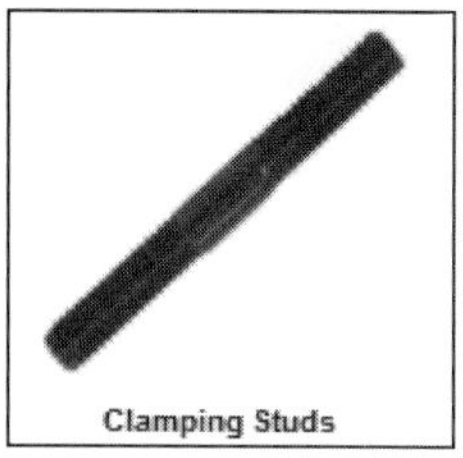

[Fig. 8.4(b)]

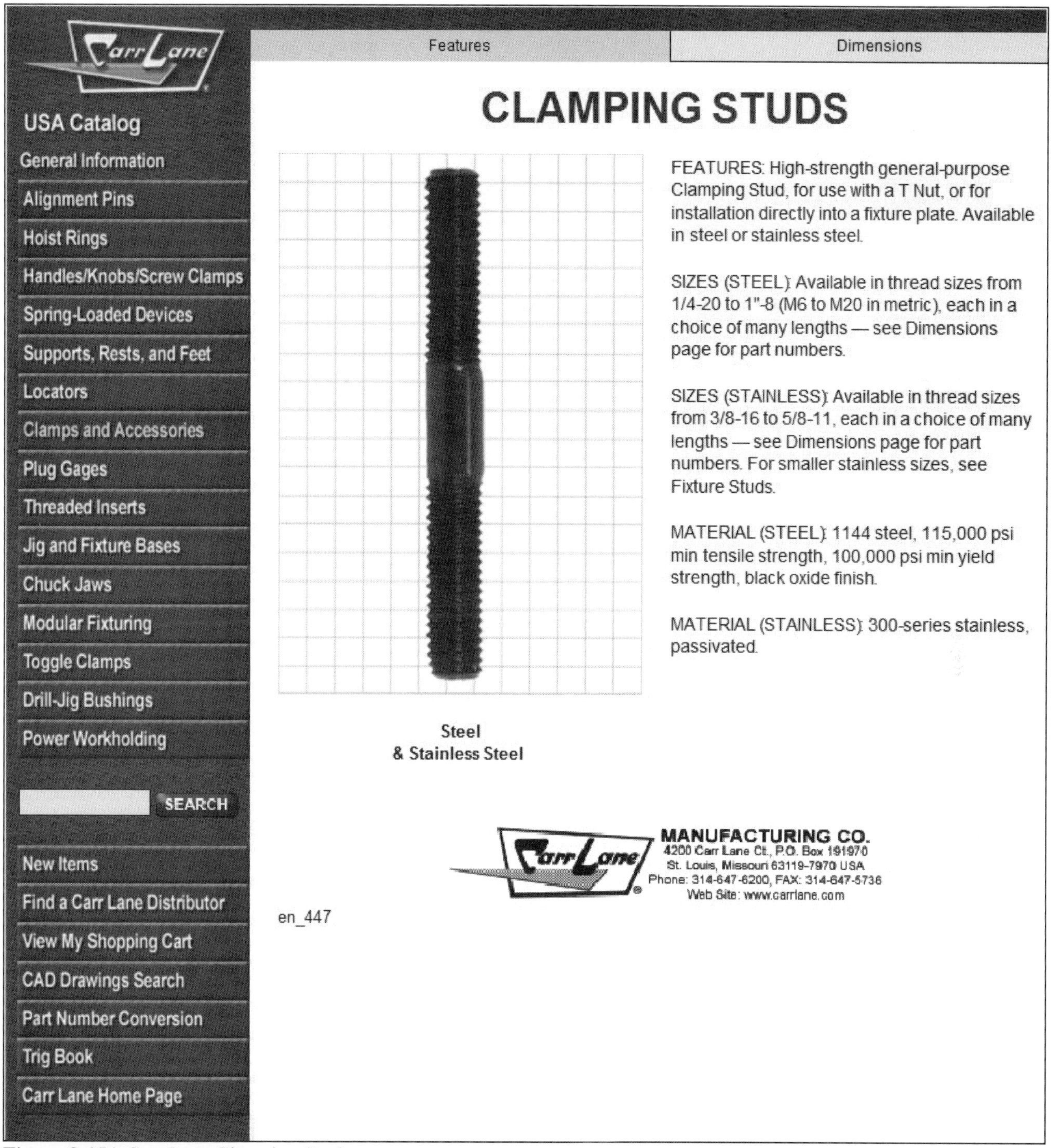

Figure 8.4(b) Carr Lane Clamping Studs

Click: **Dimensions** [Dimensions] [Fig. 8.4(c)] ⇒ select [CL-1/2-13X3.50-STUD]

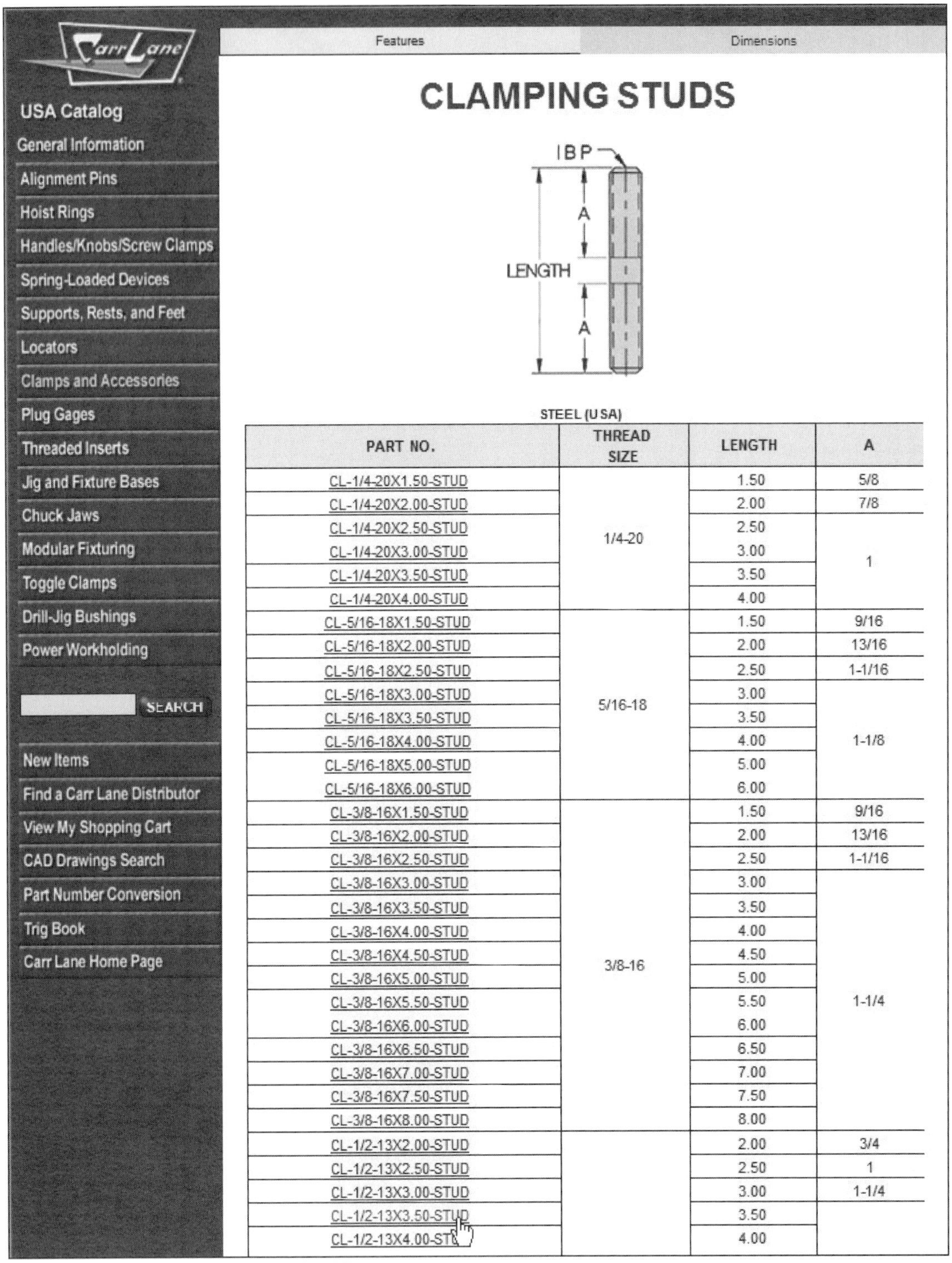

STEEL (USA)

PART NO.	THREAD SIZE	LENGTH	A
CL-1/4-20X1.50-STUD	1/4-20	1.50	5/8
CL-1/4-20X2.00-STUD		2.00	7/8
CL-1/4-20X2.50-STUD		2.50	1
CL-1/4-20X3.00-STUD		3.00	
CL-1/4-20X3.50-STUD		3.50	
CL-1/4-20X4.00-STUD		4.00	
CL-5/16-18X1.50-STUD	5/16-18	1.50	9/16
CL-5/16-18X2.00-STUD		2.00	13/16
CL-5/16-18X2.50-STUD		2.50	1-1/16
CL-5/16-18X3.00-STUD		3.00	1-1/8
CL-5/16-18X3.50-STUD		3.50	
CL-5/16-18X4.00-STUD		4.00	
CL-5/16-18X5.00-STUD		5.00	
CL-5/16-18X6.00-STUD		6.00	
CL-3/8-16X1.50-STUD	3/8-16	1.50	9/16
CL-3/8-16X2.00-STUD		2.00	13/16
CL-3/8-16X2.50-STUD		2.50	1-1/16
CL-3/8-16X3.00-STUD		3.00	1-1/4
CL-3/8-16X3.50-STUD		3.50	
CL-3/8-16X4.00-STUD		4.00	
CL-3/8-16X4.50-STUD		4.50	
CL-3/8-16X5.00-STUD		5.00	
CL-3/8-16X5.50-STUD		5.50	
CL-3/8-16X6.00-STUD		6.00	
CL-3/8-16X6.50-STUD		6.50	
CL-3/8-16X7.00-STUD		7.00	
CL-3/8-16X7.50-STUD		7.50	
CL-3/8-16X8.00-STUD		8.00	
CL-1/2-13X2.00-STUD		2.00	3/4
CL-1/2-13X2.50-STUD		2.50	1
CL-1/2-13X3.00-STUD		3.00	1-1/4
CL-1/2-13X3.50-STUD		3.50	
CL-1/2-13X4.00-STUD		4.00	

Figure 8.4(c) CL-1/2-13X3.50-STUD

Click: **3D Viewer** [3D Viewer] [Fig. 8.4(d)] (download the viewer if necessary or go directly to the next command) ⇒ **CAD Downloads** [CAD Downloads] ⇒ 3D Download Formats **Pro/E Part/Assembly (*.prt)** ⇒ **Go** [Go] [Fig. 8.4(e)]

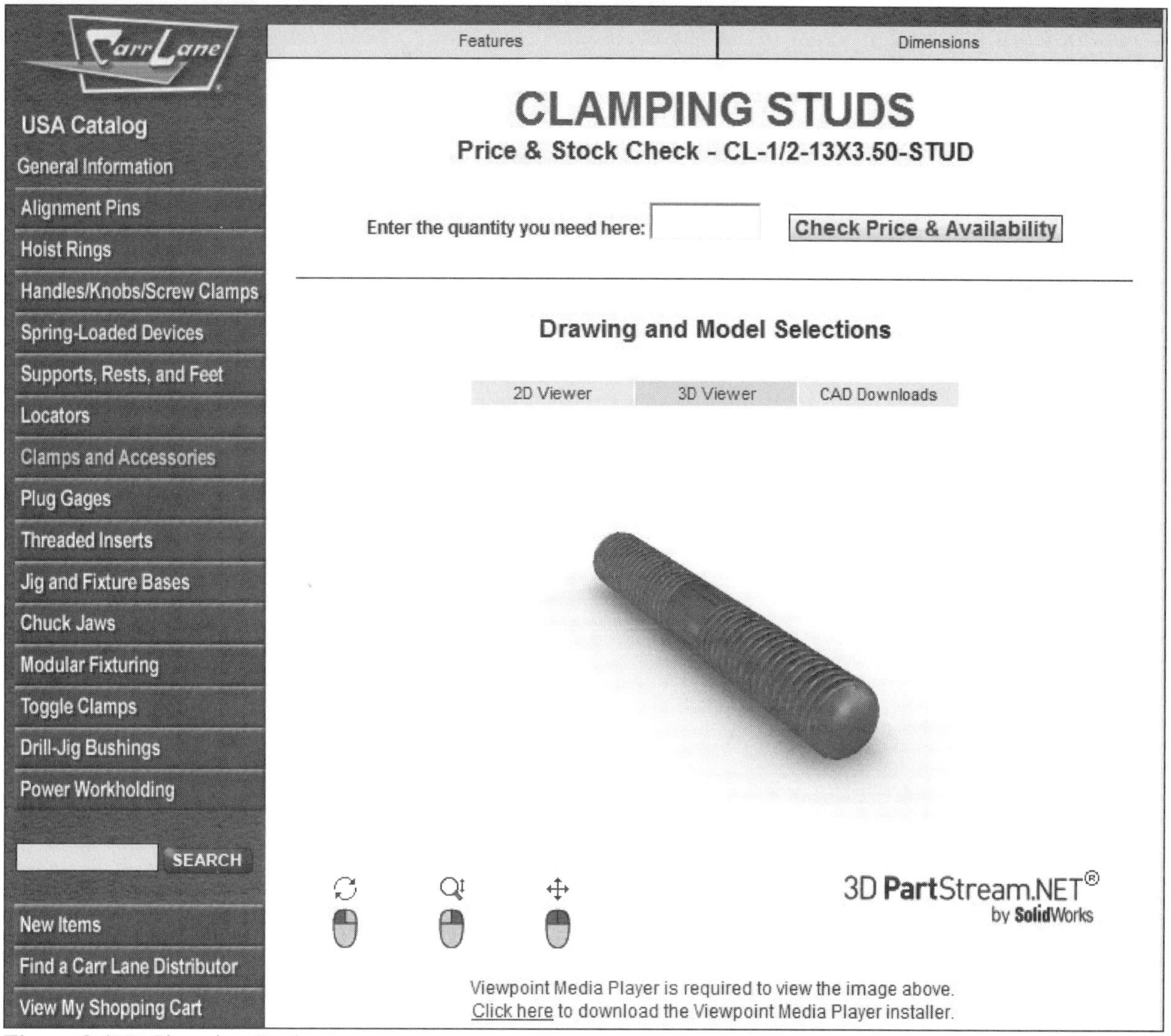

Figure 8.4(d) Clamping Studs

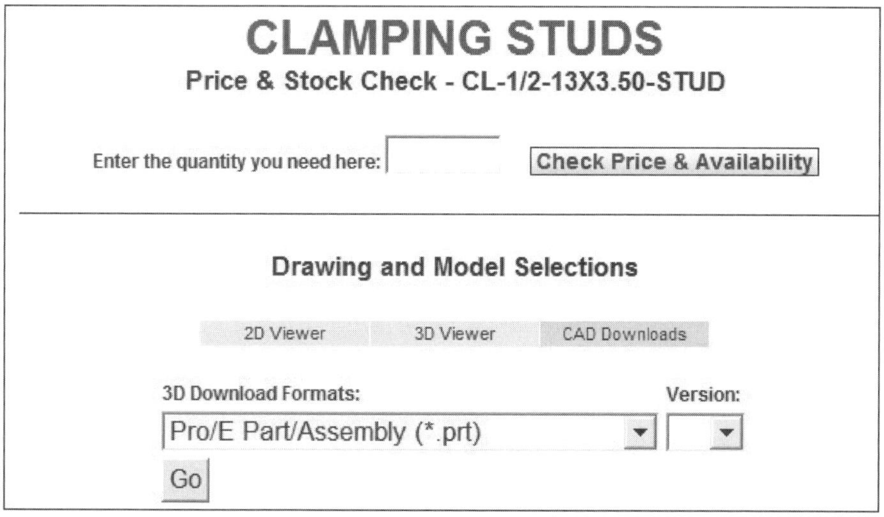

Figure 8.4(e) Pro/E Part/Assembly

Click: **Download Drawing** 

⇒ **Save** (Save As dialog box opens) *[if this does not work, right click on* 

and select "Save Target As..."] ⇒ navigate to your Working Directory ⇒ type (a new name) in File Name: **Carrlane-12-13350_STUD.prt** 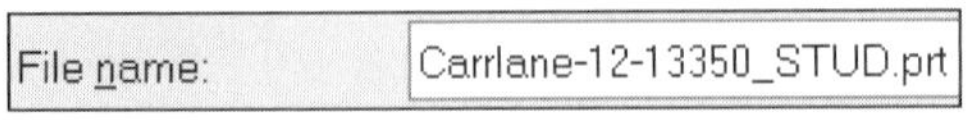

⇒ **Save** ⇒ **Close** ⇒ close the Browser shade  ⇒  **Open** [Fig. 8.4(f)] ⇒ **Carrlane-12-13350_STUD.prt** ⇒ **Open** ⇒ **View** ⇒ **Color and Appearance** ⇒ **Clear** ⇒ **Close** ⇒ change the coordinate system name to **350_STUD** [Fig. 8.4(g)] *(you may also model the stud [Fig. 8.4(h)])* ⇒ **Ctrl+D** ⇒ **Ctrl+S** ⇒ **Enter** ⇒ **Window** ⇒ **Close**

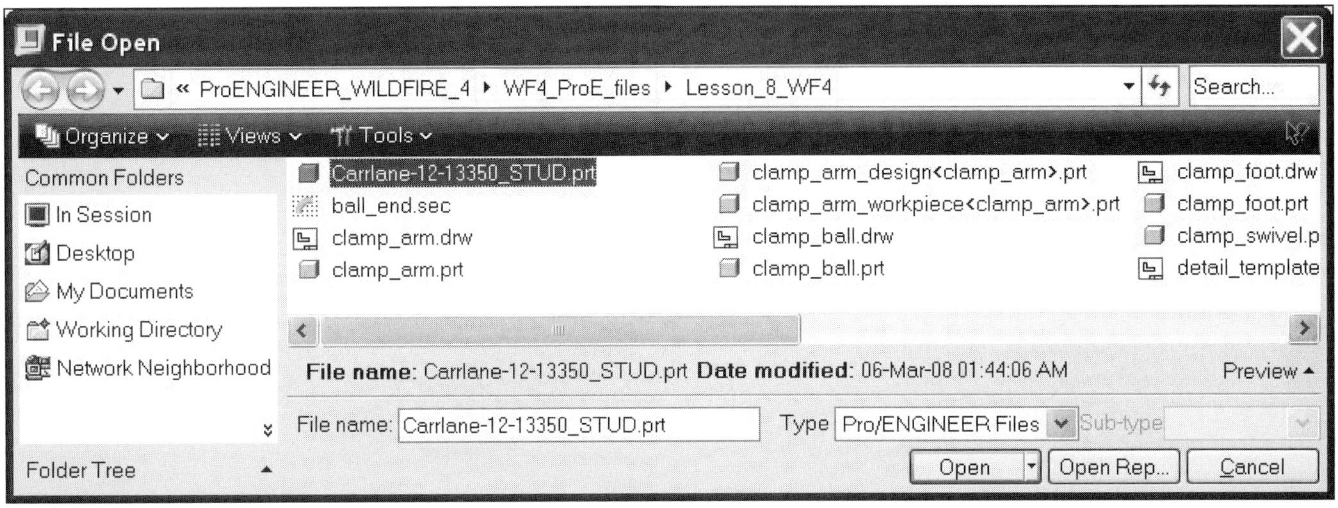

Figure 8.4(f) Pro/E Part/Assembly

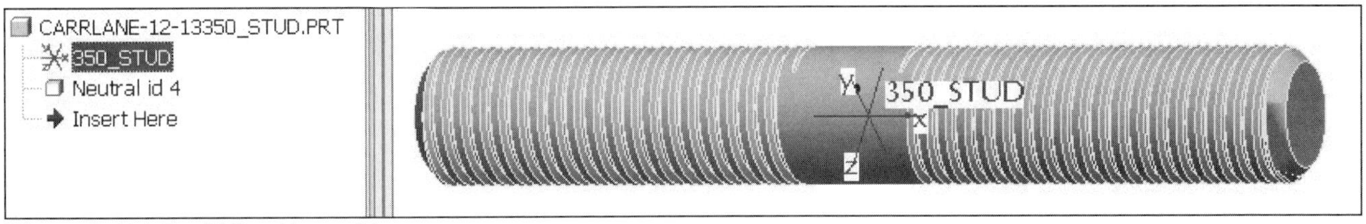

Figure 8.4(g) Pro/E Part/Assembly

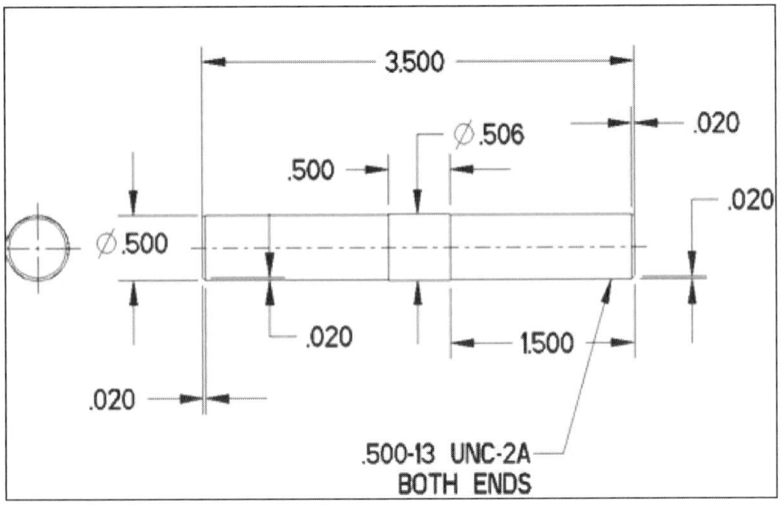

Figure 8.4(h) Model your own stud if the catalog is not available

Reopen the Browser shade ⇒ ⇒ ⇒ ⇒ ⇒ Clamping Studs Dimensions page should be active ⇒ select CL-1/2-13X5.00-STUD ⇒ **3D Viewer** ⇒ **CAD Downloads** ⇒ 3D Download Formats **Pro/E Part/Assembly** ⇒ **Go** ⇒ **Download Drawing** ⇒ **Save** (Save As dialog box opens) ⇒ navigate to your Working Directory ⇒ File Name **Carrlane-12-13500_STUD.prt** ⇒ **Save** ⇒ **Close** ⇒ close the Browser shade ⇒ **Open** [Fig. 8.4(i)] ⇒ **Carrlane-12-13500_STUD.prt** ⇒ **Open** ⇒ **View** ⇒ **Color and Appearance** ⇒ **Clear** ⇒ **Close** ⇒ change the coordinate system name to **500_STUD** [Fig. 8.4(j)] *(you may also model the stud [Fig. 8.4(k)])* ⇒ **Ctrl+D** ⇒ **Ctrl+S** ⇒ **Enter** ⇒ **Window** ⇒ **Close**

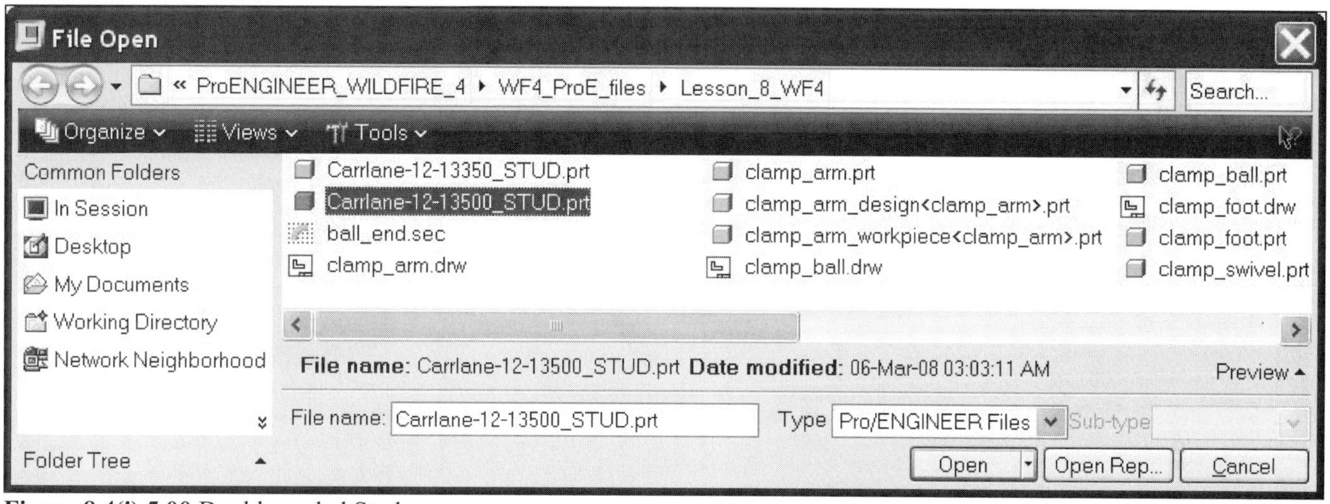

Figure 8.4(i) 5.00 Double-ended Stud

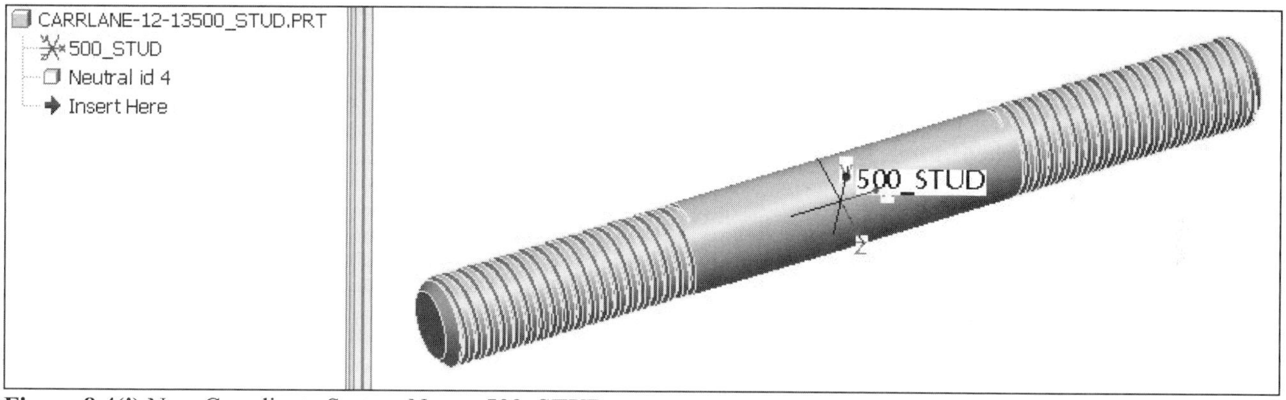

Figure 8.4(j) New Coordinate System Name: **500_STUD**

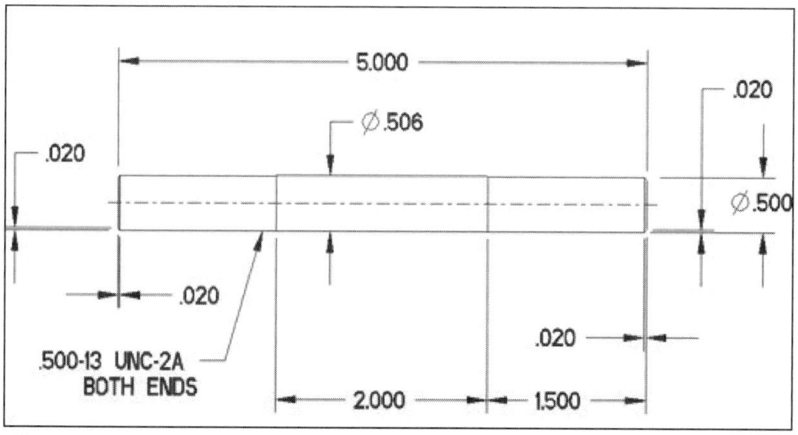

Figure 8.4(k) Double-ended Stud **∅.500** by **5.00** length

Open Browser shade ⇒  ⇒  ⇒  ⇒  ⇒  ⇒  ⇒  ⇒ Clamps and Accessories page should be active [Fig. 8.4(l)] ⇒ **Nuts** 

⇒ **Flange Nuts** 

[Fig. 8.4(m)]

Figure 8.4(l) Clamps and Accessories

Figure 8.4(m) Flange Nuts

Click: **Dimensions** [Fig. 8.4(n)] ⇒ select CL-3-FN ⇒ **3D Viewer** ⇒ **CAD Downloads** ⇒ 3D Download Formats **Pro/E Part/Assembly** ⇒ **Go** ⇒ **Download Drawing** ⇒ **Save** (Save As dialog box opens) ⇒ navigate to your Working Directory ⇒ File Name **carrlane_500_fn.prt** ⇒ **Save** ⇒ **Close** ⇒ close the Browser shade ⇒ **Open** ⇒ **carrlane_500_fn.prt** ⇒ **Open** ⇒ **View** ⇒ **Color and Appearance** ⇒ **Clear** ⇒ **Close** ⇒ change the coordinate system name to **CARRLANE_500_FN** [Fig. 8.4(o)] *(you may also model the Flange Nut [Fig. 8.4(p)])* ⇒ **Ctrl+D** ⇒ **Ctrl+S** ⇒ **Enter** ⇒ **Window** ⇒ **Close**

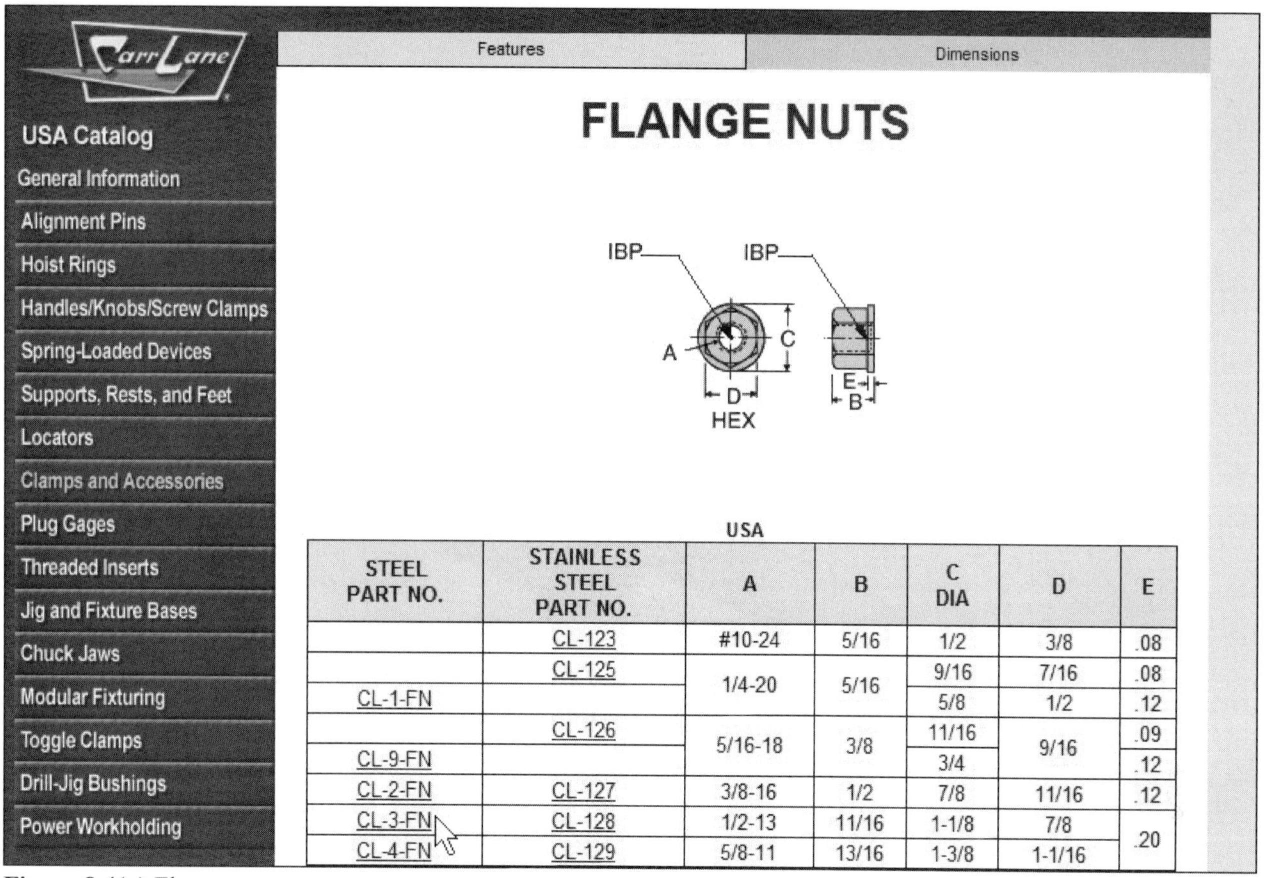

STEEL PART NO.	STAINLESS STEEL PART NO.	A	B	C DIA	D	E
	CL-123	#10-24	5/16	1/2	3/8	.08
	CL-125	1/4-20	5/16	9/16	7/16	.08
CL-1-FN				5/8	1/2	.12
	CL-126	5/16-18	3/8	11/16	9/16	.09
CL-9-FN				3/4		.12
CL-2-FN	CL-127	3/8-16	1/2	7/8	11/16	.12
CL-3-FN	CL-128	1/2-13	11/16	1-1/8	7/8	.20
CL-4-FN	CL-129	5/8-11	13/16	1-3/8	1-1/16	

Figure 8.4(n) Flange

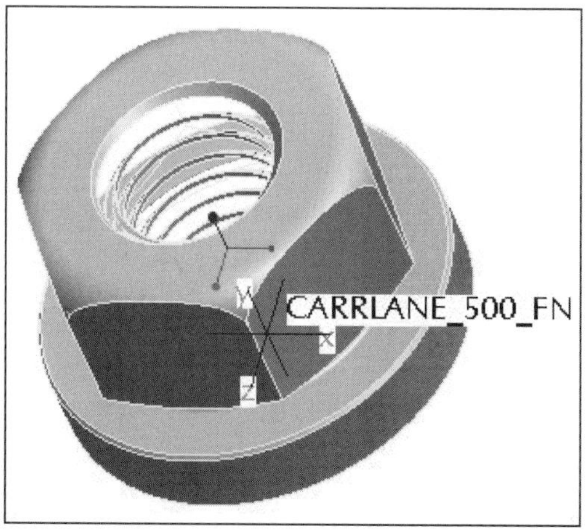

Figure 8.4(o) Flange Coordinate System Name

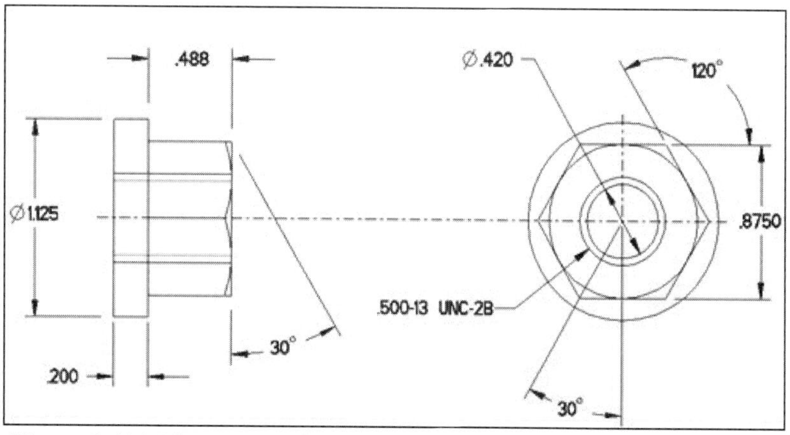

Figure 8.4(p) Flange Nut Dimensions for Modeling

Open the CLAMP_ARM_DESIGN [Fig. 8.5(a)], the Clamp_Swivel [Fig. 8.5(b)], the Clamp_Ball [Fig. 8.5(c)] and the Clamp_Foot [Fig. 8.5(d)] ⇒ review the components and standard parts for correct color, layering, coordinate system naming, and set datum planes. Close all windows. Components remain "in session".

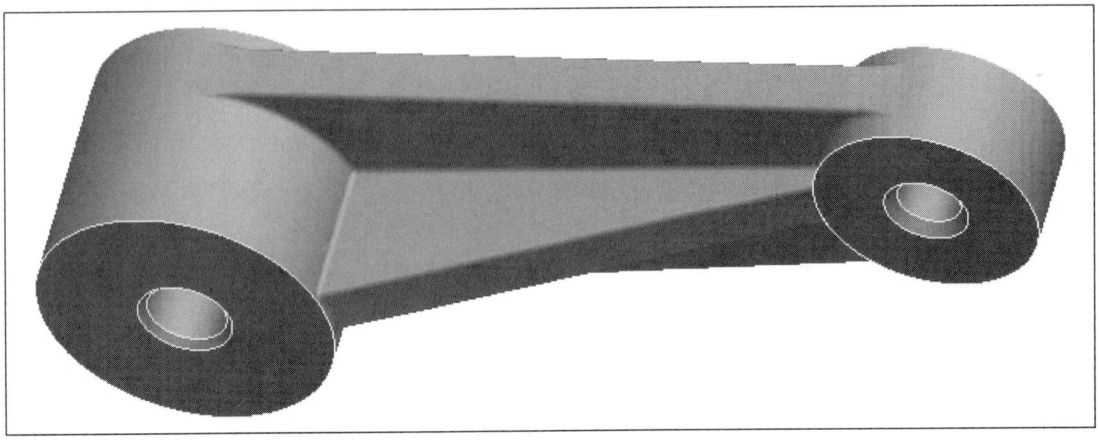

Figure 8.5(a) Clamp_Arm_Design

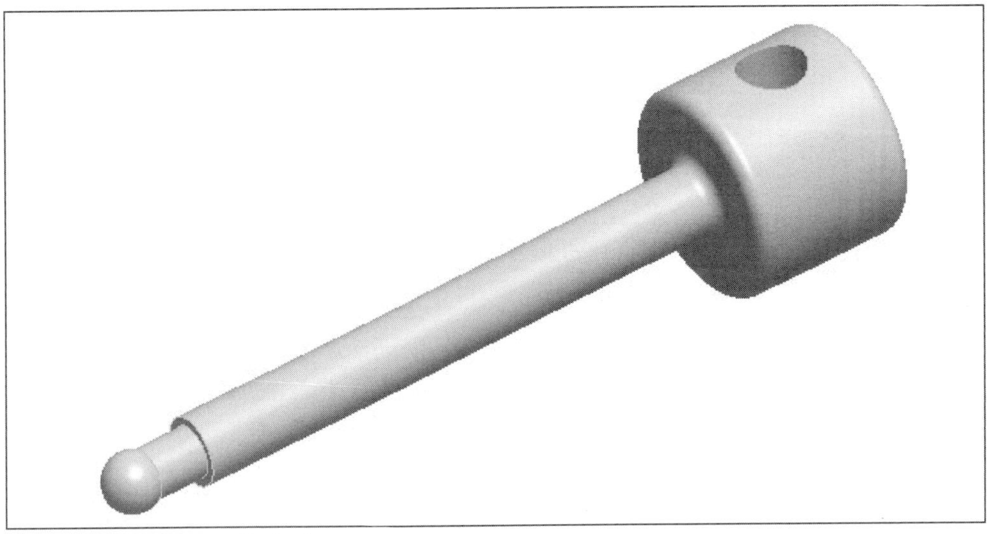

Figure 8.5(b) Clamp_Swivel

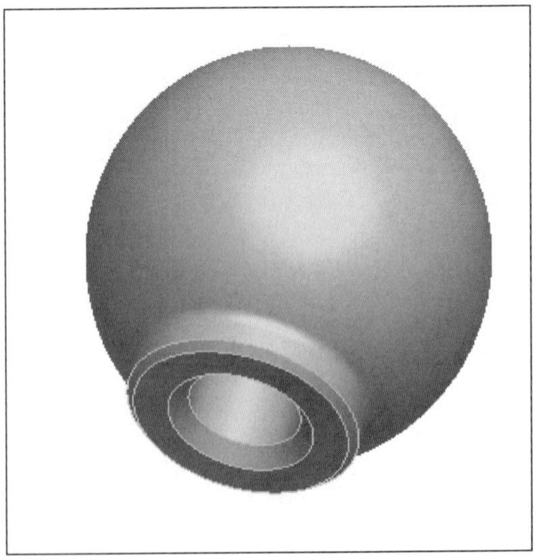

Figure 8.5(c) Clamp_Ball

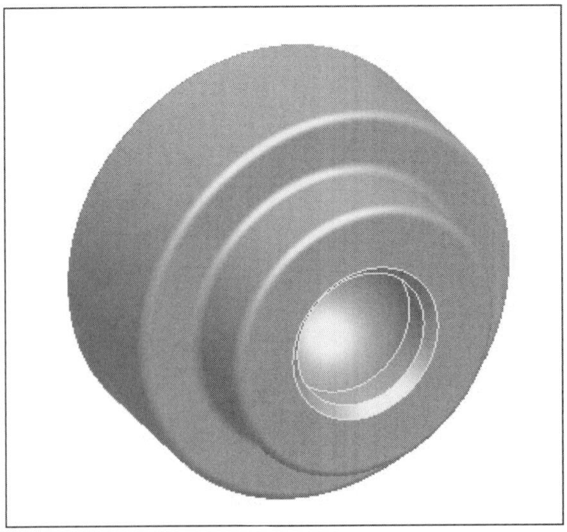

Figure 8.5(d) Clamp_Foot

You now have eight components (two identical Clamp_Ball components are used) required for the assembly. The ninth component- the Clamp_Plate (Fig. 8.6) will be created using *top-down design* procedures when you start the main assembly. *All parts must be in the same working directory used for the assembly.*

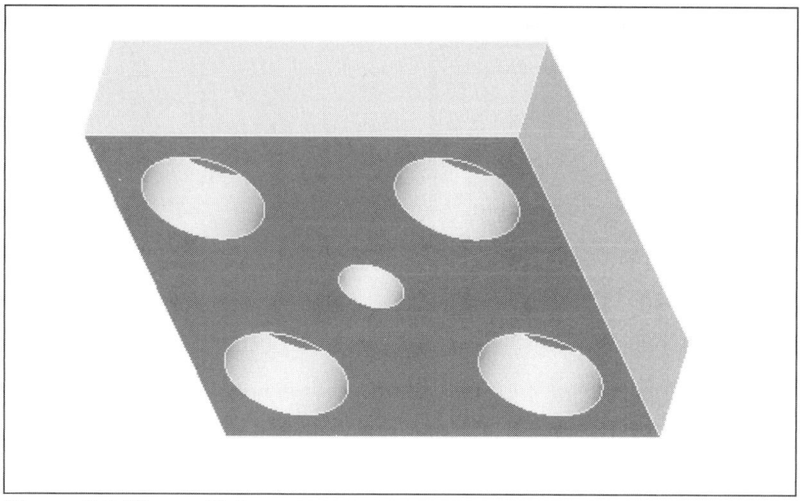

Figure 8.6 Clamp_Plate

A subassembly will be created first. The main assembly is created second. The subassembly will be added to the main assembly to complete the project. Note: the assembly and the components can have different units. Therefore, you must check and correctly set the assembly units before creating or assembling components or sub-assemblies.

Start the subassembly, click: **Create a new object** ⇒ Assembly ⇒ Name **clamp_subassembly** ⇒ Sub-type Design ⇒ Use default template [Fig. 8.7(a)] ⇒ **OK** ⇒ Template: **inlbs_asm_design** ⇒ Parameters: DESCRIPTION: type **Swing Clamp sub-assembly** ⇒ MODELED_BY ***type your name*** [Fig. 8.7(b)] ⇒ **OK** ⇒ **Edit** ⇒ **Setup** ⇒ **Units** ⇒ Inch lbm Second (Pro/E Default) ⇒ **Close** ⇒ **Done**

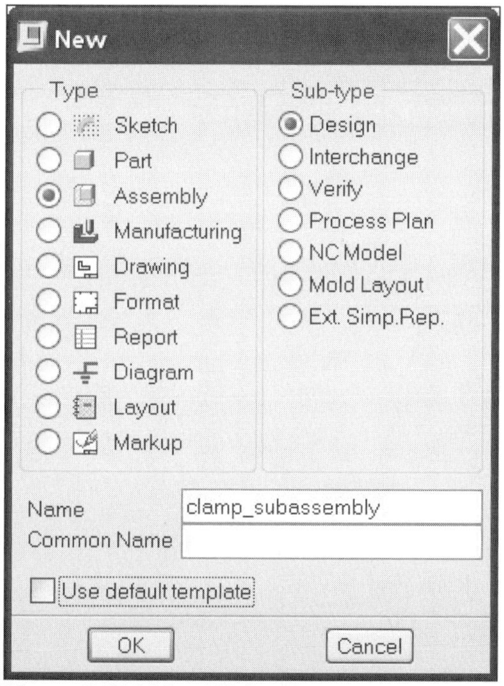

Figure 8.7(a) New Dialog Box

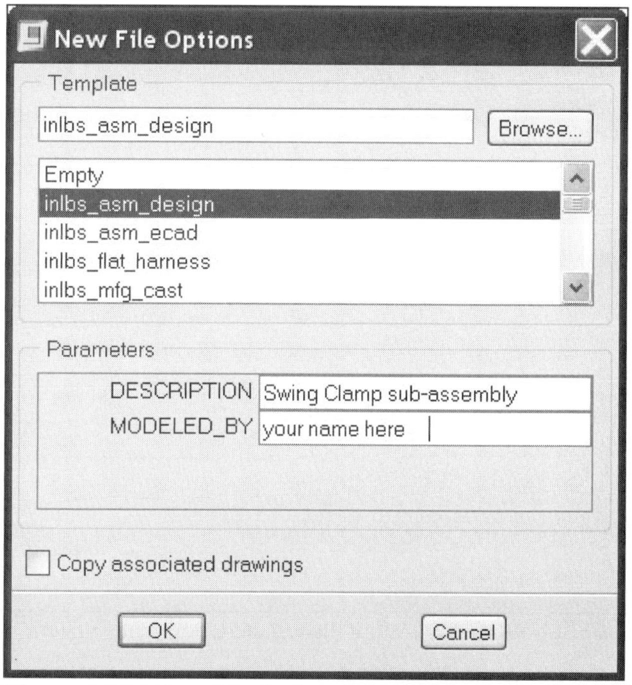

Figure 8.7(b) New File Options Dialog Box

Click: Settings from the Navigator ⇒ Tree Filters... ⇒ check all Display options on [Fig. 8.7(c)] ⇒ **Apply** ⇒ **OK** ⇒ ⇒ **OK**

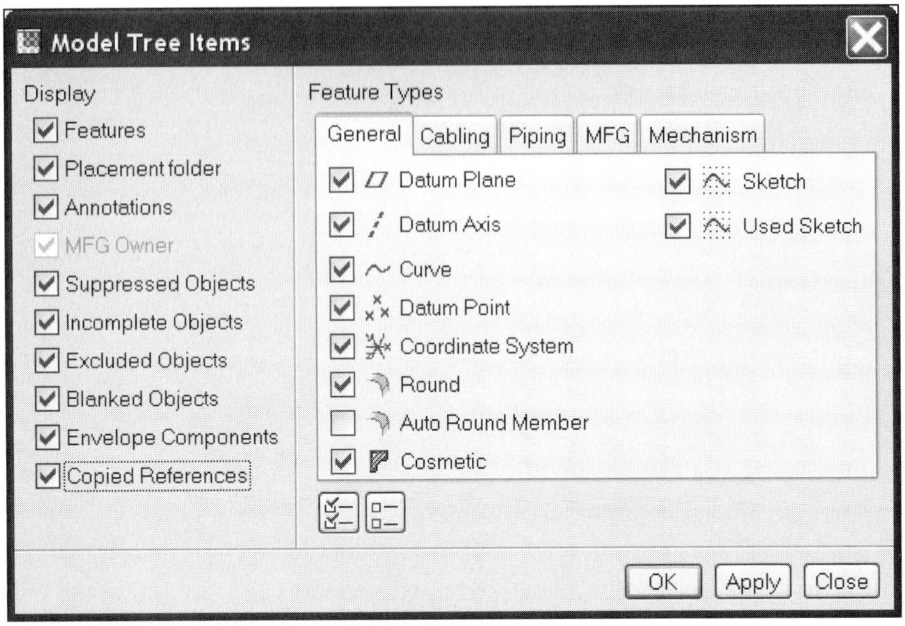

Figure 8.7(c) Model Tree Items Dialog Box

Datum planes and the coordinate system are created per the template provided by Pro/E. The datum planes will have the default names, **ASM_RIGHT**, **ASM_TOP**, and **ASM_FRONT**.

Change the coordinate system name: slowly double-click on ASM_DEF_CSYS in the Model Tree ⇒ type new name **SUB_ASM_CSYS** SUB_ASM_CSYS ⇒ **LMB** in the graphics window [Fig. 8.7(d)]

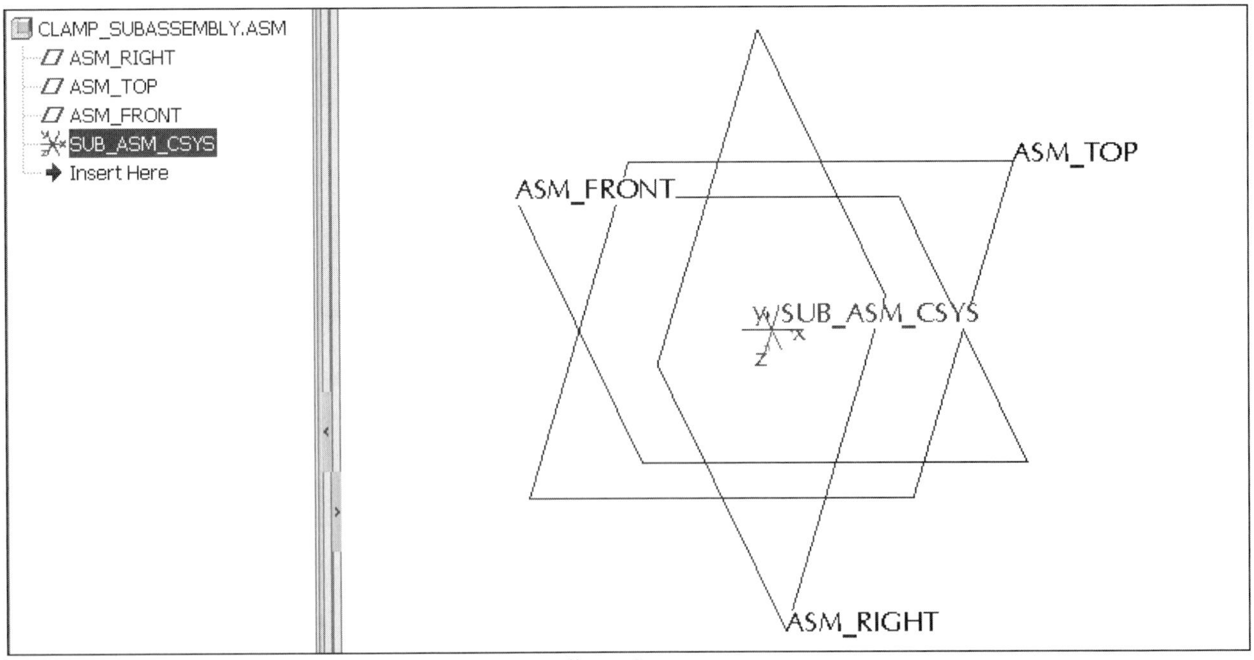

Figure 8.7(d) Sub-Assembly Datum Planes and Coordinate System

Regardless of the design methodology, the assembly datum planes and coordinate system should be on their own separate *assembly layer*. Each part should also be placed on separate assembly layers; the part's datum features should already be on *part layers*. Look over the default template for assembly layering.

Click:  ⇒  ⇒ expand the branches [Fig. 8.7(e)] ⇒  ⇒  *(or click:*  *on ⇒ click:*  *off)*

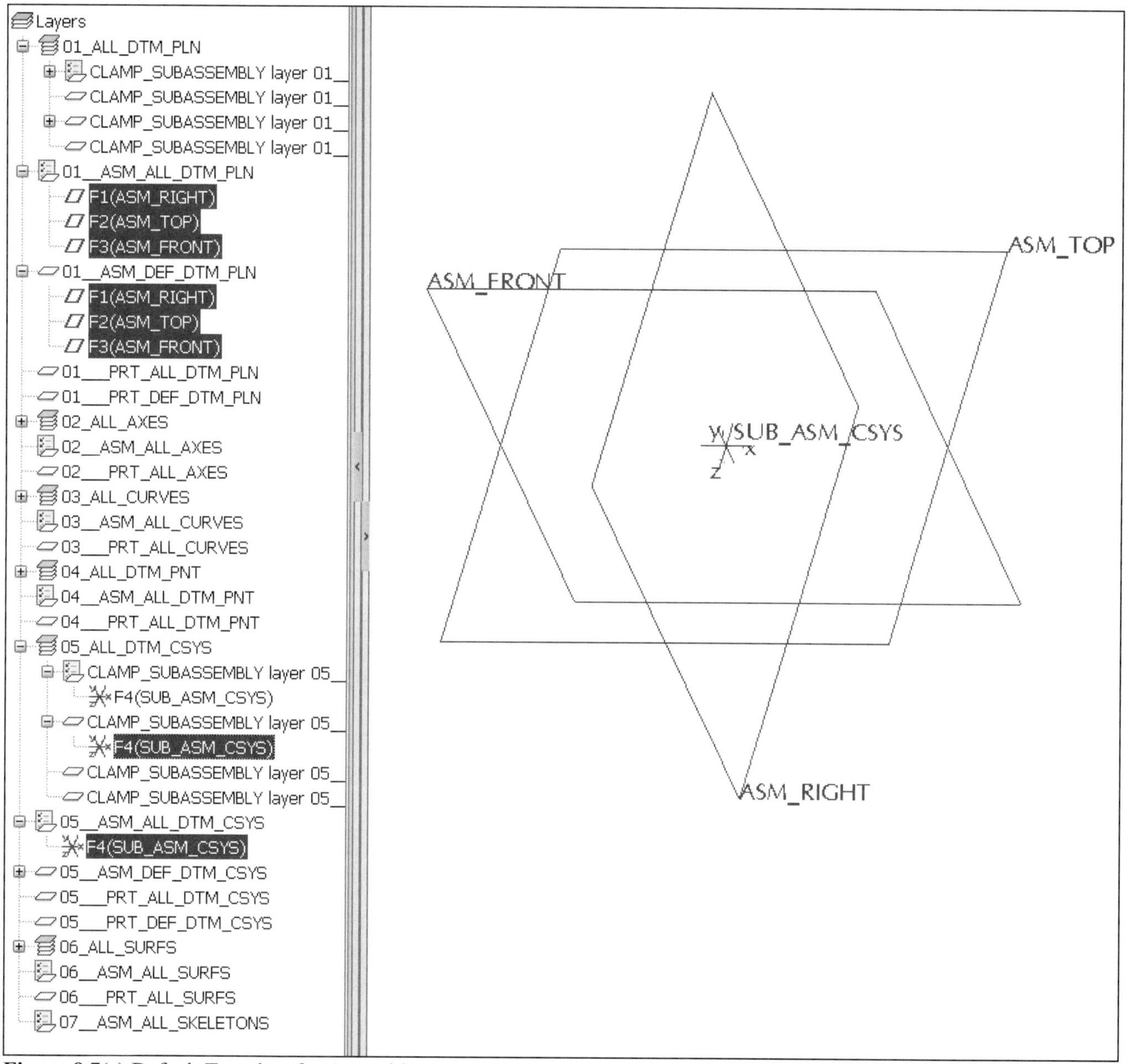

Figure 8.7(e) Default Template for Assembly Layering

 **Context sensitive help** ⇒  ⇒ **About Automatic Placement of Components** ⇒ read ⇒ 

The first component to be assembled to the subassembly is the Clamp_Arm. The simplest and quickest method of adding a component to an assembly is to match the coordinate systems. The first component assembled is usually where this *constraint* is used, because after the first component is established, few if any of the remaining components are assembled to the assembly coordinate system (with the exception of *top-down design*) or, for that matter, other parts' coordinate systems. Make sure all your models are in the same working directory before you start the assembly process.

Click: **Add component to the assembly** ⇒ select the **clamp_arm.prt** ⇒ **Preview** [Fig. 8.8(a)] ⇒ **Open** ⇒ **CLAMP_ARM_DESIGN** [Fig. 8.8(b)] ⇒ **Open** ⇒ **Show component in a separate window while specifying constraints** [Fig. 8.8(c)] *(resize and reposition your windows if needed)*

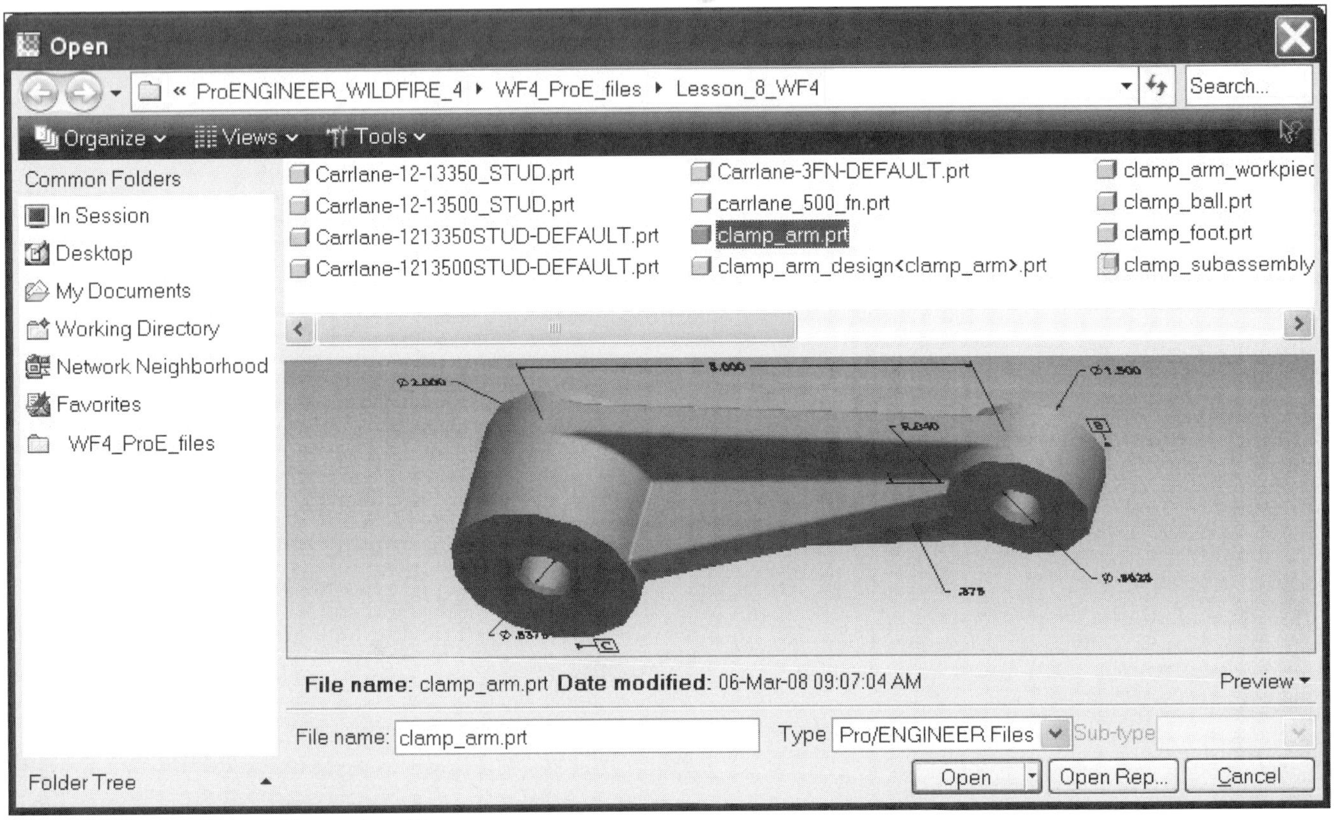

Figure 8.8(a) Previewed Clamp_Arm Component

Figure 8.8(b) Clamp_Arm_Design Instance

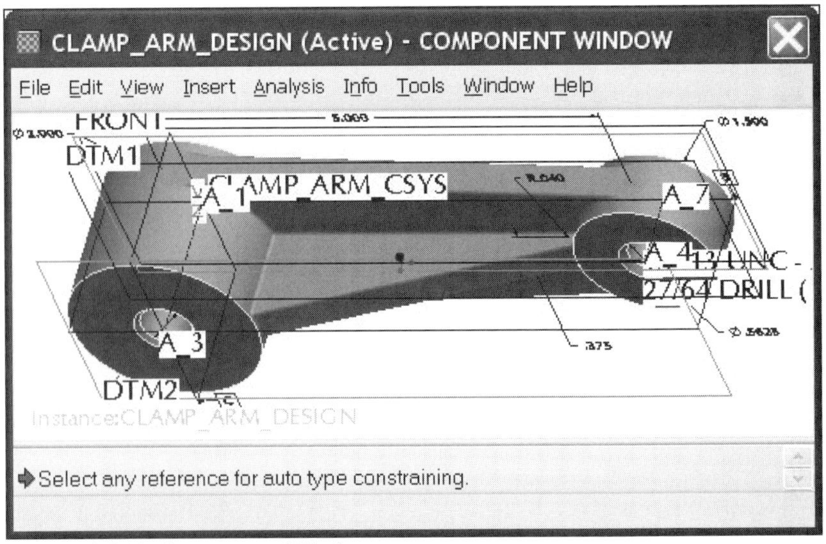

Figure 8.8(c) Component Window

When an assembly is complicated, or it is difficult to select constraining geometry, a separate window aids in the selection process. For simple assemblies (as is the Clamp Assembly and Clamp Subassembly) working in the assembly window is more convenient. Also, since this is the first component being added to the assembly, a separate window is not needed.

Click: to toggle off the component window ⇒ **RMB** ⇒ **Default Constraint** [Fig. 8.8(d)] (this option puts the component in the default position on the assembly model, which is the same as using the Coordinate System constraint) ⇒ **MMB** ⇒ **Tools** ⇒ **Environment** ⇒ Standard Orient: **Isometric** ⇒ **Apply** ⇒ **OK** ⇒ in the Model Tree, click: CLAMP_ARM_DESIGN.PRT ⇒ Placement ⇒ Set2 (your Set # may be different) ⇒ Default ⇒ off [Fig. 8.8(e)] ⇒ **Shading**

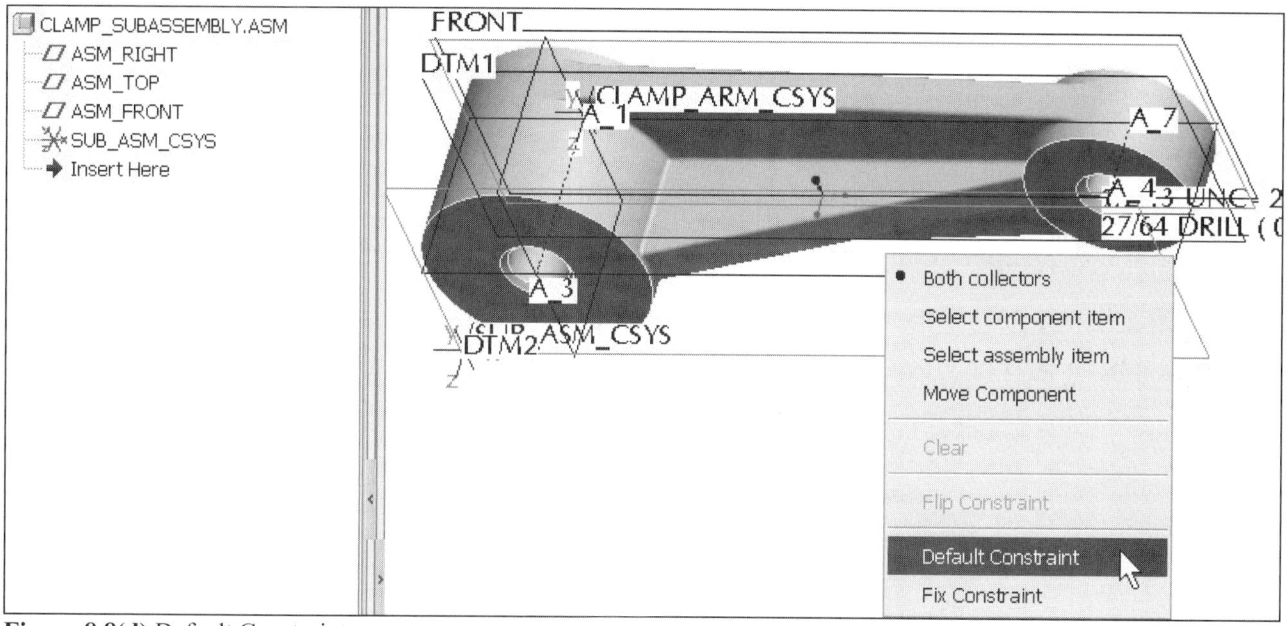

Figure 8.8(d) Default Constraint

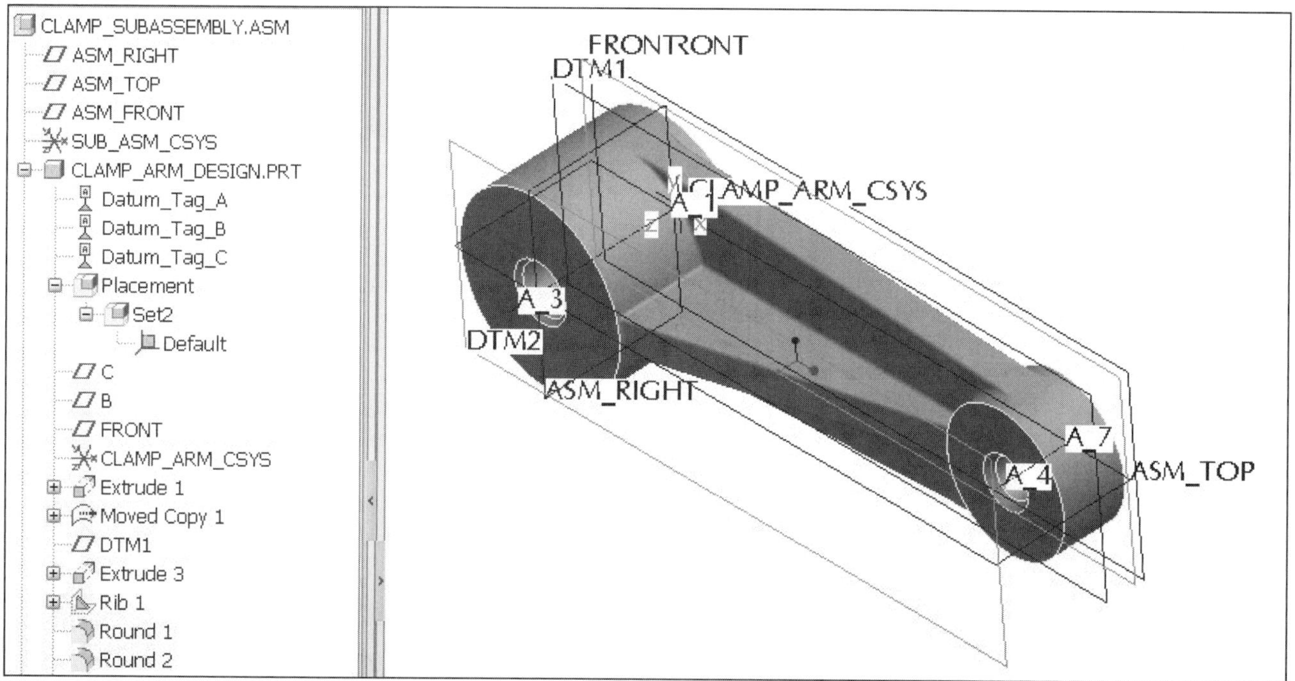

Figure 8.8(e) Expanded Placement Folder

The Clamp_Arm has been assembled to the default location. This means to align the default Pro/E-created coordinate system of the component to the default Pro/E-created coordinate system of the assembly. Pro/E places the component at the assembly origin. By using the constraint, Coord Sys and selecting the assembly and then the component's coordinate systems would have accomplished the same thing, but with more picks.

The next component to be assembled is the Clamp_Swivel. Two constraints will be used with this component: Insert and Mate (Offset). *Placement constraints* are used to specify the relative position of a *pair of surfaces/references* between two components. The Mate, Align, Insert, commands are placement constraints. The two surfaces/references must be of the same type. When using a datum plane as a placement constraint, specify Mate or Align. When using Mate (Offset) or Align (Offset), enter the *offset distance*. The *offset direction* is displayed with an arrow in the graphics window. *If you need an offset in the opposite direction, enter a negative value.*

Click: **Insert** ⇒ **Component** ⇒ **Assemble** 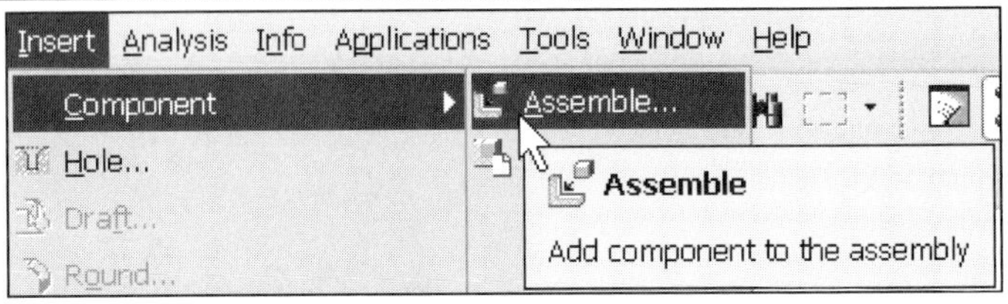

⇒ select the **clamp_swivel.prt** from the Open dialog box ⇒ **Preview** *(if not already on)* ⇒ **Open** ⇒ 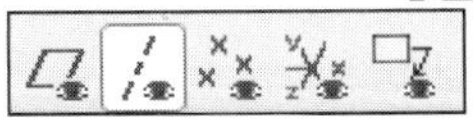 Toggle off the datum planes, datum point, coordinate systems and 3D annotations. Leave datum axis on. ⇒ **Placement** tab on [Fig. 8.9(a)] ⇒ **Placement** tab off

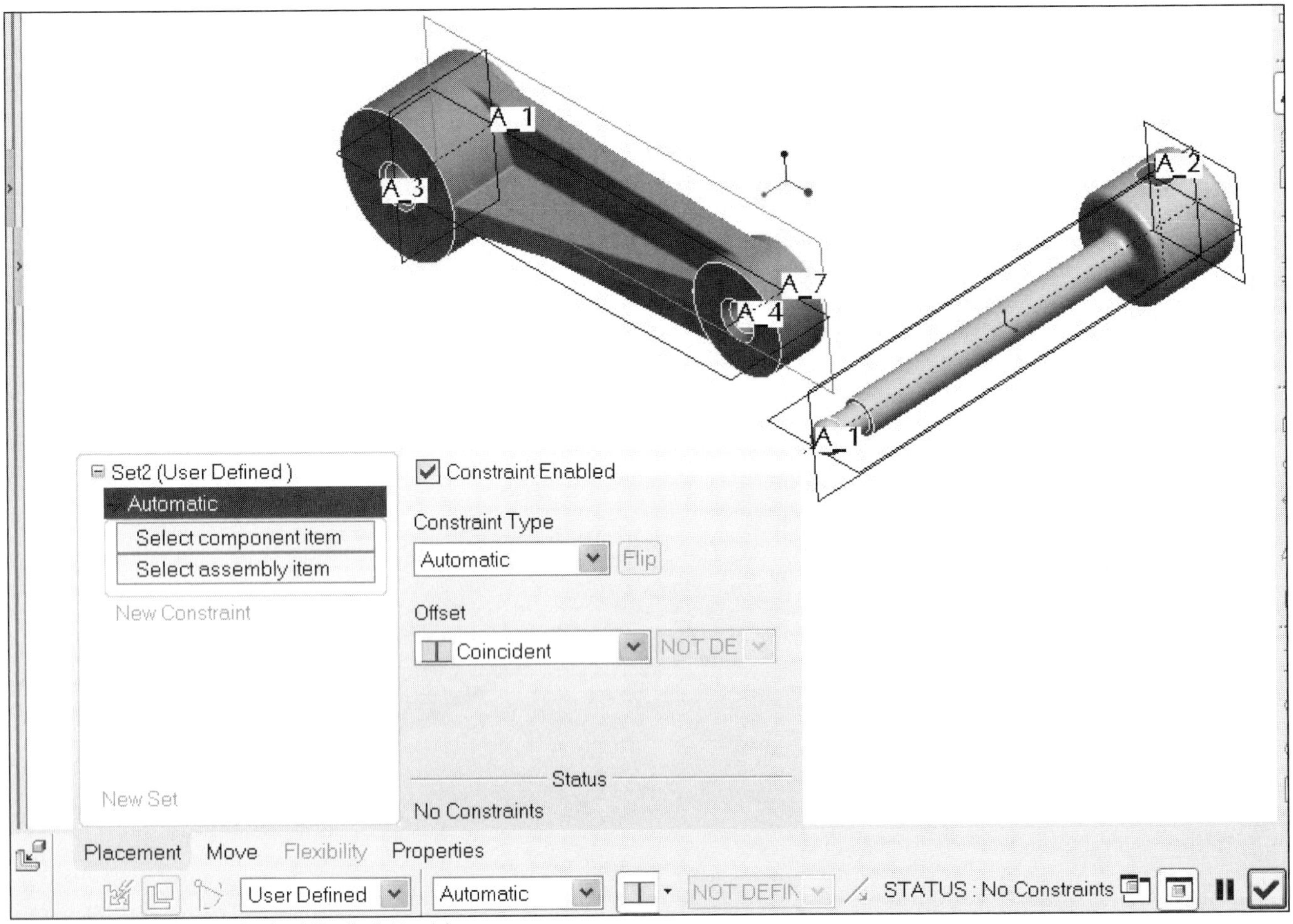

Figure 8.9(a) Clamp_Swivel Default Position

Pick the cylindrical surface of the Clamp_Swivel [Fig. 8.9(b)] ⇒ pick the hole surface of the Clamp_Arm [Fig. 8.9(c)] *constraint type becomes Insert* [Fig. 8.9(d)]

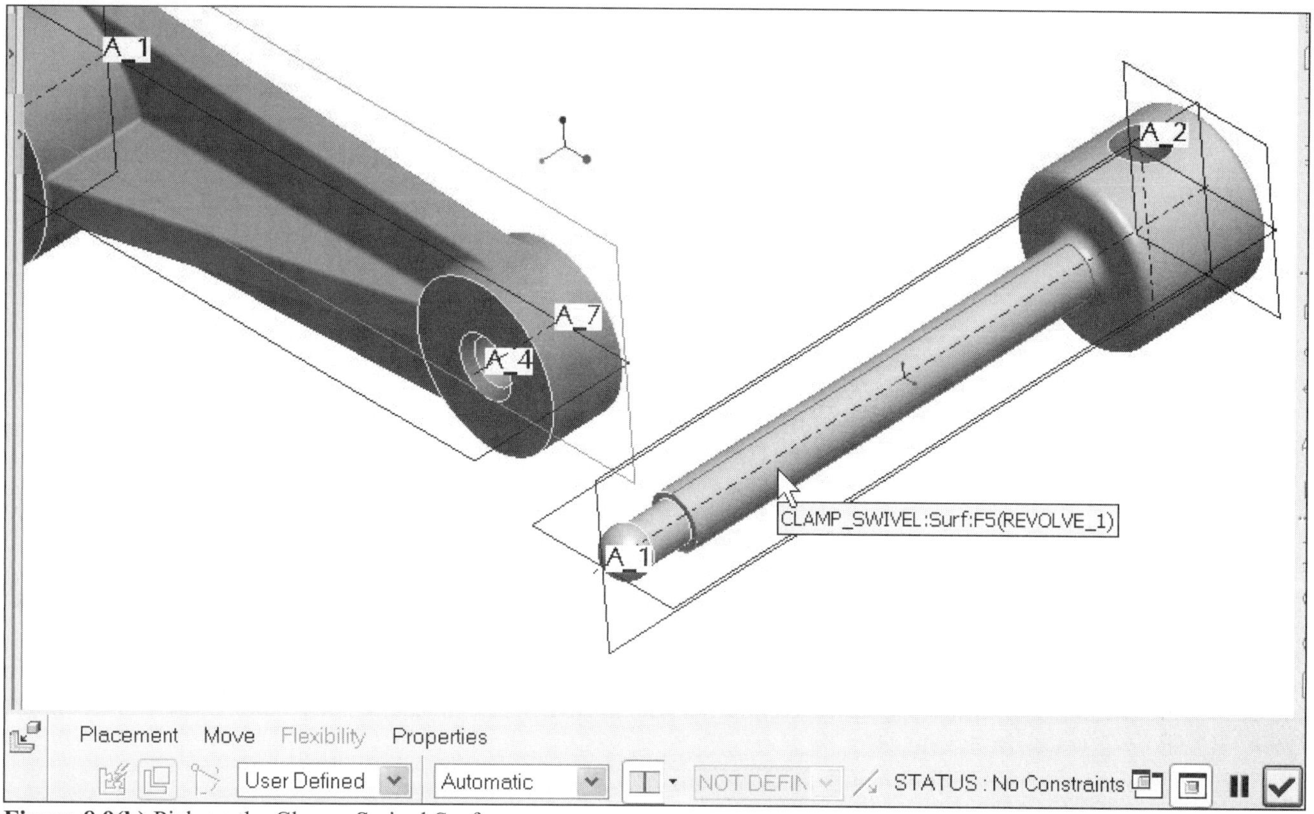

Figure 8.9(b) Pick on the Clamp_Swivel Surface

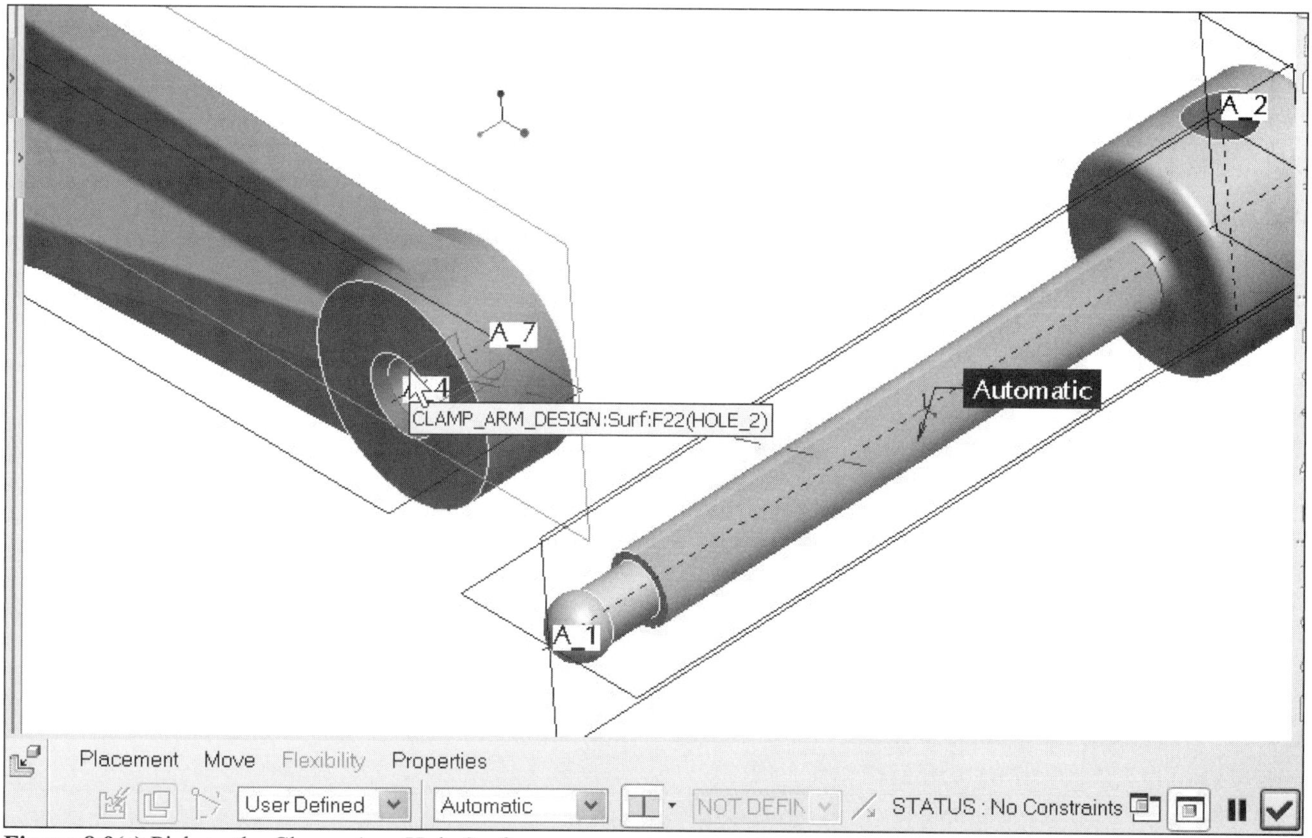

Figure 8.9(c) Pick on the Clamp_Arm Hole Surface

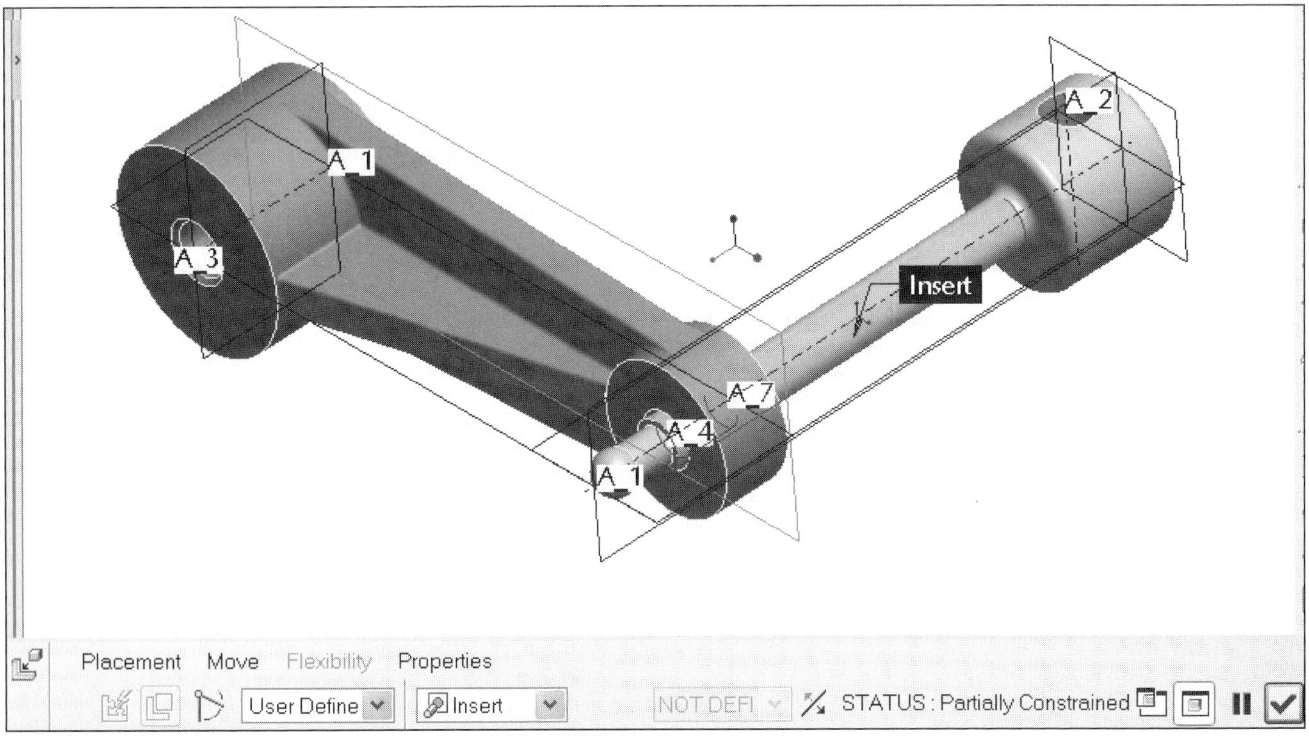

Figure 8.9(d) Insert Constraint Completed 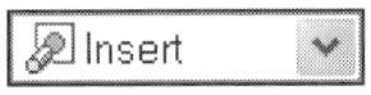

Click: **RMB ⇒ New Constraint** [Fig. 8.9(e)]

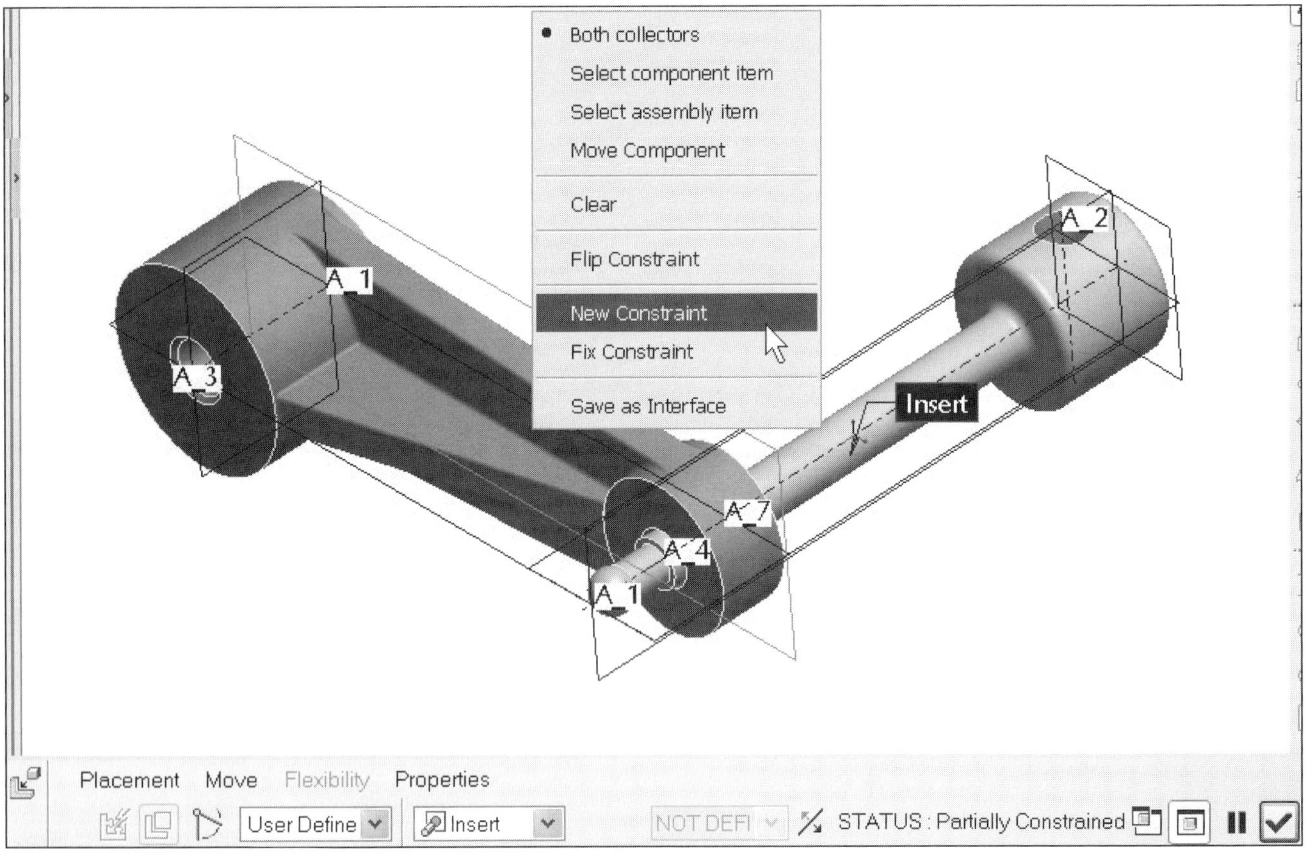

Figure 8.9(e) RMB ⇒ New Constraint

Pick the surface of the Clamp_Swivel [Fig. 8.9(f)] ⇒ pick the surface of the Clamp_Arm [Fig. 8.9(g)]

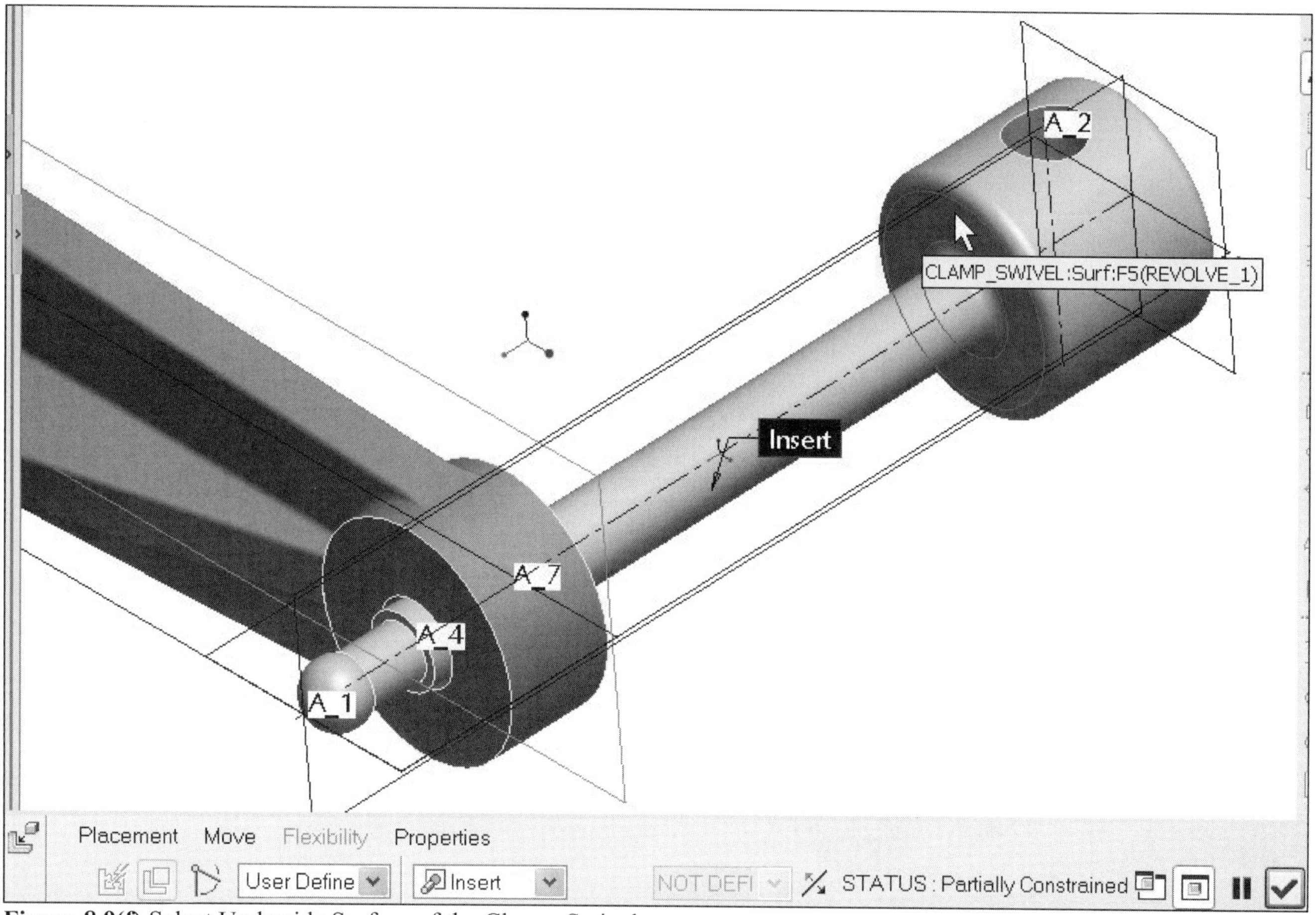

Figure 8.9(f) Select Underside Surface of the Clamp_Swivel

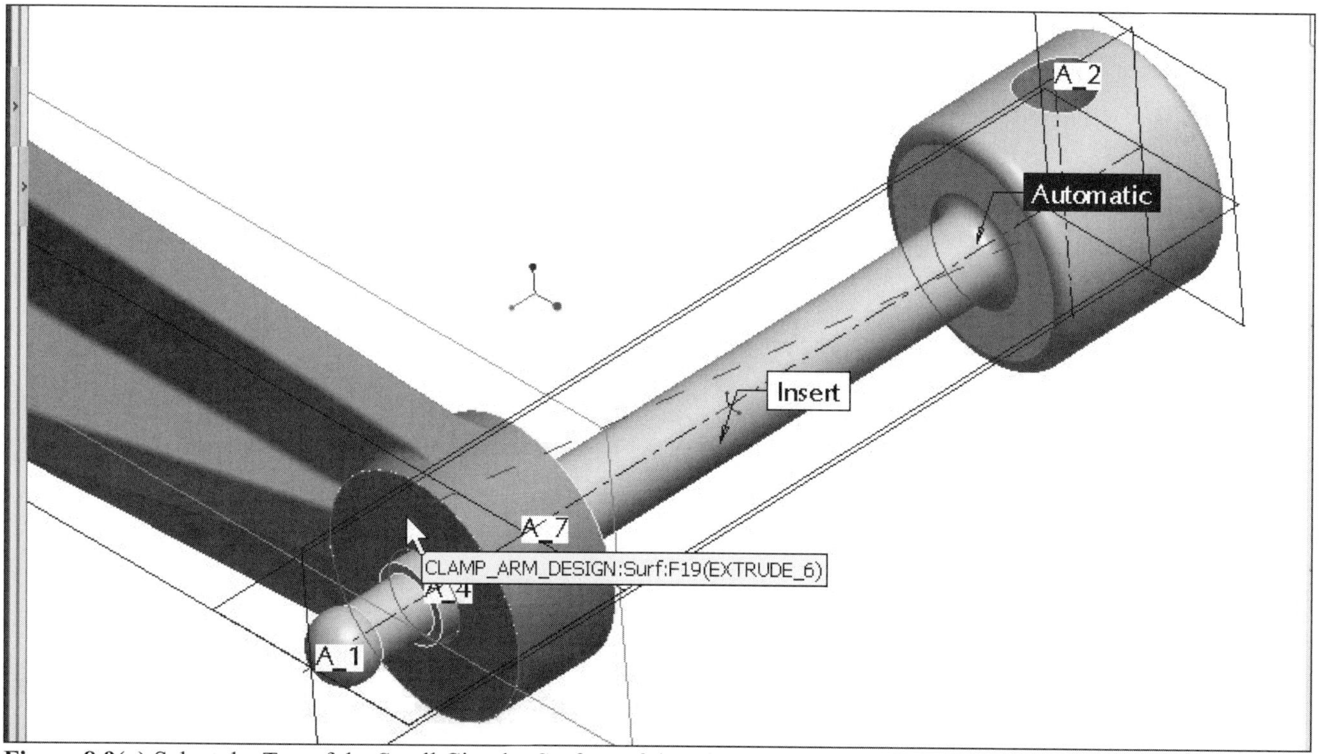

Figure 8.9(g) Select the Top of the Small Circular Surface of the Clamp_Arm

Click: **RMB** ⇒ **Flip Constraint** [Fig. 8.9(h)] *the Clamp_Swivel reverses* ⇒ 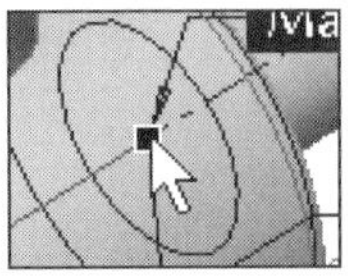 **LMB** drag the Clamp_Swivel until it is **1.50** offset from the Clamp_Arm surface [Fig. 8.9(i)]

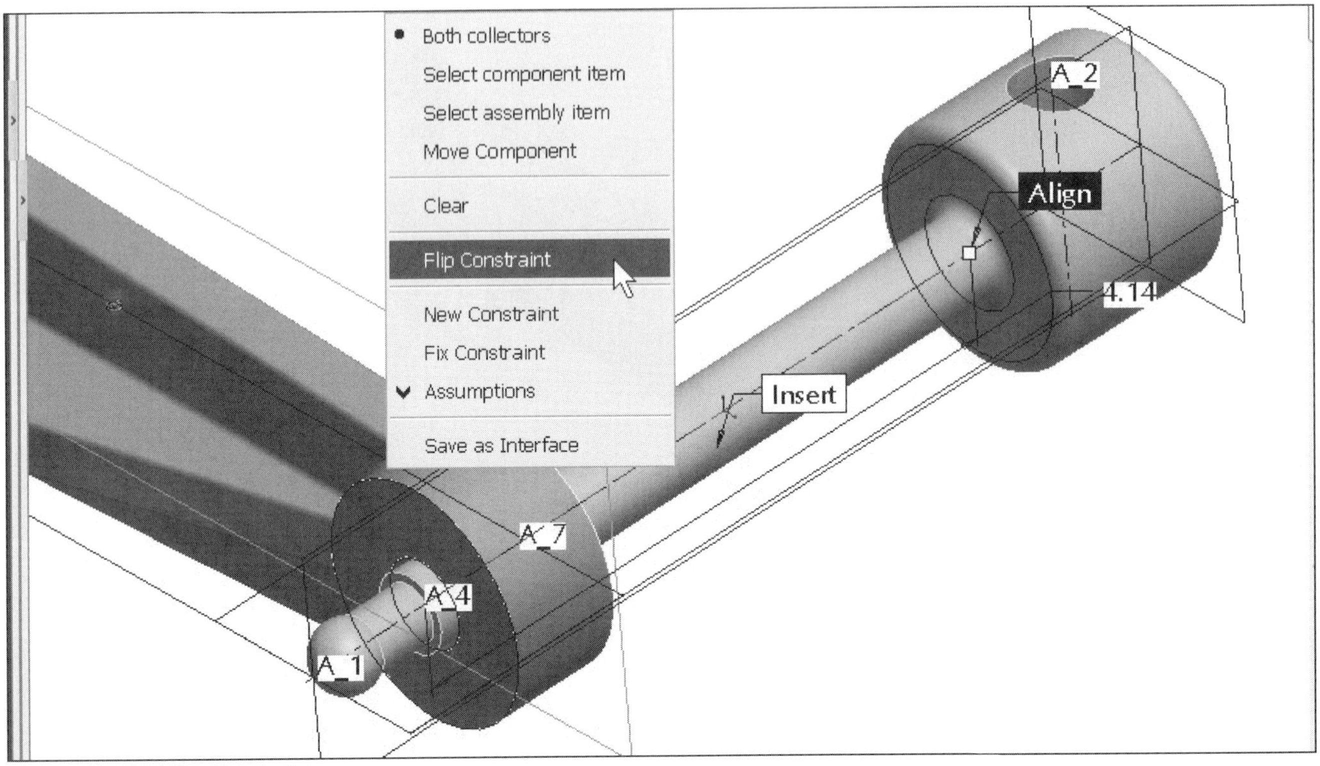

Figure 8.9(h) Flip Constraint

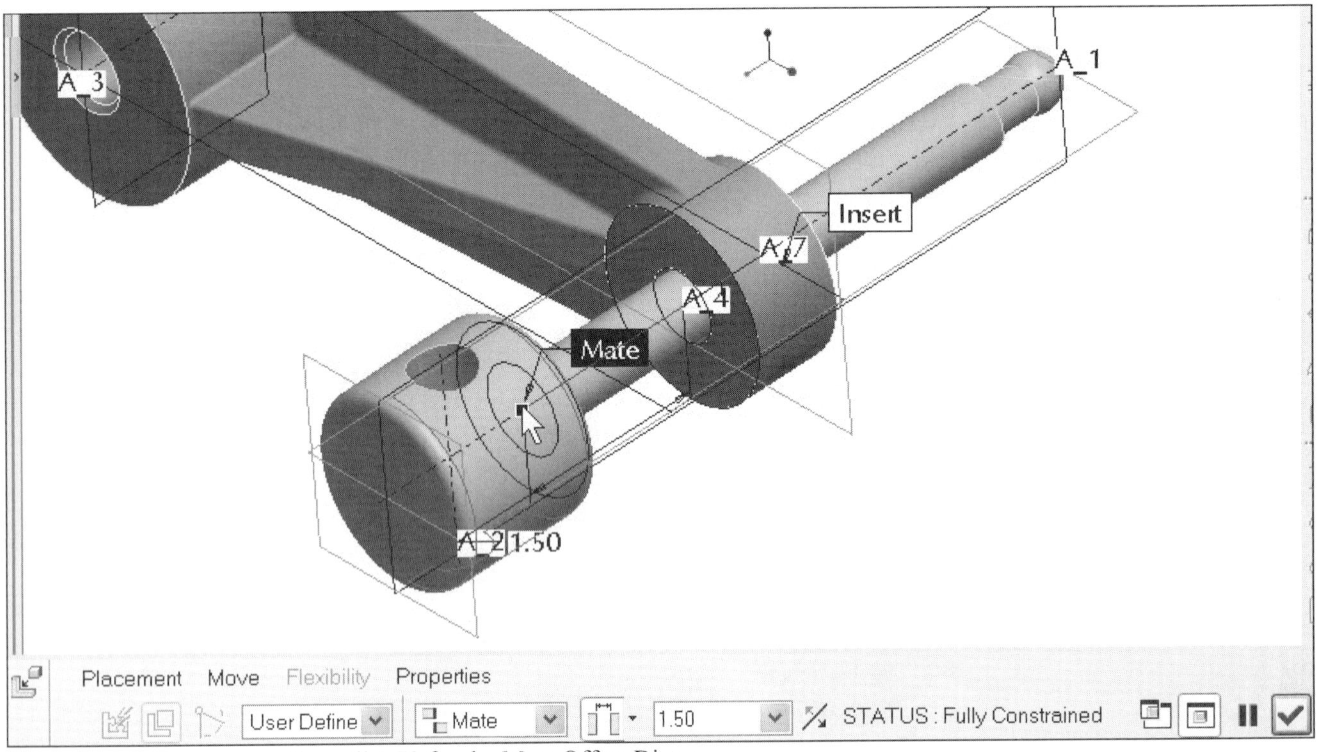

Figure 8.9(i) Drag the handle until **1.50** for the Mate Offset Distance

Click: **Placement** tab [Fig. 8.9(j)] the Placement Status shows the component is STATUS : Fully Constrained and ☑ Allow Assumptions is checked by default ⇒ ☑ ⇒ **Edit** ⇒ **Regenerate** ⇒ **View** ⇒ **Repaint** ⇒ **View** ⇒ **Orientation** ⇒ **Refit** ⇒ **Ctrl+S** ⇒ **OK** ⇒ **File** ⇒ **Delete** ⇒ **Old Versions** ⇒ **MMB**

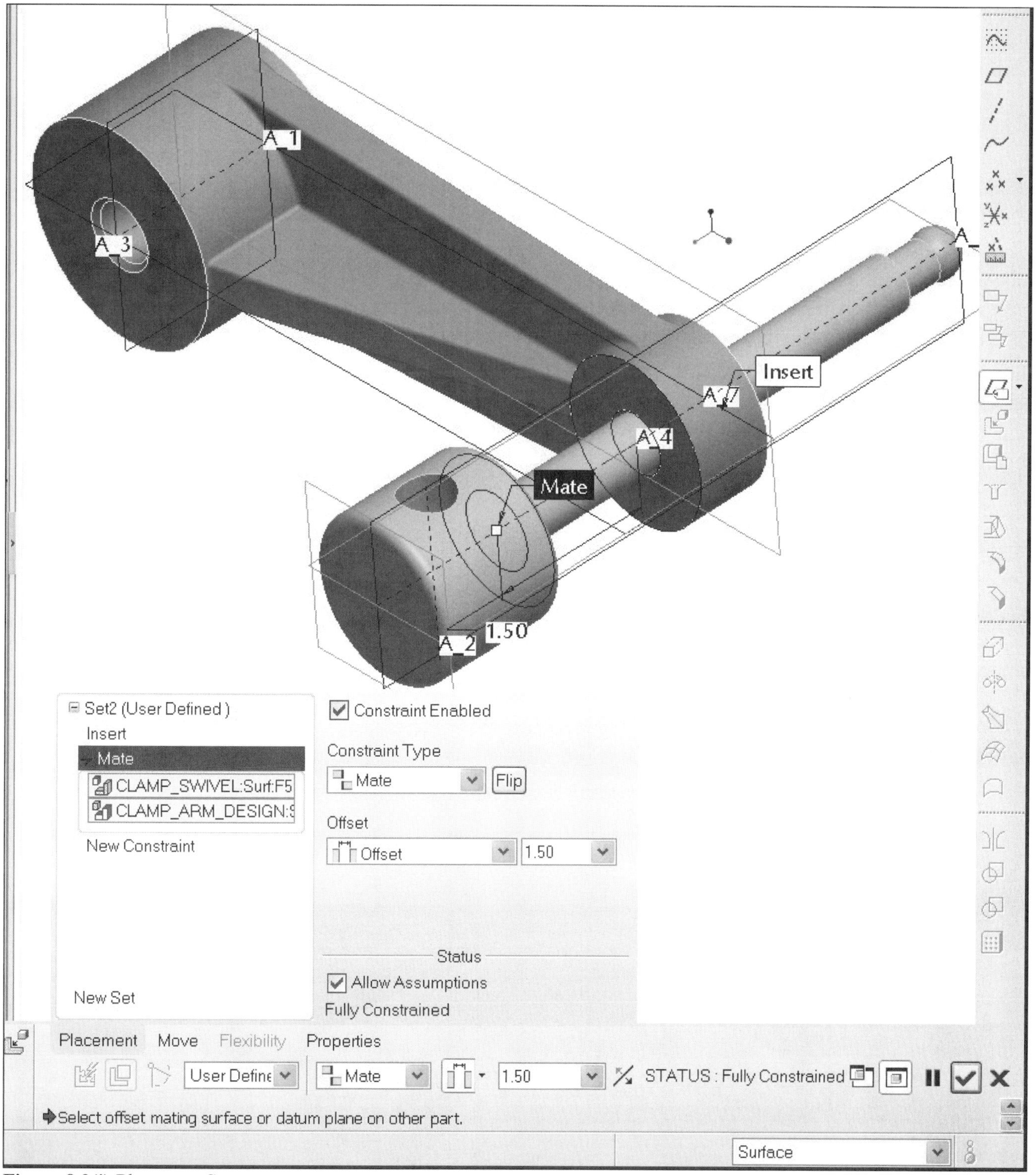

Figure 8.9(j) Placement Status

Regenerating Models

You can use Regenerate to find bad geometry, broken parent-child relationships, or any other problem with a part feature or assembly component. In general, it is a good idea to regenerate the model every time you make a change, so that you can see the effects of each change in the graphics window as you build the model. By regenerating often, it helps you stay on course with your original design intent by helping you to resolve failures as they happen.

When Pro/E regenerates a model, it recreates the model feature by feature, in the order in which each feature was created, and according to the hierarchy of the parent-child relationship between features.

In an assembly, component features are regenerated in the order in which they were created, and then in the order in which each component was added to the assembly. Pro/E regenerates a model automatically in many cases, including when you open, save, or close a part or assembly or one of its instances, or when you open an instance from within a Family Table. You can also use the Regenerate command to manually regenerate the model.

The Regenerate command, located on the Edit menu or using the icon **Regenerates Model**, lets you recalculate the model geometry, incorporating any changes made since the last time the model was saved. If no changes have been made, Pro/E informs you that the model has not changed since the last regeneration. The Custom Regenerate command (Assembly Mode), located on the Edit menu or using the icon, **Specify the list of modified features or components to regenerate**, opens the Regeneration Manager dialog box if there are features or components that have been changed that require regeneration. A column next to the Regeneration list indicates which entries you have selected to skip regeneration or to be regenerated.

In the dialog box you can:

- Select all features/components for regeneration, select Regenerate and then Select All.
- Omit all features/components from regeneration, select Skip Regen and then Select All.
- Determine the reason an object requires regeneration, select Highlight and select an entry in the Regeneration list.

Double-click on the Clamp_Swivel [Fig. 8.10(a)] ⇒ double-click on the **1.50** dimension and modify it to **1.75** [Fig. 8.10(b)] ⇒ **Enter** ⇒ **Specify the list of modified features or components to regenerate** ⇒ Regeneration Manager dialog box opens [Fig. 8.10(c)] ⇒ Regenerate ⇒ **Select All** ⇒ **OK**

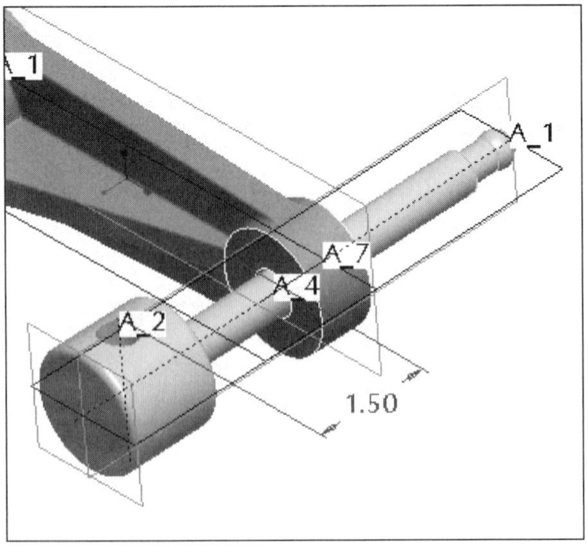

Figure 8.10(a) Double-click on the Clamp_Swivel

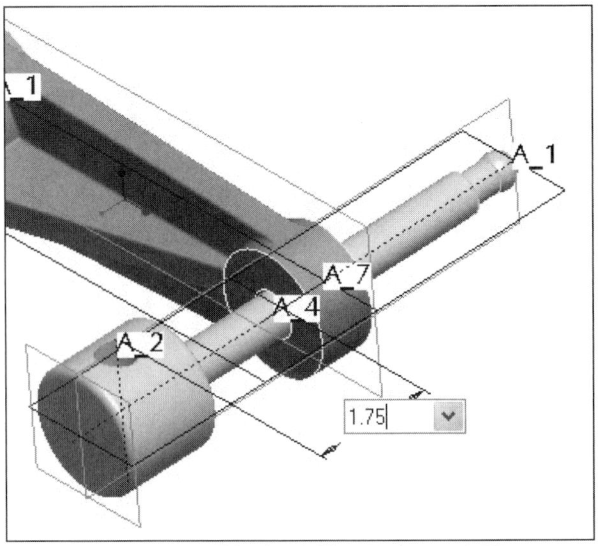

Figure 8.10(b) Modify the Offset Value to **1.75**

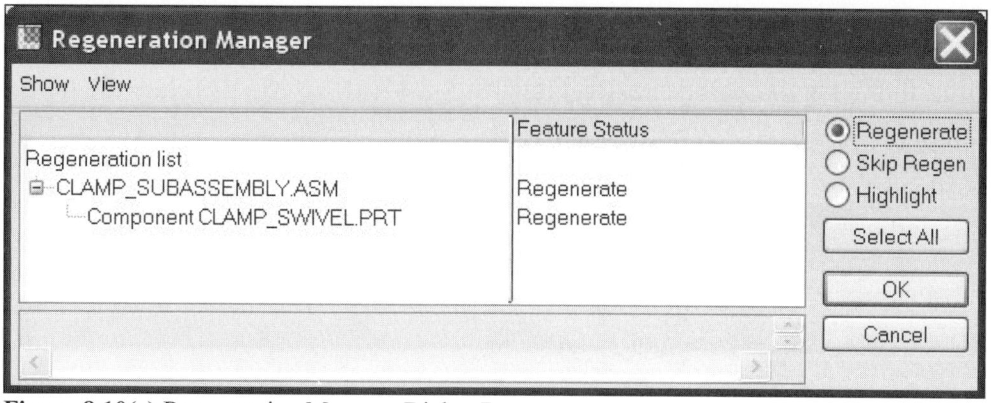

Figure 8.10(c) Regeneration Manager Dialog Box

Double-click on the Clamp_Swivel again ⇒ double-click on the **1.75** dimension and change it back to **1.50** ⇒ **Enter** ⇒ **Regenerates Model** ⇒ ⇒ ⇒ **Enter** (Fig. 8.11)

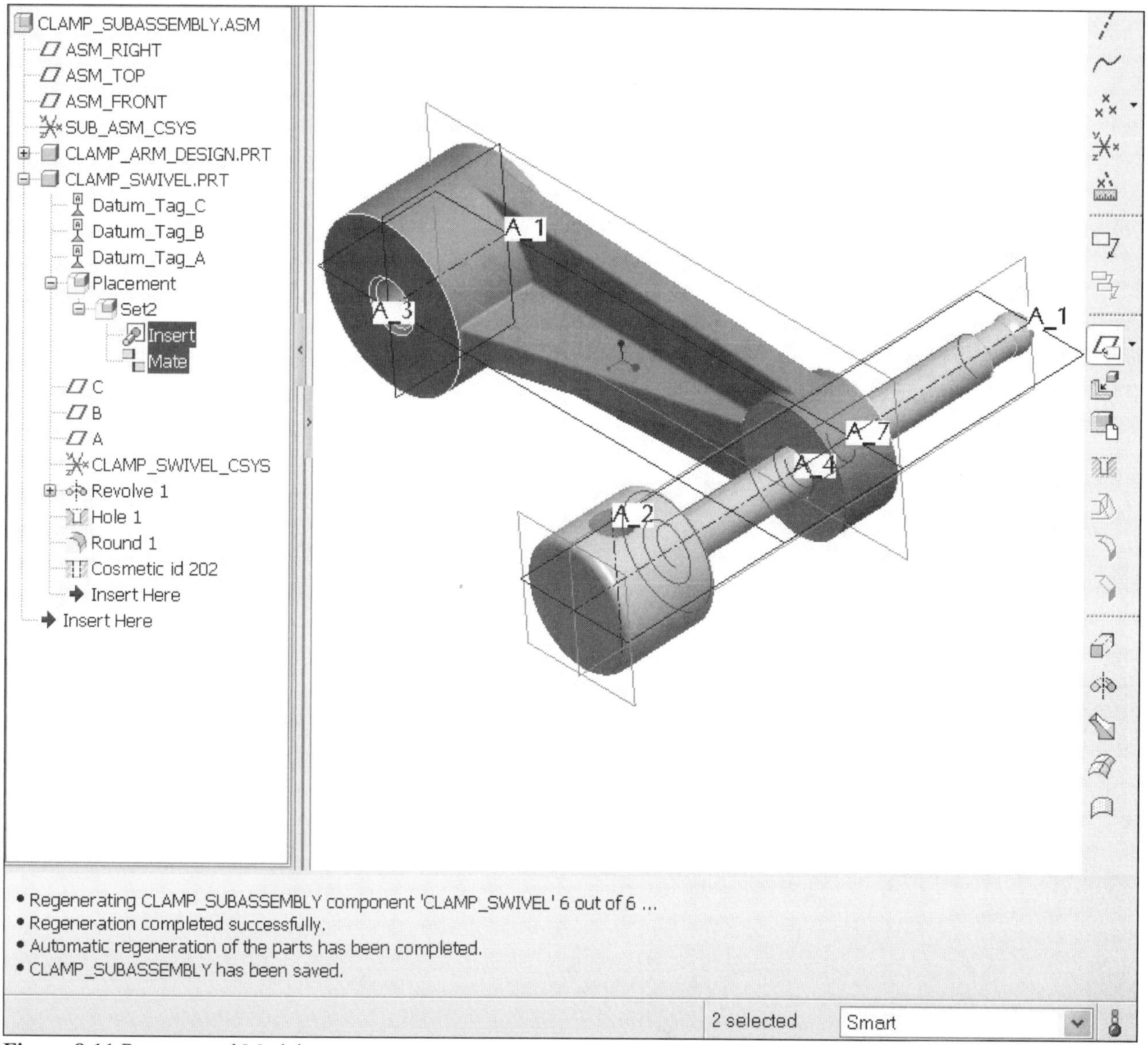

Figure 8.11 Regenerated Model

The next component to be assembled is the Clamp_Foot. Click: **Add component to the assembly** ⇒ select the **clamp_foot.prt** from the Open dialog box ⇒ **Preview** [Fig. 8.12(a)] ⇒ **MMB** rotate the model ⇒ **Open** ⇒ **Show component in a separate window while specifying constraints**

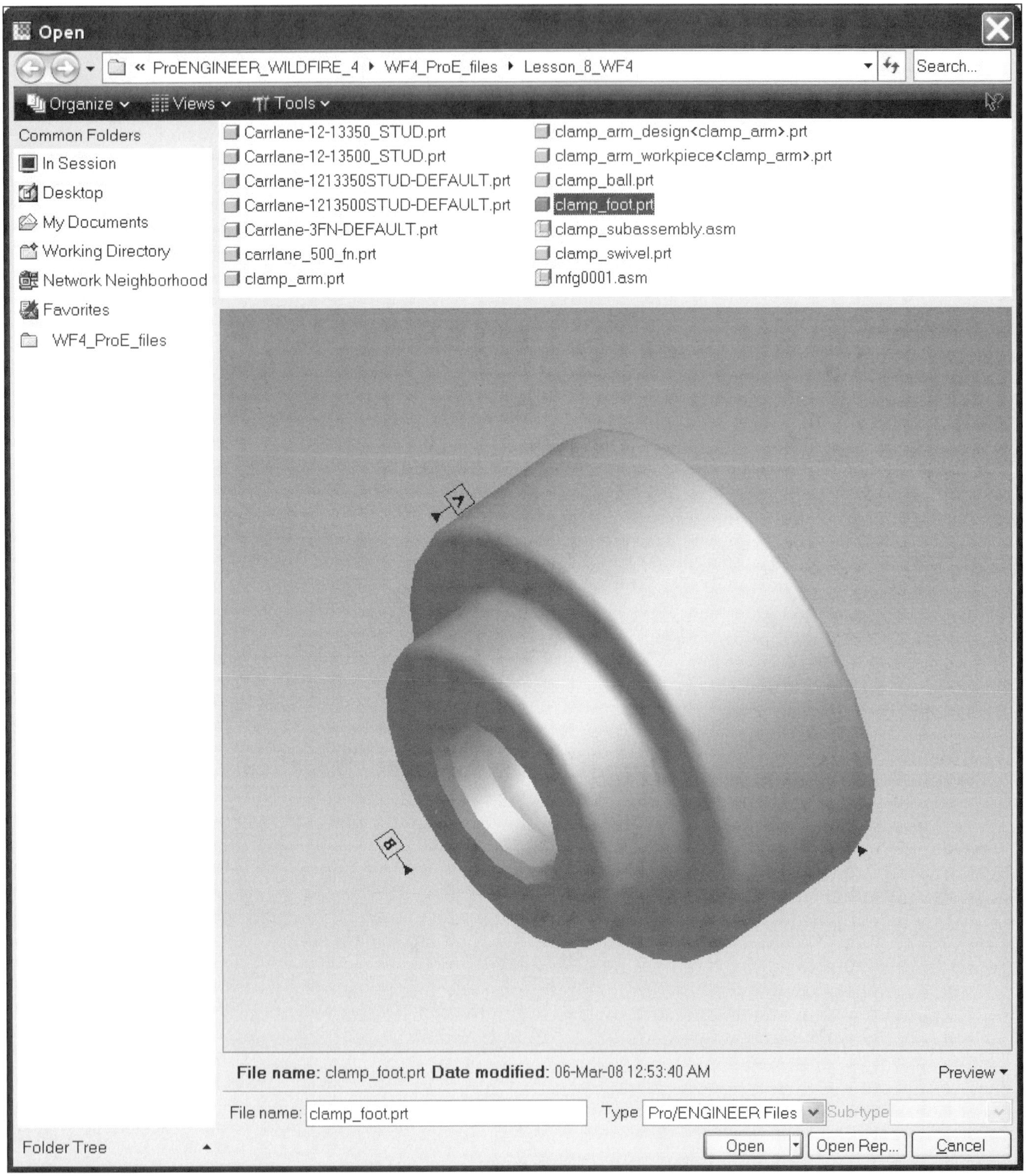

Figure 8.12(a) Clamp_Foot Preview

Spin and zoom in on your model and resize your windows as desired ⇒ Constraints [Automatic] ⇒ Pick the internal cylindrical surface of the Clamp_Foot [Figs. 8.12(b-c)]

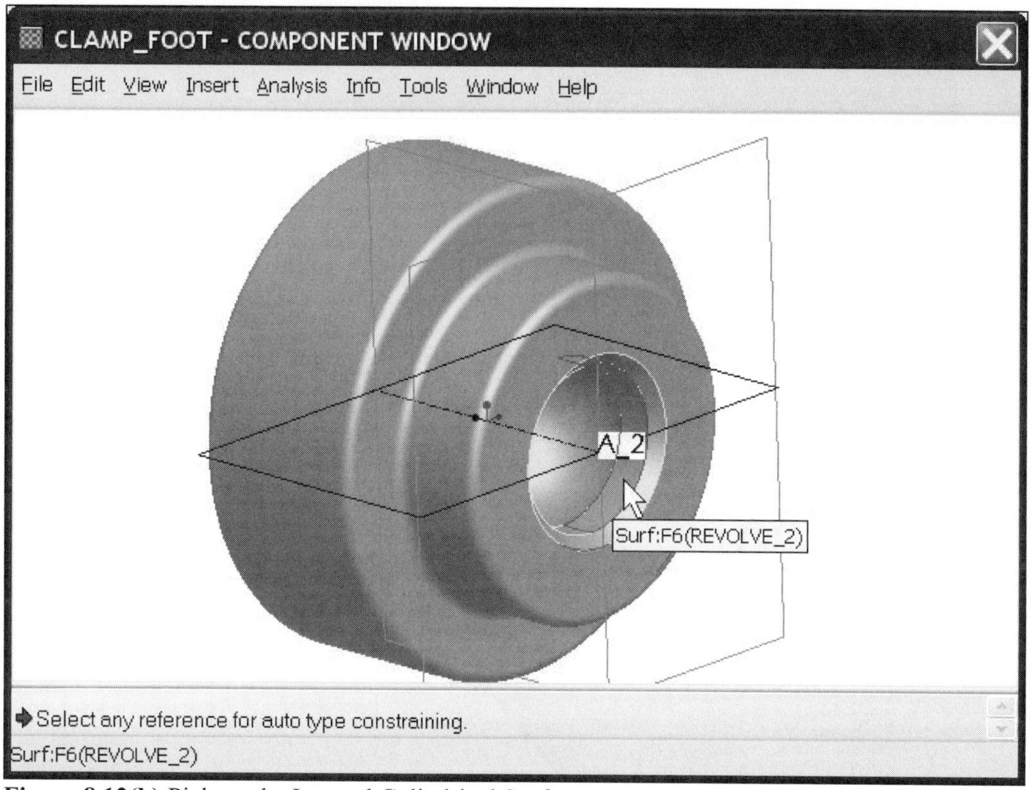

Figure 8.12(b) Pick on the Internal Cylindrical Surface

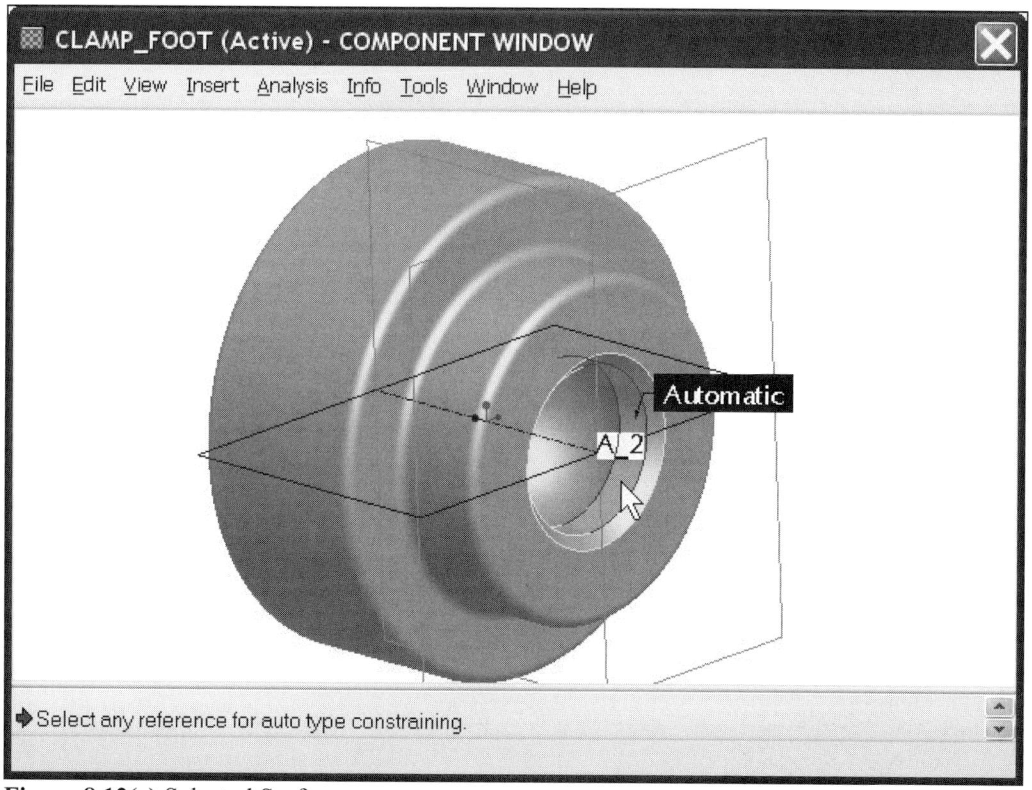

Figure 8.12(c) Selected Surface

Pick the external cylindrical surface of the Clamp_Swivel [Fig. 8.12(d)] *constraint becomes Insert* Insert ⇒ **Placement** tab Placement [Fig. 8.12(e)]

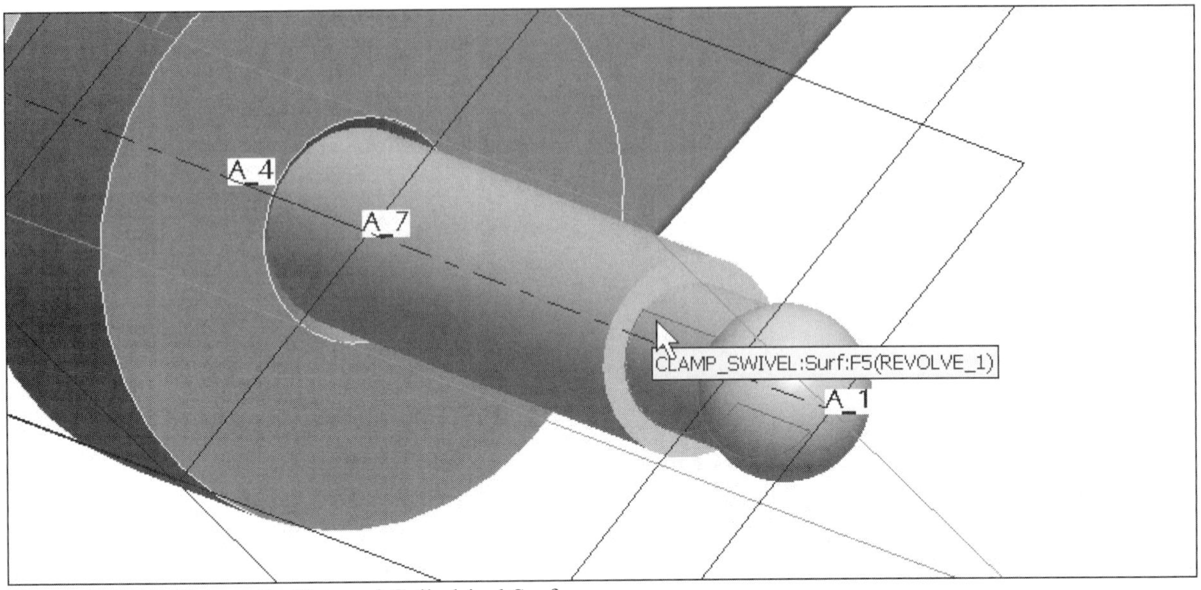

Figure 8.12(d) Pick on the External Cylindrical Surface

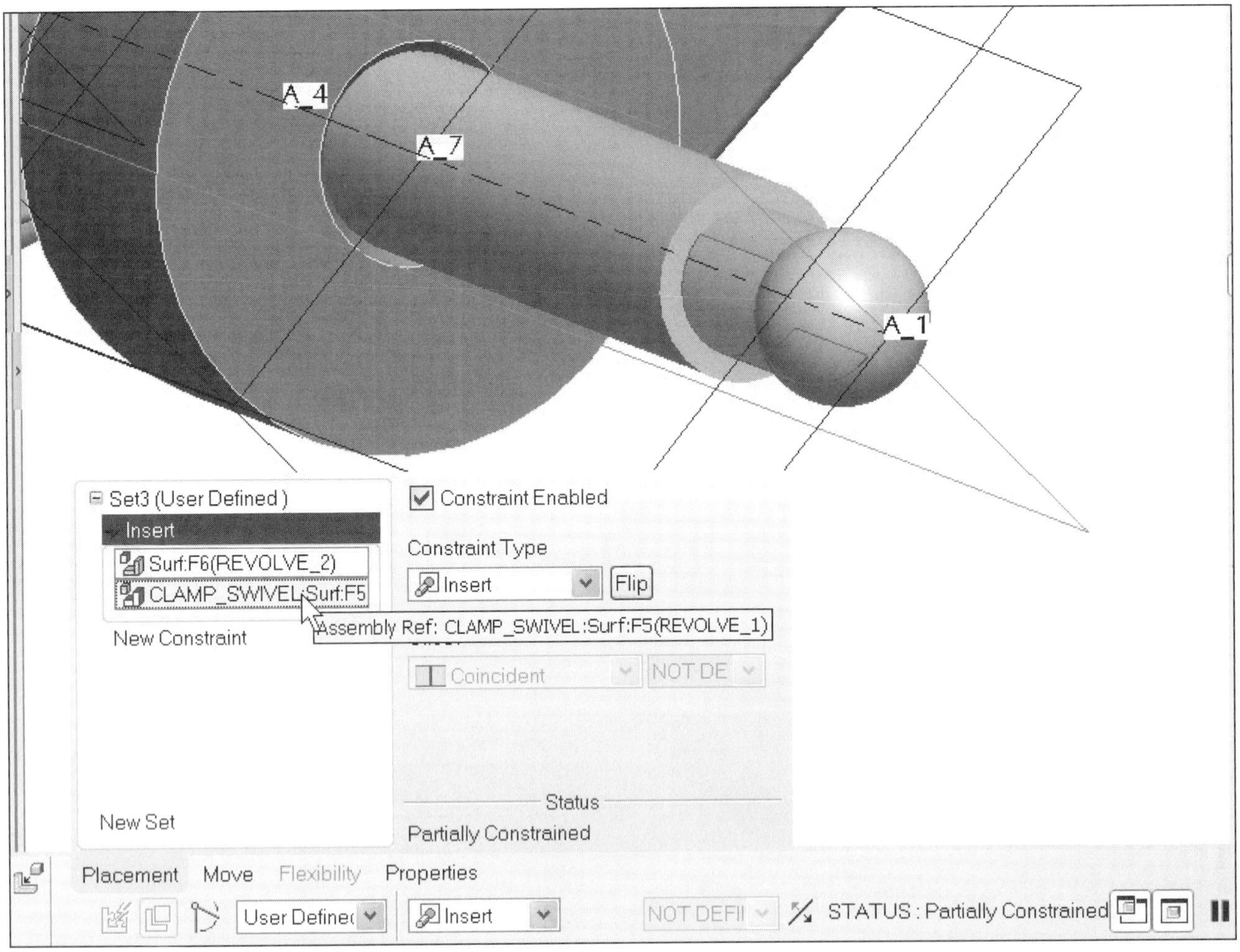

Figure 8.12(e) Constraint Type Insert

Click: **RMB** ⇒ **New Constraint** ⇒ pick the spherical end of the Clamp_Swivel [Fig. 8.12(f)] ⇒ pick the spherical hole of the Clamp_Foot [Figs. 8.12(g-h)] *constraint becomes Mate* Mate ⇒ off ⇒ **Ctrl+D** [Fig. 8.12(i)] ⇒ ⇒ ⇒ **Enter**

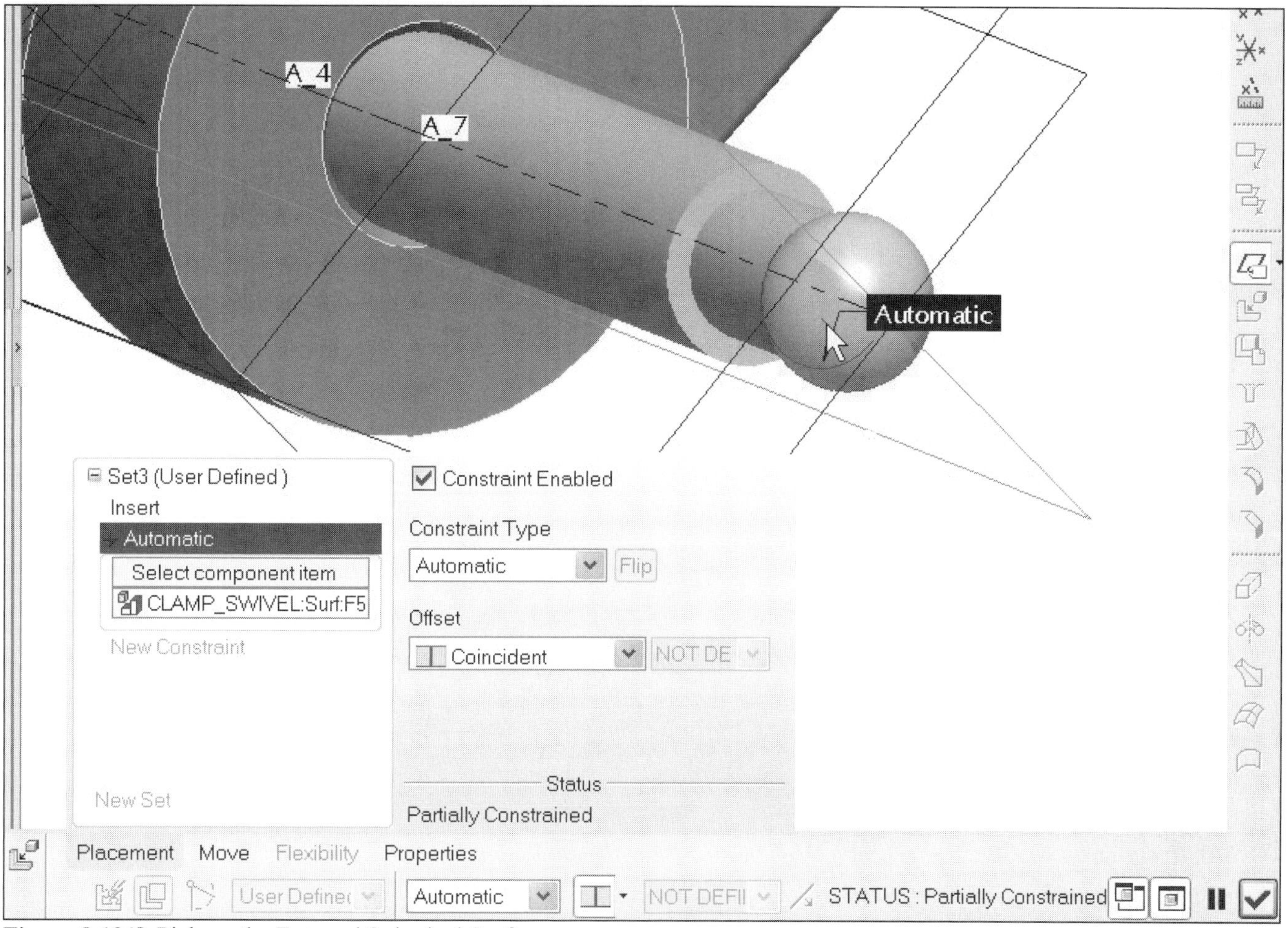

Figure 8.12(f) Pick on the External Spherical Surface

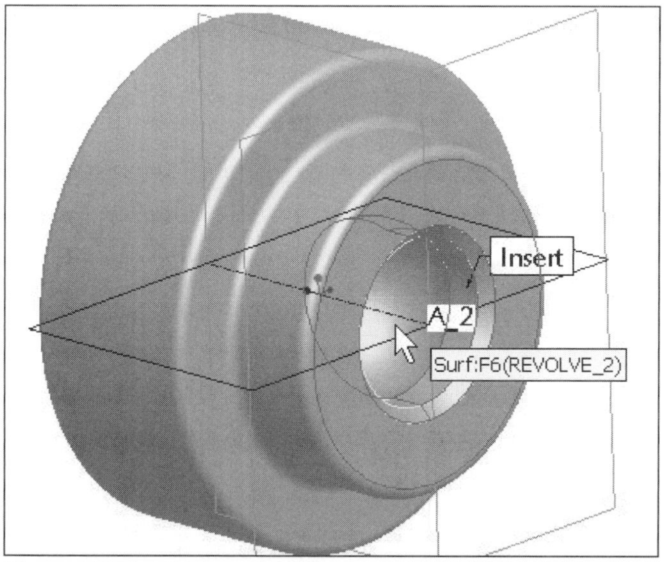

Figure 8.12(g) Pick on the Internal Spherical Hole Surface

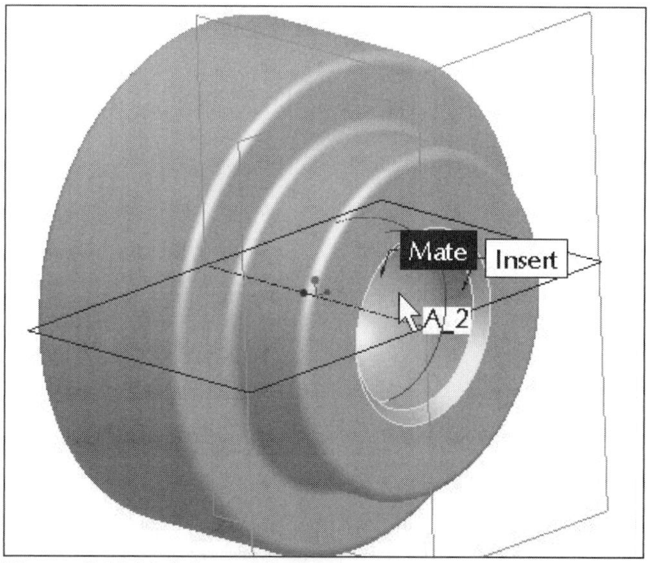

Figure 8.12(h) Selected Surface

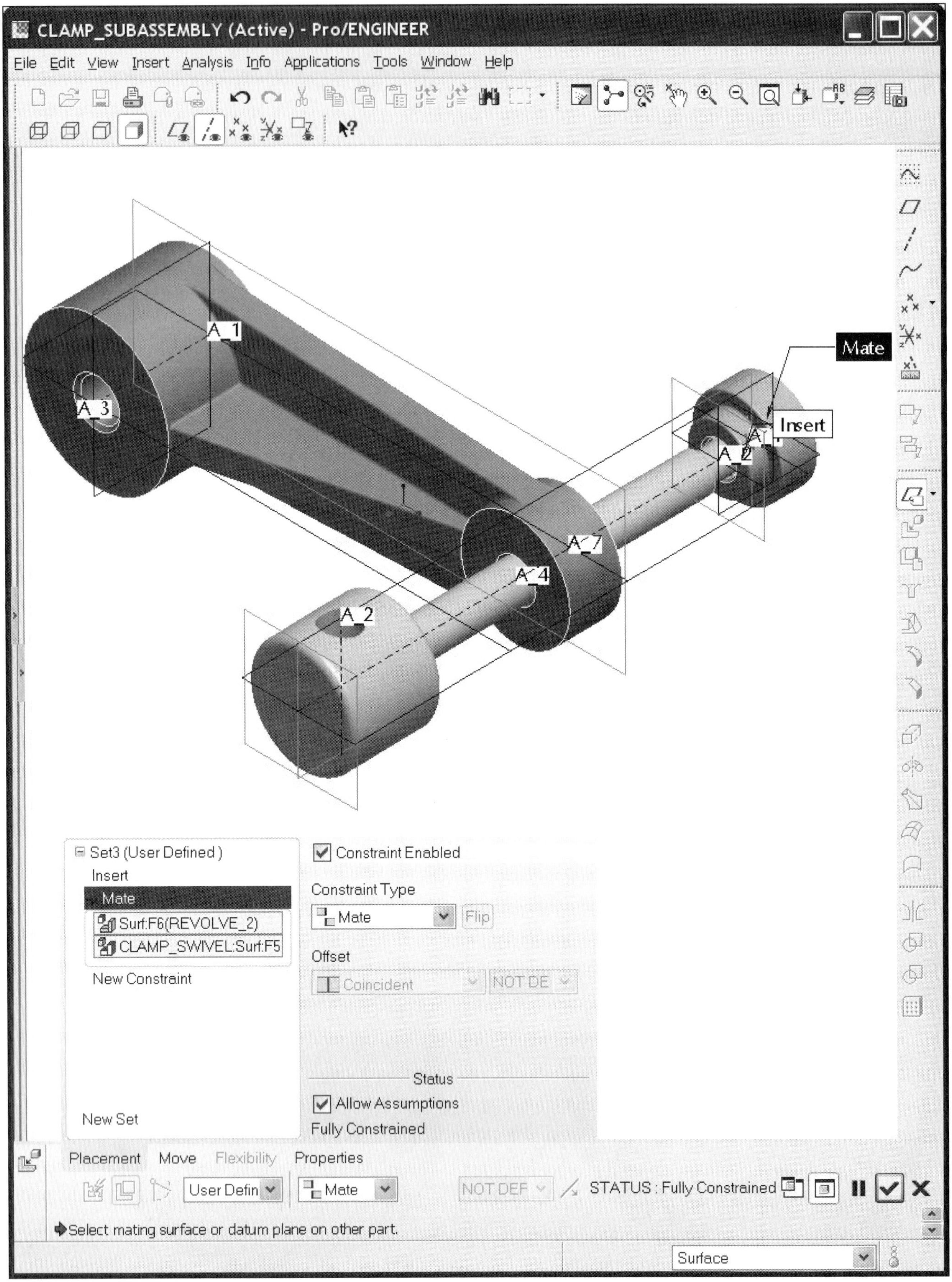

Figure 8.12(i) Fully Constrained Clamp_Foot

Rotate and zoom in on the model. The Clamp_Foot may be facing the wrong direction [Fig. 8.12(j)]. Pick on the **Clamp_Foot ⇒ RMB ⇒ Edit Definition** [Fig. 8.12(k)]

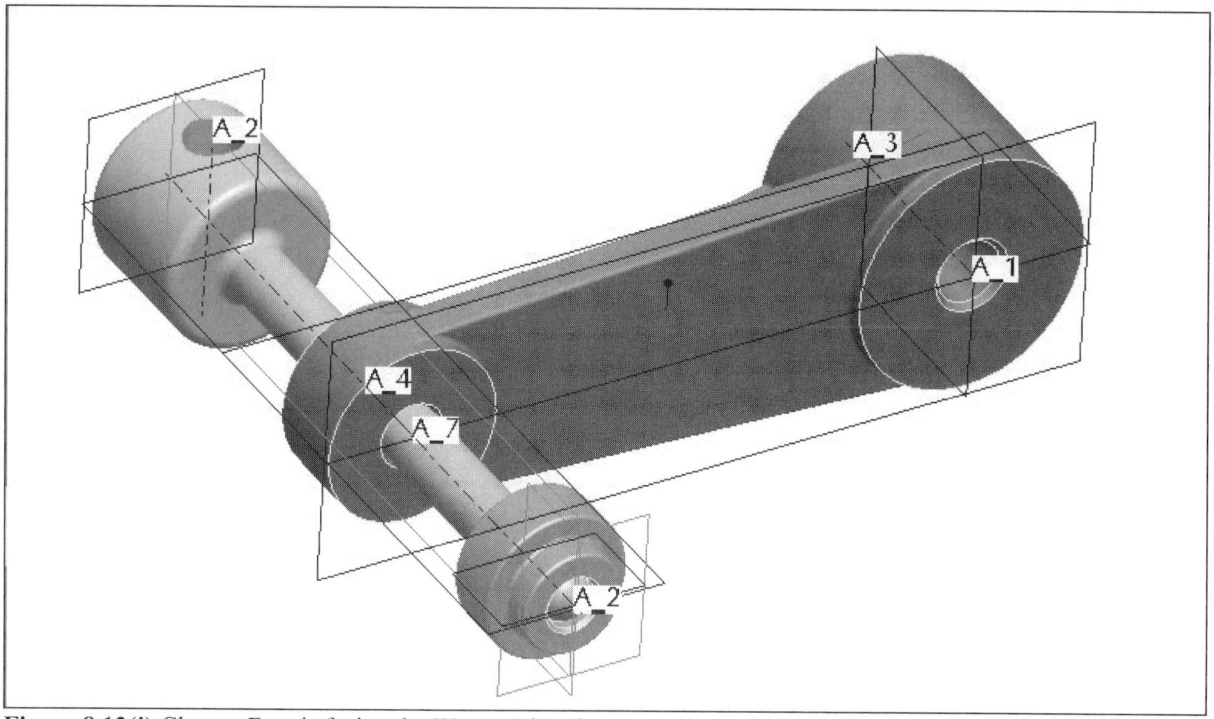

Figure 8.12(j) Clamp _Foot is facing the Wrong Direction

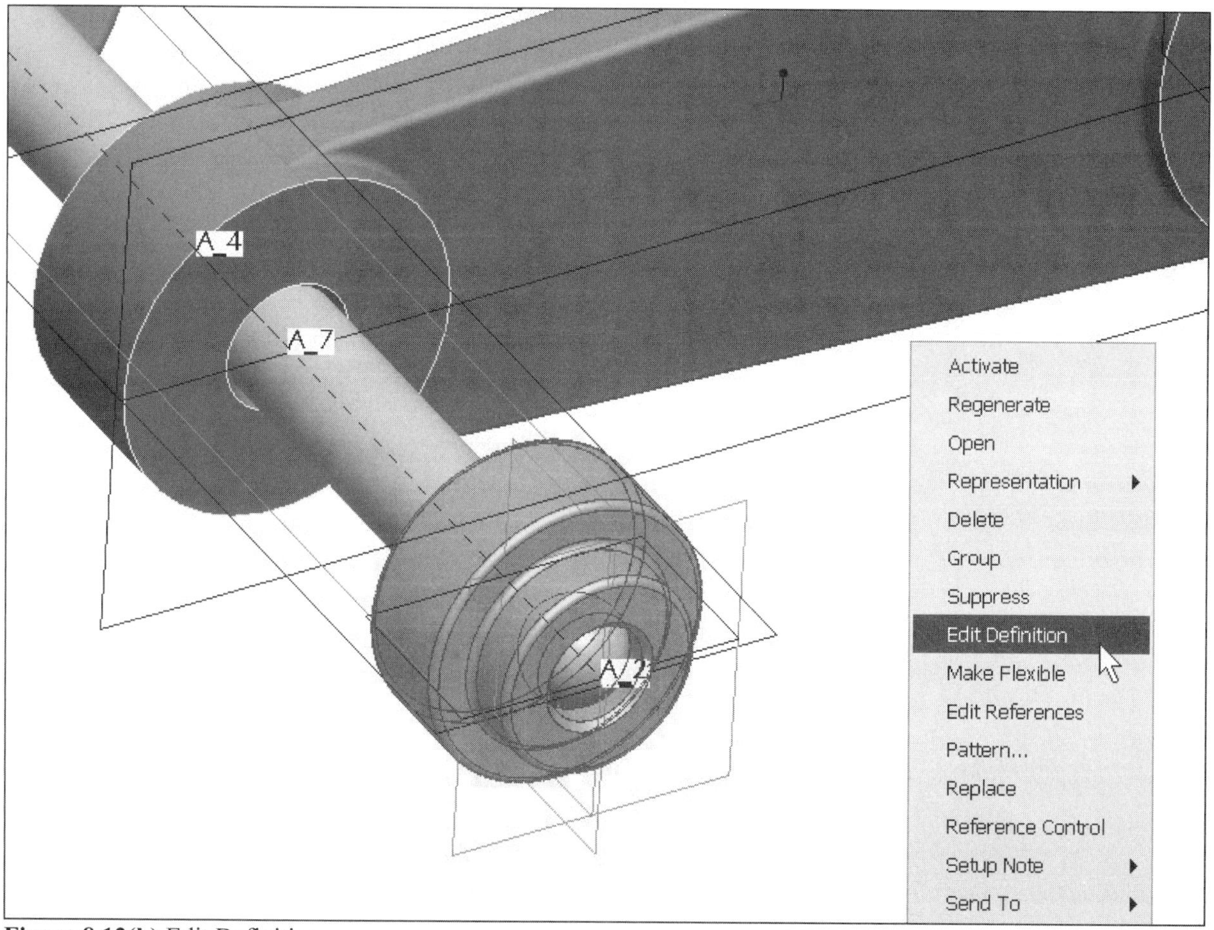

Figure 8.12(k) Edit Definition

Click: **RMB** ⇒ **New Constraint** ⇒ **Placement** tab ⇒ pick on the face of the Clamp_Foot [Figs. 8.12(l-m)] ⇒ pick on the face of the Clamp_Arm [Fig. 8.12(n)]

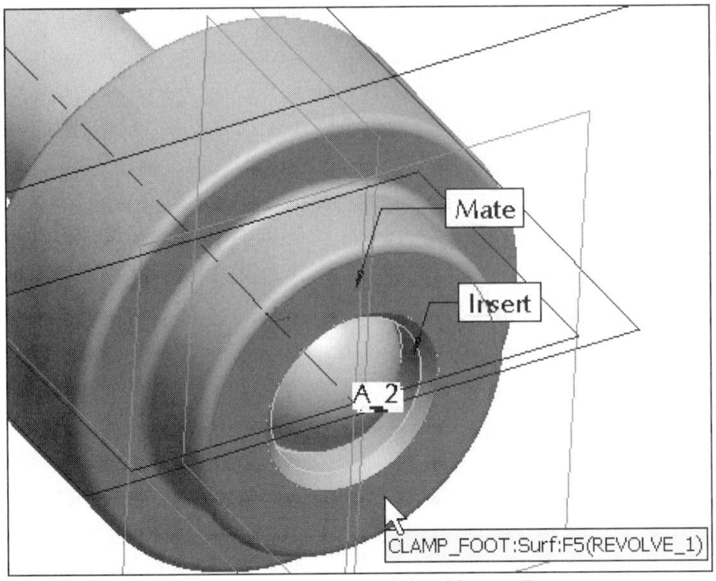

Figure 8.12(l) Pick on the Surface of the Clamp_Foot

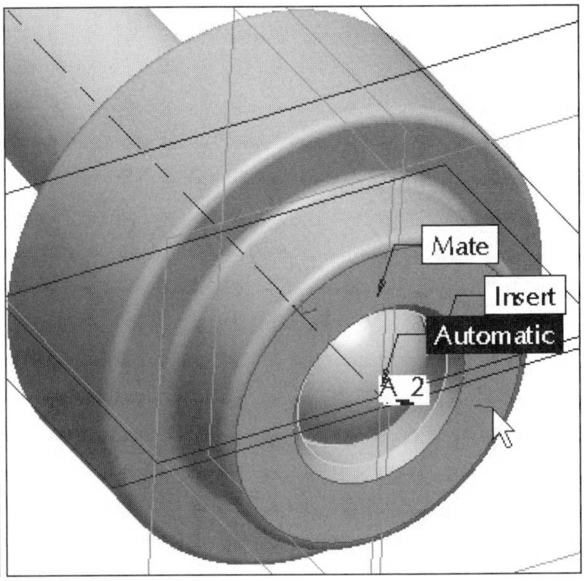

Figure 8.12(m) Selected Surface

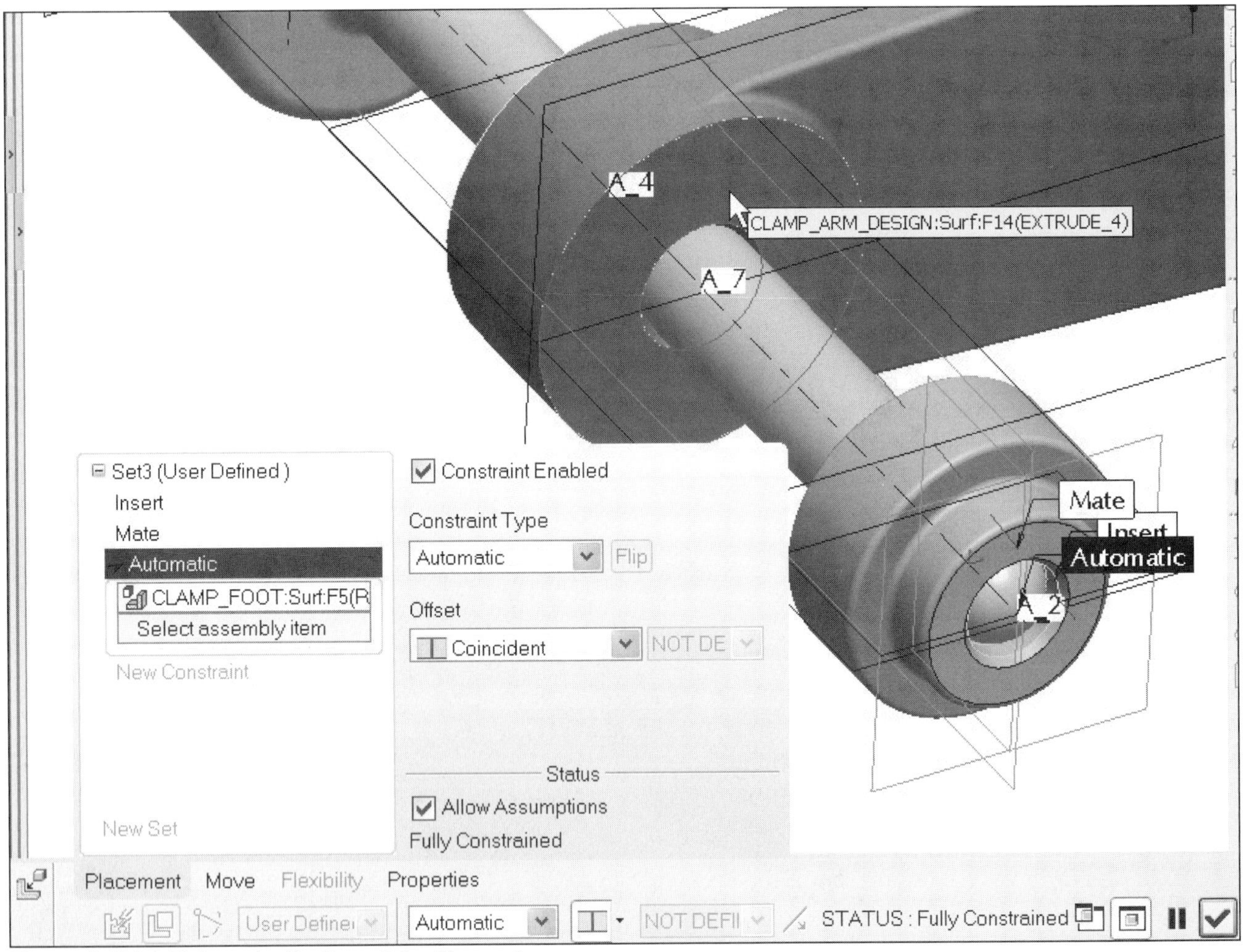

Figure 8.12(n) Pick on the Surface of the Clamp_Arm

Click: **Oriented** [Fig. 8.12(n)] ⇒ **Flip** (to Mate) [Fig. 8.12(o)] ⇒ ☑ ⇒ **Ctrl+D** ⇒ **Ctrl+S** ⇒ **OK**
(Though three constraints were used; the two Mate constraints would have been enough.)

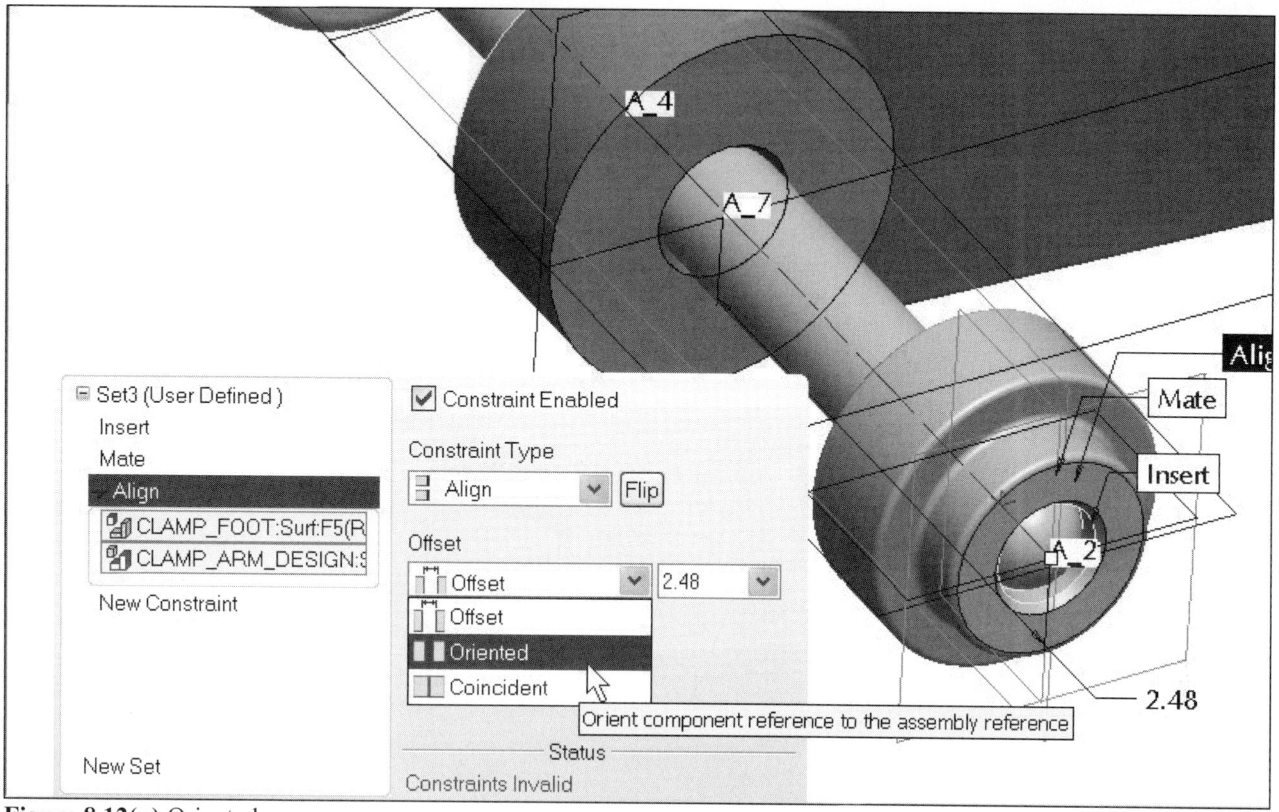

Figure 8.12(n) Oriented

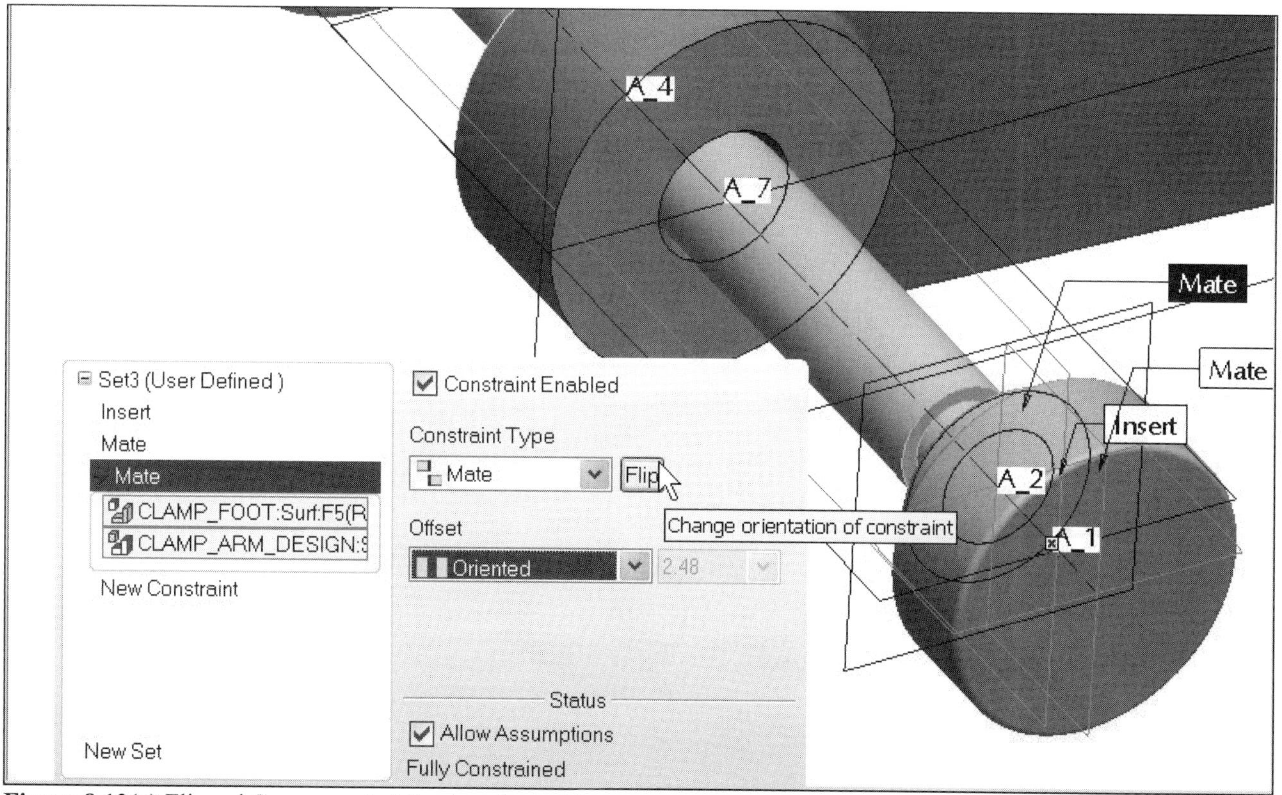

Figure 8.12(o) Flipped Component. Align toggles to Mate

Pick on the Clamp_Foot ⇒ **RMB** ⇒ **Edit Definition** ⇒ **Placement** tab ⇒ click on the Insert constraint [Fig. 8.12(p)] ⇒ **RMB** ⇒ **Delete** ⇒ ☑ ⇒ **Ctrl+D** ⇒ expand the Model Tree [Fig. 8.12(q)] *If a small box appears in front of any component shown in the Model Tree* ⊞ ▢ ᴰCLAMP *it means that it is a "packaged" component, in other words it is "partially constrained" not "fully constrained".*

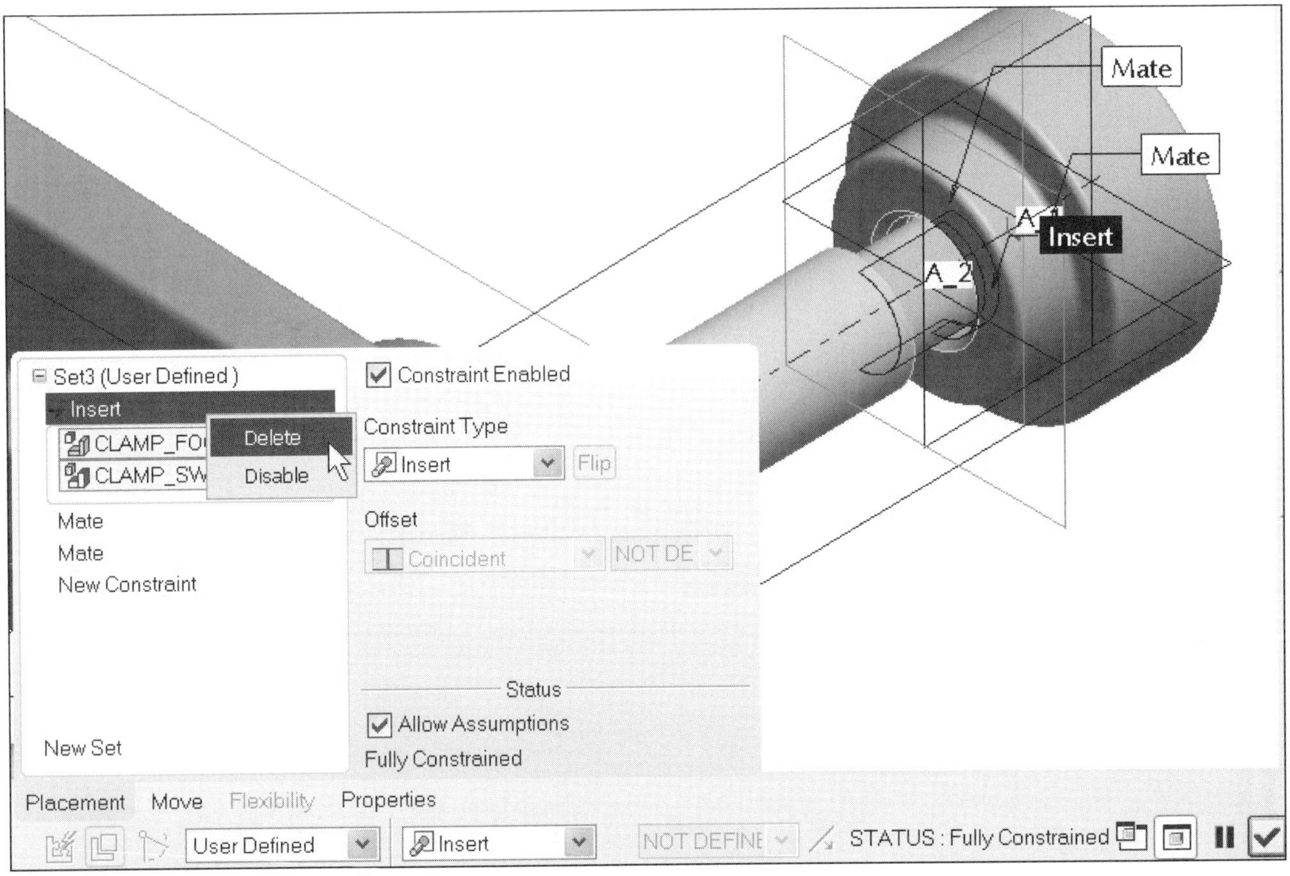

Figure 8.12(p) Delete the Insert Constraint

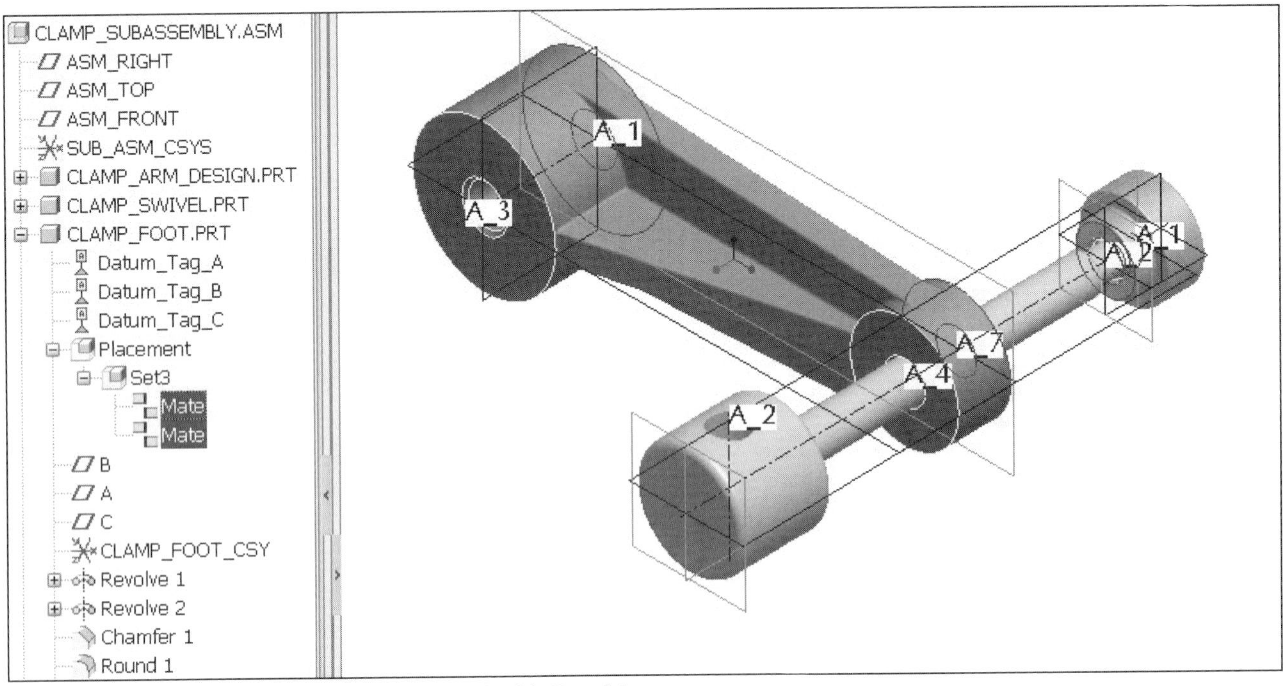

Figure 8.12(q) Clamp_Foot Fully Constrained with Two Mates

Assemble the **5.00** double-ended stud. Click: **Add component to the assembly** ⇒ select **Carrlane-12-13500_STUD.prt** ⇒ **Open** ⇒ **Show component in a separate window** off ⇒ pick the external cylindrical surface of the Carrlane-12-13500_STUD.prt [Fig. 8.13(a)] ⇒ pick the internal cylindrical surface of the hole of the Clamp_Swivel [Fig. 8.13(b)] *constraint becomes Insert* Insert

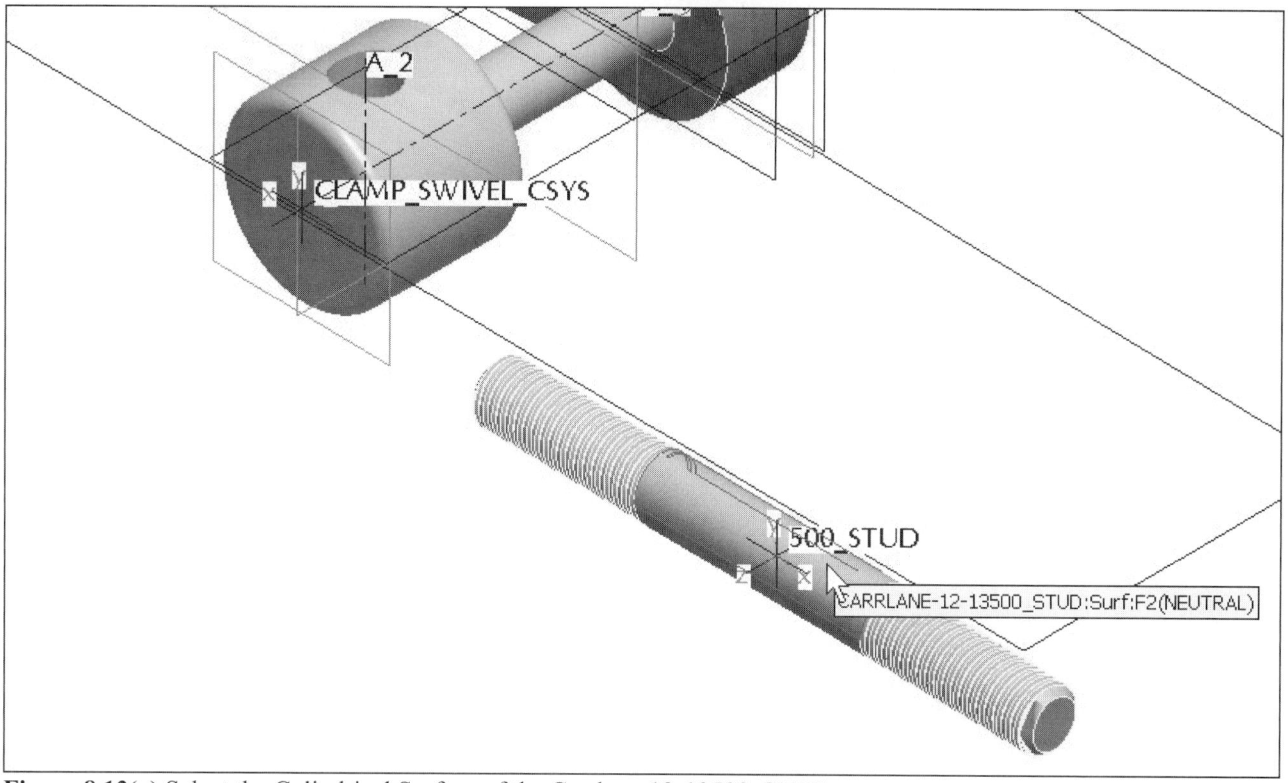

Figure 8.13(a) Select the Cylindrical Surface of the Carrlane-12-13500_STUD.prt

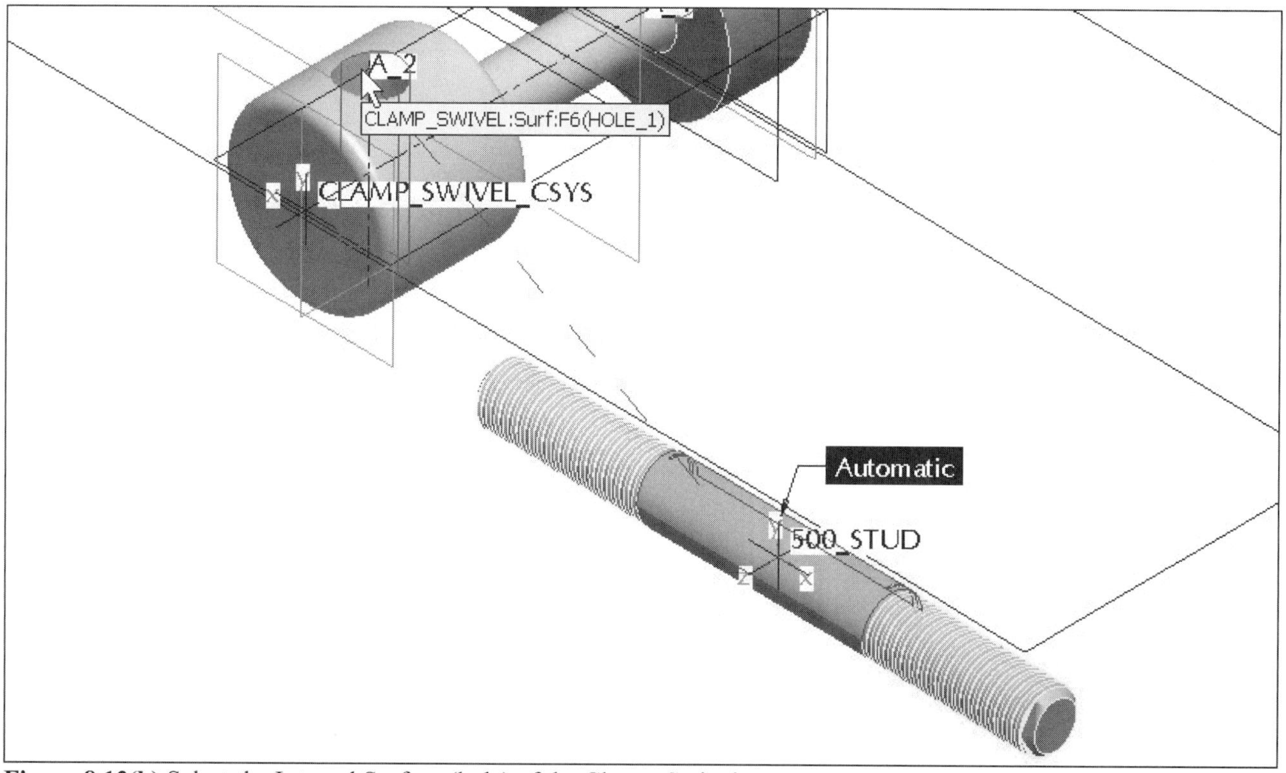

Figure 8.13(b) Select the Internal Surface (hole) of the Clamp_Swivel

Click: **RMB** ⇒ **Move Component** ⇒ pick the Carrlane-12-13500_STUD.prt (hint: release LMB) and slide the component [Fig. 8.13(c)] ⇒ **LMB** to position the component ⇒ **RMB** ⇒ **New Constraint** [Fig. 8.13(d)] ⇒  off

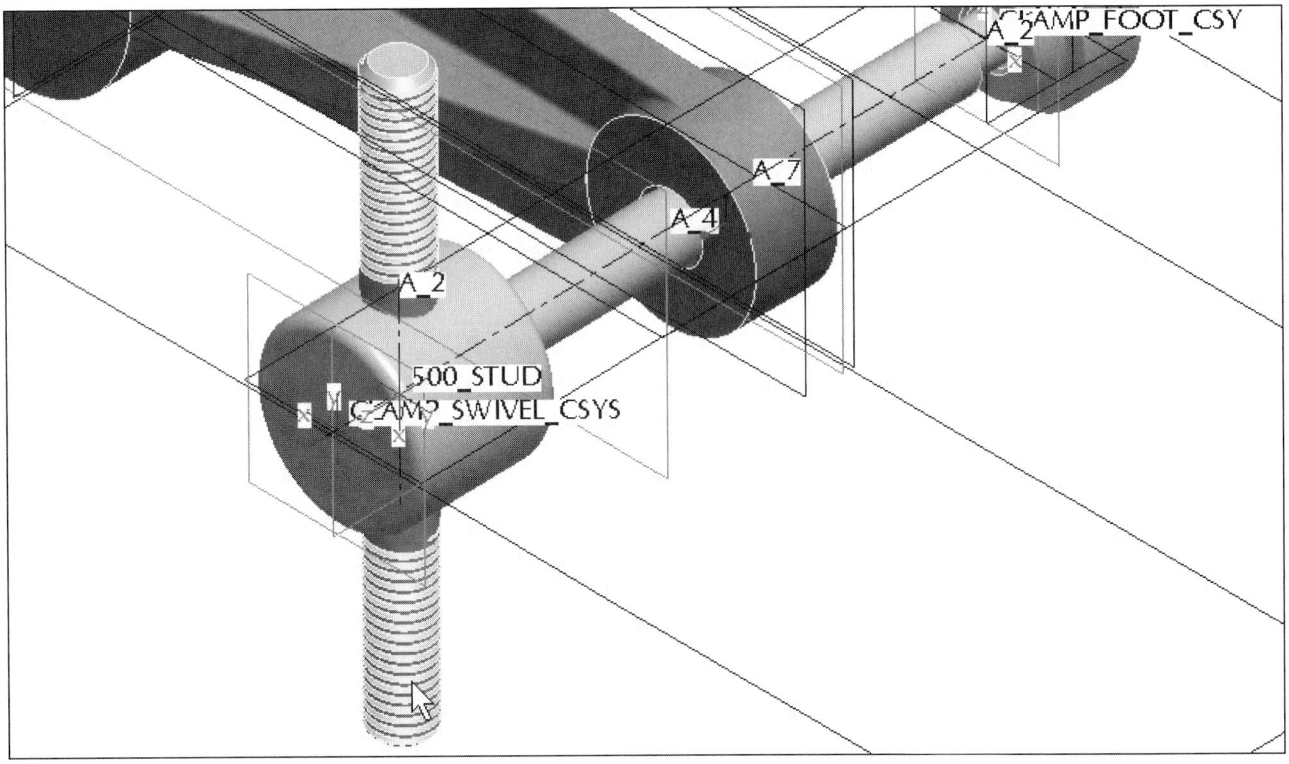

Figure 8.13(c) Move the Component

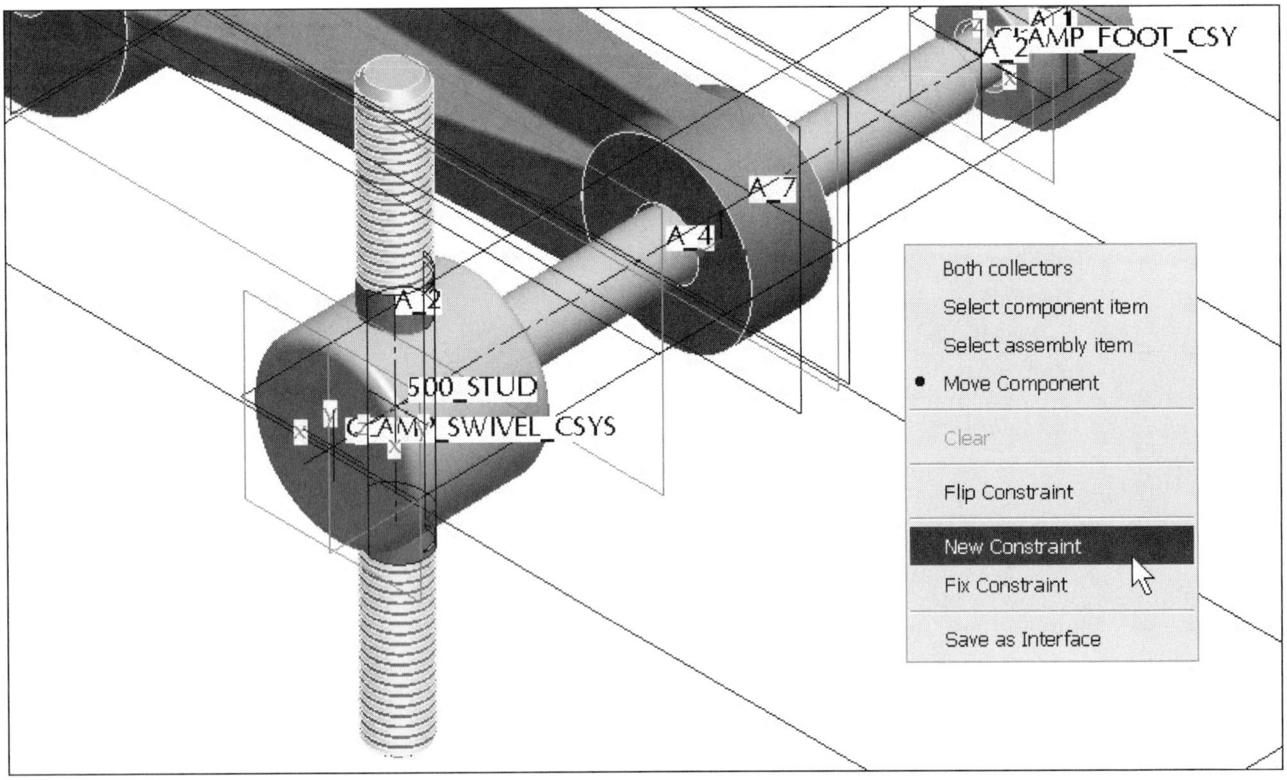

Figure 8.13(d) RMB ⇒ New Constraint

Pick the end surface of the Carrlane-12-13500_STUD.prt [Fig. 8.14(a)] ⇒ pick datum **B** of the Clamp_Swivel [Fig. 8.14(b)] ⇒ [icon] ⇒ [icon] ⇒ change offset to **2.50** ⇒ **Enter** ⇒ **Placement** tab [Fig. 8.14(c)] ⇒ [icon] ⇒ [icon] ⇒ **OK**

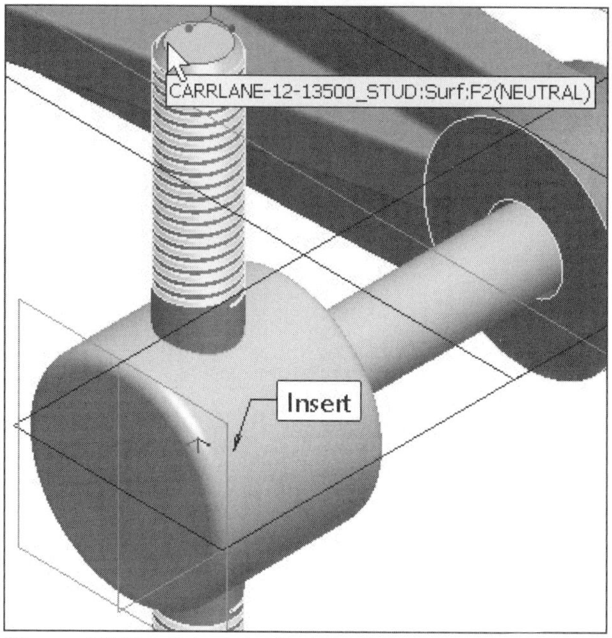

Figure 8.14(a) Select the End Surface

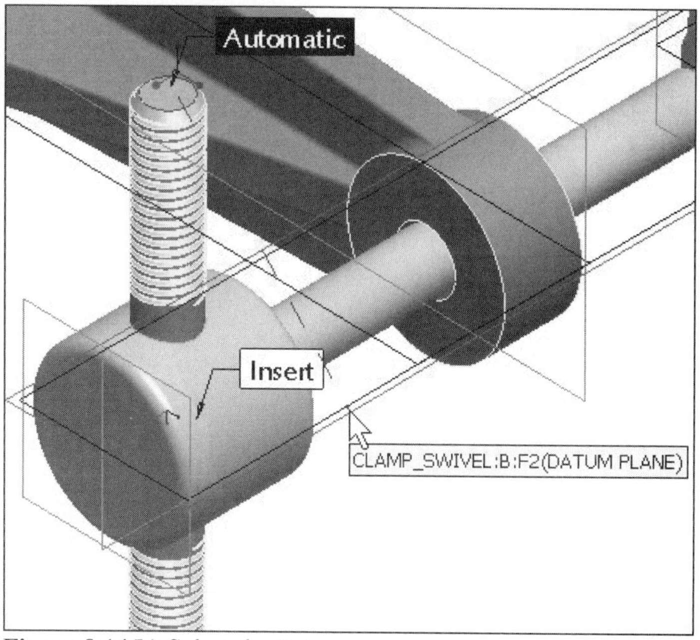

Figure 8.14(b) Select the Datum

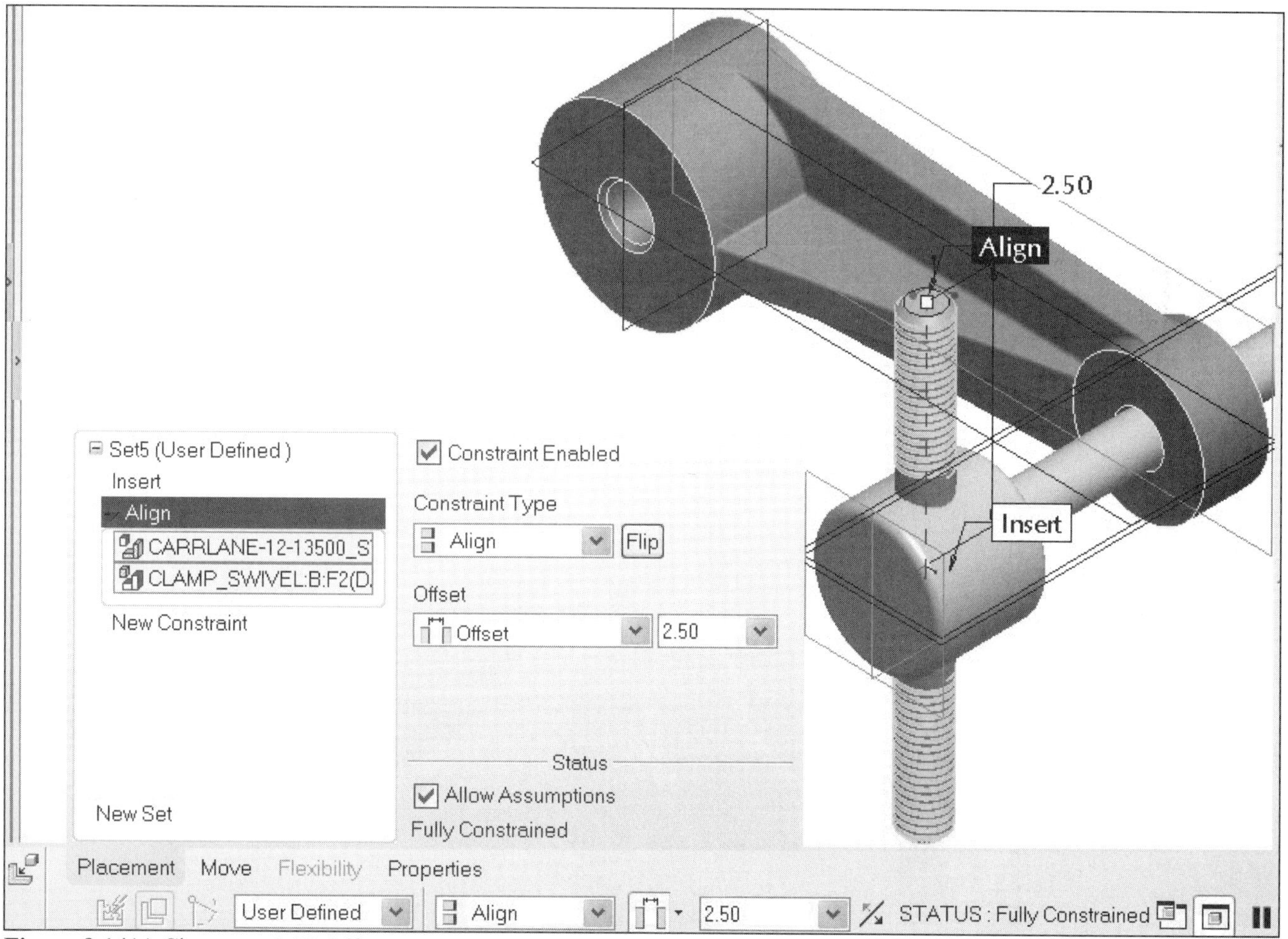

Figure 8.14(c) Change to **2.50** Offset

Spin the model as needed ⇒ Click: **Settings** from the Navigator ⇒ **Tree Columns** ⇒ use >> to add the columns names to be Displayed [Fig. 8.15(a)] ⇒ **Apply** ⇒ **OK** ⇒ resize the Model Tree column widths [Fig. 8.15(b)]

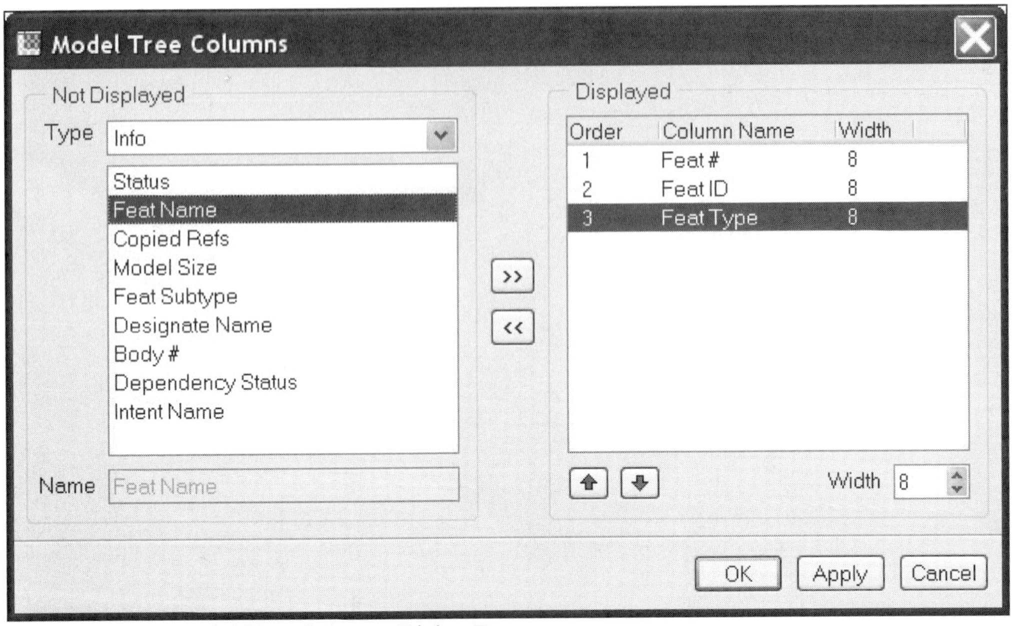

Figure 8.15(a) Model Tree Columns Dialog Box

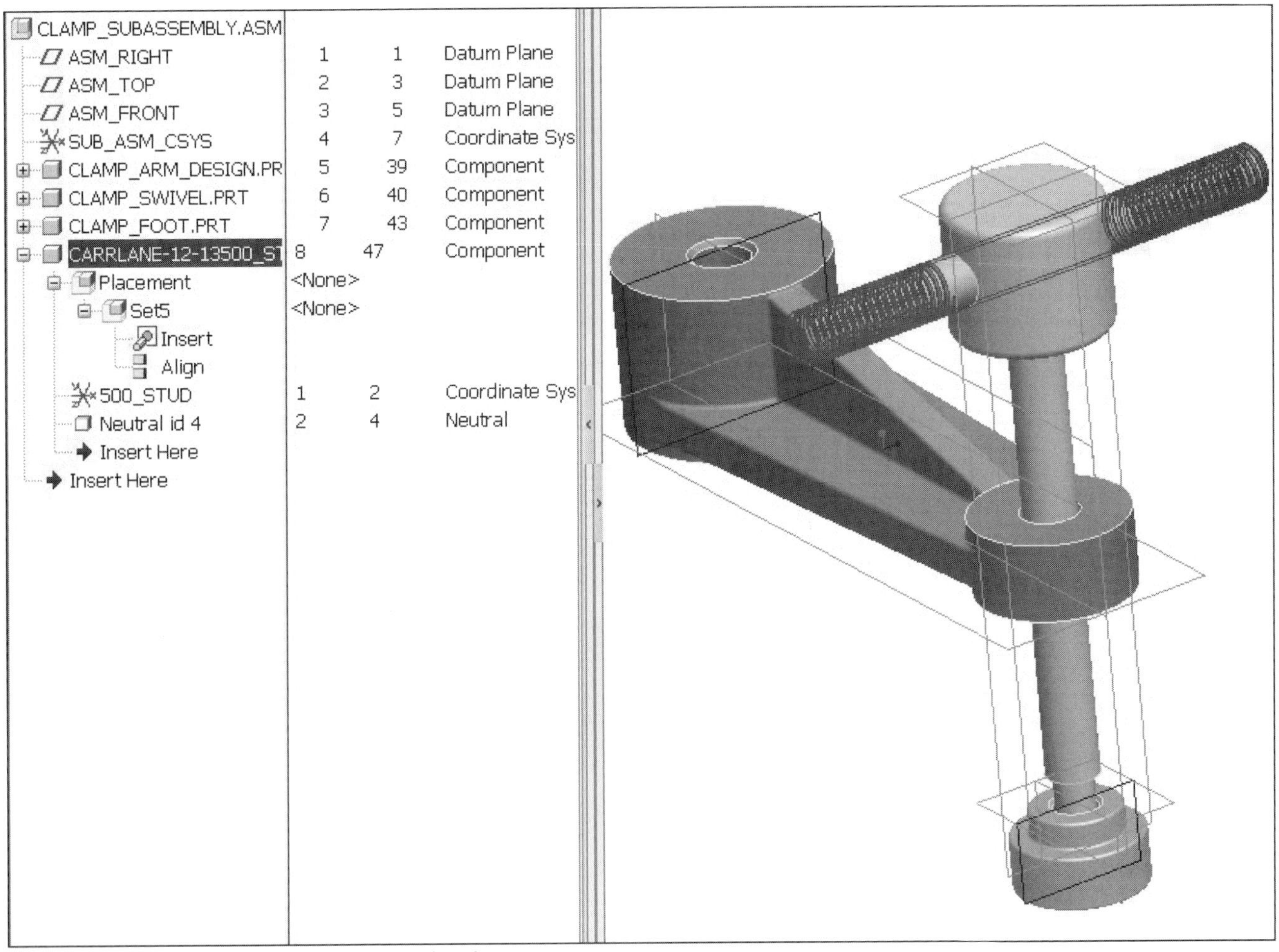

Figure 8.15(b) Model Tree with Adjusted Columns

The Clamp_Ball handles are the last components of the Clamp_Subassembly. Close the Navigator shade 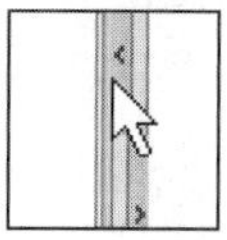 ⇒ **Add component to the assembly** ⇒ select the **clamp_ball.prt** ⇒ **Preview** [Fig. 8.16(a)] ⇒ **MMB** spin and zoom the preview model ⇒ **Open** [Fig. 8.16(b)] spin the model as shown

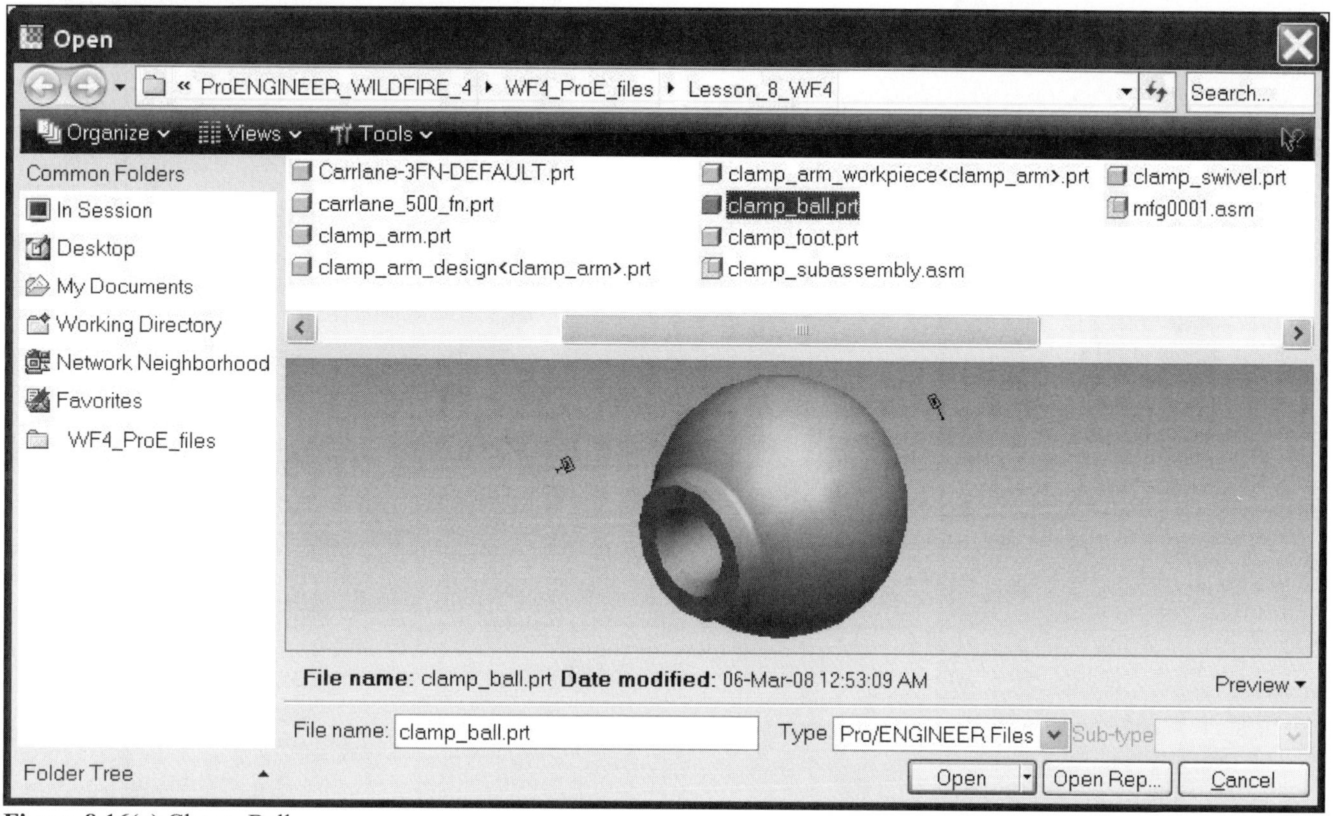

Figure 8.16(a) Clamp_Ball

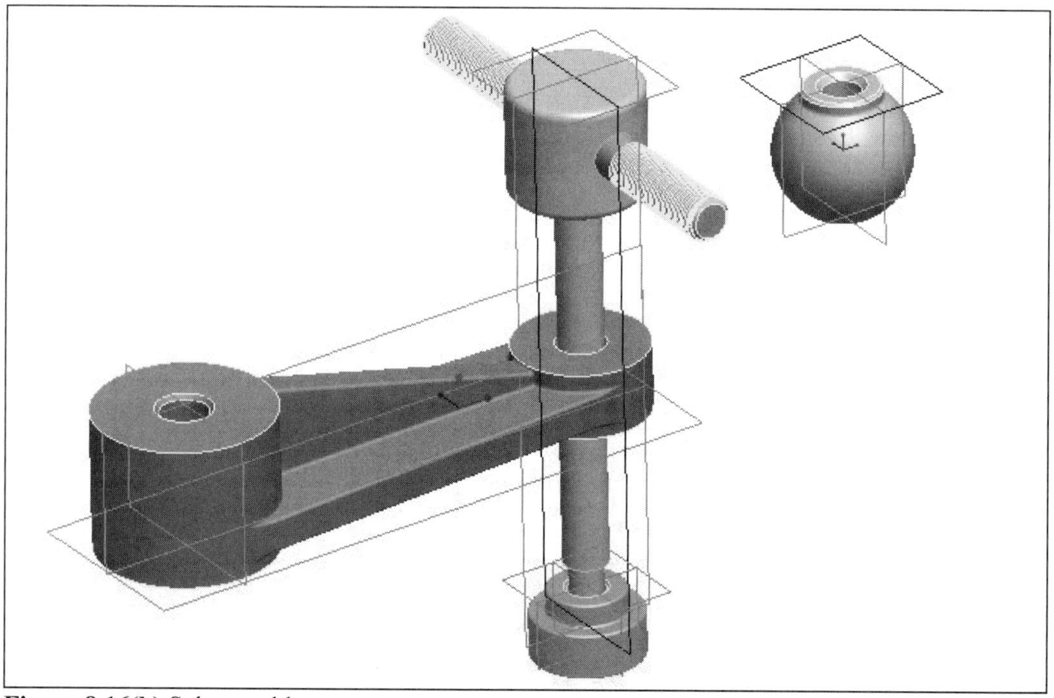

Figure 8.16(b) Subassembly

Press and hold **Ctrl+Alt** ⇒ **MMB** on the Clamp_Ball and rotate ⇒ **Ctrl+Alt** ⇒ **RMB** pan as needed ⇒ pick the internal cylindrical surface of the hole of the Clamp_Ball [Fig. 8.16(c)] ⇒ pick the external cylindrical surface of the Carrlane-12-13500_STUD [Fig. 8.16(d)] *(constraint becomes Insert)*

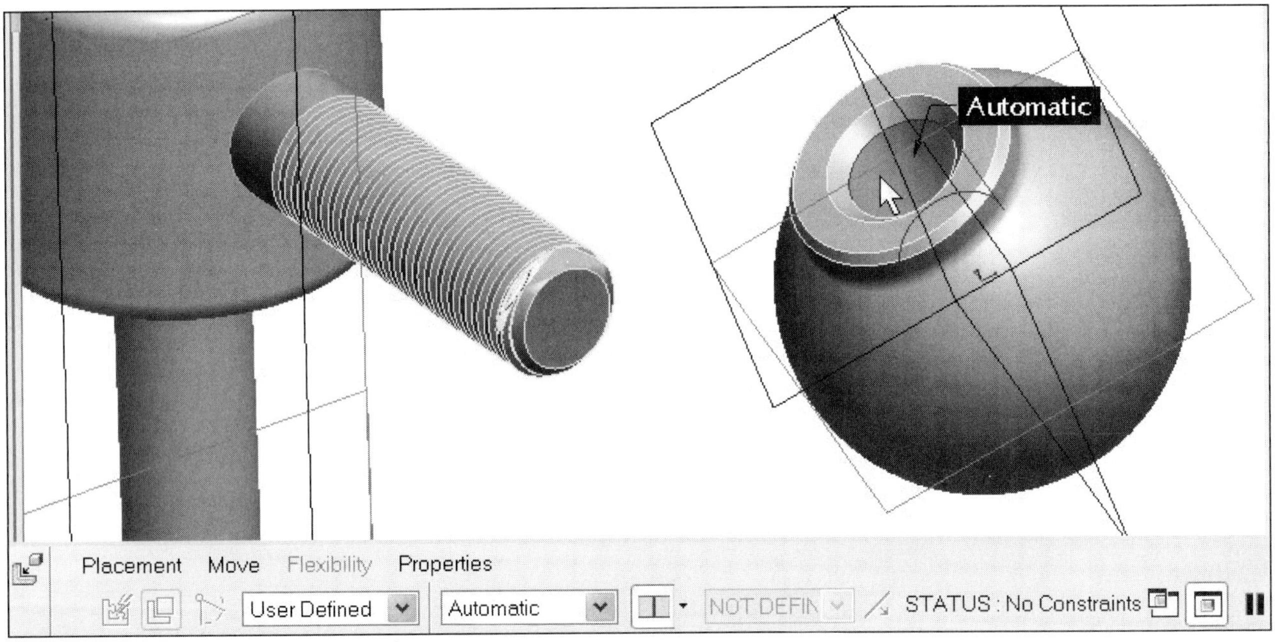

Figure 8.16(c) Select the Hole's Surface

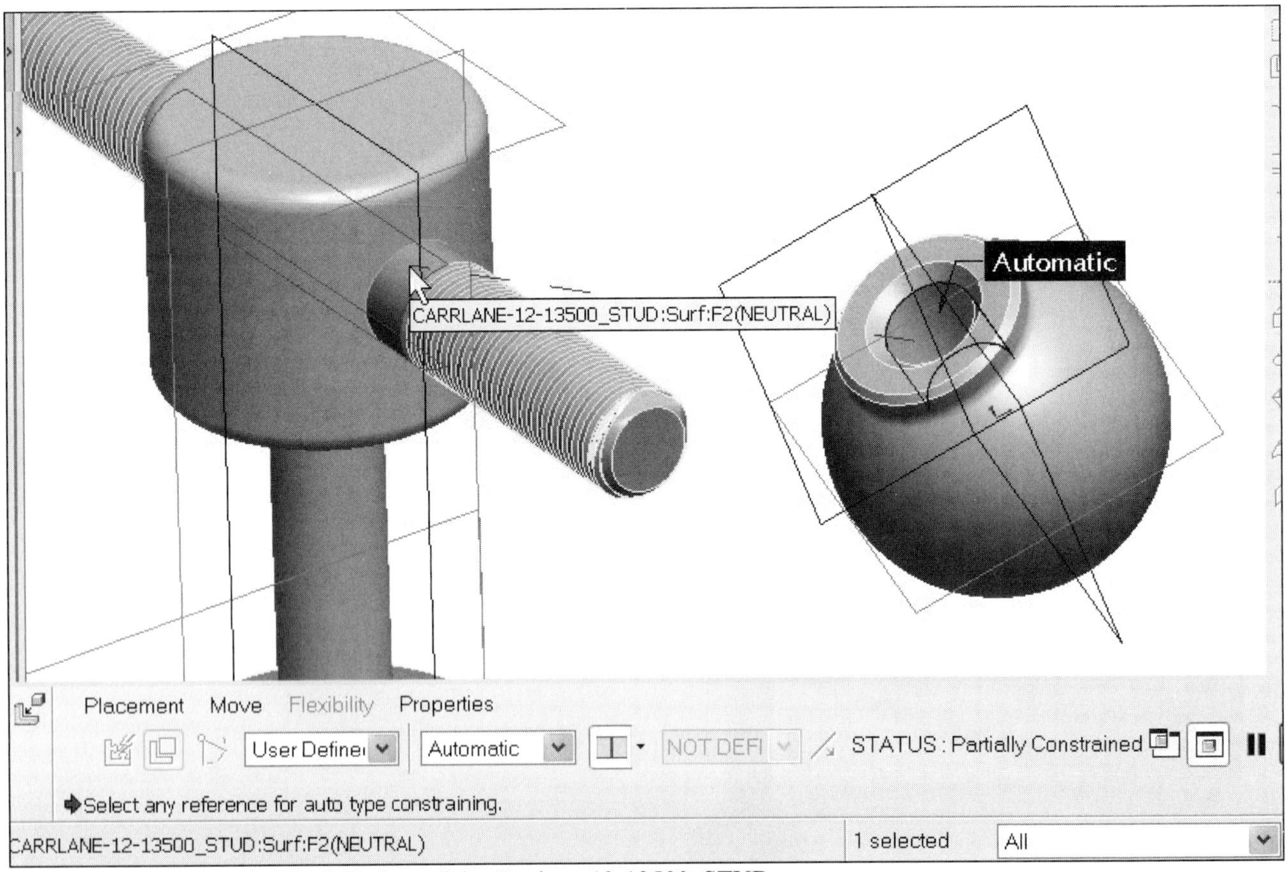

Figure 8.16(d) Select the Shaft Surface of the Carrlane-12-13500_STUD

Pick **Placement** in the dashboard ⇒ **New Constraint** ⇒ pick the end surface of the Carrlane-12-13500_STUD [Fig. 8.16(e)] ⇒ pick the flat end of the Clamp_Ball [Fig. 8.16(f)]

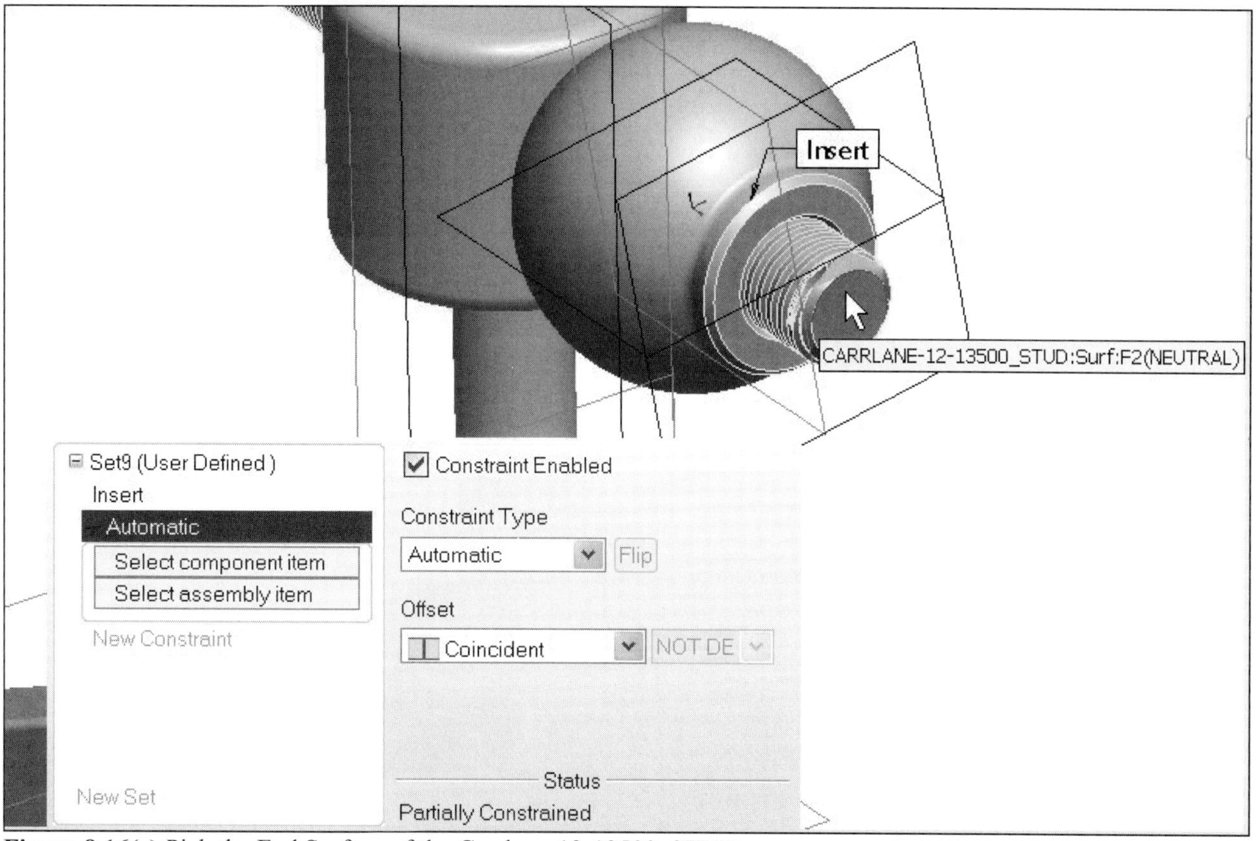

Figure 8.16(e) Pick the End Surface of the Carrlane-12-13500_STUD

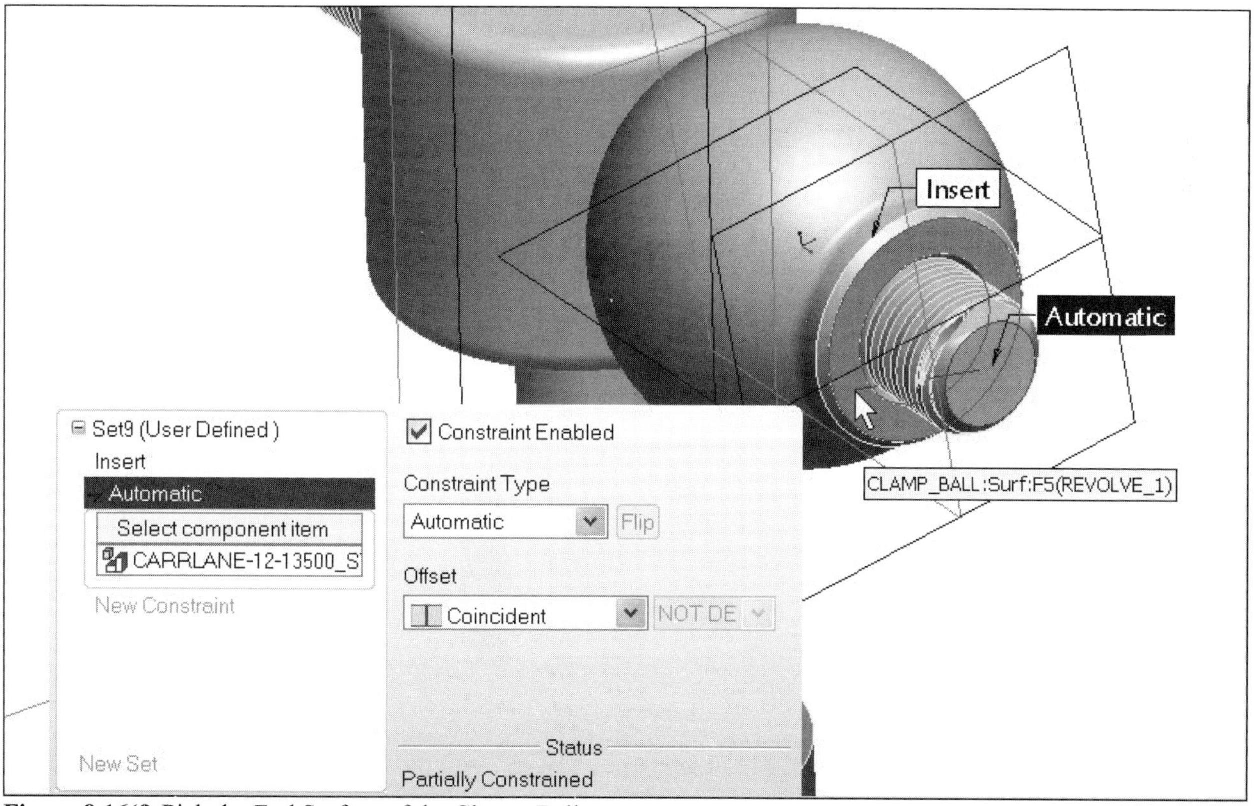

Figure 8.16(f) Pick the End Surface of the Clamp_Ball

Click: Constraint Type **Flip** (changes the constraint type from Align to Mate) [Fig. 8.16(g)] ⇒ [icon] ⇒ [icon] ⇒ drag the handle toward the Clamp_Swivel *or type -.500* [Fig. 8.16(h)] ⇒ [icon] ⇒ [icon] ⇒ **OK**

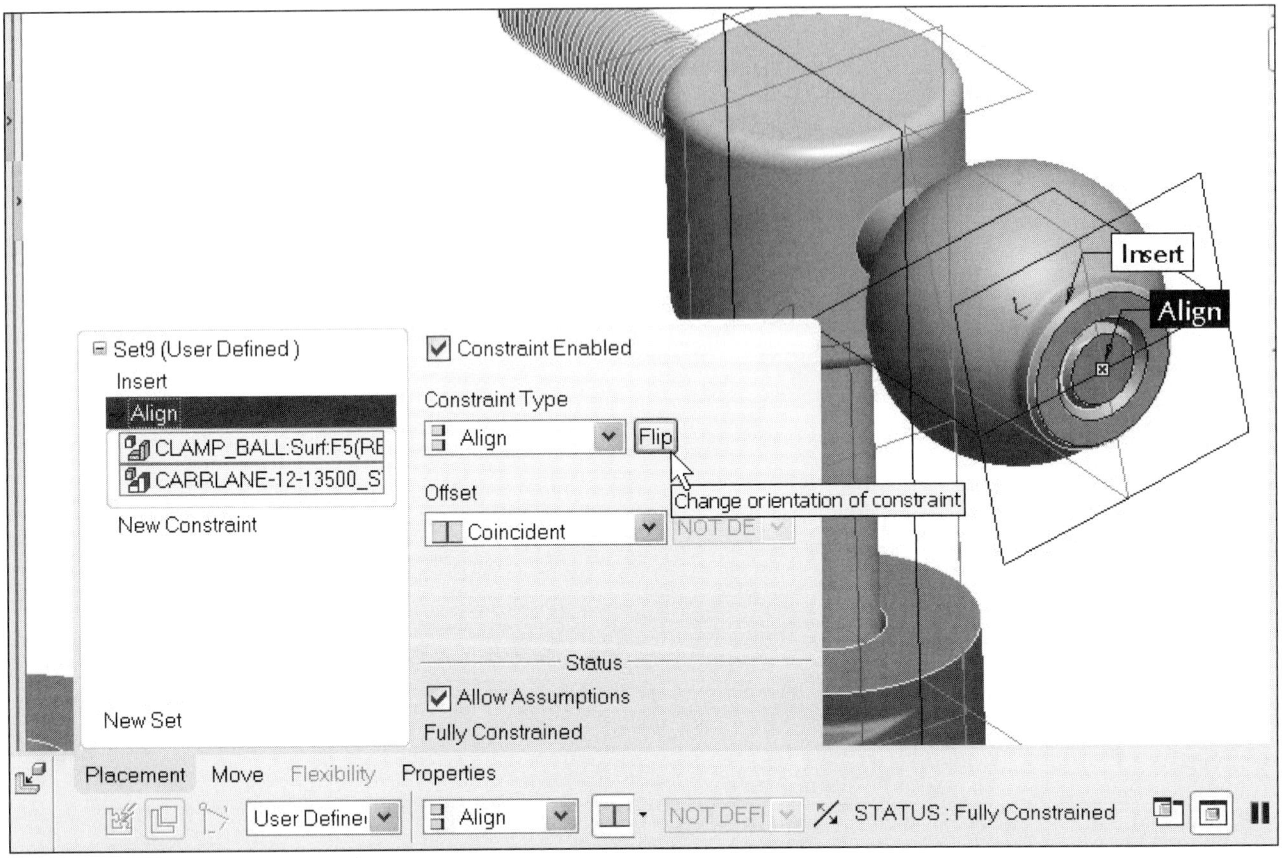

Figure 8.16(g) Fully Constrained Component

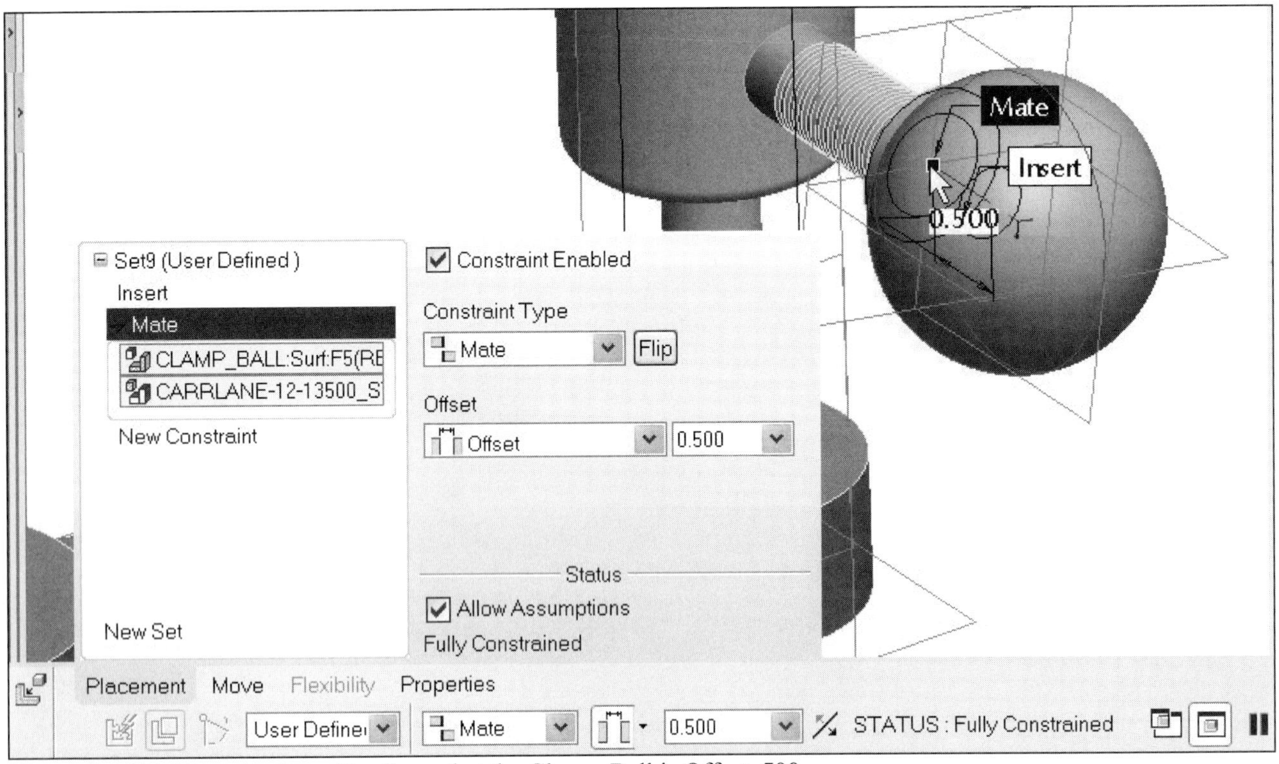

Figure 8.16(h) Move the Drag Handle so that the Clamp_Ball is Offset **.500**

Spin the model ⇒ pick on the **Clamp_Ball** component to select it *(highlights in red)* [Fig. 8.16(i)] ⇒ [icon] **Hidden line** ⇒ **Ctrl+C** ⇒ **Ctrl+V** ⇒ pick the opposite end cylindrical surface of the Carrlane-12-13500_STUD as the new reference [Fig. 8.16(j)

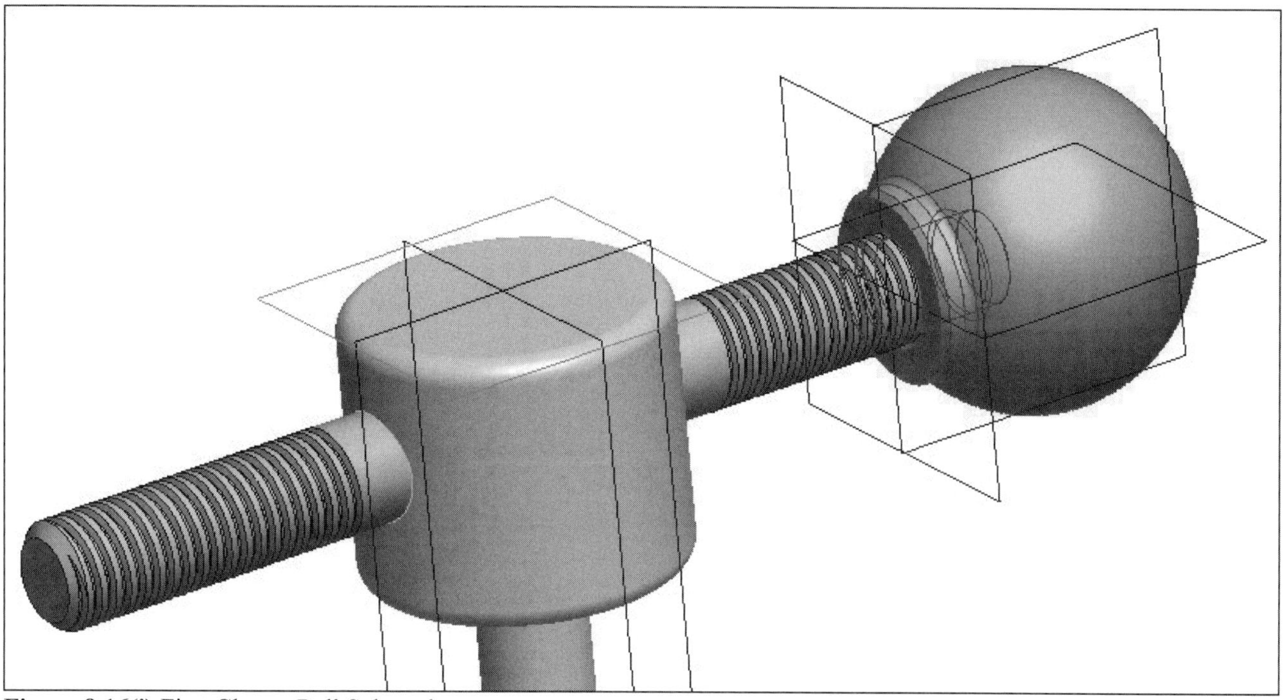

Figure 8.16(i) First Clamp_Ball Selected

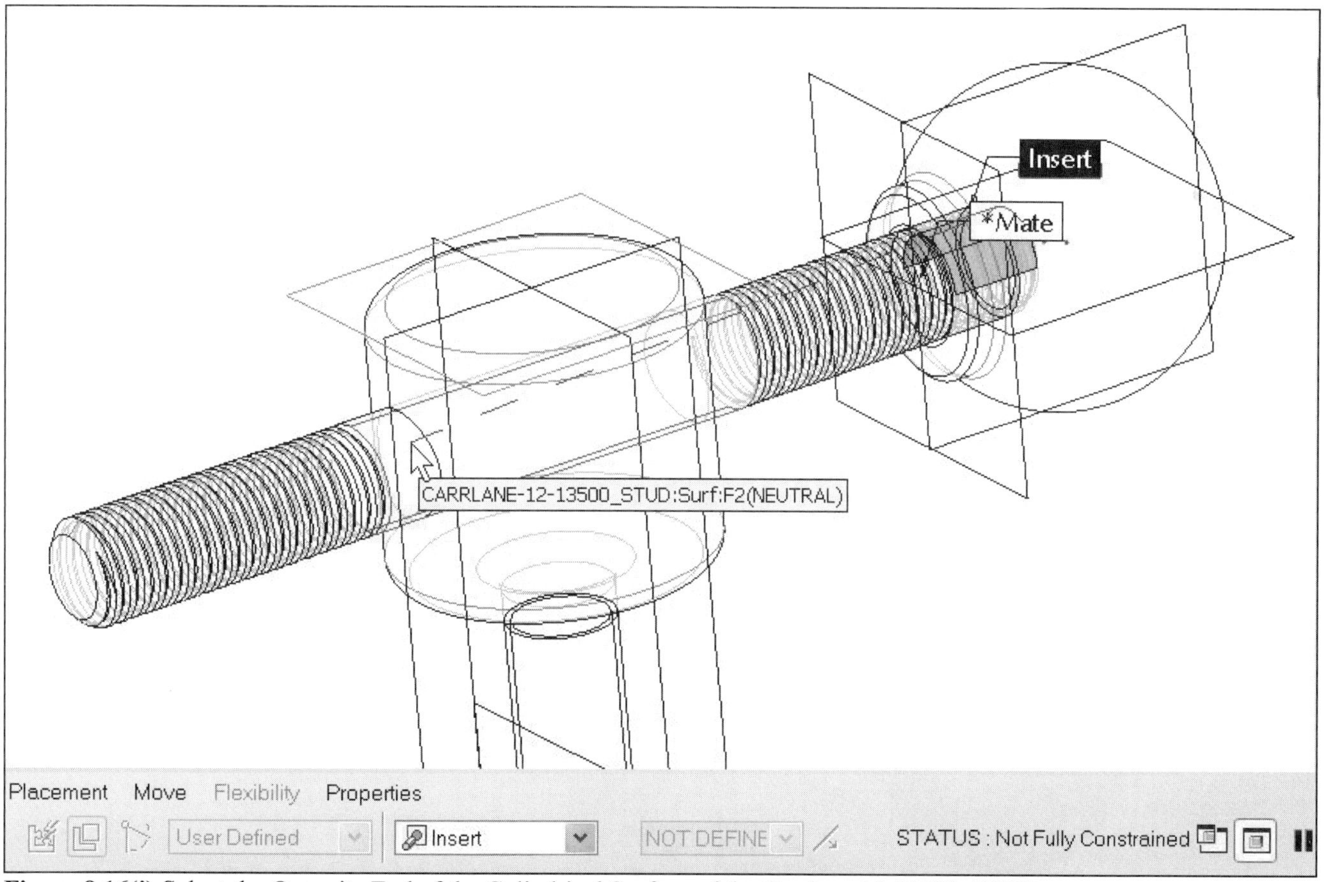

Figure 8.16(j) Select the Opposite End of the Cylindrical Surface of the Carrlane-12-13500_STUD

Pick the opposite end (flat) surface of the Carrlane-12-13500_STUD [Fig. 8.16(k)] ⇒ **Shade** ⇒ in the graphics window; double-click on **.500** dimension ⇒ type **.4375** to modify the distance from the end of the shaft so that it does not bottom-out [Fig. 8.16(l)] ⇒ **Enter** ⇒ ⇒ ⇒ **OK** ⇒ **LMB**

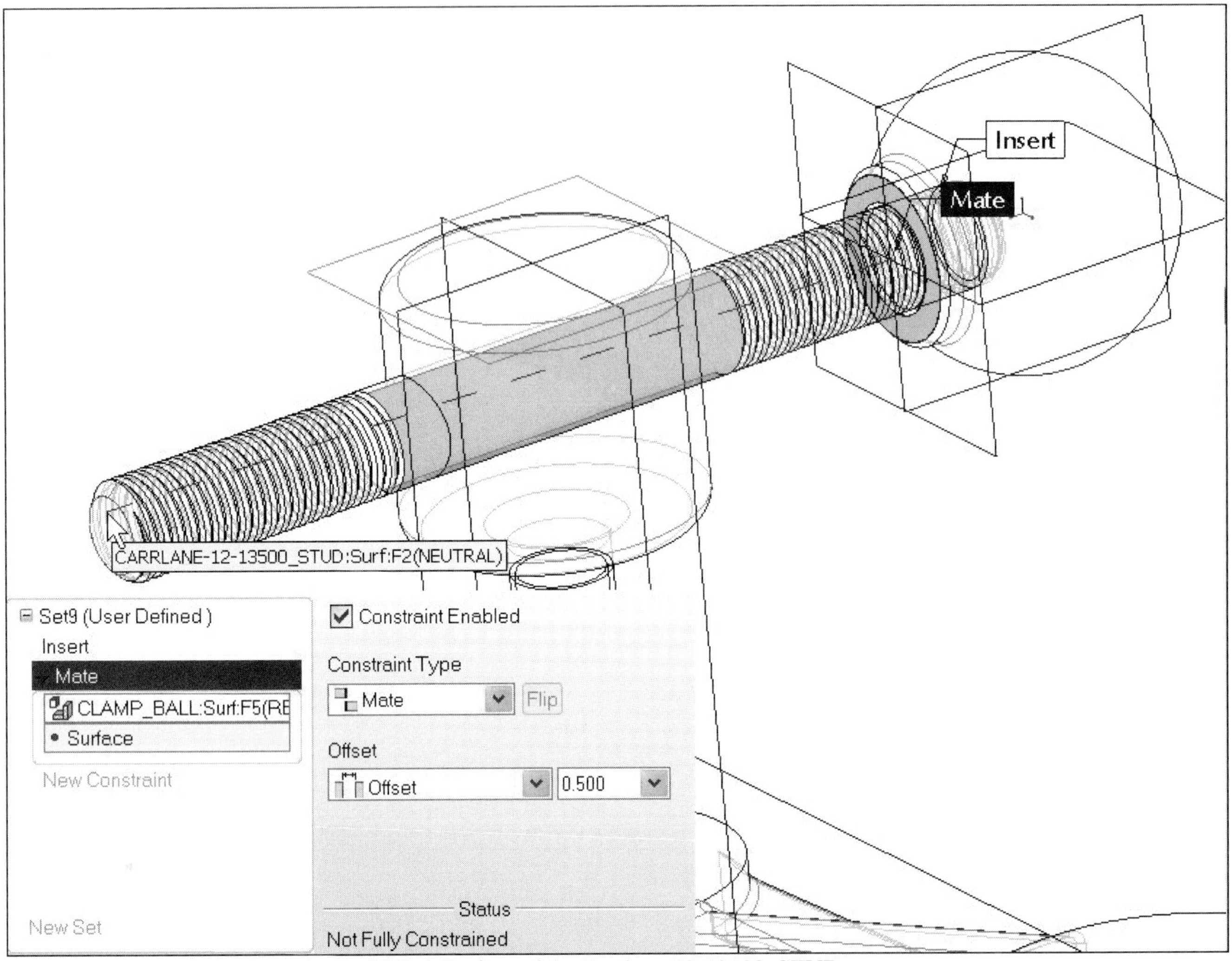

Figure 8.16(k) Pick the Opposite End (Flat) Surface of the Carrlane-12-13500_STUD

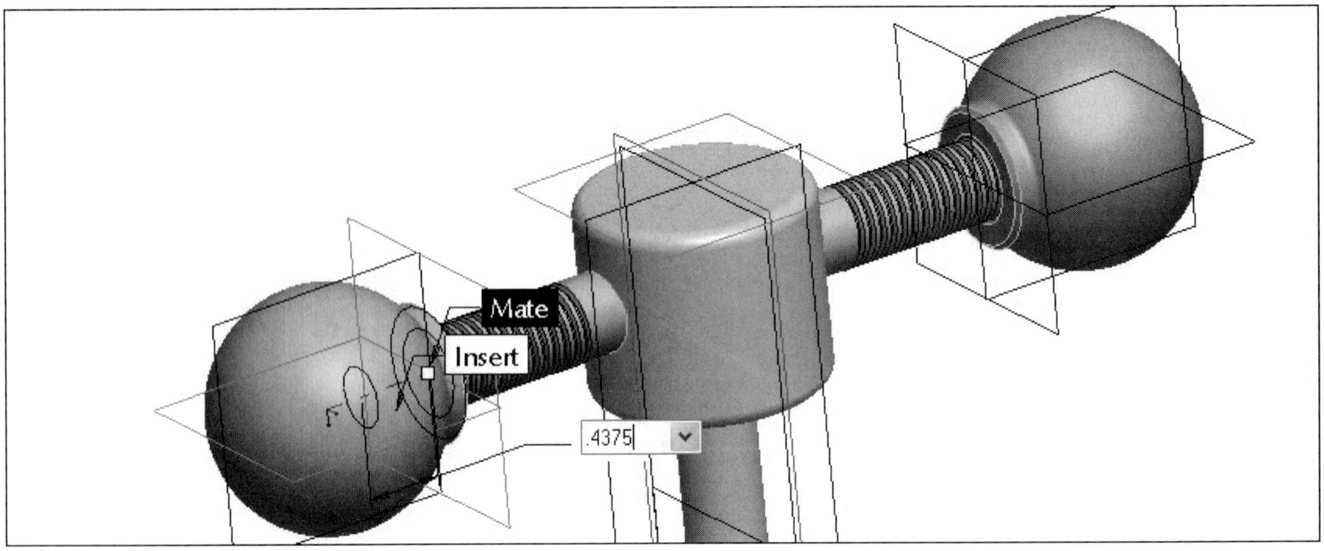

Figure 8.16(l) Offset **.4375**

Remember to change the offset distance of the first Clamp_Ball from **.500** to **.4375** and to **Regenerate**. Click: **Ctrl+D** ⇒ **View** ⇒ **Shade** [Fig. 8.17(a)] ⇒ **Info** from menu bar ⇒ **Bill of Materials** ⇒ Top Level [Fig. 8.17(b)] ⇒ **OK** [Fig. 8.17(c)] ⇒ close the Browser shade ⇒

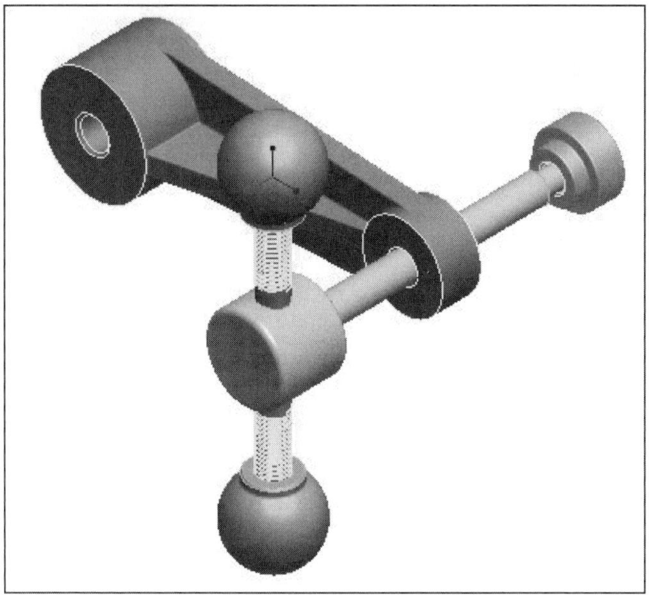

Figure 8.17(a) Completed Clamp_Subassembly

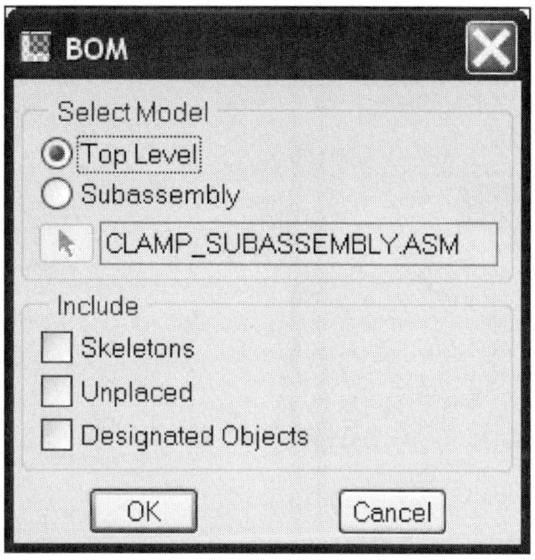

Figure 8.17(b) BOM Dialog Box

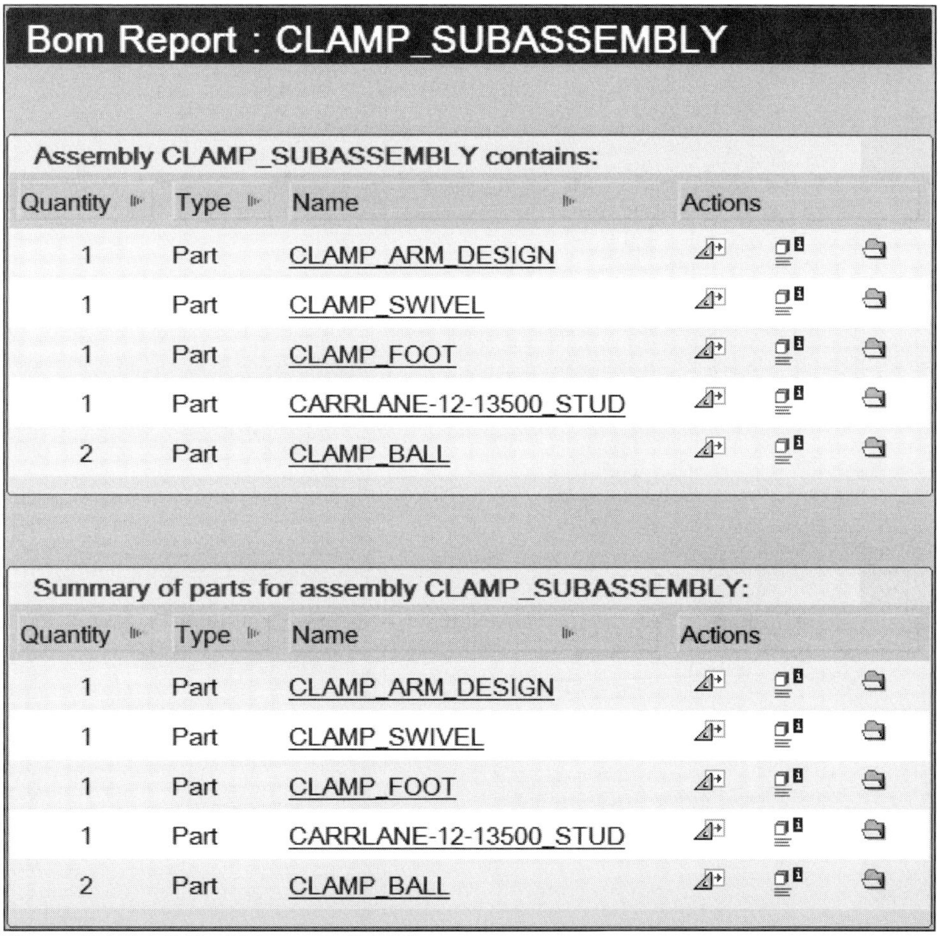

Bom Report : CLAMP_SUBASSEMBLY

Assembly CLAMP_SUBASSEMBLY contains:

Quantity	Type	Name	Actions
1	Part	CLAMP_ARM_DESIGN	
1	Part	CLAMP_SWIVEL	
1	Part	CLAMP_FOOT	
1	Part	CARRLANE-12-13500_STUD	
2	Part	CLAMP_BALL	

Summary of parts for assembly CLAMP_SUBASSEMBLY:

Quantity	Type	Name	Actions
1	Part	CLAMP_ARM_DESIGN	
1	Part	CLAMP_SWIVEL	
1	Part	CLAMP_FOOT	
1	Part	CARRLANE-12-13500_STUD	
2	Part	CLAMP_BALL	

Figure 8.17(c) Bill of Materials (BOM)

Click: **Datum planes** on ⇒ **Start the view manager** ⇒ **Xsec** tab [Fig. 8.18(a)] ⇒ **New** [Fig. 8.18(b)] ⇒ type name **A** [Fig. 8.18(c)] ⇒ **Enter** ⇒ **Model** ⇒ **Planar** ⇒ **Single** [Fig. 8.18(d)] ⇒ **Done** ⇒ **Plane** ⇒ pick datum **ASM_TOP** [Figs. 8.18(e-f)]

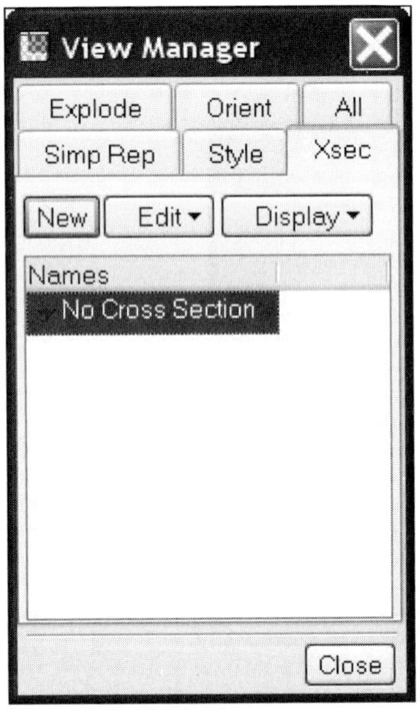

Figure 8.18(a) Xsec

Figure 8.18(b) New Xsection

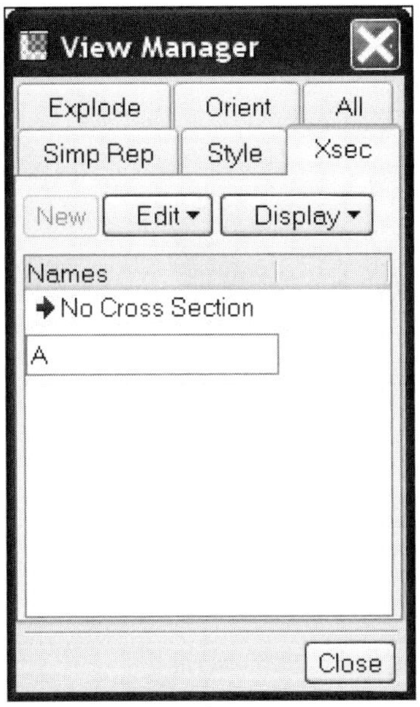

Figure 8.18(c) A

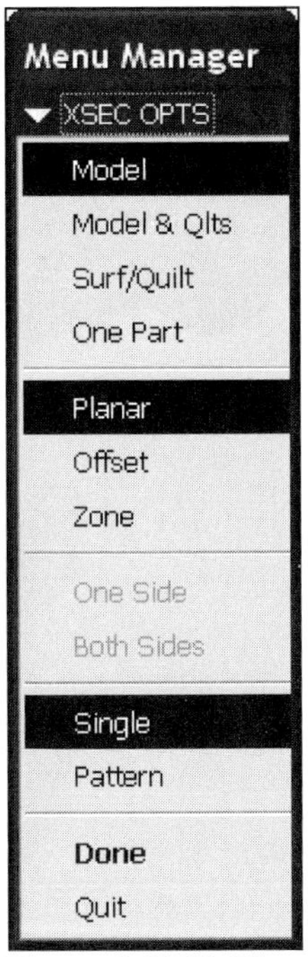

Figure 8.18(d) Planar

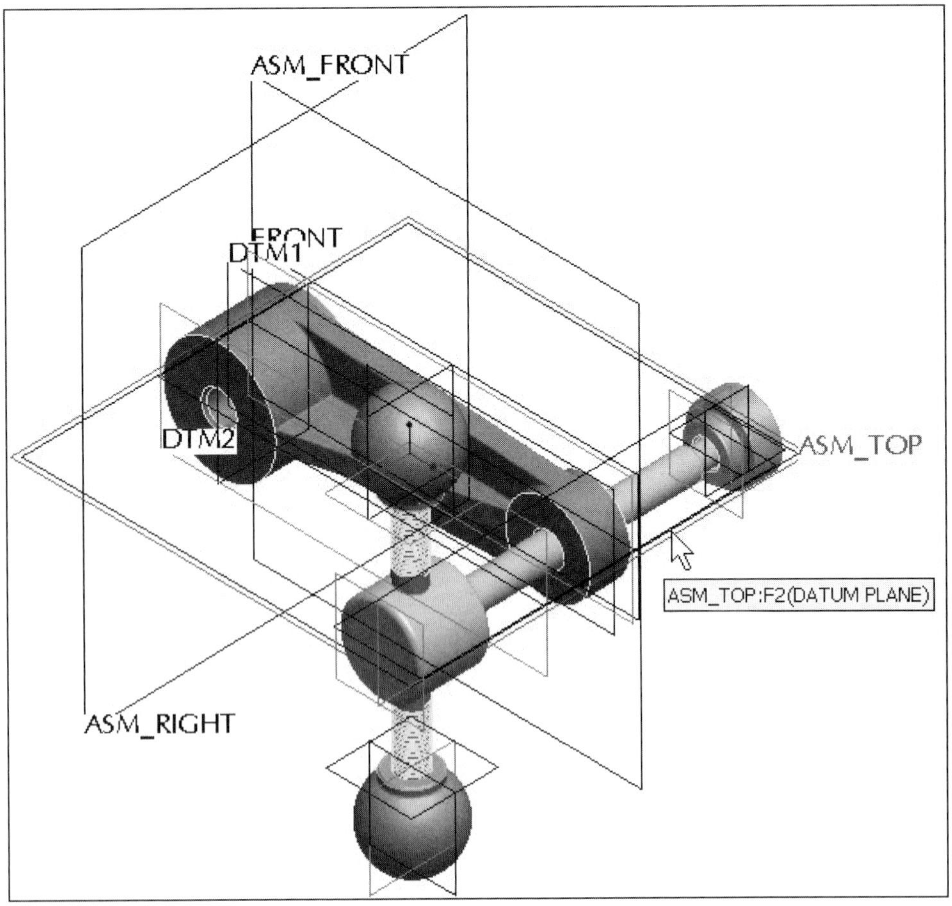

Figure 8.18(e) Select Assembly Datum Top

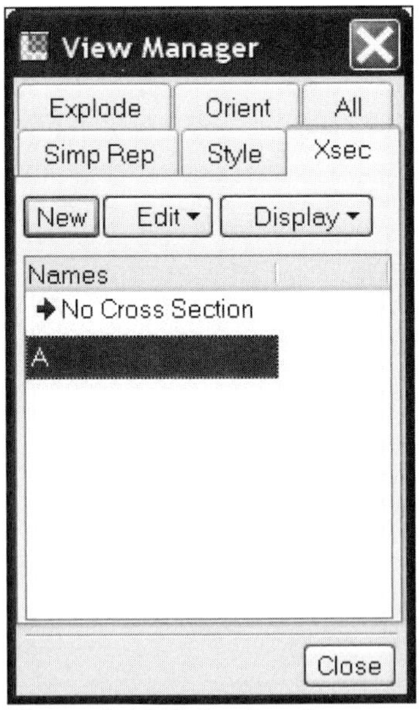

Figure 8.18(f) Section A

Click: Datum planes off ⇒ **Display** ⇒ **Set Active** [Fig. 8.18(g)] ⇒ **Display** ⇒ **Visibility** ⇒ **No Cross Section** No Cross Section ⇒ **Display** ⇒ **Set Active** [Fig. 8.18(h)] ⇒ **A** A ⇒ **RMB** ⇒ **Visibility** (toggles off) ⇒ **Close** ⇒ **Ctrl+D** ⇒ **Ctrl+S** ⇒ **OK** ⇒ **File** ⇒ **Delete** ⇒ **Old Versions** ⇒ **MMB** ⇒ **File** ⇒ **Close Window** (close the subassembly)

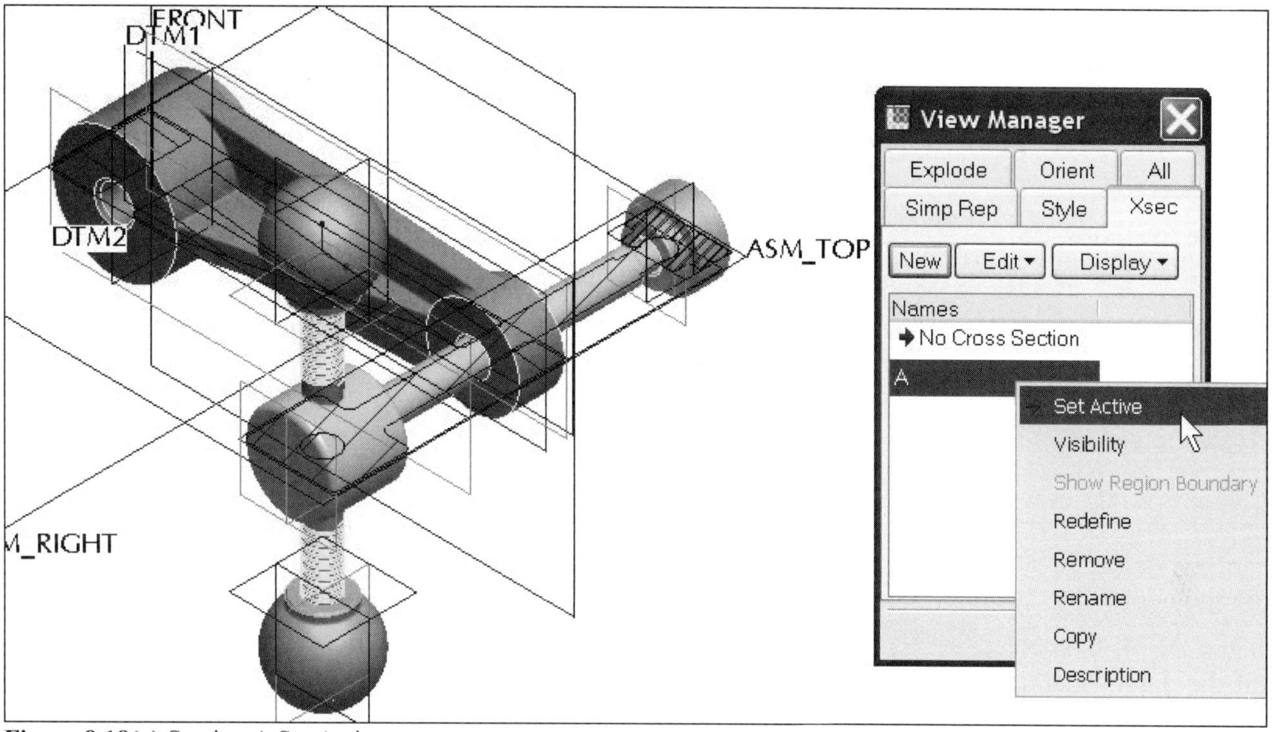

Figure 8.18(g) Section A Set Active

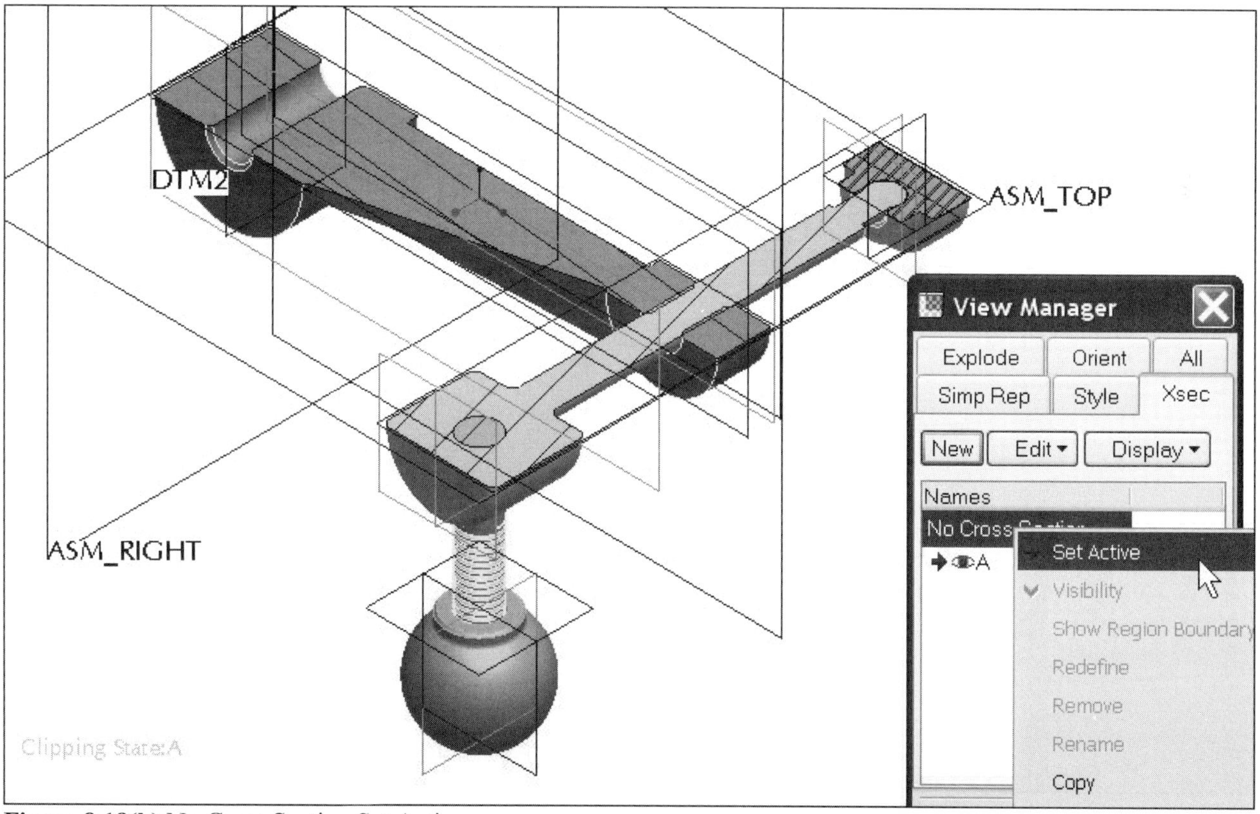

Figure 8.18(h) No Cross Section Set Active

Part Two Top-Down Design

Swing Clamp Assembly

The first features for the main assembly will be the default datum planes and coordinate system. The first part assembled on the main assembly will be the Clamp_Plate, which will be created using *top-down design*; where the assembly is active, and the component is created within the assembly mode. The subassembly is still *"in session-in memory"* even though it does not show on the screen after its window is closed. For the two standard parts of the main assembly, you will use specific constraints instead of using Automatic, which allows Pro/E to default to an assumed constraint.

Click: **File** ⇒ **New** ⇒ [Assembly] ⇒ Name **CLAMP_ASSEMBLY** ⇒ Sub-type [Design] ⇒ [Use default template] ⇒ **OK** ⇒ Template: **inlbs_asm_design** ⇒ Parameters: DESCRIPTION: *type-* **Swing Clamp Assembly** ⇒ MODELED_BY ***type your name*** ⇒ **OK** ⇒ **Tools** ⇒ **Environment** ⇒ Standard Orient **Trimetric** ⇒ **Apply** ⇒ **OK** ⇒ [datum display icons] on ⇒ **Edit** ⇒ **Setup** ⇒ **Units** ⇒ [Inch lbm Second (Pro/E Default)] ⇒ **Close** ⇒ **Done** ⇒ open the Navigator shade with the quick sash ⇒ [Settings ▾] from the Navigator ⇒ [Tree Filters...] ⇒ check all Display options ⇒ **Apply** ⇒ **OK**

Change the coordinate system name: slowly click twice on [ASM_DEF_CSYS] in the Model Tree ⇒ type new name **ASM_CSYS** [ASM_CSYS] ⇒ **LMB** anywhere in the graphics window ⇒ continue to change each of the datum identifiers in the Model Tree and add **CL_** as a prefix for each (i.e. **CL_ASM_TOP**) (Fig. 8.19) ⇒ [save icon] ⇒ **OK**

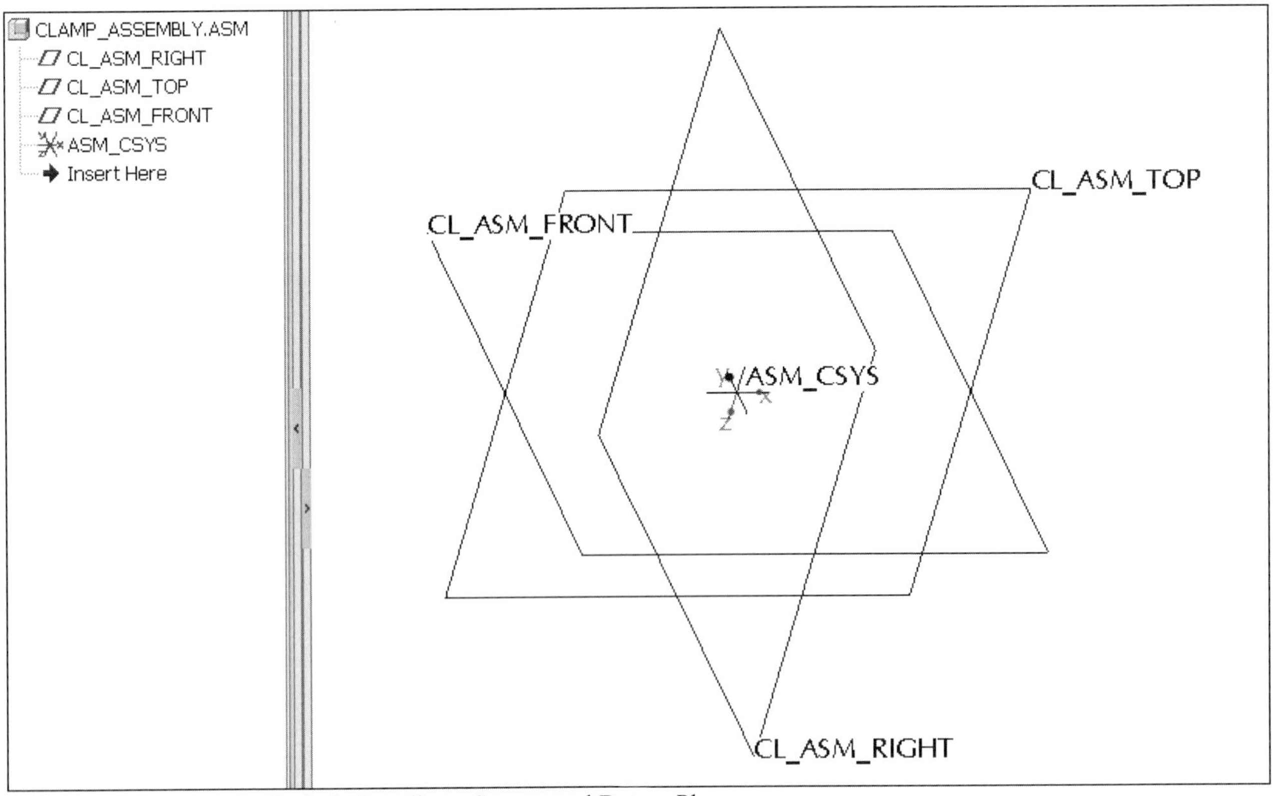

Figure 8.19 Rename the Default Coordinate System and Datum Planes

Creating Components in the Assembly Mode

The Clamp_Subassembly is now complete. The subassembly will be added to the assembly after the Clamp_Plate is created and assembled. Using the Component Create dialog box (Fig. 8.20), you can create different types of components: parts, subassemblies, skeleton models, and bulk items. You cannot reroute components created in the Assembly mode.

The following methods allow component creation in the context of an assembly without requiring external dependencies on the assembly geometry:

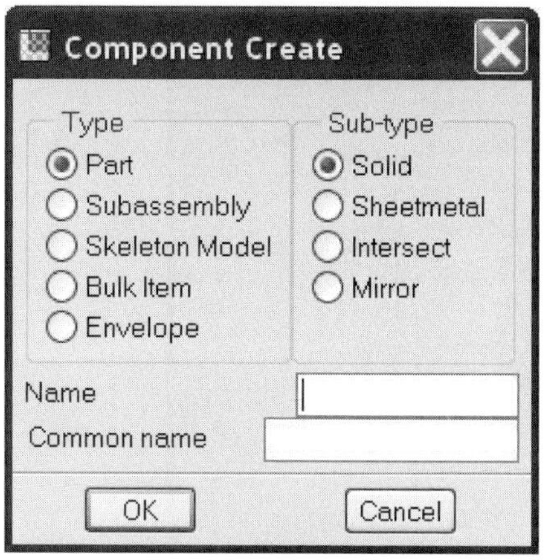

Figure 8.20 Component Create Dialog Box

- Create a component by copying another component or existing start part or start assembly
- Create a component with default datums
- Create an empty component
- Create the first feature of a new part; this initial feature is dependent on the assembly
- Create a part from an intersection of existing components
- Create a mirror copy of an existing part or subassembly
- Create Solid or Sheetmetal components
- Mirror Components

Main Assembly, Top-Down Design

The Clamp_Plate component is the first component of the main assembly. The Clamp_Plate is a new part. You will be modeling the plate *"inside"* the assembly using *top-down design*. The drawing in Figure 8.21 provides the dimensions necessary to model the Clamp_Plate.

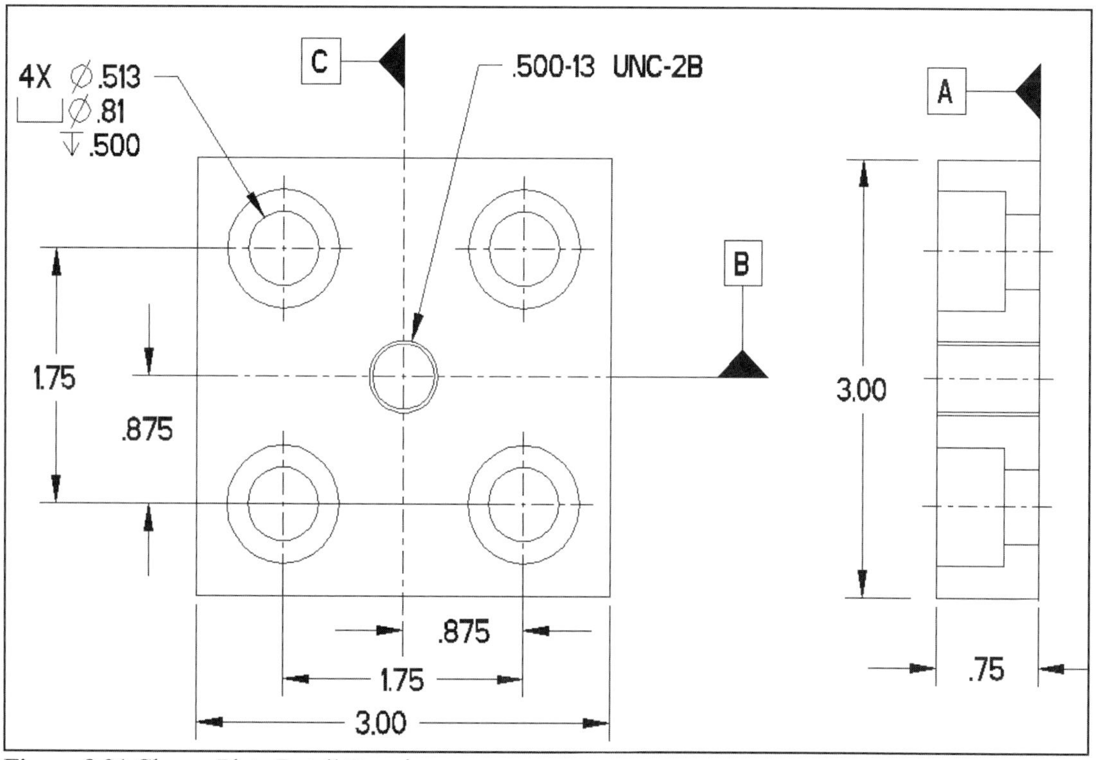

Figure 8.21 Clamp_Plate Detail Drawing

Click: **Create a component in assembly mode** Component Create dialog box displays ⇒ Type Part ⇒ Sub-type Solid ⇒ Name **CLAMP_PLATE** [Fig. 8.22(a)] ⇒ **OK** Creation Options dialog box displays [Fig. 8.22(b)] ⇒ Locate Default Datums ⇒ Align Csys To Csys ⇒ **OK** ⇒ pick **ASM_CSYS** from the graphics window, CLAMP_PLATE.PRT displays in the Model Tree [Fig. 8.23(a)], *the small green symbol/icon indicates that this component is now active* ⇒ expand and highlight the items in the Model Tree for the CLAMP_PLATE.PRT

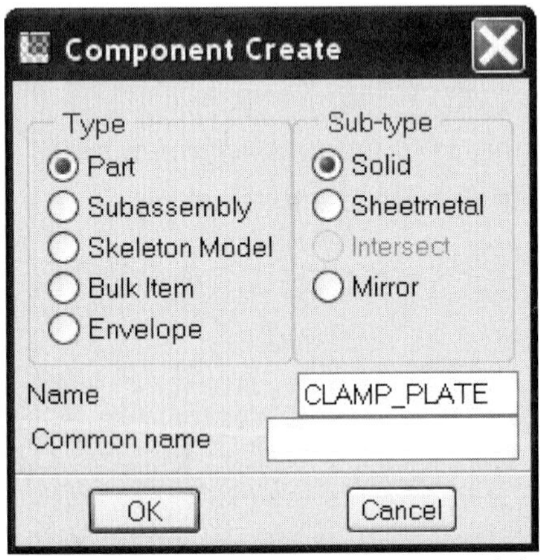

Figure 8.22(a) Component Create Dialog Box

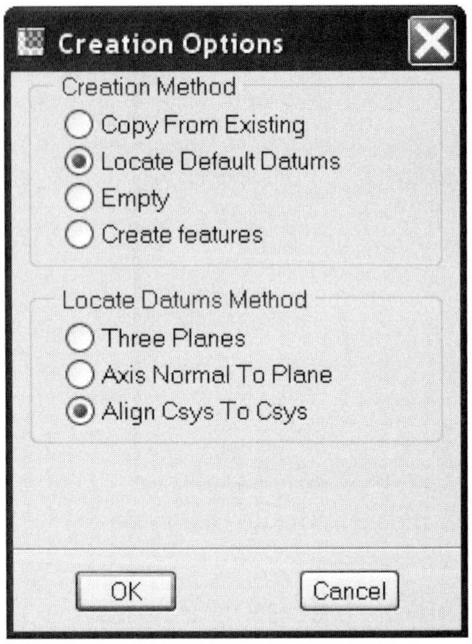

Figure 8.22(b) Creation Options Dialog Box

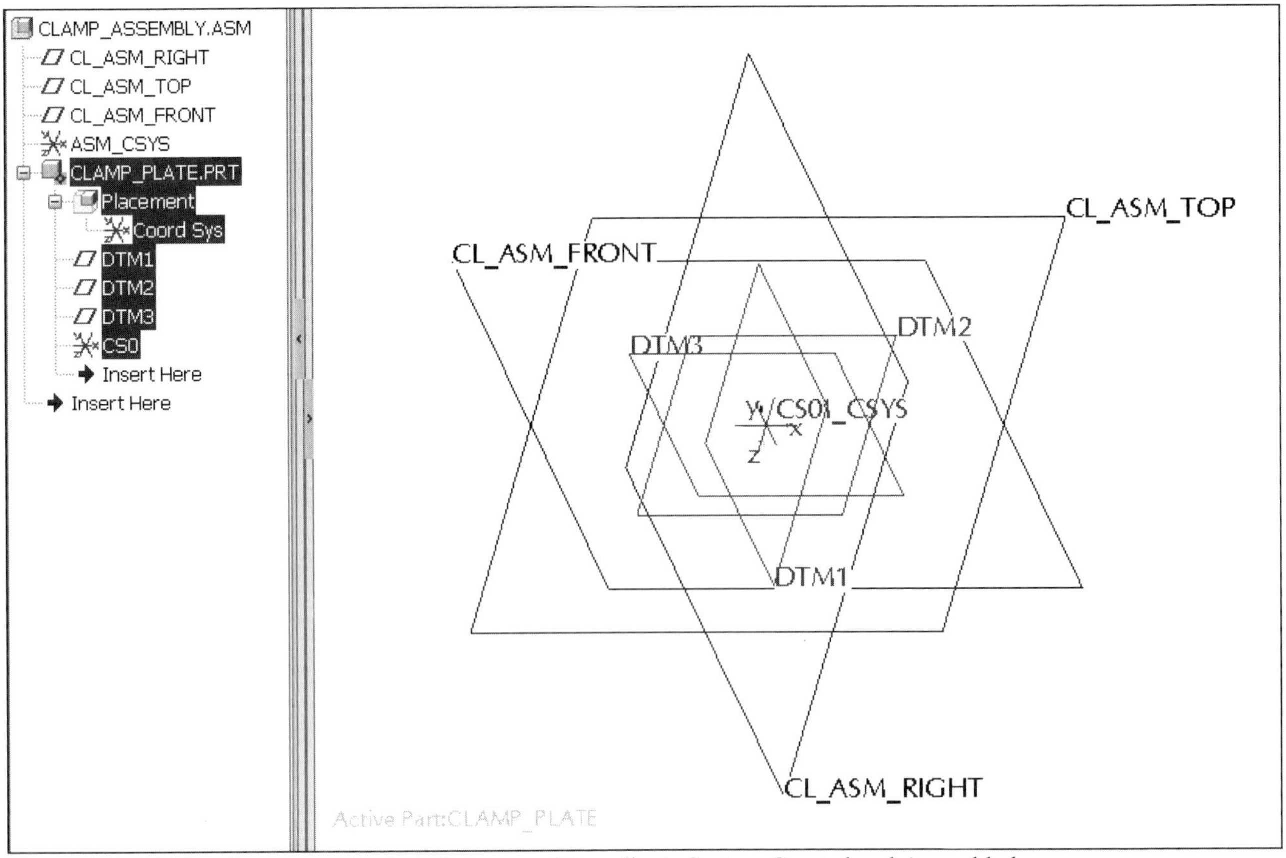

Figure 8.23(a) New Component Default Datums and Coordinate System Created and Assembled

Assembly Tools are now *unavailable* in the Right Toolchest. The current component is the Clamp_Plate. *You are now effectively in Part mode*, except that you can see the assembly features and components. Be sure to reference only part features as you model (part datum planes and part coordinate system), otherwise you will create unwanted external references. There are many situations where external references are needed and desired, but in this case, you are simply modeling a new part.

Create the protrusion, click: **Extrude Tool** ⇒ change the depth value to **.750** ⇒ **Enter** ⇒ **Placement** tab ⇒ **Define**

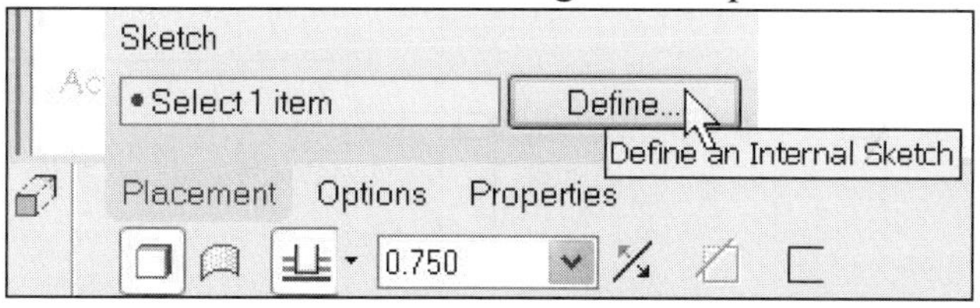

Sketch dialog box opens ⇒ Sketch Plane--- Plane: select **DTM3** [Fig. 8.23(b)] ⇒ **Sketch** button [Fig. 8.23(c)]

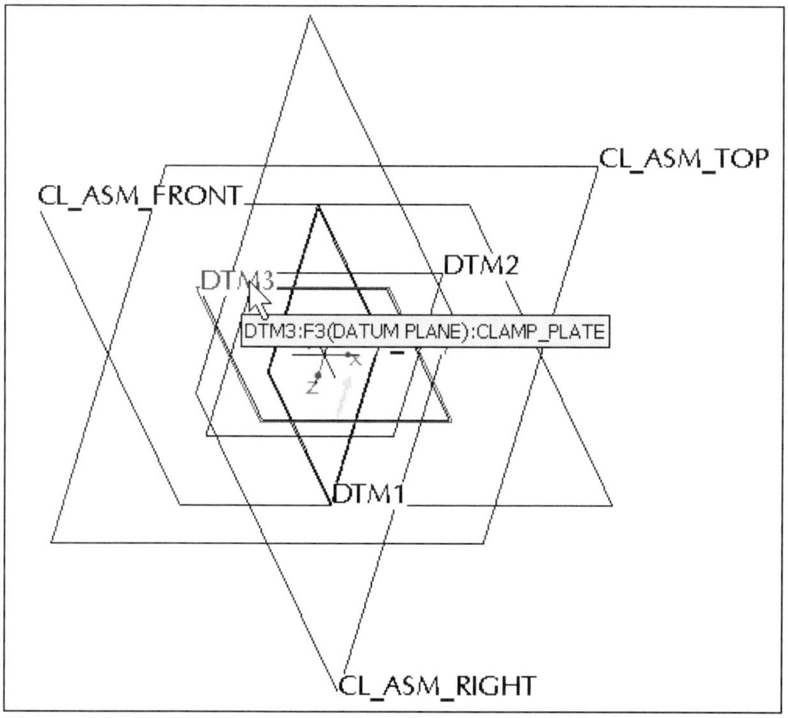

Figure 8.23(b) Sketch Plane and Reference Orientation

Figure 8.23(c) Sketch Dialog Box

Click: **RMB** ⇒ **Centerline** ⇒ sketch a vertical centerline ⇒ sketch a horizontal centerline ⇒ **Create rectangle** ⇒ pick the two corners of the rectangle *(if you select the corners carefully you will get a square with only one dimension, if not, then use the* ⇒ *constraint to make the two sides equal)* [Fig. 8.23(d)] ⇒ pick the dimension ⇒ **RMB** ⇒ **Modify** modify the value to **3.00** [Fig. 8.23(e)] ⇒ **Enter** ⇒ ⇒ **Ctrl+D** ⇒ [Fig. 8.23(f)] ⇒ [Fig. 8.23(g)] ⇒ click on the CLAMP_ASSEMBLY.ASM in the Model Tree ⇒ **RMB** ⇒ **Activate**

⇒ ⇒ **Enter** ⇒ click on the CLAMP_PLATE.PRT in the Model Tree ⇒ **RMB** ⇒ **Activate**

(the Clamp_Plate is now active CLAMP_PLATE.PRT) ⇒ **Ctrl+S** ⇒ **Enter** ⇒ **Hidden line**

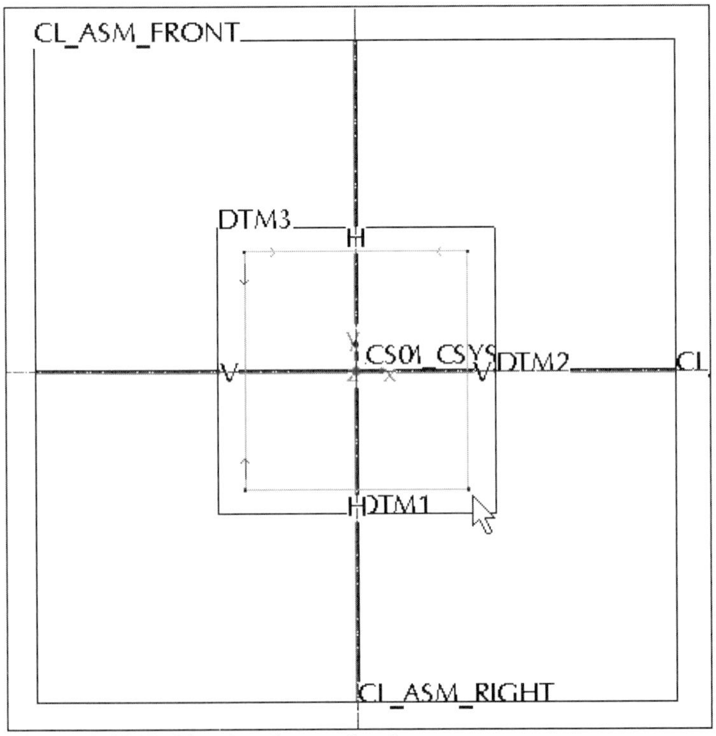

Figure 8.23(d) Sketch a Square Section

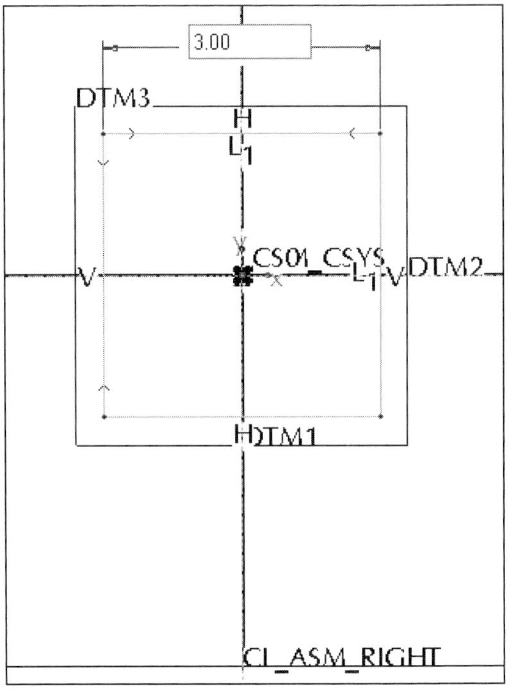

Figure 8.23(e) Modify the Value to **3.00**

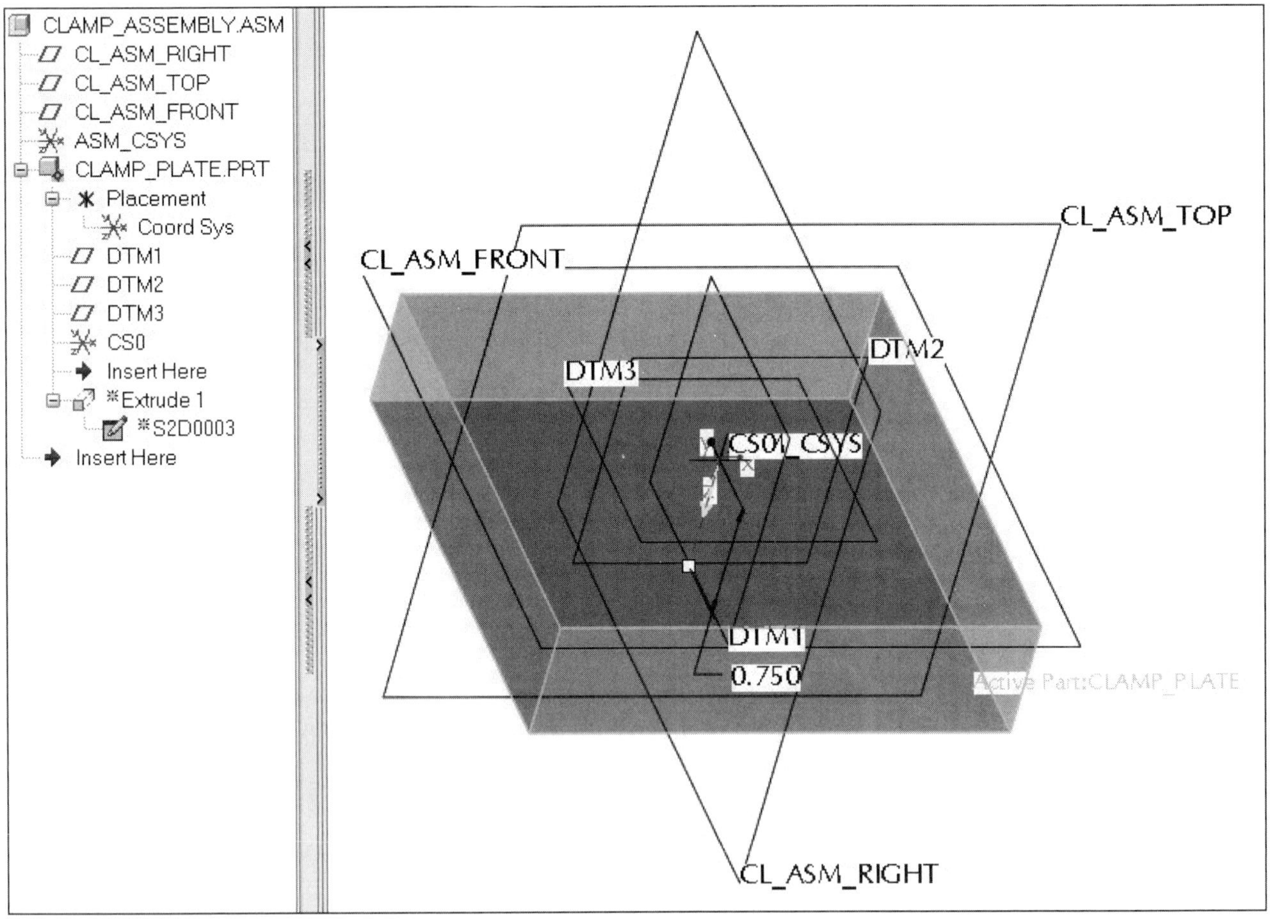

Figure 8.23(f) Depth and Direction Preview

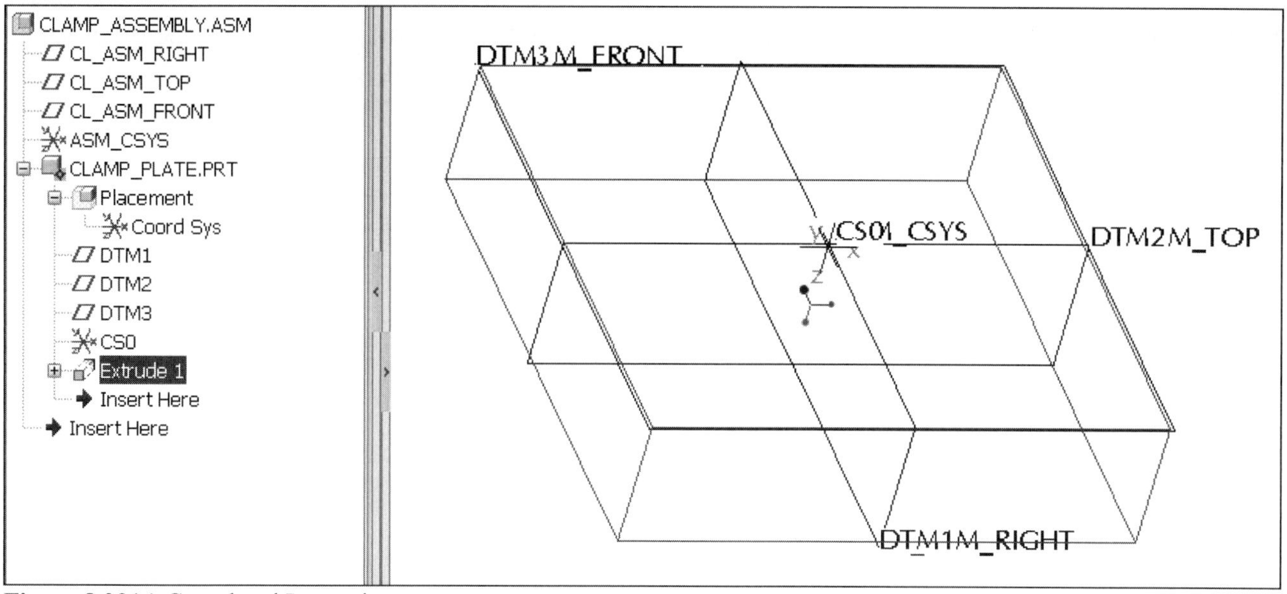

Figure 8.23(g) Completed Protrusion

Throughout the lesson, follow the steps as given. Drag the handles to the appropriate references and let Pro/E select the references automatically ***even if you think the selections are incorrect***. Later, we will address these inconsistencies, but if you correct them on your own, steps described later will not work (or be needed).

Create the tapped hole in the center of the part, click: **Hole Tool** ⇒ **Create standard hole** ⇒ toggle off ⇒ toggle off ⇒ ⇒ on ⇒ select the tap and drill size **1/2-13**

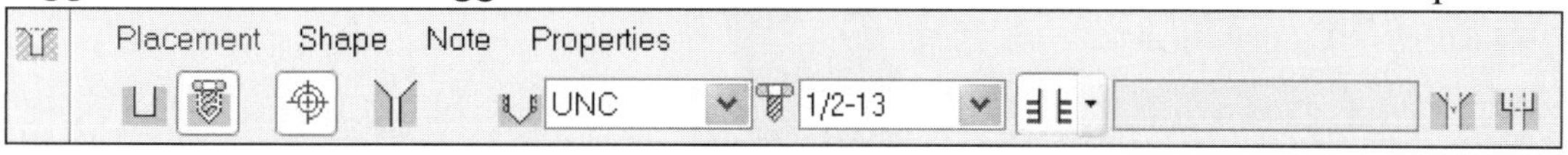

⇒ **Shape** tab [Fig 8.23 (h)] ⇒ Include thread surface ⇒ Thru Thread ⇒ **Properties** tab [Fig 8.23 (i)] ⇒ **Placement** tab ⇒ Pick on the top surface of the protrusion as the placement plane. The preview hole displays with drag handles for hole position, diameter adjustment, depth adjustment, and two reference handles for establishing the dimensioning scheme [Fig 8.23(j)].

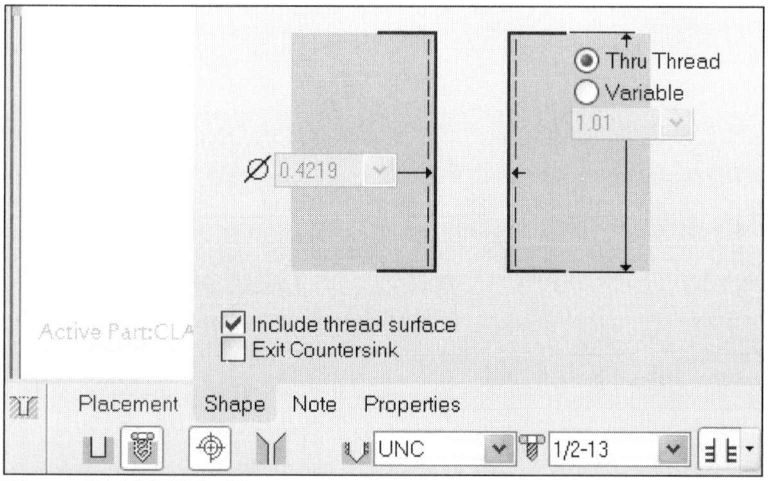

Figure 8.23(h) Hole Shape

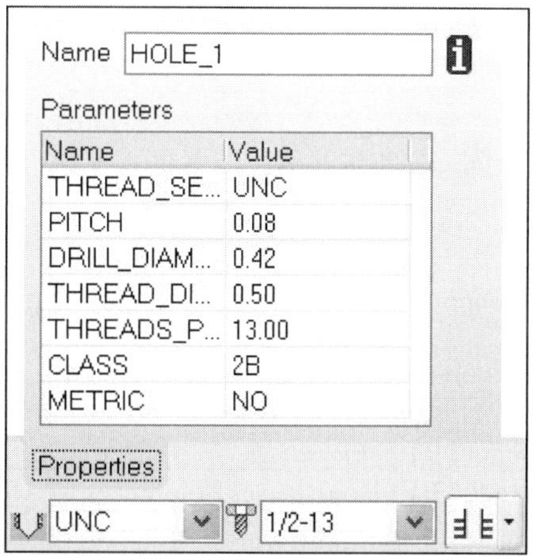

Figure 8.23(i) Hole Properties

To establish the Offset References, move one *green* drag handle to the vertical datum and the other *green* drag handle to horizontal datum respectively ⇒ change *both* linear dimensions to **Align** [Fig 8.23(k)] ⇒  ⇒ **LMB** in the graphics window to deselect ⇒ **Ctrl+D** [Fig. 8.23(l)]

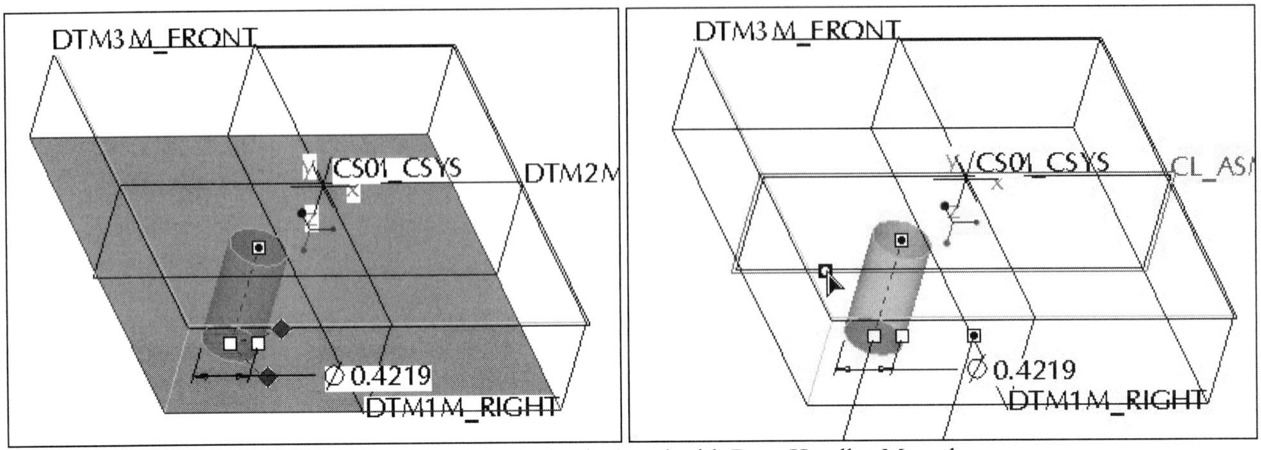

Figure 8.23(j) Initial Hole Placement (at point of selection) and with Drag Handles Moved

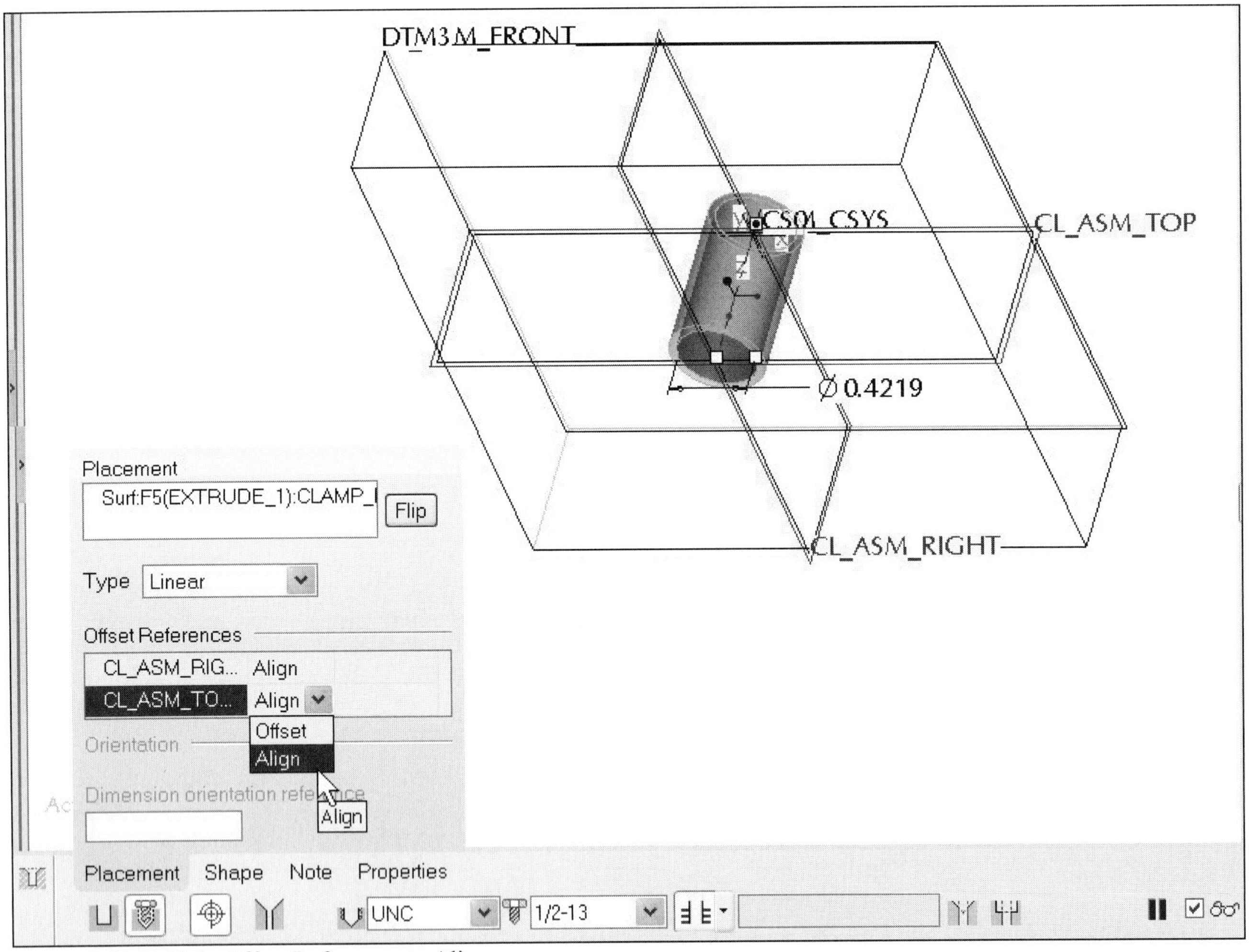

Figure 8.23(k) Hole Offset References to Align

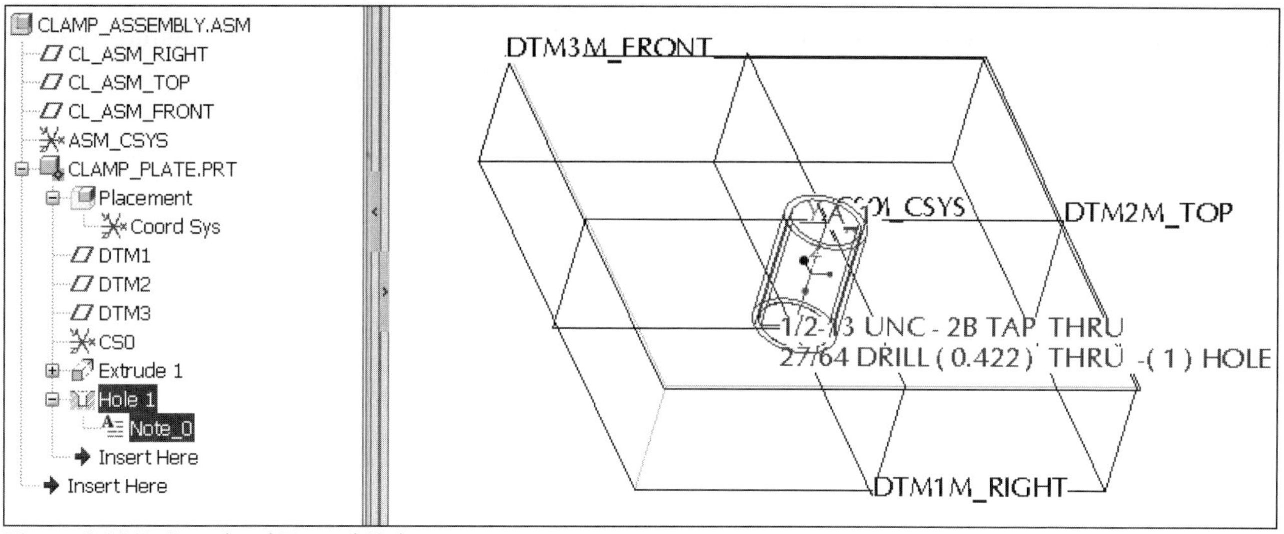

Figure 8.23(l) Completed Tapped Hole

Create a counterbore hole, click: **Hole Tool** ⇒ ⇒ toggle off ⇒ 5/8-11 ⇒ toggle off ⇒ **Adds counterbore** on ⇒ **Drill to intersect with all surfaces** ⇒ **Shape** tab ⇒ input the counterbores sizes [Fig. 8.23(m)] *the detail drawing callouts are slightly different* [Fig. 8.23(n)] ⇒ pick on the front surface to place the counterbore hole as in Figure 8.23(o).

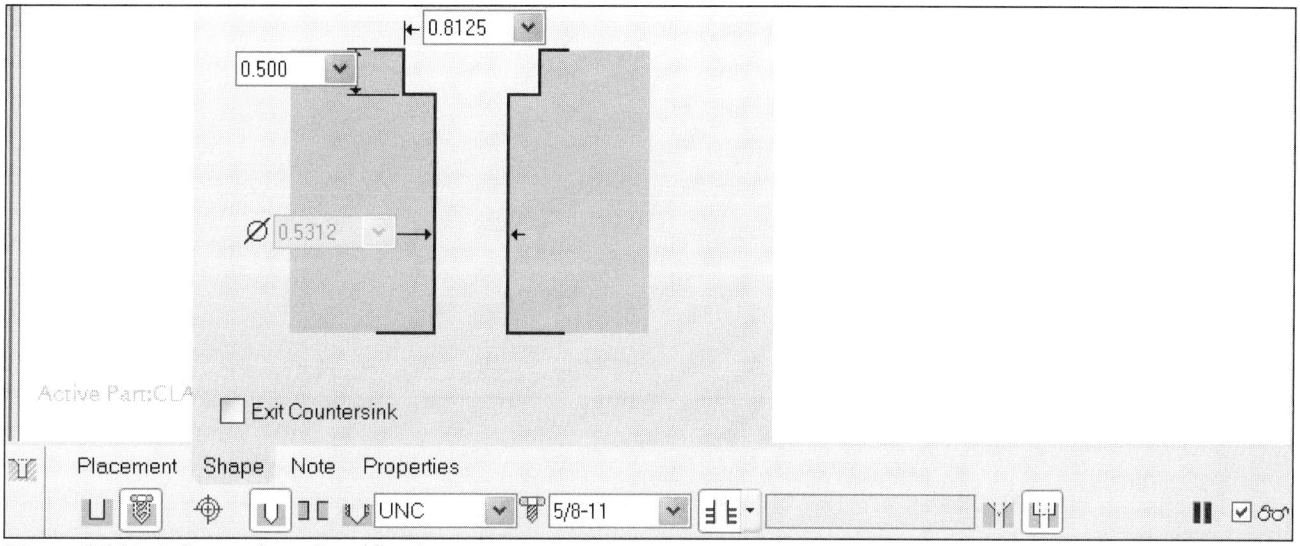

Figure 8.23(m) Counterbore Specifications

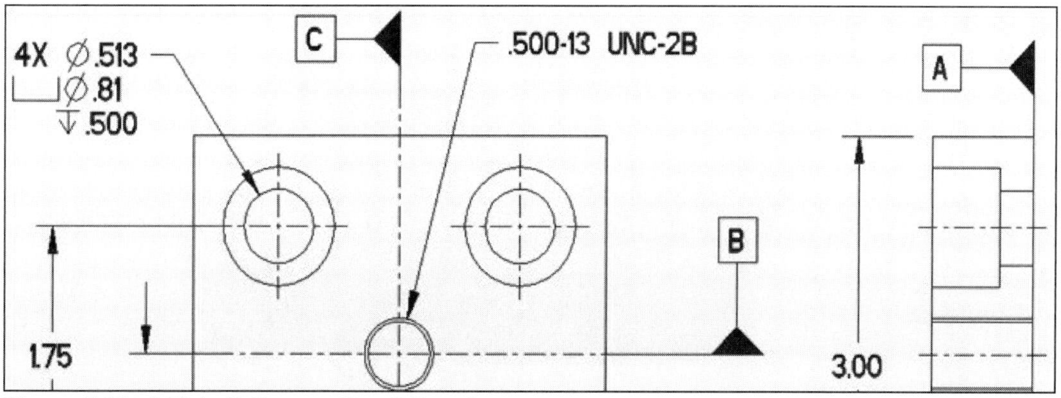

Figure 8.23(n) Hole Callout on Detail Drawing

Click: **Placement** tab ⇒ **RMB** ⇒ **Offset References Collector** ⇒ pick on the vertical datum ⇒ press **Ctrl** and pick on the horizontal datum [Fig 8.23(o)] ⇒ change both linear dimensions to **.875** *(you may need to use negative .875)* ⇒ **Enter** ⇒ ☑ ⇒ ☐ [Fig. 8.23(p)] ⇒ **LMB** to deselect

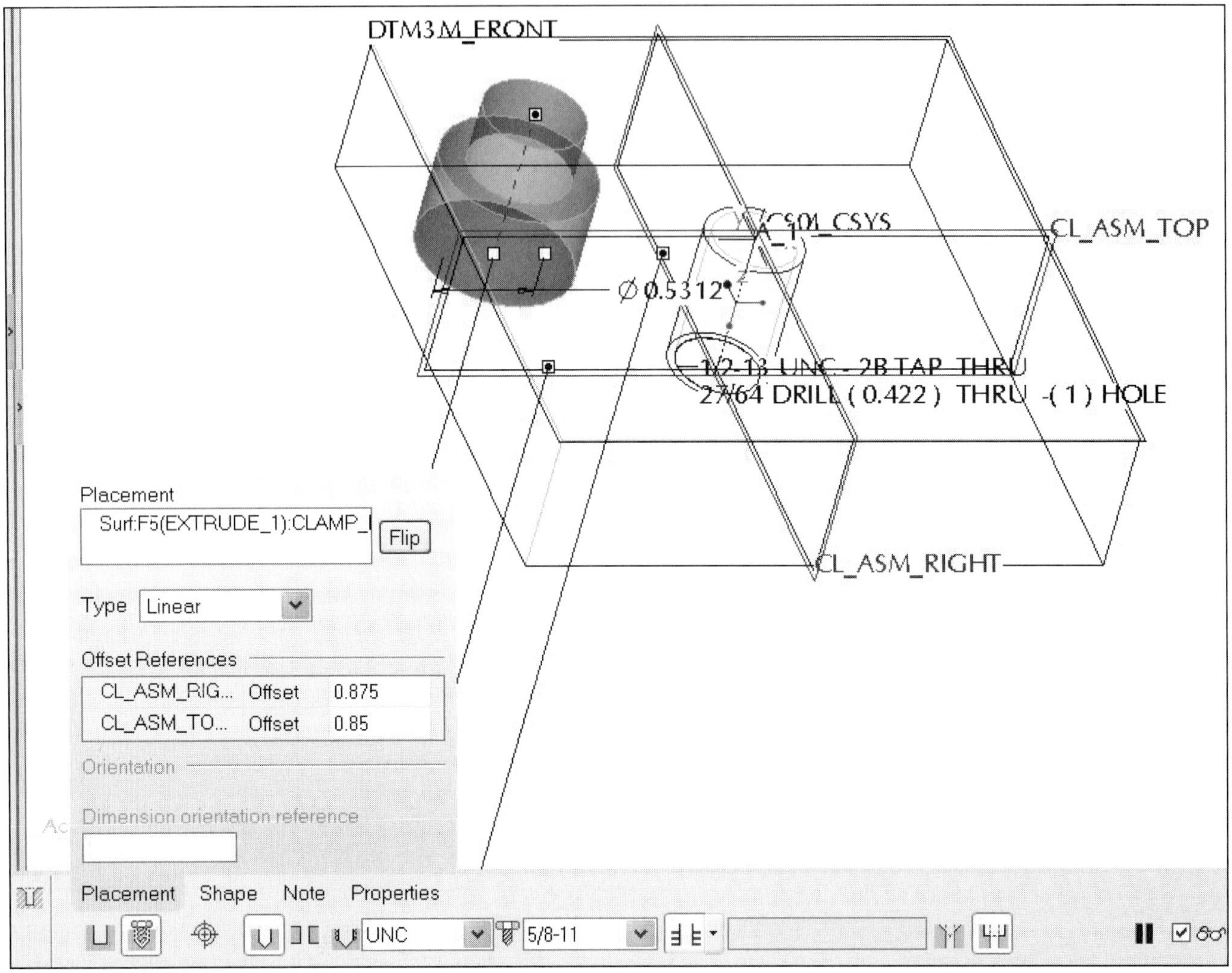

Figure 8.23(o) Offset References

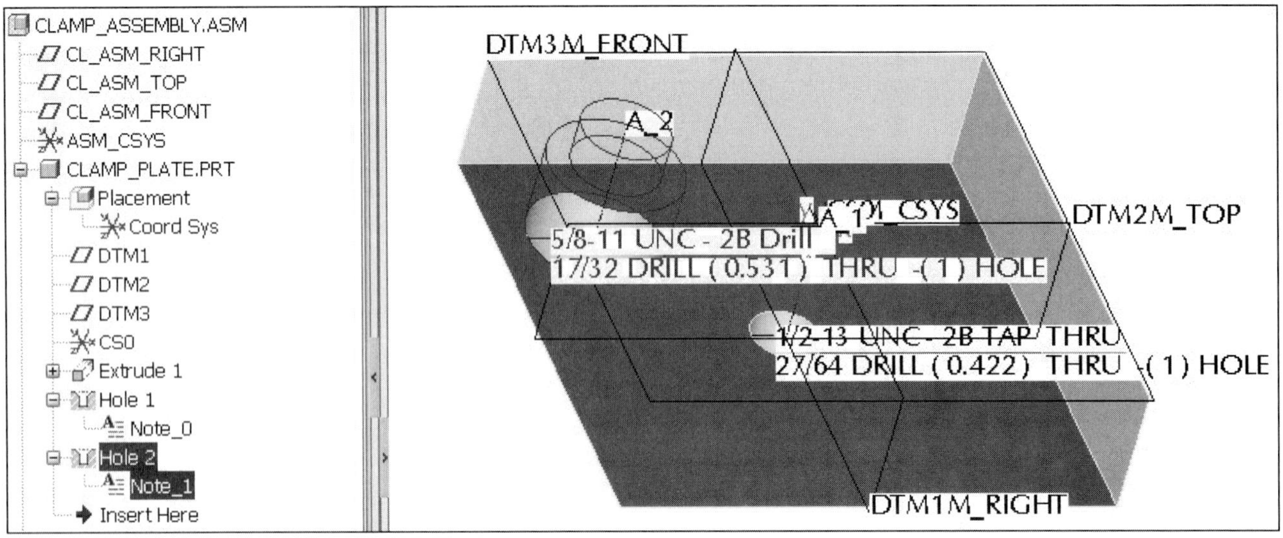

Figure 8.23(p) Completed Counterbore

The holes "seem" correct, but it is important to check for external references. Click on **Hole_2** from the Model Tree or model ⇒ **RMB** ⇒ **Info** ⇒ **Reference Viewer** ⇒ **References Filters** tab ⇒ **References** on ⇒ **Components in path** on [Fig 8.23(q)] ⇒ expand the references [Fig. 8.23(r)]

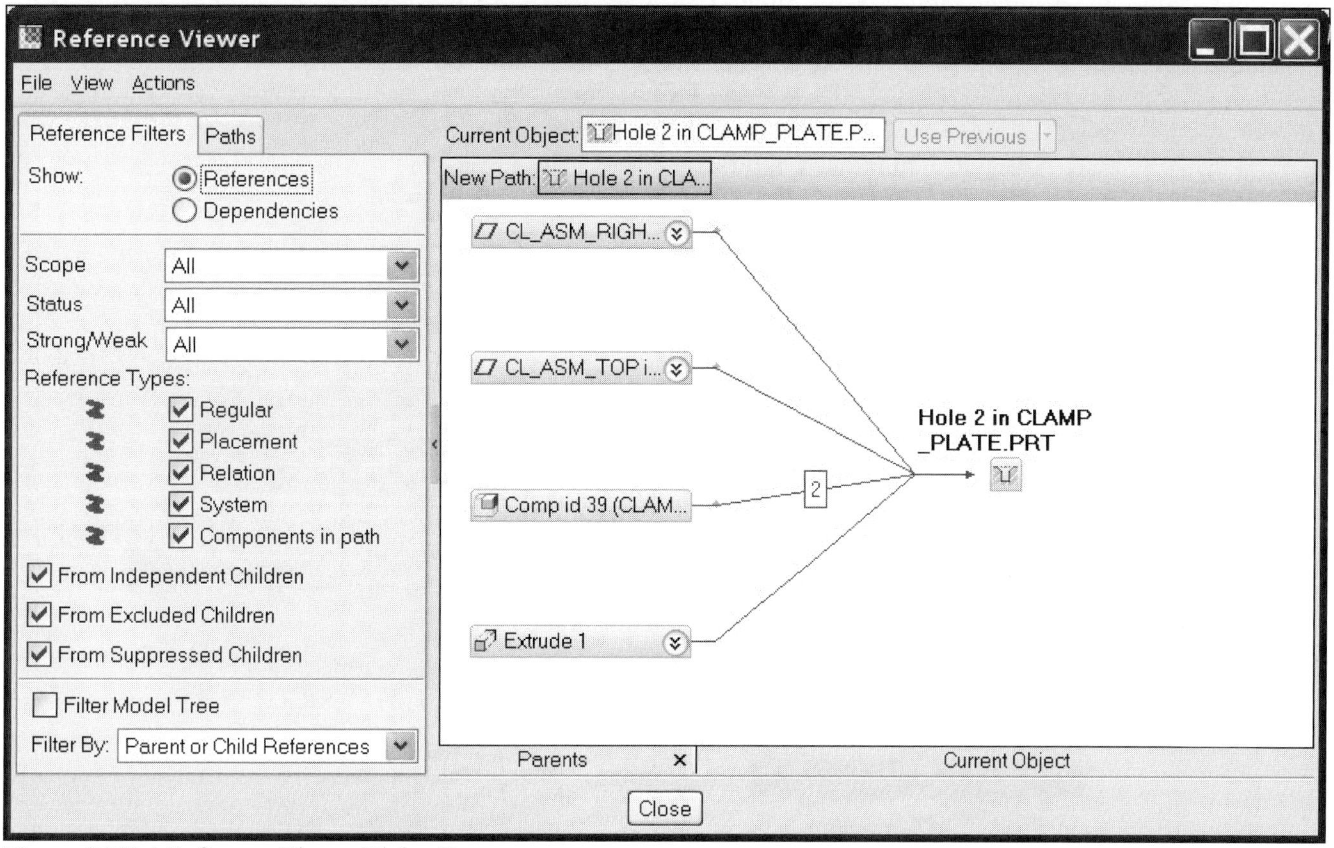

Figure 8.23(q) Reference Viewer Dialog Box

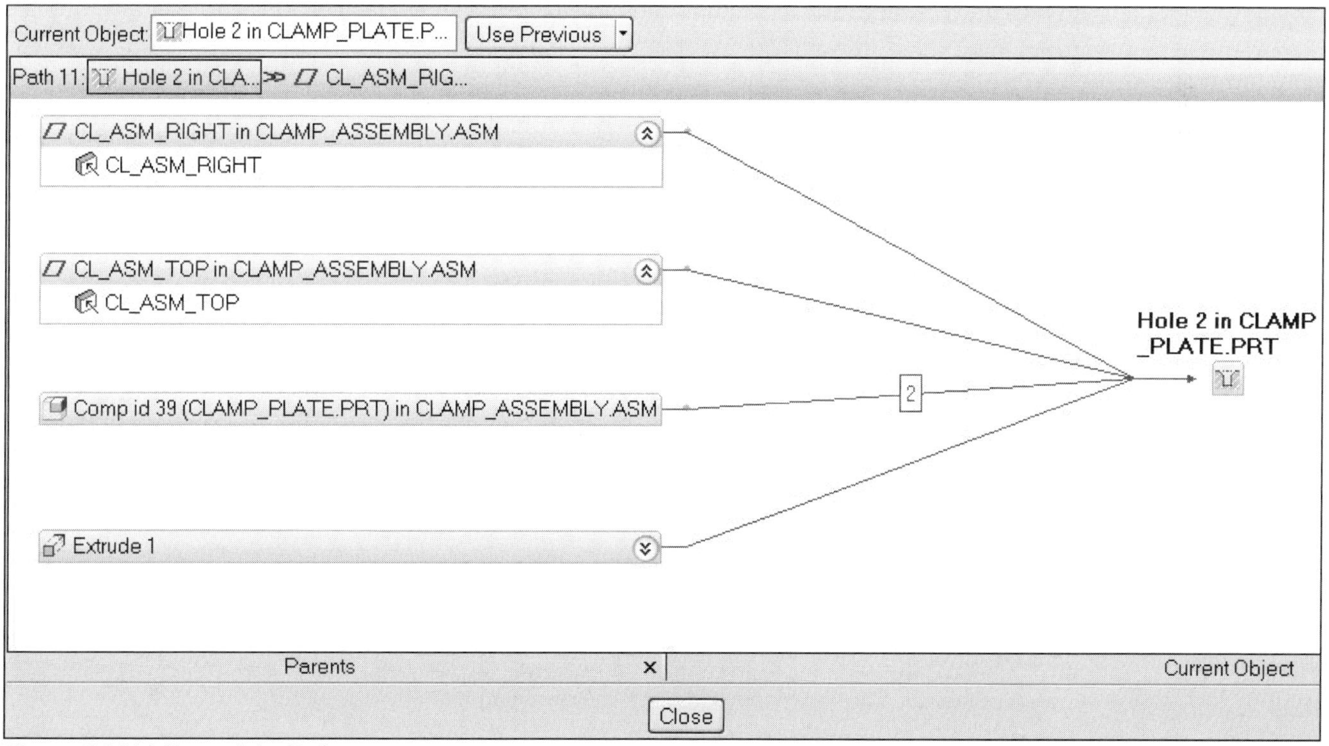

Figure 8.23(r) Expand the References

If you look back at Figure 8.23(k) and Figure 8.23(o) you will see that the assembly datums were used to place the holes instead of DTM1 and DTM2. The following commands will show you how to edit the references of one of the holes. Redo the references of the other hole using similar steps.

From the Reference Viewer Dialog box, click: **Close** ⇒ **RMB** on the hole in the Model Tree ⇒ **Edit Definition** [Fig. 8.23(s)]

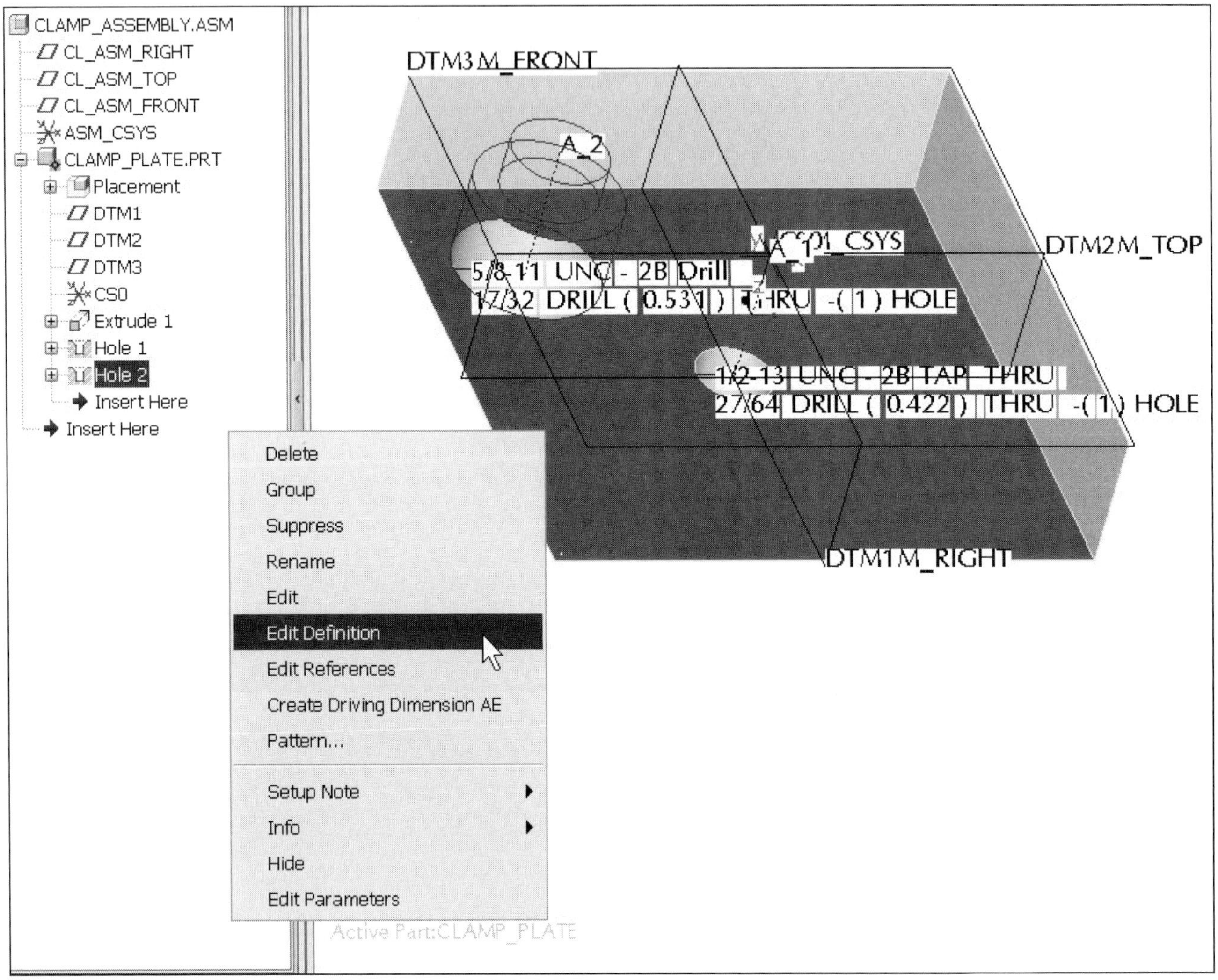

Figure 8.23(s) Edit Definition

Click: **Hidden line** ⇒ **Placement** tab ⇒ click in the Offset references field ⇒ **RMB** ⇒ **Remove All** [Fig. 8.23(t)] ⇒ place your cursor over datum CL_ASM_TOP to highlight ⇒ click **RMB** *[do not hold down the RMB, unless you want to "Pick From List" (accessing the old query select list) or choose one of the other options]* ⇒ DTM2 highlights [Fig. 8.23(u)] ⇒ pick on **DTM2**

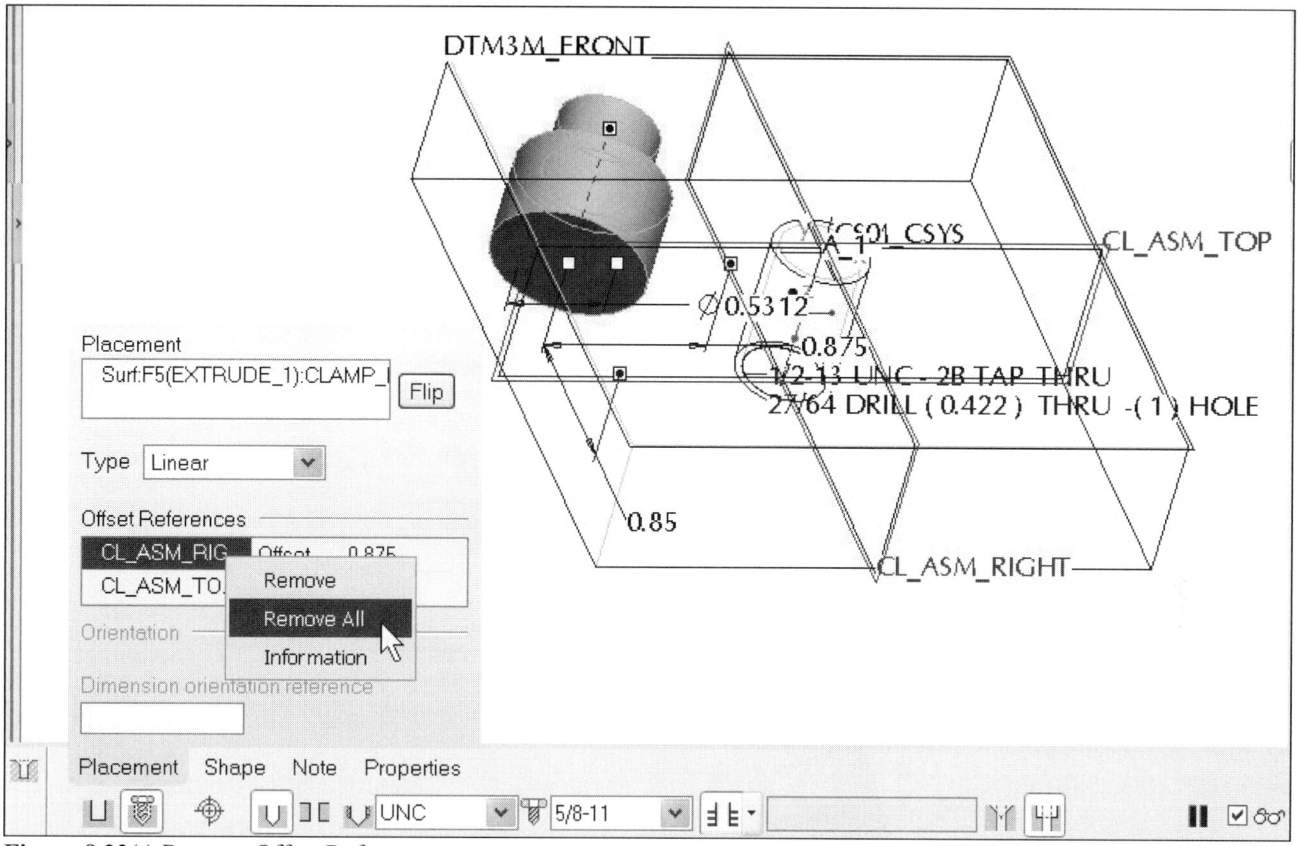

Figure 8.23(t) Remove Offset References

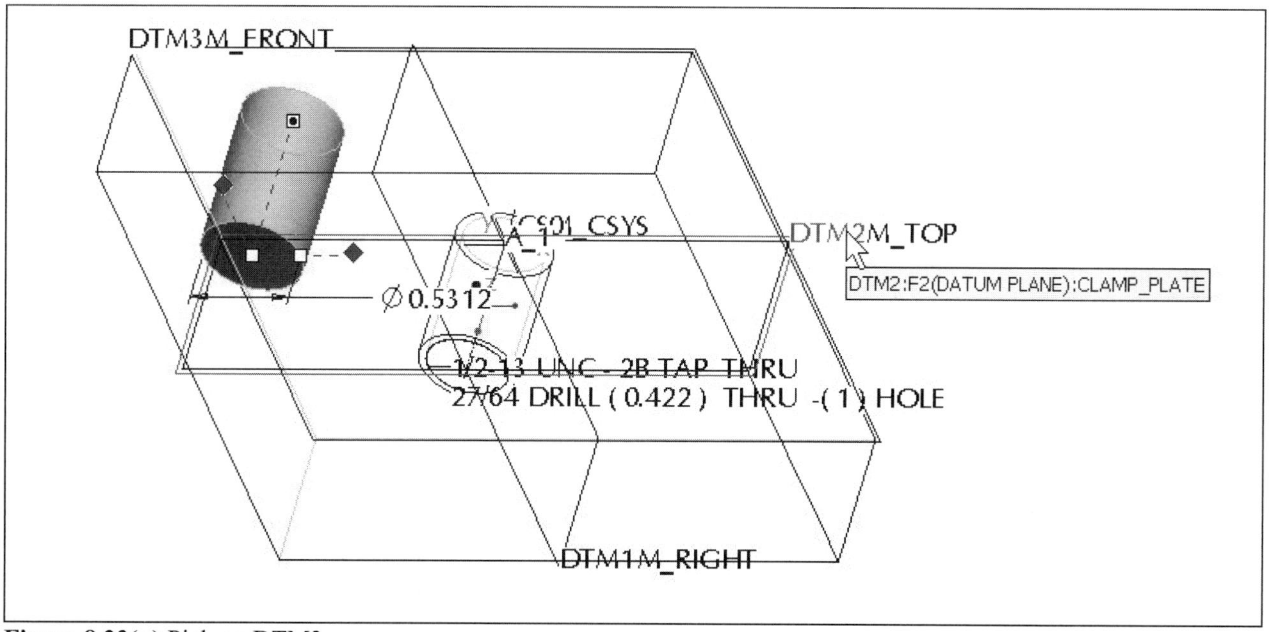

Figure 8.23(u) Pick on DTM2

Press and hold **Ctrl** key and place your cursor over CL_ASM_RIGHT (highlights) ⇒ click **RMB** (DTM1 highlights) [Fig. 8.23(v)] ⇒ pick on **DTM1** [Fig. 8.23(w)] ⇒ **.875** is still the Offset distance for both References ⇒ **Enter** to complete the command ⇒ **Ctrl+S** ⇒ **MMB** ⇒ **LMB** to deselect

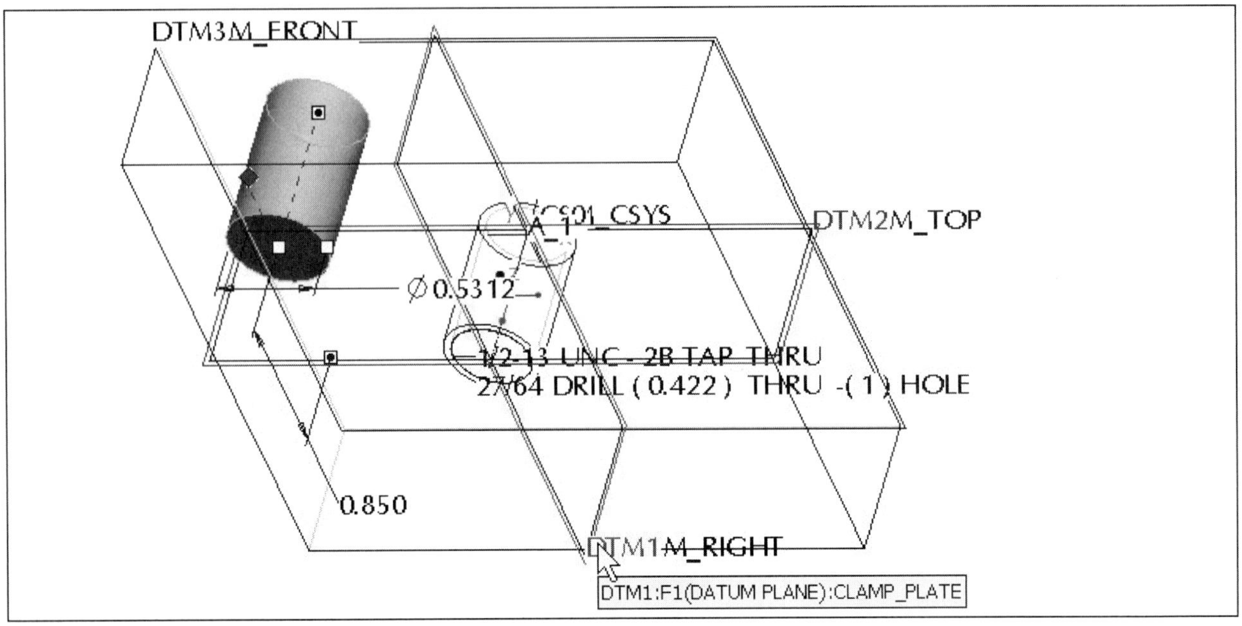

Figure 8.23(v) Pick on DTM1

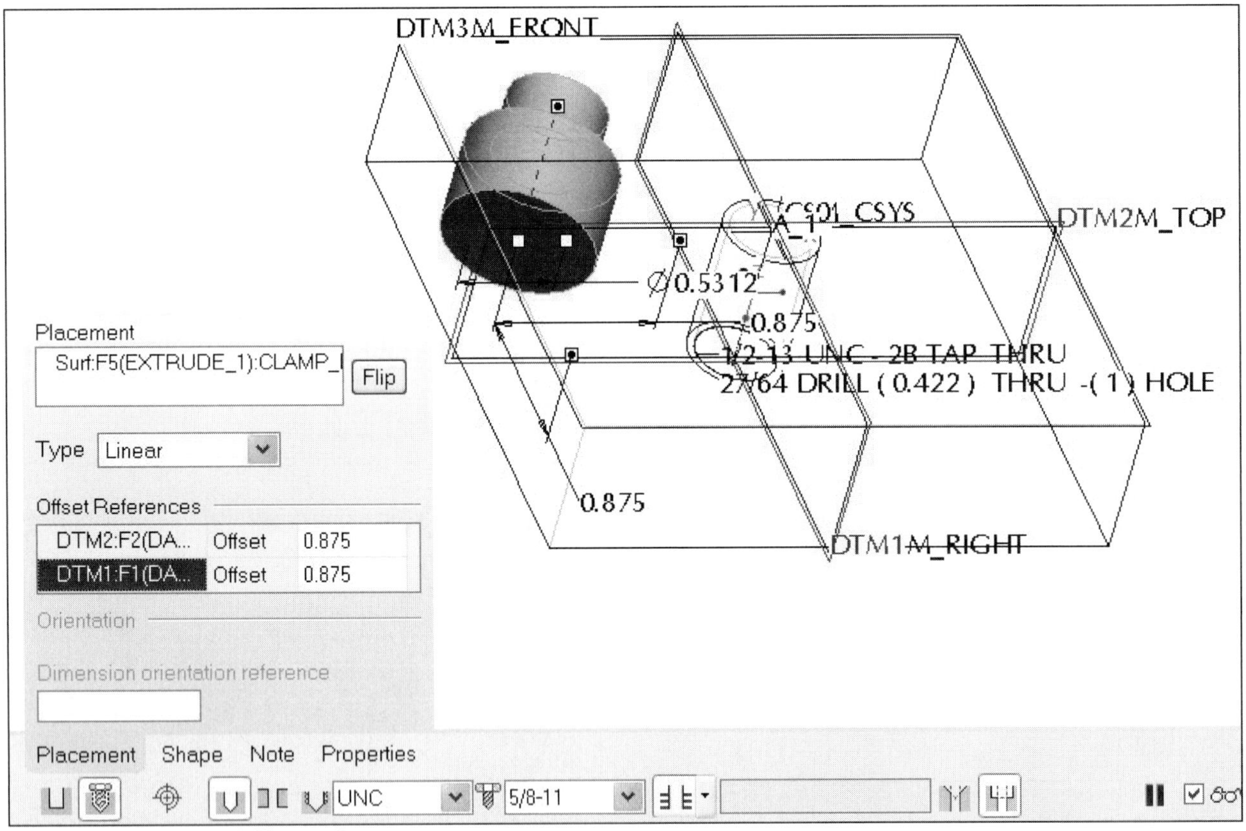

Figure 8.33(w) New Offset References

If you have difficulty with this method, simply click on the three assembly datum planes in the Model Tree ⇒ RMB ⇒ Hide. Then Edit Definition and redo the references. Unhide the assembly datums after you are finished changing the references.

Click on **Hole_2** from the Model Tree or model ⇒ **RMB** ⇒ **Info** ⇒ **Reference Viewer** ⇒ **References Filters** tab ⇒ **References** on ⇒ [icon] expand the references [Fig. 8.23(x)] ⇒ **Close**

Repeat the previous process to select the correct references (DTM2 and DTM1) for the tapped hole in the center of the part [Fig. 8.23(y)]. The hole will be aligned with both references.

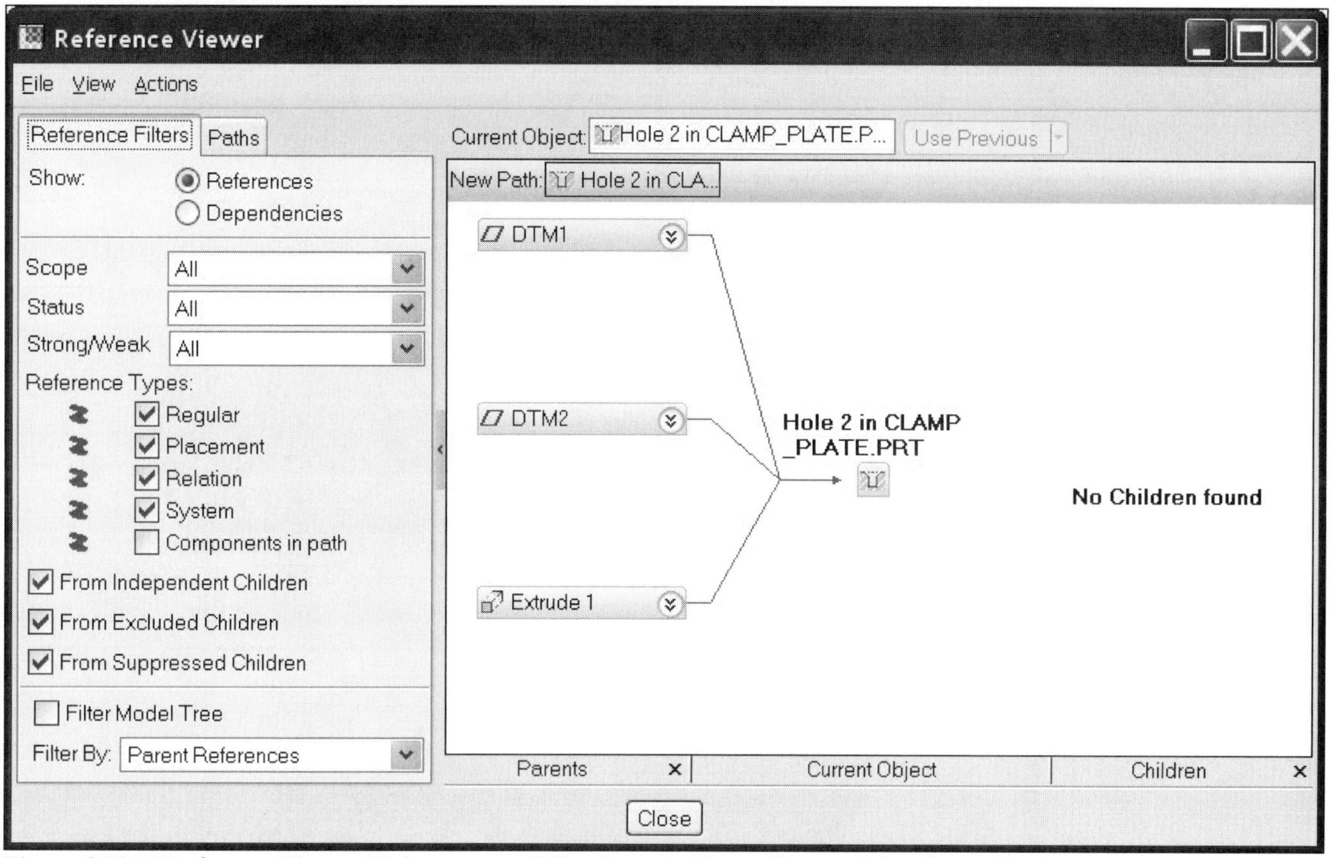

Figure 8.23(x) Reference Viewer (Holes are now Offset from the Datum Planes of the Clamp_Plate)

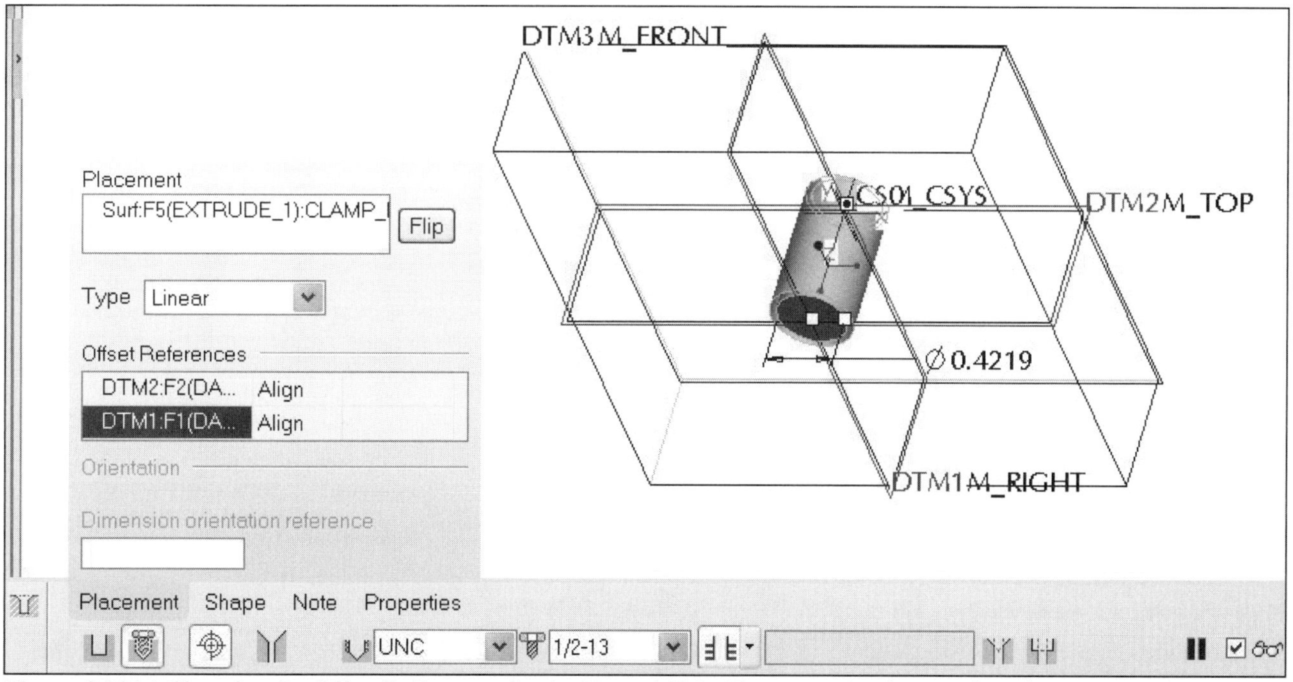

Figure 8.23(y) Changing the Offset References for the Tapped Hole (the external references have been removed)

Pattern the counterbore holes. Click: **Ctrl+D** ⇒ ⇒ off ⇒ *(if the CLAMP_PLATE.PRT is not active, click on* ***CLAMP_PLATE.PRT*** *in the Model Tree* ⇒ ***RMB*** ⇒ ***Activate)*** ⇒ click on the counterbore hole in the Model Tree ⇒ **RMB** ⇒ **Pattern** [Fig. 8.24(a)] ⇒ **Dimensions** tab ⇒ pick on the horizontal **.875** dimension (a *"combo box"* opens) ⇒ type **–1.75** [Fig. 8.24(b)] ⇒ **Enter**

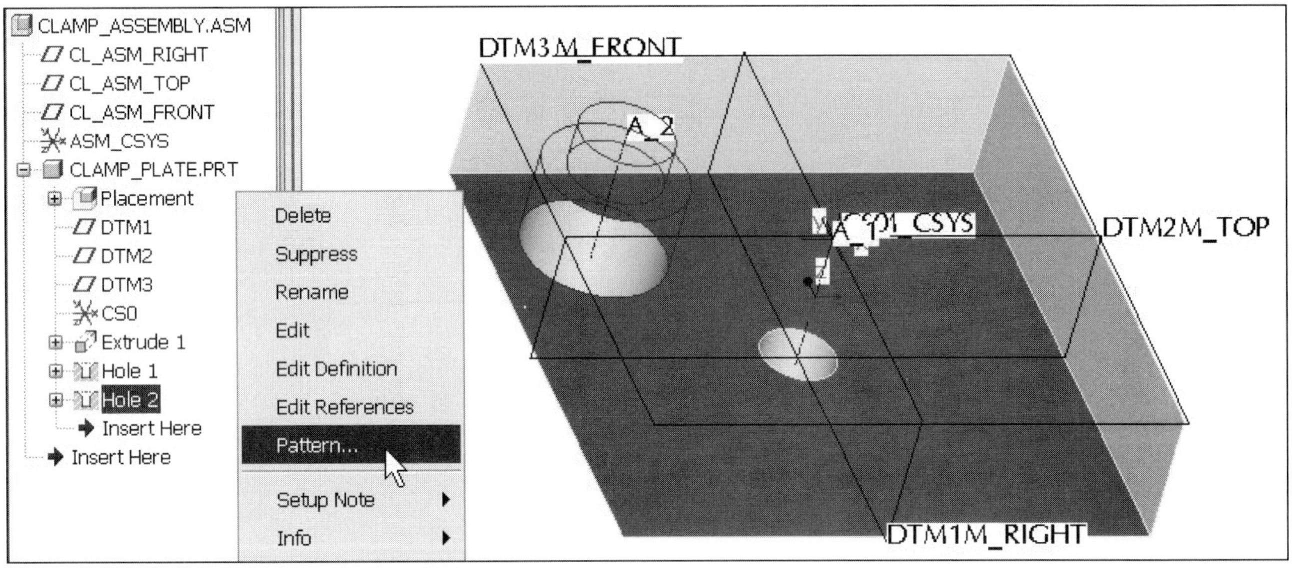

Figure 8.24(a) Pattern the Counterbore Hole

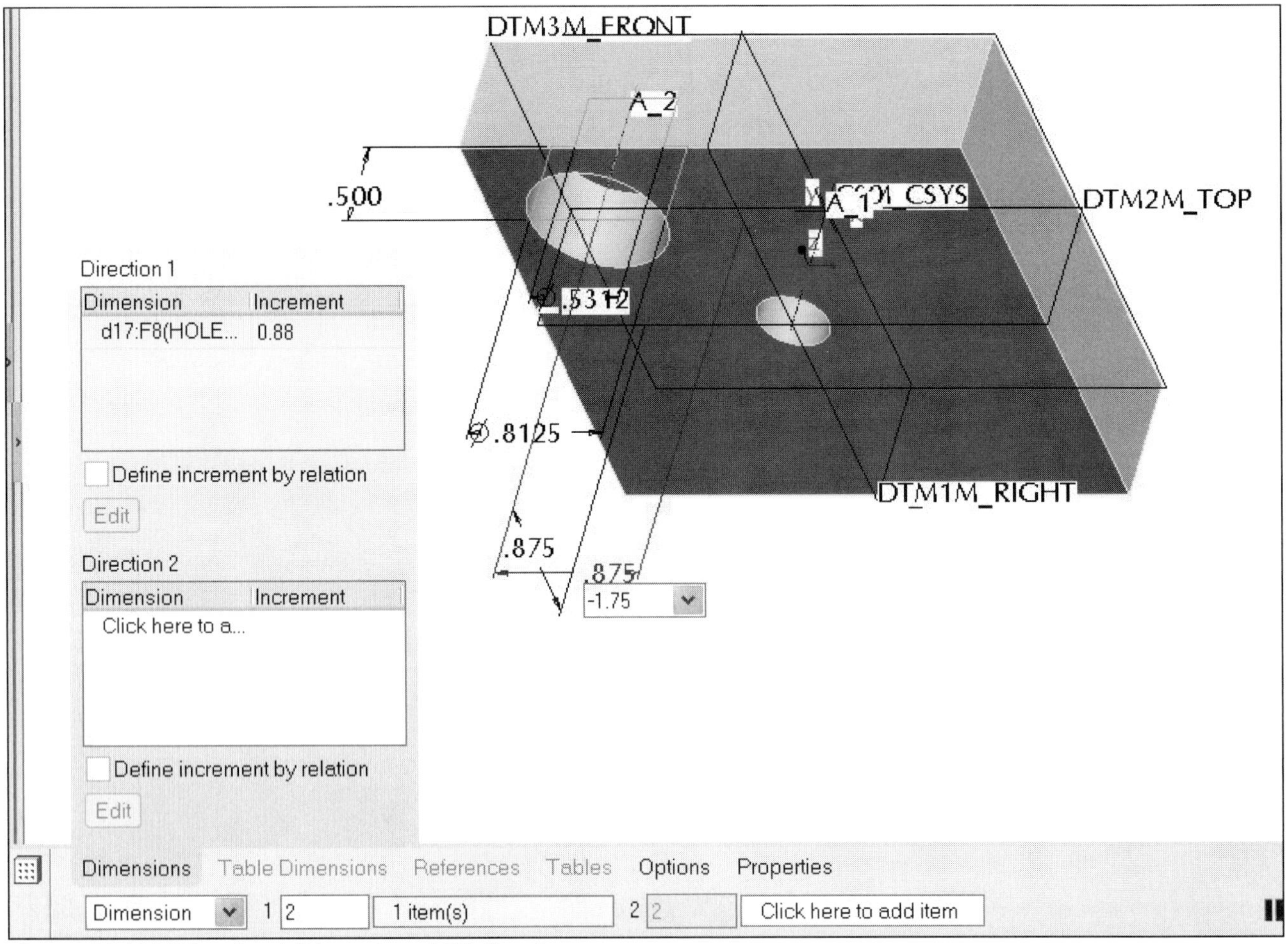

Figure 8.24(b) Direction 1 Dimension **–1.75**, Two Items

Click: **RMB** ⇒ **Direction 2 Dimensions** (*or click in Direction 2 collector box, pick* [Click here to a...]) ⇒ pick on the vertical **.875** dimension [Fig. 8.24(c)] ⇒ A "combo box" opens in the graphics window, with the dimension increment initially equal to the dimension value. Highlight the value (if needed) and type – **1.75** ⇒ **Enter** ⇒ ☑ ⇒ **Tools** ⇒ **Environment** ⇒ [Standard Orient | Isometric] ⇒ **Apply** ⇒ **OK** [Fig. 8.24(d)] ⇒ **Edit** ⇒ **Regenerate** ⇒ **Ctrl+S** ⇒ **Enter**

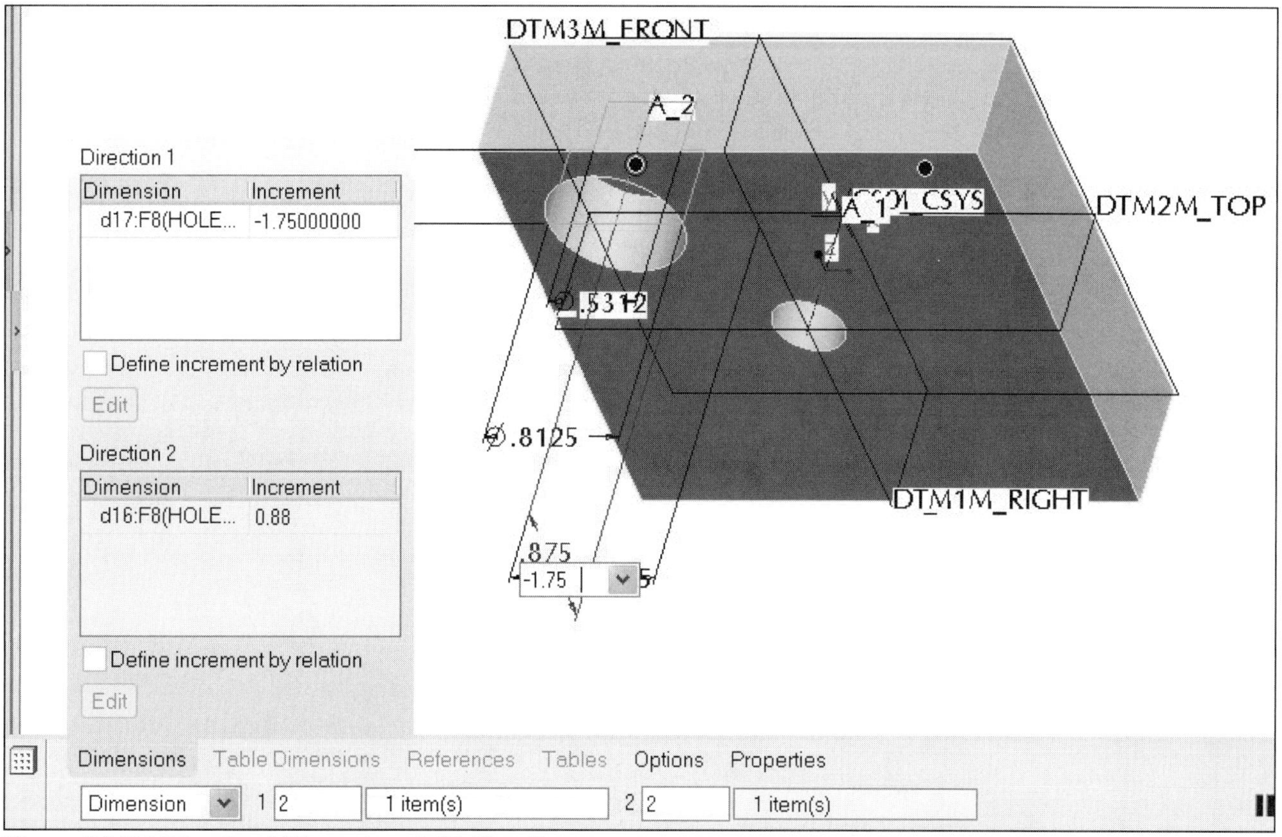

Figure 8.24(c) Direction 2 Dimension **–1.75**, Two Items

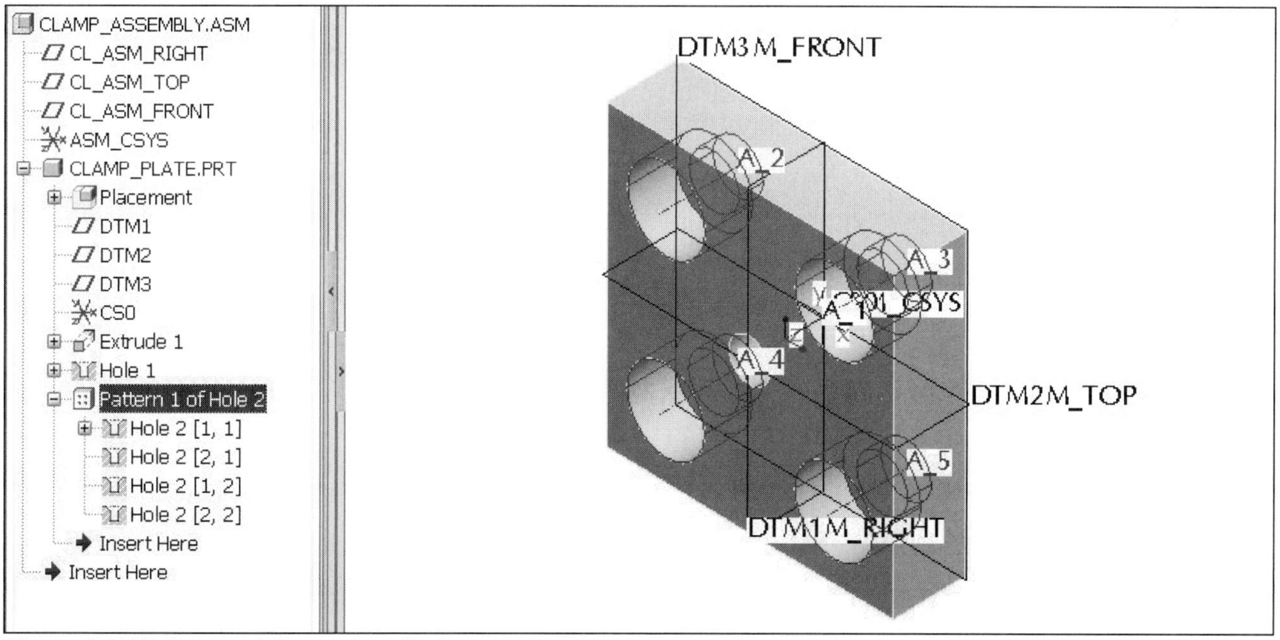

Figure 8.24(d) Patterned Counterbore Holes

Click on **CLAMP_ASSEMBLY.ASM** from the Model Tree ⇒ **RMB** ⇒ **Activate** ⇒  **Add component to the assembly** ⇒ select the **clamp_subassembly.asm** ⇒ **Preview** [Fig. 8.25(a)] ⇒ **Open**
Make sure you follow the steps exactly as provided, even if you "think" the references should be different.

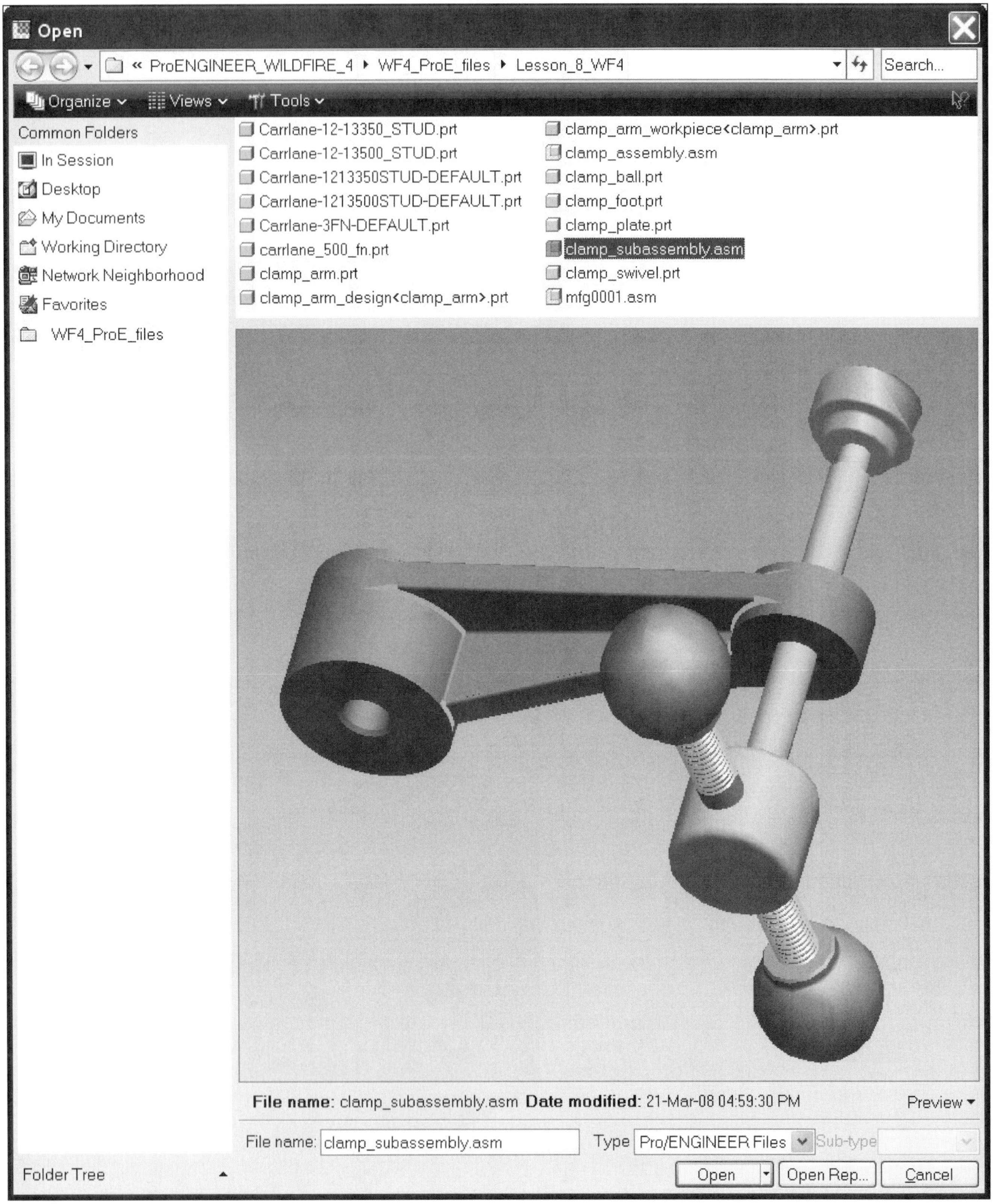

Figure 8.25(a) Preview Clamp_Subassembly

Click: [toolbar icons] off ⇒ [icon] toggle off component window ⇒ [Automatic] ⇒ **Mate** [Fig. 8.25(b)] ([Mate]) ⇒ spin the model ⇒ **View** ⇒ **Shade** ⇒ pick the bottom surface of the Clamp_Arm (of the Clamp_Subassembly) [Fig. 8.25(c)]

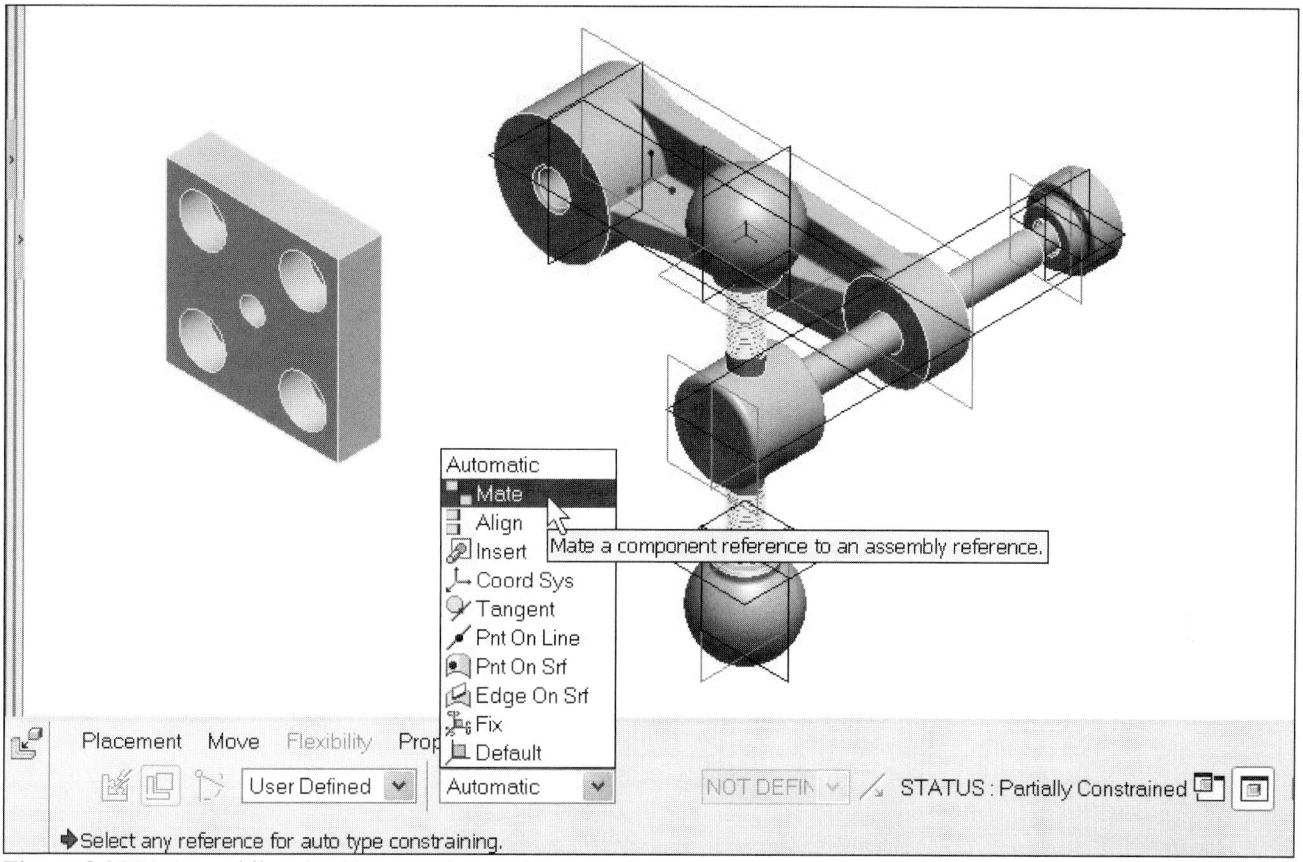

Figure 8.25(b) Assembling the Clamp_Subassembly

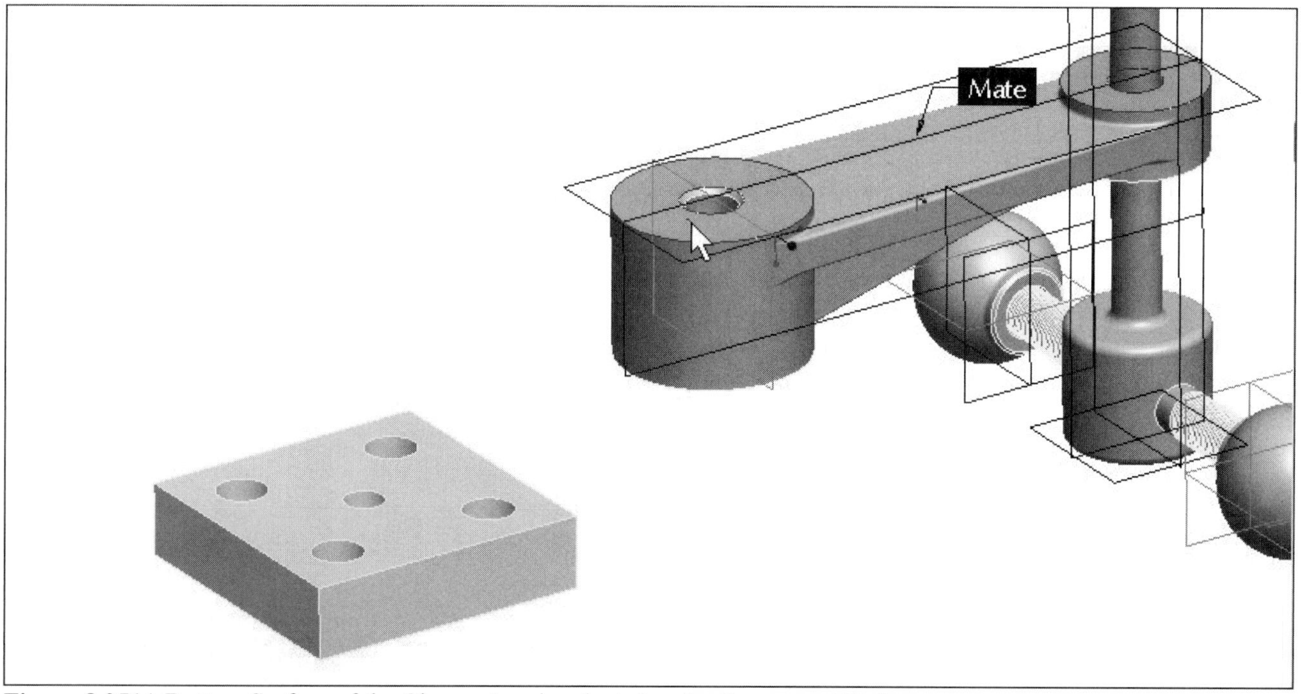

Figure 8.25(c) Bottom Surface of the Clamp_Arm is selected as the Component Reference

Click: **MMB** to spin the model ⇒ **View** ⇒ **Shade** ⇒ **Placement** tab ⇒ pick the non-counterbore surface of the Clamp_Plate [Figs. 8.25(d-e)]

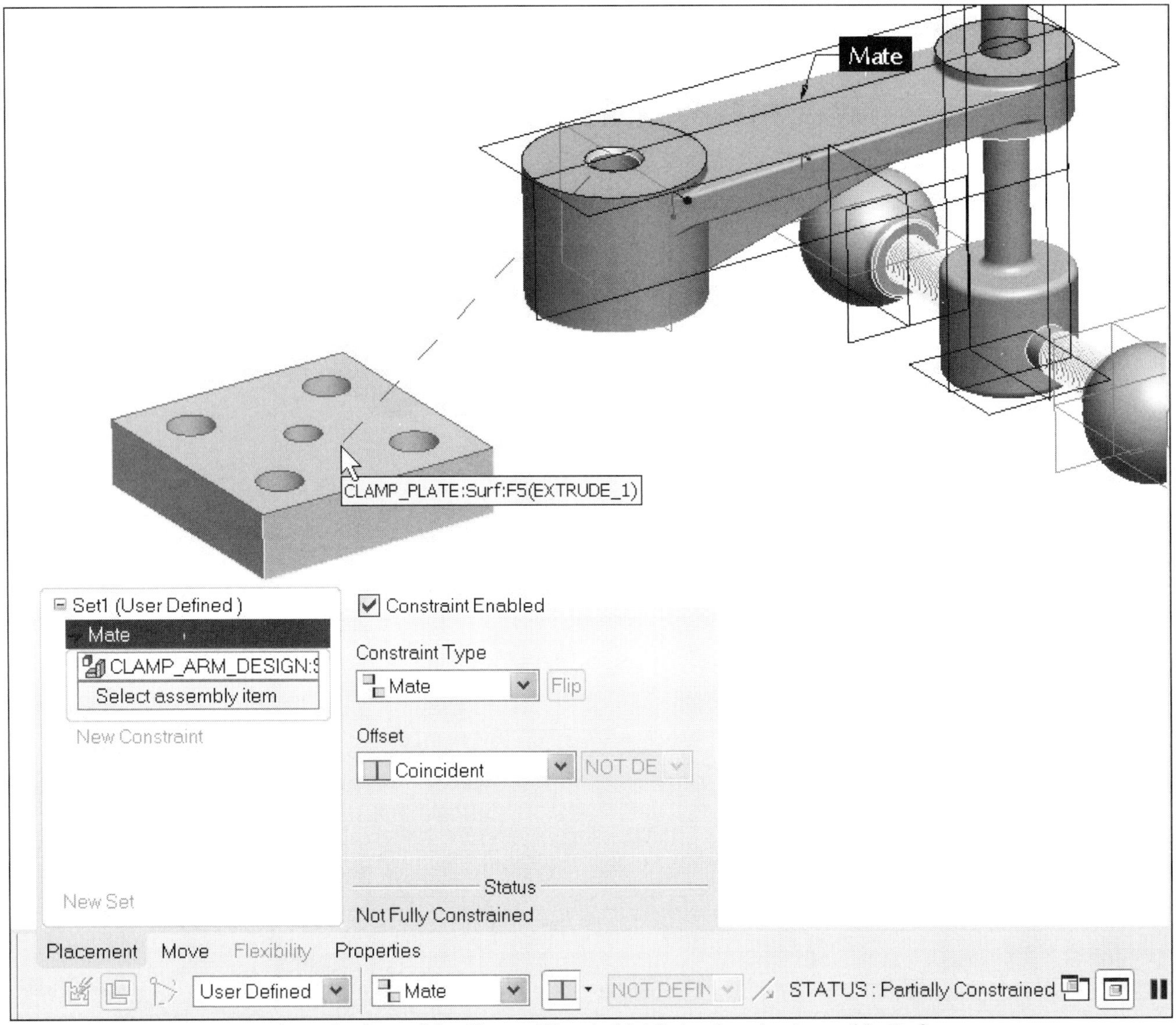

Figure 8.25(d) Non-counterbore Surface of the Clamp_Plate is highlighted as the Assembly Reference

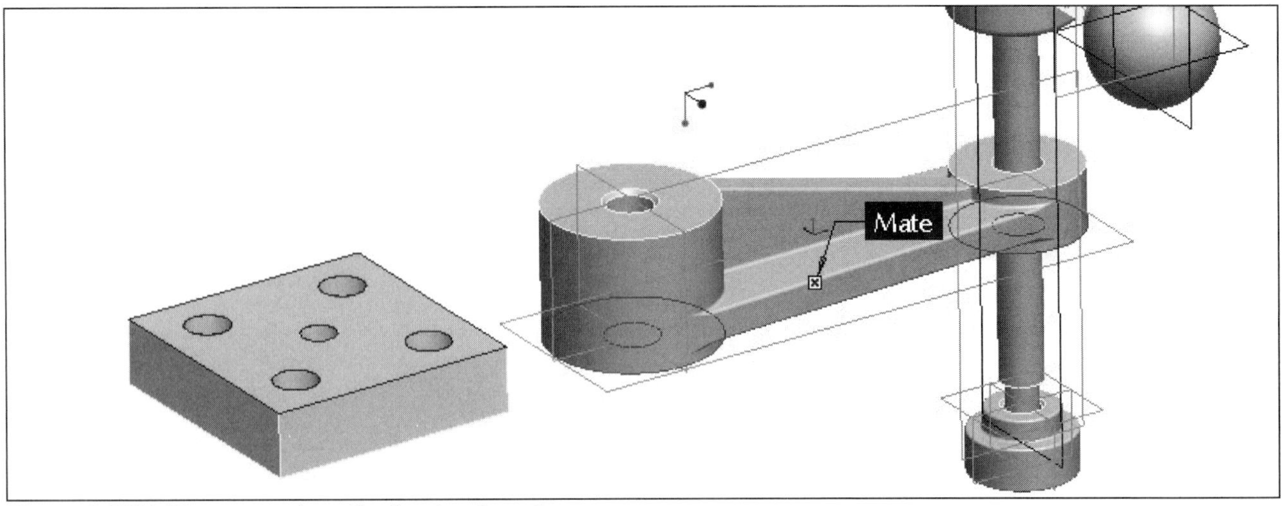

Figure 8.25(e) Non-counterbore Surface is selected

Click:  on ⇒ **RMB** ⇒ **New constraint** ⇒  ⇒ **Align** (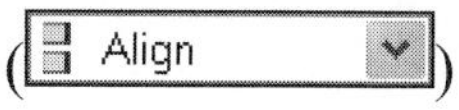) ⇒ **Ctrl+R** ⇒ pick the hole axis of the Clamp_Arm [Fig. 8.25(f)] ⇒ pick the center tapped hole axis of the Clamp_Plate [Fig. 8.25(g)]
(Note that you could have picked the two respective hole surfaces to achieve the same result using an Insert constraint.)

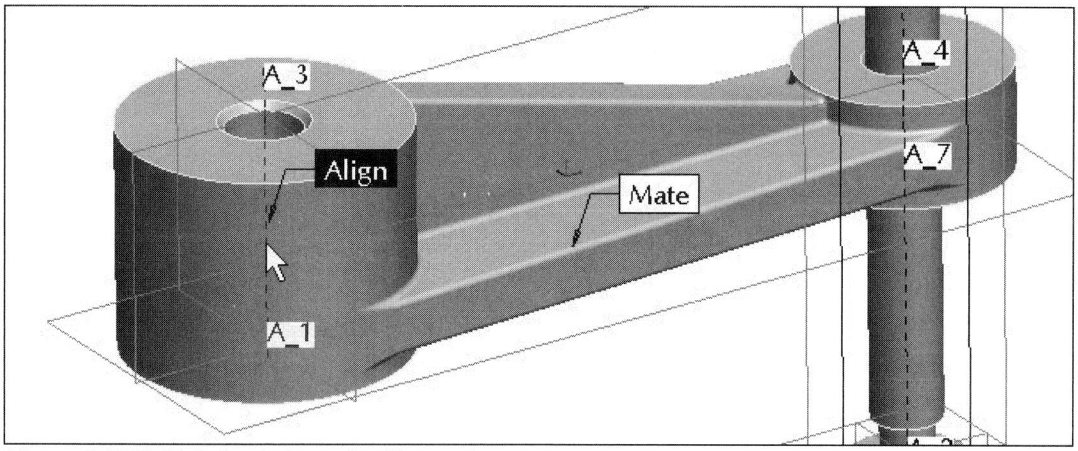

Figure 8.25(f) Pick the Axis on the Clamp_Arm

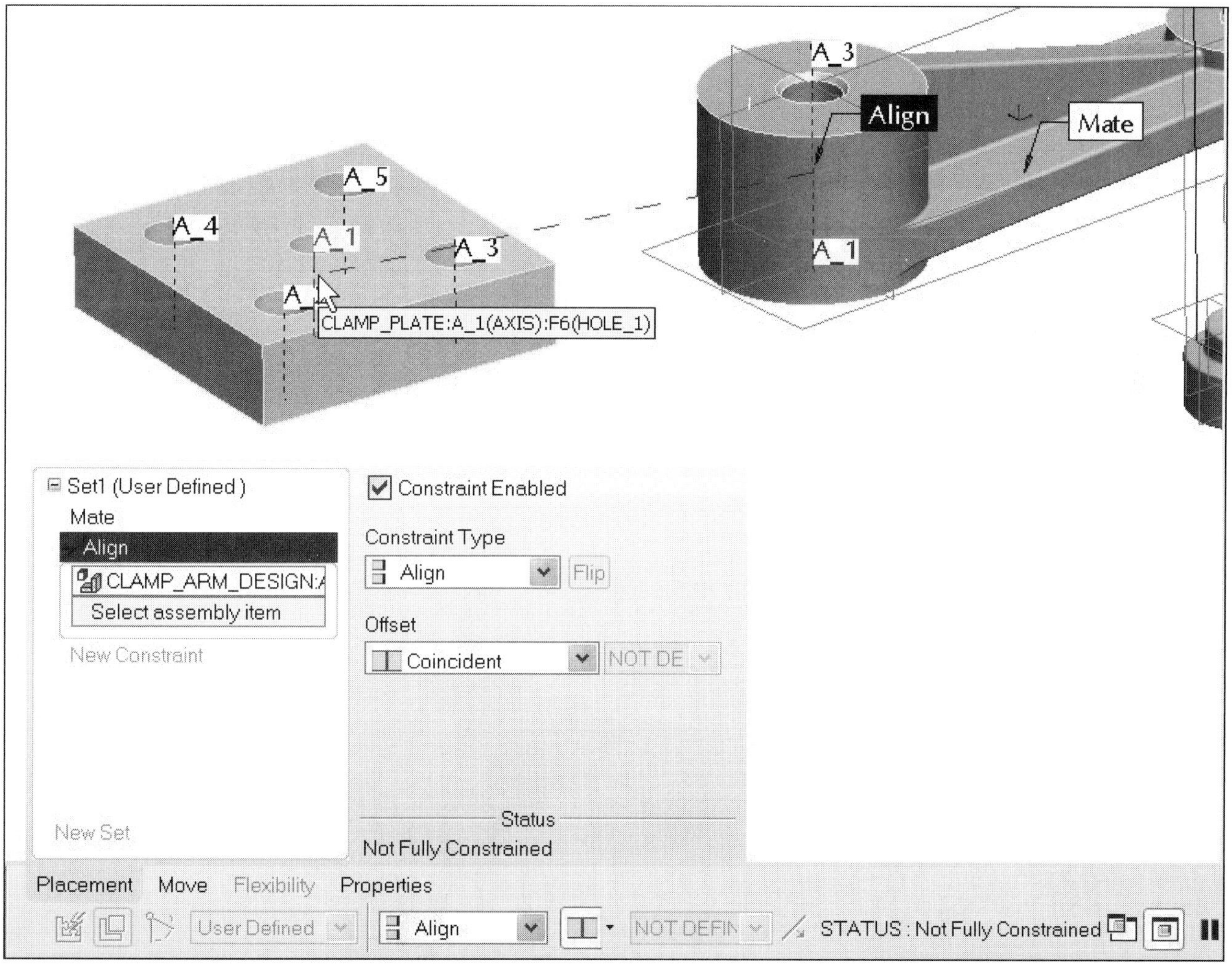

Figure 8.25(g) Pick the Tapped Hole Axis on the Clamp_Plate

Click: ☑ [Fig. 8.25(h)] ⇒ **Ctrl+D** ⇒ 🖫 [Fig. 8.25(i)] ⇒ **MMB**

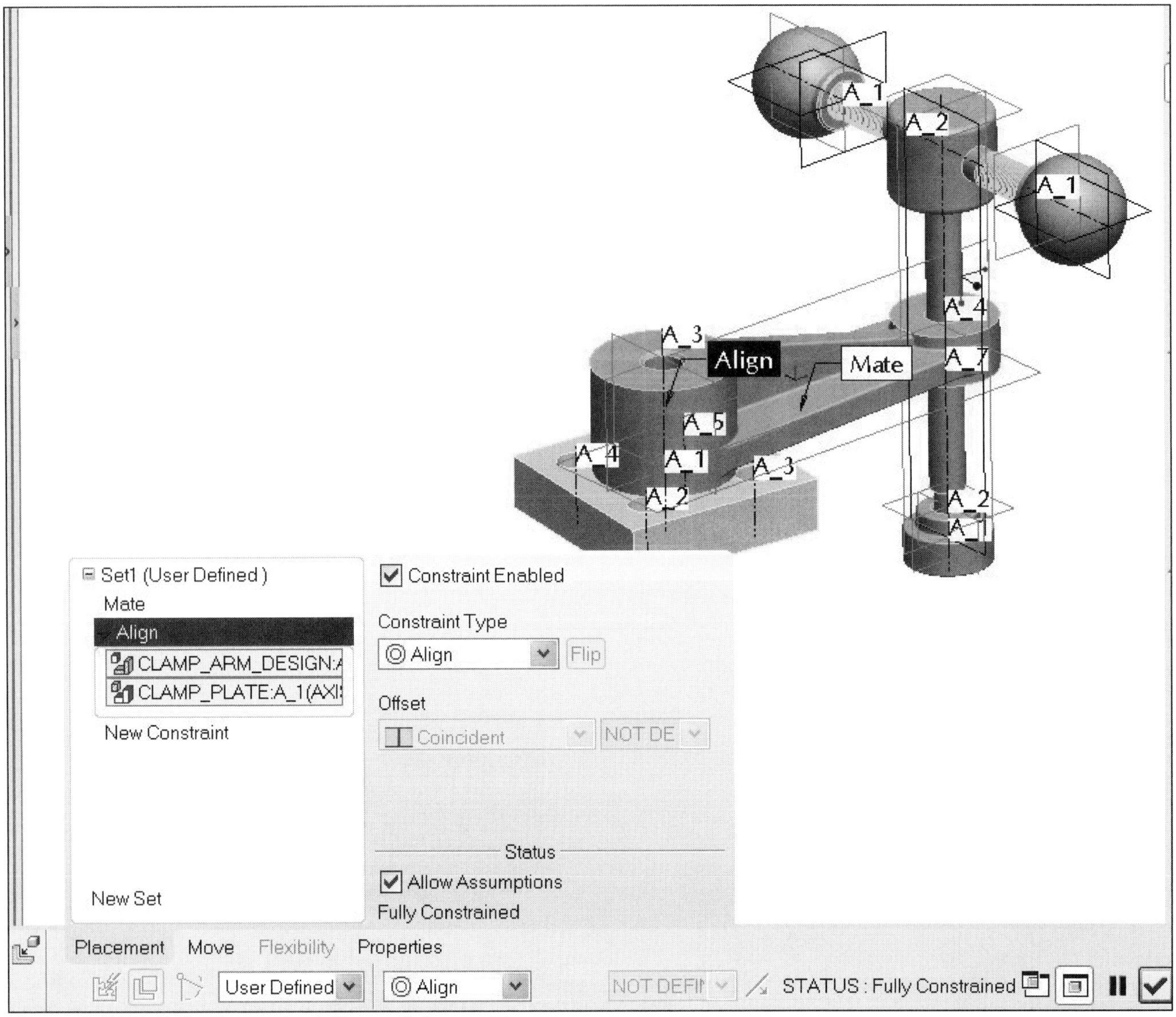

Figure 8.25(h) Fully Constrained Clamp_Subassembly

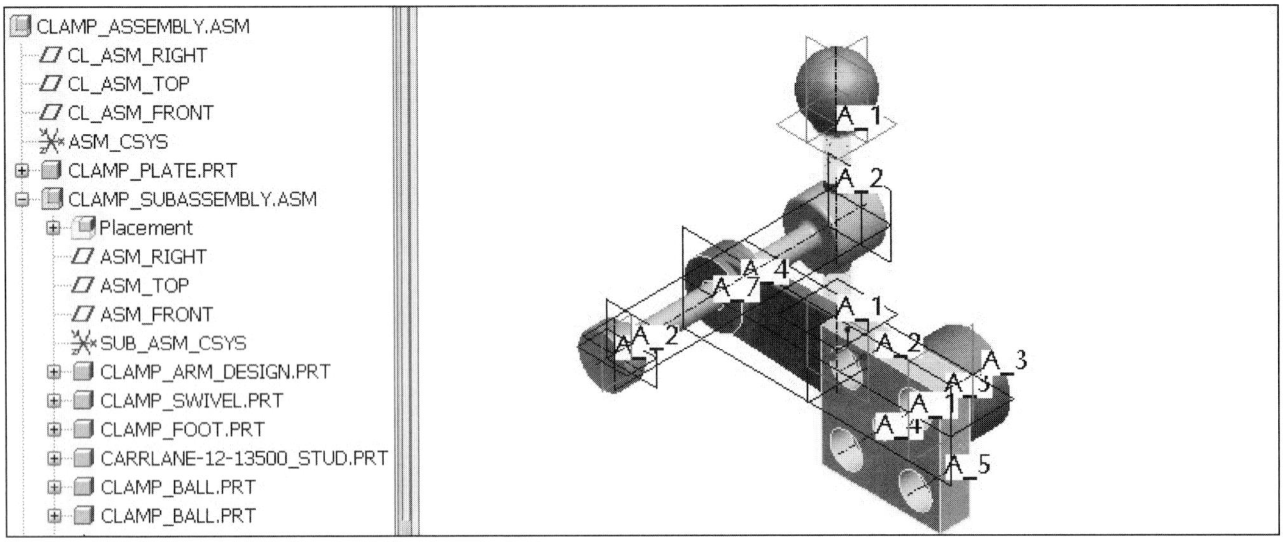

Figure 8.25(i) Assembled Clamp_Subassembly

Click: **Add component to the assembly** ⇒ pick the **Carrlane-12-13350_STUD.prt** ⇒ **Preview** [Fig. 8.26(a)] ⇒ **Open** ⇒ **Placement** tab [Fig. 8.26(b)] ⇒ Constraint Type ⇒ **Insert** Insert ⇒ press and hold **Ctrl+Alt** ⇒ using your **RMB** move the component closer to the assembly

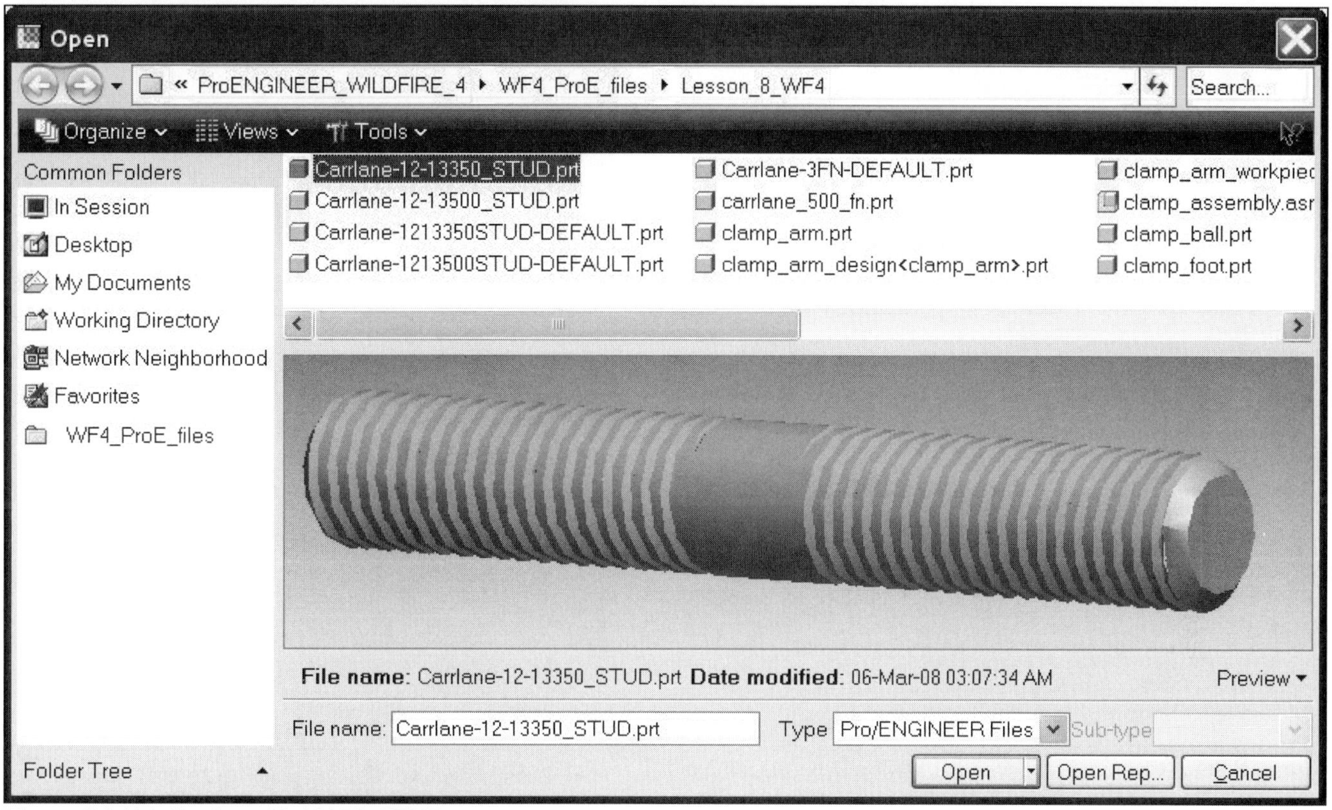

Figure 8.26(a) Carrlane-12-13350_STUD.prt

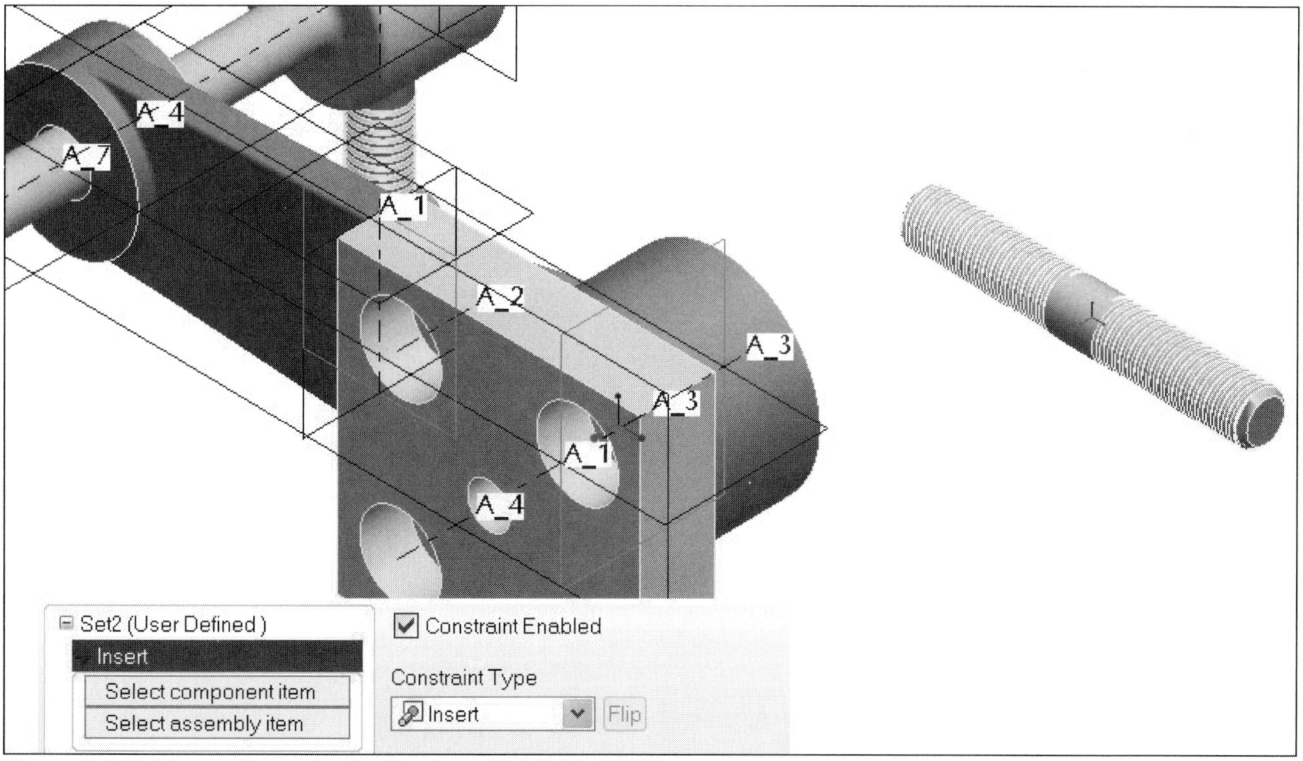

Figure 8.26(b) Insert Constraint for Carrlane-12-13350_STUD.prt

Click: **MMB** rotate the (assembly) model ⇒ pick the cylindrical surface of the Carrlane-12-13350_STUD.prt [Fig. 8.26(c)] ⇒ pick the hole surface of the Clamp_Arm [Fig. 8.26(d)]

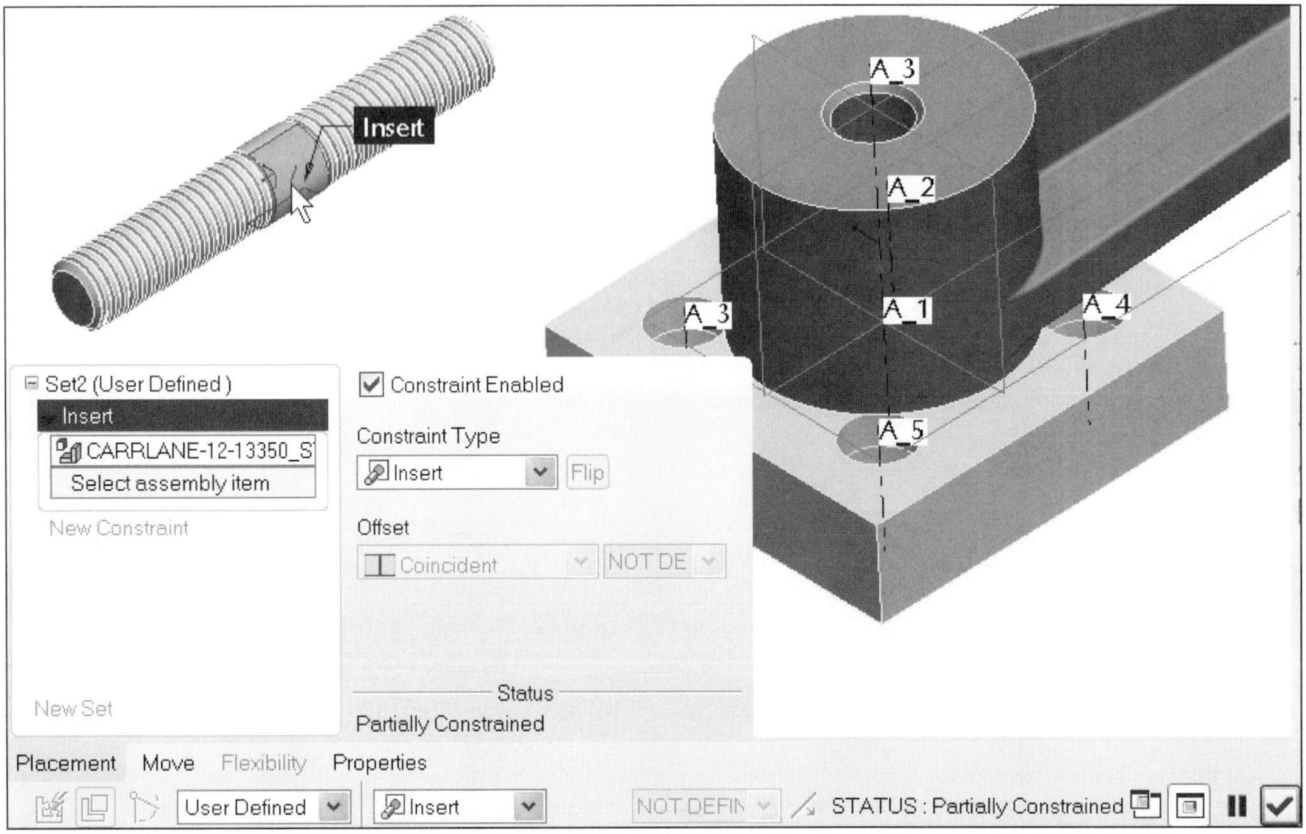

Figure 8.26(c) Component Placement Dialog Box, Placement Tab, Select Surface of Carrlane-12-13350_STUD.prt

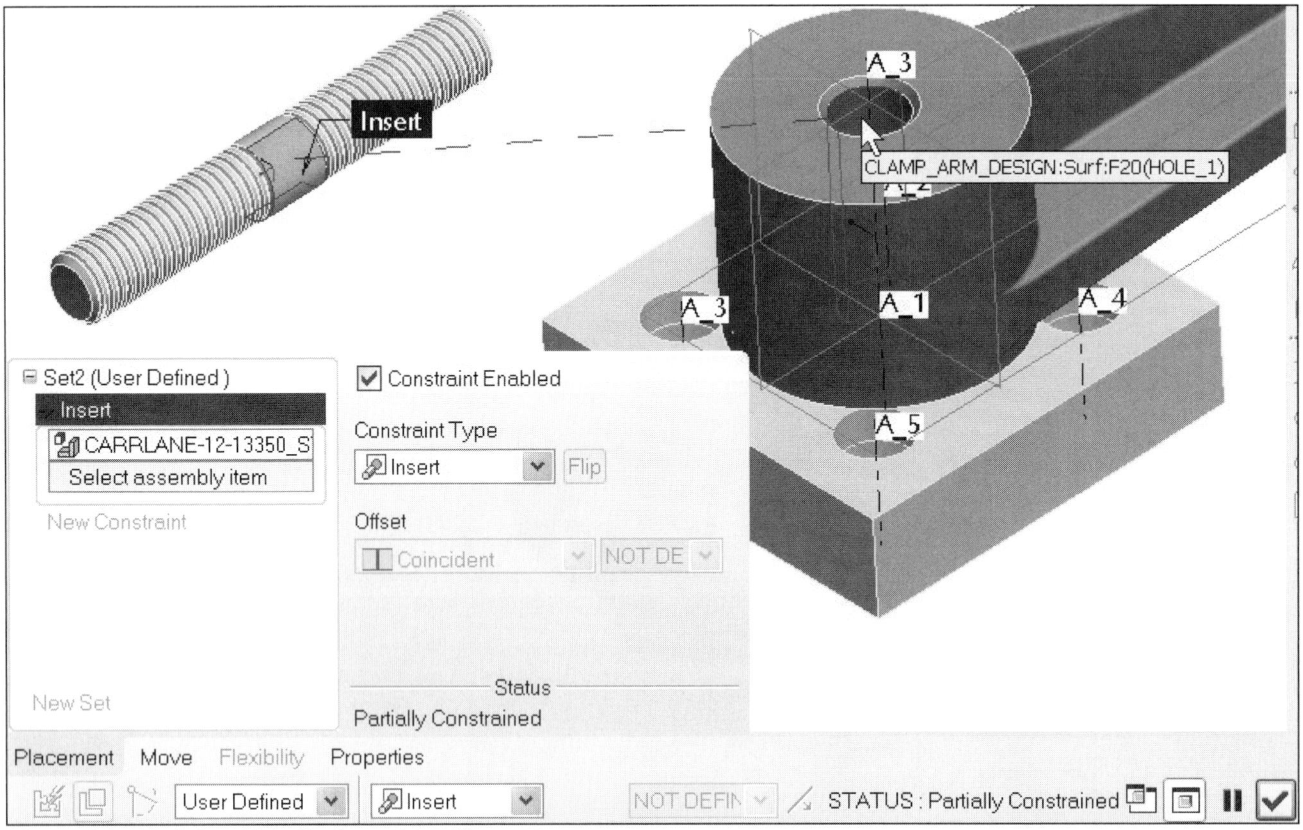

Figure 8.26(d) Select Hole Surface of the Clamp_Arm

Click: **Ctrl+D** ⇒ **Move** tab ⇒ Motion Type Translate ⇒ pick the Carrlane-12-13350_STUD.prt (hint: release LMB) and slide it deeper into the hole ⇒ **LMB** to place [Fig. 8.26(e)] ⇒ ⇒ **TOP** ⇒ **Hidden line** ⇒ pick Carrlane-12-13350_STUD.prt and translate (slide) it until it is inside the Clamp_Plate [Fig. 8.26(f)] ⇒ **LMB** to place

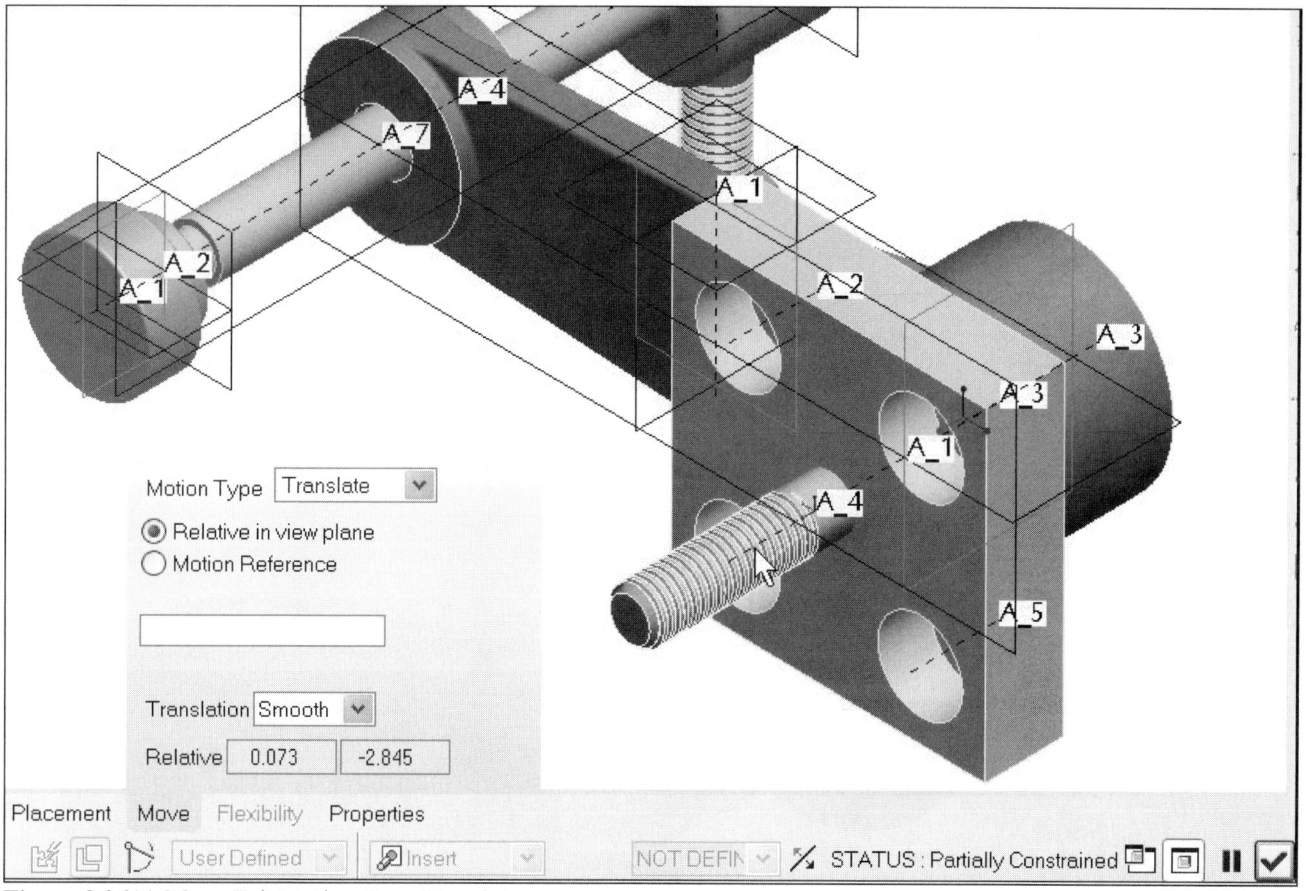

Figure 8.26(e) Move Tab Motion Type Translate

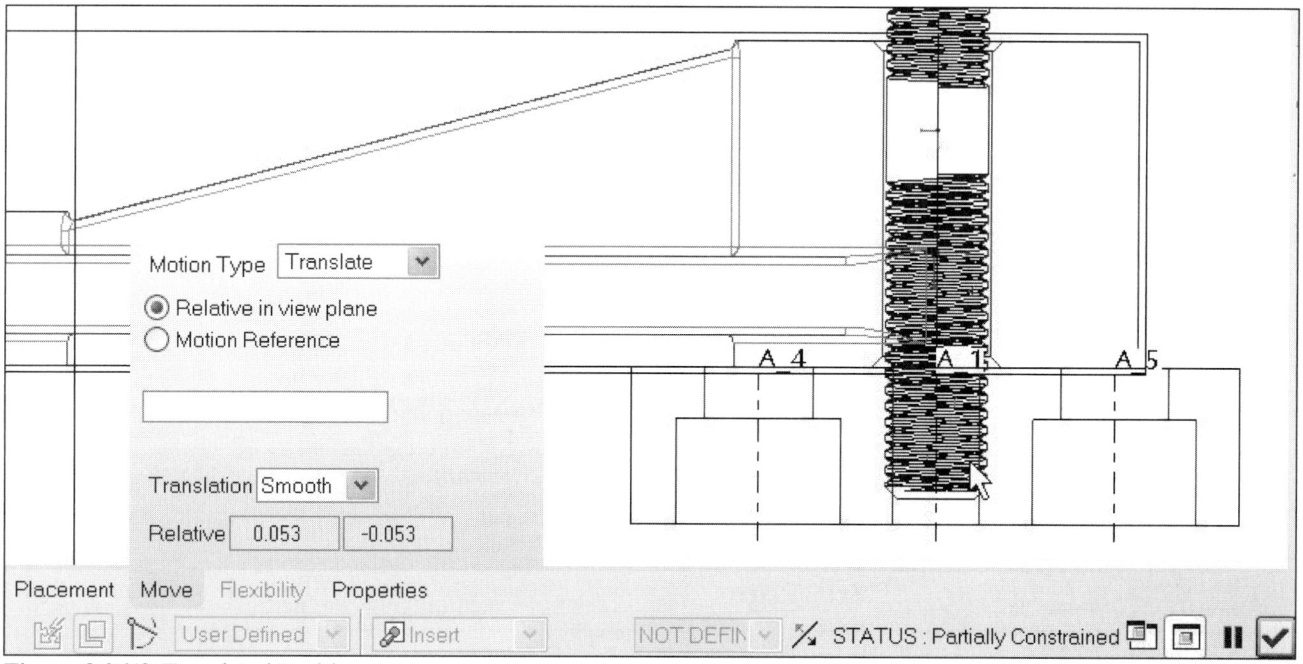

Figure 8.26(f) Translated Position

Click: [icon] ⇒ **MMB** spin the model to see the stud ⇒ **Placement** tab ⇒ **New Constraint** ⇒ Constraint Type [Fix] **Fix component to current position** [Fig. 8.26(g)] ⇒ [✓]

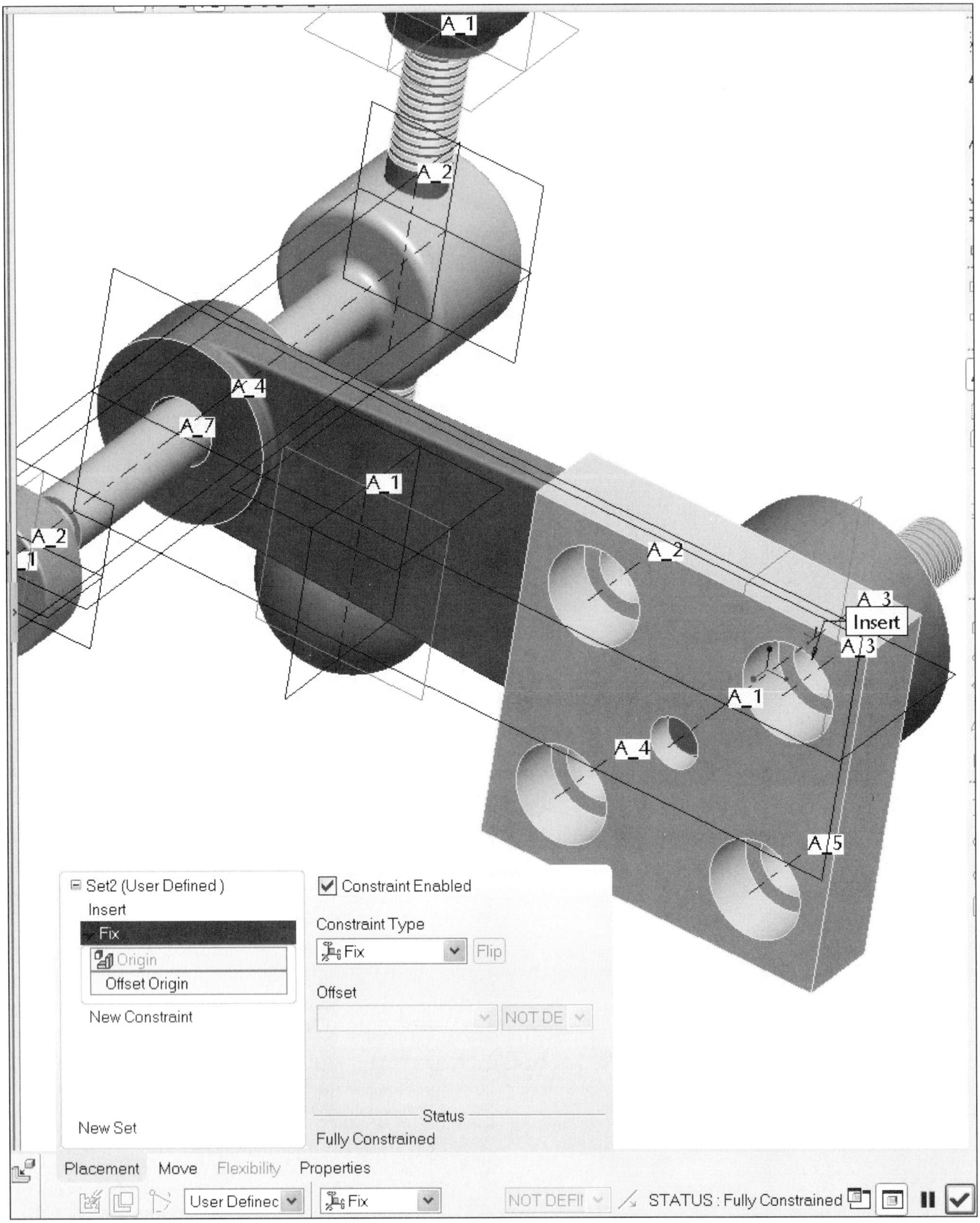

Figure 8.26(g) Fix Constraint

Click: **Ctrl+D** ⇒ **Ctrl+S** ⇒ **Enter** ⇒ **Add component to the assembly** ⇒ **carrlane_500_fn.prt** [Fig. 8.27(a)] ⇒ **Open** ⇒ **Ctrl+Alt** ⇒ **RMB** on the flange nut and move it closer to the assembly [Fig. 8.27(b)]

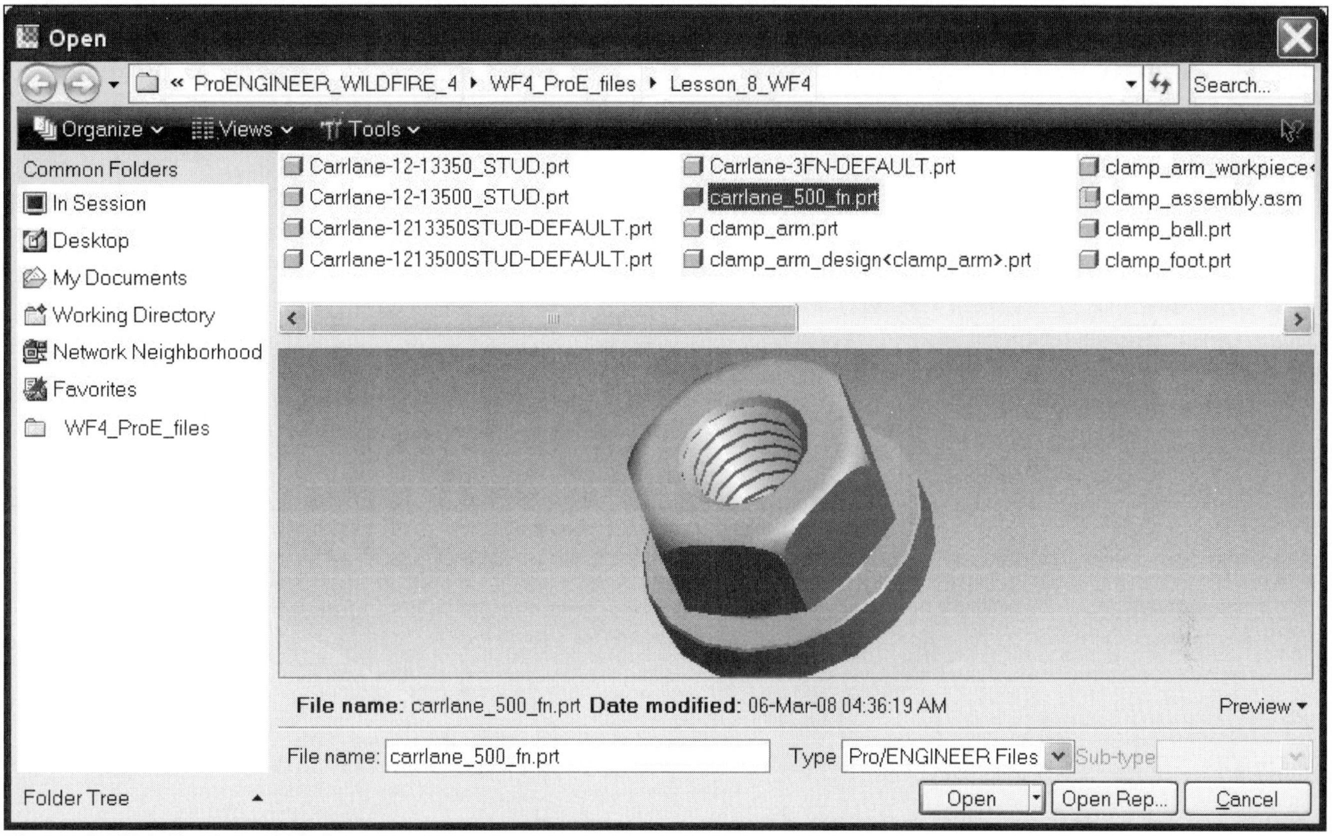

Figure 8.27(a) Carrlane_500_fn.prt

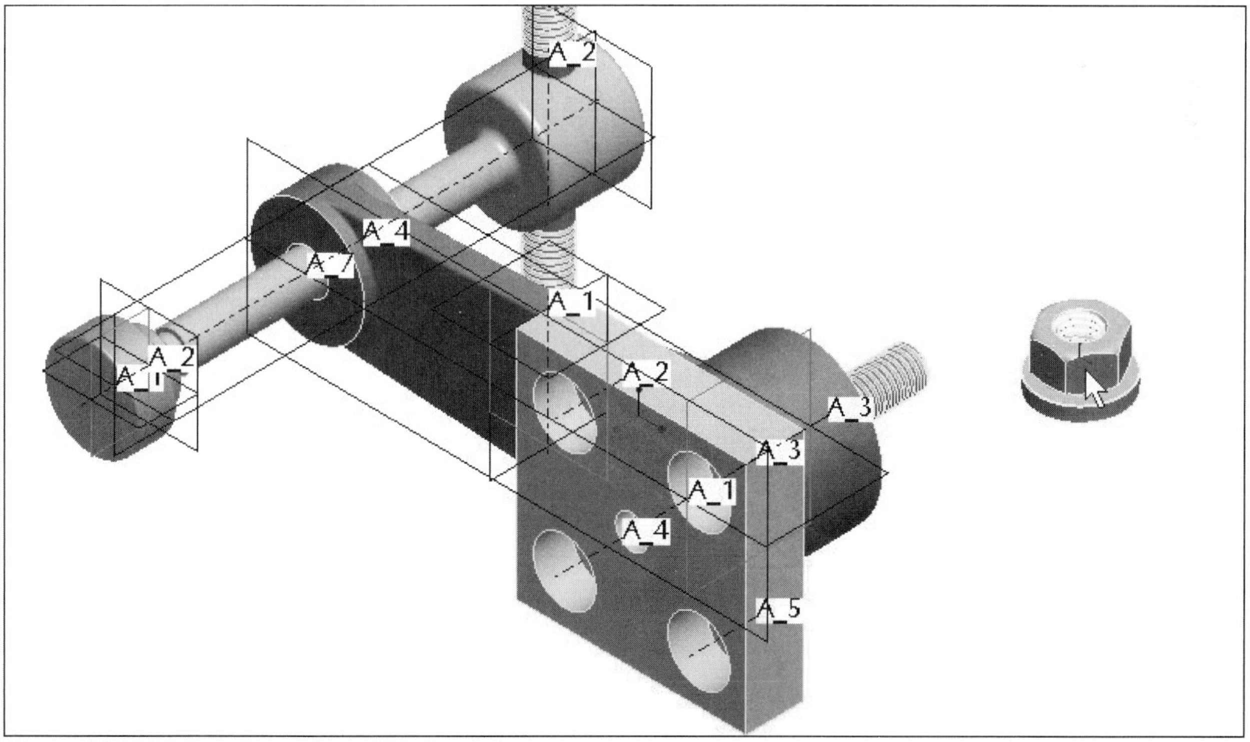

Figure 8.27(b) Move the carrlane_500_fn.prt Component

Press and hold **Ctrl+Alt** ⇒ **MMB** on the Flange_Nut and rotate [Fig. 8.27(c)] ⇒ **RMB** ⇒ **Select component item** [Fig. 8.27(d)]

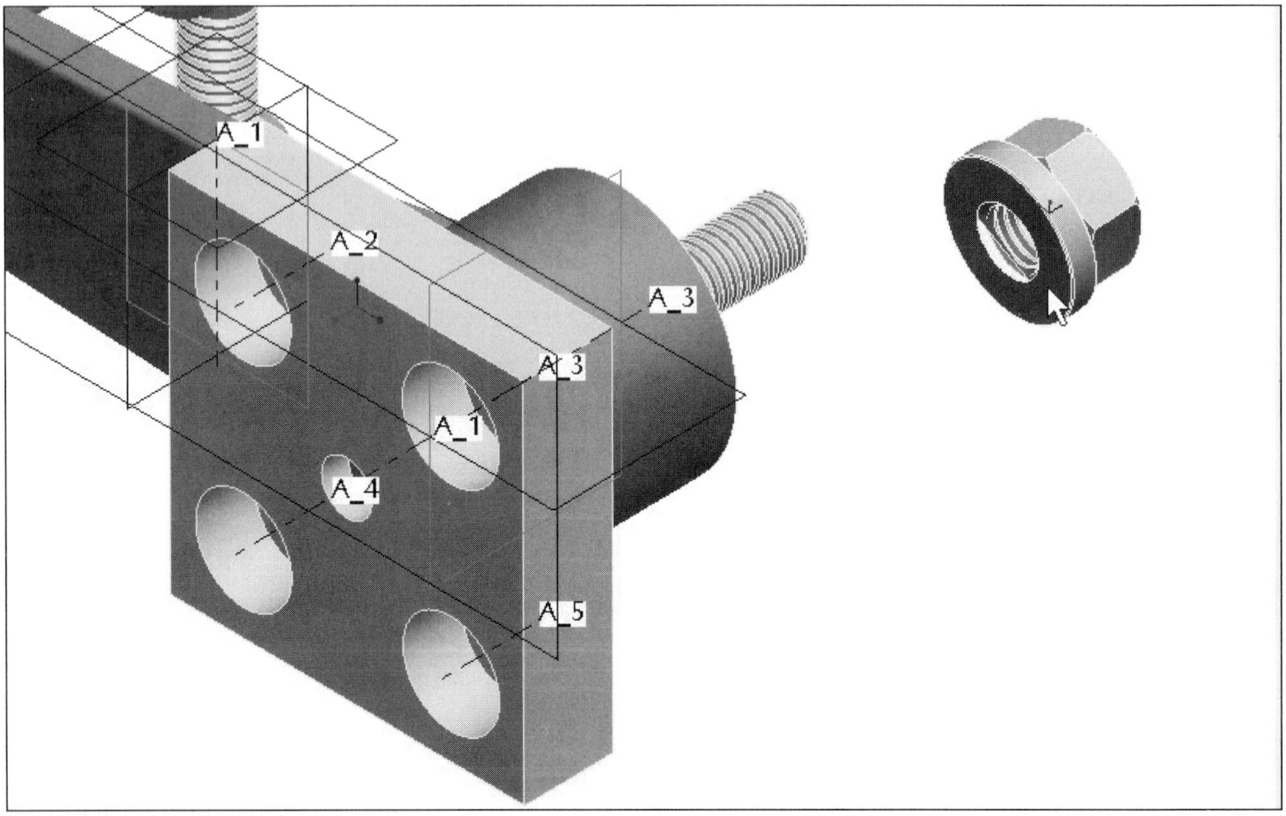

Figure 8.27(c) Rotate the Flange Nut

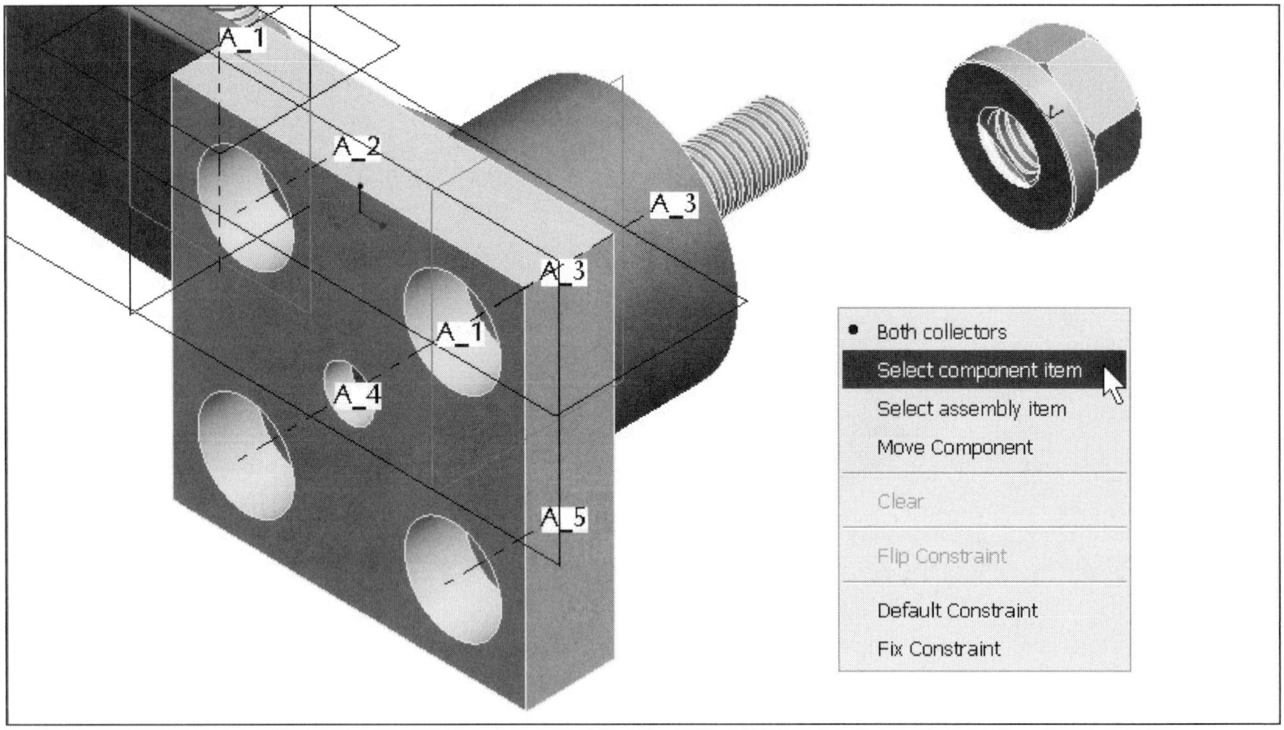

Figure 8.27(d) RMB ⇒ Select component item

Carefully pick the surface of the flange nut [Fig. 8.27(e)] ⇒ carefully pick the shaft surface of the stud [Fig. 8.27(f)]

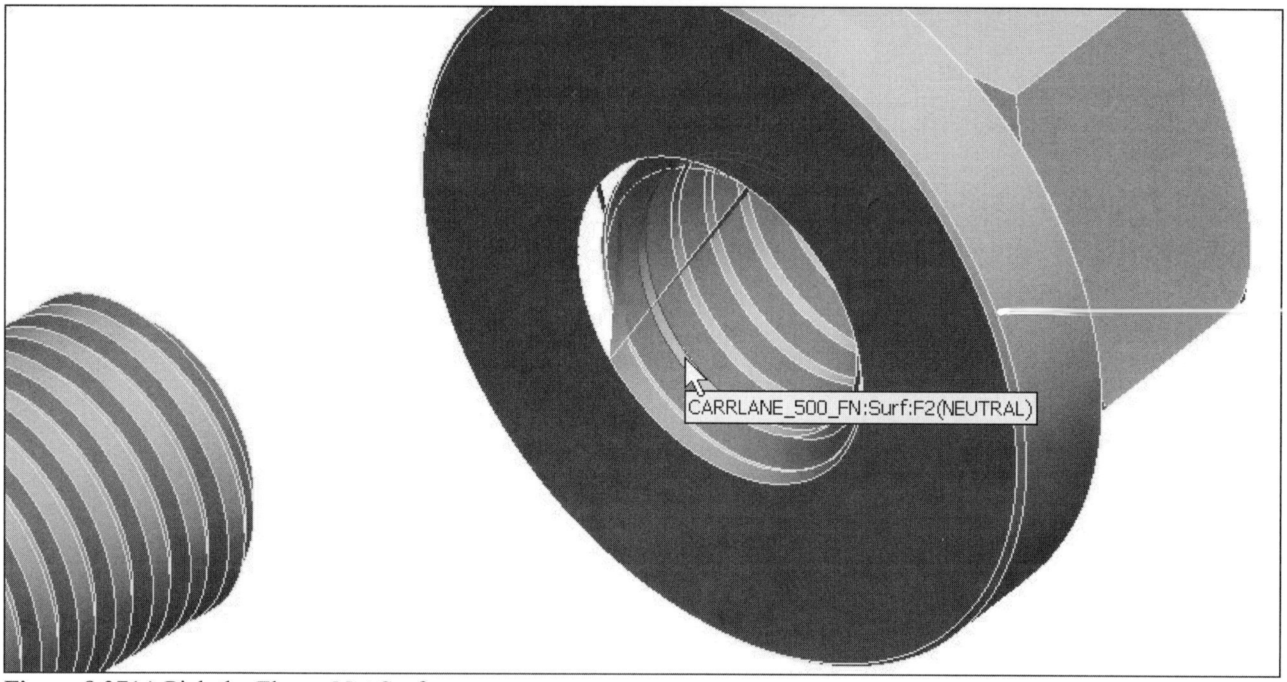

Figure 8.27(e) Pick the Flange Nut Surface

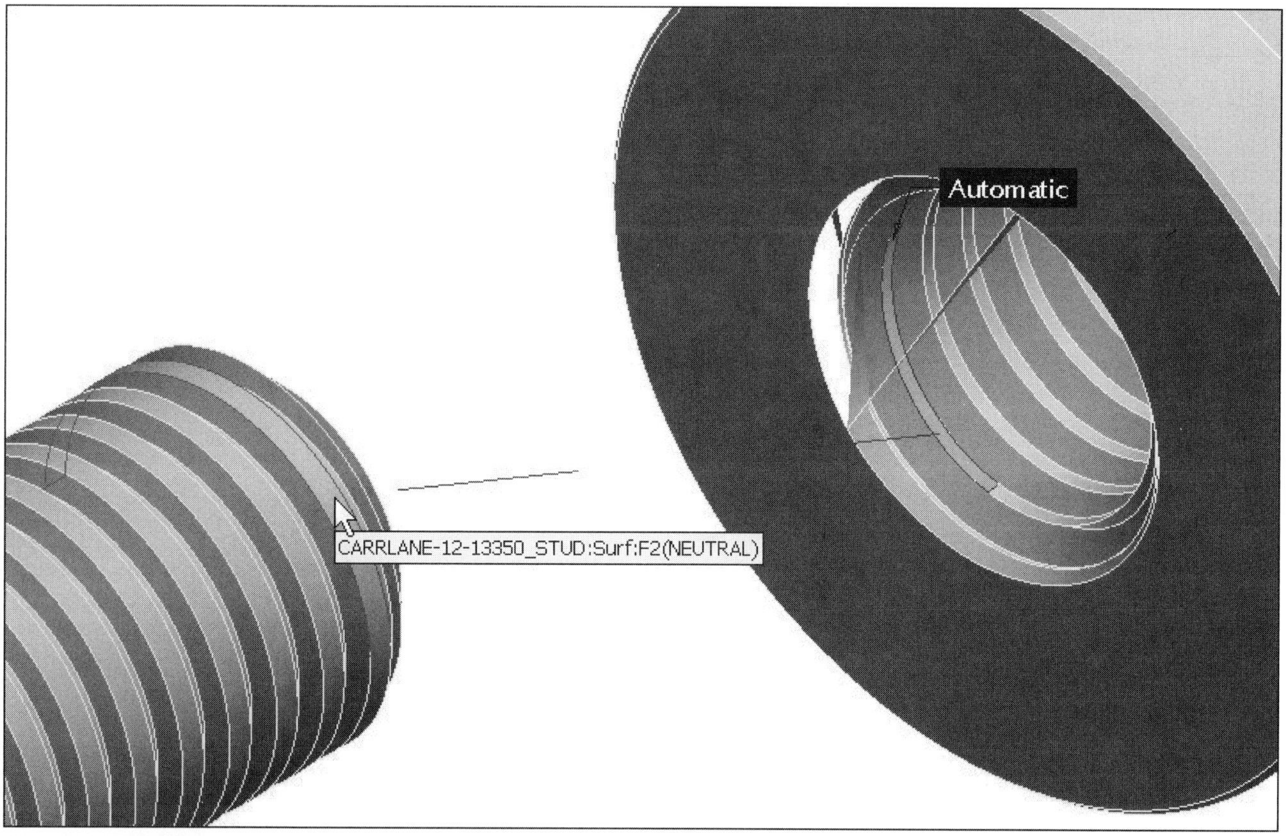

Figure 8.27(f) Pick the Stud Surface

Click: **MMB** spin the model as shown ⇒ **Placement** tab ⇒ **New Constraint** ⇒ rotate and translate the component as required ⇒ pick on the Clamp_Arm surface [Fig. 8.27(g)]

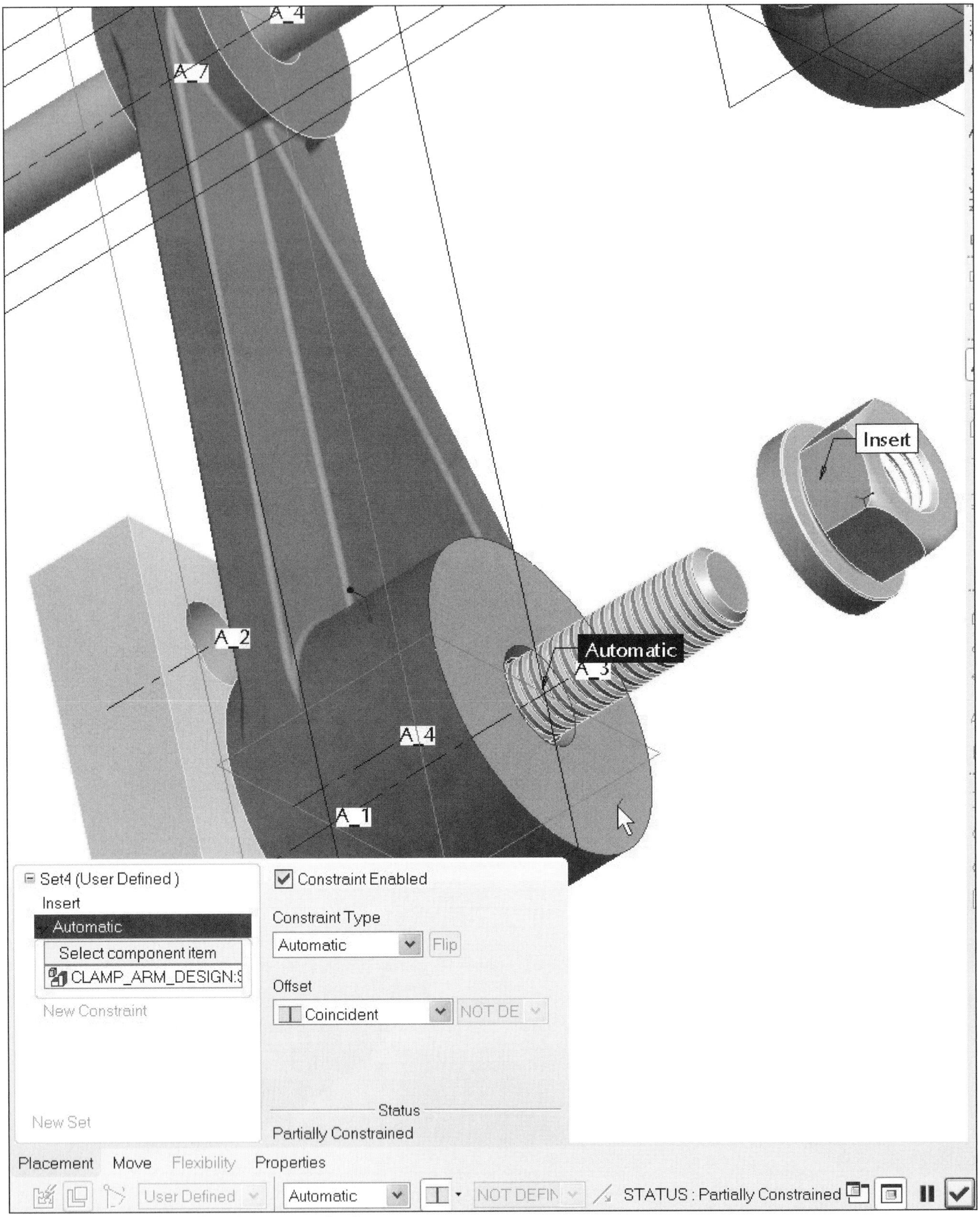

Figure 8.27(g) Pick on the Clamp_Arm Surface

Click: **MMB** spin the model as shown ⇒ pick on the flange nut bearing surface [Figs. 8.27(h-i)]

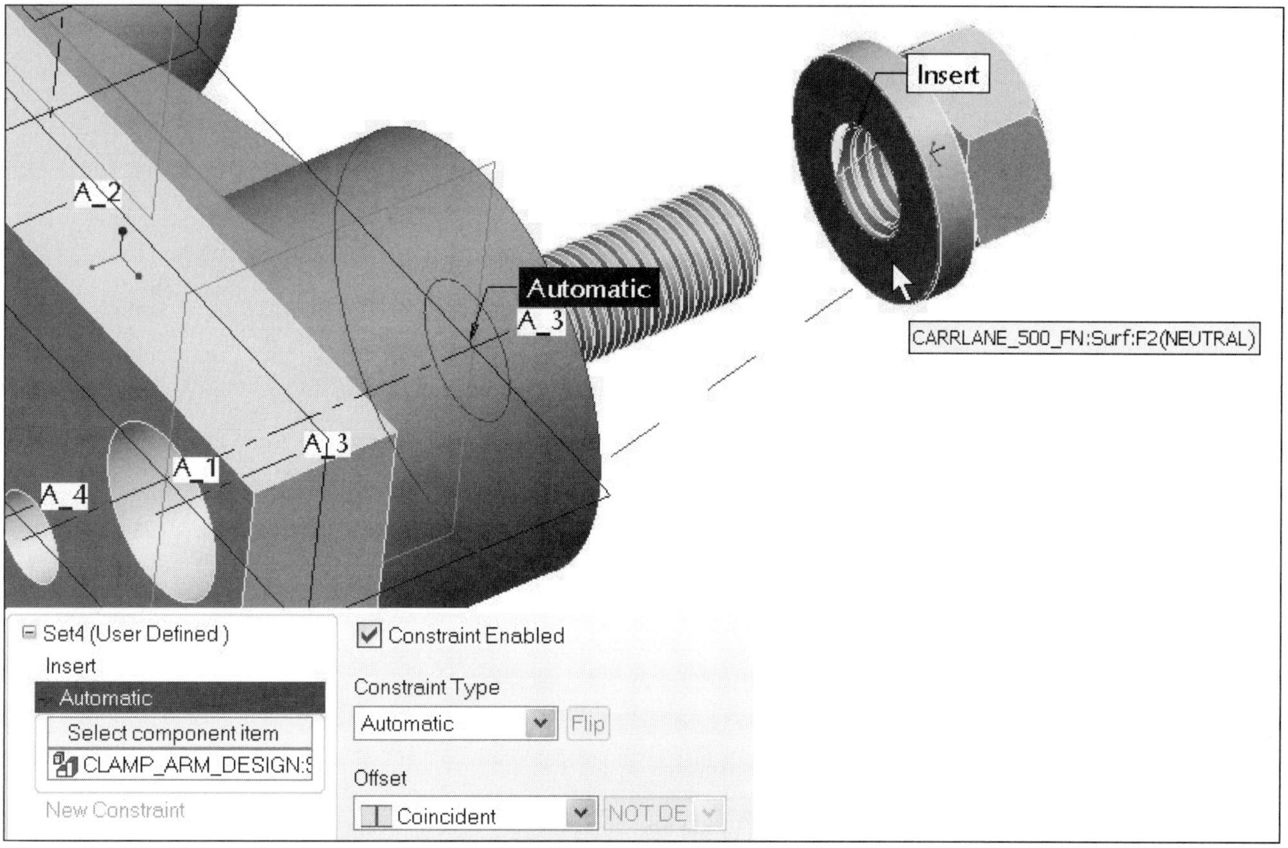

Figure 8.27(h) Pick on the Carrlane_500_fn.prt Surface

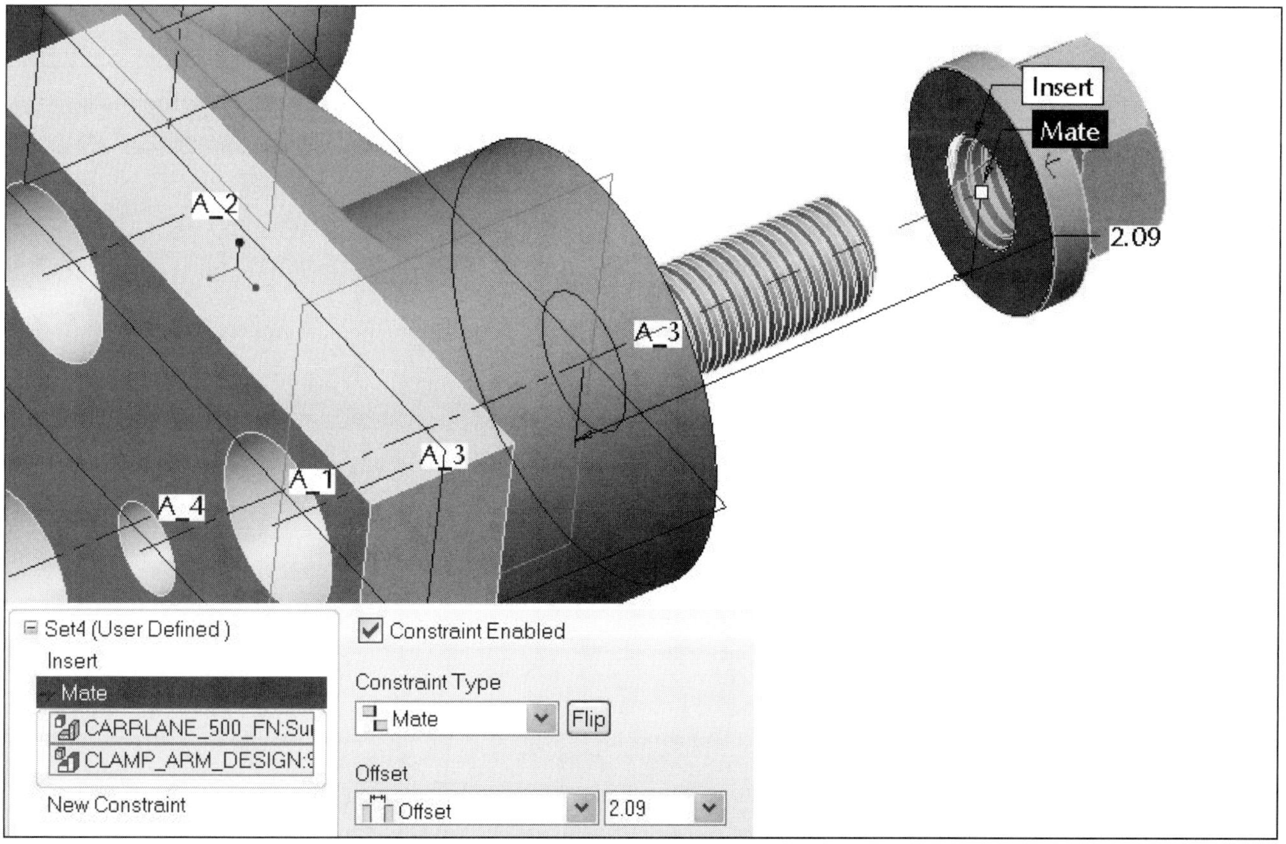

Figure 8.27(i) Flange Nut Offset

Click: 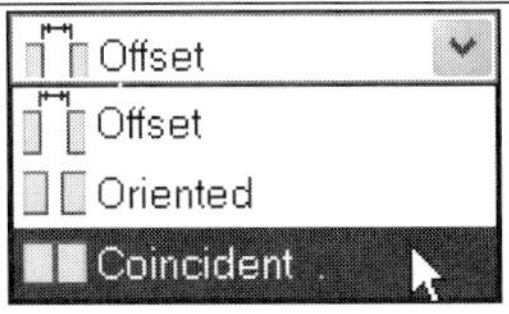

[Fig. 8.27(j)] ⇒ ✓ ⇒ **Ctrl+D** ⇒ **Ctrl+S** ⇒ **Enter** ⇒ **No hidden**

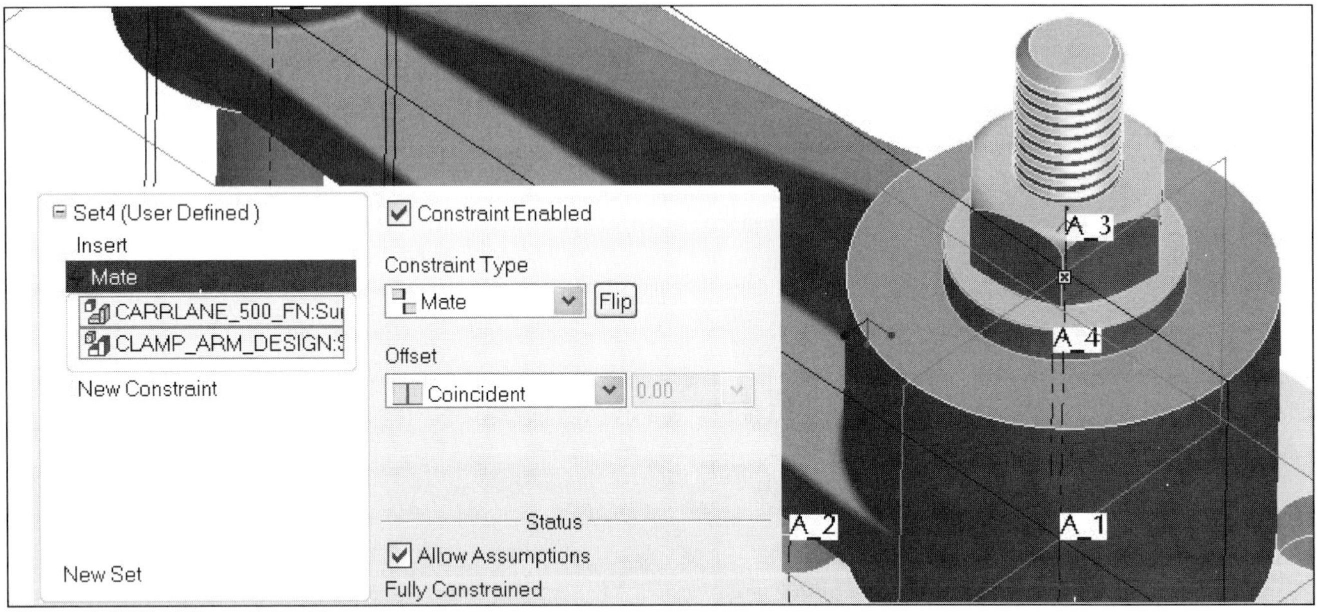

Figure 8.27(j) Offset: Coincident

Perform a check on the assembly using the **Analysis** command. Click: **Analysis** ⇒ **Model** ⇒ **Global Interference** ⇒ **Compute current analysis for preview** (Fig. 8.28) ⇒ *note the interference between components at each threaded connection* ⇒ ✓ ⇒ ⇒ **Ctrl+S** ⇒ **Enter**

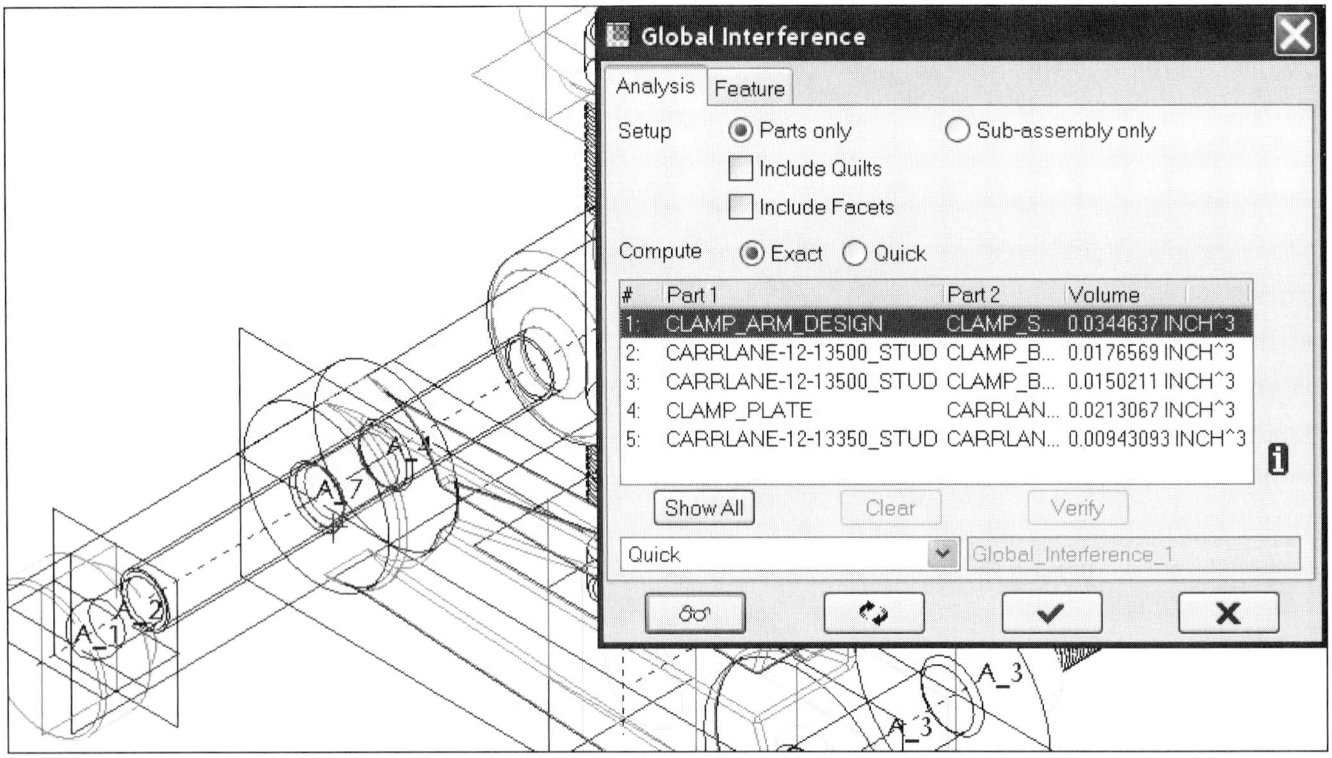

Figure 8.28 Global Interference Dialog Box

Bill of Materials

The Bill of Materials (BOM) lists all parts and part parameters in the current assembly or assembly drawing. It can be displayed in HTML or text format and is separated into two parts: breakdown and summary. The Breakdown section lists what is contained in the current assembly or part. The BOM HTML breakdown section lists quantity, type, name (hyperlink), and three actions (highlight, information and open) about each member or sub-member of your assembly:

- **Quantity** Lists the number of components or drawings
- **Type** Lists the type of the assembly component (part or sub-assembly)
- **Name** Lists the assembly component and is hyper-linked to that item. Selecting this hyperlink highlights the component in the graphics window
- **Actions** is divided into three areas:
    -  **Highlight** Highlights the selected component in the assembly graphics window
    -  **Information** Provides model information on the relevant component
    -  **Open** Opens the component in another Pro/ENGINEER window

A bill of materials (BOM) can be seen by clicking: **Info** ⇒ **Bill of Materials** [Fig. 8.29(a)] ⇒ 

⇒ check all Include options [Fig. 8.29(b)] ⇒ **OK** [Fig. 8.29(c)] ⇒ 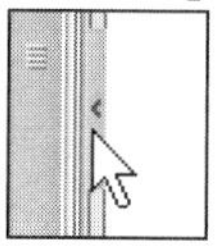 click quick sash to close the Browser

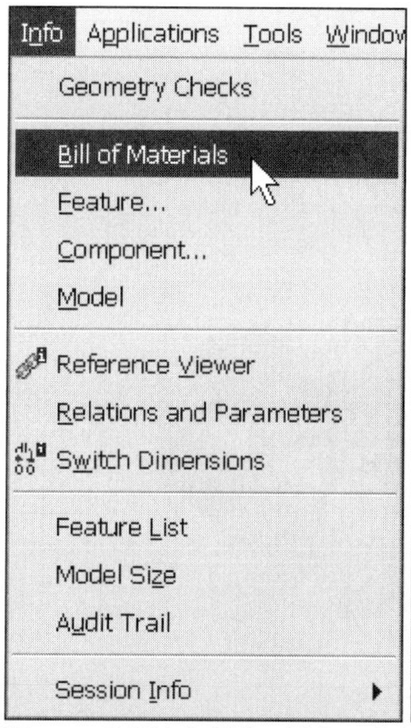

Figure 8.29(a) Bill of Materials

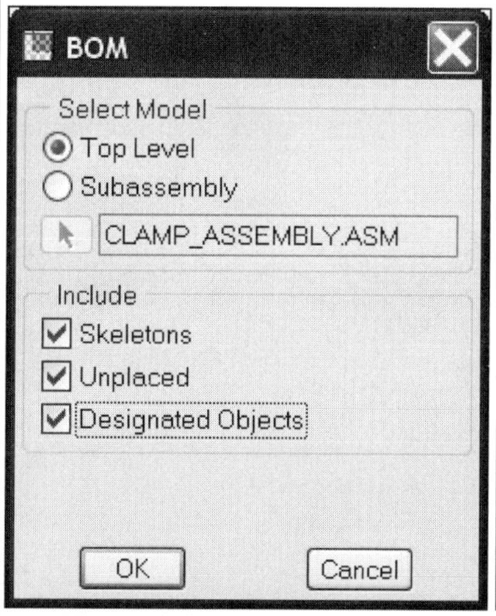

Figure 8.29(b) BOM Dialog Box

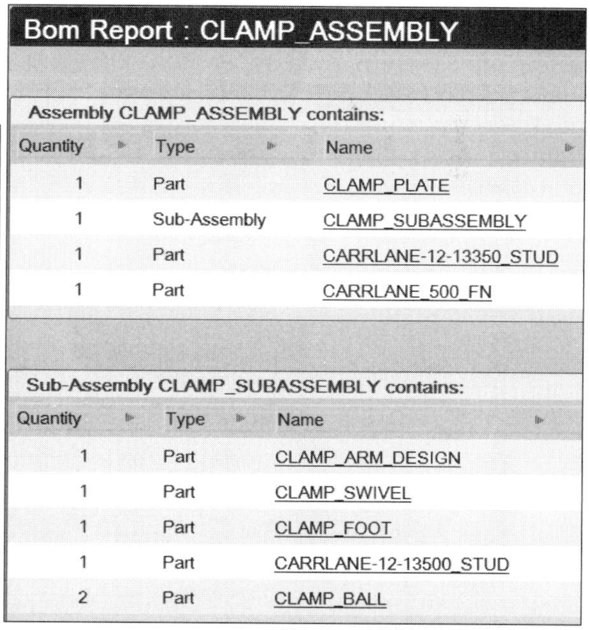

Figure 8.29(c) BOM Report

Upon closer inspection the assembly is not assembled correctly. The Clamp_subassembly is assembled on the wrong side of the Clamp_Plate. When using Pro/E, you can edit items, features, components, etc., at any time. The Edit Definition command is used to change the constraint reference of assembled components.

If necessary, Open the Clamp_Assembly, or click: **Window ⇒ Activate** (Fig. 8.30) to activate the top assembly model ⇒ click on the **CLAMP_SUBASSEMBLY** in the Model Tree ⇒ **RMB ⇒ Edit Definition** [Fig. 8.31(a)]

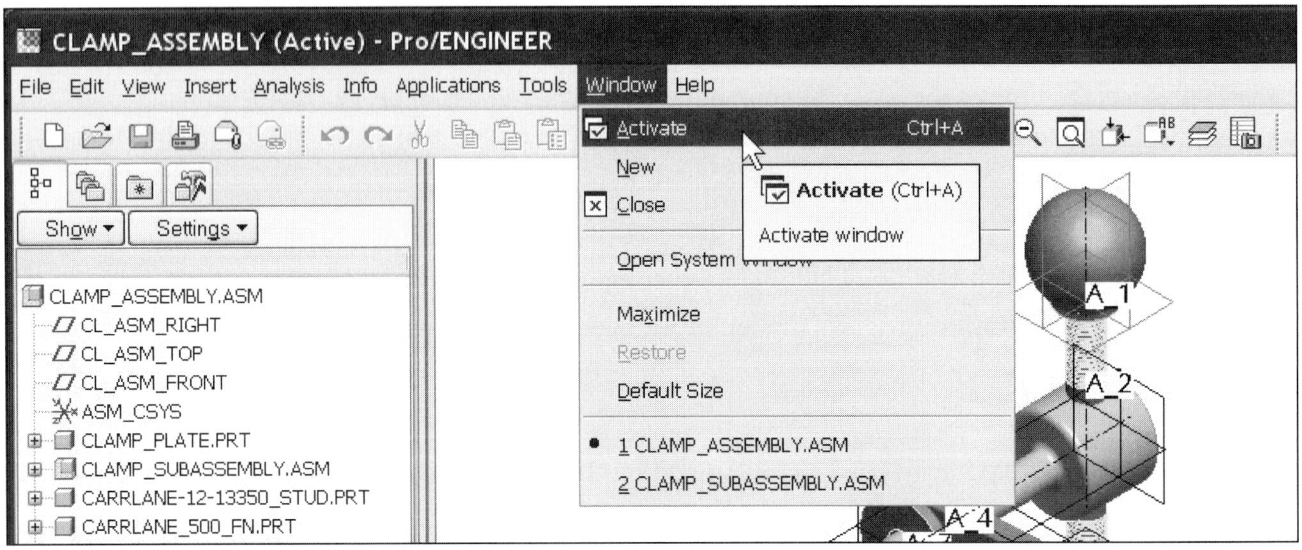

Figure 8.30 Activate the Assembly

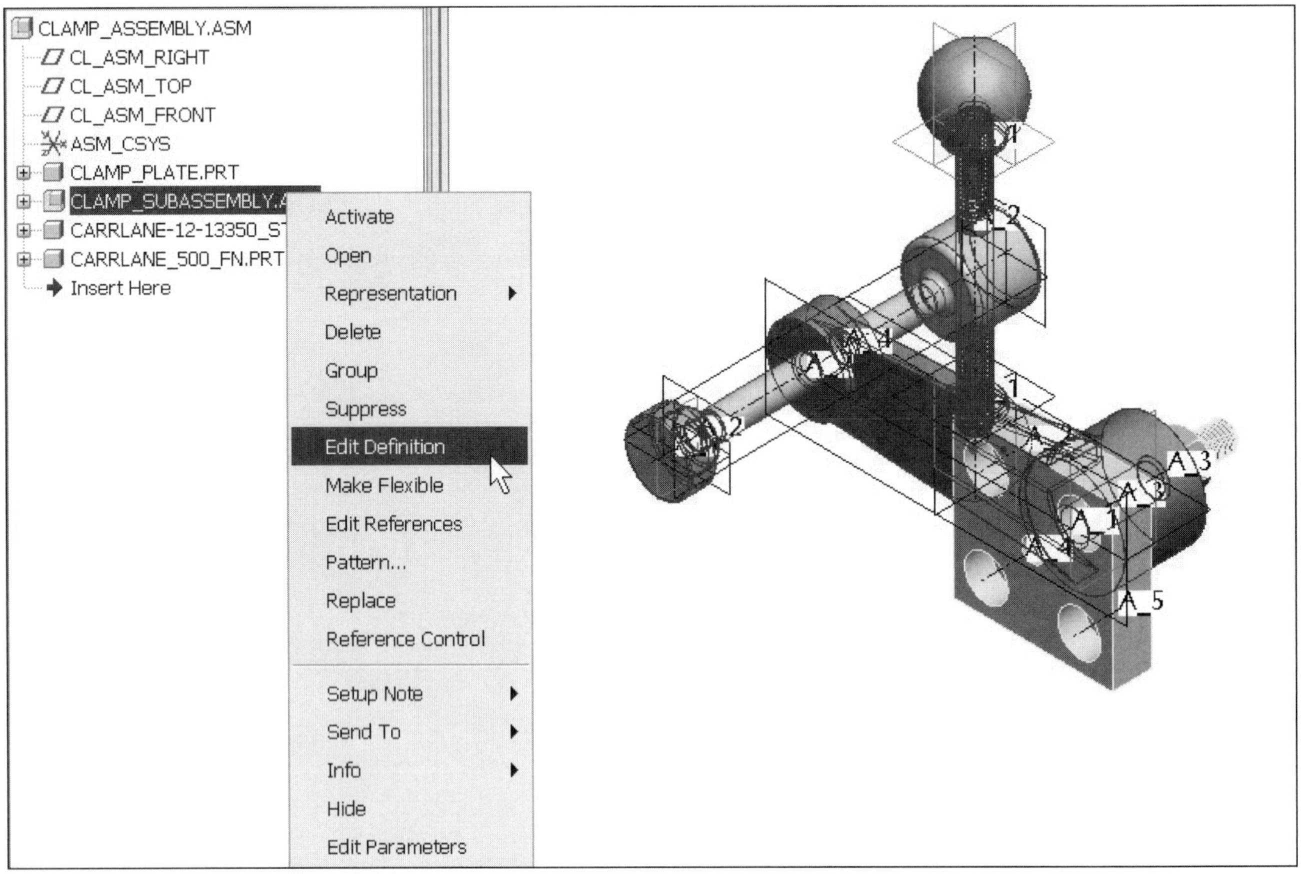

Figure 8.31(a) RMB ⇒ Edit Definition

Click: **Placement** tab, to see the existing Mate [Mate] references ⇒ pick the second (assembly) reference [CLAMP_PLATE:Surf:F5(E] from the Placement dialog box [Fig. 8.31(b)] *(turns light yellow to indicate that it is active, check the Tool Tip* [Assembly Ref: CLAMP_PLATE:Surf:F5(EXTRUDE_1)] *to make sure you selected the Clamp_Plate reference)*

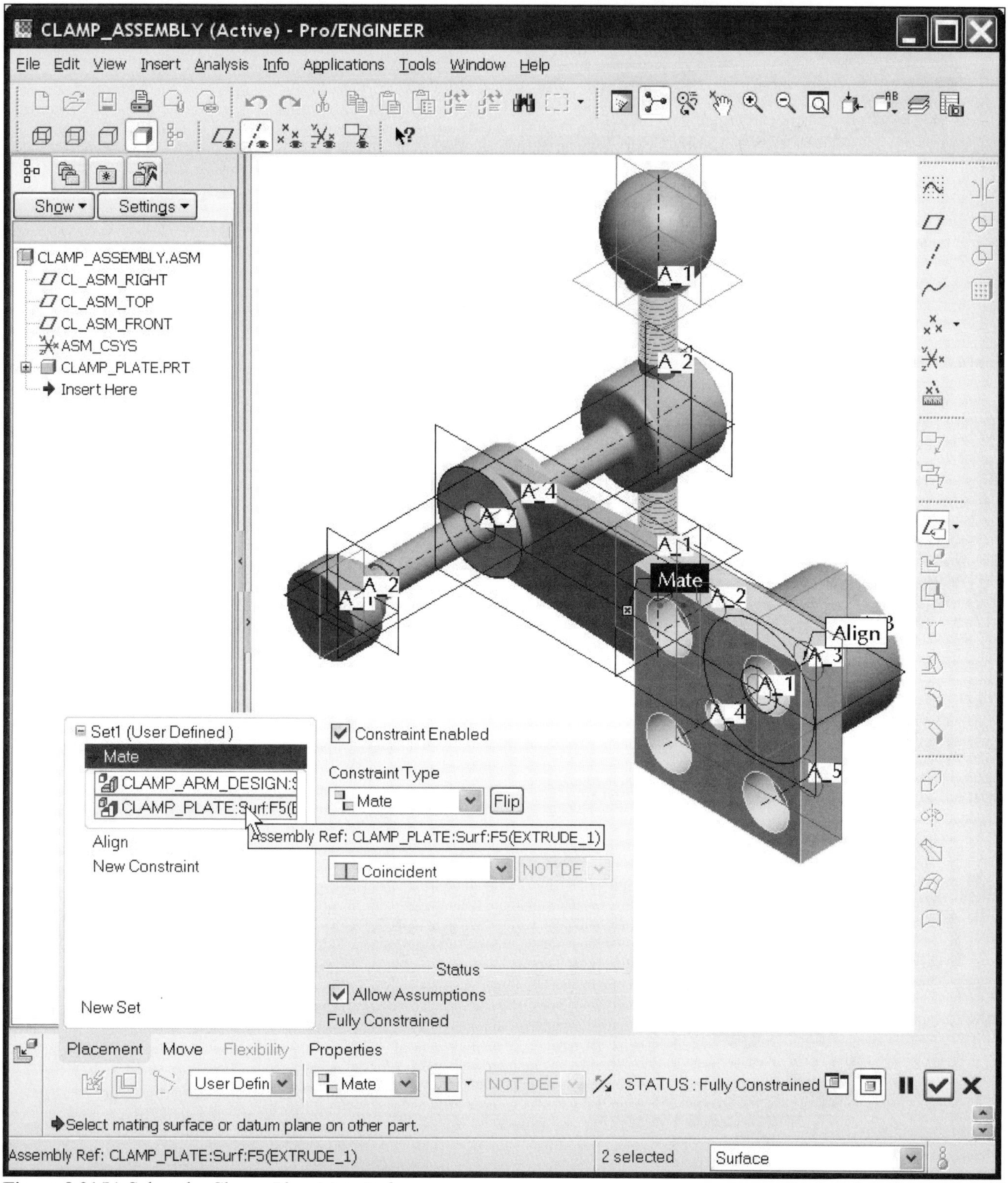

Figure 8.31(b) Select the Clamp_Plate Mate Reference

Pick the counterbore side of the CLAMP_PLATE as the new mate reference [Fig. 8.31(c)] and the subassembly reverses [Fig. 8.31(d)] ⇒ ☑ ⇒ **View** ⇒ **Shade** [Fig. 8.31(e)]

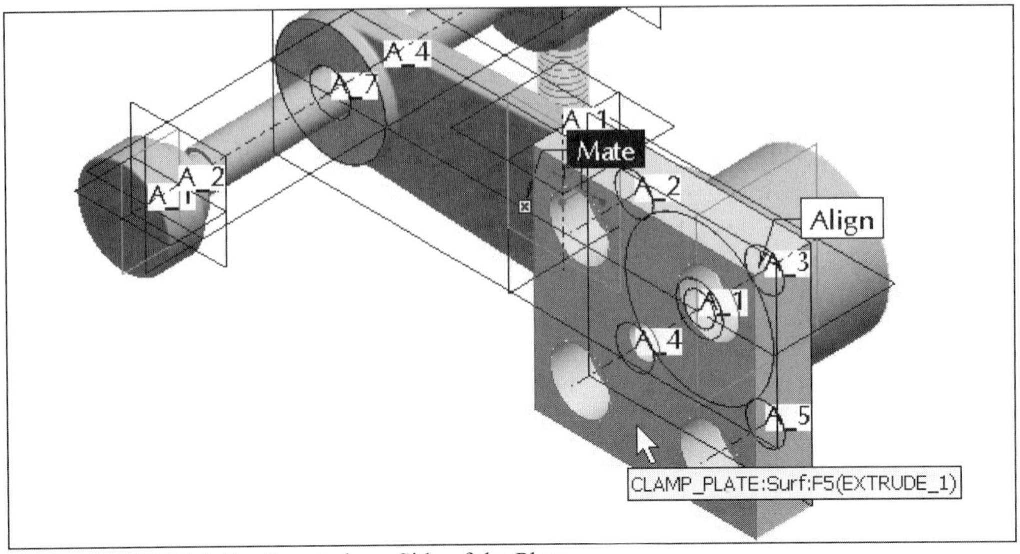

Figure 8.31(c) Pick the Counterbore Side of the Plate

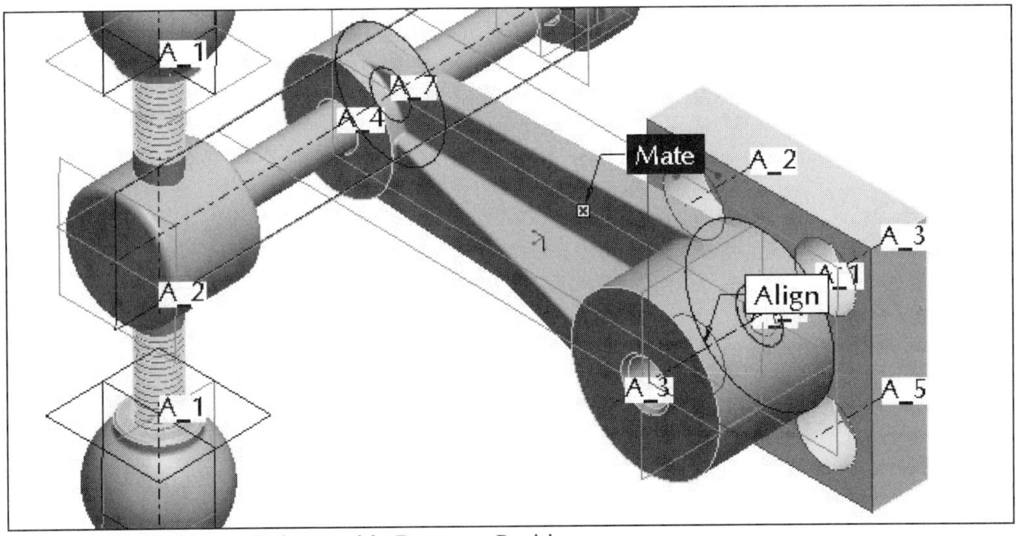

Figure 8.31(d) Clamp_Subassembly Reverses Position

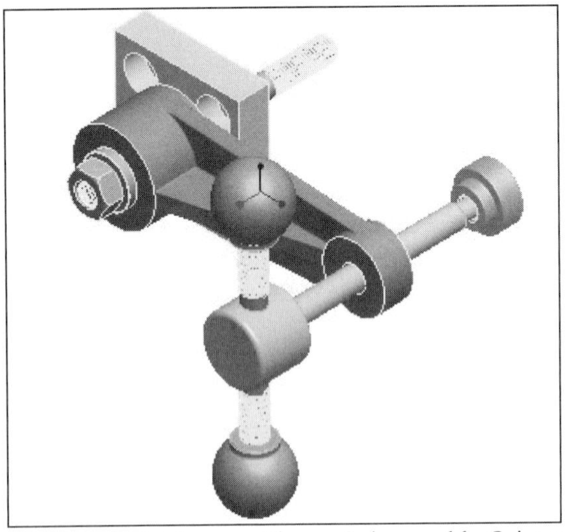

Figure 8.31(e) Correct Clamp_Subassembly Orientation

Click: **Ctrl+R** ⇒ click on the **Carrlane-12-13350_STUD.prt** in the Model Tree ⇒ **RMB** ⇒ **Edit** [Fig. 8.32(a)] ⇒ spin the model so that the offset dimension can be seen clearly ⇒ double-click on your offset dimension 2.888 ⇒ type **-2.00** [Fig. 8.32(b)] -2.00 ⇒ **Enter** ⇒ **Ctrl+G** regenerate ⇒ **Ctrl+D**

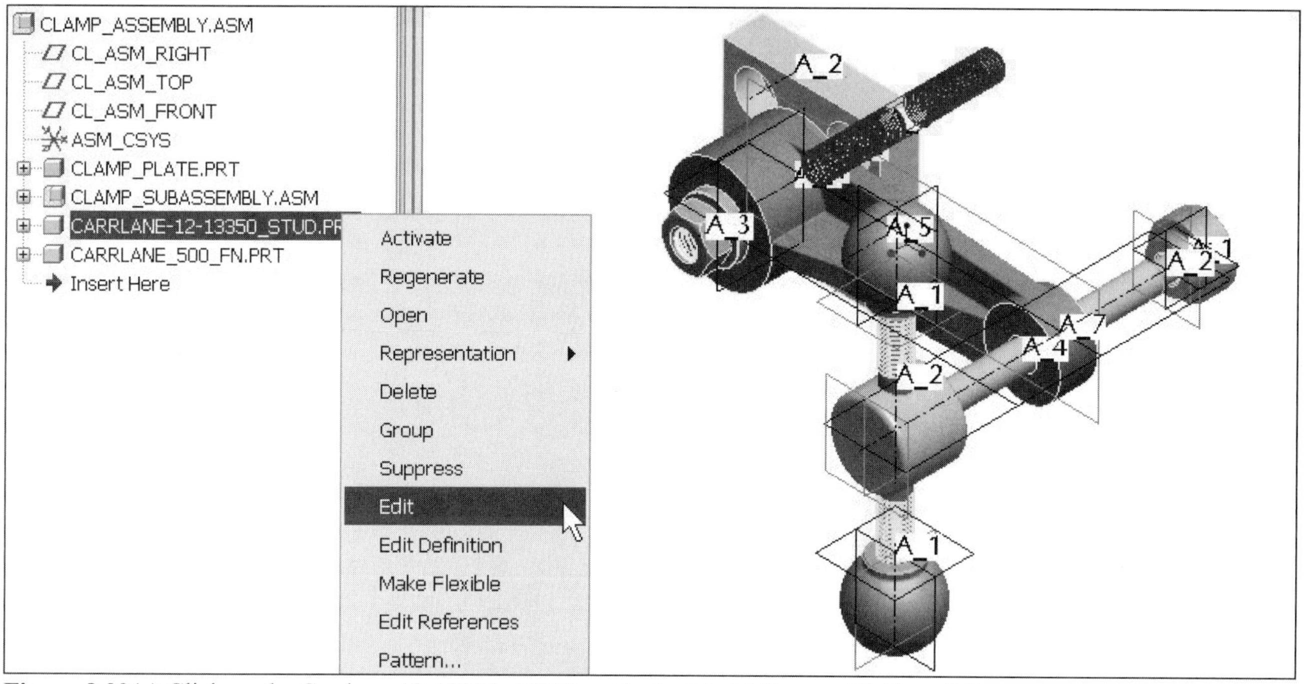

Figure 8.32(a) Click on the Carrlane-12-13350_STUD.prt in the Model Tree ⇒ RMB ⇒ Edit

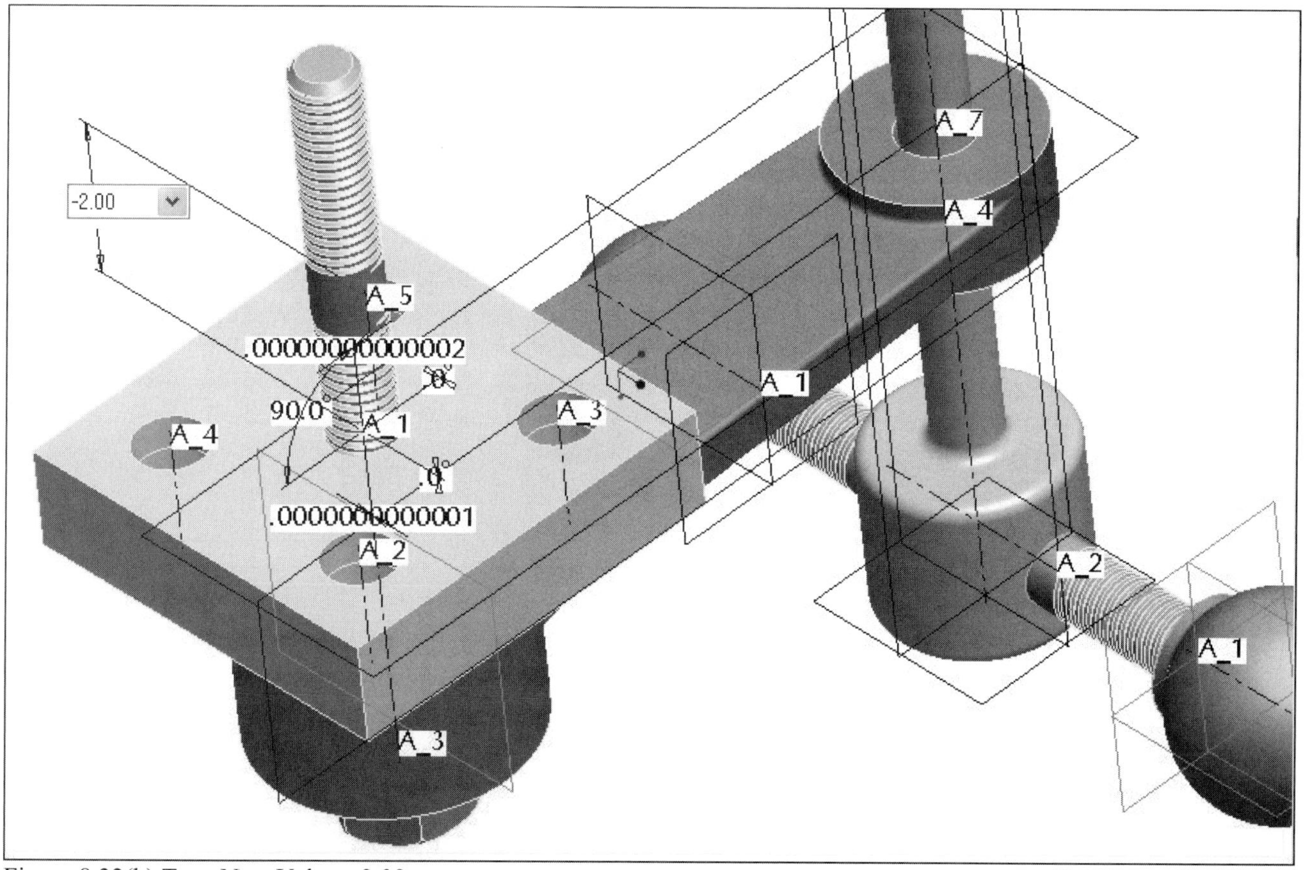

Figure 8.32(b) Type New Value: **-2.00**

Click: **Ctrl+S** ⇒ **Enter** ⇒ **File** ⇒ **Delete** ⇒ **Old Versions** ⇒ **Enter** ⇒ expand Model Tree Items ⇒ **Spin center** off ⇒ **View** ⇒ **Shade** (Fig. 8.33)

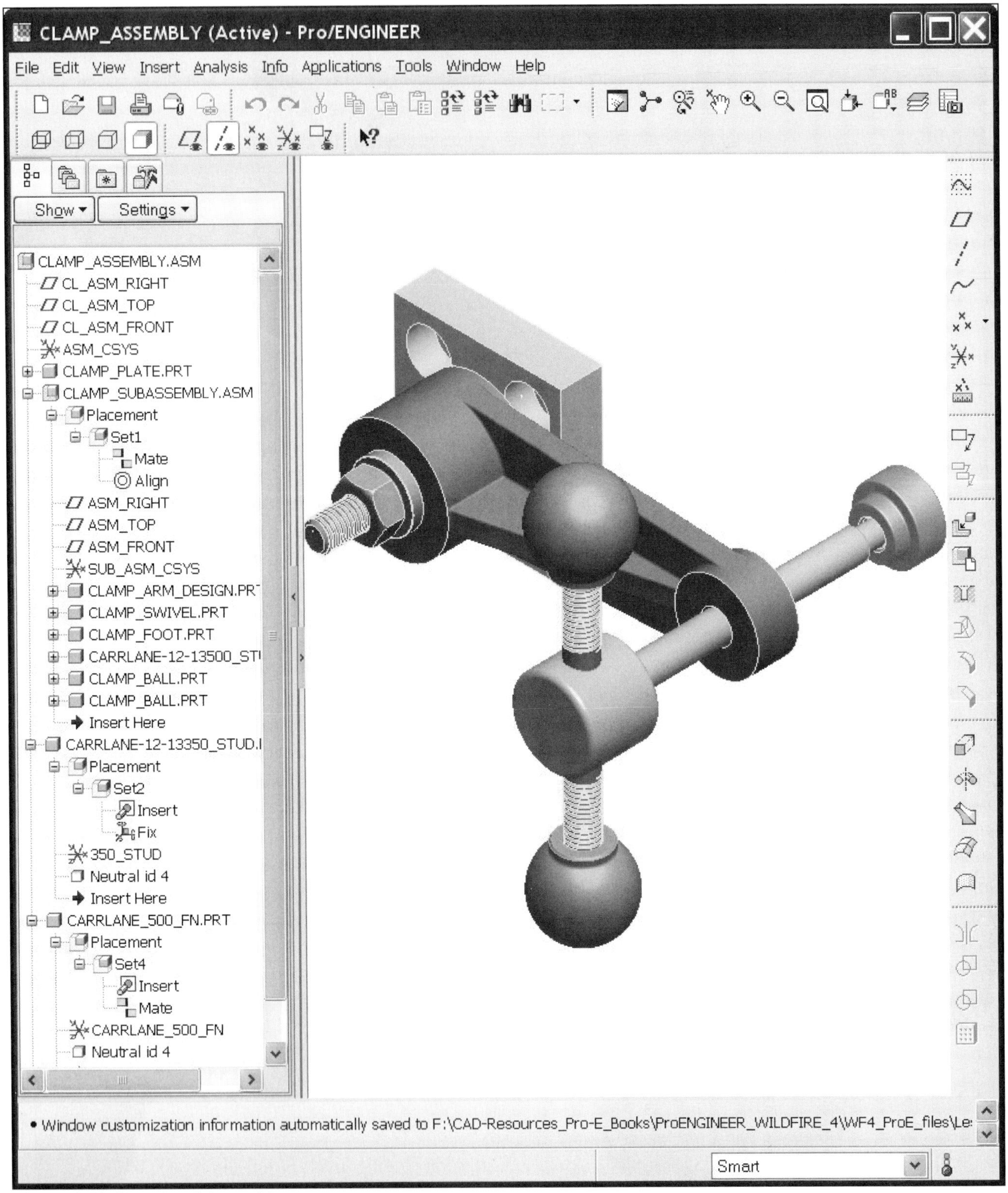

Figure 8.33 Completed Assembly

If you wish to model an assembly project (without step-by-step) instructions, a complete set of projects and illustrations are available at ***www.cad-resources.com*** ⇒ ***Downloads***.

Lesson 9 Exploded Assemblies and View Manager

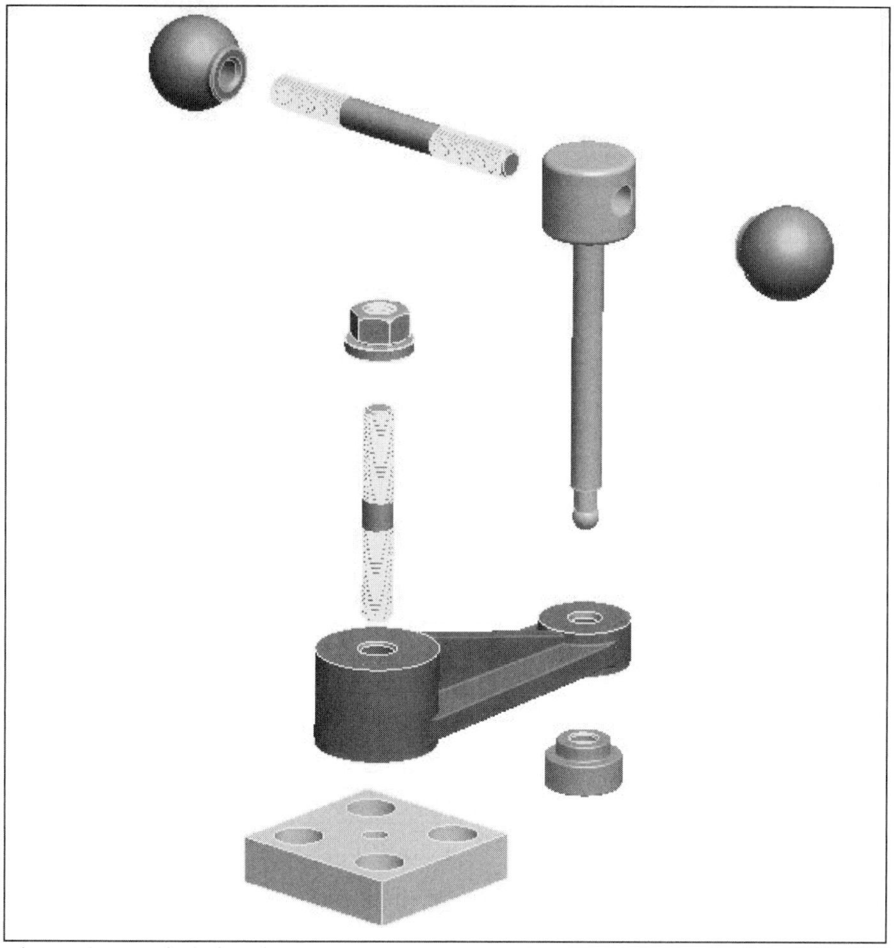

Figure 9.1 Exploded Swing Clamp Assembly

OBJECTIVES

- Create **Exploded Views**
- Create **Explode States**
- Utilize the **View Manager** to organize and control views states
- Create unique component **visibility settings (Style States)**
- **Move** and **Rotate** components in an assembly
- Add a **URL** to a **3D Note**
- Create **Perspective** views of the model
- View the **BOM** for an assembly

EXPLODED ASSEMBLIES

Pictorial illustrations, such as exploded views, are generated directly from the 3D model database (Fig. 9.1). The model can be displayed and oriented in any position. Each component in the assembly can have a different display type: wireframe, hidden line, no hidden, and shading. You can select and orient the component to provide the required view orientation to display the component from underneath or from any side or position. Perspective projections are made with selections from menus. The assembly can be spun around, reoriented, and even clipped to show the interior features. You have the choice of displaying all components and subassemblies or any combination of components in the design.

Creating Exploded Views

Using the **Explode State** option in the **View Manager**, you can automatically create an exploded view of an assembly (Fig. 9.2). Exploding an assembly affects only the display of the assembly; it does not alter true design distances between components. Explode states are created to define the exploded positions of all components. For each explode state, you can toggle the explode status of components, change the explode locations of components, and create and modify explode offset lines to show how explode components align when they are in their exploded positions. The Explode State Explode Position functionality is similar to the Package/Move functionality.

You can define multiple explode states for each assembly and then explode the assembly using any of these explode states at any time. You can also set an explode state for each drawing view of an assembly. Pro/E gives each component a default explode position determined by the placement constraints. By default, the reference component of the explode is the parent assembly (top-level assembly or subassembly).

To explode components, you use a drag-and-drop user interface similar to the Package/Move functionality. You select the motion reference and one or more components, and then drag the outlines to the desired positions. The component outlines drag along with the mouse cursor. You control the move options using a Preferences setting. Two types of explode instructions can be added to a set of components. The children components follow the parent component being exploded or they do not follow the parent component. Each explode instruction consists of a set of components, explode direction references, and dimensions that define the exploded position from the final (installed) position with respect to the explode direction references.

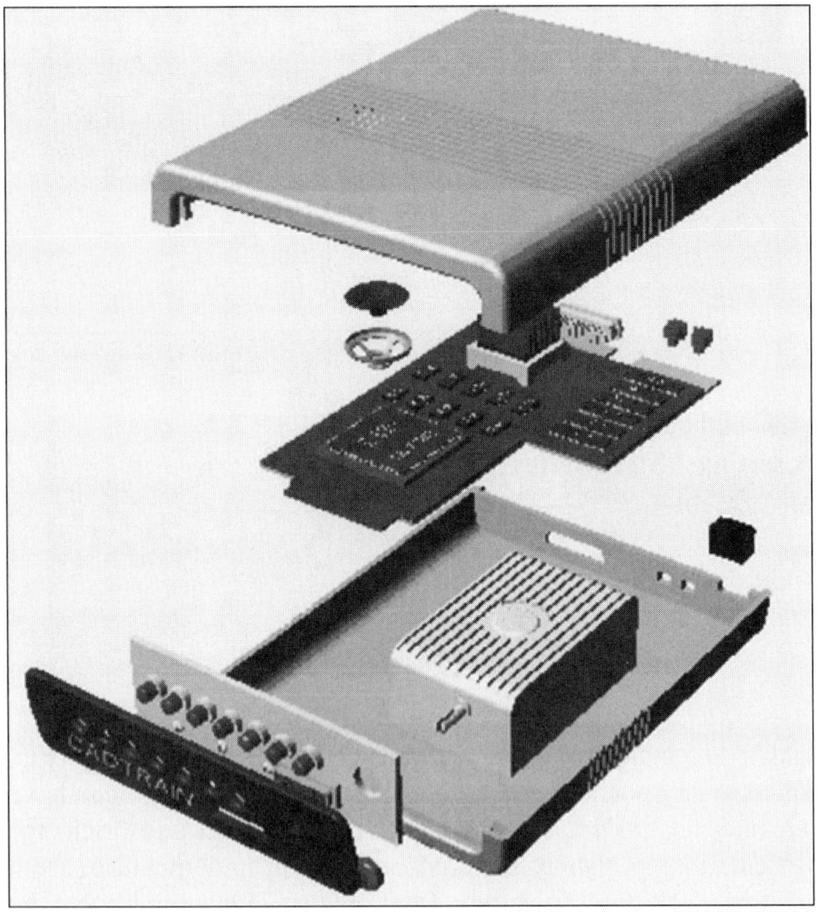

Figure 9.2 Exploded Assemblies (CADTRAIN, COAch for Pro/ENGINEER)

When using the explode functionality, keep in mind the following:

- You can select individual parts or entire subassemblies from the Model Tree or main window.
- If you explode a subassembly, in the context of a higher-level assembly, you can specify the explode state to use for each subassembly.
- You do not lose component explode information when you turn the status off. Pro/E retains the information so that the component has the same explode position if you turn the status back on.
- All assemblies have a default explode state called "Default Explode", which is the default explode state Pro/E creates from the component placement instructions.
- Multiple occurrences of the same subassembly can have different explode characteristics at a higher-level assembly.

Component Display

Style accessed through using: View ⇒ View Manager [Fig. 9.3(a)] ⇒ Style [Fig. 9.3(b)], manages the display styles of an assembly. Orient and Explode [Fig. 9.3(c)] are also available on separate tabs from this dialog. Wireframe, hidden line, no hidden, shaded, or transparent display styles can be assigned to each component. The components will be displayed according to their assigned display styles in the current style state (that is, blanked, shaded, drawn in hidden line color, and so on). The current setting, in the Environment menu, controls the display of unassigned components.

Figure 9.3(a) View Manager

Figure 9.3(b) View Manager- Style

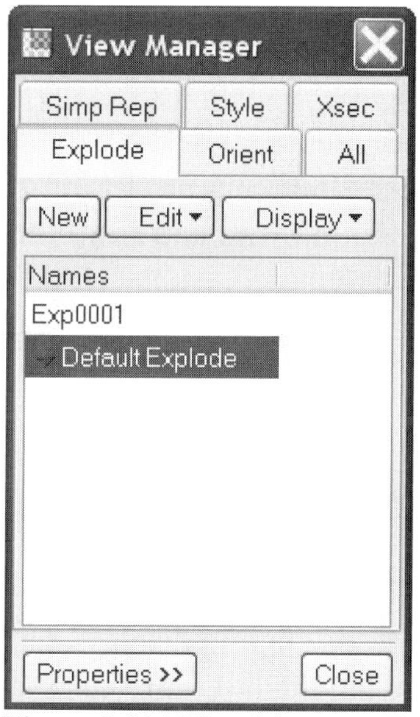

Figure 9.3(c) View Manager- Explode

Components appear in the currently assigned style state (Fig. 9.4). The current setting is indicated in the display style column in the Model Tree window. Component display or style states can be modified without using the View Manager. You can select desired models from the graphics window, Model Tree, or search tool, and then use the View ⇒ Display Style commands to assign a display style (wireframe, hidden, no hidden, shaded and transparent) to the selected models. The Style representation is temporarily changed. These temporary changes can then be stored to a new style state, or updated to an existing style state. You can also define default style states. If the default style state is updated to reflect changes different from that of the master style state, then that default style state will be reflected each time the model is retrieved.

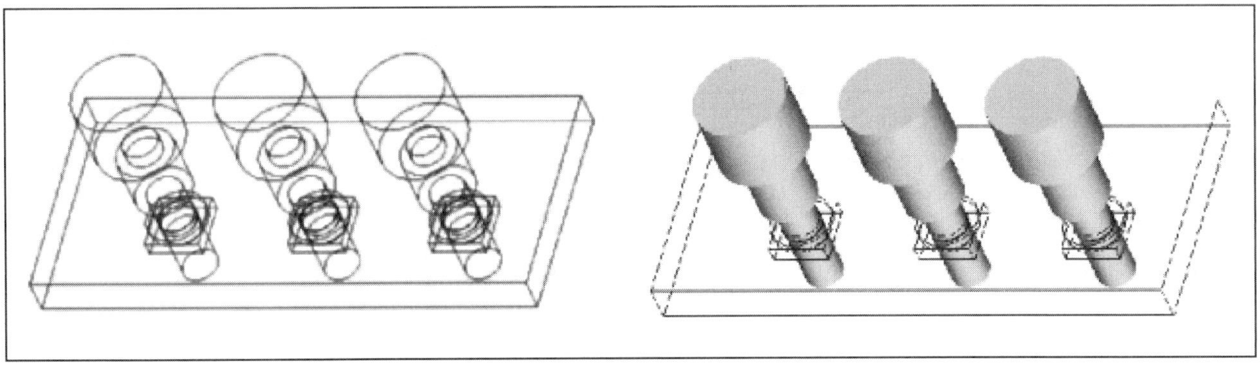

Figure 9.4 Style States

Types of Representations

The main types of simplified representations: **master**, **geometry**, **graphics**, and **symbolic**, designate which representation appears, using the commands in the View Manager dialog box.

Graphics and geometry representations speed up the retrieval process of large assemblies. All simplified representations provide access to components in the assembly and are based upon the Master Representation.

You cannot modify a feature in a graphics representation, but you can do so in a geometry representation.

Assembly features are displayed when you retrieve a model. Subtractive assembly features such as cuts and holes are represented in graphics and geometry representations, making it possible to use these simplified representations for performance improvement while still displaying on screen a completely accurate geometric model.

You can access model information for graphics and geometry representations of part models from the Information menu and from the Model Tree. Because part graphics and geometry representations do not contain feature history of the part model, information for individual features of the part is not accessible from these representations.

- The *Master Representation* always reflects the fully detailed assembly, including all of its members. The Model Tree lists all components in the Master Representation of the assembly, and indicates whether they are included, excluded, or substituted.
- The *Graphics Representation* contains information for display only and allows you to browse through a large assembly quickly. You cannot modify or reference graphics representations. The type of graphic display available depends on the setting of the *save_model_display* configuration option, the last time the assembly was saved:
    - ***wireframe*** (default) The wireframe of the components appear.
    - ***shading_low***, ***shading_med***, ***shading_high*** A shaded version of the components appears. The different levels indicate the density of the triangles used for shading.
    - ***shading_lod*** The level of detail depends on the setting in the View Performance dialog box. To access the View Performance dialog box, click: View ⇒ Display Settings ⇒ Performance.

While in a simplified representation, Pro/E applies changes to an assembly, such as creation or assembly of new components, to the Master Representation. It reflects them in all of the simplified representations (Pro/PROGRAM processing also affects the Master Representation). It applies all suppressing and resuming of components to the Master Representation. However, it applies the actions of a simplified representation only to currently resumed members, which is, to members that are present in the BOM of the Master Representation.

Lesson 9 STEPS

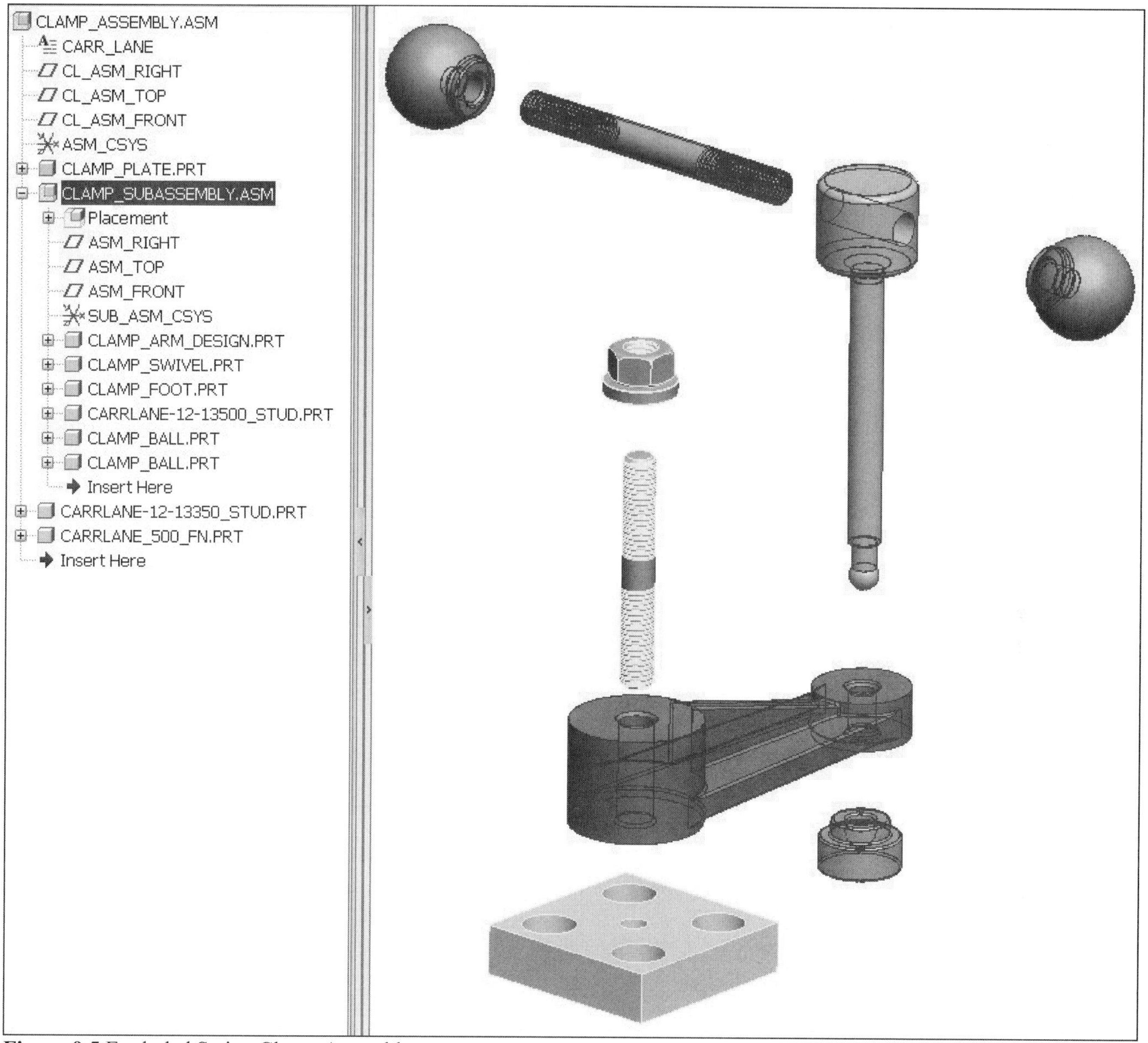

Figure 9.5 Exploded Swing Clamp Assembly

Exploded Swing Clamp Assembly

In this lesson, you will use the previously created subassembly and assembly (Fig. 9.5) to establish and save new views, exploded views, and views with component style states that differ from one another. The View Manager Dialog box will be employed to control and organize a variety of different states.

You will also be required to move and rotate components of the assembly before cosmetically displaying the assembly in an exploded state. The creation and assembly of new components will not be required. A bill of materials will also be displayed using the Info command. A 3D Note with a URL will be added to the model as the last information feature.

URLs and Model Notes

Since the assembly (Fig. 9.6) you created is a standard clamp from CARR LANE Manufacturing Co.; you could go to CARR LANE's Website and see the assembly, order it (about $70.00 U.S.) or download a 3D IGES model. Here you will create and attach a 3D Note to the assembly, identifying the manufacturer and Website.

Click: **File** ⇒ **Open** ⇒ **clamp_assembly.asm** ⇒ **Open** ⇒ **Tools** ⇒ **Environment** ⇒ Standard Orient **Isometric** ⇒ **OK** ⇒ **Insert** ⇒ **Annotations** ⇒ **Notes** ⇒ **New** ⇒ Name **CARR_LANE** ⇒ Text: type text lines (Fig. 9.7) ⇒ [✓ Place note flat to screen in model space] ⇒ URL **Hyperlink** ⇒ type **www.carrlane.com** (Fig. 9.8) ⇒ **ScreenTip** ⇒ Screen tip text: **CARR LANE Manufacturing** (Fig. 9.9) ⇒ **OK** ⇒ **OK** ⇒ **Place** ⇒ **No Leader** ⇒ **Standard** ⇒ **Done** ⇒ [Select LOCATION for note.] ⇒ [icon] pick near the assembly in the graphics window ⇒ **OK** ⇒ **Done/Return**

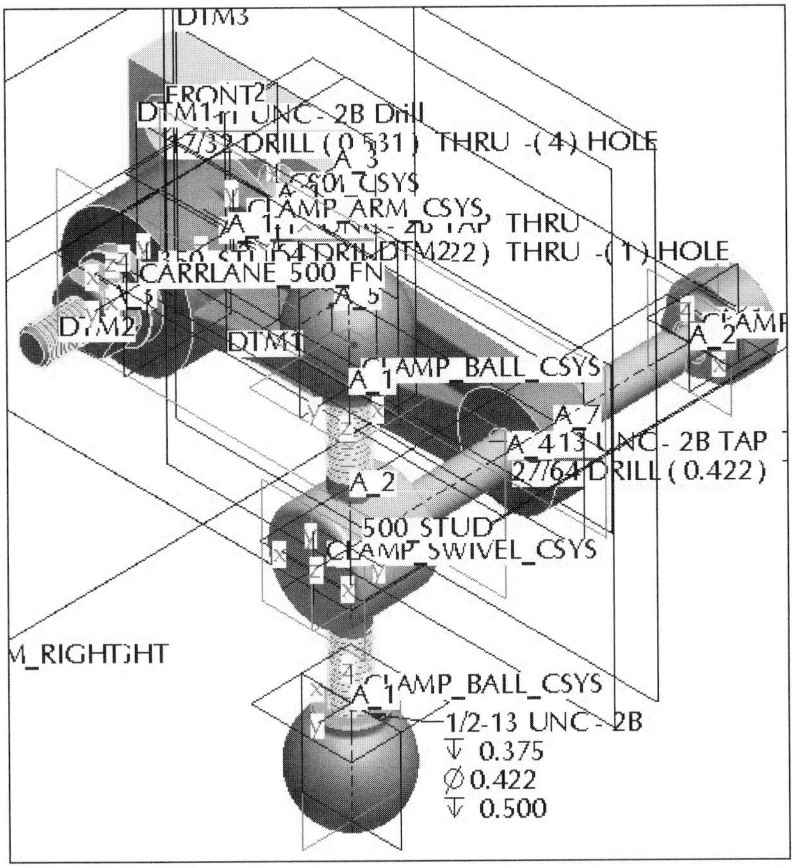

Figure 9.6 Model

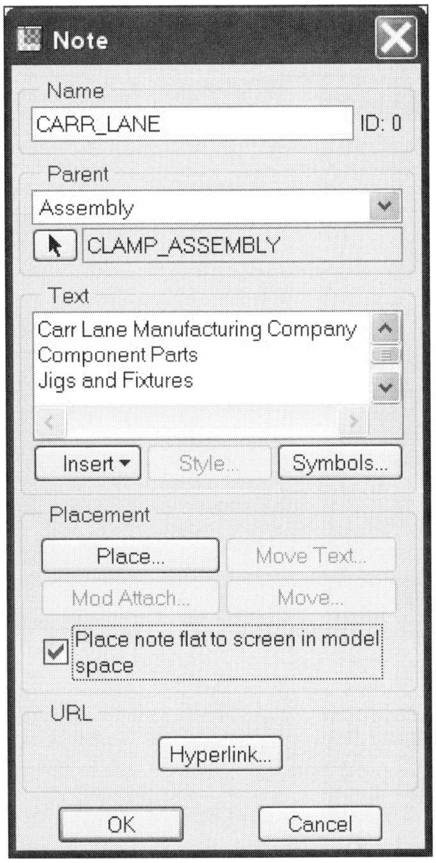

Figure 9.7 Note Dialog Box

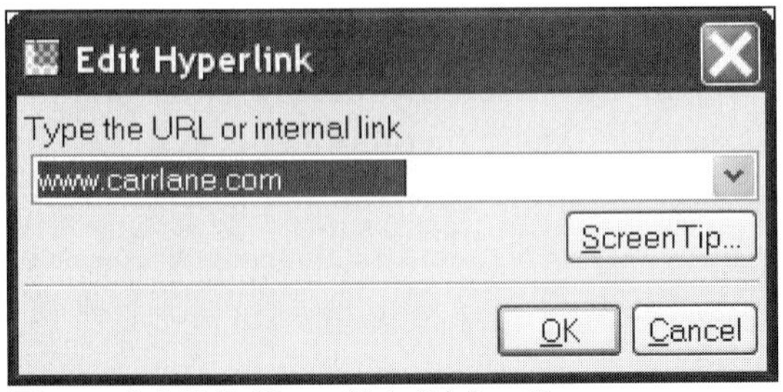

Figure 9.8 Edit Hyperlink Dialog Box

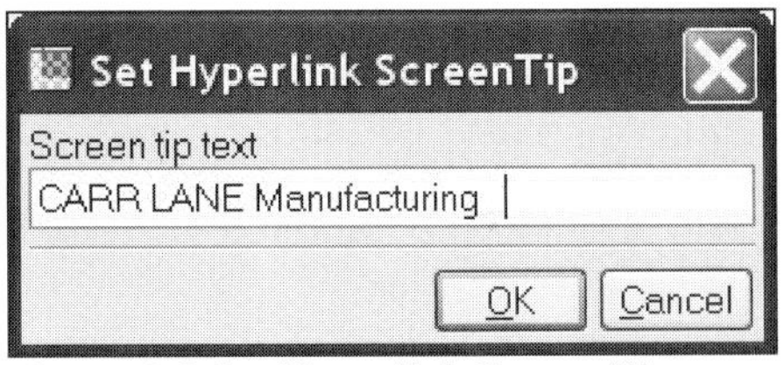

Figure 9.9 Set Hyperlink Screen Tip

Settings ⇒ **Tree Filters** ⇒ toggle all on (Fig. 9.10) ⇒ **OK** (Fig. 9.11) ⇒ **Ctrl+R** ⇒ **Ctrl+S** ⇒ **Enter**

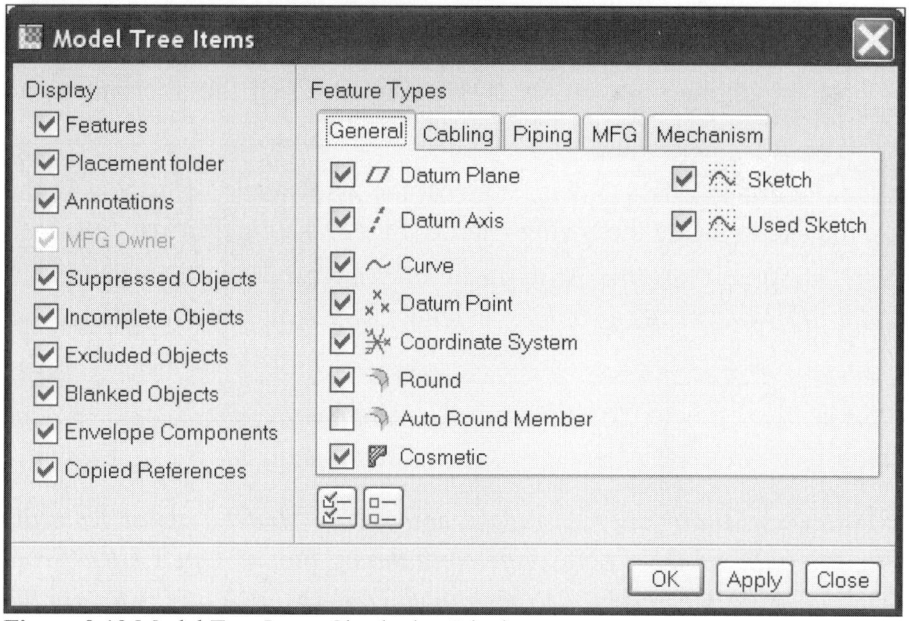

Figure 9.10 Model Tree Items Checked to Display

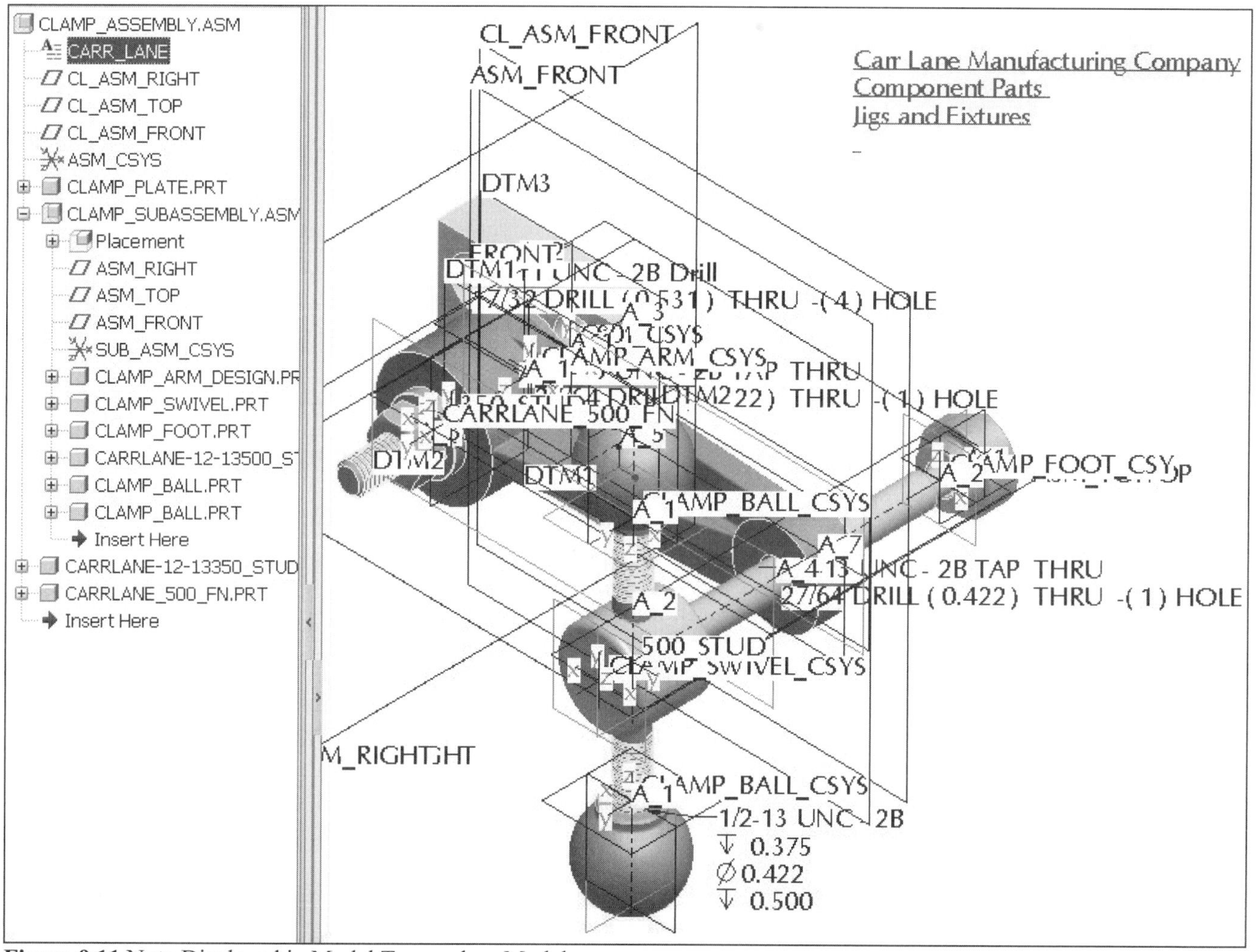

Figure 9.11 Note Displayed in Model Tree and on Model

To open a Hyperlink defined in a model note and to launch the embedded Web browser and go to a World Wide Web URL (Universal Resource Locator) associated with a model note, use one of three methods.

1. Click on the note in the Model Tree or graphics screen ⇒ **RMB** [Fig. 9.12(a)] the shortcut menu displays ⇒ **Open URL** ⇒ 

[Fig. 9.12(b)] the associated URL opens in the Pro/E embedded browser ⇒ navigate the site to locate the Swing Clamp [Figs. 9.12(c-g)] ⇒ use the scroll bars to the see the drawing [Figs. 9.12(h-i)]) ⇒ close the Browser by clicking on the quick sash ⇒  **Save** ⇒ **MMB** *(Note that to pick the 3D Note in the graphics window; you may have to change the Selection Filter to Annotation)*

*2. Move the cursor over the hyperlinked note ⇒ press **Ctrl** the cursor changes to a hand icon ⇒ click **LMB** the associated URL opens in the Pro/E embedded browser ⇒ navigate to locate the Swing Clamp*

*3. Click: **Insert** ⇒ **Annotations** ⇒ **Notes** ⇒ **Modify** ⇒ pick on the note in the Model Tree or from the screen to see the note before opening the URL ⇒ **OK** to close the Note dialog box ⇒ **Open URL** (from the Menu Manager) ⇒ pick the 3D Note from the screen or the Model Tree ⇒ navigate to locate the Swing Clamp ⇒ Done/Return from the Menu Manager*

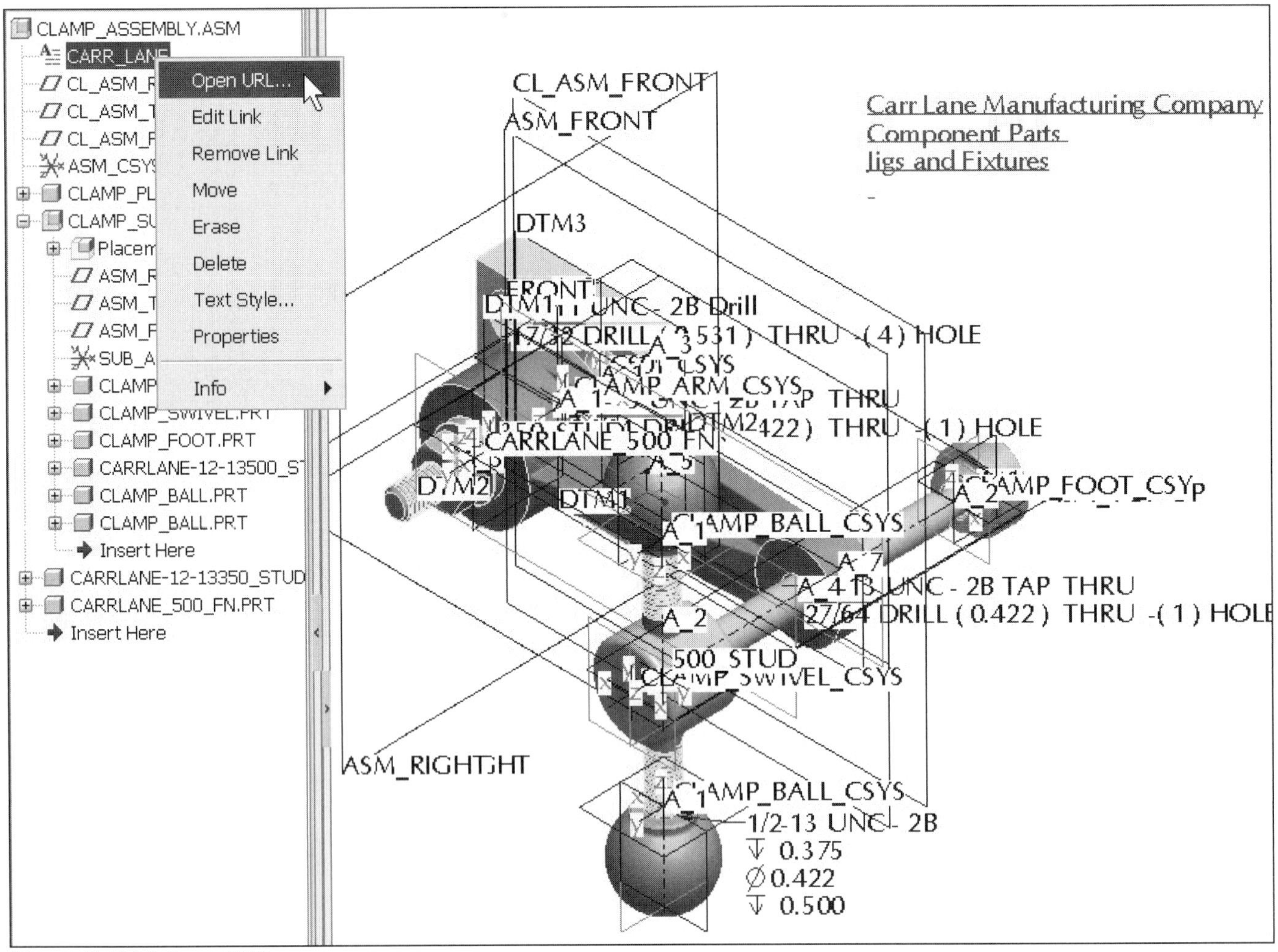

Figure 9.12(a) Open URL

Figure 9.12(b) Carr Lane Web Site 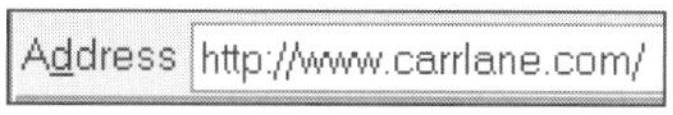

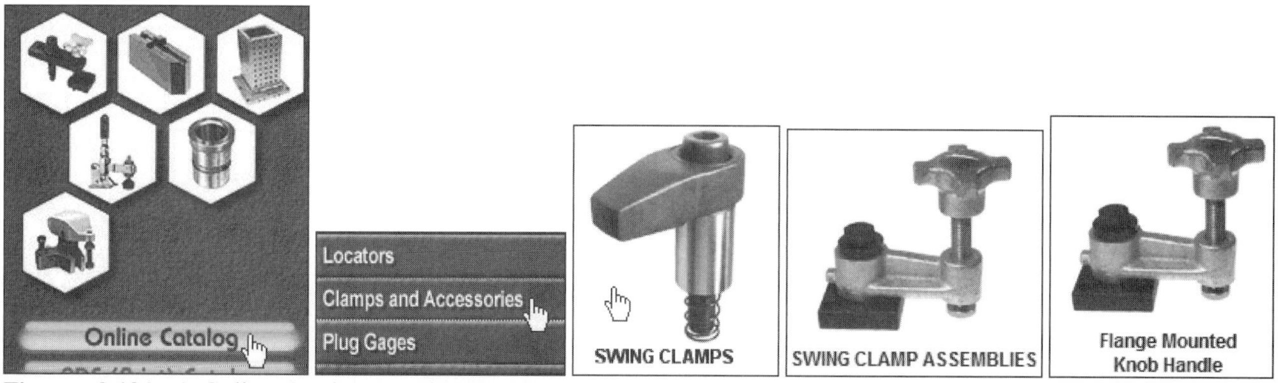

Figures 9.12(c-g) Online Catalog ⇒ Clamps and Accessories ⇒ Swing Clamps ⇒ Swing Clamp Assemblies ⇒ Flange Mounted Knob Handle

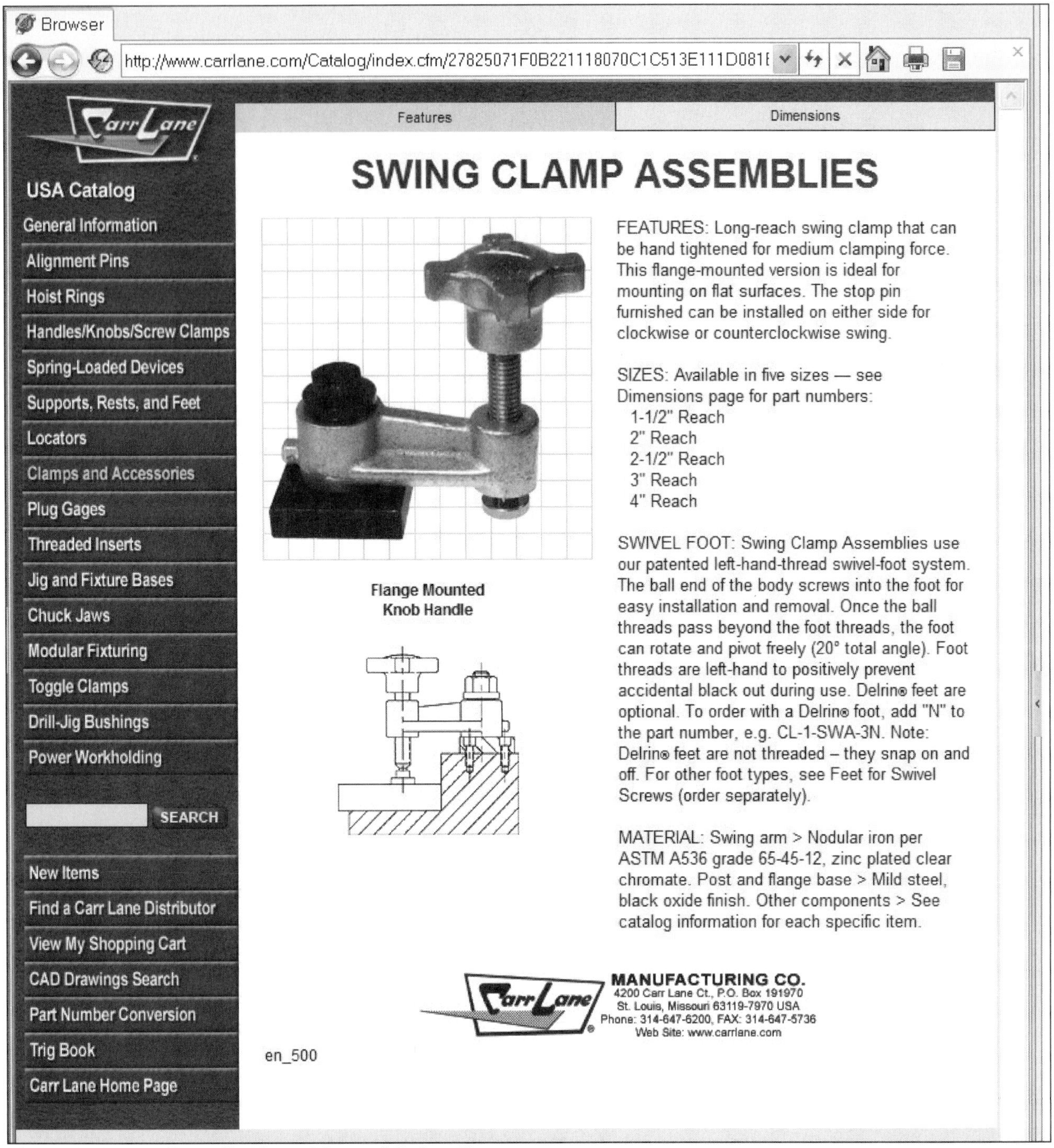

Browser

http://www.carrlane.com/Catalog/index.cfm/27825071F0B221118070C1C513E111D081E

Carr Lane

USA Catalog

- General Information
- Alignment Pins
- Hoist Rings
- Handles/Knobs/Screw Clamps
- Spring-Loaded Devices
- Supports, Rests, and Feet
- Locators
- Clamps and Accessories
- Plug Gages
- Threaded Inserts
- Jig and Fixture Bases
- Chuck Jaws
- Modular Fixturing
- Toggle Clamps
- Drill-Jig Bushings
- Power Workholding

SEARCH

- New Items
- Find a Carr Lane Distributor
- View My Shopping Cart
- CAD Drawings Search
- Part Number Conversion
- Trig Book
- Carr Lane Home Page

Features | Dimensions

SWING CLAMP ASSEMBLIES

Flange Mounted
Knob Handle

FEATURES: Long-reach swing clamp that can be hand tightened for medium clamping force. This flange-mounted version is ideal for mounting on flat surfaces. The stop pin furnished can be installed on either side for clockwise or counterclockwise swing.

SIZES: Available in five sizes — see Dimensions page for part numbers:
1-1/2" Reach
2" Reach
2-1/2" Reach
3" Reach
4" Reach

SWIVEL FOOT: Swing Clamp Assemblies use our patented left-hand-thread swivel-foot system. The ball end of the body screws into the foot for easy installation and removal. Once the ball threads pass beyond the foot threads, the foot can rotate and pivot freely (20° total angle). Foot threads are left-hand to positively prevent accidental black out during use. Delrin® feet are optional. To order with a Delrin® foot, add "N" to the part number, e.g. CL-1-SWA-3N. Note: Delrin® feet are not threaded – they snap on and off. For other foot types, see Feet for Swivel Screws (order separately).

MATERIAL: Swing arm > Nodular iron per ASTM A536 grade 65-45-12, zinc plated clear chromate. Post and flange base > Mild steel, black oxide finish. Other components > See catalog information for each specific item.

Carr Lane MANUFACTURING CO.
4200 Carr Lane Ct., P.O. Box 191970
St. Louis, Missouri 63119-7970 USA
Phone: 314-647-6200, FAX: 314-647-5736
Web Site: www.carrlane.com

en_500

Figure 9.12(h) Ball Handle Flange Mounted Swing Clamp

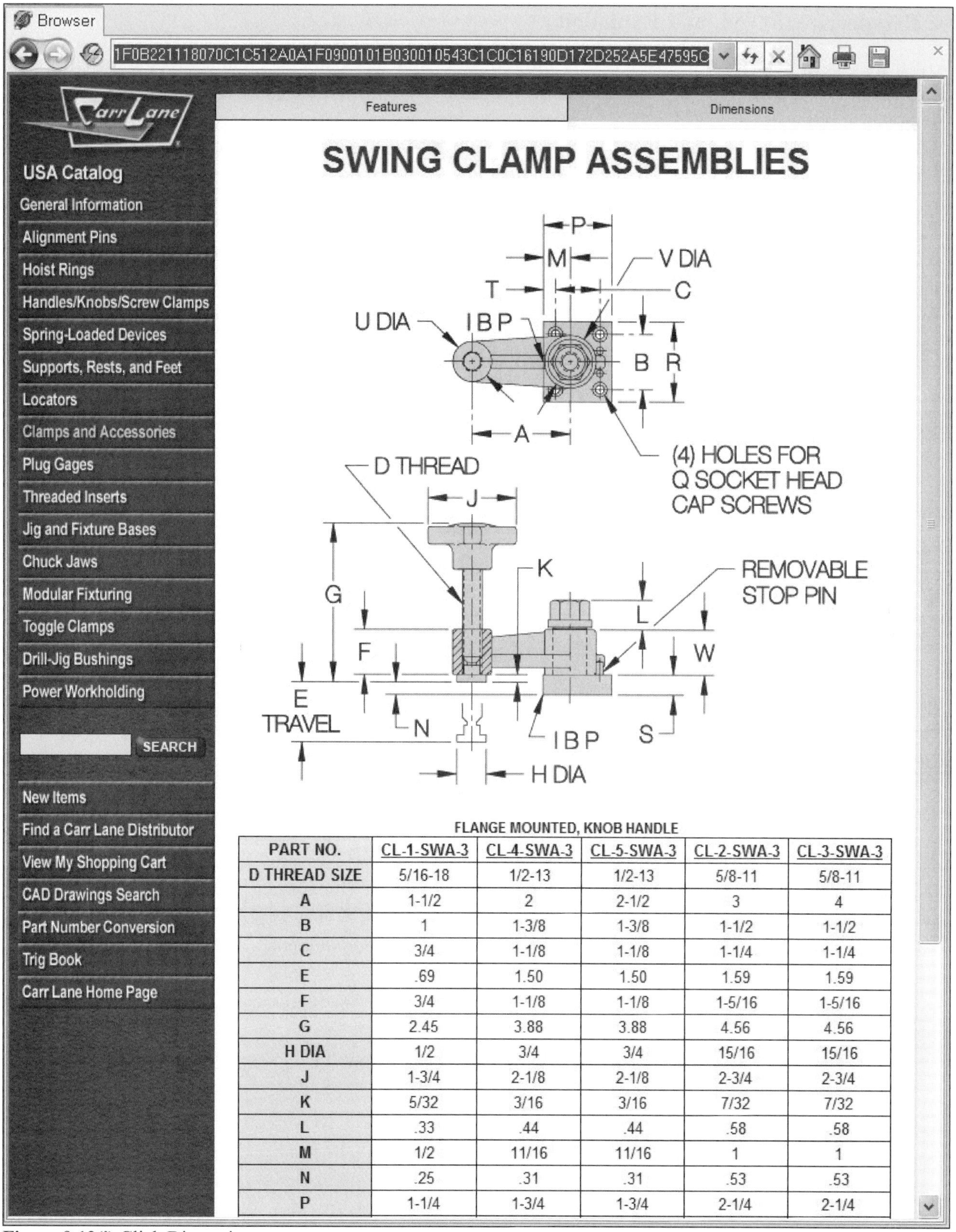

PART NO.	CL-1-SWA-3	CL-4-SWA-3	CL-5-SWA-3	CL-2-SWA-3	CL-3-SWA-3
D THREAD SIZE	5/16-18	1/2-13	1/2-13	5/8-11	5/8-11
A	1-1/2	2	2-1/2	3	4
B	1	1-3/8	1-3/8	1-1/2	1-1/2
C	3/4	1-1/8	1-1/8	1-1/4	1-1/4
E	.69	1.50	1.50	1.59	1.59
F	3/4	1-1/8	1-1/8	1-5/16	1-5/16
G	2.45	3.88	3.88	4.56	4.56
H DIA	1/2	3/4	3/4	15/16	15/16
J	1-3/4	2-1/8	2-1/8	2-3/4	2-3/4
K	5/32	3/16	3/16	7/32	7/32
L	.33	.44	.44	.58	.58
M	1/2	11/16	11/16	1	1
N	.25	.31	.31	.53	.53
P	1-1/4	1-3/4	1-3/4	2-1/4	2-1/4

Figure 9.12(i) Click Dimensions

Close the Browser shade ⇒ off ⇒ **Ctrl+D** ⇒ **Ctrl+S** ⇒ **Enter**

Views: Perspective, Saved, and Exploded

You will now create a variety of cosmetically altered view states. Cosmetic changes to the assembly do not affect the model itself, only the way it is displayed on the screen. One type of view that can be created is the perspective view.

Perspective creates a single-vanishing-point perspective view of a shaded or wireframe model. These views allow you to observe an object as the view location follows a curve, axis, cable, or edge through or around an object. To add perspective to a model view, you select a viewing path and then control the viewing position along the path in either direction. You can also rotate the perspective view in any direction, zoom the view in or out, and change the view angle at any point along the path.

Click: **View** ⇒ **Shade** ⇒ **View** ⇒ **Model Setup** ⇒ **Perspective** ⇒ adjust the slide bars (Fig. 9.13) ⇒ **Reset** ⇒ **OK** ⇒ **Saved view list** ⇒ **Standard Orientation** ⇒ ⇒ ⇒ **Help** ⇒ **Help Center**

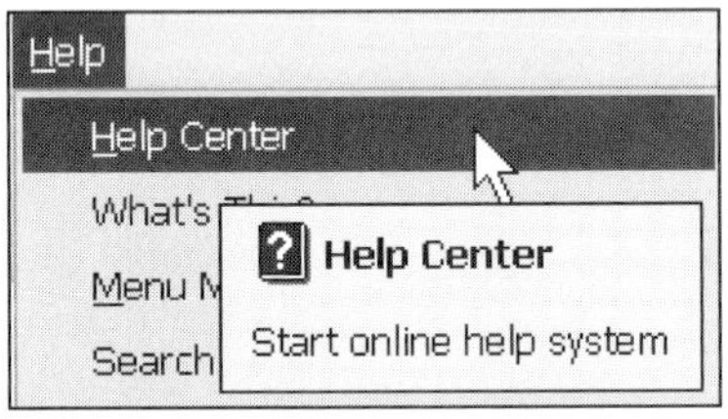

⇒ **Fundamentals** ⇒ **Pro/ENGINEER Fundamentals** ⇒ **Search** tab ⇒

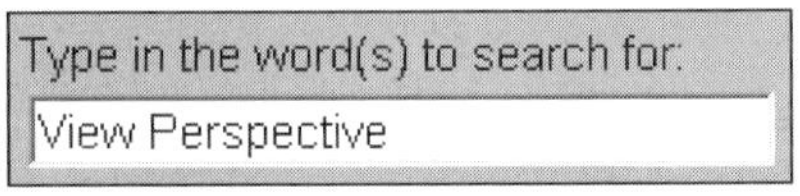

⇒ **Enter** ⇒ About Adding Perspective to a View ⇒ read the Help file ⇒

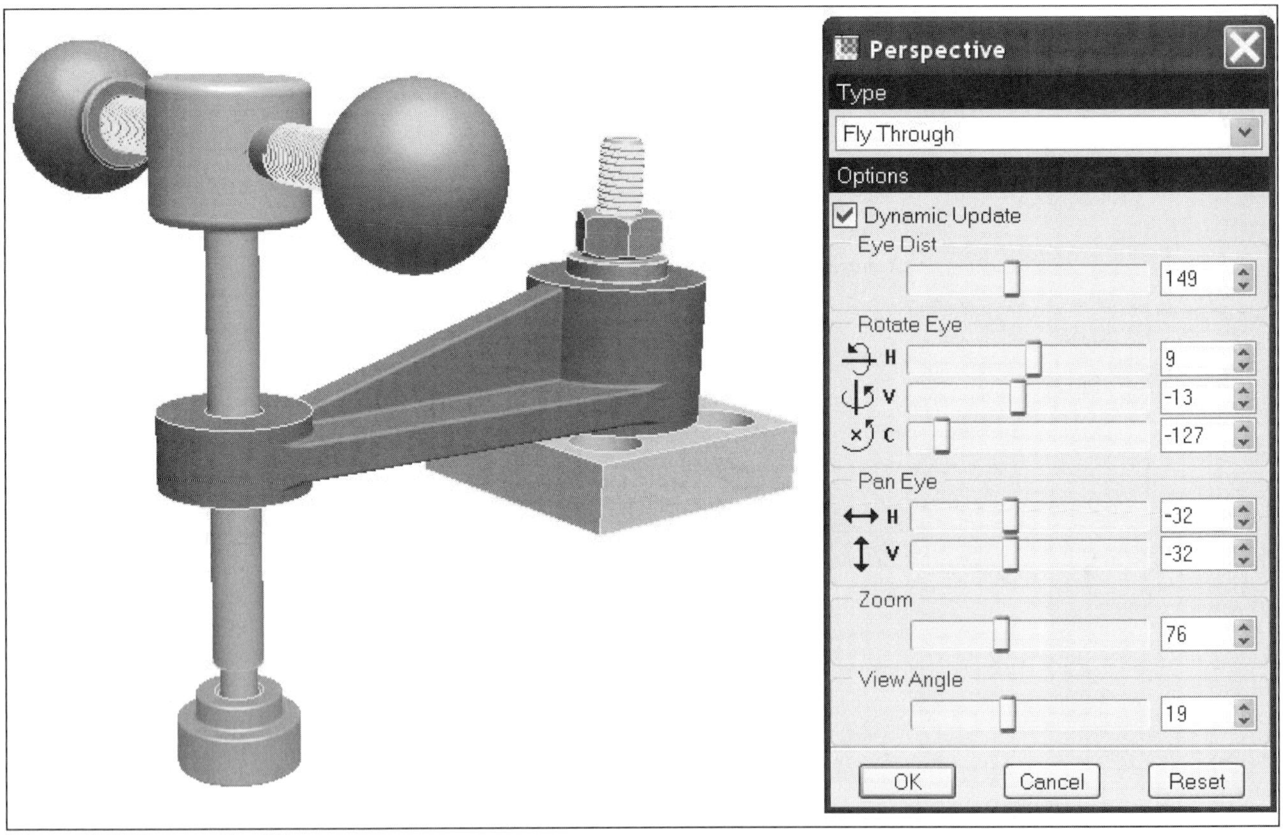

Figure 9.13 Perspective View

Saved Views

You need to create a saved view and explode state to use later on the assembly drawing. When using the View Manager to set display and explode states, it is a good idea to create one or more saved views to be used later for exploding. The default trimetric and isometric views do not adequately represent the assembly in its functional position.

Click: [toolbar icons] off ⇒ using the **MMB**, rotate the model from its default position [Fig. 9.14(a)] to one where the Clamp_Swivel is vertical [Fig. 9.14(b)] ⇒ [icon] **Reorient view** [Fig. 9.14(c)] ⇒ **Saved Views** [Fig. 9.14(d)] ⇒ Name: type **EXPLODE1** ⇒ **Save** [Fig. 9.14(e)] ⇒ **OK**

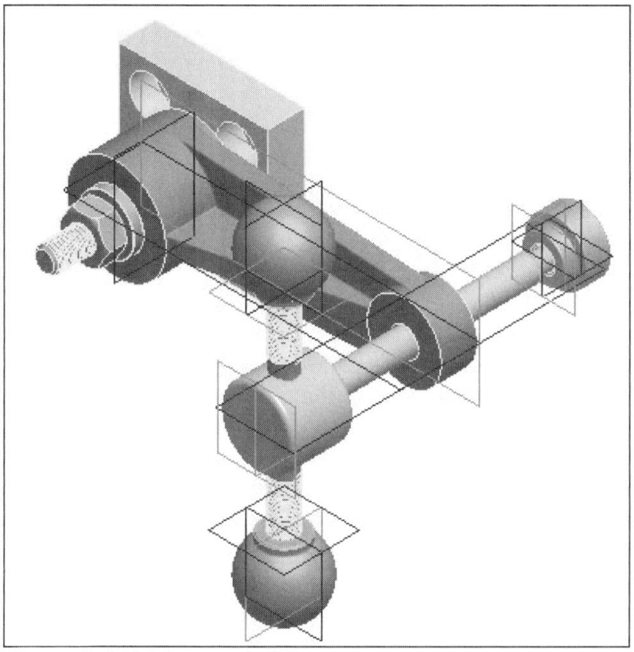

Figure 9.14(a) Isometric View

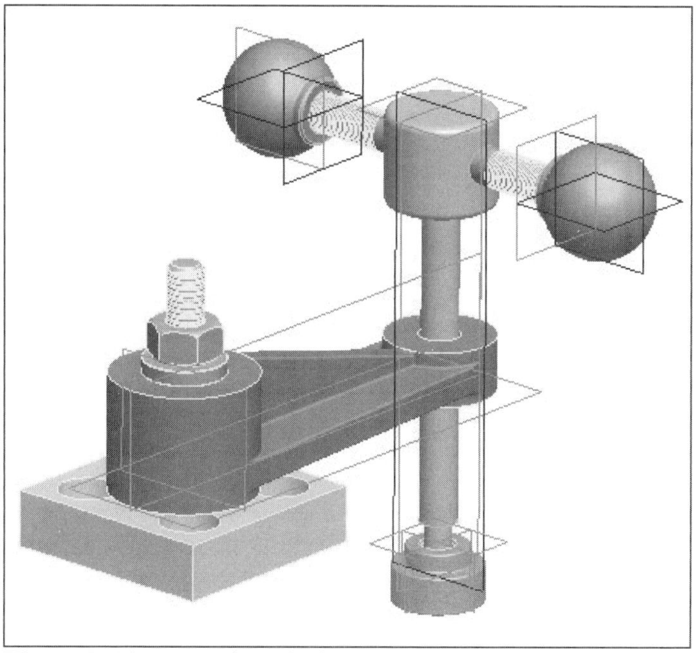

Figure 9. 14 (b) Reoriented View

Figure 9.14 (c) Orientation Dialog Box

Figure 9.14(d) Saved Views

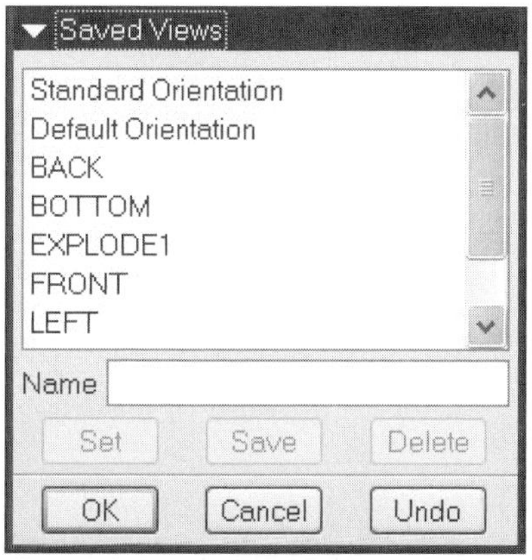

Figure 9.14(e) EXPLODE1 View

Default Exploded Views

When you create an *exploded view*, Pro/E moves apart the components of an assembly to a set default distance. The default position is seldom the most desirable.

Click: **View** ⇒ **Explode** [Fig. 9.15(a)] ⇒ **Explode View** [Fig. 9.15(b)] ⇒ **View** ⇒ **Shade** [Fig. 9.15(c)] ⇒ **View** ⇒ **Explode** ⇒ **Unexplode View** ⇒ **View** ⇒ **Orientation** ⇒ **Refit**⇒ **Ctrl+R** ⇒ **Ctrl+S** ⇒ **OK**

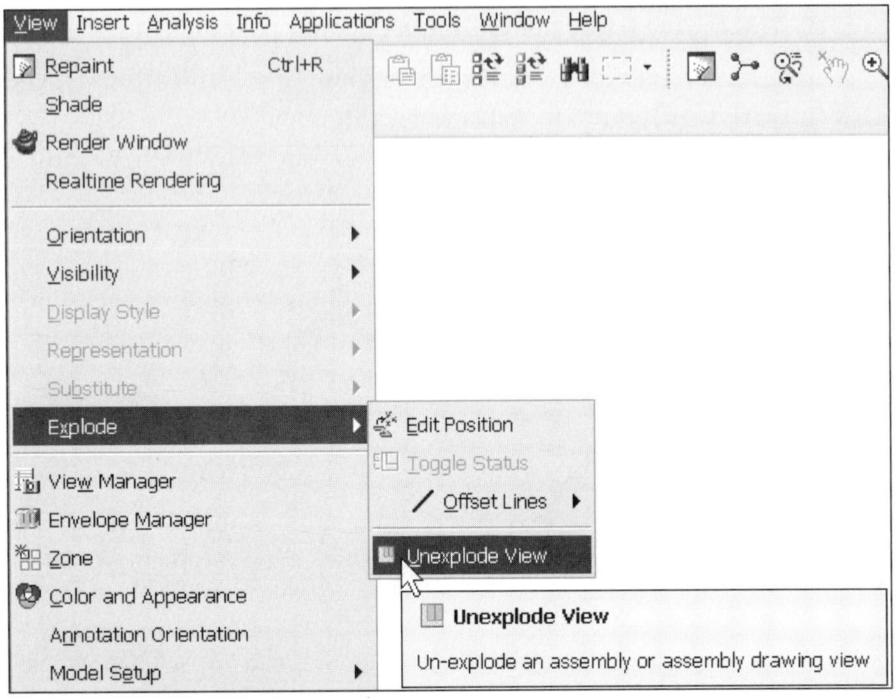

Figure 9.15(a) Explode Command

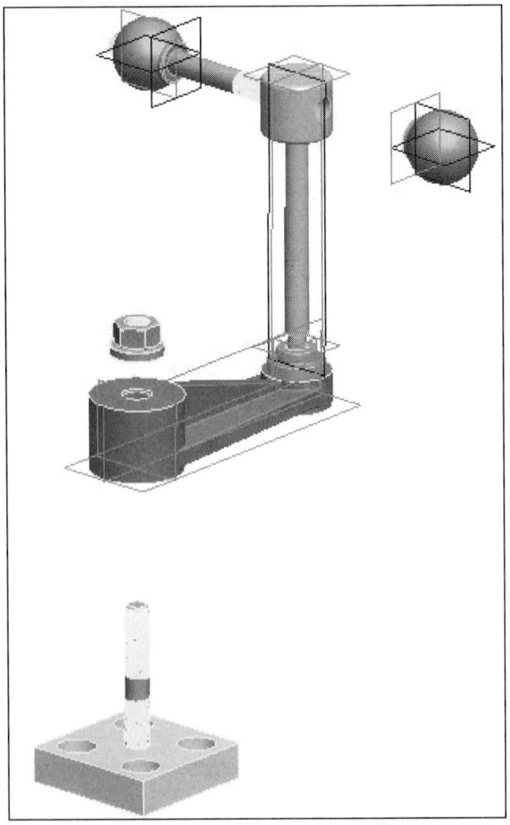

Figure 9.15(b) Default Exploded View

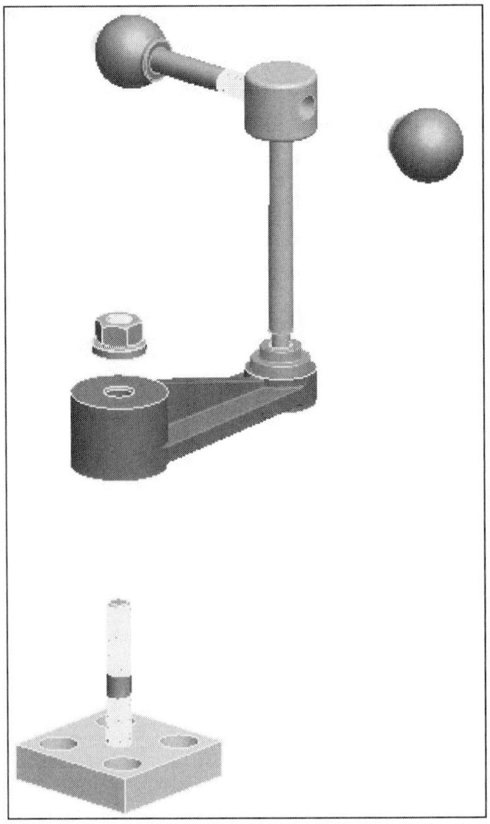

Figure 9.15(c) Shaded Default Exploded View

View Manager

The View Manager allows you to see and control the models representation [Fig. 9.16(a)], orientation [Fig. 9.16(b)], explode [Fig. 9.16(c)], style [Fig. 9.16(d)], all (combined) states [Fig. 9.16(e)], and Xsec [Fig. 9.16(f)].

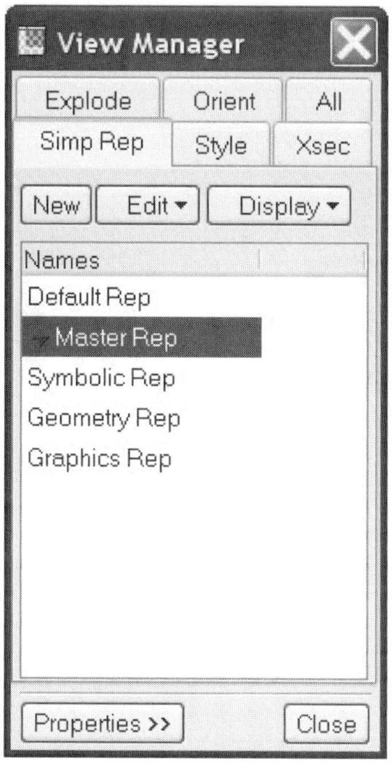

Figure 9.16(a) Representation

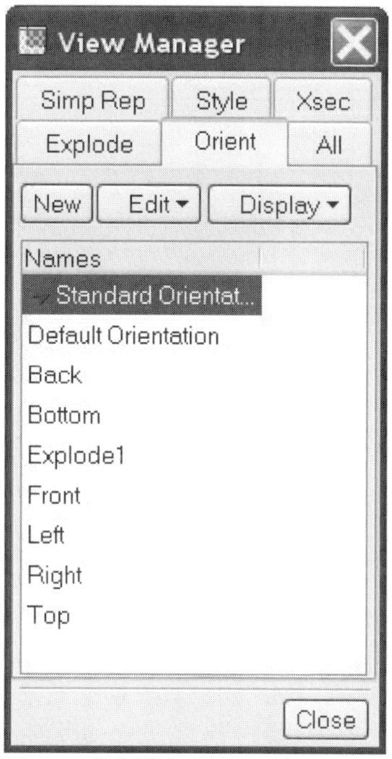

Figure 9.16(b) Orientation

Figure 9.16(c) Explode

Figure 9.16(d) Style

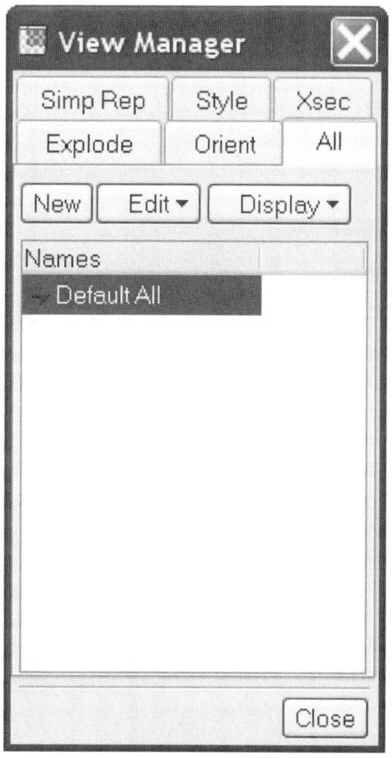

Figure 9.16(e) All (combined state)

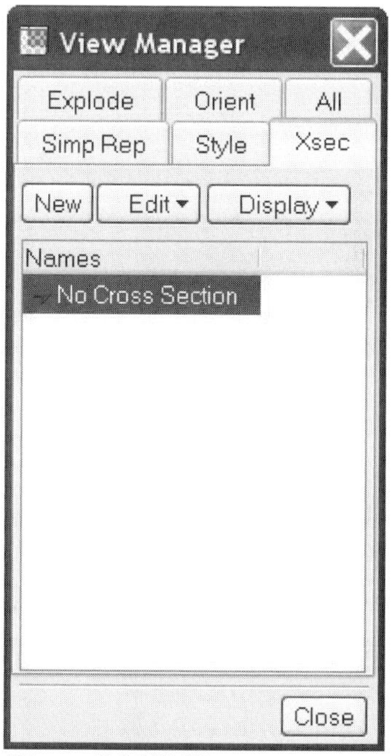

Figure 9.16(f) Xsec

Click: Start the view manager ⇒ **Simp Rep** tab ⇒ **Master Rep** ⇒ **Display** ⇒ **Set Active** ⇒ **Orient** tab ⇒ **Explode1** ⇒ **RMB** ⇒ **Save** [Fig. 9.17(a)] ⇒ **OK** ⇒ **Explode** tab ⇒ **New** ⇒ **Enter** to accept the default name: **Exp0001** [Fig. 9.17(b)] ⇒ **Properties>>** ⇒ **Edit position** [Figs. 9.17(c-d)] ⇒ on

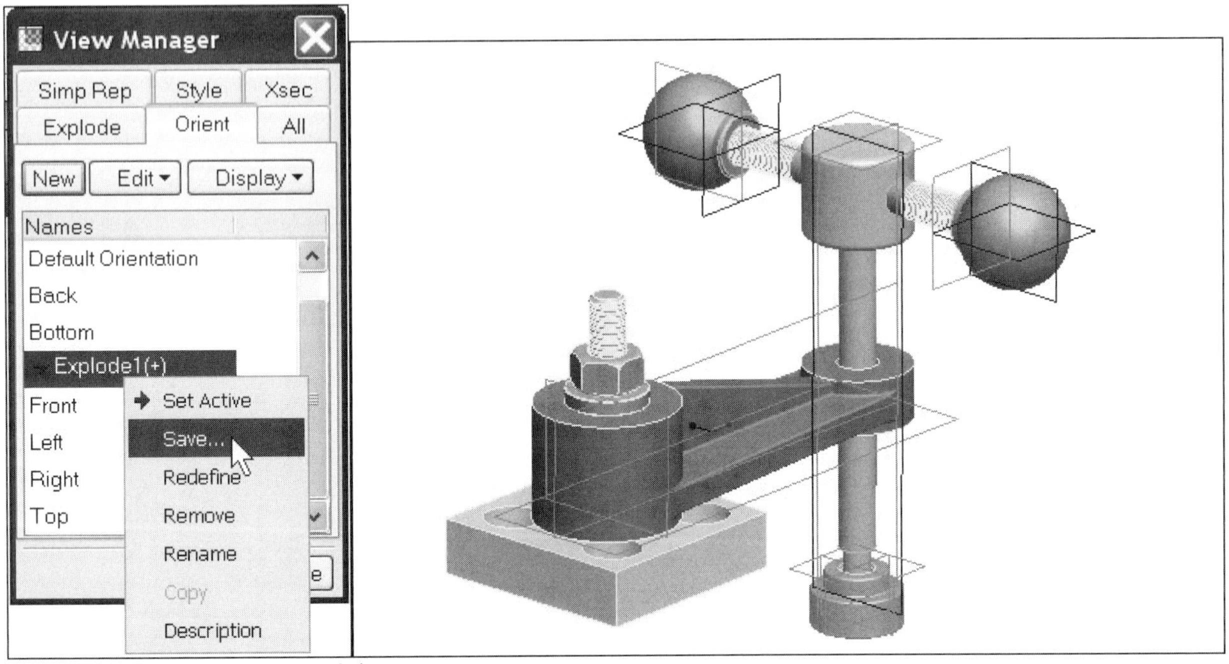

Figure 9.17(a) View Manager Orient

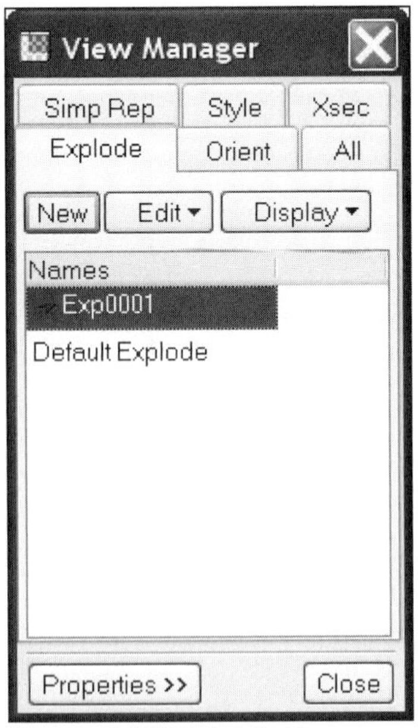

Figure 9.17(b) Set Active

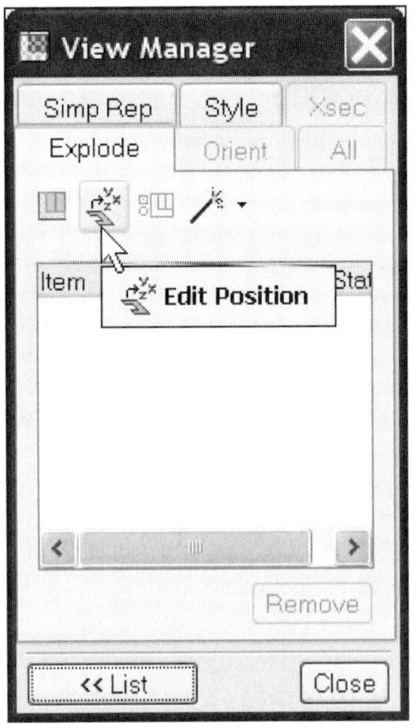

Figure 9.17(c) Edit Position

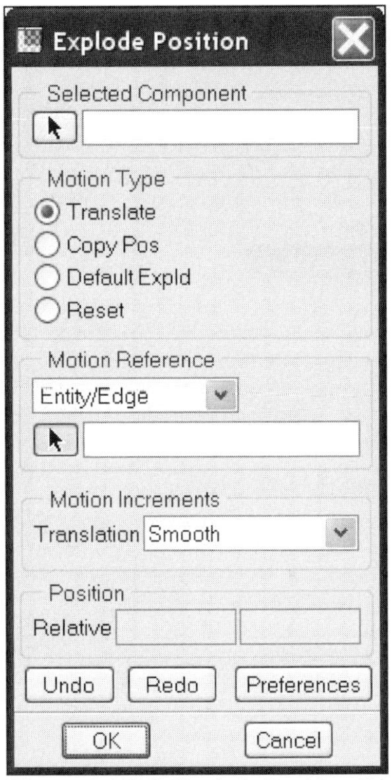

Figure 9.17(d) Explode Position Dialog

Click: Motion Reference [icon] [Figs. 9.18(a-b)] ⇒ select the vertical axis of the **CLAMP_ARM_DESIGN** [Figs. 9.18(c-d)] *or any vertical edge of the CLAMP_PLATE* ⇒ **Preferences** ⇒ **Move Many** ⇒ **Close**

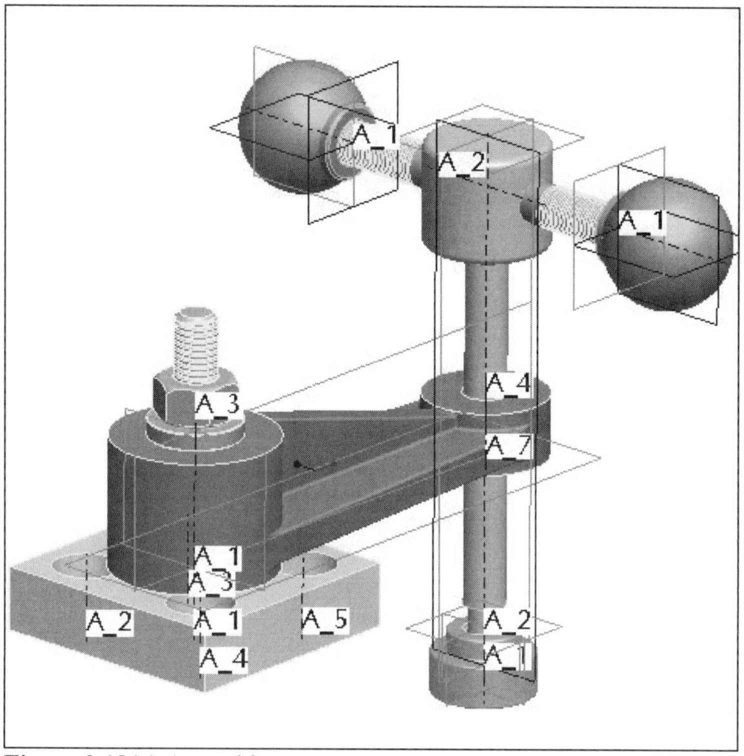

Figure 9.18(a) Assembly

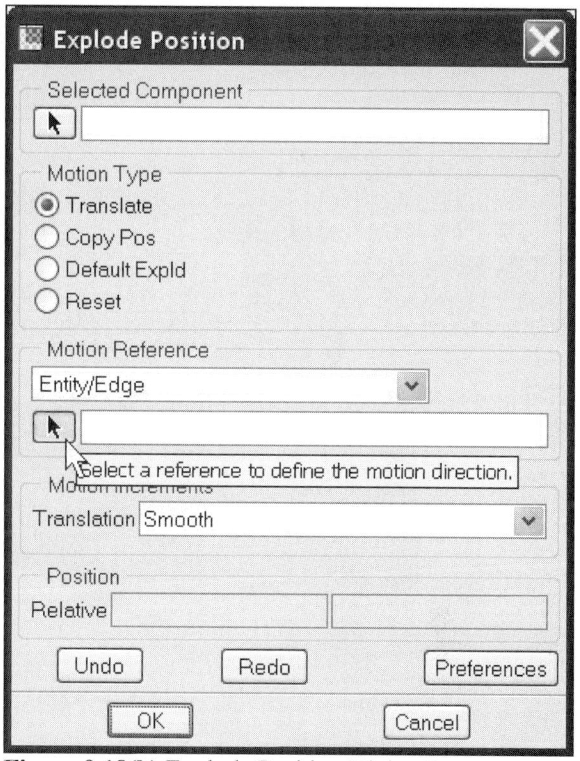

Figure 9.18(b) Explode Position Dialog Box

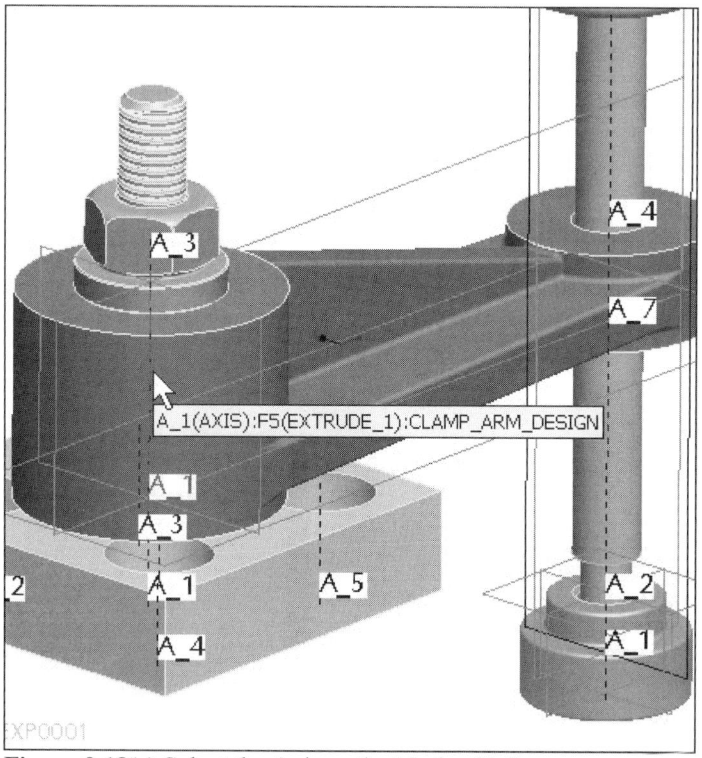

Figure 9.18(c) Select the Axis as the Motion Reference

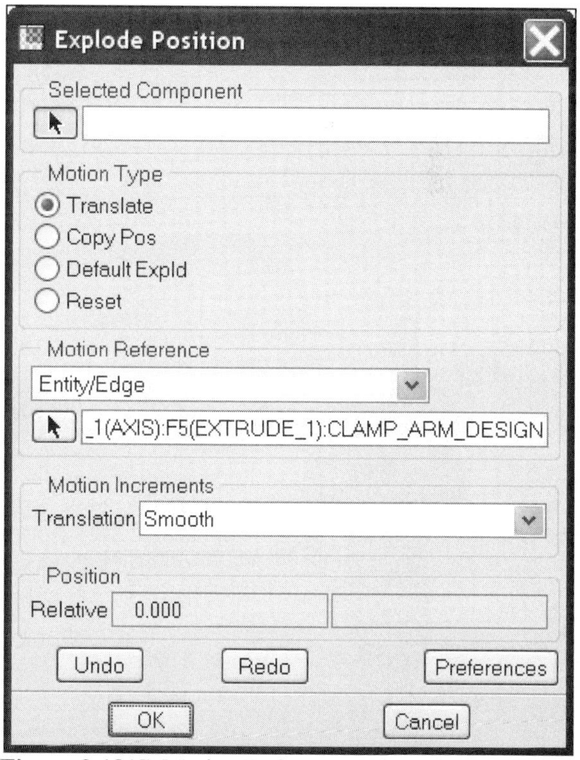

Figure 9.18(d) Motion Reference Selected

Select the **CARRLANE-12-13350_STUD** ⇒ press and hold the **Ctrl** key and select **CARRLANE_500_FN** [Fig. 9.18(e)] ⇒ **MMB** ⇒ click **LMB** in the graphics window and start dragging [Fig. 9.18(f)] ⇒ click **LMB** to place

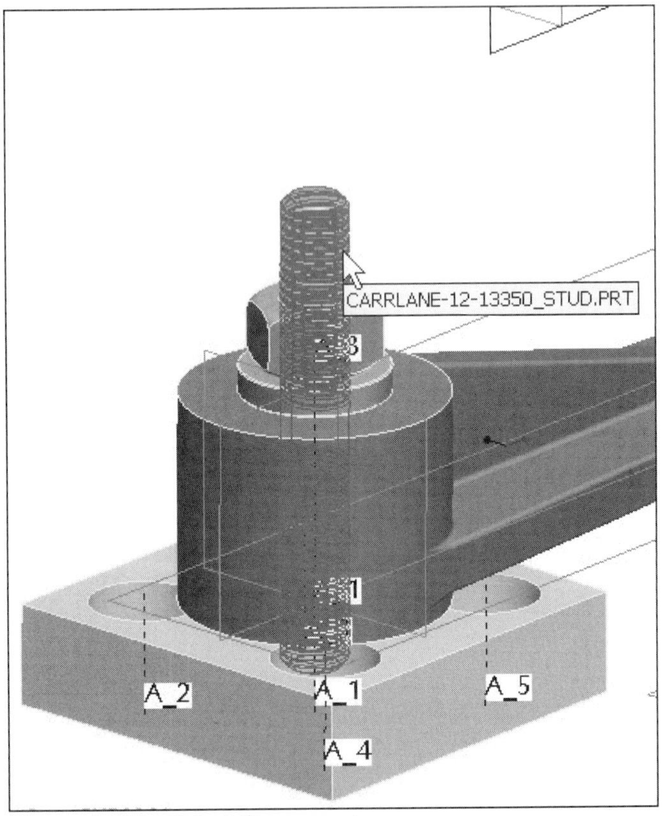

Figure 9.18(e) Select Components

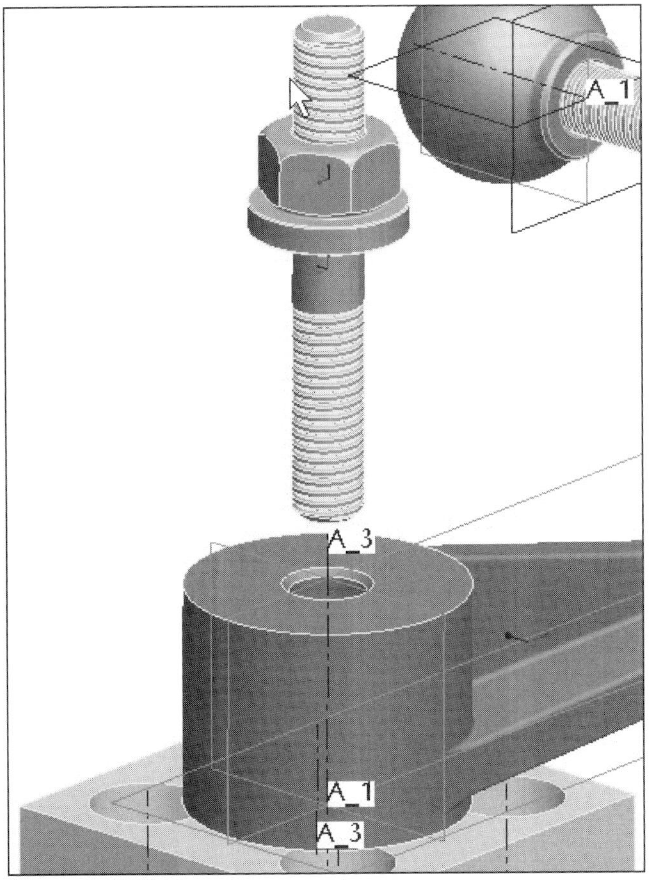

Figure 9.18(f) Drag Components

Select **CARRLANE_500_FN** [Fig. 9.18(g)] ⇒ **MMB** ⇒ click **LMB** in the graphics window and start dragging [Fig. 9.18(h)] ⇒ click **LMB** to place

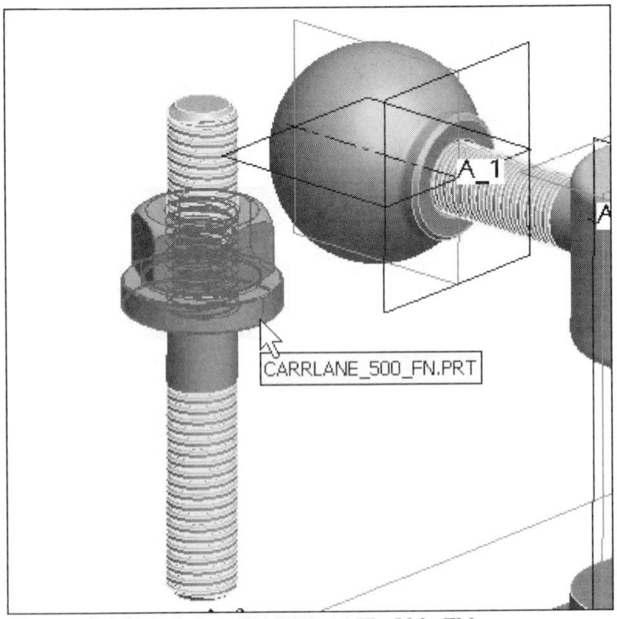

Figure 9.18(g) Select CARRLANE_500_FN

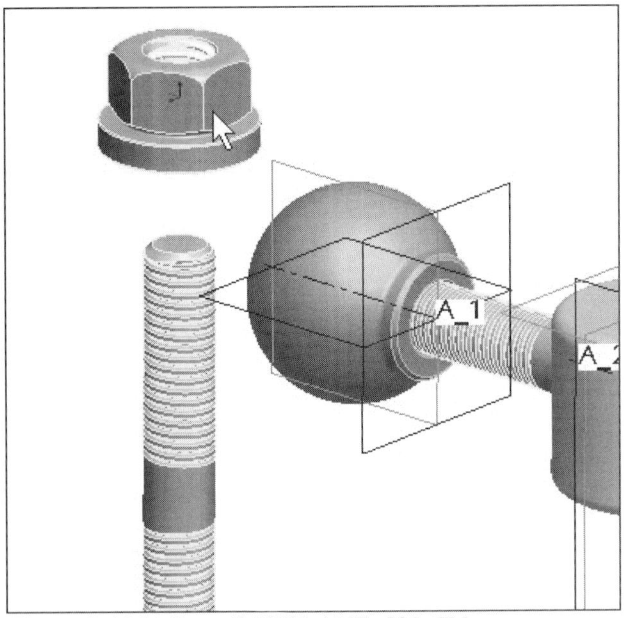

Figure 9.18(h) Drag CARRLANE_500_FN

Select **CLAMP_PLATE** [Fig. 9.18(i)] ⇒ **MMB** ⇒ click **LMB** in the graphics window and start dragging [Fig. 9.18(j)] ⇒ click **LMB** to place ⇒ **Preferences** ⇒ **Move With Children** ⇒ **Close** ⇒ select **CLAMP_SWIVEL** [Fig. 9.18(k)] ⇒ start dragging [Fig. 9.18(l)] ⇒ click **LMB** to place

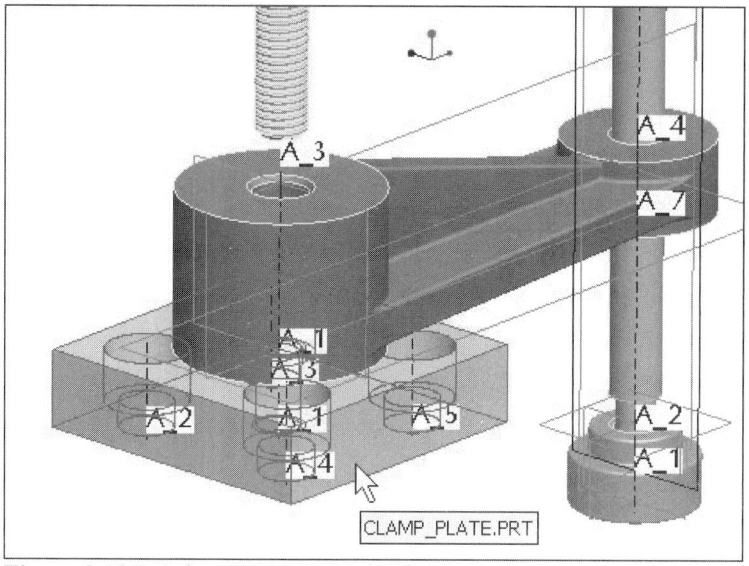

Figure 9.18(i) Select CLAMP_PLATE

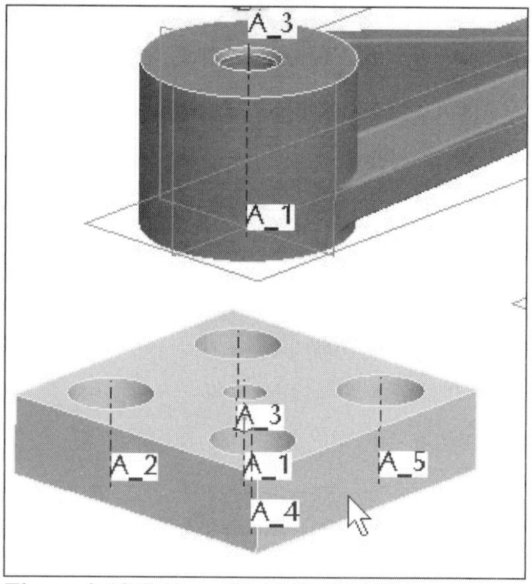

Figure 9.18(j) Drag CLAMP_PLATE

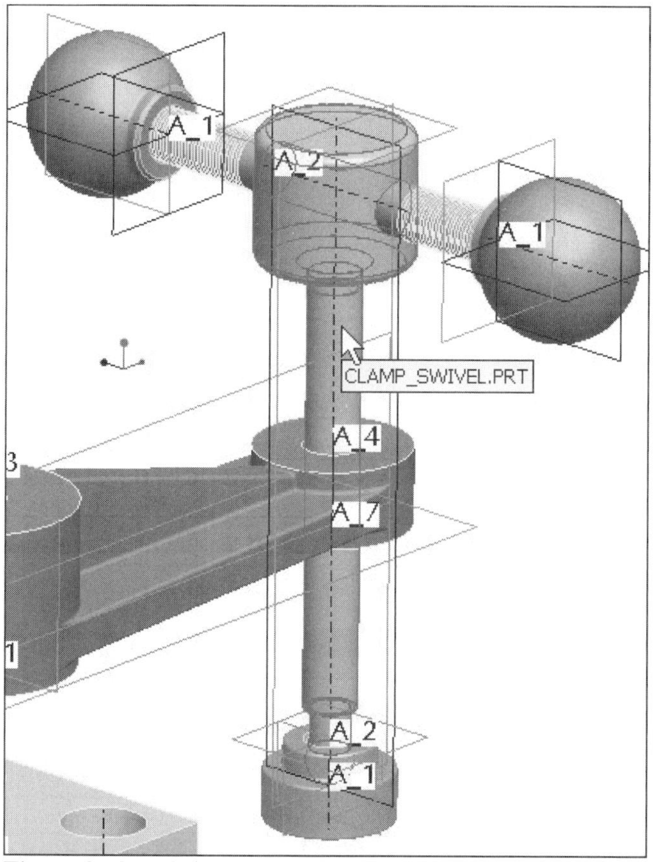

Figure 9.18(k) Select CLAMP_SWIVEL

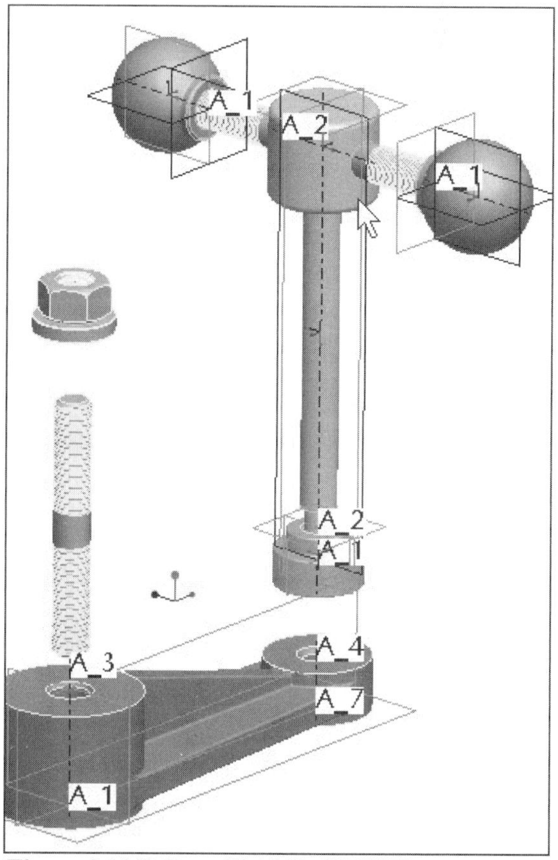

Figure 9.18(l) Drag CLAMP_SWIVEL

Click: **Preferences** ⇒ **Move One** ⇒ **Close** ⇒ select **CLAMP_FOOT** [Fig. 9.18(m)] ⇒ start dragging [Fig. 9.18(n)] ⇒ click **LMB** to place

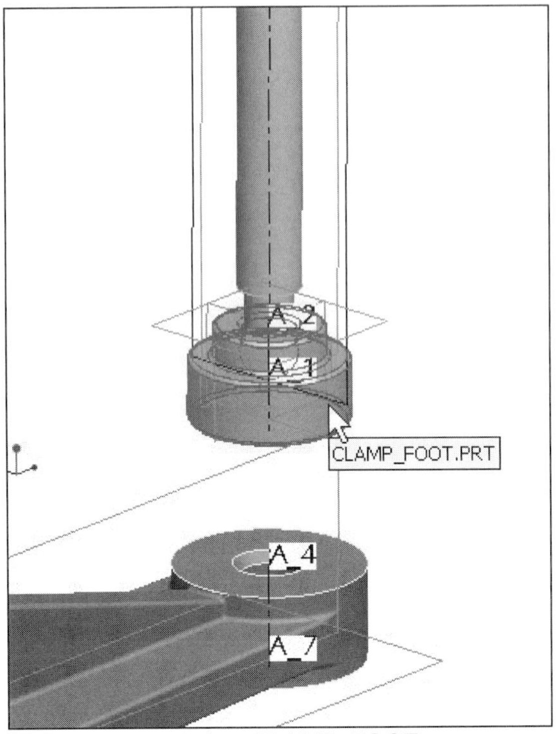

Figure 9.18(m) Select CLAMP_FOOT

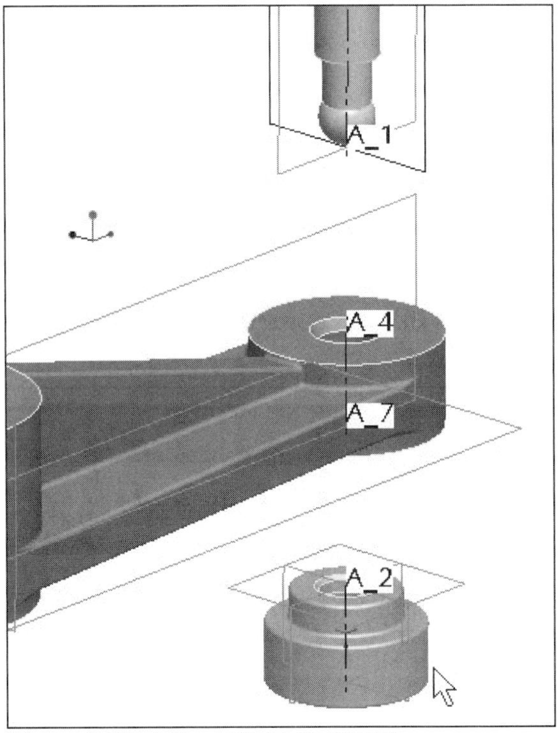

Figure 9.18(n) Drag CLAMP_FOOT

Motion Reference [icon] ⇒ select the horizontal edge of the CLAMP_PLATE [Fig. 9.18(o)] ⇒ select **CLAMP_BALL** ⇒ start dragging ⇒ click **LMB** to place [Fig. 9.18(p)] ⇒ select the second **CLAMP_BALL** ⇒ start dragging ⇒ click **LMB** to place [Fig. 9.18(q)] ⇒ select the **CARRLANE-12-13500_STUD** ⇒ start dragging ⇒ click **LMB** to place [Fig. 9.18(r)] ⇒ **OK** ⇒ [icon]

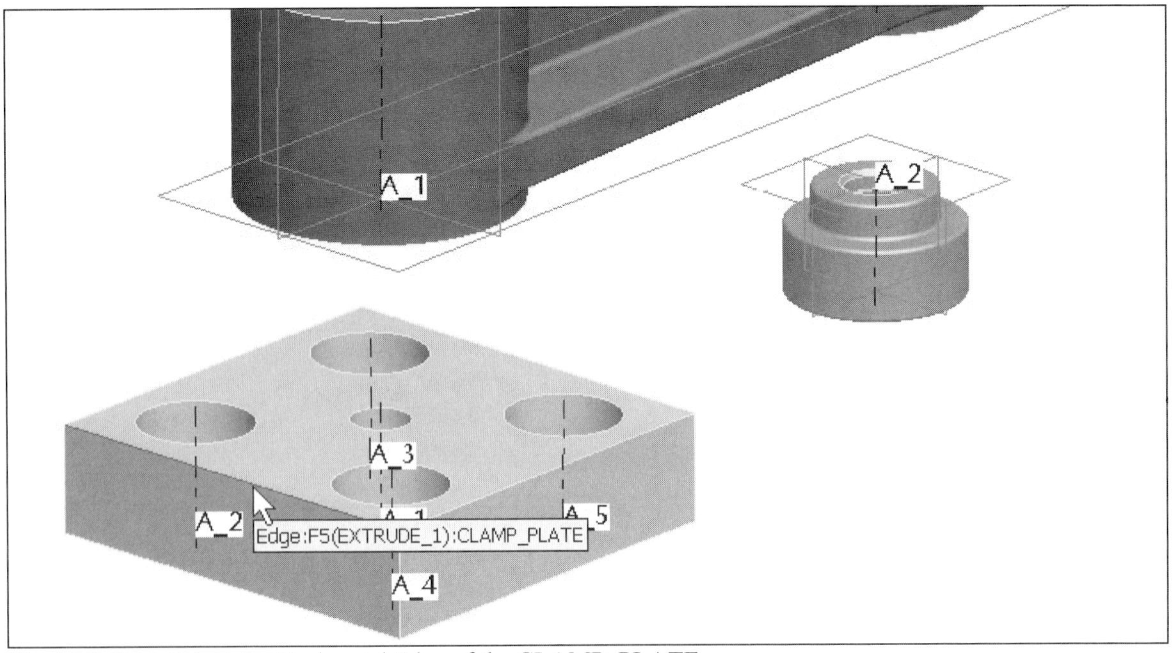

Figure 9.18(o) Select the Horizontal Edge of the CLAMP_PLATE

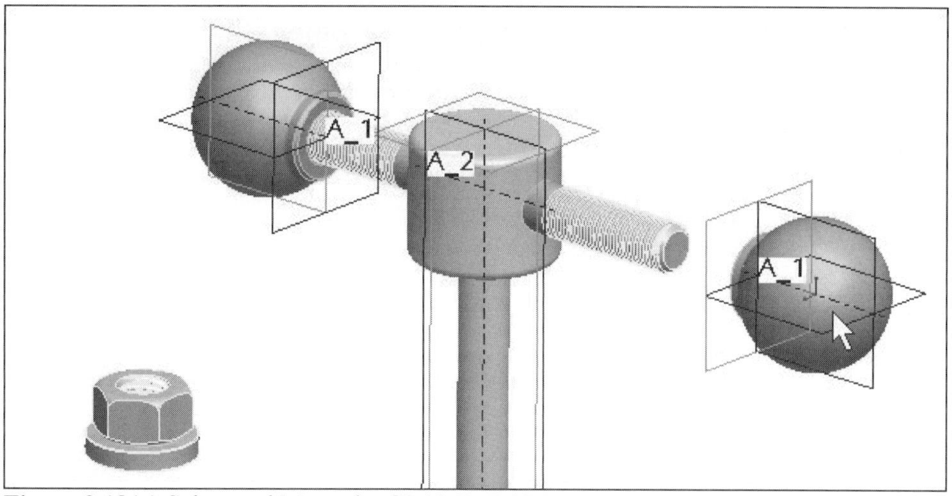

Figure 9.18(p) Select and Move the CLAMP_BALL

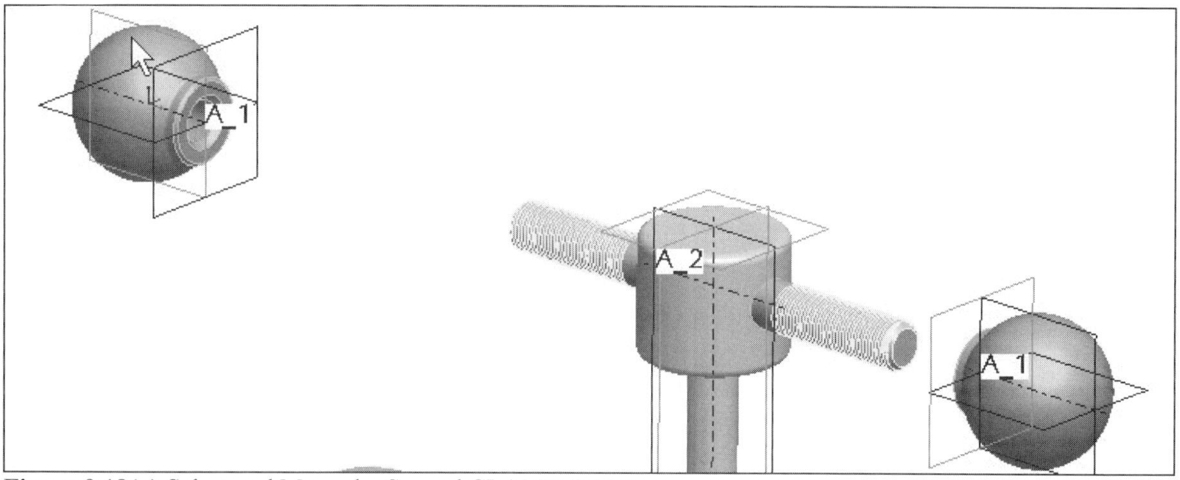

Figure 9.18(q) Select and Move the Second CLAMP_BALL

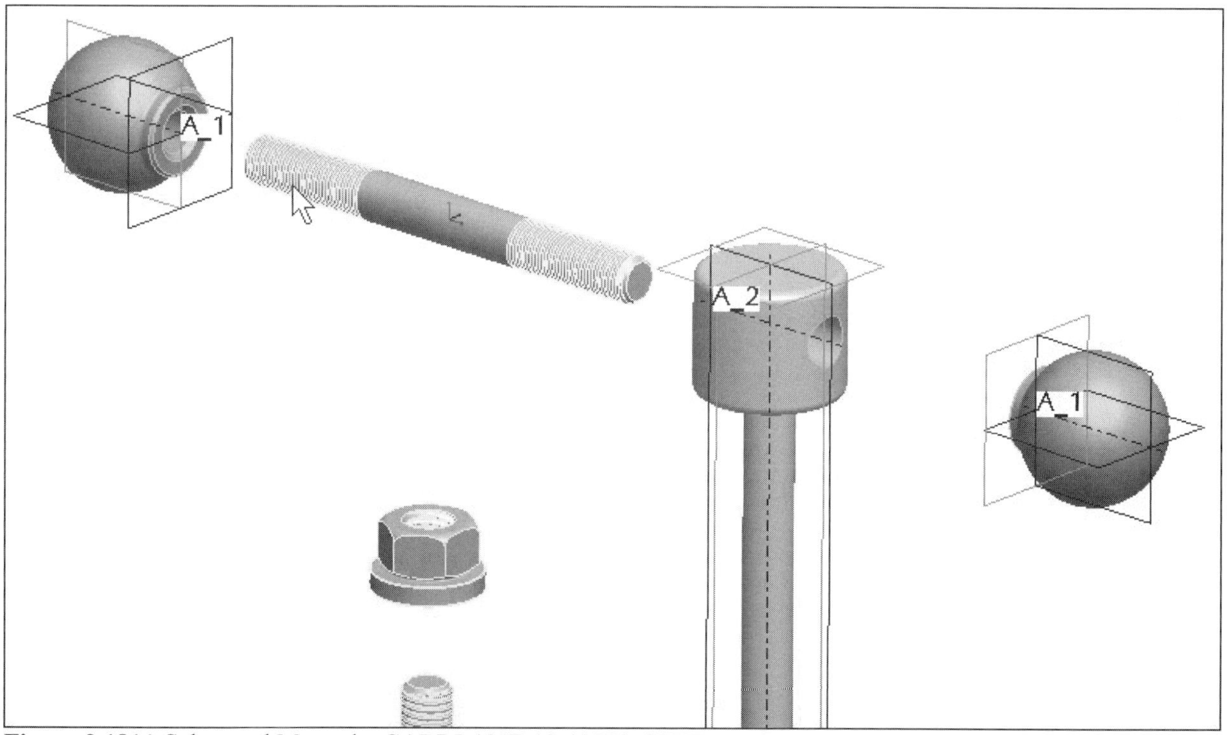

Figure 9.18(r) Select and Move the CARRLANE-12-13500_STUD

Click: <<**List** ⇒ **RMB** ⇒ **Save** [Fig. 9.18 (s)] ⇒ **OK** ⇒ **Close** ⇒ **Ctrl+S** ⇒ **Enter**

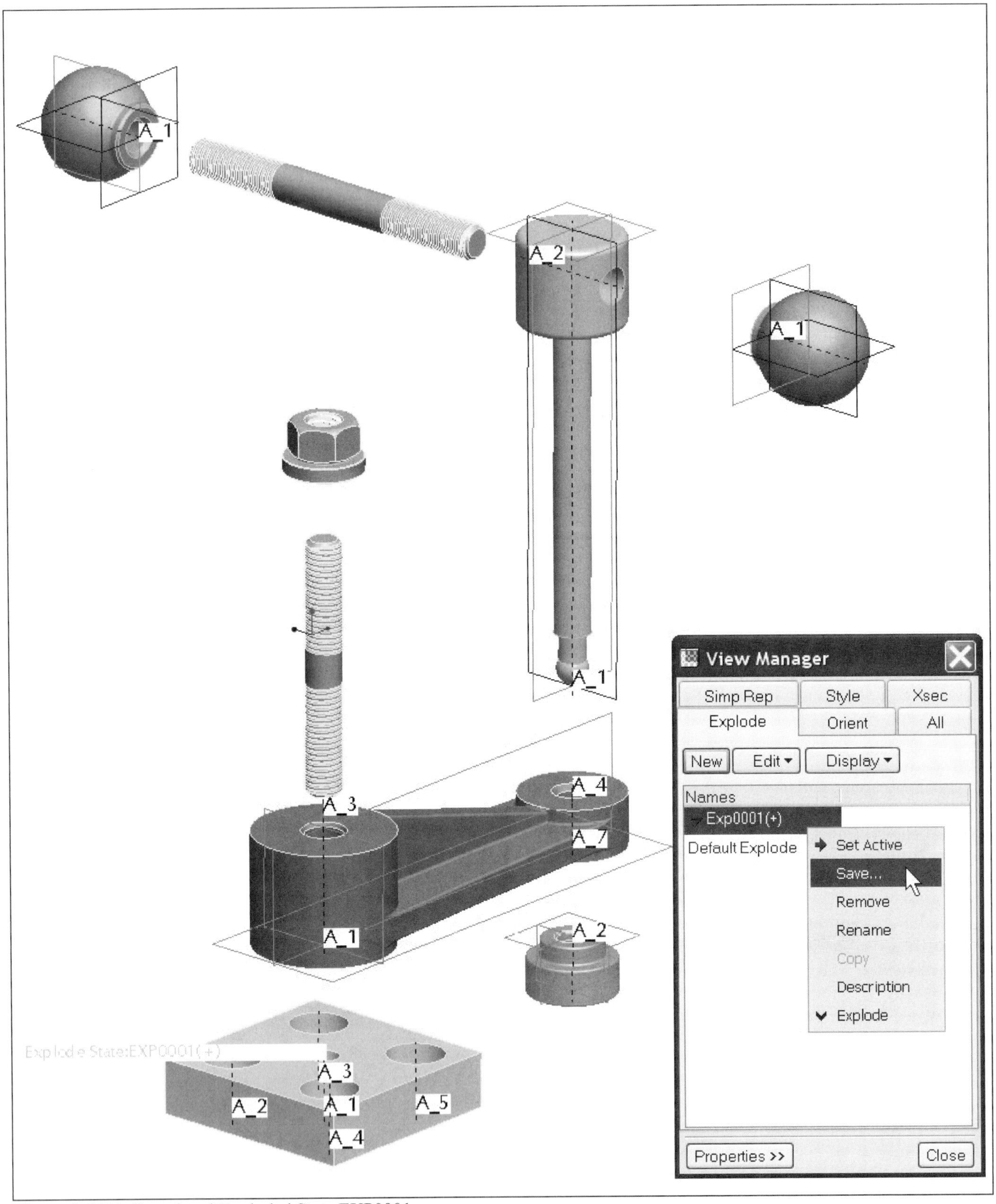

Figure 9.18(s) Completed Exploded State EXP0001

View Style

The components of an assembly (whether exploded or not) can be displayed individually with **Wireframe**, **Hidden Line**, **No Hidden**, **Shading**, or **Transparent**. Style is used to manage the display styles of an assembly's components. The components will be displayed according to their assigned display style in the current style state. The setting, in the Environment menu, controls the display style of unassigned components and they appear according to the current mode of the environment.

Click: off ⇒ **Start the view manager** from Top Toolchest ⇒ **Style** tab ⇒ **New** ⇒ **Enter** [Fig. 9.19(a)] to accept the default name: Style0001 ⇒ **Show** tab [Fig. 9.19(b)] ⇒ Wireframe ⇒ select the **CLAMP_SWIVEL** and the **CARRLANE-12-13350_STUD** ⇒ **Update**

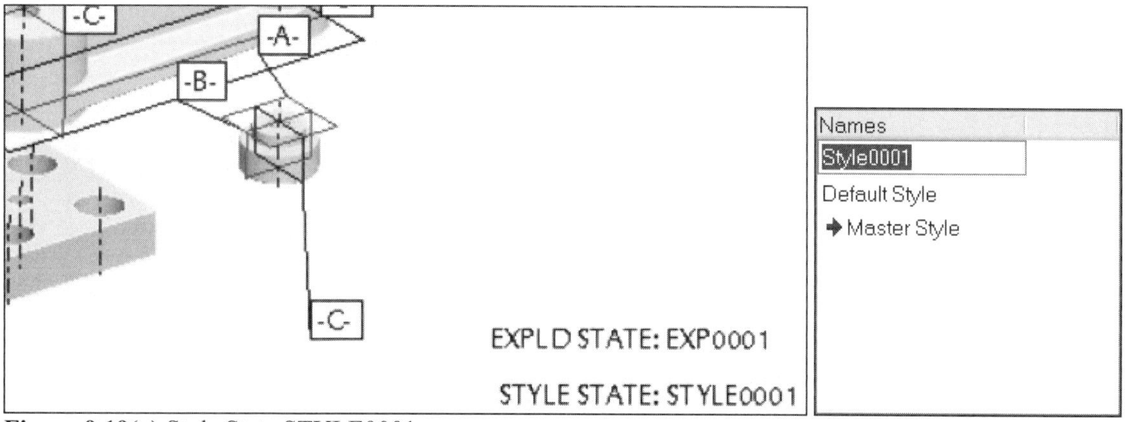

Figure 9.19(a) Style State STYLE0001

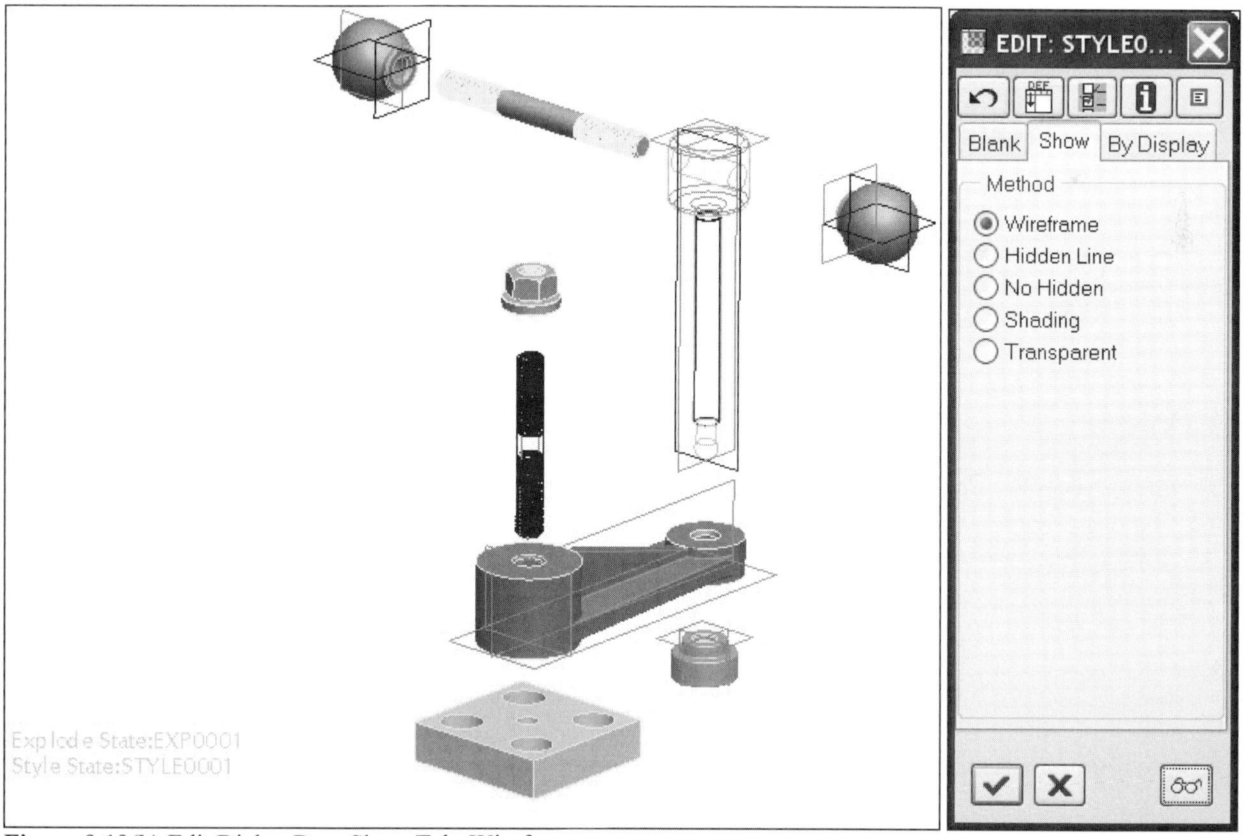

Figure 9.19(b) Edit Dialog Box, Show Tab, Wireframe

Click: Hidden Line ⇒ select the **CLAMP_PLATE** and the **CLAMP_FOOT** [Fig. 9.19(c)] ⇒  **Update Model View According To Specified Changes**

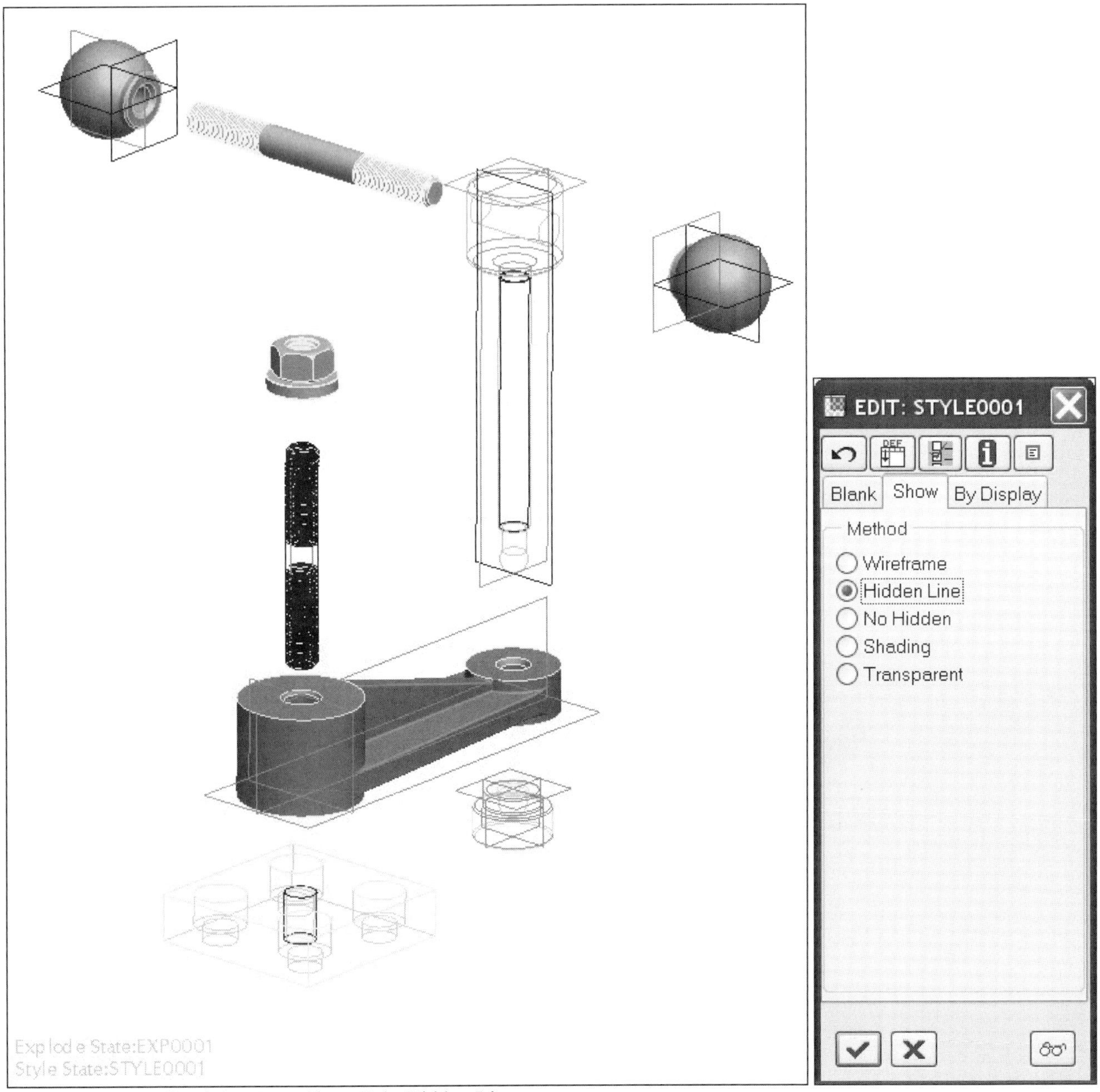

Figure 9.19(c) Edit Dialog Box, Show Tab, Hidden Line

Click: No Hidden ⇒ select the two **CLAMP_BALL** components ⇒ **Update** ⇒ Shading ⇒ select **CARRLANE_500_FN** and **CARRLANE-12-13500_STUD** ⇒ **Update Model View According To Specified Changes** [Fig. 9.19(d)]

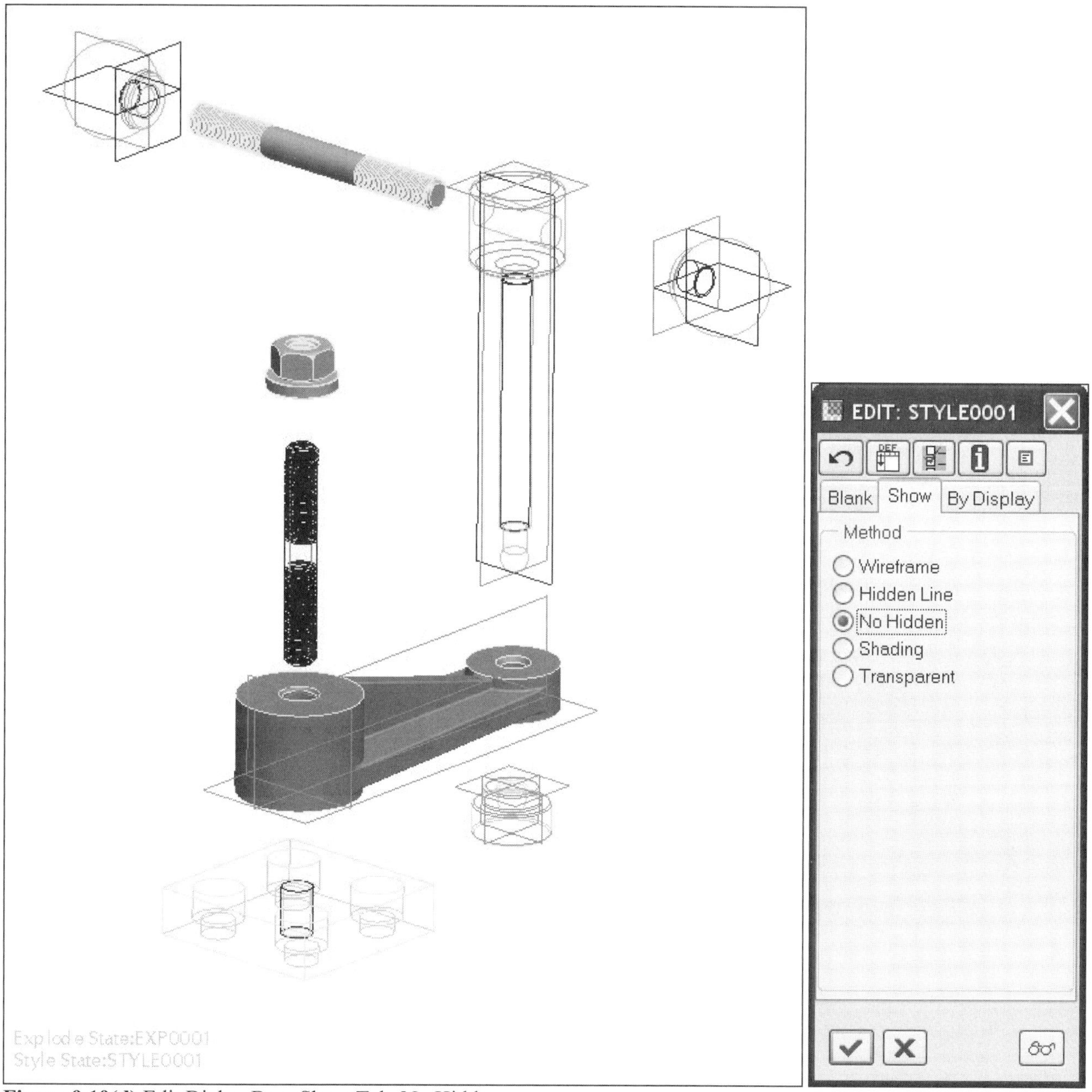

Figure 9.19(d) Edit Dialog Box, Show Tab, No Hidden

Click: 

⇒ select **CLAMP_ARM_DESIGN** ⇒  **Update Model View According To Specified Changes** [Fig. 9.19(e)] ⇒ 

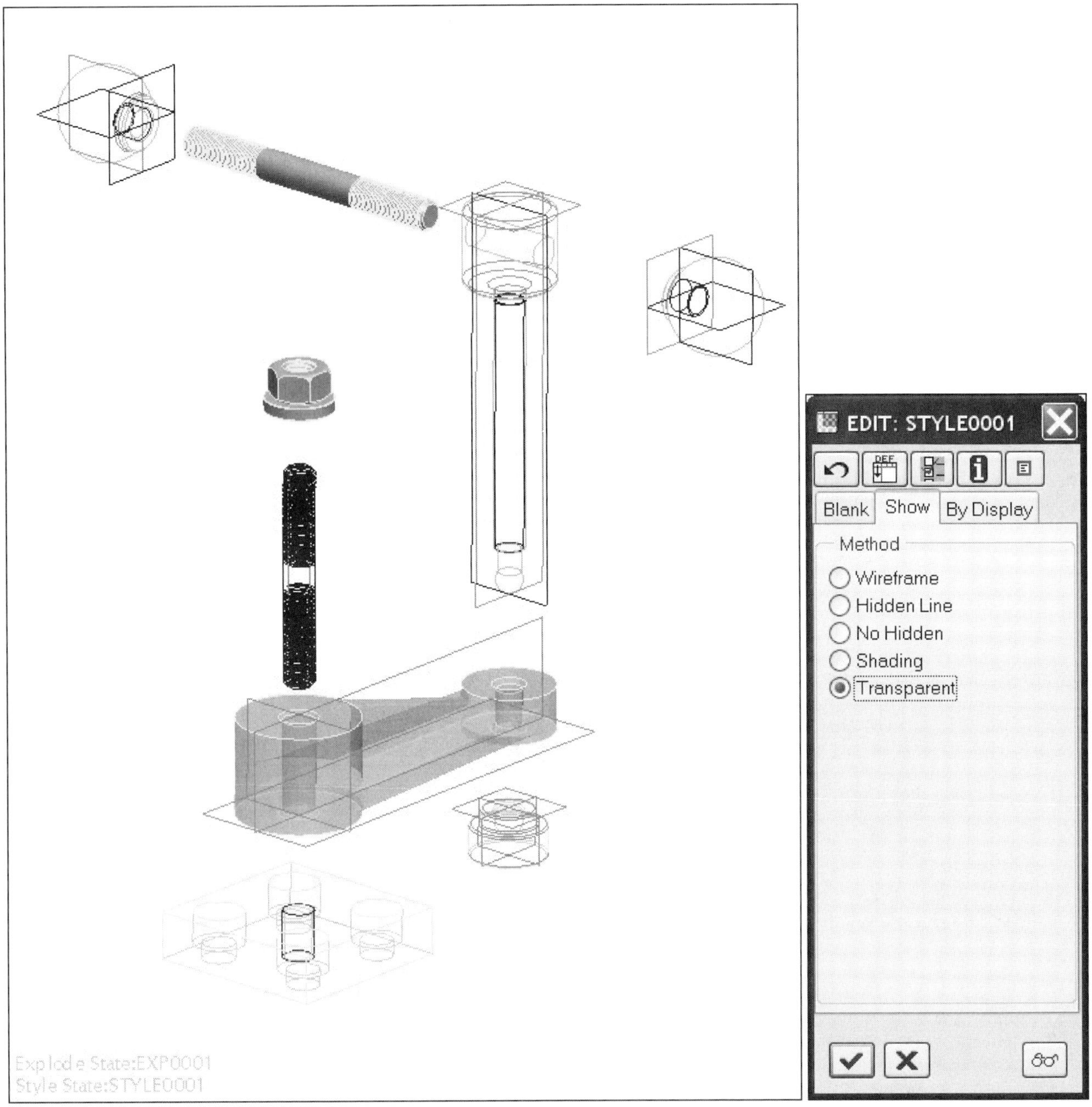

Figure 9.19(e) Completed Style State

Click: **Orient** tab ⇒ **New** ⇒ **View0001** ⇒ **Enter** [Fig. 9.19(f)] ⇒ **All** tab ⇒ **New** ⇒ **Enter** to accept the default name: **Comb0001** [Fig. 9.19(g)] ⇒ **Reference Originals** 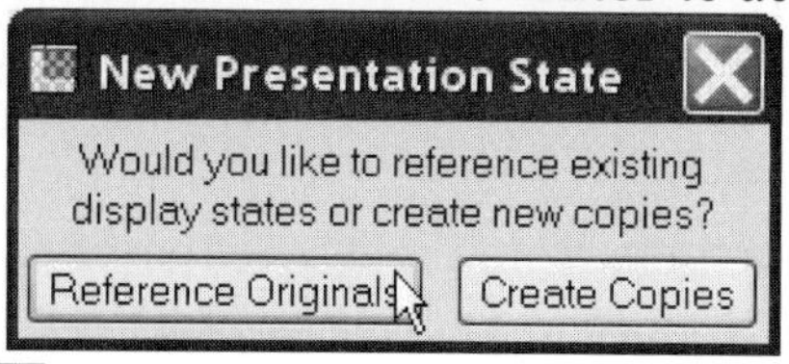

⇒ **Edit** ⇒ **Preferences** [Fig. 9.19(h)] ⇒ **OK** [Fig. 9.19(i)] ⇒ **Close** ⇒  ⇒ **MMB**

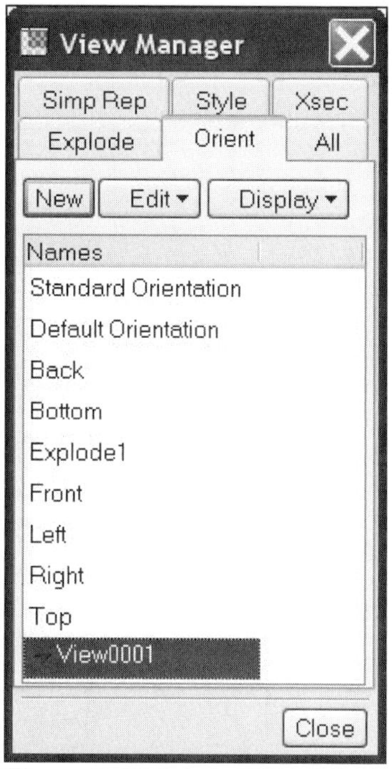

Figure 9.19(f) Orient Tab

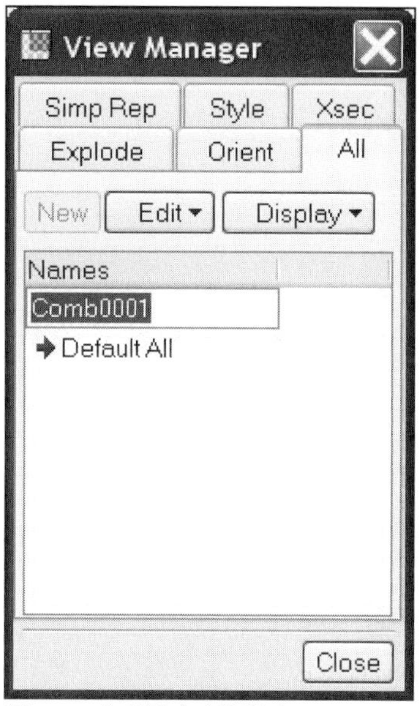

Figure 9.19(g) All Tab

Figure 9.19(h) Preferences

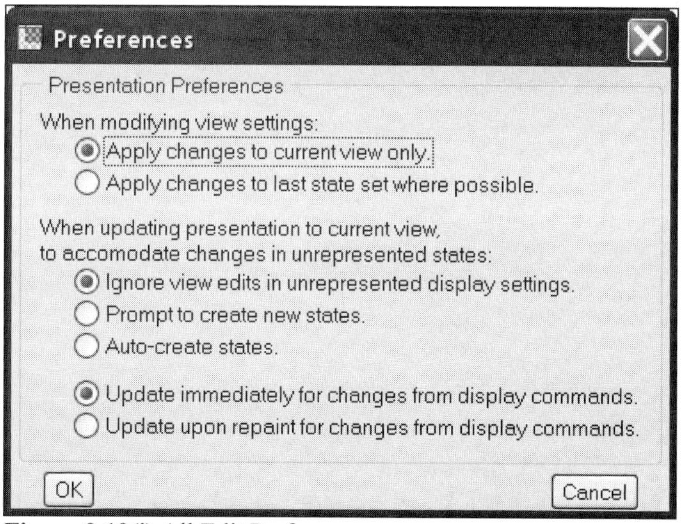

Figure 9.19(i) All Edit Preferences

Model Tree

You can display component style states, and explode states in the Model Tree.

Click: **Settings** from the Navigator ⇒ **Tree Columns** ⇒ Type: click [Fig. 9.20(a)] ⇒ **Display Styles** ⇒ **STYLE0001** ⇒ >> [Fig. 9.20(b)] ⇒ **Apply** ⇒ **OK** ⇒ adjust your column format [Fig. 9.20(c)]

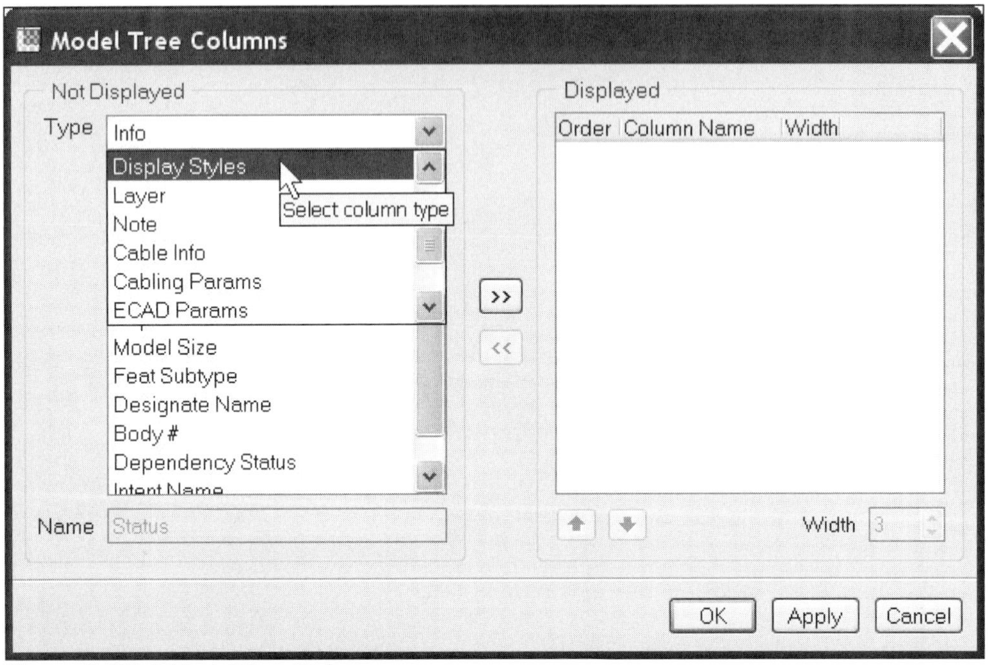

Figure 9.20(a) Model Tree Columns

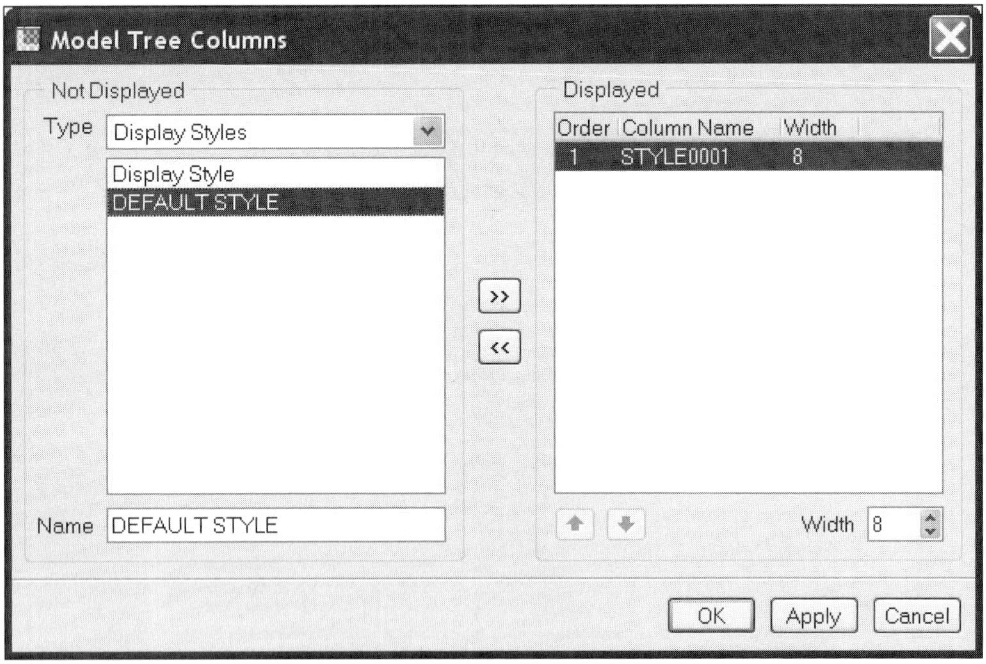

Figure 9.20(b) Adding Display Style STYLE0001 to the Model Tree Columns

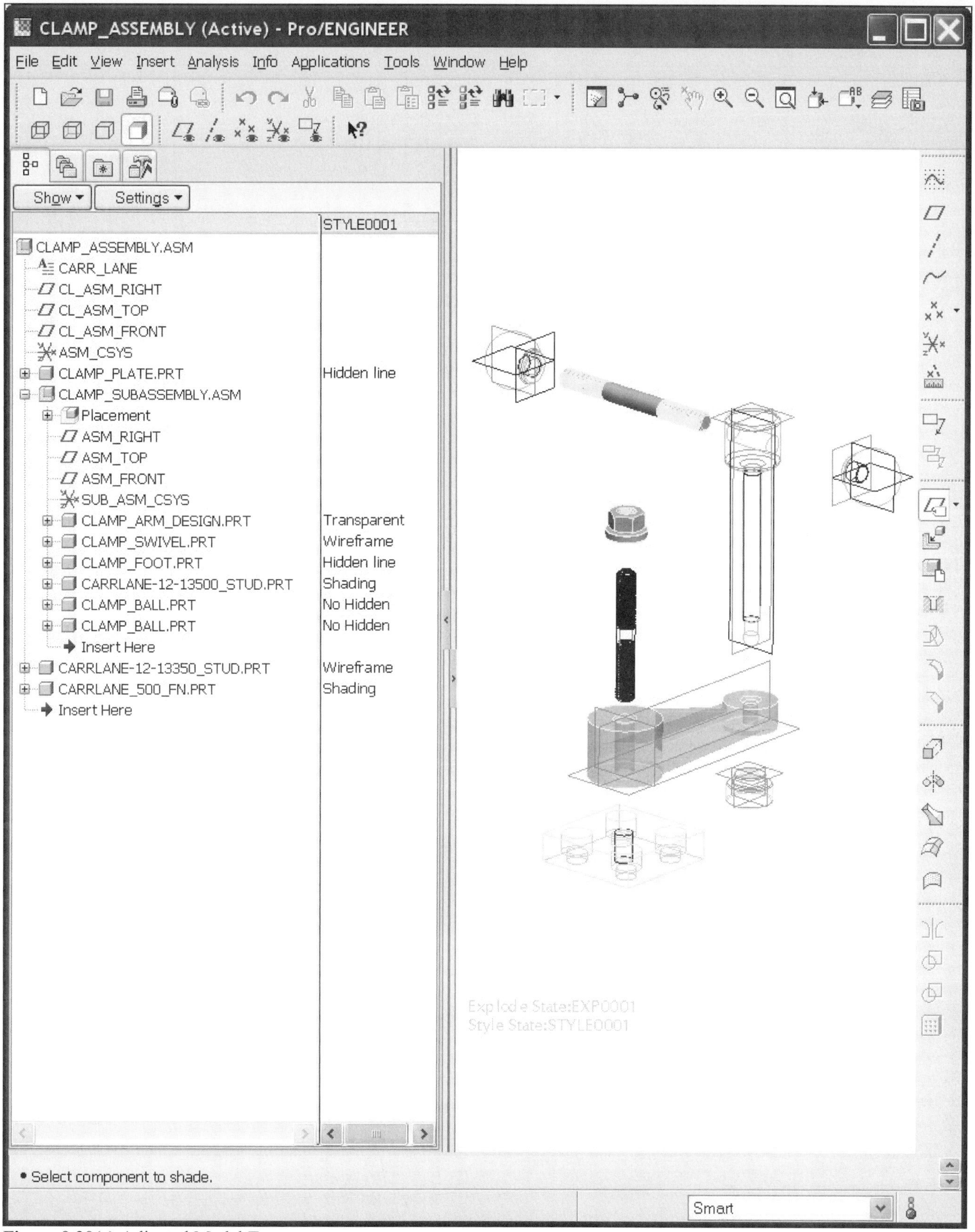

Figure 9.20(c) Adjusted Model Tree

Collapse the Navigator/Model Tree ⇒ **View** ⇒ **Shade** ⇒ **Ctrl+S** ⇒ **Enter** [Fig. 9.20(d)]

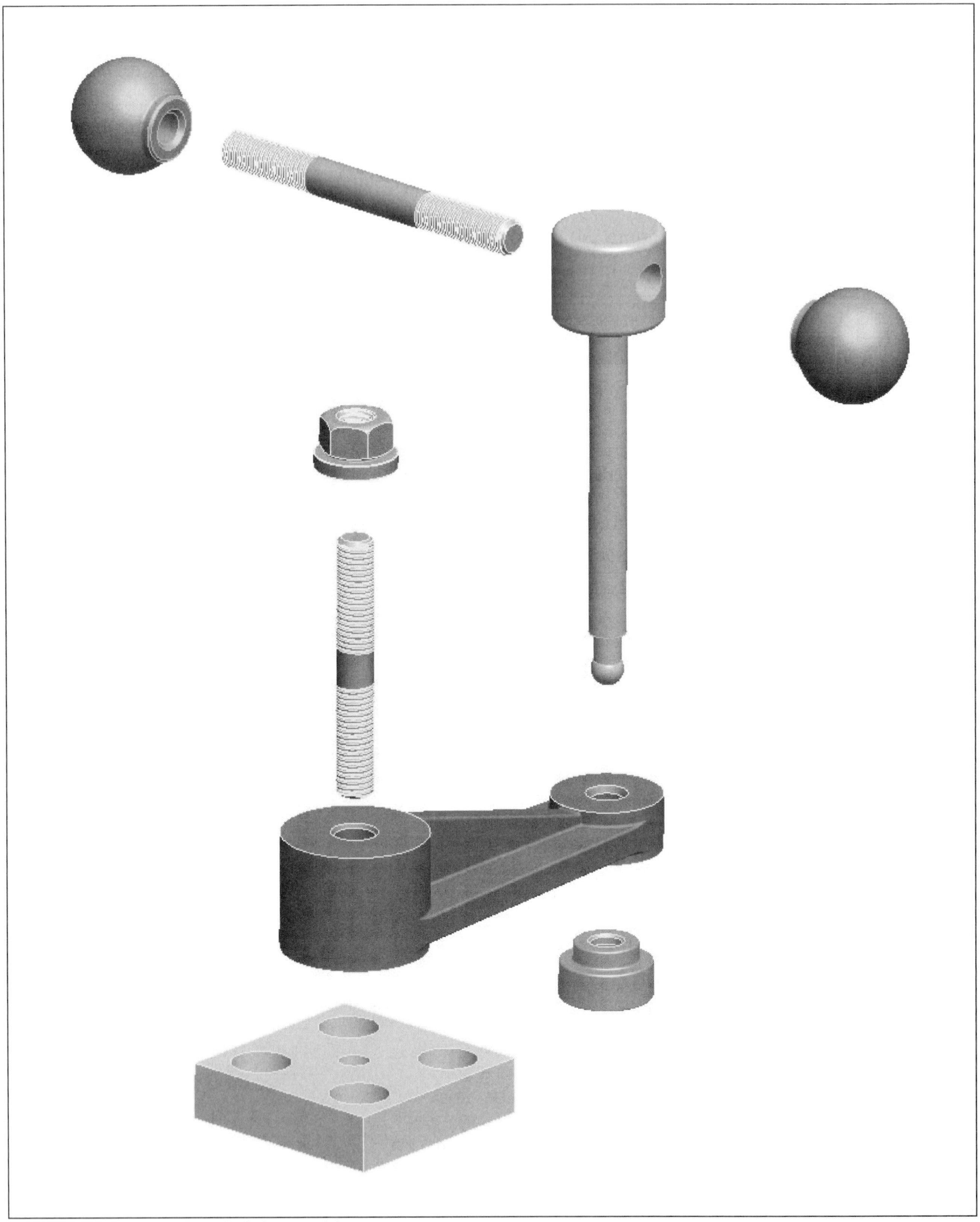

Figure 9.20(d) Shaded Exploded Assembly

Lesson 10 Introduction to Drawings

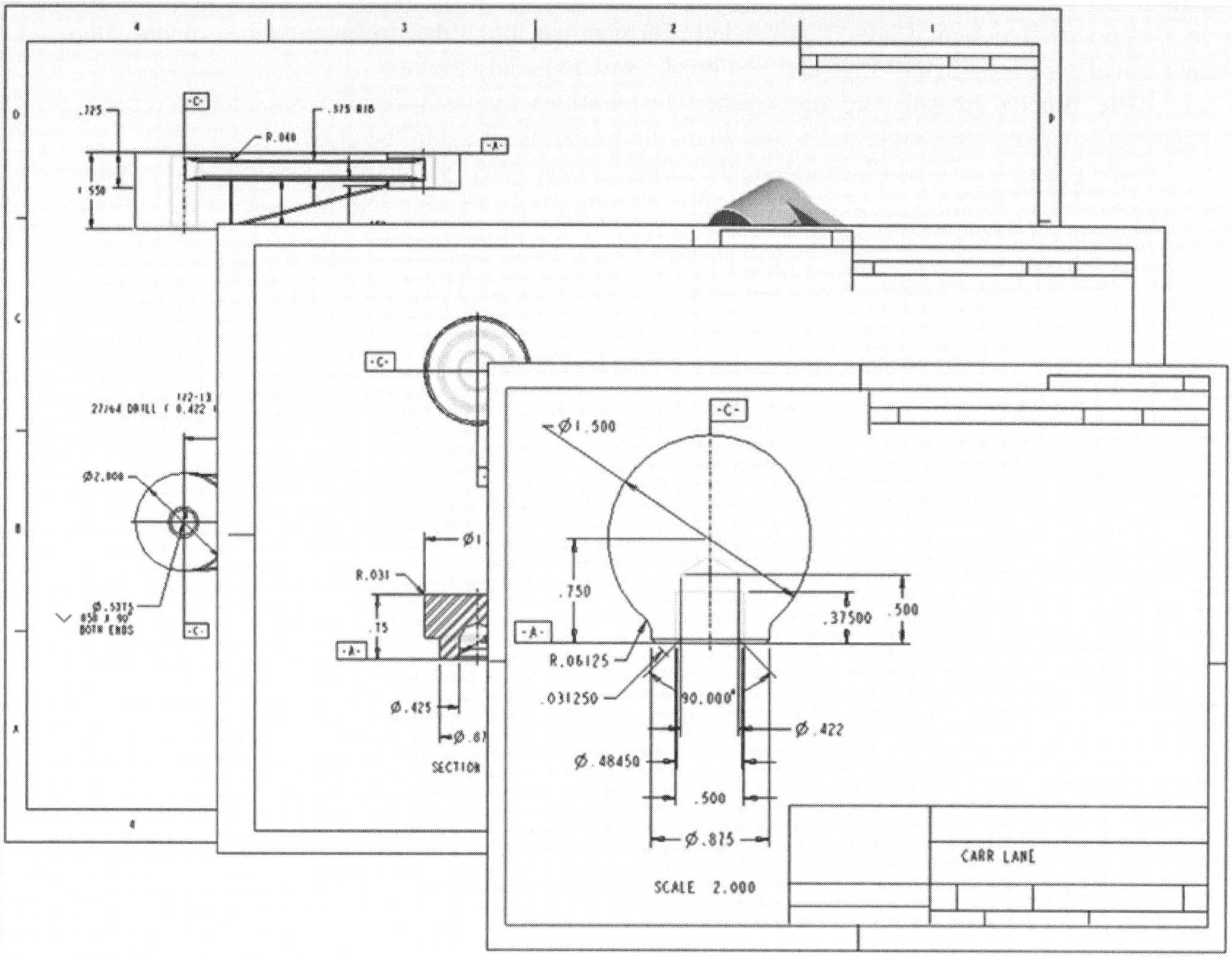

Figure 10.1 Clamp Arm Drawing

OBJECTIVES

- Create part drawings
- Display **standard views** using the template
- **Add, Move**, **Erase**, and **Delete** views
- Retrieve a standard **format**
- **Display** gtol datums, centerlines, and dimensions
- **Cleanup** dimension positions
- Specify and retrieve standard **Format** paper size and units
- Change the **Scale** of a view

Introduction to Drawings

This lesson will introduce you to the basic concepts and procedure for creating detail drawings (Fig. 10.1). Using parts modeled for the Swing Clamp Assembly, you will quickly create drawings, change formats, delete and add views, and display dimensions and centerlines. Since this lesson is meant to introduce drawing mode concepts and procedures, a minimum of cleanup, reformatting and repositioning of dimensions and centerlines will be required.

FORMATS, TITLE BLOCKS, AND VIEWS

Formats are user-defined drawing sheet layouts. A drawing can be created with an empty format and have a standard size format added as needed. Formats can be added to any number of drawings. The format can also be modified or replaced in a Pro/E drawing at any time.

Title Blocks are standard or sketched line entities that can contain parameters (object name, tolerances, scale, and so on) that will show when the format is added to the drawing.

Views created by Pro/E are identical to views constructed manually by a designer on paper. The same rules of projection apply; the only difference is that you choose commands in Pro/E to create the views as needed.

Formats

Formats consist of draft entities, not model entities. There are two types of formats: standard and sketched. You can select from a list of **Standard Formats** (**A-F** and **A0-A4**) or create a new size by entering values for length and width.

Sketched Formats created in Sketcher mode (Fig. 10.2) may be parametrically modified, enabling you to create nonstandard-size formats or families of formats.

Formats can be altered to include note text, symbols, tables, and drafting geometry, including drafting cross sections and filled areas.

Figure 10.2 Sketched Format (CADTRAIN, COAch for Pro/ENGINEER)

With Pro/E, you can do the following in Format mode:

- Create draft geometry and notes (Fig. 10.3)
- Move, mirror, copy, group, translate, and intersect geometry
- Use and modify the draft grid
- Enter user attributes
- Create drawing tables
- Use interface tools to create plot, DXF, SET, and IGES files
- Import IGES, DXF, and SET files into the format
- Create user-defined line styles
- Create, use, and modify symbols
- Include drafting cross sections in a format

Whether you use a standard format or a sketched format, the format is added to a drawing that is created for a set of specified views of a parametric 3D model.

When you place a format on your drawing, the system automatically writes the appropriate notes based on information in the model you use.

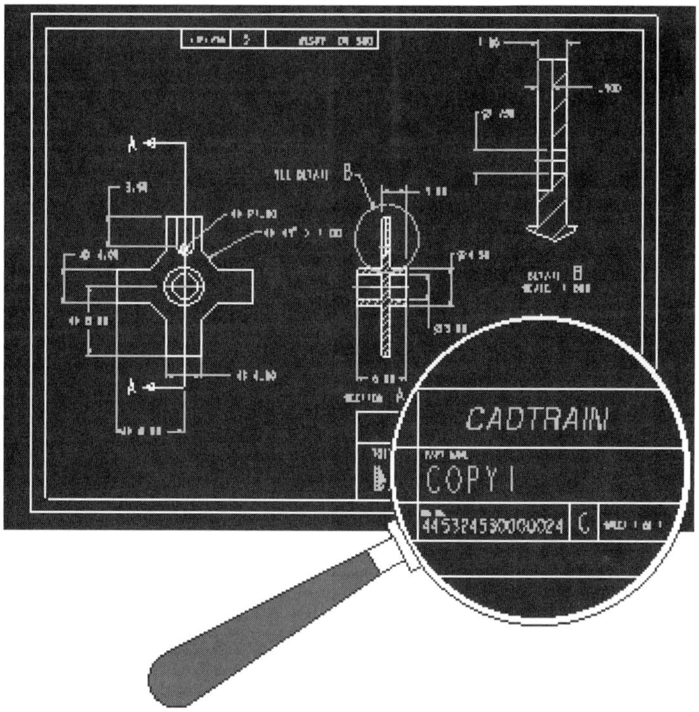

Figure 10.3 Format with Parametric Notes, Added to a Drawing (CADTRAIN, COAch for Pro/ENGINEER)

Specifying the Format Size when Creating a New Drawing

If you want to use an existing template, select a listed template [Fig. 10.4(a)], or select a template from the appropriate directory. If you want to use the existing format [Fig. 10.4(b)], using Browse will open Pro/E's System Formats folder [Fig. 10.4(c)]. Select from the list of standard formats or navigate to a directory with user or company created formats.

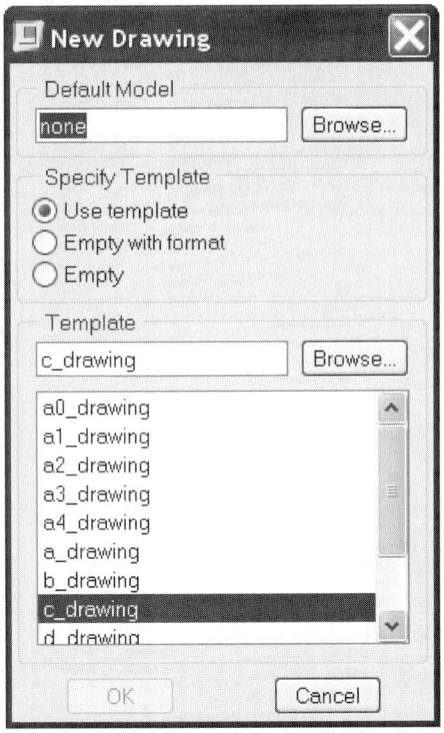

Figure 10.4(a) New Drawing Dialog Box (Use Template)

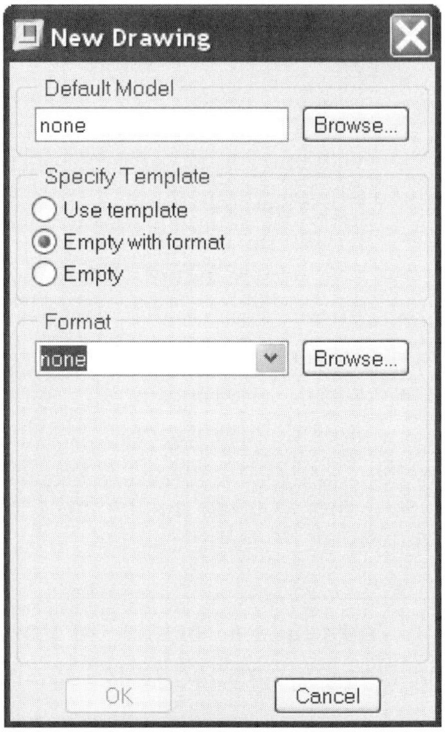

Figure 10.4(b) Empty with format

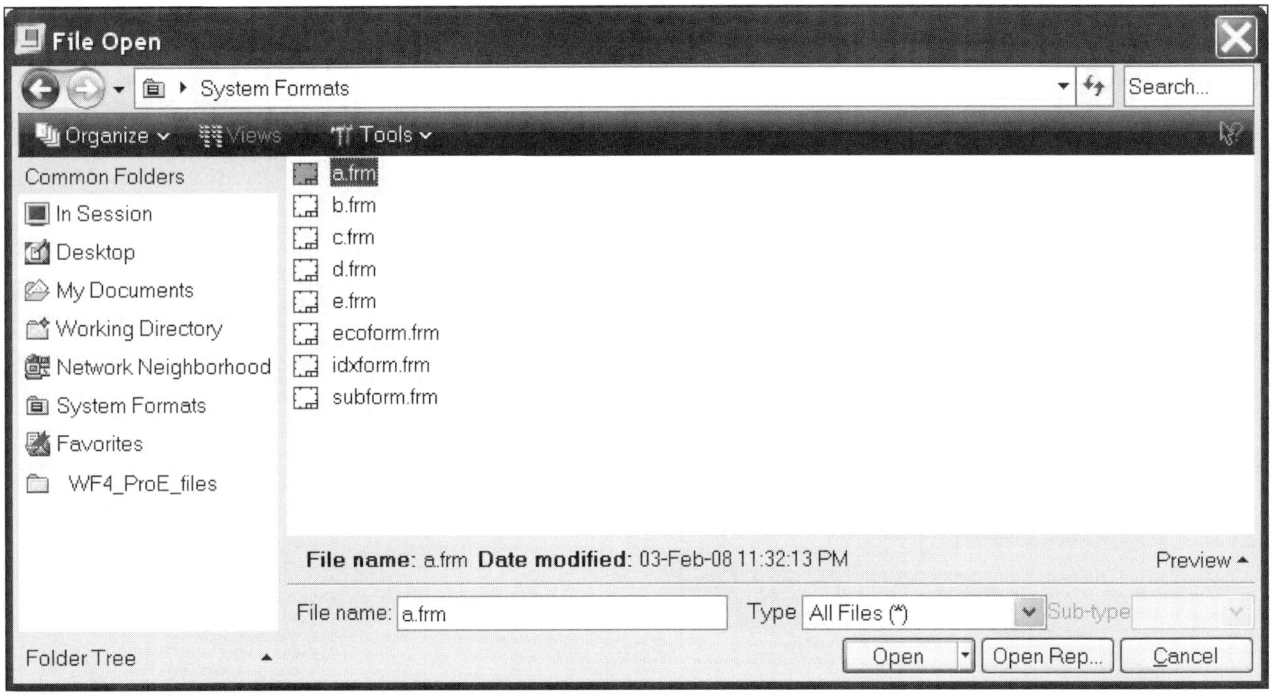

Figure 10.4(c) System Formats

If you want to create your own variable size format [Figs. 10.4(d-e)], enter values for width and length. The main grid spacing and format text units depend on the units selected for a variable size format.

The New Drawing dialog box also provides options for the orientation of the format sheet:

- **Portrait** Uses the larger of the dimensions of the sheet size for the format's height; uses the smaller for the format's width
- **Landscape** Uses the larger of the dimensions of the sheet size for the format's width; uses the smaller for the format's height [Fig. 10.4(d)]
- **Variable** Select the unit type, Inches or Millimeters, and then enter specific values for the Width and Height of the format [Fig. 10.4(e)]

A0	841 X 1189	mm	**A**	8.5 X 11	in.
A1	594 X 841	mm	**B**	11 X 17	in.
A2	420 X 594	mm	**C**	17 X 22	in.
A3	297 X 420	mm	**D**	22 X 34	in.
A4	210 X 297	mm	**E**	34 X 44	in.
			F	28 X 40	in.

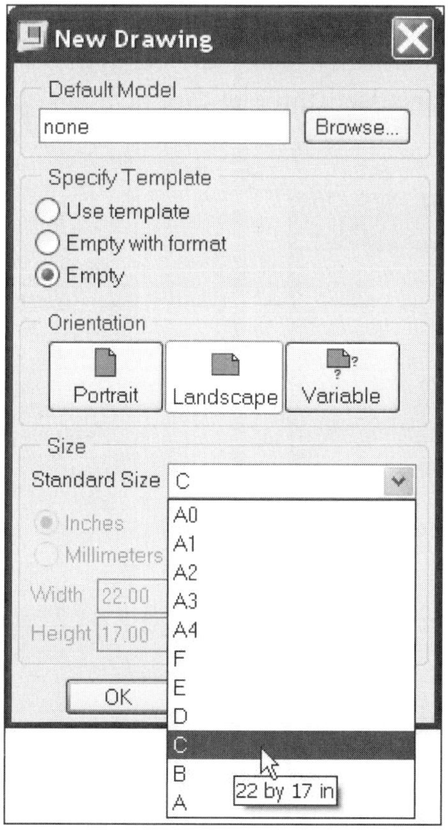

Figure 10.4(d) New Drawing Dialog Box (Empty- Landscape)

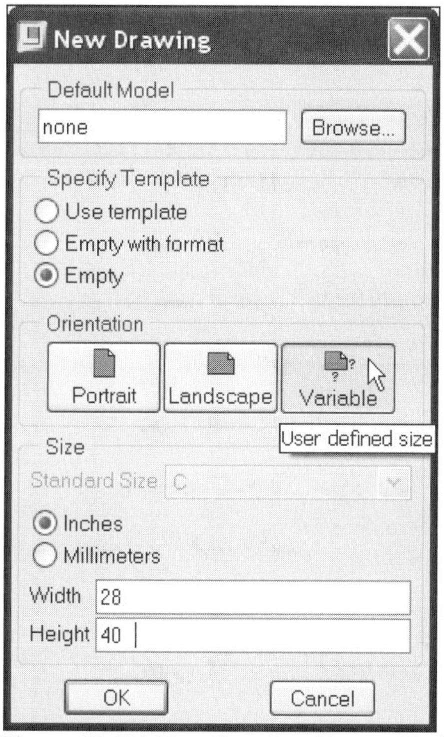

Figure 10.4(e) New Drawing (Empty- Variable)

Drawing Templates

Drawing templates may be referenced when creating a new drawing. Templates can automatically create the views, set the desired view display, create snap lines, and show model dimensions. Drawing templates contain three basic types of information for creating new drawings. The first type is basic information that makes up a drawing but is not dependent on the drawing model, such as notes, symbols, and so forth. This information is copied from the template into the new drawing. The second type is instructions used to configure drawing views and the actions that are performed on their views. The instructions are used to build a new drawing object. The third type is a parametric note. Parametric notes are notes that update to new drawing model parameters and dimension values. The notes are re-parsed or updated when the template is instantiated. Use the templates to:

- Define the layout of views
- Set view display
- Place notes
- Place symbols
- Define tables
- Create snap lines
- Show dimensions

Template View

You can also create customized drawing templates for the different types of drawings that you construct. Creating a template allows you to establish portions of drawings automatically, using the customizable template. The Template View Instructions dialog box (Fig. 10.5) is accessed through Applications ⇒ Template ⇒ Insert ⇒ Template View, when in the Drawing mode.

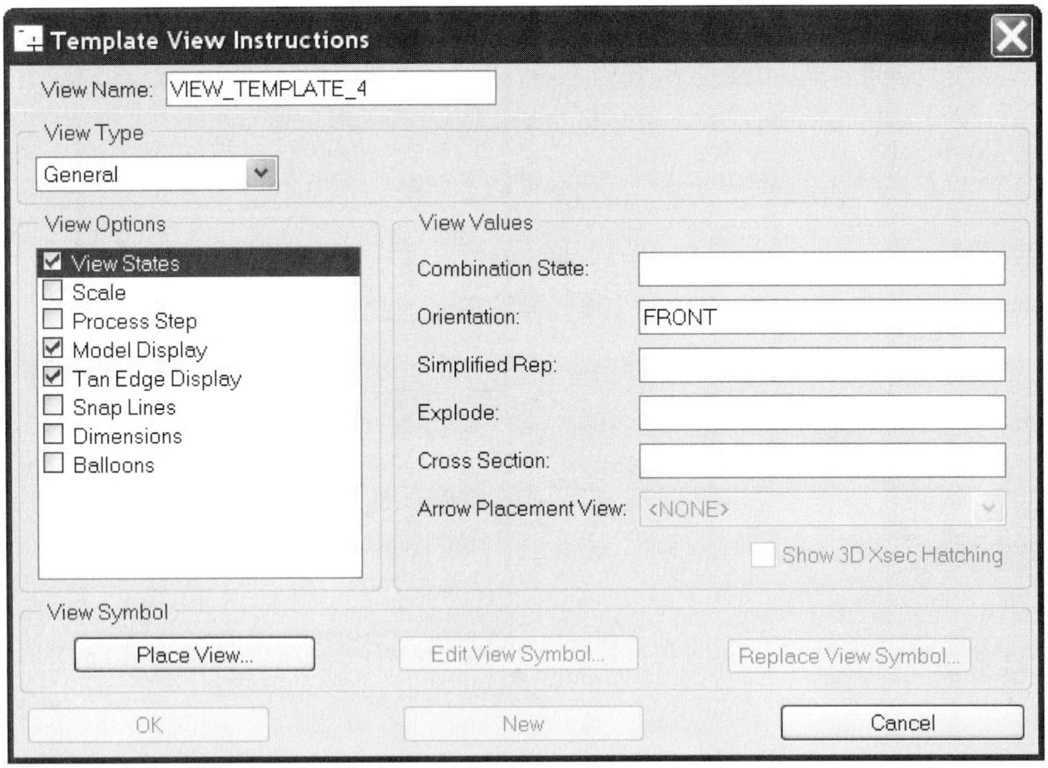

Figure 10.5 Template View Instructions Dialog Box

You can use the following options in the Template View Instructions dialog box to customize your drawing templates:

- **View Name** Set the name of the drawing view that will be used as the view symbol label
- **View Orientation** Create a General view or a Projection view
- **Model "Saved View" Name** Orient the view based on a name view in the model
- **Place View** Places the view after you have set the appropriate options and values
- **Edit View Symbol** Allows you to edit the view symbol using the Symbol Instance dialog box
- **Replace View Symbol** Allow you to replace the view symbol using the Symbol Instance dialog box

View Options	View Values
View States—Specify the type of view. By default, the **View States** check box is selected and the view state is displayed when the **Template View Instructions** dialog box opens. The value of **Orientation** is FRONT, by default. If you select **Combination State**, the **Orientation**, **Simplified Rep**, **Explode** and **Cross Section** boxes display the text Defined by Combination State. The **Arrow Placement View** and **Show X-Hatching** check boxes also become available for selection after you specify the Combination State.	Combination State, Orientation, Simplified Rep, Explode, Cross Section, Arrow Placement View
Scale—Type a new value for the scale or use the default value.	View Scale
Process Step—Set the process step for the view. You can specify the step number and set the view of the tool by checking the **Tool View** check box under **Process Step**.	Step Number, Tool View
Model Display—Set the view display for the drawing view.	Follow Environment, Wireframe, Hidden Line, No Hidden, Shading
Tan Edge Display—Set the tangent edge display.	Tan Solid, No Disp Tan, Tan Ctrln, Tan Phantom, Tan Dimmed, Tan Default
Snap Lines—Set the number, spacing, and offset of the snap lines.	Number, Incremental Spacing, Initial Offset
Dimensions—Show dimensions on the view.	Create Snap Lines, Incremental Spacing, Initial Offset
Balloons—Show balloons on the view.	

Views

A wide variety of views (Fig. 10.6) can be derived from the parametric model. Among the most common are projection views. Pro/E creates projection views by looking to the left of, to the right of, above, and below the picked view location to determine the orientation of a projection view. When conflicting view orientations are found, you are prompted to select the view that will be the parent view. A view will then be constructed from the selected view.

At the time when they are created, projection, auxiliary, detailed, and revolved views have the same representation and explosion offsets, if any, as their parent views. From that time onward, each view can be simplified, be restored, and have its explosion distance modified without affecting the parent view. The only exception to this is detailed views, which will always be displayed with the same explosion distances and geometry as their parent views.

Once a model has been added to the drawing, you can place views of the model on a sheet. When a view is placed, you can determine how much of the model to show in a view, whether the view is of a single surface or shows cross sections, and how the view is scaled. You can then show the associative dimensions passed from the 3D model, or add reference dimensions as necessary.

Basic view types used by Pro/Engineer include general, projection, auxiliary, and detailed:

- **General** Creates a view with no particular orientation or relationship to other views in the drawing. The model must first be oriented to the desired view orientation established by you.
- **Projection** Creates a view that is developed from another view by projecting the geometry along a horizontal or vertical direction of viewing (orthographic projection). The projection type is specified by you in the drawing setup file and can be based on third-angle (default) or first-angle rules.
- **Auxiliary** Creates a view that is developed from another view by projecting the geometry at right angles to a selected surface or along an axis. The surface selected from the parent view must be perpendicular to the plane of the screen.
- **Detailed** Details a portion of the model appearing in another view. Its orientation is the same as that of the view it is created from, but its scale may be different so that the portion of the model being detailed can be better visualized.

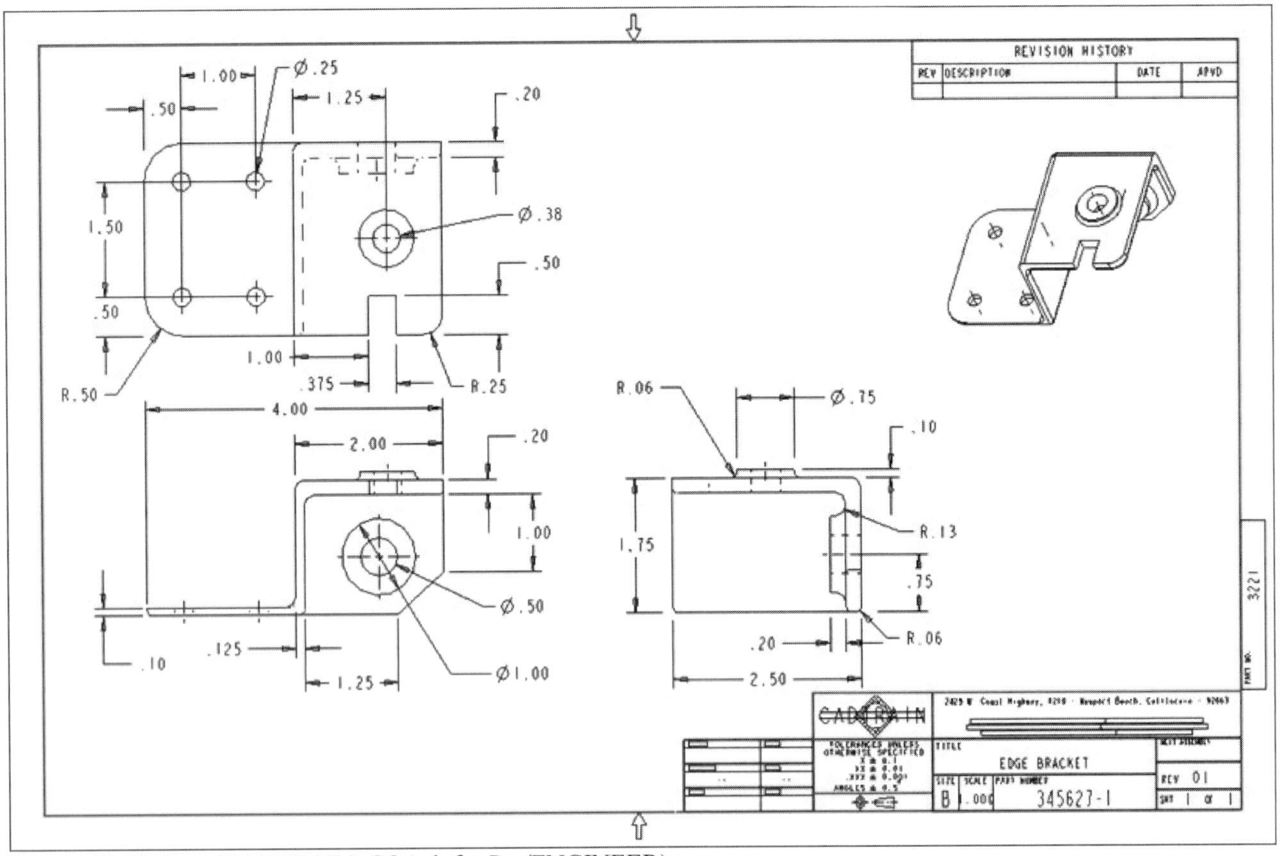

Figure 10.6 Views (CADTRAIN, COAch for Pro/ENGINEER)

The view options that determine how much of the model is visible in the view are:

- **Full View** Shows the model in its entirety.
- **Half View** Removes a portion of the model from the view on one side of a cutting plane.
- **Broken View** Removes a portion of the model from between two selected points and closes the remaining two portions together within a specified distance.
- **Partial View** Displays a portion of the model in a view within a closed boundary. The geometry appearing within the boundary is displayed; the geometry outside of it is removed.

The options that determine whether the view is of a single surface or has a cross section are:

- **Section** Displays an existing cross section of the view if the view orientation is such that the cross-sectional plane is parallel to the screen (Fig. 10.7).
- **No Xsec** Indicates that no cross section is to be displayed.
- **Of Surface** Displays a selected surface of a model in the view. The single-surface view can be of any view type except detailed.

The options that determine whether the view is scaled are:

- **Scale** Allows you to create a view with an individual scale shown under the view. When a view is being created, Pro/E will prompt you for the scale value. This value can be modified later. General and detailed views can be scaled.
- **No Scale** A view will be scaled automatically using a pre-defined scale value.
- **Perspective** Creates a perspective general view.

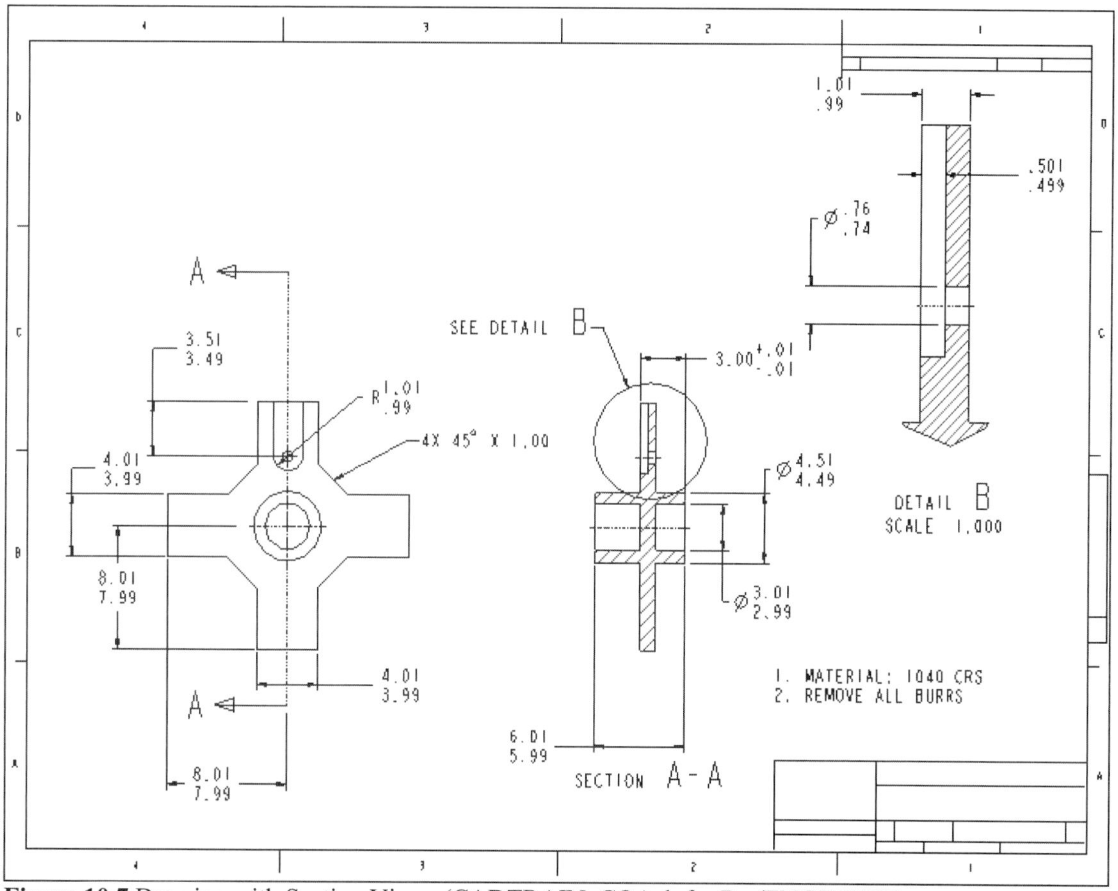

Figure 10.7 Drawing with Section Views (CADTRAIN, COAch for Pro/ENGINEER)

Lesson 10 STEPS

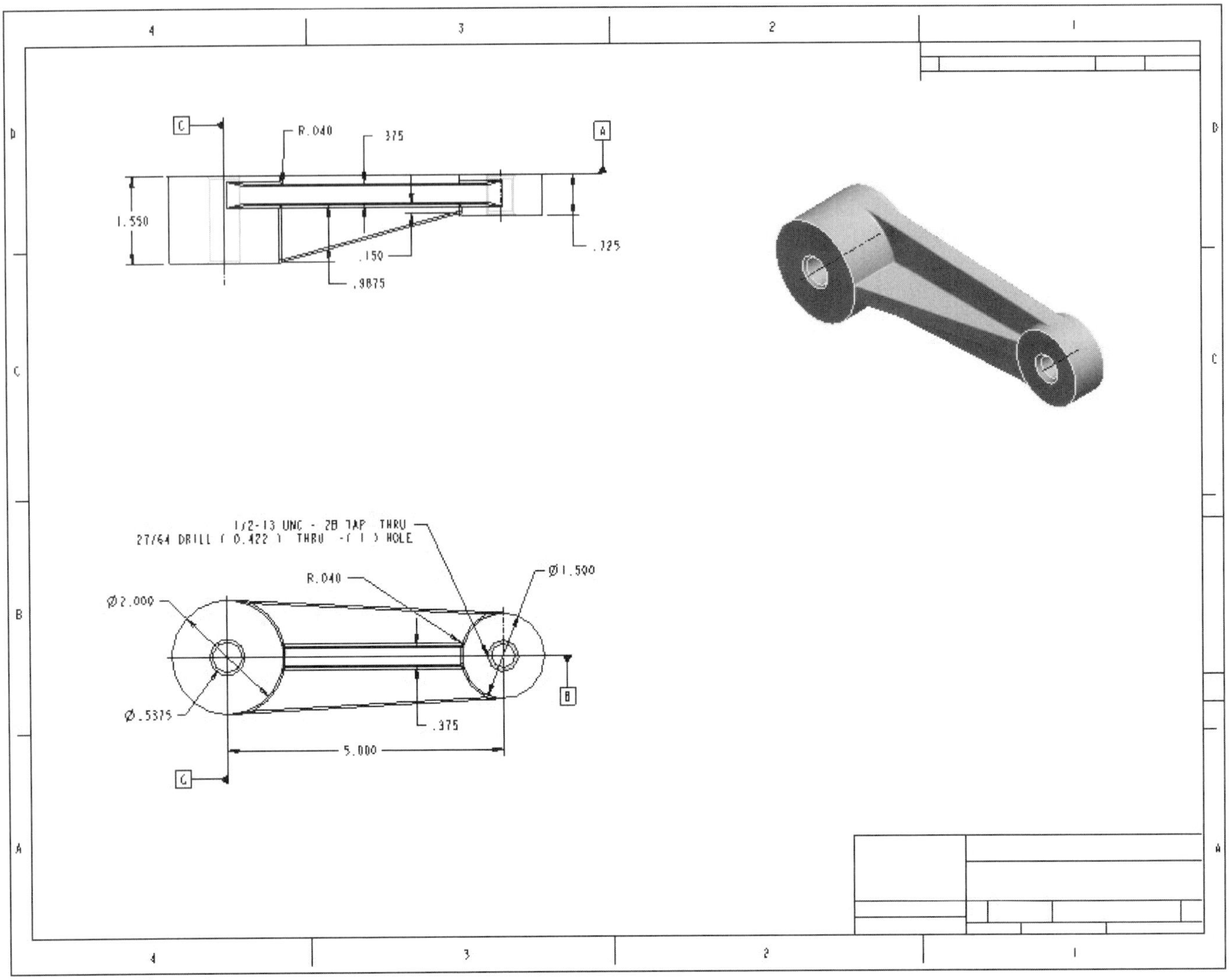

Figure 10.8 Clamp Arm

Creating a Drawing of the Clamp Arm

Before starting the drawing you will need to set your working directory to the folder that contains the Swing Clamp Assembly (and therefore all the components of the assembly). The standard three views of the Clamp_Arm (Fig. 10.8) will display automatically on the drawing, since you will use the default drawing template for this project.

Click: **File** ⇒ **Set Working Directory** ⇒ select the directory where the **clamp_assembly.asm** was saved ⇒ **OK** ⇒ **Create a new object** ⇒ Drawing ⇒ Name **clamp_arm** ⇒ Use default template [Fig. 10.9(a)] ⇒ **OK** [Fig. 10.9(b)] ⇒ Default Model **Browse** ⇒ pick **clamp_arm.prt** [Fig. 10.9(c)] ⇒ **Preview**

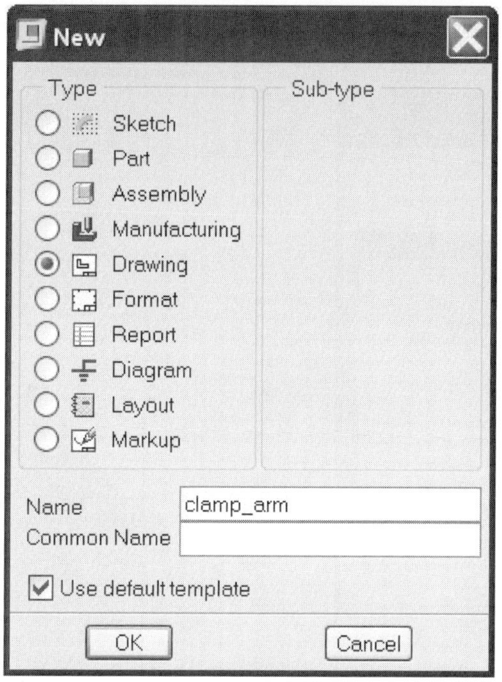

Figure 10.9(a) Type: Drawing

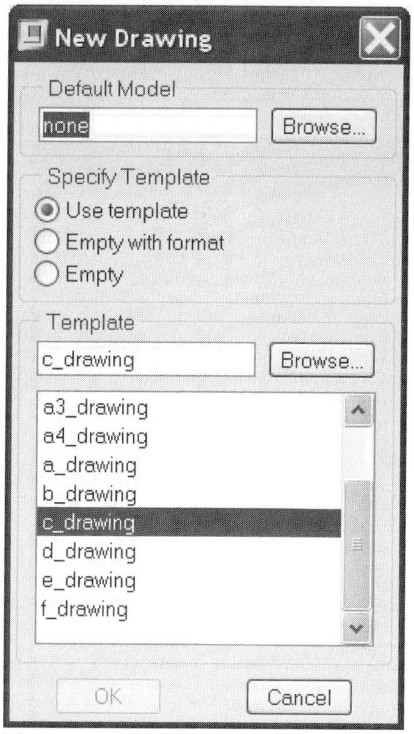

Figure 10.9(b) New Drawing Dialog

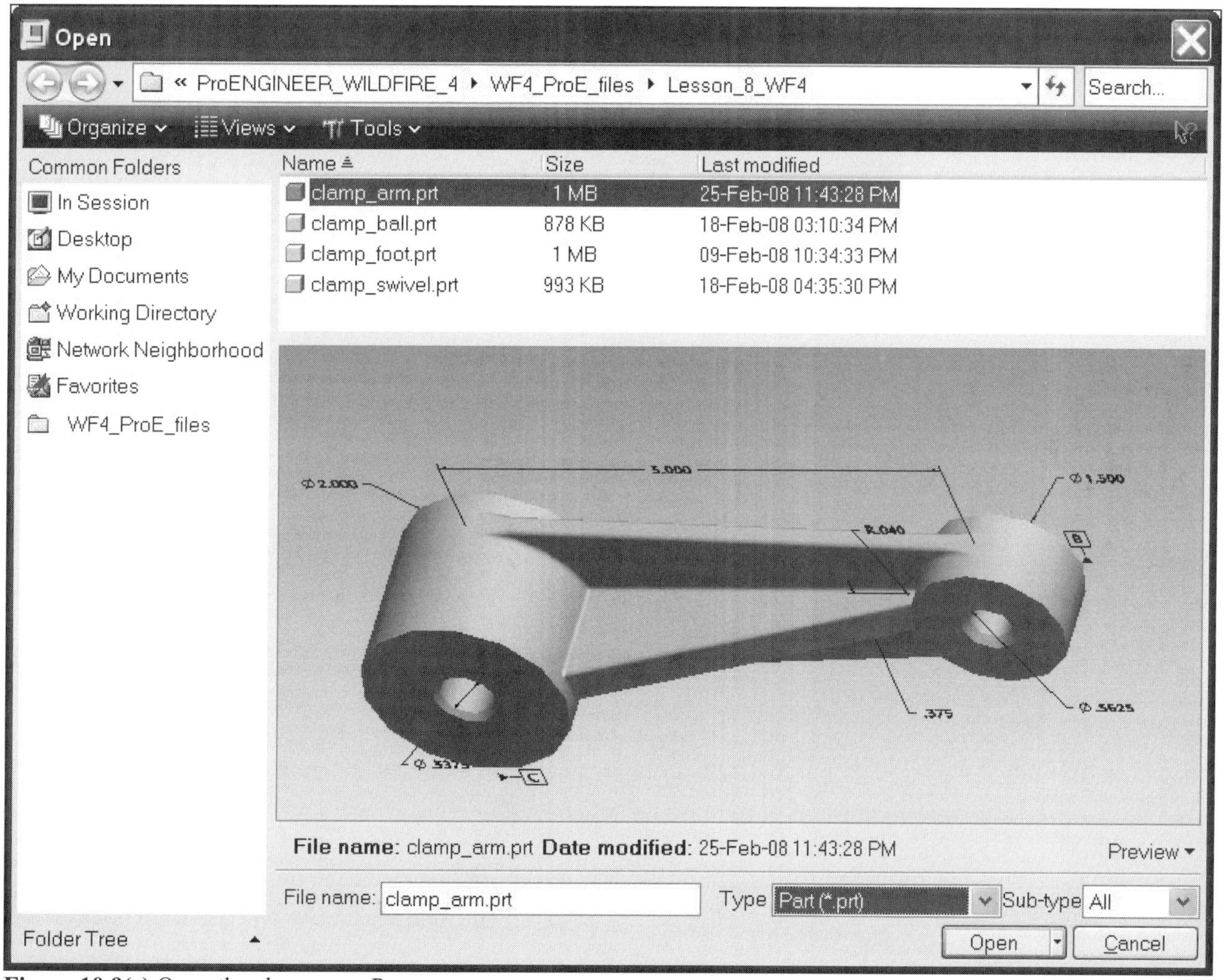

Figure 10.9(c) Open the clamp_arm Part

Click: **Open** ⇒ Template: **c_drawing** [Fig. 10.9(d)] ⇒ **OK** ⇒ CLAMP_ARM_DESIGN from the Select Instance dialog box [Fig. 10.9(e)] ⇒ **Open** [Fig. 10.9(f)]

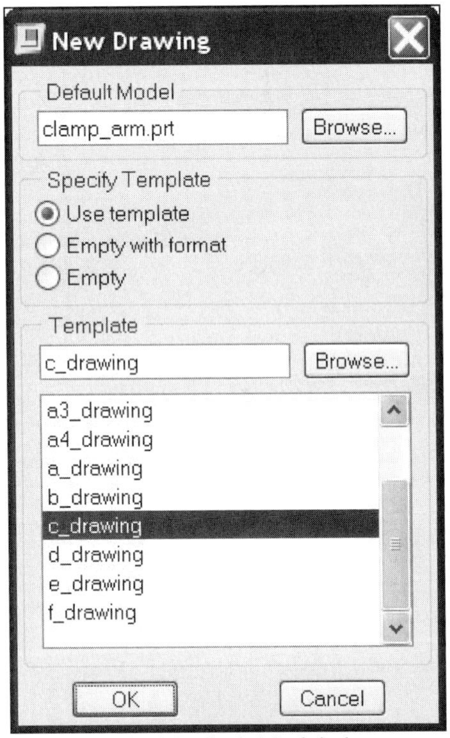

Figure 10.9(d) Default Model clamp_arm.prt

Figure 10.9(e) Select Instance Dialog Box

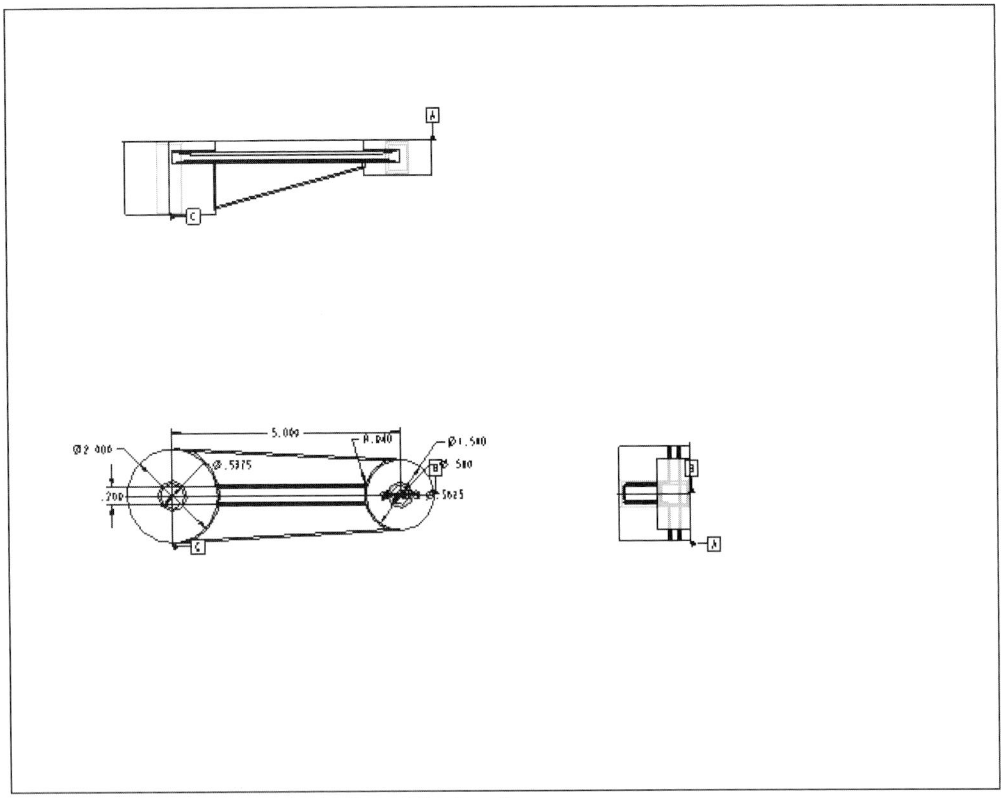

Figure 10.9(f) Three Standard Views Displayed

Click: 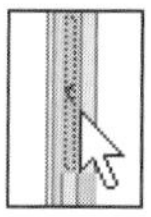 close the Navigator shade ⇒ **File** ⇒ **Page Setup** 

⇒ 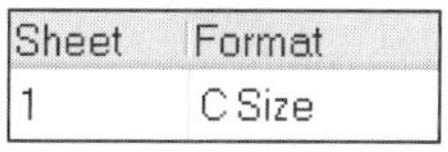

[Fig. 10.9(g)] ⇒ 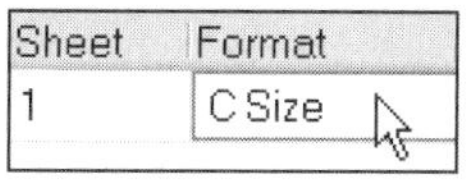

[Fig. 10.9(h)] ⇒  ⇒ **Browse** ⇒ System Formats Open dialog box opens ⇒ pick **c.frm** [Fig. 10.9(i)] ⇒ **Open** ⇒ **OK** ⇒  ⇒ **OK**

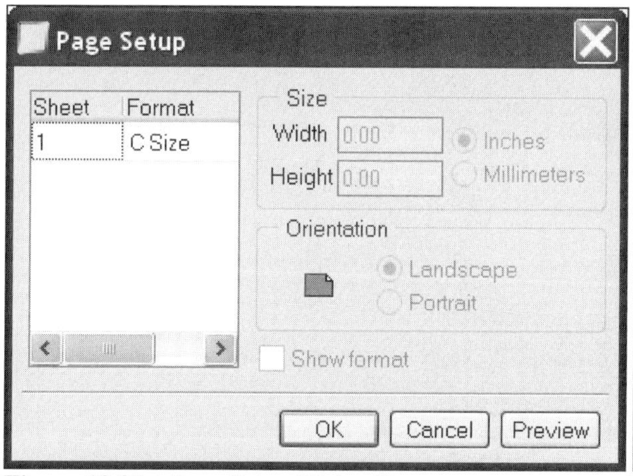

Figure 10.9(g) Page Setup Dialog Box

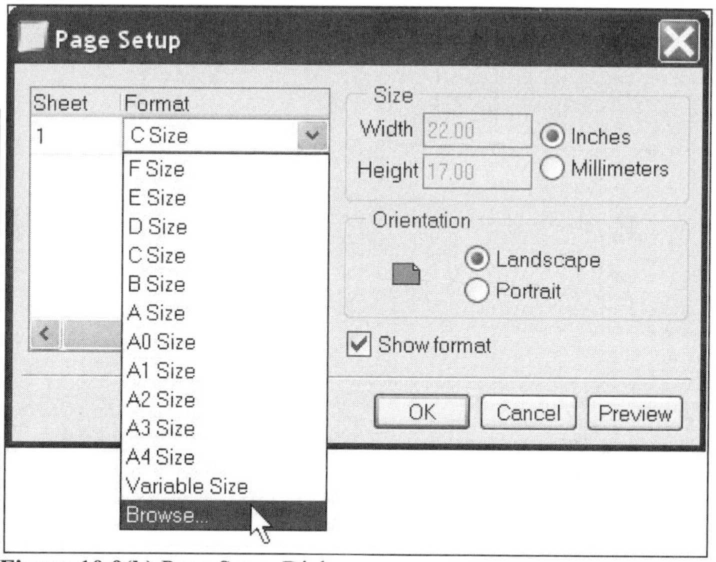

Figure 10.9(h) Page Setup Dialog

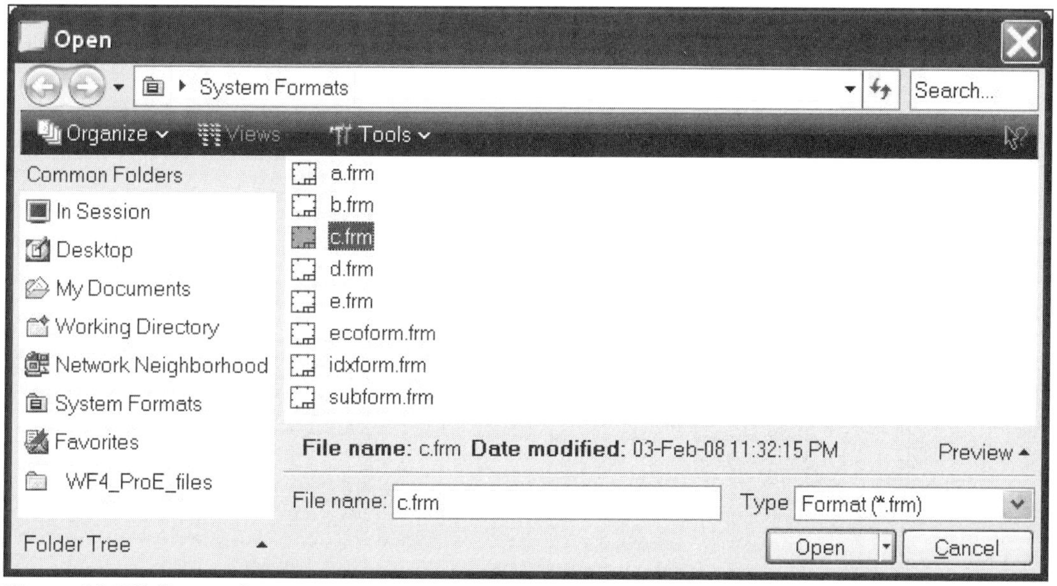

Figure 10.9(i) Open c.frm

Click: [toolbar icons] off ⇒ [icon] **Repaint** ⇒ [icon] ⇒ **OK** [Fig. 10.9(j)]

Since you created some annotations (oriented to the front view) of driving dimensions on the model, they will automatically display on the drawing along with the set datum planes.

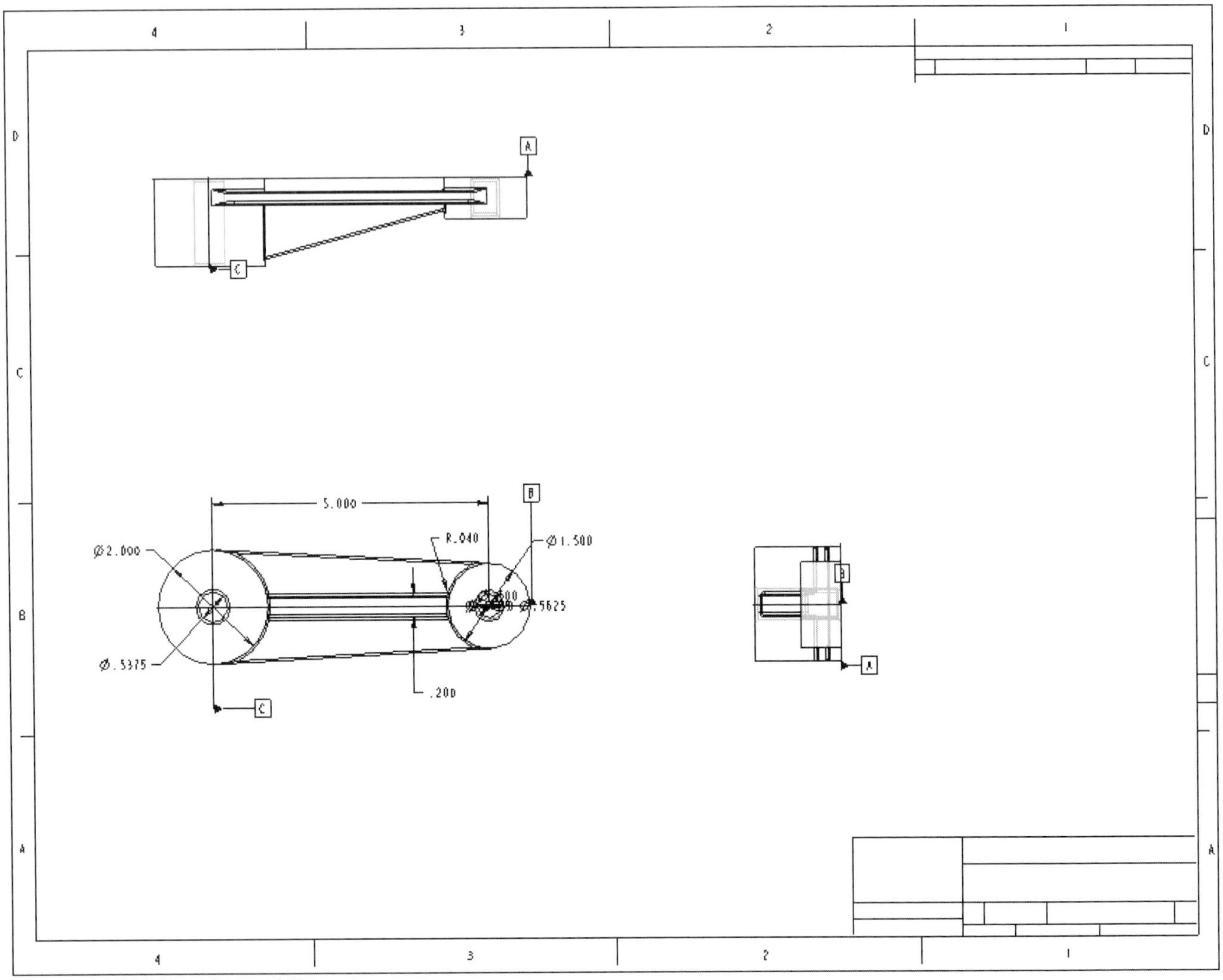

Figure 10.9(j) Drawing and Format

Click: ⇒ ⇒ **RMB** ⇒ **Insert General View** [Fig. 10.9(k)] ⇒ pick a position for the new view [Fig. 10.9(l)]

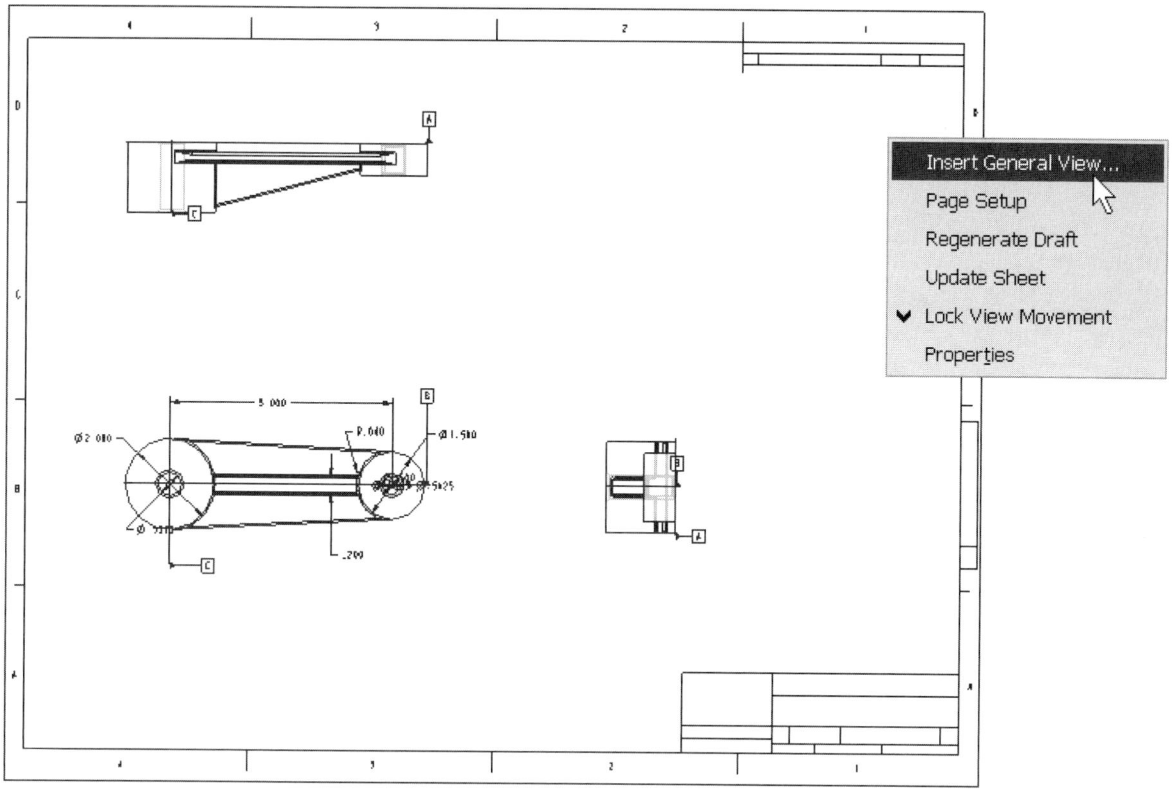

Figure 10.9(k) Insert General View

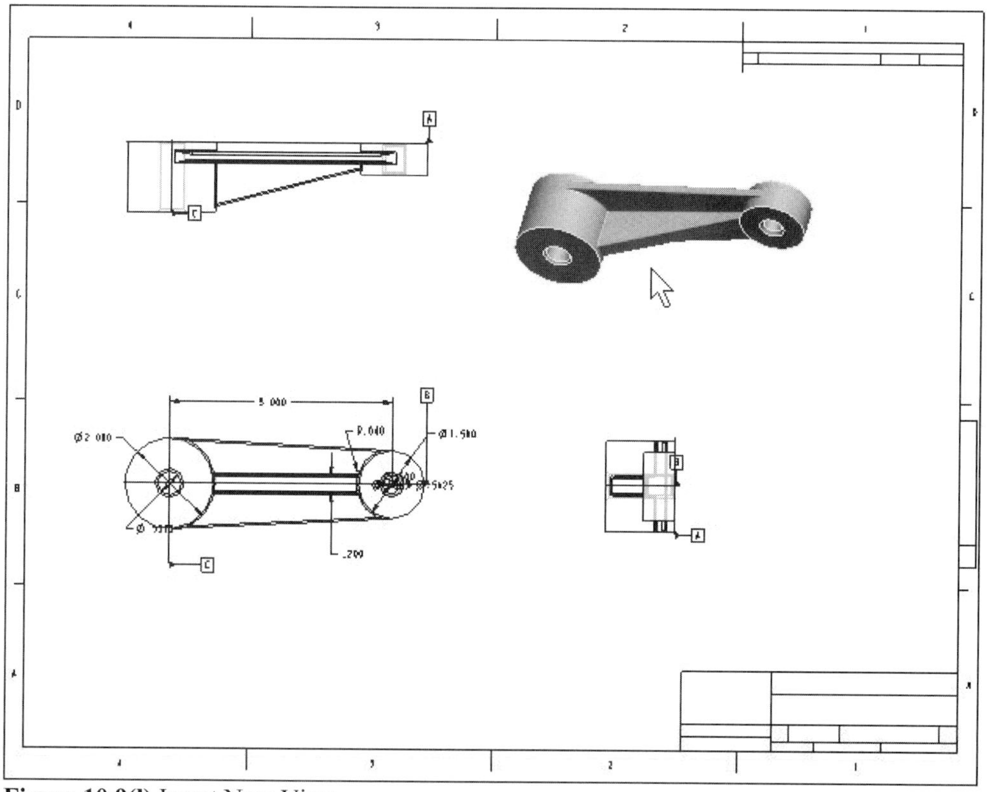

Figure 10.9(l) Insert New View

Click: Default orientation: **Isometric** [Figs. 10.9(m-n)] ⇒ **OK** ⇒ [save icon] ⇒ **OK** ⇒ **LMB** to deselect

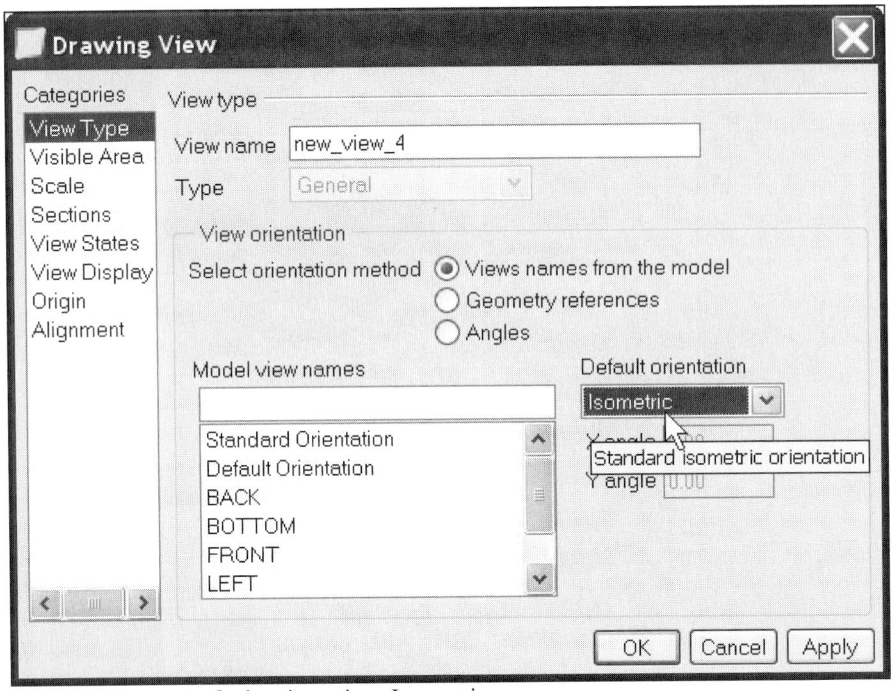

Figure 10.9(m) Default orientation: Isometric

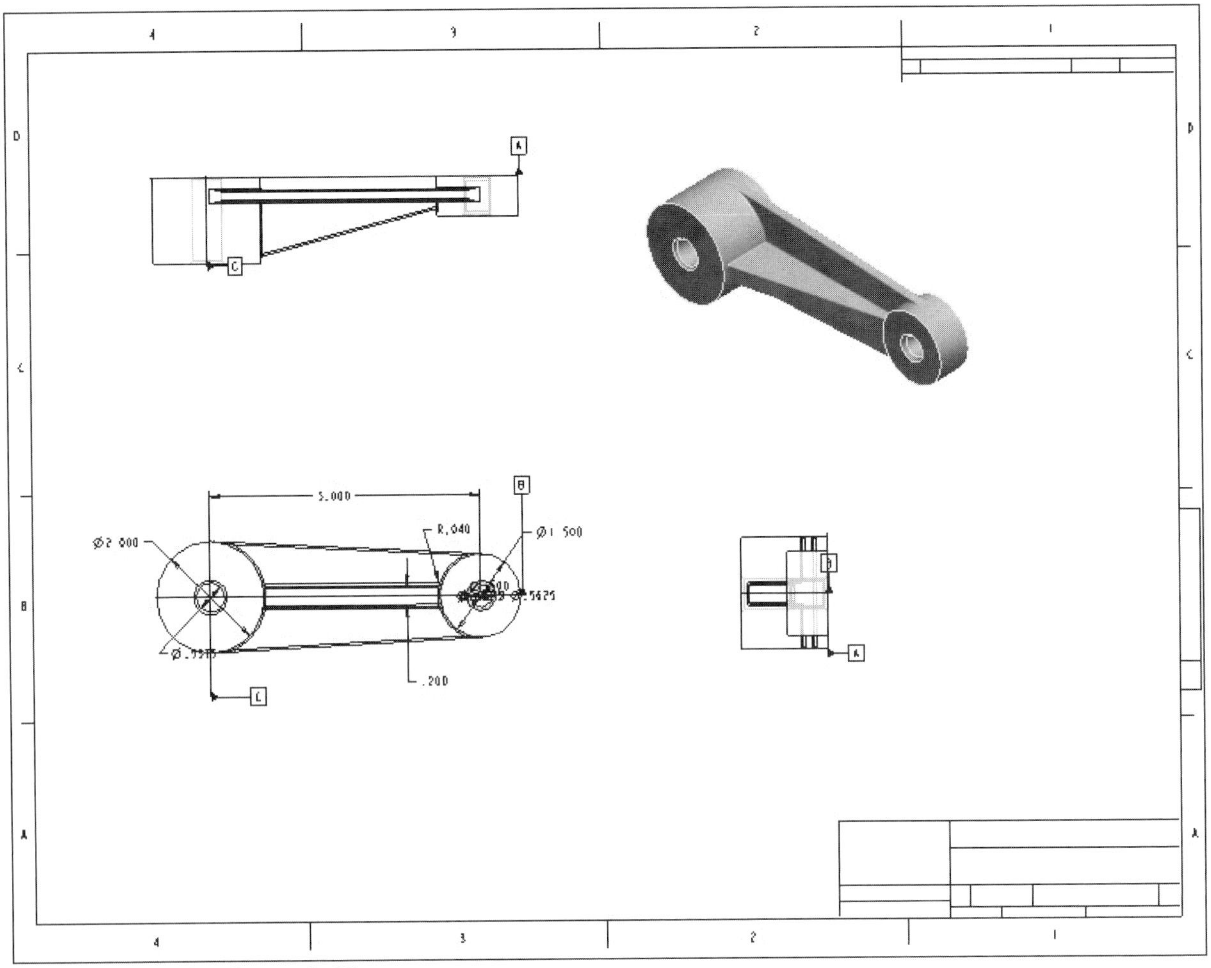

Figure 10.9(n) New Isometric View

Click:  **Update ⇒**  **Open the Show/Erase dialog box ⇒**

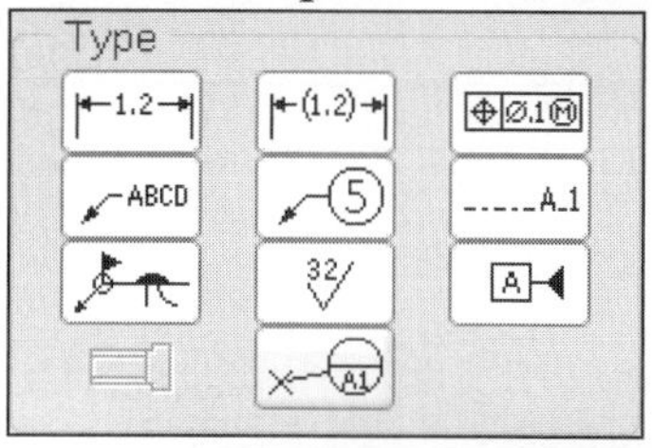

toggle on all options [Fig. 10.9(o)] ⇒ **Show All**

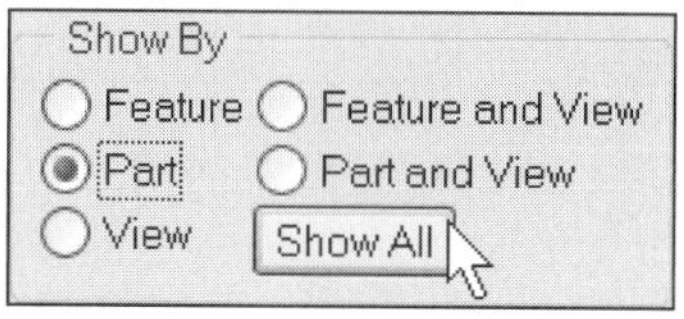

⇒ **Yes**

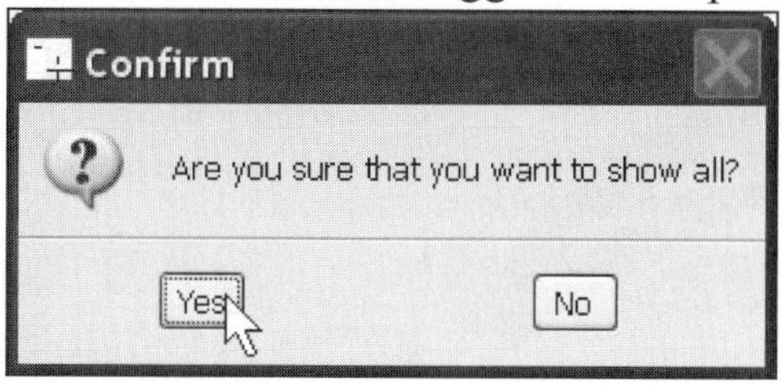

⇒ **Accept All**

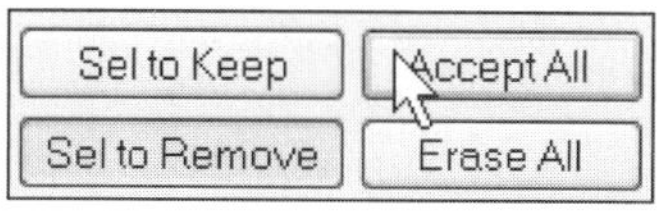

⇒ **Close**

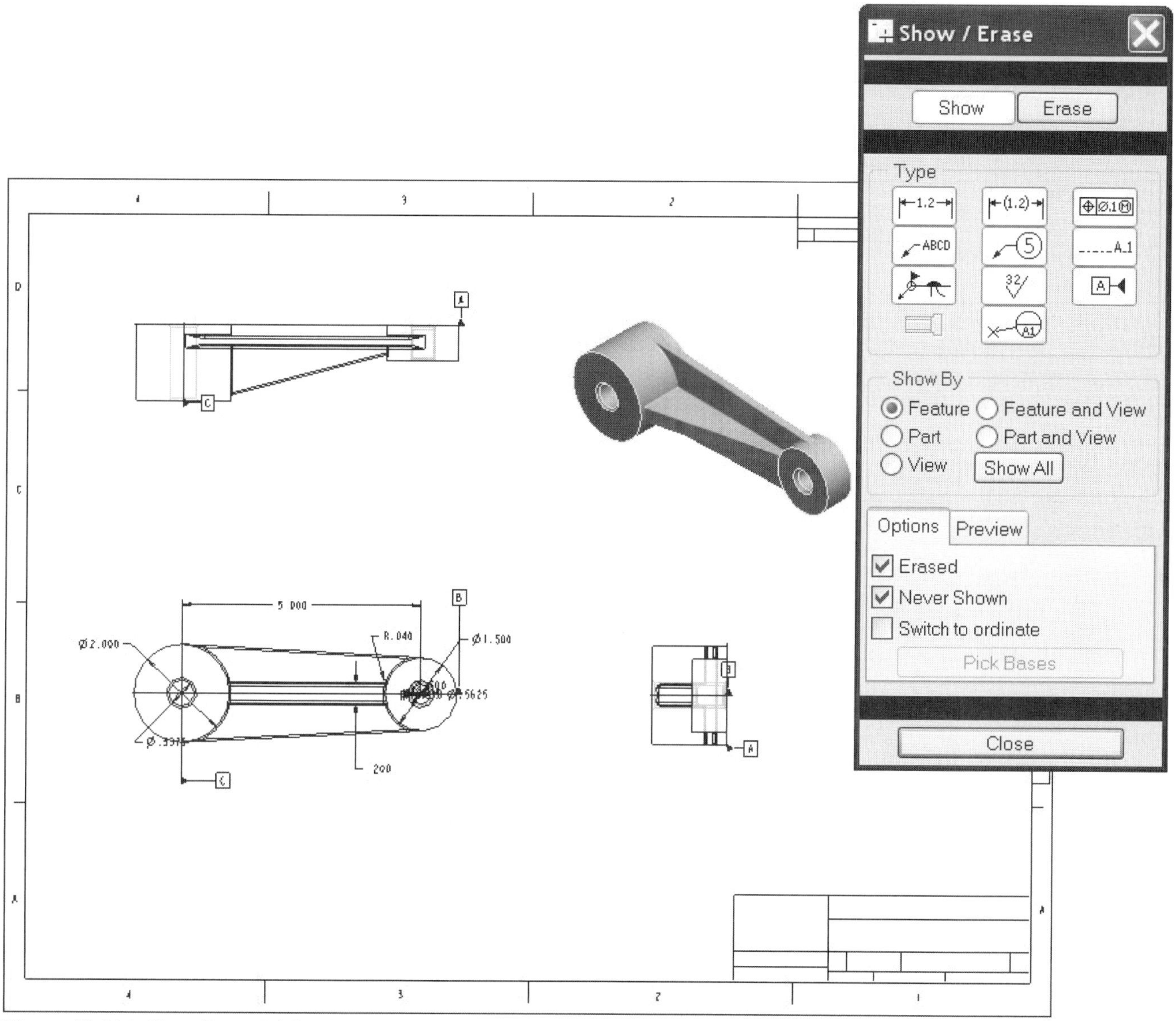

Figure 10.9(o) Show/Erase Dialog Box

With the items still highlighted, click: **RMB ⇒ Cleanup Dimensions** [Fig. 10.9(p)] ⇒ ☐ Create Snap Lines [Fig. 10.9(q)] ⇒ **Apply ⇒ Close ⇒ LMB** to deselect

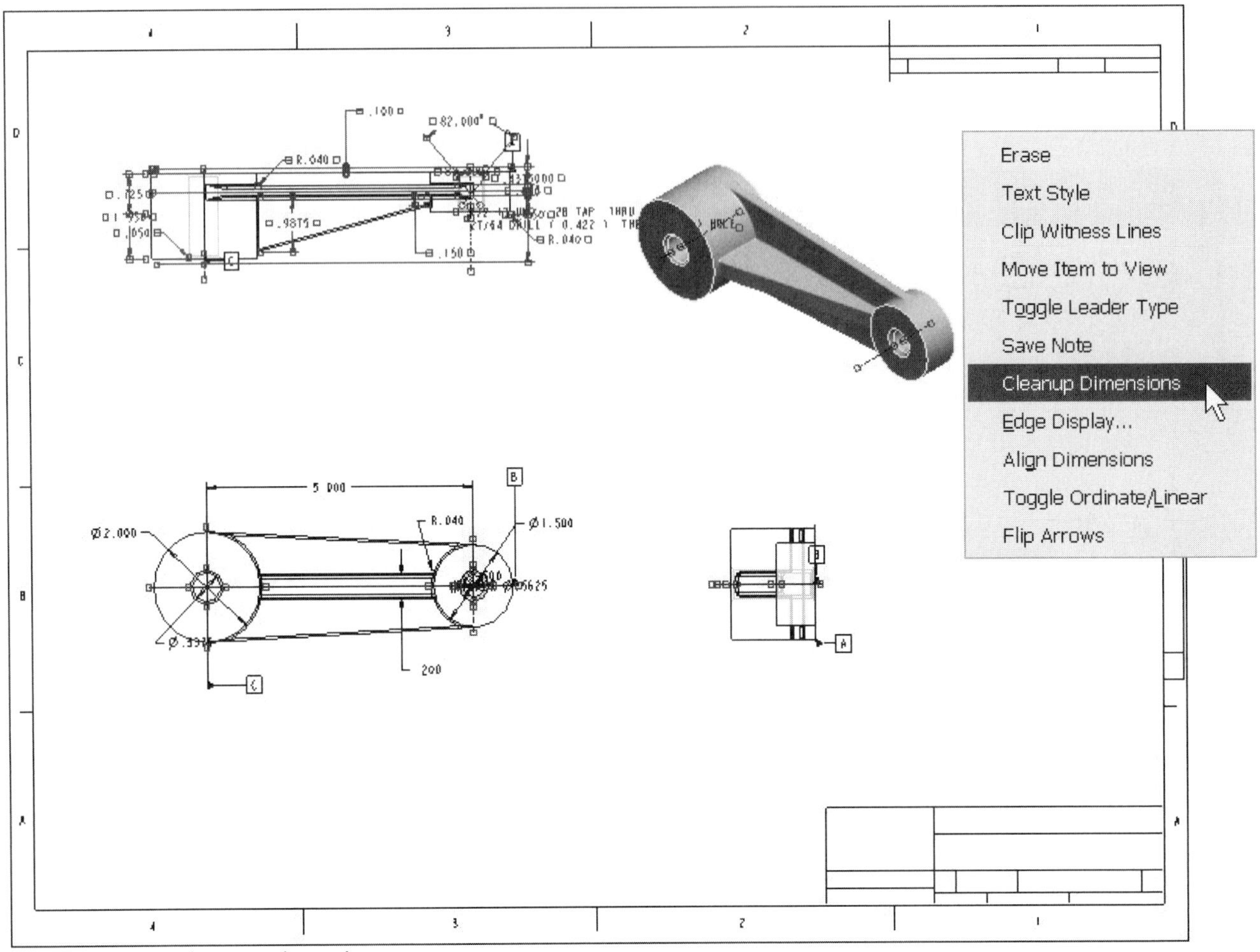

Figure 10.9(p) Cleanup Dimensions

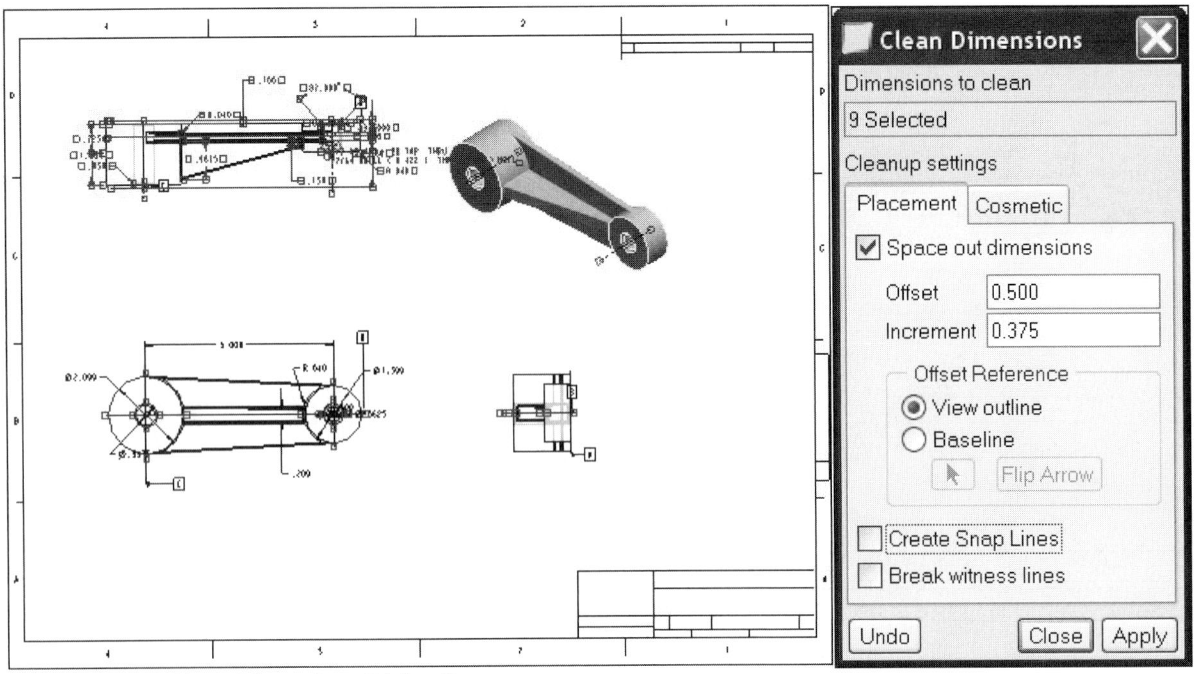

Figure 10.9(q) Clean Dimensions Dialog Box

Click: **RMB** ⇒ Lock View Movement ⇒ Lock View Movement ⇒ pick on the isometric view ⇒ [move icon] [Fig. 10.9(r)] ⇒ hold down **LMB** and move view away from the top view [Fig. 10.9(s)] ⇒ **LMB** outside of the view to set the position ⇒ **Ctrl+S** ⇒ **OK**

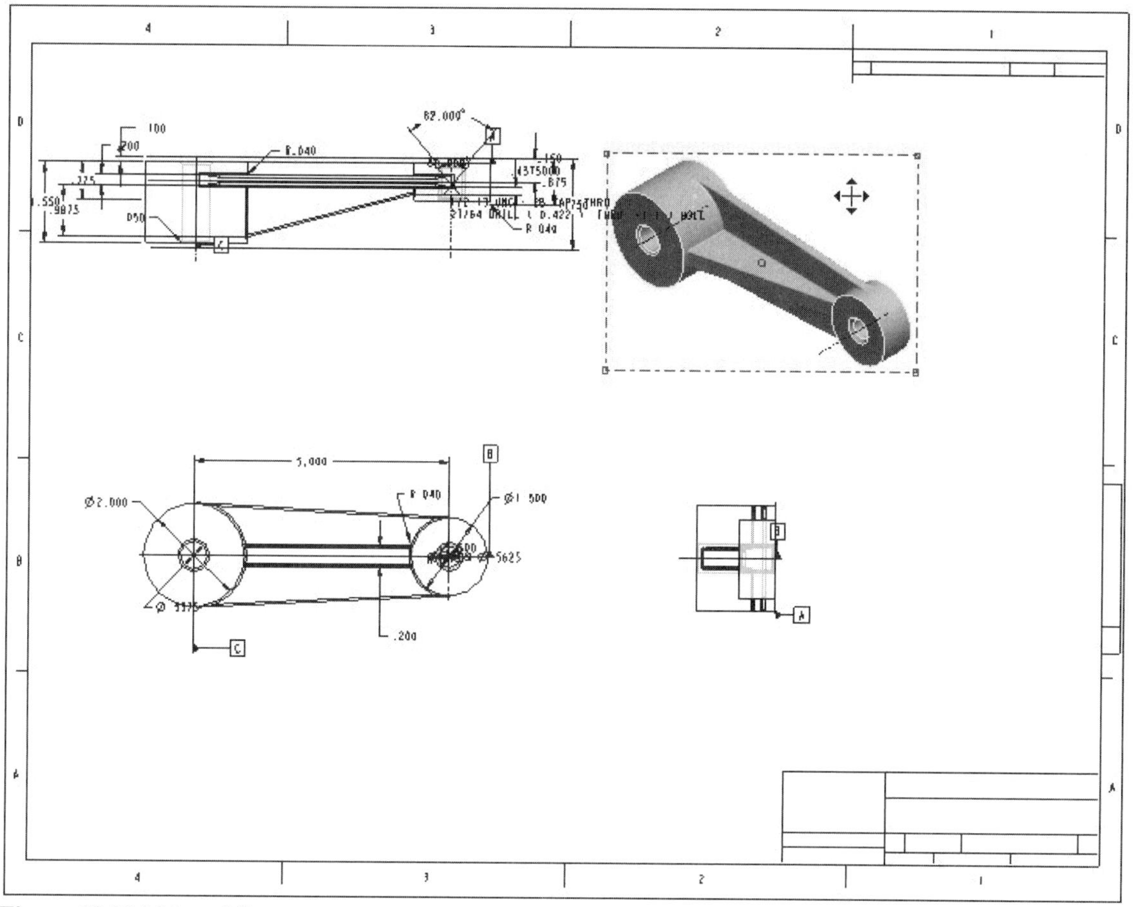

Figure 10.9(r) Move View

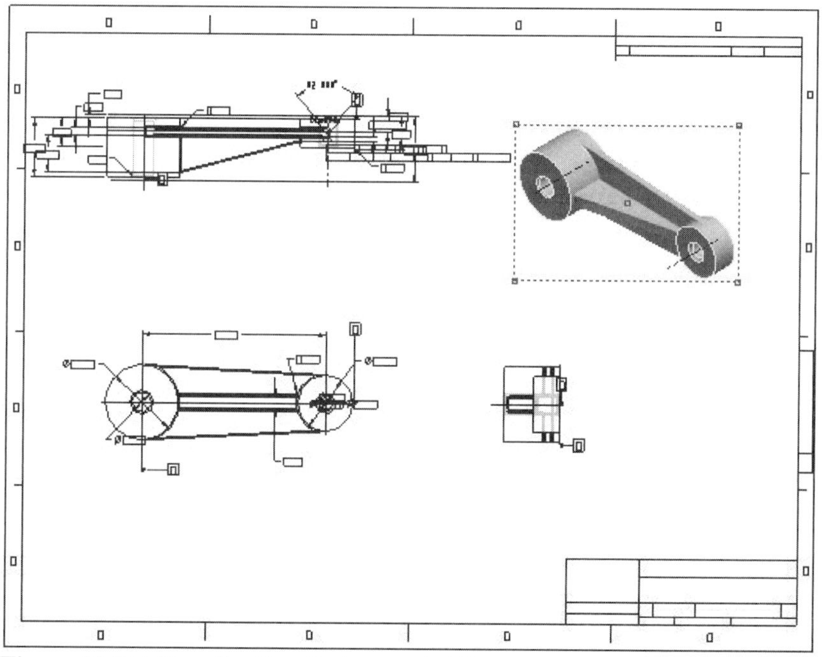
Figure 10.9(s) New Position for Isometric View

Pick on the right side view ⇒ ⇒ **RMB** ⇒ **Delete** [Fig. 10.9(t)] ⇒ **Ctrl+S** ⇒ **OK** ⇒ **Zoom In** on the front view *(pick a screen position to start the zoom box, pick another screen position to finish the zoom box)* [Fig. 10.9(u)] ⇒ **MMB** to end zoom box tool

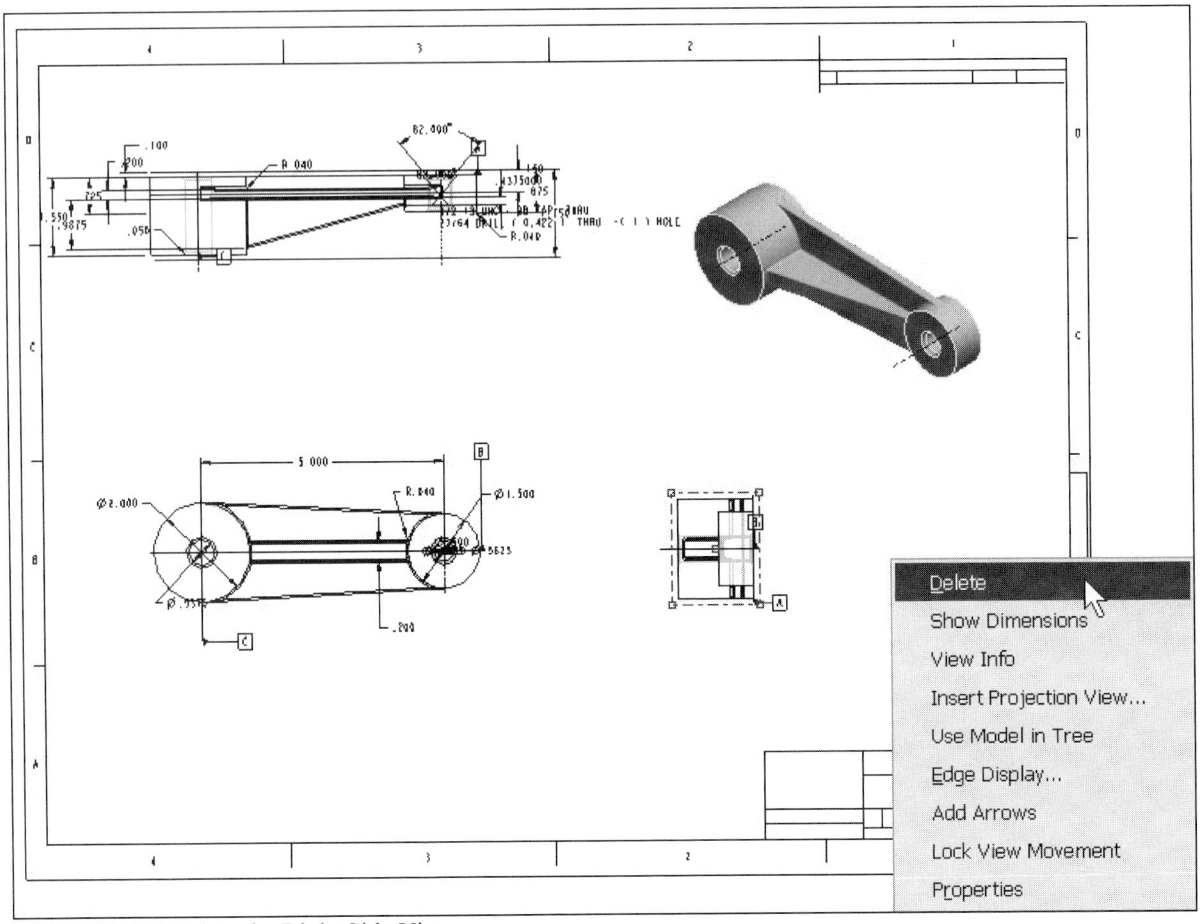

Figure 10.9(t) Delete the Right Side View

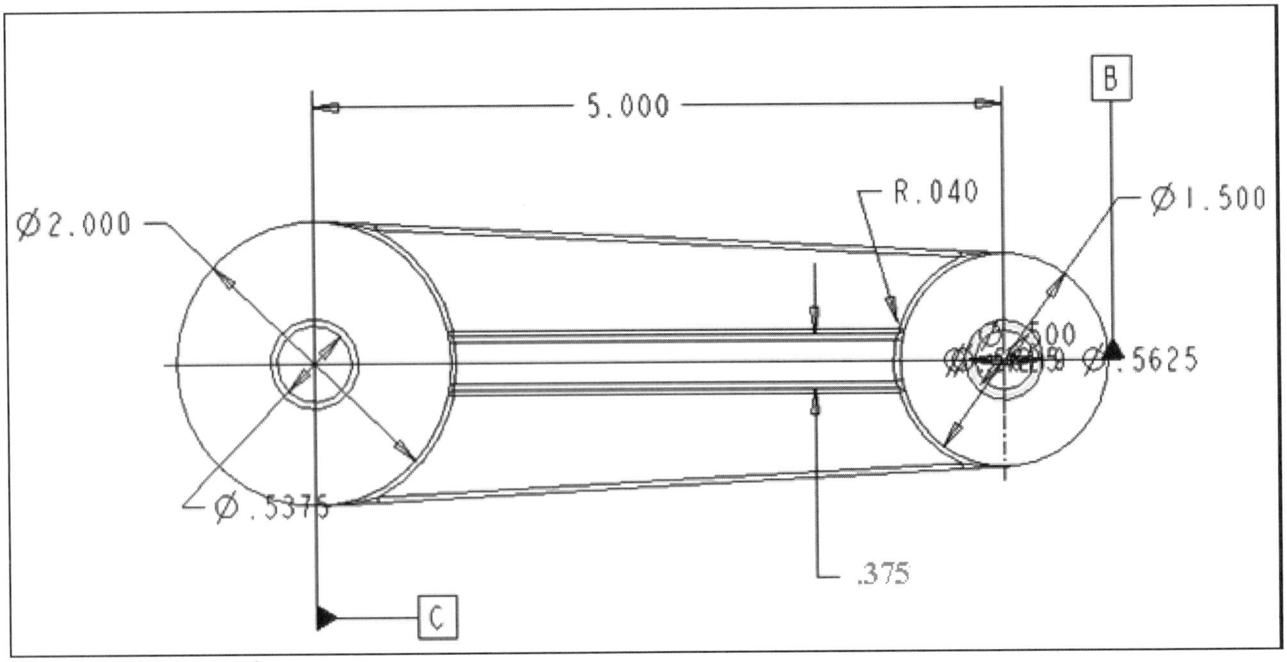

Figure 10.9(u) Front View

Pick on the hole note and move it to a new location [Fig. 10.9(v)] ⇒ reposition dimensions as needed ⇒ reposition each datum tag by picking on the item and then moving it to a new location

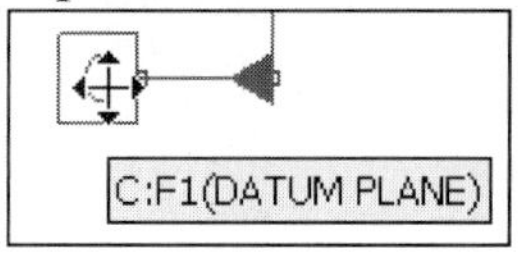

⇒ to erase an item, pick on it ⇒ **RMB** ⇒ **Erase** ⇒ **LMB** (in the graphics window) to accept ⇒ pick on the **.5375** diameter dimension ⇒ **RMB** ⇒ **Flip Arrows** [Fig. 10.9(w)]

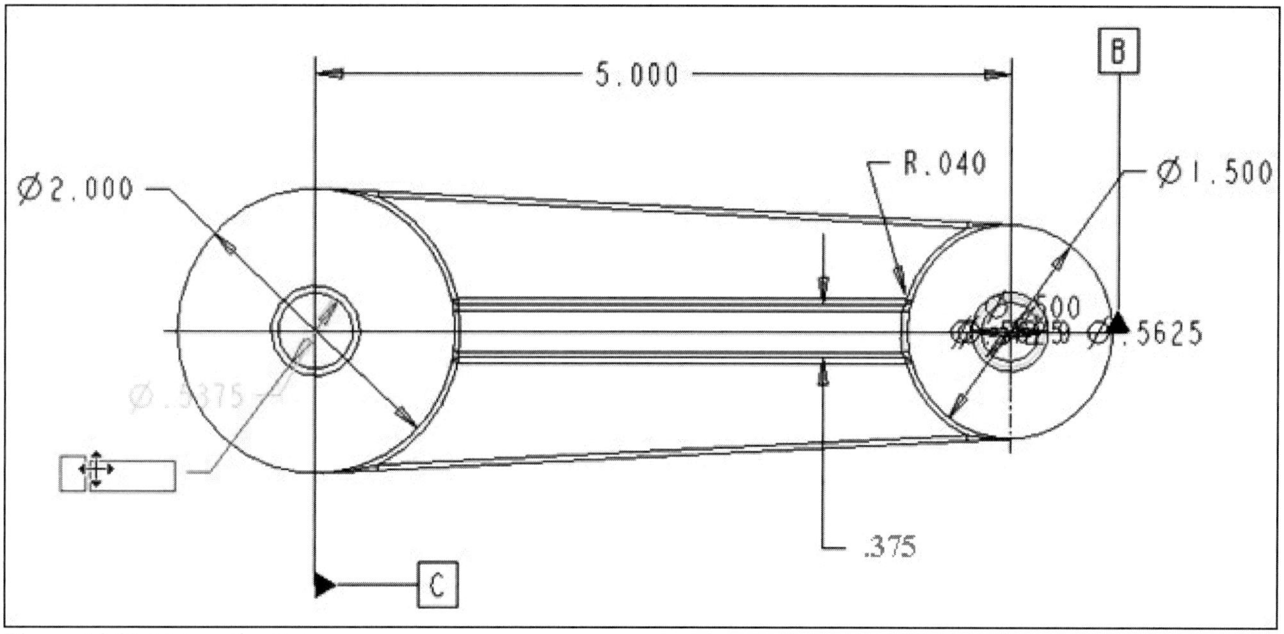

Figure 10.9(v) Move the Note

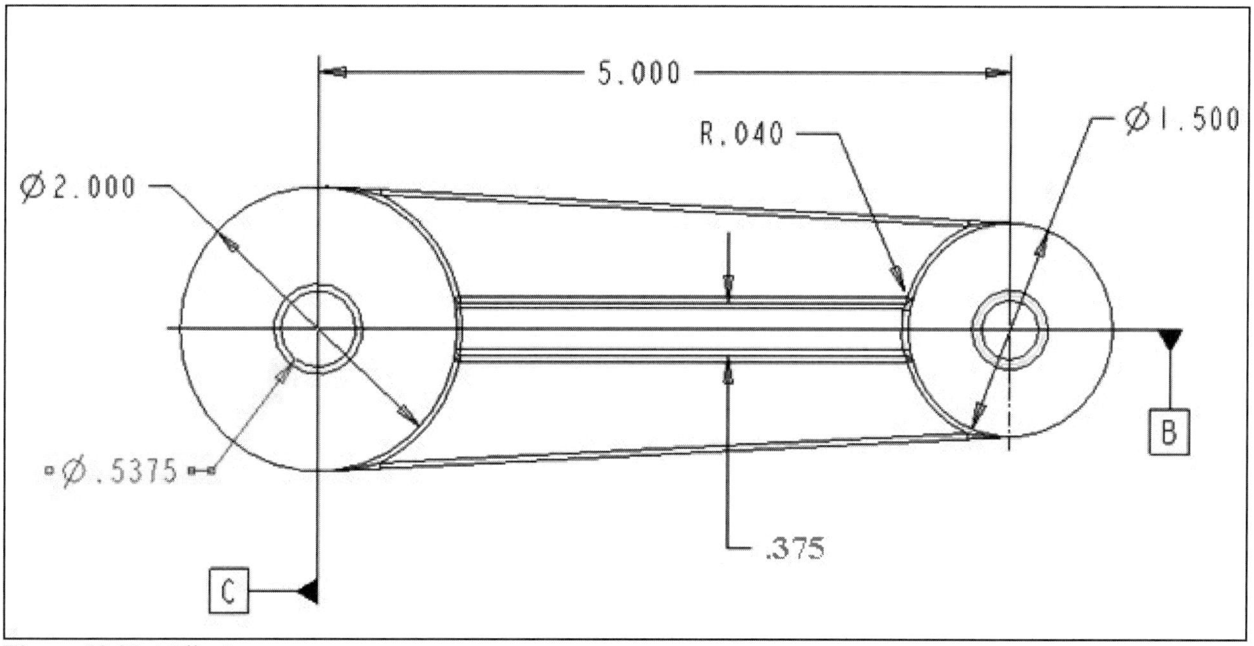

Figure 10.9(w) Flip Arrows

Erase unneeded dimensions and reposition dimensions, notes, and datum tags as needed [Fig. 10.9(x)]. Pick on the note [Fig. 10.9(x)] ⇒ **RMB** ⇒ **Move Item to View** [Fig. 10.9(y)] ⇒ pick on the front view

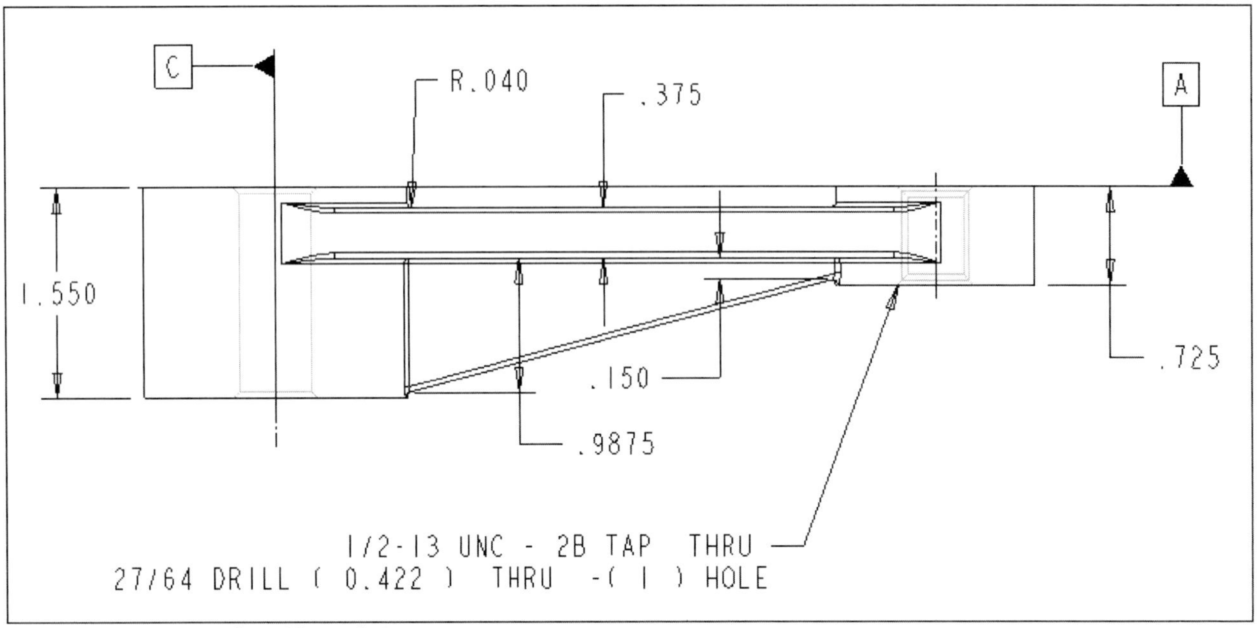

Figure 10.9(x) Top View Cleaned Up

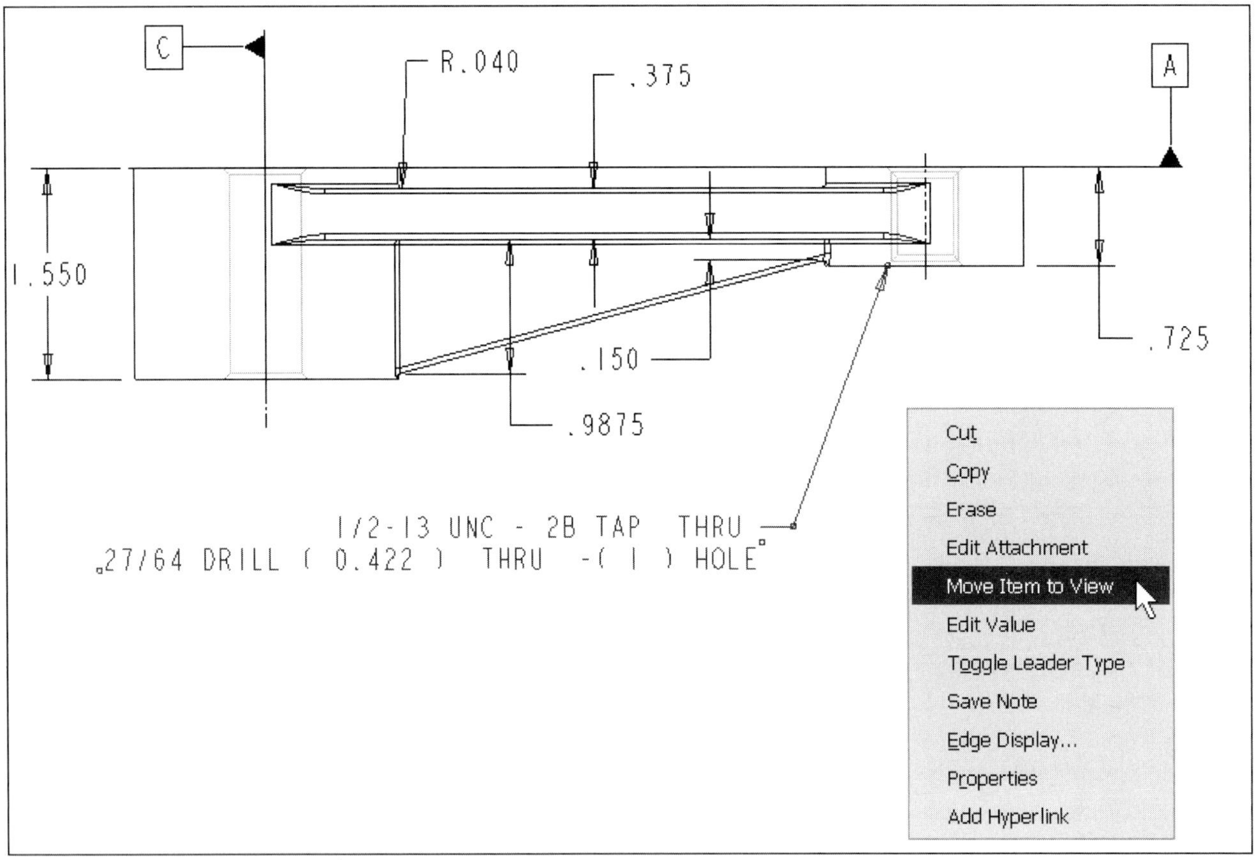

Figure 10.9(y) Move Item to View

Reposition the notes as needed [Fig. 10.9(z)]. Click: 🔍 **Refit ⇒ Ctrl+S ⇒ Enter ⇒ Window ⇒ Close**

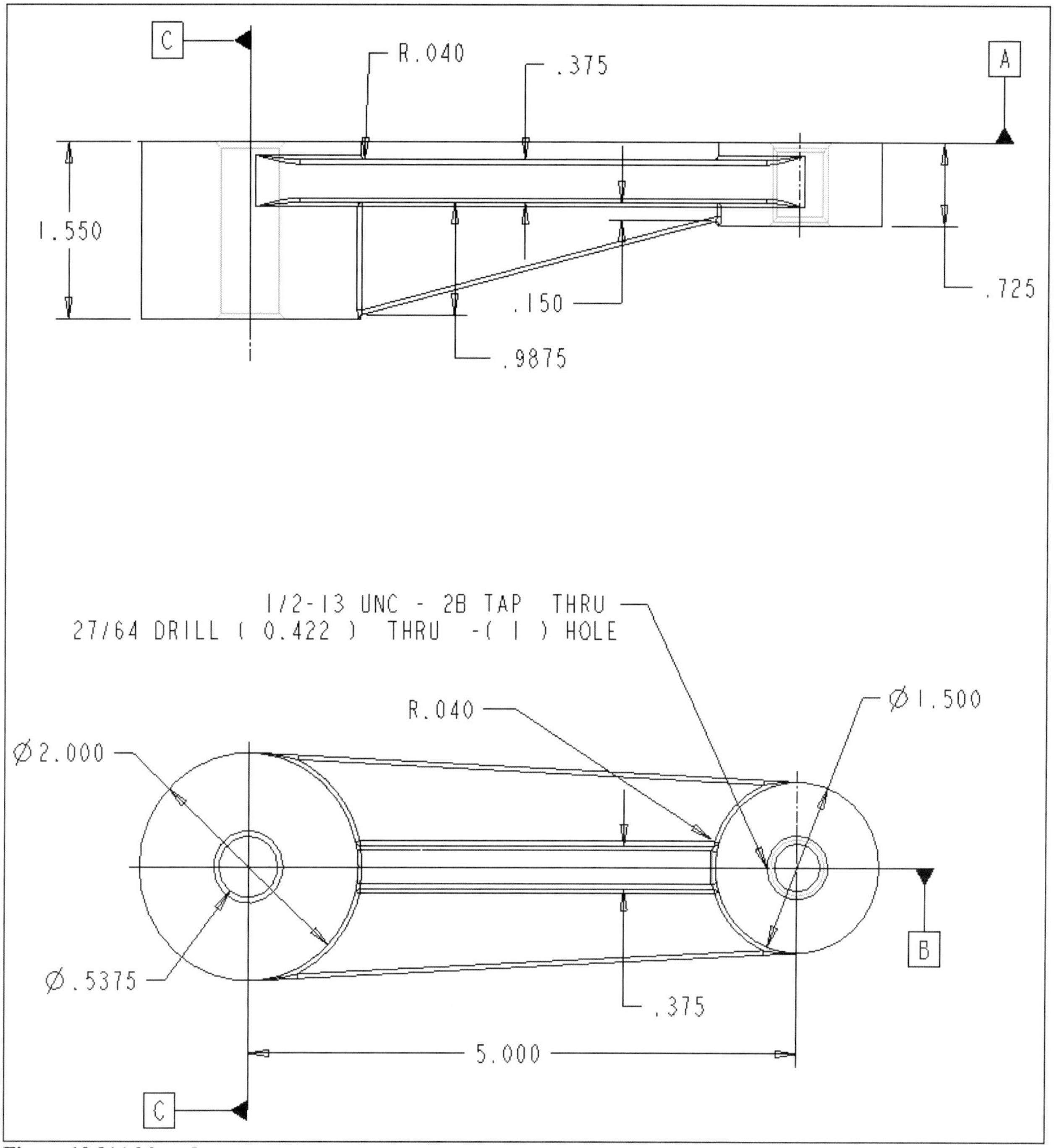

Figure 10.9(z) Move Item

Though not complete, or correct to ASME Y14.5M standards, we will leave this drawing as is. You may finalize the drawing after completing the next lesson.

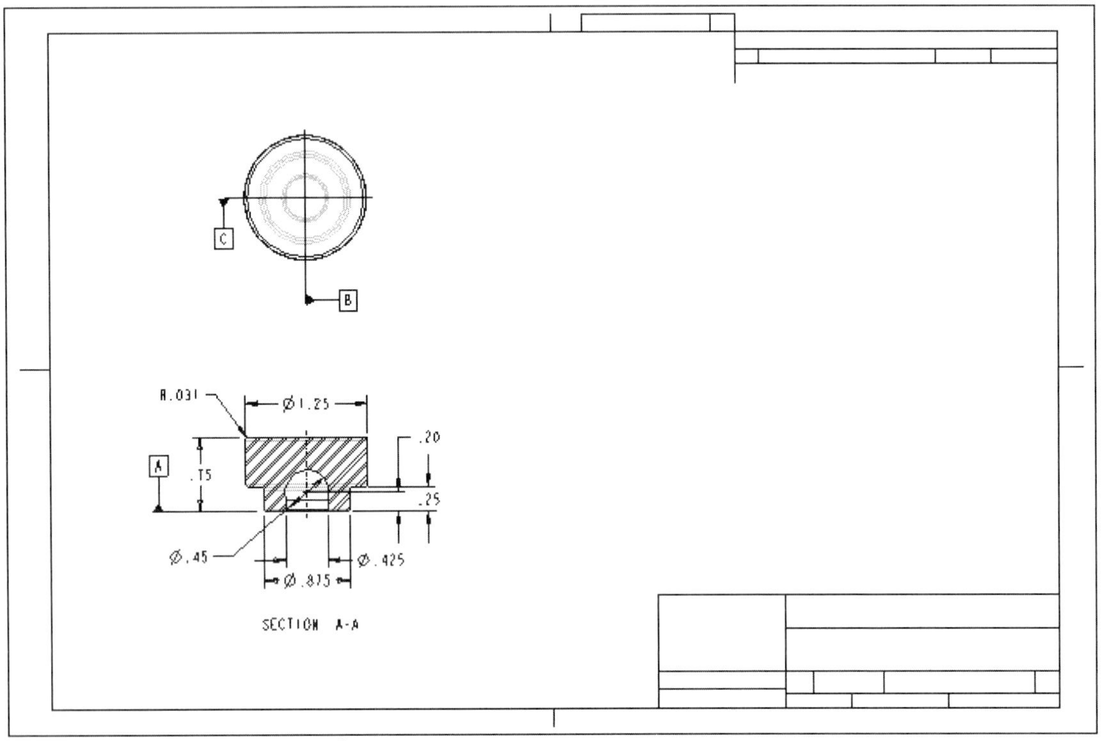

Figure 10.10 Clamp Foot Drawing

Creating a Drawing of the Clamp Foot

Before starting the drawing you will need to set your directory to the folder that contains the Swing Clamp Assembly (and therefore all the components of the assembly), and this time you will bring the Clamp_Foot (Fig. 10.10) in session by opening it. The standard three views of the model will display automatically on the drawing, since you will use the default drawing template for this project.

Click: **File** ⇒ **Set Working Directory** ⇒ select the directory where the **clamp_assembly.asm** was saved ⇒ **OK** ⇒ **File** ⇒ **Open** ⇒ **clamp_foot.prt** ⇒ **Open** ⇒ **Tools** ⇒ **Environment** ⇒ **Trimetric** (Fig. 10.11) ⇒ **OK** ⇒ **Create a new object** ⇒ Drawing ⇒ Name **clamp_foot** ⇒ Use default template ⇒ **OK** ⇒ Template: **b_drawing** [Fig. 10.12(a)] ⇒ **OK** [Fig. 10.12(b)]

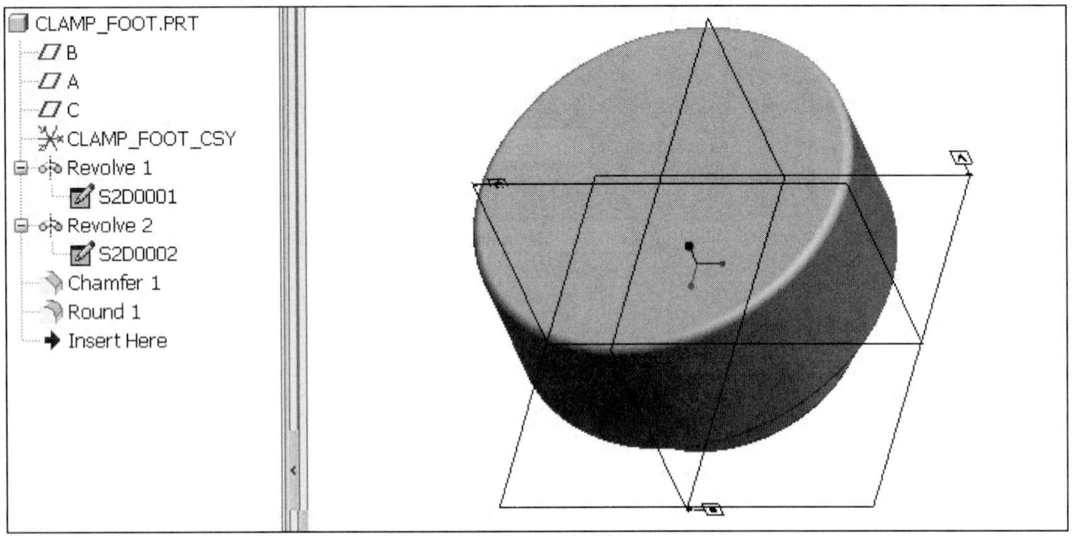

Figure 10.11 Clamp_Foot

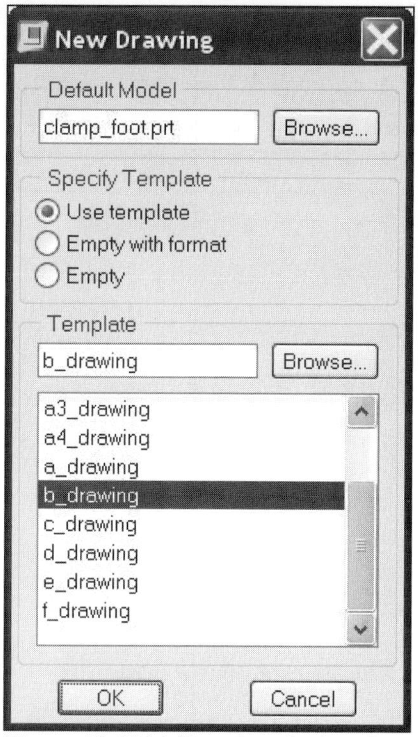

Figure 10.12(a) Template: b_drawing

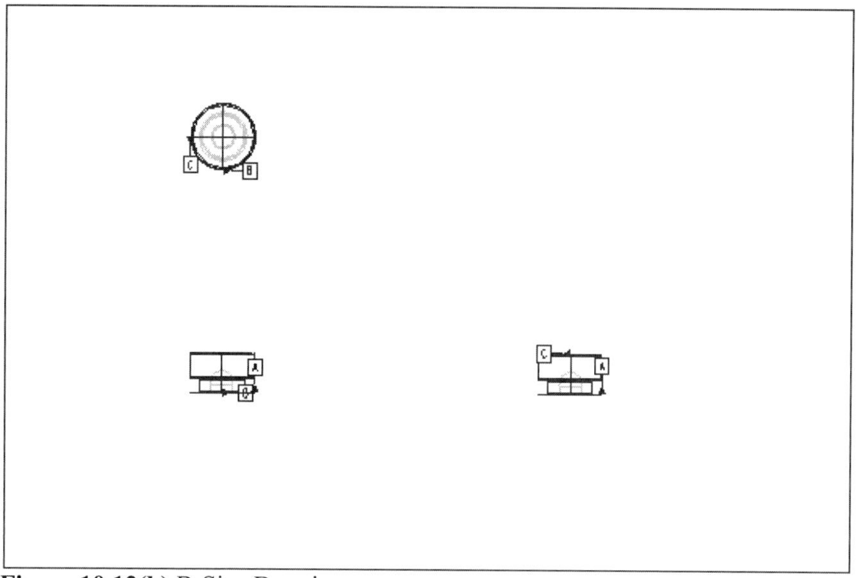

Figure 10.12(b) B-Size Drawing

Click: 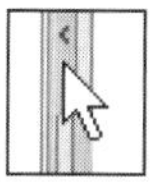 close the Navigator shade ⇒ **RMB** ⇒ **Page Setup** [Fig. 10.12(c)] ⇒ 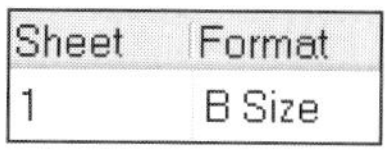

⇒ 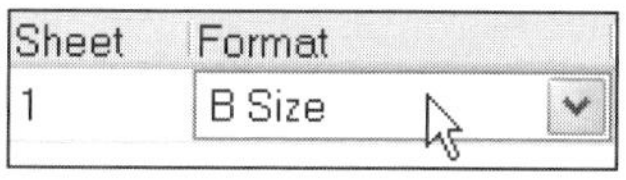

⇒  ⇒ **Browse** ⇒ System Formats Open dialog box opens ⇒ pick **b.frm** ⇒ **Open** ⇒ **OK** ⇒  ⇒  ⇒ **MMB** [Fig. 10.12(d)]

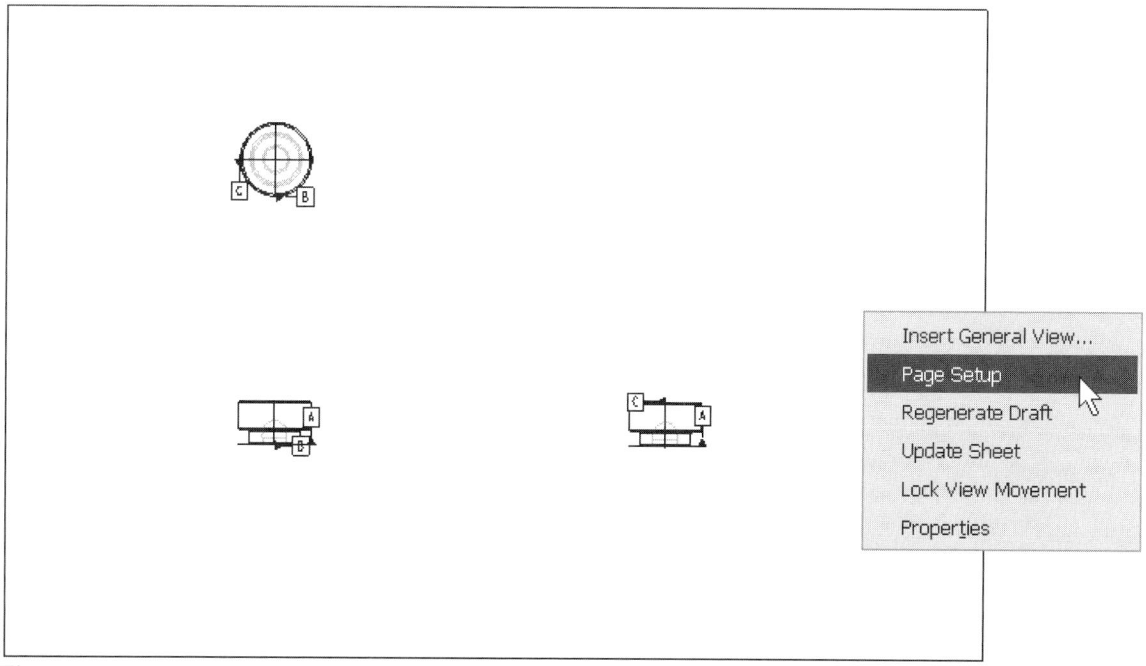

Figure 10.12(c) Page Setup

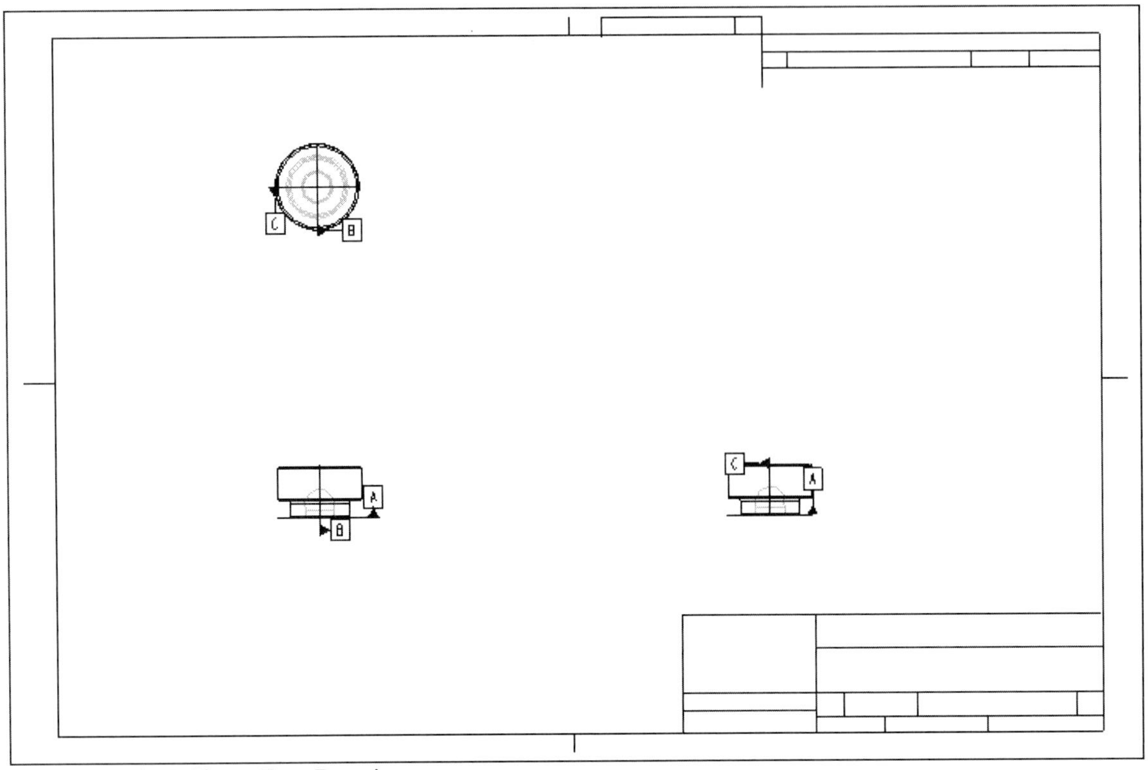

Figure 10.12(d) Clamp Foot Drawing

Double-click on **SCALE: 1.000** [Fig. 10.12(e)] ⇒ type **1.500** ⇒ **Enter** [Fig. 10.12(f)]

Figure 10.12(e) Scale **1.500**

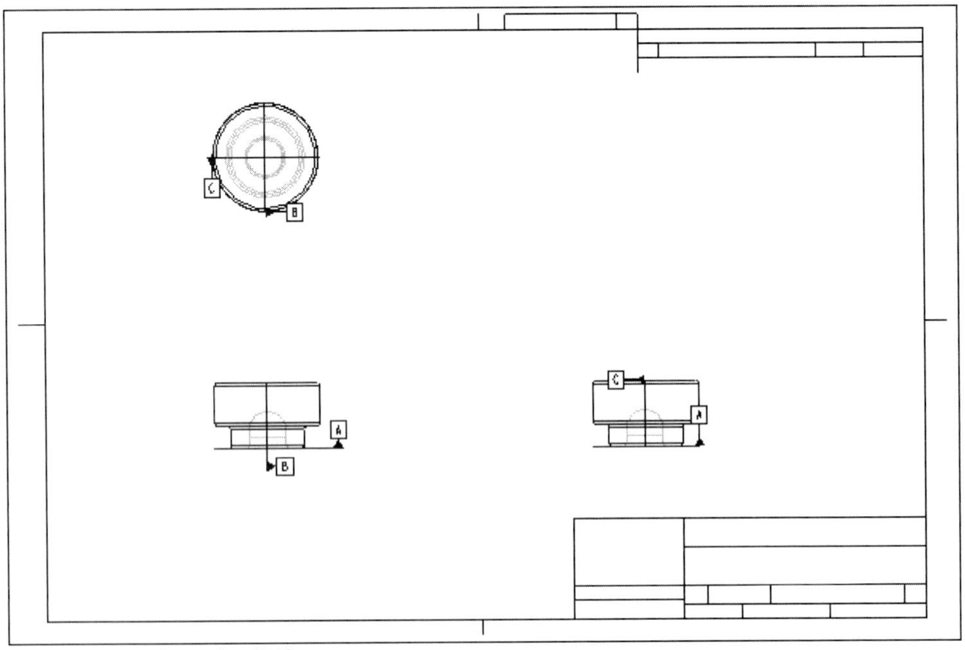

Figure 10.12(f) Resized Views

Click: off ⇒ **Update** ⇒ **Repaint** ⇒ **Open the Show/Erase dialog box** ⇒

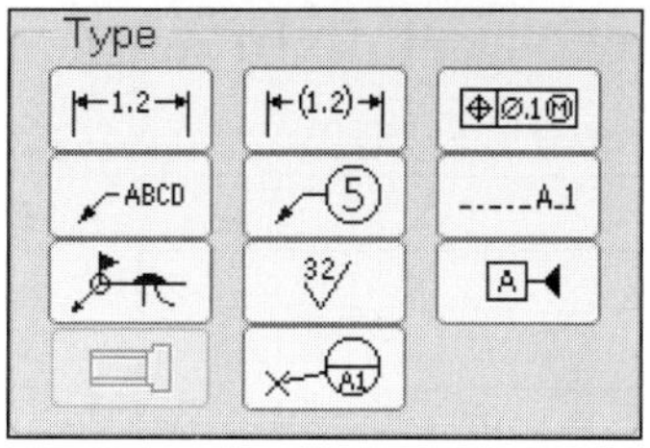

toggle on all options ⇒ **Show All** ⇒ **Yes** ⇒ **Accept All** ⇒ **Close** ⇒ **RMB** ⇒ **Cleanup Dimensions** ⇒ Create Snap Lines ⇒ **Apply** ⇒ **Close** [Fig. 10.12(g)] ⇒ pick on the right side view ⇒ **RMB** [Fig. 10.12(h)] ⇒ **Delete** ⇒ **Ctrl+S** ⇒ **OK**

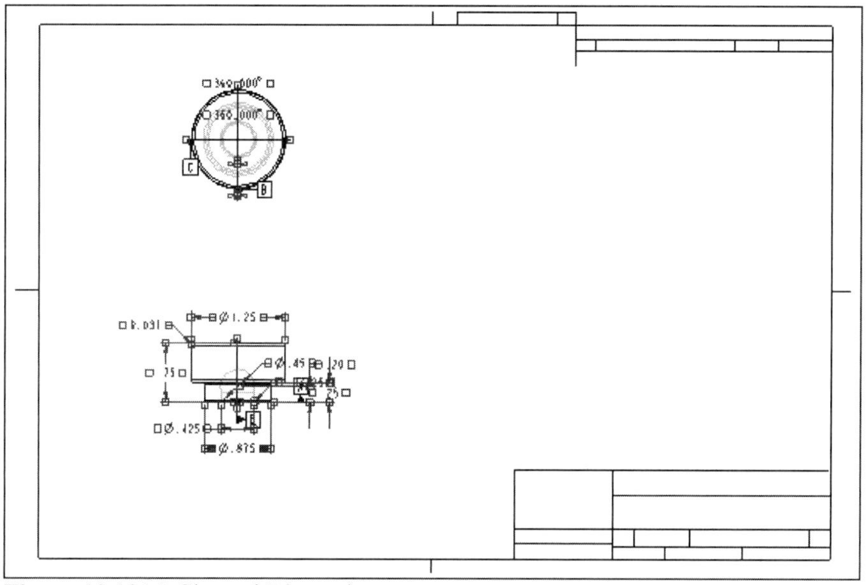

Figure 10.12(g) Cleaned Dimensions

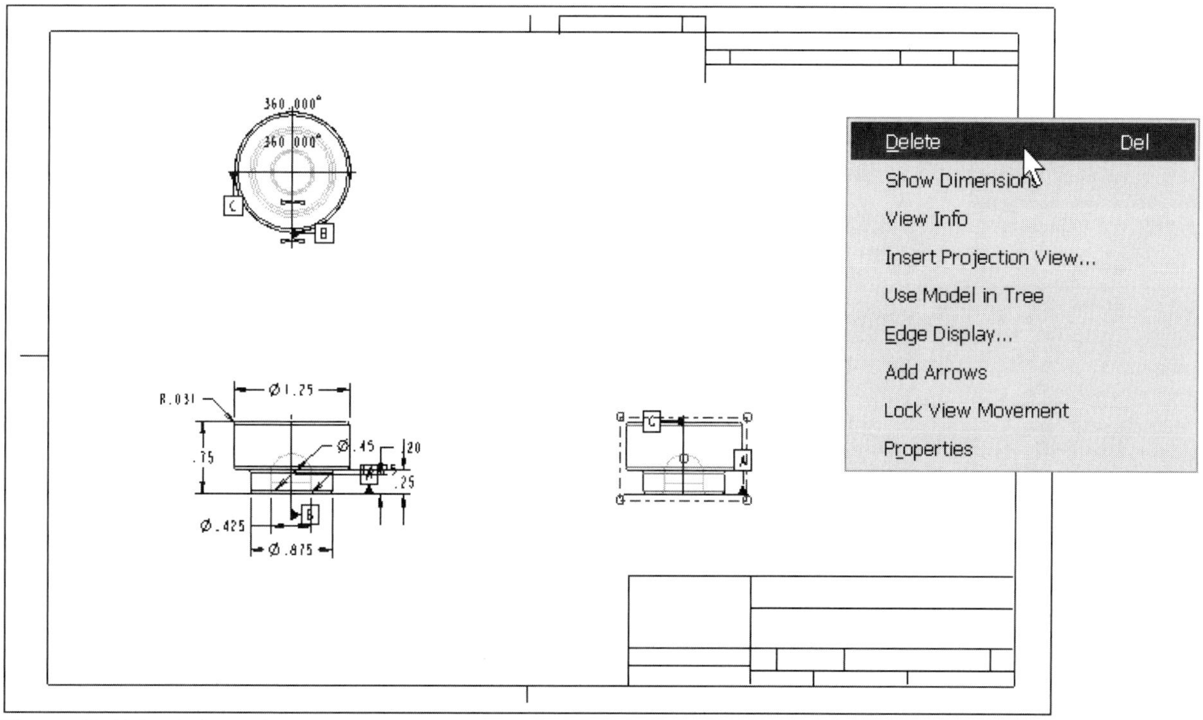

Figure 10.12(h) Delete View

Click: **Window** [Fig. 10.12(i)] ⇒ **CLAMP_FOOT.PRT** [Fig. 10.12(j)] ⇒ **Start the view manager** ⇒ **Xsec** tab ⇒ **New** ⇒ **A** [Fig. 10.12(k)] ⇒ **Enter** ⇒ **Done** ⇒ pick on datum **C** [Fig. 10.12(l)] ⇒ **Close** ⇒ **Window** ⇒ **CLAMP_FOOT.DRW** [Fig. 10.12(m)] ⇒ **Ctrl+S** ⇒ **MMB** [Fig. 10.12(n)]

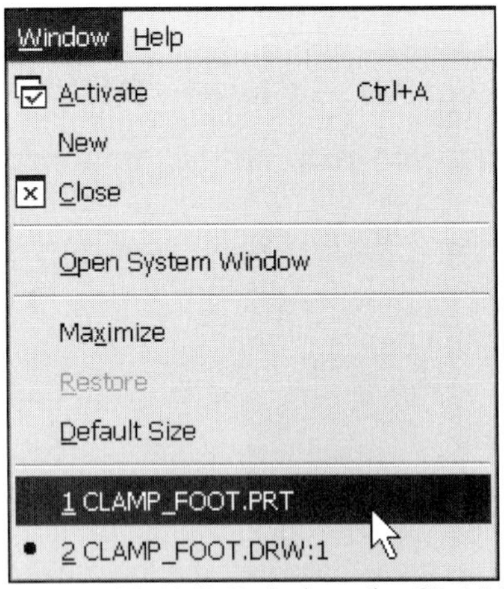

Figure 10.12(i) Switch to the CLAMP_FOOT Window

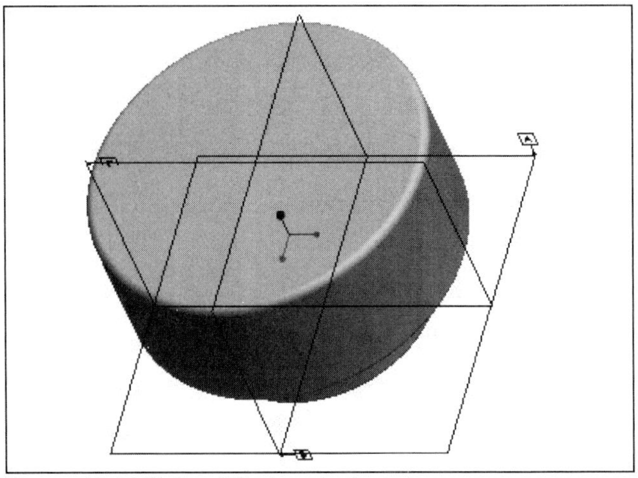

Figure 10.12(j) Part Window Activated

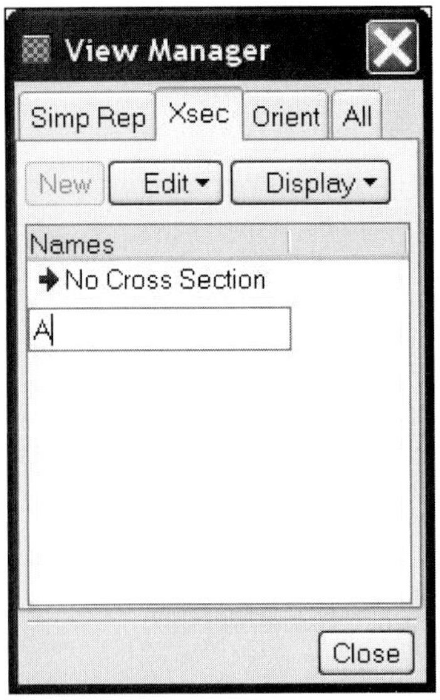

Figure 10.12(k) View Manager Dialog Box

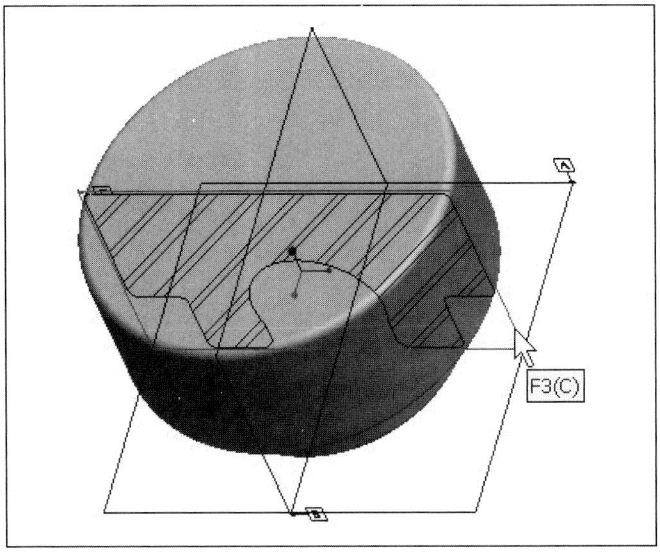

Figure 10.12(l) Part Xsection A

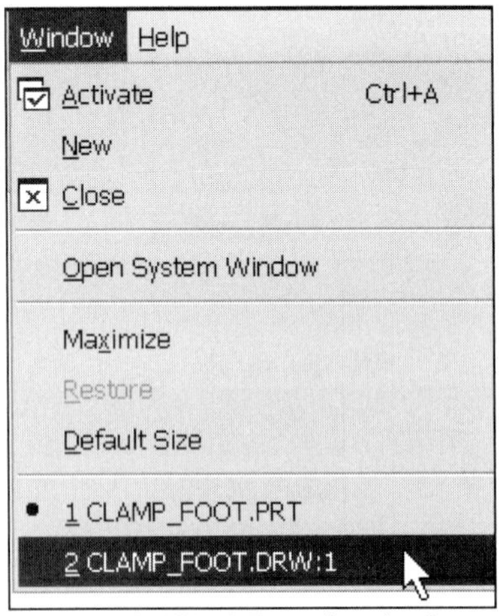

Figure 10.12(m) Drawing Active

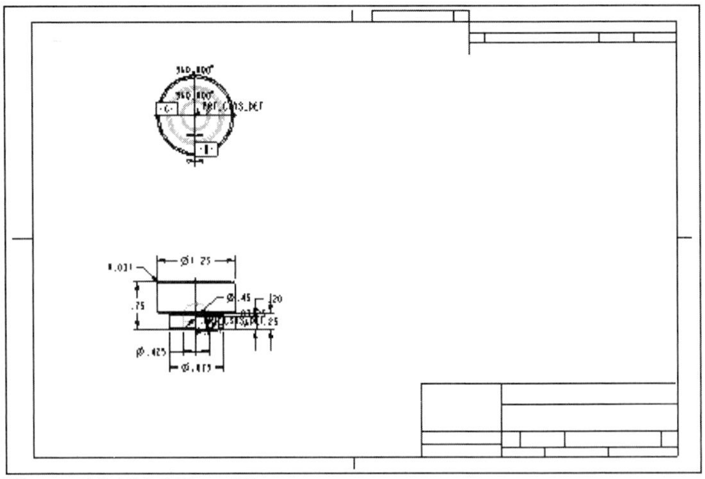

Figure 10.12(n) Two Views

Pick on the front view ⇒ **RMB** ⇒ **Properties** [Fig. 10.12(o)] ⇒ **Sections** ⇒ 2D cross-section ⇒ + ⇒ A ⇒ **Apply** [Fig. 10.12(p)] ⇒ **Close**

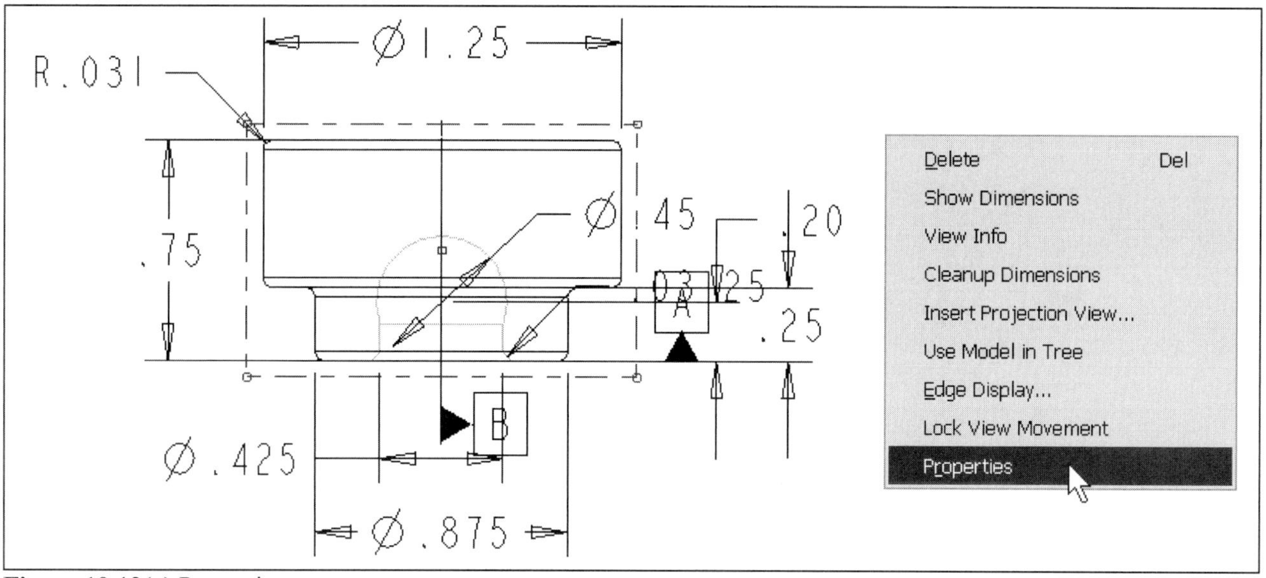

Figure 10.12(o) Properties

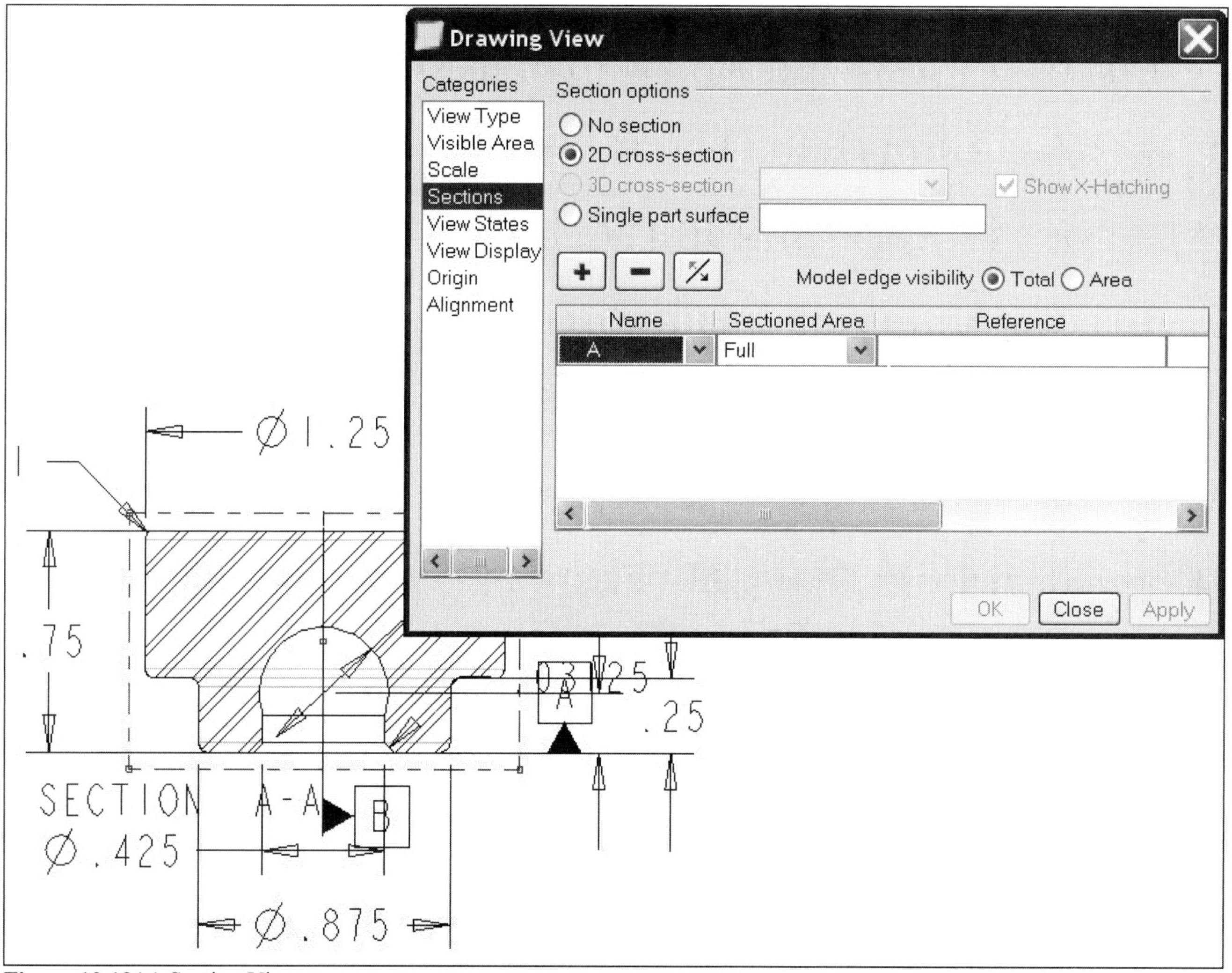

Figure 10.12(p) Section View

Cleanup the drawing views (by erasing unneeded dimensions, moving datums, etc.) [Fig. 10.12(q)].
Click: **Update** ⇒ ⇒ **Ctrl+S** ⇒ **MMB** ⇒ **File** ⇒ **Delete** ⇒ **Old Versions** ⇒ **Enter** ⇒ **Window** ⇒ **Close** ⇒ **Window** ⇒ **Close**

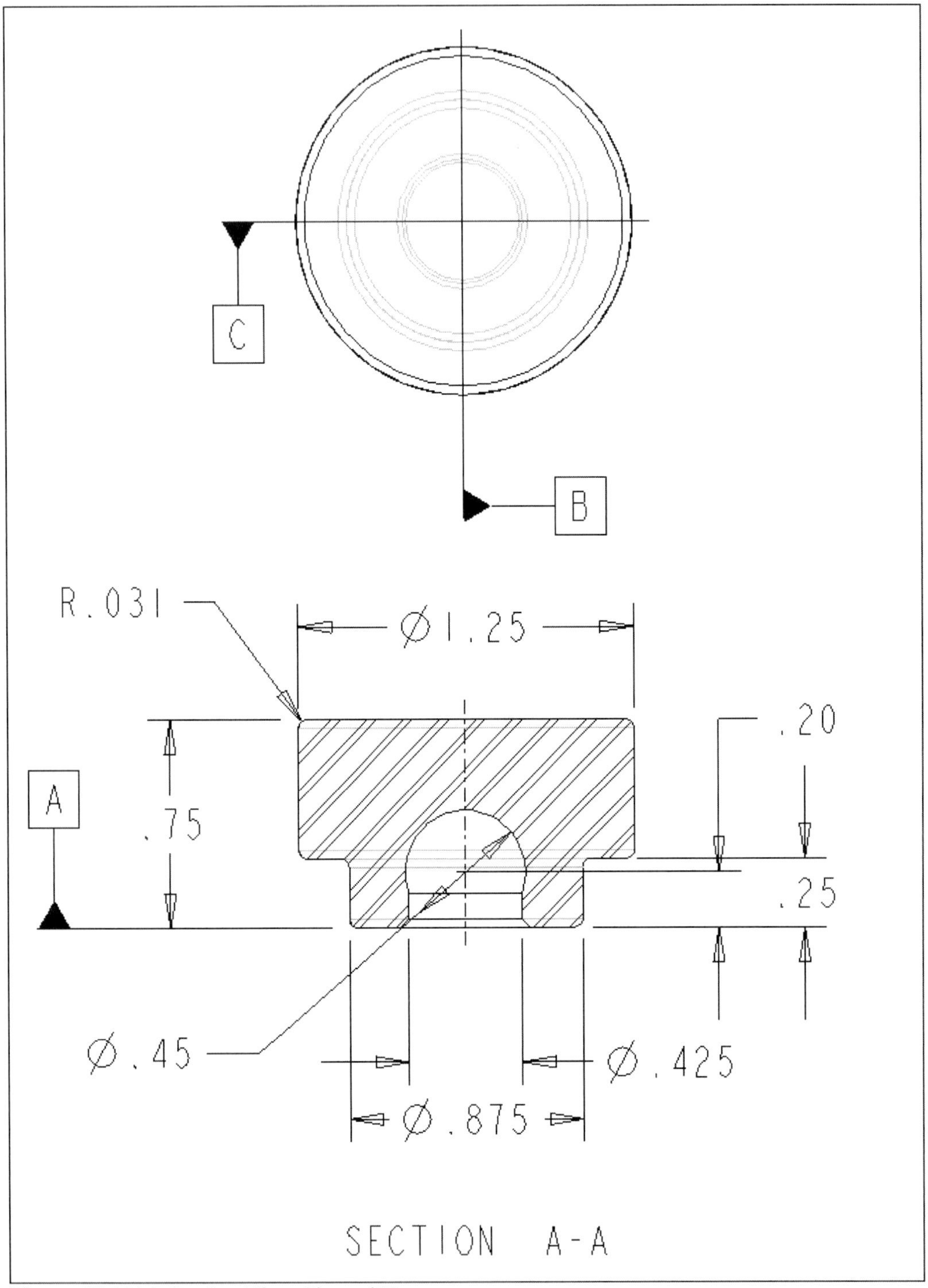

Figure 10.12(q) Section View Cleaned Up

Though not complete, or correct to ASME Y14.5M standards, we will leave this drawing as is. You may finalize the drawing after completing the detail drawing in the next lesson.

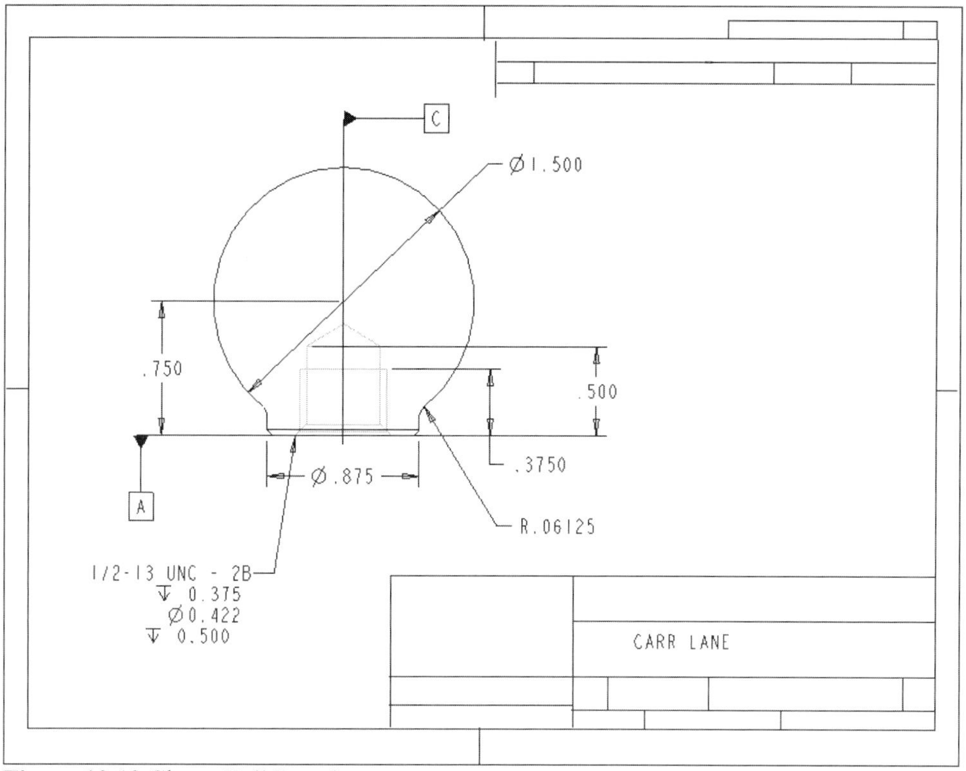

Figure 10.13 Clamp Ball Drawing

Creating a Drawing of the Clamp Ball

Before starting the Clamp Ball drawing (Fig. 10.13) for the Clamp_Ball (Fig. 10.14), you will create a template that can be used on any drawing.

Click: **File** ⇒ **Set Working Directory** ⇒ select the directory where the **clamp_assembly.asm** was saved ⇒ **OK** ⇒ **Create a new object** ⇒ Drawing ⇒ Name: **detail_template** ⇒ Use default template ⇒ **OK** ⇒ Default Model **Browse** ⇒ pick **clamp_ball.prt** ⇒ **Open** ⇒ Empty with format ⇒ Format: **Browse** ⇒ **a.frm** ⇒ **Open**

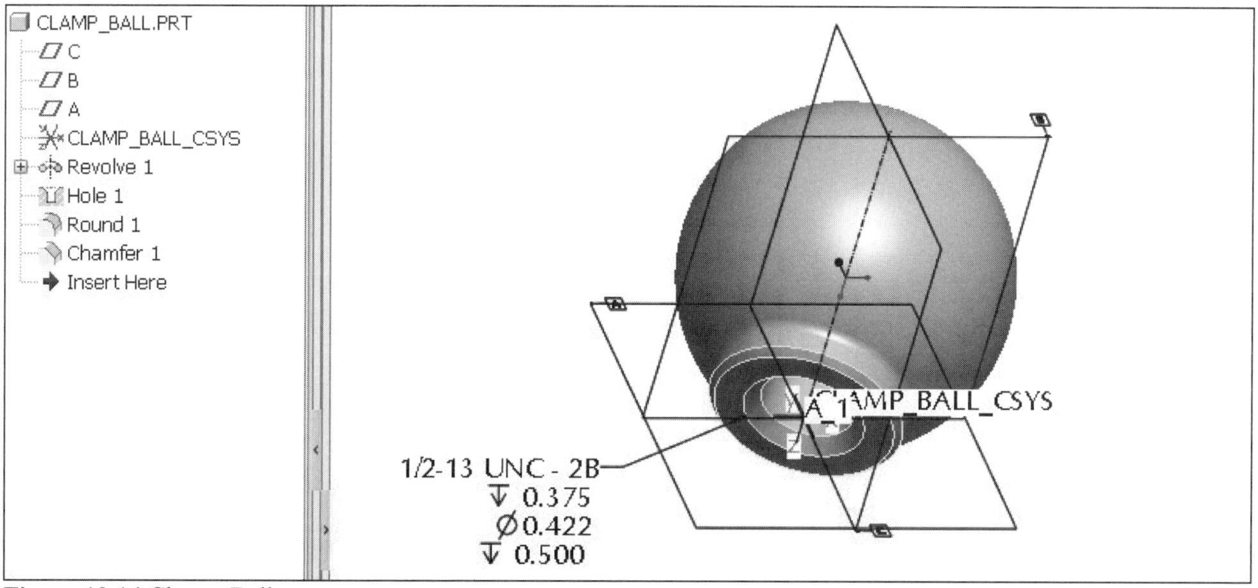

Figure 10.14 Clamp_Ball

Click: **OK** [Fig. 10.15(a)] ⇒ **Applications** from menu bar ⇒ **Template** [Fig. 10.15(b)]

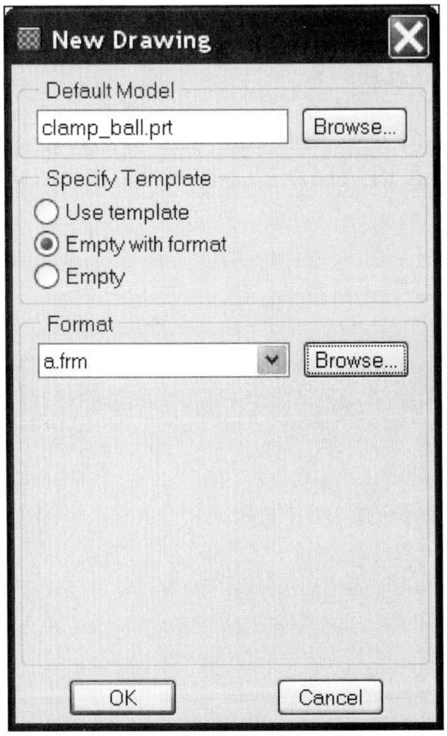

Figure 10.15(a) New Drawing Dialog Box

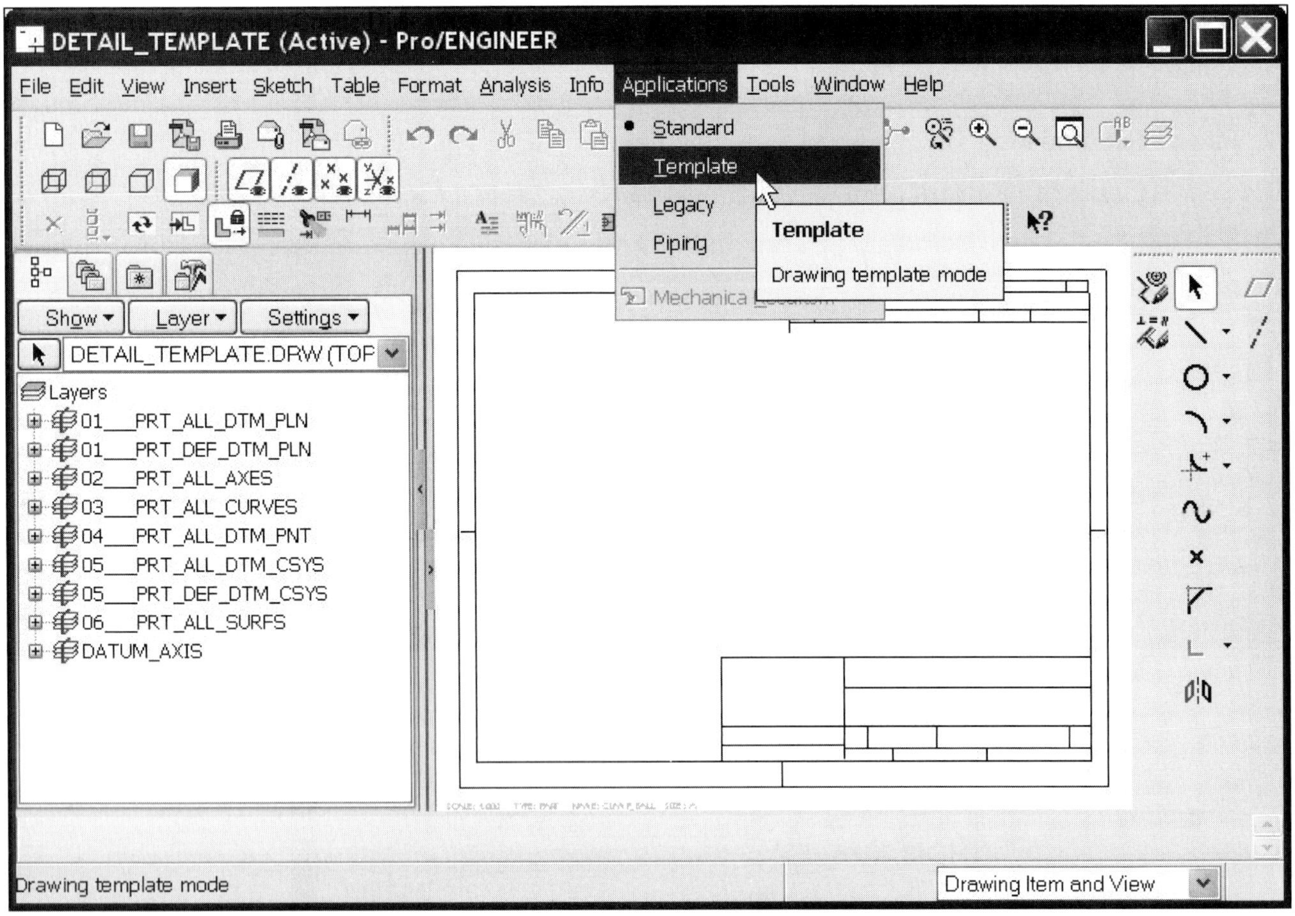

Figure 10.15(b) Applications ⇒ Template

Click: **Insert** from menu bar ⇒ **Template View** [Fig. 10.15(c)] ⇒ View Values Orientation: type **TOP**⇒ Scale ⇒ View Scale: 2.000000 ⇒ **Enter** ⇒ Dimensions ⇒ Incremental Spacing: 0.500000 Initial Offset: 0.500000 ⇒ **Enter** ⇒ Tan Edge Display ⇒ No Disp Tan ⇒ Model Display ⇒ Hidden Line ⇒ **Place View** [Fig. 10.15(d)]

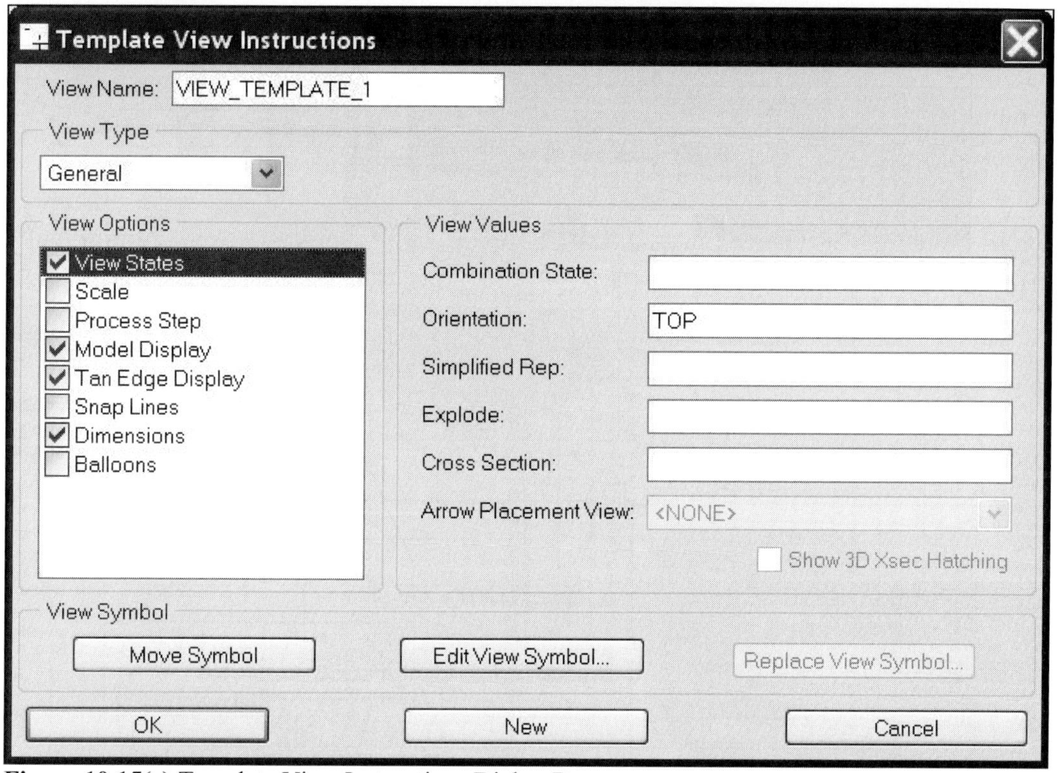

Figure 10.15(c) Template View Instructions Dialog Box

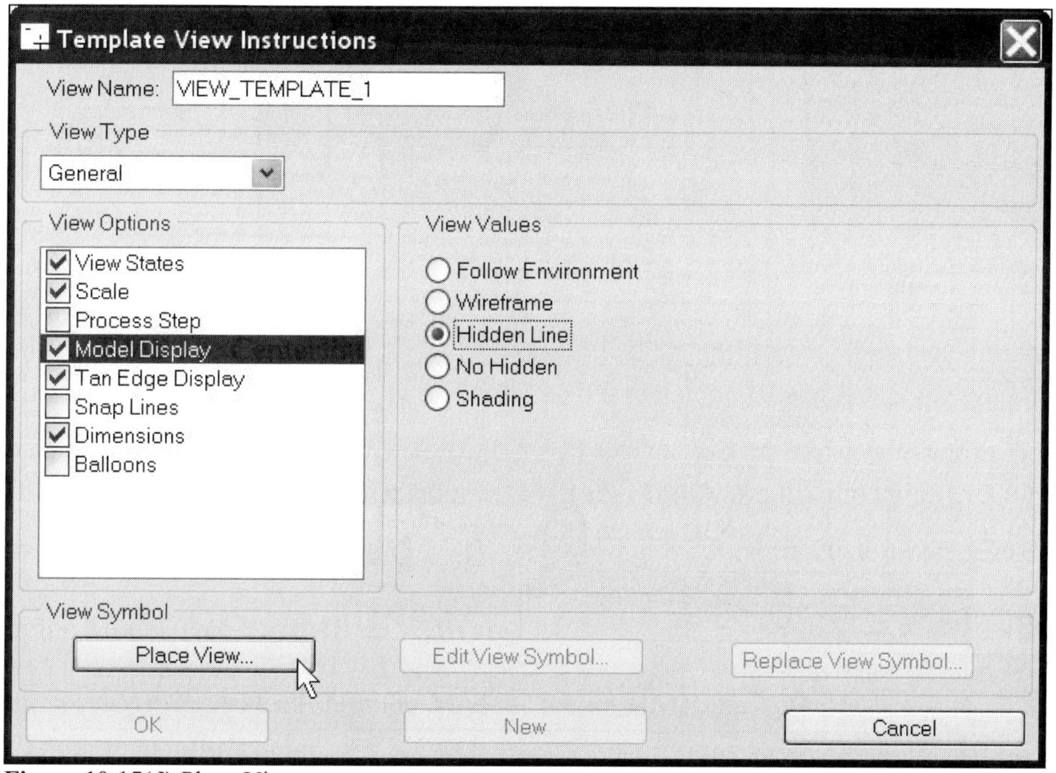

Figure 10.15(d) Place View

Pick a position for the front view [Fig. 10.15(e)] ⇒ **OK** ⇒ **Ctrl+S** ⇒ **Enter** ⇒ **File** ⇒ **Close Window** ⇒ **File** ⇒ **Erase** ⇒ **Not Displayed**

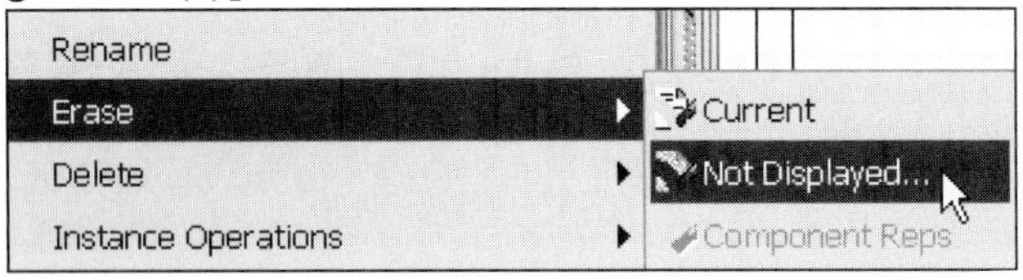

⇒ **OK** ⇒ **File** ⇒ **Open** ⇒ **clamp_ball.prt** ⇒ **Open** [Fig. 10.15(f)]

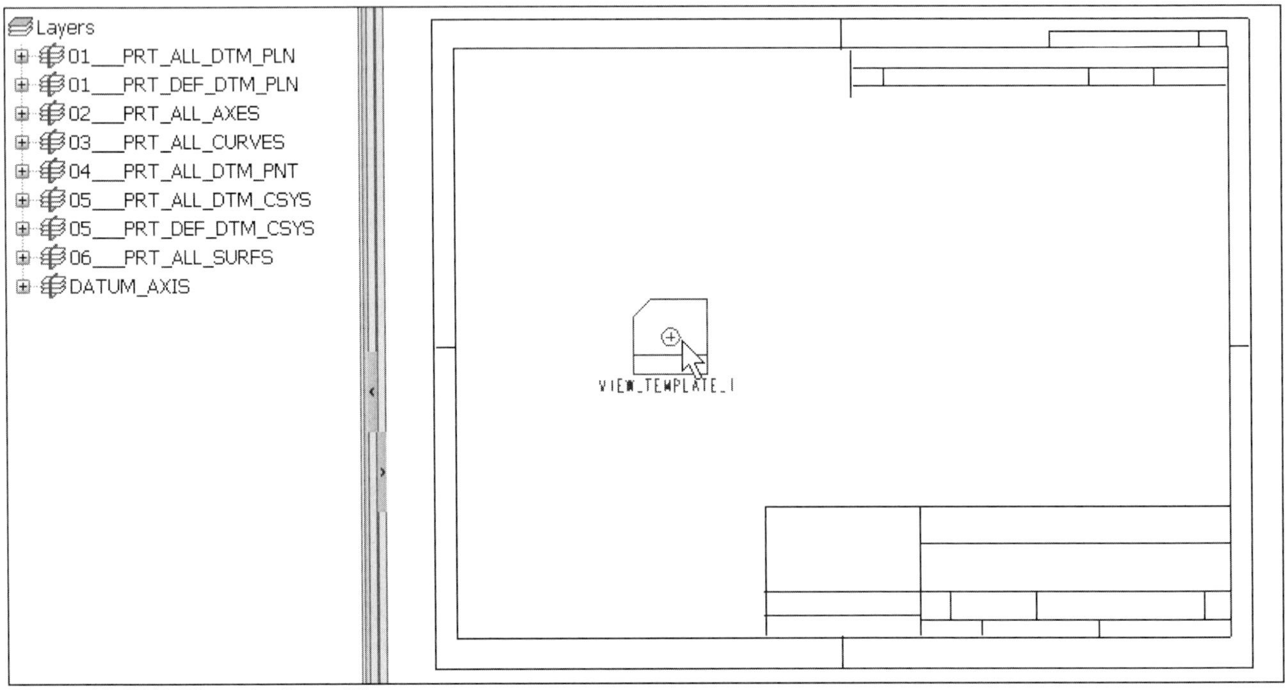

Figure 10.15(e) Place the Front View

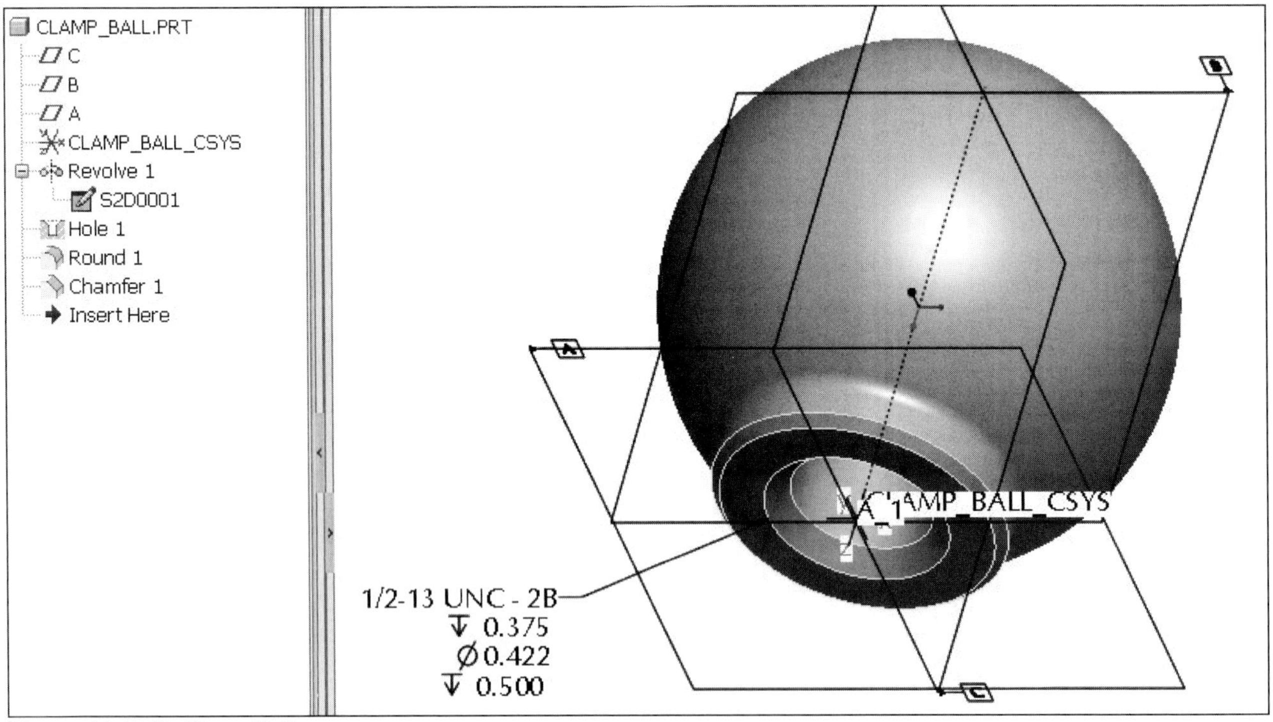

Figure 10.15(f) Clamp Ball

Click: Create a new object ⇒ Drawing ⇒ Name **clamp_ball** ⇒ Use default template ⇒ **OK** ⇒ Use template ⇒ Template **Browse** ⇒ detail_template.drw ⇒ **Open** [Fig. 10.15(g)] ⇒ **OK** [Fig. 10.15(h)] ⇒ **RMB** ⇒ **Lock View Movement** uncheck ⇒ pick on the view and move it to a better position [Fig. 10.15(i)] ⇒ ⇒ **Ctrl+S** ⇒ **MMB** ⇒ **LMB** to deselect

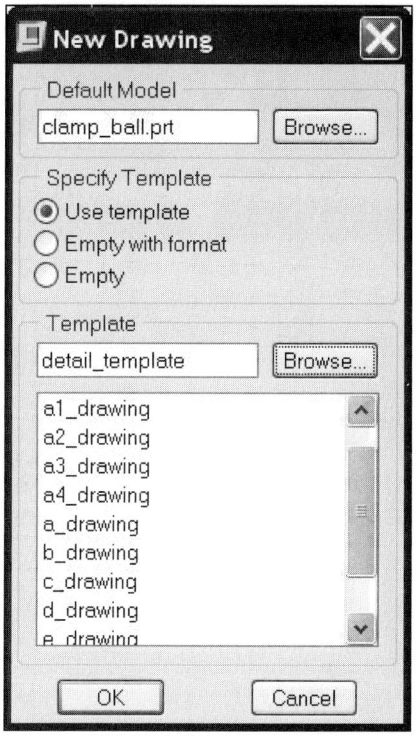

Figure 10.15(g) Template: detail_template

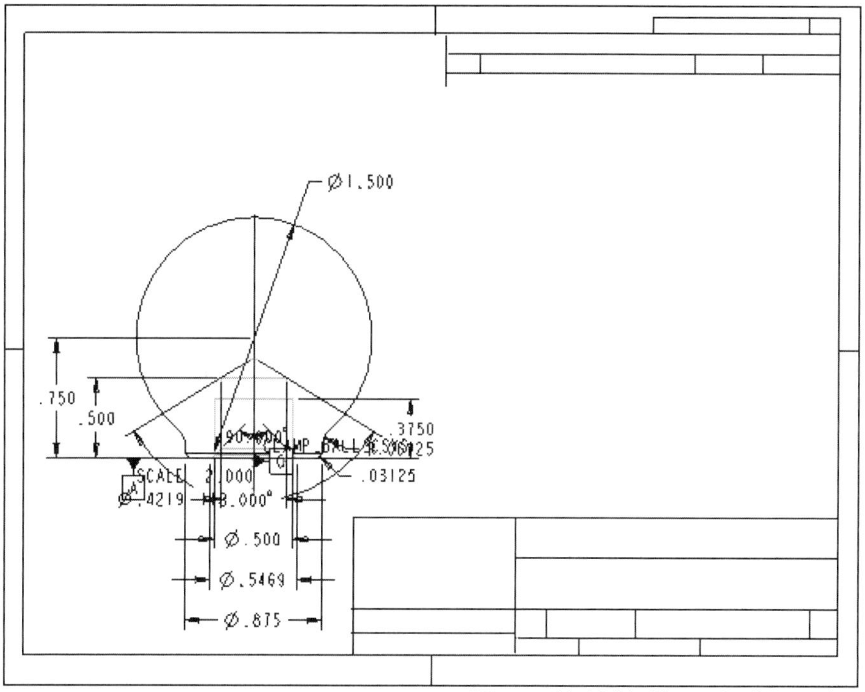

Figure 10.15(h) Clamp Ball Drawing

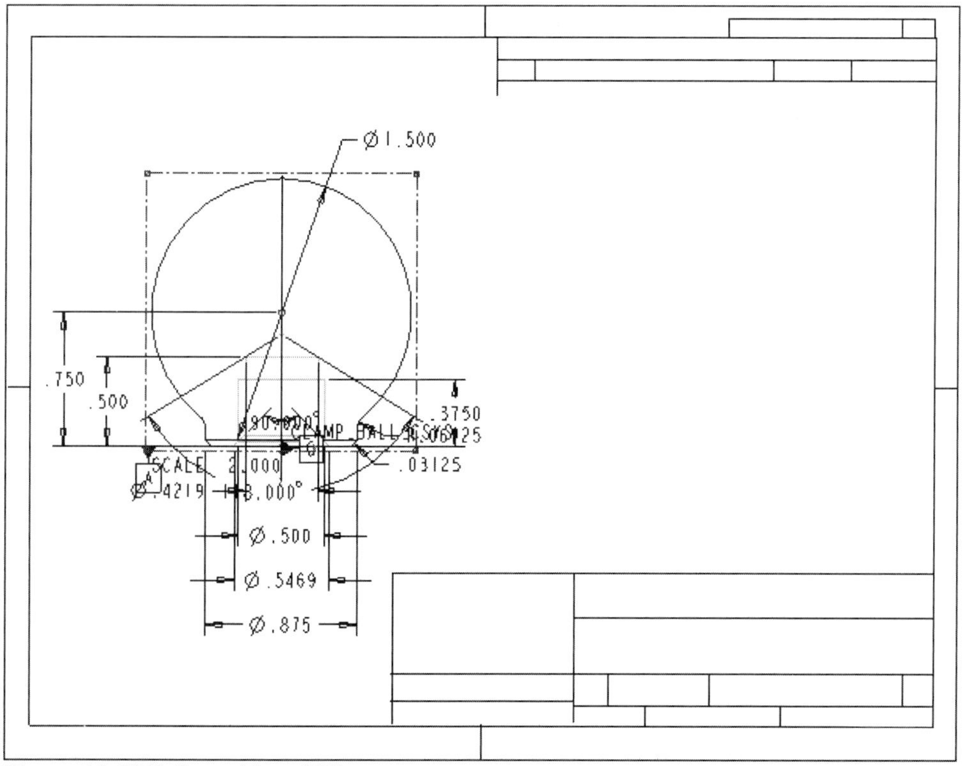

Figure 10.15(i) Repositioned View

Click: off ⇒ cleanup the drawing views by erasing unneeded dimensions, the SCALE note, moving datums, etc. ⇒ **Update** ⇒ ⇒ **Ctrl+S** ⇒ **Enter** [Fig. 10.15(j)] ⇒ **File** ⇒ **Delete** ⇒ **Old Versions** ⇒ **Enter**

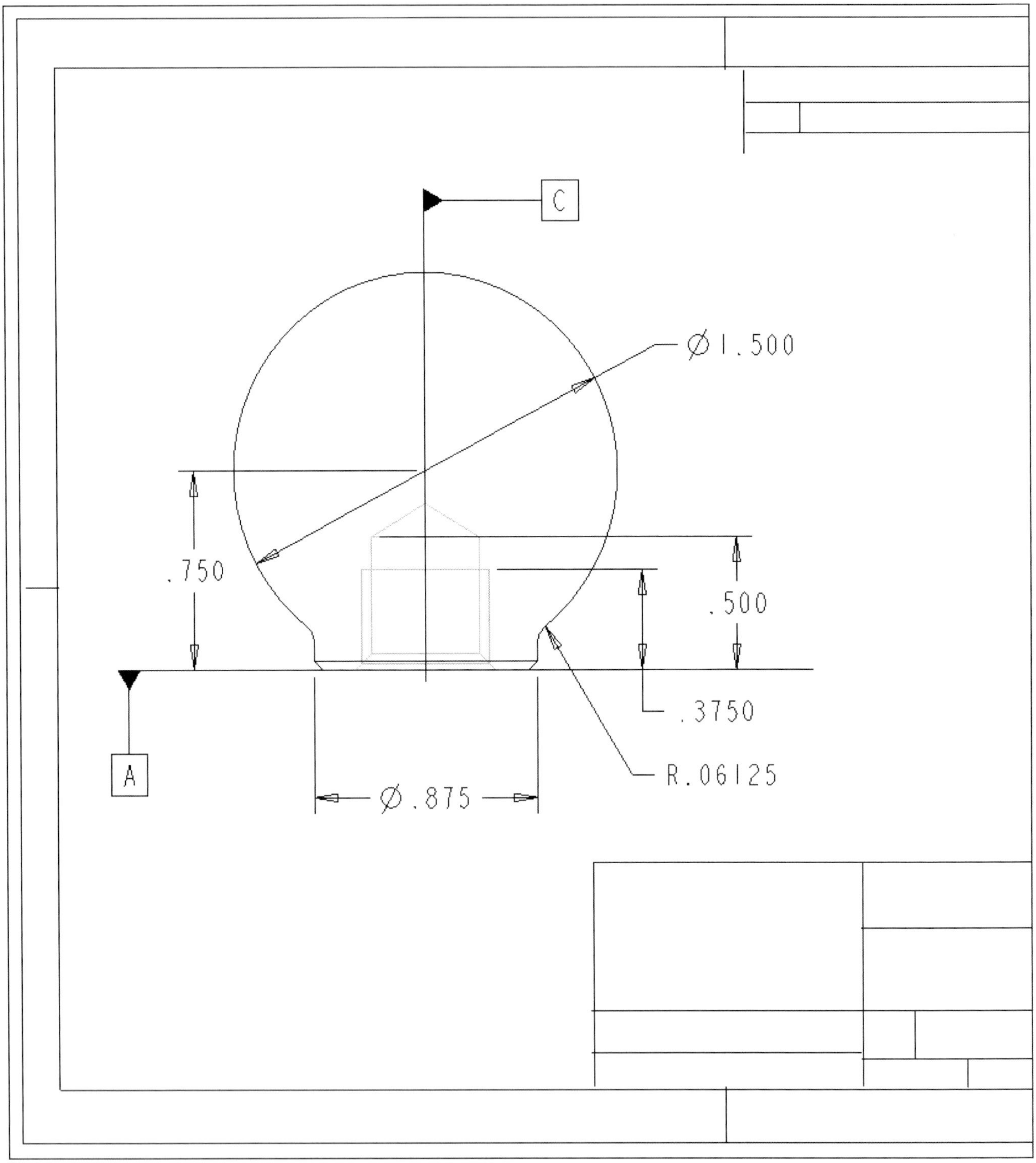

Figure 10.15(j) Axis Added to View

Click: **Create a note** ⇒ **Make Note** ⇒ Select LOCATION for note [Fig. 10.15(k)] ⇒ type Enter NOTE: CARR LANE ⇒ ⇒ ⇒ **Done/Return** [Fig. 10.15(l)] ⇒ **LMB** to deselect ⇒ ⇒ **Show** ⇒ only on ⇒ View ⇒ **Show All** [Fig. 10.15(m)]⇒ **Yes** ⇒ **OK** ⇒ **Close** [Fig. 10.15(n)]

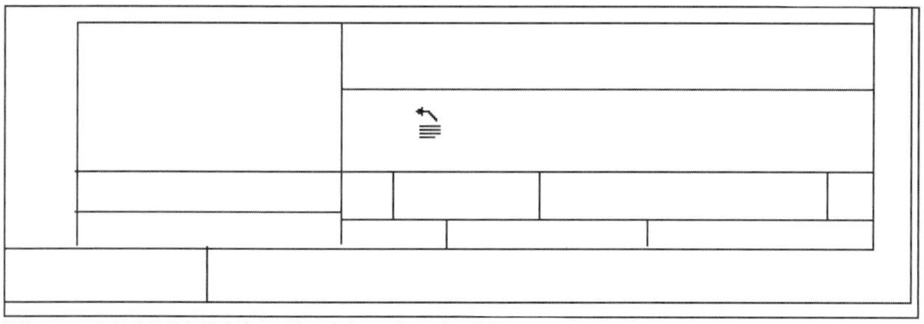

Figure 10.15(k) Pick a Position for the Note

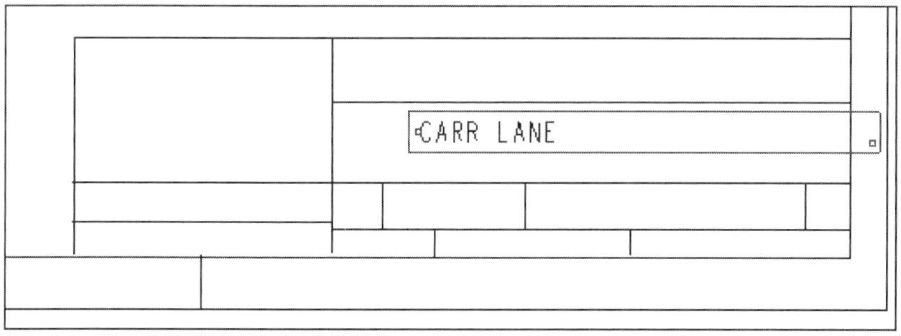

Figure 10.15(l) CARR LANE Note

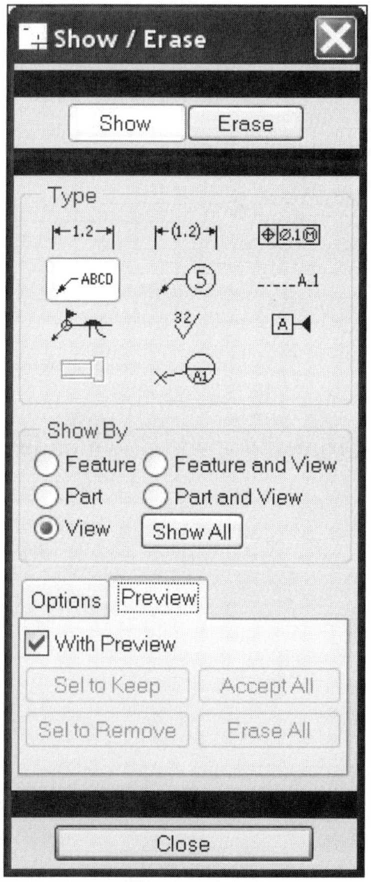

Figure 10.15(m) Show Note

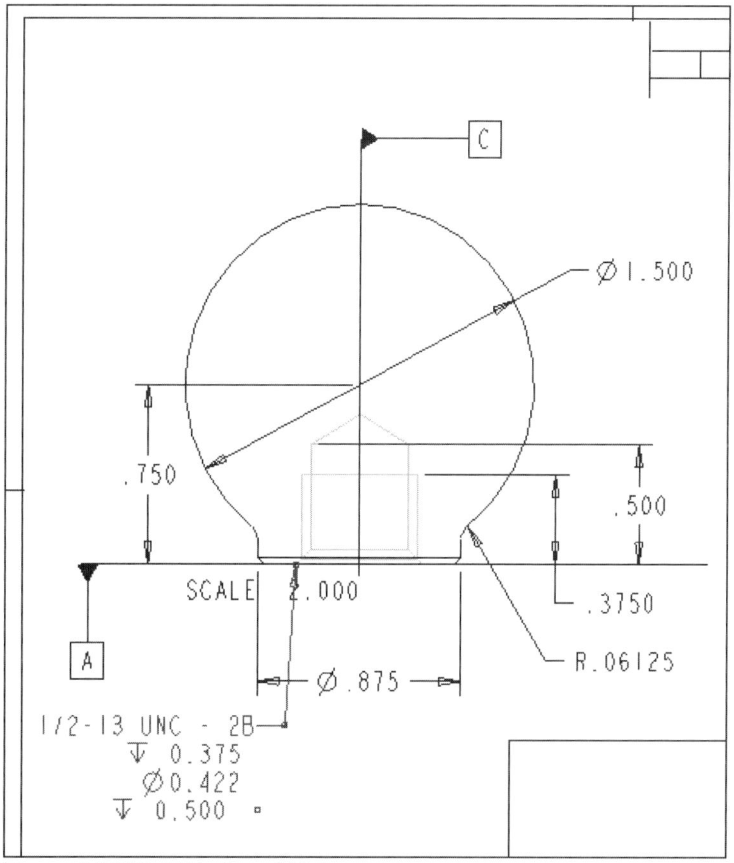

Figure 10.15(n) Displayed Note

Pick on the hole note and reposition ⇒ erase the SCALE note if it is displayed again ⇒ move the view as needed ⇒ reposition dimensions and the note to clearly present the detail ⇒ [icon] ⇒ **Ctrl+S** ⇒ **Enter** [Fig. 10.15(o)] ⇒ **File** ⇒ **Delete** ⇒ **Old Versions** ⇒ **Enter** ⇒ **File** ⇒ **Close Window** ⇒ **File** ⇒ **Close Window**

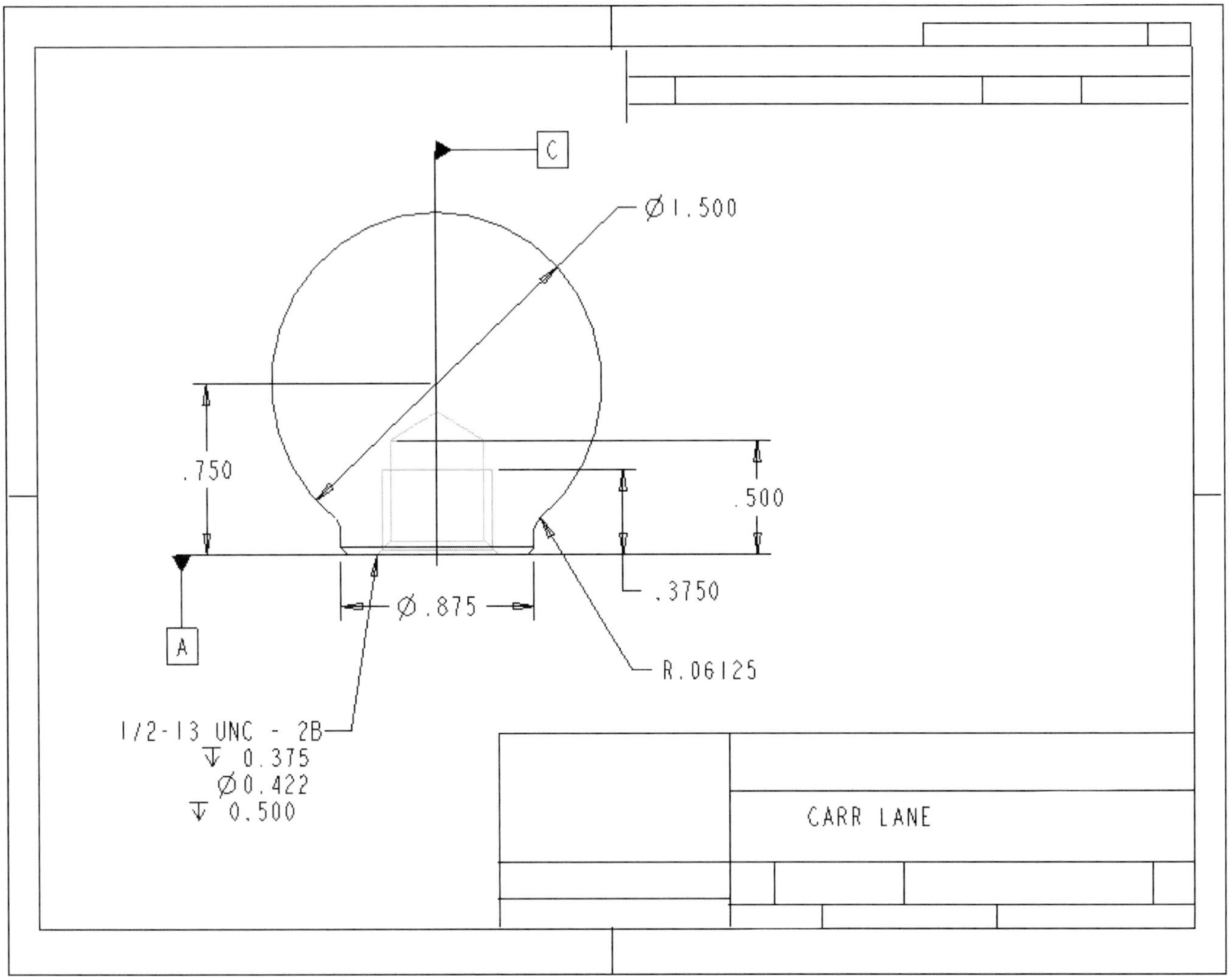

Figure 10.15(o) Completed Drawing

You may finalize the drawing after completing the next lesson. You can create drawings of the remaining assembly parts. And, a variety of part, assemblies, and drawing projects can be downloaded from the website ***www.cad-resources.com***.

Lesson 11 Part Drawings

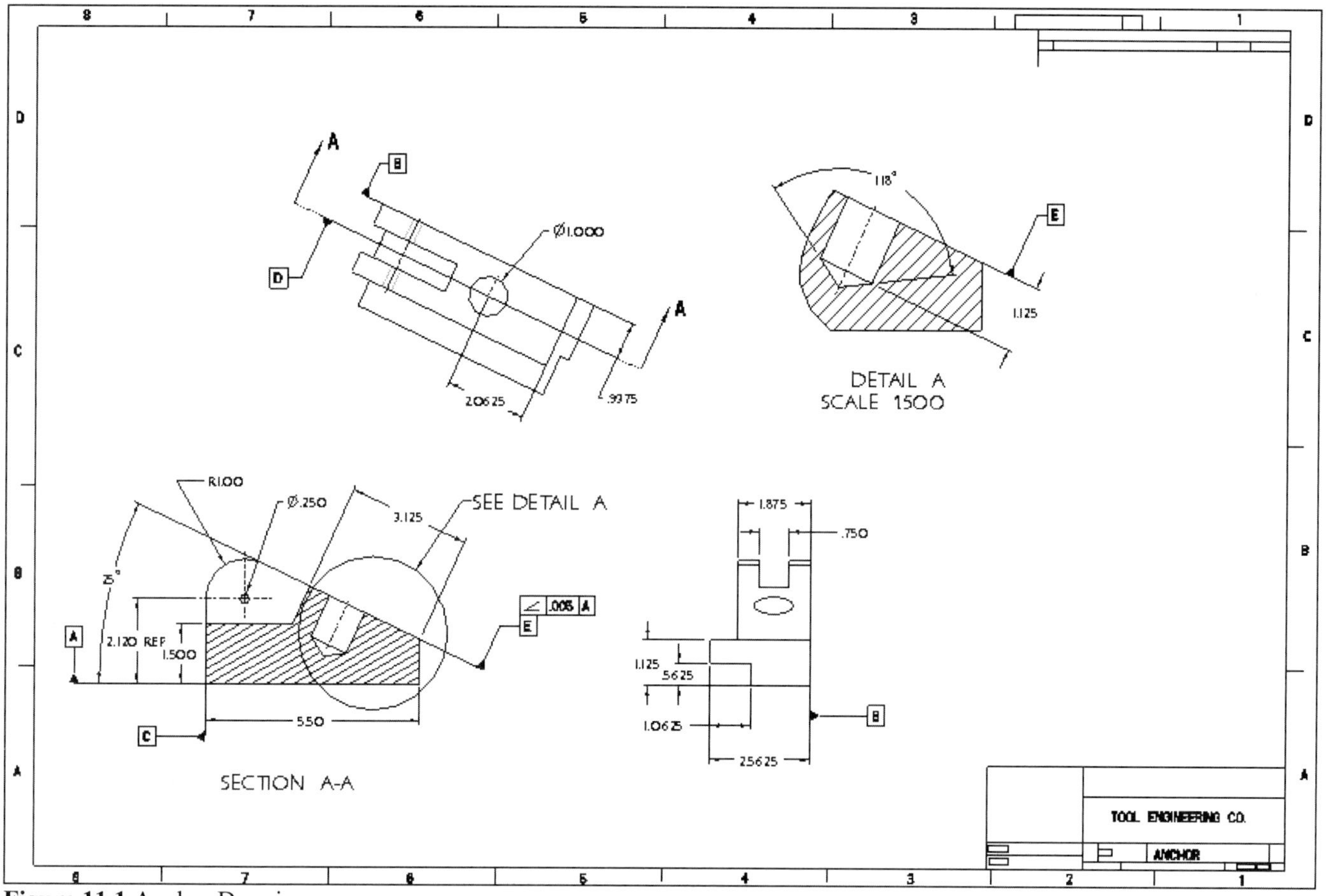

Figure 11.1 Anchor Drawing

OBJECTIVES

- Establish a **Drawing Options** file to use when detailing
- Identify the need for **views** to clarify interior features of a part
- Create **Cross Sections** using datum planes
- Produce **Auxiliary Views**
- Create **Detail Views**
- Use **multiple drawing sheets**
- Apply **standard drafting conventions** and linetypes to illustrate interior features

PART DRAWINGS

Designers and drafters use drawings to convey design and manufacturing information. Drawings consist of a **Format** and views of a part (or assembly) (Fig 11.1). Standard views, sectional views, detail views, and auxiliary views are utilized to describe the objects' features and sizes. **Sectional views**, also called **sections**, are employed to clarify and dimension the internal construction of an object. Sections are needed for interior features that cannot be clearly described by hidden lines in conventional views. **Auxiliary views** are used to show the *true shape/size* of a feature or the relationship of features that are not parallel to any of the principal planes of projection. Many objects have inclined surfaces and features that cannot be adequately displayed and described by using principal views alone. To provide a clearer description of these features, it is necessary to draw a view that will show the *true shape/size*.

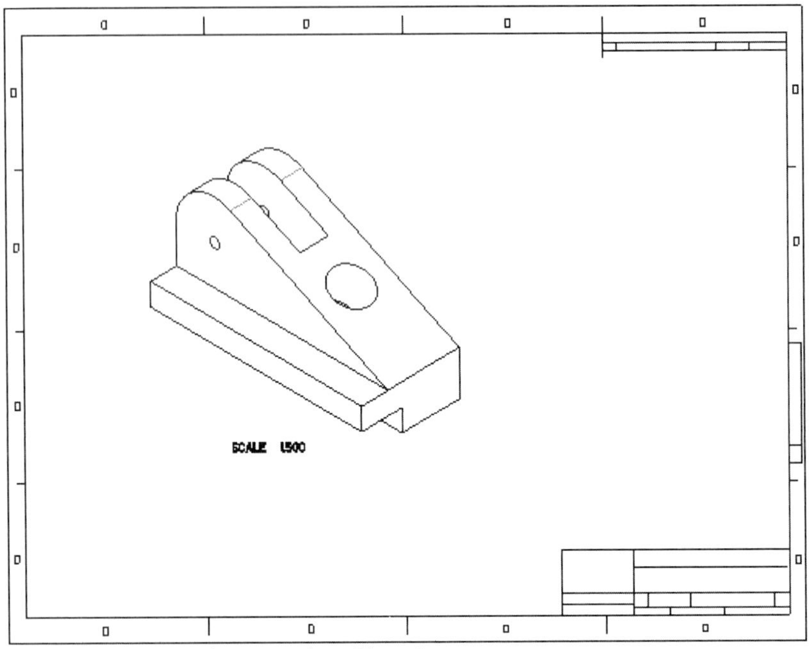

Figure 11.2(a) Anchor Drawing, Sheet 2

Anchor Drawing

You will be creating a multiple sheet detail drawing of the Anchor [Fig. 11.1 and Fig. 11.2(a)]. The front view will be a full section. A right side view and an auxiliary view are required to detail the part. Views will be displayed according to visibility requirements per ANSI standards, such as no hidden lines in sections. The part is to be dimensioned according to ASME Y14.5M. You will add a standard format. Detailed views of other parts will be introduced to show the wide variety of view capabilities.

Click: **File** ⇒ **Set Working Directory** [Fig. 11.2(b)] ⇒ select the directory where the **anchor.prt** was saved ⇒ **OK** ⇒ [New] ⇒ [Drawing] ⇒ Name **ANCHOR** ⇒ [Use default template] *(if you keep the "Use default template" checked; the Front, Top, and Right views will be automatically created for you)* ⇒ **OK**

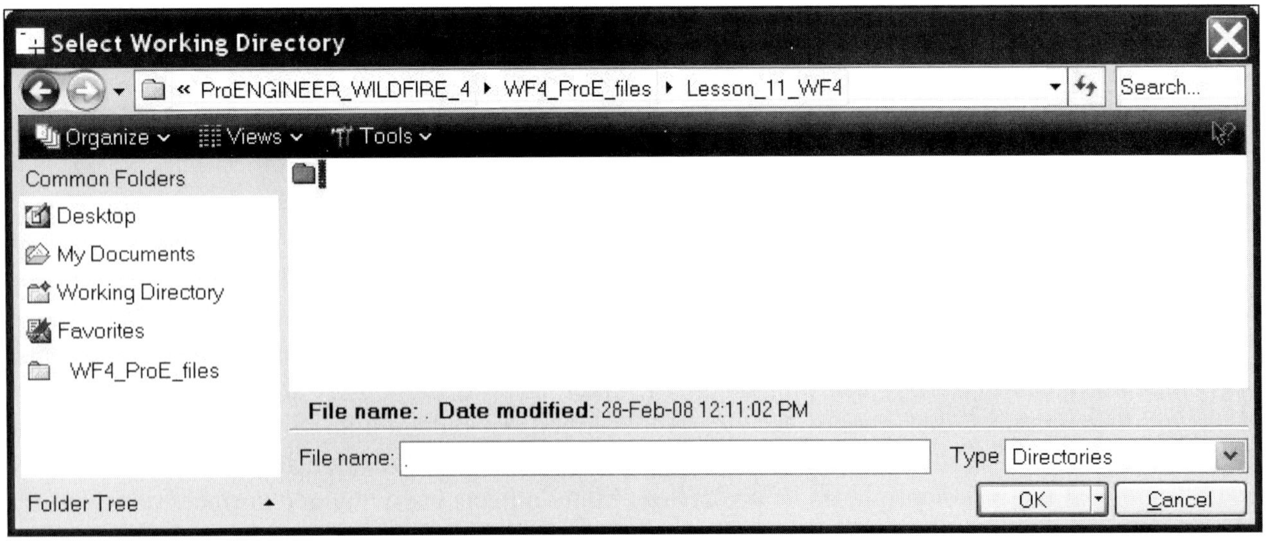

Figure 11.2(b) Anchor Drawing

Click: Default Model **Browse** ⇒ pick **anchor.prt** ⇒ **Open** ⇒ Standard Size **D** [Fig. 11.2(c)] ⇒ **OK** ⇒ **RMB** ⇒ **Page Setup** ⇒ D Size [Fig. 11.2(d)] ⇒ ⇒ **Browse** ⇒ System Formats Open dialog box opens ⇒ pick **d.frm** [Fig. 11.2(e)] ⇒ **Open** ⇒ **OK** ⇒ ⇒ **OK**

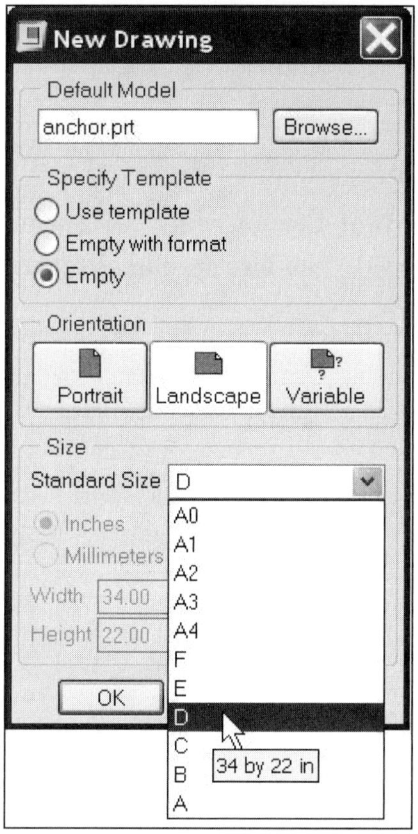

Figure 11.2(c) Standard Size D

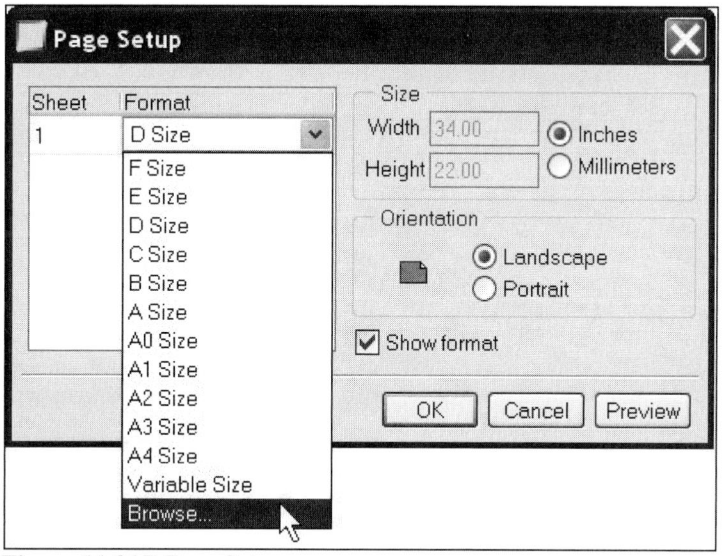

Figure 11.2(d) Page Setup

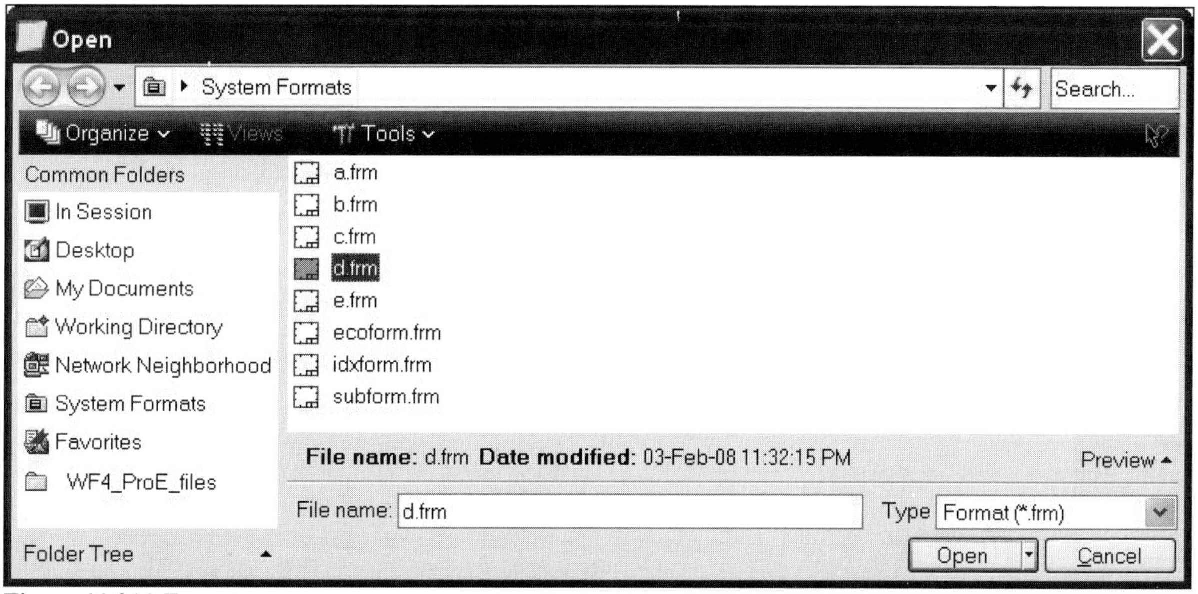

Figure 11.2(e) Formats

Click: **RMB** ⇒ **Properties** [Fig. 11.2(f)] ⇒ **Drawing Options** [Fig. 11.2(g)] the Options dialog box opens ⇒ create a new **.dtl** file, *(or*  *⇒ pick a previously saved* ***.dtl*** *from your directory list ⇒ Open)*
Change the following options to the values listed below:

Option:	***drawing_text_height***	Value:	***.25***	⇒	**Add/Change**
Option:	***default_font***	Value:	***filled***	⇒	**Add/Change**
Option:	***draw_arrow_style***	Value:	***filled***	⇒	**Add/Change**
Option:	***allow_3d_dimensions***	Value:	***yes***	⇒	**Add/Change**

Click: **Apply** ⇒ Sort: **Alphabetical** [Fig. 11.2(h)] ⇒  **Save a copy of the currently displayed configuration file** ⇒ type a unique name for your file (*draw_options*) ⇒ **Ok** ⇒ **Close** ⇒ **Done/Return** ⇒ **Ctrl+S** ⇒ **MMB**

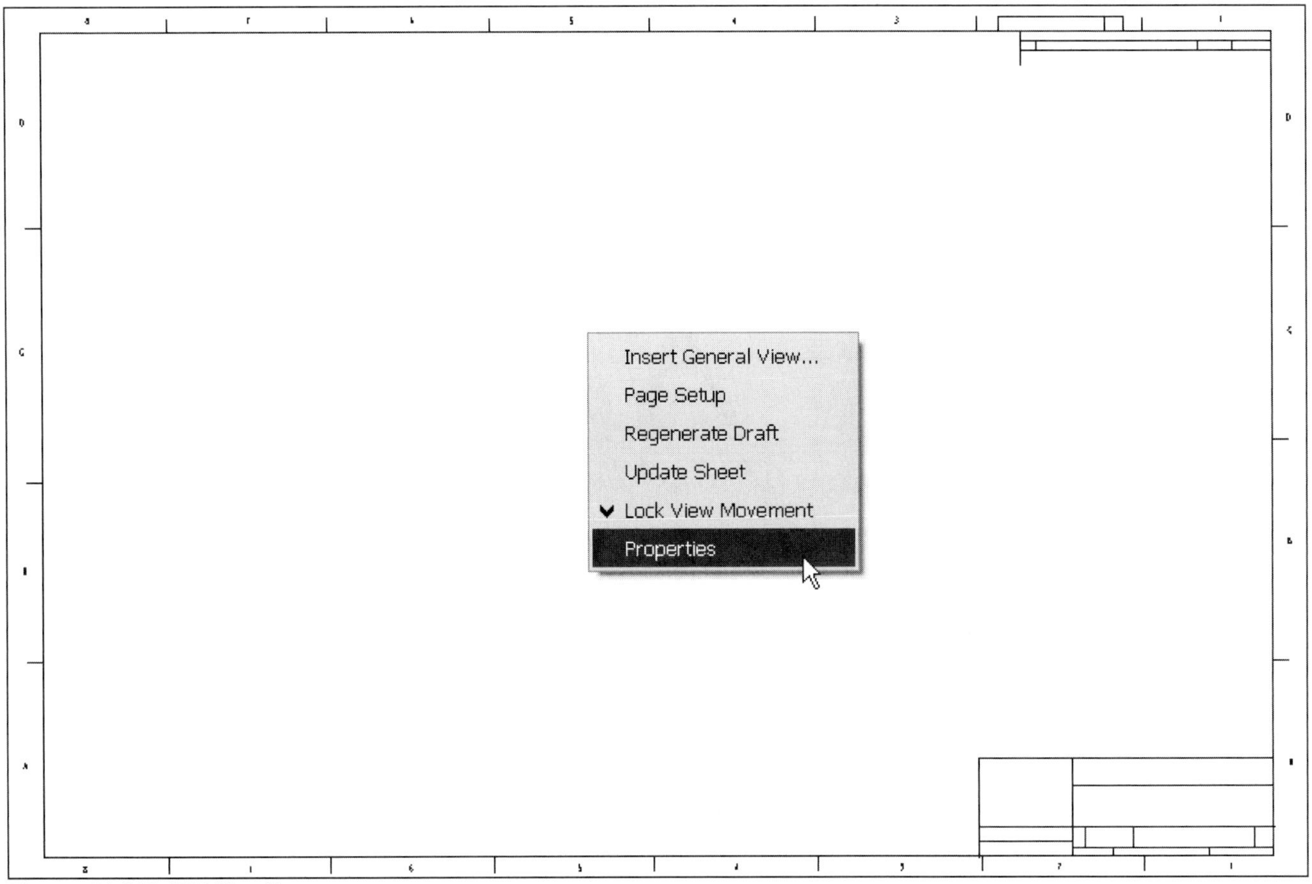

Figure 11.2(f) "D" Size Format

Figure 11.2(g) Drawing Options

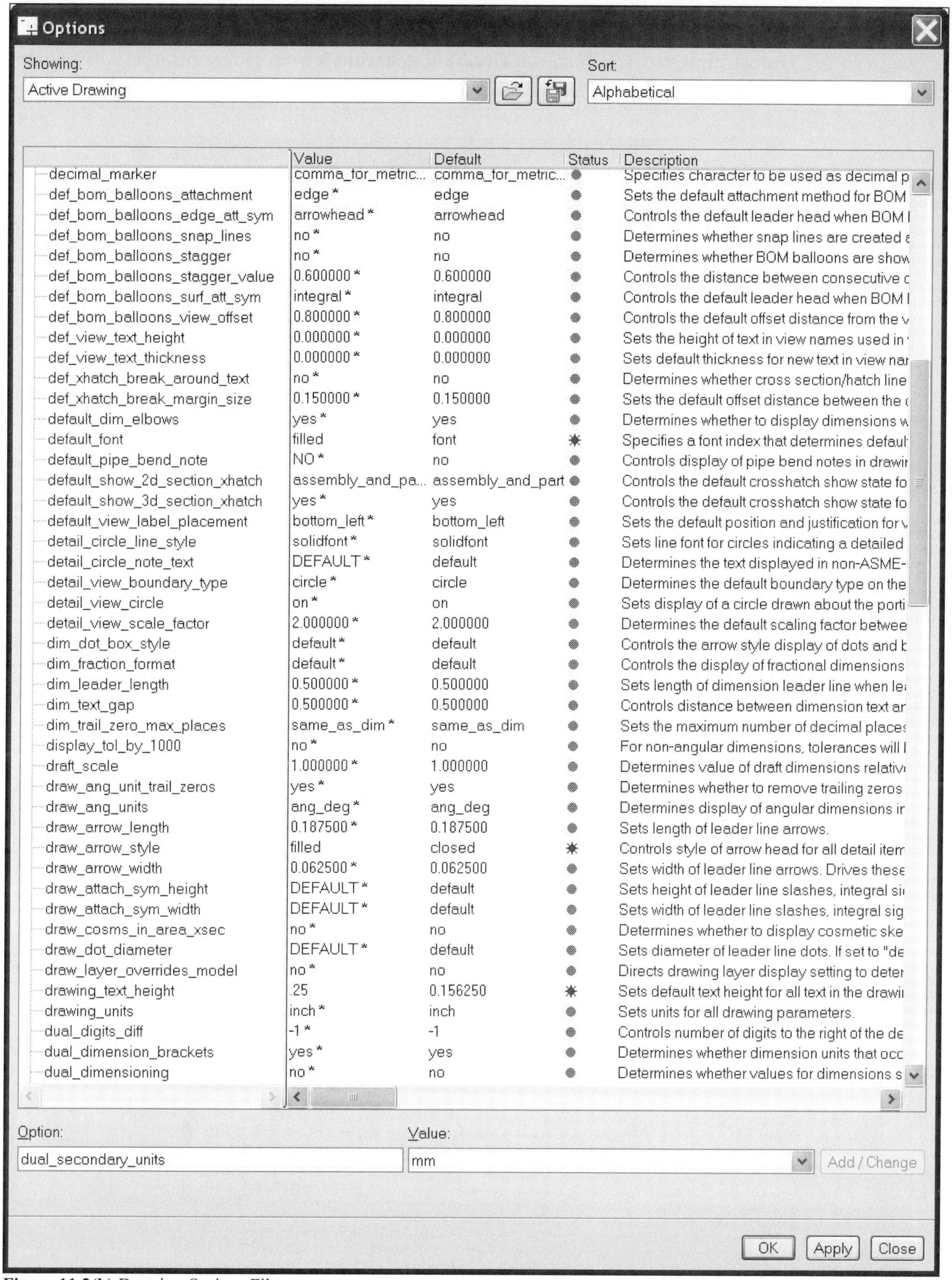

Figure 11.2(h) Drawing Options File

Click on the quick sash to close your Navigator (normally close the Navigator so you can see a larger drawing) ⇒ **Hidden line** ⇒ **Ctrl+R** ⇒ **Create a general view** ⇒ pick a position for the view [Fig. 11.3(a)] ⇒ **FRONT** [Fig. 11.3(b)] ⇒ **Apply** ⇒ **Close**

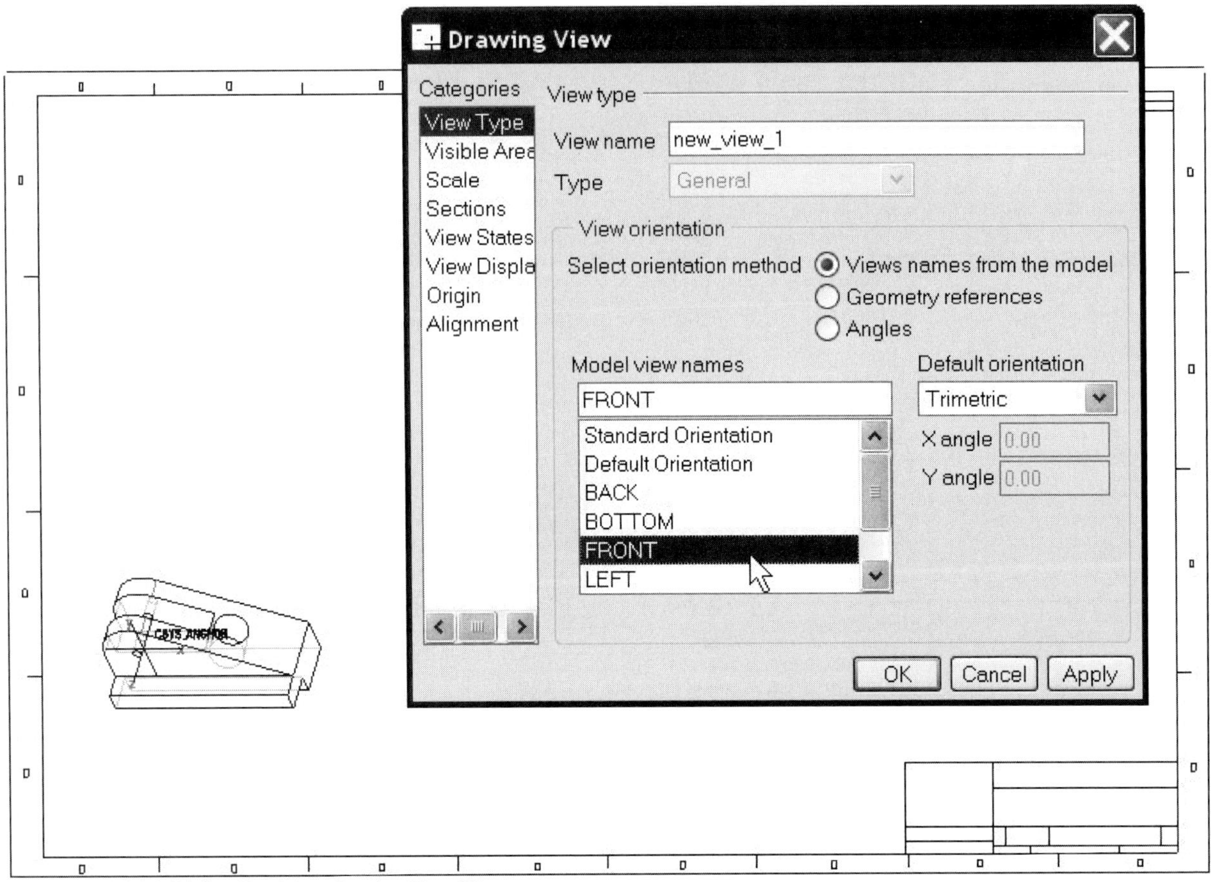

Figure 11.3(a) Pick a Position for the First View

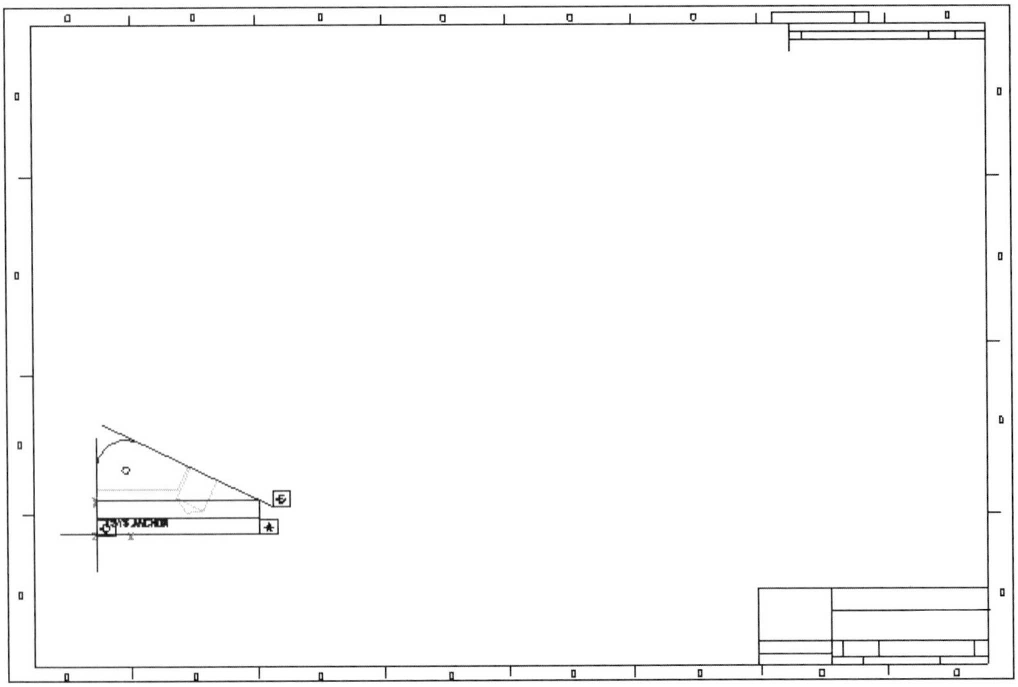

Figure 11.3(b) Front View

Click: off ⇒ Add two more views. With the Front view selected (highlighted), click: **Insert** ⇒ **Drawing View** ⇒ **Projection** ⇒ Select CENTER POINT for drawing view. pick a position for the Right side view [Fig. 11.3(c)] ⇒ **LMB** to deselect the Right View ⇒ **No hidden** ⇒ **RMB** ⇒ **Update Sheet** ⇒ pick on the Front view ⇒ **RMB** ⇒ **Insert Projection View** ⇒ Select CENTER POINT for drawing view. pick a position for the Top view [Fig. 11.3(d)] ⇒ **LMB** ⇒ **Disallow the movement of drawing views with the mouse** (unlock) ⇒ pick on a view, hold down the **LMB**, and reposition as needed ⇒ **LMB** to deselect ⇒ **Update** ⇒ ⇒ ⇒ **MMB** [Fig. 11.3(e)]

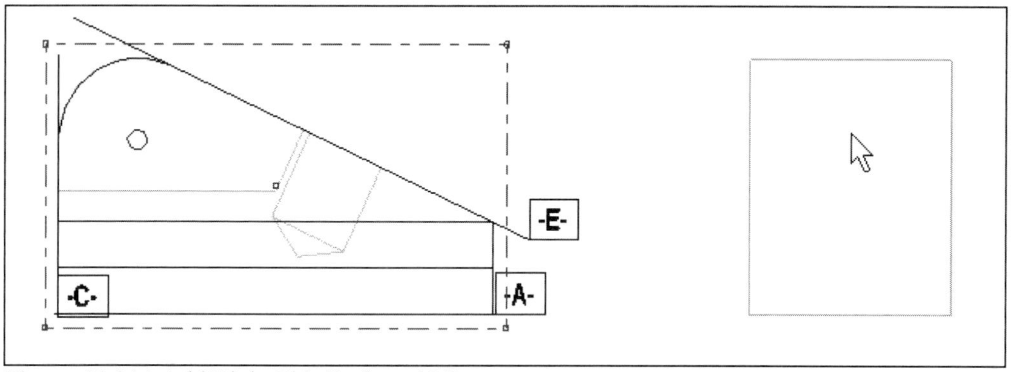

Figure 11.3(c) Add Right Side Projected View

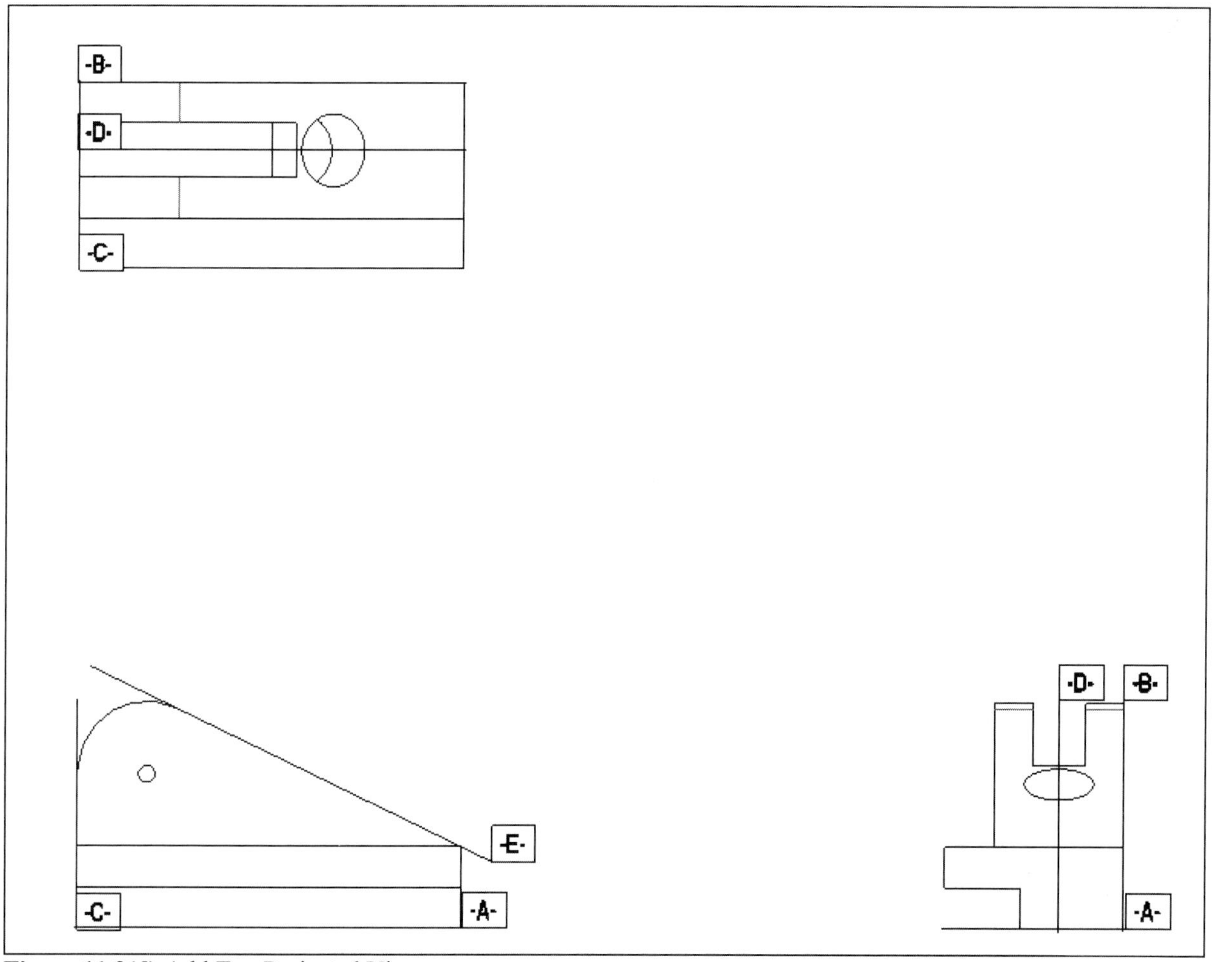

Figure 11.3(d) Add Top Projected View

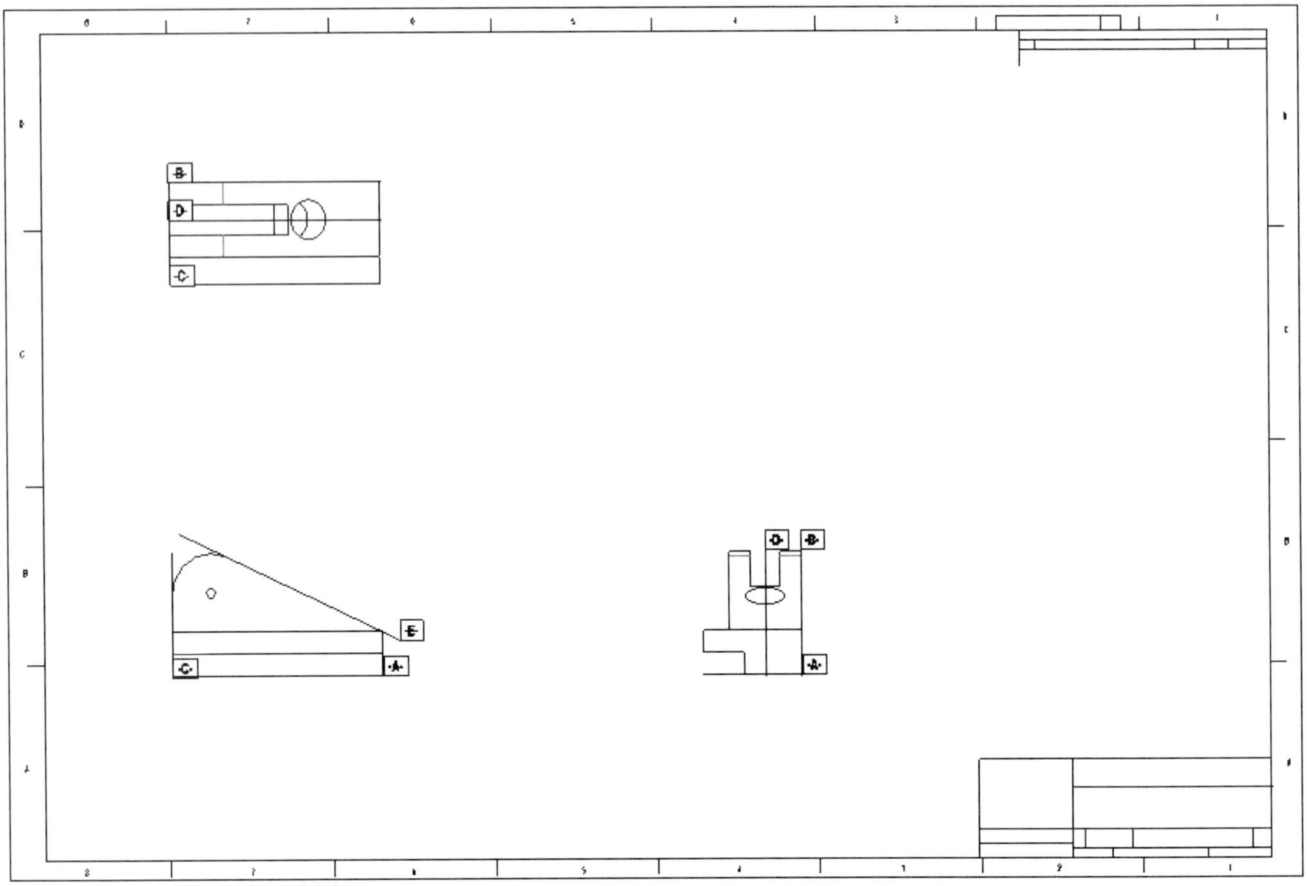

Figure 11.3(e) Repositioned Views

The top view does not help in the description of the part's geometry. Delete the top view before adding an auxiliary view that will display the true shape of the angled surface.

Pick on the Top view ⇒ **RMB** [Fig. 11.3(f)] ⇒ **Delete** ⇒ **Update** ⇒

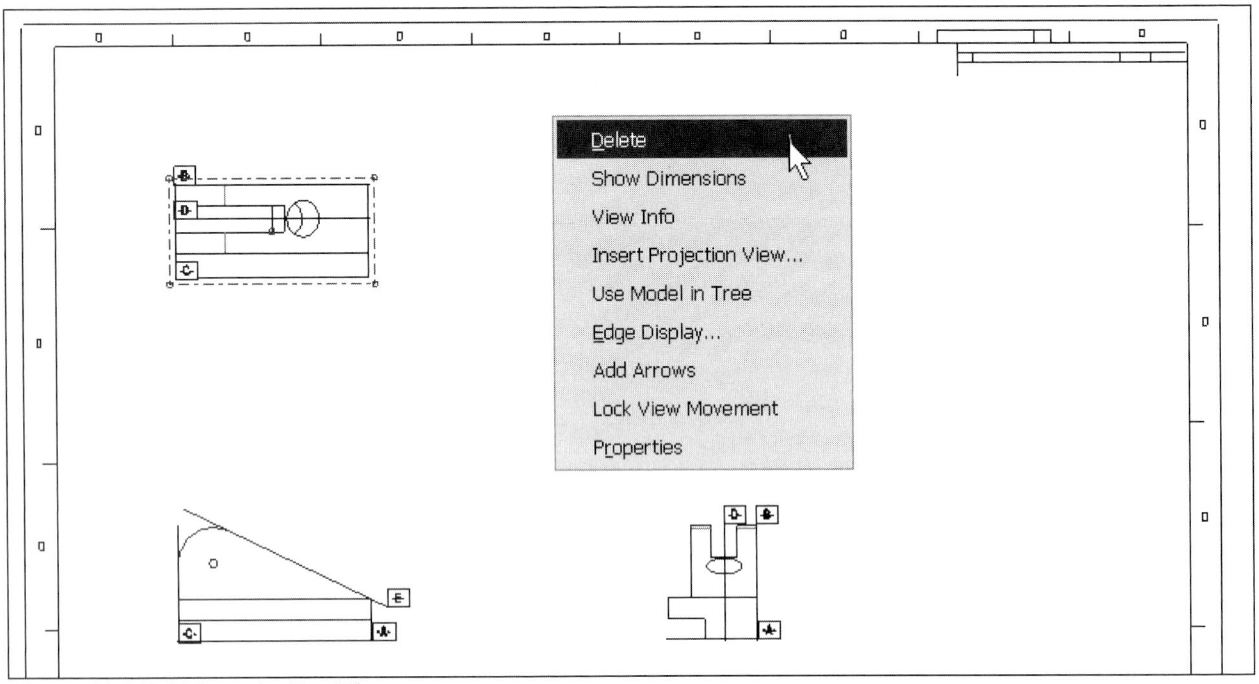

Figure 11.3(f) Pick on the Top View

Click: **Insert** ⇒ **Drawing View** ⇒ **Auxiliary** [Fig. 11.3(g)] ⇒ Select edge of or axis through, or datum plane as, front surface on main view. pick on datum **E** [Fig. 11.3(h)] ⇒ Select CENTER POINT for drawing view. [Fig. 11.3(i)] ⇒ **LMB** to place the view [Fig. 11.3(j)] ⇒ **LMB** to deselect ⇒ **RMB** ⇒ **Update Sheet** ⇒ [save icon] ⇒ **MMB**

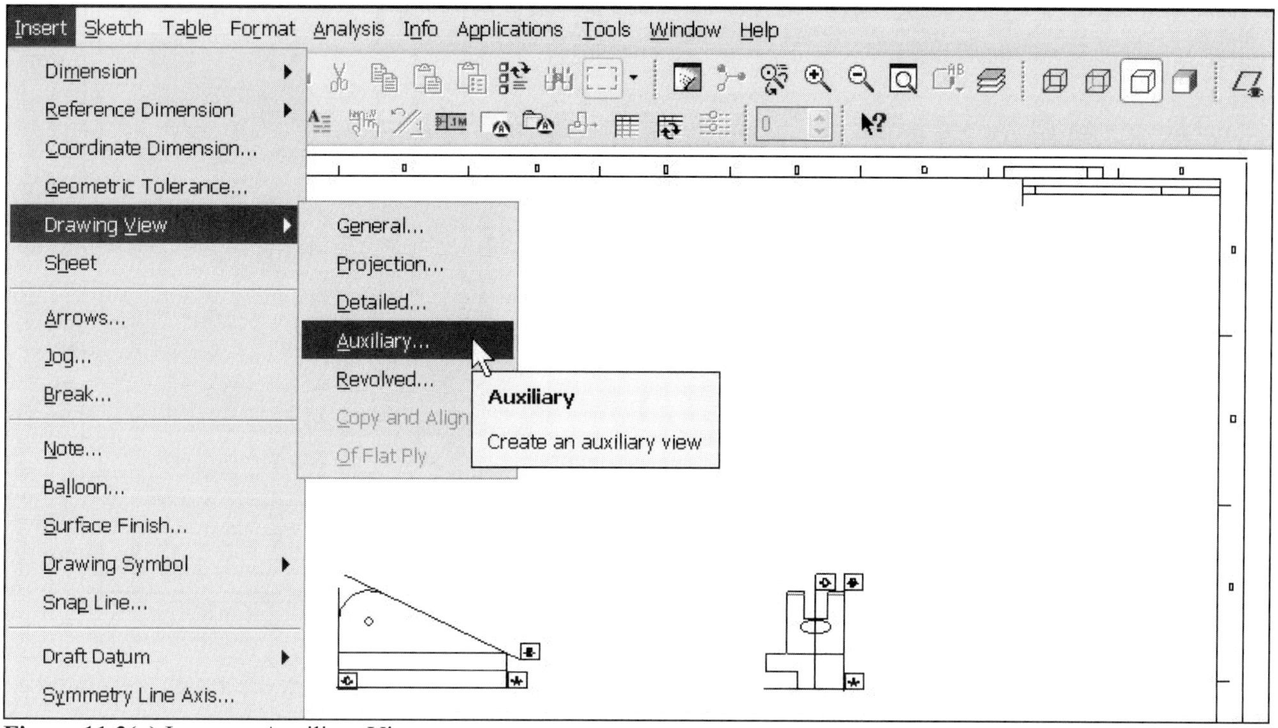

Figure 11.3(g) Insert an Auxiliary View

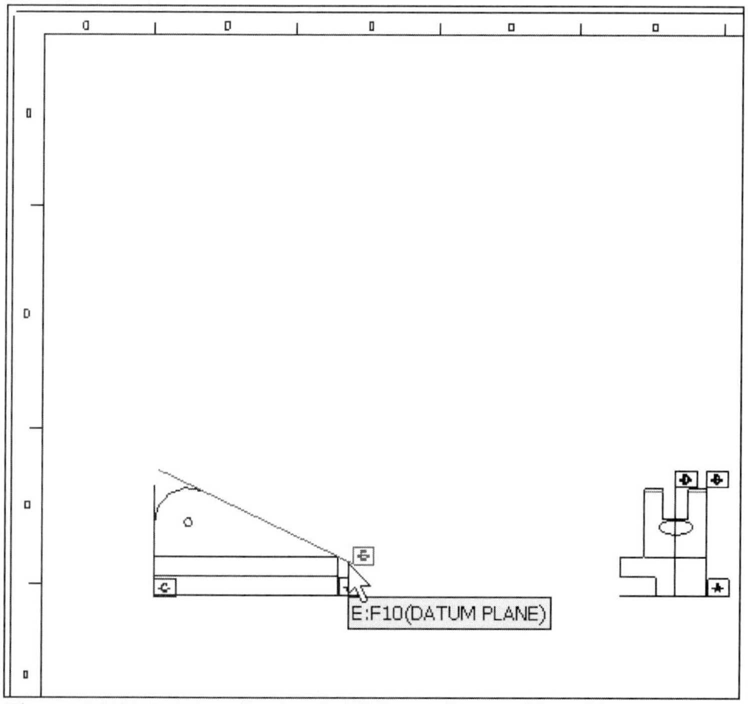

Figure 11.3(h) Select the Angled Edge or Datum Plane E

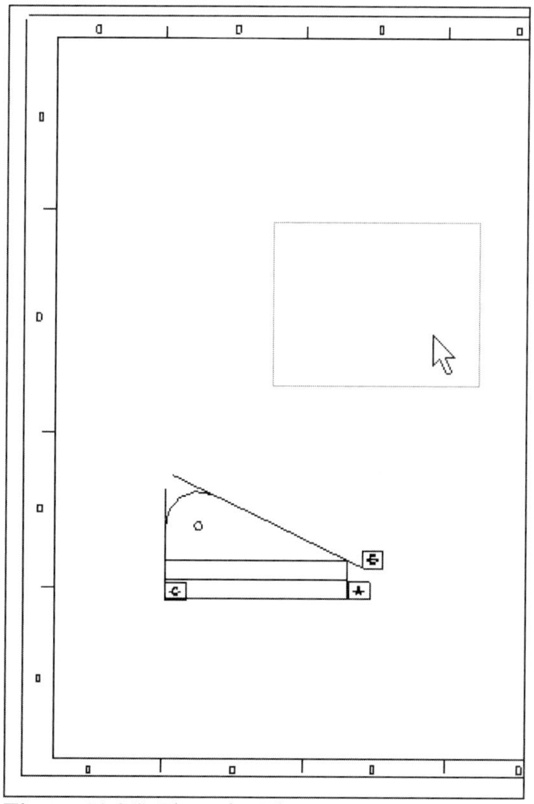
Figure 11.3(i) Place the View

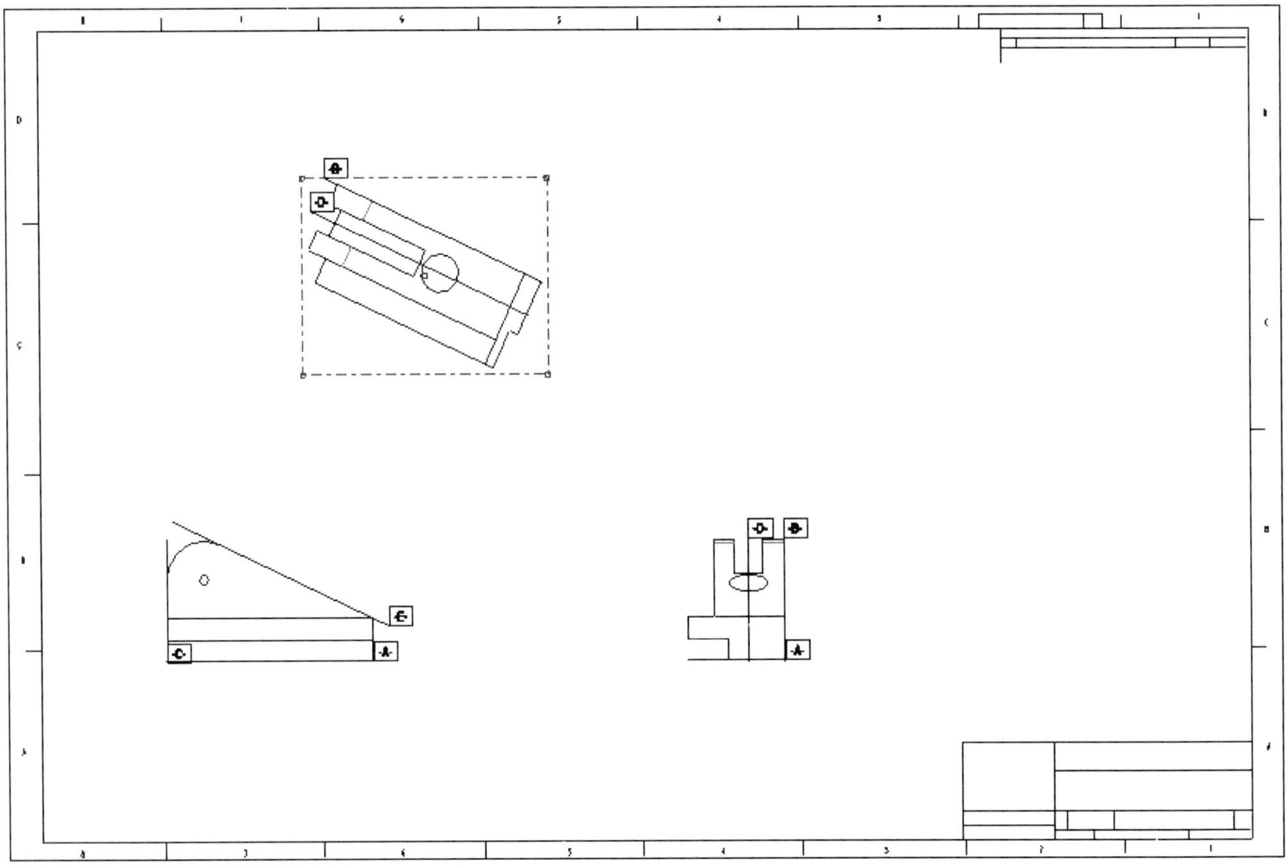

Figure 11.3(j) Auxiliary View

The correct standard was not selected when you made changes to the Drawing Options. ASME symbols for datum planes (gtol_datums) are the correct style standard used on drawings. The ANSI style [Fig. 11.3(k)] was discontinued in 1994, though retained by some companies as an "in house" standard. For outside vendors and for manufacturing internationally, the ISO-ASME standards should be applied to all manufacturing drawings.

Click: **RMB** ⇒ **Properties** ⇒ **Drawing Options** ⇒ Option: gtol_datums Value: std_asme ⇒ **Add/Change** ⇒ **Apply** ⇒ ⇒ select your previously saved file to overwrite ⇒ **Ok** ⇒ **Close** ⇒ **Done/Return** ⇒ ⇒ ⇒ [Fig. 11.3(l)] ⇒ ⇒ **OK** ⇒ **Sketch** ⇒ **Sketcher Preferences** ⇒ Grid intersection off ⇒ **Close**

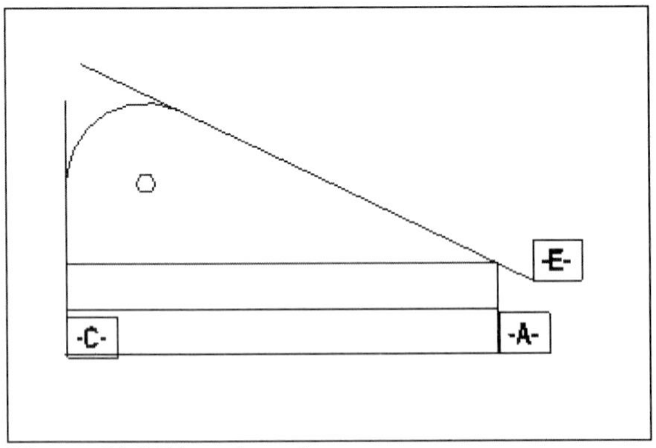

Figure 11.3(k) gtol_ansi

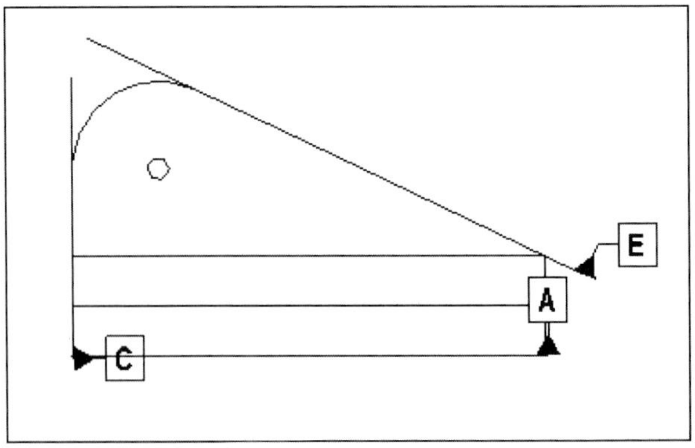

Figure 11.3(l) gtol_asme

Next, change the front view into a sectional view. Section **A** was created in the Part mode.

Pick on the Front view as the view to be modified [Fig. 11.4(a)] ⇒ **RMB** ⇒ **Properties** [Fig. 11.4(b)] ⇒ Categories **Sections** ⇒ 2D cross-section ⇒ + **Add cross-section to view** ⇒ pick section **A** from the Name list [Fig. 11.4(c)] ⇒ **Apply** [Fig. 11.4(d)] ⇒ **Close** ⇒ **RMB** ⇒ **Add Arrows** [Fig. 11.5(a)]

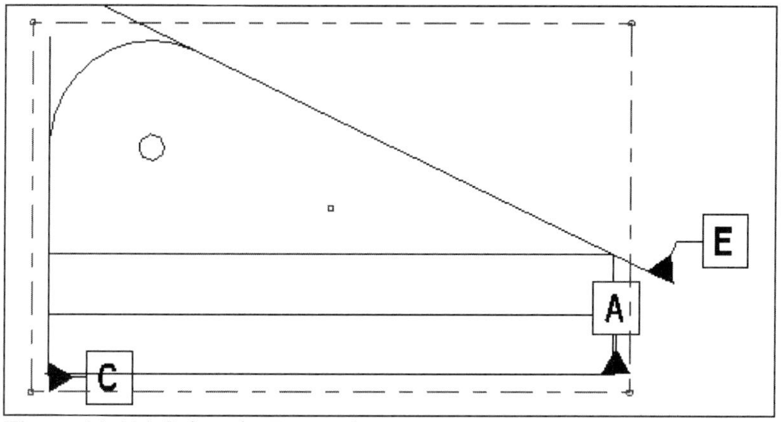

Figure 11.4(a) Select the Front View

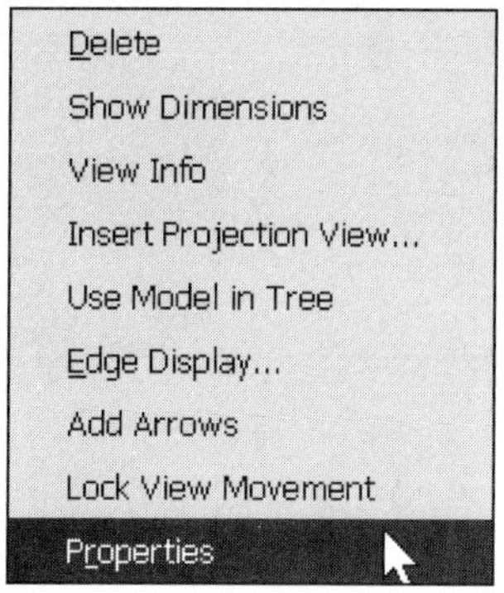

Figure 11.4(b) RMB Properties

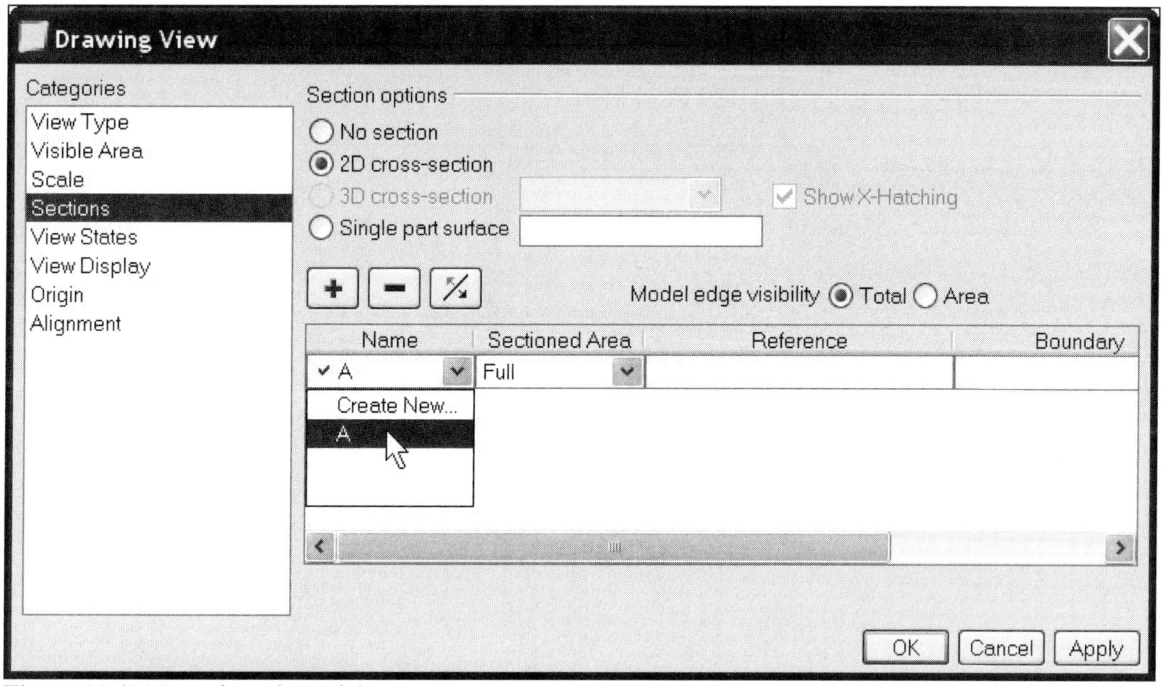

Figure 11.4(c) Drawing View Dialog Box

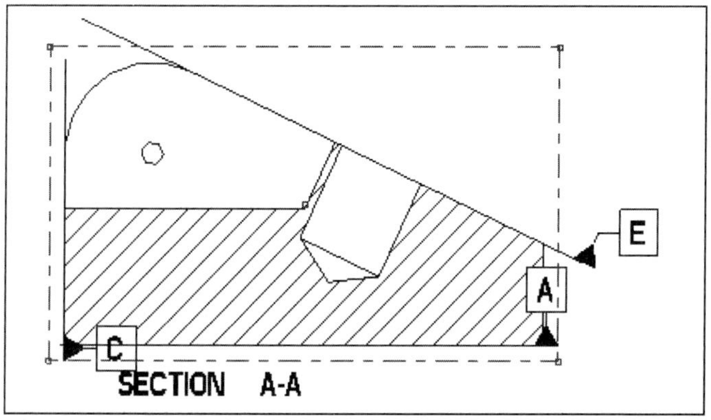

Figure 11.4(d) Section A-A

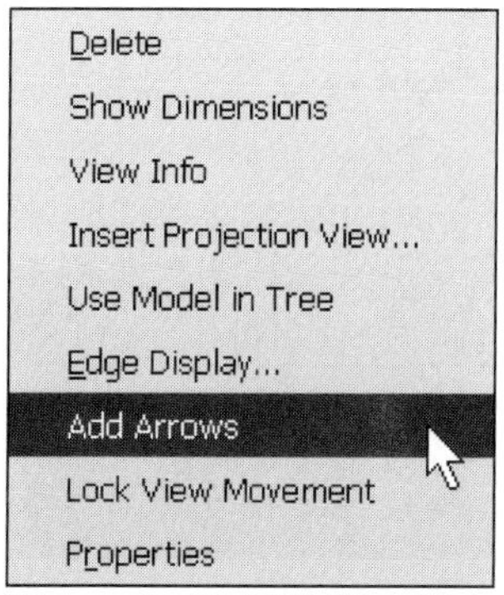

Figure 11.5(a) RMB Add Arrows

Pick the auxiliary view [Fig. 11.5(b)] ⇒ **LMB** ⇒ [icon] ⇒ [icon] ⇒ [icon] ⇒ [icon] ⇒ **OK** [Fig. 11.5(c)]

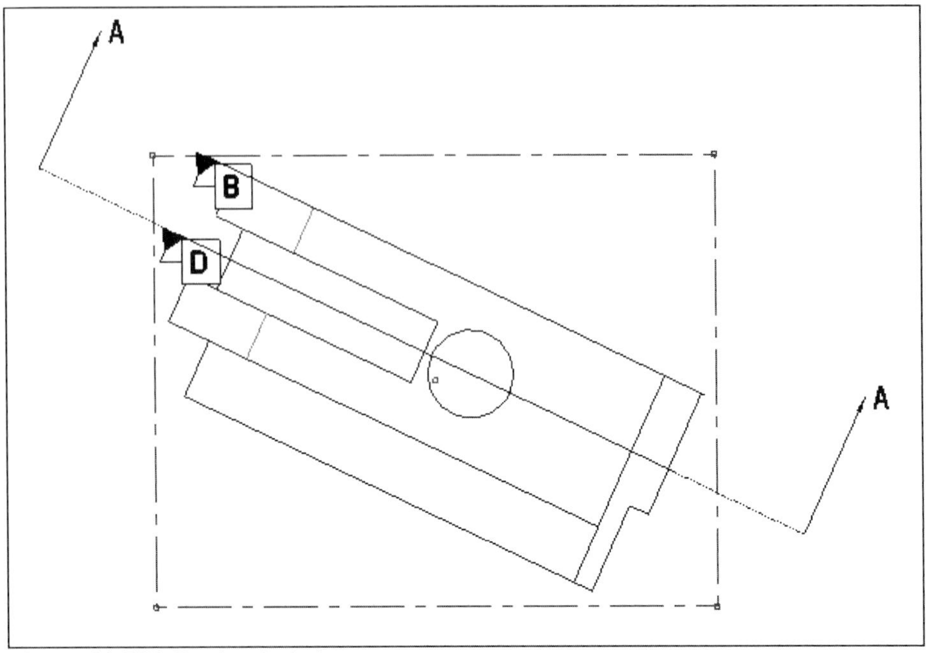

Figure 11.5(b) Section A-A Arrows

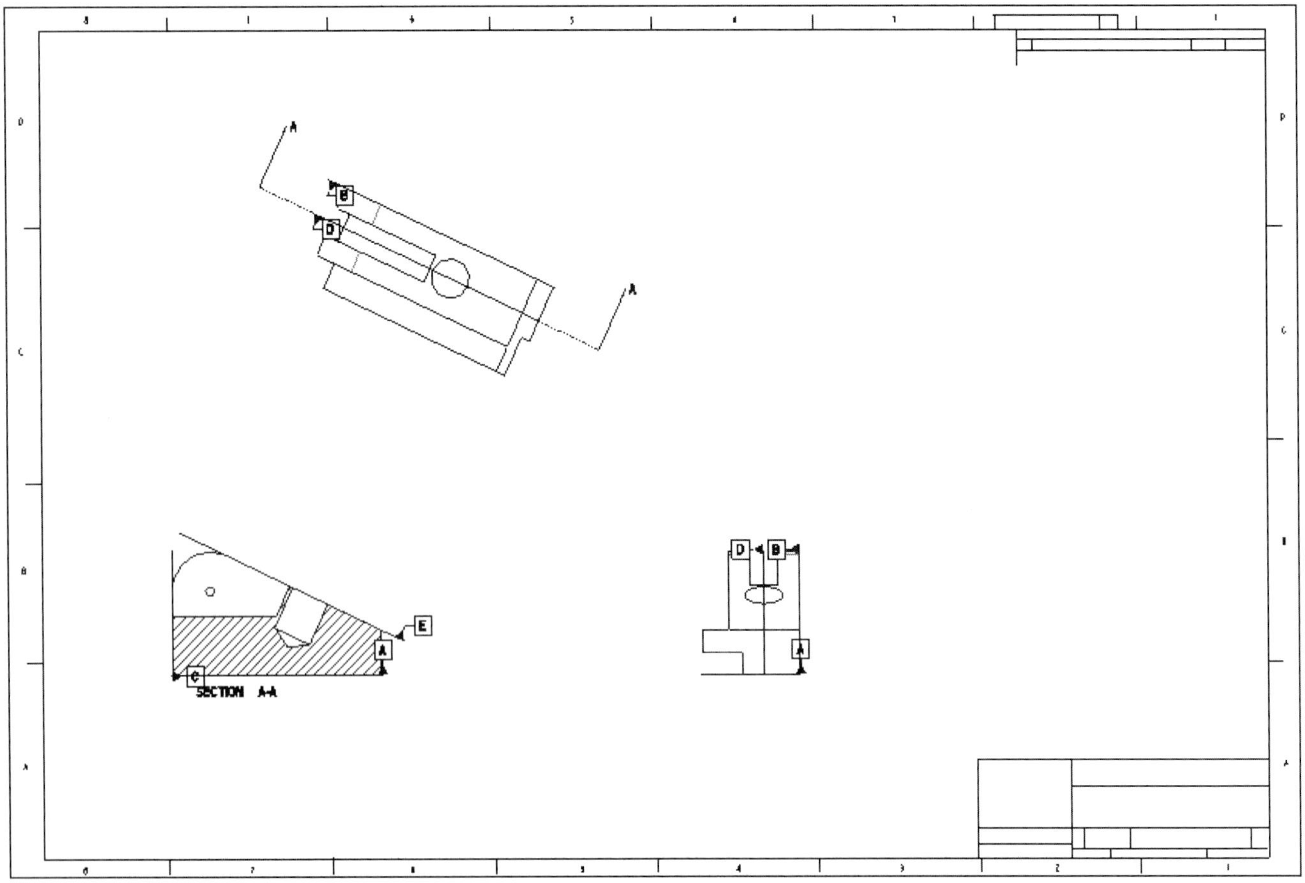

Figure 11.5(c) Section A-A

Show all dimensions and axes (centerlines), click: **Open the Show/Erase dialog box** ⇒ Show ⇒ **Dimension** **Axis** **Note** **Datum Plane** [Fig. 11.6(a)] ⇒ **Show All** ⇒ **Yes** ⇒ **Accept All** ⇒ **Close** [Fig. 11.6(b)] ⇒ **LMB** to deselect

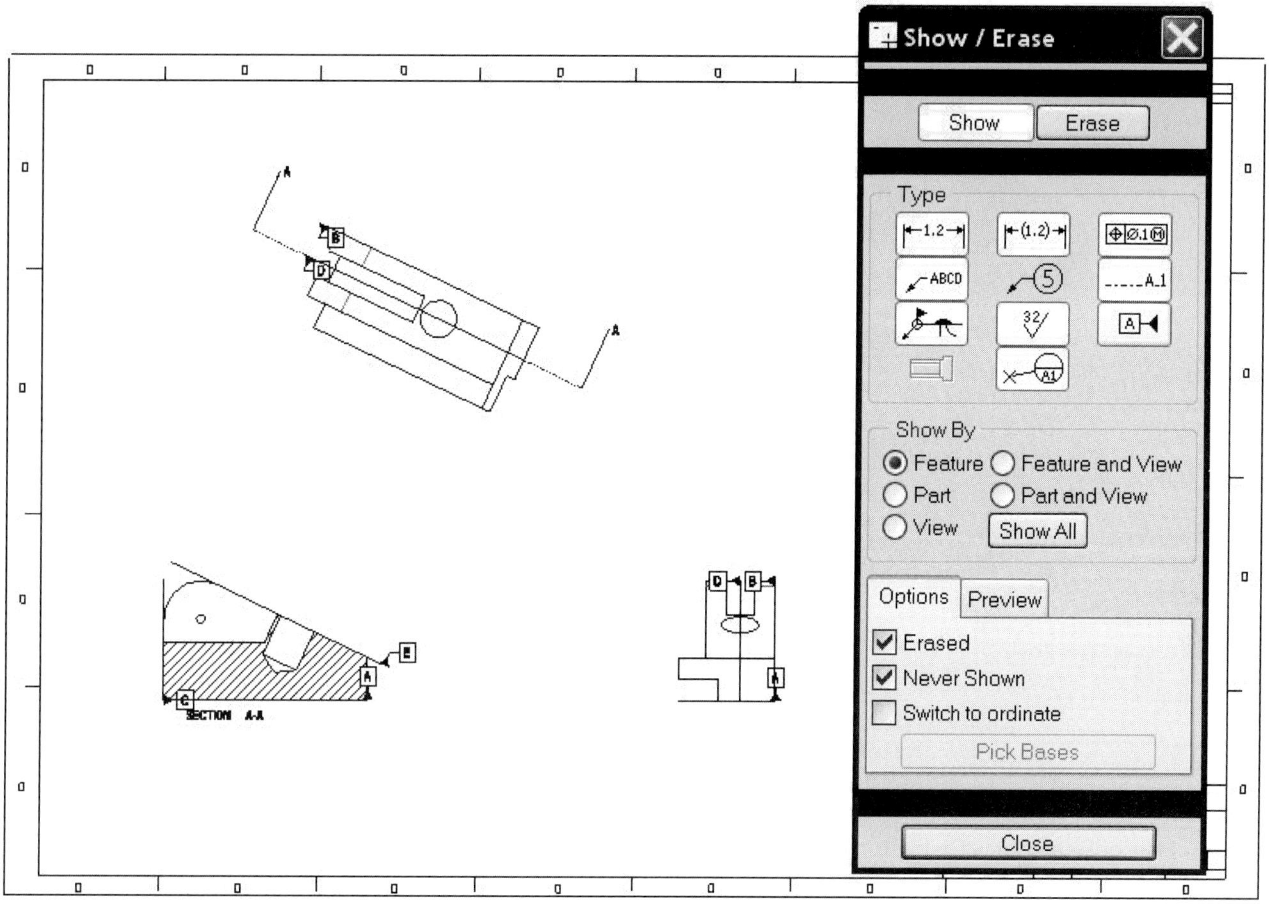

Figure 11.6(a) Show/Erase Dialog Box

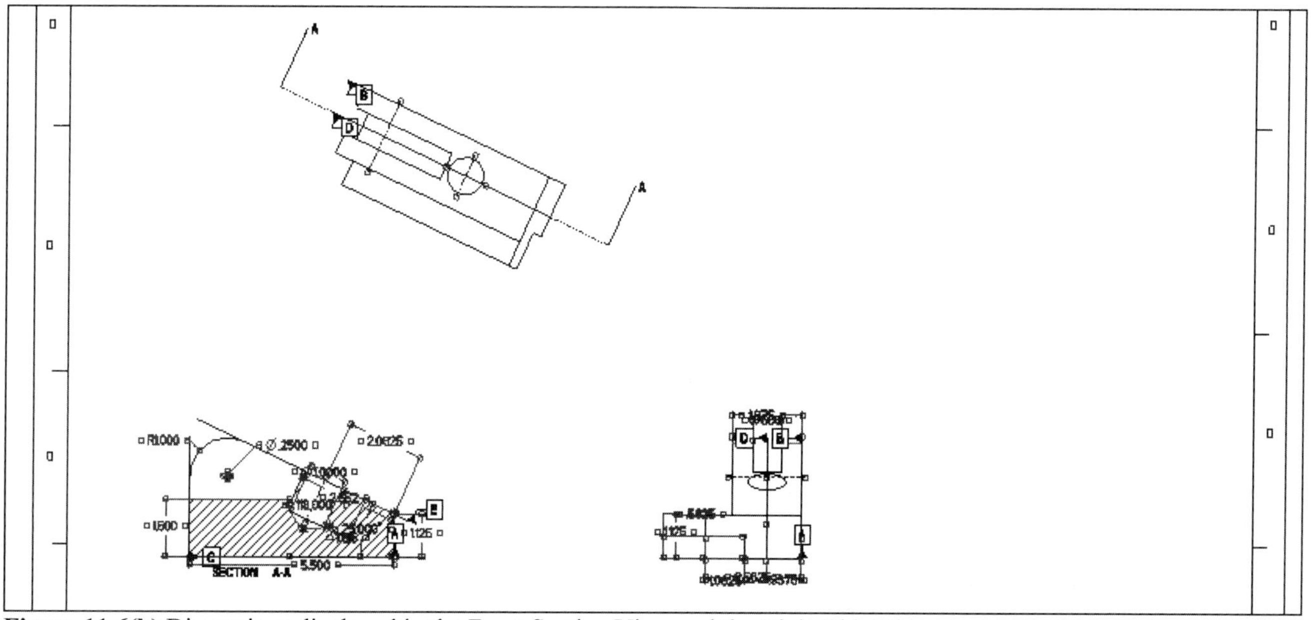

Figure 11.6(b) Dimensions displayed in the Front Section View and the Right Side View

Click: **Clean up position of dimensions around a view** *(or select all dimensions ⇒ RMB ⇒ Cleanup Dimensions)* ⇒ select all dimensions by enclosing them in a selection box ⇒ **OK** ⇒ Create Snap Lines ⇒ Increment .5 [Fig. 11.6(c)] ⇒ **Enter** ⇒ **Apply** [Fig. 11.6(d)] ⇒ **Close** ⇒ ⇒ **Ctrl+S** ⇒ **OK**

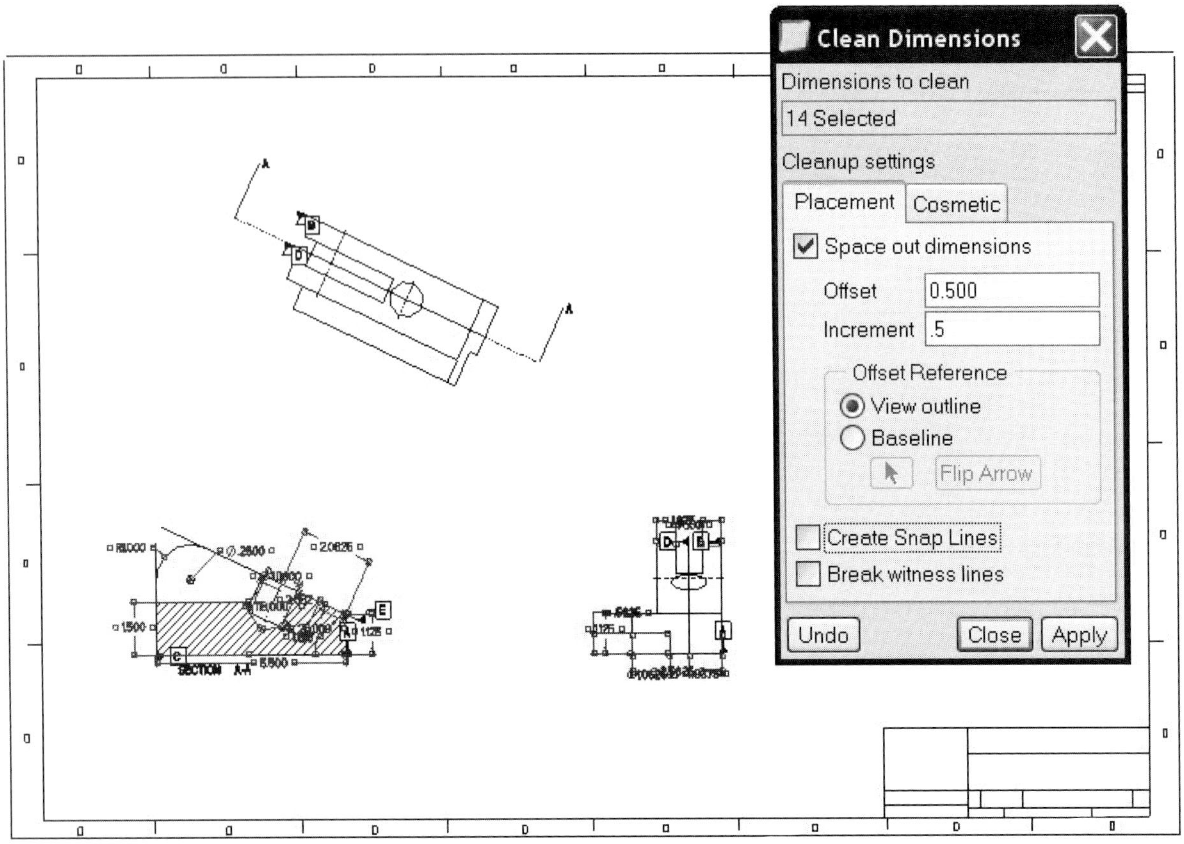

Figure 11.6(c) Clean Dimensions Dialog Box

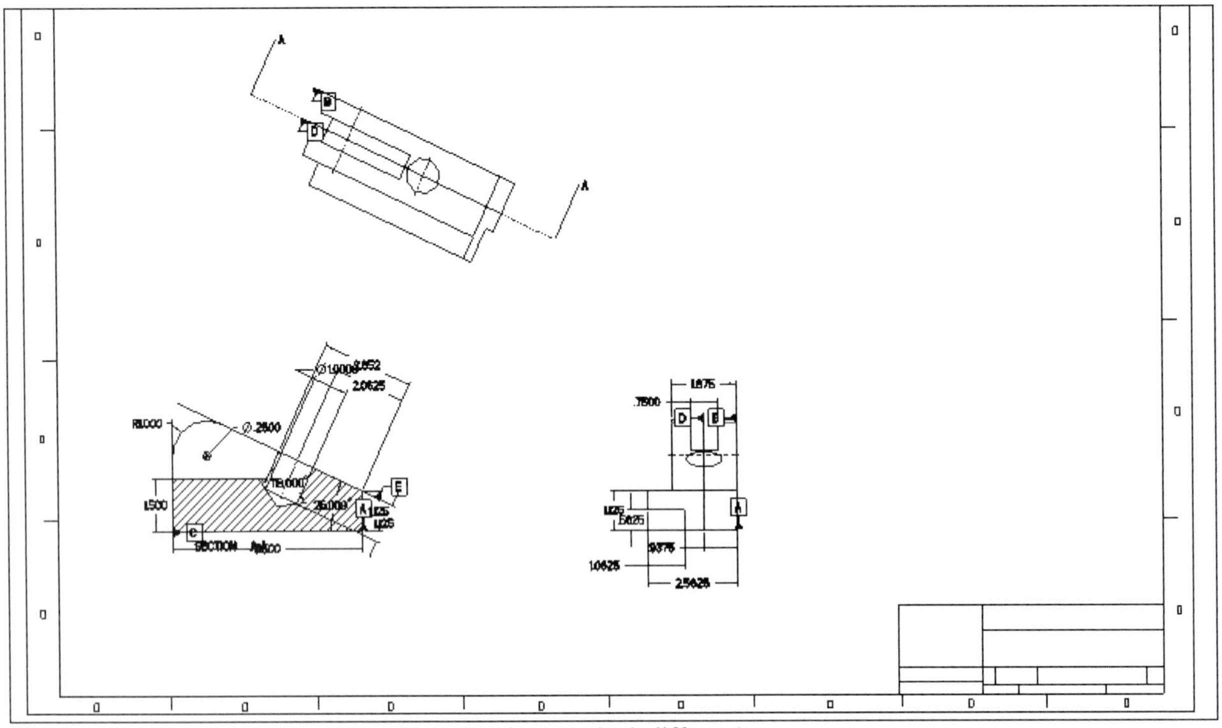
Figure 11.6(d) Cleaned Dimensions (your drawing may look different)

Pick on the **1.00** diameter hole dimension ⇒ **RMB** ⇒ **Move Item to View** [Fig. 11.7(a)] ⇒ pick on the auxiliary view [Fig. 11.7(b)] ⇒ pick on and reposition the **1.00** dimension ⇒ using the dimension handles  to reposition, move dimension text, clip extension lines, or flip arrows [Fig. 11.7(c)] where appropriate ⇒ move, erase, or clip axes 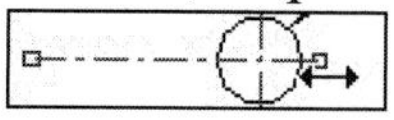 and position datums 

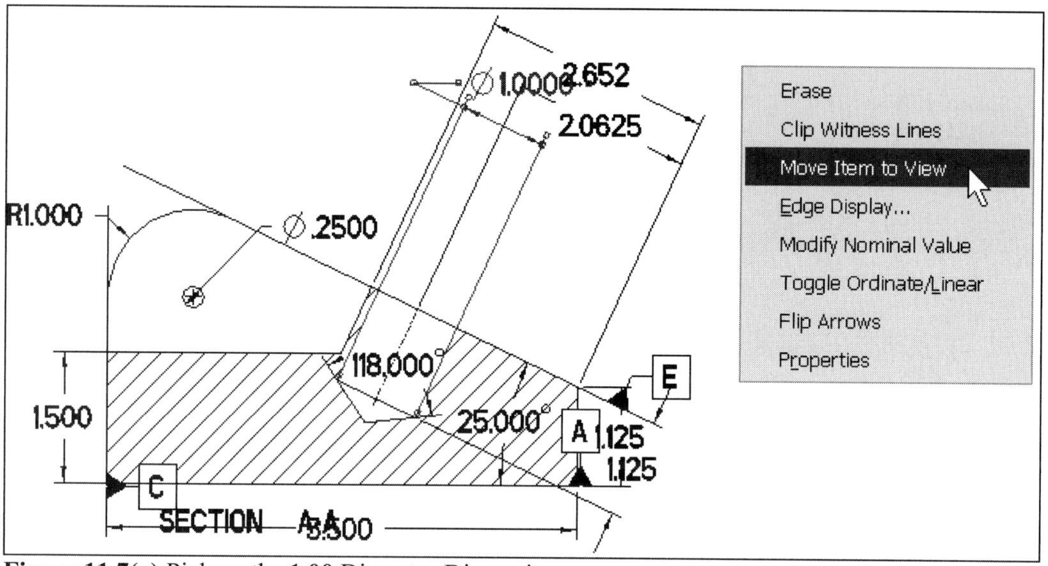

Figure 11.7(a) Pick on the **1.00** Diameter Dimension

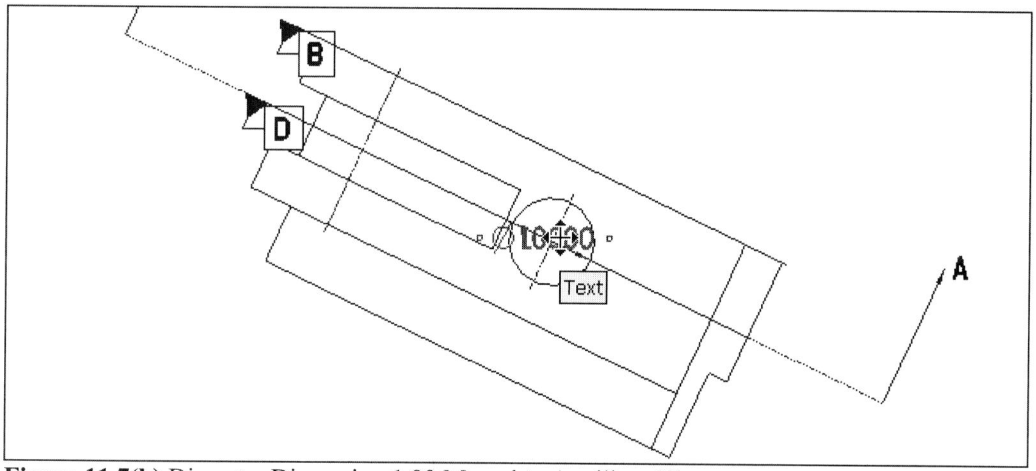

Figure 11.7(b) Diameter Dimension **1.00** Moved to Auxiliary View

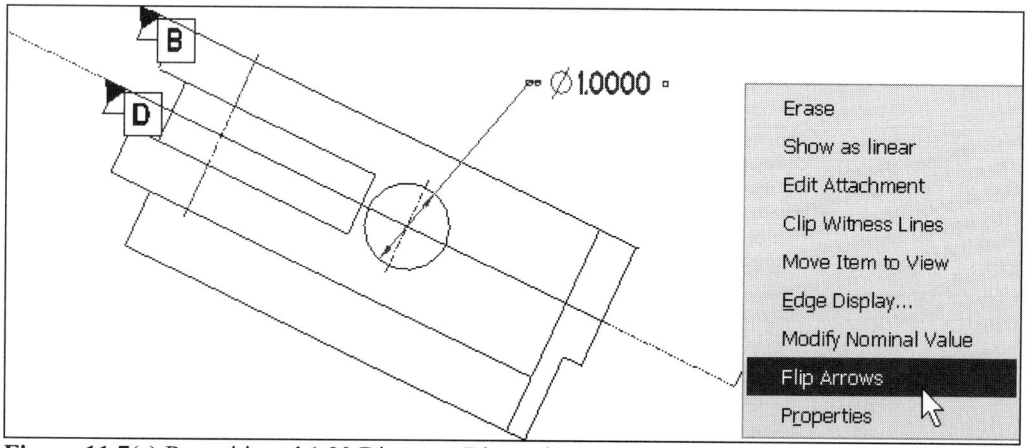

Figure 11.7(c) Repositioned **1.00** Diameter Dimension

Add a reference dimension to the small hole [Fig. 11.8(a)], click: **Insert** ⇒ **Reference Dimension** ⇒ **New References** ⇒ pick the horizontal axis of the small hole ⇒ pick datum **A** ⇒ **MMB** to place the reference dimension [Fig. 11.8(b)] ⇒ **OK** ⇒ **LMB**

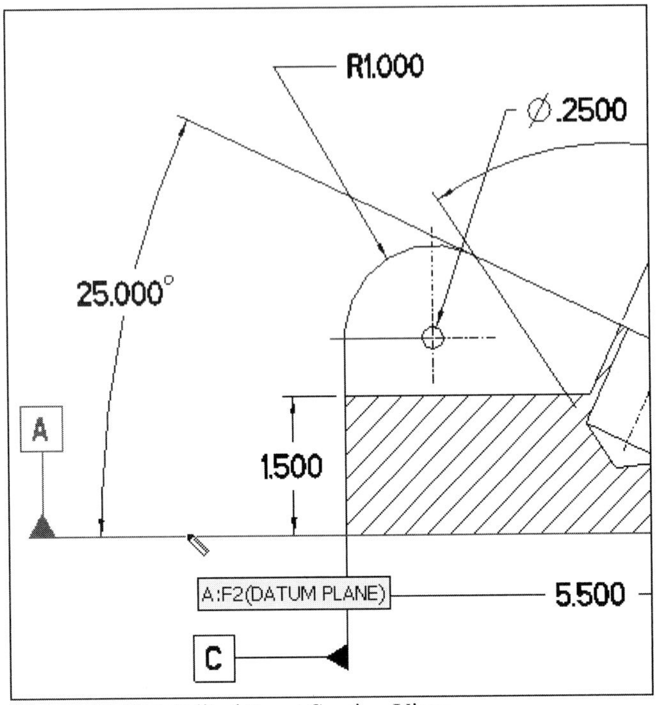

Figure 11.8(a) Edited Front Section View

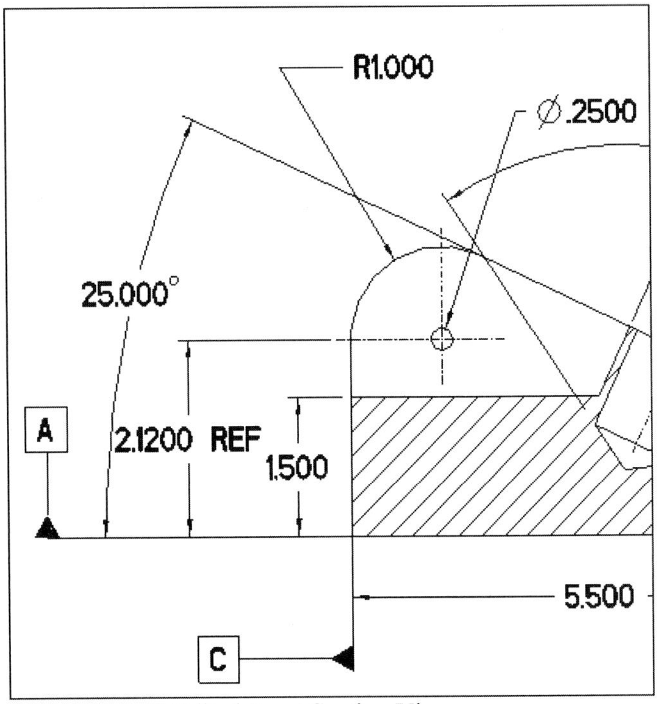

Figure 11.8(b) Edited Front Section View

Figure 11.8(c) shows the edited Auxiliary View and Figure 11.8(d) the Right side view.

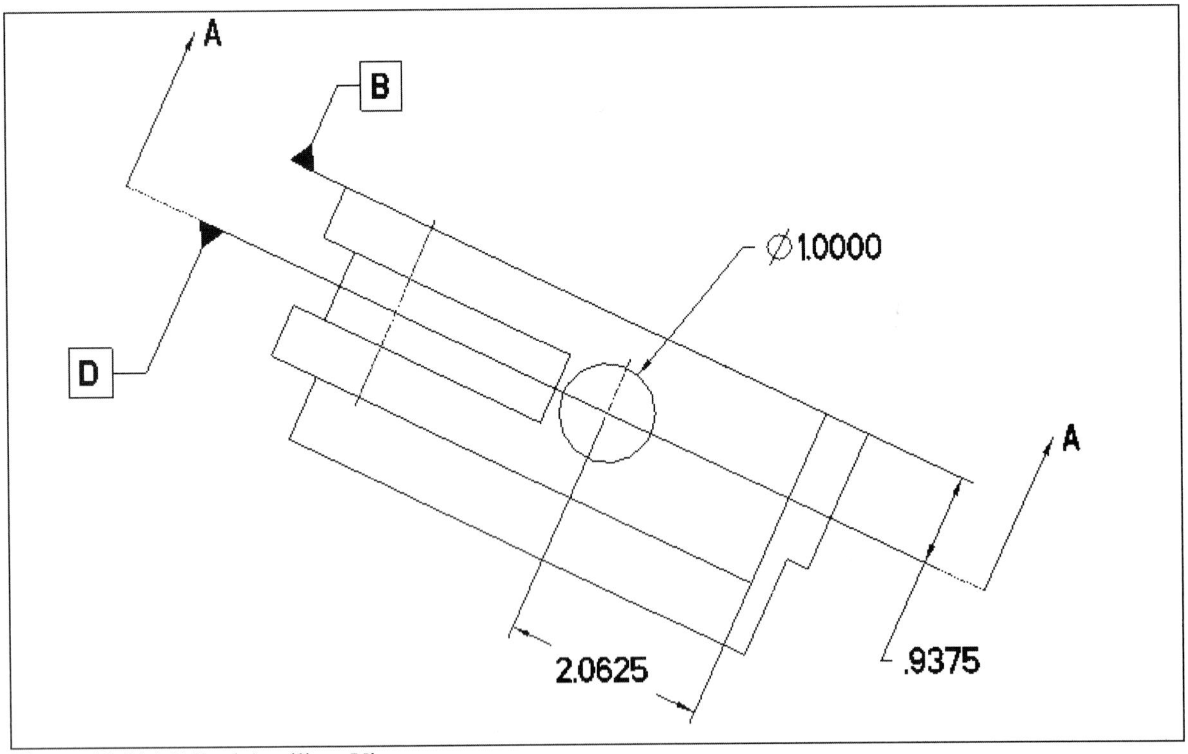

Figure 11.8(c) Edited Auxiliary View

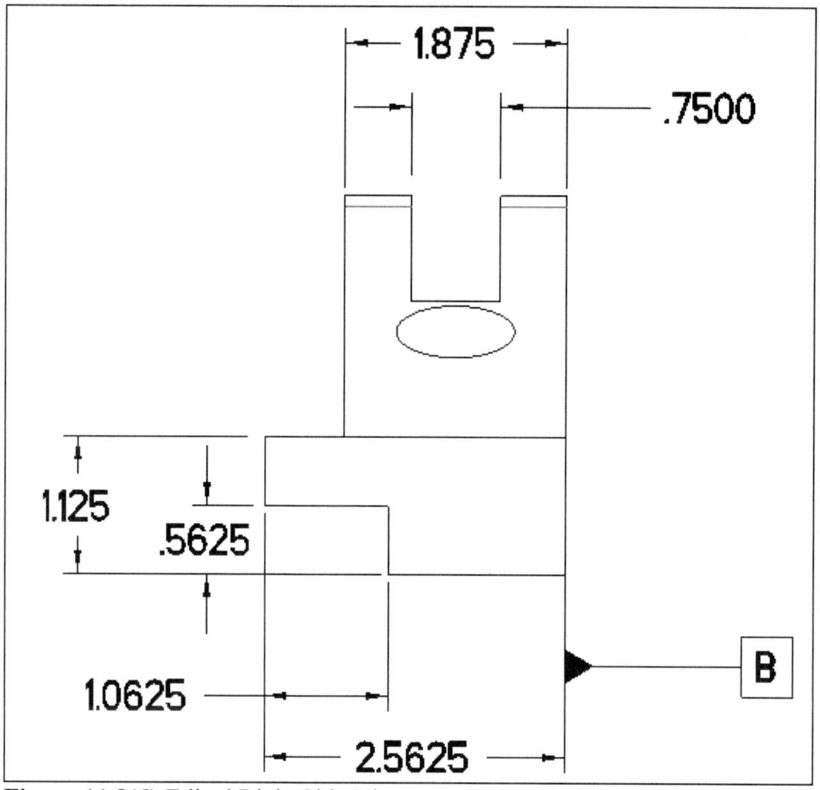

Figure 11.8(d) Edited Right Side View

Add the edges of the small through hole back into the auxiliary view. The edges will display in the graphics window as light gray, but print as dashed on the drawing plot.

Pick the **Auxiliary View** ⇒ **RMB** ⇒ **Edge Display** ⇒ **Hidden Line** ⇒ press and hold the **Ctrl** key and pick near where the small through hole would show as hidden in the auxiliary view [Fig. 11.9(a)] ⇒ pick the opposite edge [Fig. 11.9(b)] ⇒ **OK** ⇒ **Done** ⇒ ⇒ ⇒ ⇒ ⇒ **Enter** [Figs. 11.9(c)]

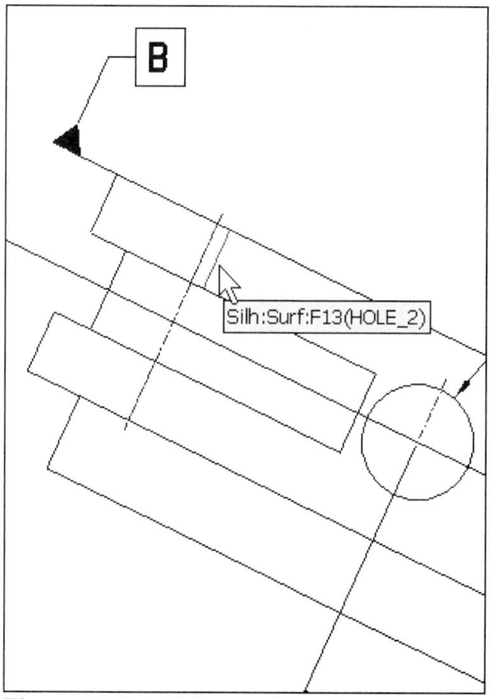

Figure 11.9(a) Pick near the Hole Edge

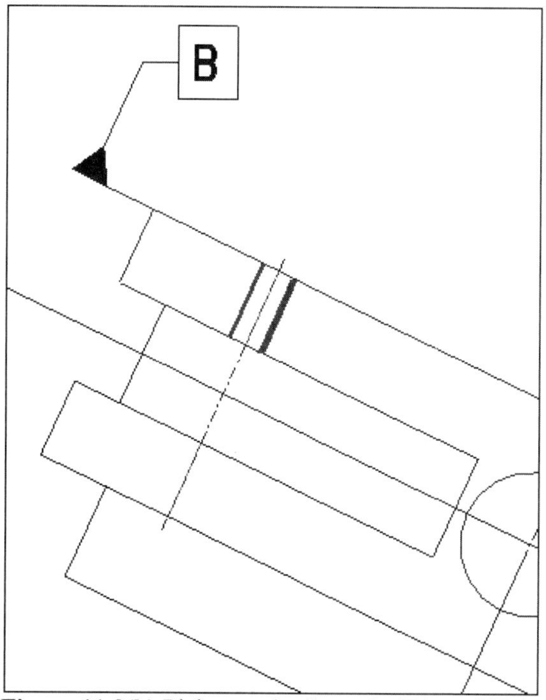

Figure 11.9(b) Pick near the Opposite Edge of Hole

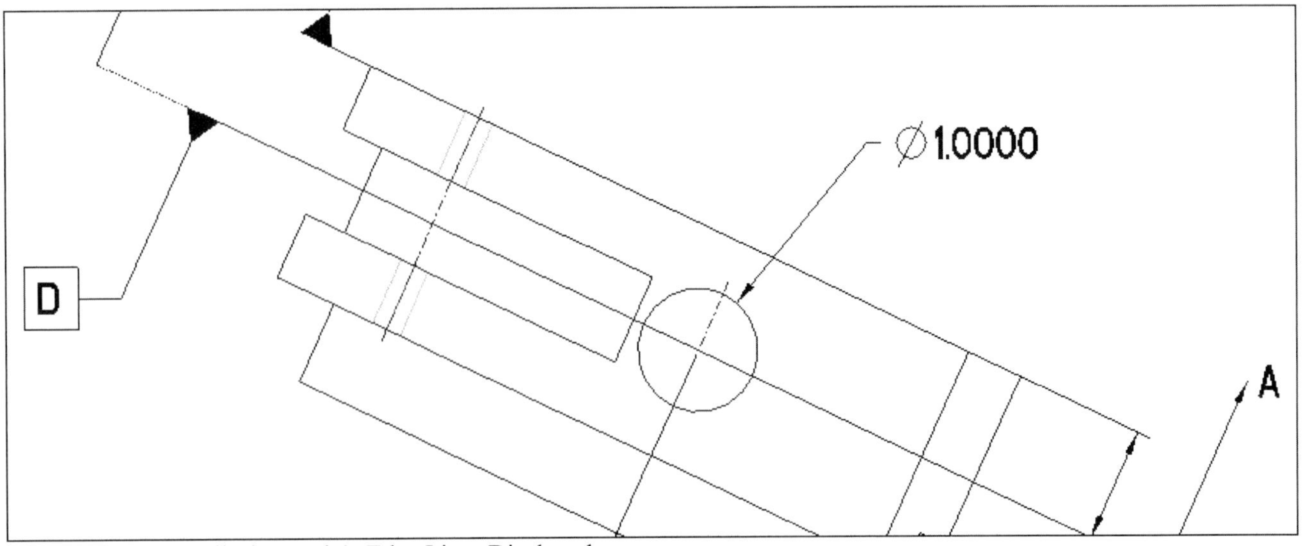

Figure 11.9(c) Small Thru Hole's Edge Lines Displayed

To increase the clarity of this drawing, you will need to master a number of capabilities. Partial views, detail views, using multiple sheets, and modifying section lining (crosshatch lines) are just a few of the many options available in Drawing mode. Create a Detail View.

Click: **Insert** ⇒ **Drawing View** ⇒ **Detailed** ⇒ [Select center point for detail on an existing view.] pick the top edge of the hole in the Front section view [Fig. 11.10(a)] ⇒ [Sketch a spline, without intersecting other splines, to define an outline.] *each **LMB** pick adds a point to the spline* ⇒ **MMB** to end the spline [Fig. 11.10(b)] *(you do not have to "close" the spline sketch)* ⇒ [Select CENTER POINT for drawing view.] pick in the upper right corner of the drawing sheet to place the view

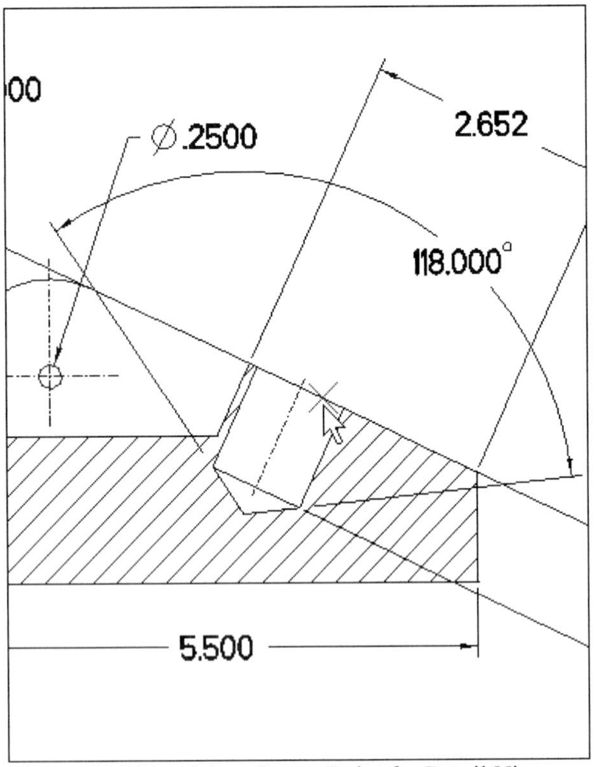

Figure 11.10(a) Select Center Point for Detail View

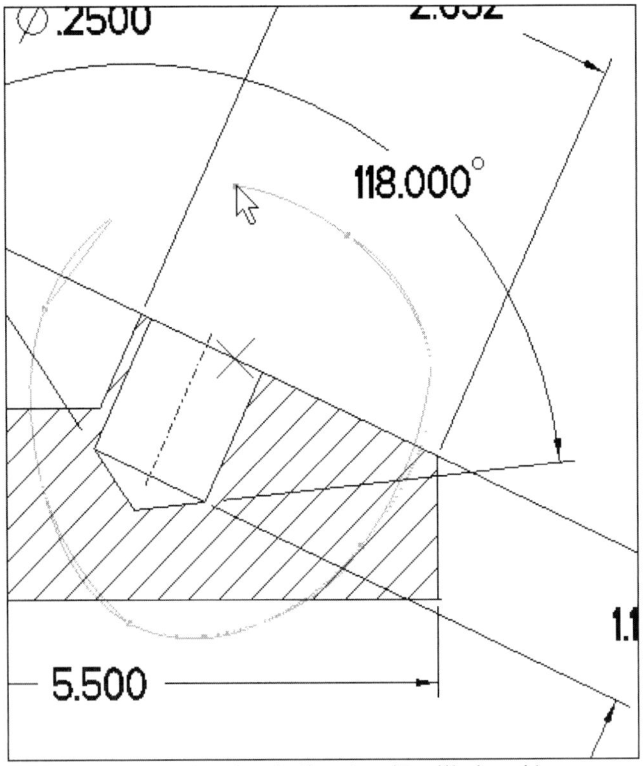

Figure 11.10(b) Sketch a Spline (Pro/E will close it)

Click: **RMB** on detailed view ⇒ **Properties** [Fig. 11.10(c)] ⇒ **Scale** ⇒ 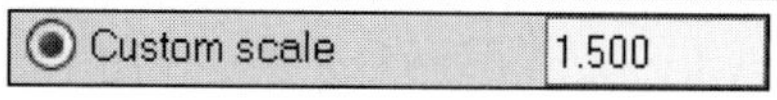

1.500 [Fig. 11.10(d)] ⇒ **Enter** ⇒ **Apply** ⇒ **Close** ⇒ **LMB** to deselect

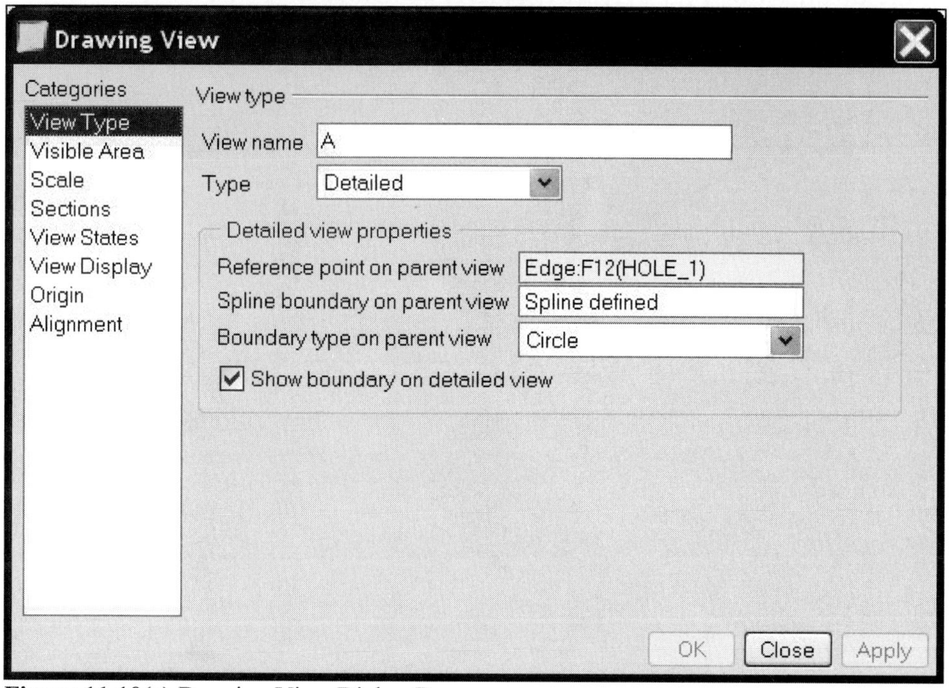

Figure 11.10(c) Drawing View Dialog Box

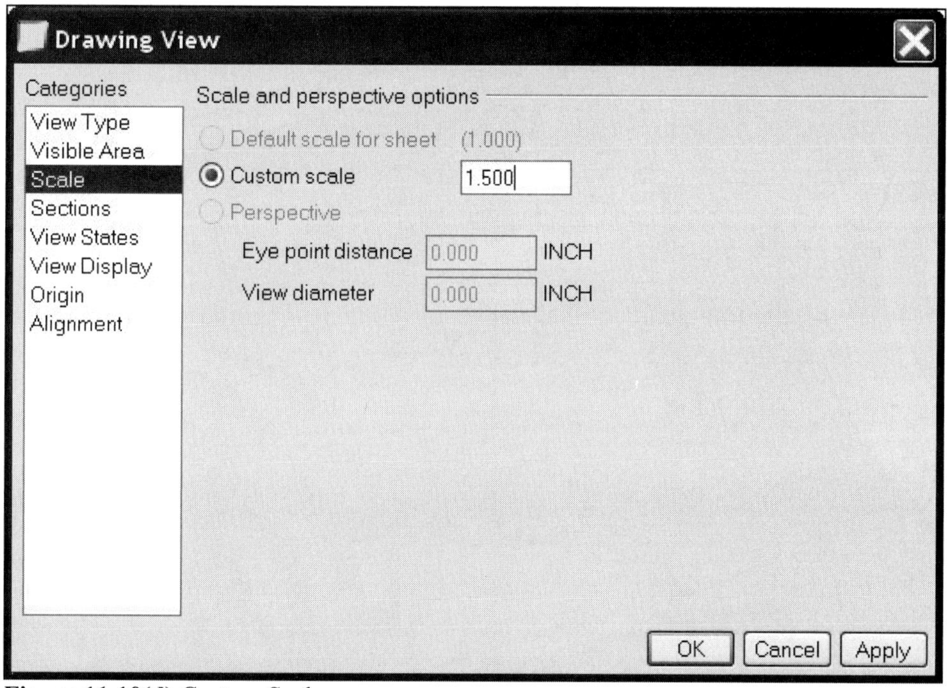

Figure 11.10(d) Custom Scale

Add an axis to the detail of the hole; click **View** from menu bar ⇒ **Show and Erase** ⇒ Show ⇒ only -----A.1 **Axis** active ⇒ Feature and View [Fig. 11.10(e)] ⇒ pick on the hole (feature) in the detail view [Fig. 11.10(f)] ⇒ **OK** ⇒ pick on the detail view [Fig. 11.10(g)] ⇒ **OK** ⇒ **Close** ⇒ **LMB**

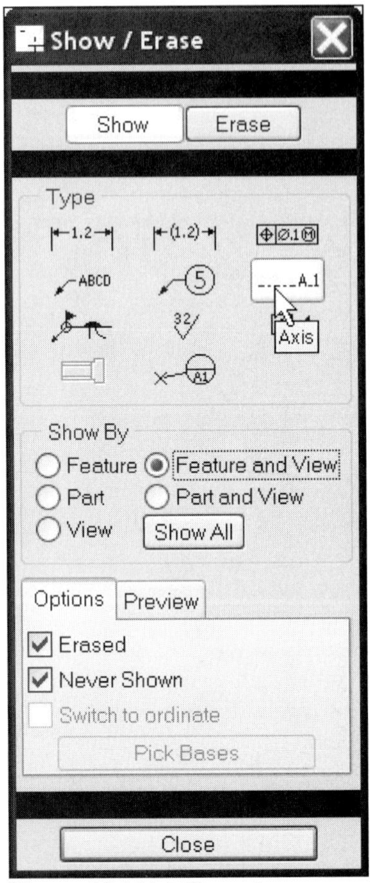

Figure 11.10(e) Show Axes

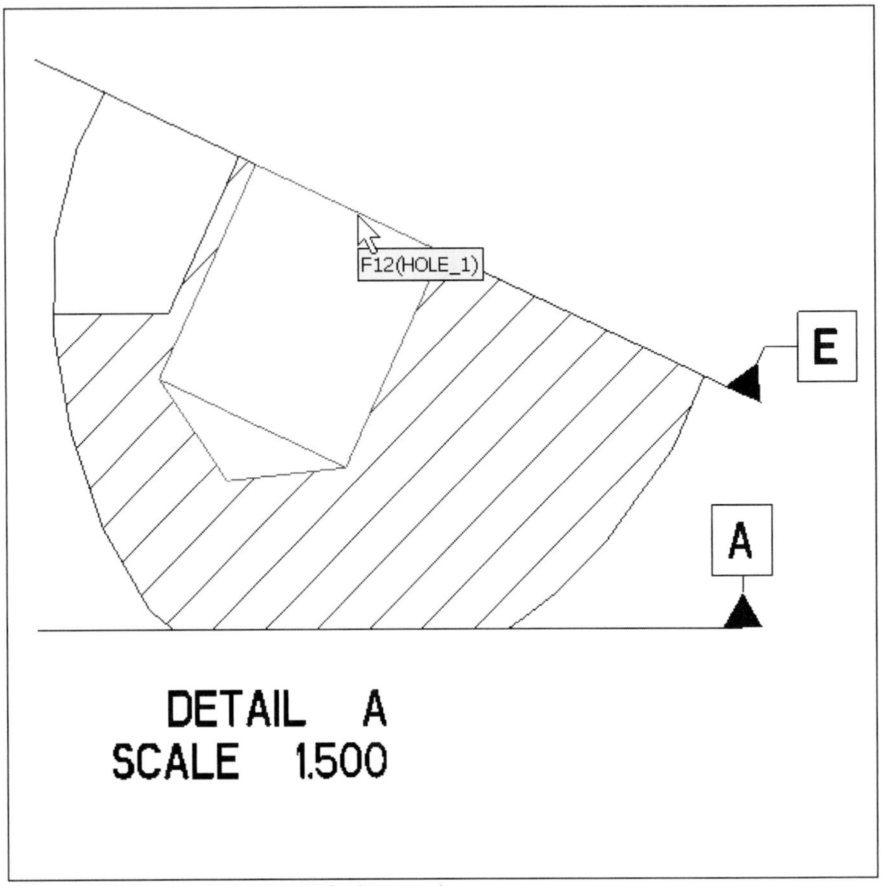

Figure 11.10(f) Pick on the Hole (Feature)

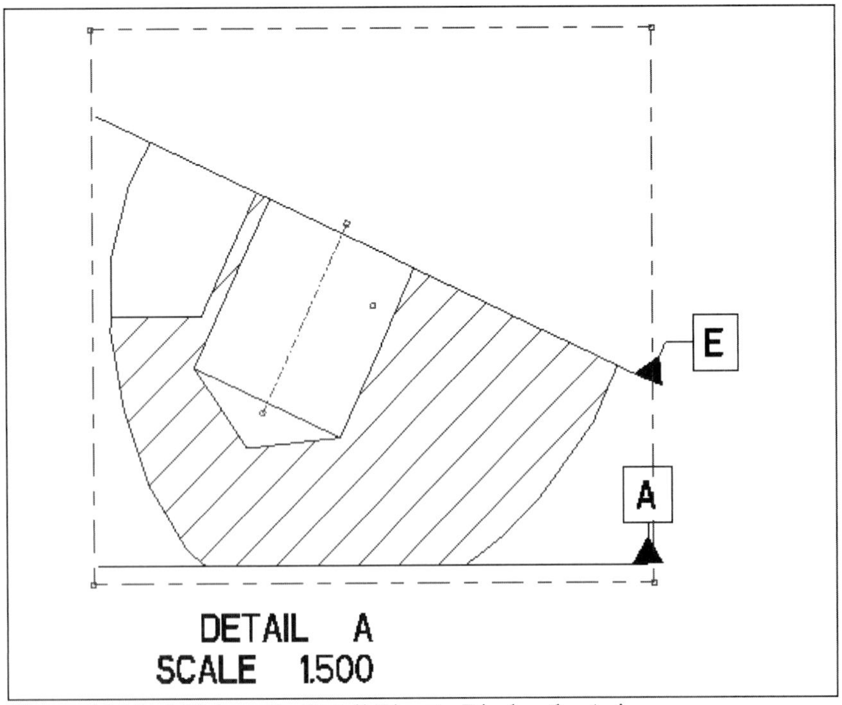

Figure 11.10(g) Pick on the Detail View to Display the Axis

In DETAIL **A**, erase datum **A** and clip datum **E**. Also, clip the axis. With the **Ctrl** key pressed, pick on the text items: **SECTION A-A**, **SEE DETAIL A** and **DETAIL A SCALE 1.500** [Fig. 11.11(a)] ⇒ **RMB** ⇒ **Text Style** ⇒ Default ⇒ Height **.375** [Fig. 11.11(b)] ⇒ **Apply** ⇒ **OK** [Fig. 11.11(c)] ⇒ **LMB**

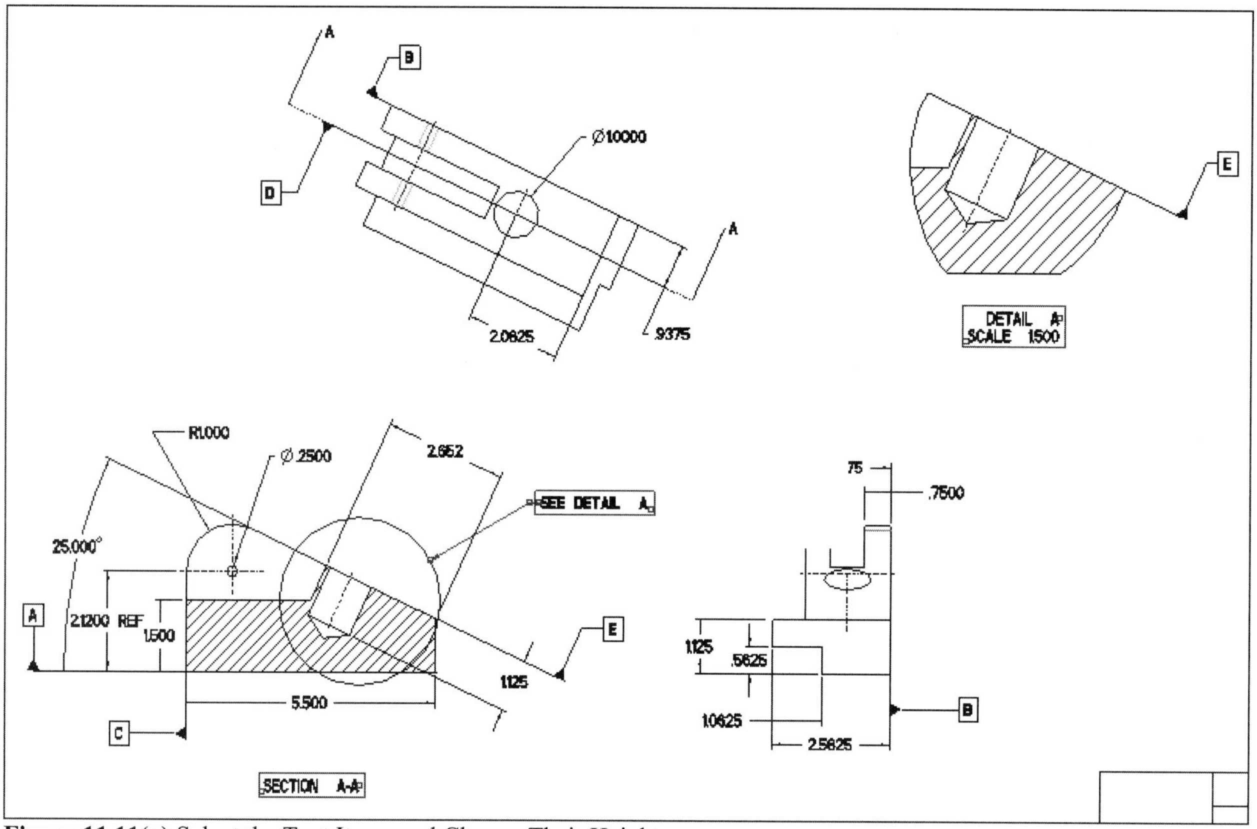

Figure 11.11(a) Select the Text Items and Change Their Height

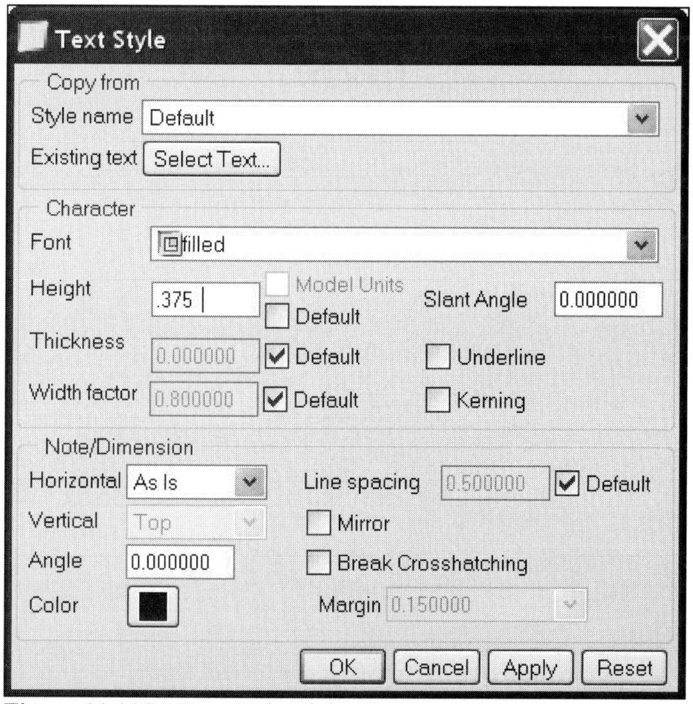

Figure 11.11(b) Text Style Dialog Box

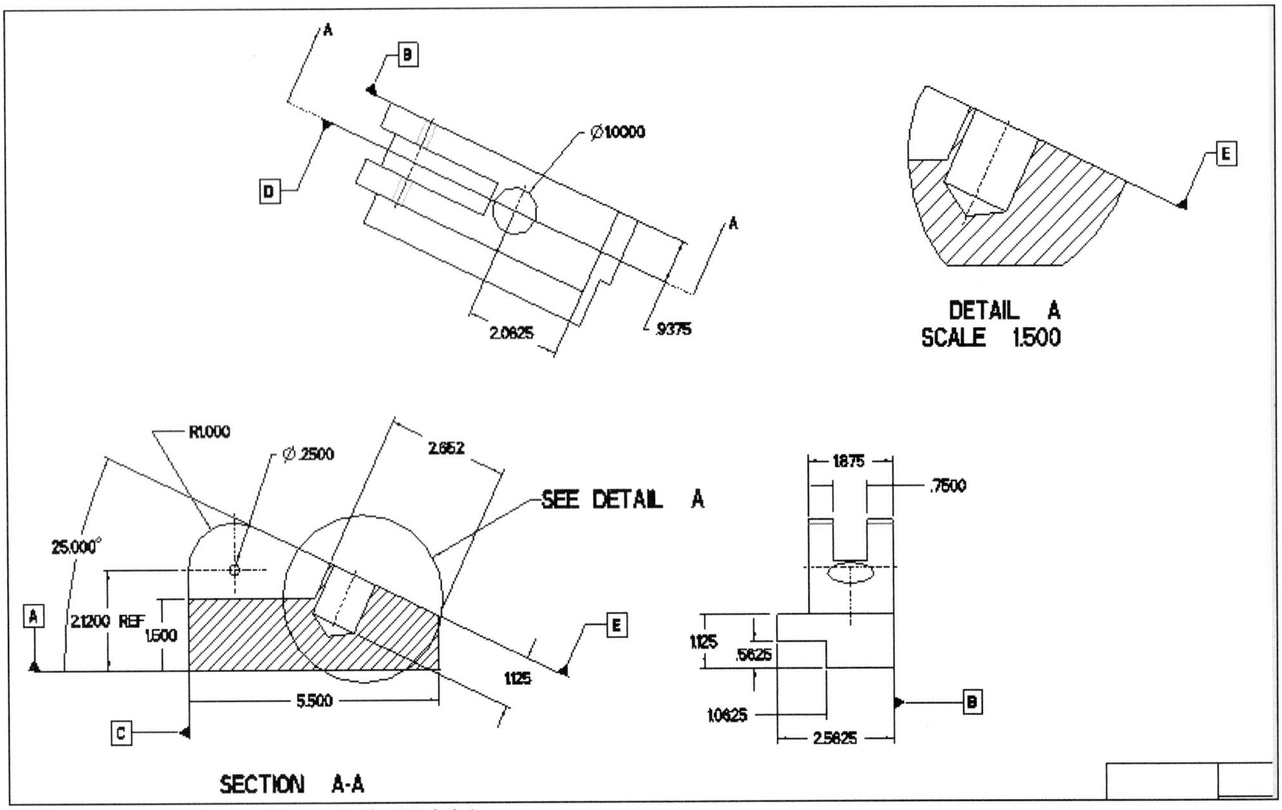

Figure 11.11(c) Changed Text Style (Height)

Change the height of the section identification lettering to **.375**, pick on "A" ⇒ **RMB** ⇒ **Text Style** [Fig. 11.11(d)] ⇒ Default ⇒ Height **.375** ⇒ **Apply** [Fig. 11.11(e)] ⇒ **OK** ⇒ **LMB**

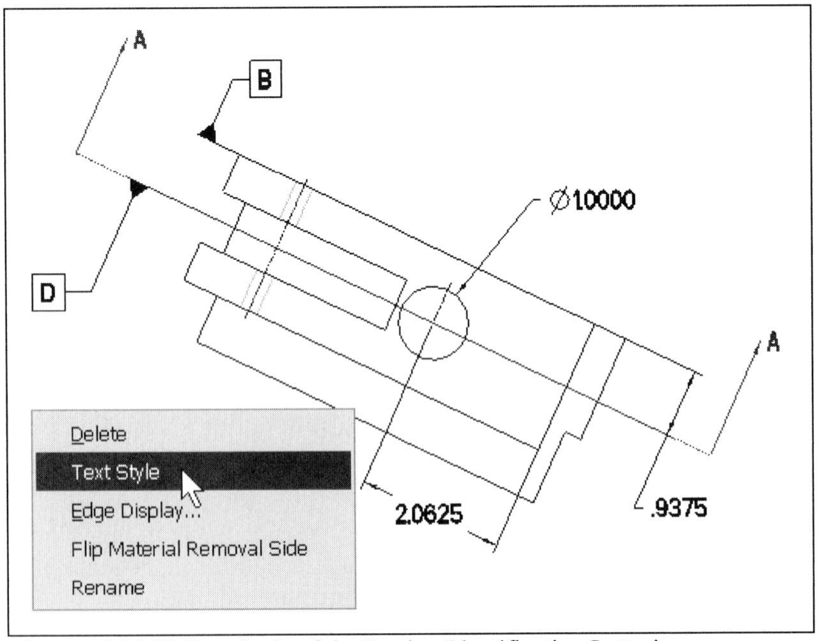

Figure 11.11(d) Text Style of the Section Identification Lettering

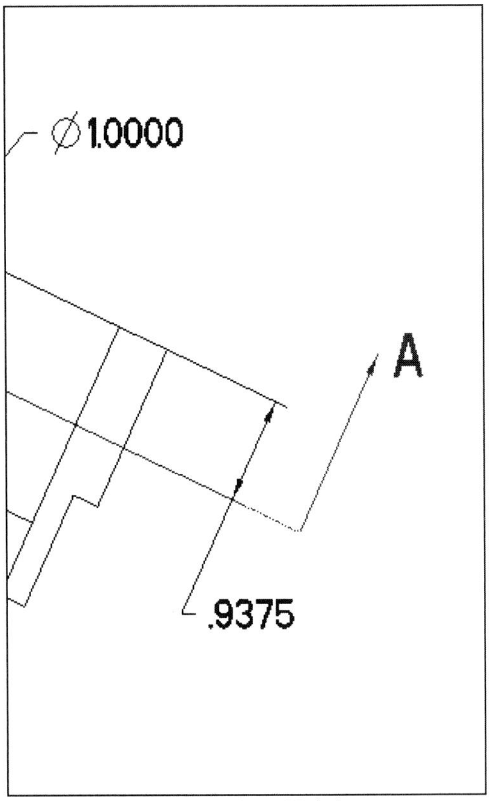

Figure 11.11(e) New Text Height

The Arrows for the cutting plane and dimensions are too small. These are controlled by the Drawing Options, click: **File** ⇒ **Properties** ⇒ **Drawing Options** ⇒ ***draw_arrow_length .25*** ⇒ **Enter** ⇒ ***draw_arrow_width .10*** ⇒ **Enter** ⇒ ***crossec_arrow_length .375*** ⇒ **Enter** ⇒ ***crossec_arrow_width .125*** ⇒ **Enter** [Fig. 11.12(a)] ⇒ **Apply** ⇒ **Close** ⇒ **MMB** ⇒ [Fig. 11.12(b)]

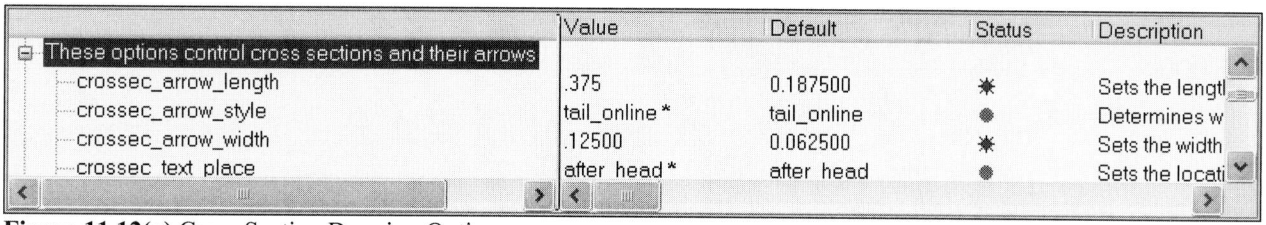

These options control cross sections and their arrows	Value	Default	Status	Description
crossec_arrow_length	.375	0.187500	✱	Sets the lengtl
crossec_arrow_style	tail_online *	tail_online	●	Determines w
crossec_arrow_width	.12500	0.062500	✱	Sets the width
crossec_text_place	after_head *	after_head	●	Sets the locati

Figure 11.12(a) Cross Section Drawing Options

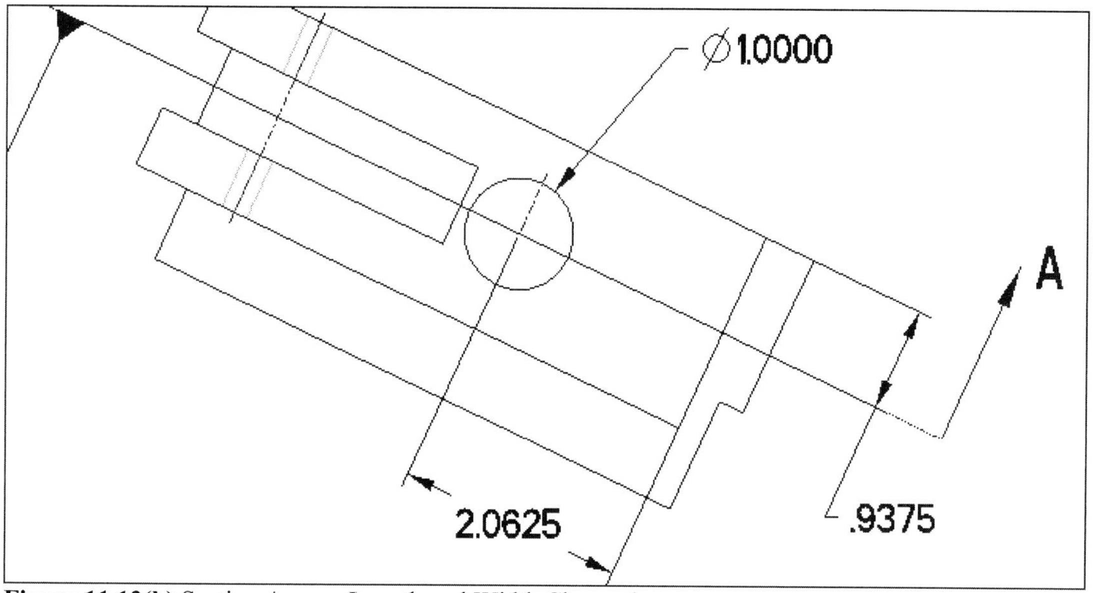

Figure 11.12(b) Section Arrows Length and Width Changed

Since the model and the drawing are associative, make a modification to the slot dimension. Pick on the **2.652** dimension ⇒ pick on the value again [Fig. 11.12(c)] ⇒ **RMB** ⇒ **Edit Value** ⇒ type **3.125** [Fig. 11.12(d)] ⇒ **Enter** ⇒ **Edit** ⇒ **Regenerate** ⇒ **Model** [Fig. 11.12(e)]

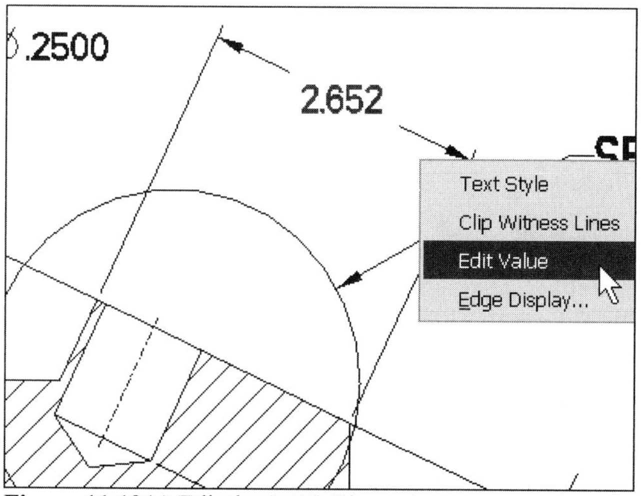

Figure 11.12(c) Edit the **2.652** Dimension

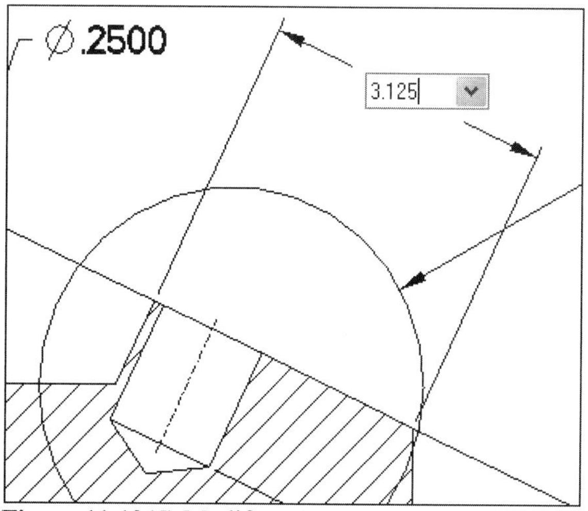

Figure 11.12(d) Modify to **3.125**

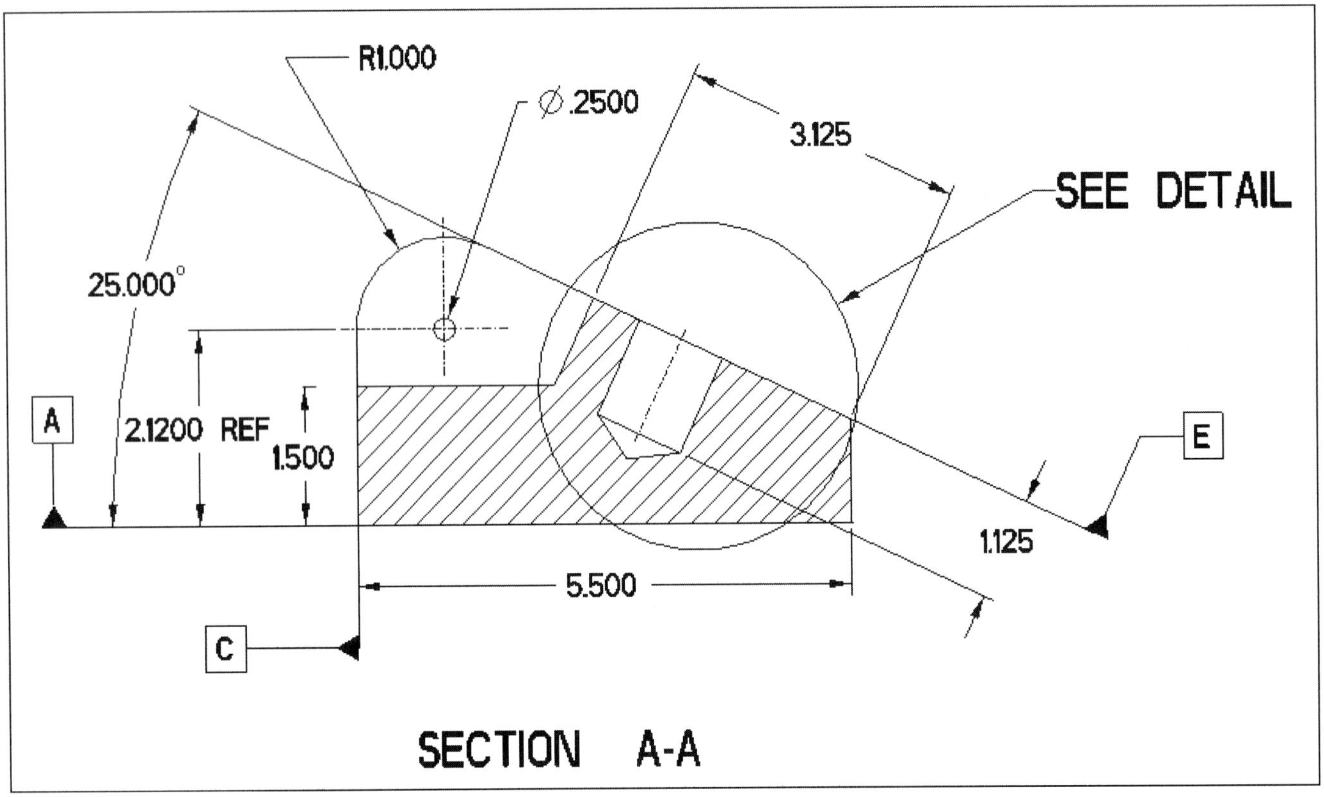

Figure 11.12(e) Regenerated Part

Pick on the **1.125** dimension in the Front section view [Fig. 11.13(a)] ⇒ **RMB** ⇒ **Move Item to View** ⇒ pick on view **DETAIL A** [Fig. 11.13(b)]

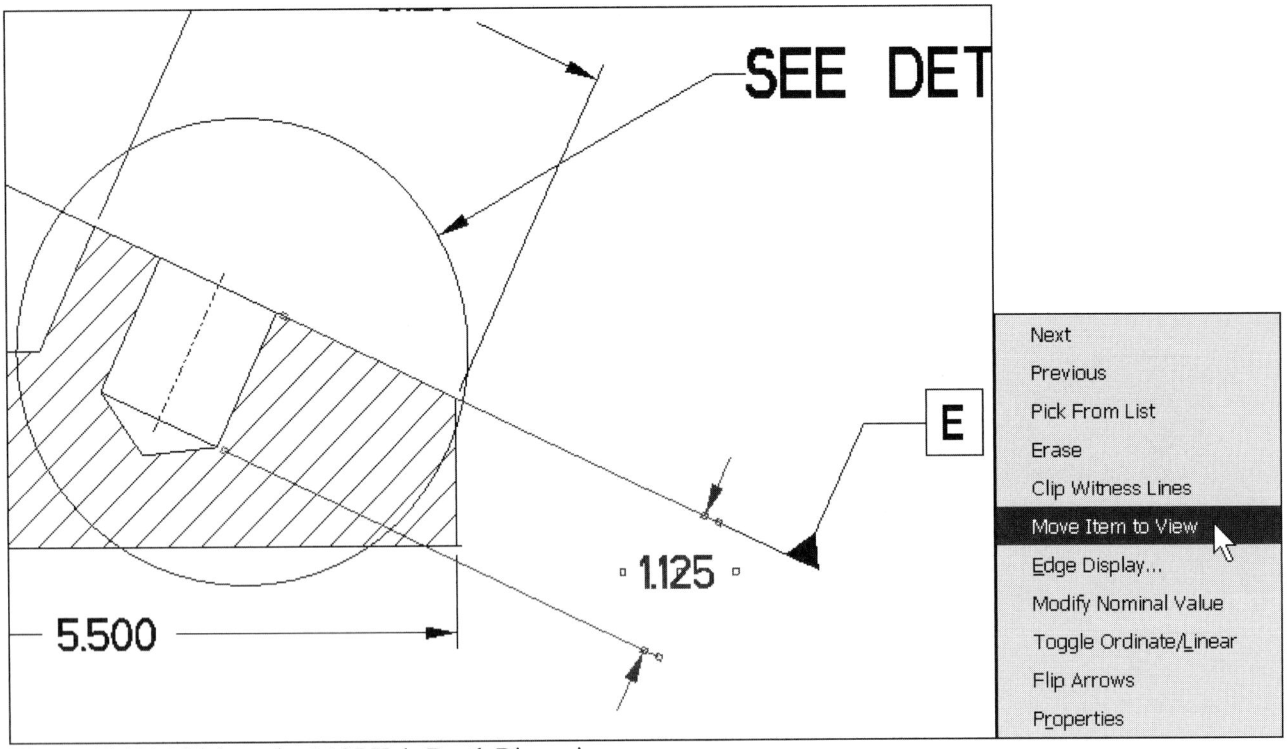

Figure 11.13(a) Pick on the **1.125** Hole Depth Dimension

Pick on the **1.125** dimension to reposition and clip as needed [Fig. 11.13(c)] ⇒ **LMB** to deselect ⇒ keeping in mind ASME standards, cleanup the drawing as needed ⇒ ⇒ ⇒ ⇒ ⇒ **MMB** ⇒ **File** ⇒ **Delete** ⇒ **Old Versions** ⇒ **MMB** (Fig. 11.14) ⇒ **LMB** to deselect

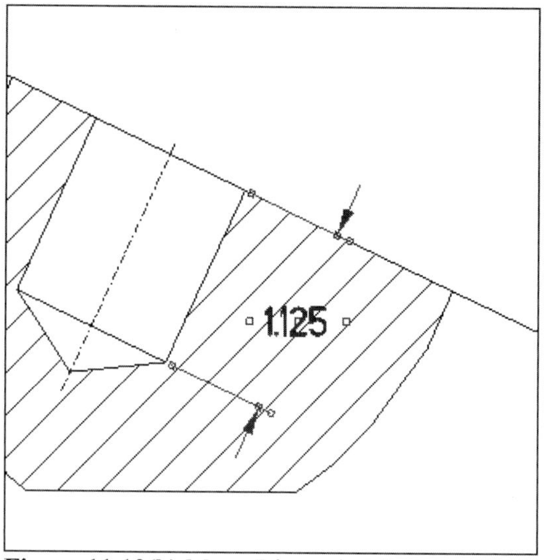

Figure 11.13(b) Move Dimension to DETAIL A

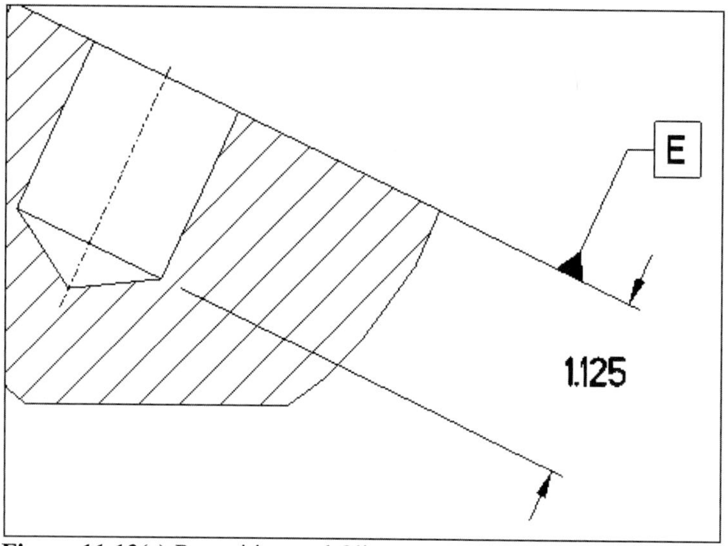

Figure 11.13(c) Reposition and Clip

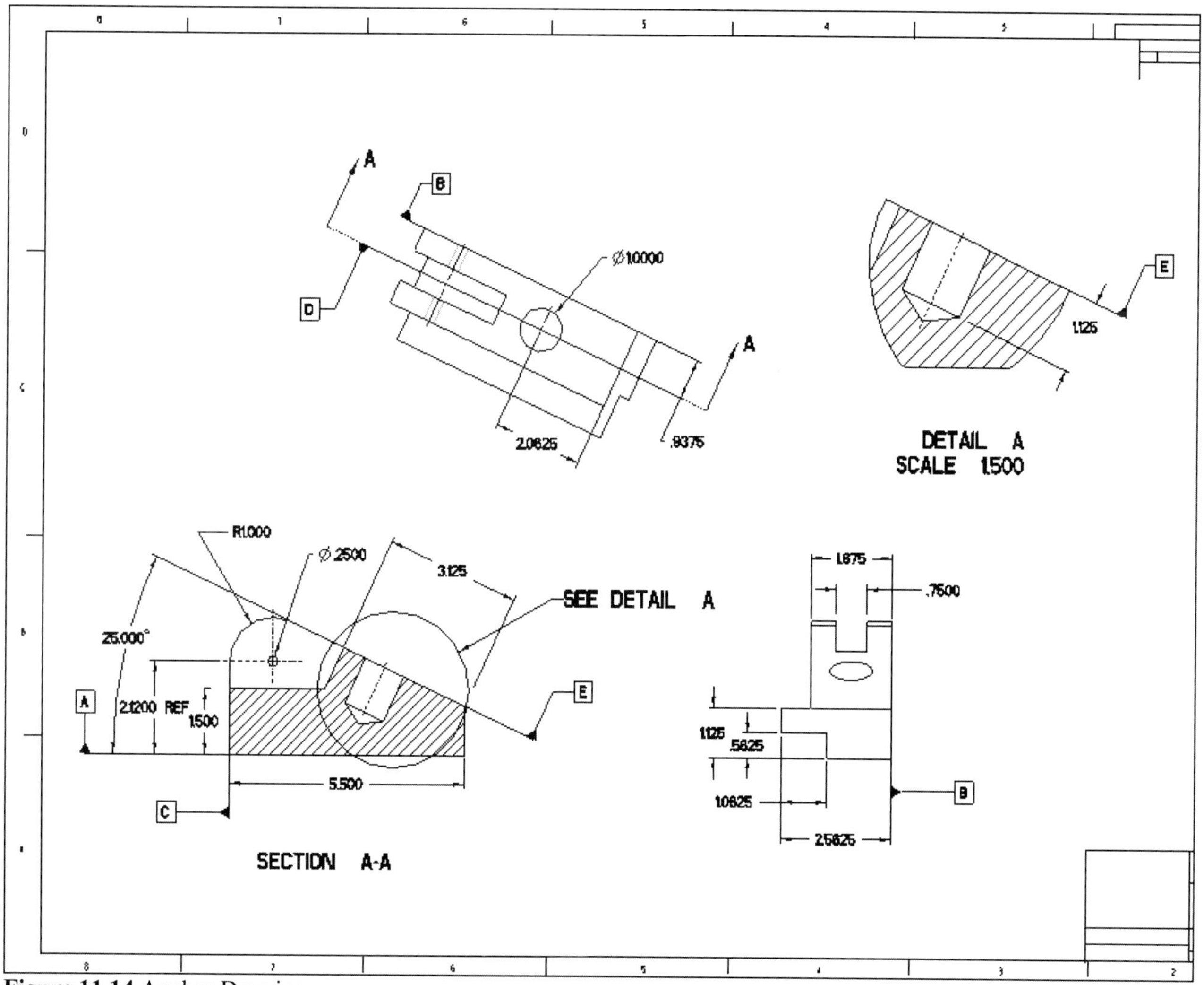

Figure 11.14 Anchor Drawing

Next, you will change the boundary of DETAIL A.

Pick on view **DETAIL A** ⇒ **RMB** (if you RMB inside the view outline, you get a pop-up list of options [Fig. 11.15(a)], whereas if you RMB outside the view outline, there are fewer options) [Fig. 11.15(b)] ⇒ **Properties** ⇒ click inside Spline area 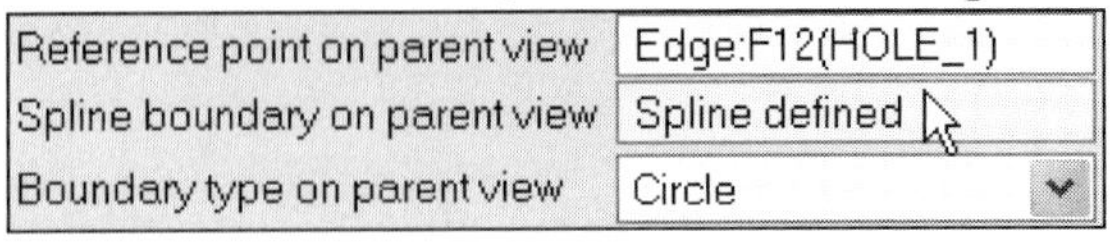

[Fig. 11.15(c)] ⇒ sketch the spline again in the Front (SECTION A-A) view [Fig. 11.15(d)] ⇒ **MMB** to end spline ⇒ **Apply** [Fig. 11.15(e)] ⇒ **Close** ⇒ **LMB** ⇒ clean up the view ⇒ [Fig. 11.15(f)] ⇒  ⇒  ⇒  ⇒  ⇒ **MMB**

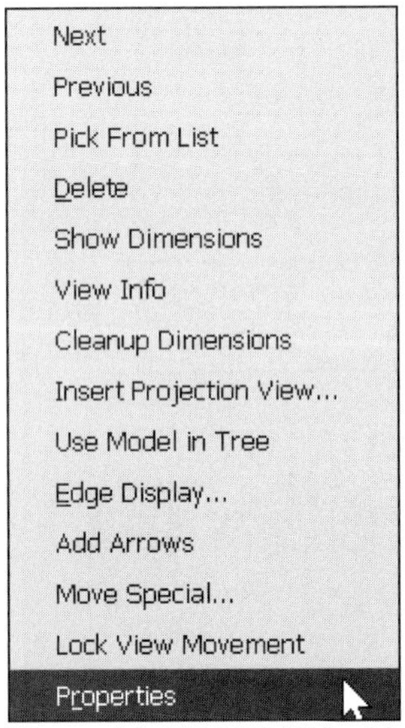

Figure 11.15(a) RMB Inside

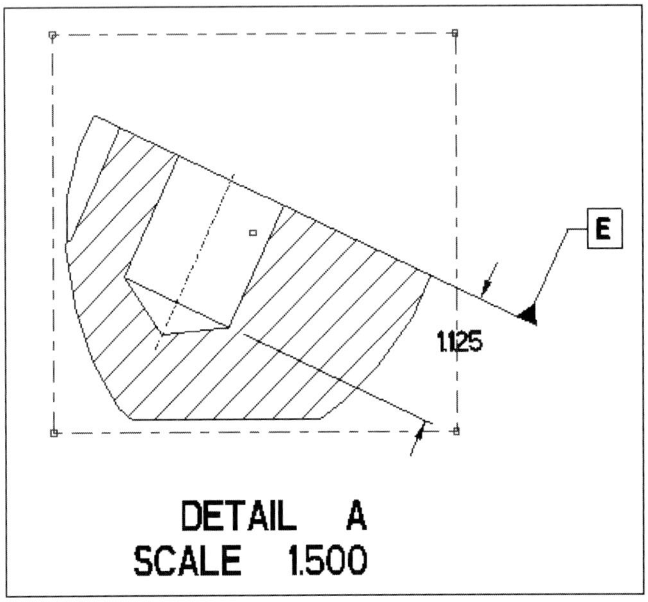

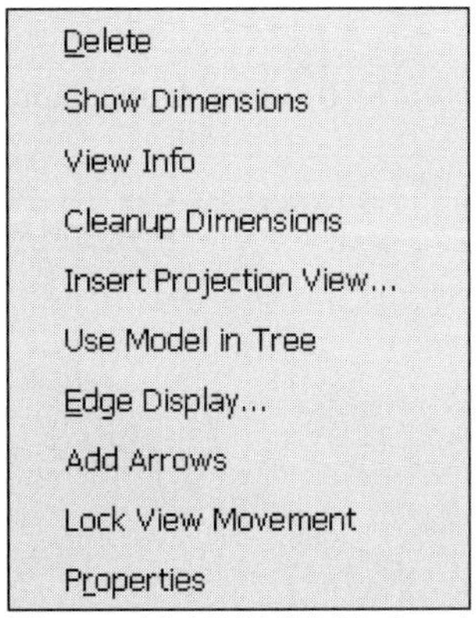

Figure 11.15(b) RMB Outside

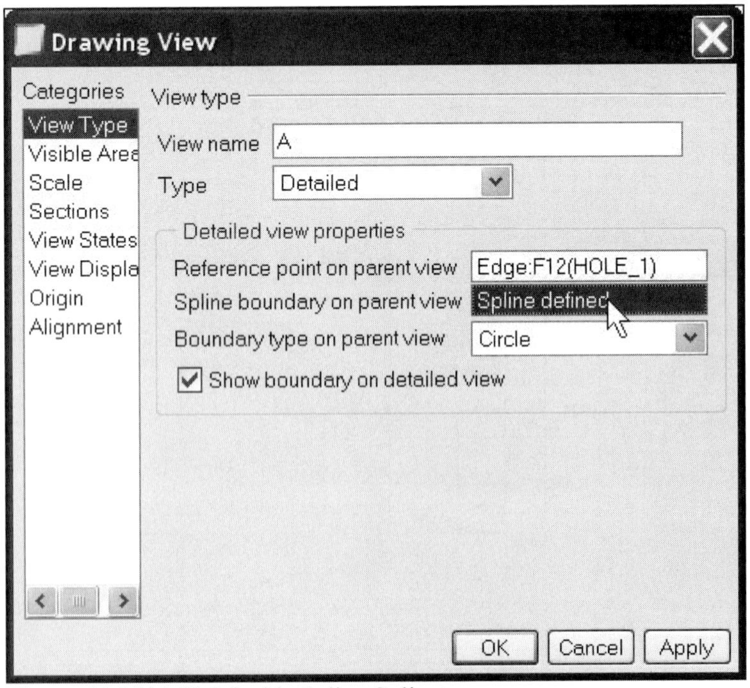

Figure 11.15(c) Pick inside Spline Collector

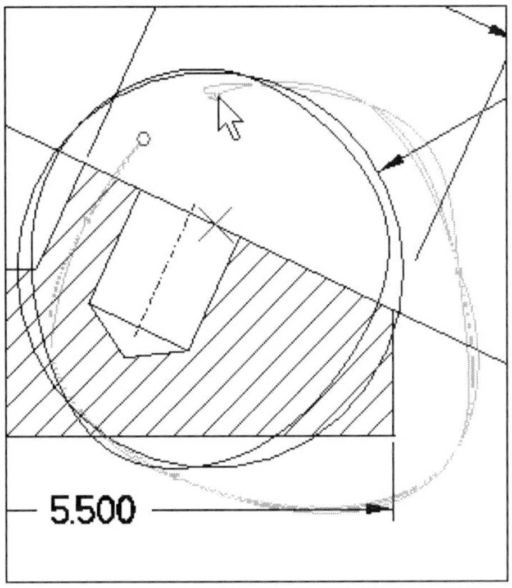

Figure 11.15(d) Sketch the Spline Again

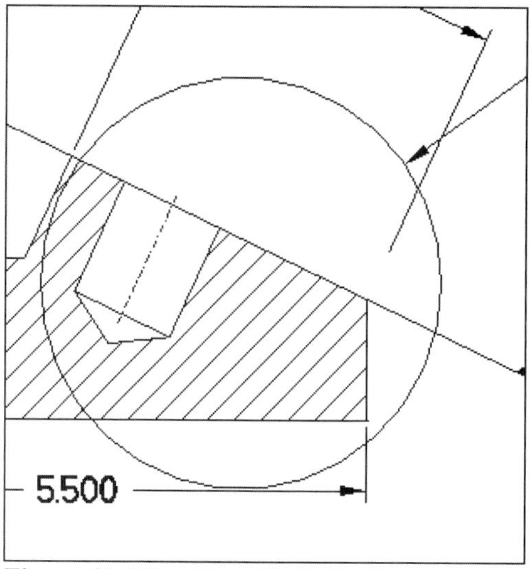

Figure 11.15(e) New Circle Position

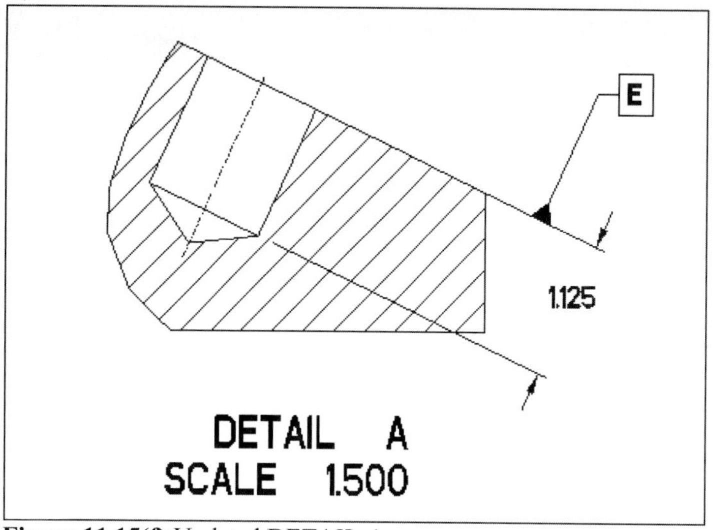

Figure 11.15(f) Updated DETAIL A

Show the (previously erased) dimension for the full angle of the hole's drill tip in **DETAIL A,** click: **View** from the menu bar ⇒ **Show and Erase** ⇒ only [1.2 icon] **Dimension** on ⇒ [Feature and View] [Fig. 11.16(a)] ⇒ pick the hole [Fig. 11.16(b)] ⇒ **MMB** to place the dimension ⇒ **OK** ⇒ **Close** ⇒ reposition and clip the dimension [Fig. 11.16(c)] ⇒ pick on the **118.000** dimension ⇒ **RMB** ⇒ **Properties** ⇒ change decimal places to **0** [Fig. 11.16(d)] ⇒ **Enter** ⇒ **OK** [Fig. 11.16(e)] ⇒ **LMB** to deselect

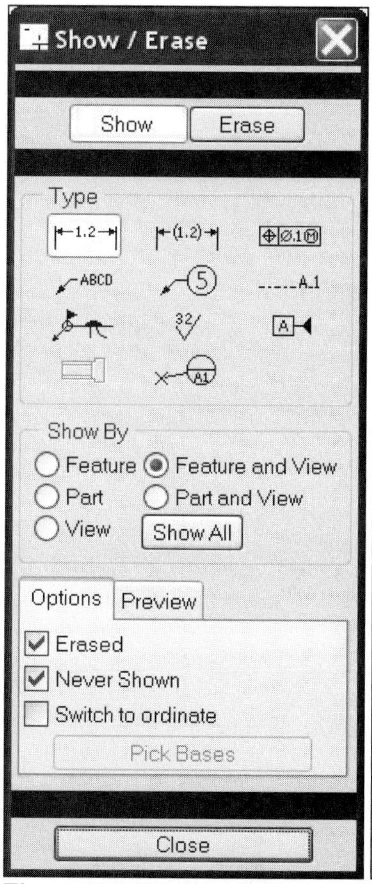

Figure 11.16(a) Show/Erase

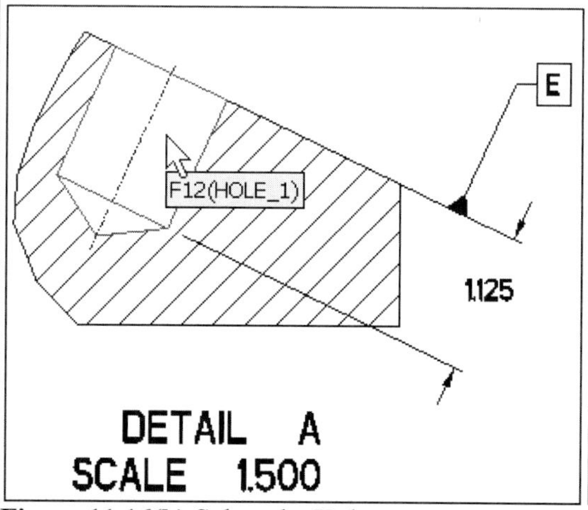

Figure 11.16(b) Select the Hole

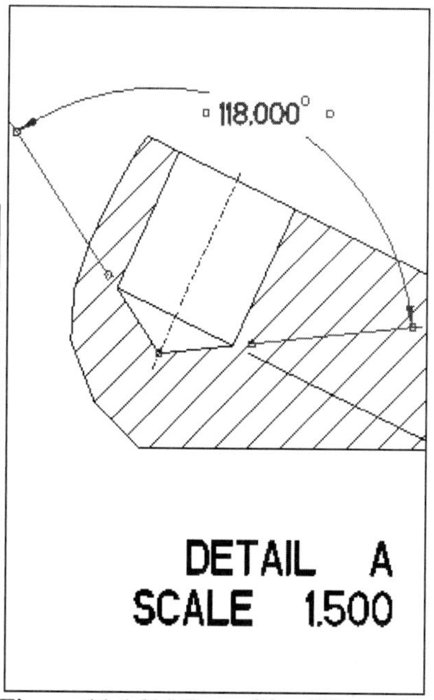

Figure 11.16(c) Reposition Dimension E

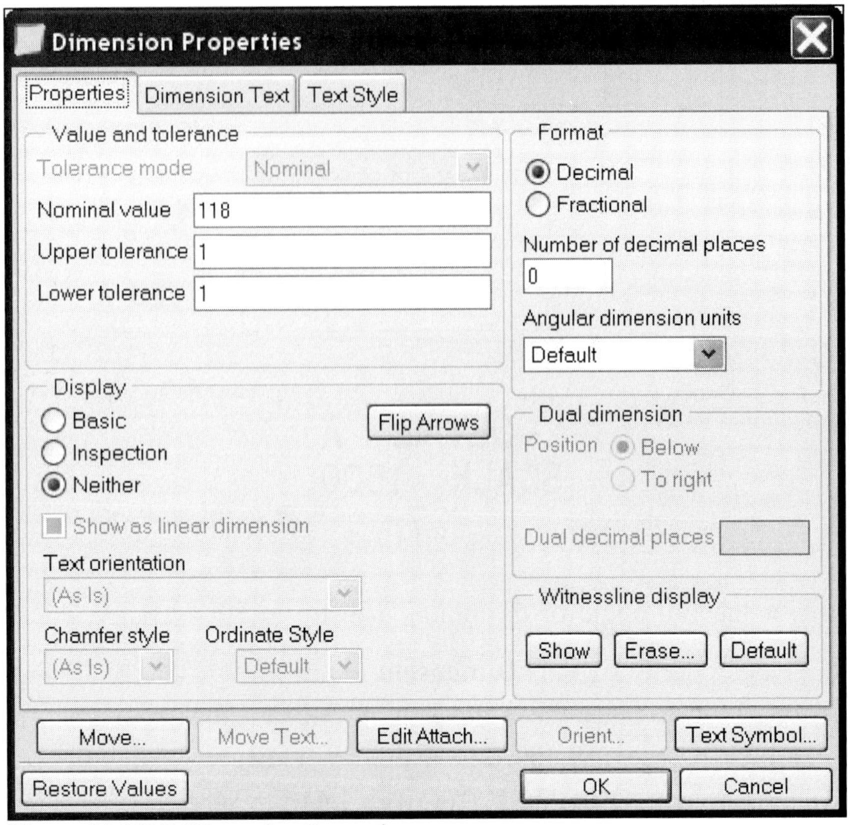

Figure 11.16(d) Dimension Properties

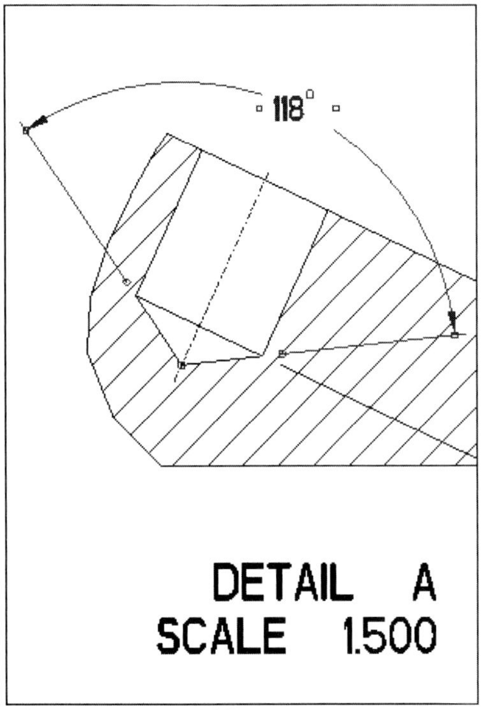

Figure 11.16(e) 118 Degrees

Change the spacing and the angle of the section lining. Double-click on the hatching in SECTION A-A [Fig. 11.17(a)] ⇒ **Angle** ⇒ **30** ⇒ **Spacing** ⇒ **Double** ⇒ **Half** ⇒ **Done** [Fig. 11.17(b)] ⇒ **LMB** ⇒ ⇒ ⇒ **View** ⇒ **Repaint** ⇒ ⇒ **MMB**

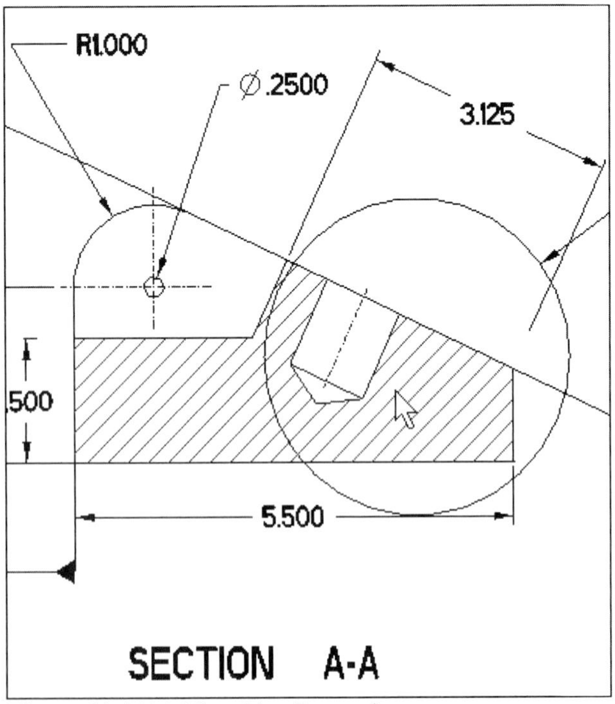

Figure 11.17(a) Xhatching Properties

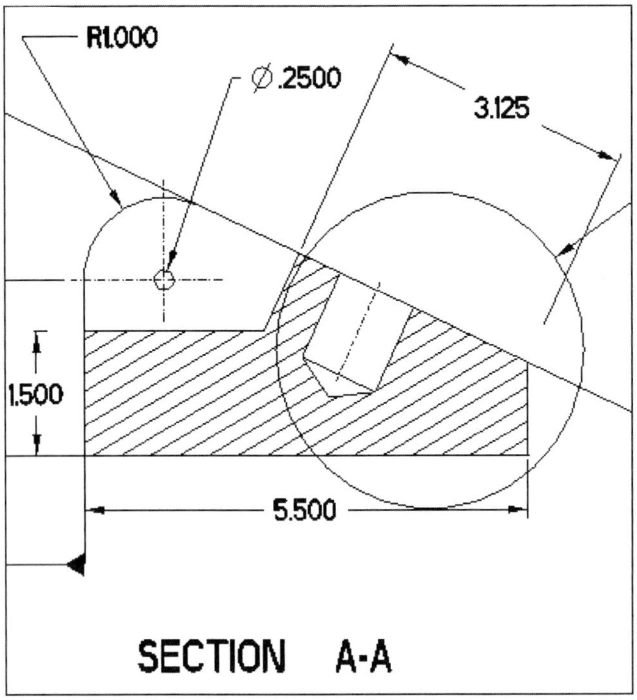

Figure 11.17(b) Hatching at **30** Degree Angle

Change the spacing and the angle of the section lining in **DETAIL A**. Double-click on the hatching in **DETAIL A** [Fig. 11.18(a)] ⇒ **Det Indep** ⇒ **Hatch** ⇒ **Spacing** ⇒ **Double** ⇒ **Angle** ⇒ **45** ⇒ **Done** [Fig. 11.18(b)] ⇒ **LMB** ⇒ **RMB** ⇒ **Update Sheet** ⇒ ⇒ ⇒ ⇒ **MMB**

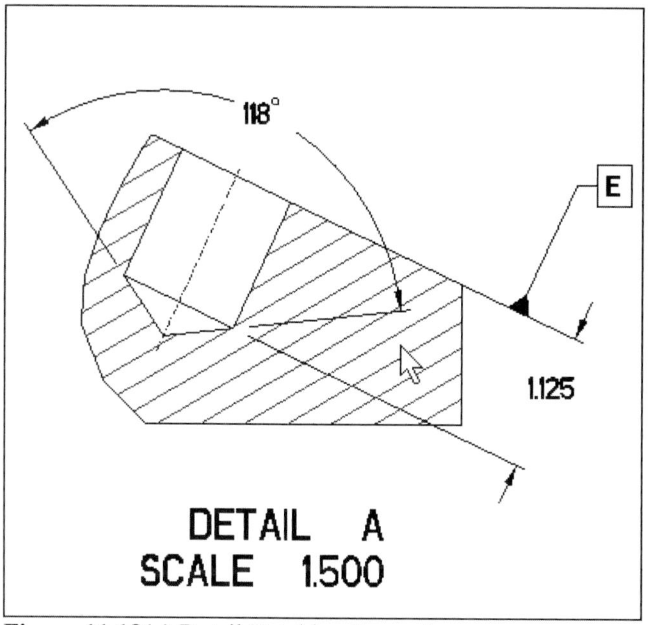

Figure 11.18(a) Detail Hatching Properties

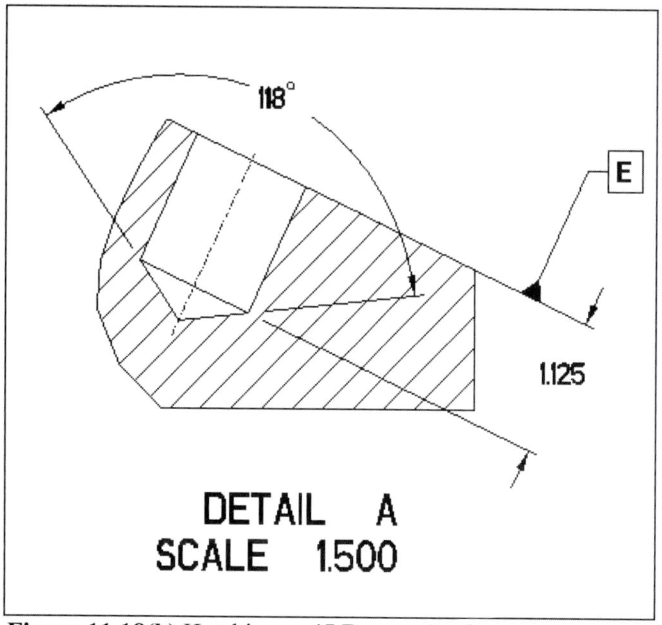

Figure 11.18(b) Hatching at **45** Degree Angle

Change the text style used on the drawing. Click: **Select items** ⇒ enclose all of the drawing text with a selection box [Fig. 11.19(a)] ⇒ **RMB** ⇒ **Text Style**

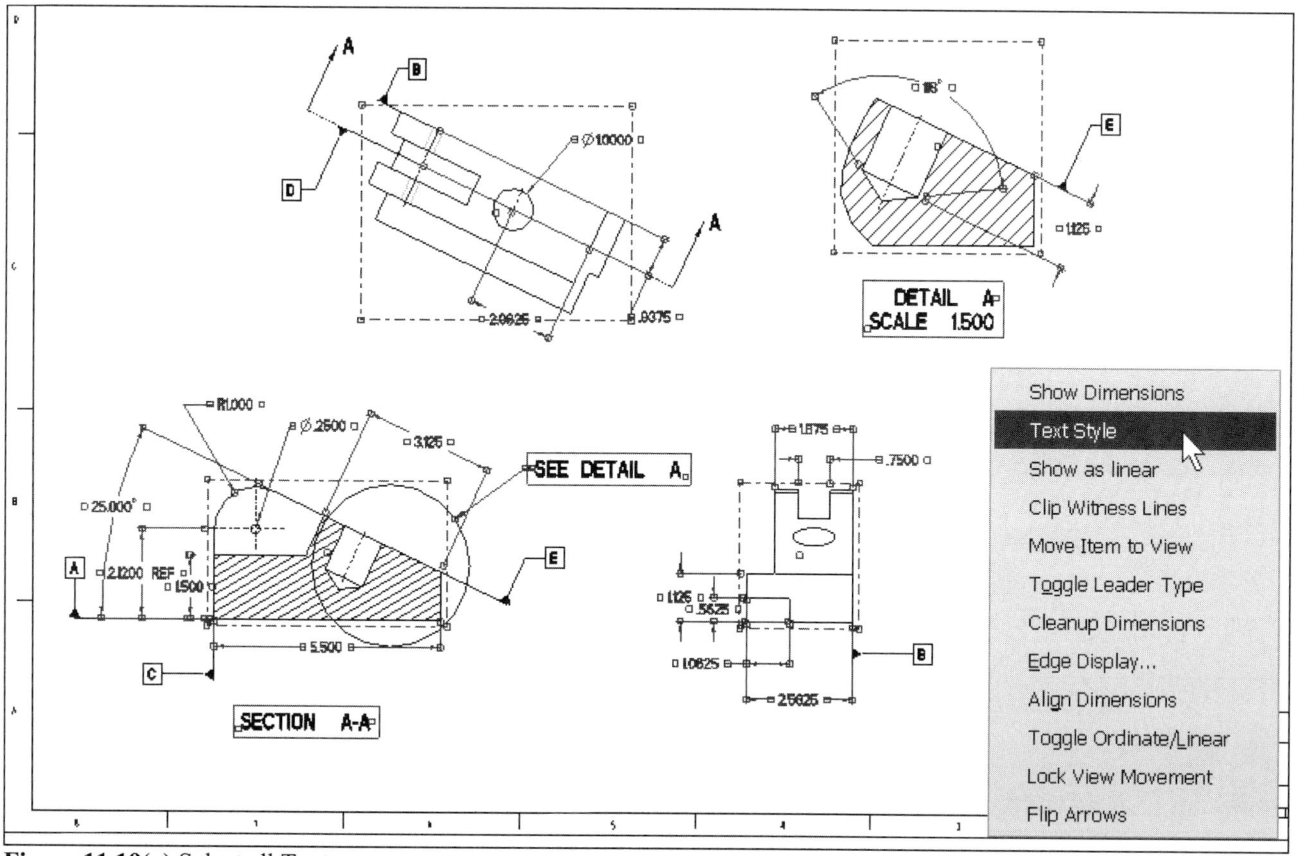

Figure 11.19(a) Select all Text

Click: Font **Blueprint MT** [Figs. 11.19(b-c)] ⇒ **Apply** ⇒ **OK** ⇒ **LMB** ⇒ change the number of decimal places for dimensions as required [Fig. 11.19(d)] ⇒ ⇒ ⇒ ⇒ ⇒ **MMB**

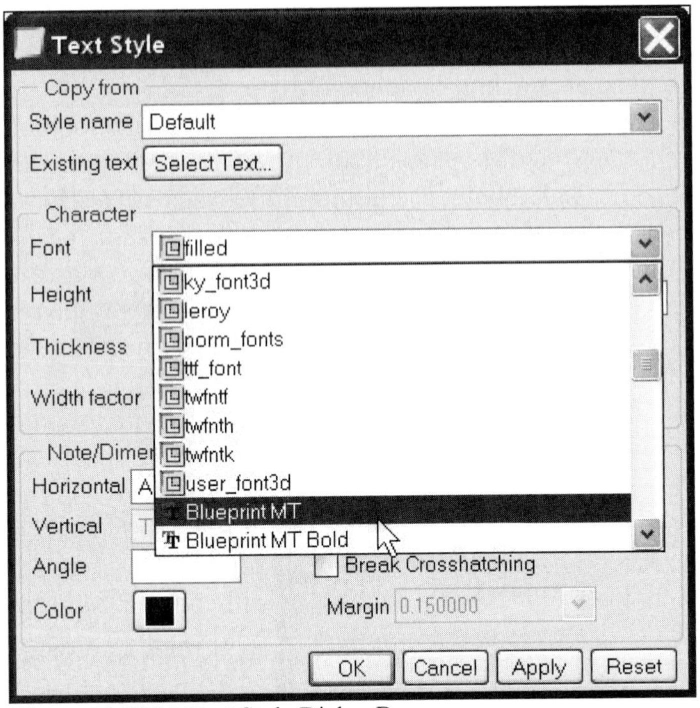

Figure 11.19(b) Text Style Dialog Box

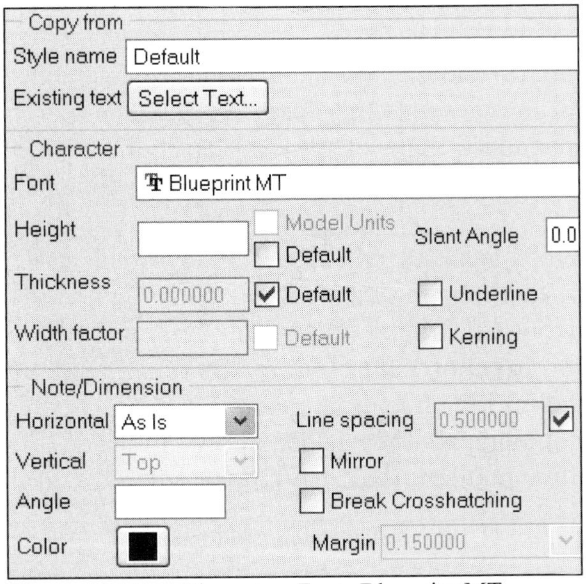

Figure 11.19(c) Character- Font- Blueprint MT

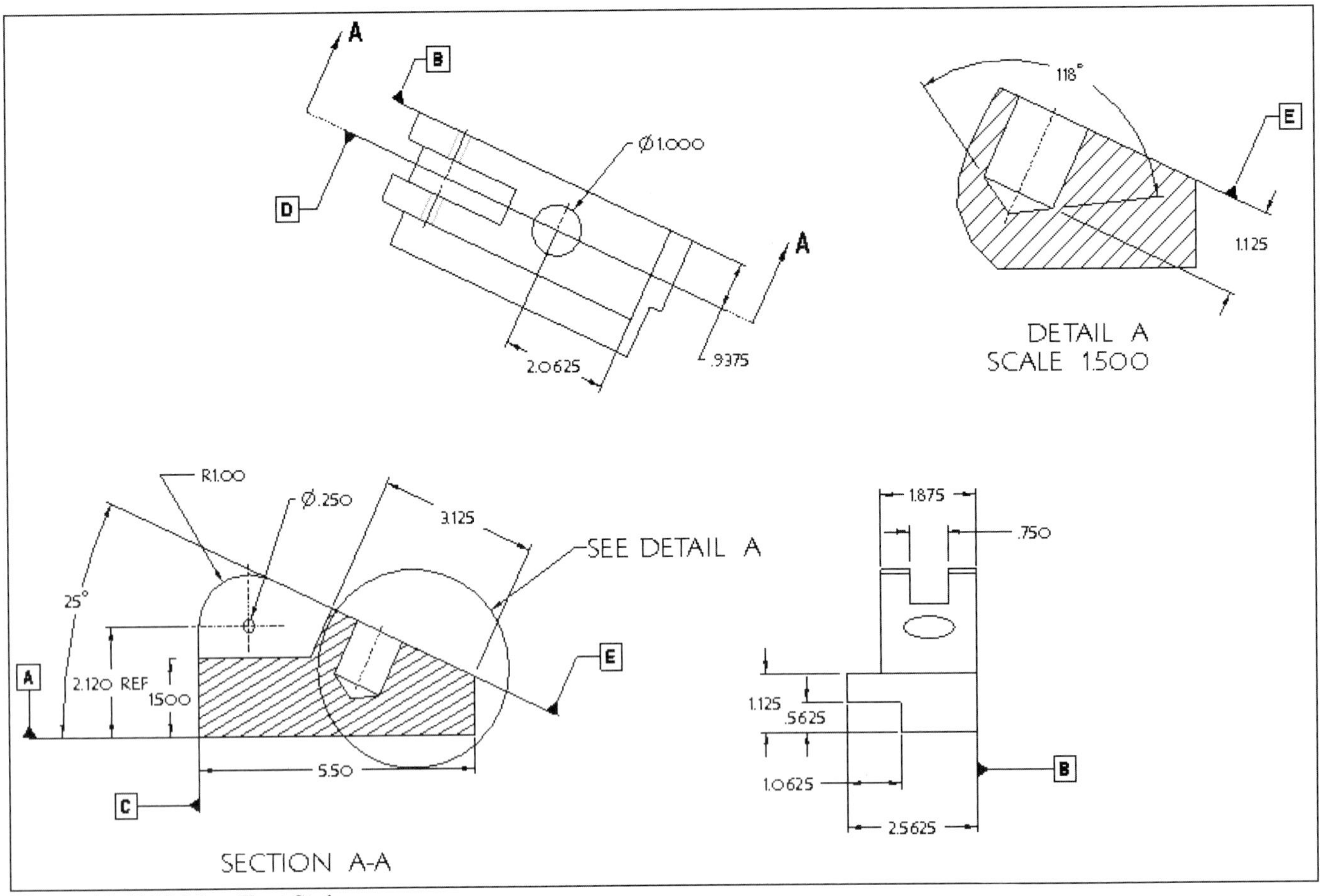

Figure 11.19(d) New Text Style

Add a geometric tolerance to the angled surface, click: **Create geometric tolerances** (*or Insert ⇒ Geometric Tolerance*) ⇒ **Angularity** ⇒ Type **Datum** [Fig. 11.20(a)] ⇒ Select Entity... ⇒ pick on datum **E** [Fig. 11.20(b)] ⇒ **Datum Refs** tab ⇒ **Primary** tab- Basic ⇒ pick on datum **A** in the Front view as the Primary Reference Datum [Fig. 11.20(c)] ⇒ **Tol Value** tab ⇒ Overall Tolerance 0.005 [Fig. 11.20(d)] ⇒ **OK** ⇒ **LMB** ⇒ **RMB** ⇒ **Update Sheet** [Fig. 11.20(e)]

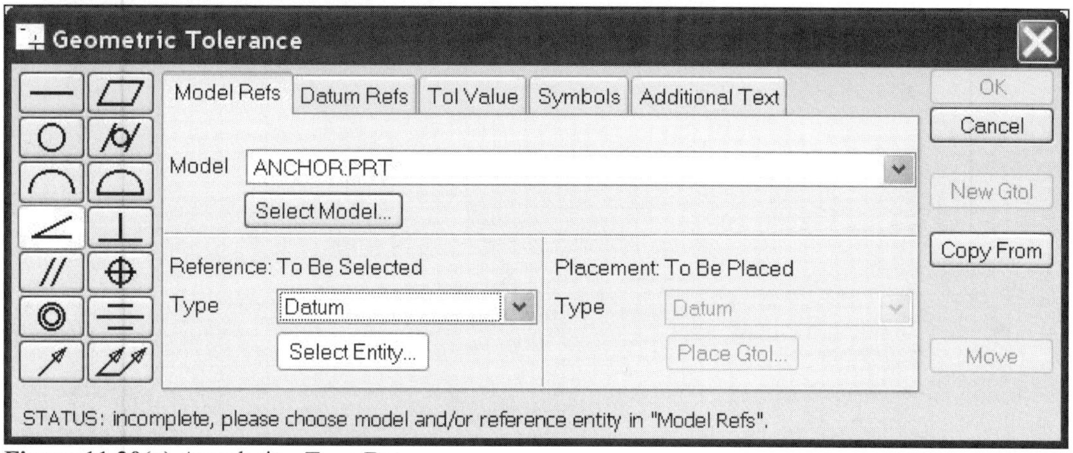

Figure 11.20(a) Angularity, Type Datum

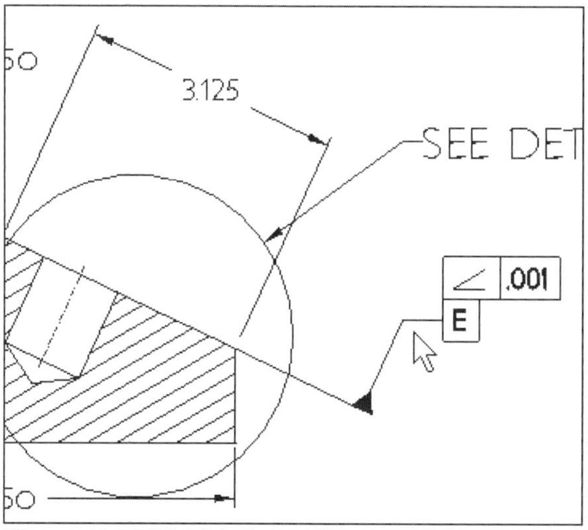

Figure 11.20(b) Pick Datum E

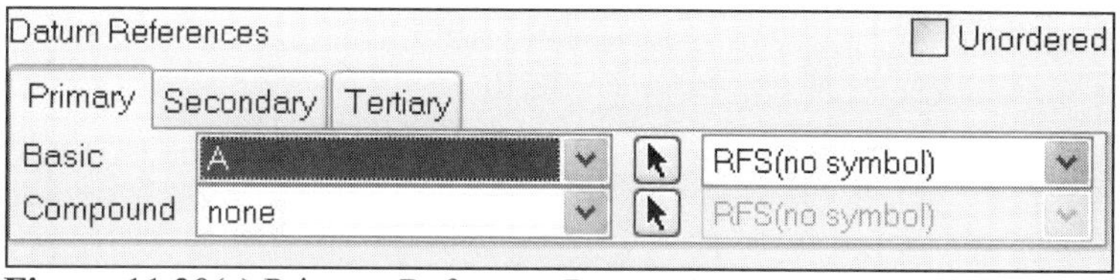

Figure 11.20(c) Primary Reference Datum A

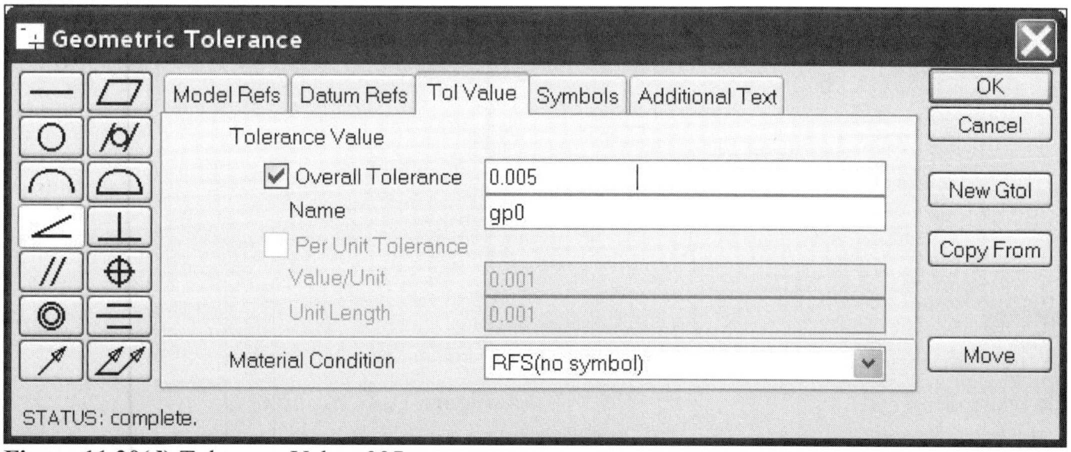

Figure 11.20(d) Tolerance Value **.005**

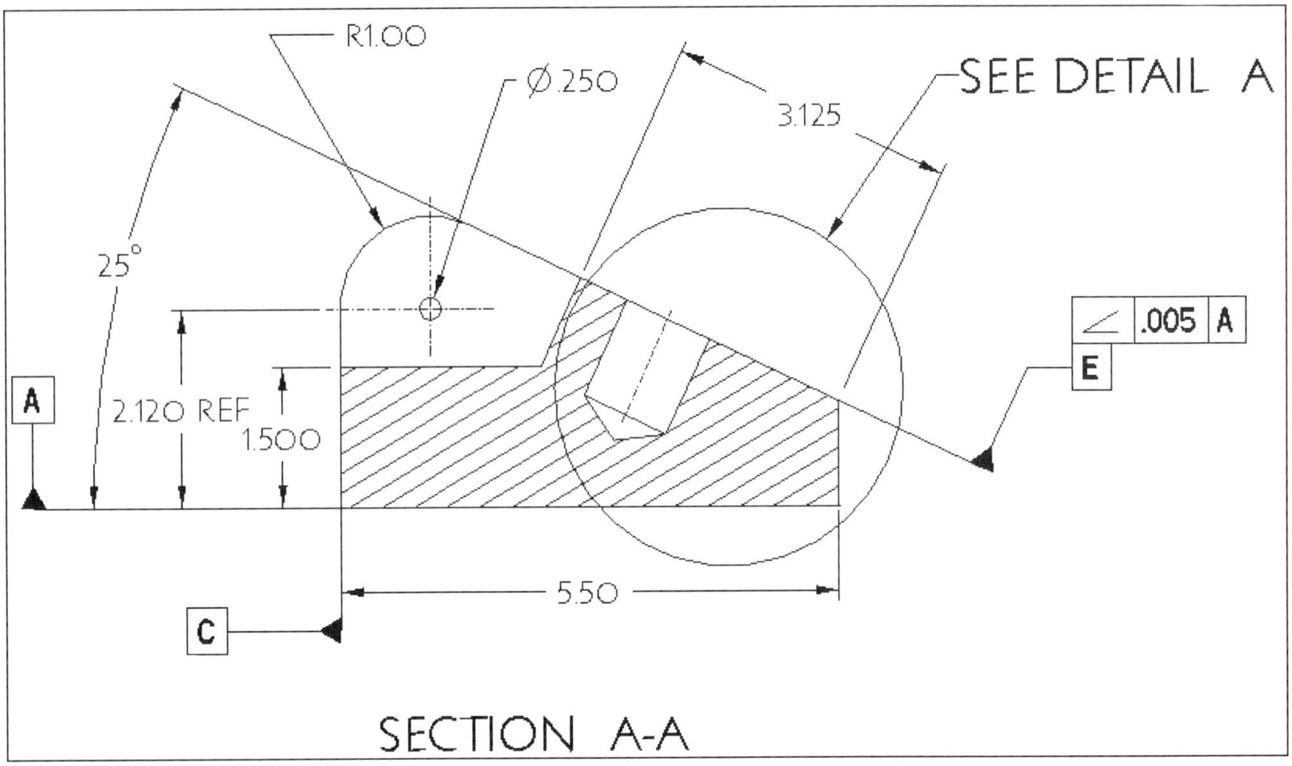

Figure 11.20(e) Geometric Tolerance

Create a *second sheet* with an isometric view of the Anchor by clicking: **Tools** ⇒ **Environment** ⇒ Standard Orient Isometric ⇒ **Apply** ⇒ **OK** ⇒ **Insert** ⇒ **Sheet** ⇒ **RMB** ⇒ **Page Setup** ⇒ D ⇒ ⇒ **Browse** [Fig. 11.21(a)] ⇒ System Formats **c.frm** [Fig. 11.21(b)] ⇒ **Open** ⇒ **OK** ⇒

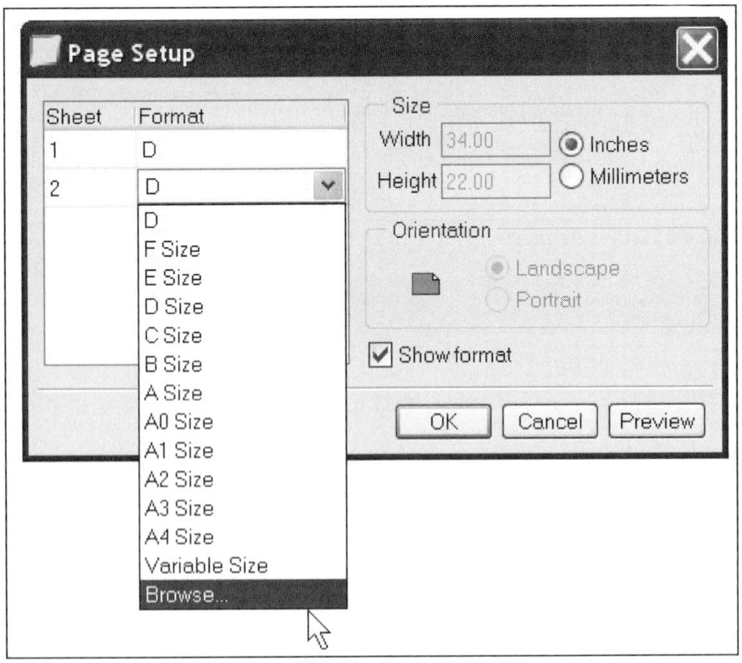

Figure 11.21(a) Page Setup Dialog Box

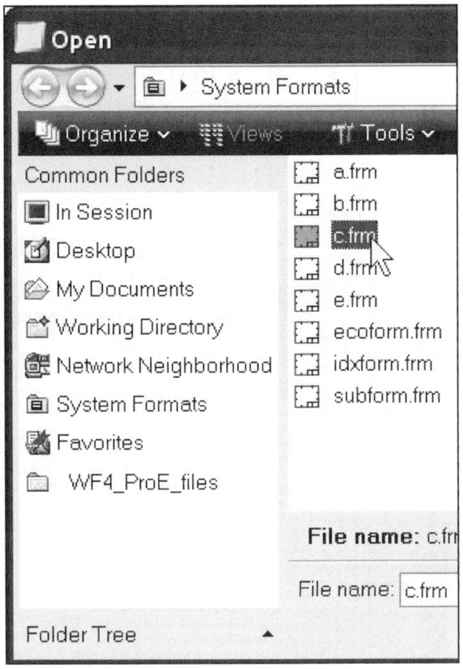

Figure 11.21(b) c.frm

Click: 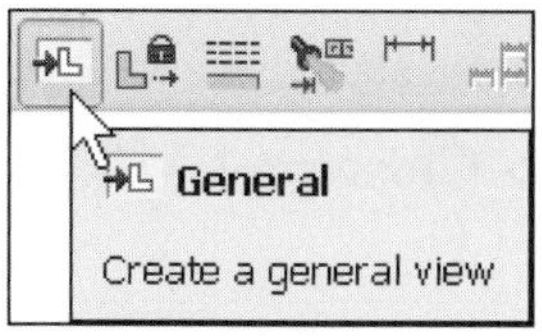

⇒ pick a center point for the view [Fig. 11.21(c)] ⇒ **Scale** ⇒ 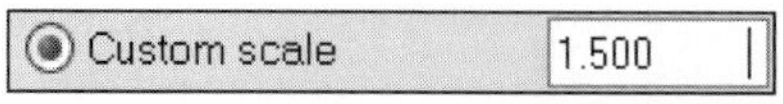

⇒ **Enter** [Fig. 11.21(d)] ⇒ **Apply** ⇒ **Close** ⇒ **LMB** to deselect

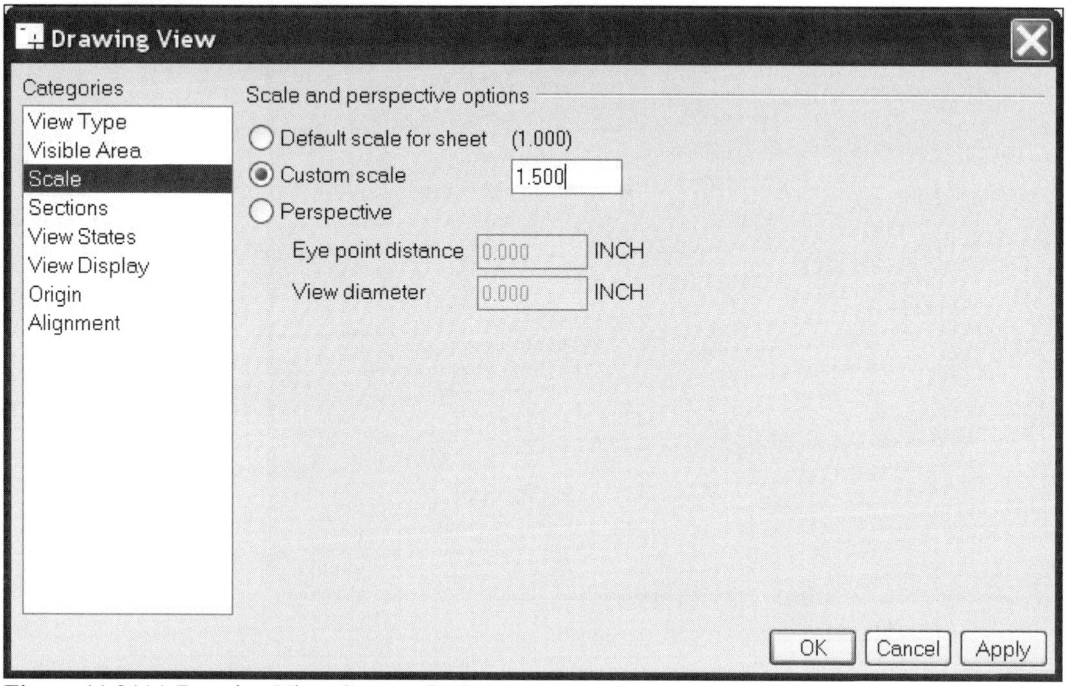

Figure 11.21(c) Drawing View Custom Scale

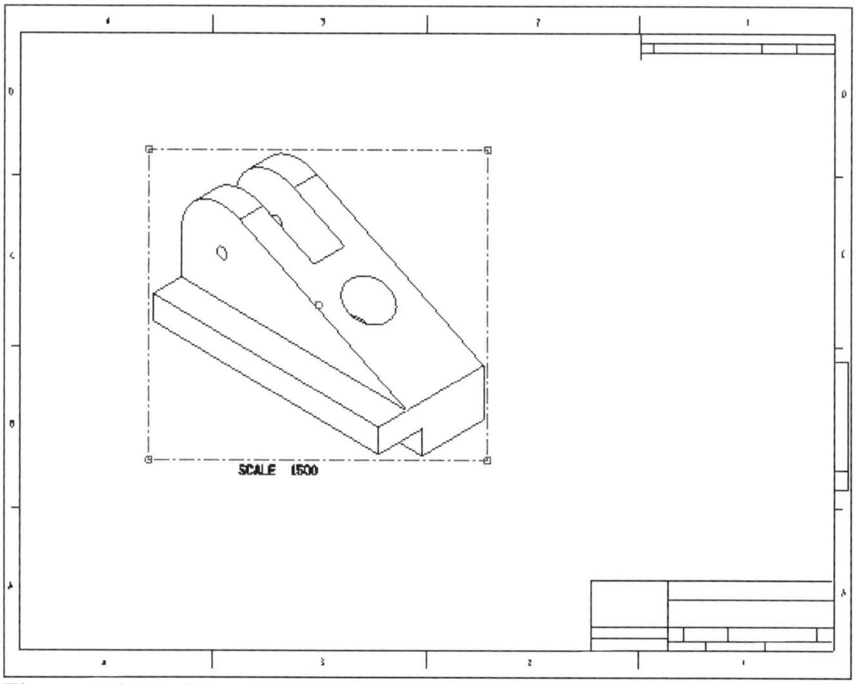

Figure 11.21(d) C Format and Isometric View

 **Context sensitive help** ⇒  ⇒ read about inserting general views ⇒  **Close**

Pick on the pictorial view ⇒ **RMB** ⇒ **Properties** ⇒ **View Display** ⇒ Display style ⇒ **No Hidden** ⇒ Tangent edges display style ⇒ **Dimmed** [Fig. 11.21(e)] ⇒ **Apply** ⇒ **Close** ⇒ **LMB** ⇒ **File** ⇒ **Save** ⇒ **MMB**

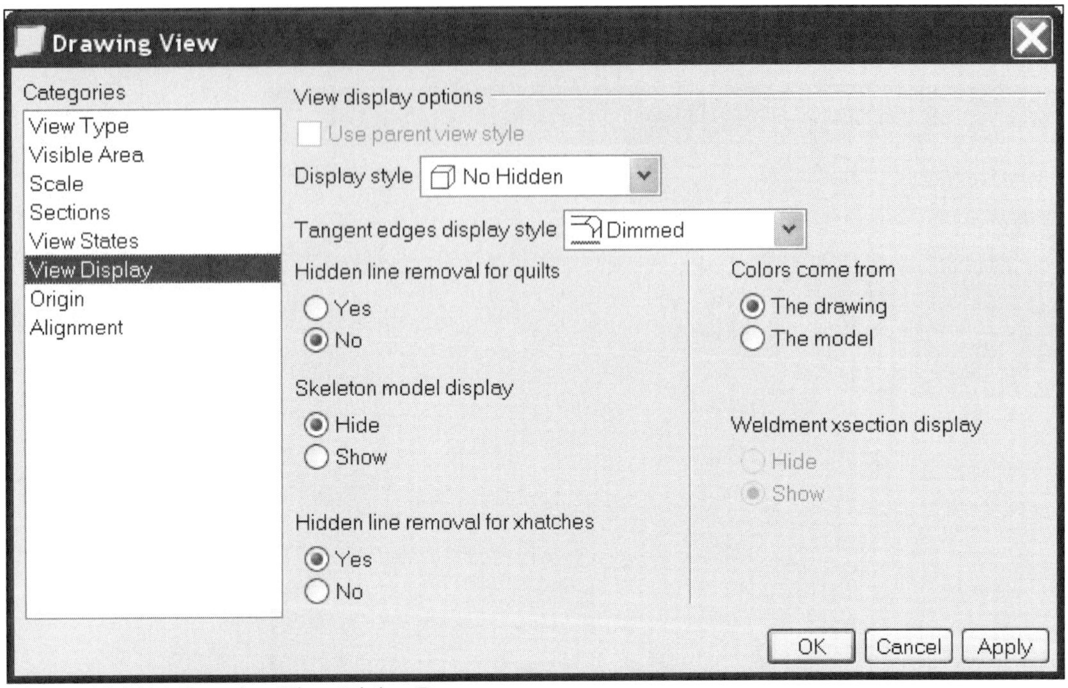

Figure 11.21(e) Drawing View Dialog Box

Click: 2 Change which sheet is currently active. ⇒ 1 ⇒ **RMB** ⇒ **Update Sheet** ⇒ ⇒ ⇒ [Fig. 11.21(f)] ⇒ ⇒ **MMB** ⇒ **File** ⇒ **Delete** ⇒ **Old Versions** ⇒ **MMB**

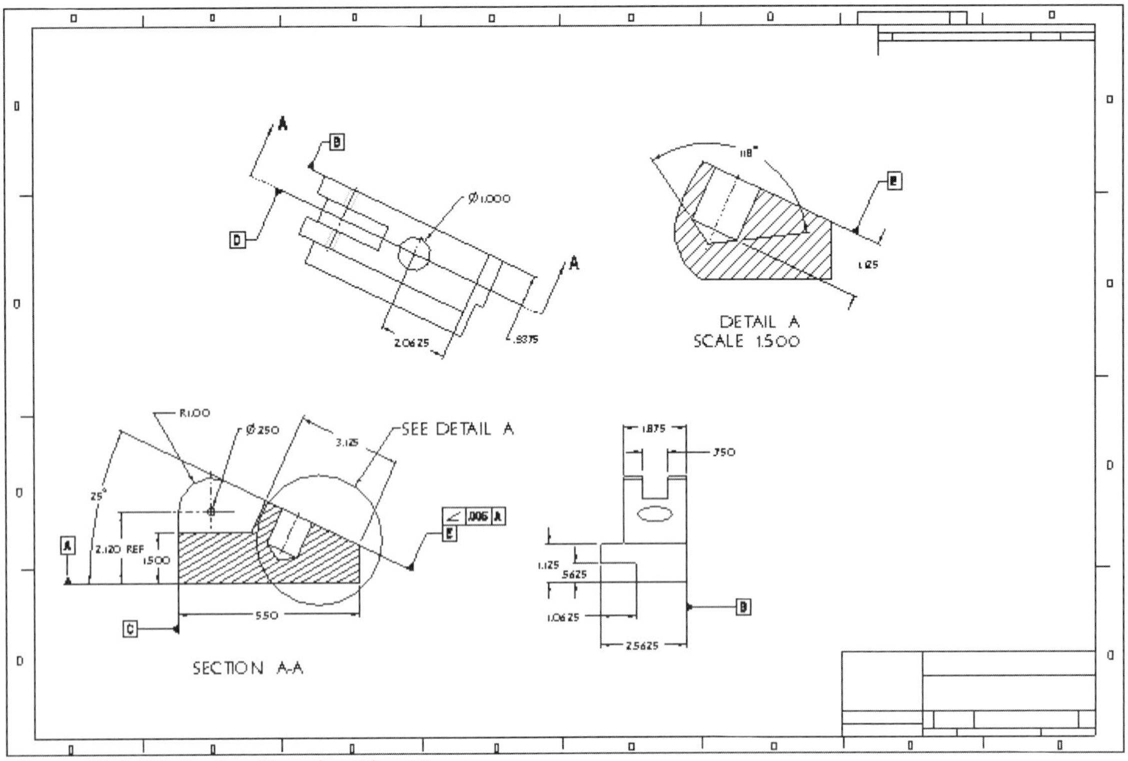

Figure 11.21(f) Anchor Drawing Sheet 1

Click: Open an existing object ⇒ **System Formats** ⇒ **d.frm** ⇒ **Open** ⇒ **File** ⇒ **Save a Copy** ⇒ type a unique name for your format: New Name **DETAIL_FORMAT_D** ⇒ **OK** ⇒ **Window** ⇒ **Close** ⇒ ⇒ pick **detail_format_d.frm** ⇒ **Open** ⇒ **RMB** in the graphics window ⇒ **Properties** Options dialog box opens ⇒ Sort **Alphabetical** ⇒ ***drawing_text_height .25*** ⇒ **Add/Change** ⇒ ***default_font filled*** ⇒ **Add/Change** ⇒ ***draw_arrow_style filled*** ⇒ **Add/Change** ⇒ ***draw_arrow_length .25*** ⇒ **Add/Change** ⇒ ***draw_arrow_width .08*** ⇒ **Add/Change** [Fig. 11.22(a)] ⇒ **Apply** ⇒ **Close** ⇒ **Ctrl+S** ⇒ **Enter**

The **format** will have a **.dtl** associated with it, and the **drawing** will have a different **.dtl** file associated with it. *They are separate **.dtl** files.* When you activate a drawing and then add a format, the **.dtl** for the format controls the font, etc. for the format only.

Active Drawing				
default_font	filled	font	●	Specifies a font index that determines default text fonts. Do not in
draft_scale	1.000000 *	1.000000	●	Determines value of draft dimensions relative to actual length of
draw_arrow_length	.25	0.187500	●	Sets length of leader line arrows.
draw_arrow_style	closed *	closed	●	Controls style of arrow head for all detail items involving arrows.
draw_arrow_width	.03	0.062500	●	Sets width of leader line arrows. Drives these: "draw_attach_sym
draw_attach_sym_height	DEFAULT *	default	●	Sets height of leader line slashes, integral signs, and boxes. If se
draw_attach_sym_width	DEFAULT *	default	●	Sets width of leader line slashes, integral signs, and boxes. If set
draw_dot_diameter	DEFAULT *	default	●	Sets diameter of leader line dots. If set to "default," uses value se
drawing_text_height	.25	0.156250	●	Sets default text height for all text in the drawing using value set f
drawing_units	inch		☒	Sets units for all drawing parameters.
leader_elbow_length	0.250000 *	0.250000	●	Determines length of leader elbow (the horizontal leg attached t
line_style_standard	std_ansi *	std_ansi	●	Controls text color in drawings. Unless set to "std_ansi," all drawi
node_radius	DEFAULT *	default	●	Sets the size of the nodes displayed in symbols.
sort_method_in_region	delimited *	delimited	●	Determines repeat regions sort mechanism. String_only - sorts a
sym_flip_rotated_text	no *	no	●	If set to "yes," then for new symbol definitions which allow rotatio
text_thickness	0.000000 *	0.000000	●	Sets default text thickness for new text after regeneration and exi
text_width_factor	0.800000 *	0.800000	●	Sets default ratio between the text width and text height. The syst
yes_no_parameter_display	true_false *	true_false	●	Controls display of "yes/no" parameters in drawing notes and ta

Figure 11.22(a) Drawing Option **.dtl** File for Format

Zoom into the title block region, and create notes for the title text and parameter text, click: **Tools** ⇒ **Environment** ⇒ ☑ Snap to Grid ⇒ **Apply** ⇒ **OK** ⇒ **View** ⇒ **Draft Grid** ⇒ **Show Grid** ⇒ **Grid Params** ⇒ **X&Y Spacing** ⇒ type **.1** ⇒ **MMB** ⇒ **MMB** ⇒ **MMB** ⇒ **Create a note** [Fig. 11.22(b)] ⇒

Make Note ⇒ pick a point for the note in the largest area of the title block ⇒ type **TOOL ENGINEERING CO.** ⇒ **Enter** ⇒ **Enter** ⇒

Make Note ⇒ ⇒ type **DRAWN** ⇒ **Enter** ⇒ **Enter** ⇒

Make Note ⇒ ⇒ type **ISSUED** ⇒ **Enter** ⇒ **Enter** ⇒

Make Note ⇒ ⇒ type **&dwg_name** ⇒ **Enter** ⇒ **Enter** ⇒

Make Note ⇒ ⇒ type **&scale** ⇒ **Enter** ⇒ **Enter** ⇒

Make Note ⇒ ⇒ type **SHEET &current_sheet OF &total_sheets** ⇒ **Enter** ⇒ **Enter** ⇒ **Done/Return** ⇒ **LMB** ⇒ **Tools** ⇒ **Environment** ⇒ ☐ Snap to Grid ⇒ **Apply** ⇒ **OK** ⇒ modify some of the text height (**.10**) and the placement of the notes as needed [Fig. 11.22(b)] ⇒ **View** ⇒ **Update** ⇒ **Current Sheet** ⇒ **View** ⇒ **Orientation** ⇒ **Refit** ⇒ **Ctrl+S** ⇒ **MMB** ⇒ **File** ⇒ **Close Window**

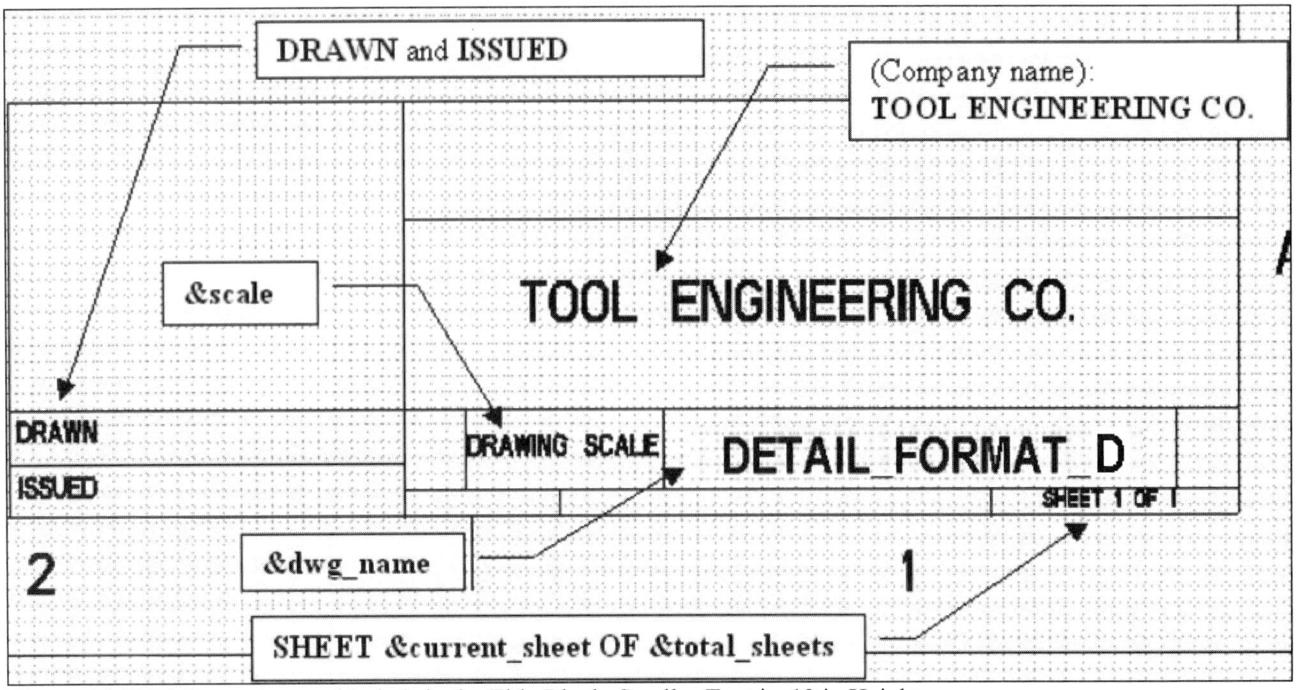

Figure 11.22(b) Parameters and Labels in the Title Block, Smaller Text is **.10** in Height

Click: **Window ⇒ Activate ⇒ RMB ⇒ Page Setup ⇒** D **⇒** **⇒ Browse ⇒ Working Directory ⇒ detail_format_d.frm ⇒ Open** DETAIL_FORMAT_D **⇒ OK ⇒** **⇒ OK** [Fig. 11.22(c)]

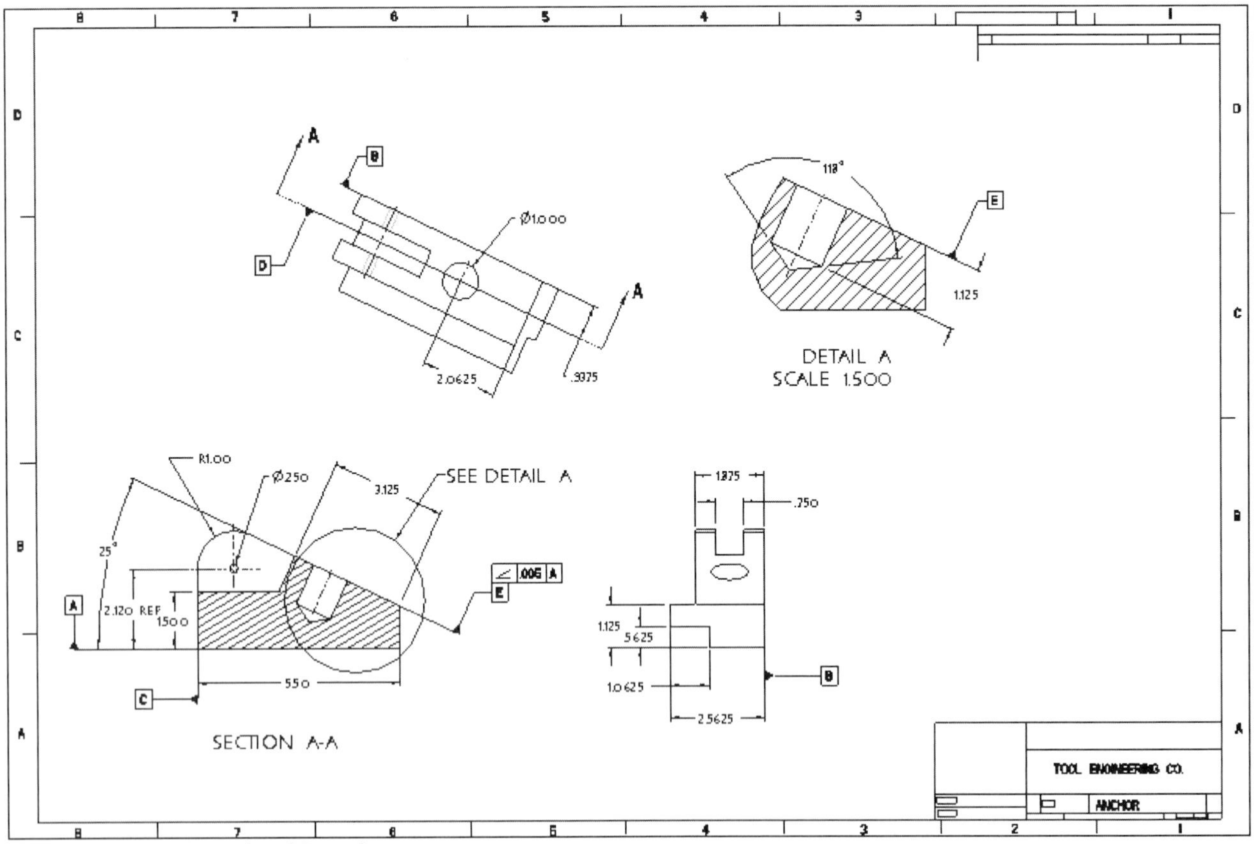

Figure 11.22(c) Completed Drawing

The lesson is now complete. If you wish to detail a project without instructions, a complete set of projects and illustrations are available at ***www.cad-resources.com*** ***⇒ Downloads.***

Lesson 12 Assembly Drawings

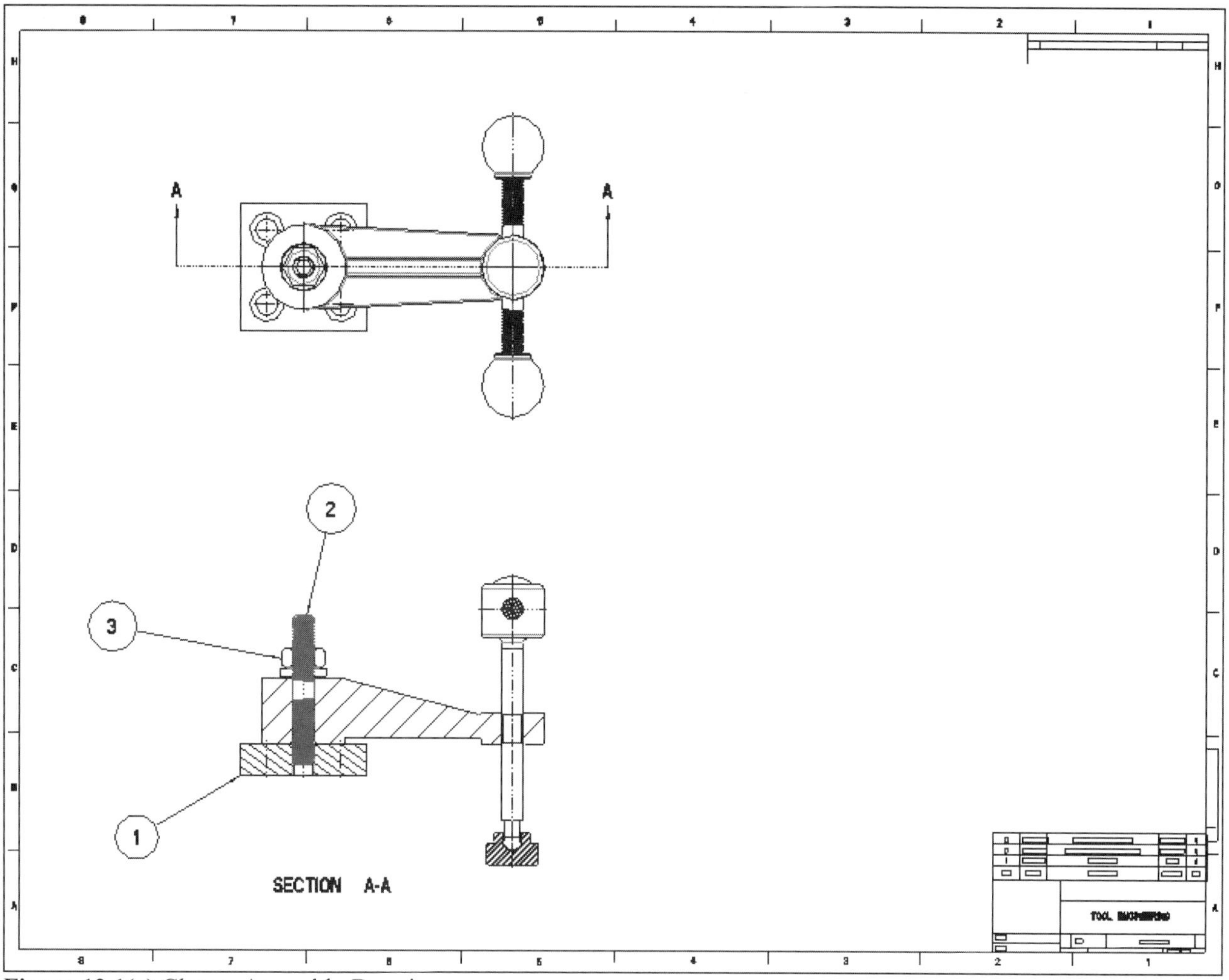

Figure 12.1(a) Clamp_Assembly Drawing

OBJECTIVES

- Create an **Assembly Drawing**
- Generate a **Parts List** from a **Bill of Materials (BOM)**
- **Balloon** an assembly drawing
- Create a **section assembly view** and change **component visibility**
- Add **Parameters** to parts
- Create a **Table** to generate a parts list automatically
- Use **Multiple Sheets**
- Make assembly **Drawing Sheets** with **multiple models**

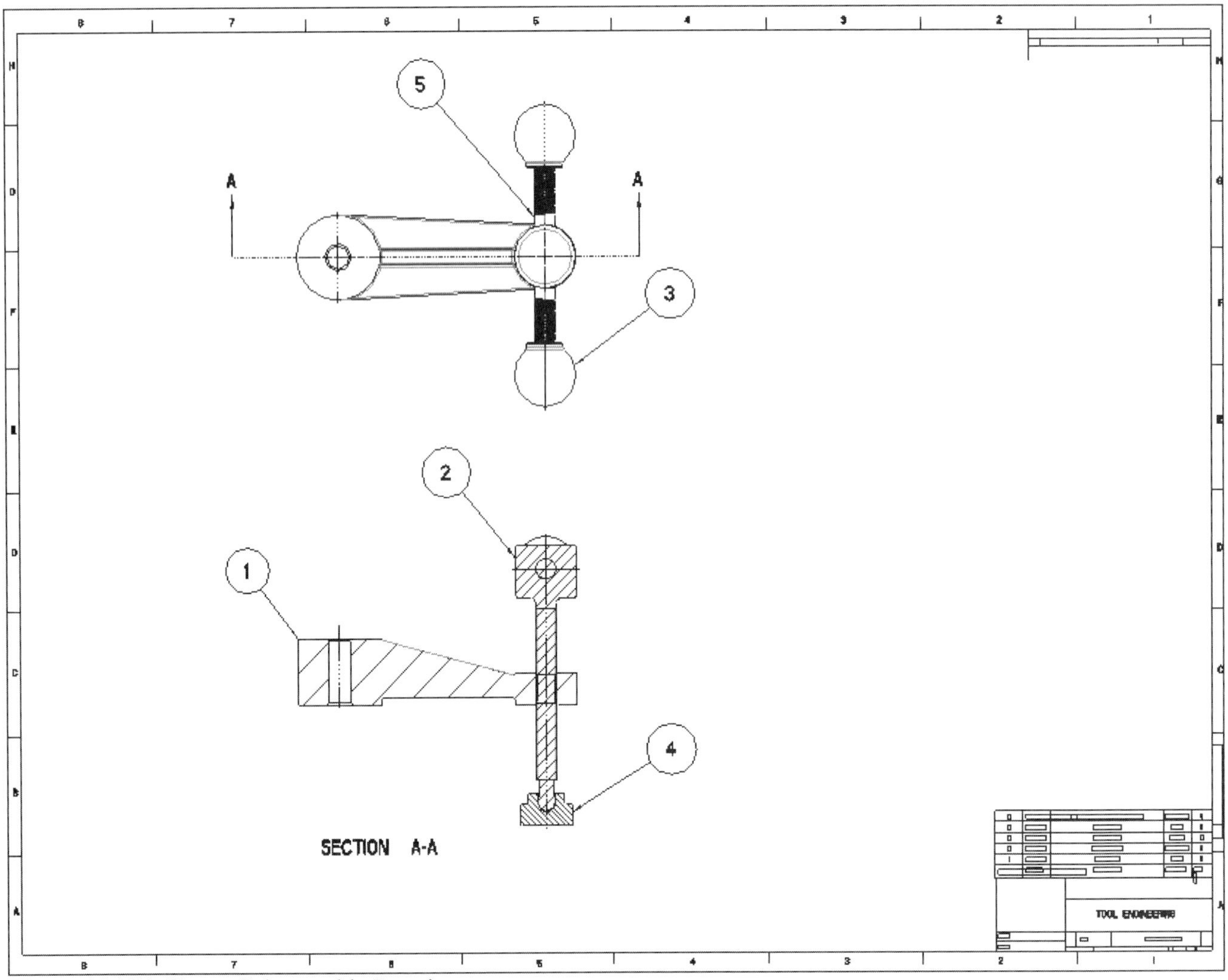

Figure 12.1(b) Clamp_Subassembly Drawing

ASSEMBLY DRAWINGS

Pro/E incorporates a great deal of functionality into drawings of assemblies [Figs. 12.1(a-b)]. You can assign parameters to parts in the assembly that can be displayed on a *parts list* in an assembly drawing. Pro/E can also generate the item balloons for each component on standard orthographic views or on an exploded view.

In addition, a variety of specialized capabilities allow you to alter the manner in which individual components are displayed in views and in sections. The format for an assembly is usually different from the format used for detail drawings. The most significant difference is the presence of a Parts List.

A parts list is actually a *Drawing Table object* that is formatted to represent a bill of materials in a drawing. By defining *parameters* in the parts in your assembly that agree with the specific format of the parts list, you make it possible for Pro/E to add pertinent data to the assembly drawing's parts list automatically as components are added to the assembly. After the parts list and parameters have been added, Pro/E can balloon the assembly drawing automatically.

In this lesson, you will create a set of assembly formats and place your standard parts list in them.

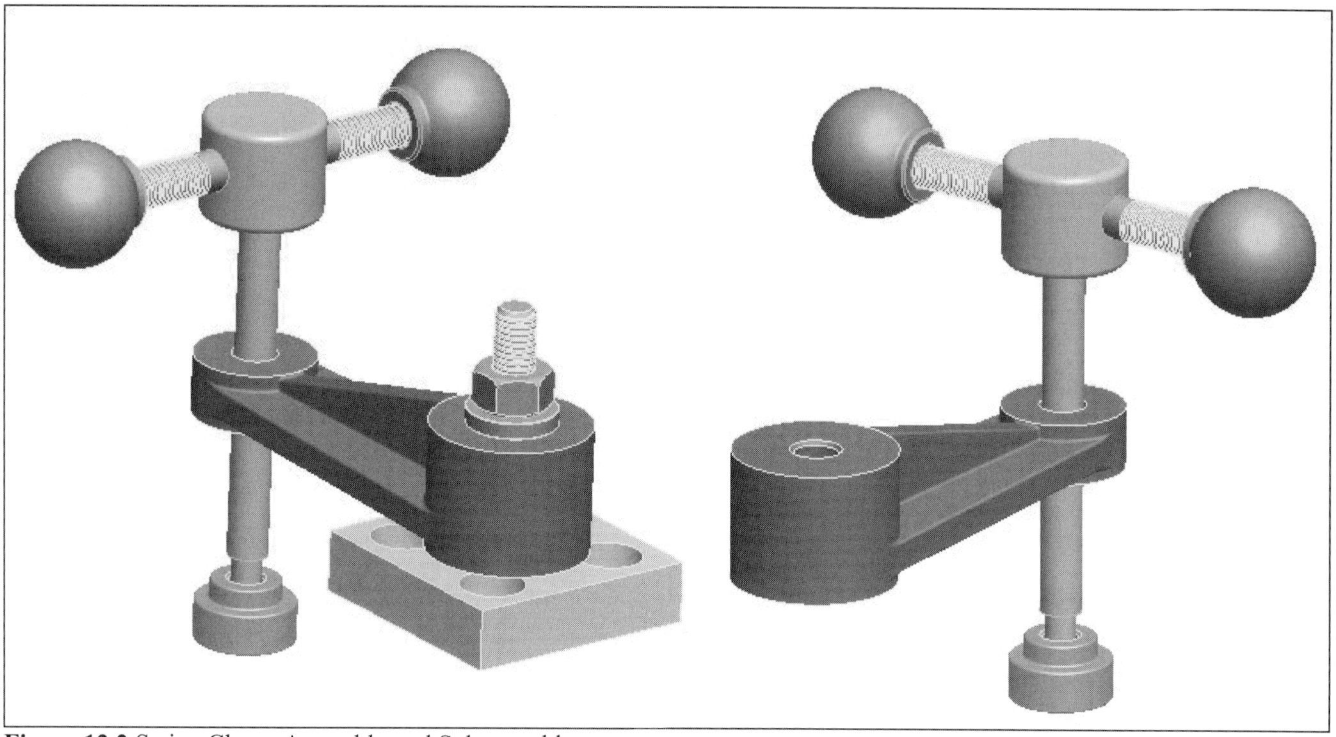

Figure 12.2 Swing Clamp Assembly and Subassembly

Swing Clamp Assembly Drawing

The format for an assembly drawing is usually different from the format used for detail drawings. The most significant difference is the presence of a parts list. You will create a standard "E" size format and place a standard parts list on it. You should create a set of assembly formats on "B", "C", "D", and "F" size sheets at your convenience.

A parts list is actually a *Drawing Table object* that is formatted to represent a bill of material (BOM) in a drawing. By defining parameters in the parts in your assembly that agree with the specific format of the parts list, you make it possible for Pro/E to add pertinent data to the assembly drawing parts list automatically, as components are added to the assembly.

After you create an "E" size format sheet with a parts list table, you will create two new drawings (each with two views) using your new assembly format. The Swing Clamp *subassembly* [Fig. 12.2(right)] will be used in the first drawing. The second drawing will use the Swing Clamp *assembly* [Fig. 12.2(left]. Both drawings use the "E" size format created in the first section of this lesson. The format will have a parameter-driven title block and an integral parts list.

Click: **File** ⇒ **Set Working Directory** ⇒ select the directory where the **clamp_assembly.asm** was saved ⇒ **OK** ⇒ **Open an existing object** ⇒ System Formats ⇒ e.frm ⇒ **Open** ⇒ **File** ⇒ **Save a Copy** ⇒ type a unique name for your format: New Name **asm_format_e** ⇒ **OK** ⇒ **Window** ⇒ **Close** ⇒ ⇒ pick **asm_format_e.frm** ⇒ **Open**

The **format** will have a **.dtl** associated with it, and the **drawing** will have a unique **.dtl** file associated with it. *They are separate **.dtl** files.* When you activate a drawing and then add a format, the **.dtl** for the format controls the font, etc. for the format only. The drawing **.dtl** file that controls items on the drawing needs to be established separately in the Drawing mode.

Click: **File** ⇒ **Properties** Options dialog box opens ⇒ ***drawing_text_height .25*** ⇒ **Enter** ⇒ ***default_font filled*** ⇒ **Enter** ⇒ ***draw_arrow_style filled*** ⇒ **Add/Change** ⇒ ***draw_arrow_length .25*** ⇒ **Add/Change** ⇒ ***draw_arrow_width .08*** ⇒ **Add/Change** (Fig. 12.3) ⇒ **Apply** (option in Status column shows Default column value active, option in Status column shows pending value for this drawing) ⇒ ⇒ File name: **asm_format_properties** ⇒ **Ok** ⇒ **Close** ⇒ **Ctrl+S** ⇒ **MMB**

Active Drawing				
These options control text not subject to other options				
drawing_text_height	.25	0.156250	✱	Sets default text height for all te
text_thickness	0.000000 *	0.000000	●	Sets default text thickness for n
text_width_factor	0.800000 *	0.800000	●	Sets default ratio between the t
draft_scale	1.000000 *	1.000000	●	Determines value of draft dime
These options control text and line fonts				
default_font	filled	font	✱	Specifies a font index that dete
These options control leaders				
draw_arrow_length	.25	0.187500	✱	Sets length of leader line arrow
draw_arrow_style	filled	closed	✱	Controls style of arrow head for
draw_arrow_width	.03	0.062500	✱	Sets width of leader line arrows
draw_attach_sym_height	DEFAULT *	default	●	Sets height of leader line slash
draw_attach_sym_width	DEFAULT *	default	●	Sets width of leader line slashe
draw_dot_diameter	DEFAULT *	default	●	Sets diameter of leader line do
leader_elbow_length	0.250000 *	0.250000	●	Determines length of leader elb
These options control tables, repeat regions, and BOM balloons				
sort_method_in_region	delimited *	delimited	●	Determines repeat regions sor
Miscellaneous options				
draft_scale	1.000000 *	1.000000	●	Determines value of draft dime
drawing_units	inch		☒	Sets units for all drawing param
line_style_standard	std_ansi *	std_ansi	●	Controls text color in drawings.
node_radius	DEFAULT *	default	●	Sets the size of the nodes displ
sym_flip_rotated_text	no *	no	●	If set to "yes," then for new syml
yes_no_parameter_display	true_false *	true_false	●	Controls display of "yes/no" pa

Figure 12.3 New Drawing Options

Click: **Context sensitive help** from Top Toolchest ⇒ click on **Insert a table by specifying the columns and rows sizes** icon ⇒ **To Create a Drawing Table** ⇒ read the help file ⇒ **About Drawing Tables** bottom of panel ⇒ read the help file ⇒ **Close** ⇒ close the Navigation area

Zoom into the title block region, and create notes for the title text and parameter text required to display the proper information.

Click: **Tools ⇒ Environment ⇒** Snap to Grid **⇒ Apply ⇒ OK ⇒ View ⇒ Draft Grid ⇒ Show Grid ⇒ Grid Params ⇒ X&Y Spacing ⇒** type **.125 ⇒ Done/Return ⇒ Done/Return ⇒** **Create a note ⇒ Make Note ⇒** pick point for the note in the largest area of the title block (Fig. 12.4) ⇒ type **TOOL ENGINEERING ⇒ Enter ⇒ Enter ⇒**

Make Note ⇒ (Fig. 12.4) ⇒ type **DRAWN ⇒ Enter ⇒ Enter ⇒**

Make Note ⇒ (Fig. 12.4) ⇒ type **ISSUED ⇒ Enter ⇒ Enter ⇒**

Make Note ⇒ (Fig. 12.4) ⇒ type **&dwg_name ⇒ Enter ⇒ Enter ⇒**

Make Note ⇒ (Fig. 12.4) ⇒ type **&scale ⇒ Enter ⇒ Enter ⇒**

Make Note ⇒ (Fig. 12.4) ⇒ type **SHEET &current_sheet OF &total_sheets ⇒ Enter ⇒ Enter ⇒ Done/Return ⇒ LMB ⇒ Tools ⇒ Environment ⇒** Snap to Grid **⇒ Apply ⇒ OK ⇒ View ⇒ Draft Grid ⇒ Hide Grid ⇒ Done/Return ⇒** modify the text height (**.10**) and the placement of the notes so that they are positioned correctly ⇒ select the appropriate text ⇒ **RMB ⇒ Text Style ⇒** Default ⇒ Height .10 (Fig. 12.4) ⇒ **Apply ⇒ OK ⇒** ⇒ ⇒ ⇒ ⇒ **Enter**

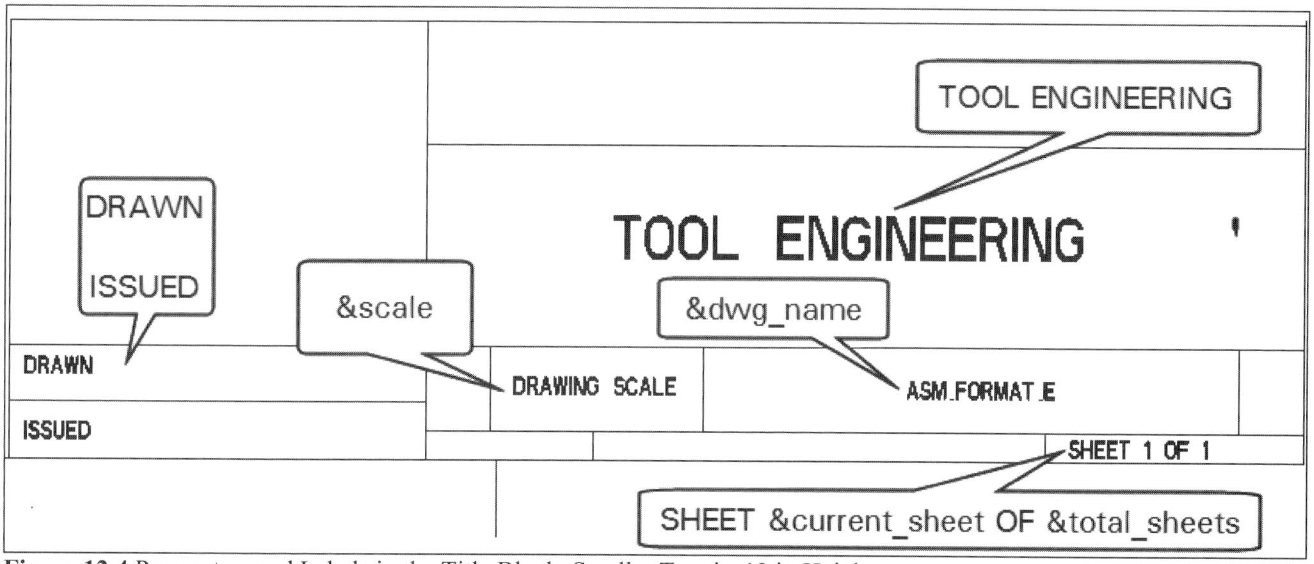

Figure 12.4 Parameters and Labels in the Title Block, Smaller Text is **.10** in Height

The parts list table can now be created and saved with this format. You can add and replace formats and still keep the table associated with the drawing. Start the parts list by creating a table.

Click: **Insert a table by specifying the columns and rows sizes** ⇒ **Ascending** ⇒ **Rightward** ⇒ **By Length** ⇒ pick a point at the upper left-hand corner of the title block [Fig. 12.5(a)] ⇒

Enter the width of the first column in drawing units (INCH) [Quit] 1.00 ⇒ **Enter** ⇒

Enter the width of the next column in drawing units (INCH) [Done] 1.00 ⇒ **Enter** ⇒

Enter the width of the next column in drawing units (INCH) [Done] 4.00 ⇒ **Enter** ⇒

Enter the width of the next column in drawing units (INCH) [Done] 1.00 ⇒ **Enter** ⇒

Enter the width of the next column in drawing units (INCH) [Done] .75 ⇒ **Enter** ⇒ **Enter** ⇒

Enter the height of the first row in drawing units(INCH) [Quit] .50 ⇒ **Enter** ⇒

Enter the height of the next row in drawing units (INCH) [Done] .375 ⇒ **Enter** ⇒ **Enter** ⇒ **LMB** [Fig. 12.5(b)]

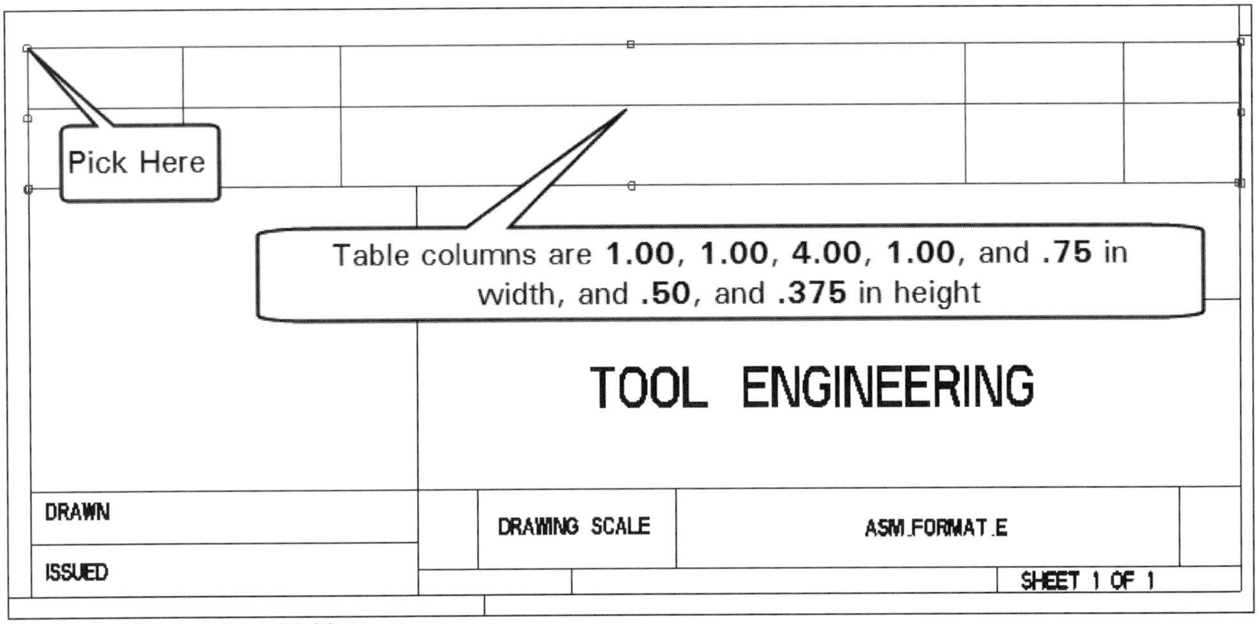

Figure 12.5(a) Inserting a Table

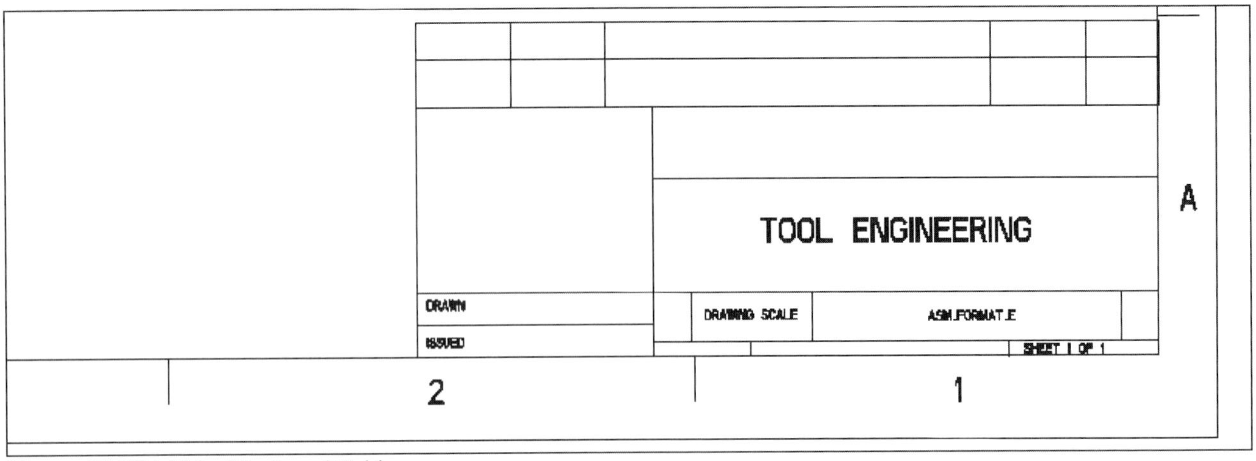

Figure 12.5(b) Title Block and Table

Click: **Ctrl+S** ⇒ **Enter** ⇒ **Table** from menu bar ⇒ **Repeat Region** ⇒ **Add** ⇒ **Simple** ⇒ pick in the left block [Fig. 12.5(c)] ⇒ pick in the right block [Fig. 12.5(c)] ⇒ **Attributes** ⇒ select the Repeat Region just created [Fig. 12.5(d)] ⇒ **No Duplicates** ⇒ **Recursive** ⇒ **Done/Return** ⇒ **Done** ⇒  **Update the display of all views in the active sheet**

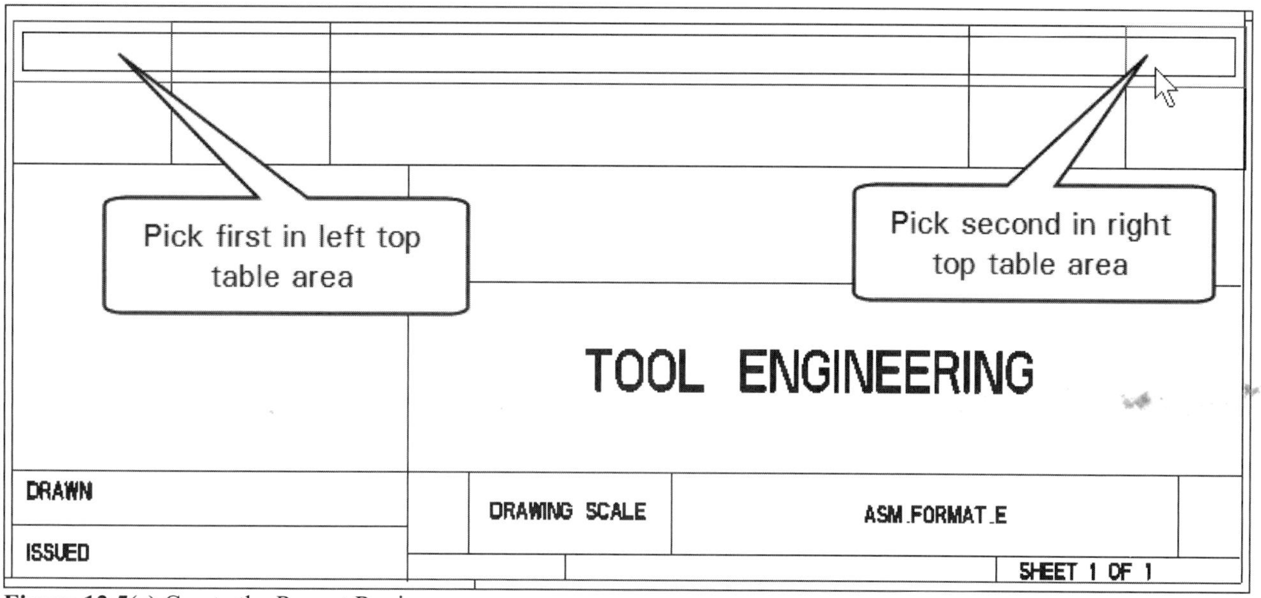

Figure 12.5(c) Create the Repeat Region

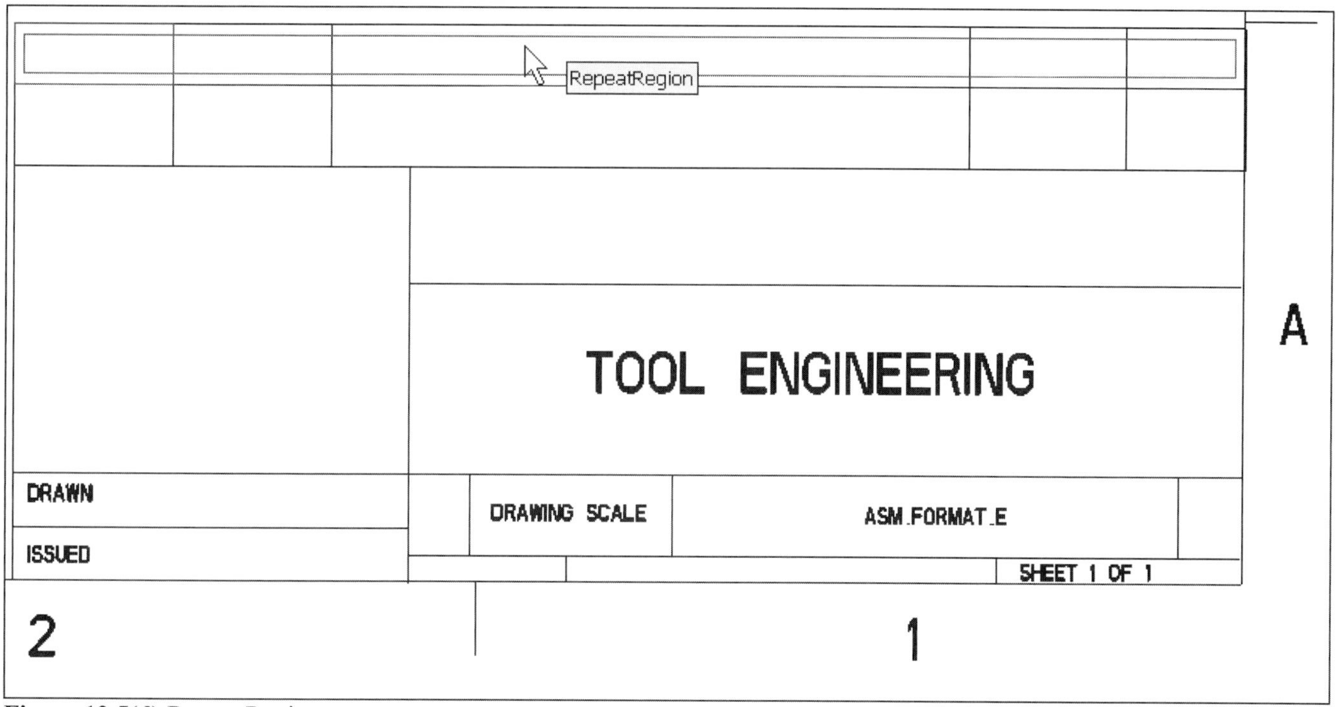

Figure 12.5(d) Repeat Region

Insert column headings using plain text, double-click on the first block [Fig. 12.6(a)] ⇒ type **ITEM** [Fig. 12.6(b)] ⇒ **OK**

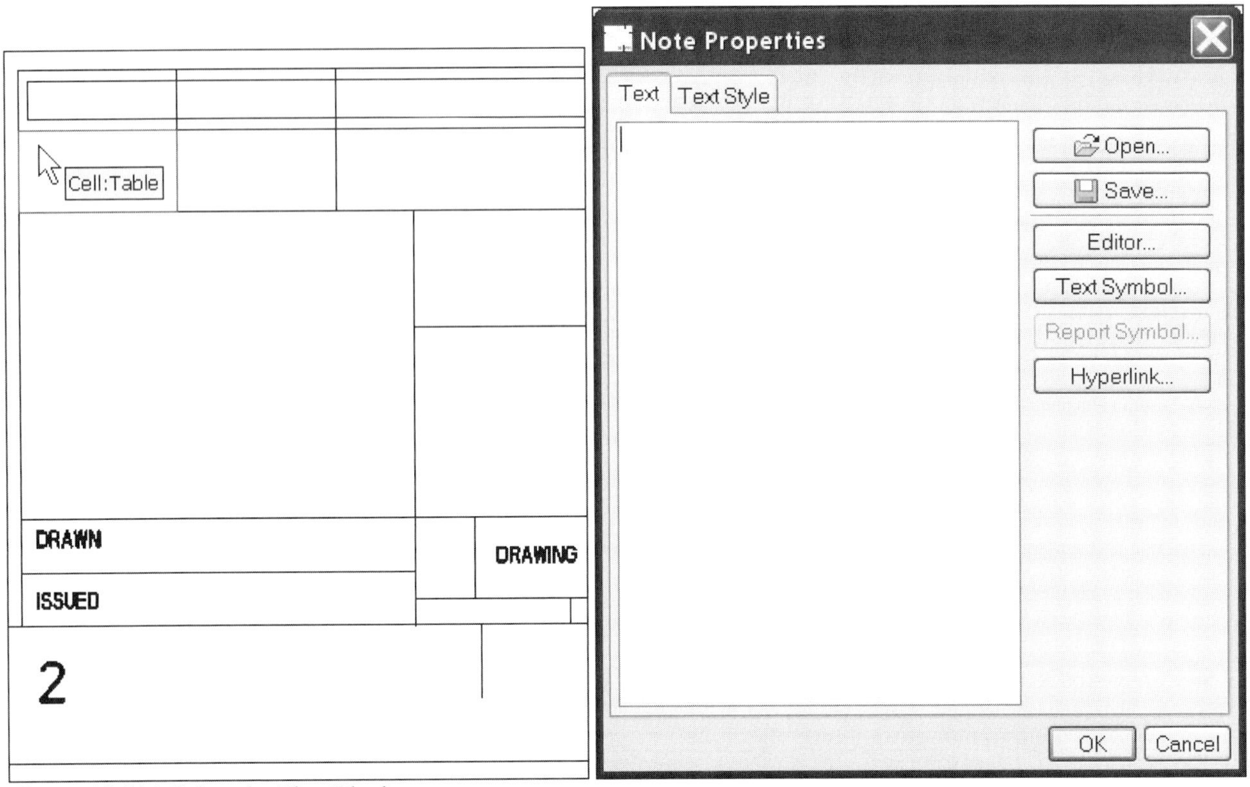

Figure 12.6(a) Select the First Block

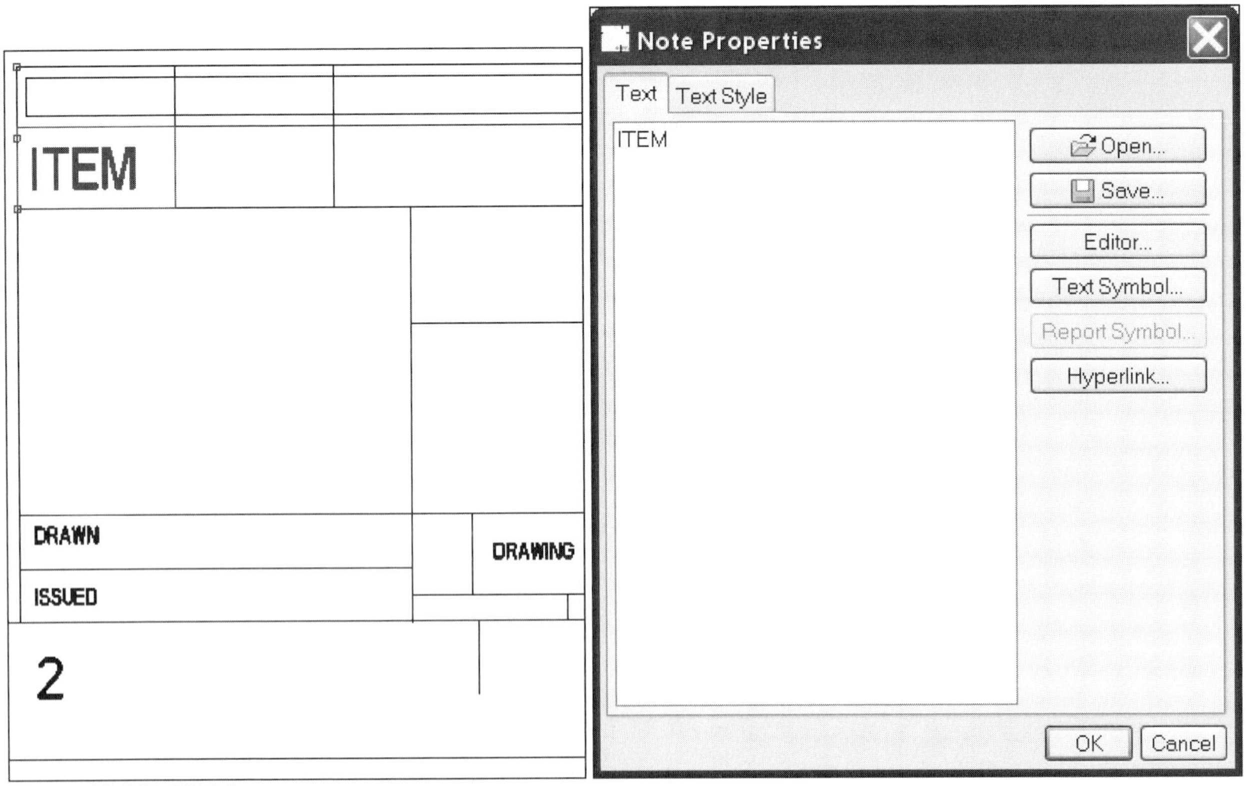

Figure 12.6(b) ITEM

The Repeat Region now needs to have some of its headings correspond to the parameters that will be created for each component model.

Double-click on the second block ⇒ type **PT NUM** [Fig. 12.6(c)] ⇒ **OK** ⇒ double-click on the third block ⇒ type **DESCRIPTION** ⇒ **OK** ⇒ double-click on the fourth block ⇒ type **MATERIAL** ⇒ **OK** ⇒ double-click on the fifth block ⇒ type **QTY** ⇒ **OK** [Fig. 12.6(d)]

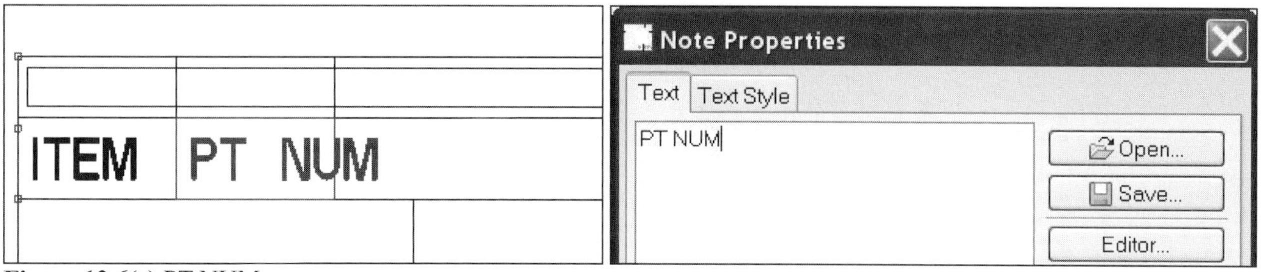

Figure 12.6(c) PT NUM

Figure 12.6(d) ITEM, PT MUM, DESCRIPTION, MATERIAL, QTY

Select all five text cells ⇒ **RMB** ⇒ **Text Style** Text Style dialog box opens ⇒ Character- 

⇒ **Apply** ⇒ **OK** ⇒ select the DESCRIPTION cell ⇒ **RMB** ⇒ **Properties** [Fig. 12.6(e)] ⇒ **Text Style** tab ⇒ Note/Dimension- 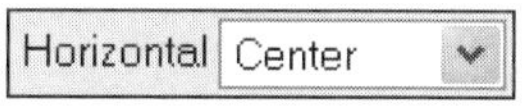

⇒ **Preview** [Figs. 12.6(f-g)] ⇒ **OK** ⇒ **LMB** ⇒ **Ctrl+S** ⇒ **Enter**

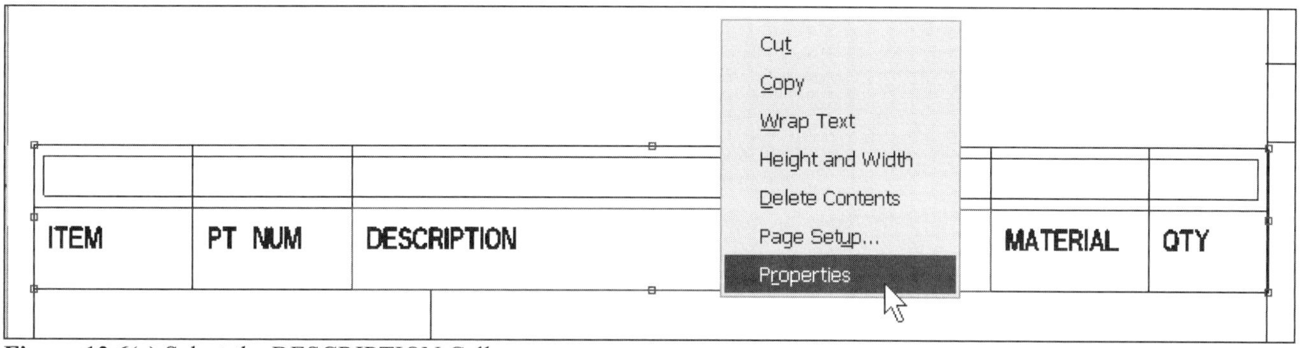

Figure 12.6(e) Select the DESCRIPTION Cell

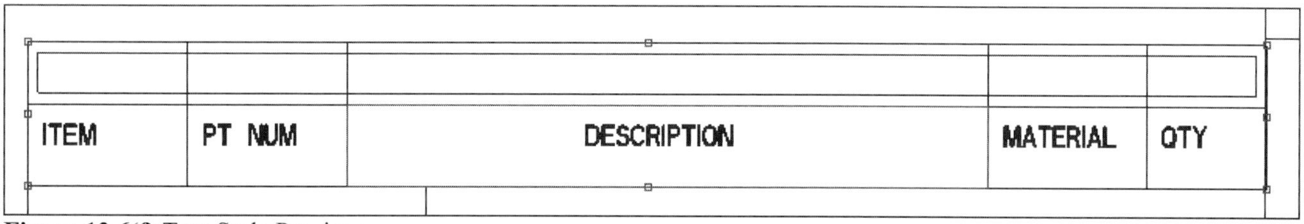

Figure 12.6(f) Text Style Preview

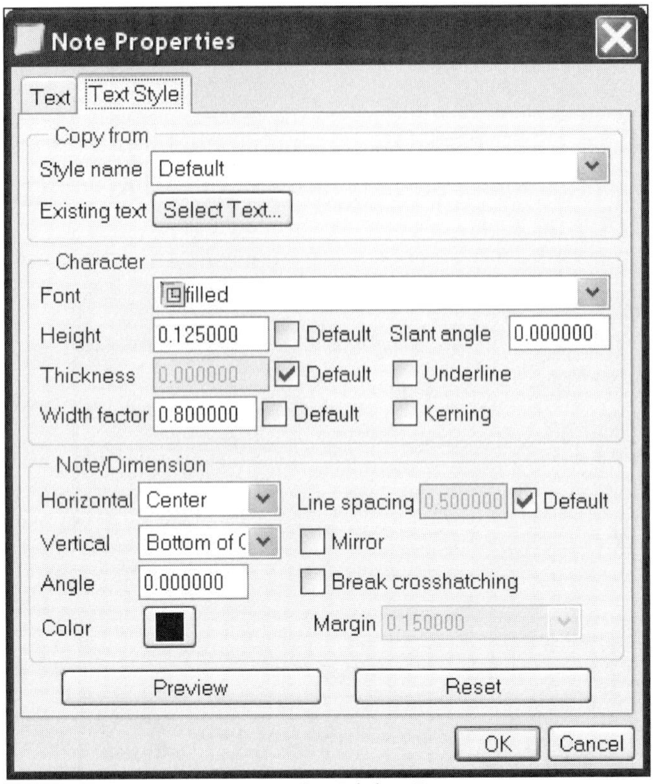

Figure 12.6(g) Note Properties Dialog Box

Insert parametric text into each appropriate repeat region block. Double-click on the first table cell of the Repeat Region [Fig. 12.7(a)] ⇒ click **rpt…** from the Report Symbol dialog box ⇒ click **index** [Figs. 12.7(b-c)]

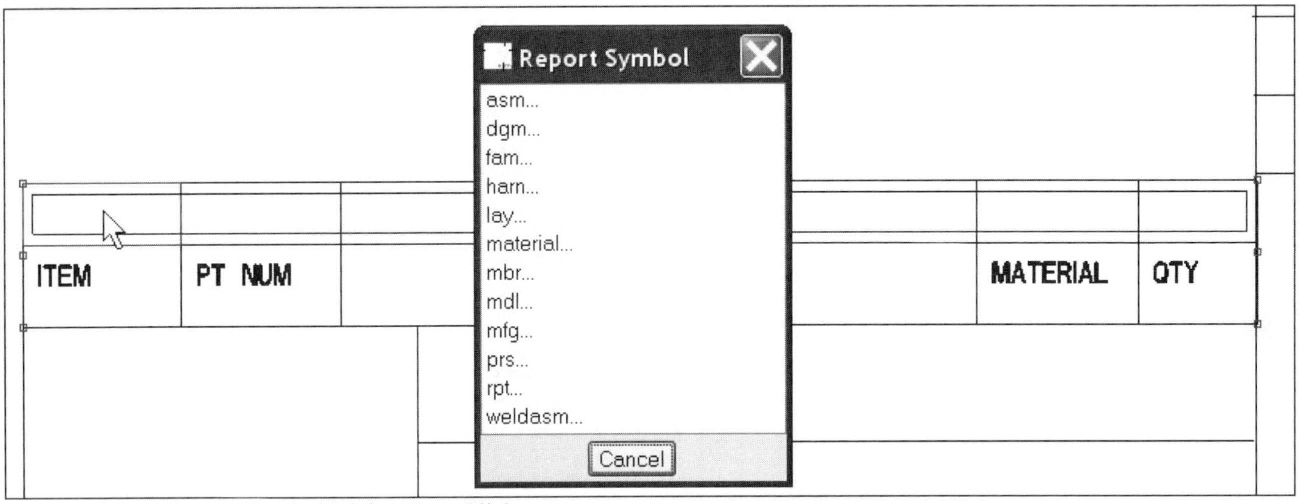

Figure 12.7(a) Report Symbol Dialog Box, Click **rpt…**

Figure 12.7(b) Click **index**

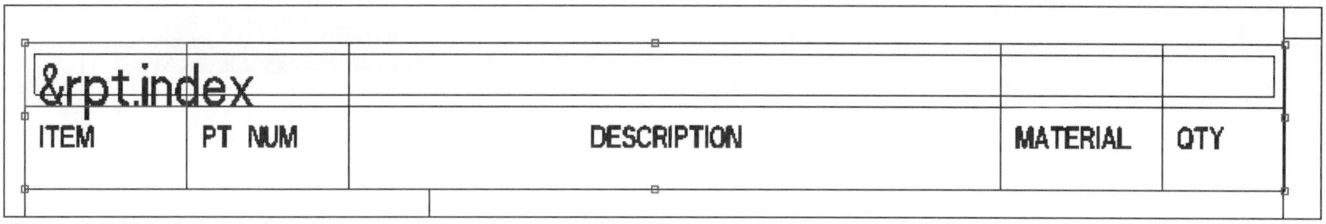

Figure 12.7(c) &rpt.index

Double-click on the fifth table cell of Repeat region ⇒ click **rpt…** [Fig. 12.7(d)] ⇒ **qty** [Figs. 12.7(e-f)]

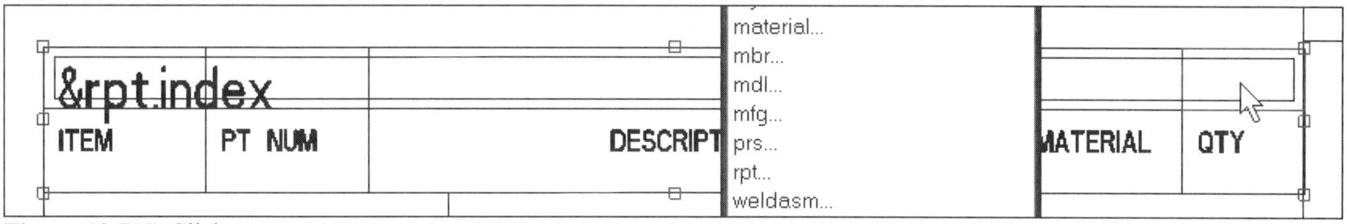

Figure 12.7(d) Click **rpt…**

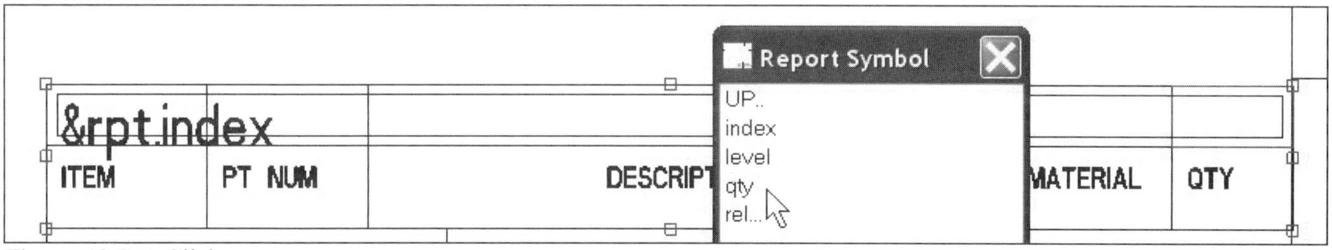

Figure 12.7(e) Click **qty**

Figure 12.7(f) &rpt.qty

Double-click on the third (middle) table cell of the Repeat Region ⇒ **asm…** [Fig. 12.7(g)] ⇒ **mbr…** [Fig. 12.7(h)] ⇒ **User Defined** [Fig. 12.7(i)] ⇒ type **DSC** 

⇒ **Enter** [Fig. 12.7(j)]

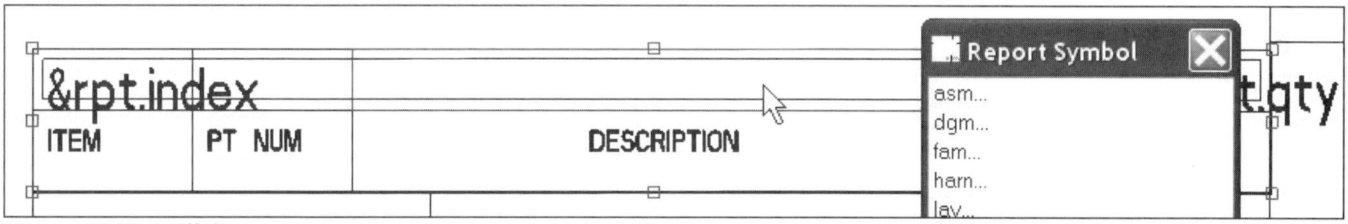

Figure 12.7(g) Click **asm…**

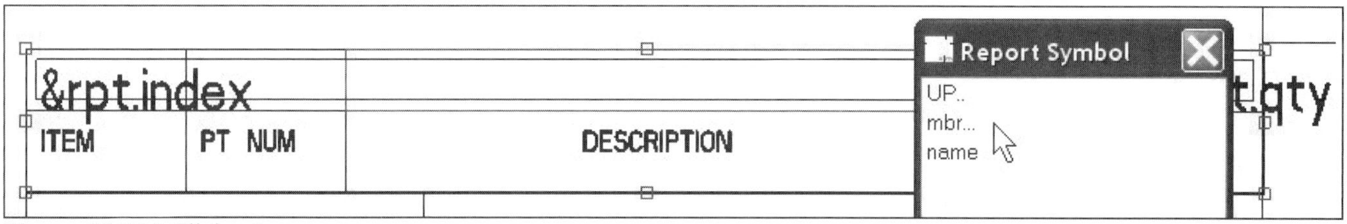

Figure 12.7(h) Click **mbr…**

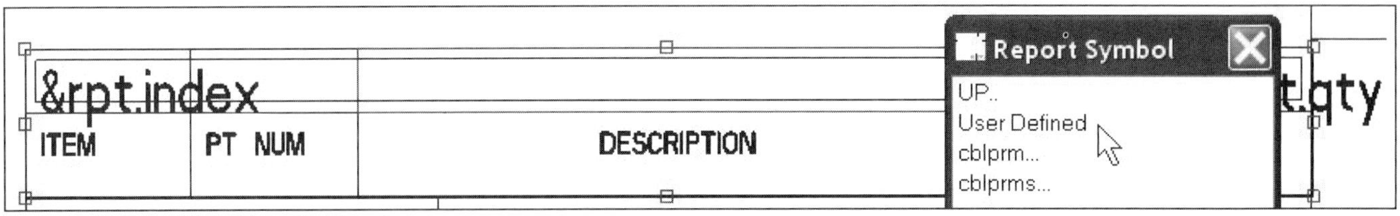

Figure 12.7(i) Click **User Defined**

Figure 12.7(j) &asm.mbr.DSC

Double-click on the fourth table cell of the Repeat Region ⇒ **asm…** ⇒ **mbr…** ⇒ **ptc_material…** ⇒ **PTC_MATERIAL_NAME** [Figs. 12.7(k-l)] ⇒ double-click on the second table cell of the Repeat Region ⇒ **asm…** ⇒ **mbr…** ⇒ **User Defined** ⇒ type **PRTNO** 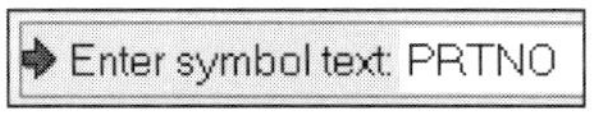

⇒ **Enter** [Fig. 12.7(m)] ⇒ **LMB** ⇒ **Ctrl+S** ⇒ **Enter**

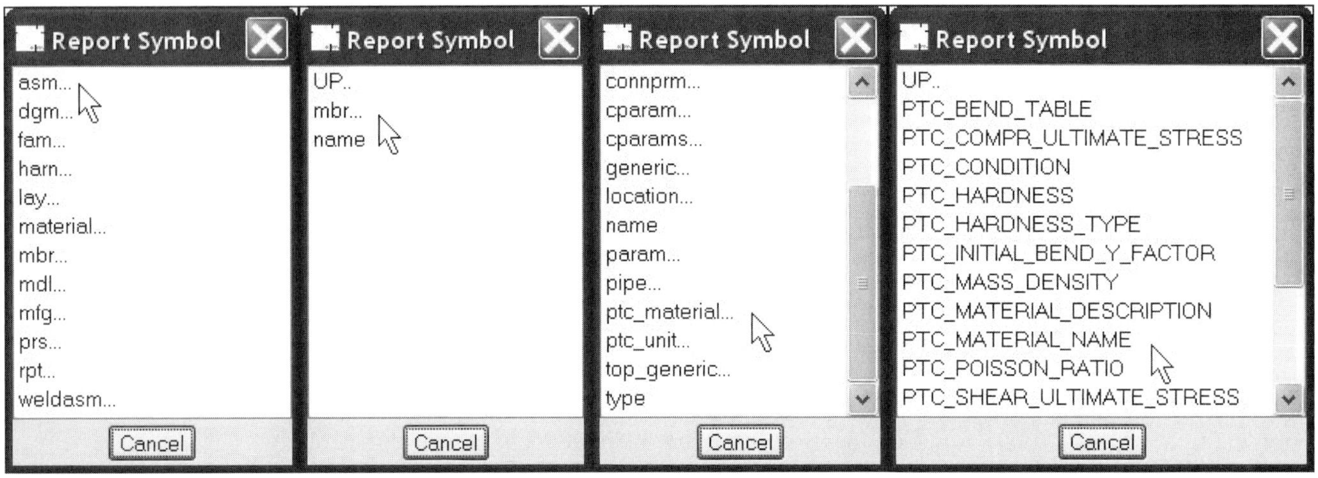

Figure 12.7(k) asm… ⇒ mbr… ⇒ ptc_material… ⇒ PTC_MATERIAL_NAME

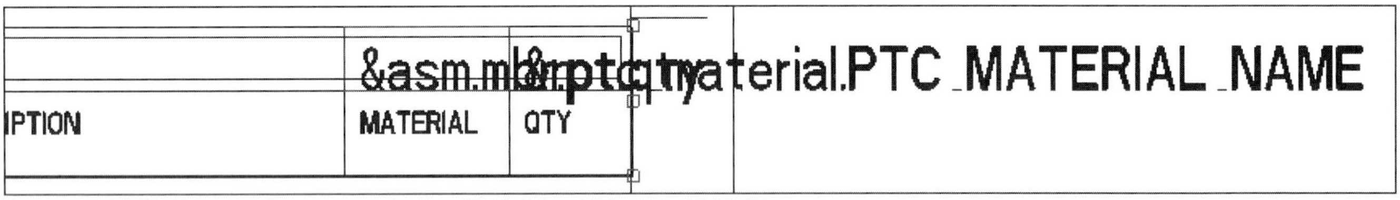

Figure 12.7(l) &asm.mbr.ptc_material.PTC_MATERIAL_NAME

Figure 12.7(m) &asm.mbr.PRTNO

Change the text height of the Report Symbols in the table cells of the Repeat Region to **.125**.

Press and hold the **Ctrl** key and select all five Repeat Region blocks ⇒ **RMB** ⇒ **Text Style** Text Style dialog box opens ⇒ Character- Height 0.125000 ⇒ Note/Dimension- Horizontal Center ⇒ Vertical Middle ⇒ **Apply** [Fig. 12.7(n)] ⇒ **OK** ⇒ **LMB** ⇒ **Update the information displayed in tables** ⇒ ⇒ ⇒ ⇒ **MMB** [Fig. 12.7(o)] ⇒ **File** ⇒ **Close Window**

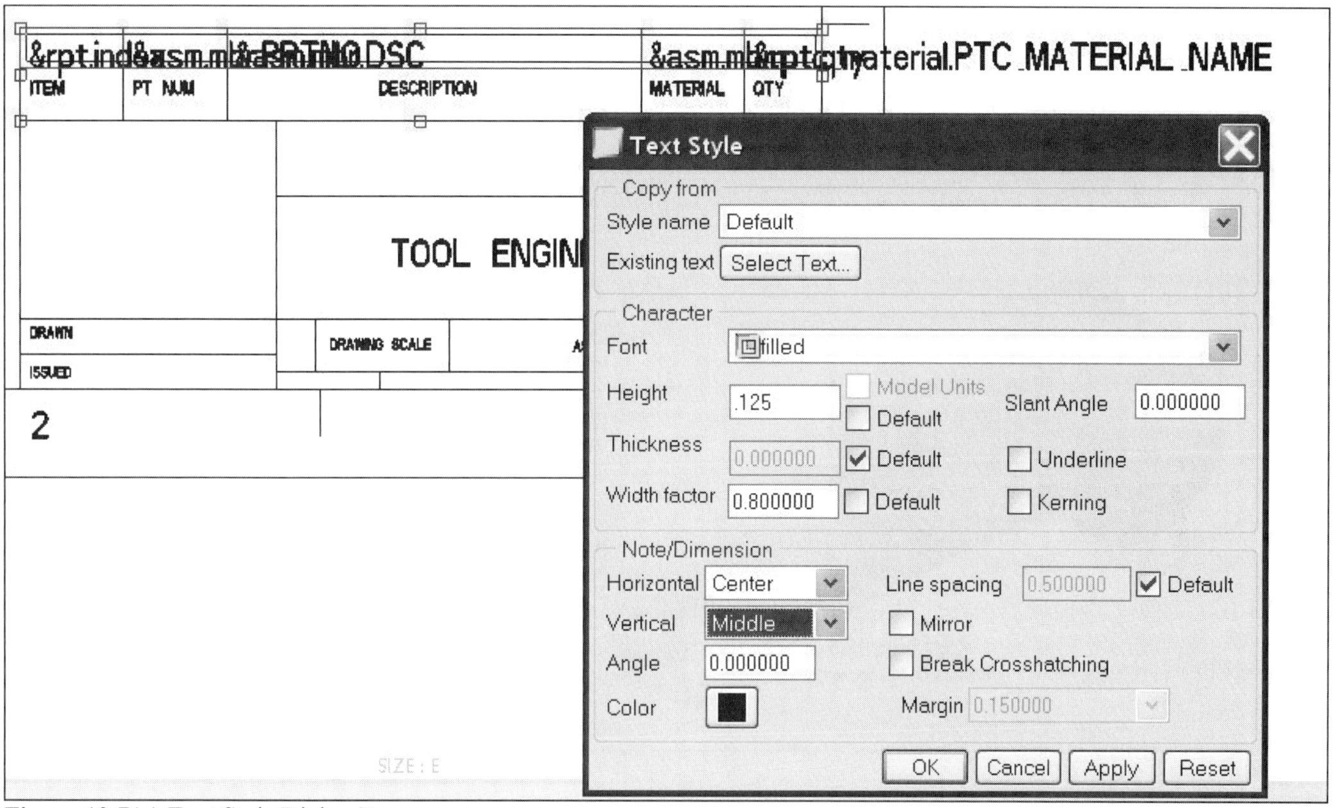

Figure 12.7(n) Text Style Dialog Box

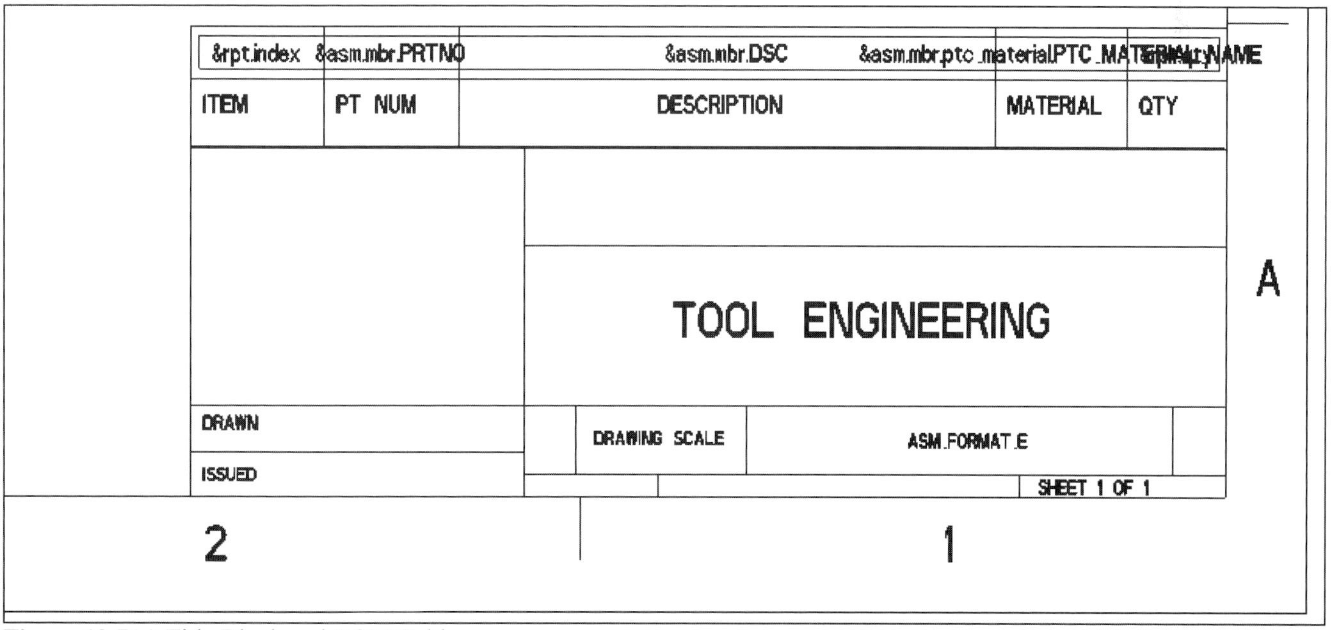

Figure 12.7(o) Title Block and BOM Table

Adding Parts List (BOM) Data

When you save your standard assembly format, the Drawing Table that represents your standard parts list is now included. You must be aware of the titles of the parameters under which the data is stored, so that you can add them properly to your components.

As you add components to an assembly, Pro/E reads the parameters from them and updates the parts list. You can also see the same effect by adding these parameters after the drawing has been created.

Pro/E also provides the capability of displaying Item Balloons on the first view that was placed on the drawing. To improve their appearance, you can move these balloons to other views and alter the locations where they attach.

Retrieve the clamp arm, click: **File** ⇒ **Open** ⇒ select **clamp_arm.prt** (The generic)⇒ **Open** ⇒ **Tools** ⇒ **Environment** ⇒ Standard Orient: **Isometric** ⇒ **OK** ⇒ ⇒ off [Fig. 12.8(a)]

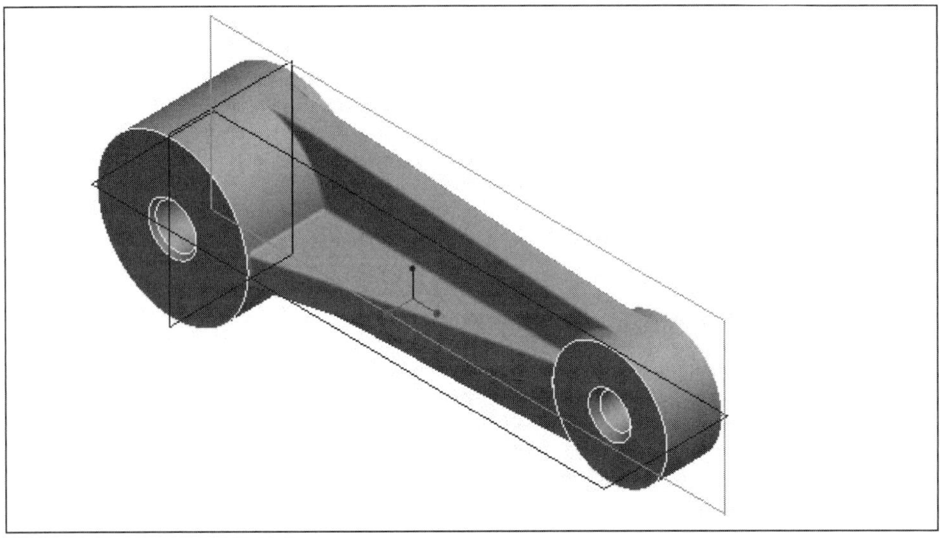

Figure 12.8(a) Clamp_Arm

Click: **Tools** from menu bar ⇒ **Parameters** [Fig. 12.8(b)]

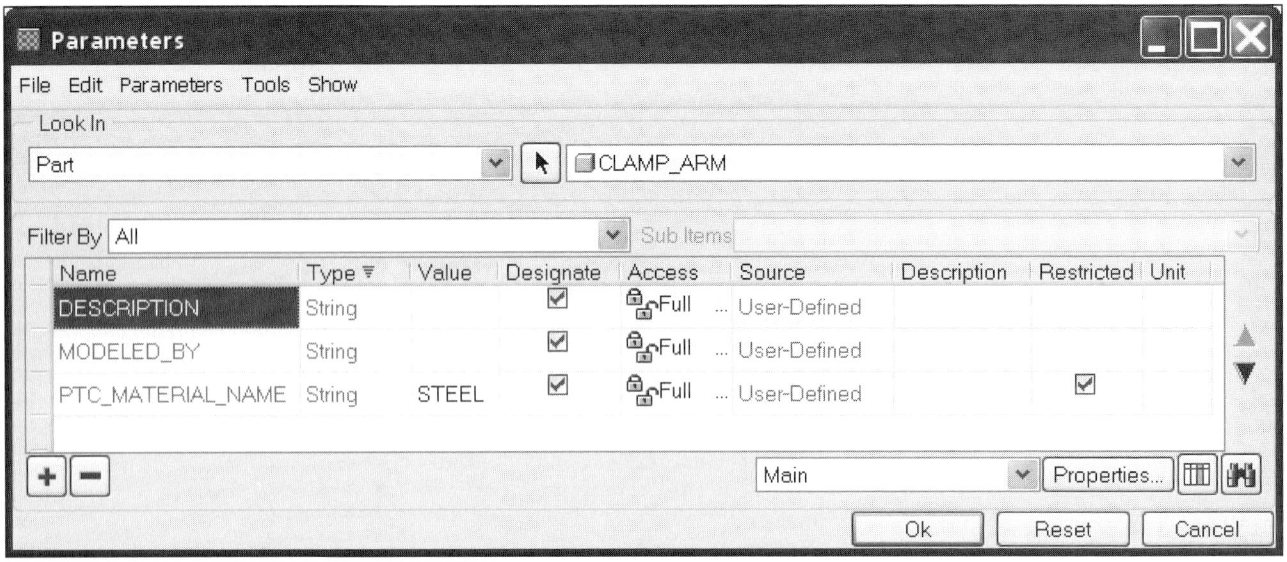

Figure 12.8(b) Parameters Dialog Box

Click: **Parameters** [Fig. 12.8(c)] ⇒ **Add Parameter** from Parameters dialog box ⇒ Name field should be highlighted (if not, click in field and highlight), type **PRTNO** [Figs. 12.8(d-e)]

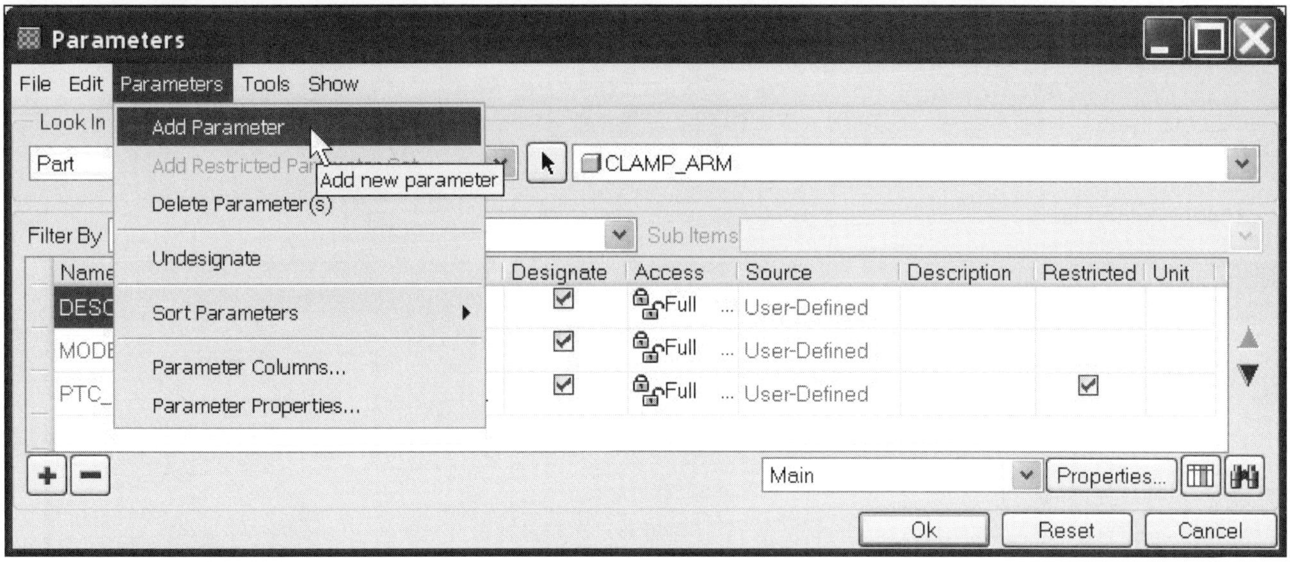

Figure 12.8(c) Add Parameter

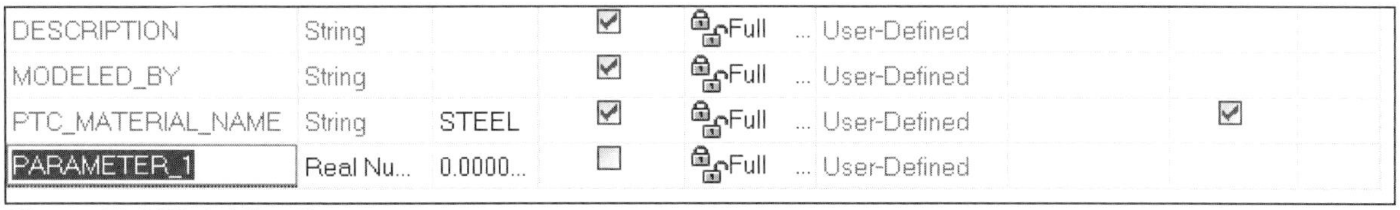

Figure 12.8(d) Adding Parameters

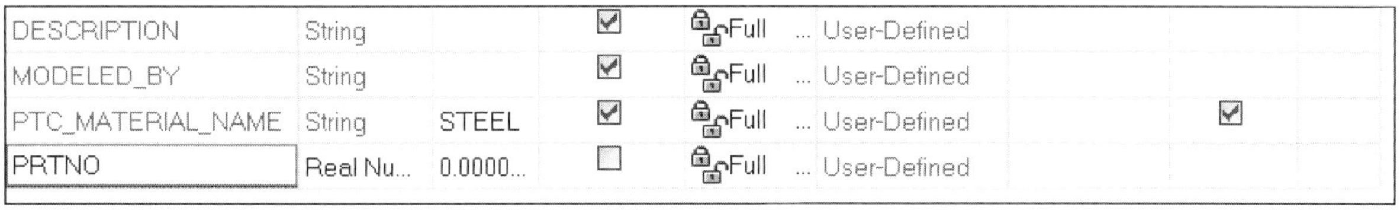

Figure 12.8(e) Name PRTNO

Click in Type field [Fig. 12.8(f)] ⇒  [Fig. 12.8(g)] ⇒ **String**

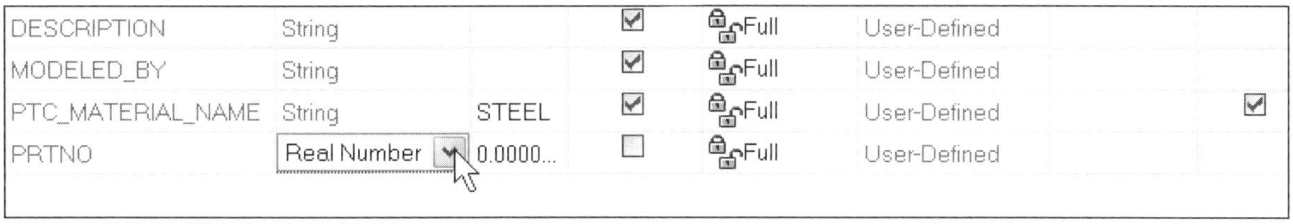

Figure 12.8(f) Click in Type Field- Real Number

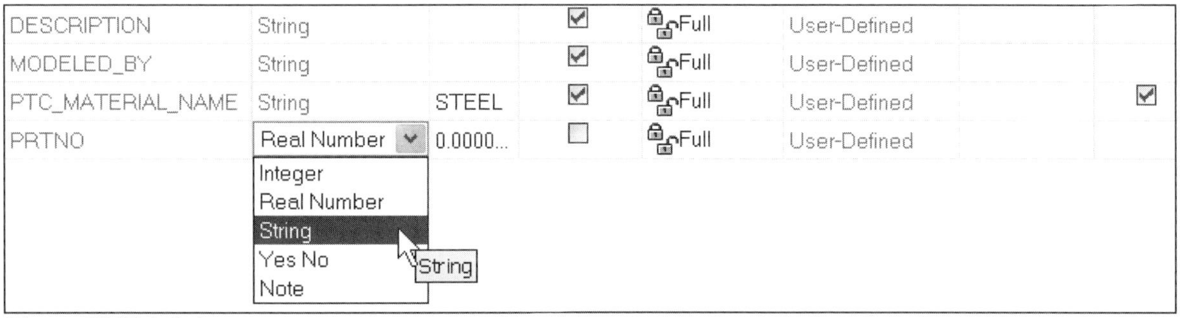

Figure 12.8(g) String

Click in Value field ⇒ type **SW101-5AR** [Fig. 12.8(h)] ⇒ Designate ☑ ⇒ click in Description field ⇒ type **part number** [Fig. 12.8(i)]

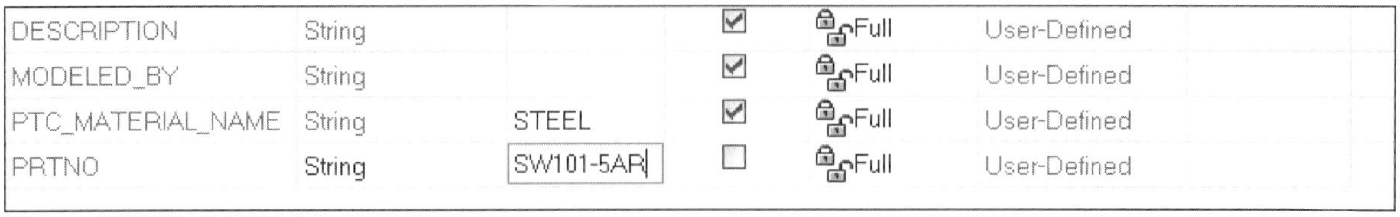

Figure 12.8(h) Add Value

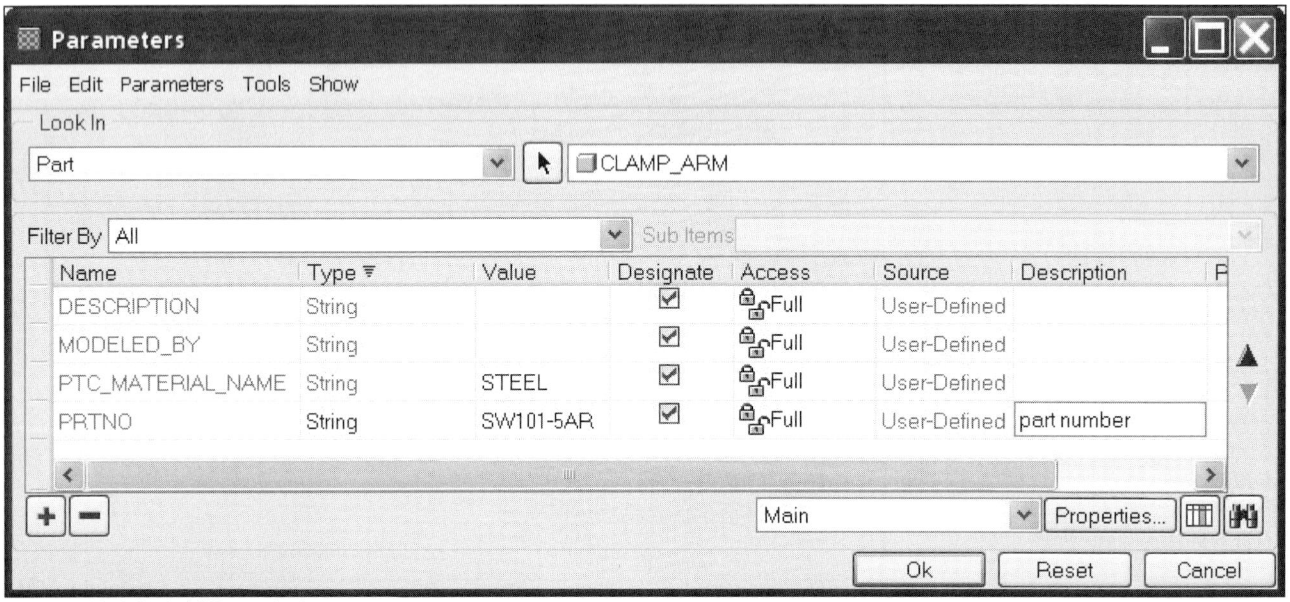

Figure 12.8(i) Add Description

Click: ⊞ **Add new Parameter** [Fig. 12.8(j)] ⇒ in Name field, type **DSC** ⇒ click in Type field ⇒ **String** ⇒ click in Value field ⇒ type **CLAMP ARM** ⇒ Designate ☑ ⇒ click in Description field ⇒ type **part description** [Fig. 12.8(k)] ⇒ **Ok** ⇒ **Ctrl+S** ⇒ **Enter**

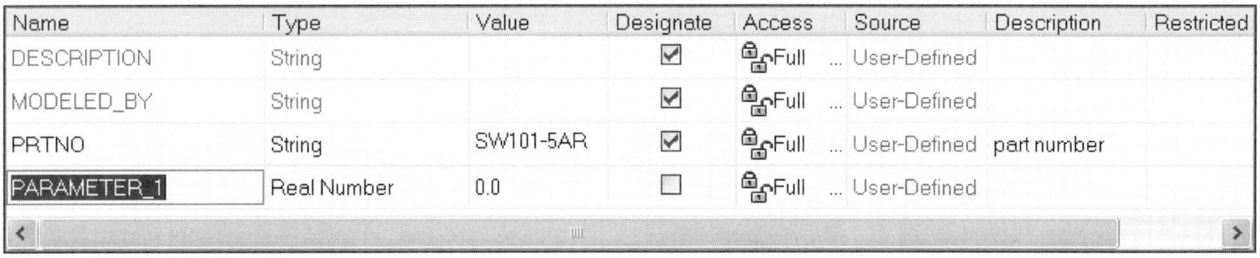

Figure 12.8(j) Add New Parameter

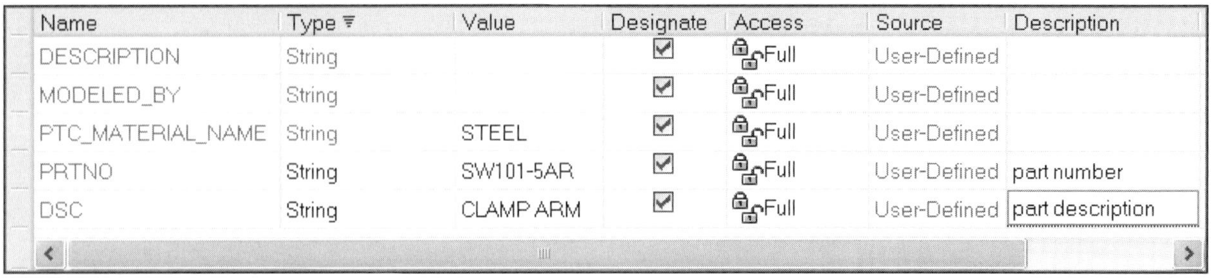

Figure 12.8(k) String

Parameters can also be displayed using the Relations dialog box. In the case of the Clamp_Arm, there was a relation created for controlling a features location. Click: **Tools** ⇒ **Relations** ⇒ **Local Parameters** [Fig. 12.8(l)] ⇒ **Ok** ⇒ **Ctrl+S** [Fig. 12.8(m)] ⇒ **OK** ⇒ **Window** ⇒ **Close**

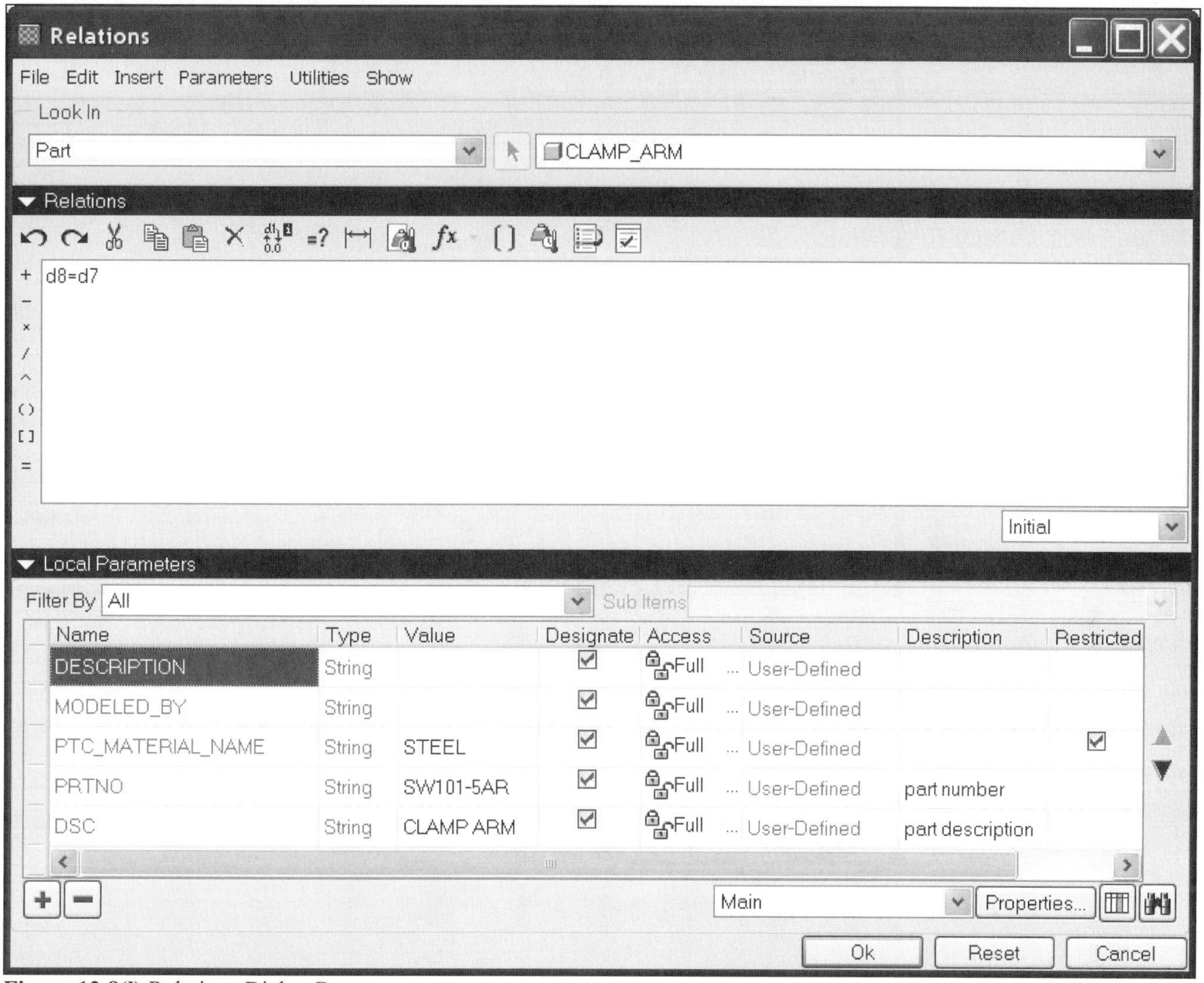

Figure 12.8(l) Relations Dialog Box

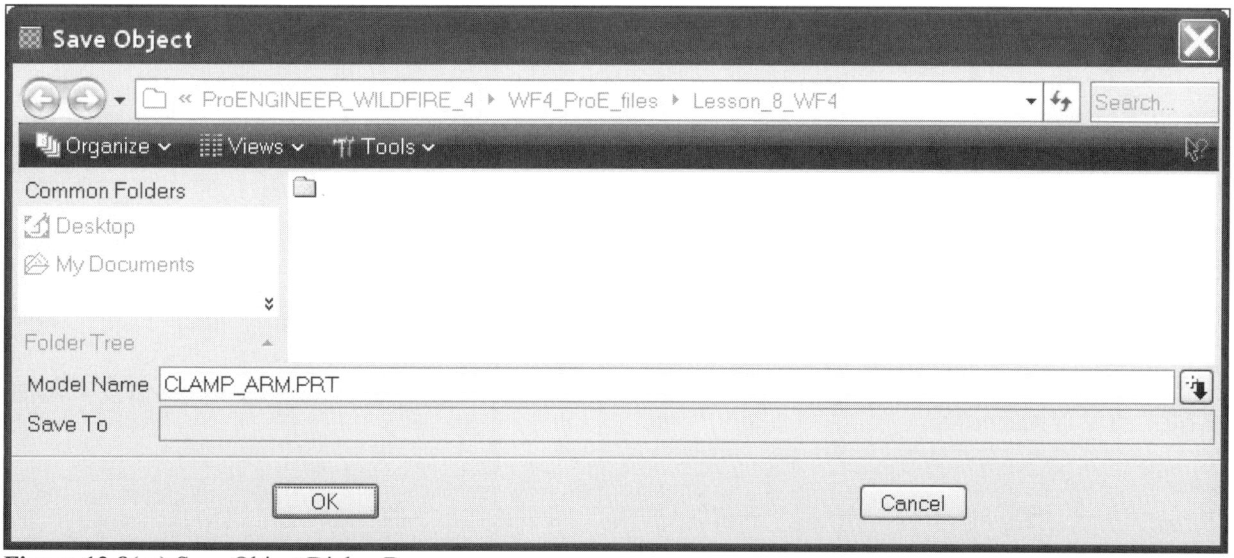

Figure 12.8(m) Save Object Dialog Box

Retrieve the clamp swivel, click: **File ⇒ Open ⇒ clamp_swivel.prt ⇒ Open ⇒ Tools ⇒ Parameters ⇒ Parameters ⇒ Add Parameter ⇒** complete the parameters as shown (Fig. 12.9) **⇒ Ok ⇒ File ⇒ Save ⇒ Enter ⇒ File ⇒ Close Window**

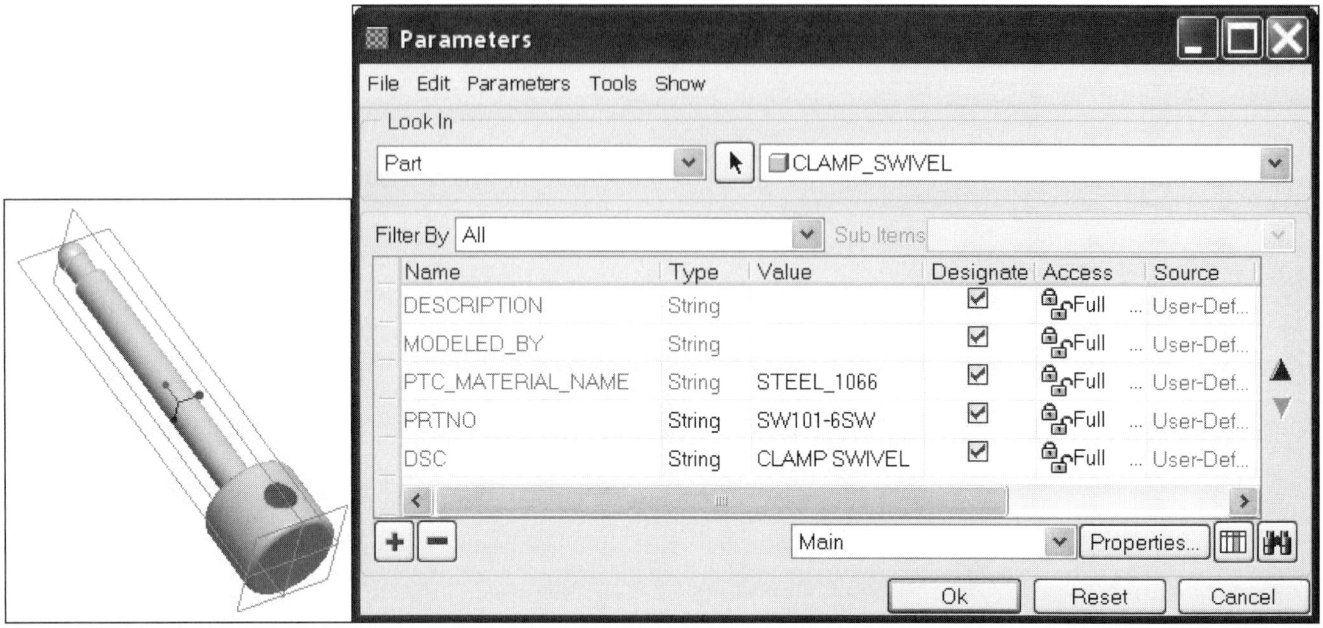

Figure 12.9 Clamp_Swivel Parameters

Retrieve the clamp ball, click: **File ⇒ Open ⇒ clamp_ball.prt ⇒ Open ⇒ Tools ⇒ Parameters ⇒ Parameters ⇒ Add Parameter ⇒** complete the parameters as shown (Fig. 12.10) **⇒ Ok ⇒ File ⇒ Save ⇒ Enter ⇒ File ⇒ Close Window**

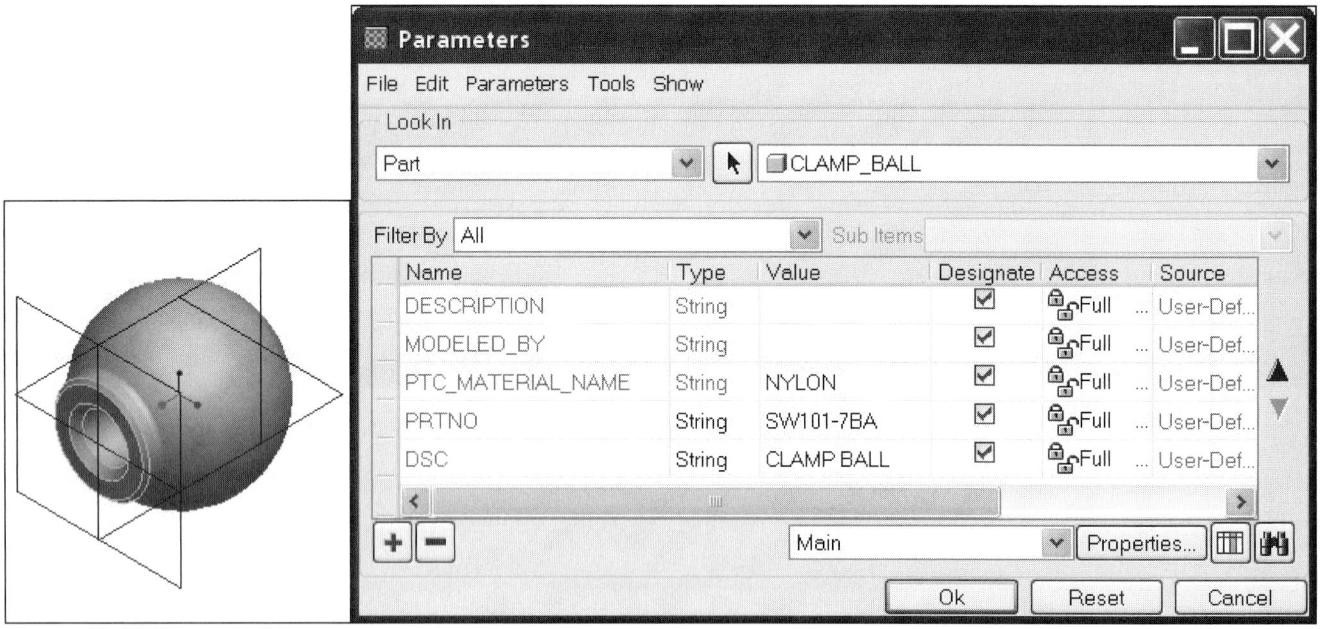

Figure 12.10 Clamp_Ball Parameters

Retrieve the clamp foot, click: **File ⇒ Open ⇒ clamp_foot.prt ⇒ Open ⇒ Tools ⇒ Parameters ⇒ Parameters ⇒ Add Parameter ⇒** complete the parameters as shown (Fig. 12.11) **⇒ Ok ⇒ File ⇒ Save ⇒ Enter ⇒ File ⇒ Close Window**

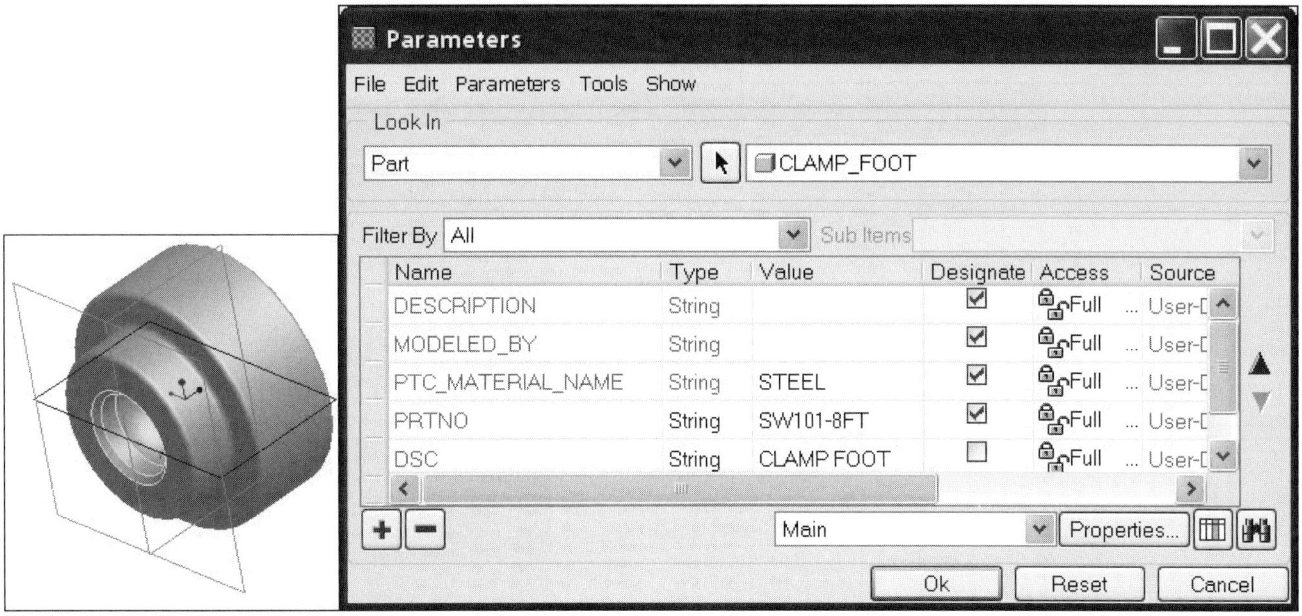

Figure 12.11 Clamp_Foot

Retrieve the clamp plate, click: **File ⇒ Open ⇒ clamp_plate.prt ⇒ Open ⇒ Edit ⇒ Setup ⇒ Material ⇒ steel.mtl ⇒**  [Fig. 12.12(a)] **⇒ OK ⇒ Done ⇒ File ⇒ Save ⇒ Enter**

Figure 12.12(a) Materials Dialog Box

Click: on ⇒ pick **DTM3** from the model ⇒ **RMB** ⇒ **Properties** ⇒ double-click in the Name field- type **A** ⇒ **Enter** ⇒ [Fig. 12.12(b)] ⇒ **OK** [Fig. 12.12(c)] ⇒ pick **DTM2** ⇒ **RMB** ⇒ **Properties** ⇒ double-click in the Name field- type **B** ⇒ **Enter** ⇒ ⇒ **OK** ⇒ pick **DTM1** ⇒ **RMB** ⇒ **Properties** ⇒ double-click in the Name field- type **C** ⇒ **Enter** ⇒ ⇒ **OK** ⇒ **LMB**

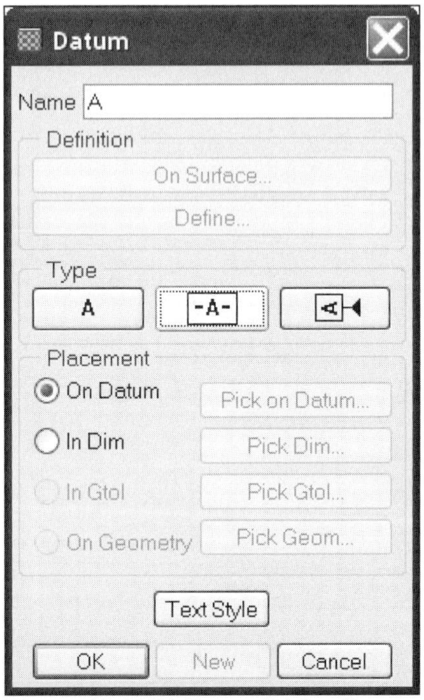

Figure 12.12(b) Datum Dialog Box

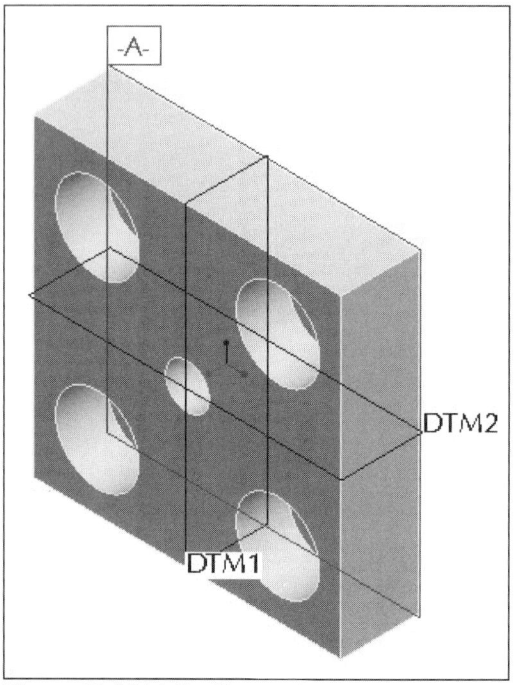

Figure 12.12(c) Set Datum A

Click: **Tools** ⇒ **Parameters** ⇒ **Parameters** ⇒ **Add Parameter** ⇒ complete the parameters as shown [Fig. 12.12(d)] ⇒ **Ok** ⇒ **File** ⇒ **Save** ⇒ **Enter** ⇒ **Window** ⇒ **Close**

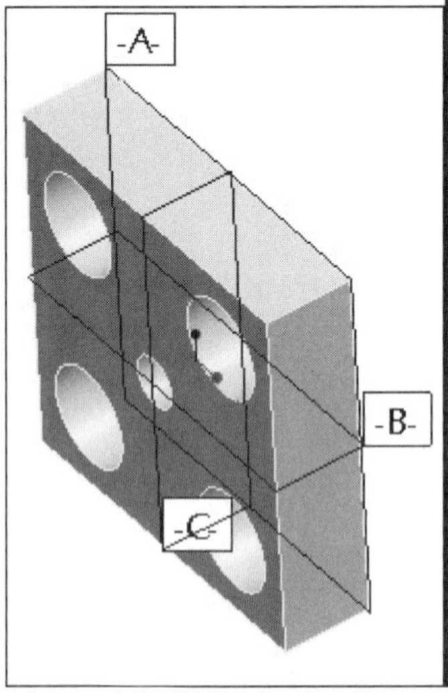

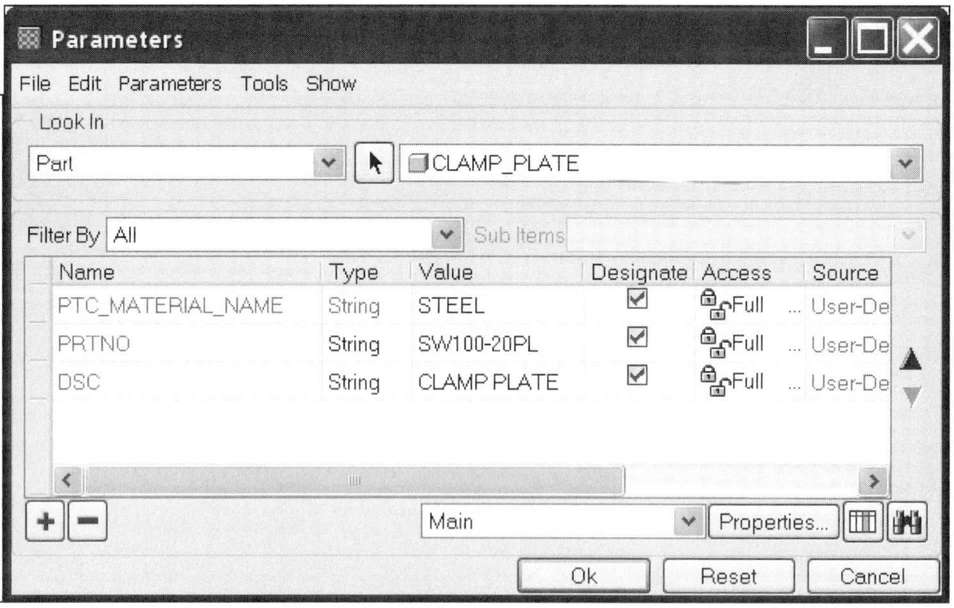

Figure 12.12(d) Clamp_Plate

Click: **File ⇒ Open ⇒ carrlane_500_fn.prt ⇒ Open ⇒ Edit ⇒ Setup ⇒ Material ⇒ File ⇒ New** [Fig. 12.13(a)] ⇒ double-click in the Name field- type **PURCHASED** ⇒ SaveTo Library... [Fig. 12.13(b)]

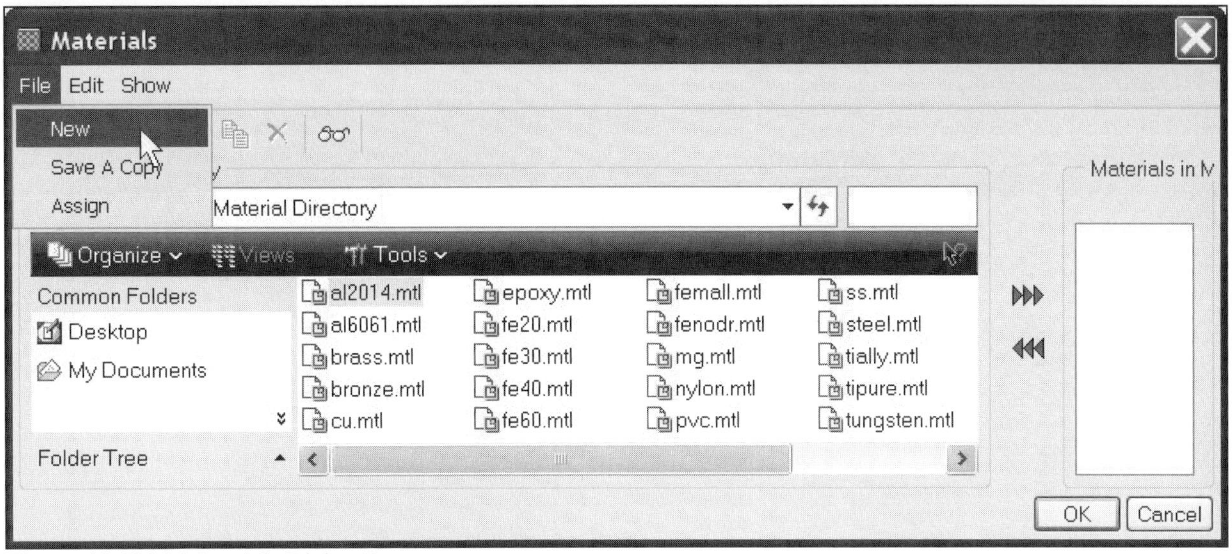

Figure 12.13(a) Materials Dialog Box

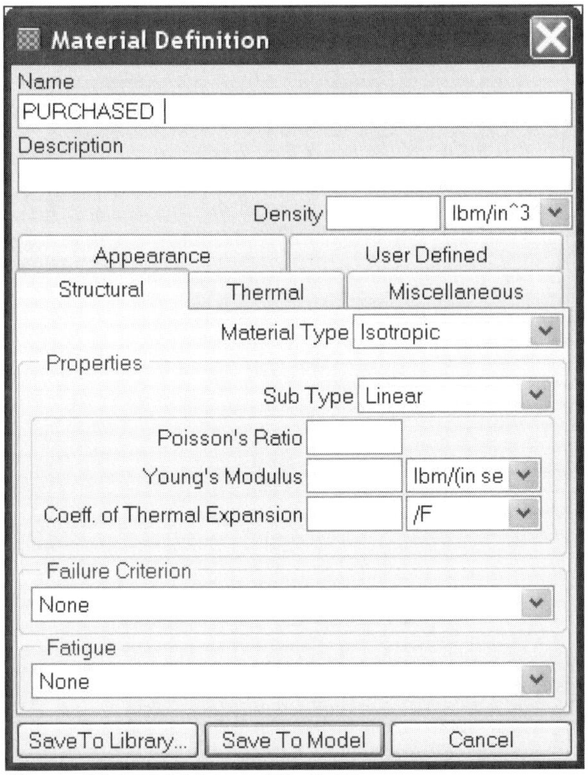

Figure 12.13(b) Material Definition

Navigate to the correct directory where you want the material to be saved, click: **OK** [Fig. 12.13(c)] ⇒ navigate to the correct directory ⇒ **purchased.mtl** ⇒ [Fig. 12.13(d)] ⇒ **OK** ⇒ **Done** ⇒ **Info** ⇒ **Model** [Fig. 12.13(e)]

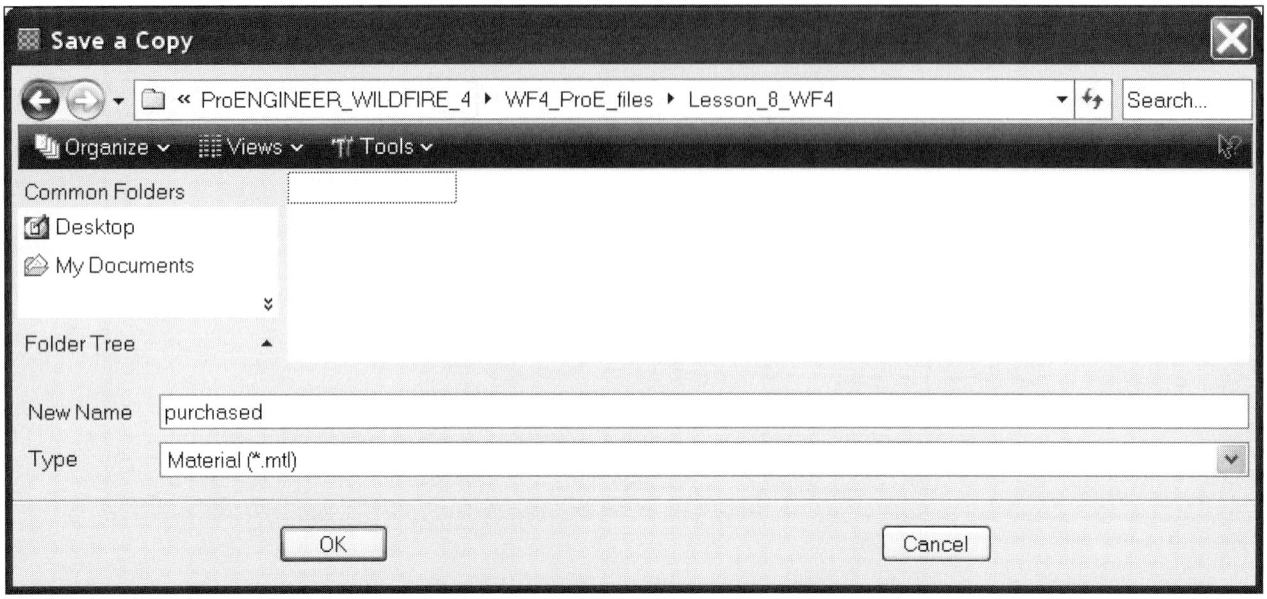

Figure 12.13(c) Save a Copy

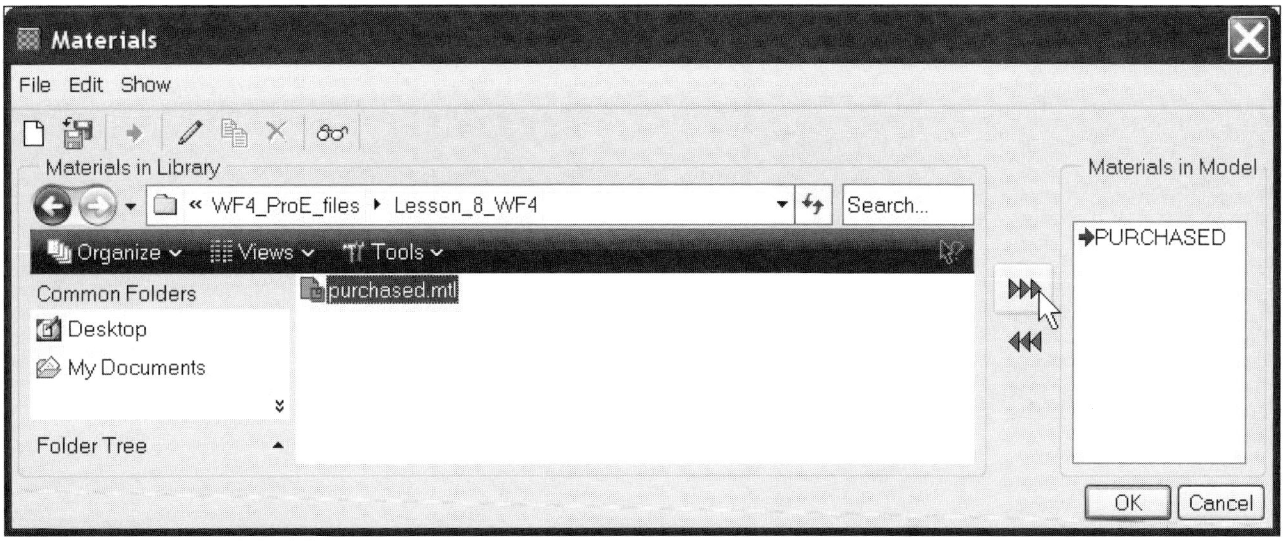

Figure 12.13(d) PURCHASED

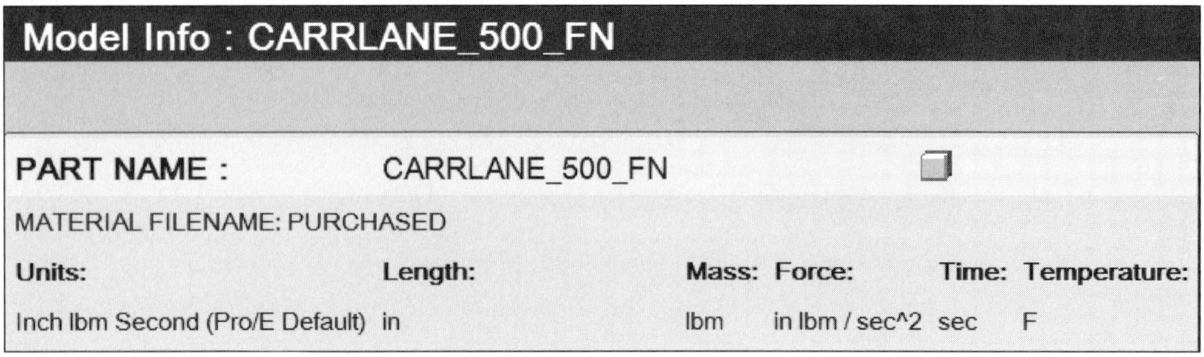

Figure 12.13(e) Model Info, MATERIAL FILENAME: PURCHASED

Click: **Tools ⇒ Parameters** [Fig. 12.13(f)]⇒ **Ok ⇒ File ⇒ Save ⇒ OK ⇒ File ⇒ Close Window**

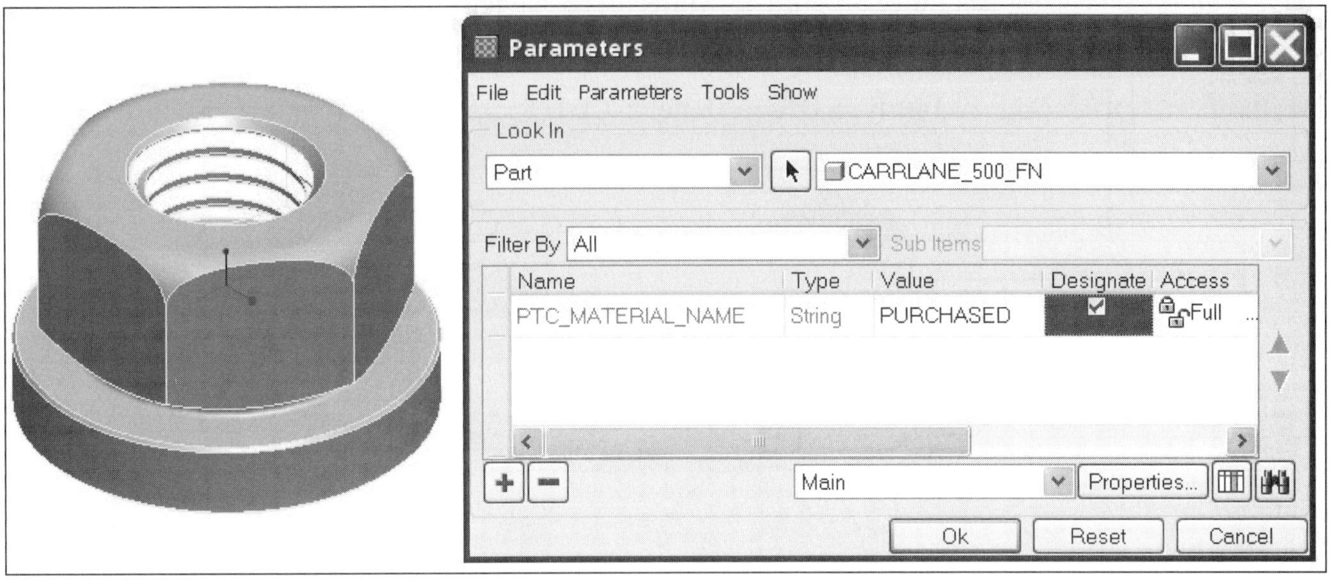

Figure 12.13(f) Parameters

Parameters can be added, deleted, and modified in Part Mode, Assembly Mode, or Drawing Mode. You can also add *parameter columns* to the Model Tree in Assembly Mode, and edit the parameter value.

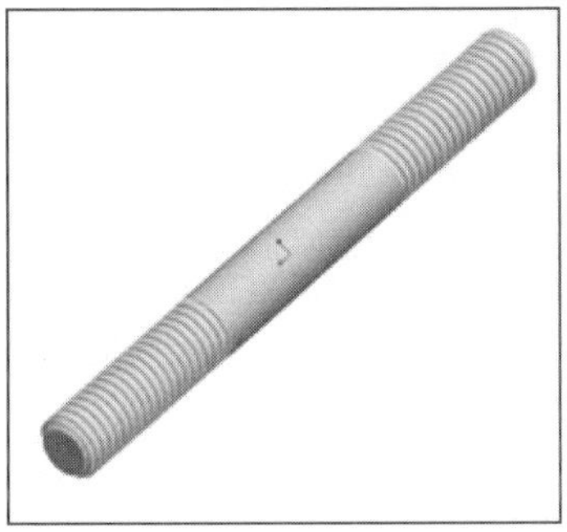

Component **Carrlane-12-13500_STUD.prt**
Part Number (PRTNO) **SW101-9STL**
Description (DSC) **.500-13 X 5.00 DOUBLE END STUD**

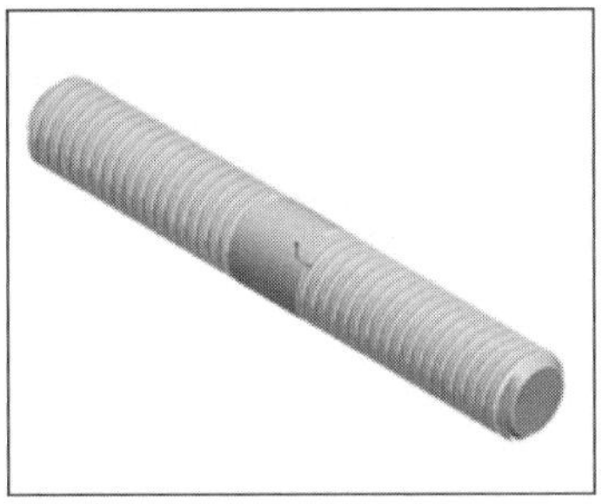

Component **Carrlane-12-13350_STUD.prt**
Part Number (PRTNO) **SW100-21ST**
Description (DSC) **.500-13 X 3.50 DOUBLE END STUD**

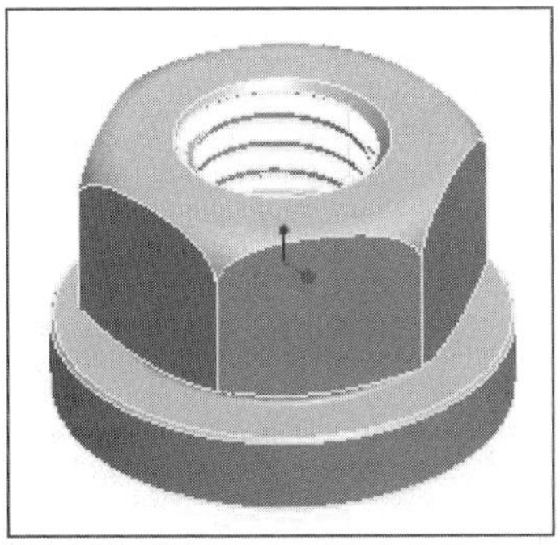

Component **Carrlane-500_FN.prt**
Part Number (PRTNO) **SW100-22FLN**
Description (DSC) **.500-13 HEX FLANGE NUT**

Click: ⇒ **Carrlane-12-13350_STUD.prt** [Fig. 12.14(a)] ⇒ **Open** ⇒ on ⇒ open the Navigator shade with the quick sash ⇒ **Edit** ⇒ **Setup** ⇒ **Material** ⇒ navigate to the correct directory where you saved the *purchased.mtl* material (you should be able to click on Working Directory) ⇒ click on the item purchased.mtl ⇒ [Fig. 12.14(b)] ⇒ **OK** ⇒ **Done** ⇒ **Ctrl+S** ⇒ **Enter** ⇒ **Window** ⇒ **Close**

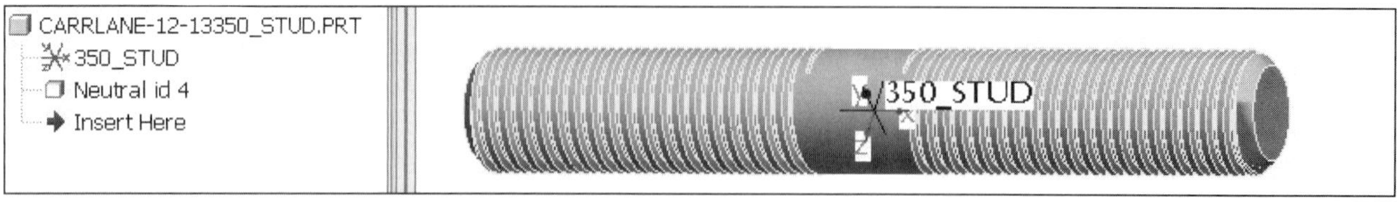

Figure 12.14(a) Carrlane-12-13350_STUD.prt

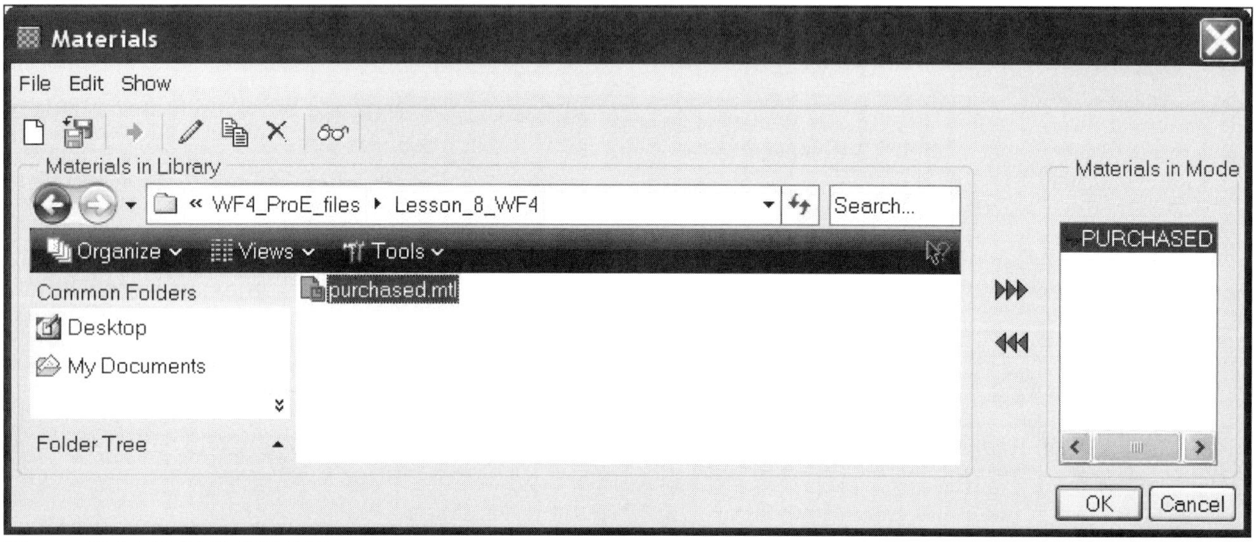

Figure 12.14(b) PURCHASED

Click: **File** ⇒ **Open** ⇒ **Carrlane-12-13500_STUD.prt** [Fig. 12.15(a)] ⇒ **Open** ⇒ **Edit** ⇒ **Setup** ⇒ **Material** ⇒ Working Directory ⇒ purchased.mtl ⇒ ⇒ **OK** ⇒ **Done** ⇒ **Ctrl+S** ⇒ **Enter** ⇒ **Info** ⇒ **Model** [Fig. 12.15(b)] ⇒ close the browser shade ⇒ **File** ⇒ **Close Window**

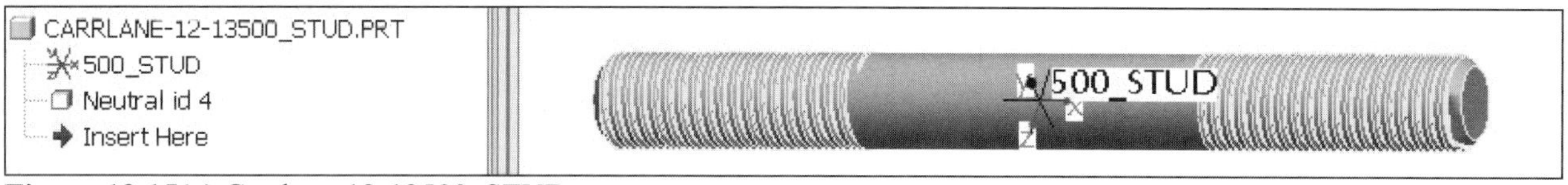

Figure 12.15(a) Carrlane-12-13500_STUD.prt

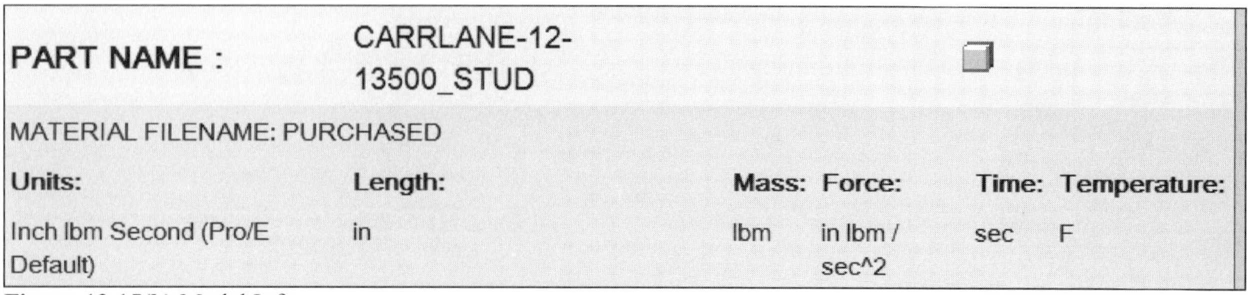

PART NAME :	CARRLANE-12-13500_STUD				
MATERIAL FILENAME: PURCHASED					
Units:	**Length:**	**Mass:**	**Force:**	**Time:**	**Temperature:**
Inch lbm Second (Pro/E Default)	in	lbm	in lbm / sec^2	sec	F

Figure 12.15(b) Model Info

Click: **File** ⇒ **Open** ⇒ **clamp_assembly.asm** ⇒ **Open** ⇒ **Settings** in the Navigator/Model Tree ⇒ **Tree Filters** ⇒ check all Display options on ⇒ **Apply** ⇒ **OK** ⇒ expand the sub-assembly, click: ⊞ next to CLAMP_SUBASSEMBLY.ASM ⇒ **Settings** ⇒ **Tree Columns** [Fig. 12.16(a)] ⇒ Type: click ⌄ ⇒ **Model Params** [Fig. 12.16(b)] ⇒ Name field, type **PRTNO** [Fig. 12.16(c)] ⇒ >> *(or **Enter**)*

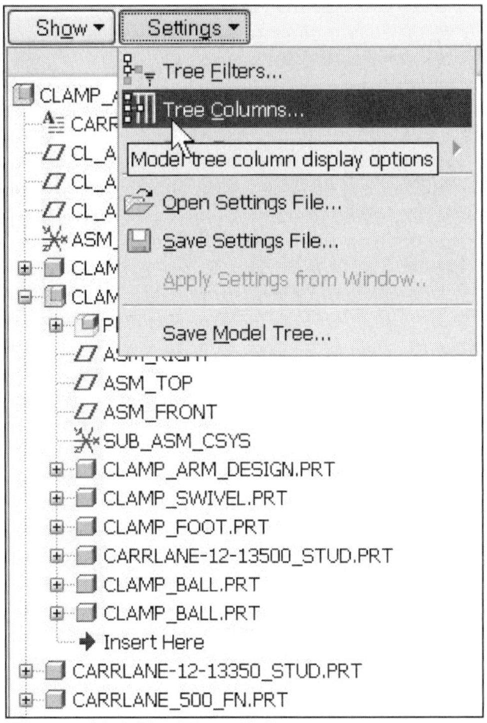

Figure 12.16(a) Tree Columns

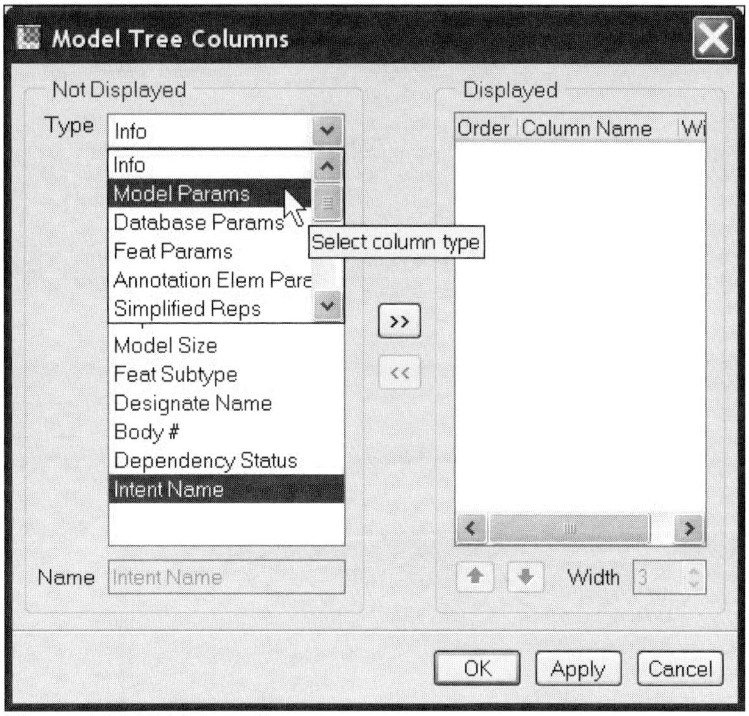

Figure 12.16(b) Model Params

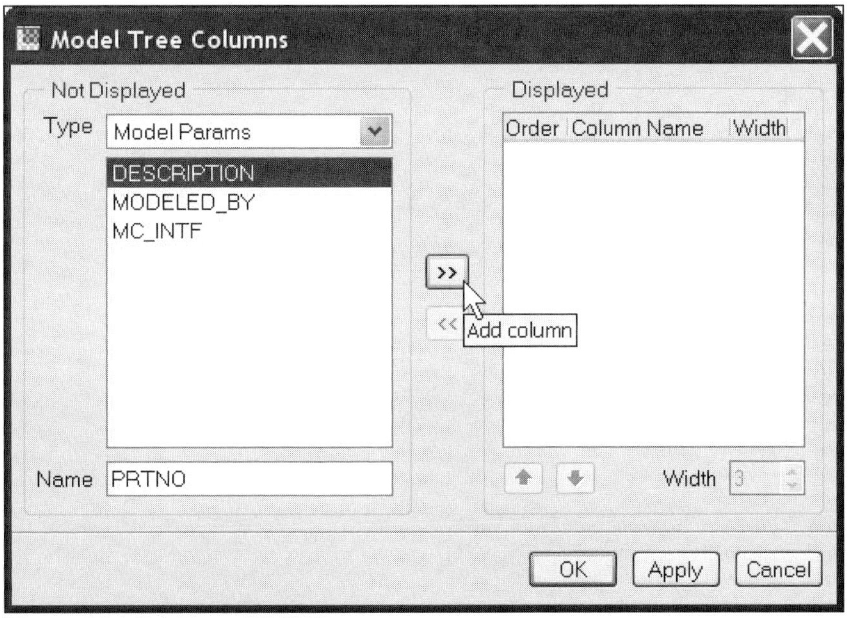

Figure 12.16(c) Name PRTNO

Click: Name field, type **DSC** [Fig. 12.16(d)] ⇒ [>>] **Add column** [Fig. 12.16(e)]

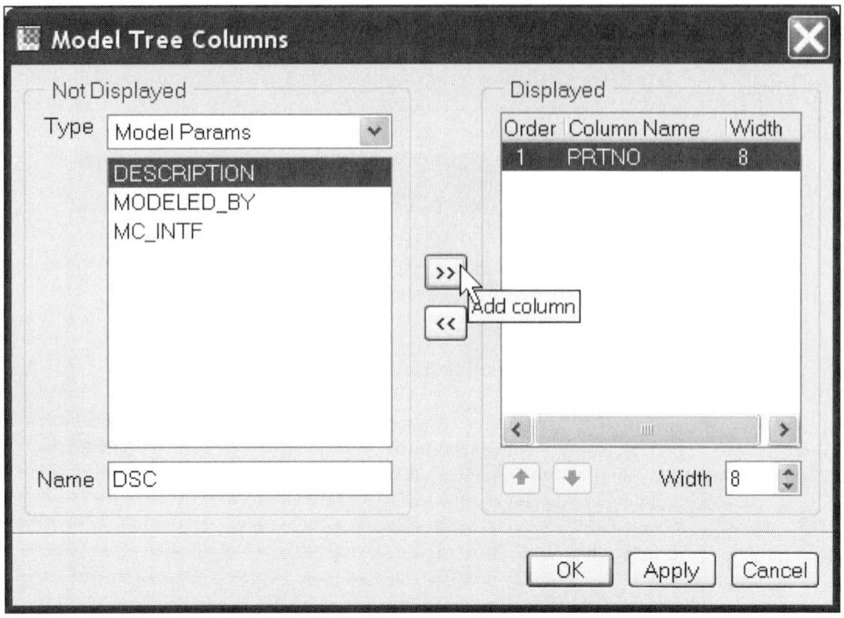

Figure 12.16(d) Name DSC

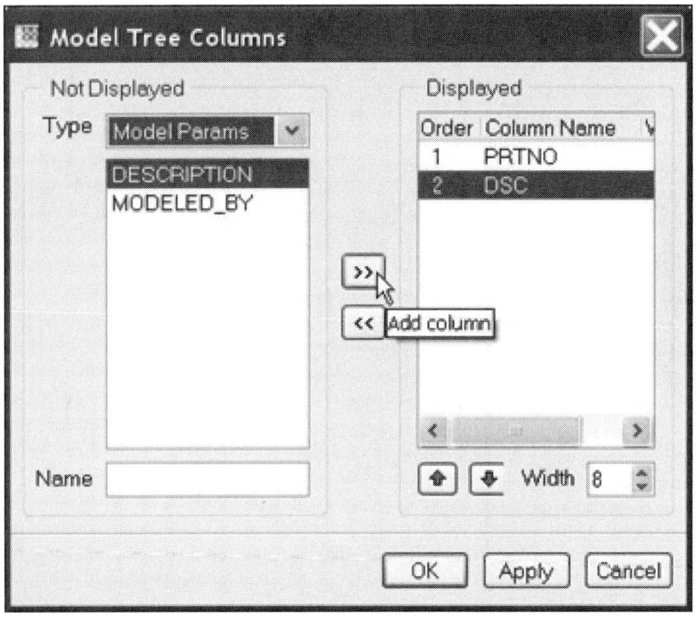

Figure 12.16(e) PRTNO and DSC

Click: Name field, type **PTC_MATERIAL_NAME** [Fig. 12.16(f)] ⇒ >> **Add column** [Fig. 12.16(g)] ⇒ **Apply** ⇒ **OK** [Fig. 12.16(h)]

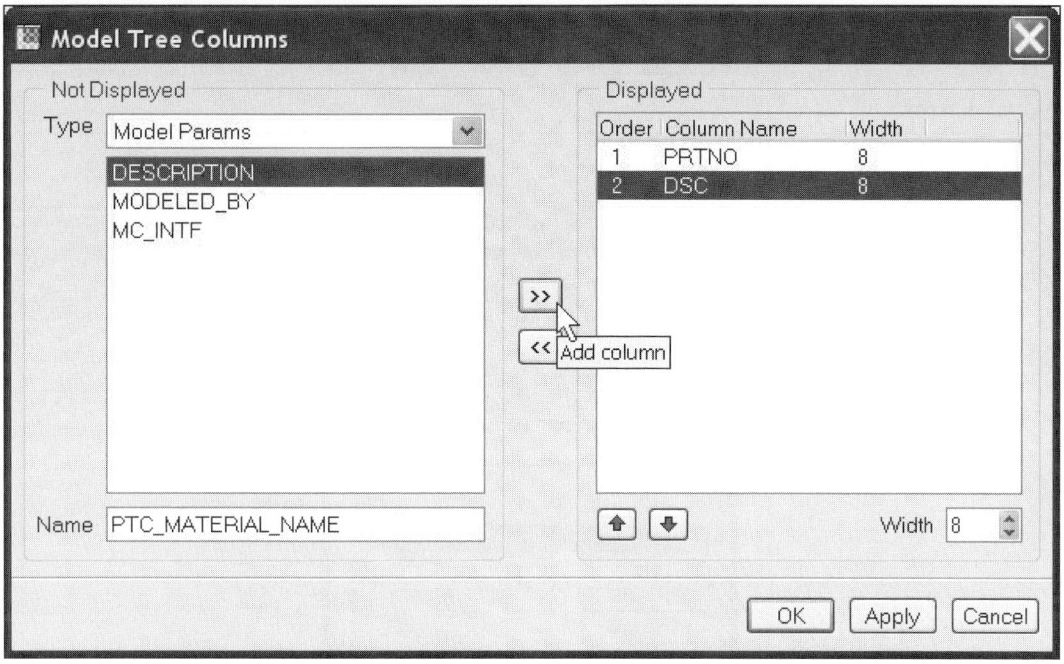

Figure 12.16(f) PTC_MATERIAL_NAME

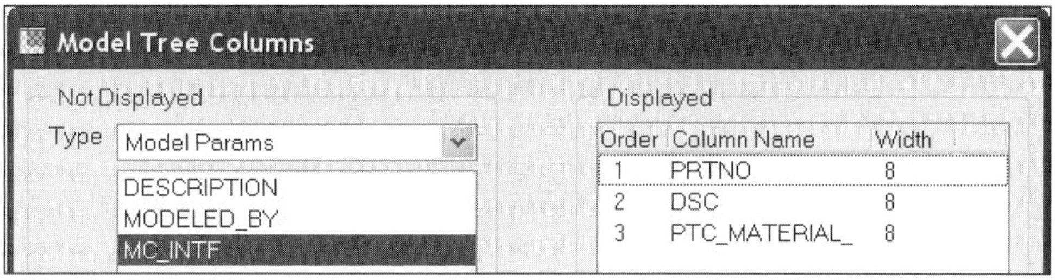

Figure 12.16(g) PTC_MATERIAL_NAME

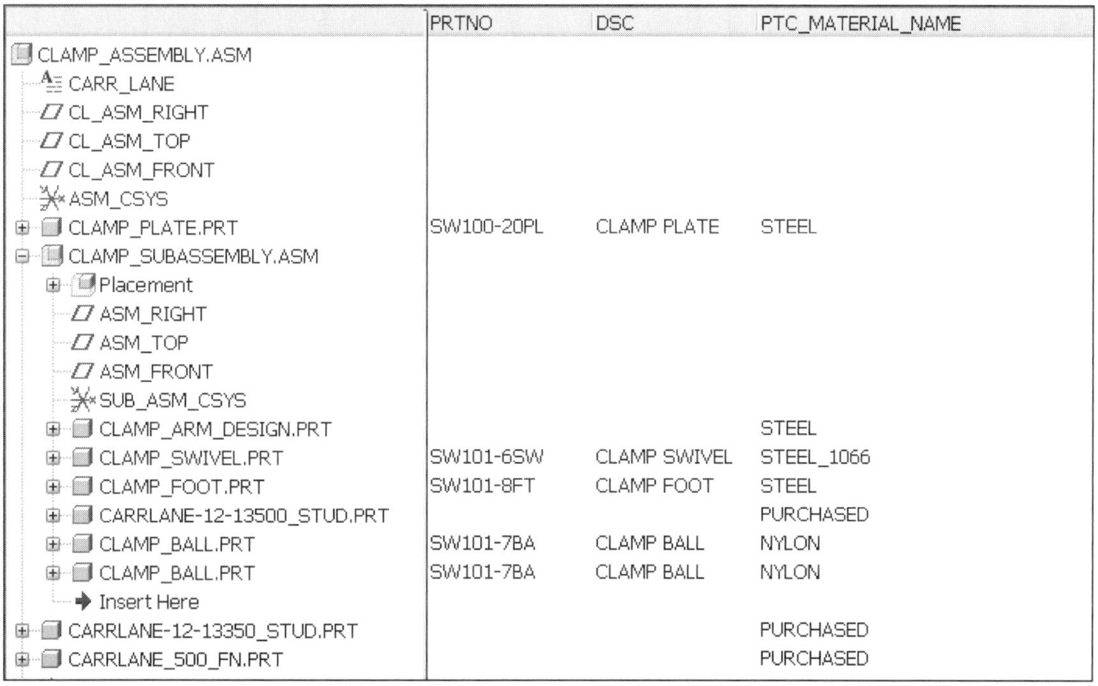

Figure 12.16(h) New Tree Columns

Resize the Model Tree and click on **CARRLANE-12-13500_STUD.PRT** ⇒ click in the **PRTNO** field [Fig. 12.17(a)] ⇒ Type: click ⇒ **String** ⇒ in Value field, type **SW101-9STL** ⇒ Designated [Fig. 12.17(b)] ⇒ **Ok** ⇒ click in the **DSC** field ⇒ ⇒ **String** ⇒ in Value field, type **.500-13 X 5.00 DOUBLE END STUD** [Fig. 12.17(c)] ⇒ Designated ⇒ **Ok** ⇒ **LMB** in the graphics window to deselect ⇒ **File** ⇒ **Save** ⇒ **Enter**

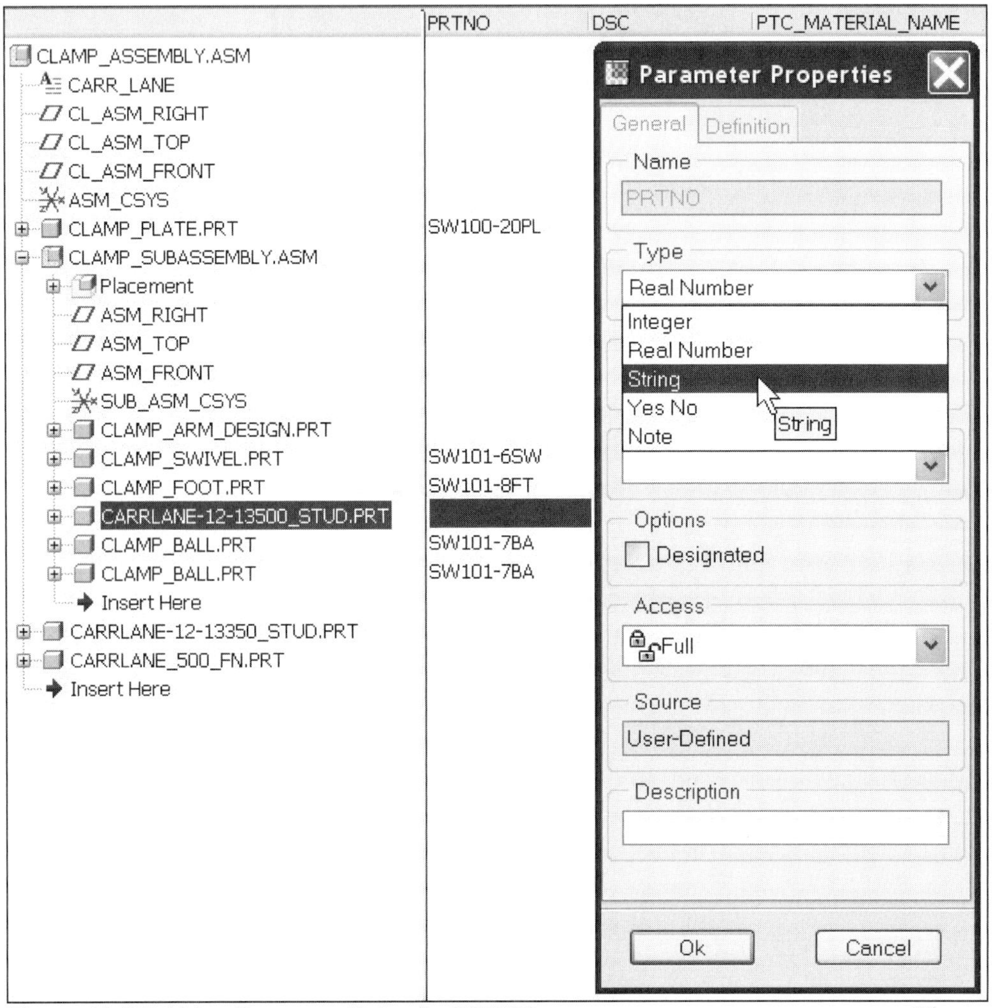

Figure 12.17(a) New Parameter String for PRTNO

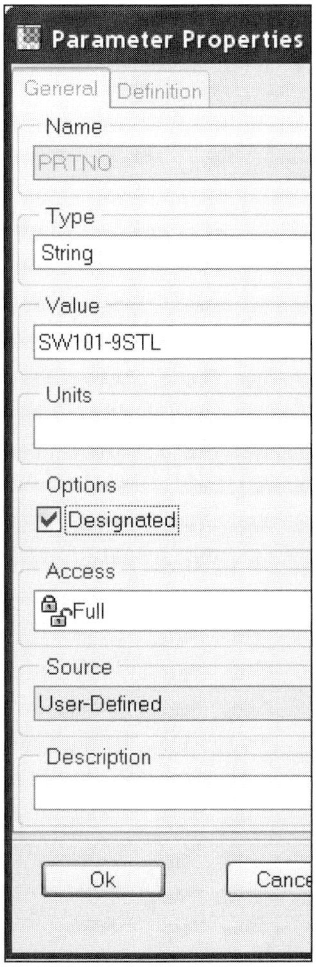

Figure 12.17(b) Value SW101-9STL

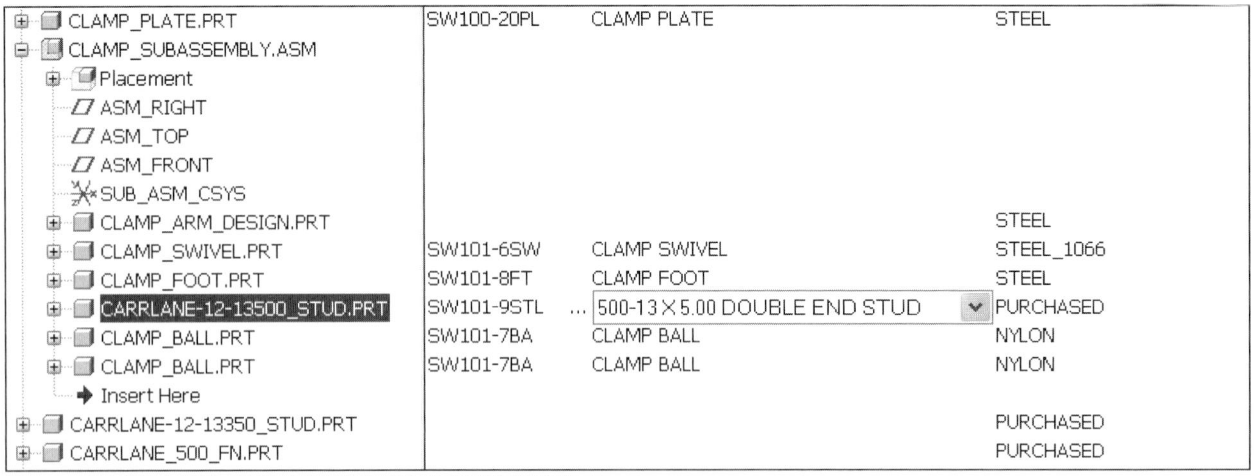

Figure 12.17(c) 500-13 X 5.00 DOUBLE END STUD

Complete the parameters for the remaining components. Where necessary modify the values for the PRTNO's to reflect what is shown here by clicking in the value field

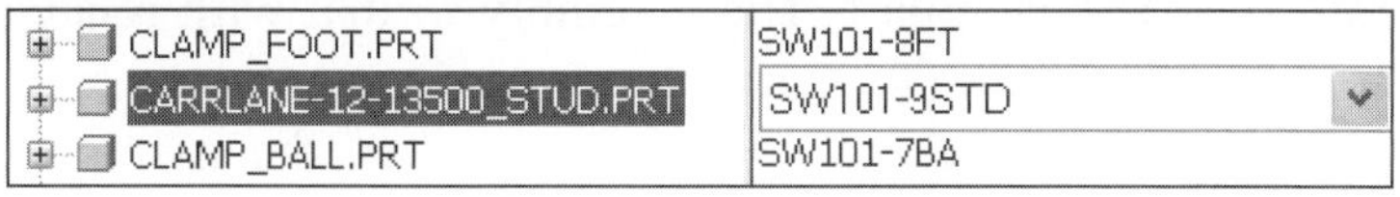

and editing the field [Fig. 12.18(a)] ⇒ **LMB** in the graphics window to deselect ⇒ **CTRL+S** ⇒ **Enter**

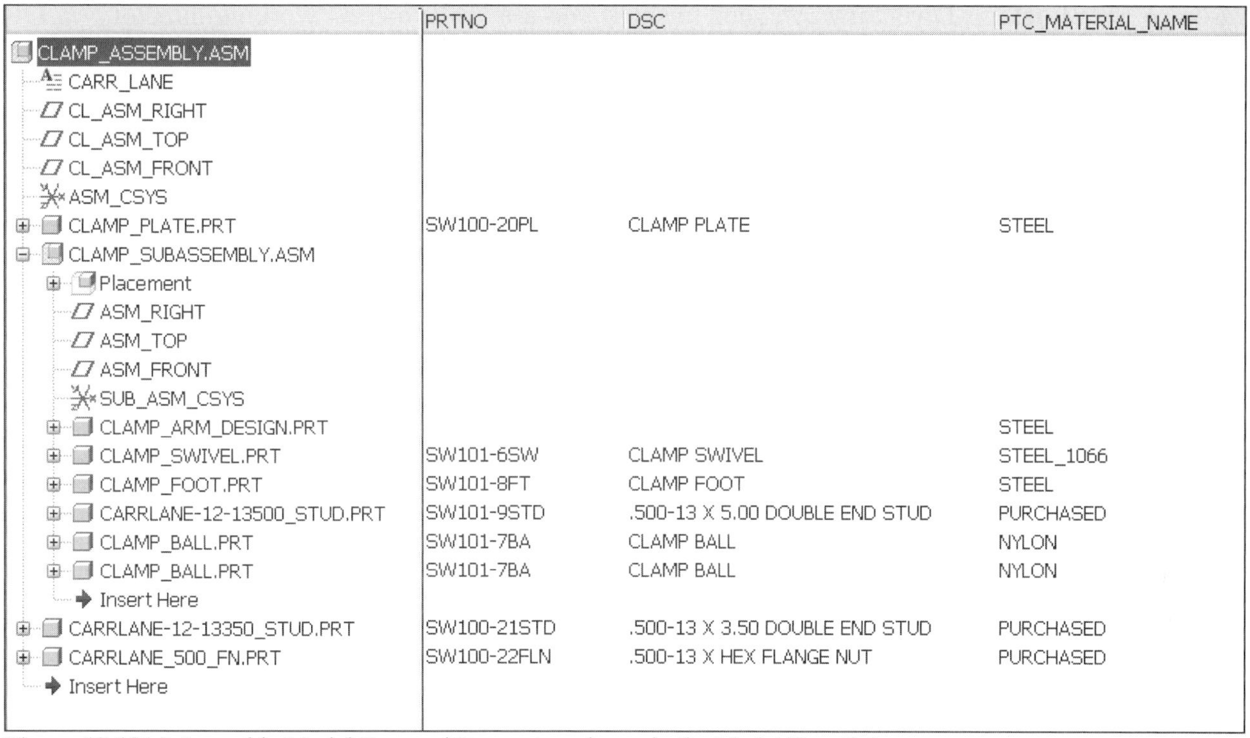

Figure 12.18(a) Assembly Model Tree and Parameters shown in the Model Tree Columns

Click: **Tools** ⇒ **Parameters** ⇒ Look In: click: [▼] ⇒ **Part** ⇒ click on **CARRLANE-12-13500_STUD.PRT** ⇒ Designate [☑] [Fig. 12.18(b)] ⇒ **Ok** ⇒ repeat for every part (including the clamp_arm.prt) to check for correct values and designation ⇒ **File** ⇒ **Save** ⇒ **OK** ⇒ **Window** ⇒ **Close**

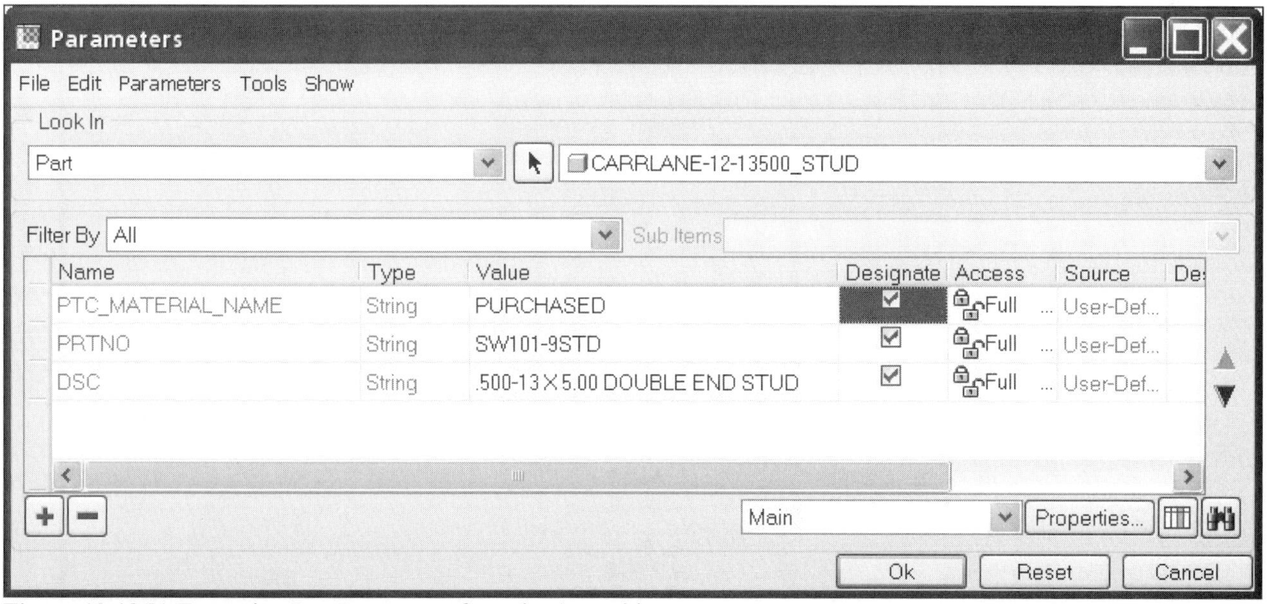

Figure 12.18(b) Extracting Part Parameters from the Assembly

Assembly Drawings

The parameters (and their values) have been established for each part. The assembly format with related parameters in a parts list table has been created and saved in your (format) directory. You can now create a drawing of the assembly, where the parts list will be generated automatically, and the assembly ballooned. The first assembly drawing will be of the ***Clamp_Subassembly.***

Click: **File** ⇒ **Set Working Directory** ⇒ check to see if you are in the correct working directory ⇒ **OK** ⇒ **Create a new object** ⇒ Drawing ⇒ Name **clamp_subassembly** ⇒ Use default template ⇒ **OK** [Fig. 12.19(a)] ⇒ Empty with format ⇒ Default Model **Browse** ⇒ **clamp_subassembly.asm** ⇒ **Open** [Figs. 12.19(b-c)]

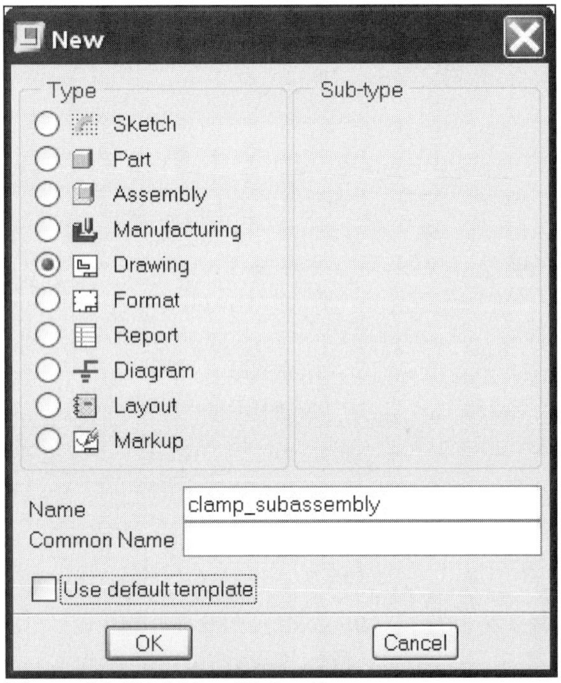

Figure 12.19(a) New Dialog Box

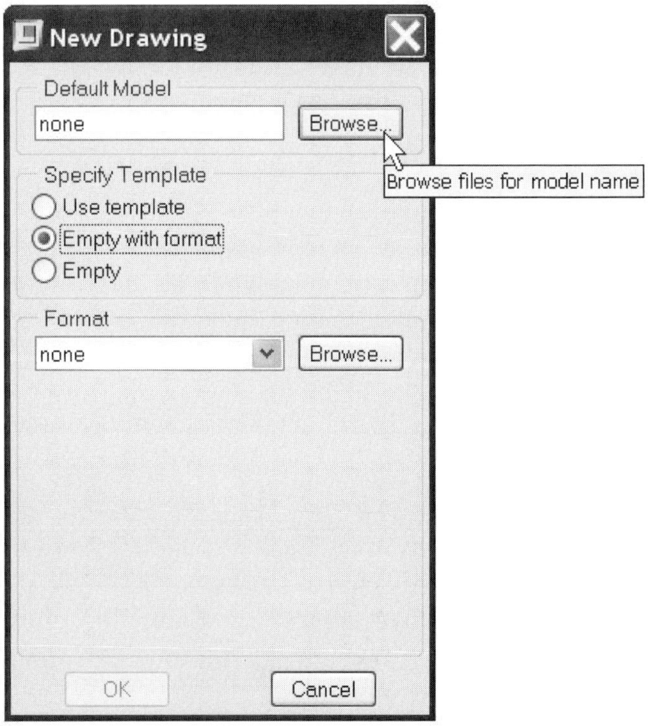

Figure 12.19(b) New Drawing Dialog Box

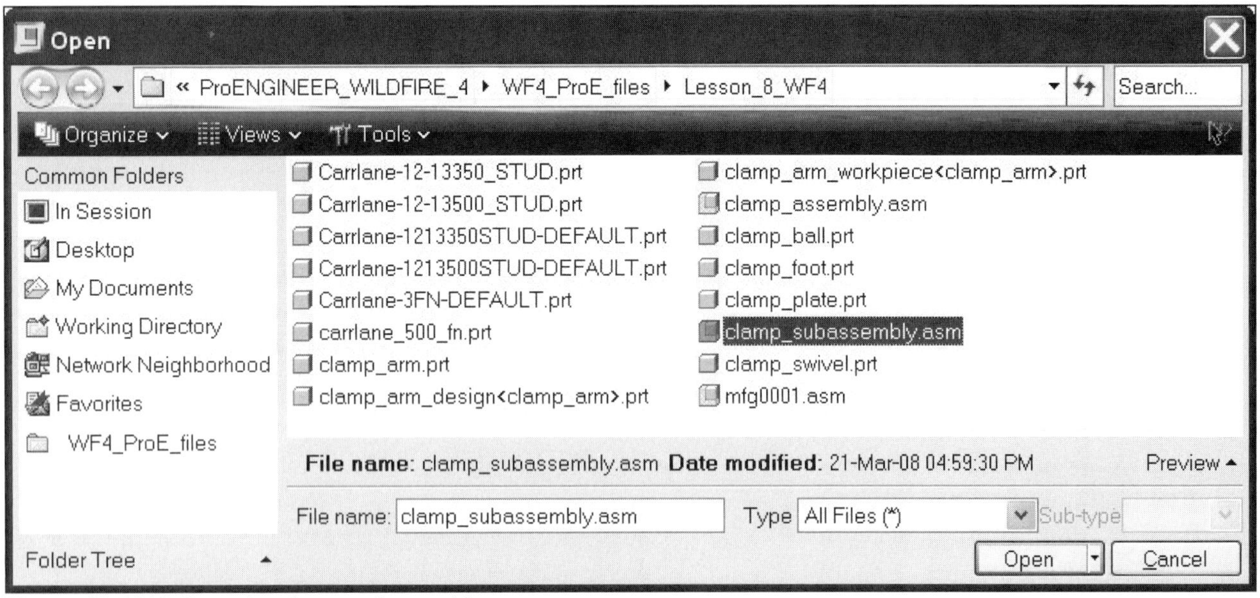

Figure 12.19(c) Open the clamp_subassembly.asm

Click: Format **Browse** [Fig. 12.19(d)] ⇒ **Working Directory** [Fig. 12.19(e)] ⇒ select **asm_format_e.frm** [Fig. 12.19(f)] ⇒ **Open** [Fig. 12.19(g)] ⇒ **OK** [Fig. 12.19(h)]

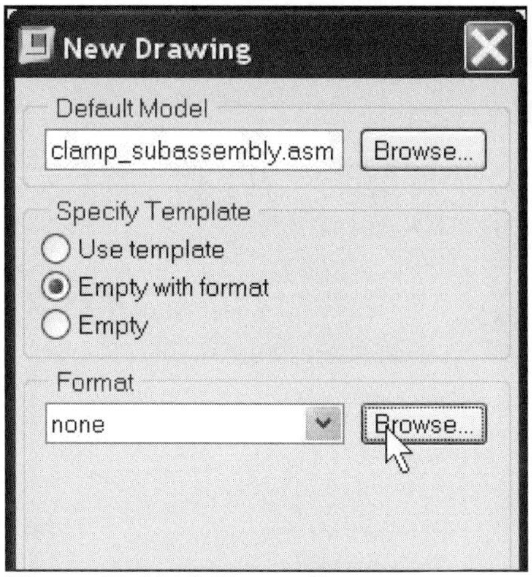

Figure 12.19(d) Format Browse

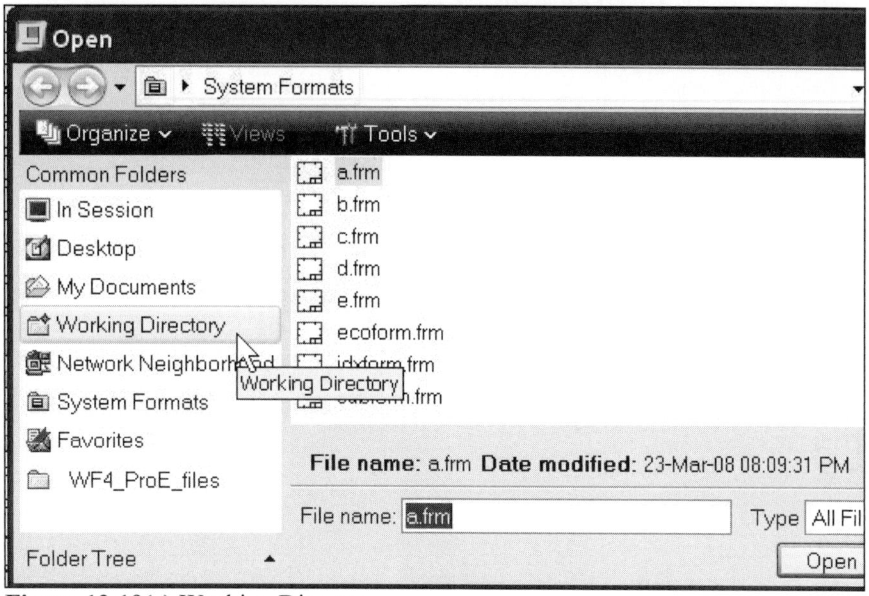

Figure 12.19(e) Working Directory

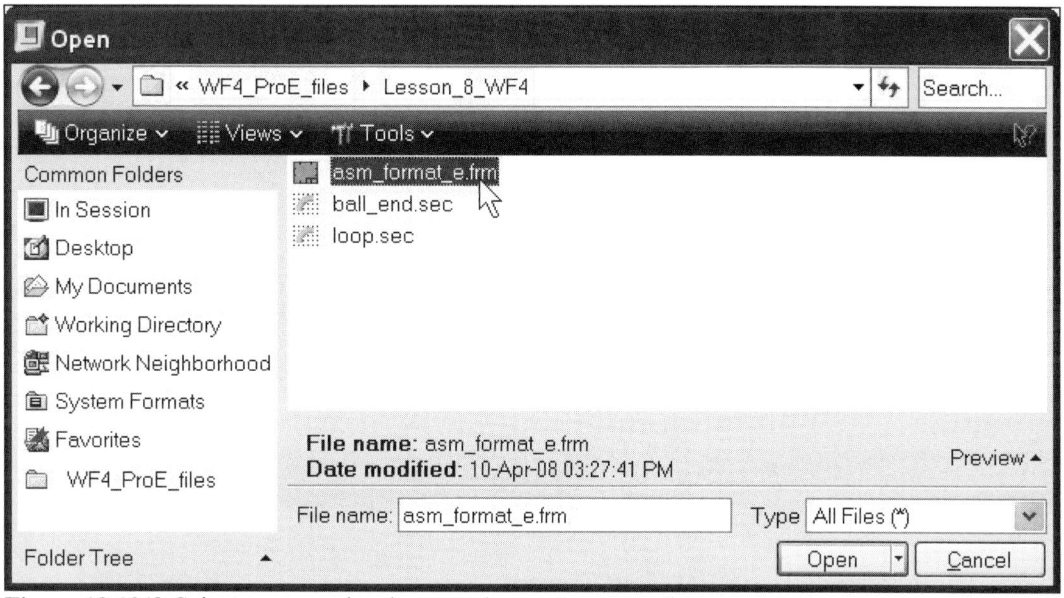

Figure 12.19(f) Select your previously created **asm_format_e.frm**

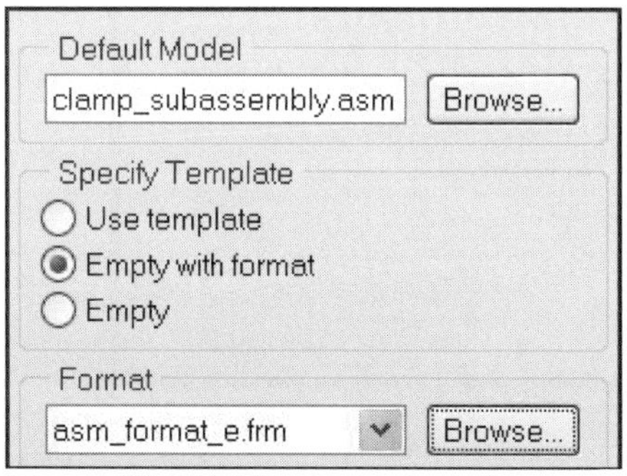

Figure 12.19(g) Format selected

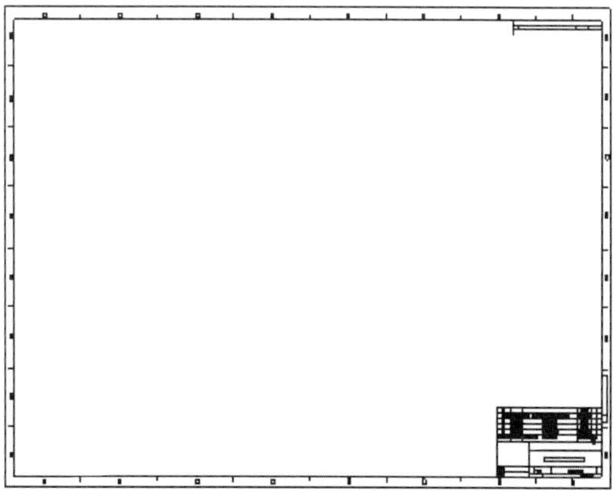

Figure 12.19(h) Drawing with Format

For the drawing to display with the correct style, modify the values of the Drawing Options.

Click: **Tools** ⇒ **Environment** ⇒ Display Style | Hidden Line ⇒ Tangent Edges | Dimmed ⇒ **OK** ⇒ **File** ⇒ **Properties** ⇒ **Drawing Options** ⇒ Sort **Alphabetical** ⇒ crossec_arrow_length | .50 ⇒ **Enter** ⇒ crossec_arrow_width | .17 ⇒ **Enter** ⇒ default_font | filled ⇒ **Add/Change** ⇒ draw_arrow_length | .375 ⇒ **Add/Change** ⇒ draw_arrow_style | filled ⇒ **Add/Change** ⇒ draw_arrow_width | .125 ⇒ **Enter** ⇒ drawing_text_height | .50 ⇒ **Add/Change** ⇒ max_balloon_radius | .50 ⇒ **Enter** ⇒ min_balloon_radius | .50 ⇒ **Enter** [Fig. 12.20(a)] ⇒ **Save a copy of the currently displayed configuration file** ⇒ File Name **CLAMP_ASM** ⇒ **Ok** [Fig. 12.20(b)] ⇒ **Apply** ⇒ **Close** ⇒ **Done/Return** ⇒ ⇒ **OK**

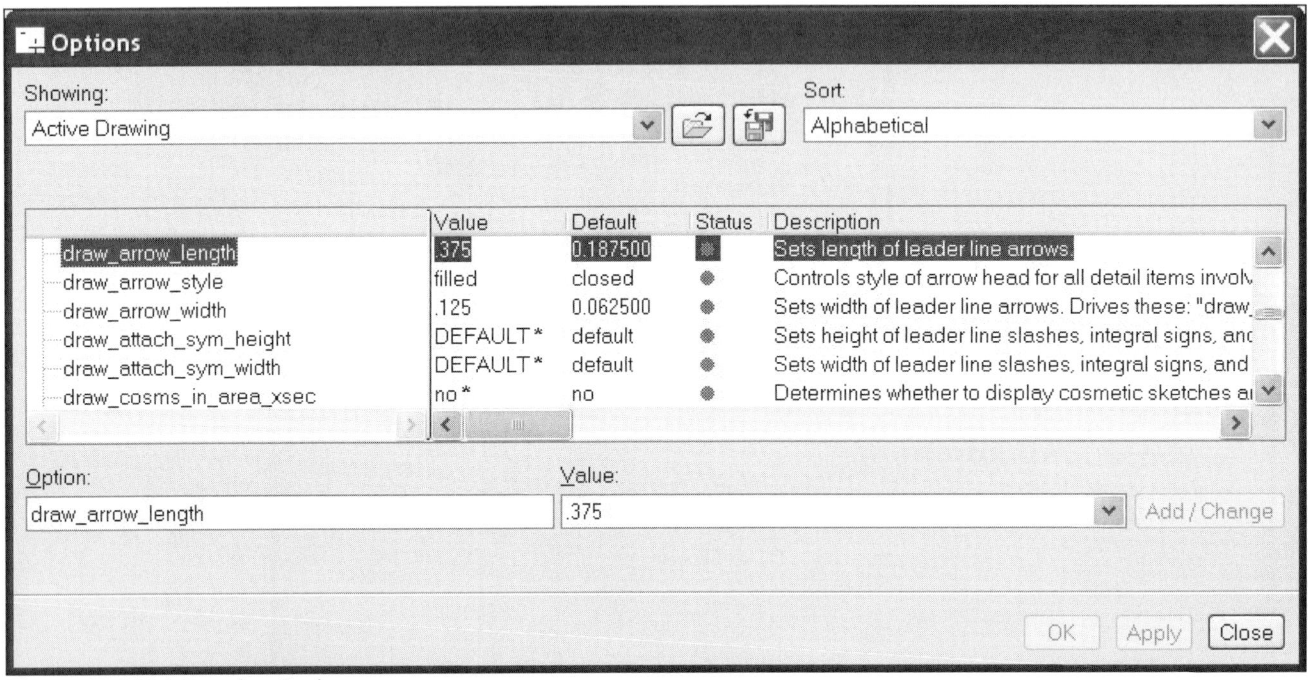

Figure 12.20(a) Drawing Options

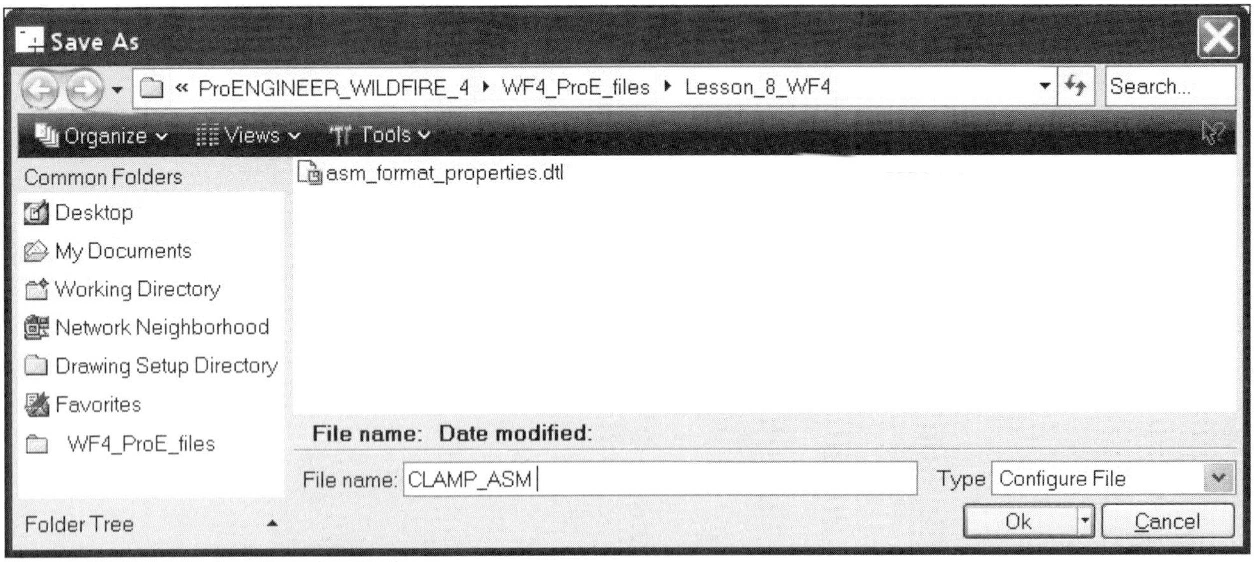

Figure 12.20(b) Saving Drawing Options

Click on the quick sash to close your Navigator ⇒ off ⇒ double-click on **SCALE: 1.000** in the lower left-hand corner of the graphics window SCALE : 1.000 TYPE : ASSEM ⇒ Enter value for scale 1.000 ⇒ type **1.50** Enter value for scale 1.50 ⇒ **Enter** ⇒ **LMB** to deselect ⇒ **Disallow the movement of drawing views with the mouse** unlock ⇒ **Create a general view** ⇒ **No Combined State** ⇒ **OK** ⇒ ⇒ Select CENTER POINT for drawing view. [Fig. 12.21(a)] ⇒ Model view names, click: **FRONT** ⇒ **Apply** [Fig. 12.21(b)] ⇒ **Close** ⇒ **Ctrl+S** ⇒ **Enter**

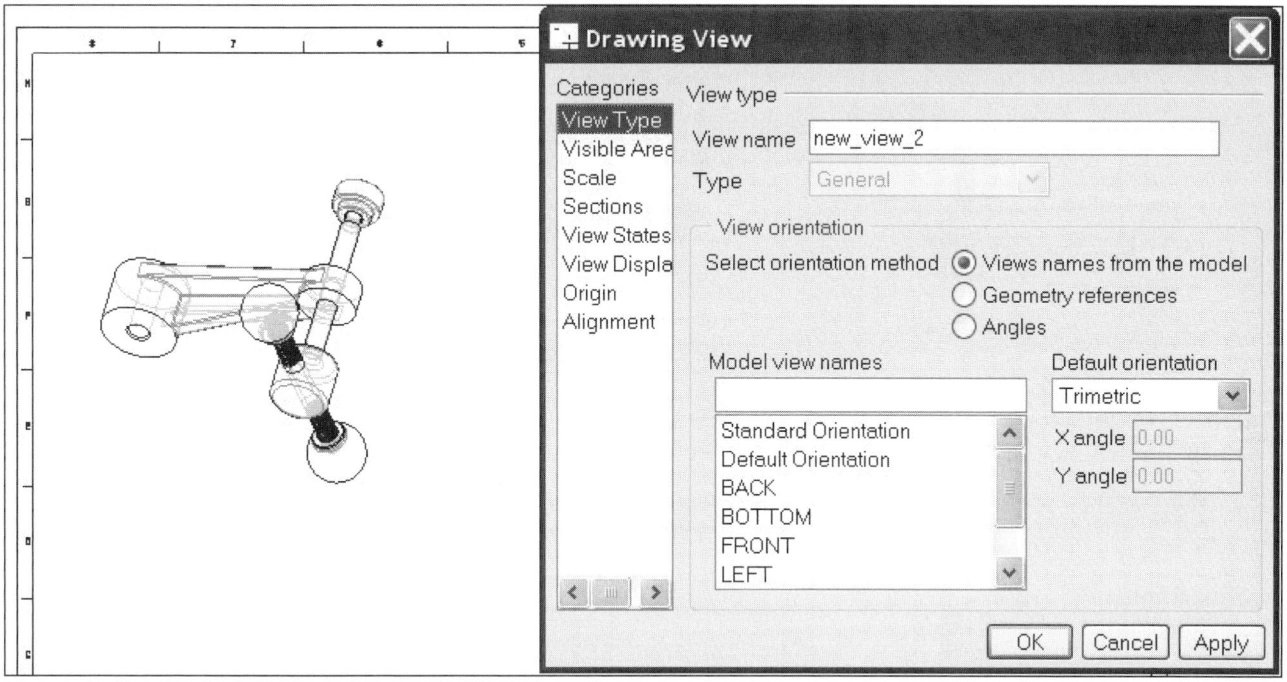

Figure 12.21(a) Select Center Point for First Drawing View

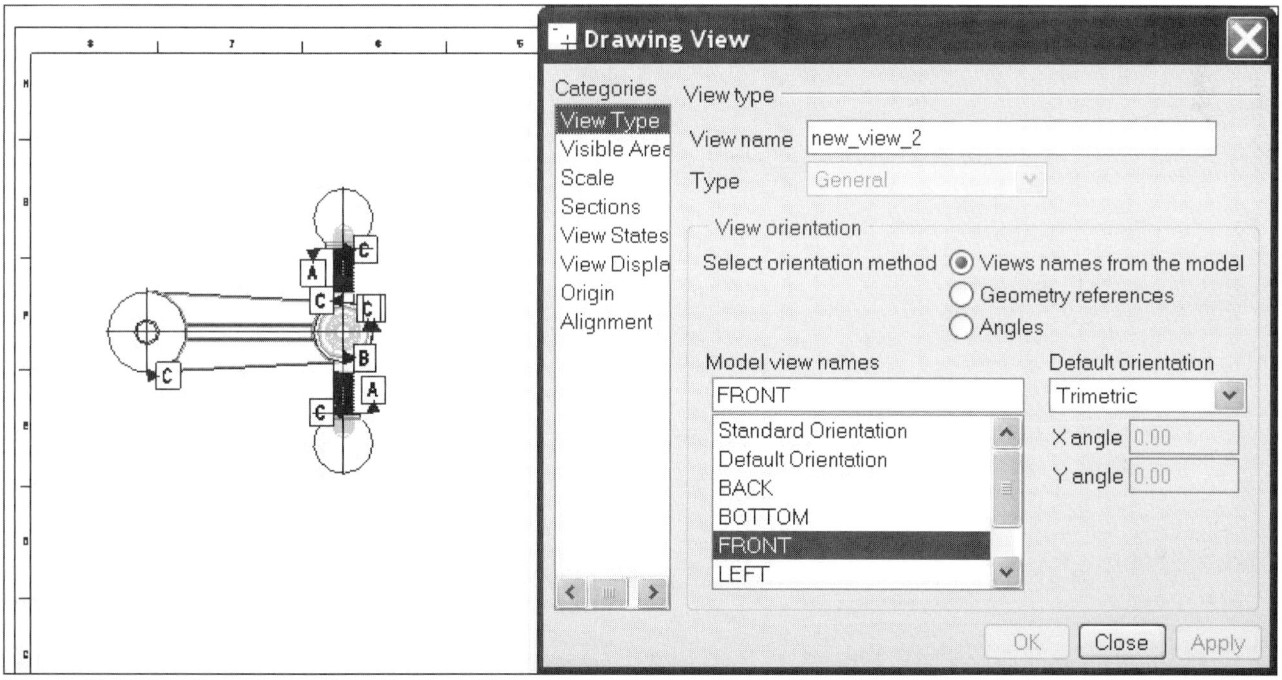

Figure 12.21(b) Views names from the model

For this drawing, the only other view needed to show the subassembly is a front section view.

With the Top view highlighted (selected) ⇒ **RMB** ⇒ **Insert Projection View** [Fig. 12.22(a)] ⇒ Select CENTER POINT for drawing view. [Fig. 12.22(b)] select below the view previously created [Fig. 12.22(c)] ⇒ **Ctrl+S** ⇒ **OK**

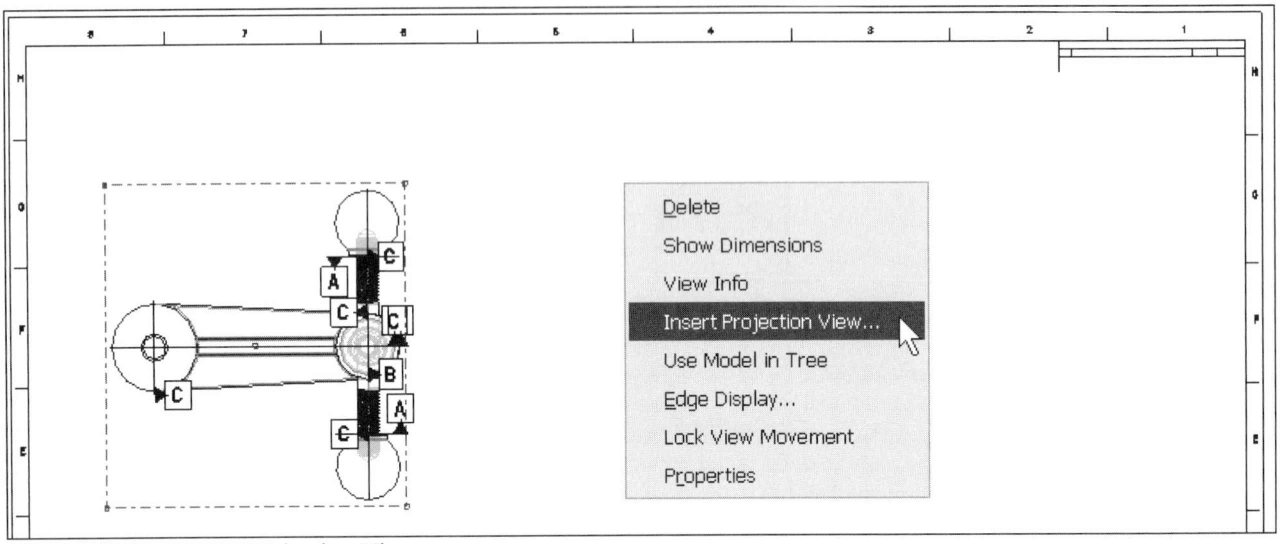

Figure 12.22(a) Insert Projection View

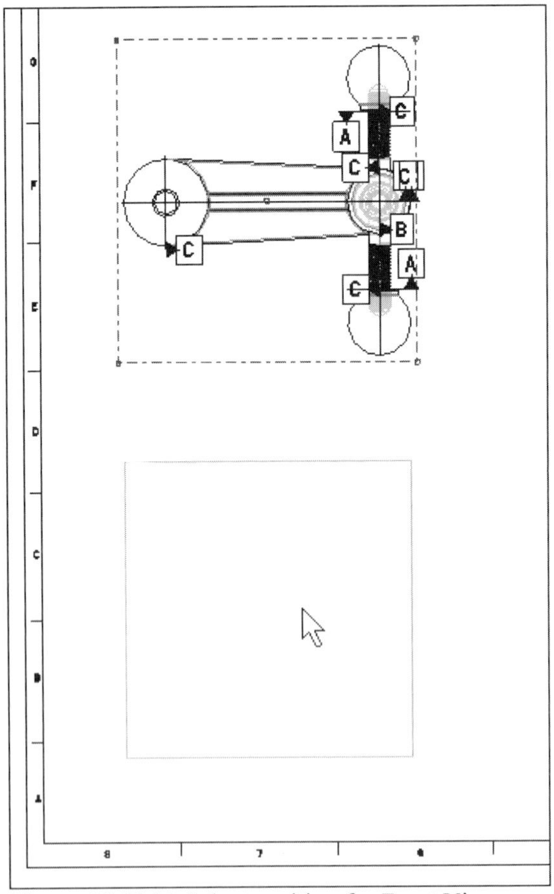

Figure 12.22(b) Select position for Front View

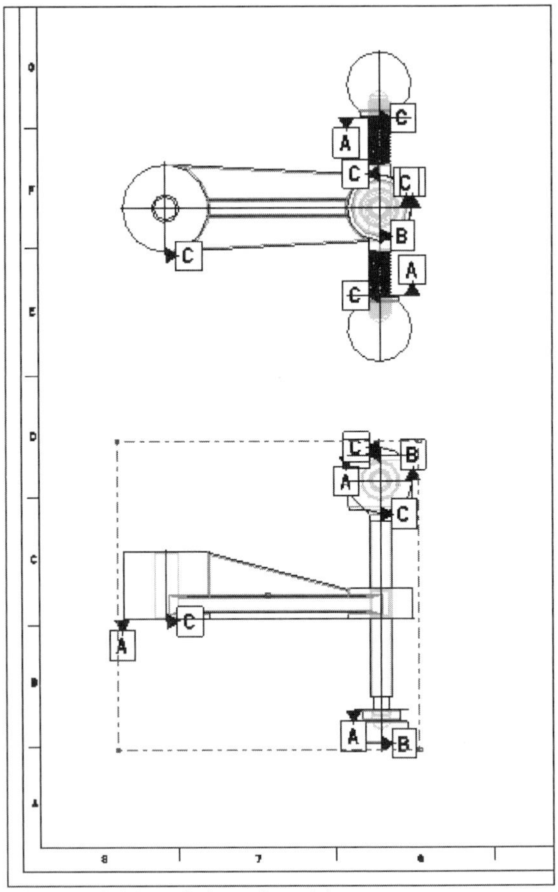

Figure 12.22(c) Selected position

With Front view still selected/highlighted, click: **RMB** ⇒ **Properties** [Fig. 12.23(a)] ⇒ **Sections** ⇒ 2D cross-section ⇒ + [Fig. 12.23(b)] ⇒ ✓ A ⇒ **OK** [Fig. 12.23(c)] ⇒ **LMB** to deselect ⇒ 

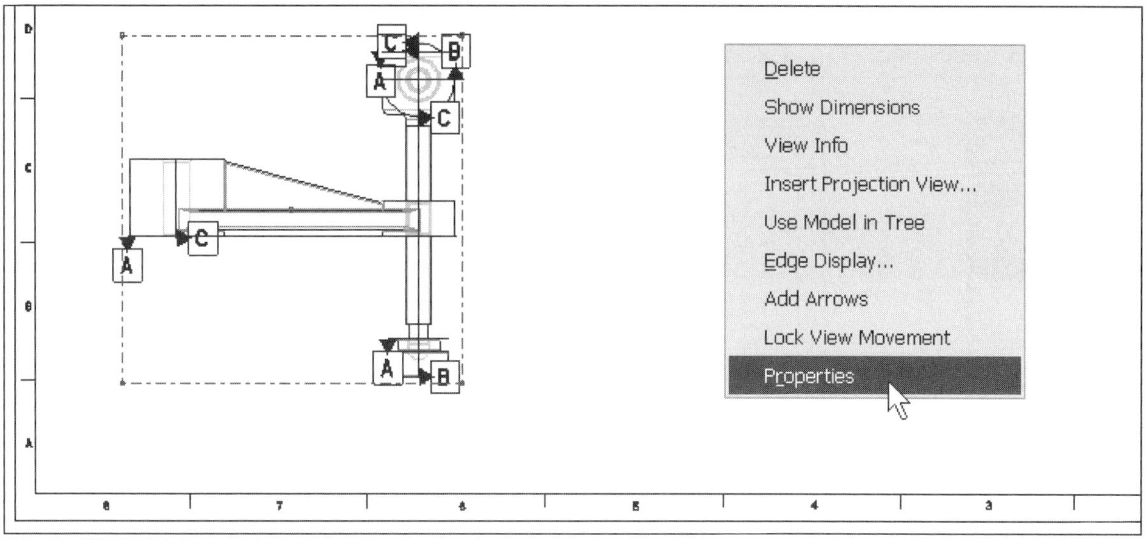

Figure 12.23(a) Properties

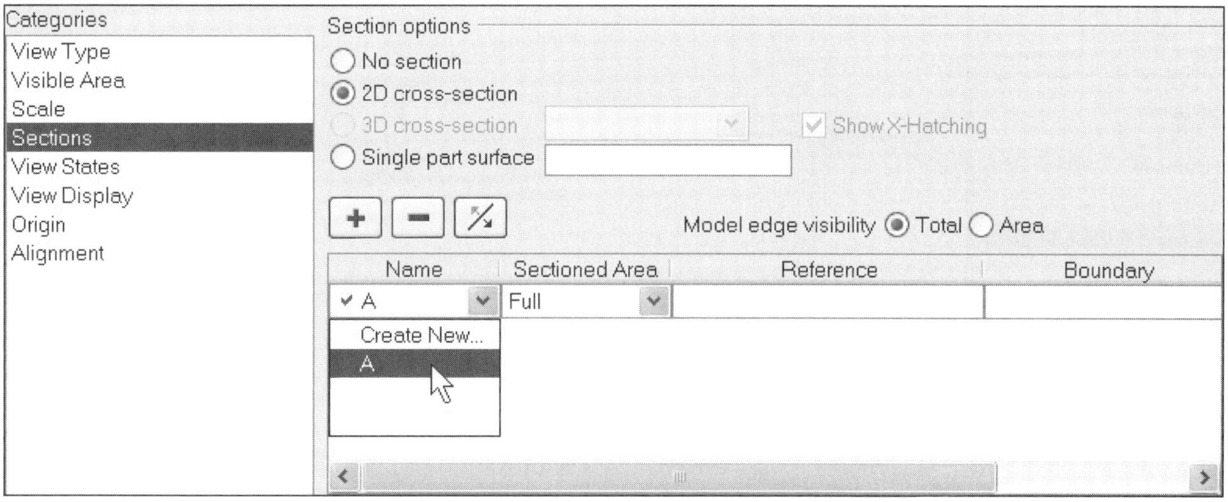

Figure 12.23(b) Drawing View 2D section (must have previously created Section A)

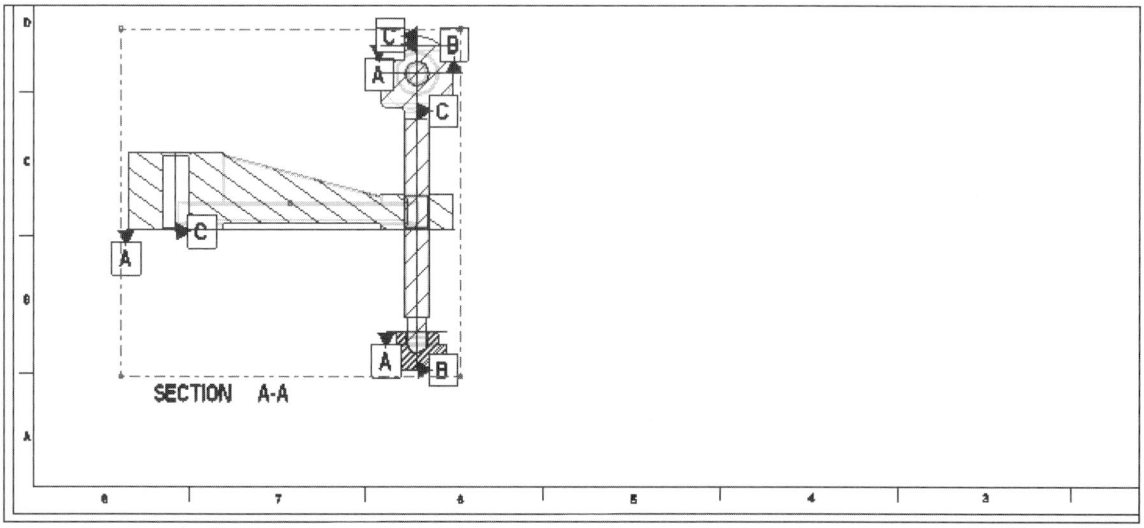

Figure 12.23(c) SECTION A_A

Click: **Open the Show/Erase dialog box** ⇒ Erase ⇒ only **Datum Plane** on ⇒ Erase All [Fig. 12.24(a)] ⇒ **Yes** ⇒ **Close** ⇒ reposition the views as needed [Fig. 12.24(b)] ⇒ ⇒ ⇒ ⇒ ⇒ **OK** ⇒ pick on the Front view to highlight/select ⇒ **RMB** [Fig. 12.25(a)] ⇒ **Add Arrows** ⇒ pick on the Top view ⇒ **LMB** to deselect [Fig. 12.25(b)]

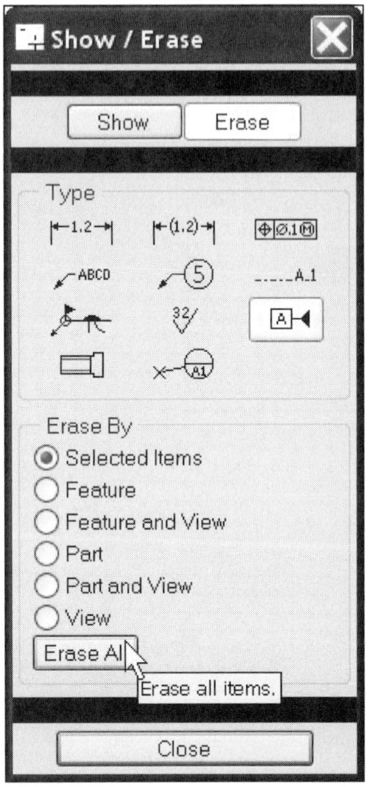

Figure 12.24(b) Erase Datum Plane

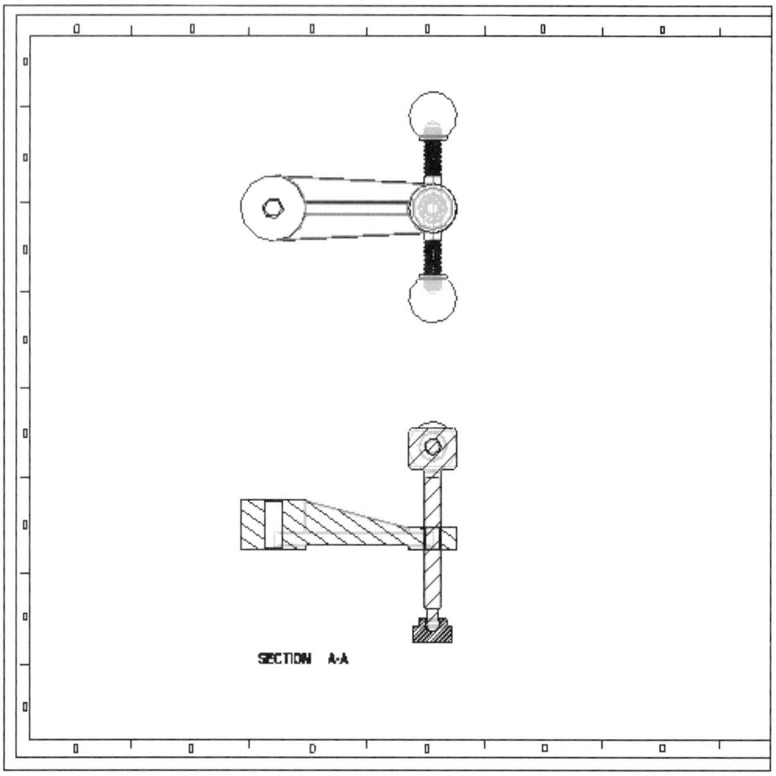

Figure 12.24(b) Reposition Views

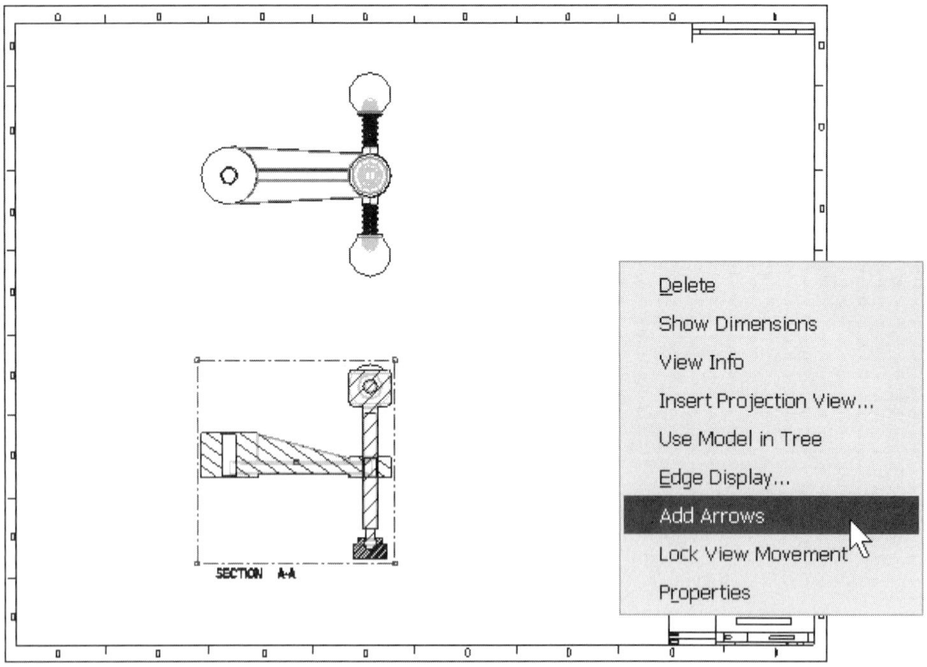

Figure 12.25(a) Add Arrows

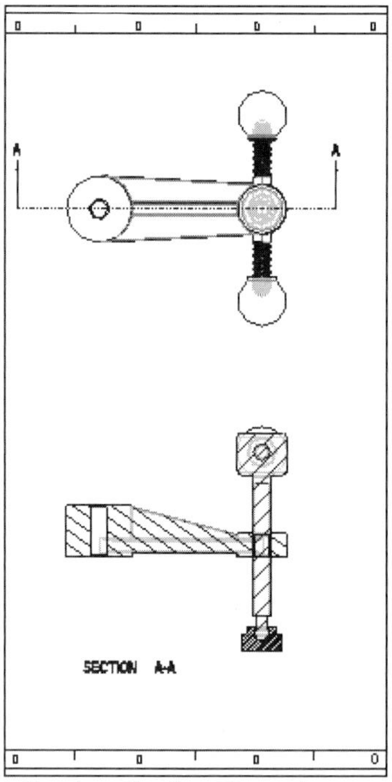

Figure 12.25(b) Section Cutting Plane Arrows

Pro/E provides tools to alter the display of the section views to comply with industry ASME standard practices. Most companies require that the crosshatching on parts in section views of assemblies be "clocked" such that parts that meet do not use the same section lining (crosshatching) spacing and angle. This makes the separation between parts more distinct. First, modify the visibility of the views to remove hidden lines and make the tangent edges dimmed. Next, show all centerlines and clip as needed.

Press and hold down the **Ctrl** key and pick on both views ⇒ click **RMB** with cursor outside of the view outlines ⇒ **Properties** ⇒ Display style [▾] ⇒ [No Hidden] [Fig. 12.26(a)] ⇒ Tangent edges display style [▾] ⇒ [Dimmed] [Fig. 12.26(b)] ⇒ **Apply** ⇒ **Close** ⇒ **LMB** in graphics window to deselect

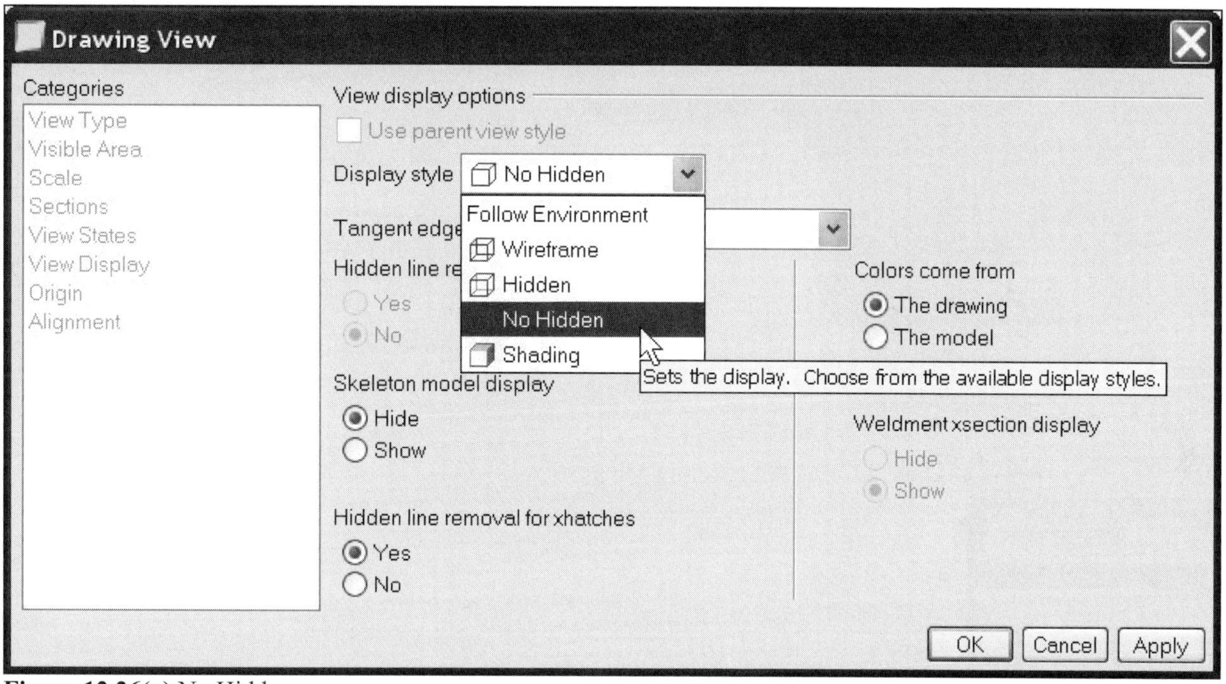

Figure 12.26(a) No Hidden

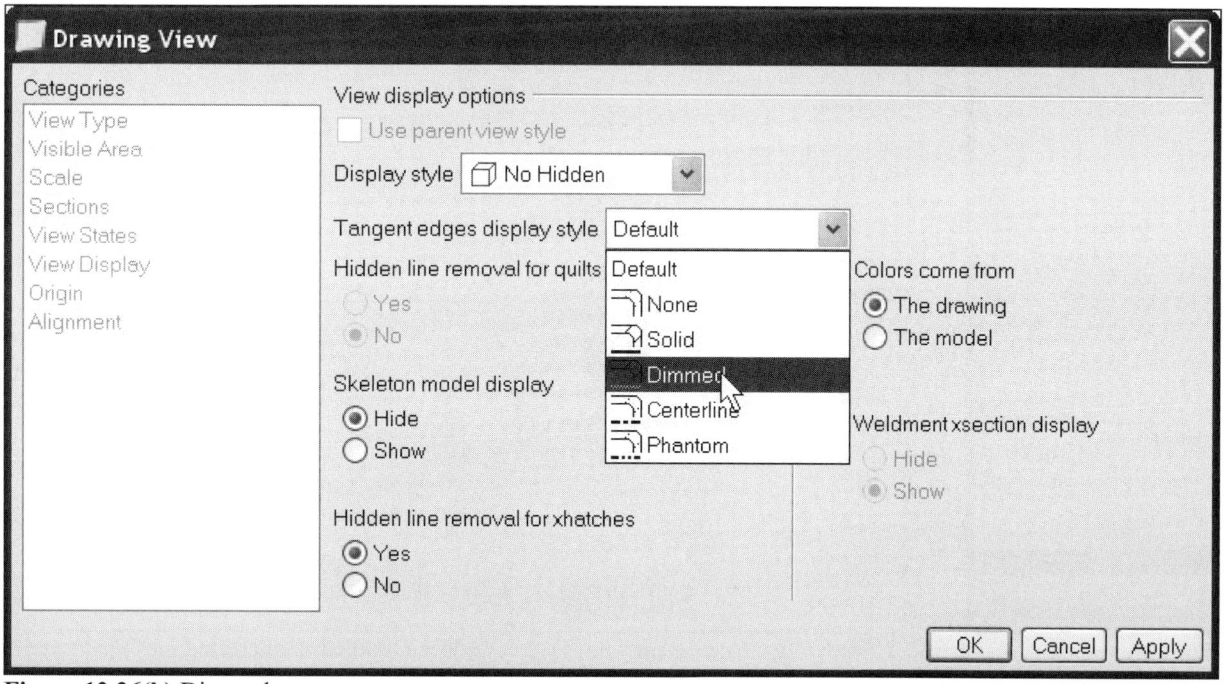

Figure 12.26(b) Dimmed

Click: Open the Show/Erase dialog box ⇒ Show ⇒ only Axis on ⇒ **Show All** [Fig. 12.26(c)] ⇒ **Yes** ⇒ **Accept All** ⇒ **Close** ⇒ **LMB** to deselect [Fig. 12.26(d)] ⇒ ⇒ ⇒ ⇒ ⇒ **MMB** ⇒ **Sketch** ⇒ **Sketcher Preferences** ⇒ Grid intersection off ⇒ **Close**

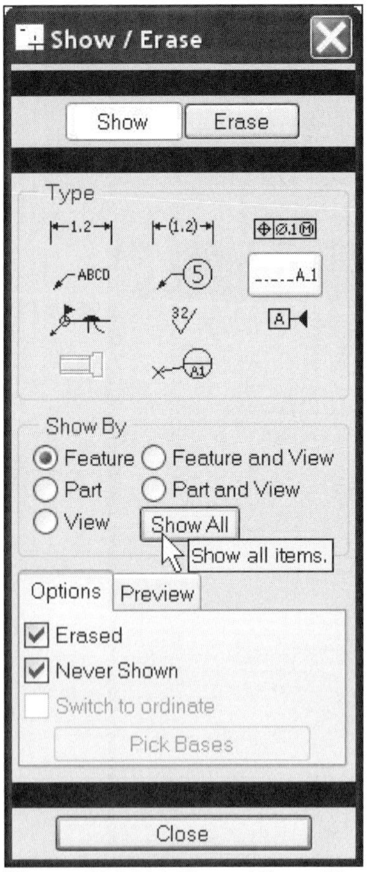

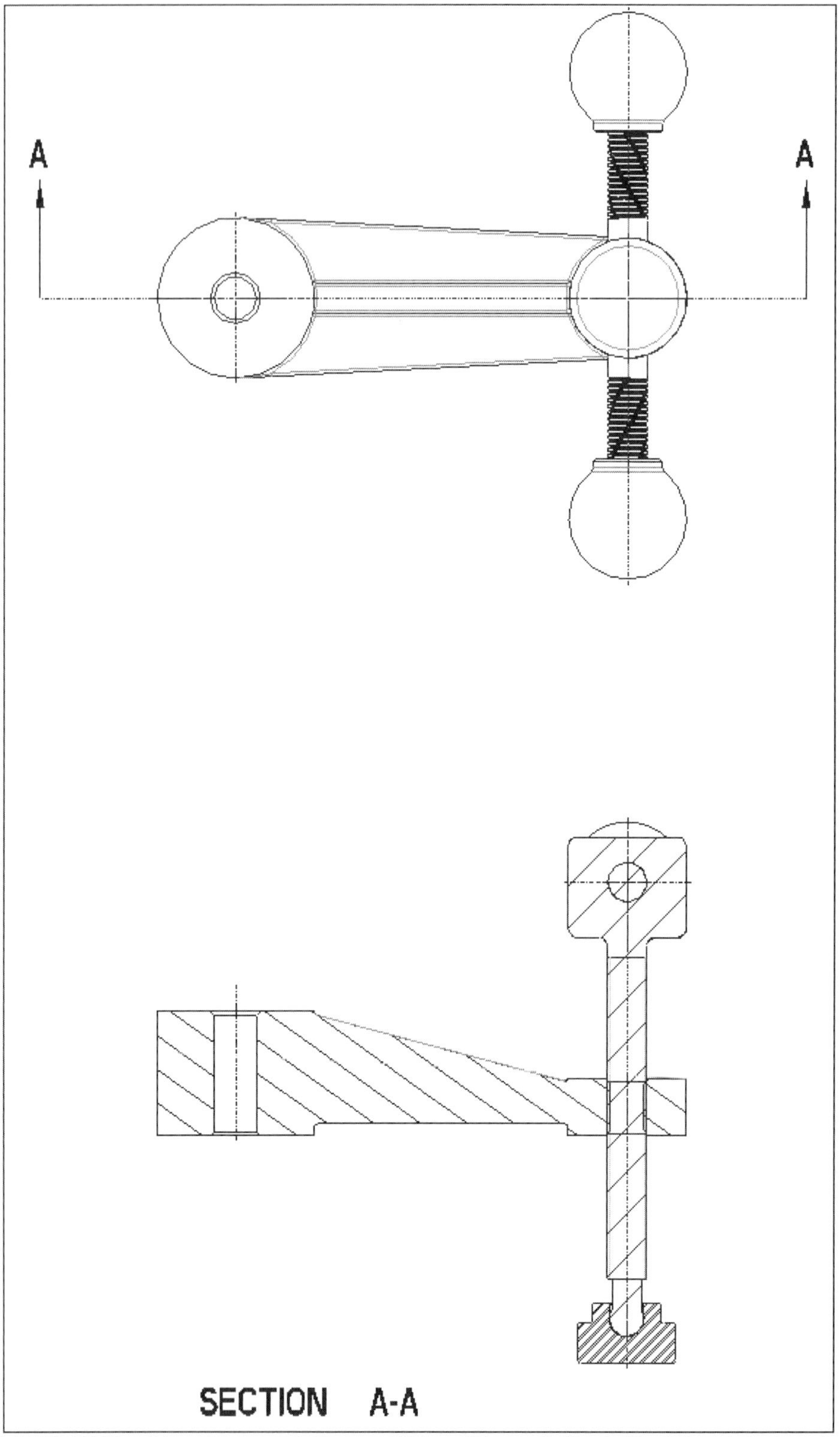

Figure 12.26(c) Show All **Figure 12.26(d)** Hidden Lines Removed, Tangent Edges Dimmed, Centerline Lines Displayed

Double-click on the crosshatching in the Front view (CARRLANE-12-13500_STUD is now active) ⇒ **Fill** [Figs. 12.26(e-f)] ⇒ click **Next** until CLAMP_FOOT is active ⇒ **Hatch** ⇒ **Angle** [Fig. 12.26(g)] ⇒ **135** [Fig. 12.26(h)]

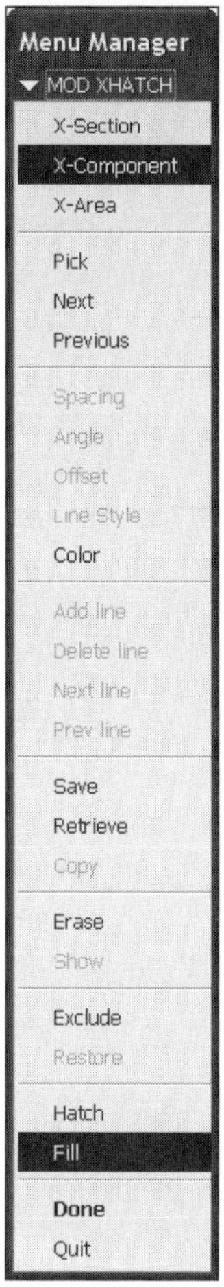

Figure 12.26(e) Modify Xhatch

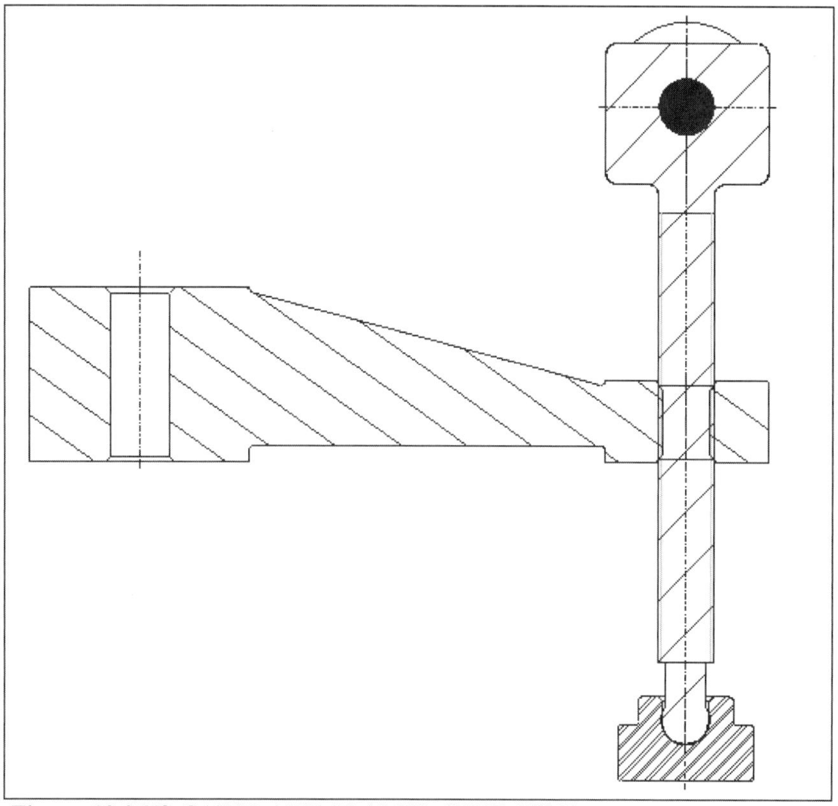

Figure 12.26(f) CARRLANE-12-13500_STUD Fill Xsec

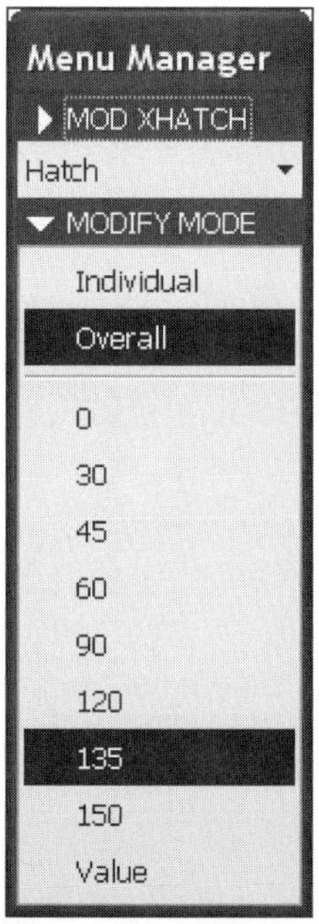

Figure 12.26(g) Angle **135**

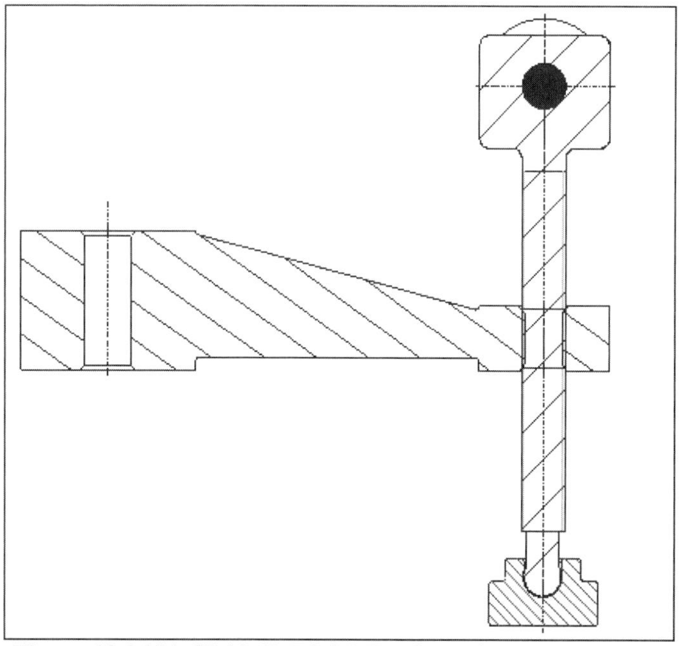

Figure 12.26(h) CLAMP_FOOT Hatch Angle **135**

Click: **Next** ⇒ **Next** CLAMP_ARM is active ⇒ **Hatch** ⇒ **Spacing** ⇒ **Half** ⇒ **Angle** ⇒ **120** [Fig. 12.26(i)] ⇒ **Previous** CLAMP_SWIVEL is active ⇒ **Spacing** ⇒ **Half** ⇒ **Done** ⇒ ⇒ ⇒ [Fig. 12.26(j)] ⇒ ⇒ **Enter**

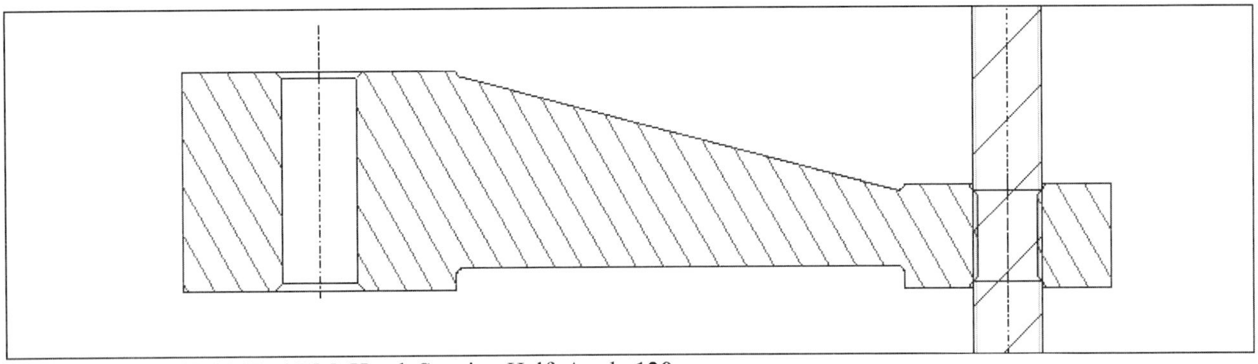

Figure 12.26(i) CLAMP_ARM, Hatch Spacing Half, Angle **120**

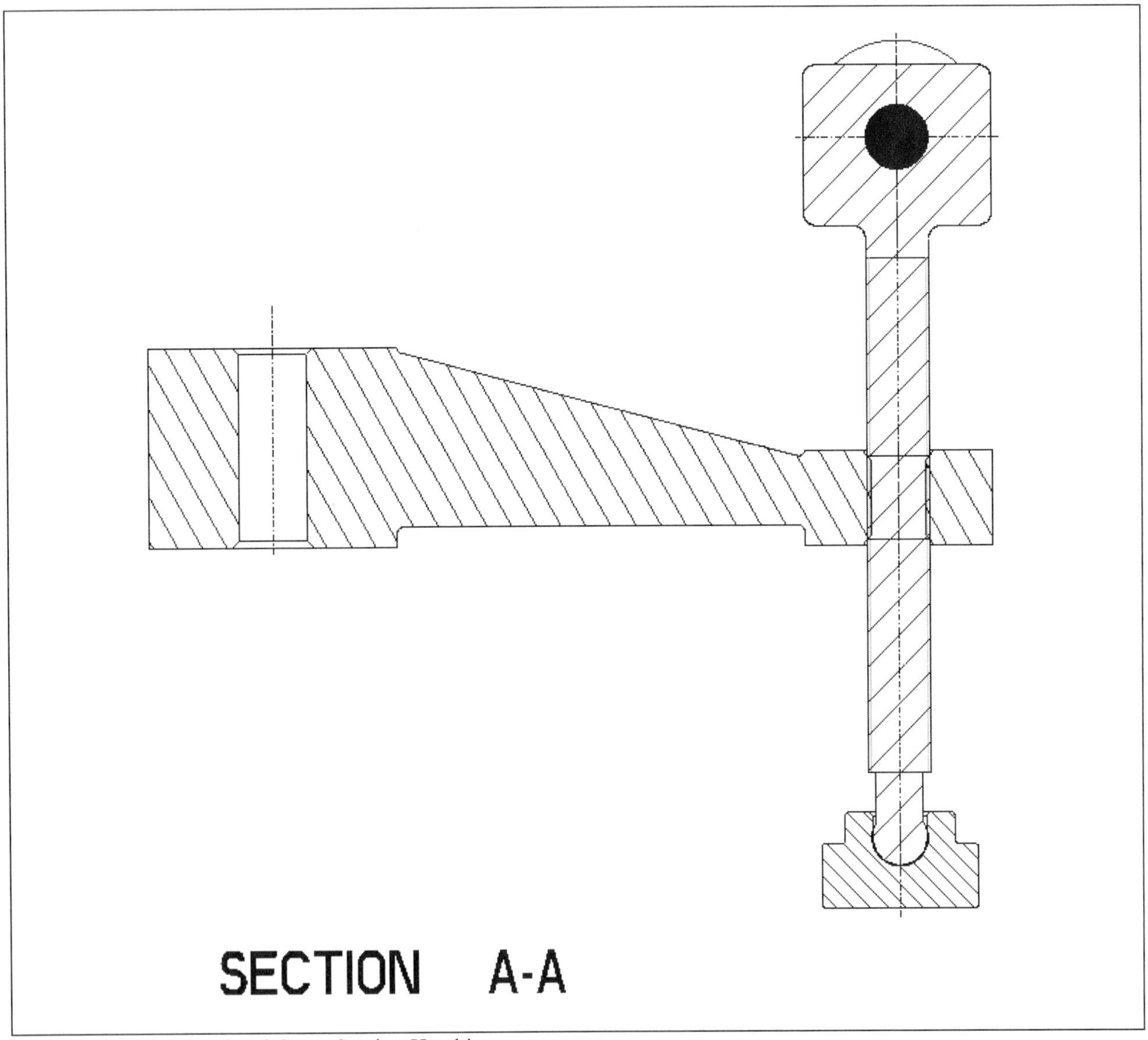

Figure 12.26(j) Completed Cross Section Hatching

Zoom in on the BOM Title Block [Fig. 12.27(a)]

5			STEEL	1
4	SW101-9STD	.500-13 X 5.00 DOUBLE END STUD	PURCHASED	1
3	SW101-8FT	CLAMP FOOT	STEEL	1
2	SW101-7BA	CLAMP BALL	NYLON	2
1	SW101-6SW	CLAMP SWIVEL	STEEL 1066	1
ITEM	PT NUM	DESCRIPTION	MATERIAL	QTY

Figure 12.27(a) BOM without CLAMP_ARM_DESIGN Description (your table may be different)

Since parameters were created in the **Clamp_Arm** component (the *generic* of a family table) not the **Clamp_Arm_Design** part (*instance*), the BOM does not reflect the correct values. Remember that the *generic* is the base model and is typically not a member of an assembly. For demonstration purposes, the next set of commands will replace the Clamp_Arm_Design part with the Clamp_Arm part. *(An alternative to replacing the component would have been to add the parameters to the family table.)*

Click: **File** ⇒ **Open** ⇒ **clamp_subassembly.asm** ⇒ **Open** ⇒ ⇒ off [Fig. 12.27(b)] ⇒ **Edit** from menu bar ⇒ **Replace** [Figs. 12.27(c-d)]

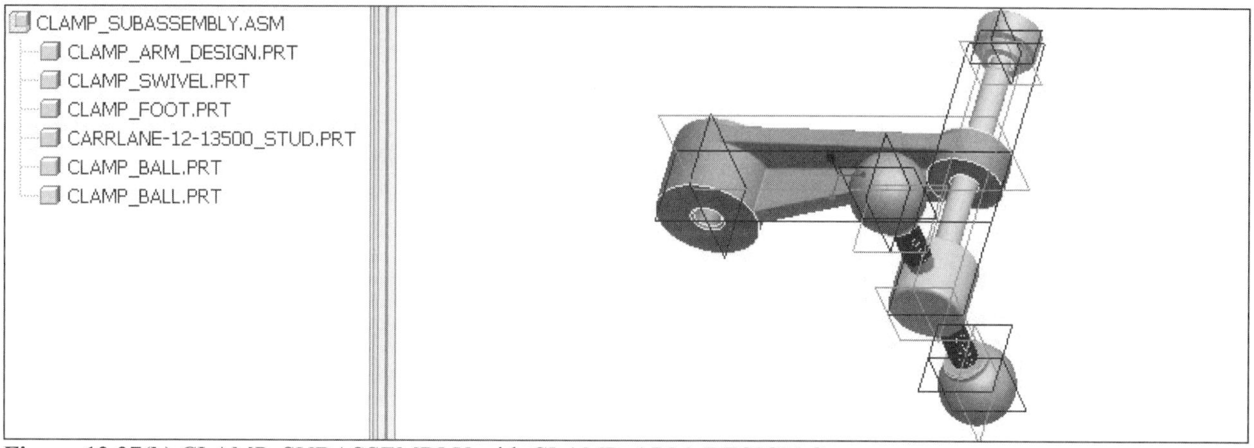

Figure 12.27(b) CLAMP_SUBASSEMBLY with CLAMP_ARM_DESIGN Component

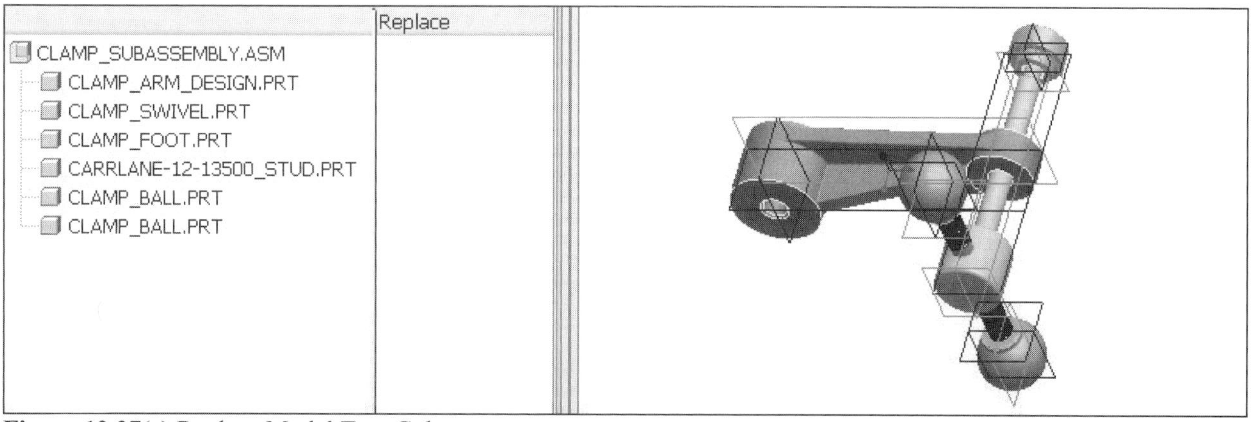

Figure 12.27(c) Replace Model Tree Column

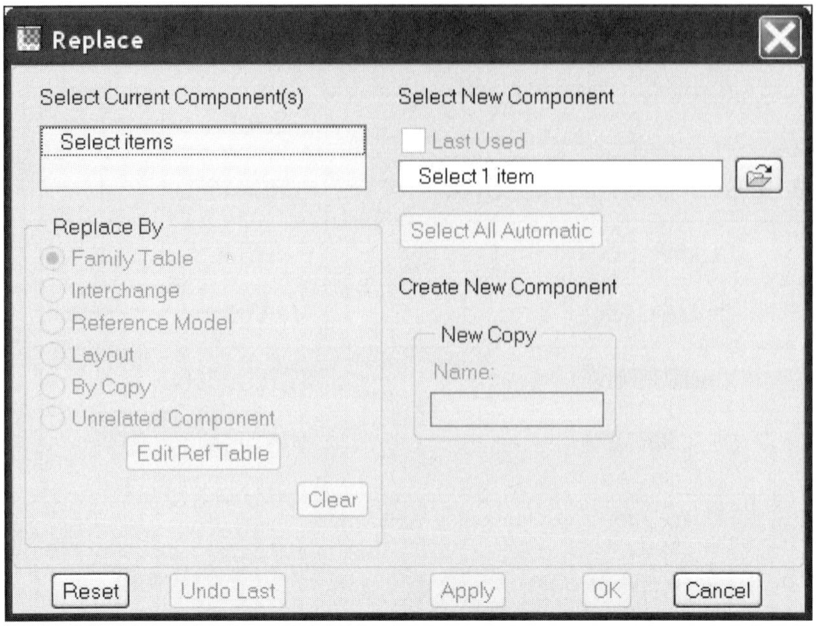

Figure 12.27(d) Replace Dialog Box

Click: Select Current Component(s): pick **CLAMP_ARM_DESIGN.PRT** [Fig. 12.27(e)] ⇒ Select New Component [Fig. 12.27(f)] ⇒ **CLAMP_ARM.PRT** ⇒ **OK** ⇒ **Apply** ⇒**OK** ⇒ **Ctrl+S** ⇒ **OK** [Fig. 12.27(g)] ⇒ **Window** ⇒ **Close**

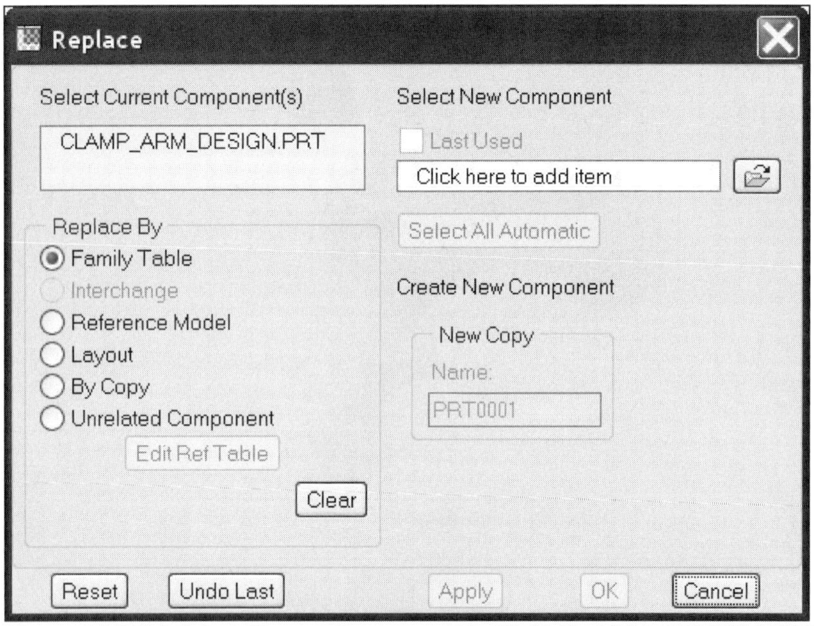

Figure 12.27(e) Select Current Component(s): CLAMP_ARM_DESIGN.PRT

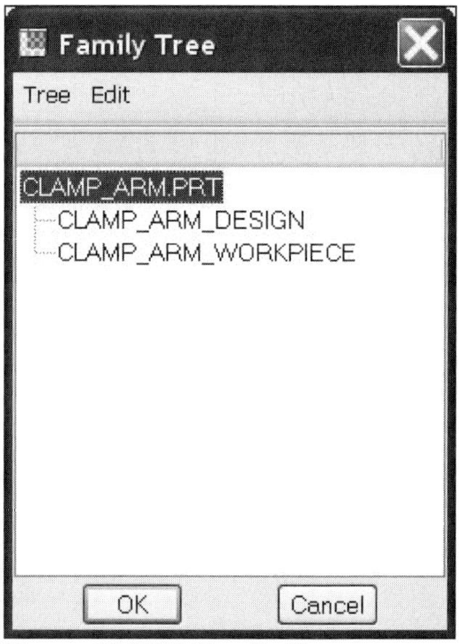

Figure 12.27(f) Family Tree

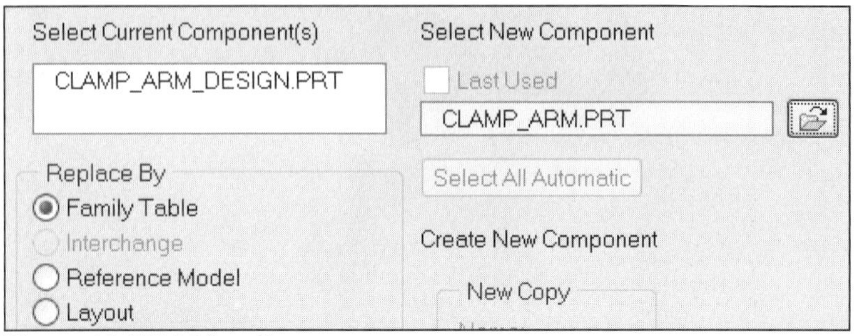

Figure 12.27(g) New Component

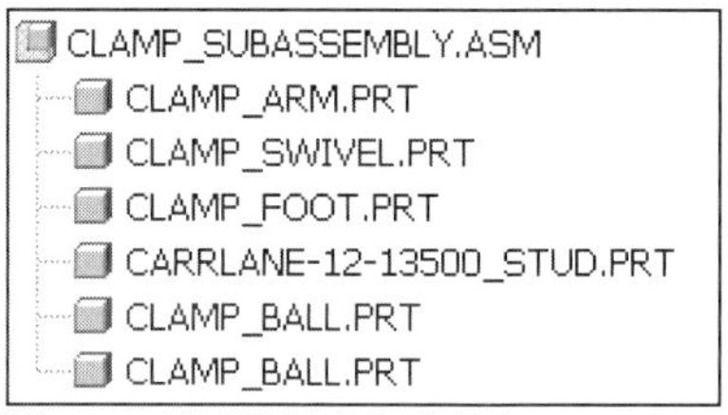

Figure 12.27(h) Model Tree

To complete the drawing, the balloons must be displayed for each component. Balloons are displayed in the top view as the default, because it was the first view that was created.

Click: **Table** ⇒ **BOM Balloons** ⇒ **Set Region** ⇒ **Simple** ⇒ pick in the BOM field [Fig. 12.28(a)] ⇒ **Create Balloon** ⇒ **Show All** [Fig. 12.28(b)] ⇒ **Done** ⇒ press and hold the **Ctrl** key and pick on the balloons for the Clamp_Arm (1), Clamp_Swivel (2) and Clamp_Foot (4) ⇒ **RMB** ⇒ **Move Item to View** [Fig. 12.28(c)] ⇒ pick in the Front view [Fig. 12.28(d)] ⇒ pick on and reposition each balloon as needed [Fig. 12.28(e)] ⇒ **LMB** to deselect ⇒ **Update** ⇒ ⇒ ⇒ ⇒ **MMB**

5	SW101-9STD	.500-13 X 5.00 DOUBLE END STUD	PURCHASED	1
4	SW101-8FT	CLAMP FOOT	STEEL	1
3	SW101-7BA	RepeatRegion CLAMP BALL	NYLON	2
2	SW101-6SW	CLAMP SWIVEL	STEEL_1066	1
1	SW101-5AR	CLAMP ARM	STEEL	1
ITEM	PT NUM	DESCRIPTION	MATERIAL	QTY

Figure 12.28(a) Set Region

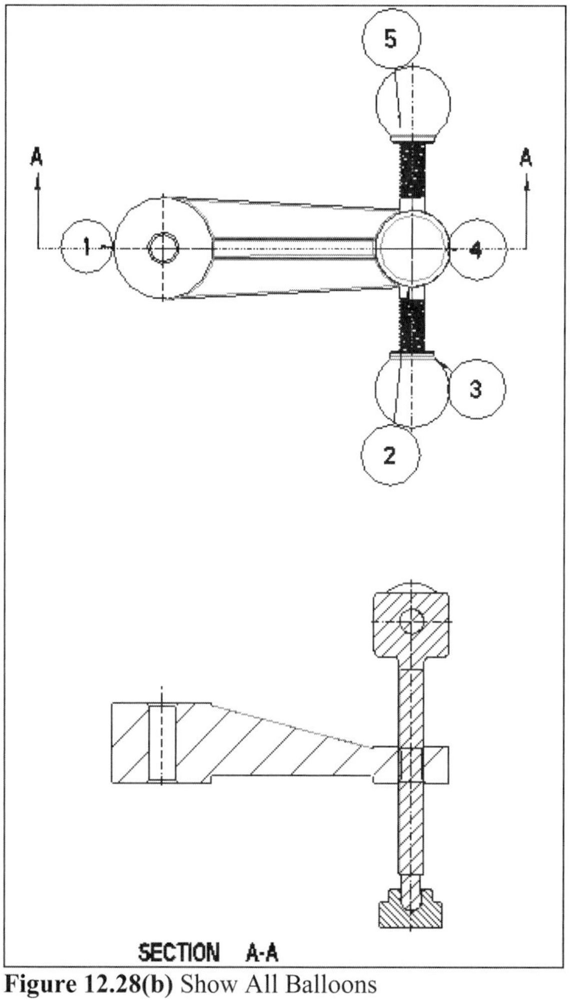

Figure 12.28(b) Show All Balloons

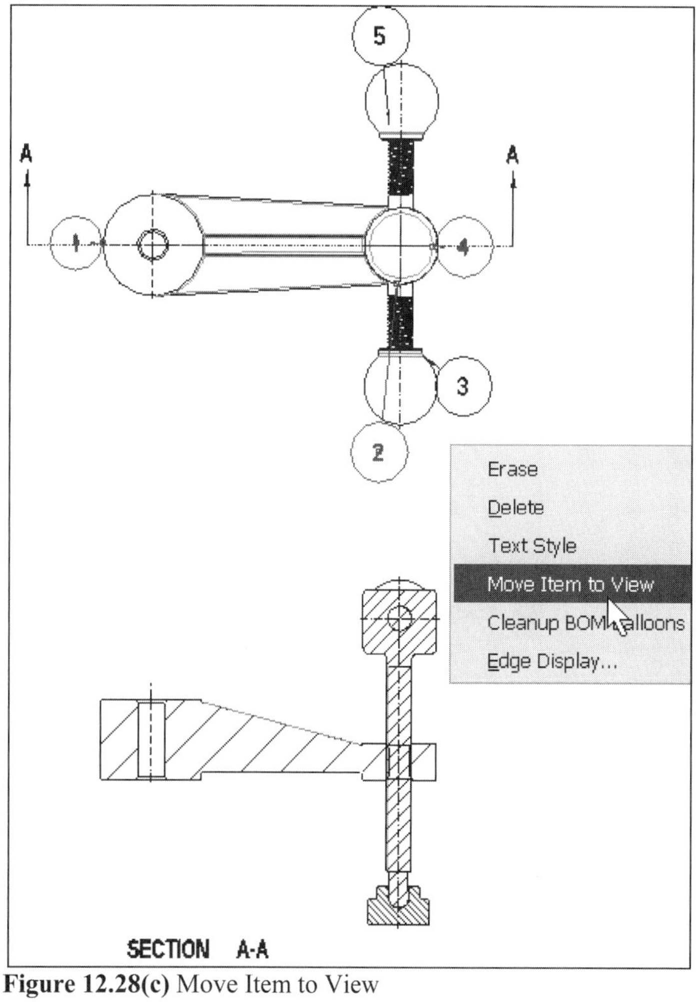

Figure 12.28(c) Move Item to View

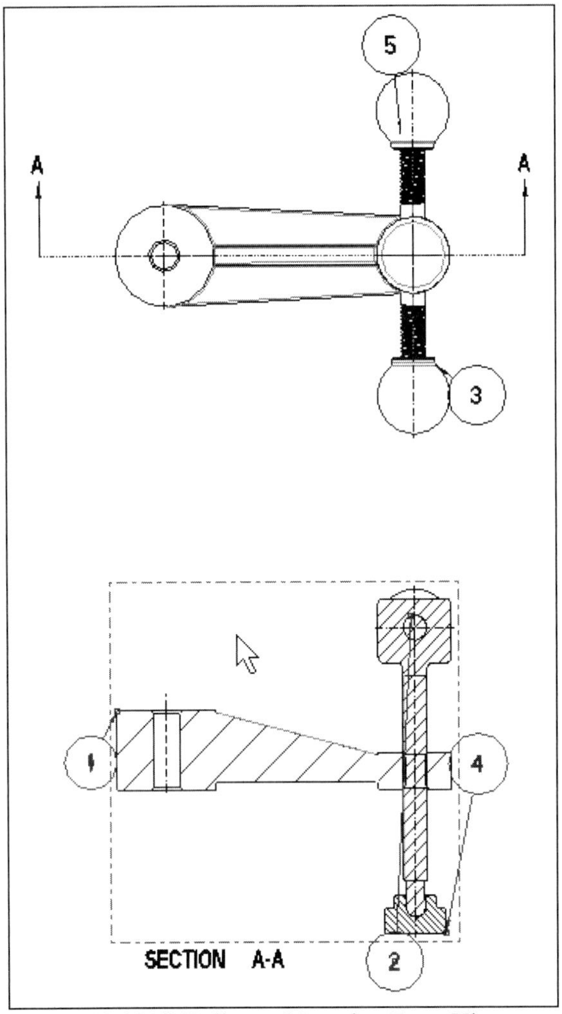

Figure 12.28(d) Balloons Moved to Front View

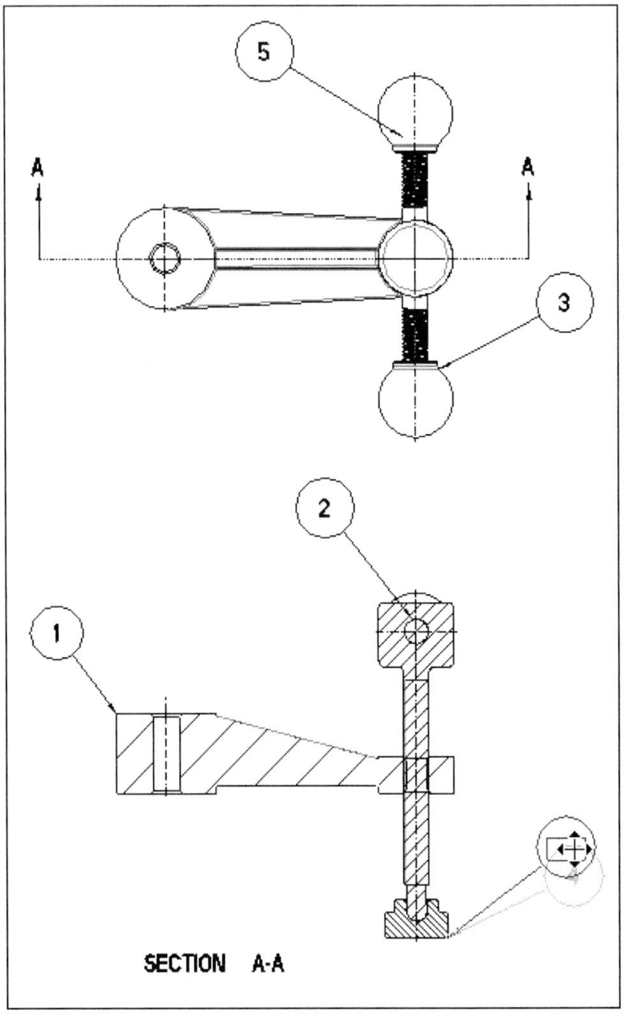

Figure 12.28(e) Reposition Balloons

Pick on balloon **5** (CARRLANE-12-13500_STUD) ⇒ **RMB** [Fig. 12.28(f)] ⇒ **Edit Attachment** ⇒ **On Entity** ⇒ pick on the edge of the CARRLANE-12-13500_STUD [Fig. 12.28(g)] ⇒ **Done/Return** ⇒ reposition balloon ⇒ **LMB** ⇒ reposition and edit the attachment of all ballons as necessary

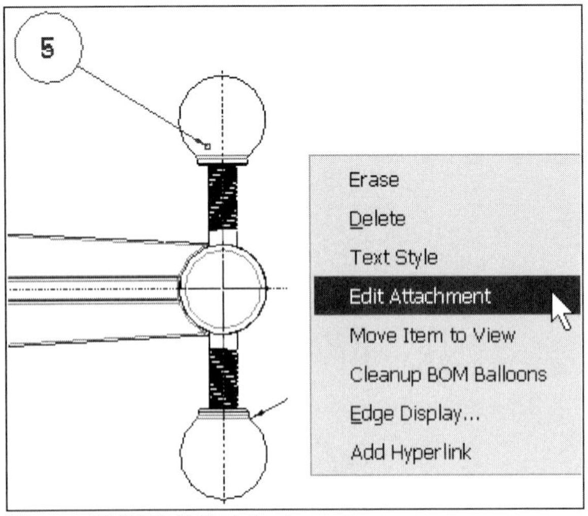

Figure 12.28(f) Edit Attachment

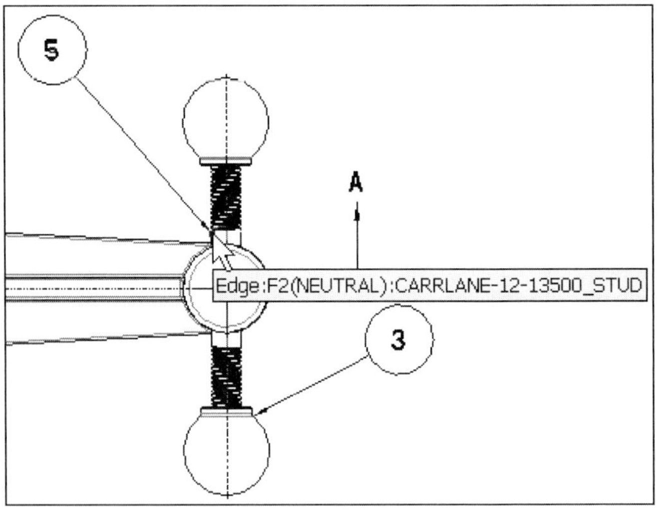

Figure 12.28(g) Pick the Edge of the STUD

*[If your balloon for item **3** (CLAMP_BALL) does not have a leader line, pick on balloon **3** ⇒ **RMB** ⇒ **Move Item to View** ⇒ pick in the Front view ⇒ **RMB** ⇒ **Erase** ⇒ **LMB** ⇒ **Table** ⇒ **BOM Balloons** ⇒ **Create Balloon** ⇒ **By Comp** ⇒ pick on the **CLAMP_BALL.PRT** [Fig. 12.28(g)] ⇒ **OK** ⇒ **Done** ⇒ pick on balloon **3** ⇒ **RMB** ⇒ **Move Item to View** ⇒ pick in the Top view ⇒ **RMB** ⇒ **Edit Attachment** ⇒ pick on the edge of the CLAMP_BALL ⇒ **Done/Return** ⇒ reposition balloon ⇒ **LMB**]* ⇒ [icon] ⇒ [icon] ⇒ **OK** ⇒ **File** ⇒ **Delete** ⇒ **Old Versions** ⇒ [icon] [Figs. 12.28(h-i)] ⇒ **Window** ⇒ **Close**

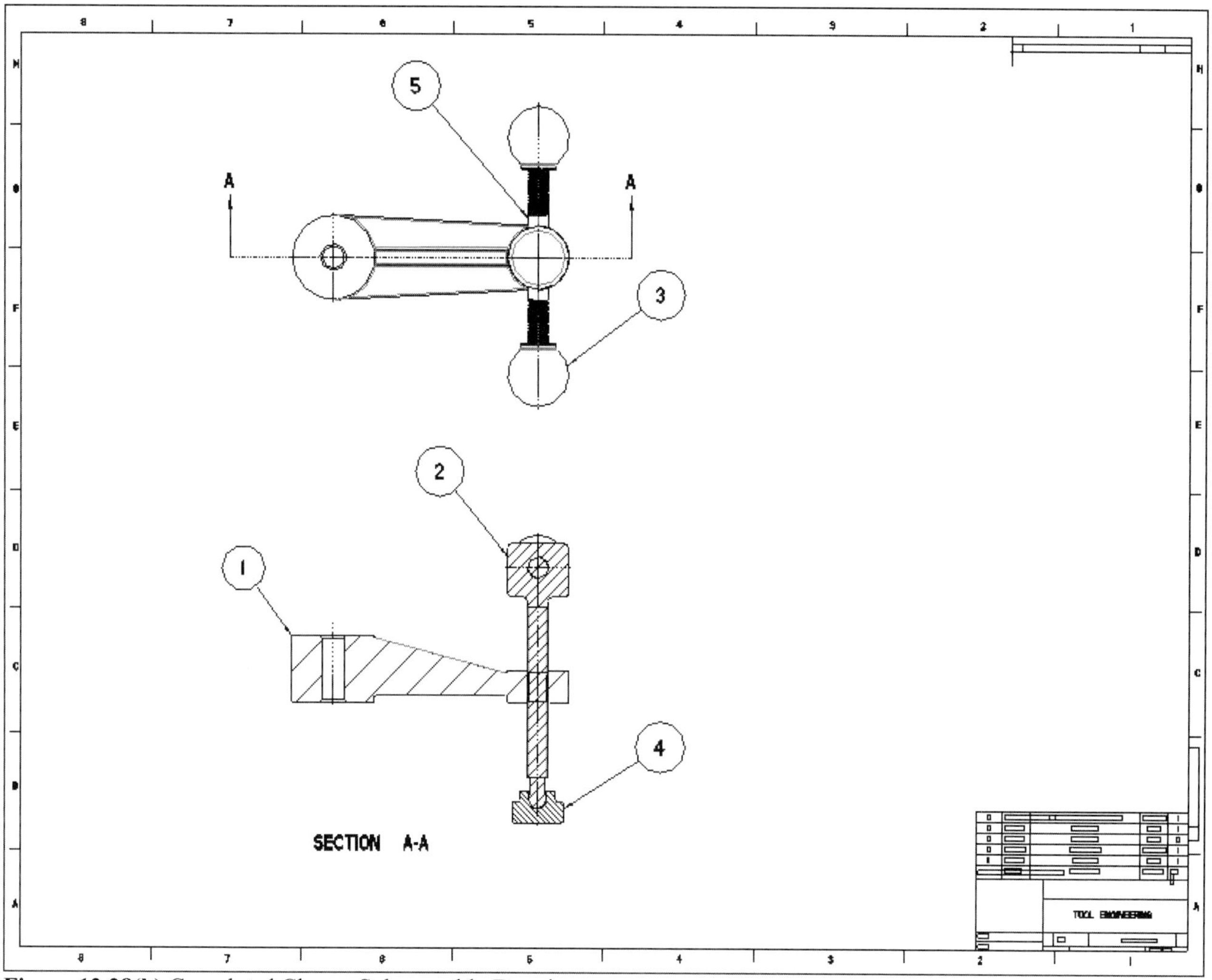

Figure 12.28(h) Completed Clamp_Subassembly Drawing

5	SW101-9STD	.500-13 X 5.00 DOUBLE END STUD	PURCHASED	1
4	SW101-8FT	CLAMP FOOT	STEEL	1
3	SW101-7BA	CLAMP BALL	NYLON	2
2	SW101-6SW	CLAMP SWIVEL	STEEL 1066	1
1	SW101-5AR	CLAMP ARM	STEEL	1
ITEM	PT NUM	DESCRIPTION	MATERIAL	QTY

Figure 12.28(i) Clamp_Subassembly BOM

Create the drawing for the Swing Clamp Assembly. This assembly is composed of the subassembly (in the previous drawing), the plate, the short stud, and the nut. The drawing will use the same format created for the subassembly. Formats are read-only files that can be used as many times as needed.

Click: open the Navigator shade ⇒ ⇒ **clamp_plate.prt** ⇒ **Open** ⇒ ⇒ click on datum A in the Model Tree or model ⇒ **RMB** ⇒ **Properties** [Fig. 12.29(a)] ⇒ [Fig. 12.29(b)] ⇒ **OK** ⇒ repeat for datum B and C ⇒ **Hidden line** [Fig. 12.29(c)] ⇒ **LMB** to deselect ⇒ ⇒ **File** ⇒ **Save** ⇒ **OK** ⇒ **File** ⇒ **Close Window** ⇒ **File** ⇒ **Erase** ⇒ **Not Displayed** ⇒ **OK**

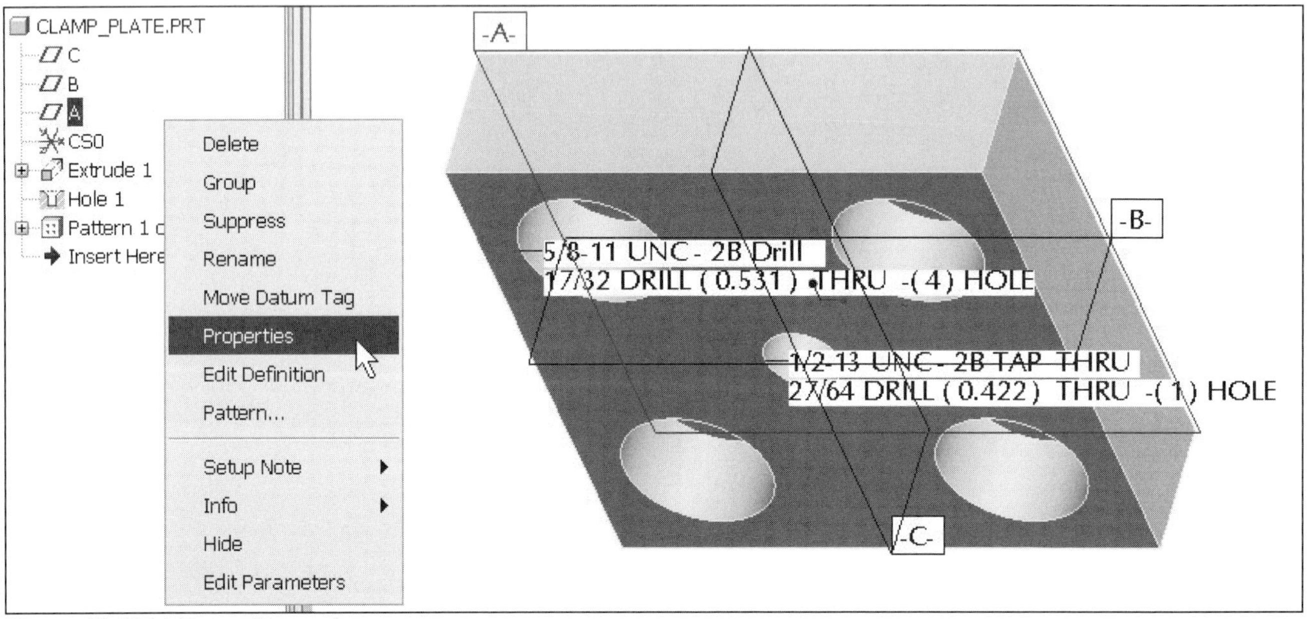

Figure 12.29(a) Datum Properties

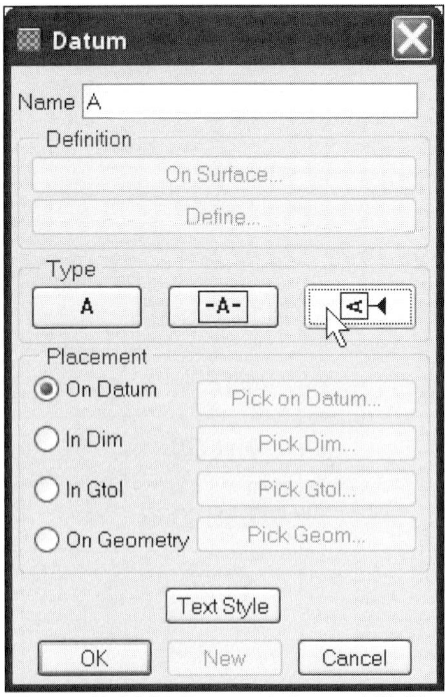

Figure 12.29(b) Datum A Set as ASME

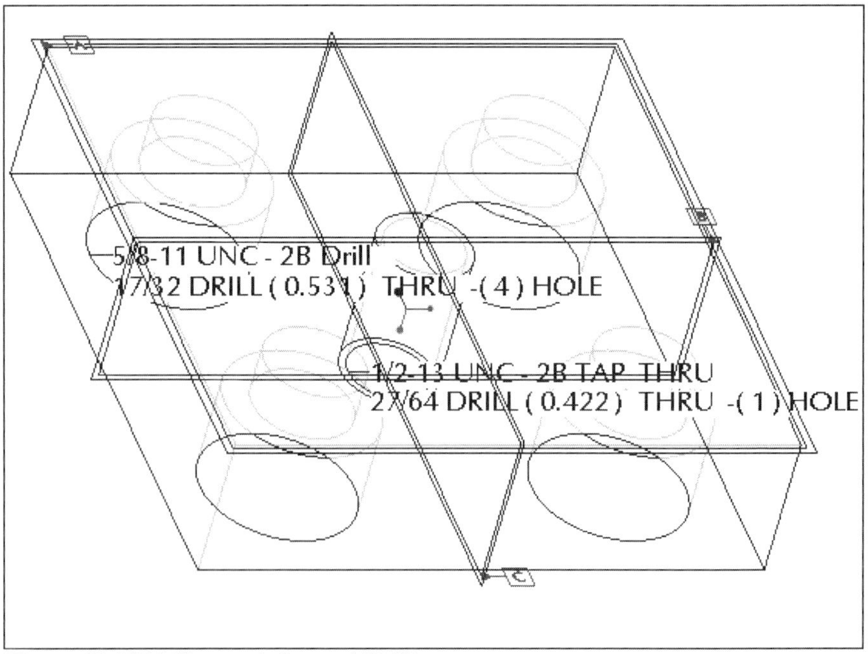

Figure 12.29(c) Datum A Set as ASME

Click: **Create a new object** ⇒ Drawing ⇒ Name **CLAMP_ASSEMBLY** ⇒ Use default template ⇒ **OK** ⇒ Default Model **Browse** ⇒ **clamp_assembly.asm** [Fig. 12.30(a)] ⇒ **Open** ⇒ Empty with format ⇒ Format **Browse** ⇒ **Working Directory** ⇒ select **asm_format_e.frm** [Fig. 12.30(b)] ⇒ **Open** [Fig. 12.30(c)] ⇒ **OK**

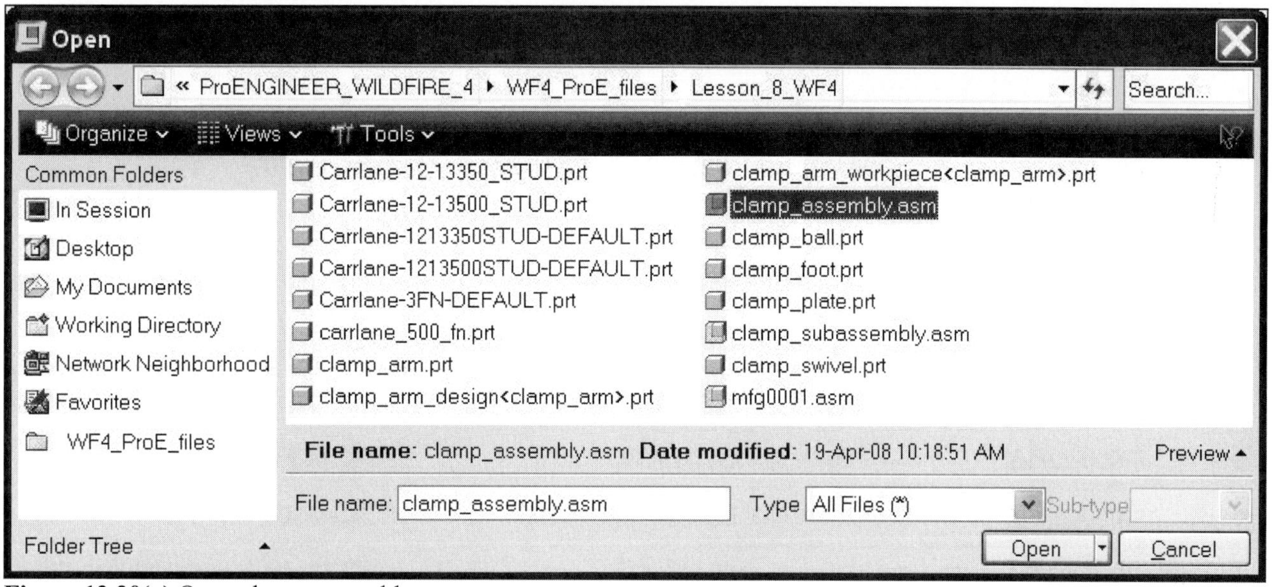

Figure 12.30(a) Open clamp_assembly.asm

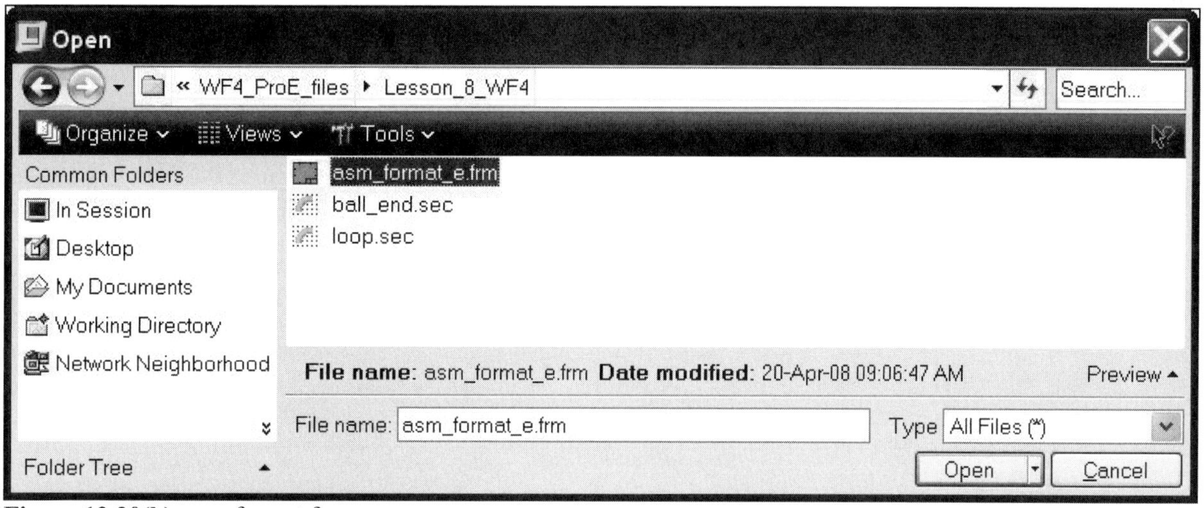

Figure 12.30(b) asm_format.frm

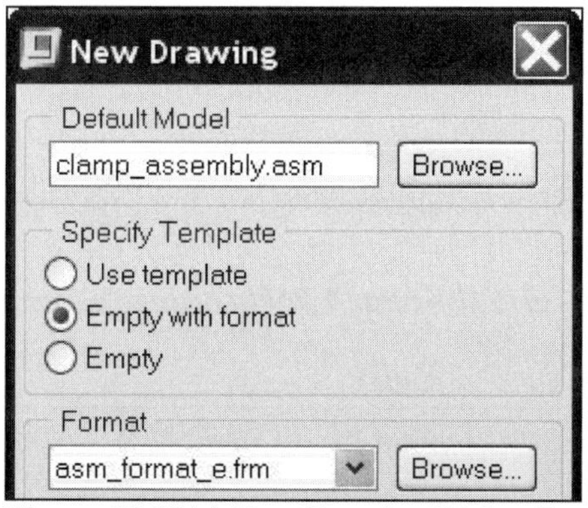

Figure 12.30(c) New Drawing Dialog Box

Click: Zoom In on the title block [Fig. 12.31(a)] ⇒ ⇒ ⇒ **RMB** ⇒ **Properties** ⇒ **Drawing Options** ⇒ **Open a configuration file** ⇒ **CLAMP_ASM.dtl** ⇒ **Open** ⇒ **Apply** ⇒ **Close** ⇒ **MMB**

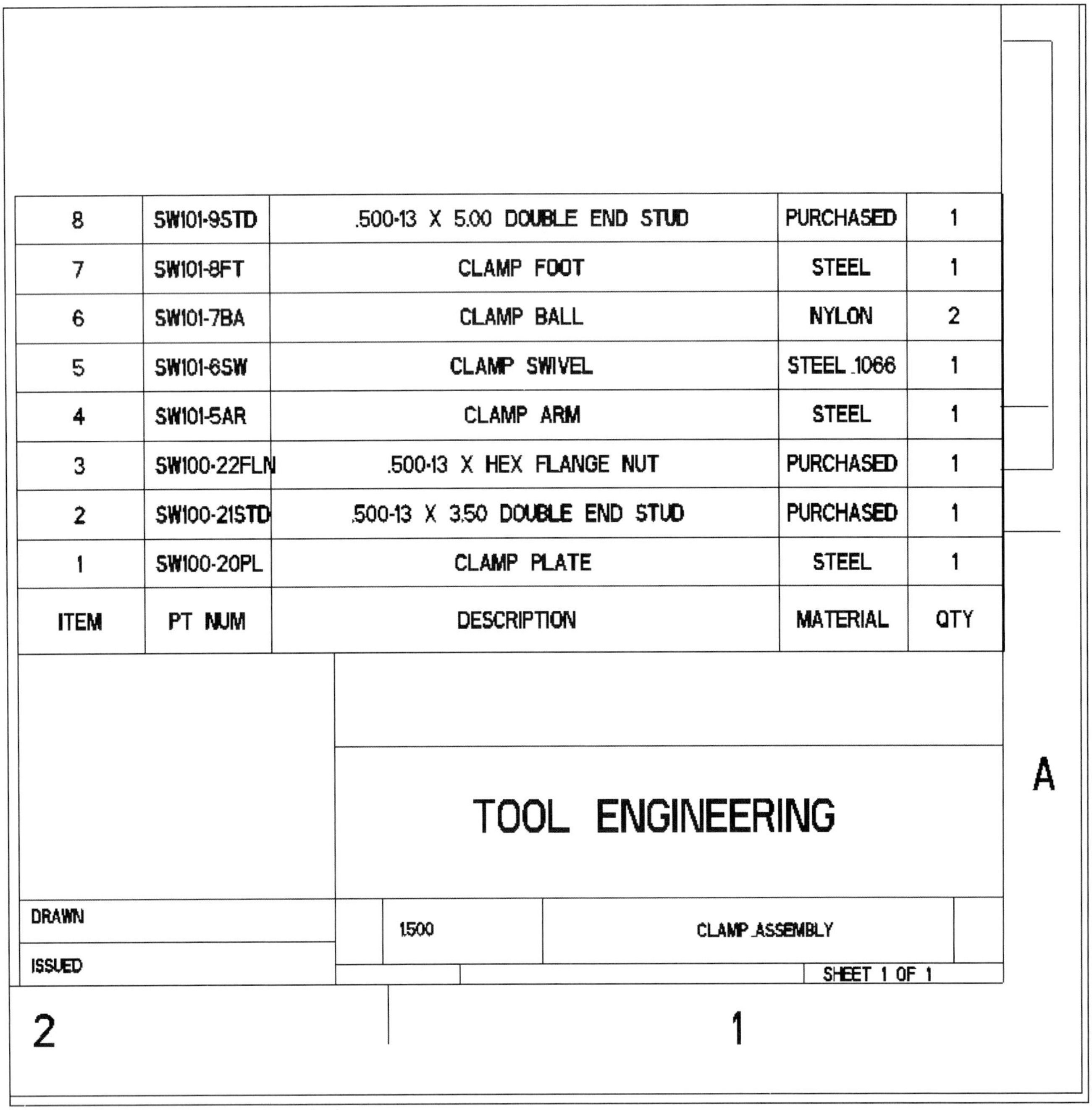

Figure 12.31(a) BOM and Title Block

Click: off ⇒ double-click on **SCALE: 1.000** in the lower left-hand corner of the graphics window ⇒ type **1.50** ⇒ **Enter** ⇒ **LMB** to deselect ⇒ **Disallow the movement of drawing views with the mouse** unlock ⇒ **Refit**

Click: Create a general view ⇒ No Combined State ⇒ OK ⇒ Select CENTER POINT for drawing view. pick where you want a Top view [Fig. 12.31(b)] ⇒ ⇒ Model view names **FRONT** [Fig. 12.31(c)] ⇒ **Apply** ⇒ **Close**

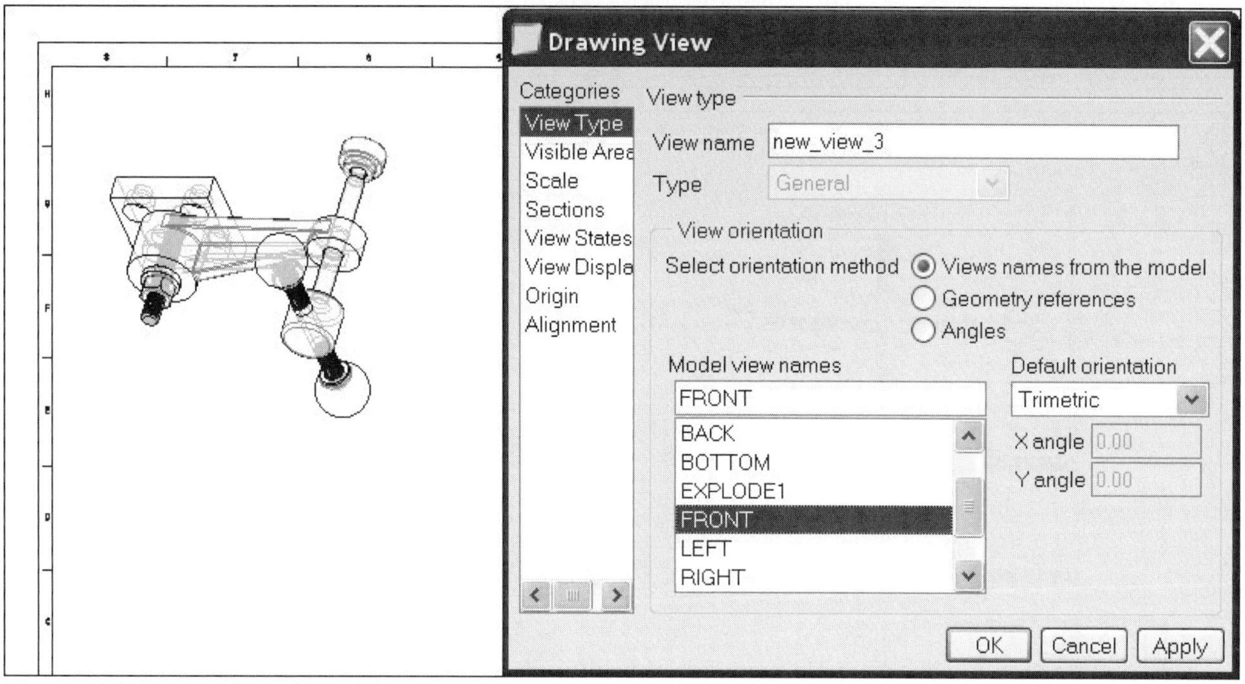

Figure 12.31(b) Drawing View Dialog Box

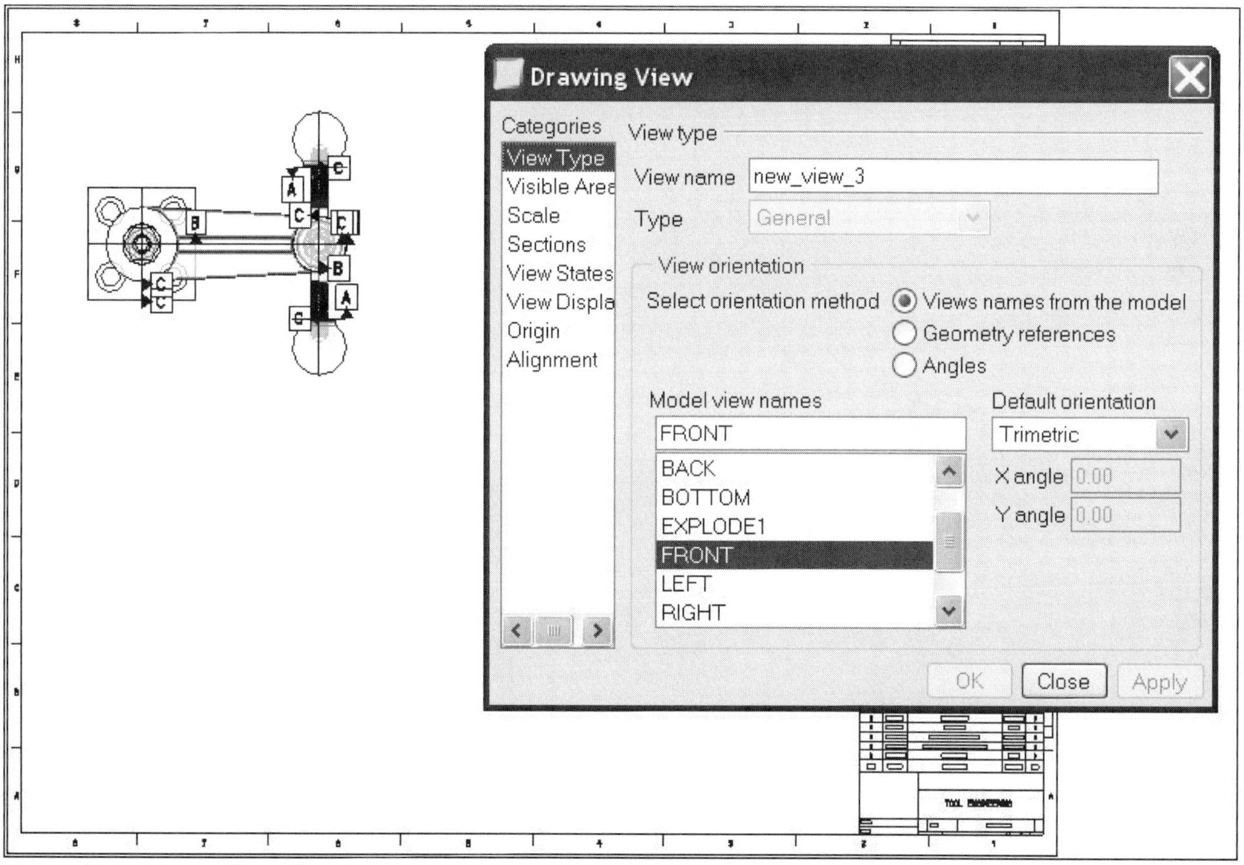

Figure 12.31(c) Top View Placed

Click: on ⇒ ⇒ **RMB** ⇒ **Insert Projection View** ⇒ pick a position for the Front view ⇒ **RMB** ⇒ **Properties** [Fig. 12.32(a)] ⇒ Categories **Sections** ⇒ 2D cross-section ⇒ ⇒ ⇒ **Create New** ⇒ **Planar** ⇒ **Single** ⇒ **Done** ⇒ type **A** ⇒ **Enter** ⇒ ⇒ Select or create an assembly datum pick on **CL_ASM_TOP** *(from the Top view)* [Figs. 12.32(b-c)] ⇒ **Apply** ⇒ **Close** ⇒ **Ctrl+S** ⇒ **Enter**

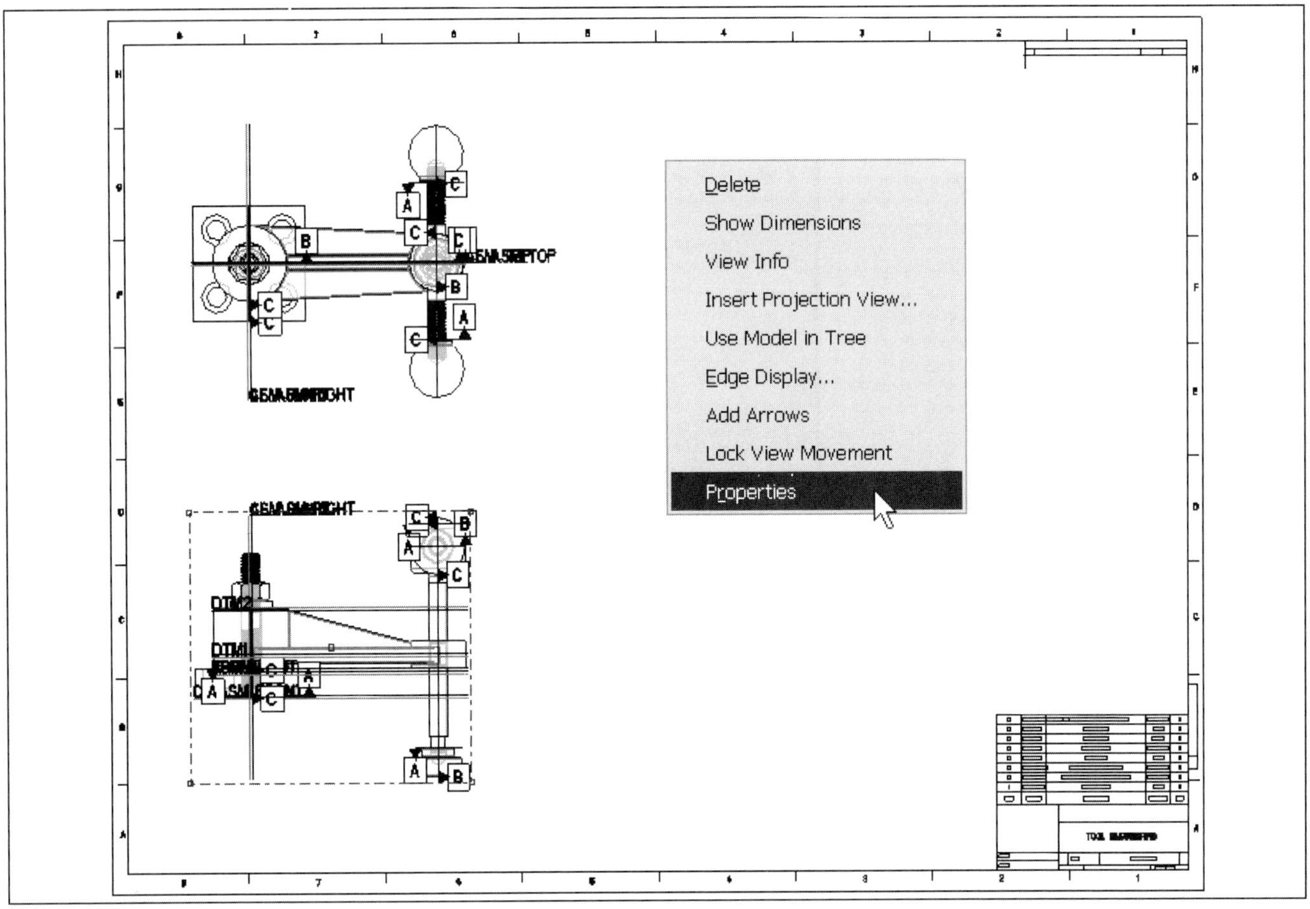

Figure 12.32(a) View Properties

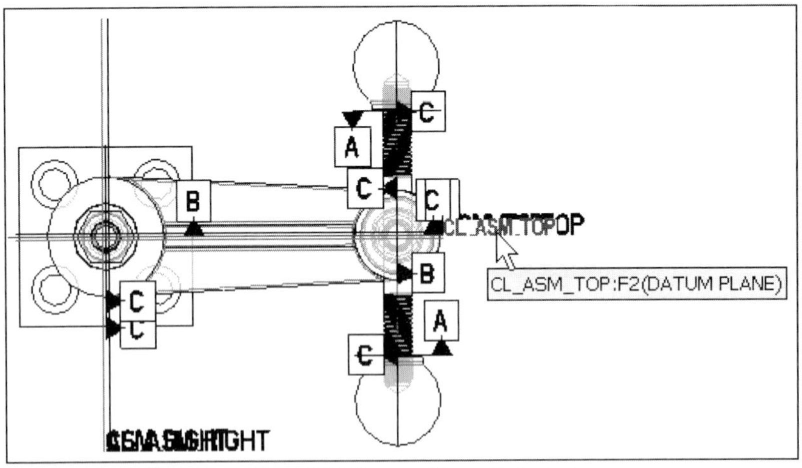

Figure 12.32(b) CL_ASM_TOP Datum Plane Selected

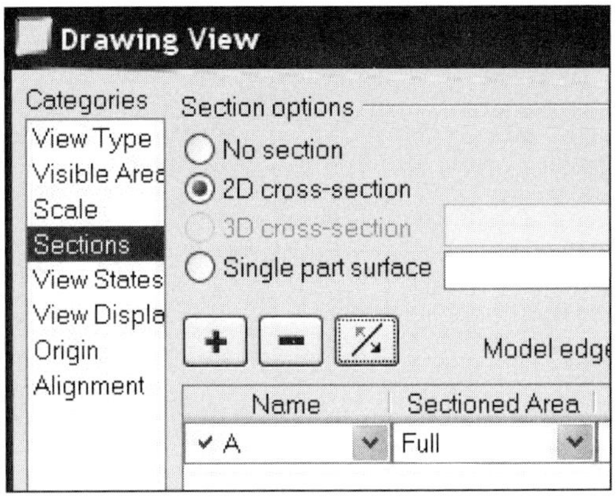

Figure 12.32(c) 2D Cross-Section

Click: **RMB** ⇒ **Add Arrows** ⇒ Pick a view for arrows where the section is perp. pick in the Top view ⇒ **Datum planes** off ⇒ **LMB** to deselect ⇒ **Open the Show/Erase dialog box** ⇒ Erase ⇒ **Datum Plane** on (everything else off) ⇒ Erase All ⇒ **Yes** ⇒ Show ⇒ **Datum Plane** off ⇒ **Axis** on ⇒ **Show All** ⇒ **Yes** ⇒ **Accept All** ⇒ **Close** ⇒ **LMB** ⇒ press and hold down the **Ctrl** key and pick on both views ⇒ click **RMB** with cursor outside of the view outlines ⇒ **Properties** ⇒ Display Style **No Hidden** ⇒ Tangent edges **Dimmed** ⇒ **Apply** ⇒ **Close** ⇒ **LMB** ⇒ ⇒ **Show** ⇒ **Model Tree** ⇒ select the three components of the assembly (not the subassembly) [Fig. 12.32(d)]

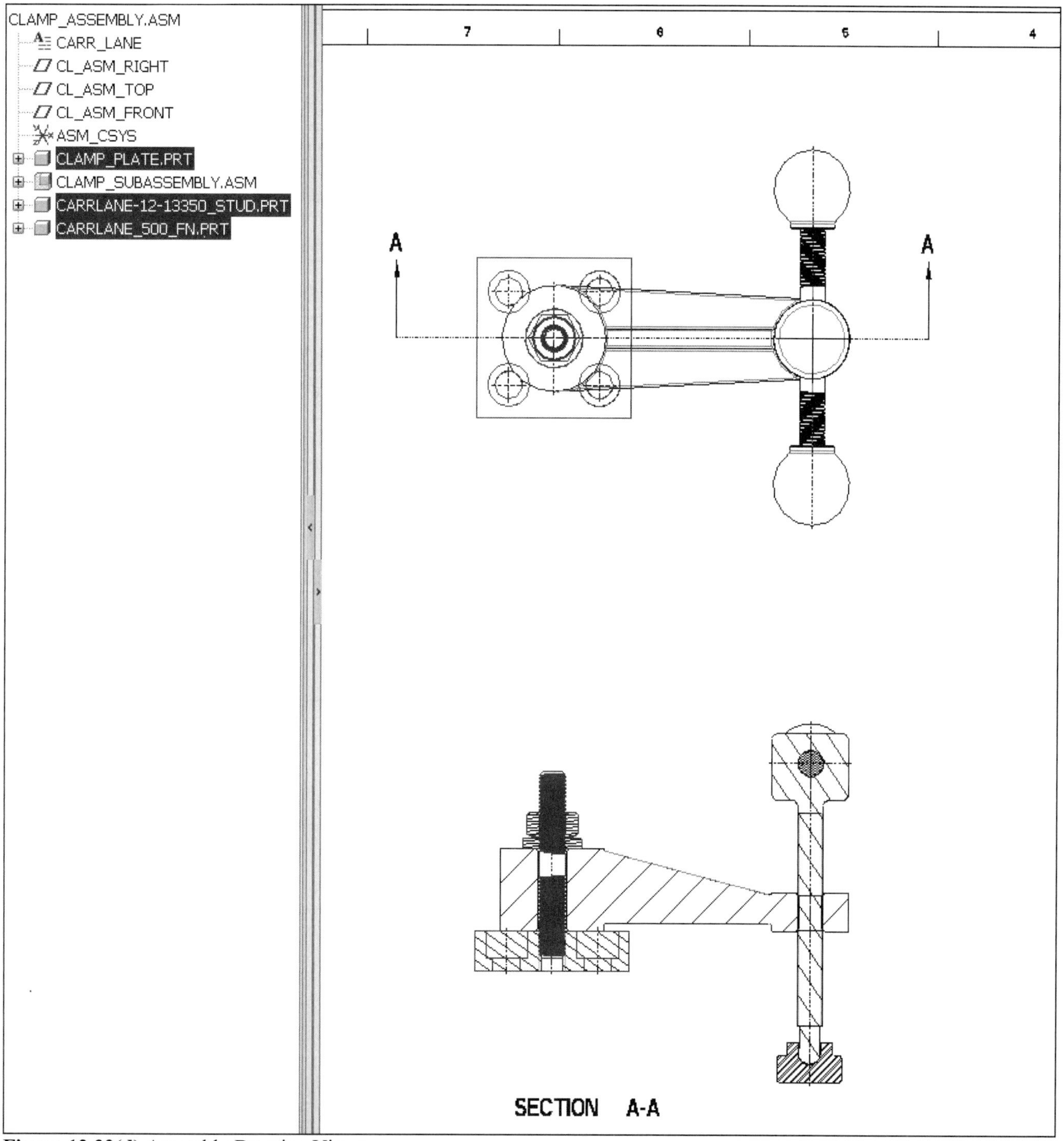

Figure 12.32(d) Assembly Drawing Views

Click: **Table** ⇒ **Repeat Region** ⇒ **Attributes** ⇒ pick in the BOM field [Fig. 12.33(a)] ⇒ **Flat** ⇒ **Done/Return** ⇒ **Done** [Fig. 12.33(b)] ⇒ **Table** ⇒ **BOM Balloons** ⇒ **Set Region** ⇒ pick in the BOM field ⇒ **Create Balloon** ⇒ **Show All** ⇒ **Done** [Fig. 12.33(c)]

8	SW101-9STD	.500-13 X 5.00 DOUBLE END STUD	PURCHASED	1
7	SW101-8FT	CLAMP FOOT	STEEL	1
6	SW101-7BA	CLAMP BALL	NYLON	2
5	SW101-6SW	CLAMP SWIVEL	STEEL 1066	1
4	SW101-5AR	CLAMP ARM	STEEL	1
3	SW100-22FLN	.500-13 X HEX FLANGE NUT	PURCHASED	1
2	SW100-21STD	.500-13 X 3.50 DOUBLE END STUD	PURCHASED	1
1	SW100-20PL	CLAMP PLATE	STEEL	1
ITEM	PT NUM	DESCRIPTION	MATERIAL	QTY

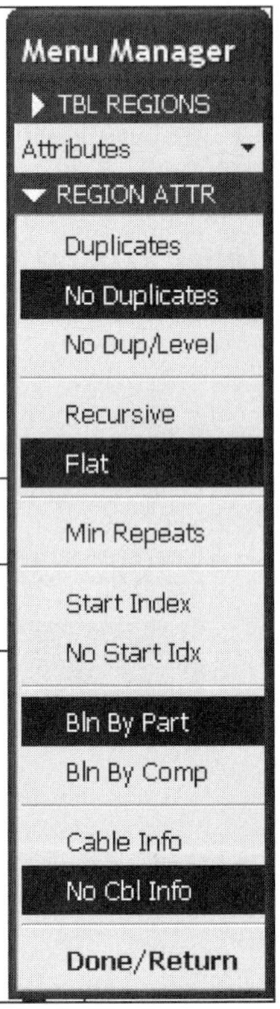

Figure 12.33(a) Showing with BOM Attribute Recursive

3	SW100-22FLN	.500-13 X HEX FLANGE NUT	PURCHASED	1
2	SW100-21STD	.500-13 X 3.50 DOUBLE END STUD	PURCHASED	1
1	SW100-20PL	CLAMP PLATE	STEEL	1
ITEM	PT NUM	DESCRIPTION	MATERIAL	QTY

Figure 12.33(b) Showing with BOM Attribute Flat

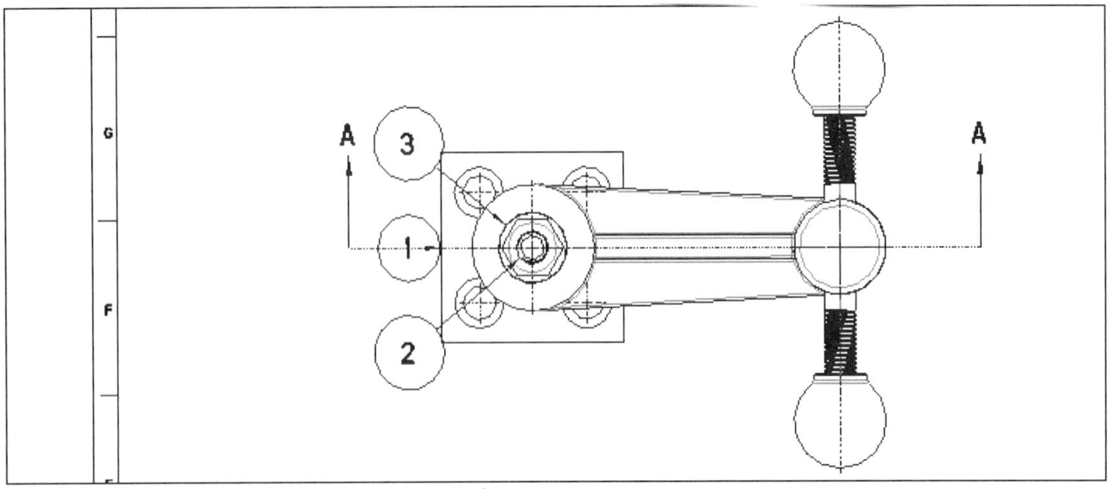

Figure 12.33(c) Balloons Displayed in Top View

While pressing the **Ctrl** key, pick on all three balloons ⇒ **RMB** ⇒ **Move Item to View** ⇒ pick in the Front view [Fig. 12.33(d)] ⇒ pick on and reposition each balloon as needed ⇒ pick on balloon **2** ⇒ **RMB** ⇒ **Edit Attachment** ⇒ **On Entity** ⇒ pick on edge [Fig. 12.33(e)] ⇒ **MMB** ⇒ **LMB**

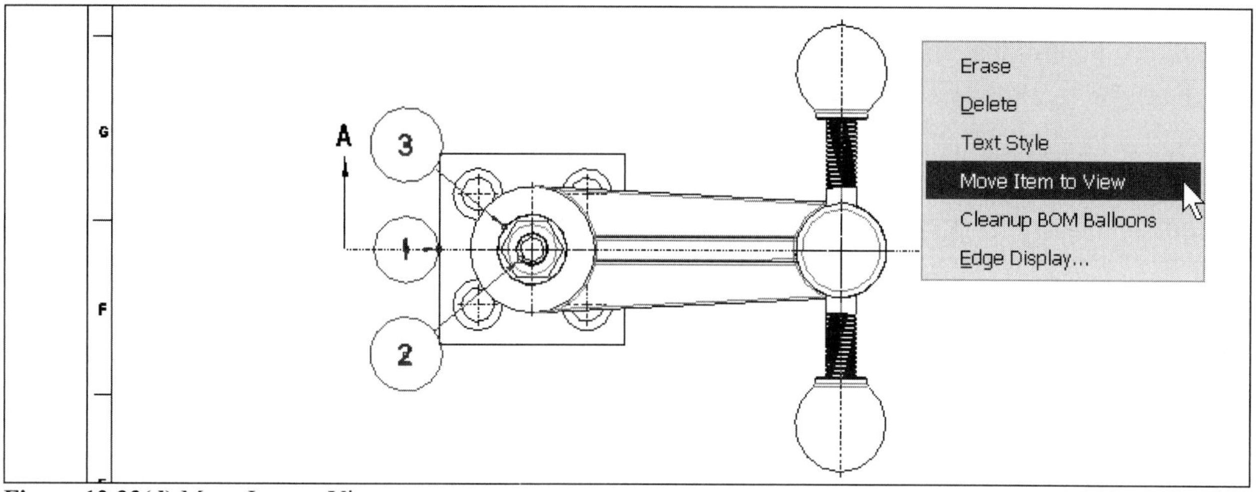

Figure 12.33(d) Move Item to View

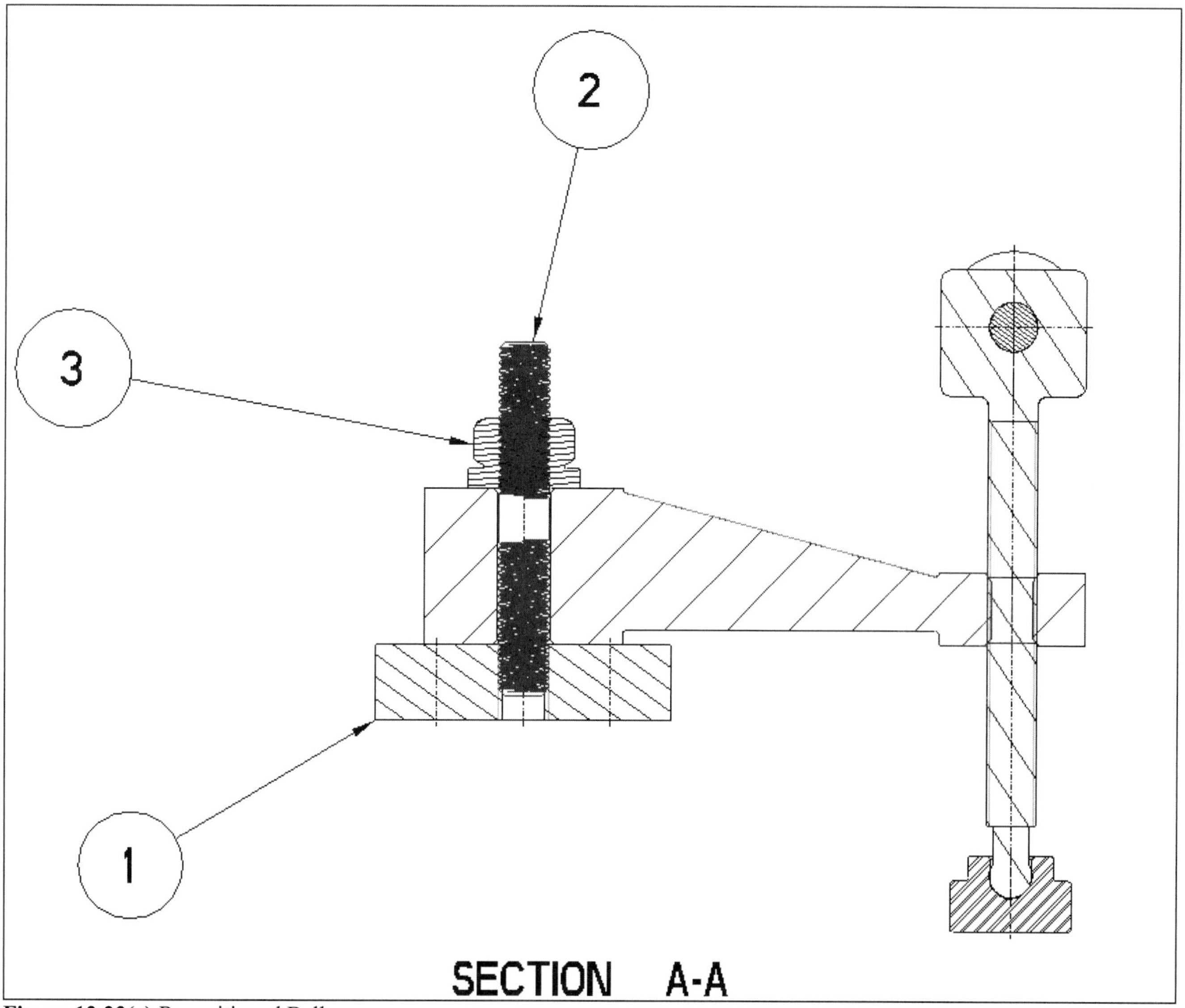

Figure 12.33(e) Repositioned Balloons

Most companies (and as per drafting standards) require that rounded purchased items, such as nuts, bolts, studs, springs, and die pins be excluded from sectioning even when the section cutting plane passes through them. Remove the section lining (crosshatching) from the CARRLANE_500_FN and the CARRLANE-12-13350_STUD in the front section view.

Double-click on the crosshatching in the Front view [Fig. 12.33(f)]. The CARRLANE_500_FN is now active. ⇒ **Exclude** to eliminate Xsec of CARRLANE_500_FN ⇒ **Next** CARRLANE-12-13350_STUD is now active ⇒ **Exclude** to eliminate Xsec of CARRLANE-12-13350_STUD ⇒ **Done** ⇒ **LMB** ⇒ ⇒ ⇒ ⇒ ⇒ **Enter**

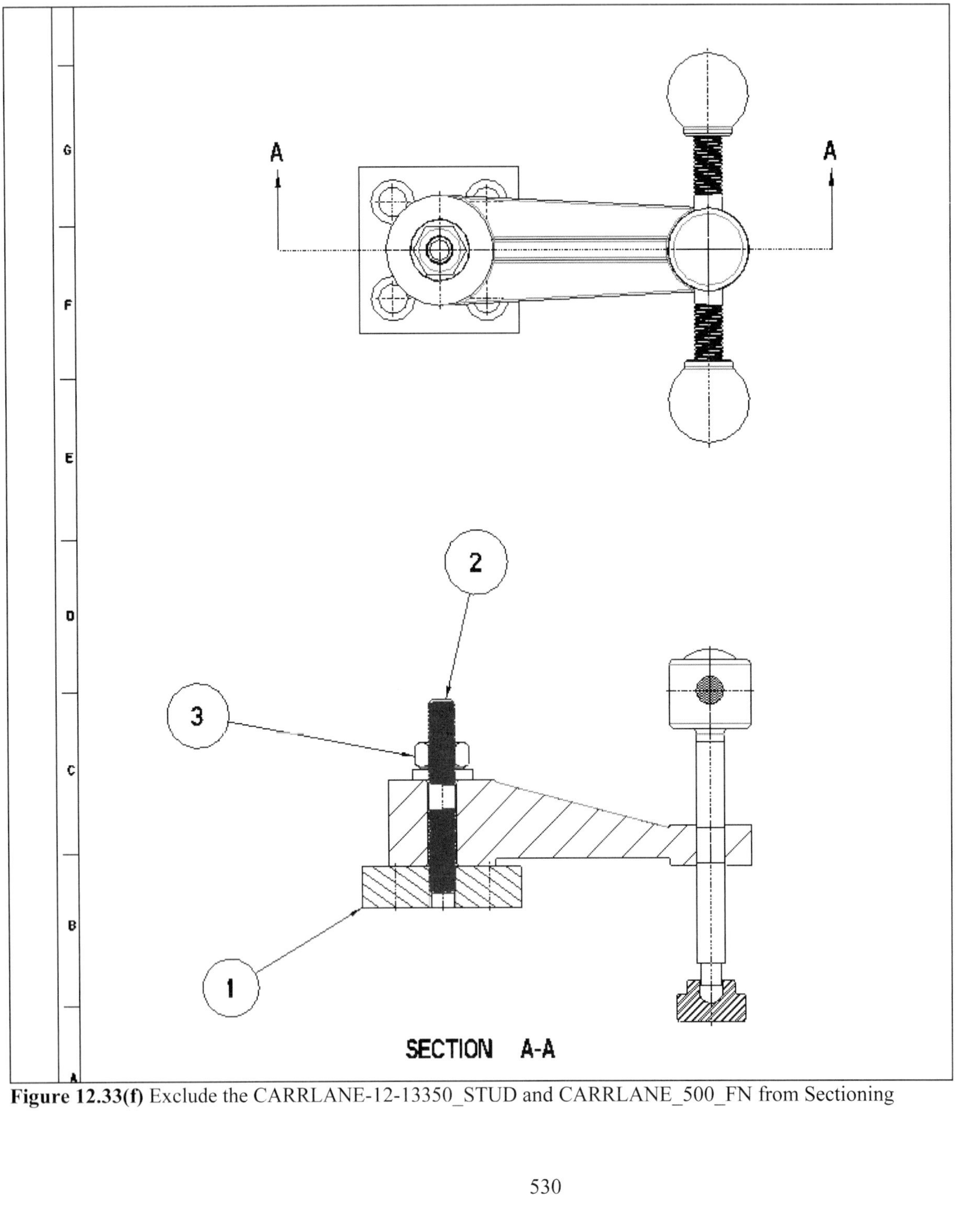

Figure 12.33(f) Exclude the CARRLANE-12-13350_STUD and CARRLANE_500_FN from Sectioning

Exploded Swing Clamp Assembly Drawings

The process required to place an exploded view on a drawing is similar to adding assembly orthographic views. The BOM will display all components on this sheet.

Click: **Insert** ⇒ **Sheet** [2] ⇒ [icon] ⇒ **RMB** ⇒ uncheck **Lock View Movement** ⇒ **Insert General View** ⇒ [No Presentation / COMB0001 / DEFAULT ALL] ⇒ **OK** ⇒ [Select CENTER POINT for drawing view.] ⇒ Categories- **View Type** ⇒ Model view names- [BACK / BOTTOM / EXPLODE1 / FRONT] ⇒ Default orientation- [▾] ⇒ [Isometric] ⇒ [OK] [Cancel] [Apply] ⇒ Categories- **View States** ⇒ Combined state- [▾] ⇒ [COMB0001] ⇒ Explode view- [☑ Explode components in view] ⇒ Assembly explode state- [▾] ⇒ [EXP0001] ⇒ **Apply** [Fig. 12.34(a)] ⇒ Categories- **View Display** ⇒ Display style- [▾] ⇒ [Display style: Shading] ⇒ **Apply** ⇒ **Close**

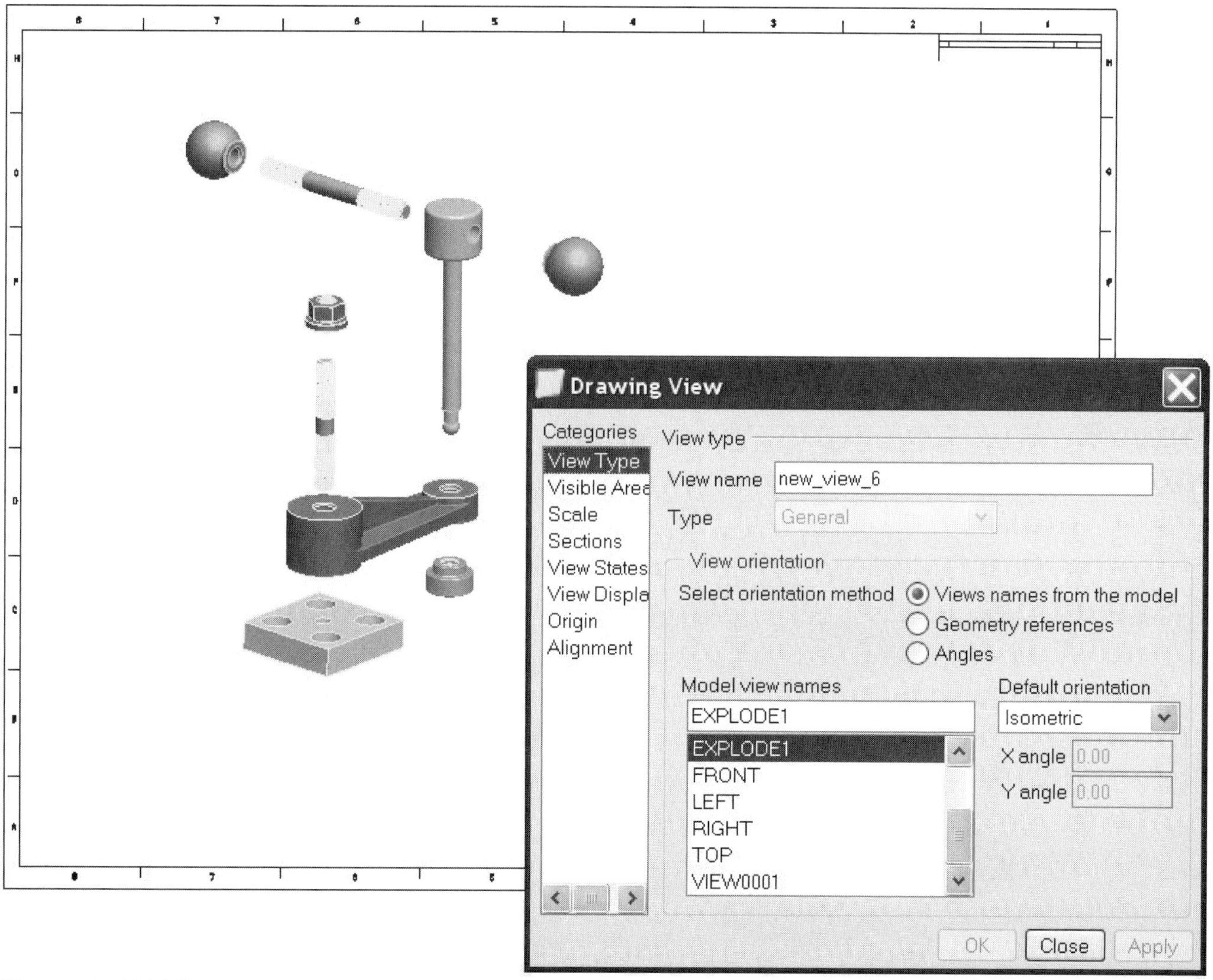

Figure 12.34(a) View Type

Click: **Open the Show/Erase dialog box** ⇒ Show ⇒ only ----A.1 **Axis** on ⇒ **Show All** ⇒ **Yes** ⇒ **Accept All** ⇒ **Close** ⇒ **LMB** [Fig. 12.34(b)] ⇒ erase extra centerlines and clip each centerline (axis) to extend between components that are in line [Fig. 12.34(c)] ⇒ ⇒ ⇒ ⇒ ⇒ **Enter**

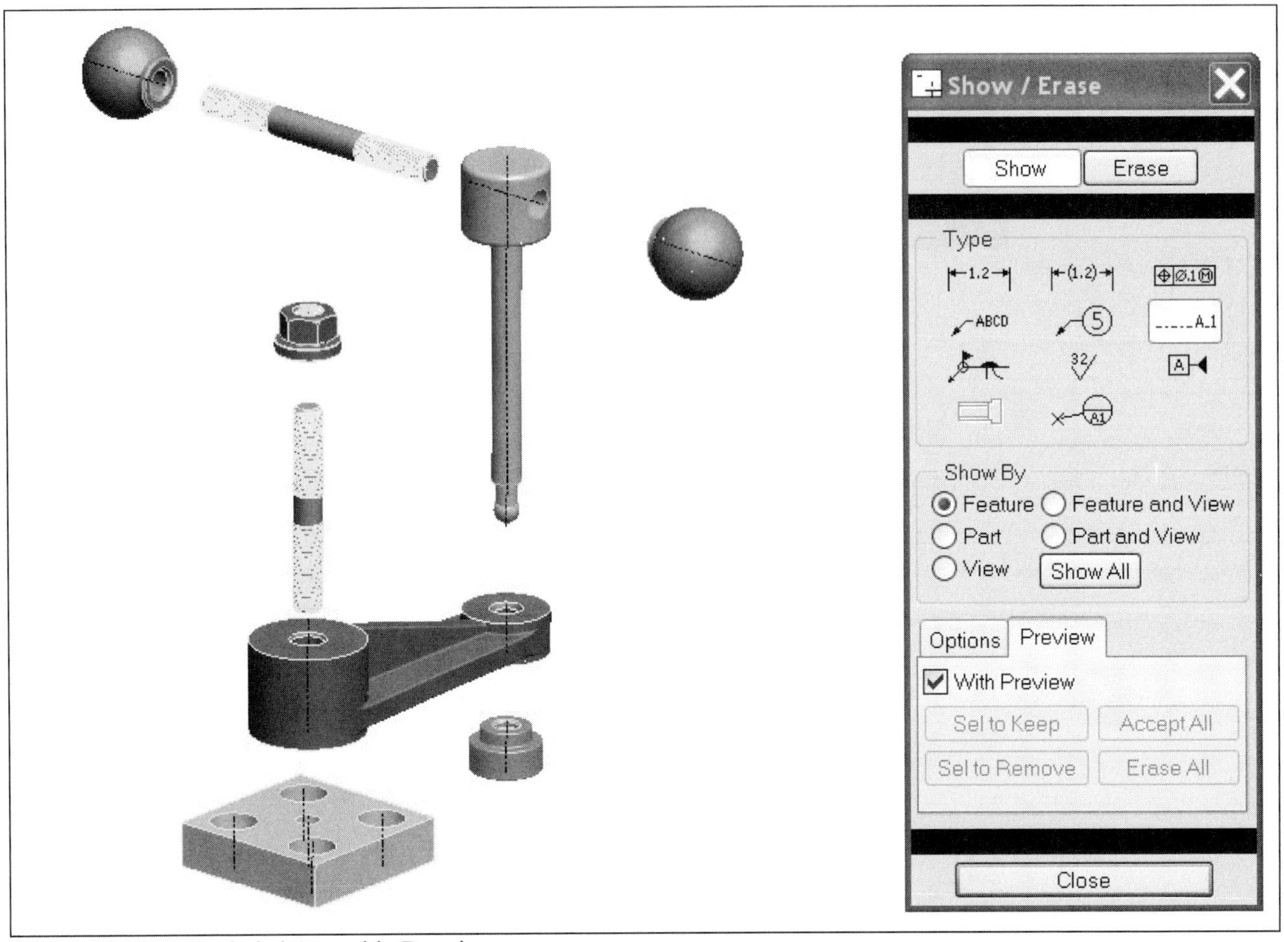

Figure 12.34(b) Exploded Assembly Drawing

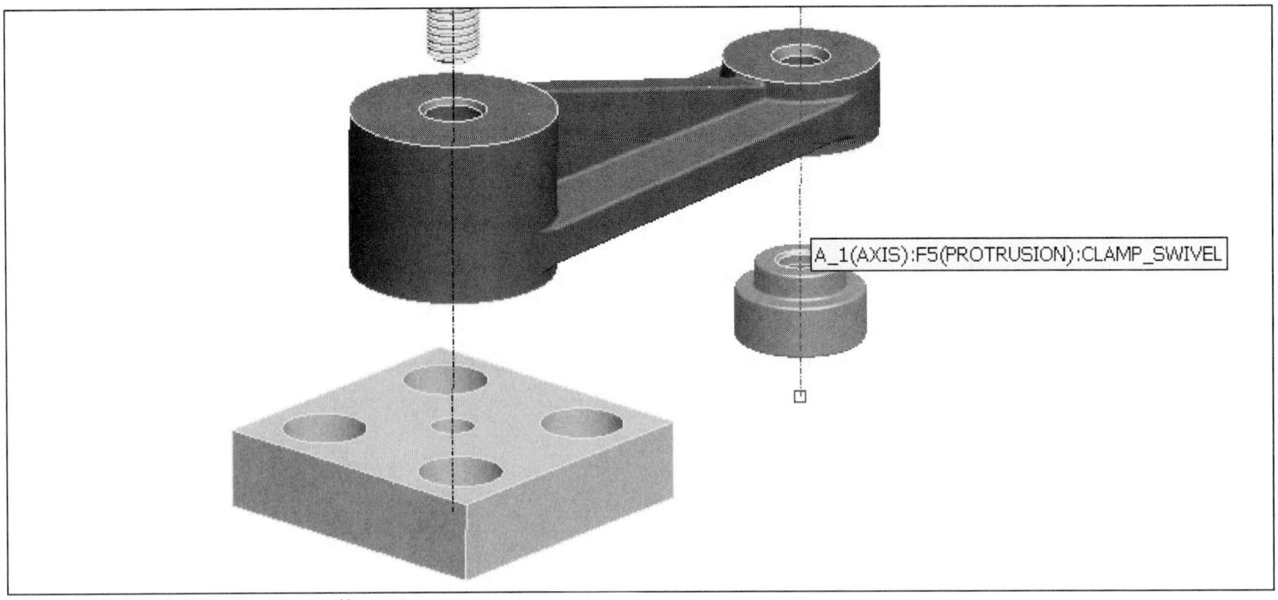

Figure 12.34(c) Clip the Centerlines Axes

Click: **Table** ⇒ **BOM Balloons** ⇒ **Set Region** ⇒ pick in the BOM field ⇒ **Create Balloon** ⇒ Set a Region: pick in the BOM field ⇒ **Show All** ⇒ **Done** ⇒ reposition the balloons and their attachment points as needed to clean up the drawing [Fig. 12.35(a)] ⇒ pick on pick balloon **6** ⇒ **RMB** ⇒ **Edit Attachment** ⇒ **On Surface** ⇒ **Dot** ⇒ pick a place on the Clamp_Ball's surface ⇒ **OK** ⇒ **Done/Return** ⇒ **LMB** ⇒  ⇒  ⇒  ⇒ **OK** [Fig. 12.35(b)] ⇒ **File** ⇒ **Delete** ⇒ **Old Versions** ⇒ **Enter** ⇒ **Window** ⇒ **Close** ⇒ **File** ⇒ **Erase** ⇒ **Not Displayed** ⇒ **OK** ⇒ Exit Pro/ENGINEER

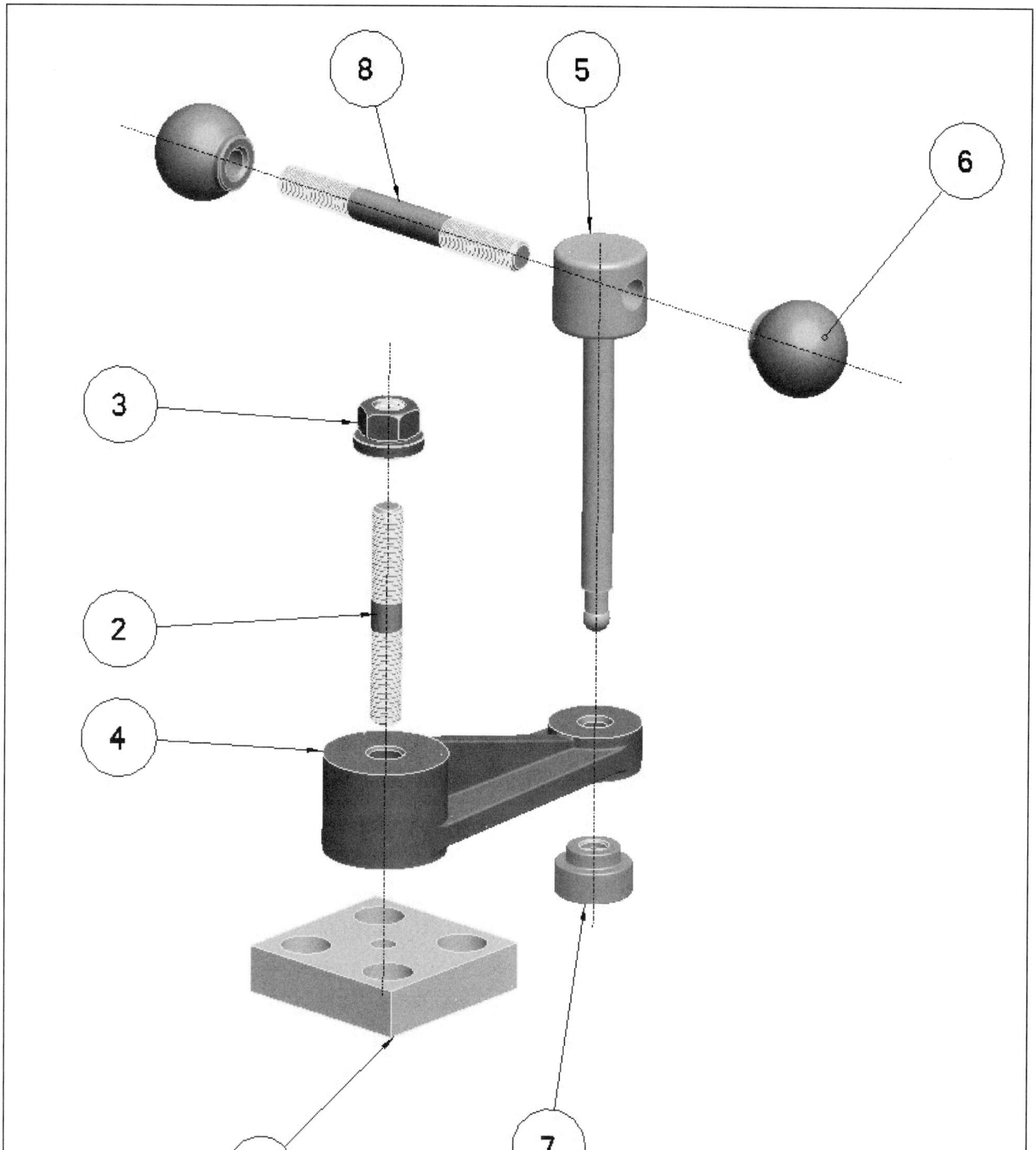

Figure 12.35(a) Ballooned Exploded View

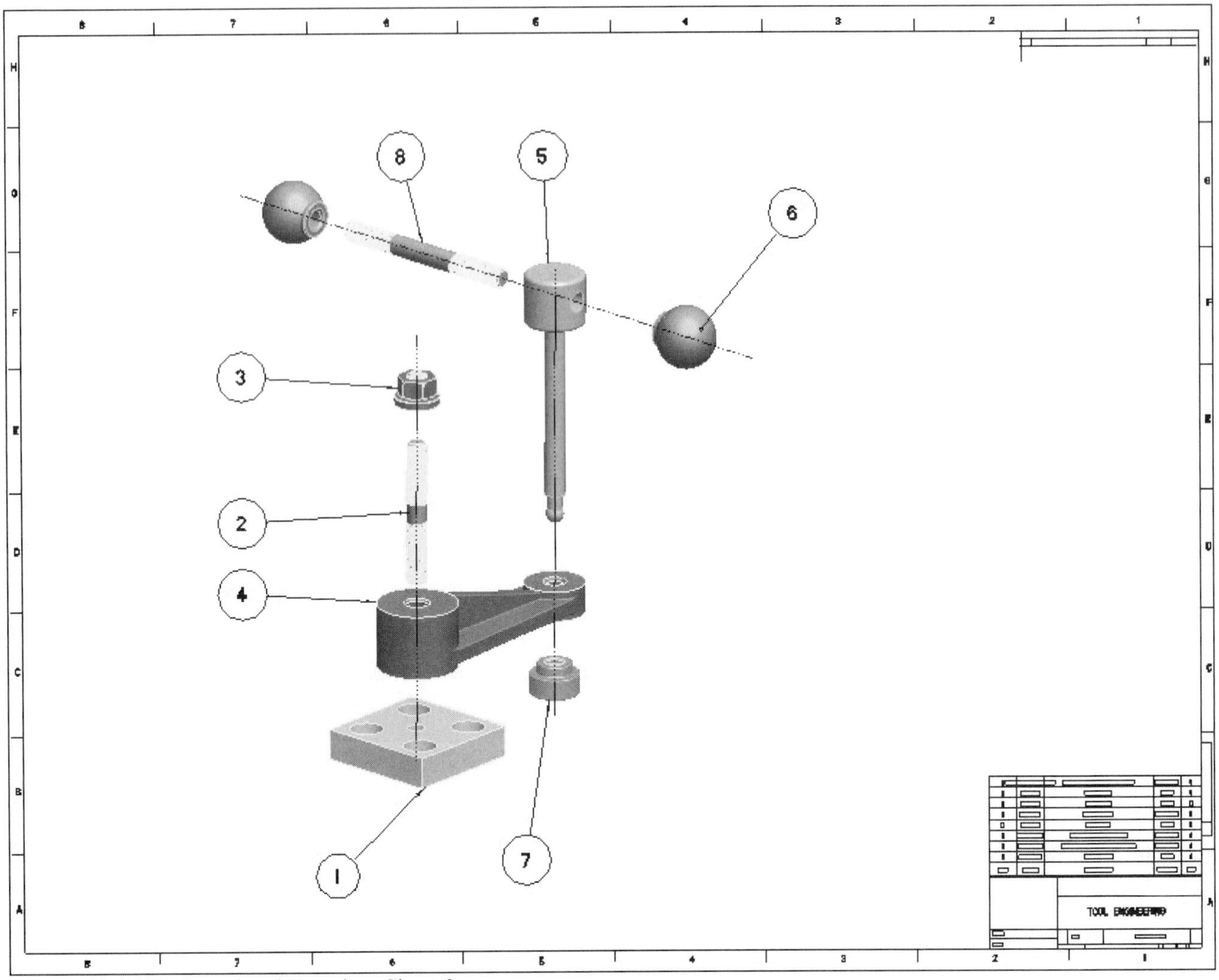

Figure 12.35(b) Completed Drawing, Sheet 2

Create a *documentation package*. A complete documentation package contains all models and drawings required to manufacture the parts and assemble the components. Your instructor may change the requirements, but in general, create and plot/print the following:

- **Part Models** for all components
- **Detail Drawings** for each nonstandard component, such as the Clamp_Arm, Clamp_Swivel, Clamp_Foot, and Clamp_Ball *(do not detail the standard parts)*
- **Assembly Drawings** using standard orthographic ballooned views
- **Exploded Subassembly Drawing** of the ballooned subassembly
- **Exploded Assembly Drawing** of the ballooned assembly

A different assembly (Coupling) is available at ***www.cad-resources.com*** ⇒ ***Downloads***.

Lesson 13 Patterns

Figure 13.1 Parts with Patterns

OBJECTIVES

- Construct a **fill pattern**
- Create an **axial pattern**
- Model a **sheetmetal** part and create a **directional pattern**
- Use the **Auto-Round Tool**

Patterns

Creating a pattern (Fig. 13.1) is a quick way to reproduce a feature. A pattern is parametrically controlled. Therefore, you can modify a pattern by changing pattern parameters, such as the number of instances, spacing between instances, and original feature dimensions. Modifying patterns is more efficient than modifying individual features. In a pattern, when you change dimensions of the original feature, Pro/E automatically updates the whole pattern.

This lesson will use direct modeling to introduce you to variations of the **Pattern Tool**.

Part Model (Plate- Fill Pattern)

Click: [New] ⇒ [Part] ⇒ **OK** ⇒ **Edit** ⇒ **Setup** ⇒ **Units** ⇒ [millimeter Newton Second (mmNs)] ⇒ **Set** ⇒ [Interpret dimensions (for example 1" becomes 1mm)] ⇒ **OK** ⇒ **Close** ⇒ **Done** ⇒ **File** ⇒ **Rename** ⇒ **Plate** ⇒ **OK** ⇒ **OK** ⇒ **Tools** ⇒ **Environment** ⇒ Standard Orient **Trimetric** ⇒ **OK** ⇒ [datum display icons] on ⇒ pick the **FRONT** datum plane ⇒ [icon] **Extrude Tool** ⇒ **RMB** ⇒ [Define Internal Sketch...] ⇒ **Sketch** ⇒ **RMB** ⇒ **Centerline** ⇒ create vertical and horizontal centerlines ⇒ [sketch display icons] on ⇒ [arc icons] **Create an arc by picking its center and endpoints** ⇒ pick the center and ends ⇒ **RMB** ⇒ **Line** ⇒ create the two horizontal lines [Fig. 13.2(a)] ⇒ **MMB** ⇒ modify the values ⇒ [select icon] ⇒ capture the sketch with a window ⇒ [icon] **Mirror selected entities** ⇒ pick on the vertical centerline [Fig. 13.2(b)] ⇒ [icon] ⇒ **Standard Orientation** ⇒ [✓] ⇒ modify the part thickness to **12mm** [Fig. 13.2(c)] ⇒ **Enter** ⇒ **MMB** to complete the extrude command ⇒ **Ctrl+S** ⇒ **OK**

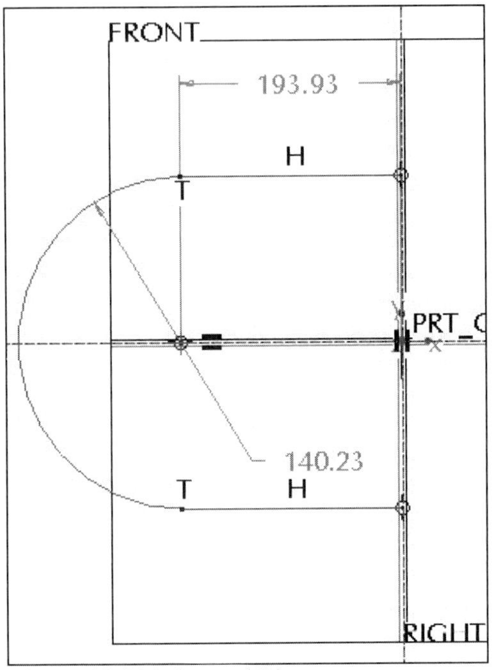

Figure 13.2(a) Arc and Lines

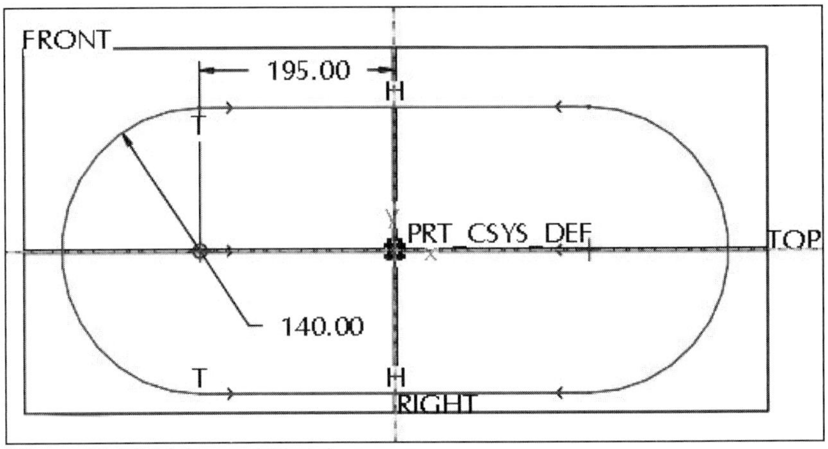

Figure 13.2(b) Mirrored Sketch

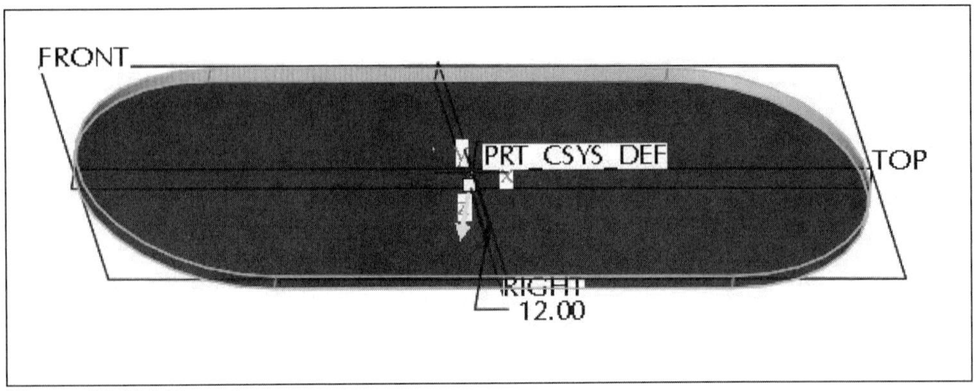

Figure 13.2(c) Extrusion

Click: **Hole Tool** ⇒ pick on the face of the part ⇒ **RMB** ⇒ Offset References Collector ⇒ pick the **TOP** datum plane ⇒ press and hold **Ctrl** key ⇒ pick the **RIGHT** datum plane ⇒ **Placement** tab ⇒ make both Offset values **0.00** [Fig. 13.2(d)] ⇒ ⇒ ⇒ change the diameter ∅ 22.00 ⇒ **Enter** ⇒ rotate the part [Fig. 13.2(e)] ⇒ from the dashboard ⇒ ⇒ **MMB** ⇒ with the Hole still selected in the Model Tree, click: **RMB** ⇒ **Pattern** [Fig. 13.2(f)] ⇒ Dimension ⇒ ⇒ **Fill** Fill ⇒ **RMB** ⇒ Define Internal Sketch... ⇒ pick the upper face [Fig. 13.2(g)] ⇒ **Sketch** ⇒ **Create an entity from an edge** ⇒ Loop ⇒ pick the face [Fig. 13.2(h)] ⇒ [Fig. 13.2(i)]

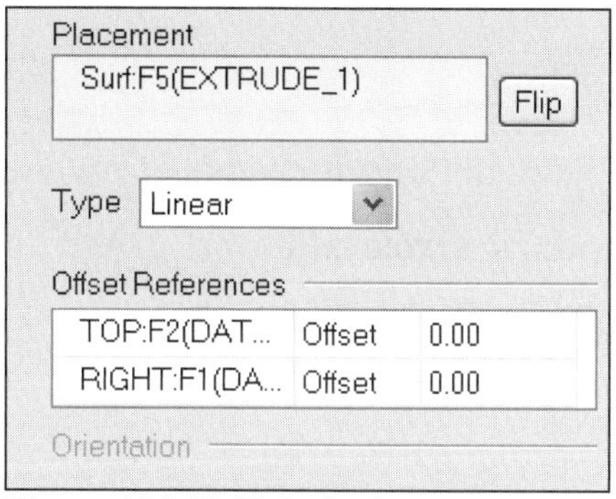

Figure 13.2(d) Hole Options

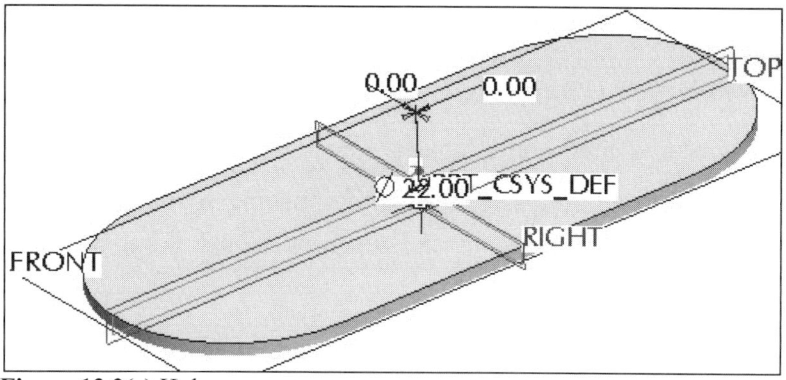

Figure 13.2(e) Hole

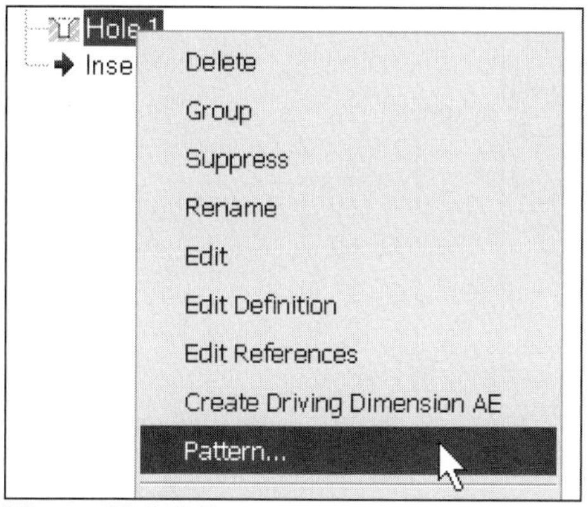

Figure 13.2(f) Pattern

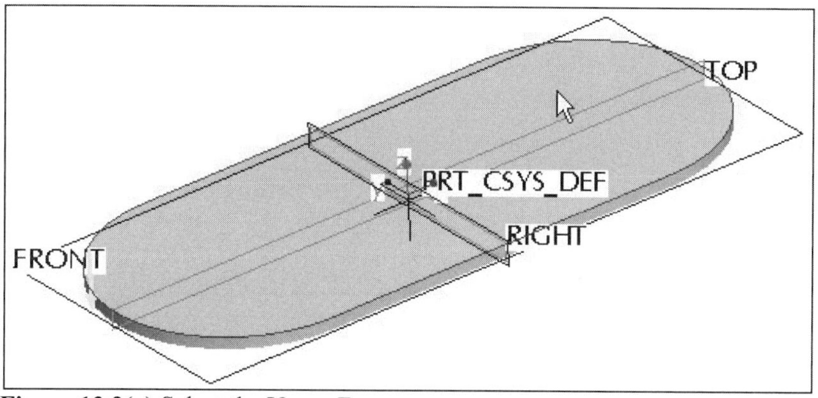

Figure 13.2(g) Select the Upper Face

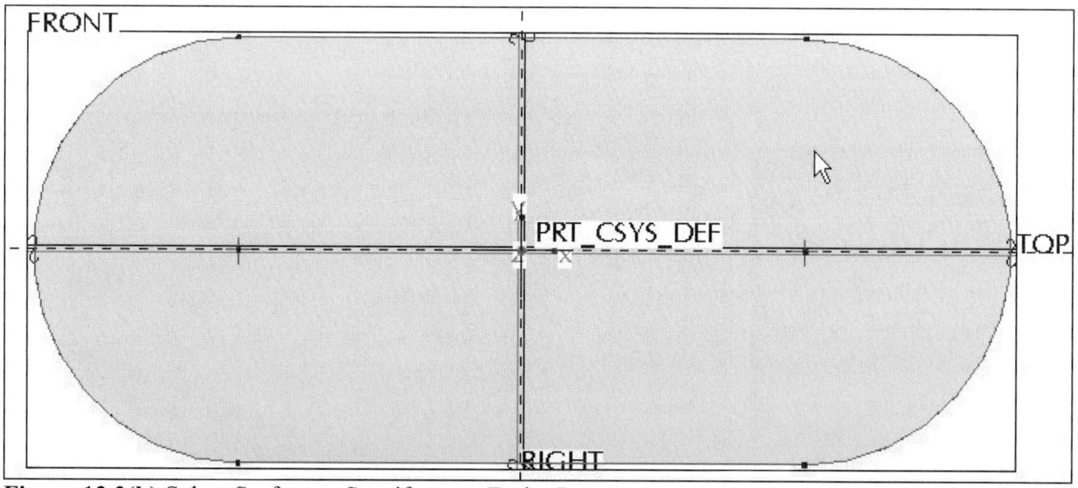

Figure 13.2(h) Select Surface to Specify as an Entity Loop

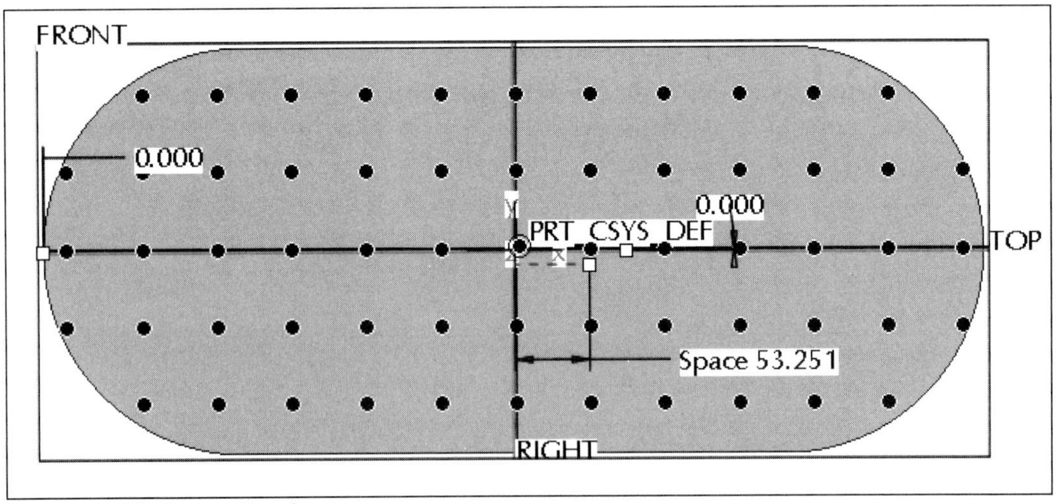

Figure 13.2(i) Pattern Preview

Pick on the unwanted copies that are near the parts edges [pick on a black dot and it changes to white (inactive)] [Fig. 13.2(j)] ⇒ modify the spacing to **53** ⇒ [icon] ⇒ **Standard Orientation** ⇒ **MMB** to complete the pattern command ⇒ **Edit** from the menu bar ⇒ **Scale Model** ⇒ **.6** ⇒ **Enter** ⇒ **Yes**

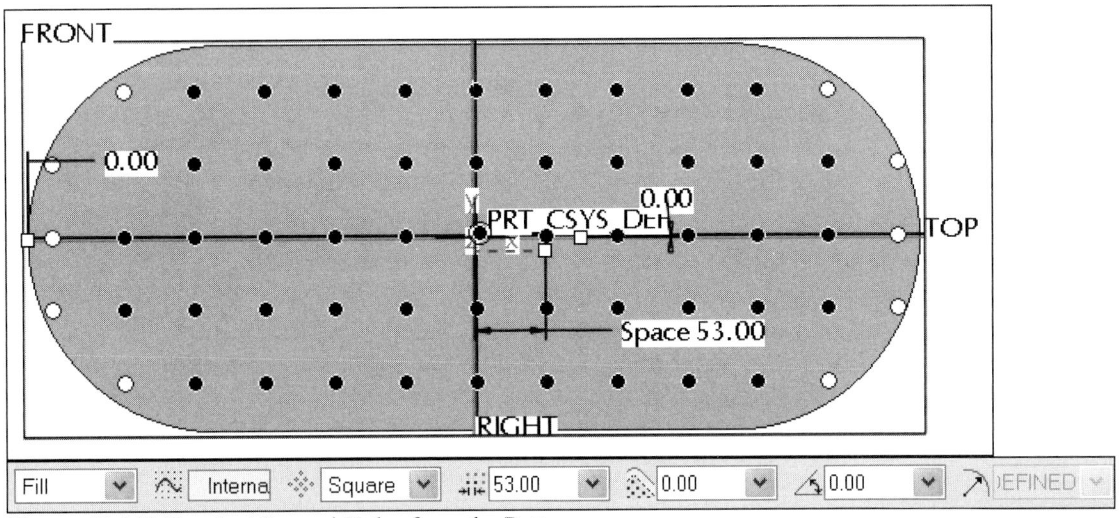

Figure 13.2(j) Deselect Several Holes from the Pattern

Click: **Insert** from the menu bar ⇒ **Auto Round** ⇒ [2.0000] ⇒ **Enter** ⇒ [✓] [Fig. 13.2(k)] ⇒ **LMB** to deselect [Fig. 13.2(l)] ⇒ **Ctrl+S** ⇒ **Enter** ⇒ **File** ⇒ **Close Window**

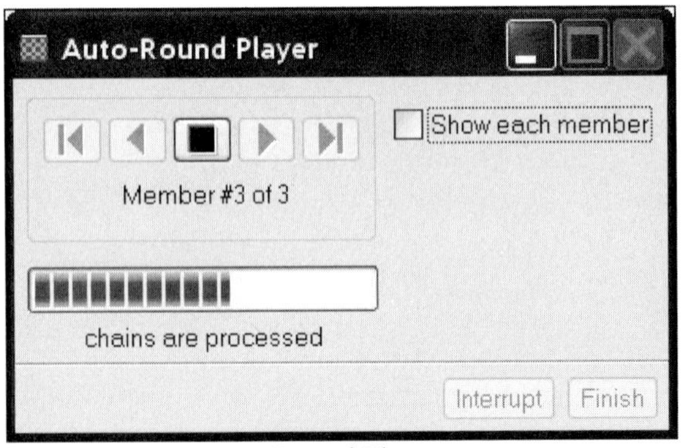

Figure 13.2(k) Auto-Round Player

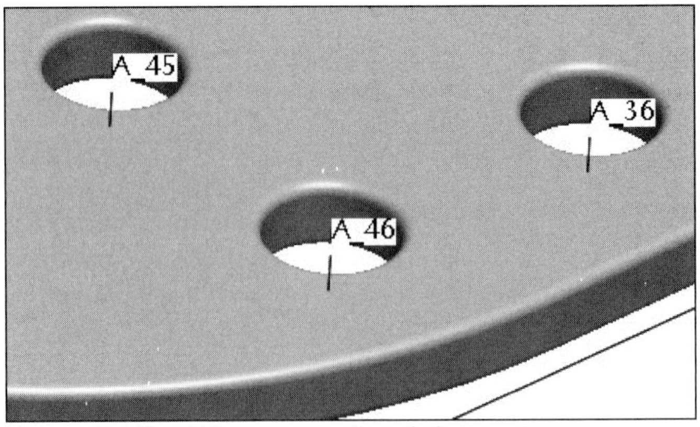

Figure 13.2(l) All Features are Rounded

Part Model (Cap- Axial Pattern)

Click: ⇒ Part ⇒ **MMB** ⇒ **Edit** ⇒ **Setup** ⇒ **Units** ⇒ millimeter Newton Second (mmNs) ⇒ **Set** ⇒ Interpret dimensions (for example 1" becomes 1mm) ⇒ **OK** ⇒ **Close** ⇒ **MMB** ⇒ pick the **FRONT** datum plane ⇒ **Extrude Tool** ⇒ **RMB** ⇒ Define Internal Sketch... ⇒ **MMB** ⇒ **RMB** ⇒ **Circle** ⇒ sketch a circle ⇒ ⇒ **Ctrl+D** ⇒ use **MMB** to rotate the model [Fig. 13.3(a)] ⇒ move the drag handle to adjust the model's height [Fig. 13.3(b)] ⇒ **MMB** ⇒ ⇒ **Reorient view** ⇒ Saved Views ⇒ Name *(type)* **PICT1** [Fig. 13.3(c)] ⇒ **Save** ⇒ **OK** ⇒ **Ctrl+D** (Standard Orientation) ⇒ ⇒ PICT1 [Fig. 13.3(d)]

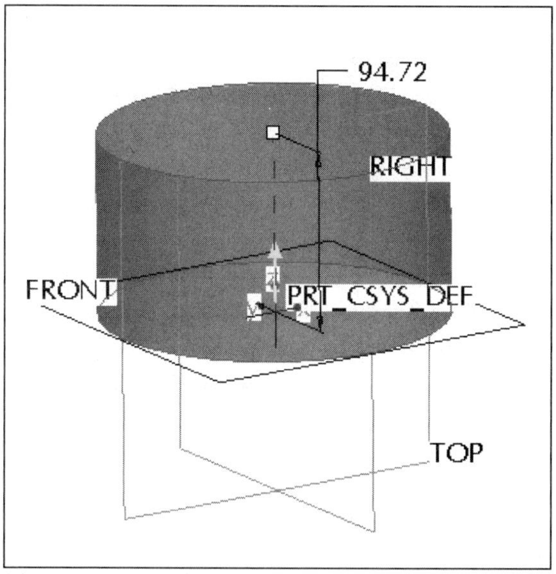

Figure 13.3(a) Circular Protrusion Preview

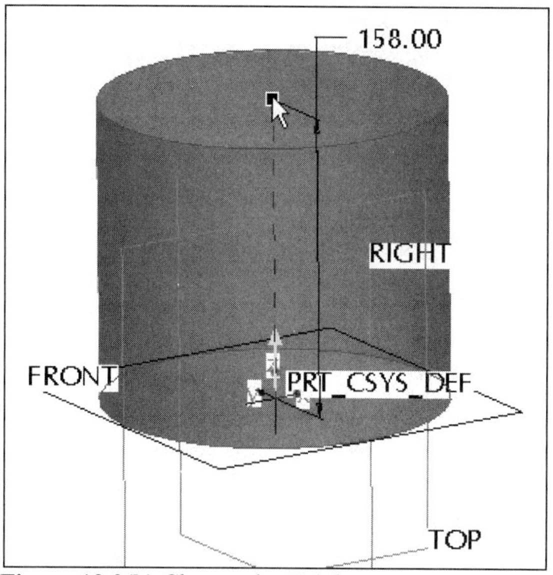

Figure 13.3(b) Change the Height

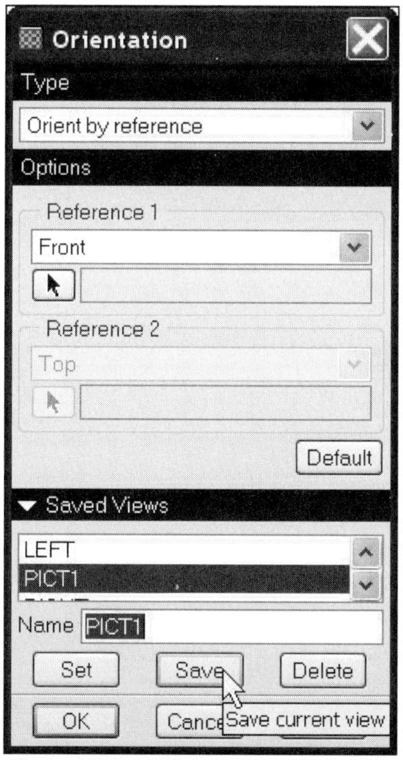

Figure 13.3(c) Orientation Dialog Box

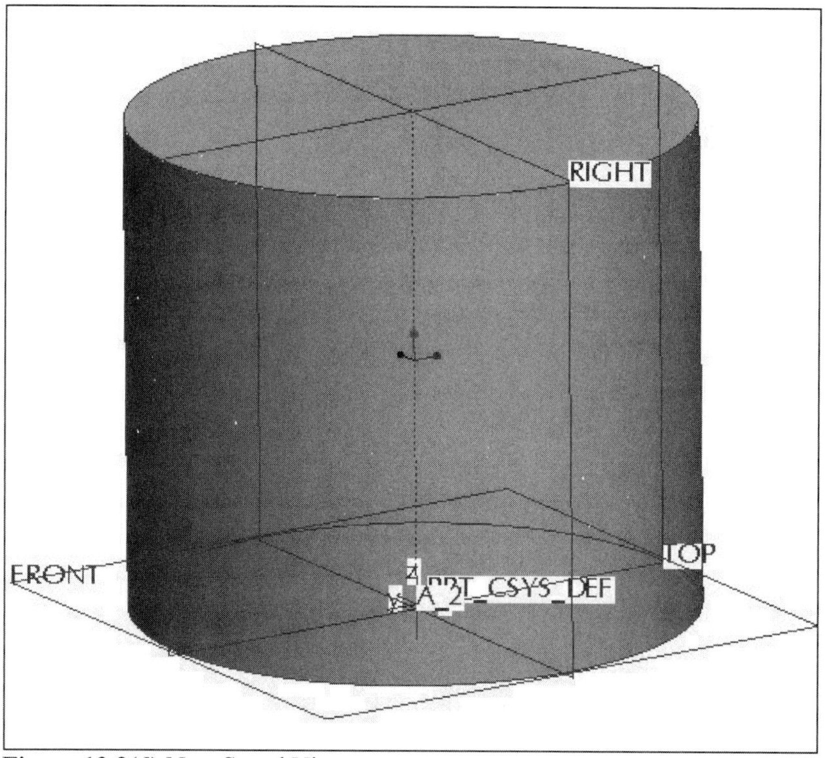

Figure 13.3(d) New Saved View

Pick on the top edge of the part ⇒ **RMB** ⇒ **Round Edges** [Fig. 13.4(a)] ⇒ **MMB** [Fig. 13.4(b)] ⇒ **Hole Tool** ⇒ pick on the vertical cylindrical face of the part ⇒ **RMB** ⇒ Offset References Collector ⇒ pick the **RIGHT** datum plane [Fig. 13.5(a)] ⇒ press and hold **Ctrl** key ⇒ pick the **FRONT** datum plane [Fig. 13.5(b)] ⇒ **Placement** tab ⇒ edit the values [Figs. 13.5(c-d)] ⇒ ⇒ **MMB** ⇒ **Ctrl+S** ⇒ **Enter** ⇒ **LMB** to deselect

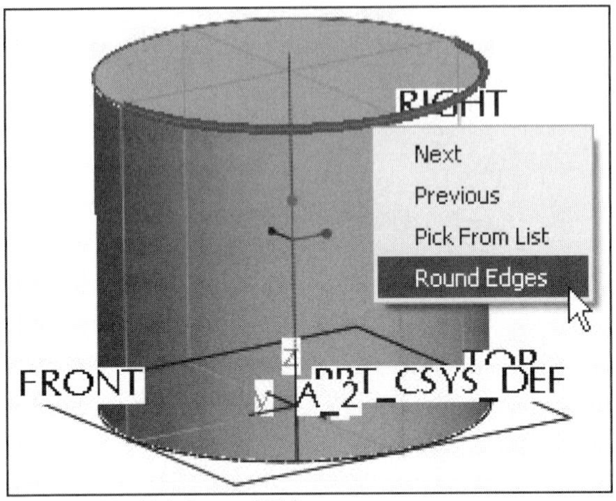

Figure 13.4(a) Round Edges

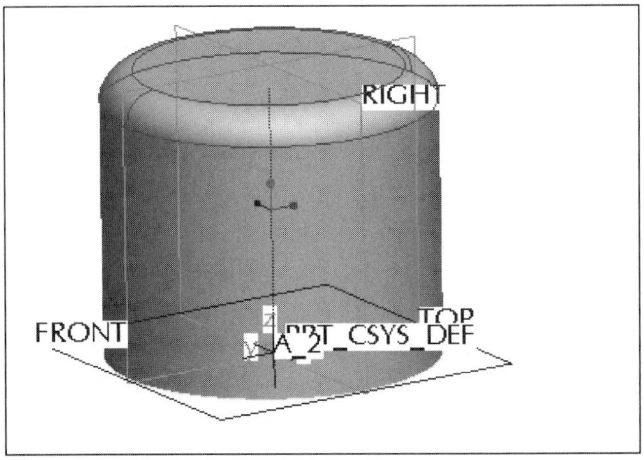

Figure 13.4(b) Completed Round

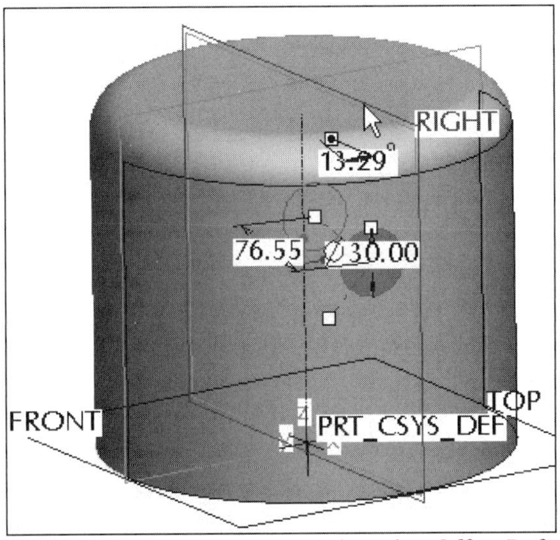

Figure 13.5(a) Right Datum Selected as Offset Reference

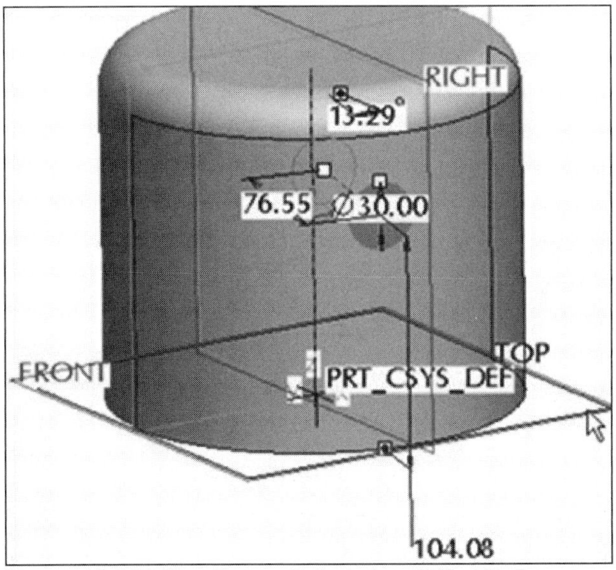

Figure 13.5(b) Top Datum Selected as Offset Reference

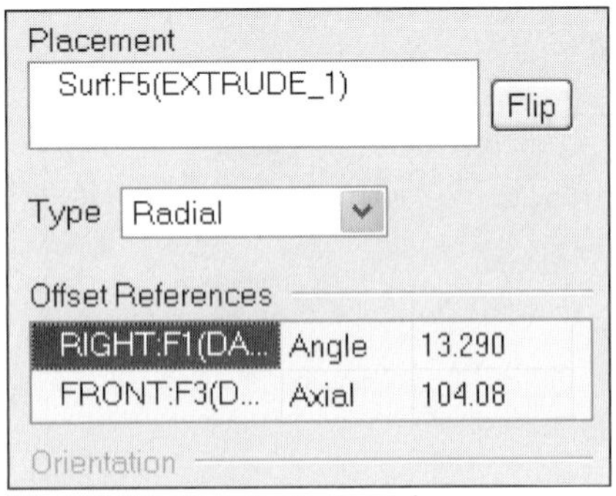

Figure 13.5(c) Default Values

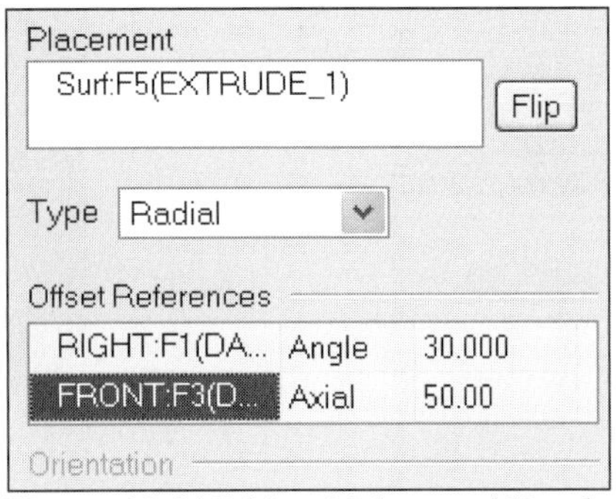

Figure 13.5(d) New Values and Depth

Pick on the hole ⇒ **Ctrl+C** ⇒ **Edit** ⇒ Paste Special... ⇒ check the options [Fig. 13.6(a)] ⇒ **OK** [Fig. 13.6(b)] ⇒ **Transformations** tab [Fig. 13.6(c)] ⇒ Direction reference- pick **Front** datum [Figs. 13.6(d-e)] ⇒ move the drag handle down [Fig. 13.6(f)] ⇒ **MMB** ⇒ **Ctrl+D** ⇒ **View** ⇒ **Orientation** ⇒ **Previous** ⇒ **LMB** to deselect ⇒ **Ctrl+S** ⇒ **Enter**

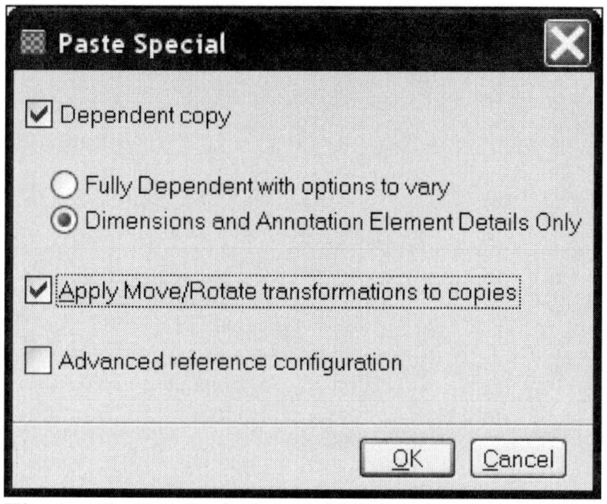

Figure 13.6(a) Paste Special Dialog Box

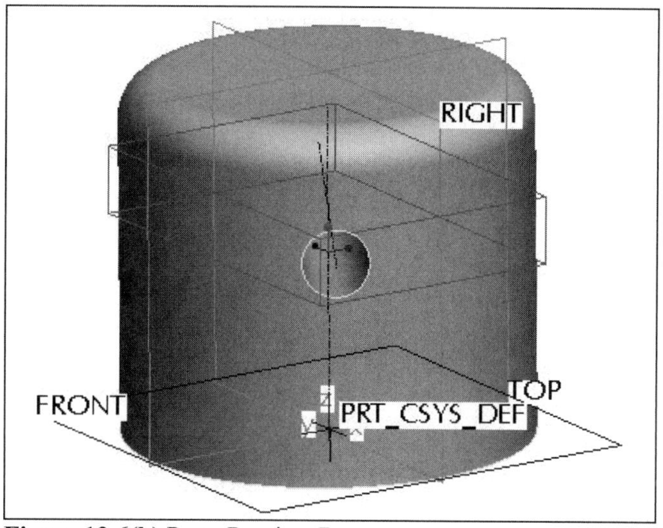

Figure 13.6(b) Paste Preview Box

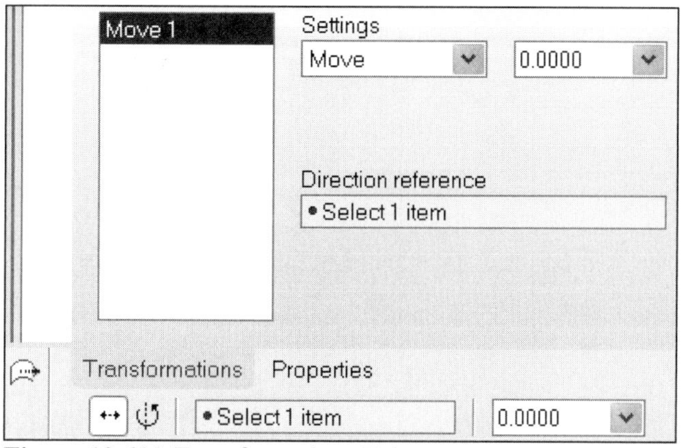

Figure 13.6(c) Transformations Tab

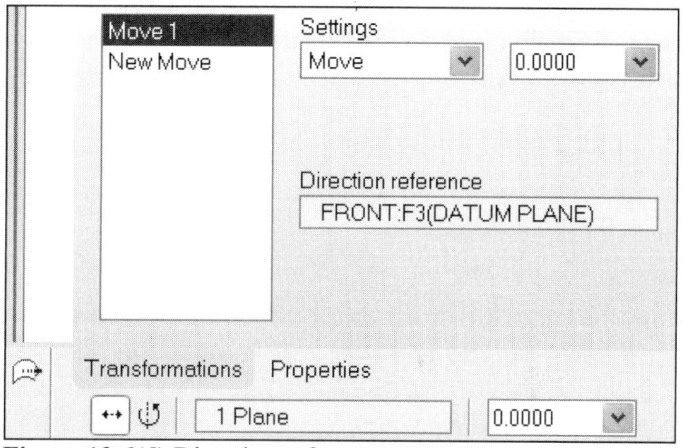

Figure 13.6(d) Direction reference FRONT

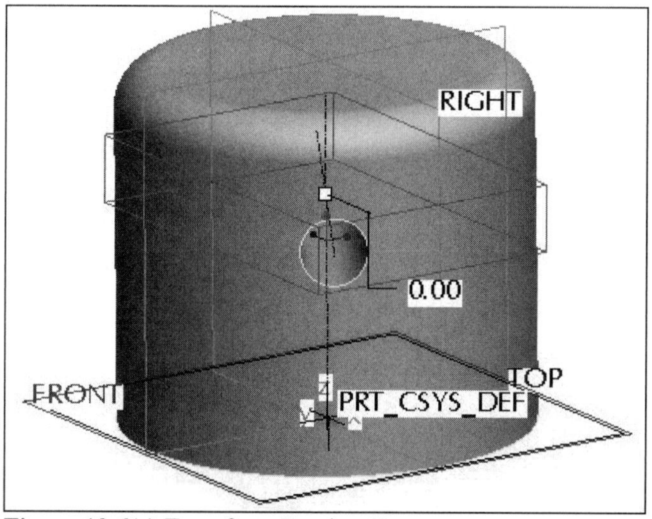

Figure 13.6(e) Transform Preview Box

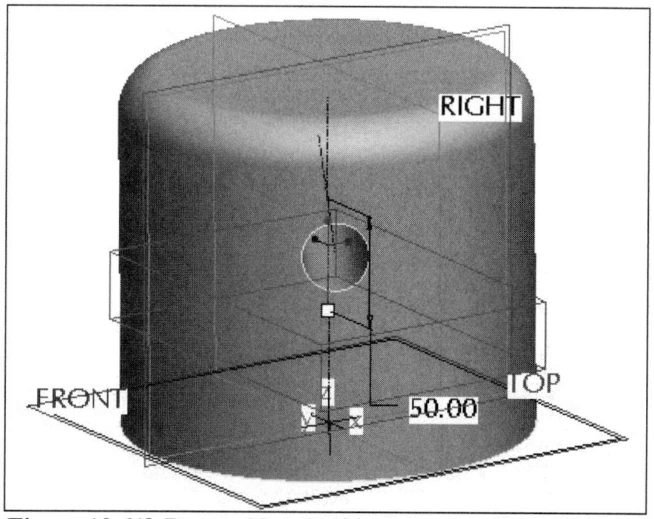

Figure 13.6(f) Drag to New Position

With the **Ctrl** key pressed, click on the **Hole** and the **Moved Copy** in the Model Tree ⇒ **RMB** ⇒ **Group** [Fig. 13.6(g)] ⇒ ⊞ ⇒ ⊞ [Fig. 13.6(h)] ⇒ **RMB** ⇒ **Pattern** [Fig. 13.6(i)] ⇒ **Dimensions** tab ⇒ **MMB** rotate the model to see the dimensions clearer ⇒ Direction 1- pick the **30°** dimension ⇒ **Enter** (to accept the default value of 30) ⇒ pattern members, type **6** [1 6 | 1 item(s)] [Fig. 13.6(j)] ⇒ **Enter** ⇒ ✔ [Fig. 13.6(k)] ⇒ ⇒ [PICT1] [Fig. 13.6(l)] ⇒ **LMB** to deselect ⇒ **Ctrl+S** ⇒ **OK** [Fig. 13.6(m)]

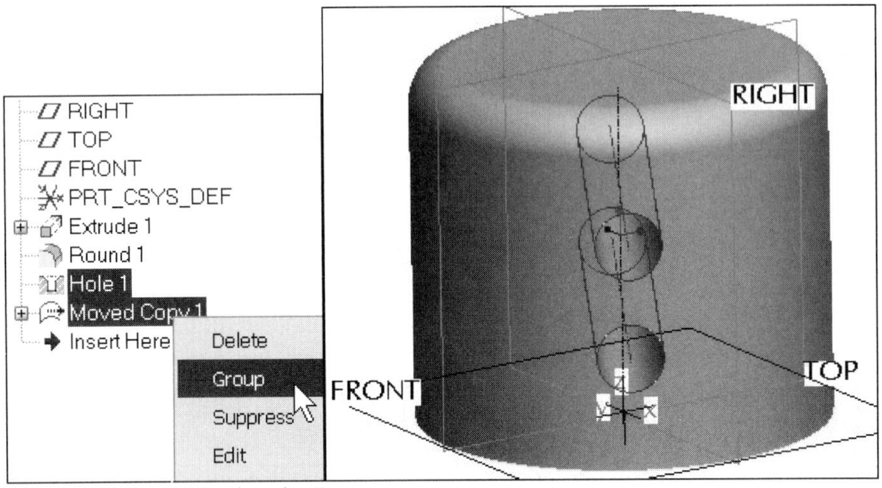

Figure 13.6(g) Grouping the Features

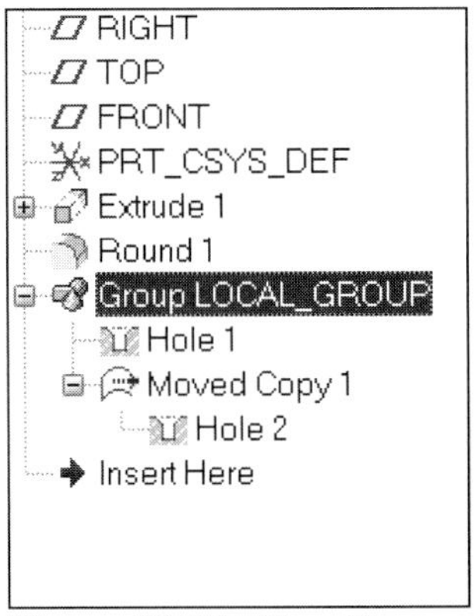

Figure 13.6(h) New Group

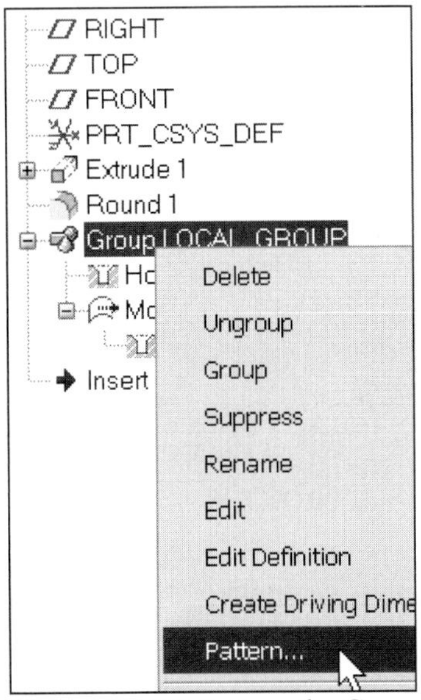

Figure 13.6(i) Pattern the Group

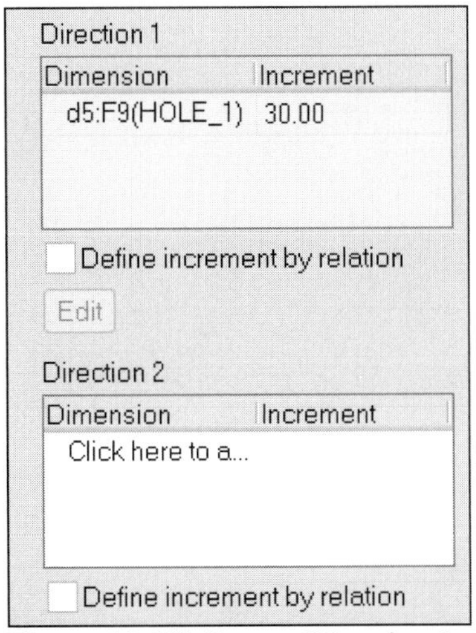

Figure 13.6(j) Pattern Direction 1

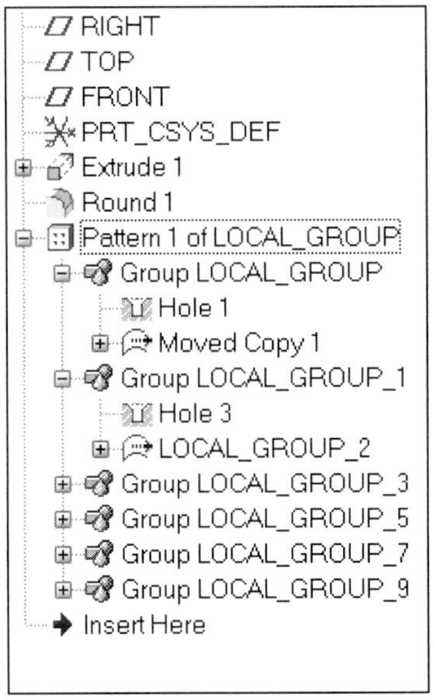

Figure 13.6(k) Pattern

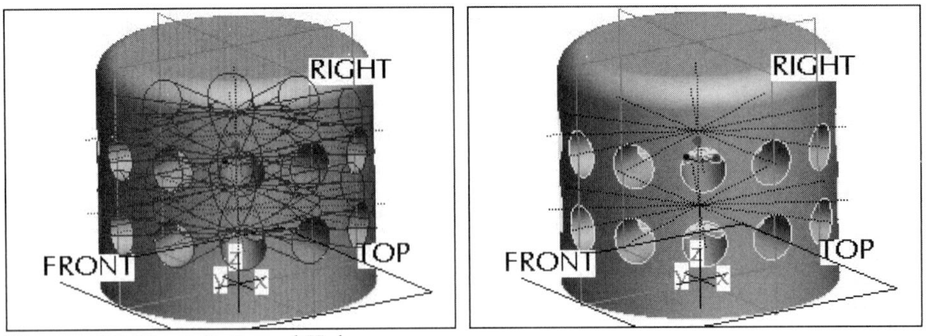

Figures 13.6(l-m) Patterned Holes

Pick on  in the Model Tree ⇒ drag **Insert Here** to a position before the **Pattern**  and drop *(this will roll back the model to a state before the pattern was created and enter the Insert Mode)* (Fig. 13.7) ⇒ rotate the model to see the bottom surface ⇒ pick the bottom surface [Fig. 13.8(a)] ⇒  **Shell Tool** [Fig. 13.8(b)] ⇒  [Fig. 13.8(c)] ⇒  ⇒ **MMB** ⇒ **Edit** ⇒ **Resume** ⇒ **Resume All** ⇒ **Ctrl+S** ⇒ **Enter** [Fig. 13.8(d)]

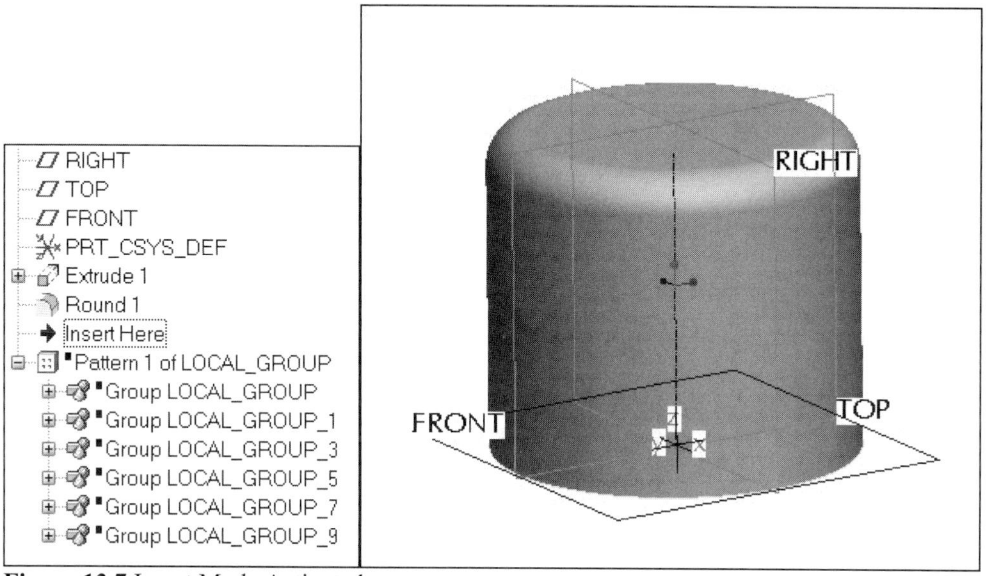

Figure 13.7 Insert Mode Activated

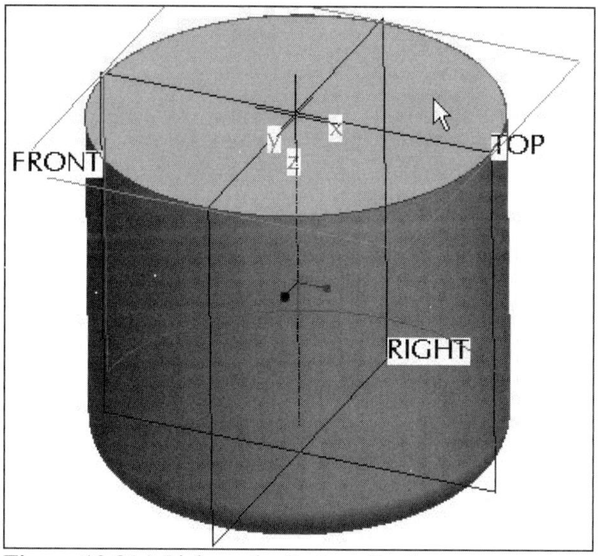

Figure 13.8(a) Pick on the Bottom Surface

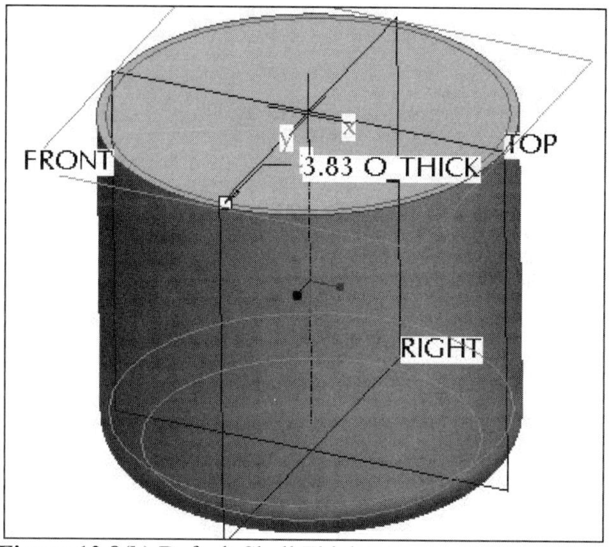

Figure 13.8(b) Default Shell Thickness

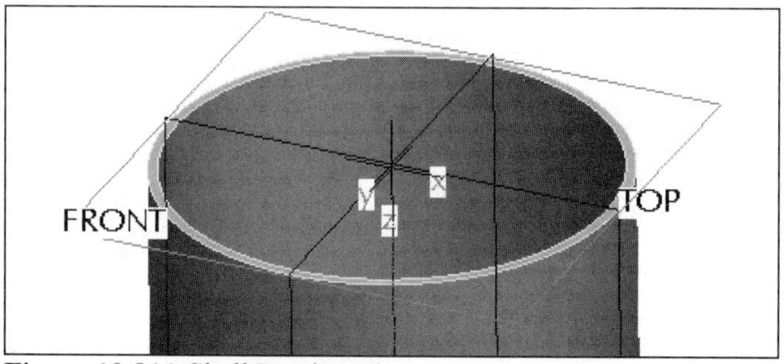

Figure 13.8(c) Shell Previewed

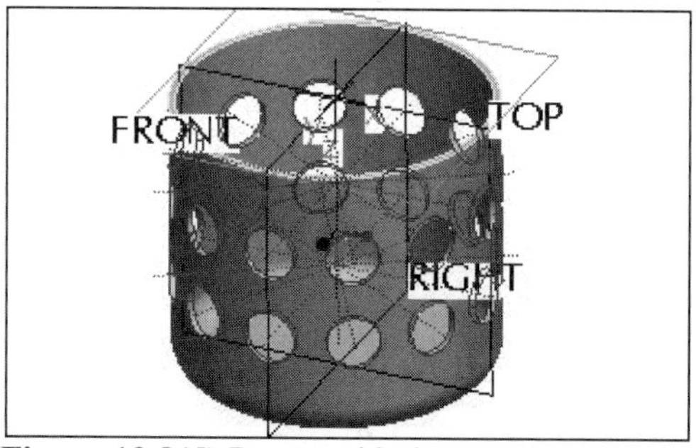

Figure 13.8(d) Resumed holes

Change the name and scale of the part, click: **File** ⇒ **Rename** ⇒ **Filter** ⇒ **OK** ⇒ **OK** ⇒ **Edit** ⇒ **Scale Model** ⇒ **.5** [Enter scale [1.0000]: .5] ⇒ **Enter** ⇒ **Yes** [Fig. 13.9(a)] ⇒ in the Model Tree, click on **Extrude** ⇒ **Shift** key ⇒ click on **Shell** ⇒ **RMB** ⇒ **Edit** [Fig. 13.9(b)] ⇒ [icon] ⇒ [PICT1] [Fig. 13.9(c)] ⇒ [icon] ⇒ **MMB** ⇒ **File** ⇒ **Delete** ⇒ **Old Versions** ⇒ **MMB** ⇒ **File** ⇒ **Close Window** ⇒ **File** ⇒ **Erase** ⇒ **Not Displayed** ⇒ **OK**

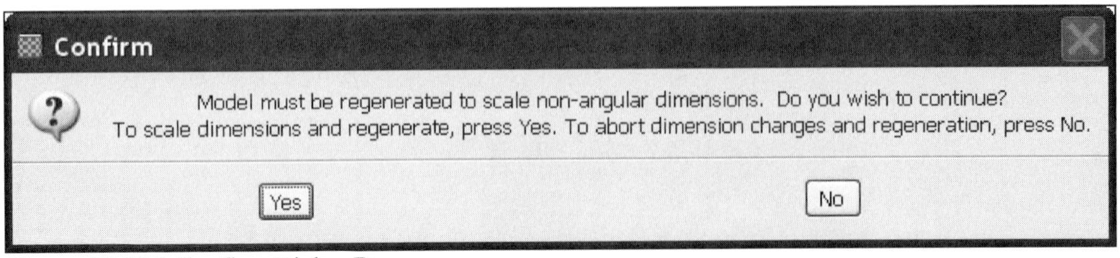

Figure 13.9(a) Confirm Dialog Box

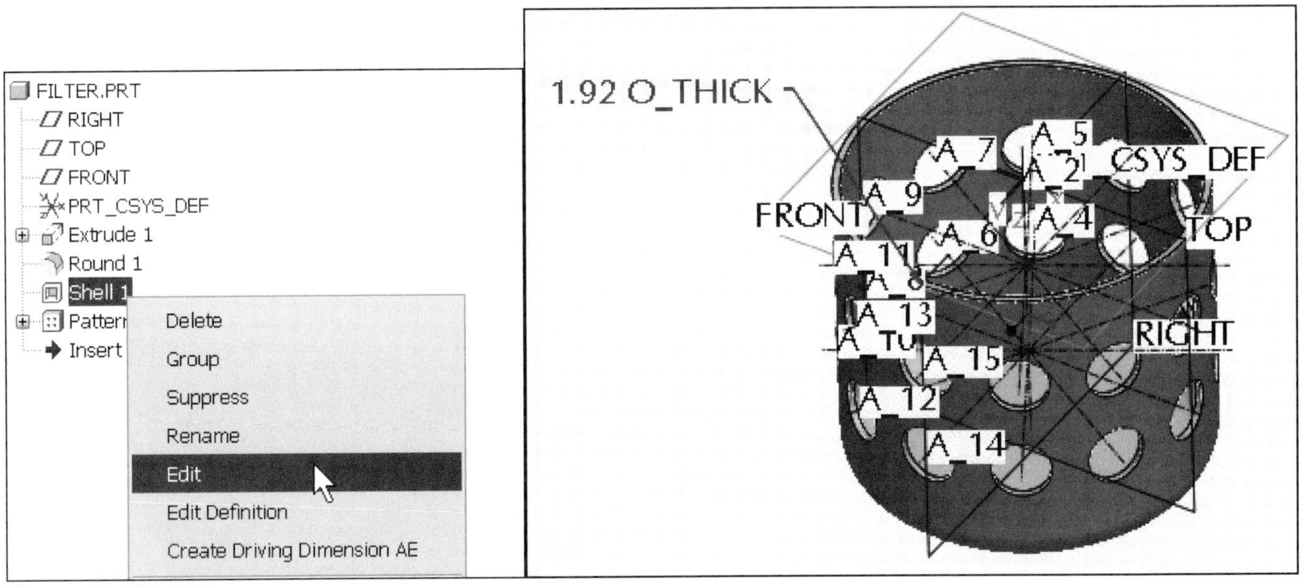

Figure 13.9(b) Edit

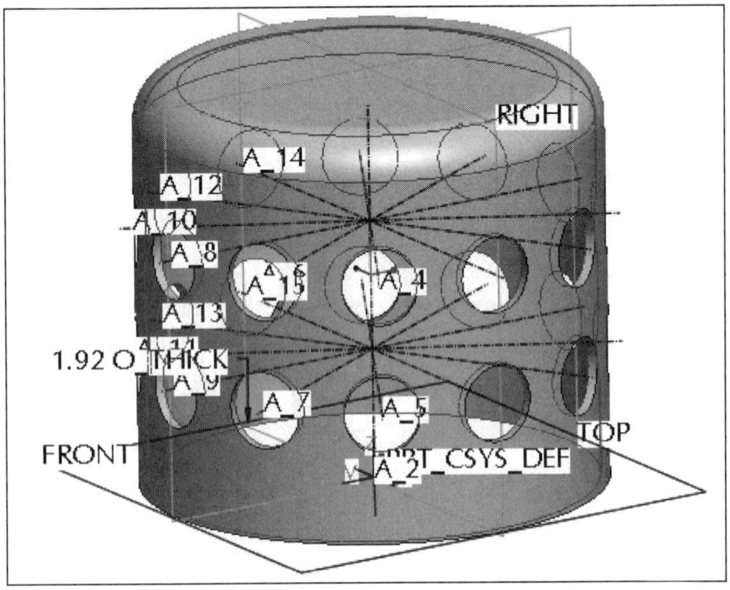

Figure 13.9(c) Completed Part

Part Model (Sheetmetal- Directional Pattern)

The Pro/E part database is different when you create parts using Pro/SHEETMETAL. All sheet metal parts are by definition thin-walled constant-thickness parts. Because of this, sheet metal parts have some unique properties, which other Pro/E parts do not have. You may convert a solid part into a sheet metal part, but not a sheet metal part into a solid part.

Click: ⇒ Part ⇒ Sheetmetal [Fig. 13.10(a)] ⇒ **OK** ⇒ **Edit** ⇒ **Setup** ⇒ **Units** [Fig. 13.10(b)] ⇒ **Close** ⇒ **Done** ⇒ **Create Unattached Flat Wall** from the Right Toolchest ⇒ .16 from the dashboard ⇒ **References** tab [Fig. 13.10(c)] ⇒ **Define** ⇒ pick the **TOP** datum plane [Figs. 13.10(d-e)] ⇒ **Sketch**

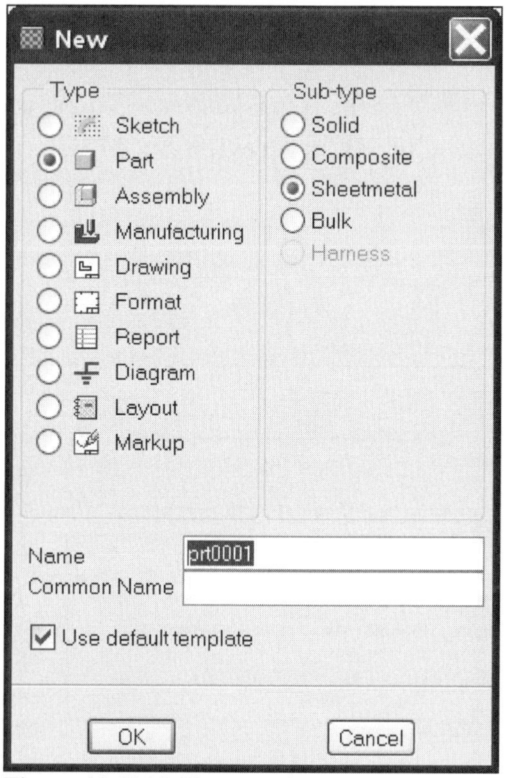

Figure 13.10(a) Sheetmetal

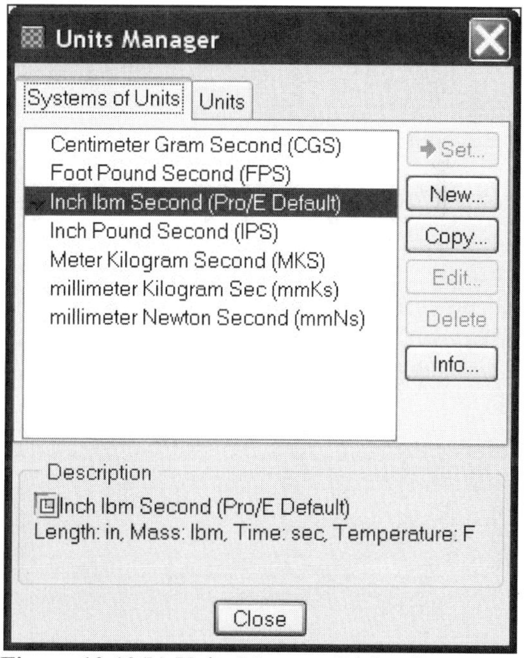

Figure 13.10(b) Units Manager Dialog Box

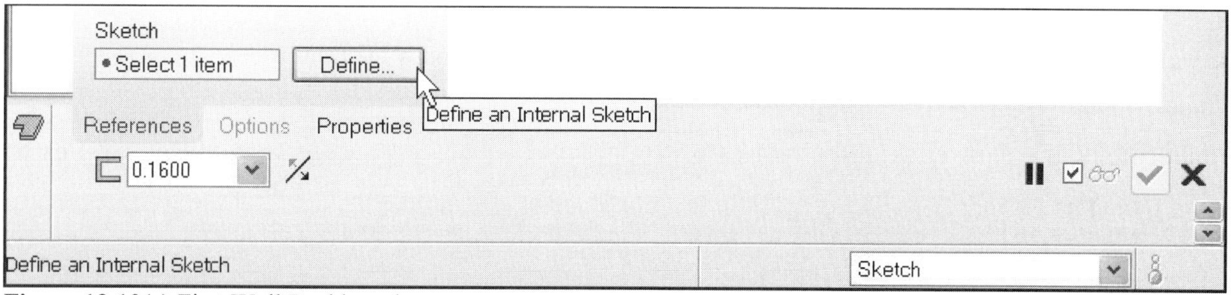

Figure 13.10(c) First Wall Dashboard

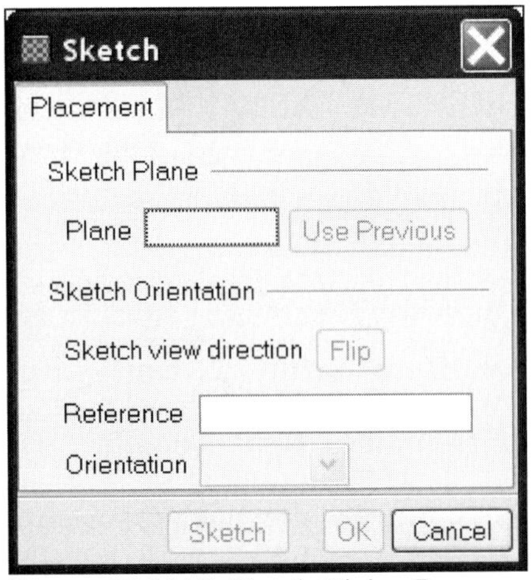

Figure 13.10(d) Sketch Dialog Box

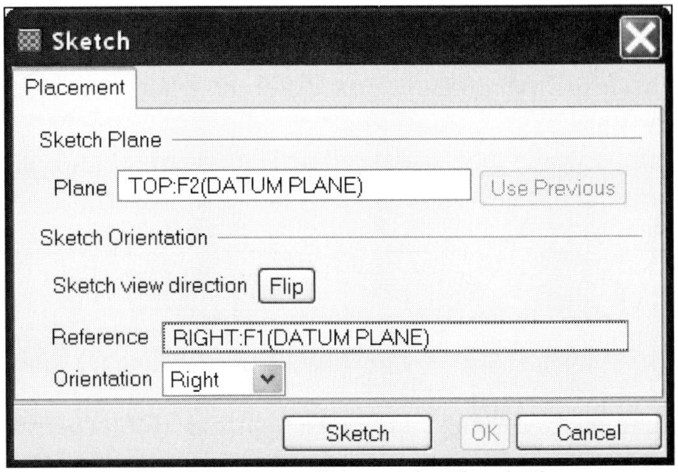

Figure 13.10(e) TOP Datum Selected for Sketch Plane

Click: **RMB** ⇒ **Centerline** ⇒ sketch a vertical and a horizontal centerline ⇒ **RMB** ⇒ **Rectangle** ⇒ sketch a rectangle [Fig. 13.10(f)] ⇒ modify the dimensions [Fig. 13.10(g)] ⇒  ⇒ **Tools** ⇒ **Environment** ⇒ Standard Orient **Isometric** ⇒ **Apply** ⇒ **Close** ⇒ **Ctrl+D** ⇒ **Preview** ⇒ **MMB** ⇒ **MMB** ⇒  ⇒ **Ctrl+S** ⇒ **Enter** ⇒ **LMB** to deselect

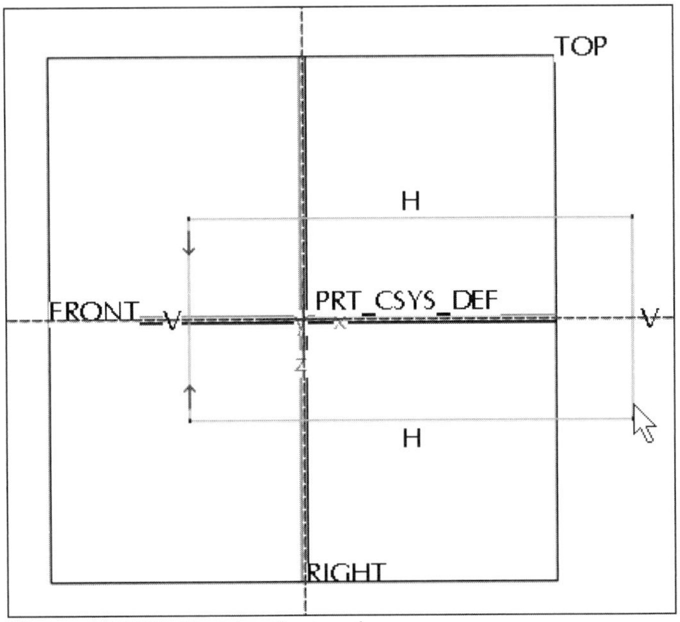

Figure 13.10(f) Sketch a Rectangle

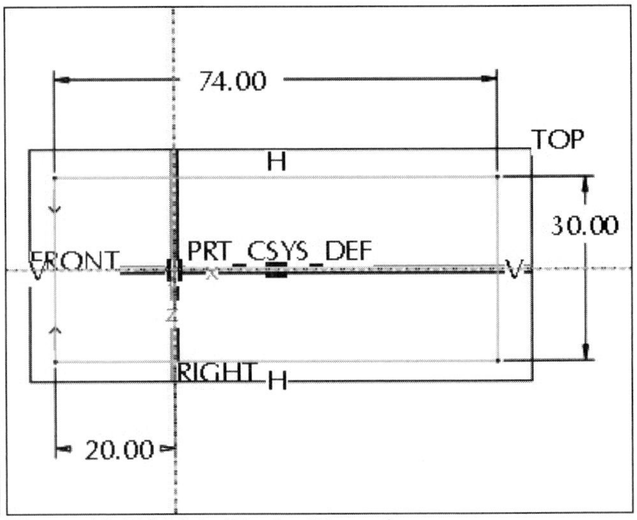

Figure 13.10(g) Modify the Dimensions

Change the scale of the part, click: **Edit** ⇒ **Scale Model** ⇒ **.2** Enter scale [1.0000]: .2 ⇒ **Enter** ⇒ **Yes** ⇒  ⇒  ⇒ **MMB**

Double-click on the wall to see its dimensions [Fig. 13.10(h)] ⇒  **Create Flat Wall** ⇒ pick on the ***lower*** edge of the first wall [Figs. 13.10(i-j)]

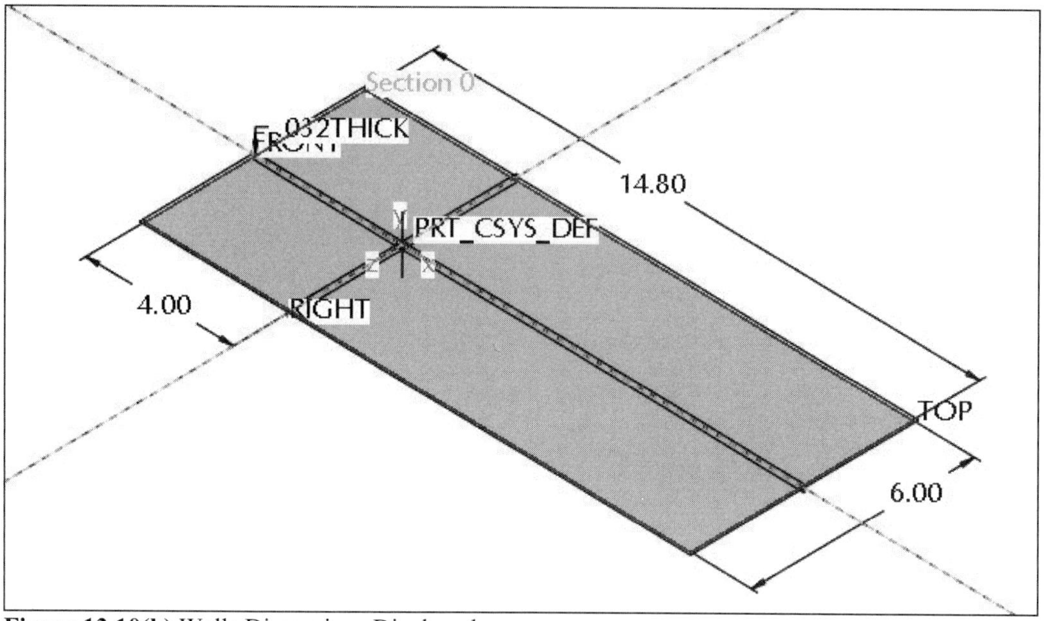

Figure 13.10(h) Walls Dimensions Displayed

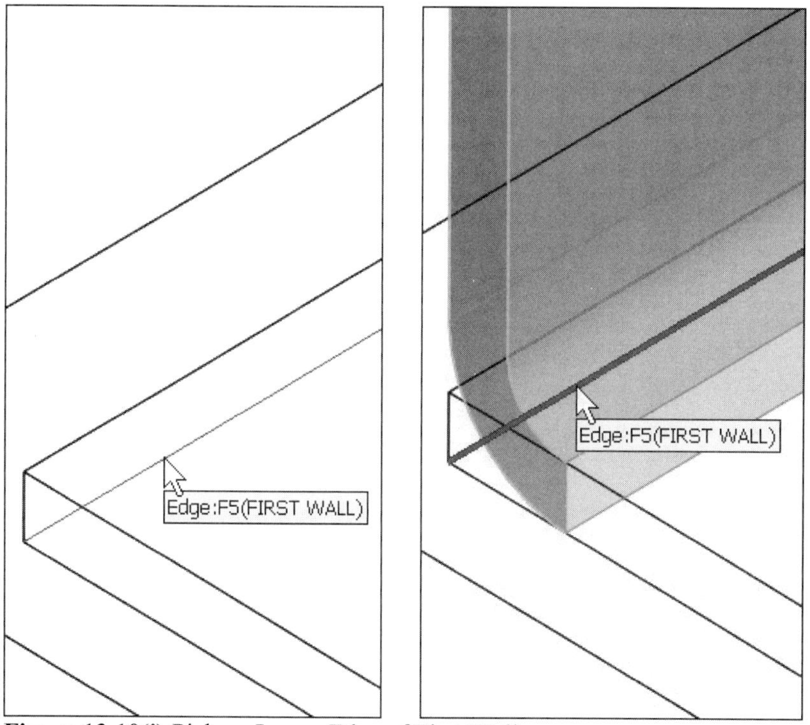

Figure 13.10(i) Pick on Lower Edge of First Wall

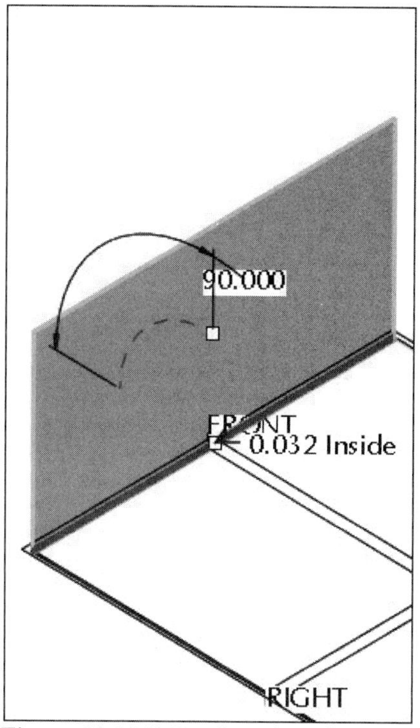

Figure 13.10(j) Previewed Wall

Click: ⇒ **Rectangle** ⇒ ⇒ **T** 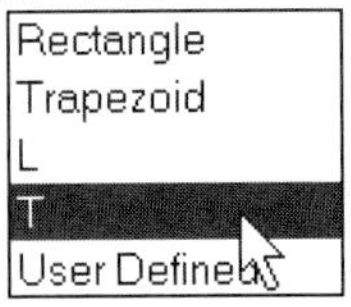

[Fig. 13.10(k)] **Shape** tab [Fig. 13.10(l)] ⇒ **MMB** ⇒ ⇒ ⇒ **MMB** [Fig. 13.10(m)]

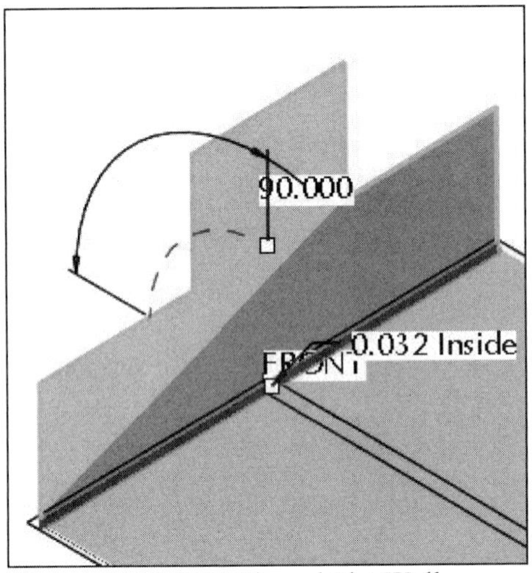

Figure 13.10(k) T-Shaped Flat Wall

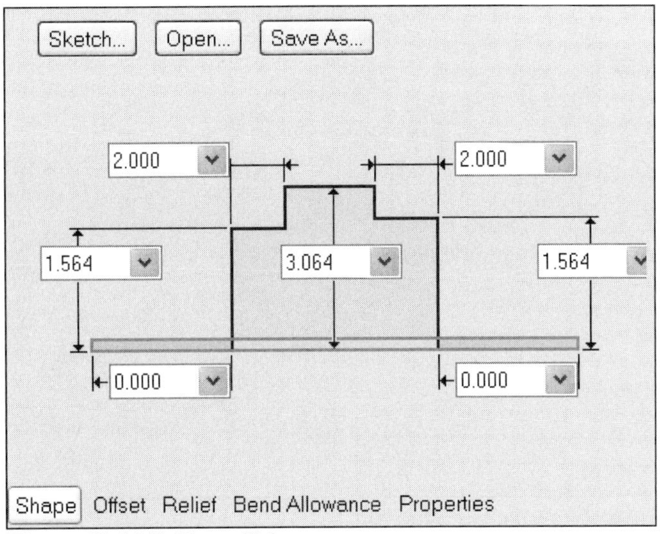

Figure 13.10(l) Shape Tab

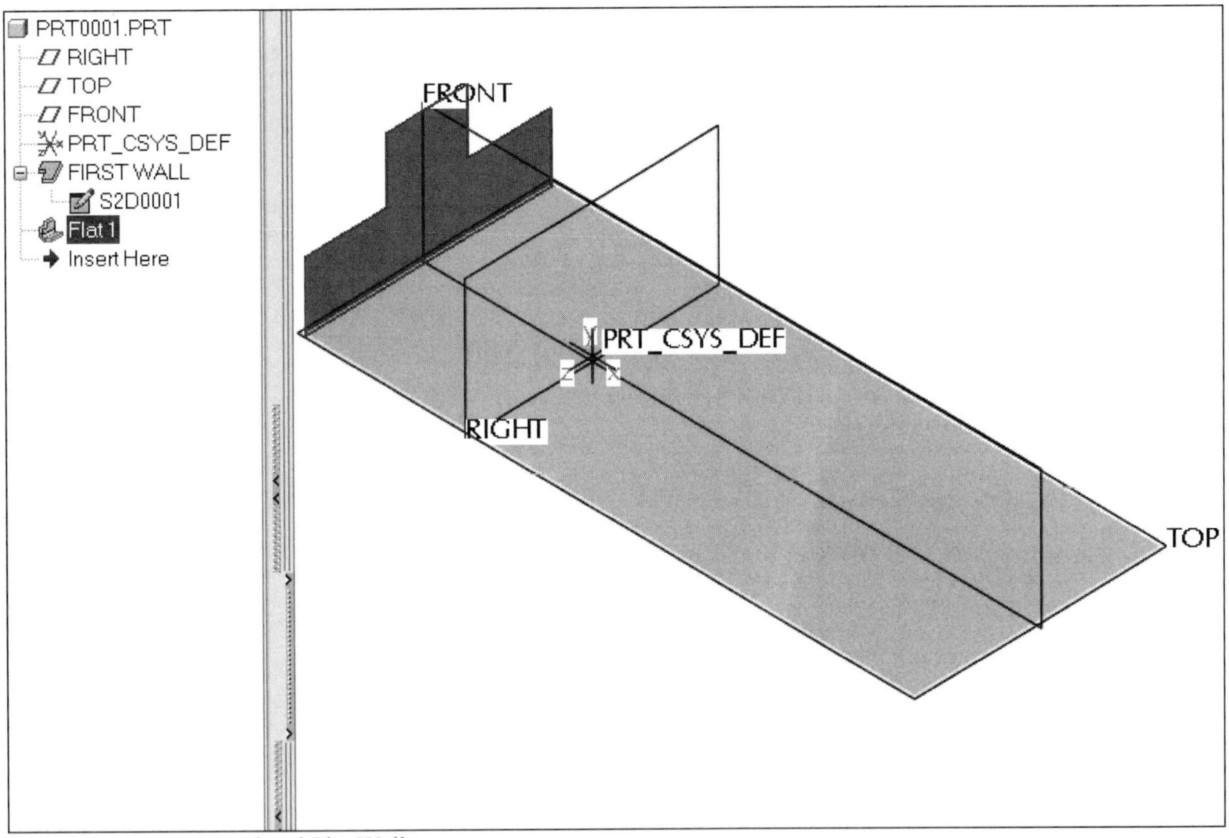

Figure 13.10(m) Completed Flat Wall

Click: Create Flange Wall ⇒ pick on the opposite edge (lower) of the first wall [Fig. 13.10(n)]

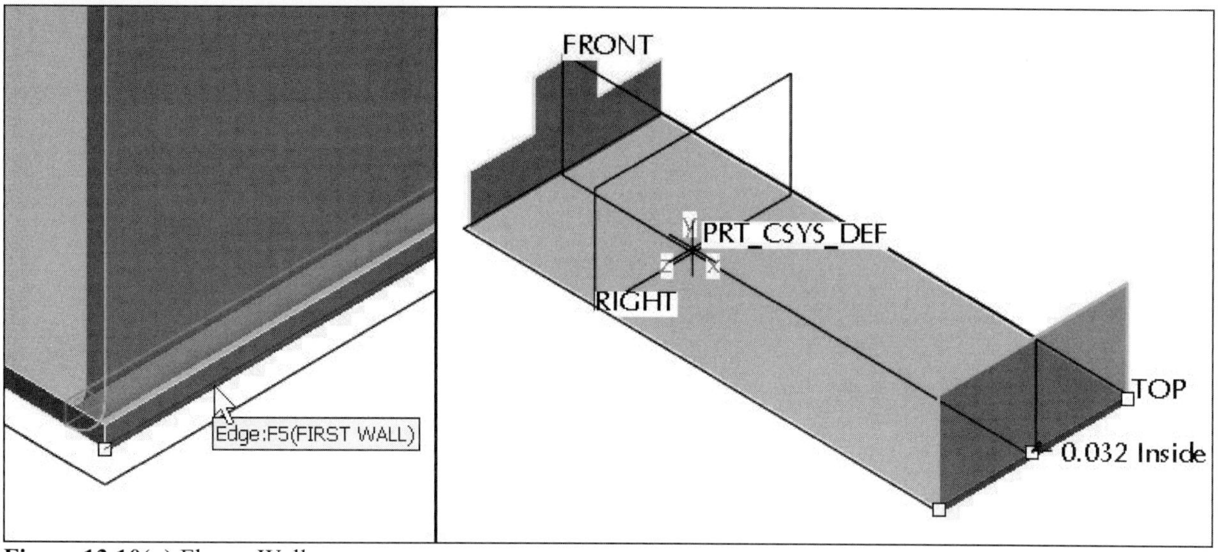

Figure 13.10(n) Flange Wall

Click: **Profile** tab [Fig. 13.10(o)] ⇒ modify the values as shown ⇒ **Enter** ⇒ rotate model [Fig. 13.10(p)]

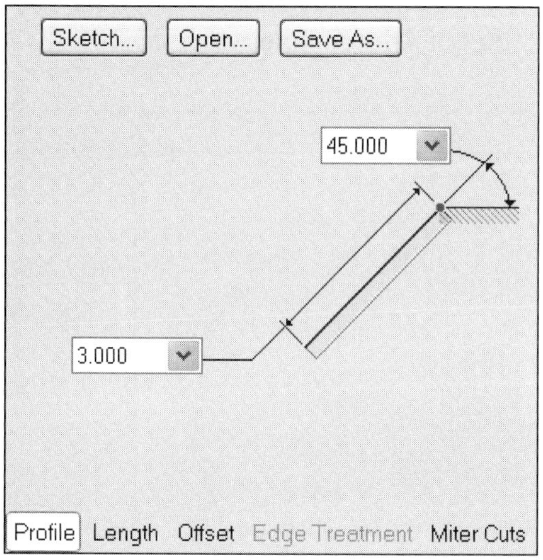

Figure 13.10(o) Profile Tab

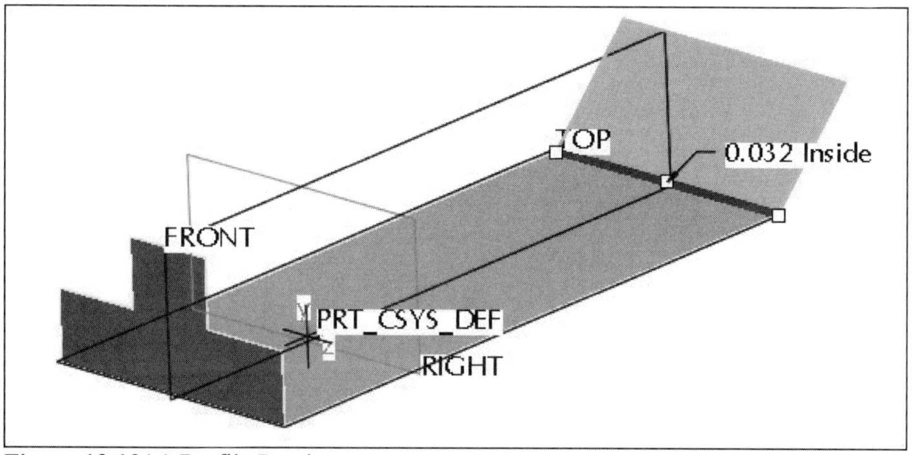

Figure 13.10(p) Profile Preview

Pick on and move a drag handle to shorten the wall [Fig. 13.10(q)] ⇒ **MMB**

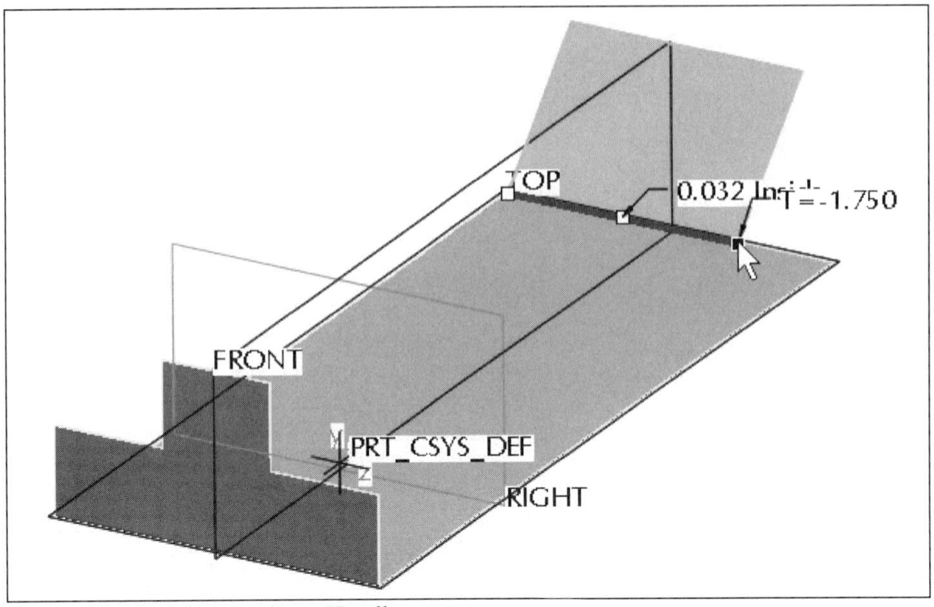

Figure 13.10(q) Move a Drag Handle

Click: **Ctrl+S** ⇒ **Enter** ⇒ **File** ⇒ **Rename** ⇒ **Sheetmetal** ⇒ **OK** ⇒ **OK** ⇒ click on the last three features in the Model Tree ⇒ **RMB** ⇒ **Edit** ⇒ **RMB** ⇒ **Show Sketch Dimensions** [Fig. 13.10(r)] ⇒ **Ctrl+S** ⇒ **OK**

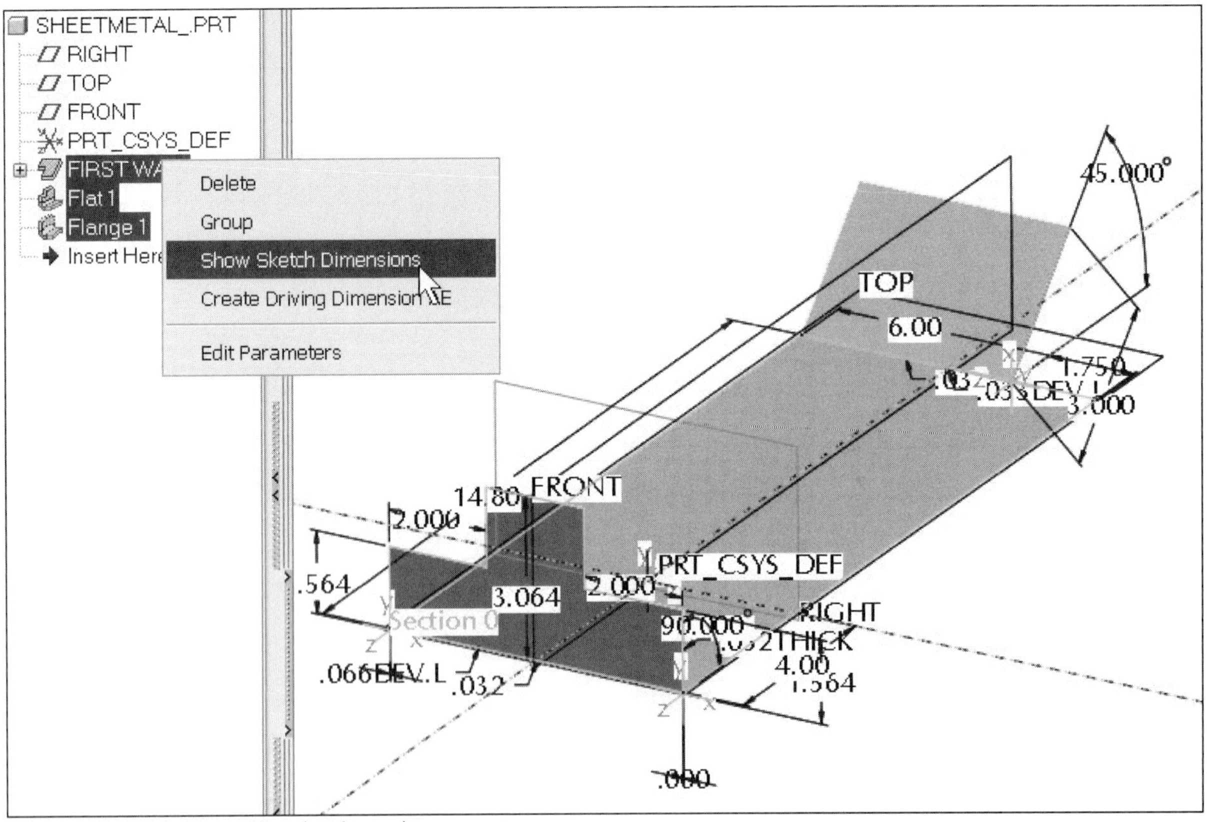

Figure 13.10(r) Show Sketch Dimensions

Click: **Insert** from the menu bar ⇒ 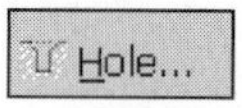

⇒ 

⇒ pick on the top surface [Fig. 13.10(s)] ⇒ **RMB** ⇒ 

⇒ pick the long edge [Fig. 13.10(t)] ⇒ press and hold **Ctrl** key ⇒ pick the short edge [Fig. 13.10(u)] ⇒ modify the dimensions [Fig. 13.10(v)] ⇒ **MMB** ⇒ **Ctrl+D** ⇒ **Ctrl+S** ⇒ **Enter**

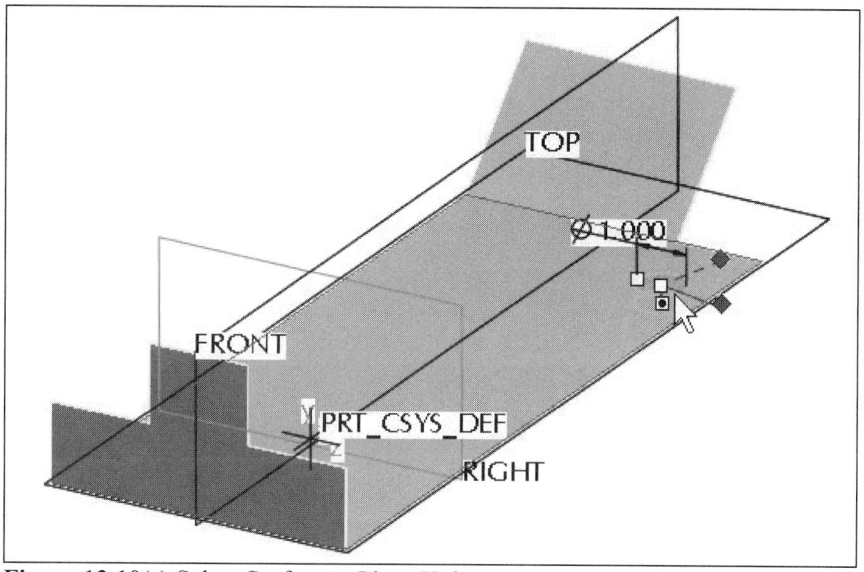

Figure 13.10(s) Select Surface to Place Hole

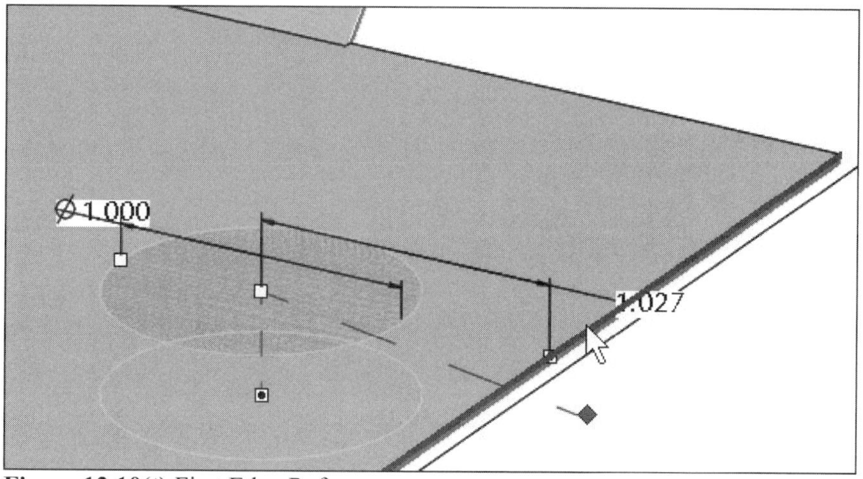

Figure 13.10(t) First Edge Reference

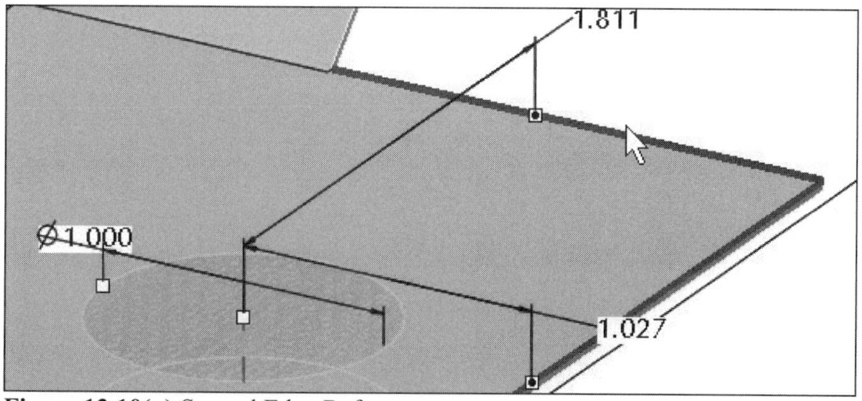

Figure 13.10(u) Second Edge Reference

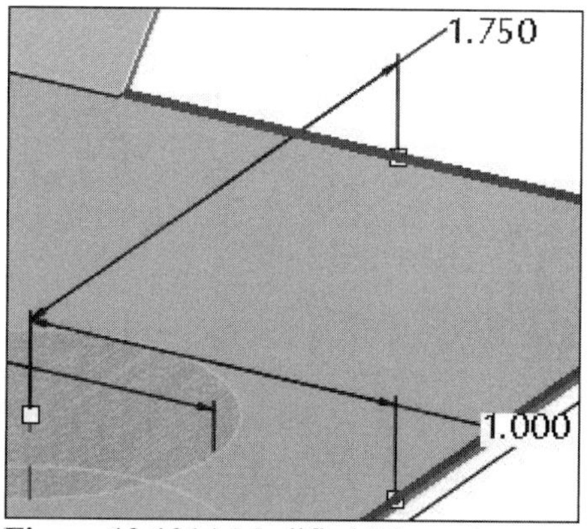

Figure 13.10(v) Modified Dimensions

With the Hole selected in the Model Tree [Fig. 13.11(a)], click: **RMB ⇒ Pattern ⇒ Ctrl+MMB** zoom in ⇒ Dimension ⇒ ⇒ Direction ⇒ pick on the same first edge used to locate the hole [Fig. 13.11(b)] ⇒ **Flip the first direction** [Fig. 13.11(c)] ⇒ type **7** as the number of holes in this direction ⇒ **Enter** ⇒ modify the dimension to **2.00** ⇒ **Enter** [Fig. 13.11(d)]

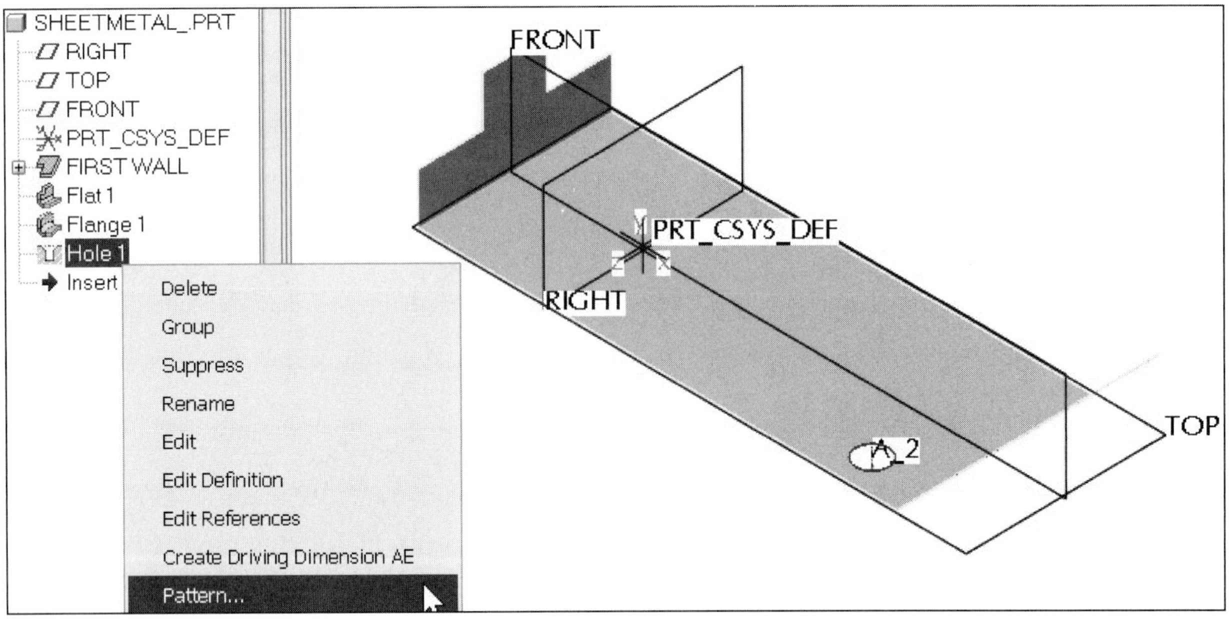

Figure 13.11(a) RMB ⇒ Pattern

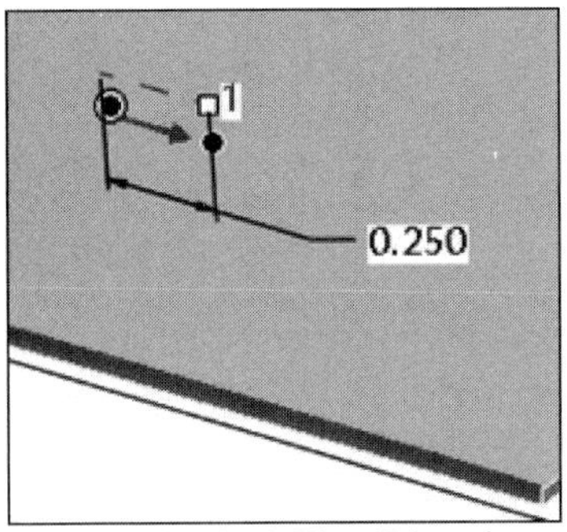

Figure 13.11(b) First Direction

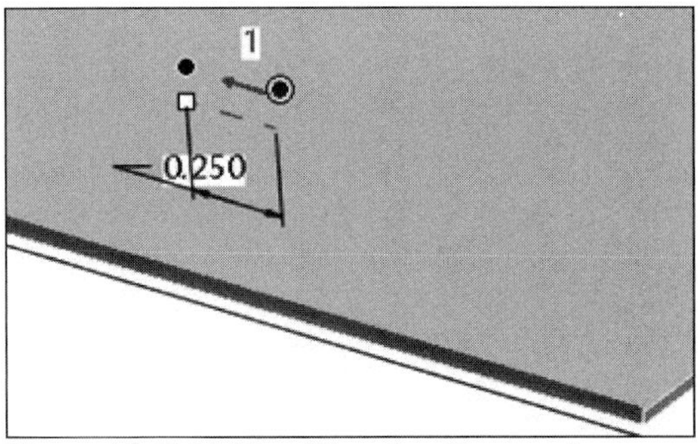

Figure 13.11(c) First Direction Reversed

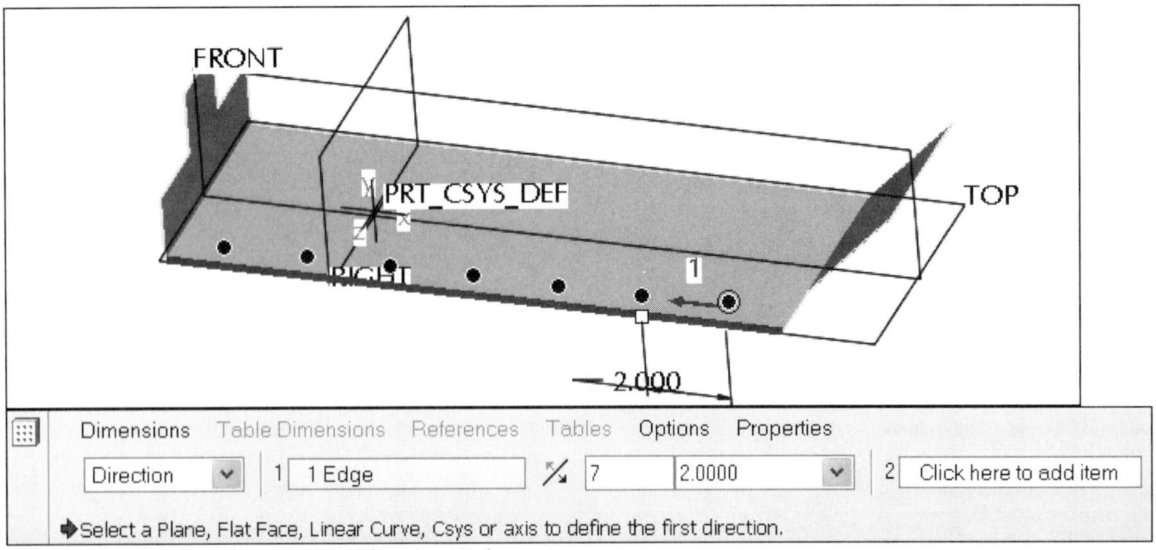

Figure 13.11(d) Previewed First Direction Holes

Click in [2 Click here to add item] ⇒ [2 Select 1 item] ⇒ pick on the same second edge used to locate the hole ⇒ modify the dimension to **4.00** ⇒ **Enter** [Fig. 13.11(e)]

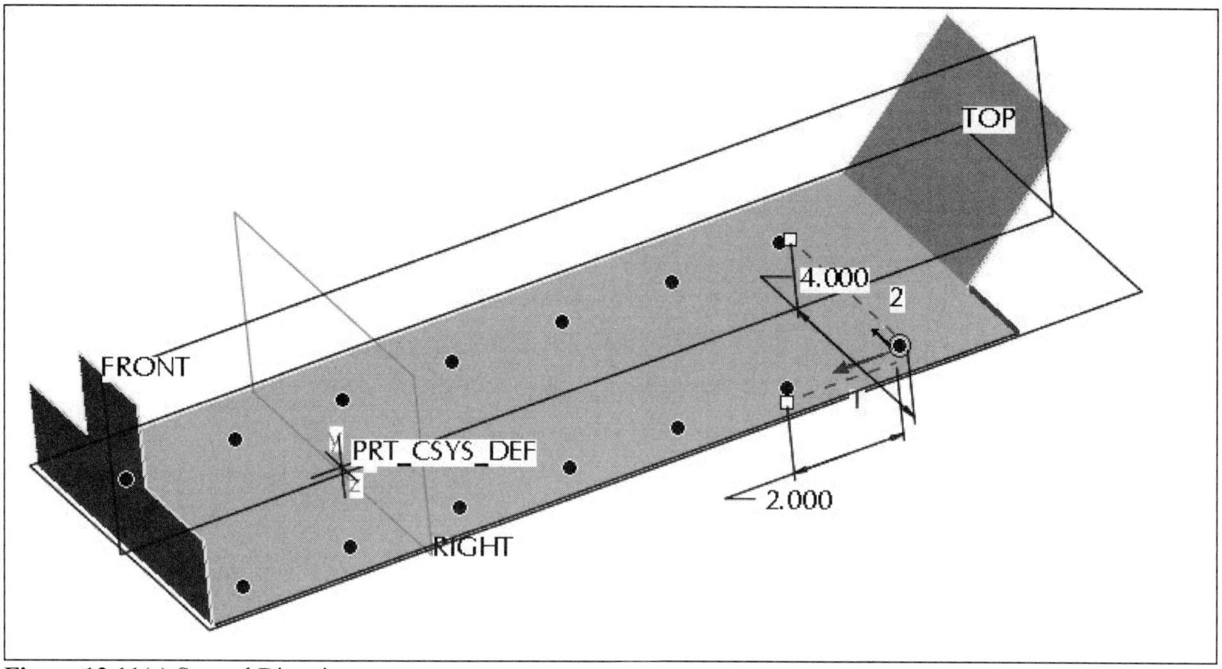

Figure 13.11(e) Second Direction

Pick on the two pattern hole preview dots (black) to remove them from the pattern (dots turn white) [Fig. 13.11(f)] ⇒ ✓ ⇒ **Ctrl+S** ⇒ **Enter** [Fig. 13.11(g)]

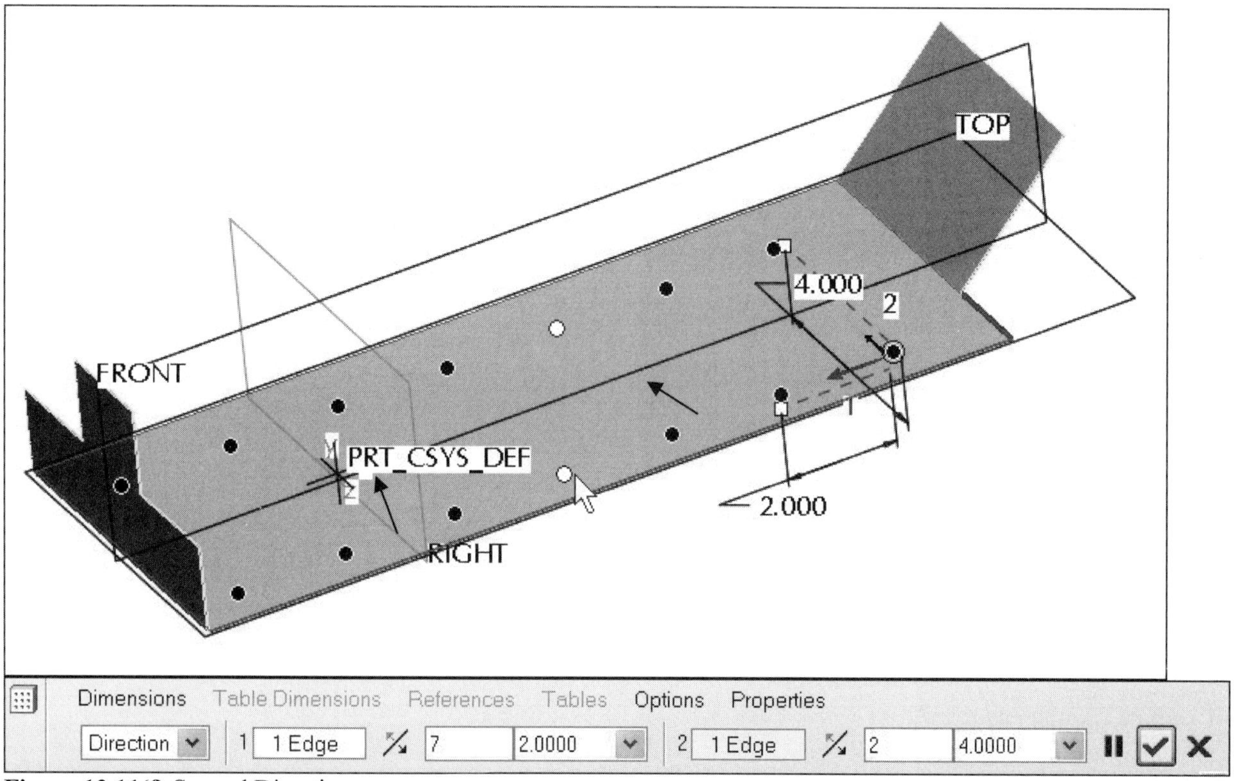

Figure 13.11(f) Second Direction

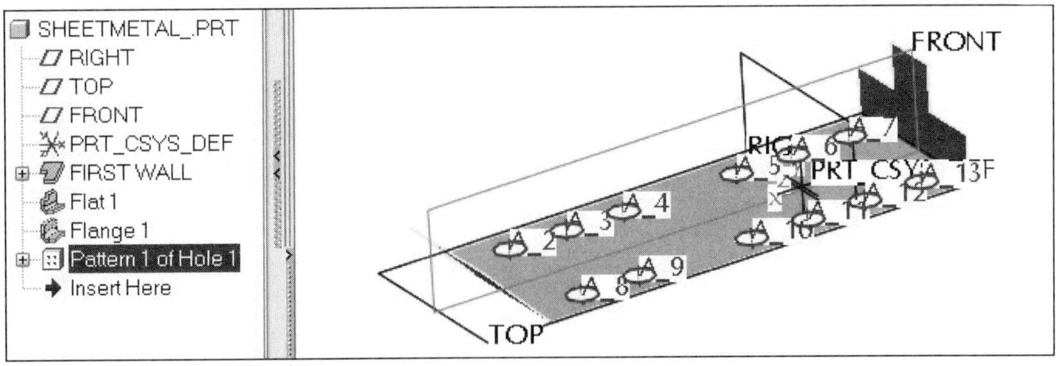

Figure 13.11(g) Completed Pattern

Click: **Refit ⇒ View ⇒ Orientation ⇒ Standard Orientation ⇒** **Create Flat Pattern ⇒** pick on the upper surface [Figs. 13.11(h-i)] ⇒ **Edit ⇒ Undo: Flat Pattern ⇒ Info ⇒ Model Size ⇒** Length of bounding box diagonal = **18.4739 INCH ⇒ Ctrl+S ⇒ Enter ⇒ File ⇒ Close Window**

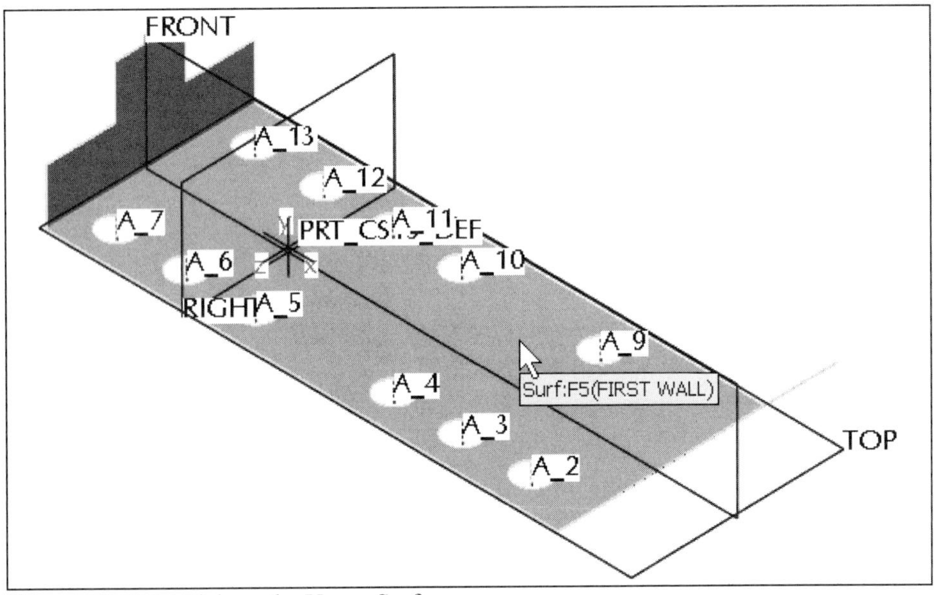

Figure 13.11(h) Pick on the Upper Surface

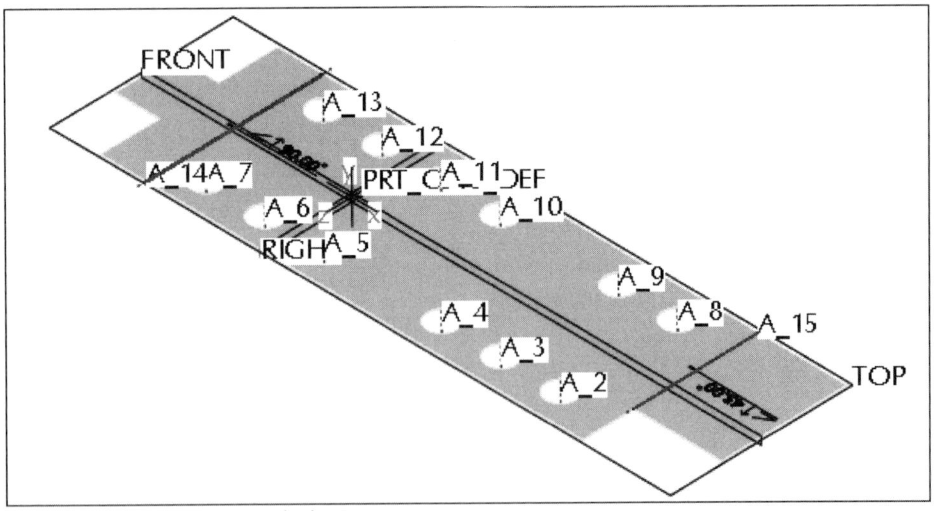

Figure 13.11(i) Completed Flat Pattern

For those who wish additional information on sheetmetal, see ***www.cad-resources.com*** ***⇒ CDI74***

Lesson 14 Blends

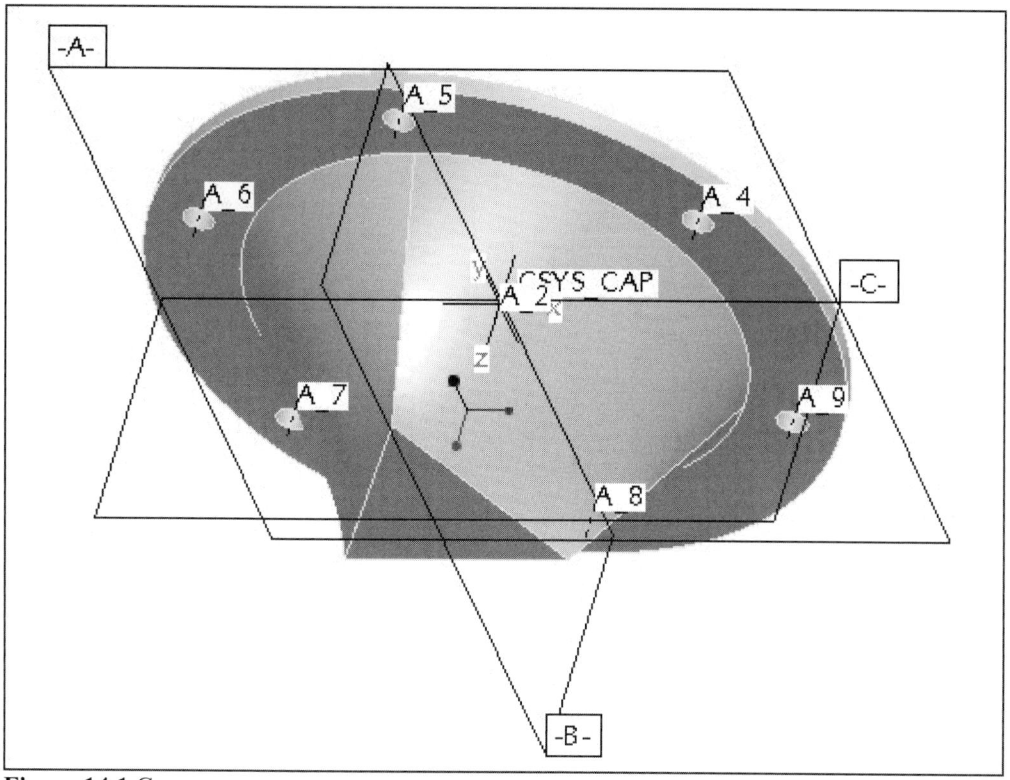

Figure 14.1 Cap

OBJECTIVES

- Create a **Parallel Blend** feature
- Use the **Shell Tool**
- Create a **Swept Blend**

BLENDS

A blended feature consists of a series of at least two planar sections that are joined together at their edges with transitional surfaces to form a continuous feature. The Cap in Figure 14.1 uses a blend feature in its design. A Blend can be created as a **Parallel Blend**, or you can construct a **Swept Blend**.

Blend Sections

Blended surfaces are created between the corresponding sections. Figure 14.2 shows a parallel blend for which the *section* consists of several *subsections*. Each segment in the subsection is matched with a segment in the following subsection; to create the transitional surfaces; Pro/E connects the *starting points* of the subsections and continues to connect the vertices of the subsections in a clockwise manner. By changing the starting point of a blend subsection, you can create blended surfaces that twist between the subsections. The default starting point is the first point sketched in the subsection. You can position the starting point to the endpoint of another segment by choosing the option Start Point and selecting the new position.

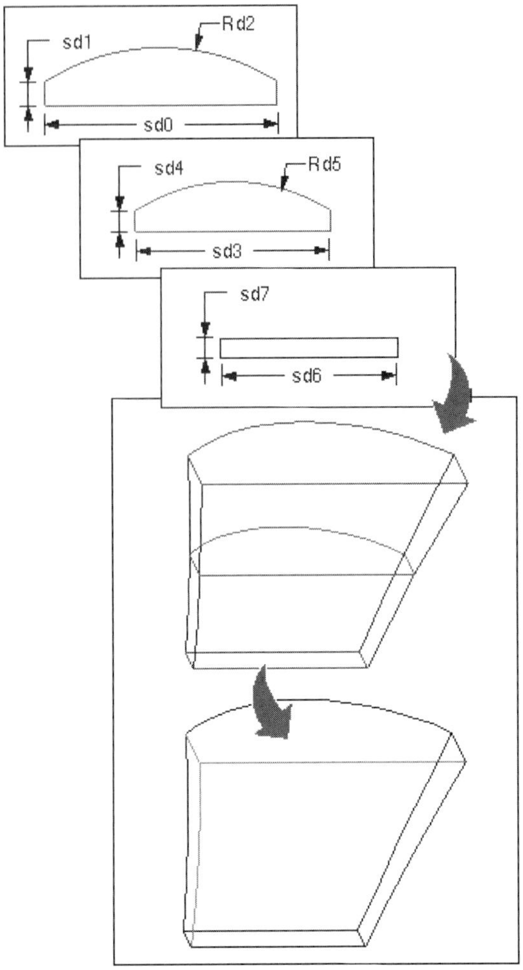

Figure 14.2 Blend Sections (CADTRAIN, COAch for Pro/ENGINEER)

Blend Options

Blends (Fig. 14.3) use one of the following transitional surface options:

- **Straight** Create a straight blend by connecting vertices of different subsections with straight lines. Edges of the sections are connected with ruled surfaces.
- **Smooth** Create a smooth blend by connecting vertices of different subsections with smooth curves. Edges of the sections are connected with ruled (spline) surfaces.
- **Parallel** All blend sections lie on parallel planes in one section sketch.
- **Rotational** The blend sections are rotated about the **Y** axis, up to a maximum of **120°**. Each section is sketched individually and aligned using the coordinate system of the section.
- **General** The sections of a general blend can be rotated about and translated along the **X**, **Y**, and **Z** axes. Sections are sketched individually and aligned using the coordinate system of the section.
- **Regular Sec** The feature will use the regular sketching plane.
- **Project Sec** The feature will use the projection of the section on the selected surface. This is used for parallel blends only.
- **Select Sec** Select section entities (not available for parallel blends).
- **Sketch Sec** Sketch section entities.

Blend Type	Section --Smooth --Straight
Parallel--All blend sections lie on parallel planes in one section sketch.	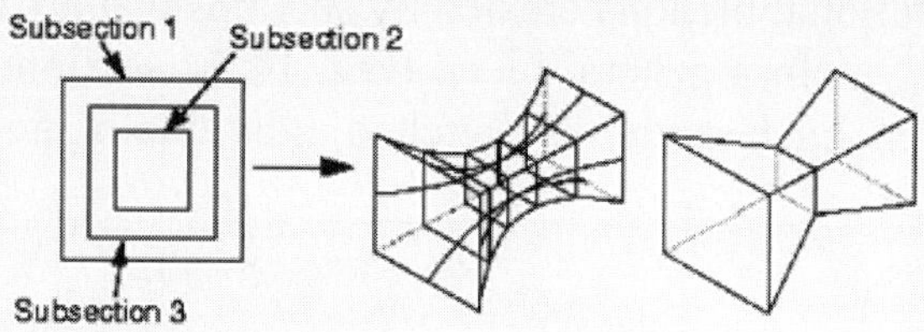
Rotational--Blend sections are rotated about the Y-axis, up to a maximum of 120 degrees. Each section is sketched individually and aligned using the coordinate system of the section.	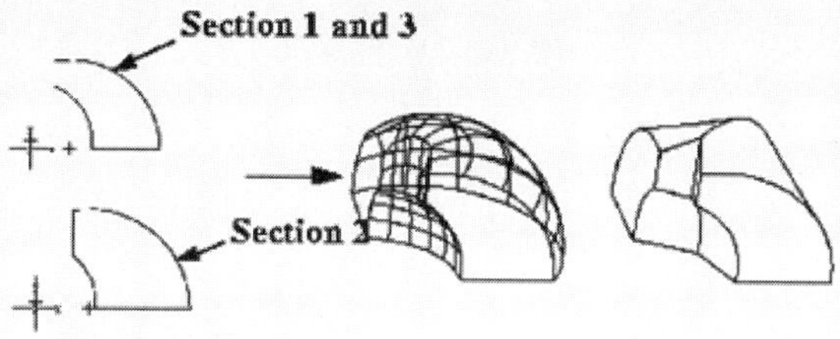
General--Sections of a general blend can be rotated about and translated along the X-, Y-, and Z-axes. Each section is sketched individually, and aligned using the coordinate system of the section.	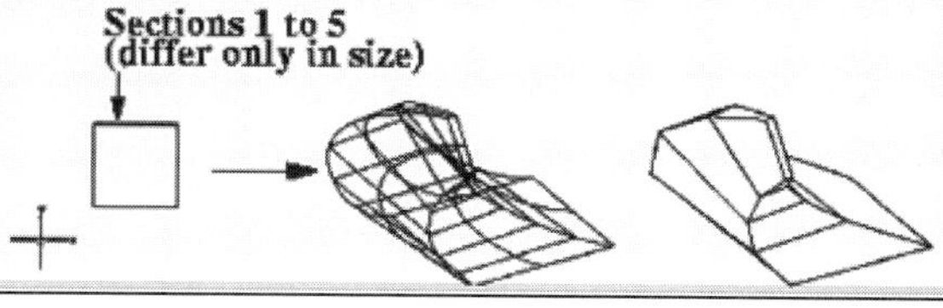

Figure 14.3 Blend Sections

Parallel Blends

A parallel blend is created from a single section that contains multiple sketches called *subsections* (Fig. 14.4). A first or last subsection can be defined as a point resulting in a blend vertex. The starting point for each subsection must be carefully selected as per the design requirements including the starting points (Fig. 14.5).

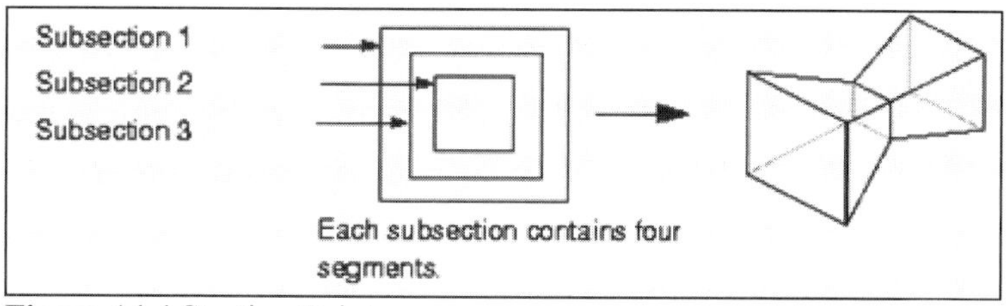

Figure 14.4 Starting Points

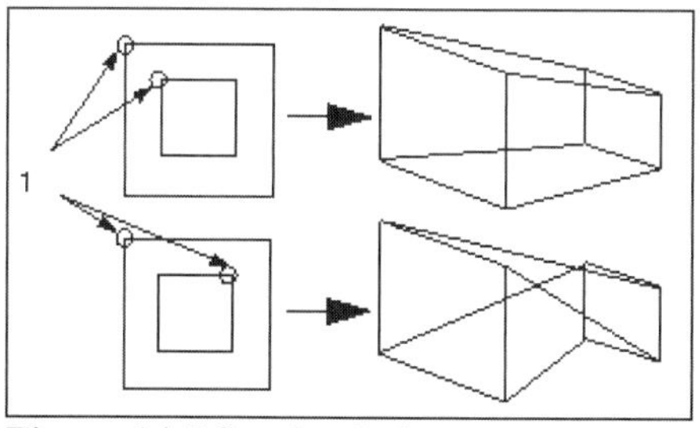

Figure 14.5 Starting Points

Rotational and General Blends

You create rotation blends (Fig. 14.6) by sketching a series of profiles which are offset by an angle. A rotational blend is created by sections that are rotated about the Y axis.

In a general blend (Fig. 14.7), sections can be rotated about and translated along the X, Y, and Z axes. Each section is sketched individually, and aligned using the coordinate system of the section.

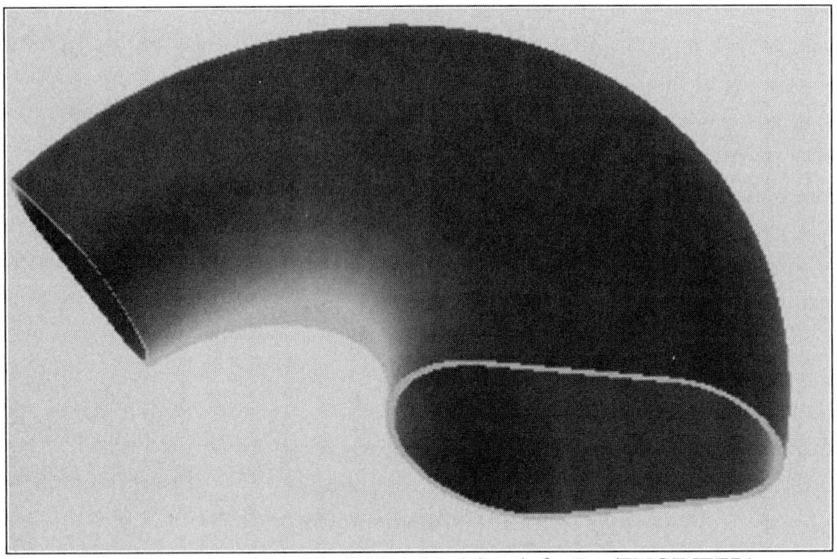

Figure 14.6 Blend Sections (CADTRAIN, COAch for Pro/ENGINEER)

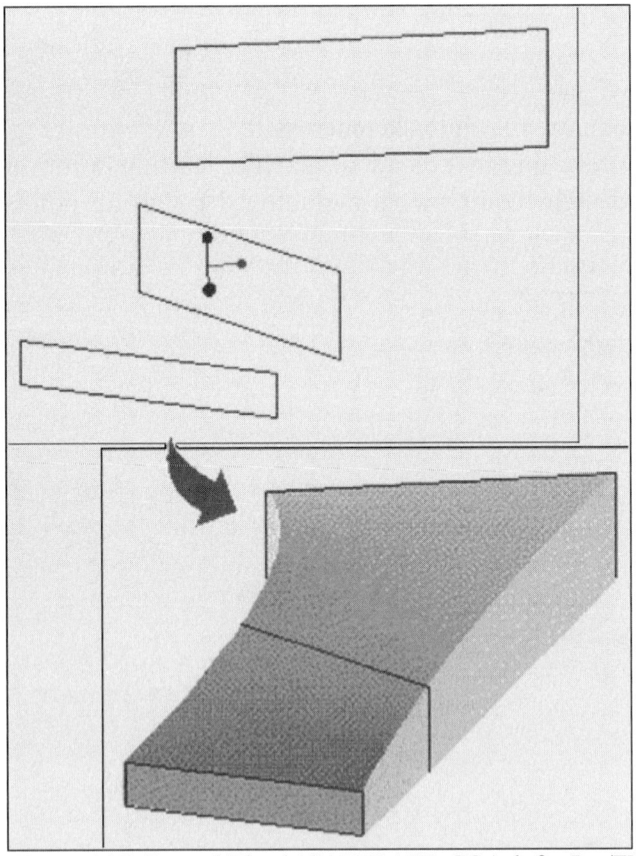

Figure 14.7 General Blend (CADTRAIN, COAch for Pro/ENGINEER)

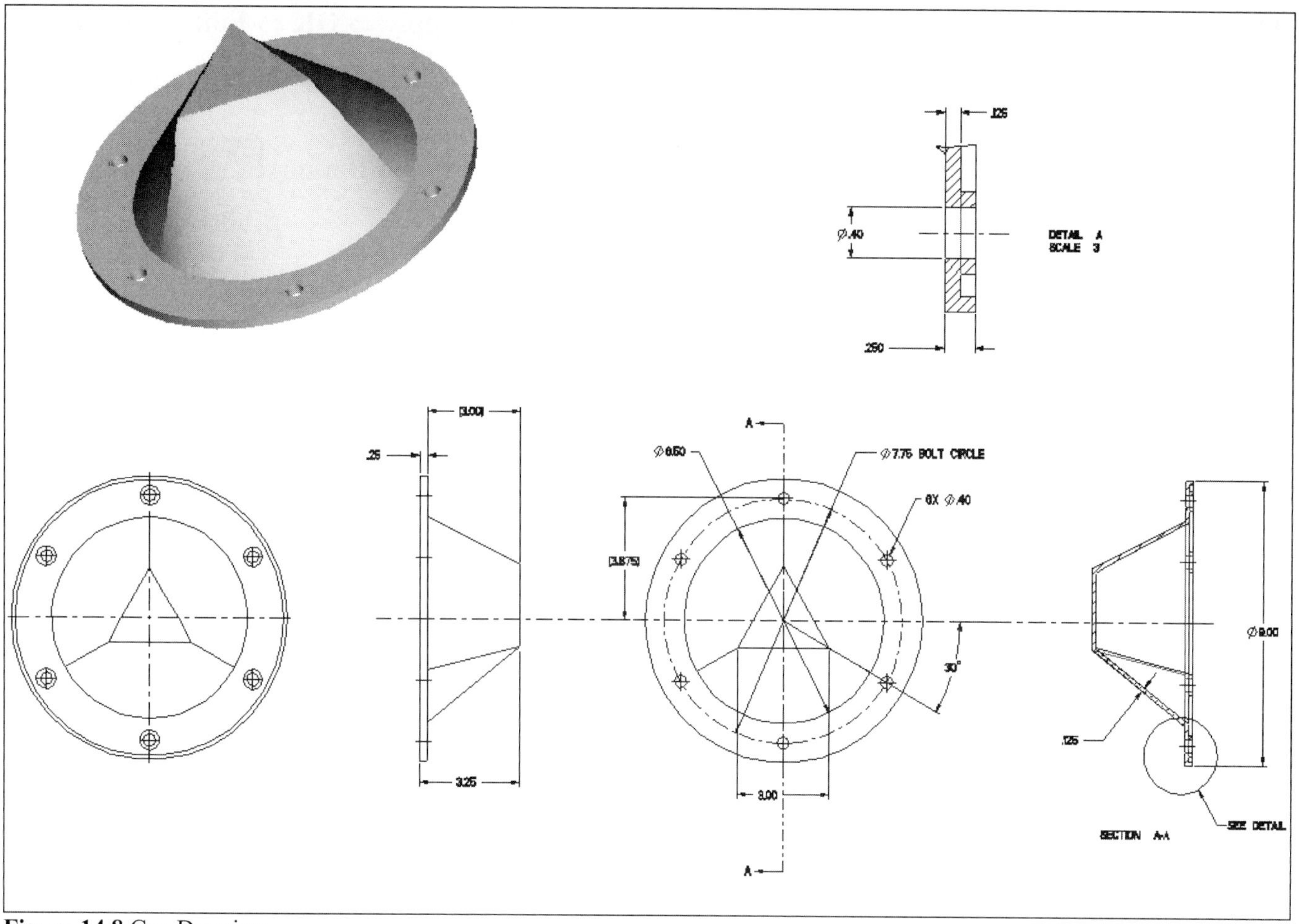

Figure 14.8 Cap Drawing

Cap

The Cap (Figs. 14.8 and 14.9) is a part created with a Parallel Blend [Figs. 14.10(a-g)]. The blend sections are a circle and a triangle. Because *the sections of a blend must have equal segments*, the "circle" is actually three equal arcs. The part is shelled as the last feature in its creation. The Shell Tool will create *bosses* around each hole as it hollows out the part. A cross section will be created in the Part mode to be used when you are detailing the Cap in the Drawing mode.

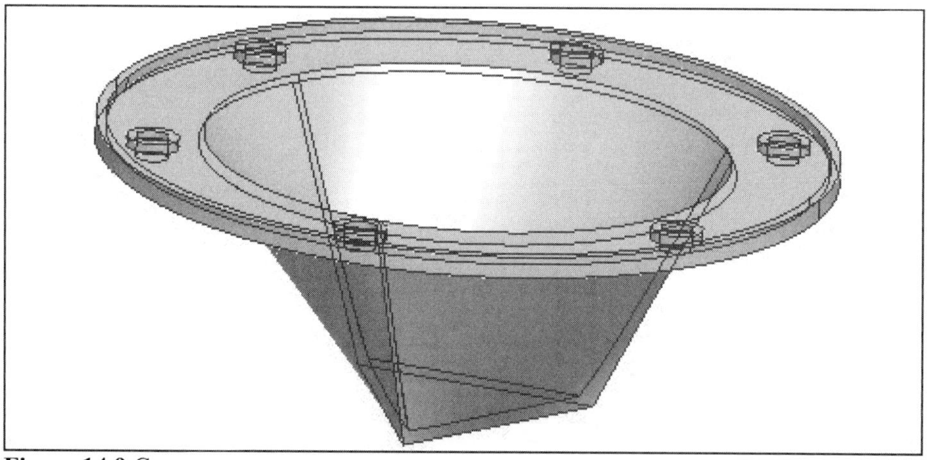

Figure 14.9 Cap

Start a new part. Click: **Create a new object** ⇒ Part ⇒ Name **CAP** ⇒ Use default template ⇒ **MMB** ⇒ **Tools** ⇒ **Environment** ⇒ Standard Orient **Trimetric** ⇒ **Apply** ⇒ **OK** ⇒ **Edit** ⇒ **Setup**

- **Material** = PVC (pvc.mtl)
- **Units** = Inch lbm Second

Set Datum (using **Insert** ⇒ **Annotations** ⇒ **Geometric Tolerance** ⇒ **Set Datum**) -A-

- Datum TOP = **C**
- Datum FRONT = **A**
- Datum RIGHT = **B**

Rename the coordinate system

- Coordinate System = **CSYS_CAP**

Tools ⇒ **Options**

- ***sketcher_dec_places*** ***3***
- ***default_dec_places*** ***3***
- ***tol_mode*** ***plusminus***

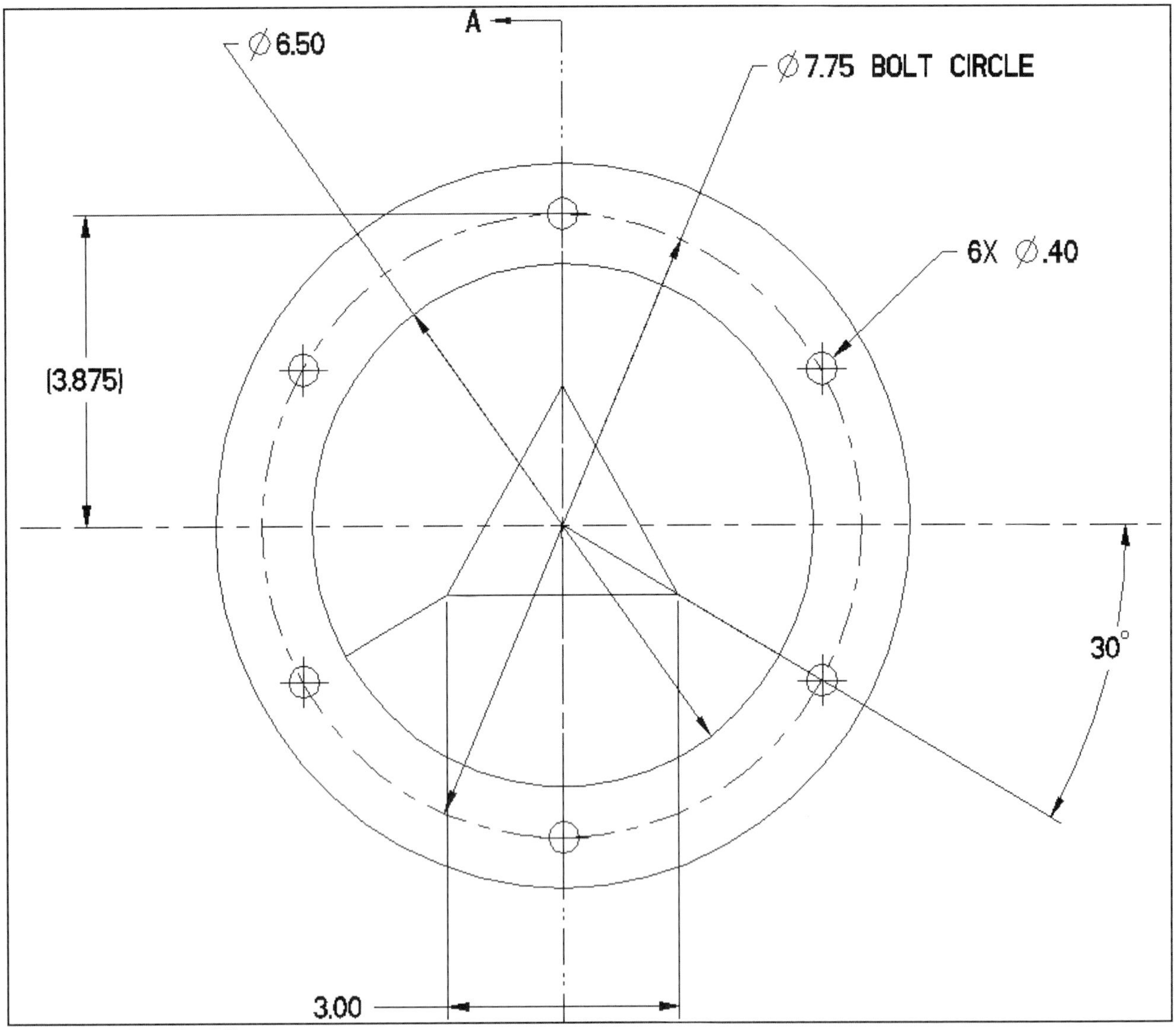

Figure 14.10(a) Top View

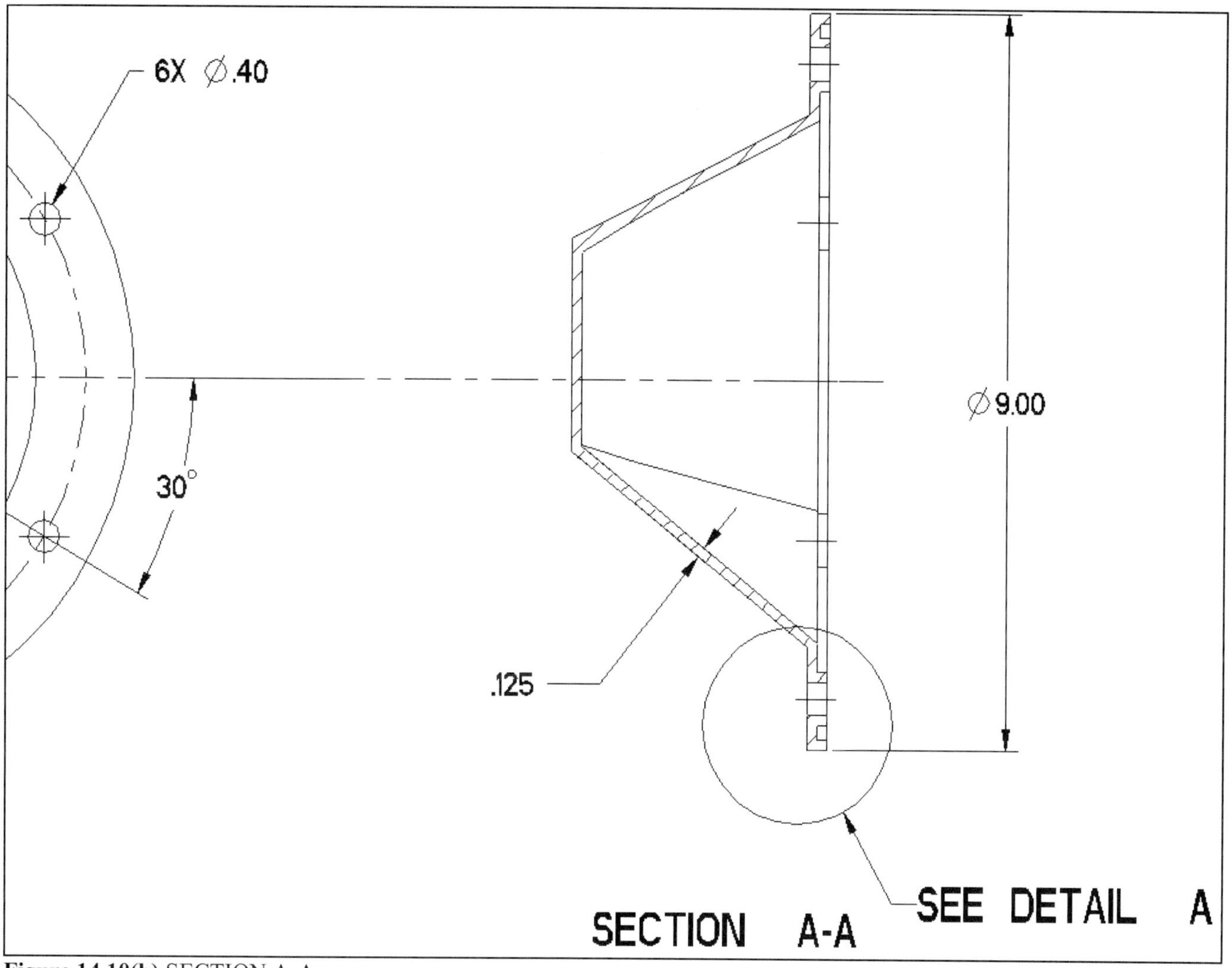

Figure 14.10(b) SECTION A-A

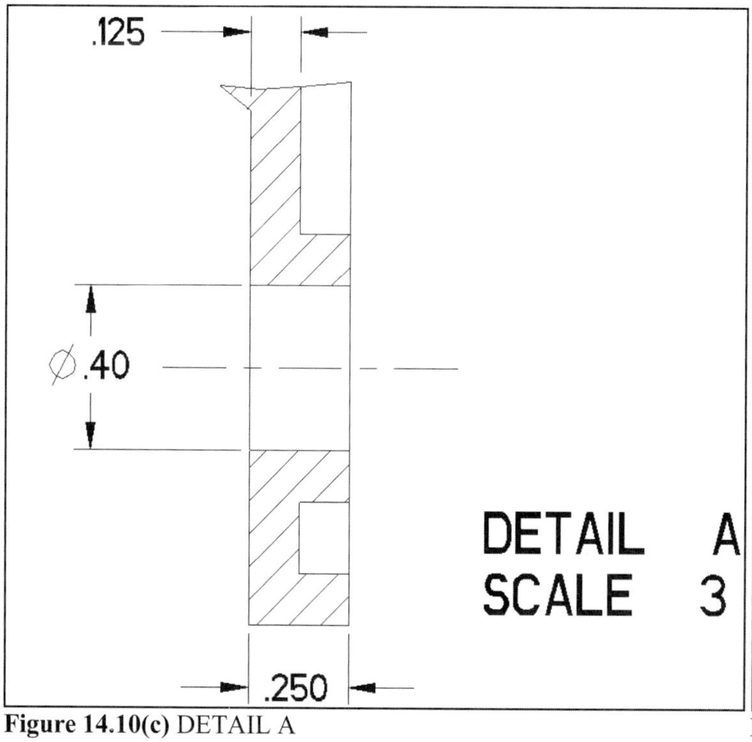

Figure 14.10(c) DETAIL A

Figure 14.10(d) Hole

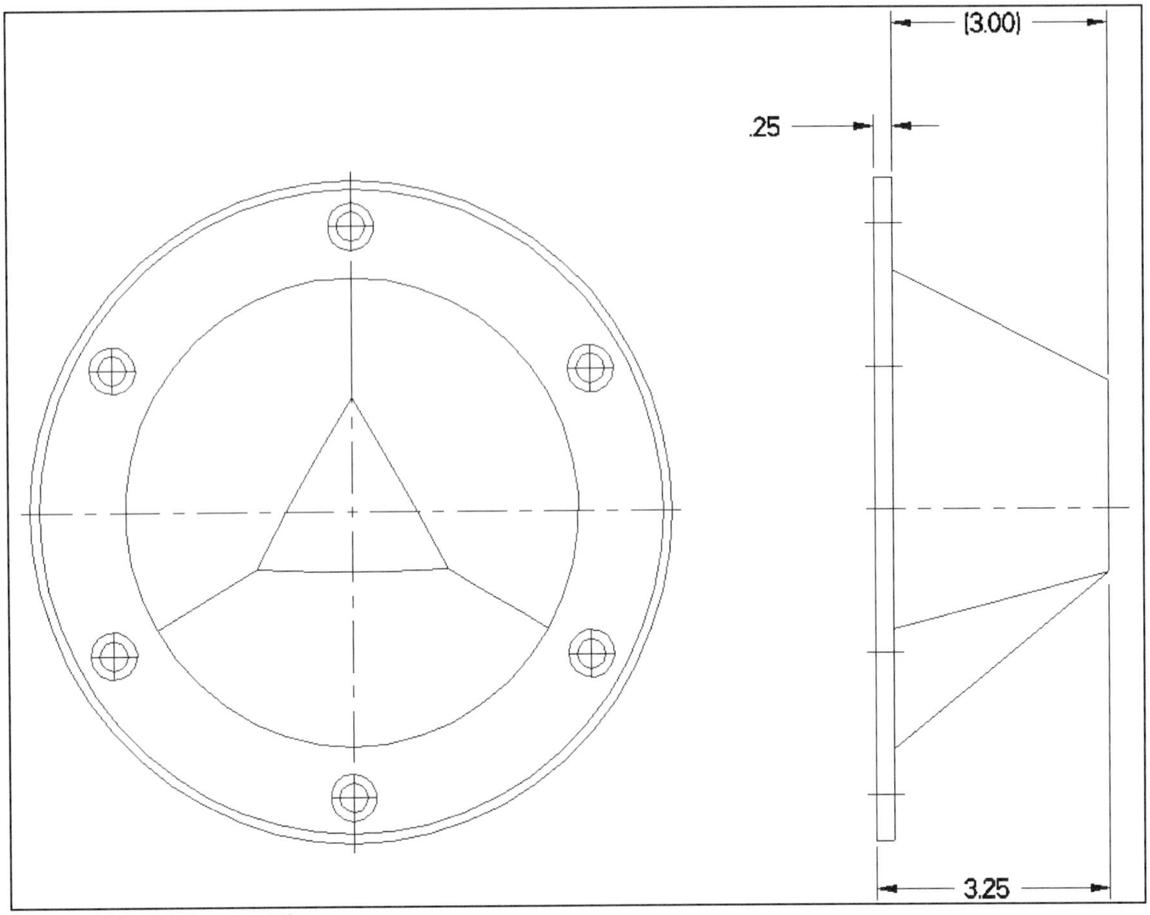

Figure 14.10(e) Bottom View

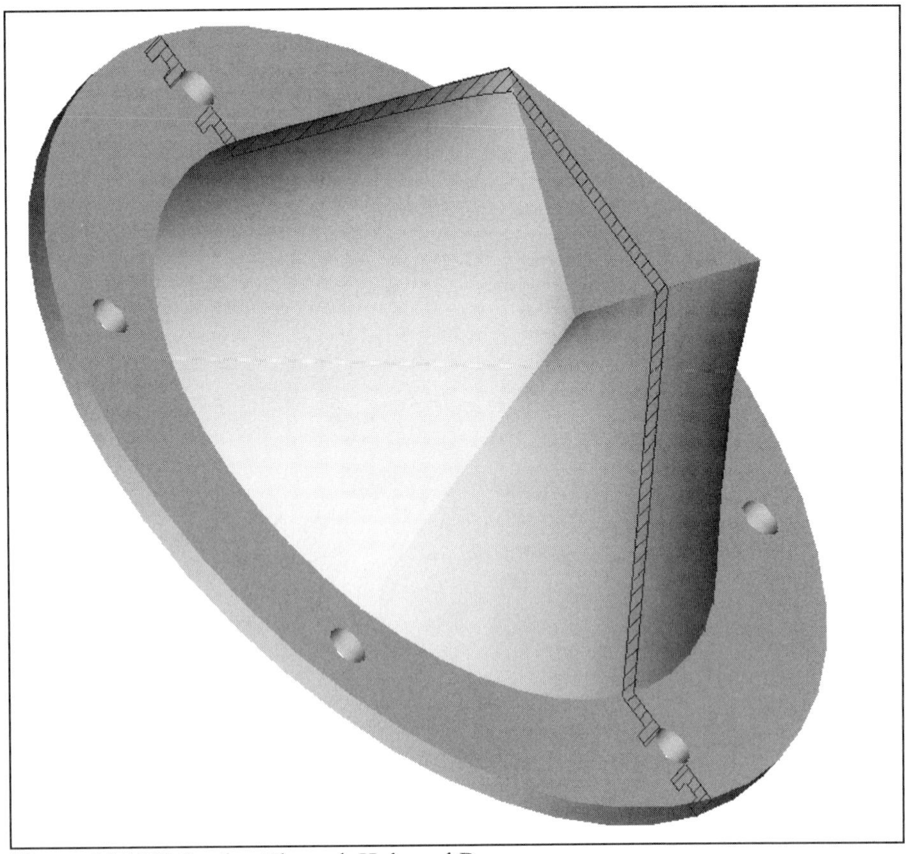
Figure 14.10(f) Section Through Hole and Boss

Figure 14.10(g) Hole and Boss

Model the circular protrusion that is ∅**9.00** by **.25** thick shown in [Fig. 14.11(a)]. Sketch the first protrusion on datum **A** (**FRONT**) and centered on **B** (**RIGHT**) and **C** (**TOP**).

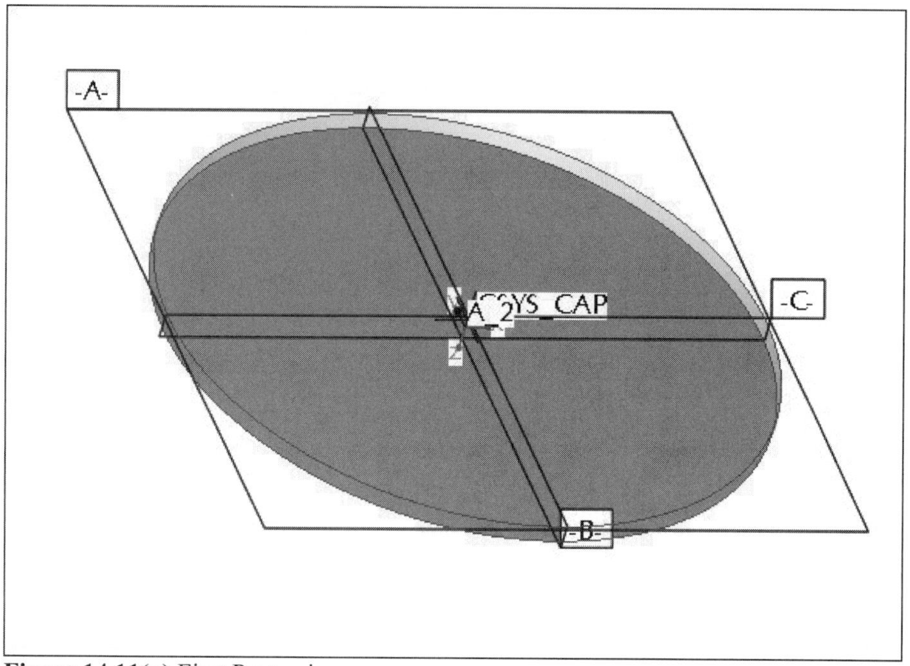

Figure 14.11(a) First Protrusion

Create the blend protrusion, click: **Insert** ⇒ **Blend** ⇒ **Protrusion** ⇒ **Parallel** ⇒ **Regular Sec** ⇒ **Sketch Sec** ⇒ **Done** ⇒ **Straight** ⇒ **Done** ⇒ pick the top surface of the first protrusion [Fig. 14.11(b)] ⇒ **Okay** to confirm direction of feature creation ⇒ **Default** for default orientation ⇒ **Toggle the grid** on [Fig. 14.11(c)]

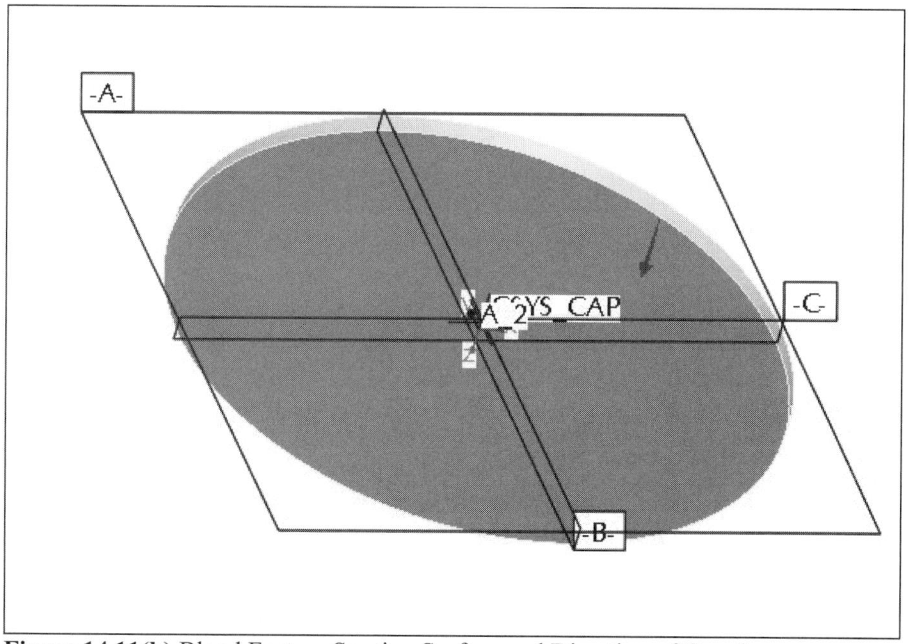

Figure 14.11(b) Blend Feature Starting Surface and Direction of Creation

Change the grid type and size, click: **Sketch** from menu bar ⇒ **Options** ⇒ **Miscellaneous** tab ⇒ Snap to Grid ⇒ **Parameters** tab ⇒ Polar ⇒ Grid Spacing ⇒ **Manual** ⇒ Radial **.50** ⇒ Angular **30°** ⇒ ✔ ⇒ **Hidden line** [Fig. 14.11(d)] ⇒ add three centerlines [Figs. 14.11(e-h)]

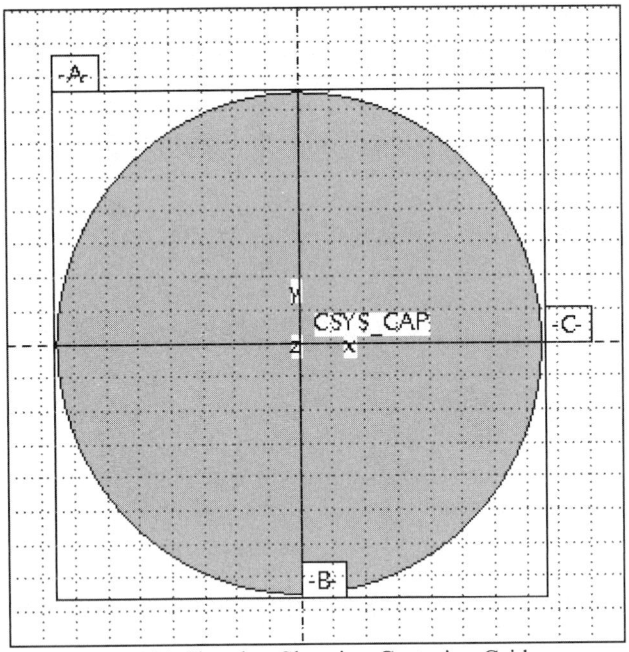

Figure 14.11(c) Sketcher Showing Cartesian Grid

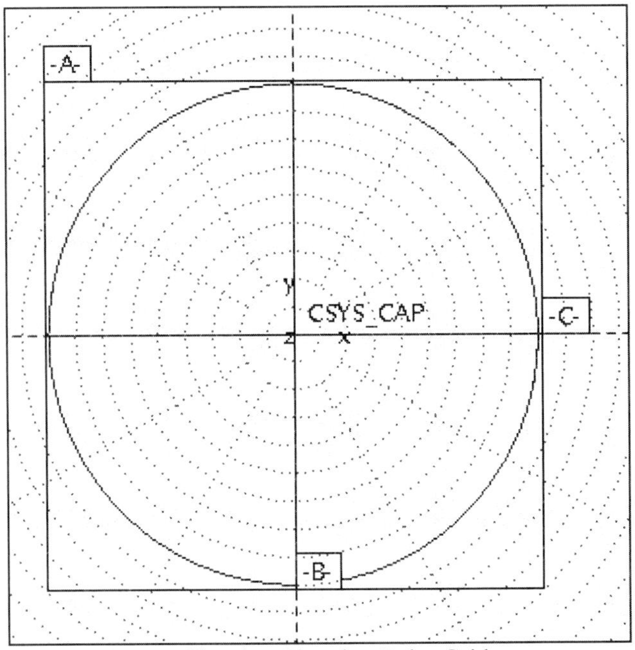

Figure 14.11(d) Sketcher Showing Polar Grid

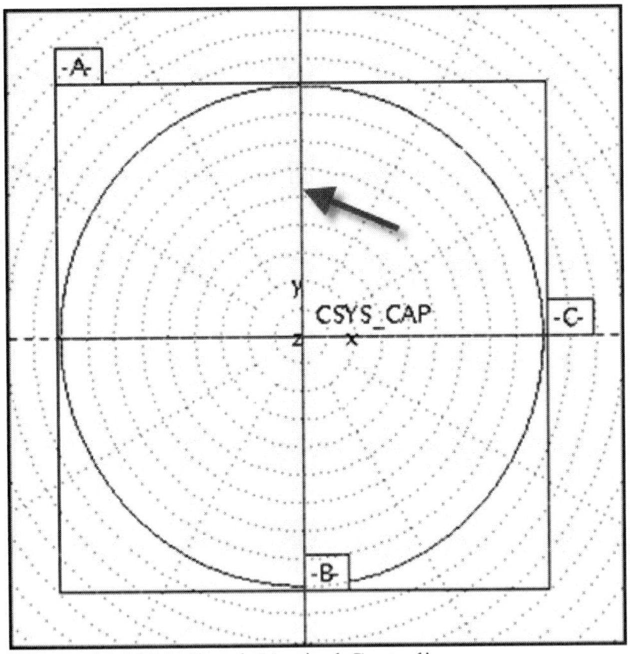

Figure 14.11(e) Sketch Vertical Centerline

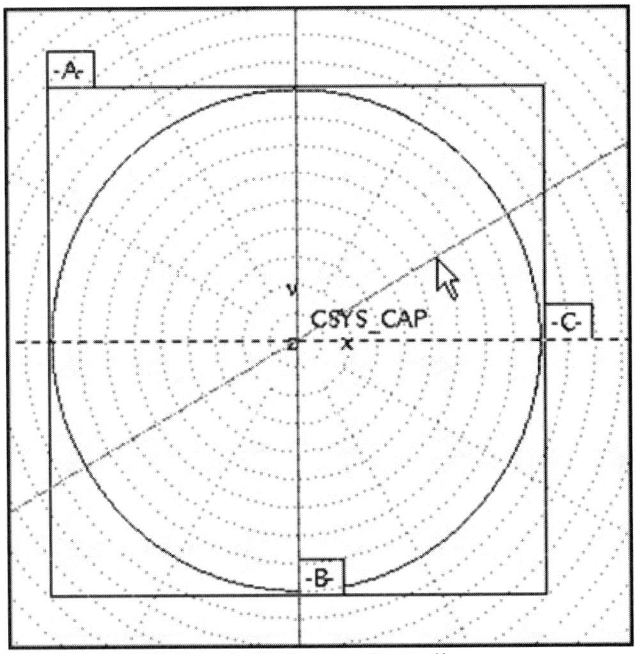

Figure 14.11(f) Sketch **30** Degree Centerline

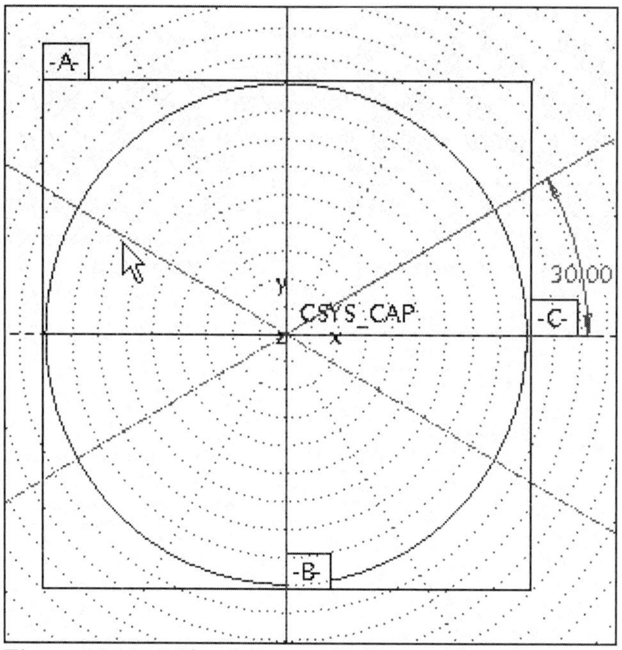

Figure 14.11(g) Sketch Second **30** Degree Centerline

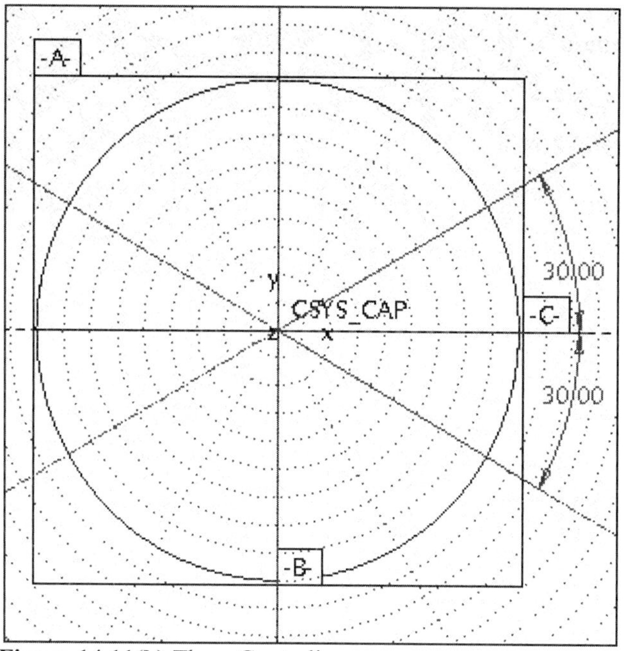

Figure 14.11(h) Three Centerlines

Sketch the *first section* of the blend by creating *three* equal **120°** arcs, click: 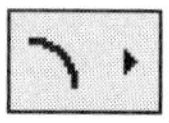 *flyout* ⇒  **Create an arc by picking its center and endpoints** ⇒ Sketch each arc in a counterclockwise direction. Note the location of the *start point* for this section. ⇒ add a *diameter* dimension [Fig. 14.11(i)]

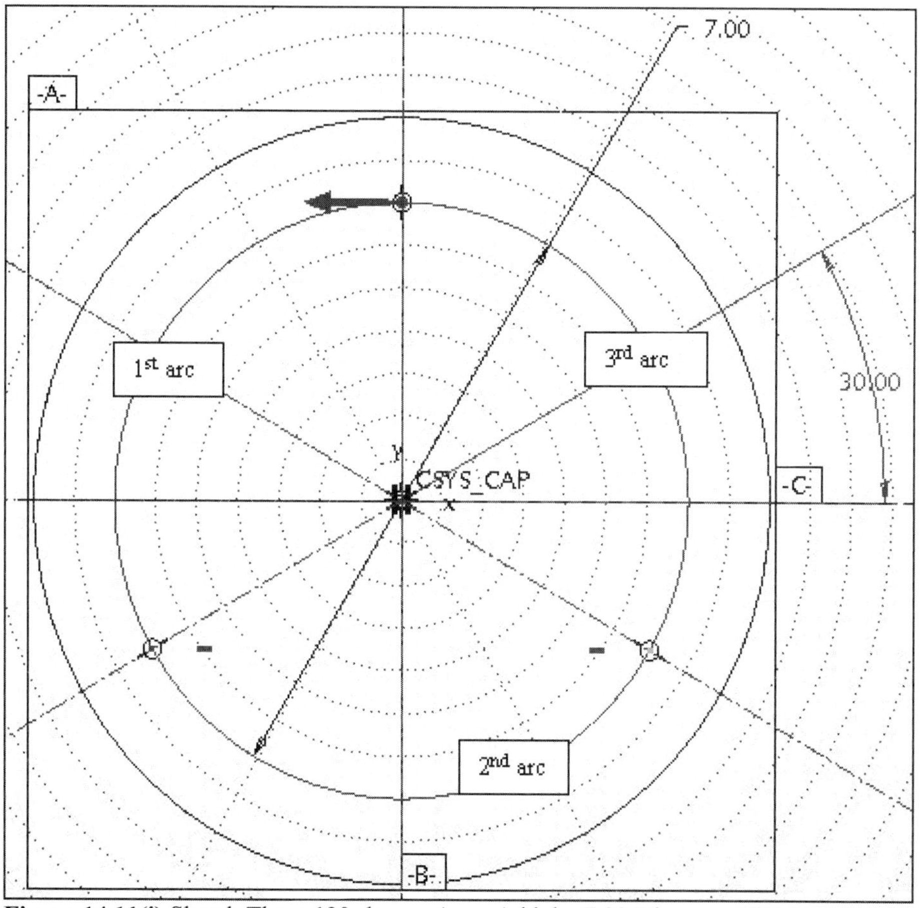

Figure 14.11(i) Sketch Three **120**-degree Arcs. Add the **7.00** Diameter Dimension

Click: **RMB** ⇒ **Toggle Section** ⇒ sketch the second parallel section (the first section is *grayed* out) ⇒ ⇒ **Create 2 point lines** ⇒ sketch the three lines of the triangle starting at the top so that the start point is near the start point of the first section and picking points *in the same direction* in which the arcs were created [Fig. 14.11(j)] ⇒ dimension and modify [Figs. 14.11(k-m)] ⇒ from the Modify Dimensions dialog box ⇒ add the two **120** degree dimensions [Fig. 14.11(n)] ⇒ **Continue with the current section**

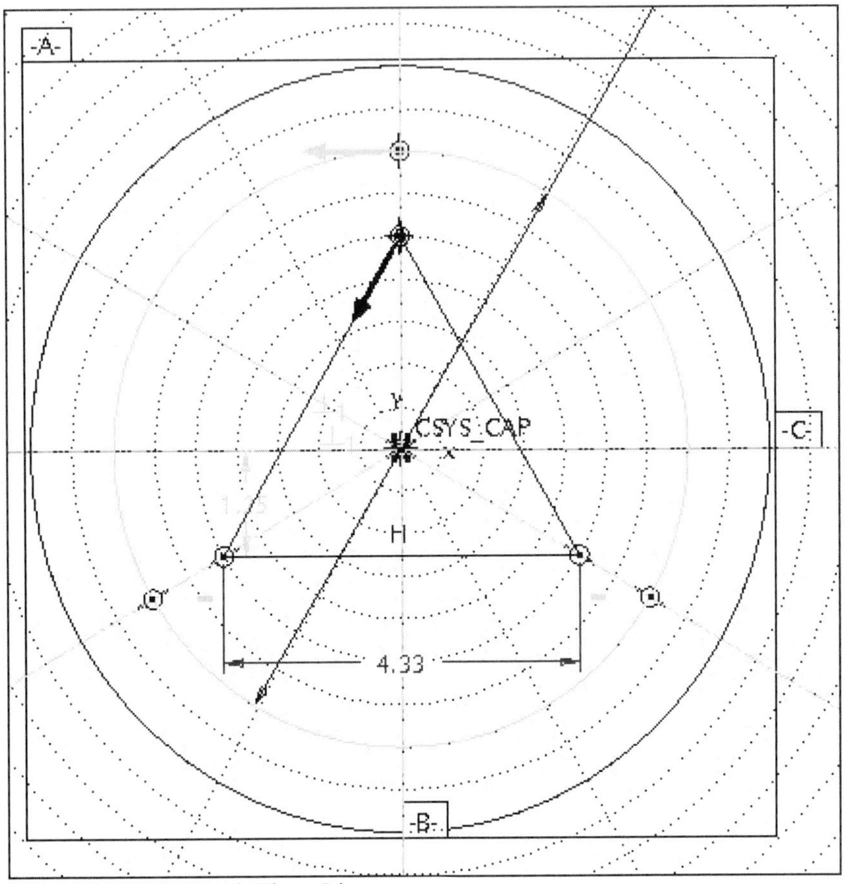

Figure 14.11(j) Sketch Three Lines

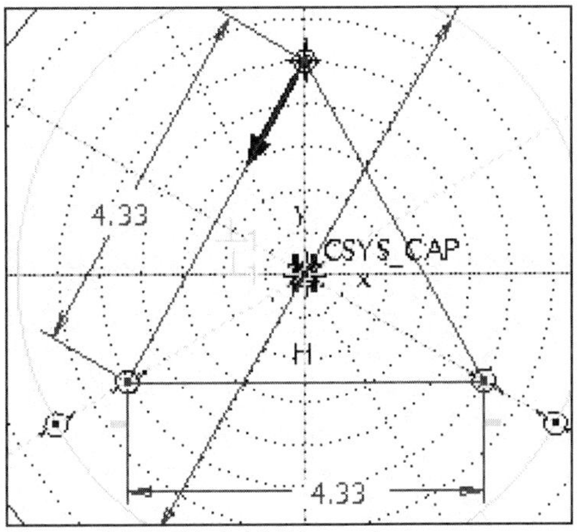

Figure 14.11(k) Add Dimensions

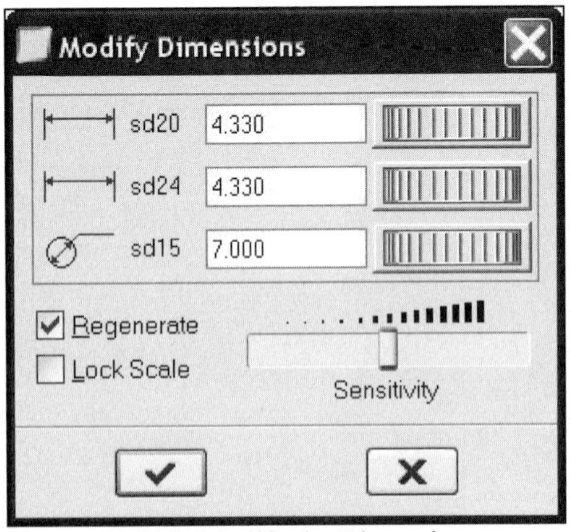

Figure 14.11(l) Modify the Dimensions

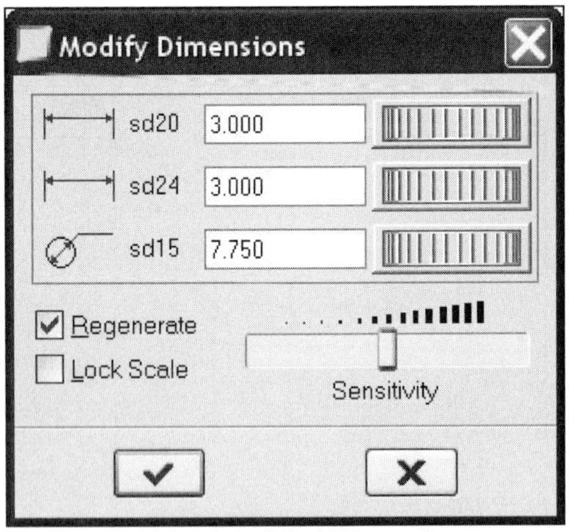

Figure 14.11(m) Modified Dimensions

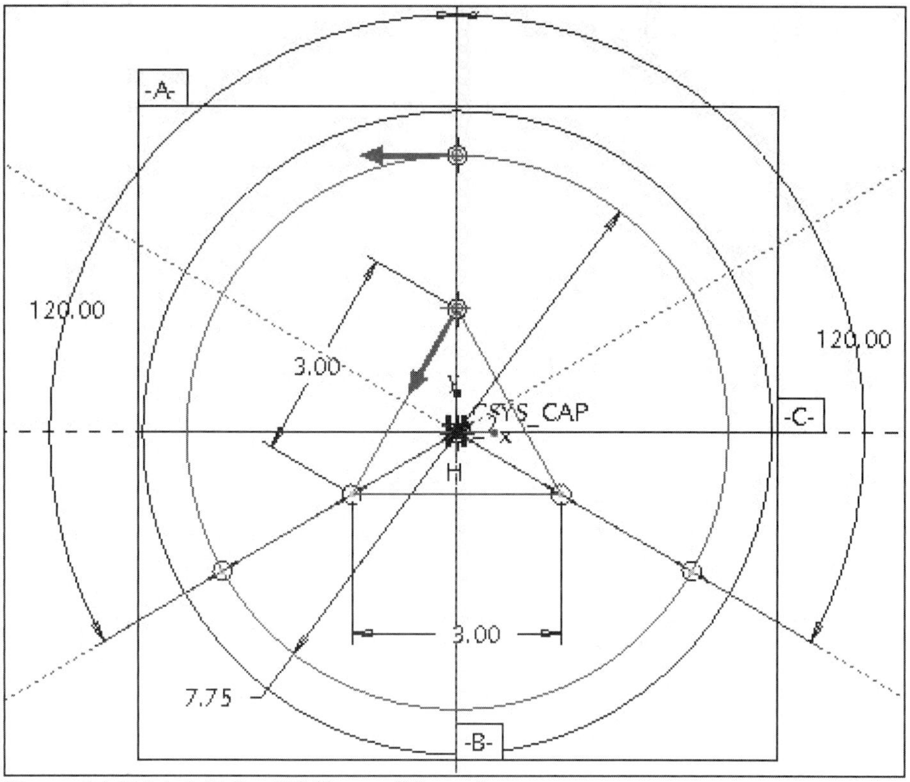

Figure 14.11(n) Add the **120** Degree Dimensions

Click: **Blind** ⇒ **MMB** ⇒ type **3.00** at the prompt ⇒ **Enter** ⇒ **Preview** ⇒ **MMB** ⇒ with the blend feature highlighted on the model and Model Tree, click **RMB** ⇒ **Edit** ⇒ double-click on the **7.75** dimension and modify the diameter to **6.50** [Fig. 14.11(o)] ⇒ **Enter** ⇒  **Regenerates Model** ⇒ **Ctrl+D** ⇒  ⇒  ⇒ **MMB** [Fig. 14.11(p)]

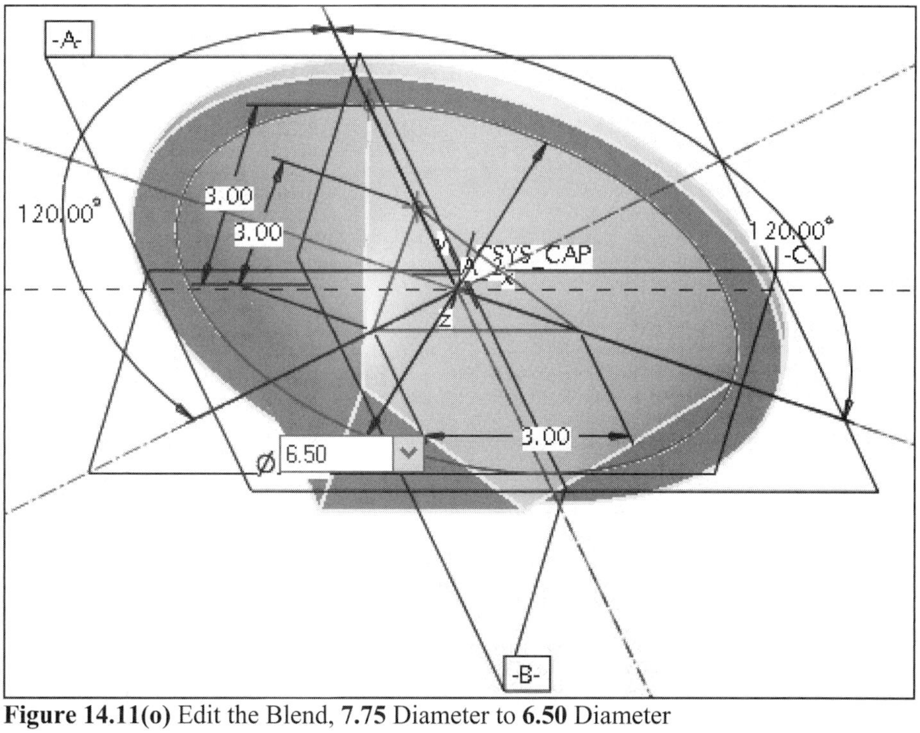

Figure 14.11(o) Edit the Blend, **7.75** Diameter to **6.50** Diameter

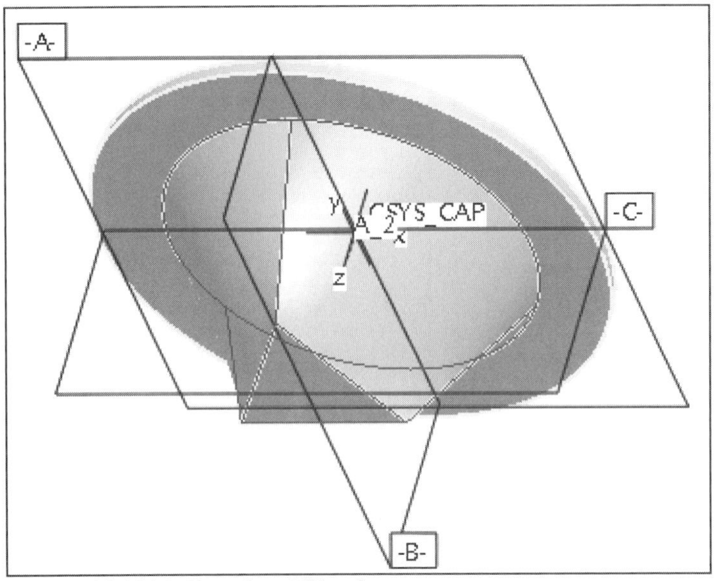

Figure 14.11(p) Completed Blend

Create and pattern the six equally spaced holes ∅**.400** on a ∅**7.75** bolt circle, click:  **Hole Tool** ⇒ complete the hole with the options and references [Fig. 14.12(a)] ⇒ **LMB** to deselect

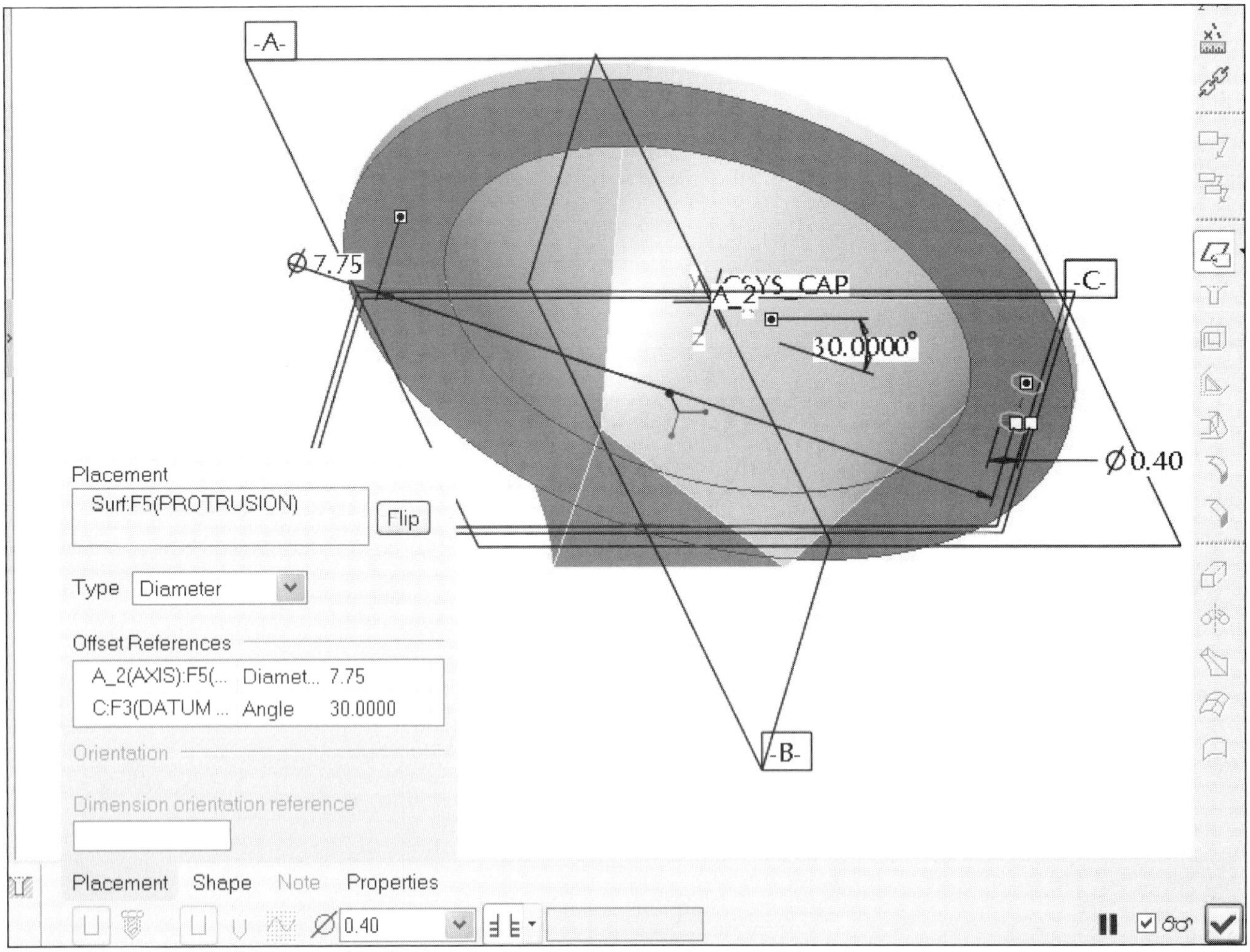

Figure 14.12(a) Hole Options and References

Click on the hole (on the model or in the Model Tree) ⇒ **RMB** ⇒ **Pattern** ⇒ complete the pattern with the options and references [Fig. 14.12(b)] *(select the **30** degree dimension to pattern)* ⇒ **LMB** to deselect [Fig. 14.12(c)]

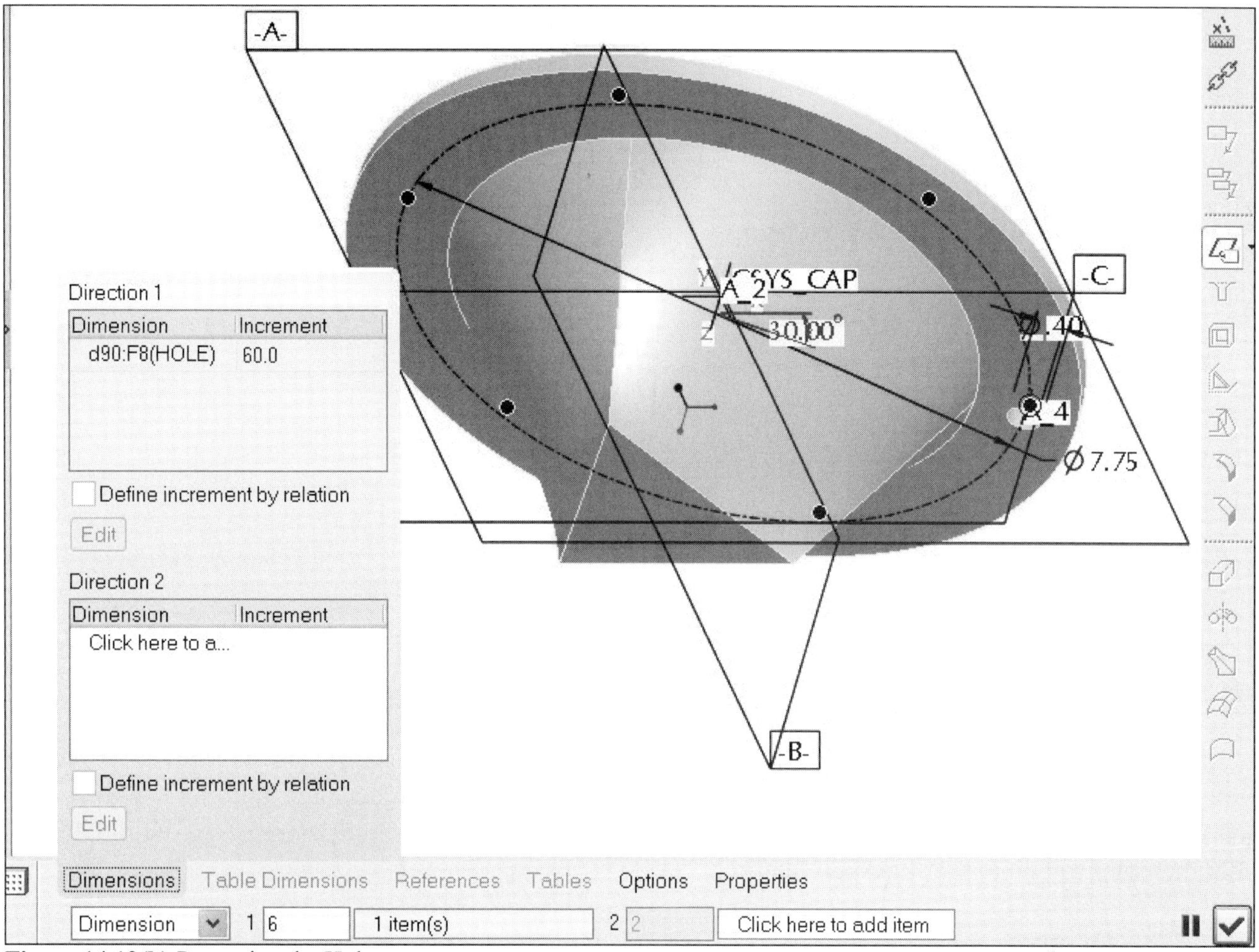

Figure 14.12(b) Patterning the Hole

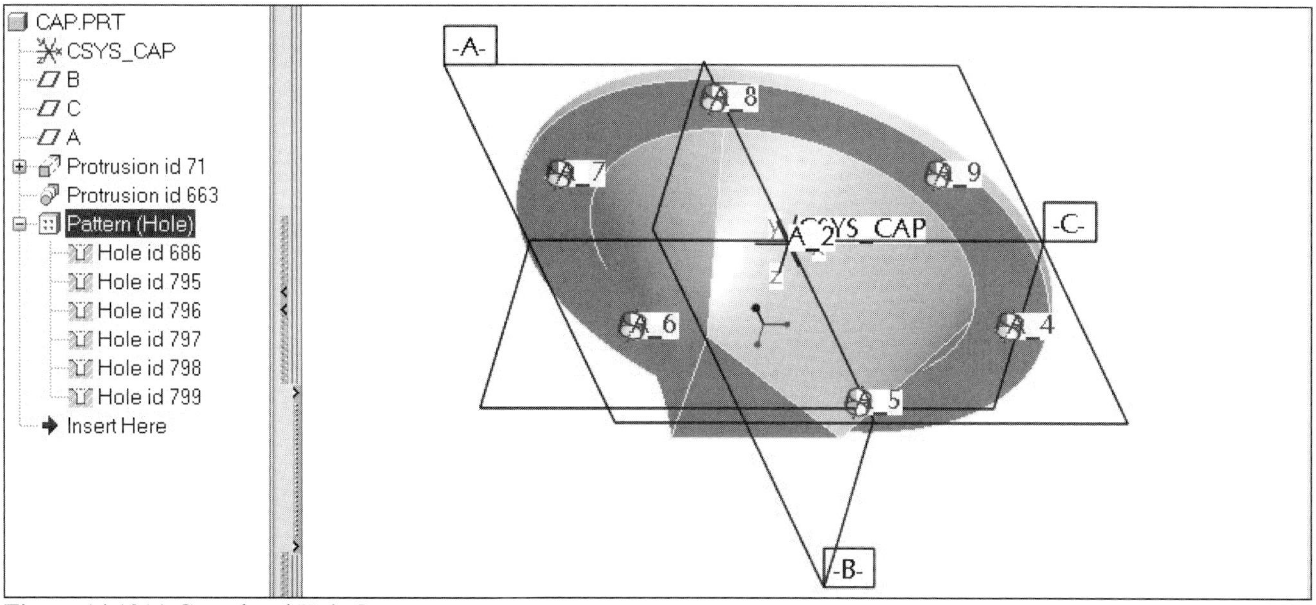

Figure 14.12(c) Completed Hole Pattern

Shell the part, click: **Shell Tool** ⇒ pick the bottom surface of the part as the surface to remove [Fig. 14.13(a)] ⇒ Thickness **.125** ⇒ **Enter** ⇒

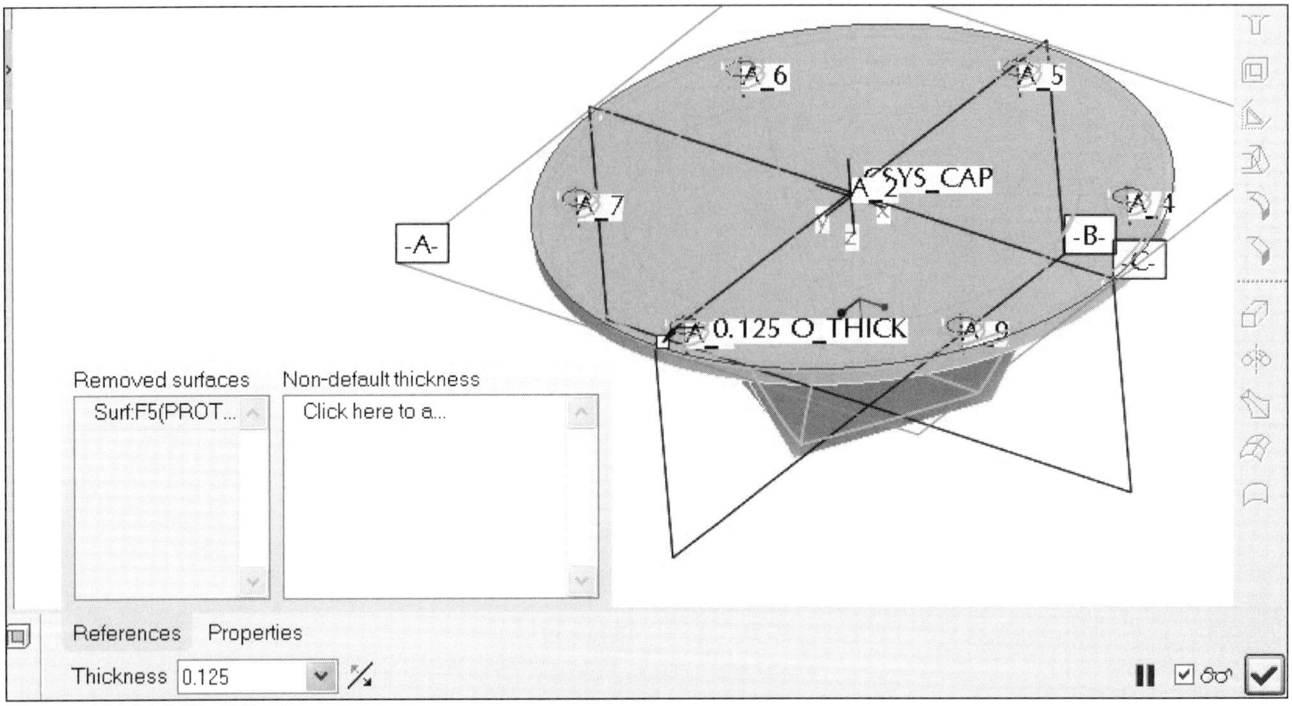

Figure 14.13(a) Shell Tool

The shell will automatically create the *bosses* [Fig. 14.13(b)], because the **.125** thickness is left around all previously created features. The *bosses* around the holes are created automatically at **.250** larger than the holes: (**.125 + .125 + .400 = ∅.650**). If the *bosses* were not desired, you would simply reorder the holes to come after the shell feature.

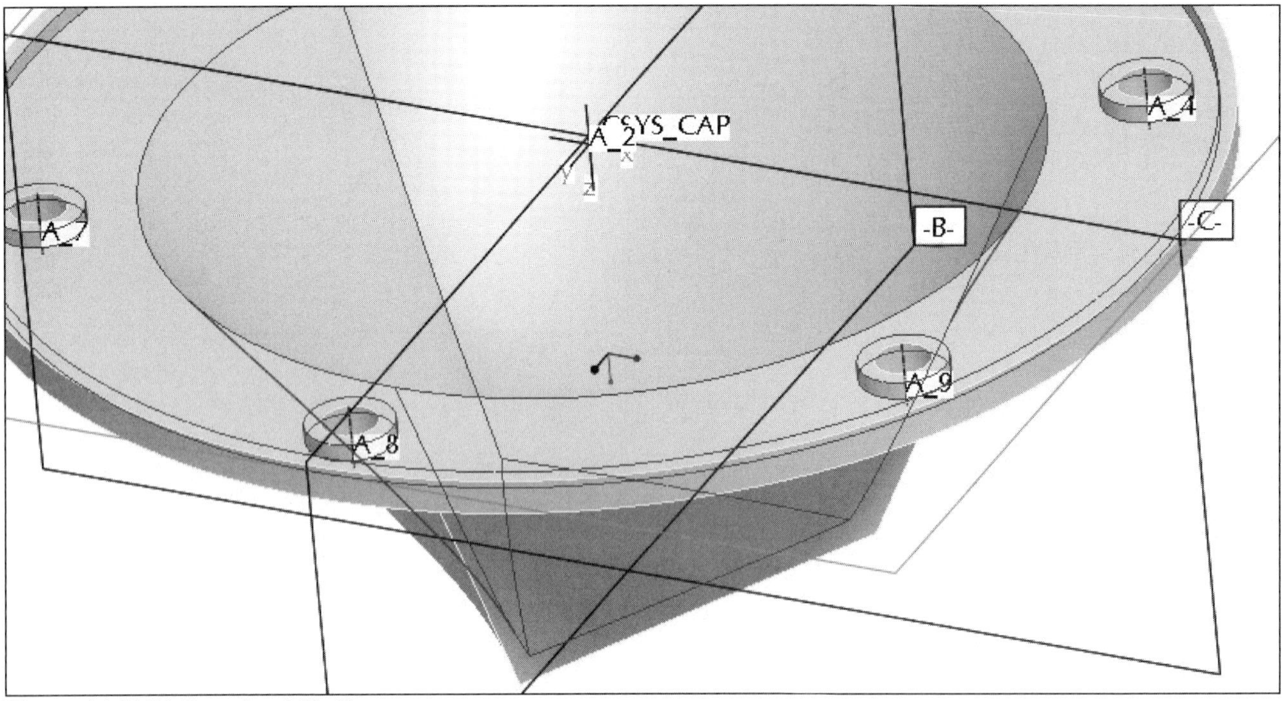

Figure 14.13(b) Completed Shell

Measure the size of the boss, click: **Analysis** ⇒ **Measure** ⇒ **Diameter** [Fig. 14.14(a)] ⇒ pick the vertical surface of a boss [Fig. 14.14(b)] ⇒ [Fig. 14.14(c)]

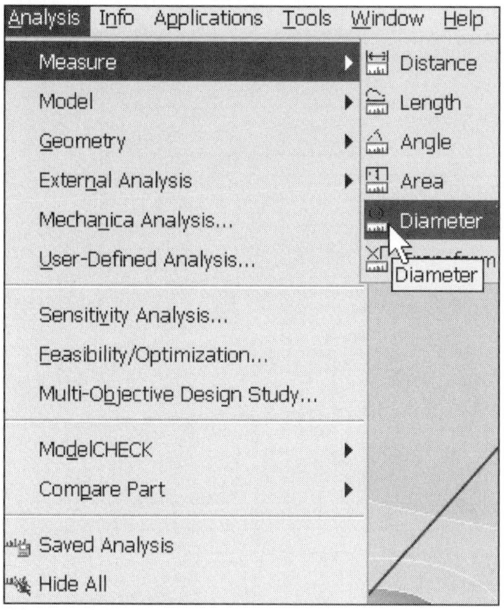

Figure 14.14(a) Measure ⇒ Diameter

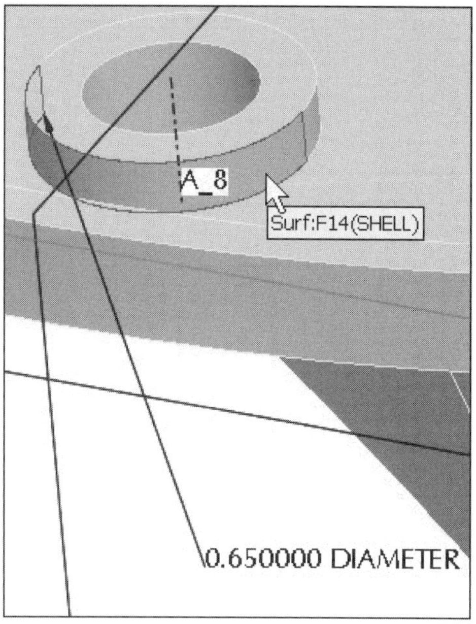

Figure 14.14(b) 0.650000 Diameter

Figure 14.14(c) Diameter Dialog

Click: **Start the view manager** ⇒ Xsec **Create Cross Sections** ⇒ **New** ⇒ type **A** ⇒ **Enter** ⇒ **Planar** ⇒ **Single** ⇒ **MMB** ⇒ pick datum **B** ⇒ **Display** ⇒ **Visibility** (Fig. 14.15) ⇒ **Close** ⇒ ⇒ **Standard Orientation** ⇒ ⇒ ⇒ **Enter**

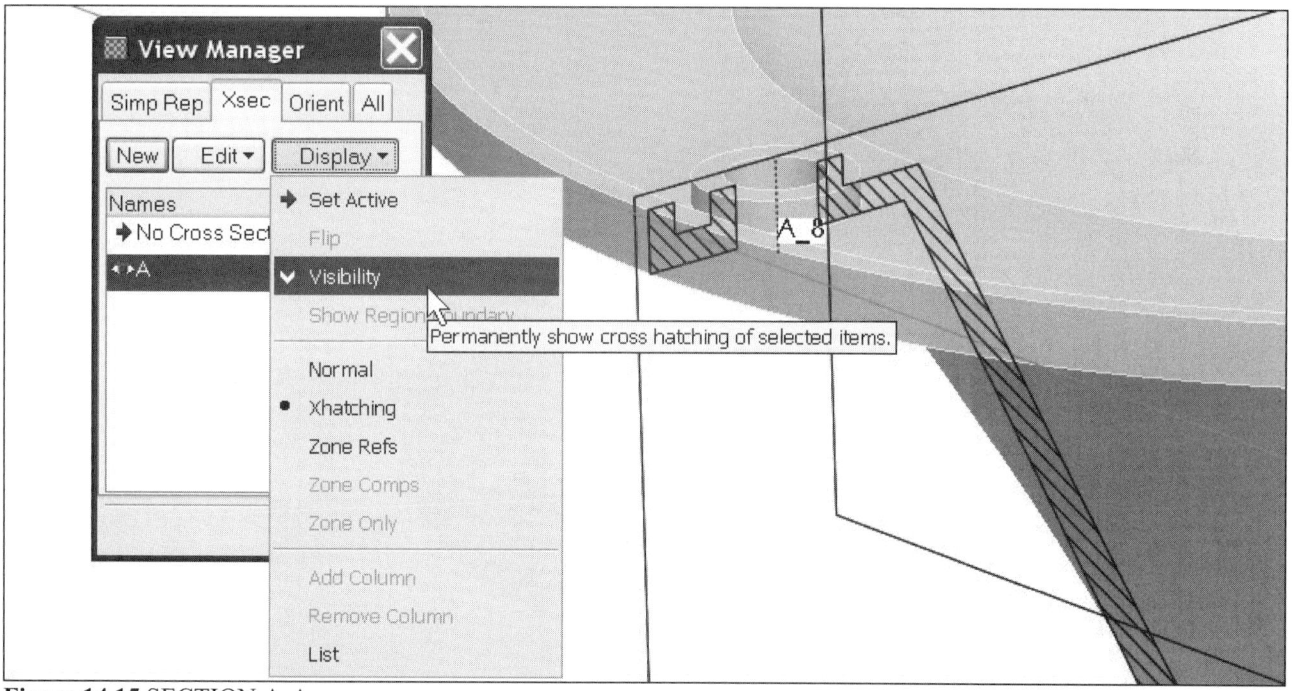

Figure 14.15 SECTION A-A

Save a copy of the Cap under a new name, and complete the **ECO** [Figs. 14.16(a-b)] to establish a second part with different dimensional sizes and features. Add **R.20** rounds to the edges. Insert the rounds to be affected by the shell.

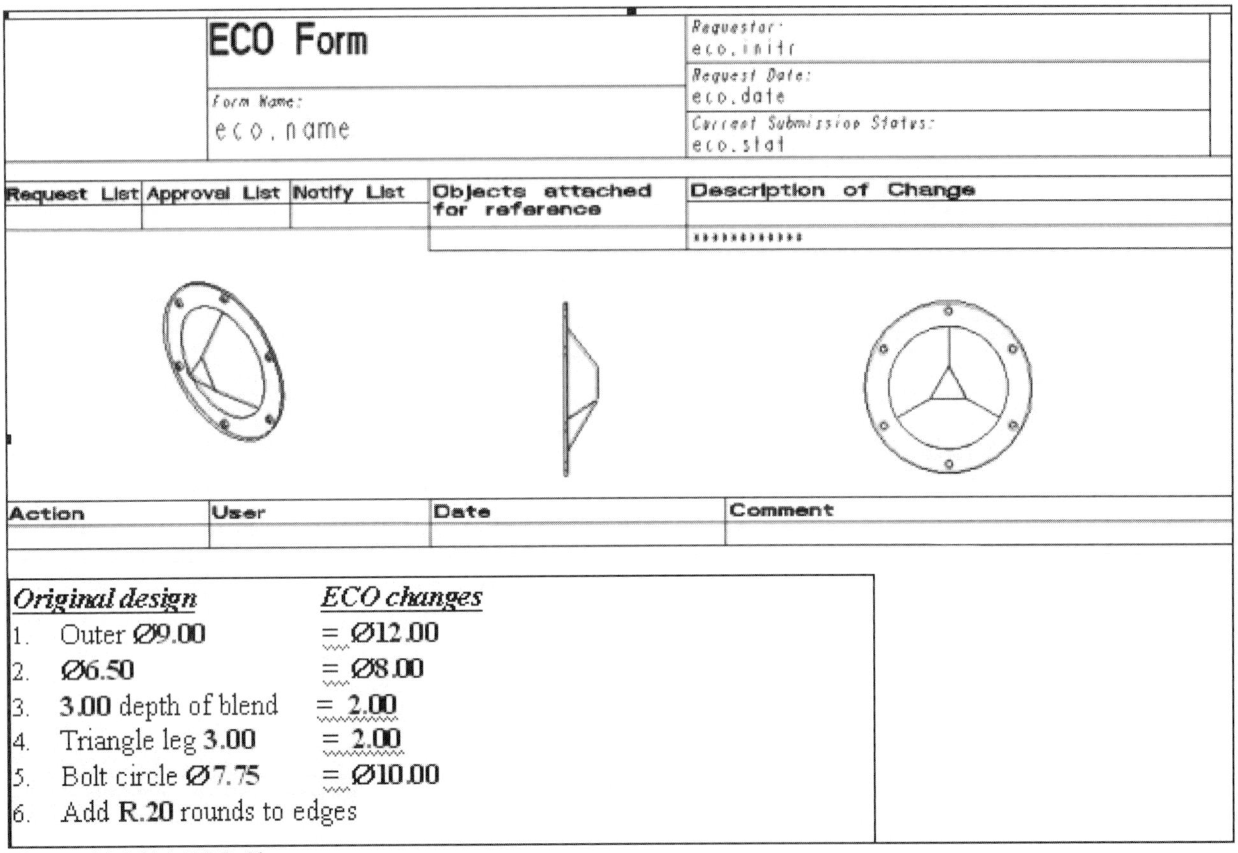

Figure 14.16(a) ECO Changes

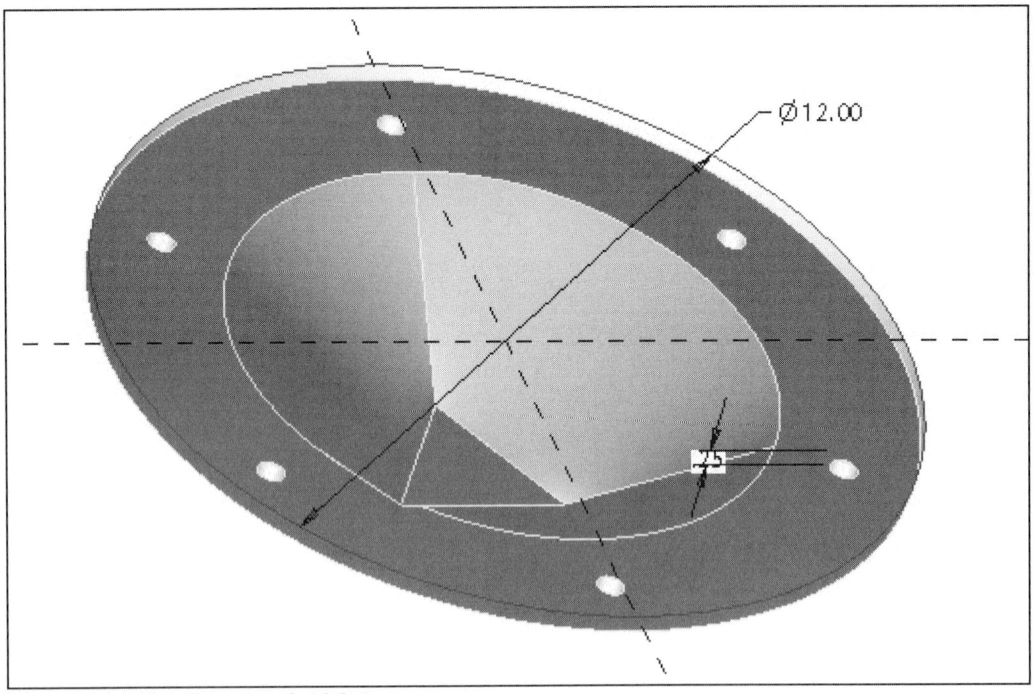

Figure 14.16(b) Completed ECO Part

For other projects see ***www.cad-resources.com* ⇒ *Downloads*.**

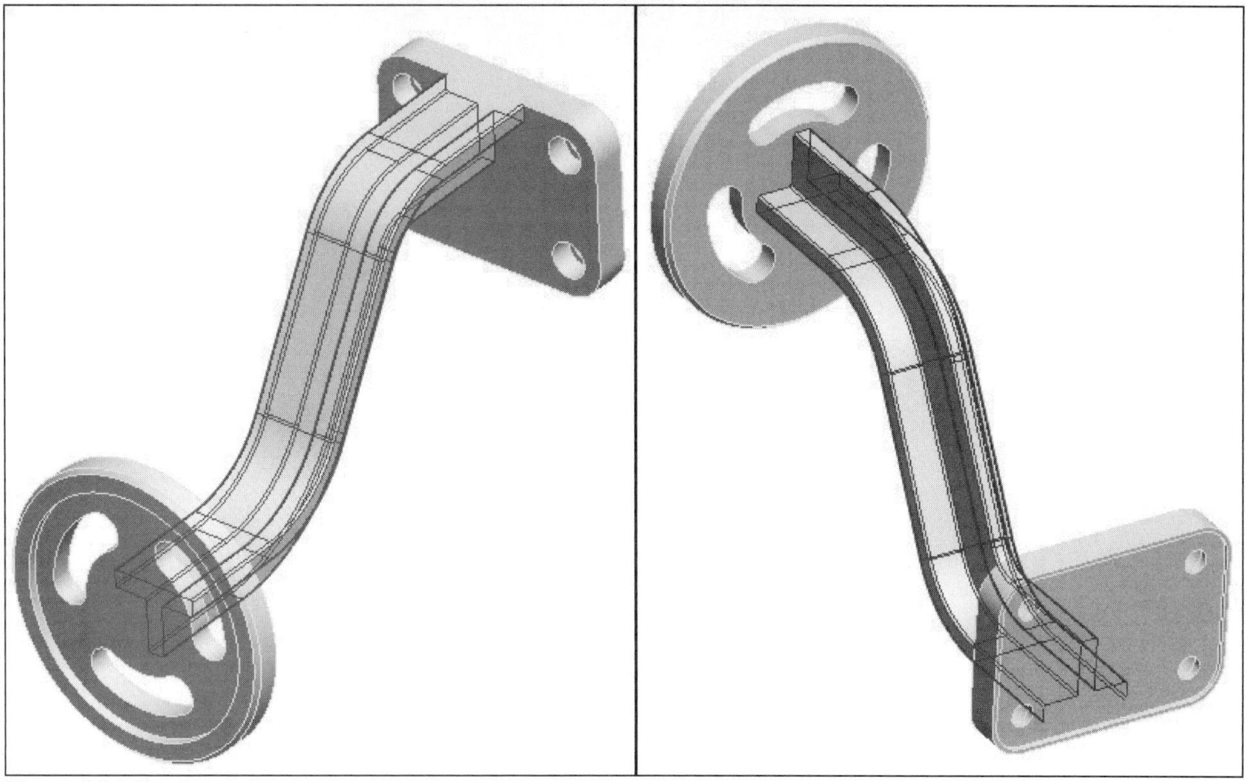

Figure 15.1 Bracket

OBJECTIVES

- Create a **constant-section sweep** feature
- Sketch a **Trajectory** for a sweep
- Sketch and locate a **Sweep section**
- Understand the difference between adding and not adding **Inner Faces**
- Be able to **Edit** a sweep

SWEEPS

A Sweep is created by sketching or selecting a *trajectory* and then sketching a *section* to follow along it. The Bracket, shown in Figure 15.1, uses a simple sweep in its design. A *constant-section sweep* (Fig. 15.2) can use either trajectory geometry sketched at the time of feature creation or a trajectory made up of selected datum curves or edges. The trajectory [Figs. 15.3 (a-c)] must have adjacent reference surfaces or be planar. When defining a sweep, Pro/E checks the specified trajectory for validity and establishes normal surfaces. When ambiguity exists, Pro/E prompts you to select a normal surface.

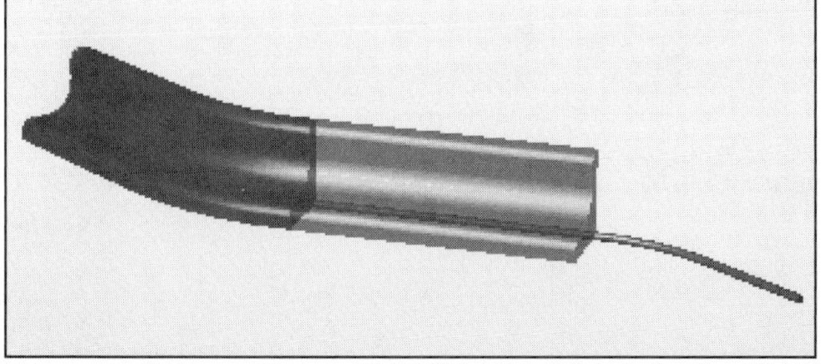

Figure 15.2 Sweep Forms (CADTRAIN, COAch for Pro/ENGINEER)

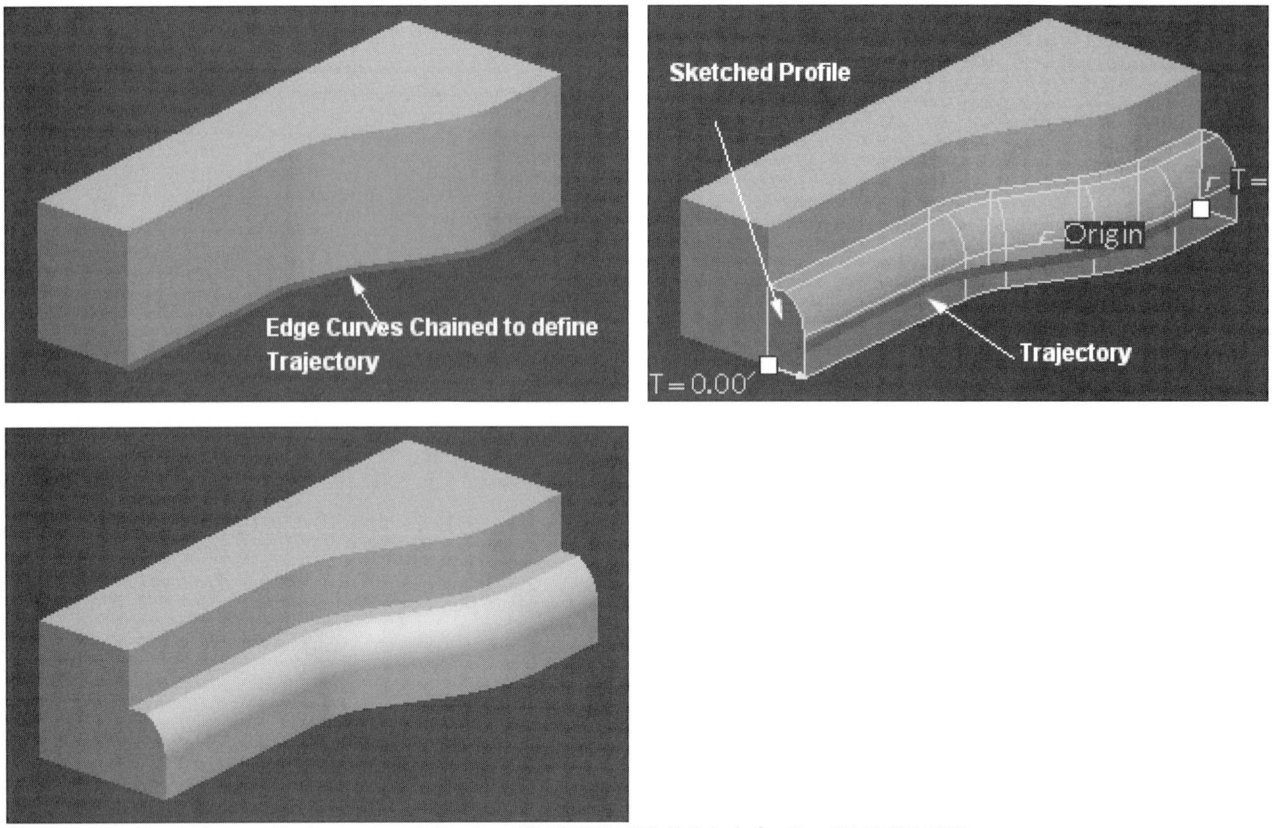

Figures 15.3(a-c) Sweep Trajectory and Section (CADTRAIN, COAch for Pro/ENGINEER)

The following options are available for sweeps:

- **Sketch Traj** Sketch the sweep trajectory using Sketcher mode
- **Select Traj** Select a chain of existing curves or edges as the sweep trajectory
- **Merge Ends** Merge the ends of the sweep, if possible, into the adjacent solid
- **Free Ends** Do not attach the sweep end to adjacent geometry
- **No Inn Fcs** For closed sections, does not add top and bottom faces [Fig. 15.4(a)]
- **Add Inn Fcs** For open sections, add top and bottom faces to close the swept solid [Fig. 15.4(b)]

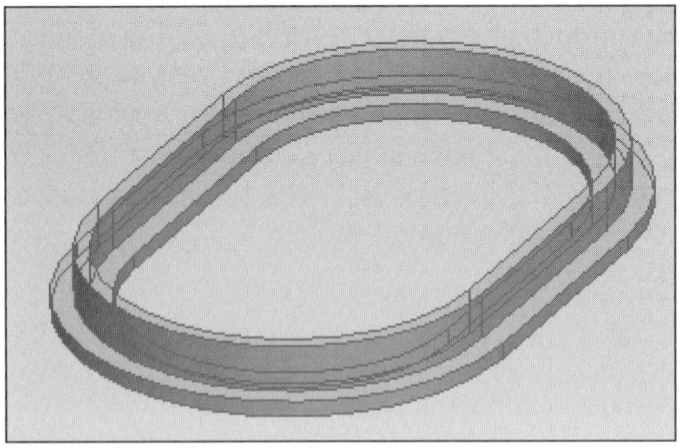

Figure 15.4(a) Sweep Attributes, **No Inn Fcs**

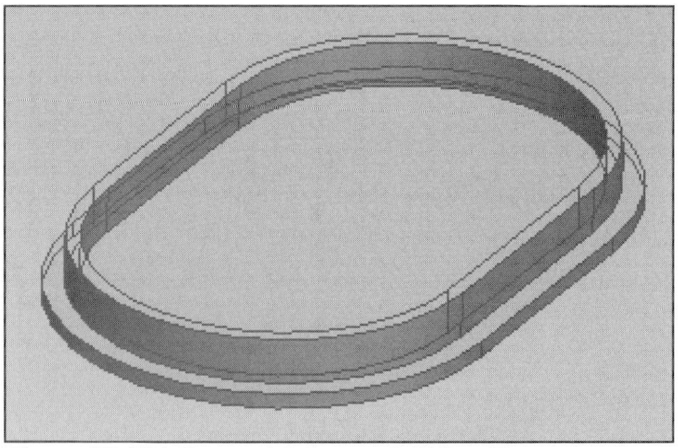

Figure 15.4(b) Sweep Attributes, **Add Inn Fcs**

Lesson 15 STEPS

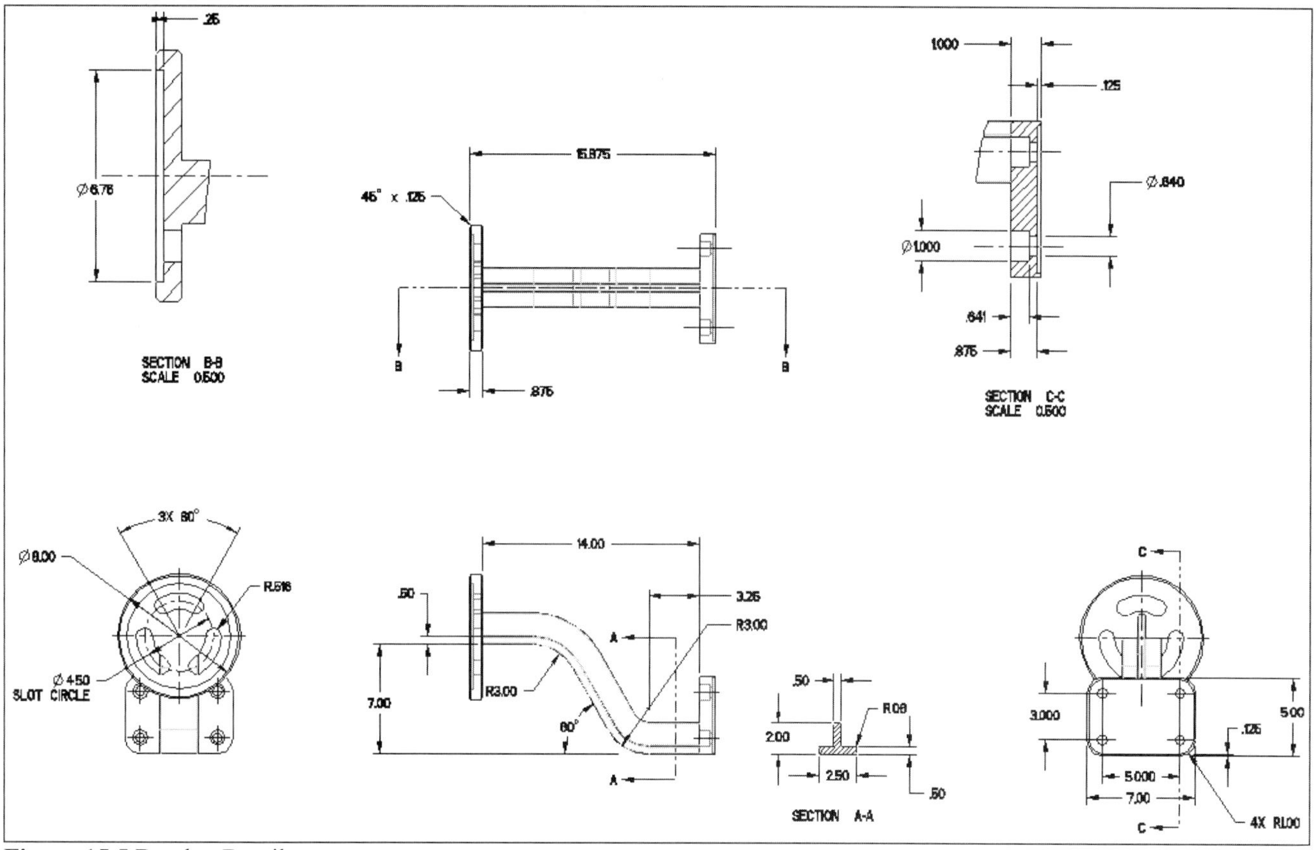

Figure 15.5 Bracket Detail

Bracket

The Bracket (Fig. 15.5) requires the use of the Sweep command. The T-shaped section is swept along the sketched *trajectory*. The protrusions on both sides of the swept feature are to be created with the dimensions shown in Figures 15.6(a-j). Systematic commands are provided only for the sweep trajectory and its cross section.

Start a new part. Click: ⇒ Part ⇒ Name **Bracket** ⇒ Use default template ⇒ **MMB** ⇒ **Edit** ⇒ **Setup**

- **Material** = al6061.mtl
- **Units** = Inch lbm Second

Set Datum -A- and **Rename** the default datum planes and coordinate system:

- Datum **TOP** = **C**
- Datum **FRONT** = **A**
- Datum **RIGHT** = **B**
- Coordinate System = **CSYS_SWEEP**

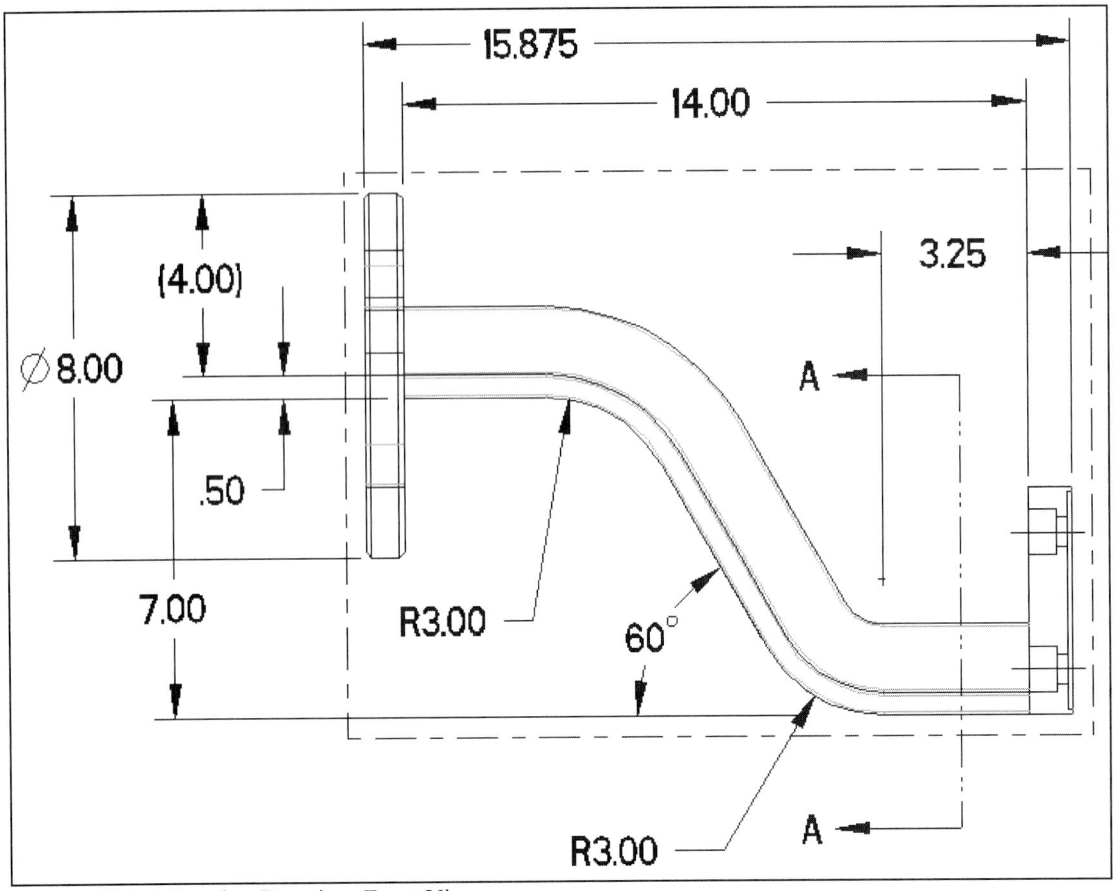

Figure 15.6(a) Bracket Drawing, Front View

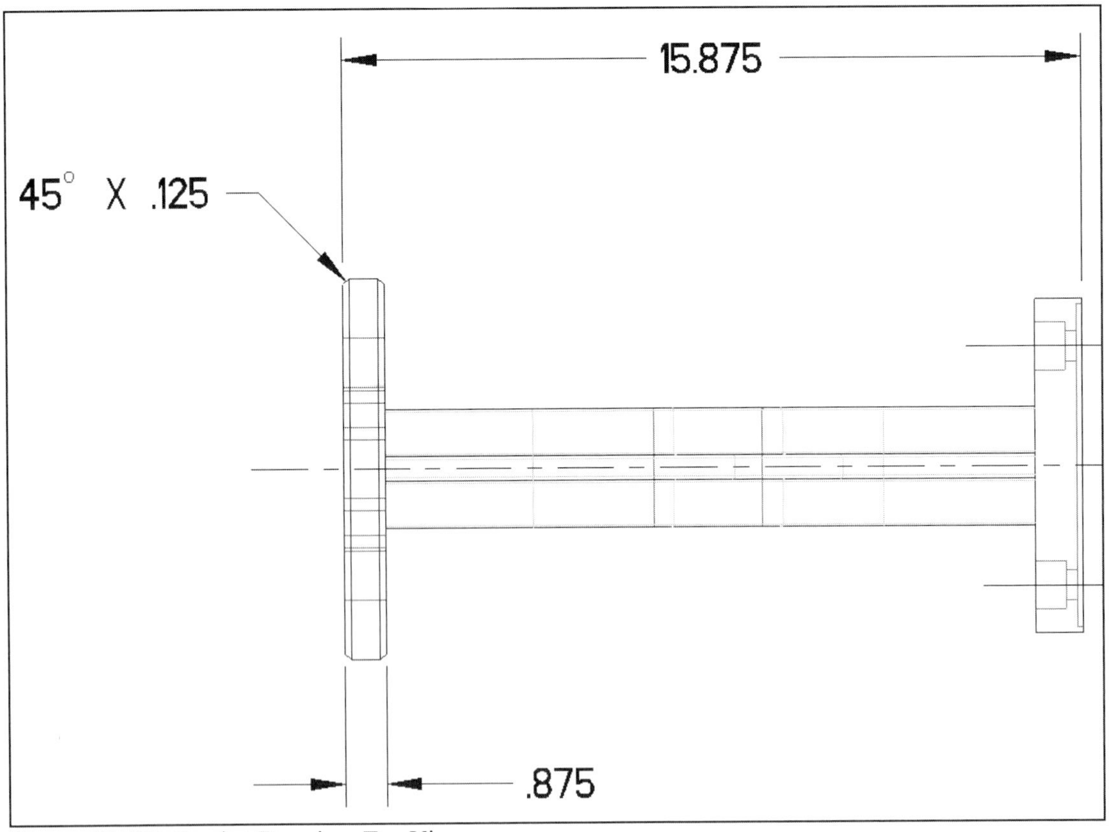

Figure 15.6(b) Bracket Drawing, Top View

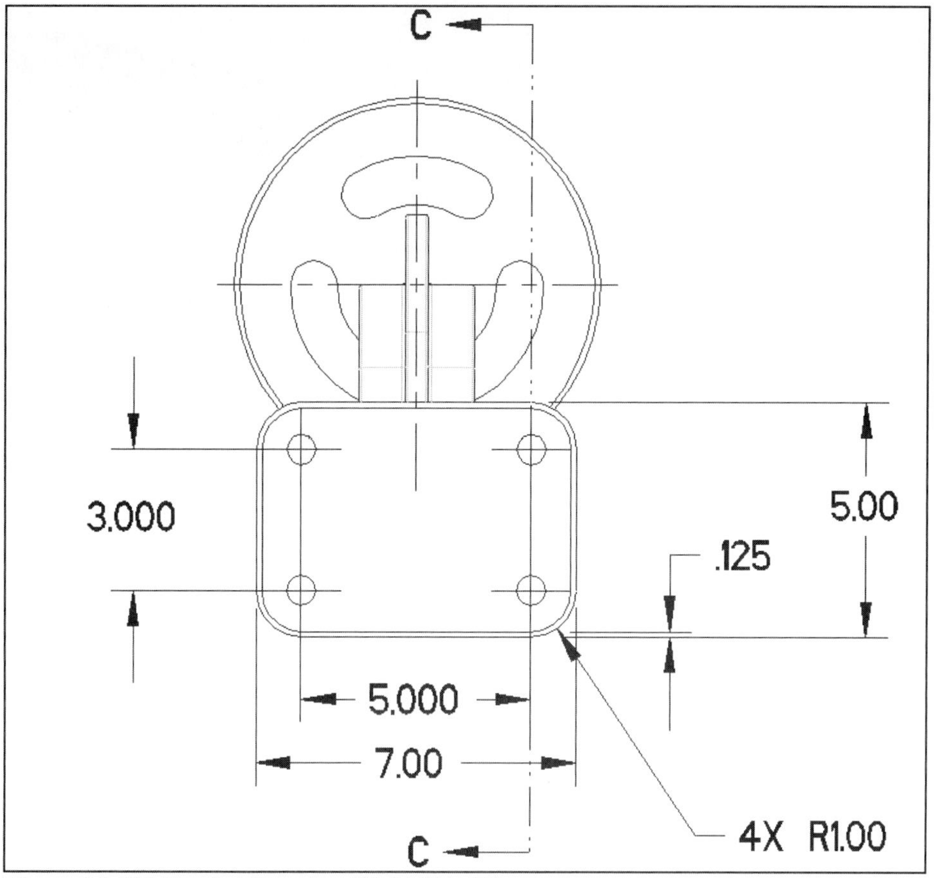

Figure 15.6(c) Bracket Drawing, Right Side View

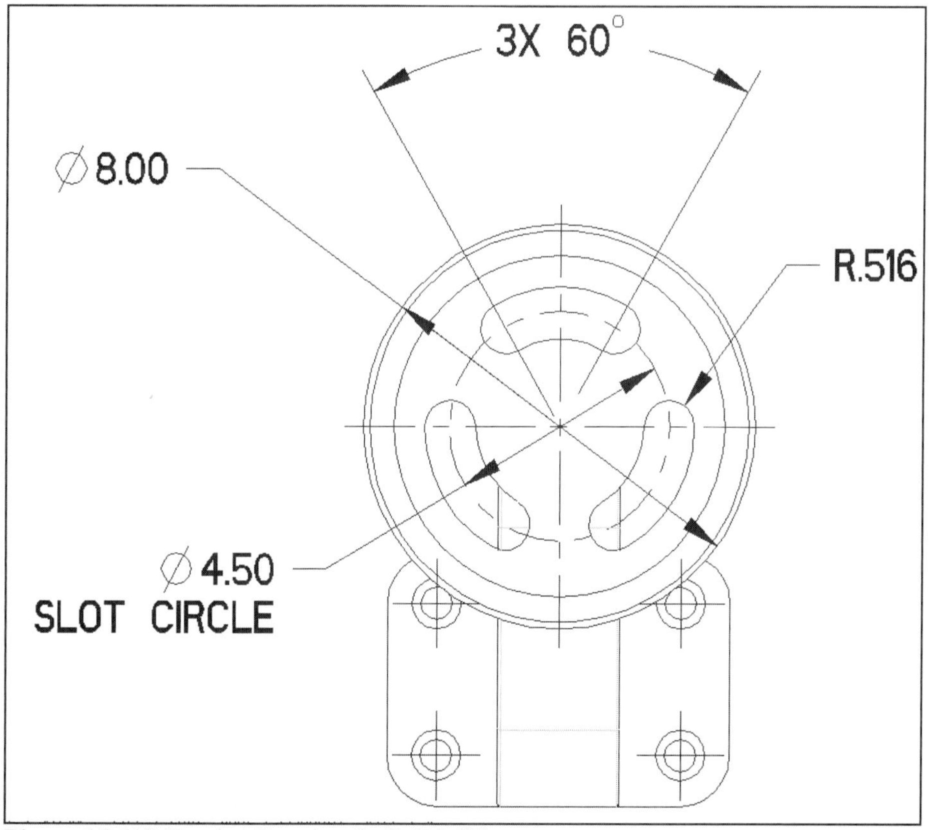

Figure 15.6(d) Bracket Drawing, Left Side View

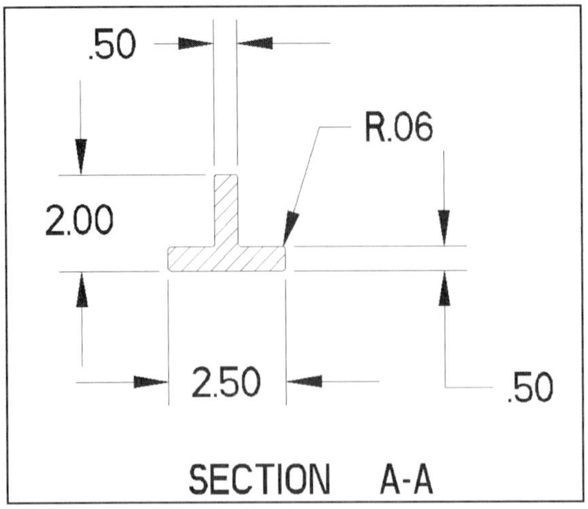

Figure 15.6(e) SECTION A-A

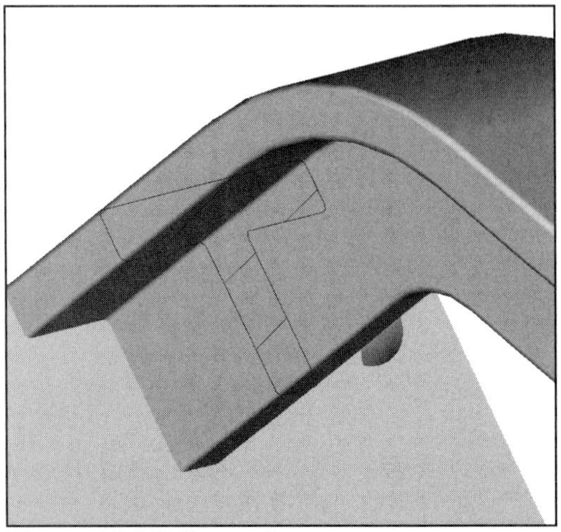
Figure 15.6(f) Swept Arm

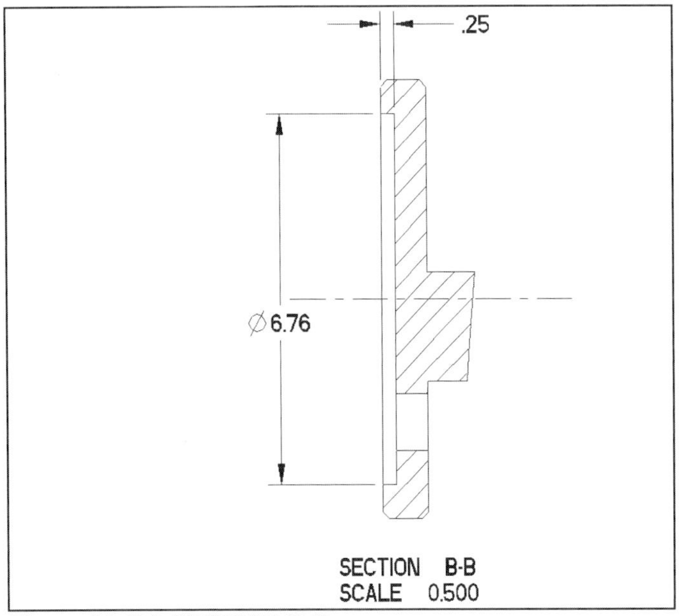

Figure 15.6(g) SECTION B-B

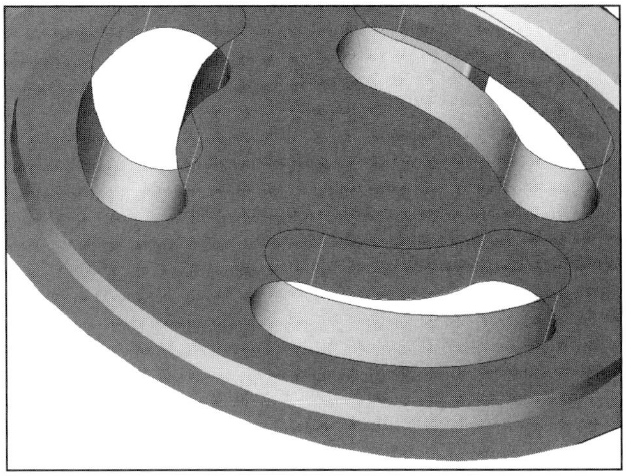
Figure 15.6(h) Cut

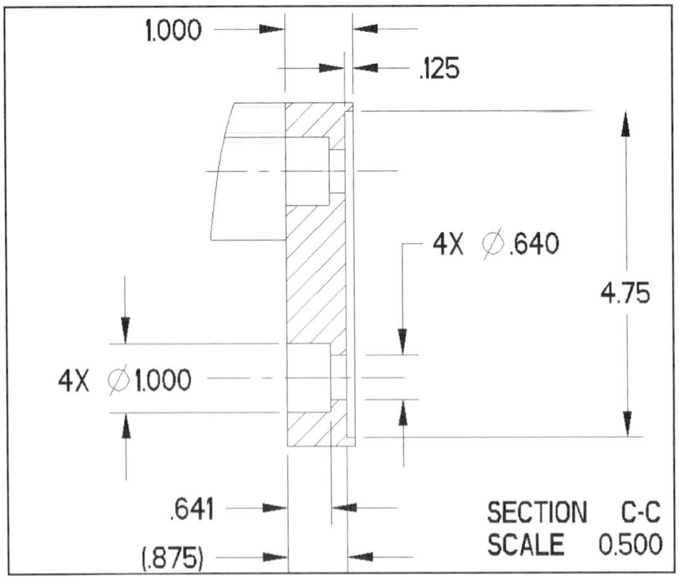

Figure 15.6(i) SECTION C-C

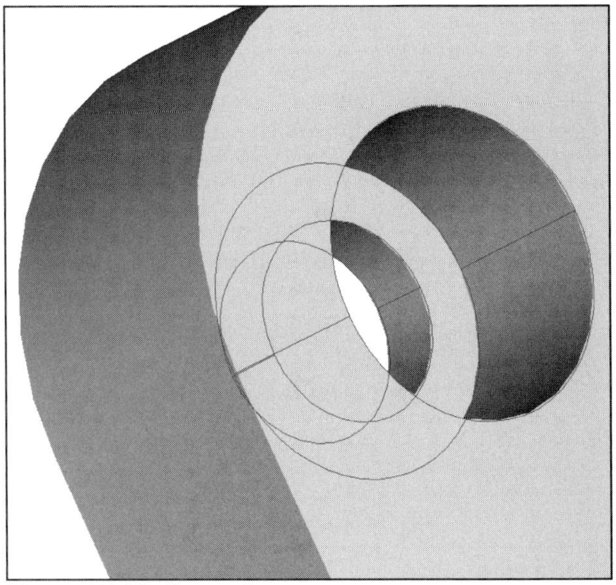
Figure 15.6(j) Counterbore

Start the Bracket by modeling the protrusion shown in Figure 15.7. This protrusion will be used to establish the sweep's position in space. Sketch the protrusion on datum **A**. The second protrusion will be the sweep [Fig. 15.8(a)].

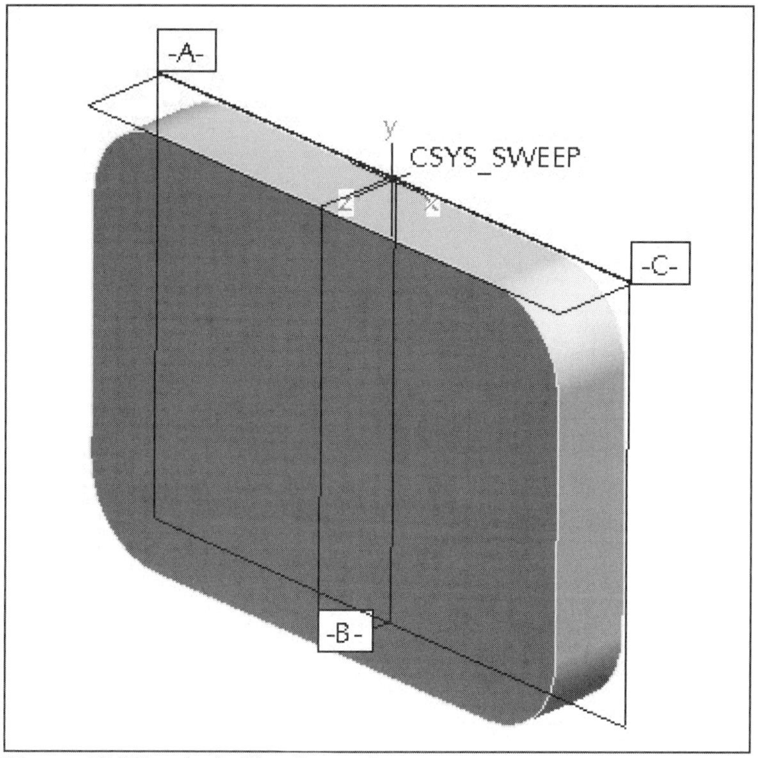

Figure 15.7 Bracket's First Protrusion

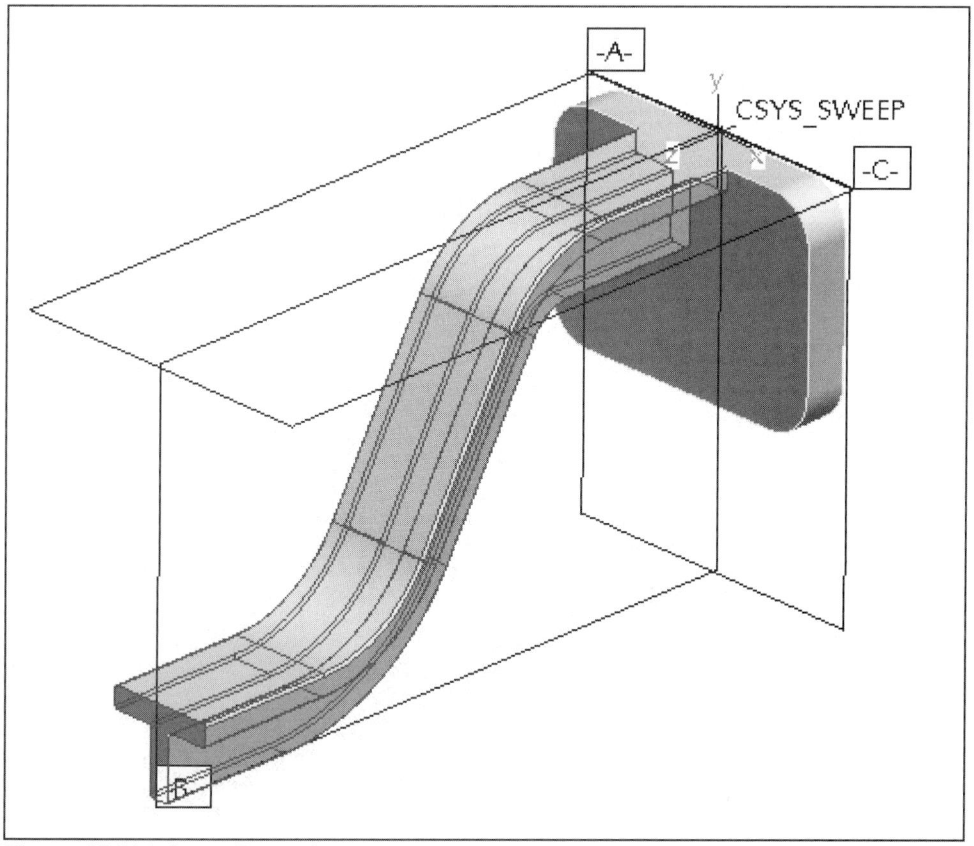

Figure 15.8(a) Swept Protrusion

Create the sweep. From the menu bar, click: **Insert** ⇒ **Sweep** ⇒ **Protrusion** ⇒ **Sketch Traj** ⇒ pick datum **B** as the sketching plane for the trajectory [Fig. 15.8(b)] ⇒ **MMB** ⇒ **Top** ⇒ pick datum **C** as the orientation plane ⇒ **Sketch** from menu bar ⇒ **References** ⇒ delete datum **A** and add the front face of the protrusion ⇒ **Close** ⇒ sketch, dimension, and modify the trajectory and complete the sweep [Figs. 15.8(c-e)] ⇒ ✓ ⇒ now you will sketch the sweep section

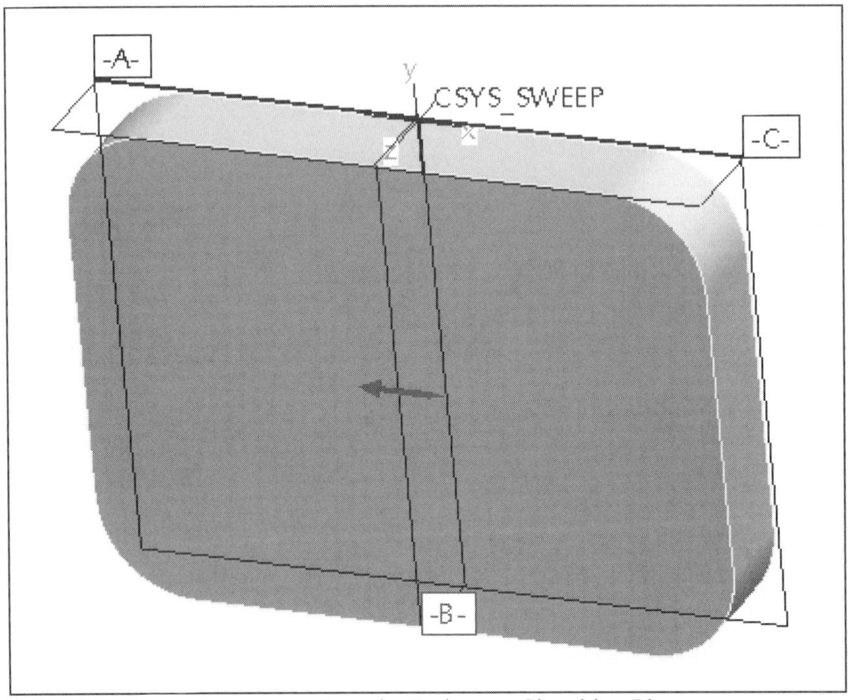

Figure 15.8(b) Select Datum B as the Trajectory Sketching Plane

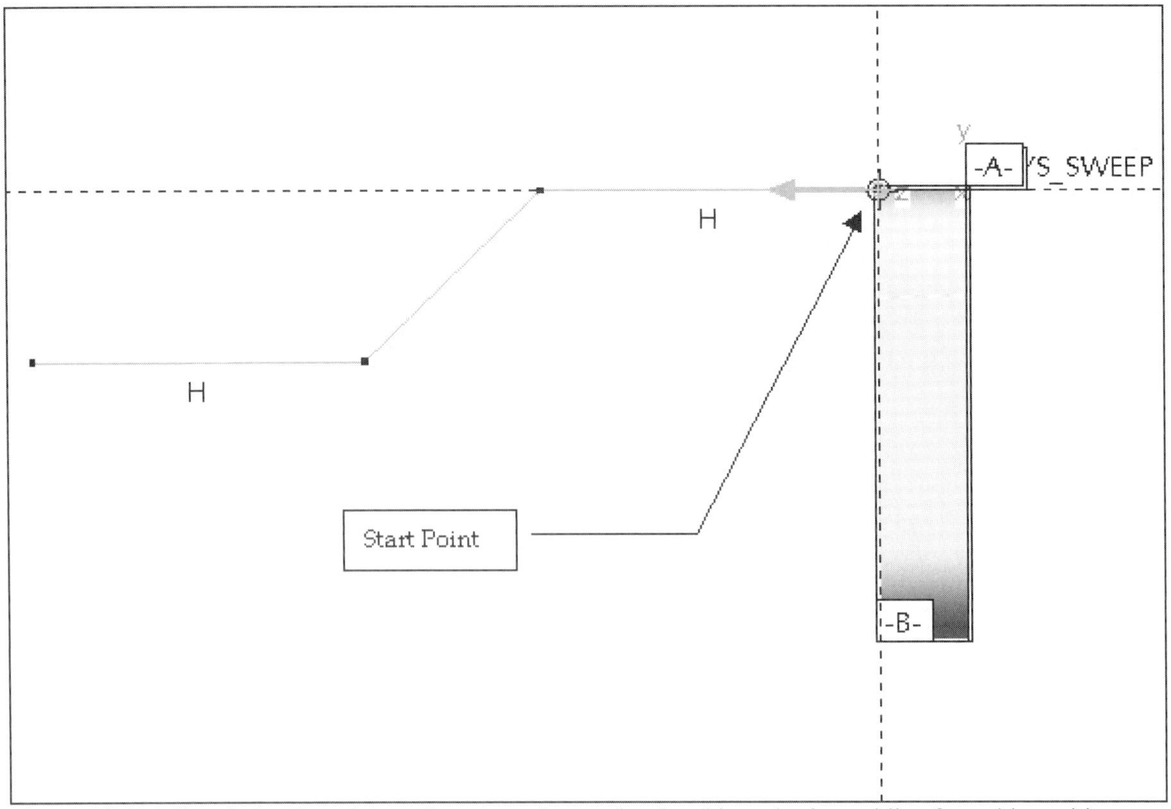

Figure 15.8(c) Sketch the Three Lines. Start the trajectory by sketching a horizontal line from this position.

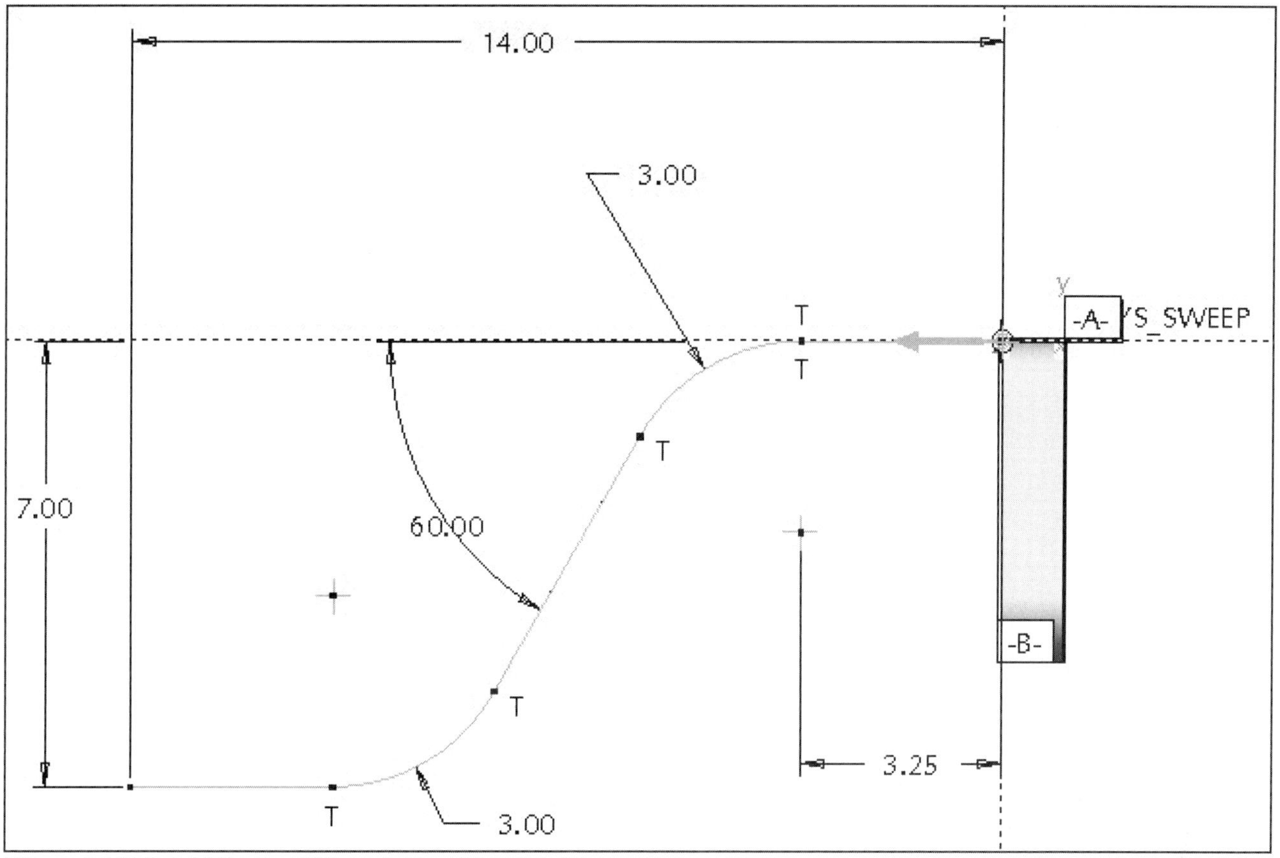

Figure 15.8(d) Add the Arc Fillets, Add and Modify Dimensions

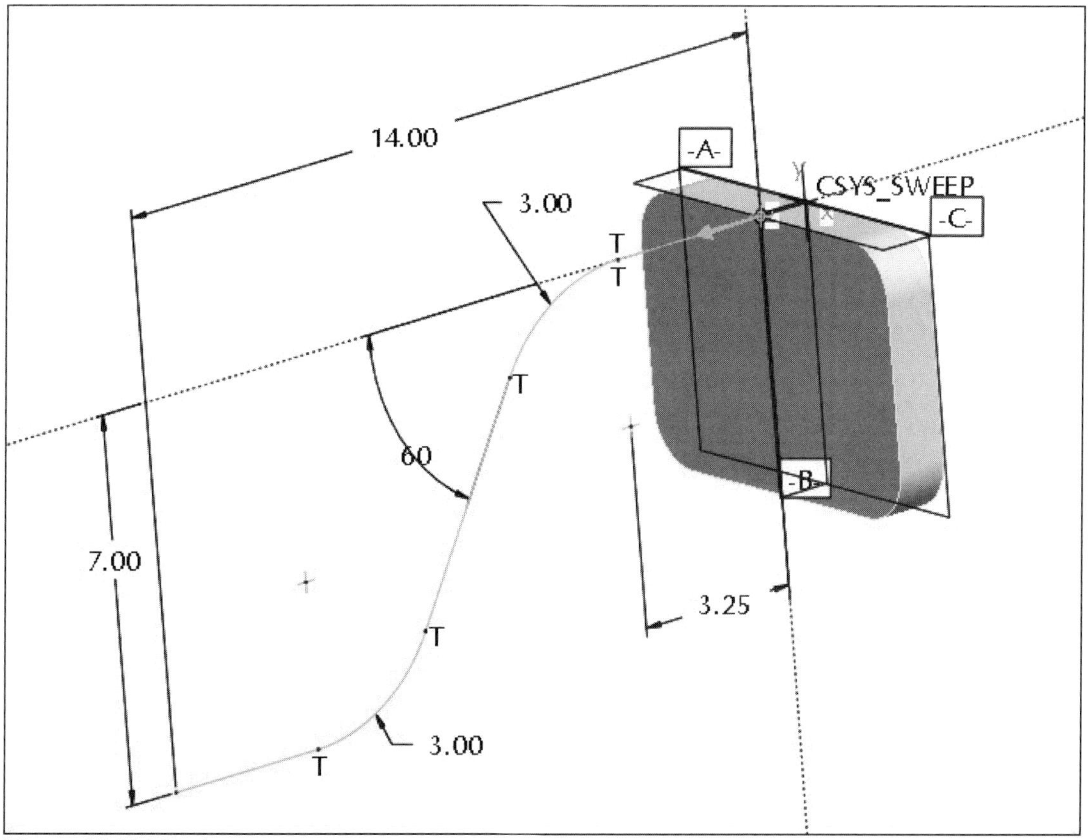

Figure 15.8(e) Trajectory

Click: **Free Ends** ⇒ **MMB** ⇒ **Sketch** (from menu bar) ⇒ **Options** ⇒ **Miscellaneous** tab ⇒ ☑ Grid ⇒ ☑ Snap To Grid ⇒ **Parameters** tab ⇒ Grid Spacing ▾ ⇒ **Manual** ⇒ ☑ Equal Spacing ⇒ **X** = **.25** ⇒ **Enter** ⇒ ✔ ⇒ sketch the *eight* lines of the section [Figs. 15.9(a-b)] ⇒ add a horizontal centerline ⇒ add *eight* fillets [Fig. 15.9(c)] ⇒ add and reposition dimensions as needed [Figs. 15.9(d-e)] ⇒ modify the section [Figs. 15.9(f-h)] ⇒ ✔ ⇒ **Preview** [Figs. 15.9(i)] ⇒ **OK** ⇒ 💾 ⇒ **MMB**

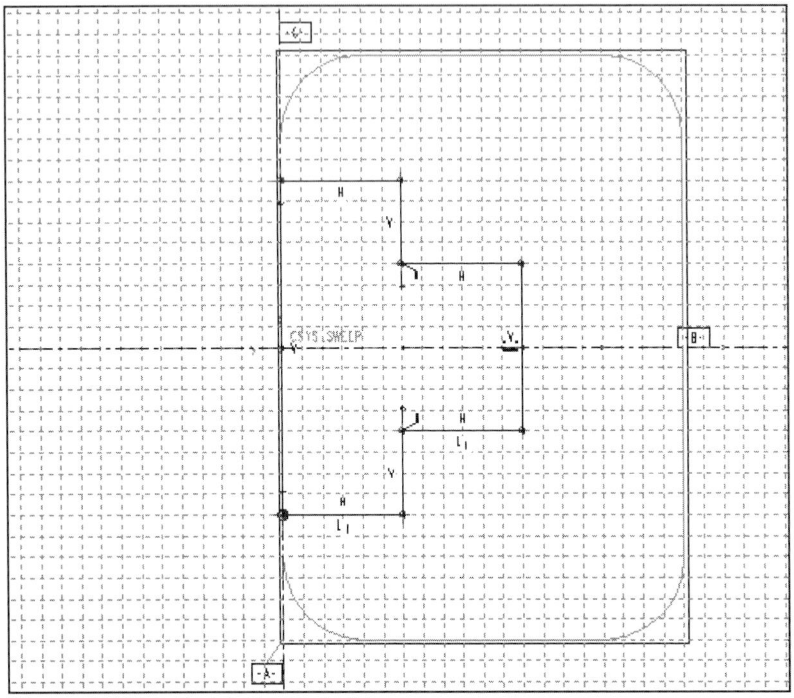

Figure 15.9(a) Sketch the Eight Lines

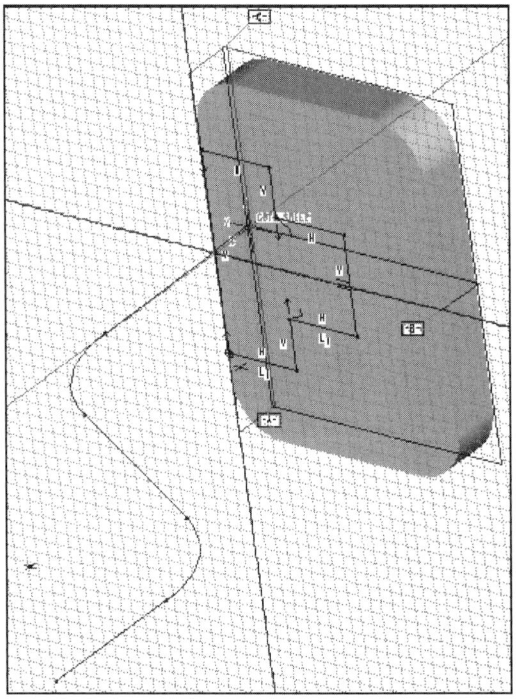

Figure 15.9(b) Sketch

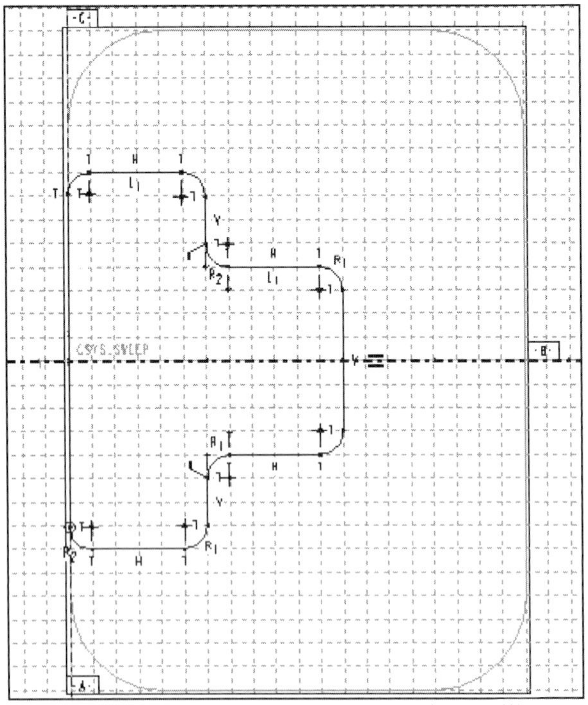

Figure 15.9(c) Sketch the Centerline and Arc Fillets

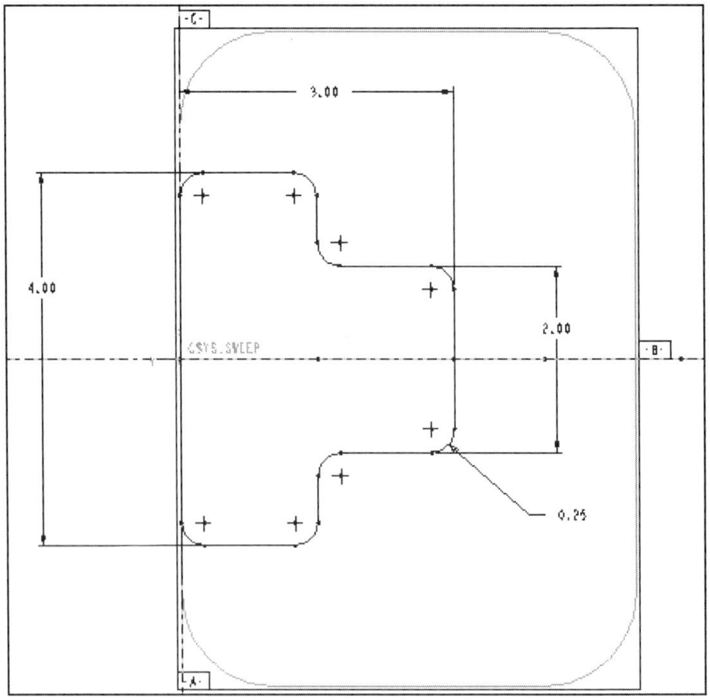

Figure 15.9(d) Add and Reposition Dimensions

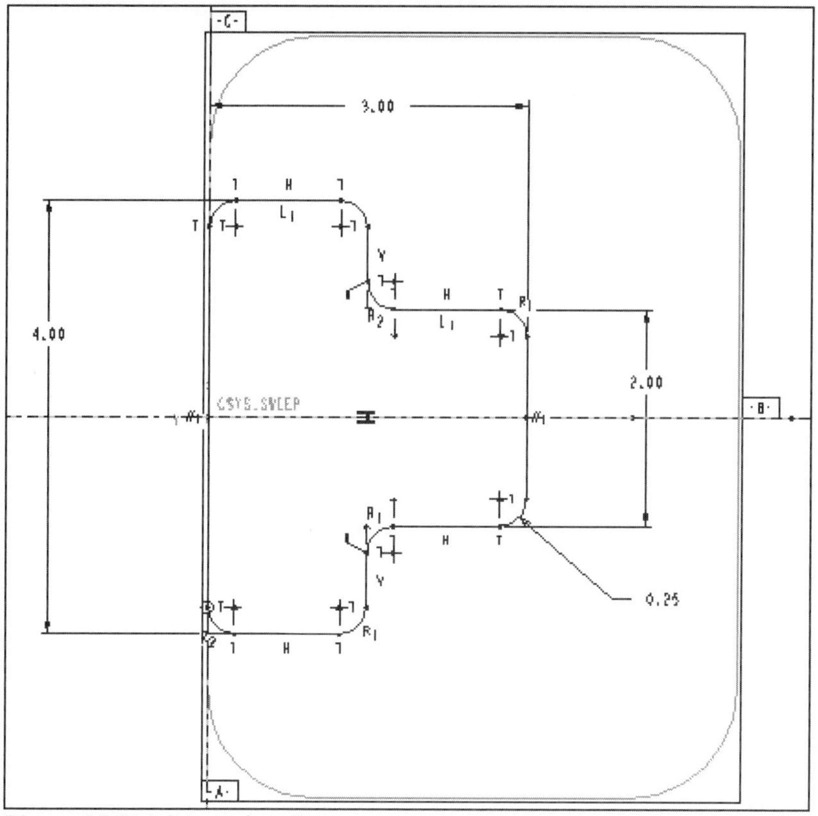

Figure 15.9(e) Constraints On

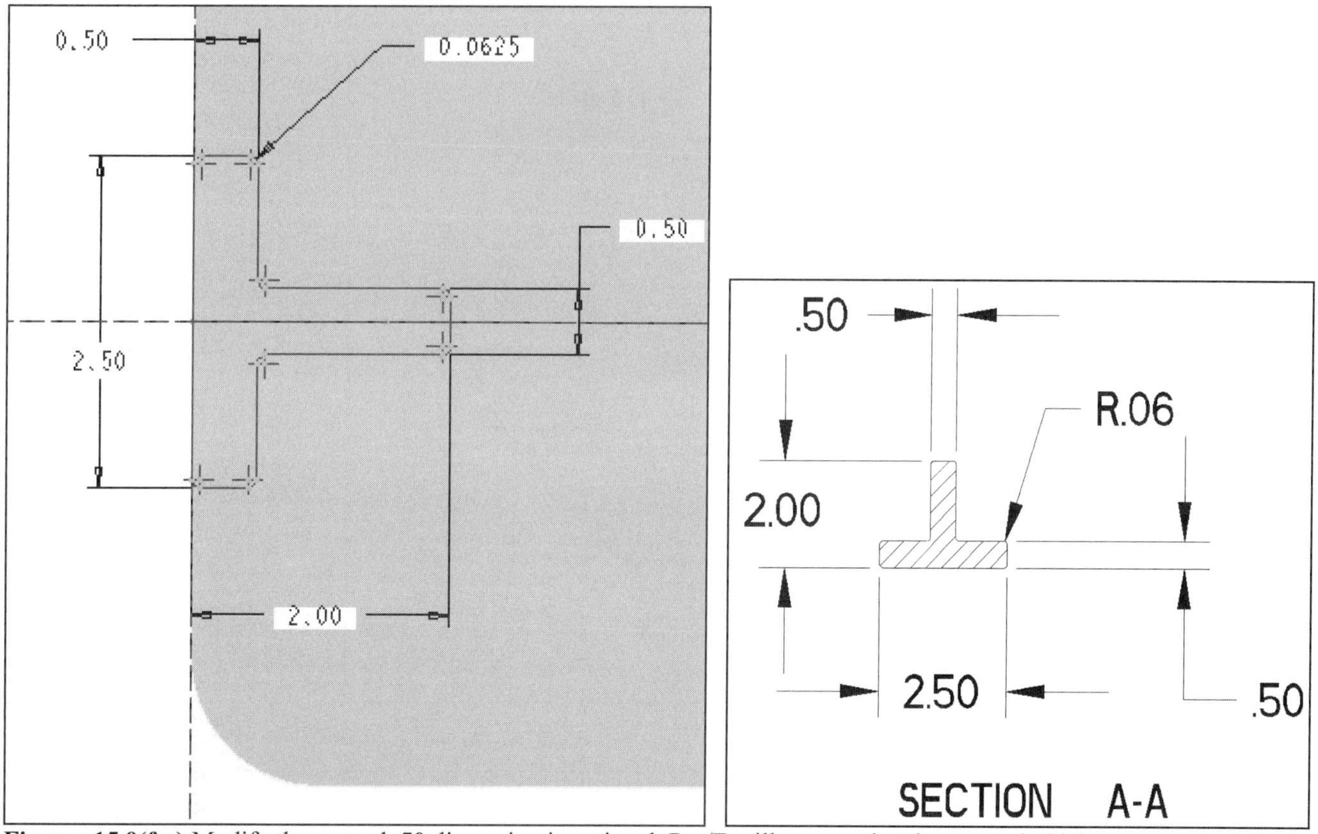

Figures 15.9(f-g) Modify the second **.50** dimension is optional. Pro/E will assume that the two web thicknesses are equal. (As per design intent, if they are both to be displayed on a drawing, then both should be on the section sketch.)

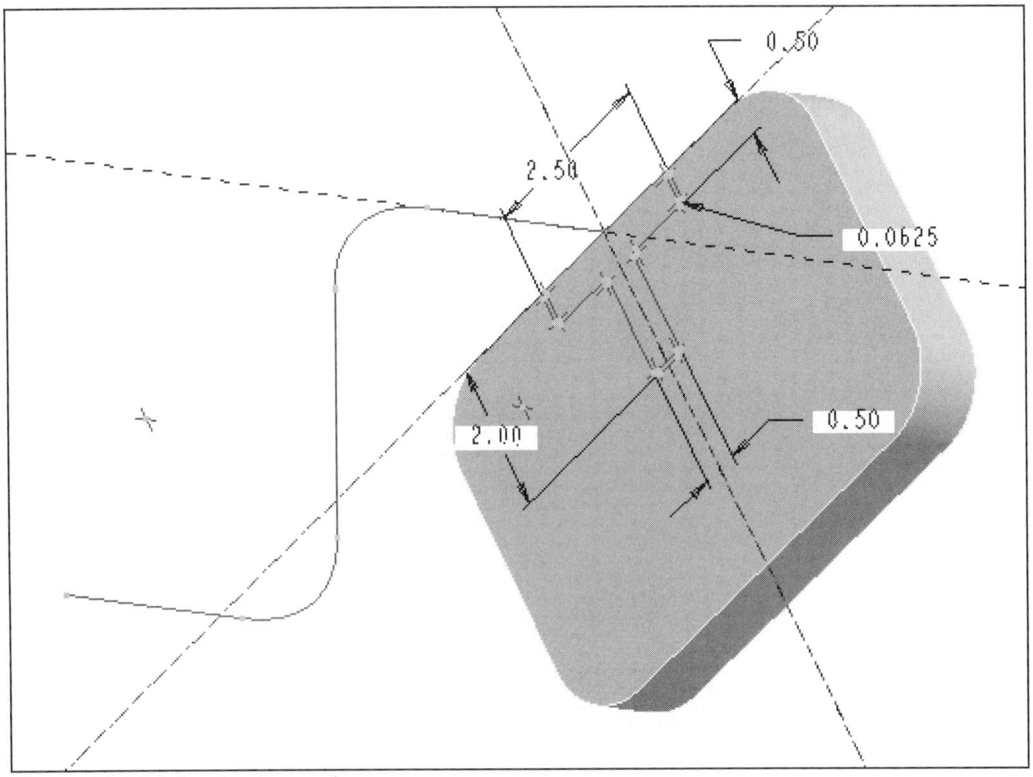

Figure 15.9(h) Section Rotated

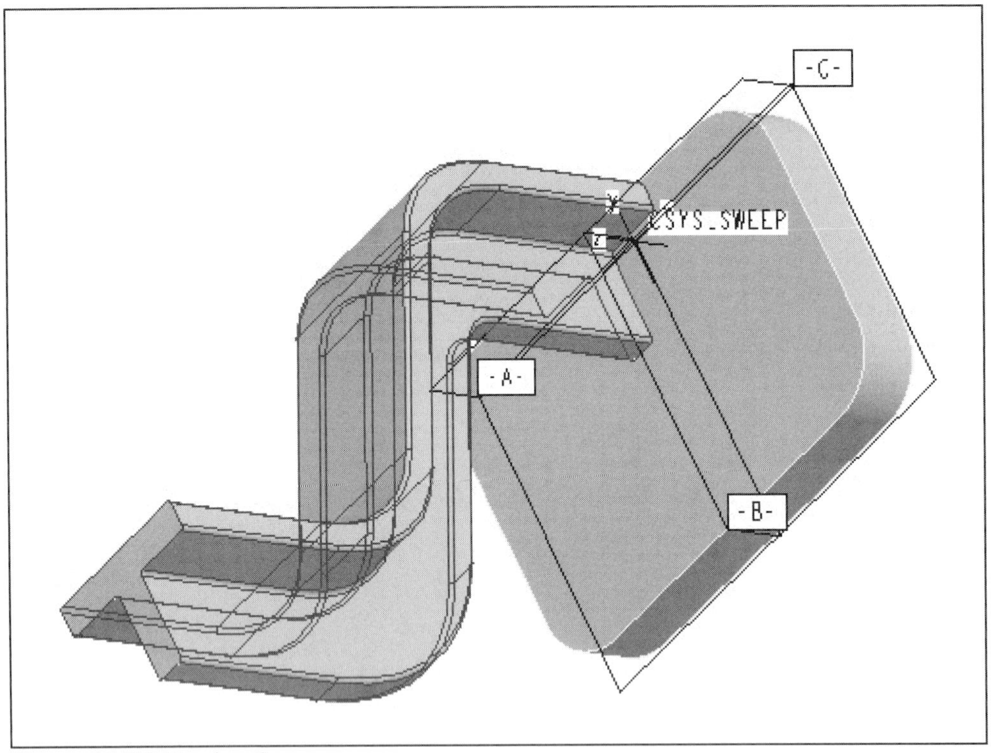

Figure 15.9(i) Preview Sweep

Complete the part by modeling the remaining features (Figs. 15.10 through 15.14).

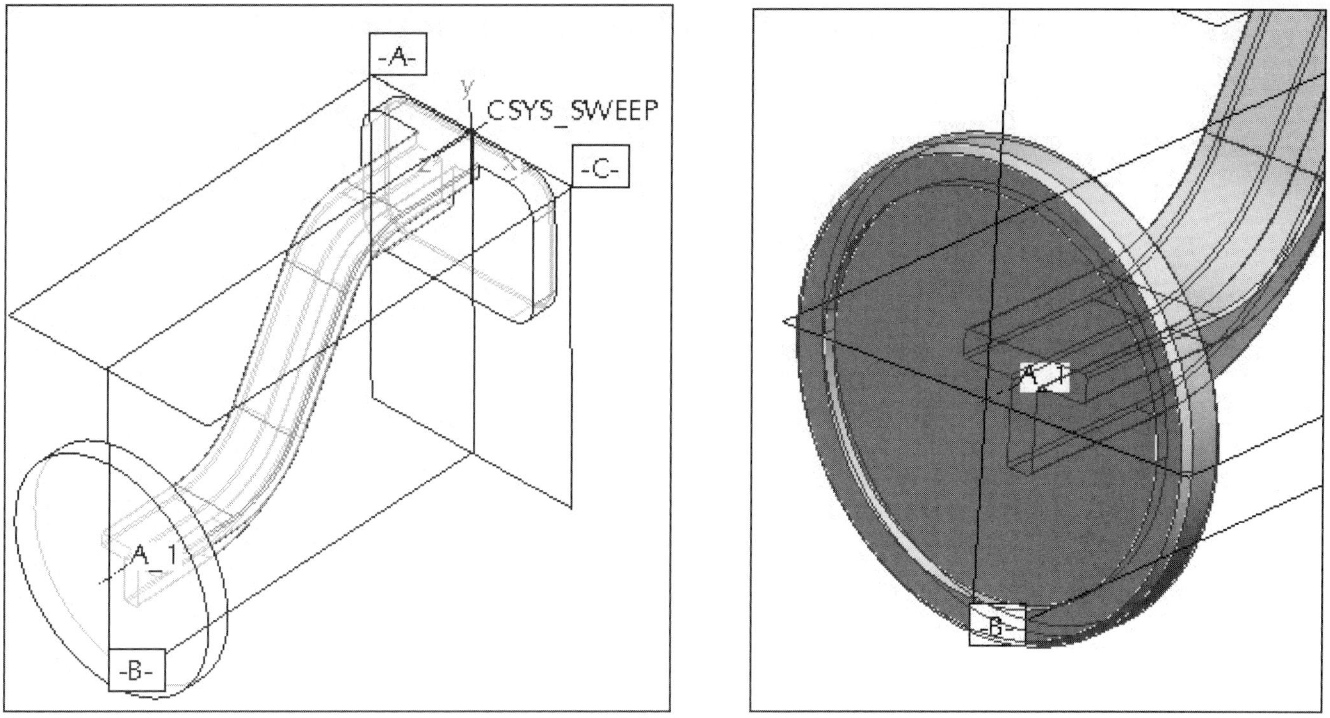

Figures 15.10(a-b) Add the Third Protrusion; Add the Cut (**∅6.76** by **.250** deep) and the Chamfers (**45° X .125**)

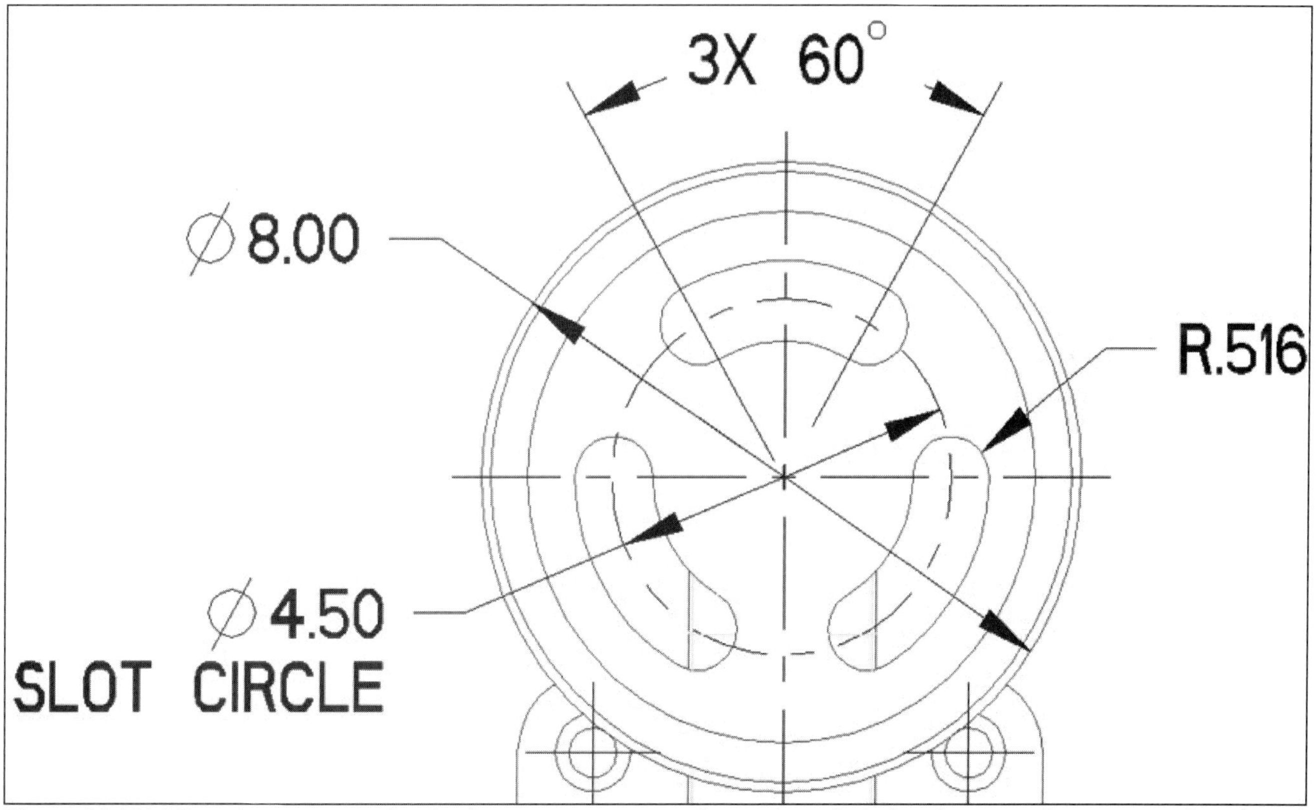

Figure 15.11(a) In order to pattern the slot feature, create a new datum through A_1 and at an angle, to use as the orientation (reference) plane.

Start the slot cut by creating a new datum through the axis and parallel to Datum C [Figs. 11(b-c)].

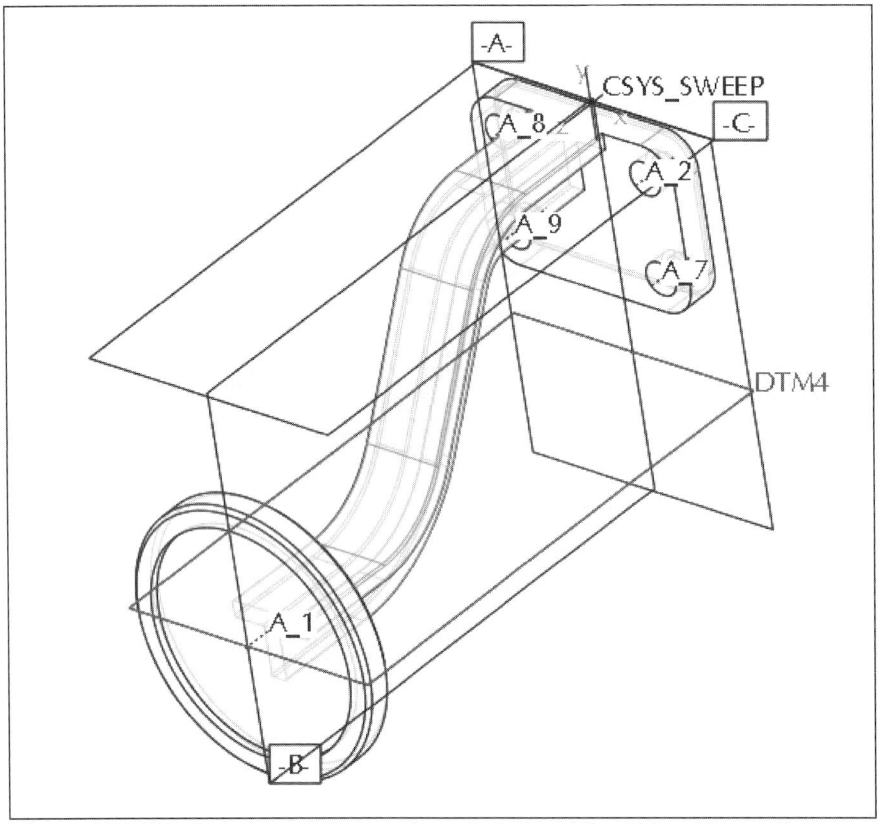

Figure 15.11(b) Create a new Datum through A_1 and Parallel to Datum C

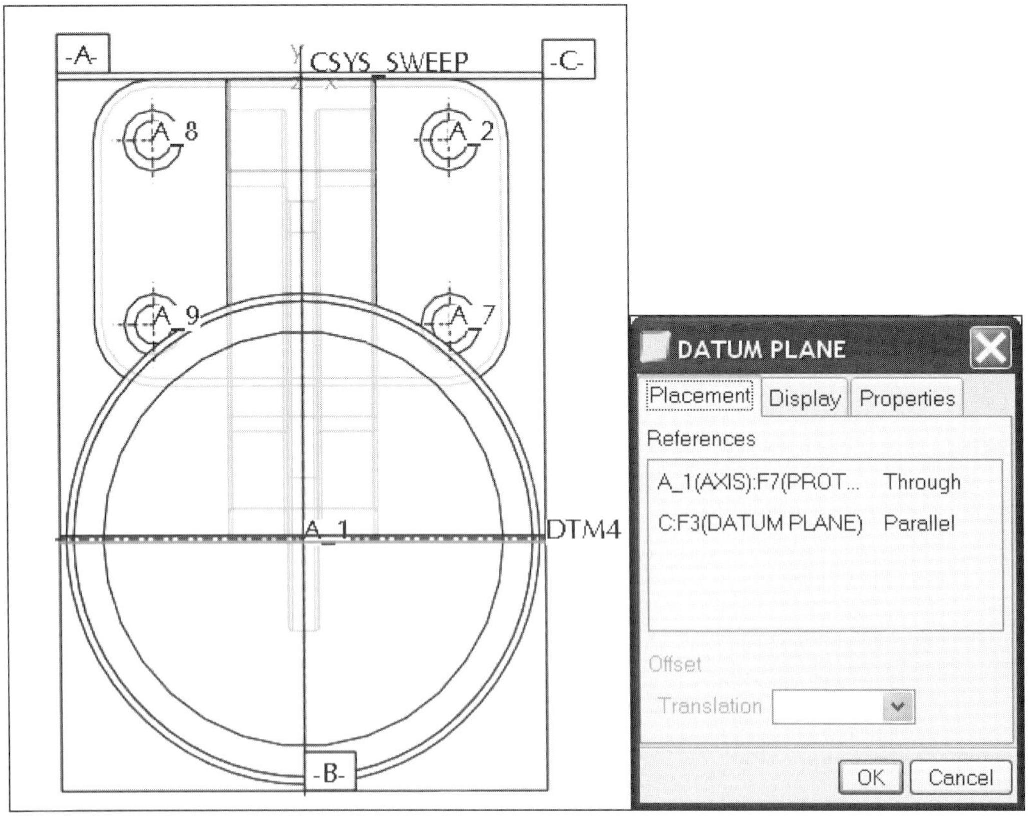

Figure 15.11(c) Datum References

Select the cut face as the sketching plane and Datum C as the Orientation Reference [Fig. 15.11(d)]

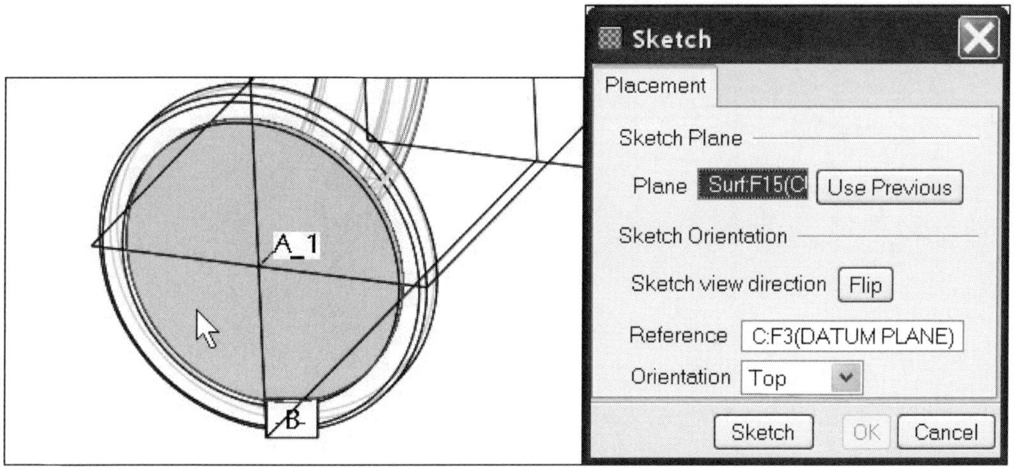

Figure 15.11(d) Sketch References

Add a Sketcher Point at the center of the round protrusion and then set the options as shown [Figs. 15.11(e-g)]. Click: **Sketch** ⇒ **Options** ⇒ **Miscellaneous** tab ⇒ set as displayed ⇒ **Constraints** tab ⇒ set as displayed ⇒ **Parameters** tab ⇒ set as displayed

Figure 15.11(e-g) Sketcher Preferences Dialog Box

Click: Grid Origin [arrow icon] ⇒ pick on the sketcher point as the **Origin** of the polar grid ⇒ [check icon] from the Sketcher Preferences dialog box ⇒ sketch the section [Fig. 15.11(h)] ⇒ complete the feature

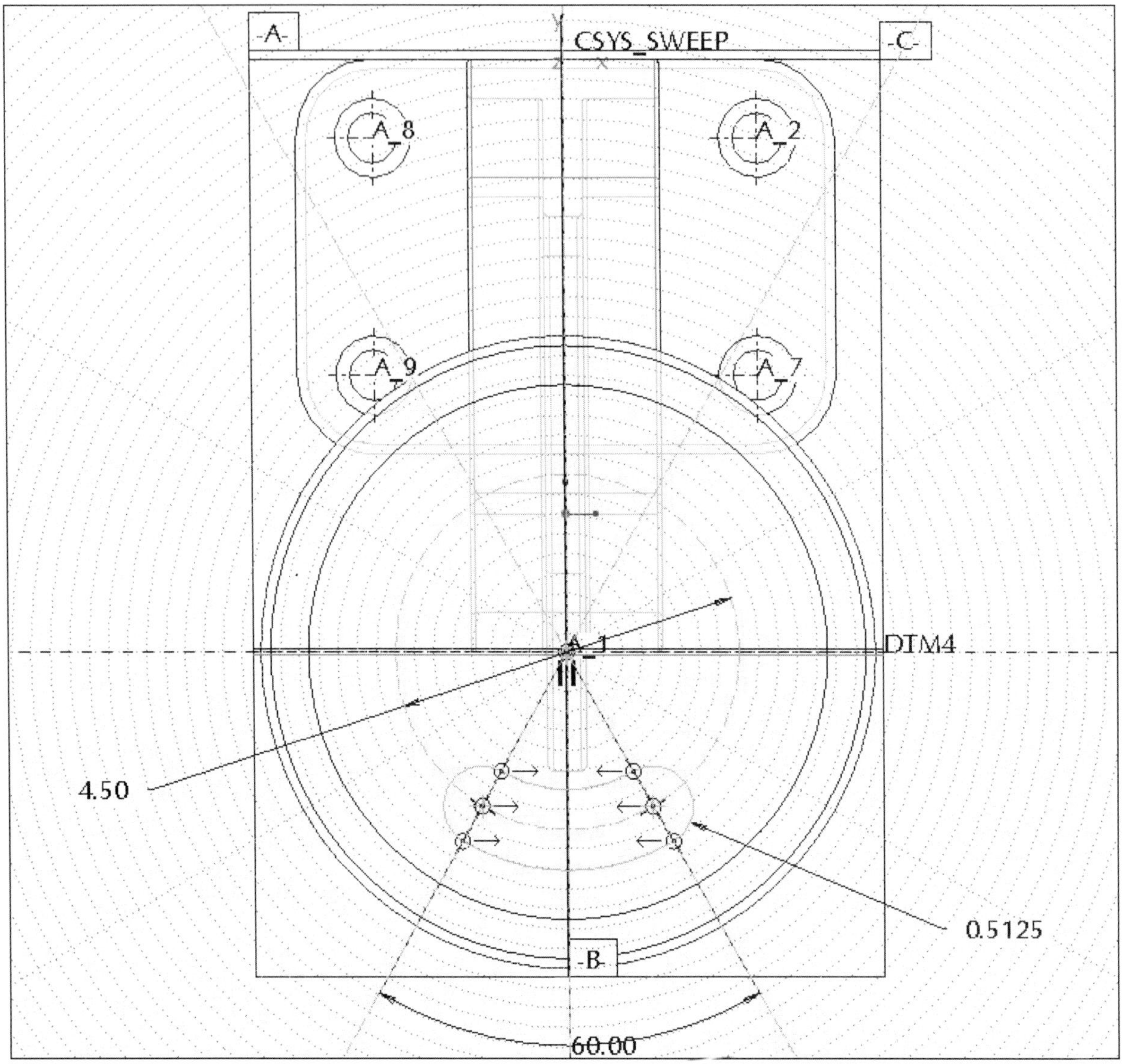

Figure 15.11(h) Sketch the Section

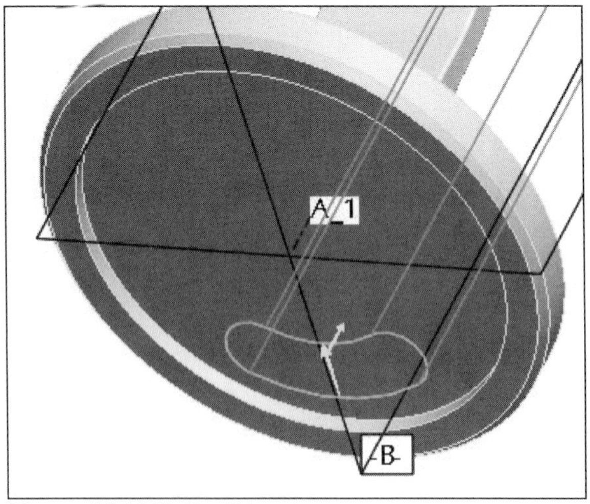

Figure 15.11(i) Cut Direction

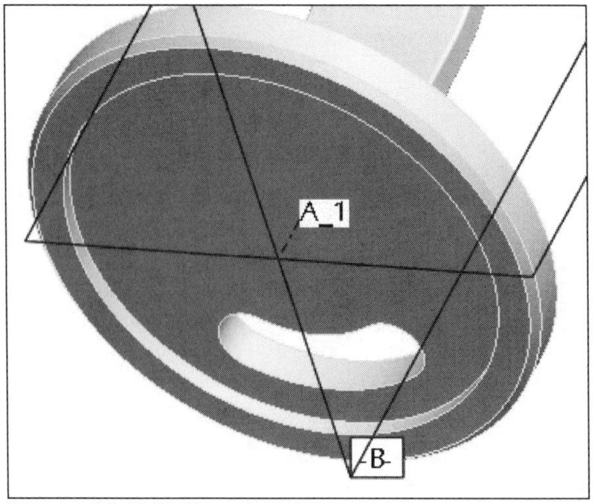

Figure 15.11(j) Completed Cut

Pattern the slot [Figs. 15.12(a-c)]

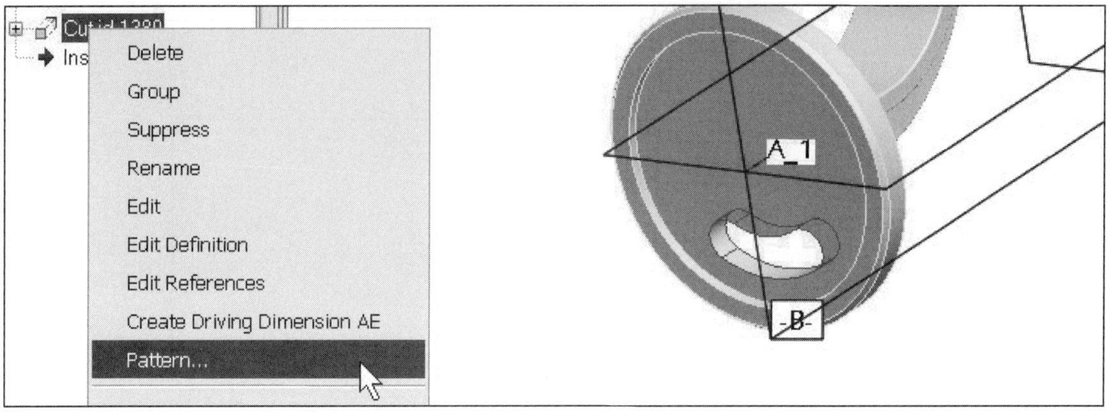

Figure 15.12(a) Pattern the Slot

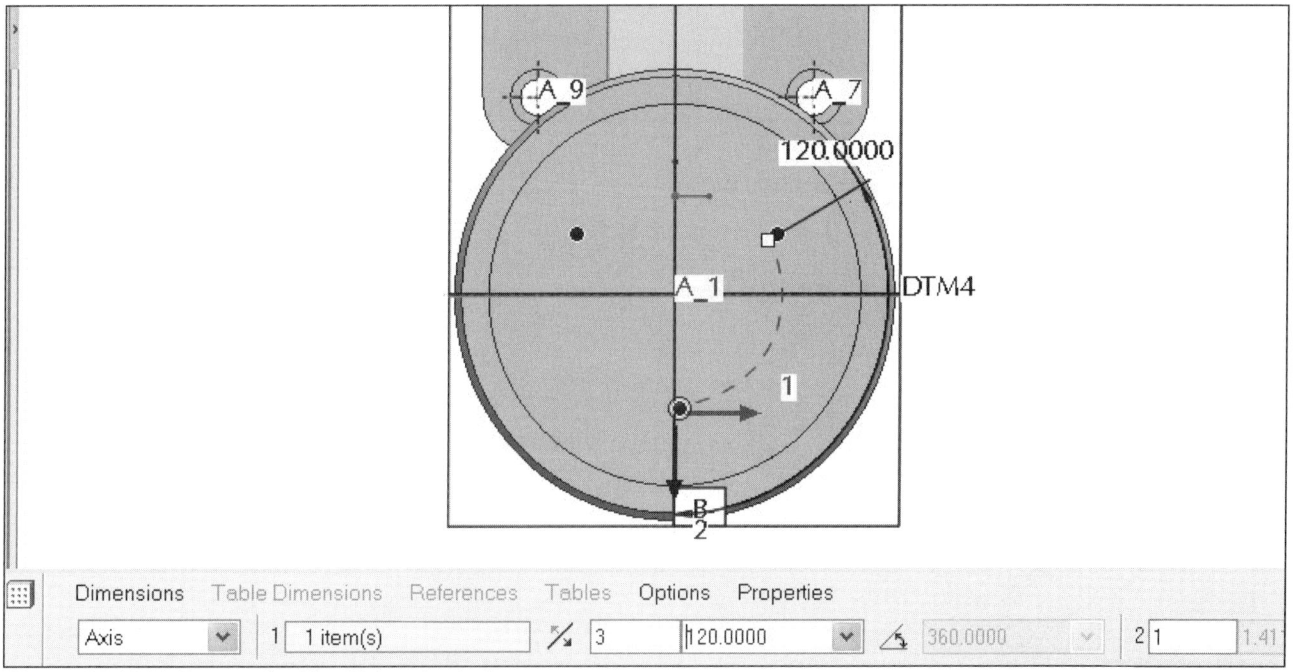

Figure 15.12(b) Pattern the Slot Using the Axis

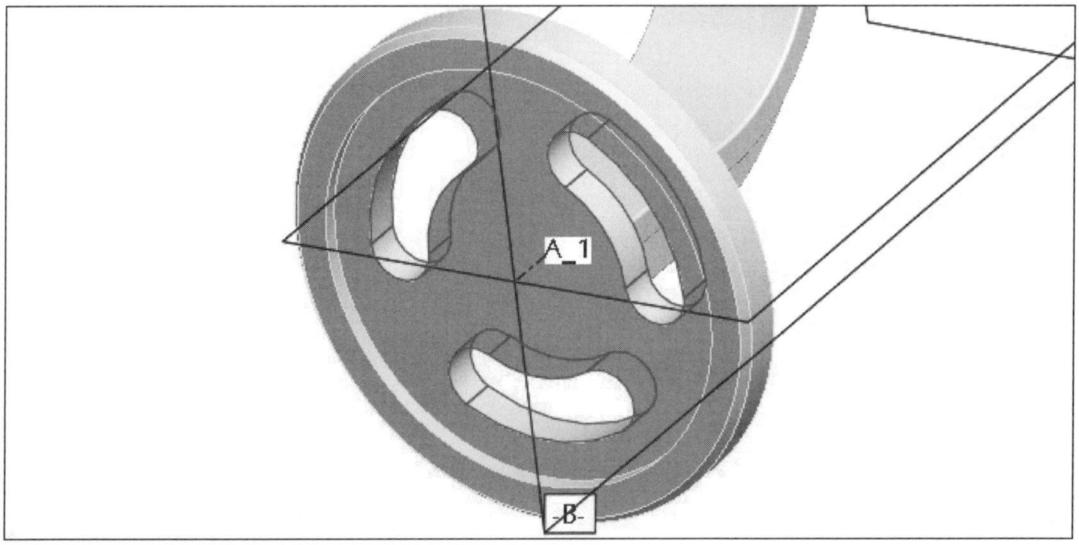

Figure 15.12(c) Patterned Slot

Model the face cut and then create and pattern the counterbore holes [Figs. 15.13(a-b)] ⇒ complete the part (Fig. 15.14)

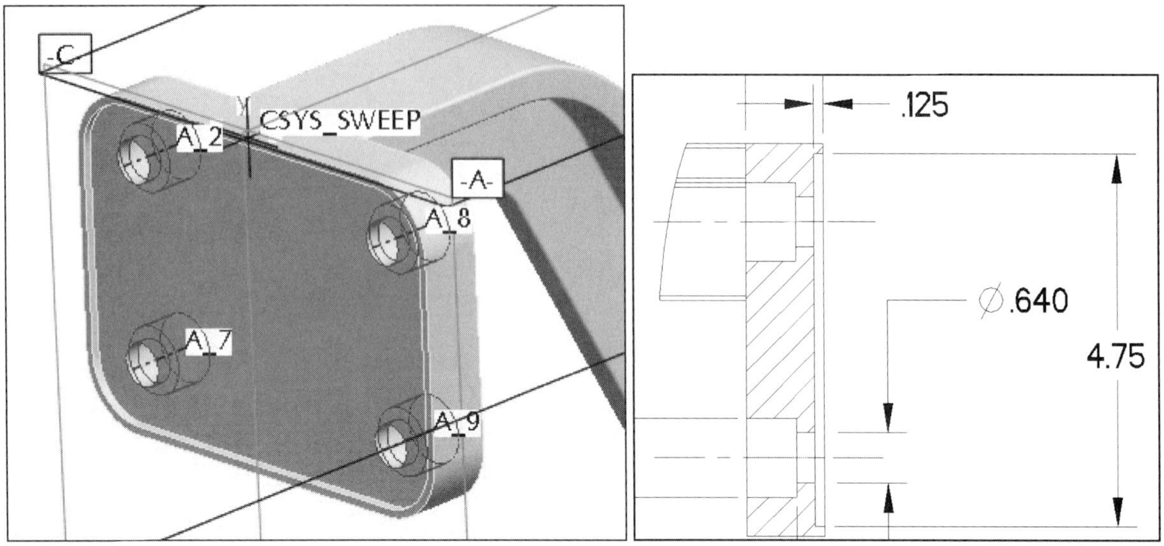

Figures 15.13(a-b) Create the Face Cut and Pattern Counterbore Holes

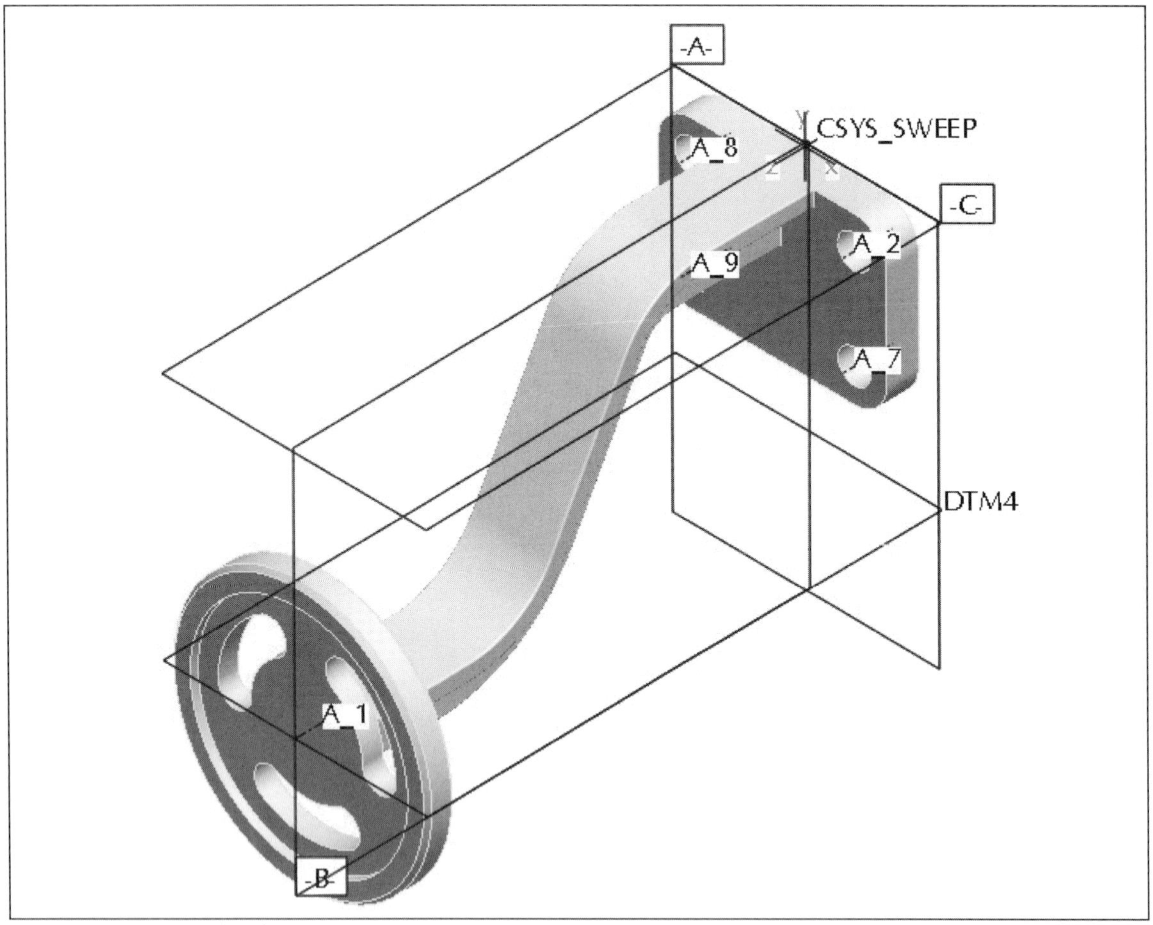

Figure 15.14 Completed Part

Download projects from ***www.cad-resources.com* ⇒ *Downloads.***

Lesson 16 Helical Sweeps, and Annotations

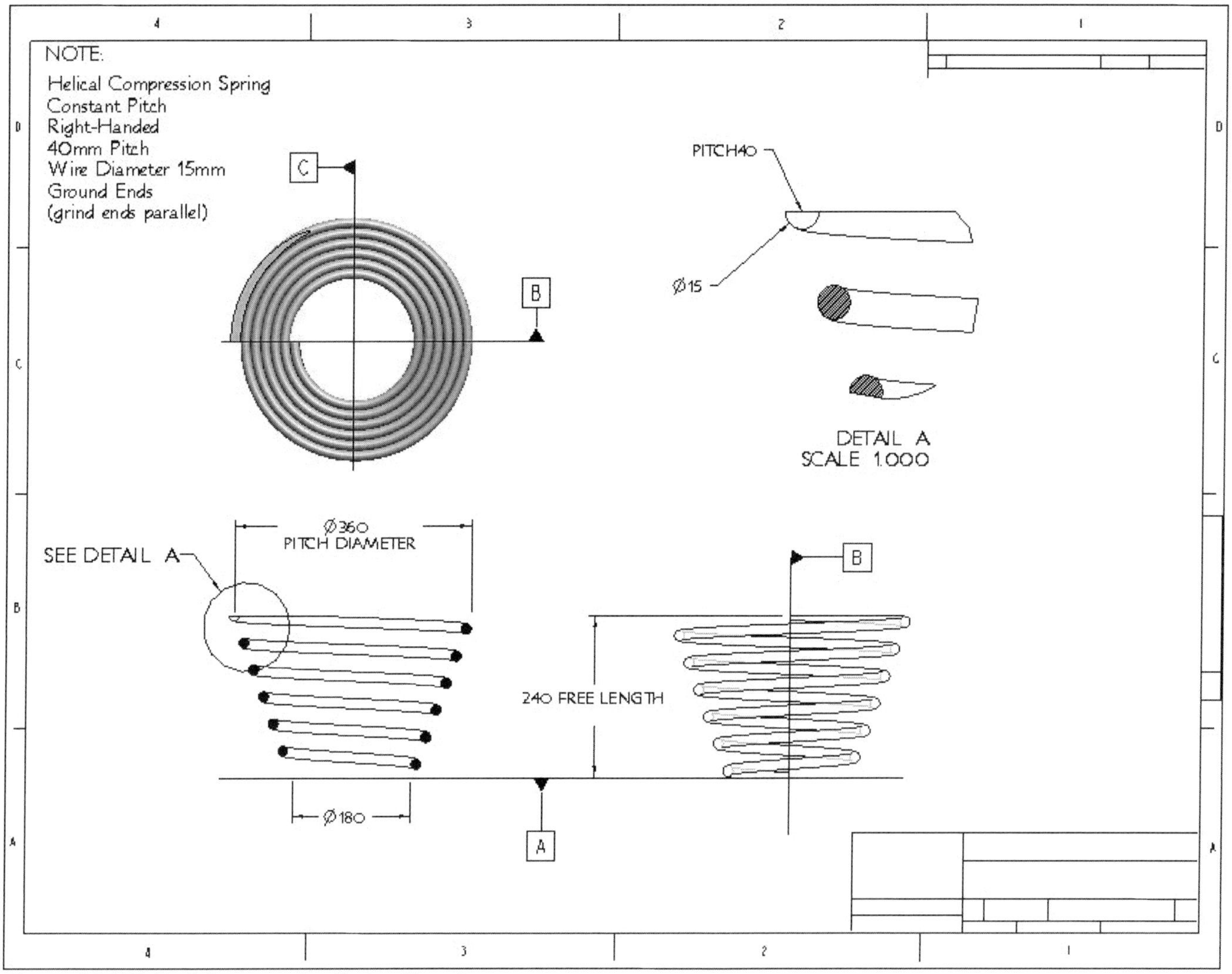

Figure 16.1 Helical Compression Spring Drawing

OBJECTIVES

- Create a **helical compression spring** with a **Helical Sweep**
- Use sweeps to create **hooks** on **extension springs**
- Create **plain ground ends** on a spring
- Create **3D Notes** and **Annotation Features**

Helical Sweeps and Annotations

A **helical sweep** (Fig. 16.1) is created by sweeping a section along a helical *trajectory*. The trajectory is defined by both the *profile* of the *surface of revolution* (which defines the distance from the section origin of the helical feature to its *axis of revolution*) and the *pitch* (the distance between coils). The trajectory and the surface of revolution are construction tools and do not appear in the resulting geometry. Model notes are pieces of text, which can contain links (URL's) to World Wide Web pages, which you can attach to objects in Pro/E. Model notes, increase the amount of information that you can attach to any object in your model. **Annotation features** are data features that you can use to manage the model annotation including surface finish, geometric tolerances, notes, and so on.

Helical Sweeps

The Helical Sweep command is available (Fig. 16.2) for both solid and surface features. Use the following ATTRIBUTES menu options, presented in mutually exclusive pairs, to define the helical sweep feature:

- **Constant** The pitch is constant
- **Variable** The pitch is variable and defined by a graph
- **Thru Axis** The section lies in a plane that passes through the axis of revolution
- **Norm To Traj** The section is oriented normal to the trajectory
- **Right Handed** The trajectory is defined by the right-hand rule
- **Left Handed** The trajectory is defined by the left-hand rule

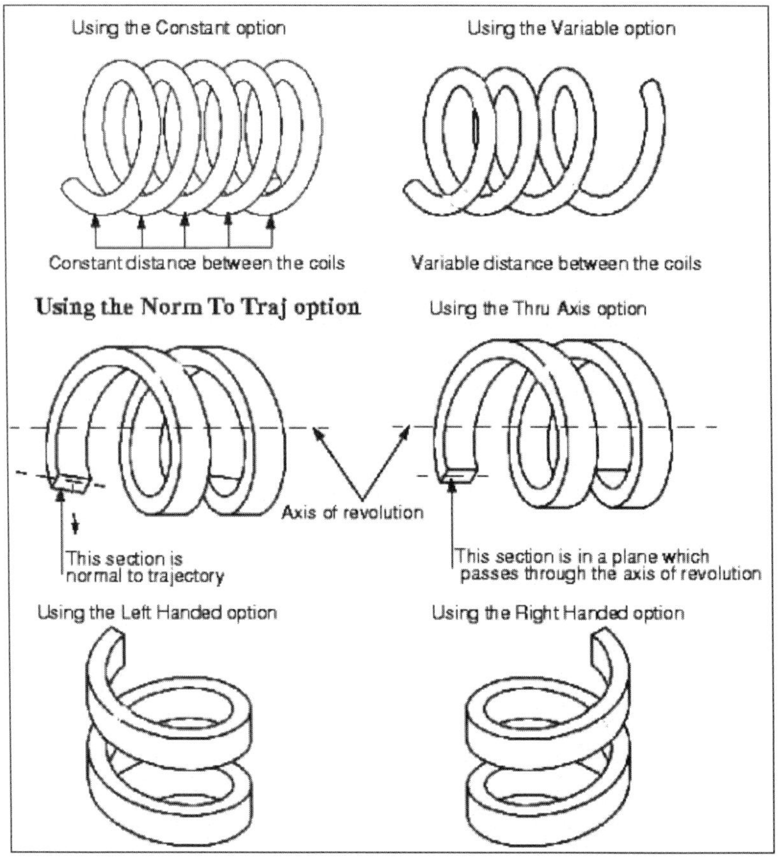

Figure 16.2 Helical Sweeps

Annotations

Model notes are text strings that you can attach to objects (Fig. 16.3). You can attach any number of notes to any object in your model. When you attach a note to an object, the object is considered the parent of the note. When you delete the parent object, all child notes are deleted with it. You can also allocate a URL to each model note. You can use model notes to communicate with members of your workgroup as to how to review or use a model, explain how you approached or solved a design problem when modeling, and explain changes that you have made to the features of a model over time.

Annotation features can also be notes, but also include: symbols, surface finish, geometric tolerance, set datum tags, ordinate baseline dimensions, driven dimension, and so on.

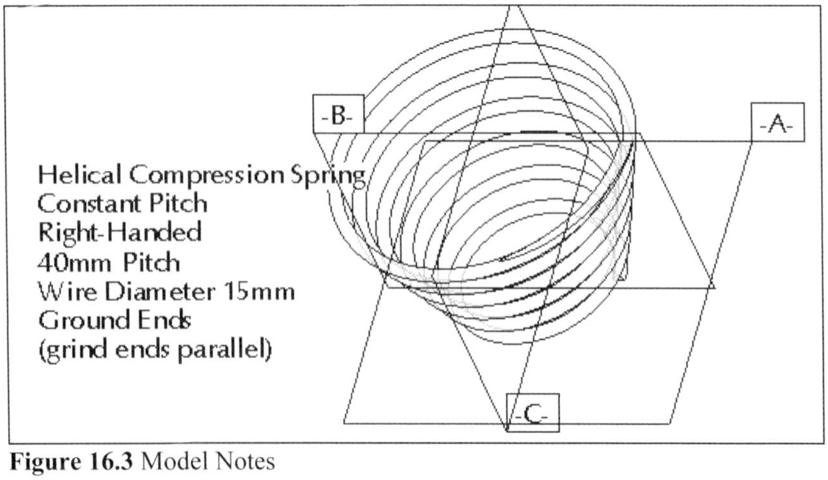

Figure 16.3 Model Notes

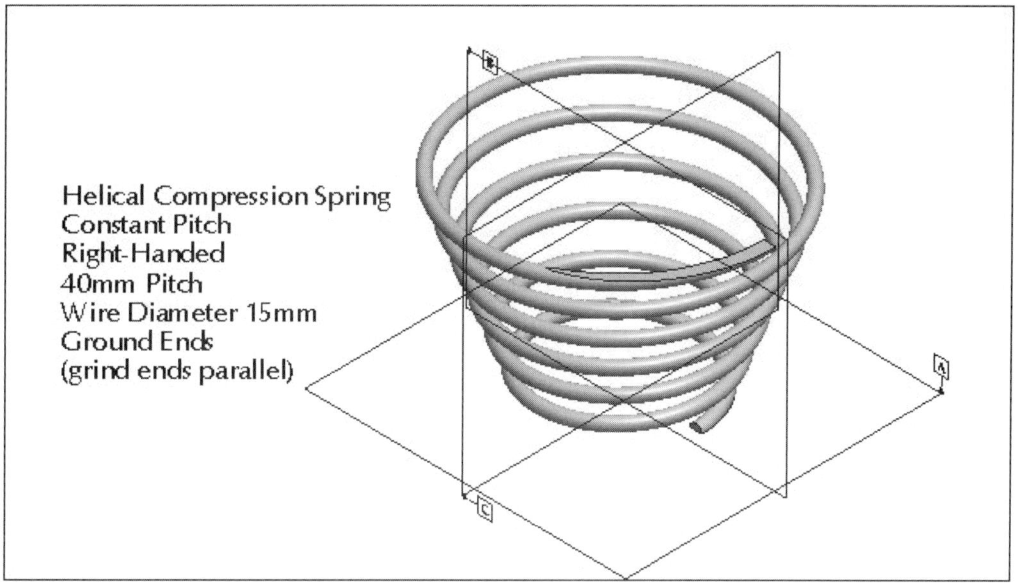

Figure 16.4 Helical Compression Spring with Datum Planes and Model Note

Helical Compression Spring

Springs (Fig. 16.4) and other helical features are created with the Helical Sweep command. A helical sweep is created by sweeping a *section* along a *trajectory* that lies in the *surface of revolution:* The trajectory is defined by both the *profile* of the surface of revolution and the distance between coils. The model for this lesson is a *constant-pitch right-handed helical compression spring with ground ends, a pitch of **40 mm**, and a wire diameter of **15 mm*** (Figs. 16.4 through 16.8).

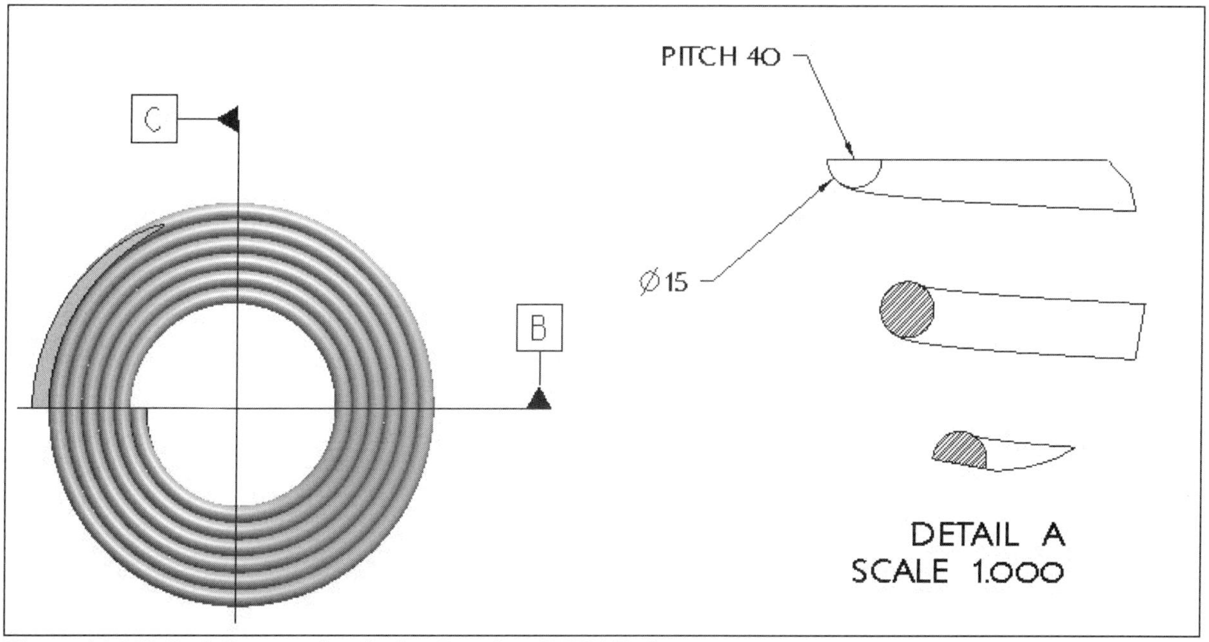

Figure 16.5 Helical Compression Spring Drawing: DETAIL A

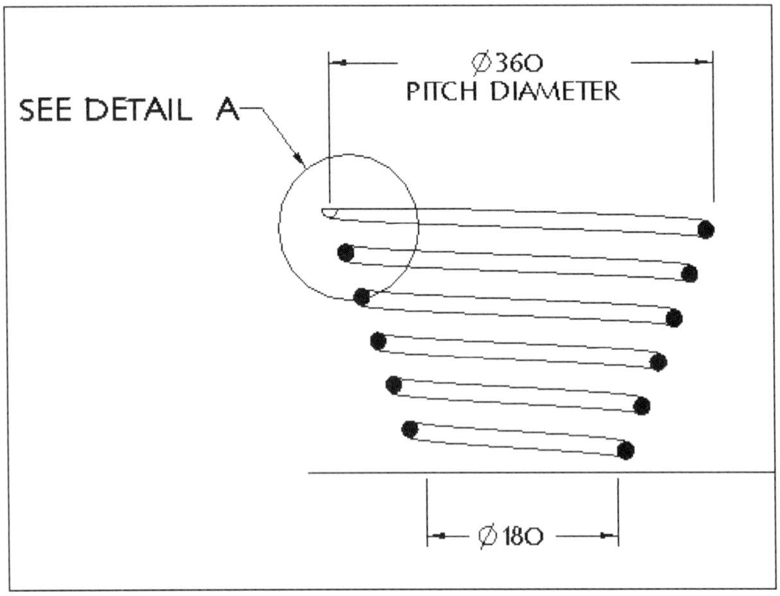

Figure 16.6 Helical Compression Spring Drawing, Section

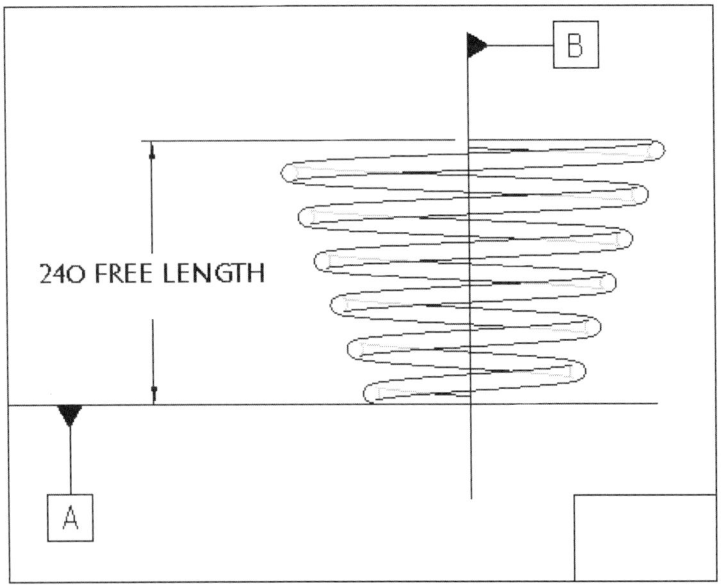

Figure 16.7 FREE LENGTH **240**

NOTE:
Helical Compression Spring
Constant Pitch
Right-Handed
40mm Pitch
Wire Diameter 15mm
Ground Ends
(grind ends parallel)

Figure 16.8 3D Note

Start a new part. Click: **Create a new object** ⇒ Part ⇒ Name **Helical_Compression_Spring** ⇒ Use default template ⇒ **MMB** ⇒ **Edit** ⇒ **Setup** ⇒ setup up your project as follows:

- **Material** = ss.mtl
- **Units** = millimeter Newton Second

Set Datum and **Rename** the default datum planes and coordinate system:

- Datum TOP = **A**
- Datum FRONT = **B**
- Datum RIGHT = **C**
- Coordinate System = **CS0**

Create the first protrusion. Click: **Insert** ⇒ **Helical Sweep** (Fig. 16.9) ⇒ **Protrusion** ⇒ **Constant** ⇒ **Thru Axis** ⇒ **Right Handed** ⇒ **Done** ⇒ pick datum **B (FRONT)** ⇒ **Okay** ⇒ **Default** ⇒ sketch a line (Fig. 16.10) ⇒ **RMB** ⇒ **Centerline** add a vertical centerline along datum **C (RIGHT)** ⇒ **RMB** ⇒ **Dimension** create the diameter dimensions and the height dimension ⇒ **RMB** ⇒ **Modify** change the values to the design sizes (Fig. 16.10)

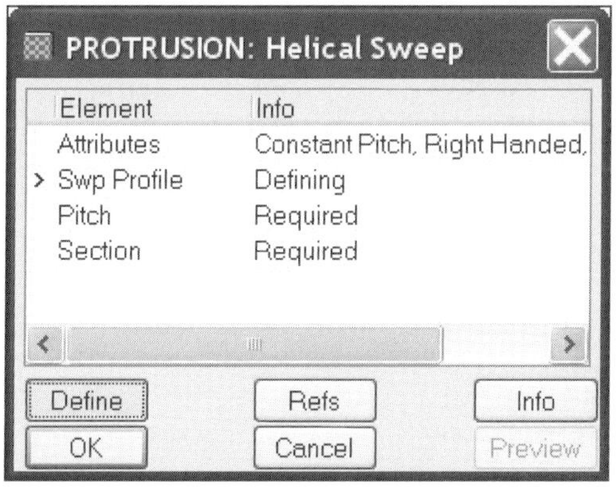

Figure 16.9 Helical Sweep Dialog Box

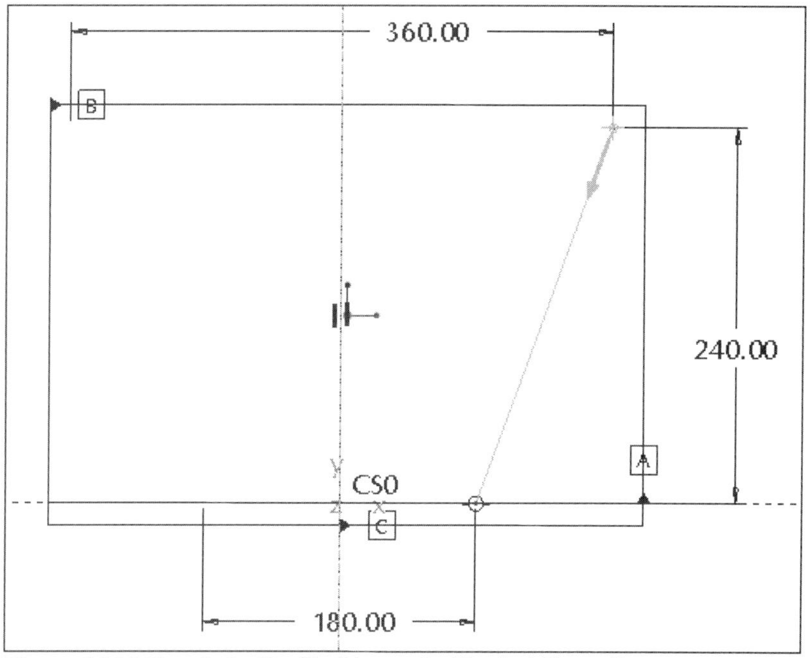

Figure 16.10 Sketch the Profile Line and a Vertical Centerline

Click: ☑ ⇒ enter the pitch value **40** at the prompt (Fig. 16.11) ⇒ **Enter**

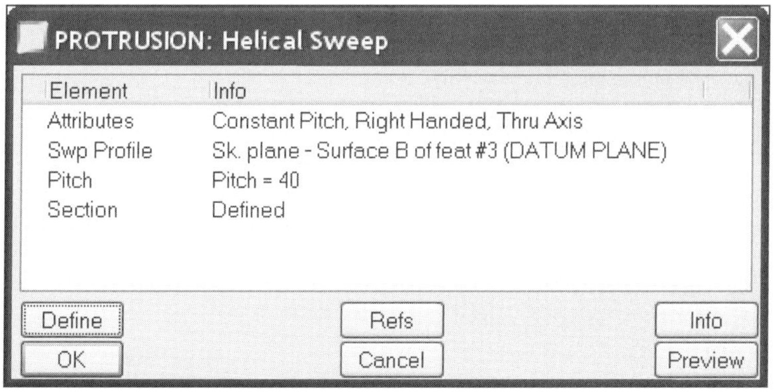

Figure 16.11 Helical Sweep Dialog Box

Sketch the section geometry of the spring at the crosshairs (a circle), click: **Create circle** [Fig. 16.12(a)] ⇒ **Modify the values of dimensions (15)** [Fig. 16.12(b)] ⇒ **MMB** ⇒ ⇒ **MMB** ⇒ **Shading** ⇒ ⇒ **MMB** ⇒ ⇒ **Standard Orientation** (Fig. 16.13)

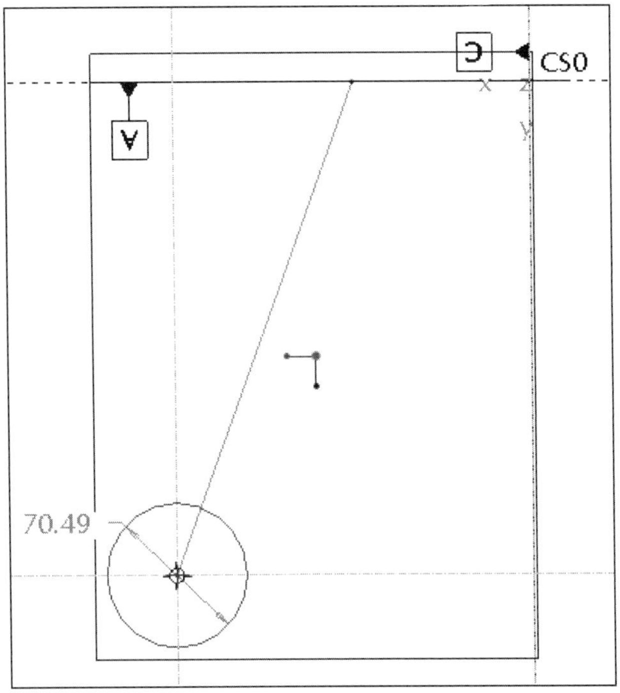

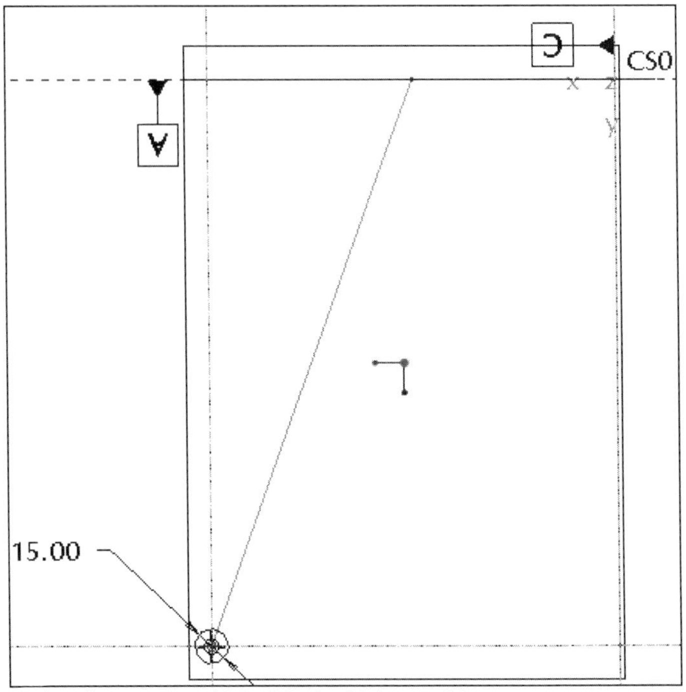

Figures 16.12(a-b) Sketch a Circle as the Section Geometry (wire diameter) and Modify the Value to **15**

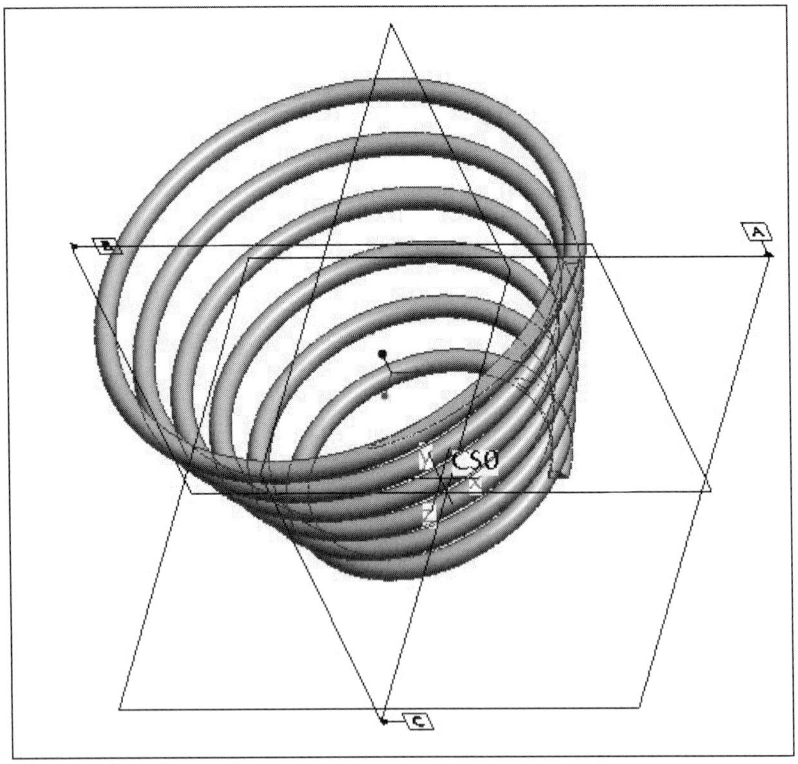

Figure 16.13 Swept Protrusion

Create the *ground ends*, click: Extrude Tool ⇒ Extrude on both sides ⇒ **RMB** ⇒ **Remove Material** ⇒ **RMB** ⇒ **Define Internal Sketch** ⇒ Sketch Plane- pick datum **C** ⇒ Reference- pick datum **A** ⇒ Orientation- **Bottom** [Figs. 16.14(a-b)] ⇒ **Sketch** ⇒ **RMB** ⇒ **Line** ⇒ create the line ⇒ modify the dimension [Fig. 16.14(c)] ⇒ spin the model as needed ⇒

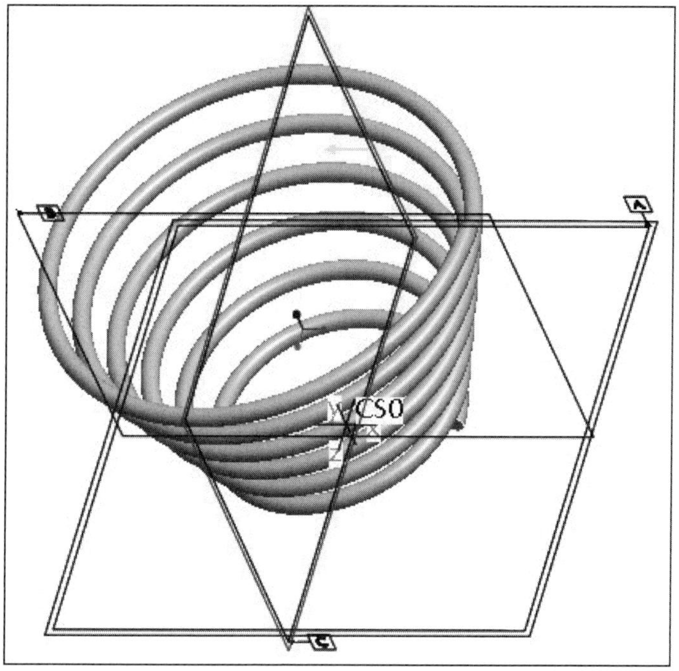

Figure 16.14(a) Cut Sketch Orientation

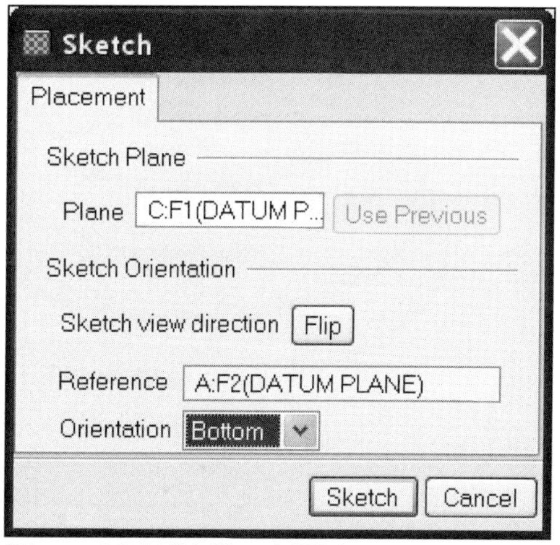

Figure 16.14(b) Sketch Dialog

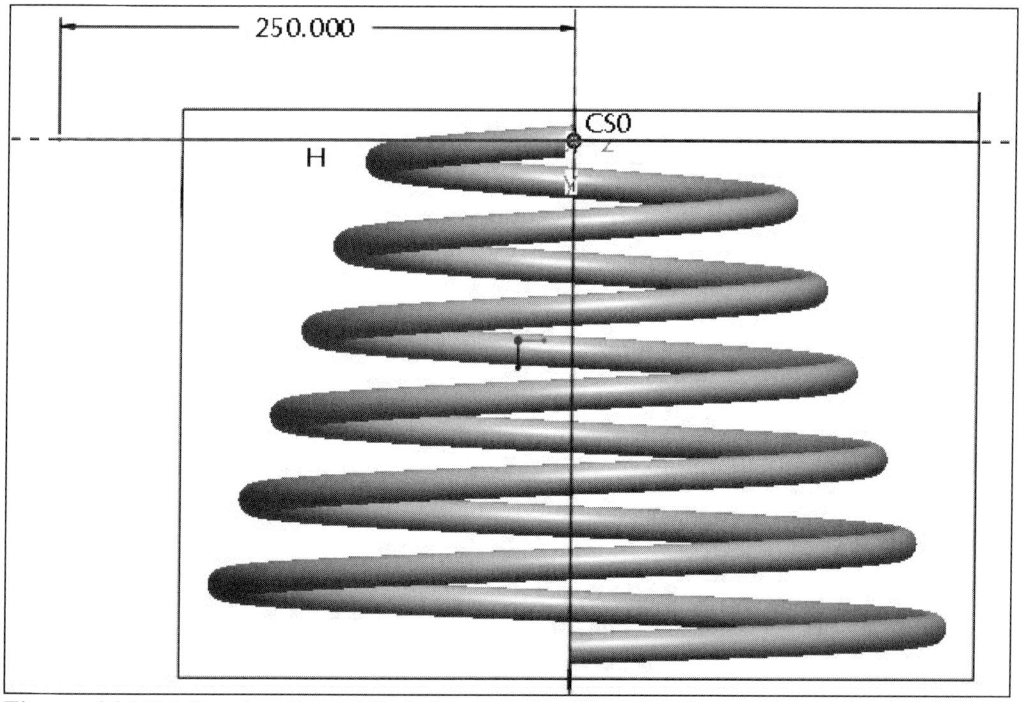

Figures 16.14(c) Creating Ground Ends (any length will work as long as it goes beyond the spring)

Extend a depth handle to include the spring [Fig. 16.14(d)] ⇒ **Change material direction** [Fig. 16.14(e)] ⇒ [Fig. 16.14(f)] ⇒ **Ctrl+S** ⇒ **Enter** ⇒ **File** ⇒ **Delete** ⇒ **Old Versions** ⇒ **Enter**

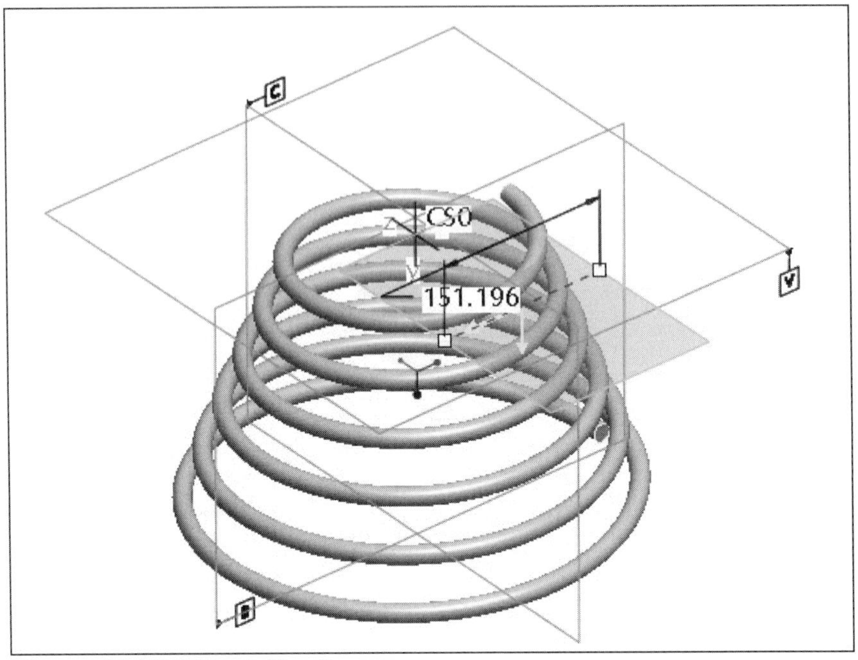

Figures 16.14(d) Depth Handle

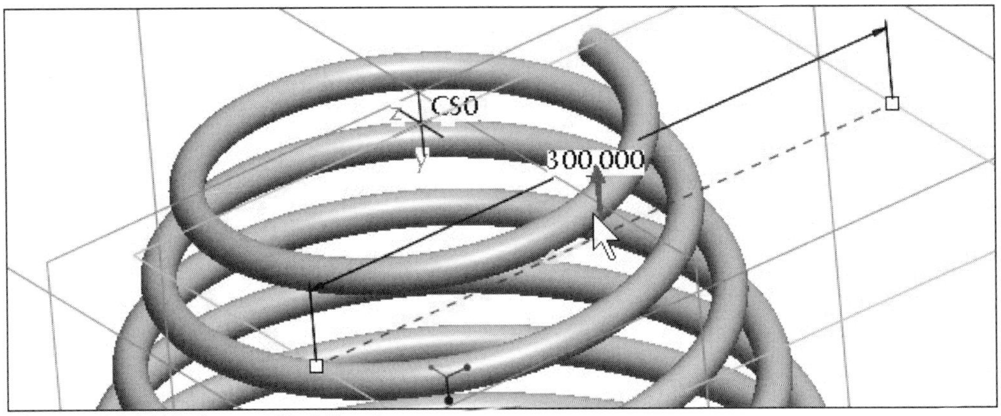

Figures 16.14(e) Ground End Depth

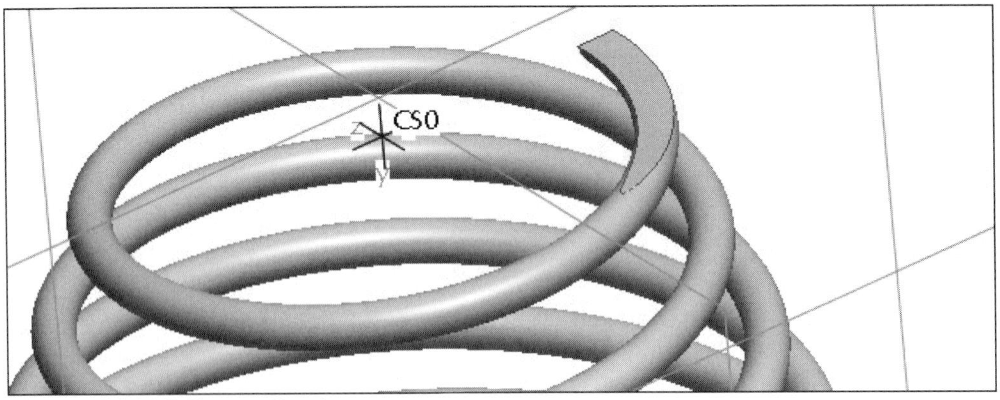

Figures 16.14(f) Feature Preview

The second ground end is created using similar commands. [Figs. 16.15(a-c)] ⇒ complete the spring ⇒ **Ctrl+D** ⇒ **Ctrl+S** ⇒ **Enter**

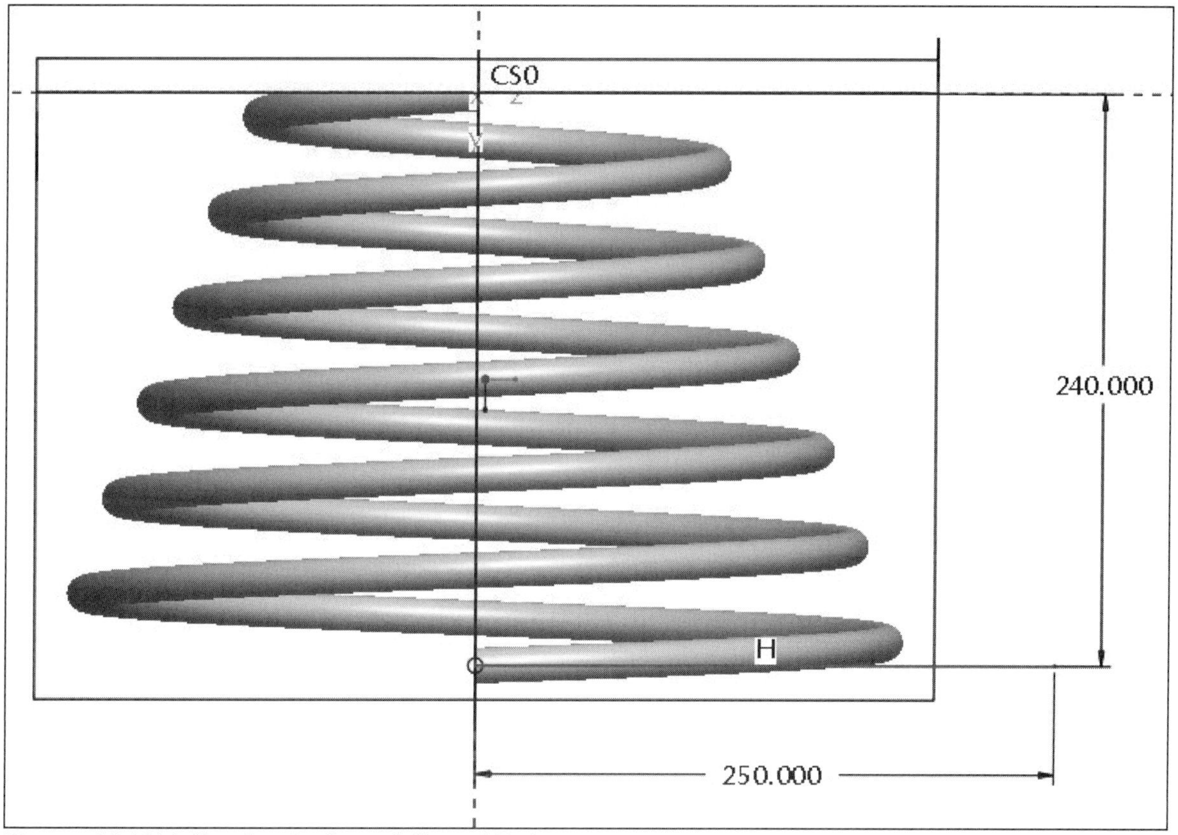

Figures 16.15(a) Creating the Second Ground End

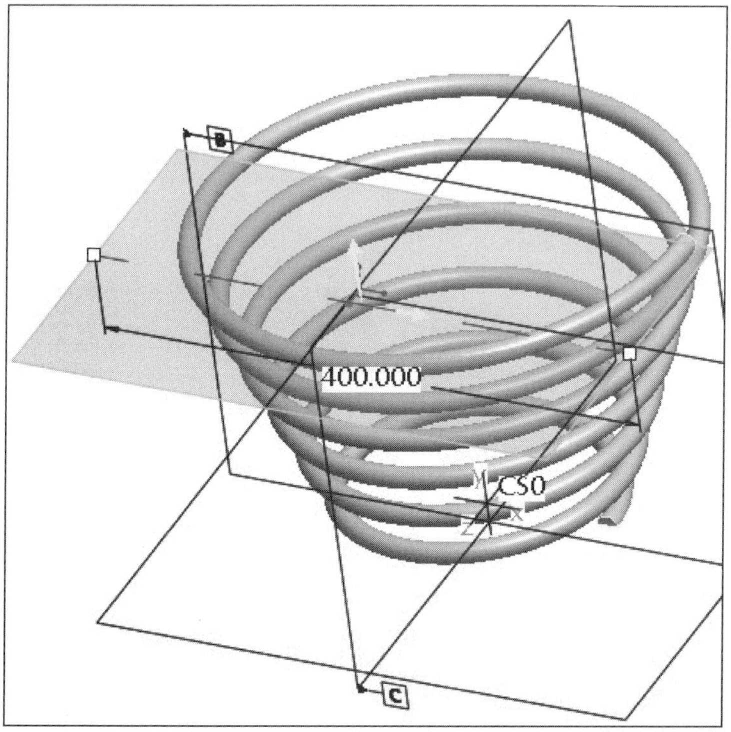

Figures 16.15(b) Cut

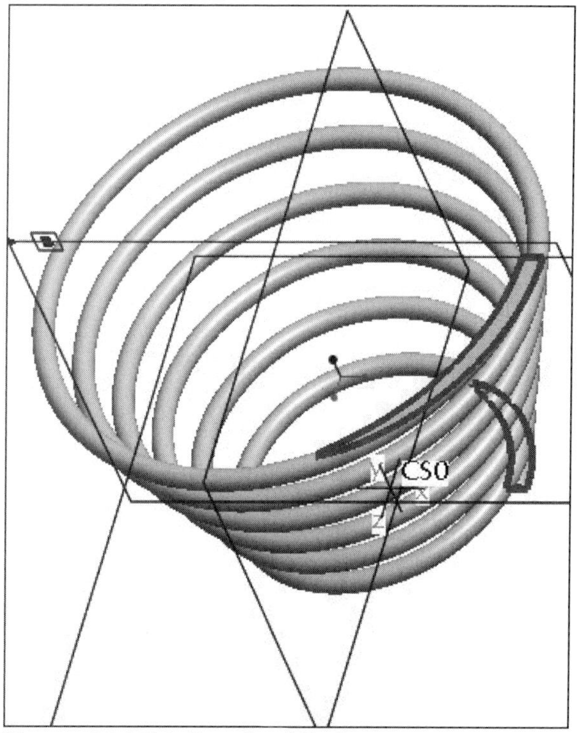

Figures 16.15(c) Completed Spring

Save a copy the Helical Compression Spring by clicking: **File** ⇒ **Save a Copy** ⇒ Type a different name—**HELICAL_COMPRESSION_SPRING_WF4**. Figure 16.16 provides an **ECO** (Engineering Change Order) for the new spring. Rename the file you are working on by clicking: **File** ⇒ **Rename** ⇒ provide a unique name such as (**HELICAL_EXTENSION_SPRING**) ⇒ delete the existing ground ends ⇒ modify the pitch to **10 mm** ⇒ change the wire diameter to **7.5 mm** ⇒ complete the extension spring (Figs. 16.17 through 16.22). The free length is to be **120 mm**. The large radius will now be **180 mm**, and the small radius will be **120 mm**.

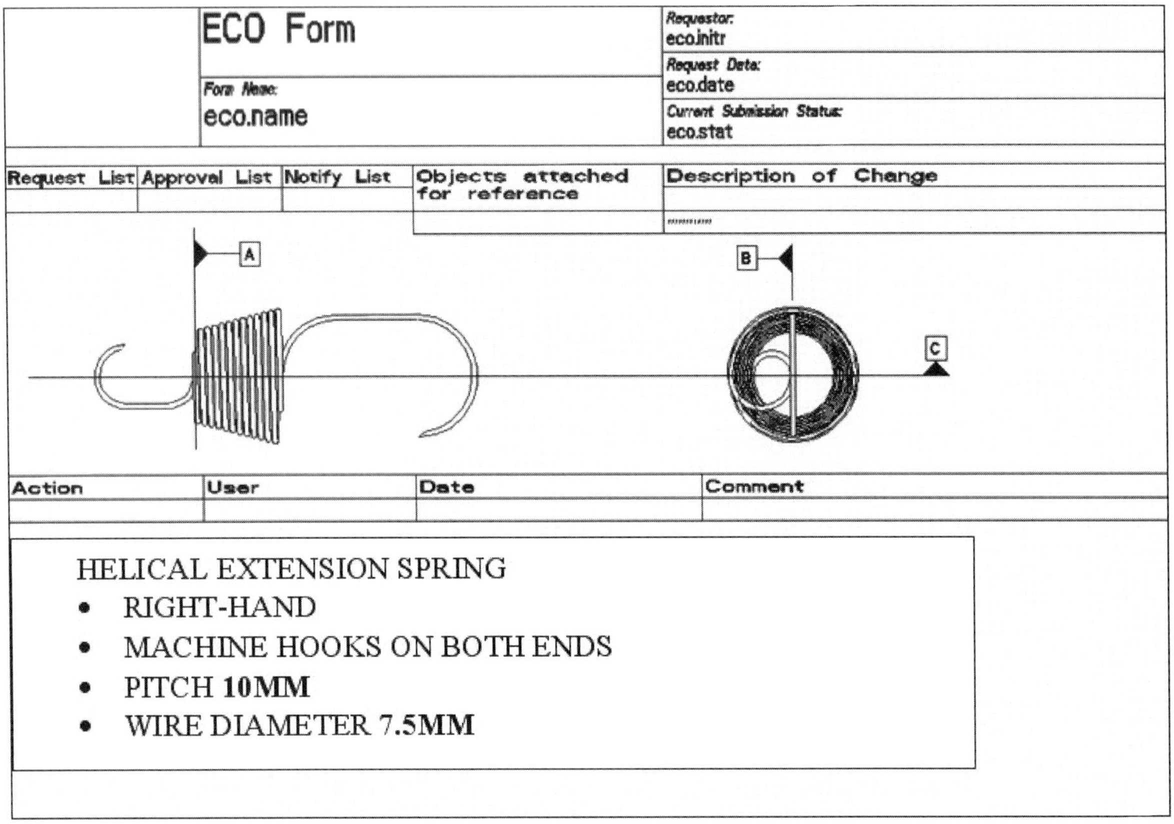

Figure 16.16 ECO to Create a Helical Extension Spring

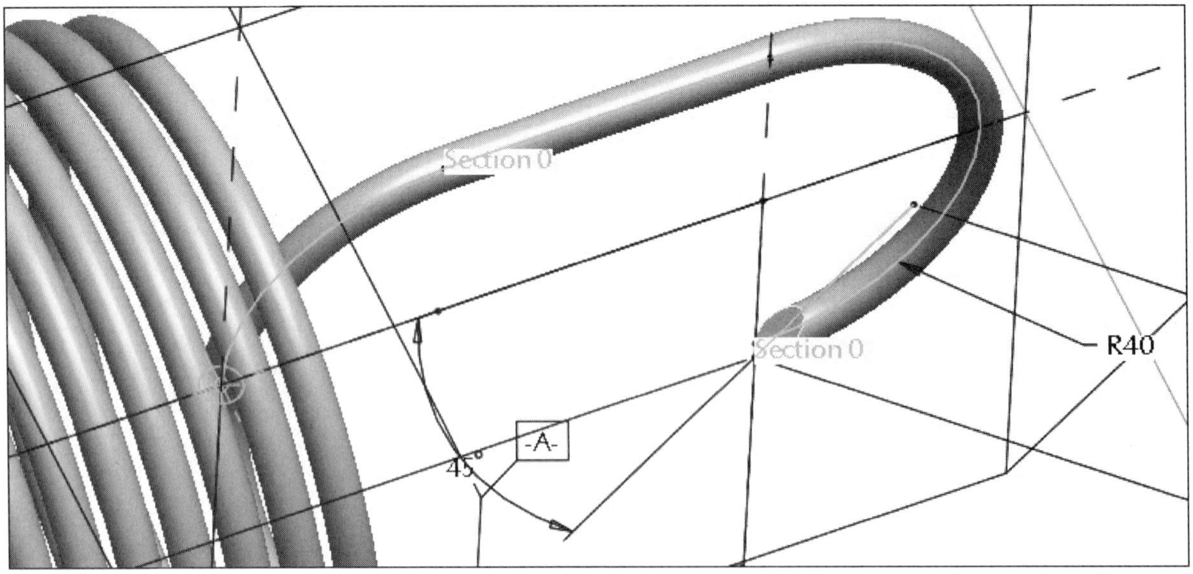

Figure 16.17 Ground End

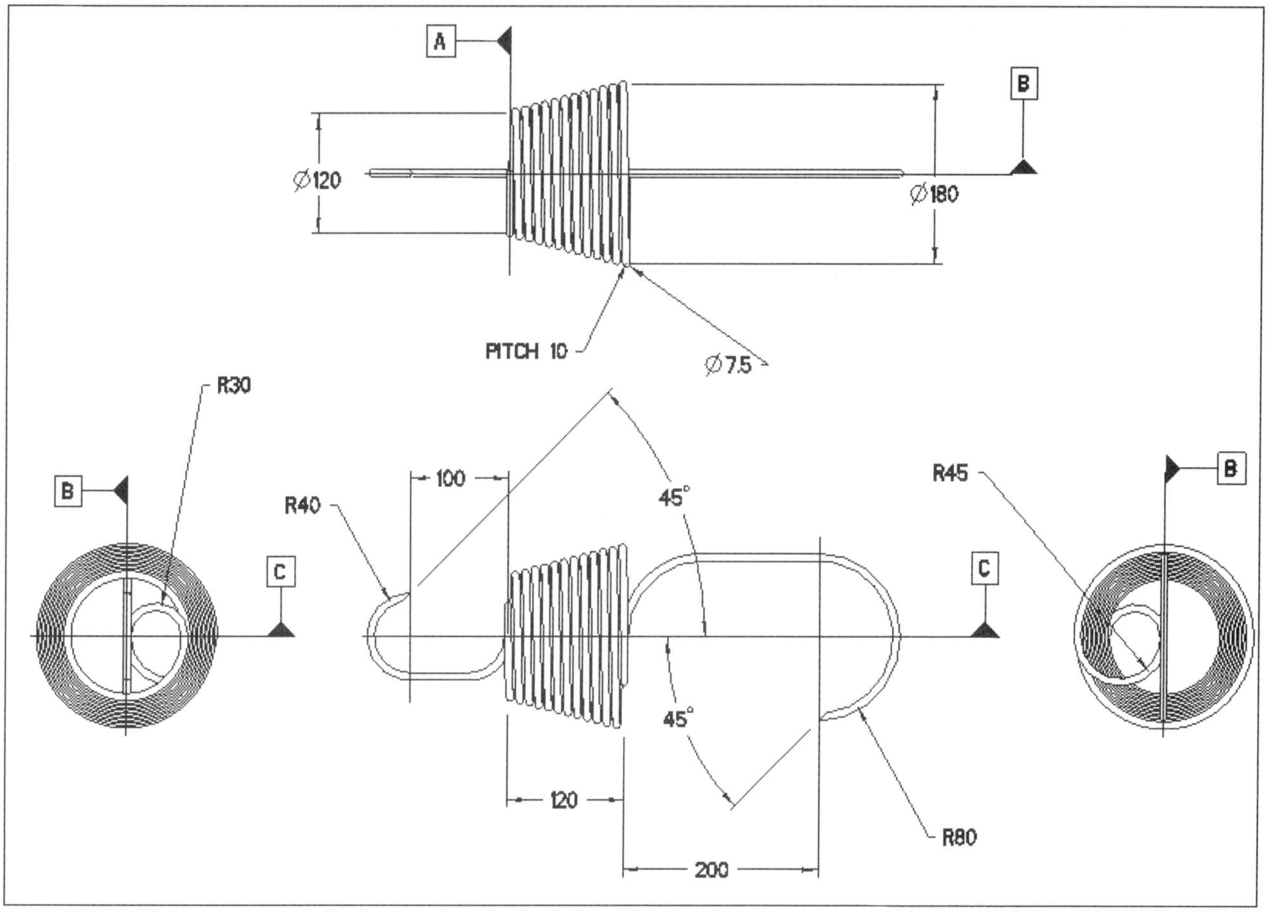

Figure 16.18 Detail Drawing of Helical Extension Spring with Machine Hook Ends

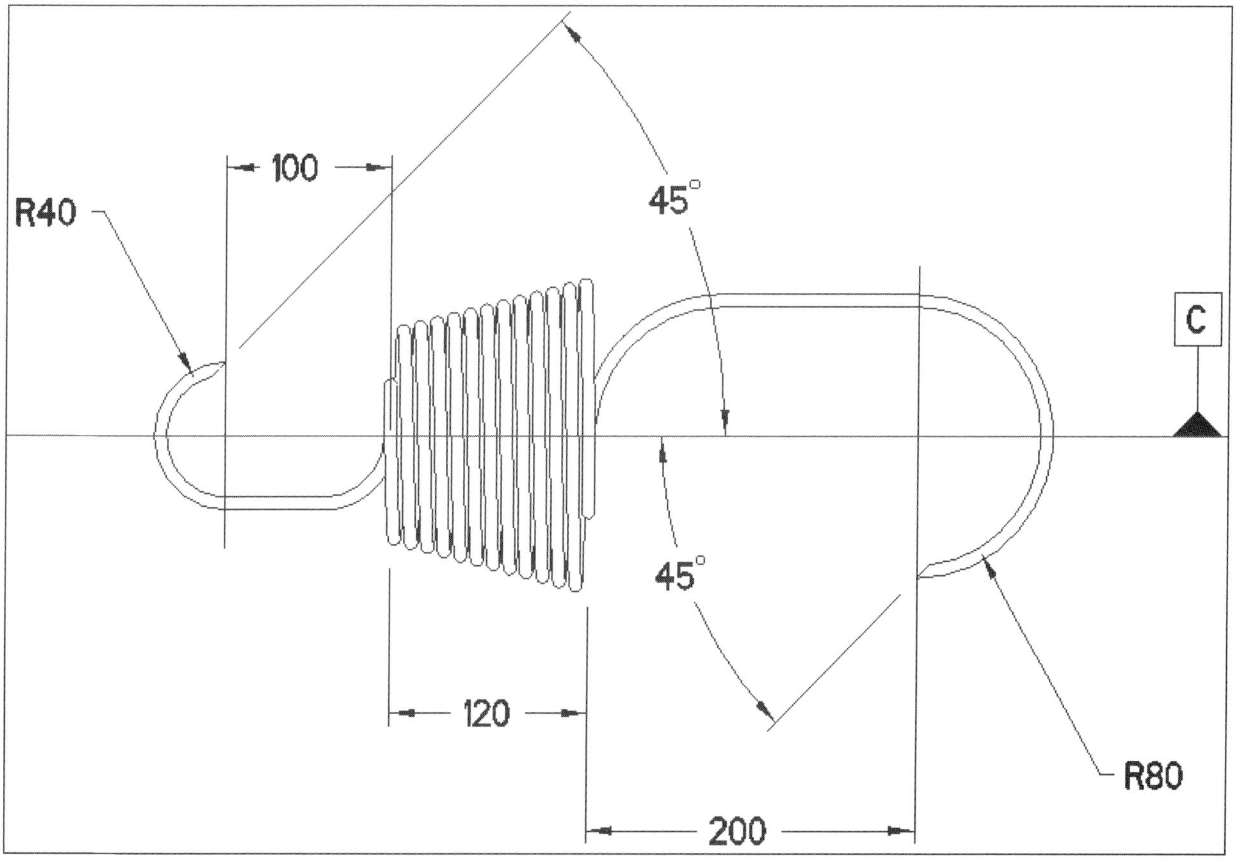

Figure 16.19 Front View

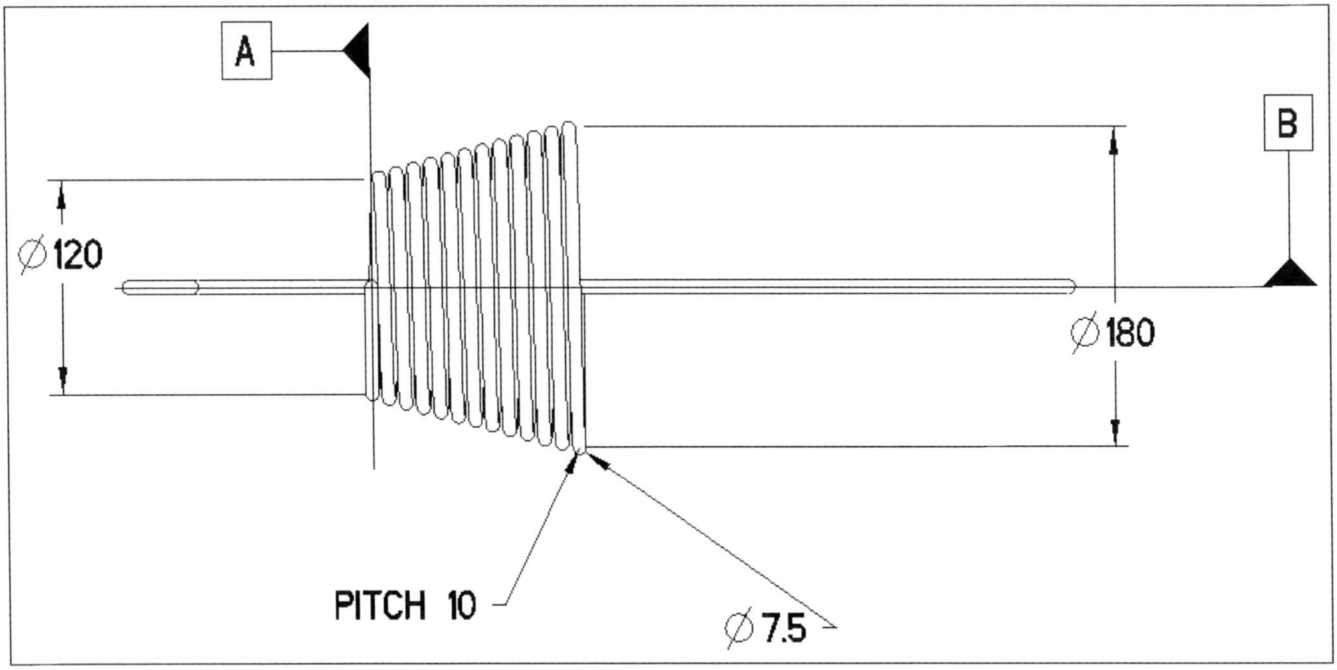

Figure 16.20 Top View

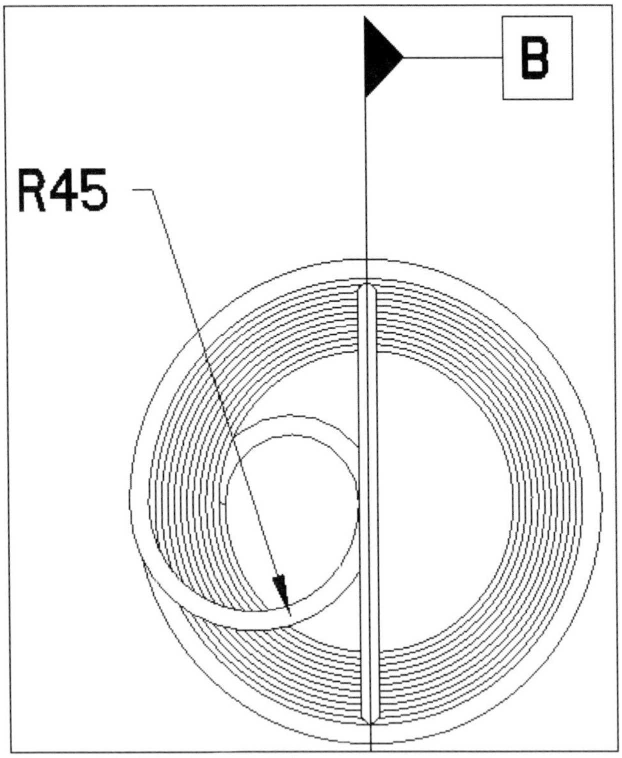

Figure 16.21 Right Side View

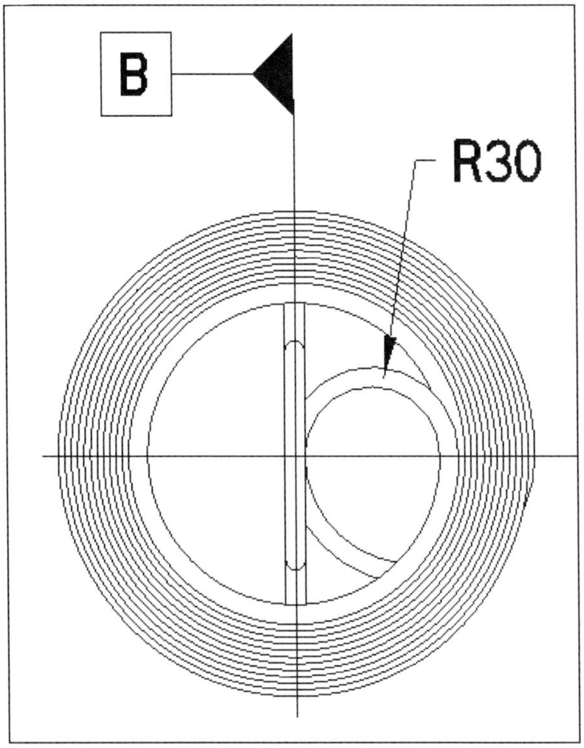

Figure 16.22 Left Side View

Create the machine hooks using simple sweeps and cuts, as shown in Figures 16.23 through 16.25.

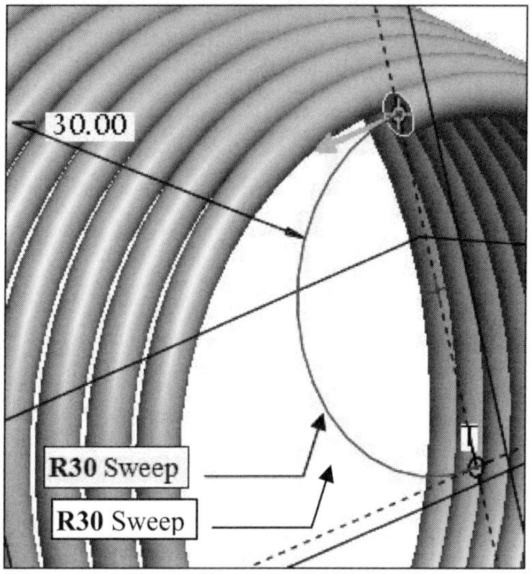

Figure 16.23(a) Sweep **R30**

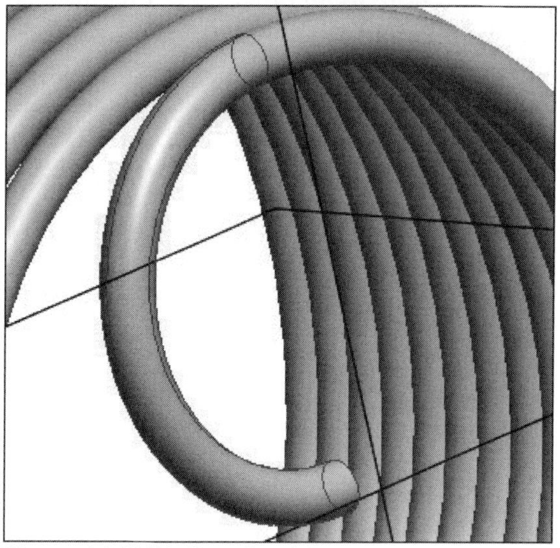
Figure 16.23(b) Completed Sweep

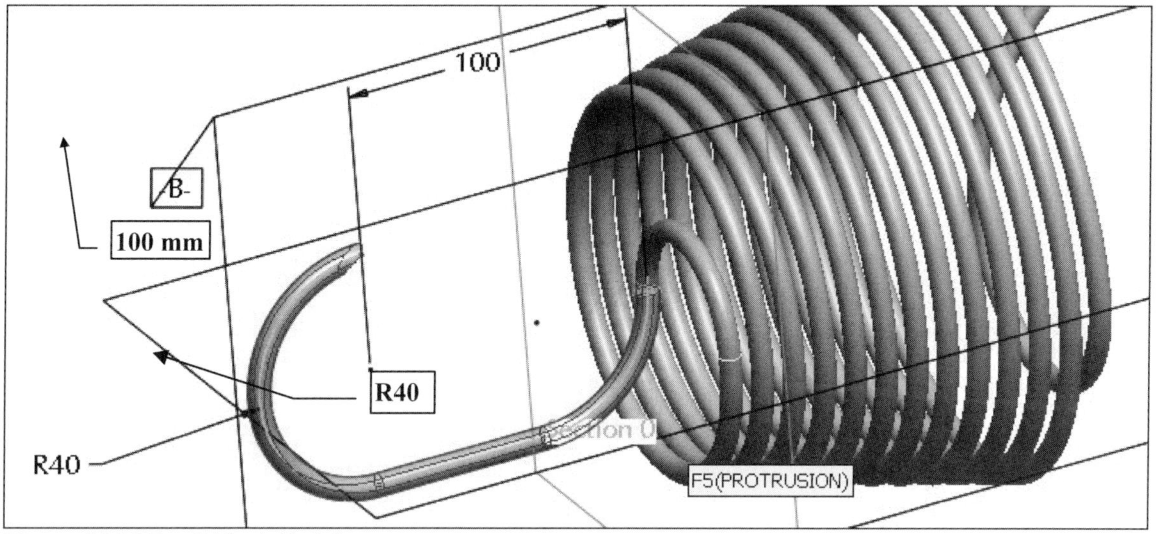

Figure 16.24 Small Hook End Sweep

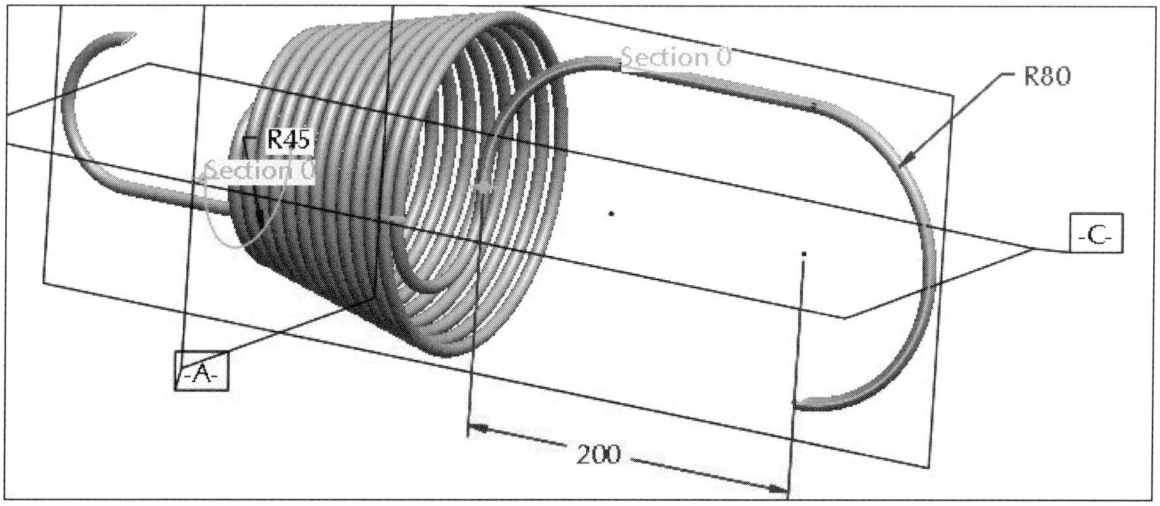

Figure 16.25 Large Hook End Sweep

Annotations

When you attach a note to an object, the object is considered the "parent" of the note. Deleting the parent deletes all of the notes of the parent. You can attach model notes anywhere in the model; they do not have to be attached to a parent. Here we will add a note to the part and describe the spring.

Open the saved spring file that has the ground ends (**HELICAL_COMPRESSION_SPRING_WF4**). Choose the following commands, click: **Insert** ⇒ **Annotations** ⇒ **Notes** [Fig. 16.26(a)] ⇒ **New** ⇒ type **Compression_Spring** as the name of the note; no spaces are allowed in the name ⇒ pick in the Text area and type the note [Fig. 16.26(b)]:

Helical Compression Spring
Constant Pitch
Right-handed
40 mm Pitch
Wire Diameter 15 mm
Ground Ends
(grind ends parallel)

Click: ☑ Place note flat to screen in model space ⇒ **Place** ⇒ **No Leader** ⇒ **Standard** ⇒ **Done** ⇒ pick a place on the screen to place the note [Figs. 16.27(a-b)] ⇒ **OK** ⇒ **Done/Return**

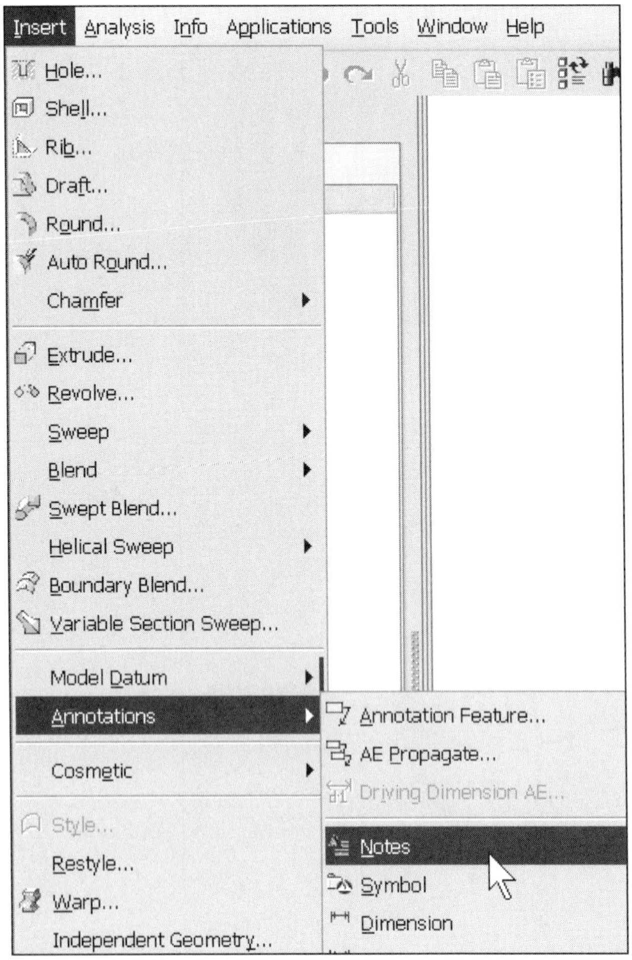

Figure 16.26(a) 3D Notes

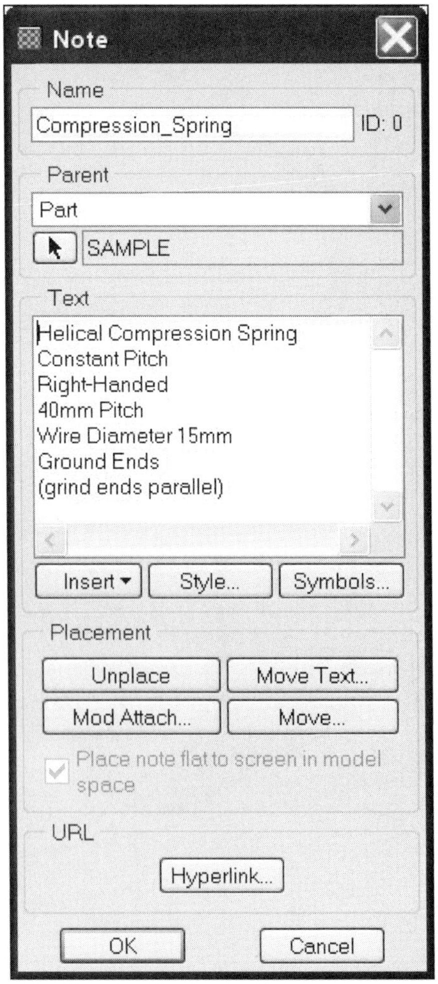

Figure 16.26(a) Note Dialog

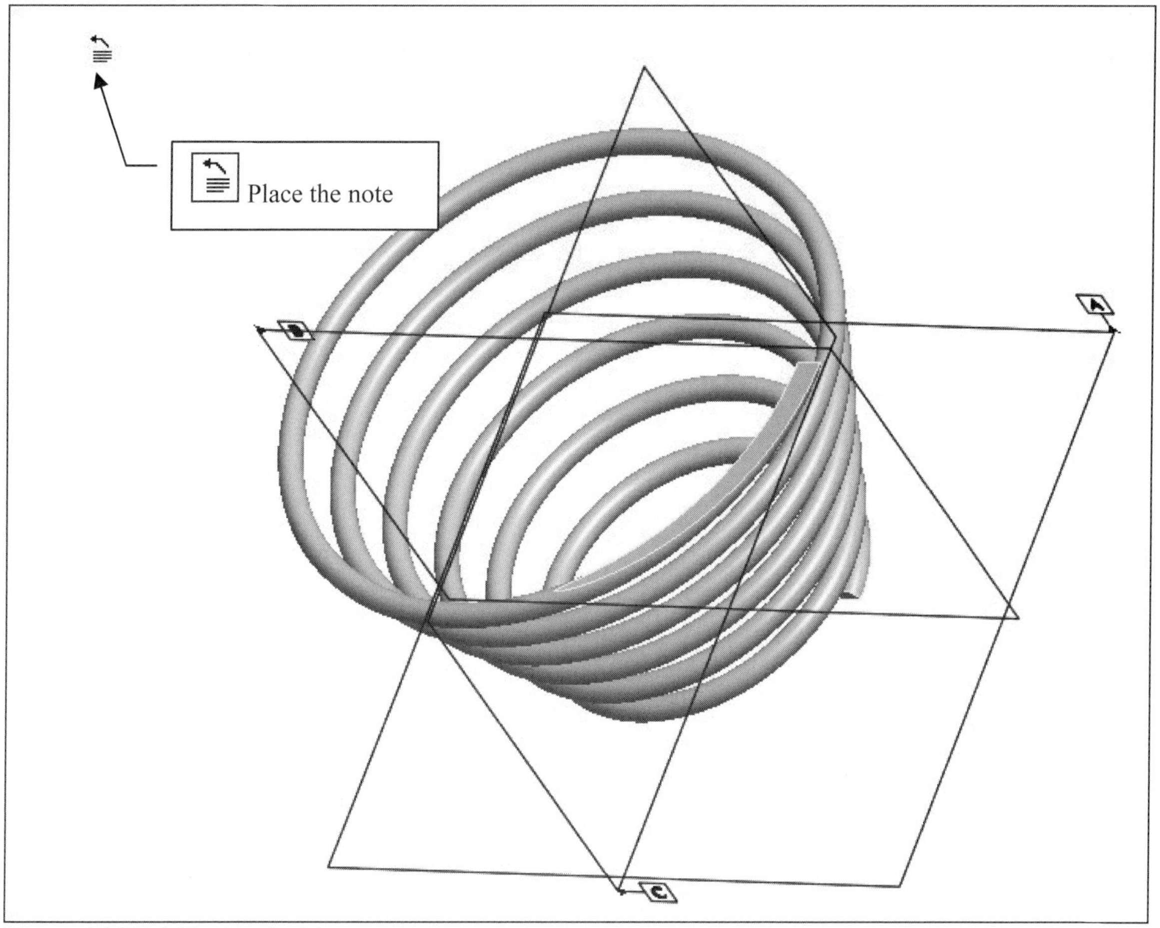

Figure 16.27(a) Placing the Note

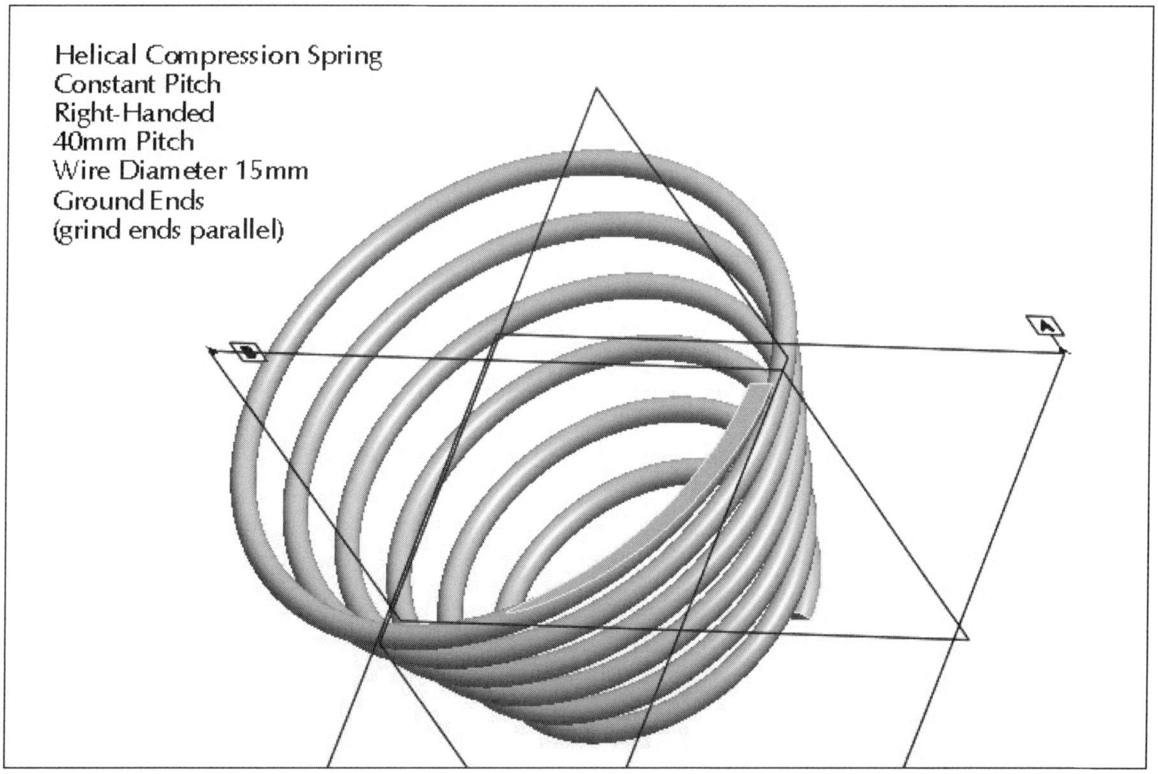

Figure 16.27(b) Completed Note

You can toggle model annotations on and off using **Turn on or off 3D annotations and annotation elements** ⇒ toggle the annotations off and on ⇒ Display the note in the Model Tree by clicking: **Settings** ⇒ **Tree Filters** ⇒ Annotations (Fig. 16.28) ⇒ **Apply** ⇒ **OK** (Fig. 16.29)

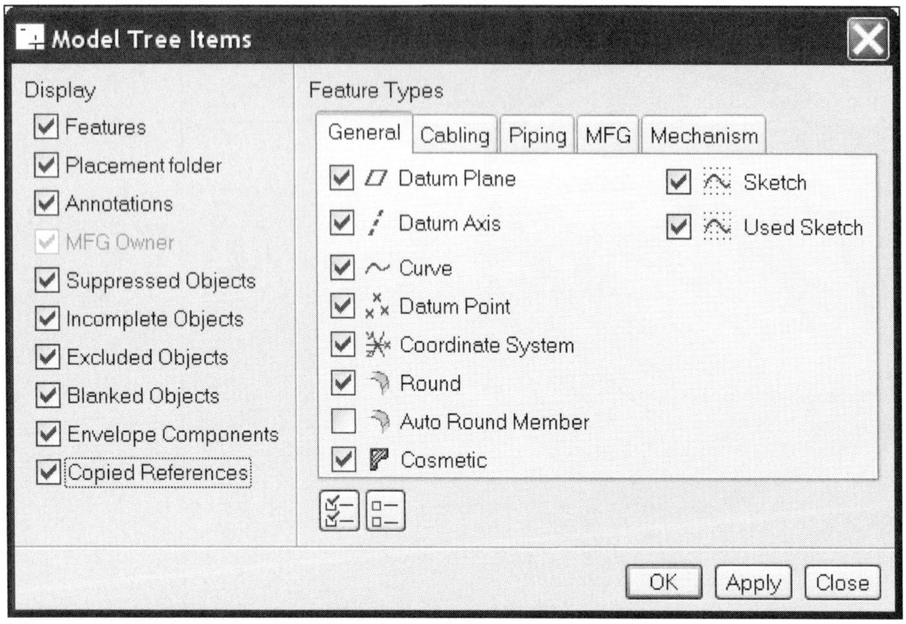

Figure 16.28 Displaying 3D Notes in Model Tree

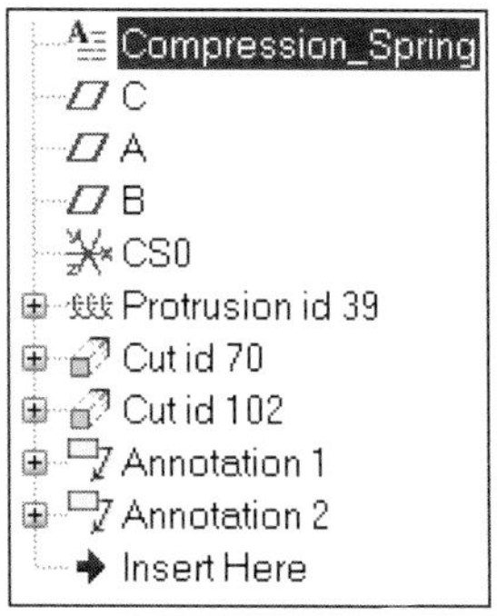

Figure 16.29 Model Tree

You can also perform a variety of functions directly from the Model Tree. Click on Compression_Spring in the Model Tree ⇒ **RMB** ⇒ **Properties** [Fig. 16.30(a)] ⇒ **Hyperlink** [Fig. 16.30(b)] ⇒ type the URL or internal link: **www.americanprecspring.com** [Fig. 16.30(c)]

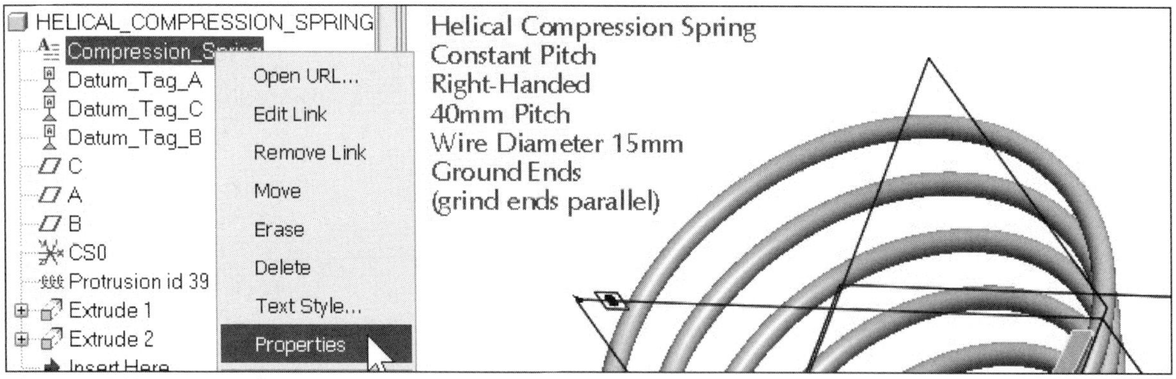

Figure 16.30(a) Properties

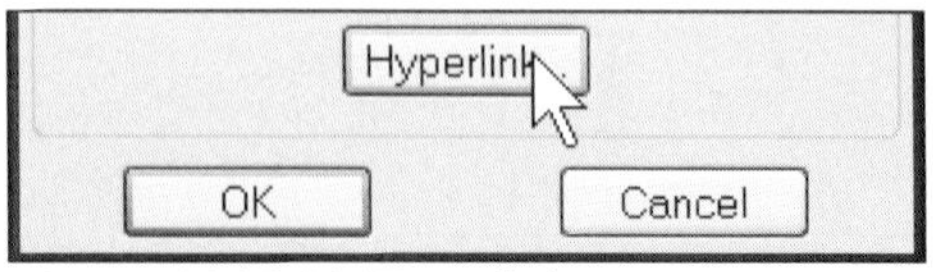

Figure 16.30(b) Hyperlink

Figure 16.30(c) URL

Create a screen tip that will display as your cursor passes over the note, click: ScreenTip... ⇒ type **SPRING COMPANY** [Fig. 16.31(a)] ⇒ **OK** ⇒ **OK** ⇒ **OK** ⇒ Smart ⇒ Annotation [Fig. 16.31(b)] ⇒ place your cursor over the note [Fig. 16.31(c)] ⇒ set your filter back to Smart

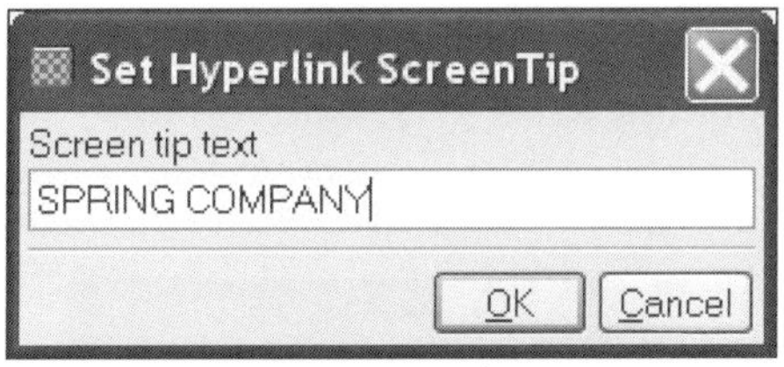

Figure 16.31(a) Screen Tip

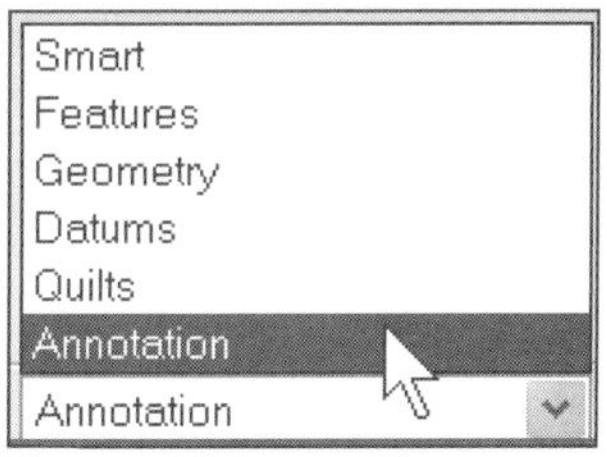

Figure 16.31(b) Filter

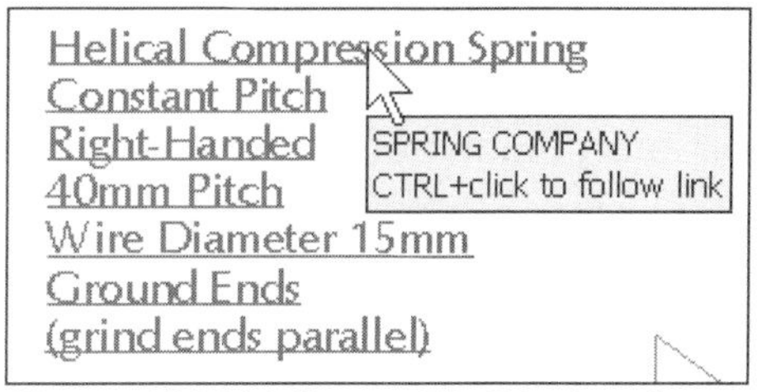

Figure 16.31(c) Screen Tip Displayed

Open the URL, click: Compression_Spring from the Model Tree ⇒ **RMB** ⇒ Open URL [Figs. 16.32(a-b)] ⇒ opens in the browser window [Fig. 16.32(c)] ⇒ close **Browser** ⇒ **LMB** to deselect ⇒ ⇒ ⇒ **Ctrl+D** ⇒ ⇒ **MMB** ⇒ **File** ⇒ **Delete** ⇒ **Old Versions** ⇒ **MMB**

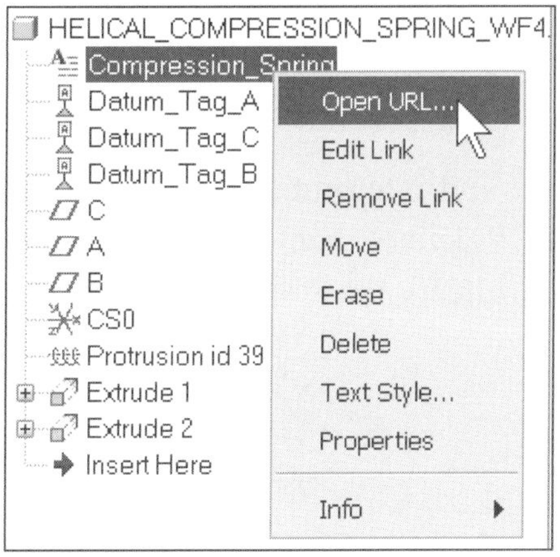

Figure 16.32(a) Open Link

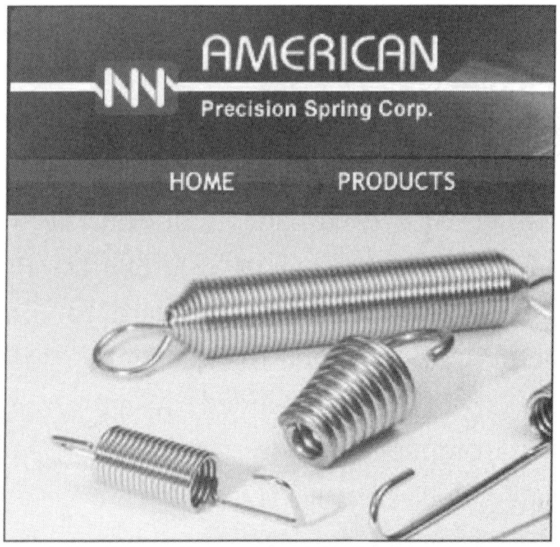

Figure 16.32(b) American Precision Spring Website

Figure 16.32(c) Spring Website

Annotation Features

3D Notes can also be added to an object using **Annotation Features**. Annotation features are data features that you can use to manage the model annotation and propagate model information to other models, or to manufacturing processes. The Annotation Feature Tool options correspond to the new ASME Y14.41 Digital Product Definition Data Practices.

An Annotation feature consists of one or more Annotation Elements. Each Annotation Element can contain one annotation item, along with associated references and parameters. You can include the following types of annotations in an Annotation Element:

- Note
- Symbol
- Surface finish
- Geometric tolerance
- Set Datum Tag
- Ordinate baseline dimension
- Driven dimension
- Ordinate driven dimension
- Chamfer dimension
- Reference dimension
- Ordinate reference dimension

Digital Product Definition Data Practices

"ASME Y14.41-2003 establishes requirements for preparing, organizing and interpreting 3-dimensional digital product images (Fig. 16.33). Digital Product Definition Data Practices, which represents an extension of the popular Y14.5 standard for 2-dimensional drawings, reflects the growing need for a uniform method of documenting the data created in today's computer-aided design (CAD) environments. The standard provides a guide for CAD software developers working on improved modeling and annotation practices for the engineering community. ASME Y14.41 sets forth the requirements for tolerances, dimensional data, and other annotations. ASME Y14.41 advances the capabilities of Y14.5, Dimensioning and Tolerancing, the standard pertaining to 2-D engineering drawings".

In the following steps you will create a single-view 3D definition of the model for manufacturing, instead of a traditional multi-view drawing.

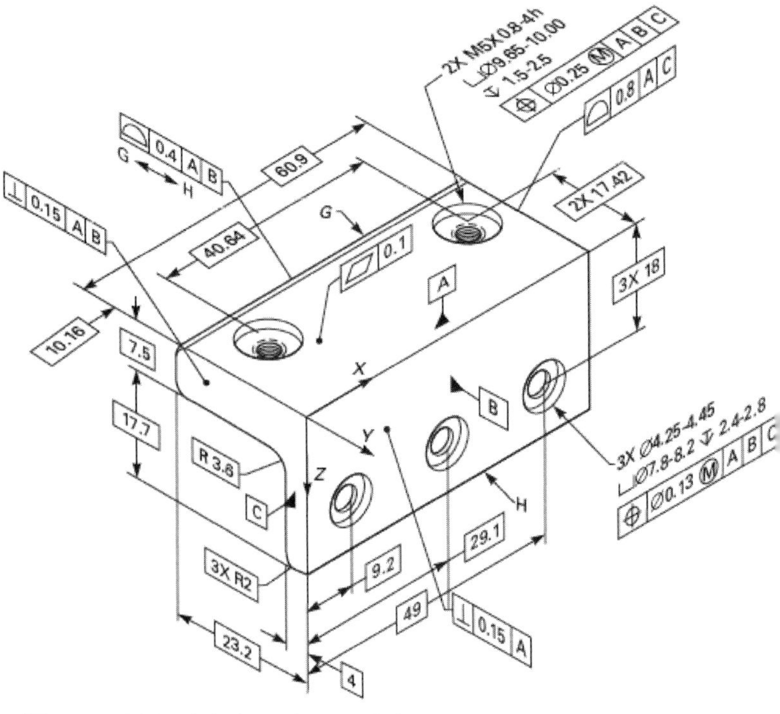

Figure 16.33 Digital Product Definition, ASME Y14.41-2003

Click [icon] **Start the view manager** ⇒ **Orient** ⇒ **New** ⇒ type **Annotation** ⇒ **Enter** [Fig. 16.34(a)] ⇒ rotate the view similar to Figure 16.34(b) ⇒ click on [Annotation(+)] ⇒ **RMB** ⇒ **Save** [Fig. 16.34(c)] *(the + sign will disappear)* ⇒ **OK** ⇒ **Close** the View Manager dialog box ⇒ **Ctrl+S** ⇒ **OK**

Figure 16.34(a) Annotation

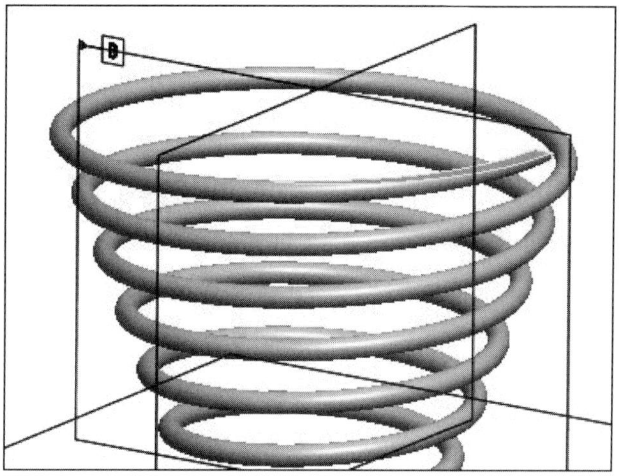
Figure 16.34(b) Reorient the Model

Figure 16.34(c) Save

Click: [Smart] ⇒ [Annotation] ⇒ pick on the 3D note [Fig. 16.35(a)] ⇒ **RMB** ⇒ **Move** ⇒ select a new position for the 3D note [Fig. 16.35(b)] ⇒ **LMB** to deselect ⇒ [Annotation] ⇒ [Smart] ⇒ **Ctrl+S** ⇒ **OK**

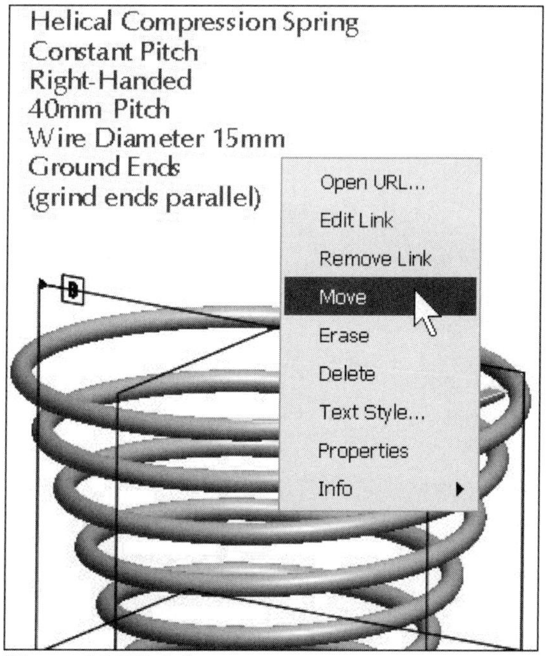

Figure 16.35(a) Save

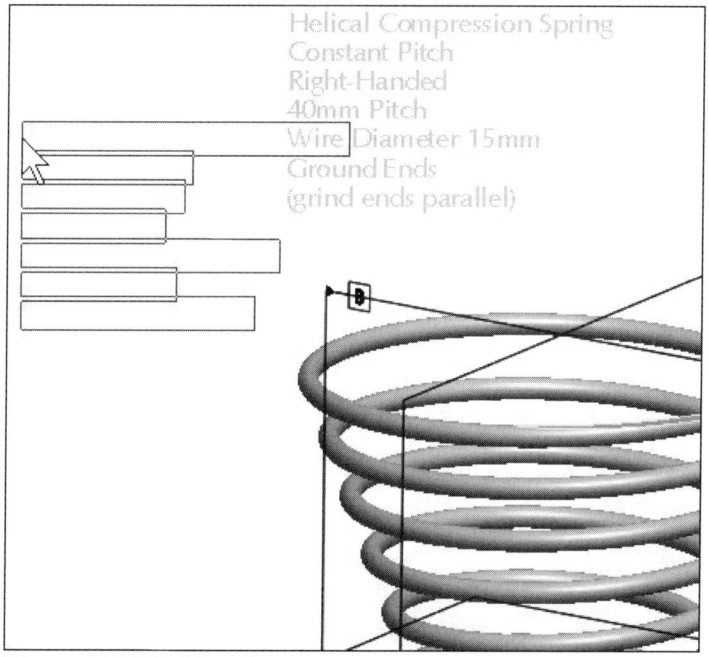

Figure 16.35(b) Place the Note

Select the first feature ⇒ **RMB** ⇒ **Create Driving Dimension AE** [Figs. 16.36(a-b)] ⇒ **LMB**

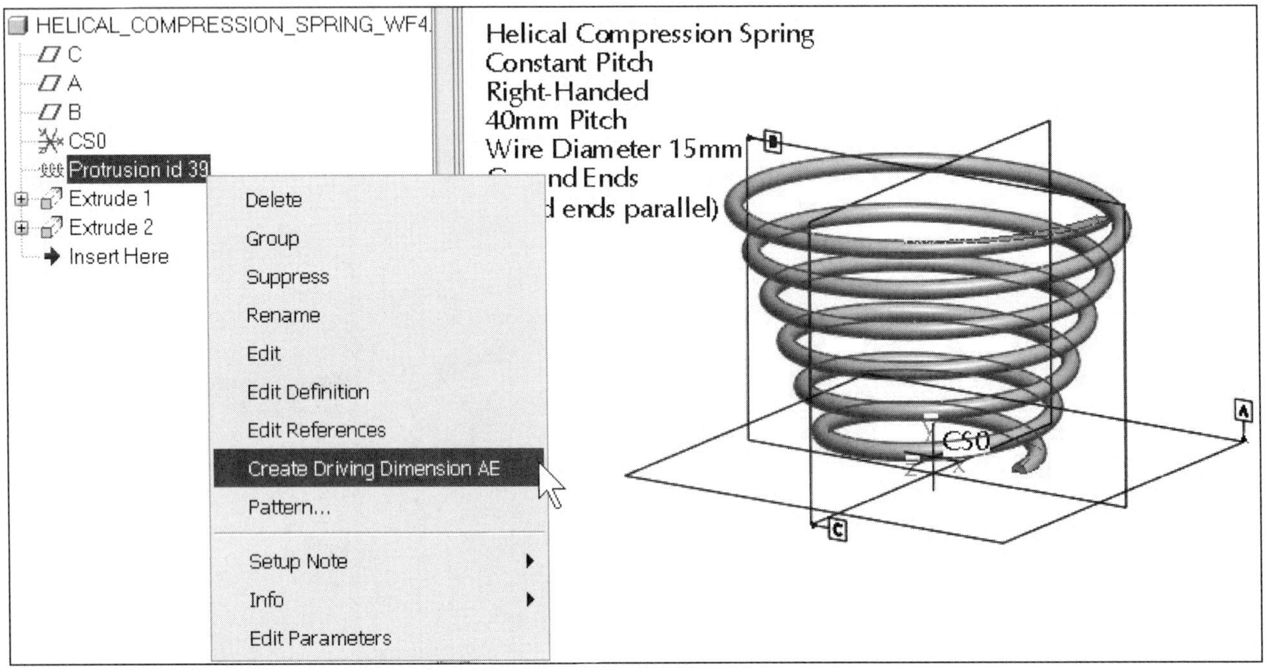

Figure 16.36(a) Create Driving Dimension AE

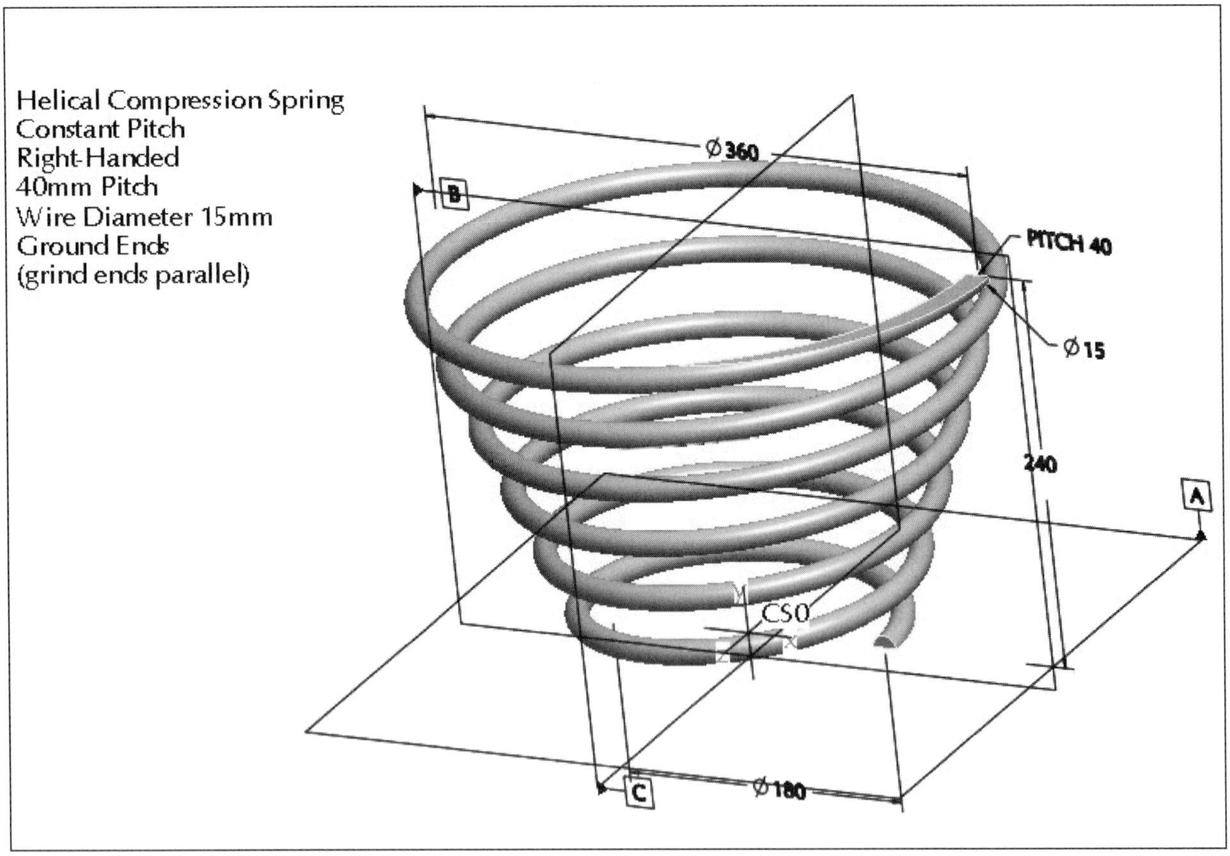

Figure 16.36(b) Displayed Driving Dimensions

Click: Smart ⇒ Annotation ⇒ pick on **Datum B** ⇒ **RMB** ⇒ **Flip** [Fig. 16.37(a)] ⇒ select an annotation ⇒ **RMB** ⇒ **Move** ⇒ pick a new location ⇒ move each to a better location ⇒ rotate the model to get the best view of all annotations ⇒ **Ctrl+S** ⇒ **OK** [Fig. 16.37(b)]

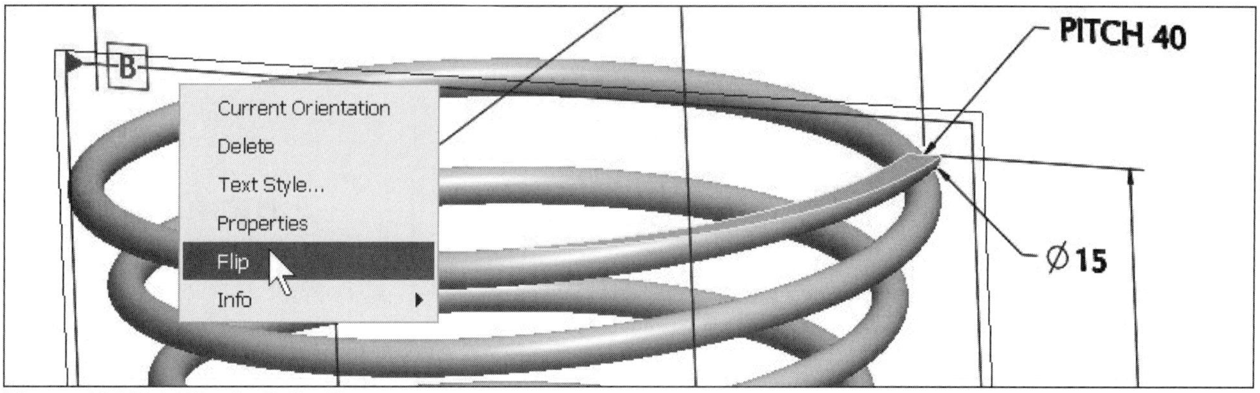

Figure 16.37(a) Flip the **B** Set Datum

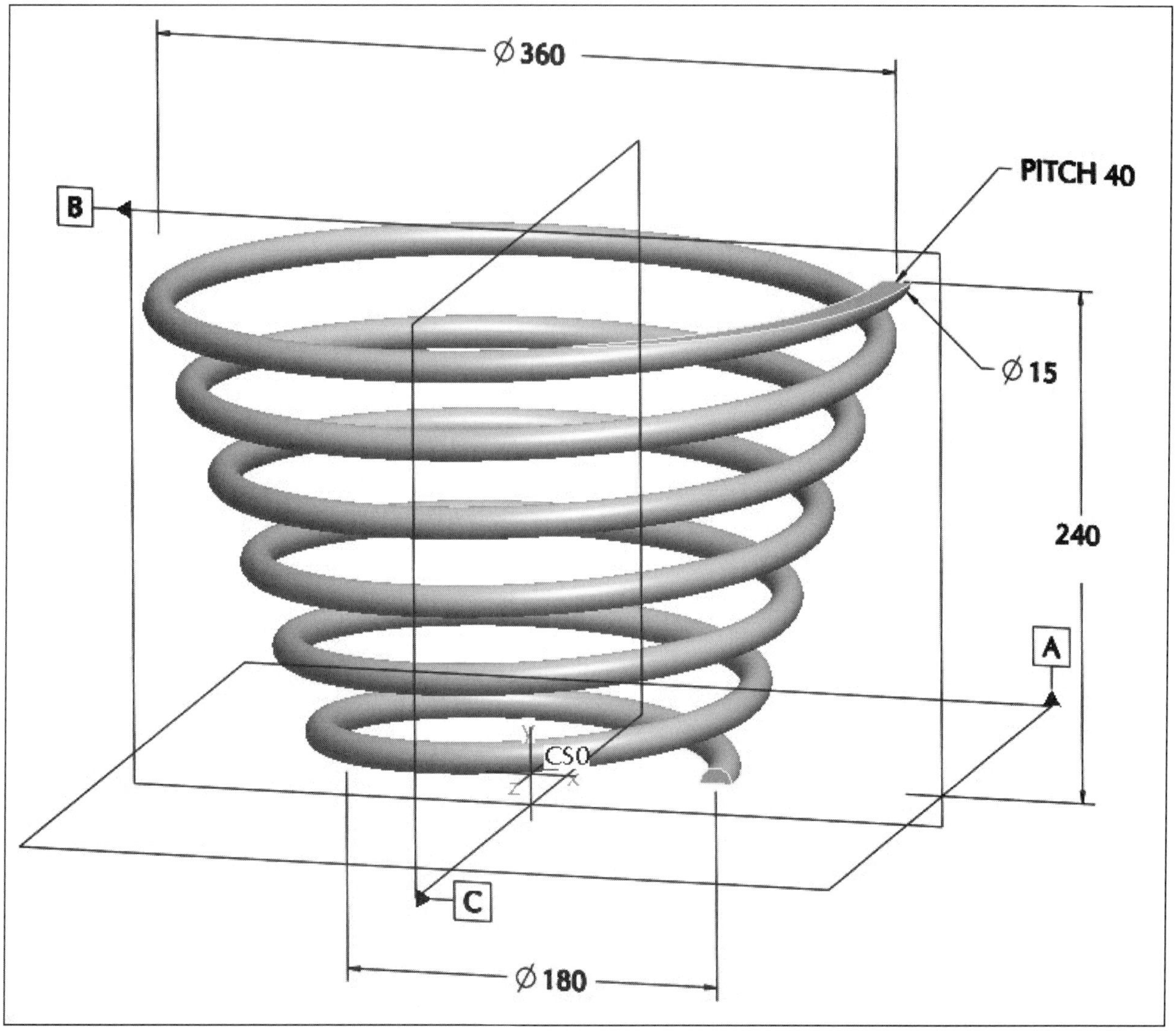

Figure 16.37(b) Repositioned Annotations

Select the **360** dimension [Fig. 16.38(a)] ⇒ **RMB** ⇒ **Properties** ⇒ **Dimension Text** tab [Fig. 16.38(b)] ⇒ type **PITCH DIAMETER** ⇒ **OK** [Fig. 16.38(c)]

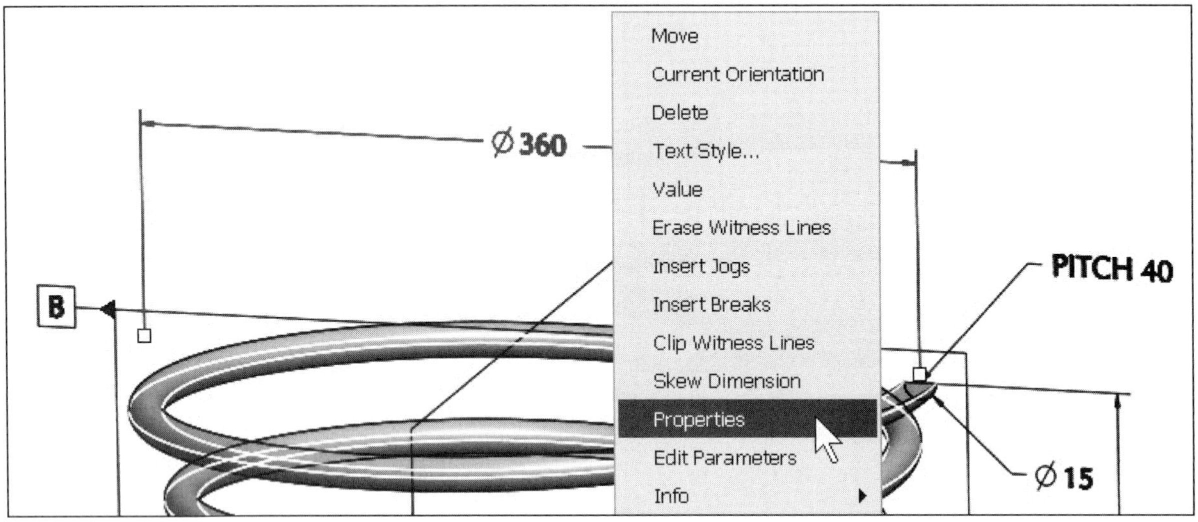

Figure 16.38(a) Dimension Properties

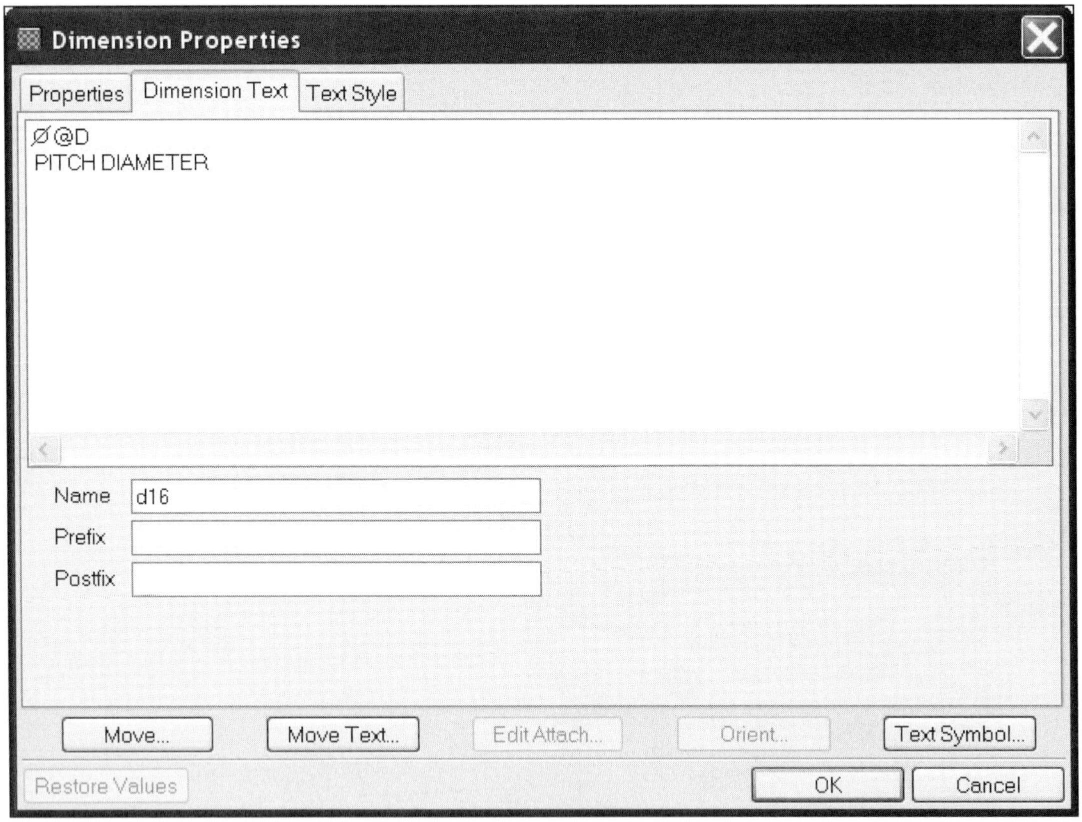

Figure 16.38(b) Dimension Text

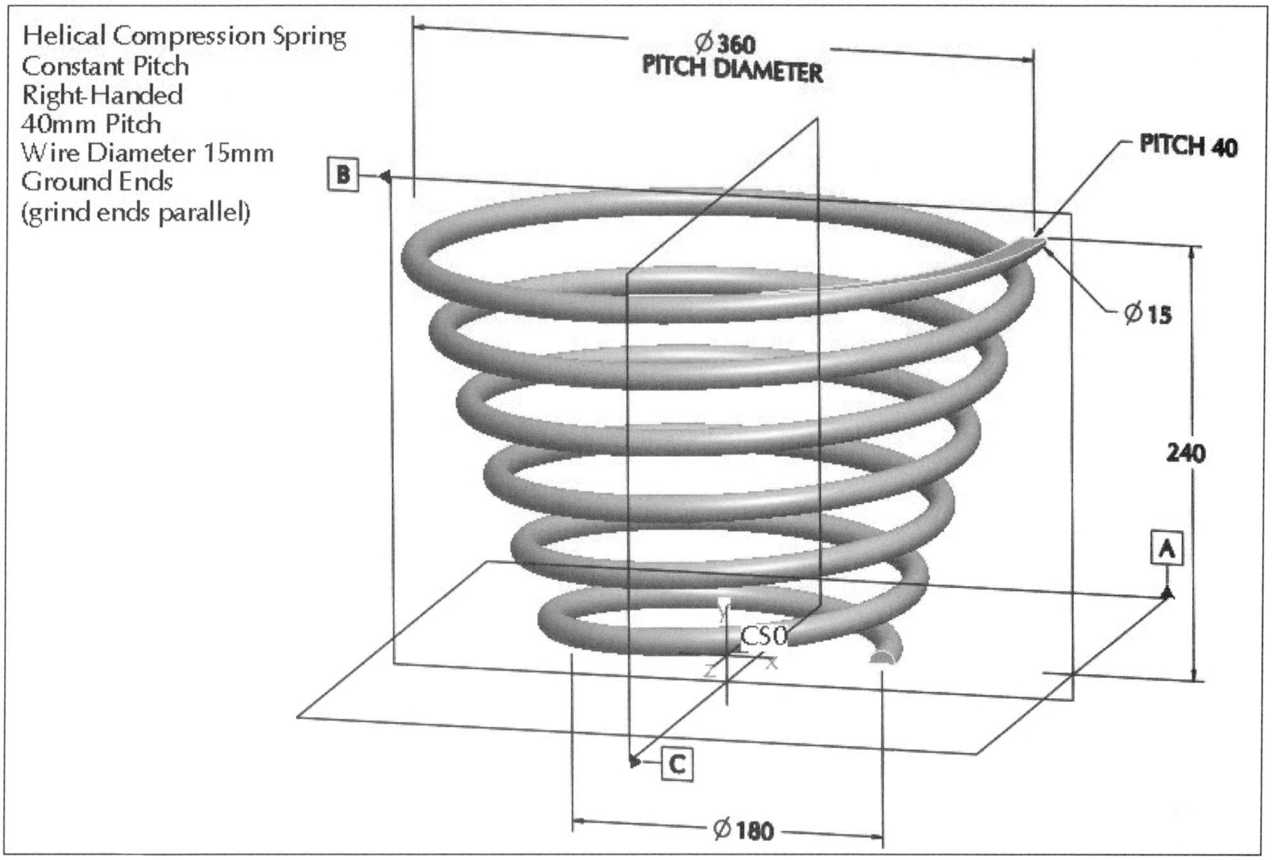

Figure 16.38(c) Dimensioned Part

Click: [icon] **Insert an annotation feature** ⇒ [Geometric Tolerance] ⇒ **OK** ⇒ [//] [Fig. 16.39(a)] ⇒ Reference: To Be Selected- Type [▾] ⇒ **Surface** ⇒ select the ground surface [Fig. 16.39(b)]

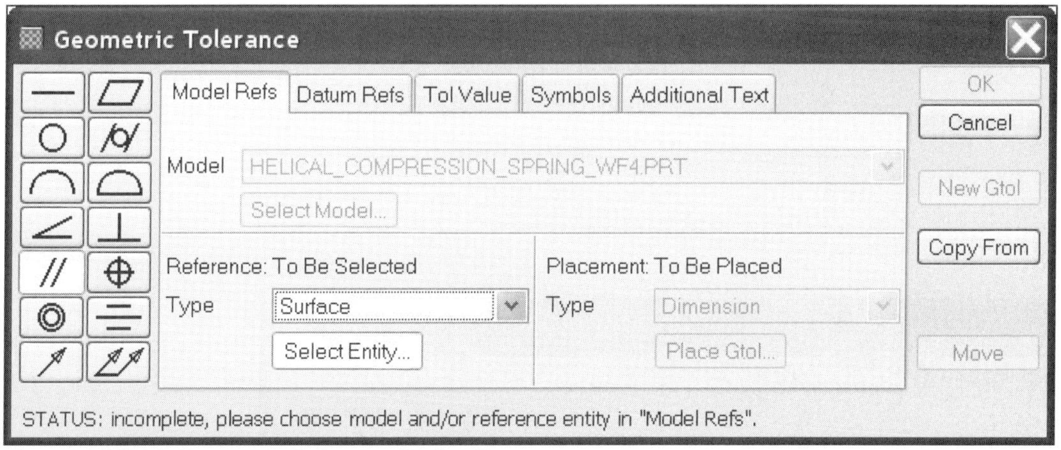

Figure 16.39(a) Geometric Tolerance Dialog Box

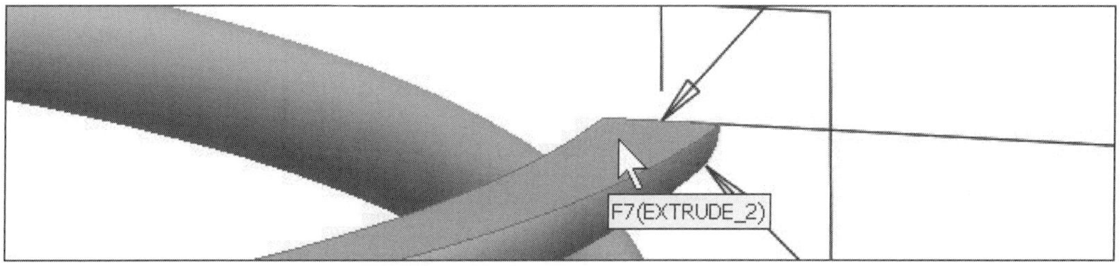

Figure 16.39(b) Select the Cut Surface

Click: Placement: To Be Placed- Type [▼] ⇒ **Dimension** [Fig. 16.39(c)] ⇒ **Place Gtol** ⇒ select the **240** dimension [Fig. 16.39(d)] ⇒ **Datum Refs** tab ⇒ **Primary** tab ⇒ select **A** [Fig. 16.39(e)] ⇒ **OK**

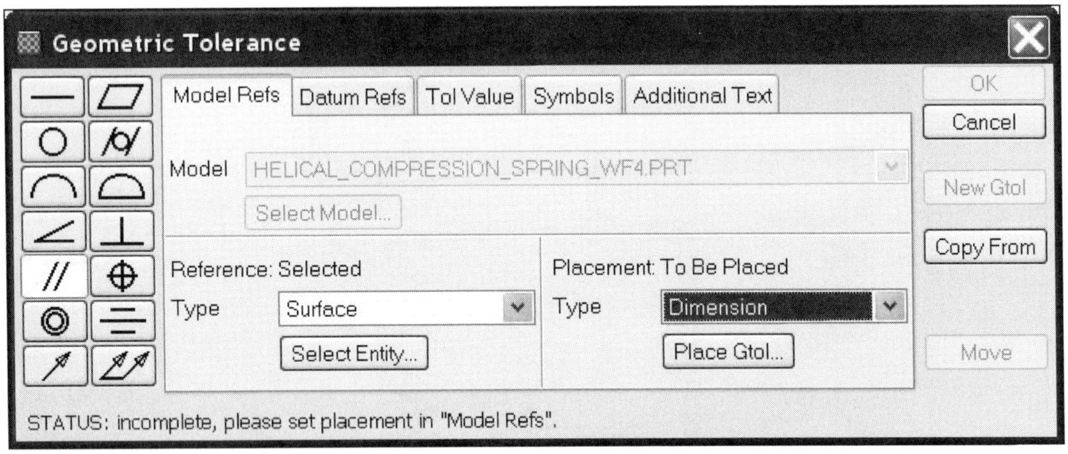

Figure 16.39(c) Type **Dimension**

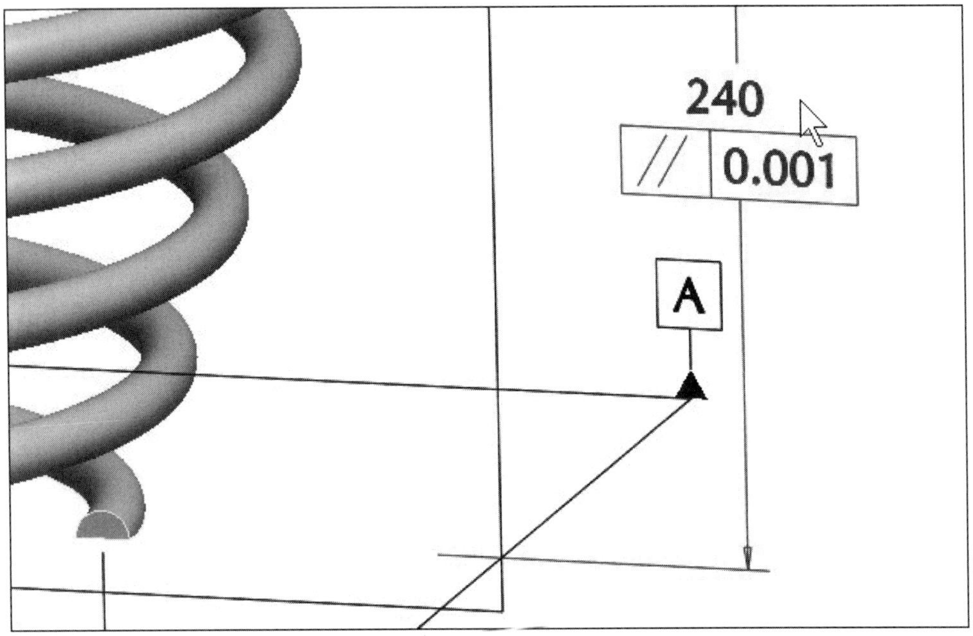

Figure 16.39(d) Select the **240** Dimension

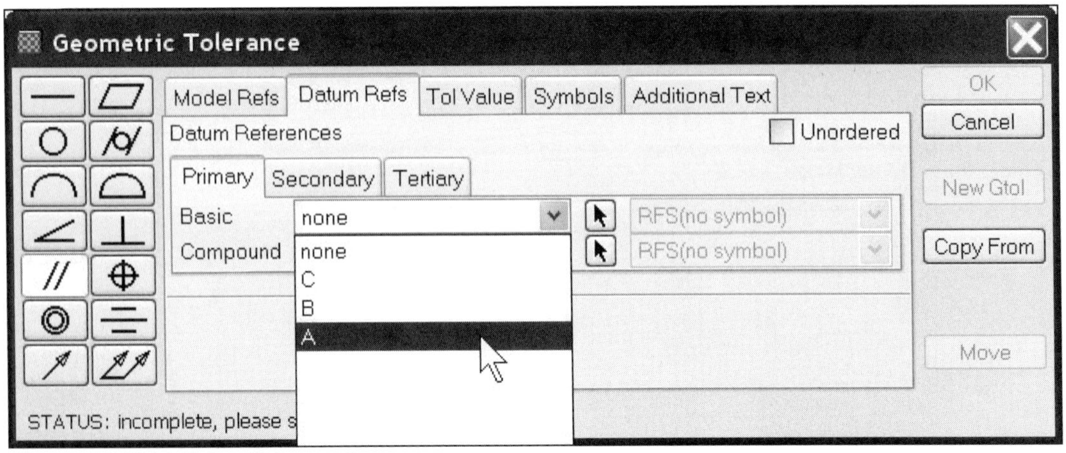

Figure 16.39(e) Select Primary **A**

Click: **OK** [Fig. 16.39(f)] ⇒ **Ctrl+S** ⇒ **Enter** [Fig. 16.39(g)]

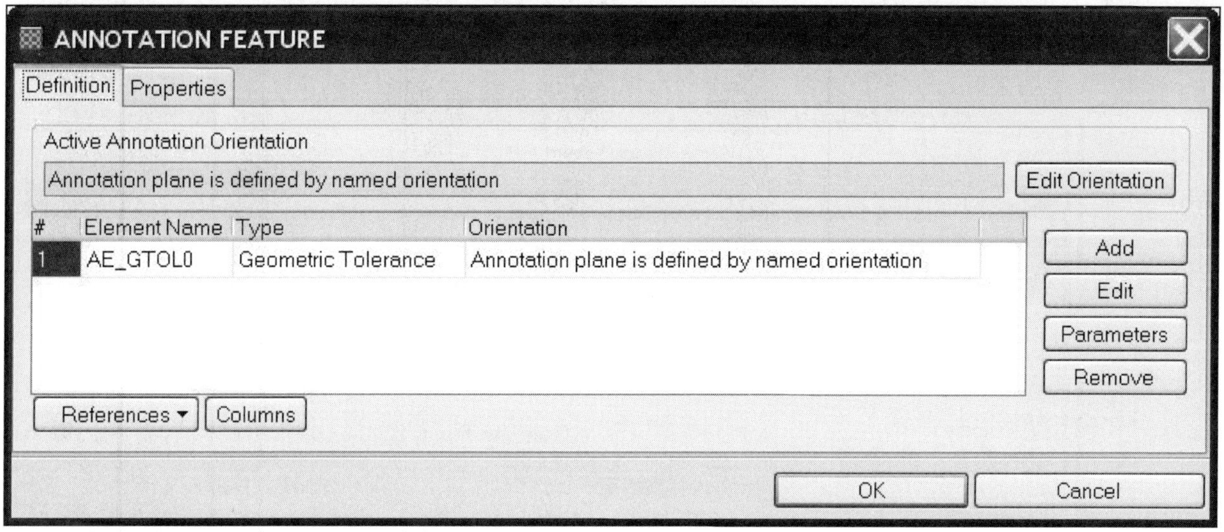

Figure 16.39(f) Annotation Feature Dialog Box

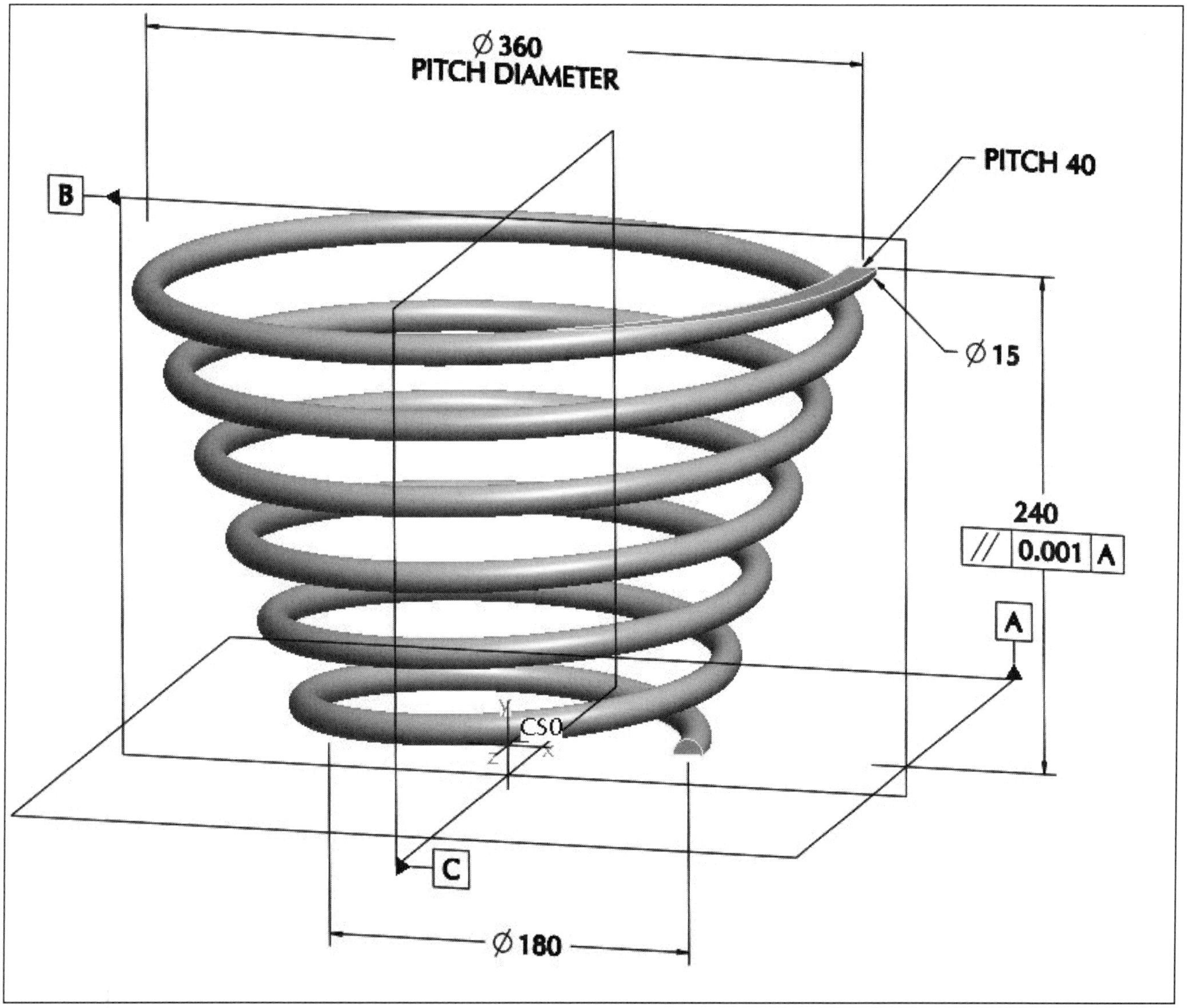

Figure 16.39(g) Annotation Feature Completed

Click: **Insert an annotation feature** ⇒ Surface Finish [Fig. 16.40(a)] ⇒ **OK** [Fig. 16.40(b)] ⇒ **Browse** ⇒ double-click on **machined** ⇒ standard1.sym ⇒ **Preview** [Fig. 16.40(c)]

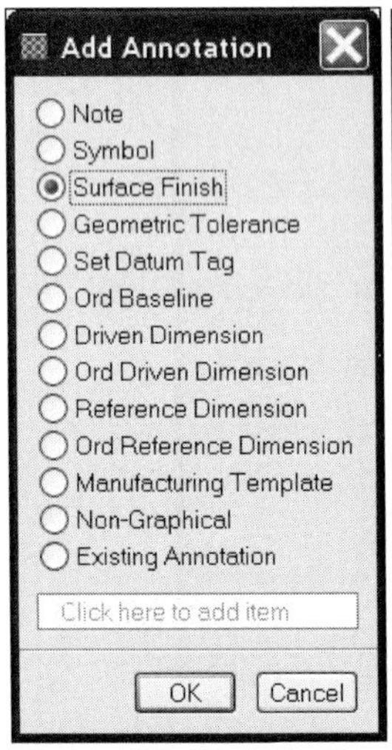

Figure 16.40(a) Surface Finish

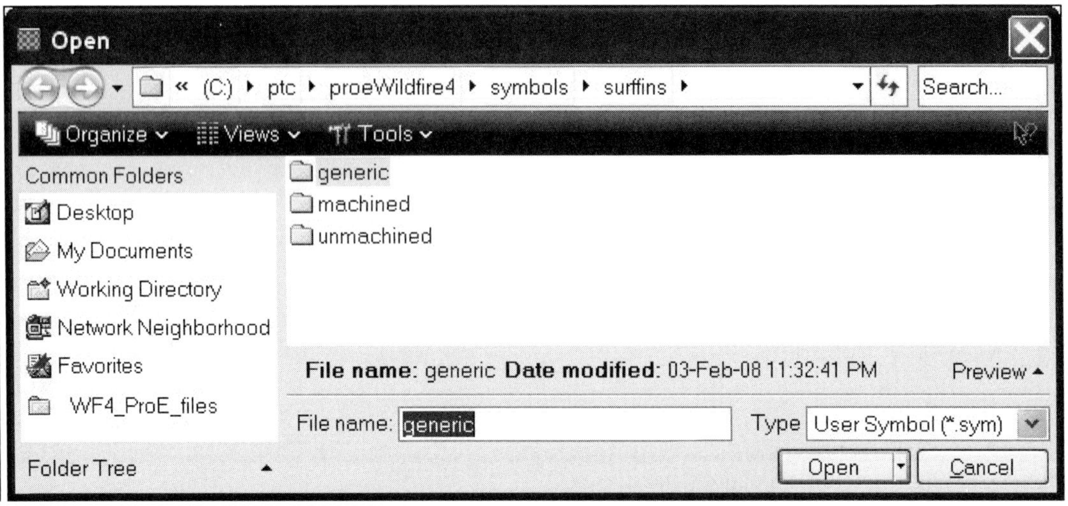

Figure 16.40(b) Surface Finish Symbol Directory

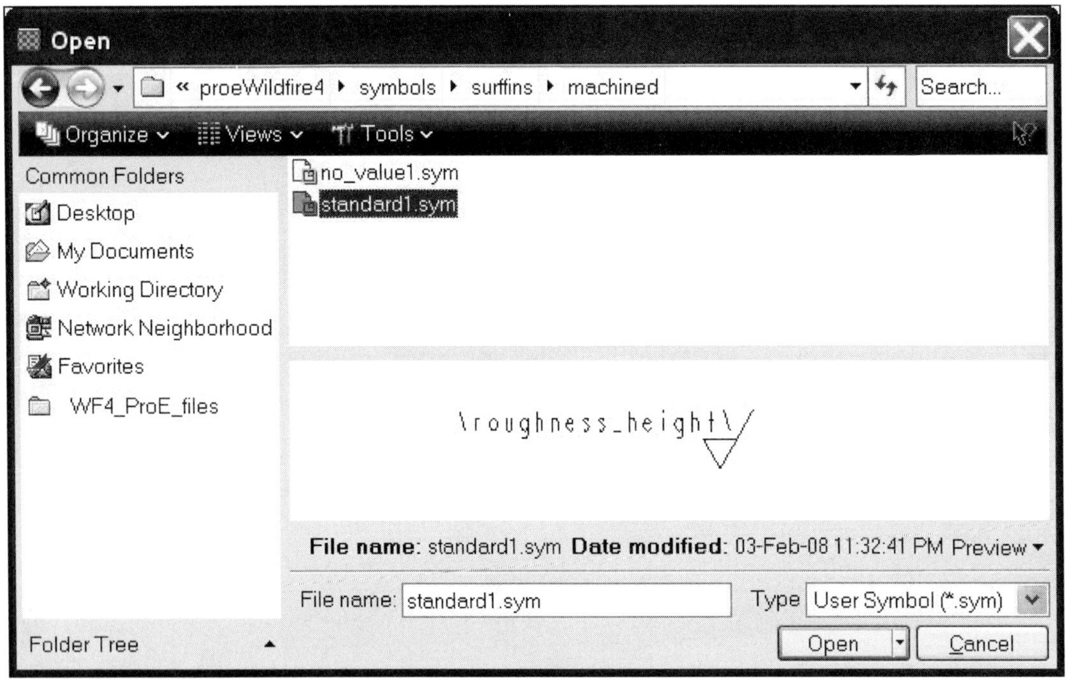

Figure 16.40(c) machined.sym

Click: **Open** [Fig. 16.40(d)] and the Surface Finish dialog box opens with its References collector active) ⇒ select the ground surface [Fig. 16.40(e)]

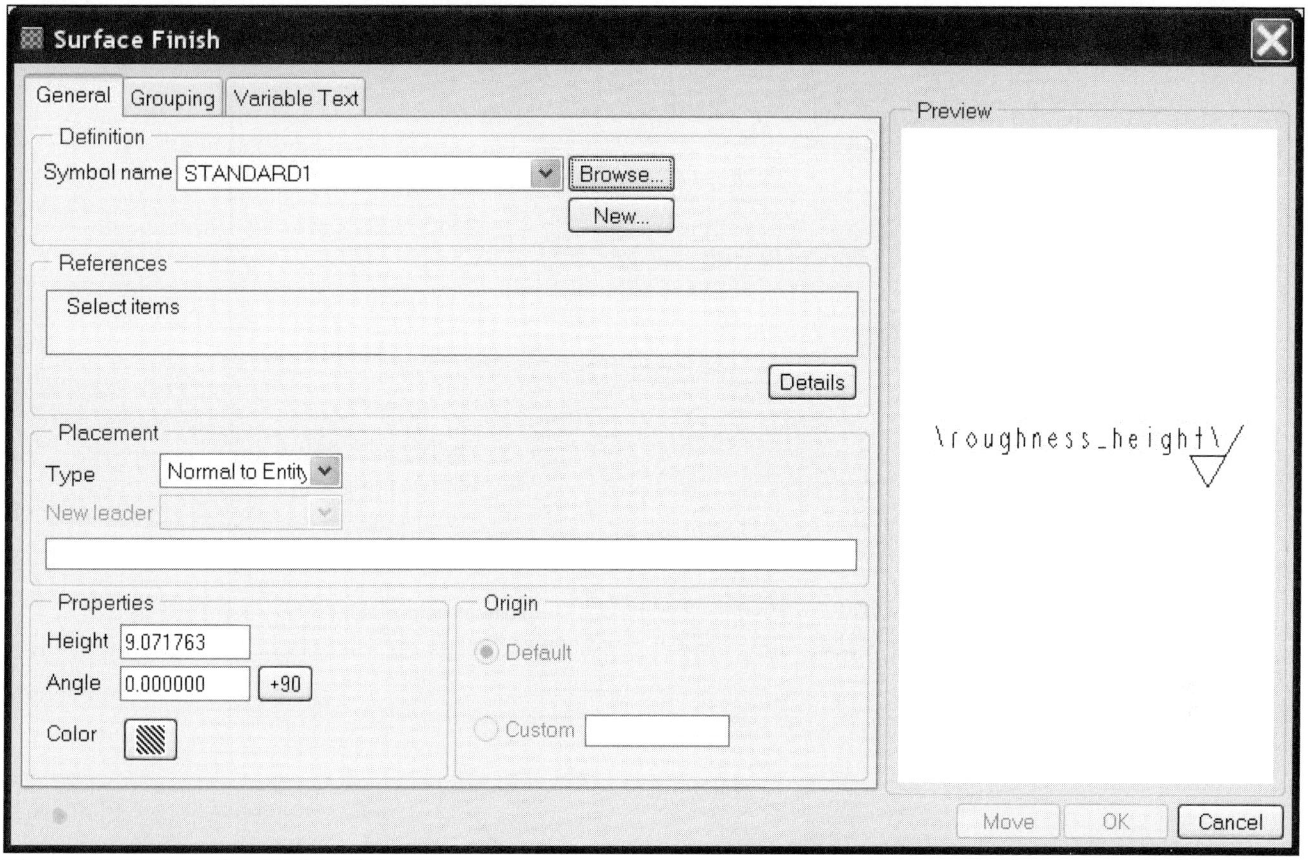

Figure 16.40(d) Surface Finish Dialog Box

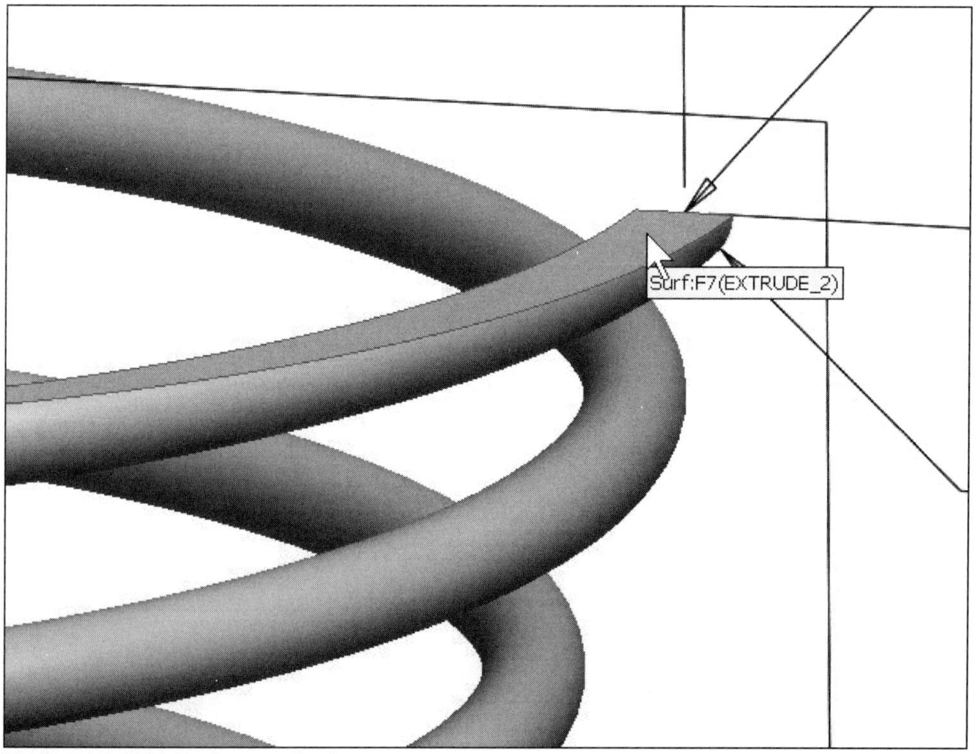

Figure 16.40(e) Reference Surface

Click inside the Placement collector for Attachment references [Fig. 16.40(f)] ⇒ pick the symbol position on the cut surface [Figs. 16.40(g-h)] ⇒ **MMB** [Fig. 16.40(i)]

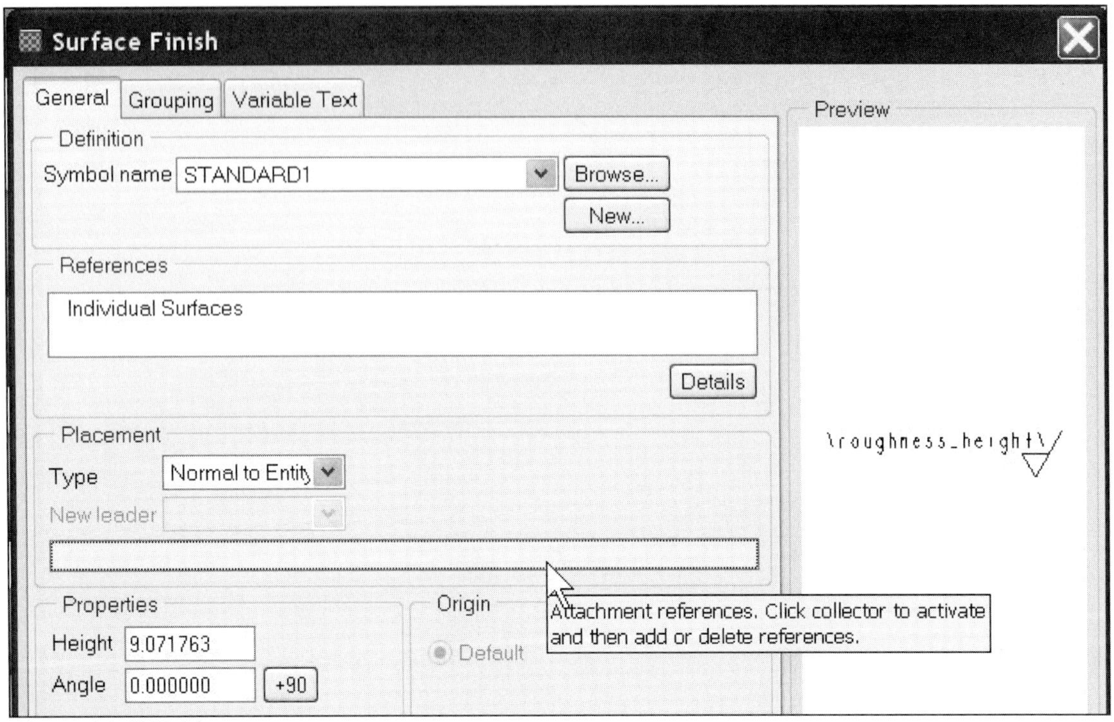

Figure 16.40(f) Placement

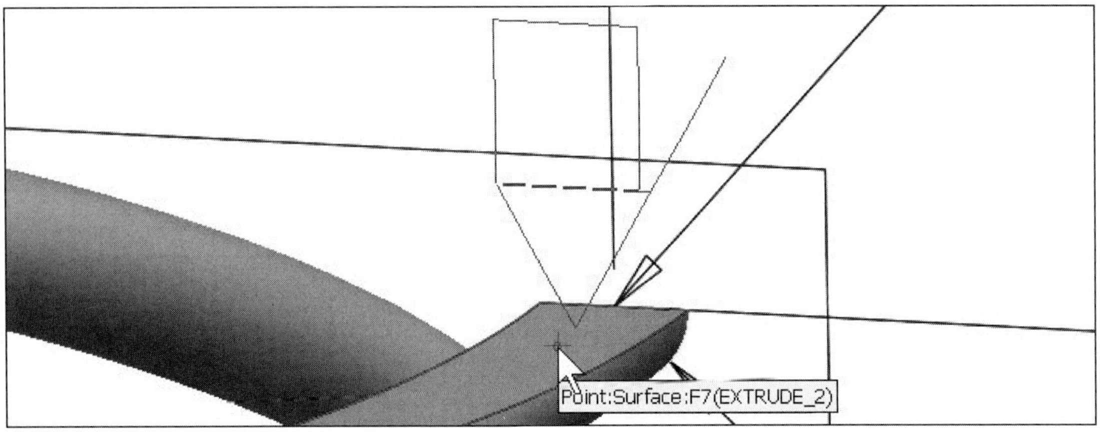

Figure 16.40(g) Pick the Surface Finish Position

Figure 16.40(h) Completed Symbol Placement

Click: **Variable Text** tab ⇒ **OK** [Fig. 16.40(j)] ⇒ **OK** [Fig. 16.40(k)] ⇒ **LMB** to deselect ⇒ repeat the process to create an annotation feature finish symbol on the opposite end of the spring

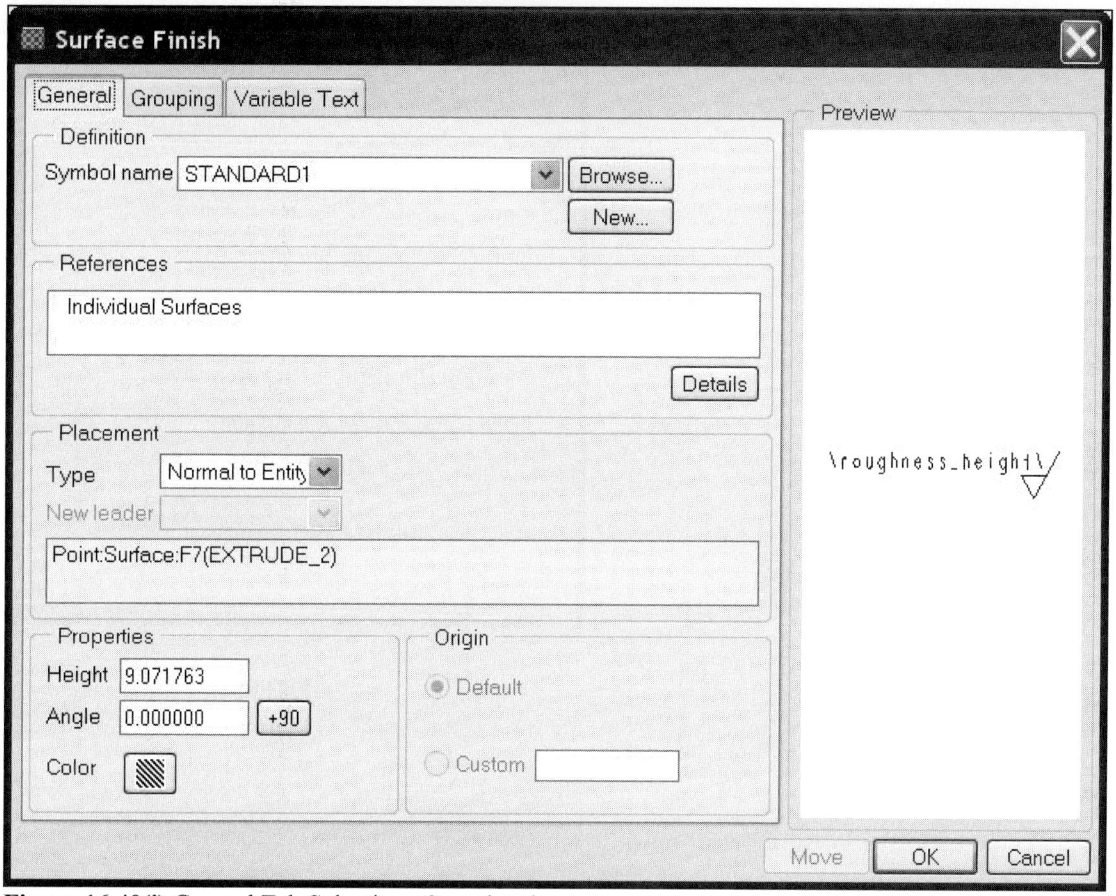

Figure 16.40(i) General Tab Selections Completed

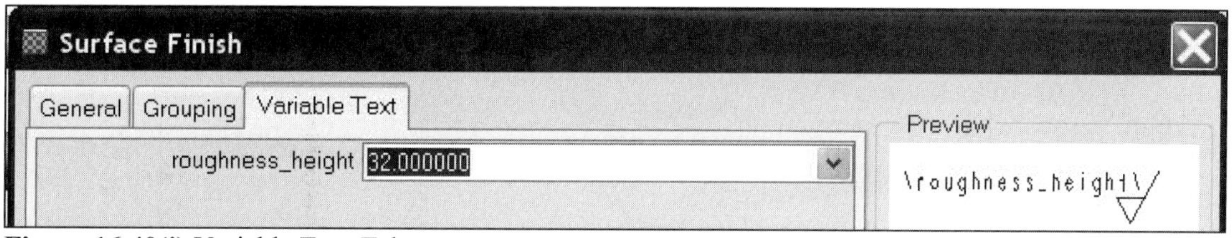

Figure 16.40(j) Variable Text Tab

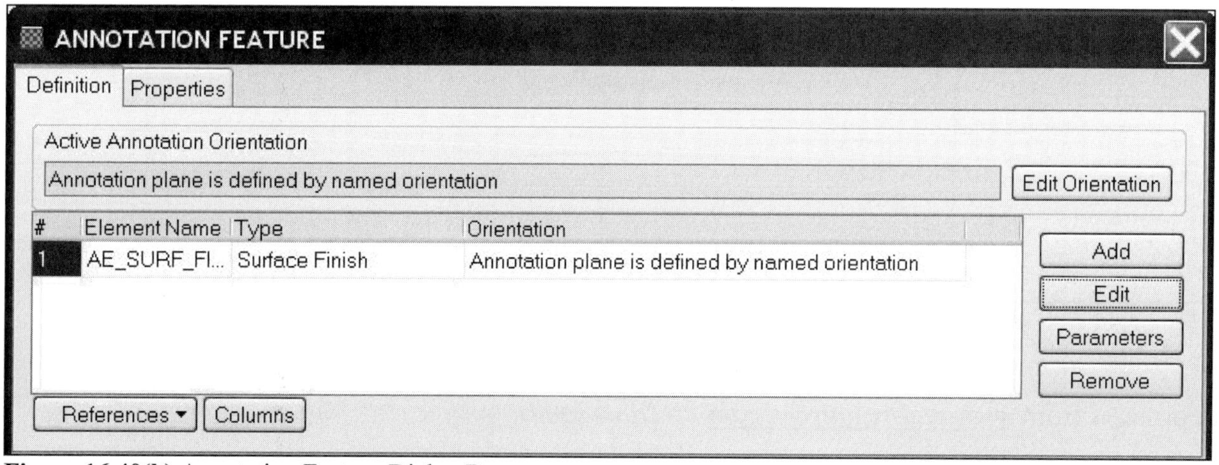

Figure 16.40(k) Annotation Feature Dialog Box

Click: **View** ⇒ **Annotation Orientation** [Fig. 16.40(l)] to see the annotation plane (green) ⇒ **Cancel** ⇒ **Ctrl+S** ⇒ **Enter** ⇒ **File** ⇒ **Close Window**

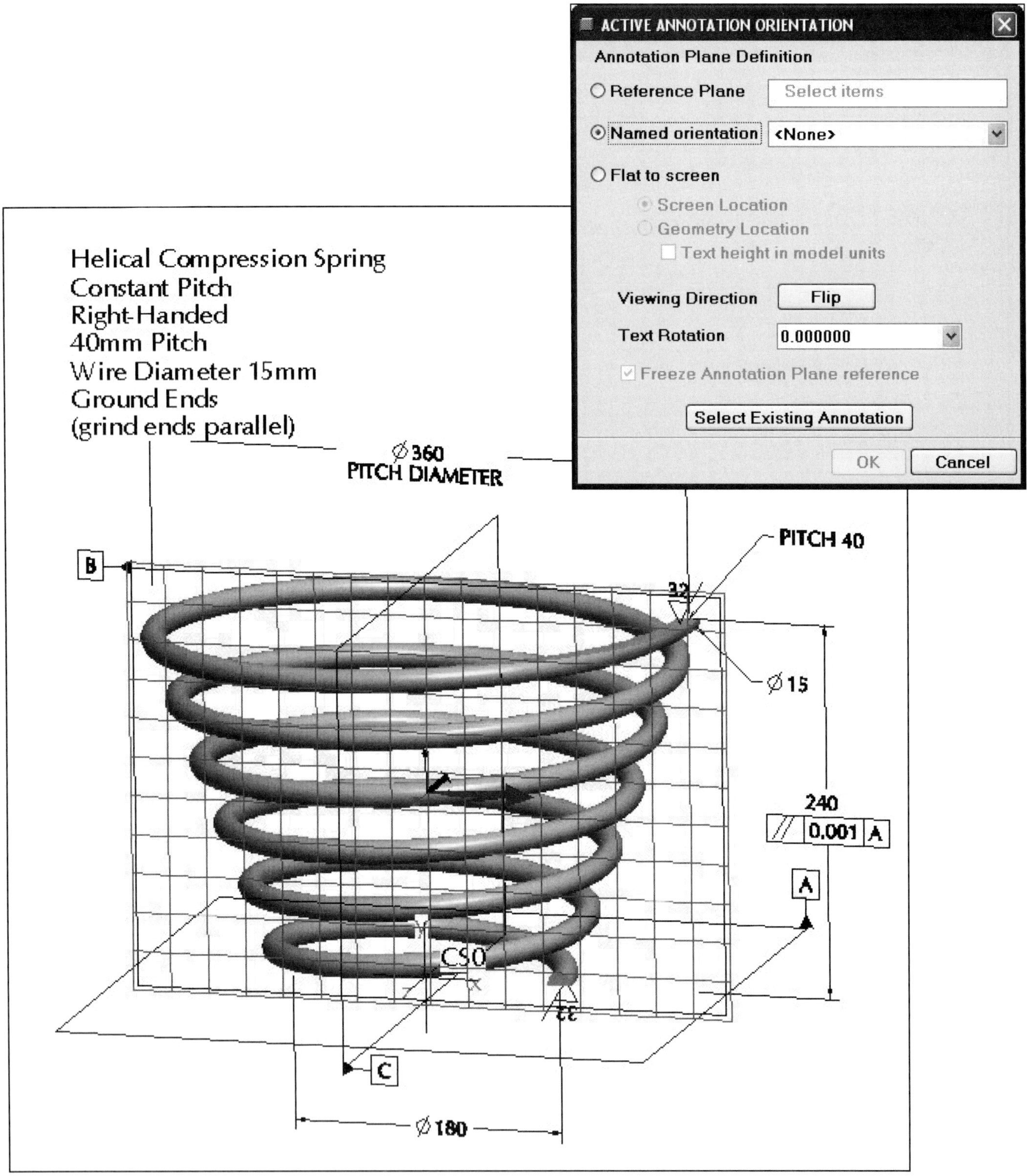

Figure 16.40(l) Active Annotation Orientation Dialog Box

Download projects from ***www.cad-resources.com*** ⇒ ***Downloads***.

Lesson 17 Shell, Reorder, and Insert Mode

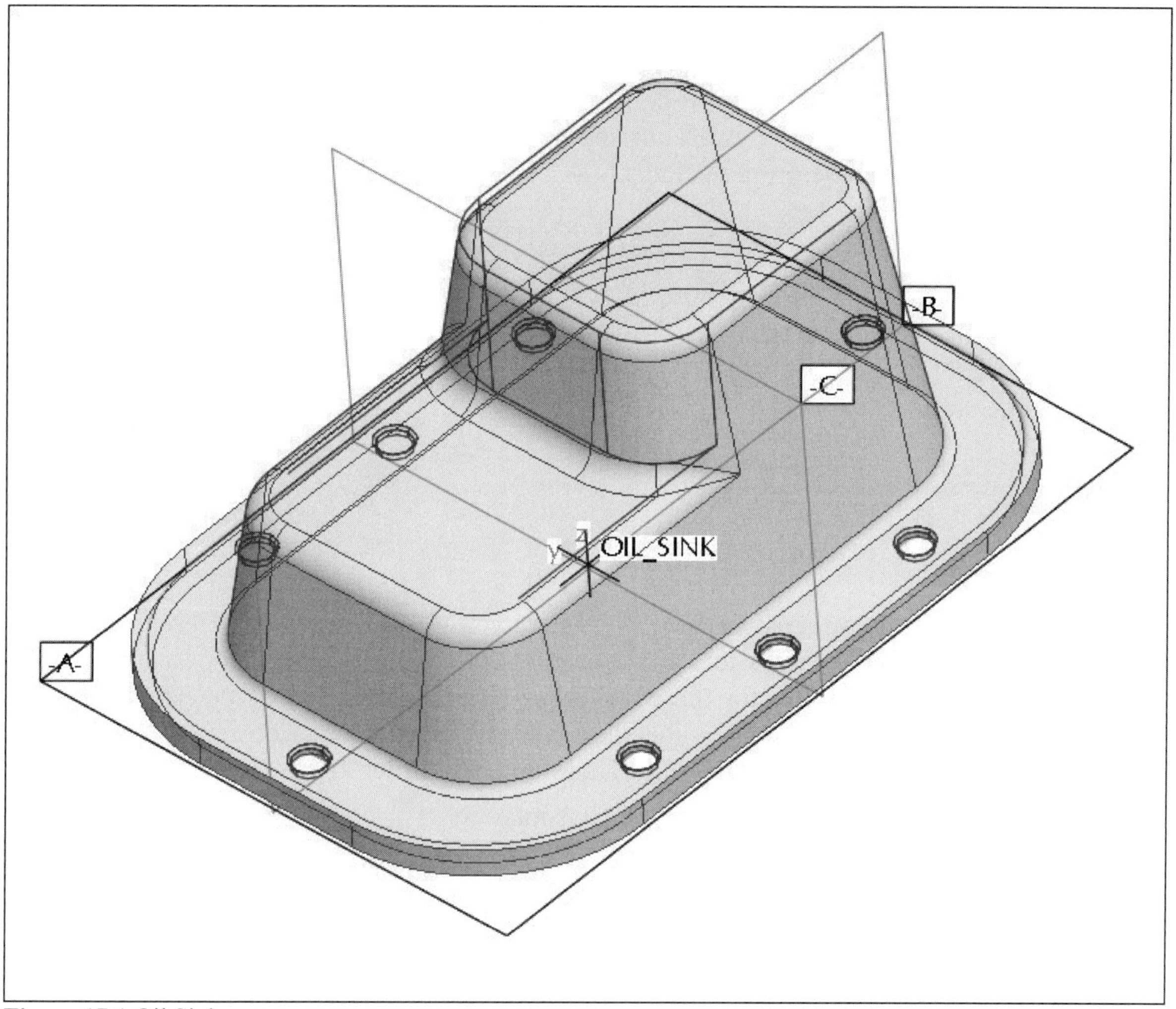

Figure 17.1 Oil Sink

OBJECTIVES

- Master the use of the **Shell Tool**
- Alter the creation sequence with **Reorder**
- **Insert** a feature at a specific point in the design order
- Create a **Hole Pattern** using a **Table**

SHELL, REORDER, AND INSERT MODE

The **Shell Tool** removes a surface or surfaces from the solid and then hollows out the inside of the solid, leaving a shell of a specified wall thickness, as in the Oil Sink (Fig. 17.1). When Pro/E makes the shell, all the features that were added to the solid before you chose the Shell Tool are hollowed out. Therefore, the *order of feature creation* is very important when you use the Shell Tool. You can alter the feature creation order by using the **Reorder** option. Another method of placing a feature at a specific place in the feature/design creation order is to use the **Insert Mode** option.

Creating Shells

The Shell Tool [Figs. 17.2(a-c)] enables you to remove a surface or surfaces from the solid, then hollows out the inside of the solid, leaving a shell of a specified wall thickness. If you flip the thickness side by entering a negative value, dragging a handle, or using the **Change thickness direction** icon, the shell thickness is added to the outside of the part. If you do not select a surface to remove, a "closed" shell is created, with the whole inside of the part hollowed out and no access to the inside. In this case, you can add the necessary cuts or holes to achieve proper geometry at a later time.

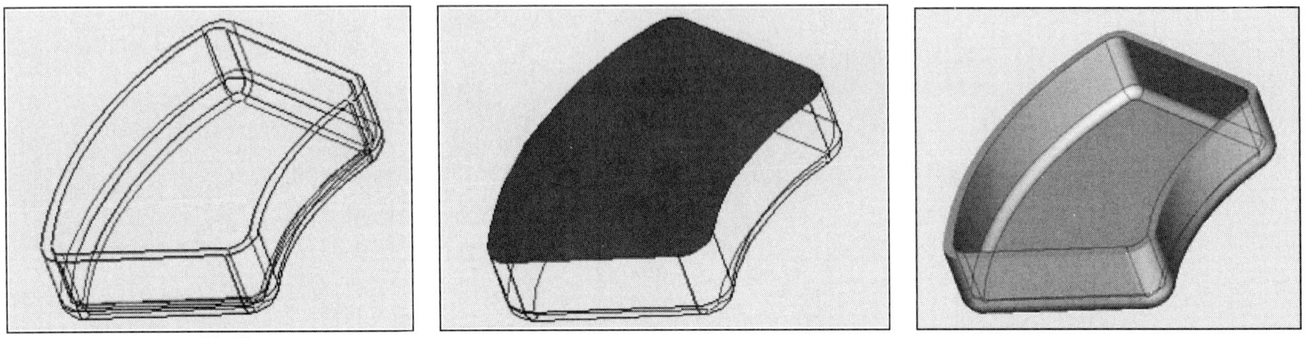

Figures 17.2(a-c) Shell

When defining a shell, you can also select surfaces where you want to assign a different thickness. You can specify independent thickness values for each such surface. However, you cannot enter negative thickness values, or flip the thickness side, for these surfaces. The thickness side is determined by the default thickness of the shell. When Pro/E makes the shell, all the features that were added to the solid before you started the Shell Tool are hollowed out. Therefore, the order of feature creation is very important when you use the Shell Tool. To access the Shell Tool, click icon in the Right Toolchest, or click **Insert ⇒ Shell** on the top menu bar. The Thickness box lets you change the value for the default shell thickness. You can type the new value, or select a recently used value from the drop-down list.

In the graphics window, you can use the shortcut menu (**RMB**) to access the following options:

- **Remove Surfaces** Activates the collector of surfaces. You can select any number of surfaces
- **Non Default Thickness** Activates the collector of surfaces with a different thickness
- **Excluded Surfaces** Activates the collector of excluded surfaces
- **Clear** Remove all references from the collector that is currently active
- **Flip** Change the shell side direction

The Shell Dashboard displays the following slide-up/down panels:

- **References** Contains the collector of references used in the Shell feature
- **Options** Contains the collector of Excluded surfaces
- **Properties** Contains the feature name and an icon to access feature information

The References slide-up/down panel contains the following elements:

- The **Removed surfaces** collector lets you select the surfaces to be removed. If you do not select any surfaces, a "closed" shell is created.
- The **Non-default thickness** collector lets you select surfaces where you want to assign a different thickness. For each surface included in this collector, you can specify an individual thickness value.

The Properties panel contains the Name text box Name SHELL_ID_200, where you can type a custom name for the shell feature, to replace the automatically generated name. It also contains the icon that you can click to display information about this feature in the Browser.

Reordering Features

You can move features forward or backward in the feature creation (regeneration) order list, thus changing the order in which features are regenerated [Figs. 17.3(a-b)]. Use **Edit ⇒ Feature Operations ⇒ Reorder** to activate the command.

You can reorder multiple features in one operation, as long as these features appear in *consecutive* order. Feature reorder *cannot* occur under the following conditions:

- **Parents** Cannot be moved so that their regeneration occurs after the regeneration of their children
- **Children** Cannot be moved so that their regeneration occurs before the regeneration of their parents

You can select the features to be reordered by choosing an option:

- **Select** Select features to reorder by picking on the screen and/or from the Model Tree
- **Layer** Select all features from a layer by selecting the layer
- **Range** Specify the range of features by entering the regeneration numbers of the starting and ending features

You can reorder features in the Model Tree by dragging one or more features to a new location in the feature list. If you try to move a child feature to a higher position than its parent feature, the parent feature moves with the child feature in context, so that the parent/child relationship is maintained.

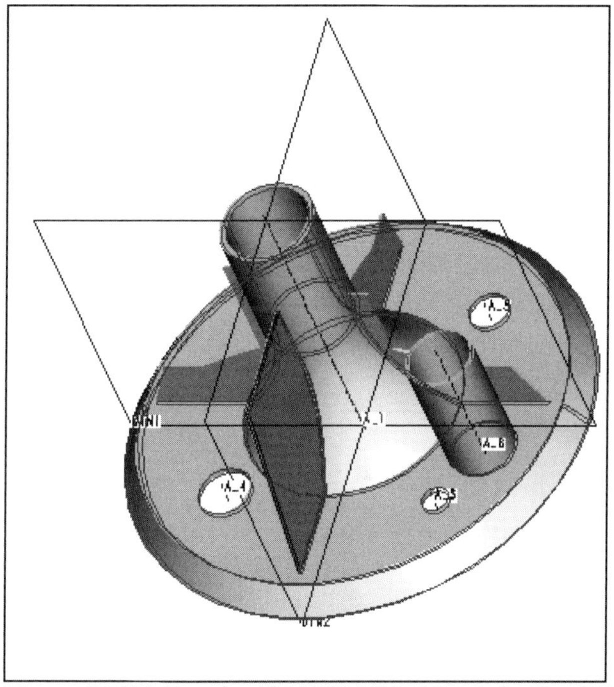

Figure 17.3(a) Reorder (CADTRAIN, COAch for Pro/ENGINEER)

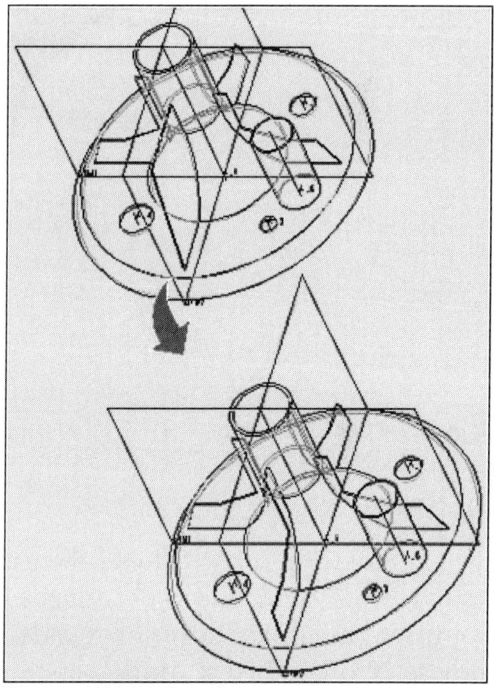

Figure 17.3(b) Reorder

Inserting Features

Normally, Pro/E adds a new feature after the last existing feature in the part, including suppressed features. Insert Mode allows you to add new features at any point in the feature sequence, except before the base feature or after the last feature. You can also Insert features using the Model Tree. There is an arrow-shaped icon on the Model Tree that indicates where features will be inserted upon creation. By default, it is always at the end of the Model Tree. You may drag the location of the *arrow* higher or lower in the tree to insert features at a different point. When the *arrow* is dropped at a new location, the model is rolled backward or forward in response to the insertion *arrow* being moved higher or lower in the tree.

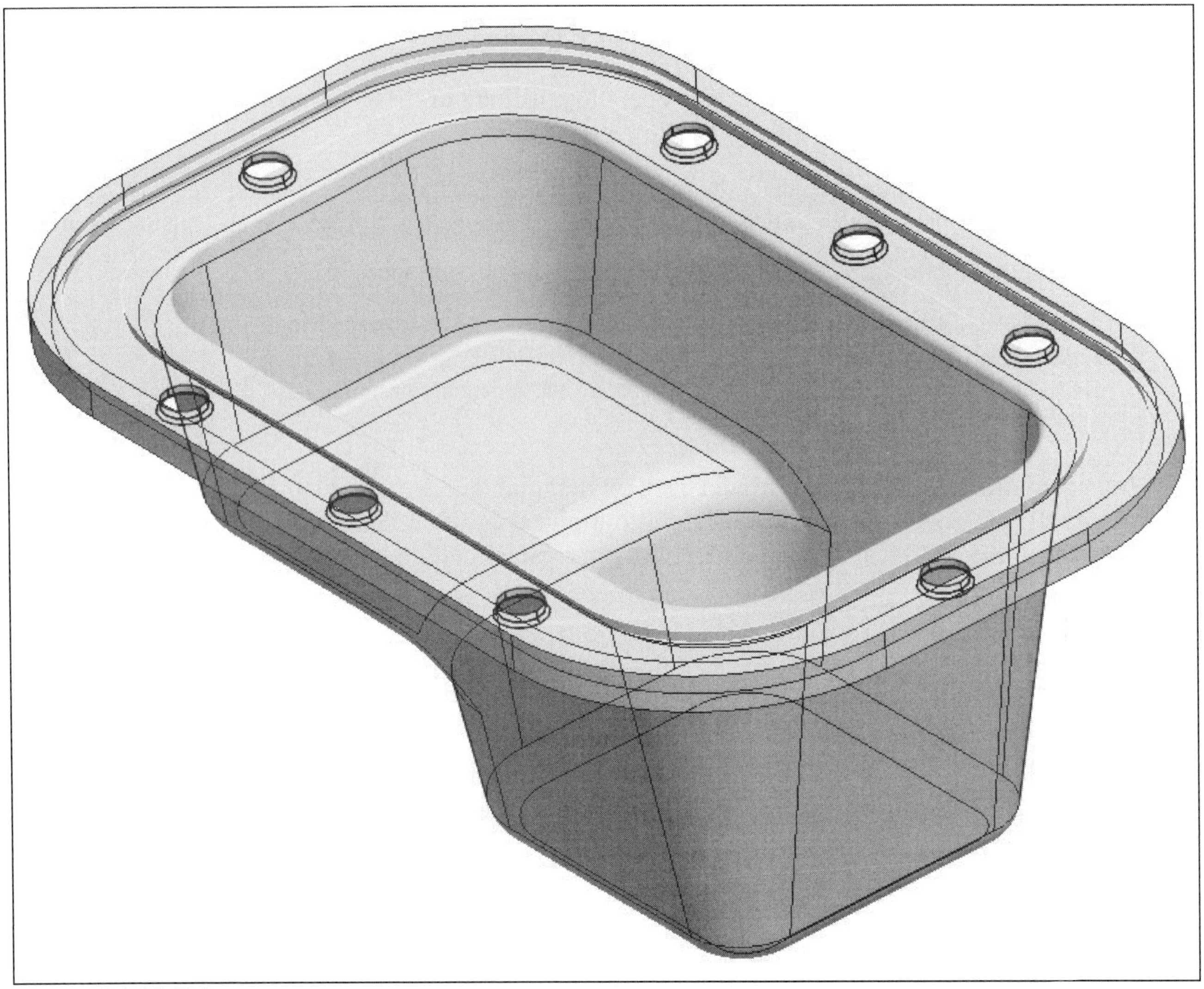

Figure 17.4 Oil Sink

Oil Sink

The Oil Sink (Fig. 17.4) requires the use of the **Shell Tool**. The shelling of a part should be done after the desired protrusions and most rounds have been modeled. This lesson part will have you create a protrusion, a cut, and a set of rounds. Some of the required rounds will be left off the part model on purpose. Pro/E's **Insert Mode** option enables you to insert a set of features at an earlier stage in the design of the part. In other words, you can create a feature after or before a selected existing feature even if the whole model has been completed. You can also *move the order in which a feature was created* and therefore have subsequent features affect the reordered feature. A round created after a shell operation can be reordered to appear before the shell, to have the shell be affected by the round.

In this lesson, you will also insert a round or two before the existing shell feature using Insert Mode. The rounds will be shelled after the **Resume** option is picked, because the rounds now appear before the shell feature. The details shown in Figures 17.5(a) through (m) provide the design dimensions.

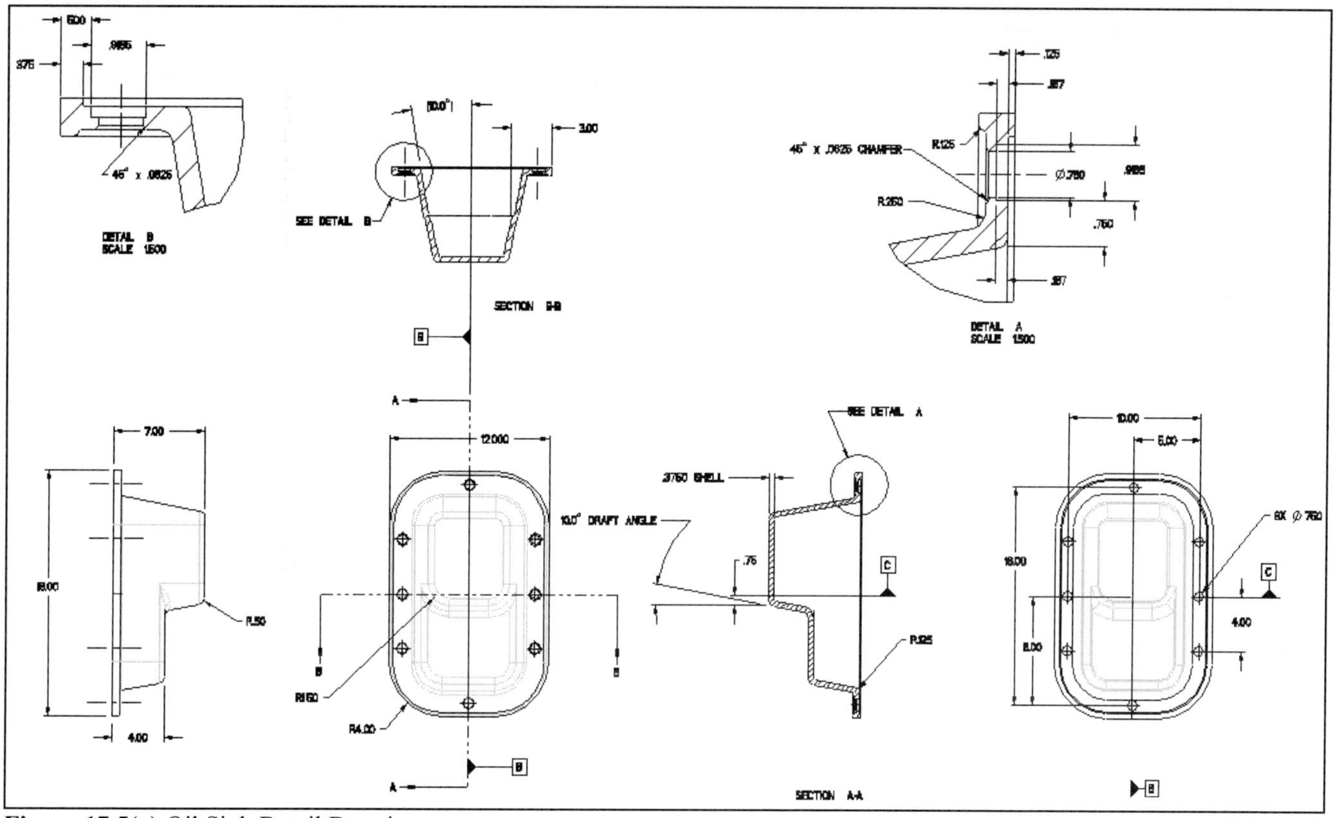

Figure 17.5(a) Oil Sink Detail Drawing

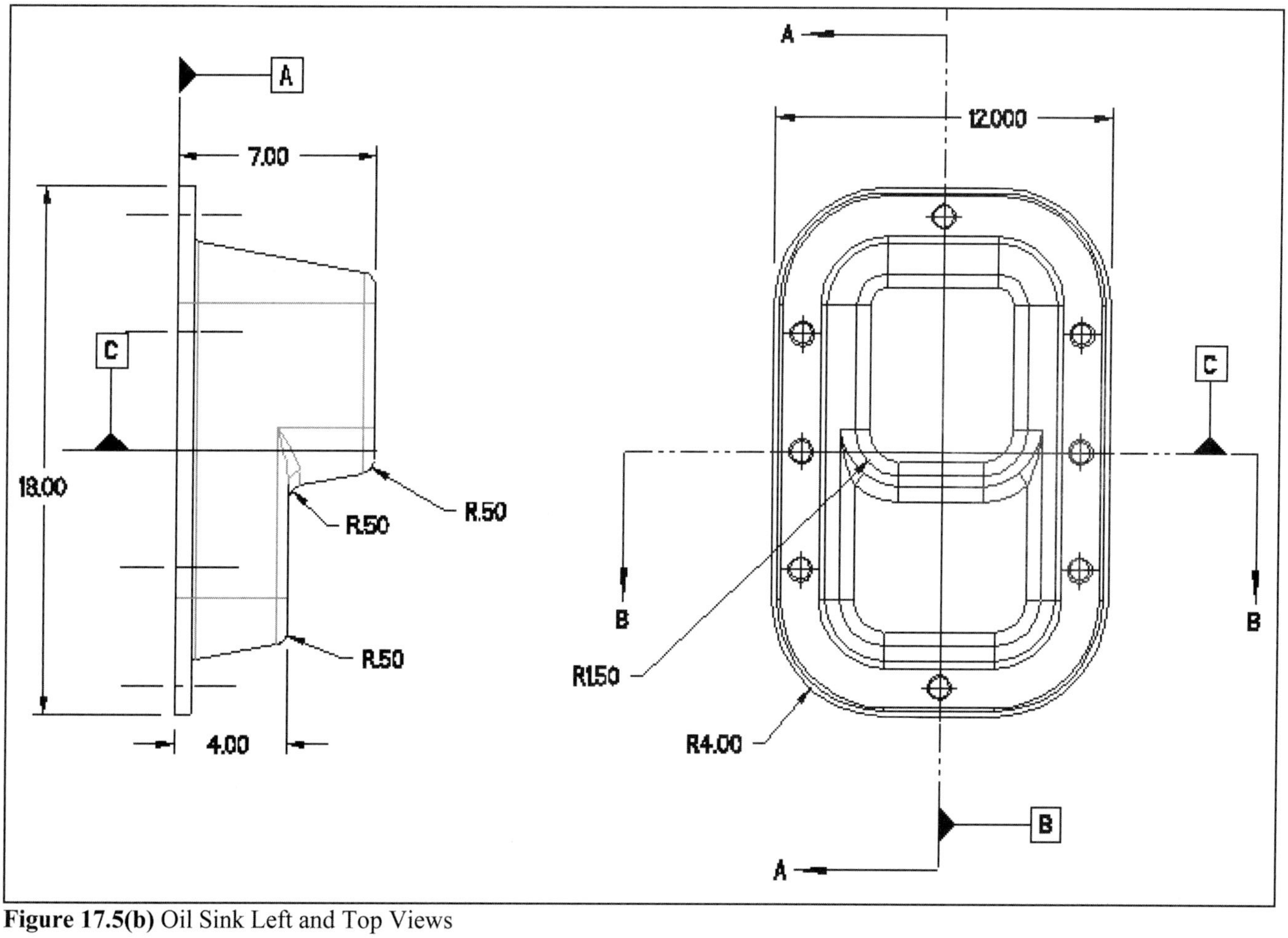

Figure 17.5(b) Oil Sink Left and Top Views

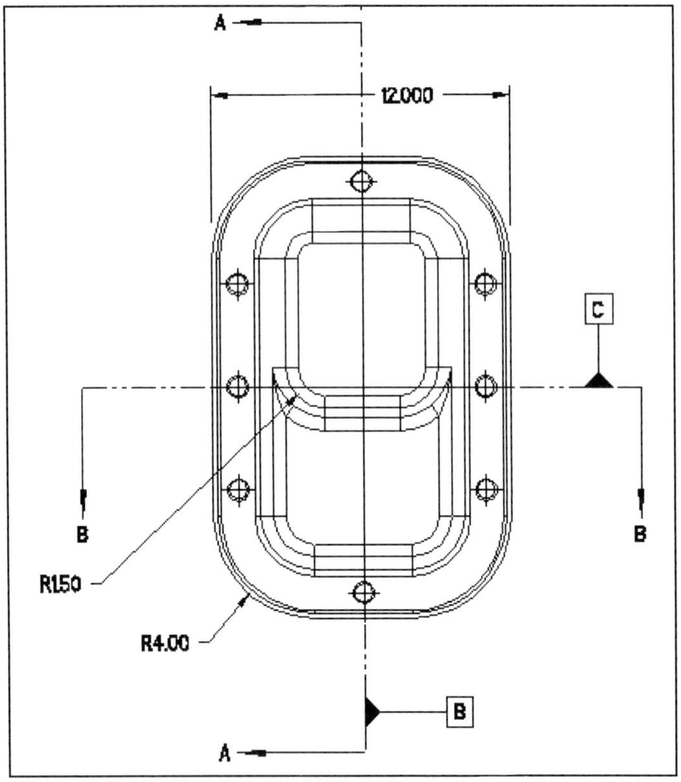

Figure 17.5(c) Oil Sink Top View Dimensions

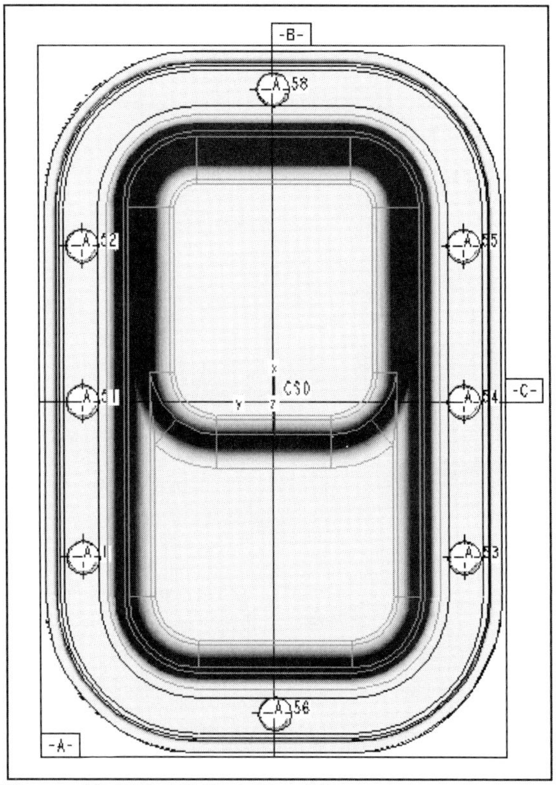

Figure 17.5(d) Oil Sink Top View

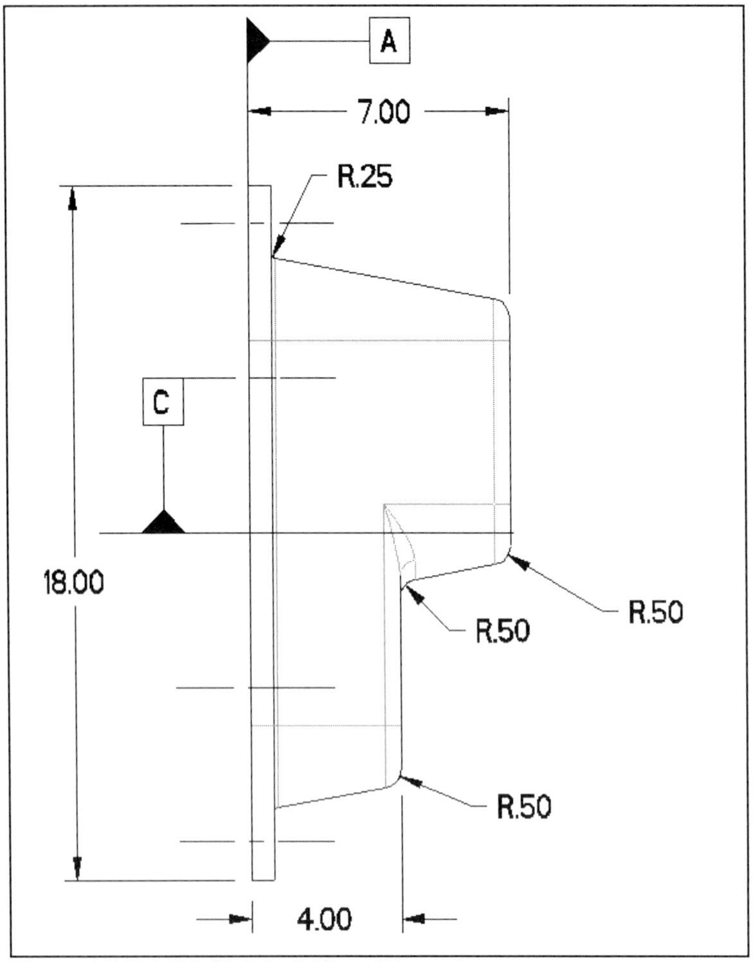

Figure 17.5(e) Oil Sink Left View Dimensions

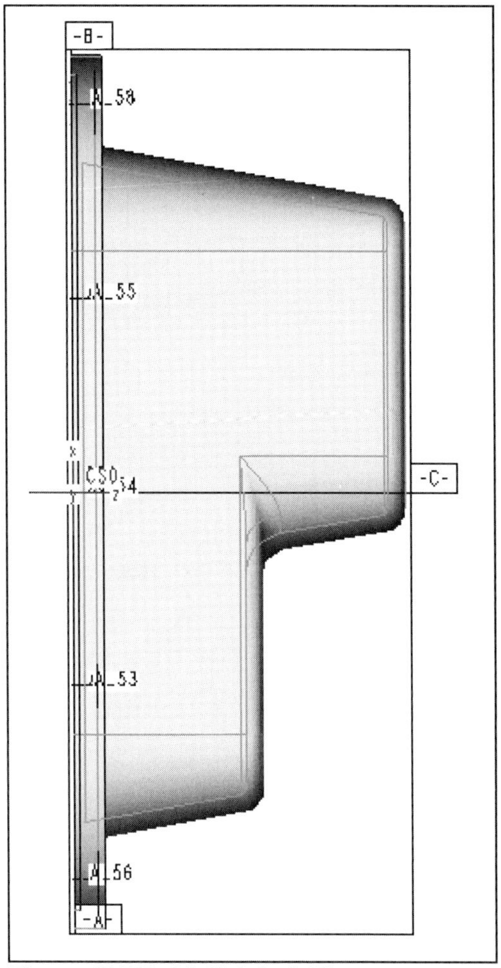

Figure 17.5(f) Oil Sink Left View

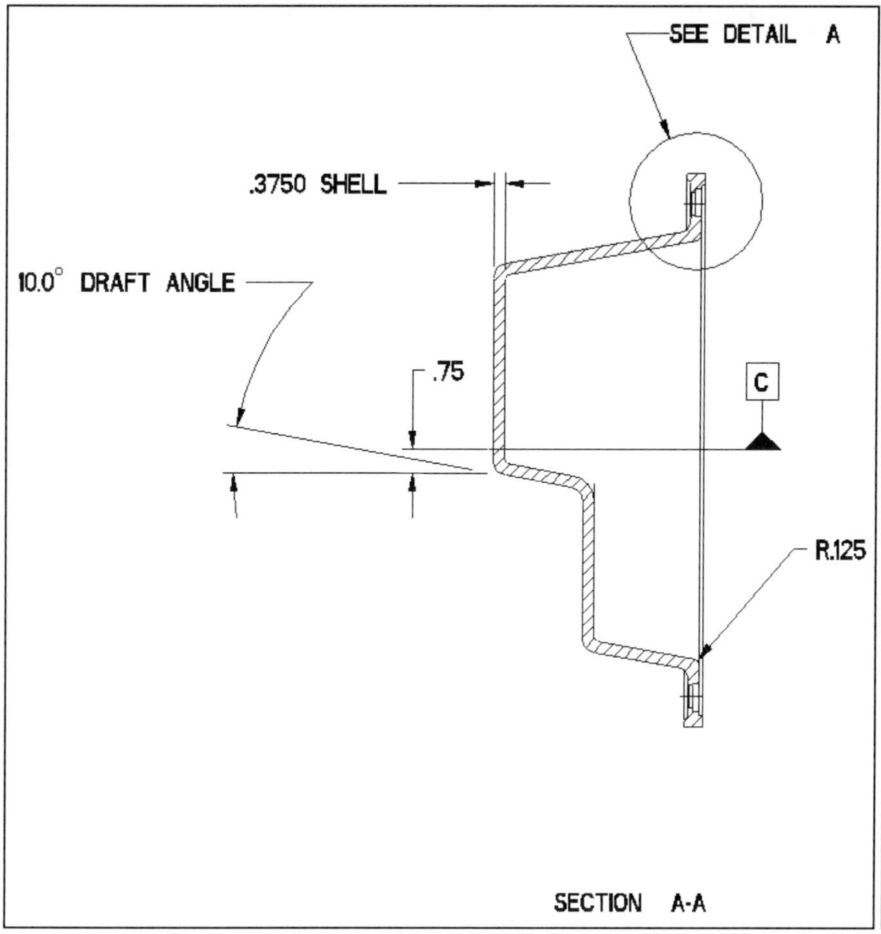

Figure 17.5(g) Oil Sink SECTION A-A

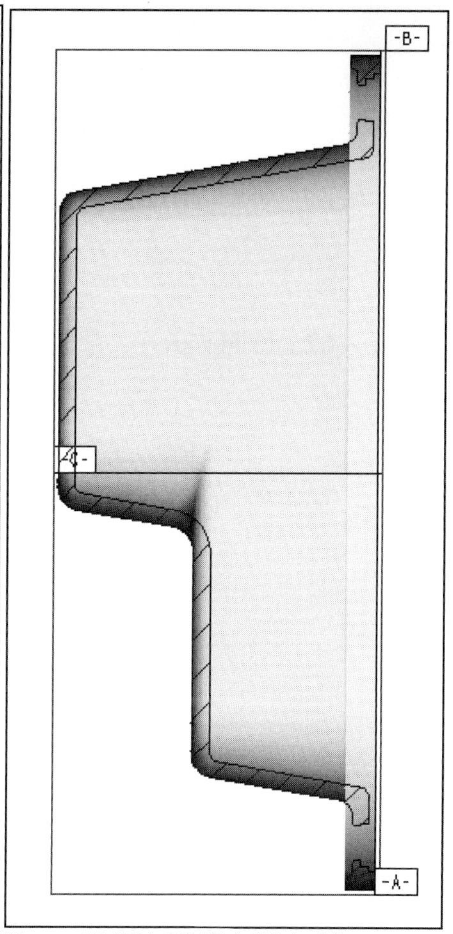

Figure 17.5(h) Oil Sink Right View

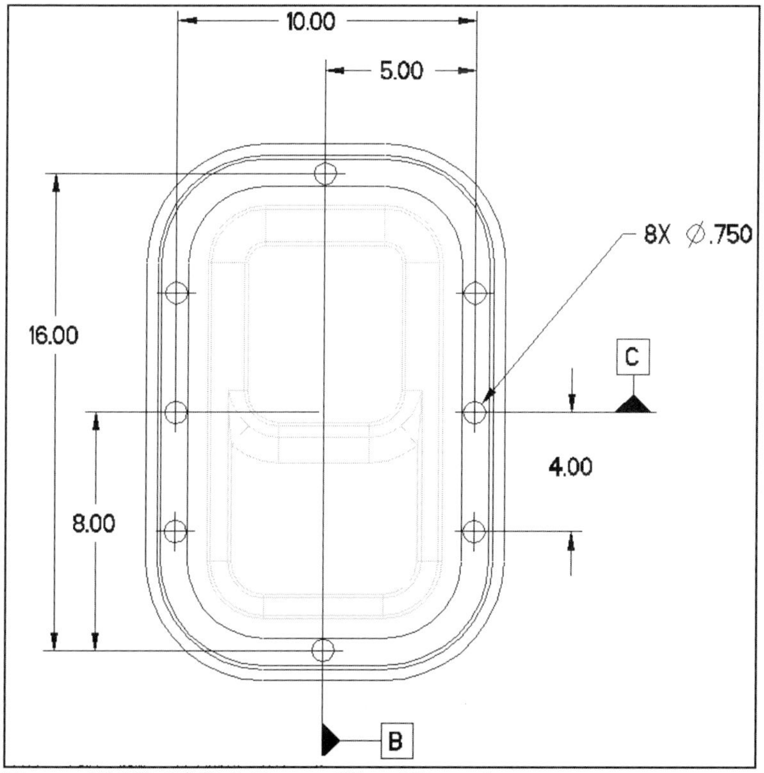

Figure 17.5(i) Oil Sink Bottom View Dimensions

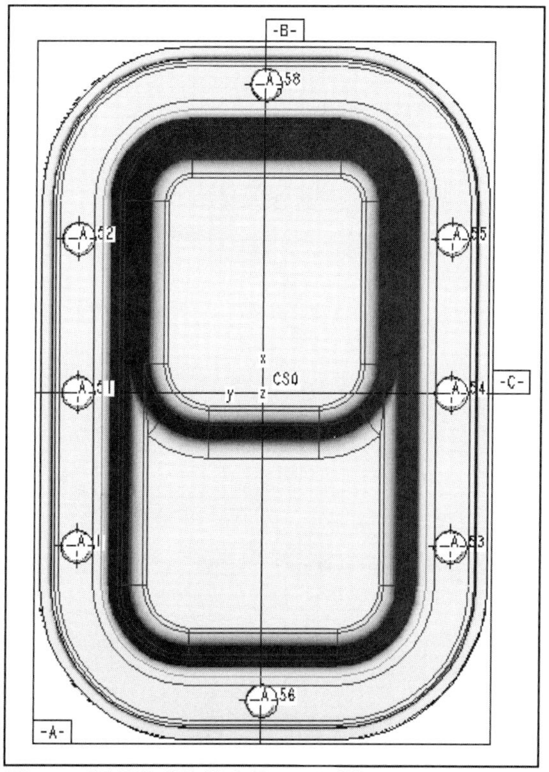

Figure 17.5(j) Oil Sink Bottom View

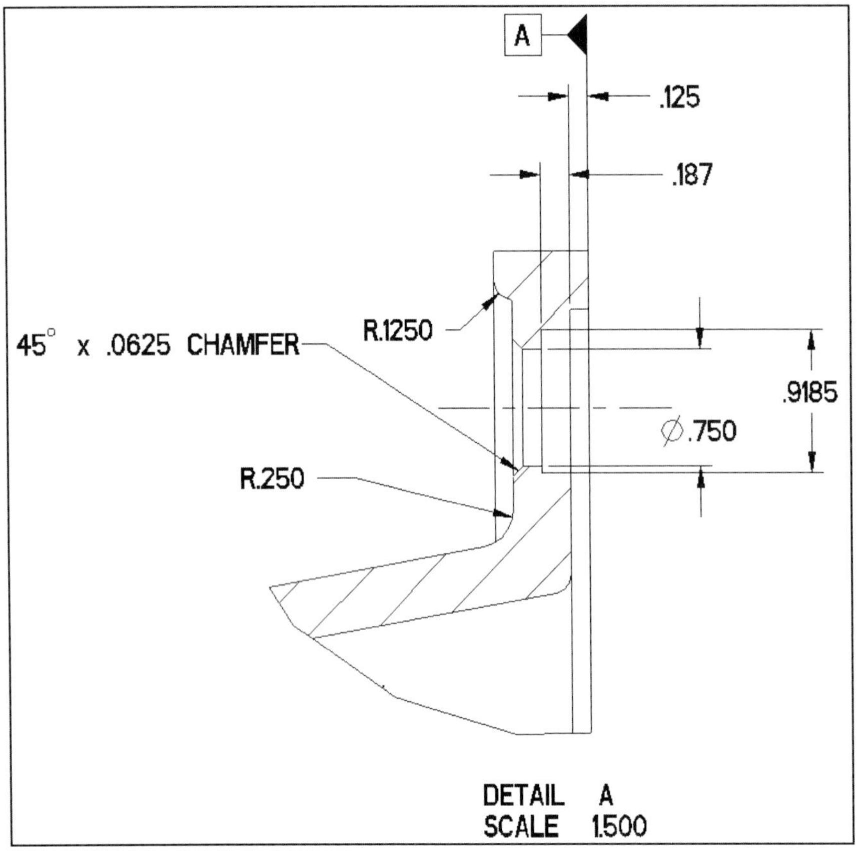

Figure 17.5(k) Oil Sink DETAIL A

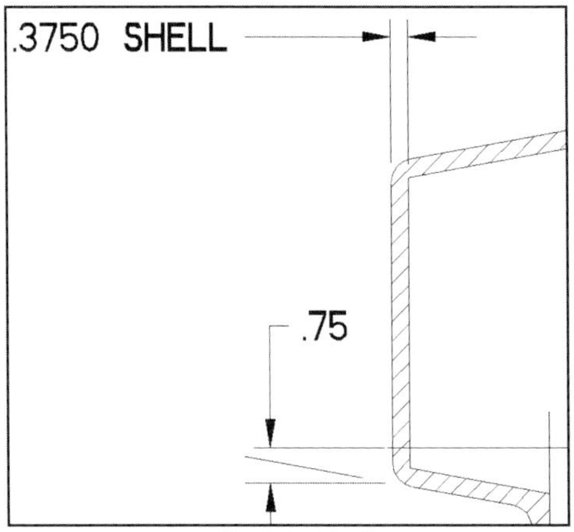

Figure 17.5(l) Oil Sink Shell

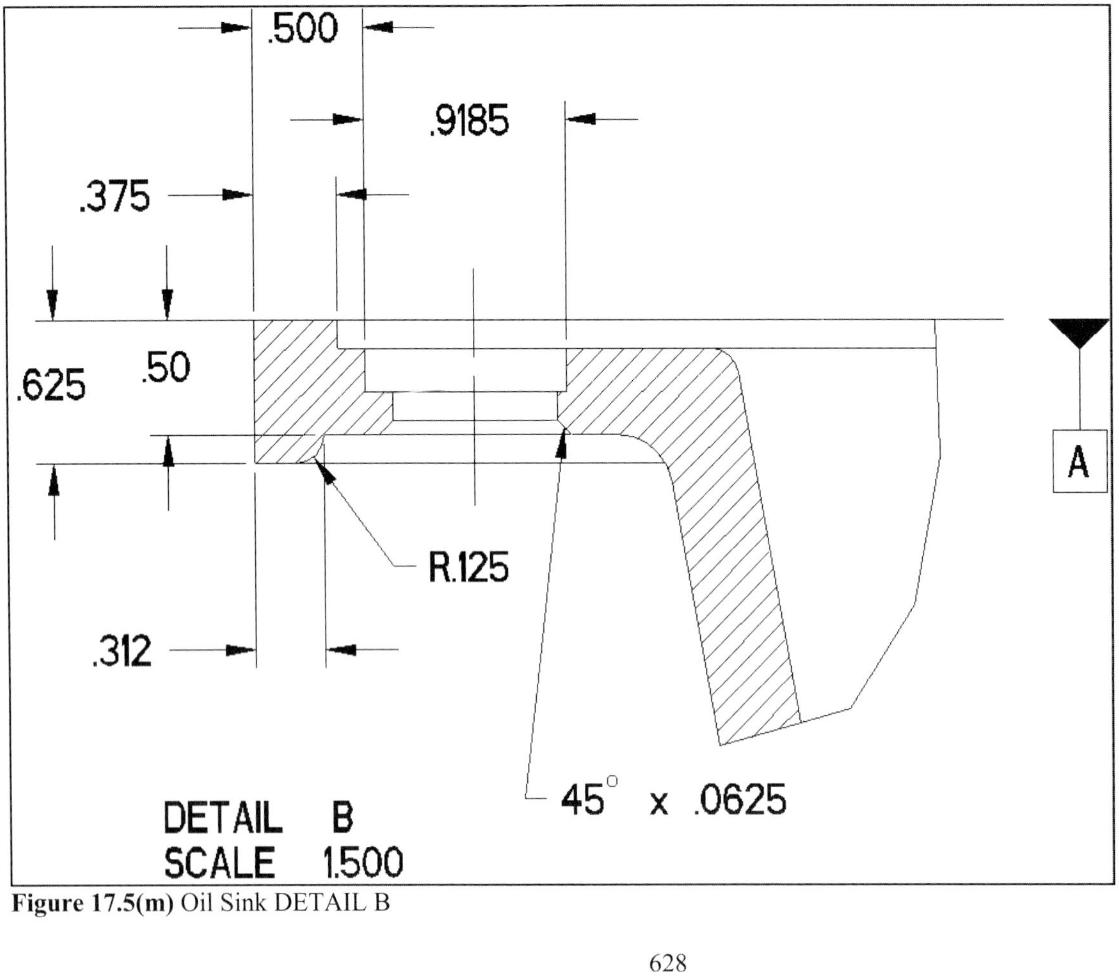

Figure 17.5(m) Oil Sink DETAIL B

Click: **File** ⇒ **Set Working Directory** select the working directory ⇒ **OK** ⇒ ⇒ Part ⇒ **OIL_SINK** ⇒ ⇒ **OK** ⇒ **Tools** ⇒ **Environment** ⇒ Snap to Grid ⇒ Use 2D Sketcher ⇒

Display Style Hidden Line ⇒ Tangent Edges Dimmed ⇒ **Apply** ⇒ **OK**

Set up the working environment and defaults: **Tools** ⇒ **Options** ⇒ Showing: **Current Session** ⇒ Option: ***default_dec_places*** ⇒ Value: **3** ⇒ **Enter** ⇒ Option: ***sketcher_dec_places*** ⇒ Value: **3** ⇒ **Enter** ⇒ **Apply** ⇒ **Close** ⇒ load your saved customization file ⇒ **Tools** ⇒ **Customize Screen** ⇒ **File** from the Customize dialog box menu bar ⇒ **Open Settings** ⇒ click on your saved file ⇒ **Open** ⇒ **OK**

Set and Assign the material: **Edit** ⇒ **Setup** ⇒ **Units** ⇒ Units Manager **Inch lbm Second** ⇒ **Close** ⇒ **Material** ⇒ **steel.mtl** ⇒ >>> ⇒ **OK** ⇒ **Done**

Change the coordinate system name and set the datums: double-click on the default coordinate system name in the Model Tree-- **PRT_CSYS_DEF** ⇒ *(type)* **OIL_SINK** ⇒ **LMB** to deselect ⇒ **Insert** ⇒ **Annotations** ⇒ **Geometric Tolerance** ⇒ **Set Datum** ⇒ pick **TOP** from the model ⇒ Name- **B** ⇒ **OK** ⇒ pick **FRONT** ⇒ Name- **A** ⇒ **OK** ⇒ pick **RIGHT** ⇒ Name- **C** ⇒ **MMB** ⇒ **MMB** ⇒ **MMB** ⇒ ⇒ **MMB** (Fig. 17.6)

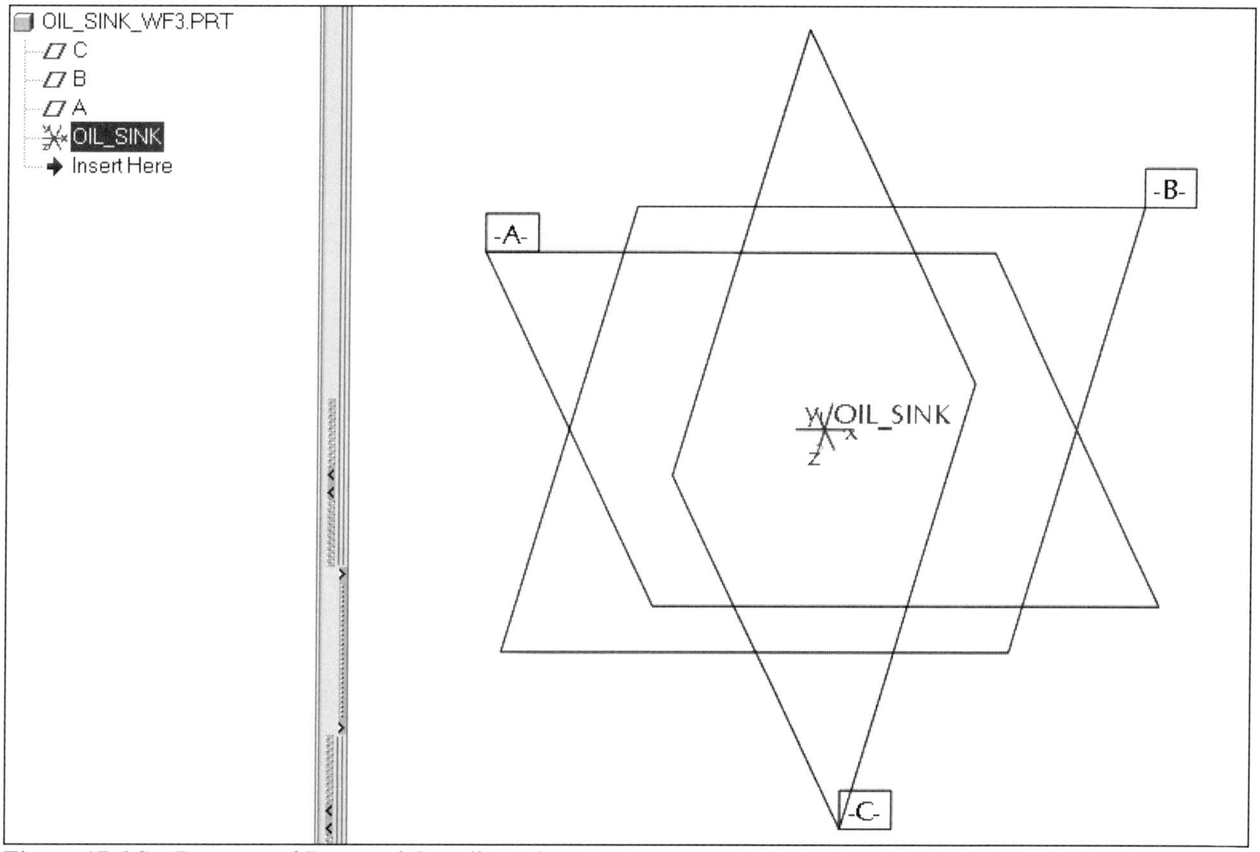

Figure 17.6 Set Datums and Renamed Coordinate System

Make the first protrusion **.50** (thickness) **X 12.00** (height) **X 18.00** (length), with **R4.00** rounds (add the fillets to the sketch). Sketch on datum plane **A**, and center the first protrusion horizontally on datum **B** and vertically on datum **C** [Figs. 17.7(a-b)].

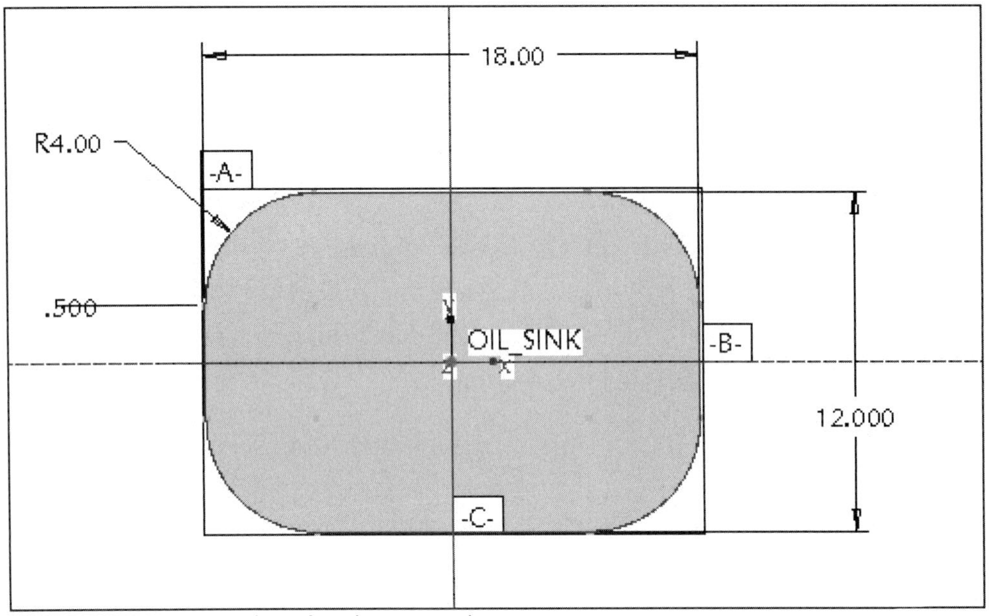

Figure 17.7(a) Dimensions for First Protrusion

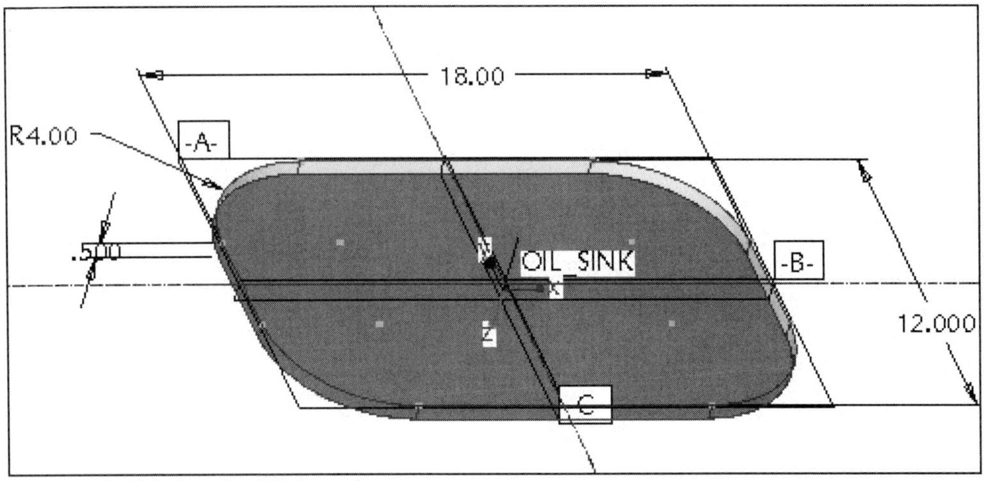

Figure 17.7(b) Standard Orientation

Make the second protrusion offset from the edge of the first protrusion **3.00**, with a height of **7.00** [Figs. 17.8(a-b)]. Sketch on top surface of the first protrusion. Then, create the cut [Figs. 17.9(a-b)].

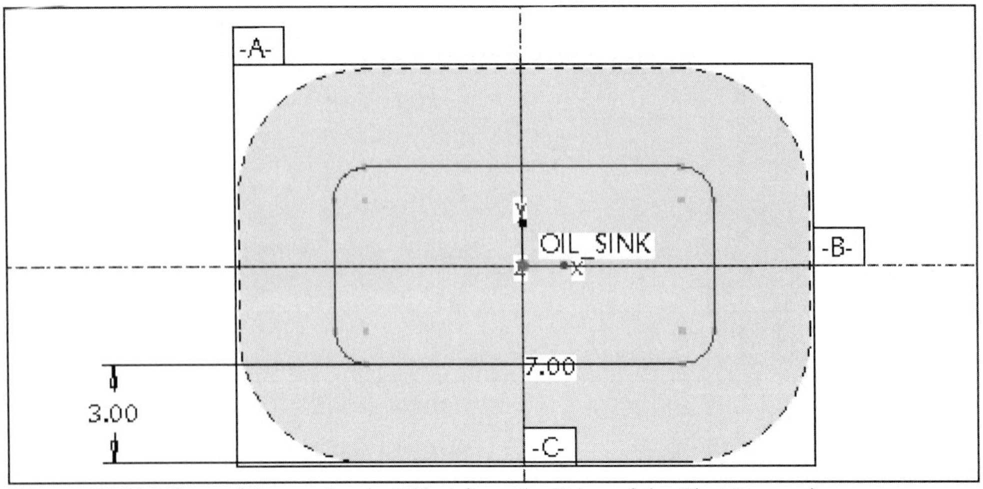

Figure 17.8(a) Second Protrusion is Offset from the Edge of the First Protrusion

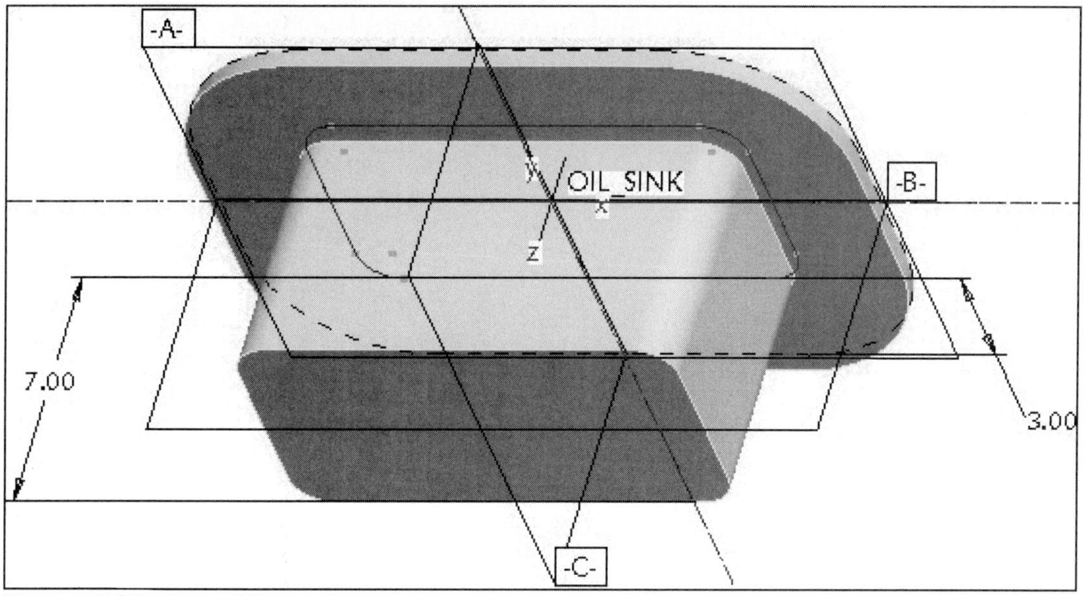

Figure 17.8(b) Second Protrusion

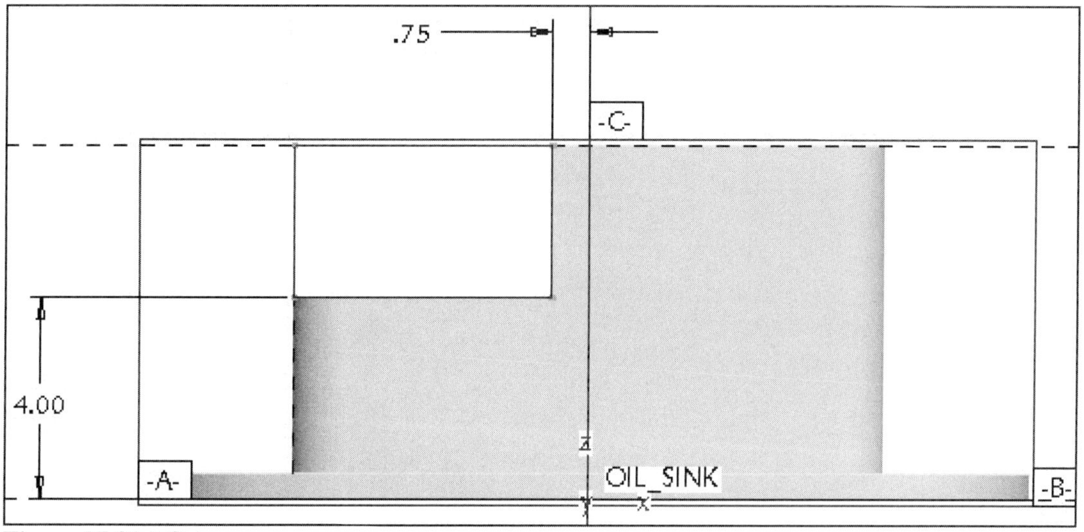

Figure 17.9(a) Cut

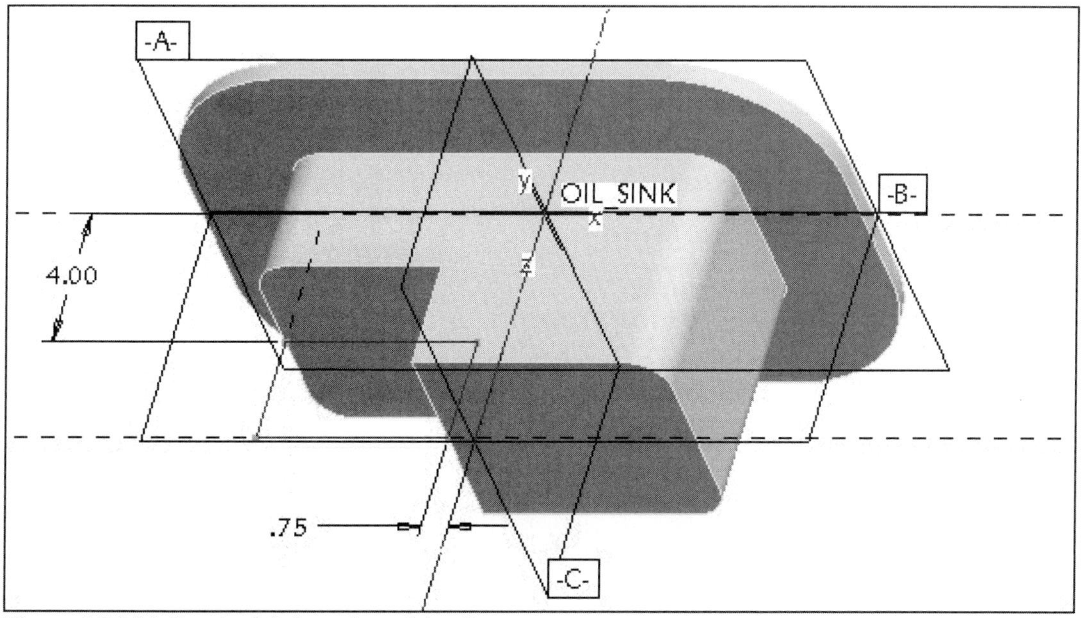

Figure 17.9(b) Standard Orientation of the Cut

Add the **R1.50** rounds [Figs. 17.10(a-b)]. Draft all vertical surfaces of the second protrusion **10** degrees. Use the top surface as the Draft hinge [Figs. 17.11(a-b)].

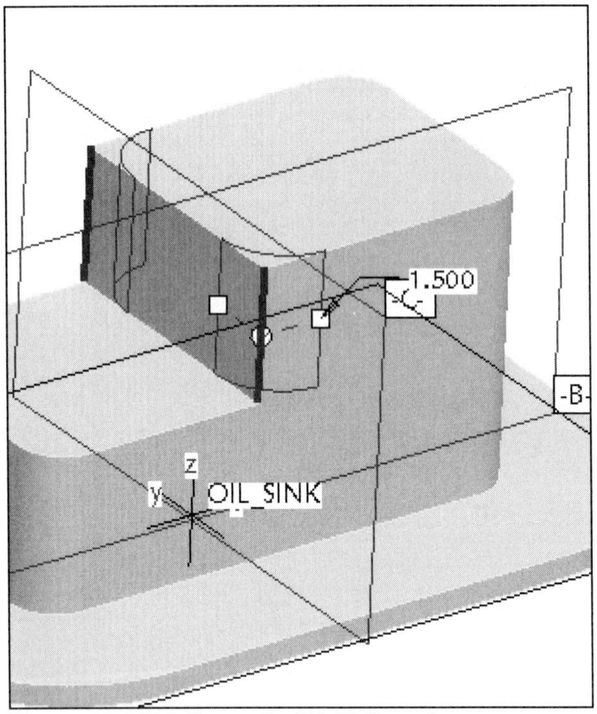

Figure 17.10(a) Create the **R1.50** Rounds

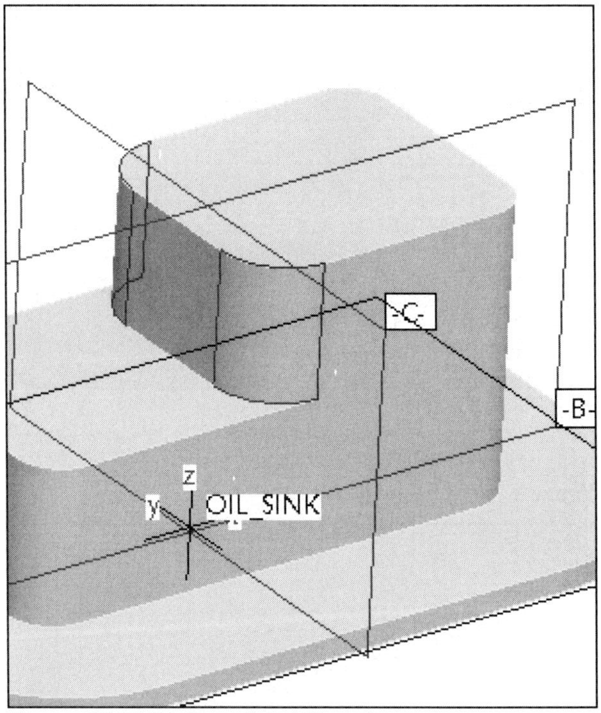

Figure 17.10(b) Completed Rounds

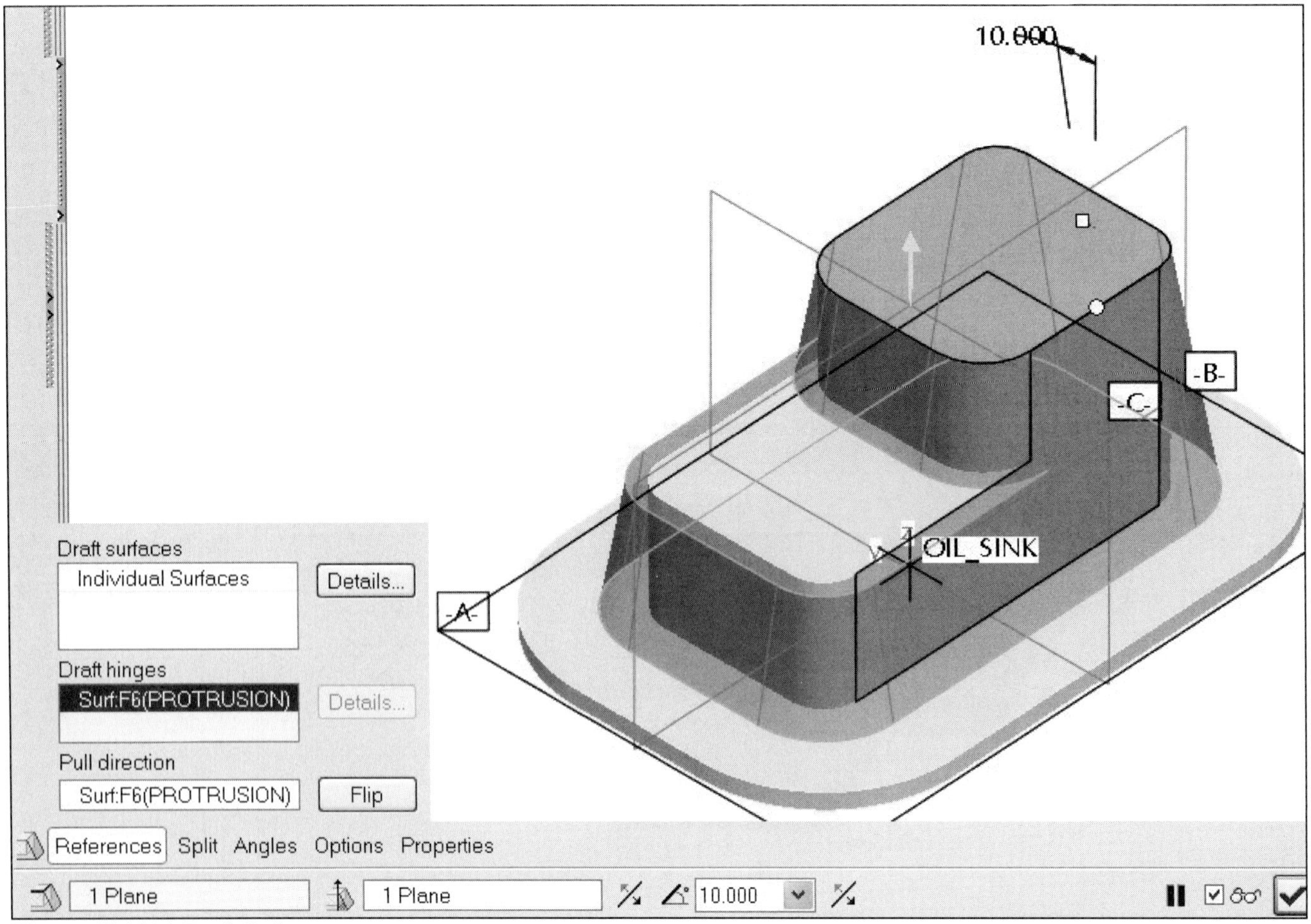

Figure 17.11(a) Draft References

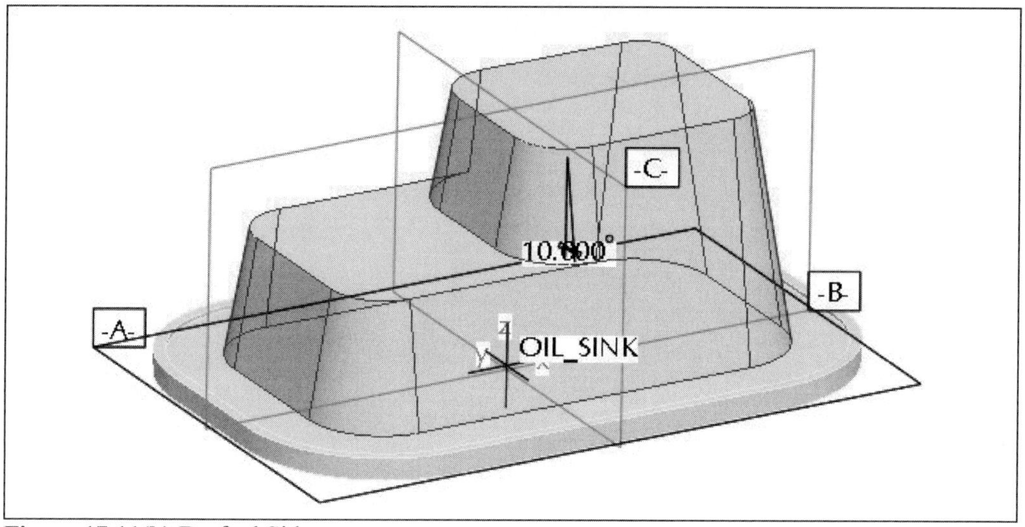

Figure 17.11(b) Drafted Sides

Click: **Context sensitive help** ⇒ ⇒ read about the Shell Tool ⇒ **Close** ⇒ **Shell Tool** ⇒ Thickness **.375** ⇒ **Enter** ⇒ spin the model ⇒ **References** tab ⇒ Removed surfaces**--** select the bottom surface of the part [Fig. 17.12(a)] ⇒ ⇒ ⇒ **MMB** ⇒ **LMB** ⇒ ⇒ ⇒ ⇒ **MMB** ⇒ **File** ⇒ **Delete** ⇒ **Old Versions** ⇒ **MMB** [Fig. 17.12(b)]

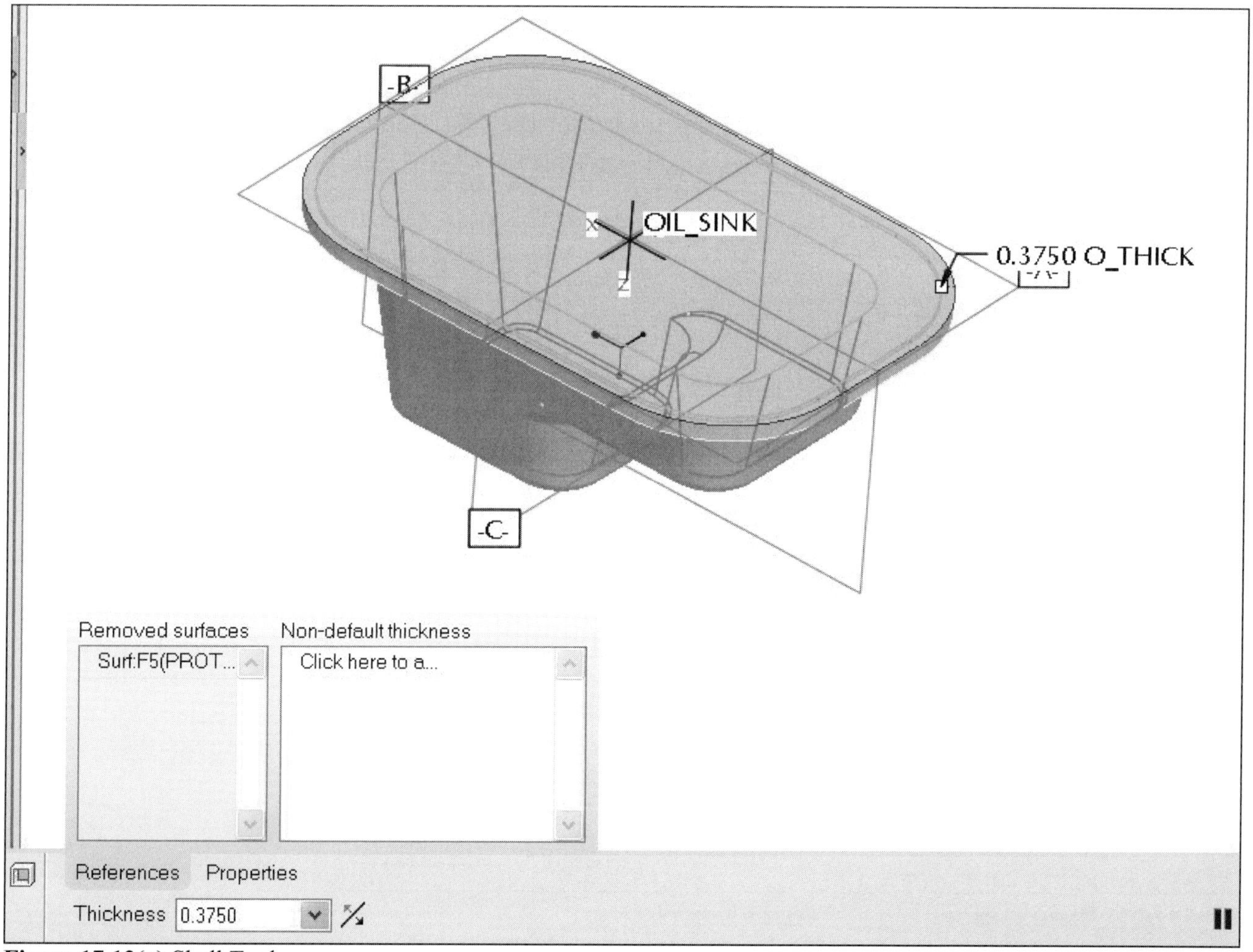

Figure 17.12(a) Shell Tool

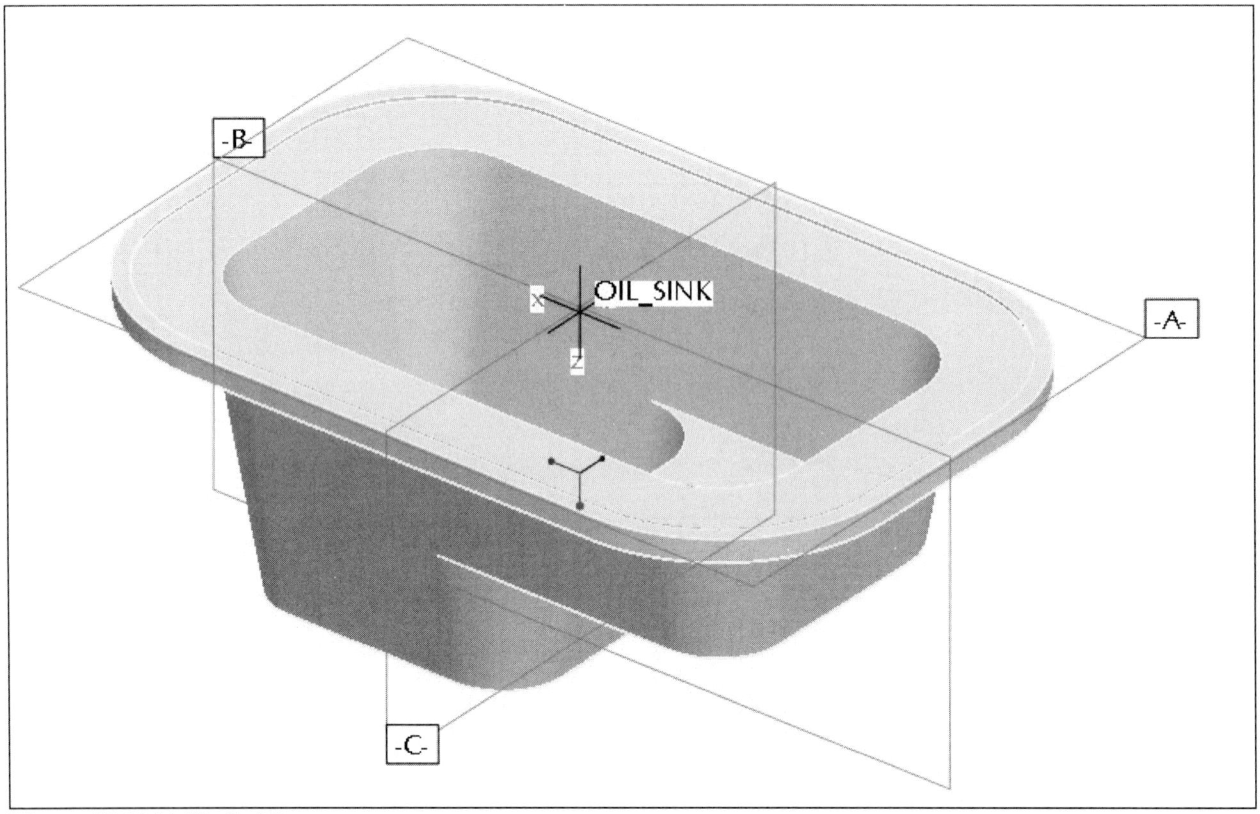

Figure 17.12(b) Shelled Part

The next feature you need to create is a *"lip"* around the part using a protrusion ⇒ Sketch two closed loops [Fig. 17.13(a)]. Use the edge of the first protrusion for the first loop and then create an offset edge (**-.3125**) for the second loop [Fig. 17.13(b)]. ⇒ The depth of the lip protrusion is **.125** [Figs. 17.13(c-d)].

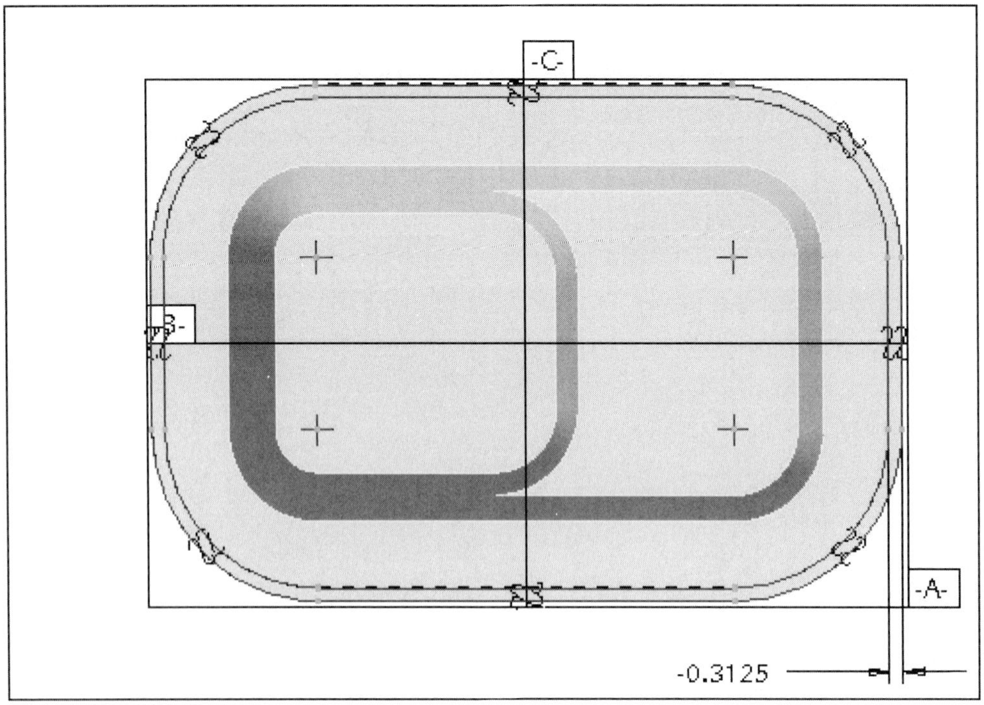

Figure 17.13(a) Sketch

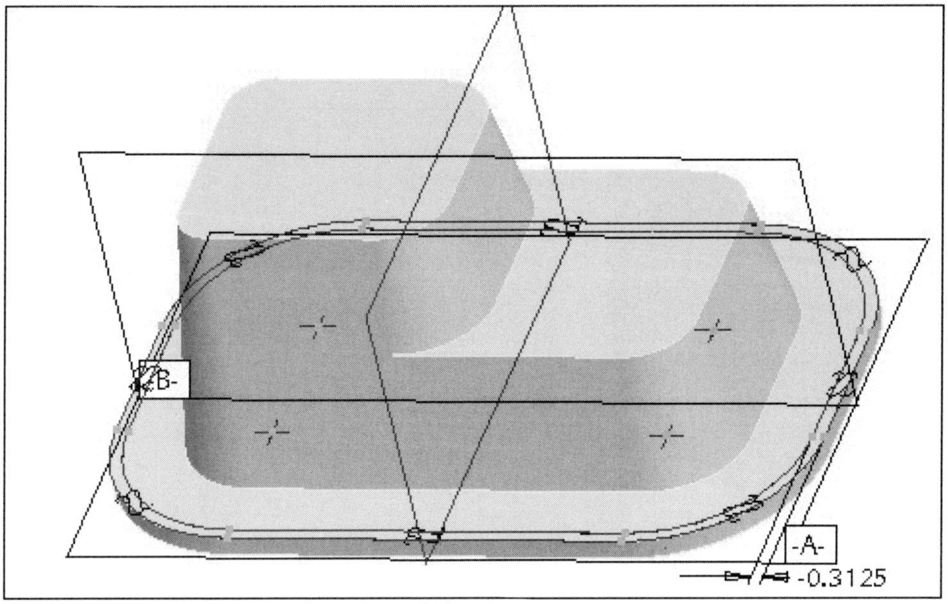

Figure 17.13(b) Standard Orientation of Sketch

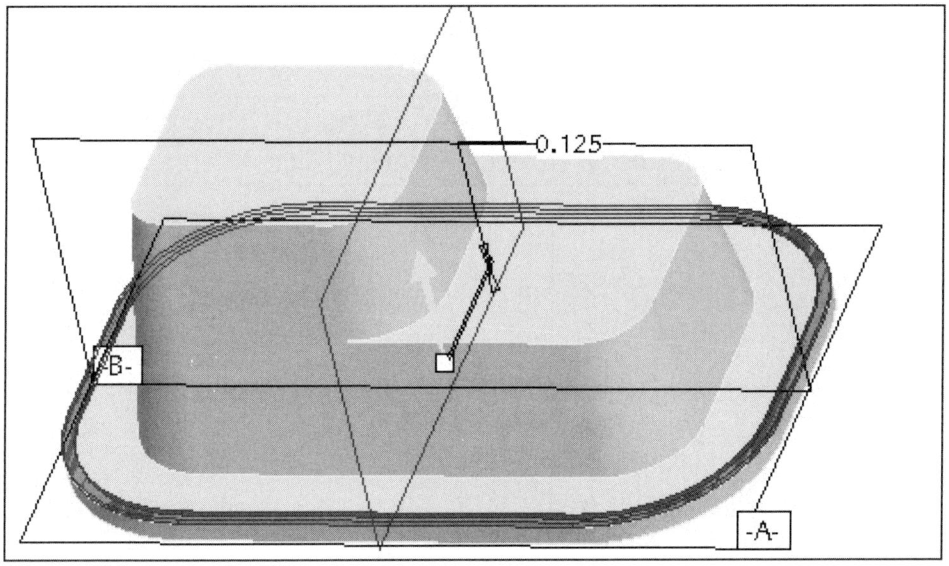

Figure 17.13(c) Depth **.125**

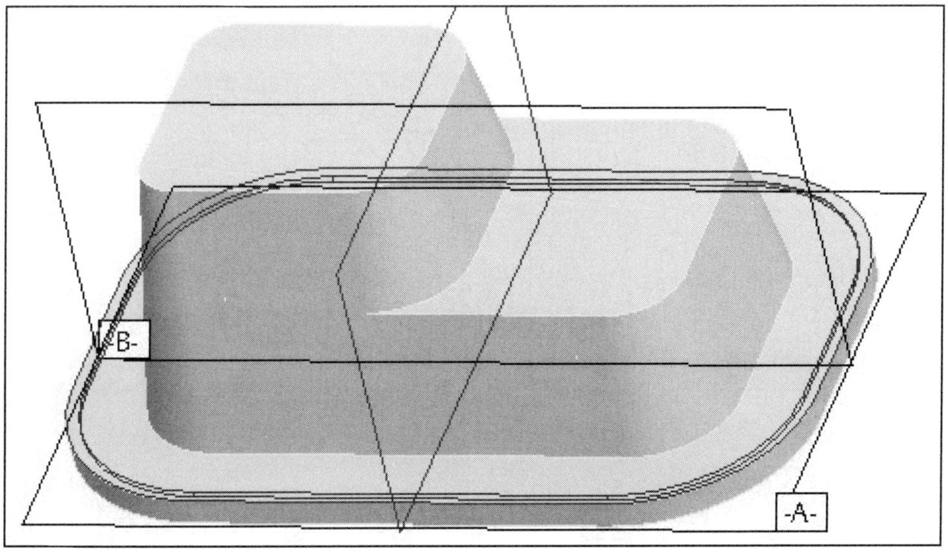

Figure 17.13(d) Completed "Lip" Protrusion

Click: **Insert ⇒ Round ⇒** add the **R.125** round to the inside of the *"lip"* (Fig. 17.14) ⇒ [save icon] ⇒ **MMB**

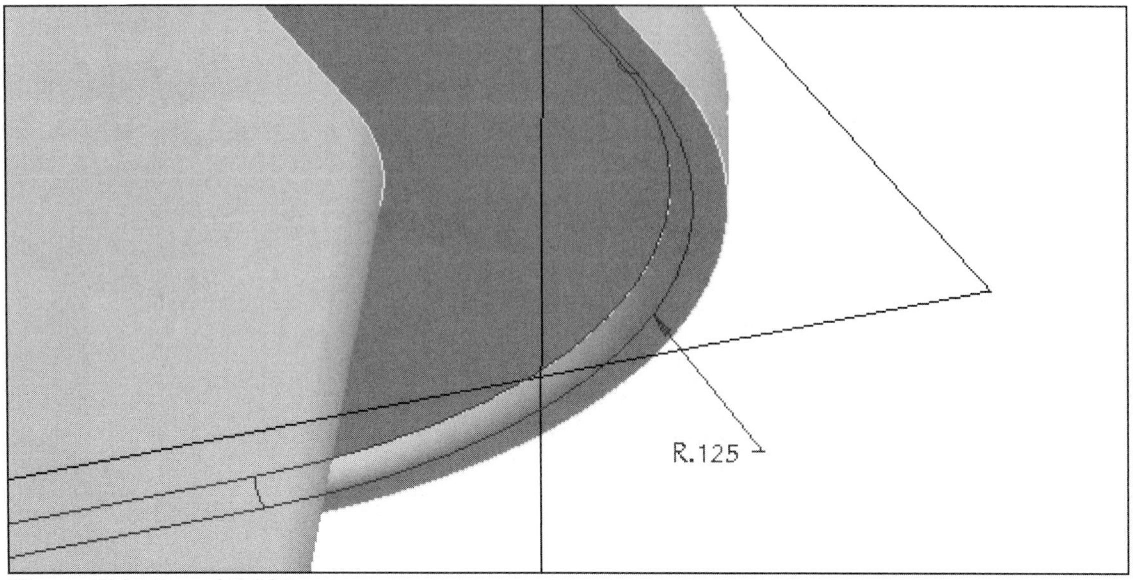

Figure 17.14 Round **R.125**

The next feature is a cut measuring **.9185** wide by **.187** deep [Figs. 17.15(a-b)]. The sketch will be composed of two closed loops as with the lip-like protrusion created previously [Figs. 17.16(a-c)].

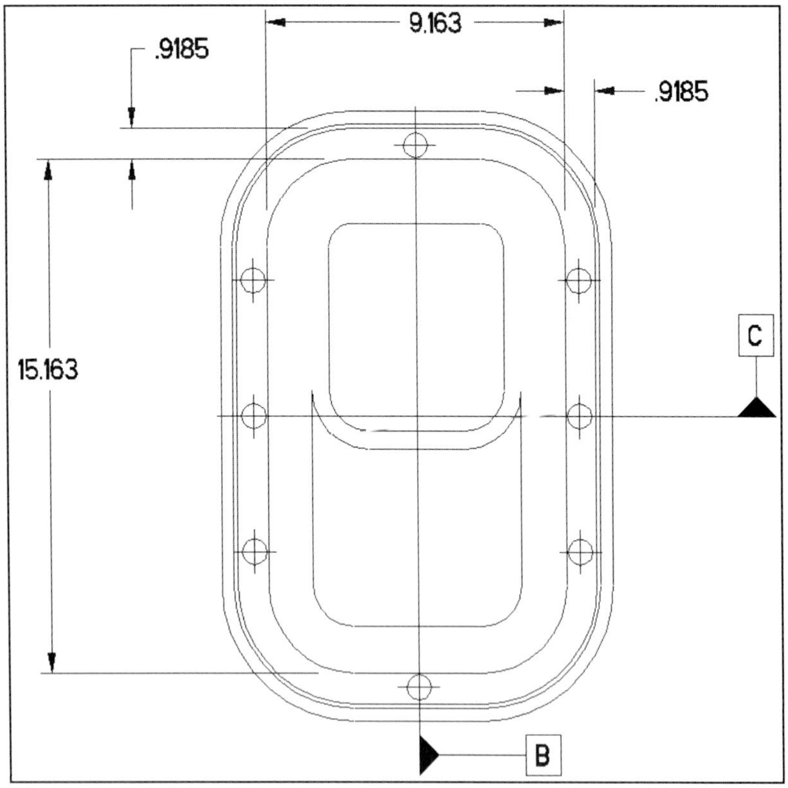

Figure 17.15(a) Cut Dimensions

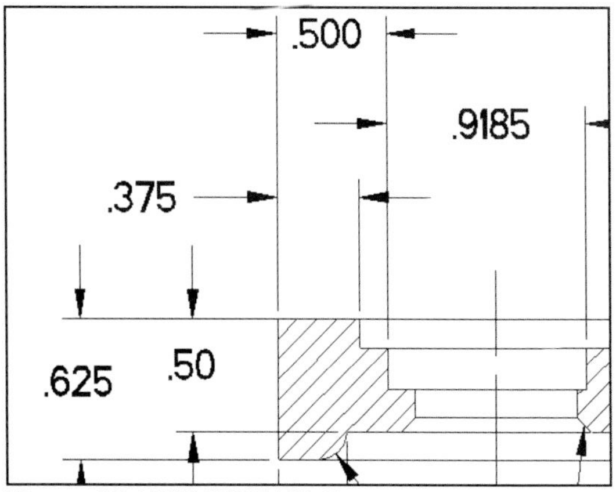

Figure 17.15(b) DETAIL B

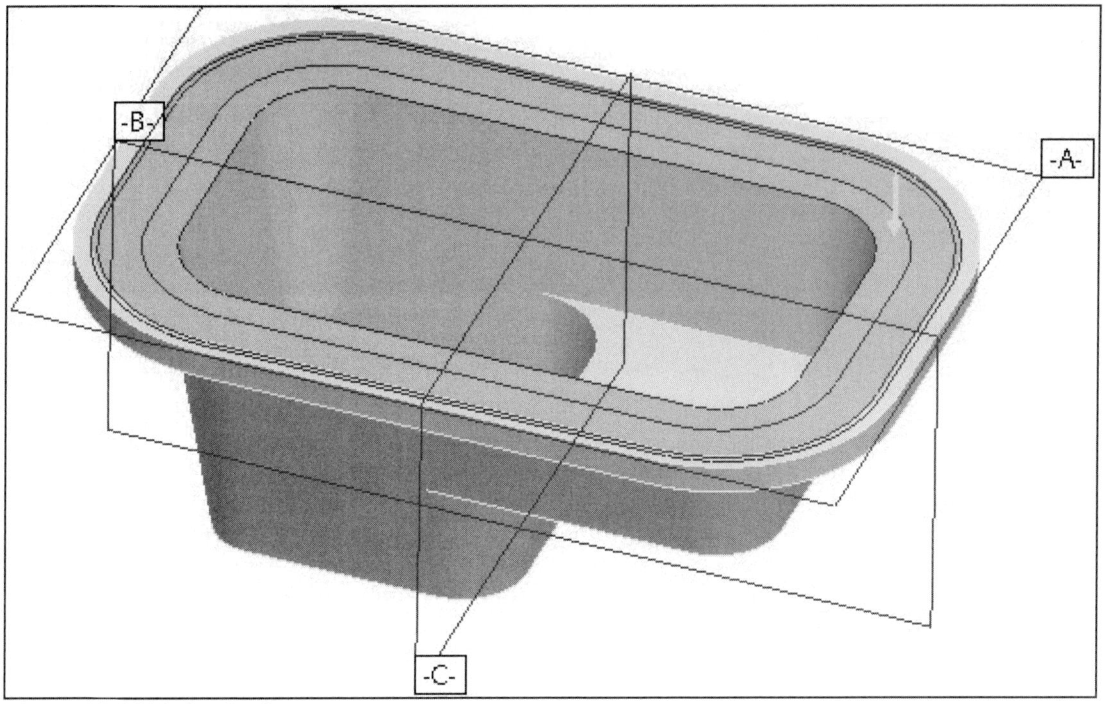

Figure 17.16(a) Cut Surface

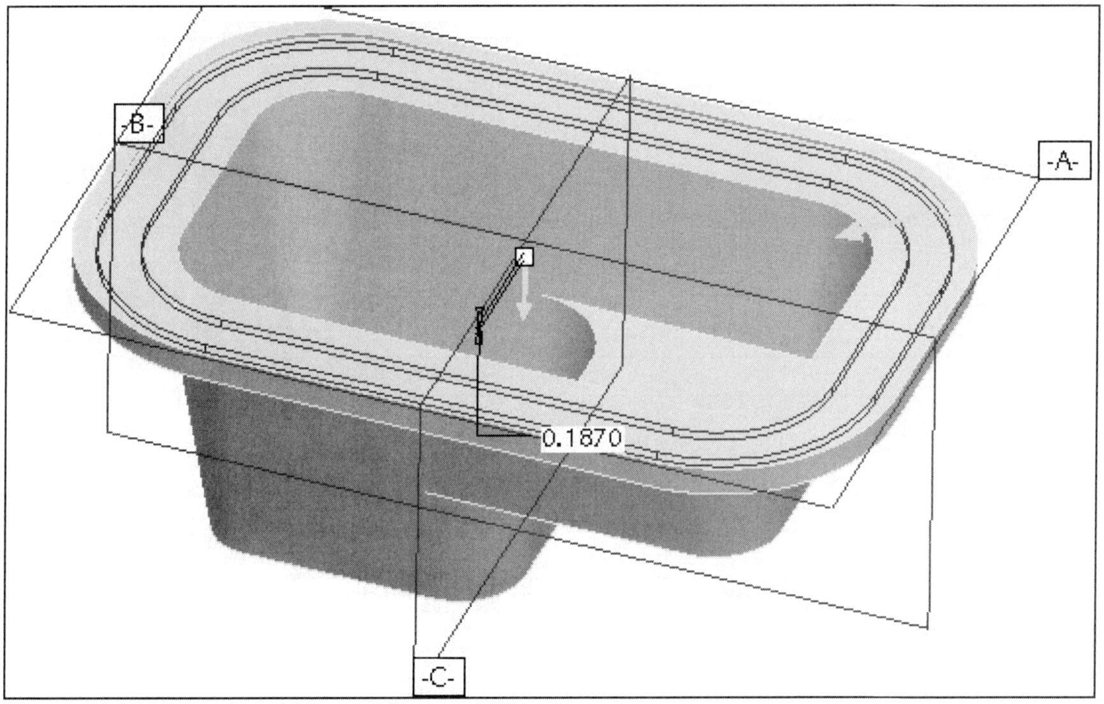

Figure 17.16(b) Depth and Material Removal Direction

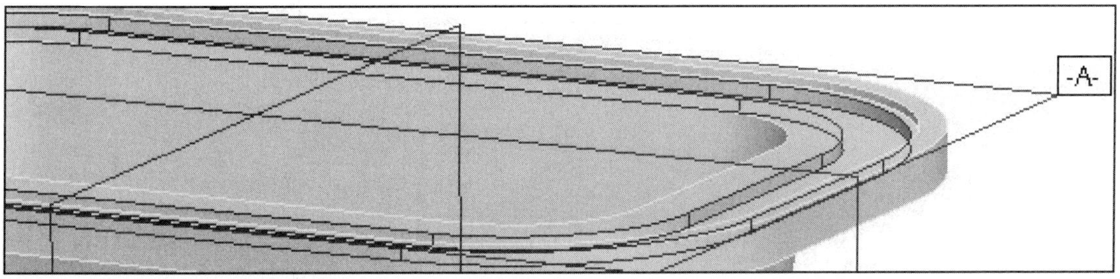

Figure 17.16(c) Completed Cut

Click: **Insert** ⇒ **Round** ⇒ add another **R.125** round to the inside edge [Figs. 17.17(a-b)] ⇒ ⇒ **MMB**

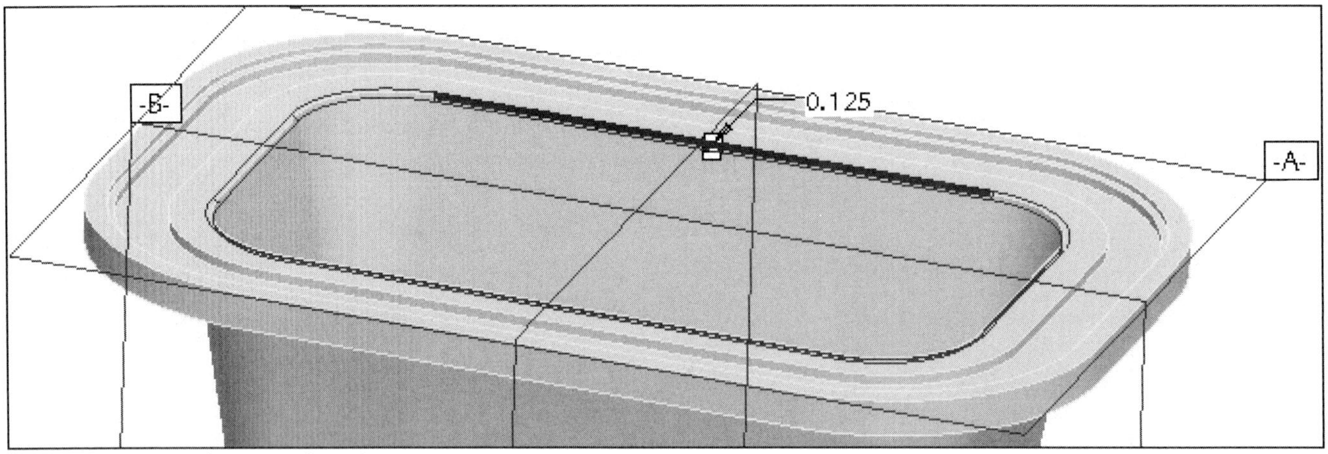

Figure 17.17(a) Edge Round **R.125**

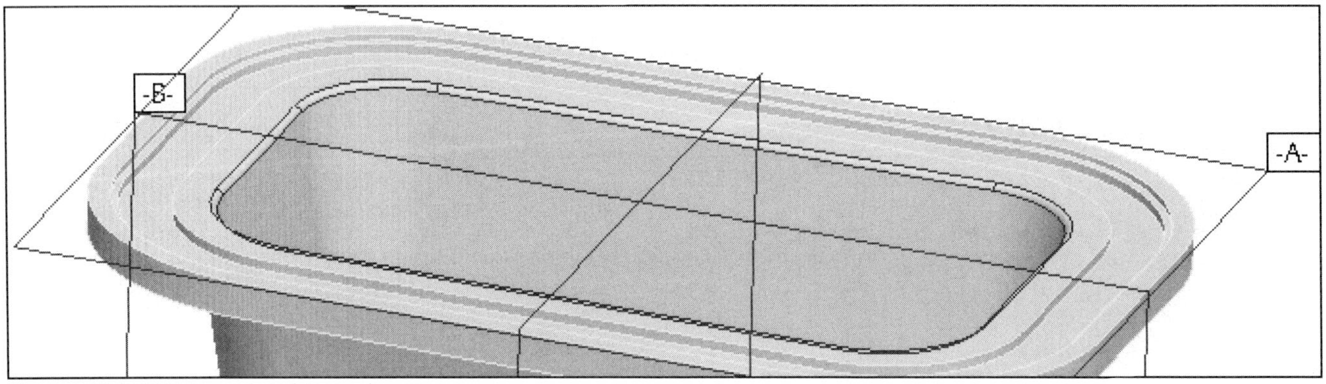

Figure 17.17(b) Completed Round

Click: **Insert** ⇒ **Round** ⇒ create a **R.250** round as shown in Figure 17.18 ⇒ **Ctrl+S** ⇒ **MMB**

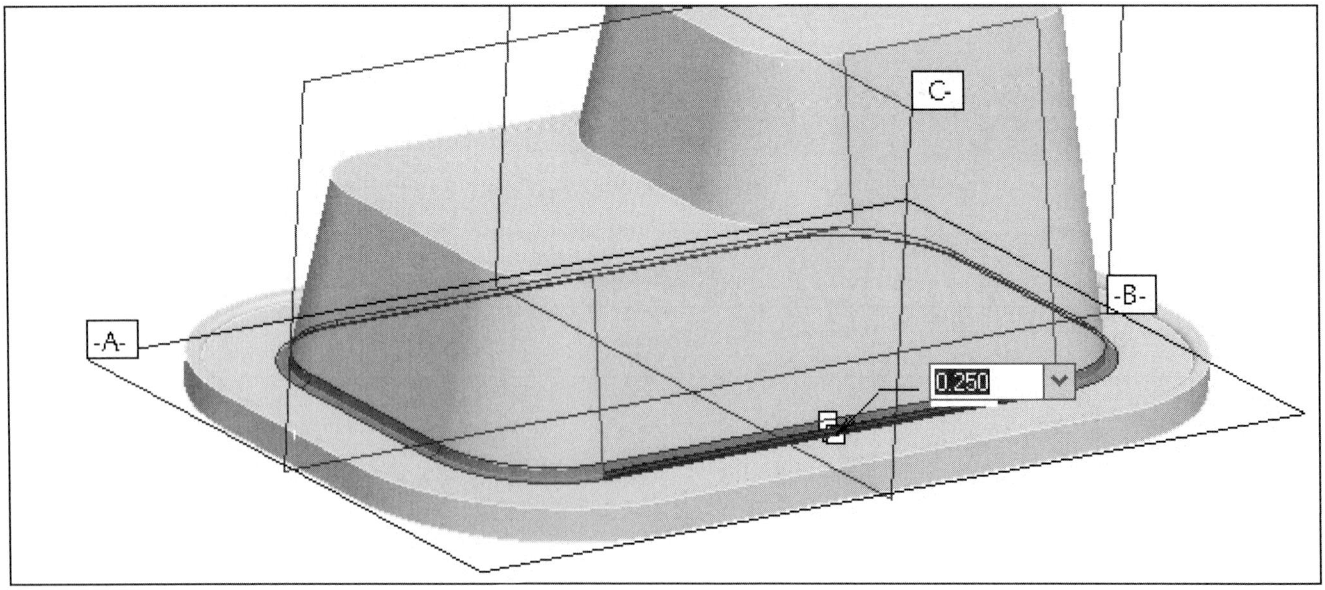

Figure 17.18 Round **R.250**

The countersunk holes will be added next [Figs. 17.19(a-b)].

Click: **Hole Tool** from Right Toolchest ⇒ spin the part ⇒ **Drill to intersect with all surfaces** ⇒ change the diameter to **.750** ⇒ **Placement** tab ⇒ select the location on the surface for hole placement [Fig. 17.19(c)]

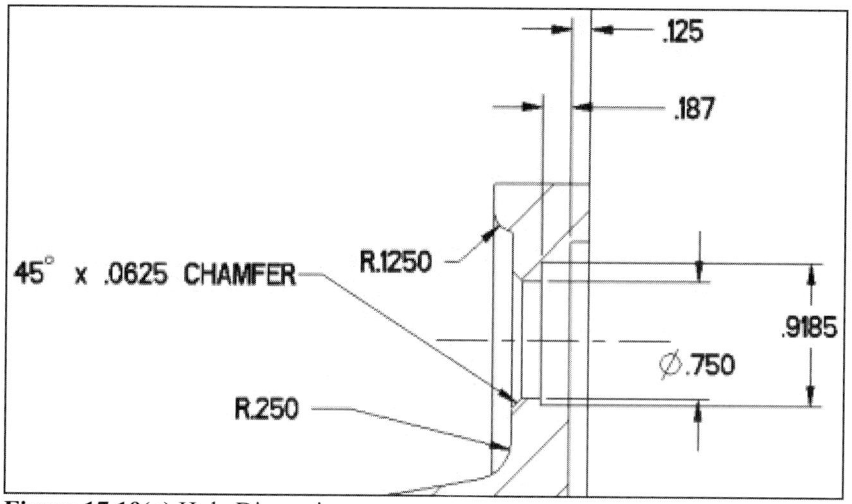

Figure 17.19(a) Hole Dimensions

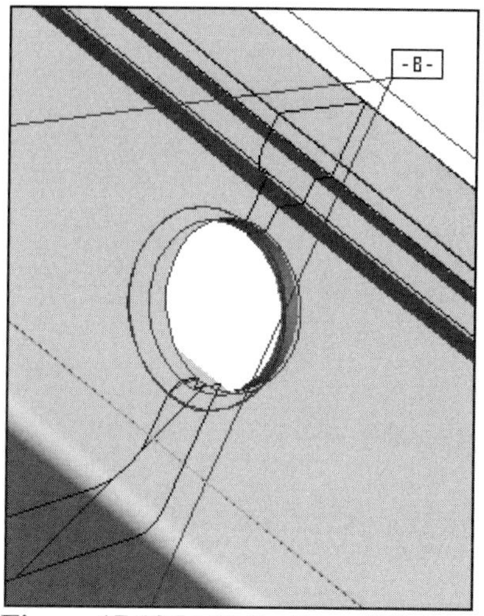

Figure 17.19(b) X-Section of Hole

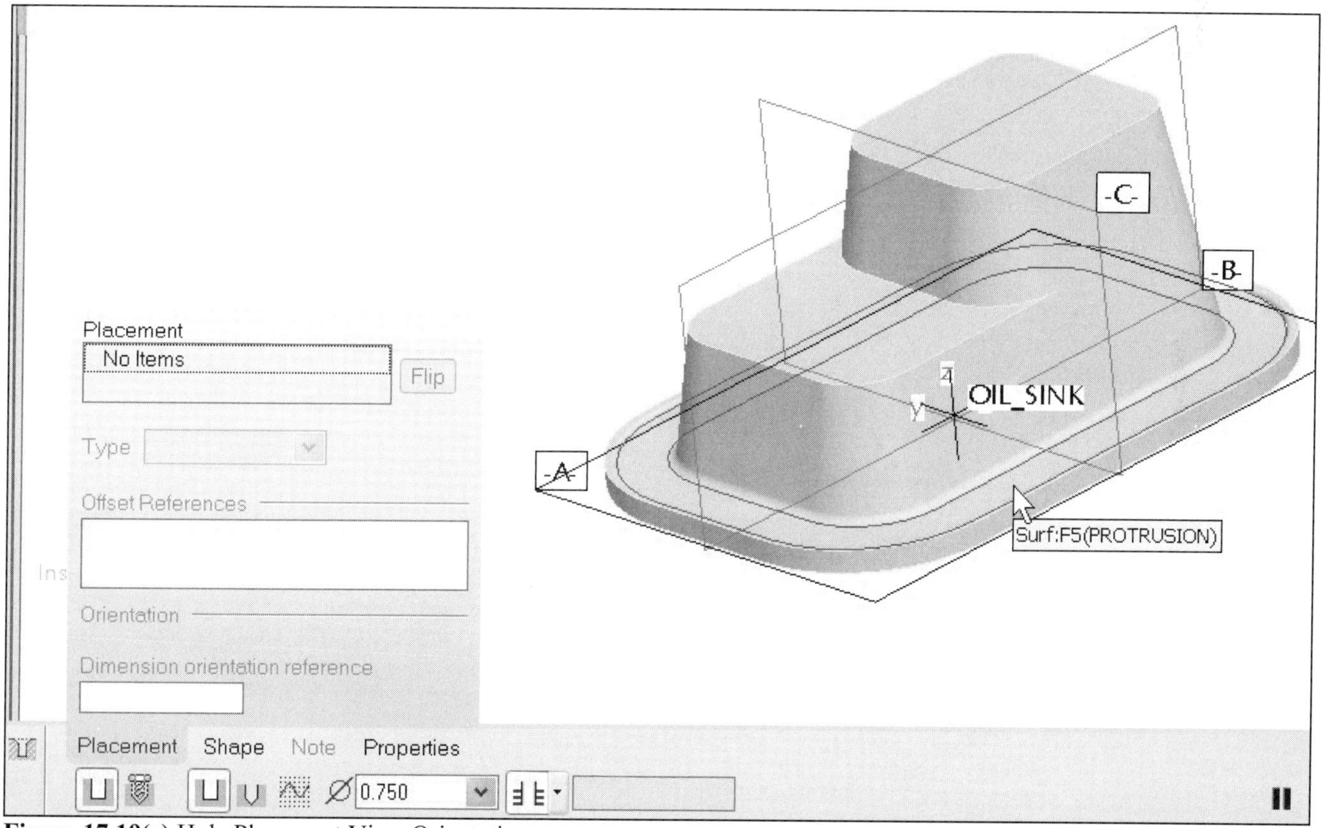

Figure 17.19(c) Hole Placement View Orientation

Pick on a drag handle *(green)* and move [Fig. 17.19(d)] to datum **C** and move the other drag handle to datum **B** [Fig. 17.19(e)]

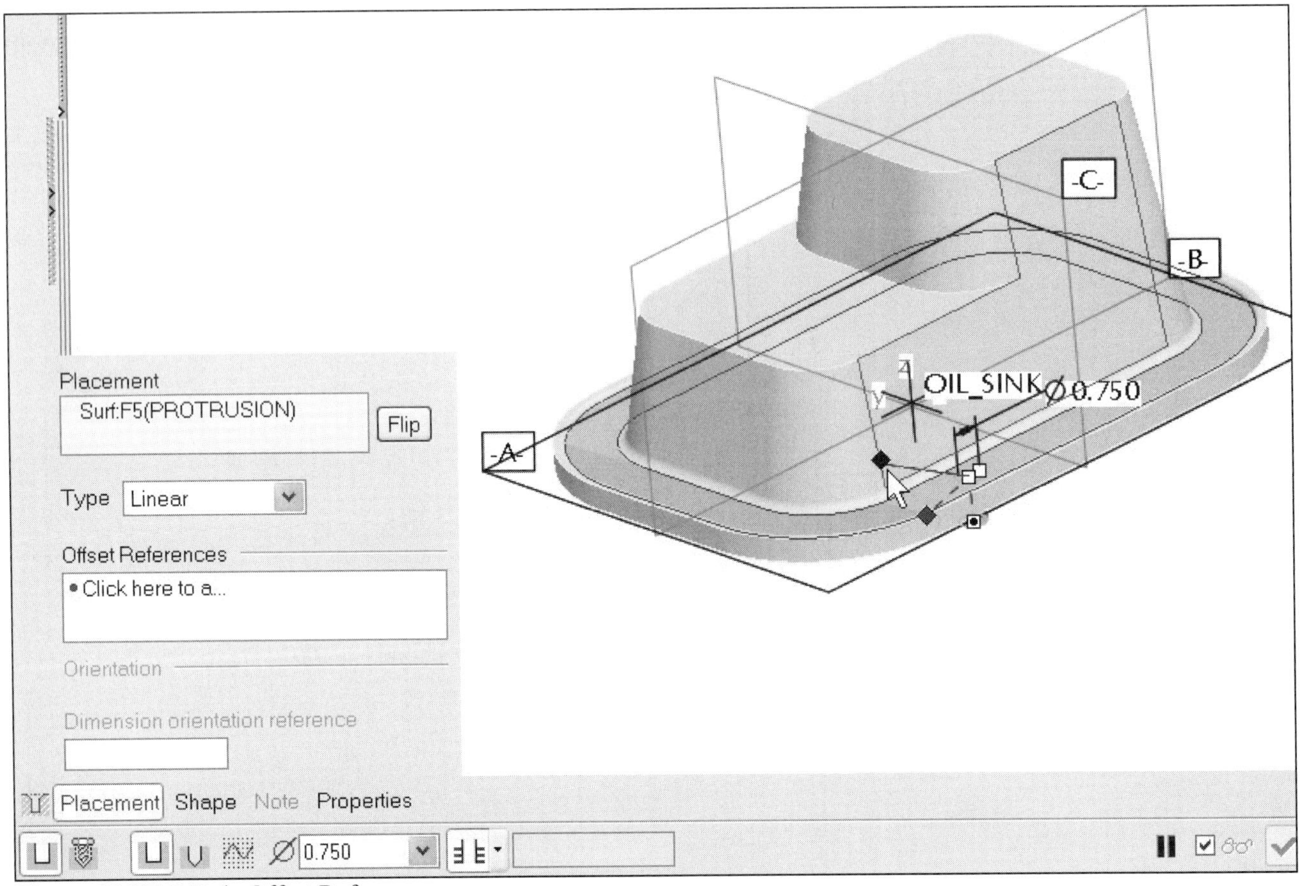

Figure 17.19(d) Hole Offset References

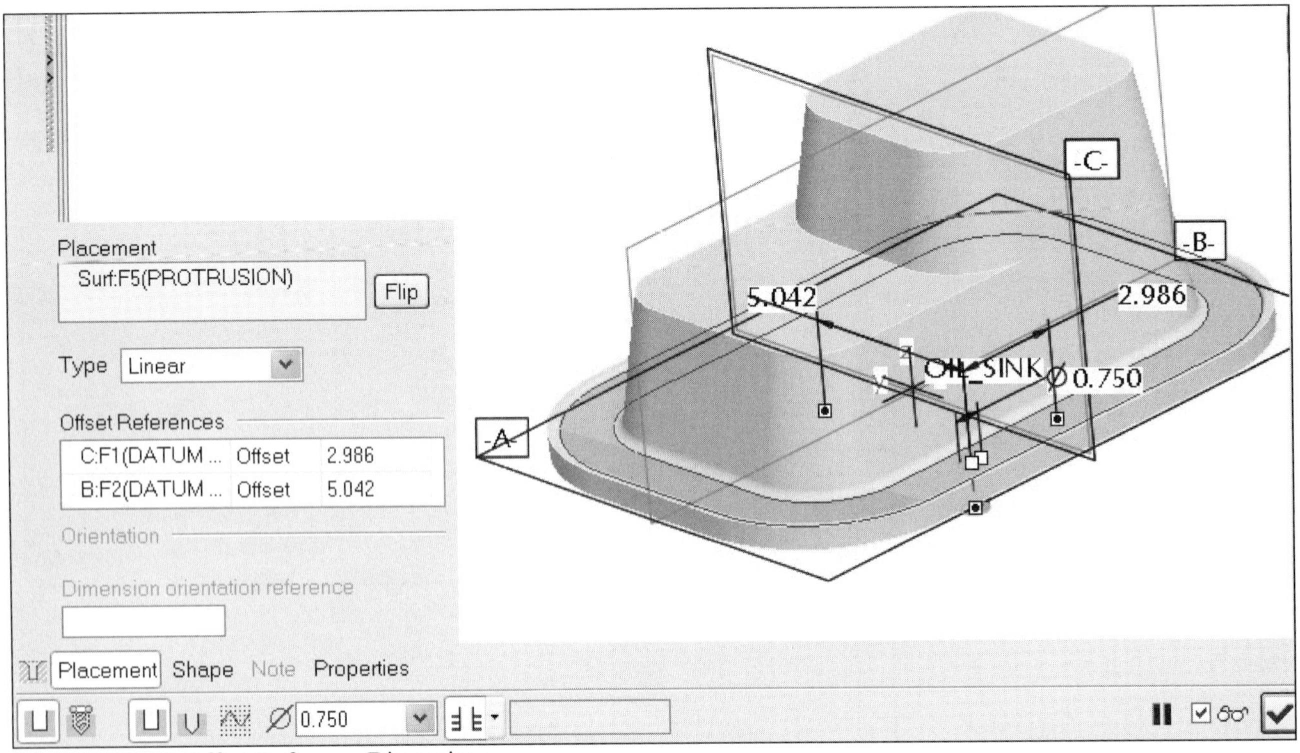

Figure 17.19(e) Offset Reference Dimensions

Modify the values to be **4.00** from datum **C** and **5.00** from datum **B** [Fig. 17.19(f)]

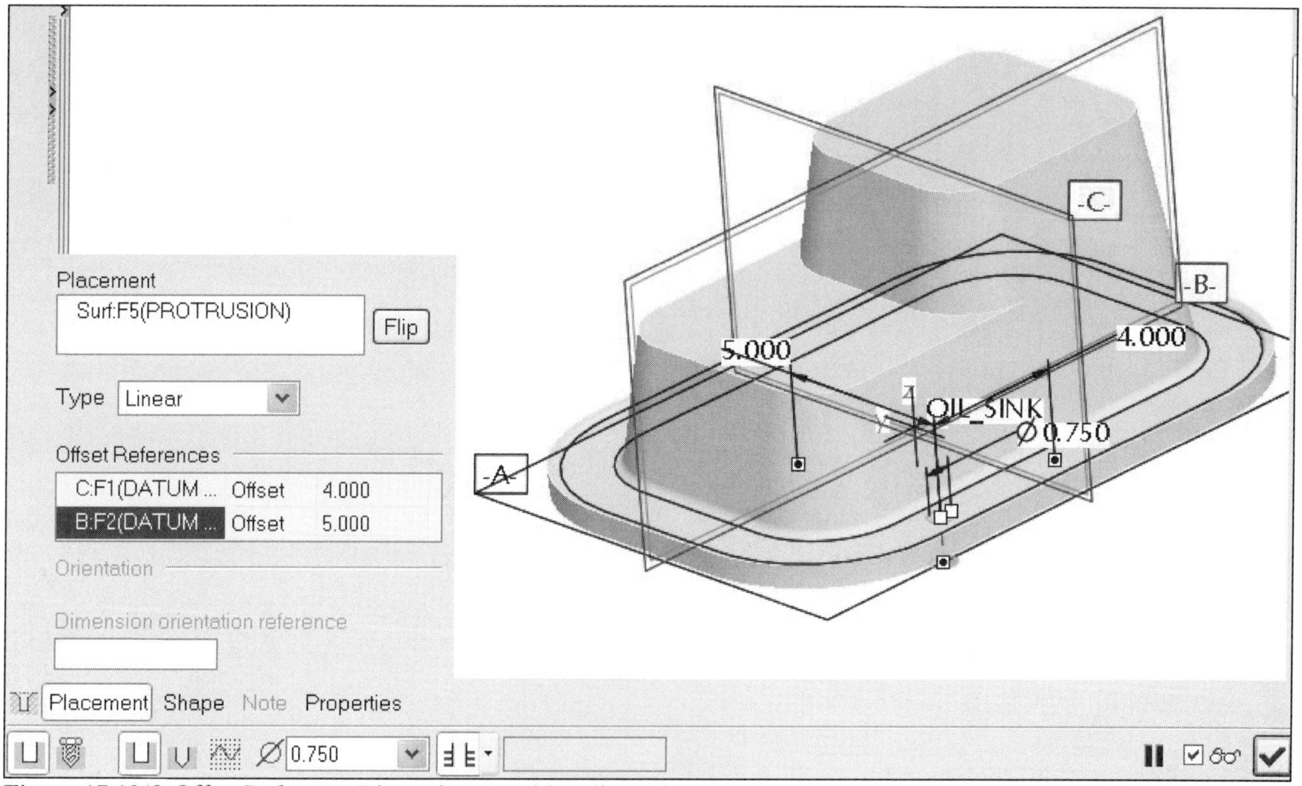

Figure 17.19(f) Offset Reference Dimensions (position dimensions)

Hole dashboard settings: 

⇒ **Shape** tab ⇒ [Fig. 17.19(g)] ⇒ ✓ ⇒ 💾 ⇒ **MMB** [Fig. 17.19(h)]

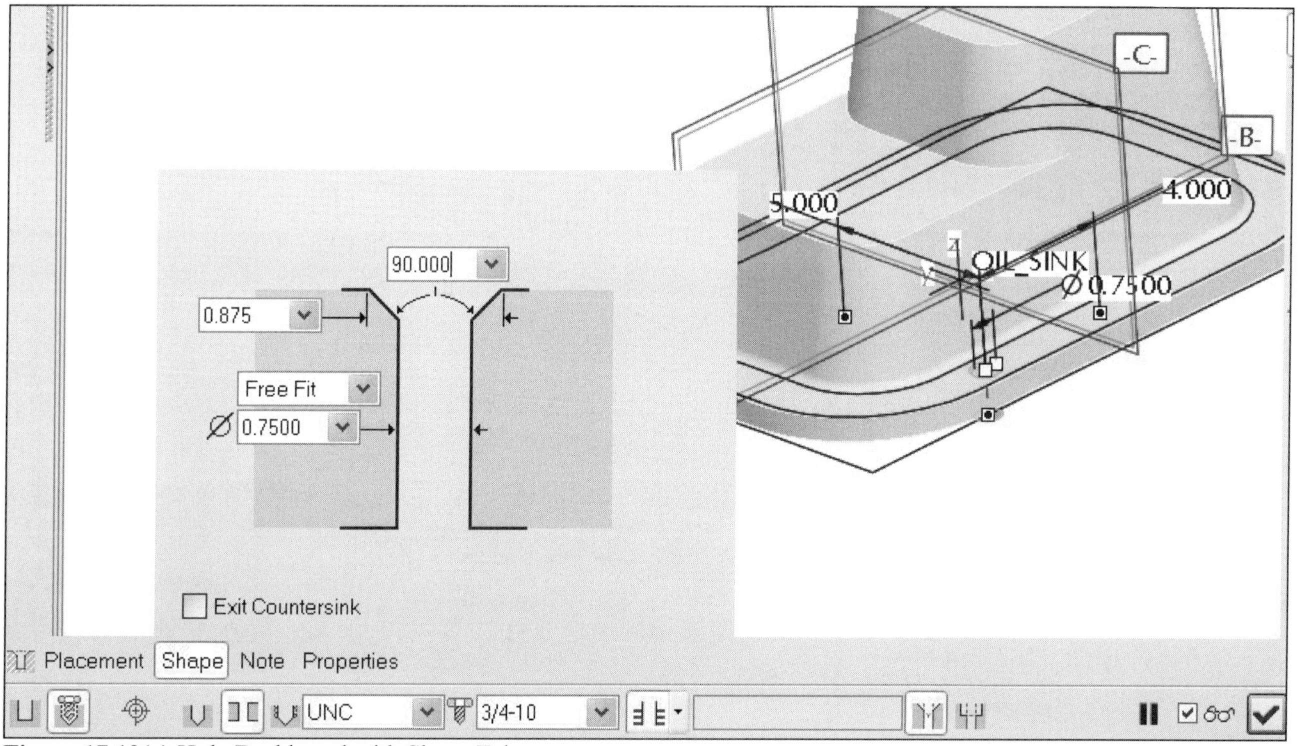

Figure 17.19(g) Hole Dashboard with Shape Tab open

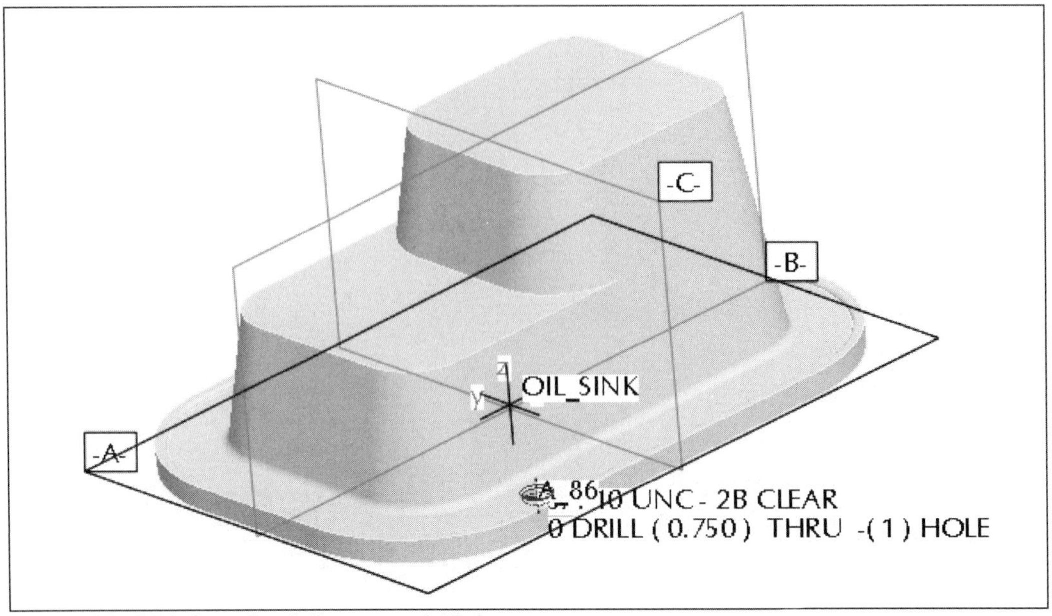

Figure 17.19(h) Completed Countersunk Hole

With the hole still selected (highlighted) ⇒ **RMB** ⇒ **Pattern** [Fig. 17.20(a)] ⇒ Dimension ⇒ ⇒ Table [Fig. 17.20(b)]

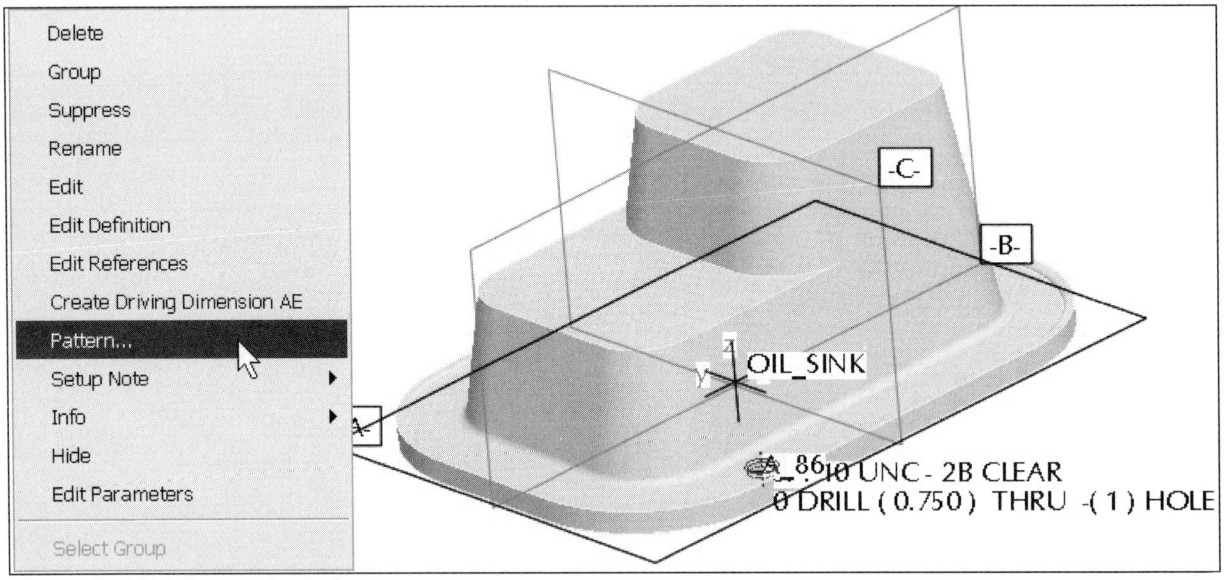

Figure 17.20(a) Pattern the Hole

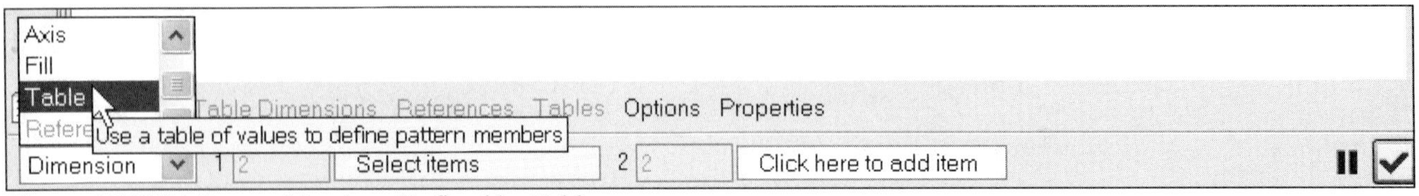

Figure 17.20(b) Pattern Members Defined by Table

With the **Ctrl** key pressed, click on **4.000** and then on **5.000** [Fig. 17.20(c)]

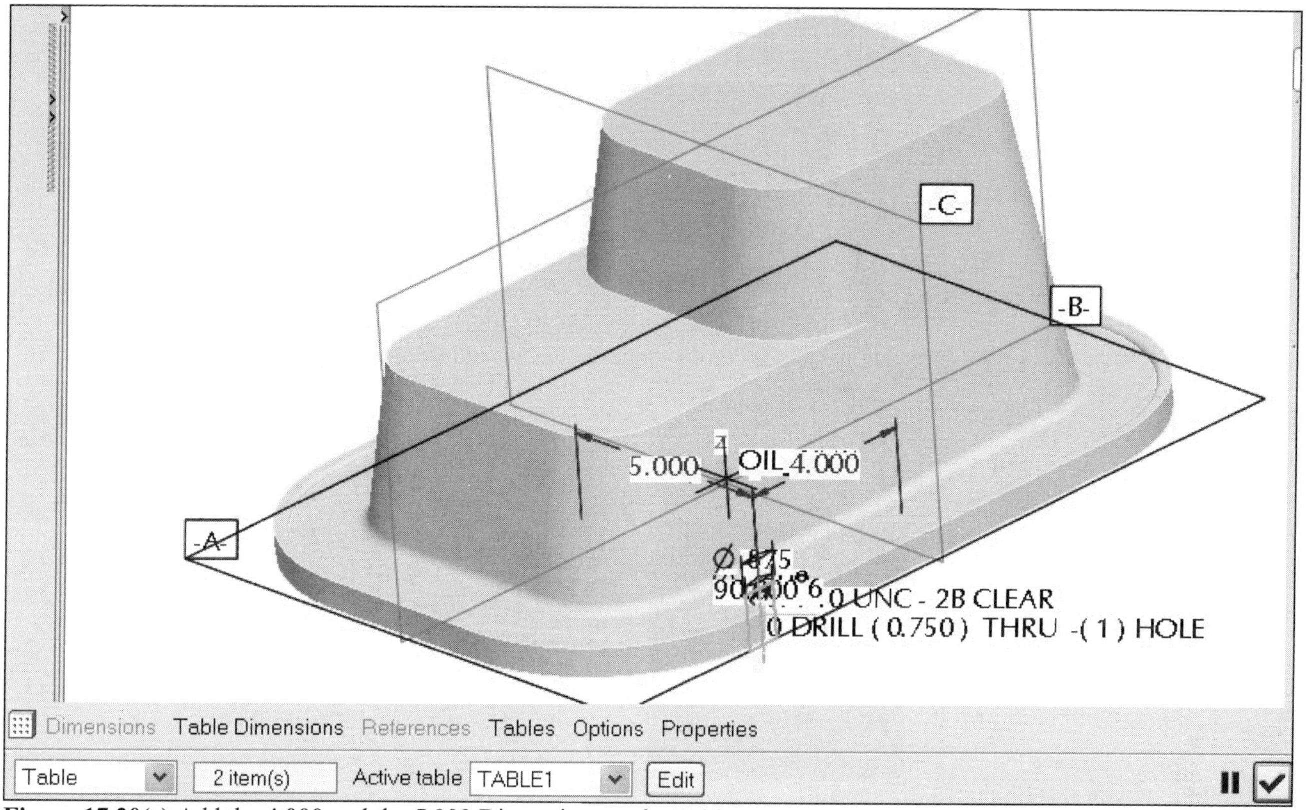

Figure 17.20(c) Add the **4.000** and the **5.000** Dimensions to the Table

Click: Edit [Fig. 17.20(d)] ⇒ add the information [Figs. 17.20(e-g)] ⇒ **File** ⇒ **Exit** [Fig. 17.20(h)]

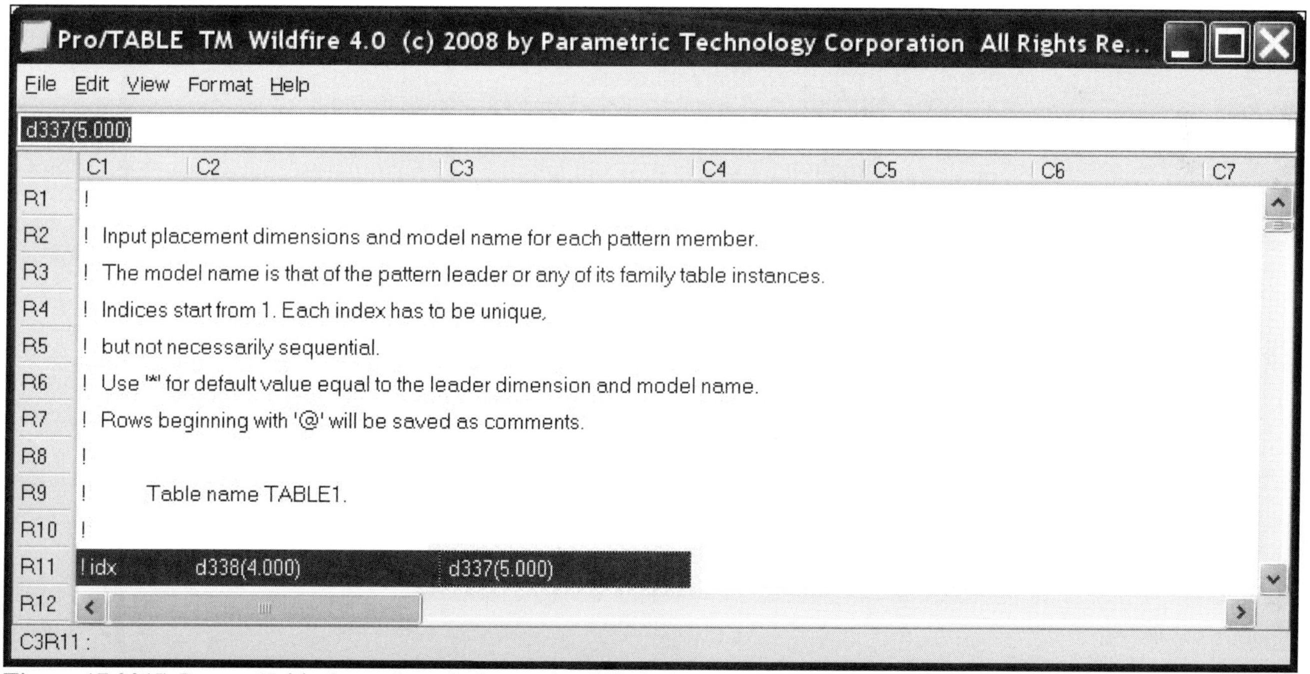

Figure 17.20(d) Pattern Table (your d symbols may be different)

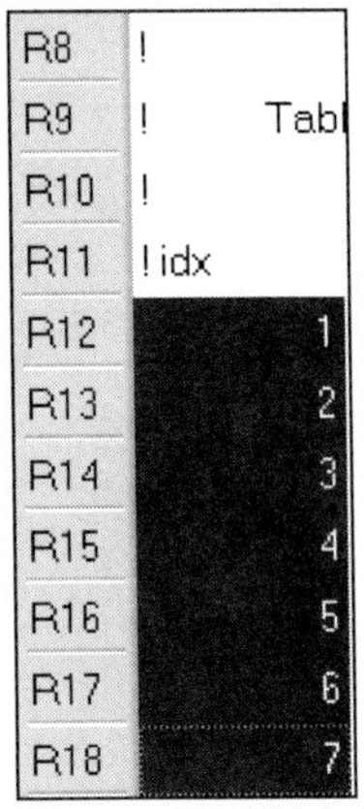

Figure 17.20(e) Add numbers 1-7

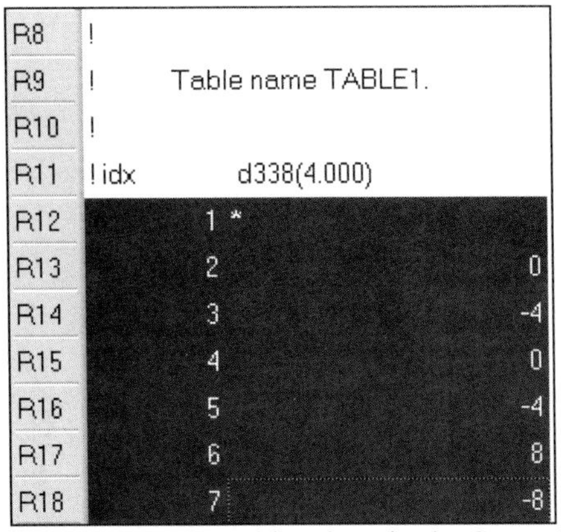

Figure 17.20(f) Add Values in the Second Column (* means identical value)

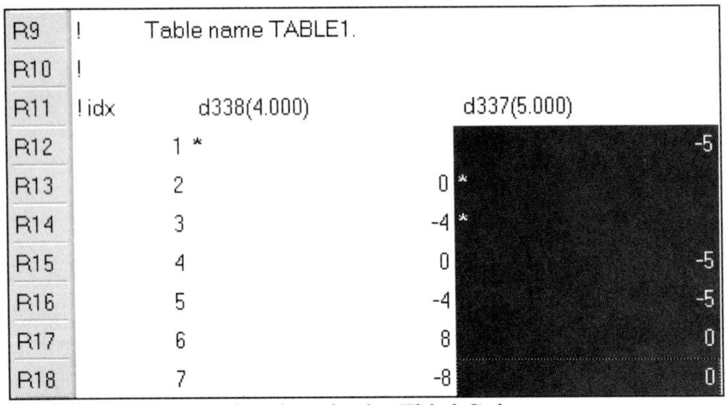

Figure 17.20(g) Add Values in the Third Column

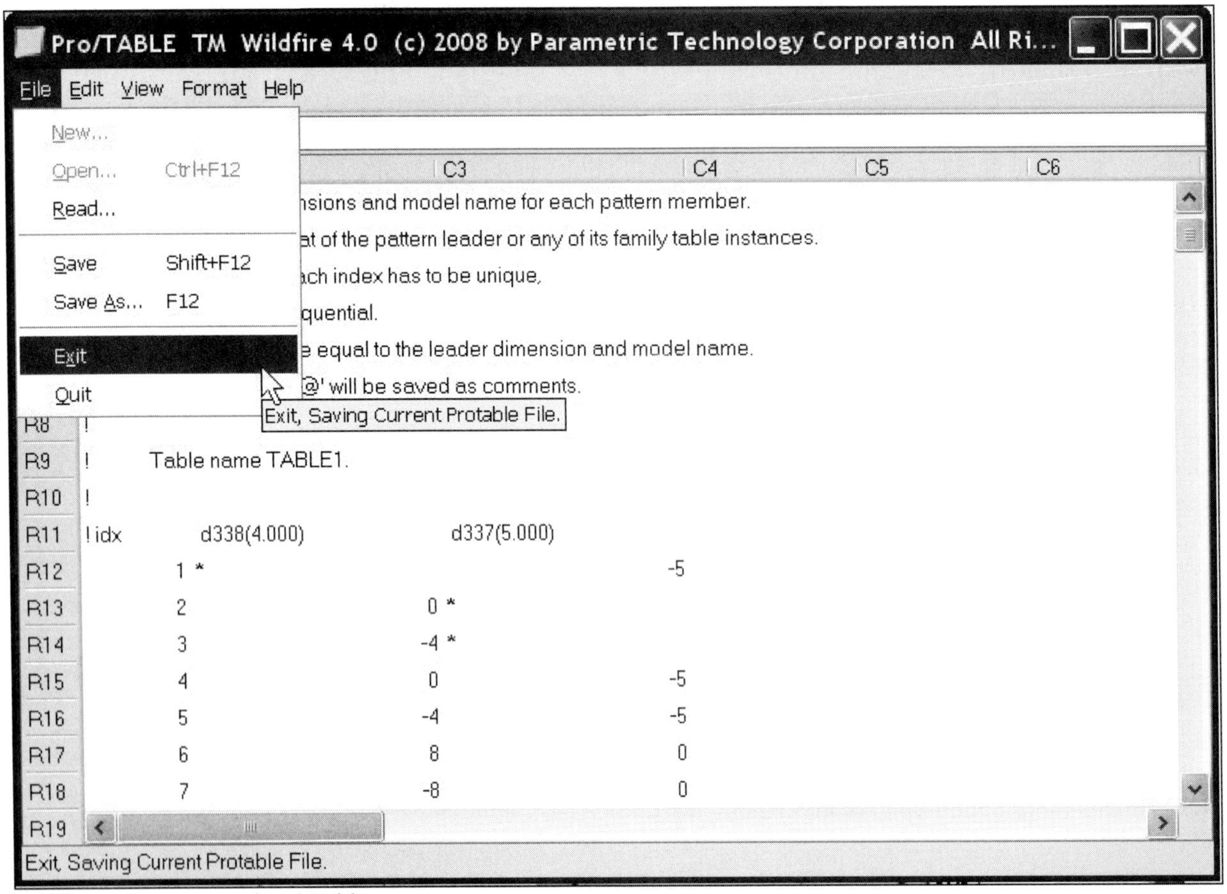

Figure 17.20(h) Completed Table

Click: ☑ [Fig. 17.20(i)] ⇒ 💾 ⇒ **MMB** [Fig. 17.20(j)] ⇒ **LMB** ⇒ check your settings in the Navigator ⇒ **Settings** ⇒ **Tree Filters** ⇒ ☑ Suppressed Objects ⇒ **OK**

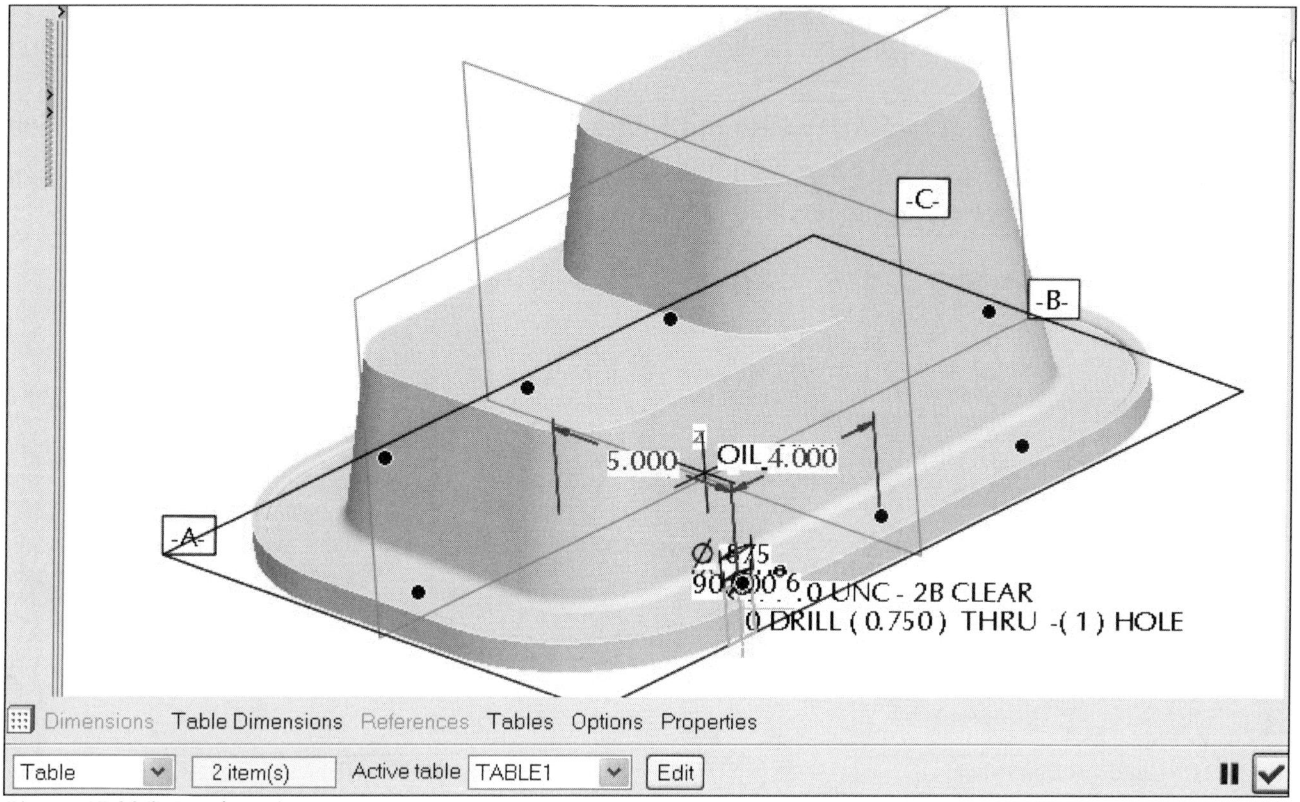

Figure 17.20(i) Previewed Pattern

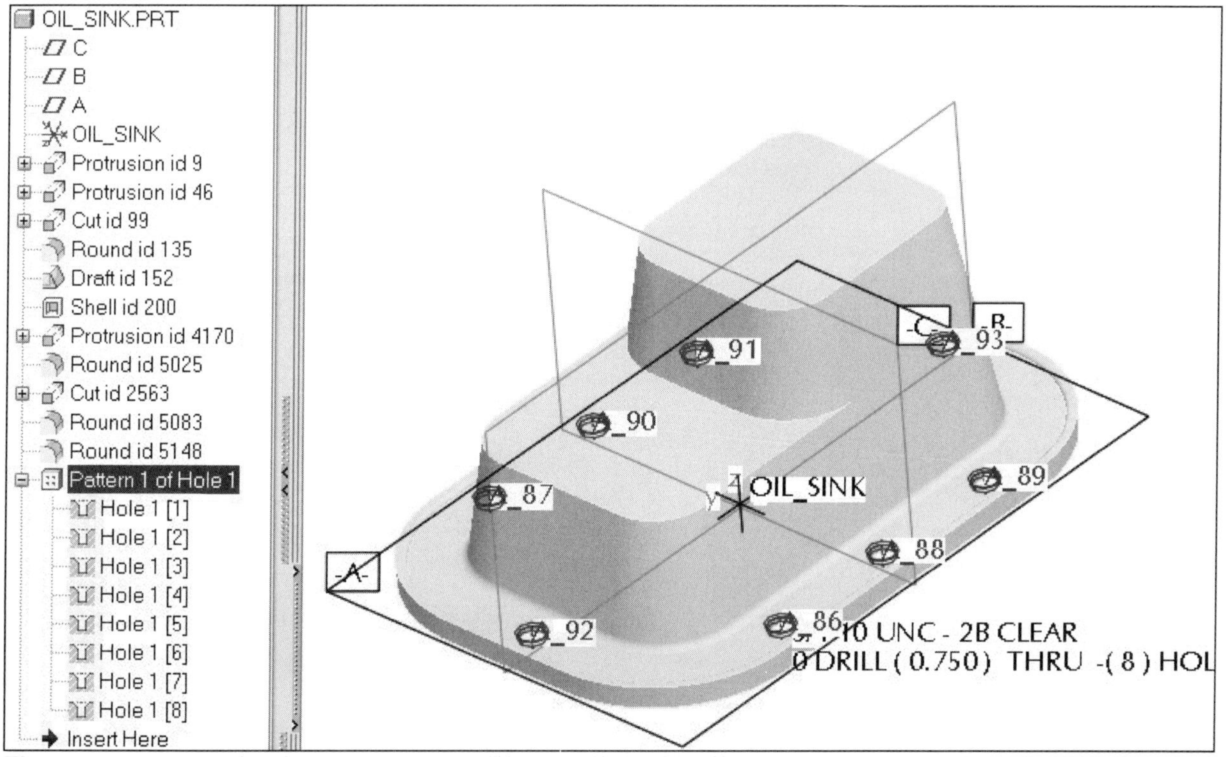

Figure 17.20(j) Completed Pattern (Your Model Tree will Look Different. It will have Extrudes instead of Protrusions)

The next series of features will be created purposely at the wrong stage in this project. You will now create the **R.50** round [Figs. 17.21(a-b)]. Because the design intent is to have a constant thickness for the part, the round should have been created before the shell. The Reorder capability will be used to change the position of this round in the design sequence. Using the Model Tree, you can pick and drag the round to a new location in the feature list. *(The Reorder capability can also be completed using Edit ⇒ Feature Operations ⇒ Reorder ⇒ pick the feature to reorder ⇒ OK ⇒ Done ⇒ select the new position).*

Pick on the top edge of the part ⇒ **RMB** ⇒ **Round Edges** [Fig. 17.21(a)] ⇒ move a drag handle to **.500** [Fig. 17.21(b)] ⇒ **MMB** [Fig. 17.21(c)] ⇒ Spin the part, and then pick on the inner surface (Fig. 17.22). Notice that the rounds do not propagate on the internal edges.

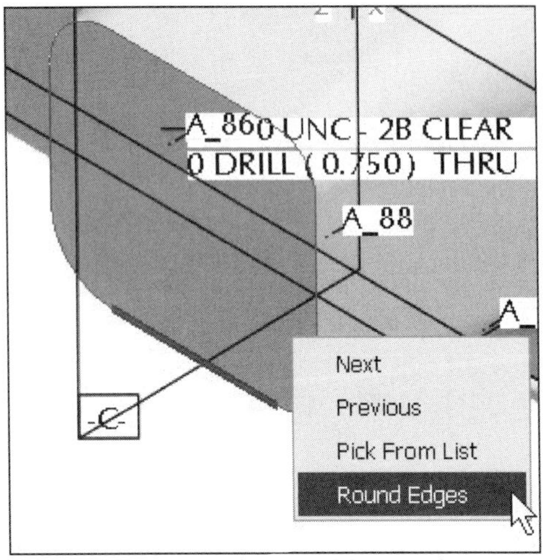

Figure 17.21(a) Add a **R.500** Round

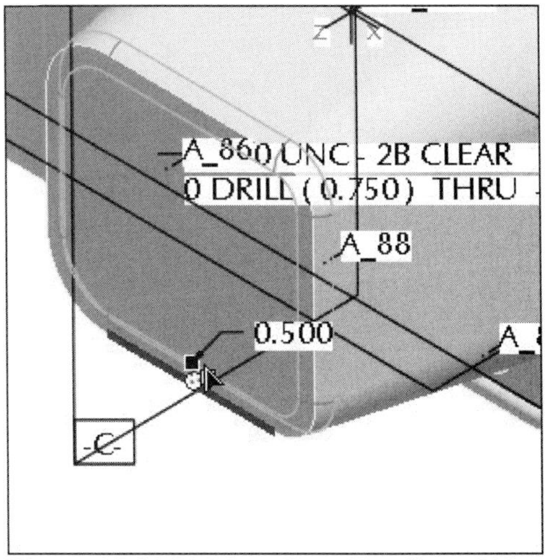

Figure 17.21(b) Move Drag Handle to **500**

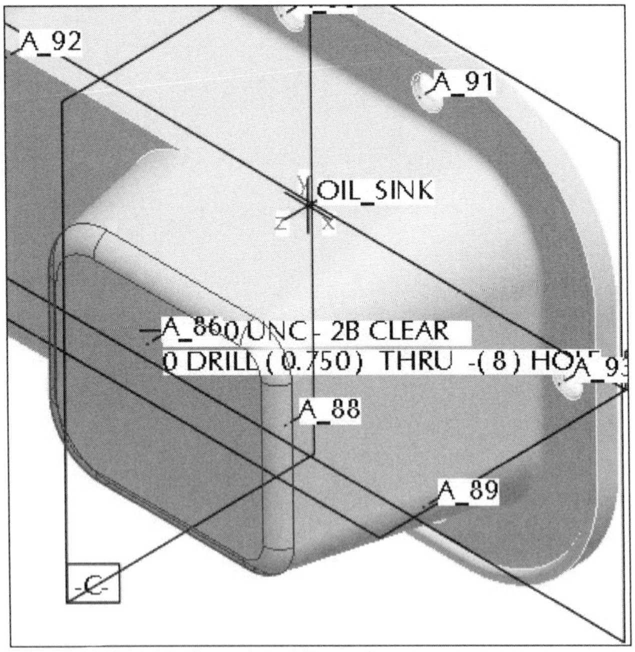

Figure 17.21(c) Completed Round

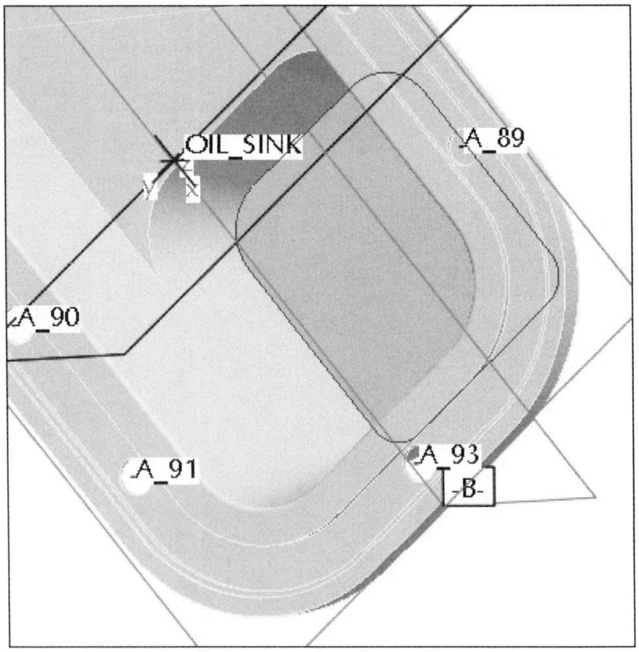

Figure 17.22 Highlighted Surface is not Rounded

Reorder the round to appear before the Shell: click on the **Round** in the Model Tree [Fig. 17.23(a)] and drag Round 1 to a position before/above the shell feature Shell id 200 [Fig. 17.23(b)] and drop [Fig. 17.23(c)] ⇒  ⇒ **MMB**

(Your Model Tree will look different. It will have Extrudes instead of Protrusions and different identifying numbers)

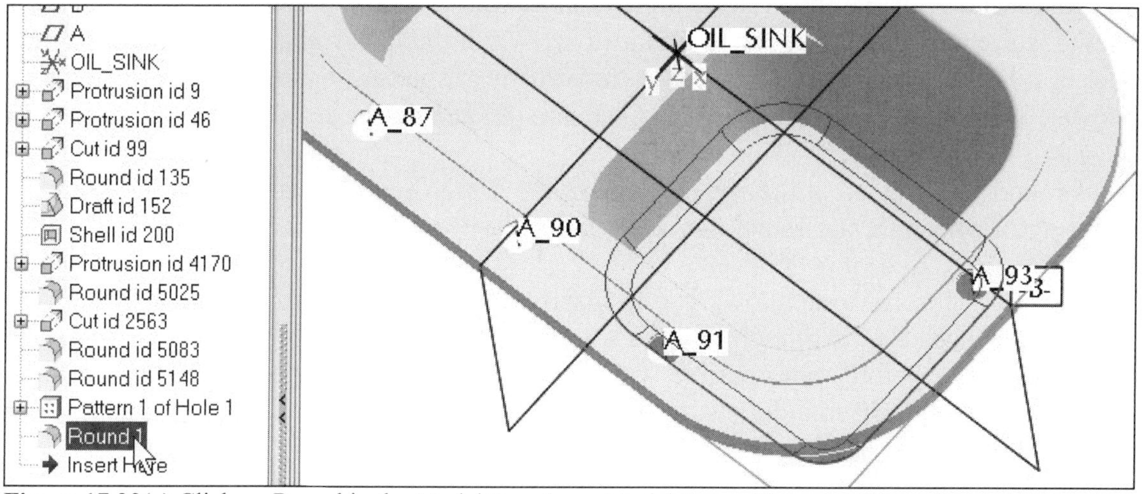

Figure 17.23(a) Click on Round in the Model Tree (your Model Tree will Look Different)

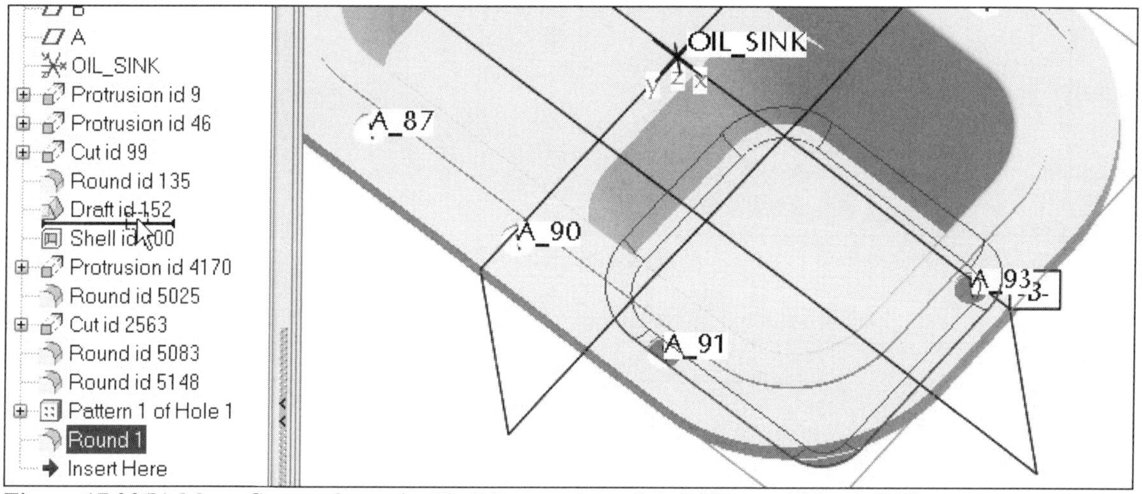

Figure 17.23(b) Move Cursor above the Shell Feature (your Model Tree will Look Different)

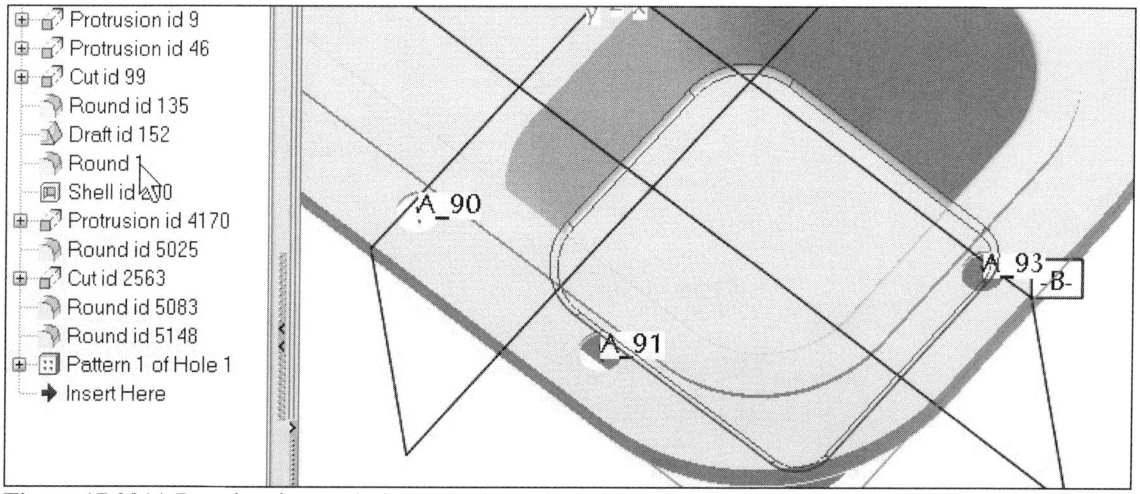

Figure 17.23(c) Reordered Round Shows on the Inside of the Part (**.500** – shell thickness)

You can also Insert new features using the Model Tree. The arrow-shaped icon [➔ Insert Here] in the Model Tree indicates where features will be inserted upon creation and is by default at the end (or bottom) of the Model Tree.

The Insert capability can also be completed using Edit ⇒ Feature Operations ⇒ Insert Mode ⇒ Activate ⇒ Select a feature to insert after ⇒ Done.

By dragging the location of the insert node higher, so that its position is before existing features, you can insert a new feature at that stage of the model history. When the *insert node* is dropped at a new location, the model is rolled backward (suppressed) or forward in response to the insertion node being moved higher or lower. The Model Tree displays a small square (■) next to the features that are not active (suppressed).

The previous round was created at the wrong stage in the design sequence and then reordered. To eliminate the reordering of a feature, the remaining **R.50** rounds will be created using Insert Mode with the Model Tree.

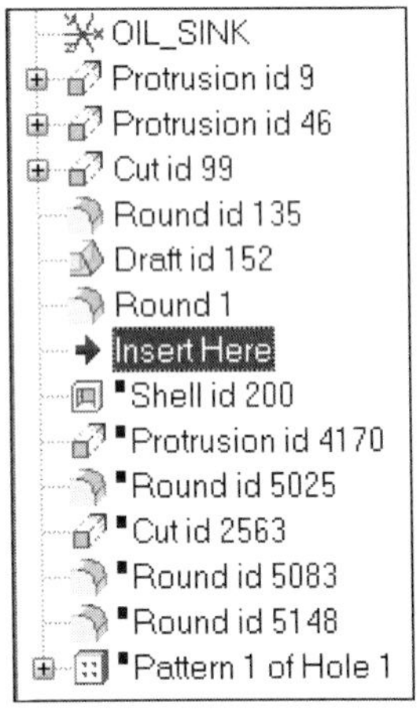

Insert Mode allows you to insert a feature at a previous stage of the design sequence. This is like going back into the past and doing something you wish you had done before--not possible with life, but with Pro/E less of a problem. Add the additional **R.50** rounds.

Your Model Tree will look different. It will have Extrudes instead of Protrusions.

In the Model Tree, click on [➔ Insert Here] [Fig. 17.24(a)] and drag it to a position before/above the shell feature [Shell id 200] and drop [Fig. 17.24(b)] ⇒ **Refit**

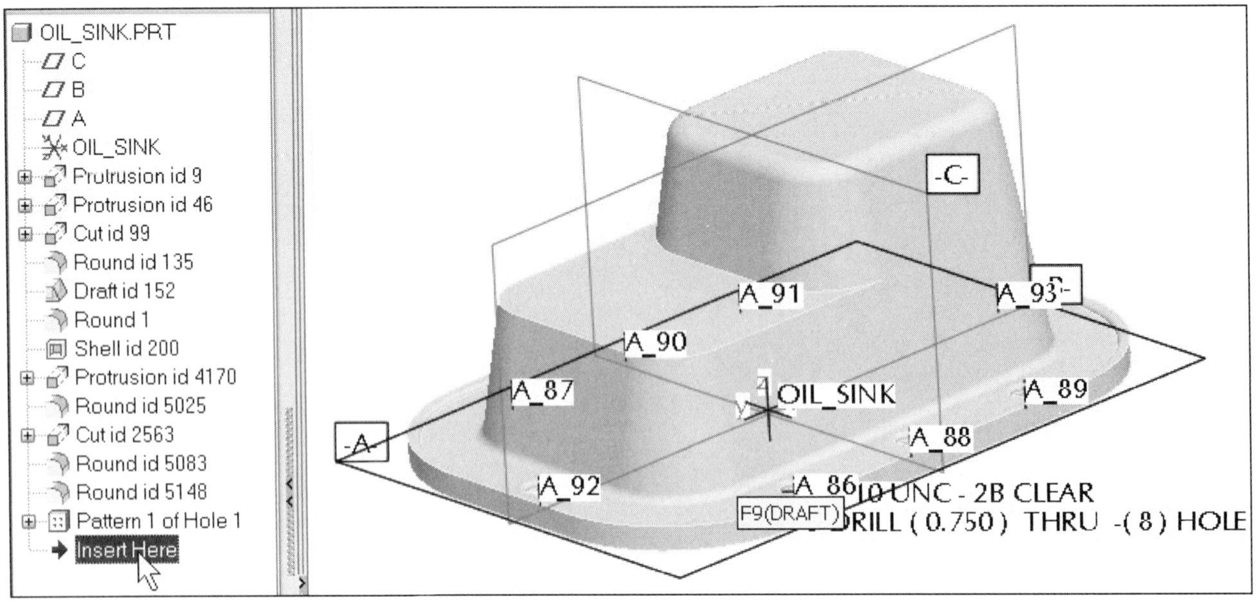

Figure 17.24(a) Insert Here Pointer [➔ Insert Here] (your Model Tree will Look Different)

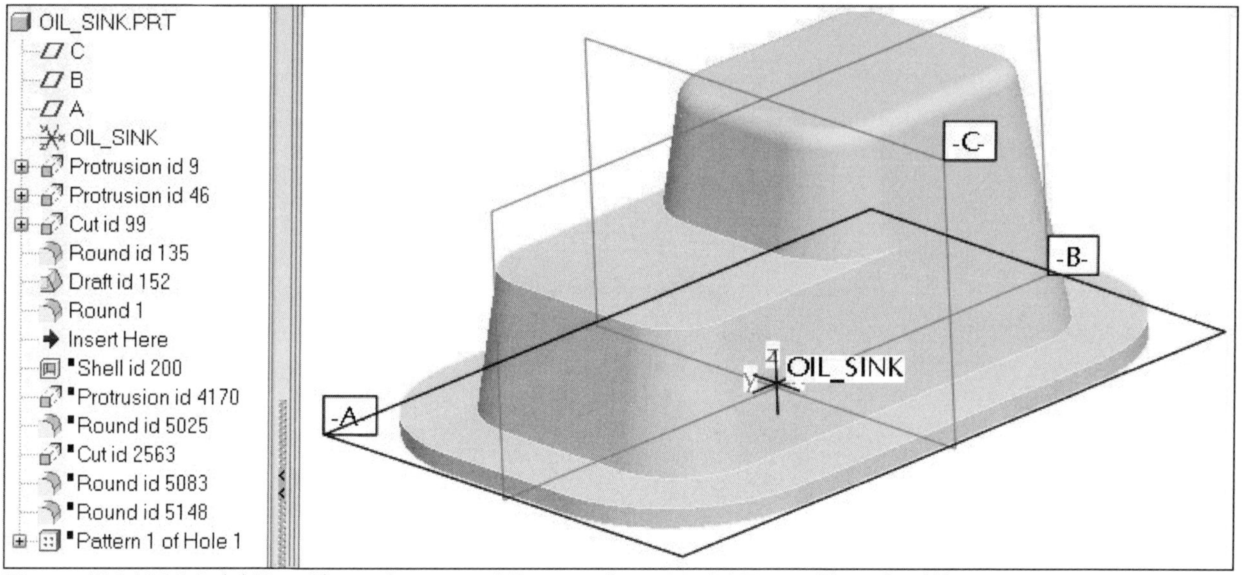

Figure 17.24(b) Model Tree Shows Suppressed Features (your Model Tree will Look Different)

Create two sets of rounds, click: **Round Tool** ⇒ **Sets** tab ⇒ pick the front edge ⇒ Radius **.50** ⇒ **Enter** ⇒ **RMB** ⇒ **Add set** ⇒ pick the second edge [Fig. 17.25(a)] ⇒ [Fig. 17.25(b)]

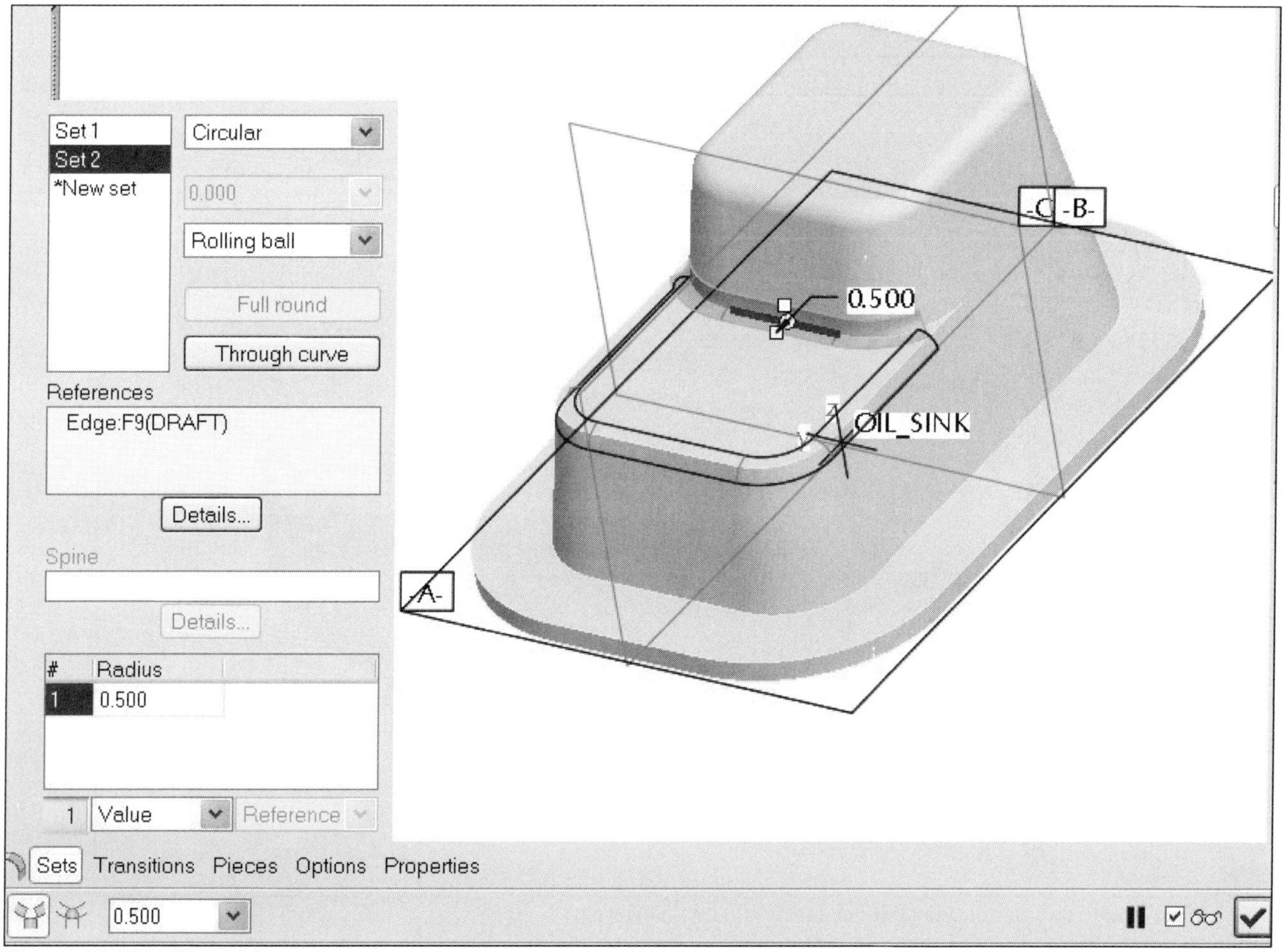

Figure 17.25(a) Round Sets

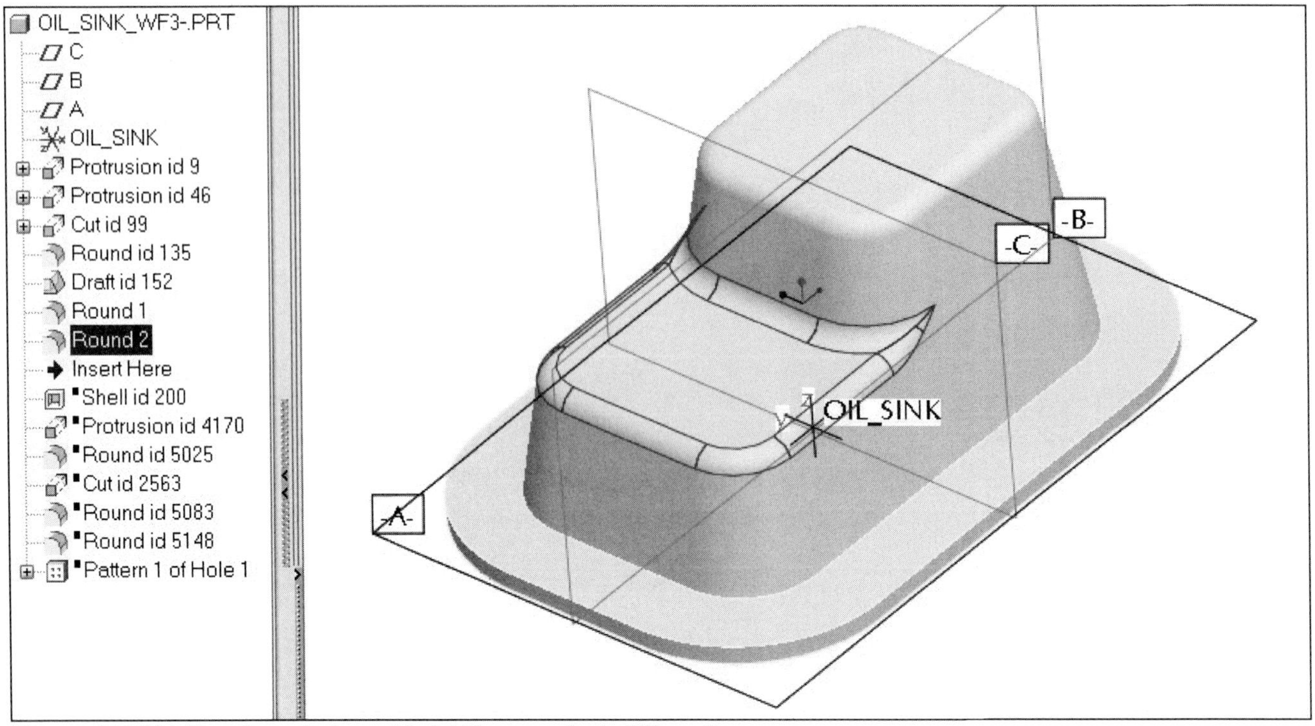

Figure 17.25(b) New Rounds (your Model Tree may look different)

Rotate the model ⇒ click on ➔ Insert Here [Fig. 17.25(c)]

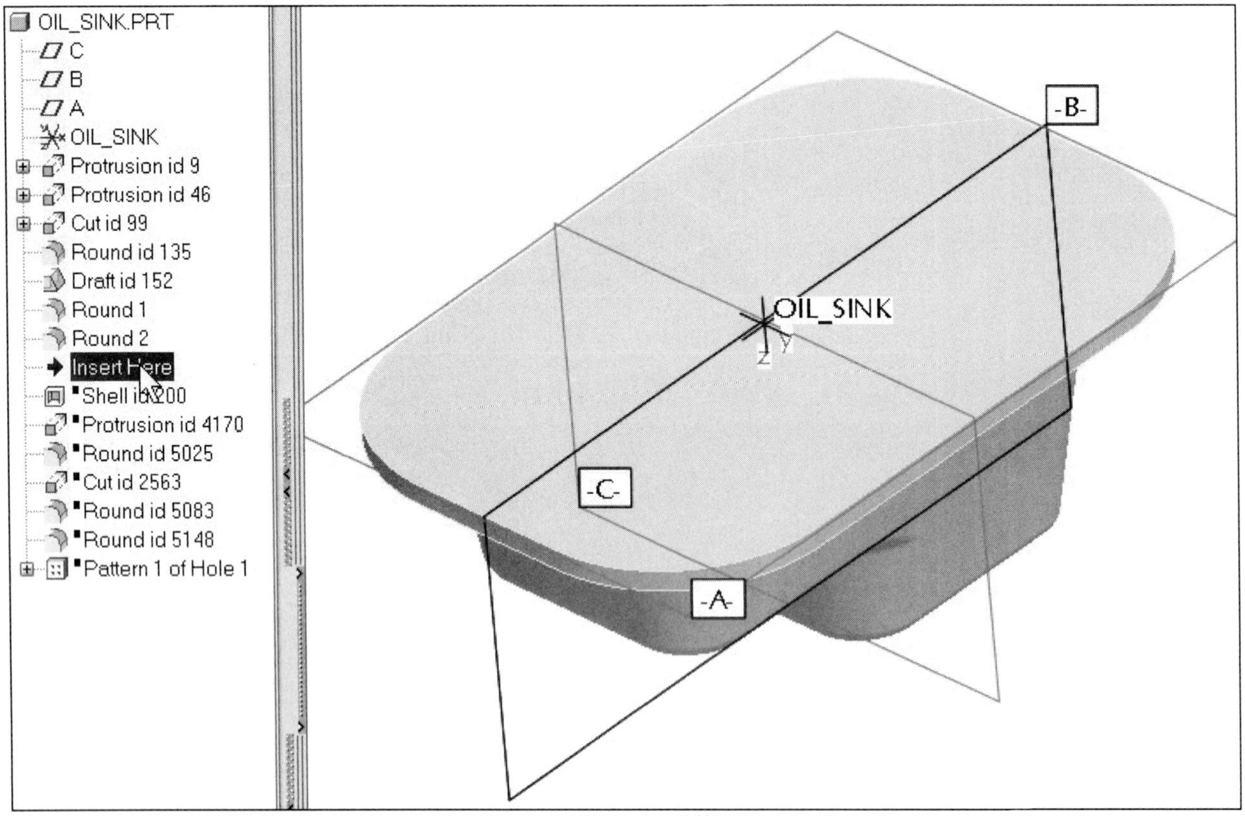

Figure 17.25(c) Rotate the Model (your Model Tree will Look Different)

Drag it to the bottom of the Model Tree list and drop [Fig. 17.26(a)] ⇒ click on the propagated internal round surfaces in the **Model Tree** ⇒ [save icon] ⇒ **MMB** [Fig. 17.26(b)] ⇒ **File** ⇒ **Delete** ⇒ **Old Versions** ⇒ **MMB** [Fig. 17.26(c)]

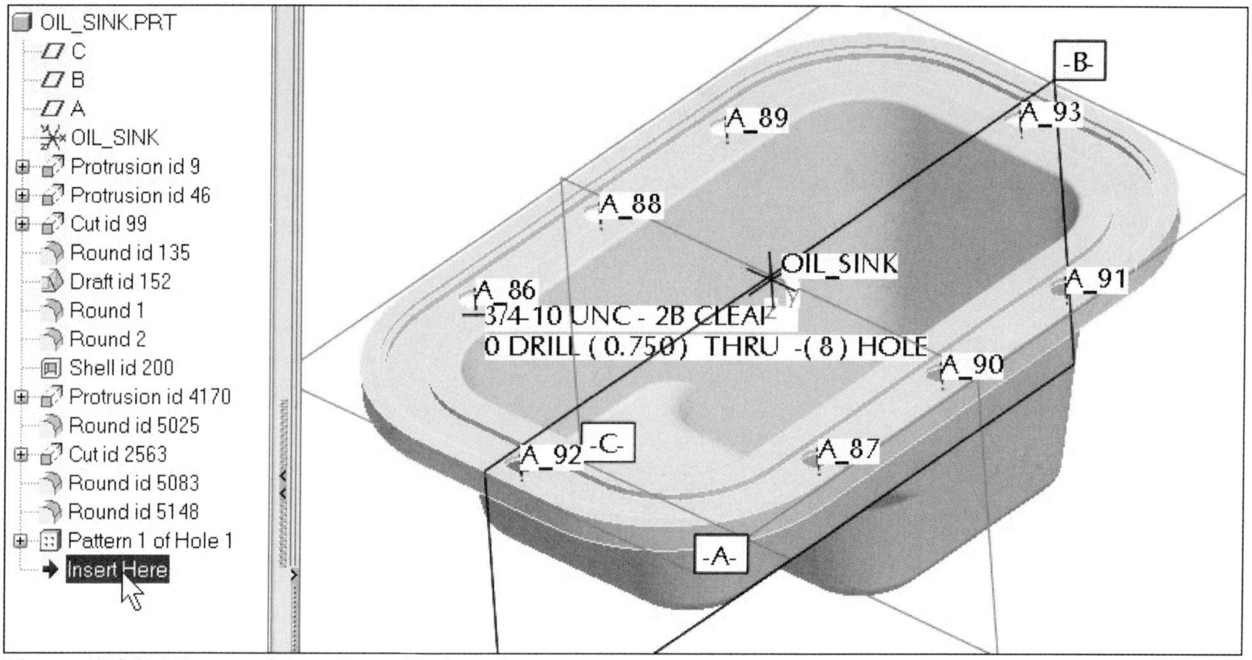

Figure 17.26(a) Drag and Drop Insert Node, All Features are Resumed

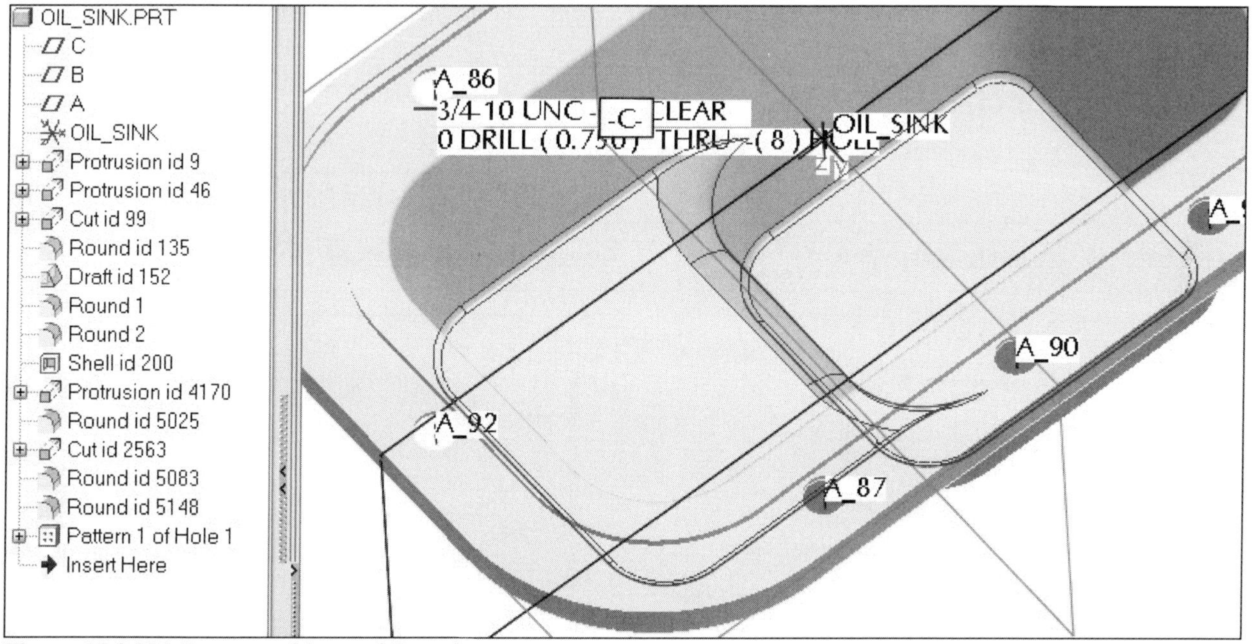

Figure 17.26(b) Propagated Internal Rounds

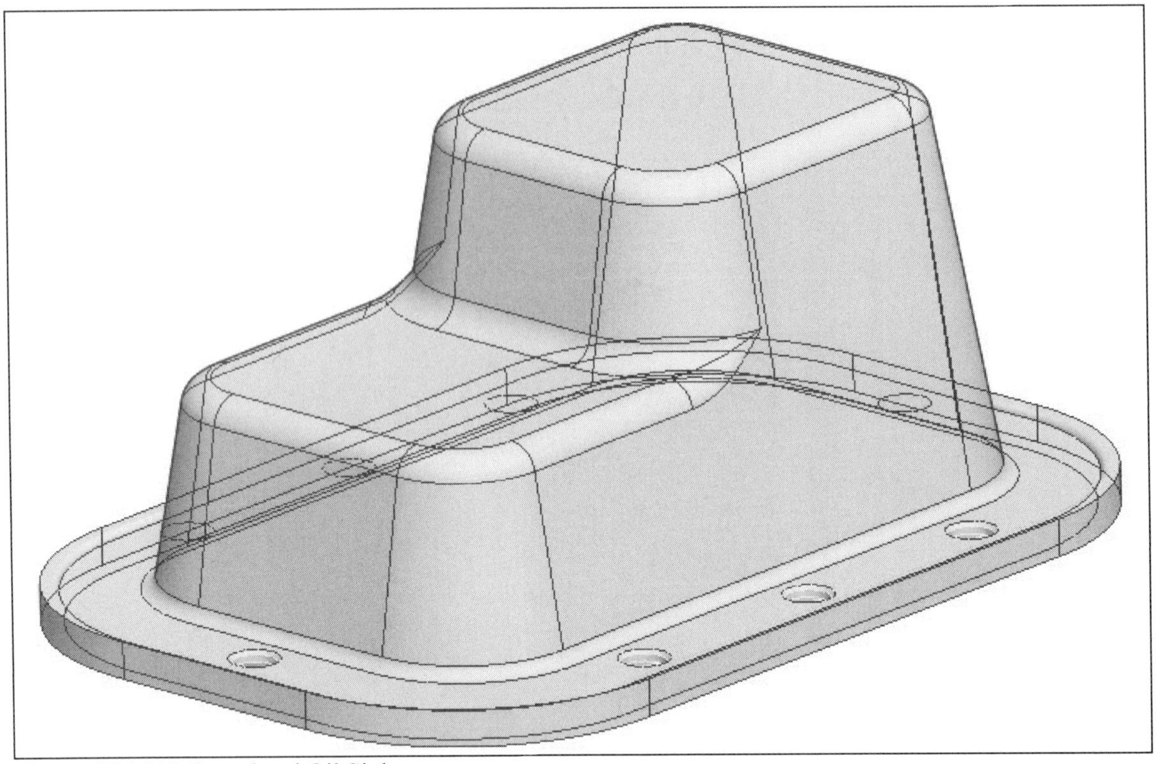

Figure 17.26(c) Completed Oil Sink

The rounds are now in the proper design sequence for the shell feature to have a constant (shell) thickness. Save the part with a new name (File ⇒ Save a Copy) and then complete the **ECO** (Fig. 17.27) using the current part (file) name. See ***www.cad-resources.com* ⇒ *Downloads*** for more part projects.

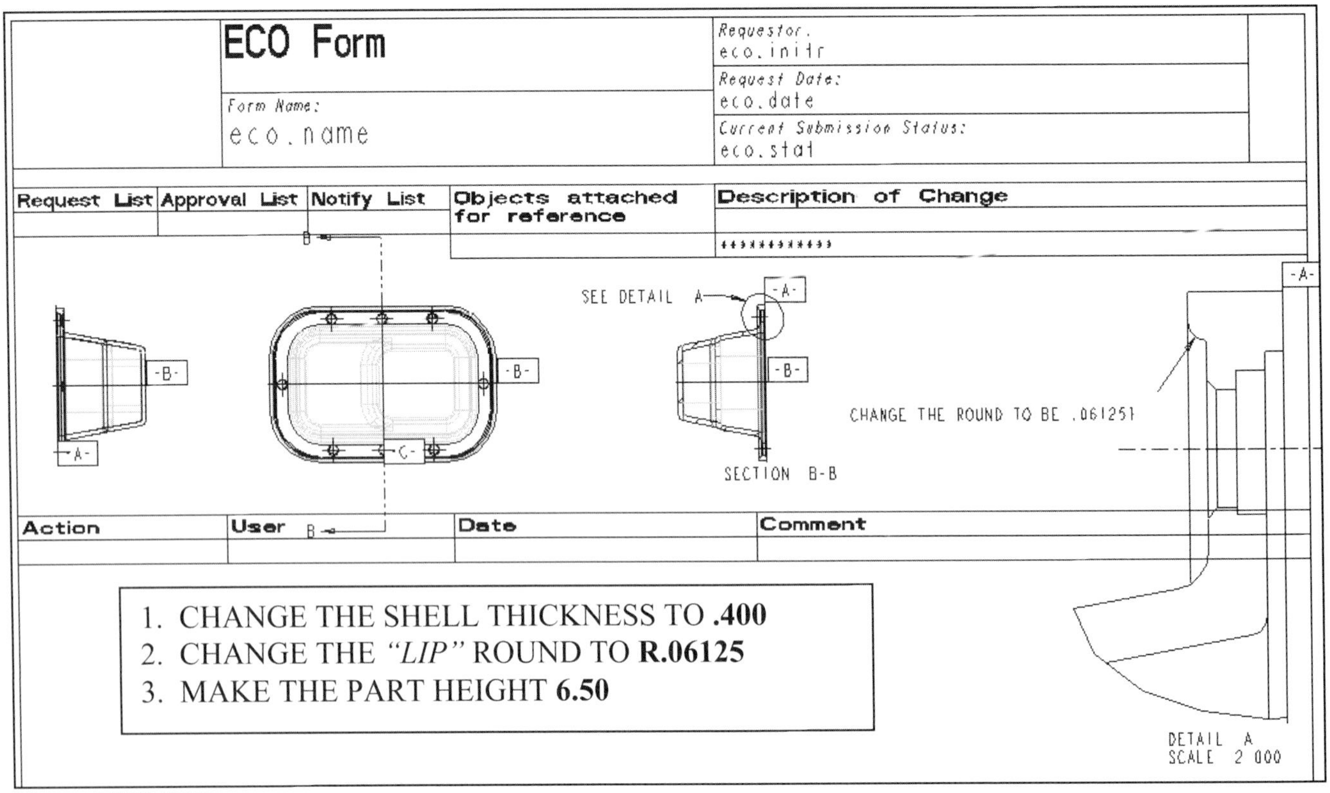

Figure 17.27 ECO

Lesson 18 Drafts, Suppress, and Text Extrusions

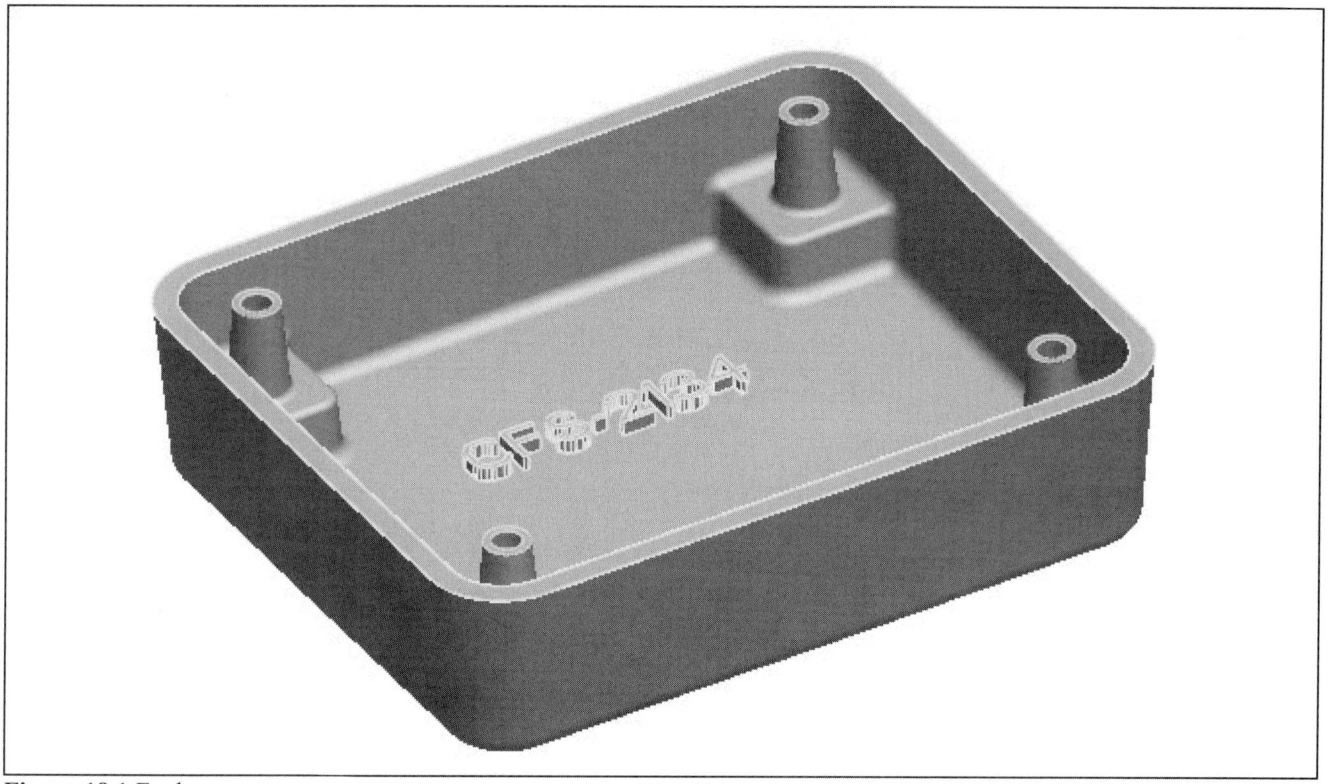

Figure 18.1 Enclosure

OBJECTIVES

- Create **Draft** features
- **Shell** a part
- **Suppress** features to decrease regeneration time
- **Resume** a set of suppressed features
- Create **Text** features on parts

DRAFTS, SUPPRESS, AND TEXT EXTRUSIONS

The **Draft** feature adds a draft angle between surfaces. A wide range of parts incorporate drafts into their design. Casting, injection mold, and die parts normally have drafted surfaces. The ENCLOSURE in Figure 18.1 is a plastic injection-molded part.

Suppressing features by using the **Suppress** command temporarily removes them from regeneration. Suppressed features can be "unsuppressed" (**Resume**) at any time. It is sometimes convenient to suppress text extrusions and rounds to speed up regeneration of the model. Suppressing removes the item from regeneration and requires you to resume the item later.

Hide is another option. Pro/E allows you to hide and unhide some types of model entities. When you hide an item, Pro/E removes the item from the graphics window. The hidden item remains in the Model Tree list, and its icon dims to reveal its hidden status. When you unhide an item, its icon returns to normal display (undimmed) and the item is redisplayed in the graphics window. The hidden status of items is saved with the model. Unlike the suppression of items, hidden items are regenerated.

Text can be included in a sketch for extruded extrusions and cuts, trimming surfaces, and cosmetic features. To decrease regeneration time of the model, text can be suppressed after it has been created. Text can also be drafted.

Drafts

The **Draft Tool** adds a draft angle between two individual surfaces or to a series of selected planar surfaces. During draft creation, remember the following:

- You can draft only the surfaces that are formed by tabulated cylinders or planes.
- The draft direction must be normal to the neutral plane if a draft surface is cylindrical.
- You cannot draft surfaces with fillets around the edge boundary. However, you can draft the surfaces first, and then fillet the edges.

The following table lists the terminology used in drafts.

TERM	DEFINITION
Draft surfaces	Model surfaces selected for drafting.
Draft Hinges	Draft surfaces are pivoted about the intersection of the neutral plane with the draft surfaces.
Pull direction	Direction that is used to measure the draft angle. It is defined as normal to the reference plane.
Draft angle	Angle between the draft direction and the resulting drafted surfaces. If the draft surfaces are split, you can define two independent angles for each portion of the draft.
Direction of rotation	Direction that defines how draft surfaces are rotated with respect to the neutral plane or neutral curve.
Split areas	Areas of the draft surfaces to which you can apply different draft angles. Split object is also a choice.

Suppressing and Resuming Features

Suppressing a feature is similar to removing the feature from regeneration temporarily. You can "unsuppress" (**Resume**) suppressed features at any time. Features on a part can be suppressed to simplify the part model and decrease regeneration time. For example, while you work on one end of a shaft, it may be desirable to suppress features on the other end of the shaft. Similarly, while working on a complex assembly, you can suppress some of the features and components for which the detail is not essential to the current assembly process.

Unlike other features, the base feature cannot be suppressed. If you are not satisfied with your base feature, you can redefine the section of the feature, or you can delete it and start over again. Select feature(s) to suppress by: picking on it, selecting from the Model Tree, specifying a *range*, entering its *feature number* or *identifier*, or using *layers*.

You can use **Suppress** and **Resume** to simplify the part before inserting features such as text extrusions. In addition, you may wish to suppress the text extrusion if there is other work to be done on the part. Text extrusions take time to regenerate, and increase the file size considerably.

Text Extrusions

When you are modeling, **Text** can be included in a sketch for extruded extrusions and cuts, trimming surfaces, and cosmetic features. The characters that are in an extruded feature use the font **font3d** as the default. Other fonts are available.

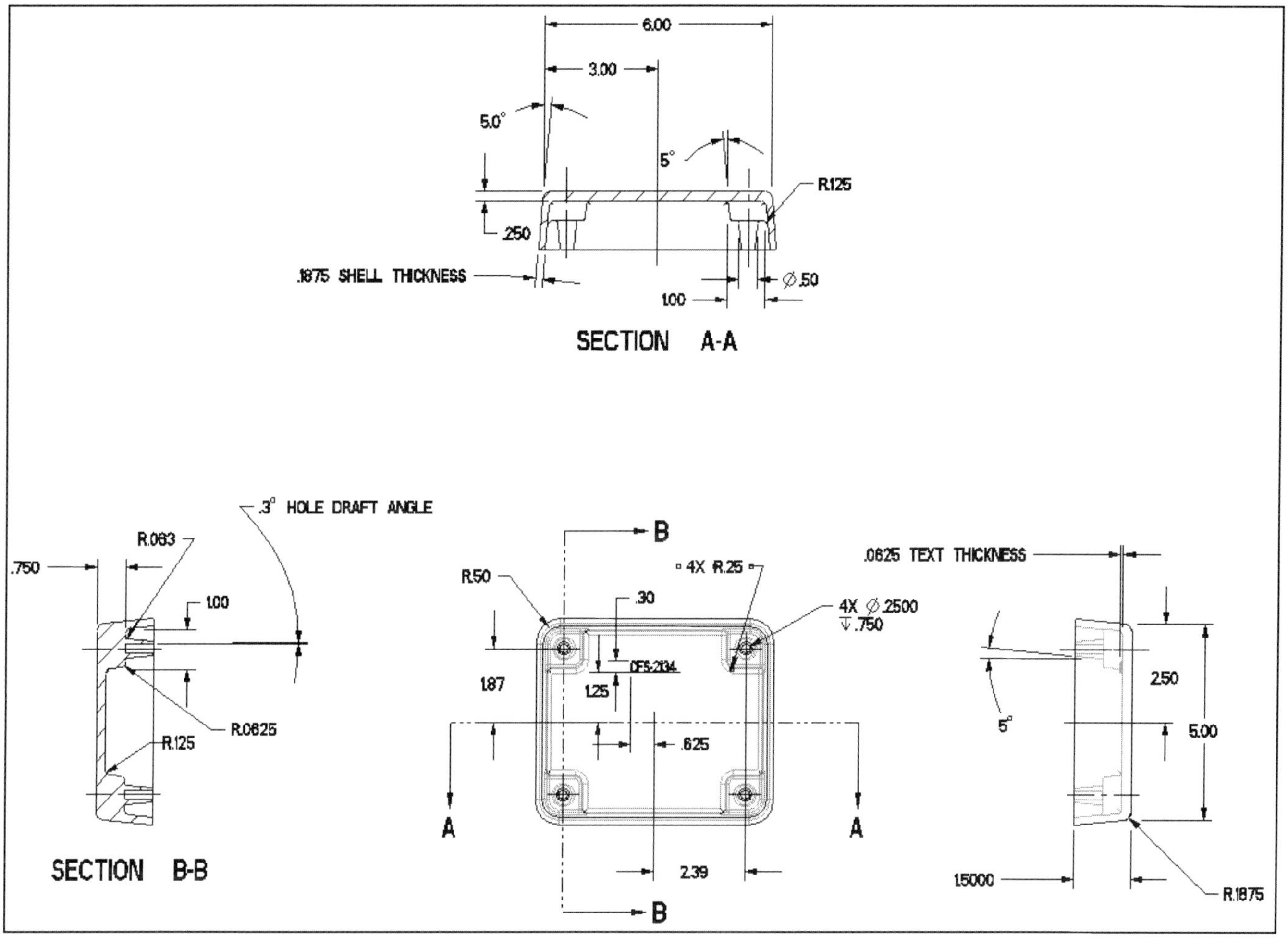

Figure 18.2(a) Enclosure

Enclosure

The Enclosure is a plastic injection-molded part. A variety of drafts will be used in the design of this part. A *raised text extrusion* will be modeled on the inside of the Enclosure, as shown in Figure 18.1. The dimensions for the part are provided in Figures 18.2(a) through 18.2(i).

Click: **File** ⇒ **Set Working Directory** select the working directory ⇒ **OK** ⇒ [icon] **Create a new object** ⇒ [◉ Part] ⇒ **ENCLOSURE** ⇒ [☑] Use default template ⇒ **OK** ⇒ **Tools** ⇒ **Environment** ⇒ [☑ Snap to Grid] ⇒ [☑ Use 2D Sketcher] ⇒ [Tangent Edges Dimmed] ⇒ **OK** ⇒ **Tools** ⇒ **Options** ⇒ Showing: **Current Session** ⇒ Option: ***default_dec_places*** ⇒ Value: **3** ⇒ **Enter** ⇒ Option: ***sketcher_dec_places*** ⇒ Value: **3** ⇒ **Enter** ⇒ **Apply** ⇒ **Close** ⇒ **Edit** ⇒ **Setup** ⇒ **Units** ⇒ Units Manager **Inch lbm Second (Pro/E Default)** ⇒ **Close** ⇒ **Material** ⇒ **fe20.mtl** (plastic) ⇒ >>> ⇒ **OK** ⇒ **Done** ⇒ double-click on the default coordinate system name in the Model Tree-- **PRT_CSYS_DEF** ⇒ type **CSYS_ENCLOSURE** ⇒ **Enter** ⇒ [icon] ⇒ **MMB** ⇒ load your customization by clicking **Tools** ⇒ **Customize Screen** ⇒ **File** ⇒ **Open Settings** ⇒ click on your saved file (only available if you created a **.win** file and saved it as instructed in a previous lesson) ⇒ **Open** ⇒ **OK**

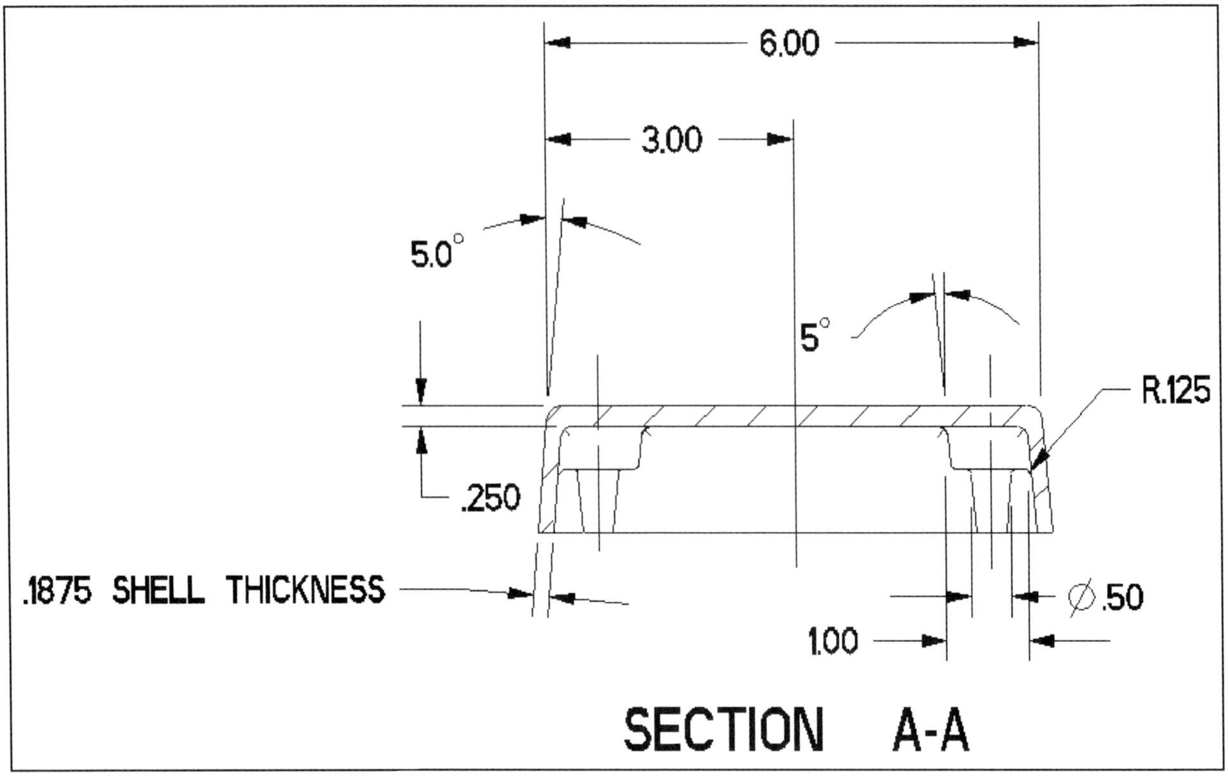

Figure 18.2(b) SECTION A-A (Top View)

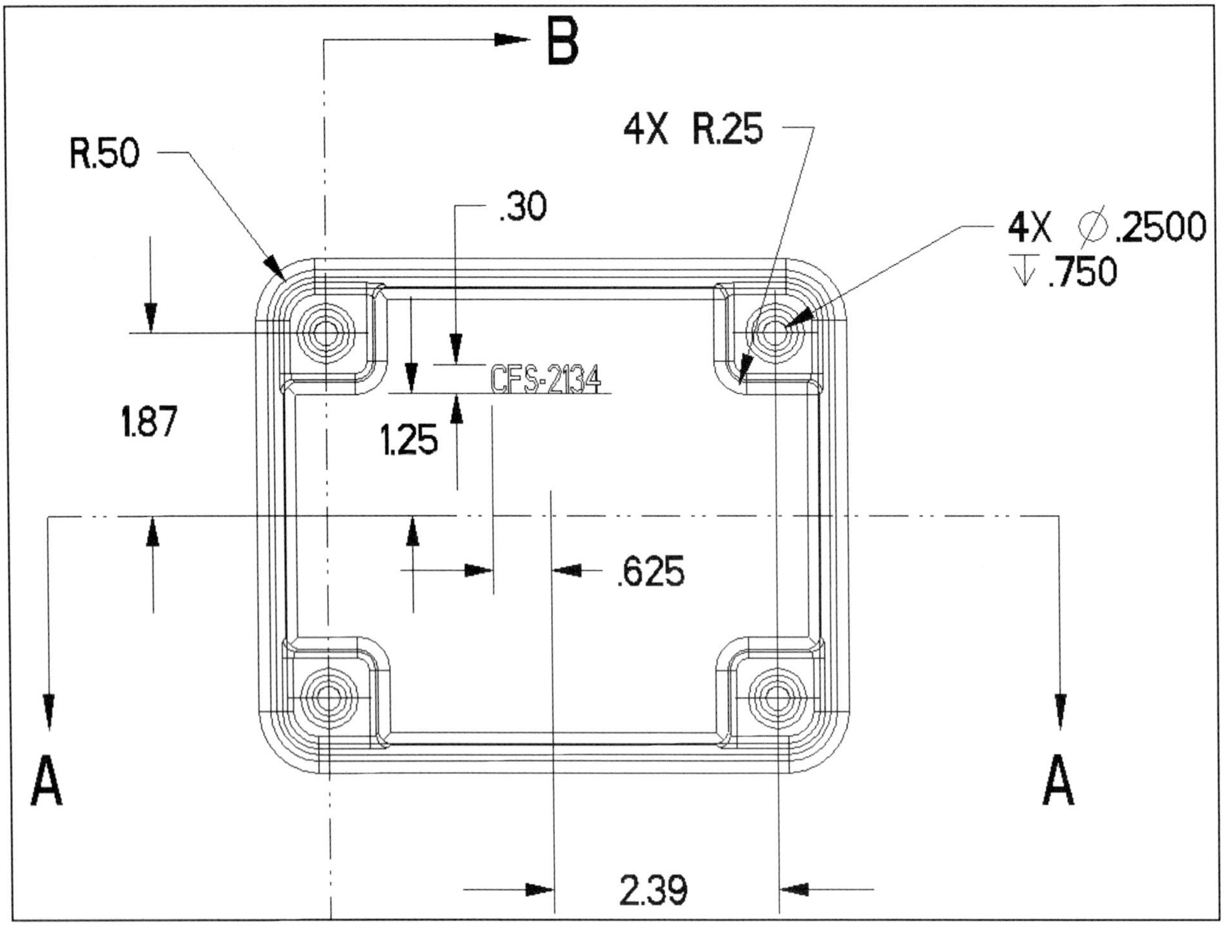

Figure 18.2(c) Front View

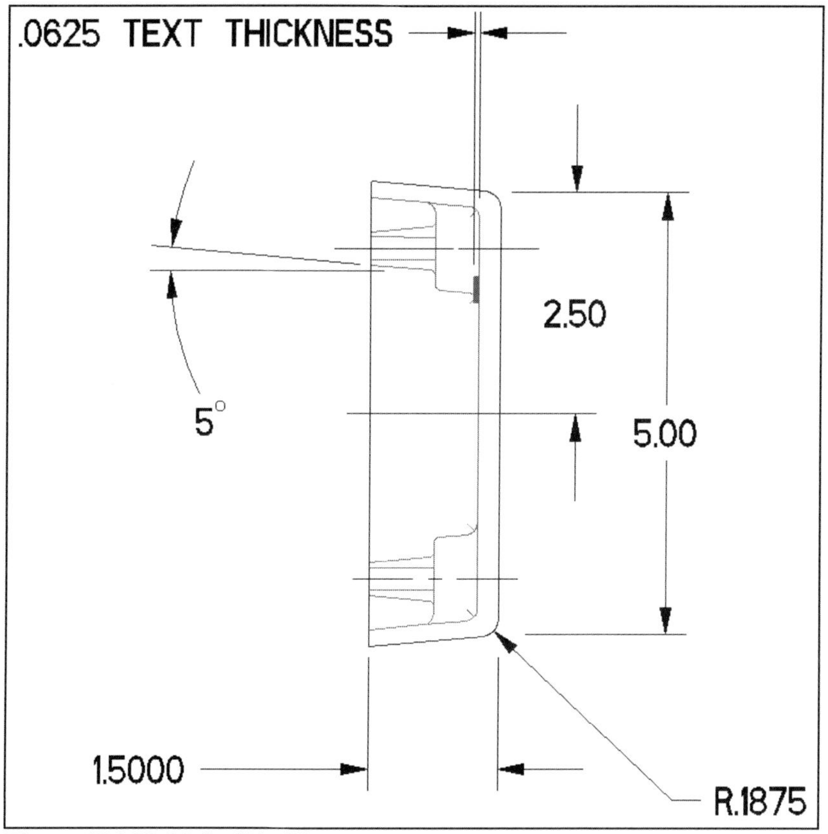

Figure 18.2(d) Right Side View

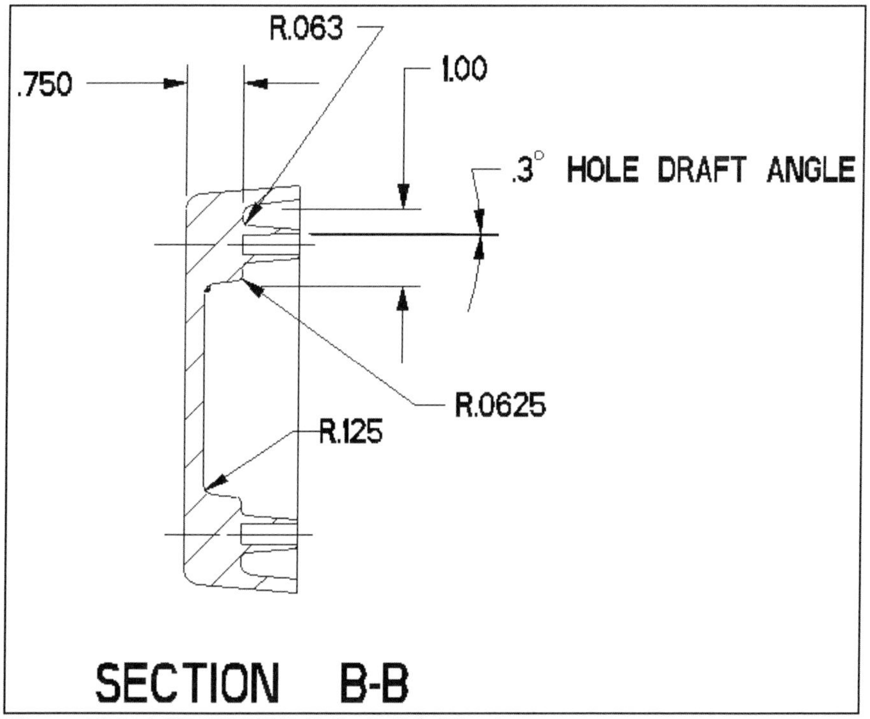

Figure 18.2(e) SECTION B-B (Left Side View)

Make the first extrusion **6.00** (width) **X 5.00** (height) **X 1.50** (depth), with **R.50** rounds. Add the fillets in the sketch instead of rounds after the first extrusion is complete. Sketch on datum **FRONT**. Center the first extrusion horizontally on datum **TOP** and vertically on datum **RIGHT**. Add constraints as needed to control your sketch geometry.

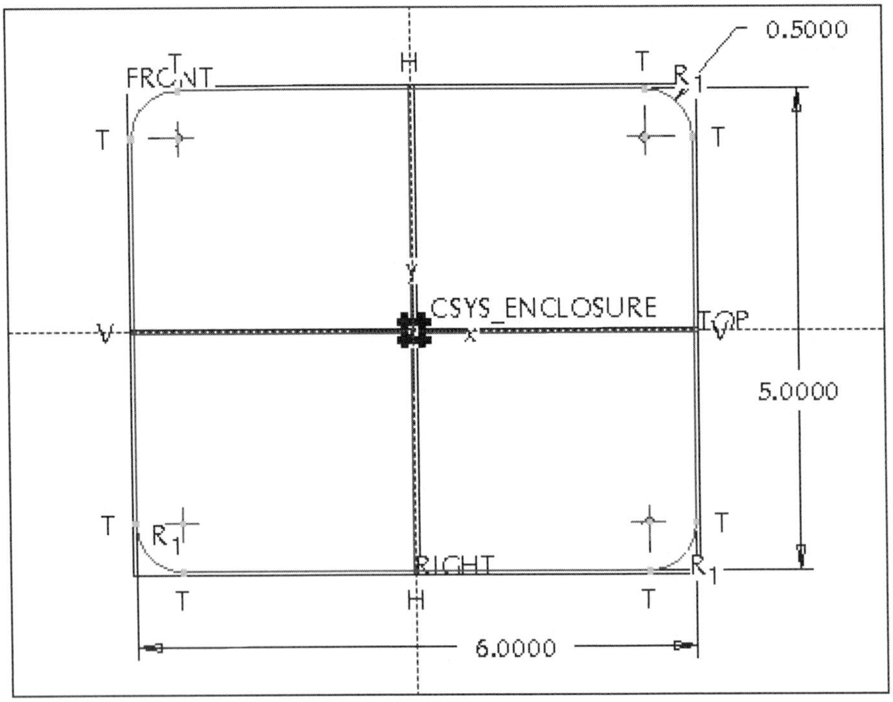

Figure 18.2(f) Sketch

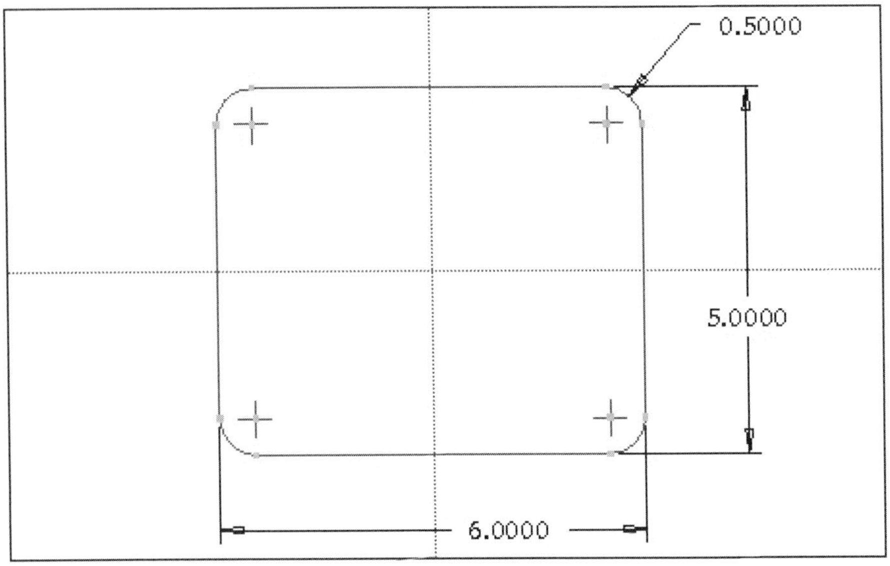

Figure 18.2(g) Dimensions

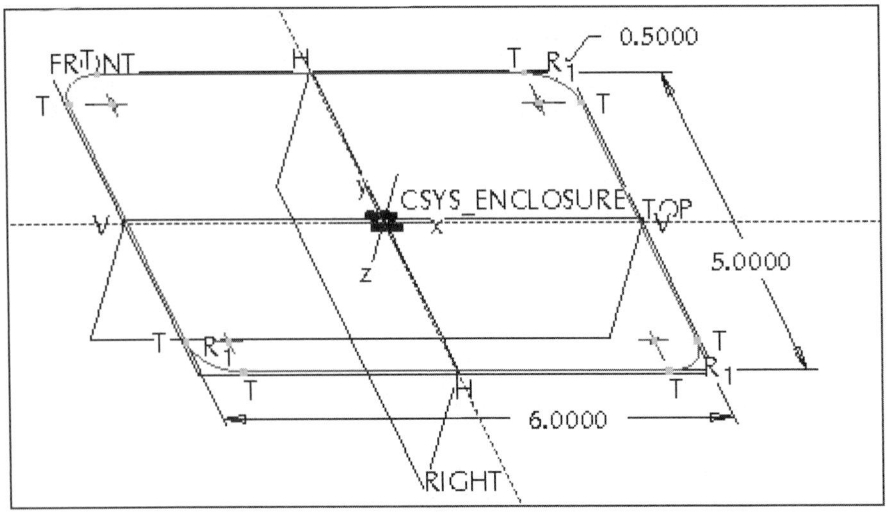

Figure 18.2(h) Standard Orientation

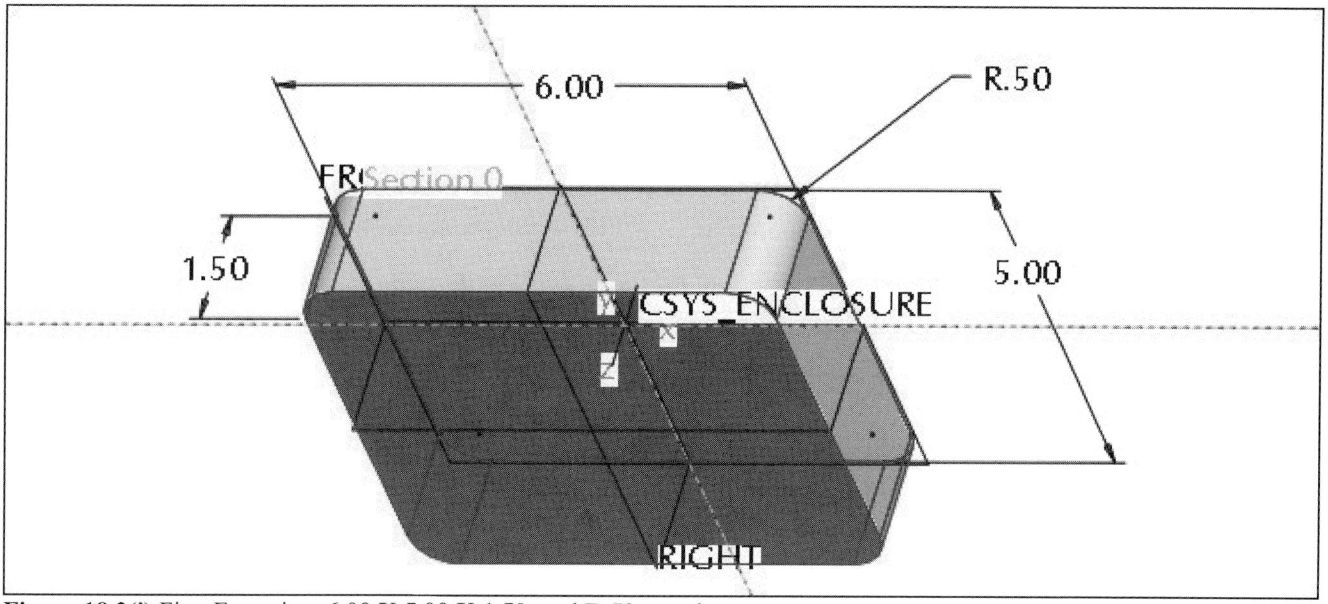

Figure 18.2(i) First Extrusion, **6.00 X 5.00 X 1.50**, and **R.50** rounds

Create the draft for the lateral surfaces of the extrusion, click: **Draft Tool** ⇒ **References** tab ⇒ select one of the lateral surfaces [Fig. 18.3(a)]

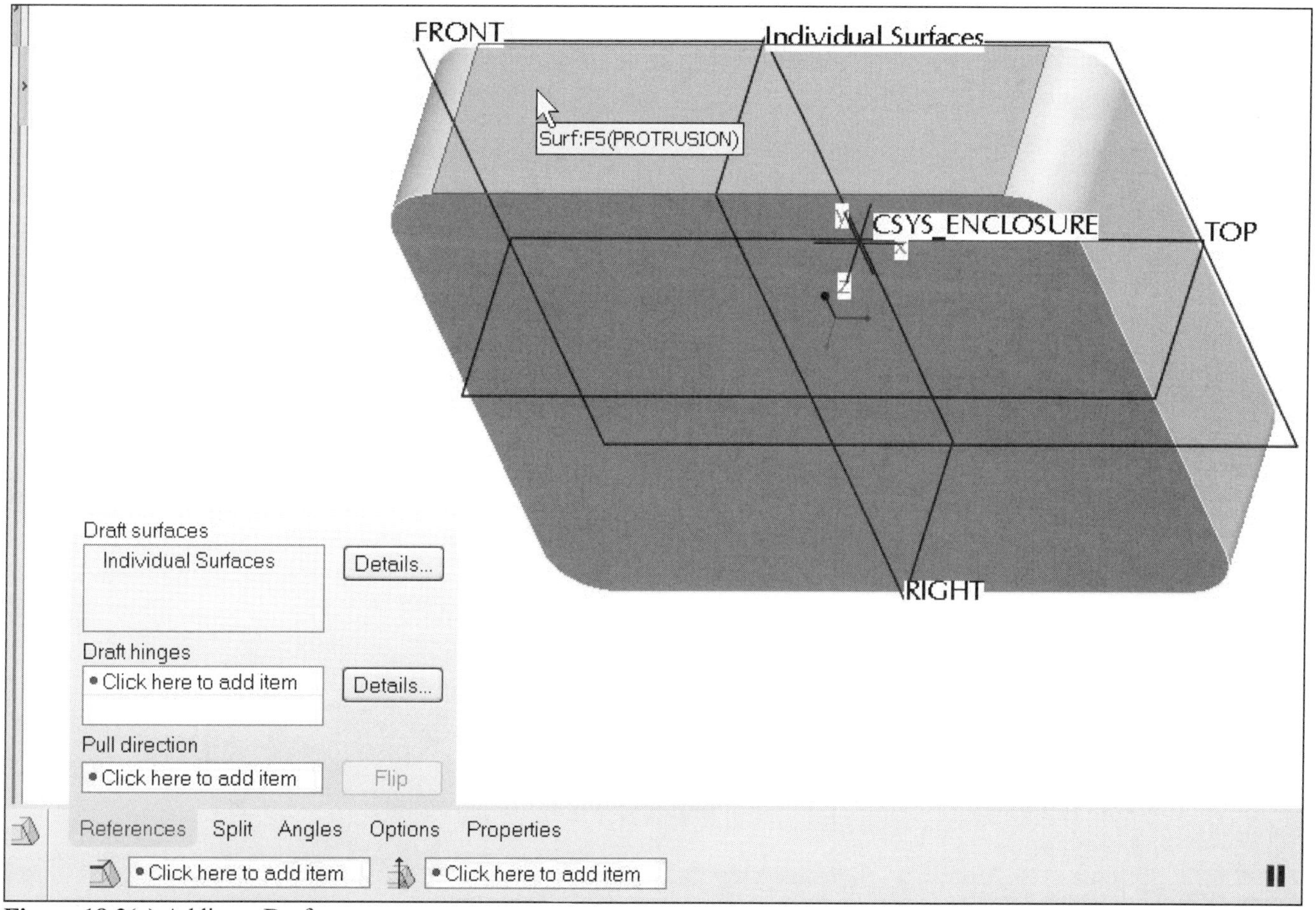

Figure 18.3(a) Adding a Draft

Click in the Draft hinges collector 

⇒ pick **FRONT** [Fig. 18.3(b)] ⇒ in the Angle field type **5** ⇒ **Enter** ⇒ ☑ ⇒ with Draft highlighted in the Model Tree, click **RMB** ⇒ **Edit** [Fig. 18.3(c)] ⇒ pick on **5.000°** ⇒ **RMB** ⇒ **Properties** ⇒ **Flip Arrows** ⇒ Number of decimal places **0** ⇒ **Enter** ⇒ **OK**

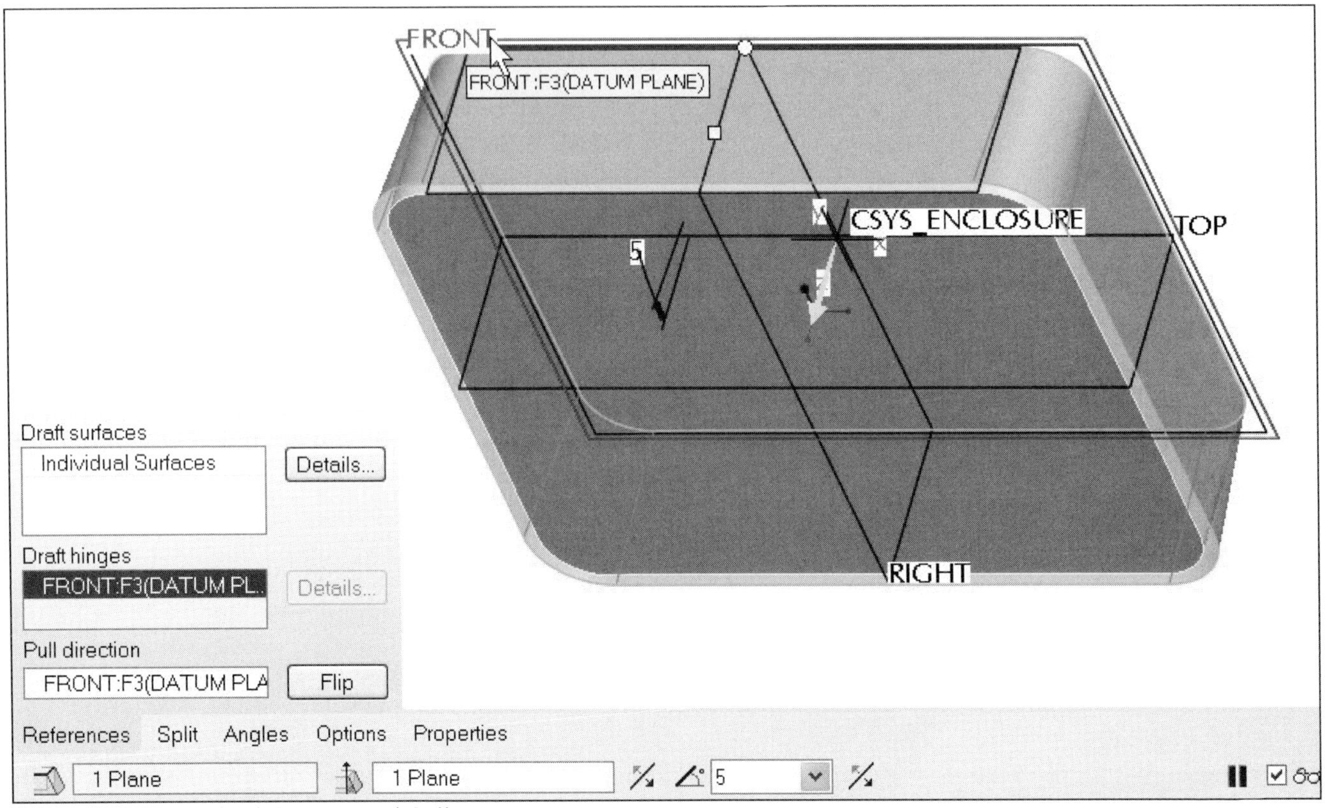

Figure 18.3(b) Draft Dialog Box and Collectors

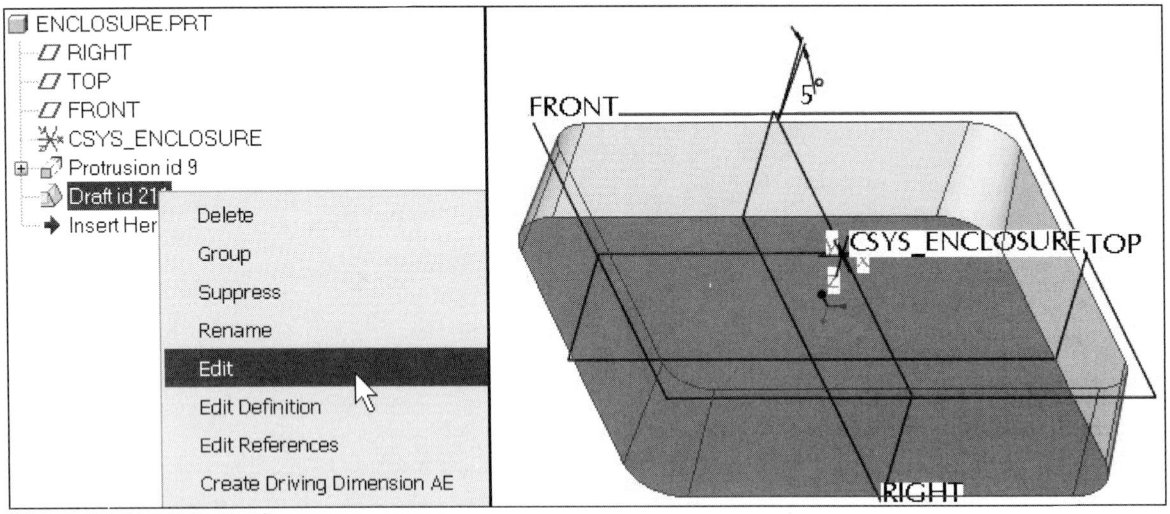

Figure 18.3(c) Edit Draft

Click: **Shell Tool** ⇒ pick the face to be removed [Fig. 18.4(a)] ⇒ type **.1875** in Thickness field Thickness 0.1875 ⇒ **Enter** ⇒ [Fig. 18.4(b)] ⇒

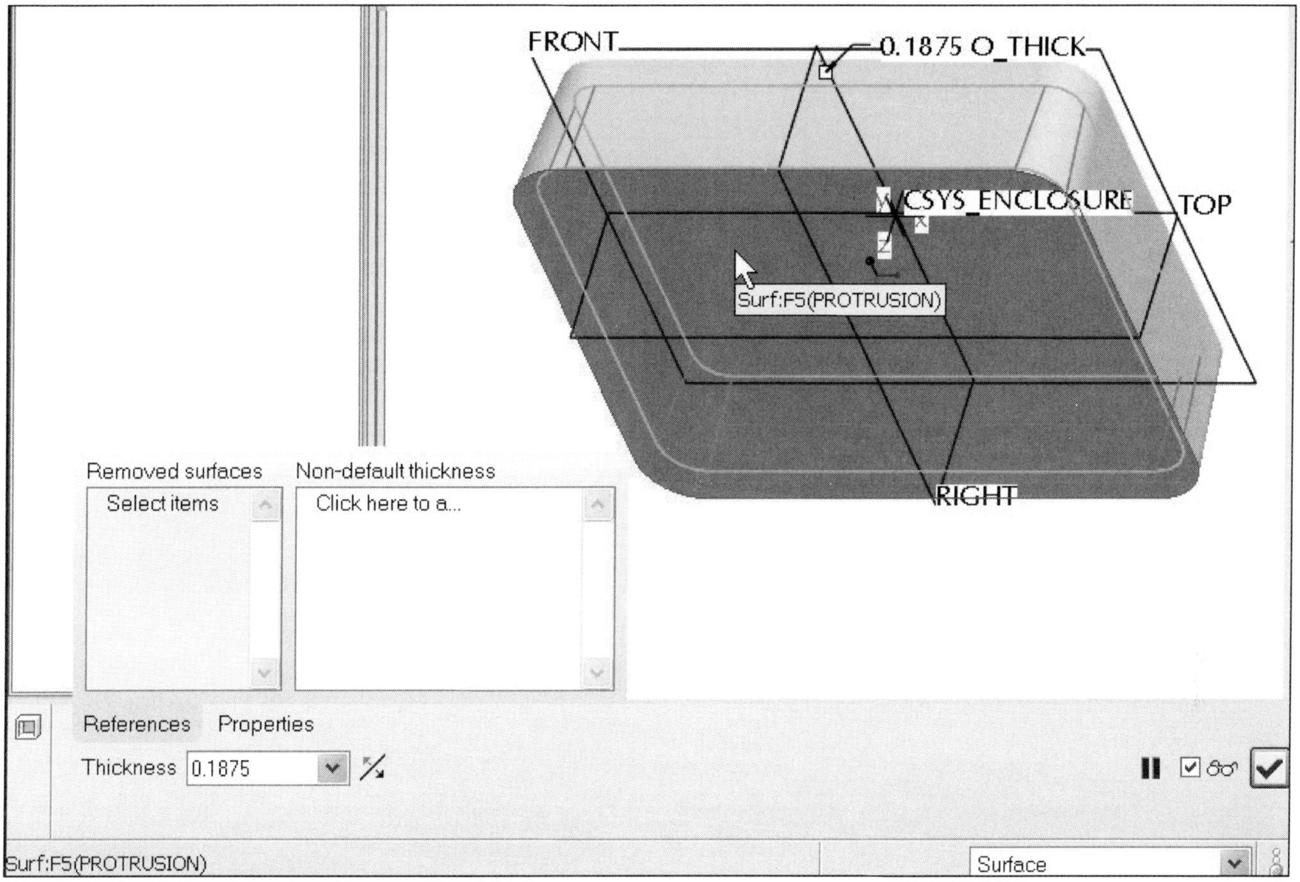

Figure 18.4(a) Using the Shell Tool

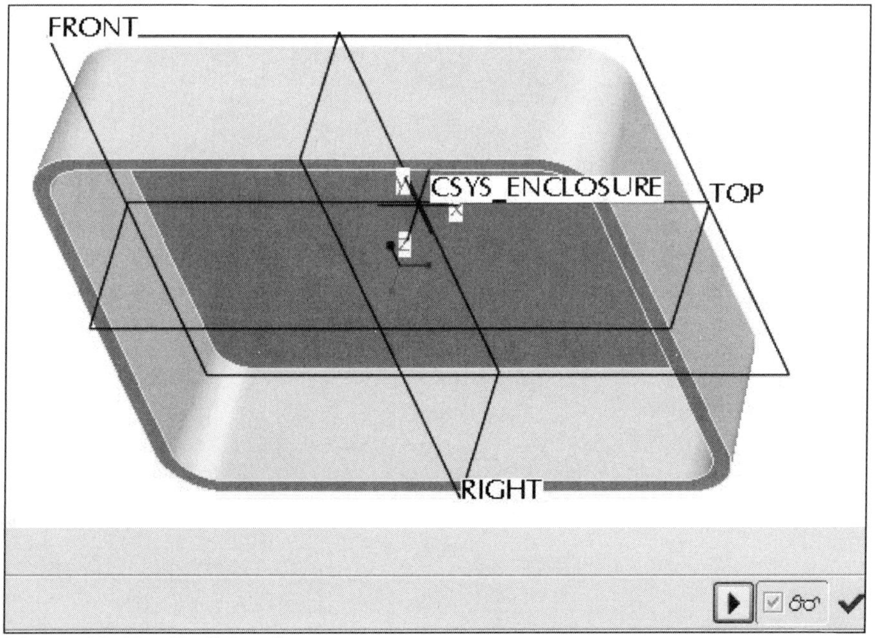

Figure 18.4(b) Shell Preview

Change the thickness of the enclosure to be **.25**, the walls will remain **.1875**. Click: **References** tab ⇒ click in the Non-default thickness collector 

⇒ pick the face [Fig. 18.4(c)] (*highlights*) ⇒ type **.250** in the Dimension field 

⇒ **Enter** [Fig. 18.4(d)] ⇒ **MMB** ⇒ **LMB**

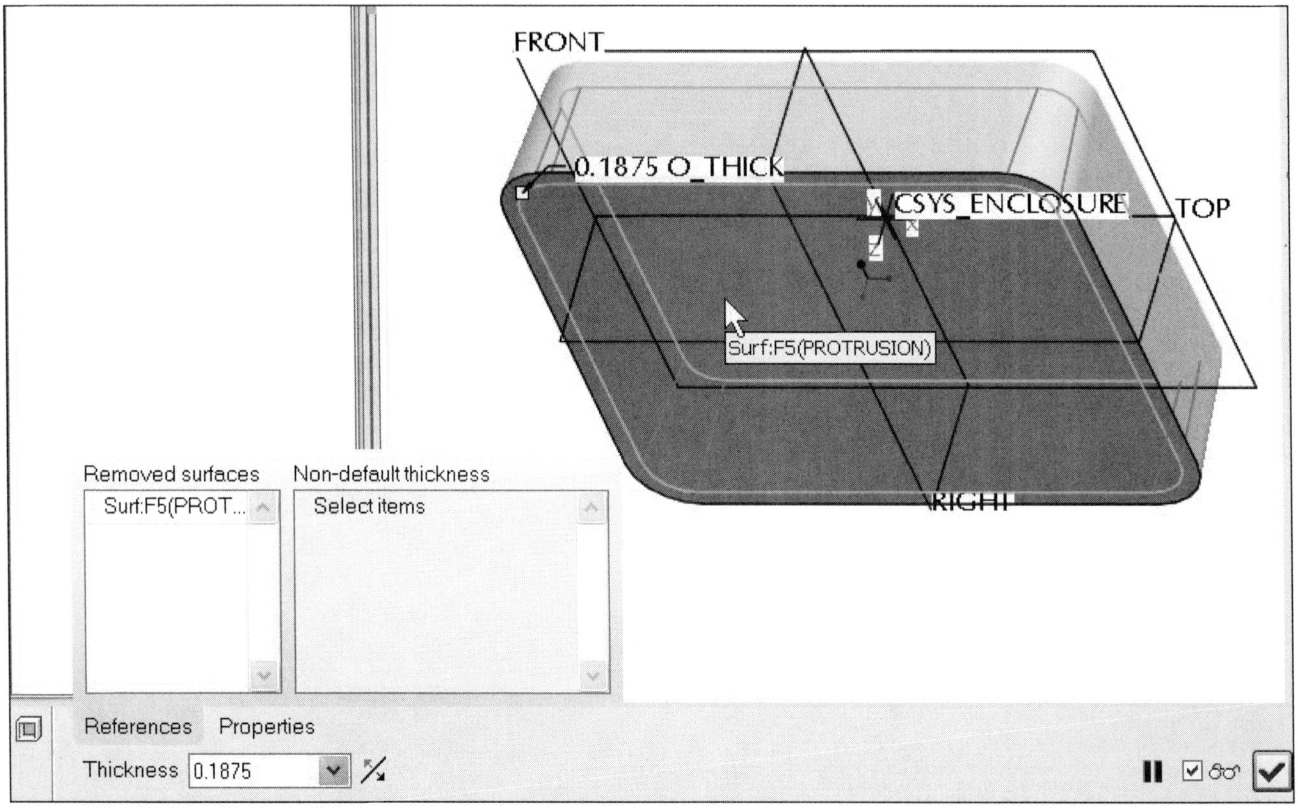

Figure 18.4(c) Non Default Thickness

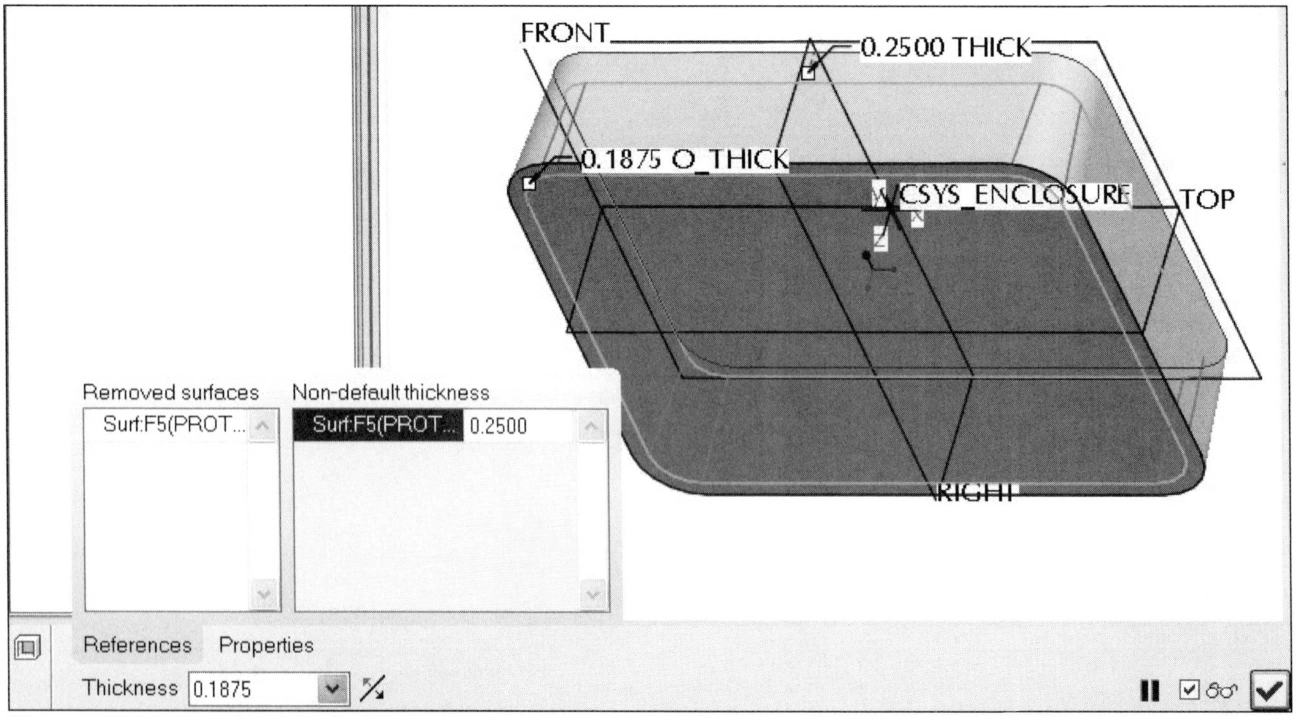

Figure 18.4(d) Completed Shelled Extrusion

Click: **File** ⇒ **Save** ⇒ **MMB** ⇒ Offset a datum plane from datum **FRONT** by **.75** [Fig. 18.5(a)]. **DTM1** will be used to control the height of the pedestal. ⇒ Create a raised pedestal-like extrusion using the inside surface as the sketching plane. ⇒ **Insert** ⇒ **Extrude** ⇒ select only the References shown in Figure 18.5(b) (delete the unneeded references) ⇒ use an existing *internal_shelled_edge* to start the section geometry [Fig. 18.5(c)] ⇒ add four lines and a fillet to complete the section [Fig. 18.5(d)] ⇒ use the dimensioning scheme in Figure 18.5(e) ⇒ [Fig. 18.5(f)]

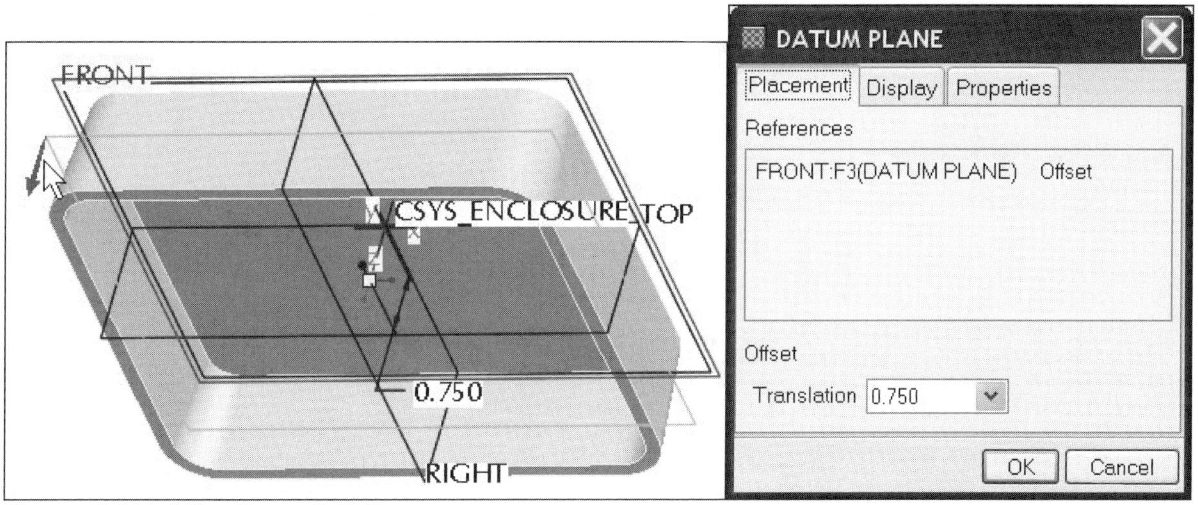

Figure 18.5(a) Offset Datum DTM1

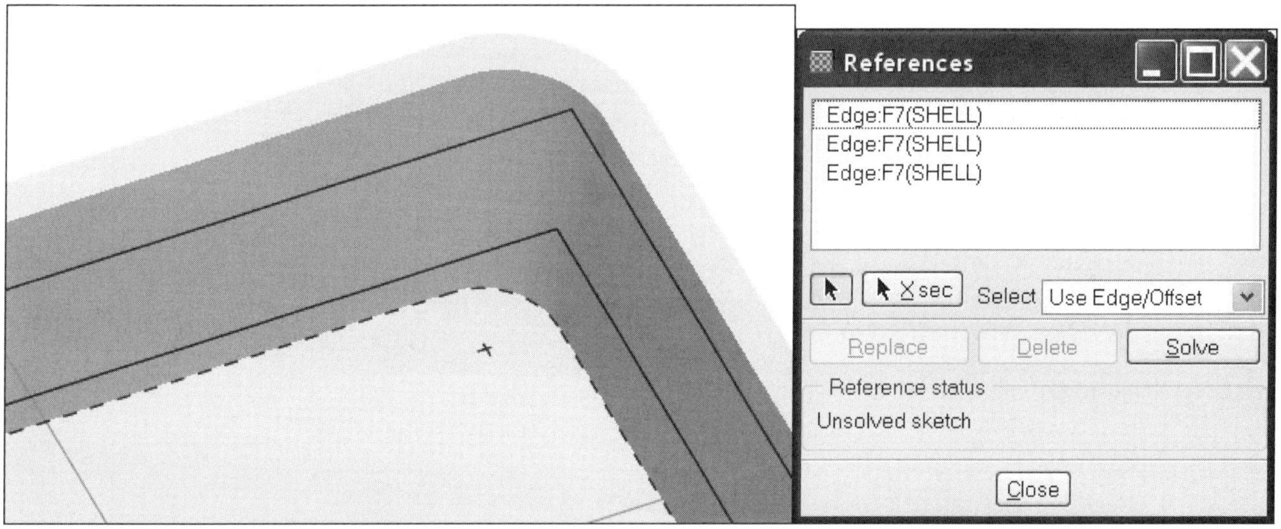

Figure 18.5(b) Extrusion References

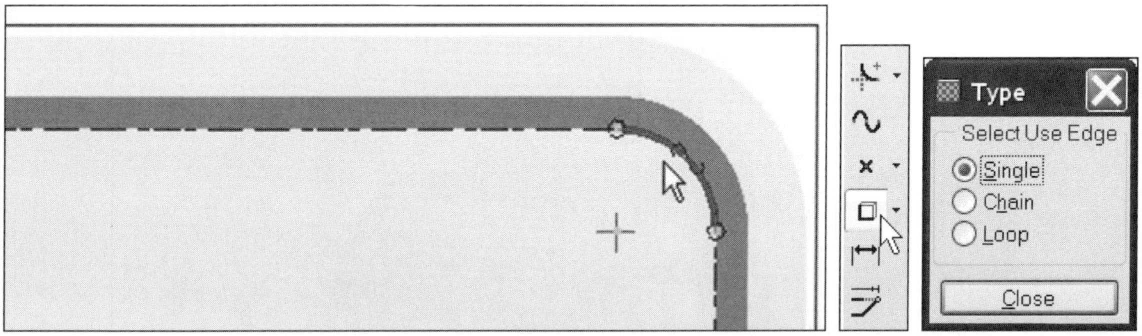

Figure 18.5(c) Create the First Entity using: **Create an entity from an edge**

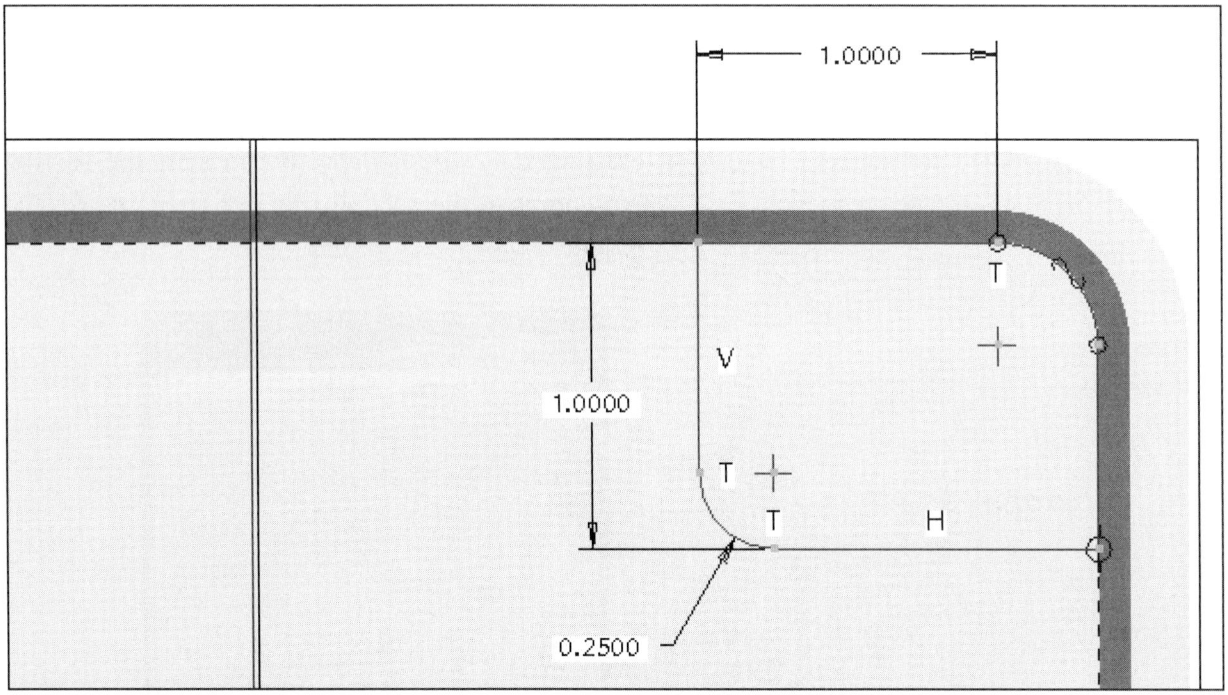

Figure 18.5(d) Section Sketch

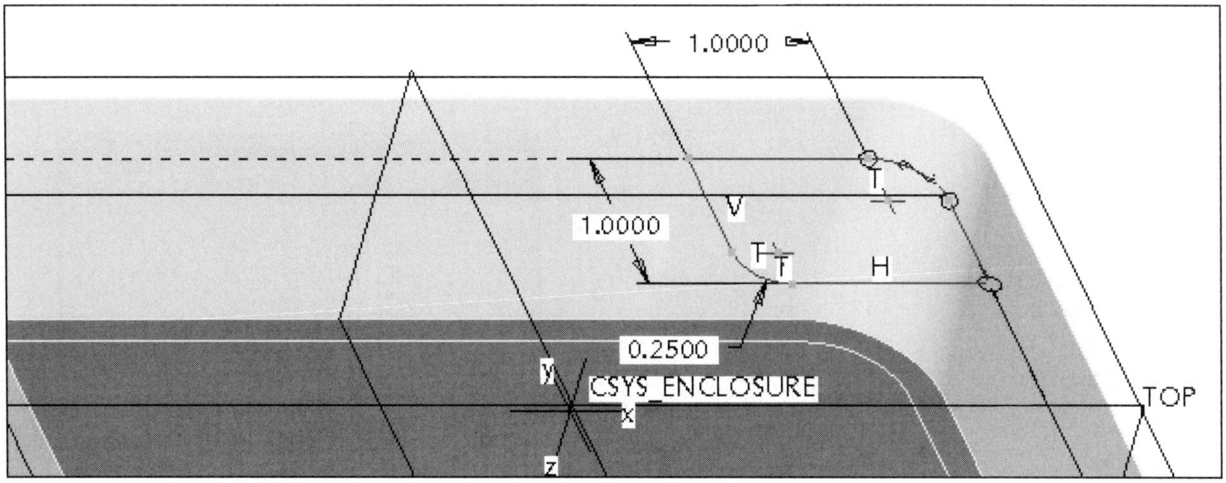

Figure 18.5(e) Standard Orientation of Section Sketch with Design Dimensions

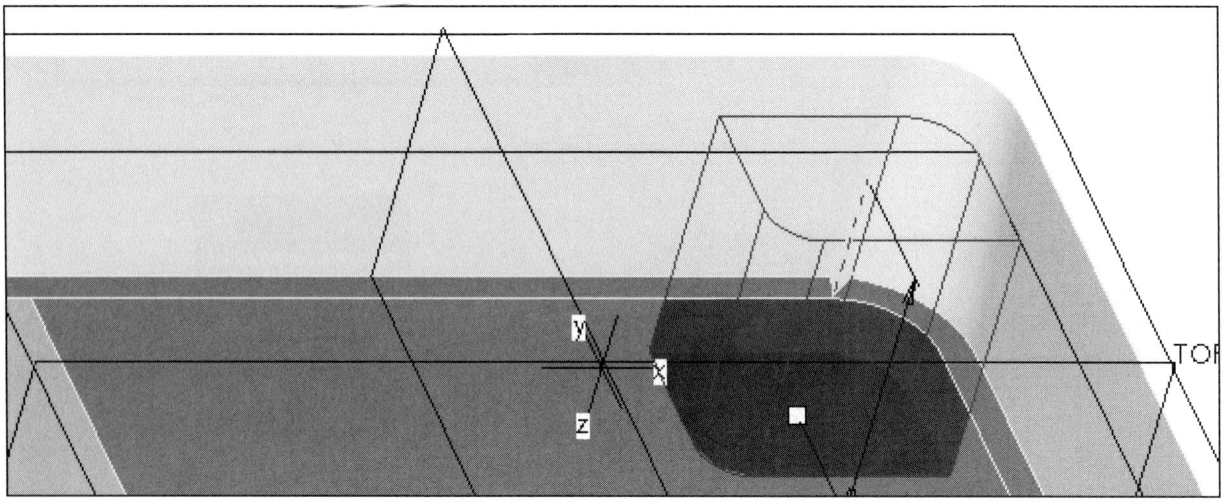

Figure 18.5(f) Previewed Extrusion Depth

Pick on the depth dimension ⇒ **RMB** ⇒ **To Selected** [Fig. 18.5(g)] ⇒ pick **DTM1** [Fig. 18.5(h)] ⇒ **MMB** ⇒ spin the model [Fig. 18.5(i)] ⇒ [save icon] ⇒ **MMB** ⇒ **LMB**

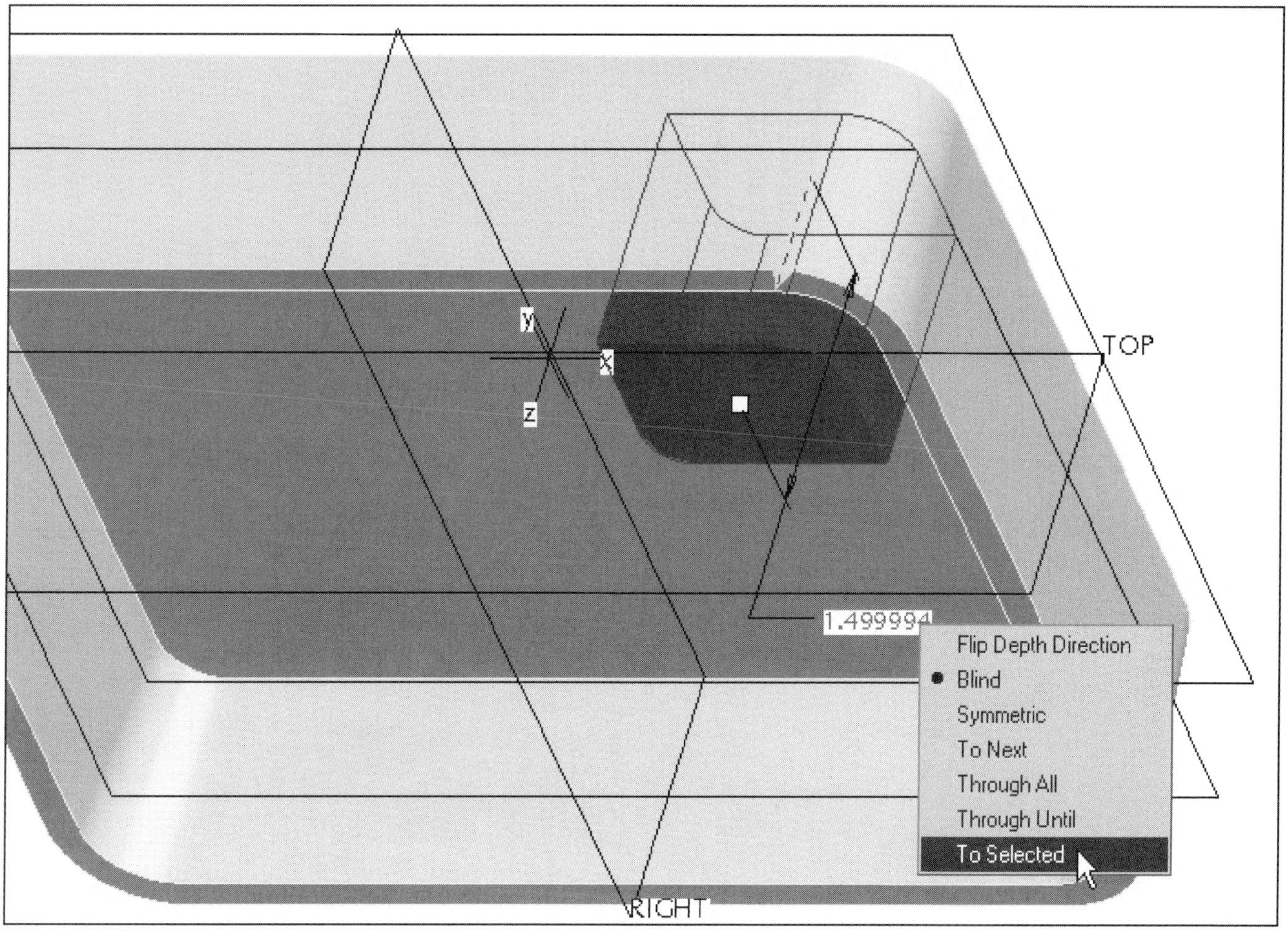

Figure 18.5(g) Determining the Depth Option

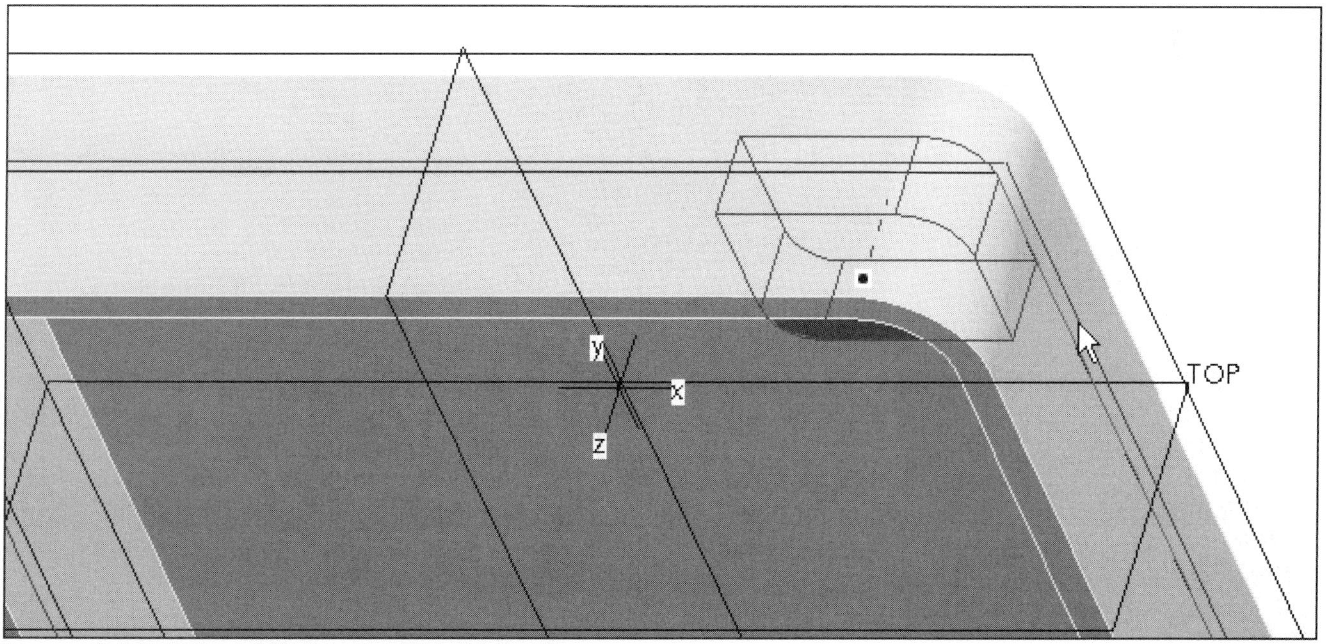

Figure 18.5(h) Pick on DTM1 to Establish the Depth

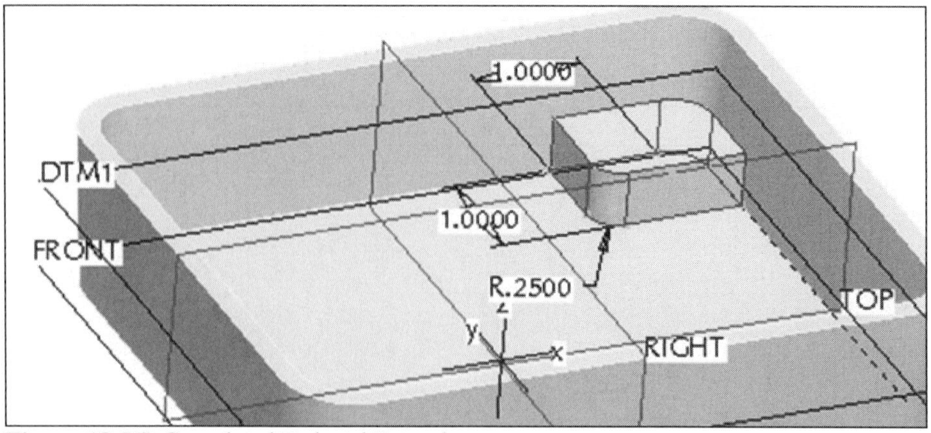

Figure 18.5(i) Completed Pedestal Extrusion

Click: **Draft Tool** ⇒ **References** tab ⇒ select one vertical surface of the pedestal ⇒ in the Draft hinges collector, click in the field labeled: [• Click here to add item] (field changes to [Draft hinges / Select 1 item]) ⇒ pick **FRONT** ⇒ type **5** in Dimension field ⇒ **Enter** ⇒ **Reverse pull direction** (Fig. 18.6) ⇒ **MMB**

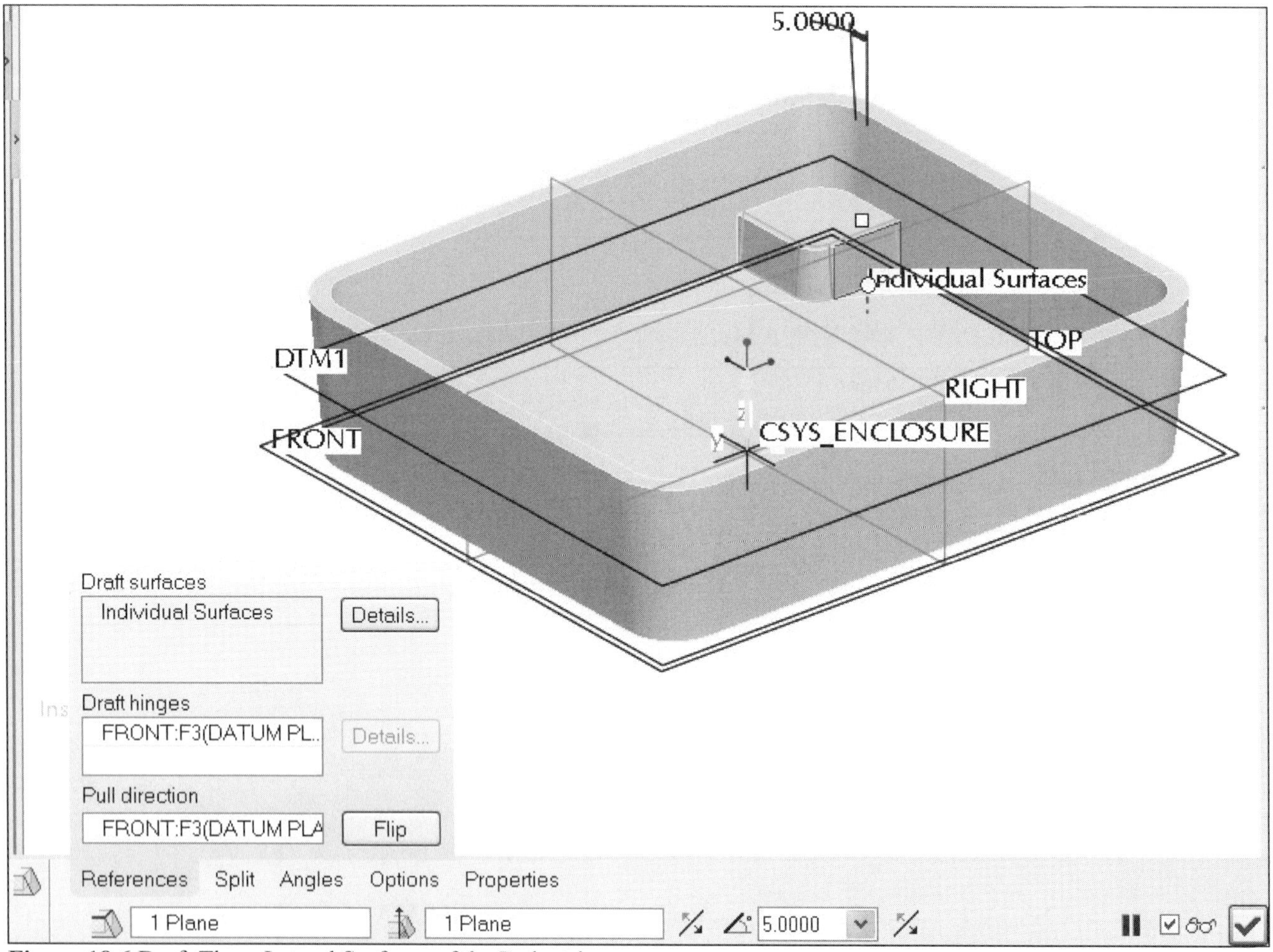

Figure 18.6 Draft Three Lateral Surfaces of the Pedestal

Model the circular extrusion using the top surface of the pedestal as the sketching plane. Click: **Insert** ⇒ **Extrude** ⇒ Keep the default references. The section consists of one circle [Fig. 18.7(a)]. ⇒ ✓ ⇒ rotate the model [Fig. 18.7(b)] ⇒ click on the depth drag handle ⇒ **RMB** ⇒ **To Selected** [Fig. 18.7(c)] ⇒ pick on the surface [Fig. 18.7(d)] ⇒ **MMB** [Fig. 18.7(e)] ⇒ **LMB** to deselect

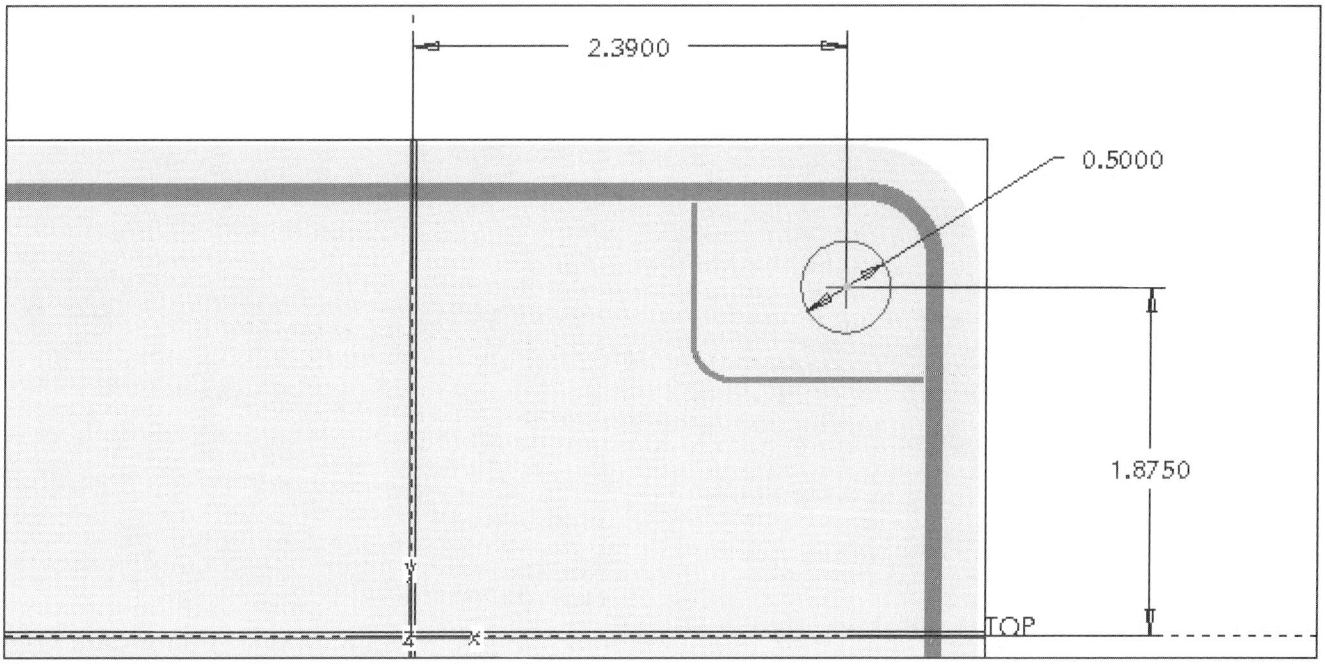

Figure 18.7(a) Section Sketch for Circular Extrusion

Figure 18.7(b) Extrusion Depth

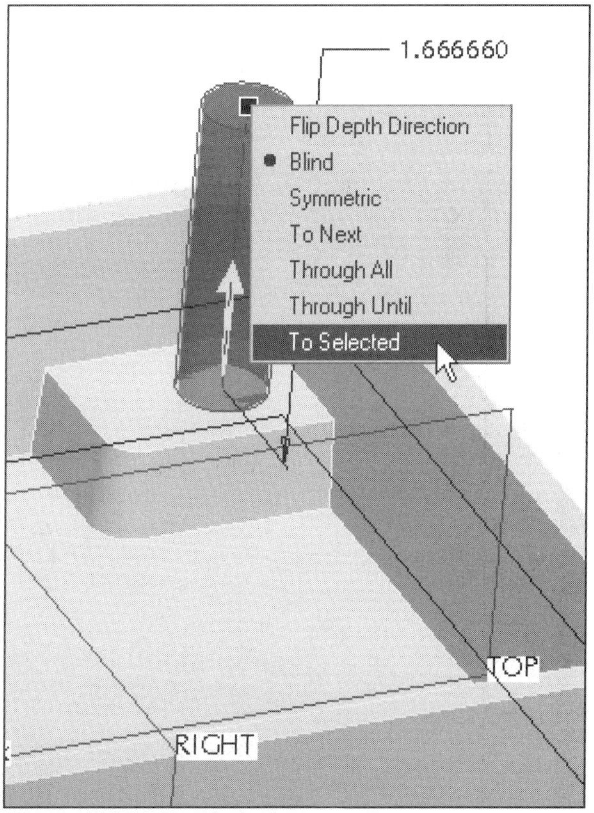

Figure 18.7(c) Depth Options

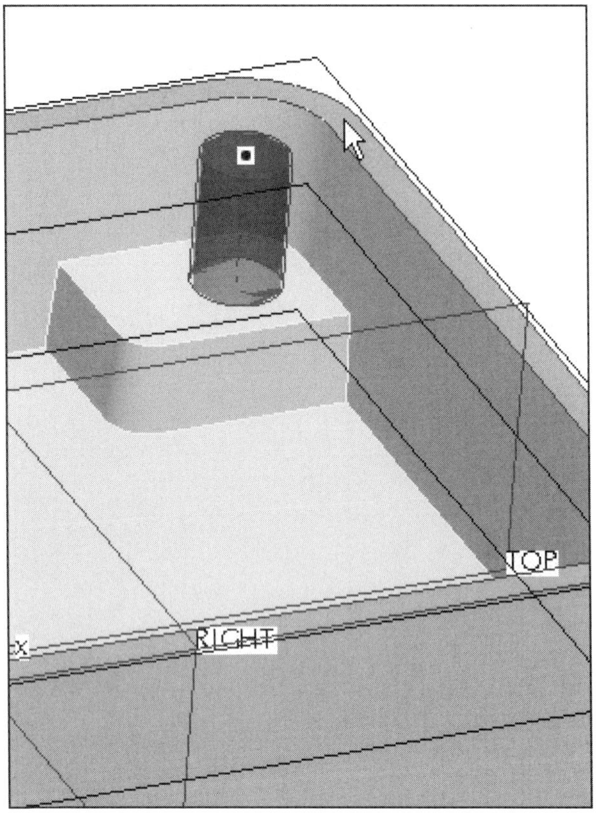

Figure 18.7(d) Select Surface

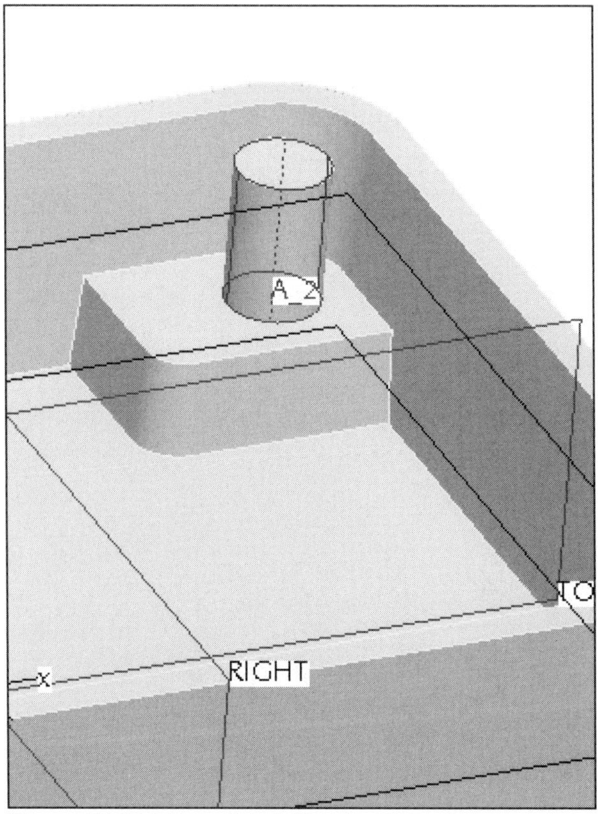

Figure 18.7(e) Completed Circular Extrusion

The circular feature looks correct, but there seems to be a problem with the pedestal.

Pick on the pedestal extrusion in the graphics window ⇒ **RMB** ⇒ **Edit Definition** [Fig. 18.8(a)]

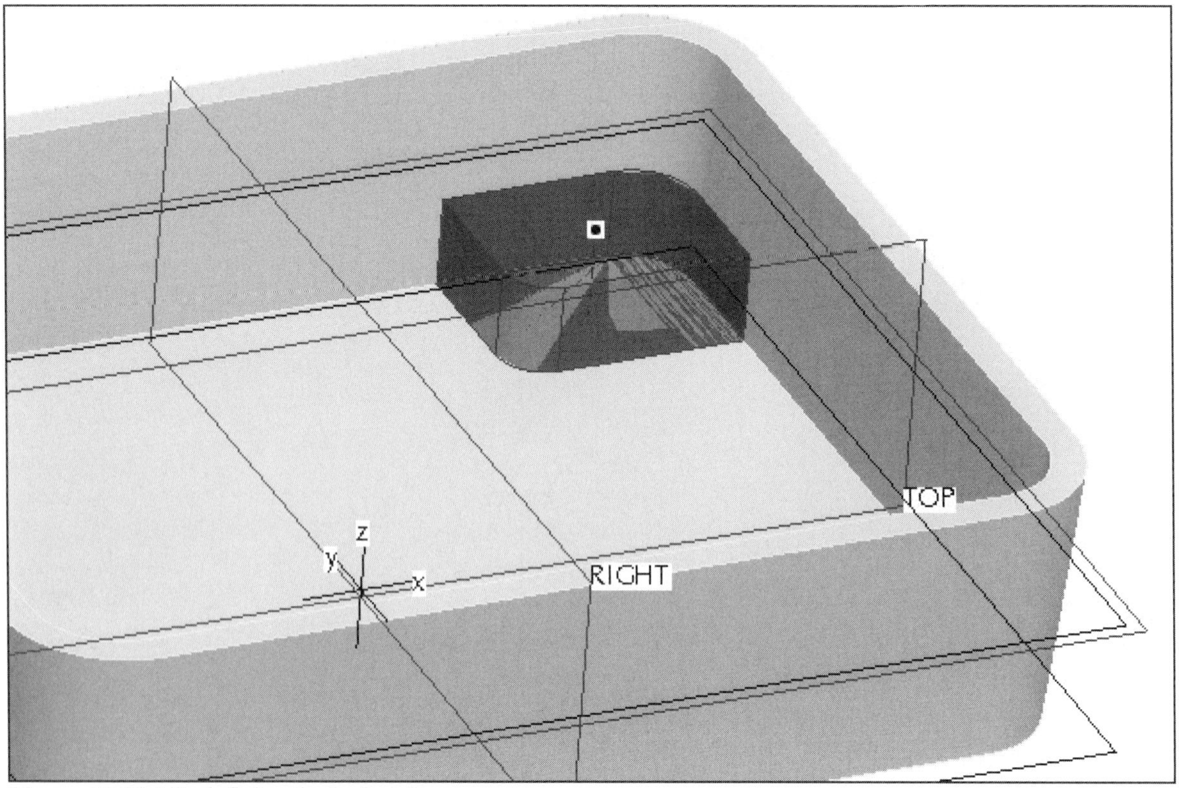

Figure 18.8(a) Redefining the Pedestal

Click: **RMB** ⇒ **Edit Internal Sketch** [Fig. 18.8(b)] ⇒ The dimension is referencing the end of the arc instead of the edge. Create a new defining dimension and modify the new dimension to **1.000** [Fig. 18.8(c)] ⇒ ✔ ⇒ ✔ ⇒ [Figs. 18.8(d-e)] ⇒ **LMB** to deselect

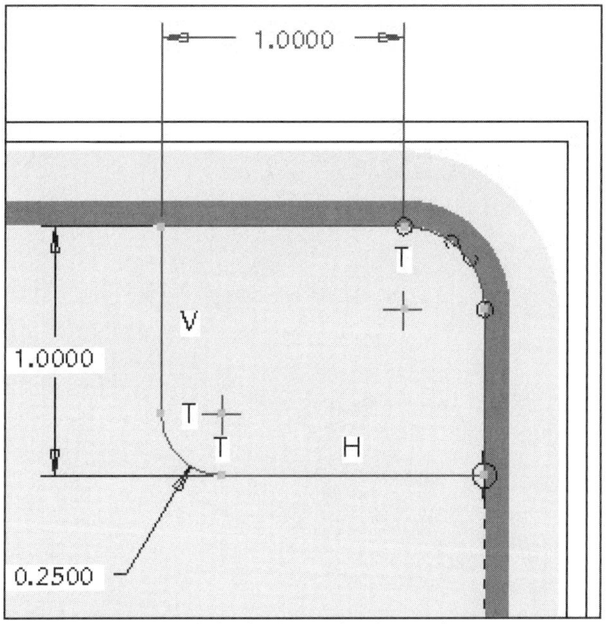

Figure 18.8(b) Original Dimensioning Scheme

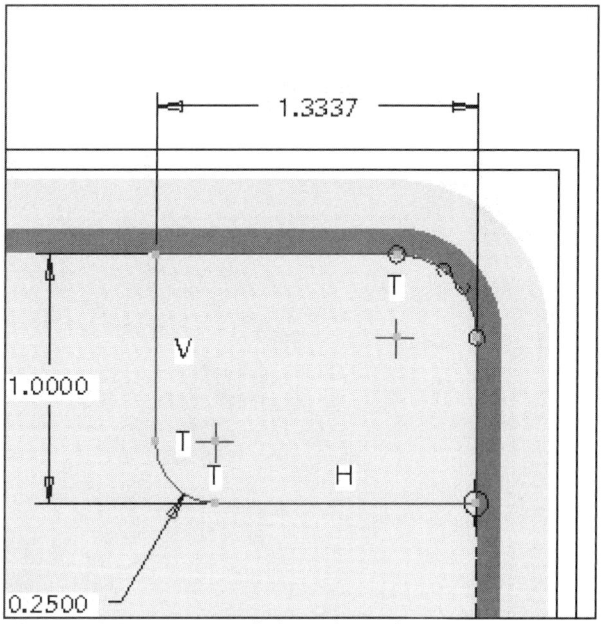

Figure 18.8(c) New Defining Dimension

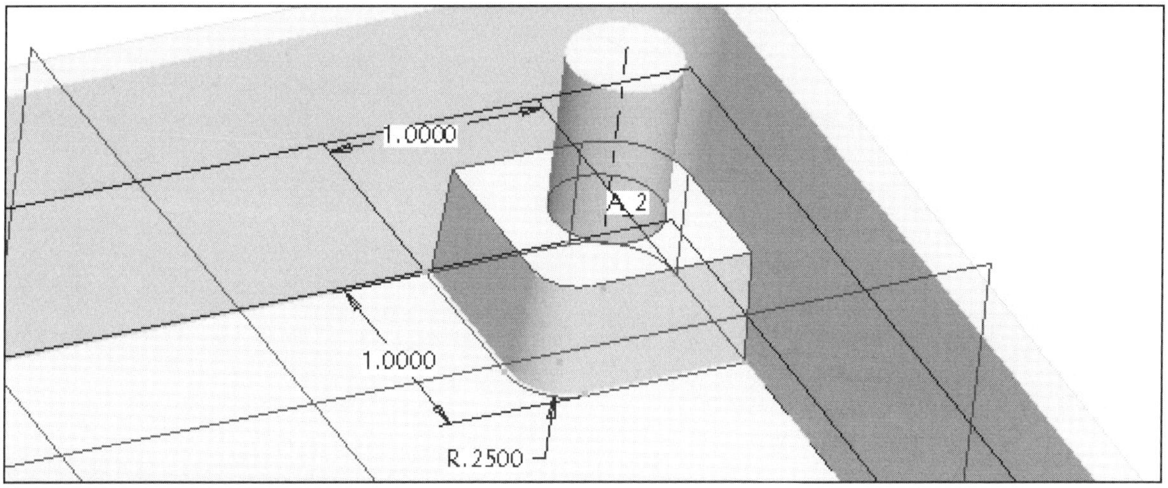

Figure 18.8(d) Redefined Pedestal

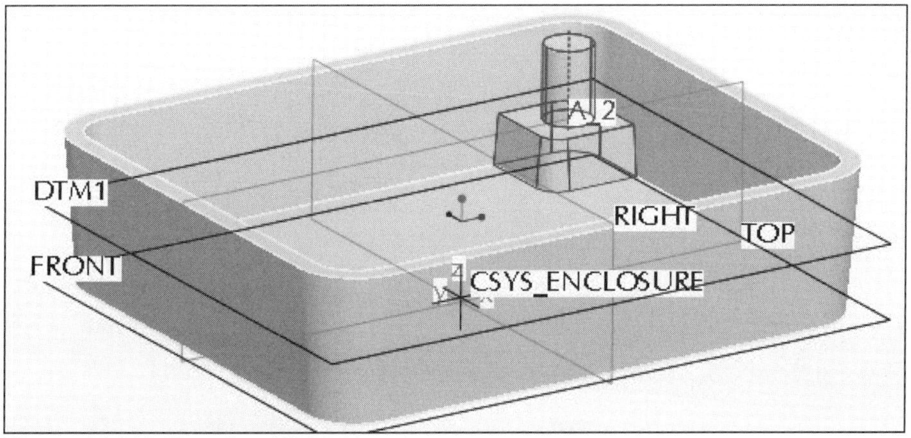

Figure 18.8(e) Completed Features

Draft the circular extrusion at **5°**. Use the upper surface of the circular extrusion as the draft hinge [Fig. 18.9(a)]. Select the drag handle. ⇒ **RMB** ⇒ **Flip Angle** ⇒ **MMB** ⇒ **RMB** ⇒ **Edit** ⇒ change the angle dimension properties to **Flip Arrows** and display **0** number of decimal places [Fig. 18.9(b)].

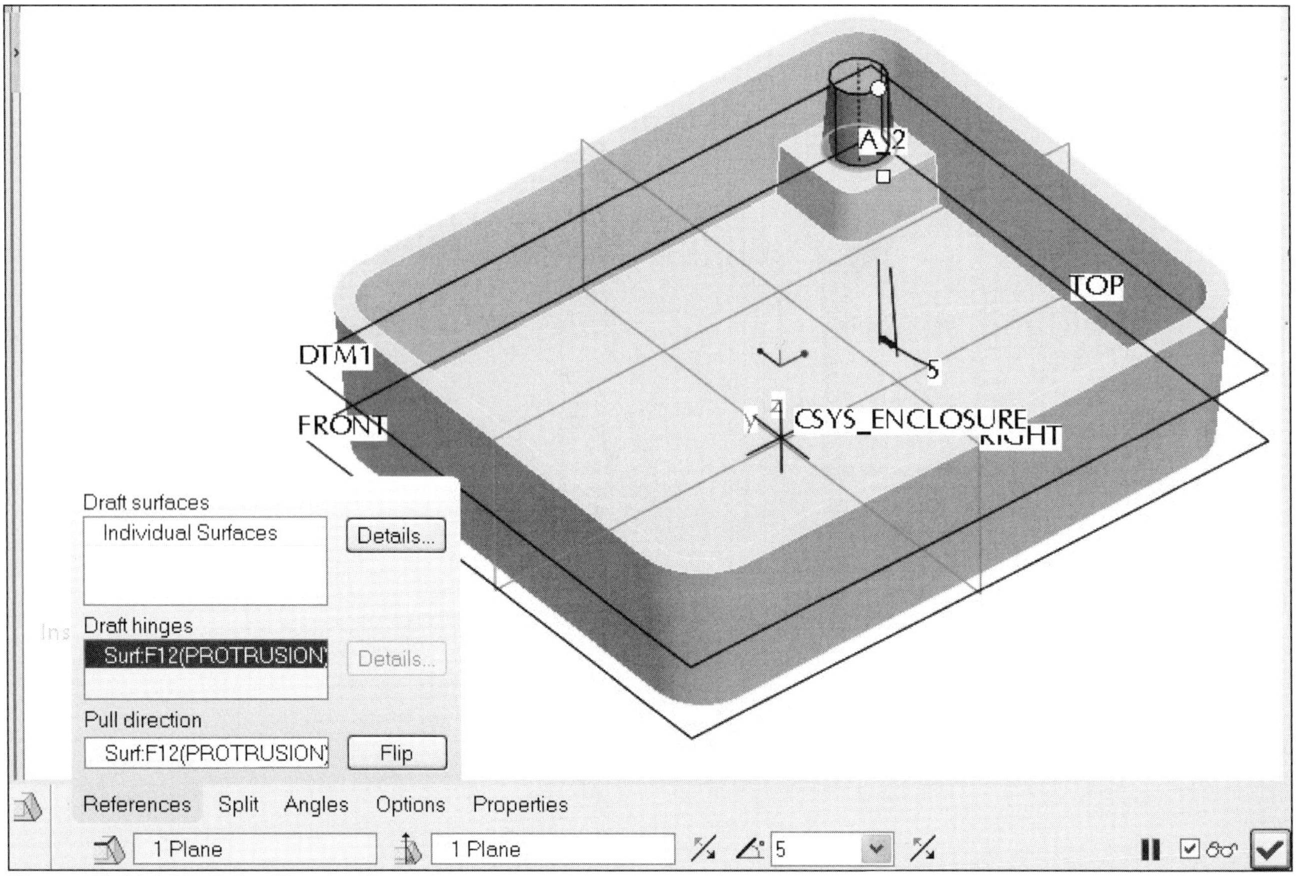

Figure 18.9(a) Draft the Circular Extrusion

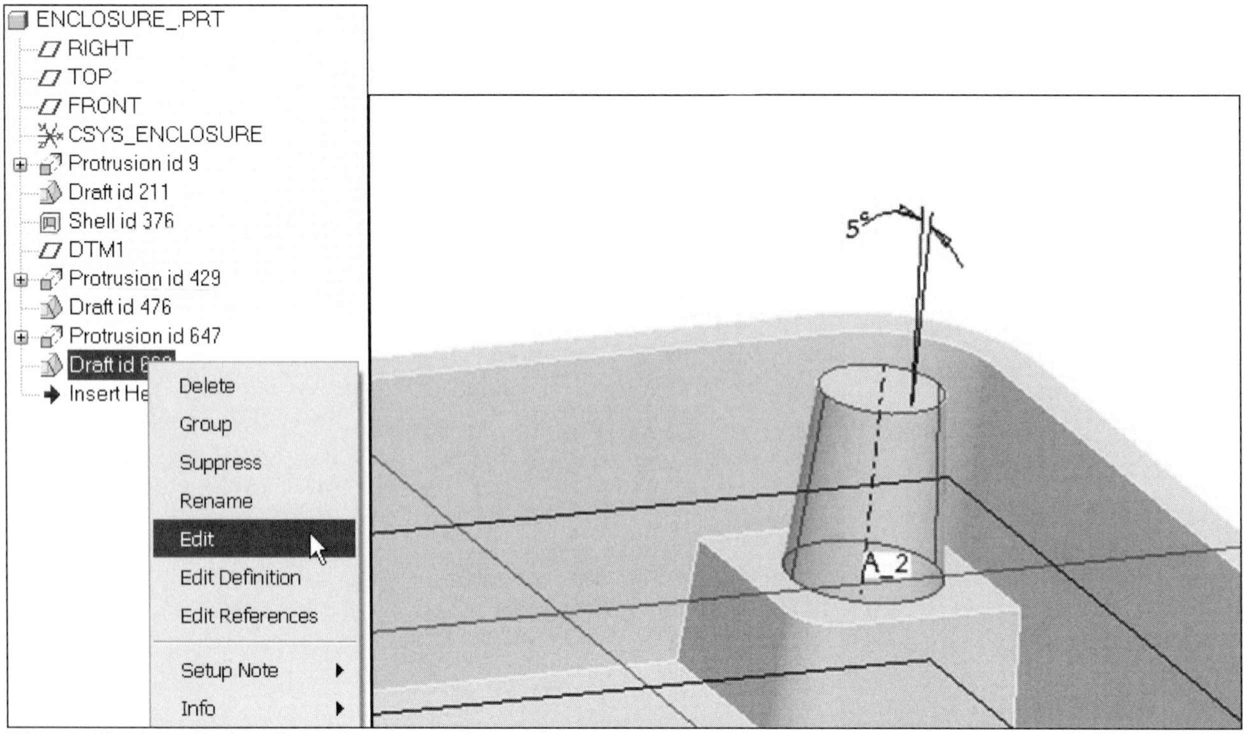

Figure 18.9(b) Edit the Dimensions Properties to Flip Arrows (properties set to 0 decimal places)

Create a **.250** diameter coaxial hole on the upper surface of the circular extrusion [Fig. 18.10(a)]. Use "To Selected" to establish the hole's depth to the top surface of the pedestal [Fig. 18.10(b)]. Complete the feature and save the part.

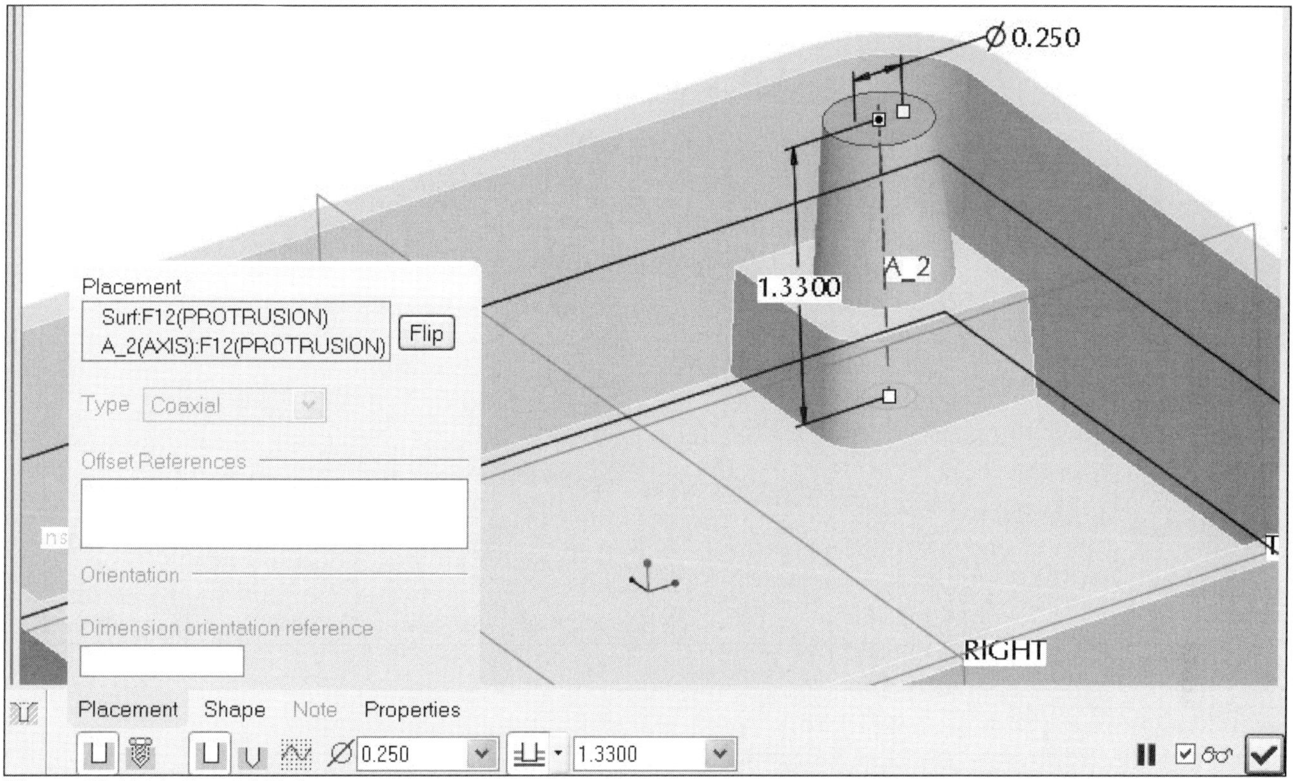

Figure 18.10(a) Coaxial Hole

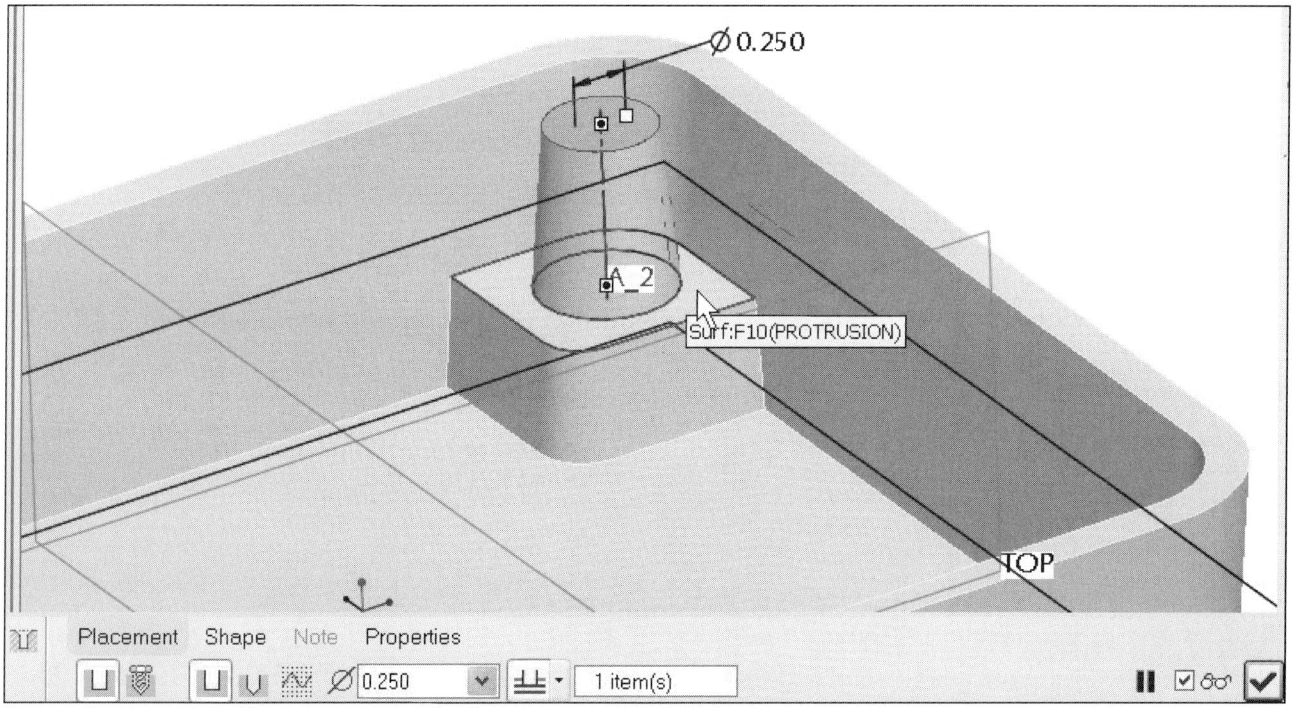

Figure 18.10(b) Hole Depth to Selected Surface

Next, add an internal draft of **.3°** to the coaxial hole (select the top surface of the cylinder as the draft hinge) (Fig. 18.11).

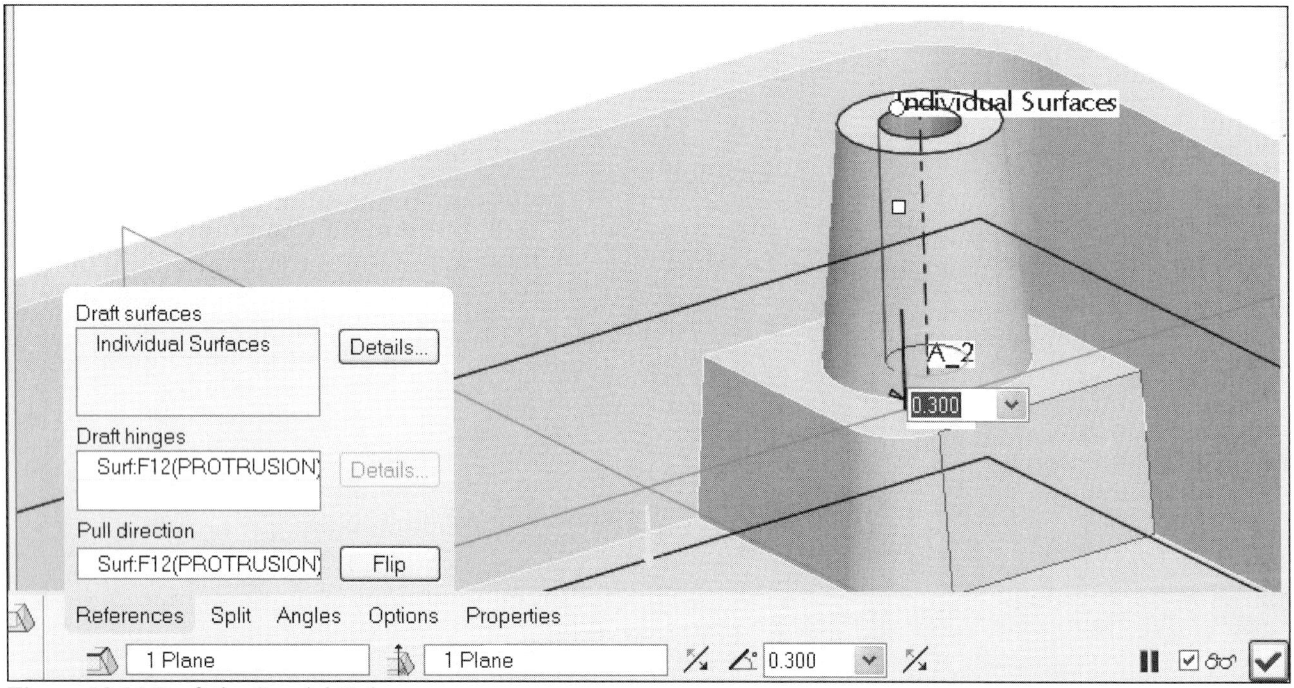

Figure 18.11 Draft the Coaxial Hole

Create the **.0625** and **.125** rounds [Figs. 18.12(a-c)].

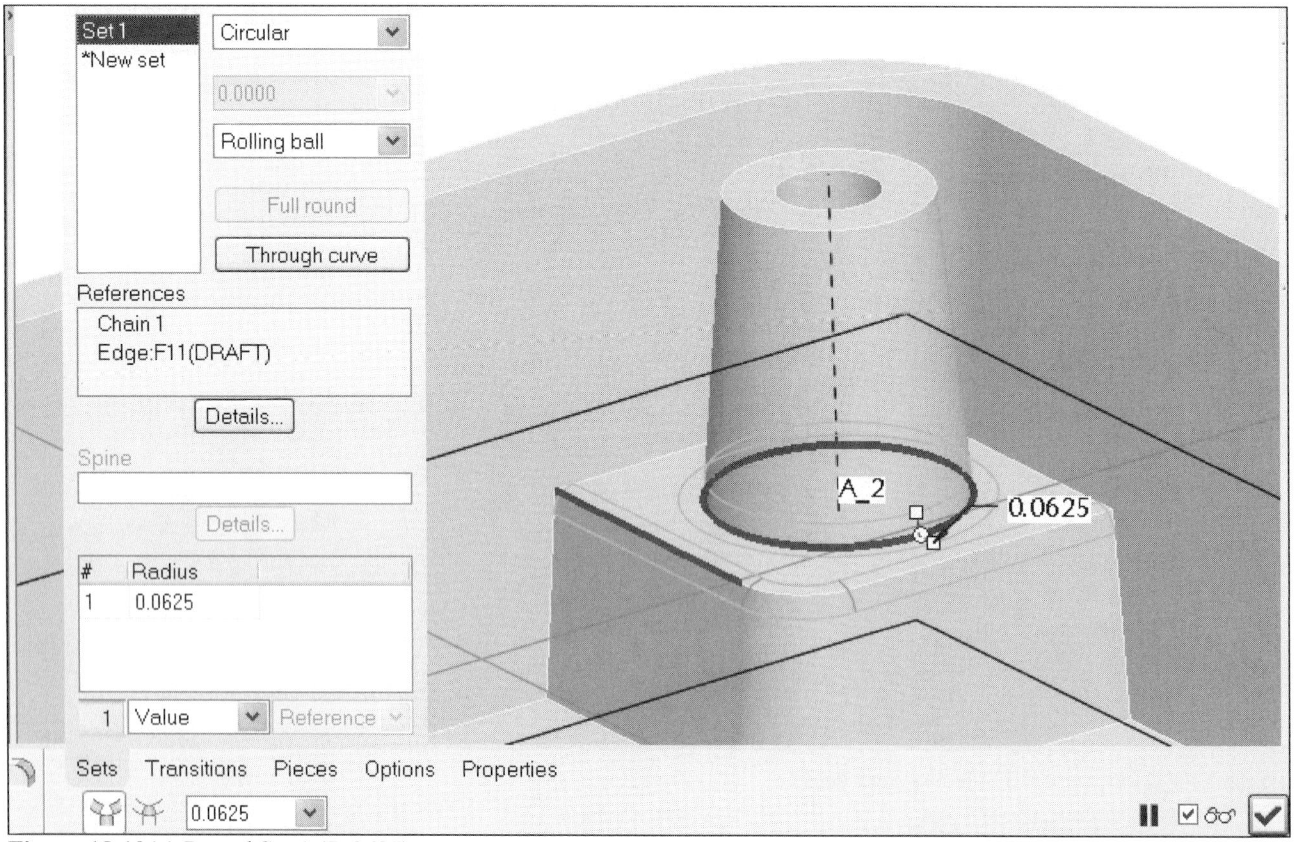

Figure 18.12(a) Round Set 1 (**R.0625**)

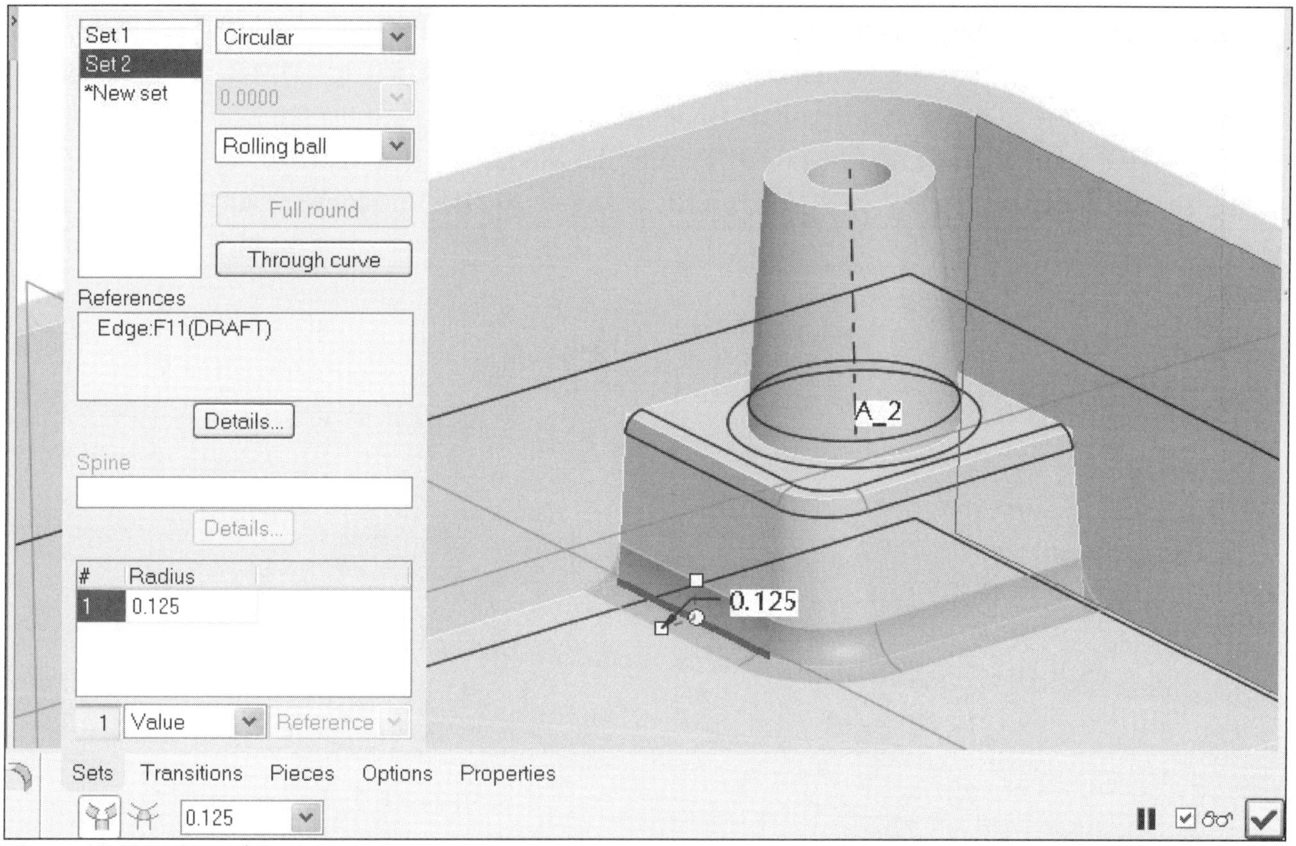

Figure 18.12(b) Round Set 2 (**R.125**)

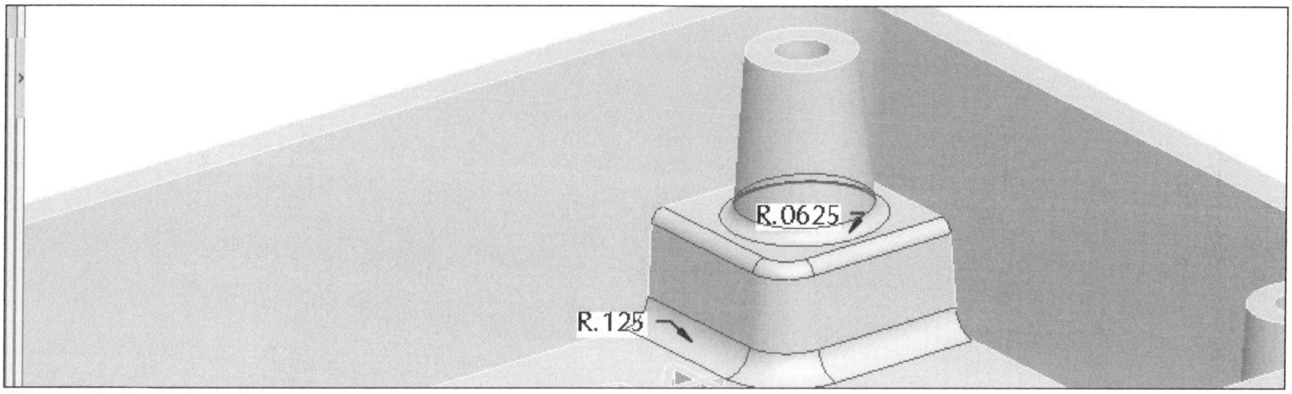

Figure 18.12(c) Completed Rounds (**R.125** and **R.0625**)

Group the extrusions, hole and rounds, click: select features ⇒ **RMB** ⇒ **Group** [Fig. 18.13(a-c)] ⇒ click on the group name ⇒ type **PED_GRP**  ⇒ **Enter**

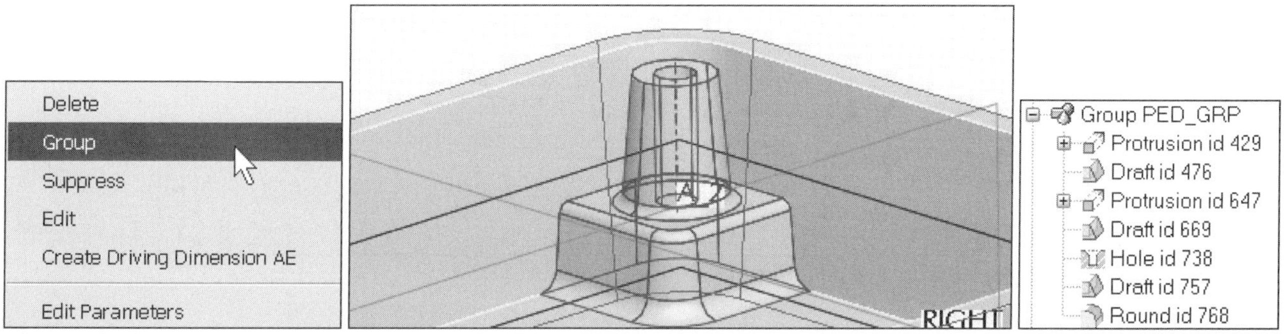

Figure 18.13(a-c) Create a Group (Your Model Tree will Look Different. It will have Extrudes instead of Protrusions)

Create three identical grouped features, click: **Edit** ⇒ **Feature Operations** ⇒ **Copy** ⇒ **Mirror** ⇒ **Dependent** ⇒ **Done** ⇒ from the Model Tree, select: Group **PED_GRP** ⇒ **MMB** ⇒ select datum **RIGHT** [Fig. 18.14(a)] ⇒ **Copy** ⇒ **Mirror** ⇒ **Dependent** ⇒ **Done** ⇒ with the **Ctrl** key pressed, select: Group **PED_GRP** and Group **COPIED_GROUP** from the Model Tree ⇒ **MMB** ⇒ select datum **TOP** [Fig. 18.14(b)] ⇒ **MMB** ⇒ **Ctrl+S** ⇒ **MMB** ⇒ **File** ⇒ **Delete** ⇒ **Old Versions** ⇒ **MMB**

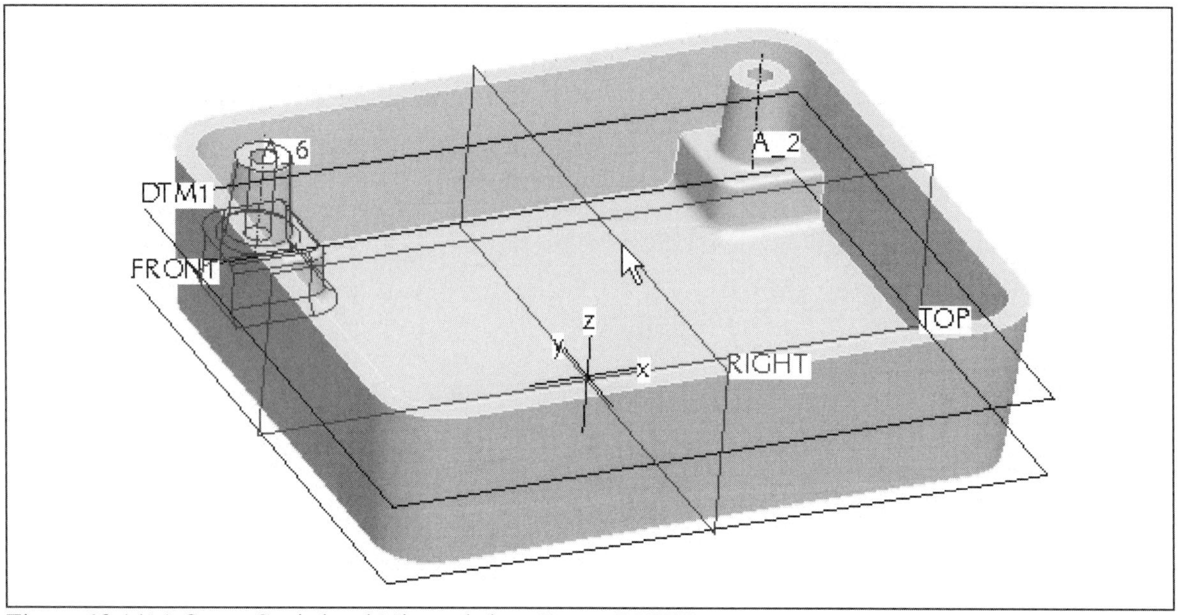

Figure 18.14(a) Group Copied and Mirrored about Datum RIGHT

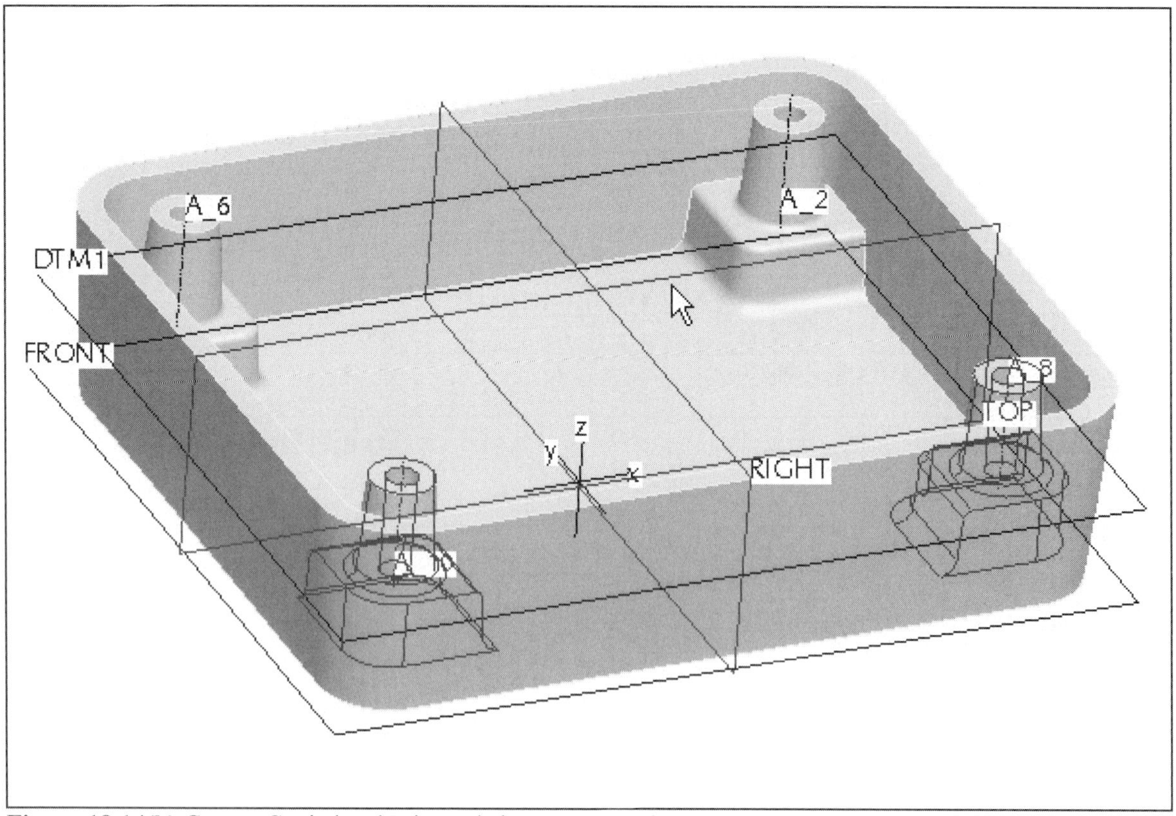

Figure 18.14(b) Groups Copied and Mirrored about Datum TOP

Create the internal round, click: **Round Tool** ⇒ type **.125** ⇒ select the inside of the shelled wall as the first reference [Fig. 18.15(a)] ⇒ **Sets** tab ⇒ press and hold the **Ctrl** key ⇒ select the top surface of the pedestal as the second reference [Fig. 18.15(b)] ⇒ **MMB** *(If you do not get a failure, slightly increase your round size and try again. You need to have the round fail to use the steps on the next few pages.)*

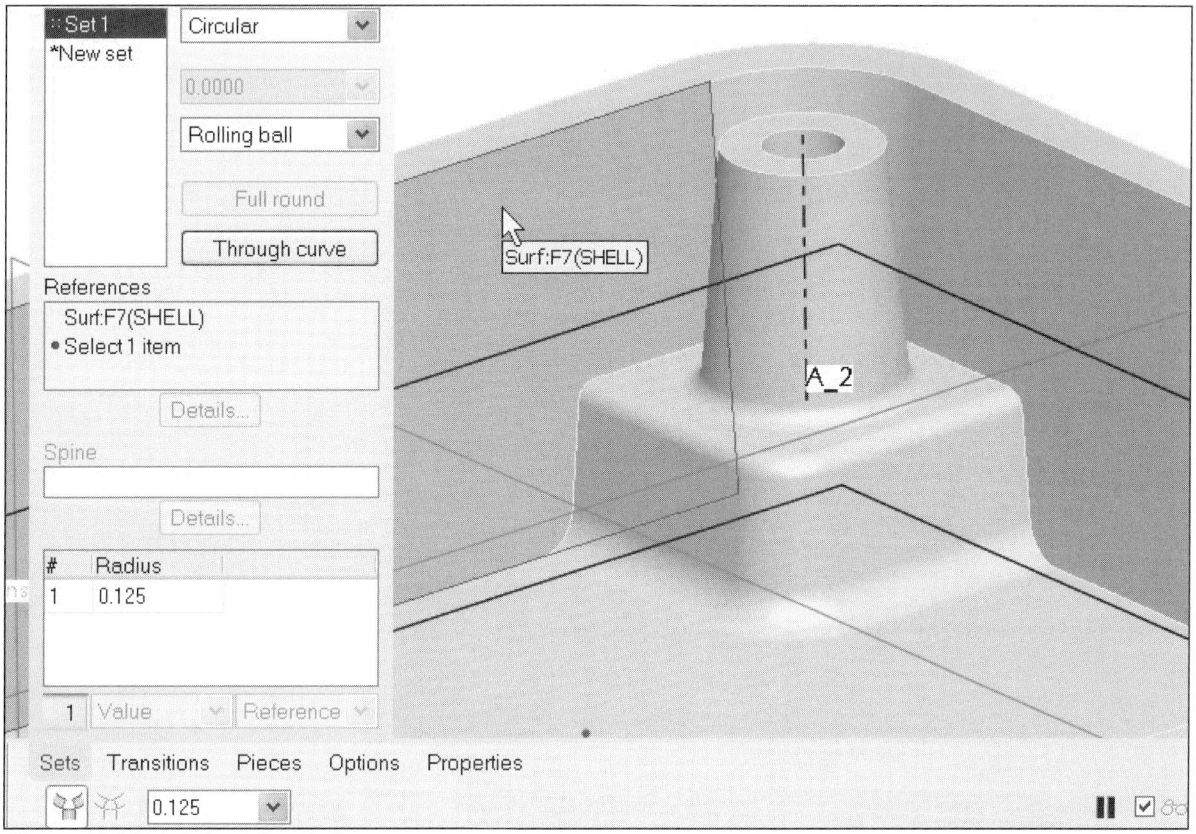

Figure 18.15(a) Select First Reference

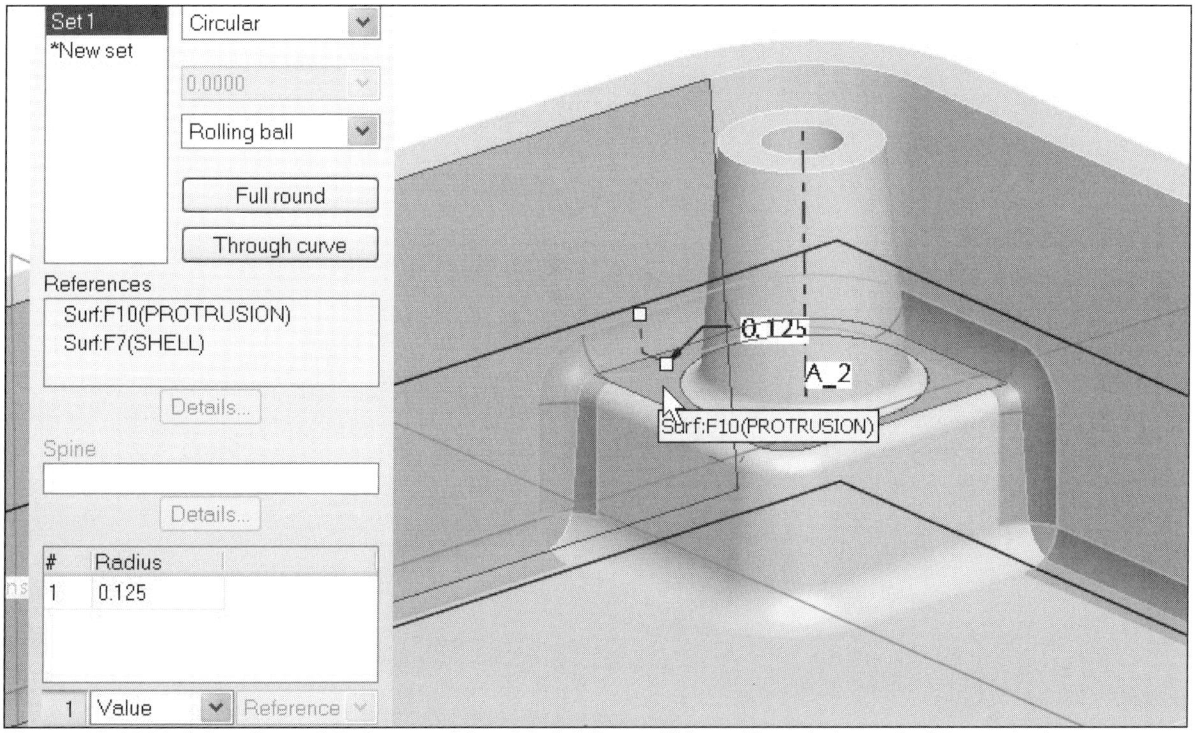

Figure 18.15(b) Select Second Reference (Your Model Tree will have Extrude instead of Protrusion)

Click: Item 2 [Fig. 18.15(c)] ⇒ Note: Item 2 ⇒ **OK** ⇒ from dashboard [Fig. 18.15(d)] ⇒ **Quick Fix** from RESOLVE FEAT menu ⇒ **Redefine** ⇒ **Confirm** ⇒ **Options** tab ⇒ **Surface** [Fig. 18.15(e)] ⇒ **Preview** ⇒ **Resume** *(if you entered a value to make the round fail and the preview does not look similar to Fig. 18.15(e), change the round size back to .125)* ⇒ **MMB**

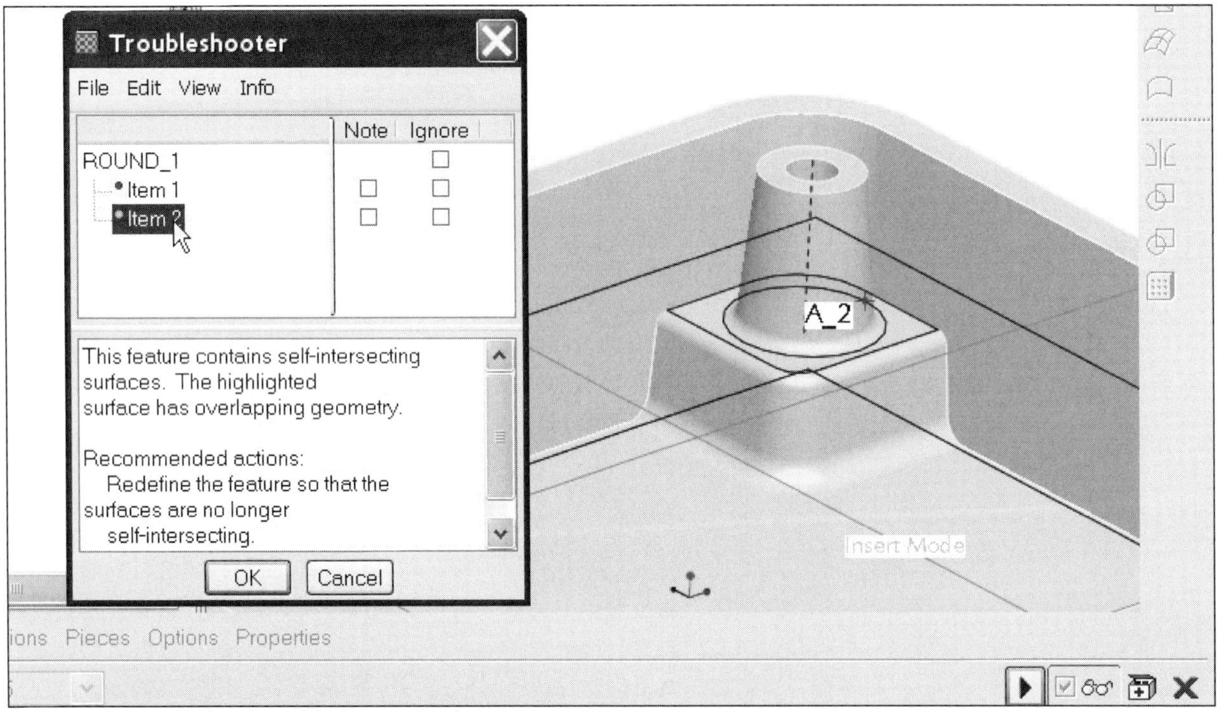

Figure 18.15(c) Select Troubleshooter Option

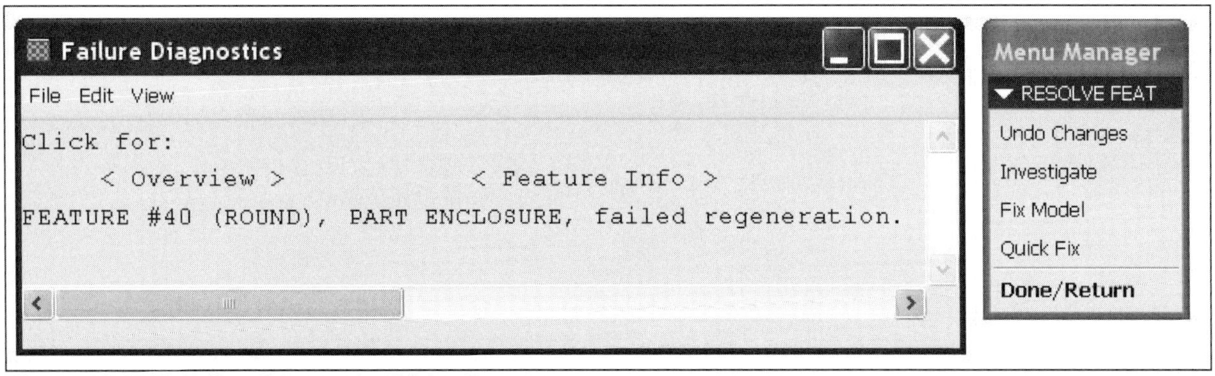

Figure 18.15(d) Failure Diagnostics Dialog Box

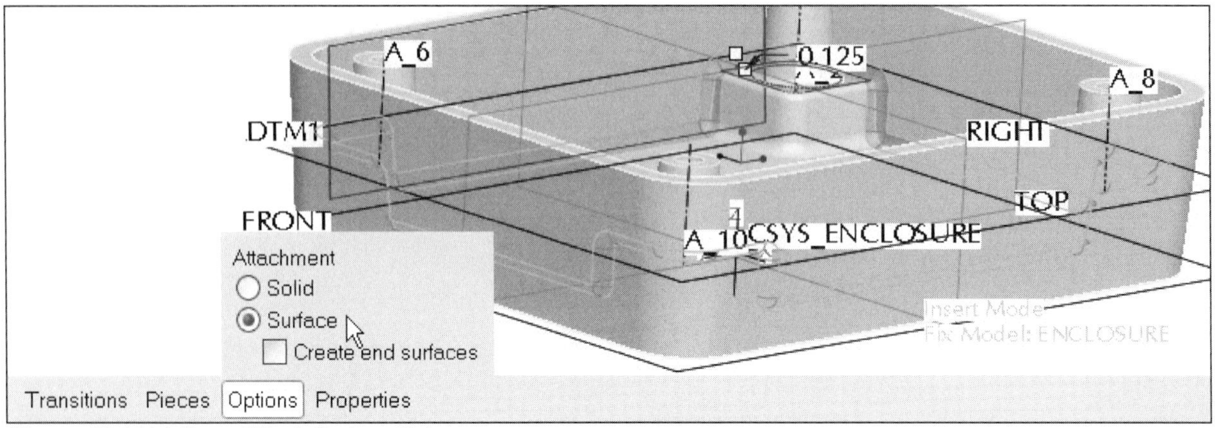

Figure 18.15(e) Round Tool Options Tab

The part's internal round may look correct, but it is not a solid round. Next, model a datum plane through the axis of two pedestals and use the datum plane to create a cross section.

Click: ▱ **Datum Plane Tool** ⇒ References: pick axis **A_2** [Fig. 18.16(a)] ⇒ press and hold the **Ctrl** key ⇒ References: pick axis **A_6** *(your id's may be different)* [Fig. 18.16(b)] ⇒ **OK** (creates DTM2)

(If you have any Troubleshooter notes remaining on the screen, click: ***Smart*** *⇒* ***Annotation*** *⇒ select the note(s) ⇒* ***RMB*** *⇒* ***Delete*** *⇒* ***Yes****)*

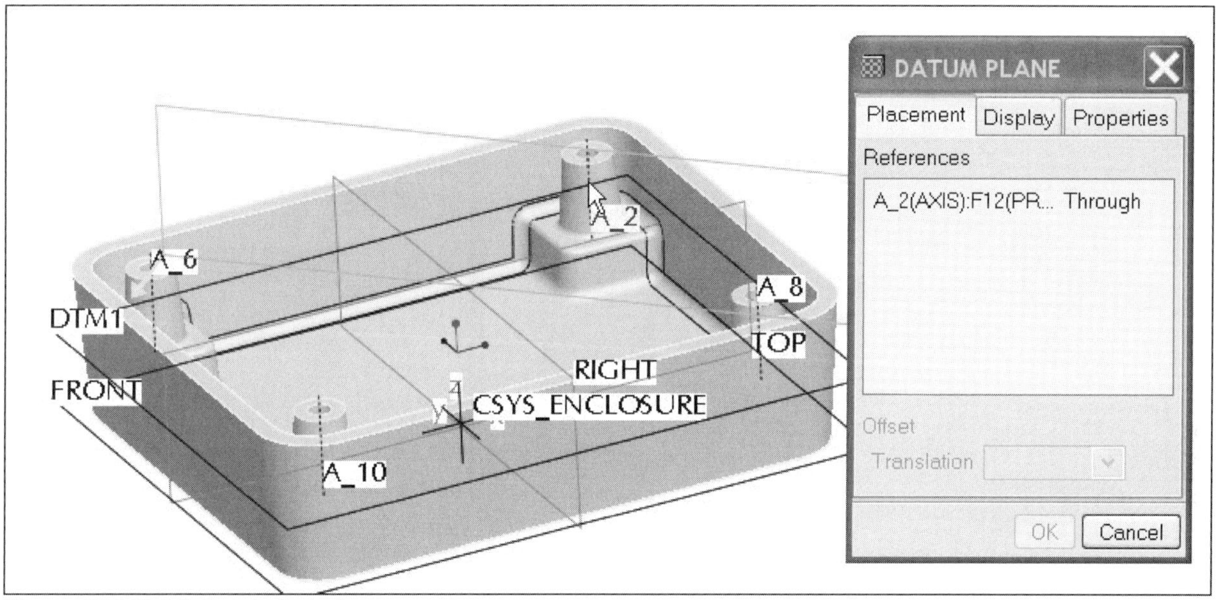

Figure 18.16(a) Select Axis A_2

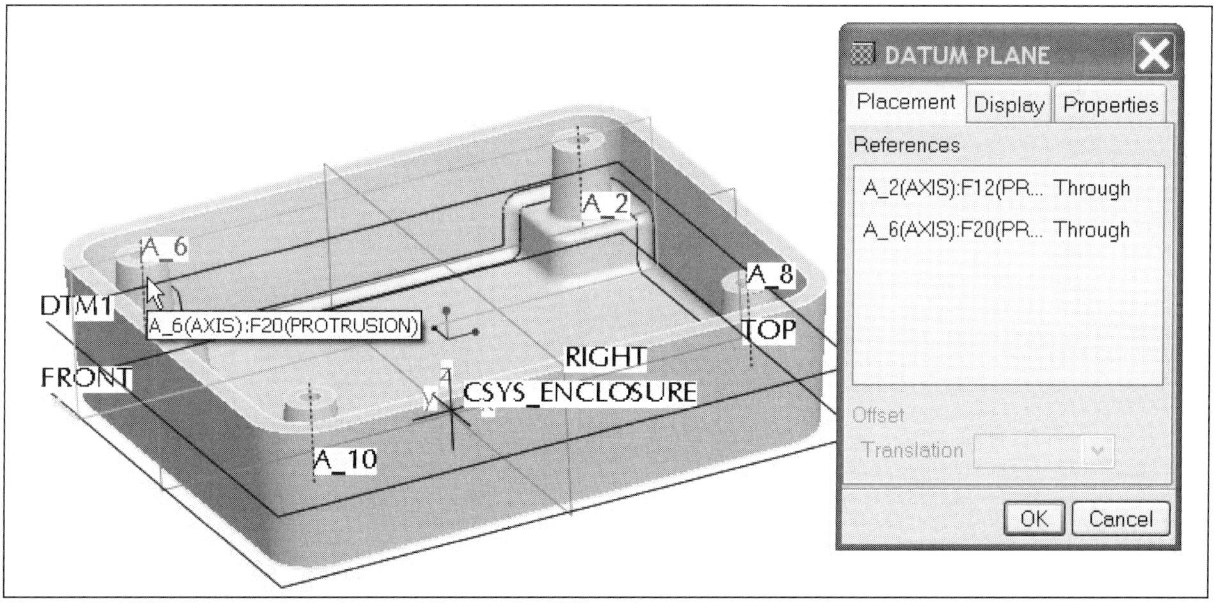

Figure 18.16(b) Select Axis A_6 (your id's may be different)

Create a cross section through the part, using datum DTM2.

Click: **Start the view manager** from Top Toolchest (View Manager dialog box displays) ⇒ **Xsec** tab ⇒ **New** ⇒ type name **A** ⇒ **Enter** ⇒ **Model** ⇒ **Planar** ⇒ **Single** ⇒ **MMB** ⇒ **Plane** ⇒ pick **DTM2** ⇒ click on **A** ⇒ **RMB** ⇒ **Visibility** [Fig. 18.17(a)] ⇒ click on section **A** ⇒ **RMB** ⇒ **Redefine** [Fig. 18.17(b)] ⇒ **Hatching** ⇒ **Fill** ⇒ **Done** ⇒ **Done/Return**

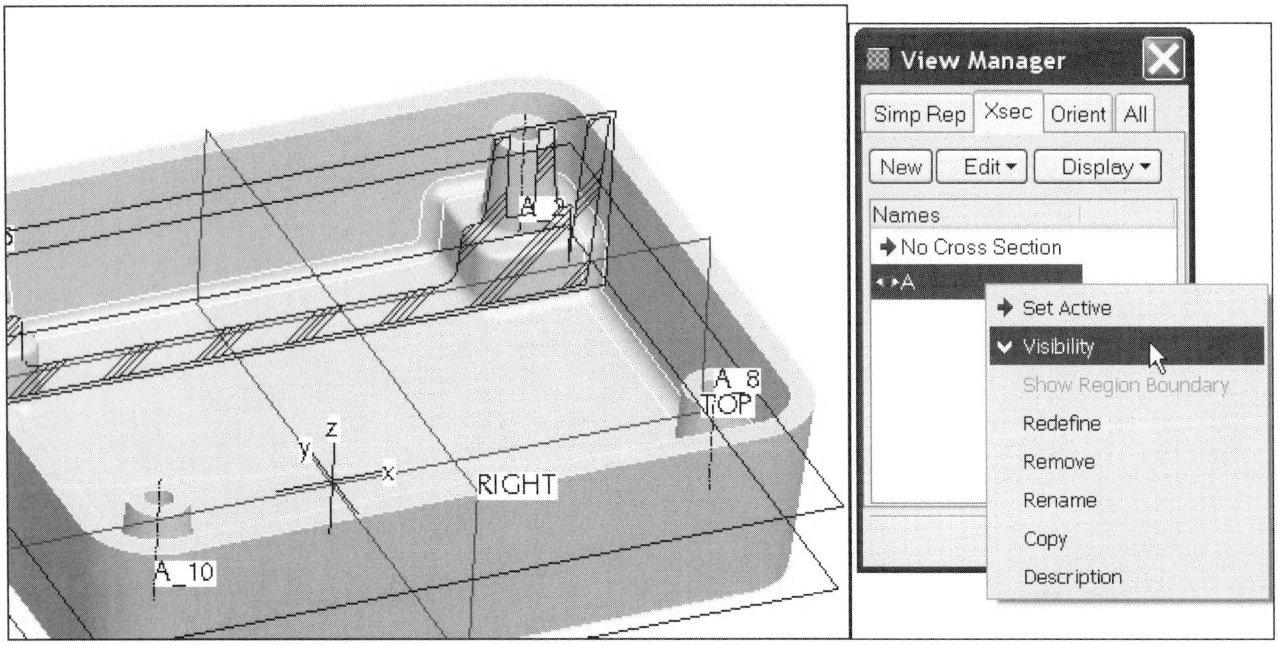

Figure 18.17(a) Show X-Section

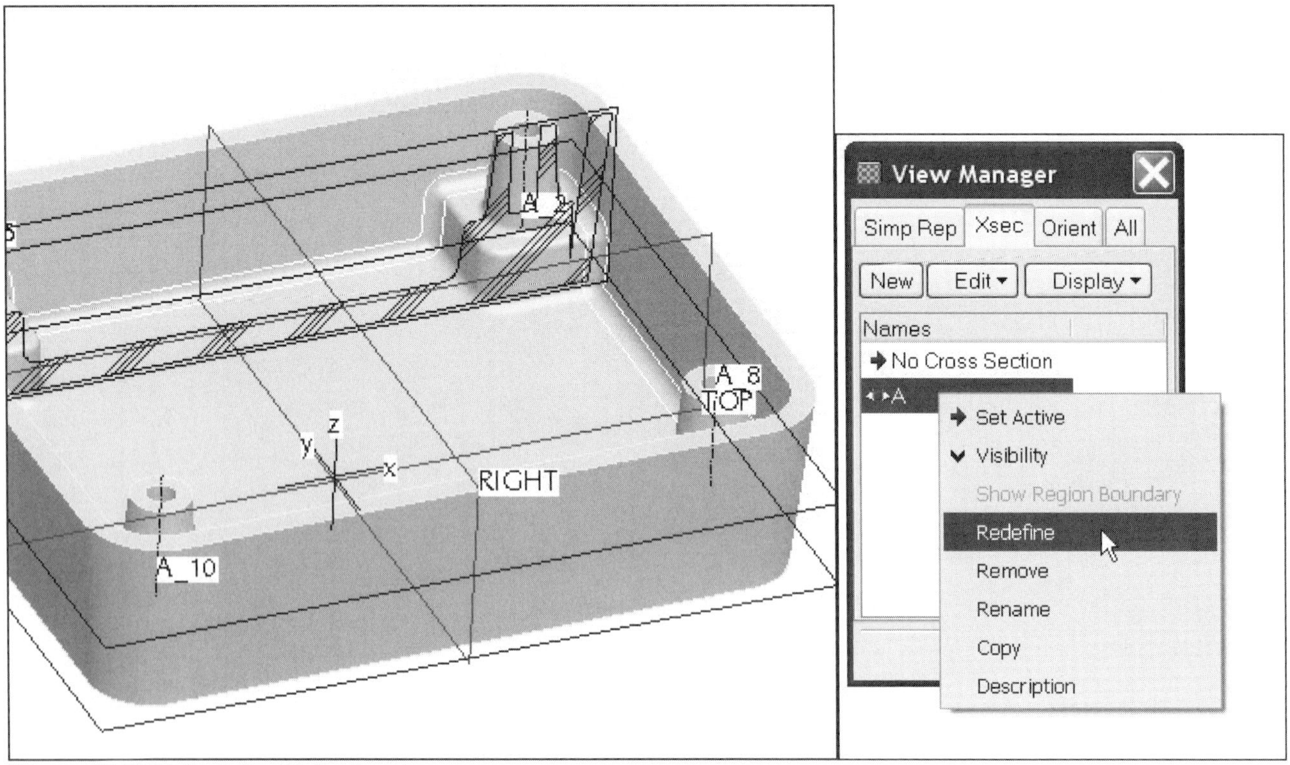

Figure 18.17(b) Redefine X-Section A

The gaps between the pedestals and the shelled walls were not filled by the Round Tool because the round was a surface round.

Click on **A** [Fig. 18.17(c)] ⇒ **RMB** ⇒ **Set Active** [Fig. 18.17(d)] ⇒ spin the model ⇒ **RMB** on No Cross Section ⇒ **Set Active** ⇒ click on **A** ⇒ **RMB** ⇒ **Visibility** (uncheck) ⇒ **Close**

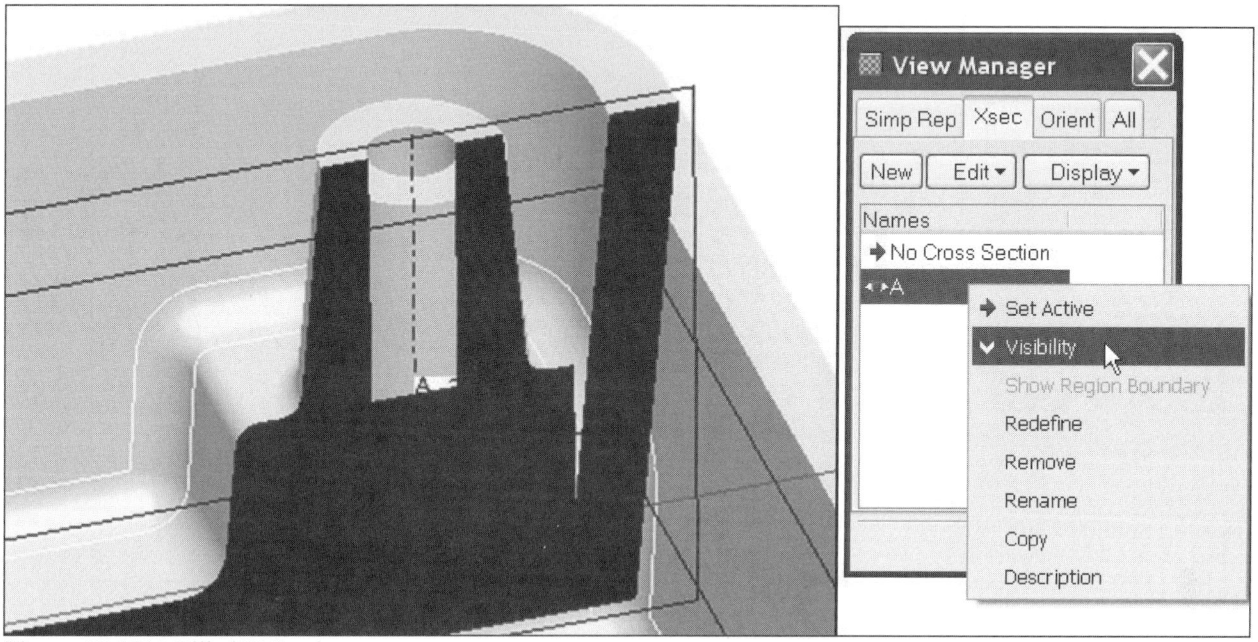

Figure 18.17(c) Filled X-Section

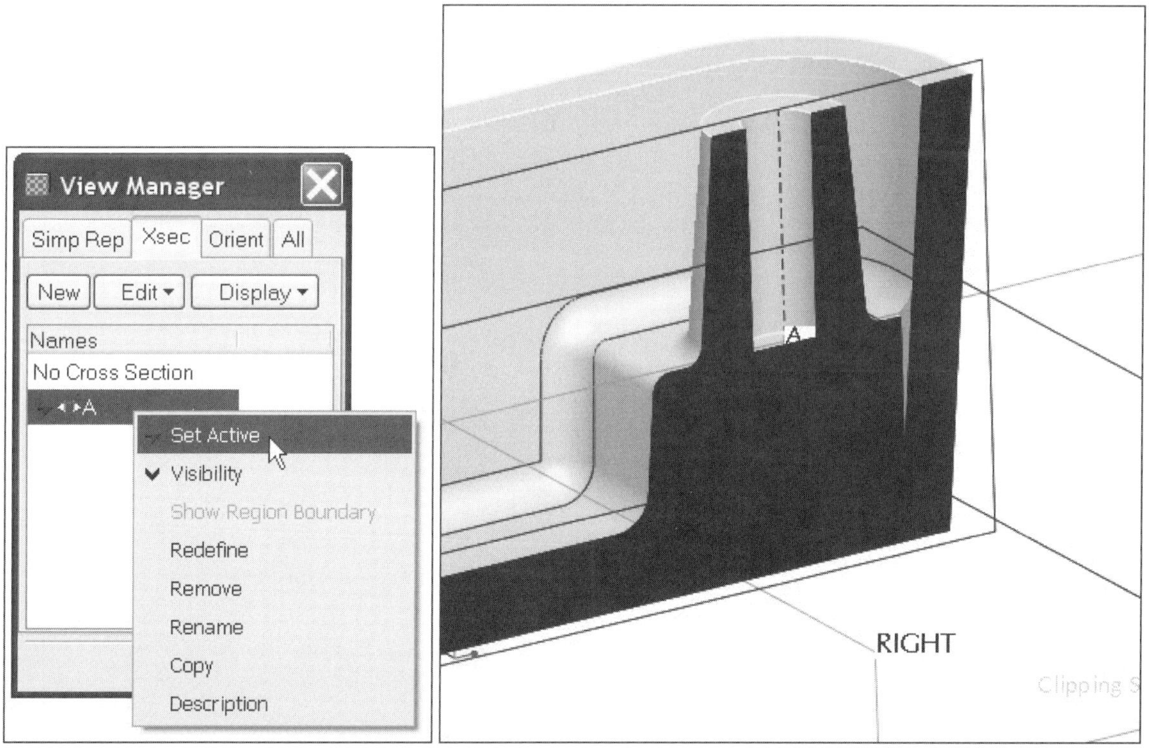

Figure 18.17(d) Xsec A Set Active

Click: on the surface round in the Model Tree (it will highlight on the model) ⇒ **Edit** from the menu bar ⇒ **Solidify** [Fig. 18.18(a)] ⇒ **References** tab [Fig. 18.18(b)]

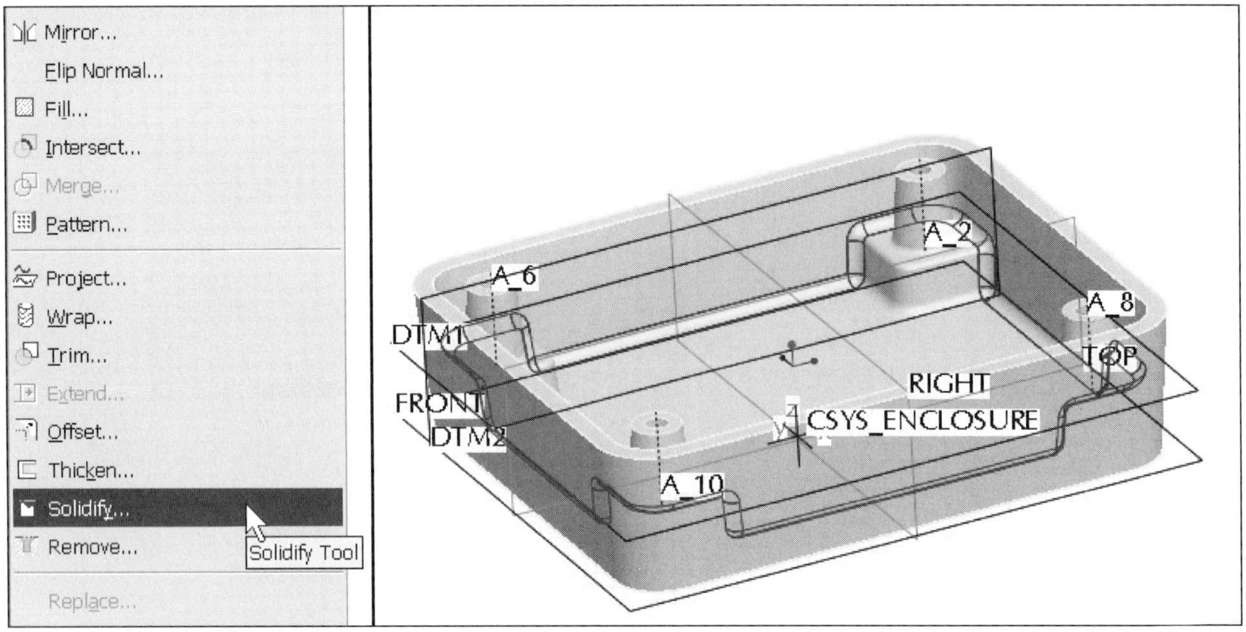

Figure 18.18(a) Solidify Tool

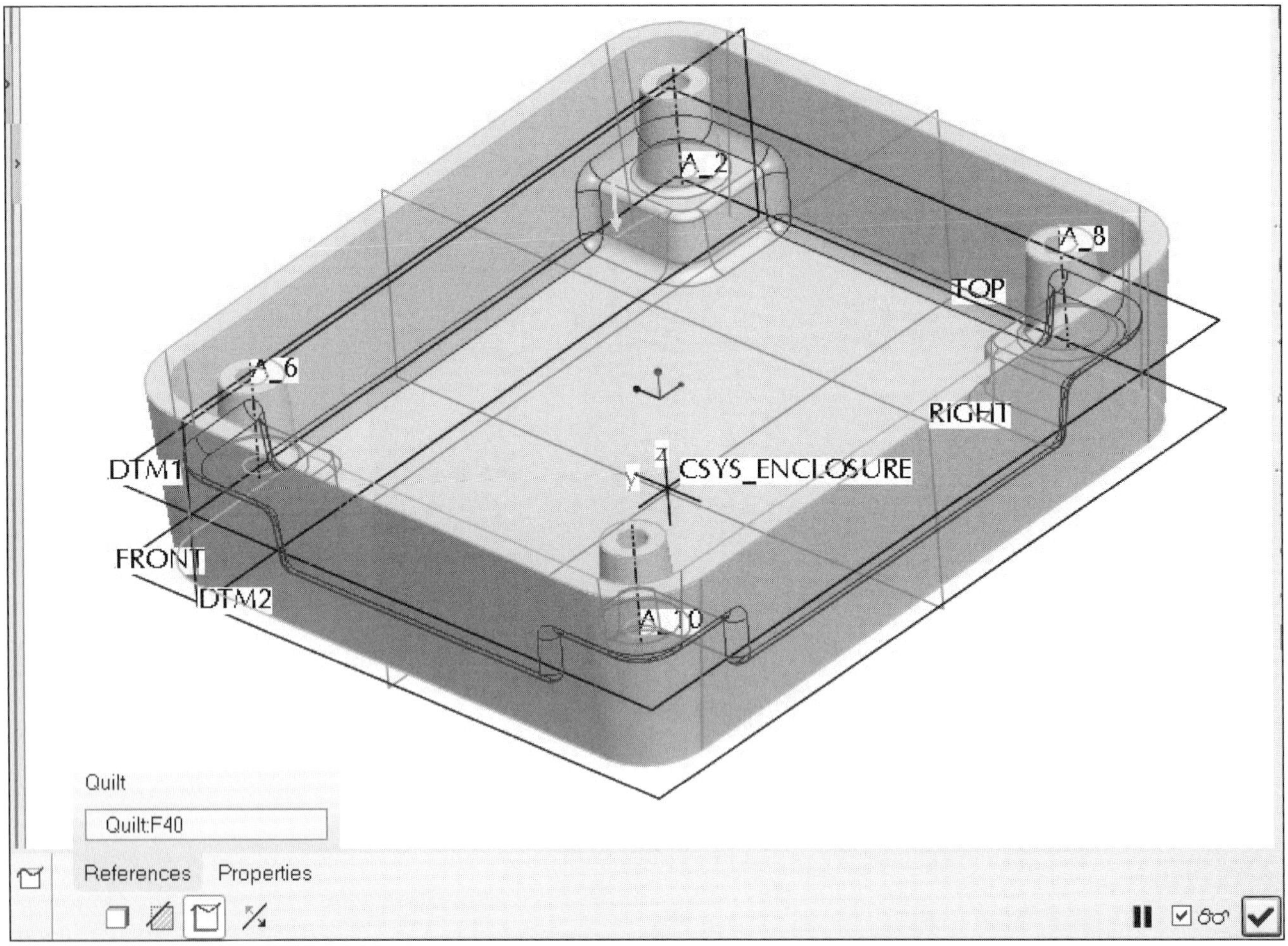

Figure 18.18(b) Solidify References Tab

Click: ▢ **Fills volume delimited by quilt with solid material** [Fig. 18.18(c)] ⇒ **MMB** [Fig. 18.18(d)] ⇒ **LMB** to deselect

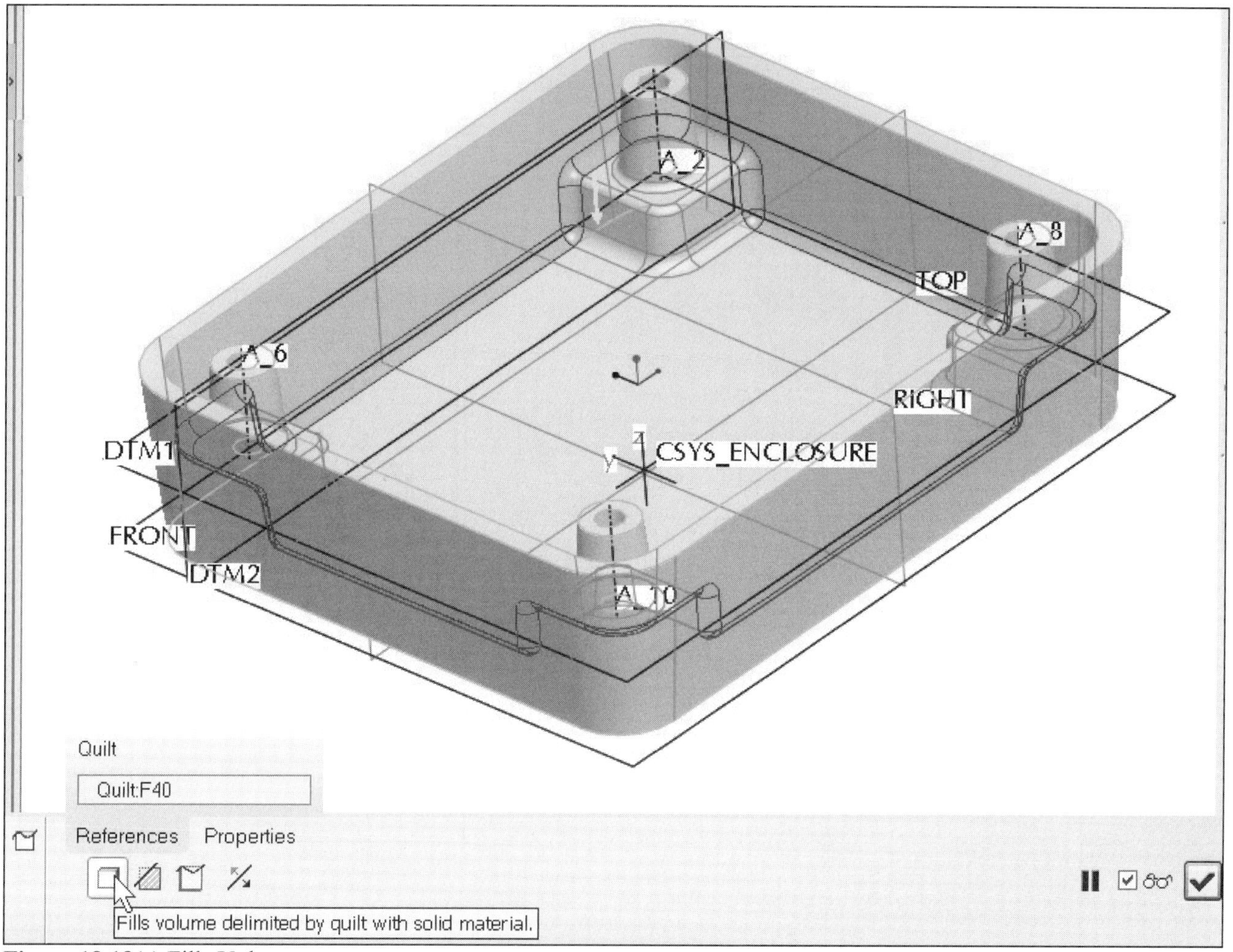

Figure 18.18(c) Fills Volume

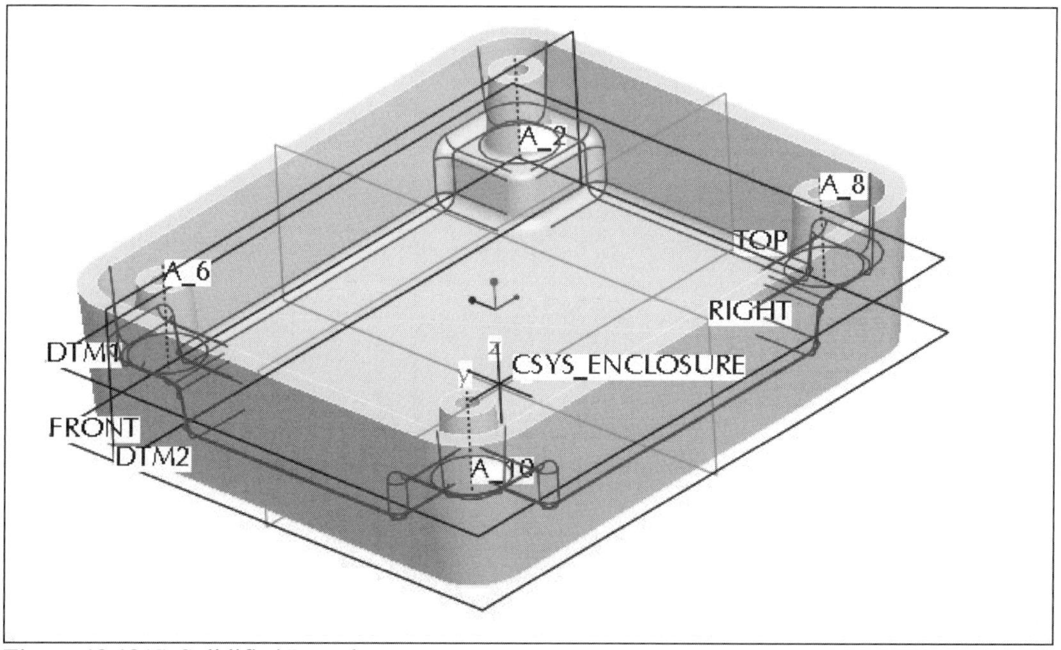

Figure 18.18(d) Solidified Round

Click: **Start the view manager** from Top Toolchest (View Manager dialog box displays) ⇒ **Xsec** tab ⇒ click on **A** ⇒ **Display** ⇒ **Visibility** (Fig. 18.19) ⇒ **RMB** ⇒ **Visibility** (uncheck) ⇒ **Close** ⇒ click on the last feature in the Model Tree to see it highlighted in the graphics window (Fig. 18.20) ⇒ ⇒ **MMB**

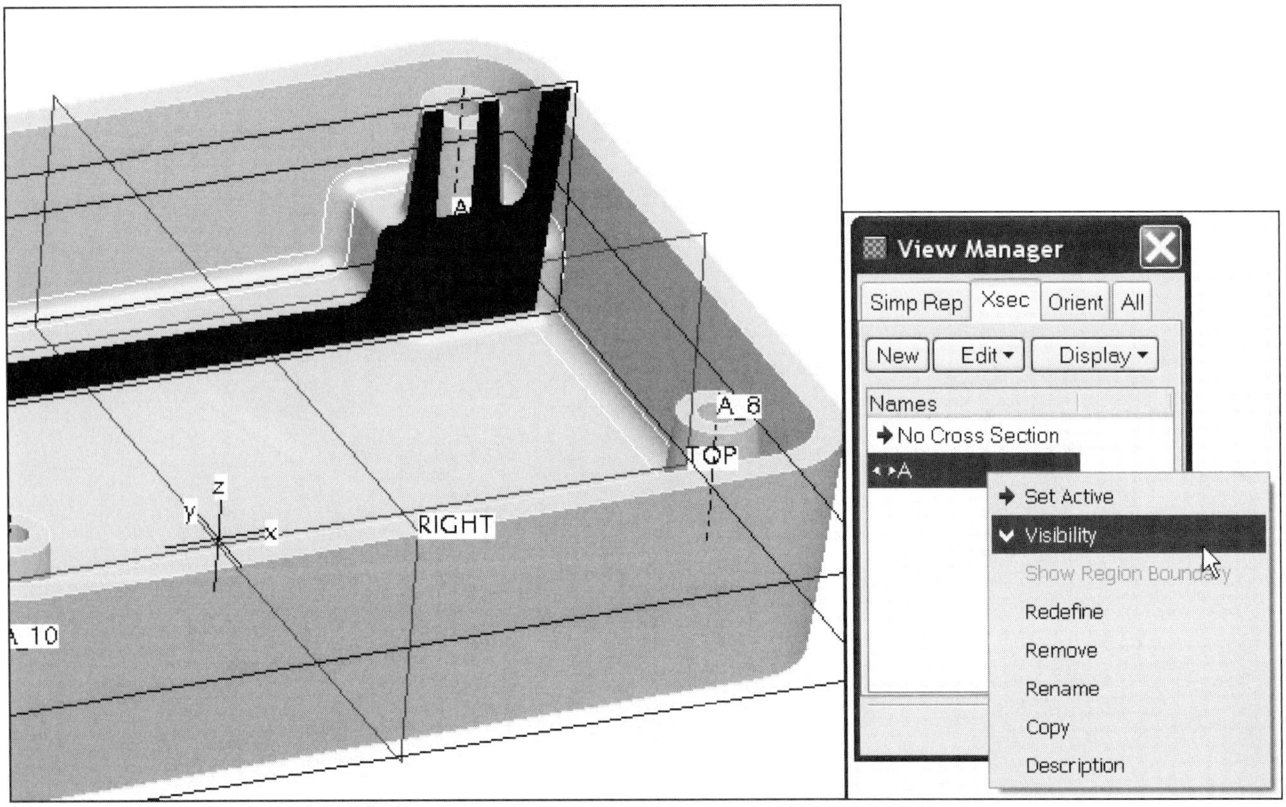

Figure 18.19 Gap has been Solidified

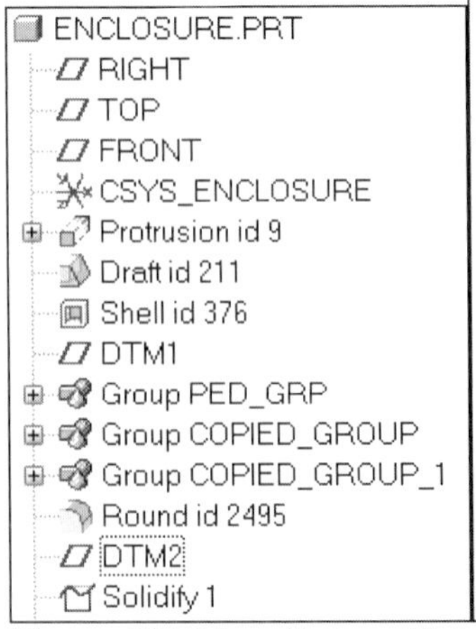

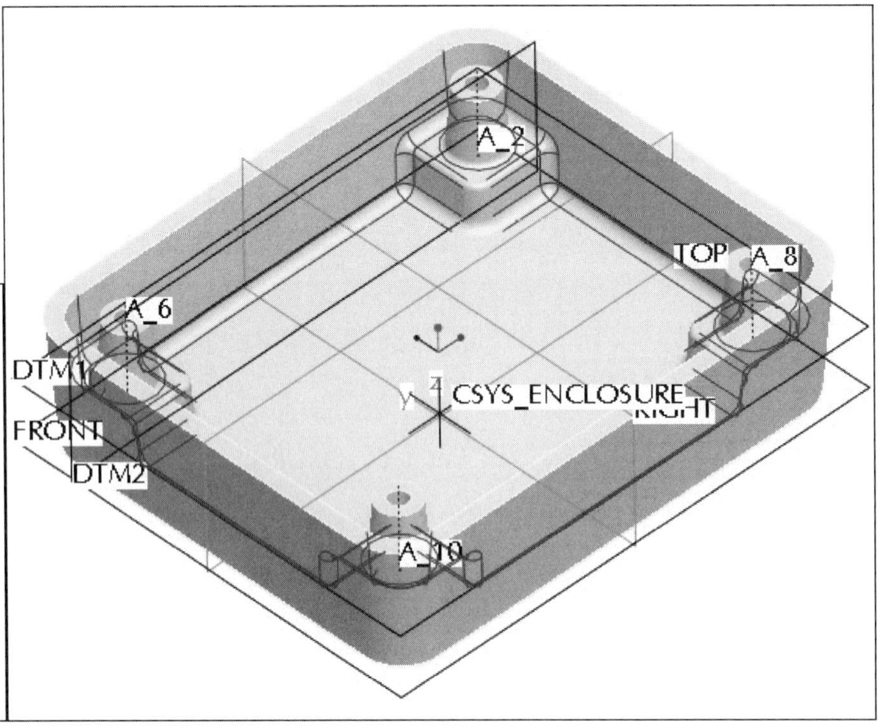

Figure 18.20 Last Feature Highlighted

Before creating the text extrusion, **Suppress** all the features after the shell command. Expand the Model Tree to include the feature number and status.

Click: **Settings** ⇒ **Tree Filters** ⇒ toggle on all options (Fig. 18.21) ⇒ **Apply** ⇒ **OK**

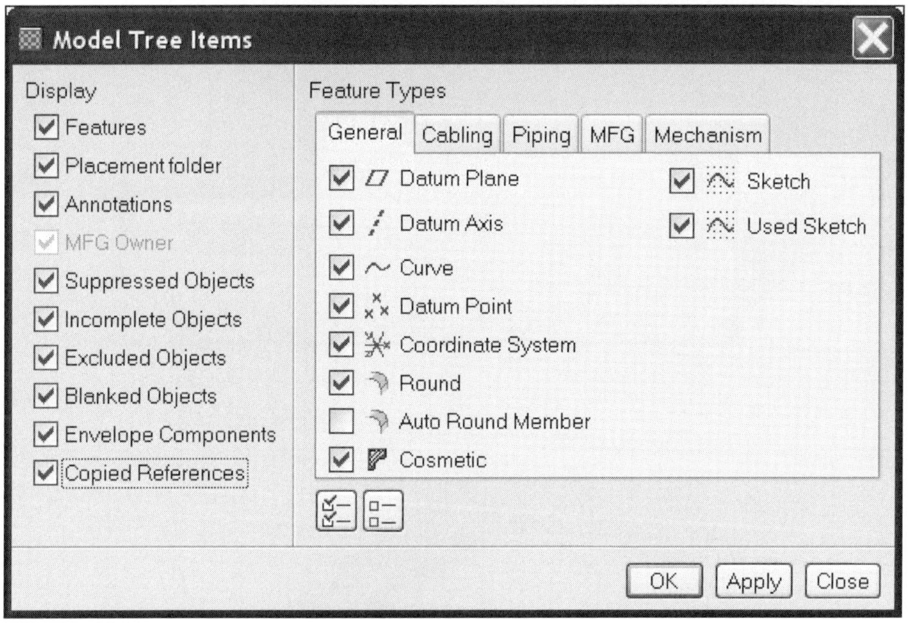

Figure 18.21 Model Tree Items Dialog Box

Click on **Group PED_GRP** in the model tree ⇒ press and hold the **Shift** key ⇒ click **Solidify_1** in the Model Tree [Fig. 18.22(a)] ⇒ **RMB** ⇒ **Suppress** [Fig. 18.22(b)] ⇒ **OK** [Fig. 18.22(c)] ⇒ **LMB** to deselect

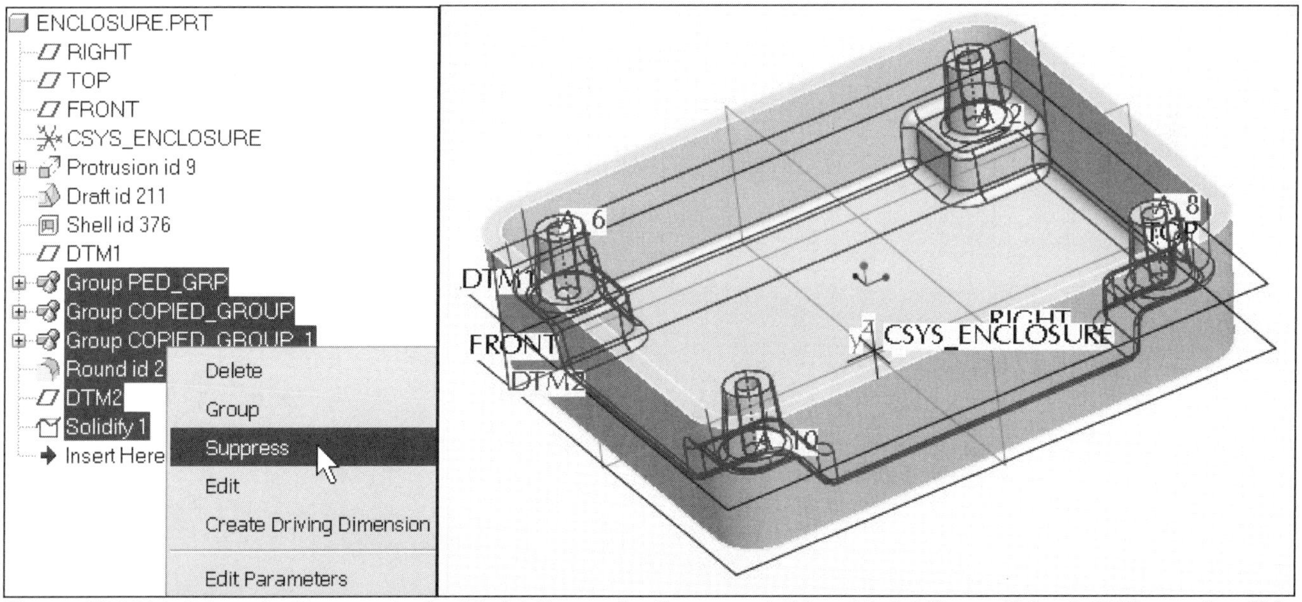

Figure 18.22(a) Select the Features in the Model Tree to be Suppressed (Your Model Tree will Look Different. It will have Extrude instead of Protrusion)

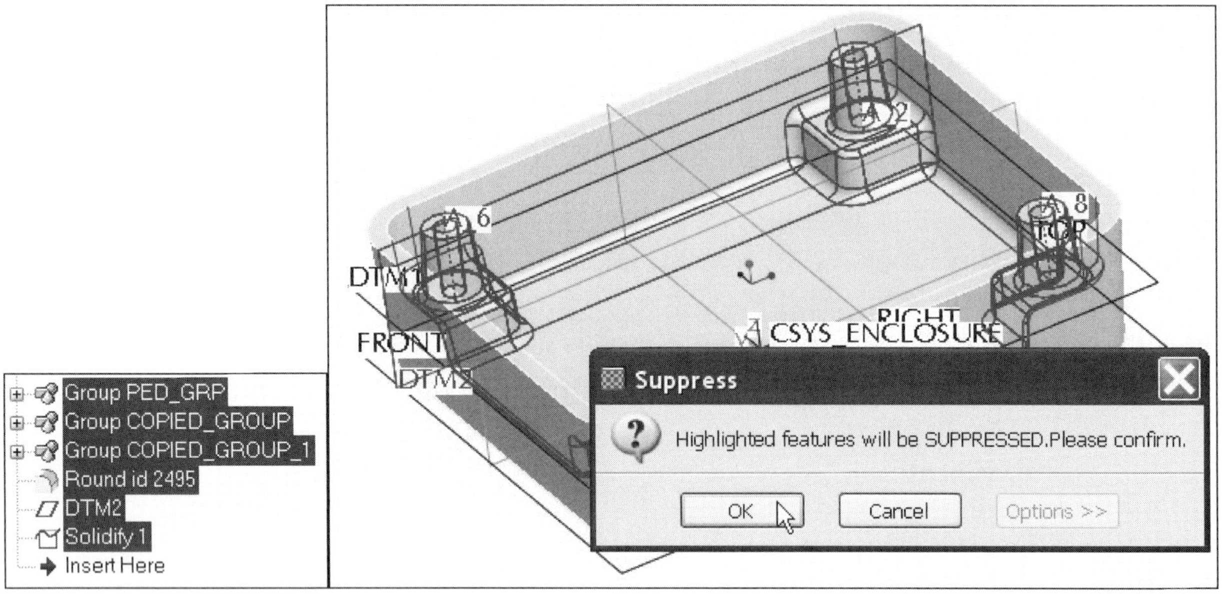

Figure 18.22(b) Highlighted Features

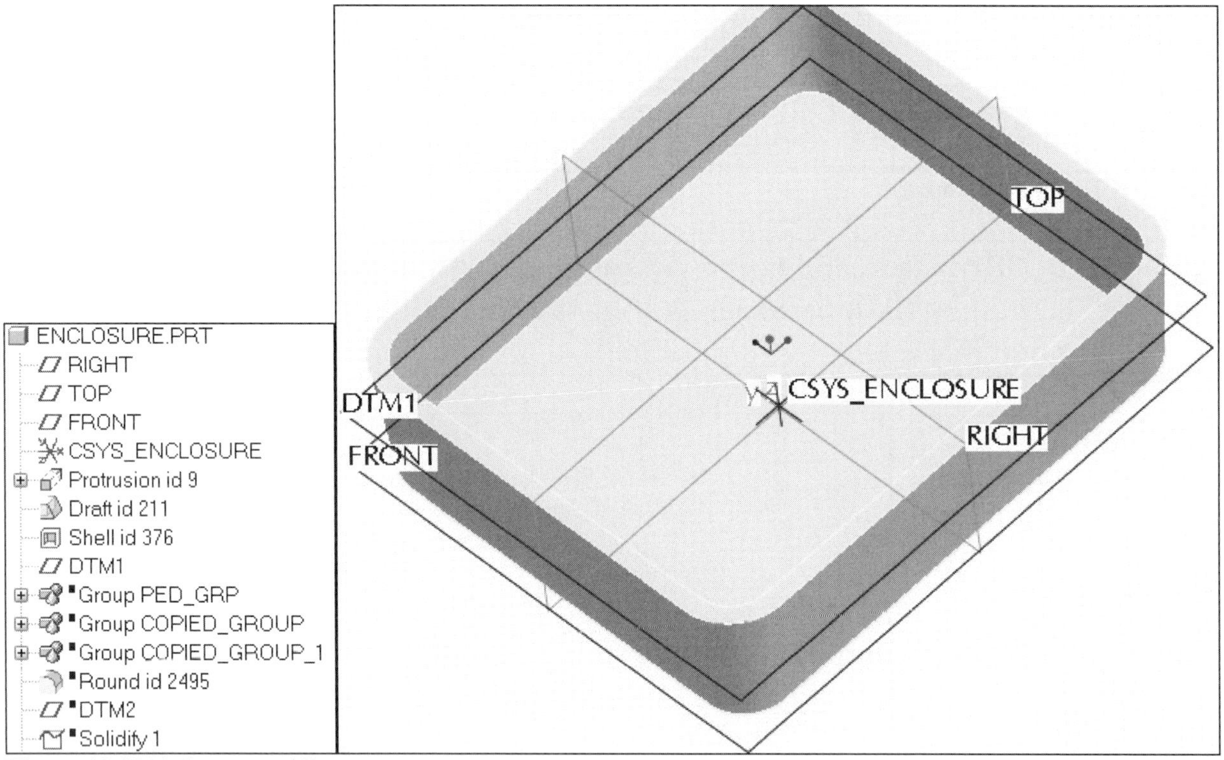

Figure 18.22(c) Suppressed Features

The regeneration time for your model will now be shorter. Next, add the text extrusion.

Click: ⇒ **Standard Orientation** ⇒ **Extrude Tool** ⇒ **RMB** ⇒ **Define Internal Sketch** ⇒ Sketch Plane--- Plane: pick the inside surface of the enclosure for the sketching plane ⇒ Reference: **TOP (DATUM)**⇒ Orientation: **Top** [Fig. 18.23(a)] ⇒ **Sketch** ⇒ **Tools** ⇒ **Environment** ⇒ Snap to Grid ⇒ **Apply** ⇒ **OK** ⇒ **Toggle the grid** on ⇒ **Create text as a part of a section** ⇒ select *start point* of line to determine text starting position [Fig. 18.23(b)] ⇒ select *second point* of line to determine text height and orientation [Fig. 18.23(c)]

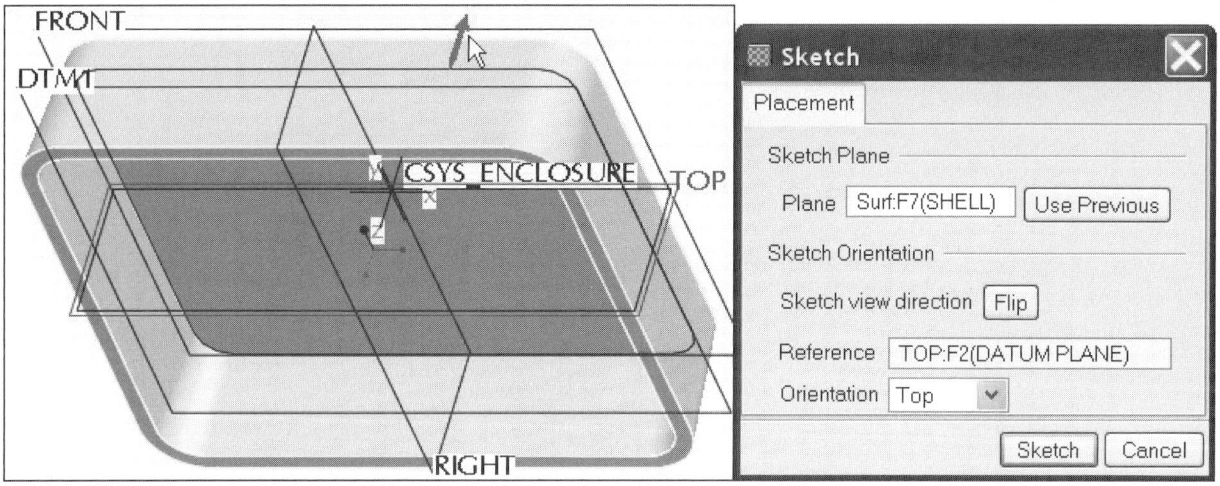

Figure 18.23(a) Sketching Plane, Inside Surface

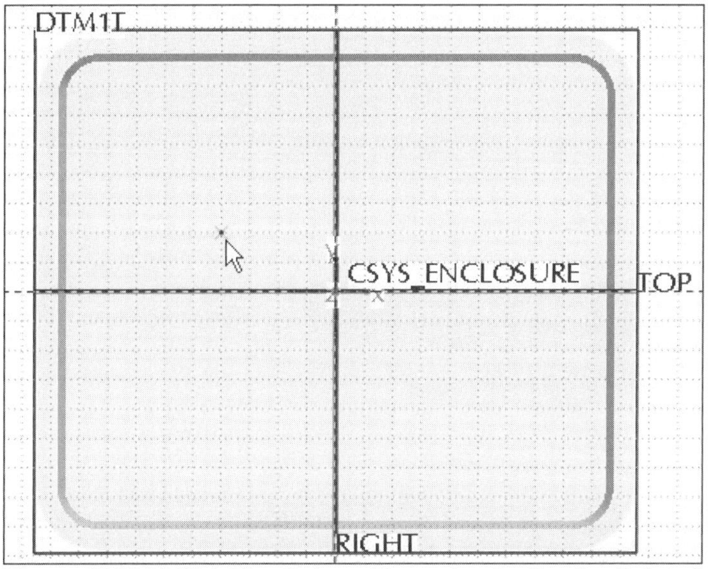

Figure 18.23(b) Pick First Point to Determine the Starting Point of the Lettering

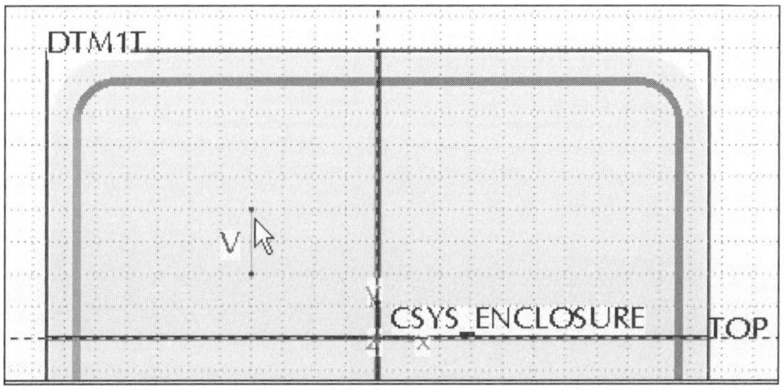

Figure 18.23(c) Pick Second Point to Determine the Height of the Lettering

Click in Text Line field- type **CFS-2134** [Fig. 18.23(d)] ⇒ **OK** ⇒ **Sketch** from menu bar ⇒ **Options** ⇒ **Parameters** tab ⇒ Num Digits ⇒ **4** ⇒ **MMB** ⇒ ⇒ window-in the sketch to capture all dimensions ⇒ **RMB** ⇒ **Modify** ⇒ modify the dimensions [Fig. 18.23(e)] ⇒ from dialog box ⇒ **Continue with the current section** from Right Toolchest

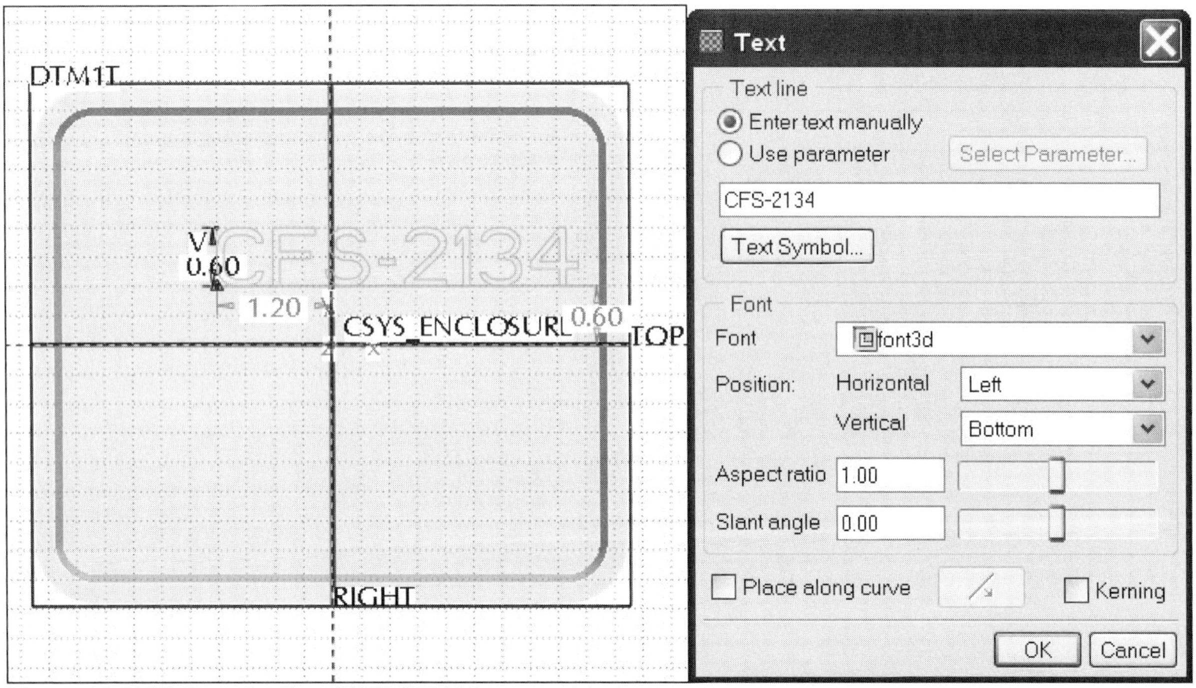

Figure 18.23(d) Type the Text "CFS-2134" (case sensitive)

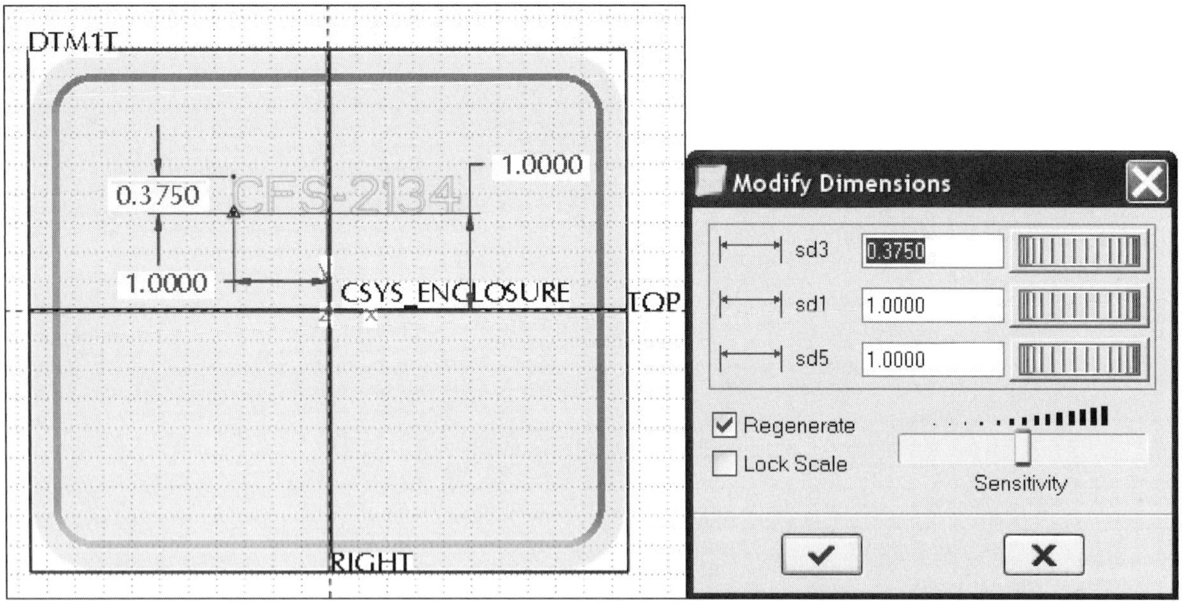

Figure 18.23(e) Modified Dimensions [dimensions are slightly different than those shown in Figure 18.2(c)]

Press and hold **MMB** to spin the model [Fig. 18.23(f)] ⇒ double-click on the height dimension and modify it to **.0625** ⇒ **Enter** [Fig. 18.23(g)] ⇒ **MMB** [Fig. 18.23(h)] ⇒ ⇒ **MMB**

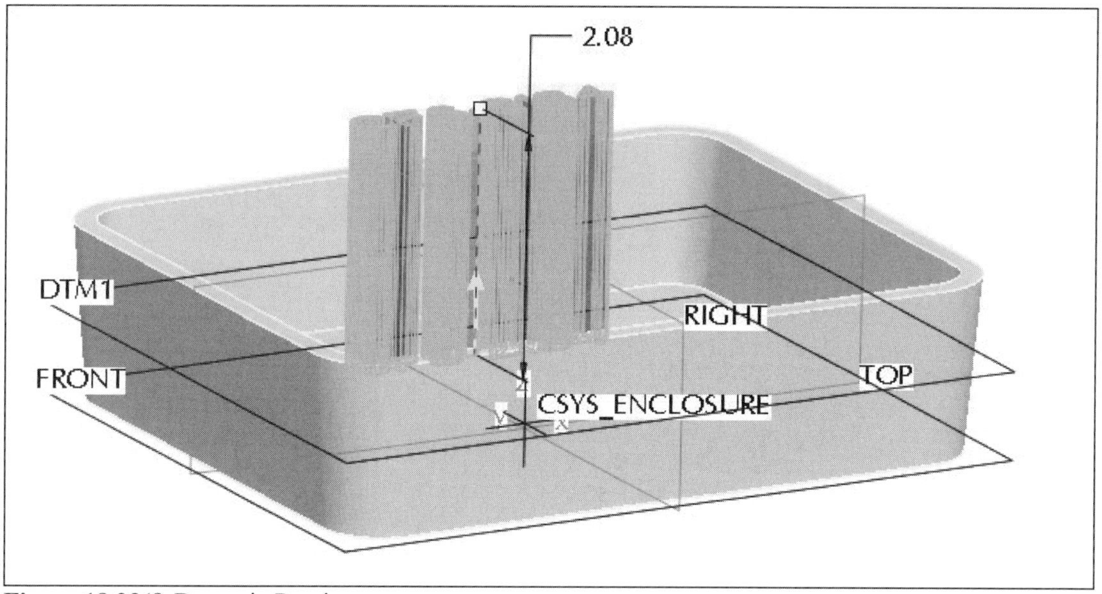

Figure 18.23(f) Dynamic Preview

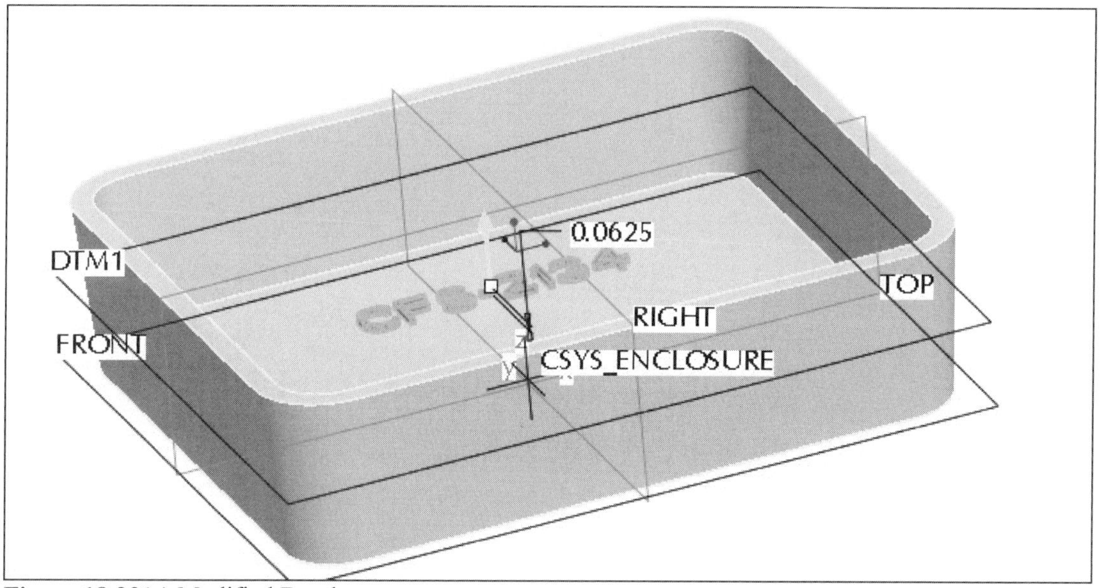

Figure 18.23(g) Modified Depth

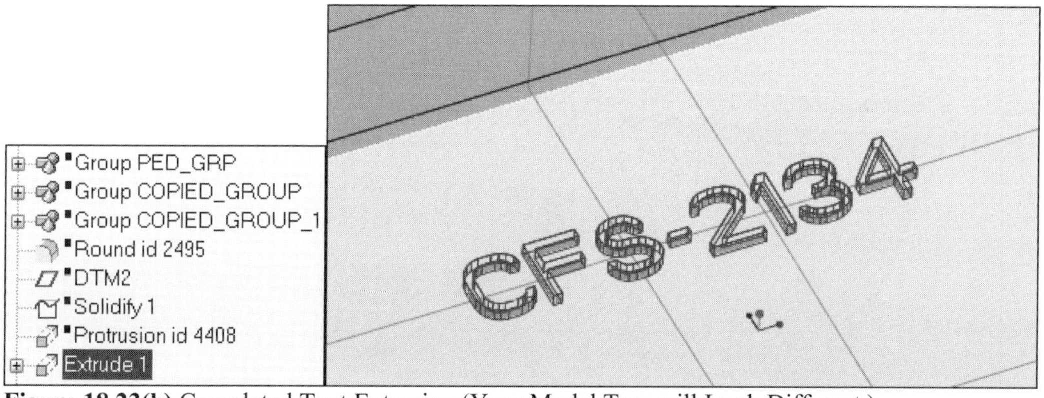

Figure 18.23(h) Completed Text Extrusion (Your Model Tree will Look Different.)

Click: **Edit ⇒ Resume ⇒ Resume All** [Figs. 18.24(a-b)] ⇒ **Ctrl+D** ⇒ **Ctrl+R** ⇒ **Ctrl+S** ⇒ **MMB** ⇒ **LMB** to deselect [Fig. 18.24(c)]

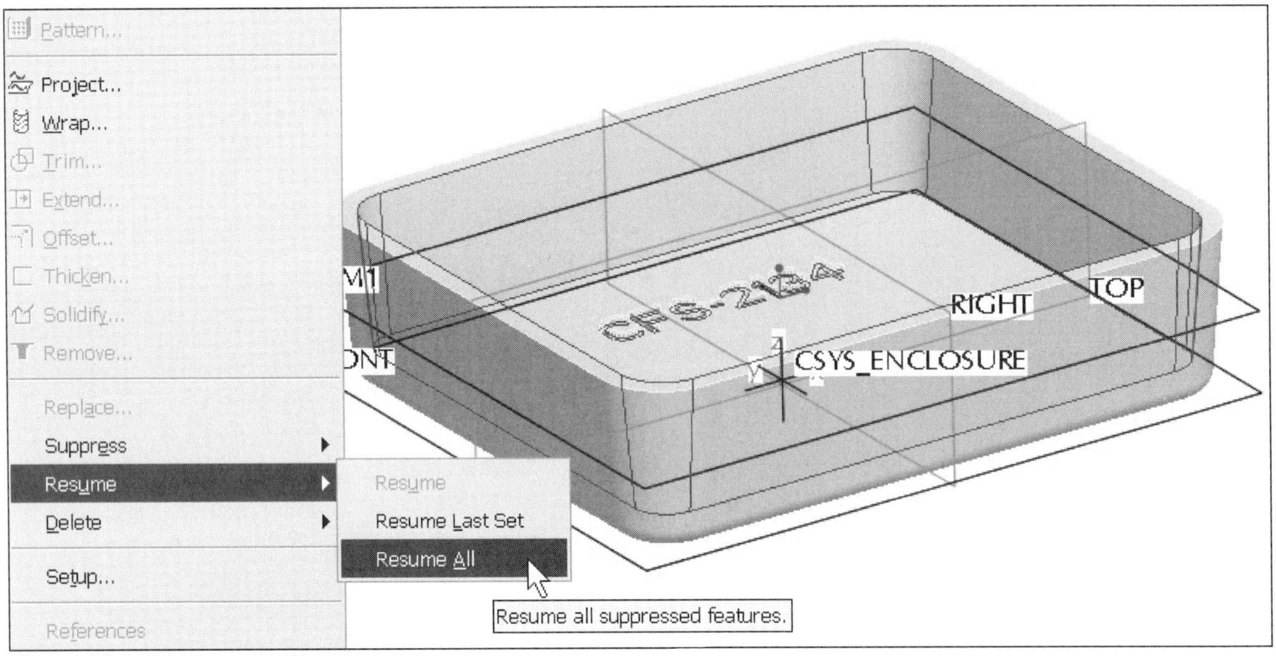

Figure 18.24(a) Resume All

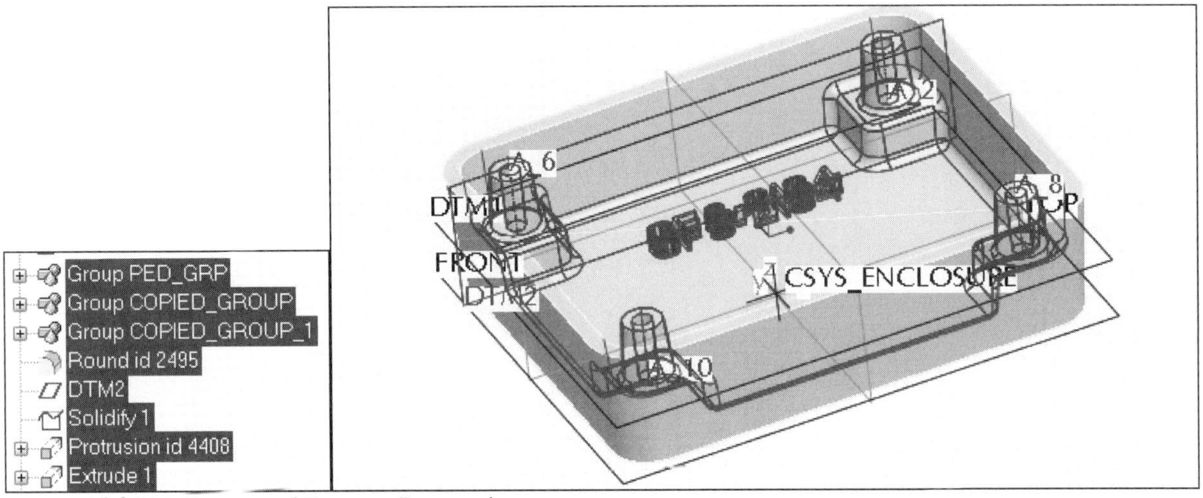

Figure 18.24(b) Suppressed Features Resumed

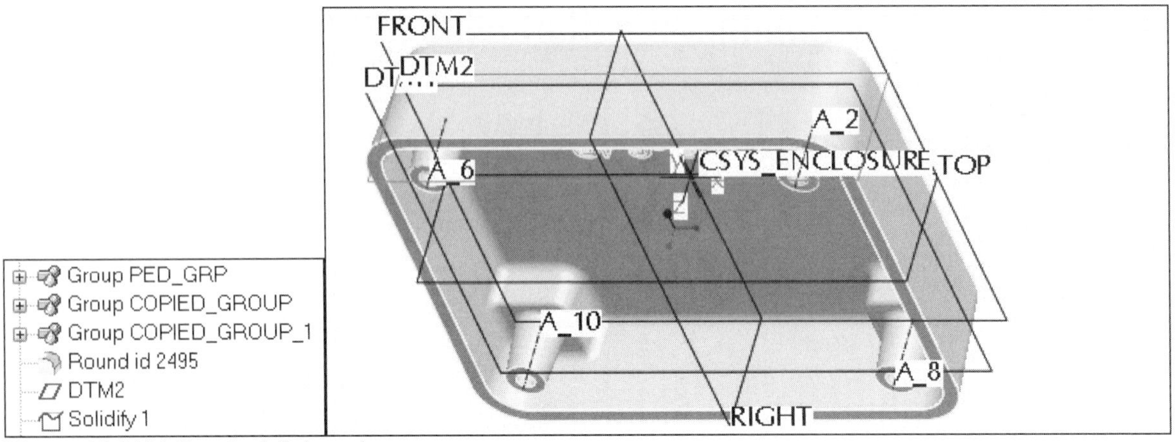

Figure 18.24(c) Standard Orientation

Spin the part ⇒ **View** ⇒ **Shade** ⇒ pick the top surface and then pick on the parts' edge until it highlights [Fig. 18.25(a)] ⇒ depress and hold **Shift** key ⇒ pick on the surface again [Fig. 18.25(b)] *[(Top surface edges will highlight, as the loop was selected. Note that this method was not necessary for this round since all the edges were tangent, but this was used to demonstrate another process of selection)]* ⇒ **RMB** ⇒ **Round Edges** [Fig. 18.25(c)] ⇒ **Ctrl+R** ⇒ modify the edge round to **.1875** [Fig. 18.25(d)] ⇒ **Enter** ⇒ **MMB** ⇒ **View** ⇒ **Shade** [Fig. 18.25(e)]

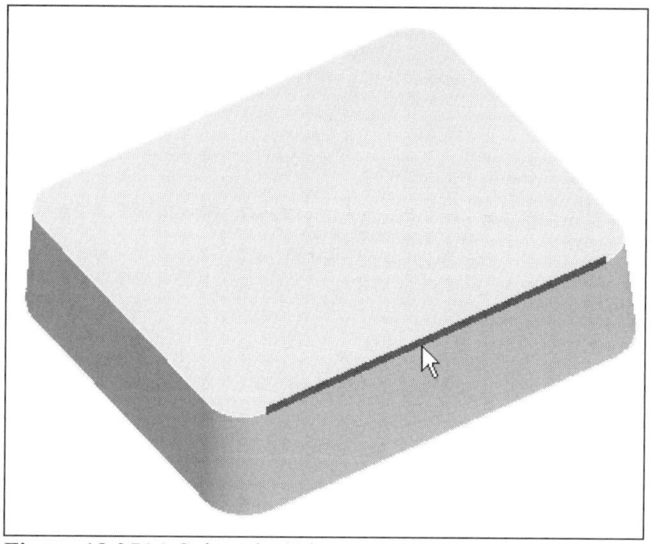

Figure 18.25(a) Select the Edge

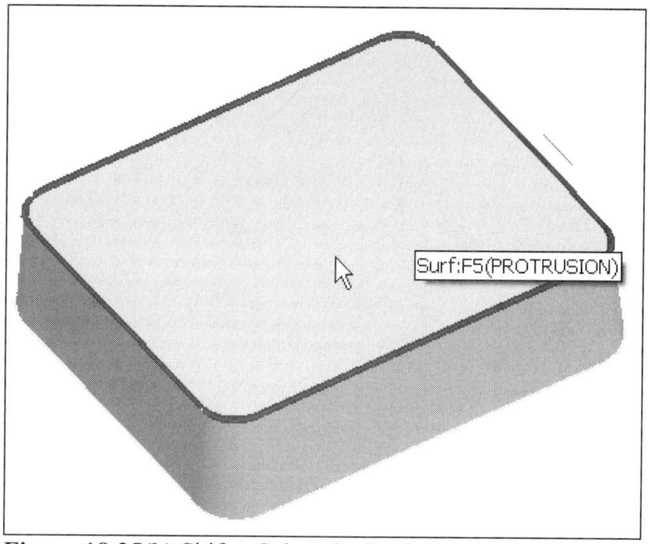

Figure 18.25(b) Shift + Select the Surface

Figure 18.25(c) Round Edges

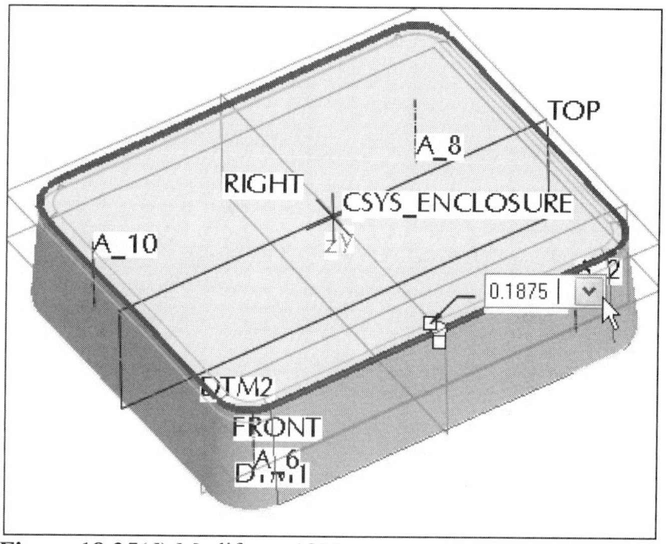

Figure 18.25(d) Modify to **.1875**

Figure 18.25(e) Completed Round

Click: **Ctrl+D** ⇒ select the text extrusion in the **Model Tree** ⇒ **Orient Mode** on ⇒ **RMB** ⇒ **Velocity** [Fig. 18.26(a)] ⇒ hold down **MMB** and move the cursor about the screen to orbit the model [Fig. 18.26(b)] ⇒ **RMB** ⇒ **Exit Orient Mode** ⇒ **LMB** to deselect

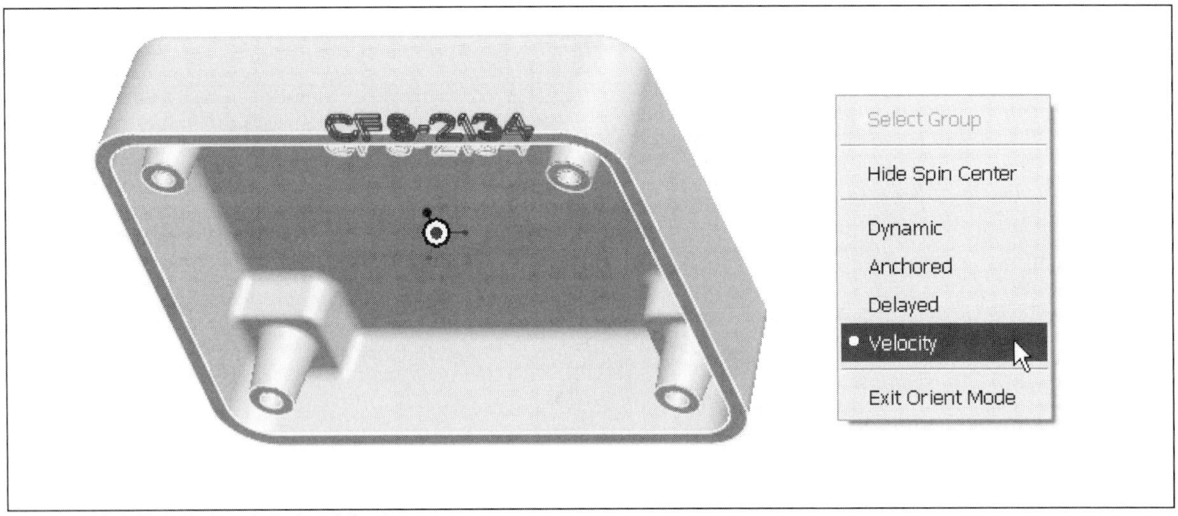

Figure 18.26(a) Orient Mode Velocity

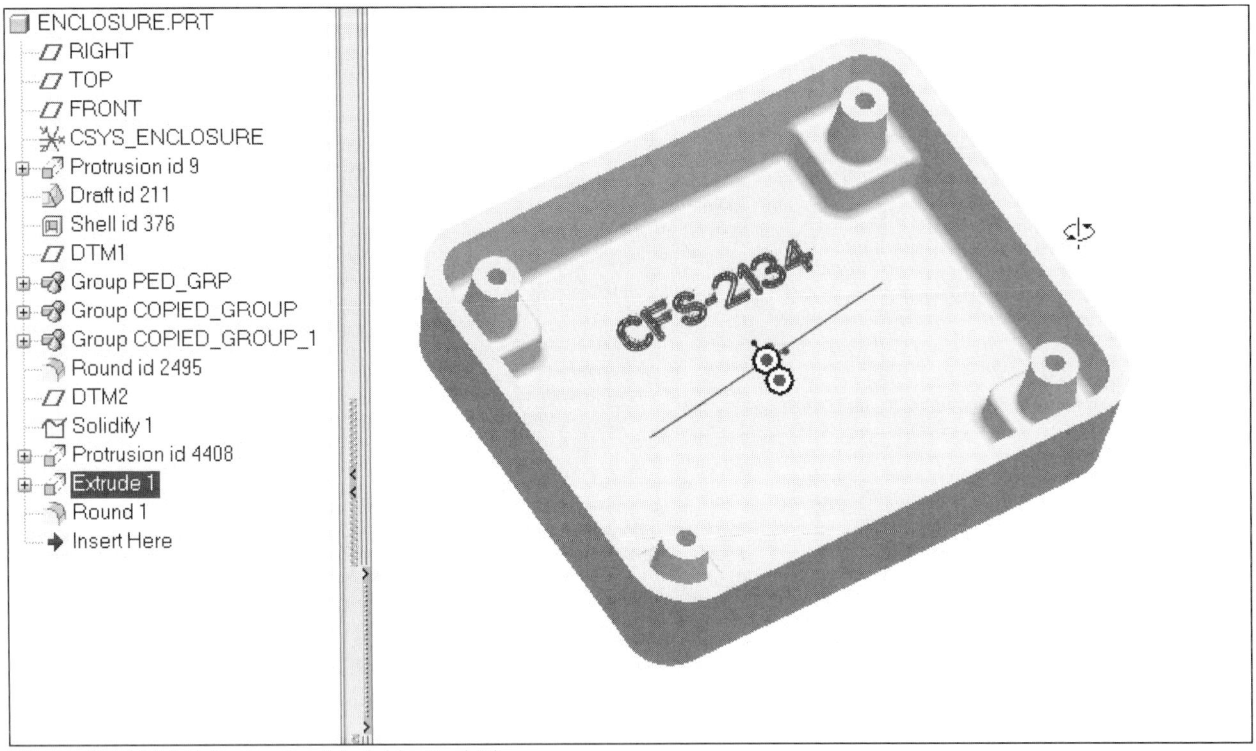

Figure 18.26(b) Spinning the Model in Real Time

Click: [icon] ⇒ double-click on the text extrusion ⇒ modify the text depth to be **.250** [Figs. 18.27(a-b)] ⇒ **Enter** ⇒ **Ctrl+G** ⇒ **Ctrl+S** ⇒ **MMB** ⇒ **File** ⇒ **Delete** ⇒ **Old Versions** ⇒ **MMB** (Fig. 18.28)

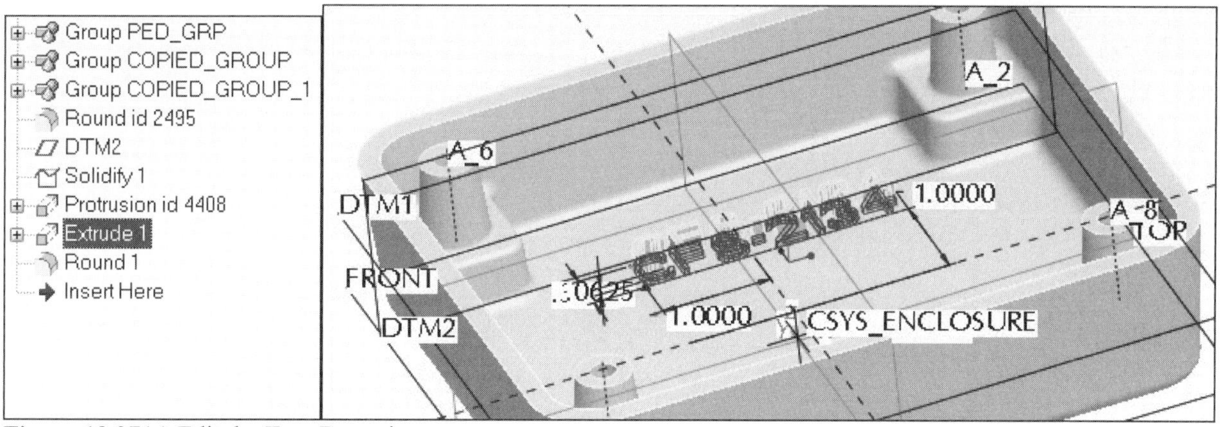

Figure 18.27(a) Edit the Text Extrusion

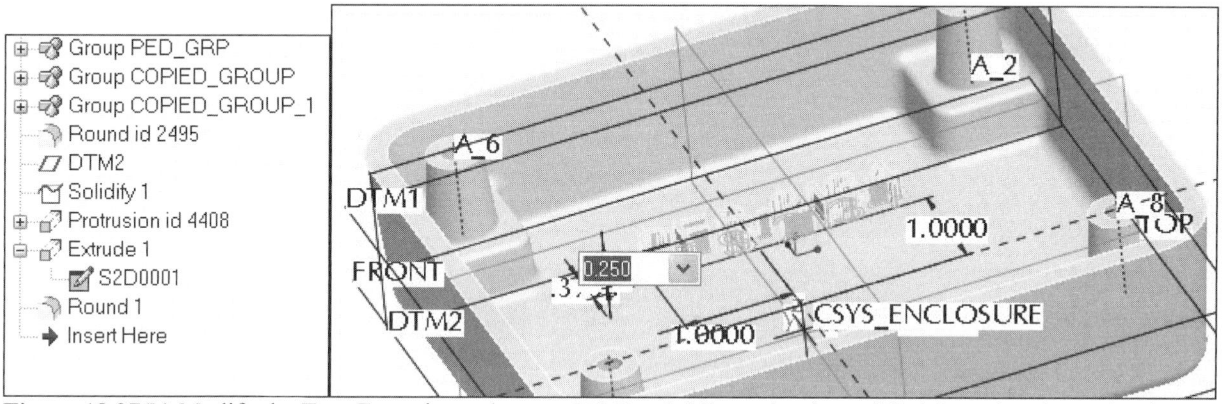

Figure 18.27(b) Modify the Text Extrusion to have a **.250** Depth [dimension is modified from that shown in Figure 18.2(c)]

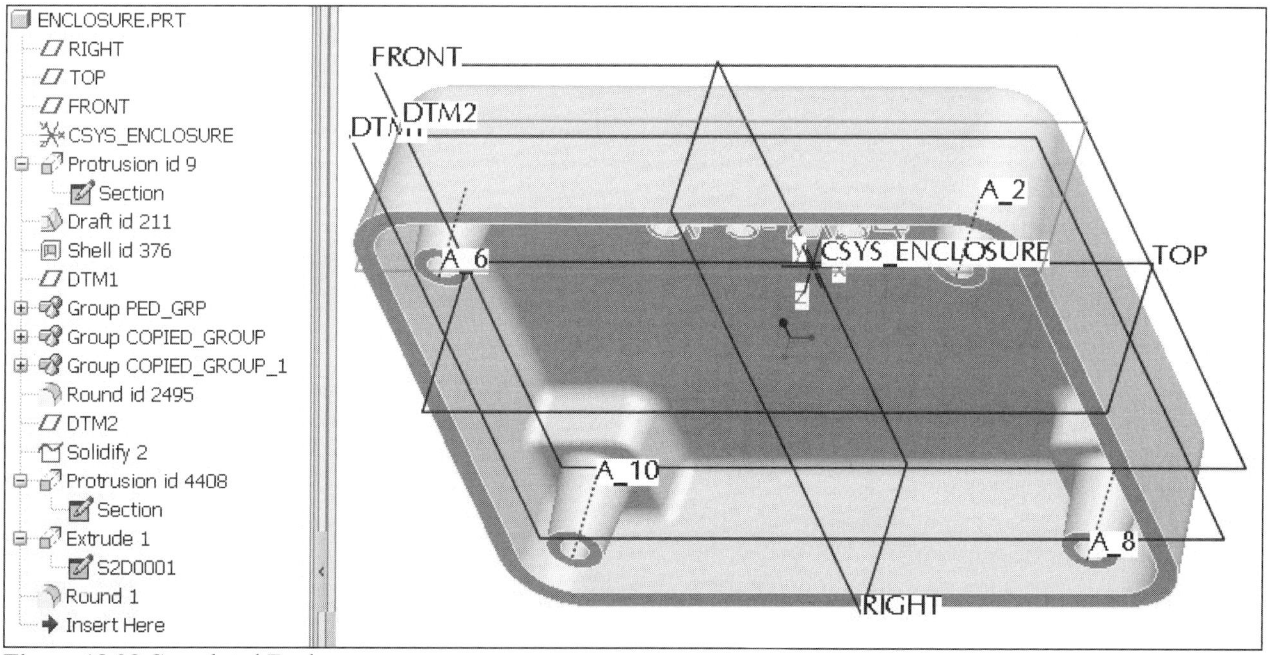

Figure 18.28 Completed Enclosure

Click: **Analysis ⇒ ModelCHECK ⇒ ModelCHECK Geometry Check** [Fig. 18.29(a)] ⇒ **OK** [Figs. 18.29(b-c)] ⇒ **Ctrl+S ⇒ MMB ⇒ Window ⇒ Close ⇒ File ⇒ Erase ⇒ Not Displayed ⇒ OK**

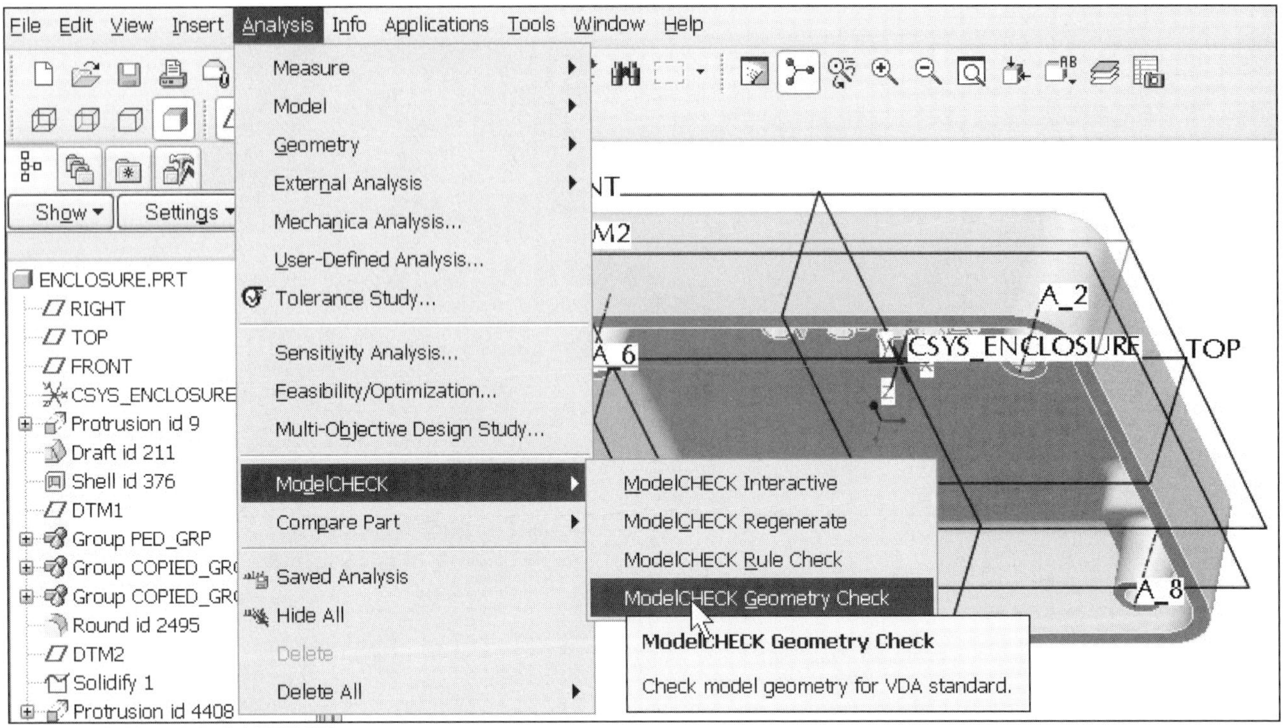

Figure 18.29(a) ModelCHECK

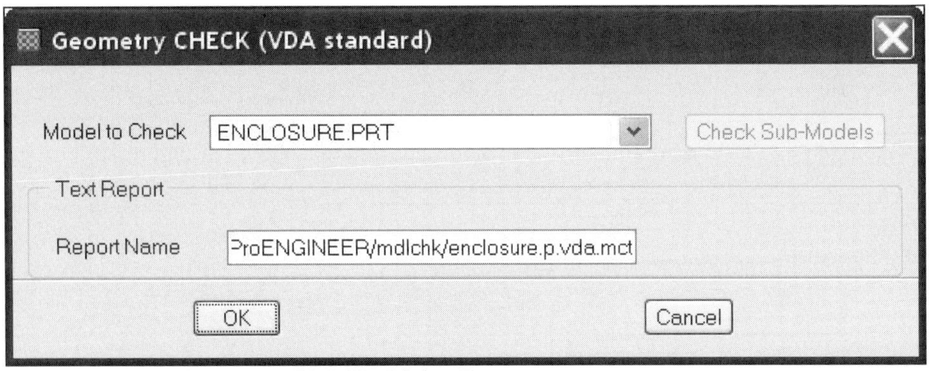

Figure 18.29(b) Geometry CHECK Dialog Box

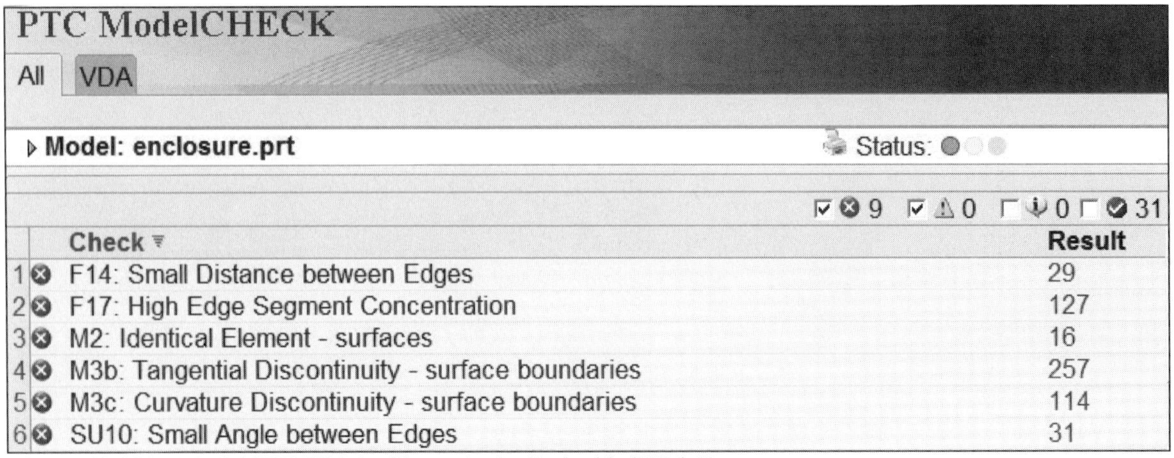

Figure 18.29(c) ModelCHECK List Shown in the Embedded Web Browser

See ***www.cad-resources.com*** ⇒ ***Downloads*** for more part projects.

Appendix

Customizing the User Interface (UI)

Customize your user interface to increase your efficiency in modeling. You can customize the Pro/E user interface, according to your needs or the needs of your group or company, to include the following:

- Create keyboard macros, called *mapkeys*, and add them to the menus and toolbars
- Add or remove existing toolbars
- Add split buttons to the toolbars (split buttons contain multiple closely-related commands and save toolbar space by hiding all but the first active command button)
- Move or remove commands from the menus or toolbars
- Change the location of the message area
- Add options to the Menu Manager
- Blank (make unavailable) options in the Menu Manager
- Set default command choices for Menu Manager menus

With a Pro/E file active, click: **Tools** ⇒ **Customize Screen** (Fig. A.1) ⇒ Customize dialog box opens with the Commands tab active (Fig. A.2) ⇒ Categories, click: **View** (Fig. A.3) ⇒ drag  **Model Colors** and drop into the Top Toolchest 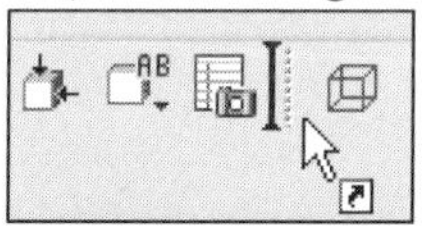 ⇒  (Fig. A.4)

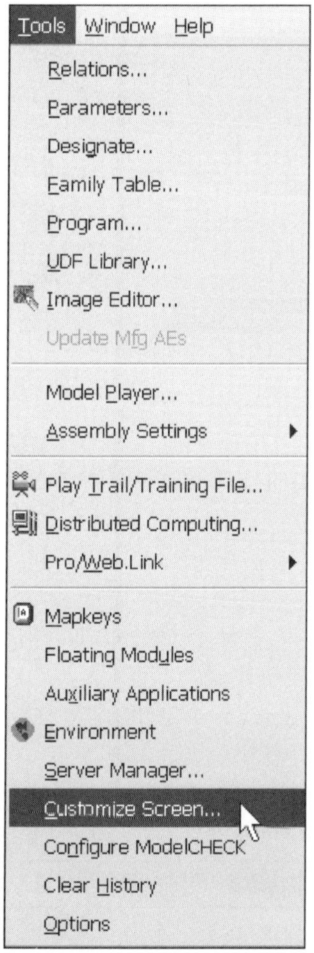

Figure A.1 Customize Screen

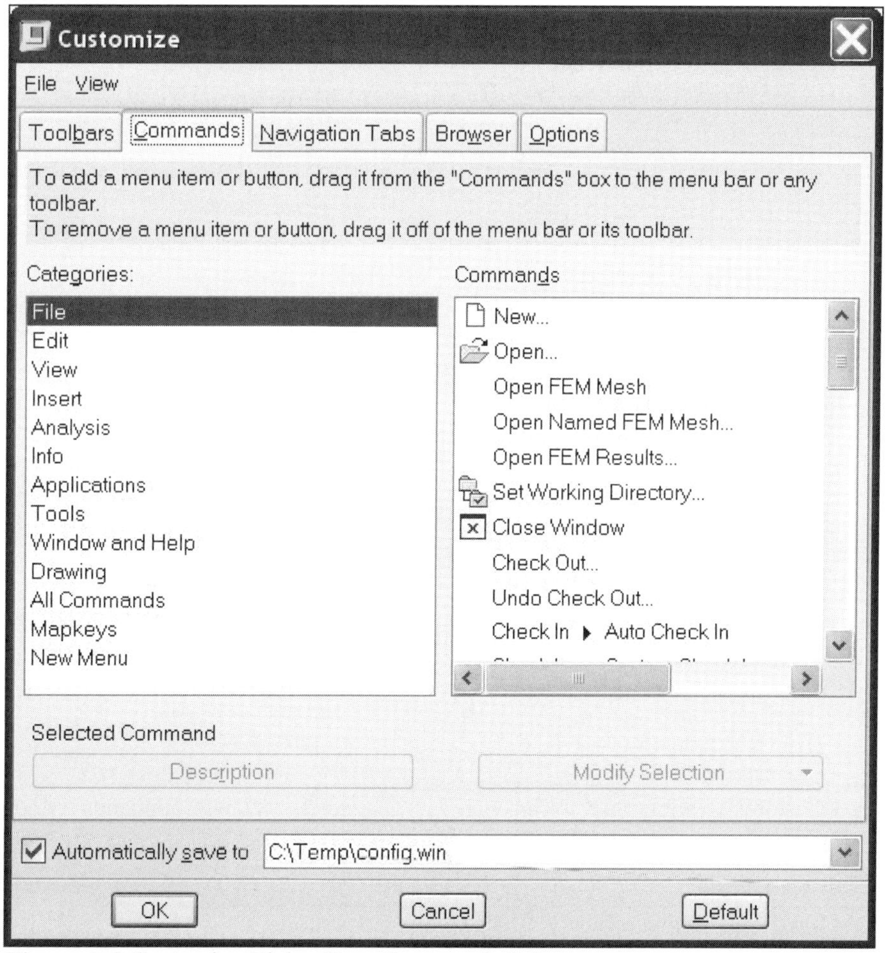

Figure A.2 Customize Dialog Box, Commands Tab

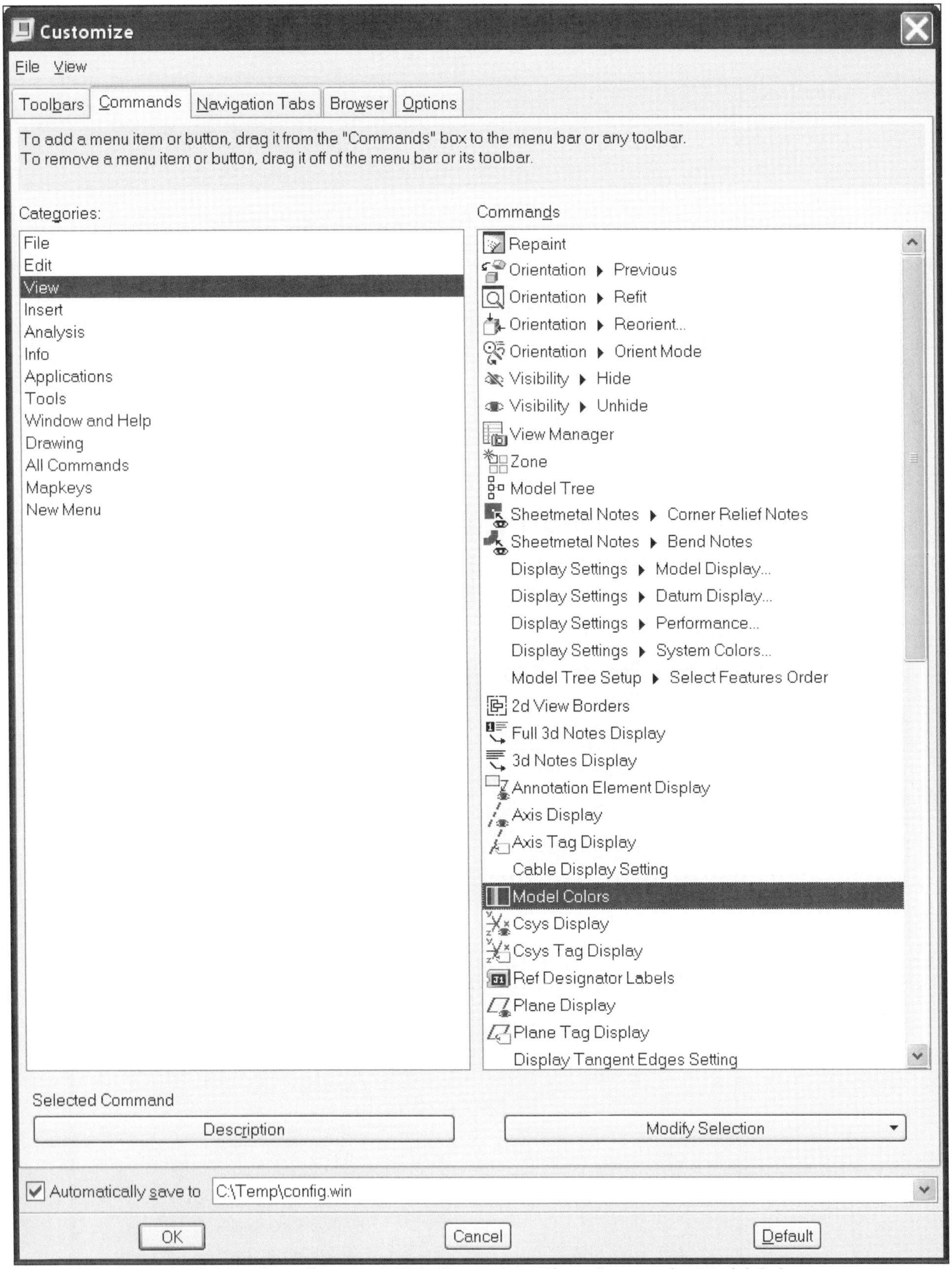

Figure A.3 Customize Dialog Box, Commands Tab, Categories– View, Commands– Model Colors

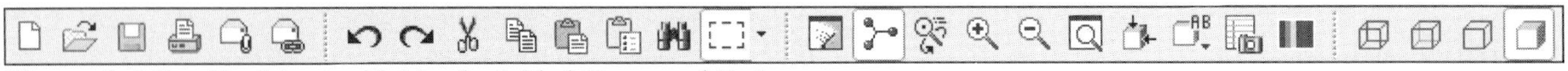
Figure A.4 Top Toolchest with Newly Added Command Button

Categories **New Menu** (Fig. A.5) ⇒ drag 

and drop into the Right Toolchest 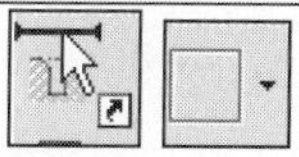 ⇒ Categories **View** ⇒ drag 

and drop into the Right Toolchest fly out 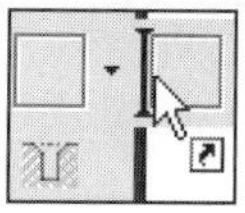 ⇒ drag 

and drop into the Right Toolchest fly out 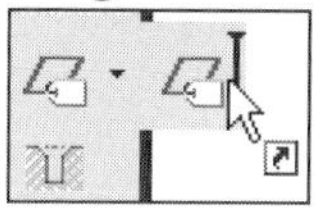 ⇒ drag 

and drop into the Right Toolchest fly out 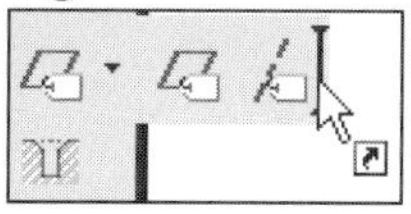 ⇒ drag 

and drop into the Right Toolchest fly out 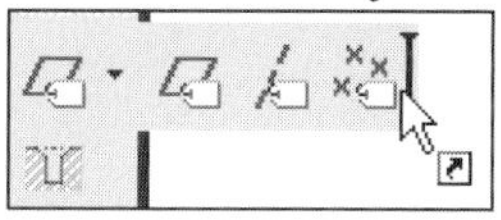 ⇒ click on  to expand the Flyout ⇒ 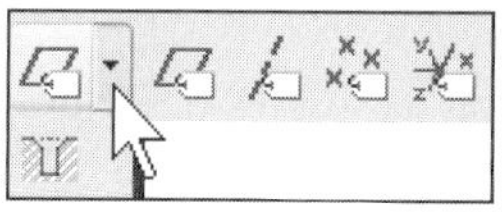 ⇒ File ⇒ 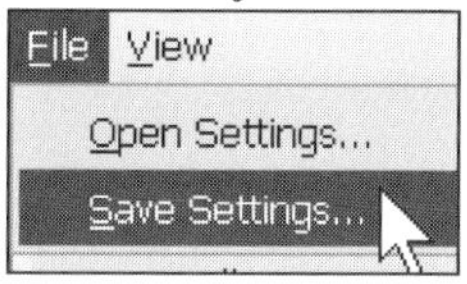

⇒ type a name for your file (i.e., **my_config.win**) ⇒ **OK** ⇒ **OK** [by default, your new Toolchest buttons are automatically saved in **your** current working directory (config.win)]

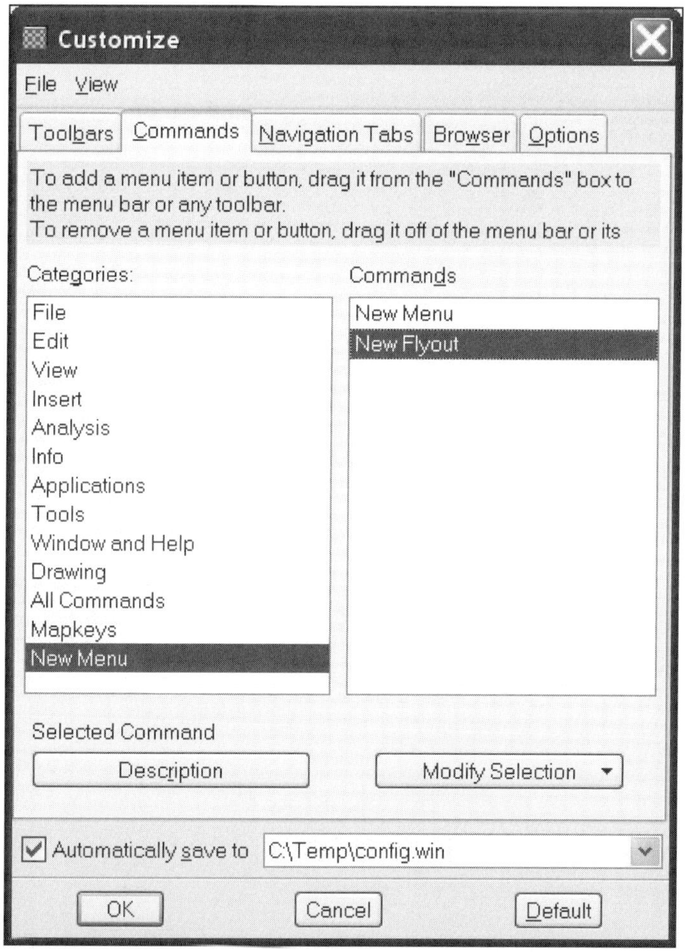

Figure A.5 New Menu Category

You can recall saved settings by clicking: *Tools ⇒ Customize Screen ⇒ File ⇒ Open Settings ⇒ select your .win file ⇒ Open.*

Buttons can be removed from the Toolchest using the exact same method, except, drag the buttons away from the Toolchest and release the mouse button. Customize your Pro/E screen with the buttons of commands you use frequently such as:

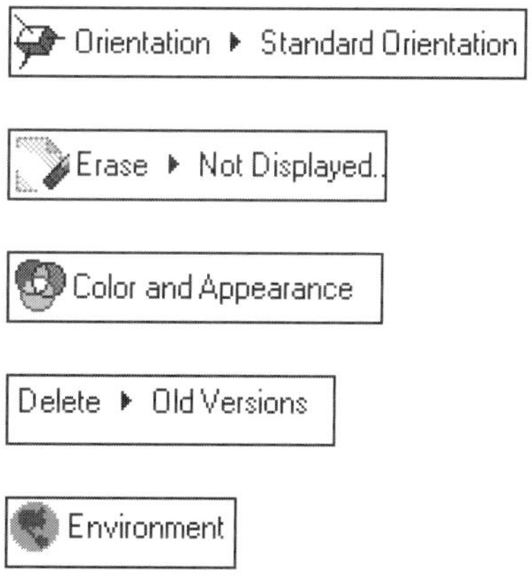

Next, add a Toolbar to the left side of the Navigator Window (Left Toolchest), click: **Tools** ⇒ **Customize Screen** ⇒ **Toolbars** tab ⇒ Tools ⇒ **Left** (Fig. A.6) ⇒ **OK**

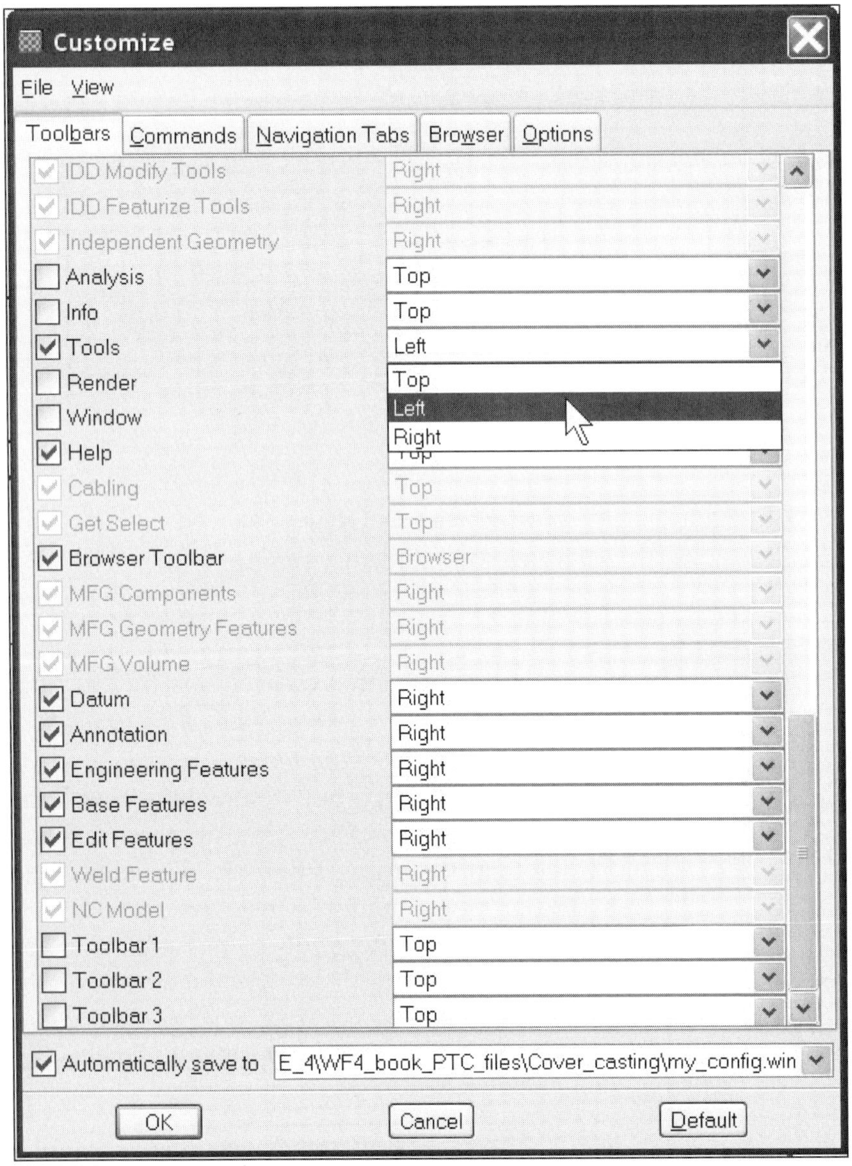

Figure A.6 Tools, Left

The Tools toolbar (Fig. A.7) includes: **Set various environment options**, **Run trail or training file**, **Create macros**, and **Select hosts for distributed computing**.

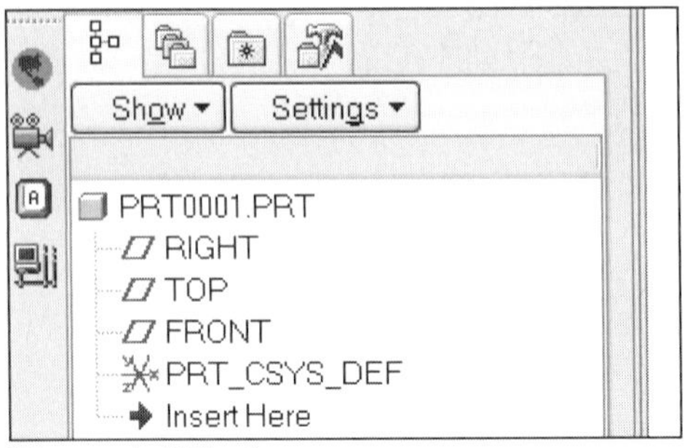

Figure A.7 Tools

The Navigation Tabs tab provides options for controlling the location of the Navigator (left or right), its width setting, and its placement in relation to the Model Tree settings.

Click: **Tools** ⇒ **Customize Screen** ⇒ **Navigation Tabs** tab and explore the settings (Fig. A.8)

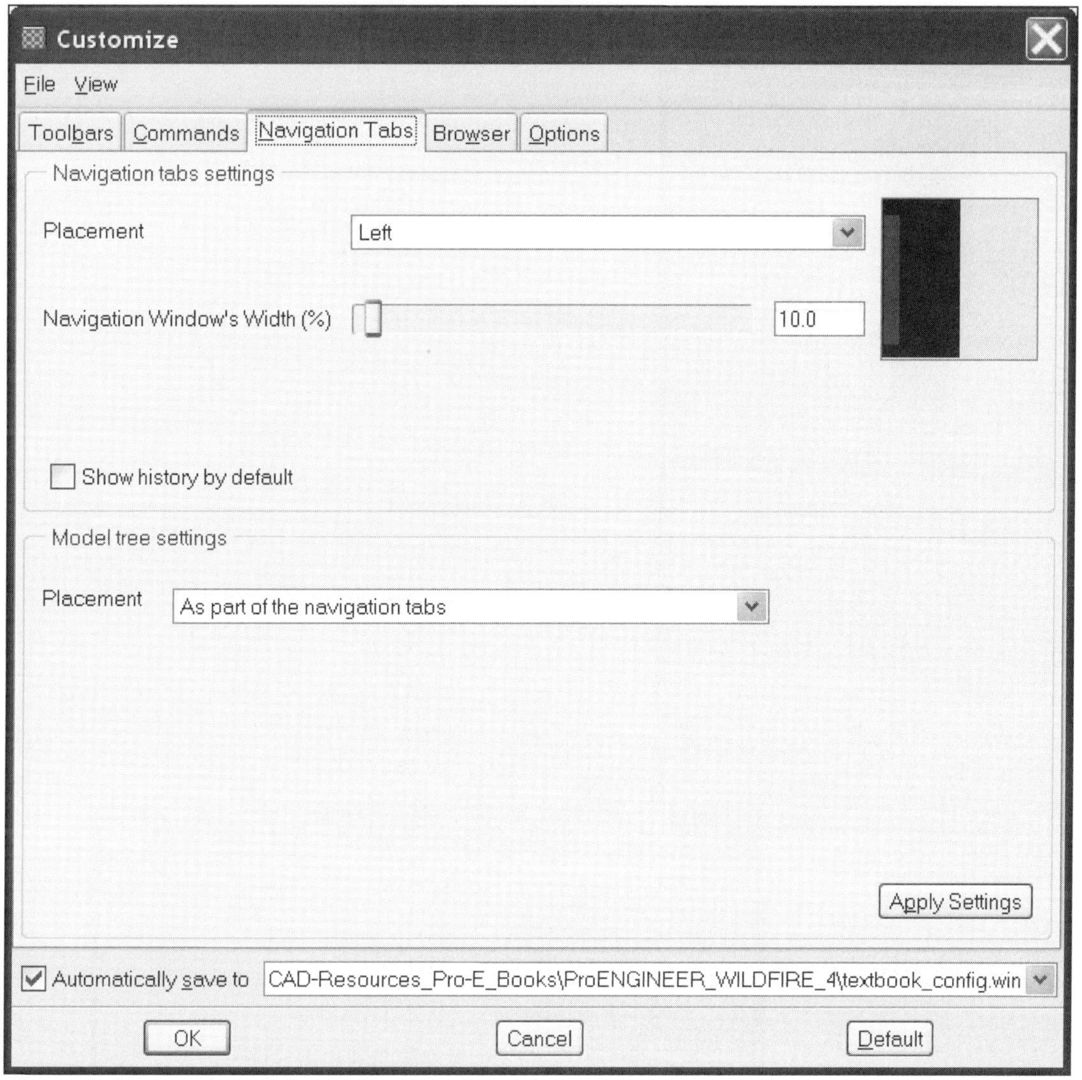

Figure A.8 Customize Dialog Box, Navigation Tabs

The Browser tab is used to control its window width and animation option. The Options tab provides settings to locate the Dashboard, Secondary Window size, and Menu display

Click: **Browser** tab and explore the options (Fig. A.9) ⇒ **Options** tab and explore its capabilities (Fig. A.10) ⇒ **OK**

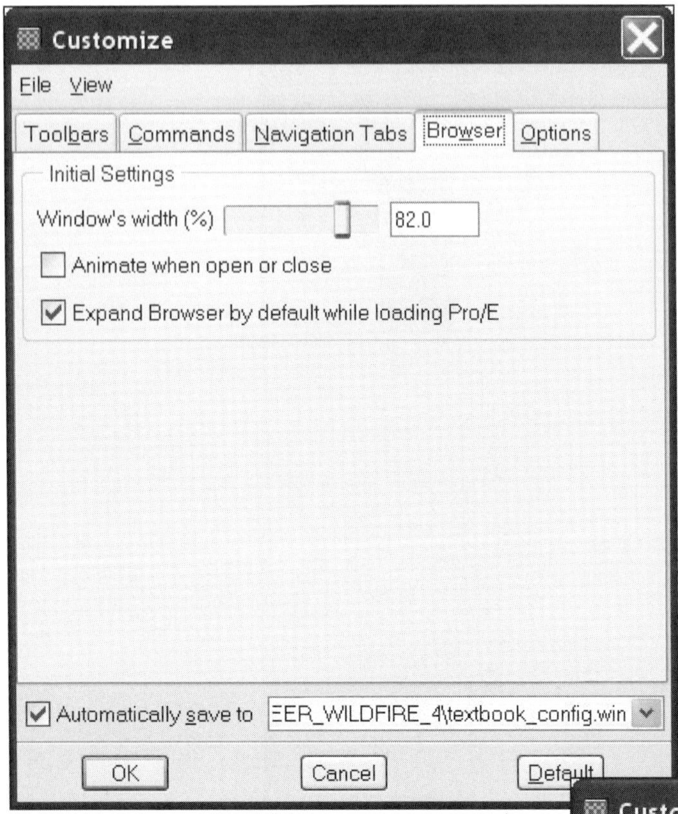

Figure A.9 Customize Dialog Box, Browser Tab

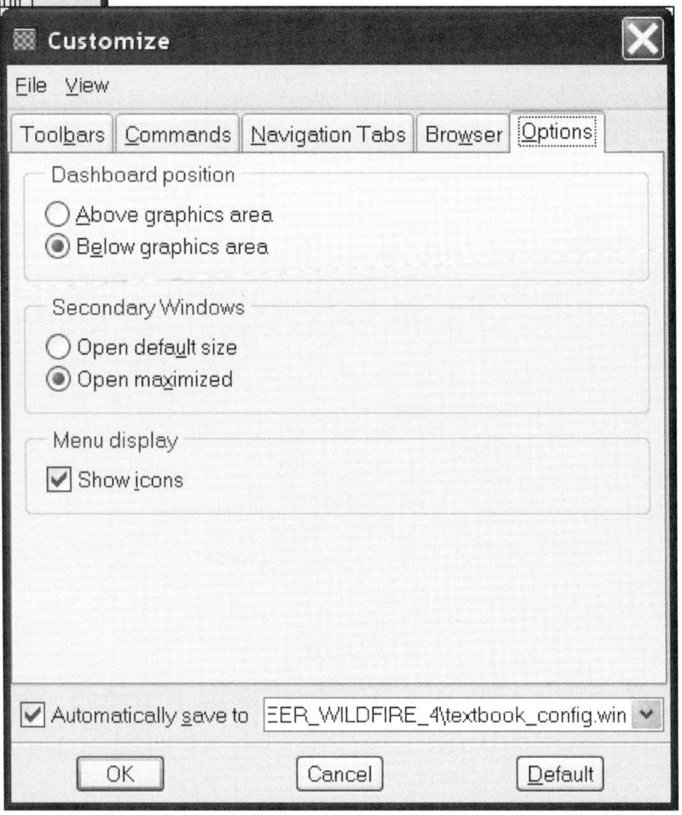

Figure A.10 Customize Dialog Box, Options Tab

Mapkeys

In Pro/E, a **Mapkey** is a macro that maps frequently used command sequences to certain keyboard keys or sets of keys. Mapkeys are saved in the configuration file, and are identified with the option *mapkey*, followed by the identifier and then the macro. You can define a unique key or combination of keys which, when pressed, executes the mapkey macro (for example, **F6** on your keyboard). You can create a mapkey for virtually any task you perform frequently within Pro/E. By adding mapkeys to your Toolchest or menu bar, you can use mapkeys with a single mouse click or menu command and thus streamline your workflow in a visible way.

Create a mapkey, click: **Tools** ⇒ **Mapkeys** Mapkeys dialog box opens (Fig. A.11) ⇒ **New** Record Mapkey dialog box opens (Fig. A.12) ⇒ Key Sequence- **$F5** ⇒ Name **COLORS** ⇒ Description **Opens Appearance Editor and starts the New Color Definition** (Fig. A.12) ⇒ **Record** ⇒ **View** from menu bar ⇒ **Color and Appearance** ⇒  **Add new appearance** ⇒ Color: **Color** button 

(Fig. A.13) ⇒ **adjust for a new color** ⇒ **Stop** ⇒ **OK** ⇒ Save Mapkeys 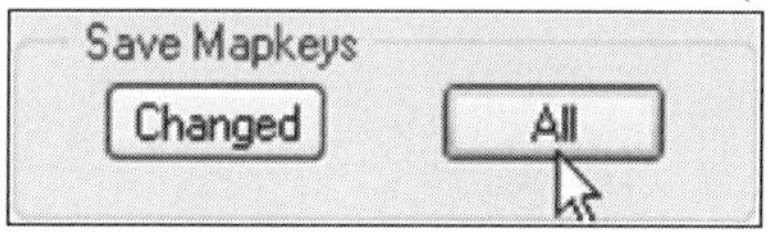

⇒ 

(Fig. A.14) ⇒ **Ok** ⇒ **Close** (Fig. A.15) ⇒ **Close** Color Editor ⇒ **Close** Appearance Editor

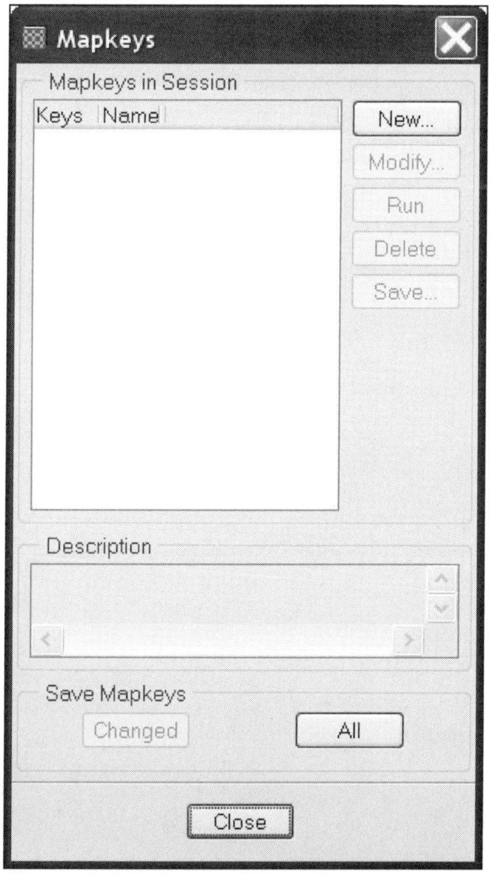

Figure A.11 Mapkeys Dialog Box

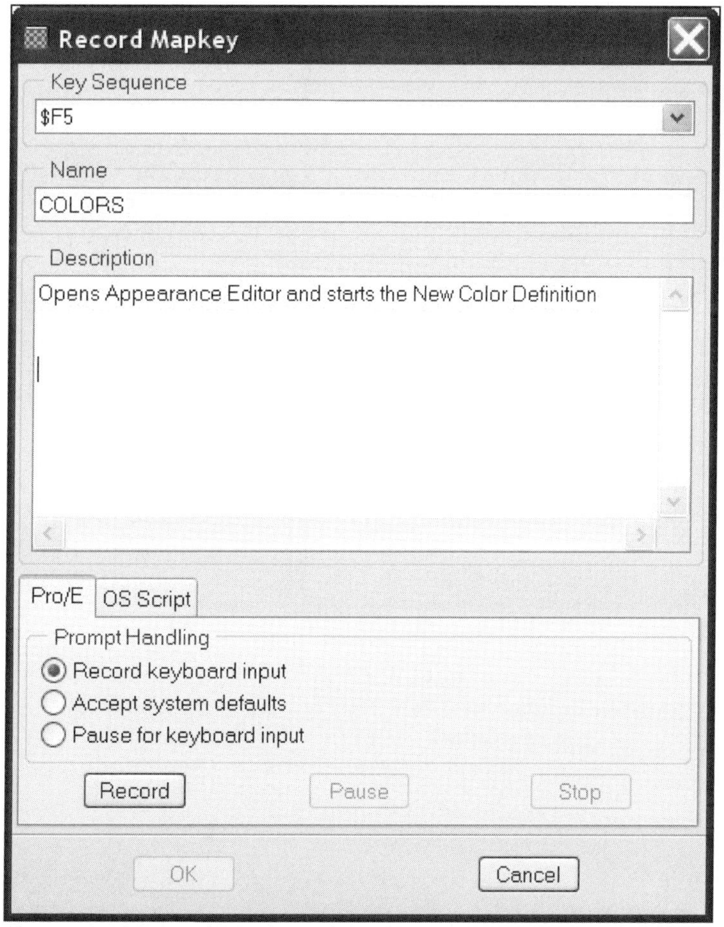

Figure A.12 Record Mapkey Dialog Box

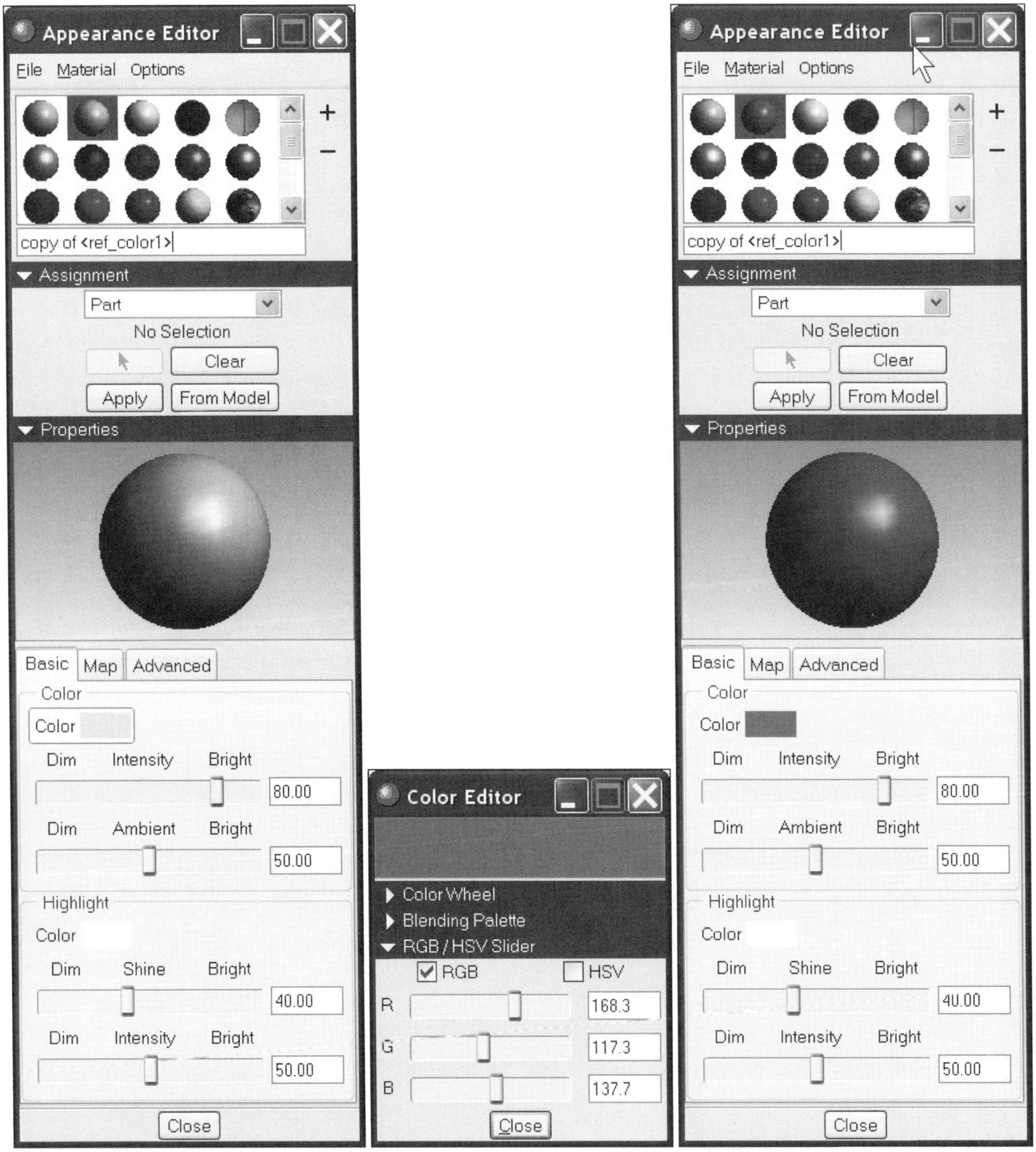

Figure A.13 Appearance Editor and Color Editor

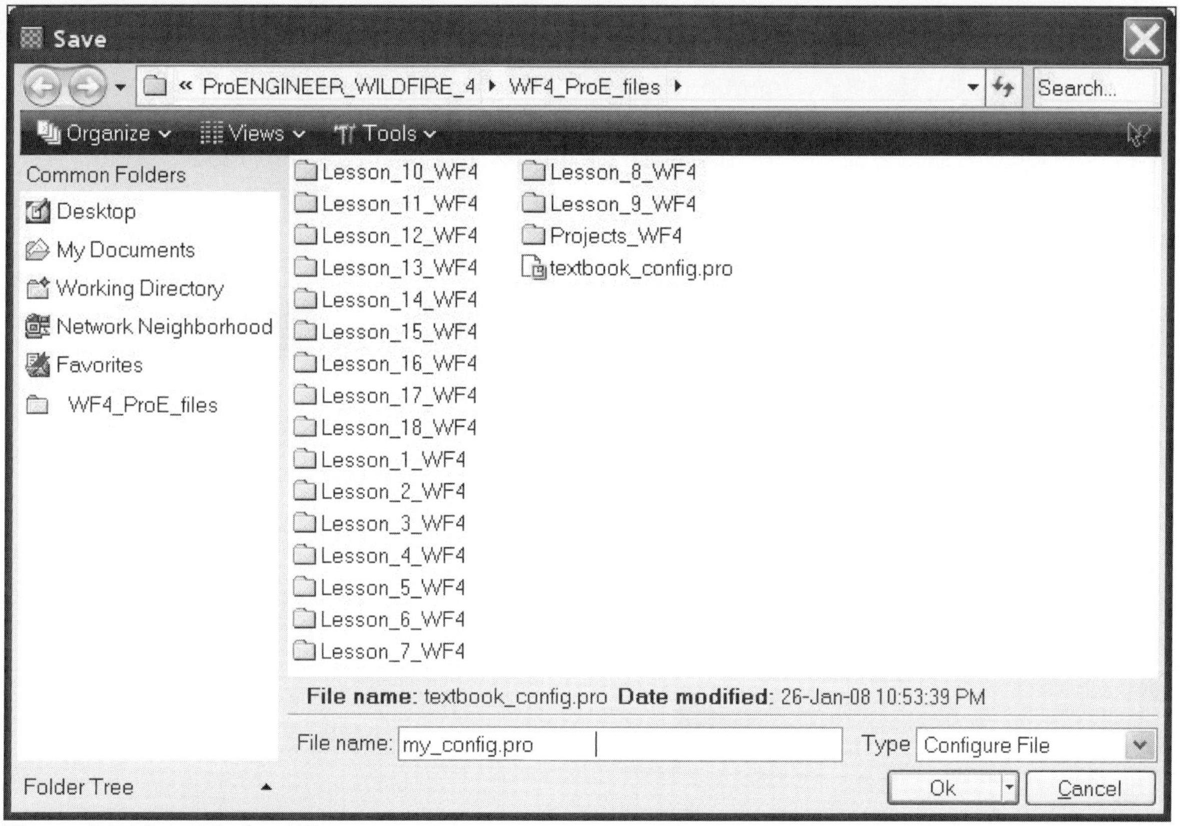

Figure A.14 Save Mapkey File

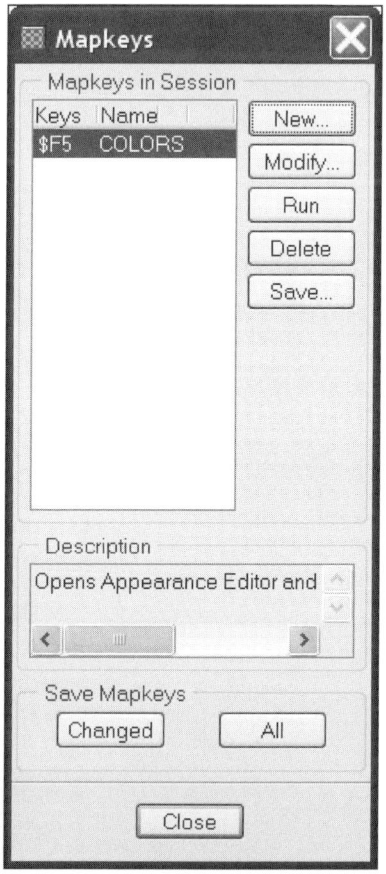

Figure A.15 Close Mapkeys

Click: **Tools** ⇒ **Customize Screen** ⇒ **Commands** tab ⇒ **Mapkeys** from the Categories list ⇒ click on the new mapkey **COLORS** 

(Fig. A.16) ⇒ **Modify Selection** ⇒ **Choose Button Image** (Fig. A.17) ⇒ select  (Fig. A.18) ⇒ move your mouse pointer over Modify Selection and click: **RMB** (Fig. A.19) ⇒ **Edit Button Image** Button Editor Opens (Fig. A.20) ⇒ click on a color block and edit the picture as you wish (Fig. A.21) ⇒ **OK**

Figure A.16 Customize Dialog Box

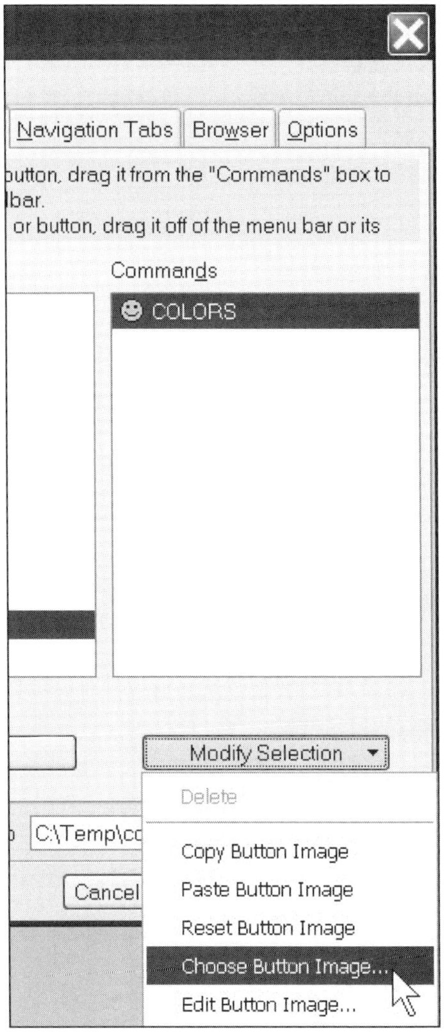

Figure A.17 Choose Button Image

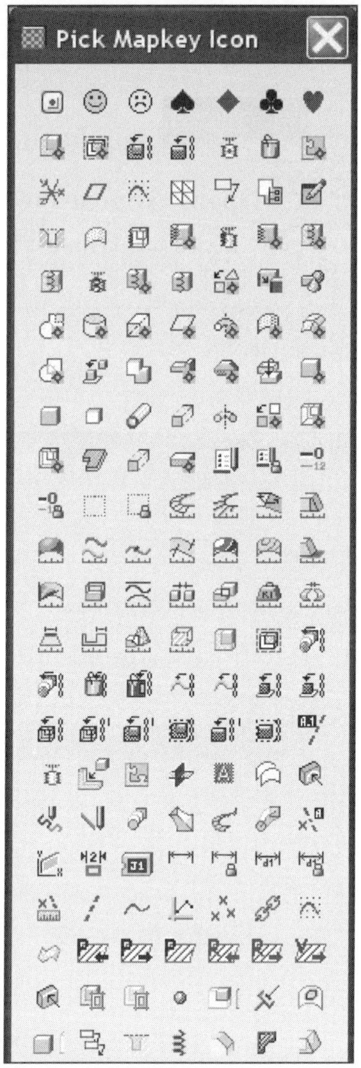

Figure A.18 Select an Icon

Figure A.19 Edit Button Image

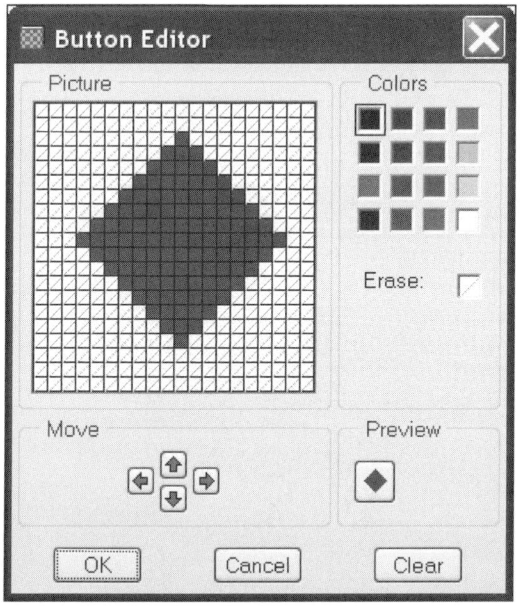

Figure A.20 Button Editor

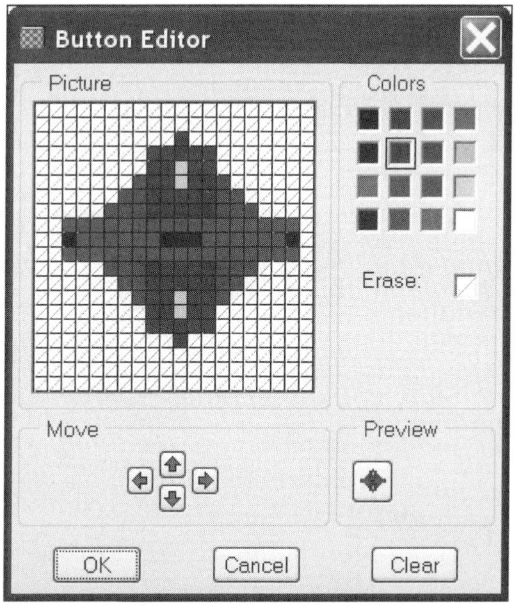

Figure A.21 Pick on Colors and Edit the Picture

Pick the new mapkey:  (Fig. A.22) ⇒ **Description** ⇒ 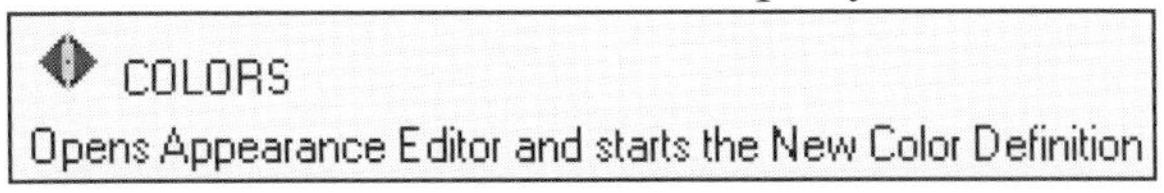

(Fig.A.23) ⇒ click on the button  and drag to the Top Toolchest and drop (Fig. A.24) ⇒ **OK** ⇒ test the new button, click:  **Opens Appearance Editor and starts the New Color Definition** (Fig. A.25), Color Editor dialog box opens ⇒ **Close** ⇒ **Close**

Figure A.22 Click on COLORS

Figure A.23 Description

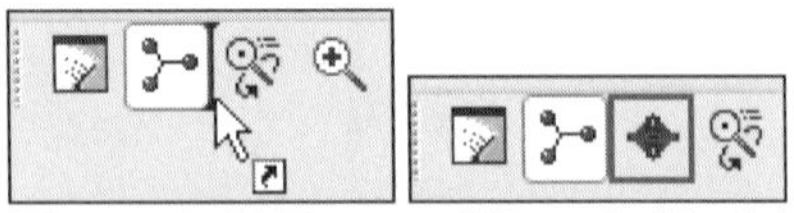
Figure A.24 Drag and Drop in Toolchest

Figure A.25 Color Editor Opens

Index